THE HANDBOOK OF SOFTWARE FOR ENGINEERS AND SCIENTISTS

Editor-in-Chief
PAUL W. ROSS

CRC Press is an imprint of the
Taylor & Francis Group, an **informa** business

First published 1996 by CRC Press
Taylor & Francis Group
6000 Broken Sound Parkway NW, Suite 300
Boca Raton, FL 33487-2742

First issued in paperback 2021

Reissued 2018 by CRC Press

Library of Congress Cataloging-in-Publication Data

The handbook of software for engineers and scientistis / editor-in-chief, Paul W. Ross
p. cm.
Includes bibliographical references and index.
ISBN 0-8493-2530-7 (acid-free paper)
1. Computer software. 2. Engineer--Data processing.
3. Science--Data processing. I. Ross, Paul W., 1938-
QA76.754.H35 1995
005.3'0245--dc20 95-6378

A Library of Congress record exists under LC control number: 95006378

Publisher's Note
The publisher has gone to great lengths to ensure the quality of this reprint but points out that some imperfections in the original copies may be apparent.

Disclaimer
The publisher has made every effort to trace copyright holders and welcomes correspondence from those they have been unable to contact.

ISBN 13: 978-1-138-10532-4 (hbk)
ISBN 13: 978-0-203-71001-2 (ebk)
ISBN 13: 978-1-138-56221-9 (pbk)

DOI: 10.1201/9780203710012

Visit the Taylor & Francis Web site at http://www.taylorandfrancis.com and the
CRC Press Web site at http://www.crcpress.com

Foreword

Although computing is only 50 years old as a discipline and barely 30 years old as a profession, it affects our lives and work as if it had existed for a thousand years. Nearly everyone in the industrialized countries works with machines that record, process, and transfer information. This revolution is one of the greatest stories of the 20th century.

In the early 1980s, Kenneth Wilson of Cornell University coined the phrase "grand challenge" to refer to very difficult scientific questions that would yield only to investigators armed with the most powerful computers and parallel algorithms. At the same time, "computational science" arose as a new mode of scientific investigation that uses computers rather than theory or experiment. In just a few years, grand challenges and computational science rose to the forefront of computing. Today, the United States government is supporting these efforts through the High Performance Computing and Communication Program, which was endorsed by Congress in 1991. A growing number of computer scientists and engineers are participating in these multidisciplinary programs. Scientists themselves no longer experience curiosity about how computers work and skepticism about their value to science; they accept computers and have a wide interest in applying them to scientific questions.

Computing has changed dramatically since Wilson's declaration. Computers and communications have merged. The international Internet has grown from four nodes in 1970 to tens of millions by the mid-1990s. Cryptographic protocols have matured from mathematical oddities to practical tools for securing privacy, exchanging money, signing documents, and authenticating users, machines, and data. Operating systems have become central to everyone's use of computers and are beginning to accommodate the design of work in organizations. A large number of parallel algorithms have been implemented and tested, demonstrating that machines consisting of thousands of computers working concurrently can operate a thousand times faster than the machines of the mid-1980s. Computer design is shifting to an approach that respects human capacities and human actions.

The future of computing will include even more changes. The breathtaking advances in hardware will continue. The Intel Pentium chip, a typical microcomputer, has more than 5 million transistors and executes over 100 million instructions per second (mips). The next generation of chips will contain more than 20 million transistors and approach speeds of 500 mips. Supercomputers will consist of hundreds or thousands of these fast processors, straining the capabilities of compilers and operating systems to support large computations. Gigabyte capacity hard disks will come in small, removable cartridges that plug into mobile computers. Data communication networks will soon carry transmissions in excess of one gigabit (one billion bits) per second, about 20 times faster than today's fastest networks. Bandwidth will be as easy to waste as computing power is. Mobile computers will become common; some will be radio-linked to the worldwide network, while others will connect occasionally and swap large files. These advances will blur the distinctions between traditional media, such as television, telephone, and data, as all transmissions become digitized. Flat-panel displays with the quality of a printed page will challenge printed-paper media. The term "multimedia" will disappear as people realize that all representations of information are encoded in the same digital medium.

The costs of computing will continue to be dominated by software, which will continue its struggle to keep up with the power of the hardware. As our reliance on computers has increased, many people have become concerned with the persistent inability of software engineers to provide software that is reliable, dependable, available on time, and within budget. A growing number of observers attribute this persistent breakdown to the underlying assumption that software production consists of deriving a program from specifications. Some designers believe that the current product-centered approach must give way to a user-centered approach, which accounts for how people work and for action in the domain in which the software will be used. Software architecture, a new approach modeled after the architecture of buildings, may provide the missing discipline.

People who use electronic mail learn to dread returning from a few days off the network. Hundreds of messages may await. Subscriptions to newsgroups and bulletin boards can demand several hours of daily reading. Yet all of this is trivial compared to the volumes of data that will soon be collected routinely by instruments or generated by computers. NASA's earth observing system will generate some trillion bits per day, enough to fill 250 compact disks. The impossibility of storing, retrieving, and comprehending such massive amounts of data challenges one of the oldest tenets of science: the tradition that all data must be saved so that others can reproduce the results. New search, compression, and information-losing storage methods will be developed for dealing with data.

Now that the telecommunications network has made people throughout the world accessible within a matter of minutes or hours, it is no longer essential that collaborators be near one another. Within a few years, telecommuting will become more common as people work at home more and visit their offices only a day or two a week. A new understanding of work as making and fulfilling commitments is emerging. Computing technologies that support work in this new sense are still in the early stages of design. A workflow management industry is already taking form and will gross over $2.5 billion by the mid-1990s. Major shifts in operating system design will be made to accommodate the new understandings of work.

The printed, archival journal or magazine is central to present-day science and engineering. The Internet and new kinds of digital media are stimulating new publication practices that, although still marginal, ultimately threaten the entire scientific print-publication enterprise. An increasing number of on-line publications distribute, free, reviewed papers within their communities. An increasing number of universities have servers on which scientists post, for free distribution, papers that have been submitted or accepted for publication. By contrast, paper publications have reviewing delays of several months cascaded with print-queue delays of 3 to 18 months. By the mid-1990s, publishers will need to offer electronic distribution, and it must be valuable enough to attract paying customers. An approach under consideration by many scientific societies is to make a transition from a series of journals to a unified database. Subscribers would be notified whenever a new entry is made on any topic that matches their interest profile. They will be entitled to extract a limited number of items at no additional charge.

All these advances would not be possible without widespread standard practices for using computers and the many software tools that they include. These basic tools include operating systems, network interfaces, window managers, document preparers, spreadsheets, graphics packages, databases, programming languages, circuit simulators, statistical analyzers, and more.

You hold in your hands a handbook that is indispensable to mastering the basic practices that will make you a player in the new, emerging world of science and engineering applications. This handbook will tell how the new tools work and how to use them. It is the Boy Scout handbook for the world of science and engineering described above: you may not be able to survive without it.

Peter J. Denning
Rosslyn, Virginia

Preface

Purpose

The Handbook of Software for Engineers and Scientists has been designed to provide, in a single volume, a ready reference on computer applications for the practicing engineer and scientist in industry, government, and academia. The book, in its comprehensive format, is divided into a number of sections that encompass the most common uses of computers in science and engineering—a veritable "Swiss Army knife" for both the beginning and advanced computer user. This handbook contains the following important carefully chosen elements:

- Tools and descriptions of computer environments, operating systems, and handy utilities for performing many common day-to-day tasks
- Descriptions of important software packages for evaluation for potential use by the practicing engineer or scientist
- Tutorial information on many packages to get you "up and running" in a minimum amount of time
- A comprehensive glossary of defining terms for the new user, or person unfamiliar with the terminology of the field
- Pertinent documents on each package or systems for study and reference
- A handy set of appendices full of useful information such as ASCII codes and common interface connections.

The goal in the development of this handbook was to provide the most up-to-date information in various areas of the continually changing field of computer applications. The wide range of topics in the handbook was chosen to include the most common computer platforms and operating system environments, languages, applications programs, and many common problem-solving tools such as statistical applications, mathematical software, and engineering tools. With the arrival of the Internet, additional extensive treatment has been devoted to the powerful telecommunications and networking tools that become the cement to bind the modern computer facility together.

Organization

The uses of computers have expanded over the last quarter century from a narrow field for specialists to a ubiquitous tool used by people in all areas of theoretical and applied science. Current statistics indicate that there are in excess of 280 installed computers per 1000 population in the United States alone. This means that more than one of every four workers have computers on their desks, or will have ready access to a computer. The exceptional decrease in the cost of computing equipment and the wide variety of inexpensive programming and application tools have led to the development of this handbook. The intent is to provide the practitioner with a combination of tutorial information coupled with such information as to evaluate easily the fitness of the application or technique for their particular problem. The handbook is organized into seven divisions.

- *Platforms and Environments*

This section contains an extensive discussion and presentation of the most common computer platforms and environments, such as MS-DOS, Windows, and the Apple Macintosh, Unix, the DEC VAX, IBM mainframes, OS/2, and NeXT.

For the novice computer user, the details of basic computer usage, such as starting applications and managing files, can pose a formidable obstacle. This section provides tutorial and descriptive material to ease this transition. The presentations in this section show users just what they will see on the screen, and what commands are commonly needed to perform a given task.

- *Structured and Object-Oriented Programming*

The elements of structured programming using various versions of the popular and powerful C and C++ languages, and object-oriented analyses and programming techniques are shown. Detailed presentations are given of the general issue of object-oriented design, as well as tutorials on the major development environments such as Eiffel and Smalltalk.

- *Application Packages*

This section presents the most important applications necessary for successful word processing, desktop publishing and presentation graphics systems, integrated packages, spreadsheets, databases, and a number of useful utilities. With the pervasive use of microcomputers in the workplace for performing routine and not-so-routine tasks, extensive tutorials on subjects such as the $\mathrm{T_EX}$ typesetting language, page makeup programs, and powerful word processors are especially valuable.

This section also presents detailed information on the use of common spreadsheet programs, the answer to any sort of simple mathematical computation that can be defined in terms of data and calculations in the form of rows and columns. The especially powerful graphical features of spreadsheets are given great emphasis.

As ancillaries to successful deliveries, computer presentation systems such as Harvard Graphics, PC Paintbrush, PowerPoint, and Freelance are illustrated. These packages are important to give the extra "punch" to successful publications and presentations.
This section also includes components on major database systems, and the ever-useful integrated package, in the form of Works for Windows, and ClarisWorks for the Macintosh. When full-functioned products are not justified for many routine tasks, integrated packages such as Works and ClarisWorks present a useful compromise between cost, ease of use, learning time, and functionality.

In addition, two groupware systems are presented, representative of this newly emerging application area for managing large and complex projects. With the tools and applications packages described in this section, the reader will be able to perform a wide range of tasks to ease their day-to-day responsibilities.

- *Mathematical and Statistical Software*

This section includes descriptions and tutorials on many powerful mathematical and statistical tools. These packages have a vast range of complex scientific and engineering applications. Analytical and statistical procedures that are beyond the realm of paper-and-pencil calculations are now easily within the reach of anyone with access to even modest computer facilities.

The mathematical software includes AXIOM, DERIVE, Macsyma, Maple, MathCad, Mathematica, and MATLAB. Examples and comprehensive descriptions are given for each package. A very useful addition to the section is an extensive listing of all major computer algebra systems, their characteristics, features, and where and how they can be obtained.

The statistical tools include presentations of the popular Minitab, SAS, STATA, S-PLUS, and WinSTAT systems.

- *Engineering Tools*

Included in this section are circuit and control simulation packages, finite element analysis tools, CAD/CAM systems such as SmartCAM, and solid modeling systems like COGO. These packages replace and supplement the drafting table or the laboratory workbench for the practicing engineer or scientist,

Tedious hours of laboratory and experimental time can be tremendously reduced with tools such as SPICE for electrical simulations for the electrical engineer, and finite element systems such as ALGOR and ANSYS for the civil and mechanical engineer.

- *Data Communications and Networking*

Computers are becoming more frequently networked in the workplace instead of being used in a simple stand-alone mode. This section describes the tools and techniques, as well as the use of various systems for creating local and wide area networks. Extensive discussions are devoted to topics such as Procomm, a telecommunications package, Novell Netware, a common local area network system, and TCP/IP, the protocol used on the Internet.

In addition, a number of applications are described for connecting computers through the dialed network and passing data between systems. Much of the material in this section is given in tutorial form, allowing the newcomer to the field to master quickly the intricacies of networking and telecommunications, so critical to functioning on the “information highway.”

- *Appendices*

This final section includes a short compendium of useful information including ASCII Codes, RS-232 and Parallel port and pinout information, and ANSI Escape Sequences. This is the sort of information users need at their fingertips at the most inopportune moment when some interconnection or printer fails or malfunctions.

Locating Your Topic

Numerous avenues of access to information contained in the handbook are provided. A complete table of contents is presented at the front of the book. In addition, an individual table of contents precedes each of the sections. Finally, each chapter begins with its own table of contents. Each chapter contains a list of defining terms, when appropriate, and references or further information on the methodology or techniques presented. Each chapter is extensively illustrated with screen dumps of important screens to provide the reader with a “roadmap” to the application at hand.

The reader should look over these tables of contents to become familiar with the structure, organization, and content of the book. The handbook is extensively and comprehensively indexed by subject.

Acknowledgments

I wish to acknowledge the invaluable assistance of the staff at CRC Press, Inc., in particular; Joel Claypool, Publisher; Robert Stern, Senior Editor; Carol Whitehead, Senior Project Editor; and Jennifer Kondras, Editorial Assistant. I also wish to thank those Advisory Board members who contributed section introductions. My employer, Millersville University, kindly made the facilities of the Internet available for managing and implementing the project. In no less a manner, I thank the contributors and the Board of Advisors for their long hours, hard work, and dedication in making this effort a success. It is truly "their" book—I merely helped direct traffic.

Paul W. Ross
Editor-in-Chief
Lancaster, Pennsylvania

Editor-in-Chief

Paul W. Ross is a professor of computer science at Millersville University, Millersville, Pennsylvania. He teaches undergraduate courses in computer science, with an emphasis on graphics, data communications, and problem-solving techniques. He earned D.E., M.E., and B.E. degrees in Electrical Engineering from Yale University, New Haven, Connecticut.

Professor Ross has an extensive background in education and industry. He was employed as a member of technical staff by the RCA Sarnoff Research Center, Princeton, New Jersey, where he did research on computer recognition of speech. He was director of the computer center at Franklin and Marshall College for a number of years, and, most recently, has taught computer science at Millersville University.

In addition, he is actively engaged in consulting activities with a variety of educational, governmental, and business organizations. He is the author or co-author of over twenty-five books, supplements, or other publications on topics as diverse as computer applications, artificial intelligence, craniofacial morphology of the cleft palate, and computer literacy. He is a member of Sigma Xi and Phi Delta Kappa honorary professional fraternities.

Advisory Board

Contributors

Paul C. Abbott
University of Western Australia
Nedlands, Australia

Salahalddin Abusalah
Florida Atlantic University
Boca Raton, Florida

Action Technologies, Inc.
Alameda, California

Alistair Adams
Integrated Systems, Inc.
Santa Clara, California

Matthew Allen
Millersville, Pennsylvania

Charles Ames
Jet Propulsion Laboratory
Pasadena, California

Tamara Artim
Integrated Systems, Inc.
Santa Clara, California

Neil R. Bauman
ComputerTalk Associates, Inc.
Blue Bell, Pennsylvania

John S. Bay
Virginia Polytechnic Institute and State University
Blacksburg, Virginia

Sean Becketti
Stata Corporation
College Station, Texas

Donald L. Byrkett
Miami University
Oxford, Ohio

Satish Chaliki
Keane, Inc.
Boca Raton, Florida

Martin Chorich
Integrated Systems, Inc.
Santa Clara, California

Debora Cole
Lotus Development Corporation
Cambridge, Massachusetts

George A. Cunningham
New Mexico Institute of Mining
Socorro, New Mexico

Ronald L. Davis
Millersville University
Millersville, Pennsylvania

Paulo Ney de Souza
University of California-Berkeley
Berkeley, California

Gregory W. Donohoe
University of New Mexico
Albuquerque, New Mexico

Mark E. Dreier
Bell Helicopter Textron, Inc.
Hurst, Texas

Dan Duricy
Miami University
Oxford, Ohio

Jonathan Edwards
University of Kenturky
Lexington, Kentucky

Carl B. Erickson
Grand Valley State University
Allendale, Michigan

Osama Ettouney
Miami University
Oxford, Ohio

Robert K. Fitch
G. Greulich Software
Staufen, Germany

Kevin Greiner
Ware, Inc.
Ephrate, Pennsylvania

James Griffioen
University of Kentucky
Lexington, Kentucky

Ashish Gupta
Integrated Systems, Inc.
Santa Clara, California

David L. Hartung Jr.
Miami University
Oxford, Ohio

Rodney Hocutt
Rubicon Technologies
Allentown, Pennsylvania

Alan Hoenig
John Jay College, CUNY
Huntington, New York

Robert "Rock" Howard
Tower Technology
Austin, Texas

David H. Hutchens
Millersville University
Millersville, Pennsylvania

M. Jamshidi
The University of New Mexico
Albuquerque, New Mexico

Roberta Jaworski
Miami University
Oxford, Ohio

David Jensen
George Mason University
Fairfax, Virginia

Elizabeth Katz
Washington Boro, Pennsylvania

James Kiper
Miami University
Oxford, Ohio

Laurie A. Knox
Millersville University
Millersville, Pennsylvania

Timothy C. Krehbiel
Miami University
Oxford, Ohio

Todd Leuthold
Todd Leuthold Consulting
Lancaster, Pennsylvania

Faustino A. Lichauco
MathSoft, Inc.
Cambridge, Massachusetts

Christina Lohr
Millersville University
Millersville, Pennsylvania

Veronica L. Longenecker
Millersville University
Millersville, Pennsylvania

T. E. Mertz
Millersville University
Millersville, Pennsylvania

Dan Moore
Rose-Hulman Institute of Techology
Torre Haute, Indiana

Thomas Murphy
Digitalk, Inc.
Santa Ana, California

W. Charles Paulsen
Algor, Inc.
Newtown, Connecticut

Richard Petti
Macsyma Inc.
Arlington, Massachusetts

Robert Pickering
Miami University
Oxford, Ohio

Rudra Pratap
Cornell University
Ithaca, New York

J. Mark Pullen
George Mason University
Fairfax, Virginia

T. M. Rajkumar
Miami University
Oxford, Ohio

William C. Rinaman
Le Moyne Çollege
Syrascuse, New York

Waldir L. Roque
Universidade Federal do Rio Grande do Sul
Porto Algre, Brazil

Paul W. Ross
Millersville University
Millersville, Pennsylvania

Rajiv Sabharwal
Bio-Rad Laboratories, Inc.
Berkeley, California

Alton F. Sanders
Miami University
Oxford, Ohio

Chr. Schmid
Ruhr-Bochum University
Bochum, Germany

Richard Sutor
American Insurance Association
New York, New York

Robert Sutor
IBM Corporation
Yorktown Heights, New York

James TenEyck
Marist College
Poughkeepsie, New York

Timothy R. Trainer
ANSYS, Inc.
Houston, Pennsylvania

Robert P. Treder
MathSoft Inc.
Seattle, Washington

Barry Walton
Millersville University
Millersvile, Pennsylvania

Judith W. Whitcomb
Northford, Connecticut

Frederick J. Wicklin
University of Minnesota
Minneapolis, Minnesota

Ali Zilouchian
Florida Atlantic University
Boca Raton, Florida

Contents

Section VI Data Communications and Networking ..... [illegible]

Appendices ..... [illegible]

Index ..... [illegible]

THE HANDBOOK OF SOFTWARE FOR ENGINEERS AND SCIENTISTS

Platforms and Environments

A REASONABLE PLACE TO START ON COMPUTER APPLICATIONS is the computer and operating system itself. This section of the handbook covers a number of the main computer hardware and software platforms that you will encounter most frequently.

- MS-DOS for IBM PCs—The classic PC operating environment on Intel 80x86 processors.
- Microsoft Windows—The most common environment for 808386, 80486, and Pentium machines.
- The Macintosh—Different variations of the Apple Macintosh hardware and system.
- UNIX—The common operating system for many workstation environments.
- DEC VAX—The hardware and operating system for Digital Equipment Corporation's major line of VAX processors.
- IBM Mainframes—The VM/CMS operating system for IBM's mainframe computers.
- OS/2 and NEXTSTEP—Two additional operating systems of interest for Intel architecture machines.

1

MS-DOS for IBM PCs

Paul W. Ross
Millersville University

Since the original IBM Personal Computer™ was introduced in 1981, MS-DOS™ (PC-DOS™ on IBM systems) has become the dominant operating system for computers that use the Intel 8088, 80286, 80386, 80486, and Pentium™ chips. More microcomputers use DOS than any other operating system, and there are thousands of application packages that run under it.

Conceptually similar in some ways to the now obsolete CP/M operating system, MS-DOS borrows concepts from the UNIX operating system. Although MS-DOS was originally designed to have the same general feel as CP/M, it was equipped with some important improvements. MS-DOS is a single-user, single-tasking operating system. Because MS-DOS runs on many different brands of computers, it is not classified as a proprietary operating system. It is, however, limited to the Intel 80x86 family of microprocessors. For this reason, MS-DOS is not a fully generic operating system either. Nevertheless, because of the continuing popularity and large installed base of IBM-compatible microcomputers, MS-DOS will remain a widely used operating system for some time.

This section is not a complete discussion of MS-DOS. You should read the reference manual that comes with your computer for more details, or use the HELP feature. This chapter reflects the major features of MS-DOS 6.0, 6.1, and 6.2.

1.1 ENVIRONMENT AND SYSTEM CONFIGURATION

Booting with a Hard Disk System

Just turn on the computer. The computer will then start, or boot up. When the computer starts up or the boot process is complete, you will see a symbol like

```
C>
```

The MS-DOS Prompt

The letter on the screen followed by the > or greater than symbol is the MS-DOS prompt. It indicates which disk drive is currently engaged. In a floppy-disk system, the first drive designated is with the letter A and the second floppy disk drive is the B

0-8493-2530-7/96/$0.00+$.50

drive. Hard drives are designated beginning with the letter C and go on through the alphabet. For example, if you have a second hard disk drive, it will be designated as the D drive. Network drives fall after the local hard drives, and CD ROM players are generally last.

If you want to instruct the computer to read information from another drive, you can switch to another disk drive by typing in B: and pressing the Enter key. Be sure to include the colon. This command shifts the computer from whatever the current drive may be over to drive B. You can shift back to the A or C drive by typing A: or C:, respectively, followed by pressing the Enter key.

Formatting Disks for a Floppy Disk System

You need to know how to format a floppy disk extensively for storing files off-line for backup purposes. Formatting completely erases any old information from the disk and lays down a skeleton structure to allow the computer to quickly find or store information on the disk.

To format a disk, follow these steps:

1. Insert the new unformatted diskette in the remaining disk drive or the A drive on a hard drive system.
2. At the MS-DOS prompt type the command:

```
FORMAT [d:] <Enter>
```

You should type FORMAT, space, the letter of the disk drive where the new, unformatted disk is inserted, and the colon. Then press the Enter key. At this point, the format process takes place. It will take a minute or so. When it is done, you will be told how much space is available on the disk and asked if you want to format another disk.

Types of MS-DOS Commands

The disk operating system, or MS-DOS, responds to a great number of commands for setting computer system parameters, manipulating data files, and executing programs. The commands can be classified as internal and external commands.

The internal commands are those that are built into the operating system itself. They are the commands that are most frequently used and that need to be executed rapidly. Whenever the operating system prompt is visible, these internal commands are available. Under certain circumstances, the commands may be available from within an applications program if it has an MS-DOS Window as is typical in many word processor and spreadsheet applications.

Less frequently used commands are known as external MS-DOS commands, or MS-DOS utility programs. They include such things as copying entire disks or sorting data files. FORMAT is an example of an external command.

Commonly Used Internal MS-DOS Commands

This section will discuss the most commonly used internal MS-DOS commands. There are other less frequently used commands that are described in the MS-DOS reference manual that comes with your computer, or through the HELP command.

What we describe and discuss here will be enough for the vast majority of situations you will come across.

Directory Command

To determine what information is stored on your disk, use the DIRectory command. Every disk contains a directory of all the files stored on it. By typing the MS-DOS command DIR after the prompt and pressing Enter, you will get a list of all the files stored on the disk in the current drive.

The General Form of the DIR Command

If you wish to see the directory for a different disk, or for a portion (subdirectory) of the files on a disk, then a more specific DIR command is needed. The complete command format for DIR is as follows:

```
DIR [d:][subdirectory][filename][.ext]
```

where

- DIR is the general Directory command.
- [d:] is the disk drive; if none is specified, then the computer will assume the default drive (the drive it is already in).
- [subdirectory] is the name of the subdirectory you want to view. Subdirectory information is optional.
- [filename] is the name of the file. If no filename is specified, then all the files in the specified drive (and path, if given) will be displayed.
- [.ext] is the extension to the filename. Files that are built into applications packages often already have extensions. With the files you create, you have the option of adding an extension if you wish. If a file has an extension, you must always use it to identify that specific file.

Files and File Names

A complete file name contains four elements:

1. The name of the disk drive. This is A or B on a floppy disk system, and is C or higher in a hard drive system. It can be omitted unless you want to refer to a file stored on a disk other than the currently logged disk drive.
2. The subdirectory name. This optional feature is a name given to a group of several similar files.
3. The name of the file. There can be no spaces in a file name. TRACKTWO is fine, but TRACK TWO is not.
4. The file extension. This is an optional extension the file name. For example, THISFILE.CUR and THISFILE.YTD might contain the current and year-to-date figures, respectively, on a given subject. The extension consists of a period followed by one to three characters. You have the option of including or not including an extension on the files you create. Most applications packages will provide an extension automatically for their purposes.

For example, if we had a subdirectory known as LETTERS on disk drive C, the DIR command to show us all the files on that subdirectory would look like this:

```
DIR C:\LETTERS or DIR C:\LETTERS\*.*
```

Both forms will have the same results. Note the use of the backslash (\) on either side of the subdirectory name. This is functionally equivalent to the / used in UNIX.

There are two global filename characters, or wildcards. The two wildcard characters are ? and *. The ? symbol can stand for any single character. The * can stand for any string or group of characters, or a null.

CLS Command

The CLS (clear screen) command clears the display screen. This command also returns the MS-DOS prompt to the top left corner of the screen.

COPY Command

The COPY command is used to make a copy of a file from a source location to a destination or target location.

The COPY command works like this:

```
COPY filename1 filename2
```

where filename1 is the source and filename2 is where it will be copied to, or the destination. If only one file name is given, it must be the name of a file on another disk or subdirectory with the same name. The general format for the COPY command is as follows:

```
COPY [d:][subdirectory]filename[.ext] [d:][subdirectory]
```

This format copies a file from one drive and/or directory to another disk, or between two directories on the same disk. The first d: in the COPY command indicates the source drive, and the second d: indicates the destination drive. You can omit the parts of the command that are not needed. For example, if you do not specify the first (source) disk drive, then MS-DOS assumes that the source is the drive you are currently logged onto. In the COPY format, the wildcards * and ? may be used in both the source and the destination file names and extensions.

REN Command

The REN (rename) command changes the name of a file to a new name. The command format for REN is as follows:

```
REN [d:][subdirectory]filename[.ext] filename[.ext]
```

You need to specify the disk drive and subdirectory only for the first file name. If no drive or subdirectory is specified, then MS-DOS will default to the current directory and subdirectory. The second file name will use the drive and subdirectory specified for the first file name. The wildcard characters * and ? may be used in the file name and extension.

DEL Command

The DEL command is used to delete an entire file you no longer need. Deleting a file recovers space on the disk for future work. It can also ward off confusion if you have earlier and later versions of the same material.

The DEL command permanently deletes one or more files from the disk. The command format is as follows:

```
DEL [d:][subdirectory][filename[.ext]]
```

If the disk drive is not specified, the default disk drive is used. Use the exact spelling for the file name, and be sure to include the extension if there is one.

The global characters * and ? may be used in the file name and extension. When using global characters, be sure you understand what you are asking for, because whole groups of files might be inadvertently deleted without warning by MS-DOS. If the content of an entire disk or directory is to be deleted, use the command DEL *.*. Since this deletes *all* files on the disk or directory, the computer first asks you to confirm your instruction. You must then enter Y or N followed by pressing the Enter key.

TYPE Command

The TYPE command displays the contents of the specified file on the screen. This is the command to use when you simply need to look and see what a file contains. The command format is as follows:

```
TYPE [d:][subdirectory]filename[.ext]
```

If the drive and subdirectory are not specified, the default drive and subdirectory are assumed. The file name may not contain the global filename characters * or ?, because only one file can be displayed at a time.

1.2 FILES AND DISK MANAGEMENT

Since a hard disk can hold a large number of files, it is important to have some way of organizing them. This is done with directories and subdirectories to produce a tree-like structure where similar files can be grouped together. An example would be to place all the files associated with a given application package in their own subdirectory.

Subdirectories are created with the MKDIR or MD command. The command format is as follows:

```
MKDIR [d:]subdirectory
```

If the command is entered without specifying a disk drive, the default drive is used. Beginning the subdirectory with a backslash (d:\) tells MS-DOS to create the subdirectory starting at the root or main directory. The absence of a leading backslash indicates that you want to create the subdirectory within the current directory. Each subdirectory may contain many subdirectories of its own. Examples of subdirectories are as follows:

```
MKDIR C:\LEVEL1
```

This command creates the subdirectory LEVEL1 on drive C starting at the root directory.

```
MKDIR LEVELN
```

This command creates subdirectory LEVELN on the default disk in the current directory.

```
MKDIR C:\LEVEL1\LEVEL2
```

This command creates subdirectory LEVEL2 on drive C within directory LEVEL1. The subdirectory LEVEL1 must exist before this command is given. Note that successive levels of subdirectories are separated by backslashes.

CHDIR Command

If we want to move to a specific directory, we use the CHDIR (change directory) command. CHDIR, or CD for short, changes the current directory of the specified or default disk drive. The command format is as follows:

```
CHDIR [[d:]subdirectory]
```

Removing a Subdirectory

The RMDIR (remove directory), or RD command removes an empty subdirectory from a disk when it is no longer needed. First, go into the subdirectory and remove all files with the DEL command. The command format for removing the empty directory is then as follows:

```
RMDIR [d:]subdirectory
```

Before specifying the subdirectory to be removed, you must go back to at least one subdirectory before the one that you want to remove. For example,

```
RMDIR \LEVEL1\LEVEL2
```

This removes the subdirectory LEVEL2 from the LEVEL1 directory on the current disk. Another example is

```
RMDIR SUBDIR
```

This command removes the subdirectory SUBDIR from the current directory.

1.3 UTILITIES

We have already discussed FORMAT, which is an external MS-DOS command. Like all other external MS-DOS commands, FORMAT is a separate program and not an integral part of MS-DOS, even though it is on the system disk. In this section, a number of useful utility programs are discussed. The MS-DOS manual or on-line HELP system should be consulted for other programs, or more specific details.

Diskcopy

The COPY command copies single files or groups of files. If you want to copy an entire floppy disk, use the DISKCOPY command. The DISKCOPY command is especially useful for making backup copies of entire floppy disks for archival purposes. The DISKCOPY command copies the contents of the specified source disk drive to the destination drive overwriting *all* the information previously on the destination diskette. It formats the destination disk at the same time, if it has not already been formatted. The command format is

```
DISKCOPY [d:] [d:]
```

The first disk drive specified is the source disk, and the second drive is the destination.

For example, DISKCOPY A: B: will copy the entire contents of disk in drive A to the disk in drive B. DISKCOPY A: A: will do a disk copy where you have only a single disk drive. You will be prompted to change disks, as necessary.

CHKDSK and SCANDISK Commands

Due to minor hardware or software problems, you can start to lose files on a disk. Some of these problems can be located and corrected with the CHKDSK (check disk) or SCANDISK (scan disk) utilities.

The CHKDSK command checks the File Allocation Table on the specified or default disk drive and produces a disk and memory status report.

The command format for CHKDSK is as follows:

```
CHKDSK [d:][filename[.ext]][/F][/V]
```

If you specify a file name, CHKDSK displays the number of non contiguous (fragmented) areas occupied by the file(s) in the current directory. The global filename characters * and ? may be used to check selected groups of files. These options may follow the filename. Note that a / is used, and not a \, as optional parameters are being specified.

/F—This option tells CHKDSK that you want to correct errors found while verifying the disk integrity. This option tells CHKDSK to gather up any lost fragments of files and gives them a name in the form FILEnnnn.CHK, where nnnn is a number. The resulting files can then be deleted so that the space can be restored to the disk for future use.

/V—This option causes each file name and directory processed to be displayed on the screen, indicating the program's progress.

After checking the disk and displaying any error messages, CHKDSK provides a status report on the disk checked and tells the amount of computer memory still available.

A more powerful utility now provided with recent releases of MS-DOS is the SCANDISK utility. This utility will attempt to repair damaged files, if possible. The syntax is simply SCANDISK <Enter>.

The DEFRAG Utility

As files are alternately created and deleted, files become distributed across the disk instead of being in contiguous locations. As a consequence, it takes longer to load or store files. This problem is known as disk fragmentation. MS-DOS provides a useful utility to reorganize the contents of the disk, placing files in contiguous locations on the disk. This utility should be run periodically. The syntax is

```
DEFRAG <Enter>
```

The utility will analyze the structure of the disk and give you some indication of the degree of fragmentation, allowing you to defragment the disk, if desired.

1.4 INTEL X86 SEGMENTED ARCHITECTURE

The processor used to support the majority of IBM or IBM clone systems is one of the Intel Corporation x86 family of processors. It is important to understand how memory is managed by this processor family. In particular, we need to primarily consider the 80286, 80386, 80486, and Pentium™ processors. In the historical development of computers based on these chips, the 8086/88 chip was used in the PC™ dual floppy machines, and the later XT™ machine that supported a hard disk system. These systems are obsolescent, but much earlier software is still used that ran on these systems. Consequently, the 8086/88 chip will be included in our discussion.

The 80286 processor chip is generally used in machines known as AT™ class machines, which are also obsolescent. The 80386, 80486, and Pentium™ chips, in various versions, are used in IBM's computer line, and various clone machines. Most newer software requires a 80386 or higher chip for proper operation. Some programs, such as those that are computationally intensive, take advantage of, or require, a math coprocessor.

The general design strategies used in memory management are based on the use of two registers. This was due to the original limitation of a 16-bit register design in the original 8086/88 processor. To access more than 64K bytes, which was the memory limitation in the earlier 8080 and 8085 processors, two registers are combined to yield 20-, 24-, or 32-bit addresses, depending on the particular processor. One register points to a 64K segment any place in memory, and the second register points to a place within this 64K segment. In the case of the 8086/88 processor, the registers are combined to yield a 20-bit address space. The 80286 processor combines the registers to give a 24-bit address space. The data path to memory is 8 bits in the case of the 8088 processor, and 16 bits in the case of the 8086 and 80286 processors.

The 80386, 80486, and Pentium processors work slightly differently, as they support a virtual memory address mode that is transparent to the user. Additionally, segments need not be 64K in size, leading to lower memory fragmentation. The complexity of the details of the complete memory management scheme for these processors is beyond the scope of this book.

All registers are 32 bits wide, instead of 16 bits wide. Software compiled for 32-bit mode must be used to take advantage of this capability. Moreover, the data bus is 32 bits wide, but can accommodate 16-bit data transfers by sensing a signal from the mother board indicating that the bus is only 16 bits wide. The address space is 32 bits.

There are two issues of which the user should be aware:

- What processor is being used (8086/88, 80286, 80386, 80486, or Pentium) by the computer system?
- What amount and kind of memory are being used by the computer system?

The earliest machines were based on the Intel Corporation 8086/88 processor. This processor is common in the IBM PC™ and XT™ class machines.

- It has a 20-bit memory address, and so is capable of addressing 1,048,576 bytes of memory
- The processor was designed to handle only a single task. With a little fiddling with the operating system, it is possible to have background tasks like a print spooler, or pop-up programs. Depending on the particular hardware configuration, some of the newer versions of MS-DOS, such as MS-DOS 6.2, may not run on these systems. MS-DOS 3.3 and 5.0 are good choices for use on these older machines.

Due to some design choices in how this memory was used, only about 640K of memory is usable for programs running under MS-DOS. This 640K memory is known as conventional memory. Initially, this amount of memory was felt to be sufficient for any program that could be conceived. As might be expected, this did not turn out to be correct. As we will explain at a later point, there are two major ways to address this problem.

Intel Corporation has developed four additional processors of note. The first is the Intel 80286 processor, used on the IBM AT™ computers. From a normal programming viewpoint, this processor is the same as the 8086/88 chip. However, it has two salient differences, besides running much faster:

- The address bus is 24 bits wide, so it can potentially address 16,777,216 bytes of memory.

It has a protected addressing mode, so that the operating system can control the integrity of memory by prohibiting a user program from accessing memory that is not specifically assigned to it. This forms the heart of the hardware support for a system capable of switching from one task to another, such as Windows 3.1, when run on the 80286 processor.

The 80386, 80486, and Pentium processors are different:

- They have a 32-bit address bus, so can potentially address 4,294,967,296 bytes of memory.
- They allow for virtual memory operation. This feature allows these processors to run more than one program or process at a time, a true multiprogramming environment. The 80286 is also capable of multiprogramming, but is generally used as a task-switching system.

Only the program or parts of programs that are running are in memory. The remaining parts are on disk or in Extended memory (in protected mode, all memory is extended memory). The user can switch between tasks, as in Standard Mode. They may also place a task, such as a file download, in the background, while another process is being run interactively in the foreground on the screen. There is also a Virtual 8086/88 mode that allows for the operation of older 8086/88 software by emulating 8086/88 hardware features, in what is known as compatibility mode.

The Intel 80486 and Pentium chips are essentially a 80386 chip with additional memory caching and a built-in math coprocessor for floating point arithmetic. The 80286 and 80386 chips can use a math coprocessor as a separate chip. There are some cut down versions of the 80486 chip that do not contain a math coprocessor. Further, there are some DX2 versions of these chips that double the clock speed on the chip, which gives enhanced processor performance.

There are a variety of ways to manage and expand memory over the original 640K memory configuration. The first type of memory to consider is what is known as Conventional memory. This is memory in the first 640 Kilobytes of memory. This memory is usable by every version of the Intel family of processors.

Above the first 640 Kilobytes of memory there is the potential in many machines to install an additional 384 Kilobytes of memory to complete a 1-Megabyte block of installed memory. This 384-Kilobyte area is generally known as the adapter segment, or high memory. This area of memory is reserved for system software (the BIOS chip), video boards, disk controllers, and other expansion boards. Some computers, such as 80286, 80386, and 80486 machines that have 1 Megabyte of memory on their motherboard may have RAM map to this address range.

However, MS-DOS applications, MS-DOS, and Windows cannot access this memory unless you are using MS-DOS 5.0 or higher. This memory is not of direct interest to us in this discussion, except for how expanded memory works. Products like Quarterdeck Office System's QEMM-386™ product or MS-DOS 5.0 and higher can make use of this memory area for loading device drivers and memory resident programs, as well as part of the operating systems.

Early in the development of PCs, the need for more memory became apparent. This led to the development of the Lotus Intel Microsoft (LIM) standard for expanded memory. This technique uses a 64-Kilobyte portion of the adapter segment memory, called a memory window, to store or retrieve 8 Megabytes (LIM Version 3.2) to 32 Megabytes (LIM Version 4). The LIM standard also provides the ability to keep program code in this expanded memory. This technique is generally no longer used, but software that requires this feature may use an emulation component provided with the operating system.

Typical programs that use expanded memory are the early MS-DOS versions of word processors from WordPerfect Corporation, or the Lotus 1-2-3 spreadsheet system. The obvious problem with expanded memory is that the appropriate page must be retrieved from, or sent to, the expanded memory system. This has the potential for slowing memory access down to a speed lower than the access a program might have to the 640 Kilobytes of conventional memory. It is rather like the restaurant that has only three tables; it can serve as many people as you wish, but only three tables full at a time.

As previously pointed out, the 80286, 80386, and 80486 processors have 24 or 32 bits in their memory addressing system, and so can potentially access memory address space above 1 Megabyte. Unfortunately, versions of MS-DOS through 4.01, and most MS-DOS applications cannot take advantage of this Extended memory. The newest releases (MS-DOS 5.0 and above) remedy this problem and allow the user to move portions of the operating system into Extended memory. This allows the user to reclaim more of the 640 Kilobytes of Real memory for applications programs. Other special portions of the system software can be placed into Extended memory, if desired, under MS-DOS 5.0 and higher. Consult your MS-DOS Reference Manual for the details of this process. It can free up a useful amount of the lower 640K of memory.

The last type of memory is known as virtual memory. This is an area of disk space used in conjunction with Real and Expanded memory to hold portions of programs that are not being used. With the 80386, 80486, and Pentium processors, and Windows 3.1, this will allow for multiprogramming, where many programs may be running at once.

The Standard mode of operation is the normal mode of operation for a 80286 computer (or higher) with at least one Megabyte of memory, or 640K of conventional memory and 256K of extended memory.

If you are using a 80386, 80486, or Pentium-based computer and experience hardware compatibility problems, do not use non-Windows applications in Windows. If you have between 2 and 3 Megabytes of memory, you should run Windows in Standard mode.

REFERENCE

Andrew S. Tannenbaum, *Structured Computer Organization*, 3rd ed. Prentice-Hall, Englewood Cliffs, NJ, 1990.

2

Microsoft Windows for IBM PCs and Compatibles

Paul W. Ross
Millersville University

2.1 THE WINDOWS ENVIRONMENT

Microsoft Windows is an increasingly popular operating environment. It serves to isolate the user from the necessity of mastering a complex set of commands in the DOS operating system. The difficulty with learning MS-DOS commands are

- They are usually not intuitive.
- There are a great number of commands. Many of them are used infrequently.
- Many of the MS-DOS commands have additional parameters that must be learned and remembered.

Microsoft Windows provides a better environment, known as a Graphical User Interface, sometimes referred to as a GUI (pronounced gooie) interface. In dealing with a textual interface, for every computer application that you use you will have to remember the necessary command or commands required to invoke the application.

The model that Microsoft Windows uses for this graphical or visual environment is that of a desktop. On a desktop, you can arrange tasks, select one at will, put it back on the desk, or put it away when you are done. Microsoft Windows uses four elements to implement this desktop-like environment:

0-8493-2530-7/96/$0.00+$.50

- A graphics mode display instead of a text mode display.
- Icons, or stylized pictures, of the tasks or applications available to you.
- Lists of commands, known as menus. These menus are usually available as a drop-down menu from a list of commands at the top of the screen.
- A mouse; a graphics pointing device used to select and perform tasks. The buttons on the mouse are used in conjunction with the mouse pointer on the screen to activate different software functions.

The key to this environment is the mouse. Execution of an application in Microsoft Windows is simply a matter of clicking the mouse button twice (called double-clicking) on the desired application icon.

There are some alternative keystroke sequences instead of mouse actions. For example, if your fingers are already on the keyboard, the sequence **Alt-F4** and **Enter** (pressing the **Alt** and **F4** function key at the same time, and then pressing the **Enter** key) to exit Microsoft Windows may be faster than selecting the Microsoft Windows close option and confirming it by pressing the **Enter** key.

2.2 BASIC MOUSE OPERATIONS

The mouse is more than simply a pointing and selection device. When you use the mouse, you will use the following mouse operations:

- *Pointing*. The position of the mouse pointer on the Microsoft Windows desktop can be controlled by moving the mouse on the actual desktop. Point to an icon on the desktop or screen by positioning the pointer over the icon.
- *Clicking*. A click of the mouse is a rapid press and release of the left button on the top of the mouse. To select an icon, point to the icon, and then click the left mouse button. If you want to abort a selection, just move the mouse pointer to an empty place on the window and release the mouse button.
- *Pressing*. The pressing operation consists of pressing and holding down the button on the mouse. The operation is normally used as part of the dragging operation.
- *Dragging*. Dragging is a frequently used operation involving one of the following:

  1. The mouse pointer, or pointer, is placed over an icon, text, or menu item.
  2. The left button on the mouse is pressed and held.
  3. The mouse is moved.
  4. When the desired location or option is presented, the mouse button is released, causing the action to take place.

Dragging is typically used in moving an icon or in selecting commands from the pull-down menus by dragging and releasing the mouse when the desired menu item is highlighted. In Windows word processors, text is selected by dragging the mouse pointer over the text.

- *Double-Clicking*. Double-clicking is performed by quickly pressing and releasing the left mouse button twice. Double-clicking is frequently used as a shortcut for other actions, such as starting an application program from an icon.
- *Shift Clicking*. Shift clicking consists of holding down the shift key while the mouse is repeatedly pointed to icons or names of files. In this way, each new item is selected and becomes highlighted even as additional items are selected.

- *Ctrl Clicking*. The items selected by Ctrl clicking are noncontiguous items on the list, in contrast to the results produced by Shift clicking. Ctrl clicking consists of holding down the Ctrl key while the mouse is repeatedly pointed to icons or names of files. In this way, each new item is selected and becomes highlighted even as additional items are selected. The Ctrl clicking option is useful in selecting multiple file names when you wish to erase a group of files, or move them to another directory with the Microsoft Windows File Manager.

If you happen to be left-handed, the mouse button operation can be reversed. Go under the Main applications group, activate Control Panel, and click on the Mouse icon. The mouse speed and reversal of the mouse buttons can be set there.

2.3 SWITCHING AND MANIPULATING WINDOWS

Windows on the desktop have a definite structure that you need to understand before you can use Microsoft Windows effectively. Figure 2.1 shows a typical applications window.

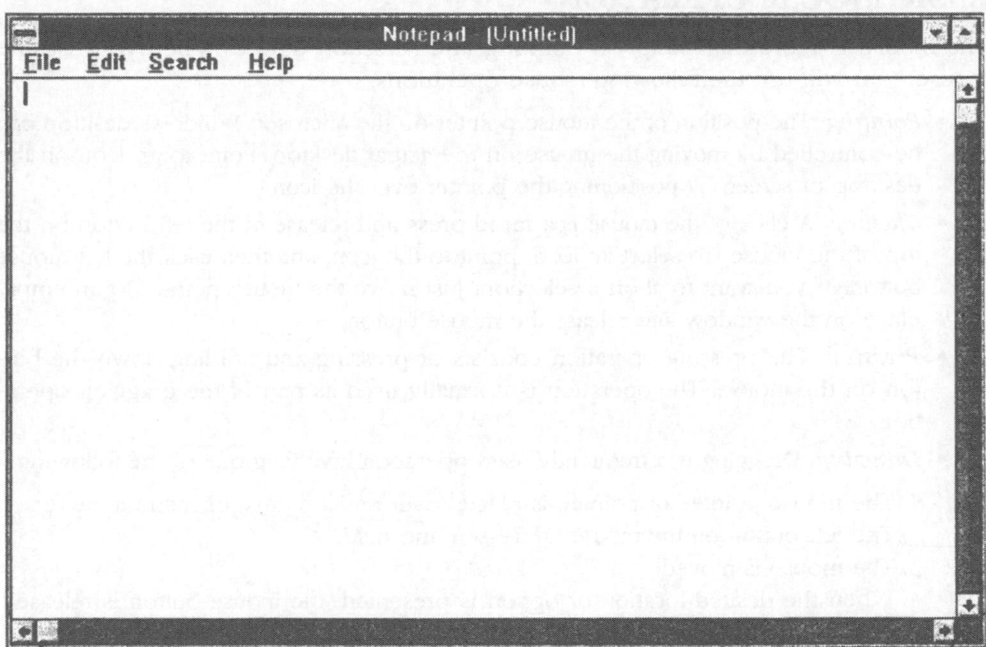

FIGURE 2.1 A typical applications window.

At the top of the screen, the name of the application or file group name appears on the title bar. The title bar contains the name of the application in use and any document used by the application.

The menu bar appears below the title bar. Selecting menu bar items activates choices within the Microsoft Windows application or system. At the upper left corner of the window is the control-menu box. The control-menu box is useful if you prefer to use your keyboard when you work with Microsoft Windows. Pressing **Alt-Spacebar** will activate this menu. Press the **Esc** key to exit from this menu.

The upper right corner of the window bears special attention. There you will find the Minimize and Maximize buttons, shown as up and down arrowheads. Clicking on the minimize button reduces the application to an icon. When the window of the previous figure is minimized, the icon on your desktop will look something like Figure 2.2.

FIGURE 2.2 A window, minimized, appearing as an icon.

Double-click on this icon to restore and resume the application. Clicking on the maximize button expands the window to full screen.

The single up arrow has changed to a combination up and down arrowhead. This means to restore to its original size. If the right button, containing two arrows, is clicked with the mouse, the window will shrink to part-screen size. The minimize button always shrinks the window to an icon. Double-click on the icon to restore it to the former size.

At the left and bottom of many applications windows you will find a scroll bar. The vertical scroll bar on the right allows you to move through a large document that you have created with one of the word processing or text editing applications of Microsoft Windows. Figure 2.1 shows a typical document with scroll bars.

The small box indicates where you are in the document. This box is called an elevator box. Dragging the elevator box up or down will allow you to move quickly through your document.

At the bottom of the screen is the horizontal scroll bar. This scroll bar allows you to move outside the margins of a word processing document, or to cells that are not visible on the window on a spreadsheet. This scroll bar will appear any time more of a window is outside your view.

Move a window by placing the mouse pointer on the title bar at the top of the screen. Drag the window with the mouse to another location. Release the mouse button to complete the process of moving the window.

Change the size of a window by placing the mouse pointer on the bottom or top edge, or either side of the window. The mouse pointer now changes to a small double-ended arrow. If you drag this arrow (hold down the left mouse button), the edge of the window will move. Release the mouse button when the edge is at the desired location.

Similarly, the four corners of a window can be moved. Placing the mouse cursor on any of the four corners allows you to change the size of the window by moving both of the edges next to the corner chosen.

If you move the corners, you can see any applications active on the desktop. You can switch to the desktop and activate another application by clicking on the portion of the desktop that shows. The application that you were working on now ap-pears behind the application now active. This is one way that you can switch applications in Microsoft Windows without having to terminate your current application.

2.4 PROGRAM GROUPS

Applications are arranged in the *Program Manager* by what are called Program Groups. Program Groups are sets of applications that have similar purposes. When Windows set-up your system several groups were made: Main, Accessories, Games, Windows Applications, Start-up, and non-Windows Applications. The table below identifies which programs typically are placed in which groups following installation.

Group	Application
Main	File Manager
	Control Panel
	Print Manager
Group	Application
	Clipboard
	DOS Prompt
	Windows Set-up
Accessories	Write
	Paintbrush
	Terminal
	Calculator
	Calendar
	Cardfile
	Clock
	Notepad
	Recorder
	PIF Editor
Windows Applications	Excel
	Word
Non-Windows Applications	Any DOS Applications you selected during installation.

To open any of the Groups, double-click on the Program Group Icon, as shown in Figure 2.3.

FIGURE 2.3 A Program Manager Group icon.

A window will open up, showing all the programs in the group. These windows are called *document windows*, and are "windows within windows."

You can create new groups, add applications to groups, and rearrange windows any way you would like. Pull down the File menu from the Program Manager, and select "New." Follow the screen prompt for creating a program group. Applications can appear in more than one group if you would like. Use the "Copy" function from the File menu under the Program Manager.

To reduce the window to an icon, click on the down arrow in the upper right corner of the window. To move the application window for a group, drag the window to where you would like it to be placed.

2.5 THE MICROSOFT WINDOWS FILE MANAGER

Directories are the mechanism that MS-DOS provides to keep your files and data organized. With a directory, you can keep all associated files in one convenient place. You can also create directories yourself from within Microsoft Windows by use of the Microsoft Windows File Manager. The File Manager icon is found in the Main program group. When Microsoft Windows 3.1 was installed on your system, Microsoft Windows applications were placed in logically similar groups, called program groups. A program group is a concept comparable to a disk directory. Instead of organizing files, it is used to organize applications programs of a similar nature. The File Manager icon is shown in Figure 2.4.

FIGURE 2.4 File Manager icon.

Double-click on the File Manager icon to start the File Manager program. After the File Manager has been started, you will see a screen somewhat like Figure 2.5.

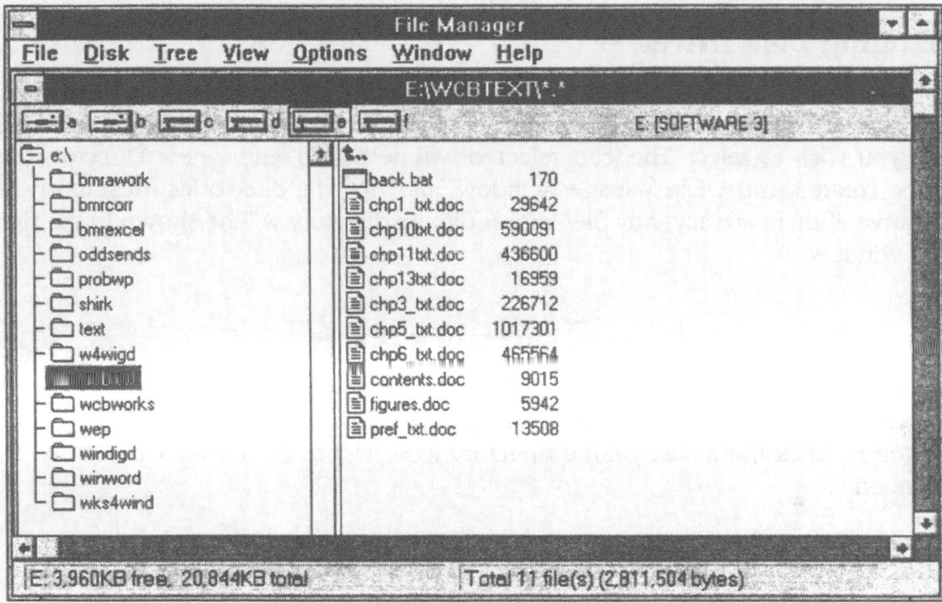

FIGURE 2.5 File Manager Directory Tree window.

The Directory Tree window contains the following parts:

- The disk-drive icons about one-third down on the screen represent the disk drives on your computer. A drive letter follows each icon, usually A: and B: for the floppy disk drives, and C: and upward for the hard drive, RAM drives, and network drives. The boxed icon and letter represent the currently logged disk drive. If you are connected to a network with your computer, the word NET will appear on the network drive icon. Any RAM drives or CD-ROM drives will also appear as drive icons on the Directory Tree window.
- The volume label is the name that identifies your disk. The volume label appears in square brackets in the status line. Volume labels are especially useful for labeling floppy disks to indicate their general contents.
- The directory path for the Directory Tree window appears on the status line below the drive icons. The directory path for a directory window appears in the Title bar for each document window.
- The current directory is the directory that is boxed in the Directory Tree window. The File Manager commands affect this directory. Clicking on a directory icon will change the current directory to the selected directory.
- The directory icons represent each directory on the currently logged disk drive. The directories will be listed in alphabetical order beneath the root directory (symbolized by characters such as **C:**). If a directory contains one or more subdirectories, a plus sign appears inside the directory icon. Each directory icon looks like a file folder.
- A scroll bar appears on the right-hand side of the document windows if there are more directories and subdirectories than will fit in the Directory Tree window.
- The right-hand side of the split screen shows all files contained in this particular directory.

If you click on the expansion arrow in the upper right corner of the window, you can expand it to full-screen size.

Changing Disk Drives

The File Manager starts by showing your currently logged disk drive. You can switch to any other disk drive shown in the Directory Tree window. Simply click on the disk drive you wish to select. The icon selected will be boxed, and a new Directory Tree will be created in the File Manager window, showing the directories present on that disk drive, if there are any. Any files present in the directory will be shown in the right-hand window.

Expanding and Collapsing Directory Levels

The initial directory tree shown in the Directory Tree window will be the first level of subdirectories. You can expand the Directory Tree to show the further levels of subdirectories. Click the mouse on the directory icon. Figure 2.6 shows a Directory Tree expanded.

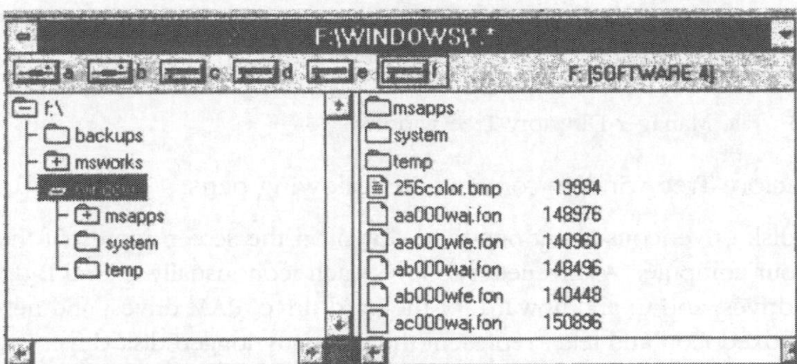

FIGURE 2.6 Expanded Directory Tree showing subdirectories.

Clicking on the root directory symbol (such as **C:**) will collapse the Directory Tree. Click on another directory symbol to show the contents of another directory. You will notice three different icons in front of the file names:

- A filing folder; this is another subdirectory. Clicking on the parent file folder will take you back to the parent directory (the one above it in the directory hierarchy).
- An icon resembling a piece of paper; this is a document or other file created by some application.
- An icon resembling a square box; this is an executable program, as the file name for executable programs or applications ends in .COM or .EXE. It is also used for .BAT (batch) files.

Launching Applications from the Directory Window

Applications can be launched or started from the File Manager window instead of double-clicking on an icon in a program group. Double-click on the icon associated with the application. The icon associated with an application is a square box. If the application is not already running, it will be launched. If the application is already running, in particular a Microsoft Windows application, you will not be able to launch it a second time. Go back to your main screen and double-click on the icon associated with the particular Microsoft Windows application to make it active.

Moving and Copying Files

You may find that a file should be in another directory. The File Manager in Microsoft Windows provides a mechanism to copy and move files. You can move files by performing the following steps:

1. Double-click on the directory icon for the source directory; the right-hand window shows the files in this directory.
2. Drag the icon of the file to be moved from the list on the right-hand side to the target directory.

If there is a file in the target window with the same file name as in the source window, it will be overwritten without warning. If you have the confirmation check boxes set (use the Options menu item at the top of the screen), Microsoft Windows will ask you to confirm each step.

Copying files from one Directory Window is done with a similar strategy. However, while you drag the file icon from the source directory window to the target directory window, hold down the CTRL key while moving the mouse. If you are copying files or directories to a different disk drive, you can simply drag with the mouse. You do not have to use the CTRL key.

Deleting Files

First, to mark the file or files to be deleted:

- Click on the name of the file to be deleted.
- Shift-click for a series of files that are ordered consecutively in the Directory Window.
- Control-click for a series of files that are not ordered consecutively in the Directory Window.
- Pull down the File menu and select Delete, or press the Delete key on the keyboard.

The Windows File Manager will now ask you to confirm the deletion. The Windows File Manager will ask you on a file-by-file basis for the confirmation of the deletion of each file.

Creating and Deleting Directories

Microsoft Windows uses directories to help keep your files organized in a logical manner. To create a directory, move to point in the Directory Tree where you wish to attach a directory. The screen might look like Figure 2.7.

To create a new directory:

- Click on the File menu
- Select the Create Directory option.

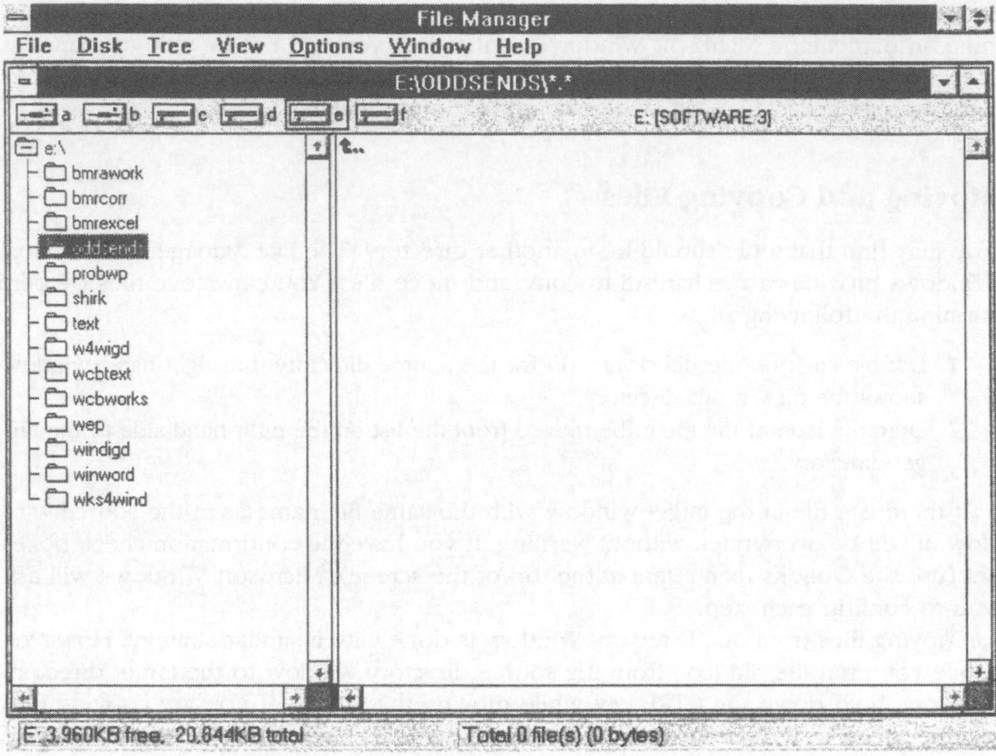

FIGURE 2.7 Directory Tree before attachment of a new directory.

You will be prompted for the name of the directory. The dialog box will look like Figure 2.8 for the creation of a new directory.

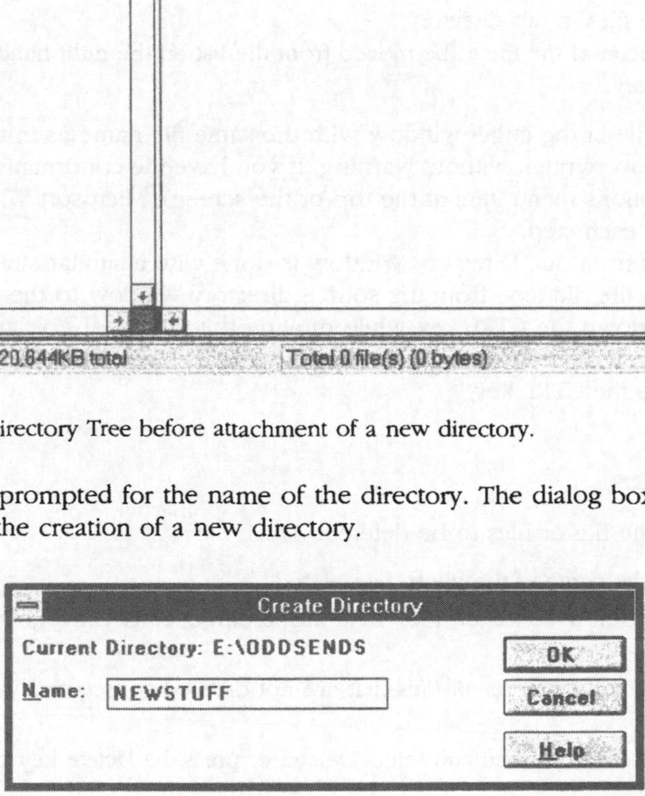

Create Directory
Current Directory: E:\ODDSENDS
Name: NEWSTUFF
OK
Cancel
Help

FIGURE 2.8 Prompt Box for creating a directory.

After accepting the name NEWSTUFF (or some other appropriate name), press the **Enter** key. The new directory will be created. When expanded, the Directory Tree looks like Figure 2.9.

If you no longer need a directory, you can remove it. The removal of a directory is a two-step process.

1. Click the directory to select it.
2. Press the Delete key or select Delete from the Files menu in the File Manager.

If you have the confirmation options activated, you will be asked to confirm the deletion of each file in the subdirectory and then the subdirectory itself.

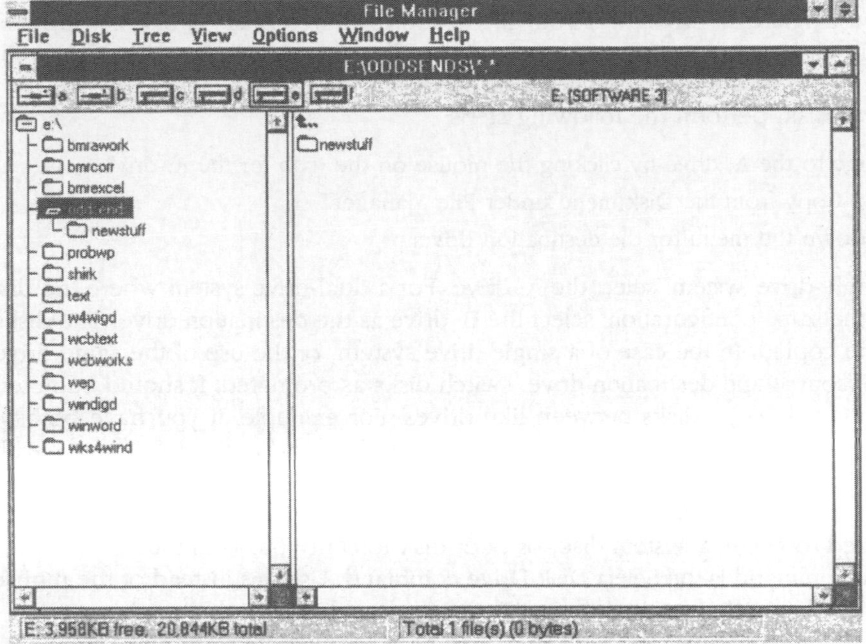

FIGURE 2.9 Directory Tree after the attachment of a new directory.

Formatting, Copying, and Labeling Floppy Disks

Since a floppy disk must be formatted before it can be used to store data, you must go through the process of formatting it. The formatting process is done through the Disk menu in the File Manager. If you pull down the Disk menu, you have five options:

1. *Copy* – to make copies of floppy disk for backup or archival purposes.
2. *Label* – to put identifying information on the floppy disk. This can also be used to change or create a label on your hard disk.
3. *Format* – to prepare a floppy disk to receive information.
4. *Make System Diskette* – to prepare a boot disk for your system.
5. *Select Drive* – to select a current logged disk drive. Use instead of a mouse.

To format a floppy disk, select Format from the Disk menu. You will be prompted for the floppy disk drive that contains the unformatted floppy disk, or the floppy disk you wish to reformat. Select the desired disk drive, either A: or B: from the drop-down menu.

If your floppy disk drive supports two recording densities, select the desired density. The Windows File Manager will then format your disk. The Windows File Manager will give you an opportunity to format additional disks.

The label feature allows you to place an appropriate name or identifying information on the disk. The label is limited to 11 characters (letters, numbers, spaces, and some special symbols).

The third command is the Copy command. The Copy command allows you to copy a floppy disk. To use the Copy command for copying floppy disk, place the source disk in the A: drive and a disk (either formatted or unformatted) in the B: drive. If you only have a single disk drive, or if your drives are for different sized disks, you must do a single disk copy be selecting the A: drive as the destination drive as well. Alternatively, the B: drive can be both the source and destination drive.

To copy a disk, perform the following steps:

- Change to the A: drive by clicking the mouse on the icon for the A: drive.
- Select Copy from the Disk menu under File Manager.
- Pull down the menu for the destination drive.

For a single-drive system, select the A: drive. For a dual-drive system where the drives are of the same configuration, select the B: drive as the destination drive. The disks will now be copied. In the case of a single drive system, or the use of the same drive as both the source and destination drive, switch disks as prompted. It should be noted that you can only copy disks between like drives. For example, if you have both 3-1/2″ and 5-1/4″ drives, you cannot copy a disk from the 5-1/4″ to the 3-1/2″ drive with the Copy command, but must copy the files one-at-a-time.

The next command is the Make System Diskette under the Files menu. This command is used to create a system disk, or boot disk to start your computer.

The final command is the Select Disk Drive command. Use this instead of the mouse to change the currently logged drive.

2.6 ALTERNATIVE MODES OF OPERATION

Windows is designed for operation on three different hardware environments. These include the 80286 processor with 640 Kilobytes of memory and optional extended memory, or the Intel 80386, 80486, or Pentium processors in 386 Enhanced Mode, 640 Kilobytes of real memory, and optional extended memory. Operation of Windows, from the systems viewpoint, is slightly different on these different hardware configurations.

Processors and Memory

It is useful to understand how memory is used by the various members of the Intel Corporation processor family. In particular, we need to consider the 80286, 80386, 80486, and Pentium processors. In the historical development of computers based on these chips, the 8086/88 chip was used in the PC dual floppy machines, and the XT machine that supported a hard disk system. Windows 3.1 will not run on computers based on the older 8086/88 processor chip.

The 80286 processor chip is generally used in machines known as AT™ class machines, and the newer 80386, 80486, and Pentium chips, in various versions are used in IBM's upper-level PS/2™ line, and various clone machines. From the normal user viewpoint, it is a matter of indifference what machine most software runs on, as only the newer software such as Windows 3.1 takes substantial advantage of the features of the 80286, 80386, or 80486 processors.

There are two issues of which the Windows 3.1 user should be aware:

- What processor is being used (80286, 80386, 80486, or Pentium) by the computer system?
- What amount and kind of memory are being used by the computer system?

The Processor and Conventional Memory

The earliest machines were based on the Intel Corporation 8086/88 processor. This processor is common in the IBM PC (TM) and XT class machines.

- It has a 20-bit memory address, and so is capable of addressing 1,048,576 bytes of memory.
- The processor was designed to handle only a single task.

Because of these design limitations, Windows 3.1 will not run on the older 8086/88 processor, but only on the newer 80286, 80386, and 80486 processors.

Due to some design choices in how this memory was used, only about 640K of memory is usable for programs running under DOS. This 640K memory is known as conventional memory.

Intel Corporation has developed three additional processors of note. The first is the Intel 80286 processor. From a normal programming viewpoint, this processor is the same as the 8086/88 chip. However, it has two salient differences:

- The address bus is 24 bits wide, so it can potentially address 16,777,216 bytes of memory.
- It has a protected addressing mode, so that the operating system can control the integrity of memory by prohibiting a user program from accessing memory that is not specifically assigned to it. This forms the heart of the hardware support for a system capable of switching from one task to another, such as Windows 3.1.

When the 80286 processor is run in protected mode, it is known as Standard Mode for Windows 3.1.

The Intel 80386, 80486, and Pentium processors go one step further.

- They have a 32-bit address bus, so can potentially address 4,294,967,296 bytes of memory.
- They allow for virtual memory operation. This feature allows the 80386, 80486, and Pentium processors to run more than one program or process at a time, a true multiprogramming environment. Only the program or parts of programs that are running are in memory. The remaining parts are on disk or in Extended memory. Not only can the user switch between tasks, as in Standard Mode, but they may place a task, such as a file downloads, in the background, while a process such as Windows Write is being run interactively in the foreground on the screen.

The Intel 80486, and Pentium chips are essentially a 80386 chip with additional memory caching and a built-in math coprocessor for floating point arithmetic. The 80286 and 80386 chips both use a math coprocessor as a separate chip.

There are a variety of ways to manage and expand memory.

Memory, Conventional, Expanded, Extended, and Virtual

It is helpful to understand how the memory in your computer is used by Windows. The first type of memory to consider is what is known as Conventional memory. This is memory in the first 640 Kilobytes of memory. This memory is usable by every version of the Intel family of processors. Windows 3.1 and other DOS applications can automatically use all the conventional memory in your computer.

High Memory

Above the first 640 Kilobytes of memory there is the potential in many machines to install an additional 384 Kilobytes of memory to complete a 1-Megabyte block of installed memory. This 384-Kilobyte area is generally known as the adapter segment, or high memory. This area of memory is reserved for system software (the BIOS chip), video boards, disk controllers, and other expansion boards. Some computers, such as 80286, 80386, and 80486 machines that have 1 Megabyte of memory on their motherboard may have RAM map to this address range.

However, DOS applications, DOS, and Windows cannot access this memory unless you are using MS-DOS 5.0 or higher. This memory is not of direct interest to us in this discussion, except for how expanded memory works. Products like Quarterdeck Office System's QEMM-386 product or MS-DOS 6.2 can make use of this memory area for loading device drivers and memory resident programs, as well as part of the operating systems.

Expanded Memory

Early in the development of PCs, the need for more memory became apparent. This led to the development of the Lotus Intel Microsoft (LIM) standard for expanded memory. This technique uses a 64-Kilobyte portion of the adapter segment memory, called a memory window, to store or retrieve 8 Megabytes (LIM Version 3.2) to 32 Megabytes (LIM Version 4.). The LIM standard also provides the ability to keep program code in this expanded memory.

To use expanded memory, you need a memory board, and software called an expanded memory manager. This will trap references to this additional memory and place the desired piece in the 64-Kilobyte window into expanded memory. The expanded memory system can be viewed as a paging system, where one 64-Kilobyte page of expanded memory is made available at a time to the program. This technique of bringing in a portion of memory when needed is often known as bank switching. Bank switching through expanded memory was the only memory expansion technique available before the development of the Intel 80286 and higher processors. It is also possible to use a value larger than a 64-Kilobyte window. In this event, the terms small and large frames are used. Generally, expanded memory schemes were used only on older PC and XT (8088 processors) class machines. Expanded memory may be used on AT and higher level machines (80286, 80386, 80486, and Pentium processors).

Typical programs that use expanded memory are the word processors from WordPerfect Corporation, or the Lotus 1-2-3 spreadsheet system. The obvious problem with expanded memory is that the appropriate page must be retrieved from, or sent to, the expanded memory system. This has the potential for slowing memory access down to a speed lower than the access a program might have to the 640 Kilobytes of conventional memory. It is rather like the restaurant that has only three tables; it can serve as many people as you wish, but only three tables full at a time.

Extended Memory

The 80286, 80386, 80486, and Pentium processors have 24 or 32 bits in their memory addressing system, and so can potentially access memory address space above 1 Megabyte. Unfortunately, versions of DOS through 4.01, and DOS applications cannot take advantage of this extended memory. The newer releases, DOS 5.0 and above, remedy these problems and allow the user to move portions of the operating system into extended memory to reclaim more of the 640 Kilobytes of real memory for applications programs. Other special portions of the system software can be placed into extended memory, if desired, under DOS 5.0, 6.0, 6.1 or 6.2.

Consult your DOS Reference Manual for the details of this process. It can free a useful amount of the lower 640K of memory for use by programs.

Windows is able to make use of this extended memory by using a memory manager that conforms to the Microsoft Extended Memory Specification (XMS). With this enhancement, all available memory becomes accessible to Windows 3.1 applications.

Virtual Memory

The last type of memory is what is knows as virtual memory. This is an area of disk space used to hold portions of programs that are not being used. With the 80386, 80486, and Pentium processors, and Windows 3.1, this will allow for multiprogramming, where many programs may be running at once. This is described later under the topic of 386 Enhanced mode operation.

Standard Mode

The Standard mode of operation is the normal mode of operation for a 80286 computer (or higher) with at least 1 Megabyte of memory, or 640K of conventional memory and 256K of extended memory.

If you are using a 80386, 80486, or Pentium-based computers and experience hardware compatibility problems, do not use non-Windows applications in Windows, you should run Windows in Standard mode. Most systems come shipped with 4 or more megabytes of extended memory, so this should not be a problem.

To force Windows to start in Standard mode, the command line should look like this:

```
WIN /S
```

386 Enhanced Mode

The final mode for starting Windows is to force 386 Enhanced mode. Your 80386, 80486, or Pentium-based computer will automatically start in 386 Enhanced mode if you have at least 640K of conventional memory and 1024K of extended memory. If you have at least 1 Megabyte of memory (640K conventional and 384K extended), you can force 386 Enhanced mode by starting Windows as follows:

```
WIN /3
```

You should note that if you have less than 2 Megabytes of memory, Windows will probably run more slowly in 386 Enhanced mode than in Standard mode. The use of mode switches is optional. If you do not specify a mode, Windows examines your system configuration and starts in the mode most appropriate for your system.

2.7 CUSTOMIZING THE WINDOWS ENVIRONMENT

Before making any changes to your Windows set-up, make sure to have a complete backup of all the files on your computer system. Use the DOS BACKUP or a similar utility program to make the backup.

Modifying the Initialization Files

Each time you start Windows, a set of files is read informing Windows of how to set various configuration options. The sets of files that define your Windows software and the way in which it works with the system are called Initialization files. Windows initialization files are ASCII text files, and have the following format:

```
[section name]
keyname=value
```

The [section name] is used to break up the files into logical groups. The keyname=value line defines the value of each setting. A keyname is the name of a setting. The value for the keyname can be an integer, a string, or a quoted string, depending on the setting. For example, the following lines from WIN.INI set the colors. Note the format of the lines in the file as defined above.

```
[colors]
HiLight=0 0 0
HilightText=255 255 255
Background=0 0 0
AppWorkspace=0 128 128
Window=255 255 255
WindowText=0 0 0
Menu=255 255 255
MenuText=0 0 0
ActiveTitle=0 64 128
InactiveTitle=193 193 193
TitleText=255 255 255
ActiveBorder=192 192 192
InactiveBorder=255 255 255
WindowFrame=0 0 0
Scrollbar=255 255 128
```

The changes that you make to the initialization files do not go into effect until you exit Windows and start the program again. You will find two initialization files in the directory where Windows was installed: SYSTEM.INI and WIN.INI.

SYSTEM.INI

The SYSTEM.INI file informs Windows about the hardware that makes up your system, and how it should interact with each component. This is where the hardware selections you made during installation of the software are filed for Windows reference. These selections define the video mode (VGA, EGA, CGA, Monochrome), keyboard type, system type, mouse, and the other parts that make up your system. This file is best modified using the setup program and not manually edited. Too many parts of this file are cryptic, and if incorrectly changed could cause Windows to misbehave or not run at all.

WIN.INI

The WIN.INI file informs Windows about the software installed on your computer system, and how it should interact with the programs. WIN.INI is where Windows saves the selections you made in the Control Panel for such things as Colors, Wallpaper, Patterns, and the mouse tracking speed.

Unlike the SYSTEM.INI file, there are interesting parameters that you might want to change or add. To make modifications to this file, use the Windows Notepad program or SysEdit, described later in this chapter. Start Notepad by double-clicking on the program icon, then from the file menu, load WIN.INI. WIN.INI is located in the WINDOWS directory.

RUN and LOAD

You can run and load programs automatically when you start Windows. To do this, find the lines in the WIN.INI file that read "LOAD=" and "RUN=." They will be near or at the top of the file.

The RUN and LOAD settings are slightly different in the way in which they function. LOAD is used to specify one or more applications that will run as icons when Windows is started. RUN starts applications in their application Window on the desktop.

2.8 SysEdit

SysEdit is a special editor that will allow you to get at most of the parameters in the .INI and other system files. SysEdit is not found in any of the Group windows of the Program Manager when you installed Windows. The program is located in the SYSTEM directory under the Windows directory:

```
C:\WINDOWS\SYSTEM\SYSEDIT.EXE
```

In this example, Windows was installed on the C:\ drive, in a directory called WINDOWS. You may want to add this program to one of the Group windows in the Program Manager. To add the SysEdit icon, go to the Program Manager. Select the File menu and the New option. Select Browse. Navigate to the C:\WINDOWS\SYSTEM subdirectory. Click on the file SYSEDIT.EXE. Accept the selection. These commands will place the SysEdit icon in the selected program group.

To start SysEdit, double-click on the Icon, if you added it to a Group. When the program starts, you will see an application that looks very similar to both the Write program, and the Notepad editor. This screen is shown in Figure 2.10.

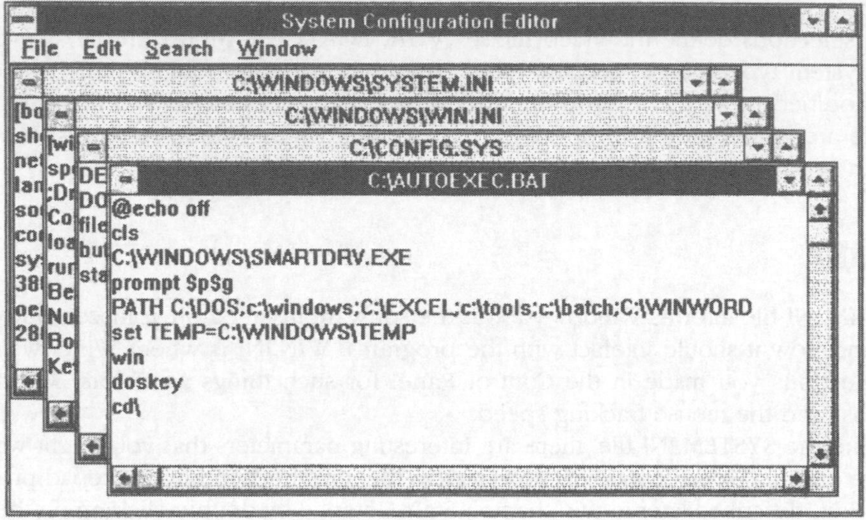

FIGURE 2.10 The SysEdit application window.

You will notice that four files are loaded at the same time each in its own document window:

- WIN.INI
- SYSTEM.INI
- AUTOEXEC.BAT
- CONFIG.SYS

SysEdit can be used to modify only one of these four files. To modify one of the files you simply click on its window and click the pointer anywhere in the file. The editor works the same as Write and Notepad applications. Most of the Notepad commands are also available from the Menus. The only additional menu of commands is Window. These commands will allow you to switch to any of the files, and arrange the files in an orderly way across the screen either tiled or cascaded.

When you have completed edits to a file, you should save the changes using the Save command from the File menu. If you try to exit the program after having made changes, you will be prompted to save the files that have been modified if you wish.

Files that are modified with SysEdit are automatically backed up in case you make a mistake. The backed up files have an extension of SYD, and can be erased when you are convinced that your changes are correct.

2.9 OPTIMIZING SYSTEM MEMORY

Optimizing your system's memory to run Windows can greatly improve the speed at which programs run, and how fast the system performs. There are a considerable number of parameters that you can adjust. For a more detailed discussion of this topic, see Chapter 1 on MS-DOS for IBM PCs.

Memory

By far the most important aspect of tuning a Windows system is making sure that memory is set up in an optimal configuration. If you are considering expansion of a system running Windows, the first component to consider must be to increase the amount of available memory.

Windows is a memory-intensive program. The more memory Windows has available the better. The first step in optimizing your system for better performance is to make sure Windows has as much memory available to run as possible. Only memory that is available when Windows is started will be available during your Windows session.

To make sure Windows has as much memory as possible check to see that your system is dealing with the following items properly:

- Memory Resident Programs—Memory resident software, also called pop-ups or TSRs, are programs that are loaded then sit in the background until needed. Typically these programs pop-up onto the screen using some kind of special keystroke. The problem with these kinds of programs is that they use memory, sometimes a significant amount. The best solution to using these programs with Windows is to run them after you have started Windows. This will give Windows some additional memory, and better manage the memory that the memory resident software uses when it is run. Often these programs are started in the AUTOEXEC.BAT file. Most of the features that are provided by typical TSRs are provided as part of the Windows environment, and will no longer be needed.
- Device Drivers—In your CONFIG.SYS file you may see several lines that start "DEVICE=" these lines are loading device drivers for your system. A device driver is a file that defines to how the system is to communicate with one of the system components. To optimize Windows in this respect, load only those that are absolutely necessary to the functionality of your system. You should exclude such items as print spoolers, RAM disks, and peripherals that are seldom used. If Windows is the only program you are using that needs or uses the mouse, you can delete any lines that load programs for it. Windows does not need these files to make use of the mouse. It knows about it from the SETUP program and the SYSTEM.INI file.

Lines to include in CONFIG.SYS:

```
BUFFERS=20
STACKS=0,0
FILES=30 (only if you are using DOS version 3.3 or greater)
```

There may also be two parameters called SMARTDRIVE and HIMEM.SYS. You should leave these two lines in the files as they are. Windows attempt to boost performance from your hard disk drive and memory, respectively. The default values Windows uses for these files will be optimal for your system.

How Much Memory?

How can you tell how much memory you system is using, and how can you tell if the changes you have made gave you additional memory? Look at the results of the Help command in the Program Manager application. Select the Help command from the

Menu Bar, then the About Program Manager command. Selection of the command will bring up a dialog box with the information. The "about" screen is shown in Figure 2.11.

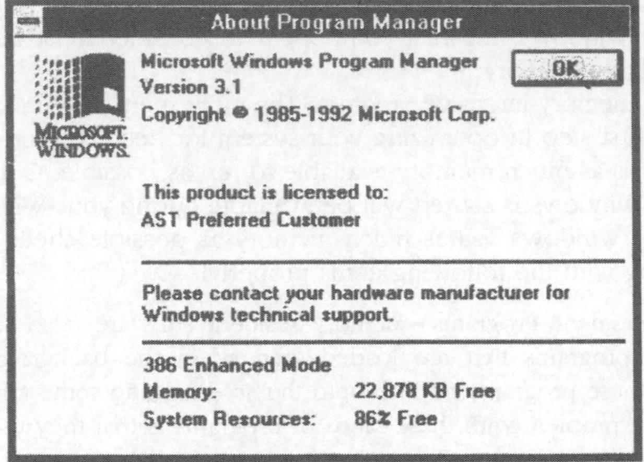

FIGURE 2.11 The About dialog box.

To check the results of your attempt at memory optimization, check these statistics before you change anything, then look at the numbers again after you have made some modifications. Be sure when you run the check that only the Program Manager is running to get a true reading on available memory; close other applications if necessary.

2.10 USING SETUP FOR MAINTENANCE

The Setup program will allow you to maintain your Windows software without having to reinstall it each time your system changes. This process will keep Windows optimized because it will run for the specific hardware you are using.

To run the application, double-click on the Setup icon in the Main Group Window. When the application starts, you will see a summary of the current hardware configuration. This includes the Display (monitor), Keyboard, Mouse, and Network (if installed). The Windows Setup application screen is shown in Figure 2.12.

To make a change to one of the settings click on the Options menu, and select the Change Systems Settings command. A dialog box will open giving you a Drop Down List Box for each of the Options. This dialog box is shown in Figure 2.13.

Windows Setup

Options Help

Display:	640x480x256 for CHIPS Super VGA 82C452
Keyboard:	Enhanced 101 or 102 key US and Non US
Mouse:	Logitech serial mouse
Network:	No Network Installed
Swap file:	Temporary

FIGURE 2.12 The Setup application window.

Change System Settings

Display: 640x480x256 for CHIPS Super VGA 82C452

Keyboard: Enhanced 101 or 102 key US and Non US keyboards

Mouse: Logitech serial mouse

Network: No Network Installed

OK Cancel

FIGURE 2.13 The Change System Settings dialog box.

To change a setting, click on the Name (Display, Keyboard, Mouse, or Network), then either use the arrow keys on the keyboard to scroll through the selections, or click on the down arrow to reveal the Drop Down List box. If you are making several changes make all of them, then click on OK. The Exit dialog box is shown in Figure 2.14.

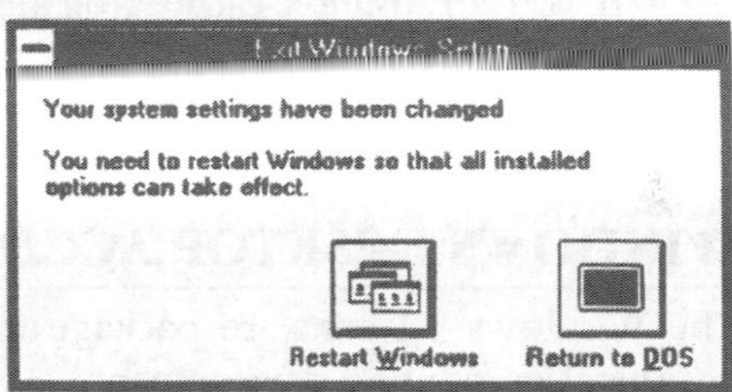
Your system settings have been changed
You need to restart Windows so that all installed options can take effect.
Restart Windows
Return to DOS

FIGURE 2.14 The Exit Windows Setup Dialog box.

When you click on OK, a final dialog box will appear asking you to restart Windows, or return to DOS. This is necessary because you have made changes to the way in which Windows is set up.

2.11 DISK CACHING

Disk caching is one way you can increase the speed with which Windows runs on your personal computer. A disk cache is a program that enhances the performance of the hard disk drive connected to your computer system. Performance is increased by holding files and information that is read frequently from the hard drive in memory.

SMART Drive

The disk caching program that comes with your Windows software is called SMART Drive. It is specifically designed for use with Windows, and is loaded in your CONFIG.SYS file.

Most likely SMART Drive was installed during the installation process. To find out, look for the following line in your CONFIG.SYS file: DEVICE=C:\WINDOWS\SMARTDRV.SYS. Even if the cache program was installed, you may be able to do some fine tuning on your own to increase system performance.

Start SysEdit, add the following line to your CONFIG.SYS file:

```
DEVICE=C:\WINDOWS\SMARTDRV.SYS 1024 256
```

Adding this line tells the system where the file is with the path and file name. Make sure this line comes after any line specifying another device called HIMEM.SYS.

The two numbers following the file name specify the amount of memory the cache can use. The first number is the normal cache size (1024 in the example above). This number specifies how much memory is used when the program is started, and the size of the cache when Windows is not running. The second number (256 in the example) is the minimum cache size. The cache size is reduced in Windows if a Windows program needs the memory.

Follow these guidelines in setting up a disk caching program (including SMART Drive):

- Caches smaller than 256K will generally be of little help in optimizing Windows. The smaller the setting, the less information that can be stored in the cache.
- Setting the cache size larger than 1024K is probably not effective use of memory. The memory you might consider allocating above 1024K would probably be of more use to Windows programs and non-Windows programs that you would like to run.
- If you are using a cache program other than SMART Drive, be sure to check the specifications to make sure it does not interfere with Windows memory management. Failure to do so may result in Windows not running.

2.12 WINDOWS DESKTOP ACCESSORIES

The Windows 3.1 software package includes several very useful applications for the graphical computing environment.

The Clipboard

To cut and paste information between applications Windows 3.1 uses an application called the Clipboard. Cutting information is selecting information or graphics within a program; pasting is taking the information being cut and placing it in another program. To start the Clipboard from the Program Manager, open the Main Group, and double-click on the Clipboard icon.

The Clipboard application is different from other Windows applications in that Windows uses it transparently. When you cut graphics or text from an application, Windows is using the Clipboard without really opening the application. The reason for this difference is that Clipboard is really a temporary holding place while information is being transferred between applications. It is not necessary to start the Clipboard application to employ its capabilities.

Window applications have Cut and Paste commands built into the application. To find the commands click on the Edit command from any Windows application Menu bar. To cut and paste information from one application to another follow the following steps:

- Select the information you want to copy. This is usually possible by dragging the cursor across information to be copied, or clicking on the graphic.
- Select Copy from the applications Edit menu, or press the keyboard combination CTRL-C.
- Open or make active the application you wish to paste the information to.
- Position the cursor where you want to paste information, then select Paste from the Edit menu. Using the keyboard, press the keyboard combination CTRL-V.
- This is a general description of how a cut and copy are made. Specific Application documentation should be consulted if you cannot get the operation to work correctly.

To view the contents of the Clipboard, open the application from the Main Group in the Program Manager. In the Clipboard's application window you will see the item that has been cut. It is a good idea to clear large items from the Clipboard after making the paste. To clear the Clipboard, open the application if it is not already, and click on the Delete command on the Menu bar.

If you would like to save the image before deleting the image from the Clipboard, select File from the menu bar, and click on Save As. At the prompt, type in a path and file name for the Clipboard file, then click on OK. Clipboard files can be retrieved and used later by opening the Clipboard, selecting File from the menu bar, and clicking on Open. A Dialog Box will open prompting for the Clipboard file to load. Once loaded thc saved image can be Pasted into your applications.

The Calculator

Included with your Windows 3.1 software is an application called Calculator. This is one of the simplest applications to use, and one of the most useful. The Calculator is a program that looks and performs just like hand-held calculators. The two modes that can be used are the Standard Calculator or the Scientific Calculator.

The Calculator has all the standard mathematical functions found on most desktop models. Entering numbers into the Calculator can be performed in three ways:

- Pasting information into the display from other applications.
- Entering the number from the Keyboard.
- Click on the numbers in the application window with the mouse.

Entering numbers or the mathematical function to use can be done with either the mouse or the keyboard. To use the keyboard, simply type in the number and functions. Using the mouse, click on the numbers and mathematical functions in the application window.

Calendar

Calendar is an organizer. With this program you can schedule appointments, meetings, and other daily events. There is also an alarm feature that serves as a reminder. Calendar events can be saved in files to be shared or used over and over again. Calendars can be saved, printed, and used in conjunction with all the other Windows applications on your system.

Cardfile

Cardfile is a simple database. The format of Cardfile is an expandable set of electronic index cards. Each card holds the information for the database entry. Cards are always sorted by the Index that is the title of the card. Cardfile databases can be set-up for many subjects.

To add a card to the stack, click on Card in the Menu Bar, then select Add from the menu. A dialog box will ask for the index for the new card. An index is the title of the card. For example, if the Cardfile database is an address list the index would most likely be the person's name. After filling in the index, click on OK. The card will be placed at the front of the stack, ready for you to type in information or paste in a picture.

To delete a card, first move the card to the front of the stack. Next, click on Card in the Menu Bar, and select Delete from the menu. A confirmation dialog box will prompt you to verify deleting the card.

After the card is created, you will see a flashing vertical line in the information area of the card. This is the insertion point. Begin typing the information just as you might write it down on an index card. To add a picture, copy the graphic from the another application. Before pasting the picture into Cardfile click on Edit in the Menu Bar, and Picture from the menu. You can then paste the graphic into the card (Edit Menu, Paste command). To move the graphic within the card drag the picture to where you want it to be. Only one picture can be placed on each card.

Clock

The clock accessory serves as an electronic version of the desktop clock. It has two modes, digital, and analog, as shown in Figure 2.15.

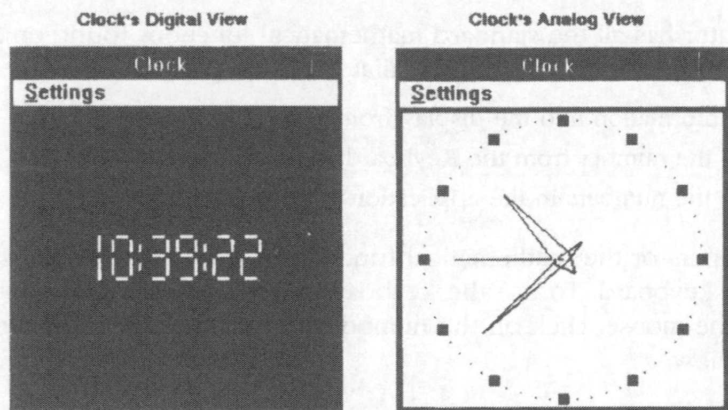

FIGURE 2.15 Clock's two Application Window views.

Change the view in the Settings Menu. The time shown on the Clock is read from your system. If necessary, reset the date and time through Time/Date in the Control Panel.

Notepad

The Notepad application is a simple editor for the Microsoft Windows environment. The files Notepad creates are in ASCII, or pure text format. ASCII, or pure text files, are different from word processing files in that the characters and text are not formatted. Some of the files that contain documentation, or the control files on your computer, contain ASCII text. Notepad is useful to create, modify, and print these files.

Text is entered into the application at the insertion point. To enter text, start typing. As you enter characters they will appear on the screen. To change Notepad so that you do not type beyond the right side, set the Word Wrap option from the Edit menu. Word Wrap will automatically return the insertion point to the left side when you reach the end of the window on the right.

To move the insertion point use the arrow keys, or click the mouse to the place in the text where you want to move. Scroll bars on either side of the Application Window can be used to move around in the file as well.

A powerful feature of most editors, including Notepad, is the ability to cut, paste, and copy some selected area of text. Notepad selects the text for these actions by dragging the mouse over the section. From this point you can use the action commands in the Edit menu or the following keyboard shortcuts:

Keystroke	Action
Ctrl-C	Copies selected text.
Ctrl-X	Cuts selected text (deleting the original).
Ctrl-V	Pastes selected text that was cut or copied.

To find text within the file use the Find command from the Search menu. Searches can match upper or lower case, or move forward or backward in the file. The Find dialog box will prompt you for this input, as shown in Figure 2.16.

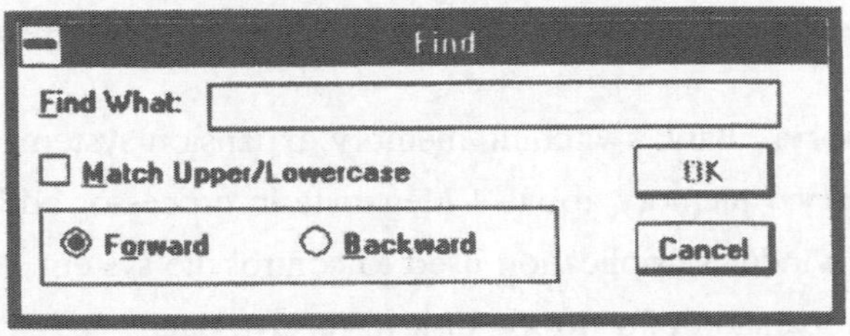

FIGURE 2.16 The Find dialog box.

To save a file, select the Save command from the File menu. Use the Save As command to save a file you have edited to a different name. Loading files into Notepad is accomplished with the Open command in the File menu. Selecting this command opens a file selection box. Use the selection box to move through drives, directories, and files to find the one you would like to edit. When you have located the file, double-click on the name in the file listing area. The file will then load into the Notepad editor. If you want to load files that do not have .TXT as the file extension, change the file extension to an * symbol.

2.13 WINDOWS 95

As this book goes to press, Microsoft, Inc. has introduced a successor product to Windows 3.1 known as Windows 95. This product is more Macintosh-like in nature. The major features of Windows 95 are: Easier installation of peripherals, such as scanners, due to the use of the Plug and Play system, which automatically recognizes and properly installs drivers for the majority of peripherals; optional longer file names, which gives the capability of having file names that are more descriptive of their contents, instead of the earlier eight-character limitation imposed by the MS-DOS operating system; and a more complete integration of the underlying operating system with the Windows environment, which leads to faster operation of Windows 95 in comparison to Windows 3.1. In addition, Windows 95 includes built-in networking capability eliminating the need for products such as Windows for Workgroups.

DEFINING TERMS

Calculator: Desk accessory providing calculator capability.

Calendar: Desk accessory providing a calendar and appointment book.

Cardfile: Simple database desk accessory.

Clicking: Pressing the mouse button.

Clipboard: Temporary storage area for cut-and-paste operations.

Clock: Desk accessory providing a desk clock capability.

Conventional Memory: The memory between 0 and 640K.

Cut and paste: Technique used to move data between or within an application.

Desktop: Paradigm used in Windows.

Desktop Accessories: Useful tools to enhance the functionality of Windows.

Disk Caching: Temporary storage of frequently used data to reduce disk access time.

Enhanced Mode: Mode of operation of processor to take advantage of Intel 386 processor.

Expanded Memory: Bank switching memory expansion system.

Extended Memory: Memory above 1 Megabyte in processor address space.

File Manager: Windows application used to control file system.

Formatting: Preparation of a floppy disk to receive data.

Initialization Files: Files used to set initial parameters for Windows.

Menu bar: Mechanism for accessing commands for Windows applications.

Menus: Lists of available commands.

Mouse: A graphics pointing device used to select and perform tasks.

Pointing: Placing mouse pointer at location for desired action.

Pressing: Pressing mouse button to select an action or application.

Program Group: Mechanism for organizing the desktop to put similar applications programs together.

Scroll bar: Mechanism to select portions of screen not visible to user.

SysEdit: Editor used for .INI files.

Virtual Memory: Using disk space to expand memory.

REFERENCES

User's Guide: Microsoft MS-DOS 6.2. Microsoft Corporation, Redmond, Washington.

Microsoft Windows Version 3.1 User's Guide. Microsoft Corporation, Redmond, Washington.

FURTHER INFORMATION

Paul W. Ross and Paul W. Ross, *Using Windows 3.0/3.1 Effectively*. Wm. C. Brown, Dubuque, Iowa, 1993. Includes Instructor's Manual.

3 Operating the Macintosh

Veronica L. Longenecker
Millersville University

In 1984, Apple Computer introduced a new type of microcomputer that was designed with the user in mind. This new technology uses a graphical user interface, better known as GUI (pronounced gooey), instead of the traditional typed command line. Users no longer need to remember commands and their parameters. With this interface, users select actions called commands, with a mouse from pull-down menus to be performed on graphics, called icons.

The Macintosh computer comes in three different models: compact, modular, and portable. The compact models have the central processing unit (CPU) and a 9- to 10-inch monitor in one case. Although their "all-in-one" style makes them easy to transport, the compact Macs have limited processing power and only one slot, if any, for expandability. Today's more popular model is the modular Macintosh, in which the central processing unit and the monitor are separate. The modular models consist of several different families: the LC, Centris, Quadra, Performa, and Power Macintosh families. Because they can use various size monitors and have one or more expansion slots, modular Macs are more versatile than the compact models. Even more portable than the compact model is the portable Macintosh. The portable models include the central pro-

0-8493-2530-7/96/$0.00+$.50

cessing unit, monitor, keyboard, and mouse in one case weighing less than 10 pounds. Currently, the portable models consist of the PowerBook and Duo families. The PowerBook and Duos run off of batteries making them easy to use just about anywhere.

No matter which model or family a Macintosh belongs to, it has several built-in features to make it user-friendly.

- The Macintosh permits file, application, and diskette names to be up to 31 characters, including spaces, in length. This flexibility allows for very descriptive names.
- Macintosh applications are very consistent in their basic functionality. Therefore, knowledge that is obtained using one application can be applied to several other software packages.
- The newer Macintosh operating system, System 7 offers an on-line help feature called Balloon Help. System 7.5 offers another on-line help feature for the general Macintosh environment called Apple Guide. Several software developers are using these help features to enhance their own software applications.
- Every Macintosh since the early Mac SEs comes with an Apple SuperDrive floppy diskette drive. This high-density floppy drive can read and write to Macintosh, MS-DOS, Windows, OS/2, and ProDOS diskettes through Macintosh PC Exchange or Apple File Exchange software, which comes standard with the operating system.
- Macintoshes can share files or devices such as printers very simply by using the standard AppleTalk port and some LocalTalk cabling to establish a network. Several of the newer Macintoshes also have an Ethernet port as a standard feature.
- Macintoshes provide a "WYSIWYG," what-you-see-is-what-you-get, environment. This environment means what you see on the screen is what you can expect when you print the document.
- Up to seven SCSI (Small Computer System Interface) devices, such as CD-ROMs, external hard drives, and scanners, can be connected to any Macintosh without additional cards or hardware.

Every Macintosh comes with the operating system preinstalled and several other very useful utilities. One of these utilities is a self-guided tour, called Macintosh Basics, of the Macintosh environment and functions. This tour is a very comprehensive tutorial for beginner users of the Macintosh computer.

3.1 TURNING ON THE MACINTOSH

The Macintosh model determines how to turn on the Macintosh. First, turn on any external devices including removable media or hard drives that are connected to the Macintosh. Then, turn on the computer according to the following instructions:

Model	Power switch
• Compact Models (SE, SE/30, Classics)	On/off toggle switch on the back left side
• LC Models (monitor may turn on separately)	On/off toggle switch on the back right side
• Modular Models (Mac II, Centris, Quadras) (monitor may turn on separately)	Power On key on the keyboard (located in upper right corner)—has a triangle pointing to the left
• AV and PowerMac Models (monitor may turn on separately)	On/off button on the front right side

For more information about turning on the monitor, please refer to the manual that came with the monitor.

After being turned on, the Macintosh looks for a startup disk containing the information it needs to start up. Usually, the startup disk is either the internal hard disk or an external hard drive connected to the Macintosh. If the Macintosh finds the startup disk, it will give a chime sound and then displays a smiling Macintosh in the middle of screen, followed by a Welcome to Macintosh window. If a startup disk cannot be found, then a diskette with a question mark appears in the middle of the screen. In this case, the operating system needs to be installed on the hard drive or a startup diskette needs to be created. When the Macintosh is finished booting up, it displays its desktop on the screen.

By looking in the upper right corner of the desktop, you can determine which operating system a Macintosh is running.

System 6–Finder Only

System 6–MultiFinder

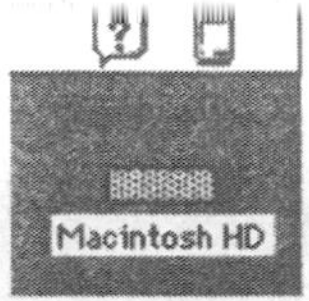

System 7

The System Folder contains the operating system consisting of the files and applications needed for the Macintosh to function. These files include fonts, device drivers, and controls over the Macintosh environment as well as the main System and Finder files that allow the Macintosh to function properly. The System Folder also contains a folder called the Startup Folder, which consists of applications, documents, folders, and sounds that are automatically opened every time the Macintosh is started. Fonts, sounds, and any configurable system resource, such as a screen saver, consume random access memory, RAM. Therefore, the more system resources desired, the more RAM is needed.

System files are loaded in the following order:

- System including resources, desk accessories, fonts, and sounds
- Control panel devices (CDEVs)
- Extensions (INITs)—extensions load alphabetically from the extensions folder, then the control panel folder, and finally the system folder itself.
- Programs in the top level of the system folder
- Finder
- Items in the startup folder

Some control panel and extension icons are displayed at the bottom of the screen during startup. To do a basic boot without all the extension files, hold down the shift key during the bootup process. The message, "Extensions off" should be displayed on the Welcome to Macintosh startup screen. Also, some of the extension icons displayed at the bottom of the screen will have an "X" through them.

To install system files into the System Folder, drag their icons over the System Folder icon (not into the System window). The system folder will recognize special files and put them in their appropriate folders within the System folder. However, items that you want in the Startup Items folder or the Apple Menu Items Folder need to be dragged to their specific folder. Also, some system software comes with an installer program to properly install files into the system folder. To use an installer program, double click on the installer application and follow the instruction on the screen.

To remove files from the system folder, drag their icons from the system folder.

3.2 TURNING OFF THE MACINTOSH

The operating system of the Macintosh is designed to go through a shutdown procedure before physically turning off the computer. To properly shut down a Macintosh, select the Shutdown command in the Special menu. The Macintosh models that started up with the Power On key on the keyboard, will automatically turn themselves off after completing the shutdown procedure. All the other Macintosh models will physically need to be downed off after completing the shutdown procedure. A message will appear on the screen stating when it is safe to physically turn off the Macintosh. After the Macintosh has been shut down, all external devices can be turned off.

3.3 RESTARTING THE MACINTOSH

If the Macintosh appears to be having some problems such as running out of memory, sometimes restarting the Macintosh will clear up these problems. The best way to restart a Macintosh is by selecting the Restart command in the Special menu. However, if the computer is "frozen," the pointer will not move, press down the command-control keys while pressing on the Power On key. If the Macintosh does not restart, then press the reset switch if one is available. Finally, if the Macintosh does not have a restart switch, simply turn the Macintosh off, wait 10 seconds, then turn it back on.

3.4 THE MACINTOSH DESKTOP

After the Macintosh starts up, the "desktop" appears on the screen. The desktop or work area consists of a menu bar across the top, an icon in the upper-right corner representing the startup disk, and the Trash can icon in the lower-right corner on a colored area. On the desktop are icons, windows, and menus. Icons or graphical images represent storage devices for more icons or information. There are several types of icons: storage (hard drives, floppy diskettes, CD-ROMs, etc.), folders, documents, and applications. Windows show what is stored "inside" the icons. For example, when a floppy diskette icon is opened, a window shows the contents of the floppy diskette. Menus give choices of commands that can be performed on selected icons or the selected contents of the icon (see diagram next page).

A process called "rebuilding the desktop" should be done on a regular basis after installing any software. "Rebuilding the desktop" helps keep track of the information on the startup disk. To rebuild the desktop, hold down the Option and the Command ⌘ keys while starting up the Macintosh. Near the end of the startup process, the message, "Are you sure you want to rebuild the desktop? All comments in the Get Info windows will be lost." will appear on your screen. Click on the OK button to rebuild the desktop or the cancel button to stop the process.

The desktop can be magnified by using the system software feature called CloseView. CloseView is a control panel that allows the screen to be black on white or white on black and magnifies the screen up to 16 times its original size. To turn on and off CloseView, choose the CloseView control panel and click on the on or off button.

3.5 THE MOUSE

The mouse controls the movements of the pointer, usually an arrow pointing towards the upper left, on the screen. With the cable attached to the mouse pointing away from you and your index finger resting on the mouse button, gently push the mouse in the direction you would like the pointer to move. If you run out of room on the mouse pad while trying to complete an action, pick up the mouse and set it back down where there is more room on the pad. Various tasks are accomplished by using five mouse actions: pointing, clicking, double clicking, pressing, and dragging.

- To point to an icon — Move the mouse until the tip of the arrow pointer is over the icon. Pointing alone does not accomplish anything, but is the first step to other actions.
- To click on an icon — Point to the icon, then press and quickly release the mouse button once. The icon becomes highlighted (darkened), which means it is selected. A selected object is affected by whatever action is chosen next.

  To select more than one icon at a time, hold down the shift key while selecting (clicking on) the various icons. This process is known as shift-clicking.
- To double-click on an icon — Point to the icon, then press and release the mouse button twice in rapid succession. Double-clicking on an icon is equivalent to opening documents, starting applications, and selecting text.
- To press on a menu — Point to a menu and hold down the mouse button without moving the mouse.
- To drag an object — Point to the object and while holding down the mouse button move the mouse. When dragging an object, an outline of the object moves with the pointer on the screen. When the mouse button is released, the object itself moves to the new location.

The tracking speed, how quickly the mouse moves across the screen, and the double-clicking speed, how short the time interval is between clicks, of the mouse can be adjusted in the Mouse control panel.

In addition to controlling the pointer with the mouse, the Macintosh system software has a feature called mouse keys. With mouse keys, the pointer is controlled using the keypad to the right of the keyboard. To turn mouse keys on, press the command-shift-clear (on the keypad) keys. To turn mouse keys off, press the clear key. With mouse keys on, all the basic mouse functions can be performed from the keypad. The 5 key is the mouse button. Press once to click; press twice to double click. The 0 key holds the mouse button down for dragging. The decimal point key or the 5 key releases the mouse button. The Easy Access control panel adjusts some of the characteristics of the mouse keys features.

3.6 ICONS

Icons are the graphical objects used by the Macintosh to represent floppy diskettes, hard drives, applications and files. They generally appear along the right side of the screen.

• To select an icon	Click on the icon. The icon will be highlighted or darkened.
• To deselect an icon	Click anywhere except on the icon.
• To select multiple icons	Hold down the shift key while clicking on the icons. Clicking on a previously selected icon while holding down the shift key deselects that icon.
• To open an icon	Double-click on the icon. The icon will be shaded or filled with a pattern while a window will open to show the contents of the icon. *or* Highlight the icon by pointing to it and clicking the mouse button once. Choose the Open command under the File menu. The icon will be shaded or filled with a pattern while a window will open to show the contents of the icon.
• To move an icon	Click on the icon and while holding down the mouse button drag the icon to where you want it.
• To duplicate an icon on the same disk	Hold down the option key and drag the icon to its new location. The item will be copied to its new location instead of being moved there. *or* Select (highlight) the icon. Choose the duplicate command under the file menu. Move the duplicated icon to its proper location.
• To copy an icon onto another disk	Drag the icon over the icon of the disk or into a window of the disk.
• To rename an icon	Click on the name (not the icon itself) of the icon. Type in the new name of the icon.
• To change an icon's appearance	Copy the new graphic to be used. Click on the icon to be changed. Select the Get Info command from the File menu. Click on the icon's graphic in the Get Info window. A box should appear around the graphic. Select the Paste command from the Edit menu. The icon's image should change to the new copied graphic.
• To restore an icon's appearance	Click on the icon to be restored. Select the Get Info command from the File menu. Click on the

icon's graphic in the Get Info window. A box should appear around the graphic. Select the Cut command from the Edit menu. The icon should return to it's default image.

- To delete or erase an icon — Drag the icon to be deleted over the Trash can icon until it is highlighted. The Trash can is bulging if there are items in it to be erased or deleted.

- To remove an icon from the Trash can — Double click on the Trash can icon to open its' window, displaying all the items in the Trash can. Drag the desired icon from the Trash can's window to the hard drive.

  or

  Select the desired the item. Choose the Put Away command from the File menu. The item will return to its original location.

3.7 WINDOWS

Windows show the contents of opened icons. Windows consisting of varying sizes can be located anywhere on the desktop.

- To make a window active — Click anywhere on the window or double-click on its grayed (shaded) icon.

- To size a window — Click on the Zoom box. The window adjusts to a size to show all of its content or as much as the screen size allows. By clicking on the Zoom box again, the window returns to its previous size.

  or

  Dragging the Size box in any direction changes the size and shape of the window. While dragging the Size box, the new size and shape are designated by the outline of the window.

- To view hidden icons — Click on the arrows in the scroll bars, or move the scroll boxes within the scroll bars to view other parts of the window, or click on the scroll bar itself to move a screen at a time. The scroll bars are shaded if there is more information to see within that window.

- To close a window — Click on the close box.

- To close all open windows — Hold down the option key while clicking on the close box in the active window.

- To move the active window — Click on the title bar and while holding down the mouse button drag the window to its new location

- To move an inactive window — Hold down the command key, click on the title bar, and while holding down the mouse button drag the window to its new location.
- To organize the icons by name in a window — Hold down the option key while selecting the Clean Up command in the Special Menu.

3.8 APPLICATIONS

Applications are software programs that accomplish a tasks. For example, word-processing applications are used to write memos, letters, reports, or any written piece of work. Most Macintosh applications function in a similar manner; therefore, techniques learned in one application can be applied to other applications. For example, the save and open windows in all Macintosh applications look identical.

Computers use special software programs to operate called the operating system or system software. The Finder is the application that creates the Macintosh desktop.

- To start an application — Double-click on the icon. The icon will be shaded or filled with a pattern while the application launches. The menu at the top of the screen will change to reflect the menus needed for the launched application.
- To open a document — To work on an existing file, double-click on that particular file. The Macintosh will automatically launch the proper application with the specified file active. If the Macintosh cannot find the application that created the file, the message "The application is missing or damaged" appears on the screen.

  or

  If the desired application is running, choose the open command under the File menu. Find the appropriate file and click on the open button.
- To choose a command — Point to the appropriate menu. Press down the mouse button to display the commands available under the selected menu. While continuing to press down the mouse button, drag the pointer over the commands. When the desired command is highlighted, release the mouse button to initiate that command.

  Many commands have keyboard shortcuts, which are combinations of keystrokes. One of the keys is always a modifier key (control, shift, option, or command), usually the command key ⌘. If a keyboard shortcut exists, it is listed to the right the command. For example, ⌘ P is the keyboard shortcut for the Print command under the File menu.

• To scroll through the list of files and folders in the open and save windows	Use the up and down arrow keys on the keyboard. *or* Click on the up and down arrows in the scroll bar. *or* Type the first couple of letters of the file name.
• To close a document	Click on the close box of the document's window or choose the close command under the File menu. Closing a document does not stop the application from running. You can tell the application is still active by the menu bar at the top and the application icon in the upper right corner.
• To start another application	Choose Finder under the application menu. Double click on the additional application icon. The number of applications that can be opened simultaneously is determined by the amount of RAM available.
• To switch between applications	Choose the desired application under the application menu. The active application will have a checkmark in front of its name. If the application is running with no windows open, then only the menu at the top of the screen will change.
• To quit (stop) an application	Choose the Quit command in the File menu. If the file created has not been saved, the message "Do you want to save this file?" appears on the screen.
• To force quit an application	If the application seems to be "frozen," try to force the application to quit by pressing the command-option-esc keys simultaneously. Although any work done in that program will be lost, any work not saved in other programs can be saved before restarting the Macintosh.

3.9 FILES/FOLDERS

File names can be up to 31 characters in length including spaces. Every character is acceptable in a file name except for the colon. Files can be organized by being placed in folders and folders can be placed within other folders.

With the use of aliases, documents, applications, or folders can be used from several locations without copying them. An alias is a pointer that points to the original file. Therefore, when you click on an alias, you are actually using the original file.

• To lock a file	Select the file. Choose the Get Info command in the File Menu. Click on the locked box in the Get Info window. An "X" will appear in the locked box. Click on the close box of the Get Info window
• To unlock a file	Select the file. Choose the Get Info command in the File Menu. Click on the locked box in the Get Info window. The "X" will be removed from the locked box. Click on the close box of the Get Info window
• To make an alias (System 7 only)	Select the desired icon that will have the alias sociated to it. Choose the Make Alias command under the File menu. The alias file will be created next to the original with the same file name plus the word "alias" at the end. The name of the alias is italicized and can be changed to anything.
• To create a new folder	Make sure the hard drive or folder where the new folder is to be created is the active window. Choose New Folder command in the File menu. Type in the new name of the folder. The default name is Empty Folder.
• To find a file/folder	At the Finder level, choose the Find command in the File menu. This command cannot find items such as fonts and sounds in the System Folder. A dialog box will appear on the screen. In the first pop-up menu, select the location (hard drive, floppy diskette, etc.) to search for the item. In the second pop-up menu, select the criteria (name, size, date, etc.) to use during the search. In the third pop-up window, select the qualifier (contains, is, starts with, etc.) to use. Type in the actual characteristic (name, size of item, etc.) to be found. Click on the FIND button.
• To find the next item matching the same criteria	Choose the Find Again command in the File menu.
• To find an item with more than one criteria	At the Finder level, choose the Find command in the File menu. This command cannot find items such as fonts and sounds in the System Folder. A dialog box will appear on the screen. In the first pop-up menu, select the location (hard drive, floppy diskette, etc.) to search for the item. In the second pop-up menu, select the criteria (name, size, date, etc.) to use during the search. In the third pop-up window, select the qualifier (contains, is, starts with, etc.) to use. Type in the actual characteristic (name, size of item, etc.) to be found. Click on the more button and select the

proper criteria and qualifier in the pop-up menu. Fill in the actual criteria for doing the search Click on the FIND button.

To add more criteria to help narrow the search, continue to click on the More Choices button.

3.10 MENUS

Every applications including the Finder has a menu consisting of commands at the top of the screen. Some menus have submenus associated with them.

- To select a command — Point to the menu. While holding down the mouse button, drag the pointer across the commands until the appropriate command is highlighted, then release the mouse button.

  or

  Some commands can also be executed from the keyboard using their "quick key" equivalent. If a "quick key" equivalent is available, it is listed to the right on the command in the menu. Usually the "quick key" is the command ⌘ in combination with another key.

A Macintosh system software feature called Sticky Keys allows the "quick key" equivalent to be typed as two separate keystrokes instead of simultaneously pressing the two keys. To turn on or off Sticky Keys, press the Shift key five times in a row without moving the mouse. When Sticky Keys is turned on, the "quick key" equivalent is typed by pressing the qualifier key (control, option, command) and then the character key. Some of the characteristics of the Sticky Keys can be set with the Easy Access control panel.

The Apple Menu

The Apple Menu always remains in the upper left-hand corner of the screen. The selections in the Apple Menu are always listed in alphabetical order. Items can be added to the Apple Menu by placing aliases, applications, sounds, documents, folders, and desk accessories in the Apple Menu Items folder in the System folder (System 7) or using the Font/DA Mover utility (System 6). Selecting an item from the Apple Menu is equivalent to double-clicking on that icon. The command "About this Macintosh" in the Apple Menu describes how much random-access memory is available and which version of the operating system is installed.

- To have an item appear in the Apple Menu — Drag the item (file, folder, application, etc.) or its alias into the Apple Menu Items Folder in the System Folder.
- To remove an item from the Apple Menu — Drag the item (file, folder, application, etc.) or its alias out of the Apple Menu Items Folder in the System Folder.

Chooser

The Chooser is the part of the system software that determines which printer to use. After selecting a printer in the Chooser, that printer will remain the active printer until another printer is selected.

- To activate a printer connected to a network — Choose the Chooser command under the Apple Menu. Click on the printer icon on the left side of the window that represents the desired printer. Names of available printers will appear on the right side of the chooser window. Highlight the name of the desired printer. Close the Chooser window by clicking on the close box.
- To activate a printer connected directly to the Macintosh — Choose the Chooser command under the Apple Menu. Click on the printer icon on the left side of the window that represents the desired printer. Click on the icon that represents the port that is being used by the printer. Be sure the word "Inactive" is selected next to the "AppleTalk" option. Close the Chooser window by clicking on the close box.

3.11 MEMORY

The term memory often refers to Random Access Memory (RAM). RAM is the memory available for applications and their documents to function. This memory is different from disk storage or hard disk space.

- To determine how much memory is installed — Choose About this Macintosh command in the Apple Menu. A window will appear stating which operating system is running and how much total memory is installed in the Macintosh.
- To adjust how much memory an application uses — Select the application by clicking on it once. Choose the Get Info command from the File menu. The memory requirements for the application will appear in the lower right-hand corner of the Get Info window. Type in the new minimum or preferred size. Applications run most effectively with the suggested or higher amount of memory.

The amount of memory set aside to help applications work more efficiently is called disk cache. This memory cannot be used to open documents or start applications. Therefore, the size of the disk cache directly affects the number of applications that can be running and the number of files that can be opened at the same time.

- To set the disk cache — Choose the memory control panel. Click on the up or down arrow to select the desired disk cache size. The Macintosh must be restarted for the new size to take effect.

Some Macintosh models can use virtual memory or disk space to increase the amount of memory available. The disk space that is being used to enhance the amount of memory in the Macintosh cannot be used to store information.

- To turn on virtual memory — Choose the memory control panel. Click on the on button in the Virtual Memory section. Using the pop-up menu, select the hard drive to use as virtual memory. Using the up and down arrows, select the amount of hard disk space to use as virtual memory. The Macintosh must be restarted for the changes to take effect.
- To turn off virtual memory — Choose the memory control panel. Click on the off button in the Virtual Memory section and restart the Macintosh.

If a Macintosh has more than 8 Megabytes of memory installed, the 32-bit addressing option in the memory control panel should be turned on to take full advantage of all the available memory.

- To turn on 32-bit address — Choose the memory control panel. Click on the on button in the 32-bit Addressing section and restart the Macintosh.
- To turn off 32-bit address — Choose the memory control panel. Click on the off button in the 32-bit Addressing section and restart the Macintosh.

3.12 DISKETTES

Macintosh computers use 3-1/2" floppy diskettes to both store and load information or data. Diskettes come in a double-sided, double-density (low-density) format, which holds approximately 200 pages of information, or a high-density format, which holds approximately 300 pages of information. Besides the additional information that can be stored on a high-density diskette, it also has an extra hole in the upper left corner and the letters "HD" printed somewhere on the diskette.

When an unformatted diskette is inserted to a Macintosh, the message "This diskette is unreadable, do you want to initialize it?" appears on the screen. If the diskette should be formatted, click on the Initialize button for high-density diskettes and the Two-sided button for low-density diskettes. If the diskette should not be formatted, click on the Eject button. If the two-sided button or Initialized button was selected, then another dialog box stating "The process will erase all information on this disk," appears on the screen. This dialog box gives you one last chance to cancel the initialization process. When prompted, enter the name of the diskette. The default name is Untitled.

- To insert a floppy diskette — Insert it with the metal protective shutter away from you, and the circular metal drive hub facing the bottom. Push it gently into the floppy drive until it snaps into place.
- To eject a floppy diskette — While holding down the command key, press the letter "E." This option leaves a "ghost" image of

	diskette icon. The Macintosh remembers information about the ejected diskette.
	or
	Choose the Eject Disk command from the Special Menu. This option leaves a "ghost" image of diskette icon. The Macintosh remembers information about the ejected diskette.
	or
	Drag the disk icon over the Trash icon until the Trash icon is highlighted. With this option, there is no "ghost" image of the diskette on the desktop and the Macintosh no longer knows about the diskette.
	or
	While holding down the command and shift keys, press the number "1." This option leaves a "ghost" image of the diskette icon. The Macintosh remembers information about the ejected diskette.
	or
	Hold down the mouse button, while rebooting the Macintosh.
	or
	Choose the Put Away command from the File menu. With this option, there is no "ghost" image of the diskette on the desktop and the Macintosh no longer knows about the diskette.
• To erase all the items on a disk	Open the diskette icon. From the window displaying the contents of the diskette, drag all the icons to the Trash can.
	or
	Select the diskette icon. Choose the Erase disk command in the Special menu.
	or
	Hold down the command, option, and tab keys while inserting a diskette. A message asking whether to completely erase the diskette appears on the screen. You have the option to click on OK and erase the diskette or click on cancel and leave the diskette alone.
• To lock a diskette	Move the switch on the backside of the floppy diskette to the up position. There should be a hole

in this corner of the diskette. Information may be read off the diskette, but no additional information can be saved to the diskette.

3.13 MISCELLANEOUS TIPS

- To prevent the internal hard from being mounted — Hold down the command, option, shift, and drive delete keys while starting up the Macintosh. Be sure a floppy diskette with a valid operating system is inserted.
- To boot from the CD-ROM drive — Hold down the letter "C" key, while starting up the Macintosh. Be sure a CD-ROM with a valid operating system is inserted.
- To take a snapshot of the screen — Hold down the command and shift keys, while pressing the number "3" key. A "shutter" sound is made by the Macintosh when it takes the snapshot of the screen. The image is saved in a graphics format that can be opened by a variety of software packages. The files are named "Picture 1," "Picture 2," etc.
- To empty the Trash can — "Deleted" items will remain in the Trash can until the Trash can is emptied. Select the Empty Trash command in the Special menu. After completing this command, the Trash can will no longer be bulging and the items can no longer be retrieved.

  Under System 6, the Trash can automatically empties itself every time the Macintosh is restarted or shut down. However, in System 7, the Trash can is not automatically emptied by the system during the shutdown process.

3.14 THE MACINTOSH HELP SYSTEM

Apple Guide

Apple Guide is an on-line help system available at the Finder level. The Apple Guide has a menu identified by a ? in the upper right-hand corner of the screen.

- To start the Apple Guide — At the Finder level, choose the Macintosh Guide command under the ? menu. The Macintosh Guide window will appear.
- To find information in Apple Guide — Click on the topic button to choose from a list of available topics. Then click on the desired topic on the left side of the window. Click on a specific item pertaining to that topic on the right side of the window. Click on the OK button. A small window appears with instructions on how

to complete the desired task. To get more help with the instructions window, click on the Huh? button. To return to the Apple Guide menu, click on the Topics button.

or

Click on the Index button to choose from an alphabetical list of subjects. Click on the desired subject on the left side of the window. Then click on the specific item pertaining to that subject on the right side of the window. Click on the OK button. A small window appears with instructions on how to complete the desired task. To get more help with the instructions window, click on the Huh? button. To return to the Apple Guide menu, click on the Topics button.

or

Click on the Look for button to search for a key word. Click on the arrow pointing right to activate that key word text box. Type in the key word. Click on the search button. A small window appears with instructions on how to complete the desired task. To get more help with the instructions window, click on the Huh? button. To return to the Apple Guide menu, click on the Topics button.

- To quit Apple Guide — Click on the close box in the upper left-hand corner.

Balloon Help

Through "thought balloons" Balloon Help describes the function or purpose of icons, menus, commands, and other features that are being pointed to on the Macintosh desktop.

- To turn balloon help on — Choose the Show Balloons command under the Help menu. As you point to various items, balloons with explanations will appear next to the item.
- To turn balloon help off — Choose the Hide Balloons command under the Help menu.

3.15 UTILITIES

Apple HD SC Setup

This utility provided by Apple Computer initializes and tests Apple internal or external SCSI hard drives.

- To initialize a hard drive

Start Apple HD SC Setup by double clicking on its icon. Click on the drive button until the name of the disk to be initialized appears in the window. Click on the initialize button. Click on the INIT button to confirm the hard drive to be initialized. If a message appears about naming the hard drive, type in a name and click the OK button. Click on the Quit button to exit the HD SC Setup application.

- To test a hard drive

Start Apple HD SC Setup by double-clicking on its icon. Click on the drive button until the name of the disk to be tested appears in the window. Click on the test button. If any problems are found, sometimes they can be fixed with Disk First Aid. When the tests are completed, click on the Quit button to exit the HD SC Setup application.

Disk First Aid

This utility provided by Apple Computer verifies and repairs some type of disk damage.

- To repair a disk

Start Disk First Aid by double clicking on its icon. Click on the disk at the top of the window to be fixed. Click on the repair button. The startup disk or the disk where Disk First Aid resides cannot be repaired, but can be verified. When finished repairing any disk damage, choose Quit from the File menu.

If problems still occur try to repair the disk again. Sometimes, to complete repair of a disk, Disk First Aid needs to be run two or three times.

- To verify a disk

Start Disk First Aid by double clicking on its icon. Click on the disk at the top of the window to be verified. Click on the verify button. The startup disk or the disk where Disk First Aid resides cannot be repaired, but can be verified. When finished repairing any disk damage, choose Quit from the File menu.

DEFINING TERMS

The reader should consult Chapter 2 on Windows for details pertaining to the terms used in Macintosh applications, as they are quite comparable to Windows applications in most respects.

4

HyperCard®

Elizabeth E. Katz

4.1 OVERVIEW

HyperCard® is a multipurpose, multimedia, hypertext application that models information as a stack of virtual index cards. The cards may contain text, graphics, sound, and links to other cards. HyperCard provides a language, HyperTalk, for scripting the actions of stack elements.

The scientist or engineer might use HyperCard to present ideas to colleagues, introduce visitors to a project, store and display data from experiments, represent an interactive or animated view of an apparatus, maintain a personal calendar, draw simple bitmapped illustrations, or explore new research ideas. Its greatest features are its accessibility and flexibility. It provides an unusual mix of capabilities with moderate performance.

The icons for the HyperCard application and a stack (a HyperCard document) are shown below. This chapter's details were checked with HyperCard 2.1, but most concepts apply to all versions of HyperCard since 1.2.

Hypercard Stack

HyperCard is already installed on many Macintoshes. If it is not installed on your machine, copy the application program and the stack called Home to a folder on the hard disk. The installation disks also include example stacks. The examples are optional, but each machine running HyperCard must have exactly one stack called Home as described later in this chapter. Opening the HyperCard application places the user at the first card of the Home stack as shown in Figure 4.1.

4.2 STACK STRUCTURE

HyperCard models information as a stack of virtual index cards. In early versions, the cards were 340 by 512 pixels (the size of the compact Macintosh screen), but versions 2.0 and above allow the cards to be varying sizes. However, all cards in a stack are the same size.

0-8493-2530-7/96/$0.00+$.50

The stack is a sequence of cards. When a stack opens, the first card appears. Going to the next card displays the second one. The card after the last one is the first. However, one of HyperCard˛s features is its support of user-directed navigation. The user can take many different paths through the stack.

For example, in the Home stack shown in Figure 4.1, the user could press the left or right pointing buttons at the bottom of the card to go to the previous or next cards, respectively. Alternatively, the user could press the Appointments button to open an appointment book stack, the Addresses button to open a stack containing addresses, or the More button to learn more about HyperCard. The user decides what to explore and which path to take.

Each card in a stack contains elements such as buttons (active areas to press), fields (containers for text), and pictures (bitmapped graphics). Pictures are bitmapped (refer to the section on MacPaint) and may be created or changed within HyperCard with the painting tools.

```
Press buttons and type in fields with the browse tool.
Command-tab chooses the browse tool.
```

Buttons usually perform actions when pressed. For example, the Appointments button opens an appointment book. Buttons often display an icon or picture suggesting their purpose. For example, Figure 4.1 shows left and right arrows on the buttons at the bottom of the card. These buttons move the user to the previous and next cards in the stack. Buttons may also be transparent, opaque, or rectangular boxes showing their names. Button actions can be simple, such as going to the next card, or complex, such as extending a calendar into a new month or year.

Fields contain text. The letters in a field may be a mixture of fonts and styles, but they do not have paragraph formatting. Long sections of text will be in scrolling fields. Locked fields can react much like buttons. In Figure 4.1, the large boxed area on the left is a field.

These concepts are best illustrated with an example. As an extended example for this chapter, consider a stack describing the stars in the Great Circle of Orion. The first card of the stack is shown at the right of Figure 4.2. The picture shows all the stars of

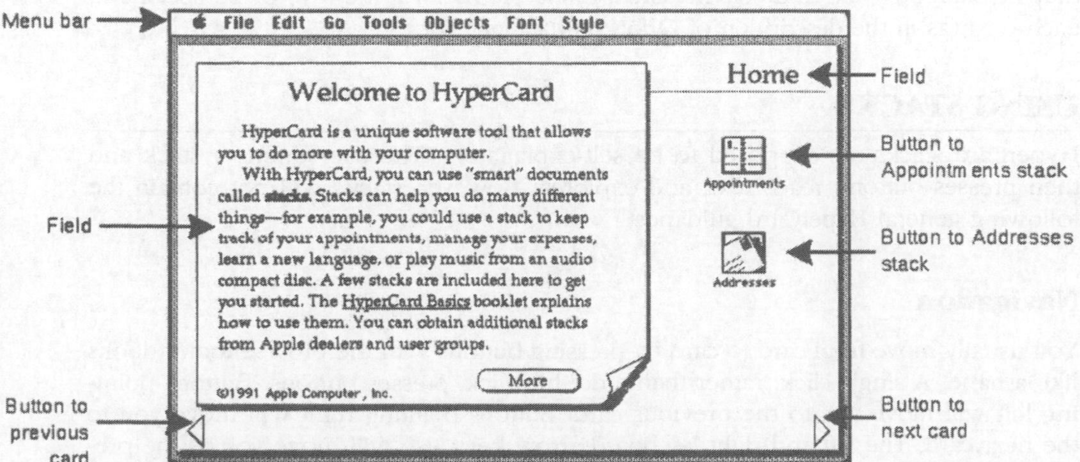

FIGURE 4.1 First card of Home stack in HyperCard 2.1.

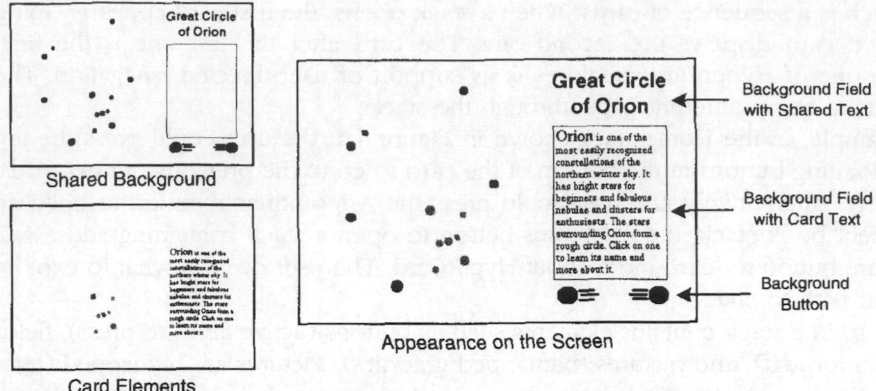

FIGURE 4.2 Background, unique card elements, and card appearance.

the circle and major stars in Orion. Each star will have a separate card describing it and highlighting its position.

Figure 4.2 also displays the use of a shared background in HyperCard. To provide a uniform feel to stacks, stack authors place common elements on a background layer. Cards that want to share the elements share the background. Therefore, the screen display is really a composition of the elements unique to the current card (lower left of Figure 4.2) and the elements on the background that may be shared by other cards (upper left of Figure 4.2). There may be many backgrounds in a stack, but each card has exactly one background. Making changes to a background changes the appearance of all cards sharing that background. For example, if the author erased one of the star pictures shown in black, it would no longer appear on any cards sharing the background.

The Great Circle stack has two different types of buttons. The next and previous buttons have custom icons (comets). There are transparent buttons over the stars painted on the background. These buttons will go to the corresponding star's card.

In the Great Circle stack, the rectangular areas on the right are background fields. The field˛s format and style are consistent across cards. The text for a background field may be shared by the cards (as in "Great Circle of Orion" at the top) or be specific to each card (as in the description of Orion on the right side).

4.3 USING STACKS

HyperCard stacks are supposed to be self-explanatory. The user opens a stack and then presses buttons, reads text, and explores. However, stacks will not contain the following general HyperCard guidance.

Navigation

You usually move from card to card by pressing buttons with the browse tool. It looks like a hand. A single click, rather than a double-click, presses buttons. Buttons pointing left will move you to the previous card. Buttons pointing right will move you to the next card. The left and right keyboard arrow keys also will move you to the previous and next card. (Hold down the Option key to use the keyboard arrows if you have the text arrows preference set in your Home stack.)

`If you change your mind about pressing a button, move the browse tool off the button before releasing the mouse.`

The author might place invisible buttons on pictures or parts of a map. If you do not see any buttons, hold down the Command and Option keys to see outlines of all the buttons. Holding down the Command, Option, and Shift keys will outline the fields.

The Go menu, shown in Figure 4.3, provides another way to navigate through the stack. First, Prev, Next, and Last will take you to the appropriate card in the stack. Back will go to the most recently viewed card. Home will go to the Home stack. Help will go to the Help stack (if it is installed). Recent will show you small pictures of the 42 most recently viewed cards regardless of which stack they are in. Click on one of the small pictures to go to its corresponding card.

Go		take user to . . .
Back	⌘~	most recently viewed card
Home	⌘H	Home stack
Help	⌘?	Help stack
Recent	⌘R	view of 42 recent cards; click one to go to it
First	⌘1	first card of this stack
Prev	⌘2	previous card in this stack
Next	⌘3	next card in this stack
Last	⌘4	last card in this stack
Find...	⌘F	display Find dialog box to find text in fields
Message	⌘M	display Message box
Scroll	⌘E	display scroll bars for a stack with a large window
Next Window	⌘L	go to window of the next stack when viewing multiple stacks in multiple windows

FIGURE 4.3 The Go menu and its uses.

Home

The Home stack contains information unique to this machine's HyperCard installation. There should be exactly one. It contains references to help HyperCard find information on the machine. For example, one card in the Home stack lists the full path names for stacks you have opened. This card, shown in Figure 4.4, tells HyperCard where to look for a stack the next time you want to open one by giving its name.

You can add frequently used buttons, pictures, and other data to cards in your Home stack for easy reference. If you want a message to be handled a certain way whenever it is not explicitly handled by a card or stack element, you would add a handler to the Home stack script. Messages move up a hierarchy and will go to Home before passing on to HyperCard. The message passing hierarchy is described later.

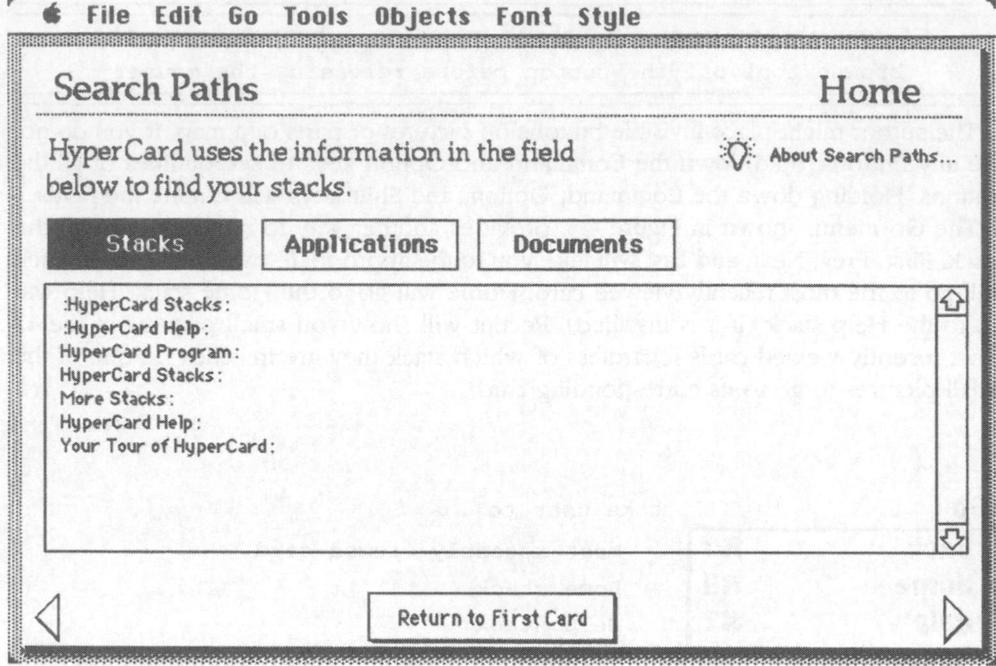

FIGURE 4.4 The Stack Search Path card of the Home stack.

The last card of the Home stack, shown in Figure 4.5, is a graphical place to set some HyperCard preferences. Turning on the Arrow Keys in Text preference allows you to use the keyboard arrow keys to move around in fields. Turning on Blind Typing allows you to type in the Message Box (see below) without seeing it. Turning on Power Keys lets you use keyboard shortcuts when painting.

The user level preference determines which activities HyperCard allows. Browsing permits clicking on buttons. Typing allows changing fields. Painting permits using the painting tools to modify pictures. Some types of animation require painting permission. Authoring allows creation of new cards, buttons, and fields as well as linking of cards. Scripting permits script creation and modification. Higher permissions, such as scripting, allow all the previous permissions.

Menu Bar

The menu bar at the top of the screen may disappear if the stack author hides it. If you are using a compact Macintosh, you may want to hide it to see more of the screen. Pressing the Command key and the space bar will toggle the hiding and showing of the menu bar.

```
Command-space toggles the menu bar on and off.
```

The menu bar changes when the painting tools are selected. This provides additional commands that apply when painting. The menu bar also displays only those commands available at the current user level. For example, if the user level is browsing, the authoring menus and commands are not shown.

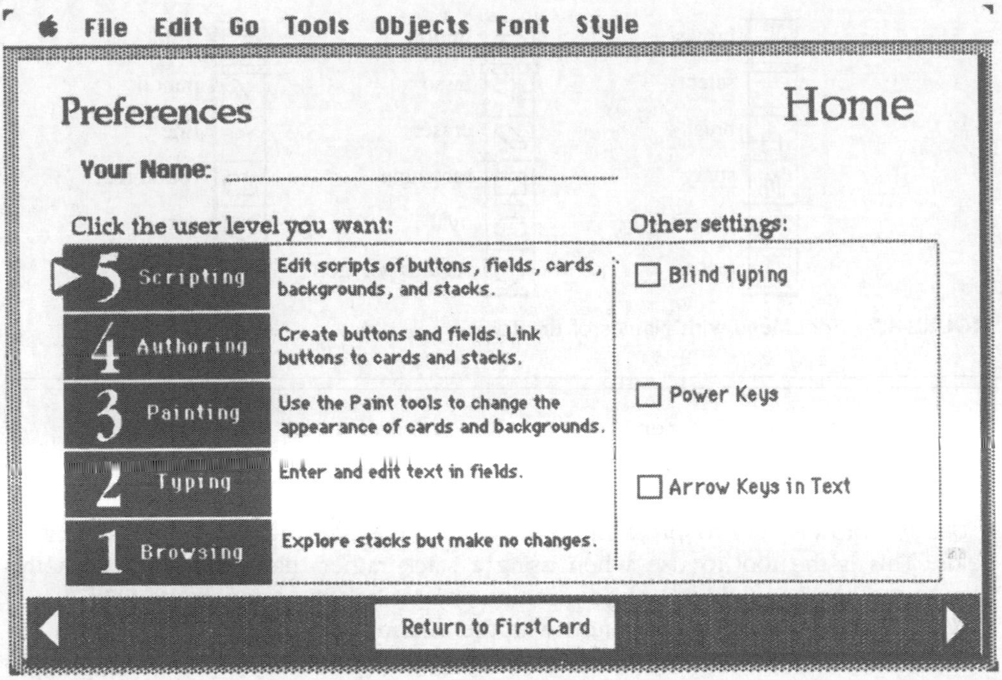

File Edit Go Tools Objects Font Style

Preferences

Home

Your Name:

Click the user level you want:

5 Scripting — Edit scripts of buttons, fields, cards, backgrounds, and stacks.

4 Authoring — Create buttons and fields. Link buttons to cards and stacks.

3 Painting — Use the Paint tools to change the appearance of cards and backgrounds.

2 Typing — Enter and edit text in fields.

1 Browsing — Explore stacks but make no changes.

Other settings:

Blind Typing

Power Keys

Arrow Keys in Text

Return to First Card

FIGURE 4.5 The Preferences card of the Home stack.

Message Box

While browsing a stack, you may want to type simple HyperTalk commands. One line commands can be typed in the message box, shown in Figure 4.6, chosen from the Go menu. For example, if you want to change the user level to allow painting (level 3), instead of setting the level on the Home stack, you could type

```
set the userLevel to 3
```

in the message box.

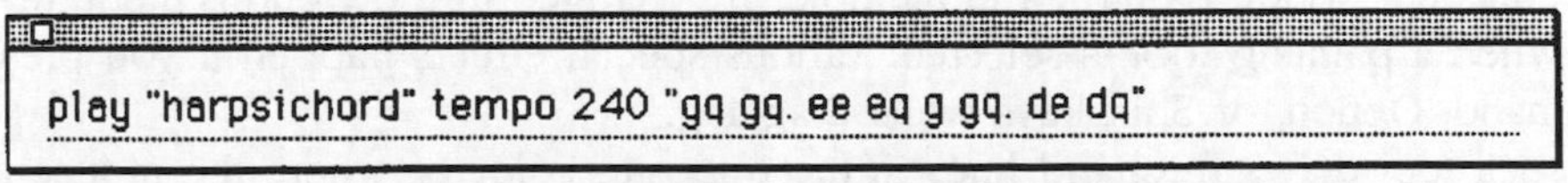

```
play "harpsichord" tempo 240 "gq gq. ee eq g gq. de dq"
```

FIGURE 4.6 The message box with a Music command.

Any **put** commands without an explicit destination will place the result in the message box.

Tools

Figure 4.7 displays and names HyperCard,s tools available on the Tool menu. This menu will tear off and remain displayed if smoothly dragged off the menu bar. The top row contains tools to use and modify buttons and fields. The lower rows contain painting tools.

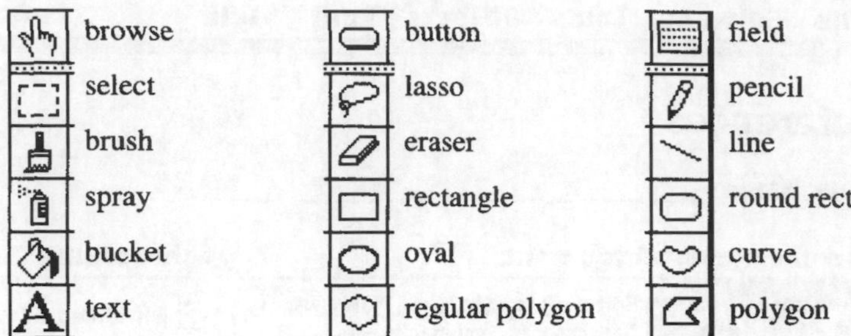

FIGURE 4.7 Tool Menu with Names of the Tools.

`A detached tool menu shows the current tool at a glance.`

Use the browse tool to press buttons, type in unlocked fields, and select text in fields. This is the tool to use when using a stack rather than authoring. Click the browse tool inside a field to get an insertion pointer before typing in the field.

When authoring, click on a button with the button tool to select it. Click with the field tool to select fields. Drag selected objects by their middles to move them. Delete selected objects by pressing the Delete key.

Painting

Along with its authoring capabilities, HyperCard includes simple painting tools. The painting is bitmapped; therefore, once something is painted, it becomes part of the picture and cannot be removed. This is in contrast to drawn objects that retain their identities.

Remember the distinction between background and card while painting. Each has its own picture. In the Great Circle stack, the stars are all painted on the background picture in black. Each card's picture has that card's star painted in a lighter pattern.

The rectangle, oval, curve, line, and polygon tools perform as expected. Select them by clicking them once on the menu, press down in the picture, and drag. Their properties, such as line width, pattern, and multiple, are changed in the Options menu that appears when a painting tool is selected. Various special effects happen if you press the Command, Option, or Shift keys while dragging.

The pencil tool draws freehand lines in the opposite color of where the mouse is first pressed. If it starts on a black pixel, it will draw in white. Single clicks will flip individual pixels. Hold down the Command key and click with the pencil to go into FatBits. This is an enlarged view of the area around the clicked point. The normal size view will appear in a small window. Click in that small window to return to the regular view. FatBits is also available as a menu item.

`Command-clicking with the pencil opens FatBits on that point.`

The eraser erases bits. Drag it over the portions to be removed. Double-clicking on it will erase the entire picture.

The text tool places painted letters on the picture at the point where the tool is clicked. The font and style are set by the Text Style command on the Edit menu. Text typed with the text tool is part of the painting. It cannot be edited after the mouse is clicked elsewhere. If changes of wording, font, size, or style are expected, use a field. Create patterned letters by carefully pouring paint from the bucket into individual letters.

The brush, spray, and bucket tools paint large areas with the selected pattern. The pattern menu tears off the menu bar. With the spray tool, holding down the mouse sprays heavier paint. The shape of the brush tool is set in Options. Hold down the mouse button and drag to paint. The bucket pours paint from the lower tip of the spout into an enclosed area. The area must be entirely enclosed in the current picture. Paint will pour out of a pixel-width hole. Paint poured on a card picture into an enclosed area of the background picture will spread over the entire card.

When accidents happen, choose undo immediately. The keyboard shortcut is Command-Z. Undo remembers only a single action. To save a picture before a series of painting actions, choose Keep. While using painting tools, Revert usually will go back to the kept picture. Another approach to saving before tricky painting is to select the entire picture, copy it, and paste it in the scrapbook for safekeeping.

`Undo with Command-Z. This undoes a single action.`

The select and lasso tools are very useful. To select a rectangular area, drag the select tool from one corner to the opposite corner. This selects the white area as well as painted pixels. To select an irregular area or only the painted pixels, drag the lasso around the perimeter of the desired area and cross the ends of the loop. The lasso will collapse around the painted pixels. The selected pixels will flicker to indicate selection.

Move the selection around the screen by dragging a pixel within the selection. Dragging while holding the Option key creates a clone. Dragging the corner of a rectangular selection while holding the Command key resizes the selection. Pressing the Delete key will remove the selection. Additional effects such as flipping horizontally or vertically, rotating 90 degrees, and inverting selected bits are available on the Paint menu.

Sound

Many HyperCard stacks use sound. Use the Control Panel under the Apple menu to set the volume. After HyperCard starts a sound, other commands may execute while the sound plays. If a stack starts to play a long or undesirable sound, open the message box and type

```
play stop
```

and the sound will stop. The other actions in the sound-generating script will continue unless they were scripted to depend on the sound playing. For example, a loop may

```
repeat until the sound is "done"
```

The sounds must exist within the stack, HyperCard, or Home. The play command can play individual notes or sampled sounds. Refer to a HyperCard reference for more details.

Type play stop in the message box to stop the current sound.

Printing

HyperCard supports printing of entire stacks, single cards, single fields, and reports summarizing selected fields throughout a stack. The dialog boxes for printing are extensive and are self-explanatory. For example, the dialog box for printing a stack is shown in Figure 4.8. Note that it uses multiple images of the current card to show the layout of a printed page. The crosshair handles on the sample page change the margins.

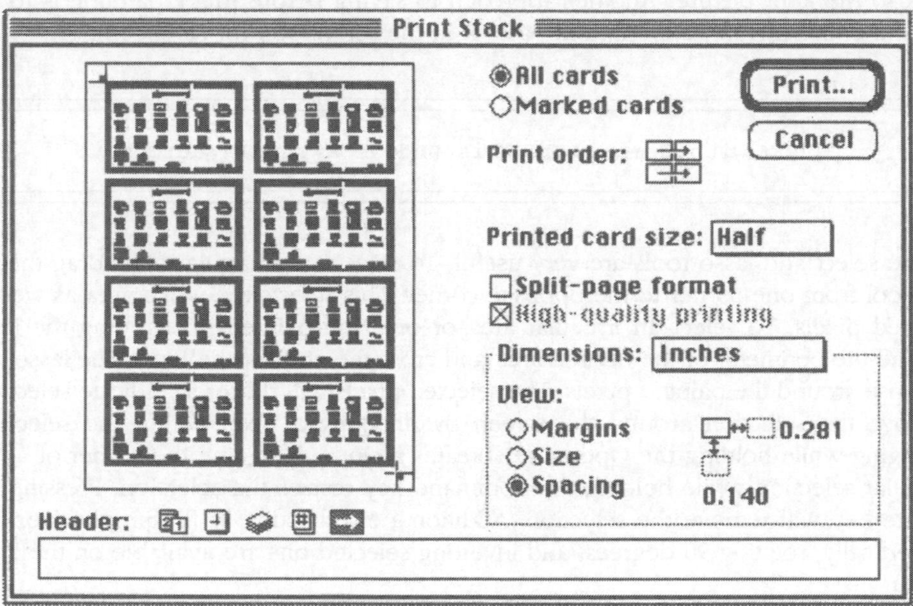

FIGURE 4.8 The Print Stack dialog box.

Saving

HyperCard saves the stack file continuously as a stack changes. This is contrary to most applications and is not an option. It is a good idea to save a copy before making major changes. The Save a Copy command creates a copy of the current stack but leaves you working on the original.

4.4 CREATING STACKS

Apple initially expected most people to use stacks rather than create them. However, its authoring features convert many users to authors. Some authors start with an existing stack and copy its background, buttons, and fields. Other authors start from an empty stack. In either case, authors must be aware of backgrounds, cards, buttons, and fields. Scripts and message passing are important if anything unusual is needed.

Most elements in HyperCard have an information dialog box for setting their names and additional properties. Invoke a button or field info box by selecting the element with the appropriate tool and either double-click or choose from the Objects menu. For cards, backgrounds, and stacks, choosing from the Objects menu brings up the info box for the current element.

Every element has a unique identification number, but meaningful names help authors refer to elements in scripts. Names may contain letters, numbers, and some symbols as well as spaces. They should be unique for each kind of element on a card. For example, the Great Circle stack has buttons with names such as “Rigel” and “Capella” but cards also have these names.

To avoid confusion, authors might need to specify whether a button or field is on the card or background. For example,

```
put empty into card field "Co-op Contacts"
```

All action in HyperCard is a response to a message. HyperCard sends messages to elements when events occur, and the elements may act in response to those messages. To respond to a message, an element has a script containing a message handler defining what should happen when that message arrives. For example, a button would have a mouseUp handler scripting its response to the mouse being released over it. An example action would be to go to the next card. Message handlers are written in HyperTalk and are described later in this chapter.

Stacks and Backgrounds

In HyperCard, each card has exactly one background. That background may be unique to that card or be shared with other cards. Figure 4.2 gives an example. Elements that will appear on several cards often belong on the background. This is a design decision. Changes made to a background will affect all cards sharing that background. Therefore, if a background button must be moved, that move is made once. If the button were on each card, it would have to be moved once for each card.

Many stacks have a single background shared by all the cards in the stack. For example, an address book or card catalog would have similar information on each card. The data will be different, but will have a consistent look. The Great Circle stack will start as this type of stack. All the cards will show the stars in the background but will have the card's star painted in a lighter pattern on the card. The field will contain information about the star.

Other stacks have disparate kinds of information. Some aspects may be similar and others different. In this case, either use multiple backgrounds or have little information on the background. Ideally, HyperCard would support layers of backgrounds that could be included on each card as needed, but it does not. The Great Circle stack could be extended to include additional cards showing the distances of the stars from Earth, their magnitude, or their spectral class. Perhaps Orion and the other constellations could be presented in more detail. Different backgrounds would distinguish these different types of information.

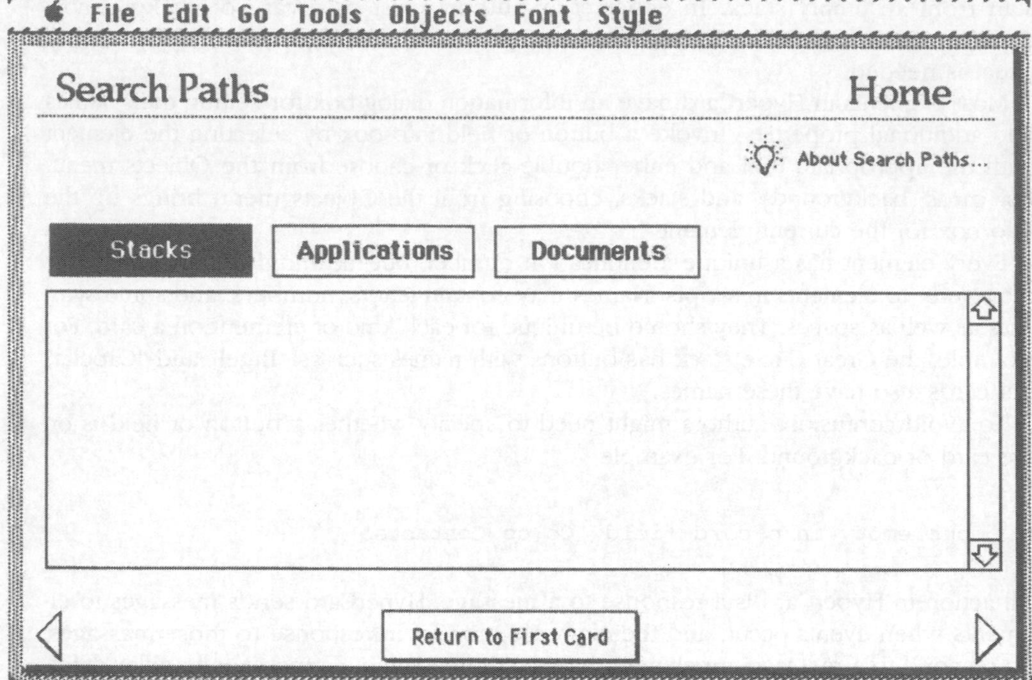

FIGURE 4.9 A Background in the Home stack.

The menu bar has a laced or hashed appearance whenever the background is selected. The Background command near the bottom of the Edit menu toggles between the background and the card. The hashed menu and another example of shared elements are shown in Figure 4.9. This is the background shared by some cards in the Home stack including the card shown in Figure 4.3.

To create a new stack based on an existing one, choose New Stack and select Copy Current Background. This creates a new stack with one background and one card. To create a new empty stack, do not select Copy Current Background. If a new background is needed within an existing stack, choose New Background from Objects. This creates an empty background and a new card.

Cards

Individual cards will contain information unique to the card. For an address book, that would be the name, address, and phone number of one person or company. For a card catalog, it would be the title, author, location, and other information. Cards might hold pictures or specialized buttons or fields. For the Great Circle stack, it will be paint to highlight the card's star and the text for the description field.

Create new cards with the current background by choosing New Card. If substantial portions of the current card's information are needed for the new card, copy the current card and paste. New or pasted cards are placed after the current card.

Cards may be rearranged by cutting and pasting. Stacks open to their first card. To add that first card later, go to the current first card, paste the desired card (or create a new card), go back to the first, cut it, and then paste. This technique could be used to add a title card to the Great Circle stack.

Buttons

Buttons are elements the user presses to cause an action. They may link to other cards or start an animation. They may have an icon or a name. They may be transparent to create "hot spots" on a map. The author chooses the most appropriate form and creates a script defining what action should be performed when the button receives a certain message.

For example, the Next button for the Great Circle stack has a custom icon rather than an arrow. The button˛s action when pressed should be to go to the next card. By convention, a button usually responds to the mouseUp message HyperCard sends it when the mouse button is released over it after it has been pressed over it. This allows the user to stop the mouse click by moving off the button before releasing the mouse. The handler for this action is

```
on mouseUp
        go next card
end mouseUp
```

The handler is placed in the script for the button. When the button is pressed with the browse tool, HyperCard will display the next card.

The button tool selects, moves, and resizes buttons. Clicking on a button with the button tool selects it. It will have a flickering border. Drag its middle to move it. Drag a corner to change its size. Drag holding the Option key to create a clone. Double-click to get its info box.

`Option-drag with the button or field tool to clone a button or field.`

Common button actions are

- Navigate serially through cards. Use go next card and go previous card in the handler.
- Go to a particular card. HyperCard will create this type of link without scripting when the Link button in the button dialog box is pressed. Press Link, navigate to the desired card, and click the This Card button in the dialog box to make the connection. HyperCard builds a handler with the command such as go card id 6745. A cleaner approach is to give the card a name, such as "Aldebaran," and then use go card "Aldebaran". It is good practice to put all names in quotes. The quotes are needed if the name has spaces.
- Go to the card with the same name as the button. This creates named links and is particularly useful in creating a directory where the user chooses a destination card from a list. For the Great Circle stack, a transparent button over a star picture has the star's name. The card describing the star also has that name. The command for the handler is

```
go card (the short name of me)
```

- Open another application. Sometimes it is appropriate to open another application, such as Microsoft Excel, for some calculation or graphing. This command suspends HyperCard while the user is in the other application. When the user quits that application, HyperCard resumes showing the same card as when it suspended. The command is

```
open "Star Distances" with "Microsoft Excel"
```

- where the file name (and with) may be omitted. HyperCard uses one of the lists in the Home card to determine where to find the files and applications. If it cannot find them, it will ask the user to find the file.

Fields

Fields contain text that is not lost when HyperCard quits. The text may have varying fonts, may be shared across many cards, and may be quite long (almost 30,000 characters). The field may be bordered, have scroll bars, or be transparent. If it is unlocked, the text may be changed with the browse tool. If it is locked, words and phrases may be hypertext links to other cards.

Fields are manipulated with the field tool much as buttons are manipulated with the browse tool. However, cloned fields do not include the contained text. The text must be copied and pasted separately. Set the overall font and style in the field's info box. Select individual words and set their font and style with the Font and Style menus. Stacks will be viewed on the screen; therefore, use the screen fonts with city names for best effect.

Creating hypertext links with field text is easy. The user will be able to click on a word or phrase in the field and go to the card with that name. More elaborate actions such as defining the word are also possible, but this example links to named cards.

Create a field and then type the text with the browse tool. Select words or phrases that will be links and apply the Group style to them. Users will identify special words more easily if they are in italics or boldface. Individual words need not be grouped. Lock the field (in the info box), and create the following script:

```
on mouseUp
    go card (the clicktext)
end mouseUp
```

HyperCard goes to the card with the name corresponding to the text (the clicktext) that the user clicked. If no such card exists, HyperCard stays on the current card. The field must be locked for this to work. Otherwise, clicks put the insertion pointer in the field for typing.

4.5 SCRIPTING STACKS

HyperCard elements are controlled by HyperTalk scripts containing message handlers. Each element may have a script, and that script can contain one or more handlers. The handlers may appear in any order within the script.

Each handler consists of the keyword on and the name of the message handled. The HyperTalk commands for the actions that should occur when the message is received follow. The handler concludes with the keyword end and the message name. The HyperCard script editor will indent the code appropriately. When the script is complete, either choose Save and Quit from the menu or press the Enter (not Return) key.

Enter saves the current script and closes the script editor.

Comments start with dash dash and continue to the end of the line. Continue long commands to the next line with Option-Return anywhere within a command's white space (spaces or tabs, not within literals). This places a NOT symbol (¬) at the end of the current line. Upper and lower case are equivalent. However, most HyperTalk authors capitalize the first letter of imbedded words.

An example script is shown below.

```
--This handler shows each card in the stack while
--also showing its name in the message box. It shows
--the cards from first to last and then returns to
--the current card. Put it in the stack's script.
--Type showCards in the message box to execute it.
on showCards--a user-defined message
  push card--remember this card
  put the number of cards into numCards
  repeat with c = 1 to numCards
    go card c
    put the short name of this ¬
    card into message
    wait 20 ticks--a tick is 1/60 second
  end repeat
  pop card--go back to remembered card
  put empty into message
end showCards
```

HyperTalk is an extensive language. It was designed to be readable, but it is more precise than natural language. There are many commands, functions, properties, and predefined values. Some of them are described below, but refer to one of the references for more details.

Variables

Variables such as numCards and c are local to the handler where used. They are not declared, are dynamically typed, and lose their contents when the handler finishes. If information must be retained between handlers, use a field and hide it. If necessary, use a global variable. It must be initialized when HyperCard starts. Every handler using the global variable, for example `myInfo`, must declare it with `global myInfo`.

HyperCard has many predefined variables and functions that seem to be variables. These are preceded by the word `the` as in `the number of cards` or `the clicktext`. Properties of elements are also available. For example, `the location of bg button "Procyon"` would give the horizontal and vertical offset in pixels from the top left corner of the card to the middle of the button.

Two special variables are `it` and the `result`. Many commands place their results into it. For example, the answer command creates a dialog box containing a question and buttons for answer choices. When the user presses one of the buttons, the answer is placed in `it`. `The result` will contain an error code or message if a command fails. `The result` may be checked for empty to determine whether a command succeeded.

The following mouseUp handler would ask the user for an instrument to play a short song. It also asks for a tempo.

```
on mouseUp--ask user for instrument and play song
  answer "Which instrument?" with "flute" or ¬
  "harpsichord" or "boing"
  put it into instrument--it is the result of answer
  answer "How fast?" with "fast" or "slow"
  put it into tempo
  if (tempo is "fast") then
    play instrument tempo 300 "cq d e e d c e"
  else
    play instrument tempo 100 "cq d e e d c e"
  end if
end mouseUp
```

Two other special variables are me and `the target`. Me refers to the element containing the current handler. For example, `put the short name of me` places the short name of the element (without the type of element or its stack's name) into the message box. `The target` is the element that originally received the message being handled. This is useful when one handler processes messages from several elements. For example, in the earlier example, each star button had a separate mouseUp handler with the command

```
go card (the short name of me)
```

Alternatively, the background or stack could have a handler such as

```
on mouseUp--go to star's card if star button is clicked
  if (the name of the target contains "button") then
    go card (the short name of the target)
  end if
end mouseUp
```

An advantage to this approach is that any changes can be made in one script rather than in several different scripts.

Control Structures

The example handlers above demonstrate some of the control structures available in HyperCard. If-then and if-then-else statements provide selection. These statements may be nested and should end with `end if`. One-line versions may omit the `end if` but are more difficult to read. For example, `if the mouseClick then play "boing"` will play a sound when the mouse is clicked.

There are many looping structures. The showCard example uses a counting loop similar to for loops in other languages. There is also a repeat while condition is true loop, a repeat until condition is true loop, and a repeat a given number of times loop. Each of these loops should end with `end repeat`. They may be nested.

HyperCard also provides for user-defined functions. Functions may be defined in the current script or anywhere up the message hierarchy. Parameters are dynamically typed. Function calls must use parentheses even if there are no parameters. For example, the function sumCubes is defined by

```
function sumCubes x, y
  put x^3 into xCubed
  put y^3 into yCubed
  return xCubed + yCubed
end sumCubes
```

It might be called by the following handler that asks the user for two numbers and then prints the result:

```
on mouseUp--ask for two numbers, sumCube them, and print an-
swer
  ask "Enter two numbers separated by a space"
  put it into theNumbers
  put word 1 of theNumbers into m
  put word 2 of theNumbers into n
  put sumCubes(m,n) into theAnswer
  put m & " and " & n && ¬
  "cubed and summed are" && theAnswer
end mouseUp
```

This example also shows how to obtain specific words from variables. HyperCard calls these chunks. Words are separated by spaces, so this code works with real numbers. HyperCard also defines chars that are individual characters, items that are separated by commas, and lines that are separated by returns.

This example also shows the concatenation operators & and &&. The single ampersand concatenates without adding a space. The double ampersand concatenates adding a space between the two strings. For example, if the user entered 3 and 2, the message box would contain

```
3 and 2 cubed and summed are 35
```

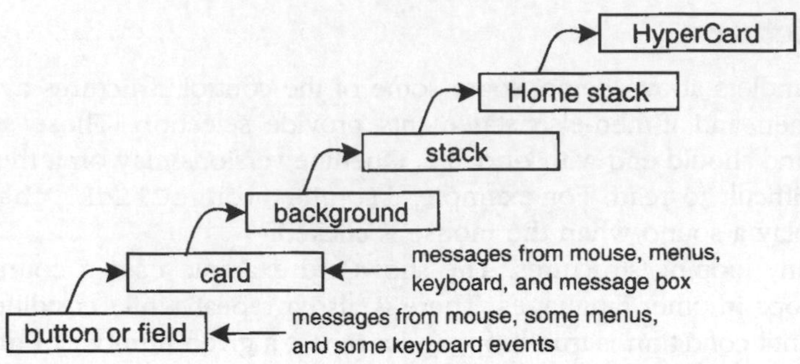

FIGURE 4.10 The message hierarchy.

Message Passing

HyperCard elements respond to messages. They may also send messages to communicate. HyperCard sends messages to elements when certain events occur. Some examples are mouseUp, openCard, mouseWithin, and newCard. If an element receives a message and does not have a handler for it, that message is passed up the message hierarchy until a handler is found or it passes through HyperCard. Figure 4.10 shows this hierarchy. Messages generated by HyperCard can pass through HyperCard without mention. HyperCard will give an error message if it receives an unhandled user created message, such as showCards.

Send messages to elements with the send command. For example, create a simulated button press by typing

```
send "mouseUp" to card button "Play Music"
```

in the message box.

Files

It may be useful to read from and write to text files from HyperCard. For example, lines from a database could become cards in a stack. HyperCard reads ASCII files character by character. The file must be opened before reading or writing and then closed. For example, to open and read a file named "Hard Disk:Alumni:The Names" and put its contents into a card field named "Card Names" use:

```
put empty into card field "Card Names"
put "Hard Disk:Alumni:The Names" into filename
open file filename
repeat--grab blocks until entire file is read
  read from file filename for 1024
  if it is empty then exit repeat
  put it after card field "Card Names"
end repeat
close file filename
```

The handler uses 1024 character blocks. HyperCard has a limit of 16,384 characters per read. To obtain single lines of data, use `read from file filename until return`.

To create and write a file named "Star Info" use:

```
open file "Star Info"
write bg field "Star Data" to file "Star Info"
close file "Star Info"
```

The first write replaces any previous contents. Subsequent writes before closing will append to the file. If the file does not exist, it is created.

4.6 BUILDING AN EXAMPLE STACK

This section describes how to build the Great Circle stack. The stack helps people explore the winter night sky.

Building the Visible Background

Start with an empty stack and empty background. Make sure you have the user level set to scripting. Refer to the upper left of Figures 4.2 and Figure 4.11 for details. The star names will not be on your picture.

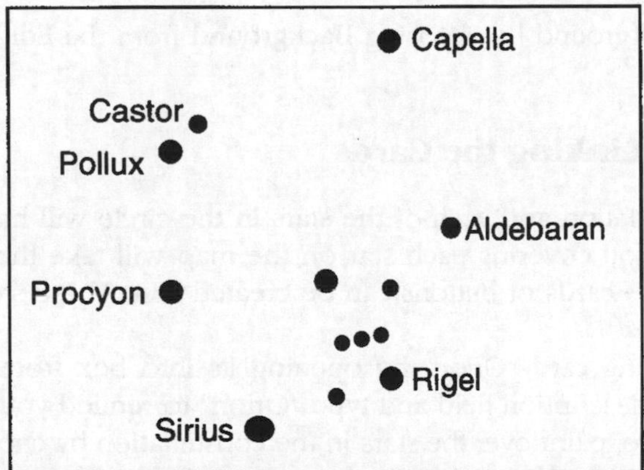

FIGURE 4.11 The star locations and names.

Name the background "Sky Picture." You will add other backgrounds later. Go to the background by choosing Background from the Edit menu. The menu bar will have hash marks around it.

Paint the stars with the brush. Use a dark pattern or black. Use the eraser for mistakes.

Create the Previous button by choosing New Button from the Objects menu. Double-click on the button to get its info box. Name it "Previous" and tell it to not show its name. Click Icon to see your icon choices. Choose something that points left and click OK. (You can create a custom comet icon later, if you wish.) Drag the corners of the button to change its size to match its icon. Drag the button by its middle to the lower right corner. Double-click the button again to get back to the info box. Make the button transparent. Enter the script editor by clicking the Script button. For the script, enter

```
on mouseUp
   go previous card
end mouseUp
```

Press Enter to save the script and quit the script editor.

Create a Next button by repeating these steps or by creating a clone of the Previous button (drag from the existing button while holding the Option key) and modifying the appropriate properties. The script should have go next card.

Save a copy of what you have done so far. HyperCard saves as you make changes, so this is a backup copy.

Create the title field with New Field from the Objects menu. Drag it by its middle to the upper right. Double-click it to see its info box. Name it "Title." Choose Shared Text. Make it transparent. Click on Font. Make the text Centered 18 point Geneva. Choose the browse tool, and click in the upper part of the field. The insertion I-beam should appear. Type "Great Circle of Orion."

Create the description field similarly. Do not choose Shared Text because each card will have its own text. The font should be left justified 12 point New York. The field style is rectangle.

Get off the background by choosing Background from the Edit menu.

Creating and Linking the Cards

The Orion constellation and each of the stars in the circle will have their own cards. A transparent button covering each star on the map will take the user to the matching card. Either the cards or buttons can be created first. These steps create the cards first.

Name the existing card "Orion" by opening its info box from the Objects menu. Also, click in the description field and type "Orion" to remind you of the card's name. On the card picture, paint over the stars in the constellation by choosing a lighter paint pattern and then using the brush.

Choose New Card from the Edit menu. Name this card "Aldebaran" and type its name in the description field. Repeat for Capella, Castor, Pollux, Procyon, Sirius, and Rigel. Paint over the appropriate star with light paint on the card picture. Rigel is both part of Orion and a separate star in the Circle.

Go to the background. Create a new button that is transparent, does not show its name, and is named "Orion." Drag and shape it to roughly cover the constellation. Clone the button by option-dragging. Make it smaller, but not too small, and name it "Aldebaran." Place it over the appropriate part of the picture. Clone this smaller button to cover the other named stars. Name each one appropriately. The Rigel button should be on top of the Orion button. It will be if you create it after the Orion button.

To ease changes, the star buttons have a centralized mouseUp handler in the background script. Get the background's info box and go to its script. Type the following handler:

```
on mouseUp--go to star's card if star button is clicked
  if (the name of the target contains "button") then
    go card (the short name of the target)
  end if
end mouseUp
```

Save the script. To test the buttons, choose the browse tool. Click on the star pictures and see that you go to the matching card. If none of the buttons work, check the background script for typing mistakes. If a button goes to the wrong card, check the button's name and the card's name. They must match exactly. An extra space after a name changes it. Upper and lower case differences are ignored. If the buttons are on only one card, either build them again on the background or carefully cut them from the card and paste them on the background.

Adding a Title Card

Go to the first card of the stack. Create a new background by choosing New Background from Objects. Create a title field and add a title. Create another field that describes this stack. Create a button named "Sky Picture" and have its action be

```
go first card of background "Sky Picture"
```

The title card should be the first card your users see when the stack opens. Therefore, go to the current first card, cut it (the title card will appear), and paste it (the old first card is now the second card). Go to the first card and check that it is the title card.

Cleaning Up and Extensions

Your basic stack is complete. Save a backup copy. Remove excess space using Compact Stack from the File menu. Then save your backup again. You might want to extend the stack with more descriptions of the stars, animation, diagrams, or sound. Refer to the references for ideas.

If you plan to create stacks to share with others, remember that you are making a presentation. Look at some examples before you start, plan your approach, and consider who your audience will be. The stack design guidelines book listed in the references has many hints.

DEFINING TERMS

Background: Layer containing elements shared by many cards in a stack.

Browse Tool: Looks like a hand ready to press a button. Use it to press buttons and type in fields. Select it from upper left of Tool menu or Command-Tab.

Button: Element on a background or card that responds to actions, usually mouseUp.

Card: Layer containing elements that are unique. Exactly one card from a stack is displayed at any one time.

Field: Element of a background or card that holds and displays text. Background fields may share text with all cards or have unique text on each card.

Home: Unique stack that holds preferences, common handlers, and other information used by HyperCard.

Message: Command sent to an element. HyperCard sends messages to elements when events occur. Additional messages may be sent by handlers or form the message box.

Message Box: Place to type one-line commands to HyperCard.

Message Handler: Program specifying action to be performed when a message is received.

Script: Collection of message handlers associated with an element.

Stack: HyperCard document. Contains data for a sequence of cards as well as any resources (e.g., sounds or icons) required to display and execute the stack.

User Level: Identifies the actions allowed with HyperCard. Browsing, typing, painting, authoring, and scripting correspond to values 1 to 5. Higher levels allow actions at lower levels. Default set in user preferences in Home stack.

REFERENCES

Apple HyperCard Script Language Guide: The HyperTalk Language. Addison-Wesley, Reading, MA, 1988.

Apple Computer, Inc., *HyperCard Stack Design Guidelines*. Addison-Wesley, Reading, MA, 1989.

George Beekman, *HyperCard in a Hurry*. Wadsworth, Belmont, CA, 1992.

Danny Goodman, *The Complete HyperCard 2.2 Handbook*, 4th ed. Random House, New York, 1993.

Mimi Jones and Dave Myers, *Hands-On HyperCard: Designing Your Own Applications*. John Wiley, New York, 1988.

Lon Poole, *HyperTalk*. Microsoft Press, Redmond, WA, 1988.

Dan Shafer, *The Complete Book of HyperTalk 2*. Addison-Wesley, Reading, MA, 1988.

Dan Winkler, Scott Kamins, and Jeanne DeVoto, *HyperTalk 2.2: The Book*. Random House, New York, 1994.

5

Basics of the Unix System

Alton F. Sanders
Robert Pickering
Miami University

5.1 HISTORY

It is virtually mandatory that a description of Unix begin with a brief history of the system. Part of the rationale for such a beginning is that the history is a significant part of the Unix culture and the Unix mystique. It can also be helpful, since an understanding of the evolution of Unix can sometimes help to make Unix commands seem less cryptic.

First Version in 1969 on DEC PDP 7

The first version of Unix was created on a Digital Equipment Corporation PDP 7, which was a small machine. The idea originated with Ken Thompson of Bell Laboratories who reportedly wanted a better system to execute a 'space wars' program. While that is not the whole story, it is at least partially true. The name Unix is a pun based on the Multics system, which Thompson considered unsatisfactory. Actually the name adopted was UNIX®, which is now a trademark of AT&T. The system was developed and refined by Thompson, Dennis Ritchie, and Rudd Cadaday and the first "real Unix" system for a DEC PDP 11 was completed in 1971. Two years later, in October 1973, the landmark paper describing the system was presented at the Fourth ACM Symposium on Operating Systems Principles and the following year appeared in the July issue of *Communications of the ACM.* Although the system described in that paper was to become one of the most influential in the history of computing, that first paper appeared to have little initial impact. Unix continued to be developed at Bell Labaratories and eventually version 6 was released to the public. At that time, Unix became available to universities at nominal cost, but no support was to provided. That decision had a tremendous effect on the nature of the system and

0-8493-2530-7/96/$0.00+$.50

on Unix installations. The users had no "big brother" to help them with their problems of maintenance, system modifications, and ports to new platforms. They consequently came to rely upon each other by sharing information and software. The rise of Unix paralleled the growth of what would become the Internet, which increasingly facilitated the sharing of software. Both Bell Laboratories (which became AT&T) and universities (particularly the University of California at Berkeley) continued to develop and use Unix until it was soon the most commonly used and understood operating system in universities throughout the United States.

In the late 1970s and most of the 1980s, there were two other factors that contributed to Unix's rise. One was the explosion of hardware vendors and the other was the severe shortage of trained talent in the computing field. That combination imposed economic advantages of adopting Unix as the system for new workstations on the market. During the time when many new platforms were entering the market, it was comparatively easy to implement Unix on those platforms and to recruit systems programmers who were already familiar with the system due to their experiences at universities. Thus Unix proliferated greatly during that period.

Philosophy

Unix was developed originally on and for small machines. If Unix represents any single overriding philosophy, it would be summarized by "small is beautiful." Unix was to be highly modular, highly efficient, and highly flexible. The modularity was largely achieved and did provide a great deal of flexibility. The efficiency was partially achieved. The high degree of modularity was undoubtedly a significant factor in Unix's rise as an operating system. Commands performed a single function and had a standard interface. There were to be something close to the minimal number of commands necessary to do all of the things a user would ever want to do. The resulting command set, with its short, cryptic names, came remarkably close to achieving that goal. In some cases, understanding the minimalist philosophy of the designers can help one understand the command set. For example, Unix has no command "type" as in MS-DOS, but the at first glance mysteriously named command "cat" appears to do the same thing. Actually, "cat" (short for "concatenate") writes the concatenation of several files into an arbitrary output file. If there is only one file to be "concatenated" and the output file happens to be the standard output file that appears on the user's screen, then the command behaves like a "type" command in MS-DOS. Thus there is no need for an extra "type" command and it was not included. It is true that Unix lets you rename commands to suit your preferences, but that is often insufficient for novices who do not know what commands to rename. Once they learn, the value of having alternative names is greatly reduced. Typically new users view Unix as user unfriendly. But Unix is actually quite friendly to the advanced experienced user and that is one of the reasons for its tremendous success.

Innovations Introduced by Unix

The manner that modularity was achieved in Unix was innovative. As mentioned above, most of Unix's commands do very little. Hence, to accomplish the complex tasks available in other systems, you must be able to combine the simple commands. Unix introduced operators (pipes, tees, and redirections) that served as the glue with which users could combine commands conveniently.

Unix was implemented in the high level language C. It was not the only system to be implemented in a high level language (e.g., the system Multics was implemented largely in PL/1), but certainly it was the first small system so implemented. The command user interface constituted a complete programming language. This allowed the development of "shells" that enveloped the system and allowed users to create new command sets more to their liking. Users were even able to change shells (and hence command sets and syntax) as they wished and could have a number of shells active at any given time. The user's view of Unix is as a set of concentric circles where the innermost circle is the system kernel and each of the outer circles is a user shell. The user may exit a shell, thus erasing the circle, or may enter a shell, thus drawing a new outer circle. A single command allows one to enter a new shell and work within that framework for a while. Upon exit, the user does not exit the system but merely reverts to the original shell. In fact, users seldom do much switching between shells, but there are occasions when it can be convenient because of special features of a particular shell. We shall discuss shells further in Chapter 6.

Since the original AT&T (Bell labs) version of Unix, numerous variants have come into existence. First the original version continued to be developed through 6 versions. Version 6 was released to the public and served as the basis for several important variants. Microsoft developed a version for PCs and called it Xenix. The University of California at Berkeley developed a variant that they called BSD. Meanwhile, AT&T added enhancements and developed a new version called UNIX 2.0. They continued to develop new systems, creating a version called System III that eventually became System V. The second release of that system was called System V, release 2, or SVR2. That was followed by SVR3 and SVR4 and, most recently, SVR4.2.

At Berkeley, work also continued through a series of new versions until 4.3 BSD. It was a combination of 4.3BSD and SVR3 that became IBM's AIX system. The 4.3 BSD version was the last to come out of Berkeley as they decided to terminate their development systems.

The most common versions of Unix for PCs is SCO Unix (based on SVR3.2) and Xenix, which is now also marketed by SCO. Linux is a recently developed system that is available for free via a variety of bulletin boards. Linux is also sold by a variety of vendors who have packaged the variations of the system in various ways along with documentation. The system itself is free, but the vendors charge for the media, documentation, and any enhancements they may have added. There is also an independently developed version for PCs called Coherent by Mark Williams Co. and a Unix-like toolkit that runs along with DOS from Mortis Kern Systems, Inc.

In addition to the variants already mentioned, Hewlett-Packard has their own version HP-UX, and DEC developed a version they called Ultrix. There is also Solaris, which runs on SPARC-based machines and on the new PowerPC.

5.2 THE FILE SYSTEM

The creators of Unix considered the system's most important role to be that of providing a file system. Virtually everything in Unix is a file. This simple design principle did a remarkable job of simplifying the system and increasing the power of the command set. From the users' viewpoint, there are three kinds of files: ordinary disk files, directories, and special files.

Ordinary files contain whatever the user chooses to put in them. Most ordinary files are expected to be text files, which are simply strings of characters. Binary files are generally those created by compilers or linkers and contain executable forms of programs.

Directories may be thought of as a table that maps the names of files to their attributes. One of those attributes is, of course, the physical location of the file, but there are other attributes in the table as well. Such additional information includes the type of file (directory, text file, binary file, special file), date of creation, etc. Note that a file in a directory may be another directory, thereby creating a hierarchy of directories and other files. If a directory C is in contained in another directory P, then we refer to C as the child of P or a subdirectory of P. We refer to P as the parent of C. Each directory always has at least two entries. The name '.' , which does not include the quote marks and is often pronounced "dot," refers to the directory itself so that a program may read the current directory by using the name '.' as an alias to the complete path name of the directory, as defined below. The other entry that is in each directory is '..', which may be used as the name of the parent directory. The "topmost" directory that has no parent is called the root directory and in that directory, '.' and '..' each refers to the same directory, namely the root.

The same file name may appear in several directories but will always have a unique path name. The directory hierarchy determines the path name for a file. The full path name of a file consists of a sequence of directory names (the root has no name) separated by slashes. Thus the path name /tom/dick/harry represents a file named harry contained in the directory dick which in turn is contained in the directory tom. The directory tom is contained in the root directory. The file harry may be an ordinary file, a directory, or a special file. The path name for the root directory is simply '/'.

5.3 GETTING STARTED ON A UNIX SYSTEM

Logging on and off a Unix system is quite simple. Typically, when assigned an account for a Unix system, a user is given a user identification (userid) and an initial password. Once connected to the system, you simply type your userid at the prompt and then your password when prompted. Usually, the first order of business is to then change your password using the "passwd" command. Note that on some systems, there may be a need to use a different command to change one's password. If such is the case, that would most likely have been make clear when the account was assigned. To log off, the command is "logout" rather than "logoff." Also, CONTROL-D will log off a user if typed at the system prompt of the outermost shell (see below). There may be some additional actions that should be completed before logging off (see discussion below). In particular, it is sometimes necessary to terminate additional processes that have been created during the session. Once logged on, the user is ready to begin issuing commands.

5.4 UNIX COMMANDS

Basic Format

The basic format of Unix commands is straightforward. We shall use the "ls" command to demonstrate the format of the commands as well as our conventions for describing them.

```
ls List contents of directory
```

Form:

```
ls [-aClRx] [file .. ]
```

where the options are

a	List all entries: usually entries whose names begin with a period are not listed
C	Multicolumn output, sorted down the columns
l	Gives "long listing," which includes the type of file, permissions, link/directory count, owner, group, size and date of last modification
R	Recursively list subdirectories encountered
x	Multicolumn output, sorted across the columns

The square brackets delimit optional portions of the commands and may be omitted. Thus, all of the following are legitimate "ls" commands:

```
ls
ls -a
ls -aCl *.doc readme.*
```

The letters following the hyphen are command options. In the above example, as in all of our examples here, there are actually many more options than we have indicated. We have included only the most commonly used options and, since not all command options are identical across all Unix systems, only those that are most commonly available.

Wildcards

The asterisks in the last example above constitute "wildcards" or "metacharacters." They are special symbols, interpreted by the shell, that can be used to represent a number of files. If you are a MS-DOS user, you are already familiar with wildcards. The wildcards of Unix are quite similar to those of MS-DOS, but those of Unix are more flexible. Since wildcards are interpreted by the shell, they could be different for different shells, but all of the major shells use the same wildcard conventions (there is one exception noted below for C shell users).

There are three types of wildcards that may be used in filenames (Table 5.1): the asterisk (*), which represents any string of characters (including the string of no characters), the question mark (?), which represents any single character, and the character set ([...]), which represents any one of the characters included within the square brackets. Within the square brackets, there are further notations that can be used. Hyphens can be used to represent a range of letters. Thus a–g is short for abcdefg. An excla-

TABLE 5.1 Wildcards

File Name with Wildcards	Will Match
*	Any file name
a*	Any file name beginning with a
a*k	Any file name beginning with a and ending with k
a	Any file name containing the letter a
a??	Any file name consisting of exactly three letters beginning with a
*[xyz]	Any file name ending with one of the letters x, y, or z
*[!xyz] (not valid for C shell)	Any file name ending with a character other than x, y, or z

mation mark may be used after the first square bracket to indicate that the character set represents any character except for the characters listed. Thus, [!abc] would represent any character other than a, b, or c. Note that the exclamation point has a different special meaning for the C shell and so it must be preceded by a backslash (\). In the C shell, we would type [\!abc] to represent any character other than a, b, or c. Note that not all implementations of the C shell implement this negation feature and so it may not work on your system. If not, you may change to another shell as explained in Chapter 6.

As a second example, we will consider the "cat" command.

```
cat [-v] [file ...]
```

Concatenates files

Where the options are:

```
-v Lists nonprintable characters
```

The "cat" command is also used to view files as explained above. For example, suppose the file named "start" contains the text

```
Four score and seven years ago
our fathers brought forth on this continent
```

and the file named "middle" contained the text

```
a new nation conceived in liberty and dedicated
```

and the file named "finish" contained the text

```
to the proposition that all men are created equal.
```

Then the command

```
cat start middle finish
```

would cause the following to appear on your screen

```
Four score and seven years ago
our fathers brought forth on this continent
a new nation conceived in liberty and dedicated
to the proposition that all men are created equal.
```

Redirection and Pipes

If we wished to save the text of the previous example into a separate file we could issue the command

```
cat start middle finish > gettysburg.address
```

That would create a new file (or add to the contents of an existing file) named gettysburg.address that would contain the text as displayed above. If we wished to view it for confirmation we might issue the command

```
cat gettysburg.address
```

which would display the file on our screen.

The character ">" is the redirection symbol that directs the system to place the output into the file "gettysburg.address" rather than the standard output.

Had we put the entire Gettysburg Address into our file instead of just the first sentence, then when the "cat" command displayed the file, most of it would run off the screen before we could view it. We could stop the display by striking CONTROL-S and then restart by striking CONTROL-Q, but we would have to be quick with our fingers to avoid missing any part of the text. Better would be to use an alternate command called "more."

```
more gettysburg.address
```

Then, each time the screen filled up, the system would generate a pause. Striking the space key will cause the next page of text to be displayed.

Pipes

Commands may be combined in Unix using the pipe operator (|), which causes the output of one command to be the input (or first input if there is more than one) of the next command. For example, if we wanted to view the three files "start," "middle," and "finish" with the "more" command, we could use three commands

```
more start
more middle
more start
```

or simply the single command

```
cat start middle finish | more
```

"tee" Command

We can go further. If we wished to concatenate the three files "start," "middle," and "finish" into a single new file and then view the results, we could use the "cat" command and then the "more" command as we did above or we could employ the "tee" command. The "tee" command is used in pipe sequences to save intermediate results into a file. Thus another way to do what we did above would be

```
cat start middle finish | tee gettysburg.address | more
```

which would save the concatenated file as well as display the results one page at a time.

Special Keys

There are several special keys beyond the CONTROL-S and CONTROL-P mentioned above. The particular keystrokes will vary with different systems, but Table 5.2 gives the most common. You can determine the keystrokes by consulting with your system administrator or by using the "stty" command.

Quotations

Certain characters have special meaning within a shell; different shells have different sets of special characters. We have already seen that the backslash can be used before a character to remove its special meaning. For strings of characters, quotation marks can do something similar. In particular, enclosing a string in single quotes will, with one exception, remove the special meaning from all of the characters in the string. The one exception is the exclamation point (called "bang") within the C shell. That particular symbol must be preceded by a backslash even when quoted. When using the Bourne or Korn shells, there is no problem with the exclamation point. Enclosing a string within double quotation marks will remove the special function of all symbols except for dollar signs, back quotes, and backslashes (and, for the C shell, exclamation points). Unix also uses the back quote or grave character, but we shall defer a discussion of that until the next section.

5.5 A BASIC MINIMAL COMMAND SET

There are several commands that can help you find your away around Unix. One of the most valuable (ultimately) is the "man" command. This will give you the pages in the Unix manual that describe any given command. Recall that our descriptions of commands do not include all of the options and capabilities of commands. The "man" command will enable you to obtain all the information you want. In fact, frequently, you get considerably more than you want. For example, if you give the command

```
man man
```

you will see approximately 11 pages of detailed information about the man command. With a little practice, you learn how to scan through the material you do not want or need, but is useful to have some sort of abbreviated command reference. Many systems will implement a simple help command that summarizes the command set, or possibly summarizes the help facilities. We will illustrate methods of combining Unix commands to do your own automatic editing a little later.

Another command that is often used is "apropos," which returns a list of the commands with a particular keyword in their title.

TABLE 5.2 Special Keys

Key	Function
CONTROL-S	Pauses the display of the standard input stream
CONTROL-Q	Resumes displaying of the standard input stream
CONTROL-C (or DELETE)	Interrupts the current command
CONTROL-D	Exits the current shell. If the current shell is the outermost shell, then logs off the system
CONTROL-H	Backspace

Copying Files

```
cp source_file target_file
```

Copies the file *source_file* into a (possibly new) file *target_file.* If *target_file* already exists, it will be overwritten (but retains its permissions). If the *target_file* does not exist, it will be created with the user as the owner.

```
cp source_file [file ..] target_directory
```

Copies a list of files into a *target_directory.* The names in the new directory will be the same as in the old and, as in the previous example, any existing file with the same name will be overwritten.

```
cp [-R] source_directory target_directory
```

Will copy all files in the *source_directory* into the *target_directory*

The option to the "cp" command is

R Recursively copies subdirectories so that entire directory subtree is copied.

Removing (Deleting) Files

```
rm [-iR] file [file ..]
```

Removes all of the files listed. If a given file is a directory, then all files in that directory are deleted.

The options to the "rm" command are

i Interactive mode. Interactively confirms each deletion.
R Recursively deletes all subdirectories of any directory listed.

Moving or Renaming Files

mv *source_file [file ..] directory*

Moves *source_file* and other listed files, if any, into *directory.* This is usually equivalent to copying the designated files and then deleting them. However, there can be a subtle but important difference because "moving" a file does not alter its permissions as copying sometimes does.

Creating Directories

```
mkdir directory-name
```

Creates a new directory named "directory-name" that will be a subdirectory of the current directory.

Moving to a New Directory

```
cd directory-name
```

The cd commands make "directory-name" the new current directory. If no directory name is specified, the user's home directory becomes the new current directory.

Setting and Changing a File's Permissions

```
chmod mode file [file..]
```

The permissions of the listed files are changed to *mode*. If a directory is listed, then all files within that directory have their permissions changed as specified by *mode*. There are two major ways to express the desired mode: symbolic and numeric. For brevity, we will discuss only the symbolic format here. In the symbolic mode format, *mode* c consists of up to three symbols that represent a group of users, an operator, and a permission (Tables 5.3, 5.4, and 5.5).

Examples

```
chmod a+rwx myfile
```

Gives full rights for everything to everything.

```
chmod g+rw,o=r mydirectory
```

Gives read and write privileges to all members to of the owner's group and read-only privileges to all other users. Note that if "mydirectory" is a directory, then the privileges applies to all files in the directory.

TABLE 5.3 Specification of user group

u	Owner or current user of the file
g	Members of same group as the user
w	All other users
a	All users

TABLE 5.4 Operators

+	Add the permission
–	Remove the permission
=	Set the permission to be . . .

TABLE 5.5 Permissions

r	Read
w	Write
x	Execute

Searching for Strings

The three commands grep, fgrep, and egrep are all used to search one or more files for lines matching a pattern. The patterns are given as regular expressions (defined below). All three commands do the same thing; the three versions exist only for efficiency. The most general command of the three is "egrep" and may be used exclusively. The fastest is "fgrep" but it can be used only when searching for specific literal strings. If you are searching for a simple pattern, "grep" can be used and will execute somewhat faster than "egrep." Since "egrep" is the most general, all of our examples will use it.

Regular Expressions

Regular expressions are patterns for a set of strings (Table 5.6). Regular expression include ordinary characters and special operators called metacharacters (Table 5.7). Ordinary characters are those that are not special operators. The metacharacters may have special meaning depending on their context, or may behave like ordinary characters. Note that some metacharacters (e.g., '*' and '?') that are treated as wildcards by the shell do not have the same meaning when used as metacharacters.

TABLE 5.6 Special Operators for Regular Expressions

.	Any single character
*	Zero or more occurrences of preceding regular expression
+	One or more occurrences of preceding regular expression
^	Beginning of line
$	End of line
[...]	Any one of the characters in . . . (ranges such as a–z may be used)
[^...]	Any character other than the characters in . . . (ranges such as a–z may be used)
\	Causes the following character to lose its special meaning (if any)
(...)	(Valid in egrep only) used to group regular expressions

TABLE 5.7 Examples of Regular Expressions

Expression	Will Match
a.z	Any three-letter string beginning with a and ending with z
a-z	Any single lower-case letter
^a.z	Any three-letter string at the beginning of the line that begins with a and ends with z
[a-zA-Z]*	Any string (including the empty string) consisting entirely of alphabetic characters
^[a-zA-Z]+$	Any line consisting entirely of alphabetic characters (excluding blank) that contains at least one character
^([a-zA-Z]* +)*$ (egrep only)	Any line consisting entirely of alphabetic words separated by blanks (note that the line may not begin with a blank and the last word must be followed by at least one blank)
^([a-zA-Z]* +)*[a-zA-Z]?$ (egrep only)	Any line consisting entirely of alphabetic words separated by blanks (note that the line may not begin with a blank)

5.6 CREATING AND EDITING FILES

Most user created files are, of course, text files, which can be created and modified by text editors. There are three major editors that are included in Unix enironments. There is a line editor called "ed." Since line editors are not particularly popular these days, we will omit any discussion of the details. We will, discuss briefly the stream editor "sed," which is based on "ed," in Section 5.7. Below we will discuss the standard Unix editor, "vi."

vi

The editor vi (pronounced vee-eye) is a powerful editor with a full set of commands for manipulating simple text. In many ways it epitomizes Unix—it is heavily used and loved by a large number of Unix advocates and genuinely hated by a different large group of Unix users. Those who love it admire its power and its convenience to users who have mastered its commands. Those who hate it resent the inconvenience of having to master a set of commands that, to them, is complex, cryptic, and counterintuitive. Novices who are accustomed to working with editors and word processors on PCs are particularly uncomfortable with the vi editor initially. We recommend developing at least a passing familiarity with vi since it is sometimes the only full screen editor available on a Unix system.

We will consider only a small subset of vi's full command set here, but that should be sufficient to get you started. As with all Unix commands, full details can be gotten by using the man command. To envoke vi, issue the command

```
vi file
```

where *file* is the name of the file to be created (or a previously created file to be edited). As is usual for Unix commands, there are a variety of options that can be employed but we shall discuss only this simple invocation. Some versions of Unix have a version of vi specifically designed for new users (e.g., vedit on System V) that supplies additional information of the command set.

When you first enter vi, you see a nearly empty screen. There will be a cursor in the upper left corner of the screen with a column of tilde (~) symbols each of which represents a blank line. The editor initially is in *command mode*, which means the first key you strike will be interpreted as a command to the editor. For many people, this is a departure from the usual setup. To actually type the text you want in your file, you need to issue a command that will allow you to enter *input mode*. Once in input mode, everything you type is input for your file. There are six commands that will allow you to enter input mode (Table 5.8). To exit insert mode and return to command mode, strike the ESCAPE key.

To navigate around your file, you most likely will be able to use the arrow keys on your keyboard as usual. However, Unix was designed to work with many terminals,

TABLE 5.8 Commands that Enter Input Mode

a	Appends text to the right of the current cursor position
A	Appends text to the end of the current line
i	Inserts text to the left of the current cursor position
I	Inserts text to the left of the first non-white character of the current line
o	Opens a new line inserted below the current line. Text is added at the beginning of that line
O	Opens a new line inserted above the current line. Text is added at the beginning of that line

and so there is a set of movement commands that allow users to navigate even on "dumb" terminals (Table 5.9). Each of the commands in Table 5.9 may be preceded by a number for multiple applications. For example, the command 5k will move up five lines in the same column (Table 5.10).

Deleting Text

Text that is deleted in vi is saved into an allocated area of memory called a *buffer*. It is possible to retrieve (or put) text from the buffer as described below. There are a number of delete commands in vi. The most useful are given in Table 5.11.

Copying and Moving Text

One copies and moves text much the way one does it in various Windows applications. You either copy (yank) or delete text, which will cause it to be put into a buffer. You then paste (put) the text from the unnamed or named buffer into the current buffer. The commands are given in Table 5.12.

TABLE 5.9 Cursor Commands

h	Moves left one position on current line
j	Moves down one line in same column
k	Moves up one line in same column
l	Moves right one position on current line

TABLE 5.10 More Cursor Movement Commands

nG	Moves to line n (where n is a number). If n is missing, moves to last line. In each case, the cursor moves to the first non-white character of the line.
0	Moves left to first character on current line
^	Moves left to first nonwhite character of current line
$	Moves right to the last character of current line. If preceded by a number n, will move to last character of the nth next line.

TABLE 5.11 Commands That Delete Text

x	Deletes the character at the current cursor position. If preceded by a number, n, then the n characters beginning at the current cursor position and moving right are deleted. If the end of line is encountered, then the deletions stop
X	Deletes the character to the left of the current cursor position. If preceded by a number, n, then the n characters beginning immediately to the left of the current cursor position and moving left are deleted. If the beginning of the line is encountered, then the deletions stop
d$	Deletes remainder of the current line
dd	Deletes the entire current line
dG	Deletes all the lines from the cursor to the end of the current buffer
d1G	Deletes all the lines from the cursor to the beginning of the current buffer

Exiting and Saving Your Work

There is a set of special commands in vi that begin with a colon (:) (Table 5.13). These commands are quite sufficient, albeit not always optimally convenient, for creating and editing files. There are a large number of vi commands not presented here that greatly increase the power and convenience of using vi. Once you are comfortable with the basics of vi, you may wish to consult man to learn how to move the cursor over words or sentences, search for text using regular expressions, work with multiple files, define macros, and many other operations.

5.7 FILTERS

A filter is a command that extacts, inserts, or rearranges its input stream. The most used filters in Unix are "sort," "sed," and "awk."

Sort Command

```
sort [-m][-d][-f][-r][-b] file [file ...]
```

The options for the sort command are:

-m	Merges presorted files
-d	Only letters, digits and blanks or used for the sort.
-f	Ignores case
-r	Reverses the sort
-b	Ignores leading blanks
+keybegin	Identifies keybegin (a positive integer) as the beginning column of the key
-keyend	Identifies keyend (a positive integer) as the ending column of the key

The sort command sorts one or more files. Each line of a file may be subdivided into fields that are separated by a delimiter (default a blank) so that sorts may be performed specific fields. As a simple exampe consider the file "fruit," which has apparently been typed hastily and has inconsistencies with captals and lower case letters. Suppose the file contains

TABLE 5.12 Commands for Copying Text

y$	Yanks (copies) the characters from the cursor to the end of the current line
yy	Yanks (copies) the entire current line
yG	Yanks (copies) all the lines form the cursor to the end of the buffer
y1G	Yanks (copies) all the lines from the cursor to the beginning of the buffer

TABLE 5.13 Commands for saving and Quiting

:w	Saves the current edit buffer by overwriting the current file name
:w <filename>	Saves the current edit buffer into the file namedd <filename>
:wq	Write the edit buffer into the current file name and quit vi
:q!	Quit vi without saving file.

```
Fruit      Price    Unit     Comment
pears      $7.50    box      Only until July 15
Apples     $3.75    box      Apple boxes contain only 6 apples
Oranges    $8.50    Box      From Florida
Grapes     $2.50    Bag      Mixed red and green
Grapes     $6.00    Basket   Red only
Apples     $22.50   Barrel   Each apple wrapped individually
Oranges    $14.70   Basket   4 basket styles available
Oranges    $5.50    Bag      overnight delivery
Grapes     $12.25   box      Mixed red and green
Pears      $5.00    bag      Only until July 15
Apples     $2.00    bag      bag contains only 4 apples
```

We may sort the the lines of the file with the simple command

```
sort fruit
```

to get

```
Apples     $2.00    bag      bag contains only 4 apples
Apples     $22.50   Barrel   Each apple wrapped individually
Apples     $3.75    box      Apple boxes contain only 6 apples
Fruit      Price    Unit     Comment
Grapes     $12.25   box      Mixed red and green
Grapes     $2.50    Bag      Mixed red and green
Grapes     $6.00    Basket   Red only
Oranges    $14.70   Basket   4 basket styles available
Oranges    $5.50    Bag      overnight delivery
Oranges    $8.50    Box      From Florida
Pears      $5.00    bag      Only until July 15
pears      $7.50    box      Only until July 15
```

Let us imagine that we wish to sort on the prices. Then we could issue

```
sort -b -n +1.1 fruit
```

which would ignore leading blanks and specifies a numeric fileld that is found by skipping the first field and the first column of that field. The result will be

```
Fruit      Price    Unit     Comment
Apples     $2.00    bag      bag contains only 4 apples
Grapes     $2.50    Bag      Mixed red and green
Apples     $3.75    box      Apple boxes contain only 6 apples
Pears      $5.00    bag      Only until July 15
```

```
Oranges  $5.50   Bag     overnight delivery
Grapes   $6.00   Basket  Red only
pears    $7.50   box     Only until July 15
Oranges  $8.50   Box     From Florida
Grapes   $12.25  box     Mixed red and green
Oranges  $14.70  Basket  4 basket styles available
Apples   $22.50  Barrel  Each apple wrapped individually
```

sed

"Sed" is a stream editor. It accepts a file (or files) as input and applies one or more editing commands to each line with the file. It is particularly useful for doing global editing of several files. Sed uses the command set from the line editor ed. For example, suppose I wish to substitute the word "sack" for "bag" in file "fruit." I could issue the command

```
sed 's/bag/sack/g' fruit
```

which would yield

```
Fruit    Price   Unit    Comment
pears    $7.50   box     Only until July 15
Apples   $3.75   box     Apple boxes contain only 6 apples
Oranges  $8.50   Box     From Florida
Grapes   $2.50   Bag     Mixed red and green
Grapes   $6.00   Basket  Red only
Apples   $22.50  Barrel  Each apple wrapped individually
Oranges  $14.70  Basket  4 basket styles available
Oranges  $5.50   Bag     overnight delivery
Grapes   $12.25  box     Mixed red and green
Pears    $5.00   sack    Only until July 15
Apples   $2.00   sack    sack contains only 4 apples
```

The string in single quotes constitutes an editor command. The first s indicates that a subsititution is desired and the word "bag" indicates the word to be replaced by "sack." The final g indicates that all occurrences of "bag" in the line should be replaced. Commonly a set of editing commands are put into a "script" file so that a number of different commands can be applied to the same file. For example, we could create the following script in the file called "scriptfile" to substitute "sack" for "bag" or "Bag" and "case" for "box" or "Box." We will create "scriptfile" as follows:

```
s/bag/sack/g
s/Bag/sack/g
s/box/case/g
s/Box/case/g
```

then we can issue

```
sed -f scriptfile fruit
```

to produce

```
Fruit    Price   Unit       Comment
pears    $7.50   case       Only until July 15
Apples   $3.75   case       Apple casees contain only 6 apples
Oranges  $8.50   case       From Florida
Grapes   $2.50   container  Mixed red and green
Grapes   $6.00   Basket     Red only
Apples   $22.50  Barrel     Each apple wrapped individually
Oranges  $14.70  Basket     4 basket styles available
Oranges  $5.50   container  overnight delivery
Grapes   $12.25  case       Mixed red and green
Pears    $5.00   container  Only until July 15
Apples   $2.00   container  container contains only 4 apples
```

There is a full set of editing commands that can be used with sed to perform complex editing tasks.

Awk

The awk program is a "pattern/action processor" that examines every line of a file (or files) for those that match a particular pattern. When a line matches the given pattern, then the specified action is performed. As with "sort," each line is divided into fields separated by delimeters. For example,

```
awk '$1 == "Apples" {print $1, $2, $3}' fruit
```

will produce

```
Apples $3.75 box
Apples $22.50 Barrel
Apples $2.00 bag
```

The string within the single quotes constitute the pattern/action pair, with the action enclosed within curly brackets. The pattern is this case is that $1, which stands for the first field in the line, is equal to the string "Apples." The action is to print the first three fields.

Awk actually constitutes a complete programming language and can be a very powerful tool for the experienced user. To illustrate its utility, we present several simple "one liners."

```
awk 'END {print NR}' file
```

will print the number of lines in a file. "END" is a special pattern that represents the end of file and "NR" is a built in variable that represents the number of records.

```
awk '$0 > 40' {print $0}' file
```

will print every line that is greater than 40 characters long. The variable "$0" represents the entire line.

```
awk '$1 = "keep" {print $2, $3}' file
```

will print only the second and third fields of each line that contains the word "keep" as the first field.

```
awk '/[aA]pples/ {print $0}' fruit
```

will print only those lines that contain strings that match the regular expression "[aA]pples." Thus, it would print any line that contains either "apples" or "Apples."

DEFINING TERMS

File Permissions: Information that is stored with a file that determines what privileges various users have. Privileges are read, write, and execute. Users are divided into the owner, the owner¸s group, and all other users.

Filter: A command that extracts, inserts, translates, or rearranges the data of its input stream.

Man Pages: The discription of a command in the online manual that is accessible via the "man" command

Metacharacter: A character that has special meaning to the shell or within a regular expression. Usually represents a set of strings.

Pipe: A mechanism, represented by "|" that sends the output of one command to another command as input.

Redirection: Changing the standard input (via "<") or the standard output (via ">") of a command to come from are go to a specified file.

Regular: An expression that represents a set of strings. It is used in "ed," "vi,"

Expression: and "grep" commands.

Shell: A command processor.

REFERENCES

Alfred V. Aho, Brian W. Kernighan, and Peter J. Weinberger, *The AWK Programming Language*. Addison-Wesley, Reading, MA, 1988.

Prabhat K. Andleigh, *UNIX® System Architecture*. PTR Prentice-Hall, Englewood Cliffs, NJ, 1990.

Maurice J. Bach, *The Design of the UNIX® Operating System*. Prentice-Hall, Englewood Cliffs, NJ, 1986.

Kaare Christian and Susan Richter, *The UNIX Operating System*, 3rd ed. John Wiley, New York, 1994.

Dale Doughery, *sed & awk*. O'Reilly & Associates, Sebastopol, CA, 1991.

Berny Goodheart and James Cox, *The Magic Garden Explained: The Internals of UNIX® System V Release 4, An Open Systems Design*, Prentice Hall of Australia Pty Ltd, Sidney, Australia.

Chris Hare, Emmett Dueaney, George Eckel, Steven Lee, and Lee Ray, *Inside UNIX®*. Riders, Indianapolis, IN, 1994.

Brian W. Kernighan and Rob Pike, *The UNIX® Programming Environment*. Prentice-Hall, Englewood Cliffs, NJ, 1984.

Samuel J. Leffler, Marshall Kirk McKusick, Michael J. Karels, and John S. Quarterman, *The Design and Implementation of the 4.3BSD UNIX Operating System*. Addison-Wesley, Reading, MA, 1989.

Dennis M. Ritchie and Ken Thompson, The UNIX time-sharing system. *Commun. ACM* 17(7):365–375, 1974.

Kenneth H. Rosen, Richard R. Rosinski, and Douglas A. Host, *Best UNIX Tips Ever*. Osborne McGraw-Hill, Berkeley, CA, 1994.

Peter H. Salus, *A Quarter Century of UNIX*. Addison-Wesley, Reading, MA, 1994.

Sams Development Team, *UNIX® Unleashed*. Sams, Indianapolis, IN, 1994.

Alan Southerton, *Modern UNIX™*. John Wiley, New York, 1993.

Alan Southerton and Edwin C. Perkins, Jr., *The UNIX and X Command Compendium*. John Wiley, New York, 1994.

Douglas W. Topham, *Portable UNIX™*. John Wiley, New York, 1992.

6

Typical Workstation Environments

Alton F. Sanders
Miami University

6.1 INTRODUCTION

The Unix environment can be tailored and personalized for the individual user. In this chapter, we will discuss the ways users create and maintain their own personal environment. We will explain the function of shells along with major differences of the three most popular shells. We will also identify software that is not actually part of Unix, but is commonly present in Unix environments.

6.2 SHELLS

The purpose of a shell is to provide an interface to the Unix system. The shell function most obvious to the user is the interpretation of the command line. How the shell interprets wildcards, pipes, and redirection symbols will have a direct effect on even novice users. Shells, however, do considerably more than that. They provide a programming facility that allows the user to combine and repeat commands in various and to control the computing environment in general.

There are three major shells for Unix systems. The first was the Bourne shell developed by Stephen Bourne of Bell Labs in 1979. It is available on virtually every Unix system. Less than 2 years after the Bourne she l was introduced, the C Shell was developed by Bill Joy who was then at Berkeley, but later became a co-founder of Sun Microsystems. The C shell was so called because of the resemblance of its shell language to the programming language C. The Korn Shell developed by David Korn, also of Bell Labs, is the most recent of the three and combines many of the better features of the previous two. If one were going to learn only one, the Korn shell would probably be the shell of choice if it were universally available. Unfortunately, AT&T chose to market the Korn shell separate from the Unix system and thus not all systems provide the Korn shell. We shall confine our examples primarily to the Bourne shell since it is available virtually everywhere.

0-8493-2530-7/96/$0.00+$.50

Variables

Each shell is capable of storing data for later use in named areas of memory called variables. Data are assigned to a variable via an equal sign. For example, at the command prompt of either the Bourne or Korn shell we may type:

```
WK=test/wk
```

or within the C shell

```
set WK=test/wk
```

to set the value of a variable that we are calling WK to the particular string "test/wk." We could examine the current value by using the echo command, which simply echoes to the screen the value of its arguments after processing by the shell. Thus in this example if we typed

```
echo $WK
```

we would see

```
test/wk
```

Notice that the dollar sign is used to retrieve the contents of the variable. Thus, if we typed

```
ls $WK/*.doc
```

it would be equivalent to typing

```
ls test/wk/*.doc
```

In our example, we used all capital letters in the variable name. That is not necessary, but it is a common convention since it makes it easy to identify variable names.

Shell Scripts

A shell script is a text file of shell commands. Instead of typing in each command in the file separately, we store them in a file and execute them as a package. As a very simple example, suppose your documentation files are always named with one of the file extensions, "doc," "bak," "wp," or "tex." Now for any given directory, you could simply type the appropriate ls command to see what files of those type are stored there, or you could create a file that assumes that the current directory of interest has been stored in the variable WK. Thus we may create the file "ls.script" that contains the single line

```
ls -l *.doc *.bak *.wp *.tex
```

Then instead of typing a long command line to see the files we want, we may type

```
sh ls.script
```

to accomplish the same thing. The command "sh" initiates a new copy of the Bourne shell ("csh" and "ksh" are the corresponding commands for the C and Korn shells, respectively) and, if a file name is specified, executes the commands in the specified file and exits.

We might want to make our script more flexible by including the particular directory of interest as part of our script. We might, for example, try making use of variables by changing our script file to

```
ls -l $WK/*.doc $WK/*.bak $WK/*.wp $WK/*.tex
```

The idea is a good one but contains a problem. A new subshell does not "remember" the values of the previous shell so we must "export" the variable $WK:

```
export WK
```

and now the command

```
ls.script
```

will work as expected (in the Bourne shell) provided the variable WK has been set to the directory path of interest.

Shell Programming

Shell programs are really just another name for shell scripts. We are creating an artificial distinction here to separate those shell programs that use control logic in the script language. Each shell has its own script language and we shall not attempt to explain them all here. As a simple example to illustrate the possiblities, consider the following Bourne shell script.

```
for i in 'ls'
do
   cat $i
done
```

The above script will set the variable i to each word in the output of the ls command and then display the contents of each file on the screen. The first line defines a "for loop," which will perform all the commands between "do" and "done" (called the body of the loop) varying the value of the variable i. The phrase

```
i in 'ls'
```

uses the back quotes around ls to indicate that it is the result of the command "ls." that is desired rather than the literal string "ls." While our example is a bit too simple to be realistic (a simple "cat*" would have worked as well), a small change in the body of the loop can produce a script that may be used for a variety of purposes. A very similar script may be used, for example, by a system manager to change users' file ownerships by changing the "cat" command in the loop body to an appropriate "chown" command.

6.3 CUSTOMIZING YOUR ENVIRONMENT

Each shell allows a certain amount of customizing by the user. We shall follow our convention of using the Bourne shell as the primary example. When you log on Unix and the interactive shell first starts executing, independent of which shell it is, it executes the commands in the file ".profile" in your home directory. The commands in that file define your terminal conventions via the "stty" command and establish your values for a set of standard system variables (Table 6.1).

The users may modify their ".profile." file for permament changes, or simply alter variable values at the command line for temporary modifications.

Aliases in the C Shell and Korn Shell

Both the C Shell and the Korn Shell allow for additional customization by using an "alias" command, which is not actually part of the Unix kernel, but a feature of those shells. Thus the command

```
alias dir='ls -CF'
```

would define a command "dir" that would be an "ls" command with the C and F options set. In the Korn shell, you would put the "alias" command in your ".profile" file, but in the C shell, you would probably put them in your ".cshrc" file. The ".cshrc" file is executed by the C shell each time a new C shell is initiated, so that new invocations of the C shell would retain the values for your aliases.

TABLE 6.1 System Variables in the Bourne Shell

$HOME	Name of your home directory
$PATH	Name of the search path used by the shell for executing commands
$CDPATH	A list of directories searched on "cd" command
$IFS	Internal fiels separators (default is space, tab, and new line)
$LOGNAME	Login name
$MAIL (or $MAILPATH)	Used by shell to notify you of mail
$MAILCHECK	Frequency of mail checks
$SHELL	Name of the login shell
$SHACCT	Name of file for accounting information
$PS1	Primary prompt string (default is "$")
$PS2	Secondary prompt string for multiline commands (default is ">")
$TERM	Name of terminal

6.4 EMACS EDITOR

The Emacs editor is not a part of Unix, but it is often available on Unix systems. It is an extremely powerful and flexible editor with a powerful macrolanguage that allows the user to extend the editor's flexibility even more. It is available for many platforms and GNU Emacs is available free from the Free Software Foundation. That fact is, no doubt, a major contributing factor to its popularity. The command to start Emacs varies from system to system and may be "emacs," "gnuemacs," or something else entirely.

Emacs is a modeless editor (almost), so when you enter you are ready to begin typing text at once. Commands are issued by using the CONTROL key along with other keys or the "meta" key along with other keys. The "meta" key varies with the terminal, but often is the ESCAPE key. The major terminal emulators of PCs usually define the meta key as the ALT key on the PC keyboard. In the Emacs documentation, "C-" represents the CONTROL key and "M-" represents the "meta" key. Hence, C-x represents CONTROL-x.

If you have no documentation handy, you may have trouble discovering how to get out. Your need to strike the sequence C-x followed by C-c to exit. If you forget that, and it is easy to forget at first, you may consult the help facility by striking C-h. C-h, like C-x, is a "prefix key" that expects something to follow. Initially, if you do not have a command list available, you will be unsure of what should come next so simply strike C-h again. If it still is not clear, strike it yet again. The third time, you will see a table displayed that will give you 17 different options (Table 6.2). Armed with that information on the help command and the quick reference table (Table 6.3) which gives you an adequate beginning command set, you should be able to deal with Emacs satisfactorily.

6.5 OTHER SOFTWARE

There are several programs in addition to Emacs that are commonly available in Unix environments. These programs will not be discussed in detail here, but as a Unix user, you should be aware of their existence.

TABLE 6.2 Emacs Help Options

Option	Meaning
A	Apropos—you type a word and it gives you all commands that contain that word in their title
B	List all key mappings
C	You type a key sequence and it gives you the function performed
F	You type a function name and it gives you documentation
I	Initiates a documentation browser
K	You type a key sequence and it displays full documentation
L	Shows the last 100 characters you typed (even if you have since deleted them)
M	Describes current mode
S	Displays annotated syntax table
T	Starts the tutorial (a good idea for a beginner)
V	You type the name of a variable, and it displays its value and function
W	You type a command name and it gives you the keystrokes
C-c	Displays copyright information
C-d	Displays ordering information
C-n	Displays log of recent Emacs changes
C-w	Displays warranty

TABLE 6.3 Emacs Commands

Command	Function
C-x C-c	Exit Emacs
C-h C-h C-h	Present 17 major help options
C-z	Suspends Emacs
C-l	Clear and redraw screen
C-f	Move forward one character
C-b	Move back one character
C-n	Move to next line
C-p	Move to previous line
M-f	Move forward one word
M-b	Move backward one word
M-a	Move back to beginning of sentence
M-e	Move forward to end of sentence
M-<	Move back to beginning of file
M->	Move forward to end of file
C-d	Delete character to right of point
Delete key	Delete character to left of point
C-k	Kill forward to the end of the line
M-d	Kill to the end of the next word
M-DEL	Kill back to the beginning of the previous word
C-x u	Undo the most recent change
C-y	Yank back killed text
C-x C-s	Save file

Perl

Perl is a programming language that is intended to handle the kind of problems that awk handles, only better. Actually, it does not use the same paradigm as awk, but it is a powerful programming language with many special features for handling text, user input, and features of Unix. It is relatively easy to learn and is available free. It is rapidly becoming one of the most popular languages for Unix users.

TeX and LaTeX

TeX (pronounced *tech*) is a text processing program developed by Donald Knuth that is also available free. In fact, Knuth has published all of the original source code for the entire program. It is a very powerful text formatter and can handle complicated mathematical notations as well as large varieties of fonts and layouts. It does that by imbedding special commands within the text. One of its values is that the text, with the TeX commands, can still be stored and transferred as plain ASCII text. Although powerful, it is not trivial to learn how to use TeX proficiently—it has been called the assembly language of text formatting. Thus Leslie Lamport introduced LaTeX, which is essentially a set of macro definitions for TeX that make it easier to use. Like TeX, LaTeX is powerful and can be stored and transferred as plain ASCII. Most Unix systems (and many other systems, for that matter) will have TeX and LaTeX available.

DEFINING TERMS

Bourne Shell: The first and simplist shell program.

C Shell: A popular shell with a C-like shell language.

Emacs: A full screen editor availabe from the Free Software Foundation.

LaTeX: A set of macros for TeX that make TeX easier to use.

Perl: A programming language tailored for many of the same problems addressed by awk.

.profile: The script executed by the shell upon login.

Shell Program: Another term for shell script. Used here as a more complex shell script.

Shell Scripts: A file of commands executed by the shell.

REFERENCES

Morris I. Bolsky and David G. Korn, *The Kornshell: Command and Programming Language*. Prentice Hall, Englewood Cliffs, NJ, 1989.

Chris Hare, Emmett Dueaney, George Eckel, Steven Lee, and Lee Ray, *Inside UNIX®*. Riders, Indianapolis, IN, 1994.

Donald E. Knuth, *The TeXbook*. Addison-Wesley, Reading, MA, 1986.

Leslie Lamport, *LaTeX: A Document Preparation System*. Addison-Wesley, Reading, MA, 1986.

Sams Development Team, *UNIX® Unleashed*. Sams, Indianapolis, IN, 1994.

Alan Southerton and Edwin C. Perkins, Jr., *The UNIX and X Command Compendium*. John Wiley, New York, 1994.

Richard Stallman, *GNU Emacs Manual*, 6th ed. Free Software Foundation, Cambridge, MA, 1987.

Larry Wall and Randal L. Schwartz, *Programming Perl*. O'Reilly & Associates, Sebastopol, CA, 1990.

7

X Windows

Alton F. Sanders
Miami University

7.1 INTRODUCTION

Modern users strongly prefer the convenience of a graphical user interface (GUI), which is missing in Unix. MS-DOS, Windows, and Apple users are accustomed to having pull-down menus, radio buttons, and the like readily available in their applications. Many users find it cumbersome to work with an interface consisting of a command line only. In some ways, GUIs are antithematic to Unix, since Unix was created originally to be functional on the full range of terminals, from the most sophisticated to the dumbest. Graphical interfaces are highly dependent on the terminal type and so violate one of the significant design principles of Unix. The X Window System is surely the graphical interface system that is most compatible with the Unix philosophy of simplicity. Actually, the X Widow System is not really a GUI. Rather it is the foundation upon which a user interface, in the form of a "window manager," can be built. Unix supports the X Window System and the two most popular window managers: Open Windows (the popular version of Open Look), from Sun Microsystems, and Motif (a window manager that functions somewhat like Microsoft Windows), from the Open Software Foundation (OSF). The window managers are built on the X Windows Systems but are not a part of it. In this section, we will discuss the X Window System and give a brief summary of the nature of Open Look and Motif. In the next section, we will cover Motif in somewhat greater detail.

The X Window System, also referred to X Windows, X, X11R5, or X11, is widely available for Unix systems and has become something of a de facto standard for a Unix windowing system. However, if you look at a several systems, you will probably find that they do not look very much alike and you may wonder whether X should be called a standard at all. Part of the reason for these different looks is that X, like Unix, is a highly flexible systems that allows a great deal of customizing. Second, the look of a system is largely determined by the window manager, typically either Open Windows or Motif. Finally, the window managers themselves vary substantially from system to system. Paradoxically, it is the great variability of Unix and X Windows that have made them "standards."

0-8493-2530-7/96/$0.00+$.50

7.2 OVERVIEW OF X WINDOWS

The basic idea of X Windows is simple. Since terminals vary so much, particularly with the details of graphical I/O, the best approach to standardize graphical interfaces is to establish a standard set of functions that is implemented by both the user's machine and Unix's machine. Then a terminal can become an X terminal by providing those standard functions. If the functions are the right ones, we have a powerful, flexible graphical interface system that is reasonably easy to implement and hence easy to port to new platforms. Of course, the realization of that idea was far from simple, but X Windows, while certainly imperfect, appears to be close to an optimal solution.

Client/Server Architecture

X is designed using a client/server architecture. Logically, there are two machines: the user's workstation, which is viewed as a terminal to a larger central Unix machine. There may actually be only one machine that fills both roles, but it is easier to think of terms of a network environment where the user's workstation will be a terminal connected to a larger, faster, central machine. In X, the roles of client and server are reversed from the usual configuration. The user's workstation functions as the server and the larger, faster machine is the client.

There are three major components of X:

1. Server. The server is the software that controls the devices on the user's workstation. That includes the display, the keyboard, and pointing device such as a mouse or graphics pad. Note that the user's display is not necessarily synonymous with the user's screen. An X display could extend over more than one monitor screen. The server software runs on the workstation itself.
2. Clients. The clients are the application programs that may execute on the user's workstation or elsewhere.
3. Communication Link. The communication link constitutes the vehicle by which messages are exchanged between clients and the server.

Server

Every workstation capable of functioning as an X terminal must be capable of providing the services required of an X server. It must handle input from clients who send requests for graphical tasks. Such tasks include creating, configuring, and deleting windows, providing the requested fonts, colors, line styles, etc., and displaying bitmapped graphic images on demand. Clients also communicate with each other, and the server functions as the central message center for such communication. It must also read the input from the user via the keyboard or mouse. Finally, the server manages any network connections.

The server makes very few decisions for its clients. It does not, for example, determine what windows are currently visible. If a window becomes hidden because another window was drawn on top if it, it is the client that must be aware of its status and request that it be redrawn. Similarly, it is the client that must (via messages to the server with the proper requests) resize windows, determine colors and fonts, and determine when a window should be deleted. The server actually deletes the window, but only when told to do so by the client.

Similarly, although the server is responsible for collecting input from the user, it does not interpret it. All interpretation of user input is done by the client. Nor does the server interpret the messages that clients use to communicate with each other. Rather, it maintains sets of named locations called "properties." Clients may request that data be stored in a certain property or be retrieved from a certain property. They may also request that data in a certain property be sent to another client. In all cases, the server does not interpret the data, but functions as an intermediary.

Finally, although the server deals directly with the user, it does not include the user interface. All user interface code is executed by clients. There were good reasons for this design decision. First, the creators of X felt that user interface design was not a task best undertaken by computer scientists and systems analysts. Their view was that human factors experts were best able to design user interfaces and so X Windows should provide enough flexibility for them to do so. Second, the standards for user interfaces are rapidly changing as the technology advances. Interfaces that were quite acceptable to users 5 years ago are now considered primitive and inadequate. The designers of X wanted a foundation that could remain stable with the changing state of the art. Finally, the relative merits of different user interfaces tend to be emotional issues. The best way to avoid the controversy that goes with such issues is to avoid the issues themselves.

The decision to exclude user interfaces from the system required leaving out everything that was not absolutely necessary. That minimalist philosophy is quite consistent with Unix and was certainly a significant factor in the success of X. However, there are some drawbacks to such a decision. The situation is similar to the early development of programming languages with respect to I/O. Using similar reasoning, the makers of Algol 60 chose to leave I/O of the language with the result that different implementations of the language had I/O that differed significantly. As mentioned in the introduction, we have a similar situation with respect to graphical user interfaces today. We currently have two major window managers, but there are at least 18 other window managers in use. Since any single window manager may vary substantially, we have many different user interfaces that are in use on Unix systems. It is not yet clear whether the evolution of interfaces will lead us to fewer alternatives and eventual standardization or to greater diversity and more confusion.

At this point, however, there is no doubt that X has been a tremendous success. There are many advantages that the X system offers. First, X is nonproprietary so that it is widely available at low cost. It minimizes the dependence of the system on the particular workstation hardware, the clients' platform, and the network. Finally, as discussed above, it is "politically neutral" with respect to the "best" policies for user interfaces.

7.3 STARTING X

If you are fortunate, X will automatically be available on your system through the shell script "xdm." In that case, when you connect to the system you will see a window saying "Welcome to Xdm" along with places for your user name and password. A window manager such as Motif or Open Windows will be started for you. Otherwise, you will have to enter "mwm" (for Motif) or "olwm" (for Open Windows) at the "xterm" prompt. If X does not start automatically, you will have to initiate it from the command line once you have signed on the system. That will involve using the script "xinit." We recommend that you consult with your system manager about how to get started. The flexibility of X provides for great variation among different systems and so it is quite impossible to provide you with a "cookbook" approach to setting up X if it is not already provided by your system. Ultimately you will wind up executing a shell script (either "xdm" or "xinit") but running X will require proper values to 30 or more environment variables so you will most likely need assistance from your system manager for the proper setup. Moreover, it is unlikely that you will want to run X without a window manager and so the more appropriate question for your system manager is "How do I run Motif?" or "How do I run Open Windows?" It is unlikely that you will have any reason to deal with "raw" X Windows. If you do find that necessary, you will require more information than we can reasonably supply here.

7.4 WINDOW MANAGERS

Window managers implement window policies, that is, where buttons or "hot spots" are on windows, which mouse button one presses and focus policy (which window is the one that will receive user input, and how do you change it). The windows of Motif and Open Windows are similar in appearance, although Open Windows often uses rounded corners or marks corners with a metalic look (to indicate the resizing area) while Motif uses rectangles almost exclusively. Both, for example, button in the upper left corner of each window that you may use to close the window. There are differences, however. For example, in Open Windows the user may "pin" a menu anywhere on the screen. Pinning a window to a location causes that window to stay there until unpinned by the user. There are also some significant differences in terminology between the two systems. In Motif, for example, "close" means to exit an application and "minimize" is used to iconify a window. In Open Windows, "close" means to iconify and "quit" is used to exit.

The left button is the default selector for Motif while the right one is the default for Open Windows. Each allows the user to reverse the mouse button order. The two managers also differ in focus policy (the focus determines which window receives input typed in by the user). Open Windows changes the focus simply by moving the mouse pointer to the window, whereas Motif requires the user to click the mouse before the focus is changed. The Open Windows policy is faster, but more error prone. The user must remain alert to mouse movements, even if such movements are due to an accidental bump.

DEFINING TERMS

Client: The applications that request various graphic tasks to be performed by the server.

Client/Server Architecture: An architecture that divides (logically) the machine requesting service from the one that provides that service.

Communication Link: The pathway through which clients and the server exchange messages.

Focus Policy: The protocol for determining and changing the window that is to receive input.

Iconify: To replace a window with an icon. In Motif's terminology, to *minimize* the window and in Open Windows's terminology, to *close* the window.

Motif: One of the two most popular window managers (Open Look is the other).

Open Look: One of the two most popular window managers (Motif is the other).

Open Windows: The popular implementation of Open Look

Pin (a window): To fix the window in place. A capability available in Open Windows.

Server: The software that fulfills the requests for graphic tasks received from clients.

xdm: One of two shell scripts for starting X (xinit is the other). Using xdm is normally considered easier to use than using xinit.

xinit: One of two shell scripts for starting X (xdm is the other). Using xinit is normally considered more cumbersome than using xdm.

REFERENCES

Chris Hare, Emmett Dueaney, George Eckel, Steven Lee, and Lee Ray, *Inside UNIX®*. Riders, Indianapolis, IN, 1994.

Eric F. Johnson and Kevin Reichard, *X Window Applications Programming*. MIS Press, New York, 1992.

Niall Mansfield, *The Joy of X*. Addison-Wesley, Reading, MA, 1993.

Levi Reiss and Joseph Radin, *X Window Inside & Out*. Osborne McGraw-Hill, Berkeley, CA, 1988.

Kenneth H. Rosen, Richard R. Rosinski, and Douglas A. Host, *Best UNIX Tips Ever*. Osborne McGraw-Hill, Berkeley, CA, 1994.

Sams Development Team, *UNIX® Unleashed*. Sams, Indianapolis, IN, 1994.

Alan Southerton and Edwin C. Perkins, Jr., *The UNIX and X Command Compendium*. John Wiley, New York, 1994.

8

OSF/Motif

Alton F. Sanders
Miami University

8.1 WHAT IS A WINDOW MANAGER

When computing was a young industry, all of a user's interaction with the computer took place in one window. The window usually encompassed the entire screen of whatever display device was being used. The user had little or no control over the display options of that window, except perhaps for the number and width of the lines being displayed. As computing progressed it became possible to have more than one process running at the same time. This led the user to interact with more than one process at a time, thus having one window for each process, and the invention of a window manager. A window manager allows the user to display, alter, and interact with more than one window at a time. The window manager's primary responsibility is to keep track of all aspects of each of the windows being displayed. Today, windows can be overlapped, put side by side, tiled, cascaded, hidden, stretched, reduced, iconified (reduced to an icon), and otherwise altered to the user's tastes and needs (Fig. 8.1).

To keep track of all of this modification and customization, let alone the functions of keeping data streams and their window assignments together, we need a window manager. Motif is such a window manager.

8.2 WHAT IS OSF/MOTIF

OSF/Motif is a window manager for the X-windows system. Motif was created by the Open Software Foundation (OSF), which is a consortium originally consisting of Digital Equipment Corporation (DEC), Hewlett-Packard (HP), and International Business Machines Corporation (IBM). It is important to note that Motif can mean many different things to different people. "Motif" can refer to

- The *Style Guide* for applications' look-and-feel as implemented in IBM's OS/2 and Microsoft Windows.
- A widget set, C library, for building *Style Guide* compliant applications.

0-8493-2530-7/96/$0.00+$.50

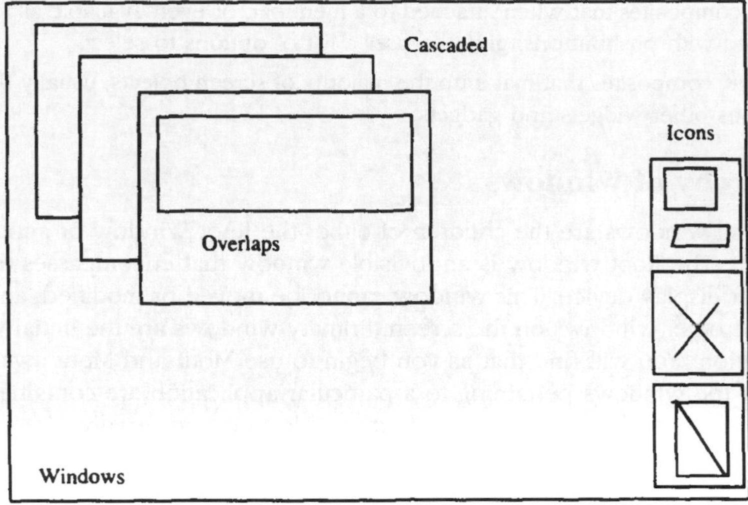

FIGURE 8.1. Cascaded and overlapped windows with icons.

- A window manager, mwm.
- A User Interface Language (UIL) interpreter, which allows the user to put all of the user interface code into an interpreted file.

This chapter will use "Motif" to mean only the window manager itself.

Components That Make up the Motif Window Manager (mwm)

A typical Motif session is made up of several objects, which are called widgets. A widget is defined as an object or group of objects that provides default behaviors to the user and programmer. It consists of a window along with a set of procedures and data structures that define the attributes and behaviors of the window. For the user, widgets provide a consistent interface into Motif. Regardless of the instance of the widget, the user knows that it will behave in a specific way. For the programmer, widgets come in classes that can inherit behaviors from parent classes of widgets, making the programming of a user interface much simpler. Widgets can be combined to form new, more complex, widgets called *composites*. Composites are therefore widgets with more than one component. Widgets that cannot be subdivided into components are called *primitives*, or *gadgets*. Strictly speaking, gadgets are actually not widgits, but rather objects without windows that are otherwise like widgets. We shall consider them to be windowless widgets. Some examples of different types of widgets are

- *Pushbuttons*, gadgets that provide the user interface with the ability to select an option or toggle a setting.
- *Scroll Bars*, composites that provide the user interface with the ability to process large amounts of data, while displaying only a small range.
- *Menu Bar*, a composite that, by itself, has no appearance on the screen, but can give the user a list of menus to choose from when filled appropriately.

- *Menus*, composites that when attached to a menubar, or even available as a pop-up, and filled with pushbuttons, give the user a list of options to select.
- *Windows*, composites that make up the majority of screen objects, usually filled with numerous other widgets and gadgets.

The Hierarchy of Windows

Within Motif all windows are the children of either the Root Window or another window (Fig. 8.2). The Root window is an invisible window that encompasses the entire screen of your display device. This window cannot be moved or modified, and it contains all of the other windows on the screen. Primary windows are the initial windows of an application. You will find that as you begin to use Motif and Motif aware applications, all of the windows pertaining to a particular application are contained within the primary window of that application. Primary windows are usually direct children of the Root window, although they can be children of other windows. Secondary windows are the subwindows of a Motif application. These are usually windows that contain information for the user, or that prompt for user input. Secondary windows include dialog boxes, status windows, and file selection boxes, to name a few.

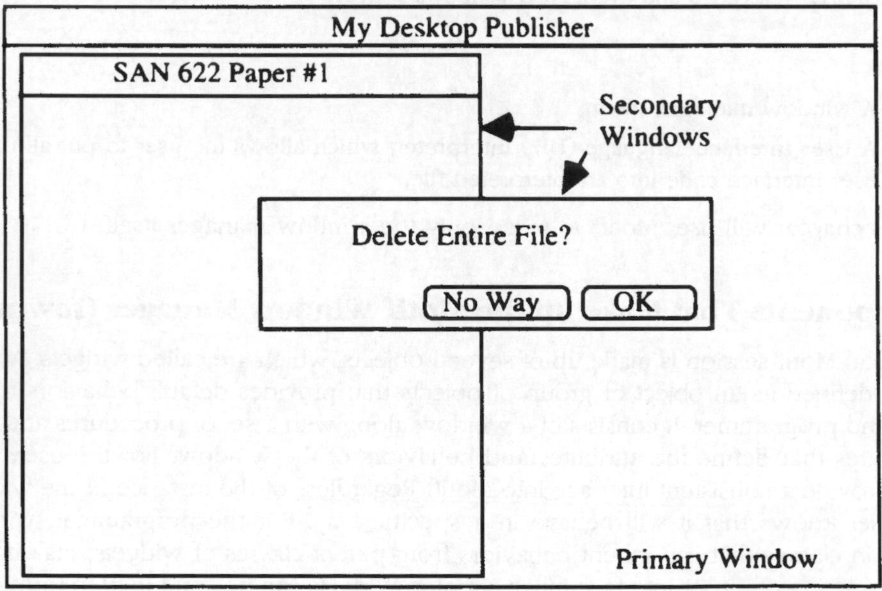

FIGURE 8.2. A description of different window types.

The Geometry of Windows

All windows on a display have (x, y) coordinates. The coordinates are usually measured in pixels and begin with (0, 0) in the top left corner of the window. As you move to the right the x coordinate grows, and as you move down the y coordinate grows. There are also two sets of coordinates for each pixel in a window. The global coordinates are in relation to the Root window, which covers the entire display screen. The local coordinates are in relation to the child window in which the pixel is currently found (Fig. 8.3). It is important when programming with Motif to make

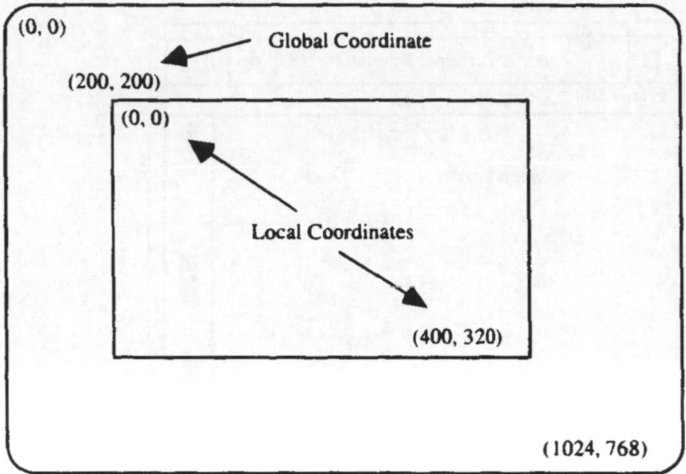

FIGURE 8.3 Local versus global coordinates.

sure you are using the proper coordinates; however, for the user, who is not programming with Motif, this knowledge is supplementary.

Window Decorations

All windows within Motif have certain decorations. These decorations, and their behavior, can be modified or removed (Fig. 8.4). The default decorations for all windows under Motif are

- *Window border*, encloses the entire window, setting it apart from the background color and other windows.
- *Title bar*, contains the name of the current window and is found at the top of each window.
- *Maximize button*, makes the window the largest size that will fit on the screen. Found in the upper right of the window.
- *Minimize button*, reduces the window to an icon on the desktop. Found to the left of the maximize button.
- *Resize areas*, allows the user to click and drag them to change the window dimensions.
- *Window menu*, contains a menu that allows certain modifications, including closing, to the window, as well as providing the functionality of the other window decorations. Found in the upper left of the window.

Look and Feel of Motif

The visual aspect of Motif is perhaps the most notable to users. It is this aspect of Motif that gives the applications their three-dimensional look and feel (Fig. 8.5). This three-dimensional look and feel is achieved through the use of colors and shading. For ex-

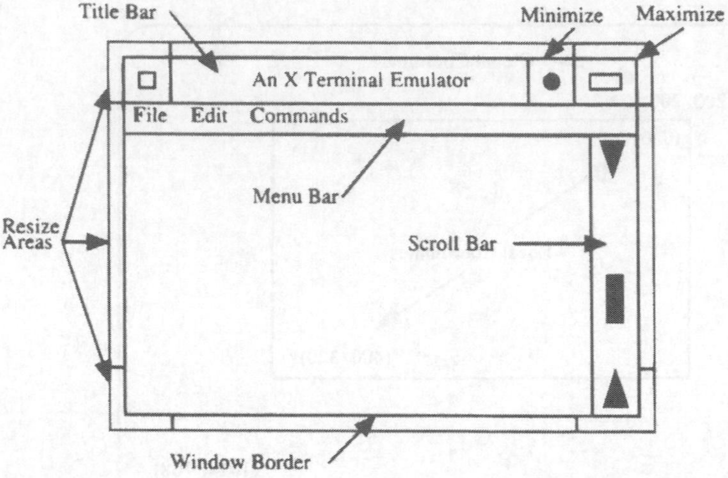

FIGURE 8.4 The parts of a Motif window.

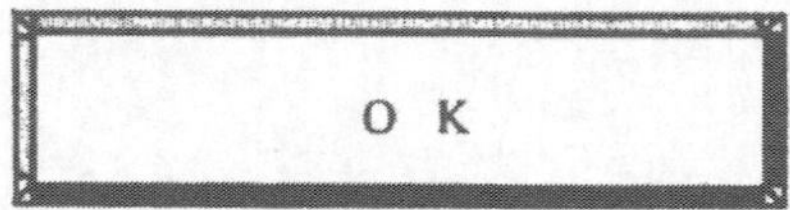

FIGURE 8.5 A three-dimensional pushbutton.

ample, on a button, the look of being "up" is achieved by having the button's shadow on the top and the left be a lighter color than the shadow on the bottom and right. To give the look of being "down" the button's shadow colors are reversed from the "up" shadow colors.

Starting a Motif Session

Most systems will be configured to allow the user to use X-windows and Motif automatically upon logging into the system. Instructions for initiating a Motif session are given here in case of problems or in case the system is not so configured. These instructions are Unix specific and may or may not work under your operating system or configuration. If they fail to work, contact your system administrator to find out the proper procedure.

The first step to starting a Motif session is logging into the computer that has Motif. The different paths that can be taken to perform this step are too numerous to list here, and also too specific to each installation. Therefore, if you are not sure how to perform this step contact your system administrator. The most important aspect of this step is that the display sitting in front of you must be running, or capable of running, X-windows. This means that you cannot run Motif on an ASCII/dumb terminal, like a

Digital VT-XXX terminal, and if you are sitting in front of a personal computer it must be running additional software, like MacX for the Apple Macintosh. Once you are logged into the computer you can proceed with the next step.

Once you are logged into the computer, if you find your display is functioning as a single terminal window, chances are that you are not yet running X-windows. On most Unix systems there is a program designed for initiating an X-windows session. This program is usually found in the **/usr/bin/X11** directory. Under AIX it is found under **/usr/lpp/X11**, and is called **startx.** The best place to start is by reading the Unix manual page for **startx.** This will tell you of any special requirements for running **startx** on your platform.

```
% man startx
```

One of the things that the manual page will tell you is that you must have the X11 binaries directory, **/usr/bin/X11 or /usr/lpp/X11** under AIX, in your search path.

```
% set path = ( /usr/bin /etc /usr/ucb /usr/bin/X11 )
```

However, you should be able to run the program from the command line with no additional options.

```
% startx&
```

This program should initiate an X-windows session and execute some initial applications for you, a terminal emulator at the very least. Once X-windows is running you might not be able to move the windows around on the screen or close the windows. You will also notice that there is no border around the windows and no title bars for the windows. This is due to the fact that Motif (the window manager) is not yet running. To execute the Motif window manager manually you must have the pointer in the same window as the following command:

```
% mwm&
```

Once you have X-windows and Motif running you can begin to run applications that require Motif.

Stopping a Motif Session

Most systems will terminate all open applications and kill off the X-windows and Motif programs whenever you log out from the initial terminal window. However, if this is not the case you can use the following procedure to log out of a system and kill off your Motif applications and the X-client for your session. First find out all the processes running on the system that are owned by you.

```
% ps axu | grep myusername
```

This will give you output similar to

```
pickerin 201  0.0  3.9  2566  443  ?  I   2:10/usr/bin/mwm
pickerin 219  0.0  3.2  2011  78   ?  I   3:05/bin/csh
pickerin 211  0.0  0.9  2806  77   ?  I   0:09/usr/bin/dxue
pickerin 210  0.0  2.3  2884  257  ?  I   4:42/usr/bin/dx
                                           term -ls
```

The first column of output is the user name and the second column is the process id. You should be able to stop all of your running processes and log yourself out. Make sure that you stop your shell process last (usually **/bin/xsh**).

```
% kill -9 201; kill -9 211; kill -9 210; kill -9 219
```

This should stop all of your processes and log you out of the computer. However, it is always a good idea to ask your system administrator about proper log-out procedures.

Resetting the Default Behavior of Motif

Throughout this chapter you will learn about how to modify Motif and customize its functionality. However, *it is possible to get Motif into a state in which you can no* longer function adequately, such as having your text color the same as your background color. To restore all of the customizations to the default try pressing down the <ALT>, <CNTL>, <SHIFT>, and <!> keys at the same time. To change the settings back to the customized settings, press the keys again. If your system does not respond to this or if you do not have all of those keys you can restore the defaults from the Root menu.

8.3 HOW TO MOVE WITHIN MOTIF

To use the Motif window manager to its fullest extent users must understand where their typing will be displayed (input focus), how to move between windows, and how to manipulate those windows once the rest is achieved.

Input Focus

There are two different policies of input focus. Under bare X-windows (X-windows without a window manager running), you have only the implicit input focus. The keyboard input is sent wherever the pointer is on the screen. So, to type into a window you must first move the mouse pointer so that it rests within the window. Explicit input focus is used when you have a window manager running. The location of the pointer does not determine destination of keyboard input. You must click within the window that is to receive the input, making it the active window. Under Motif, the active window and the inactive windows have different colors attributed to them, which are user configurable.

How to Move between Windows

The input focus policy in effect for a Motif session is configurable by the user. However, the default behavior for Motif is to have an explicit input focus policy. Thus, you must select the window to receive keyboard input by clicking within it and making it the active window. The keyboard can also be used to select a new active window, <ALT> <TAB> moves to the next window in the window stack and <SHIFT> <ALT> <TAB> moves to the previous window.

Window Manipulation

Earlier we discussed the different window decorations. Now, we will discuss how to use each of the decorations to manipulate the window.

The *window border* is an area of each window that separates the window from the background and from other windows on the screen. This border usually appears as a slightly different color at the edge of each window and extends into and around the title bar.

The *title bar* is an area of the window that displays the current title for a window. The title bar is also where the maximize, minimize, and window menus are found. Generally, you click and drag on the title bar to move a window to another location on the screen. You might also be able to double-click on the title bar to iconify a window; however, this is a user-configurable option (see next section).

The *maximize button* appears in the upper right-hand corner of the title bar and appears to be a raised box within a slightly larger box. The maximize button, when pressed, will expand the active window to the largest size possible for that window, usually the size of the display being used. When pressed again, the window will return the size that was being displayed before the maximize button was activated.

The *minimize button* appears in the upper right-hand corner of the title bar to the left of the maximize button. The minimize button provides a way to iconify a window. When a window is iconified, the window is closed and a small icon that represents the window appears on the root window. The icon can include a picture of the currently running application, or it might just be the label showing the title of the window.

Resize areas are some of the most-used decorations on a window. The directional resize areas are found just outside of the window border at the top, bottom, left, and right of the window, appearing as slightly rounded areas at the edge of the window. You know that you are on top of a resize area because your cursor changes into a small bar with an arrow pointing at it. These directional resize areas allow you to resize the window only on one axis at a time. The left and right resize areas allow resizing in the horizontal direction, while the top and bottom resize areas allow resizing in only the vertical direction. Additionally, there are multidirectional resize areas found in each corner of the window that allow you to resize along any axis. The cursor also changes shape when you are on a multidirectional resize area; it appears as a corner with an arrow pointing into it.

Finally, the *window menu* is a button found at the upper left-hand side of the title bar. This menu includes choices to make all of the same modifications available from the different window decorations and, additionally, a menu choice for closing the window itself; it is very useful menu if you have opted to eliminate some or all of the other window decorations.

8.4 BASIC MOTIF CUSTOMIZATION

Through the use of customization, an individual user can tailor her Motif session to provide the most productive and pleasing environment possible. The customization abilities of Motif are due in part to a series of variables, called resources, that specify customization information and are stored in a resource database. The resource database is created from a series of resources specified in configuration files for each application. This database then specifies all of the customization information for a given application. Here are the common configuration files that you will be looking at and modifying.

```
~/.Xdefaults
~/.mwmrc
/usr/lib/X11/system.mwmrc
/usr/lib/X11/rgb.txt
```

Note: The ~ above can be used on many systems to represent the path into your home directory; if it does not work, you will have to type out the entire path to your home directory.

The Resource Database

Motif achieves its customization abilities through the resource database. To understand how this operates we must first understand file search paths. A file search path is an ordered list of paths. If the application cannot find the information it is looking for, it moves to the next path in the list, until it either runs out of possible paths or finds the information.

The resource database is actually a composite of a number of different resource files and command line options. Each of these resource files and command line options can override ones that were used previously. Therefore, it is important to note the order of precedence that is taken by an application when creating its resource database.

- Command line options specified for the application by the user.
- An *environment* resource file created by the user on the local host.
- Resources for the specific screen being used.
- A resource property found on the server or a user preference resource file found on the local host.
- Application-specific resource files found on the local host.
- Application-specific resource files found on the server.

Each of these areas is searched when an application is first started, and a resource database for that application is then created.

Resource Specification

Before we cover some of the different types of customization you first need to know how to specify the customization. Resource specification has a common syntax that applies across all applications that you will encounter under X-windows and Motif. A resource is specified using the following general syntax:

```
Application*resource: value
```

An example of the use of resource specifications would be to customize part of the **xclock** application by putting the following lines in your **.Xdefaults** file.

```
Xclock*background: white
Xclock*foreground: CornflowerBlue
Xclock*hands: wheat
```

The next time you execute the **xclock** application it would have a background of white, a foreground of cornflower blue, and the hands of the clock would be displayed in a wheat color. You may have noticed that there is an asterisk present in the resource definition above. Asterisks are used within resource specifications to represent a number of Motif components found between the application itself and the resource. Some of these components could be **pushButton, window1, statwin**, or anything else that the application programmer called her widgets. With this knowledge you are now ready to begin some simple customization of the Motif Window Manager (mwm).

Customizing Colors

One of the most performed, and simplest, customization is changing your default colors. Here is an example of some color modifications that you can make:

```
Mwm*background: BlanchedAlmond
Mwm*foreground: MistyRose
Mwm*activeBackground: MistyRose
Mwm*activeForeground: BlanchedAlmond
Mwm*menu*foreground: DodgerBlue
Mwm*menu*background: LawnGreen
```

These colors would perform two functions: they would allow your active window to stand out, since it would have colors that are exactly opposite of those for the inactive windows. It would also specify the colors of any menus that are generated within the window manager. Motif and X-Windows allow for a rich suite of colors to use for your customizations. However, you must be careful about specifying colors, since you also will have to view these customizations on a daily basis. All of the colors available on your system can be found in the **/usr/lib/X11/rgb.txt** file. This file lists all of the color names, as well as their numeric representations, that can be used to customize aspects of your window manager. Here is a sample of an rgb.txt file as found under AIX.

255 250 250	snow
248 248 255	ghost white
248 248 255	GhostWhite
245 245 245	white smoke
245 245 245	WhiteSmoke
220 220 220	gainsboro
255 250 240	floral white

255 250 240	FloralWhite
253 245 230	old lace
253 245 230	OldLace
250 240 230	linen
250 235 215	antique white
250 235 215	AntiqueWhite
255 239 213	papaya whip
255 239 213	PapayaWhip
255 235 205	blanched almond
255 235 205	BlanchedAlmond
255 228 196	bisque
255 218 185	peach puff
255 218 185	PeachPuff
255 222 173	navajo white
255 222 173	NavajoWhite
255 228 181	moccasin

Customizing Windows

Everything that you see within Motif is contained in a window. Therefore, it might be important that all of your windows have a specific set of customization options.

```
Mwm*keyboardFocusPolicy: Explicit
Mwm*clientDecoration: resizeh border menu title
Mwm*resizeBorderWidth: 10
Mwm*moveOpaque: False
Mwm*wMenuButtonClick2: False
```

Each of these resources specifies a behavior for all windows under Motif. The keyboard focus policy was covered earlier, and here you can specify either implicit or explicit. The client decorations allow you to specify each of the decorations that you wish to have on your windows, **all** for all of them, or **none** for none of them. The resize border width measures how many pixels to make the resize border for each window. Move opaque allows the user to determine whether or not the entire window is drawn when it is being moved. Finally, the menu button double-click is a Boolean value that when set to true allows the user to double-click on the window title to iconify the window. These resources are just a few that are available to configure your window behaviors.

Customizing Icons

The last area of basic customizations we will cover is performed on icons. Icons are small graphic representations of objects on the screen. An icon can represent an application, a file on disk, or a window that has been iconified. Like all other aspects of Motif, icons' decorations and colors can be customized.

```
Mwm*iconImageForeground: CadetBlue
Mwm*iconImageBackground: white
Mwm*iconImageAutoShade: True
```

```
Mwm*iconDecoration: activelabel label image
Mwm*iconPlacement: Bottom Right
Mwm*useIconBox: True
Mwm*iconBoxTitle: Icon box
```

These customizations would first set the foreground and background colors for any icons displayed on the screen. Icon decorations are specified including whether or not a label or an image is displayed within the icon (the icon image is determined by the type of application that is being represented by the icon). Finally, there is a icon box that can be displayed on the screen. This icon box is a window in the screen that contains an icon for each of the running applications. A nice feature of the icon box is the fact that it displays a grayed image of the icon for each of the applications that are currently displayed on the screen, making it much easier to manage a large number of windows.

8.5 ADVANCED MOTIF CUSTOMIZATION

Once you have mastered the resource files and understand the resource database for applications, it is time to move beyond changing colors and fonts and into modifying behaviors of applications and changing the choices available under the different menus as well as creating your own menus.

Configuring Application Resources

Configurations are possible for each application that is running on your computer under X-windows and Motif and are not restricted to only the window manager. Normally, there are always the standard resources that can be modified for an application, resources such as colors, fonts, window sizes, and input focus. However, most applications also have additional behaviors and decorations that can be modified. But how do you find out how to modify them? Most applications under X-windows and Motif have what is called an app-defaults file. This file is usually shipped with the application and specifies all of the default decorations, appearances, and behaviors of the application. This file needs to be placed into a special location on your system so that each time the application is executed it can find the resource file and load it. You can override those defaults in your **.Xdefaults** file in your personal account. The app-defaults file for an application should be placed in the **/usr/lib/X11/app-defaults** directory and have the same name as the application, except with the first letter of the name being capitalized. Now, by looking at the app-defaults file for your application you can determine all of the resources that are available to customize, in addition to the window manager customizations that will affect the application.

Altering Menus

One of the most exciting configurations to make is altering an existing menu or even creating a menu of your own within Motif. First, let us take a look at modifying some existing menus within the Motif window manager (while this can be done under applications other than the window manager, we will not be covering it here).

The default format for a menu within the window manager is:

```
Menu menuName
{
choice1 [mnemonic] [accelerator] function [argument]
choice2 [mnemonic] [accelerator] function [argument]
.
.
.
choiceN [mnemonic] [accelerator] function [argument]
}
```

A mnemonic is a character (or group of characters) that represent the menu choice. Within the menu itself these appear as underlined characters within the menu choice, and pressing that key on the keyboard while the menu is being displayed will select that menu item. An accelerator is a keyboard equivalent of selecting that menu choice from the menu. Accelerators can be used even if the menu is not currently being displayed. The function is the action that will be taken when the menu choice is selected (or the mnemonic is pressed, or the accelerator is activated). Occasionally the function requires one or more arguments. For the Motif window manager there are a number of defined functions that must be used for the different menu choices. The most useful are

```
f.exec
f.kill
f.menu
f.separator
f.title
```

Executes the shell command defined in [argument]
Kills/closes the current window
Executes the menu named in [argument]
Creates a separator bar (dotted or solid line across menu)
Sets the title for the menu

There are several other functions available; see the *OSF/Motif Programmer's Reference* for more information.

The Default Window Menu

Previously, we discussed the window menu that is present in the upper left corner of each window displayed under Motif. Now, we will present how to make modifications to that menu.

First, you must make a copy of the system **.mwmrc** file. This file contains the default configurations for many of mwm's behaviors and colors.

```
% cp /usr/lib/X11/system.mwmrc ~/.mwmrc
```

Once you have a copy of the **.mwmrc** file you can make modifications to your copy to override the defaults in the system **.mwmrc** file. Edit the file using the editor of your choice.

```
% /usr/ucb/vi.mwmrc
```

The standard definition of the default window menu is as follows:

```
Menu DefaultWindowMenu MwmWindowMenu
{
"Restore"     _R   <Alt><F5>    f.normalize
"Move"        _M   <Alt><F7>    f.move
"Size"        _S   <Alt><F8>    f.resize
"Minimize"    _n   <Alt><F9>    f.minimize
"Maximize"    _x   <AltF<F10>   f.maximize
"Lower"       _L   <Alt<F3>     f.lower
no-label                        f.separator
"Close"       _C   <Alt><F4>    f.kill
}
```

Notice the menu choices, mnemonics, accelerators, and functions that we discussed earlier. You can now make any changes you want to the menu structure including changing a menu choice name, adding a menu choice, deleting a menu choice, or just rearranging the order of the menu choices. You may also decide to change either the mnemonics, accelerators, or both. Let us get rid of the move, minimize, maximize, and lower items and add a menu item to display a submenu of minimize and maximize. Here is what our new menu will look like:

```
Menu DefaultWindowMenu MwmWindowMenu
{
"Growth"    _G   <Alt><F2>   f.menu GrowMenu
"Restore"   _R   <Alt><F5>   f.normalize
"Size"      _S   <Alt><F8>   f.resize
no-label                     f.separator
"Close"     _C   <Alt><F4>   f.kill
}
```

However, we are not finished yet. We now need to create an entire new menu called GrowMenu. Here is the definition for it:

```
Menu GrowMenu MwmWindowMenu
{
"Minimize"  _n   <Alt><F9>    f.minimize
"Maximize"  _x   <Alt><F10>   f.maximize
```

You have now not only made modifications to your default window menu, but you also know how to create submenus for use elsewhere. **Note**: You must restart mwm to have your new changes take effect. You can usually perform this from the Root menu.

8.6 TROUBLESHOOTING A MOTIF SESSION

As in life, things do not always work as planned. Sometimes there are problems with using a system, and you need quick, accurate solutions to those problems. In this section we outline some of the common problems users encounter with Motif and ways to correct those problems.

Mouse Does Not Respond

A graphical user interface is not very useful if you do not have a mouse or pointer device that allows you to specify where input should be sent and allows you to move windows that are overlapping other windows. Usually, when a mouse does not respond it is because it was not plugged into the computer when it was first started. However, you have probably fired up your entire X-session before realizing that you do not have any cursor control. Fortunately, Motif allows many functions to be performed with only the keyboard. Here are a few keystrokes that should help you navigate through Motif and allow you to exit your session so that you can reboot your computer and regain mouse control.

`<Alt><Tab>`	Brings up next window in the stack and moves input focus into it (if explicit input focus is configured)
`<Alt><Esc>`	Brings up next window, but does not move input focus
`<Alt><Space>`	Pops up the window menu if the input focus is properly set
`<Alt><F6>`	Navigates between a primary window and its secondary windows
`<Shift><F10>`	Displays a pop-up menu (such as the root menu)

Window Is Larger than the Screen

Normally, this problem is caused because the default geometry for an application window is larger than the geometry of the display. There are several possible solutions to the problem; however, none of them is guaranteed.

1. Attempt to resize the window using one of the resize areas. Then move the window into the display so that the hidden portions of the window can be viewed.
2. Close the offending window using the keyboard command: **<Alt><F4>**.
3. Many windows have some form of default response awaiting a user's input. The solution to the problem could be as simple as hitting the <Return> key to answer <OK> to a dialog box.

Colors Conflict

Usually the only way that colors will create an unusable environment is if background and foreground colors are set to the same value, or a very close value. An easy fix for this problem is to restore the default color definitions for your windows using the following key sequence: **<ALT><CNTL><SHIFT><!>**. This will normally restore the default colors and behavior for the entire Motif session.

8.7 BEHIND THE SCENES WITH MOTIF

This section is devoted to C programmers who are interested in getting started with writing their own Motif-aware applications, and for those users who are curious about how it all works together.

Inheritance of Classes

Motif is a class-driven programming environment. What this means to the programmer is that many of the objects used to develop applications inherit their properties from other classes of objects. The root class of objects that can have windows within X-windows and Motif is the Core class. Core includes basic properties such as geometry, colors, and input focus policies. Composite is a subclass of Core and is the fundamental class for all objects that can have children. Composite manages a list of all of the children and their associated geometries. Constraint is a subclass of composite and manages additional information (stored in constraint resources) about children. Shell is a base class for all widgets that envelope other widgets. Shells are what allow windows to interact with the Motif window manager. Shells are a subclass of the Constraint class. There are several types of shell widgets that are all subclasses of the Shell class.

```
OverrideShell:
WMShell:
VendorShell:
TransientShell:
TopLevelShell:
ApplicationShell:
```

Shell class that contains all widgets that the window manager should ignore, such as menus

Superclass of shell widgets that need to interact with the window manager

WMShell subclass that implements toolkit specifics

VendorShell subclass for widgets that appear on the screen briefly, such as dialog boxes

VendorShell subclass for different top-level application components

TopLevelShell subclass that represents the top-level widget for an entire application

Widgets and Their Uses

There are a multitude of different widgets available to the Motif programmer. All widgets fall into one of two categories: primitive and manager. Primitives are widgets that cannot have children; managers are widgets that can have children. There are several types of primitive widgets:

- *Label* widgets, designed to display static text or a pixmap (graphic).
- *Separator* widgets, group objects by displaying a horizontal line.
- *Button* widgets, probably the most used widget, the button widget allows the user to activate it with a mouse click and perform an action. The button widget comes in several different types, such as Cascade, Pushbutton, Togglebutton, and Drawbutton.
- *ScrollBar* widgets, allow the user to control the portion of a total window that is being displayed.
- *List* widgets, display textual items in a list that allows the user to select items either one at a time or in a continuous manner.
- *Text* widgets, display and possibly allow editing of text in either a single or multiple line mode.

Manager widgets are also found in numerous types:

- *Frame* widget, surrounds a child with a shadow and a margin.
- *Scale* widget, appears as a scroll bar without arrows on the ends. The user can then control a slider that changes the values of the scale.
- *PanedWindow* widget, arranges all of its child widgets in a vertical column and forces their widths to all be the same. Additionally, the PanedWindow manager inserts a control sash that allows the user to control the height of each individual pane.
- *ScrolledWindow* widget, a window that implements a scroll bar and viewport. The application must manage the updates of the window in response to movement of the scroll bar thumb.
- *MainWindow* widget, a subclass of ScrolledWindow that can act as the primary window for an application. Optionally implements a title bar, command window, and message window.
- *RowColumn* widget, contains other widgets to create either a menu or nonmenu area, usually created using PushButton widgets.
- *BulletinBoard, Form, MessageBox, SelectionBox*, dialog boxes that present information to the user and retrieve responses to that information.
- *Drawing Area* widget, used as a canvas for creating graphic objects within a window.

Motif Program Structure

There are several steps that the application programmer must take to create, display, and remove widgets within an application; however, the Motif toolkits handle all of the communication between the user and the application.

Including the Proper Header Files

In addition to any header files that are required by the program's main logic, you must include header files for X-windows, the Motif toolkit, and every widget that is used within the program. For example,

```
#include <Xm/Xm.h>
#include <Xm/Label.h>
#include <Xm/PushB.h>
```

The Xm.h file is the header file that brings in the Motif toolkit, as well as all of the necessary X-windows header files and libraries. The second two includes allow us to use a Label widget and a PushButton widget within our program. It is not uncommon for a Motif program to have up to 25 or 30 Xm includes since there are over 90 different header files.

Initialize the X Intrinsics

Through a call to **`XtAppInitialize()`** the Motif program performs several initialization steps necessary for a Motif program to operate properly, including creating a top-level application shell widget.

Create Program Widgets

Widgets are the heart of any Motif program. To create a widget, the programmer makes a call to the **`XtCreateWidget()`** function and passes in the name, class, and parent of the widget to be created. This step must be repeated for each widget that is created within the application program.

Add Widget Callback Routines

When a widget is activated on the display, or the user makes a modification to a widget, the application program needs to be made aware of the change to the environment. This is handled by assigning a callback routine to each widget with which the user has interaction. After creating a callback function to handle the user interaction, the programmer attaches the callback function to the widget using a call to the procedure **`XtAddCallback()`**. Then, when the widget is interacted with, the callback will be activated to handle the user's actions. This procedure must also be performed for each widget that requires a callback function.

Make the Widgets Visible

When Motif creates a widget it does not display it on the computer screen immediately. This would cause an effect where different aspects of a program interface appeared at different times while the application was initializing. To prevent this, Motif keeps all of the widgets invisible until a call to **`XtRealizeWidget()`** is made. This function accepts a parent widget as its argument and makes that parent and all of its children visible. Normally, the realize widget function is given the top level parent as its argument so that the entire application appears at once on the user's display.

Enter the Main Event Loop

Now all the work for creating a Motif environment is completed. It is now up to the application to maintain that environment and perform whatever tasks the application was written to perform. This work is all performed within the main event loop after a call to **`XtAppMainLoop()`** is made. This loop handles all user input to the application by dispatching that action to the proper event handler. The main event loop is an infinite loop and does not provide a way to exit; this is left up to the application programmer to provide.

A Simple Motif Program

The following is a simple "Hello World" program written in X-windows and Motif. The various aspects of the program should now be clear.

```
/* Hello World */
#include <Xm/Xm.h>
#include <Xm/Label.h>
#include <Xm/PushB.h>

XtAppContext context;
XmStringCharSet defChar=XmSTRING_DEFAULT_CHARSET;

/* First create the callback, this is what we do when the
"Hello        World" button is pressed */
void qCallback( Widget widget,
caddr_t clientData,
caddr_t callData ) {
Display *display;

display = XtDisplay( widget );
XtCloseDisplay( display );

exit (0);
}
main ( int argc, char argv[] ) {
Arg args[10];
int n;

Widget toplevel, label, quitWidget;
/* Initialize the X-intrinsics */
toplevel =
XtAppInitialize (&context, " ", NULL,0,
&argc, argv, NULL, NULL, 0);

/* Set the defaults for our application*/
n=0;
XtSetArg (args[n], XmNlabelString,
XmStringCreateLtoR ("Hello World!",
defChar)) ;n++;
XtSetArg(args[n], XmNwidth, 100 ); n++;
XtSetArg (args[n], XmNheight, 100); n++;
/* Create our quit button, called
"Hello World!" */
```

```
quitWidget = XtCreateManagedWidget ( "Quit",
xmPushButtonWidgetClass,
toplevel, args, n );
/* Attach our callback to the "Hello World"
button we created. */
XtAddCallback( quitWidget,
XmNactivateCallback,
qCallback, NULL);

/* Display the program window */
XtRealizeWidget (toplevel);

/* Move into the main loop */
XtAppMainLoop (context);
}
```

This program can be compiled using the following command:

```
% cc -o hello hello.c -lXm -lXt -lX11
```

There are three libraries being included above. The Xm library is the X/Motif library that contains all of the information for Motif. The Xt library is the X-Intrinsics library that contains information about manipulating X. And the X11 library contains information about X-windows itself. You can then run the "Hello World" program by typing **`hello`** at the shell prompt.

DEFINING TERMS

Binaries: Unix executable programs.

Cascaded: Overlapping but down and to the right, so that the title bar can still be seen.

Cursor/Pointer: The representation of the user's mouse or input device on the screen.

Emulator: Representing something else, as a terminal emulator running on a computer represents the functions of a terminal within the computer.

Gadget: A device on a window that performs an action, cannot have children.

Iconify: Closing the open window and representing it with an icon.

Inheritance: The ability to take on all of the attributes of a parent process or class.

Interaction: As with a computer, all actions that a user makes of which the computer must be aware.

Pixels: Picture elements on a display screen.

Process: An executing program in a multiprogramming environment.

Shell: The Unix command line parser environment, commonly the Korn Shell (`/bin/ksh`) or the C-Shell (`/bin/csh`).

Toolkit: A group of programs or routines that allow a programmer to more easily generate application programs

Widget: A device on a window that performs an action, can have children.

Window Stack: A group of windows on a screen, ordered by most recently used.

X-client: Under X-windows the client/server relationship seems reversed. The X-client runs on the host machine while the X-server runs on the computer in front of the user.

REFERENCES

Eric F. Johnson and Reichard, Kevin, *Power Programming . . . Motif*, 2nd ed. MIS Press, New York, 1993.

Marshall Brain, *Motif Programming*. Digital Press, Burlington, MA, 1992.

OSF, *OSF/Motif Programmer's Guide*. Prentice-Hall, Englewood Cliffs, NJ, 1993.

OSF, *OSF/Motif User's Guide*. Prentice-Hall, Englewood Cliffs, NJ, 1993.

OSF, *OSF/Motif Programmer's Reference*. Prentice-Hall, Englewood Cliffs, NJ, 1993.

Alan Southerton, *The Shell Hacker's Guide to X and Motif*. John Wiley, New York, 1994.

9

DEC VAX

R. L. Davis
L. A. Knox
T. E. Mertz
Millersville University

9.1 THE VAX/VMS ENVIRONMENT

VAX/VMS (Virtual Address eXtentsion/Virtual Memory System) is a computer system made by Digital Equipment Corporation. VAX describes the architecture of the machine; VMS is the operating system, which allows users to share the resources of the computer.

The VAX System

The VAX System is designed to execute many different kinds of jobs concurrently. The jobs consist of one or more processes, each of which can be executing a program image that interacts with on-line users. The system enables the users to write, compile, and test programs interactively or in batch mode.

User Accounts and Passwords

To be able to use the VAX computer you must have an account. Accounts are set up by the System Manager to authorize the use of the system. When you are given an account, you will be provided with a user name and a password. Whenever you log into the VAX, you identify yourself as an authorized user by entering your user name and password, as prompted.

You may change your password at any time. Type the words SET PASSWORD at the prompt, and follow the directions on the screen. Passwords should be at least six or more characters in length, up to 31 characters, and may consist of alphabetic, numeric, the dollar sign ($), and underscore (_). Passwords cannot contain blank spaces. Uppercase and lowercase characters are treated the same. Passwords are checked against a dictionary and a history list, to prevent the use of common words and the reuse of old passwords.

To logout, type LOGOUT at the system prompt.

0-8493-2530-7/96/$0.00+$.50

9.2 BASIC DCL TECHNIQUES

Users communicate with the VAX system through the operating system's command language, which is called Digital Command Language (DCL). It enables the user to log into the system, manipulate files, develop and test programs, and obtain system information. Users can also extend or redefine commands using DCL.

DCL is used to communicate with the VAX/VMS operating system through DCL commands. These commands are words (generally verbs) that describe the functions they perform. They may be typed in upper or lowercase. (They will appear in uppercase in this guide for easier identification by the reader.)

DCL has its own vocabulary and rules of grammar: the vocabulary consists of commands, parameters, and qualifiers, while the grammar provides rules for using these components.

Parameters define what the command acts upon, while *qualifiers* modify how the action will occur. The addition of qualifiers is usually optional. For example, the PRINT command requires a parameter to indicate what is to be printed:

```
$ PRINT myfile.lis <Return>
```

However, the /COPIES qualifier can be used to indicate how many times the file is to be printed:

```
$ PRINT /COPIES=2 myfile.lis <Return>
```

The line above is known as a *command string*: a command with its required parameters and any qualifiers that might have been used. Command strings must always be followed by <Return>.

If you do not provide all necessary parameters on a command line, you will be prompted for them. For example, the COPY command requires two parameters:

```
$ COPY <Return>
_From: MY.PAS <Return>
_To: YOUR.PAS <Return>
```

DCL commands may be abbreviated to no less than three characters, as long as the abbreviation is unique.

The HELP Command

HELP is a DCL command that is useful for looking up command formats on the VAX system. This command supplies the user with a list of possible help topics:

```
$ HELP <Return>
```

In response to the above command, the system will display a list of HELP topics, and prompts the user for a specific topic. If help on a particular command is needed, it can be specified as a parameter to the HELP command:

```
$ HELP PRINT <Return>
```

or

```
$ HELP PRINT /COPIES <Return>
```

To leave the HELP command, which continues to prompt for topics, hit <Return> until the dollar sign prompt is displayed again.

Displaying Previously Entered Commands

During an interactive session, you may enter many commands. To see these command strings after they have been processed, you can do one of the following:

- press the up-arrow key
- enter the DCL command RECALL

These commands allow you to display a previously entered command. Then you may edit and process the command again by pressing <Return>.

Add the /ALL qualifier to the RECALL command to display the last 20 commands you have entered. The most recently entered command is number 1.

```
$ RECALL/ALL <Return>
```

You can reenter a command by typing its number with the RECALL command.

```
$ RECALL 14 <Return>
```

This will display the fourteenth command. If you wish to process this command press <Return>.

9.3 USING SYMBOLS AND COMMAND PROCEDURES TO SAVE TIME

Abbreviating DCL Commands with Symbols

DCL commands can be abbreviated to the first three characters of the command. However, DCL commands also can be denoted by abbreviations of your own choice. Then you can invoke DCL commands by using those abbreviations. For example, the DCL command "SHOW USERS" can be denoted by the abbreviation "SU" as follows:

```
$ SU="SHOW USERS" <Return>
```

You can use the symbol SU in place of SHOW USERS as follows:

```
$ SU <Return>
```

You can use the DCL command SHOW SYMBOL to see the value of any symbol you create as follows:

```
$ SHOW SYMBOL SU <Return>
```

Note: Any symbols you assign will disappear when you log out unless you put them in a LOGIN.COM file.

When defining a symbol, care should be taken to avoid using abbreviations of existing DCL commands. For example,

```
$ ST = "SHOW TIME"
```

ST could also be used to abbreviate the START and STOP commands.

A symbol to modify a DCL command should look like the DCL command it represents:

```
$ DEL = DELETE/ERASE/CONFIRM
```

You can check for possible conflicts with DCL commands by doing a HELP on the symbol you want to use. If any help information appears about a DCL command, that name should probably not be used.

Command Procedures

A command procedure is a file that contains a sequence of DCL commands. A command procedure is created by creating a file and filling it with DCL commands. When the command procedure is invoked, the commands are executed beginning with the first and continuing consecutively to the end of the procedure.

The default file type of a command procedure is COM. To execute a command procedure, precede its name with an at sign (@):

```
$ @tryit <Return>
```

The LOGIN.COM File

When you log in to the system, it automatically searches for a LOGIN.COM file and executes the commands within this file. You can place the global assignment statements you use for command synonyms in a separate file called SYNONYM.COM. Your LOGIN.COM file would contain the following line:

```
$ @SYNONYM
```

If this file contains abbreviations of commands, use two equal signs ("==") when making the symbol assignment:

```
$ SU == "SHOW USERS"
```

You can also include the SET commands that you want to be in effect for every login session. You can change the VMS prompt from the dollar symbol ($) to a "VMS==" prompt like this:

```
$ SET PROMPT = VMS==
```

To return to the default prompt ($), type SET PROMPT.
Note: Your prompt cannot contain blanks.

Submitting Batch Jobs

A command procedure can be submitted as a batch job, so that it can be executed as its own process. This allows you to use your terminal interactively for other tasks as the command procedure executes. To submit a command procedure as a batch job, use the SUBMIT command. A procedure CLEAN.COM in the [ACCOUNT] directory can be submitted to the system as a batch job as follows:

```
$ SUBMIT [ACCOUNT]CLEAN <Return>
```

Use the /PARAMETERS qualifier to pass any needed parameters. The following command passes two files, BILLS.DAT and RECEIPT.DAT, as parameters:

```
$SUBMIT [ACCOUNT]CLEAN/PARAMETERS=(BILLS.DAT,RECEIPT.DAT)
```

When the batch job is finished, a log file containing the output from the command procedure is printed on the line printer. After the log file is printed, it is deleted. If you want to save the log file, use the /KEEP qualifier with the SUBMIT command. Using a combination of /NOPRINT and /KEEP will prevent the log file from printing, but will retain a copy in your directory.

```
$ SUBMIT/KEEP
[ACOUNT]CLEAN/PARAMETERS=(BILLS.DAT,RECEIPT.DAT)
```

When a log file is kept, it is given the same name as the command procedure with a file type of LOG, and it is placed in your directory.

Looking at Jobs in the Batch and Print Queues

To display the list of jobs in a queue, use the following command:

```
$ SHOW QUEUE queue-name <Return>
```

The queue name is provided in a message when the batch/print job is entered. The SHOW QUEUE command provides the names, job entry numbers, owners' names, and status of the jobs in the indicated queue. If the SHOW QUEUE command is used without specifying a queue name, information about each available batch and print queue will be printed.

Removing a Job from the Batch or Print Queues

To remove a job from the batch/print queue before it starts executing/printing, enter the following command:

```
$ DELETE/ENTRY=job-entry-number queue-name <Return>
```

9.4 CONTROL KEY SEQUENCES

Control key sequences are produced by holding down the <Control> key and pressing the indicated letter. Following is a list of useful control key sequences, with an explanation of their functions.

Control Key Sequence	Function
<Control/B>	Displays previously entered commands
<Control/U>	Deletes the current line, providing a fresh prompt (useful for canceling a line containing many mistakes)
<Control/S>	Suspends terminal output until <Control/Q> is pressed
<Control/Q>	Resumes terminal output suspended by <Control/S>
<Control/C>	Cancel an entire command, or interrupt the system
<Control/Y>	while it executes a command
<Control/W>	Refreshes the screen (image dependent)
<Control/R>	
<Control/T>	Provides a single line of information about the current process (helps to identify if the system is currently active)
<Control/Z>	Within a process, acts the same as "Exit": returning to VMS after saving the work
<Control/O>	Turns off messages written to SYS$OUTPUT until another <Control/O> is issued.

9.5 VAX/VMS FILES

What Is a File?

A file contains information for the computer to process. It can contain text, data, or program statements written in any computer language (i.e., FORTRAN, PASCAL, C). The user can create as many files as he or she wants, subject to disk quota limitations.

File Identification

A file is identified mainly by file name and file type as follows:

```
Filename.Type
```

A filename can have from one through 39 characters chosen from the letters A through Z, the numbers 0 through 9, and an underscore (_) or a dollar sign ($). A filetype can be from 0 to 39 characters long and must be preceded by a period.

The system recognizes many default file types that are used for special purposes. Some of the more common default file types are listed below.

File Type	Use
COM	Command procedure
DAT	Data file
DIS	Distribution list for MAIL
EDT	Start up command file for EDT editor
JOU	Journal file used by the EDT editor
LIS	Output listing file
MAI	MAIL message file

File Type	Use
OBJ	Object module file output from a compiler or assembler
MAR	Input source file for the VAX MACRO assembler
PAS	Input source file for the VAX PASCAL compiler
C	Input source file for the VAX C compiler
ADA	Input source file for the VAX Ada compiler
COB	Input source file for the VAX COBOL compiler
MOD	Input source file for the Modula-2 compiler

Every file also has a version number that the system assigns to it when the file is created or revised. When you initially create a file, the system assigns it a version number of 1. When you edit a file and then save it, even without making any changes to the file contents, the system saves this file with the old version number automatically increased by one. The version number follows the file type in a file specification:

```
Filename.type;1
```

If you reference a file without specifying a version number, the system assumes the highest version.

Using Wildcard Characters

A wildcard character is a symbol that can be used with many DCL commands to apply the command to several files at once, rather than specifying each file individually. The two wildcard characters are

1. Asterisk (*)
2. Percent symbol (%)

The asterisk (*) may be used to match any sequence of characters (other than period or semicolon) such as the file name, file type, or version number.

Examples	Explanation
Test.dat;*	All the versions of Test.dat file
.dat;	All the versions of all the files whose file type is DAT
Test.*;*	All the versions of the file whose file name is Test and all the file types
.;*	All the files in a directory

The percent sign (%) allows you to specify all the files containing any single character in the position that the percent sign occupies in the file specification.

Example	Explanation
Test%.dat;5	All the fifth versions of the files with the file type of DAT whose file names consist of five characters, whereas the first four characters are Test and the fifth character can be any other character Examples: Test1.dat;5, Test2.dat;5, Test3.dat;5

How to Delete a File

The DELETE command deletes a specific file. When you use the DELETE command, you must specify the file name, file type, and version number. However, wildcard characters can also be used in the file specification. You can also enter more than one file specification in a command string, separating them with commas.

Command	Results
$ DELETE test.dat;2	Deletes the file named test.dat;2
$ DELETE *.lis;*	Deletes all the versions of all the files of file type LIS
$ DELETE test.dat;3, t3.lis;2	Deletes the two files named test.dat;3 and t3.lis;2

How to Purge Files

The PURGE command allows you to delete all but the most recent version of a file. Therefore, no version number is specified in the PURGE command. The PURGE command is entered as follows:

```
$ PURGE test.dat <Return>
```

Use the /KEEP qualifier with the PURGE command to specify that you want to keep more than one version of a file. To keep the two highest versions of a file, type the following command:

```
$ PURGE/KEEP=2 test.dat <Return>
```

To clean up your entire directory, enter the PURGE command without any parameters. After this command, your directory will have only the highest version of each file:

```
$ PURGE <Return>
```

How to Display a File at Your Terminal

The TYPE command allows the user to scroll through a file. The user cannot edit the file while scrolling through it.

```
$ TYPE test.dat <Return>
```

Press <F1> or <Control/S> to stop scrolling. Press <F1> or <Control/Q> to resume scrolling.

To display a file one page at a time on the terminal, use the /PAGE qualifier:

```
$ TYPE /PAGE test.dat <Return>
```

You must press the <Return> key to view the next page. Press <Control/Z> to exit the TYPE command.

How to Print a File

To make a hard copy of a file, type the following command:

```
$ PRINT filename.type;version <Return>
```

If you do not give the version number, the PRINT command will print the highest version of that file.

How to Copy a File

The COPY command is used to make copies of files in your default directory, to copy files from one directory to another directory, to copy files from other devices, or to create files consisting of more than one input file.

The following COPY command copies the contents of the file PAYROLL.TST to a file named PAYROLL.OLD.

```
$ COPY payroll.tst payroll.old <Return>
```

If a file named PAYROLL.OLD exists, a new version of that file is created with a higher version number.

To copy a file from the directory [MALCOLM] to the subdirectory [MALCOLM.TEST-FILES], and give it a new name, OLDFILE.DAT, enter the following command string:

```
$ COPY newfile.dat [malcolm.TESTFILES]oldfile.dat <Return>
```

When you copy files from devices other than your default disk, you must specify the device name with the COPY command. The following command copies a file from your default directory onto DMA1 (the logical name of the disk).

```
$ COPY payroll.tst dma1: <Return>
```

If the output file specification did not include a directory, file name, or file type, the COPY command uses the same directory, file name, and file type as the input file (by default).

How to Rename a File

To change the identification of a file, type the following:

```
$ RENAME old-file-name.type new-file-name.type <Return>
```

The rename command can also move a file from one directory to another directory.

```
$ RENAME [old-directory-name]file-name.type [new-directory-
  name]
```

Notes: Do not type a blank between the old directory name and the file name. The RENAME command does not change the ownership of a file; if you move one of your files to a directory that you do not own, the file itself will still be owned by you, and will not be accessible to the owner of that directory.

How to Protect a File

To prevent other users from gaining unauthorized or undesired access to your files, use the SET FILE/PROTECTION command specifying user categories with access types.

The following table lists the four user categories and the four access types.

User Category	Type of User
SYSTEM	All users who have system privilege (SYSPRV) or low group numbers
OWNER	The user who created the file
GROUP	The users, including the owner, who have the same group number (or group identifier) in their user identification codes (UIC) as the owner of the file
WORLD	All users

Access Type	Type of Access
READ	The right to examine, print, or copy a file
WRITE	The right to modify or write a file
EXECUTE	The right to execute a file that contains executable program images (of type EXE) or command procedure files (of type COM)
DELETE	The right to delete a file

READ and EXECUTE mean something different for directories than for files. READ permission on a directory means that the accessor can

1. Read files under the directory, and
2. Use DIR to list files in the directory.

EXECUTE permission on a directory allows the accessor to read files under the directory but not to use DIR to list files in the directory. File names must be specified exactly; wildcard characters cannot be used.

The following table shows the abbreviations for user categories and access types to be used in the SET FILE/PROTECTION command:

User Category	Access Type
S—SYSTEM	D—DELETE
O—OWNER	R—READ
G—GROUP	W—WRITE
W—WORLD	E—EXECUTE

By system defined default, when you create a file, it is protected in the following way:

- Both the Owner and the System Manager have total access, or the ability to READ, WRITE, EXECUTE, and DELETE the file.
- All other users (i.e., the GROUP and WORLD categories) have no access.

Examples of File Protection

```
$ SET FILE/PROTECTION=(O:RWE,G:R,W) names.lis
```

The SYSTEM protections are left unchanged.
The OWNER has READ, WRITE and EXECUTE access.
The GROUP has READ access only.
The WORLD has no access.

```
$ SET FILE/PROTECTION=(G:R,W)
```

The SYSTEM and OWNER protections are left unchanged.
The GROUP has READ access only.
The WORLD has no access.

9.6 DIRECTORY STRUCTURE

Full File Specification

Example:

```
DRACUL::VAMP:[MALCOLM]STORIES.TXT;7
```

Node:	DRACUL
Device:	VAMP
Directory:	[MALCOLM]
Filename:	STORIES
Filetype:	TXT
Version:	7

Node: When computer systems are linked together, they form a network. Each system in a network is called a node. When you access a file on your own local node, you do not need to specify the node.

Device: The device name identifies the physical device on which the file is stored. If you omit the device name from the file specification the system assumes the file is on the default device. This disk is called the default disk. (The default disk is established at login time and can be changed using the SET DEFAULT command.)

Directory: Each user of a given disk has a directory that catalogs all the files belonging to him or her on that device. The directory file contains the names and locations of files in a format that only the system understands. If you omit the directory name, the system assumes your default directory (the directory in which you are currently working).

How to List Files in a Directory

To see the list of files in your default directory, type

```
$ DIRECTORY <Return>
```

To list the files in a particular directory, specify the name of the directory preceded by the device name if it is required.

```
$ DIRECTORY device:[directory_name] <Return>
```

To see how many versions of a particular file currently exist in your directory, type the command as follows:

```
$ DIRECTORY filename.type <Return>
```

You can also use wildcard characters (*, %) to display a selected group of files.

If you want to see detailed information about the files in your directory, type the following command:

```
$ DIRECTORY/FULL <Return>
```

Note: You can abort the screen output by typing <Control/C>.

Understanding Subdirectories

The System Manager provides each user with one directory to maintain files, which is called the main directory. When a user first logs in, their default directory is their main directory. Your default directory is the directory in which you are currently working. The name of your default directory is listed after the colon (:) on the first line displayed after issuing a DIRECTORY command.

You may create your subdirectories under the main directory and place files in these subdirectories; this allows you to organize your files hierarchically. For example,

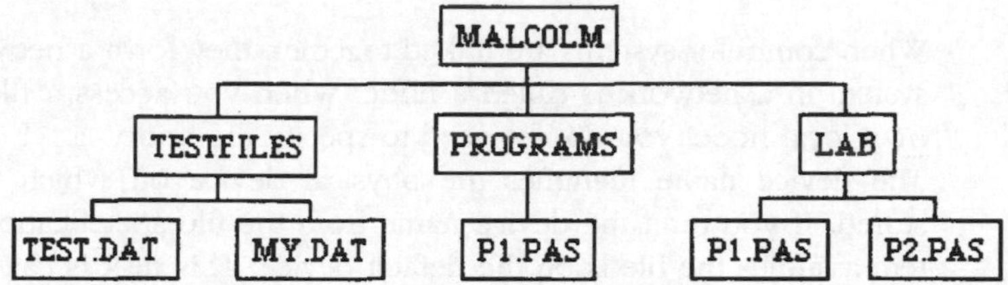

In the example, MALCOLM is the name of the main directory. TESTFILES, PROGRAMS, and LAB are subdirectories under the main directory; TEST.DAT, MY.DAT, and P2.PAS are files. There are two files named P1.PAS, and they are not necessarily the same.

How to Create a Subdirectory

Use the CREATE/DIRECTORY command to create a subdirectory. The following command creates the subdirectory file TESTFILES.DIR in the directory [MALCOLM], resulting in a subdirectory with the name [MALCOLM.TESTFILES].

```
$ CREATE/DIRECTORY [MALCOLM.TESTFILES] <Return>
```

If your default directory is MALCOLM, then the following command would produce the same result:

```
$ CREATE/DIRECTORY [.TESTFILES] <Return>
```

How to Change the Default Directory

Use the SET DEFAULT command to change your default directory as follows:

```
$ SET DEFAULT [MALCOLM.TESTFILES] <Return>
```

If your default directory is [MALCOLM], you can change your default directory to [MALCOLM.TESTFILES] as follows:

```
$ SET DEFAULT [.TESTFILES] <Return>
```

To move "upward" in the directory hierarchy, you can use a minus sign, which is a special directory wildcard symbol. If your default directory is [MALCOLM.TESTFILES], you can change your default directory to [MALCOLM] with the following command:

```
$ SET DEFAULT [-] <Return>
```

You can also use SET DEFAULT to change your default disk.

```
$ SET DEFAULT PILOT: <Return>
```

Note: The changes you make with the SET DEFAULT command remain in effect until you either issue another SET DEFAULT command or log out.

How to Delete a Subdirectory

Subdirectories show up in a DIR listing as a directory file, that is, a file with the type DIR.

To delete a subdirectory, use the following process:

1. DELETE all contents of the directory.
2. Change the protection on the directory file, giving DELETE permission to the owner (see sections on "How to Protect a File").
3. Delete the directory file.

Directory Protection

For directories, the protection codes WRITE and DELETE refer to what the accessor can do to files under that directory. For example, if a directory has no WORLD:WRITE access, then other users would not be able to alter files under the directory, even if the individual files have WORLD:WRITE access.

In other words, in recalling the file hierarchy, by removing access at any node in the tree (e.g., a directory) you disallow that access to any descendant of that node.

READ and EXECUTE mean something different for directories than for files. READ permission on a directory means that the accessor can

1. Read files under the directory, and
2. Use DIR to list files in the directory.

EXECUTE permission on a directory allows the accessor to read files under the directory but not to use DIR to list files in the directory. Filenames must be specified exactly; wildcard characters cannot be used.

Therefore, you can allow others to access your directory in a limited way by giving it EXECUTE access. They then can access only files whose names you have given them. If you wish to remove all access to your directory, remove both READ and EXECUTE permissions.

How to Use Logical Names

A logical name is used to reduce the typing done when specifying files or directories that you refer to frequently. For example, the following command assigns the logical name ZAP to the file specification **DRACUL::DOC1:[MALCOLM]TEST.DAT;21**

```
$ DEFINE ZAP DRACUL::DOC1:[MALCOLM]TEST.DAT;21 <Return>
```

After issuing the above command, the user can display the same file using the logical name in the TYPE command as follows:

```
$ TYPE ZAP <Return>
```

Use the SHOW LOGICAL command to display a logical name and its equivalent as follows:

```
$ SHOW LOGICAL ZAP <Return>
```

Several default logical names follow:

Name	Use
SYS$COMMAND	The initial file (usually your terminal) from which DCL reads input
SYS$INPUT	The default input stream from which the system reads commands and your programs read data
SYS$OUTPUT	The default device or file to which DCL writes error, warning, and informational messages
SYS$ERROR	The default device to which the system writes all error and informational messages
SYS$DISK	Your default disk device, which can be changed with the DCL command SET DEFAULT
SYS$LOGIN	The default device and directory when you first log in to the system

The following example of a logical name assignment demonstrates how to make a hard copy of the on-line HELP output for the DCL command SHOW LOGICAL:

```
$ DEFINE SYS$OUTPUT HELPFILE.LIS <Return>
$ HELP SHOW LOGICAL EXAMPLES <Return>
```

The system response will be Topic? Press <Return>

```
$ DEASSIGN SYS$OUTPUT <Return>
```

All the system responses are captured in the file HELPFILE.LIS that can then be printed by issuing the following PRINT command:

```
$ PRINT HELPFILE.LIS <Return>
```

Another example would be to capture the output from an executing program:

```
$ DEFINE SYS$OUTPUT MYOUTPUT <Return>
$ RUN MYPROGRAM <Return>
$ DEASSIGN SYS$OUTPUT <Return>
```

The results of running the executable image MYPROGRAM would be stored in the file MYOUTPUT, which could then be printed on the printer with the PRINT command.

9.7 MAIL

The MAIL utility provides electronic mail service for the exchange of messages and files. The MAIL prompt is MAIL>.

Command	Results
$ MAIL	Invokes the MAIL utility
EXIT or CTRL-Z	Exiting from MAIL
QUIT or CTRL-Y	If you change your mind and do not want to delete those messages that you have marked for deletion
HELP	On-Line help
READ (or just press RETURN)	Reading and selecting messages
BACK	Reading and selecting messages
CURRENT	Reading and selecting messages
FIRST	Reading and selecting messages
LAST	Reading and selecting messages
NEXT	Reading and selecting messages
<message number>	Reading and selecting messages
DELETE	Deletes the message currently being read

The MAIL utility automatically creates files in your directory, which start with "MAIL$," followed by letters and numbers, and ending in the type ".MAI." These files contain MAIL messages that have either been saved or have not yet been read. You should NOT delete these files with the DCL DELETE command. These files will go away when messages have been read and saved messages have been deleted in the MAIL utility.

Exchanging Messages

Send a new message:

```
SEND (or MAIL)
```

Reply to the current message:

```
REPLY (or ANSWER)
```

You will be prompted for the recipient, subject, and text of your message. In the case of REPLY, you will be prompted only for the text. To end the message and send it, type CTRL-Z.

You may also SEND or REPLY by specifying a file name:

```
SEND mymessage.txt
REPLY mymessage.txt
```

Distribution Lists

Suppose that you wish to send a message or a file to a group of users such as your department. You may create a distribution list by placing the user names of the recipients, one user name per line, in a file whose type is DIS. Do it like this:

```
SEND TO: @DEPT <return>
```

Where DEPT.DIS is the file containing the list of department members, in this example.

Examining and Deleting Messages

Command	Results
DIRECTORY	Numbered listing of your current messages
13	Make the 13th message the current message
DELETE	Delete the current message
DELETE 25	Delete message number 25

9.8 PHONE

The PHONE utility permits you to exchange messages with users who are currently logged on to the system. It simulates a real telephone with the hold button, conference calls, and telephone directories. If you do not want to be bothered by phone calls, you can "take your phone off the hook" by including the following command in your LOGIN.COM file:

```
$ SET BROADCAST=NONE
```

The PHONE utility is invoked by the VMS command

```
$ PHONE
```

Summary of PHONE Commands

ANSWER	HELP
DIAL	HOLD
DIRECTORY	MAIL
EXIT (OR CTRL-Z)	REJECT
FACSIMILE	UNHOLD
HANGUP	

Examples of the Use of Phone

1. You want to call Jane Q. Public if she is currently logged on.

```
$ PHONE
% > DIRECTORY
% > DIAL jqp25617m012
```

2. You receive a message that Jane Q. Public is calling you.

```
$ PHONE ANSWER
```

3. You need help on how to use the PHONE utility

```
$ PHONE HELP
```

9.9 THE EDT EDITOR

EDT is an interactive text editor. You can use EDT to edit many kinds of text files—letters, memos, or complex computer programs. With EDT you can create new files, insert text into them, and edit that text. You can also edit text in existing files.

Invoking EDT

To invoke EDT, enter the EDIT command after the $ prompt, followed by the file name. If you do not enter the file name, the system will prompt you to enter the file name.

```
$ EDIT MEMO.TXT <Return>
```

Note: On VAX/VMS Version 6.x systems, enter the following command to access the EDT editor, as the default for 6.x is the TPU editor.

```
$ EDIT/EDT MEMO.TXT <Return>
```

To automatically access the EDT editor when using the EDIT command, place the following line in your LOGIN.COM file:

```
$ EDIT := EDIT/EDT
```

See the section on LOGIN.COM files for detail about the use of the LOGIN.COM file.

If you are creating a new file, the following message will be displayed:

```
Input file does not exist
[EOB]
```

If you are editing a file that already exists, the following will appear:

```
1 This is the first text line in this file.
*
```

Once you enter EDT, you can choose among three modes:

1. Line mode: Line mode allows you to edit line by line using line numbers to refer to each individual line in the file.
2. Keypad mode: Keypad mode allows you to move the cursor to the text you want to edit and perform most editing functions by pressing keypad keys.
3. Nokeypad mode: Nokeypad mode allows you to move the cursor directly to the text you what to modify and enter nokeypad commands.

One difference among the three modes is that you press keys to edit text in Keypad mode whereas you enter commands to edit text in Line mode and Nokeypad mode.

By default, EDT puts you into Line mode. You will know that you are in Line mode when you see the asterisk prompt (*).

Terminating EDT

At the asterisk prompt, type EXIT to save the file and leave EDT. EDT saves the file with the same file name and type, but the version number is increased by one.

```
$ EDIT FUN.DAT <Return>
.
.
.
*EXIT <Return>
$
```

If you want to store the edited file under a different name, then enter the new file name after EXIT as follows:

```
*EXIT mem.dat <Return>
```

If you do not want to save your changes when you end an editing session, use the QUIT command. All the changes you made to the file will be lost and no output file will be created.

```
*QUIT <Return>
```

How to Use EDT Line Mode

EDT assigns line numbers and indents the text 12 spaces. These line numbers and indentation are not part of the text and are not kept when you end the editing session.

Accessing Help from Line Mode

The HELP command can be typed after the asterisk (*) prompt for information about EDT commands. Information about specific commands can be obtained by typing the command after HELP:

```
*HELP INSERT <Return>
```

For help on a qualifier, type the qualifier with the command:

```
*HELP EXIT/SAVE <Return>
```

Inserting Text

When you want to insert text, enter the INSERT command after the asterisk prompt (*), and press <Return>. The cursor will indent 12 spaces and wait for you to start typing. Indentation appears on the screen only, and does not become part of the edited text.

```
*INSERT <Return>
```

The following keys are useful while inserting text:

Keys	Function
<Control/U>	Deletes a portion of a line, from the cursor to the left margin
<Control/Z>	To exit INSERT mode and return to the line editing prompt (*)
<Return>	To end a line. However, when you end a line with <Return>, you can no longer delete characters from that line in the INSERT mode

EDT always inserts text in front of the current line. If you want to insert text after the fifth line, make the sixth line the current line and type INSERT at the EDT prompt (*). Type as many lines as you want and press <Control/Z>. Use the TYPE WHOLE command to see the whole file. The lines you inserted will be numbered like this: 5.1, 5.2, 5.3, 5.4. . . . You can renumber the lines by using the command RESEQUENCE at the EDT prompt:

```
* RESEQUENCE
```

How to Display a File on the Screen

The following is the list of commands to display a part or the whole file on the screen:

Commands	Explanation
TYPE	Displays the current line
TYPE 35	Displays the 35th line of text
TYPE "string"	Displays the first line containing the string
TYPE BEGIN	Displays the first line of the current buffer
TYPE END	Displays the [EOB] mark
TYPE LAST	Displays the most recent line of text displayed
TYPE WHOLE	Displays every line in the current buffer
TYPE BEFORE	Displays the group of lines in the current buffer starting with the first line and ending with the line just before the current line

Commands	Explanation
TYPE REST	Displays the group of lines in the current buffer starting with the current line and ending with the last line in the buffer
TYPE 3,6,9	Displays 3rd, 6th, and 9th line
TYPE 3 AND 9	Displays 3rd and 9th line
TYPE 25:"sam"	Displays the group of lines starting with line number 25 and ending with the line containing 'sam'
TYPE. THRU 5	Displays all the lines in the group starting with the current line and ending with line number 5
TYPE 4#3	Displays line 4 and the three lines following line 4
TYPE. FOR 8	Displays the current line and the next 8 lines
TYPE BEGIN +5	Displays the 6th line of the current buffer
TYPE-3 THRU	Displays the current line and the three lines preceding it

Line Numbers

Line numbers have the following characteristics:

- They are assigned to each line in a buffer in every editing session, including newly inserted lines and text added with the INCLUDE command.
- They start with 1 and are numbered 1,2,3. . . .
- They are decimal numbers if newly inserted (e.g., 3.1 and 3.2).
- They are removed by the EXIT command when you save the file.
- They can be renumbered in increments of 1 or more with the RESEQUENCE command.

If you want to resequence the line numbers in the whole file (with an increment of 1) after executing an INSERT or DELETE command, type the following:

```
*RESEQUENCE <Return>
```

If you want to resequence the lines after a specific line number, then use the /SEQUENCE qualifier followed by the starting line number and the increment. The following command resequences the line numbers, starting at line number 6, in increments of 4.

```
*RESEQUENCE/SEQUENCE:6:4 <Return>
```

Deleting Text

You can delete individual lines or groups of lines by using the DELETE command. After a deletion operation, EDT displays a message stating the number of lines deleted and displays the line following the last line deleted; this is the new current line. You will not see the actual line numbers of the text that has been deleted by the DELETE command. So, if you want to resequence the line numbers, you will have to issue the RESEQUENCE command.

The syntax of the DELETE command is the same as that of the TYPE command, which has been discussed earlier.

If you enter the DELETE command and do not specify a range or the line number, EDT deletes the current line. You can use the /QUERY qualifier if you want EDT to prompt you before deleting each line of a specified range. The prompt is a question mark (?) and the possible answers to the prompt are

Y	(Yes)	Delete this line.
N	(No)	Do not delete this line.
A	(All)	Deletes all remaining lines in the specified range.
Q	(Quit)	Quit the delete operation.

Substituting Text

Use the SUBSTITUTE command for replacing one string with another. The general syntax of this command is

```
*SUBSTITUTE/old-string/new-string/[range] [/QUERY] <Return>
```

Examples of Substitution

To substitute the word GOOD for BAD throughout a file, type

```
*SUBSTITUTE/BAD/GOOD/WHOLE <Return>
```

Note: If the qualifier WHOLE is omitted, the substitution is done only on the current line.

Most nonalphanumeric characters can be used as delimiters instead of the slash (/). The following command substitutes the string A/3 for A/2 in the current line, using a dollar sign ($) as the delimiter:

```
*SUBSTITUTE$A/2$A/3$ <Return>
```

To get EDT to prompt you before each substitution, use the /QUERY qualifier with the SUBSTITUTE command:

```
*SUBSTITUTE/ in / at /WHOLE/QUERY <Return>
```

EDT prompts you for one of the following responses before each occurrence of the search string:

Y	(Yes)	do the substitution
N	(No)	do not do the substitution
A	(All)	do the rest of the substitution without query
Q	(Quit)	terminate the command

Moving and Copying Text from One Location to Another

The MOVE and COPY commands have the same syntax. The only difference between the two commands is that the COPY command does not delete the text from its original location, whereas the MOVE command does. The following is the syntax for the MOVE and COPY commands:

```
*MOVE first-range TO second-range/QUERY <Return>
*COPY first-range TO second-range/QUERY <Return>
```

The /QUERY qualifier is used to verify each line to be inserted.

Examples of Move and Copy

To move lines 43 through 56 above line 13, enter the following command:

```
*MOVE 43 THRU 56 TO 13 <Return>
```

To move the current line above line 65, type

```
*MOVE TO 65 <Return>
```

You can use the /DUPLICATE qualifier with the COPY command if you want to insert the range of text more than once.

To copy the first line, placing it just above line 65 and repeating the operation 9 times, enter the following command:

```
*COPY BEGIN TO 65/DUPLICATE:9 <Return>
```

The EDT response will be

```
1 line copied 9 times
*
```

Replacing Text

The REPLACE command combines the DELETE and INSERT functions in one command. You can use REPLACE when you need to delete a block of text and want to type new text in that location. The syntax for the REPLACE command is

```
*REPLACE range <Return> text <Control/Z>
```

The following REPLACE command deletes the current line and shifts to insert mode:

```
*REPLACE <Return>
```

Type the text and then press <Control/Z> to save the inserted text and get out of Insert mode.

How to Use EDT Keypad Mode

Keypad editing is available on the VT200 Series, VT100, and VT52 terminals. This is a full screen editor. The contents of a file are displayed on the screen as you edit. You can see the changes you make to a file as they take place.

The Layout of the Numeric Keypad on VT200 Series Terminals

PF1 **GOLD**	PF2 HELP	PF3 FNDNXT **FIND**	PF4 DEL L **UND L**
7 PAGE **COMMAND**	8 SECT **FILL**	9 APPEND **REPLACE**	- DEL W **UND W**
4 ADVANCE **BOTTOM**	5 BACKUP **TOP**	6 CUT **PASTE**	, DEL C **UND C**
1 WORD **CHNGCASE**	2 EOL **DEL EOL**	3 CHAR **SPECINS**	ENTER **SUBS**
0 LINE **OPEN LINE**		. SELECT **RESET**	

Accessing Help From Keypad Mode

Pressing the HELP key while you are in Keypad mode will provide a diagram of the keypad functions. Once this diagram is displayed, information for a key can be obtained by pressing the key in question. Pressing a keypad key gives information about the main function of the key, and its function in a Gold key sequence (if any). For information on a control key sequence, press the desired sequence.

Creating a File and Inserting Text

At the VMS prompt ($), type the following:

```
$ EDIT file-name <Return>
```

The EDT prompt (*) will appear on the screen. Type C or CHANGE at the asterisk prompt:

```
*C <Return>
```

Now you are in Keypad mode. Type the text, using the <Return> key to go to the next line. When finished, press <Control/Z>. This will take you out of Keypad mode and put you in the EDT line mode at the asterisk prompt. Type EXIT to save the file as in line mode. All the keys on the numeric keypad behave as function keys in Keypad mode. These keys have been explained above.

The following is the list of functions assigned to each key on the numeric keypad.

Key	Functions	Functions Performed with GOLD Key
PF1	GOLD key	
PF2	HELP	
PF3	FNDNXT	FIND
PF4	DEL L	UND L
0	LINE	OPEN LINE
1	WORD	CHNGCASE
2	EOL	DEL EOL
3	CHAR	SPECINS
4	ADVANCE	BOTTOM
5	BACKUP	TOP
6	CUT	PASTE
7	PAGE	COMMAND
8	SECT	FILL
9	APPEND	REPLACE
–	DEL W	UND W
,	DEL C	UND C
.	SELECT	RESET
ENTER	SUBS	

Using the GOLD Key

<PF1> is called the GOLD key in Keypad mode and is used for two purposes:

1. To use the alternate of the two functions on a keypad key. All the alternate functions are in boldface on the keypad layout shown on the previous page.
2. To execute a function a specified number of times.

To use an alternate function, press the GOLD key followed by the desired keypad key. All the functions typed in boldface on the keyboard map are to be used in conjunction with the GOLD key.

To execute a function a specified number of times, follow these instructions:

1. Press the GOLD key (<PF1>).
2. Type in the number of times to repeat the function.
3. Press the desired function.

Moving the Cursor

You can move the cursor in any direction using the four arrow keys. You can also change the direction of the cursor movement.

The following table explains the keys on the numeric keypad for cursor movement.

Key(s)	Function
< 0 >	Moves the cursor one line
< 1 >	Moves the cursor word by word
< 2 >	Moves the cursor to the end of line
< 3 >	Moves the cursor character by character
< 4 >	To set the cursor movement in the forward direction
< 5 >	To set the cursor movement in the backward direction
< PF1 >< 5 >	Moves the cursor to the top of file
< PF1 >< 6 >	Moves the cursor to the bottom of the file
< 8 >	Moves the cursor 16 lines up or down the screen

Deleting and Undeleting Text

You can delete text by character, word, and line. The deleted text is stored in three buffers. The character buffer contains the last character deleted, the word buffer contains the last word deleted, and the line buffer contains the last line deleted. Only the most recent deletion can be restored, but you can restore this unit of text numerous times to any location. You can use the following keypad functions to delete text:

Function	Key(s)	What It Does
DELETE	< DEL >	Deletes the preceding character
DEL C	< , >	Deletes the current character
DEL W	< _ >	Deletes the current word
DEL L	< PF4 >	Deletes the current line
DEL EOL	< PF1 >< 2 >	Deletes from the cursor to the end of line
CTRL/U	< CTRL/U >	Deletes from the beginning of the line to the cursor

The following keypad functions can be used to undelete text:

Function	Key(s)	What It Does
UND L	<PF1><PF4>	Restores the last deleted line at the cursor position
UND W	< PF1 >< _ >	Restores the last deleted word at the cursor position
UND C	< PF1 >< , >	Restores the last deleted character at the cursor position

Locating Text

Use the FIND keypad function to locate strings of text as follows:

1. Press <PF1> and then press <PF3>. The system response will be: Search for:
2. Type the string of text you are looking for and press one of the following keys.

< 4 > (ADVANCE)	To set the search in the forward direction and search in that direction
< 5 > (BACKUP)	To set the search in the backward direction and search in that direction
<ENTER>	To search in the direction already set
<DO>	To search in the direction already set

3. If you want to find the next occurrence of the same string, press <PF3> to invoke the FNDNXT function of the Keypad mode.

Note: Remember the following three items when you are entering a search string:

1. EDT ignores the cases of letters while making searches, unless you enter the SET SEARCH EXACT command.
2. <Delete> is the only key you can use to edit an incorrectly typed search string.
3. To cancel a search string, press <Control/U>.

Moving Text

You can move text using three groups of keypad functions:

1. RETURN and OPENLINE (<PF1><0>)
2. DEL L (<PF4>), UND L (<PF1><PF4>), DEL W (<_>), UND W (<PF1> <_>), DEL C (<,>), UND C (<PF1><,>)
3. CUT (<6>) and PASTE (<PF1><6>) (or <REMOVE> and <INSERT HERE>)

Use the following six steps to move text:

1. Place the cursor at the beginning of the text you want to move.
2. Press <.> (SELECT) to mark the starting location.
3. Move the cursor to the end of the desired range.
4. Press <6> (CUT) to delete the text from its current position.
5. Move the cursor to the character just beyond the point at which you want the text inserted.
6. Press the GOLD key (<PF1>) followed by <PF6> (PASTE).

Substituting Text

The following are the steps for substituting text:

1. Press <.> (SELECT) and enter the replacement text.
2. Press <6> (CUT). (The select range will disappear into the delete buffer.)
3. Press the GOLD key (<PF1>) followed by <PF3> (FIND). Type in the search string, then press <ENTER>.
4. Press <PF1><ENTER> (SUBS) to exchange the existing text for the replacement text.

Five More Keys to Use with the GOLD (<PF1>) Key

To use any of the following keypad functions, you must press the GOLD key (<PF1>) first.

Keypad Function	Key	What It Does
COMMAND	< 7 >	Enables you to enter a line mode command from Keypad mode
CHNGCASE	< 1 >	Reverses the case of letters in your text Uppercase letters become lowercase; lowercase letters become uppercase
FILL	< 8 >	Recognizes the select range so that the maximum number of whole words can fit within the current line width
RESET	< . >	Changes the following conditions of your editing session: Cancels an active select range Empty the search buffer so that there is no current search string Sets EDT's current direction to advance
SPECINS	< 3 >	Enables you to insert any characters from the DEC Multinational Character Set into your text using the character's decimal equivalent value

Moving between Modes

EDT has three modes for editing text:

1. Line mode
2. Keypad mode
3. Nokeypad mode

The following table lists the commands to move between modes:

From	To	Command
Line mode	Keypad mode	*CHANGE or *C
Line mode	Nokeypad mode	*SET NOKEYPAD *CHANGE
Keypad mode	Line mode	<CTRL/Z>
Keypad mode	Nokeypad mode	<CTRL/Z> *SET NOKEYPAD *CHANGE
Nokeypad mode	Line mode	E X
Nokeypad mode	Keypad mode	E X *SET KEYPAD *CHANGE

DEFINING TERMS

Account: Authorization and space allocation necessary to access the DEC VAX.

Command Procedures: Files containing DCL files. Similar to .BAT files in DOS, or shell scripts in UNIX. Takes the form filename.COM.

Control Keys: Key sequences implemented by holding down the CTRL key and at the same time pressing another key. Extends the functionality of the system.

DCL: Digital Command Language.

EDIT: Full-screen VAX text editor (/EDT option gives you a line editor).

File Protection: Controls access to your files by other users.

Logical Name: Macro implementation to give you an alias for commonly used files or directories.

Password: User-selected identification code used to verify authorization to use the system. Comparable to the PIN number you use with your ATM card.

Purge: Removal of all but most recent version of a file.

Wildcard Characters: The characters * and %, which stand for any sequence or any single character, respectively. Useful in managing groups of files.

REFERENCES

DEC VAX/VMS Reference Manuals. Digital Equipment Corporation, Maynard, MA.
DEC VAX/VMS On-Line Help System.

FURTHER INFORMATION

Paul C. Anagnostopoulos, *VAX/VMS: Writing Real Programs in DCL*. Digital Equipment Corporation, Maynard, MA, 1989.

James F. Peters III and Patrick Holmay, *An Introduction to VAX/VMS*. Digital Equipment Corporation, Maynard, MA, 1990.

Ronald M. Sawey and Troy T. Stokes, *A Beginner's Guide to VAX/VMS*. Digital Equipment Corporation, Maynard, MA, 1989.

10

IBM Mainframes

Christina Lohr
Barry Walton
Millersville University

10.1 VM/CMS

Overview of VM/CMS

VM/CMS (Virtual Machine/Conversational Monitor System) is an IBM operating system that is available for use on IBM mainframes. General users use two of the components of VM/CMS:

> *CMS (Conversational Monitor System)* CMS deals with creating, updating, and managing files. It is also used for compiling, loading, and running programs. CMS is essentially a tool for communicating with the system via a terminal.
>
> *CP (Control Program)* CP is necessary for CMS to operate. CP handles hardware devices required to store files (disks and tapes), work on files (terminals and memory), and output files (printers, terminals, punches, and readers).

CMS and CP work closely together to perform all of the tasks deemed necessary for users.

CMS Files

VM/CMS stores information in CMS files and retrieves and manipulates information with CMS and CP commands. A CMS file is identified by a unique three-part *fileid*. The *fileid* consists of a *file name*, *file type*, and *file mode* in respective order. Examples of *fileids* are:

STATPROG	SAS	A
NOTEBOOK	ALL	A
FALL1990	DATA	A

file name can be any combination of alphanumeric characters up to eight characters long. *file type*, like *file name*, can also be any combination of alphanumeric characters up to eight characters long. However, certain filetypes must be used in conjunction with certain mainframe applications. For example,

0-8493-2530-7/96/$0.00+$.50

SAS	Should be used with SAS programs
PASCAL	Should be used with Waterloo Pascal programs
SPSSX	Should be used with SPSSx programs
FORTRAN	Should be used with VS FORTRAN programs

file mode is generally a single letter. This letter identifies the minidisk on which the file is stored. The default filemode is A. Normally, when you create a file, it is stored on your A-disk.

Logging On and Off VM/CMS

Before a user can use the VM system, he or she should obtain a valid *userid* and *password* from the system adminstrator. When logging in from a terminal, a user will usually have to press the <ENTER> key to clear an introductory system screen before logging on to VM/CMS. After doing so, a user will have a blank screen with a prompt in the lower right hand corner indicating CP READ. To logon, the user should type

```
L userid <ENTER>
short for LOGON
```

The computer will then prompt the user for a password. The user should type it in and then press <ENTER>. The password will not be displayed on the screen for security reasons. If the user is successfully logged on, VM will respond with the following *Ready*; message:

```
READY; T=0.02/0.02 08:05:53
```

short for TIME, followed by time taken by VM to log the user on, followed by the time of day.

If a user has finished working on the terminal then he/she must LOGOFF. *Switching the terminal off does not log off the user.* Anyone using the terminal after the user could gain access to the user's files. To LOGOFF, the user should be at the READY; prompt, type **LOG**, and press the <ENTER> key. VM will display a logoff message stating the time that the user logged off.

Using the FILEList Screen

A popular utility used on most VM/CMS systems is FILELIST. FILELIST is used to manage the user's files on the system. At the READY; prompt, users can type **FILELIST** or **FILEL** to bring up the screen shown in Figure 10.1.

Appearing at the bottom of the FILEList screen in Figure 10.1 are a series of numbers followed by equal signs and commands (1= Help, 2= Refresh, and so on). These numbers refer to the commands performed when a particular PF key (Program Function key) is pushed on a hard-wired terminal or 3270. The definition of these keys can be changed by your system administrator, so the screen shown in Figure 10.1 may be slightly different than those on your system. To exit the FILEList screen, as with most VM/CMS screens, press <PF3> or type **QUIT**.

The arrow prompt (====>) at the bottom of the FILEList screen is an area where VM commands can be entered as they would normally be entered at the READY;

prompt. Other commands can be entered under the **Cmd** column located at to the left side of the listed files. These commands, when entered on the same line that a file is displayed, performs the command on that file:

`DISCARD`	Erases the file
`RENAME / newfn newft newfm`	Renames a file to the new filename (*newfn*), new filetype (*newft*), and new filemode (*newfm*)
`COPYFILE / newfn newft newfm`	Makes an identical copy under the new filename, new filetype, and new filemode
`PRINT`	Prints a file to the user's default line printer

```
 ACAD2    FILELIST A0  V 108  Trunc=108 Size=4 Line=1 Col=1 Alt=3
Cmd   Filename Filetype Fm Format Lrecl    Records      Blocks  Date     Time
*     FALL1990 DATA     A1 F          80         4           1 11/07/93 17:18:26
      PROGRAM  PASCAL   A1 V           3         3           1 11/07/93 17:16:51
      STATPROG SAS      A1 F          80         1           1 11/07/93 16:33:18
      PROFILE  EXEC     A1 F          80         3           1  9/15/93 11:13:37

1= Help       2= Refresh  3= Quit   4= Sort(type)  5= Sort(date)  6= Sort(size)
7= Backward   8= Forward  9= FL /n 10=             11= XEDIT      12= Cursor

====> _
                                                          X E D I T  1 File
4A                          Aa   A Session1     R 23 C 7   o-o__ 17:18   11/7/93
```

FIGURE 10.1 FILEList screen.

Some of the commands, when typed in the **Cmd** column, are long and will appear to "overwrite" the characters listed for a file's filename. The user should not be concerned with this, for when the <ENTER> key is pressed to perform the command, the command will disappear and the screen will be updated accordingly.

The above commands can also be entered at the arrow (====>) prompt on the FILEList or the Ready; prompt under VM. However, the user must now supply the input *fileid* information for these commands. For example, if a user would like to rename the file MYPROG SAS A1 that appears in Figure 10.1, at the arrow prompt on the FILEList screen or at the READY; prompt under VM he can type:

```
RENAME MYPROG SAS A1 NEWFILE SAS A1
```

Likewise, the user can copy a file from the READY; prompt or FILEList arrow prompt as well. The following command would copy the file MYPROG SAS A1 to BACKUP SAS A1:

```
COPYFILE MYPROG SAS A1 BACKUP SAS A1
```

Note that the slash (/) normally used under the **Cmd** column of the FILEList screen is not part of the command line version of these commands.

When entering command-line instructions (i.e., READY; prompt or FILEList arrow ====> prompt commands), groups of files can be referred to via the wildcard character (*). For example, if a user is at the READY; prompt and would like the FILEList utility to just display all those files with the SAS filetype on his A-disk, he would enter

```
FILEL * SAS A <ENTER>
```

Likewise, the equal sign (=) used in an output fileid will retain the same fileid information as the input fileid for certain commands. For example,

```
RENAME TEMP SAS A PERM = = <ENTER>
```

will rename the file TEMP SAS A to PERM SAS A (the filetype and filemode are kept the same as the original, as indicated by the equals signs).

Communicating with Other System Users

CMS has commands to send files and messages to other users on the same mainframe. To send a brief message to someone that is logged onto the system use the TELL command at the READY; prompt:

```
TELL JSMITH WHEN CAN WE MEET? <ENTER>
```

If JSMITH is logged on, the following message will appear on his screen:

```
MSG FROM MJONES : WHEN CAN WE MEET?
```

To send the file MYPROG SAS A1 to user JSMITH, at the READY; type:

```
SENDFILE MYPROG SAS A1 TO JSMITH <ENTER>
```

Files can be received via the RDRList (Readerlist) under CMS. The RDRList is an area where all incoming mail (files and e-mail) is put to be processed (see Fig. 10.2). At the READY; prompt, the user types **RDRL** to access the RDRList. The RDRList appears similar to the FILEList, however, some of the PF key options are different than those of the FILEList.

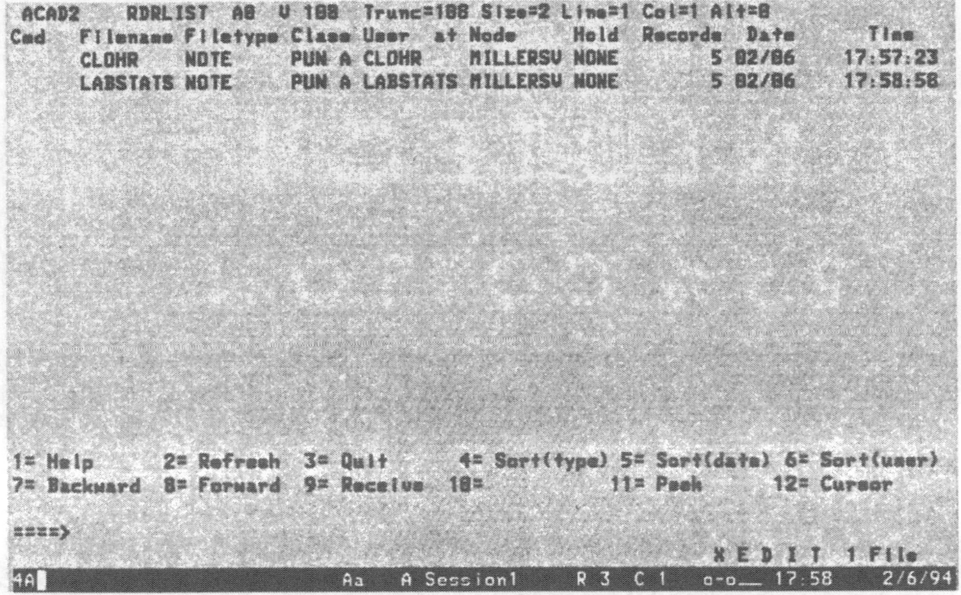

FIGURE 10.2 Readerlist.

A mail facility is also provided for e-mail via the NOTE command under CMS. To send a note to JSMITH, the user would type

```
NOTE JSMITH <ENTER>
```

Entering the NOTE command puts the user in XEDIT mode and a screen similar to the screen shown in Figure 10.3 would appear (see more information on how to use XEDIT in Section 10.2).

PF keys are defined at the bottom of the NOTE screen to perform operations such as send, move backward and forward in a longer file, invoke help, and quit the utility. When a user sends or receives a file using the NOTE command, a log of these transactions is put in a file called ALL NOTEBOOK A0 on the user's disk. Different NOTEBOOKs can be created to help organize notes (refer to the *VM/SP CMS User's Guide* for more information).

10.2 USING XEDIT

The XEDIT Editor

While different editors are available for use on IBM mainframes under VM/CMS, one of the most popular is XEDIT. XEDIT is a line-oriented editor that provides a means of editing text and data files to be used for programming, data storage, and, sometimes, electronic mail.

Note: The actual layout of an XEDIT screen can vary from system to system, as VM/CMS is flexible enough to let a user change the layout of the screen to put the command line, line numbers, column indicators, and file information almost anywhere

```
ACAD2   NOTE    A0  V 132  Trunc=132 Size=10 Line=10 Col=1 Alt=0
* * * Top of File * * *
OPTIONS: NOACK   LOG   SHORT   NOTEBOOK ALL



Date: 7 November 1993, 18:12:10 ET
From: ACAD2    at MU2
To:   JSMITH
Subject:

* * * End of File * * *





1= Help     2= Add line  3= Quit   4= Tab      5= Send       6= ?
7= Backward  8= Forward   9= =     10= Rgtleft 11= Spltjoin  12= Power Input

====>
                                                        X E D I T  1 File
4A                        Aa   A Session1    R 23 C 7   o-o__ 18:11   11/7/93
```

FIGURE 10.3 Note screen.

on the screen. One possible layout, which is very clear to read, appears in this section of the book and will be used in the figures, but realize that this is just one way of customizing the XEDIT screen.

Creating an XEDIT File

After logging onto the system, at the READY; prompt, a user types XEDIT followed by the name of the file he wants to create.

```
XEDIT PARKING MUD A1
```

The above command creates a file called PARKING MUD A1 and invokes XEDIT. The screen shown in Figure 10.4 shows the XEDIT screen and the now empty PARKING MUD A1 file, which is now ready for text entry. At the top of the screen appears an arrow prompt (====>) at which XEDIT editor commands are typed. Line numbers may appear on the left hand side of the screen to indicate where the body of the file is. **`Top of File`** and **`Bottom of File`** also indicate where the file begins and ends.

Inputting Data into an XEDIT File

After using the XEDIT command under VM/CMS to create and enter text into an empty XEDIT file, to input the file's initial data, the user would type **`INPUT`** at the arrow prompt at the top of the screen. The line numbers disappear, and the user can begin typing text or data (see Fig. 10.5). It is important to note that the user presses the <RETURN> key as opposed to the <ENTER> key when completing a line of text.

```
 PARKING  MUD       A1  F 80  Trunc=80 Size=0 Line=0 Col=1 Alt=0
====> _
      |...+....1....+....2....+....3....+....4....+....5....+....6....+....7...
00000 * * * Top of File * * *
00001 * * * End of File * * *

PF: 1=HLP 2=SPLIT 3=FILE 4=TOP 5=BOT 6=EQ 7=UP 8=DWN 9=QUIT 10=LFT 11=RT 12=CUR
4A                          Aa   A Session1    R 2  C 7   o-o___ 18:22   11/7/93
```

FIGURE 10.4 Blank XEDIT screen for PARKING MUD A1.

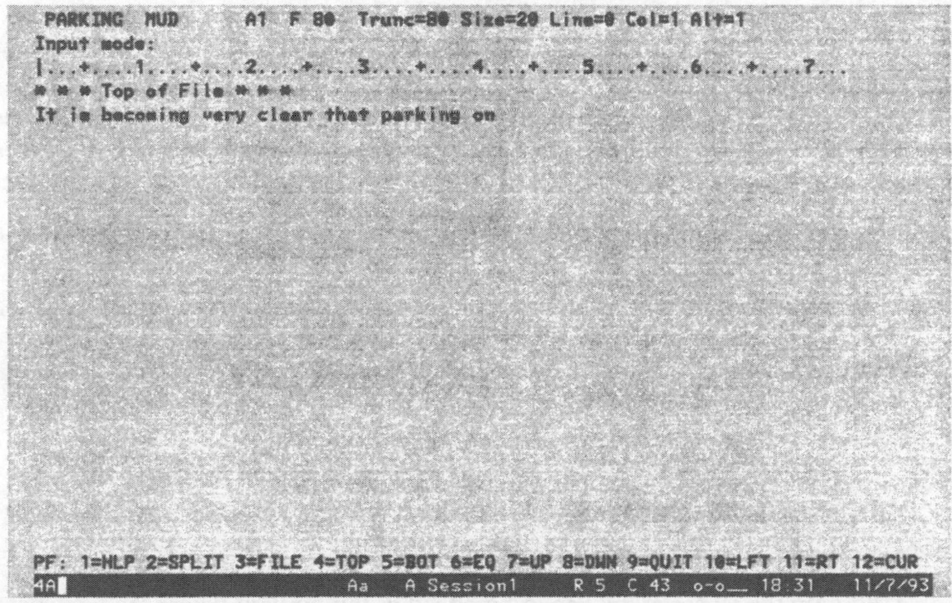

FIGURE 10.5 XEDIT input screen.

```
 PARKING  MUD       A1  F 80  Trunc=80 Size=4 Line=0 Col=1 Alt=5
====>
      |...+....1....+....2....+....3....+....4....+....5....+....6....+....7...
00000 * * * Top of File * * *
00001 IT IS BECOMING VERY CLEAR THAT PARKING ON THIS CAMPUS IS GOING TO BE A
00002 PROBLEM FOR MANY YEARS TO COME.  HOWEVER, CONTRIBUTING TO THE PROBLEM IS
00003 THE FACT THAT MANY "STUDENT-CREATED" PARKING LOTS ARE NOTHING MORE THAN
00004 MUDDY FIELDS IN WHICH STUDENTS (AND UNFORTUNATELY SOME STAFF) MUST PARK.
00005 * * * End of File * * *

PF: 1=HLP 2=SPLIT 3=FILE 4=TOP 5=BOT 6=EQ 7=UP 8=DWN 9=QUIT 10=LFT 11=RT 12=CUR
4A                     Aa   A Session1     R 2  C 7    o-o__  18:35    11/7/93
```

FIGURE 10.6 XEDIT editor screen after input.

When the user has finished entering data, the <ENTER> key is pressed twice to indicate the completion of input. The user then leaves the input screen and is returned to the XEDIT editor screen (Fig. 10.6).

Inserting/Deleting Characters in a Line

When using XEDIT, the default setting for insertion of characters is via typeover mode, i.e., as the user types characters they replace any characters that appear at the cursor position. However, the following steps can be followed to insert characters in the middle of a line of text:

- Position the cursor at the line to be edited at the position where text is to be inserted.
- Press the *insert* key. On hard-wired terminals and 3270s, this appears as an â on a key. If using a different keyboard or other terminal emulation, consult your system adminstrator to determine which key has been mapped to provide character insertion.
- Begin typing the text to be inserted. The text will be inserted in the line at the cursor, and the characters that appear after the cursor will be pushed to the right.
- Press <ENTER> to register the changes and exit the insert mode.

To delete characters on a line:

- Position the cursor at the line to be edited at the position where the first character is to be deleted.

- Press the *delete* key until all of the characters to be deleted disappear. On hard-wired terminals and 3270s, this appears as an Á on a key. If using a different keyboard or other terminal emulation, consult your system administrator to determine which key has been mapped to provide character deletion.
- Press <ENTER> to register the changes.

Using XEDIT Prefix Commands

When using XEDIT to edit a file, at the left of the text that is in the file appears either a five-digit line number (e.g., 00001) or five equal signs (=====). This is referred to as the line *prefix*. Any command typed in the prefix area is called a *prefix command*. The following letters are used in prefix commands:

`I`	for Inserting line(s)
`C`	for Copying line(s)
`D`	for Deleting line(s)
`M`	for Moving lines
`P`	for Preceding (used in conjunction with Copy or Move)
`F`	for Following (used in conjunction with Copy or Move)
`B`	for Before (same as Preceding)
`A`	for After (same as Following)

The above letters can be typed anywhere in the prefix area. Users should just type over the line numbers or equal signs in the prefix area—they will be restored after the action is performed. Insert, Copy, Delete, and Move can be used in conjunction with Preceding/Following, Before/After, and numbers (e.g., 6) to perform particular tasks. Several examples of each are shown below with the ===== prefix area, with an explanation as to the action performed when the prefix command is entered (the user should press <ENTER> after entering a prefix command to perform the action).

Inserting Lines

`====I`	Inserts one blank line after this line
`I====`	Inserts one blank line after this line
`==6I=`	Inserts six blank lines after this line
`===I5`	Inserts five blank lines after this line

Deleting Lines

`====D`	Deletes this line
`==D==`	Deletes this line
`===D5`	Deletes this line and the next four lines

To delete a block of lines type "DD" in the prefix area at the first line of the block and the last line of the block to be deleted. For example, Figure 10.7 shows the deletion of lines **00005** through **00009**. The lines disappear from the file after the <ENTER> key is pressed.

```
PARKING  MUD       A1  F 80  Trunc=80 Size=11 Line=0 Col=1 Alt=12
====>
        |...+....1....+....2....+....3....+....4....+....5....+....6....+....7...
00000 * * * Top of File * * *
00001 IT IS BECOMING VERY CLEAR THAT PARKING ON THIS CAMPUS IS GOING TO BE A
00002 PROBLEM FOR MANY YEARS TO COME.  HOWEVER, CONTRIBUTING TO THE PROBLEM IS
00003 THE FACT THAT MANY "STUDENT-CREATED" PARKING LOTS ARE NOTHING MORE THAN
00004 MUDDY FIELDS IN WHICH STUDENTS (AND UNFORTUNATELY SOME STAFF) MUST PARK.
0  05 THE APPROPRIATE ADMINISTRATORS HAVE BEEN MADE AWARE OF THE PROBLEM, BUT
00006 SEEM RELUCTANT TO TAKE ANY ACTION ON THE MATTER.  NEVER BEFORE HAVE I
00007 SEEN SUCH APATHY TOWARDS THE HORRENDOUS SITUATION THAT THE STUDENTS AND
00008 STAFF MUST ENDURE DAY IN AND DAY OUT.  AS THEY CONTINUE TO PARK IN THE
000   CAR-SWALLOWING MUD BOGS, I WONDER IF THERE WILL EVER BE A DAY WHERE THE
00010 APPROPRIATE ADMINISTRATORS WILL BE FORCED TO PARK IN ONE OF THESE SWAMPS
00011 AS WELL.
00012 * * * End of File * * *

PF: 1=HLP 2=SPLIT 3=FILE 4=TOP 5=BOT 6=EQ 7=UP 8=DWN 9=QUIT 10=LFT 11=RT 12=CUR
4A                  Aa    A Session1     R 13 C 7    o-o__ 18:43    11/7/93
```

FIGURE 10.7 Deletion of block.

Copying Lines

The copy prefix must be used in conjunction with the Preceding or Following prefix (or Before/After prefix)

`====C`	Copies this line
`====F`	Following this line

A number of lines can be copied by specifying a number with the Copy command:

`==B==`	Before this line
`===C5`	Copy this line and the next four lines

To copy a block of lines, type "CC" in the prefix area of the first line of the block and the last line of the block. Specify the destination using P or F (or B or A). For example, in Figure 10.8, lines **00001** through **00003** will be copied as a block and placed after line **00005**.

Moving Lines

The Move prefix must be used in conjunction with the Preceding or Following prefix (or Before/After prefix) and is used much like the Copy prefix. The selected lines are moved from their original place and put in another place.

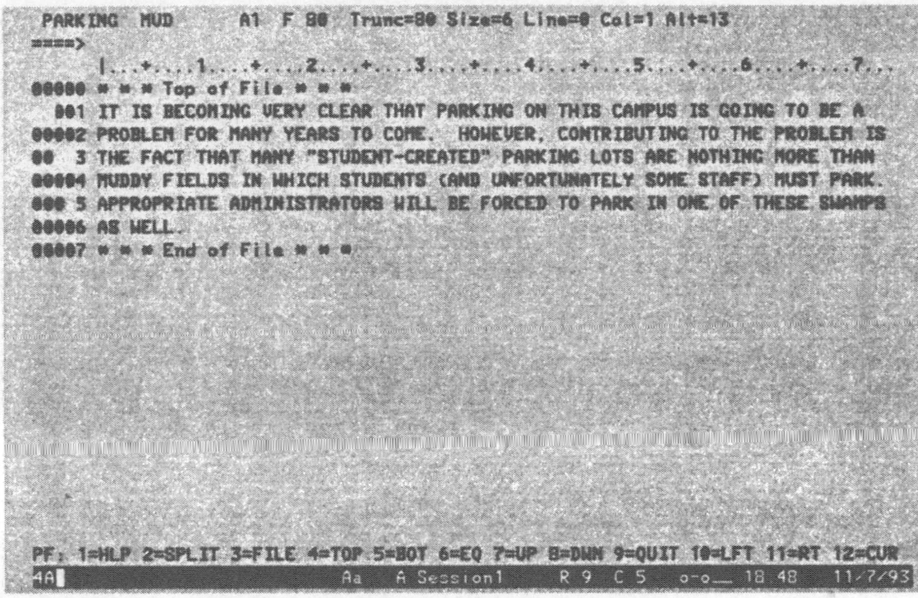

```
 PARKING  MUD      A1  F 80  Trunc=80 Size=6 Line=0 Col=1 Alt=13
====>
     |...+....1....+....2....+....3....+....4....+....5....+....6....+....7...
00000 * * * Top of File * * *
  001 IT IS BECOMING VERY CLEAR THAT PARKING ON THIS CAMPUS IS GOING TO BE A
00002 PROBLEM FOR MANY YEARS TO COME.  HOWEVER, CONTRIBUTING TO THE PROBLEM IS
00  3 THE FACT THAT MANY "STUDENT-CREATED" PARKING LOTS ARE NOTHING MORE THAN
00004 MUDDY FIELDS IN WHICH STUDENTS (AND UNFORTUNATELY SOME STAFF) MUST PARK.
000 5 APPROPRIATE ADMINISTRATORS WILL BE FORCED TO PARK IN ONE OF THESE SWAMPS
00006 AS WELL.
00007 * * * End of File * * *

PF: 1=HLP 2=SPLIT 3=FILE 4=TOP 5=BOT 6=EQ 7=UP 8=DWN 9=QUIT 10=LFT 11=RT 12=CUR
4A                 Aa    A Session1      R 9  C 5    o-o__  18 48     11/7/93
```

FIGURE 10.8 Block copy.

`====M`	Moves this line
`====F`	Following this line

A number of lines can be moved by specifying a number with the Copy command:

`==B==`	Before this line
`===M5`	Move this line and the next four lines

To move a block of lines, type "MM" in the prefix area of the first line of the block and the last line of the block. Specify the destination using P or F (or B or A). For example, in Figure 10.9, lines **00006** through **00008** will be copied as a block and placed before line **00003**.

Using the PF Keys in XEDIT

At the bottom of the XEDIT screen the PF keys are defined to perform very basic functions for the user when editing a file (again, this is an option usually set up by a systems administrator and may be different from system to system). These keys will provide the user with basic functions, including on-line help, saving, paging up and down, paging left and right, and others.

Using the Command Line in XEDIT

A wealth of commands are also available under XEDIT to assist the user with the editing process. These commands are typed in at the arrow (====>) prompt while in XEDIT. To get a complete list of the available commands and their various parameters, simply type **HELP** at the arrow prompt and press <ENTER>.

```
 PARKING  MUD       A1  F 80  Trunc=80 Size=9 Line=0 Col=1 Alt=14
====>
       |...+....1....+....2....+....3....+....4....+....5....+....6....+....7...
00000 * * * Top of File * * *
00001 IT IS BECOMING VERY CLEAR THAT PARKING ON THIS CAMPUS IS GOING TO BE A
00002 PROBLEM FOR MANY YEARS TO COME.  HOWEVER, CONTRIBUTING TO THE PROBLEM IS
00 03 THE FACT THAT MANY "STUDENT-CREATED" PARKING LOTS ARE NOTHING MORE THAN
00004 MUDDY FIELDS IN WHICH STUDENTS (AND UNFORTUNATELY SOME STAFF) MUST PARK.
00005 APPROPRIATE ADMINISTRATORS WILL BE FORCED TO PARK IN ONE OF THESE SWAMPS
  006 IT IS BECOMING VERY CLEAR THAT PARKING ON THIS CAMPUS IS GOING TO BE A
00007 PROBLEM FOR MANY YEARS TO COME.  HOWEVER, CONTRIBUTING TO THE PROBLEM IS
  008 THE FACT THAT MANY "STUDENT-CREATED" PARKING LOTS ARE NOTHING MORE THAN
00009 AS WELL.
00010 * * * End of File * * *

PF: 1=HLP 2=SPLIT 3=FILE 4=TOP 5=BOT 6=EQ 7=UP 8=DWN 9=QUIT 10=LFT 11=RT 12=CUR
4A                  Aa   A Session1      R 7  C 4   o-o__ 18 53   11/7/93
```

FIGURE 10.9 Block move.

10.3 REXX (RESTRUCTURED EXTENDED EXECUTOR)

Overview of REXX

Restructured Extended Executor (REXX) language is a command programming language that allows the user to combine useful sequences of commands into programs. REXX combines easy access to the operating system (CP and CMS) with a complete set of control instructions and control structures, similar to high-level languages such as Pascal, Fortran, and PL/1. Unlike most high-level languages, REXX programs are interpreted line by line, like BASIC.

Creating a REXX Program

REXX programs are created using any system editor, such as XEDIT (see Chapter 6.2). The file name of a REXX program is like any other CMS file, however, the file type of the file must be EXEC. In naming the EXEC file, care must be taken to avoid conflicts between your EXEC file and system files of the same name. Care should also be taken to avoid conflicts between the name of a user-written REXX program and CMS commands. For example, if you write a REXX program called COPYFILE EXEC A (which has the same name as the CMS command COPYFILE), you will not be able to use the CMS command COPYFILE to duplicate a file. Rather, when you type COPYFILE, the system will attempt to execute your REXX program called COPYFILE.

The first line of a REXX program *must* be a comment. REXX comments consist of a sequence of characters delimited by /* and */. In addition to the required first line comment, comments should be used throughout the program to explain and document it. Comments may be alone on a line (as is required for the first line), or be appended to the right of program statements:

```
/* Example of a comment taking up one line */
A=2  /* Appended comment */
```

Executing a REXX Program

To run a REXX program, first be sure you are in the READY; mode. Then simply type in the file name of the program to be run and press <ENTER>. For example, to run the program MYPROG EXEC A, type **MYPROG** and press <ENTER>.

Basic REXX Symbols and Terms

Each line of a REXX program may contain one or more of the following: strings, numbers, and symbols.

Strings

A string is a sequence of characters delimited by either single or double quotes, provided identical delimiters are used on a given line (Example: 'string', not "string"). Use two quotes to obtain one quote inside a string (as in Fortran). Some examples of valid strings are:

```
"Soccer"  'Don't worry'  '124'
```

Numbers

A number is a series of decimal digits with an optional decimal point. A number can be prefixed with a plus or minus sign, and/or written in exponential (i.e., scientific) notation. Some examples of valid numbers are

```
17  98.6  -15  +29.003  12.2E6  -.031
```

Symbols

Symbols are similar to strings, except they are not delimited by quotes. The can consist of any combination of letters, numbers, or the following characters:

```
@  #  $  !  .  ?
```

Note that symbols cannot start with a number or a period.

When REXX finds a symbol (a string that is not in quotes), it checks to see if the symbol is the name of a variable, that is, whether it has previously been given a value by an assignment statement. If it has been assigned a value, the REXX interpreter substitutes the value for the variable name before executing the line. Otherwise, the interpreter translates the symbol to uppercase and uses it literally. The following are examples:

```
   Program Statement                  Statement passed to VM/CMS

Erase x data a            -->         ERASE X DATA A
x = 'prog1'
Erase x data a            -->         ERASE PROG1 DATA A
'Erase' 'x data a'        -->         ERASE X DATA A
```

Booleans

Booleans are variables that can have one of two values: TRUE or FALSE. REXX uses a numeric 1 for TRUE and 0 for FALSE. A sample program segment follows:

```
/* USE OF BOOLEANS */
X = 1                    /* X is TRUE */
IF X THEN SAY 'HELLO'    /* Displays "HELLO" on screen */
Y = 0                    /* Y is FALSE */
IF Y THEN SAY 'BOO'      /* "BOO is NOT displayed on screen*/
A = 3                    /* Variable A gets the value 3 */
Z = (A = 3)              /* TRUE : Z gets the value 1 */
IF Z THEN SAY 'A IS 3'   /* Displays "A IS 3" on screen */
```

Arrays

Arrays in REXX are fairly easy to use. Like simple variables, they do not have to be declared as arrays prior to use. The period (.) is the subscript delimiter; for example, LIST.3 would denote the third member of the array LIST—equivalent to LIST[3] in Pascal or LIST(3) in Fortran and BASIC. Multidimensional arrays are allowed—for example, BOARD.I.J is equivalent to BOARD[I,J] in Pascal or BOARD(I,J) in Fortran and BASIC.

One handy feature of REXX arrays is the use of a stem to refer to all members of an array. A stem is an array variable name that ends with a period. Stems are particularly useful when one wants to initialize all the members of an array at the beginning of a program. For example, to set all the members of the array LIST to a blank, use the following code:

```
LIST. = " "
```

Variables and Data Types

REXX variables are symbols, as in other high-level languages. Unlike Pascal, variables do not have to be declared as a particular type; in fact, one variable can hold both character and numeric data (at different times) within the same program.

The assignment statement names a variable and gives it a value. This statement can be used to give an initial value to a variable or to change its value. The equals sign ("=") is the assignment operator in REXX. The following is a simple REXX program that uses assignment statements and demonstrates the fact that one variable can hold several different data types:

```
/* First line must be a comment */
A = 10              /* A contains the value 10 */
B = 2               /* B contains the value 2 */
C = A * B           /* C contains the value 20 */
A = 'XYZ'           /* A now contains the string XYZ */
C = A               /* C now contains the string XYZ */
```

Expressions and Operators

REXX variables are combined with operators in expressions, which are computed. Similar to expressions in other high-level languages, REXX expressions follow normal algebraic style and are evaluated from left to right according to their priority (modified by parentheses). The following is a list of the more common REXX operators, in order of their priority:

Symbol	Explanation	Sample Expr.	Value
+	Uniary plus (optional)	+31	31
-	Uniary minus (required)	-29	-29
**	Exponentiation	3**2	9
* /	Multiplication, division	3*4/2	6
%	Integer divide (Pascal DIV)	7%2	3
/ /	Remainder (Pascal MOD)	7//2	1
+ -	Addition, subtraction	3+4-2	5
| |	Concatentation	"A" | | "B"	AB
< > = < > =	Comparison operators	0 > 2	0(FALSE)
< > -< -> =		2 < 3	1 (TRUE)
¬	Logical NOT	-(A = 3)	0 if A = 3
&	Logical AND	1 & 0	0 (FALSE)
|	Logical OR	1 | 0	1 (TRUE)

[**Note**: The equals sign (" = ") serves as both an assignment operator (Example: A = 3) and a comparison operator (IF A = 0 . . .). The logical NOT ("¬") may be used in combination with any of the other comparison operators (Example: IF B ¬ = A . . .) or as a prefix of an entire expression (Example: ¬ (A > B | C = D).]

REXX Input and Output

SAY and PULL Commands

Most I/O in REXX programs is done via the terminal. There are two primary mechanisms for terminal I/O in REXX. One is the SAY instruction, which displays output from the program on the screen (similar to the Pascal WRITELN). The other one is the PULL instruction, which accepts data from the keyboard (similar to the Pascal READLN). One important difference between PULL and READLN is that the PULL instruction automatically converts input into uppercase; you must use the PARSE PULL command if you want to keep the input in the same case it was typed in. A simple example of these instructions follow:

```
/* REXX I/O EXAMPLE */
SAY "Please enter your name"      /* Prompts for name */
PULL name                         /* User types Paul */
SAY "HELLO" name                  /* Displays HELLO PAUL */
SAY "Please enter your name"      /* Prompts for name */
PARSE PULL name                   /* User types John */
SAY "GOODBYE" name                /* Displays GOODBYE John */
```

It is also possible to selectively accept input. If you list two or more variables after the PULL keyword, REXX assigns the first word (i.e., string of characters separated by one or more blanks) to the first variable, the second word to the second variable, etc., and the remainder of the line (one or more words) to the last variable. If there are more variables following the PULL keyword than words typed in response to the command, the remaining variables are assigned the null string (" "). See the following examples:

Program line:	`PULL FIRST SECOND THIRD`
User enters:	`Winken Blinken Nod`
Variable assignments:	`FIRST = "WINKEN"`
	`SECOND = "BLINKEN"`
	`THIRD = "NOD"`
Program line:	`PULL FIRST SECOND THIRD`
User enters:	`This is the time to remember`
Variable assignments	`FIRST = "THIS"`
	`SECOND = "IS"`
	`THIRD = "THE TIME TO REMEMBER"`
Program line:	`PULL FIRST SECOND REST`
User enters:	`Joe Bob`
Variable assignments:	`FIRST = "JOE"`
	`SECOND = "BOB"`
	`REST = " "`

This last program segment suggests a good way to be sure the right number of words are typed in. If two words are expected, and the user types in too many, REST will not equal the null string (" "), and this can be easily checked. If the user types in too few values, the value of SECOND will be the null string (" "), and this can also be checked. We can combine these concepts into one program which will accept three and only three words as input:

```
/* ACCEPTS THREE AND ONLY THREE WORDS AS INPUT */
OK = 0
DO UNTIL OK
```

```
        SAY "PLEASE ENTER EXACTLY THREE WORDS"
        PULL FIRST SECOND THIRD REST
        IF THIRD = " " THEN SAY "NOT ENOUGH WORDS"
        ELSE
                IF REST ¬ = " " THEN SAY "TOO MANY WORDS"
                ELSE
                        OK = 1
END /*DO UNTIL */
```

The ARG Command

Another way to get input into a REXX program is to use the ARG command. Standing for ARGument, this command takes values typed in after the file name (when the program is executed) and assigns those values to the corresponding variables listed after the ARG statement. For example, consider the following program ADD EXEC that adds two numbers:

```
/* ADD TWO NUMBERS */
ARG NUM1 NUM2
SUM = NUM1 + NUM2
SAY "THE SUM OF " NUM1 " AND " NUM2 " IS: " SUM
```

To execute the program, Type **ADD 4 5**. The system response will be

```
THE SUM OF 4 AND 5 IS: 9
```

REXX Control Language Statements

REXX provides several logical control statements to allow structured programs to be written. These statements include the logical IF / THEN / ELSE keywords, the iterative DO statement, the WHILE and UNTIL statements, and the SELECT statement.

Logical IF / THEN /ELSE

The logical IF instruction with an optional ELSE functions the same way as similar constructs in other languages. The general format of the IF instruction with only one instruction following the THEN and ELSE keywords is as follows:

```
IF <boolean expression> THEN
        instruction    /* executed if boolean exp. is true */
ELSE
        instruction    /* executed if boolean exp. is false */
```

If more than one expression is to be executed conditionally, the DO and END keywords are required. They perform the same function as the BEGIN and END in Pascal, or the { and } in C:

```
IF <boolean expression> THEN DO
        instruction 1a
        .
        .
        .
END
ELSE DO
        instruction 1b
        .
        .
        .
END
```

Iterative DO Loops

To execute the same block of commands repeatedly, use an iterative DO loop. This performs the same function as the Fortran DO loop or the Pascal FOR. The simplest form of the iterative DO will execute a block of code (delimited by the keywords DO and END) a given number of times. The number of times is given by the value of an expression (number or formula) following the keyword DO:

```
DO <numerical expression>
        instruction 1
        .
        .
        .
END
```

For example, the code:

```
DO 10
        SAY "HELLO THERE"
END
```

will display 10 lines of "HELLO THERE" on the screen.

A more useful version of the DO loop uses a control variable that is automatically incremented during each pass through the loop:

```
DO variable = first TO last
        instruction
        .
        .
        .
END
```

For example, to print out the first 10 integers on the screen, use the following code:

```
DO I = 1 to 10
        SAY I
END
```

Notice that the control variable (I in the last example) was incremented by 1. This is the default. If you want to increment the control variable by a value other than 1, use the BY keyword in the DO instruction:

```
DO variable = first BY increment TO last
```

Here, *increment* gives the number that will be added to the DO loop variable during each iteration of the loop. For example, to list the odd integers from 1 to 10, use the following code:

```
DO I = 1 BY 2 TO 10          /* 2 is the increment */
        SAY I
END
```

Conditional Looping

In addition to iterative (i.e., DO) loops, REXX provides conditional looping, where the loop executes as long as a Boolean expression evaluates to "TRUE" or "FALSE" (i.e., 1 or 0). There are three kinds of conditional loops in REXX. The first one is the DO WHILE instruction, where the decision whether or not to continue executing the loop is made at the top of the loop—similar to the Pascal WHILE statement:

```
DO WHILE <boolean expression>
        instruction
        .
        .
        .
END
```

In this loop, the instructions between the DO WHILE and END statements are executed as long as the Boolean expression is true.

The second conditional loop is similar to the Pascal REPEAT in that the decision to iterate or leave the loop is made at the bottom; hence the loop will execute at least once. This is the DO UNTIL statement:

```
DO UNTIL <boolean expression>
        instruction
        .
        .
        .
END
```

With this type of loop, the instructions between the DO UNTIL and END statements are executed until the Boolean expression becomes true.

The third type of conditional loop allows for an "endless loop" type of construct. In this case, the decision to continue looping is made within the body of the loop (as opposed to the top or bottom). The REXX keyword LEAVE is used to tell the interpreter to stop looping, and control passes to the statement following the END statement of the current loop:

```
DO FOREVER      /* Endless loop */
      instruction 1
      .
      .
IF <boolean expression> THEN LEAVE
      instruction n
      .
      .
      .
END
```

This loop executes until the Boolean expression in the body of the loop becomes TRUE.

Select

The SELECT statement performs the same function as the Pascal CASE statement. Here, a series of Boolean expressions is listed, each preceded by the keyword WHEN. REXX evaluates the expression from top to bottom and executes the instruction or block of instructions (which *must* be delimited by DO and END) that follow the first true Boolean expression. The instruction or block of instructions following the reserved word OTHERWISE (similar to the DEFAULT of the C language) is executed if the program "falls through" the other choices (i.e., all of the boolean expressions are false). The following is an example of the SELECT statement:

```
/* Example SELECT statement */
SELECT
    WHEN AGE < 6 THEN STATUS = 'PRESCHOOLER'
    WHEN AGE < 12 THEN STATUS = 'CHILD'
    WHEN AGE < 20 THEN DO /* Execute block of commands */
        STATUS = 'TEENAGER'
        TEENFLAG = 'TRUE'
        TCOUNT = TCOUNT + 1
END      /* Block of commands */
WHEN AGE < 30 THEN STATUS = 'YOUNG ADULT'
OTHERWISE          /* Default - do if all other */
```

```
                /* statements are false */
        STATUS = 'ADULT' /* Block-no DO/END here */
        ADULTFLAG = 'TRUE'
END     /* SELECT statement and OTHERWISE */
```

Note that the OTHERWISE statement is required if all the preceding Boolean expressions (WHEN...) evaluate to FALSE. Also, the DO..END construct allows a block of instructions to be evaluated if a WHEN condition is met. DO..END is not required for a block of commands following the OTHERWISE statement—the final END delimits the block as well as the SELECT statement itself. When the SELECT statement is executed, the statement or block of statements (delimited by DO..END) after the first true Boolean expression (WHEN...) is executed, and then processing continues with the statement following the last END (i.e., the end of SELECT) statement. In other words, the SELECT statement executes only one statement (or block of statements) following the first true expression after WHEN.

Reserved Words (Keywords) in REXX

The following is a list of reserved words in REXX. These words CANNOT be used as variable names:

```
ARG       ELSE      ITERATE   PULL      SELECT
CALL      EXIT      LEAVE     PUSH      THEN
DO        FOREVER   MAKEBUF   RETURN    UNTIL
DROPBUF   IF        PARSE     SAY       WHILE
```

Loop Control Keywords

There are three important keywords in REXX that control looping within the body of the loop. One has already been discussed: LEAVE. This exits the innermost loop, with execution continuing with the statement following the first END encountered after the LEAVE statement. A loop with a LEAVE statement in it is executed until a certain condition is met. This is particularly useful if you wish to accept data until a sentinel data value is read.

If, however, you want to skip to the end of the loop (bypassing the remaining statements in the loop) use the ITERATE keyword. This is equivalent to a GOTO statement that transfers control to the end of the loop. If the loop exit conditions are not met, a new iteration of the loop begins (unlike LEAVE, which halts looping). The ITERATE statement is useful when you want to read a list of numbers and process only the numbers that are valid:

```
DO FOREVER
        PULL NUMBER
        IF (NUMBER < 0) THEN ITERATE
                                    /* process non-negative */
                                    /* values only */
                                    /* otherwise, drop to */
```

```
                                 /* end of loop and read */
                                 /* new number */
        instruction
        .
        .
        .                        /* process data */
    END
```

The third loop control keyword, EXIT functions much like STOP in Fortran. This can be used anywhere in a REXX program, and causes immediate termination of the program. This can be used in conjunction with a sentinel value or error flag to tell the program to stop execution.

REXX Functions and Procedures

Like many high-level languages, REXX has functions and subroutines. There are a wide variety of types: they may be built-in (i.e., a library routine such as absolute value), appended to the end of the program (i.e., user-defined), or even external (i.e., contained in a separate file).

String Functions

REXX string functions are built-in (i.e., do not have to be declared) and provide powerful tools for string manipulation. The table below summarizes some of the more useful string functions available in REXX:

Routine	Explanation	Example	Result
SUBSTR(S,I1,L)	Substring of S, beginning at position I1 and of length L	SUBSTR('CONRAIL',4,2)	'RA'
LENGTH(S)	Length of string S	LENGTH('AMTRAK')	6
COPIES(S,I1)	I1 copies of string S	COPIES('HO',3)	'HOHOHO'
LEFT(S,L1)	String of length L1 with left justification	LEFT('TON',5)	'TON'
POS(S1,S2)	Position of first character of S1 as a, substring of S2	POS('ANT','GIANT') POS('XX','TEN')	3 0 (not found)
OVERLAY(S1,S2, I1,I2)	Overlays S1 onto S2 beginning at position I1, for a length of I2	OVERLAY ('XYZ', '123456', '12XY56' 3,2)	

Routine	Explanation	Example	Result
COMPARE(S1,S2)	Position of the first character in S1 that doesn't match S2	COMPARE('TRACK','TRAP')	4
		COMPARE('TMI','TMI')	0

Arithmetic Functions

In addition to built-in string functions, REXX features built-in arithmetic functions, which operate on and return integers and real numbers. Some of the more useful arithmetic functions are summarized in the following table:

Function	Explanation	Example	Result
ABS(R1)	Absolute value of R1	ABS(-25.0)	25.0
SIGN(R1)	-1, 0, or 1, depending on whether R1 is negative, zero, or positive, respectively	SIGN(4.04)	1
		SIGN(-0.35)	-1
		SIGN(0.00)	0
TRUNC(R1,I1)	Integer part of number and I1 decimal places	TRUNC(11.5)	11
		TRUNC(1.456,2)	1.45
RANDOM(MIN,MAX, SD)	Non-negative random number between MIN and MAX, with seed SD	RANDOM()	305
		RANDOM(5,8)	7
		RANDOM(1,10,.1)	3
MAX(R1,R2, . . .)	The largest of R1, R2, . . . (up to 10 arguments)	MAX(-5,7,9)	9
MIN(R1,R2, . . .)	The smallest of R1, R2, . . . (up to 10 arguments)	MIN(-5,7,9)	-5

NOTE: No blanks are allowed between the name of the function and the left parenthesis; otherwise, an error results. Thus ABS (N) would be illegal, while ABS(N) and ABS(N) are valid.

Internal Functions

Internal functions are user-written functions defined in the same program, as in Fortran and Pascal. They are similar to functions in other high-level languages in syntax and structure. Functions are defined at the end of a program, and an EXIT statement (analogous to the STOP / END in Fortran) is required as the last statement in the main program (immediately before the first function definition).

The function is defined with the name of the function, followed by a colon (:). The REXX keyword ARG is used to pass the arguments, while the keyword RETURN is used to return the calculated value of the function to the main program. Finally, the function is invoked in the main program by giving its name, followed by a list of the arguments enclosed in parentheses. A sample program with a function (named AREA) follows:

```
/* PROGRAM TO CALCULATE THE AREA OF A RECTANGLE */
SAY "ENTER HEIGHT AND WIDTH"
PULL HEIGHT, WIDTH
X = AREA(HEIGHT,WIDTH)
SAY "AREA OF RECTANGLE IS " X
```

```
EXIT                /* required */
AREA:               /* begin function definition */
     ARG HT,WD      /* HT -> HEIGHT; WD -> WIDTH */
     AR=HT*WD       /* calculate area of rectangle */
RETURN AR           /* return calculated area to main prog */
```

External Functions

If you have a function used by several different programs, it may be more convenient to have the function in a separate file, analogous to INCLUDE files in Pascal. In this case, the main program calls the function by specifying its file name. Otherwise, external functions operate in the same fashion as internal functions. The following is the same AREA function used in the previous example, implemented as an external function:

```
/* MAIN PROGRAM: FILENAME IS 'MAIN EXEC A' */
SAY "ENTER HEIGHT AND WIDTH"
PULL HEIGHT WIDTH
X = AREA(HEIGHT,WIDTH)
SAY "AREA OF RECTANGLE IS: " X
EXIT                 /* last line of main program */
/* AREA FUNCTION: FILENAME IS 'AREA EXEC A' */
ARG HT,WD
AR = HT * WD
RETURN AR
```

Procedures

Procedures have a format very similar to functions; the main difference is that procedures do not simply return a calculated value. Instead, they are executed as program segments, and can read in data, write out data, and, in general, do the same tasks as regular REXX programs.

A procedure is invoked from the main program by the keyword CALL, followed by the procedure name, and arguments (if any). The procedure itself has a syntax nearly identical to that of a REXX function, except that the last statement must be RETURN. As was the case with REXX functions, an EXIT statement is required between the main program and the procedure definition; otherwise the procedure will always be invoked when the main program terminates. The following example illustrates an AREA procedure similar to the AREA function illustrated previously:

```
/* EXAMPLE OF A PROGRAM WITH A PROCEDURE */
SAY "ENTER HEIGHT AND WIDTH"
PULL HEIGHT WIDTH
CALL AREA HEIGHT WIDTH
/* AREA is a procedure; HEIGHT and WIDTH are arguments */
EXIT    /* End of Main program */
AREA:   /* Beginning of procedure definition */
        ARG HT WD
```

```
        AR = HT * WD
        SAY "AREA OF RECTANGLE IS: " AR
    RETURN
```

Procedures can be internal, as in the previous example, or external, just like functions.

Issuing Commands to CMS and CP

One of the most powerful features of REXX is its ability to issue commands to the CMS and CP operating systems. To issue most commands, simply type them in as lines in your REXX program. When your program executes, anything not recognized by the computer as a REXX keyword (e.g., CALL), as an assignment (e.g., X=2), or a function/procedure call [e.g., AREA (HEIGHT,WIDTH)] is evaluated and passed to the CMS/CP operating system.

CMS Commands

As previously mentioned, the fact that an expression is evaluated before being passed to CMS/CP should be kept in mind. It is easy for commands to be passed to the operating system that were not intended to be passed by the programmer. Consider the following example:

```
A = "STUFF" /* variable A is assigned the value "STUFF" */
.
.
.
ERASE TEMP WORKFILE A /* erase the file 'TEMP WORKFILE A' */
```

Because REXX evaluates each line before passing it to the operating system, "STUFF" is substituted for "A." This happens because "A" has been declared as a REXX variable with a value of "STUFF" earlier in the program. Thus, the command "ERASE TEMP WORKFILE STUFF" is passed to the operating system, and an error message results, since there is no file mode STUFF.

Another problem that can arise concerns REXX operators. Suppose you wanted to erase all files with a file name of TEMP on your "A" disk, so you place the following line in your REXX program:

```
ERASE TEMP * A
```

Although seemingly correct, this line will cause an error. REXX attempts to multiply TEMP by A, since the asterisk ("*") is interpreted as a multiplication operator (as in 2 * 3 = 6).

To avoid these problems, it is standard practice to enclose all symbols that are NOT variable names in quotes. This will prevent inadvertent substitutions when your REXX program runs. Anything in quotes is interpreted literally, and passed directly to the CMS and CP operating system (OS). It is also legal to combine literal quoted strings with variable names:

```
  = "TEMP" /* X is a variable with the value "TEMP" */
"ERASE" X "DATA A" /* "ERASE TEMP DATA A" is passed to OS */
```

It is often desirable to suppress messages generated on the screen when CMS commands are executed in a REXX program. The CMS command "SET CMSTYPE HT" will halt message generation, while "SET CMSTYPE RT" will restore it:

```
"SET CMSTYPE HT"    /* halt typing */
"ERASE" FN FT FM    /* erases without displaying message */
"SET CMSTYPE RT"    /* resume typing */
```

One problem with using SET CMSTYPE HT to suppress system messages is how to detect and correct error conditions that may arise during the execution of a REXX program. For example, suppose your program tries to erase a file called MYFILE TEXT A that does not exist. If you issue this command outside of REXX (i.e., directly from the READY; prompt), a message similar to the following will appear:

```
FILE MYFILE TEXT A NOT FOUND.
Ready(00028); T = 0.01/0.01 Ready
```

The number in parentheses following the R is of special significance: it is the return code from the CMS command. A return code of zero usually means that the command executed correctly. A nonzero return code (28 in this case) usually indicates an error condition. Thus, it would be useful to save the return code from a CMS command issued in a REXX program, and warn the user if it were a number other than zero.

REXX provides a way to do just that by using a special variable: RC. When a CMS command is executed in REXX, RC is given the value of the return code. A simple IF-THEN statement can then be used to alert the user to the error and/or take the corrective action:

```
"ERASE MYFILE TEXT A"    /* file does not exist */
IF RC = 28 THEN          /* return code will be set to 28 */
       SAY "WARNING—FILE MYFILE TEXT A DOES NOT EXIST"
```

The problem with this segment as it is written is that a warning message will be displayed on the screen from CMS, as well as from the SAY command. If we use SET CMSTYPE HT and SET CMSTYPE RT, they will alter (i.e., overwrite) the value of RC from the erase command, since they are themselves CMS commands. So, we must save the value of RC in another variable and test that variable:

```
/******************/
"SET CMSTYPE HT"         /* RC is set to zero */
"ERASE NOFILE TEXT A"    /* does not exist—RC is set to 28 */
RCSAVE=RC                /* save RC (28) in variable RCSAVE */
"SET CMSTYPE RT"         /* RC is set to zero */
IF RCSAVE = 28 THEN
       SAY "WARNING—FILE DOES NOT EXIST"
```

Note: You may not use SAY commands after issuing the command SET CMSTYPE HT (unless you subsequently give the command SET CMSTYPE RT). SAY commands are ignored when CMS messages are suppressed.

CP Commands

Many CP commands can be written into REXX programs, just like CMS commands. Remember to use quotes around all symbols that are not REXX variables. The following program segment shows a list of users currently logged on:

```
/* Show currently logged on users */
"QUERY NAMES"        /* a CP command */
```

It is possible to suppress the messages issued by some CP commands—just prefix the command with "EXECIO 0 CP (STRING." The number zero indicates that no lines are to be returned from the command. To change the color of messages from the operating system (and messages from the SAY command in your program) without having the "COMMAND COMPLETE" message appear on the screen, issue the following command in you REXX program:

```
"EXECIO 0 CP (STRING" "CP SCREEN VMOUT RED NONE"
```

Capturing Output from CMS and CP Commands

It is possible to capture output from CMS and CP commands and assign them to variables. Consider the program on the next page which saves each line of output from the CP command QUERY NAMES:

```
/* Program to save list of users in array USERLST */
I = 0
MAKEBUF
"EXECIO * CP (STRING" "CP QUERY NAMES"
DO WHILE QUEUED () > 0  /* do while lines remain on stack */
       I = I + 1
       PULL USERLST.I   /* pull line off stack and */
                        /* assign to USERLST */
END /* do while */
DROPBUF
"EXECIO 0 CP (STRING" "CP SCREEN VMOUT WHITE NONE"
SAY "CURRENTLY LOGGED ON USERS:     " /* in white */
"EXECIO 0 CP (STRING" "CP SCREEN VMOUT RED NONE"
DO J = 1 TO I
       SAY USERLST.J   /* list username in red */
END /* do J */
"EXECIO 0 CP (STRING" "CP SCREEN VMOUT GREEN NONE"
            /* subsequent output will be in green */
```

The MAKEBUF command creates a temporary storage area, similar to a program stack, where output from the command QUERY NAMES is placed. The output is put in this area (called a buffer), rather than being displayed on the screen. The actual CP command is invoked by the command line "EXECIO * CP (STRING" "CP QUERY NAMES," where the asterisk (*) specifies that all lines returned by the command are to be placed in the buffer. QUEUED() is a special REXX function that returns the number of lines currently in the buffer; we want to process them until the buffer is exhausted (QUEUED() = 0). PULL USERLST.I assigns each line from the buffer to an element in the array USERLST, and QUEUED() is automatically decremented as lines are pulled from the buffer. Once all the lines are pulled, DROPBUF discards the now-empty buffer. Finally, the DO loop at the end displays each line (element) of USERLST, which contains a list of currently logged-in users.

Note that the use of the "EXECIO 0 CP (STRING" commands to change the color of the lines displayed on the screen by the SAY commands. In this example, the first such command—"EXECIO 0 CP (STRING" "CP SCREEN VMOUT WHITE NONE"—causes the heading line "CURRENTLY LOGGED ON USERS" to be displayed in white. Subsequent commands cause the actual list of users to be displayed in red, and any output after the end of the program to be displayed in green.

To capture output from CMS commands, use the STACK option, together with MAKEBUF and DROPBUF:

```
/* Capture output from CMS commands */
MAKEBUF
'QUERY DISK A (STACK'  /* put results in buffer (stack) */
PULL HEADER
PULL LABEL ADDR DISKLET STAT CYL TYPE BLKSIZE NUMFILES REST
DROPBUF
SAY 'Number of files on A-disk: ' NUMFILES
```

The above example captures the number of files from the CMS command QUERY DISK A and displays that number on the screen in an easily read format.

REXX File Processing

Input can be read from CMS files, and output can be written to files:

```
/* REXX program to read a file and display it on screen */
SAY
SAY "ENTER THE FILENAME, FILETYPE, AND FILEMODE OF THE"
SAY "FILE TO BE DISPLAYED: "
PULL FILENAME FILETYPE FILEMODE /* user inputs filed */
MAKEBUF     /* create buffer */
DO FOREVER  /* endless loop */
     "EXECIO 1 DISKR" FILENAME FILETYPE FILEMODE
                                   /* read a line from the file */
     IF (RC = 2) THE LEAVE        /* end of file reached */
```

```
        IF (RC ¬ = 0) THEN LEAVE   /* error condition */
        PULL ALINE                 /* get a line from the buffer */
        SAY ALINE                  /* display line on screen */
    END /* DO */
    DROPBUF /* discard buffer */
    "FINIS" FILENAME FILETYPE FILEMODE /* close file */
```

As in the previous example, MAKEBUF sets up a buffer to store lines read in from the file. The corresponding DROPBUF command near the end of the program discards the buffer after processing is complete.

Within the DO loop, EXECIO 1 DISKR is a CMS command that reads one line at a time from the file (which is specified by the variables FILENAME, FILETYPE, and FILEMODE). The first IF statement (IF RC = 2) tests for the end-of-file condition (return code 2), and exits the loop (LEAVE) if that condition is reached. The second IF statement (IF RC ¬ = 0) tests for an error condition (nonzero return code), and also leaves the loop if the condition is met. Note that this IF statement could be enhanced to display a warning on the screen if a processing error occurs:

```
    IF (RC ¬ = 0) THEN DO
        SAY "ERROR IN PROCESSING FILE"
        SAY "RETURN CODE WAS " RC
        LEAVE
    END /* IF */
```

Finally, the FINIS command closes the file being processed.

Writing output to files in REXX is as easy as reading them. Consider the sample program below that reads file FNAME1 FTYPE1 FMODE1 and writes to file FNAME2 FTYPE2 FMODE2 (i.e., copies the first file to the second file in a line-by-line fashion):

```
    /* REXX program to copy a file */
    /* FNAME1 FTYPE1 FMODE1 --> FNAME2 FTYPE2 FMODE2 */
    SAY
    SAY "COPY WHAT FILE?"
    PULL FNAME1 FTYPE1 FMODE1
    SAY
    SAY "COPY TO WHERE?"
    PULL FNAME2 FTYPE2 FMODE2
    MAKEBUF
    LINECOUNT = 0
    DO FOREVER
        "EXECIO 1 DISKR" FNAME1 FTYPE1 FMODE1
        IF (RC = 2) THEN LEAVE
        IF (RC ¬ = 0) THEN LEAVE
        PULL ALINE              /* Get a line from input file */
```

```
        SAY ALINE              /* Display line on screen */
        PUSH ALINE             /* Place line on buffer */
        LINECOUNT = LINECOUNT + 1
        "EXECIO 1 DISKW" FNAME2 FTYPE2 FMODE2 LINECOUNT
    END /* do */
    DROPBUF
    "FINIS" FNAME1 FTYPE1 FMODE1
    "FINIS" FNAME2 FTYPE2 FMODE2
```

In the above sample program, lines added to the previous program are **boldfaced**.

First, you will notice the addition of a new variable—LINECOUNT. This variable is required to tell the second EXECIO command which line is being written to the file. Notice that a line is PULLed from file 1, displayed on the screen ("SAY ALINE"), and then pushed on the buffer so that it can be written.

EXECIO 1 DISKW is a CMS command that takes a line from the buffer and writes it to the given file (FNAME2 FTYPE2 FMODE2). The value of LINECOUNT given at the end of the EXECIO 1 DISKW command determines which line number in the new file to write (i.e., if LINECOUNT = 3, the line pulled off the buffer will become the third line in the output file). Finally, two FINIS instructions are needed to close both the input and output files.

DEFINING TERMS

CP: Control Program; provides access to mainframe hardware devices.

File: Under CMS, storage for data on disk, identified by a unique file name, file type, and file mode.

FILEL: Utility for file management under VM/CMS (file directory).

Filemode: Third identifier of a CMS file; usually is the letter of the disk on which the file resides.

Filetype: Second identifier of a CMS file; can be up to 8 characters in length and usually describes the kind of file.

Filename: First identifier of a CMS file; can be up to 8 characters in length.

Keywords: Reserved words in REXX that cannot be used as variable names.

NOTE: e-mail system under VM/CMS.

Password: User-selected alphanumeric up to 8 characters long to protect userid.

PF key: Key pressed on an input terminal to perform a specific function ('Program Function' key).

Prefix: Under XEDIT, an area to the left side of file contents where line-oriented commands are entered.

RDRL: Utility for incoming e-mail management under VM/CMS.

Ready: Command prompt under VM/CMS.

REXX: Restructured Extended Executor—programmable command language.

Userid: Account name as provided by a systems administrator to permit access to a mainframe.

VM/CMS: Virtual Machine/Conversational Monitor System, an IBM mainframe operating system.

XEDIT: VM/CMS file editor.

REFERENCES

VM/SP CMS User's Guide. IBM Corporation, Programming Publications Dept., Endicott, NY 13760.

VM/ESA Procedures Language VM/REXX Reference Release 1.1. IBM Corporation, Programming Publications Dept., Endicott, NY 13760.

VM/ESA Procedures Language VM/REXX User's Guide Release 1.1. IBM Corporation, Programming Publications Dept., Endicott, NY 13760.

11

OS/2

James Griffioen
Jonathan Edwards
University of Kentucky

11.1 INTRODUCTION

IBM introduced the IBM PC in 1981. The team responsible for its design accomplished their task in 2 years. This is a remarkable achievement quickly when measured against IBM's standards for product development cycles at the end of the 1970s.

The original IBC PC architecture reserved 384 Kilobytes of the machine's 1-Megabyte address space for system support functions (video memory, adapter addresses, etc.). This left 640 Kilobytes of memory for the operating system (DOS) and applications. In short, the hardware capabilities of the IBM PC were severely limited when compared to mainframe computers of that period. Thus PCs were unable to support the more advanced features of large machines such as virtual memory. Moreover, because the PC targeted single users, there seemed to be no reason to include multitasking support in the operating system. Consequently, DOS can run only one application at a time.

Advances in microprocessor development in the mid-1980s revolutionized the power and capabilities of the PC. Intel's 80286 microprocessor has an address space of 16 Megabytes. Its successors (the 80386 and its descendants) have an address space of 4 Gigabytes. The 80386 and compatible microprocessors also have all the hardware capabilities of earlier mainframe computers, such as virtual memory, privileged and nonprivileged execution modes, and high-speed computing power. But the DOS operating system takes no advantage of these features. Using an 80386-based PC to run DOS is very much akin to using a Learjet to taxi down the street to the corner grocery store.

OS/2 version 2.x exploits all the features of modern microprocessors. The address space available to applications is greatly increased, multiple applications can run simultaneously without interfering with one another, and the operating system is protected from malfunctioning and malicious programs. OS/2 version 2.x is a well-con-

0-8493-2530-7/96/$0.00+$.50

ceived system that incorporates recent advances in operating system design and exhibits the flexibility required to capitalize on future developments in computer architecture. OS/2 will therefore be a strong contender in the operating system arena for years to come. But OS/2 also preserves one's investment in software: most DOS and Windows applications run under OS/2 as well as or better than they do under their "native" operating systems.

11.2 THE HISTORY OF OS/2

IBM and Microsoft recognized that a replacement for DOS was necessary and in 1985 signed an agreement to create the next generation PC operating system. The ensuing development project was extraordinarily broad in scope, with programmers in four states and the United Kingdom. The initial version of OS/2, OS/2 1.0, first shipped in December 1987. It did not include a graphical user interface. OS/2 1.0 was designed to run on 80286-based machines. The 80286 does not have the hardware features necessary to create multiple virtual 8086 machines, so OS/2 1.0 provided only limited support for DOS programs.

The Presentation Manager, OS/2's first graphical user interface, was added in November 1988, when OS/2 1.1 was released. Further improvements were made in OS/2 1.2 (which added the High Performance File System and was released in late 1989) and OS/2 1.3 (which reduced the amount of memory required to run OS/2 and appeared in late 1990). However, all of these versions could run on 80286-based machines and so none of them addressed one of the major shortcomings of OS/2: its relatively weak support for running DOS programs.

Work on OS/2 2.0 began in the middle of 1989. OS/2 2.0 adopted many of the best features of OS/2 1.x, including multitasking, protected address spaces, and a modern file system, but it replaced the Presentation Manager with a new graphical user interface and greatly improved DOS support. OS/2 2.0 was released in 1991. OS/2 2.0 exploits hardware features of the 80386 that are not present in the 80286 and thus will not run on 80286-based PCs.

In late 1990, Microsoft withdrew from OS/2 development to focus on Windows and Windows NT. However, IBM has continued to support OS/2. The latest versions, OS/2 2.1, requires fewer resources than OS/2 2.0 and incorporates a number of internal enhancements. New developments continue. IBM recently demonstrated a version of OS/2 2.x that runs on a symmetric multiprocessor. This version is slated to be released before the end of 1994.

11.3 KEY FEATURES

OS/2's primary strengths are

- *Memory management*—OS/2 eliminates the 640K address space limit. Each OS/2 application executes in a 512-Megabyte virtual memory space. Consequently, OS/2 applications may be several orders of magnitude larger than DOS applications. Furthermore, OS/2 uses a "flat" memory model consisting of a single large region. DOS partitions its 640K address space into 64K "segments," making it difficult to manipulate large data structures, since they must be spread across multiple segments. OS/2 also uses the microprocessor's virtual memory hardware to ensure that one application or device driver cannot accidentally (or maliciously) corrupt the memory used by another application. Anyone who has found it necessary to experiment with options on DEVICE statements in a DOS CONFIG.SYS file in order to get a machine to boot properly will appreciate this feature.

- *Multitasking*—OS/2 allows several applications to execute simultaneously. This translates into increased productivity. For example, someone using a CAD program that takes some time to perform its computations can read and send electronic mail while waiting for the CAD program to complete its work. OS/2's multitasking capabilities are superior to those offered by Quarterdeck's DesqView and Microsoft's Windows. Such systems do not provide true multitasking because they are built on top of DOS, which does not support multitasking and cannot process more than one request at a time. Thus, when one program running under DesqView or Windows makes a DOS call, no other program can use DOS services until the first program's call has returned. OS/2 is "reentrant," which means more than one system call can be active at a time. This increases overall throughput and productivity.
- *Multithreading*—OS/2 is also "multithreaded," which means a single program can be designed as a collection of cooperating threads of execution. This model of computation makes it easier to implement powerful applications that must perform several operations concurrently on shared data. For example, a multithreaded word processor can spawn a thread that checks the spelling of each word as it is typed. Multimedia applications can spawn separate threads to handle video, audio, and user input.
- *High Performance File System*—The original IBM PC had no hard disk; its diskette drives could store 180 Kilobytes on a 5.25-inch floppy diskette. DOS's file system (the File Allocation Table or FAT file system) was designed with this in mind. Recent versions of DOS have extended the FAT file system to handle large hard disks, tree-structured directories, etc., but its origins have certainly imposed unwelcome constraints on its design that significantly limit its performance.

  OS/2 supports the FAT file system but also includes a new file system: the High Performance File System or HPFS. As its name implies, HPFS is generally faster than the FAT file system. (FAT uses a linear list to keep track of where on disk each piece of a file is located. HPFS uses a balanced binary tree, which can be searched to find a particular piece more quickly.) HPFS also uses a smaller block size (512 bytes) than the FAT file system (2 Kilobytes), resulting in more efficient use of disk space. (Every file, no matter how small, occupies at least one block. On average, half of the last block in a file contains no data and is wasted.)
- *Support for Multiple File Systems*—Although HPFS is a significant improvement over FAT, OS/2 is not inextricably bound to it. New OS/2 file systems can be "installed" using a statement in the CONFIG.SYS file. This means that file systems developed in the future can easily be integrated into an OS/2 environment.
- *Graphical User Interface*—OS/2 includes a modern graphical user interface: the WorkPlace Shell (WPS). The WPS is easier to learn than the DOS commandline. It also differs in subtle ways from older GUIs such as Windows.

The differences are the result of an extensive effort in human-factors, graphics design, and beta testing.

11.4 THE WORKPLACE SHELL

The WorkPlace Shell provides windows, icons, and a mouse-directed cursor as the user interface. The WPS displays these items against a background known as the "desktop." Figure 11.1 shows the desktop with three windows: and OS/2 command

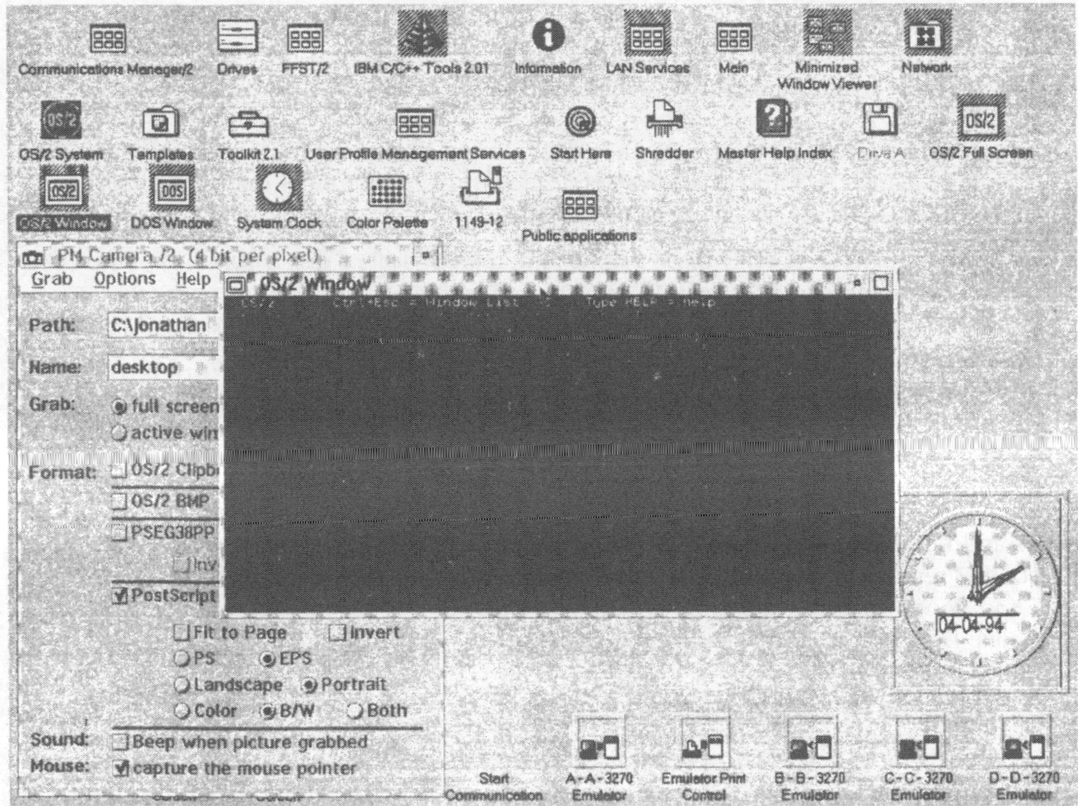

FIGURE 11.1 The OS/2 WorkPlace Shell desktop.

session, a clock, and an application called PM Camera. Several icons appear at the top and bottom of the desktop.

Actions are performed by moving the mouse pointer to an icon or window and clicking or double-clicking on one of the mouse buttons. The mouse buttons are designated "mouse button 1" and "mouse button 2" rather than "left" and "right" because the WorkPlace Shell can be customized for right-handed or left-handed mouse operation. Mouse button 1 is the button your index finger controls. The WorkPlace Shell is far too rich an environment to cover in this section. Fortunately, OS/2 provides a tutorial that quickly brings new users up to speed. The WPS tutorial automatically runs after OS/2 is installed. The tutorial can also be launched by moving the mouse pointer over the "Tutorial" icon and double-clicking mouse button 1. To get to the tutorial icon it may be necessary to open the "Information" icon by moving the mouse pointer over the icon and double-clicking mouse button 1. Both icons are shown in Figure 11.2. Several of the references at the end of this section discuss the WorkPlace Shell and its capabilities.

FIGURE 11.2 The OS/2 Information and Tutorial icons.

11.5 OS/2 AN INTEGRATION PLATFORM

One argument frequently used against OS/2 stems from a widely propagated misconception that there are almost no applications that run on OS/2. This could not be further from the truth. There is certainly no lack of software for OS/2.

More than 1000 applications have been written explicitly for OS/2. The *OS/2 Applications Directory*, a 270-page catalog containing information on these applications, is available from the publisher, Miller Freeman, Inc. Copies may be ordered by calling 415-905-2728.

Moreover, the vast majority of DOS and Windows programs will run just fine under OS/2. In fact, many DOS and Windows applications run better under OS/2 than they do under DOS! There are several reasons for this:

- Many programs (especially those that perform a great deal of disk I/O) actually run more quickly under OS/2 than under DOS or Windows.
- OS/2's memory management capabilities prevent an application from crashing the system—the application itself may fail, but OS/2 will not.
- OS/2 makes it easy to run DOS and Windows applications whose configurations are mutually incompatible. Each application can in effect have its own CONFIG.SYS and AUTOEXEC.BAT.

Because OS/2 2.x is such a good environment in which to run DOS and Windows programs and because there are so many DOS and Windows applications available, we expect that readers of this book who use OS/2 will use it primarily as a platform for DOS and Windows applications. Accordingly, the rest of this section discusses how to customize DOS and Windows settings in OS/2.

11.6 RUNNING DOS PROGRAMS UNDER OS/2

OS/2 provides three levels of support for running DOS programs:

1. *VDMs*—The Intel 80386 and its successors support a "virtual 086" mode specifically designed to allow the processor to run multiple DOS programs simultaneously. OS/2 takes advantage of this facility to implement multiple Virtual DOS Machines (VDMs). Each VDM provides an environment that looks very much like DOS running on an 80386-based PC. Most DOS services are provided in the same manner as in native DOS, extended and expanded memory are available, the 80386 instruction set is supported, and programs may be loaded into upper memory blocks (UMBs) or the high memory area (HMA). Under OS/2 several VDMs may be active simultaneously. Furthermore, device drivers take up almost none of a VDM's address space (they reside in the OS/2 system space) so more memory is available to applications.
2. *Booting native DOS inside OS/2*—Some programs that rely on certain undocumented DOS services or that require direct access to specific hardware adapters will not run in a VDM. It is possible to boot a native version of DOS *inside* OS/2. OS/2 essentially sets up a "bare" virtual 8086 and runs the native DOS on it. The major difference between a VDM and a native DOS session is that the native DOS session can bypass OS/2's virtual device drivers and access physical devices in some circumstances. This means that the device cannot be used by other programs running at the same time (including VDMs and applications running in other native DOS sessions), but it does preserve the user's existing investment in software.
3. *Boot Manager*—In some rare cases it may be necessary to boot DOS rather than OS/2 to run an application. OS/2 has an optional feature that allows a user to install multiple operating systems and to choose which one is activated at boot time. The choices need not be limited to OS/2 and DOS. A user could install Windows NT or a Unix clone (assuming the hard dirk is big enough) and have the option to boot any of the installed systems. When using HPFS, keep in mind that other operating

systems (including DOS) probably will not be able to access HPFS files. Because this option does not use OS/2's DOS emulation features, we will not discuss it further. See the references at the end of this section for a complete discussion of the Boot Manager.

Virtual DOS Machines

OS/2's default installation procedure creates two icons that can be used to start a VDM. The "DOS Full Screen" icon starts a full-screen VDM and "DOS Window" icon starts a VDM running in a window on the WorkPlace Shell desktop. Each VDM acts like a PC running DOS. Creating a new VDM icon is like configuring a new PC. To create a new VDM icon:

1. Open the "Templates" folder by moving the mouse pointer to the icon for the folder and double-clicking mouse button 1. The icon for the folder is shown in Figure 11.3 and the contents of the folder once opened are shown in Figure 11.4.
2. Associated with each WPS icon is a menu that can be opened by moving the mouse pointer to the icon and clicking mouse button 2. We will call this menu the "object menu." Open the object menu for the "Program" icon. The object menu for this icon is shown in Figure 11.5.

Templates

FIGURE 11.3 The OS/2 Templates folder icon.

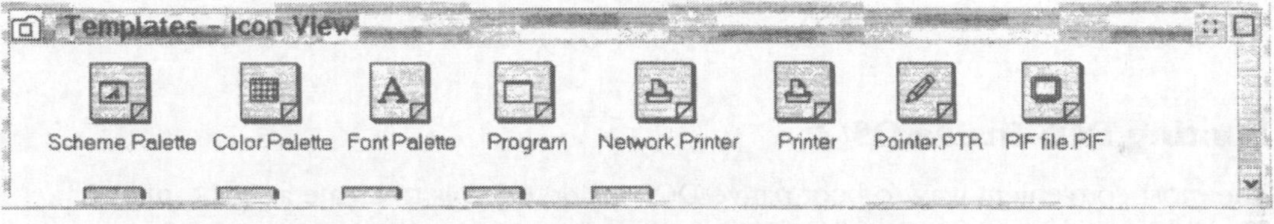

FIGURE 11.4 The OS/2 Templates folder icon (opened).

3. Select the "Create Another" option by moving the mouse pointer to the option and clicking mouse button 1. A window will open that prompts for the name to be given to the new icon and the location (folder or desktop) where the new icon should reside.
4. After the new template has been created, a settings notebook will be opened. The first page of the notebook prompts for the name of the program to run in the new VDM. Enter the full pathname of the program or '*' to run DOS. Then select the "Session" tab by moving the mouse pointer to the tab and clicking mouse button 1. The Session page of the settings notebook is shown in Figure 11.6.

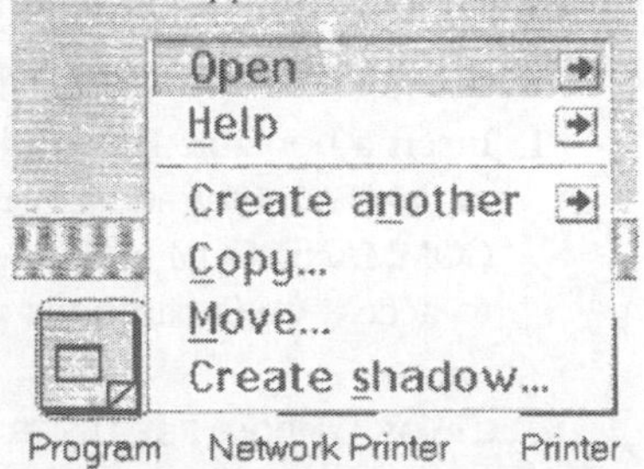

FIGURE 11.5 The OS/2 Program template and object menu.

5. Click in the "DOS full screen" radio button to create a new DOS full-screen VDM or in the "DOS window" radio button to create a new DOS windowed VDM.
6. Select the "DOS Settings" button by moving the mouse pointer to the button and clicking mouse button 1 to set options for the new VDM as described in Configuring DOS Sessions.

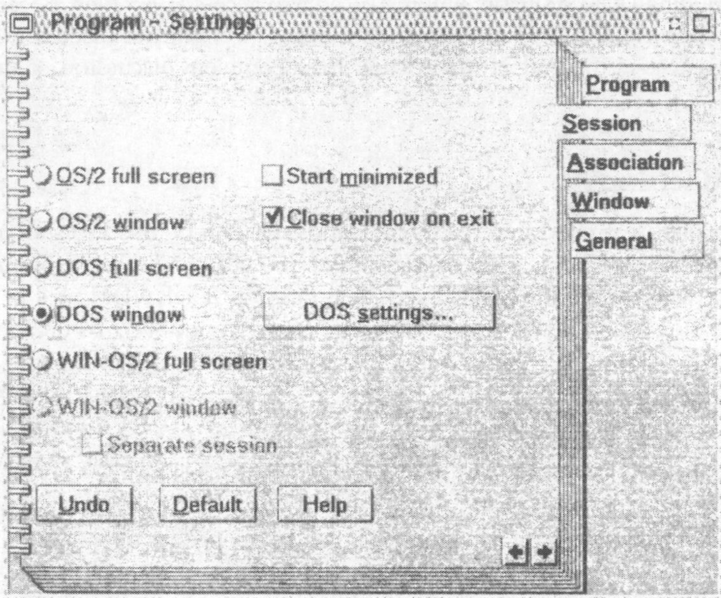

FIGURE 11.6 The OS/2 settings notebook—Sessions tab.

Booting DOS Inside OS/2

The most convenient way to boot native DOS inside OS/2 is to create a "boot image" file. A boot image file is an OS/2 file that is used to obtain the necessary DOS boot files when booting native DOS inside OS/2. Conceptually, the boot image file can be though of as an archive of the files found on a bootable DOS diskette. In fact, OS/2 uses a bootable DOS diskette to create a boot image file. OS/2's VMDISK command creates a boot image file as explained below:

1. Insert a bootable DOS diskette into a diskette drive. (We will assume the diskette is in drive A: in the subsequent discussion.) The diskette must contain the native DOS's COMMAND.COM and an editor that runs under native DOS. To preserve the ability to access OS/2 partitions on your hard drive, copy the OS/2 file

   ```
   C:\OS2\MDOS\FSFILTER.SYS
   ```

   onto the diskette.
2. Open an OS/2 window or fullscreen session and enter the command

   ```
   vmdisk a: bootfile
   ```

   where **`bootfile`** is the name of the boot image file to create.
3. When the VMDISK command completes, create a new DOS icon for the native DOS session following the instructions given in Virtual DOS Machines above.
4. Select the DOS_STARTUP_DRIVE option from the "DOS Settings" menu and enter the full pathname of the boot image file created in step 2.

5. Save the settings and open the new DOS icon by moving the mouse pointer to the icon and double-clicking mouse button 1. A new session begins. The session is either full screen screen or windowed, depending on how the icon was created. A native DOS session is now running. Note that the current drive is A: and that it is not possible to switch to C: or any other hard drive. Also note that the DIR command reports that the contents of the A: drive are identical to the diskette from which the boot image file was created.
6. To make the hard drive visible to the native DOS session, add the line

```
DEVICE = FSFILTER.SYS
```

to the native DOS CONFIG.SYS using the editor from the boot image file. To use EMS or XMS memory, add the lines

```
DEVICE = C:\OS2\MDOS\HIMEM.SYS
```

and

```
DEVICE = C:\OS2\MDOS\EMM386.SYS
```

as well (they must come *after* the line that loads FSFILTER.SYS).
7. The next step is to make further changes to the CONFIG.SYS and the AUTOEXEC.BAT to suit a specific environment. To restart the native DOS session, close it using the window menu for a windowed session or the task list for a fullscreen session—the EXIT command does not work because as far as the native DOS session is concerned there is nowhere to exit to—and double-click on its icon again.

Files in the boot image file cannot be examined from anywhere but the native DOS session that boots from the file. Accordingly, it make sense to make the boot image file as small as possible (CONFIG.SYS, AUTOEXEC.BAT, COMMAND.COM, editor, and FSFILTER.SYS) and put any other files for the native DOS session in a subdirectory. You will of course have to make the proper changes to the native DOS session's CONFIG.SYS and AUTOEXEC.BAT to ensure the other files are found.

A native DOS session thinks the A: drive is the boot image file. If a native DOS session needs to use the real A: drive, issue the command

```
fsaccess a:
```

in the native DOS session. All future accesses to A: will be directed to the real (physical) diskette drive. This means that the native DOS session can no longer reference files in the boot image file. Thus it is vital to set the session's parameters (PATH, COMSPEC, SHELL, etc.) so that the session does not need to reference the boot image file after it has booted.

Configuring DOS Sessions

To examine or modify the settings for a VDM or native DOS session:

1. Open the object menu for the icon associated with the VDM or native DOS session by moving the mouse pointer to the icon and clicking mouse button 2).

2. Select the "Open → Settings" option: move the mouse pointer to the arrow to the right of the "Open" option, click mouse button 1, move the mouse pointer to the "Settings" option on the secondary menu, and click mouse button 1 again.
3. Select the "Session" tab on the settings notebook (move the mouse pointer to the tab and clicking mouse button 1.
4. Select the "DOS Settings" button by moving the mouse pointer to the button and clicking mouse button 1.

Mouse button 1 can now be used to select a specific option or to move the scrollbar to show additional options.

A brief description of each option follows. Options that affect both VDMs and native DOS sessions are marked with an asterisk (*). Options without an asterisk affect only VDMs.

- COM_DIRECT_ACCESS—This option tells OS/2 to give applications running in a VDM direct access to the machine's COM port.
- COM_HOLD(*)—This option tells OS/2 to give applications running in a VDM or native DOS session exclusive access to the machine's COM ports. Programs running in other sessions (OS/2, VDM, or native DOS) will not be allowed to use the COM ports.
- COM_RECEIVE_BUFFER_FLUSH—This option tells OS/2 what to do with data arriving on a COM port when an application running in a VDM enables the "received data interrupt" or when a VDM is switched from the background to the foreground. Some applications may wish to discard data when one of these events occurs (e.g., because the data are destined for another session); others may not.
- COM_SELECT—This option specifies which ports an application running in a VDM can use. (Some DOS applications try to take control of all available ports regardless of which ones they actually use; this option prevents such applications from grabbing a port they do not need and preventing other programs from using it.)
- DOS_AUTOEXEC—This option specifies the full pathname of a .BAT file that OS/2 uses as the AUTOEXEC.BAT file for a VDM. That is, the .BAT file specified by the DOS_AUTOEXEC option is executed after the VDM is started but before any program run in the VDM receives control. (A native DOS session has its own AUTOEXEC.BAT.)
- DOS_BACKGROUND_EXECUTION(*)—This option determines whether applications running in a VDM or native DOS session are given CPU timeslices when the VDM or native DOS session does not have the focus. In other words, this turns multitasking of those applications on and off.
- DOS_BREAK—This option tells OS/2 whether to pass Ctrl-Break and Ctrl-C key sequences to the application running in a VDM or to interpret them as "break" signals. It is equivalent to the "BREAK =" statement in the DOS CONFIG.SYS.
- DOS_DEVICE—This option specifies device drivers that are to be loaded in a VDM. DOS device drivers can also be loaded in OS/2's CONFIG.SYS. DOS device drivers loaded in this manner are loaded in every VDM. (A native DOS session has its own CONFIG.SYS.)
- DOS_FCBS—This option tells OS/2 how many file control blocks (FCBs) to allocate for a VDM. It is equivalent to specifying m in the "FCBS = m, n" statement in the DOS CONFIG.SYS.
- DOS_FCBS_KEEP—OS/2 may decide to close FCBs to keep the number open at or below the limit specified by the DOS_FCBS option. The DOS_FCBS_KEEP option

tells OS/2 how many FCBs must be kept open when OS/2 decides to close them. It is equivalent to specifying *n* in the "FCBS = *m,n*" statement in the DOS CONFIG.SYS.

- DOS_FILES—This option tells OS/2 the maximum number of files that can be open simultaneously by an application running in a VDM. It is equivalent to the "FILES =" statement in the DOS CONFIG.SYS.
- DOS_HIGH—This option tells OS/2 to load part of the DOS emulation code in high DOS memory in a VDM. DOS_HIGH frees about 24K of conventional RAM for DOS programs. It is equivalent to the "DOS = HIGH" statement in the DOS 5.0 and later CONFIG.SYS. XMS memory must be enabled for this option to work correctly (see "XMS_MEMORY_LIMIT" below).
- DOS_LASTDRIVE—This option identifies the last disk drive an application running in a VDM can access. It is equivalent to the "LASTDRIVE =" statement in the DOS CONFIG.SYS.
- DOS_RMSIZE(*)—This option specifies the amount of memory (in Kilobytes) that OS/2 allocates for the program area in a VDM or native DOS session. The default (and maximum) is 640K. If you use a VDM or native DOS session to run a specific application that requires less than 640K of memory, you can reduce the resources the application uses by specifying a smaller value for this option.
- DOS_SHELL—This option specifies the full pathname of the command processor to use in a VDM. It is equivalent to the "SHELL =" statement in the DOS CONFIG.SYS.
- DOS_STARTUP_DRIVE(*)—This option tells OS/2 to boot a native version of DOS in the session in which it is specified. The value of this option is the full path of the boot file that contains the native version of DOS to boot. (The boot file is created with the VMDISK command as explained above in Booting DOS Inside OS/2.)
- DOS_UMB—This option tells OS/2 to allow a VDM to load applications or device drivers in an Upper Memory Block (e.g., using the DOS LOADHIGH command). It is equivalent to the "DOS = HIGH/LOW,UMB" statement in the DOS 5.0 and later CONFIG.SYS.
- DOS_VERSION—This option tells OS/2 how to respond when an application running in a VDM asks which version of DOS is running in the VDM. By default, OS/2 returns version 20.10. This may confuse some applications; others may require a specific value. The syntax is

  ```
  progname, major, minor, count
  ```

  where **`progname`** is the name of the application, **`major`** is the DOS major version number, **`minor`** is the DOS minor version number, and **`count`** is the number of times to respond in this manner (after which OS/2 returns version 20.10). If **`count`** is 255, the specified version is always returned. For example, **`lotus.exe,3,30,255`** tells OS/2 to always say the version is 3.30 when lotus asks.
- DPMI_DOS_API(*)—This option tells OS/2 whether an application running in a VDM or native DOS session uses DPMI services to access extended memory and, if so, whether the application performs the necessary translations itself or requires assistance from the operating system.
- DPMI_MEMORY_LIMIT(*)—This option specifies how many Megabytes of DPMI memory OS/2 provides to a VDM or native DOS session. The default is 4M. This memory does not consume resources unless it is actually used by an application.

- DPMI_NETWORK_BUFF_SIZE(*)—This option specifies how many Kilobytes of RAM OS/2 allocates for the network translation buffer for programs running in a VDM or native DOS session. This buffer is used by programs that use DPMI memory and transfer data across a network.
- EMS_FRAME_LOCATION(*)—This option specifies where to place a 64K page frame used by LIM expanded memory. If the applications to be run in a VDM or native DOS session do not use LIM expanded memory, LIM support may be disabled (and the 64K used for something else) by setting this option to "none."
- EMS_HIGH_OS_MAP_REGION(*)—This option specifies an additional amount of memory (in Kilobytes) to use as EMS memory in a VDM or native DOS session. This essentially expands the size of the EMS page frame.
- EMS_LOW_OS_MAP_REGION(*)—Some applications that use EMS memory can "remap" conventional memory to increase the amount of EMS memory available. This option tells OS/2 how much of the conventional memory available to a VDM or native DOS session may be remapped by such applications.
- HW_NOSOUND(*)—This option tells OS/2 to suppress sounds generated by an application running in a VDM or native DOS session.
- HW_ROM_TO_RAM(*)—This option tells OS/2 to copy the contents of BIOS ROM to RAM. Code in ROM typically cannot be accessed as quickly as code in RAM. Setting this option on may make applications in a VDM or native DOS session run more quickly than they otherwise would.
- HW_TIMER(*)—This option tells OS/2 to give applications running in a VDM or native DOS session direct access to the hardware timer. It is intended to be used with programs that have a critical need for precise timing functions.
- EMS_MEMORY_LIMIT(*)—This option specifies how many Kilobytes of LIM expanded memory OS/2 provides to a VDM or native DOS session. The default is 2M (2048K). This memory consumes resources even if it is not used by an application in the VDM or native DOS session, so EMS_MEMORY_LIMIT should be set to 0 if expanded memory is not needed.
- IDLE_SECONDS(*)—This option controls how quickly OS/2 will decide to decrease the resources dedicated to a VDM or native DOS session that does not appear to be doing useful work. The default value is 0, so OS/2 normally takes resources from an idle DOS application immediately. Some games or timing-dependent programs may actually be doing something constructive by waiting; IDLE_SECONDS can be increased for such programs to give them more resources during this activity.
- IDLE_SENSITIVITY(*)—This option controls how OS/2 decides that an application in a VDM or native DOS session is not doing useful work. Specifically, OS/2 monitors such applications to determine whether they are "polling" the keyboard (i.e., twiddling their thumbs until the user hits a key). If IDLE_SENSITIVITY is *n*, OS/2 decides an application is idle if the rate at which it is polling is greater than *n*% of the maximum possible polling rate. Some programs (notably communications programs such as Procomm) may appear to be idle when they are in fact doing useful work (e.g., downloading files). IDLE_SENSITIVITY should be set to 100 for such programs.
- INT_DURING_IO(*)—This option tells OS/2 to allow an application in a VDM or native DOS session to receive interrupts while disk I/O is in progress. This essentially allows the application to perform useful work at the same time data are being written to or read from the disk. Multimedia applications can derive particular benefit from this option.

- KBD_ALTHOME_BYPASS(*)—This option tells OS/2 to pass the Alt-Home key sequence (which normally switches a fullscreen session to windowed and vice versa) to the application running in a VDM or native DOS session. This option is intended to be used with applications that use Alt-Home for their own purposes.
- KBD_BUFFER_EXTEND(*)—This option tells OS/2 to buffer keystrokes to an application running in a VDM or native DOS session. This effectively increases the size of the typeahead buffer available. Some applications have their own typeahead buffers and may not be affected by this option.
- KBD_CTRL_BYPASS(*)—This option tells OS/2 to pass the Ctrl-Esc key sequence (which normally brings up the Workplace Shell task list) or the Alt-Esc key sequence (which normally switches to the next active task) to the application running in a VDM or native DOS session. This option is intended to be used with applications that use Ctrl-Esc or Alt-Esc for their own purposes.
- KBD_RATE_LOCK(*)—This option tells OS/2 whether to allow an application running in a VDM or native DOS session to change the rate at which keystrokes repeat when a key is held down.
- MEM_EXCLUDE_REGIONS(*)—This option tells OS/2 to prevent a VDM or native DOS session from using specific regions of upper memory as an EMS page frame or as program memory.
- MEM_INCLUDE_REGIONS(*)—This option tells OS/2 to make specific regions of upper memory available to a VDM or native DOS session for use as an EMS page frame or as program memory. It is often possible to include regions that would have to be excluded when running DOS outside of OS/2 because the regions are occupied by adapters.
- MOUSE_EXCLUSIVE_ACCESS(*)—This option tells OS/2 to let an application running in a VDM or native DOS session control the mouse. It is intended to be used with applications such as WordPerfect that do not cooperate well with the OS/2 mouse management code.
- PRINT_SEPARATE_OUTPUT(*)—This option tells OS/2 whether to treat printer output from two or more applications running in a VDM or native DOS session as separate print jobs.
- PRINT_TIMEOUT(*)—This option tells OS/2 how long to wait for an application running in a VDM or native DOS session to generate an end-of-job code before releasing spooled output to the printer. This in turn affects how soon something an application has printed begins to appear at the printer.
- VIDEO_8514A_XGA_IOTRAP(*)—This option tells OS/2 to give applications running in a VDM or native DOS session direct access to the 8514/A or XGA video adapter. This may allow games and other graphics-intensive programs to run more quickly.
- VIDEO_FASTPASTE(*)—This option controls how OS/2 uses its clipboard to transfer information between two applications running in VDMs or native DOS sessions. OS/2 normally transfers data from the clipboard to the keyboard buffer (so that the target application thinks the data have been entered by a very fast typist). Applications that have their own keyboard buffers or that manipulate the keyboard buffer in other ways may require that this option be turned off.
- VIDEO_MODE_RESTRICTION(*)—This option determines whether an application running in a VDM or native DOS session can use VGA graphics or is restricted to CGA graphics and text. Setting this option to CGA frees about 96K of upper memory block RAM for applications. (Some applications do not work well with this option for reasons that are not entirely clear.)

- VIDEO_ONDEMAND_MEMORY(*)—This option tells OS/2 whether to allocate a complete video buffer for a VDM or native DOS session before starting the session or to wait until an application accesses the video buffer before allocating it. Setting this option to ON can reduce the amount of memory and other resources in use.
- VIDEO_RETRACE_EMULATION(*)—Some older CGA adapters would display "snow" if an application attempted to change the contents of the screen while the picture was actually being drawn. Some applications therefore poll the retrace status port and update the screen only while the electron beam is moving from the end of one scan line to the beginning of another. EGA, VGA, and other modern adapters do not suffer from this problem, so programs that wait to change the screen run more slowly than they need to. The VIDEO_RETRACE_EMULATION option causes all polls of the retrace status port to report it is safe to update the contents of the screen.
- VIDEO_ROM_EMULATION(*)—Many video adapters keep data or service routines in ROM. Data in ROM typically cannot be accessed as quickly as data in RAM. This option tells OS/2 to emulate standard video calls using routines in RAM, which improves performance. Some applications use undocumented video services or rely on features provided by a specific video adapter and require that this option be set off.
- VIDEO_SWITCH_NOTIFICATION(*)—This option tells OS/2 to notify the VDM or native DOS session when the session is switched from fullscreen to windowed or vice versa. An application running in the VDM or native DOS session can use this information to decide when to redraw its screen and how much of the screen to redraw. VIDEO_SWITCH_NOTIFICATION is particularly useful when an IBM 8514 adapter (or equivalent) is present. It may also help when running applications that use nonstandard video modes that OS/2 does not support.
- VIDEO_WINDOW_REFRESH(*)—This option determines how often the screen for a VDM or native DOS session is redrawn. If an application frequently changes the contents of the screen, setting this option to a higher value may improve performance and reduce flicker by reducing the number of times OS/2 actually updates the screen image.
- XMA_HANDLES(*)—This option specifies the number of handles OS/2 reserves for a VDM or native DOS session to use to reference XMS memory.
- XMS_MEMORY_LIMIT(*)—This option specifies how many Kilobytes of LIM extended memory OS/2 provides to a VDM or native DOS session. The default is 4M (4096K). XMS memory consumes resources even if it is not used by an application in the VDM or native DOS session, so should be set to 0 if extended memory is not needed. XMS memory is used to provide more conventional memory to applications by loading DOS in the high memory area. At least 64K XMS memory is needed to do so.
- XMS_MINIMUM_HMA(*)—This option tells OS/2 the smallest amount of memory an application running in a VDM or native DOS session must request before it is given any memory in the high memory area HMA. This option can be used to prevent an application that needs only a small amount of HMA from reserving all of it.

11.7 RUNNING WINDOWS PROGRAMS UNDER OS/2

Most Windows applications can be run under OS/2. There are two basic decisions you must make:

1. A Windows session can be run in fullscreen mode or in a window (the latter is often called "seamless" Windows because the Windows session behaves in the same manner as any other windowed application on the WorkPlace Shell desktop). This choice

is made in the "Session" page of the settings notebook for the icon used to launch the Windows session. (The VWIN.SYS device driver must also be loaded in the OS/2 CONFIG.SYS to use seamless Windows sessions.) Graphics operations are somewhat faster in a fullscreen session.

2. Each Windows application can be run in a separate Windows session or several Windows applications can be run in a single Windows session. Running several Windows applications in a single session mimics the way native Windows handles multiple applications. The Windows task list is used to switch control among the applications. Running multiple applications in a single session has advantages and disadvantages: there is no need to switch back to the WorkPlace Shell desktop to move between applications, but because all the applications run in a single address space an error in one application can cause the others to crash. In other words, running multiple applications in a single session removes the protection OS/2's memory management offers.

   The "Session" page of the settings notebook for the icon used to launch a Windows session has a checkbox with the title "Separate sessions." If this box is checked, opening the icon launches a separate Windows session to run the application associated with the icon. If the box is not checked, opening the icon causes the application associated with the icon to be run in the same Windows session as all other applications whose checkboxes are not checked. (We will refer to this as the "default" Windows session below.) Note that more than one application may be run in any Windows session, regardless of whether the checkbox is checked.

Configuring Windows Sessions

The configuration of a Windows session is examined and changed in almost exactly the same way the configuration of a VDM is manipulated. The only difference is that the button that leads to the options is labeled "WIN-OS/2 Settings" rather than "DOS Settings."

All of the options that apply to VDMs also apply to Windows sessions. There are three additional options that are used only by Windows sessions.

- WIN_CLIPBOARD—This option determines whether the clipboard is public (shared between Windows programs and windowed OS/2 and DOS sessions) or private to a single Windows session.
- WIN_DDE—This option determines whether dynamic data exchange (DDE) is public (shared between Windows programs and OS/2 sessions) or private to a single Windows session.
- WIN_RUN_MODE—This option specifies whether applications running in a Windows session run in standard or enhanced mode.

OS/2's Windows emulation also adds a number of items to the standard Windows SYSTEM.INI file:

- display.drv—This option specifies the display driver to use in conjunction with fullscreen Windows sessions.
- MAVDMAPPS—This option lists applications that are started automatically when the default Windows session is launched.
- os2shield—This option specifies the program that manages the interaction between a Windows session and the remainder of OS/2. WINSHIELD.EXE is the default.

- SAVDMAPPS—This option lists applications that are started automatically when a Windows session that is not the default is launched.
- sdisplay.drv—This option specifies the display driver to use to allow seamless Windows sessions to appear on the WorkPlace Shell desktop.
- useos2shield—This option is used to enable loading of the program named by the os2shield option. Useos2shield should be set to 0 in Windows sessions that are used to launch applications that must be the first program to run in a session (e.g., the Norton Desktop).
- WAVDMAPPS—This options lists applications that are started automatically when a seamless Windows session is launched.

Caveats

Windows applications that use Win32s calls or VxDs will not run under OS/2 2.1. Such applications are being designed to run under "Chicago," successor to Windows 3.x. One prominent example is Math Cad 4.0. If you are unsure whether an application uses either of these features, contact the application vendor.

11.8 SUPPORT AND REFERENCES

OS/2 includes a large amount of online help (including descriptions of the DOS and Windows options) and a tutorial. The online help can be accessed via the "Start Here" icon, the "Master Help Index" icon, or the "Information" icon; the tutorial can be accessed via the "Information" icon. The "Information" icon appears in Figure 11.2 the other icons appear in Figure 11.7.

FIGURE 11.7 The OS/2 Start Here and Master Help Index icons.

Several books are available that contain more detailed information on OS/2's capabilities and how to make the most of them:

- Tyson and Oliver, *10-Minute Guide to OS/2 2.1*. Alpha Books, 1993.
- Dvorak, Whittle, and McElroy, *Dvorak's Guide to OS/2 Version 2.1: Learn to Navigate the Operating System of the Future*. Random House, New York, 1993.
- Minasi et al., *Inside OS/2 2.1*, 3rd ed. New Riders Publishing, 1993.
- Moskowitz and Kerr et al., *OS/2 2.1 Unleashed*. SAMS Publishing, 1993.
- Brown and Howard, *Using WorkPlace OS/2: The Power User's Guide to IBM OS/2 Version 2.1*. Van Nostrand Reinhold, New York, 1993. ISBN 0-442-01590-9.
- Tyson, *Your OS/2 2.1 Consultant*. SAMS Publishing, 1993.
- Levenson and Hertz, *Now That I Have OS/2 2.0 On My Computer, What Do I Do Next?* Van Nostrand Reinhold, New York, 1992. ISBN 0-442-01227-6.
- Tyne, *OS/2 and the WorkPlace Shell: A User's Guide and Tutorial for OS/2 2.0*. Computer Information Associates, 1992.
- Nance, *Using OS/2 2.0*. Que Corp., 1992.

(The books published in 1992 cover OS/2 version 2.0; there are no substantial differences between its user interface and OS/2 version 2.1, so they may still be of interest.)

More information on the internal design of OS/2 can be found in *OS/2 2.1 Unleased* and in Deitel and Kogan, *The Design of OS/2*. Addison-Wesley, Reading, MA, 1992.

There are several online avenues for support for OS/2. Official IBM support can be obtained on CompuServe (GO IBMOS2, GO OS2USER, GO OS2SUPPORT) or Prodigy (JUMP OS2 CLUB). IBM also offers fee-based support through its IBM TalkLink/OS2BBS facility; call 800-547-1283 for information on cost and enrollment.

The IBM Personal Computer Company maintains a free bulletin board at 919-517-0001 that includes several areas specific to OS/2 (general questions about OS/2 version 2.x are in conference 6). This support is "semiofficial." A complete set of bugfixes (IBM calls them "CSDs," for Corrective Service Diskettes) and interesting utilities are available, but questions are answered on a "best-effort" basis.

Unofficial support is available on America OnLine (Keyword OS2) and in several Internet newsgroups. The list As of March 20, 1994 includes

- comp.os.os2.bugs
- comp.os.os2.misc
- comp.os.os2.multimedia
- comp.os.os2.networking
- comp.os.os2.programmer.misc
- comp.os.os2.programmer.porting
- comp.os.os2.setup

DEFINING TERMS

Balanced Binary Tree: A binary tree data structure in which all leaf nodes are the same distance from the root.

Flat Memory Model: A design in which the address space in which an application runs consists of a single (usually large) region. See segmented memory model.

Multithreading: The ability to design a single program as a collection of cooperating "threads" of execution. The multithreaded model of computation is well suited to powerful applications that must perform several operations concurrently on shared data.

Segmented Memory Model: A design in which the address space in which an application runs consists of several (usually fairly small) regions or "segments". See flat memory model.

Settings Notebook: A collection of customized settings for a WorkPlace Shell object in OS/2 2.x.

VDM: Virtual DOS Machine. An environment provided by OS/2 2.x that behaves very much like DOS running on an 80386-based PC.

Virtual 8086 Mode: A mode of operation available on the Intel 80386 and its successors that allows the chip to emulate several 8086s running in real mode simultaneously. Virtual 8086 mode uses the 386's paging hardware to enforce separation of each virtual 8086's address space.

12

NEXTSTEP

David H. Hutchens
Millersville University

12.1 OVERVIEW

NEXTSTEP™ is an operating system from NeXT™ Computer Inc. It is based on BSD Unix™ running on a Mach kernel. Nearly all interaction is done through a graphical user interface (GUI). The GUI uses a unified PostScript™ imaging model for displaying text and graphics on the screen or on the printer. It runs on several types of hardware ranging from industry standard PCs to high end workstations from HP. Versions for other hardware are being announced regularly. A version called OPENSTEP™ is planned for use on Sun™ workstations running Solaris™. It is distributed in two parts: NEXTSTEP User and NEXTSTEP Developer.

Information about available ports and hardware requirements may be obtained by sending Internet e-mail to nextanswers@next.com with the subject HELP. Alternatively, visit the Web page at http://www.next.com.

NEXTSTEP's primary advantages for scientists and engineers are the very consistent user interface, multitasking, networking, and the ease with which new applications can be built using the developer's kit.

12.2 THE WORKSPACE

NEXTSTEP is designed as a multiuser operating system. To use NEXTSTEP, you must first log on. The system is shipped with two users; one called *root*, the other called *me*. If you have not set any passwords, the system logs on as *me* on power up. This is not normally a good thing and the system administrator should immediately provide a password for the *me* and *root* accounts and provide new accounts (userids and passwords) for everyone who should be using the system. Once this has been done, to log in type your userid and password in the fields provided, pressing the return key after each one.

Once logged in, you should see the File Viewer, the Workspace Menu, and the Dock (see Fig. 12.1). Each of these is described in the following paragraphs.

0-8493-2530-7/96/$0.00+$.50

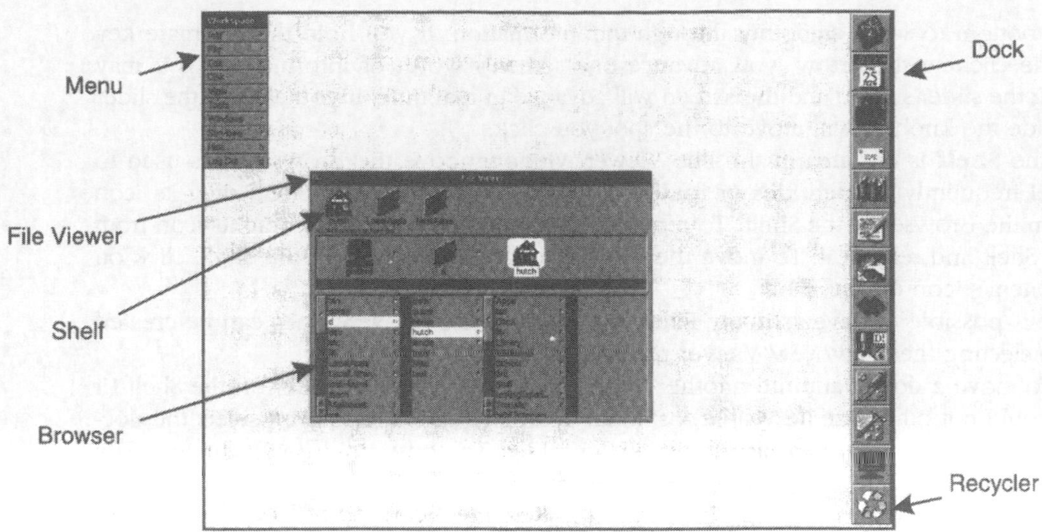

FIGURE 12.1 The Workspace.[1]

The **Workspace Manager™** hereafter referred to as the Workspace, is an application in NEXTSTEP. The purpose of the Workspace is to launch other applications and to provide for normal file system activities such as moving, copying, and deleting files. Like all applications, it has a main menu that is displayed in the upper left portion of the screen. This menu is used to set preferences and invoke many of the functions provided by the workspace.

The **File Viewer** is used to navigate the file system. It has several display modes that may be selected in the View submenu. The most common one is browser mode, which will be described here.

Since NEXTSTEP is a multiuser operating system, each user needs a home directory or folder. This folder is pictured in the file browser with a house icon. Typically, a user keeps all personal files in this folder or its subfolders. At any time, some files and/or folders are selected. The enclosing folders are displayed in the browser. To select a file or folder, click on its name with the mouse. To select more than one, click on each additional one while holding down the shift or alternate key. All selected items must be in a single folder.

A **Document** is a file or folder containing a logically connected group of information that is manipulated by an application. Each document has a default application determined by the name suffix. For example, all documents that end with .ps are taken to be PostScript files and are to be opened by default in the PostScript Previewer. To open a document with the default application for that document, double-click the document name or icon. To open an application without an initial document, double-click on the application. For more information on launching applications, see the discussion of the Dock.

Notice that the Browser presents the file and folder names in **scrolling views**. This is the standard scrolling view in NEXTSTEP. It is used by most applications to display information that does not all fit on the screen at one time. You may use the arrows at

[1]Computer Screen shots of NEXTSTEP software (Copyright Next Computer, Inc., 1988–1994) contained in this chapter are reproduced with permission of NeXT Computer, Inc.

the bottom to scroll smoothly through the information. If you hold the Alternate key while clicking the arrow, you advance one screen's worth of information. You may drag the slider's knob and the screen will advance in real time. If you click in the slider beside the knob, it will move to the spot you click.

The **Shelf** is the area in the File Viewer window above the Browser. It is used to hold frequently needed files or folders. To add something to the Shelf, drag its icon from the Browser to the Shelf. To remove something from the Shelf, drag its icon from the Shelf and release it. To move the Browser to an item that is on the Shelf, click on the item's icon on the Shelf.

It is possible to have as many File Viewers as is desired. New ones can be created by selecting the View/New Viewer menu item.

To move a document into another folder, place the receiving folder on the Shelf (if it would not otherwise be visible) or find it in an alternate File Viewer, select the document, and drag its icon into the receiving folder. To copy, hold down the alternate key while dragging.

To delete a document, drag its icon to the **Recycler** icon in the Dock (see below). This actually moves the document to the Recycler's folder. You may empty the Recycler's folder by selecting File/Empty Recycler from the menu. Until you empty the Recycler you may retrieve the document by double-clicking on the Recycler icon. Another way to delete a document is to select the document and then select File/Destroy from the menu. In this case, you cannot retrieve the document from the Recycler.

The **Dock** is the collection of icons down the right hand side of the screen (as shown in Fig. 12.1). These icons should represent frequently used applications. The top icon is the Workspace icon. The bottom icon is the Recycler. The user may fill the rest of the Dock with any applications desired. This is done by finding the application in the browser and dragging the icon to the dock. Applications may be removed from the dock by dragging the icon away. The Workspace Info/Preferences menu may be used to select any of the Dock applications for automatic launching at the time the user logs in.

To activate a launched application, click on its icon. This will also unhide the application if you have previously selected the Hide command from the menu. Note that the icons for applications that are on the Dock have three little dots in the lower left of the icon to signify that the application is not launched. The icon has no dots if the application is launched. Launching an application that is not on the Dock will place its icon along the bottom portion of the screen.

Each application has a main **Menu**. This menu should be visible in the upper left of portion of the screen whenever the application is the active application. Like any window in NEXTSTEP this menu may be moved to some other location by dragging its title bar. To select a menu item, click on it. To select a submenu, click on the submenu name. This will cause the submenu to be displayed to the right of its parent menu. To close a submenu, click on the name in the parent menu. A submenu may be torn off its parent menu and moved to another place on the screen. To close a torn off menu, click in its close box.

The **Services** menu provides a mechanism for invoking other programs to perform tasks for the current application. Any application that includes the Services menu has access to all the services provided by any application currently installed on the system. For example, if you have Webster's dictionary installed (it is included on the installation CD), whenever you select a piece of text in a field you may select the

Services/Define in Webster menu to bring up Webster and look up the selection. There are a variety of services provided by NEXTSTEP applications.

It is common practice among NEXTSTEP applications to provide an Inspector panel for examining the properties of objects supported by the application. Consider the Workspace Inspector panel (Fig. 12.2). It may be displayed on the screen by selecting the Tools/Inspector menu. Once displayed, it provides information about the currently selected document. There are several types of information that may be displayed by using the pop-up menu at the top of the inspector. These are described below.

The **Attributes** panel shows

- size
- owner
- date changed
- access permissions

The **Contents** panel shows the contents of certain types of files. For example, a tiff picture file will be displayed inside this panel.

The **Access Control** panel allows the owner of the file to change its access permissions. This panel is used to provide or deny access to other users on the system.

Many NEXTSTEP applications support the standard Drag and Drop protocol. For the user this means that to include the contents of a document in an application, all that

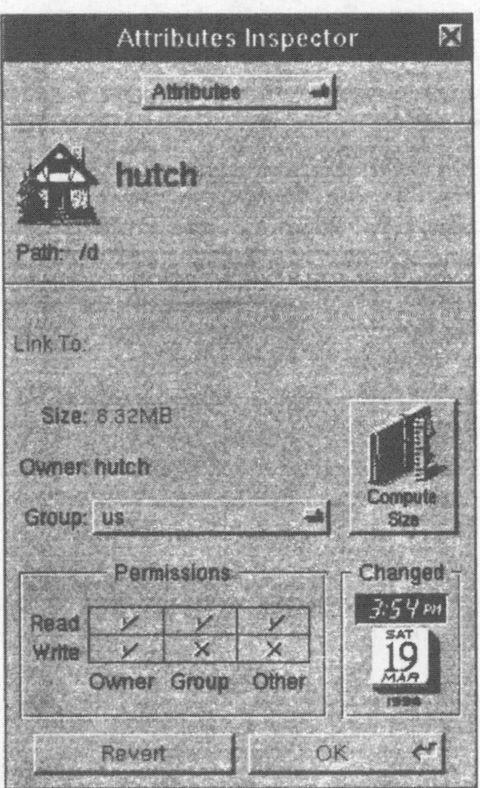

FIGURE 12.2 The Workspace Inspector.

need be done is find the document in the browser and drag its icon over the target application's window and drop it in. For example, this is a standard way to put pictures into documents.

12.3 THE HELP SYSTEM

Most applications for NEXTSTEP provide a Help system. The Help system may be invoked by selecting the Info/Help menu, which brings up the applications Help panel. Figure 12.3 shows the Workspace Help panel. At the top of the panel are buttons for finding occurrences of keywords, locating the index, and backtracking to previous information. Below the buttons is a field containing the titles of the section of the document. The user may select a section by clicking on its title. Below the titles is a field that contains the currently selected section of the document.

Context Sensitive Help is provided by holding down the Help key and clicking on the item for which help is desired. This will bring up the Help panel positioned on the section describing the item selected. The Help key is system dependent. In some computers, the F1 key acts as the Help key. On computers with no F1 key, holding down the Control and Alternate keys together invokes context-sensitive help. When invoked, the mouse pointer will change to a question mark.

12.4 USEFUL SYSTEM FOLDERS

While browsing the file system, a user will encounter several folders that are not part of the standard UNIX distributions. Some of the more important ones are described below.

/**NextApps** is the folder that contains most of the standard applications that are provided with the system. See the section on Standard Applications for more detail.

/**NextAdmin** contains the applications that are used by the system administrator to maintain the system. This includes programs to install new applications, set up network connections, configure the system's device drivers, initialize disks, and create, modify, and destroy user accounts.

/**NextLibrary** contains the system documentation and standard information used by various applications on the system. This information includes fonts, colors, printer configurations, Webster's dictionary, and much more.

/**LocalApps** is the folder where the system administrator installs applications that are not part of the standard NEXTSTEP distribution. Its contents are entirely determined locally

/**LocalLibrary** normally contains generally available information that is installed by the local system administrator. It may, for example, contain fonts that are not part of the standard NEXTSTEP distribution or the documentation for applications found in LocalApps.

/**NextDeveloper/Apps** contains the programs that are provided on the NEXTSTEP Developer CD. These applications include Project Builder, Interface Builder, and Icon Builder.

/**NextDeveloper/Demos** contains several demonstration applications and games.

12.5 STANDARD APPLICATIONS

Much of the power of NEXTSTEP derives from the rich set of standard applications that are provided with it. This section will describe several of them.

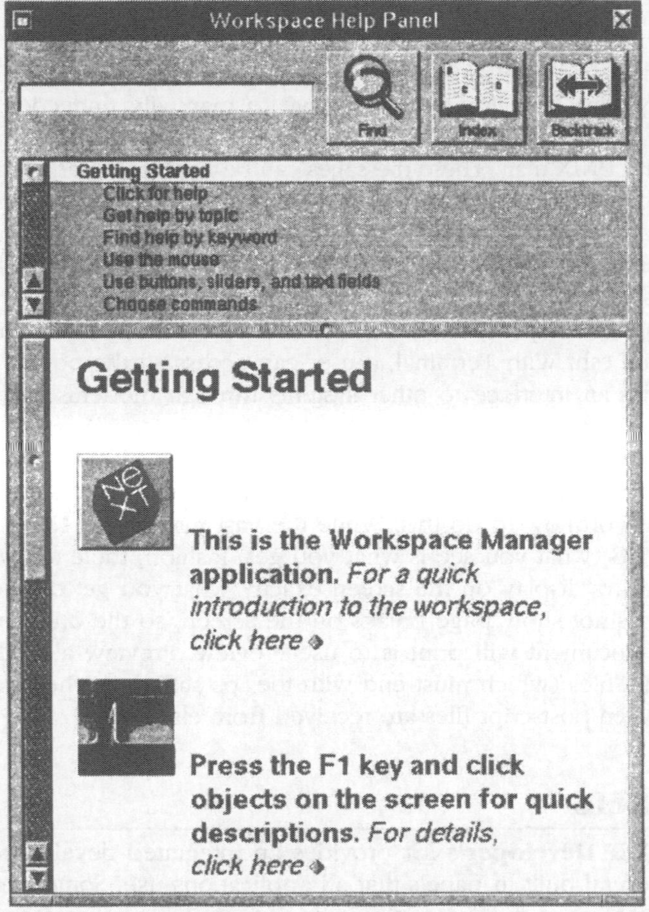

FIGURE 12.3 The Help Panel.

The **Preferences** application is used to set preferences that apply across applications. Things that can be set include mouse speed, click delay, right or left handed mouse, keyboard repeat speed, automatic dimming or screen saver activation time, background color, volume, language (English, German, French, etc.), system beep sound, default fonts, time/date (might be restricted to the administrator), password, and others. A UNIX expert should select the UNIX Expert preference so that the File Viewer will display all the files. In its default mode, the File Viewer hides several of the files and folders associated with UNIX. A user may also add additional preferences modules to the Preferences application.

The **Digital Librarian™** application is mainly used for searching the on-line documentation, although it can be used to find information in other folders as well. By typing keywords in the search field and pressing the search button, you can find all the documents in the target folder that contain those keywords.

The **Mail** application is used to send electronic mail to other users. If the system is connected to a network, the Mail application can also be used to send mail to the users of the other systems on the network. A button in the compose window selects between the two kinds of mail supported by the Mail application.

One is NeXT mail. NeXT mail is allowed to contain documents and sound. The documents can be added to the message by Drag and Drop. The sound may be recorded directly from the Mail application. NeXT mail can easily be read on other NEXTSTEP systems. On other systems, a recipient would have to manually undecode, uncompress, and untar, the message.

The other is standard UNIX mail. These messages can be read by anyone who can receive them, whether on a UNIX system or not, as they contain standard ASCII characters.

The **Edit** application provides the ability to read and write standard ASCII files. It is also capable of reading and writing RTF (Rich Text Format) files and RTF files with embedded tiff or eps pictures.

The **Terminal** application provides the standard UNIX command line interfaces (shells) such as sh and csh. With Terminal, a user can access all the power of UNIX. Terminal also provides an interface to other systems through modems and network protocols such as rlogin and ftp.

The PostScript **Preview** application allows the user to see exactly what will be printed before sending a document to the printer. See the discussion of the Print Panel below for more details on how to do this. While the vast majority of NEXTSTEP operates in the WYSIWYG (what you see is what you get) fashion, there are some programs that do not always display on the screen exactly what you get on the printer. For example, Edit does not show page breaks on the screen, so the only way to see exactly how an Edit document will print is to use Preview. Preview also allows the user to view postscript files (which must end with the .ps suffix) on the screen. This can be very handy when postscript files are received from elsewhere.

12.6 STANDARD PANELS

Because the **NEXTSTEP Developer's kit** provides an integrated development environment, there are several built in panels that all applications use. Some of these are described below. For the user, this integration means a very consistent user interface.

The **Print Panel** (Fig. 12.4.) is obtained by selecting the Print menu. The menu contains the options: Save, Preview, Fax, Cancel, and Print. It also allows the user to select the number of copies and which pages of the document are to be printed. In some cases there may be choices for Resolution, Paper Tray, and other options. From this panel, the user can direct the print to a file (Save), to the PostScript Preview application (Preview), to a modem for Faxing (Fax) if there is a Fax modem available, or to the Printer (Print). Every application that allows printing uses this panel, or a derivative of it, so all of those options are available consistently.

The **Save Panel** (Fig. 12.5) is displayed whenever a user wishes to select a place to save a document. It is normally obtained by the Save As command from the File or Document menu. Once obtained, the user is expected to use the browser in the panel to select a destination for the save. There are several ways to manipulate the browser. One may select the target folder in the same way as the File Viewer browser, drag a folder onto the panel to have it jump there instantly, or type the path name from the current folder in the field. The new document name must also be provided.

The **Open Panel** is similar to the Save Panel except you must specify an already existing document. The application may limit the types of documents that you may specify.

The **Font Panel** (Fig. 12.6) is provided by any application that allows text to be edited with different fonts. It allows the selection of the font family (e.g., Times Roman), its size (in points), and the specific typeface (e.g., Italic).

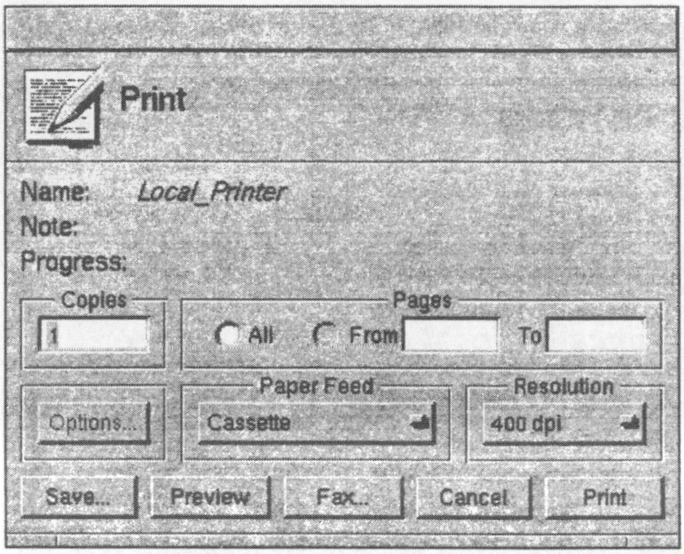

FIGURE 12.4 The Print Panel.

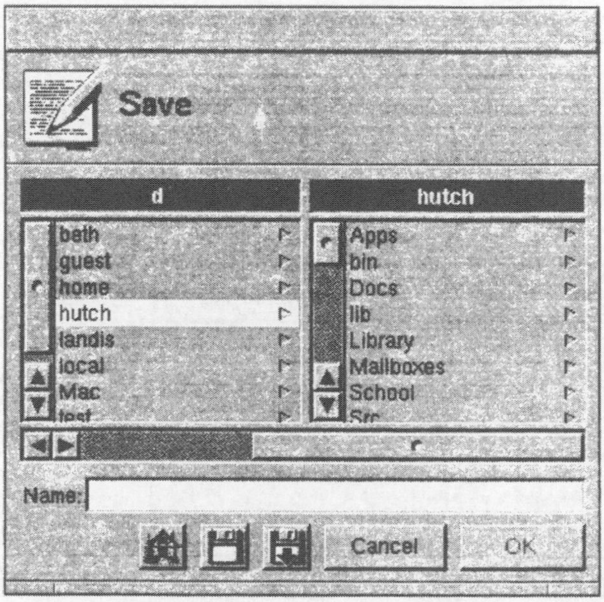

FIGURE 12.5 The Save Panel.

The **Color Panel** (Fig. 12.7) is provided by applications that allow the user to select color. The color panel allows the selection of color in several different modes. These include the standard color wheel, sliders for the Red, Green, and Blue components, CMYK sliders (shown in Fig. 12.7), and Pantone Color by number. The user is allowed to save selected colors in the color wells at the bottom of the panel for later use in either the current application or a different one.

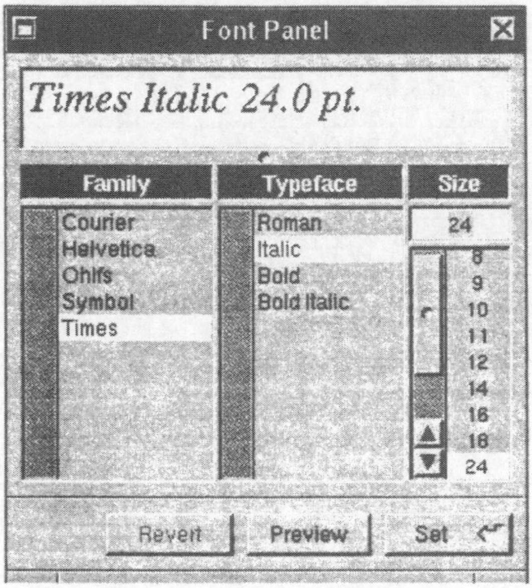

FIGURE 12.6 The Font Panel.

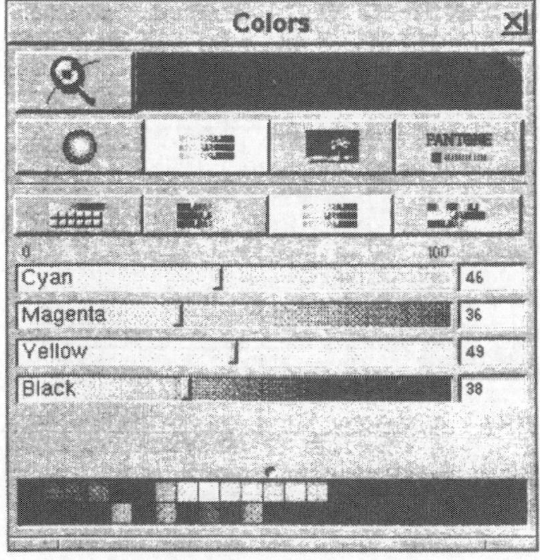

FIGURE 12.7 The Color Panel.

12.7 ADVANCED FEATURES

This section will provide a brief overview of some of the advanced features of NEXTSTEP, particularly the developer applications Interface Builder and Project Builder. These applications are available as part of the NEXTSTEP developer's CD.

Project Builder (Fig. 12.8) is an application that controls the construction of a project. It maintains information about the files that are used in the implementation and builds the Make files (see UNIX) that build the project. The projects may be applications, bundles, or palettes. Applications are stand alone programs that may be run by the workspace.

Bundles are collections of code that may be dynamically linked into a running application. Many commercial applications for NEXTSTEP draw significant power by allowing the inclusion of bundles created by the user. For example, a spreadsheet might be enhanced by providing additional capabilities though bundles such as the ability to monitor a database and update the cells when the database changes.

Palettes are code chunks, essentially special bundles, that may be added to **Interface Builder** (Fig. 12.9). In this way a programmer can extend the types of objects that may be added to the user interface through simple drag and drop mechanisms. Interface Builder already provides most of the common objects such as buttons, menus, windows, and fields. It also provides access to the NEXTSTEP object library containing objects that are not views and hence cannot be dropped in a window.

Interface Builder provides the ability to build a user interface through dragging objects from palettes and dropping them into windows. The objects can then be connected to other objects by doing a control drag from one to the other.

The next several paragraphs will lead you though the building of a simple project that demonstrates some of the power of these development tools.

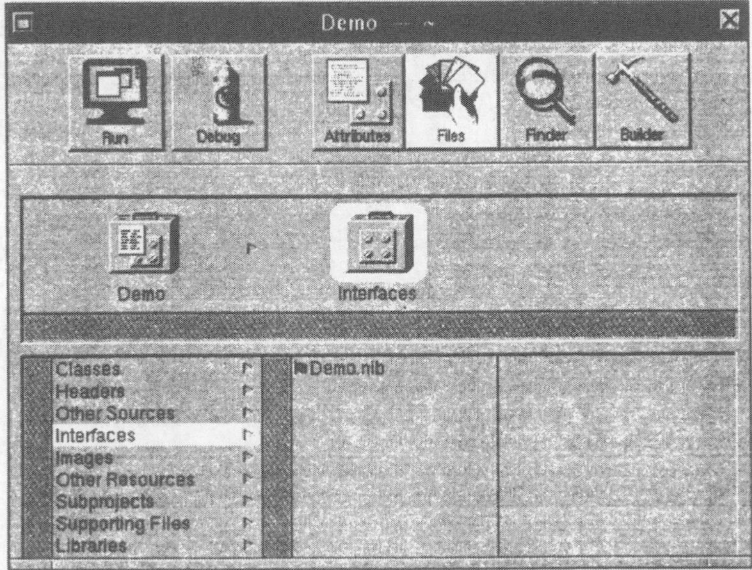

FIGURE 12.8 The Project Builder Window

First, create an empty folder to hold the project. Then open Project Builder by double-clicking it in /NextDeveloper/Apps. Then create a new project with the Project/New menu and tell the save panel to use the folder you created. You should see the window in Figure 12.8.

In the Files panel, obtained by pressing the Files button, you will see an interfaces group. Inside this will be the default interface file for your project. Interface files end with the suffix .nib (Next Interface Builder). Double-click the name and Interface Builder will be invoked on that nib file. You should see the some of the windows in Figure 12.9.

Use the tools menu to bring up the Inspector panel. If you click in the window titled My Window, you will see the Window Inspector. From this inspector you can change the behavior of this window object.

Drag a Button from the Palettes Inspector to the window. It will appear with **Handles**, little dots around the object. The button may be resized or moved by dragging the handles or the middle with the mouse. Hold down the Alternate key and drag the right center handle out until 3 buttons appear. You now have a matrix of buttons. Double-click on the left button. Its inspector should appear in the inspector panel. Double-click again and the title will highlight. You may now rename it. Change its name to Play. Change the middle buttons name to Record and change the right buttons name to Stop.

Select the Classes file in the lower left. Move the slider to the left so that you see the first object called Object. Click on Sound in the right browser panel. Now use the

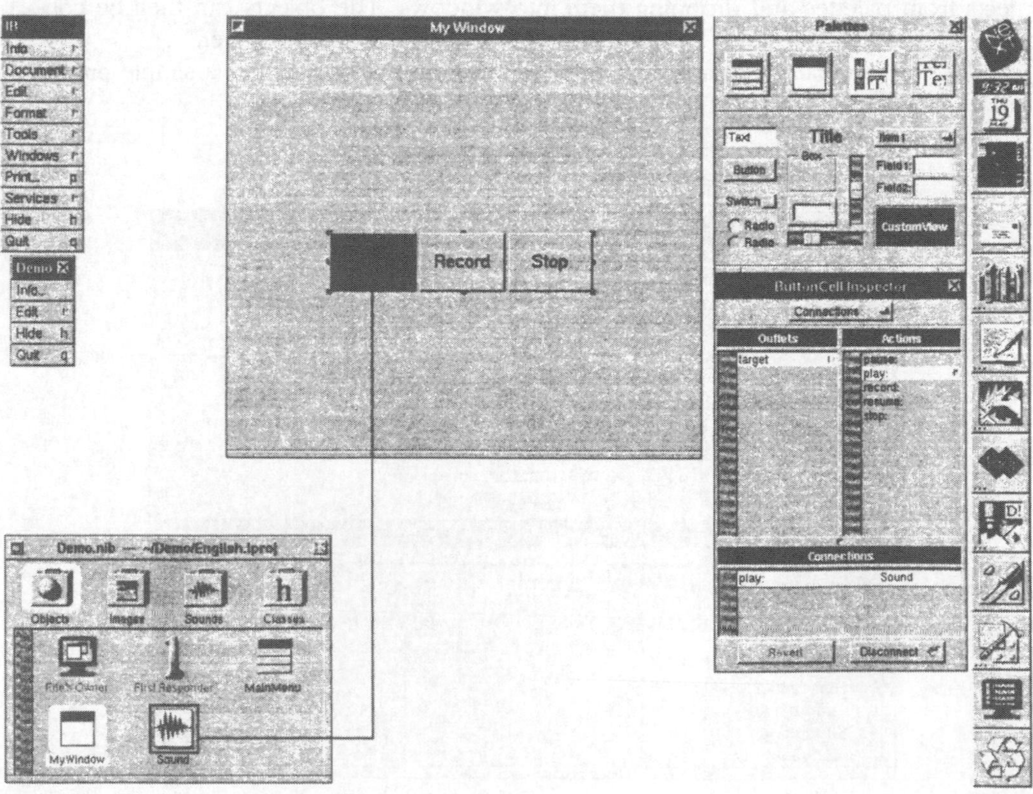

FIGURE 12.9 Interface Builder

pop-up menu to select instantiate. You have added a sound object to your project and you should see it in the Objects file.

Now select the Play button in the window and hold down the control key while dragging from the button to the sound object's icon in the lower left window. A line should be made connecting the two objects. In the Inspector panel, the connections panel will show the methods of Sound that you may connect to be the target of the Play button. Select the Play: method by double-clicking its name. Similarly connect the Record and Stop buttons.

You now have an application that can record and play sound, assuming your hardware is capable. To test the interface, select the Document/Test Interface menu. You should now be able to press the record button, speak, press the stop button, and hear the recording by pressing the play button.

To compile the program, return to Project Builder select the Builder button, and press Build. The application will be built in the project directory. To install the application in your home Apps directory, select the pop-up menu entry Install, and press Build. Your application is now ready for use.

There is nothing special about the Sound object. It is an object supplied by the NEXTSTEP object library. You may build your own objects using Objective-C, inform Interface Builder of their existence, instantiate them, and connect them to the other objects in the interface through control-drag actions.

A full description of the power of Interface Builder is beyond the scope of this book. For more information, see the on-line documentation and the books listed at the end of this chapter.

Trademark Information

NeXT, the NeXT logo, NEXTSTEP, OPENSTEP, Digital Librarian, and Workspace Manager are trademarks of NeXT Computer Inc. PostScript is a trademark of Adobe Systems, Inc. Sun and Solaris are trademarks of Sun Microsystems, Inc. All other trademarks are the property of their respective owners.

DEFINING TERMS

Application: A program that is run under NEXTSTEP. It has an icon on the screen.

Browser: A display tool for navigating tree structured data; in particular, the browser in the File Viewer navigates the file system.

Directory: The standard UNIX term for a Folder.

Dock: The column of icons found at the right of the screen.

Document: A file or folder that contains information used by an application as a unit.

File Viewer: A window of the Workspace that is used for finding documents in the file system.

Folder: A collection of folders and files in the file system (see also directory).

Menu: The buttons provided in the upper left that are used to control an application.

Panel: A window that provides some special functionality. It cannot normally be resized.

Recycler: The document disposal represented by the icon at the bottom of the Dock.

Services: Capabilities that are provided by one application to another.

Shelf: The space at the top of the File Viewer where frequently used files and folders may be referenced.

Window: A rectangular area of the screen that often may be independently moved, resized, or closed. It often contains a view of a document.

Workspace: The program that manages the display, launches applications, and provides File Viewers to manage the file system.

REFERENCES

Simson L. Garfinkel and Michael K. Mahoney, *NEXTSTEP Programming Step One: Object-Oriented Applications*. Springer-Verlag, New York, 1993.

Alex Duong Nghiem, *NEXTSTEP Programming: Concepts and Applications*. Prentice-Hall, Des Moines, IA, 1993.

Michael B. Shebanek, *The Complete Guide to the NEXTSTEP User Environment*. TELOS: Springer-Verlag, New York, 1993.

User's Guide. NEXT Computer Inc., Redwood City, CA, 1992–1993.

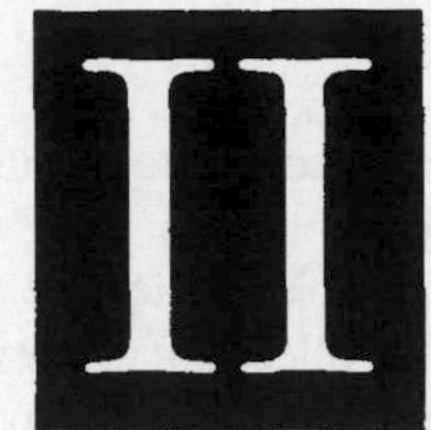

Structured and Object-Oriented Programming

Before the widespread use of higher level programming languages, the only data types were what was directly implemented in the hardware of a computer. These would be integers, floating-point numbers, and some sort of field representing characters. Even fundamental ideas such as strings or arrays were only in the programmer's mind; a contiguous block of items in memory.

Early programming languages reflected little more than these data types. Their main purpose was to allow programming in a more natural manner, and to simplify and standardize the management of looping and subroutines.

As programming languages evolved, so did abstractions of data into data types. One particularly important development was the record or structure, which was a data type consisting of one or more other data items, which could, in turn, be of mixed types. A record could be created that contained a character string for a name field, a number to hold account information, and another string for address. The operations on the fields in this record structure were the usual ones supported in the language. There would be minor modifications used to select and operate on a field from a record.

If we can bundle the data and the operations that represent and manipulate a new type, then we call that an object. We think in terms of telling the object what we basically want, and it is the responsibility of the object to perform these operations customized to its own design. Some early object-oriented languages referred to message-passing as their fundamental operation.

Objects lend themselves to object hierarchies. The new class of objects derived from a previous class of objects inherits all the properties of its parent class and may add some new data or operations. It might also replace some of the operations of its parent class.

Object-oriented systems accomplish two goals. You can express new data types to be filled out in new ways. Because each object is self-contained, the management of the information and operations the object represents can be updated without affecting other parts of the system. This approach helps in the management of large, complex information systems.

This section of the handbook is divided into two discussions. The first covers the conventional strategies of programming in C, and the second expands these ideas to encompass some of the major techniques of using object-oriented programming tools.

13

C Programming

James Kiper
Miami University

Charles Ames
Jet Propulsion Laboratory

13.1 CURRENT PROGRAMMING PRACTICE

The C programming language is currently one of the most popular procedural programming languages. In the first 10 sections of this chapter, we will present the major features of C and illustrate these features with several sample programs. In the last section, we will discuss various C programming environments. Before presenting the specifics of C, we first provide a brief summary of current programming practice.

Development of procedural software generally centers around **structured programming**. Historically, an awareness of the need for structure in software arose in the 1950s and 1960s as programs grew larger and more complicated. Dijkstra's classic article "Goto's Considered Harmful" (Dijkstra, 1968) was the catalytic event that brought this issue to consensus among researchers and practitioners. It became apparent that the indiscriminate use of the "goto" statement produced software that was exceedingly difficult to understand, verify, or modify. Even newer paradigms, e.g., object-oriented programming, use structured programming as their basis.

Stated in a positive way, the essence of structured programming is that structured control constructs are used without exception. A **control construct** is a syntactic unit of a programming language that determines the sequencing of instructions. (A "goto" statement is a control construct that causes control to branch to some other arbitrary point in the program.) A "structured" control construct is one that has a single entry point and a single exit point. That is, there is one statement that is the first in the group to be executed; when this group of instructions is finished, the next instruction to be executed is invariant.

0-8493-2530-7/96/$0.00+$.50

Some examples can help to understand the meaning of this definition of structured programming constructs. Three important structured control constructs are if-then-else, while loop, and functions. The if-then-else is a control structure for encapsulating a decision. The flow chart fragment in Figure 13.1 illustrates the "if-then-else" construct.

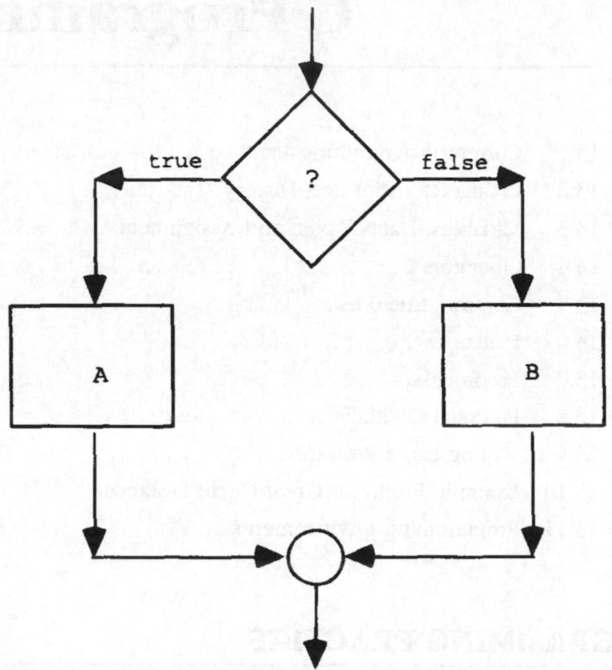

FIGURE 13.1 Flow chart for if-then-else.

If the condition specified in the box labeled with a "?" is true, then the statements in the box labeled A are executed; if this condition is false, those statements in the box labeled B are executed. Notice that the statements in box A or B may be quite complicated. In fact, there may be another structured control construct nested in these statements. Furthermore, note that the condition is always the first of these statements to be executed, i.e., a single entry point and that, regardless of which branch was taken, true or false, control comes back together when these are finished, i.e., a single exit.

The flow chart for the while loop (Fig. 13.2) has the same properties: single entry, single exit.

A function is a way of organizing a program into smaller pieces to aid understandability. A function is a section of code that has a name, and that can be called multiple times. The flowchart in Figure 13.3 shows the effect of a function call.

There are several advantages of structured programming. The first is established by the "Structured Programming Theorem" as proved by Bohm and Jacopini (1966). The essence of this theorem is that any program, i.e., "effective procedure," that can be represented algorithmically can be written with three primary structured control con-

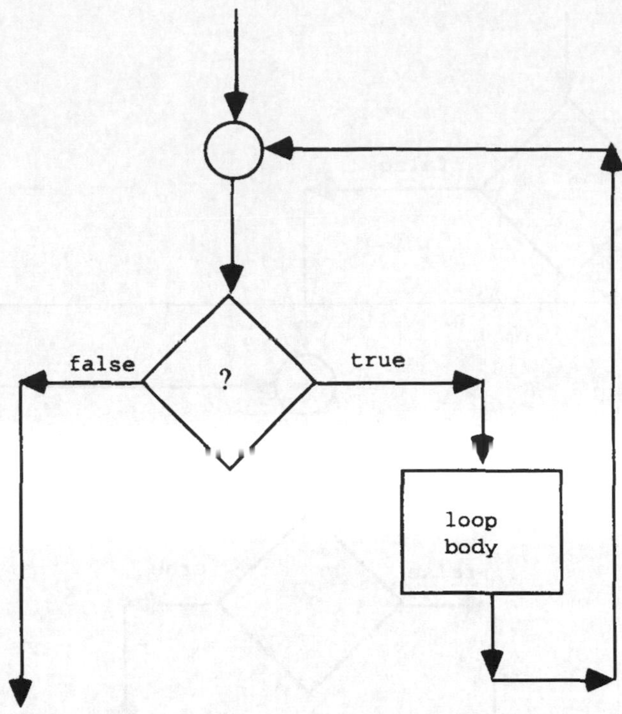

FIGURE 13.2 Flow chart for the while loop.

structs: sequencing, selection, and repetition. Sequencing means the default control in which one statement is executed after another. This is obviously structured. The if-then-else described above is an example of selection; the while loop implements repetition. The Structured Programming Theorem establishes that these are the only forms of control necessary. Other constructs are often useful, but are not necessary. In particular, jumps to arbitrary program locations are not necessary. This theorem is accepted with little controversy now. However, it was quite important historically to motivate practitioners to move to this programming paradigm.

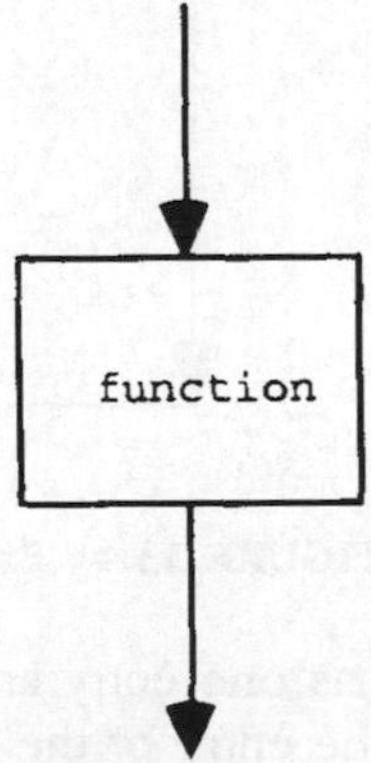

FIGURE 13.3 Flow chart for function call.

Again, you should notice that each of these control constructs has a single entry and a single exit.

One of the important properties of structured control constructs is that they maintain their structure property when nested. For example, a while loop can be nested inside an if-then-else as illustrated in Figure 13.4. Or an if statement, function calls, and a while loop can be nested inside a separate while loop.

The value of structured programming lies in the fact that each control component used in a program can be constructed and tested independently of the remainder. These pieces can be combined subsequently in interesting ways. Since each construct

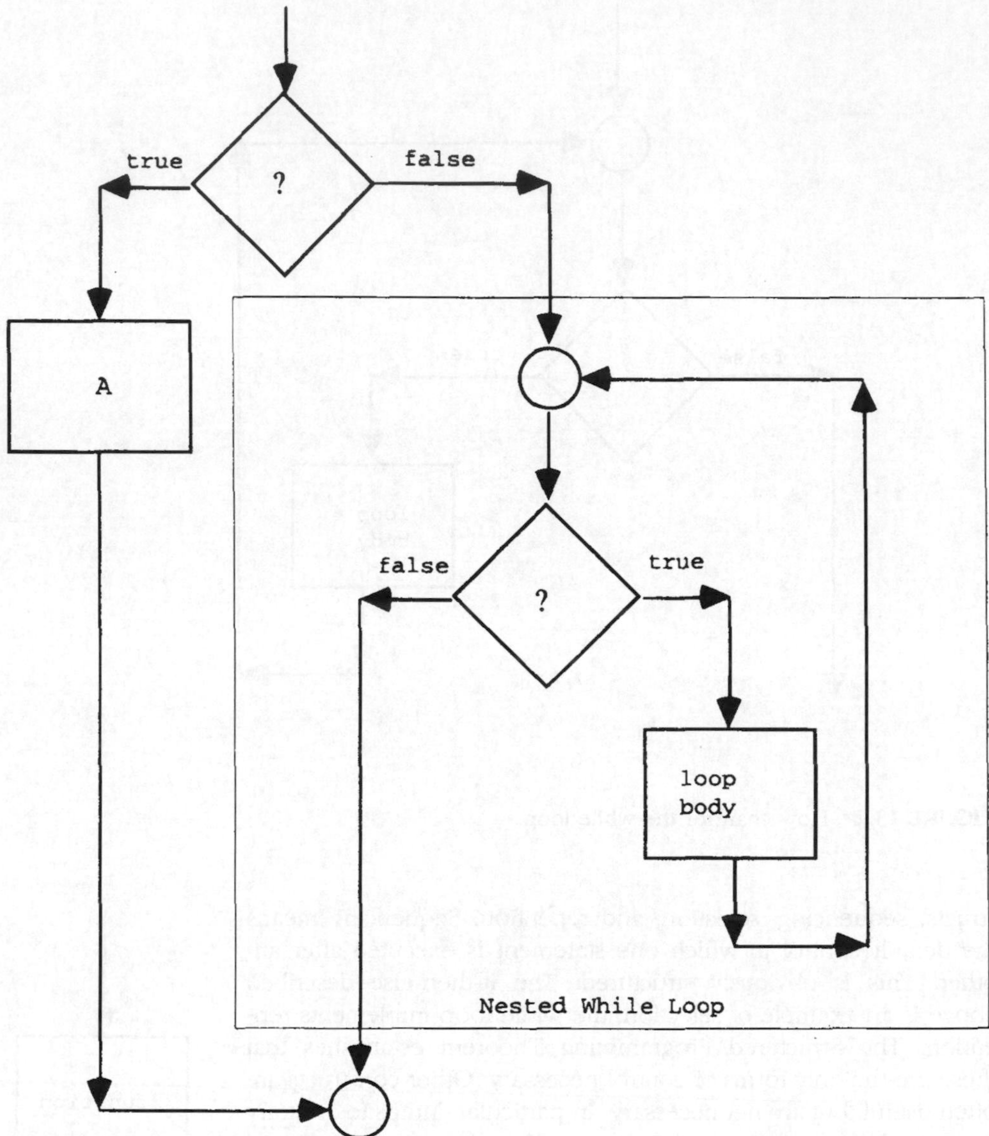

FIGURE 13.4 Example of nested constructs.

has one entry and one exit, they can be sequenced by connecting the exit of one to the entry of the next. Furthermore, they can be nested as illustrated above. Most importantly, either of these forms of combination produces a new control construct that is also structured. A mathematical theory of structured programming has been produced by Linger, Mills, and Witt (1979).

In summary, structured programming has proven to be an important paradigm for developing software in procedural programming languages that is sufficiently powerful to capture any algorithm, and that can be used to produce understandable, verifiable, and modifiable code for complex algorithms. The C programming language is one of a number of programming languages that directly support structured programming.

The field of software engineering has seen further developments since the advent of structured programming. One of the more effective of these is the use of **abstract data types** (ADT) (Guttag, 1977; Liskov and Zilles, 1974). This refers to the encapsulation of data and related operations into a single syntactic unit. This paradigm does not preclude the use of structured programming for the functions and operations necessary for each ADT. The use of ADTs is another level of organization applied to the program.

Object-oriented programming (OOP) has become quite popular in recent years. This paradigm centers program development on "objects." An object is synonymous with representation of an ADT, that is, a collection of logically associated data and related operations. OOP adds another layer of organization. Each object is an implementation of a "class" that specifies the data and operations for all objects in that class. Classes can be combined in various ways, e.g., composition and inheritance. Objects communicate with one another via "messages" (Booch, 1986, 1994; Coad and Yourdon, 1991). OOP is a significant paradigm shift in software design and implementation. However, OOP builds on the power of structured programming rather than replacing it. C can be use to program with ADT's or in an OOP style, although it provides no direct support for either paradigm. Other languages, e.g., C++ and Objective C, provide syntactic and semantic extensions to C in these areas.

13.2 C CHARACTERISTICS AND HISTORY

C has quickly evolved to become one of the most commonly used programming languages from its birth at Bell Labs in the late 1960s. C was developed by Dennis Ritchie as an outgrowth of an internal language called BCPL. It was not intended to be a commercial product, but rather a language to fit a need at Bell Labs. Its use as the base language for the Unix operating system was the start of a synergistic relationship that established both as industry standards.

C continued to evolve at Bell Labs as its influence outside expanded. The popularity and widespread use made the need for a standard apparent. An ANSI standard for C was approved in 1989 (ANSI, 1989).

C's initial popularity was enhanced by its utility in systems programming. Early versions of C had an easily used interface to the Unix operating systems. C provides several features that aid systems programmers: bitfields, shift operators, etc. Current uses of C have expanded from systems programming to general applications programming of all types.

One of the central characteristics of C is that it is a typed language. All identifiers have a **type** assigned to them, either by the language itself in the case of reserved identifiers, or by the user in the case of user-defined names, like variable names or function names. In general, a type is a description of the kinds of values that a particular identifier can hold, and the operations that can be performed on those data. The philosophy of types in C is somewhat less rigorous than this general definition. The fact that all data are stored as strings of bits is more obvious in C. C types give alternative ways of viewing those bit strings. For example, a string of bits can be viewed as an integer, or can be interpreted as the ASCII code corresponding to that integer. Types must be associated with an identifier prior to the first use of that identifier. In the case of variables, this association occurs during a declaration. We will discuss C's types in the next section.

As described in the previous section, C supports structured programming through its control constructs: sequencing as a default, decisions through the if and if-else structures, and repetitions through for loops, while loops, and do loops.

C is considered to be a **block** structured programming language. A block of statements, i.e., a sequence of statements surrounded by braces { }, can appear anywhere that a single statement is legitimate. Variables can be declared at the beginning of such a block. These are then local variables to that block, and cannot be referenced outside the block.

Another important characteristic of C is its dependence on the ANSI C **library.** A library is a collection of prewritten and translated functions to provide commonly needed actions. For example, there are no input or output statements built into C. All input and output occurs through library functions. The ANSI C standard defines library functions for many purposes: standard input and output functions, math functions, string manipulation function, various time functions, etc. To be an effective C programmer, it is not sufficient merely to know the statement types provided by C; it is vital to understand C's use of library functions and to know what type of functions are available in it.

C has been used with considerable effectiveness in the development of large programs. (The development of large programs is not only quantitatively, but also qualitatively different from developing small programs.) C allows a single program to be divided across several files allowing teams of programmers to work independently on sections of that program. These files containing portions of a program can be compiled separately to eliminate syntax errors before being combined during the link phase. (There are some restrictions on what portions of a program can appear in a single file and be successfully compiled.) Section 13.8 discusses this topic in more detail.

C is an expression-oriented rather than statement-oriented language. Most programming languages require a sequence of complete statements to form a syntactically valid program. In C, a sequence of expressions can form a valid program. An example of a C statement is "sum = sum + test," which adds together the value of variable sum and variable total and returns the result to variable sum. An example of a C expression is "students < 100," which compares the contents of variable *students* to 100 but changes no variables. The differences between a statement-oriented and an expression-oriented language are subtle. However, this is the source of some of the power of C, and is also a cause of some of the potential crypticness of C programs.

The remaining sections of this chapter will present an introduction to the C programming language. A complete and comprehensive coverage of C would require much more space. The interested reader is referred to one of the many C programming books, the classic of which is authored by Kernighan and Ritchie (1988; see also Troy and Kiper, 1989).

13.3 VARIABLES, DATA TYPES, AND ASSIGNMENT

C Format

C is free format, that is, statements can begin at any position on a line, can span several lines, or can be placed on the same line as another statement. White space (blanks, tabs, or newlines) can be placed anywhere in a statement except in the middle of a keyword or identifier.

Comments

Comments are begun with the characters /* and ended with */ and can be placed in a C program at any point that white space is permitted.

Variable Names

Variable names in C are unlimited in length although some compilers recognize only the first 32 characters of the name. Names must begin with a letter, which may be followed by a sequence of letters, digit, and underscores. C is case sensitive, so the variable names **sum**, **Sum**, and **SUM** are all valid and are distinct names. Other examples of valid variable names are **total4**, **final_result**, **initialVoltage**, **a1234**, **R54ET**. Of course, some of these are better than others from the point of view of style and understandability.

Basic Types

As discussed above, all variables have to be associated with a type, i.e., declared, before being used. C has several built-in *simple* types, and has the capability of user-defined types. The simple types are (1) *int*, *short*, *long*, *char*, (2) *float*, *double*, *longdouble*, and (3) *pointer*.

Integer Types

The first group discussed here is the collection of integer types. Each of these signifies that variables of this type will hold an integer value and can be combined with the integer operations. (Operations are described in a subsequent section.) The various varieties indicate the amount of space in the computer's memory to be reserved for each such variable, and, consequently, the potential range of values that can be represented. The most common is the type *int*. The amount of space reserved for a variable of type *int* is generally the size of a computer word on the underlying operating system and hardware, which is typically 16 bits (binary digits) on a personal computer and 32 bits on a larger machine (workstation or mainframe). The other varieties either have fewer bits (*short* or *short int*) or more bits (*long* or *long int*). The type *char* is used to indicate that it represents a single character value. This type is considered an integer type in C because the value stored in a *char* variable is actually the integer code representing that character, typically the ASCII code. The C standard prescribes the following relationships as given by the predefined function *sizeof*( ):

```
sizeof( char ) <= sizeof(short int) <= sizeof (int) <= sizeof
(long int)
```

Figure 13.5 gives valid declarations of integer variables in C. Notice that each of these declarations ends with a semicolon. In fact, every statement in a valid C program ends with a semicolon. These examples also demonstrate that multiple declarations can be given in a single statement. In the second line of this example, both **initialNumber** and **final_Number** are declared to be of type *int*.

```
int count;
int initialNumber, finalNumber;
char ch, answer;
short int small_number;
long int large_number;
```

FIGURE 13.5 Valid integer declarations in C.

The example declarations in Figure 13.6 show some other alternatives:

```
unsigned int employeeCount = 0;
unsigned short int answer = 7;
char answer = 'Y';
```

FIGURE 13.6 Alternate forms of integer declarations in C.

An integer type can be preceded by the keyword "unsigned" to indicate that it should allow only positive integers. These examples also demonstrate that variables can be initialize. The variable **employeeCount** is given the initial value 0; the *char* variable **answer** is given the initial value 'Y'. This latter initialization actually places the integer code for the letter 'Y' in that variable's memory location. A character literal such as 'Y' is always surrounded by single quote marks (apostrophes).

Floating Point Types

The next class of simple types is the floating point group that represents real numbers: *float*, *double*, and *long double*. Variables with one of these types can hold values that contain a decimal point. This includes numbers represented in scientific notation. In the computer's memory these are recorded by storing an exponent and a fractional part (mantissa). A variable of type *double* has a larger mantissa than a *float*; a variable of type *long double* has a larger exponent and a larger mantissa than a *float*. Figure 13.7 shows some valid declarations of floating point types, including some initializations. The last example shows the form used for scientific notation. Figure 13.8 gives some examples of assignment to these variables.

```
float length;
float radius = 5.056;
double pi = 3.14159;
long double nationalDebt = 3.00e12;
```

FIGURE 13.7 Valid float declarations in C.

```
float area;
area = pi * radius * radius;
nationalDebt = nationalDebt * 2;
length = 49.45;
```

FIGURE 13.8 Valid float variable assignments.

Pointer Type

One of the most powerful features of C, and one that is most difficult to master, is the **pointer** type. A variable that is declared as a pointer is one that holds the address of another variable. Thus, the pointer variable "points" to another memory location. Although in C every pointer variable requires the same amount of space in the computer's memory, each is declared to point to a particular type of value. That is, we can declare a variable to be a pointer to an integer, or a pointer to a float, etc. A pointer declaration is indicated with an asterisk *. Figure 13.9 gives examples of pointer variable declarations.

```
int * ptr_int;
float * ptr_float;
unsigned long int * longPtr;
```

FIGURE 13.9 Valid pointer variable declarations in C.

Pointers form a central part of C because of their close relationship to arrays as discussed below. Pointers are also useful to build generic types and are necessary in construction and manipulation of dynamic data structures.

Assignment of values to pointers is complicated by the fact that the assignment operator can be used to change the address stored in a pointer variable, and it can be used to change the value that the pointer variable points to. In using pointer variables, the address operator **&** is often useful. This operator gives the address of the variable it modifies.

The statement "pi = &i;" causes the address of variable **i** to be placed in pointer variable **pi**. Thus **pi** "points to" variable **i**. In Figure 13.10, the box labeled "Memory Contents 1" diagrams a possible memory configuration that would result from these first three statements.

The assignment statement "j= *pi;" causes the value of the location to which **pi** points to be copied into **j**. The asterisk * here is the dereferencing operator that obtains the contents of the memory location pointed to by the pointer variable. This has

Variable Name	Memory Location	Memory Contents 1	Memory Contents 2
	998		
i	1000	7	13
j	1002	0	7
pi	1004	1000	1000
	1006		

```
int i = 7;        j = *pi;
int j= 0;         *pi = 13;
int * pi;
pi = &i;
```

FIGURE 13.10 Pointer variable assignment.

the effect of giving **j** the value 7 since **pi** points to **i** and **i** has the value 7. The last statement "*pi = 13;" causes the value 13 to be placed into the location that **pi** points to. Since **pi** points to the location of **i**, this causes **i**'s value to become 13. The result of these final two assignments are displayed as "Memory Contents 2." This brief discussion can give you an understanding of the complications that pointers can present to a novice programmer.

Initialization

Initialization is a useful way of giving a value to a variable when it is declared. One way to change this value when the program is run is through the assignment operator =. This operator appears to the right of a single variable (called an **l-value**). The value of the expression to the right of the assignment operator is placed in the memory location of the **l-value**. The expression on the right can be a simple constant or variable, or a complicated expression of values, variables, functions, and operators. (Functions and operators will be presented in subsequent sections.) Consider the simple examples in Figure 13.11.

```
int increment = 2; /*declarations and initialization */
int count;
char answer = 'Y';
double pi = 3.14159;
int * countPtr = &count;
count = 7;               /* assignments */
count = count + increment;
answer = 'N';
```

FIGURE 13.11 Assignments in C.

Notice that the variables **increment**, **answer**, **pi**, and **countPtr** are given initial values in their declaration statement. The variable **countPtr** is an integer pointer that is made to point to the memory location occupied by variable **count**. The first assignment statement "count = 7;" places the value 7 into the memory location for the *int* variable **count**. The next assignment statement then adds 2 to this value so that 9 is stored as count's value. Then, the *char* variable **answer** is given the value 'N'.

Aggregate Types

In addition to the simple types, C allows *aggregate* types to be built. The two most important internal aggregate types are *arrays* and *structures*.

Arrays

Most procedural programming languages allow the use of **arrays**—a collection of storage locations with a single name whose components can be referenced by indexing. In C, every array declaration contains three pieces of information: the array's name, its size, and its base type. Square brackets indicate the declaration of an array (Fig. 13.12).

```
float salaries[100];
char name[30];
long int lengths[1000];
```

FIGURE 13.12 Valid array declarations in C.

The first example declares an array called **salaries** that has 100 elements each of which is a **float**. If floats are represented in 4 bytes (32 bits) as they are on many machines, then this array occupies 400 bytes. That is, *sizeof(salaries)* == 400. The next example declares an array of 30 characters called **name**; the last is an array of 1000 long integers called **lengths**. Thus, *sizeof(name)* == 30 since characters occupy one byte; *sizeof(lengths)* == 8000 if a *long int* requires 8 bytes.

Multidimensional Arrays and Array Initialization

C allows multidimensional arrays and array initialization as demonstrated in Figure 13.13.

```
float voltages[5] = { 0.0, -1.0, 0.5,-0.5, 1.0 };
char husbandName[30] = { 'G', 'e', 'o', 'r', 'g', 'e', ' ', 'W',
                'a', 's', 'h', 'i', 'n', 'g', 't', 'o', 'n', '\0'};
char wifeName[30] = "Martha Washington";
float grades[100] [3];
double dimensions[4][4][4] = { {1.0, 1.0, 1.0, 1.0}
                                 { 2.0, 2.0, 2.0, 2.0}
                                 { 3.0, 3.0, 3.0, 3.0} };
```

FIGURE 13.13 Array declarations and initializations.

The first example declares an array of 5 *floats* called **voltages** and initializes these 5 values. The next initializes a character array to the string "George Washington" by specifying each character one at a time. In this example there are fewer initial values than the size of the array. This is valid; the remaining array values are left uninitialized.

Strings in C

When using a character array, it is common practice in C to make this array a **string**. A string in C is an array of characters whose last value is the null character '\0'. The null character is a special character whose underlying (ASCII) code has the value 0. This special value is used to indicate the end of the string. In this way, this array can be used to hold any string value whose length is less than 30, the length of the array. [Special character values are represented in C by means of the backslash\. There are several other special values that will prove useful to you: '\n' represents end of line, '\t' represents a tab. (Table 13.1).]

The third line of Figure 13.13

TABLE 13.1 Escape Characters

\0	Null byte
\n	Newline
\t	Tab

```
char wifeName[30] = "Martha Washington";
```

gives an alternative and easier method of initializing a character array to a string. A sequence of characters surrounded by quotes is defined as a string. It is automatically null-terminated.

The next declaration

```
float grades[100] [3];
```

demonstrates the declaration of a two-dimensional array called **grades**. You should think of this as a table with 100 rows and three columns. The last example above declares a three-dimensional array and initializes the entire array. Notice that the initialization is bracketed into three rows with four values in each row. These could be listed in one long list with no intervening braces, although that form is less readable.

Arrays and Pointers

An array name is identical to a pointer to that array with one exception: the name of an array is not an l-value. This similarity to pointers has many implications that will be pointed out in more advanced discussions of C. The fact that an array name is not an l-value has an important implication in value assignment. An array name cannot appear alone on the left side of an assignment operator. That is, a single assignment operator cannot be used to give an entire array values at one time. (Note that the use of the = sign in a declaration is for initialization, not assignment.) However, the assignment operator can be used to assign values to individual elements of an array.

Array Indexing

Individual elements of an array are referred to by means of the array indexing operator [ ]. Array indices in C always begin with 0. Thus, valid indices for the array **voltages** above are 0, 1, 2, 3, and 4. The C programmer should be warned that C does not check indices to determine if they are out of range. It is possible to index beyond an array's allocated space although the results can be disastrous. Figure 13.14 gives valid assignments of values into arrays as declared previously. Notice particularly the last two, which are assignments into the first and last elements of the two-dimensional array called **grades**. The third line indicates that an array index can be an expression whose value is calculated. This assignment statement gives **voltage[2]** a new value. An array index can be any valid expression that evaluates to an *int*.

```
int index = 1;
voltage[0] = 2.0;
voltage[index + 1] = voltage[0] - 0.5;
grades[0][0] = 95.0;
grades[99][2] = 100.0;
```

FIGURE 13.14 Assignments to array elements.

Structures

Like an array, a **structure** is a collection of data. It differs from an array in two ways. First, a structure can contain items of data that are not all the same type. Second, the components of a structure, i.e., fields, are referred to by name rather than through an index. A structure is declared with the keyword "struct" as shown in the following Figure 13.15.

```
struct empRec
{
   char name[30];
   float salary;
   long int id;
} employee;
```

FIGURE 13.15 A valid structure declaration in C.

In this example, **employee** is a *variable* that is declared as a structure with three fields. If a character occupies one byte (8 bits), a *float* four, and an *int* two bytes, as is typical, then the *sizeof (employee)* gives the value 36. The identifier "empRec" in the first line of the declaration is the *structure tag*, which is optional, but useful for declaring other structures with this same type. For example, "struct empRec emp2;" declares another structure **emp2** that has the same type as the variable **employee.**

To refer to individual fields, the dot notation is used. The salary field is accessed with the notation "employee.salary." These field names are not variable names, and cannot be used alone. They are valid only when used in the dot notation with the appropriate structure name. A structure name is an l-value, which means that it can be used on the left side of an assignment statement to give the entire structure a value at one time. Thus, the statement "emp2 =employee;" is valid. This result in the values of all three fields being copied from the variable **employee** to the corresponding fields of **emp2**. This type of assignment is possible only when the two structures have the same type.

Structures can be nested and combined with arrays as shown in Figure 13.16.

```
struct addressRec
{
   char street[30];
   char city[30];
   char state[3];
   long int zipCode;
} employee;
struct empRec
{
   char name[30];
   struct addressRec address;
   float salary;
   long int id;
};
struct empRec company[1000];
```

FIGURE 13.16 Nested structure declaration.

Here, there is one variable, **company**, that is an array of 1000 **empRec** structures. The **empRec** structure has four fields. One of these, **address**, is a structure type itself. The following are all valid references to various portions of this array:

```
company[0],
company[i].name
company[i+1].address,
company[999].address.street
company[1].address.state[0]
```

Type Modifiers

Typedef

C type modifiers can be used to adapt types in some useful ways. The "typedef" keyword is used to create a new name for an existing type. The first example in Figure 13.17 allows the name "boolean" to be an alias for *int*. (C has no boolean type.) Then the variable "flag" is declared of type boolean. This variable is actually an *int*, but it is clear that the programmer intends to use this to store true and false values. The second *typedef* example creates a new name for this structure. This new name is then used to declare the variable **largeRectangle**.

```
typedef int boolean;
boolean flag;
typedef struct
{
    float length;
    float width;
    double area;
} rectangle;
rectangle largeRectangle;
```

FIGURE 13.17 Examples of type def.

Static and Extern

The keyword "static" is used before a local variable to give that variable **program duration**. This means that this variable will retain its value between activations of the block that contains that variable. The other alternative is "auto" for automatic, which means that a fresh copy of a local variable will be created each time that block is entered. The default is automatic so that this keyword is rarely used.

A **file variable** is one that is declared outside any block or function. (These are sometimes called global variables.) Such a variable can be used in multiple files by giving it **external linkage** with the keyword "extern." The defining declaration is placed in one file outside any functions. Then, in any other files that need to refer to that variable, an identical declaration is placed with the keyword "extern" attached prior to that declaration. A file variable can be given **internal linkage**, meaning that it cannot be used outside this file, by preceding it with the keyword "static." This allows a certain amount of information hiding. (Notice that the keyword "static" has a different meaning for local, i.e., block, variables than for file variables. In the first case, it changes to program duration; in the second, to internal linkage.)

Register

C allows the keyword "register" as a suggestion to the compiler that it keep this value in a hardware register for faster operation. This is just a suggestion, and may be ignored by a compiler.

Const

The ANSI standard introduced the keyword "const" to create named constants (Fig. 13.18). In the next example, the variable **true** is assigned the value 1, and **false** is assigned the value 0. These values cannot be modified—that is, these variables are not l-values.

```
const boolean true = 1;
const boolean false = 0;
```

FIGURE 13.18 Valid constant declarations in C.

Type Casts

Generally, C requires that variables involved in an assignment statement, or in an expression as operands of operators (see section 13.5) have the same type. However, in certain cases, the compiler automatically converts from one of the integer types to float or double. This is an implicit **cast**. The programmer can explicitly change from one type to another by means of an explicit cast. An explicit cast is the name of a type in parentheses that precedes a variable. The example in Figure 13.19 explicitly converts the integer value –13 to its floating point equivalent. In Figure 13.20 there is a loss of accuracy. After the assignment, i has the value 7.

```
int i = -13;
float f, g;
f = (float) i; /* explicit cast */
g = i; /* implicit cast */
```

FIGURE 13.19 Cast to a float.

```
f = 7.8
i = (int) f;
```

FIGURE 13.20 Cast to an *int*.

13.4 OPERATORS

C has a rich set of **operators** for manipulating data in various forms. This section will present the most important operators in four categories: arithmetic, relational, logic, and assignment.

Arithmetic Operators

The arithmetic operators include those that are common among procedural languages: addition +, subtraction –, multiplication *, and division /. When two integers are divided with / the result is an *int*. That is, the fractional part of the result is lost. For example, 15/4 is 3 and 15.0/4.0 has the value 3.75 since these are floating point literals. The *mod* operator % can be used only on integer types, and produces the remainder of the division so that 22 % 4 gives 2. There is no exponentiation operator in C, although there is a standard mathematical function *pow()* in the standard library.

Unary Operators

The **increment** and **decrement operators** are two unique characteristics of C. The increment ++ operator adds 1 to the value of an integer variable; decrement – – subtracts 1 from that value. For example, if the *int* variable **x** has the value 6 and *int* variable **y** has the value 1994, then after **x++**, **x** has the value 7, and after **y––**, **y** has the value 1993. Each of these operators can be used in two forms: either before or after the variable, e.g., **++x** or **x++**. There is a difference between these two when the value is used in a larger expression. For example,

```
x = 7;
y = ++x;
```

gives both variables the value 8. However,

```
x = 7;
y = x++;
```

assigns the value 7 to **y** before incrementing **x**.

These operators can be used to combine any of the integer types or to combine any of the *float* types (except for %, which can be used only with *int* types). When used in a mixed type expression, e.g., int + float, the integer value is automatically converted (cast) to *float*, and the result is type *float*.

Relational Operators

The relational operators are used to compare values of basic types: <, >, >=, >=, !=, ==. The value of expressions using these operators is 0 if their result is false, and 1 if true. (In general, a value of 0 is interpreted as false, and any non-zero value is true in C.) If variables x and y have type *int*, with x having the value 7 and y the value 20, then the relational operators will have the values given in Table 13.2.

These operators work in analogous ways for *float*-type values. They can be used on expressions, i.e., constants, variables, and function calls combined with arithmetic, relational, or logical operators. Thus, with the declarations "float f, g, h; int i, j;," the following are all valid uses of relational operators:

```
2*f +3.4/g < 34.56 * ( i - j)
-f + sqrt( g*g - 2.0*f*h )/ 2.0 > 0
```

Each of these expressions has either a true (1) or false (0) value depending on the values of these variables.

Logic Operators

Logic operators are logical and &&, logical or ||, and logical not!. These operators can be combined to form more complicated logical conditions. Such logical expressions are most often used to form conditions in control structures. The next section will provide several examples of these. The semantics of logical and (&&) are that the result is false (0) if either or both of the operands have a false value; if both operands are true, i.e., nonzero, then the logical and gives true. The logical or operator (||) is a binary one whose value is true if either or both of its operands are true, and is false when both are false. Logical not (!) is a unary operator that converts false to true or true to false.

Assignment Operators

C provides a set of useful **assignment operators** to give a new value to a variable. The most common form of assignment is =. This has the effect of copying the value from the expression on the right-hand side of this operator to the variable on the left-hand side. This leftside variable must be an l-value. (The only kind of variable that we have discussed that is not an l-value is the array.)

Other assignment operators are formed by combining the basic assignment operator with one of the arithmetic operators: +=, –=, *=, /=, %=. The meaning of these is illustrated in Figure 13.21. For each line of this example, the right statement and the left have the same effect.

TABLE 13.2 Comparison operators in C (with x ==7, y == 20)

Operation Name	Operation	Resulting Value
Less than	x < y	1
Less than or equal	x <= y	1
Greater than	x > y	0
Greater than or equal	x >= y	0
Equal	x == y	0
Not equal	x ! = y	1

```
x += 2;            x = x + 2;
f -= 34.567;       f = f - 34.567;
x *= 3*y - 2;      x = x * (3*y - 2);
f /= 7.7;          f = f / 7.7;
x %= y + 3;        x = x % (y + 3);
```

FIGURE 13.21 Examples of C assignment operators.

The address operator & and the dereferencing operator * were described in the previous section. Note that the symbol * is used for multiplication when it is a binary operator and dereferencing when it is unary.

These operators follow the normal **precedence** and **associativity** generally used by programming languages. Table 13.3 list these operators in precedence order from highest to lowest.

13.5 CONTROL STRUCTURES

C permits the use of structured programming by providing structured control constructs as described in Section 13.1. The default control construct is sequencing—meaning that statements are normally executed in the order that they are written.

Selection (if) Control Structures

Selection or decision control structures are the *if*, the *if-else*, and the *switch* structures.

If Statement

The if control statement consists of two major pieces: the condition and the statement.

```
if ( condition ) statement;
```

The condition is a parenthesized logical condition that is, in the case of C, an expression that has an integer value. When this statement evaluates to true (nonzero) then the statement is executed. If the condition is false, this statement is skipped. Consider the following example: "if (x > 0) z = y / x;" The condition is (x > 0) and the statement is "z = y/x;." The condition can be a simple condition as shown, or a complex

TABLE 13.3 Operator Precedence and Associativity

Highest to Lowest Precedence	Associativity
&, * (dereference)	Right-to-left
! ++ --	Right-to-left
* / %	Left-to-right
+ –	Left-to-right
== !=	Left-to-right
&&	Left-to-right
||	Left-to-right
< <= > >= != = += *= –= /= %=	Right-to-left

expression that even has side-effects. The statement can be a compound statement which is a list of statements surrounded by brackets. Figure 13.22 shows the conversion of lower case vowel to upper case by using the library function *toupper()* and prints the letter out with the library function *putchar()*. (These library functions will be discussed in the next section.)

```
char ch;
ch = 'a';
if ( ch == 'a' || ch == 'e' || ch = 'i' || ch == 'o' || ch = 'u'
)
{
  ch = toupper( ch );
  putchar( ch );
}
```

FIGURE 13.22 C conditional statement (if).

If-Else Statement

The if-else control statement adds another piece to the if statement. If the condition is false, the statement or compound statement given after the keyword "else" is executed. In Figure 13.23 **average** is computed by dividing the variable **sum** by **count**. However, if **count** has the value 0, division will result in an error. In this case, **average** is given the value 0.

```
float average, sum;
int count;
if ( count != 0) average = sum / count;
else average = 0.0;
```

FIGURE 13.23 C conditional statement(if-else).

Switch Statement

The switch statement allows a choice among multiple conditions. The value of the condition (given in parentheses after the keyword "switch") is evaluated and matched with the value of one of the cases. Execution begins at the matching case. Each case value must have type *int*; as before the condition is any *int* valued expression, i.e., a logical expression.

```
switch( condition )
{
  case value : statement;
  case value : statement;
...
  default : statement;
```

The example of Figure 13.24 uses the output function *printf()* and the input function *scanf()*. If the user enters the value 2 when this program segment is run, then the function *modify()* is called. Notice that in each condition the *break* statement is used to terminate. This statement causes control to leave the case statement and go to the next statement that follows the closing brace. The C switch statement has the interesting characteristic that once a case has been selected, control progresses through all remaining cases. Thus, it is quite common to end each case with a "break." The "default" condition at the bottom of the structure is the case that is chosen if none of the other cases is matched.

```
int choice;
printf("Menu\n");
printf("1. Insert record\n");
printf("2. Modify record\n");
printf("3. Delete record\n");
printf("4. Quit\n");
printf("Choose 1 - 4");
scanf("%d", &choice)
switch (choice)
{
   case 1: insert(); break;
   case 2: modify(); break;
   case 3: delete(); break;
   case 4: break;
   default: printf("Error in choice.\n");
}
```

FIGURE 13.24 C conditional statement (switch).

Repetition Statement

There are several repetition statements in C. The two most common are the *while loop* and the *for loop*.

While Loop

The while loop has the following form:

```
while (condition) statement;
```

The condition is a logical condition, i.e., *int* valued expression similar to that described for the if statement. The statement can be replaced by a statement block. This statement, or statement block, is repeatedly executed as long as the condition expression has a true value. To be precise, the condition is checked. If it is false, the body is skipped entirely and execution continues at the next statement. If true, the body is executed. Then the condition is evaluated again. The body should have changed some values in the condition. If the condition is now false, the loop is finished. If true, the body is executed again. This process continues as long as the condition evaluates to true.

```
int x, sum;
x = 100;
sum = 0;
while (x > 0)
{
   sum += x- -;
}
```

FIGURE 13.25 While loop example.

The loop in Figure 13.25 finds the sum of the series 100 + 99 + 98 + 97 + . . . + 4 + 3 + 2 + 1. It repeats as long as **x** has a positive value. Each time the loop statement (the loop body) executes, it decrements the value of **x** by 1.

It is quite common to use a *while* loop to read from a file (Fig. 13.26). This loop uses input/output functions from the standard C library to open a text file, and to read its contents character by character, printing these characters to the screen until the file is empty.

```
FILE * fp;
char ch;
fp = fopen("test.dat", "r");
while(!feof(fp))
{
   ch = getc(fp);
   putchar(ch);
}
```

FIGURE 13.26 While loop file read example.

There is an alternative form of the while loop: the *do* loop. The semantics of the do loop differ from those of the while in that the test is applied after each iteration of the loop, rather than prior to the loop. The syntax can be seen in the following specification:

```
do
   statement
while (condition);
```

For Loop

The other commonly used loop in C is the *for* loop. The first line of the *for* statement consists of three parts: initialization, test, and increment. These three sections are separated by semicolons. The following specifies the general form:

```
for (initialization ; condition ; increment)
     statement;
```

In Figure 13.27 the expression "i = 100" is the initialization, "i > 0" is the test, and "i– –" is the increment. This *for* loop has the same function as the first *while* loop of Figure 13.25.

```
int i;
for (i = 100; i > 0; i–)
{
  sum += i;
}
```

FIGURE 13.27 For loop example.

The initialization section is evaluated once at the beginning of this statement's execution. Then, the test or condition is evaluated. If this gives a true value, then the loop body is executed. After execution of this body, the increment section statement is performed.

```
FILE * fp;
char ch;
fp = fopen("test.dat", "r");
for (; ch = getchar(fp); ch != EOF)
  putchar(ch);
```

FIGURE 13.28 For loop file read example.

The *for* loop example in Figure 13.28 is equivalent to the file reading *while* loop in Figure 13.26. Notice that the initialization portion of this for loop header is missing. In this case, no initialization is necessary since the *fopen()* function initializes the FILE pointer.

13.6 FUNCTIONS

Functions are the primary unit of modularity in C. There are two types of C functions: predefined and user-defined.[1] We will first examined some important predefined functions. These functions are not part of the C language itself, but are provided in the accompanying standard library.

[1] There is no real difference between predefined and user-defined functions. Both are C functions that have the same syntax. The predefined functions are those functions that have been created previously and packaged in the standard C library with the compiler.

```
statements:
int i = 7;
float pi = 3.14159;
char name[30] = "Abraham Lincoln";
printf("Hello World!\n");
printf("The value of i is %d.\nAnd 2 * i = %d\n\n", i, 2 * i);
printf("pi = %f\n", pi);
printf("%s\n", name);

corresponding output:
Hello World
The value of i is 7
And 2 * i = 14


pi = 3.141590
Abraham Lincoln
```

FIGURE 13.29 Examples of the use of *printf()*.

Standard Input/Output Functions

The standard input/output library functions provide many functions for input from the keyboard, output to the screen, and file input and output (from either text or binary files) (Table 13.4).

The *printf()* function allows the value of a list of variables to be displayed on the screen. This function requires at least one parameter: a format specification string. This string can merely consist of a message to be printed. It can also contain special characters that tell how to display any variables that follow. The symbol "%d" specifies that a variable's value be displayed as an integer; "%f" as a float; "%c" as a character; and "%s" as a string of characters. Consider the examples in Figure 13.29.

As discussed in a previous section (see Table 13.1), the special character '\n' is used to represent a new line character. The presence of this character in the second *printf()* causes the number 14 and its caption to be printed on a new line. The two new lines at the end of this *printf()* specification produce a blank line after the 14 is printed. There are many other output functions including *putchar()* that cause a single character to be displayed. The example in Figure 13.30 causes the letter 'K' to be displayed.

TABLE 13.4 C Library Input/Output Functions

printf (formating string, variable list)	Formatted printing
scanf (formatting string, pointer variable list)	Formatted input
putchar (*char*)	Single character output
getc()	Returns a single input character from keyboard (standard input)

```
char ch = 'K';
putchar(ch);
```

FIGURE 13.30 Character output function *putchar()*.

The most versatile keyboard input function is *scanf()*. Like *printf()*, this function requires a format specification string as its first parameter to describe the number and type of variables to be input. The specifications have the same form: "%d" for *int*, "%f" for *float*, "%c" for *char*, and "%s" for *string*. The input statement "scanf("%d%d", &i, &j);" reads two integer values into the *int* variables *i* and *j*. Note that the addresses of these values are used as the parameters rather than the variables themselves. Failure to pass the address will cause an error when the program is run. This is a very common source of programming errors among novice C programmers. Consider the valid uses of *scanf()* given in Figure 13.31.

Notice that reading a string into the character array "name" does not require the use of the address operator. This is because an array variable is a pointer to that array.

The *getc()* function is analogous to the *putchar()* function described above. This function reads a single character from the terminal, which it returns as its value. It is typically used as shown in Figure 13.32. When any of these functions are used, you should insert the preprocessor command "#include <stdio.h>" at the beginning of that file. This command will be explained in a subsequent section.

```
float f, g;
int i, j;
char ch;
char name[30];
scanf("%f %d %f", &f, &i, &g);
scanf("%c %d", &ch, &j);
scanf("%s", name);
```

FIGURE 13.31 Input examples using scanf().

```
char ch;
ch = getc();
```

FIGURE 13.32 Character input using *getc()*.

Standard Mathematics Functions

The C library provides the typical mathematical functions. To use any of these functions use the preprocessor command "#include<math.h>." The functions given in Table 13.5 work for either *int* or *float* values.

Standard Character Functions

The character functions in the standard library ("#include <ctype.h>") provide those functions that manipulate characters in various ways. Table 13.6 lists and briefly explains some of the most commonly used of these.

TABLE 13.5 C Library Mathematical Functions

Function	Semantics (Value Returned)
sin(x)	Sine of x (angle in radians)
cos(x)	Cosine of x (angle in radians)
tan(x)	Tangent of x (angle in radians)
exp(x)	e raised to the power x
log(x)	Logarithm of x base e
log10(x)	Logarithm of x base 10
pow(base, exp)	base raised to the power exp
sqrt(x)	Square root of x
ceil(x)	Round x up to next integer
floor(x)	Round x down

TABLE 13.6 C Library Character Functions

Function	Semantics (Value Returned)
isalpha(ch)	True if ch is an alphabetic character
isalnum(ch)	True if ch is alphabetic character or a digit
isdigit(ch)	True if ch is a digit
islower(ch)	True if ch is a lower case character
isupper(ch)	True if ch is an upper case character
isspace(ch)	True if ch is a space, tab, or new line
toupper(ch)	Converts ch to upper case if it is lower case
tolower(ch)	Converts ch to lower case if it is upper case

Standard String Functions

A string in C is a sequence of characters that is terminated with the special character '\0'. There is a set of functions that is helpful when working with such strings. In each case in Table 13.7, these lists of functions are not complete (see Kernighan and Ritchie, 1988; Troy and Kiper, 1989).

Function Definition

Every function has to be defined before it is used. We have described calls to some predefined i/o, math, string, and character functions above. These functions have to be defined somewhere. Their definitions appear in the standard libraries that come with every ANSI compiler. The user can define and use functions also. The syntax for a call to a user-defined function has the same form as a call to a predefined function. Let us first examine the form of a definition of such a function.

A function definition has four parts: the return type, the function name, the formal parameter (or argument) list, and the function body. In Figure 13.33, the function name is "cube()," the return type is "double," and the formal parameter[2] is called **x** with type *double*. The function body computes the cube of the input parameter and then returns this as the value of the function.

[2]A formal parameter is a variable used to represent a value in a function definition. An actual parameter, or argument, is the corresponding variable or expression used in the function call.

TABLE 13.7 C Library String Functions

Function	Semantics
strcpy(dest, source)	Copies the characters in string *source* to the string *dest*
strcat(dest, catStr)	Adds the string in *catStr* to the end of the string in *dest*
strcmp(str1, str2)	Returns >0 if *str1* > *str2*; <0 if *str1* < *str2*; returns 0 if *str1* ==*str2*

```
double cube(double x)
{
  return x * x * x;
}
```

FIGURE 13.33 Sample function declaratiion.

This **cube()** function is called twice in Figure 13.34. Figure 13.35 takes two parameters and returns no value. The keyword "void" as the return type indicates that this function returns no value. (In some programming languages, this form of module would be called a "procedure" rather than a function.) If the return type is left blank, rather than *void*, the C compiler assumes that a function returns *int*. That is not what is intended in this case.

```
double x, y;
x = 5.3;
y = cube(x);
printf("%f\n", y);
printf("%f\n", cube(2.1*(x+7.1));
```

FIGURE 13.34 Example function calls.

```
void printAverage(char name[], float grades[])
{
  int i;
    float average;
    float sum = 0.0;
    sum = grades[0] + grades[1] + grades[2] + grades[3];
    average = sum/4.0;
    printf("The average grade for %s is %f %\n", name,
      average*100);
}
```

FIGURE 13.35 Example of a function definition with array parameters.

Every valid C program must have a function called "main()." This is the first function to be executed. Typically, the **main()** function is written first in the program file, followed by definition of any other user-defined functions. Function definitions cannot be nested in C. Some or all of these user-defined functions can be placed in other files that can be separately compiled into separate object files. However, these separate object files must be linked together with the object file for the **main()** to form a single executable file. (The details of separate compilation and subsequent linkage are compiler dependent.)

```
void main(void)
{
   char studentNames[10][30];
   float studentGrades[10][4];
   int i;
   for (i = 0; i < 10; i++)
   {
     printf("Please enter the student's name: ");
     scanf("%s", name[i]);
     printf("Now enter this student's 4 test scores: ");
     scanf("%f %f %f %f", &studentGrades[i][0],
        &studentGrades[i][1], &studentGrades[i][2],
        &studentGrades[i][3]);
     printAverage(studentNames[i], studentGrades[i]);
   }
} /* end of main */
```

FIGURE 13.36 Example program to call functions.

Argument Passing Techniques

Notice that the parameters in the function parameter list for the function **printAverage()** are declared as arrays with no size specified. These parameters, **name[]** and **grades[]**, are the *formal* parameters. When an array is used as a formal parameter, the size is typically not specified. However, when the function is called, the *actual* parameters, **studentNames** and **studentGrades** in this case, must be declared with a size so that space is allocated for that array (Fig. 13.36).

All parameters in C are passed **by value**. This means that when a function is called, a copy of the actual parameter is made as the value of the formal parameter. This formal parameter value can then be modified without affecting the value of the corresponding actual parameter. This protects the actual parameter from changes to the formal parameter. However, there are times when the value calculated in a function needs to be passed back to the calling function. If a single value needs to be calculated and sent back, the function's return value can be used. But this will not suffice to return multiple values. Changes to parameter values can be passed back by use of pointer variables. For example, to create a **swap()** function that interchanges the value of two variables, we use pointer types for both parameters (Fig. 13.37).

```
void swap(float * x, float * y)
{
    float temp;
    temp = *x;
    *x = *y;
    *y = temp;
}
```

FIGURE 13.37 Pointers as parameters.

We call this function by passing the address of the actual parameters. That is, given two float variables f and g, the call "swap(&f, &g);" will interchange the values of these variables. Notice also that there is no ***return*** statement in this function. The return is optional when the function returns no value.

Function Prototypes

Many programmers place the **main()** function first in the file and place other function definitions later in the file. This can potentially lead to a problem. Remember that a general principle of C is that identifiers should be declared before they are used. That principle holds for the name of functions also. However, if the **main()** function calls a user-defined function whose definition comes later in the file, this principle is violated. C allows this violation, but makes some assumptions about such a function. It assumes that it returns an *int*, and that the types of the actual parameters are the correct types of formal parameters. These assumptions may be incorrect, and may lead to some run-time errors in the program. The ANSI standard provides a solution to this potential problem: **function prototypes**. A function prototype is a declaration of the return type, function name, and parameter type with no function body. If a function prototype is placed at the beginning of a file, then the full function definition can be placed at the end of the file after some of the function calls. The compiler will have the necessary information to correctly translate function calls. Figure 13.38 gives function prototypes for some of the function examples given above.

```
void printAverage(char [], float []);
void swap(float * , float *);
double cube(double);
```

FIGURE 13.38 Functioin prototypes.

Header Files

When using predefined functions, it is good practice to include function prototypes for these functions also. These function prototypes are incorporated in special files called **header files**. For example, prototypes for the standard i/o functions are supplied in the file "stdio.h". This file can be copied into your file by using the preprocessor #include command: "#include <stdio.h>".

13.7 PREPROCESSOR

The ANSI standard describes a set of **preprocessor commands** that every standard C compiler must support. These preprocessor commands are distinguished from C statements by the fact that they begin the character # and that they generally do not end with a semicolon. The preprocessor executes these commands prior to the translation of the C program.

#include Command

The two most common such commands will be discussed here: ***#include*** and ***#define***. We have already seen a use of the #include in previous sections. This command is followed by the name of a file. The preprocessor inserts the contents of this file into the program's file at the point of the #include command. This is frequently used to include files that contain function prototypes or definition of constants or types when these definitions or prototypes are needed in several files. If the file name is surrounded by quotes, the system looks for the file in the user's default directory. If the name is surrounded by < and >, then the preprocessor looks for this file in special library directories. Thus, in the previous section the preprocessor command "#include <stdio.h>" searches for this header file in the standard library. The command "#include "list.h" causes the preprocessor to look for this file in the user's default directory.

#define Command and Macros

The "#define" preprocessor command allows the user to create new symbols for fixed values. For example, the command "#define PI 3.14159" creates asymbol called PI with value 3.14159. The preprocessor replaces all occurrences of this symbol PI with the number 3.14159 from the point of definition to the end of the file. When the C compiler gets the file, the symbol PI is gone—having been replaced. Macros do not apply to string literals. Here are some valid uses of the preprocessor define.

```
#define BUFSIZE 512
#define DEBUG 0
#define EOF (-1)
```

A commonly used style is to use capital letters for such symbols, although this is just a convention, not a requirement. A common error is to end a symbol definition with a semicolon. This will cause the semicolon to become part of the value of the symbol, and to be inserted at inappropriate places throughout the file.

The #define is sometimes used to create macros. These are pseudo-functions that are expanded by the preprocessor. For example,

```
#define SQUARE( x ) (x) * (x)
#define TRIANGLEAREA( h, b ) 0.5 * (h)*(b)
```

If the preprocessor finds the symbol "SQUARE(v – 2 * w)," it expands this to become "(v – 2 * w) * (v – 2 * w)." The parentheses are not necessary although they can be useful in avoiding some errors. For example, consider the expansion that would result in this case if the parentheses were not present. The expression TRIANGLEAREA(sqrt(height), base – 0.5) would be expanded to 0.5 * sqrt(height) * base – 0.5.

Generally, the #define should be avoided. It can lead to a programs that are cryptic and for which the compiler can provide little error identification. These problems are caused by the fact that symbols defined in this way are not known to the C compiler. They do not exist at compile time, they have no type, and they are not restricted in scope. The C *const* construct provides a better way to create symbols with constant values; and C functions can be used very easily in place of #define macros.

13.8 POTENTIAL C PITFALLS

C has become a very popular language over the past decade. That is not to say that it is free of problems as programming language theorists will quickly point out. An awareness of some of these problems can help a programmer avoid them. We discuss a few of these problems here.

C is an expression-oriented language rather than a statement-oriented one. This leads to some flexibility in coding that can be useful; however, it can also permit erroneous statements to be written and compiled with no compiler error message. One of the most common errors of this variety is the use of the assignment statement rather than an equality test in the condition of an if or while statement. For example, the statement "if (a = 1) . . ." will compile with no errors, but is probably not what the programmer intended. This if condition assigns the variable **a** the value 1; since the value of this expression is 1, it treats this as a true condition and then executes the statement part of the if. (Some compilers produce a warning for this type of situation.)

C has types, but is not strongly typed. For example, in the absence of a function prototype, a C compiler will assume that a function whose call appears before its definition returns an *int*. If this function actually returns a *double*, the compiler without this knowledge will have interpreted the return value incorrectly. The compiler produces no error when a prototype is missing; it silently makes an assumption. (ANSI compilers produce a warning in this and similar situations.) The consistent use of function prototypes and the inclusion of appropriate header files (containing function prototypes) for standard library functions will avoid this type of problem.

Function definitions cannot be nested. In many programming languages, nested function definitions are used to achieve some degree of structure and "information hiding." That is, the details of a complicated algorithm can be organized into submodules whose details are hidden away inside the function. In this way, users of the algorithm will not be affected by any changes to the implementation details of these submodules. C can achieve some of the same benefits from the use of nested blocks, and through separate files. Remember that a block in C is a collection of statements that is surrounded by braces. A statement block can appear anywhere that a statement is valid. Thus, blocks can be nested. Variables can be declared at the beginning of a block. Their scope is from the point of declaration until the closing brace of that block. However, blocks are not named and cannot have parameters.

Information hiding can be achieved via separate files. A collection of related functions that might be nested in some languages can be isolated in a separate file. This file can then be compiled individually before being linked with the other program files. Any file level variables, i.e., variables declared in this file but outside of any function, can be marked with the keyword *static*, which restricts its scope to this file.

The fact that the name of an array is a pointer to that array is a source of convenient flexibility in C, but also a source of confusion. Although an array name is a pointer, there are some important differences. The declaration of an array as, for example, "int A[100]," creates a pointer to a location in memory that is sufficient to store 100 integers. The declaration "int *B" creates a pointer to an integer, but allocates no space for such an integer; however, it can still be used as if space had been allocated. Furthermore, this pointer **B** can now be used as an l-value, although the array name **A** cannot. This is complicated further by the fact that the name of a structure is an l-value. It is important for C programmers to understand these differences so that they know when each should be used and when they are interchangeable.

A common error in the use of pointers inside a function is to attempt to return a pointer to a local variable. When the function returns, its local variable space is recovered and reused by the system. Thus, the address returned now points to an invalid location in memory. The solution is to return a copy of the value rather than a pointer to that location.

13.9 WRITING LARGE PROGRAMS

Writing large programs is qualitatively different from writing small programs. The techniques described in this section will help you organize large programs to make them easier to write, understand, test, and maintain.

Separate Compilation

The simplest way to manage complexity of large C programs is to exploit C's support for **separate compilation**. C programs can be partitioned across any number of files, which can be compiled individually, then linked together to make an executable program. This helps reduce compile time since only files that are modified need be recompiled.

However, the primary advantage of breaking large programs into multiple smaller files is that smaller files are generally easier to understand and modify than larger files. Since C imposes almost no constraints on the way programs can be split into multiple files, it is up to the programmer to do so in a way that makes sense. The following guidelines describe how to take advantage of the program structuring mechanisms that C does provide.

Organize Files into Modules

Group related functions together in separate .c files. For example, if your program needs to create and manipulate lists, it would be good practice to group a set of functions for creating and manipulating lists together in a separate file. This functional grouping will help provide readily observable gross level structure to your program. In addition, this approach allows you to focus on list functionality exclusively while you are working on this module, which will help you create a more robust and general set of routines. These routines can then be treated like standard library functions when writing other modules that use list routines.

Think of Your Program in Terms of Layers

Another helpful guideline is to think of your program in terms of layers, where functions in higher (or outer) layers call functions in lower (or inner) layers. This will help you decide how to group functions, and when to redefine functions that do not seem to fit anywhere.

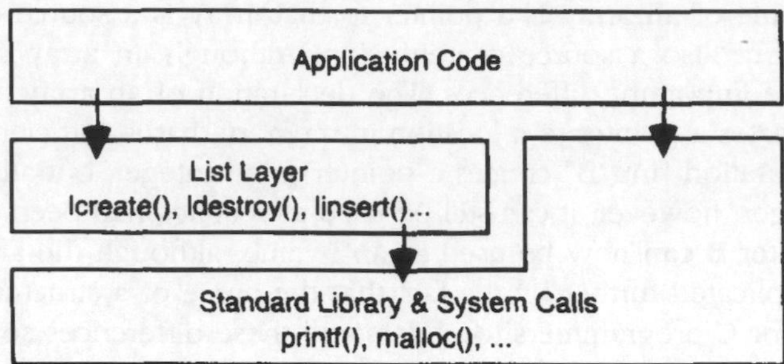

FIGURE 13.39 Programming in layers.

In Figure 13.39, functions in the List layer would call functions from the Standard Library and System layers, e.g., *lcreate()* would call *malloc()*. The application code would call functions from both lower layers. Thus higher layers are implemented in terms of the functionality of lower layers. Breaking the application into layers helps provide clean interfaces for testing, in addition to promoting reuse.

Thinking in terms of layers will also help avoid circular dependencies among files, i.e., having functions in file A that call functions in file B *and* functions in file B that call functions in file A. While this can be useful in some instances (e.g., "callbacks"), it is generally easier to understand and maintain code with readily observable, unidirectional dependencies.

The previous two guidelines will help decide how to organize programs into files corresponding to cleanly defined modules. The next few sections describe some techniques for solidifying module interfaces.

Use 'Static' to Hide Private Functions and Data

By default, variables and functions declared outside of any block, i.e., at the "file" level, are globally visible, meaning they can be referenced from anywhere in the program—even from another file. The 'static' keyword can be used to encapsulate global variables and functions within the file in which they are declared. File level variables and functions declared with 'static' are visible to all statements that appear after their declaration in the same file, but not visible to any statement outside that file.

```
int global_var;       /* visible to any statement in the pro-
gram*/
static int private_var;/* not visible outside this file */
```

The 'static' keyword may be applied to any declaration. In addition to reducing visibility of the item being declared, it also modifies the storage class. This means that variables declared within a function with the 'static' keyword will hold their values after the function returns. Since variables declared at the file level already have static lifetime, only the scope reduction effect of 'static' is observed.

Use Header Files to Export Functions and Data to Other Modules

Where 'static' can be used to explicitly hide functions and data from other modules, header (.h) files can be used to explicitly "export" data and functions to other modules. Function prototypes and type definitions that will be used outside a module should be put in a header file to be #included by the client module. For example, Figure 13.40 contains a partial listing of type definitions and prototypes from the header file of a list package.

Header files should only contain non-storage-allocating definitions, i.e., function prototypes, struct definitions (but not declarations), type defs, and #define macros. If global variables are required, they should be declared in the module's .c file and exported by an *external* declaration in the header file. This guideline helps prevent "multiple definition" errors caused when a header file containing storage-allocating declarations is included in multiple files in the same program.

Another useful trick for preventing "multiple definition" errors is to use #define to set a flag that prevents a header from being evaluated more than once during a single compilation.

```
/* list.h–typedefs and prototypes for a genericlist package */
#ifndef LIST_H
#define LIST_H
struct listElt_struct {
    void *data;
    struct listElt_struct *prev;
    struct listElt_struct *next;
    ...
};
typedef struct listElt_struct *ListElt;
    struct list_struct {
    struct listElt_struct *head;
    struct listElt_struct *tail;
    ...
    };
typedef struct list_struct *List;
List lcreate(void);                /* create and destroy Lists */
void ldestroy(List l);
List linsert(List l, void *data); . ./* insert and delete ele-
ments */
List ldelete(List l, ListElt e);
ListElt lfirst(List l);           /* traverse Lists */
ListElt lnext(ListElt e);
ListElt llast(List l);
...
#endif
```

FIGURE 13.40 List package header file.

In Figure 13.40, the preprocessor directive '#ifndef LIST_H' checks to see if the macro 'LIST_H' has already been defined. If not, the next statement, '#define LIST_H', defines it. If 'LIST_H' has already been defined, e.g., if list.h was included in another .h file that was included by the same file that included list.h, then the preprocessor scans forward, ignoring all input until it find the '#endif' at the end of the file, thus preventing the contents of list.h from being evaluated more than once.

Using header files to define the external interface to a module increases the level of encapsulation. Users of the module need only refer to the header file (and not the module's implementation) to use the module. This helps limit the effect of later changes to the module, as long as the interface published by the header file is kept constant.

Use 'Extern' to Subscribe to Data Defined in Other Files

The 'extern' keyword is to variable declarations what prototypes are to functions. Declaring a variable with 'extern' allows the compiler to continue as if that variable were declared in the current file, but requires that the linker supply a reference to the actual variable declaration at link time. External declarations are often used in header files to explicitly export data items defined in the module implementation. The 'extern' keyword can also be used to explicitly import data items that are declared in other files. Using 'extern' in this way is less desirable than explicitly exporting the data item in a header, but is a useful workaround when such declarations do not exist.

13.10 EXAMPLE PROGRAM: GREAT CIRCLE DISTANCES

```
/* This program determines the great circle distance from
pairs of locations on the surface of the earth. These loca-
tions are specified by giving the latitude and longitude of
each location specified in degrees.
  Input:             Latitude and longitude of two locations
                     on earth.
  Output:            Distance in statute miles.
  Accuracy:          This program makes the assumption that
                     the earth is a sphere with radius 3963
                     miles.
  Programmer:        James D. Kiper
  Acknowledgments: The great circle distance algorithm is
                     due to Roger Noe, Department of Computer
                     Science, University of Illinois.
*/
#include <stdio.h>
#include <math.h>
/* function prototypes */
```

```
void inputData(double *, double *, double *, double *);
double determineSubtendedAngle( double, double, double,
double );
double convert2Radians(double);
const double radius = 3963.0;
const double pi = 3.14159;
void main( void )
{
  FILE * source, * dest;
  double lat1, lat2, long1, long2;
  double angleCos, angle, distance;
  char answer[4];
  do {
    inputData( &lat1, &long1, &lat2, &long2);
    lat1 = convert2Radians(lat1);
    lat2 = convert2Radians(lat2);
    long1 = convert2Radians(long1);
    long2 = convert2Radians(long2);
    angleCos = determineSubtendedAngle( lat1, long1, lat2,
               long2);
    angle = acos(angleCos);
    printf("\nangle = %6.2f degrees\n", angle*180/pi);
    distance = angle * radius;
    printf("Distance is: %8.2f miles.\n", distance);
    printf("More?");
    scanf("%s", answer);
  } while( answer[0] == 'y-' );
}
void inputData( double * lat1Ptr, double * long1Ptr,
                double * lat2Ptr, double * long2Ptr )
{
  printf("Enter longitude and latitude of point 1 in
degrees: ");
  scanf("%lf %lf", long1Ptr, lat1Ptr);
  printf("Enter longitude and latitude of point 2 in
degrees: ");
  scanf("%lf %lf", long2Ptr, lat2Ptr);
}
double convert2Radians(double lat)
{
  return lat/180.0*pi;
```

```
double determineSubtendedAngle(double lat1, double long1,
                               double lat2, double long2)
{
  return sin(lat1)*sin(lat2) + (cos(lat1)*cos(lat2)*
                  cos(long2 - long1) );
}
```

13.11 PROGRAMMING ENVIRONMENTS

In this section will discuss **programming environments** that are commonly used to create, run, test, and maintain C programs. First we discuss typical components of C compilers. Then we examine several of the most commonly used compilers and their environments.

Compiler Components

A compiler is not used alone, but is supported by a set of related tools to enter and modify the program, to coordinate the multiple files that often comprise large C programs (the "make" facility), and to debug programs, as well as for version control. Most, if not all, commercial C programming environments include an ANSI C compiler. Some include switches that allow the compiler to also accept other dialects of C. But the environment that surrounds this compiler varies from product to product. The functions described above (an editor, a make facility, a debugger, and configuration manager) are present in each. The primary difference among environments is the degree of integration of these various tools. Environments for the MacIntosh or Window OS typically present a single interface to the user that integrates all the functionality of editor, compiler, debugger, make, and configuration management. The Unix philosophy is to provide small, independent tools that can be "pasted" together, through pipes or i/o redirection, in interesting ways.

Program Editor

Editors are common enough to require little or no explanation. Some products, particularly those designed for MacIntosh or Windows computers, have a menued editor that supports a mouse or similar input device with the usual functions of opening and closing files, copying and pasting, etc. A variety of editors are available for Unix machines although *vi* and *emacs*, both keystroke-driven editors, are the most commonly used.

Make

One of the reasons that C has been so successful is that it handles large programs well as described in Section 13.9. The primary facility for this is the division of a large program into multiple files that can then be individually compiled. The collection of compiled files is then linked together to create an executable program. When the number of files becomes large, remembering which files need to be recompiled when a change is made, and recalling the names of all files that need to be linked together to form the file product can be quite difficult. Most C programming environments give some support to the user in this task. We will refer to this function as the "make" task. This name is that of one of the first such tools—the *make* utility of Unix.

On-Line Debugger

Most modern C programming environments allow the user to obtain information about a program as it is running to aid the process of finding errors. A "breakpoint" is a location in a program at which the program pauses. The user can set a breakpoint at any executable statement. When the program pauses there, the state of the program can be determined by displaying values of active variables. These values can be reset to other values if desired. Then the program's execution can be resumed from that point. The value of user-specified variables can be "watched" to see their values change during program execution. The flow of the program can be "traced" by displaying function calls as they occur.

Revision Control

One of the major tasks in the development of a large computer program is to track the various versions of files, and of the program as a whole. Programs are generally developed incrementally. When multiple people are working a single project, it is important for each to know what version of each C file is the current one. This is complicated when multiple versions of the program are desired. For example, a team may want first to create a program with minimal functionality. This functionality is then expanded step by step until the total functionality is present. However, the team will want to be able to recover previous, more stable versions as they create subsequent versions. Configuration management tools aid this process. Environments targeted for personal computers (MacIntosh or Windows operating systems) are notoriously weak in this area since they are intended for small to medium sized projects constructed by an individual or small team. Unix environments have much stronger tools in this area.

In the next four sections, we discuss some of the most popular C programming environments. The first, Borland C++, is a product for DOS or Windows operating systems. The next, Semantec Think C, is a MacIntosh-based product. The final one is the GNU C compiler and Unix tools. You will notice that the first product is a C++ compiler. The C++ programming language, discussed in Chapter 14, is a superset of C; thus, most compiler companies have chosen to market a single product that compiles either C or C++ program. The compiler differentiates between these by an option set in the environment, a command line flag, or the file extension. Semantec has a related product, Semantec C++, that shares the same style of use and many menus. The GNU C compiler for unix, called *gcc*, can also compile C++ programs. These products all support ANSI C (in addition to C++ and some other versions of C). A large majority of the programming environment options for editing, debugging, version control, and make in all of these products are generic to both C and C++.

Borland C++

This product runs under DOS or Windows, and has a windowed, menued interface for both operating systems. (Cilwa, 1994; Ezzell, 1992; Pappas, 1994). Menu options can be selected by a mouse or by typing the *Alt* key to move the cursor to the menu bar, then using the arrow keys to choose the appropriate option. The main menu options that appear across the top of the screen or window are **File**, **Edit**, **Search**, **View**, **Project**, **Debug**, **Tool**, **Option**, **Window**, and **Help**. The "File" menu is typical for a Windows tool. Options include **New**, **Open**, **Save**, **Save as**, **Print**, **Print setup**, and **Exit**. The choice "New" creates an initial window below the menu bar that provides a canvas for the user to type a C program. Multiple windows can be open simultaneously by choosing "New" more than once, or by opening multiple files. A window generally corresponds to a file allowing modularization of large programs. The Window menu choice gives several options for display of multiple windows: "Cascade"—windows are displayed as a stack of papers with only the top one entirely visible; "Tile" (horizontal or vertical)—windows are reduced in size so that they are all visible on the screen simultaneously.

The "Edit" menu choice provides typical editor functions: **Cut**, **Copy**, **Paste**, **Undo** (remove the last editor modification), **Redo**, etc.

The "Cut" and "Copy" commands operate on the currently selected item. (Text is selected with a mouse using the normal click and drag technique.)

The "Search" menu choice has options for finding text, replacing it with alternatives, locating a particular function, and moving from one error message to the next or previous one.

The option to compile the C code in the current window is under the "Project" main menu choice. This menu choice is used to create a "Project"—that is, a collection of files that forms a program. The "New Project" option creates a project; "Open Project" opens an existing project. The choice "Make" all causes all files (windows) to be compiled if there has been a change since the last modification. "Build all" forces all files to be compiled.

The Borland C++ on-line debugger is accessible via the Debug menu option. From this menu, a program can be run (it will be compiled if needed before running), a stopping point in a program can be set ("Toggle breakpoint"), single statement can be executed by the commands "Step over" or "Trace into" ("Step over" considers a called function to be a single step), an on going watch on the value of a particular variable can be set up ("Add watch"), and the current value of a variable at a breakpoint can be inspected and/or modified ("Evaluate/modify").

The "Options" menu choice provides a means of customizing the environment. "Help" is provided in several forms including a table of contents to guide the request and a keyword search.

Symantec Think C

Think C is a MacIntosh product that makes good use of the MacIntosh interface style. (Symantec, 1993; Parker, 1993). Windows and dialog boxes in this style are used effectively in addition to the menued interface. You begin to program in the Think C environment by creating a project. Double-clicking on the icon for "THINK Project Manager" displays a dialog box that allows you to open an existing project or to create a new one. (The convention is to use the file extension π for the name of a project.)

After creating or opening a project, the menu choices displayed are **File**, **Edit**, **Search**, **Project**, **Source**, and **Windows**. The "File" menu has the usual choices for opening an existing file, creating a new one, saving, printing, and quitting. The "Edit" menu offers the normal MacIntosh choices to undo, cut, copy, paste, etc., and some other useful choices to set tabs and fonts, to shift left or right for appropriate indentation of a program, and to check that parentheses and/or braces are balanced. The "Search" menu has various search and replace options; the ability to go to a particular line number, to the next or the previous error, etc.

When you have entered C code into a file and saved it, it is necessary to add this file to the project. This is accomplished through the "Source" option and its choices of **Add**, **Add files**, and **Remove**. The "Add" option causes the currently active file to be added; "Add files" gives a typical MacIntosh file selection dialog box. Files that have been added are displayed with their sizes in a special project window.

Think C requires that any necessary library files be explicitly added to the project. For example, if you use input and output functions from the standard library, you must explicitly add the ANSI library to the project. If the Think C installation has been standard, the ANSI standard library is in the file called "ANSI" stored in the folder "Standard Libraries" that is inside the "Think C" folder. A library file is added in the same way as a user file—i.e., through the "Add files" option of the "Source" choice.

The "Source" option is the location of the commands to compile a file or make a project. The make option compiles those files that have not been compiled since their last modification. This menu also allows you to check syntax (no object code generated), to run the preprocessor only, and even to disassemble assembly level routines. The command to run a program is found under the project menu choice. Before a program is run, the "Run" choice saves and compiles any necessary files (per user instruction).

To debug a program in Think C, the debug option (under the "Project" menu) must be selected. After this choice is made, the choice "Run" (under "Project") causes the debugger to be entered. This opens a debug window that includes buttons to "Go," "Step," "In," "Out," "Trace," and "Stop." It also changes the main menu options at the top of the window: **File**, **Edit**, **Debug**, **Source**, **Data**, and **Windows**. The "Debug" option includes all the options listed as buttons on the debugger window and has the same effect as the corresponding button. The debugger window has a copy of the source code with an arrow beside the executable source statement that is the next to be executed. The "Go" button runs the program; "Step" causes the current statement to be executed. (This button treats any functions called as a single statement.) When the next statement is a function call, the "In" button causes execution to pause inside that function. "Out" causes execution to continue until the current function is finished.

The "Debug" menu has a few additional choices: "Go Until Here" that causes execution to continue until it reaches the current cursor position in the debug window (a temporary breakpoint), and "Exit to Shell" that returns from the debugger to the project manager menu options.

The "Source" menu is the location of commands to set and clear breakpoints. The choice "Set Breakpoint" causes a breakpoint to be entered for the current location of the cursor in the debugger window. Breakpoints are visible in the left column of the debugger window. Every executable statement has a diamond in this column. Those that have a breakpoint are filled diamonds. "Clear Breakpoint" removes a breakpoint from the statement that is currently selected. Breakpoints can also be set or cleared by clicking on the appropriate diamond. When the "Go" button is selected, execution continues to the first breakpoint. Subsequent "Go" commands cause execution to proceed to the next breakpoint to be encountered.

When the debug window is opened as a result of running in debug mode, another window, the "data" window, is created. When execution is suspended, at a breakpoint or as the result of a step command, data may be examined in the "data" window and through the "Data" menu. The data window has an input box positioned at the top. The current values of variable names that are typed into this input box are displayed in a lower pane of this window. The type of each such variable, i.e., the lens through which its value is to be viewed, can be selected from the "data" menu. For example, a value can be displayed as decimal (signed or unsigned), character, hexadecimal, pointer, or C string.

Of the tools described thus far, Think C has the best support for configuration management. The "Source Server" choice of the "Source" menu has options for checking in and out files from a data base of file versions. To use this utility, first choose "Launch Source Server" from the "Source Server" submenu of the "Source" menu. This runs the Source Server tool. (It can also be launched from the MacIntosh Finder by double clicking on the "Source Server" file in the Tools folder that is in the THINK C folder.) Now choose "New ProjectorDB," which creates a new database to contain project file revisions. You will be asked to enter a name for this new database, and will be given the opportunity to type in explanatory comments about the purpose of this project database. Now, select "Mount ProjectorDB" from the "Source Server" submenu. This database now is the current project; files can be checked in and checked out. When you create a new file for the project, or have finished with a revision to an existing file, the choice "Check In. . ." in the "Source Server" submenu enters this in the database. You will be asked to enter a comment describing the changes that you have made to this version of the file. For subsequent modifications to a file, you will have to check it out with the "Check Out. . ." option. The database keeps track of the history of changes to these files, and allows you to return to any version that you desire.

Unix Tools and GNU C

The Unix operating system provides a rich set of tools for the development of C programs. Standard Unix tools include a C compiler (**cc**), several editors (**vi** and **emacs**), a symbolic debugger (**adb** or **dbx**), the original **make** tool, and a revision control system (**SCCS** or **RCS**). Most versions of Unix also provide a complete set of document preparation tools for both on-line and printed documentation. Unix also provides a rich set of tools for text processing and "scripting"—rapid development of ad hoc applications.

The **GNU** set of Unix tools that are made available by the Free Software Foundation of Cambridge, Massachusetts, are becoming the standard and preference for many Unix users because of their high quality. In this section we will briefly discuss some of the features of these tools using the GNU versions where possible. Each of these is a sophisticated tool with many options. More thorough coverage is generally available in the appropriate manuals which are generally available on-line through the "man" command (e.g., "man gcc" presents the manual pages for the GNU C compiler).

The two most common command editors for Unix are **vi** and **emacs**. Both are full screen editors that allow the user to view and edit text files. The primary difference is the **vi** has two modes, command and insert, whereas **emacs** is modeless using special key combinations (e.g., control-x or alt-y), to specify commands. Unix editors are discussed in the Unix chapter of this book. We briefly described these two editors here.

The heritage of **vi** can be traced to earlier line editors, such as **ex**. Therefore, most **vi** commands apply to a single line of text. **Vi** has two modes: "insert" mode and "command" mode. "Insert" mode is used to add lines of characters into a file; "command" mode is used to navigate and edit existing lines of text. Most **vi** commands are a single keystroke (e.g., 'i' means "enter insert mode", 'x' means "delete the character under the cursor", and 'u' means "undo the last edit operation"). Modifiers can be applied to many commands to effect their scope (e.g., 'dw' means "delete the word the cursor is on", '2dw' means "delete 2 words starting with the word the cursor is on"). By default, most **vi** commands apply only to the line the cursor is on. However, addresses (ranges of lines) can precede most commands to apply the command to each line in that range (e.g., 's/foo/bar' means "replace the first string matching the regular expression 'foo' on the current line with the string 'bar'", and '1,3s/foo/bar/g' means "replace all strings matching the regular expression 'foo' on lines 1 through 3 with the string 'bar'"). The '/g' at the end of the command means "apply the preceding command globally within the current line."

Emacs is a fully programmable, full screen editor. All **emacs** commands are invoked by some combination of shift keys and a character (e.g., <control>-x or <alt>-y). Thus, 'x' by itself means "enter the character 'x' at the cursor position" while '<control>-x' is a command. Emacs is programmable via a Lisp derivative called "Elisp," and scripts written in Elisp can be "bound" to keystrokes to create new commands. The functionality of these new commands can range from "insert my standard memo header" to complete applications. Examples of applications that have been written in Elisp include Mail and News readers, source code browsers, and even a World Wide Web browser.

Most vendor's stock C compilers will compile only a single dialect of C, most often K&R (i.e., pre-ANSI) C. ANSI C, as well as C++ and other language compilers, must be purchased and installed separately. The GNU C compiler, when properly configured, can compile K&R C, ANSI C, C++, and Objective C, as well as a dialect of C that includes "GNU extensions." Additional language "drivers" can be added to GNU C to compile Fortran 77 and ADA as well. Finally, GNU CC can be configured as a cross compiler—to produce binaries to be executed on a different kind of computer than the one on which the binary was compiled—for a number of host and target architectures.

The GNU C compiler is initiated by the command "gcc." This compiler command **gcc** is really just the compiler driver: it calls the preprocessor, compiler, and linker in the proper order to turn input source files into executable binaries. The general convention is to use '.c' as the file extension for each source file to be compiled. Thus, the simplest use of this compiler is to compile a single file that does not need to be linked with any library functions. If the file "sample.c" contains a complete C program, the command "gcc sample.c" will compile this program including the preprocessing step, and will link the program as necessary to produce an executable program. The executable will be stored in the file "a.out."

There are many compiler options available for the GNU C compiler. These options are indicated by giving them on the same line as the "gcc" command preceded by a hyphen. To tell the compiler to place the executable in another file (rather than a.out), use the '-o' option. For example, the command "cc -o hello hello.c" will produce the file "hello" which is a compiled and linked executable. The command "cc -c -o hello hello.c" will stop after the compile step (-c== "compile only") and produce the object file "hello.o." You can then manually invoke the linker to produce an executable using the command "ld -o hello hello.o." If you use functions that are included in standard libraries, you will need to indicate that these should be linked by use of the '-l' option. For example, to link in the math library functions (e.g., sgrt, sin, pow, etc.), use the option '-lm' at the end of the command line. You should check the documentation on your system to determine the available libraries and their names.

It is common for a large C program to be spread across several files. These can all be compiled by listing them on the command line (e.g., "gcc sample.c otherFunctions.c moreFunctions.c -o sample"). Note that these file names are listed separated by spaces not commas. It is often useful during program development to compile a single file even though that file cannot be linked to create a executable since it contains only part of the program. The '-c' option described above is useful here since it will cause a file to be compiled with its object code saved in the same file name with extension '-o', but not linked. Thus "gcc otherFunctions.c -c" compiles the functions in this file and saves the object code in the file 'otherFunctions.o'.

When errors are discovered in the compilation process, these are reported to the terminal by giving the line and a brief description of the error. With this information, the source file can be edited (using vi, emacs, or your favorite text editor) to correct these errors. The gcc compiler compiles to the ANSI standard (although switches are available to adjust this to older C styles.)

The GNU on-line debugger is initiated with the command "gdb" followed by the name of an executable program and, optionally, the name of a core file. For example, the command "gdb sample" would allow you to debug the program compiled above with the command "gcc sample.c -o sample." The core file option is used to help determine what happened in a previous execution of the program in question. When a Unix C program fails during execution, a core file is created. This core file is a copy of the contents of memory when that program failed.

This **gdb** debugger operates in interactive mode by allowing a user to enter commands at the prompt that govern execution of the attached program. Important commands are **break**, **run**, **bt**, **print**, **c**, **next**, **step**, **help**, and **quit**. The command "break" sets a stopping point (break point) in a program at the function whose name follows. Thus, the command "break calculate" causes a break point to be set at the function called "calculate." When this function in the program is reached in a subsequent execution of the program, execution is paused and control passed back to **gdb** for further user commands. The command "run" starts execution of the program. It can be followed by a list of argument if the program requires command line arguments. The simple command "c" causes execution of the program to continue after it was stopped as at a break point. The **gdb** commands "next" and "step" are used to execute just the next statement after a break point. The "next" command treats a function call as a single statement and steps over it. The "step" command goes into any called functions. The "print" command can be used to print out the value of any expression using variables that are in scope at that particular stopping point. This can be used to print out the value of simple variables to determine if the correct calculations are occurring, to check input or parameter values, or to determine the value of complicated arithmetic or logical expressions. The "bt" command (backtrace) displays the values that are currently on the program stack. You exit **gdb** with the "quit" command.

Large C programs may be comprised of many C source files and libraries. The build process for such programs can be automated with the "make" utility. Once the rules for building a given program are specified in a "Makefile," then the program can be built with the command "make." (Note: "make" with no arguments will look for a file in the current directory called "Makefile" and use its contents to decide what to do; nondefault names can be specified on the command line, e.g., "make<filename>".)

In addition to saving typing, **make** also minimizes compile time by re-compiling only those parts of a program that depend on source files that have changed since the last compilation. This notion of "dependency" is the basis for the rules that make up a Makefile. Make rules have the form:

```
<target>:<source>[<source>...]
<command>
[<command>...]
```

where the specified commands are those that must be applied to given source(s) in order to produce <target>. For example, consider a very simple Makefile for the venerable hello.c:

```
hello:hello.c
      gcc -o hello hello.c
```

Makefiles can contain cascading rules which automate individual steps of the program building process, and rules can contain special variables which can be used make the rules more generic:

```
hello:hello.c
      gcc =c -o $@hello.c


hello:hello.o
      ld -o$@hello.o
```

In the above example, modifying hello.c will make hello.o "outdated," so Make will apply the command specified in the first rule to update hello.o. This will in turn make "hello," which depends on "hello.o" outdated, so Make will apply the command specified in the second rule to update the target "hello." The special symbol "$@" is replaced by the name of the target for that rule before the command is invoked.

Consider the example of Figure 13.41:

```
sample:          sample.o otherFiles.o moreFiles.o sample.h
      gcc sample.o otherFiles.o moreFiles.o
sample.o:        sample.c sample.h
      gcc sample.c -c
otherFiles.o:    otherFiles.c sample.h
      gcc otherFiles.c -c
morefiles.o:     moreFiles.c
      gcc moreFiles.c -c
```

FIGURE 13.41 Example Makefile.

When the command "make" is given, this Makefile is consulted. If the file "otherFiles.c" is the only one that has been modified since the last compilation, then the file "otherFiles.o" will have to be reconstructed with the command "gcc otherFiles.c -c." Then the file executable file "sample" can be constructed with the command "gcc sample.o otherFiles.o moreFiles.o." Notice that the files "sample.o" and "moreFiles.o" do not have to be reconstructed.

SCCS (Source Code Control System) is a tool to help maintain and control various versions of software as it is developing (see Peek et al., 1993). When a team of several programmers is working on various pieces of a large software system, it is frequently the case that there are various versions of each component file. For example, there may be the files for an old version of the system; others for the current version; and still others for future versions that are under development. SCCS provides a set of commands to assure that only one person is editing a given file, and that older versions of a file can be recovered. This is accomplished in an efficient manner by storing changes from on version of a file to the next rather than storing complete versions of all files. These so-called "deltas" allow any specific version to be recovered when needed.

The first step in using SCCS is to create a subdirectory called SCCS. For any file that you want to place under revision control, insert the characters "%W% %G%" somewhere in that file, probably inside a comment. Now give the SCCS command "sccs create filename" where "filename" is the name of the file that you want to place under control.

To access the file now use the command "sccs get filename" or "sccs edit filename." The first gives a read-only copy of the file; the second gives an editable copy. More than one team member can check out a file in read-only mode; but only one is allowed to edit the file at a time. When finished editing this file, it can be returned with the command "sccs delta filename". Each use of this delta command causes a new version of the file to be created with a corresponding version number.

The command "sccs unedit filename" is used when you decide that you do not want to edit the file. The file is returned and no new version is created. The command "sccs check" gives a list of all files currently checked out.

DEFINING TERMS

Abstract Data Types (ADT): The encapsulation of data and related operations into a single syntactic unit.

Array: A collection of storage locations with a single name whose components can be referenced by indexing.

Assignment Operator: A C operator that gives a new value to a variable: =, +=, –=, *=, /=, %=.

Associativity: The order of evaluation of two operators in a single expression that have the same precedence level.

Block: A sequence of statements surrounded by braces { }.

By Value Parameter Passing: When a function is called, a copy of the actual parameter is made as the value of the formal parameter.

Cast: A change of a memory location's type either implicitly or explicitly.

Control Construct: A syntactic unit of a programming language that determines the sequencing of instructions.

Decrement Operator: Subtracts 1 from the value of an integer variable.

External Linkage: A category of variables in which a variable can be referenced outside its defining file.

File Variable: One that is declared outside any block or function.

Function: A syntactic unit that encapsulates a section of code. Functions are the primary unit of modularity in C.

Function Prototype: A declaration of the return type, function name, and parameter type of a function with no function body.

Header Files: Files that contain function prototypes to be included in other files where these functions are used.

Increment Operator: Adds 1 to the value of an integer variable.

Internal Linkage: A category of variables in which a variable can be referenced only in its defining file.

Library: A collection of prewritten and translated functions to provide commonly needed actions.

l-Value: An identifier that stores a value that can be changed during program execution.

Operator: A syntactic element that designates data manipulation in one of four categories: arithmetic, relational, logic, and assignment.

Pointer: A variable that holds the address of another variable.

Precedence: The order of evaluation of operators in an expression.

Preprocessor Commands: Instructions that are executed prior to the translation of the C program.

Program Duration: A category of variables in which a variable retains its value between activations of the block that contains that variable.

Programming Environment: An integrated collection of tools that is commonly used to create, run, test, and maintain programs.

separate compilation: The capability of breaking a single C program into several files, each of which can be compiled by itself.

String: An array of characters whose last value is the null character '\0'.

Structure: A collection of data of different types whose components, i.e., fields, are referred to by name.

Structured Programming: A method of constructing computer programs in which structured control constructs are used without exception.

Type: A description of the kinds of values that a particular identifier can hold, and the operations that can be performed on those data.

EFERENCES

American National Standard for Information Systems Programming Language—C. American National Standards Institute, New York, approved December 14, 1989.

C. Bohm and G. Jacopini, Flow diagrams, Turing machines and languages with only two formation rules. Commun. ACM 9(5):366–371, 1966.

Grady Booch, Object-oriented development. *IEEE Trans. on Software Eng.* SE-12(2):211–221, 1986.

Grady Booch, *Object-Oriented Analysis and Design with Applications*, 2nd ed. Benjamin/Cummings, Redwood City, CA, 1994.

Paul S. Cilwa, *Borland C++ insider*. John Wiley, New York, 1994.

Peter Coad and Edward Yourdon, *Object-Oriented Design*. Yourdon Press, Englewood Cliffs, NJ, 1991.

Edsger W. Dijkstra, GoTo statement considered harmful. *Commun. ACM*, 11(3), 1968.

Ben Ezzell, *Borland C++ 3.0 programming*, 2nd ed. Addison-Wesley, Reading, MA, 1992.

J. Guttag, Abstract datatypes and the development of data structures, *Commun. ACM* 20(6):396–405, 1977.

Brian W. Kernighan and Dennis M. Ritchie, *The C Programming Language*, 2nd ed. Prentice Hall, Englewood Cliffs, N 1988.

Richard C. Linger, Harlan D. Mills, and Bernard I. Witt, *Structured Programming, Theory and Practice*. Addison-Wesley, Reading, Ma, 1979.

B. Liskov and S. Zilles, Programming with abstract data types. *ACM SIGPLAN Notices* 9(4):50–59, 1974.

Chris H. Pappas, *Borland C++ Handbook*, 4th ed. Osborne McGraw-Hill, Berkeley, CA, 1994.

Richard O. Parker, *Easy Object Programming for the Macintosh Using AppMaker and THINK C*. Prentice-Hall, Englewood Cliffs, NJ, 1993.

Jerry Peek, Tim O'Reilly, and Mike Loukides, *UNIX Power Tools*, O'Reilly and Associates, Sebastopol, CA, 1993.

Symantec THINKC/Symantec C++ User's Guide. Symantec Corporation, 1993.

Douglas A. Troy and James D. Kiper, *The C Programming Language: Including ANSI C, Portability, and Software Engineering*. Scott Foresman, Glenview, Il, 1989.

14

C++ and Objective C

arl B. Erickson
rand Valley State University

This chapter is written for C programmers who want to know what the common object-oriented C hybrids (C++ and Objective C) add to their base language. Readers unfamiliar with C should begin with Chapter on ANSI C. Some familiarity with object-oriented (OO) ideas is also assumed. Readers for whom OO languages are entirely new should first read Chapter 15 on OO analysis and programming.

Why are there two commercially successful, hybrid OO languages based on C? The answer lies in the popularity, flexibility and performance of their base language. As a general purpose language, C has much to offer as a base upon which to build: it runs on nearly every architecture, there are many C programmers, it is fast and efficient, and existing C programs will work with the new compilers with either no or only minimal modification. When C++ and Objective C were being developed, a common criticism of the pure OO languages was their performance. By choosing a hybrid approach, the developers of these languages sought, among other goals, to find a compromise between speed and support for OO programming.

This chapter is organized into five parts: this introduction section, a brief background on some important ideas from C, a section on the C++ language, a section on the Objective C language, and finally a comparison and summary of the two. While the focus of the language sections is on the support for OO programming, the C++ section has additional coverage of the non-OO features of the language. These sections conclude with coverage of advanced language features.

14.1 COMMON IDEAS

This short section considers common issues of code organization, declaration and definition, scope, binding, and typing. These important ideas are used later in the language sections.

Files and C

C programming involves three file types: header files, with .h extensions; source files with .c extensions, and libraries, with various extensions (.lib and .a are both common). Header files usually contain variable, type and function declarations and pre-

-8493-2530-7/96/$0.00+$.50

processor macros. These declarations are included in the source files which need them with a preprocessor directive, the #include. Source files are compiled to object files (often a .o extension), and the object files are linked with the appropriate libraries to make an executable program file. The same system of files is used in C++ and Objective C.

Object-oriented programming makes a fundamental distinction between interface and implementation. A common way of writing the code for a custom class mimics this dichotomy; a header file contains the interface to a class and a source file contains the implementation of the class methods. Class hierarchies can be placed in libraries to which header files serve as the roadmap.

Declaration and Definition

Understanding the difference between declaration and definition is essential to understanding C programs. A declaration of a variable or function is simply a statement of its type and name, and, in the case of functions, its parameter list. Declarations allow the compiler to do type checking. Something may be declared as many times as is necessary to inform the compiler about the thing being declared. Multiple declarations are implicitly part of using the #include preprocessor facility, since multiple source files may #include the same declarations. With ANSI C, the compiler must see a declaration for a thing before it can use that thing. The declaration of a function is known as its prototype.

In contrast to declarations, a thing (variable, function, or object) must be defined once and only once. Multiply defined things result in errors at compile-time. When the compiler sees the definition of a thing, it allocates memory for it. The definition of a function is the body of code for that function.

Scope

Scope is the area of a program in which the name of a variable, function, or object is visible. C++ and Objective C offer three kinds of scope: global, local, and class. Global scope is done just as in C, by defining the variable outside of any function or class scope. The static modifier can be used to limit global scope to a single file. Local scope is provided for automatic variables (those defined within functions). Class scope is provided for data members of a class, and encompasses all of the methods of the class.

Typing and Binding

Typing refers to the association of a variable with a data type. Strongly typed languages are those that enforce the assignment of values that match the declared type of a variable. The advantage of strong typing is that the compiler can find errors at compile time by checking that, for instance, function arguments match the type declaration of the function, and that function return values are assigned to compatible type variables. The original C language was not strongly typed. ANSI C is more strongly typed than C. The other consideration of the typing issue is when checking should be performed. Static type checking is done at compile-time. Dynamic type checking is done at run-time. Dynamic type checking results in error messages at run-time when a type error is encountered.

Binding refers to the association of a function call or message to the actual code to execute. Object-oriented languages fall into two camps, static binding and dynamic binding. Dynamic binding means deferring the decision about what code is executed for a given message until run-time. The class of the receiving object is not determined until the time the message is sent, so the code that is executed can be determined by the class of the receiver. Static binding determines the code to execute at compile time, based on the type of the variable used to message the receiving object. Pointer variables allow the type of the receiving object to differ from the type of the pointer used to send the message. Static binding is faster and safer, since it resolves the binding question at compile-time, rather than run-time. Dynamic binding is more flexible, and is better able to support polymorphism, a fundamental OO principle.

An object-oriented language may provide either form of typing (static or dynamic) and either form of binding (static or dynamic), which provides four possibilities. Traditionally, non-OO high level languages use static typing and static binding. Pure OO languages use dynamic typing and dynamic binding.

14.2 THE C++ LANGUAGE

C++ is an immensely rich language. This richness is both a blessing and a curse: a blessing, because of the expressive power and support for several programming paradigms; a curse because this richness means complexity, and there is much to master. C++ is a language to grow with, one for which each experience can teach new features or better understanding. Since each of C++'s features may interact with the others, learning C++ feels like gradually filling in a not-so-sparse matrix of knowledge formed by the cross-product of the C++ feature vector with itself. No serious use of the language should be undertaken without good references at hand. The end of this section has several suggestions for more complete sources of C++ knowledge, along with brief commentary on the strengths of each reference.

C++ was developed by Bjarne Stroustrup at Bell Laboratories in the 1980s for his own use in writing complex simulations (Stroustrup, 1991). His stated design goals for the language were

- to improve upon C,
- to better support data abstraction, and
- and to support object-oriented programming.

In addition to these goals, compatibility with C was to be maintained, and C's applicability to demanding low-level system programming problems retained. The support for object-oriented programming was inspired by Simula67.

In 1989 interest in C++ was sufficiently widespread to warrant the initiation of an ANSI/ISO standardization effort. The ANSI X3J16 committee is expected to issue the first C++ language standard by the end of 1996. Until that time, the de facto standard for the language is the ANSI committee baseline document, ***The Annotated C++ Reference Manual*** by Ellis and Stroustrup (1990), known as the "ARM."

The original C++ "compiler" was actually a translator that preprocessed C++ into plain C, and compiled the result. The cfront translator is currently in Release 3.0. Many other C++ compilers are now available, including the free (via the Internet) GNU C++ compiler from the Free Software Foundation, which has been ported to many architectures and operating systems. The portability of C++ code between compilers tends to depend on the usage of recent, or advanced, features of the language. The more features used, the less likely it is that the code will be portable across many compilers. In general, C++ is more portable between the many UNIX-based compilers than it is from one of the UNIX-based compilers to a DOS/Windows-based compiler.

As planned, ANSI C is nearly a perfect subset of C++. Most ANSI C programs will compile without change using a C++ compiler. Books such as Lippman, (1991) or Stroustrup, (1991) describe the few areas in which C++ and ANSI C differ. Older Kernighan and Ritchie (K&R) C programs require a little more work, chiefly in the area of function declarations.

Improvements over ANSI C

The features described in this section represent those that were added to C++ to improve some aspect of the C language. Many of them answer criticisms of C (e.g., weak type checking, syntax for pass-by-reference parameters) and others were included to improve the representation and use of abstract data types. The features described here represent the most important non-object-oriented extensions that C++ adds to ANSI C, and are loosely classified into five groups: comments, types and declarations, functions, memory management, and I/O.

Comments

The traditional C comment delimiters (/* and */) have been supplemented with a new, to-the-end-of-line delimiter, the double-slash, //. One major advantage of the new style comment is that they may be nested, whereas the old style may not. The three examples below show typical uses of C++ comments. The third code fragment shows a use for the new style that could not be achieved with the old style alone.

```
int numberOfEmployees; // number of employees in DB
   // for all employees
for(i=0; i < numberOfEmployees; i++)
   /* commented out for debugging
   if(salary > 10000.0){
     print_employee(i); // display emp info first
   */
```

Types and Variable Declarations

The names of structures are types, so that the keyword struct may be be omitted in the definition or declaration of structures, or when describing structures as parameters. This eliminates the need for typedefs in many situations.

```
struct Process {
   int pid;
   int ppid;
   int priority;
};
struct Process proc1; // old style C
Process proc2; // ok in C++
extern void printProcess(Process);
```

Automatic (i.e., local) variables may now be defined anywhere in blocks, rather than strictly at the top of the function. This allows variables to be defined at their point of use, rather than at the top of the function, potentially well before their actual use.

```
int main(void)
{
   for(int i=0; i < <max; i++){
     float salary = salaryForDept(i);
   }
}
```

A new type modifier, **const**, allows truly constant-valued variables, rather than having to use the preprocessor to this effect.

```
#define MAX 10 /* old way */
const int max = 10;// new way
```

Variables that are defined const must be initialized, since they may not be assigned to. The compiler will enforce the meaning of the constant modifier. This is different than a #define macro "constant," since the const variable is a real variable, with space allocated for it by the compiler.

C++ provides an alternative to the type cast means of converting from one type to another. A data type may be applied to a variable with a function-like syntax:

```
int count, sum;
float avg;
avg = (float) sum / (float) count; // old way
avg = float(sum) / float(count); // new way
```

C++ supports anonymous unions (unions without names). This reduces the level of access operators that must be applied to access an element of a structure containing a union.

```
struct VarData {
  int type;
  union {
    int integerData;
    float floatData;
    char charData;
  };
};
VarData d;
d.integerData = 2; // no union name
                   // so only 1 dot operator
```

Functions

Many improvements were made in the support for functions. The first, and arguably the most important, is the improved type checking of function parameters and return values. This improvement is shared by ANSI C, but is important enough to describe here. Old style C allowed use of a function before its definition. The compiler assumed the return value of the function was an integer, and did not attempt to check whether the function was invoked with the correct number and type of parameters. In C++, the compiler must be given a prototype declaration of a function before it is used. The function prototype consists of three things: the name of the function, the type of the return value, and the type of each parameter. Function prototypes are usually placed in header files, which are #include'd by the source file where the function is used. With the information from the prototype, the compiler can check each function invocation to see if the parameters are of the proper type and number, and if the return value is stored into a variable of the proper type. Function prototypes must agree with the definition of the function. The actual names (but not the types) of the parameters are optional in a prototype.

```
   // prototype (declaration)
void printProcessInfo(Process);
   // name of param is optional
void printProcessInfo(Process p);
   // definition
void printProcessInfo(Process p)
{
   // body of func
}
```

C++ allows the **overloading** of function names. This means that more than one function may have the same name, as long as the compiler can distinguish the like-named functions by their return type and parameter list. For example, ANSI C would require three different names for three functions that printed variables of different types.

```
void printFloat(float);
void printInt(int);
void printChar(char);
```

But C++ allows a single name to be overloaded, and used for each of these functions. The compiler knows which function is being invoked by the type of the parameter being passed.

```
void print(float);
void print(int);
void print(char);
float area;
print(area); // invokes the correct function
```

Default parameters allow for the invocation of functions with no values for those parameters that have defaults. Default values are specified for parameters in the function declaration. A parameter can have its default value specified only once in any given file. The header file that contains the function prototype is preferred for default values over the function definition.

```
float washerArea(float id = 0.5, float od = 1.0);
washerArea();          //  area of 1/2" washer
washerArea(0.2);       //  area of "fat" washer
washerArea(0.75, 1.5);//  area of 3/4" washer
```

The positional matching of values with parameters restricts which parameters may be allowed to default. Every parameter to the right of the first default-value parameter must also default. In the example above, there is no way to specify a function invocation of washerArea, which has a default value for od, but not id.

A notation for denoting that a function takes an unknown number of unknown type parameters is available in C++. The classic example for the need for such a notation (previously done with the C preprocessor in varargs.h) is the function printf. After the first argument to printf (a char * format string), the number and type of the arguments depend on the specific use of the function. In C++ we could declare the printf function like this:

```
int printf(char *fmt, ...);
```

The ellipsis turns off the argument checking normally done by the compiler.

Function parameters in C are passed by value. To have another function operate on a variable local to the scope where the function is invoked, the address of the local variable is passed to the function. This is a common use of the pointer in C. For example,

```
struct Process {
   int pid;
};
void enterProcessInfo(Process*, int*);
int main (void)
{
   Process proc;
   int done;
   enterProcessInfo(&proc, &done);
   if( done )
      // take some action
}
void enterProcessInfo(Process* proc, int* flag)
{
   scanf("%d", &proc→pid);
   if( proc→pid > 10)
```

```
      *flag = 1;
   else
      *flag = 0;
   }
```

The disadvantage of passing addresses as parameters to the function enterProcessInfo is that it requires us to use a different syntax to access these parameters in the function. The dereferencing operators → and * must be used to read or write the parameters. We also must use the & operator in the function invocation to send the address of our data objects, and not their value.

C++ has another means of passing a parameter by reference. Reference parameters to functions may be treated in the function as a "normal" variable, i.e., no dereferencing operators are necessary. In addition, the address of variables is automatically passed at invocation, so the & operator is also unnecessary. With the reference parameter, we could rewrite the above code fragment to

```
struc Process {
   int pid;
};
void enterProcessInfo(Process&, int&);
int main(void
{
   Process proc;
   int done;
   enterProcessInfo(proc, done);
   if( done )
      // take some action
}
void enterProcessInfo(Process& proc, int& flag)
{
   scanf("%d", &proc.pid);
   if( proc.pid > 10 )
      flag = 1;
   else
      flag = 0;
}
```

Reference variables may also be used, like pointers, to create another name for a data object:

```
int x;
int& rx = &x;
x = 2;
rx = 2; // same thing as setting x directly
```

Reference variables must be initialized to the address of a variable of compatible type. Their referent cannot be changed at run-time.

In terms of the output of the compilation process (an executable program file), the use of functions to organize and reuse code represents a tradeoff between space and time. By having multiple function invocations to the same code, the size of the program is reduced, as the code for the function is not replicated at the point of each use of that function. There is a cost in time, since the passing of parameters, and the change of control flow takes some time for the CPU to perform. The presence of an instruction cache and a heavily pipelined CPU tend to make this cost relatively greater on high performance computers. C++ provides a means of controlling the tradeoff between time and space more precisely. By declaring a function with the keyword **inline**, the compiler will replace the usual assembly language code to jump to the function with the actual code of the function.

```
inline int aSmallFunction(int);
```

By putting the actual function code inline with its invocation, the size of the program is increased, but the program will execute faster. For very small functions, both code size and time may be saved. The inline specifier is only a hint to the compiler. The compiler gets to decide whether the function can be "inlined" at each point of invocation (it may be too complex, or the circumstances may not allow it). Novice C++ programmers inline too readily. Inlining should be considered only after careful performance analysis of the program under execution.

Memory Management

C++ offers a new model of memory management to replace C's malloc and free. The operators **new** and **delete** allocate and release memory from the free store. Memory allocated by new will be available to the programmer until it is returned to the free store with delete. See the earlier discussion on object lifecycle issues for more information on this important subject. The example below is typical of the use of new; a node in a tree is being dynamically allocated then released.

```
struct TreeNode {
  int id;
  TreeNode* child;
  TreeNode* parent;
};
TreeNode* newNode(int id)
{
  TreeNode* tnp = new TreeNode;
  tnp→id = id;
  return tnp;
}
```

The function newNode dynamically allocates a node by calling new. The new operator returns a pointer to the newly available memory. The function returns this pointer so that the new TreeNode structure object can be used elsewhere in the program.

Eventually, when the program no longer needs this TreeNode object, the delete operator is called with the address of the TreeNode structure. This returns the memory to the free store.

```
delete node;        // where "node" is the value returned
                    // by the function newNode
```

The new and delete operator have a slightly different syntax for allocating arrays of objects.

```
int* arrayOfInt = new int [10];
delete [] arrayOfInt;
```

It is important that you use the delete [ ] operator when you are deleting something that was allocated with new [ ] (a vector), and use delete when you are deleting something that was allocated with new (a scalar).

The malloc/free facility is still available, though new and delete are preferred. Mixing the two facilities in the same program is dangerous.

Input and Output

The standard input/output facility of C continues to work in C++. An improved facility, known as iostream, takes advantage of the improved support for data abstraction and operator overloading. To fully understand the iostream facility requires knowledge of the class concept. This issue is revisited in more detail later in this section.

The standard input of a C++ program is reached via the object **cin**. Similarly, the standard output is available at **cout**. The operators << and >> are overloaded to work with cin and cout and produce formatted input and output. Consider the classic "hello, world" program in C++:

```
int main(void)
{
  cout << "hello, world!\n"
  return 0;
}
```

The << operator may be chained with other << operators, so that

```
int sum, count;
cout << "count is " << count << " and sum is " << sum;
```

is equivalent to

```
count << "count is ";
cout << count;
cout << " and sum is ";
cout << sum;
```

and neither output line will be flushed to the standard output until either

```
cout << "\n";
```

or

```
cout << endl;
```

is performed. The corresponding input operator, >>, works with cin to read the standard input. Notice the use of the reference operator in the input function below.

```
void queryForMaxMin(int& max, int& min)
{
  cout << "please enter max and min ";
  cin >> max >> min;
}
```

The standard error is available to C++ programs as cerr.

Advice for C Programmers

The following advice for C programmers converting to C++ was gleaned from Stroustrup (1991), Lippman (1991), and Meyers (1992):

- convert #define macros to const data types, or inline functions
- use the const modifier whenever possible
- prefer C++ style comments (//) over C style comments (/* */)
- use the iostream library, rather than stdio
- define variables where they are first used, rather than at the top of a function
- use new and delete, not malloc and free
- use the improved data abstraction and OO features of C++ to avoid void*, pointer arithmetic, arrays, and type casts
- think in terms of objects messaging each other, rather than data structures operated on by functions.

C++ code may be linked with C code provided certain steps are taken. For each straight C function that is to be linked into a C++ program, the C++ program must have a special external declaration. To properly link a C function from an existing library,

```
void straight_c_func(int);
```

the C++ program must have the following extern symbol:

```
extern "C" void straight_c_func(int);
```

C++ Support for OO Programming

One of the major design goals for C++ was to support object-oriented programming. The strong compile-time typing of C++ adds complexity not found in pure object-oriented languages such as Smalltalk. The need to retain the efficiency and speed of C also influenced the support for OO programming. This section is organized around the fundamental concepts of object-oriented programming, describing how each is implemented in C++. A final section addresses some of the features of the language that allow a programmer to subvert the principles of OO programming.

Classes

The concept of class is the principal means of data encapsulation and abstraction in C++. You can think of a class as a structure with functions. In fact, in C++ struct is only a special case of class. The relation of class to object is that of a data type to a variable. A class consists of members: data and methods. Instance variables (sometimes abbreviated as "ivars") are the data members of a class. Functions and operators are the methods of a C++ class. This section will use a common convention for the names of instance variables, member functions, objects, and classes. Names are concatenated, descriptive words. In the case of classes, the first letter of each word is capitalized, like Employee, GraphNode, Animal. For ivars, functions, and objects the first letter of each word *except the first word* is capitalized, like printFirstName(), fedTaxWithholding, and newNode.

Consider representing an employee with a C++ class. Perhaps you are writing an application for a human resources department. You encapsulate everything you need to know about employees in this class with instance variables. The declaration of a simple employee class might look like this:

```
class Employee {
  int        empId;
  char*      fName;
  char*      lName;
};
```

Classes differ from C structs by also having methods. These member functions and operators define what an object of this class can do, whereas the data members define what the object knows. Suppose an employee object can do three things: tell you its id number, calculate its length of service, and print its name in a company standard format. We can declare this functionality in our Employee class like this:

```
class Employee {
  int        empId;
  char*      fName;
  char*      lName;
  int        idNumber(void);
  int        lengthOfService(Date d);
  void       printName(void);
};
```

The definition of the member functions is usually in a separate source code file (an implementation file, e.g., Employee.cc) whereas the declaration for a class is in a header file (an interface file, e.g., Employee.h). The convention of separating the implementation from the interface promotes the sharing and reuse of classes.

The implementation file for the Employee class would have three functions defined in it, and would include the class header file, as well as any other header files needed for the function implementations.

```
#include        "Employee.h"
#include        <iostream.h>
// Return the employee id number.
int Employee::idNumber(void)
{
  return empId;
}
// Return the length of service in years
// on the date passed as a parameter.
int Employee::lengthOfService(Date d)
{
  Date hDate = hire_date_from_DB(empId);
  if(d < hDate)
    return 0;
  else
    return d - hDate;
}
// Print name to the stdout in a standard format
void Employee::printName(void)
{
  cout << lName << ", "
       << fName << endl;
}
```

The member functions of a class have full access to the instance variables of that class. The scope operator symbol,::, is used to identify a function as a member function of a particular class.

Access to the members (both data and functions) of a class can be controlled. The keywords public, protected, and private are used to control access to the members of a class. Private members are limited to access by the member functions of the class. They are hidden from the outside world. Public members represent the interface offered to the outside world by this class. Protected members are accessible to the class itself and subclasses of the class. To achieve true data encapsulation, no data members should be in the public interface of a class. Instead, the public interface is composed of those functions that users of the class may ask objects of this class to perform. The public functions may themselves access private data members, but the details of this knowledge remain hidden from outside the class.

Here is a more OO declaration of our simple Employee class:

```
class Employee {
public:
  int    idNumber(void);
  int    lengthOfService(Date d);
  void   printName(void);
private:
  int    empId;
  char*  fName;
  char*  lName;
};
```

Protected access is described more fully in the section on inheritance.

C++ supports class members (i.e., data and methods shared by all objects of the same class) through the static modifier. For example, declaring a static int count variable in the private portion of the Employee class would result in all Employee objects sharing a single integer variable called count. This is in contrast to instance variables, where each object has its own personal copy. A static data member provides a means of having a "class global" variable while preserving data encapsulation.

A member function may also be declared static and be shared by all instances of a class, providing that the function accesses only static data members of the class, and not regular ivars or regular member functions.

A C++ class declared in a header file is defined the first time the header file is seen by the compiler. One common problem with placing class declaration/definitions in a header file is that there are situations in which it is easy to have multiple inclusions of the same header file. This results in multiple definitions for that class. For example, suppose you have a Circle class and a Square class, which are both subclasses of Shape. In a particular source file, say main.c, you are using both Circle and Square objects, so you #include Circle.h and Square.h. But each of those header files #includes Shape.h, since they are subclasses of Shape. Now, in your source file, you have multiple inclusions of Shape.h, via Circle.h and Square.h. Figure 14.1 illustrates the problem of #including Shape.h twice, and hence the compiler trying to define class Shape twice.

There are various solutions to this problem, but a commonly used one in the C++ world is to use the preprocessor directive, which allows for conditional compilation. A macro is reserved for each class header file. If this macro has not yet been defined, the class declaration is included, otherwise it is not. The reserved macro is defined the first time the class declaration is included. Here is what the header file for the Shape class would look like using this approach:

```
#ifndef SHAPE_H
#define SHAPE_H
/* declaration of Shape class interface */
#endif
```

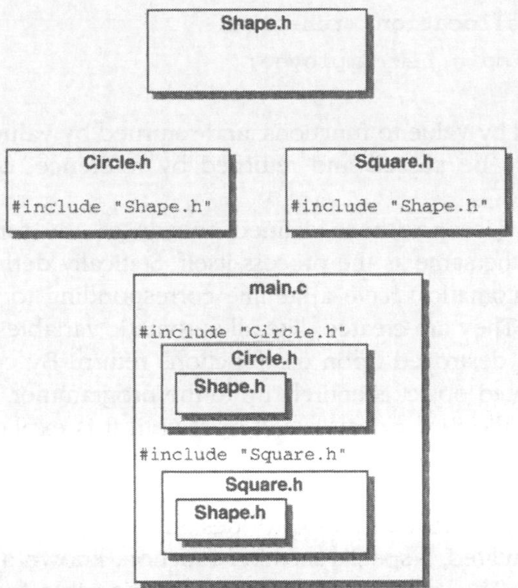

FIGURE 14.1 The problem of multiple inclusions of a class header file.

Objects

Classes play the role of data type to objects. Classes are passive; they do not get your work done. By contrast, objects are active. An object definition causes the allocation of memory by the compiler. An object is instantiated from, or is an instance of, its class. Objects get your work done.

Objects get work done by messaging each other. Each message has at least two components: a receiver and a method. The method must be a member of the receiver's class definition. The optional third component of a message is the set of parameters required by the method being invoked. The messaging operators in C++ are identical to the structure access operators (. and →) in C. The general form is

```
// an object
Class    object;
// a pointer to an object
Class*   pobj;
// invoke function method with two args
object.method(arg1, arg2);
pobj→method(arg1, arg2);
```

Objects, like their built-in cousins ints, chars, and floats, may be created statically, at compile-time, or dynamically, at run-time, using the new operator.

```
   // static allocation (compile-time)
Employee president, secretary;
```

```
    // dynamic allocation (run-time)
Employee* newEmp = new Employee;
```

C++ objects are passed by value to functions, and returned by value from functions by default. They may also be passed and returned by reference, or by address, as a pointer.

Objects with external scope (those defined outside of any function, or globally) have a lifetime that is the same as the process itself. Statically defined objects that are local to a function (automatics) have a lifetime corresponding to the duration of the scope of the function. They are created, like all automatic variables, upon entry to the function, and they are destroyed upon the functions return. By contrast, the lifetime of a dynamically allocated object is entirely up to the programmer. An object allocated from the free store by the new operator will exist until it is explicitly destroyed with the delete operator. The management of object lifetimes is a central part of the correct and robust solution to a problem in an OO paradigm. In C++ it is also the potential source of many problems and weaknesses of design.

When an object is created, a special member function, known as a **constructor**, is invoked automatically. The constructor function is responsible for the correct initialization of the object being created. Constructors, like any function in C++, may be overloaded to provide multiple interfaces for the creation of objects. All constructors for a class are given the name of the class. We could add a constructor declaration to our Employee class by putting the following lines in the public section of the class declaration:

```
class Employee {
public:
        // constructors
    Employee();
    Employee(int id);
    Employee(int id, const char *first, const char *last);
        // accessor methods
    int idNumber(void) const;
    void setIdNumber(int i);
    const char* firstName(void) const;
    void setFirstName(const char* n);
    const char* lastName(void) const;
    void setLastName(const char* n);
        // calculation methods
    int lengthOfService(Date d);
    void printInfo(void) const;
private:
    int         empId;
    char*       fName;
    char*       lName;
};
```

This new declaration of our Employee class shows the constructors, providing three distinct interfaces for creating Employee objects, additional methods (known as accessor or get/set methods) for accessing the instance variables of the class, and a const keyword following certain class functions. The const keyword is applied to functions that are safe to perform with const Employee objects. These functions are guaranteed not to alter the instance variables of the object.

The three constructors can be distinguished by their parameter list. In the first case, no values are passed with which to initialize the object. In the second case, the employee id could be initialized from the integer argument. The last case provides values for employee id and name. Using the ability of C++ to specify default values for function parameters, these three constructors could be replaced by a single constructor that provides default values for id, first name and last name. The declaration of this single constructor would look like:

```
Employee(int id = 0,
  const char *first = 0,
  const char *last = 0);
```

An Employee object that was created with no arguments would have an employee id of 0 and no first and no last name. We could implement this rather flexible constructor in the following way:

```
Employee::Employee(int id,
                   const char *first,
                   const char *last)
{
  empId = id;
  // allocate space, whether or not we have an arg
  fName = new char[(first ? strlen(first): 0) + 1];
  // if there was an arg, use it for init value
  if( first )
    strcpy(fName, first);
  else
    fName[0] = '\0';
  lName = new char[(last ? strlen(last): 0) + 1];
  if( last )
    strcpy(lName, last);
  else
    lName[0] = '\0';
}
```

This constructor demonstrates the access to instance variables that member functions of a class have; it directly sets empId. It also shows a typical memory management issue faced in classes. The constructor dynamically allocates enough character storage to hold the names to which the object's ivars are being initialized. In a production class it would be more likely to see the details of character array allocation, and the char* ivars replaced by String objects.

A special type of constructor is generated automatically by the C++ compiler. The compiler-provided default copy constructor does a member-by-member copy of the instance variables. The copy constructor is used when an object is passed by value, returned by value, or initialized with another object of the same class through assignment. The prototype for the copy constructor for Employee would look like this:

```
Employee(const Employee& emp);
```

and the implementation might be

```
Employee::Employee(const Employee& emp)
{
   empId = emp.idNumber();
   fName = new char[strlen(emp.firstName()) + 1];
   strcpy(fName, emp.firstName());
   lName = new char[strlen(emp.lastName()) + 1];
   strcpy(lName, emp.lastName());
}
```

The copy constructor copies the values of the ivars from the parameter object into its own ivars. Here are some examples when each type of constructor is invoked, as well as the equivalent values for the parameters:

```
// Employee::Employee(0, " ", " ");
Employee vp2;
// Employee::Employee(512, " ", " ");
Employee vp2(512);
// Employee::Employee(33, "Biggus", "Wiggus");
Employee vp3(33, "Biggus", "Wiggus");
// copy constructor, vp3 passed as arg
Employee vp4 = vp3;
```

To understand the importance of writing a custom copy constructor for the Employee class, rather than just rely on the default implementation provided by the compiler, consider the effect of doing a member-by-member copy on the ivars of the Employee class. For empId there would be no problem, simply copying the ivar by value is sufficient. However, copying the char* fName and lName by value would mean that object vp4 would have pointers to the same character arrays as vp3. If vp3 were deleted, then vp4's fName and lName ivars would point to reclaimed memory. Figure 14.2 illustrates the result of the default copy constructor.

The custom copy constructor given earlier correctly allocates storage for the name of the employee and then initializes this space from the object passed by reference as a parameter to the constructor. This is illustrated in Figure 14.3.

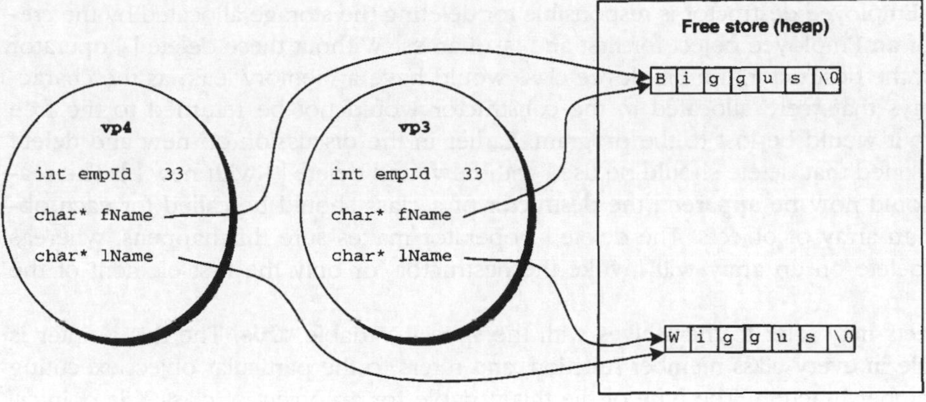

FIGURE 14.2 Effect of default copy constructor.

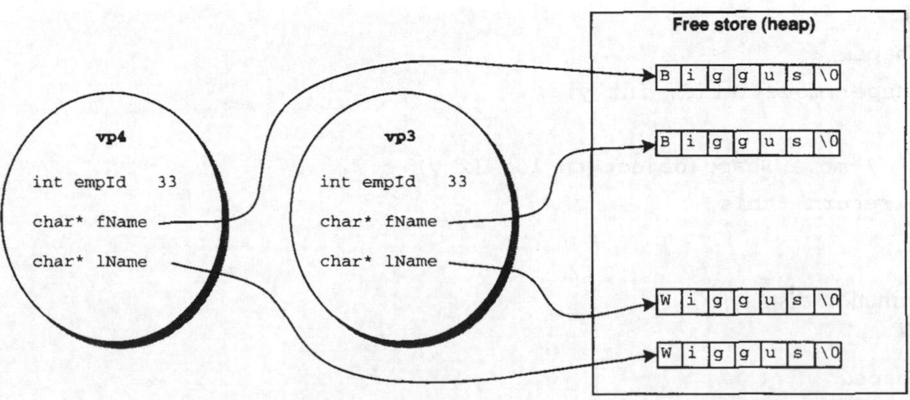

FIGURE 14.3 Effect of custom copy constructor.

If a constructor is concerned about the birth of an object, then a **destructor** is concerned about its death. A destructor function of a class is invoked whenever a dynamically allocated object is being deleted, or when a statically allocated object is going out of scope. The name of the constructor function is the name of the class preceded by a tilde.

```
// destructor declaration
~Employee();
// destructor implementation
Employee::~Employee()
{
   delete [] lName;
   delete [] fName;
```

The Employee destructor is responsible for deleting the storage allocated by the crè-ation of an Employee object for first and last names. Without these delete [ ] operator calls in the destructor, the Employee class would have a memory leak, as the character arrays that were allocated in the constructor would not be returned to the free store, but would be lost to the program. Earlier in the discussion on new and delete it was stated that delete should be used with new, and delete [ ] with new [ ]. The reason should now be apparent; the destructor of a class should be called for each object in an array of objects. The delete [ ] operator makes sure this happens, whereas using delete on an array will invoke the destructor for only the first element of the array.

Objects may refer to themselves with the special variable, **this**. The this pointer is available in every class member function, and refers to the particular object executing the member function. The type of the this variable for an object of class X is pointer-to-X. One use of the this pointer is as a return value for class member functions. Returning the contents of this allows for class functions to be nested. For example, if a class to represent geometric shapes that are drawn in a window had a function move,

```
Shape&
Shape::move(int x, int y)
{
   // move Shape object to loc (x,y)
   return *this;
}
```

and a function display,

```
Shape&
Shape::display()
{
   // display shape at current location
   return *this;
}
```

then a Shape object, s1, could be sent consecutive messages to move and display like this:

```
Shape s1;
s1.move(0,0);
s1.display();
```

or it could be written in a nested fashion, like this:

```
s1.move(0,0).display();
```

The this pointer is also commonly used where memory management issues arise, as in adding elements to or removing elements from linked lists.

Inheritance

One of the most distinguishing means of representing relationships between abstractions in an OO program is with inheritance. Inheritance means forming subclass/superclass relationships between classes, also known as a **base class** and **derived class** relationship. A subclass inherits members from its superclass. Many classes together form an inheritance, or a class, hierarchy. Classes at the top of a hierarchy represent the data and functionality common between the classes that inherit from them. Classes at the top tend to be more abstract; objects are usually not instantiated from such classes. Classes at the bottom of the hierarchy are more concrete, and are more likely to be instantiated into objects. Classes that inherit from only one superclass exhibit single inheritance. Classes that inherit from more than one superclass exhibit **multiple inheritance**. Rich language that it is, C++ supports both single and multiple inheritance.

Continuing with our Employee class example, suppose that you needed to represent a variety of people in your programs, and not simply employees. You might have a Customer class, for example. Both customers and employees are people; the commonality between Employee and Customer class can be pushed upward to an abstract Person class. Employee and Customer can inherit functionality and data from Person, rather than each having to define it themselves.

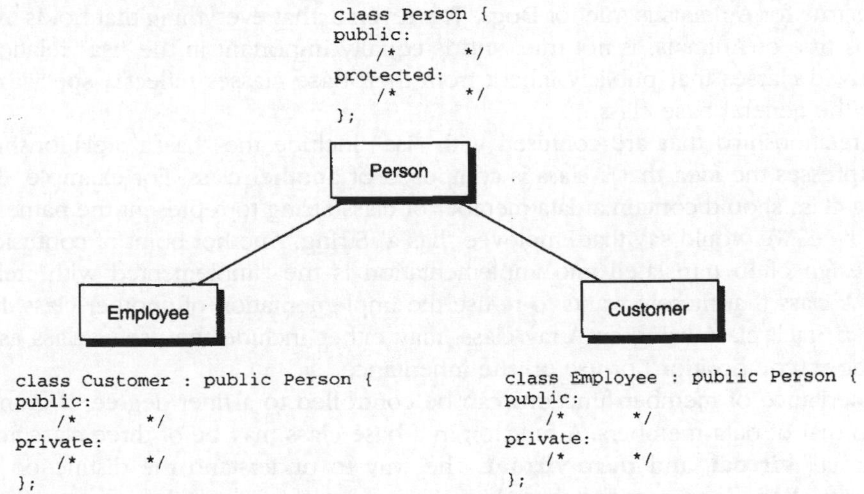
class Person {
public:
/* */
protected:
/* */
};
Person
Employee
Customer
class Customer : public Person {
public:
/* */
private:
/* */
};
class Employee : public Person {
public:
/* */
private:
/* */
};

FIGURE 14.4 Inheritance relationship between Person, Employee, and Customer classes.

Figure 14.4 illustrates the base relationship that Person has to the derived classes Customer and Employee. The skeleton of the class declarations shows the syntax for indicating the base class in a derived class declaration. Since both employees and customers have names, the data members fName and lName of the Employee class could be removed from Employee and put instead in Person. The member functions to get/set these ivars would also be placed in Person. Objects instantiated from Employee and Customer inherit the data (names) from Person, and the ability to get/set those names. Consider the issue of access control to the fName and lName ivars in Person. They should be hidden outside of member functions to preserve data encapsulation. With the Employee class we achieved this level of protection by declaring them private. If we declare them private in Person, however, then objects of type Employee and Customer will not be able to use those ivars. The derived classes do not inherit the private members of the base class. The compiler will detect illegal uses of private ivars in derived classes. The solution to this problem is to declare fName and lName as protected. Protected members are public to derived classes, but private to the rest of the world.

Separate from the issue of access control for individual class members just discussed is the issue of how a derived class inherits from a base class. Inheritance may be either public, protected, or private. Table 14.1 summarizes the possibilities from the perspective of the derived class. Public inheritance means that the public and protected members of the base class retain their status for the derived class. Protected inheritance from a base class means that public and private members of the base class are protected in the derived class. Private inheritance means that public and protected members of the base class are private in the derived class. Private inheritance can be used to ensure that a public interface inherited from a base class is not available (i.e., not public) to the users of the derived class. Or, put another way, the fact that the derived class uses the base class via inheritance is hidden from the world by the derived class.

Meyers says the single most important thing to know about inheritance in C++ is that public inheritance means "isa." A derived class "isa" base class (and hence should publicly inherit from the base) if everything that is true about the base class is also true about the derived class. For example, a Dog class "isa" Animal, and everything that holds true for Animals is true for Dogs. The reverse, that everything that holds true of Dogs is true of Animals, is not true, and is equally important in the "isa" relationship. Derived classes that publicly inherit from their base classes reflect a specialization over the general base class.

Other relationships that are confused with "isa" include the "has a" relationship, which expresses the idea that a class is composed of another class. For example, the Employee class should contain a data member of class String to represent the name of the employee. We would say that Employee "has a" String. Another point of confusion in OO design often translated into implementation is the "implemented with" relationship. A class that merely wants to re-use the implementation of another class, for example, a Stack class using an Array class, may either include the useful class as a data member (composition) or use private inheritance.

The inheritance of member functions can be controlled to a finer degree of granularity than that of data members. A function in a base class may be of three classifications, normal, **virtual**, and **pure virtual**. The way to understand the distinction of these classifications is to separate the inheritance of interface from the inheritance of implementation. The data in Table 14.2 summarizes the use of these inheritance classifications. Normal functions are those that are neither declared virtual nor pure virtual. Such functions specify an interface that is inherited by derived classes, as well as

TABLE 14.1 Access Control to Class Members from the Perspective of a Derived Class

	Inherited from Base Class		
Declared in Base Class	Public "Is a"	Protected "Is a for subclasses"	Private "Is implemented with"
Public	Public	Protected	Private
Protected	Protected	Protected	Private
Private	Private	Private	Private

TABLE 14.2 Meaning of the Inheritance of Functions by How They Are Declared in Base Class

Declared in Base Class	Meaning in Derived Class		
	What's inherited	Override in derived?	OOD term
Normal	Interface and implementation	Possible, but should not	Invariant over specialization
Virtual	Interface and default implementation	Optional	Specialization
Pure virtual	Interface only	Mandatory	Required specialization

a mandatory implementation for the function. While a derived class may in fact implement its own version of such a function, and the scope operator can be used to access the base class function within the derived class, doing so violates the spirit of public inheritance as representing an "isa" relationship. When a derived class overrides a function from the base class by reimplementing it, the derived class is in effect saying that what is true for the base class (the function in dispute) is *not* true for the derived class, and needs to be implemented differently. This represents an exception to the generalization/specialization relationship between the base and derived classes. On a very practical note, the next section on messaging and polymorphism discusses a very confusing behavior of C++ related to this issue of the inheritance of normal functions.

Virtual functions are made so by prefacing their declaration within a base class by the keyword virtual. These functions are designed to be implemented by a derived class in a way meaningful to the derived class (i.e., a specialization over the general base class). A virtual function may optionally have an implementation in the base class. If so, and if the derived class does not implement this function, then the implementation in the base class serves as a default implementation for the derived class.

A pure virtual function is declared by "initializing" the function in the base class declaration with the value 0. To derive a class from a base class with a pure virtual function, the derived class *must* implement the inherited pure virtual function. The example below illustrates the possibilities of function inheritance.

```
class Base {
public:
   void fundamental(void);
   virtual void depends(void);
   virtual void mandatory(void) = 0;
};
class Derived {
public:
   void depends(void); // optional to implement
   void mandatory(void);        // must implement
};
```

The function Base::fundamental() should not be reimplemented in Derived. The function Base::depends() may or may not be. If it is not, and if Base defines it, that implementation will serve for Derived. The function Derived::mandatory() *must* must be implemented in order for Derived to inherit from Base.

As an example of the possible uses of function inheritance, consider the possibilities faced when Employee was generalized to Person, and the ivars that hold the name of an employee were moved to this more abstract class. What should become of the function Employee::printInfo()?

```
void
Employee::printInfo(void) const
{
  cout << empId << ": "
       << (lName.length() ? lName.value() : "<none>")
       << ", "
       << (fName.length() ? fName.value() : "<none>")
       << ", "
       << endl;
}
```

In Employee this function printed the employee id, then the first and last names of the employee object. Migrating the functionality for printing names up the class hierarchy offers two advantages: this code need not be reimplemented in Customer, and every Customer and Employee object would print their name in a single, standard fashion. If the means of printing names were indeed rigorously standardized, then Employee::printInfo() could be taken apart and become a new, non-virtual function Person::printName(). By being nonvirtual, every derived class of Person would inherit not only the ability to print names, but a mandatory, standardized manner of doing it. The new Employee::printInfo() and the Person::printName() functions would look like this:

```
class Person {
public:
  void printName(void);
};
void
printName(void)
{
  cout << (lName.length() ? lName.value() : "<none>")
      << ", "
      << (fName.length() ? fName.value() : "<none>");
}
void
Employee::printInfo(void) const
{
  cout << empId << ": ";
```

```
    Person::printName();
    cout << endl;
}
```

The scope qualifier of Person:: is unnecessary in the printInfo function, since Employee only has one version of this function, but it serves to point out exactly how an Employee object goes about printing its information.

As a design alternative, consider that since the printing of names may be highly context dependent, it might be better to allow the subclasses of Person to override this functionality. In this case Person::printName() would be more appropriate as a virtual function.

```
class Person {
public:
  virtual void printName(void);
  // ...
};
class Employee {
public:
  void printName(void);
  void printInfo(void);
  // ...
};
void
Employee::printName (void)
{
  //print name in special Employee fashion
{
void
Employee::printInfo(void) const
{
  cout << empId << ": ";
     // use the default version
  Person::printName();
     // OR
     // use the custom version
  printName();
}
```

If the designer of Employee chose to implement a printName( ) function as above, then this function would be used on Employee objects, unless an explicit scope qualifier forced use of the inherited function. If no printName( ) function were implemented in Employee, then the default behavior, inherited from Person, would be used on Employee objects.

Inheritance complicates the initialization of objects. When an Employee object is allocated the constructors for both Employee and Person must be invoked, since the Employee object has instance variables from both Employee and Person. The syntax for doing this is

```
Employee::Employee(int id,
                    const char *first,
                    const char *last)
  : Person(first, last)
{
   empId = id;
}
```

The name of the base class is placed in the initialization list of the derived class constructor. Parameters to the constructor may be passed to the base class constructor as shown. Constructors in a multilevel class hierarchy are executed from the top of the hierarchy down. In this case the constructor for Person will execute before the constructor for Employee.

Polymorphism

Polymorphism is the idea that the code that is executed when a message is sent to an object depends on both the object's class and the name of the method in the message. In traditional procedural languages, the code that is executed by a function call is uniquely determined by the name of the function.

Before the idea of virtual functions was adopted, support for polymorphism in C++ was limited by the **static binding** that is done for nonvirtual functions. Static binding results in faster executing programs, since the alternative, dynamic binding, leaves the decision about what code is actually executed when a message is sent to be determined at run-time. The drawback of static binding is that it violates the idea of polymorphism, that the code executed should depend on the receiver's class. When the receiver's class is not known at compile-time, a static binding cannot be made, and hence true polymorphism is not possible. Why would the class of a receiver not be known at compile-time? When an object is dynamically allocated, it is messaged via a pointer to the object. The classic example of this is a List class. Consider the graphics class hierarchy in Figure 14.5.

Each class derived from Shape must implement a draw function to display itself. Clearly, how a shape is drawn is quite class dependent, and so must be implemented in the subclasses of Shape. We can now do things like

```
Circle          circ;
Square          sq;
circ.draw();    // draws the circle
sq.draw();      // draws the square
```

and see polymorphism in action as each figure correctly draws itself.

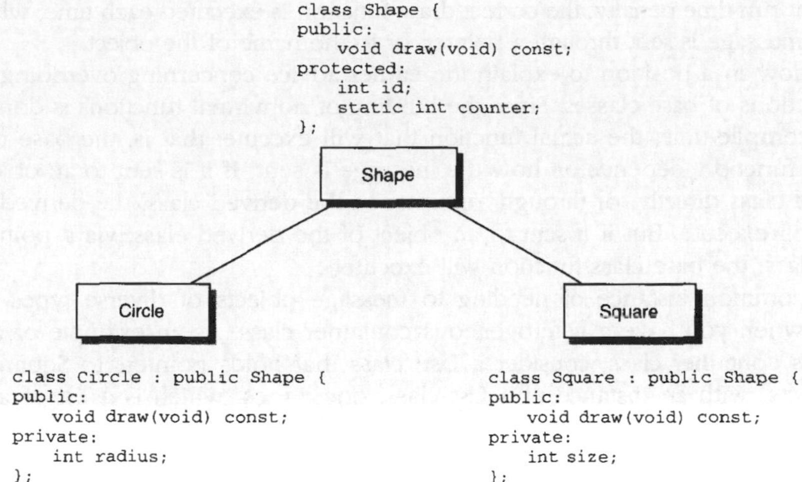
class Shape {
public:
void draw(void) const;
protected:
int id;
static int counter;
};
Shape
Circle
Square
class Circle : public Shape {
public:
void draw(void) const;
private:
int radius;
};
class Square : public Shape {
public:
void draw(void) const;
private:
int size;
};

FIGURE 14.5 Inheritance relationship between Shape, Circle, and Square classes.

Now suppose you needed to message the Circle or Square object via a pointer. For example,

```
Circle*          circP = &circ;
Square*          sqP = &sq;
circP->draw();   // still draws a circle
sqP->draw();     // still draws a square
```

and polymorphism is still working for us. But what if you needed to message Circles and Squares, via the same pointer? For example, if you had a List class that was able to store heterogeneous objects, then you could walk through this list asking each object to draw itself. But what pointer type would you use? Since both Circle and Square are subclasses of Shape, we can address objects of both these classes via a pointer to their superclass Shape:

```
Shape*           genericShapeP;
genericShapeP = &circ;
genericShapeP→draw();
// executes Shape::draw(), not Circle::draw()
genericShapeP = &sq;
genericShapeP→draw();
// executes Shape::draw(), not Square::draw()
```

The problem is that the compiler statically binds the Shape::draw() method to the above two messages, based on the data type of the pointer. Having to worry about which actual function will be executed depending on which way an object is messaged does not seem very object-oriented. The solution to this problem is to declare the Shape::draw() function virtual. This defers the binding of the code to the message until run-time. With only a slight run-time penalty, the correct draw function is executed each time, whether or not the message is sent through a pointer, or to the name of the object.

We are now in a position to explain the earlier advice concerning overriding nonvirtual functions of base classes. Since the binding of nonvirtual functions is done statically, at compile-time, the actual function that will execute, that is, the base or derived class function, depends on how the message is sent. If it is sent to an object of the derived class directly, or through a pointer to the derived class, the derived class function will execute. But if it sent to an object of the derived class, via a pointer to the base class, the base class function will execute.

A very common instance of needing to message objects of diverse types via a pointer is when you have a heterogeneous container class. As an example of a heterogeneous container class, consider a List[1] class that holds pointers to Square and Circle objects, with an instance of a List class, docShapes, which is a list of all the shapes currently known to the program. The List class could store pointers to void, and a type cast could be performed to message the contents of the List. Suppose you want to ask each of these objects to draw themselves (as you might, for example, to refresh the window in a drawing tool). The following code seems plausible:

```
List docShapes;
for(int i=0; i<docShapes.count(); i++)
   ((Shape*) docShapes.objectAt(i))->draw();
```

At compile-time the object that is being messaged with the draw() method has type pointer-to-Shape; the compiler therefore statically binds the code Shape::draw() for these messages. The result is that the individual Circle and Square objects do not draw themselves, but rather execute the default implementation of draw(), which they inherit from Shape. Making the Shape::draw() function virtual solves the problem. Type casting is generally frowned upon in C++. Scott Meyers puts it this way: "Casts are to C++ programmers what the apple was to Eve." Meyers describes several means of eliminating type casts, depending on the problem the cast is used to solve.

Virtual functions provide dynamic binding, and hence complete polymorphism, only for classes that share a common base class. In the example above, it was possible to message a Circle object and a Square object only via a Shape pointer because Circle and Square derived from Shape. Virtual functions do not allow a message to be sent to every object in a general heterogeneous list of objects.

One final note about the Shape class concerns the static data member, count. If each object of classes derived from Shape were to have a unique integer identity, then it makes sense to have the Shape class manage this information. Each time a Circle or a Square is instantiated, the constructor for Shape will be invoked. By simply incrementing a counter stored in the Shape class, and assigning this counter to the id of the object being created, we guarantee that each object will have a unique identity.

```
int MyShape::counter = 0;
MyShape::MyShape()
{
   id = counter++;
}
```

[1]This List class is not type-safe. A real List class is a perfect candidate for a template.

Operator overloading for user-defined classes is another means of support for polymorphism in C++. The traditional C operators, such as =, +, –, /, *, [ ], <, == may be overloaded and defined for a class. Class operators are declared and implemented in the same way that member functions are declared; the syntax is a bit less traditional.

```
class String {
public:
    String (const char* str);
    String& operator+=(const String& str);
    int operator!=(const String& str1, const String& str2);
private:
    // ...
};
```

The above incomplete declaration of a String class shows two overloaded operators, += to concatenate and != to test for equality. The implementation of these operators would perform the necessary steps (perhaps using the C library) for adding one string to another (+=) and comparing two strings for equality (!=). We could now use these operators like this:

```
String s1("mary had a");
String s2("little lamb");
if(s1 != s2)
    s1 += s2;
```

Almost all existing operators may be overloaded. New operators cannot be made up. Operators new and delete may be overloaded for a class that wishes to manage its own memory.

Defeating OO Principles

One aspect of the richness of C++ is the ability to support more than simply the OO paradigm of software development. Coupled with the desire to make C++ fast and efficient, several language features were created that can be (mis-)used in the implementation of an OO design. Experienced C++ programmers learn to make judicious and profitable use of these features. Novice C++ programmers often make a mess with them.

Free functions are functions that are defined outside the scope of any class. They may be used by class member functions. Library functions are examples of free functions in C++. Looking in the index of a C++ textbook will not usually turn up anything under free function, since a free function is simply a normal function in C. In the purer world of object-oriented thinking, where "everything's an object," free functions stand out.

If free functions are only passively un-OO, friend functions are actively so. A free function may be declared to be a "friend" to a class in the declaration of that class. Doing this grants access to the nonpublic parts of the class to the friend function. Throwing data hiding in the trash should not be done lightly, but there are times when breaking the rules can be enormously handy. Friend status may also be granted to other classes.

Support for global data (i.e., objects with global scope) is similar in C++ and C. The use of global data should be similarly limited. Complications arise when global objects are initialized with other global objects, particularly if the objects are in different files. Refer to a good book on C++ if you need to do this. Meyers has a lengthy discussion on this topic.

Data members in the public interface of a class subvert the data hiding intent of encapsulation. Access to data members should be controlled by member functions in the public interface.

Advanced Features

The nature and length of this chapter preclude an exhaustive description of C++. A nonscientific sampling of the length of C++ books reveals page counts of 660, 670, 980, and 420. While the features of the language discussed in this section are of vital importance to flexible and robust C++ class design and use, their complete description must be left to other references. These features are also likely to vary greatly between compilers.

Exception Handling. Despite the best efforts of the C++ compiler to enforce strong typing with compile-time checking and static binding, run-time errors can still occur. These include misuse of user-defined classes (e.g., giving an Employee a negative id), operating system limitations (e.g., running out of free store), and mathematical operations like dividing by zero. C++ refers to these sorts of events as exceptions. The verb "to throw" is used to describe the event of raising an exception. The exception handling facility of C++ allows the programmer to "catch" exceptions, execute a block of code, and exit. To be caught the exception must be thrown in a try block, or in a function called directly or indirectly within a try block. Try blocks are associated with one or more catch blocks. The following example shows throwing an exception when the free store has been exhausted:

```
try {
   String* name = new String("Joe");
   if(name == 0)
      throw "out of memory";
}
catch (const char* msg) {
   cerr << "Exception raised: " << msg << endl;
   exit(1);
}
```

Upon being thrown, control flow is transferred to the appropriate catch block, if any. After executing the catch block, control flow does *not* return to the point of the exception throw.

Templates. Templates allow you to parameterize a class by type. A good way to judge whether a template is an appropriate solution to defining a new class (versus subclassing an existing class, for instance) is by whether or not the semantics of the class depend on the type of data being operated on. If the semantics of the class are independent of the type of data, then a template is a good choice.

A classic example of use of a template is the Stack class. Suppose you have already written an IntegerStack class, implementing push and pop functions, and constructors and destructors for the class. Your IntegerStack class is written to handle integers; it is completely type safe, and so cannot be used for other data types. If you eventually need a FloatStack class, you have two choices: you can try to create a generic Stack base class, from which both IntegerStack and FloatStack inherit, or you can make a template class of Stack. A Stack template will allow you to create, as needed, Stack classes for any data type. The operations of the stack are entirely independent of the type of data being manipulated.

Here is the declaration for the interface of a template Stack class that implements a stack as a linked list of StackElement objects. The StackElement class is defined within the Stack class (StackElement is nested within Stack):

```
template<class Type> class Stack {
private:
   struct StackElement {
     Type value;
     StackElement *next;
     StackElement(const Type& newValue, StackElement *nextElem):
value(newValue), next(nextElem) {}
   };
   StackElement *top;
public:
   Stack();
   ~Stack();
   void push(const Type& val);
   Type pop();
};
```

And the skeleton implementation of the template Stack class would look like

```
#include "Stack.h"
template <class Type>
Stack<Type>::Stack(): top(0)
{
}
template <class Type>
Stack<Type>::~Stack()
{
   // implement destructor here
}
template <class Type>
void
```

```
Stack<Type>::push(const Type& val)
{
   // implement push here
}
template <class Type>
Type
Stack<Type>::pop()
{
   // implement pop here
}
```

In order to use the Stack template, a type must be provided to instantiate the Stack class for a particular type of Stack, for example:

```
Stack<int> integerStack;
Stack<float> floatStack;
```

Multiple Inheritance. The necessity of multiple inheritance (MI) is widely disputed. The advantage of multiple inheritance is that a class hierarchy can be constructed that more closely resembles the domain being modeled. Some OO languages support it, others do not. C++ offers multiple inheritance.

Detractors of MI claim it is not necessary, and have other techniques for modeling what may be done with MI (e.g., delegation, forwarding, and protocols in Objective C). Whether or not MI is necessary, there is general agreement on the complexity of using MI. As Meyers (1992, p. 157) puts it,

> The one indisputable fact about MI in C++ is that it opens up a Pandora's box of complexities that simply do not exist under single inheritance.

Multiple inheritance should be avoided by novice C++ programmers. Experienced programmers should use MI judiciously, and should study the sections on MI in several texts closely before doing so.

RTTI. One of the most recent features of the C++ language is support for RTTI, or Run Time Type Information. This allows an object to introspect at run-time to determine its class. This sort of introspection is commonly used in more dynamic OO languages, such as Objective C and Smalltalk.

DEFINING TERMS: C++

Base Class: A class that is inherited from, or subclassed by another class.

cin, cout, cerr: The I/O stream replacements for stdin, stdout, stderr.

const: A modifier to make a data object constant.

Constructor: An initialization function that is executed automatically when an object is created.

Delete: The equivalent of C's free. Returns dynamically allocated storage to the heap.

Derived Class: A class that inherits from a base class.

Destructor: A function that is executed automatically when an object is deleted.

Inline: Asks compiler to avoid function call, place copy of function body directly at point of invocation.

Multiple Inheritance: The ability to inherit from more than one base class.

New: The equivalent of C's malloc. Gets a dynamically allocated data object from the heap.

Overloading: Using the same name for more than one function or operator. They must be distinguished by parameters and return type.

Pure Virtual Function: A virtual function must be implemented by a subclass. Specifies inheritance of interface only.

Static Binding: Code executed by a message is determined at compile-time, based on how message is sent.

This: An object's self-referential pointer.

Virtual Function: A base class function that may be implemented by each subclass that inherits it. Allows for inheritance of interface and default implementation.

14.3 THE OBJECTIVE C LANGUAGE

Objective C is an object-oriented extension to the ANSI C language developed by Brad Cox at The Stepstone Corporation (Cox, 1991). The singular design goal for Objective C was to add support for object-oriented programming to C. This is in marked contrast to the multiple goals for C++ (see earlier section). The inspiration for the OO support in Objective C was Smalltalk. The influence of Smalltalk can be seen in fundamental ways, such as the dynamic nature of the language, and in less important ways, such as the messaging and class definition syntax.

The differences between Objective C and C++ can be summarized very simply. Compared to C++, Objective C is

- less feature-rich, less complex
- more dynamic

These differences have both good and bad implications. For instance, a language that is simple is also easier to learn and use. On the other hand, a rich language is more likely to have exactly the one feature that happens to be a perfect solution to a particular programming problem. Philosophically, the largest difference between the two is the static nature of C++ versus the dynamic nature of Objective C. Both philosophies have their strengths and weaknesses. The final section of this chapter directly addresses the question of C++ versus Objective C.

There are currently four Objective C compilers commonly available: the original product from Stepstone including a library of foundation classes, the compiler and class libraries for NEXTSTEP development from NeXT Computer Inc. (NeXT, 1993), the Free Software Foundation's GNU C compiler, which is also an Objective C compiler, and the Berkeley Productivity Group product, which can be used with the Borland C compiler to support Objective C in a Microsoft Windows environment. There is no standardizing effort underway for the Objective C language. Each compiler supports different features. The NeXT and GNU compilers are very similar, as NeXT started from GNU C, then returned their Objective C extensions to the Free Software Foundation.

A summary of the differences between compilers can be found in the advanced features section below.

Relation to ANSI C

ANSI C is a subset of Objective C. Nothing special need be done to use ANSI C functions within Objective C programs. Objective C does not improve upon ANSI C in any way other than adding support for OO programming.

Objective C Support for OO Programming

As a C programmer familiar with OO concepts, learning Objective C is a very simple matter. As a C programmer without much OO experience, Objective C is considerably easier to learn than C++, as it minimizes the number of new language features needed for OO programming. The extensions made to Objective C are in two areas: support for messaging and a syntax for class definition.

Classes

The concept of class is the principal means of data encapsulation and abstraction in Objective C. The relation of class to object is that of a data type to its variable. A class consists of members: data and methods. Instance variables (sometimes abbreviated as "ivars") are the data members of a class. Objective C methods may be of two types, class or instance. This section will use a common convention for the names of instance variables, methods, objects, and classes. Names are concatenated, descriptive words. In the case of classes, the first letter of each word is capitalized, like Employee, GraphNode, Animal. For ivars, methods, and objects the first letter of each word *except the first word* is capitalized, like printFirstName, fedTaxWithholding, and newNode.

Objective C class definitions consist of two parts, an **interface** section, and an **implementation** section. In general, the interface to a class is stored in a header file, while the implementation is in a separate file with a .m extension. Following the example used in the C++ section of this chapter, here's our minimal Employee class interface declaration:

```
@interface Employee : Object
{
    int         empId;
    char*       fName;
    char*       lName;
}
@end
```

Objective C compiler directives begin with an @ symbol. Since every Objective C class is part of a hierarchy with the root class known as **Object**, custom classes which are "standalone" (i.e., not subclasses of some other class in the hierarchy) are subclasses of Object. We declare the interface to the functionality of our Employee class by adding prototypes for the methods of the class.

```
@interface Employee : Object
{
    int             empId;
    char*           fName;
    char*           lName;
}
- init: (int)eid;
- (int) empId;
- (int) lengthOfService: date;
- (void) printName;
@end
```

The prototype of each instance method is preceded by a "–." Much like the default type in C is int, the default type for Objective C is **id**. The id data type is used to hold the identity of an Objective C object. Methods that do not return type id, or have arguments that are not of type id, must be explicitly denoted as such with a type cast notation.

The implementation file for this Employee class would have four methods defined in it, and would include the header file for the class, as well as any other header files the implementation required. Objective C preprocessors support a smart version of #include called #import, which will not include a file twice. This means the programmer does not have to worry about and guard against this possibility.

```
#import    "Employee.h"
#import    <stdio.h>
@implementation Employee
- init: (int)eid
{
    hireDate = [[Date alloc] init];
    [hireDate setDate: hire_date_from_DB(eid)];
    return self;
}
- (int) empId
{
    return empId;
}
- (int) lengthOfService: date
{
    if( [hireDate isBefore: date] )
        return 0;
    else
        return [hireDate differenceInDays: date];
}
-(void) printName
```

```
{
   printf("%s, %s\n", lName, fName);
}
@end
```

Instance methods have full access to the instance variables of the class. Control over instance variable access is provided via the @public, @protected, and @private directives. Instance variables declared with the @protected directive are accessible to instances of the class and subclasses of the class. This is the default level of protection. Instance variables that are declared @private are not inherited by subclasses. Instance variables declared @public are accessible to the world. Since @public variables defeat the idea of encapsulation, they are almost never used. The following declaration of the Employee class interface would make empId public, and would hide fName and lName from subclasses of Employee.

```
@interface Employee : Object
{
@public
   int          empId;
@private
   char*        fName;
   char*        lName;
}
```

Objective C does not support class data members (i.e., data shared by all objects of the same class), however, they can be emulated through the use of the static modifier. For example, declaring a static int count variable in the implementation file of the Employee class would result in all Employee objects sharing a single integer variable called count. This is in contrast to instance variables, where each object has its own personal copy. A static data member provides a means of having a "class global" variable while preserving data encapsulation.

Objective C does support class methods. Each class has a class object associated with it that is constructed by the compiler. The class object does not have instance variables, but can be messaged just like any other object. Class methods, also known as factory methods, may be part of a message sent to the class object, rather than the instances of a class. Since the main role of a class object is to create instances of its class type, class methods most often have to do with the creation of objects. The declaration of class methods in a class interface is distinguished from instance methods by being preceded by a "+", instead of a "–."

Objects

Classes play the role of data type to objects. In Objective C, however, classes are not passive; they may be messaged to perform such tasks as object creation. Even with class methods and active classes, a class does not get all your work done. Objects are active. An object is instantiated from, or is an instance of, its class. Objects get your work done. Objective C added the data type id to represent objects. A variable of type id can hold the identity of an object of any class.

Objects get work done by messaging each other. Each message has at least two components: a receiver and a method. The method must be a member of the receiver's class definition, or it must be inherited from a superclass. The optional third component of a message is the set of parameters required by the method being invoked. The syntax for messaging is

```
[receiver method: arg];
```

where receiver is the identity of an object, method is the name of a method that the receiver responds to, and arg is the parameter supplied to method.

Parameters to the method, if any, are separated by colons with optional descriptive names for each parameter. The colons are part of the actual method name, and may distinguish like-named methods from each other. For example, a method init with no arguments is distinct from a method init: which requires one parameter. In the examples below, myObject, window, employee, and square are all variables of type id, which have been initialized to objects:

```
[window display];
[myObject aMethod: 1 andSecondArgument: "two"];
[employee calculateEarningsFor: 1995];
[square moveTo: 0.0 :0.0];
```

Messaging in Objective C has a clean syntax because of the id data type and dynamic binding. In effect, id is equivalent to the void* type in ANSI C; it can hold a pointer to an object of any type. The class of the receiver of a message is not determined until run-time. Messages may be nested, provided the return value of all but the outermost message is the identity of an object:

```
[[square moveTo: 0.0 :0.0] display];
```

The following example shows a typical message sent to a class object. The class method alloc asks the Employee class to allocate a new Employee object:

```
[Employee alloc];
```

The GNU and NeXT compilers do not allow for static object allocation; the Stepstone compiler does. All Objective C compilers support dynamic object allocation, most often using the class method alloc. The return value of the alloc method is the identity of the newly allocated object. This return value is then used as the receiver in a second message, where the init method initializes the new Employee object. Since an uninitialized object is a dangerous object, it is safer to nest the init message:

```
[[Employee alloc] init: 132];
```

The init method is responsible for initializing the instance variables of the object. Since an object inherits instance variables from its superclass, the first line of code in an init method is usually a message to the superclass init method:

```
- init
{
    [super init];
    // initialize ivars here
    return self;
}
```

Objects can refer to themselves with the special instance variable self. This allows an object to message itself, which facilitates the decomposition of complex methods into smaller pieces. In this example, the parsing of a line of input (oftentimes ugly, detailed C code) is isolated in another method to simplify the takeActionFor: method and possibly share the parsing algorithm with other methods in the class:

```
- takeActionFor: (const char* line)
{
    int command;
    command = [self parseLine: line];
    // take action for command
    return self;
}
```

To facilitate the nesting of messages, methods often return self when they do not have anything else to return. Objects may also invoke methods in their superclasses, even when their own class has overridden an inherited method. An object does this by messaging super, for example,

```
[super draw];
```

Unlike self, super is not an instance variable, and thus cannot be changed at run-time. Super is a flag to the compiler that changes where the run-time search begins for a method to execute.

Objective C objects can be of either method, class or global scope. For compilers that support them, statically allocated objects may be scoped to the method in which they are defined. Such objects are created automatically when a method is entered, and destroyed when the method returns. As with all instance variables, objects declared as class members have the scope of that class. The dynamic creation of objects allows objects of global scope; they are visible in any portion of the program at which their identity is available, and they are not automatically destroyed when the method in which they were created returns. Objects created dynamically must be explicitly freed to return their memory to the free store:

```
id date = [[Date alloc] init];  // new Date object
[date free];  // return memory to free store
```

The NeXT Foundation Kit class library provides a means of semiautomatic garbage collection via reference counting. This scheme represents a compromise between performance and programmer convenience.

Inheritance

One of the most characteristic means of representing relationships between abstractions in an OO program is with inheritance. Inheritance means forming subclass/superclass relationships between classes. A subclass inherits members from its superclass. Many classes together form an inheritance, or a class, hierarchy. Classes at the top of a hierarchy represent the data and functionality common between the classes which inherit from them. Classes at the top tend to be more abstract; objects are usually not instantiated from such classes. Classes at the bottom of the hierarchy are more concrete, and are more likely to be instantiated into objects. Classes that inherit from only one superclass exhibit single inheritance. Classes that inherit from more than one superclass exhibit multiple inheritance. In keeping with its clean and simple design, Objective C does not support multiple inheritance, though features have been developed to replace some of the functionality provided by multiple inheritance (see run-time section below).

The root of the Objective C class hierarchy is the Object class. While this name causes endless confusion in teaching neophyte programmers about OO programming, it is descriptive of the primary responsibilities of this class. At first, the idea of a common class from which all classes inherit is odd. After all, what could possibly be common to every class ever defined? The answer lies in the name: support for creating, initializing, releasing, copying, and comparing objects and the interface of objects to the run-time system are the main responsibilities of this class.

The superclass of a class defines where in the hierarchy a class fits in. The syntax for declaring a class' superclass is simply a colon, followed by the name of the superclass in the class interface declaration. The skeleton declarations shown in Figure 14.6 are illustrative.

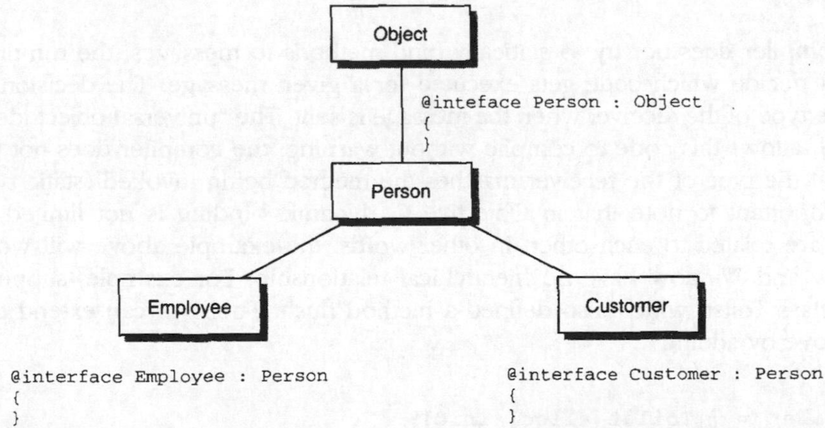

FIGURE 14.6 Inheritance relationship between Person, Employee, and Customer classes.

All class and instance methods of a class are inherited by subclasses of that class. Unlike the compiler directives for instance variables, there is no control over the inheritance of methods. One convention for private methods is simply to not declare them in the interface of the class. Such "private by convention" methods may still be used by other methods of the class (i.e., internally), but they will not be visible to the programmer clients who use the class, either directly as an object or indirectly via subclassing. If, however, a client learns the name of these private methods, they may be used; this form of access control is not supported by the compiler or run-time, it is merely a statement that a method is not considered part of the public interface of the class.

Polymorphism

Polymorphism is the idea that the code that is executed when a message is sent to an object depends on both the object's class and the name of the method in the message. In traditional procedural languages, the code that is executed by a function call is determined by the name of the function only. Objective C's support for polymorphism is simple: the type of the receiver of a message is not determined until run-time, so the method that is executed is not decided until run-time, and that decision is based on the receiver's class.

For example, suppose Window is a subclass of View, and they both implement a method called flush.

```
id   window = [[Window alloc] init];
id   view = [[View alloc] init];
id   anObject;
[view flush];     // the flush method in View
[window flush];   // the flush method in Window
anObject = view;
[anObject flush];        // the flush method in View
anObject = window;
[anObject flush];        // the flush method in Window
```

Since the compiler does not try to statically bind methods to messages, the run-time system must decide which code gets executed for a given message. The decision is based on the type of the receiver when the message is sent. The "universal object identity" type, id, allows this code to compile without warning; the compiler does not try to check that the type of the receiver matches the method being invoked (static typing). It is important to note that in Objective C, dynamic binding is not limited to classes that are related to each other. In other words, the example above will work even if View and Window have no hierarchical relationship. For example, suppose you had a class Toilet, which also defined a method flush. Then we can extend the example above by adding

```
id toilet = [[Toilet alloc] init];
anObject = toilet;
[anObject flush];// the flush method in Toilet
```

and the correct method is still invoked.

The id data type in Objective C is a good match to dynamic binding, since it can represent the identity of objects of any class, and the compiler does not attempt to statically bind code to messages. In this scheme the programmer loses the advantage of having the compiler check that the method being sent to an object is one that the object responds to. Performing this check is known as static type checking. Objective C can support static type checking by declaring variables that represent objects to be of type pointer to the object's class, rather than id. For example, in the following code fragment, the compiler could check that the method display was understood by the Window class:

```
Window *help;
[help display];
```

If the variable help had been defined as id type, the compiler could not do this static type check. Declared as a pointer to Window, the compiler will allow the variable help to be assigned only to the identity of a Window class object, or an object of a type that is a subclass of Window. However, the compiler will generate a warning only when an assignment is done to another statically typed object. If the variable help is assigned the value of an id variable, the compiler will not (cannot) check that the types are compatible. Static typing does not affect dynamic binding, so if help was assigned the identity of an object of a subclass of Window at run-time, the code for the subclass' display method would still be executed.

Objective C does not support the operator overloading form of polymorphism.

Run-time Features

The run-time system in Objective C is a critical part of the language. The run-time provides the behavior of dynamic binding, as well as some other very powerful language features, such as dynamically loading classes into a running program, providing for persistence of objects, and supporting some of the features of multiple inheritance. The capabilities and implementation of the run-time component of Objective C tend to vary between compilers more than other features of the language.

Dynamic type checking means that it is possible for a message to be sent to an object with a method that the object does not implement. The run-time system will detect when this will occur, and will instead send a forward:: message to the object. The forward:: method is part of the Object class, and is thus understood by all classes. The two arguments to forward:: are the name of the method originally invoked and the arguments to that method. By default, the implementation in Object class of the forward:: method is simply to print an error message and exit. But a programmer can override this behavior and do any of a number of things. For example, all messages sent to an object with an unknown method may be caught and sent to an error handling object. Another use of forwarding is the ability to have one object perform some action for another. This can be used to mimic multiple inheritance. For example, suppose a Professor class object is sent a method that it does not implement, say writeRealCode. A Professor object may know of an ex-student Programmer object working in industry who writes real code and is willing to help him out. Objects of the Programmer class implement the writeRealCode method. The Professor object can then implement a forward:: method to have a Programmer object execute the method on its behalf. The Professor forward:: method might look like this:

```
- forward:(SEL)aMethod :(marg_list)args
{
  if([exStudent respondsTo: aMethod])
     return [exStudent performv: aMethod : args];
  else
```

```
    return [super forward: aMethod :args];
}
```

As implemented above, the forward:: method first checks to see if the exStudent object responds to the writeRealCode method. It does this by asking the exStudent object about itself. The respondsTo: method is one of several methods in Object that are concerned with the capabilities of the class of an object. If this particular Professor object has no exStudent to turn to, exStudent will be null, and the message is a no-op. If exStudent is not null, and if it responds to the writeRealCode method, then exStudent will perform the writeRealCode method for the Professor object. The performv:: method has exStudent execute the writeRealCode method just as if it had been messaged directly. The Professor class has used the writeRealCode method from the Programmer class just as it is defined in the Programmer class. If a particular Professor object does not have a willing Programmer object in the guise of exStudent, then it will instead invoke the inherited behavior for forward::, which, if not overridden in a superclass, would be the default implementation in Object. Professors without anywhere to turn to write real code would print an error message and exit.

How does forwarding replace multiple inheritance? Suppose Programmer and Professor are both subclasses of Person, the Professor class has a single method, teach, and Programmer has a single method writeRealCode. Figure 14.7 illustrates the inheritance hierarchy. They both inherit common functionality from Person.

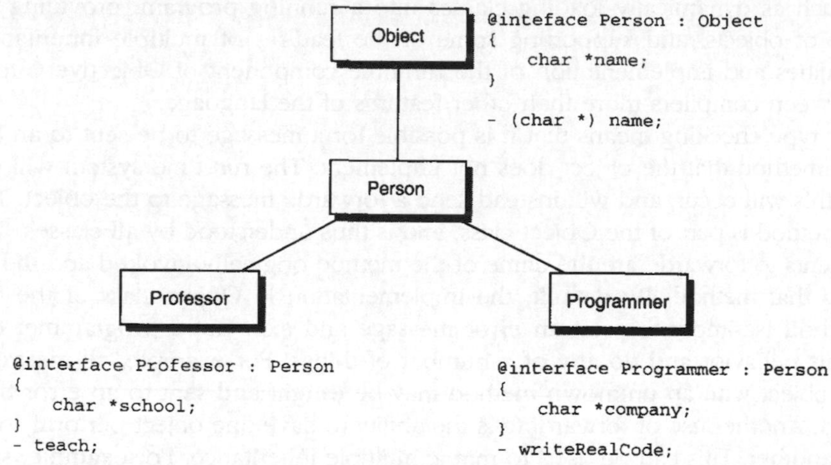

FIGURE 14.7 Inheritance relationship between Person, Professor, and Programmer classes.

Suppose we want to give some Professor objects the ability to write real code, in addition to the ability to teach. One way to do this would be to have the Professor class inherit from the Programmer class. But this violates the "isa" relationship between super and subclass. If every professor was not a programmer, then it is not true that every instance of Professor "isa" Programmer. Another way to have Professor objects know how to write real code would be to use multiple inheritance; a Professor could inherit both from Person and Programmer, as in Figure 14.8a. This might make Professor a complex class, since it would inherit everything (instance variables and methods) defined by Programmer. Additionally, there is immediate ambiguity concerning the methods and ivars that would be inherited by Professor from Person via two paths. For example, the ivar char *name declared in Person would be inherited twice by Professor. Multiple inheritance would make the otherwise simple Objective C language complex. A third strategy is required. The forward:: method of the Object class and support for it in the Objective C run-time provide this third way. With forwarding, the Professor class need only use what functionality it wants from the Programmer class. Note that, as written above, the Professor class is not selective about what it forwards to Programmer; all unknown methods would be forwarded. Figure 14.8b illustrates the forwarding solution.

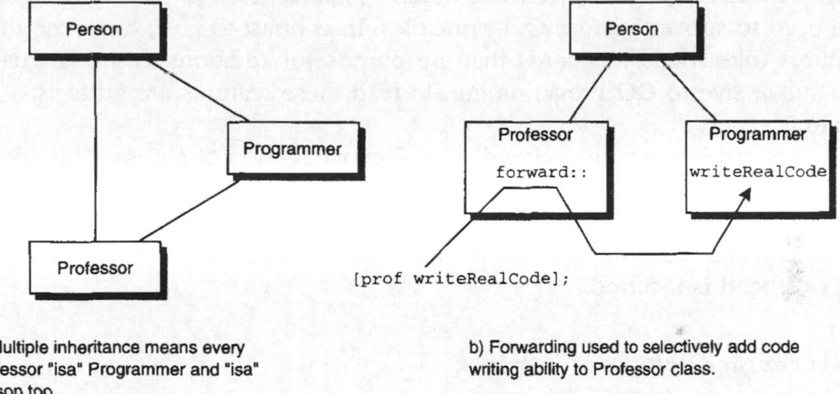
Person
Programmer
Professor
Person
Professor
forward::
Programmer
writeRealCode
[prof writeRealCode];
a) Multiple inheritance means every Professor "isa" Programmer and "isa" Person too.
b) Forwarding used to selectively add code writing ability to Professor class.

FIGURE 14.8 Two possible solutions to the dilemma of professors who can also write code.

The multiple inheritance solution suggested for Professor objects who can code requires that every Professor "isa" Programmer. The forwarding solution allows for some Professor objects to employ the services of a Programmer object, or for some Professor objects to also be programmers, via composition.

One consequence of the dynamic nature of Objective C is the ability to dynamically load classes at run-time. This capability offers incredible flexibility to programmers, as they can construct programs that use classes not even anticipated when the program is written. The original programmer provides a framework, and subsequent programmers can supply classes that are dynamically loaded, extending the framework in some unforeseen fashion. The Preferences application in NEXTSTEP is a good example of dynamic class loading. The Preferences application is used to customize various environmental factors to suit a particular user. There are modules in Preferences.app to tailor the speed of the mouse, the flavor of the keyboard, the color of the screen background, etc. When launched, the Preferences application searches in known locations for dynamically loadable modules. If any are found, they are seamlessly integrated with the application. The custom classes dynamically loaded from a module are no different at run-time than the classes compiled into the Preferences.app itself. The author of Preferences specified only an interface; subsequent programmers added functionality to Preferences.app never anticipated by the author by extending the Objective C class hierarchy.

In the NeXT compiler, object persistence is supported by the run-time capability of archiving an object's instance variables to a stream. This allows the state of an object to be saved and later restored. The Object class defines read: and write: methods to restore an object from an archive, and to write an object to an archive. Archiving is also useful for the exchange of objects via a pasteboard, or a network. The read:/write: methods must be overriden for custom classes, since Object obviously cannot anticipate the form of the ivars in every class. Each class that declares instance variables (and wishes to be archiveable) must implement read: and write:. These methods first invoke their superclass' corresponding method, so that archiving actually happens from the top of the class hierarchy down to the class of the object being archived.

Defeating OO Principles

As a hybrid OO language, Objective C inherits features from its base language, which can be used to subvert various OO principles. In contrast to C++, however, there are no features (like friend functions) that are purposeful additions to the language and can be subversive to OO programming. Instead these features are artifacts of its hybrid nature.

Instance variables may be declared in an @public section of a class. This allows statically typed references to objects of such a class to access the instance variables directly, with the C structure pointer dereference operator. The only possible motivation for this sacrifice of data hiding is the increased efficiency of access to the ivar; no messaging overhead is incurred.

```
@interface MyClass : Object
{
@public
   int wideOpen;
}
MyClass *myObject;
myObject→wideOpen = 1; // no messaging, no hiding
```

Functions not part of the class hierarchy may be used within methods. Using the C library is a common example of such free functions. Nothing special needs to be done to compile or link standard C functions. While these functions do not directly attack any OO principles, they are also not part of the object-orientedness of a solution.

The final two techniques, getting and using the address of a function and getting and using the data structure behind a class, can be used to defeat dynamic binding and encapsulation. The syntax for these techniques is tortuous and is not covered here.

Advanced Features

Since there is no standard for the Objective C language, there are differences, some small, others not so, between the compilers. The features discussed in this section are highly compiler dependent. A summary of differences is shown in Table 14.3.

Categories. Categories provide a means of extending the functionality of a class without subclassing, and without access to the class source code. For example, suppose you would like every object of every class in a program to respond to a debug message. When sent such a message, an object might print its identity with a time stamp. If you control the root of the class hierarchy, this is easy. But if what if you do not have the source code for Object? A **category** of Object can solve your problem.

TABLE 14.3 Comparison of features of three Objective C compilers.

Compiler	Language Feature					
	statistic allocation and binding	protocols	categories	forward class declarations	distributed objects	class library
NeXT	No	Yes	Yes	Yes	Yes[a]	Yes
GNU	No	Yes	Yes	Yes	Yes[a]	No[b,c]
Stepstone	Yes	No	No	No	No	Yes[b]

[a]Not intercompatible.
[b]Distributed with source code.
[c]Evolving very quickly—a work in progress.

```
#import <Object.h>
@interface Object (DebugExtensions)
- (void) debug;
@end
@implementation Object (DebugExtensions)
- (void) debug
{
   printf("%s: %u\n", time_stamp(), (unsigned) self);
}
```

Categories may not declare new instance variables for a class, they may only add methods to the class. Categories are also convenient means of organizing large classes by grouping related methods together. Each category can be implemented with a separate header and implementation file. Private methods, discussed in the section above on inheritance, provide a good example of the use of categories for organization. The private methods of a class can be gathered together and placed in a ClassPrivate category, clearly separating and documenting their private nature.

Distributed Objects. The NeXT and GNU compilers support, in a mutually incompatible fashion, what are known as distributed objects. A distributed object is one that can be messaged by another object outside of its address space. Lifting this restriction on address space means that an object can message another object residing in a different application, perhaps on a different computer altogether. Distributed objects are a means for extending OO programming to client/server applications. Objects associated with user interface elements can reside in the client, while objects associated with database resources, or heavy computational demands, can be placed on special purpose servers. Rather than connecting the two sets of objects via a traditional interprocess communication mechanism, e.g., sockets or RPCs, distributed objects allow for the standard OO messaging model to work between the client and server objects. Distributed objects also provide the infrastructure for applications to provide services to other applications via application program interfaces (APIs). For example, a data graphing application could provide the service of graphing and presenting collections of data for custom programs via a distributed object interface.

Protocols. Inheritance allows for the reuse of interface and implementation. Single inheritance limits this reuse to a tree structure. Every class can appear at only one location in the class hierarchy. That location defines what is inherited by the class. The inheritance hierarchy is not good at imposing interface requirements on a group of unrelated classes. Protocols allow for an interface specification (a set of methods) to be defined that can be adopted and adhered to by otherwise unrelated classes. Protocols have nothing to do with sharing implementation; they strictly define an interface. A formal protocol is declared with the @protocol directive. Consider a protocol to define the interaction between people over the telephone.

```
@protocol TelephoneEtiquette
- saySomething: (char *)msg;
- respond: (char *)msg;
- initiateGoodbye;
- acknowledgeGoodbye;
@end
```

A class could adopt this protocol by including it in the @interface declaration for the class. Suppose objects of the Customer class should be able to talk on the phone:

```
@interface Customer : Object <TelephoneEtiquette>
{
  // ivars
}
// methods
```

The compiler will now warn you if you fail to implement all of the methods in the protocol. Clients of the Customer class are guaranteed to be able to send the messages defined in the protocol to Customer class objects. There is no guarantee about the implementation of the protocol methods, simply that an implementation is provided. Protocols are natural means of documenting APIs, as they define the methods a class implements without revealing anything else about the class. In the reverse direction, a protocol can serve to document the requirements of a custom class to interact with a given object. For example, a module in the NeXT Preferences application (see earlier discussion in the section on run-time) can be developed by anybody, providing they agree to meet a protocol that the Preferences application defines for communication with the modules.

Mixed Objective C and C++. The NeXT compiler supports the simultaneous use of C++ and Objective C. Features of the two languages may be used together in almost any manner. For example, C++ functions can send messages to Objective C objects. Objective C methods can create and message C++ objects. C++ retains its strong typing and static binding and Objective C its dynamic nature.

DEFINING TERMS: OBJECTIVE C

Category: A means of extending the functionality of a class by adding methods.

Distributed Object: An object running in a different address space that can be messaged.

Dynamic Binding: Code executed by a message is determined at run-time, based on class of receiver.

Interface: The declaration of the interface of an Objective C class. Starts with @interface directive.

Implementation: The implementation of an Objective C class. Starts with @implementation directive.

id: A data type that can represent the identity of an object of any class.

Object: The root class of the Objective C class hierarchy.

Protocol: The declaration of a set of methods implemented by a class that is independent of class hierarchy.

Self: An object's self-referential instance variable of id type.

Super: Reference to a class' superclass.

14.4 SUMMARY AND COMPARISON

Every significant feature of C++ and Objective C can be considered to be either a weakness or a strength, depending on your perspective. The main philosophical difference between the two is the static nature of C++ versus the dynamic nature of Objective C. For the C programmer new to OO programming, the choice is primarily between the feature-rich and complex C++, with its wide industry acceptance, or the easier-to-learn, more flexible, but less widely used Objective C.

The interesting characteristics of C++ are summarized in Table 14.4. This summary is ordered by generality, with specific language features considered last.

The interesting characteristics of Objective C are summarized in Table 14.5. This summary is ordered by generality, with specific language features considered last.

TABLE 14.4 Positive and Negative Aspects of Major C++ Characteristics

Characteristic	Positive Aspects	Negative Aspects
Based on C	Compatible with legacy code, programmers, and tools Retains efficiency	Old, fairly low-level language Easy to continue non-OO habits
Feature set	Accommodates more programming paradigms, applications	Complex to learn, use well Differences between compilers
Industry acceptance	Many compilers, tools, people	
Static typing/binding	Shifts errors to compile-time	Hard to do some things easily Limited extensibility at run-time
Messaging syntax	None	Does not nest cleanly; looks too much like C structure access and function invocation
Virtual functions	Adds more complete polymorphism	Limits reuse via subclassing if functions not made virtual Polymorphism limited to common base class
Operator overloading	Extends basic operations in very consistent ways to user-defined types	Can be confusing when applied to complex user types like classes
Reference types	Disguises pointer syntax	Can lead to errors based on returning references to limited-scope data objects
Multiple inheritance	Helps in OO modeling	Many complexities introduced to implementation

TABLE 14.5 Positive and Negative Aspects of Major Objective C Characteristics

Characteristic	Positive Aspects	Negative Aspects
Based on C	Compatible with legacy code, programmers, and tools	Old, fairly low-level language
Based on Smalltalk	Good for OO influence	Less efficient
Industry acceptance		Relatively few compilers, tools, people Differences between compilers
Feature set	Easier to learn	Fewer ways for solving problems
Dynamic typing/binding	Easier to learn Faster to develop Highly extensible after development Dynamic loading	Shifts errors to run-time Less efficient than static binding
Messaging syntax	Simple, nests well Distinct from other operators	
Protocols	API specification and reuse Some aspects of MI	Cannot provide default implementation
Forwarding	Useful aspects of MI	Must have object to forward to
Rooted class hierarchy	Allows for common features of all classes	None

REFERENCES

C++

Tom Cargill, *C++ Programming Style*, Addison-Wesley Reading, MA, 1992.

H.M. Deitel and P.J. Deitel, *C++ How To Program*, Prentice-Hall, Englewood Cliffs, NJ, 1994.

Margaret Ellis and Bjarne Stroustrup, *The Annotated C++ Reference Manual*, Addison-Wesley, Reading, MA, 1990.

Neill Graham, *Learning C++*, McGraw-Hill, New York, 1991.

Stanley B. Lippman, *C++ Primer*, Addison-Wesley, Reading, MA, 1991.

Scott Meyers, *Effective C++*, Addison-Wesley, Reading, MA, 1992.

Bjarne Stroustrup, *The C++ Programming Language*, 2nd ed. Addison-Wesley, Reading, MA, 1991.

Objective C

Timothy Budd, *An Introduction to Object-Oriented Programming*, Addison-Wesley, Reading, MA, 1991.

Brad Cox, *Object Oriented Programming: An Evolutionary Approach*, Addison-Wesley, Reading, MA, 1991.

NeXT Computer, Inc. *Object-Oriented Programming and the Objective C Language*, Addison-Wesley, Reading, MA, 1993.

Lewis J. Pinson and Richard S. Wiener, *Objective C: Object-Oriented Programming Techniques*, Addison-Wesley, Reading, MA, 1991.

FURTHER INFORMATION

C++ Sources

The *Annotated C++ Reference Manual*, by Ellis and Stroustrup, and *The C++ Programming Language*, by Stroustrup, are the ultimate sources for determining the meaning of C++ features. Neither makes a good first book to learn C++. For a first book, Graham or Lippman are better choices. Lippman is much more thorough; Graham covers the essentials. Deitel and Deitel offer complete coverage, with useful examples and tips.

Once basic familiarity with C++ is gained, the book by Meyers is an invaluable tool for understanding how to really use the language. Organized around 50 items of advice for C++ programming, with clear and detailed explanations of each, this book can improve your programs as well as your understanding of the language. Cargill's book is similar to Meyers' book in providing understanding beyond a simple description of features.

Useful sources of information on C++ are available on the Internet. An excellent place to start is with the C++ FAQ (Frequently Asked Questions):

```
ftp://sun.soe.clarkson.edu/pub/C++/FAQ
```

The following document is a collection of references to tutorials, code collections, white papers, etc. for learning C and C++:

```
http://vinny.csd.mu.edu/learn.html
or
ftp://rtfm.mit.edu/pub/usenet/news.answers\
     /C-faq/learn-c-cpp-today
```

An award winning, Web-based course, "Object-Oriented Programming Using C++," is available at

```
http://uu-gna.mit.edu:8001/uu-gna/text/cc/index.html
```

Objective C Sources

The inventor of Objective C, Brad Cox, also wrote one of the first books on OO programming. Cox describes Objective C and uses it to teach the principles of OO programming. This is not a language tutorial.

The NeXT Computer Inc. book is the best tutorial and reference for the language. Some of the information is specific to the NEXTSTEP compiler.

Budd has an excellent book on OO programming, as well as a comparison and examples of several OO languages, among them Objective C.

Useful sources of information are available on the Internet. A Web browser (e.g., Mosaic, NetScape, OmniWeb) can be used to browse the following resources:

The Objective C frequently asked questions (FAQ) document:

```
http://csld.ucr.edu/NeXTSTEP/objc_faq.html
```

A collection of Objective C resources, questions, and general information:

```
http://www.symnet.net/~dekorte/Objective-C/objc.html
```

A short course on OO and Objective C programming:

```
http://www.cs.indiana.edu/classes/c304/oop-intro.html
```

15

Object-Oriented Programming

James TenEyck
Marist College

15.1 THE OBJECT-ORIENTED PROGRAMMING PARADIGM

With the advent of LSI and VLSI computer hardware has

- dramatically decreased in size and price,
- become much more reliable, and
- has become increasingly easier to use.

These gains in performance and price have come about because of the application of sound engineering principles, such as

- modularity of construction,
- standardization of components and design, and
- large scale production of standard (general purpose) parts

which have led to the construction of machines with a rather standard set of capabilities from "off the shelf" parts. These machines can also be readily upgraded by adding additional standard components for memory enhancement, disk storage, graphics capability, or networking. Over the same period of time, the cost of software and the productivity of programmers have not shown any dramatic change, while the demand for software products continues to grow.

It is a goal, therefore, of the computer science community to develop a software construction methodology that utilizes sound engineering principles like those that have achieved success in hardware production, and to support this methodology with compatible languages. The difficulty in achieving this goal lies in the difficulty in answering the questions:

- How does one modularize a system into its components and what is the best approach to modularization, and how does one identify the specific components when constructing the system?

0-8493-2530-7/96/$0.00+$.50

- Does the methodology support the maintenance (upgrading) of a system by adding or removing (standard) components?
- How does one promote the reusability of software components in diverse application?

The object-oriented programming methodology provides a basis for answering the first question (around objects), adds features (primarily inheritance) that support reusability, and is supported by a number of object-oriented programming languages.

The Development of Programming Methodology

Object-oriented programming methodology and object-oriented programming languages that support this methodology have evolved together with the concept of what a program is, the platform upon which programs are mounted, and the language support that programmers are accorded. It recognizes that the identified objects, which encapsulate data and functionality, within a software system are less prone to change as the system evolves through successive maintenance than the function that such a system provides. These objects, therefore, become the basis for modularizing the system.

The object-oriented programming paradigm adds new emphasis and some new features to earlier programming paradigms. These programming paradigms include

- Functional Decomposition
- Data-Driven Decomposition
- Object-Oriented Methodology

Functional Decomposition

Functional decomposition was the first programming methodology to emerge in the early years of digital computers. It originally featured the flow chart. In flow chart methodology, lines with directional arrows connected simple geometric shapes that represented either processing stages, decision branch points, input/output activities, or state changes. The directed line segments in the diagrams represented the sequencing of activities used to accomplish a specific task. Arbitrary feedback (and feedforward) loops within the diagrams were realized by programming languages of the day, such as FORTRAN, that not only supported but encouraged the use of GOTOs in program construction.

An early attempt to discipline the functional decomposition approach to program construction was described by Niklaus Wirth as a Structured Programming methodology. Its main tenets were top down design with stepwise refinement. The flow chart was replaced by a hierarchy of descriptions written in the style of a programming language with control flow constructs like those found in PASCAL. Successive refinement of the higher level tasks continued until pseudocode description of each subtask emerged. Structured programming produced a decomposition of the program into the procedures that provided the needed functionality.

The early computing environment was constrained by

- limited user/machine interaction capability and
- limited computer memory.

A functional decomposition approach to program construction reflected the nature of the early applications, which was to solve a specific, computationally intensive task. Without severe constraints on user interaction capability and memory, today's programs might best be considered as an interactive environment in which a multiplicity of tasks can be performed. Functional decomposition is still appropriate for designing procedures and algorithms that are task specific, but it does not scale up very well to use for designing large systems.

Data Driven Approach

The evolution in the concept of a program led to an emphasis on the data that were being transformed rather than on any single task that the system might be required to do. The structured programming methodology that emphasized top down design with stepwise refinement now embraced the primacy of data in the specification of a program. Data flow diagrams, that are one product of this methodology, have a superficial resemblance to earlier flow charts, but depict an environment that contains data sources and sinks, data stores, data transfers, and processes that transform data. The diagrams are developed from an initial high level context diagram through successive refinements where intermediate data stores and more detailed process descriptions are included. Data flow diagrams show all of the possible computational paths that are available within the specified system. They are not meant to convey a particular data path and sequence of processes for instances of program execution. Control flow information is modeled separately, as is the relationship between the identified data entities within the specified system.

The data dictionary is an important tool in the analysis and design process. Every term that is used in constructing models of the application should appear in some central repository along with its kind or type (class, attribute, association, etc.) and a brief description of how or why this identifier appears in the model. The data dictionary is extremely important in any effort to cross-check models for consistency and completeness, and it provides an essential record for any programmer charged with maintaining the system.

The data flow approach to structured analysis and design leads to a decomposition of the program according to the principal data structures and the operations that are performed upon them. Good design stresses the decomposition of the system into "units" that are highly cohesive and loosely coupled. This means that the data are encapsulated with all of the operations that act upon those data and only those operations. The interaction between "units" is limited to what is allowed in the specification. "Loosely coupled" requires that a change in the implementation of one "unit" have no effect upon any other "unit" in the program.

Languages such as Modula-2 and Ada support the structured approach to program construction with the following features:

- Separately compiled specification and implementation "units." Specification can be done independently of implementation, and a consistent interface can be established before implementation begins.
- Private data. Access to private data is limited to those operations that are provided in the specification "unit." This ensures that the coupling between "units" is well defined and limited.

Object-Oriented Approach

The object-oriented paradigm is a refinement of the data-driven structured approach that attempts to produce a more stable decomposition of the system under construction, and to provide for the greater reuse of software components. It differs from the structured approach in two main ways:

- The data flow model is the principal viewpoint of the system to be modeled in the structured approach. The object model that examines the relationship between entities in the problem environment is the principal viewpoint in the object-oriented approach.
- The structured approach either to function or data is top down and the object-oriented viewpoint of stepwise refinement is more bottom up. Data flow diagrams emerge from a successive refinement of the context diagram. The object model is refined by the generalization of classes with common features within the object model.

The object model is similar to the Entity-Relation (E-R) diagram of the data-driven approach, but the classes in the object model include a listing of the features that objects of those classes possess. In the E-R diagram, an entity is a data item. Any attributes of of the entity are depicted as bubbles attatched to the entity box, and operations are not considered.

Many of the language features that support structured programming also support reusability and may be found in object-oriented languages. In addition to private data and separate compilation, these features include

- Name and operator overloading—the meaning of the name or operator that is referred to is determined by context during compilation.
- Genericity—a stack, for instance, has the same function regardless of the data type being stacked, but nongeneric languages require separate type declarations such as intstack and charstack with their associated operations.

Two additional features characterize object-oriented languages:

- Inheritance—promotes the construction of new classes that are subclasses of ones already constructed. A proper subclass inherits features from its parents. This adds stability to the structure of the system, and may limit the amount of new coding that is necessary.
- Polymorphism—allows references to objects of a parent class to refer at run-time to instances of subclasses. For instance, an entity referring to a rectangle object could refer at run-time to an instance of a square if square is defined as a subclass of rectangle.

Chapters 16 and 17 discuss popular object-oriented languages that incorporate many of these features.

15.2 OBJECT-ORIENTED ANALYSIS (OOA)

The goal of an analyst/programmer is to produce a software system that is "properly" modularized into highly cohesive, loosely coupled software components that may be, in part, already available, and to produce a system that is easy to maintain. The organization of the system into objects facilitates an appropriate modularization. During the analysis phase, the software developer trics to identify the real world objects and their associations that model the essential structure of the problem domain. The analysis phase progresses by ascribing attributes and some operations to each of the classes. The design phase builds upon the analysis by modeling the solution domain in a similar fashion. New solution specific classes may be added at this time. Specific data structures are chosen and algorithms for implementing operations are also selected. The designer must particularly consider the communication that is required between objects from the various classes, and model the object interface that describes how this communication is implemented. In large part, however, the design phase is a continuation of the effort begun in the problem domain of associating attributes and operations or methods to the classes. The class structure that emerges becomes the basis for the modularization of the system.

The task of the system analyst has two main components:

- develop an abstraction or model of the problem domain and
- then consider the form and substance of the client's input and output requirements for this system.

A distinction between these two aspects of the task is made here because a premature concentration on the menu and report screens that form the user interface produces an analysis that imposes a particular solution before the problem domain is thoroughly examined. Concentration on the client's input/output requirements forms the basis of the structured approach to program construction that emphasizes the job to be completed. The object-oriented paradigm emphasizes the environment in which the job is to be performed.

Two methodologies that provide a formal framework for doing object-oriented analysis and design are

- The Object Modeling Technique and
- The Fusion Method.

The Object Modeling Technique is more mature and has a relatively inexpensive CASE tool supporting it. The Fusion Method is now being popularized and its CASE support is more embryonic. It borrows heavily from the Object Modeling Technique during the analysis phase, and integrates techniques used in other object-oriented approaches into a coherent structure of analysis and design that is likely to be widely accepted.

Object Modeling Technique (OMT)

The object modeling technique (OMT) introduces a graphical notation for expressing abstractions of an application or problem domain. OMT breaks description of the application domain into three essentially orthogonal viewpoints.

- the object model, which describes the static or structural relationship between the objects in the problem environment;
- the dynamic model, which describes the temporal behavior of the system; and
- the functional model, which describes the transformational aspects of the system.

The primacy of the object model description of the application domain is what distinguishes the object-oriented approach to analysis and design from the earlier structured approach. The relative utility of the three viewpoints in constructing an abstract description of the problem domain will vary with the nature of the application.

OMT is an approach to object-oriented analysis and design that was developed at the General Electric Research and Development Center in Schenectady, and promoted in the text, *Object-Oriented Modeling and Design*, by Rumbaugh et al. (1991). It is supported by a CASE product called OMTool that is available from the Martin-Marietta Corp. (1992). OMT provides the analyst with a formalism and notation for constructing models, but it provides no additional formal support to the designer. The distinction between the analysis phase and the design phase is blurred, with the designer composing and refining previous models. The Fusion Method of Coleman et al. (1994) provides additional design models that will be briefly examined in the next section.

OMTool comes in a UNIX version for SUN and Hewlett-Packard workstations and a PC version that runs under Windows 3.1. The PC version has pop-up menu features that are very similar to those found in the Windows tool, Paintbrush. The documentation is clear and concise, and the tool is very easily learned. OMTool provides the analyst with the ability to construct graphic models of each of the three viewpoints and annotate any of the features within a model. The tool can be used to generate a data dictionary and C++ code from the annotated object model.

OMT provides a small set of formal graphical constructs that are used to depict classes and the relationship between classes. This formal notation will be briefly introduced in this subsection, then used in the remainder of the discussion of analysis and design in this chapter to construct models for an example application.

A rectangular box will be used to denote a class. Inside the box one lists the class name and features. Features are of two types: attributes and operations. An object is an instance of a class and is depicted as a box with rounded corners. OMTool, however, does not provide the analyst with the ability to draw an object. The class and not the specific objects is the usual focus of model construction.

Note that age may be considered as an attribute of the player, a derived attribute that is denoted with a slash before the name, or an operation that computes the difference of two dates. Meyer (1988) recommends that such an implementation decision be deferred, and the term feature not distinguish between attribute and operation. The association between classes is shown by connecting the boxes with a solid line. A label adjacent to the line will describe the association, and the multiplicity of the association—the number of instances of one class that may relate to a single instance of another—will be indicated with a ball in front of the class with multiple instances. A solid ball will be used to indicate one or more instances, and a hollow ball will indicate a multiplicity of zero or one. Figure 15.1a reflects the fact that a group of tellers will be served from a common waiting line.

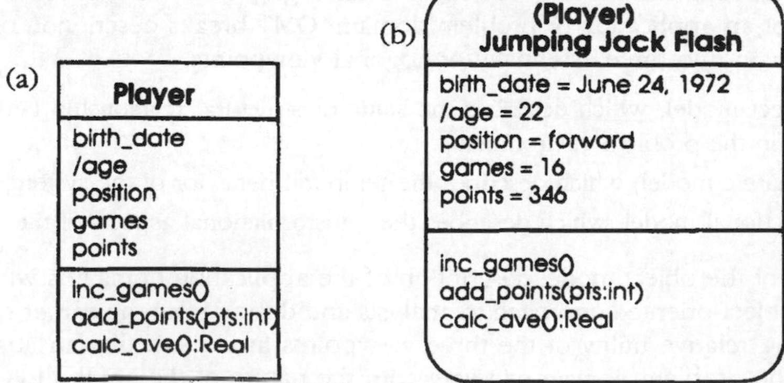

FIGURE 15.1 **(a)** Class Player. **(b)** Object of Class Player.

Classes should not merely reflect roles, but the role that the individual object plays in a link may be written at the end of the association as shown in Figure 15.2b.

Sometimes the one-to-many or many-to-many link can be refined by use of qualification. The qualifier distinguishes among the objects that are at the many end of such

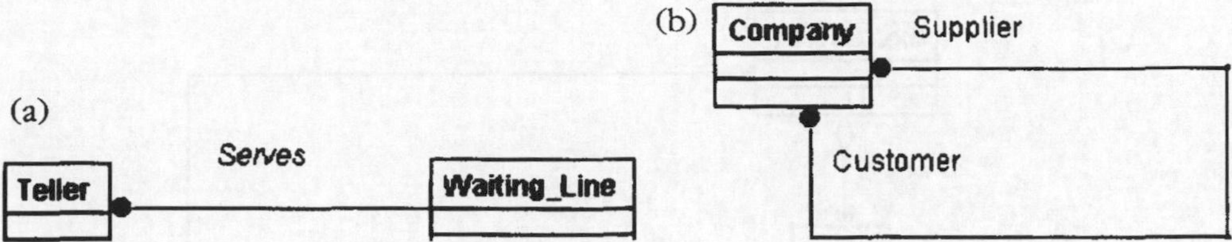

FIGURE 15.2 (**a**) Multiplicity of tellers with a common queue. (**b**) Roles of objects.

an association. For instance, in Figure 15.3a, the many-to-many association indicates that students take more than one course, and courses have more than one student enrolled. The course attribute, catalog number, distinguishes from among the course objects that link to a particular student object.

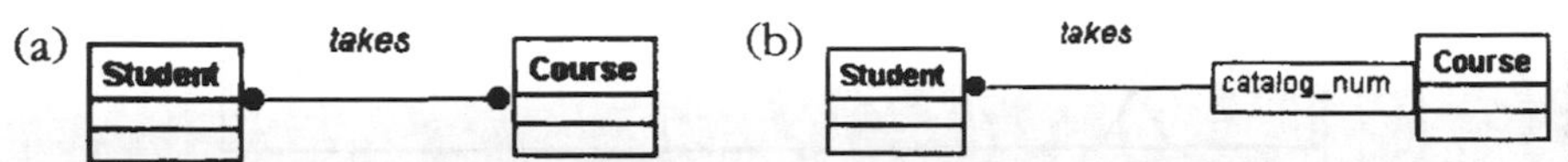

FIGURE 15.3 (**a**) Unqualified association. (**b**) Same association qualified.

Aggregation and generalization are two special associations between classes (Fig. 15.4). Aggregation is used when the relationship between classes is "part-of." A car is an aggregate of its chassis, engine, transmission, and exhaust system. Recognition of an aggregation association becomes important when there is an operation in the aggregate that is performed by doing the corresponding operation in each of the parts. Aggregation is denoted by a small diamond next to the aggregate class in an association.

Generalization denotes an "is-a" association. A subclass object "is-a" object of the super class with additional properties. For example, a rectangle "is-a" quadrilateral with two pair of parallel sides that intersect at right angles. If perimeter, area, and color are features of quadrilateral, they are also features of rectangle. The method for calculating the area and perimeter of a rectangle might be changed to take advantage of any computational or conceptual simplification, but the meaning of the term must be the same in each class. The term generalization suggests that the analyst is looking to relate similar classes to a common ancestor. It is often useful to create an abstract class with deferred operations that will have no run-time instances, but will provide a template for the real subclasses that have some common properties and behaviors. Generalization can be used to simplify some of the associations that occur in earlier versions of an analysis, and produce a more robust structure. It is depicted by a triangle as shown in Figure 15.5. Subclasses inherit features from the superclass. Multiple inheritance is indicated by filling in the triangle.

The Fusion Method

The Fusion Method was developed by Derek Coleman and others (1994) to provide a more formal object-oriented framework for the process of analysis and design. It attempts to supply the analyst and designer with a set of techniques and tools that supports a consistent progression from analysis through design. This method also features the object model as the centerpiece of the analysis effort, but downplays the impor-

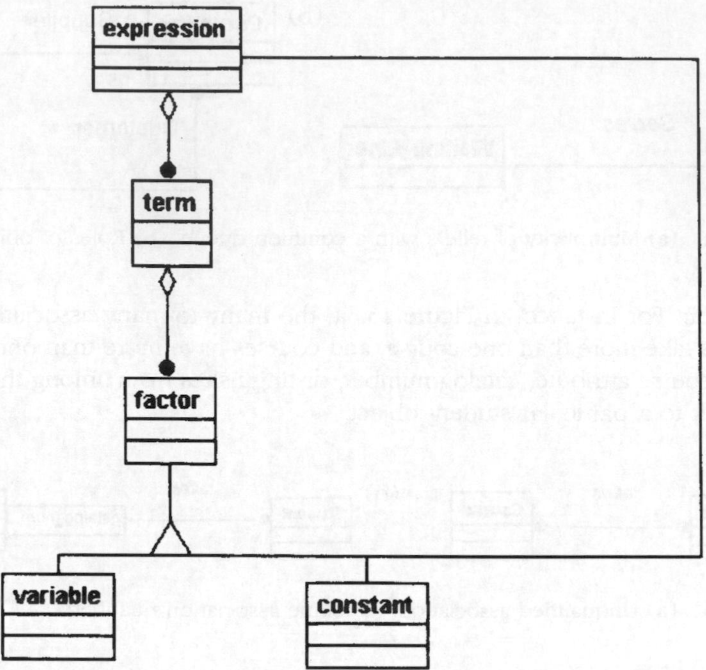

FIGURE 15.4 Aggregation with generalization to show recursion.

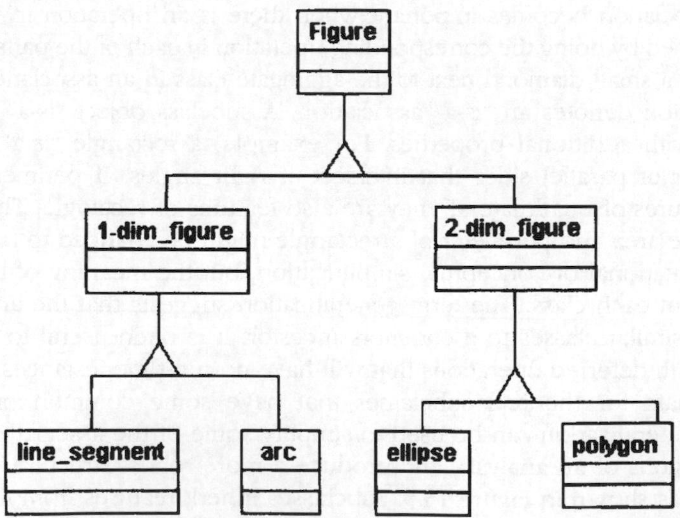

FIGURE 15.5 An example of generalization.

tance of the dynamic and transactional models. Instead it determines the context of the system to be constructed—the boundary between the classes in the object model that are part of the system itself and those that are only in the environment of this system. Once this determination is made, one regards the classes that are part of the environment to be agents that interact with the system proper. They are the source of events that cause the state of the system to change and may introduce data into the system. These events initiate both the dynamic and transactional behavior of the system, and are called system operations. Coleman et al. (1994) provides a schemata for describing system operations that stresses specification of the pre- and postconditions of the system. System operations are identified by investigating scenarios that detail the sequence of events that occurs when the various agents interact with the system. The formal description of each system operation, and the life cycle model of the system, that indicates formally the order in which each of the system operations is allowed to occur, supplement the object model in the analysis phase of the Fusion Method. An example of how these models are constructed is contained in the next subsection.

Example of OOA

As an example, let us suppose that we are asked by a client to build a system for stimulating a queuing network. The system we are modeling should be sufficiently general to allow for the simulation of many diverse applications such as a bank, a restaurant, a computer system, a computer network, an assembly line, or a distribution network. The first task of the analyst is to identify the common elements of these applications. They all have service centers with an arbitrary number of servers. These service centers will have waiting lines associated with them with a queuing discipline such as FIFO (first in first out), LIFO (last in first out), or priority service. In addition to the active servers, the queuing network may have passive resources, such as computer memory or tables in a restaurant, that are limited and must be allocated and deallocated. These passive resources may also have waiting lines. A third type of resource that may exist in a queuing network might be called transport resources, which could include trucks or tankers in a distribution problem, conveyor belts, or simply a scheduled delay for whatever is being processed. That thing being processed will be called a customer such as a customer at a bank, but this "customer" may also be considered a party at a restaurant, a job in a computer system, a packet on a network, or a product on an assembly line. Our "customers" are scheduled to arrive at entry points in the network by selecting values from a statistical distribution that the client can define before run-time. Customers visit the various queuing resources in an order that is preconfigured by the client, but have the ability to choose client-specified alternate paths based on the state of the system at the time. Service time and transport time may also be statistically determined according to client specification before run-time, and certain servers may be scheduled to "arrive" or "leave" at certain times in the simulation.

The client will want a report from the simulation run that will detail statistics for such things as server utilization, queue length, customer time in service, and number of customers in the system. The client would like to be able to include a histogram for each customer averaged statistic provided in the report. The client would also like to be able to animate sections of the queuing network, and be able to view the dynamic behavior of this selected portion of the system during run-time.

The above statement of the problem will serve this example as a rudimentary specification of the problem domain. A more complete specification would be made and refined by the analyst through a series of discussions with the client, and through his or her own research of the problem area. The analyst will next use the specification to construct models of the problem domain. In OOA it is the static relationship between the objects that forms the structure upon which the system will ultimately be built, and, therefore, constructing the object model is the first and most important task. However, model building is very much a discovery process, and one's original conception of the object model will often be clarified and refined when modeling the system from the other viewpoints.

The first step in building an object model is to identify the classes of objects that are part of the problem environment. One technique for finding classes is to list the distinct nouns in the problem specification and then decide which of these refer to classes in the problem environment. Our brief specification yields the following candidates:

system	applications	bank assembly line
service centers	waiting lines	queuing disciplines
(FIFO..priority)	passive resources	computer memory
tables	customer	transport resource
path	client	report
histogram	queuing network	utilization..number in system
statistical distribution		

Nouns such as system are rejected as being too vague or having no clearly defined boundary. Classes must be specific.

Applications and the list of example applications are irrelevant. We are constructing a generic queuing network simulation system that can be used for a variety of applications. The actual choice of application is irrelevant to the structure of the system being modeled. Actual examples of passive resources such as computer memory or tables are also irrelevant to our model.

Names such as server utilization and number in system will not be regarded as independent classes but attributes of the report. Nouns that do not have features associated with them should not be independent classes.

In other situations, names that refer to roles, such as owner or employer, and names that refer to implementation constructs, such as procedure or interrupt, should be rejected as inappropriate class names.

After reviewing the list of candidate nouns, we will retain **service center, server, waiting line, (FIFO line, etc.), passive resource, transport resource, path element, customer, client, report, statistical distribution package, histogram, and queuing network** as potential classes. From our knowledge of the problem domain we may recognize that other classes implicitly belong in the model, and additional classes may emerge from a process of generalization that we will apply to our model. It is also possible that some of our initial choices may be dropped after more careful consideration.

Two additional tasks remain in the construction of our model:

- identify the associations between the selected classes, and
- identify the features—the attributes and operations of each of the classes.

The two tasks may be regarded as respectively adding bones and flesh to the classes of the system under construction.

We will find many of the associations between classes by returning to the problem specification and listing the verb phrases. An action that connects a subject class with its target class specifies the association between these two classes. Verb phrases that refer to nouns that were rejected as candidate classes can themselves be rejected. In the description of the simulation system problem domain, we encountered the following verb phrases:

service centers	HAVE	(arbitrary number of) servers
service centers	HAVE	waiting lines
passive resources	HAVE	waiting lines
customers	VISIT	service centers
client	CONFIGURES	queuing resource order
client	RECEIVES	report
passive resources	ARE ALLOCATED & DEALLOCATED	

Each of these fragments except the last will denote an association between classes. The last phrase indicates two operations on passive resources. The indication that both service centers and passive resources have waiting lines will lead us to investigate whether a generalization of these two classes might be appropriate. There may also be implicit relationships between classes that we are able to discern from our study of the problem environment.

A second set of relationships between classes results from generalization. The analyst should look for classes that have common features or common associations and attempt to abstract out the common behavior and characteristics of such classes. In the example we are considering, the passive resource center must allocate and deallocate the resources that it manages. Customers that arrive at such a center when insufficient resources are available must join a waiting line until additional resources are reallocated. At a service center or a transportation center, customers wait until a server or transporter is available, then an end of service event can be scheduled. These events can be considered analogous to allocating and deallocating a resource. Path elements may also be considered a resource that needs to be managed. A simple path between two successive service centers is trivially entered and left by the customer, but the path may also contain decision points where alternate paths are chosen either randomly or based on waiting line or server status. At such a point we may say that a path element is allocated (and then deallocated when the customer arrives at the next resource center). We will therefore construct the abstract class **QResourceCtr** that has real subclasses: **Active_ResCtr**, **Passive_ResCtr**, and **Path_Element** (Fig. 15.6). An **Active_ResCtr** itself consists of **ServiceSts** or **TransportSts**. The abstract class is used as a single target for associations with **Waiting_Line**s and **Customer**s that makes for a cleaner model and ultimately a better design. The three subclasses of Waiting Line (FIFO,LIFO,and priority) form another fairly obvious generalization.

We have observed from the specification and the subsequent discussion that customers visit the abstract **QResourceCtr** class. During these visits service is provided, or resources may be allocated, and waiting may occur. In short, all of the behavior that one wants to record and analyze takes place during these visits. Therefore, let us say that a transaction occurs during the visit of a customer to a queuing resource center. The only other dynamic behavior of interest to the client is the arrival and departure of customers and servers to and from the system. (Transporters may also be put in and removed from service at certain times during the simulation run.) These arrivals and departures will constitute a second type of transaction. **Transaction**s may have a location, and duration, and cause changes of state at **QResourceCtr**s. They are an additional class of objects that belong in the model of this problem domain (Fig. 15.7).

We have identified a set of classes that produces objects that are active in the queuing network, **QNetwork**. They arrive, depart, and visit **QResourceCtrs**. These active elements, or **Participants** will consist of **Customer**s, **Server**s, and **Transporter**s; and

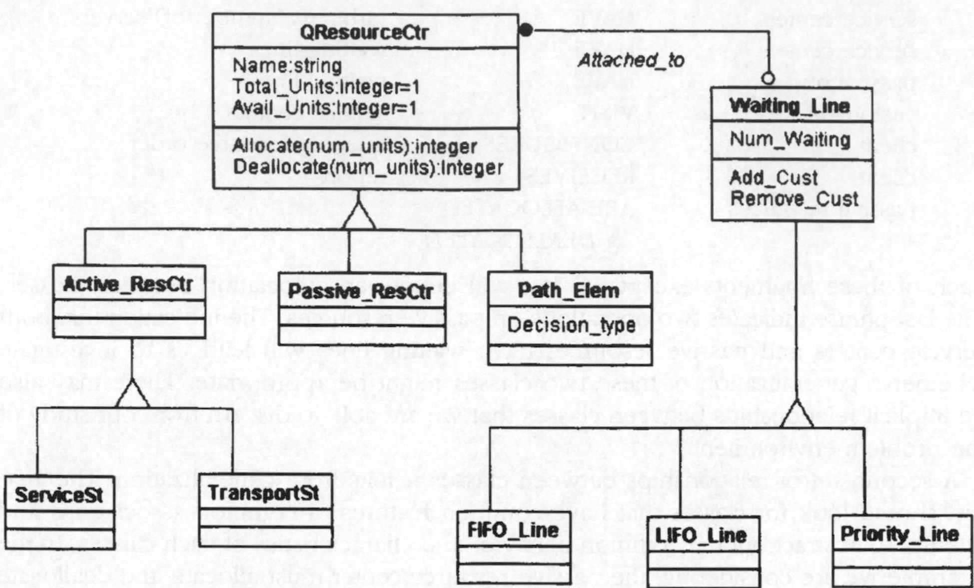

FIGURE 15.6 QResourceCtr and related classes.

may experience either or both **QResourceCtr** visitation transactions and Arrival/ Departure type transactions during their "life span" within the system (Fig. 15.8).

The previous discussion leads to the construction of the object model that is depicted in Figure 15.9. It is a first or partial model because it is an effort to understand the queuing system itself before analyzing more thoroughly the simulation of this system. In modeling a simulation system we are in effect constructing a model of a model. The model diagram is the product of numerous revisions and refinements. Constructing an abstract model of any complex problem domain is a challenging endeavor that methodology and tools can only partially ameliorate. The model depicted also represents one experienced individual's abstraction of the problem domain. A consensus model from a team of analysts might produce a better conceptualization.

The object model in Figure 15.9 depicts the static relationship between objects in the problem domain. It will be the basis for constructing a modularized, hierarchical system in the object-oriented paradigm. The dynamic and transactional behaviors of the objects are two additional descriptions of the system that are needed to develop this model more thoroughly. When the information contained in these models should be collated differs in the approach of various authors. The boundary between the analysis phase and the design phase is not clearly defined. Since state transition diagrams and data flow diagrams that depict the dynamic and transactional behavior of the problem domain have long been a part of the data-driven approach to structured programming, they will not be considered in detail here, but an approach that identifies the agents in the system environment and describes precisely their interaction with the system under construction can be very useful for many applications.

The client has been included as a class in the object model of the simulation system. He or she is part of the environment in which the application runs, but clearly not one of the classes that the designer needs to implement. The client is the source of input events that causes the system to change state and perhaps produce output

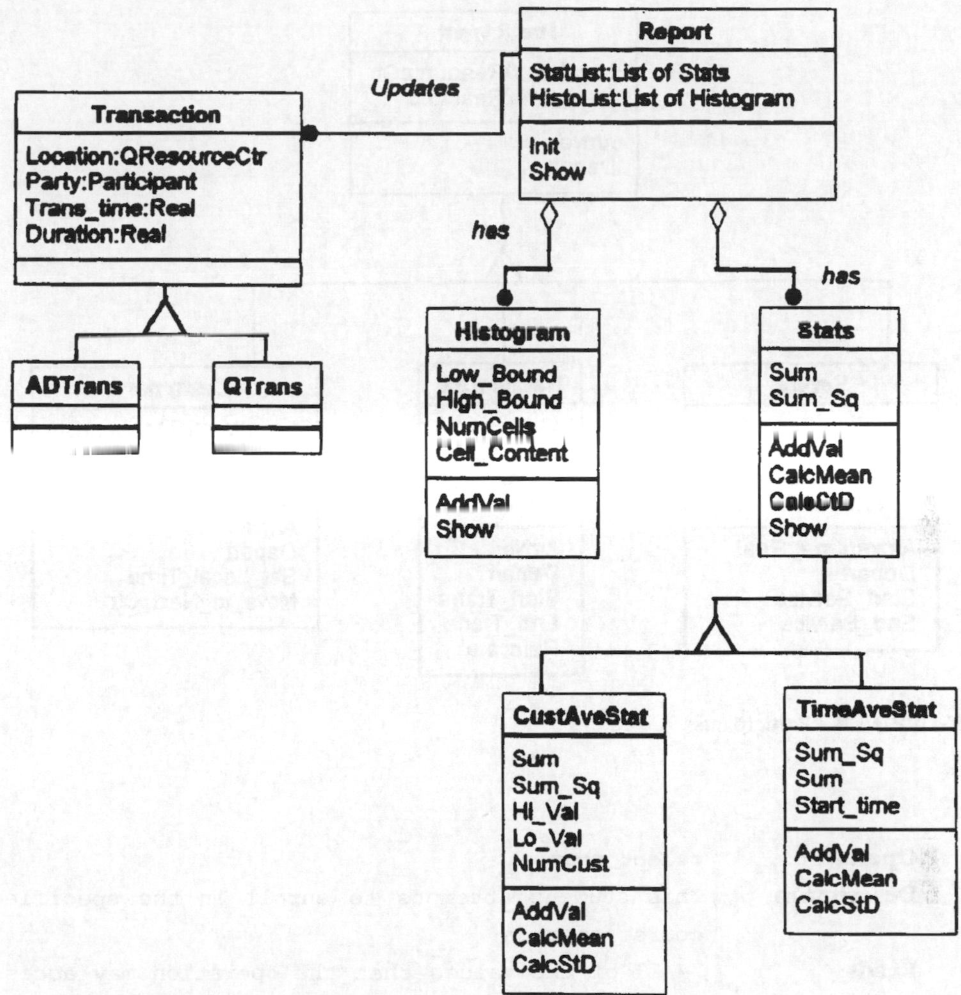

FIGURE 15.7 Transaction and related classes.

events. The client in our example is also the destination for any such output event. The boundary between the classes that need to be implemented as part of the system proper and those that are only agents within the environment of the system must be determined. In our example the client is the only agent. The client configures a queuing network during setup and receives a report of the transactions at the end, but has little interaction with the system during the course of the run itself.

A better illustration of the interaction between a system to be built and the agents in the environment that interact with it is a college registration system. At first sight, the agents in this environment include the students who register for courses, the registrar who manages the system, and the bursar who bills the students for each course taken. The system operation, **select_course**, is initiated by the student, and is one of the input events that needs to be identified and modeled by the analyst. The registration system responds only to system operations such as **select_course** that are initiated by the various agents. To model a system operation (Coleman et al., 1994) suggests the following schemata:

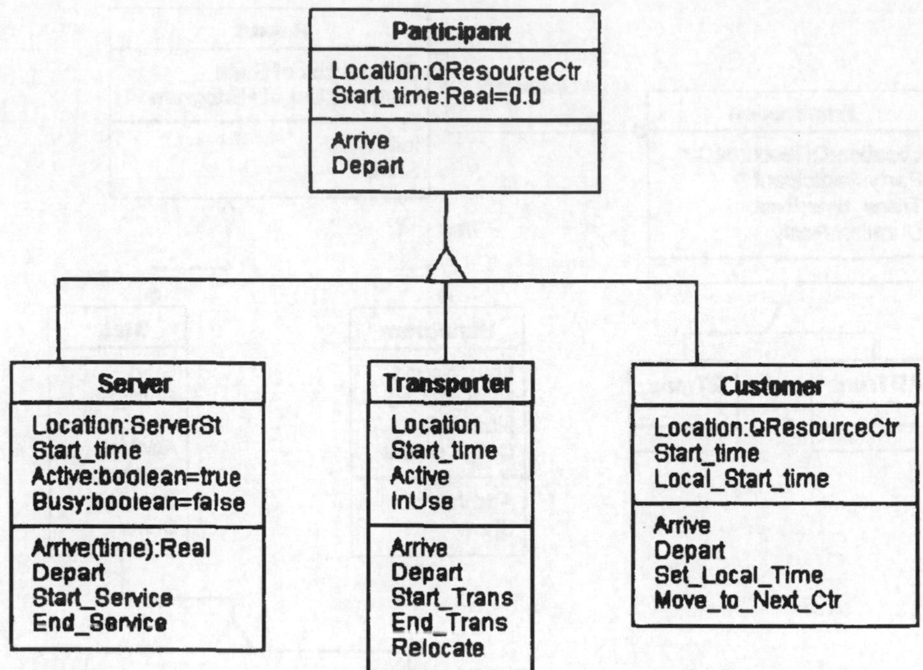

FIGURE 15.8 Participants.

Operation:	**select_course**
Description:	This student attempts to enroll in the specified course section.
Reads:	//All of the values that the operation may access but not change.
	supplied name: student_name
	supplied ID: student_ID
	supplied: catalog_num:course_ID
	Course:prereqs
Changes:	Course: class_size
	Course: class_list
Sends:	Student:(confirmation)
Assumes:	//Precondition for the operation to be defined
	student_ID is valid //student currently registered
Result:	if (Course:class_size < max_size) & Course:prereqs are satisfied
	then inc(Course:class_size) and add(Student: student_name and Student:student_ID) to Course: class_list and confirmation = enrolled
	else confirmation = not_enrolled

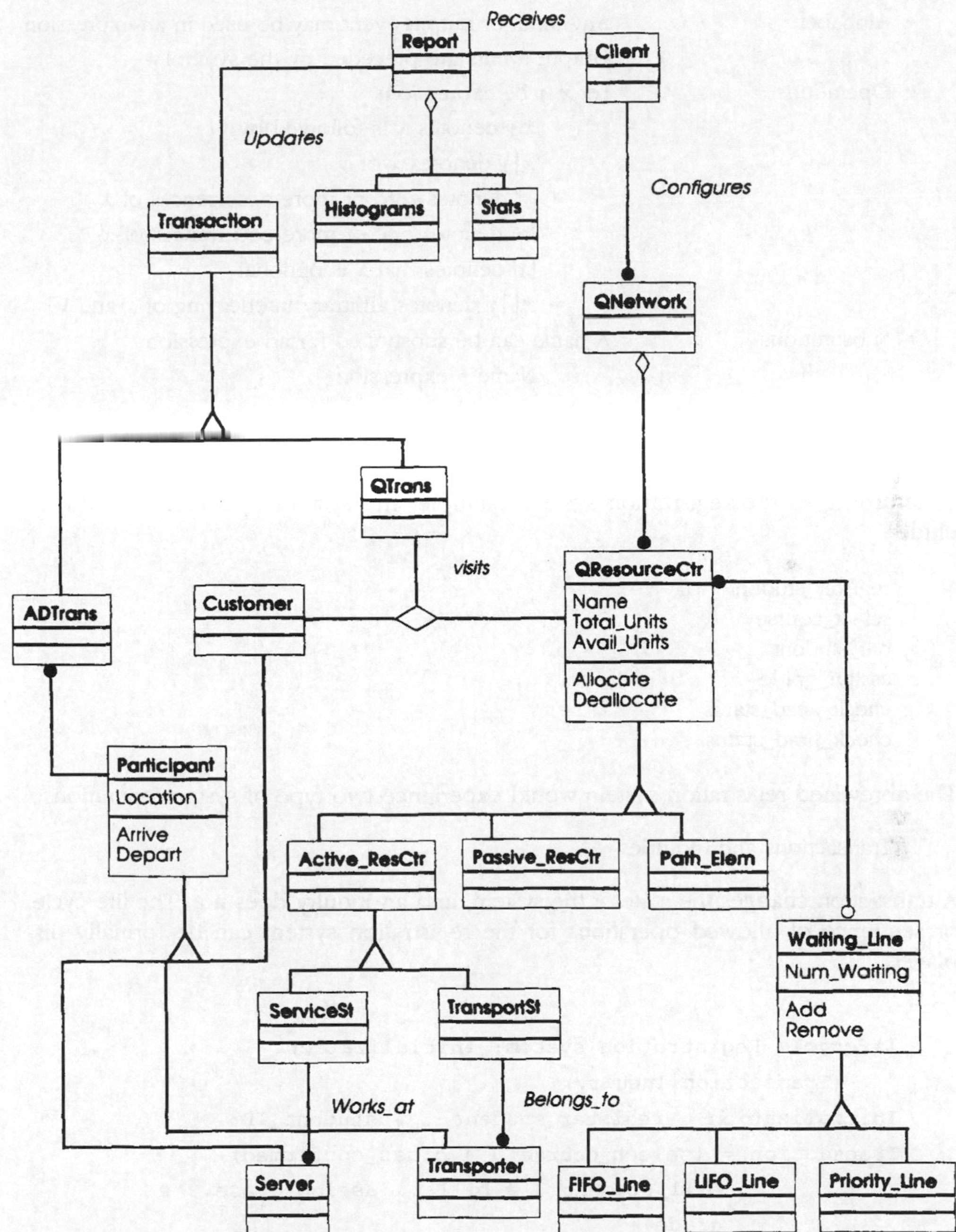

FIGURE 15.9 The (partial) object model of the queuing system.

Each system operation that each agent initiates is similarly modeled. The sequence in which these operations can be initiated may not be arbitrary, but deterministic.

Coleman et al. (1994) again provide a formalism for displaying the sequence of system operations that may be performed. The notation is borrowed from the syntax of specifying formal grammars.

- Alphabet — Any input or output event may be used in an expression. Output events are preceded by the symbol #
- Operations — Let *x*, *y* be expressions
  - $x.y$ denotes x is followed by y
  - $x|y$ denotes x or y
  - x^* denotes zero or more occurrences of x
  - $x+$ denotes one or more occurrences of x
  - $[x]$ denotes that x is optional
  - $x||y$ denotes arbitrary interleaving of x and y
- Substitutions — A name can be substituted for an expression
  - Name = expression
- Operator Precedence — [ ], *, +, ., |, ||

Returning to the registration system example, the system operations would include

register_student
select_course
bill_student
assign_grade
check_acad_status
check_grad_status

The abreviated registration system would experience two type of system operations:

Transactions and Inquiries

A transaction changes the state of the system, and an inquiry does not. The life cycle, or sequence of allowed operations for the registration system can be formally displayed.

```
lifecycle Registration system: Initialization.
    (Transaction|Inquiry)*
Initialization = register_student . # student_ID
Transaction = (select_course . # class_confirmed)+ .
            bill_student . # bill . (assign_grade . #
            grade)+
Inquiry = (check_acad_status . (# none //ave > 2.0 | #
        academic_warning) | check_grad_status. #
        remaining_requirements)
```

Modeling the system operations and system life cycle in this formal manner captures much of the information that is available to the analyst in the dynamic and transactional models of OMT, and it emphasizes the use of pre- and postconditions to establish the criteria for the robustness of the system.

15.3 OBJECT-ORIENTED DESIGN (OOD)

A good analysis should produce models of the problem domain without concern for implementation. The objects that are identified should be real world things or concepts. The designer, on the other hand, must be concerned with implementation detail from the general down to the specific. He or she must create "solution domain" realizations of real world objects. The designer must also consider solution domain constraints such as performance criteria, space limitation, cost, and delivery schedule. The analysis models will be the starting point of a design process that ultimately produces a structure of the solution domain, filled in with implementation specific detail.

The OMT Design Philosophy

In OMT the two aspects of the design problem are

- System Design and
- Object Design.

These may be characterized as first decomposition of the problem domain system and allocation of resources, then specification of the individual solution domain components.

The object of an analysis and design effort is to produce a hierarchical organization that limits the interaction between elements of a given hierarchy. The object-oriented paradigm suggests classes and modules of related classes as the basis for decomposing a system into appropriate components. A well engineered hierarchical system should be easier to build, test, and maintain. In system design, the first task is to break the problem domain model into appropriate subsystems. These subsystems are usually identified by the service they provide. For example, an operating system provides memory management, file system management, process control, and accounting services. The grouping into subsystems should encapsulate functionality and minimize interaction.

The designer may choose to implement a particular subsystem in hardware, software, or otherwise. For instance, in a magazine subscription system, sorting letters into separate categories such as new subscriptions, renewals, and inquiries can be done by the Postal Service if different PO Box numbers are provided for each category. Concurrency is another feature that the designer might seek to exploit. It may exist within the system if objects can be identified that have activities to perform at the same time with infrequent need to communicate. Independent subsystems may be assigned to different processors with minimal communication overhead. Also, a database management system (DBMS) may be employed to manage certain data stores. In general, the application may be decomposed into subsystems, not all of which need to be implemented with newly written software. The designer is encouraged to exploit the use of existing resources to the fullest extent.

The designer is concerned with implementing a solution using data structures and algorithms that access, transform, and display data. The design model describes a class structure that encapsulates data with appropriate function. Its classes will mimic, augment, or possibly refine the classes that were identified in the analysis model, but their features will describe structured data and the operations that may be performed upon it. The design model class structure should reflect three main concerns:

- It should accurately represent the structure depicted in the OOA object model. This model describes the natural structure of the problem domain and suggests the strongest, most stable framework for an implementation;

- It should affect concern for performance and cost criteria; and
- It should take advantage of opportunities for reuse of existing code.

Once the design object model is developed, the task is to design each of the individual classes. A thorough job of design will make coding routine. The three models from the analysis phase are used together to identify the operations that must be performed, and the class to which each should be assigned. When the designer is comfortable that the specification of attributes and operations for a particular class is both appropriate and complete, he or she will select the structure that best organizes the data and algorithms that most efficiently perform the operations. New classes that manage intermediate results may be appropriate.

Besides the features that are assigned to classes to implement the dynamic and transactional behavior of the system, the associations from the object model have to be implemented as part of a class interface. The implementation of an association will depend on the multiplicity of the association, and whether it may have additional attributes. An association creates a link between two, or possibly more, objects. For instance, a many-to-one association, **Works_for**, between objects of classes Person and Company might have an instantiation where Phyllis Stein **Works_for** Rambo Films Inc. If the association is traversed only in one direction, it may be implemented by Miss Stein having a reference to the Company object as one of her attributes. In the other direction, Rambo Films would need to have a set of references to each of its employees. The two-way association provides the greatest degree of generality, and allows for the greatest flexibility in future maintenance, but complicates the implementation. If Phyllis left Rambo, the dissociation would have to be made in both of the objects.

The Fusion Method

The weakness of the OMT methodology is that it does not provide the designer with any new tools or methods to construct his or her model. The design of objects becomes largely a task of identifying and fleshing out the features that the class should contain. The Fusion Method, by Derek Coleman and other authors (1994) incorporates an analysis process very similar to OMT with methodologies that provide additional support to the designer, chiefly by enabling him or her to develop models that describe the object interface and communication paths in the system.

Objects, unlike classes, have a dynamic relationship with one another. They have a life span of their own, and during this life span they interact with other objects by sending and receiving messages. In essence, they collaborate to provide the function expected from the system. The extent of awareness and collaboration between objects can be modeled in a number of different ways:

- Object Interaction Graph—For each function that the system is to provide, responsibility for the implementation of that operation must be delegated to a set of objects. For each system operation one object is designated the controller and initiates the message passing between the other objects collaborating in this task. The sequence of messages between the collaborating design objects defines the functional behavior of the system for the particular operation. The Fusion Method provides a graphical syntax for representing this collaboration.
- Visibility Graphs (Booch, Coleman)—Client and Server classes are identified, and the nature of the relationship is described. The client's reference to the server may be on-going (permanent) or dynamic, and the objects may exist only together or inde-

pendently. The reference may also be either constant or mutable (changed to reference a new object of the same class). The nature of the relationship can be expressed graphically.

- Inheritance Graphs—The task of the designer is twofold and opposite. It is to identify opportunities for generalization that were either missed during analysis or have developed in the design model, and to look for situations where the class hierarchy is unnecessarily deep. In the simulation system example, the designation of an ActiveResCtr class provided pleasing symmetry and useful abstraction, but this second level of abstraction adds little or no additional definition to its two subclasses, only additional overhead to the run-time system. It should be eliminated from the design model. The nature of the inheritance in each instance where it is present can be investigated by constructing separate inheritance graphs. Inheritance is an often abused privilege that the object-oriented paradigm extends. The is_like association is treated as a proper is-a by ignoring the bothersome features. This can lead to an improperly structured model, and maintenance problems. The common features of the two is_like classes should be generalized into a new abstract superclass. The inheritance graphs can help ascertain that the depicted class hierarchy is both necessary and proper.

The Fusion Method integrates system design and object design into a common process. The Object Interaction Graph is an especially convenient tool to allow the designer to explore alternatives by choosing different feasible classes as controller for a particular system operation. The Object Interaction Graphs and Visibility Graphs display the coupling between system components, and can be used to develop an appropriate hierarchical structure.

The individual class descriptions are developed by collating the information contained in each of the models. From the Object Model comes the class structure and the data attributes that each class contains. The Visibility Graphs provide the object references and the Object Interaction Graphs provide the methods. The Inheritance Graphs identify any superclass from which this class may inherit.

Design of the Queuing Network Simulation System

The object model diagram for the simulation system environment that was developed in the section on the Object-Oriented Analysis provides a framework for building a computer model of the application. It was constructed with proper disregard for any implementation specific detail. Several classes that properly belong to a simulation system were left out of the first analysis model because they are part of the description of the simulation of a queuing network, not of the queuing network itself. Even the question of whether this system should employ an event-driven approach or process interaction approach is still open.

It would be appropriate at the start of the design phase for the designer to write an implementation-directed specification of the system under construction. It should consider "solution domain" objects and constructs such as files, icons, hardware devices, and appropriate data structures, and reflect the designer's detailed plan for implementing the analysis phase model. This specification would augment the previous "problem domain" specification with implementation-specific detail, not supersede or substantially ignore it. Separately specifying the design of selected subsystems serves to reinforce the original conceptualization of the "problem domain" and provide an implementation-sensitive criteria for identifying an appropriate decomposition of it into subsystems. Since it is the intention in this discussion to illustrate only design concerns and methodology, not to produce a complete design of the example problem, only fragments of the design specification will be given when needed. These fragments will be identified with surrounding quotation marks.

The first excerpt describes the setup procedure for configuring the queuing network. It illustrates the need to refine a class that has more than one set of compatible behaviors.

"The client has expressed an interest in obtaining a general purpose simulation system that allows her to construct a model with a standard set of capabilities without having to write her own code. The queuing network model is built by using a mouse to click on one of a set of icons that represents resources or path elements and then adds that icon to the model diagram that is being drawn. Each chosen icon is added in turn to a sheet and has a coordinate position and orientation on the sheet. As each icon is added to the model, appropriate pop-up menus will direct the model builder to add the detail necessary to describe the element. For a ServiceSt element, the name, number of servers, and identity and discipline of the waiting line will be among the items that are requested. Each element may have an identification number that has a meaning only within the solution domain. At the completion of the model building, the person performing the simulation can select some portion of the model to lie within a window for viewing. The service and path elements will be represented by depiction of the appropriate icons. Servers will be represented by green (busy), red (idle), and amber (unavailable) dots. Customers will move from location to location, and be represented by black dots. Customers will arrive at and depart from designated arrival and departure path elements points."

When the simulation model has been constructed, four data structures will be indicated:

- An adjacency list that indicates the ID of the successor(s) to each **QResourceCtr** element,
- An accounting table properly configured to collect each statistic of each identified **QResourceCtr** element,
- A table that provides the information needed to construct each of the **QResourceCtr** objects to an interpreter.
- A sheet that contains graphical information such as icon type and location in "world" coordinates for each **QResourceCtr** and **Participant** object.

The configuration of a **QNetwork** requires multiple data elements, and therefore suggests that additional classes are necessary. It is not only an aggregate of **QResourceCtr** objects. It also must provide routing information, and information for animation details. The routing information will be stored as an adjacency list. Operations on the list will include the trivial **Find_Successor**, and the more difficult to implement, **Add_Element**. Operations **Add_Element** or **Remove_Element** will necessitate changes in each of the aggregate of class objects of **QNetwork** and in the accounting table of **Report**. They are a good example of operations that are performed in the aggregate by performing portions of the operation in each of the contributing classes. The cohesion of these aggregate classes imposes a framework for the decomposition of the system into subsystems.

The next example illustrates the use of Object Interaction and Visibility Graphs. Except for during setup, there is no client interaction with the system aside from the passive observation of the animation (and directed abortion of the run). Therefore, during a simulation run there are no proper system operations. Design aid is available, however, if we consider subsystem operations. In an event-driven simulation, events are repeatedly removed from an ordered list and processed. The variety of event types will constitute the (sub)system operations in this example.

"**QResourceCtr** objects are constructed from the user's specifications during setup. They persist for the duration of the simulation run, and are visited by Customers that enter and leave the **QNetwork** at various times during the course of the simulation run. When a **Customer** visits a **QResourceCtr**, a transaction occurs. If the **Customer** were to visit a **ServiceSt**, a free server, if any, would be assigned to the customer and a start of service event would commence. The server would be made busy, and a duration of service would be calculated from a statistical distribution associated with this server. The server process would then be paused, and an end of service transaction scheduled. When the end of service time arrives, the server is reactivated, and either removes a waiting customer from his queue or, if no new customer is present, becomes idle. Statistical details of the transaction are recorded in the appropriate **Report** category, and the next **QResourceCtr** that the customer visits is identified. A new transaction between customer and **QResourceCtr** is initiated.

Transactions identify **Participants** with resource elements. They have a start time and an association. A schedule of transactions, ordered by increasing start time, provides the sequence of events that occurs in the simulation run. The transaction schedule is another instance of a priority queue, and it may be implemented by a heap."

The designer must establish the communication path between objects that collaborate in the execution of the various functions that the system provides. Object Interaction Graphs are one technique that assist the designer to model the collaboration between objects and to readily investigate alternatives. For each system operation, one object is selected as the controller that receives the operation and then initiates the message passing between the other cooperating objects. A relatively simple graphical notation that uses boxes for objects and directed line segments between the boxes to denote the communication describes the sequence of messages exchanged by the collaborating objects. The Object Interaction Graphs should be accompanied by descriptive narrative in the data dictionary.

As an example of an Object Interaction Graph, consider the operation, Start_Service. The controller is chosen to be an object of type ServiceSt. The time sequence of the messages is indicated numerically above each message; a conditional is expressed by a prime after the number of one of the alternatives, such as: if x then (3) else (3′); and iterated messages are denoted with an asterisk after their number (Fig. 15.10).

```
Description
   Operation ServiceSt:start_service(c:Customer,tnow:Real)
      Check for available server
      if available server then
         Decrement # of available servers
         Select sever from active servers list
         Set server to busy and record
         Find end of service time
         Schedule end of service event
      else
         Add customer to queue and record
```

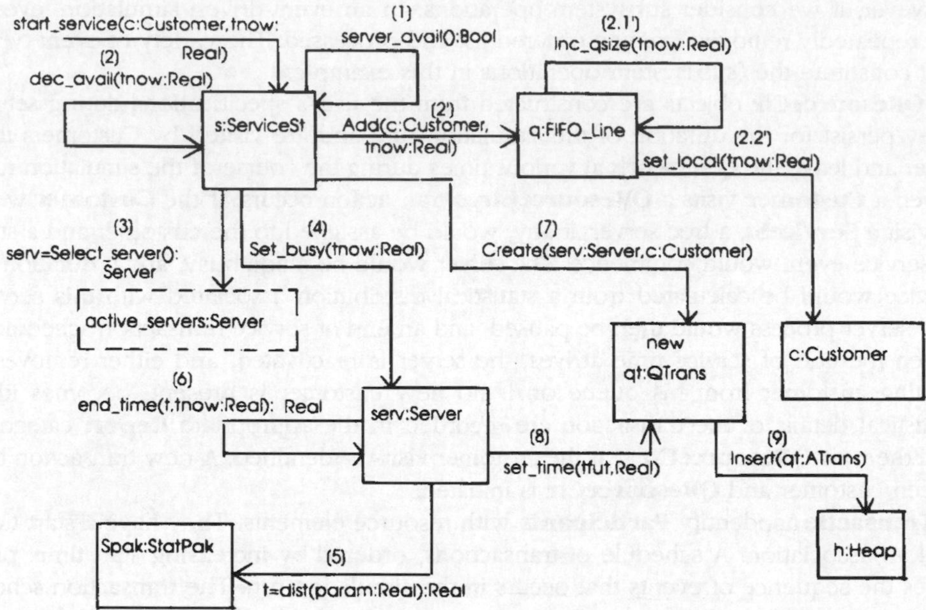

FIGURE 15.10 Object Interaction Graph for start_service operation.

```
method select_server0:Server
  if active_servers list is not empty then
    Remove first server from list
  else
    Error
```

Note that the nature of the accounting activity for the generation of the final report is one area where design alternatives need to be investigated. One could store statistical information in each of the affected objects (as the object interaction graph currently indicates), or one could attach statistical gathering objects to each **QResourceCtr** and **Participant** object, or one could record all of the necessary information on a **Transaction** sheet that is passed to a central **Report** accounting object. Recording accounting information with an individual **Server**, for instance, simplifies the interaction interface, but it adds function that is not particularly appropriate to the class **Server** and would require a change in the structure of the object model.

When object interaction graphs are composed for all of the system operations, a Visibility Graph that indicates which objects reference which other objects, and the duration of that reference, is developed to further specify the object interface. From Figure 15.10, it is seen that a **ServiceSt** object, which persists for the duration of the simulation run, has a dynamic reference to a **QTrans** object, and a permanent reference to an **Active_Servers_List**. The permanent reference is indicated with a solid line; the dynamic reference, with a dashed line. The list of Servers is indicated by a box with dashed lines. Exclusive reference is indicated with a double border, and those objects that do not exist once they are dereferenced from some other object are said to be bound to that object and can be drawn residing inside of the box depicting the object referencing them. Some visibility graphs for the objects in Figure 15.10 are shown in Figure 15.11a–c.

When the designer has completed building the object interaction and visibility graph models, he or she is ready to construct the individual class descriptions. The Fusion Method provides some formal support for both system and object design, but, regardless of the tools, design like analysis is still very much a craft learned through experience. A few general observations may be appropriate to close.

- Graphical tools are useful, but they do not scale up to very large problems conveniently. Too many boxes and lines in the object model are ugly and indicate that either the analyst did not capture the essential simplicity of the underlying structure or the underlying structure of this model is truly ugly. In the latter case, the analyst may separately model individual subsystems, but then the task of finding appropriate subsystems is shifted to the analysis phase for the convenience of the model builder. This may or may not produce an appropriate design.
- The usefulness of the paradigm is problem dependent. The OOP paradigm adds a new emphasis and some new tools to the methods developed in the data-driven structured approach. Individual applications fit this paradigm like clothes fit the individual: sometimes very well and sometimes only with extensive tailoring. The underlying OOP approach or way of thinking about the problem, however, is useful even when the specific tools provide little new support.

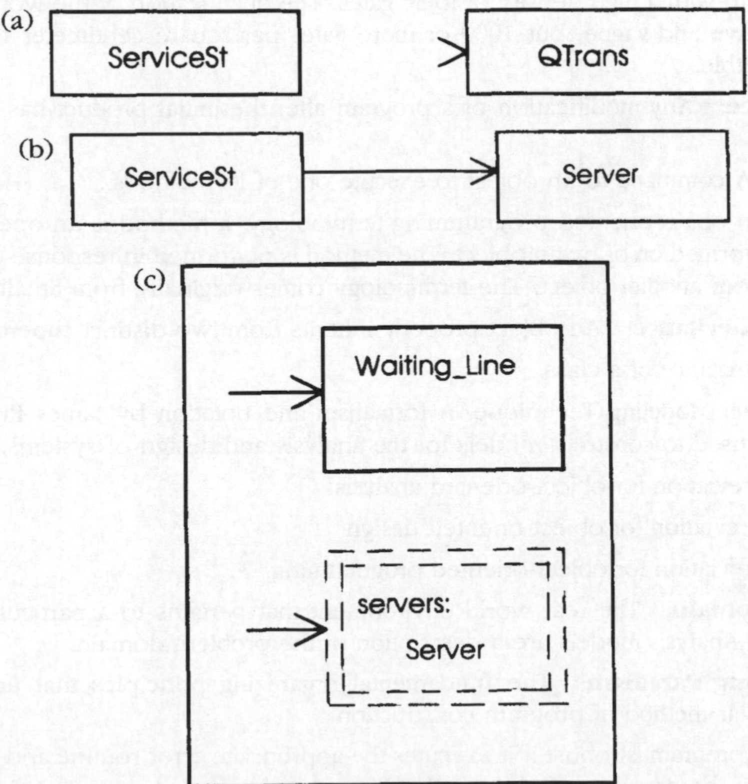

FIGURE 15.11(a) Dynamic reference of **ServiceSt** object to **QTrans** object. **(b)** Permanent reference of **ServiceSt** object to server object. **(c)** Lifetime of client and server objects are bound.

DEFINING TERMS

Adjacency List: A data structure used where a graph represents the underlying relationship between the entities that are depicted as vertices. In the adjacency list each indexed row contains the name (given by its number) of all of the other vertices directly reachable by an edge from the vertex whose number is that of the index.

Attribute: Property of an object. For example, diameter would be an attribute of an object of class circle.

Class: A type, or a template that describes the features that objects, which are instances of the class, possess. It is more than just a data type. It encapsulates both data and function.

Context Diagram: The initial data flow diagram that defines the bound between the system and the actors outside the system that are the sources and sinks of data. The context diagram displays the input/output requirements of the system.

Data Dictionary: A list of all of the identifiers, together with a description of their kind, location, and function, that are found in a software project.

Data Structure: A format by which memory allocated to instances of a data type is organized.

Feature: A property of an object—either an attribute or an operation.

LSI: Large Scale Integration. The technology capable of producing an integrated circuit chip with a high density of logic gates. The qualification of the word high is subjective and vague, but 1000 or more gates per square centimeter would be reasonable.

Maintenance: Any modification of a program after the initial product has been released.

Message: A command to an object to execute one of its methods.

Method: In object-oriented programming terminology, a method is an operation to be performed on or by an object. The method is performed in response to a message from another object. The terminology comes originally from Smalltalk.

Multiple Inheritance: An object properly inherits from two distinct superclasses.

Object: An instance of a class.

OMT: Object Modeling Technique. A formalism and notation by James Rumbaugh that is used to construct models for the analysis and design of systems.

OOA: Abbreviation for object-oriented analysis.

OOD: Abbreviation for object-oriented design.

OOP: Abbreviation for object-oriented programming.

Problem Domain: The real world environment that pertains to a particular application. Analysis models are a description of the problem domain.

Programming Paradigm: The fundamental organizing principles that underlie a particular method of program construction.

Robust: A program is robust if it executes the appropriate error routine and does not crash in the presence of aberrant input.

Solution Domain: The environment appropriate for constructing a "computer" model of the application.

System Operation: An input event initiated by an agent in the system environment.

Units: A term used here to refer to any separately compiled entity within a program. In modula-2 these will be modules; in Ada, packages; in Turbo Pascal 5.5 or higher, they will also be called units.

VLSI: Very Large Scale Integration. An inflationary term reflecting the continued advance in integrated circuit technology after LSI was already in use.

REFERENCES

Derek Coleman, Patrick Arnold, Stephanie Bodoff, Chris Dollin, Helena Gilchrist, Fiona Hayes, and Paul Jeremaes, *Object-Oriented Development—The Fusion Method*. Prentice-Hall, Englewood Cliffs, NJ, 1994.

OMTool User Guide, Martin Marietta Corp., Advanced Concepts Center, King of Prussia, Pennsylvania, 1992.

James Rumbaugh, Michael Blaha, William Premerlani, Frederick Eddy, and William Lorenson, *Object-Oriented Modeling and Design*. Prentice Hall, Englewood Cliffs, NJ, 1991.

FURTHER INFORMATION

Grady Booch, *Object-Oriented Design with Applications*. Benjamin Cummings, Redwood City, CA, 1991.

Peter Coad and Edward Yourdon, *Object-Oriented Analysis*, 2nd ed. Yourdon Press Computing Series, Prentice-Hall, Englewood Cliffs, NJ, 1991.

Bertrand Meyer, *Object-Oriented Software Construction*. Prentice-Hall International Ltd., Hertfordshire, UK, 1988.

Object-Oriented CASE Technology, *Paradigm Plus Case Tool*. ProtoSoft Corp., Houston, Texas, 1994.

16

Eiffel: A Language for Object-Oriented Software Engineering

Robert "Rock" Howard
Tower Technology Corporation

The Eiffel programming language was designed to allow the creation of large-scale software applications that exhibit correct behavior and general robustness even in the face of constant improvements brought on by changing requirements. Eiffel supports and relies on the object-oriented approach to programming and provides many advanced software engineering capabilities that can be used to yield benefits such as improved time to market, a manageable maintenance cycle, easily understood code, and reusable software components. Leading practitioners have built domain-specific Eiffel frameworks that serve as strategic assets for the continual development of specialized software solutions.

This chapter introduces the basic concepts of the Eiffel language. It also provides an overview of the common tools and mechanisms that are used to support Eiffel development projects. A look at advanced project management techniques that have been used successfully along with Eiffel is also included.

16.1 WHAT IS EIFFEL?

Eiffel was designed for creating software applications that work correctly—and continue to work correctly even after dozens of new features are added by people other than the original team. To accomplish this remarkable feat, Eiffel combines solid software engineering ideas together with object-oriented programming principles. The result is a rich and powerful programming language that is fun to use and surprisingly simple to learn.

Eiffel supports Responsibility-Driven Design (RDD), which is an important object oriented (OO) software engineering discipline. RDD centers on capturing key assumptions about object behavior by the simple technique of placing boolean state-

0-8493-2530-7/96/$0.00+$.50

ments called "assertions" at strategic places in the code. These assertions act as contracts between objects that forbid invalid or erroneous cases, thus obviating the need for "defensive programming." Eiffel has been shown to dramatically decrease the time spent tracking down bugs. Furthermore, Eiffel allows the lead programmers for a project to better leverage their design skills while at the same time preventing junior programmers from creating code with subtle problems that cause cascading defects.

Eiffel allows a greater degree of software extensibility and reuse than is possible with most common software development approaches. As we shall demonstrate, this is accomplished by parameterization and inheritance—two of Eiffel's easy and powerful mechanisms for reuse. Also important is the clean syntax of Eiffel and the aforementioned assertions that make it possible to create software that is easy to understand. In fact, Eiffel is often used to great effect as a specification tool that just happens to be able to create executable programs. The fact that Eiffel includes tools that can extract documentation directly from program source code is an additional reason why Eiffel is great for building reusable software.

While Eiffel can be utilized as a stand-alone development solution, most large-scale software projects rely on its ability to interface with other technologies. Eiffel can be used along with popular OO methodologies and OO CASE tools, OODBMS, RDBMS, GUI and graphics systems, distributed object technologies, operating system libraries and tools, and more. There are supported Eiffel libraries and tools that mechanize many of these interfaces. Most Eiffel compilers provide interfaces with the C language. Some Eiffel systems also provide interfaces to C++ and Objective C. The various Eiffel implementations also exhibit a high degree of interoperability with each other—especially when compared to implementations and libraries that are available for other leading OO language systems.

16.2 EIFFEL'S ROLE IN OO SOFTWARE ENGINEERING

Object-Oriented Programming (OOP) is popularly identified with GUI systems and rapid prototyping, but recently OOP has been touted as a technology that can reduce the complexity of large programming projects. Doing this successfully requires a software engineering mindset. Of the nonproprietary OO languages, Eiffel best exemplifies this mindset while fully supporting the OO methodology. Perhaps more than any other language, Eiffel removes unnecessary complexities from software engineering without limiting the ability to deal with the actual complexity that is inherent in modern software systems.

Software engineers need to be fully grounded in basic OO concepts to utilize Eiffel properly. Fortunately, Eiffel itself is an excellent tool for introducing and mastering these concepts. Accordingly there are several informative texts that can be used to learn both Eiffel and OOP.

Proper OO analysis and design are important for creating software systems with Eiffel. The standard OO methodologies and techniques including AMD3/4, Booch, BON, Coad-Yourdon, CRC, Fusion, MOSES, OMT, OOSE, SOMA, and Shlaer-Mellor can all work effectively as front-ends for Eiffel implementation. As Eiffel directly supports Responsibility-Driven Design (RDD) (also known as Design-By-Contract), you may want to check out CRC, BON, MOSES, and other methodologies that emphasize RDD. Larger projects will find use-case scenarios and object interaction diagrams to be helpful organizational tools for ensuring that important aspects of the system design are not accidently ignored.

As stated, Eiffel is commonly used as a specification language. It is excellent for this, particularly as the specification can be compiled and checked for correctness. Simulated or stubbed systems can be generated directly from the specification.

Eiffel is also is an excellent tool for creating an integration layer above existing software systems. It can be used to encapsulate legacy systems and thus define an unequivocal interface that can be targeted by an eventual replacement system.

Eiffel's most important role in software engineering is as a preferred mechanism for defining and building large-scale OO frameworks that can be used to address highly complex application domains. Example usages for Eiffel include network management systems for telecom applications; risk analysis, market simulation, and portfolio tax calculations in financial services; business process specification, tracking, and simulation; factory floor control software; market demand and load forecasting in the travel industry; advanced multimedia, virtual reality, and games with multiplayer and/or advanced simulation capabilities, to name just a few.

16.3 A QUICK INTRODUCTION TO EIFFEL

Eiffel is a "pure" object-oriented language in the sense that every value is either an object, a reference to an object or Void. (Void specifically denotes a reference that is not currently attached to an object.) Eiffel syntax is familiar as it is loosely derived from the Algol/Pascal/Ada family of languages. Eiffel is not a hybrid, however, as it was designed from the ground up as a class-based statically typed object-oriented language. In particular:

- All code in Eiffel must be expressed within a class.
- All routine calls are relative to some object.
- All attributes are relative to some object—there are no global variables.
- There is inheritance including repeated and multiple inheritance.
- There is automatic garbage collection.
- There is strong encapsulation with fine-grained control over visibility.

Conceptually, all class attributes and routine arguments are thought of as object references. The compiler may decide that it can implement certain feature calls in a more direct and efficient manner, but generally these details do not concern the Eiffel programmer. The uniformity in calling conventions simplifies the class interfaces in Eiffel when compared to languages in which there are choices between value and reference semantics (e.g., C, C++).

16.4 EIFFEL TERMINOLOGY

Like every other object-oriented language that emerged in the late 1980s, Eiffel has its own set of terminology. Some of Eiffel's terms are a bit long. Fortunately the Eiffel syntax itself is simple. If a term confuses you, look it up in the "Defining Terms" section at the end of this chapter.

Eiffel is a "class-oriented" language as programming centers on class definitions. A "class" in Eiffel is the collection of code (called "routines") and data (called "attributes") that describes the run-time behavior of "objects"—also called "instances." (Routines and attributes are closely aligned in Eiffel, so collectively they are also known as "features.") Objects are created and then interact with each other during the execution of the program to perform work. An Eiffel object always knows its class type.

The grand simplifying assumption in Eiffel is CLASS = TYPE = MODULE. This means that classes not only define how objects behave, they also indicate how objects are related to each other, i.e., which objects can stand in for other objects in certain situations. Further the "type" information is used for checking that ensures that an object's features are always called properly. In addition, each class must reside in its own file. Thus there is only one class per module.

Related Eiffel classes are gathered together in directories called "clusters." To build an Eiffel application, you create a simple specification file called an "Ace" file. This is like a "makefile," but in an Ace file it suffices to list the clusters where classes might be found. The Eiffel compiler does full dependency analysis on your behalf. Also required is the name of the class that will produce the "root" object for the run-time system, as well as the name of the first routine that will be run for the root object. (When the code for this feature ends, the program terminates.) An example of an Ace file is given later in the chapter.

Figures 16.1–16.4 show the code for an Eiffel system. These examples introduce most of the elements of the language. (The code shown is not from any standard Eiffel library since standard libraries tend to have dozens of routines instead of the handful included in these classes.) We recommend that you look over the code and the associated annotations before returning to the following explanations.

Parameterized Classes

In Figure 16.1 we note that the class name is defined as SIMP_LIST[Element]. The bracketed name designates this class as "parameterized." This means that it can be specialized when used. For example, a class can have two SIMP_LISTs that are parameterized in different ways:

```
my_list  : SIMP_LIST[NUMERIC];-Will hold instances of class
                              -NUMERIC or subclasses of same
                              -(e.g. INTEGER, REAL, etc.)
my_names : SIMP_LIST[NAME];   -Will hold instances that
                              -conform to the class NAME.
```

Trying to add an INTEGER to 'my_names' or a NAME to 'my_list' would be a caught as an error by the compiler.

The most typical use of parameterized classes is for containers classes such as the SIMP_LIST example. Many other uses are also possible. In fact, advanced Eiffel designers will often use parameterized types to unify entire reusable frameworks. For example, a framework for solving genetic algorithms can be parameterized by a GENE class for which any number of different implementations might exist.

We can see an example of this in class SIMP_LIST itself. Note that there is an attribute defined near the bottom of the page with the name 'list_head'. This attribute is designated to be of type NODE[Element]. As we shall see shortly, the NODE class is itself parameterized. Due to being defined with 'Element,' which is the same generic parameter that SIMP_LIST itself uses, NODE will always match up with SIMP_LIST in terms of what types of objects are allowed for use. Similarly, almost all of the other features of SIMP_LIST also use the parameter 'Element' to specify either an input argument or a function result.

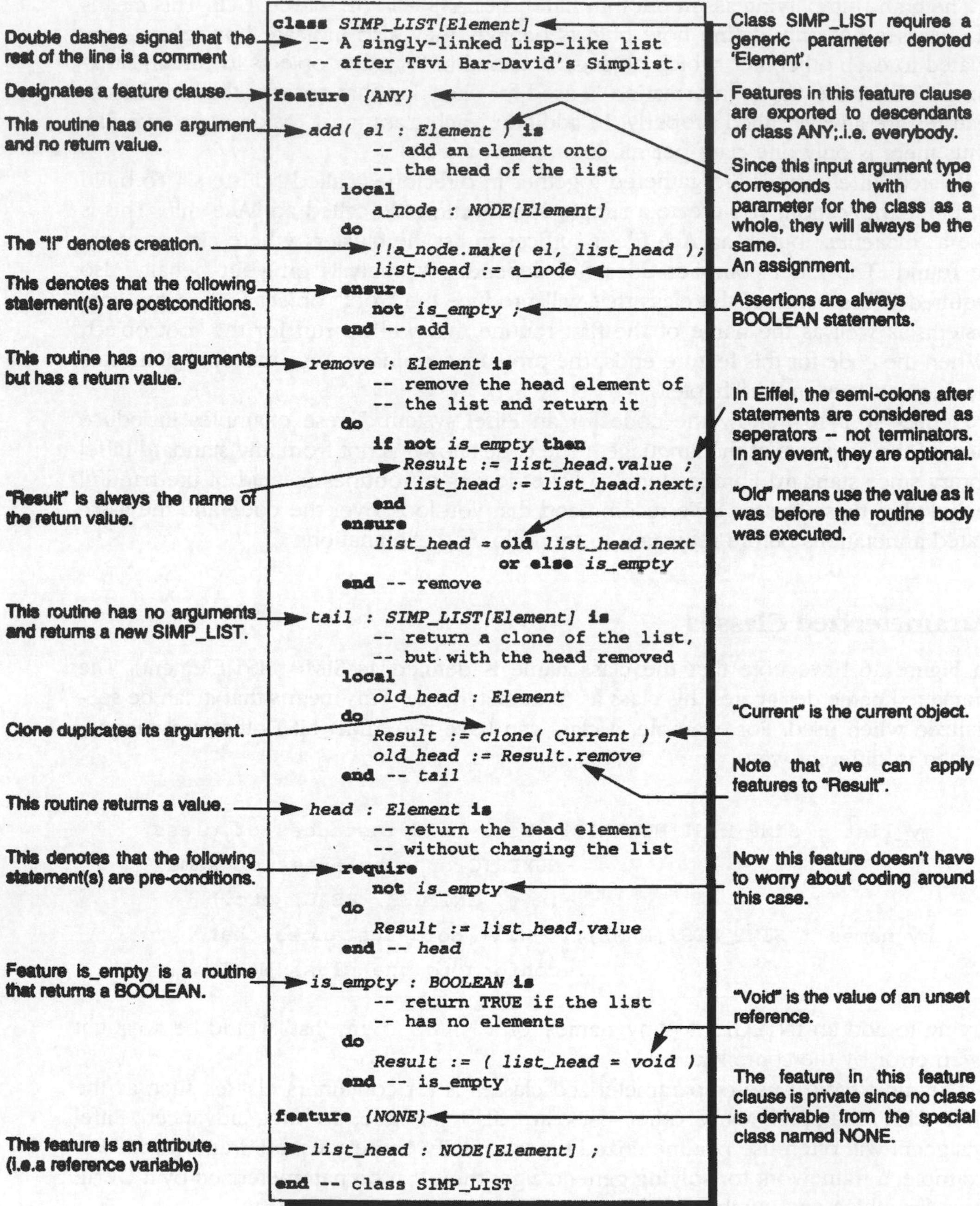

class SIMP_LIST[Element]
-- A singly-linked Lisp-like list
-- after Tsvi Bar-David's Simplist.
feature {ANY}
add(el : Element) is
-- add an element onto
-- the head of the list
local
a_node : NODE[Element]
do
!!a_node.make(el, list_head);
list_head := a_node
ensure
not is_empty ;
end -- add
remove : Element is
-- remove the head element of
-- the list and return it
do
if not is_empty then
Result := list_head.value ;
list_head := list_head.next;
end
ensure
list_head = old list_head.next
or else is_empty
end -- remove
tail : SIMP_LIST[Element] is
-- return a clone of the list,
-- but with the head removed
local
old_head : Element
do
Result := clone(Current) ;
old_head := Result.remove
end -- tail
head : Element is
-- return the head element
-- without changing the list
require
not is_empty
do
Result := list_head.value
end -- head
is_empty : BOOLEAN is
-- return TRUE if the list
-- has no elements
do
Result := (list_head = void)
end -- is_empty
feature {NONE}
list_head : NODE[Element] ;
end -- class SIMP_LIST
Double dashes signal that the rest of the line is a comment
Designates a feature clause.
This routine has one argument and no return value.
The "!!" denotes creation.
This denotes that the following statement(s) are postconditions.
This routine has no arguments but has a return value.
"Result" is always the name of the return value.
This routine has no arguments and returns a new SIMP_LIST.
Clone duplicates its argument.
This routine returns a value.
This denotes that the following statement(s) are pre-conditions.
Feature is_empty is a routine that returns a BOOLEAN.
This feature is an attribute. (I.e.a reference variable)
Class SIMP_LIST requires a generic parameter denoted 'Element'.
Features in this feature clause are exported to descendants of class ANY;.i.e. everybody.
Since this input argument type corresponds with the parameter for the class as a whole, they will always be the same.
An assignment.
Assertions are always BOOLEAN statements.
In Eiffel, the semi-colons after statements are considered as seperators -- not terminators. In any event, they are optional.
"Old" means use the value as it was before the routine body was executed.
"Current" is the current object.
Note that we can apply features to "Result".
Now this feature doesn't have to worry about coding around this case.
"Void" is the value of an unset reference.
The feature in this feature clause is private since no class is derivable from the special class named NONE.

FIGURE 16.1 Class SIMP_LIST: an annotated example of Eiffel source code.

Feature Clauses and Encapsulation

All features in an Eiffel class are defined inside of "feature clauses." There are two of these in SIMP_LIST. The first includes all but one of the features. It defines the export status for the included features—in this case ANY—which means that any Eiffel object can utilize these features. The other feature clause at the bottom has only one feature

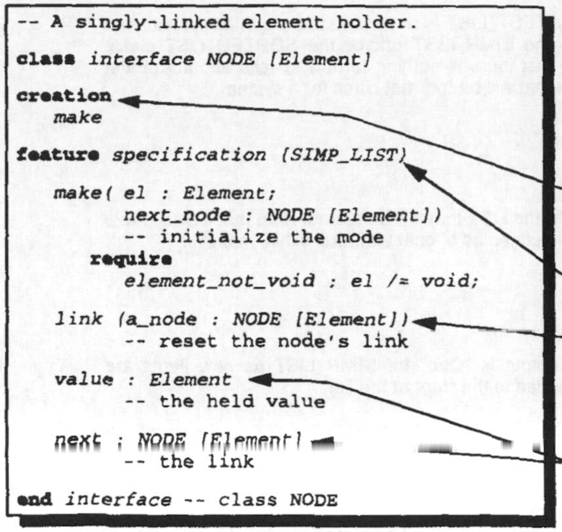

```
-- A singly-linked element holder.

class interface NODE [Element]

creation
    make

feature specification {SIMP_LIST}

    make( el : Element;
          next_node : NODE [Element])
          -- initialize the mode
      require
          element_not_void : el /= void;

    link (a_node : NODE [Element])
          -- reset the node's link

    value : Element
          -- the held value

    next : NODE [Element]
          -- the link

end interface -- class NODE
```

You may have noticed that class SIMP_LIST used a class NODE to hold elements. This is the automatically generated "short" form of the NODE class. It includes the publicly exported features including their signatures, their major comment and their pre and post-conditions. This is precisely what is needed in order to use a class instance in a client/server relationship.

The creation clause is used to denote procedures that are used to create new instances and/or re-initialize existing instances.

The features of this class are only available for use by the class SIMP_LIST (and its descendants.)

Note how the NODE class shares the same generic parameterization designation as the SIMP_LIST class. This technique is an important aspect of Eiffel's support for the safe construction of patterns or frameworks where work is accomplished by cooperating classes.

It is safe to allow the attributes to be visible since outside instances are not allowed to assign directly to them. Only the object containing the attributes enjoys that privilege.

FIGURE 16.2 Class NODE presented via automatic documentation generation.

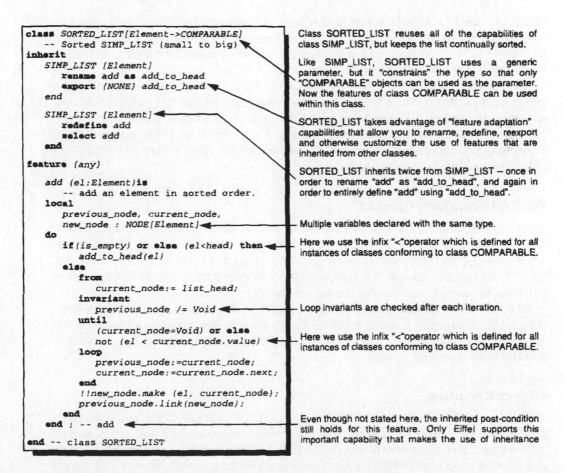

```
class SORTED_LIST[Element->COMPARABLE]
  -- Sorted SIMP_LIST (small to big)
inherit
   SIMP_LIST [Element]
      rename add as add_to_head
      export {NONE} add_to_head
   end
   SIMP_LIST [Element]
      redefine add
      select add
   end
feature {any}
   add (el:Element)is
      -- add an element in sorted order.
   local
      previous_node, current_node,
      new_node : NODE[Element]
   do
     if(is_empty) or else (el<head) then
        add_to_head(el)
     else
        from
           current_node:= list_head;
        invariant
           previous_node /= Void
        until
           (current_node=Void) or else
           not (el < current_node.value)
        loop
           previous_node:=current_node;
           current_node:=current_node.next;
        end
        !!new_node.make (el, current_node);
        previous_node.link(new_node);
     end
   end ; -- add
end -- class SORTED_LIST
```

Class SORTED_LIST reuses all of the capabilities of class SIMP_LIST, but keeps the list continually sorted.

Like SIMP_LIST, SORTED_LIST uses a generic parameter, but it "constrains" the type so that *only* "COMPARABLE" objects can be used as the parameter. Now the features of class COMPARABLE can be used within this class.

SORTED_LIST takes advantage of "feature adaptation" capabilities that allow you to rename, redefine, reexport and otherwise customize the use of features that are inherited from other classes.

SORTED_LIST inherits twice from SIMP_LIST -- once in order to rename "add" as "add_to_head", and again in order to entirely define "add" using "add_to_head".

Multiple variables declared with the same type.

Here we use the infix "<"operator which is defined for all instances of classes conforming to class COMPARABLE.

Loop invariants are checked after each iteration.

Here we use the infix "<"operator which is defined for all instances of classes conforming to class COMPARABLE.

Even though not stated here, the inherited post-condition still holds for this feature. Only Eiffel supports this important capability that makes the use of inheritance

FIGURE 16.3 Extending SIMP_LIST using constrained genericity.

```
-- A simple test driver for SIMP_LIST
-- and/or SORTED_LIST classes.

class TEST_LIST

creation
   make

feature

   test_list : SIMP_LIST[STRING] ;
       -- Can be changed to SORTED_LIST

   make is
       -- create a list and exercise it
     do
       !!test_list ;
       test_list.add( "Four" ) ;
       test_list.add( "Three" ) ;
       test_list.add( "Two" ) ;
       test_list.add( "One" ) ;

       io.putstring( the_list.head ) ;
     end

end -- class TEST_LIST
```

Class TEST_LIST is a root class for an Eiffel system that tests the SIMP_LIST and/or the SORTED_LIST class. Note that there is nothing terribly special about an Eiffel class that will be the root class for a system.

This is the LIST that will be tested. Note that it could have been defined as a local attribute within 'make'.

The output is "One" for SIMP_LIST as new items are appended to the front of the list.

The output is "Four" for SORTED_LIST as the items are sorted in alphabetical order.

When the designated feature of the root class is done, the program terminates.

FIGURE 16.4 Test driver for the annotated example.

that is exported to NONE—meaning that only the current object can access this feature. You can also provide a list of class or classes that the features in a feature clause are visible to. This allows detailed control of the class encapsulation. This technique is utilized in class NODE shown in Figure 16.2.

Attributes in Eiffel are strongly encapsulated—they can only be modified by the object instance that contains them. This means that features like 'get_value' are almost entirely unnecessary. You need not specifically protect attributes in fear that a separate object that can view them might change them and thus break encapsulation. Conversely you *must* supply a 'set_value' style feature for cases where you want other objects to be able to alter an attribute value.

Built-In Standard Classes

Eiffel implementations come with a variety of built-in kernel classes that handle basic types including integers, reals, floats, booleans, strings, arrays, and bits. The compiler knows about these types and thus can produce optimized code when they are used. Other kernel classes include IO, standard files, and exceptions. The Eiffel Consortium (NICE) is working to standardize the kernel classes.

All Eiffel classes implicitly inherit from class ANY and thus include copying, cloning, comparison, and other features included in this class. Eiffel supports a clean mechanism whereby programmers can easily add their own universally available capabilities into the class hierarchy as well.

Object Creation

Some kernel classes including INTEGER, REAL, FLOAT, CHARACTER, and BOOLEAN are "basic types." This means that they are created automatically and set to default values when they are used. (It also means that the compiler will optimize their use by not creating object references for them until and unless a specific feature call reqires the object in a reference form.)

Nonbasic types, which encompass the vast majority of supplied and user-defined classes, must be specifically created before they can be used. We see the syntax for this ("!!") in the feature 'add' in class SIMP_LIST. Note that there is a feature call to 'make' associated with the creation. This feature is specifically marked in class NODE as a "creation routine." It is expected to initialize the object.

Creation routines are optional. If there is no creation routine, then you just create the object by using "!!," but tacking on no routine call. A class can also have more than one creation routines. If there is one or more creation routine, you are obliged to use one of the creation routines to create the object.

Eiffel routines that return values (called "functions") do so with an automatically specified local attribute called "Result." The type of Result is the same as the return type of the function. Note that if Result is not a basic type, then it too must be created (using "!!" as per normal) if you need for it to return a non-Void object reference. (A nice aspect of using Result as an automatic return value for Eiffel functions is that you can create it and use it in any manner that you like. The value that Result has when you end the function body is the value returned. This is more powerful than the mechanism in Pascal, for instance.)

If you are creating all these objects, you eventually are going to have to consider their destruction as well. Fortunately this is all managed automatically for you by Eiffel's garbage collector. This is a major advantage of programming in Eiffel. It dispenses with most of the defects that are commonly associated with memory management. Case studies indicate that this can reduce defects by 20 to 50%.

There are many kinds of garbage collectors for Eiffel. They vary in design, speed, and flexibility. They all can generally be controlled by the application to some degree. (At least turned on and off.) Some can perform incrementally. This reduces the possibility of long pauses while the garbage collector works at the expense of some additional overhead. Real-time garbage collectors also exist and they can be set up to guarantee a required response time for an application.

If an object points to an external resource (usually via a C pointer passed in from an external call to C), it can define a feature called 'dispose' that the collector calls to free the external resource when the Eiffel object is reclaimed by the garbage collector. This standard Eiffel capability is known as "finalization."

Eiffel Assertions and Design-By-Contract

The most unique aspect of Eiffel is shown within the clauses marked 'require' and 'ensure.' The boolean statements within these clauses are the "preconditions" and "postconditions" of the routines and are forms of Eiffel assertions. (Class invariants, loop invariants, loop variants, and general checks are also supported.) The preconditions require that provided arguments are acceptable and that the object is in the proper state for handling the feature call. Postconditions ensure that a feature performed correctly and left the object in a correct state.

Since passing uninitialized references is one of the most common programming mistakes, preconditions checking for non-Void references are quite common. (Note that Eiffel can also insert optional checks in the application that trap any feature call made upon a void reference.)

When a run-time error occurs, the Eiffel system exception handling capability kicks in. Assuming no rescue clause is defined to handle a given assertion failure, it is handled by the Eiffel run-time system, which dumps traceback and other information to the developer. Note that precondition failures always denote errors in the *caller* while postconditions always denote errors in the *callee*. The run-time knows this and thus errors are immediately isolated across the critical class boundary. The liberal use of assertions usually suffices to isolate defects quickly and their use is a huge net time saver.

The key point to the assertions is that they are respected even when a new Eiffel class inherits and redefines a given feature. Thus pre- and postconditions are parts of a binding contract that holds not only for objects of a given class, but for all descendent classes as well. This capability is uniquely supported in Eiffel and is actually the touchstone for OO software engineering of large systems. Not only do assertions directly support the well-known design-by-contracting metaphor, they also aid code comprehensibility, interact with Eiffel's exception handling capabilities, and serve to improve overall confidence in system reliability.

Class invariants are another important and unique kind of Eiffel assertion. The invariant is a set of BOOLEAN statements that must be true before and after every nonlocal call to an Eiffel feature. This means that except for calls that are within a single object, the invariant of the callee must be correct upon entry to and exit from the call. Invariants are often used to collect checkable statements about the logically allowable states of an object.

Now you might be starting to worry about the effect of all this checking on the performance of your program. You are correct to worry as the checking can impose a lot of overhead. Fortunately the compiler makes it easy to turn the checking on and off. You can control the checking on a per class basis and can specify different amounts of checking (called "assertion levels"). For example, "preconditions only" is a popular setting for well-tested code. After all, just because you have tested a class thoroughly does not preclude someone else from calling a given feature incorrectly.

Note that most Eiffel systems will generate optional trace information and/or allow run-time control of the assertion levels. Some even have functions that will read in files specifying detailed assertion levels. Run-time setting of assertions is particularly useful when used in conjunction with an Eiffel source-level debugger.

Class Abstraction

Figure 16.2 is an example of an automatically abstracted class interface. The example was created automatically by a tool that is referred to as "short" as it generates a short form of the Eiffel class. The short form includes the visible features of a class including the signature (which includes the calling parameters and the return type), the main comment and the pre- and postconditions.

Class abstractors also include the ability to "flatten" a class definition. This means that the entire interface for a class is presented, including all inherited features. We will revisit class abstractors later and take a look at additional capabilities.

16.5 SUPPORT FOR SOFTWARE REUSE IN EIFFEL

Eiffel supports the reuse of classes via three main mechanisms—inheritance, parameterization, and composition.

The inheritance mechanism is quite well known as a mechanism for reuse in OOP. Eiffel allows renaming and redefinition of features by subclasses so that they can be specialized easily. Multiple and repeated inheritance are also allowed. In Eiffel, inheritance can be used for many purposes including refinement, specialization, reuse of an abstract interface, and even simple code reuse.

We have already explored parameterization, which is also known as "genericity." Note that class parameters can be "constrained." This means that you can specify a class from which the generic parameter must inherit either directly or indirectly. By doing so, it becomes possible to use the features of the constraint class within the enclosing parameterized class. This results in enhanced opportunities for the implementation of high level abstract data types.

Using classes via composition simply means that we have attributes of a given class that we use in a client-supplier manner. This type of reuse is probably the most common and is aided by assertions that clearly elucidate the assumptions of this relationship. (This is also known as "containment.")

Eiffel actively supports the reuse process. Its clear syntax, assertion technology, and flexible and straightforward type system all combine to make reuse practical in day-to-day programming. Eiffel even includes an indexing clause to help locate and track reusable classes. Here is an example:

```
indexing
    module:          "$RCSfile: hp82000_pin.e, v $"
    author:          "$Author: wilson $, Amazing Devices"
    version:         "$Revision: 1.4 $"
    original:        "Original:14/3/94 5:22pm."
    last_modified:   "Last Modified $Date: 1994/10/26 10:43:37 $"
    class_cluster:   "CVS Module: vector_hub/hp82000"
    keywords:        "pin, vector, hp82000, translation"
```

This indexing clause is optionally defined at the beginning of an Eiffel class. Note that information includes revision control information. Another important use of the indexing clause is to provide relevant keywords or other designations that can help specify or differentiate the unique capabilities of a given class.

Eiffel's powerful support for reuse results in real time savings. Field reports from large projects indicate that nontrivial Eiffel programs tend to have 25 to 30% fewer lines of code than equivalent programs in C++. The support for correctness reduces the time spent on bug fixes and increases the ease of most software enhancements. Users report that major new capabilities in actively supported Eiffel applications can be added and tested in shorter development cycles.

A Practical Example of Reuse

In Figure 16.3 we present a class called SORTED_LIST that is an extension of class SIMP_LIST from Figure 16.1. The SORTED_LIST class shows how Eiffel's powerful mechanisms for reuse can be combined to great effect. In particular, the class uses constrained genericity, multiple and repeated inheritance, feature renaming, and feature redefinition. This sounds quite daunting, but the example shows how easy this can be with Eiffel.

What we are trying to do in SORTED_LIST is to extend SIMP_LIST, which is an extremely simple linked list implementation, such that the elements are kept in order, from greatest to smallest, according to some criteria that we do not want to know the details about. The strategy is simple. We will change the 'add' feature, which is the only feature for adding a new element to the list, so that it adds the element according to some checkable criteria instead of to the head of the list.

Changing 'add' is easy enough. First we inherit from SIMP_LIST and indicate that 'add' will be redefined.

```
inherit
   SIMP_LIST [Element]
      redefine add
   end
```

Now we just define the new 'add' feature in the same manner as if it was a brand new feature and we are done. (Note that the postcondition from the definition of 'add' in SIMP_LIST is automatically inherited. If we needed to, we could have added additional postconditions as well.)

So we start to write the code for 'add' in SORTED_LIST and we notice that the old version of 'add' would be handy to call. Unfortunately calling 'add' from 'add' would recursively call the very routine we are writing. This is not what we want at all! To get at the old definition of 'add' from SIMP_LIST we use an expediency that seems funny at first, but is actually a standard idiom in Eiffel. What we do is inherit from SIMP_LIST again and rename 'add' to some other name so that we have a way to call it.

The approach of inheriting twice (which is the simplest possible form of repeated inheritance) may seem extravagant, but the compiler will recognize that all of the features and attributes that have not been redefined are actually the same. They are combined such that there is no additional overhead. (Note that this is handled in all cases of code sharing via inheritance, not just the simple case of direct repeated inheritance shown here.)

So now we are inheriting from SIMP_LIST twice, but we are not quite done. We note that the renamed version of 'add' should not be available to be called from outsiders. (If we allow it to be called, we can no longer guarantee that the list will be sorted.) This is handled by reexporting the renamed feature to {NONE}. You can see this done in the first inheritance clause in the example in Figure 16.3. In fact, all inheritance done for the sake of code reuse should be reexported to {NONE} so that the visible class interface is not altered.

At this point, the compiler will indicate a small remaining detail. If an object of type SORTED_LIST should become attached to a reference of type SIMP_LIST, which version of 'add' should be invoked—the redefined one or the renamed one? Clearly we want the redefined one to be used and we need to indicate this to the compiler via the 'select' statement as has been done in the code example.

The fine-tuning of inheritance just performed is called "feature adaption." This example showed most, but not all, of Eiffel's feature adaption mechanisms. These mechanisms give tremendous latitude in the utilization of inheritance for reuse and system design.

Now there is just one remaining problem—how do we compare the elements of the list so that we know where to insert them into the list? Given a feature to do that, it is easy to loop through the list and figure out where the element belongs.

The mechanism we use to solve this problem is called constrained genericity. The notation [Element →COMPARABLE] at the top of the SORTED_LIST class shows this construct in action. Remember that we encountered generic parameters before in both SIMP_LIST and NODE, but this time we see the "→" mark and this indicates that the parameter must conform to a constraint class—namely the class COMPARABLE. The class COMPARABLE defines the features named "<" and ">," which are both infix operators. These features are just what we need to use in the new version of 'add' to write the code we need. The problem is solved. Note that anyone using the SORTED_LIST class may designate only parameters that inherit from COMPARABLE.

16.6 STANDARD PROGRAMMING CONSTRUCTS

Eiffel provides many of the standard programming constructs that are familiar from structured programming languages. These include if statements (if . . . elseif . . . else . . . then . . . end), switch statements (inspect . . . when . . . then . . . else . . . end), and loops (from . . . until . . . loop . . . end). There are no break statements nor are there provisions to return from routines prematurely. The philosophy of the Eiffel language is to keep the language clean and simple by providing a minimal set of powerful constructs for looping and branching.

Another common construct is equality testing via "=" and "/=." These constructs test the equality of references—not the underlying object values. Features for doing comparisons of object values are universally available as well. They include 'is_equal,' 'equal,' and 'deep_equal.'

Assignment is designated via ":=." The compiler insists that type conformity be respected when assignments are done. There is an interesting case where this is not sufficient. Say that you got an Eiffel object from a file or from a separate Eiffel process. When you get it you do not know what type it is and yet you want to assign it to a reference in your Eiffel system. For this we use "?=" as in

```
a : STRING ;
a ?= b ;
```

This is called an "assignment attempt." What happens is that the assignment is attempted. If the object conforms, then the assignment occurs. This means that if the object referred to by 'b' in the example happens to be a STRING or a class that inherits from STRING, then 'a' is set equal to 'b.' Otherwise, 'a' gets set to Void.

Eiffel and C

Eiffel supports an interface to external languages. All Eiffel systems support calls to C. Some handle other languages as well. The basic construct is

```
c_call is
external C: <include_me.h>
alias "c_function"
end
```

This means that the feature is called from Eiffel using the name 'c_call' but that the call maps to the underlying call to the c function named 'c_function.'

Notice that external routines can have pre- and postconditions just like a normal Eiffel routine. It can also be redefined, renamed, and so forth. There are also ways to map calling arguments and return values. These mechanisms can either utilize or override ANSI prototypes that may exist for the external C function. Some Eiffel vendors supply "inlining" as an option to reduce unnecessary overhead when utilizing C functions. Support for calling C macros is also available.

Eiffel implementations include a POINTER class for holding C pointers. It is used chiefly for managing C pointers that are returned from external C calls. The C pointer must be passed back to the external C code for dereferencing. At the Eiffel level the POINTER object can be checked for equality versus another pointer object or versus NULL, but that is all that can be done.

Calling Eiffel from C is generally done through a C library called CECIL that allows dynamic lookup of Eiffel class and feature names. TECIL is a faster calling construct from Tower Technology that supports direct polymorphic feature calls.

Tower also includes direct interfaces to C++ and Objective-C. These mechanisms are similar to calling "inline" C, but take advantage of Tower's ability to generate code for C++ or Objective C. These interfaces allow encapsulation of C++ or Objective C objects such that calls through to the respective features have zero overhead (no "wrappers" are required). The interface also make it possible to use Eiffel's garbage collector to trigger the destruction of the underlying C++ or Objective C object, thus extending the benefits of automatic memory management.

Other Eiffel Constructs

Eiffel has a few constructs that are reasonably unique. One is the notion of a "once" function. A "once" function is a normal feature that has the property that it behaves like a normal feature call the first time that it is called, but for every subsequent call to the same feature it simply returns the same Result (if any) as the first time that it was invoked. This is an unusual but highly useful construct for initializing subsystems and arranging for sharing among objects.

Another important construct is that features can be marked as 'deferred.' This means that they are fully defined features in terms of having their interface defined—and even pre- and postconditions can be optionally included—but there is no code body. Classes with one or more deferred features are known as deferred or "abstract" classes. They are a useful organization tool.

Remember class COMPARABLE? This class from the Eiffel kernel library is deferred since the "<" feature is deferred. This means that you are not actually allowed to create an object of type COMPARABLE. Every creatable descendent of comparable must supply a definition for the "<" feature. Of course, this is precisely the semantics we need for the constrained parameter of SORTED_LIST.

This concludes our brief introduction to the mechanisms and constructs of the Eiffel language. As you can probably see already, there is a minimum of syntax but a lot of available power. Our experience is that the language constructs have a synergy that makes the whole much more powerful than the sum of the parts.

16.7 EIFFEL TOOLS AND SYSTEMS

Development Automation

A strong software engineering approach such as that provided by Eiffel is only part of an overall development solution. Equally important for success is to take advantage of opportunities to automate the development process. Eiffel systems include important capabilities in this area. Furthermore, the relative simplicity of Eiffel makes it easy to create tools that generate or otherwise interact with Eiffel systems. A number of such tools are available.

System Builds

Relying on programmers to create dependency analysis information in the form of makefiles is a needless waste of time and a constant source of problems. Also, requiring the maintenance of separate class interface files, including complex include file invocation order, is another time waster that is unnecessary with Eiffel. Management of the build process in Eiffel is accomplished via a system configuration file commonly referred to as Ace file.

Figure 16.5 shows the Ace file that corresponds with the code example shown in Figures 16.1–16.4. As you can see, the Ace file designates

- the executable to create—in this case "simp_list_test"
- the first object to create and first routine to call—test_simp_list and 'make'
- the clusters in which to search for relevant Eiffel source code.

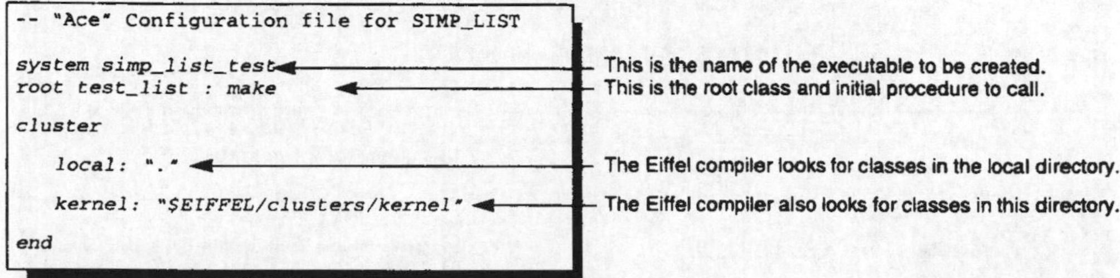

-- "Ace" Configuration file for SIMP_LIST
system simp_list_test
root test_list : make
cluster
local: "."
kernel: "$EIFFEL/clusters/kernel"
end
This is the name of the executable to be created.
This is the root class and initial procedure to call.
The Eiffel compiler looks for classes in the local directory.
The Eiffel compiler also looks for classes in this directory.

FIGURE 16.5 Example configuration (Ace) file.

There are many other things that can be handled in Ace files. These include the designation of assertion levels, compiler options, linker options, options to choose or set up the garbage collector, options to link in .o files or system libraries, call outs to make or shell scripts at strategic points during the build cycle, commands to create or utilize precompiled libraries, and more. Most of these options take effect for an entire system. Some can be designated for individual clusters or even for individual classes. The Ace file is a sophisticated control mechanism that serves as the front line for automating the creation of Eiffel systems.

Eiffel Compilers

Eiffel compilers check for correct type usage, without imposing a confusing or limited type system on the developer. As mentioned previously, subclassing and subtyping are one and the same in Eiffel, making the use of inheritance easy and powerful. Eiffel chooses to support the more natural covariant type system that allows classes to be easily specialized in concert. This gives the programmer a powerful mechanism for subclassing while the compiler does all the work of ensuring type safety.

Eiffel compilers take on most of the burden of performing system-wide and local optimizations. For example, practically all Eiffel features are "virtual" as far as the Eiffel programmer is concerned. The Eiffel compiler locates all opportunities for safely eliminating dynamic binding or for in-lining code. The compiler can also eliminate dead code that is never used in the application. This work permits the programmer to focus on choosing proper and efficient data structures and designing effective patterns for object interactions instead of worrying about trade-offs concerning reusability versus performance.

Looking at Figure 16.6 we can see that a typical Eiffel compiler provides a host of options covering error handling, creation of static or shared libraries, number of processes to utilize, memory utilization, as well as types of warning messages and other feedback. The library options are particularly useful. Eiffel compilers allow the creation of precompiled libraries. These can be used to cut the link time and the dependency analysis time for stable Eiffel clusters. This is important for the effective support of scalable systems.

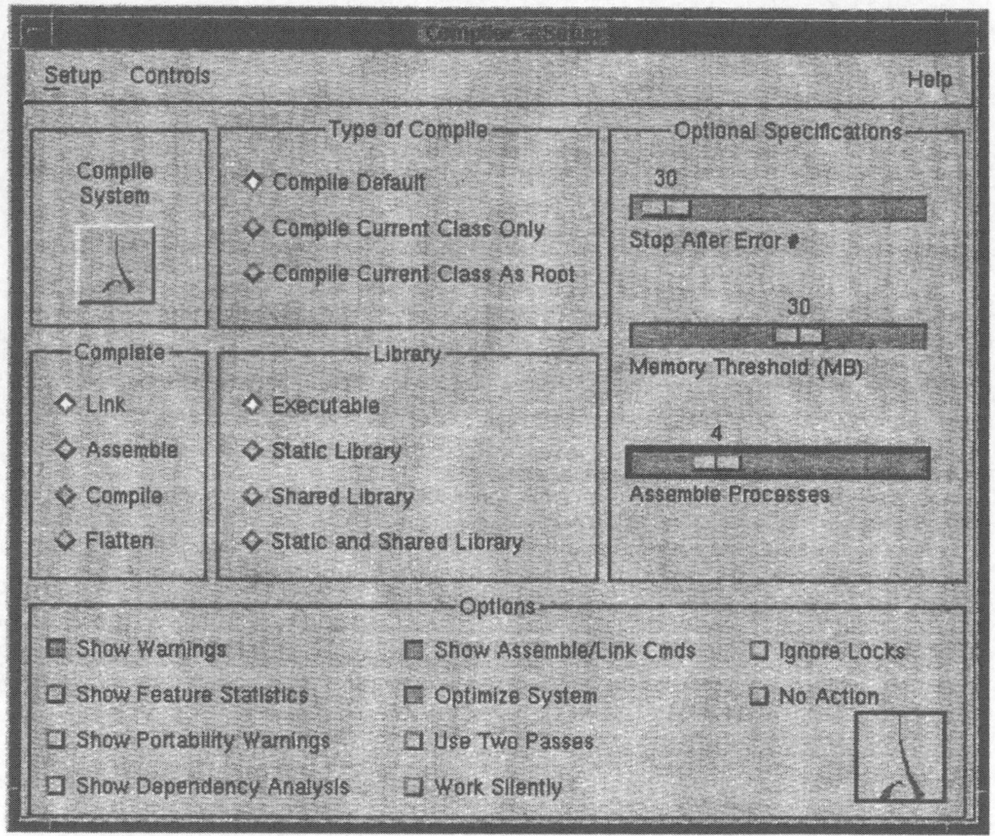

FIGURE 16.6 Control panel for an Eiffel compiler.

Other advanced compilation capabilities that are available with various Eiffel compilers today include mixed interpretation/compilation systems, compiler/database integration for aiding browsing and incrementality, and support for cross-compilation. Additional capabilities are expected over time.

Short Tools

As was noted in the discussion of Figure 16.2, Eiffel systems generally provide tools for extracting the interface of classes. These "short" tools are used to create and present up-to-the-minute information on classes that are under development. For more stable Eiffel classes and clusters, these same tools can be used to extract and generate class documentation in a publishable format. For example, Figure 16.7 is a page from a TowerEiffel library reference manual that was generated directly from program source code.

Figure 16.8 shows the controls for Tower's class abstractor, which is typical "short" tool. Notice the support for different output formats and the level of detail included in the output.

Browsers

Large systems require tools to aid understanding of the overall system. A typical tool for this is shown in Figure 16.9. This tool allows clusters and classes to be found and examined quickly. For ease of updating, browsers are typically integrated with a programmer's editor of choice.

Other browsing options that are usually available include tools for seeing the ancestors and/or descendents of a given class, for revealing the client/supplier relationships between classes and for finding the usages of parameterized classes. Some tools can display the results and interact with the Eiffel programmer using a graphical display. Some tools not only allow browsing, but also support analysis and design activities and thus qualify as full-fledged OO CASE tools.

Debuggers

While it is possible to accomplish a great deal using Eiffel without recourse to a debugger, no software professional can afford to be without such a tool. Eiffel debuggers vary in their ability to provide the full range of features that most programmers are accustomed to such as single stepping and watch points, so check this out when you consider which system to use. See Figure 16.10 for a look at a Motif-based debugger in action.

Other Tools

Other tools that are available for Eiffel include GUI building tools, database schema generators for marrying Eiffel systems with OO databases or RDBMS, proxy generation tools for hooking up Eiffel objects with other languages, Eiffel-oriented CASE tools for handling design and analysis issues, and Eiffel interfaces to simulation systems. There are also plenty of Eiffel libraries for GUI, database, lexing, parsing, data structures, simulation, graphics, expert systems, networking, mathematics, and more.

16.8 PROJECT MANAGEMENT FOR EIFFEL

Academic Experiments with Eiffel Engineering

There have been enough projects done with Eiffel now to report some interesting, findings. First, Eiffel is not magic. It does not erase the myraid problems associated with software development. Nor does Eiffel ensure that code written is high quality, object oriented, readable, or reusable. Eiffel is a tool that needs to be properly mastered before high quality results will ensue.

In an experiment in advanced project development and management with Eiffel at a major University, one professor had the interesting experience of creating a large-scale application in a 2-year time frame with a group of seven students who had very

Support BC_PAIR

24.1 BC_PAIR

NAME

class *BC_PAIR [Key_T, Value_T]*

DESCRIPTION

BC_PAIR is a key/value pair.

ANCESTORS

BC_PAIR [Key_T, Value_T]

FEATURE SUMMARY

```
change_key (new_key : Key_T)
change_value (new_value : Value_T)
copy (a : like Current)
is_equal (a : like Current) : BOOLEAN
key : Key_T
make (new_key : Key_T; new_value : Value_T)
out : STRING
value : Value_T
```

CREATORS

```
make
```

FEATURES *{ANY}*

```
key : Key_T

value : Value_T

make (new_key : Key_T; new_value : Value_T)
                -- Create (or reinitialize) a new pair.
        require
                new_key_not_void : new_key /= void;

change_key (new_key : Key_T)
                -- Change the key.
        require
                new_key_not_void : new_key /= void;

change_value (new_value : Value_T)
                -- Change the value

out : STRING
                -- A string which represents the pair

is_equal (a : like Current) : BOOLEAN
        require
                other_not_void : a /= void;

copy (a : like Current)
        require
                other_not_void : a /= void;
                conformance : a.conforms_to (Current);
        ensure
                result_is_equal : is_equal (a);
```

CLASS INVARIANT

```
key_not_void : key /= void;
```

TowerEiffel **Booch Components**

FIGURE 16.7 An automatically generated page from an Eiffel manual.

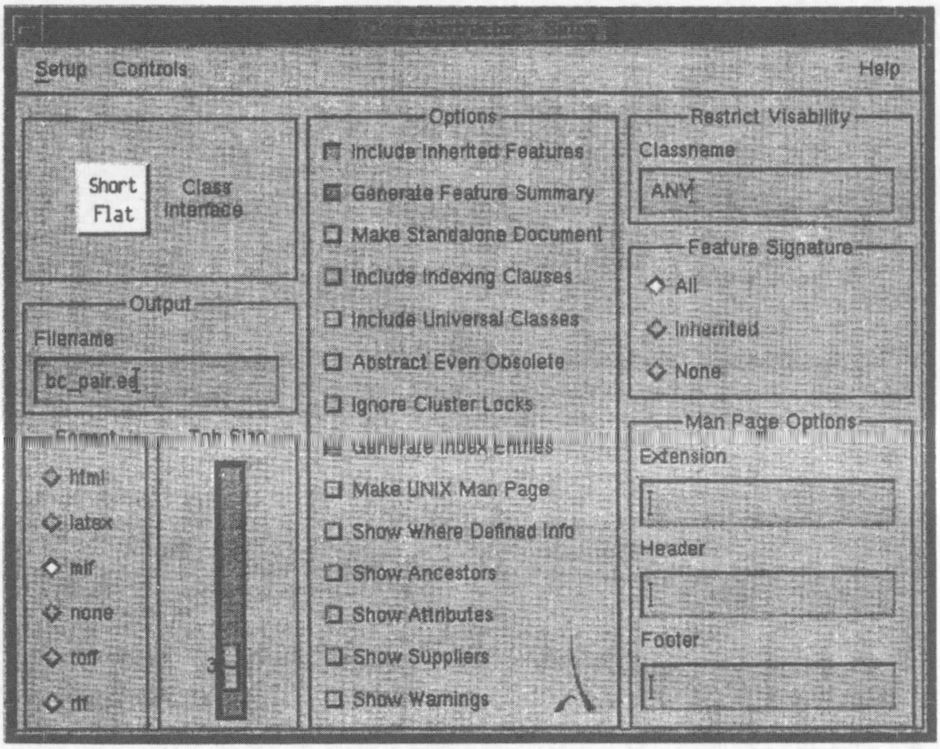

FIGURE 16.8 Control panel for an Eiffel class abstractor.

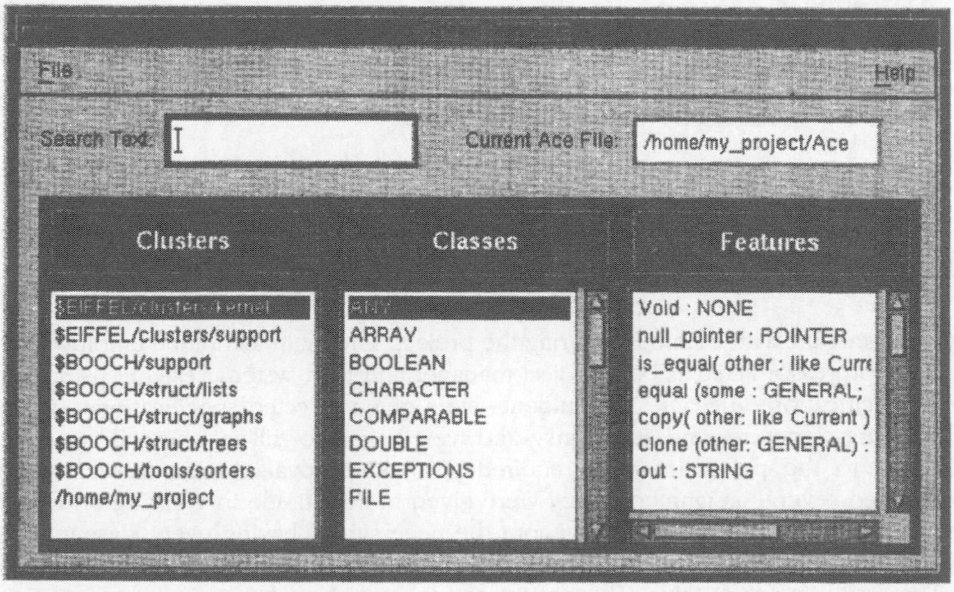

FIGURE 16.9 An Eiffel class browser.

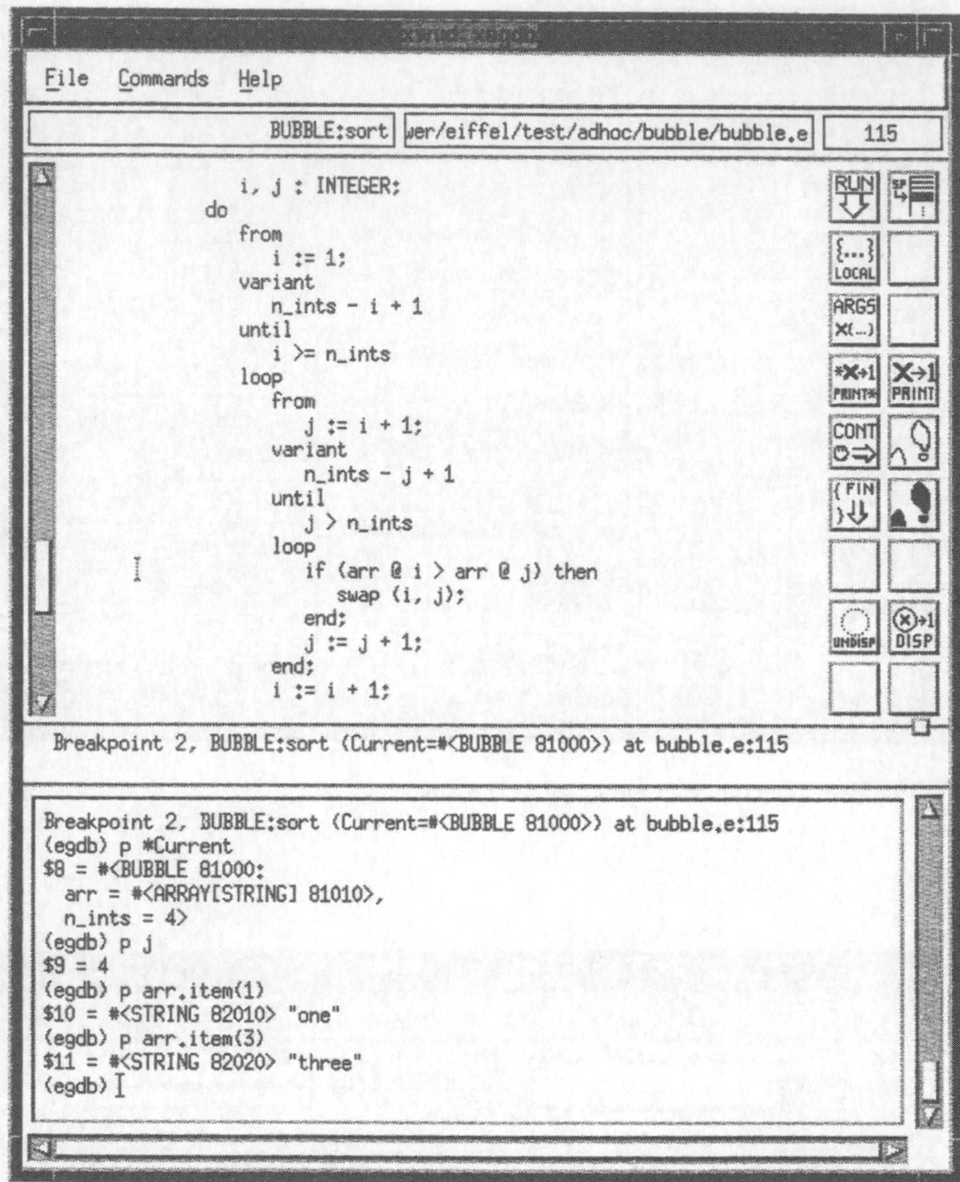

File Commands Help
BUBBLE:sort
ver/eiffel/test/adhoc/bubble/bubble.e
115
i, j : INTEGER;
do
from
i := 1;
variant
n_ints - i + 1
until
i >= n_ints
loop
from
j := i + 1;
variant
n_ints - j + 1
until
j > n_ints
loop
if (arr @ i > arr @ j) then
swap (i, j);
end;
j := j + 1;
end;
i := i + 1;
RUN
LOCAL
ARGS
PRINT
CONT
FIN
UNDISP
DISP
Breakpoint 2, BUBBLE:sort (Current=#<BUBBLE 81000>) at bubble.e:115
Breakpoint 2, BUBBLE:sort (Current=#<BUBBLE 81000>) at bubble.e:115
(egdb) p *Current
$8 = #<BUBBLE 81000:
arr = #<ARRAY[STRING] 81010>,
n_ints = 4>
(egdb) p j
$9 = 4
(egdb) p arr.item(1)
$10 = #<STRING 82010> "one"
(egdb) p arr.item(3)
$11 = #<STRING 82020> "three"
(egdb)

FIGURE 16.10 An Eiffel source level debugger.

little interaction with each other during the project. The main technique employed was that the professor acted as both project manager and lead system designer using Eiffel as a specification language. The students were given selected specifications—including feature signatures and assertions—and were asked to "fill in" the code to match. Changes to the public interface required special approval. As the better students completed several assignments, they were given more latitude in working on extending selected areas of the system without the necessity of having every class interface change specifically approved.

The results were positive. The project was a success and the resulting application worked quite reliably—which is a rarity for a student-developed large-scale project.

Furthermore, the project was stable throughout its development cycle. The team members remained coordinated despite the lack of communication between team members.

This experiment was a graphic example of the most important lesson for Eiffel project management. This is to differentiate the programming team not only by convenient demarcations in the design of the overall system, but also by skill level. Choose the most skillful designer(s) to oversee the entire design until the point in time where other proven Eiffelists are available to oversee selected areas of the project. Managing the project in a way that maximizes the impact of the top design staff is a key factor in successful Eiffel projects.

The flip side of the same coin is that the junior members on the project are always handed well-defined chunks of work that they could easily perform. This allows them to get their "sea legs" with Eiffel while absorbing design concepts through repeated exposure. They also learned about the importance of the class interface. They find that when they got their code right it "just worked" because it had been designed in advance to be preintegrated into the overall system.

Commercial Success with Eiffel Engineering

The most notable commercial success with Eiffel thus far was achieved by a LAN equipment manufacturer. Here a brand new class of hardware device was launched whose major value was predicated on complex software created and delivered using Eiffel. The project manager readily admits that while the team experienced many false starts and blind alleys in getting used to Eiffel, the ability to use Eiffel for concentrating on quality was the key to success.

The success was predicated by the creation of a solid design for the overall system. The interfaces between the subsystems were especially important as was the technology for distributing the Eiffel objects as required between the subsystems. With this design work done, selected subsystems were concentrated on while the rest were "stubbed." The result was a minimal system that actually worked in very little time.

The minimal system exhibited enough robustness for the team to target an early release of a low functionality system. This early release was critical for gaining market interest and for learning more from early customers about the true requirements that the system would need to fulfill. Due to the excellent quality of the releases, the enhancement requests got the majority of the attention and the team did not get bogged down in handling bug reports.

The main point is that Eiffel can be used to reduce the "time to quality" and thus allow early releases that are robust. Eiffel's support for reuse and extension then becomes the crucial mechanism for meeting new requirements as they arrive. While redress occurs, there is no need to completely "start over" every few releases.

The Forefront of Eiffel Engineering

The current forefront of Eiffel project management is with an application that is a huge and ambitious next generation options trading system with risk analysis and a slew of front office and back office capabilities. Eventually it will grow to serve as the complete software backbone for an investment bank. From checking accounts to hedge accounts to simulation to cash flow, all will be united with a single tightly integrated system. For this to work, the designers must deliver a design that can grow and adapt for many years.

The project manager choose several highly experienced Eiffel experts to serve as the anchor for his team. Several additional experienced programmers are also on board as well as a person entirely devoted to making sure all documentation is clean and that the system is always up to date. The rest of the team is made up of domain experts—actual traders and banking experts who are not only defining the system requirements but are doing a significant portion of the coding. Eiffel's clean syntax and understandability are key for this plan to work.

So far the project is on track having amassed a tremendous amount of technology in a relatively short period of time. With over 2000 classes and 200,000 lines of code, it is easy to imagine that such a project might reach some scaling problems soon. To combat this, the project leaders must stay focused on reusability and maintainability. Careful guidelines for the design of clusters and module interfaces are important for keeping the system continually expandable.

A recent technology review of this project by Arthur Anderson came back with positive report. The system itself has been installed for preliminary use. It will be interesting to see what additional project management ideas for Eiffel software engineering emerge from this successful large-scale project.

Summary Thoughts about Eiffel

Eiffel's support for assertions, abstraction, and parameterization combines to allow the construction of robust, reusable software components. The powerful yet clear syntax coupled with the elucidation of assumptions via assertions makes it easy to share these components with confidence. The result is that the benefits of OO programming can be more fully realized.

Finally, Eiffel is fun. Many of the nasty and tedious problems that programmers face every day are handled automatically. Eiffel eliminates the possibility of entire classes of errors, including memory management errors, unintended side effects due to poor encapsulation, bad links from erroneous makefiles, and improper routine dispatching due to type errors. The freedom from trivial matters coupled with error prevention techniques reduces mundane work. This leaves more time to use the power of Eiffel to attack the challenging and interesting problems that face modern software engineers.

DEFINING TERMS

Ancestor: A class that has been inherited directly or indirectly.

Assertion: A run-time boolean check.

Attribute: A variable associated with a class.

Class: The description of an object written in Eiffel source code.

Client: An instance that has a reference to the current class.

Constrained Genericity: Restricting parameterization to a specified type in order to have access to the capabilities of the type.

Creation: The act of allocating and initializing a new instance.

Creation Procedure: A procedure defined in a class to initialize a new instance.

Deferred: Not effective.

Deferred Class: A class with one or more deferred routines.

Deferred Routine: A routine with a full signature, but no implementation.

Descendent: A class that inherits from this class either directly or indirectly.

Effective: A feature or class that is fully defined.

Export: Make a feature visible to instances of specified classes or their descendents.

Feature: An attribute or routine.

Feature clause: A group of features that share the same export status.

Function: A routine that returns a value.

Garbage Collection: Automated memory management whereby instances that no longer have references stemming from a root instance are freed and their memory released back into the system at large.

Generic Class: A class that is parameterized by one or more other classes.

Genericity: The ability to make a generic class that is parameterized by other classes.

Inheritance: Designating a class to which the current class conforms. This process also obtains the features of the inherited class for the current class.

Instance: A created object. The run-time personification of a class.

Invariant: An assertion that is checked upon entry and exit from every exported routine.

Multiple Inheritance: The ability to inherit from more than one class at a time.

Parameterization: Another word for genericity.

Polymorphism: The ability of related classes to define different implementations for the same feature.

Postcondition: An assertion that is attached to a routine and is invoked after the code body of the routine has been completed, but before the routine returns control to the callee.

Precondition: An assertion that is attached to a routine and is invoked before the code body of the routine is run.

Procedure: A routine that does not return a value.

Reference: A handle for an instance. Eiffel's form of safe pointer.

Repeated Inheritance: The ability of the compiler to deduce the proper dispatching for routines even in the face of multiply inherited classes that share some common ancestor.

Signature: The definition of a routine including the feature name and, if present, the name and types of arguments, the type of the return value, the feature comment, preconditions and postconditions.

Supplier: The class type of a reference to an instance attribute within a class.

Void: The value of an uninitialized reference.

REFERENCES

Journals

The Eiffel Outlook Journal, Tower Technology Corporation (bimonthly journal).

The Journal of Object Oriented Programming, SIGS Publication (monthly journal).

Books

Brian Henderson-Sellers and Julian Edwards, *BOOKTWO, The Working Object*. Prentice Hall, Englewood Cliffs, NJ, 1994.

Bertrand Meyer, *Eiffel: The Language*. Prentice Hall, Englewood Cliffs, NJ, 1992.

Bertrand Meyer, *Object Oriented Software Construction*. Prentice Hall, Englewood Cliffs, NJ, 1988.

Bertrand Meyer, *Object Success*. Prentice Hall, Englewood Cliffs, NJ, 1995.

Bertrand Meyer and Jean-Marc Nerson, *Object Oriented Applications*. Prentice Hall, Englewood Cliffs, NJ, 1993.

Robert Rist and Robert Terwilliger, *Object Oriented Programming in Eiffel*. Prentice Hall, Englewood Cliffs, NJ, 1995.

Robert Switzer, *Eiffel: An Introduction*. Prentice Hall, Englewood Cliffs, NJ, 1993.

Kim Walden and Jean-Marc Nerson, *Seamless Object-Oriented Architecture*. Prentice Hall, Englewood Cliffs, NJ, 1995.

Richard Wiener, *Software Development Using Eiffel*. Prentice Hall, Englewood Cliffs, NJ, 1995.

Rebecca Wirfs-Brock, *Designing Object Oriented Software*. Prentice Hall, Englewood Cliffs, NJ, 1990.

17

Small but Mighty: Smalltalk

Thomas Murphy
Digitalk, Inc

17.1 INTRODUCTION

Few teams have so widely influenced the world of computing as the XEROX Palo Alto Research team led by Alan Kay. They gave us the workstation with a graphical interface, ethernet cards, the mouse, "portable" computers, and Smalltalk. Smalltalk is a pure object-oriented language that truly began a revolution in the way we think about software development. But Smalltalk is much more than a language.

17.2 HISTORY OF SMALLTALK

Beginnings at Xerox

Alan Kay had a vision in the early 1970s of a portable hand-held device he termed the Dynabook. This computer would interact with touch and contain the data for 90% of our needs and has an interface that was simple to use. While working at Stanford's Artificial Intelligence research group, a window of opportunity opened with the foundation of the XEROX Palo Alto Research Center. Gathering a small group of creative minds, Smalltalk was envisioned to make programming simple—something a child could understand. The beginnings were humble, a 1000-line BASIC program that in 1972 evaluated 3 + 4. From this humble beginning a humble language was born. Kay has stated: "I called the language Smalltalk so that if anything interesting was accom-

0-8493-2530-7/96/$0.00+$.50

plished with it the developer would be pleasantly surprised." Building on the ideas of Simula, Smalltalk is the first fully object-oriented programming language. With much hard work, the first fully operational Smalltalk was finished and named Smalltalk-72.

Smalltalk evolved in a process of design, implement, test, evaluate, and repeat again. Indeed an important feature of Smalltalk is its dynamic nature that encourages exploration and rewards with instant feedback. This first Smalltalk system had three main problems: performance of token lookup, a complicated and inefficient message stream, and classes were not first-class objects with the now common inheritance. In 1974 a new version appeared and fixed these difficulties. It also included the now common *BitBlt* as its major bitmap graphics operation. This release also contained the first implementation of virtual memory.

In 1976 the next major release of Smalltalk came to life. This version made classes real objects, contained a class hierarchy, featured a much streamlined messaging system, and performed 4 to 100 times faster than the previous model. This version was more widely circulated within PARC but is was still a somewhat of a secret from the world at large.

With Smalltalk maturing Xerox PARC began to work at delivering a Smalltalk system in a "portable" computer. The original concept called for a hand-held device known as the Note Taker. This system was a great change from the bitmapped ALTOS workstation that had been the location of previous Smalltalk systems. Much of this work led to the current implementations of Smalltalk and prepared the way for Smalltalk-80 (Krasner, 1984).

In 1986 XEROX spawned ParcPlace® Systems to make a commercial Smalltalk. ParcPlace continues to develop Smalltalk for a wide variety of platforms. In 1994, ParcPlace became a publicly traded company.

Formation of Digitalk

Inspired by an issue of *Byte* magazine, Jim Anderson, George Bosworth, and Barbara Noparstak formed Digitalk in 1985 to bring Smalltalk to the PC. All prior Smalltalk systems developed at PARC were realized on specialized workstations placing the power of Smalltalk beyond the reach of common developers. The first Digitalk release, Methods, ran on Intel 8086 based MSDOS machines with 512K of RAM.

In 1986, Digitalk released the first commercial Smalltalk for personal computers—Smalltalk/V. This system is a full implementation of Smalltalk for DOS based machine that works in 512K of RAM and allows the use of virtual memory. It features the standard Smalltalk environment with code browsers, overlapping windows, multitasking, and incremental compilation. As the PC industry pushed forward, Digitalk moved its system to the now predominant GUI platforms. Digitalk merged with ParcPlace to form ParcPlace-Digitalk in August of 1995.

17.3 SMALLTALK IMPLEMENTATIONS

Today there are several Smalltalk implementations available. Each offers many similarities, but each has unique capabilities.

ParcPlace-Digitalk

With the merger, ParcPlace and Digitalk will bring their product lines together to enable high-performance client/server applications. ParcPlace's tools have great strength in portability and support for a wide variety of platforms. Digitalk's tools offer tight host integration with the Windows and OS/2 interfaces and by combining, customer will be able to build distributed applications.

IBM

IBM has been a large user of Smalltalk for several years. It is IBM who created the proposed ANSI draft standard for Smalltalk and thereby catalyzed the standardization effort. 1994 marked IBM's entry into the Smalltalk tool market. IBM's product offering, VisualAge, is similar to the Digitalk PARTS Workbench product. It allows developers to create components in Smalltalk and then assemble instances of these in a visual workbench. IBM also provides a standalone Smalltalk system. The underlying Smalltalk engine for IBM is provided by OTI. Along with their core Smalltalk and VisualAge products, IBM also sells several "wrappers" for use with VisualAge that allow easy use of communication protocols, legacy COBOL, and Multimedia.

Easel

Easel entered the Smalltalk market through the purchase of Enfin. This was first produced as a Smalltalk like system and was marketed more like a 4GL than a Smalltalk language product. As time went on, Easel (who bought Enfin) realized that having a standard language was valuable to customers. The Easel *ObjectStudio* is a complete line of Smalltalk-based development tools including CASE tools, GUI designers, and language products. In 1995, Easel was purchased by VMark Systems.

QKS

Quasar Knowledge Systems manufactures a modified Smalltalk system called Smalltalk Agents. It features extensions to the base language to support constructs such as CASE statements and the ability (using the External Code Linking Toolbox) to make inline calls to routines written in C, Pascal, FORTRAN, and assembly. Smalltalk Agents comes with a built-in GUI builder and is designed to be fully thread safe. Support is also provided for connectivity to a variety of SQL databases and all Macintosh Toolbox functionality. Currently Smalltalk Agents is available on the Macintosh with Windows due in late 1995.

OTI

OTI entered the world with ENVY Developer—a set of development tools similar to Digitalk's Team/V in theory but much different in execution. Envy uses a simple object database to make the image changes persistent across multiple client connections. Envy also includes advanced code analysis tools on some supported platforms. In addition OTI has produced several Smalltalk systems primarily for real-time applications. Building upon their success and experience in Smalltalk, OTI has recently entered the Smalltalk language market and partners together with IBM on the VisualAge and IBM Smalltalk products.

GNU

GNU Smalltalk is a public domain implementation of Smalltalk that is part of the GNU Software Foundations suite of tools. It is unlike other implementations in that it has no development tools except for an interface to the GNU Emacs editor. Versions in beta at this time add extensions for full support of the Motif library and traditional development tools.

Hewlett Packard

Building on top of ParcPlace's VisualWorks, Hewlett Packard produced HP Distributed Smalltalk. This product is a CORBA compliant and allows various objects distributed over a system to send message to each other. The system can also make use of HP's request broker to communicate with non-Smalltalk objects in the system.

17.4 KEY TERMS AND CONCEPTS

Smalltalk is a simple language with only five real concepts to learn: class, instance, message, object, and method. These concepts are all interrelated and drive the entire system. As will be seen later, Smalltalk has a very simple syntax.

The first grouping of terms is class, instance, and object. All three are about the main concept in object-oriented programming—objects. They just describe different ways of looking at things in the world or actually different states. Classes are abstract definitions; they are the things you are modeling and are the textual entities that the system works with to define systems. Object has two meanings. One is to represent what it is you model. Systems such as businesses are composed of various entities or objects. In the Smalltalk environment there are many objects such as browsers, windows, and text streams. Instances are almost the same thing. They are objects that have been realized from your classes. You may have several instances of classes. There will be many instances of the Character class that represent character objects in any text window.

The second group is method and message. A method represents a ***behavior*** that an object exhibits. These are the things that objects can tell you about or do. A bank account can tell you its balance or it can carry out a withdrawal transaction. Messages are the way that methods are invoked. Objects send messages to other objects to request information or action. Traffic lights send the messages, go, stop and prepare to stop to drivers. In Smalltalk, all messages are two way. When a message is sent, a value is always returned. If a specific value is not specified the object itself is returned. This allows the sender to know that the message was received—something that would be useful in traffic control.

There are three types of messages: unary, binary, and keyword. Unary messages are messages with no arguments sent to an argument. They are often used as accessor methods to select and return the value of an instance variable. Examples are

```
circle circumference.
anAngle sin.
```

Binary messages are normally used for numeric computations and for comparison operations. They consist of two objects and one operation message. Examples are

```
3 * 2.
balance > minimunBalance.
```

Keyword selector messages allow for the sending of messages with arguments. Message selectors with an ending colon are messages with arguments. The simplest use of keyword messages is to set the value of instance variable such as

```
circle circumference: 9.3.
```

Selectors for setting and accessing can have the same names and only differ by the use of the colon. They also typically have the same name as the instance variable they are acting upon. More complex keyword selectors allow you to pass more arguments to an expression. Arguments are not mutated in this process. An example is

(person accounts) withdraw: 50.00 from: 'checking'.

In this sequence the accounts for a person are retrieved and then this list is sent the message to withdraw money from the checking account.

17.5 SMALLTALK SYNTAX

Smalltalk therefor has a simple syntax. One could look upon the name Smalltalk as readily defining a small, simple language based on the concept of objects "talking" to each other. Smalltalk was designed to be simple so that developers could spend time thinking about design, not developing verbose texts but crafting literate solutions.

EBNF

```
1. method = messagePattern [temporaries] [primitiveNumber]
   expressionSeries ["."].
2. messagePattern = unarySelector | binarySelector
   variableName | keyword vaiableName {keyword
   variableName}.
3. primitiveNumber = "<" "primitive:" number ">".
4. temporaries = "|" {variableName} "|".
5. expressionSeries = expression {"." [expression]} ["."
   "^" expression]] | ["^" expression].
6. expression = {variablename ":="} messageExpression {";"
   cascadeMessage].
7. messageExpression = unaryExpression} binaryExpression}
   keywordExpression.
8. unaryExpression = primary {unarySelector}.
9. primary = variableName | literal | block | "("expression")".
10. block = "[" ′ {":" variableName} "|"] expressionSeries "]".
11. bianryExpression = unaryExpression binaryMessage
    {bianryMessage}.
12. binaryMessage = binarySlector unaryExpression.
13. keywordExpression = (unaryExpression |
    binar4yExpression) keywordMessage {keywordMessage}.
14. keywordMessage = keyword (unaryExpression |
    binaryExpression).
15. cascadeMessage = unarySelector | binaryMessage |
    keywordMessage
16. literal = number | string | characterConstnat | "#" (sym-
    bol | array_.
```

```
17. keyword = identifier ":".
18. variableName = identifier.
19. unarySelector = identifier.
20. binarySelector = "-" | selectorCharacter
    [selectorCharacter].
21. symbol = identifier | binarySelector + keyword {keyword}.
22. array = "(" {number | string | symbol | array |
    characterConstant} ")".
23. number = (digits "r" ["-"] bigDigits ["." bigDigits] |
    ["-"] digits ["." digits) ["e" ["-"] digits].
24. digits = digit {digit}
25. bigDigits = bigDigit {bigDigit}.
26. bigDigit = digit | captialLetter.
27. string = " ′ " {character | " ' ' " | " " "}" ′ ".
28. characterConstant = "$" (character | " ′ " | " " ").
29. comment = " " " {character | " ′ " | " "" "} " " ".
30. identifier = letter{letter | digit}.
31. character = selectorCharacter | letter | digit | "[" | "
    "{" | "}" | "(" | ")" | "^" | ";" | "$" | "#" | ":" | "." |
    "_" | "'".
32. selectorCharacter = "," | "+" | "/" | "\" | "*" | "~" |
    "-" | "<" | ">" | "=" | "@" | "%" | "|" | "&" | "?" | "!".
33. letter = capitalLetter | "a" | "b" | "c" | "d" | "e" | "f"
    | "g" | "h" | "i" | "j" | "k" | "l" | "m" | "n" | "o" | "p" |
    "q" | "r" | "s" | "t" | "u" | "v" | "w" | "x" | "y" | "z".
34. capitalLetter = "A" | "B" | "C" | "D" | "E" | "F" | "G" |
    "H" | "I" | "J" | "K" | "L" | "M" | "N" | "O" | "P" | "Q" |
    "R" | "S" | "T" | "U" | "V" | "W" | "X" | "Y" | "Z".
35. digit = "0" | "1" | "2" | "3" | "4" | "5" | "6" | "7" | "8" | "9".
```

A Simple Example

Figure 17.1 shows the beginning of a set of classes to describe bank accounts. The listing is broken up into sections. The first section defines the classes. A class definition contains information about what the super or parent class is, instance variables and class variables, and usage of pool dictionaries. Thus the first class defined is *Account*, which a subclass of *Object*. Instance variables are values that are unique to each instance of this class. A key object concept is that data and function are encapsulated together. A class can make use of instance variables that belong both to the class itself and its super classes. Thus class *CheckingAccount* has an instance variable *clearedChecks* and can also make use of its parent's instance variables: *accountNumber* and *balance*. All classes must have leading capitals. Methods normally have lower-case leading characters.

Class variables are values that are shared across all instances of a class. In this case, none is required. Pool dictionaries are groups of constants stored in key value pairs. These can contain things such as system character constants and color maps.

Just as variables are grouped into either class or instance variables, methods are also grouped. Class methods allow for the creation of new instances of a class. Instance variables allow operations to be performed on a specific instance of a class. Class methods may also be used in setting up linkage between events and various components.

Different chunks of code—class definitions and instance and class method definitions—are separated by apostrophes. Sections are introduced by headings such as *!Account methods!*. In addition each definition ends with an apostrophe. In normal Smalltalk development, code is never seen in this format. Instead browsers are employed. These allow you to focus your thoughts on the immediate task at hand.

```
Object
      subclass: #Account
      instanceVariableNames: 'accountNumber balance'
      classVariableNames: ''
      poolDictionaries: ''!

Account
      subclass: #CheckingAccount
      instanceVariableNames: 'clearedChecks'
      classVariableNames: ''
      poolDictionaries: ''!

!Account methods !

accountNumber
      ^accountNumber!

accountNumber: anID
      accountNumber := anID!

balance
      ^balance!

balance: anAmount
      balance := anAmount!

deposit: anAmount
      self balance: self balance + anAmount!

withdraw: anAmount
      self balance: self balance - anAmount! !

!CheckingAccount methods !

canCover: aCheck
      ^aCheck <= self balance!
```

```
stopPayment: aCheck
     "add to cleared list, don't debit"
     self clearedChecks add: aCheck.!
clearCheck: aCheck
     "if check hasn't been cleared previously and is less
     than the available balance then clear it."
     clearedChecks cleared: aCheck
     ifTrue: [self canCover: aCheck
          ifTrue: [self clearedChecks add: aCheck.
                    self withdraw: (aCheck amount)]]! !

!Account class methods !

COMMENT
"Abstract Class to describe basic account attributes, this
class will have children for various types of accounts such
as savings, types of checking, etc." ! !

!CheckingAccount class methods !

COMMENT
"Basic checking account—has children like interestBearing to
modify the basic behavior." ! !
```

FIGURE 17.1

The constant goal of Smalltalk is to reduce complexity. This is accomplished with the simple syntax, iterative development, simple messaging system, and inheritance and encapsulation.

All Smalltalk interaction is accomplished by sending message to objects. A message requests a service from an object. If the object does not have the service itself, the parent classes will be checked back to the root class *Object*. If a message is not handled, an error will occur. In the *CheckingAccount* class the simple unary message *balance* can be sent to instances of the class. In the following code fragment a new instance of the class is defined and the *balance* message is sent to it:

```
MyAccount := CheckingAccount new.
MyAccount balance
```

This also illustrates line separations. Each line in Smalltalk ends with a period. A semicolon can end a line in which you want to send successive messages to the same object. In Figure 17.2 a portion of a window is defined. Various attributes of the window are set by sending messages to it.

In Smalltalk, all code is strictly evaluated from left to right. Evaluation order can be forced by the use of parenthesis. Thus, if you evaluate 4 + 3 * 2 the answer will be 14. This is because the object 4 will be sent the message + with the argument 3; then the resulting object 7 will be sent the message * 2 resulting in the object 14. In addition unary messages (messages without arguments) will be evaluated before binary messages. The complete order of evaluation for messages is unary messages, binary messages, and then keyword messages.

```
addSubpane: (
                    TextPane new
                        owner: self;
                        framingBlock: (FramingParameters new
                        iDUE: 1143
    @ 280; lDU: 0 r: #left; rDU: 0 r: #right; tP: 52/133; bDU:
    44 r: #bottom);
                        paneName: 'codeView';
                        startGroup;
                        yourself
                );
```

FIGURE 17.2

Code can also be grouped together into blocks. Blocks are generally used in control structures such as loops. They represent a set of deferred actions such as: **`[MyAccount clearCheck. MyAccount updateBalance.]`**. Every time this block is executed, each of the messages will be sent in order. For example, say you have a block that is going to iterate through all elements of some collection of objects. In this case you could do the following:

```
#('te' 're' 'do' 'me' 'so') do:[:each leach become: (String
new)]
```

This example create a new array of strings and then iterates through each one and replaces it with a new blank string. The :each is the local block variable.

Like any language, Smalltalk has several structures for looping. Fixed repetition can be determined with the following: ***Number*** **`timesRepeat: [ block statements]`**. Conditional loops allow a test to be performed to control the execution of a block. Often you need local variables inside a block. These are introduced to a block after the opening bracket.

Smalltalk also allows users to make use of true/false condition testing. This is illustrated in the **`clearCheck: aCheck`** method. Smalltalk does not make use of case statements. Large sets of conditional logic mean that object-oriented principles are not being properly used. Rather than trying to maintain a non-object-oriented case statement structure of conditional checking you should create state variables and methods.

As illustrated in listing one, Smalltalk allows you to refer to the instance of a class by the key word **self**. In addition you can refer to the parent class with the term **super**. By referring to **super** the system will make use of a parent's definition of a method even if you have a version defined for your class.

17.6 THE SMALLTALK SYSTEM

There are three essential pieces in a Smalltalk development system: the virtual machine, the image, and the class hierarchy. The virtual machine is the engine that makes it all go. If you spend much time in a Smalltalk system you will find that sending a message will result in a whole chain of messages being sent, and finally you will get to a point where the virtual machine takes over and tells the system a set of instructions to execute. Smalltalk execution is traced in the Smalltalk stack that allows you to trace execution. Figures 17.3–17.5 show the result of beginning the execution of 3 factorial. The factorial is computed recursively (see Figure 17.3) and the real work is when the system does the subtraction that is accomplished by negating the argument and adding as shown in Figure 17.4. At this point a primitive is performed and "real" computing is done. Of course much real computing is done in Smalltalk to get to this point.

```
factorial
    "Answer the factorial of the receiver."
  self> 1
    ifTrue: [^(self-1) factorial *self].
  self< 0
    ifTrue: [^(self error: 'negative factorial'].
  ^1
```

FIGURE 17.3 Factorial method.

```
+ aNumber
    "Answer the sum of the receiver and aNumber."
  <primitive: 21>
  ^aNumber + self
```

FIGURE 17.4 SmallInteger >>–.

When you compile a set of code in a Smalltalk system the following happens. First the system checks for syntax problems in you class definition or method. Next it checks to see if the class name (in the case of creating a new class) already exists. This is done by checking a system dictionary called the SymbolTable to see if it already contains the symbol in question. If the symbol does not exist, a new entry is made in the dictionary and the code is compiled to byte codes.

When you execute a set of code—run your application—these bytecodes are changed into machine code and executed. This allows systems to achieve a high level of portability. Byte codes can either be interpreted or compiled. Compiled systems compile the byte codes and store them in a memory space called the code cache. This allows the code to be executed much faster the second time around.

The Smalltalk image is the collection of all currently defined classes and globals in your environment. This includes representatives of all the existing classes. Because the system knows about every defined class and method, Smalltalk systems can easily be queried to return information. For instance, to determine the number of defined symbols in the system you can simply execute the a piece of code like **`SymbolTable`**

```
SmallInteger(Integer)>>factorial
UndefinedObject>>Doit
[] in MethodExecutor>>evaluate
ProtectedFrameMarker(Context)>>setUnwind:
ZeroArgumentBlock(Context)>>ensure:
MethodExecutor>>evaluate
MethodExecutor class>>execute:for:
CompilerInterface>>evaluate:withReceiver:
CompilerInterface class>>evaluate:in:to:notifying:ifFail:
TextPane>>evaluate:ifError:
TextPane>>doIt:
[] in TextPane>>doIt
ProtectedFrameMarker(Context)>>setUnwind:
ZeroArgumentBlock(Context)>>ensure:
CursorManager>>changeFor:
TextPane>>doIt
TranscriptWindow(ViewManager)>>doIt
TranscriptWindow(Object)>>perform:
```

FIGURE 17.5 Example of stack during execution of 3 factorial.

size. This provides a way in which you can perform static metrics on your code base by counting things like average lines of code in all methods.

Classes are arranged in a hierarchical relationship. This helps group functionality and aides in the documentation and understanding of a system. Basic Smalltalk systems come with more than 1000 predefined classes and thousands of methods for use with these classes. Classes also provide for code reuse by allowing you to specify general behavior and then add special behaviors as your move deeper into the hierarchy.

Memory Management

Smalltalk systems automatically manage memory. This system is know as garbage collection and provides system level management of the programs heap allocations. The Smalltalk memory space is actually two spaces, the *new* space and the *old* space. When a new object is created the system places it into the new space.

The first part of garbage collection is flipping. The flipper runs through new space and looks for objects that are no longer referenced and objects that have existed for a period of time. Objects without reference can be thrown away. Objects that have been around for a long enough period of time are flipped or tenured into the old space.

When old space becomes full two things can happen, you can grow the old space (if you have room) or you can clear out old objects. Just as in new space the transitive closure is computed and objects without references are thrown out. After memory has been released, the system then compacts the remaining objects and returns memory to the system.

In this system you typically have a fairly small new space and a larger old space. Old space unlike new space is not statically sized though. Garbage collectors can be adjusted to change the tenuring policy, and the sizes or limits of the two memory spaces.

Many Smalltalk systems also make use of *finalization*. Finalization provides a system by which you can detect when objects have changed, the garbage collector has run, and by which objects can fade from existence. For instance, if you are connected to a database and delete a record, finalization will allow the database to be notified of your change so that the database can take care of any work it needs to do. This also allows both the database to release resources and Smalltalk to release the resources allocated for the database.

ANSI Effort

In 1993, IBM released an IBM Redbook entitled *Smalltalk Portability: A Common Base*, and launched an effort to form and ANSI standardization committee for the Smalltalk language. This committee has been formed, ANSI X3J20, and is filled with representatives of the major Smalltalk vendors. The IBM Redbook served as the foundation for the standards effort detailing the language syntax and a base set of common functionality. The committee hopes to have a draft standard submitted for review by 1996.

Since Smalltalk is much more than a syntactical description the ANSI standard will cover specified classes and behaviors. While covering basic behavior, the standard will not say how the class hierarchy should be shaped, thus a vendor could provide a class in a different location of the hierarchy than a competitor and, as long as the behavior of each met the specification, still be conformant.

The standard also will not cover all classes in a Smalltalk system. For instance, the classes used to specify user interface and database frameworks are not covered. Overall the standard will strive not to impact legacy code or constrain current or future implementations.

17.7 THE BASIC CLASS HIERARCHY

As Smalltalk developed and gained inheritance, classes were arranged in a hierarchy. This hierarchy groups classes logically into functional sets. A hierarchy allows a parent to define generic behaviors that are inherited by all subclasses or children. The children may then add their own new behaviors or modify inherited ones.

Smalltalk's rich hierarchy provides a rich foundation for assembling applications. Classes provide support for basic data structures, numerical manipulations, reading and writing data, and creating graphical interfaces.

Smalltalk is a singlely inherited language. This aids in understanding relationships and dependencies but also requires more discipline in design. Single inheritance also provides Smalltalk with efficiency and eliminates the problems associated with precedence of method implementation that arise when multiinherited parents both provide methods with the same name and protocol. Smalltalk's system also makes it easy to build powerful tools for organizing and browsing code.

Object

The top of the Smalltalk food chain is Object. This class defines common behaviors shared by all classes in the system. This includes the ability to create new objects, perform basic comparisons, and other basic behaviors. Object contains many testing methods for testing the "type" of an object.

Collections

Collection classes provide structures for dealing with ordered and unordered collections of data. Ordered collections such as arrays and dictionaries allow values to be stored and retrieved by a key or index value. Nonordered collections such as set and bags simply hold groups of data. Strings are also types of Collections.

All Collections have the ability to add, find, and remove elements as well as the ability to return information such as the size of the collection. Collections can also be asked to return new objects of their various children. Thus a Collection can be asked to return a new OrderedCollection. Most Collections are able to hold generic objects. An exception to this is the class ByteArray, whose instances are a fixed size indexable collection of bytes.

Smalltalk itself makes great use of Collections. When a new class is defined in the system it is placed in a Collection called Smalltalk. Smalltalk is an instance of the Dictionary class and contains a set of symbols. Each class then has Collections or arrays to hold methods, instance variables, and pools. Most of a class's attributes are also types of Collections.

This makes it easy to query the system about itself and develop metrics. For instance by evaluating the message **(Smalltalk at: #Collection) allSubclasses size** you can count the children of Collection. You can also look for things like the implementors and senders of various messages such as **(Smalltalk implementorsOf: #add:)**.

Streams

Streams control the flow of data to and from an application. This includes transactions such as read or writing data from a file, working with databases, or any other device providing a flow of information. The stream methods make it simple to deal with data in files. Methods are provided to assist with parsing streams allowing you to jump to the next word or letter, strip out linefeeds, or other characters, and return information like the size of the stream.

The most common use for streams is reading and writing data from and to files. You can quickly send information out to a file with code such as shown in Figure 17.6.

```
aFileStream := File pathName: 'c:\hello.txt'.
aFileStream nextPutAll: 'Hello World'.
aFileStream close
```

FIGURE 17.6

Magnitudes

Magnitudes provide ways to measure and compare entities. Common Magnitudes are the various classes of numbers such as integers or reals. Magnitudes can be compared with operations such as greater than and combined or manipulated with operations such as +.

Time is also a type of Magnitude. Smalltalk systems contain mechanisms for timing chunks of code: **Time millisecondsToRun: [block to time]**. Other

Magnitude subchildren include Date, Association, and Character. These allow you to quickly compare various entities and provide the basis for being able to sort strings, for instance.

GUIs

The last major system in a Smalltalk system is the classes used to build and manipulate GUIs. These class structures greatly simplify the creation of GUI-based applications. For instance, to create the classic Hello World application for Microsoft Windows in C, one would write code similar to that of Figure 17.7 (Petzold, 1991). The same can be accomplished in Smalltalk/V with the code in Figure 17.8. Indeed the message-based structure of Smalltalk lends itself to the development of GUI applications. Little wonder since Smalltalk is where it all started.

Building more complex GUI interfaces than a simple Hello World window does require more work, mainly to specify all the various parameters for colors, labels, placement, and the sizing behavior. The key is that much work is going on behind the scenes.

Smalltalk implementations have employed several mechanisms to separate the building of GUI interfaces from the under lying application logic. The idea is that code developed in this manner is more reusable. This is because the underlying operations or business logic are unlikely to have rapid change but the way people want to view the result of this logic is continually changing.

Frameworks

Frameworks provide abstractions in which systems can be organized and assembled. The father of frameworks is the MVC or Model View Controller framework developed by ParcPlace. Frameworks describe the way in which application logic, views of data, and flow of control are organized. It is desirable that application or business logic is separate from the presentation logic. GUI frameworks provide this separation. The necessary components of a good framework are graphic building blocks, support for the domain model, and loose coupling.

Model View Controller. The Model View Controller framework divides applications into three sections. Business logic is built as Model code, presentation logic and GUI layout are described by the View, and the GUI is loosely coupled to the Model through the Controller. This provides a simple three-part architecture that allows you to easily change the GUI front-end independent of the non-GUI elements. The various pieces are connected by inputs being queued up by an inputSensor in the controller. The MVC architecture is also used by QKS, IBM, and Liant for their C++ Views product.

ApplicationCoordinator. The ApplicationCoordinator is Digitalk's latest framework. It allows multiple views and multiple controllers. The previous frameworks at Digitalk included ApplicationWindow and ViewManager. ApplicationCoordinator provides loose coupling by allowing views to be assembled and attached to coordinator objects through events. The coordinator is then connected to the domain model. This allows application control to be designed around a flow of communications between the various interfaces and the domain through a coordinating object. As in much of Smalltalk, many of the differences in a framework are basically naming conventions. This is because the names used in Smalltalk for classes and methods signify various attributes such as the style of control. For instance, in MVC, a cen-

C example - Hello World

```
#include<windows.h>

long FAR PASCAL WndProc (HWND, WORD, WORD, LONG);

int PASCAL WinMain(HANDLE hInstance, HANDLE hPrevInstance, LPSTR lpszCmdParam, int
nCmdShow)
{
static char szAppName[] = "Hello World";
HWND    hwnd;
MSG     msg;
WNDCLASS        wndclass;

if(!hPrevInstance)
        {
        wndclass.style = CS_HREDRAW | CS_VREDRAW;
        wndclass.lpfnWndProc    = WndProc;
        wndclass.cbClsExtra     = 0;
        wndclass.cbWndExtra     = 0;
        wndclass.hInstance      = hInstance;
        wndclass.hIcon = LocadIcon(NULL, IDI_APPLICATION);
        wndclass.hCursor        = LoadCursor(NULL, IDC_ARROW);
        wndclass.hbrBackground = GetStockObject(WHITE_BRUSH);
        wndclass.lpszMenuName =NULL;
        wndclass.lpszClassName = szAppName;

        RegisterClass(&wndclass);
        }
hwnd = CreateWindow(szAppName,
        "Hello World Program",
        WS_OVERLAPPEDWINDOW,
        CW_USEDEFAULT,
        CW_USEDEFAULT,
        CW_USEDEFAULT,
        CW_USEDEFAULT,
        NULL,
        NULL,
        hInstance,
        NULL);
ShowWindow(hwnd, nCmdShow);
UpdateWindow(hwnd);

while(GetMessage(&msg, NULL, 0,0))
        {
        TranslateMessage(&msg);
        DispatchMessage(&msg);
        }
return msg.wParam;
}

long FAR PASCAL WndProc(HWND hwnd, WORD message, WORD wParam, LONG lParam)
{
HDC     hdc;
PAINTSTRUCT ps;
RECT    rect;

switch(message){
        case WM_PAINT:
                hdc = BeginPaint(hwnd, &ps);
                GetClientRect(hwnd, &rect);
                DrawText(hdc, "Hello, World From Windows",-1,&rect, DT_SINGLELINE |
DT_CENTER | DT_VCENTER);
                EndPaint(hwnd, &ps);
                return 0;
        case WM_DESTROY:
                PostQuitMessage(0);
                return 0;
        }
return DefWindowProc(hwnd, message,wParam, lParam);
}
```

FIGURE 17.7

Smalltalk Example - Hello World

```
TextWindow windowLabeled: 'Hello World' frame:(Rectangle leftTop:
0 @ 0 extent: 300@100)
```

FIGURE 17.8

tral element is the Controller and in ApplicationCoordinator is a Coordinator. This implies in MVC a central object controls the flow of operations whereas in the ApplicationCoordinator the coordinator delegates jobs to other intelligent objects.

Handling Exceptions

The Smalltalk system dynamically checks for many common error conditions such as array bounds checking. When errors do occur most systems provide facilities for handling exceptions. The most basic of these is the ability to ensure that a section of code is executed. This is useful for making sure that system handles are released and for other cleanup tasks such as closing files. In Digitalk's dialect of Smalltalk this block of code contains a *try* block and an *ensure* block and looks like this **`["try this code"] ensure: ["make sure this is done even if try block fails"]`**.

In addition, general exception handling blocks can be written. An example of this is division by zero or coming to the end of a stream before you expect. When a defined exception occurs a message is sent to the system. If the exception is handled it may result in either a continuable or noncontinuable state. A continuable exception allows the system to attempt to correct the condition and resume operations where the exception occurred.

Exceptions are objects, and by defining Exception classes you can define the default actions to occur when a particular exception occurs. The standard action is to open a walkback window or the system debugger. You can provide default handling for the error at this point to tell the system whether the error condition is resumable or not and do any necessary cleanup.

Exception Handlers allow you to go a step further. By providing a handler you can catch the exception and potentially fix the problem without bringing it to the user's attention or allowing the user to provide a correction. For instance, if you had a section of code that calculated the ratio of two values and the denominator was zero you would have an error condition and the ZeroDivide (in Smalltalk/V) exception would be raised. This exception can be caught and processed with the following (Figure 17.9):

```
[x/y]
  on: ZeroDivide
  do:
    [anError |
    y := (Prompter prompt: 'Can not divide by zero, enter
new value:' default: '1') asInteger
    failure retry]
```

FIGURE 17.9

If a process is not resumable, or you do not ask the system to *retry* the result, the *do*: block will be returned to the calling process.

17.8 TALKING TO OTHER LANGUAGES

Modern Smalltalk systems do not enjoy the luxury of dedicated environments such as the XEROX Altos. They must share the space with underlying operating systems and applications developed in a wide variety of languages. The major Smalltalk vendors all provide various mechanisms for talking to the outside world.

Indeed this is the way that the system itself accomplishes a large bulk of what it needs to do such as creating windows. The process by which system primitives are issued is the same process used by the system to talk to other languages. By allowing the user to create new primitives the system can be extended to accomplish various tasks. In this manner if you wanted to accomplish a complex numeric manipulation you could write a module in FORTRAN or C and then transfer the data between this module and your Smalltalk application.

There are two pieces required by Smalltalk to accomplish this task. The first is to provide the actual method for making the call to you routine. This looks like

```
myFunctionName: anArgument
   <api: MyFunctionName argType returnValue>
   ^self invalidArgument
```

The second piece is your objects to represent in Smalltalk the external structures you have defined. You may have a C structure to define a customer as (Figure 17.10):

```
typedef customer struct{
     name char*;
     age int;
     height int;
     id char [6];
}
```

FIGURE 17.10

You will need to define a similar class in Smalltalk to move data to and from Smalltalk to non-Smalltalk memory. Thus you may have a class such as the following:

17.9 THE SMALLTALK DEVELOPMENT ENVIRONMENT

Smalltalk is much more than syntax and a compiler. It is a complete environment that includes a variety of browsing and inspecting tools, a large class hierarchy, and other language tools all combined into a dynamic environment that allows you to trace and dynamically change and compile your system easily from any view. Most programmers spend their time in hierarchy or class browsers and workspaces.

A workspace is a simple editing window that allows you to enter and evaluate expressions. When you evaluate an expression you can either *do-it* or *show-it*. Doing

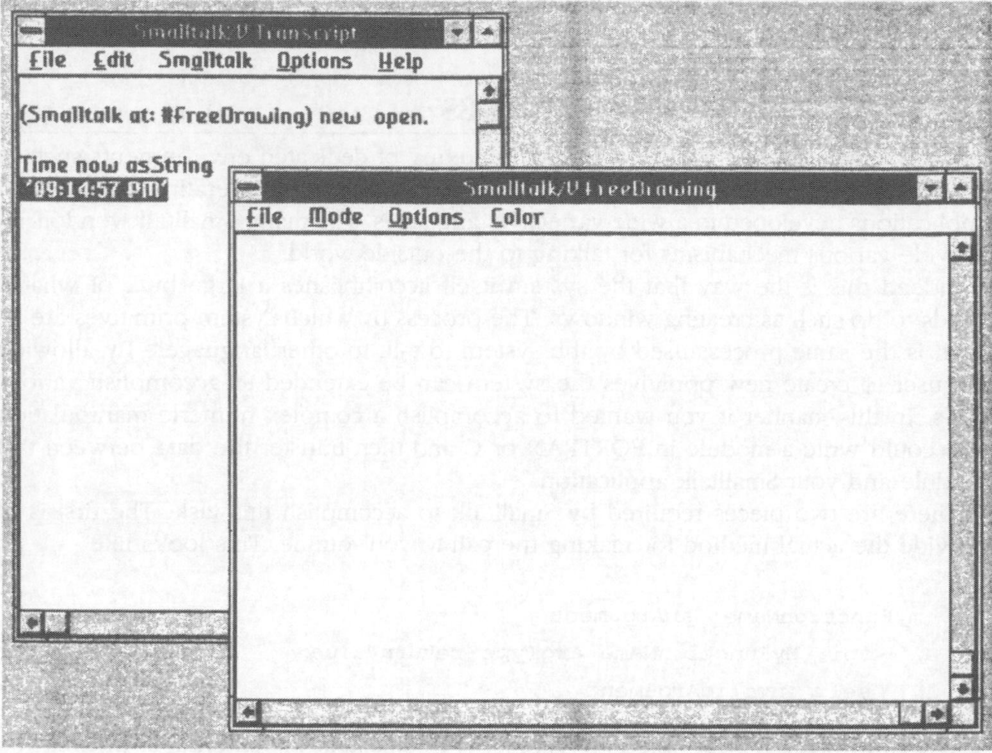

FIGURE 17.11 Use-it and show-it.

something is like pressing a button and seeing what happens; showing something asks the system to return any information to you. In Figure 17.11 you can see the result of issuing a do-it to open an application and a show-it to evaluate the current time.

The Smalltalk system of browsers aids in organizing application composition and the reuse of base classes. Figure 17.12 shows a basic Smalltalk hierarchy browser. In the left-most pane, classes are listed in a hierarchical fashion. All Smalltalk classes descend from the root class—Object. This browser also contains information about variables and methods and displays code and definitions in the lower pane. The user can switch between viewing class methods (ways to create instances of classes) and instance methods (operations on instances). This system provides structure at the same time it promotes an iterative explorational style of development. The browser also makes it easy to maintain a perspective on the system you are developing while concentrating in at the task (current method) at hand.

As you work in Smalltalk, it keeps track of all of your work in a *change log*. This is a text file that allows you to recover from system crashes and crrors. Smalltalk also has a unique idea called the *image* that creates a persistent programming environment. The image contains all your currently defined classes and instances of those classes. This includes things like instances of various browsers and their state. Thus you can be editing a set of code with several browsers opened on various classes and methods, snapshot your image by saving it to disk, and exit the system. When you restart this image you return to the exact state that you had defined when you did the save.

Delivering applications in this traditional development style environment is a matter of removing unneeded classes and objects from your system after adding the classes that make the image have the behavior you need. Some of the modern systems have taken an approach of building applications by assembling components. In this fashion you still use an image to keep the advantage of a persistent development environment but create applications by creating component files that can be dynamically loaded and removed by the run-time system. These advances are helping to improve the memory efficiency of Smalltalk applications. This also aids in the traditional configuration management roll of reproducing a version of an application by use of a build script.

Debugging

While Smalltalk offers developers many advantages, errors still occur. Like all major programming languages, Smalltalk provides full-featured debugging environments. Unlike static, compiled languages, Smalltalk debuggers are just like any other code browser and allow full exploration, modification, and compilation of code. A debugging window is illustrated in Figure 17.13. In this window you can see the execution stack in the upper left window, currently defined object instances in the center, and their value in the right top pane. The bottom pane allows you to view the source code for the currently selected method in the stack. Thus you can trace back through the stack to where the error occurred, view the various local state information, and modify the code inside the debugger and be off and running.

You can automatically open a debugger by placing the message **`self halt`** in your code. In addition most systems allow you to set break points or grab an active Smalltalk process. While in a debugging environment, you can also perform standard debugging operations such as single stepping.

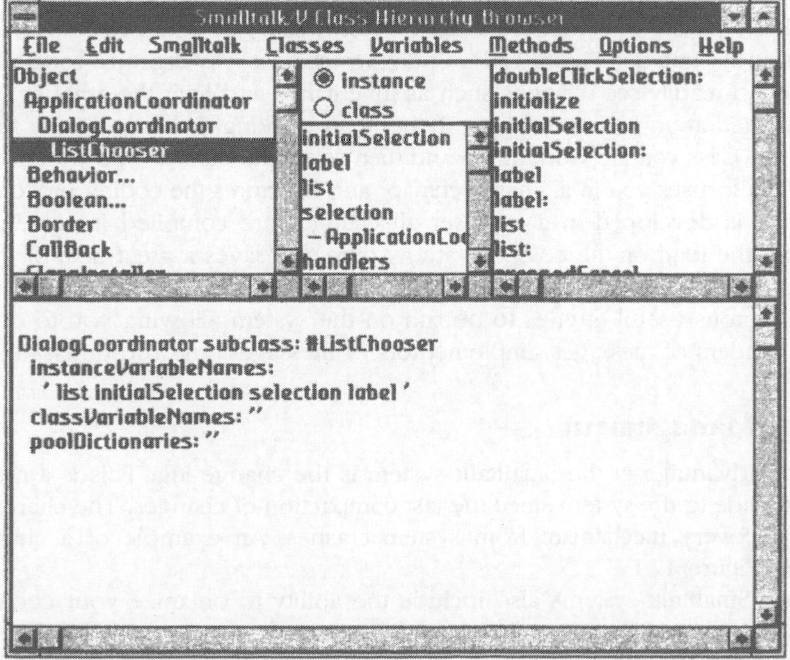

FIGURE 17.12 Class hierarchy browser.

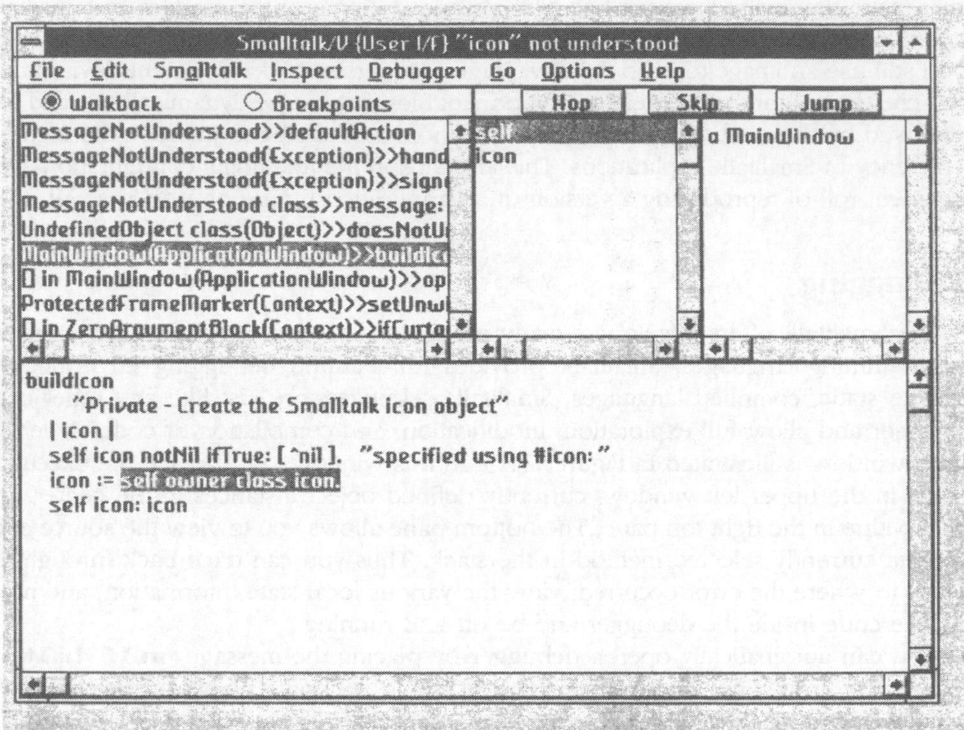

FIGURE 17.13 Smalltalk debugger.

Browsing Code

Smalltalk is typically written inside a browser. Browsers give you a view of a class and its methods as illustrated in Figure 17.14. Code browsers visually organize code allowing the developer to readily see relations such as inheritance and limit the amount of information presented at any given moment. Rather than working with a large text file, you select first the class you are working on and then select the method you want to modify or create. This focuses you in a single behavior and structures the coding process.

As code is developed in a browser all changes are compiled incrementally. This minimizes the load on the compiler at any time and saves a great deal of time allowing for a more experimental development style. In addition, the Smalltalk system supports numerous useful queries to be run on the system allowing you to quickly find classes, senders of messages, implementors of messages, and run quick metrics.

Change Management

A unique advantage of the Smalltalk system is the change log. This is a record of all changes made to the system since the last compaction of changes. The change log provides a recovery mechanism from system crashes. An example of a change log is shown in Figure 17.15.

Modern Smalltalk systems also include the ability to organize your code into projects and apply version control and configuration management tools. These systems also benefit teams of developers by providing resource locking, conflict resolution, and difference browsers.

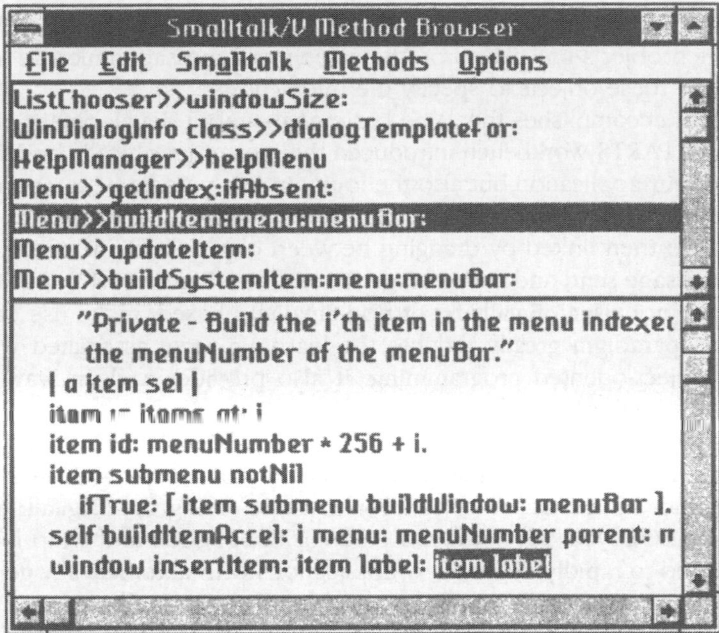

FIGURE 17.14 Method browser.

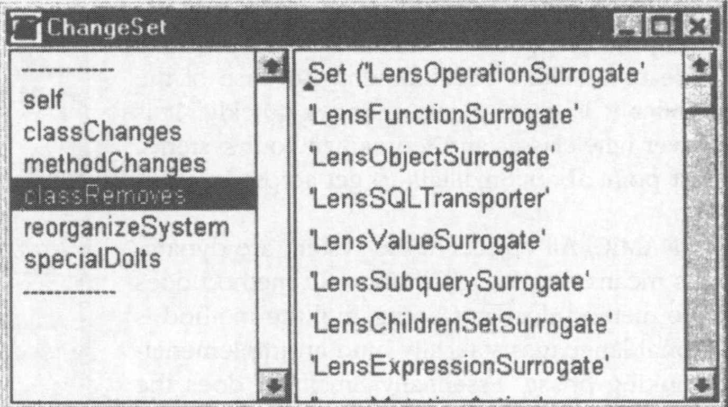

FIGURE 17.15 Working with changes.

One of the goals of Smalltalk and object-oriented development is the componentization of code. This allows developers to create autonomous units that interact with each other through defined interfaces. Team/V's organizational tools allow developers to version packages and build configurations called clusters. OTI's Envy manages applications and then automatically creates the executable package.

Visual Development

Smalltalk truly has its roots in the development of the modern graphical user interface. As systems have progressed, they have become much more visual in nature. This allows interfaces to be assembled more rapidly and also can allow much of the interface's logic to be easily specified. There are two systems in use for this type of development. The first is a class-based approach in which interfaces are painted and then code is attached to the pieces to make everything work. The second is instance based, in which objects are placed on the screen and message links are also visually drawn between these objects to specify the interactions.

Either system accomplishes the same basic goal—rapid development of GUI interfaces. Digitalk's PARTS Workbench introduced the idea of graphically building not only the layout of your application but also the logic. In this paradigm, components are assembled by dragging them off a palette (see Figure 17.16) onto a workspace. Components are then linked by dragging between objects and specifying what event triggers the message send and the message to send (see Figure 17.17). All components, whether visual, nonvisual, Smalltalk, or non-Smalltalk based, make use of a standard interface. This paradigm greatly reduces the learning curve associated with learning Smalltalk or object-oriented programming. It also provides a clean way to integrate new and legacy components.

Since components have a uniform interface, they become easy to document and share with others. It also means that pieces from very non-object-oriented systems can be wrapped into an extensible object. For instance, using the Digitalk Wrapper for MicroFocus COBOL, you can point at a piece of Cobol code and wrap it. This object can then be used to rapidly deploy new applications with automatically generated GUI interfaces that can share data and function with other Smalltalk code, SQL database, and other systems.

17.10 THE SMALLTALK DEVELOPMENT PROCESS

One of the most important aspects of Smalltalk development is the iterative nature of Smalltalk development. Because of the incremental compiler it is easy to test changes quickly. It is also easy to discover new classes and reorganize your systems. This is the hardest point about Smalltalk to get across in a static text document.

Smalltalk is DYNAMIC. All objects in the system are dynamically bound—this means that the resolution of a method does not occur until the method does not occur until the method is executed. Traditional languages statically bind an implementation during the linking phase. Essentially Smalltalk does the same thing, only linking occurs at the point of execution.

This nature makes Smalltalk an excellent prototyping environment because new ideas and changes can be incorporated and tested rapidly. Developers can experiment with different algorithms and designs because there is not a large time cost to create and test. Dynamic binding also provides an easy system for creating componentized software and dividing work up between teams.

Smalltalk is also different because it is a semipersistent system. As you develop code you are modifying the actual system, adding classes and behaviors until the system performs as you want. Then you bundle your set of objects up with the run-time, tell it how to start, and you are finished.

Combining this persistent system of an object collection with dynamic binding, teams can interact together on a project, changing the implementation while keeping the public inter-

FIGURE 17.16

face the same. As other members of the team work with the same object collection, they will always resolve to the current implementation.

17.11 APPLICATIONS

Smalltalk has been applied to a wide variety of problems. During the early years at Xerox PARC, each new system was used to build applications and the experience gained then was fed back into the next iteration of the environment. But in these early days Smalltalk needed more memory and processing power than was commonly available. This hampered the early adoption of Smalltalk and limited its use to prototyping, simulation, and experimentation.

Fortunately, hardware has advanced rapidly since the introduction of the PC and machines are now equipped with the processing power and memory required to make Smalltalk efficient. In addition, great advances have been made in Smalltalk technology. The best Smalltalk systems today have dynamic compilation rather than interpretation and fast garbage collection schemes allowing Smalltalk programs to function nearly as efficiently as most 3GL languages. Combining quick performance with rapid development and a simple syntax, Smalltalk is now rapidly gaining favor as the object language of choice.

The greatest interest in Smalltalk has come from the corporate sector. As corporations look to distributed application solutions and invest in business process reengineering, Smalltalk proves to be a ready and able vehicle with its benefits of object orientation and iterative development. In addition, the modern Smalltalk implementations offer the ability to easily integrate legacy solutions with new development. This allows corporations that ability transition into new systems over time.

Smalltalk applications today run a wide variety of the corporate world's operations: controlling inventories, shipping products, billing and problem resolution, and all aspects of customer service. Because of the ease with which Smalltalk can be interfaced to other languages, it is able to accomplish a very broad variety of tasks. Smalltalk allows the developer to focus on the problem domain and not worry about the complexity of the solution space. It is has been shown in commercial projects that Smalltalk provides development time and gets products to market faster than traditional methodologies.

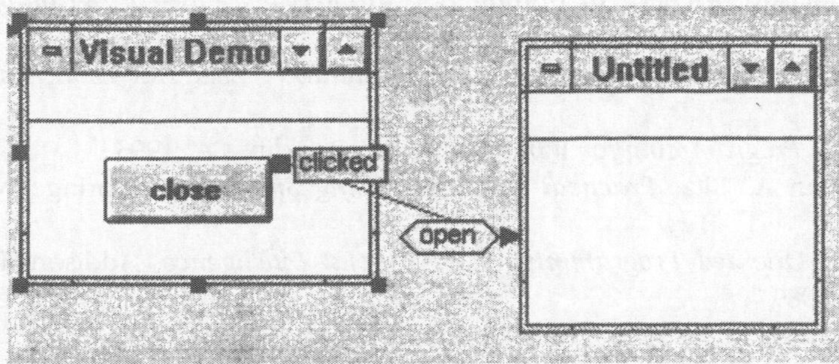

FIGURE 17.17

17.12 CONCLUSION

Smalltalk is an easy to learn but powerful pure object-oriented language that has grown from the vision of an individual to a dedicated team to be a leading language in a wide variety of applications. The benefits of Smalltalk are simple language constructs, a rich class hierarchy, and reduction of complexity through this library and garbage collection. As Smalltalk continues to evolve it will continue to allow developers to concentrate on getting their job done—building reliable, working applications.

REFERENCES

Smalltalk is covered in a wide variety of books. The most famous series is that produced by the Xerox PARC team. These cover the language, implementation, benchmarking, historical errata, etc. In addition there are many books on programming with Smalltalk. A magazine, *The Smalltalk Report*, is dedicated to Smalltalk and the issues of design and coding style. The definitive history of the genesis of Smalltalk is contained in Alan Key, "An Early History of Smalltalk," in *History of Programming Languages Conference* (HOPL-11), April 20–23, Cambridge, MA, published in *ACM SIGPLAN Notices*, 20(3), March 1993.

Adele Goldberg, *Smalltalk 80 The Interactive Programming Environment*. Addison-Wesley, Reading, MA, 1984.

Adele Goldberg and David Robson, *Smalltalk 80 The Language*. Addison-Wesley, Reading, MA, 1989.

IBM, *Smalltalk Portability: A Common Base*. IBM Document Number GG24-3903, Boca Raton, FL, 1993.

Glenn Krasner, *Smalltalk 80 Bits of History, Words of Advice*. Addison-Wesley, Reading, MA, 1984.

Wilf Lalonde, *Discovering Smalltalk*. Addison-Wesley, Reading, MA, 1993.

Wilf Lalonde and John Pugh, *Inside Smalltalk*, Volume I. Prentice-Hall, Englewood Cliffs, NJ, 1991.

Wilf Lalonde and John Pugh, *Inside Smalltalk*, Volume II. Prentice-Hall, Englewood Cliffs, NJ, 1991.

Mark Mullen, *Rapid Prototyping for Object Oriented Systems*. Addison-Wesley, Reading, MA, 1990.

Charles Petzold, *Programming Windows*.

Lewis Pinson and Richard Wiener, *An Introduction to Object Oriented Programming & Smalltalk*. Addison-Wesley, Reading, MA, 1988.

Dusko Savic, *Object Oriented Programming with Smalltalk/V*. Ellis Horwood, London, 1990.

Dan Shafer, *Smalltalk Programming for Windows*. Prima, Rocklin, CA, 1993.

Dan Shafer and Dean A. Ritz, *Practical Smalltalk: Using Smalltalk/V*. Springer-Verlag, Berlin, 1992.

Ernest Tello, *Object Oriented Programming for Artificial Intelligence*. Addison-Wesley, Reading, MA, 1989.

The Smalltalk Report. Sigs Publications, New York.

Stephen Travis Pope, *The Well Tempered Object*. MIT Press, Cambridge, MA, 1991.

Rebecca Wirfs-Brock, Brian Wilkerson, and Lauren Wiener, *Designing Object-Oriented Software*. Prentice-Hall, Englewood Cliffs, NJ, 1990.

Application Packages

A few decades ago, computing power was so expensive that computer access was limited to a few key applications. In those days, scientists and engineers used computers to perform complex computations, accountants used them to prepare periodic reports, and the need for efficiency meant working through a technical expert. The availability of inexpensive microcomputers has changed the rules. Today, almost anyone who wants a computer can have one, and parameters such as ease of use and functionality have replaced efficiency as the primary focus of modern application software.

This part covers several popular application packages. The tasks they perform do not fit the usual image of scientific and engineering applications, but these packages have the potential to significantly enhance personal productivity.

Word processing software (Chapters 18–21) simplifies the task of creating memos, reports, documentation, and other printed material. Once you learn to use a word processor, you will never again be satisfied with a typewriter or a secretarial pool. An understanding of word processing conventions can also help to clarify e-mail (see Part VI) and similar applications.

Word processing is concerned with preparing text. Desktop publishing (Chapters 22–25) is concerned with preparing camera-ready pages that integrate text, graphics, tables, and other elements prepared by various application packages. A basic knowledge of desktop publishing is invaluable to those scientists and engineers who regularly submit papers to professional journals, write formal proposals, or prepare finished documentation.

A spreadsheet is an accountant's tool that can also be used to hold and manipulate tables of scientific and engineering data. Today's spreadsheet packages (Chapter 26) support sophisticated mathematical and statistical computations and allow the user to prepare graphs and charts literally at the push of a button. Additionally, spreadsheets have become a de facto standard for data collection, data definition, and data distribution, much as FORTRAN FORMAT statements were the data definition standard for a previous generation of statistical and mathematical software.

Database software (Chapter 27) defines the new standard for storing, retrieving, and managing large volumes of data. This chapter and Chapter 28 cover the most common database packages: dBASE IV and Microsoft Access 2.0.

Integrated packages (Chapters 29 and 30) incorporate the basic features of word processing, spreadsheet, database management, and presentation graphics software. Although the more advanced features of the individual applications are missing, an integrated package is an excellent choice for the scientist or engineer who needs only the basic features and lacks the time to learn multiple programs.

As the name implies, presentation graphics software (Chapter 31) is used to prepare presentations. The user types text, enters tabular data, selects artwork, and prepares charts and graphs on the screen. The software then converts the information to professional-looking screen images. The resulting presentation can be played back on the computer or converted to transparencies or even color slides for more conventional presentations.

Chapter 32 describes several utility programs that can help manage and protect data, improve the efficiency of a computer system, and recover from system failures. Chapters 33 and 34 discuss the new area of Groupware that allows the efficient management and exchange of data for large groups of people working with complex projects.

18

WordPerfect 5.1

)avid L. Hartung Jr.
liami University

18.1 WORD PROCESSING CONCEPTS

Word processing, one of the earliest applications developed for microcomputers, is the process of creating documents using a computer and a word processing software package. The advantages of word processing over typing include the following:

- The ability to make multiple revisions to documents without having to retype the entire page.
- The ability to store documents for easy filing and retrieval.
- The ability to correct spelling errors.
- The ability to change the physical appearance of the text.

WordPerfect is one of the most popular word processing software packages in use today.

18.2 SYSTEM OVERVIEW

Major Features

WordPerfect offers the following major features:

- Creating, deleting, and changing text.
- Controlling the physical appearance of the text by changing the font and the point size.
- Support for a wide variety of printers.
- A spell checker and thesaurus.
- Automatically timed backup of the current document.
- An option that allows for the creation of tables and multiple columns of text.
- A merge capability for creating form letters and mailing labels.
- The ability to construct mathematical formulas.

)-8493-2530-7/96/$0.00+$.50

Starting/Loading WordPerfect

If you have a system with a hard drive, skip the section for loading WordPerfect on a two-disk system.

Loading WordPerfect on a Two-Disk System

1. Start DOS by turning on the computer.
2. Insert the WordPerfect 1 disk into drive A and the data disk into drive B.
3. At the A prompt type **B:** and press the ENTER key. This will change the default drive to B.
4. At the B prompt enter **A:WP** and press the ENTER key to start WordPerfect.
5. When prompted, remove the WordPerfect 1 disk and replace it with the WordPerfect 2 disk, then press any key. After several seconds, the WordPerfect screen will appear.

Loading WordPerfect on a Hard Disk System

1. Start DOS by turning on the computer.
2. At the C prompt type **CD** *directory name* (directory name is the name of the subdirectory where the WP.EXE file is located) and press the ENTER key.
3. At the new prompt, enter **WP** and press the ENTER key to start WordPerfect.
4. After several seconds, the WordPerfect screen will appear.

The WordPerfect Screen

The WordPerfect screen (Fig. 18.1) is designed to simulate a blank sheet of paper. The blinking dash at the top of the screen is the cursor. The status line, found at the bottom of the screen, displays the document number and the cursor's current page number, line number, and position.

When you first begin to use WordPerfect, you do not need to be concerned with initial settings for margins, page size, tab stops, etc. These values have been preset. Later in the chapter you will learn how to change these settings. For a complete list of the default settings, please refer to the appropriate appendix in the WordPerfect manual.

Executing WordPerfect Commands

WordPerfect is a command-driven package. This means that commands are executed by pressing a function key by itself or by pressing a function key in combination with the ALT, SHIFT, or CTRL key. The WordPerfect template displays the various functions and the associated key combinations. A template that lays over the function keys is provided when you purchase WordPerfect. The template also may be viewed by accessing WordPerfect's HELP function. Press **F3** and the WordPerfect help menu appears. Press **F3** again to display the Word Perfect Template. To exit the help function press ENTER or the space bar.

If you accidentally get into a function, you can always back out by pressing **F1** (Cancel).

Cursor Movement

Moving around the document can be done using the cursor movement keys alone or in combination with other keys. Below is a summary of keys that control cursor movement.

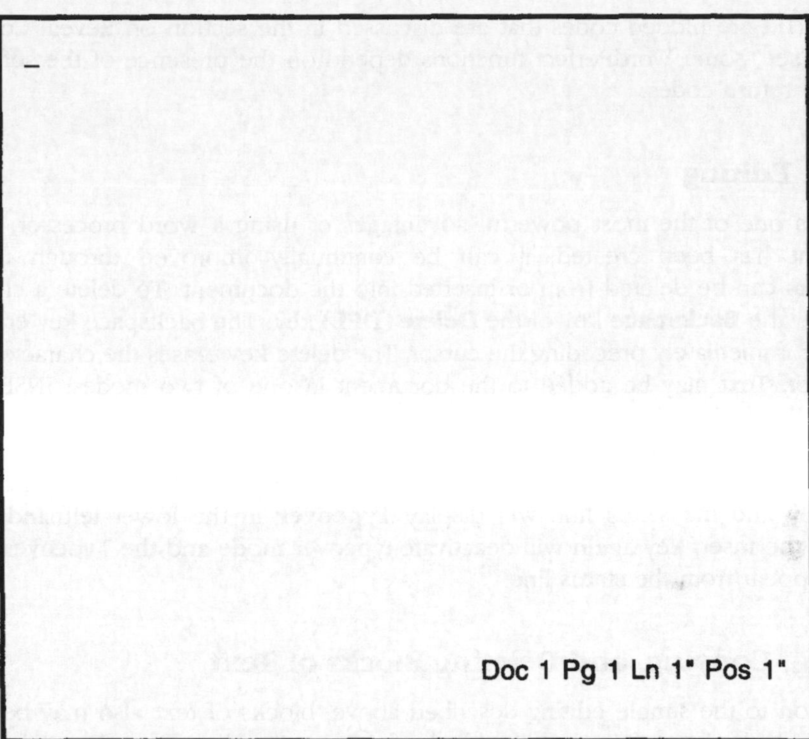

FIGURE 18.1 Initial WordPerfect screen.

Keystroke	Cursor Movement
Up Arrow	One line up
Down Arrow	One line down
Left Arrow	One character to the left
Right Arrow	One character to the right
CTRL and Left Arrow	One word to the left
CTRL and Right Arrow	One word to the right
END or HOME and Right Arrow	End of the line
HOME and Left Arrow	Beginning of the line
HOME and Up Arrow	Top of the screen
HOME and Down Arrow	Bottom of the screen
HOME, then HOME and Up Arrow	Top of the document
Home, then HOME and Down Arrow	Bottom of the document

18.3 ELEMENTARY COMMANDS/FUNCTIONS

Entering Text

To create a document in WordPerfect simply begin to type after accessing the initial WordPerfect Screen. Upon reaching the right margin, WordPerfect automatically advances the cursor to the beginning of the next line. This word processing feature is known as **wordwrap**. Do not press the ENTER key at the end of the line to advance to the next line. ENTER is used to end a paragraph or to insert a blank line. When the text wraps around to the next line, a soft return code, SRt, is placed at the end of the line. Pressing the ENTER key places a hard return code, HRt, at the end of the line.

SRt and HRt are hidden codes that are discussed in the section on Reveal Codes. As you will see, some WordPerfect functions depend on the presence of the soft return and hard return codes.

Simple Editing

Editing is one of the most powerful advantages of using a word processor. Once a document has been created, it can be continually improved through revision. Characters can be deleted from or inserted into the document. To delete a character, use either the **Backspace** key or the **Delete (DEL)** key. The backspace key erases the character immediately preceding the cursor. The delete key erases the character above the cursor. Text may be added to the document in one of two modes: INSERT and TYPEOVER. In INSERT mode, text is added at the current location of the cursor and text following the cursor automatically is pushed to the right. TYPEOVER mode replaces existing text with new text. To activate TYPEOVER mode, press the **Insert (INS)** key and the status line will display **Typeover** in the lower lefthand corner. Pressing the Insert key again will deactivate typeover mode and the Typeover display will disappear from the status line.

Moving, Copying, and Deleting Blocks of Text

In addition to the simple editing described above, blocks of text also may be edited. To edit a block of text, first mark the block of text that is to be edited. To mark the text, place the cursor under the first character in the block and press either the **ALT** and **F4** keys or the **F12** key. This turns the block on, indicated by the **Block on** message flashing in the lower left-hand corner of the screen. Move the cursor to the end of the block of text. This can be done by using the left or right arrow keys to block one character at a time, the up and down arrow keys to block one line at a time, or the Page-Up and Page Down Keys to block an entire page at one time. After the text is blocked, press the **CTRL** and **F4** keys. This displays the move menu as follows: **Move: 1 Block; 2 Tabular Column; 3 Rectangle: 0**. Select **1 Block**. The next menu displays the following selections: **1 Move; 2 Copy; 3 Delete; 4 Append: 0**. Moving text will delete the blocked text from its current position and store a copy in temporary memory; it then can be inserted elsewhere in the document. Copying text stores a duplicate copy of the text in temporary memory. While the original text remains in place, a copy can be inserted elsewhere in the document. Delete removes the blocked text from the document. Select Move, Copy, or Delete. After selecting Move or Copy, position the cursor at the location the text should appear and press the ENTER key.

Saving Your File

There are two ways to save a document: **F10**—save and continue editing or **F7**—save and exit. To save and continue editing press F10. Enter a name for the document as prompted. A document name consists of one to eight characters followed by an optional extension containing one to three characters. If an extension is used a period must be placed between the document name and the extension. Press ENTER; the file will be saved to disk and its name will appear on the left side of the status line. At this point you may continue editing the document.

To save and exit the document, press F7. Enter the document name as described above. After the name is entered, Wordperfect will give you the option of exiting the software or creating a new document. Press "Y" to exit the software or press "N" to create a new document.

When saving a file that already has been saved on disk, WordPerfect will suggest the same file name and ask if you want to replace the file. Press "Y" to replace the contents of the old file with the version currently appearing on the screen. To save the current document under a different name, enter the new name and press the ENTER key.

Printing a Document

To print a document, press the **SHIFT** and **F7** keys. This will access the print menu (Fig. 10.2).

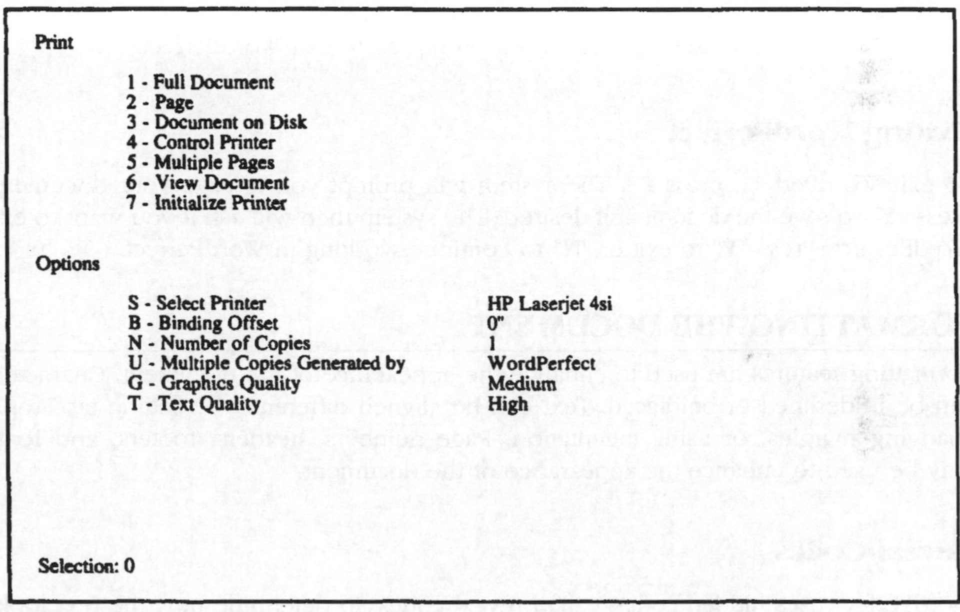

Print

1 - Full Document
2 - Page
3 - Document on Disk
4 - Control Printer
5 - Multiple Pages
6 - View Document
7 - Initialize Printer

Options

S - Select Printer	HP Laserjet 4si
B - Binding Offset	0"
N - Number of Copies	1
U - Multiple Copies Generated by	WordPerfect
G - Graphics Quality	Medium
T - Text Quality	High

Selection: 0

FIGURE 18.2 Print menu.

The following selections are available to print a document:

- **Full Document**: Prints all the pages in the active document.
- **Page**: Prints the page on which the cursor is currently located.
- **Document on Disk**: Prints the document from a file stored on disk.
- **Control Printer**: Controls printing functions, including stopping the printing of a document.
- **Multiple Pages**: Prints a specified range of pages.
- **View Document**: Previews how the document will appear when printed.
- **Initialize Printer**: Initializes the printer.

To change any of the selections below the OPTIONS heading on the Print Menu, press the letter preceding the option. For example, if the printer displayed to the right of "Select Printer" is incorrect, press "**S**" and use the cursor to select from the list of available printers.

Retrieving a Document

There are two ways to retrieve a document from a disk: pressing **Shift** and **F10** or **F5**. If the name of the document is known press SHIFT and F10. When prompted, enter the WordPerfect file name and press RETURN.

If the name of the document is not known or you wish to view a list of files on your disk, press the **F5** (List Files) key. WordPerfect displays the default disk drive and directory, for example, C:\WP51*.*. The drive and/or directory may be altered as necessary. When the desired disk drive and directory appear, press the **RETURN** key. All the files on your disk will be listed on the screen. Position the cursor on the document you wish to retrieve and select **1 Retrieve**.

Note: Exercise caution when retrieving files. WordPerfect will not automatically remove the current file from memory (screen). If file is retrieved when another file is in memory (screen), the files will be combined. It is a good idea to clear memory(screen) before moving on to the next document. To clear the screen, press **F7**. Save the document if desired and then type "**N**" when prompted to exit WordPerfect.

Exiting WordPerfect

To exit WordPerfect, press **F7**. The system will prompt you to save your document. Press "**Y**" to save the document if desired. The system then will ask if you want to exit WordPerfect. Press "**Y**" to exit or "**N**" to continue working in WordPerfect.

18.4 FORMATTING THE DOCUMENT

Formatting features are used to enhance the appearance of the document. Characters can be underlined or boldfaced. Text can be aligned differently by setting tab stops, changing margins, or using indentation. Page numbers, headers, footers, and fonts may be used to enhance the appearance of the document.

Reveal Codes

WordPerfect uses hidden codes, called reveal codes, to determine how the text looks on the screen and when it is printed. All WordPerfect formatting features have an associated reveal code. To view these codes press the **ALT** and **F3** keys or the **F11** key. The screen is divided in half displaying the document on top and the reveal codes on the bottom. Misplaced or unwanted codes can cause problems with a document. Fortunately, codes can be edited. Unwanted codes can be deleted by placing the cursor in the code and pressing the DELETE key. Misplaced codes can be moved or copied by following the rules for moving blocks of text described in the section on Moving, Copying, and Deleting Blocks of Text. To turn off the reveal codes, press Alt and F3 or F11.

Underlining Text

Underlining text may be accomplished two ways: underlining as you type or underlining text that already has been entered. To underline as you type, press **F8** to activate the underline feature. Enter the text. As you type, the text appears in reverse video. This indicates that the text is underlined. When finished entering the text, once again press **F8** to turn off the underline feature. F8 acts as a toggle switch, first turning on the underline feature and then turning it off.

Underlining text that already has been entered is a two step process. First, block the text. This is done by first placing the cursor at the beginning of the text and pressing **ALT** and **F4** or **F12**. Second, move the cursor to the end of the text and press F8. The highlighted text will be underlined.

Bolding Text

As with underlining, bolding text also may be accomplished two ways: boldfacing as you type or boldfacing text that already has been entered. To boldface as you type, press **F6** to activate the boldface feature. Enter the text. As you type, the text appears in boldface. When finished entering the text, press **F6** again to turn off the boldface feature. F6 acts as a toggle switch, first turning on and then turning off the boldface feature.

Boldfacing text that already has been entered is a two step process. First, **block** the text. This is done by first placing the cursor at the beginning of the text and pressing **ALT** and **F4** or **F12**. Second, move the cursor to the end of the text and press F6. The highlighted text will be boldfaced.

Centering Text

Centering text may be accomplished two ways: centering text as you type or centering text that already has been entered. To center text as you type, press **SHIFT** and **F6**. The cursor will move to the center of the screen. The text is automatically centered as you type. Enter the text. Press ENTER to turn off the centering feature.

To center text that already has been entered, place the cursor on the first character in the word or phrase that is to be centered and press **SHIFT** and **F6**. The center function stops when it reaches a hard return code (HRt).

Indenting Text

With WordPerfect, the following three types of indentation may be done:

1. Indenting the first line of a paragraph.
2. Indenting a paragraph from the left margin.
3. Indenting a paragraph from both the left and right margins.

To indent the first line of a paragraph, place the cursor at the beginning of the line and press the **TAB** key. Each time the TAB key is pressed the line will shift one tab stop from the left margin (Tab stops are preset every 0.5 inches). To indent an entire paragraph, place the cursor at the beginning of the paragraph and press the **F4** key. This will indent the paragraph one tab stop from the left margin. Each time the F4 key is pressed the paragraph is indented an additional tab stop. To indent a paragraph from both the left and right margins, also known as double indenting, press the **SHIFT** and **F4** keys. This will indent the paragraph one tab stop from both the left and right margins. As with indenting, each time this procedure is repeated, the paragraph is indented an additional tab stop from both margins.

Indenting shows the importance of the word-wrap feature. The indent and double indent functions will work only from the position indentation is turned on until it reaches a hard return. If a hard return is placed at the end of every line, each line would have to be indented individually.

The Format and Line Submenus

The following topics are all found on the format menu. Some fall under the heading **PAGE** commands and others fall under the heading **LINE** commands. The difference between these commands is that page commands affect the entire page while line commands affect only that part of the document where they are set. For page commands to work properly they must appear in the reveal codes **before** the line command reveal codes. The Format menu is shown in Figure 18.3 and the Line and Page submenus are shown in Figure 18.4.

Setting Tabs

Tab stops in WordPerfect are preset at half-inch intervals. To change the settings press **SHIFT** and **F8** to access the format menu. Select **`1 - Line`**. Select **`8 - Tab Set`**. The ruled line shown at the bottom of the screen displays the current tab settings, indicated by the letter "L" (left align). To clear the current settings press the **CTRL** and **END** keys. The current settings will disappear. To set the new tab stops, type the position of the tab stop in decimal form, for example, 1.5. Press the **ENTER** key and an indicator will appear at that position. Continue to enter additional tab settings as described above. When finished, press **F7** to save the new tab settings. Press the ENTER key twice to return to the document.

To set the tabs back to the original settings, repeat the above commands. Instead of typing each individual tab stop, all tab stops can be set in one step. Again access the tab set screen. Enter the position in which the tab stops are to begin, followed by a comma. Then enter the increment desired for the tab settings. For example 0.0,0.5. This will start the tabs at position 0.0 and insert a tab stop every half inch.

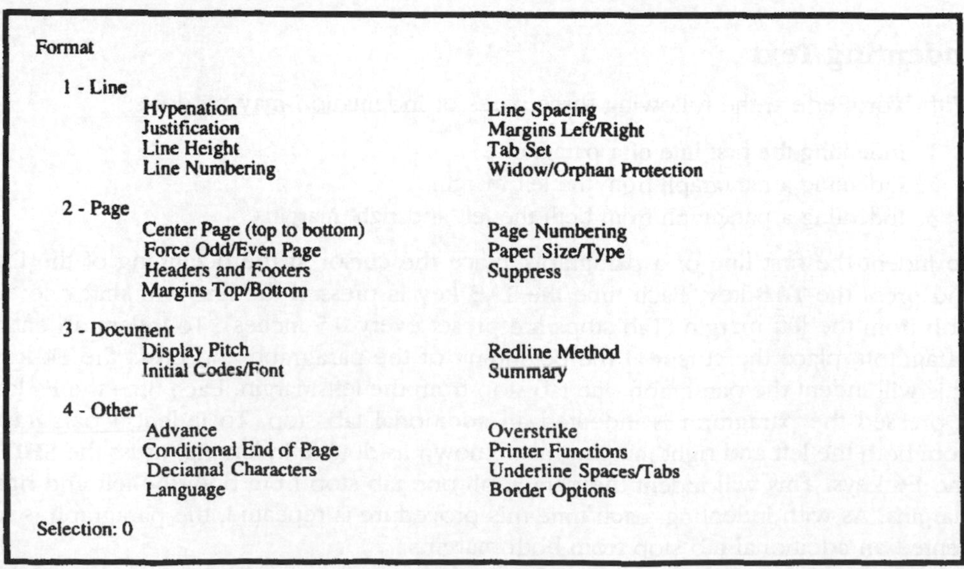
Format

1 - Line

Hypenation	Line Spacing
Justification	Margins Left/Right
Line Height	Tab Set
Line Numbering	Widow/Orphan Protection

2 - Page

Center Page (top to bottom)	Page Numbering
Force Odd/Even Page	Paper Size/Type
Headers and Footers	Suppress
Margins Top/Bottom	

3 - Document

Display Pitch	Redline Method
Initial Codes/Font	Summary

4 - Other

Advance	Overstrike
Conditional End of Page	Printer Functions
Deciamal Characters	Underline Spaces/Tabs
Language	Border Options

Selection: 0

FIGURE 18.3 Format menu.

```
Format: Line
   1 - Hypenation                        No
   2 - Hypenation Zone - Left   10%
                        Right   4%
   3 - Justification                     Full
   4 - Line Height                       Auto
   5 - Line Numbering                    No
   6 - Line Spacing                      1
   7 - Margins - Left           1"
                 Right          1"
   8 - Tab Set                           Rel: -1", every 0.5"
   9 - Widow/Orphan ProtectionNo

Selection: 0
```

FIGURE 18.4 Line menu.

Line Spacing

To change line spacing in the document, place the cursor at the position you want the change in line spacing to take effect. Press **SHIFT** and **F8** to access the format menu. Select **`1 - Line`**. Select **`6 - Line Spacing`**. The default value is single-spaced text. To change to double-spaced text, enter the number 2. After entering the desired spacing, press the **ENTER** key. Then press the **ENTER** key to return to the document. The chosen spacing is in effect until it is changed.

Justification

Full justification is the default in WordPerfect. This means the text is evenly spread out between the right and left margins. Text that is left-justified has an even left margin and a ragged right margin. Center justification centers the text on the page. Text that is right justified has a ragged left margin and an even right margin. To change justification, access the format menu by pressing **SHIFT** and **F8**. Select **`1 - Line`** and **`3 - Justification`**. The following menu is displayed:

```
1 Left; 2 Center; 3 Right; 4 Full: 0
```

After selecting the desired justification, press the **ENTER** key until you are returned to the document.

Widow/Orphan Protection

A Widow occurs when the first line of a paragraph appears by itself on the last line of a page. An Orphan occurs when the last line of a paragraph appears by itself on the first line of a page. To avoid this situation use the line command Widow/Orphan Protection. Press **SHIFT** and **F8** to access the Format menu. Select **`1 - Line`**. Select **`9 - Widow/Orphan Protection`**. Type "**Y**" to activate this feature. Press Enter several times to return to the document.

Setting Margins

To set margins, either left and right or top and bottom, press **SHIFT** and **F8** to access the format menu. Changes to the left and right margins can be made under the **line** command menu; changes to the top and bottom margins must be made under the **page** command menu. Default margins settings are one inch from all sides of the page. To change the left and right margins select **7 - Margins** from the line command menu and enter the new value for the margin using the decimal form. To change the top and bottom margins select **5 - Margins** from the page command menu and enter the new margin values. Press the ENTER key after entering the new margins. Press ENTER several times to return to the document.

Page Submenu

The Page submenu is shown in Figure 18.5.

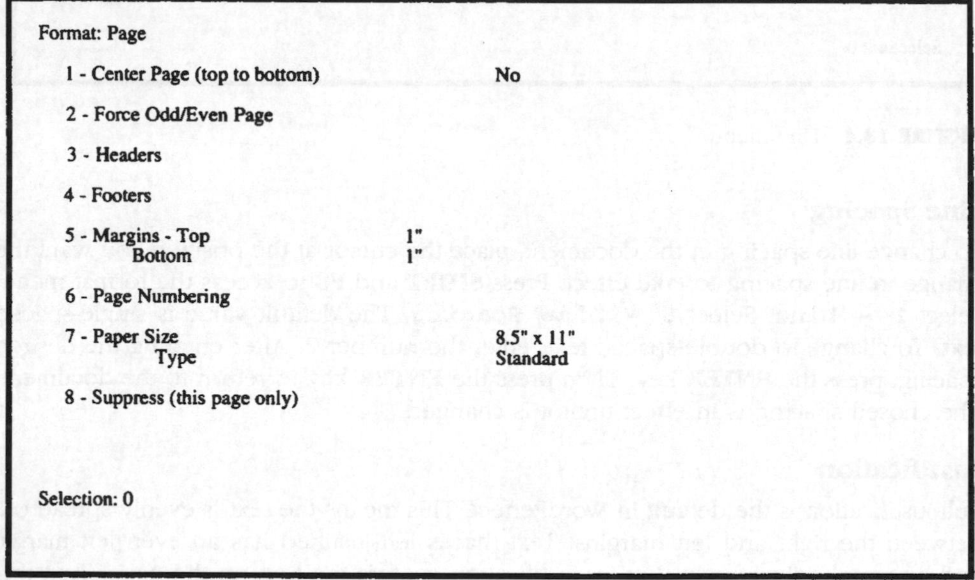

Format: Page
1 - Center Page (top to bottom) No
2 - Force Odd/Even Page
3 - Headers
4 - Footers
5 - Margins - Top 1"
Bottom 1"
6 - Page Numbering
7 - Paper Size 8.5" x 11"
Type Standard
8 - Suppress (this page only)
Selection: 0

FIGURE 18.5 Page menu.

Page Headers and Footers

Headers and footers are used to insert text at the top or bottom of a page. The user controls the frequency with which the header or footer is printed. The following choices are available: Discontinue, Every Page, Odd Pages, and Even Pages. To insert a header or a footer in a document, access the format menu by pressing **SHIFT** and **F8**. Select **2 - Page** from the menu. Selection 3 is **Header** and selection 4 is **Footer**. Upon selecting Header or Footer, a menu is displayed with choices for Header/Footer A or Header/Footer B. You may have two different headers or footers in the document. After selecting A or B, select the frequency with which you want the header or footer to appear. An additional selection, EDIT, also is displayed. Use EDIT to revise a header or footer. After selecting the frequency, enter the text for the header or footer just as you would any other text. When finished entering text, press F7 two times to return to the document. Headers and Footers do not show on the screen. To view their appearance on the printed page, use the VIEW function from the print menu.

Page Numbering

To add page numbers to the document, access the Format menu by pressing **SHIFT** and **F8**. Select **`2 - Page`** and then **`Page Numbering`** from the Page menu. Select **`4 - Page Number Position`** from the Page Numbering menu. WordPerfect then will display an image showing page number placement options. Selecting 6, for example, will insert the page number at the bottom center. After selecting the position, press F7 until you get back to the document.

Fonts

Font refers to the design of the printed characters. It is possible to change the font and the text attribute. To change the font, position the cursor in the document and press the **CTRL** and **F8** keys. Select **`4 Base Font`**. A list of fonts is displayed. To view available fonts, scroll through the fonts using the arrow keys. Select the desired font by placing the cursor on the font and pressing the ENTER key. Enter the pitch size when prompted and press the ENTER key

To change the text attribute, block the text using **ALT** and **F4** or **F12**. Press the CTRL and F8 keys. The following menu is displayed: **`Attribute: 1 Size; 2 Appearance: 0`**. Select 1 or 2. The size menu contains such choices as superscript, subscript, and a number of selections for varying the size of the printed text. The appearance menu contains selections such as boldface, underline, and italicize. Select the appropriate size and/or appearance number from the menu.

The ability to change fonts depends on what fonts your printer has available. If the printer has only one font, then all documents are printed in that font.

18.5 SPELL CHECKER AND THESAURUS

The Spelling Checker

One of the advantages of using a wordprocessor instead of a typewriter is the ability to check spelling. The WordPerfect spelling checker can be used to check a word, a page, a document, or a block of text. However, the spelling checker will not detect a word that is spelled correctly but used incorrectly. When using the spell checker, you may have to use a separate speller disk. Check the requirements for your particular hardware configuration. Before invoking the spell checker, it is a good idea to first save the file by pressing the F10 key. To bring up the WordPerfect spell checker press the **CTRL** and **F2** keys. The following menu will be displayed:

```
Check: 1 Word; 2 Page; 3 Document; 4 New Sup. Dictionary; 5
Look Up; 6 Count: 0
```

The spelling selections available are as follows:

Word: Checks the word on which the cursor is located.
Page: Checks the page on which the cursor is located.
Document: Checks the entire document.
New Supplementary Dictionary: Changes the supplementary dictionary.
Look Up: Looks up a word in the main dictionary.
Count: Displays a count of the number of words in the block, page, or document.

After the amount of text to spell check has been selected, the WordPerfect spelling checker starts at the top of the text and scans each word until it comes to a misspelling. When the checker encounters a misspelled word, a list of suggested spellings is displayed on the lower half of the screen. If WordPerfect does not have any suggested spellings, the message "Not Found" appears. If one of the suggested spellings is the correct form of the word, press the letter corresponding to the word. The selected word will replace the incorrect word in the document. If none of the suggested spellings is correct the following options are available from the menu displayed at the bottom of the screen:

```
Not Found: 1 Skip Once; 2 Skip; 3 Add; 4 Edit; 5 Look Up; 6
Ignore Numbers: 0
```

Skip Once: Skips this occurrence of the word, but will flag it as an error if it is encountered later in the document.
Skip: Skips all occurrences of the word in the document.
Add: Adds the word to the supplementary dictionary.
Edit: Allows you to edit the word and resume spell checking.
Look Up: Looks up a word in the main dictionary.
Ignore Numbers: Ignores words containing numbers.

When the checking is completed, a count of the number of words is displayed. Press any key to exit the spell checker.

The Thesaurus

WordPerfect has a thesaurus that displays related words (synonyms) and antonyms for words selected from the text or words entered from the keyboard. To use the thesaurus, position the cursor on the word you wish to look up. Press the **ALT** and **F1** keys. WordPerfect displays a list of words that can be used in place of the current word. A menu at the bottom of the screen contains the following options:

```
1 Replace Word; 2 View Doc; 3 Look Up Word; 4 Clear Column: 0
```

Replace Word: Replaces the selected word with a related word or antonym from the list of alternatives by entering the letter corresponding to the word.
View the Document: Allows you to view different parts of the document.
Look Up: Looks up the selected word; displays a list of related words and antonyms.
Clear Column: backs up one level in the thesaurus.

To replace the word selected in the text with an alternative from the list select **`1 - Replace Word`** and then select the letter corresponding to a word in the list. Additional suggestions can be viewed by pressing the letter corresponding to a word in the list.

18.6 FEATURES

Search and Replace

WordPerfect has the ability to search for and replace text strings. A text string is defined as any sequence of characters. The search and replace feature is an excellent way to search for and replace multiple occurrences of words or phrases without having to be concerned about missing one of the occurrences. To begin the search, press the **ALT** and **F2** keys. The search will begin at the current location of the cursor. When prompted, enter "Y" to replace with confirmation or "N" to replace without confirmation. If replace with confirmation is chosen, WordPerfect will stop at each occurrence of the word for confirmation of replacement. At the prompt, → **Srch:**, enter the string to be replaced. Press F2 after the string has been entered. DO NOT press ENTER at this point unless the search includes a hard return. Enter the replacement string at the prompt, **Replace with:**. Press F2. Again, only press the ENTER key if a hard return is to be inserted. The search now begins. If the confirmation option was selected, type "Y" to replace the string or "N" to skip the string each time it is encountered.

Footnotes

It often is necessary to incorporate footnotes into documents. Numbered footnotes automatically are renumbered if there are additions, deletions, or other changes that affect the numbering. To add a footnote to a document, first position the cursor at the location the footnote reference number is to appear. Press the **CTRL** and **F7** keys. The following menu is displayed:

```
1 Footnote; 2 Endnote; 3 Endnote Placement: 0
```

Select **1 Footnote**. From the next menu select **1 Create**. A blank screen will be displayed. Enter the text for the footnote on this screen. When finished, press **F7** (Exit). A footnote reference number automatically is placed in the text at the location of the cursor.

If the footnote needs editing, press **CTRL** and **F7**. Select **1 Footnote**. Then select **2 Edit**. When prompted, enter the number of the footnote that is to be edited and press ENTER. When finished editing, press F7 (Exit) again.

Mail Merge

Mail Merge is a feature that allows you to customize documents by inserting variable data into identical letters or forms. Two files are needed for the mail merge: the primary file and the secondary file. The primary file contains fixed information and markers to indicate where the variable information from the secondary file will be placed. The secondary file contains the variable information to be inserted or merged into the primary file. The most common application for this feature is sending the same letter to different addresses. The primary document contains the letter and the secondary document contains the addresses.

The Secondary File

Creating the secondary file consists of defining the fields that will be inserted into the primary document. A field may consist of a single line of text or multiple lines of text. The fields are combined to form a record. When running the merge option, one document (i.e., letter) is created from each record in the secondary file. To define the fields, start with a blank screen. Type the text that is to appear in the first field. After entering the text, press the **F9** key. This will mark the entry with the end field marker, **{END FIELD}**. Repeat this process for each field in the first record. Once the first record is complete, the end of the record should be marked. To insert the end of record marker, press the **SHIFT** and **F9** keys. The following menu is displayed:

```
1 Field; 2 End Record; 3 Input; 4 Page Off; 5 Next Record;
6 More: 0
```

Select **2 End-Record** from the menu. This will insert an **{END RECORD}** marker and a double dashed line marking the end of the record. Repeat this process for each record you want included in the secondary file. Use **F10** to save this file when finished. Figure 18.6 shows the fields in three records of a secondary file.

```
Mr.{END FIELD}
Mac{END FIELD}
Smith{END FIELD}
123 Campus Street
Oxford, Ohio 44444{END FIELD}
{END RECORD}
================================================================
Mrs.{END FIELD}
P.C.{END FIELD}
Jones{END FIELD}
456 Happy Court
Oxford, Ohio 45056{END FIELD}
{END RECORD}
================================================================

{END FIELD}
Xain{END FIELD}
Yuning{END FIELD}
9876 International Blvd.
New York, NY 11111{END FIELD}
{END RECORD}
================================================================
```

FIGURE 18.6 The secondary file.

The Primary File

The primary file consists of a document (i.e., form letter) and the appropriate field markers indicating where the fields from the secondary file are to be inserted. To create the Primary File, start with a blank screen or a previously created document. To insert the field markers, position the cursor at the location in the document where the field should appear. Press the **SHIFT** and **F9** keys. Select **1 Field** from the menu. At the prompt, **Enter Field:**, enter the number of the field that is to appear at this location and press the **ENTER** key. To position field one, place the cursor at field one's location, press **Shift** and **F9**, enter the number of the field, "1," and press the **ENTER** key. The field marker, **{FIELD}1~**, will appear in the document. Continue this process until all the desired markers have been inserted into the primary file. Using F10, save and name this file when finished. Figure 18.7 is an example of a primary document with field markers placed in appropriate places.

Helpful hints:

- Fields may be used multiple times in a document by using the same field number when placing the marker.
- Field markers can be imbedded in the document to insert any variable text.

January 1, 1994

{FIELD}1~ {FIELD}2~ {FIELD}3~
{FIELD}4~

Dear {FIELD}1~ {FIELD}3~:

This is a sample of a form letter used as a Primary file. When the primary and secondary files are merged, the marked fields above will be replaced by the text from the corresponding fields in the Secondary File. Notice fields 1 and 3 above. During the merge of the two documents, the text placed in these fields is repeated. A field can also be imbedded in the text. On the line below, the text from field 2 will appear again, centered on the line.

{FIELD}2~

Sincerely,

Mrs. P. C. Perfect

FIGURE 18.7 The primary file.

The Merge

To merge the two documents, first clear the screen. Press the **CTRL** and **F9** keys and the following menu is displayed:

```
1 Merge; 2 Sort; 3 Convert Old Merge Codes: 0
```

Select **1 Merge** from the menu. When prompted, supply the name of the primary file and press the ENTER key; then supply the name of secondary file and press the ENTER key. After a few seconds, the merged documents will appear on the screen. These merged documents then can be edited, printed, and/or saved to disk.

Adding the Date to the Document

It often is customary to add the date to correspondence and other documents. WordPerfect provides an easy way to add the current date to a document. Place the cursor at the location the date is to be inserted and press the **SHIFT** and **F5** keys. Selection one, **Date Text**, will insert the current system date in the document at the cursor's current position. Selection two, **Date Code**, will insert the current date in the document, however, a code is inserted rather than the date text. Each time the document is retrieved, the date is updated with the current system date. The format of the date can be changed by selecting the third choice, **Date Format**, from the menu.

Undelete

If text is accidentally deleted, it can be recovered by using the undelete function. Press **F1** to access the undelete function. The most recently deleted text will appear at the current location of the cursor and the following menu is displayed:

```
Undelete: 1 Restore; 2 Previous Deletion: 0
```

Select **1 Restore** to restore the text or **2 Previous Deletion** to restore previously deleted text.

Creating Equations

Often in technical documents, it is necessary to include mathematical equations. WordPerfect offers an easy way to include these equations in a document. Place the cursor at the location the formula is to appear and press the **ALT** and **F9** keys. The following menu is displayed:

```
1 Figure; 2 Table Box; 3 Text Box; 4 User Box; 5 Line; 6
Equation: 0
```

Select **`6 Equation`**. From the next menu, select **`1 Create`** to display the equation definition screen shown in Figure 18.8.

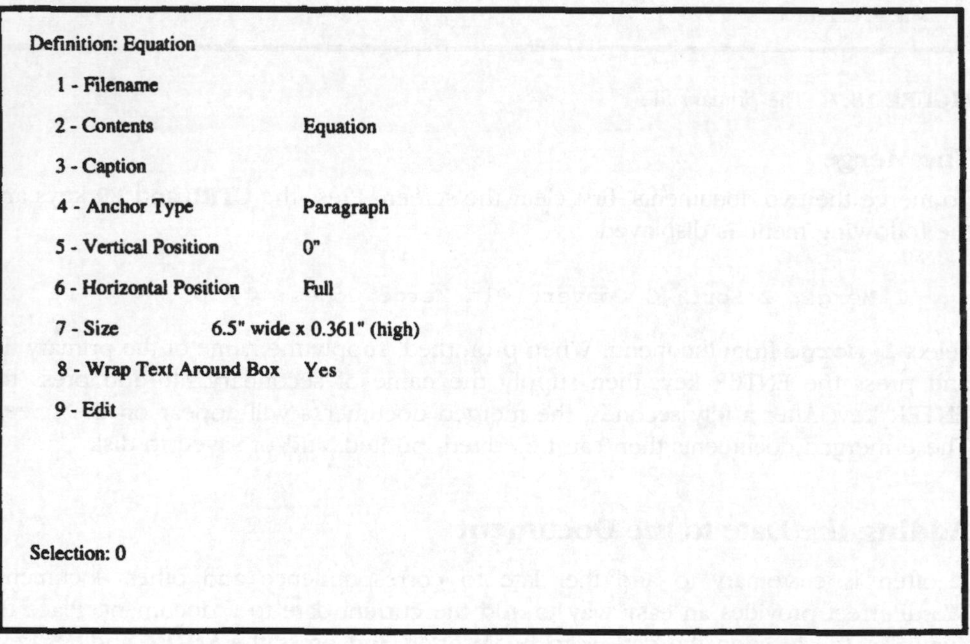

Definition: Equation
1 - Filename
2 - Contents Equation
3 - Caption
4 - Anchor Type Paragraph
5 - Vertical Position 0"
6 - Horizontal Position Full
7 - Size 6.5" wide x 0.361" (high)
8 - Wrap Text Around Box Yes
9 - Edit
Selection: 0

FIGURE 18.8 The equation definition screen.

Select **`9 - Edit`** from the equation definition screen to access the Equation Editor. The equation editor screen is divided into three parts: the Editing window (bottom left screen), the Equation Palette (right side of screen), and the Display window (top left screen). By selecting commands and symbols from the Equation Palette and typing characters from the keyboard, an equation is created in the Editing window and then can be displayed in the Display window as it will appear in the document. The Equation Palette provides the choice of various mathematical symbols, commands, and functions. For example, typing *x* over *y* creates the following equation:

$$\frac{x}{y}$$

To view the equation press the **SHIFT** and **F3** keys or the **F9** key and the equation will appear in the Display window. If an equation such as the following, requires more complex symbols, the equation palette may be used.

$$\mu_A = E(y) = \beta_0 + \beta_1 x_1 + \beta_2 x_2$$

To access the equation palette press **F5**. The **PAGE UP** and **PAGE DOWN** keys may be used to view different palettes. To make a selection from one of the palettes, use the arrow keys to highlight the command or symbol and press the **ENTER** key. When the equation is completed, press **F7** until you return to the document.

The equation feature of WordPerfect has many options. Consult the WordPerfect manual for a more extensive explanation.

Columns and Tables

Often, text presented in a tabular or columnar format enhances the readability of the document. This section demonstrates several options for displaying text in columns or tables.

Cellular Table

A cellular table displays text in a spreadsheet type format using rows and columns. To create a cellular table, move the cursor to the left margin of the line where you want the table to appear. Press the **ALT** and **F7** keys to access the Columns and Tables menu and select **2 `Tables`**. The following menu will appear:

```
Table: 1 Create; 2 Edit: 0.
```

Select **1 `Create`** to create a new table. When prompted, enter the number of columns and rows desired in the table. Figure 18.9 displays a table containing five columns and four rows.

FIGURE 18.9 A cellular table.

At this point, the table is displayed on the screen in **table editing** mode. Note that there are two modes for a table on the screen. In **normal editing** mode, text may be entered in the cells and all WordPerfect features may be utilized. In **table editing** mode, the physical appearance of the table may be changed, but text cannot be entered. To exit table editing mode and access normal editing mode, press the **F7** key.

In table editing mode, it is possible to alter the size and appearance of the table. To change the width of the columns use the **CTRL** and **left** or **right arrow keys**. To add and/or delete rows and columns, use the **Insert** or **Delete** keys. To change the borders of the table and/or cells, select **Lines** from the menu. Consult the WordPerfect manual for a comprehensive explanation of the available options.

Newspaper Columns

Newspaper columns allow text to flow continuously from column to column, as it does in a newspaper article. To define and create newspaper columns, first place the cursor at the location the columns are to begin. Press the **ALT** and **F7** keys to access the Columns and Tables menu and select **`1 Columns`**. The following menu will appear:

```
Columns: 1 On; 2 Off; 3 Define: 0
```

The first step is to define the columns. Select **`3 Define`** and the column definition screen (Fig. 18.10) will appear.

Text Column Definition

1 - Type **Newspaper**

2 - Number of Columns **2**

3 - Distance Between Columns

4 - Margins

Column	Left	Right	Column	Left	Right
1:	1"	4"	13:		
2:	4.5"	7.5"	14:		
3:			15:		
4:			16:		
5:			17:		
6:			18:		
7:			19:		
8:			20:		
9:			21:		
10:			22:		
11:			23:		
12			24:		

Selection: 0

FIGURE 18.10 Column defintion screen.

The menu options are defined as follows:

Type: establishes the type of columns (Newspaper, Parallel, or Parallel with Block protection).

Number of Columns: sets the number of columns

Distance between Columns: sets the distance between columns; defaults to .5″ unless otherwise specified.

Margins: the left and right margins are adjusted automatically depending on the number of columns entered. These default values may be changed if desired.

After the columns are defined press the **Enter** key to save the settings.

Select the **Type** of columns (Newspaper) preferred and enter the **Number of Columns** desired. The margins and the distance between columns automatically will be adjusted; however, these can be changed if desired. To enable columns, select **`1 On`** from the columns menu. As text is typed, it will fill the first column. Upon reaching the bottom of the column, the cursor automatically advances to the top of the next column. When the bottom of a page is reached the cursor automatically advances to the top of the first column on the next page and text continues in column form. To move to the top of the next column before reaching the end of the current column, press the **CTRL** and **ENTER** keys.

When finished entering text in column mode, it is necessary to turn the column feature off. Press the **ALT** and **F7** keys to access the Columns and Tables menu and select **1 Columns**. To turn the column mode off, select **2 Off** from the Columns menu. To repeat the same column feature elsewhere in the document, select **1 On** from the columns menu.

Parallel Columns

Parallel columns are used when unequal blocks of text are to be printed side by side on a page. The steps for setting up parallel columns are identical to those for setting up Newspaper columns (see instructions and figure above) with one exception; when selecting the type of column, select **Parallel** instead of Newspaper. Complete the definition process by entering the number of columns, distance between columns, and the margins, as necessary. After the columns are defined press the **Enter** key to save the settings. To enable columns select **1 On** from the columns menu.

After enabling the columns, begin to enter text. When the end of a column is reached, press the **CTRL** and **ENTER** keys simultaneously. This will position the cursor at the top of the next column so the next entry can be made. When the end of the last column is reached, again press the **CTRL** and **ENTER** keys simultaneously; the cursor will be placed at the beginning of the first column, directly under the first entry. Figure 18.11 is an example of parallel columns.

When finished entering text in column mode, it is necessary to turn off the column feature. Press the **ALT** and **F7** keys to access the Columns and Tables menu and select **1 Columns**. To turn off column mode, select **2 Off** from the Columns menu.

TITLE	DESCRIPTION	SALARY
Systems Analyst	Responsible for the day-to-day management of the system	$50,000
Analyst	Responsible for analysis and design of system enhancements	$40,000
Programmer	Responsible for coding and testing programs	$30,000

FIGURE 18.11 Parallel columns.

DEFINING TERMS

Block: A section of highlighted text.

Cursor: A blinking bar or shaded box on the screen that indicates where the next character will be displayed.

Directory: A table of contents maintained by DOS of the information on a disk.

DOS: An acronym for the Disk Operating System.

Extension: An optional one to three character name that may be added to the file name. The file name and extension are separated by a period. The extension may consist of letters of the alphabet, numbers one through nine, and/or the special characters $ # & @ ! % ( ) - { } ‘` _ ~^.

File Name: The first part of a name given to a file stored on a disk. The file name may be one to eight characters in length and consist of letters of the alphabet, numbers one through nine, and/or the special characters $ # & @ ! % ( ) - { } ‘` _ ~ ^.

Font: The shape of the characters in a document.

Pitch: The size of the font.

Prompt: Appears on the screen after DOS has been loaded into memory; it is usually a letter followed by a greater than sign (>). The letter indicates the current disk drive. DOS commands are entered following the prompt.

Subdirectory: Additional directories created by the user on the disk.

Wordwrap: A feature that senses when you have gone beyond the right margin and automatically moves text and the cursor to the next line.

19

WordPerfect for Windows 5.2

David L. Hartung Jr.
Miami University

19.1 WORD PROCESSING CONCEPTS

Word processing, one of the earliest applications developed for microcomputers, is the process of creating documents using a computer and a word processing software package. The advantages of word processing over typing include the following:

- The ability to make multiple revisions to documents without having to retype the entire page.
- The ability to store documents for easy filing and retrieval.
- The ability to correct spelling errors.
- The ability to change the physical appearance of the text.

WordPerfect is one of the most popular word processing software packages in use today.

19.2 SYSTEM OVERVIEW

Major Features

WordPerfect offers the following major features:

- Creating, deleting, and changing text.
- Controlling the physical appearance of the text by changing the font and the point size.
- Support for a wide variety of printers.
- A spell checker and thesaurus.
- Automatically timed backup of the current document.
- An option that allows for the creation of tables and multiple columns of text.

0-8493-2530-7/96/$0.00+$.50

- A merge capability for creating form letters and mailing labels.
- The ability to construct mathematical formulas.

Starting/Loading WordPerfect

To access WordPerfect for Windows, you must first initiate the windows program. Assuming that windows is installed on the C drive, use the following steps to load the software onto the computer.

1. Start DOS by turning on the computer.
2. At the C prompt type **CD\WINDOWS** to change to the WINDOWS directory and press the ENTER key.
3. Type **WIN** and press the ENTER key.

The windows software is loaded into the memory of the computer. A screen similar to Figure 19.1 will appear.

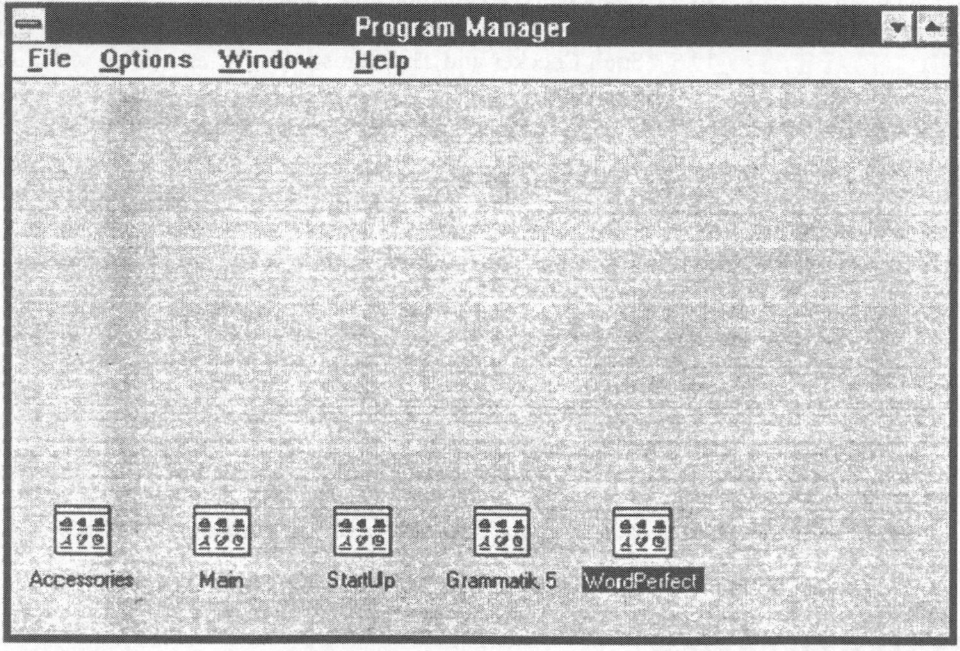

FIGURE 19.1 Initial Windows screen.

Mouse operations needed to use WordPerfect for Windows include the following:

Pointing: Moving the mouse to position a pointer over items; the mouse pointer is shaped like an arrow when pointing to selections in the menu or scroll bars and like an I-beam when pointing to text.
Clicking: Pressing then releasing the mouse button.
Double-clicking: Quickly pressing the mouse button twice then releasing.
Dragging: Pressing and holding down the mouse button while moving the mouse.

To access WordPerfect for Windows software, position the mouse pointer on the **WordPerfect** icon and double-click the left mouse button. The screen shown in

Figure 19.2 will appear. Point to the **WPWin** icon and double-click to load the WordPerfect for Windows software into memory.

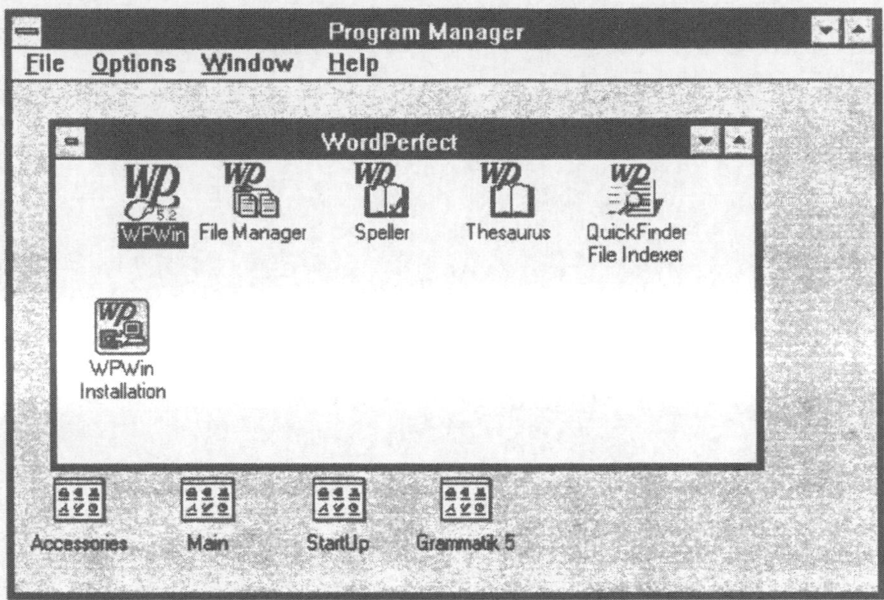

FIGURE 19.2 WordPerfect icon.

The WordPerfect Screen

The WordPerfect screen is designed to simulate a blank sheet of paper. The blinking line at the top of the screen is the cursor. The parts of the screen, starting from the top, are shown in Figure 19.3.

Title Bar: Contains the title of the application window and the name of the current document.
Menu Bar: Contains the Main Menu selections used in creating a document.
Status Bar: Displays the cursor's current page number, line number, and position.

When you first begin to use WordPerfect, you do not need to be concerned with initial settings for margins, page size, tab stops, etc. These values have been preset. Later in the chapter you will learn how to change these settings. For a complete list of the default settings, please refer to the appropriate appendix in the WordPerfect manual.

Executing WordPerfect Commands

Commands in WordPerfect for Windows may be executed by using the mouse or the keyboard. To use the mouse to select items, position the mouse in the menu bar until it points to the desired selection; when the arrow touches the desired selection, click the left mouse button.

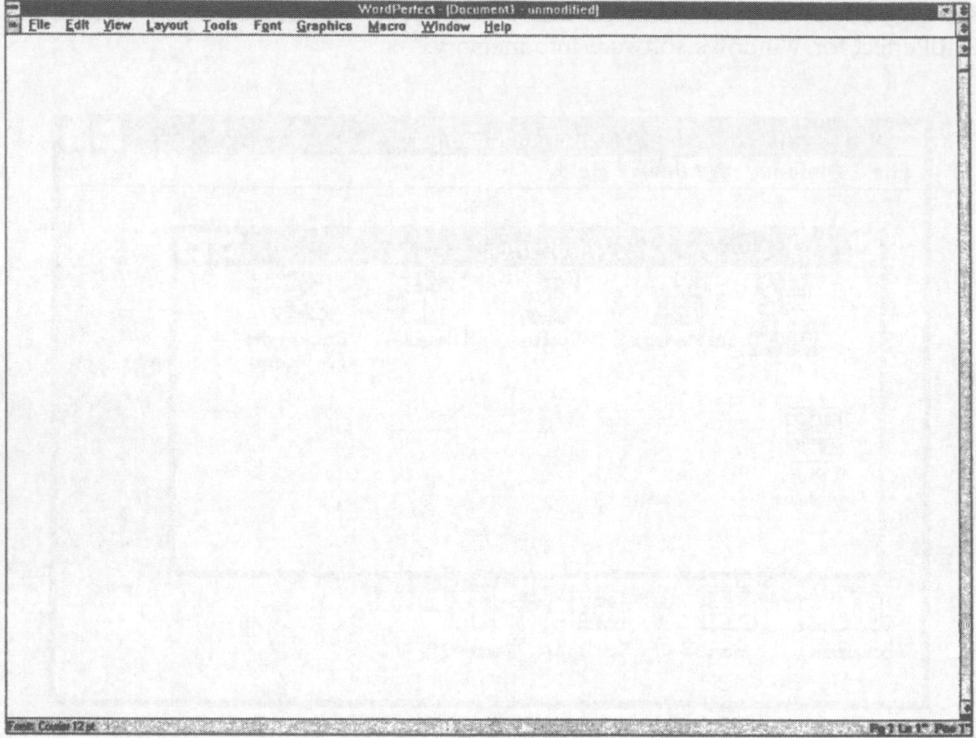

FIGURE 19.3 WordPerfect screen.

Another way to execute WordPerfect commands is to use the keyboard to select menu items by pressing the ALT key in conjunction with the underlined letter from the menu.

If you accidentally get into a menu, exit the menu by positioning the I-beam anywhere inside the document area and clicking the left mouse button or pressing the ESC key on the keyboard.

Cursor Movement

Moving around the document can be done using the cursor movement keys alone or in combination with other keys or using the mouse. Below is a summary of keys that control cursor movement.

Keystroke	Cursor Movement
Up Arrow	One line up
Down Arrow	One line down
Left Arrow	One character to the left
Right Arrow	One character to the right
CTRL and Left Arrow	One word to the left
CTRL and Right Arrow	One word to the right
END	End of the line
HOME	Beginning of the line

19.3 ELEMENTARY COMMANDS/FUNCTIONS

Entering Text

To create a document once you have accessed the WordPerfect for Windows screen, simply begin to type. Upon reaching the right margin, WordPerfect for Windows automatically advances the cursor to the beginning of the next line. This word processing feature is known as **wordwrap**. Do not press the ENTER key at the end of the line to advance to the next line. ENTER is used to end a paragraph or to insert a blank line. When the text wraps around to the next line, a soft return code, SRt, is placed at the end of the line. Pressing the ENTER key places a hard return code, HRt, at the end of the line. SRt and HRt are hidden codes that are discussed in the section on Reveal Codes. As you will see, some WordPerfect for Windows functions depend on the presence of the soft return and hard return codes.

Simple Editing

Editing is one of the most powerful advantages of using a word processor. Once a document has been created, it can be continually improved through revision. Characters can be deleted from or inserted into the document. To delete a character, use either the **Backspace** key or the **Delete (DEL)** key. The backspace key erases the character immediately preceding the cursor. The delete key erases the character in front of the cursor. Text may be added to the document in one of two modes: INSERT or TYPEOVER. In INSERT mode, text is added at the current location of the cursor and text following the cursor automatically is pushed to the right. TYPEOVER mode replaces existing text with new text. To activate the TYPEOVER mode, press the **Insert (INS)** key and the status line will display **Typeover** in the lower left-hand corner. Pressing the Insert key again will deactivate the typeover mode and the Typeover display will disappear from the status line.

Moving, Copying, and Deleting Blocks of Text

In addition to the simple editing described above, blocks of text also may be edited. To edit a block of text, first mark the block of text that is to be edited. To mark the text, place the mouse pointer in front of the first character in the block. Holding down the left mouse button, drag the mouse pointer across the text. When the text is **blocked**, it appears in reverse video (highlighted). Release the mouse button. After the text is blocked, move the mouse to the Menu Bar and click on **Edit**. The Edit Menu shown in Figure 19.4 will be displayed.

To move or delete blocked text, select **Cut** from the Edit Menu. The text disappears from the screen and is stored in temporary memory. To move the text, position the cursor at the location the text is to reappear, click on **Edit** from the Main Menu, and select **Paste**; the text reappears in the new location. To copy text, select **Copy** from the Edit Menu. A copy of the text remains on the screen and a copy also is stored in temporary memory. Position the cursor at the location the text should appear again, click on **Edit** from the Main Menu, and select **Paste**. A copy of the text is inserted at the new location.

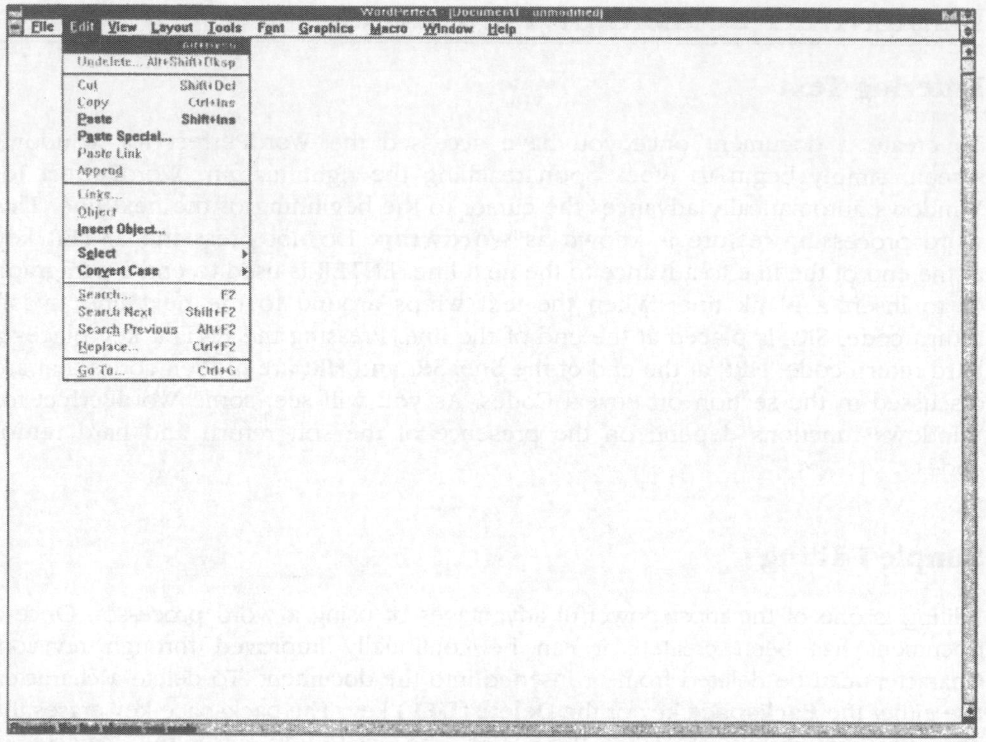

FIGURE 19.4 Edit menu.

Saving Your File

WordPerfect for Windows has two save commands: **File Save As** and **File Save**. Use the File Save As command the first time you save and name a file or any time you want to change the file name or directory. Use the File Save to save the file under the old name, replacing the old document contents with the new document contents. To SAVE AS, click on **File** from the menu bar. Move the mouse pointer to **Save As** and click. The dialog box shown in Figure 19.5 appears.

The current directory will appear below the Save As text box. To change the directory, move the mouse pointer to one of the directory selections and click. Enter the document name in the Save As text box. A document name can have up to eight characters followed by an optional three character extension. If an extension is used, a period must be placed between the name and the extension. Once the directory and name are entered, click on **Save** and the file is saved to the corresponding directory.

To SAVE a document, click on **File** from the Menu Bar. Move the mouse pointer to **Save** and click. The current version of the document is saved, replacing the old document.

Printing a Document

To print a document, move the mouse pointer to **File** on the Main Menu and click the left mouse button one time. Move the pointer to **Print** and click the left mouse button. The Print dialog box appears as shown in Figure 19.6.

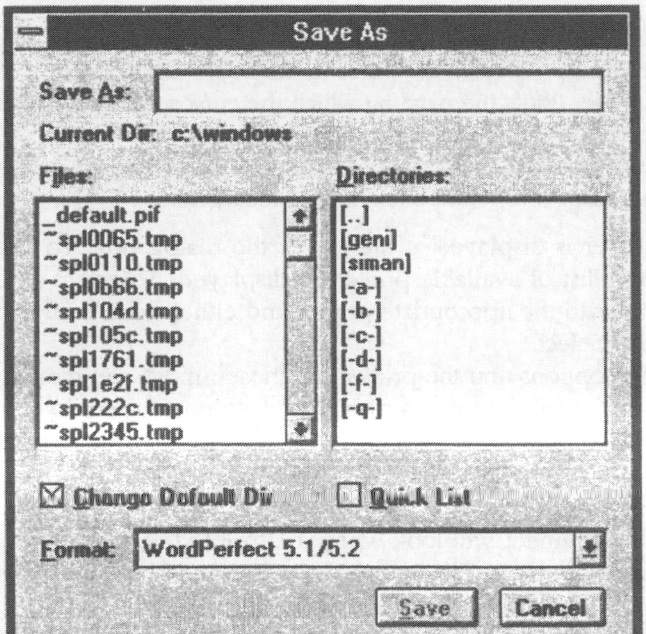

Save As

Save As:

Current Dir: c:\windows

Files:

_default.pif
~spl0065.tmp
~spl0110.tmp
~spl0b66.tmp
~spl104d.tmp
~spl105c.tmp
~spl1761.tmp
~spl1e2f.tmp
~spl222c.tmp
~spl2345.tmp

Directories:

[..]
[geni]
[siman]
[-a-]
[-b-]
[-c-]
[-d-]
[-f-]
[-q-]

Change Default Dir

Quick List

Format: WordPerfect 5.1/5.2

Save

Cancel

FIGURE 19.5 Save As dialog box.

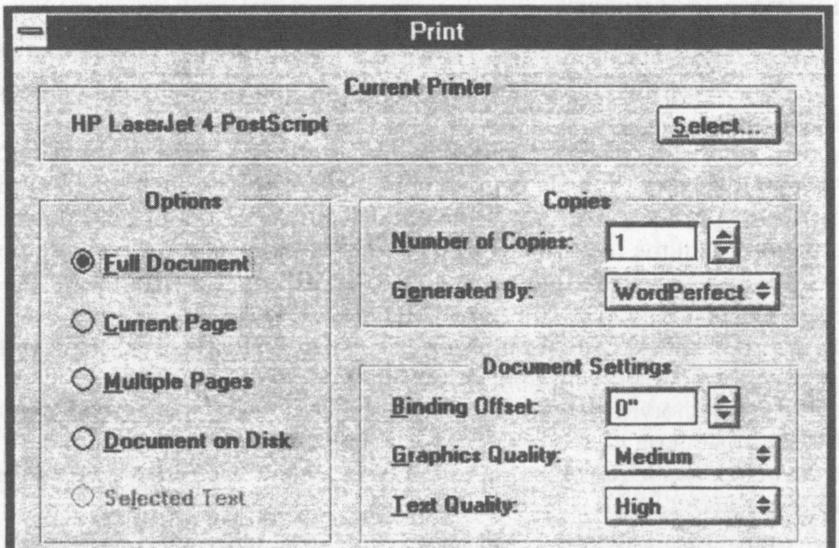

Print

Current Printer

HP LaserJet 4 PostScript

Select...

Options

Full Document

Current Page

Multiple Pages

Document on Disk

Selected Text

Copies

Number of Copies: 1

Generated By: WordPerfect

Document Settings

Binding Offset: 0"

Graphics Quality: Medium

Text Quality: High

Initialize Printer...

Print

Close

FIGURE 19.6 Print dialog box.

The following selections are available to print a document:

Full Document: Prints all the pages in the active document.
Current Page: Prints the page on which the cursor currently is located.
Multiple Pages: Prints a specified range of pages.
Document on Disk: Prints the document from a file stored on disk.
Selected Text: Prints marked text in the document.

The current printer is displayed at the top of the dialog box. To change the printer, click on Select. A list of available printers is displayed. To select a new printer move the mouse pointer to the appropriate printer and either double-click or click one time and then choose Select.

Once the print options and the printer are chosen move the mouse pointer to **Print** at the bottom of the dialog box and click. The document is sent to the specified printer and a dialog box containing the status of the print job is displayed on the screen.

Print Preview

To view how a document will look when it is printed, use the Print Preview option. Place the mouse pointer on **File** in the Menu Bar and click the mouse button. Select **Print Preview** from the File Menu. This selection produces a screen that shows you what the document will look like when it is printed. At this point a number of options are available. By choosing one or more options from the Button Bar on the left of the screen, it is possible to view full pages, facing pages, previous pages, or the next page. In addition, portions of the document may be viewed by selecting 100%, 200%, or by "zooming" in and out on a page. After you are satisfied with the appearance of the document, select **Print** from the Button Bar to produce a hard copy of the document or **Close** to exit Print Preview and return to the document.

Retrieving a Document

To retrieve a document from a disk, move the mouse pointer to **File** and click the mouse button. Move the mouse pointer to **Retrieve** and press the mouse button. The dialog box shown in Figure 19.7 will appear.

Disk files are displayed in the Files box. Locate the the name of the file you want to retrieve in this box. If the file name is not visible, scroll through the files (place the mouse pointer on one of the arrows to the left of the box and hold down the left mouse button) until the file name appears in the window. Move the mouse pointer to the file name and click the mouse button to **highlight the file**. Move the mouse pointer to **Retrieve** and click the mouse button to retrieve the file from disk.

Note: Exercise caution when retrieving files. WordPerfect will not remove the current file from memory (screen) until you tell it to do so. If you retrieve a file when another file is in memory (screen), the files will be combined. It is a good idea to close a file before moving on to another document. To close a file, move the mouse pointer to **File** and click the mouse button. Save the document if desired. Select **Close** from the File Menu and a blank screen appears.

Exiting WordPerfect for Windows

To exit WordPerfect for Windows, move the mouse pointer to **File** and click the mouse button. From the File Menu select **Exit**. This will exit the WordPerfect software.

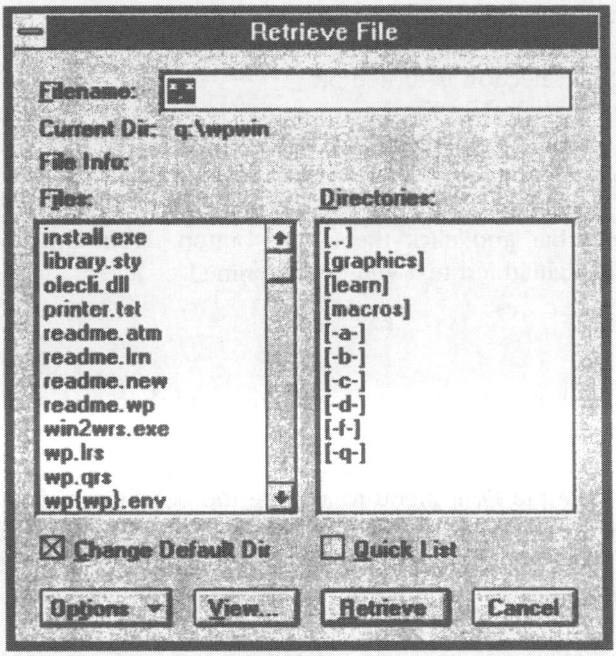

FIGURE 19.7 Retrieve File dialog box.

19.4 FORMATTING THE DOCUMENT

Formatting features are used to enhance the appearance of the document. Characters can be underlined or boldfaced. Text can be aligned differently by setting tab stops, changing margins, or using indentation. Page numbers, headers, footers, and fonts may all be used to enhance the appearance of the document.

Reveal Codes

WordPerfect for Windows uses hidden codes, called reveal codes, to determine how the text looks on the screen and when it is printed. All WordPerfect for Windows formatting features have an associated reveal code. To view these codes move the mouse pointer to **View** and click the mouse button. Select **Reveal Codes** from the View menu. The screen is divided in half displaying the document on top and the reveal codes on the bottom. Misplaced or unwanted codes can cause problems with a document. Fortunately, codes can be edited. Unwanted codes can be deleted by placing the cursor on the code and pressing the DELETE key. Misplaced codes can be moved or copied by following the rules for moving blocks of text described in the section on Moving, Copying, and Deleting Blocks of Text. Repeat the instructions to turn off the reveal codes.

Underlining Text

Underlining text may be accomplished two ways: underlining text as you type or underlining text that already has been entered. To underline as you type, move the mouse pointer to the **Font** Menu selection and click the left mouse button. Select **Underline** from the Font Menu to activate the underline feature. Enter the text. As you type the text appears underlined. When finished entering the text, once again select

Underline from the Font Menu to turn off the underline feature. A check mark appearing next to a menu item indicates the feature has been activated; the check mark disappears when the selection is turned off.

Underlining text that already has been entered is a two step process. First, **Block** the desired text by placing the cursor at the beginning of the text and holding down the mouse button while dragging the mouse pointer over the text. When all of the text to be underlined is highlighted, release the mouse button. Move the mouse pointer to **Font** on the Menu Bar and click the mouse button. From the Font Menu select **Underline** and all highlighted text will be underlined.

Bolding Text

As with underlining, bolding text also may be accomplished two ways: boldfacing as you type or boldfacing text that already has been entered. To boldface as you type, move the mouse pointer to the **Font** Menu selection and click the left mouse button, then select **Bold** from the Font Menu to activate the boldface feature. Enter the text. As you type the text appears in boldface. When finished entering the text, once again select Bold from the Font Menu to turn off the boldface feature. A check mark appearing next to the word Bold on the menu indicates the feature has been activated and disappears when it is turned off.

Boldfacing text that already has been entered is a two-step process. First, **Block** the text by placing the cursor at the beginning of the desired text and holding down the mouse button while dragging the mouse pointer over the text. When all of the text to be boldfaced is highlighted, release the mouse button. Move the mouse pointer to **Font** on the Menu Bar and click the mouse button. From the Font Menu select **Bold** and all highlighted text will be boldfaced.

Centering Text

Centering text may be accomplished two ways: centering as you type or centering text that already has been entered. To center text as you type, move the mouse pointer to the **Layout** Menu and click the left mouse button. Select **Line** from the Layout Menu and then select **Center**. The cursor will move to the center of the screen. The text is automatically centered as you type. Press ENTER to turn off the centering feature.

To center text that already has been entered block the word or phrase that is to be centered, move the mouse pointer to the **Layout** Menu and click the left mouse button. Select **Line** from the Layout Menu and then select **Center**. The blocked text will appear centered on the page.

Indenting Text

With WordPerfect for Windows, the following three types of indentation may be done:

1. Indenting the first line of a paragraph.
2. Indenting a paragraph from the left margin.
3. Indenting a paragraph from both the left and right margins.

To indent the first line of a paragraph, place the cursor at the beginning of the line and press the **TAB** key. Each time the TAB key is pressed the line will shift one tab stop from the left margin (Tab stops are preset every 0.5 inches.) To indent an entire paragraph, place the cursor at the beginning of the paragraph and move the mouse pointer to **Layout** and click the left mouse button. Select **Paragraph** from the Layout Menu and then select **Indent**. This will indent the paragraph one tab stop from the left margin. Each time this procedure is repeated, the paragraph is indented an additional tab stop. To indent a paragraph from both the left and right margins, also known as double indenting, select **Paragraph** from the Layout Menu and then select **Double Indent**. This will indent the paragraph one tab stop from both the left and right margins. As with indenting, each time this procedure is repeated, the paragraph is indented an additional tab stop from both margins.

Indenting shows the importance of the wordwrap feature. The indent and double indent functions work only from the position indentation is turned on until it reaches a hard return. If a hard return is placed at the end of every line, each line would have to be indented individually.

The Format Menu

All the topics discussed in this section are located within the **Layout** Menu. Some of the topics are found under the **PAGE** commands and others are found under the **LINE** commands. The difference between these commands is that page commands affect the entire page while line commands affect only that part of the document where they are set. For page commands to work properly they must appear in the reveal codes **before** the line command reveal codes. Figure 19.8a shows the **Layout** menu and the Line submenu and Figure 19.8b shows the Layout and the Page submenu.

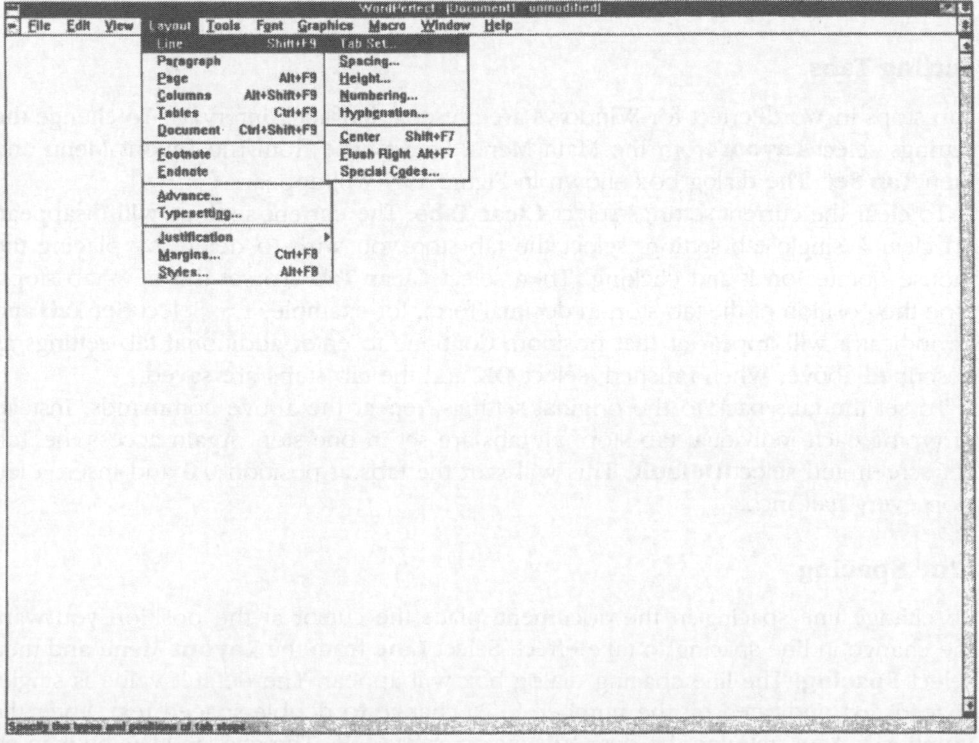

FIGURE 19.8a Layout with Line submenu.

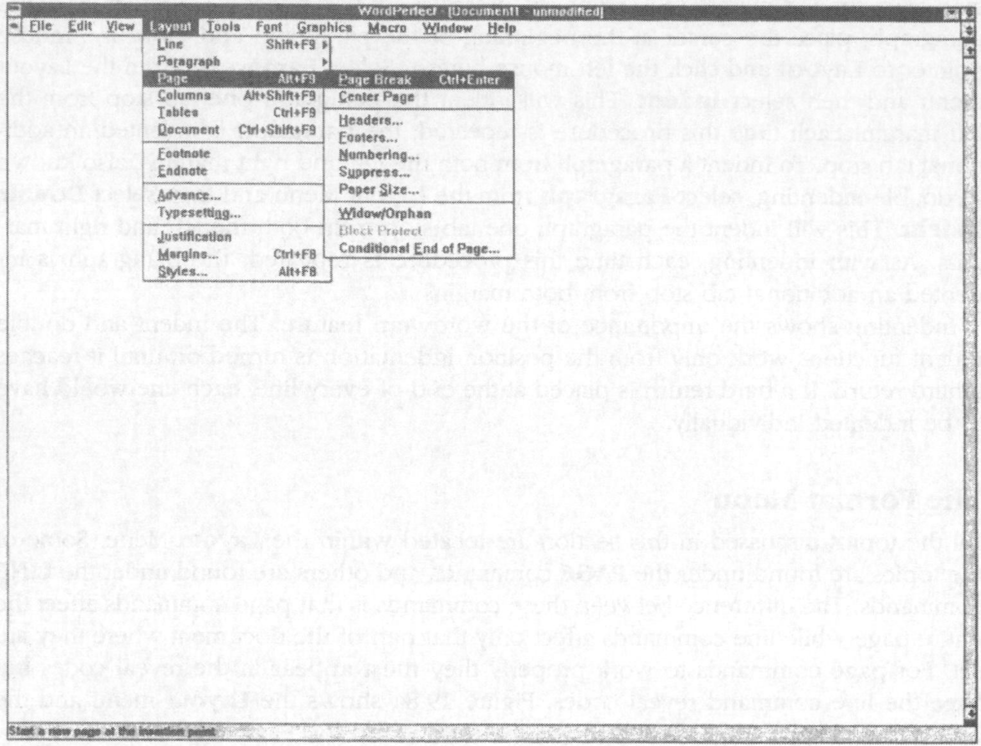

FIGURE 19.8b Layout menu with Page submenu.

Setting Tabs

Tab stops in WordPerfect for Windows are preset at half-inch intervals. To change the settings select **Layout** from the Main Menu. Select **Line** from the Layout Menu and then **Tab Set**. The dialog box shown in Figure 19.9 will appear.

To clear the current settings select **Clear Tabs**. The current settings will disappear. To clear a single tab setting, select the tab stop you wish to delete by placing the mouse pointer on it and clicking. Then select **Clear Tab**. To set the new tab stops, type the position of the tab stop in decimal form, for example, 1.5. Select **Set Tab** and an indicator will appear at that position. Continue to enter additional tab settings as described above. When finished, select **OK** and the tab stops are saved.

To set the tabs back to the original settings, repeat the above commands. Instead of typing each individual tab stop, all tabs are set in one step. Again access the Tab Set screen and select **Default**. This will start the tabs at position 0.0 and insert a tab stop every half inch.

Line Spacing

To change line spacing in the document, place the cursor at the position you want the change in line spacing to take effect. Select **Line** from the **Layout** Menu and then select **Spacing**. The line spacing dialog box will appear. The default value is single-spaced text, indicated by the number 1. To change to double-spaced text, enter the number 2. After entering the desired spacing, select **OK**. The chosen spacing is in effect until it is changed.

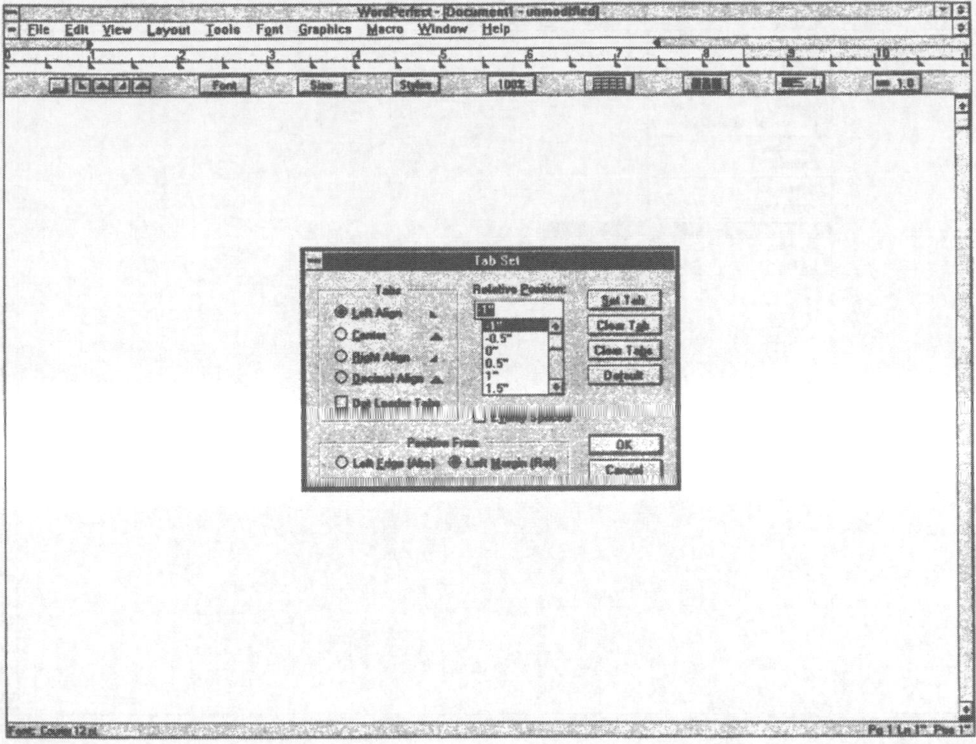

FIGURE 19.9 Tab Set dialog box.

Justification

Full justification is the default in WordPerfect for Windows. This means the text is evenly spread out between the right and left margins. Text that is left-justified has an even left margin and a ragged right margin. Center justification centers the text on the page. Text that is right justified has a ragged left margin and an even right margin. To change justification, access the **Layout** Menu and select **Justification**. The menu shown in Figure 19.10 is displayed. After selecting the desired justification, you are returned to the document.

Widow/Orphan Protection

A Widow occurs when the first line of a paragraph appears by itself on the last line of a page. An Orphan occurs when the last line of a paragraph appears by itself on the first line of a page. To avoid this situation use the Page command Widow/Orphan Protection. Select **Page** from the **Layout** Menu. Select **Widow/Orphan** to activate this feature.

Setting Margins

To set margins, either left and right or top and bottom, access the **Layout** Menu and select **Margins**. Default margin settings are 1 inch from all sides of the page. To change the margins, place the mouse pointer on one of the margin boxes and press the left mouse button. Enter the new value for the margin using decimal form. Follow this procedure for all margins. When all margins are set, click on **OK** and the margins are set. Changes to the left and right margins will take effect from the position in the document where they are set. Top and bottom margins affect the entire page.

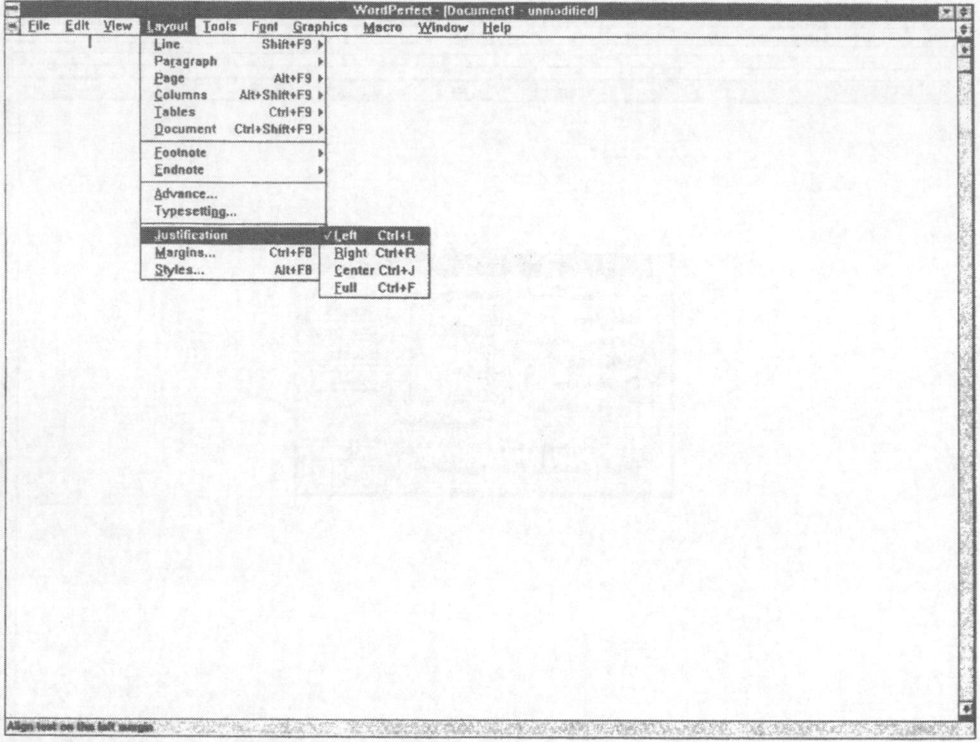

FIGURE 19.10 Justification menu.

Page Headers and Footers

Headers and footers are used to insert text at the top or bottom of a page. The user controls the frequency with which the header or footer is printed. The following choices are available: Discontinue, Every Page, Odd Pages, and Even Pages. To insert a header or a footer in a document, access the **Layout** Menu and select **Page**. The Page Menu contains selections for both the Header and Footer. Upon selecting Header or Footer, a menu is displayed with the choice of Header/Footer A or Header/Footer B. You may have two different headers or footers in the document. Select either A or B. Additional selections include Discontinue, Edit, and Create. Select **Edit** to revise a header or a footer and **Discontinue** to discontinue its use. Select **Create**, then choose the frequency with which the header or footer is to appear by selecting **Placement**. After selecting the frequency of appearance, enter the text for the header or footer on the screen, just as you would any other text. When finished entering text, select **Close**. Headers and Footers do not show on the screen. To view their appearance on the printed page, use the **Print Preview** function under the **File** Menu.

Page Numbering

To add page numbers to a document, access the **Layout** Menu and select **Page**, then select **Numbering**. A dialog box showing page number placement options will be displayed. To select the placement for page numbers, place the mouse pointer on the box beside **Position** and hold down the left mouse button. Drag the pointer to the desired placement for the page numbers and release the button. An image of the page will show a sample of where the number will be placed. When the page number position is set, select **OK**.

Fonts

Font refers to the design of the printed characters. It is possible to change the font and the text attribute. To change the font, position the cursor in the document and select **Font** from the Main Menu. The selection of Font appears again under the Font Menu. Select **Font** again. A list of fonts is displayed. To view available fonts, scroll through the fonts using the arrows to the right of the font box. Select the desired font by pointing to your choice and clicking the left mouse button. After selecting the font, select the pitch size. Examples of the font are displayed in a window below the font selection box.

To change the text attribute, block the text by placing the mouse pointer at the beginning of the text and dragging the mouse across the text. When the text is highlighted, release the mouse button. Select **Font** from the Main Menu. The selection of Font appears again under the Font Menu. Select **Font** again. On the right side of the dialog box are selections for **Size** and **Appearance**. The Size Menu contains such choices as superscript, subscript, and a number of selections for varying the size of the printed text. The Appearance Menu contains selections such as boldface, underline, and italicize. Select the appropriate size and/or appearance by clicking on the box preceding the selection. Click on **OK** when finished.

The ability to change fonts depends on what fonts your printer has available. If the printer has only one font, then all documents are printed in that font.

19.5 SPELL CHECKER AND THESAURUS

The Spelling Checker

One of the advantages of using a wordprocessor instead of a typewriter is the ability to check spelling. The WordPerfect for Windows spelling checker can be used to check a word, a page, a document, or blocks of text. However, the spelling checker will not detect a word that is spelled correctly but used incorrectly. When using the spell checker, you may have to use a separate speller disk. Check the requirements for your particular hardware configuration. Before invoking the spell checker, it is a good idea to first save the file. To bring up the WordPerfect for Windows spell checker, select **Tools** from the Main Menu and select **Speller**. The speller dialog box shown in Figure 19.11 is displayed. Click on **Check** to establish the breadth of the search. The spelling selections available are as follows:

Word: Checks the word on which the cursor is located.
Page: Checks the page on which the cursor is currently located.
Document: Checks the entire document.
To the End of the Document or the Page: Checks the spelling to the end of the document or the end of the page.
Selected Text: Checks a block of text.

After selecting the amount of text to spell check, click on **Start**. The WordPerfect for Windows spelling checker starts at the top of the text and scans each word until it comes to a misspelling. When the checker encounters a misspelled word, a list of suggested spellings is displayed in the suggestion box. If WordPerfect for Windows does

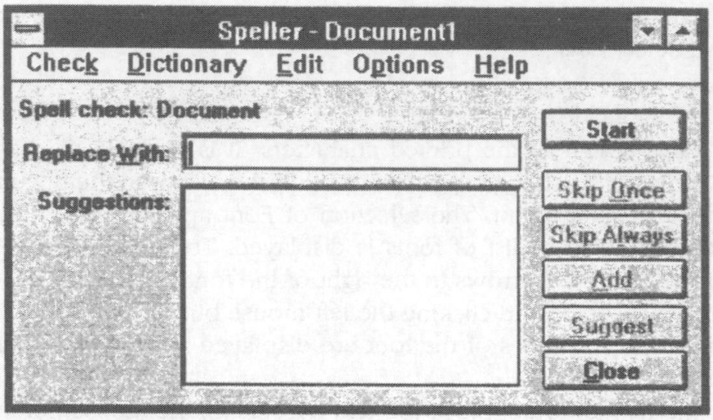

FIGURE 19.11 Spell Check dialog box.

not have any suggested spellings, the message "None Found" appears. If one of the suggested spellings is the correct form of the word, click on the word and then select **Replace**; the selected word will replace the incorrect word in the document. If none of the suggested spellings is correct, the following options are available:

Skip Once: Skips this occurrence of the word, but will flag it as an error if it is encountered later in the document.
Skip: Skips all occurrences of the word in the document.
Add: Adds the word to the supplementary dictionary.
Edit: Allows you to edit the word and resume spell checking.
Ignore Numbers: Ignores words containing numbers.

When the checking is completed, a dialog box appears. Select **Yes** to exit the spell checker or **No** to continue using the spell checker.

The Thesaurus

WordPerfect for Windows has a thesaurus that displays related words (synonyms) and antonyms for words selected from the text or words entered from the keyboard. To use the thesaurus, position the cursor on the word you wish to look up. Select **Tools** from the Main Menu, then select **Thesaurus**. The thesaurus dialog box appears and the selected word is identified in the lower part of the dialog box in the Word window. The selected word also appears above a list of alternative words. Buttons at the bottom of the dialog box include the following options:

Replace: Replaces the selected word with a related word or antonym from the list of alternatives and exits the thesaurus dialog box.
Look Up: Looks up the selected word; displays a list of related words and antonyms; the dialog box remains active.
Close: Exits the thesaurus dialog box.

To replace the word selected in the text with an alternative from the list or look up one of the suggested words from the list, first click on the word to select it. The newly selected word will appear in the Word window. Then click on one of the options above. Words also can be entered into the Word window from the keyboard.

19.6 FEATURES

Search and Replace

WordPerfect for Windows has the ability to search for and replace text strings. A text string is defined as any sequence of characters. The search and replace feature is an excellent way to search for and replace multiple occurrences of words or phrases without having to be concerned about missing one of the occurrences. To begin the search, select **Edit** from the Main Menu, then select **Replace** from the Edit Menu. The dialog box shown in Figure 19.12 is displayed.

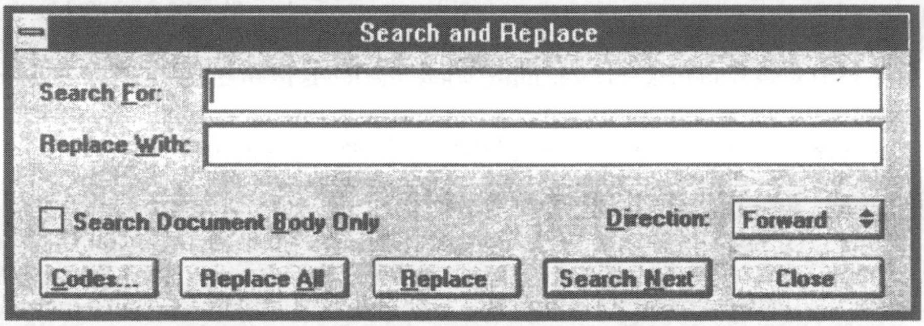

FIGURE 19.12 Search and Replace dialog box.

Enter the string for which you are searching in the space next to **Search for** then enter the string that will replace the original text in the space next to **Replace with**. The search will begin at the current location of the cursor and the default is to search forward through the document. To initiate a backward search, select **Direction** and change from forward to backward. At the bottom of the dialog box, select **Replace All** to replace all occurrences of the word, select **Replace** to replace only one occurrence of the word, or select **Search next** to find and replace the next occurrence of the word.

Footnotes

It often is necessary to incorporate footnotes into documents. Numbered footnotes automatically are renumbered if there are additions, deletions, or other changes that affect the numbering. To add a footnote to a document, first position the cursor at the location the footnote reference number is to appear. Select **Layout** from the Main Menu, then select **Footnote** from the Layout Menu, and finally, select **Create** from the Footnote submenu. A blank screen will be displayed. Enter the text for the footnote on this screen. When finished, select **Close**. A footnote reference number automatically is placed in the text at the location of the cursor.

If a footnote needs editing, select **Edit** after selecting **Footnote** from the Layout menu. When prompted, enter the number of the footnote that is to be edited and select **OK**. When finished editing, select **Close**.

Mail Merge

Mail Merge is a feature that allows you to customize documents by inserting variable data into identical letters or forms. Two files are needed for the mail merge: the primary file and the secondary file. The primary file contains fixed information and markers to indicate where the variable information from the secondary file will be placed. The secondary file contains the variable information to be inserted or merged into the primary file. The most common application for this feature is sending the same letter to different addresses. The primary document contains the letter and the secondary document contains the addresses.

The Secondary File

Creating the secondary file consists of defining the fields that will be inserted into the primary document. A field may consist of a single line of text or multiple lines of text. The fields are combined to form a record. When running the merge option, one document (i.e., letter) is created from each record in the secondary file. Figure 19.13a shows the fields in three records of a secondary file (19.13a) and a form letter (primary document, 19.13b) with field makers indicating where the fields from that file will be inserted.

To define the fields, start with a blank screen. Type the text that is to appear in the first field. After entering the text, select **Tools** from the Main Menu. Select **Merge** from the Tools Menu and then select **End Field**. This will mark the entry with the end field marker **{END FIELD}**. Repeat this process for each field in the first record. Once the first record is complete, the end of the record should be marked. To insert the end of record marker, select **Merge** from the Tools Menu and then select **End Record**. This will insert an **{END RECORD}** marker and a double dashed line marking the end of the record. Repeat this process for each record you want included in the secondary file. Using the **Save As** option, save and name this file when finished.

The Primary File

The primary file consists of a document (i.e., form letter) and the appropriate field markers indicating where the fields from the secondary file are to be inserted (Fig. 19.13b). To create the Primary File, start with a blank screen or a previously created document. To insert the field markers, position the cursor at the location in the document where the field should appear. Select **Tools** from the Main Menu. Select **Merge** from the Tools Menu and then select **Field**. Enter the number of the field that is to appear at this location and select OK. To position field one, place the cursor at field one's location, select **Field** from the Merge Menu, enter the number of the field, "1," and select OK. The field marker, **{FIELD}1~**, will appear in the document. Continue this process until all the desired markers have been inserted into the primary file. Using the **Save As** option, save and name this file when finished.

Helpful Hints

- Fields may be used multiple times in a document by using the same field number when placing the marker.
- Field markers can be imbedded in the document to insert any variable text.

The Merge

To merge the two documents, first clear the screen. Select **Tools** from the Main Menu. From the Tools Menu select **Merge**, then select **Merge** again from the Merge Menu. The dialog box shown in Figure 19.14 appears. Supply the name of the primary and secondary files in the dialog box and select **OK** to run the merge. After a few seconds, the merged documents will appear on the screen. These merged documents then can be edited, printed, and/or saved to disk.

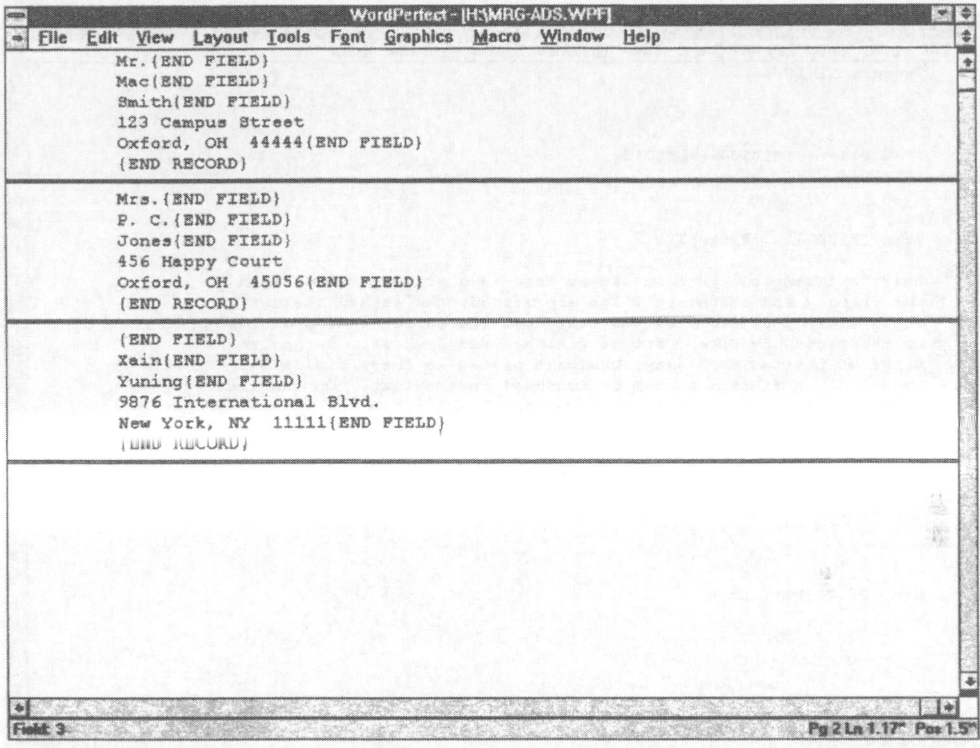

FIGURE 19.13a Merge documents (secondary).

Adding the Date to the Document

It often is customary to add the date to correspondence and other documents. WordPerfect for Windows provides an easy way to add the current date to a document. Place the cursor at the location the date is to be inserted. Select **Tools** from the Main Menu, then select **Date**. Selection one, **Text**, will insert the current system date in the document at the cursor's current position. Selection two, **Code**, will insert the current date in the document, however, a code is inserted rather than the date text. Each time the document is retrieved, the date is updated with the current system date. The format of the date can be changed by selecting the third choice, **Format**, from the menu.

Undelete/Undo

If text is accidentally changed or deleted, the error can be reversed by using the undo or the undelete function. UNDO reverses the most recently completed command or function; UNDELETE restores previously deleted text only. To access either function, select **Edit** from the Main Menu, then select **Undelete** or **Undo**. Selecting Undo automatically reverses your most recent action. However, when selecting Undelete, the most recently deleted text will appear at the current location of the cursor and the dialog box shown in Figure 19.15 will appear. Select **Previous** or **Next** to view previously deleted text; select **Restore** to restore the desired text.

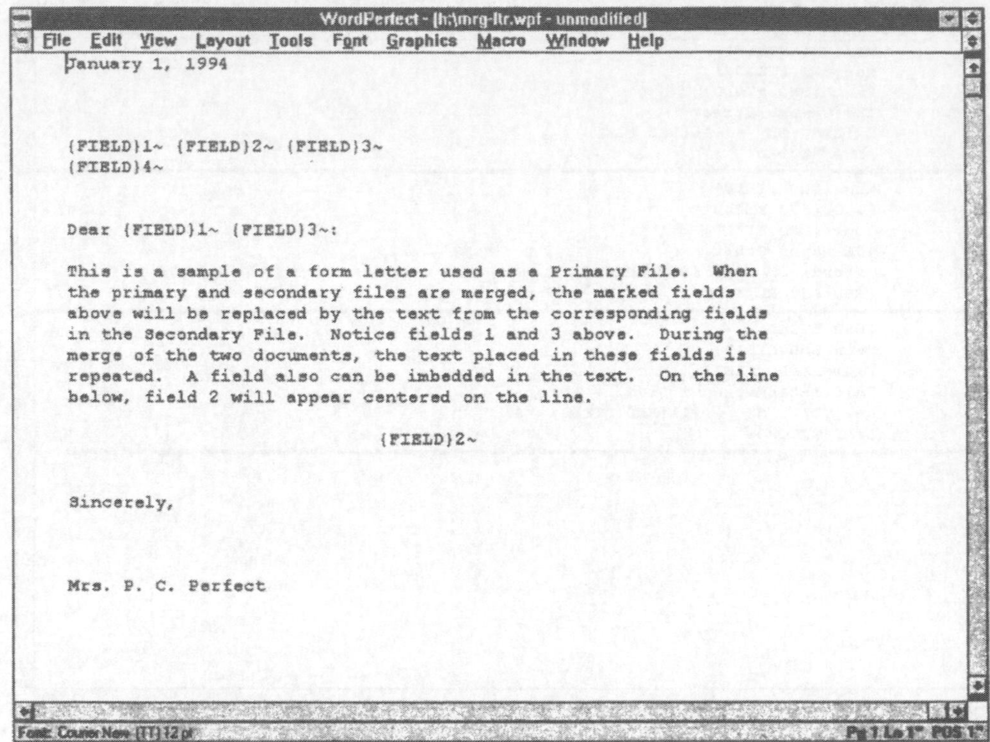


```
January 1, 1994


{FIELD}1~ {FIELD}2~ {FIELD}3~
{FIELD}4~


Dear {FIELD}1~ {FIELD}3~:

This is a sample of a form letter used as a Primary File.  When
the primary and secondary files are merged, the marked fields
above will be replaced by the text from the corresponding fields
in the Secondary File.  Notice fields 1 and 3 above.  During the
merge of the two documents, the text placed in these fields is
repeated.  A field also can be imbedded in the text.  On the line
below, field 2 will appear centered on the line.

                        {FIELD}2~


Sincerely,



Mrs. P. C. Perfect
```

FIGURE 19.13b Merge documents (primary).

Merge
Merge Files
Primary File:
Secondary File:
ASCII Delimited Text (DOS)
OK Cancel

FIGURE 19.14 Merge dialog box.

Creating Equations

Often in technical documents, it is necessary to include mathematical equations. WordPerfect for Windows offers an easy way to include these equations in a document. Place the cursor at the location the formula is to appear and access the **Graphics** Menu in the Main Menu. The menu shown in Figure 19.16 is displayed.

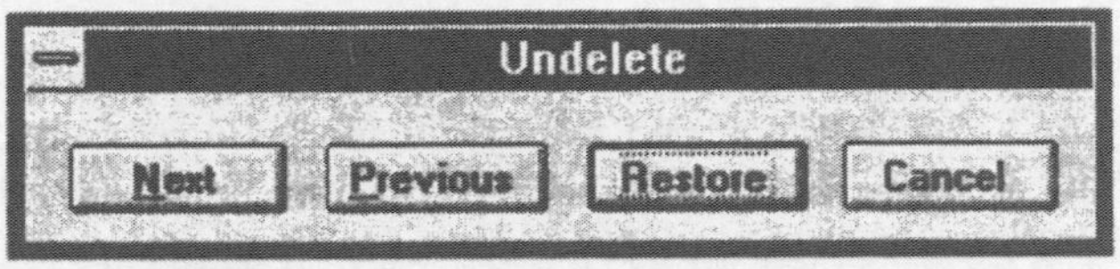

Undelete

Next Previous Restore Cancel

FIGURE 19.15 Undelete dialog box.

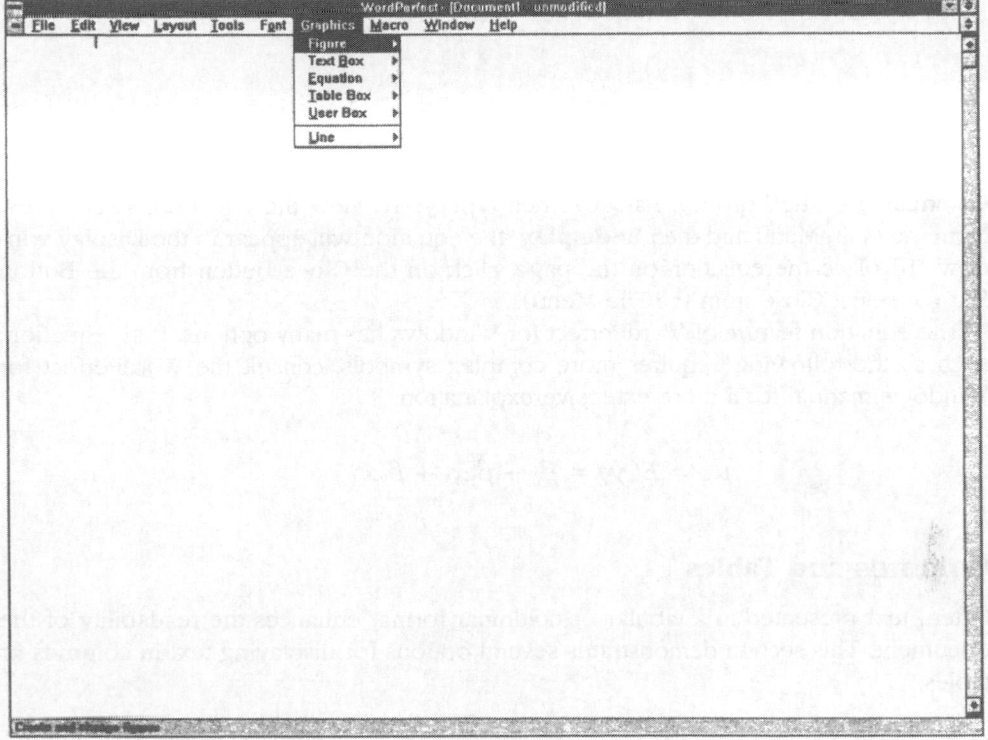

FIGURE 19.16 Graphics menu.

Select **Equation** from the Graphics Menu. After selecting Equation a number of menu options are displayed. These options include the following:

> **Create**: Allows you to create an equation.
> **Edit**: Allows you to edit an existing equation.
> **Position**: Positions the equation on the page.
> **Caption**: Allows you to add a caption to the equation.

Select **Create** or **Edit** to access the Equation Editor. The equation editor screen is divided into three parts: the Equation Palette (left side of screen), the Editing window (top right screen), and the Display window (bottom right screen). By selecting commands and symbols from the Equation Palette and typing characters from the keyboard, an equation is created in the Editing window and then can be displayed in the Display window as it will appear in the document.

The Equation Palette provides the choice of various mathematical symbols, commands, and functions. To choose one of the palettes, place the mouse pointer on the button above the palette window, hold down the left mouse button, and drag the mouse to one of the selections found on the menu. A different palette is displayed for each menu choice. To make a selection from one of the palettes, point to the symbol, function, or command and double-click.

The equation can be displayed in the Display window for viewing. For example, to create and place the following equation on a page, position the cursor where the formula is to appear on the page.

$$\frac{x}{y}$$

From the Main Menu select **Graphics**, then click on **Equation** and double-click on **Create**. The cursor automatically will be positioned in the Editing window. Type **x** from the keyboard, double click on the word **OVER** (the first menu item under Commands in the Equation Palette), then type **y**. To view the equation, select **View** from the Main Menu and then **Redisplay**; the equation will appear in the Display window. To place the equation on the page, click on the **Close** Button from the Button Bar (or select Close from the File Menu).

The equation feature of WordPerfect for Windows has many options. If an equation, such as the following, requires more complex symbols, consult the WordPerfect for Windows manual for a more extensive explanation.

$$\mu_A = E(y) = \beta_0 + \beta_1 x_1 + \beta_2 x_2$$

Columns and Tables

Often, text presented in a tabular or columnar format enhances the readability of the document. This section demonstrates several options for displaying text in columns or tables.

Cellular Table

A cellular table displays text in a spreadsheet type format using rows and columns. To create a cellular table, move the cursor to the left margin of the line where you want the table to appear. Select **Layout** from the Main menu. From the Layout menu select **Tables** and then select **Create** from the Tables submenu. The dialog box shown in Figure 19.17 will appear.

Enter the desired number of rows and columns in the appropriate boxes, then move the mouse pointer to **OK** and click the left mouse button. The table will appear in the document. Figure 19.18 below displays a table containing five columns and three rows.

After the table is created, text may be entered in the cells and all WordPerfect for Windows features may be utilized. To edit the physical appearance of a cell, place the cursor in any cell and access the

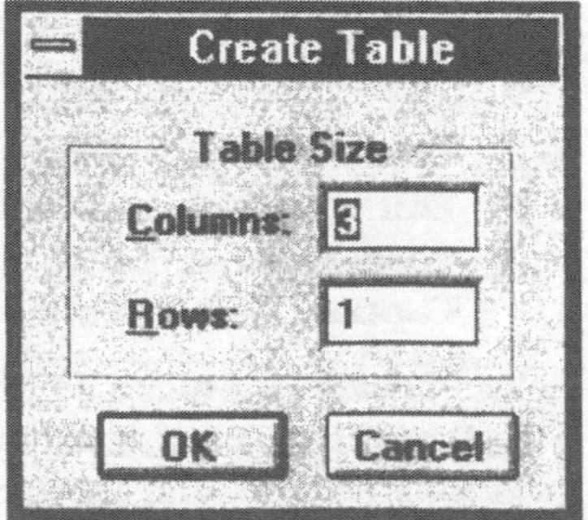

FIGURE 19.17 Create Table dialog box.

FIGURE 19.18 Cellular table.

Tables menu. To edit the physical appearance of the entire table, position the I-beam in the upper left cell (cell A1), hold down the left mouse button, and drag the mouse to the bottom right cell (cell E3); the entire table must be highlighted. The Tables submenu includes the following selections:

> **Options**: Allows you to add rows and columns or change the positioning on the page.
> **Insert and Delete**: Allows you to insert or delete rows and/or columns.
> **Cell, Column, and Row**: Allows you to change the format of the cell, column, or row.
> **Lines**: Allows you to change the borders of the table or cells.

Consult the WordPerfect for Windows manual for a comprehensive explanation of the available options.

Newspaper Columns

Newspaper Columns allow text to flow continuously from column to column, as it does in a newspaper article. To define and create newspaper columns, first place the cursor at the location the columns are to begin. Select **Layout** from the Main Menu. From the Layout Menu select **Columns** and then select **Define** from the Columns submenu. The screen shown in Figure 19.19 will appear.

The options are defined as follows:

> **Type**: Establishes the type of columns (Newspaper, Parallel, or Parallel with Block Protection).
> **Number of Columns**: Sets the number of columns.

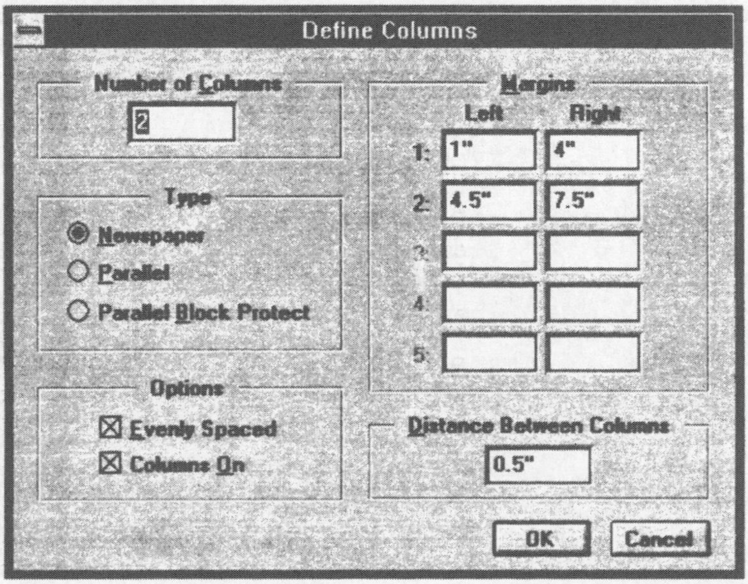

Define Columns
Number of Columns
2
Type
Newspaper
Parallel
Parallel Block Protect
Options
Evenly Spaced
Columns On
Margins
Left Right
1: 1" 4"
2: 4.5" 7.5"
3:
4:
5:
Distance Between Columns
0.5"
OK Cancel

FIGURE 19.19 Define Columns dialog box.

Distance between Columns: Sets the distance between columns; defaults to .5 inches unless otherwise specified.

Margins: The left and right margins are adjusted automatically depending on the number of columns entered. These default values may be changed if desired.

Options: Evenly Spaced (automatically spaces the columns evenly on the page); Columns On (automatically turns on the column feature when it has been defined).

Select the **Type** of columns (Newspaper) preferred and enter the **Number of Columns** desired. The margins and the distance between columns automatically will be adjusted; however, these margins can be changed if desired. After entering all information, the columns are defined and the column feature is enabled by clicking on **OK**. As text is typed, it will fill the first column. Upon reaching the bottom of the column, the cursor automatically advances to the top of the next column. When the bottom of a page is reached, the cursor automatically advances to the top of the first column on the next page and the text continues in column form. To move to the top of the next column before reaching the end of the current column, press the **CTRL** and **ENTER** keys.

When finished entering text in column mode, it is necessary to turn the column feature off. Select **Layout** from the Main Menu. From the Layout Menu select **Columns** and then select **Columns Off** from the Columns submenu. To repeat the same column feature elsewhere in the document, select **Columns On** from the Columns submenu.

Parallel Columns

Parallel columns are used when unequal blocks of text are to be printed side by side on a page. The steps for setting up parallel columns are identical to those for setting up newspaper columns (see instructions and Figure 19.19) with one exception; when selecting the **Type** of column, click on **Parallel** instead of Newspaper.

After enabling columns by entering all information in the Define Columns dialog box and clicking on **OK**, begin to enter text. When the end of a column is reached, press the **CTRL** and **ENTER** keys simultaneously. This will position the cursor at the top of the next column so the next entry can be made. When the end of the last col-

TITLE	DESCRIPTION	SALARY
Systems Analyst	Responsible for the day-to-day management of the system	$50,000
Analyst	Responsible for analysis and design of system enhancements	$40,000
Programmer	Responsible for coding and testing programs	$30,000

FIGURE 19.20 Parallel columns.

umn is reached, again press the **CTRL** and **ENTER** keys simultaneously; the cursor will be placed at the beginning of the first column, directly under the first entry. Figure 19.20 is an example of parallel columns.

When finished entering text in column mode, it is necessary to turn off the column feature. Select **Layout** from the Main Menu. From the Layout Menu select **Columns** and then select **Columns Off** from the Columns submenu.

DEFINING TERMS

Block: A section of highlighted text.

Clicking: Pressing then releasing the mouse button.

Cursor: A blinking bar or shaded box on the screen that indicates where the next character will be displayed

Directory: A table of contents maintained by DOS of the information on a disk.

DOS: An acronym for the Disk Operating System.

Double-Clicking: Quickly pressing the mouse button twice then releasing.

Dragging: Pressing and holding down the mouse button while moving the mouse.

Extension: An optional one to three character name that may be added to the file name. The file name and extension are separated by a period. The extension may consist of letters of the alphabet, numbers one through nine, and/or the special characters $ # & @ ! % ( ) - { } ‘` _ ~ ^.

File Name: The first part of a name given to a file stored on a disk. The file name may be one to eight characters in length and consist of letters of the alphabet, numbers one through nine, and/or the special characters $ # & @ ! % ( ) -{ } ‘ ` _ ~^.

Font: The shape of the characters in a document.

Icon: Small pictures that represent objects such as software or documents.

Pitch: The size of the font.

Pointing: Moving the mouse to position the pointer over items; the mouse pointer is shaped like an arrow when pointing to selections in the menu or scroll bars and like an I-beam when pointing to text.

Prompt: Appears on the screen after DOS has been loaded into memory; it is usually a letter followed by a greater than sign (>). The letter indicates the current disk drive. DOS commands are entered following the prompt.

Subdirectory: Additional directories created by the user on the disk.

Window: A defined work area on the screen.

Wordwrap: A feature that senses when you have gone beyond the right margin and automatically moves the text and the cursor to the next line.

20

Word for Windows

Paul W. Ross
Millersville University

20.1 SYSTEM OVERVIEW

Microsoft Word®, a product of Microsoft Corporation, is available on the Macintosh, MS-DOS, and Windows platforms. This article will discuss the Windows implementation of the product. The Macintosh version of the program is quite similar to the Windows version. The MS-DOS version of the program has been generally displaced by the Windows version of the product. Some individuals may prefer the MS-DOS version, if their systems are unable to support a Windows environment and the fairly large disk storage requirements of Word for Windows.

Word for Windows is a word processor with capabilities approaching that of many desktop publishing packages, and can be used for tasks that might be considered appropriate to desktop publishing software. Word for Windows allows the inclusion of art from any image format supported by Windows that can be transferred to a BMP or comparable format. It also includes an equation editor, a picture editing program, a macro language processor, and other utilities to extend the basic functionality of a word processor.

Word for Windows has the usual drop-down menus, which we will explore later. Besides the menus, Word for Windows has an extensive set of toolbars, two of which are shown at the top of the screen. These toolbar features are simply started by clicking on the icon representing the desired function. Approximately 80% of the commonly used features are available through the toolbars, saving a great deal of time and energy in using Word for Windows. This section will give a presentation of the major features of Word for Windows. For more details, consult the Word for Windows reference manual. There are a number of sample files and clip art selections provided as part of the installation process. These samples can be found in the various subdirectories under the WINWORD directory where Word for Windows is installed. If a custom installation, such as might be done for a laptop computer with limited storage, some of this sample material may be missing.

20.2 FEATURES AND FUNCTIONS

Creating Files and Entering Text

To start Word for Windows, simply double-click on the Word for Windows icon. When

0-8493-2530-7/96/$0.00+$.50

Word for Windows is installed, it will normally be installed in its own program group. The initial window looks like that shown in Figure 20.1.

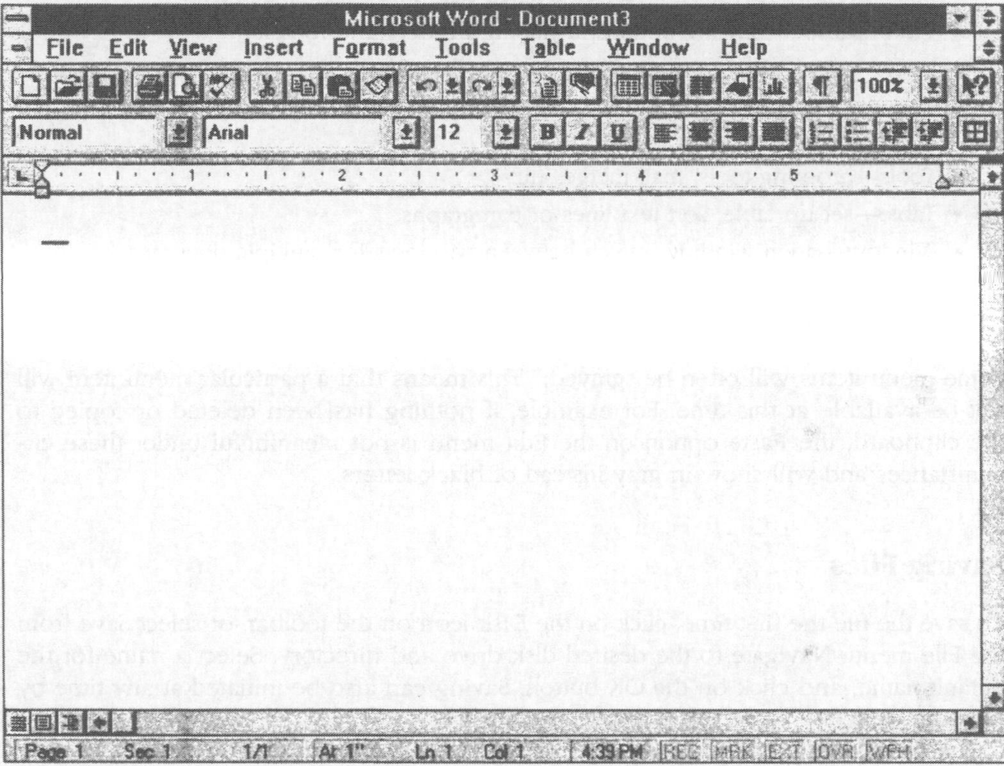

FIGURE 20.1 Initial Word for Windows screen.

Like all Windows applications, the close button is in the upper left hand corner, the sizing buttons are in the upper right corner, and the scroll bars allow you to move through the document. For a general discussion of the Windows environment, see Chapter 2 on Windows.

A default file name of Document1 is assigned to the file, as shown at the top of the screen. When you save it the first time, you can name the disk and directory where the file will be stored, and the name you wish to assign to your word processing file.

To enter text on the new word processing document, simply type it. However, there is one feature of a word processor that is different from what you may be used to on a typewriter. When you reach the end of a line, just keep on typing. Do not press the Enter key until you reach the end of a paragraph or wish to insert an extra blank line between paragraphs.

Use of the Menu System

Word for Windows is a menu-driven and mouse-driven system. The various features of Word for Windows are accessed through a series of menus at the top of the screen. This may be done by selection with the Alt key or by clicking the mouse on the menu item.

The nine menus at the top of the screen perform the following functions:

- File—create, save, open and close files and windows, print, select printer options, and exit from Word for Windows to Windows.
- Edit—move, copy, delete, restore deleted text, find and replace text.
- View—various viewing options, including headers and footers.
- Insert—insert page number, footnotes, art, date fields, etc.
- Format—appearance control with boldfacing, italics, centering, etc. Most of the important ones are on the toolbar.
- Tools—proofing tools, mail merge, etc.
- Table—set up table, sort text lines or paragraphs.
- Window—open window, switch between windows for multiple document.
- Help—an extensive set of Help information. Press the F1 function key at any time to make the Help system active.

Some menu items will often be "grayed." This means that a particular menu item will not be available at this time. For example, if nothing has been deleted or copied to the clipboard, the Paste option on the Edit menu is not meaningful under these circumstances and will show in gray instead of black letters.

Saving Files

To save the file the first time, click on the Disk icon on the toolbar, or select Save from the File menu. Navigate to the desired disk drive and directory. Select a name for the default name, and click on the OK button. Saving can also be initiated at any time by pressing CTRL-S.

The process of loading an already existing file is much the same as saving the file. Click on the Open file icon on the toolbar, select Open File from the File menu, or press CTRL-O.

Moving the Insertion Point

Move the insertion point by using the up, down, left, and right cursor movement keys or click the mouse pointer at the location where you wish to place the insertion point.

For larger movements, such as an entire screen, the following keystrokes can be used:

Up one screen	Page Up
Down one screen	Page Down
To the end of a line	End
To the beginning of a line	Home
To the end of the document	CTRL-End
To the beginning of the document	CTRL-Home
To the beginning of word to right	CTRL-Right Arrow
To the beginning of word to left	CTRL-Left Arrow
To the top of the screen	CTRL-Page Up
To the bottom of the screen	CTRL-Page Down

Deleting, Copying, and Inserting Text

To alter or delete text, drag the mouse over the text in question. Press the Delete key, or type in the new material. The Scissors icon on the toolbar does the same thing, as does the keystroke combination CTRL-X. The CTRL-C keystroke combination will copy text to the clipboard. The icon with the two pages on it on the toolbar does the same thing.

Insert text by pressing CTRL-V, selecting Insert from the Edit menu, or using the Clipboard icon on the toolbar. Any material that is currently on the clipboard, such as an image, can be inserted in the same way. To scale images, click on the image and use the handles to scale or change the dimensions of the image.

Setting Margins

To set the margins, use the mouse to move the small markers on the ruler line at the top of the screen. Drag them to the desired position, and release the mouse button when it is at the desired location. The margin stops on the left end of the ruler line are different. Instead of a single triangle, the triangle is broken down into two pieces. The upper one sets the margin for the first line in a paragraph. The lower triangle sets the left margin for succeeding lines.

Setting Tab Stops

To use the tabulation feature in Word for Windows, press the tab key. Initially, Word for Windows uses a set of predefined tab stops set for every 0.5 inch. To change the location of tab stops, click on the Ruler bar where you want a tab stop to appear.

Centering and Aligning Text

To center or align text, highlight the text (line or lines) using the mouse. Then click the mouse on the appropriate alignment symbol just above the ruler line.

Printing the Document in Word

Select Print from the File menu at the top of the screen, or press CTRL-P. You can get an idea of what your document will look like by selecting the Print Preview option. You can also set the number of copies in the Print dialog box. For a single copy, just click on the Printer icon on the toolbar.

Finding Text

The procedure to locate or find text is as follows:

- Position the insertion point somewhere before the location where you expect to find the desired text. CTRL-Home will position you at the beginning of your word processing document.
- Choose Find (CTRL-F) from the Edit menu at the top of the screen.
- Type in the text you wish to find, either one word or a phrase. Press the Enter key to start the search

- Optionally, select the Match Case, if upper case and lower case letters are important in the search. Select Match Whole Word Only if you do not want the search to be satisfied for just a part of a word.
- Click on the OK to start the search, or Cancel to abort the search.

If the search is successful, the insertion point is automatically moved to the first occurrence of the target word or phrase and highlighted. If the word or phrase cannot be found, you will receive an error message telling you that no match is found.

Finding and replacing text is similar to Find. Choose Replace from the Edit menu, or press CTRL-H. Follow the prompts on the screen.

Text Attributes and Appearance Issues

Text attributes refer to the font or design of type available and how it appears, such as boldface, italics, etc. In addition, you can center text, make the left, right, or both margins justified, and create superscripts and subscripts for mathematical or chemical formulas.

The variety of fonts available to you depends upon the printer you have on your computer and the fonts available within Windows itself. Check what fonts are currently installed by pulling down the Fonts menu at the top of the Word processing screen.

The text attributes such as boldfacing, etc., available in Word for Windows' include plain text, italics, boldface, and underline.

- Select the desired text by dragging the mouse over the text.
- Select the desired attribute by clicking on the B, I, or U buttons on the toolbar at the top of the screen.

Justification and Centering

You can perform justification and centering in four ways:

- Left—all lines are vertically aligned on the left side, but not the right side.
- Right—all lines are vertically aligned on the right side, but not on the left side.
- Justified—all lines are vertically aligned on both the left and right sides.
- Centered—equal spaces on both the left and right.

Highlight the desired text and click on the alignment icons on the toolbar.

Proofing Tools

Word for Windows contains three powerful proofing tools:

- Spelling checker
- Thesaurus
- Grammar Checker

First, make sure to save your document. For spell checking, click on the "ABC" icon on the toolbar, or select Spelling (F7) from the Tools menu. Simply follow the prompts on the screen to make any correction or add words to your custom dictionary. A number of options can be elected, including detecting and removing double words, and optionally suggesting a word to replace a misspelled word.

Drag the mouse over the desired text, and select Thesaurus (Shift-F7). Follow the prompts on the screen to select an alternate word.

The grammar checker is specially useful. Select Grammar from the tools menu. Edit lines as necessary, and select Change to change the line in question. Alternately, go on to the next sentence, or ignore the rule.

DEFINING TERMS

The reader should consult Chapter 2 on Windows for more details pertaining to the terms used in all Windows applications.

Drop-Down Menus: Menu items that stay displayed if single-clicked with the mouse. Contrast with pull-down menus in the Apple Macintosh environment.

Proofing Tools: Spelling checker, thesaurus, and grammar checker.

View Menu: Menu giving various viewing options, such as full page as will be printed, normal mode, outline mode, and headers and footers.

REFERENCES

Word for Windows on-line Help screens.

Word for Windows Reference Manuals. Microsoft, Inc.

21

Ami Pro 3.0

Todd Leuthold
Todd Leuthold Consulting

21.1 INTRODUCTION

Using Ami Pro V3.0 is straightforward. Start the program by double clicking on the Ami Pro 3.0 icon. Usually, unless you are typing a specific type of document or form, you can just start typing and print out the end result, all without making any changes in Ami Pro.

The initial start-up screen (Fig. 21.1) shows a page with the margins markedout with a blue background, which helps to identify the printable portion of the page. You can use the cursor keys or the mouse to move the character indicator (the item that shows where your next character will appear) where you want it, then begin typing. To move the indicator with the mouse, move your mouse indicator (the item that shows where your mouse pointer is located) to the desired location and click the left mouse button. The character indicator will then move to that location.

In most word processors, your type starts at a varying distance from the outside edge of the page. With Ami Pro, your type begins right at the edge of the printable area. The entire area of both blue and white comprises the actual page. However, only the white area is meant for typing.

21.2 FEATURES AND FUNCTIONS

Margins

If you need to change the margins, use the mouse to move the mouse indicator into the blue area, then click with the right mouse button. A screen will pop up, allowing you to change a number of things, from margin settings to tab settings. Make the necessary changes and choose OK. This will make the changes effective right away. If you change the margin settings, the entire page will be affected, no matter what physical location the mouse indicator is in. If you decide that you do not want to make any changes, click the Cancel button and any changes will be discarded.

SmartIcons

The default opening screen has an Icon Bar, with SmartIcons. These SmartIcons allow you to perform certain tasks without having to navigate the menu structure. If you see a SmartIcon without a descriptive enough picture, put your mouse pointer

0-8493-2530-7/96/$0.00+$.50

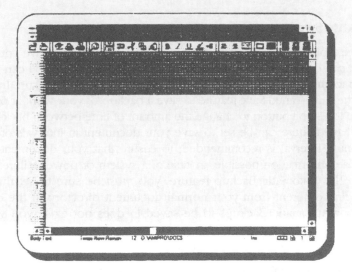

FIGURE 21.1 Initial start-up screen.

on it and hold down the right mouse button. The title bar (the topmost bar on the screen that has the program name in it) will show a brief description of what that icon does.

Once you have started the Ami Pro program, you have the option of beginning a new document, or calling up an old document. The first icon on the bar has a picture of a file folder on it, with an arrow pointing up and to the right. This represents removing an existing file from a folder. This can be done by clicking your mouse pointer on this icon.

When you click this SmartIcon, a window pops up showing all the Ami Pro files available for editing. You can also look for other types of files by clicking on the button in the lower portion of the screen. There are 39 different types of files that Ami Pro can read directly, including several different types of spreadsheets. You can also check the file without loading it, to see if it is the one that you want. Click the Preview button to try this option. Once you have selected a file by choosing the OK button, you can then edit it, using any of the numerous tools at your command.

Saving

After typing or editing your document, you have a number of options. However, the most important is that you save what you have done to this point. To do this, you can choose the second SmartIcon on a default setup, which is a file folder with an arrow pointing down and to the right. This indicates placing a file into a folder, or saving. If you are working on a new document and have not yet saved it, you will be asked a few questions about the document, such as name and file type. Once you have saved your document, any further editing of that document will allow the Save SmartIcon to save your document without any questions. If you wish to save your changed document under a different name, in a different location, or in a different file type, you will have to choose the File menu option and select Save As. This will give you the options to save your document with different options than the original.

Backup/AutoSave

While you are typing or editing your document, you may want to save your work occasionally. You can do this by either clicking the Save SmartIcon, or you can use the Tools menu option to change the User Setup options for an Automatic Backup. In Tools menu, you can set the Auto-Timed Save feature to save a backup of your work at predetermined intervals. Use the scroll button to change the amount of time between backup saves. The Auto Timed Save feature can be set to save your document in intervals of 1 to 99 minutes. A 5-minute interval is recommended, to ensure that your document contains the most up-to-date information possible, in case of a system or power failure.

If you use the Automatic Backup feature, you *must* be sure to set the Backup directory as being different from your normal document directory. If the directory that you want your Automatic Backups to be saved in does not exist, you will first have go to the File Manager to create it. Once it is created, you can then choose the Paths button in the setup screen to set a path for the Automatic Backups to be placed in.

Error Correction

If you make a simple error in your document, you can use the Backspace key to delete the error. You can also move the character indicator to the location and use the delete key. Another method is to highlight the text with your mouse indicator and press the delete key. Using this feature, you can delete large blocks of text. For obvious reasons, you should exercise caution when using this feature.

If you accidentally delete too much, you can use the Undo feature, which you can get from the default SmartIcon set. It is the button with the two blue arrows on it making a reverse turn. In the default mode, the undo level is set to one. This means that you can only undo the last thing that you did. If you wish to have a little more protection, use the User Setup option on the Tools menu to change it. You can set it up to four levels deep, meaning that you can undo up to the last four changes.

You can use the Undo feature to see what will happen to a certain section of text, or the body of a selection. If you do not like it, you can Undo the change, putting things back the way they were. As you make new changes, the oldest change in the list (depending on how many levels you have active) will be removed, and the new change will be put at the top of the list. As you use the Undo feature, the changes are undone in the reverse order they were put in, or Last In, First Out.

Of course, if you make a number of changes and cannot repair them with the Undo feature, you also have the option of closing the document and reloading the original. To do this, you will use the CLOSE function under the *File* menu selection. When the drop down box is displayed, choose Close or New. Either of these options will allow you to get rid of the current document. When you select one of these choices, another box will pop up reminding you that the file has changed. You have three options: Yes (Save), No (Don't save), and Help (get information on your options and their results). If you want to start over, choose the No option and then reload the original document.

Following is a brief description of the function keys and what they can provide.

Function Keys

1. **F1** provides on-line help.
2. **F2** will give you a Body Text style sheet.
3. **F3** will give you a Body Single style sheet.
4. **F4** will give you an Asterisk Bullet.

5. **F5** will give you a Diamond Bullet.
6. **F6** will give you a Number-Style Bullet.
7. **F7** will make the following text (up to the next Hard Return) into a Subheading.
8. **F8** will make the following text (up to the next Hard Return) into a Title.
9. **F9** will make the following text into a Header.
10. **F10** activates the Menu field.

11. Access Styles Box	**Ctrl + Y**
12. Bold text	**Ctrl + B**
13. Center text	**Ctrl + E**
14. Copy	**Ctrl + Ins**
	Ctrl + C
15. Cut	**Shift + Del**
	Ctrl + X
16. Delete next word	**Ctrl + Del**
17. Delete previous word	**Ctrl + Backspace**
18. Draft/layout view	**Ctrl + M**
19. Exchange the selected paragraph with the paragraph above it	**Ctrl + Up Arrow**
20. Exchange the selected paragraph with the paragraph below it	**Ctrl + Down Arrow**
21. Fast format	**Ctrl + T**
22. Find and Replace	**Ctrl + F**
23. Full page/previous layout view	**Ctrl + D**
24. Go To	**Ctrl + G**
25. Go To next item	**Ctrl + H**
26. Insert glossary record	**Ctrl + K**
27. Italicize text	**Ctrl + I**
28. Justify text	**Ctrl + J**
29. Left align text	**Ctrl + L**
30. Modify a paragraph style	**Ctrl + A**
31. Normal text	**Ctrl + N**
32. Open document	**Ctrl + O**
33. Paste	**Shift + Ins**
	Ctrl + V
34. Print document	**Ctrl + P**
35. Right align text	**Ctrl + R**
36. Save	**Ctrl + S**
37. Show/Hide set of SmartIcons	**Ctrl + Q**
38. Underline text	**Ctrl + U**
39. Undo previous action	**Alt + Backspace**
	Ctrl + Z
40. Word underline text	**Ctrl + W**

Menus

If you do not see what you need on the SmartIcon bar, then you can choose one of the menu options. To choose a menu option, either move your mouse pointer on it and click, or hold down the Alt key and press the underlined letter of the menu option that you would like.

At this point, a drop down list appears, showing the different functions available under that menu option. You can either press the underlined letter, move the highlight bar with the cursor keys, or use your mouse pointer to point at the desired choice and then click the left mouse button to activate. These options require only a single click, rather than a double click.

If a menu option is followed by three dots (. . .), this indicates that there is a dialog box with choices for that option. If no dots follow, or if there is a control key combination displayed (^X, where "X" is the letter shown), then the action will be executed as soon as you select that option. If there is a black triangle (shown as > in this discussion), there will be another menu window available for this option. All the following menu definitions will have a letter underlined. This means that you can press this letter when the menu window is dropped down, to activate this option.

Note: If you see an option to use Ctrl-x, where the x stands for any particular key, the Ctrl key is the Control Key. It is usually found on the right and left lower edges of the typing key area, just below the Shift keys. To use the key combination, press and hold the Control Key, while tapping the second key shown, then release both keys. If you see an ALT option, then you would hold down the ALT key and press the desired key to get the desired option.

The menu options at the top of the screen offers a complete set of tools to help you with your document. Here is a breakdown of the various options under each menu choice.

File

- New . . .

  Provides the opportunity to start with a clean slate. A screen pops up to allow you to choose the style sheet that you want, with the most frequently used options highlighted. If you do not want to make a change, just choose OK.

- Open . . .

  Allows you to open a previously created document. A box will pop up, allowing you to type in a file name, select a file from a scroll box, change drives or directories, or change the type of file that you want to work with. Using Ctrl-O will also perform this task.

- Close

  Closes the current document in memory.

- Save

  Allows you to save the current document in memory. If you have not previously saved this document, you will get a pop-up box that allows you to give the file a name, as well as selecting where you want to save the file and what file format to use. If you have previously given the file a name, the old file will be overwritten with the new changes. Use with caution. Ctrl-S will save your document without any other prompts, if it already has a name. If not, the normal dialog box will appear.

- Save As

  Allows you to select a new file name different from the original file name. This is normally used when you want to make changes to the document, but do not want to lose the original.

- Revert To Saved

  Returns you to the last saved version of the current file that you are working on. This is another method that can be used if you have made too many changes to fix with Undo. Use with caution, since you will erase any changes that you have made.

- Import Picture . . .
This will allow you to import a graphic into your document. The document formats directly supported are AmiDraw, AmiEquation, AutoCAD, CGM, DrawPerfect, EPS, Freelance, HPGL, Lotus PIC, PCX, TIFF, Windows Bitmap, and Windows Metafile. Do not worry about recognizing the various graphic file types. You need to know which ones only if you want to import one of them. You can select the file type when you choose the Import Picture option. A box will pop up, allowing you to type in the file name and/or select the file type, which will then give you all files that fit the category that you select.
- Doc Info . . .
This option allows you to get additional information on the file that you currently have in memory. You can specify additional fields, if necessary.
- File Management . . .
This is a small version of the File Manager in Windows. It is a bit more convenient than the ones that comes with Windows, though.
- Master Document . . .
Allows you to generate a Master Document as well as include any other documents. You can make any document the Source Document, including a New document (no document loaded). If you want to use another document as the source document, load that document into an active window and have that window active when you choose the Master Document option. You can also use the option to generate a Table of Contents. Since a Master Document generates pages numbers, be sure to include any documents in the order that you want them to be referenced. This option is good if you want to write different parts of a book or manual, but want to do the chapters as separate files.
- Merge . . .
This allows you to merge a data file into a *Template* file. The template can be a new file that you have just created, an existing file, or a file that you have edited.
- Print Envelope . . .
Will search your document for a name and address. It will then use this information to put into the mailing address portion of an envelope. A pop-up box will appear, allowing you to choose the type of envelope that you want to use (including the ability to set up a custom envelope description) and the type of printer that you wish to print the envelope with.
- Print . . .
Allows you to print your document. Options include which pages to print, what order to print them in, and what options to use in setting up your printer before printing. If you wish to use a different printer, you will have to choose Printer Setup first. Ctrl-P will start printing right away, using the defaults. It is recommended that you choose the Print option from the menu, just in case there are things that need setting.
- Printer Setup . . .
Allows you to select a printer for printing your document.
- Exit
Quits the Ami Pro program. If you have a file open, you will be given the opportunity to save it before quitting. You can also quit using Alt-F4, just as in other Windows programs.
- 1, 2, 3, 4, 5
You will see a list of zero (0) to five (5) files at the bottom of the FILES menu. These are a list of files that are available for editing. You can increase or decrease the num-

ber of files shown here by making changes in your User Setup. The minimum number of files displayed can be zero (0) and the maximum can be set to five (5).

Edit

- Undo

  This will undo the last changes that you have made, up to the limit that you have set. The default number of changes that can be undone is set at 4. This can be changed in the User Setup area. You can also use Ctrl-Z to activate this option.

- Cut

  This will remove a highlighted portion of text from your document. To highlight an area, put your mouse pointer just before the first character of the area you want to cut. Press and hold the left mouse button, and move the mouse pointer to just after the last character that you want to cut. Release the left mouse button. Choose Cut from the Edit menu, or use Ctrl-X, to put the highlighted area into a temporary buffer. This will also remove it from your document. If you wish, you can then put it in another location in the current or another document. If you do not do anything with the area stored in the temporary buffer, it will be erased when you place something else there, or when you exit the program. It is also important to note that this option is active only when you have an area highlighted. If you do not have an area highlighted, the option will appear gray and will not respond.

- Copy

  This option works much the same as Cut, but it does not remove the material from its current location. Instead, it makes a copy and puts it into the temporary buffer. You can then proceed just as you did with Cut. You can also activate this option with Ctrl-C.

- Paste

  This option allows you to place material in the temporary buffer into your document. This operation does not erase the buffer, so you can make more than one insertion of the same material. The shortcut for this option is Ctrl-V.

- Paste Link

  This option allows you to put a special "mark" into your document that will allow you to place information/data from another Windows program into your document. The option can set up either a DDE (Dynamic Data Exchange) or OLE (Object Linking and Embedding) link. See the chapter on DDE and OLE for a more in-depth explanation of what these are.

- Paste Special . . .

  This allows you to embed data from another Windows program into your Ami Pro document. You can also use this option to format data that you want to paste at a particular location.

- Link Options . . .

  Allows the user to display information on existing DDE and OLE links, or to update, unlink, deactivate, edit, or create a link anywhere in the document.

- Find and Replace . . .

  Allows the user to find any character or string of characters, including specialized characters, such as those that are in special fonts, etc., as well as styles. You can then replace the item or items with another set of items, including special fonts and styles.

- Go To . . .

  Allows the user to go to anywhere in the document, including headers and footers and styles, as well as bookmarks. You can also Go To a New Object, which will allow you to use a picture or a spreadsheet area created in another Windows program, into a frame in your Ami Pro document.

- Insert >
  Provides options for inserting various items into your document. When this option is chosen, another menu of options will drop down, providing several options. Each of these options will have several options of their own, except Note, which allows the user to attach a note to a particular location in your document.
- Power Fields >
  Allows the user to Insert, Update, Update All, go to the Next Power Field, and go to the Previous Power Field anywhere in your document, including footers, headers, tables, cells, etc.
- Mark Text >
  Allows the user to mark a character or string of text for special purposes, including use as an Index Entry, Table of Contents Entry, Glossary entry, or Revision Insertion. It also allows the user to mark specific area as Protected, so that the text cannot be easily accessed with the cursor keys or the mouse.
- Bookmarks . . .
  Allows the user to insert a Bookmark at the specified location. The name of the Bookmark can consist of up to 17 characters. The name can consist of any combination of letters and numbers, but it cannot start with a number or consist solely of numbers.

View

- Full Page
  Allows the user to see the entire page at the same time. This lets you see what the page looks like as you type. Of course, this display is too small to do any real typing, but it helps you get a feel of the entire layout. You can also get this option directly by using Ctrl-D.
- Custom xx%
  The xx% indicates what percentage of a full page you would like to have your display set at. I personally chose 91%, because the text and fonts are large enough to see, while also allowing you to see your margins and sufficient text above and below to keep track of where you are. You can set this option to whatever is the most comfortable for you by selecting the last option in this menu.
- Standard
  This option will push the margins off of the screen and just show you the typing area. The text gets larger and you cannot see as much on the screen as with smaller displays.
- Enlarged
  This option will enlarge the print on the display. If you have trouble seeing the small print on a standard display screen, this could be the perfect option for you. It also allows you to see more detail on the screen.
- Facing Pages
  This shows you two pages, side-by-side, in very small print. Though you cannot use the screen for editing, you can use it to see the overall layout.
- Layout Mode
  This option allows you to see what will be printed as you are typing/editing the document. It will show you the page breaks, formatting, etc.

- Outline Mode
  This options lets you see the layout of your page, by showing you the various options chosen at different parts of the document. It shows special (unprinted) characters that indicate whether or not something is a Header, Bullet, etc.
- Draft Mode
  This option can also be chosen with Ctrl-M. This option puts you in plain typing mode, without concern for layout, positioning, page breaks, or anything else. It is good when you are just typing in information and it will be formatted at a later time.
- Hide SmartIcons
  Using Ctrl-Q will activate this option, too. This will remove the SmartIcons from the screen.
- Show Clean Screen
  This option will just show you the page and the borders, without any of the scroll bars, menus, icons, etc. To return to the standard display, choose the icon shown at the bottom right of your screen, or use Alt-V, N, to Hide Clean Screen. This item toggles itself from one mode to the other. When you choose it, the option will change to allow you to switch back to the way it was.
- Show Ruler
  You can use this option to display a ruler at the top of the screen, just below the SmartIcon bar. This might be helpful if you need to know the positioning of a particular item on the document. This item is also set up to toggle itself.
- Show Styles Box
  You can put a small box on your screen where you can pick styles directly from. You can move this box to any convenient location on your screen. This option is a toggle.
- Show Power Fields
  If you use Power Fields, this option will display them on the screen, among the text. This is helpful if you need to keep track of their position. This option is a toggle.
- View Preferences
  This option allows you to set a number of things about the screen that you see. You can also set the percentage of magnification for the Custom option, as well as the things that will remain when you choose the Clean Screen option. You can also turns features on and off by checking or clearing the check boxes for various options.

Text

- Font . . .
  This option allows to select the font that you wish to use (from all available fonts) by scrolling down the list of choices and picking one. You can also change the size of the font by selecting your choice from the scroll box marked Size. If the specific size that you want is not shown, you can enter the point size directly from the Points scroll box, or use the arrows to select a point size. You can also choose a text color from this menu. However, only color printers will take advantage of this. You can also see a sample of what your text will look like while using this font. Just look in the box at the bottom of the window.
- Alignment >
  This item will allow you to align your text on the page. The options are Left Align, where your text is smooth on the left side of the page, Center Align, where each line

of text is centered on the page according to its length, Right Align, where the text is smooth along the right side of the page, and Justify, which will attempt to make the text smooth on both sides of the page, by adding extra spacing between words to fill out a line.

- Indentation . . .
  This dialog box will allow you to set the start of indentation and the amount of indentation when you use it. You can enter the numbers directly, or use the scroll buttons to change the values.
- Spacing . . .
  This dialog box will allow you to select how much space is between each line of text. You can select one of the preset values, or use the scroll box to set custom spacing. You can also use this option to go back to the original style that you were using when you started the document, or since the last style change that you made.
- Normal
  You can choose this option with Ctrl-N, also. This puts text back to normal. This is useful if you have several options defined, such as Italics, Underline, and Bold all at the same time.
- Bold
  You can activate this by using Ctrl-B, as well. This option will darken the text that it is applied to, allowing it to stand out from surrounding text. This option is a toggle.
- Italic
  Using Ctrl-I will activate this option. It will italicize the text. You can either turn it on for text that you are about to type, or you can highlight some text, then choose this option to italicize just the selected text. This option is a toggle.
- Underline
  Using Ctrl-U will activate the Underline mode. You use this as a toggle to turn underlining on and off. This will allow you to underline a single letter or character, or to underline a highlighted block of text, or just one section of your sentence.
- Word Underline
  Use Ctrl-W to toggle this option. This will quickly underline an entire work, beginning with the next word that you type. It will skip the spaces between words, allowing you to just underline the words themselves.
- Caps >
  This option will drop down another menu that allows you to choose between four different types of capitalization. They are Upper Case, Lower Case, Initial Caps, and Small Caps.
- Special Effects . . .
  This option will allow you to Superscript, Subscript, Double Underline, and Strikethrough text. You can also select Overstrike, which allows you to choose the character to use.
- Fast Format
  Provides quick formatting of your document. Highlight an area of text that has the formatting that you wish to use. Choose Fast Format and the rest of your document will then use that formatting. You can also use it to format text in a table, or a text frame.

Style

- Create Style . . .
  This option allows you to create a new style, based on other styles already present. You can save the new styles and use them again at a later time.
- Modify Style . . .
  Use Ctrl-A to access this option. You can modify the current style that you are using. You can change the font, color, size, points, typeface, and attributes of the current style, then save it as a totally new style, or permanently modify the current style.
- Define Style
  This option uses the current text enhancements in use to modify the current paragraph style.
- Outline Styles . . .
  Use this option to change the way Outlines are set up. You can change the numbering, the levels, and separators, as well as the reset options.
- Use Another Style sheet . . .
  Allows you to select a different style sheet from a list of options.
- Save as a Style Sheet . . .
  Allows you to create a style, then save it, along with a description, for later use. You can also set it up to run a macro (which you can choose). You have the option to save the contents of the style sheet, also.
- Style Management . . .
  This option will provide you with the tools to modify styles based on their content. You can select the different styles available, then add or remove elements of the style. You can also change the "F-Key" designation for a given style.
- Select a Style
  Use Ctrl-Y to quick-start this option. A scroll box will drop down to allow you to choose a style from a list. You can resize the box and position it anywhere you wish on the screen, for easy access.

Page

- Header/Footer . . .
  This option allows you to create a header or footer, with a variety of options. You can choose to place the header/footer, on just the current page, on all pages of the document, or on just specific pages. Click the question mark in the upper right corner of the dialog box to see what options are available.
- Insert Page Layout >
  This option drops down a second menu to allow you to insert a layout, remove a layout, or revert the current layout to the previous layout.
- Modify Page Layout . . .
  This options provides for changing or setting all page layout options, including tabs, margins, lines, headers, footers, gutter width, columns, and on which pages it will all take effect.
- Ruler >
  This option allows you to add or remove a ruler at the left side of your document.
- Page Numbering . . .
  This option allows you to set the page numbering for your document. You can start any page or number, and you can even add leading text. You can also se-

lect the style of page numbering (using numbers or letters) from a drop-down box.

- Line Numbering . . .
  Provides the option of numbering your lines. You can select to number all lines, every other line, or every fifth line, or you can specify how often to number lines. You can also number just text lines, or all lines on a page. There is also an option to reset the line numbering at the beginning of each page.
- Breaks . . .
  Provides a means to insert or remove a page or column break. You can also use this option to center the text on a page between page breaks.

Frame

- Create Frame . . .
  This option provide a method of creating frames in your document. You can specify the type of frame, the size of the frame, and its position on the page. You can also create a Manual frame. Click on the question mark in the upper right corner of the dialog box to get help on this option.
- Modify Frame Layout . . .
  Use this option to change the features and positioning of the frame that it active.
- Graphics Scaling . . .
  Use this option to change the size of a picture in a graphics frame. You can fill the entire frame, or adjust it to fit your needs.
- Group
  Use this option to specify two or more frames, if you need to delete, copy, or move them. This will allow you to change several frames at the same time.
- Bring to Front
  This will bring the active frame to the front of all other frames at that particular location.
- Send to Back
  This will send the active frame behind all other frames at that particular location.

Tools

- Tables . . .
  This option will create a table at the current location. You can specify how many rows and columns you needs, as well as specifying the particular layout you desire.
- Image Processing . . .
  This option will allow you to enhance the gray scales in a scanned image. You can import an image in, then lighten or darken it to suit your needs.
- Drawing . . .
  This options starts the built-in drawing program for Ami Pro. Use this program to design a drawing of your choice, or to import a graphic and then modify it. When you are finished, the results will be placed into a graphics frame in your document. You can use ESCape to stop drawing and return to your document.
- Equations . . .
  This option starts the Equation Editor built into Ami Pro. Use this option to insert equations into your document. There are a wide variety of symbols available, as well as options for placement and usage.

- Charting . . .
  This option allows you to add charting features to you document. You should have data in the clipboard to insert into the chart, but you can also enter data directly into the chart upon creation.
- Footnotes . . .
  This option is used to insert or edit a footnote. You can set up your footnote with other options, such as resetting the numbering on each new page, the indentation, and whether to set the length at the margins, or to set it to a custom setting that you provide. You can also make it an end note and you can select what number to start the number sequence at.
- Revision Marking . .
  This option allows you to use the document as a markup to make or indicate revisions. You can also set options for strike-through and overstrike. Changing colors here will allow you to add text comments that can be later removed. You can also accept or reject some or all revisions.
- Doc Compare . . .
  This option allows you to compare the current document to another document on disk.
- Sort . . .
  This option allows you to sort either your entire stream of data, or just a selected text portion. If you just choose the option, you will be given a chance to stop the sort and try again.
- TOC, Index . . .
  This option allows you to create/modify a Table of Contents or an Index. You can also save the results in separate files. Options are the same as those available for Outlining.
- SmartIcons . . .
  This allows you to add or remove icons from the SmartIcon bar at the top of your document. You can also edit an icon and save or delete an entire set of icons to meet your needs.
- User Setup . .
  This is where you make your changes to the way things are set up in your application. You can set the frequency and types of backups, what macros to run on startup and closing, what colors to use, the levels of Undo available, and a large number of other things. Click the question mark in the upper right corner for specific details on all the options available.
- Macros >
  This option will drop down a submenu with options to Quick Record or Playback, or Normal Record or Playback a macro, as well as allowing you to edit existing macros.

Window

- New Window
  This will open a new document window for editing or creating a new document, while leaving your current document in memory.
- Tile
  This option will set your open windows up so that you can choose one to work on.

- Cascade
  This option will overlay each window with the next most-current one, so they all fit in a smaller area. You can select the one you wish by clicking on its exposed area.
- 1 xxxxx.xxx
  This shows the available editing windows active. You can select the window you wish to edit from this list by either pressing the number or clicking on the desired document name.

Help

This option provides on-line help for the various features of Ami Pro. It is recommended that you select Contents the first few times you go in, so that you can find the information that you are looking for as quickly as possible. This is set up to show the different categories of information available, from which you can choose the topic that you need information on. You can also get help by pressing the **F1** key, or by clicking on the question mark in the upper right corner of many dialog boxes.

DEFINING TERMS

Dialog Box: A box containing information about an option. You will also be given a Yes or No option, or be informed of something pertaining to your current operation or task.

File Type: The format that you wish the document to be saved in. Each manufacturer's program usually creates a file format unique to itself. Ami Pro can convert its own file type to that of another program, making it easier to use in the other program.

Import: To add a graphic or text originating in another program to your current file.

Master Document: The "Home Base" of a collection of documents. This is the document that all other documents are attached to. This is a useful feature when creating chapters as separate files.

Merge: To join two files together in a special manner. A "template" file is used to define what information will be taken from the second file and how it will be arranged.

Style Sheet: A collection of specific settings, predesigned, or designed on-the-fly, to provide a certain look to the document.

Word Processor: A program that assists the user in creating a document.

22

Desktop Publishing

Neil R. Bauman
ComputerTalk Associates, Inc.

Paul W. Ross
Millersville University

22.1 PRINTERS

Paul W. Ross

CRT terminals suffice for many purposes, but often a permanent record on paper is needed. To fill this need, many different printer technologies have been developed. This article is not meant to be comprehensive, but attempts to outline the major capabilities of the various technologies available to most computer users.

Types of Printers

Printers may be divided into two general classifications: impact, and nonimpact. Impact printers can be characterized by mechanisms where a font or comparable device is inked, and pressed against a piece of paper. An ordinary typewriter is a simple example of an impact printer. A laser printer is an example of a nonimpact printer.

Two major varieties of impact printers are encountered:

- Daisy Wheel
- Dot Matrix

The Daisy Wheel Printer

The daisy wheel printer is most commonly encountered in electronic typewriters. The mechanism is quite simple. The fonts are formed of plastic or metal and placed on thin strips much like the petals of a flower, hence the name, daisy wheel printer. An inked ribbon, or single-strike carbon ribbon is interposed between the print wheel and the paper. An electrically driven hammer is activated to bring the type face in contact with the ribbon, pressing it against the paper to form the letter. The type wheel is rotated by a precision stepper motor to select the desired letter.

0-8493-2530-7/96/$0.00+$.50

This printer produces a quality comparable to hand-set type, but suffers from being noisy, slow (about 35 characters per second, maximum), and requiring that the type wheel be changed every time a different font is required. Consequently, is has fallen into disuse, but may still be encountered from time-to-time. It is not capable of producing graphical output. Paper can be fed a sheet at a time, from a bin feeder, or from continuous forms from a tractor feed mechanism.

The Dot Matrix Printer

The other commonly encountered impact printer is the dot matrix printer. The letters are formed by activating a series of pins (typically 9 or 24), which form the letter as the print head is moved across the paper. This printer is also noisy, but much faster than the daisy wheel printer. The 24 pin printers are capable of forming letters and graphics that are almost indistinguishable from nonimpact printing technologies, such as ink jet and lower resolution laser printers. With the falling costs of ink jet and laser technology, the dot matrix printer is being displaced. However, because of its rugged quality and reliable performance, the printer continues to be used. This printer produces graphical output, if appropriate software is used. The paper can be fed a sheet at a time, or from continuous forms with a tractor feed mechanism.

Nonimpact Printers

Nonimpact printers are characterized by various technologies that do not involve pressing an inked font or pins against a piece of paper. Consequently, they can be faster and quieter, and are often capable of better resolution than impact printers. The three most common nonimpact technologies are thermal, ink jet, and laser.

Thermal Printers

Thermal printers require a chemically treated, heat-sensitive paper. Instead of pins to form the letters, small heater elements cause the paper to darken. This printer technology is most commonly found in low-cost facsimile machines. The paper is fairly costly, difficult to handle, and does not produce a permanent image.

Ink Jet Printers

The technology of choice for those on a budget is probably the ink jet printer. The basic image formation strategy is similar to that of a dot matrix printer. The dots are formed by small globules of ink projected through minute holes in a diaphragm. The printers are exceptionally quiet, have excellent resolution (300 dots per inch, typically), and print at moderate speeds (two pages per minute, typically). The printers are capable of color printing, which is important in the production of presentation quality graphics and charts.

The paper used is common photocopier paper, fed from a bin, a sheet at a time. Single sheets or envelopes can be hand fed. The only drawbacks, in comparison to a laser printer, is the low resolution, slower printing rate, and darker print. The ink is a water-based dye in most cases, so has a slight "gray" tinge. However, the output photocopies well. There may be some streaks in large dark areas. The water-based ink may cause the paper to curl or warp, if thin stock is used. However, for moderate volume, high quality work, it represents an appropriate and useful technology. Moderate quality facsimile machines often use ink jet printer technology.

Laser Printers

Laser printers, once quite expensive, have fallen below $1000 for entry-level units. The technology of the laser printer is that of a photocopier. The image is formed on a photosensitive drum, discharging an electric charge selectively. The remaining charge is used to electrostatically attract a toner powder, which is transferred to paper stock. The toner powder, which contains carbon black and wax, is then fused to the paper with a built-in heat lamp. Resolution in excess of 600 dots per inch is available, along with printing speeds exceeding a dozen sheets per minute. For desktop publishing, or high-volume work, the laser printer is the most desirable technology. Laser printers are also used in high-quality facsimile machines.

Printer Driver Software

To this point we have addressed only the mechanical issues of printer. The problem is made more complex by the vast variety of printers, with all sorts of special features, produced by a wide variety of manufacturers.

Initially, the situation was fairly simple. The available printers responded to ASCII text and control characters. For each character sent to the printer, a letter was produced. Advancement to the next line was made by means of a carriage return and line feed character. If overprinting was needed for boldface effects, only a carriage return character was sent, and the line was reprinted. Typewriter-like printers were also available, which responded to a backspace character, so boldfacing and underlining were easily accomplished.

As printer technology advanced with the development of dot-matrix printers, the capability for font changes and graphical options became feasible. The outcome of this was that word processors needed a set of software for each available printer, as each manufacturer had its own ideas of which features might be useful, and implemented them in a different way. With each new printer make or model, a new set of software, called driver software, was needed. Suffice to say, this produces an awkward problem for both the user of the word processor printer and word processor vendors.

As the need for more elaborate printing capabilities grew, a number of solutions arose. The heart of the problem is where to put the "intelligence" for the printing process. There are three choices:

1. **In the word processor itself**. The word processor becomes somewhat awkward to use unless a fast CPU and good interface are available. Among other things, this leads to a proprietary file format for the word processing document. It becomes awkward to move highly formatted documents between different platforms or word processors, unless a common data format can be used, such at RTF or DCA. If no common formatted data structure can be found, it may be possible to transfer the files in "generic" format, where a carriage return/line feed is used only at the end of a paragraph. It will then be necessary to reformat the file.
2. **In the printer**. This leads to what are known as page description languages. The most common of these is one known as PostScript. The printer contains a processor and program to interpret the commands contained in the PostScript output file. The PostScript output file is an ASCII text file that can be edited manually, if necessary.

Other proprietary page description and control languages are also available, such as those in the Hewlett-Packard line of laser printers. The postscript implementation is activated by buying and installing a PostScript language chip for the printer.

3. **Another alternative**. A final alternative, intermediate to these two, is a typesetting program such as TeX (see Chapter 25) that creates a file containing the typesetting commands. This file is then converted into a device-independent format that must be passed through a driver program for the particular printer.

Graphics support is implemented in two general ways:

1. **Essentially a graphics "dump" of the word processing screen** Again, the necessary printer driver is needed for the printer being used. It is also possible to imbed the art within the document itself, such as in Microsoft Word.
2. **An "encapsulated" PostScript file**. This contains the code created by the drawing program in a format to be interpreted. An Encapsulated PostScript file can also contain a preview image of the picture in a raster format, such as TIFF, or vector format, such as Windows Metafile for simple but direct screen manipulation of the image.

Again, which method is chosen is a matter of personal preference, and the particular implementation available.

22.2 DESKTOP PUBLISHING—BACKGROUND

Neil R. Bauman

We are all busy. Most of us, too busy. And we have too much to read. Further, we are constantly barraged with more to read. Much of what we are given to read, either by colleagues or through the mail, we will throw away. Not because we don't *need* or *want* to read it but because we do not have enough time to read all that we should.

Because of limits on our time, we actually look for excuses to not read a particular document. As a disseminator of information, in a world competing for the reader's attention, you need to do what you can to improve the chances that your document will not be discarded.

This section will provide you with the knowledge you need so that your written document not only looks good but is easy to build and modify using state-of-the-art desktop publishing tools and, more importantly, is appealing to the reader as well as easy to read.

We Put Things Aside That Are Hard to Read

If a given piece of literature puts up unnecessary roadblocks for the reader, the reader will simply disregard the document. If the reader cannot determine a (perceived) benefit, almost immediately, and including all of the following:

- what the document is about (subject),
- why it is important to them and why they need to read it (interest, relevancy), or
- what they will, or can expect to, get from it (knowledge gained), and
- how (organization) to quickly get information from the document,

the reader will not read the item.

On the other hand, when we come across things that are of interest or relevant *and* easy to get into *and* we quickly learn something (probably by just scanning the captions, subheads, or something else very accessible) chances are we will pick the document up and will start to read it. You want your document to be one of those picked up.

Is There a Way to Measure a Well-Designed Document?

There is. You know something is well designed if you learned something quickly. You probably did not notice any design. You did notice the topic and then a phrase or sentence or two. Then you decided to start to scan it. Maybe you then read the first paragraph. Then maybe the last. Then maybe you study a chart or figure. Before too long you were learning something that was worth your time.

The simple fact is the document was

- of interest and
- easy to read.

Readability Is the Single Key

Documents are very readable when one can read blocks of words with one glance (this is actually how we all read, if given the chance). If eye movement is controlled in a comfortable, subtle way the reader can, and does, skip around the document all the while picking up concepts or interesting little tidbits

What Is the Measure of Good Readability?

When a reader can learn the maximum number of things in the minimum amount of time.

How Can You Create Readable Material?

Before we explore the "how to's" we need to understand the "hows" and "why's." Let us start by studying how readers read things.

A reader will read things in the order that he or she sees them. Simply put, the reader will be attracted to the easy-to-read items. For most adequately designed documents a reader will read and/or notice things in the following order:

1. The title
2. Pictures and their captions
3. Other pieces of art such as graphs and tables
4. The lead-in (called the abstract in technical journals, the deck in a magazine article)
5. Callouts or shouts (also called pull quotes)
6. Subheads or paragraph titles
7. The first and/or last paragraph of the document
8. Bold, italicized, or otherwise emphasized words.

It should be your goal to use these eight points to your advantage. That is, what you need to do is to control your reader through your material, superficially at first, and then hopefully, at some length. To control your reader you must first organize your messages in some hierarchical way. The title, for instance, *must* be catchy and clever. Your goal here should be to make the potentially interested reader curious while simultaneously stimulating thought. You will need to spend a lot of time on your title.

Next, the reader will look at any pictures that appear. Knowing this, make sure your captions explain more than just the obvious. Try to bring the reader in. Here you are presented with your second chance to explain something about the document. Again, try to provoke curiosity but also try to teach the reader something.

If you do not have pictures, graphs, or tables you will need to employ another trick to entice your reader: contrast.

Contrast is the key to creating interesting looking things. And what is interesting to look at is often interesting enough to start to read. The art and science of creating and implementing contrast in your document (which we will explore in detail later) is necessary if your reader is to continue on with items (4) through (8) above. Fortunately, each of these, by necessity, will stand out. One cannot avoid noticing an abstract or a lead-in. However, keep in mind that these five items require that the reader start reading your document. They must be easy to read, interesting (from the very start) to read, and either pose questions the reader cannot answer or provide information (quickly, readily, and easily) the reader did not know.

Four simple mechanisms are used by publishers to create contrast:

- Larger type vs. smaller surrounding type
- Longer lines vs. shorter lines of the surrounding type (or vise versa)
- More space rather than less space between lines (for the surrounding type)
- Different color type (which can be achieved, for example, by either making the type bolder and blacker, printing the type in a nonblack (second) color, or putting the type in a tinted or colored box).

Each of these techniques works because the surrounding environment is different than the item being emphasized. That is, the important phrase or sentence clearly stands out because it is visually different from the rest of what is seen.

Before we begin to provide you with the tools and capabilities of implementing contrast, you need to understand several of the terms and concepts used by designers and typographers. Knowing these terms is fundamental to using the desktop publishing packages.

22.3 TERMINOLOGY

Neil R. Bauman

Points

The world of publishing uses its own measurement system. The basic unit is the point. All type, line, and paragraph spacing is described in points.

Since the advent of PostScript, a page-description programming language used to place very small dots on a computer-generated page, the size of the point has become 1/72 of an inch. Prior to PostScript (which was developed in the early 1980s), there were 72.27 points per inch. However, for practical and technical reasons, the developers of PostScript changed the size of the point and rounded it down so that 72 would exactly be equal to 1 inch. Much to everyone's surprise, PostScript quickly became the standard within the desktop-computing and publishing industry and has since become the standard within the graphics arts industry.

Picas

There are, and have always been, 12 points to a pica. Again, however, the exact value of the pica changed when the size of the point changed.

Typeface

Typeface is the term for a particular font and all of its related family members. Probably, the two most known (and commonly used today) typefaces are Times Roman and Helvetica. The Times Roman family has four variants: the plain (also called regular), the italicized version, a boldface version, and an italicized bold. The four variants, or fonts, that make up the entire Times Roman typeface are shown here:

abcdefghijklmnopqrstuvwxyz
abcdefghijklmnopqrstuvwxyz
abcdefghijklmnopqrstuvwxyz
abcdefghijklmnopqrstuvwxyz

Like Times, Helvetica comes in the same four ways (regular, italics, bold, and bold italics). However, unlike Times Roman, the Helvetica family has several additional variants: light, condensed, light condensed, bold condensed, black, and black condensed.

Regular—abcdefghijklmnopqrstuvwxyz
Italic—abcdefghijklmnopqrstuvwxyz
Bold—abcdefghijklmnopqrstuvwxyz
Bold italic—abcdefghijklmnopqrstuvwxyz
Light—abcdefghijklmnopqrstuvwxyz
Light italic—abcdefghijklmnopqrstuvwxyz
Light condensed—abcdefghijklmnopqrstuvwxyz
Light condensed italic—abcdefghijklmnopqrstuvwxyz
Condensed—abcdefghijklmnopqrstuvwxyz
Condensed italic—abcdefghijklmnopqrstuvwxyz
Bold condensed—abcdefghijklmnopqrstuvwxyz
Bold condensed italic—abcdefghijklmnopqrstuvwxyz
Black—abcdefghijklmnopqrstuvwxyz
Black italic—abcdefghijklmnopqrstuvwxyz

Probably the most distinguishing difference between Times Roman and Helvetica is that the individual letterforms of Times have "hickeys," called serifs, on each letter, as shown in Figure 22.1. Helvetica, on the other hand, is an example of a typeface from a class of typefaces called sans serifs (sans is Latin for "without").

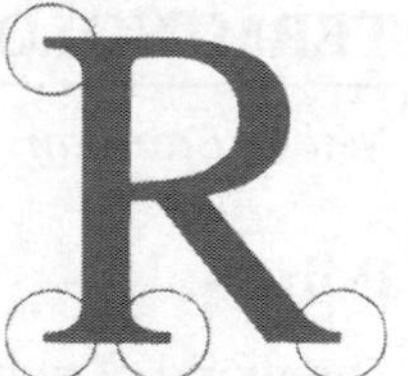

FIGURE 22.1
Serifs.

As we have seen, a typeface is made up of a family of related fonts. The individual fonts are all designed to work together within the typeface family, just like the individual letters of one font should be related. For instance, the bowl (the enclosed looping part of the letterform) of the letters R, P, and B should all be identical within one font. From light to black (called type weight) the loops, axes of the curvature of the letters, length of the serif, etc. (all those things that distinguish one font from another) should be very similar. This similarity is what binds the individual fonts into a typeface family.

Leading

Leading (pronounced "ledding") is the typographic term for line spacing. When using a typewriter, we only have a few options: single space, double space, triple space. When we set type, either traditionally or with desktop publishing systems, we have many more choices, most of which are far more subtle than just single or double space.

Leading is best described within the context of the specific type being set or measured. Type, like just about everything in publishing, is measured in points. Unfortunately, the size of a typeset letter is not measured exactly. Believe it or not, a 24-point letter set in 10 different faces will present itself in 10 different sizes. The basic explanation for this is that type was originally (and still is!) created by artists, not mathematicians. Originally, a font size was labeled 24 points if the vertical distance measured from the top of any capital letter to the bottom of any lowercase descender (like a "g" or a "y") was 24 points. (Actually, the measurement was taken from the top of a lowercase ascender, like a "k" or "h"). The problem is that not all lowercase ascenders necessarily have the same topmost point, for example. Since each typeface is designed by an individual artisan, there are no rules or laws that must be strictly adhered to when creating a font.

Although typeface sizes are not precise, leading is. It is the actual measured distance between typeset lines. The measurement starts from the baseline, which is an artificial (but clearly defined and fixed) line that the type stands on. Because type is an artful expression, not every letter sits *exactly* on the baseline. The type designer wants the reader to believe that every letter neatly "just touches" the baseline; however, this is achieved only by making every letter *appear* to be on the baseline.

By slightly moving some letters up, and some letters down, the type designer can create the illusion that every letter sits exactly on the baseline, as shown in Figure 22.2. This is a fairly common example (in this case, the font is Optima Regular): centered-curved-at-the-base letters (like C, G, O, Q, S, U) sit lower on the baseline than letters that have a stem touching (A, F, H, I, K, M, N, P, R, T), which sit just a bit lower than letters that have a segment of the letterform flattening out (D), or whose arm or stroke is the base of the letterform (E, Z). Optically, when read together as words, the letters all appear to sit exactly on the baseline.

FIGURE 22.2 The baseline.

It is from the baseline that we measure leading. If the space between any (and hopefully every) two lines in a single paragraph, for instance, is exactly 12 points (or one pica), the leading is 12. Graphic art-types, as well as typographers, call this "on twelve" and it is often written as a fractional denominator ("/12"). Continuing with this example, if the type size is 10 points, we describe this type and leading as "ten on twelve" (or "10/12"). When the type size and the leading are the same, typographers call this "set solid." This is similar to single spacing lines on a typewriter. Ten on twenty (10/20) would be double spacing. (By the way, characters produced on a typewriter approximate 12-point typeset type.)

Justification

Justification is the term used to describe how the lines of a paragraph line up. There are four possibilities: flush (meaning aligned) left, ragged (meaning random edges) right—like a mechanical typewriter would do; flush right, ragged left; centered—each line would be ragged and the paragraph would have an easily discernible center axis; and justified—both flush left and right. We tend to assume that justified type is professionally set and the easiest to read. Both assumptions can be wrong.

With ragged type, the space between words in a paragraph is always the same. When a paragraph is set justified, the space between words will vary, line by line. Similarly, with ragged type, the space between letters is always the same. Again, with justified type, the space between letters will vary.

The traditional typesetter or desktop software user can set the parameters for both of these spacing options, and in the case of letter spacing, can force the letter spacing so that like ragged type, the space between the letters stays fixed. However, word spacing will always vary when type is justified.

When justifying, sometimes one word (or more depending on the line length) does not consume enough of a line to be spaced out to create the proper flushing (both left and right). The term "force justify" is used to describe what is done to force a line of type to justify. It is achieved by adding space between whatever words and/or letters there are in the line.

The Fixed Spaces: Em, En, Thin

The em (a most handy word when used by scrabble players) is a spacing concept. Its actual definition is a nonprinting square whose side measures the number of points of the type with which it is set. For instance, if a line of type is set with 10-point type, the em space would be 10 points tall and wide. The fact that height is included in the definition is useless as leading settings take care of proper spacing between lines. The em is commonly used as the initial spacing for a new paragraph. When new paragraphs are typed out on a typewriter, we usually put five spaces (called a spaceband when using typesetting equipment or software) before the first letter of a new paragraph. This is what the em space was originally designed for.

The en space is exactly half (the width) of the em space. If set as 10-point type, the en space is 5 points wide. The en space is most practically used when putting a fixed amount of space between a bullet and the bulleted copy. The width of the bullet plus the en space then is usually the left indent for the remaining bulleted copy:

- Now is the time for
  all good men
  to come to the aid

To describe the thin space you must, once again, realize that artists, not physicists, are behind the scenes. The thin space, by definition, is either 1/3 or 1/4 of an em space. It can vary, but rarely does, within these two bounds. The thin space also defines the width of the typeset numerals zero through nine, plus several other number-type characters (the plus sign, minus or dash, dollar symbol, etc.). In some fonts these characters are thicker than in other fonts (accounting for the variability of the thin space). For most fonts, the thin space width is also the same as the spaceband (ragged, unadjusted word-space use) width.

In addition to the nonprinting uses of the em, en, and thin spaces, there are three corresponding, identical width, types of dashes: the em and en dashes and the hyphen. The em dash is the longest and is used in place of the two typewritten dashes used when writing parenthetically, for example. The en dash is used primarily for specifying ranges. For example, ten to twelve as 10–12. The hyphen is used as a single dash for hyphenating words and as a minus sign (although it is preferable to use the en dash for minus signs when not in a tabular form).

Measure

This is the term used to define the length of the typeset line. Unless setting just one column of numbers, it is measured in picas and points and written: number, the lowercase letter p, and another number. The first number refers to the number of picas and the last number is the fractional part of one pica (expressed as points). For example, 13p6 is 13 picas and 6 points, which also happens to be 13.5 picas. Generally, decimals are not used because most of us do not think quickly enough to convert twelfth's to points and back again. Also, which is easier: 13p5 or 13.41666 picas?

A general rule of good typography is that the longer the line (i.e., the greater the measure), the more leading one should use. Studies have shown that blocks of long lines of set type can be hard to read (we often reread the same line over and over). The problem is that your eye has a hard time going backward over 6 inches worth of copy and then picking up the next line, particularly when trying to read phrase at a time, rather than word at a time. To make the journey easy for the eye, when setting long lines of type, keep the space or leading between lines larger than usual. This way, the eye can quickly back track using the baseline of the line just read.

Margin

This is the space on the page between the copy and the edge of the paper. Since copy never bleeds (the graphic-arts term for something, usually art or a photo, appearing to go beyond the edge of the paper) the lines of type define the boundary for the columns. The margin space from the edge of a column to the binding is called the gutter. If a multipage printed document is glued together (called perfect binding) like many journals are, the gutter space should be slightly larger than the outer margin.

22.4 DESIGN TECHNIQUES

Neil R. Bauman

This section will attempt to lay out a (simplified) formula to follow when designing and putting together a scientifically oriented published document. As we go along, traditional guidelines to follow and traditional "do's" and "don'ts" will be addressed as well as a generic approach (or a quick-and-dirty solution) to many common problems and situations will presented.

Before any typesetting starts, some fundamental layout options need to be decided. In addition, some physical appearance decisions need to be made. Although these decisions can be changed, and often are, keep in mind that arriving at "what is right (or best)" for the copy will entail a trial-and-error formulation.

How Many Pages Will the Printed Document Be?

For instance, if you are about to put together a résumé, you should try to get the copy to fit in one page. If you have three typewritten pages to fit on one page you now have some guideline as to what you must do: use a condensed typeface and/or use relatively small type (9 or 10 points—remember, 12 points is the typical typewritten size character), and/or if the copy can be nicely broken into sections (as all résumés can), consider more than one column for the type. (To really pack a lot of copy on one page, publishers sometimes go to five or more columns. This is out of the question for a résumé, however.)

If the document is going to be multiple pages, do you have a limit? Do you want to set a limit? The answer is always yes since no one wants to read 1,000 pages. However, you do not always have to use small and condensed type. The most important thing is readability. If you know, for instance, that the finished document will be two pages, as you set the type, plan on going larger—from maybe 10 or 11 to 12 point type—if the copy falls somewhat short on our second page.

One limiting factor, not to be ignored, is how the document will be initially generated and then finally reproduced. If you do not have access to a laser printer (or some other high quality, reasonable resolution output device) 9-point type, and sometimes even 10-point type, may not be very legible for the reader. Similarly, if the document is to be photocopied, rather than printed in the traditional plate and offset way, be careful using small, light weight, or very condensed typefaces: they often do not reproduce well from a copy machine.

How Many Columns of Type Do I Want to Present?

Single column approaches to copy, like a paperback book or a typewritten letter, do not offer the designer a lot of maneuverability. Without design options, our ability to create contrast is limited. However, single-column typeset situations do not eliminate all of our options—at all! Although visual interest is somewhat reduced in one-column settings, any two-to-three word subheads will stick out like a sore thumb.

The other side, of course, is layout difficulty. The more columns, the greater the challenges and the greater the effort. Keep it simple. Two or three columns offer practically all the flexibility inherent in a multicolumn design. Plan on using four columns when you have a particular problem to solve, like a too much copy for too few pages.

Further, studies have shown that people read best with about 40 to 50 letters (or 8 to 10 words) on a line. Given the size of the type you may need to use, and given a certain look to the line spacing, the number of columns may be obvious.

Figure 22.3 depicts two-column and three-column grid templates. Each page has a 3/4 inch margin (all around), and each column is separated by 1/6 inch.

Depending on the actual copy, two columns might actually present themselves as four columns to the reader, as shown in Figure 22.4. When typesetting data like this, consider using one or two columns, as shown. If you go the one-column route, consider extra spacing between each entry or a rule between each entry.

What Typeface Should I Use? Serif or Sans Serif?

Certain typefaces will only work with certain output devices. For example, on a fairly generic PostScript laser printer (300 dpi, Canon engine), try to use Adobe PostScript fonts because

- they are universally available (and, indeed the standard) at all PostScript-based service bureaus, and
- they come with special printing algorithms that produce a consistently excellent look on 300 dpi printers.

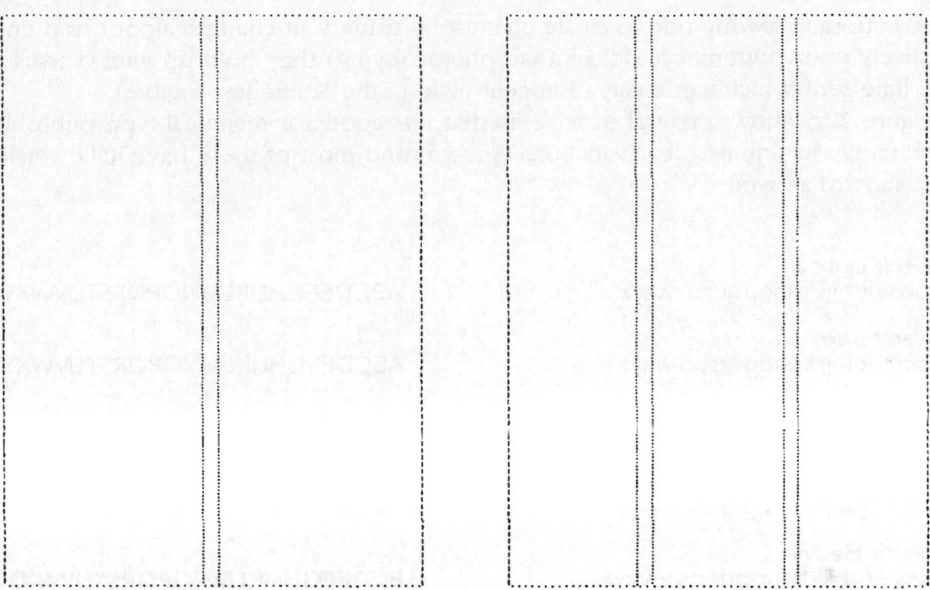

FIGURE 22.3 Two- and three-column templates.

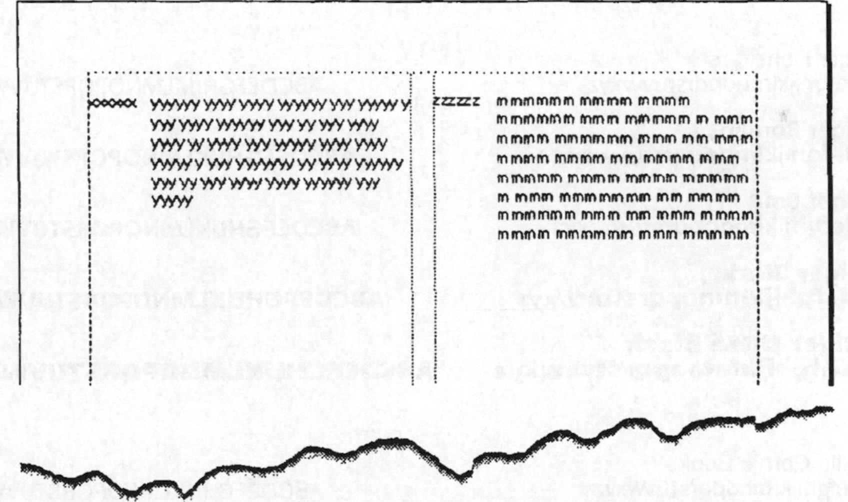

FIGURE 22.4 Indented copy in two columns really looks like four columns.

For a lot of word-based textual material (more than two pages), we recommend using a serif typeface like Times Roman, Palatino, or Century Schoolbook. All three are readily recognized (subconsciously) by the reader (hence, a reasonably high comfort level), are available for hundreds of printers and all major operating systems, and are very legible. It is generally believed, by those who study such things, that serif typefaces are easier to read when reading a lot of words. The hickeys (serifs) on each letter tend to push our eye toward the next letter. It appears that serif-based copy is easier to read when one is reading phrase by phrase (which is how all of us tend to read).

However, for technical documents, résumés, charts, or numeric data (where reading is naturally slow) we recommend sans serif faces. Many have rich typeface families to choose from (allowing one to create contrast by using font changes alone), and under relatively poor printing conditions (like photocopying) they hold up well (versus the tiny little serif which can easily disappear making the letters less legible).

Figure 22.5 shows several sans serifs that are good for technical-type publishing. Each family has quite a few variations (shown) and most of these have italic versions (not shown) as well.

Avenir Light
abcdefghijklmnopqrstuvwxyz ABCDEFGHIJKLMNOPQRSTUVWXYZ

Avenir Book
abcdefghijklmnopqrstuvwxyz ABCDEFGHIJKLMNOPQRSTUVWXYZ

Avenir Roman
abcdefghijklmnopqrstuvwxyz ABCDEFGHIJKLMNOPQRSTUVWXYZ

Avenir Medium
abcdefghijklmnopqrstuvwxyz ABCDEFGHIJKLMNOPQRSTUVWXYZ

Avenir Heavy
abcdefghijklmnopqrstuvwxyz ABCDEFGHIJKLMNOPQRSTUVWXYZ

Avenir Black
abcdefghijklmnopqrstuvwxyz ABCDEFGHIJKLMNOPQRSTUVWXYZ

Frutiger Light
abcdefghijklmnopqrstuvwxyz ABCDEFGHIJKLMNOPQRSTUVWXYZ

Frutiger Roman
abcdefghijklmnopqrstuvwxyz ABCDEFGHIJKLMNOPQRSTUVWXYZ

Frutiger Bold
abcdefghijklmnopqrstuvwxyz ABCDEFGHIJKLMNOPQRSTUVWXYZ

Frutiger Black
abcdefghijklmnopqrstuvwxyz ABCDEFGHIJKLMNOPQRSTUVWXYZ

Frutiger Ultra Black
abcdefghijklmnopqrstuvwxyz ABCDEFGHIJKLMNOPQRSTUVWXYZ

Franklin Gothic Book
abcdefghijklmnopqrstuvwxyz ABCDEFGHIJKLMNOPQRSTUVWXYZ

Franklin Gothic No. 2 Roman
abcdefghijklmnopqrstuvwxyz ABCDEFGHIJKLMNOPQRSTUVWXYZ

Franklin Gothic Demi Bold
abcdefghijklmnopqrstuvwxyz ABCDEFGHIJKLMNOPQRSTUVWXYZ

Franklin Gothic Heavy
abcdefghijklmnopqrstuvwxyz ABCDEFGHIJKLMNOPQRSTUVWXYZ

Franklin Gothic Condensed
abcdefghijklmnopqrstuvwxyz ABCDEFGHIJKLMNOPQRSTUVWXYZ

Franklin Gothic Extra Condensed
abcdefghijklmnopqrstuvwxyz ABCDEFGHIJKLMNOPQRSTUVWXYZ

FIGURE 22.5 Sans serif samples.

The thickness of the individual characters is called the typeface's weight. Figure 22.6 shows the gamut: from light to ultra bold. The spectrum of weights usually goes like this (from thinnest to thickest): light, book, regular, medium, demi bold, heavy, bold, extra bold, black, ultra bold. No one typeface family has all of these weights, but several have many of these.

If you use a typeface that has four different weights (light, book, medium, and bold are four common options), plan on using the book or regular version for most of the copy of your document. Use the book italic or regular italic to emphasize a word or two within the copy or when referring to a book title. (The rule is, whenever you would

Futura Light
abcdefghijklmnopqrstuvwxyz ABCDEFGHIJKLMNOPQRSTUVWXYZ

Futura Book
abcdefghijklmnopqrstuvwxyz ABCDEFGHIJKLMNOPQRSTUVWXYZ

Futura Regular
abcdefghijklmnopqrstuvwxyz ABCDEFGHIJKLMNOPQRSTUVWXYZ

Futura Heavy
abcdefghijklmnopqrstuvwxyz ABCDEFGHIJKLMNOPQRSTUVWXYZ

Futura Bold
abcdefghijklmnopqrstuvwxyz ABCDEFGHIJKLMNOPQRSTUVWXYZ

Futura Extra Bold
abcdefghijklmnopqrstuvwxyz ABCDEFGHIJKLMNOPQRSTUVWXYZ

Eras Light
abcdefghijklmnopqrstuvwxyz ABCDEFGHIJKLMNOPQRSTUVWXYZ

Eras Book
abcdefghijklmnopqrstuvwxyz ABCDEFGHIJKLMNOPQRSTUVWXYZ

Eras Medium
abcdefghijklmnopqrstuvwxyz ABCDEFGHIJKLMNOPQRSTUVWXYZ

Eras Demi Bold
abcdefghijklmnopqrstuvwxyz ABCDEFGHIJKLMNOPQRSTUVWXYZ

Eras Bold
abcdefghijklmnopqrstuvwxyz ABCDEFGHIJKLMNOPQRSTUVWXYZ

Eras Ultra Bold
abcdefghijklmnopqrstuvwxyz ABCDEFGHIJKLMNOPQRSTUVWXYZ

Akzidenz Grotesk Light
abcdefghijklmnopqrstuvwxyz ABCDEFGHIJKLMNOPQRSTUVWXYZ

Akzidenz Grotesk Roman
abcdefghijklmnopqrstuvwxyz ABCDEFGHIJKLMNOPQRSTUVWXYZ

Akzidenz Grotesk Bold
abcdefghijklmnopqrstuvwxyz ABCDEFGHIJKLMNOPQRSTUVWXYZ

Akzidenz Grotesk Black
abcdefghijklmnopqrstuvwxyz ABCDEFGHIJKLMNOPQRSTUVWXYZ

FIGURE 22.6 Typeface weight samples.

underline on a typewriter, use italics when setting type.) The medium weight would be most appropriate for a subhead or caption. The boldest font would be used for the title (although not necessarily because the title, often set large and placed on top, will be the first thing seen anyway), lead-in (if not too many words), or anyplace else where you want to hit your reader on the head with a hammer. The eye will always go to the darkest thing on the page and, therefore, words set in bold will be the first words seen.

In addition to weight, many faces have variations based on the condensing and expanding of the characters. The condensed faces are the most commonly available. Expanded faces are used more as "display type" (decorative or type used primarily for titles and other places where the reader has to read only a few words). You can occasionally use expanded faces for one word subheads or descriptors on a résumé.

Condensed faces are most useful when setting a lot of type in tabular formats. Generally, when there is with too much copy, tabular data, or just too many words for the space allocated, think "condensed." However, condensed faces are slightly harder to read than regular faces, so use the condensed typefaces when necessary, not by design.

What Point Size(s) Do I Use? How Much Bigger Should My Subheads or Titles Be Than the Body Copy?

Unless setting a disclaimer or trademark/copyright notice, never use type smaller than 8 point. Generally, you would rarely want to set copy smaller than 9 point. It is simply too hard for a reader to see. Furthermore, only set type in 9 point when space is very tight; better to think 10 point condensed. Ten point is a good readable size and you should not hesitate in setting body with this size type. If space is not a problem, try 11 point. And if you have plenty of space, and fairly sparse copy, try to use 12 point for the body.

Subheads should be around 20%± larger than the body copy, callouts and lead-ins possibly 40%± larger, and the title anywhere from twice as large to four times larger.

Prior to computerized and digitized type, typesetters would buy fonts in prepackaged sizes. As a result, certain fixed sizes became industry standards. These standard sizes are still good to work with when setting all sorts of type, regardless of the application: 9, 10, 11, 12, 14, 18, 24, 36, 48, and 60 points. Generally speaking, the smaller four are good for body copy, the larger four for titles, and the two (or so) in the middle for intermediate uses (like subheads).

What Is the Right Amount of Line Spacing (Leading) to Employ?

The answer is relative and is a function of measure (length of the line); point size of the type dictates the absolute value. Some specific examples with guidelines will serve to explain:

- With a single column of copy (where the copy probably is 6.5 to 7 inches in length) the leading should be *at least* 30% over solid. Some practical examples: 10/13, 10/13.5, 10/14. Fifty percent would be good also. Example: 12/18.
- With two columns of copy (each with a measure of 4 to 5 inches, or so) 20 to 30% over solid would be a reasonable minimum. Some practical examples: 10/12 (20%), 12/15 (25%), 11/14 (27%).
- With three columns of copy (each with a measure around 2 to 3 inches) solid through 20% would be a reasonable way to treat the line spacing.

Headlines, subheads, callouts, or other typographic displays can be treated individually. For example, if the title were all caps (the term for all capital letters), negative leading might look best and be easiest to read as well. An example of negative leading (for a title, for instance) might be 60/51 (20 on 17 shown here):

THIS IS MY TITLE
AND IT IS QUITE CLEVER!

On the other hand, Callouts as an example can be interesting with a lot of extra line spacing. In a document with a lot of relatively small size type (and thus a dark looking page), this type of callout would really stand out:

This is the first

now the second

and finally, the

fourth line of a callout.

What Is Easier to Read: Fully Justified Type or Ragged Right Type?

This is a question typographers and designers have been arguing over for decades. There have been studies (done in the 1960s) that concluded that right ragged was easier to read because the studied readers were plowing through the copy faster. More recent scientific studies seemed to prove nothing, however. It is now generally believed that either right ragged or justified columns do not, by themselves, improve or impede readability. Other factors, however, clearly do such as font, a poor leading and measure combination, design of the overall page, poorly spaced lines, and left ragged or centered copy to name a few.

Due to the way type is justified (i.e., by adding or removing space between letters and words) justified type is almost always more compact and more condensed appearing. And word-for-word, justified type does consume fewer lines. It also has a more formal, even polished, look. If you want your copy to feel breezy, use right ragged. As an historical note: the original Gutenberg Bible was set justified, as was just about most type in the early years of typesetting.

However, use what seems appropriate for the copy. For some guidelines:

- Do not justify copy if it is just a few lines long (although a disclaimer or copyright notice, in small type, at the bottom might be fine).
- Do not use justified type for a lot of short-measure lines: the variable amount of spacing between words and/or letters of the words could easily become distracting.
- Do not justify if the type is to be in a single column. Your typical hard cover book, set this way, offers no visual appeal. On the other hand, application permitting, when space is a problem, justifying copy is one tool to use to minimize the space the type will consume.

How Much Space Should I Put Between Paragraphs?

Chances are, when you learned to type on a typewriter, you were told to put five spaces at the start of each paragraph. When setting type you should now know to indent one em space. Another option, however, rather than indenting, would be to put some extra space between paragraphs.

Generally, when you are setting justified type, extra space between each paragraph is the way to go. There is a good reason: justified type already is a bit heavy and compact and the extra space opens the copy up and makes it less dense looking (and visually, more appealing). Picture, if you will, a typical hard cover novel: fully justified, one em indents, no extra spacing between paragraphs. Now picture the page, with extra space between the paragraphs. It is clearly easier to put down the latter and pick up the book later knowing exactly where you stopped. Of course, extra space between paragraphs consumes more space and, when setting a book, more pages.

How much extra space between paragraphs is right? The answer is a function of leading, mostly. (You can use your eye too, making sure the page you are producing is appealing.) If the leading is solid, or within 20% of solid, the extra space between the paragraphs should be half the leading value. For example, if the type is set 10/12 (ten on twelve), six points between paragraphs would be good. However, if you are trying to create a very open look by setting the type 10/15, then probably a full line of extra space (in this case, 15 points) would be good. The goal always is readability. Print out the page and look at it from this standpoint. Are the individual paragraphs easy to spot? Or does the page look or feel cluttered? Are you having a difficult time (no matter how minute) moving from word to word, or from line to line, or from paragraph to paragraph?

Subheads should be kept reasonably close to the copy to which they pertain. The leading can be the same as the paragraph, even if they are bigger in type size. However, in this case, you will need extra space before the subhead so that the previous paragraph does not collide with the subhead.

How Can I Make a Subhead Stand Out Without It Overwhelming the Surrounding Copy?

Some white space above and below the subhead will make the subhead easy to pick up when just scanning the copy. And this is important as a subhead can reveal much about your copy. Obviously, interesting or good summarization's when writing the subheads will get a lot of information across to the reader in a minimal amount of time. If the subheads appear to describe copy that readers will find useful they are apt to read the copy!

In addition to leaving some space around a subhead, one should make the subhead type larger and somewhat different than the copy. You want your subheads to be easy to read, no matter how many pages and regardless to the number of subheads per page. Of course, if you go overboard with too many subheads, it will be hard for the reader to see anything else on the page.

Using type that is 10 to 20% bigger should be enough. Or keep in the same size (or just slightly larger, say 5 to 10%) and boldface the subhead. Another variation on the same theme would be to use all caps (again only slightly larger, if at all). Other typographic variations are discussed in the next section on contrast.

22.5 CREATING CONTRAST

Neil R. Bauman

The single most important element in creating both easy to understand and inviting copy is contrast. The contrast, or difference, that exists between things helps us perceive what we see.

The sky is noticeably blue because the clouds are white, the grass is green, and the road is black. When looking at the leaves in a dense forest it is hard to imagine that *each* trees' leaves are slightly different in color than all the other tress. But this is true! However, if you could remove any two trees from the forest and put them against a bright blue background with nothing else around you would detect the two different shades of green. The slight and subtle difference that makes each tree unique is simply not noticeable without creating more contrast. Contrast was created by placing two isolated trees against a completely different colored background.

This is what you must do with the important words and phrases in your copy. Otherwise, all your words will look like trees in a forest—too much to look at all at once. Not only is this not appealing to a reader but you convey by this presentation that there is nothing of importance amongst the words!

What the Experts Do

Deliberate designers create contrast by placing various elements (type, photos, drawings, logos, etc.) in an unusual way and treating the elements of the page somewhat differently. The most obvious way to create contrast is make one element big and all the other elements small. This is the typical treatment of most print ads you see in magazines. The photo is on the top and usually consumes half the page. The rest of the stuff needed in the ad (like the copy) goes underneath.

But if you take a closer look, the material underneath consists of more than just a few paragraphs. There is the headline, probably a logo (identifying the seller), price, unique feature of the product, or benefit to be gained if you use or buy the product.

Given all this material, the designer needs to create a reason for you to want to decipher all this. This is typically done with a clever one liner (headline), tying the picture to the concept. If there is to be any chance that you will read the copy, you must take note of the one liner—at about the same time that you figure the picture out. And of course, all of this must be accomplished by the reader in the 3 to 5 seconds most of us spend looking at the average page in a magazine. Unless of course, something interesting slows us down!

What You Must Do

You are now faced with this dilemma as the creator and disseminator of a document: your potential reader will somehow come across your document and decide, in a small finite number of seconds, whether or not to read on further. Chances are, the headline alone will not yield a committed reader. First, the headline and then either the lead-in (or abstract), call-outs, or subheads will be scanned quickly, *if* the reader can clearly see them all. The reader's ability to quickly decipher these additional important keys to your copy—and to quickly understand much of what you intend to teach—will be the critical path if you expect the reader to go further into your copy.

Obviously the headline needs to be the largest typographic object on the first page (and all subsequent pages, too). The lead-in, or some other description or reason to read on, must be the next most visible thing, followed by subheads or callouts. All of these things should be on the first page, in addition to some copy.

Five Simple Tricks

How can you make all of these things not only readable but appealing? Contrast. One kind of type style will contrast with another when the following five effects are used in conjunction with each other:

- light typeface/heavy (or bold) typeface
- condensed type/expanded type
- very tight letter spacing/very loose letter spacing
- black type on white background/knocking-type out (i.e., white type in a dark area)
- all capital letters/caps and lowercase letters.

Type Is Not Black and White

Lay on the floor several different types of printed black and white pages. When comparing them you will notice that some pages are darker than others. None of the pages will appear all black (even though the type is all black) and none of the pages will be totally white. Each page will be somewhere in between: each will look gray: either dark gray, light gray, or somewhere in between.

Each of the five typographic mechanisms described in the previous section is an attempt at creating near opposite color effects. For each one of these contrasting pairs one effect is dark and the other effect is light. The expanded type and the loose letter spacing are much lighter when near condensed type and very tight letter spacing, respectively. Observe:

abcdefg hijklmn opqrst
abcdefg hijklmn opqrst

22.6 THINGS TO KEEP IN MIND

Neil R. Bauman

How the Eye Moves

When we read a page—or a spread (two facing pages)—excluding other influences, we always start at the top of the page. Furthermore, we also tend to examine a page from left to right, and from the outside to the inside. Of course, the designer can alter this by drawing the eye to a specific element or two. One thing that draws the eye, and will affect where the reader starts looking, is a photograph. Everyone loves to look at pictures. Another thing that draws the eye is a dark object in a lighted area. To prove the point, find some object that is very dark (nearly black). Put the object somewhere out in the open in a room filled with things. Now ask someone to tell you what is the first thing they notice in the room. Try it on yourself. You can't help it: your eye will always be drawn to the darkest thing.

The down side is that we read very dark lettering, very light lettering, and most other typographic effects (condensed type, expanded type, very tight or very loose letter spacing, reversed-out type, and all capital letters) more slowly than normal type. This side effect, however, can be very useful. For instance, sometimes we want the reader to read a phrase slowly, like a subhead. And if the title is not made up of too many words, here is the perfect place where a slow read may actually improve the chances that any subtleties will be fully comprehended.

However, do not slow the reader down *too* much. Use typographic tricks sparingly. By that I mean when employing any of the above effects, apply the effect only to a small number of words. A lengthy sentence will be too much if the effect is very strong.

Going back a bit, remember that the eye moves to dark things. Knowing this, we can control readers and direct them to see the important words on a page, helping them decide whether to read on. If the title captures a lot of the meaning and purpose of your document, make it the darkest thing on the page. If the title is not, in your judgment, a sufficiently strong enticement, make it less dark. The reader will still read it anyway. Since you have concluded the title will not convince the reader to read on, can the lead-in or subheads do the job? If they can, make them bold but surrounded by a lot of white space. (Surrounded by a lot of space will make them appear, by contrast, darker.) If a particular callout or phrase really captures the value of your document, try to make it the darkest text on the page. Do not make it difficult for the reader to figure out what your document is about and why he should read it. **LET THE KEY JUMP OUT**.

Eight Simple Do's and Don'ts with Type

1. **Never use two space bands between sentences or after a colon**. As a matter of fact, never use two space bands together at all! It makes the copy choppy and slows the reader down.
2. **Use italics sparingly**. They are also harder to read than the roman or regular version of a typeface. They should be used to denote titles of books and to emphasize *single* words.
3. **Do Not Use An Initial Capital Letter For Each Word In A Sentence**. The up and down effect will give your readers hiccups. It also forces them to read each word, one at a time, rather than phrase at a time, again, unnecessarily slowing them down.
4. **For a really nice effect hang punctuation, especially quote marks**:

   “Now is the time for all
good men to come to
to the aid of their
country.”

5. **Try to keep the space between the same elements the same**. That is, if you use a half-line extra space between one paragraph, do it for all the paragraphs. The same is true for subheads. Different spacing between copy confuses the readers because the document looks confused and disorganized. Simple consistency gives the readers a quick opportunity to figure out how you have your document organized. Once they understand what one subhead looks like they will instantly recognize all subheads. Also, this tells the reader that all subheads are equal in importance. Which should be true except for that *one* special callout, lead-in, sentence, word, etc. that you want the reader to see right away.

By the way, this is the one advantage to ragged right copy (as opposed to justified copy): the space between all the letters and all words is exactly the same, sentence to sentence, paragraph to paragraph.

6. **Always hyphenate copy**. By turning off hyphenation, even in justified copy, the location of line endings vary to much. The copy looks like a jumbled mess. Besides, we are all used to reading hyphenated words. And 98% of the time when you see a hyphenated word at the end of a line you know what the balance of the word, on the next line, will be.
7. **When in doubt about the right type size, set it bigger**.
8. **When setting long lines** (if you must), **adjust the type size so that you keep the number of words per line below fourteen.**

Eight Simple Do's and Don'ts When Designing a Printed Document

1. **Be consistent**. Whatever pattern you develop, stick to it. Do not change the pattern so that *all* the columns can end at exactly at the same place.
2. **Be simple**. Clutter does not improve readability. Your objective is for the reader to focus on, and to be able to easily read, your words—not on deciphering where the information is.
3. **Figure out what is the most important concept, and make sure it gets across to the reader quickly**. Do not make your reader figure everything out. Your are writing about something that you have studied. Give your reader the advantage of your superior understanding of the subject. And give extra thought to how to express the important words and phrases. When making a point, use a specific example. People relate better to the experiences of others rather than some esoteric point or pontification.
4. **Figure out, for your reader, what all the secondary points should be**. Make them your subheads.
5. **Align things (paragraphs, pictures, anything) optically**. That is, use your eye when deciding if things are lined up correctly. When aligning things, line them up along the top or along a left or right edge. Try not to line things along the bottom.
6. **Use more white space than you think you need**. Heavy copy tends to turn readers off right away. Plus, white space allows for contrast. When designing a page, think of white space as your friend.
7. **Use design elements judiciously**. Examples are boxes, rules, bullets, leaders (those repeated dots between a topic and a page number), drop caps, and an extra large initial letter. They simply get in the way of the words. A fancy bullet or dingbat will not impress a reader as much as a phrase or concept that holds the promise to teach something.
8. **Use the right tool or mechanism to show relationships**:

To show	Use a
A group of related items with no specific order	Bulleted list
Group of related items with a specific order	Numbered list
Relationships and steps involved in a process	Flow chart or process diagram
An evaluation of items against several criteria	Rating table
A comparison of several things in relation to one variable	Bar chart
Concepts	Illustration

22.7 SOME GENERAL DON'TS: A REVIEW

Neil R. Bauman

Many other elements draw attention and tend to slow the reader down. However, most of these you do not want the reader to notice. If he does, you will have lost a reader:

- Hard to read type
- Type that is too **big** or too small
- TOO MANY WORDS IN ALL UPPER CASE (capital letters)
- Capitalizing Each Word In A Sentence. If You Do This For An Entire Paragraph You Can Forget Keeping Someone's Attention.
- Underlining words is very distracting
- **Making more than just a word or two bold; however, when used sparingly says**, "Hey, I'm an **important** word or concept."
- *Making more than just a word or two italic; however, when used sparingly says*, "Hey, I'm an *important* word or concept."
- A lot of special characters like * # $ % @ &.

Type set without understanding some of the basic rules of good typography can create quite unreadable material. Some of these basic "don'ts" are

- Putting two spaces between sentences. (Yes, this is how we type letters using a typewriter where all characters have the same width. But when setting type for printing out on an ink-jet or low resolution laser printer, never space twice.)
- Poor character spacing (either too tight or too open).
- Poor word spacing (either too much space between words or too little space between words.
- Poor line spacing (either too little or too much space between each line). In either case, it can be hard for the reader to pickup (and therefore read) a group of words at a time.
- Poor paragraph spacing (generally, too much space between each paragraph).
- Inconsistent use of anything. (Particularly, varying the space between paragraphs or between lines.)
- Mixing different character sets (also called typefaces)

22.8 DESIGN WITH PAGEMAKER

Neil R. Bauman

Designing a Résumé

A résumé can be a challenging document to design because we like them to fit on one page. Many, like the one that follows, has more information than can fit on a page if typeset conventionally. The biggest problem here is that we have two and a half pages of publication references.

Andy. B. Candy

1111 Homeystreet Road
Anytown, Pennsylvania 15555
(717) 555-1212

Department of
Computer Science
Sometown University
Sometown, PA 16666
(717) 666-1212

Education

B.E., Yale University, 1960—Electrical Engineering, with additional studies in Physics and Mathematics.

M.E., Yale University, 1961—Electrical Engineering.

D.E., Yale University, 1963—Electrical Engineering. Supported by NSF Research Grant. Thesis on Traveling Wave Parametric Amplification. Major course work was in applied mathematics and circuit design as applied to computer technology.

Employment and Professional Experience

1978 to present: Professor, Department of Mathematics and Computer Science, Sometown University, Sometown, Pennsylvania.
Teaching undergraduate courses in advanced programming, systems analysis and programming, computer architecture, graphics, software engineering and microcomputer systems.

1970 to 1978: F and G College, Maintown, Pennsylvania. Director of Academic Computing. Responsibility for coordinating all academic computing activity for faculty and students with the computer center staff. This included extensive in-house consulting activity in various research areas with both faculty and students. In addition, I directed a grant under the auspices of the National Science Foundation College Science Improvement Program from 1971 through 1974.

In addition, I taught a number of courses and seminars on programming and systems design. From 1970 through 1972, I was the director of an NSF/COSIP grant for the Middle Atlantic Educational and Research Center, a regional computer center based at Franklin and Marshall College. This project was directed toward the training of college faculty members in computer use in the sciences. The consortium included four colleges, one local school district, and a private preparatory school.

1969 to 1970: RCA Electronic Components Division, Lancaster, Pennsylvania. My major activity was to head a design team for the installation of a computer controlled automatic production and testing facility for the manufacture of electronic components. In addition, I taught a series of in-plant courses on digital computers and control systems.

1963 to 1969: RCA Laboratories, Princeton, New Jersey. Various assignments as a Member of Technical Staff associated with the recognition of speech by computer. This included activity in the following areas:

Development of a speech-to-computer interface. A team "Outstanding Achievement" award was received for this work in 1967.

Design of an adaptive speech recognition system with a restricted vocabulary.

Summer, 1959, 1960: Technician; Edgerton, Germeshausen and Grier, Inc., Boston, Massachusetts. Various assignments dealing with the development of high-speed pulse circuits and time-domain reflectometry systems for the testing and simulation of nuclear devices.

In addition, I am a Member of the Society of Sigma Xi and Phi Delta Kappa, honorary professional societies. I hold both a First Class Commercial Operators license and a Class C Amateur Radio Operator's License, W3FIS. I also do extensive consulting in computer applications for business, private industry, and governmental agencies.

Further, I serve as a consulting editor and reviewer for various computer science textbook publishers. I have received recognition for this editorial work in over twenty major text books in the computer science and related fields. I actively consult with business, government, and industry in computer applications problems.

Publications

1. A. B. Candy and J. G. Skalnik, "An Analysis and Experimental Investigation of the Binistor," IRE Trans. ED-9, No. 2, pp. 153–161, March, 1962.
2. A. B. Candy, "A Limited-Vocabulary Adaptive Speech Recognition System," J. Audio Engineering Society, pp. 414–418, October, 1967.

(for more than a page)

The way to start any design process is to figure out what are the important points, in order of importance. This, of course, is usually subjective. In my opinion, there are only three things that are *really* important. The first is that Andy Candy has a doctorate; the second is that he currently is employed as a full professor; third is that he is well published. Let us bring the file into Adobe PageMaker 5.0 and see how we can design Dr. Candy's résumé.

The first step is to create a new PageMaker file, as shown in Figures 22.7, 22.8, and 22.9. We are now looking at an 8 1/2 by 11 inch blank page with a 3/4 inch margin on all four sides.

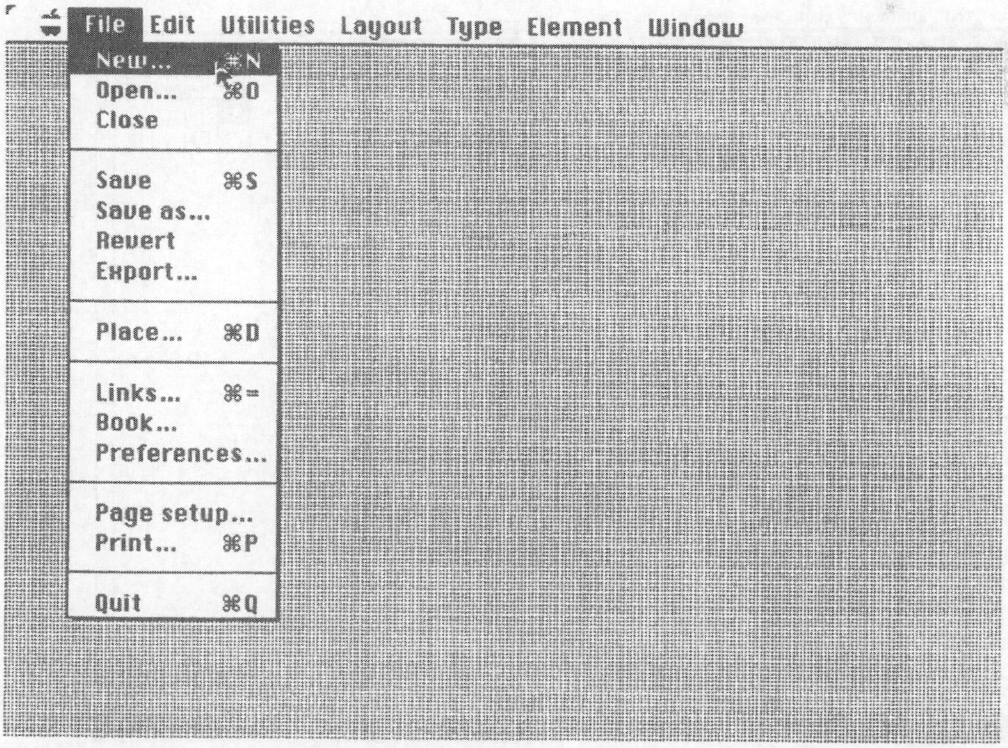

FIGURE 22.7 A new PageMaker file is created.

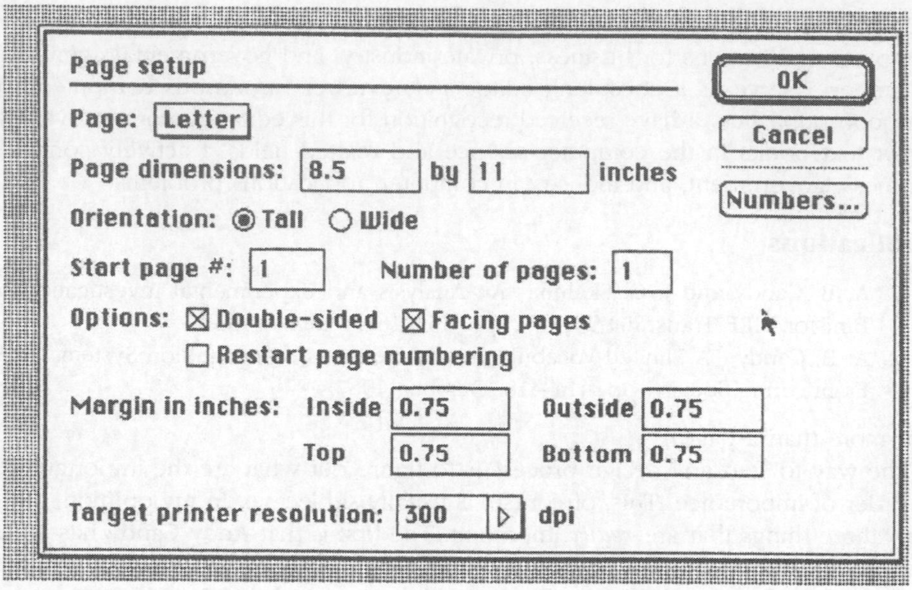

Page setup

Page: Letter

Page dimensions: 8.5 by 11 inches

Orientation: ◉ Tall ○ Wide

Start page #: 1 Number of pages: 1

Options: ☒ Double-sided ☒ Facing pages

☐ Restart page numbering

Margin in inches: Inside 0.75 Outside 0.75

Top 0.75 Bottom 0.75

Target printer resolution: 300 dpi

OK

Cancel

Numbers...

FIGURE 22.8 The Page setup dialog box.

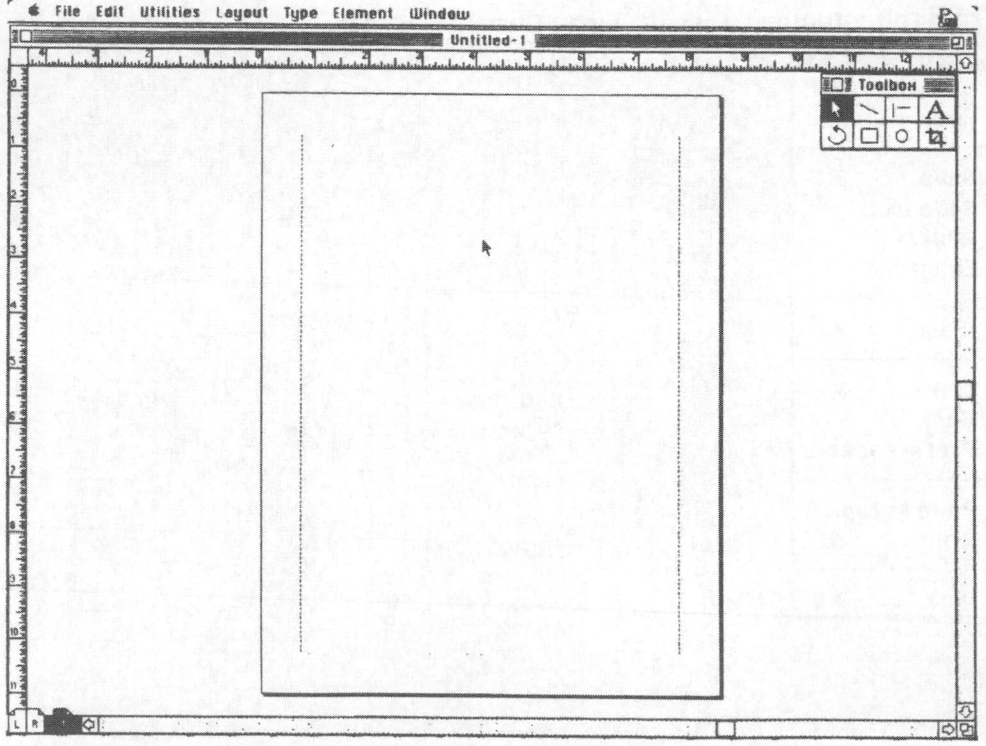

FIGURE 22.9 After hitting "OK," a new, untitled document is available.

If we are going to attempt to fit all of this on one page the publication citations will have to be in a rather small and in a condensed typeface. In addition, rather than use the traditional one-column measure of a typical résumé, for these citations we will use three columns. By using both condensed faces and three columns, we can get a lot of copy in as little of space as possible. However, taken all together, they are important so we will try to fit all of them in less than half but more than one-third of the entire space.

We start by creating a three-column grid on the page and place the citation copy within the grid, as shown in Figures 22.10, 22.11, and 22.12. Figure 22.12 also shows a guideline pulled down from the ruler at the top. It is this general area on the bottom of the page that we plan to put the citations.

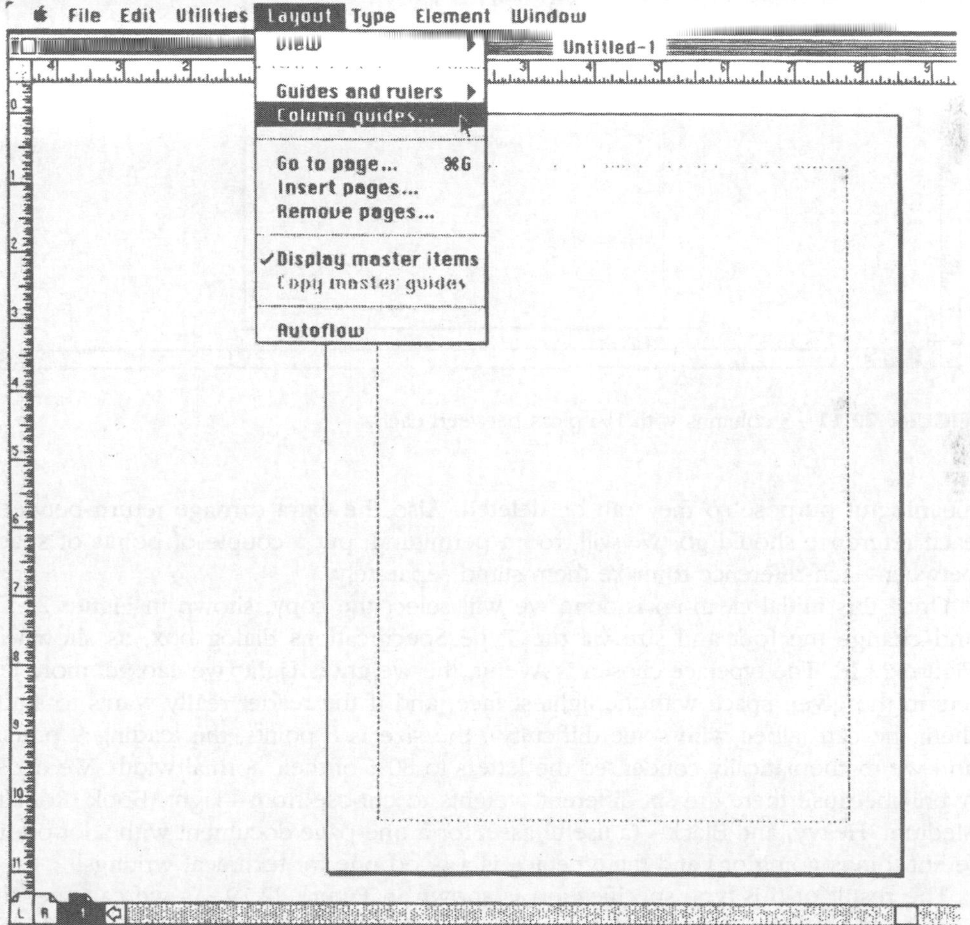

FIGURE 22.10 "Layout column guides" is used for setting up the number and size of columns you want.

Figures 22.13 and 22.14 show the steps involved in selecting the file to place down. Once we are ready to place the text, as shown in Figure 22.15, you click near the intersection of two guidelines and the copy starts flowing in. After we have placed the text in the approximate space we want to allocate, as shown in Figure 22.16, we will have some work to do. First, the actual numbers before each reference serve no

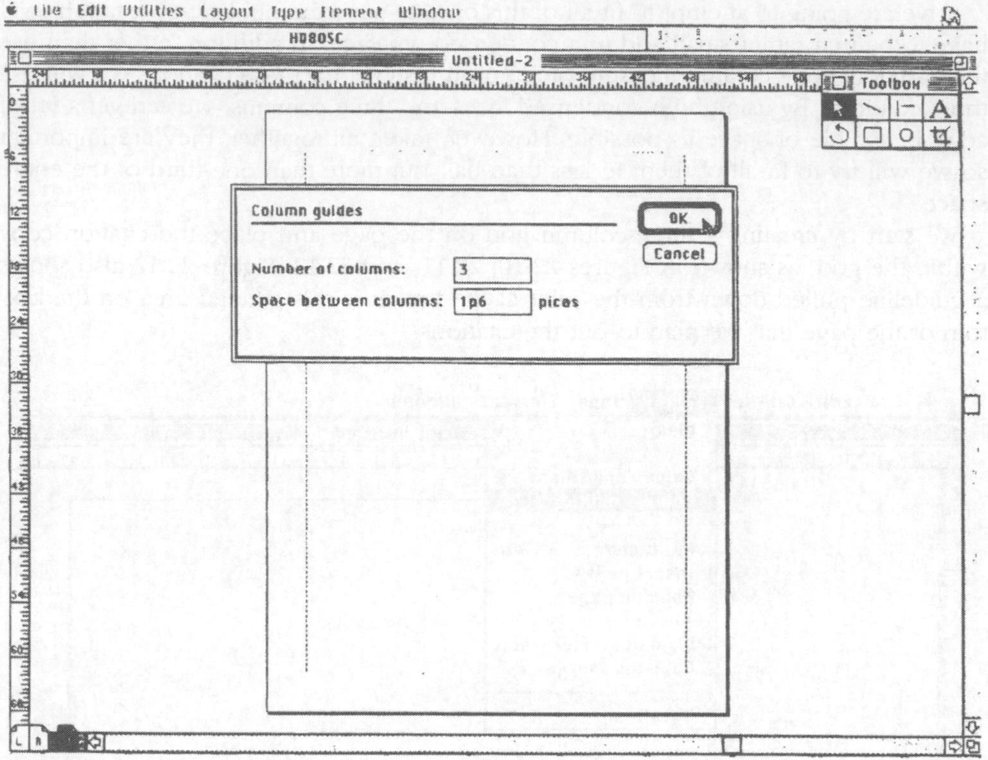

FIGURE 22.11 3 columns with $1\frac{1}{2}$ picas between each.

meaningful purpose so they can be deleted. Also the extra carriage return between each reference should go. We will, room permitting, put a couple of points of space between each reference to make them stand separately.

Once this initial clean-up is done we will select the copy, shown in Figure 22.17, and change the font and size via the Type Specifications dialog box, as shown in Figure 22.18. The typeface chosen is Avenir, the weight is Light (we can get more letters in the given space with the lightest face, and if the reader really wants to study them, he can, albeit with some difficulty), the size is 8 points, the leading 9 points, and we mathematically condensed the letters to 80% of their normal width. We chose Avenir because there are six different weights to choose from—Light, Book, Roman, Medium, Heavy, and Black—(a useful asset for a one-page document with a lot of different things going on) and the typeface is a good one for technical writings.

The result of this type specification is shown in Figure 22.19. As we can see, the copy is about four column inches too long. After printing out the page, it was clear: we could go a bit smaller with the size of the type; the leading could be a little tighter; we can reduce the amount of the hanging indent in the hope of picking up a couple of lines; and, we can move the copy up on the page a little and let the columns go lower a bit. The result of 7.8 on 8.3 with a 1 pica indent now yields about two extra inches of copy. We now grab all three text blocks and physically move them up about 1/2 inch (see Fig. 22.20). Grabbing the plus icon at the bottom of each text block and dragging down lengthens the text block (see center text block, Fig. 22.21).

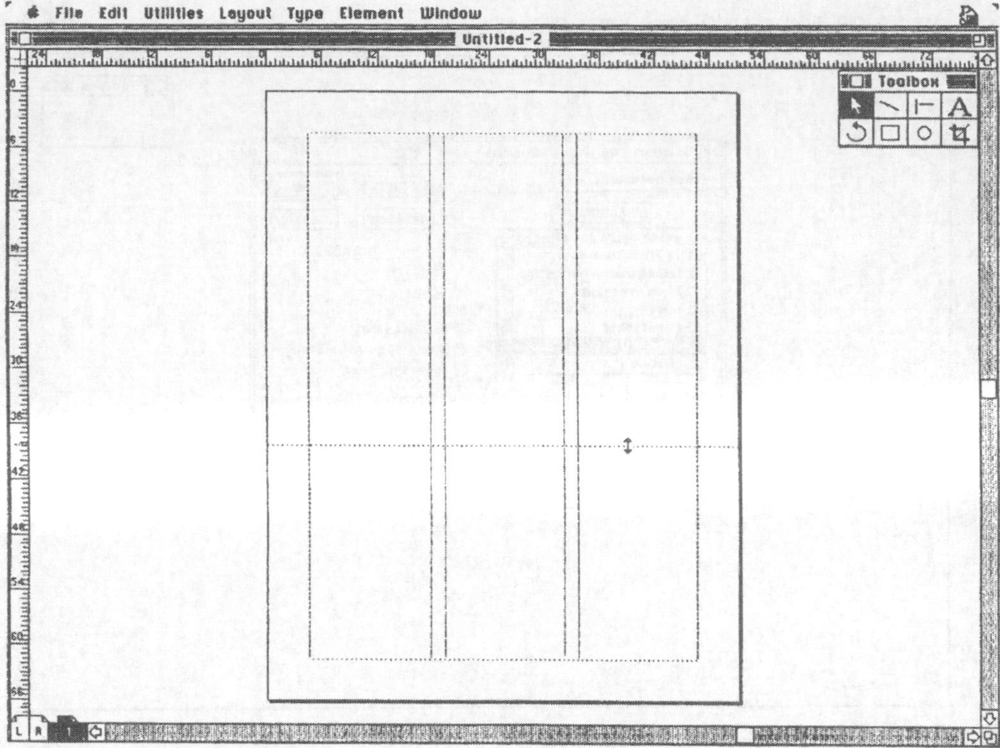

FIGURE 22.12 Guide is brought down from top rule.

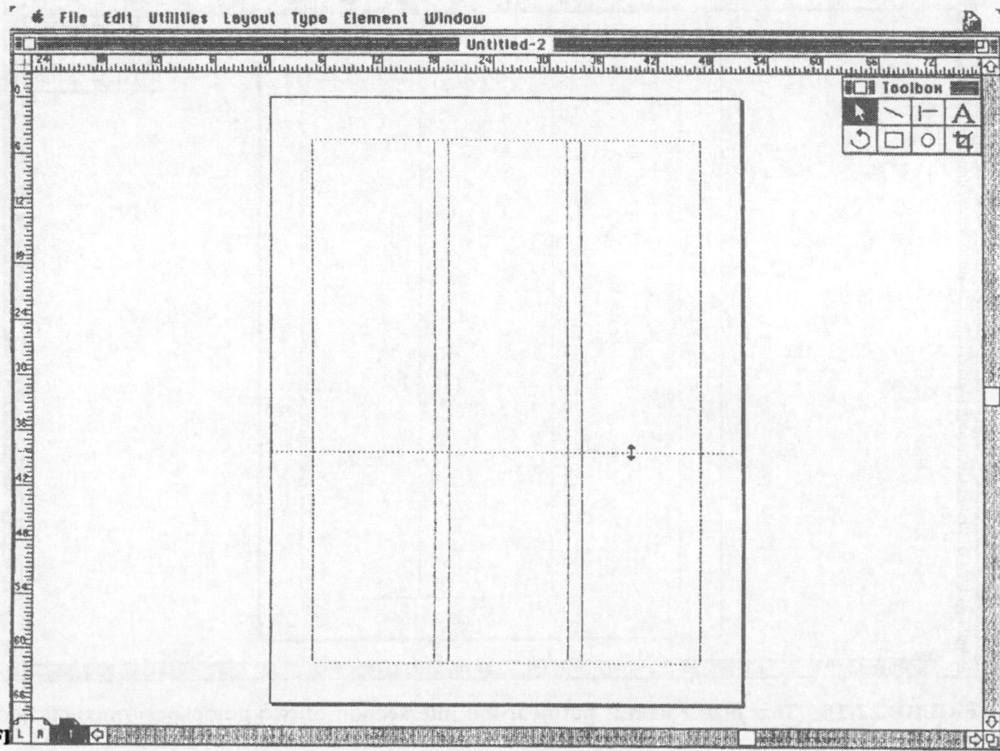

FI

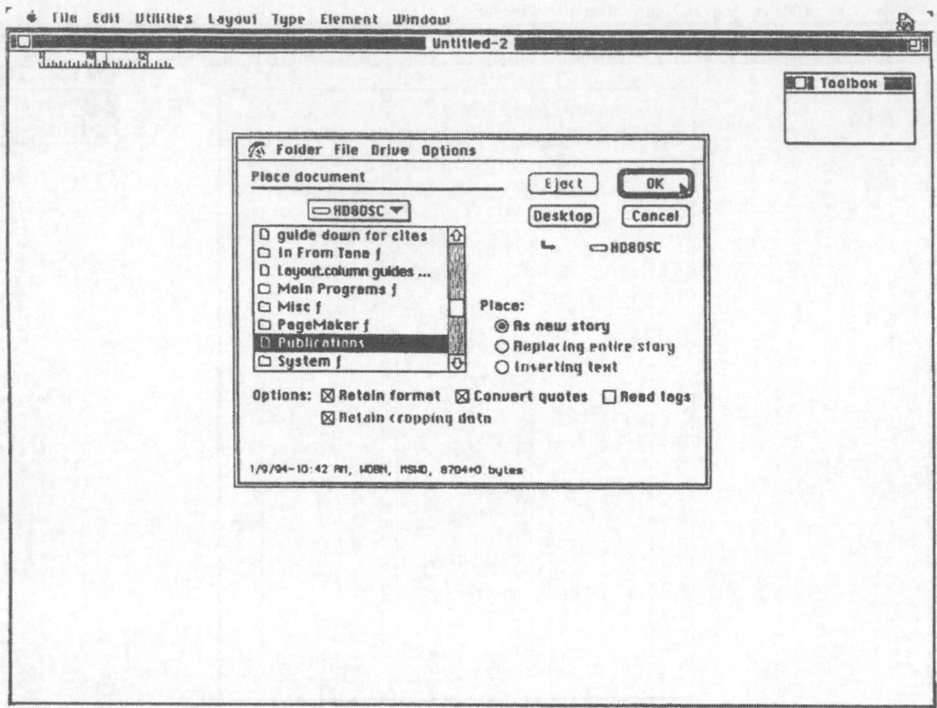

FIGURE 22.14 The file called "Publications."

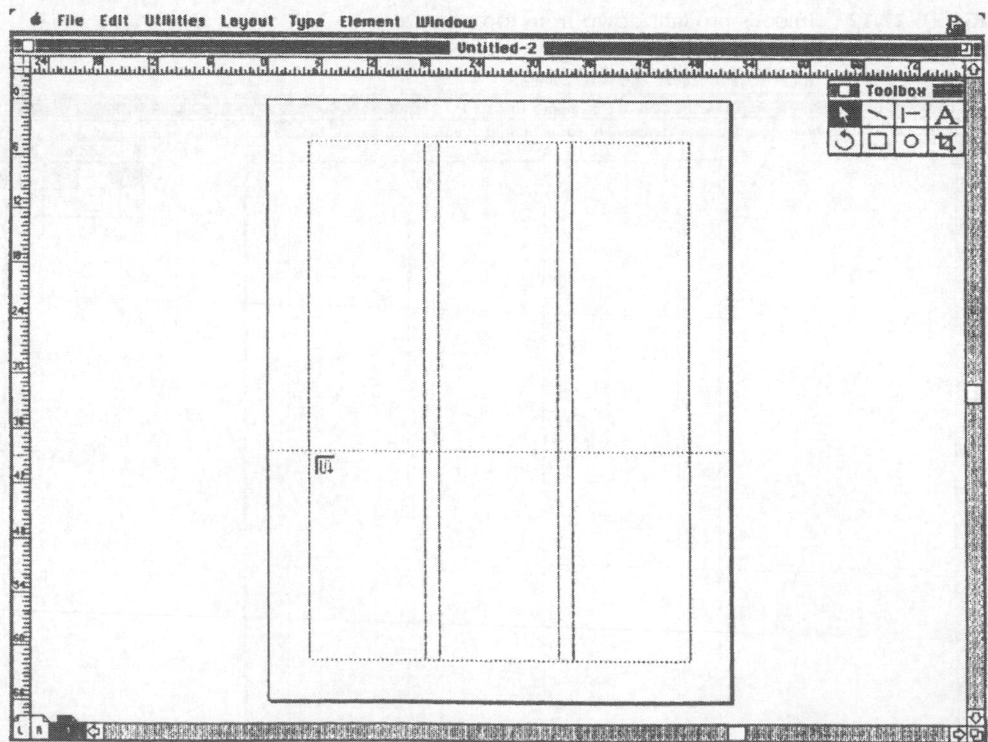

FIGURE 22.15 Text placer icon is put near the intersection of two guidelines; once clicked, the text will flow in.

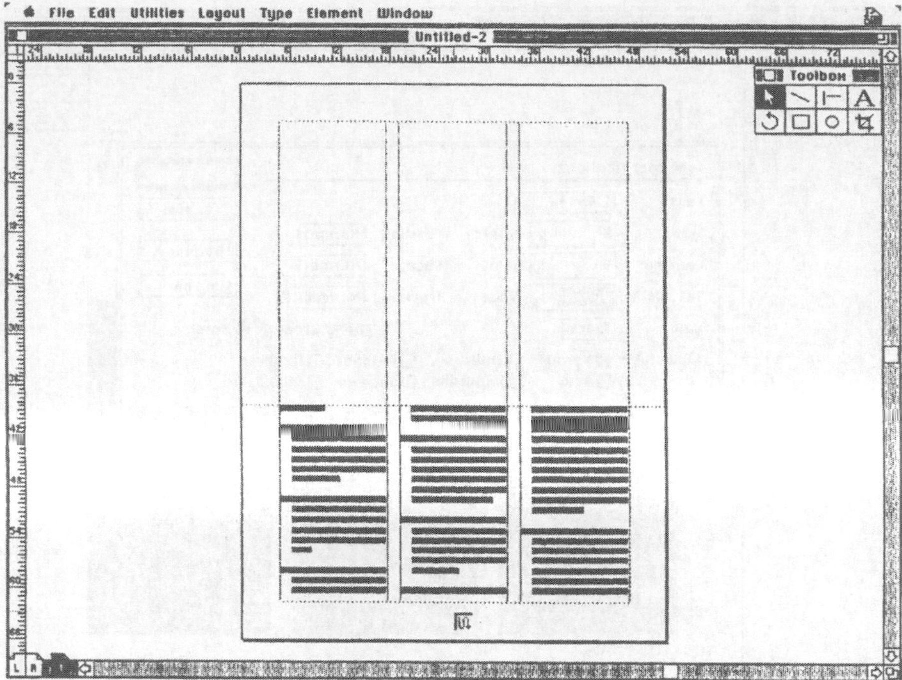

FIGURE 22.16 Initial laydown.

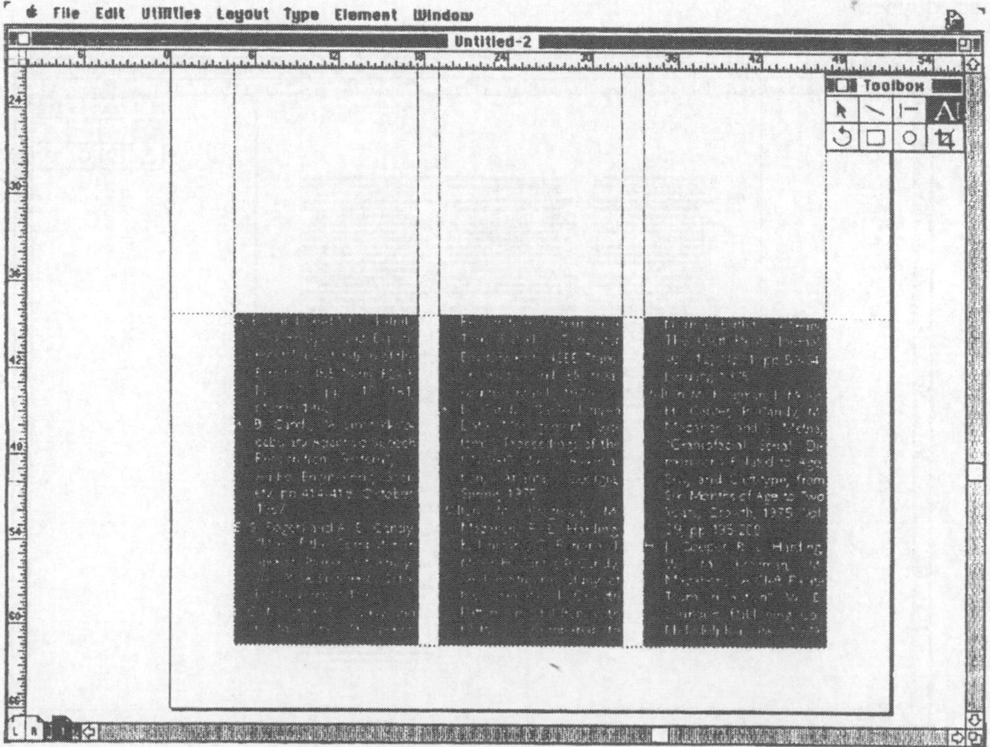

FIGURE 22.17 From the Edit pull-down menu, Select All is chosen.

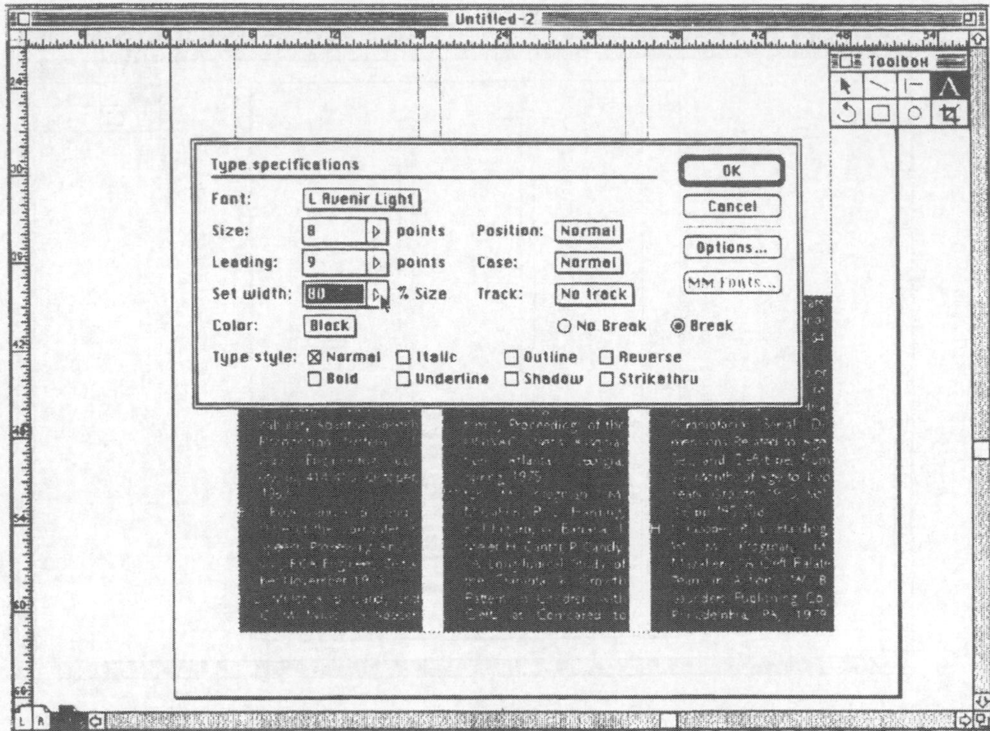

FIGURE 22.18 The type specifications dialog box is activated to change the specs of the selected type.

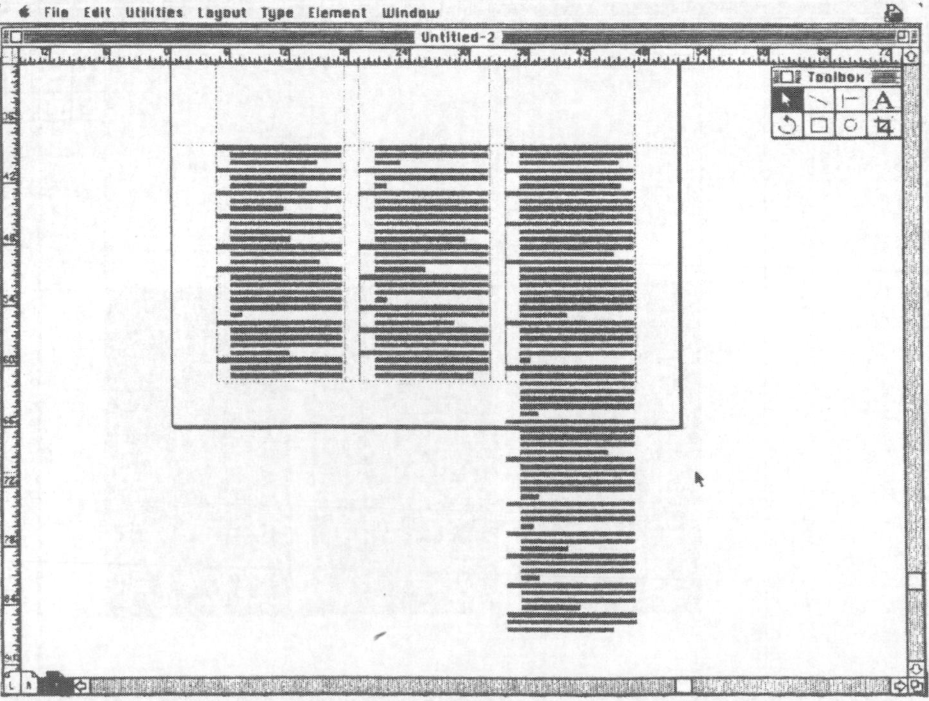

FIGURE 22.19 The result of the change after hitting the "OK" button from above.

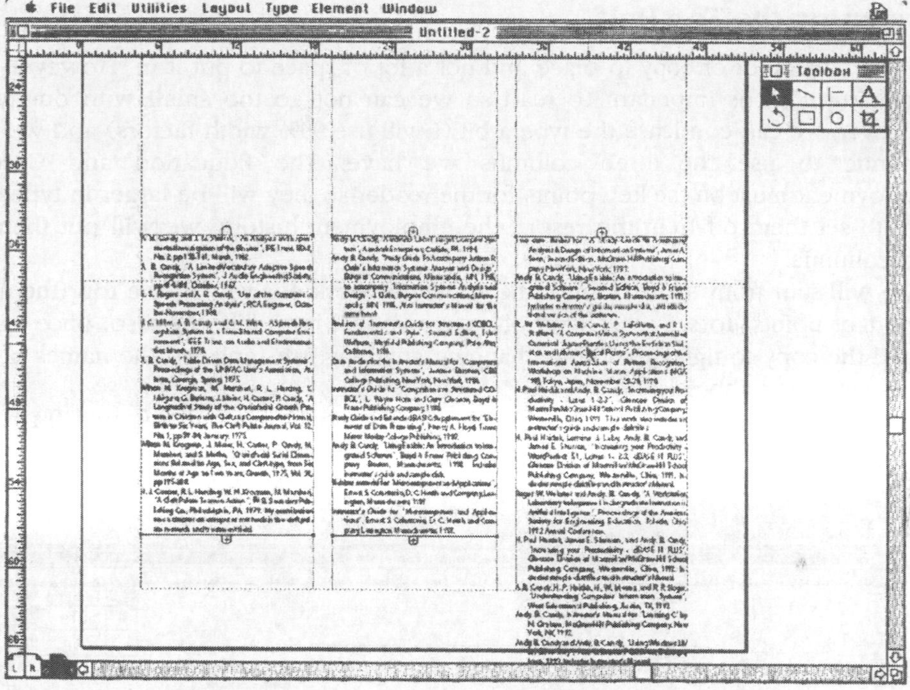

FIGURE 22.20 The three text blocks are selected by holding down the shift key and then clicking on each block. Once selected, they can be dragged up on the page.

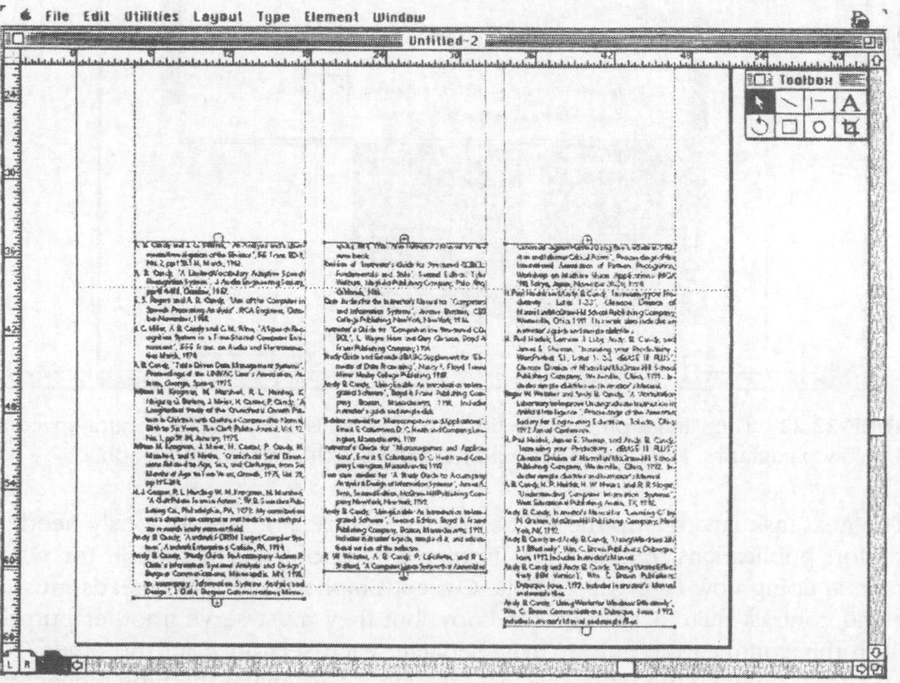

FIGURE 22.21 The handles at the bottom of columns 1 and 2 are pulled down to even the columns out.

Designing the Top Half

We still have a lot of copy to place and not a lot of space to put it in. However, the remaining copy is important to read so we can not go too small with our type. However, we can condense the type a bit (I will use 90% width factors) and we will continue to use the three columns we have. The Education and Current Employment must be the key points for the reader so they will be larger in type size and, to set them off from the rest of the employment history, we will put them in two columns.

We will start from the bottom of the copy and work our way to the top (the most important points) to see what kind of space we will have left. We also, once again, edited the copy to tighten without changing anything: we replaced the names of the states with their abbreviations.

To get this copy to stand out from the publications material, we will flush left and put one-half line of spacing between each paragraph, as shown in Figure 22.22.

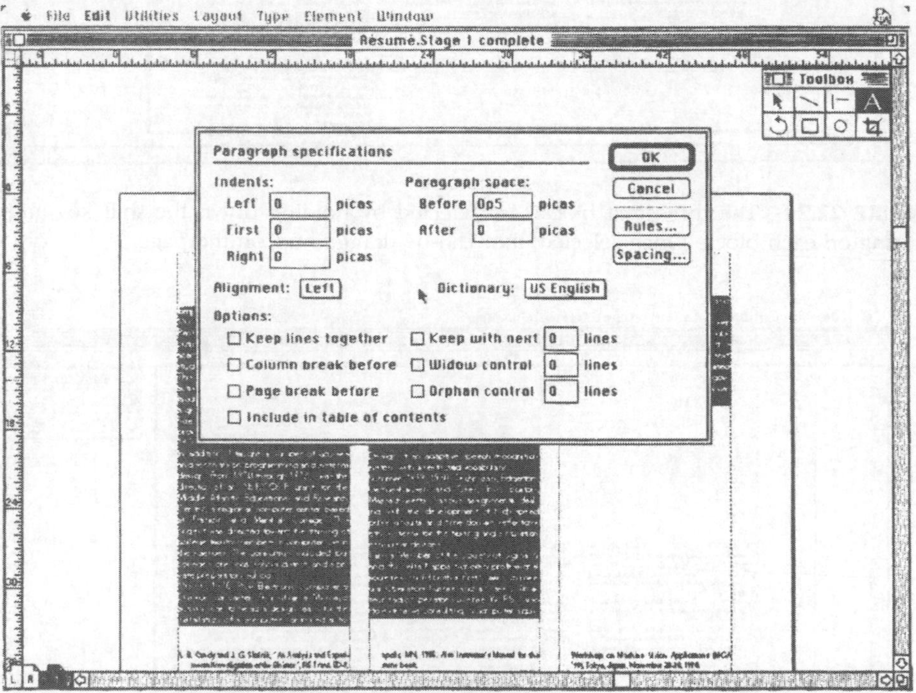

FIGURE 22.22 The Paragraph Specifications dialog box is used to place extra space above and below paragraphs. Here we put five points before each selected paragraph.

The next task is to set up the subheads (see Fig. 22.23). We obviously need a sections for: publications, employment history, education, and a section for what the person is doing now (we will call this Current Emphasis). The subheads must stand out and contrast quite a bit with the copy, but they must serve another purpose as well in this confined page: they must separate each section from the other. To create contrast, open up the letter spacing, (see Fig. 22.24) make the type bigger, darker, and expanded, (see Fig. 22.25) and place a rule over the text the subhead will cover.

Within the Employment and Professional Experience section we need to get each

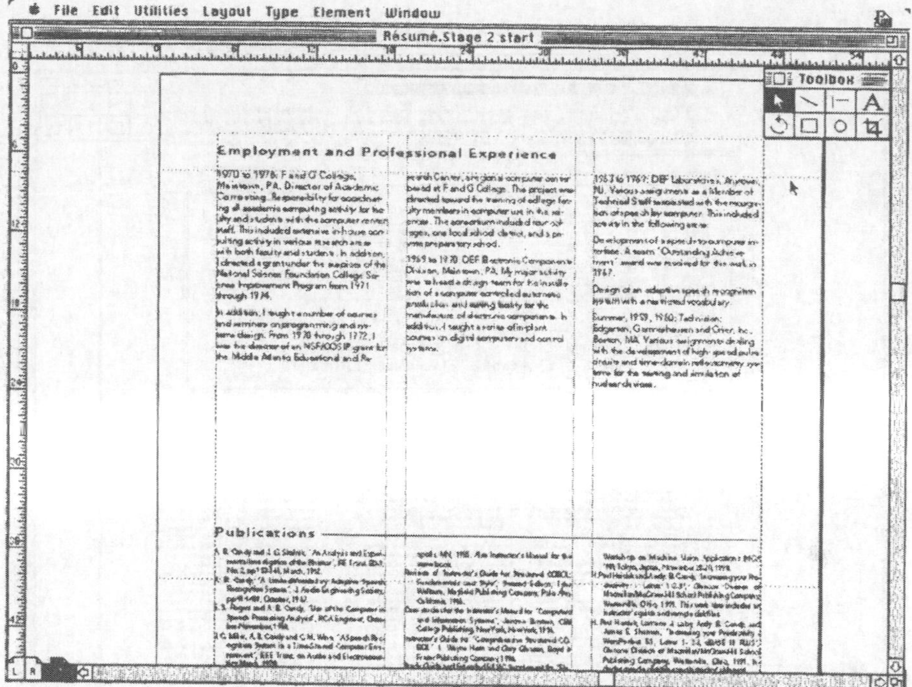

FIGURE 22.23 Subheads starting to take shape.

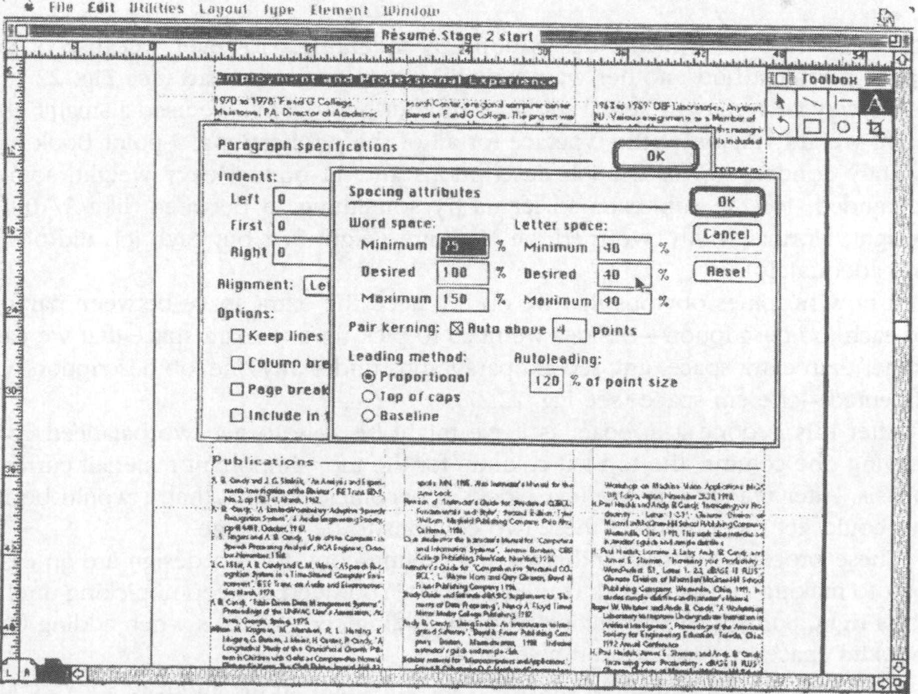

FIGURE 22.24 The spacing attributes sub-dialog box of the Paragraph Specifications dialog box allows us to add extra letter spacing.

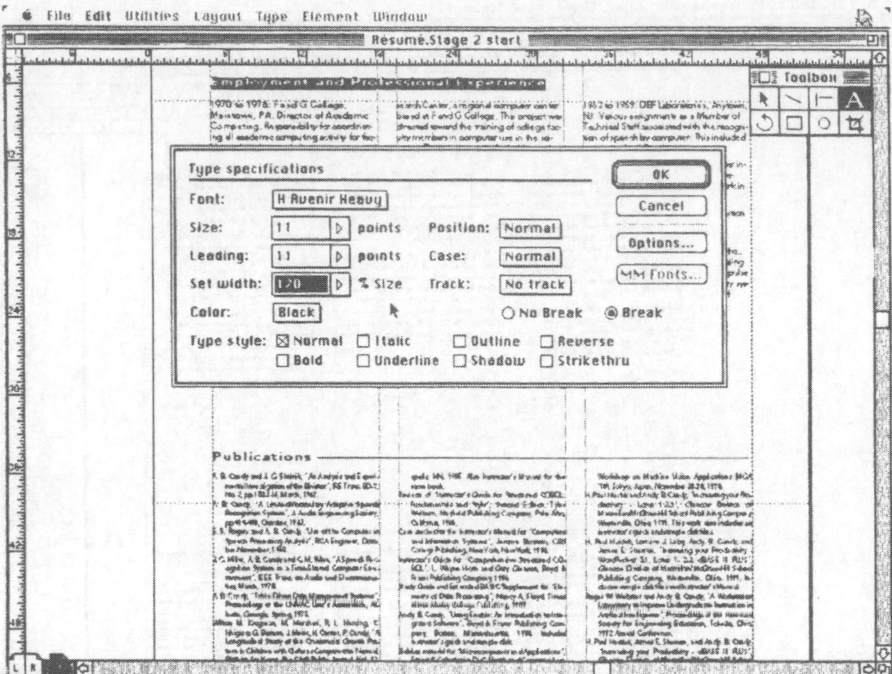

FIGURE 22.25 Type specifications for the subheads: 11/11 H Avenir set at 120% normal width.

job to stand out. Typically this is done by indenting. However, because of limited space we must find another way. We will create a subsubhead (see Fig. 22.26) that is halfway between our actual subhead and our body copy. We used a small (7.8 size) Light weight version of the typeface for all of the publications, 9-point Book weight, slightly condensed, for the job descriptions and 11-point Heavy weight, somewhat expanded, for the subheads so let us try something in between, like 9/10 Heavy weight, Normal width (we tried the Medium weight first but each job did not stand out adequately).

It now becomes obvious that we do not need the extra space between paragraphs of each job description—besides we need to pick up any extra space that we can. So, rather than extra space, any second paragraphs under any one job description will get indented—one em space (see Fig. 22.27).

After this is done it appears as if we might be able to get two balanced columns leaving one column, the top left column, for the most important material: current emphasis. After rearranging the text blocks it becomes obvious that it would be nice if we could get the lengths of these two columns to be the same.

These processes of trial and error and editing copy to fit a design are an essential part to making a design work (see Fig. 22.28). You need to keep nitpicking until it just feels right. Sometimes the editing requires adding copy. This is when adding Callouts or extra space everywhere can help.

After placing the Education (we put Yale University in the subhead so we could remove the redundancy from each degree) we now had another spacing problem: the consulting comment line did not fit in this column with the education and emphasis also. So, we created another text block, italicized the type, and made three very long

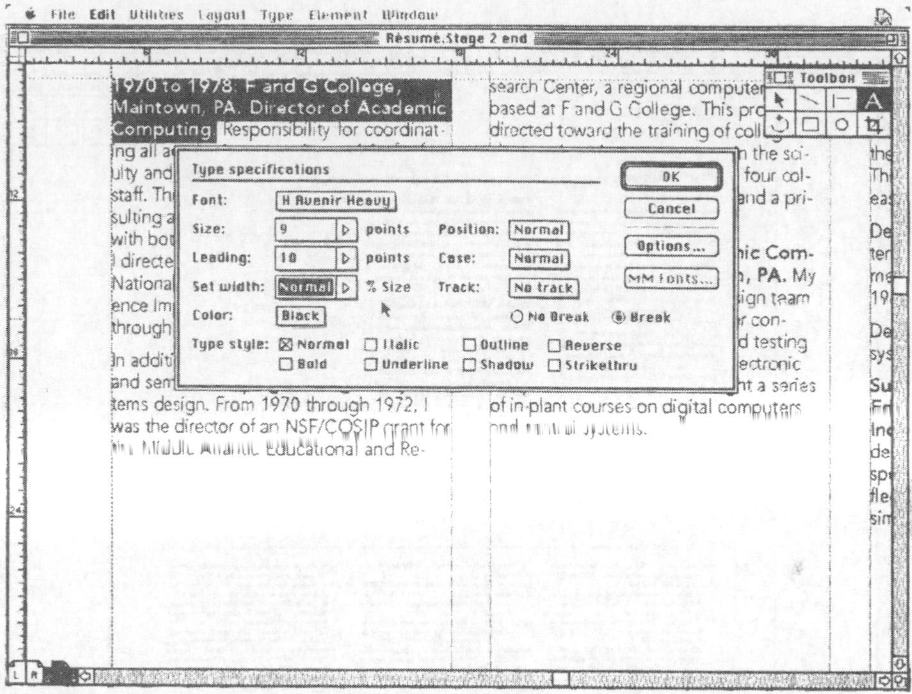

FIGURE 22.26 Each job will be highlighted in 9-point Heavy Avenir.

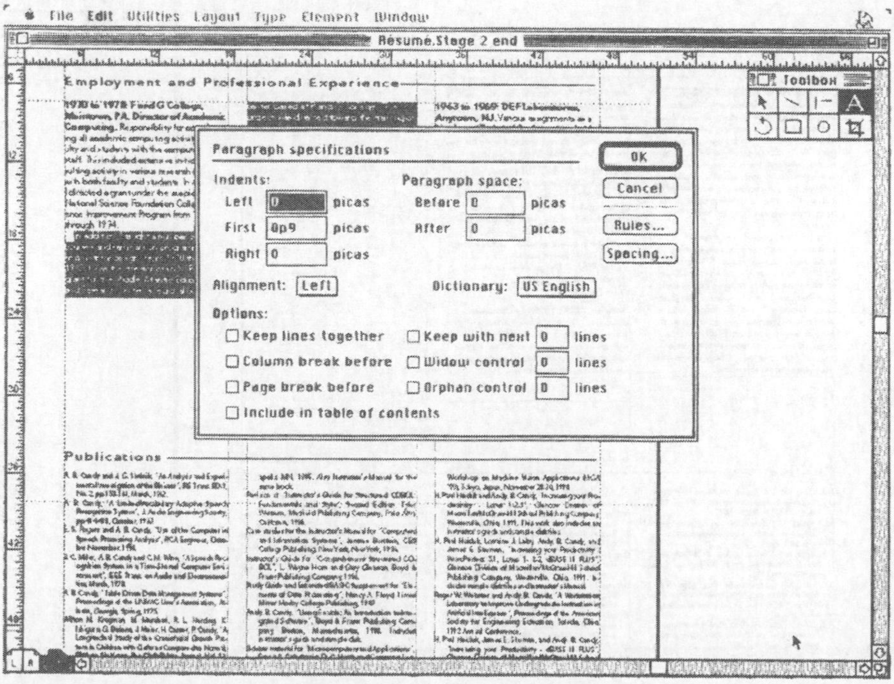

File Edit Utilities Layout Type Element Window

Resume.Stage 2 end

Toolbox

Employment and Professional Experience

Paragraph specifications

Indents: Left 0 picas; First 0p9 picas; Right 0 picas

Paragraph space: Before 0 picas; After 0 picas

Alignment: Left

Dictionary: US English

Options: Keep lines together; Keep with next 0 lines; Column break before; Widow control 0 lines; Page break before; Orphan control 0 lines; Include in table of contents

OK / Cancel / Rules... / Spacing...

Publications

FIGURE 22.27 For the em space, 9-point type gets a 9-point indent.

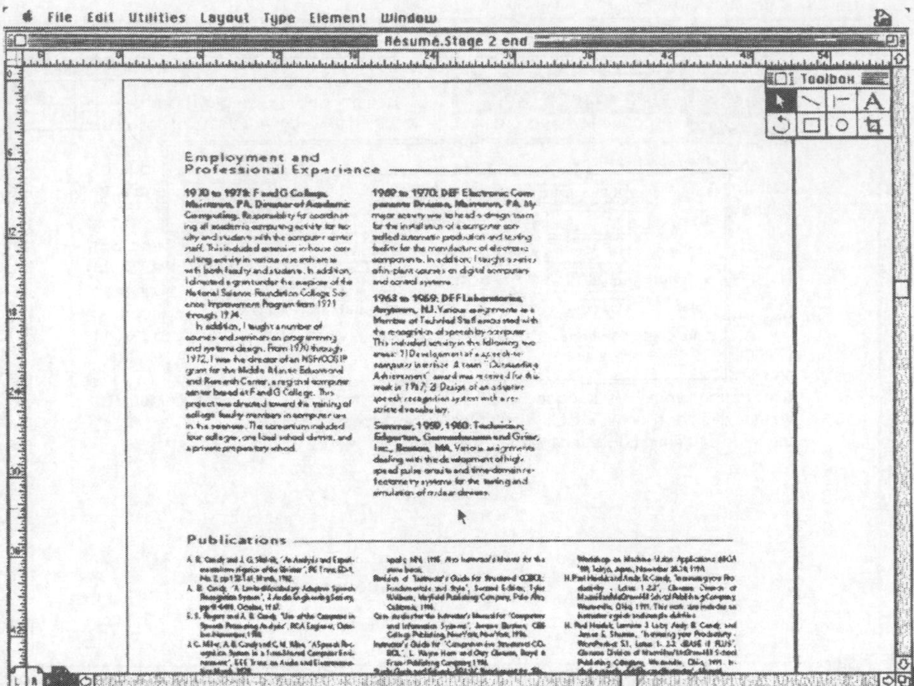

FIGURE 22.28 The trial-and-error process is starting to yield the solution.

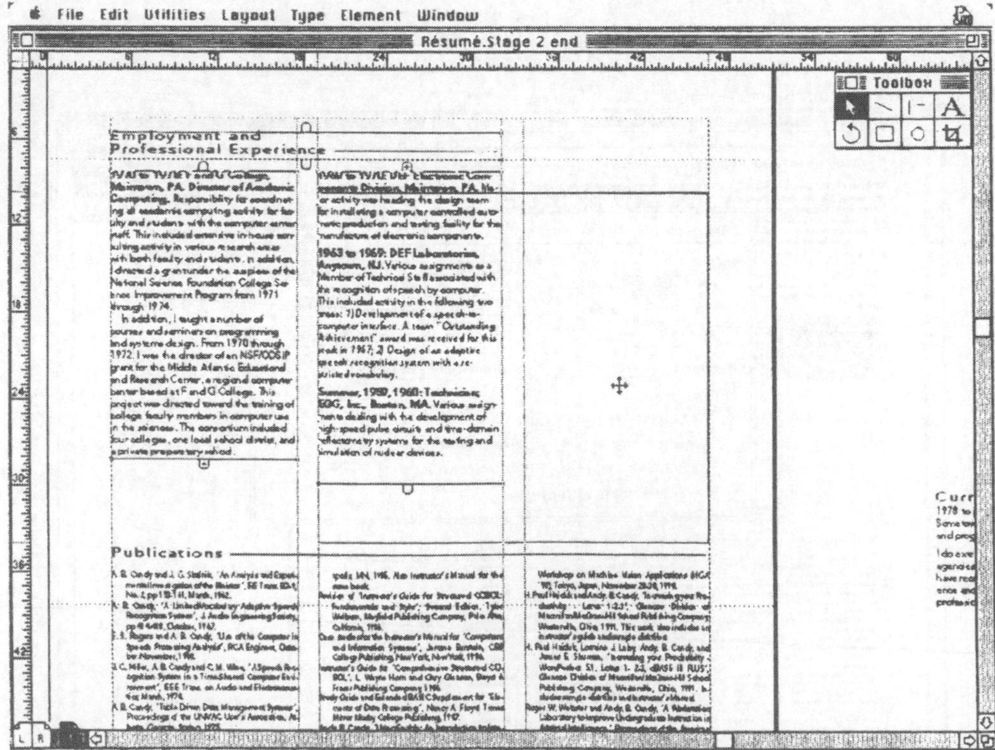

FIGURE 22.29 When moving blocks of type, their outline appears as a box which you can see as it is moved.

lines. Normally, one would not put so many words on a single line; however, since we only have three lines the readability is not a problem. Further, they add a little flair, serve to separate the three sections above from the detail publications below, and while not as important as the education and current emphasis, their importance now stands out (see Figs. 22.29, 22.30 & 22.31).

We are not sure than each subhead, when seen all together, are as clear as they could be. We will make them even heavier by using the Black weight of Avenir.

The Finishing Touch

To separate the person's name and address from the actual résumé we'll use slightly bigger type (9.5 point), somewhat darker type (Roman weight), lighten the effect by generous leading (on 11), and italicize both name and address.

The final result is shown in Figure 22.32.

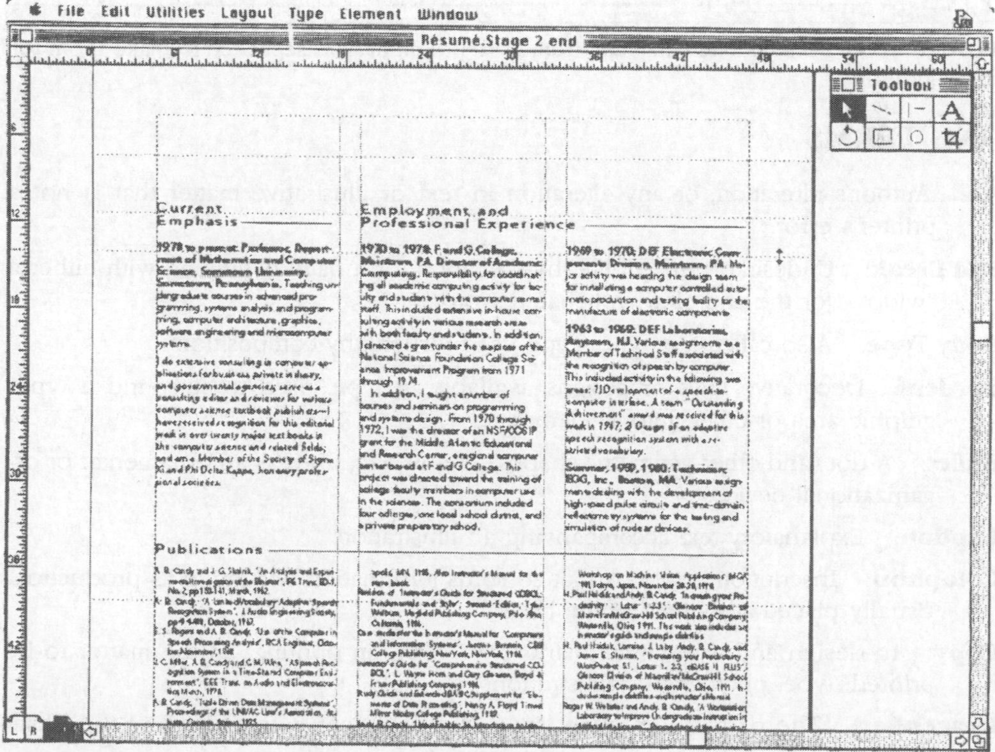

FIGURE 22.30 Another guideline is brought down from the ruler to make certain all the columns line up.

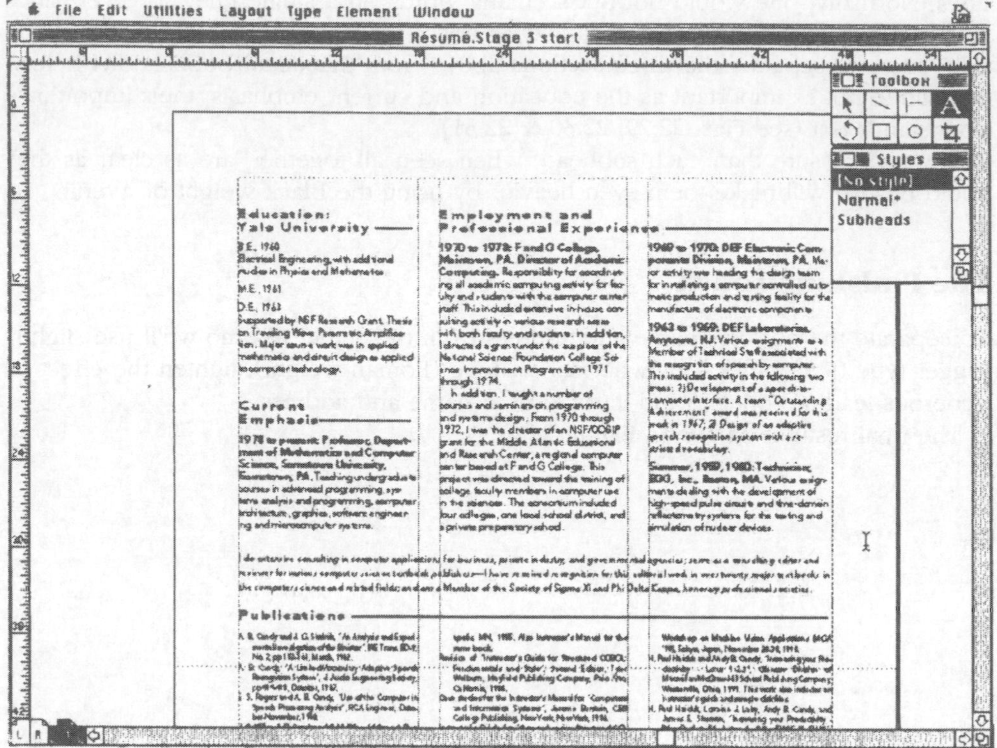

FIGURE 22.31 The second paragraph from Current Emphasis is separated to make room for the Education.

DEFINING TERMS

AA: Author's alteration, or any alteration in text or illustrative matter that is not a printer's error.

Bad Break: Undesirable end-of-line hyphenation, or a page beginning with either a "widow" or the end of a hyphenated word.

Body Type: Also called text type. Type used for lengthy composition.

Borders: Decorative lines or designs available in type, used to surround a typographic area or other graphic elements.

Bullet: A dot (and other geometric shapes) of any size, used as an ornamental or organizational device.

Caption: Explanatory text accompanying an illustration.

Colophon: Inscription in a book that contains information relating to its production. Usually placed at the end of the book.

Copy: In design and typesetting, manuscript copy. In printing, all the matter to be printed: type, photographs, illustrations, etc.

Descender: The part of the letters "g, h, y," etc., that extends below the baseline.

Display Type: Type which, by its size or weight, is used to attract attention, usually 14 point or larger.

ANDY. B. CANDY, D.E.

1111 Homeystreet Road
Anytown, Pennsylvania 15555
(717) 555-1212

Department of Computer Science, Sometown University
Sometown, PA 16666 (717) 666-1313

Education: Yale University

B.E., 1960
Electrical Engineering, with additional studies in Physics and Mathematics.

M.E., 1961

D.E., 1963
Supported by NSF Research Grant. Thesis on *Traveling Wave Parametric Amplification.* Major course work was in applied mathematics and circuit design as applied to computer technology.

Current Emphasis

1978 to present: Professor, Department of Mathematics and Computer Science, Sometown University, Sometown, PA. Teaching undergraduate courses in advanced programming, systems analysis and programming, computer architecture, graphics, software engineering and microcomputer systems.

Employment and Professional Experience

1970 to 1978: F and G College, Maintown, PA. Director of Academic Computing. Responsibility for coordinating all academic computing activity for faculty and students with the computer center staff. This included extensive in-house consulting activity in various research areas with both faculty and students. In addition, I directed a grant under the auspices of the National Science Foundation College Science Improvement Program from 1971 through 1974.

In addition, I taught a number of courses and seminars on programming and systems design. From 1970 through 1972, I was the director of an NSF/COSIP grant for the Middle Atlantic Educational and Research Center, a regional computer center based at F and G College. This project was directed toward the training of college faculty members in computer use in the sciences. The consortium included four colleges, one local school district, and a private preparatory school.

1969 to 1970: DEF Electronic Components Division, Maintown, PA. Major activity was heading the design team for installating a computer controlled automatic production and testing facility for the manufacture of electronic components.

1963 to 1969: DEF Laboratories, Anytown, NJ. Various assignments as a Member of Technical Staff associated with the recognition of speech by computer. This included activity in the following two areas: *1) Development of a speech-to-computer interface. A team "Outstanding Achievement" award was received for this work in 1967; 2) Design of an adaptive speech recognition system with a restricted vocabulary.*

Summer, 1959, 1960: Technician; EGG, Inc., Boston, MA. Various assignments dealing with the development of high-speed pulse circuits and time-domain reflectometry systems for the testing and simulation of nuclear devices.

I do extensive consulting in computer applications for business, private industry, and governmental agencies; serve as a consulting editor and reviewer for various computer science textbook publishers—I have received recognition for this editorial work in over twenty major text books in the computer science and related fields; and am a Member of the Society of Sigma Xi and Phi Delta Kappa, honorary professional societies.

Publications

A. B. Candy and J. G. Skalnik, "An Analysis and Experimental Investigation of the Binistor", IRE Trans. ED-9, No. 2, pp 153-161, March, 1962.

A. B. Candy, "A Limited-Vocabulary Adaptive Speech Recognition System", J. Audio Engineering Society, pp 414-418, October, 1967.

E. S. Rogers and A. B. Candy, "Use of the Computer in Speech Processing Analysis", RCA Engineer, October-November, 1968.

J. C. Miller, A. B. Candy and C. M. Wine, "A Speech Recognition System in a Time-Shared Computer Environment", IEEE Trans. on Audio and Electroacoustics, March, 1970.

A. B. Candy, "Table Driven Data Management Systems", Proceedings of the UNIVAC User's Association, Atlanta, Georgia, Spring, 1975.

Wilton M. Krogman, M. Mazaheri, R. L. Harding, K. Ishiguro, G. Bariana, J. Meier, H. Canter, P. Candy, "A Longitudinal Study of the Craniofacial Growth Pattern in Children with Clefts as Compared to Normal, Birth to Six Years, The Cleft Palate Journal, Vol. 12, No. 1, pp 59-84, January, 1975.

Wilton M. Krogman, J. Meier, H. Canter, P. Candy, M. Mazaheri, and S. Metha, "Craniofacial Serial Dimensions Related to Age, Sex, and Cleft-type, from Six Months of Age to Two Years, Growth, 1975, Vol. 39, pp 195-208.

H. J. Cooper, R. L. Harding, W. M. Krogman, M. Mazaheri, "A Cleft Palate Team in Action", W. B. Saunders Publishing Co., Philadelphia, PA, 1979. My contribution was a chapter on computer methods in the cleft palate research and treatment field.

Andy B. Candy, "Aardvark FORTH Target Compiler System", Aardvark Enterprises, Carlisle, PA, 1984.

Andy B. Candy, "Study Guide To Accompany Judson R. Ostle's Information Systems Analysis and Design", Burgess Communications, Minneapolis, MN, 1985, to accompany "Information Systems Analysis and Design", J. Ostle, Burgess Communications, Minneapolis, MN, 1985. Also Instructor's Manual for the same book.

Revision of "Instructor's Guide for Structured COBOL Fundamentals and Style", Second Edition, Tyler Welburn, Mayfield Publishing Company, Palo Alto, California, 1986.

Case studies for the Instructor's Manual for "Computers and Information Systems", Jerome Burstein, CBS College Publishing, New York, New York, 1986.

Instructor's Guide for "Comprehensive Structured COBOL", L. Wayne Horn and Gary Gleason, Boyd & Fraser Publishing Company, 1986.

Study Guide and Extended BASIC Supplement for "Elements of Data Processing", Nancy A. Floyd, Times Mirror Mosby College Publishing, 1987.

Andy B. Candy, "Using Enable: An Introduction to Integrated Software", Boyd & Fraser Publishing Company, Boston, Massachusetts, 1986. Includes instructor's guide and sample disk.

Sidebar material for "Microcomputers and Applications Ernest S. Colantonio, D. C. Heath and Company, Lexington, Massachusetts, 1989

Instructor's Guide for "Microcomputers and Applications", Ernest S. Colantonio, D. C. Heath and Company, Lexington, Massachusetts, 1989.

Two case studies for "A Study Guide to Accompany Analysis & Design of Information Systems", James A. Senn, Second Edition, McGraw-Hill Publishing Company, New York, New York, 1989.

Andy B. Candy, "Using Enable: An Introduction to Integrated Software", Second Edition, Boyd & Fraser Publishing Company, Boston, Massachusetts, 1991. Includes instructor's guide, sample disk and educational version of the software.

R. W. Webster, A. B. Candy, P. LaFollette, and R. L. Stafford, "A Computer Vision System that Assembles Canonical Jigsaw Puzzles Using the Euclidean Skeleton and Isthmus Critical Points", Proceedings of the International Association of Pattern Recognition, Workshop on Machine Vision Applications (MGA '90), Tokyo, Japan, November 28-30, 1990.

H. Paul Haiduk and Andy B. Candy, "Increasing your Productivity - Lotus 1-2-3", Glencoe Division of Macmillan/McGraw-Hill School Publishing Company, Westerville, Ohio, 1991. This work also includes an instructor's guide and sample disk files.

H. Paul Haiduk, Lorraine J. Laby, Andy B. Candy, and James E. Shuman, "Increasing your Productivity - WordPerfect 5.1, Lotus 1- 2-3, dBASE III PLUS", Glencoe Division of Macmillan/McGraw-Hill School Publishing Company, Westerville, Ohio, 1991. Includes sample disk files and Instructor's Manual.

Roger W. Webster and Andy B. Candy, "A Workstation Laboratory to Improve Undergraduate Instruction in Artificial Intelligence", Proceedings of the American Society for Engineering Education, Toledo, Ohio, 1992 Annual Conference.

H. Paul Haiduk, James E. Shuman, and Andy B. Candy, Increasing your Productivity - dBASE III PLUS Glencoe Division of Macmillan/McGraw-Hill School Publishing Company, Westerville, Ohio, 1992. Includes sample disk files and Instructor's Manual.

A. B. Candy, H. P. Haiduk, H. W. Means, and R. R. Sloger, "Understanding Computer Information Systems West Educational Publishing, Austin, TX, 1992.

Andy B. Candy, Instructor's Manual for "Learning C" by N. Graham, McGraw-Hill Publishing Company, New York, NY, 1992.

Andy B. Candy and Andy B. Candy, "Using Windows 3.0/3.1 Effectively", Wm. C. Brown Publishers, Dubuque, Iowa, 1993. Includes Instructor's Manual.

Andy B. Candy and Andy B. Candy, "Using Works Effectively (IBM Version)", Wm. C. Brown Publishers, Dubuque, Iowa, 1993. Includes Instructor's Manual and sample files.

Andy B. Candy, "Using Works for Windows Effectively" Wm. C. Brown Communications, Dubuque, Iowa, 1993. Includes Instructor's Manual and sample files.

FIGURE 22.32 The trial résumé for Andy B. Candy, D.E.

Em: The square of a given point size of type.

Folio: Page number.

Font: The letter, numerals, punctuation marks, and special characters that constitute a complete character set of a given style of typeface. Prior to digital type, each size was a unique font as well.

Headline: Usually the most prominent element of type in a piece of printing: that which attracts the reader to read further or summarizes at a glance the content of the copy that it accompanies.

Kern: To space two letters closer together than customary to create visually consistent spacing between all letters.

Layout: Preliminary plan of the basic elements of a design shown in their proper positions.

Markup: In typesetting, to mark type specification on the layout and copy for the typesetter

Mechanical: Camera-ready pasted-up assembly of all type and design elements mounted in exact position and containing instructions, either in the margins or on an overlay, for the printer.

Modern: Term used to describe a type style developed in the late eighteenth century.

Outline Characters: Open characters made from solid ones by putting a line on the outside edge of a letter.

Pica: A measure of type equal to 12 point or approximately 1/6 of an inch. Derived from an old term for metal type of that size.

Point: Basic increment of typographic measurement, equal to 0.0138 inch. Twelve points equal a pica.

Ragged (Unjustified): The setting of text type with an irregular appearance on either one or both margins, such as ragged right or ragged left. In ragged setting, interword spaces are not varied. Ragged setting is the opposite of flush setting in which even margins are achieved on one or both sides of the text.

Roman: Name often applied to the Latin alphabet as it is used in English and European languages. Also used to identify upright, as opposed to italic or cursive alphabet designs.

Runaround: Type set to fit around an illustration box or irregular shape.

Running Head: A book title or chapter head repeated at the top of every page in a book.

Serif: A line crossing the main strokes of a character.

Stet: Proofreaders' mark indicating copy marked for correction should stand as it was before the correction was marked.

Text: The body copy in a book or on a page, as distinct from the headings.

White Space Reduction: The reducing of space allocated to the characters.

Widow: The single-line end of a paragraph or of a column of reading matter that is undesirably short: a single, short word: or the end of a hyphenated word, as in "ing."

REFERENCES

David Holzgan, *PostScript Programmer's Reference Guide.* Scott, Foresman, 1988.

Roger C. Parker, *The Aldus Guide to Basic Design,* 2nd ed. Aldus Corporation, Seattle, WA,.

Glen C. Reid, *PostScript Language Program Design.* Addison-Wesley, Reading, 1988.

Stephen F. Roth, editor, *Real World PostScript.* Addison-Wesley, Reading, MA, 1988.

Jan White, *Editing by Design,* 2nd ed. R. R. Bowker, New York, 1982.

Jan White, *The Grid Book, a Guide to Page Planning,* Letraset, Paramus, NJ, 1989.

23

Corel Ventura Publisher

Richard Sutor
American Insurance Association

The chapter that follows is not a how-to manual, but rather a description of what Corel Ventura Publisher does and how it does it.

This article was based on a late beta copy[1] of Ventura Publisher 5.0, which is a complete rewrite of this venerable desktop publishing program. Although the implementation of some features may be different in the released version the philosophy behind the program remains the same.

23.1 WHAT YOU GET IN THE PACKAGE

You Are Not Alone

Ventura Publisher is a powerful page design and typesetting program that the Corel Corporation sells alone or as part of a suite of graphic programs. The suite includes

- CorelDraw, a popular and powerful drawing program
- Corel Ventura Publisher
- Corel Photo-Paint, a bit map manipulation program, similar to, though not quite as powerful as, Adobe Photoshop
- CorelChart, a graphing and charting program
- CorelMove, which allows you to do simple animation
- CorelShow, a presentation program
- Corel Mosaic, an image compression and storage program. A group of graphic files can be compressed and placed in a "library file" either on your hard disk or on a floppy disk. Mosaic creates a low resolution "thumbnail" image of each of the files so you can see at a glance what is in a given library.
- Corel Trace, a program that will trace bit maps and turn them into images that can be manipulated by CorelDraw

[1]A beta version of a program is one that is still under development but has been released to a group of "civilian" testers who do not work for the software company.

0-8493-2530-7/96/$0.00+$.50

- CorelKern, a kerning[2] program for Type 1 fonts
- 825 fonts (TrueType and Type 1), and 22,000 clip art images and symbols.

System Requirements

The package includes the program on floppy disks and CD ROMs. Only a few of the fonts and clip art pieces that are on the CD ROMs are also included on the floppy disks so you will need a CD ROM drive to get at them.

Installing the whole package will take up about 50 Megabytes of hard disk space, but, of course, you can install only those programs you feel that you will need.

Corel Ventura Publisher runs under Windows Version 3.1 or higher. It, along with CorelDraw and most of the other programs in the package, are large and powerful and need a large amount of memory and a powerful computer to run them. A 33 MHz, 486-based computer with a 250-Megabyte hard drive and at least 8 Megabytes of memory is the minimum configuration I would recommend.

23.2 OVERVIEW

Do You Need Corel Ventura Publisher?

If you produce documents that are sent off to be typeset, use a word processor. If you produce documents that are going to be reproduced (by photo copy or offset) just as they look when they came out of your computer, you have a more difficult decision. For a majority of people a modern word processor, such as WordPerfect, Microsoft Word for Windows, or Ami Pro, will be all they need. These programs are powerful text editing tools, allow WYSIWYG (What You See Is What You Get) placement of graphics and tables and have a host of features such as style sheets and indexing that bring them very close to page layout programs such as Ventura Publisher.

What Corel Ventura Publisher offers is document and layout structure, precision, and flexibility. If you need to control every aspect of your document, from color separations and halftone screen settings for graphics to page breaks and kerning for text, then consider Ventura. With a word processor you can write a book; with Ventura you can make a book.

What You Can Do with Ventura

Venture was originally created to produce long documents, but over the years, and especially with the newest version, it has evolved into a versatile program that can be used to design just about anything, from one page advertisements using spot color and romantic photographs, to technical manuals with headers, footers, tables, graphs, footnotes, and indexes. Ventura gives you the tools to design page layouts for single or multipage documents with precise and easily controlled formatting.

This range is obtained by two complementary sets of tools that I shall call *local control* and *document control.*

[2]Kerning adjusts the spacing between certain letters or numbers so that they look better together. For instance "A" and "V" often look too far apart.

Local Control

Local formatting is familiar to anyone who has used a word processor—you select something on a page and do something to it—change a paragraph into nine-point Helvetica and indent its first line. You must go through all the steps again if you want to do the same thing to another paragraph. This works for short "design as you go" documents, especially ones that have more graphics than text. Ventura provides a powerful set of local formatting tools for these tasks. See Figure 23.1 for an example of a document created using local formatting.

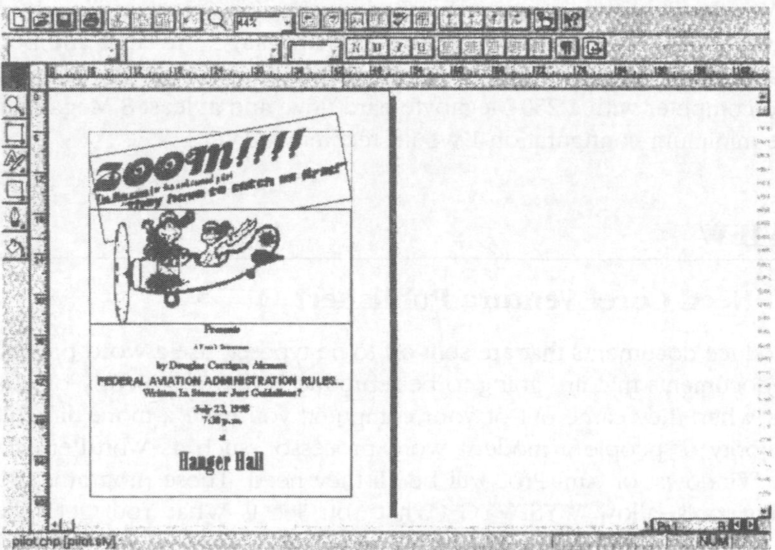

FIGURE 23.1 Corel Ventura Publisher 11.2.

For longer documents, local formatting can quickly get out of hand—how *was* that bulleted list on page 45 of Chapter 8 formatted?

Because the reader of this book is probably more interested in producing long documents than one page flyers, I will emphasize document formatting in the following discussion. Local and document controls, however, are by no means mutually exclusive; they are tools that are meant to be used together. How they are used, and in what proportion, depends on the job that has to be done and how you want to do it.

Document Control

Ventura calls its document controls *publications, chapters, frames, style sheets,* and *tags.* At first these terms can be confusing, but once understood they will make long document production much easier. The terms identify controls that allow writers to set up a framework for a document and then plug in graphics and pour in text previously prepared in a word processor.

Ventura has many other features that make long document production easier, such as its ability to generate a table of contents, an index, cross-references, and variable text, all of which will be discussed below.

Publications. Unless you are planning to write a 98-page super-duper self-help book with a catchy title that sells millions of copies and makes its author rich before her time, any technical book you are planning to produce is going to take organization. One organizational method is to produce each chapter as a single unit and then print them all together as a publication at the end.

The one chapter, one unit approach allows great flexibility: chapter order can be changed without cutting and pasting large quantities of text and graphics; the unit that you are working with is smaller and, I find at least, that it improves my concentration having smaller, less threatening, units to work with.

"Ah," I can hear you saying, "that is all well and good, but I have 17 beautifully crafted and illustrated chapters tucked here and there in my computer and, after a tense series of e-mail messages with my editor, we have decided that it would make much more sense if Chapter 12 were moved between Chapters 6 and 7, Chapter 3 between Chapters 9 and 10, and three of the four photographs of my cat (illustrations number 5, 30, and 74) deleted entirely. He would like to see the changes next week. I would like to know how I'm going to explain this to my cat, much less renumber the pages and illustrations?"

Piece of cake.

The top organizational level of a Ventura document is a "publication" that keeps track of the structure of the entire document. When you open a publication with one chapter that chapter opens automatically. When you open a publication with many chapters you are presented with a list so you can jump to any one you want. Chapters can be added and subtracted from the publication, brought in from other publications, their order changed, and page, chapter, and illustration numbers made to start with any number at any place in the chapter or flow sequentially from the first chapter to the last. Numbers may be roman or arabic and may be shown numerically or spelled out. The publication also keeps track of the index setup and the table of contents data, the document version number, and a the date/time stamp.

You can fulfill your editor's requests and still have time to start that book on your dog you have been thinking about.

You do not have to tell your cat.

Chapters. Just as a Ventura publication is a way to organize parts of a document, including chapters, a Ventura chapter is a way to organize text files and graphics. Having said that, let me backpeddle a bit. If you are writing a chapter of a book, the chapter probably consists of one text file, and may or may not contain graphics. In this case the "collection" consists of one text file, perhaps all by itself, perhaps with a few graphics. If you are putting a newsletter together, then you will have a collection of many text files and graphics. Either way, Ventura allows you to bring them into one place to be edited, and Ventura calls that document a chapter.

A chapter, in Ventura, is where text and graphics are united on the computer screen ready to be edited, formatted, and turned into a section of a book, newsletter, brochure, or almost anything else you can imagine.

The Ventura chapter, which looks so seamless on the screen, is, way down deep, a file that contains a list of "pointers" to the location of the text, graphics, and style sheet (see below) used in that chapter. Text and graphic files created in other programs and imported into Ventura are kept in their original directories and can be modified at any time by those programs. The changes will show up the next time the chapter is loaded.

Text, but not graphics, is also completely editable in Ventura and is saved back into its original format and file.

Frames. In Ventura everything on a page—all text, graphics and tables—is contained in a frame. A frame is a discrete box created either by Ventura (the page it self is a special kind of a frame called the page frame) or drawn by the layout designer using the frame tool. A drawn frame is called a free frame. A free frame can be sized and placed by eye or by entering numbers into a menu (Ventura will accept inches, centimeters, picas, and points) for the height, width, and distance from the top and left side of the page. A free frame can also be moved anywhere in the document, or anchored to the page or to a paragraph.

A free frame that has been placed on a page frame can have text running around it and you can specify how much space there will be between the top and/or sides of the frame and the text. You can also tell Ventura that you want the page frame text to flow over the free frame, useful if you want to "ghost" a graphic.[3] Text can also be made to run around the edges of an irregularly shaped object.

In many ways Ventura treats free frames like page frames. Free frames can have margins and columns that are distinct from the margin and column setting of the page or other free frames. They can also have attached caption frames on the bottom, top, left, or right side. Free frames can be drawn with ruling lines on the top and/or bottom, or all around and filled with shades of gray or with color.

When imported text is placed directly into the page frame, Ventura will generate as many pages as needed to hold the text. Text placed in a free frame will be cut off if the frame is too small to hold it, but the text may be made to continue into other free frames.

In a linear document such as a book you would probably want to place the text on the page frame and let Ventura control its flow from page to page. In a nonlinear document such as a newsletter or newspaper where stories are often continued from one page to another, you would place the text into free frames so that you can control its flow.

Style Sheets. If the publication is the basic structure in Ventura, then style sheets are the blueprints that, once perfected, make the building go up quickly and permit the construction of other buildings just like it.

A style sheet is a file, linked to a chapter, that holds detailed directions about page, document, paragraph tag, and frame-tag-formatting information.

Every chapter must have a style sheet attached to it, but different chapters can share the same style sheet. For example, if you are writing a book and have divided it into separate chapter files, (see "Chapters" above), you will want to make very sure that the formatting is the same from chapter to chapter (too many variations in type style or paragraph indents may give readers the impression that your grasp of your subject is equally in doubt). Once you have developed a workable style sheet for one chapter, you can use it for all other chapters, and your reader's confidence is assured, at least typographicaly.

[3]A ghost is a graphic that has text running over it. For the text to be seen the graphic is rendered in a very light shade of gray.

Paragraph Tags. If one characteristic of the human race is its need to name everything in sight (or out of sight, for that matter), then Ventura is an intensely human program. Every paragraph in Ventura has a tag or name, whether it is the default tag Body Text, or a name given it by the layout designer, and every paragraph with the same tag has the same typographic attributes unless the local formatting tool has been used to change it.

If you change a paragraph's formatting using the paragraph formatting tool then every paragraph with the same tag name will change in the same way throughout the whole chapter (or the collection of chapters in a publication), if they are all using the same style sheet.

A paragraph, in Ventura, is anything that ends with a carriage return. It could be a single number, or letter, or as many words as your editor will let you get away with.

Unless tagged in a word processor, all paragraphs brought into Ventura are tagged as Body Text. The paragraphs can then be retagged or new tags can be created for them.

Frame Tags. If you are creating a document that has a lot of identical free frames, you can create a frame that has the features you want: frame size, margins, columns, background color, ruling line above, below, or around, and tag it just like a paragraph. All frames tagged with the same name will be identical and if one is changed, they all will change.

Other Good Things

The following is a listing, with a brief description, of some of Ventura's other features.

Templates. One of Venture's underlying themes is "don't reinvent the wheel." If you have designed a document, such as a newsletter, that has the same layout from issue to issue, Ventura makes it easy to use it as a template for the next issue. Ventura lets you choose which text and graphic files to include when you load the document. For a newsletter, for instance, you would want to leave in everything that repeats for every issue, and take all the other files out. Ventura leaves the empty frames that contained the text and graphics in place, ready to be used again.

Type Assist. When the engineers at Corel designed Ventura 5.0 they obviously took a long look at Microsoft Word for Windows' elegant interface and liked what they saw. One of the many similarities between Ventura and Word is Ventura's Type Assist, which is similar to Word's Auto Correct. Type Assist can be set up to correct misspelled words and capitalization mistakes as well as change straight quotes (" ") to "curly" or typographic quotes (“ ”), capitalize the days of the week and the first word of a sentence, all as you type. If you find that you fingers have decided that "all" should be spelled "lal" then you can tell Type Assist to substitute "all" for "lal" every time you type it.

Another use for Type Assist's substitution feature is being able to type a shorthand code and have Type Assist expand it to the word or phrase you want. If you find, for instance, that you frequently type "My client refuses to answer on the grounds that he may incriminate himself," you can type it once into Type Assist and use "fa" (for Fifth Amendment, you can use any code you want) as the code. Every time you type "fa" and a space Ventura will type out the whole phrase.

Table of Contents and Index. Ventura provides a way to mark text in your document so that it can later generate a file that is a list of the text and its location. The paragraphs in the generated file are given tag names (see Tags, above) and the list is saved as a separate file. The list can then be brought into an empty chapter and formatted as a Table of Contents or Index and placed in the publication.

Cross-Reference and Variables. "Continued on page three" and "See table number seven on page forty-nine" are probably the simplest example of a cross-references. You are telling the reader that the story is continued, or more information can be found, in another place. One of the problems is keeping track of the "other place." As you edit a document the story that was continued on page 3 might end up on page 4, or you might not be sure just where the table on page 49 has gone to, or what number it should be. Ventura's cross-reference feature can keep track of all this information and can automatically generate new cross-reference numbers whenever you want.

The cross-reference feature can also be used to insert variable text into a document. If, for instance, you (Smith) and a colleague (Jones) are working on a theory and you have not yet decided to call it the Smith Unification Theory or the Jones Unification Theory you can insert a marker where the name will go and update it with the correct name, after you both have decided.

Tables. A table is a way of organizing information in columns and rows. Every block of information in a table is called a cell. A spreadsheet is a table and so is a train schedule. Ventura's powerful table feature allows great flexibility in the creation and formatting of tables:

- Tables can be inserted in a page frame or a free frame. Putting the table in a free frame allows you to caption and number it by adding a cross-reference counter (see "Cross-Reference and Variables" above) so that Ventura can update the numbering as needed.
- Tables in free frames can be moved anywhere in the document, or anchored to the page or to a paragraph.
- Columns and/or rows in a table can be added or subtracted at any time.
- Border lines around the table or individual cells can be added or eliminated and their color and thickness changed.
- Cells can be shaded with a gray or a color.
- The contents of cells (numbers and/or text) can be tagged and formatted like any other text.
- Column widths may be resized by eye or by entering numbers into a menu.
- Information can be imported from a database and Ventura will build a table to hold it.
- Cells can be merged or split.

Equations. With Ventura's Equation Editor you can insert an equation as simple as $^3/_4$ or as complicated as

$$\sum_{i=1}^{\infty} \frac{x_i}{x_i + 1}$$

directly in the text of your document.

Equations are built using commands and special words. The above equation, for instance, was constructed by inserting the following text string in Ventura's equation editor:

```
sum from {i ~ = ~1} to inf ~~ {x sub i over {x sub I ~ + ~ 1}}
```

Once built, the equation can be edited or cut, copied and pasted anywhere in the document.

Spell Check and the Thesaurus. Ventura provides a spell checker, with it you can check spelling for the entire document, selected text, footnotes, individual text files, or text that has been changed or added since the document was opened or last saved.

Ventura's thesaurus enables you to look up synonyms for a selected word and replace the word with its synonym.

Find and Replace. Using Corel Ventura Publisher's find and replace commands you can find and optionally replace text, character attributes (such as bold and italic), and paragraph tags. Find can also be used to locate frame anchors, footnote references, and index entries.

Halftone Screen Controls. Halftoning converts an image that contains shades of gray (a grayscale or continuous tone image), such as a photograph, into a pattern of black dots, so that the image can be printed on a device that is capable of producing only solid dots, such as an offset printing press.

Ventura provides a way to set the halftone screen type and angle and the lines per inch in a grayscale image. It should be noted that producing computer-generated halftone screens is much more of an art then a science and that immediate success, even using a sophisticated graphics program, is by no means assured. In other words, do not try this on a deadline and practice, practice, practice!

DEFINING TERMS

Beta Version: A beta version of a program is one that is still under development but has been released to a group of "civilian" testers who do not work for the software company and will test the program under field conditions.

Byte: A byte equals one character of data. Used for measuring computer memory and disk storage.

Clip Art: Collection of photographs and/or illustrations that can be "clipped" from a book, floppy disk, or CD ROM and used in publication. The art is sometimes, but not always royalty free and great care should be taken to use clip art legally..

CD ROM: Means Compact Disk Read Only Memory. Identical to a CD music disk but information (up to 550 million characters) is stored rather than music. The information can be text, computer programs, clip art, or almost anything else.

Chapter: One of the building blocks of Corel Ventura Publisher. All documents have at least one chapter. One or more chapters make up a publication file.

Footers: Text that is repeated on the bottom of each page in a document. A page number is a simple example of a footer.

Frame: In Ventura a frame is a container used to hold text and graphics. Ventura automatically creates page frames to hold the main body of document text, but "free frames" can be treated to hold illustrated, etc.

Ghost: A ghost is a graphic that has text running over it. For the text to be seen the graphic is rendered in a very light shade of gray.

Headers: Text that is repeated on the top of each in a document. The chapter name is a simple example of a header.

Jumped Text: Mostly used in newspapers, magazines, and newsletter, jumped text is a story that starts on one page (usually the first) and "jumps" to or is continued on another page.

Kerning: Adjusts the spacing between certain letters or numbers so that they look better together. For instance "A" and "V," when one follows the other, often look too far apart.

Megabyte: One million bytes.

Publication: Ventura is extremely hierarchical and a publication is at the top of the hierarchy. It contains one or more chapters, which, in turn, contain style sheets.

Spot Color: Sport colors are chosen from a swatch book that will tell the printer what proportions of basic colored inks should be mixed to get the desired color.

Style Sheet: A Ventura file that contains information about tags, columns, margins, page orientation and other formatting.

Tags: A Ventura tag defines text attributes such as fonts used, paragraph alignment, and spacing. Once a tag is defined and named it can be applied to any paragraph. Frames can also be defined and tagged.

WYSIWYG: What You See Is What You Get. A WYSIWYG program will attempts to allow you to create a document or illustration on the computer screen that will be identical to the document that comes out of the printer or off the press. One should remember, particularly when using color, that what you see will often bear no resemblance to what you get.

24

QuarkXPress

Rajiv Sabharwal
Bio-Rad Laboratories, Inc.

24.1 THE QUARKXPRESS INTERFACE

QuarkXPress is electronic publishing software for the desktop. You can use it to create any kind of publication, from a black-and-white business card to a multi-color magazine. Its powerful typographical controls, word processing capabilities, page layout features, advanced color support, and sophisticated graphics have made it the preferred software program among professional designers. The program gives you the freedom to be more creative and more productive in any publishing situation.

The Document Window

You can see the QuarkXPress interface in Figure 24.1, which displays an open QuarkXPress document window. The Title bar of QuarkXPress contains the QuarkXPress logo followed by the program's menu bar along the top with nine menu titles and document window(s), which contain the document or file name; you can have multiple documents open at the same time.

Setting Program and Document Defaults

QuarkXpress refers to these as preferences or default values and settings that you can modify to enable, disable, and customize the way in which many automatic features work. QuarkXpress provides four dialog boxes for specifying preferences: the **Application Preferences** dialog box, the **General Preferences** dialog box, the **Typographic Preferences** dialog box, and the **Tool Preferences** dialog box. Each

0-8493-2530-7/96/$0.00+$.50

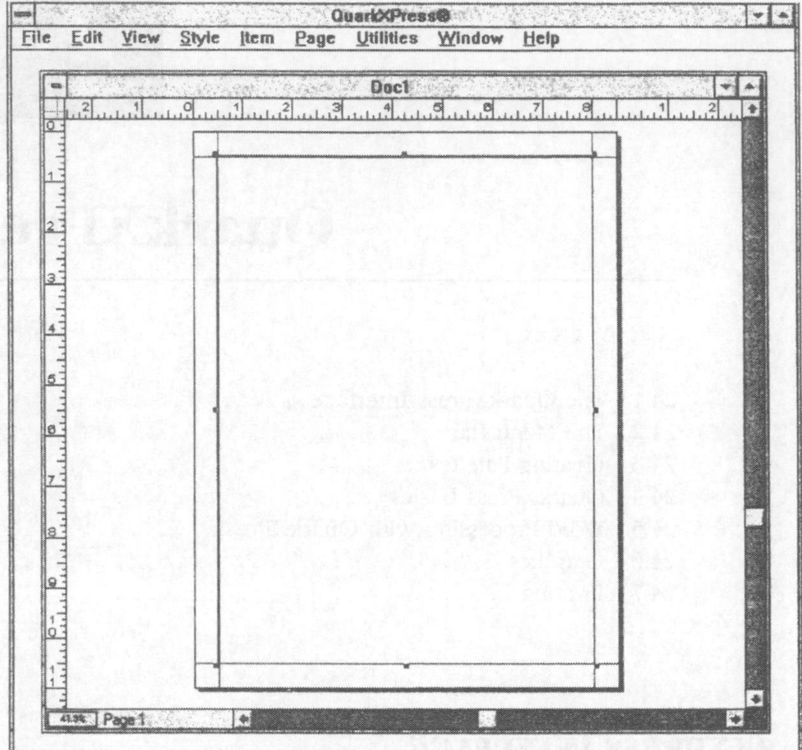

FIGURE 24.1 The QuarkXPress interface—an open document window.

of these dialog boxes may be accessed under the (**Edit→Preferences**) menu. You can modify these preferences for a specific document by making changes in each dialog box when the document is active. You can modify the preferences for all newly created documents by making changes in the dialog boxes when no documents are open.

24.2 THE MENU BAR

The menu bar appears below the Title bar and contains nine pull-down menu titles—**File**, **Edit**, **View**, **Style**, **Item**, **Page**, **Utilities**, **Window**, and **Help**—running across the top of the screen, which can be accessed by clicking on each menu title.

The **File** menu is dedicated to opening, closing, and saving documents, importing images or text, setting document preferences such as page size and margins, and printing. It includes commands that relate to entire documents or files. You can make copies of a document and retrieve the original document you started before you made those unwanted changes. You can instantly change a document's size and control the way in which it will print.

The File menu includes the following commands:

New	Creates a new document or library
Open	Opens a document, template, or library
Close	Closes the frontmost window
Save	Saves a QuarkXPress document
Save as	Saves a QuarkXPress document

Revert to Saved	Restores a document from disk
Get Text	Imports text
Get Picture	Imports pictures
Save Text	Saves a QuarkXPress story
Save Page as EPS	Saves an EPS page image
Document Setup	Changes document specifications
Printer Setup	Controls how a document is printed
Print	Prints a QuarkXPress document
Exit	Ends a QuarkXPress session

The **Edit** menu is used to manipulate text and pictures, to control the Clipboard, and to set program defaults. QuarkXPress has added its own selections to this menu for specifications that apply to the entire document, such as style sheets, color choices, hyphenation and justification and typographic, page layout, and tool customization.

The Edit menu includes the following commands:

Undo/Redo	Reverses the last action performed
Cut	Removes items, text, or pictures and places them on the Clipboard
Copy	Copies items, text, or pictures to the Clipboard
Paste	Copies items, text, or pictures from the Clipboard
Paste Special	Controls the way in which objects are placed in a document
Paste Link	Places a linked object in a document
Delete	Deletes items, text, or pictures
Select All	Highlights story or activates all items
Links	Controls object links in a document
Clipboard	Displays/hides the clipboard window
Find/Change	Searches for and modifies text
Preferences	Edits QuarkXPress default settings
Style Sheets	Edits style sheets
Colors	Edits program/document colors
H&Js	Edits program/document hyphenation & justification specifications

The **View** menu allows you to control what you see on the screen and how items (text boxes, picture boxes, lines, and groups of these) and pages are displayed. You can specify the size or magnification of the document, control the way in which visual layout aids such as guides and rulers appear, and display or hide palettes that contain tools, fields, and icons for working with documents.

The View menu includes the following commands:

Fit in Window	Fits page in document window
50%	Zoom 50%
75%	Zoom 75%
Actual Size	Zoom 100%
200%	Zoom 200%
Thumbnails	Different scales for viewing documents
Guides	Displays/hides guides
Baseline Grid	Displays/hides grid

Snap to Guides	Makes pointer snap to guides
Rulers	Displays/hides document rulers
Invisibles	Displays/hides special characters
Tools	Displays/hides the Tool palette
Measurements	Displays/hides the Measurements palette
Document Layout	Displays/hides the Document Layout palette
Style Sheets	Displays/hides the Style Sheets palette
Colors	Displays/hides the Colors palette
Trap Information	Displays/hides the Trap Information palette

The **Style** menu is used to control the character attributes and paragraph formats applied to text, to add color and special effects to pictures, and to make stylistic changes to lines. The operation of the Style menu depends on the kind of item that is active (text box, picture box, or line).

Style Menu for Text

If a text box is active, the Style menu can be used to vary the appearance of highlighted text or text that is entered at the text insertion point.

The first section of the Style menu for text controls attributes of individual characters. It includes the following commands:

Font	Specifies the font for characters
Size	Specifies the point size of characters
Type Style	Specifies the type style of characters
Color	Specifies the color of characters
Shade	Specifies the shade of characters
Horizontal/Vertical Scale	Specifies the width of characters
Kern/Track	Specifies intercharacter spacing for a character pair or a range of highlighted characters
Baseline Shift	Specifies the offset of characters from their normal baseline
Character	Displays a dialog box that contains all character attributes

The last section of the Style menu for text controls formats of individual paragraphs:

Alignment	Specifies the alignment of paragraphs
Leading	Specifies the spacing between lines of paragraphs
Formats	Displays a dialog box that contains most paragraph formats, including drop caps, H&Js, and widow/orphan controls
Rules	Specifies lines that are anchored to paragraphs
Tabs	Specifies tab stops for paragraphs
Style Sheets	Applies style sheets to paragraphs

Style Menu for Pictures

If a picture box is active, the Style menu can be used to add color or special effects to the picture. Check marks indicate the formatting that is currently in effect. You can choose from Color, Shade, Negative, Normal Contrast, High Contrast, or Posterized.

Color	Colorizes a bitmap-based black and white or grayscale picture
Shade	Shades a bitmap-based black and white or grayscale picture
Negative	Inverts a bitmap-based black and white or grayscale picture
Normal Contrast	Retains original contrast of bitmap-based black and white/grayscale pictures
High Contrast	Displays a bitmap-based grayscale picture in two levels of gray
Posterized	Displays a bitmap-based grayscale picture in six levels of gray

Style Menu for Lines

If a line is active, the Style menu can be used to change the style of the line (solid, dashed, dotted, etc.), to add endcaps (arrowheads and/or tail feathers) to the line, or to change the width, color, or shade of the line.

Line Style	Displays a submenu containing various line styles
Endcaps	Displays a submenu containing various endcaps
Width	Indicates the line width that is currently in effect. Line widths of hairline, 1, 2, 4, 6, 8, and 12 points can be chosen directly; other line widths from 0 points (hairline) to 504 points in 0.001-point increments can be specified in the **Other** entry
Color	Displays a submenu containing various colors
Shade	Indicates the shade that is currently applied. Shades from 0 to 100% in 10% increments can be chosen directly; shade values from 0 to 100% in 0.1% increments can be specified in the **Other** entry

The **Item** menu is used to perform operations on items (text boxes, picture boxes, lines, and groups). You can change attributes, place frames on boxes, control the way in which text flows around items, delete items, create duplicates or groups of items, prevent items from being moved or resized accidentally, change the stacking order of items on a page, control spacing and alignment, change the shape of picture boxes, and reshape polygon picture boxes.

The Item menu includes the following commands:

Modify	Changes item attributes
Frame	Places a border around a box
Runaround	Controls text flow around items and pictures
Duplicate	Makes a copy of an item
Step and Repeat	Makes multiple copies of an item
Delete	Deletes an item
Group	Makes an association between items
Ungroup	Disassociates items
Constrain	Restricts/removes restrictions on item movement
Lock/Unlock	Prevents/allows item changes with the mouse
Send Backward	Moves an item one layer back in the stacking order

Send to Back	Moves an item to the bottom of the stacking order
Bring Forward	Moves an item one layer forward in the stacking order
Bring to Front	Moves an item to the top of the stacking order
Space/Align	Arranges items
Box Shape	Changes the shape of a text or picture box
Reshape Polygon	Changes polygon editing mode

The **Page** menu contains commands for arranging pages in a document and for navigating quickly through a document, allowing you to insert, delete, and move pages easily. You can modify the placement of page guides, change the numbering system of a document or a range of pages in a document. You can go to any page in a document and display a master page or a document page.

The Page menu includes the following commands:

Insert	Inserts pages in a document
Delete	Deletes pages from a document
Move	Moves pages within a document
Master Guides	Specifies master page margin and column guides
Section	Creates independently numbered ranges of pages within a document
Previous	Scrolls to the previous page
Next	Scrolls to the next page
First	Scrolls to the first page
Last	Scrolls to the last page
Go to	Scrolls to any page in the document
Display	Switches between document pages and master pages

The **Utilities** menu includes commands for checking spelling and hyphenation, for tracking and kerning, and for listing fonts and pictures used in a document. You can create custom spelling dictionaries, obtain suggested hyphenation for words, and even create a table of hyphenation exceptions. You can list and change all fonts and pictures used in a document and edit tracking and kerning information for fonts—a special feature of this program.

The Utilities menu includes the following commands:

Check Spelling	Checks the spelling of a word, story, or document
Aux. Dictionary	Creates/opens supplements to the QuarkXPress dictionary
Edit Auxiliary	Allows editing of an auxiliary dictionary
Suggested Hyphenation	Displays syllable breaks
Hyphenation Exceptions	Customizes syllable breaks
Font Usage	Shows fonts used in a document
Picture Usage	Shows pictures used in a document
Tracking Edit	Customizes automatic tracking
Kerning Table Edit	Customizes kerning pairs for fonts

The **Window** and **Help** menus are standard Windows menus that contain commands for viewing multiple windows or documents and accessing QuarkXPress' on-line help system.

These descriptions give you an idea of what the menus offer. The menus change depending on the different conditions of your document and some selections are dimmed or disabled if you try to perform an action that the selection is unable to perform.

24.3 FLOATING PALETTES

Palettes, which float on top of the screen (they can actually be placed outside the boundaries of the QuarkXPress interface), function like mini windows in that you can move one by dragging on the title bar or close one by double-clicking on its close box. Instead of accessing the pull-down menus to find an appropriate command, you can click on these icons for that same command. There are four kinds of QuarkXPress palettes:

1. Tools palette
2. Measurements palette
3. Document Layout palette
4. Library palette.

Tools Palette

Almost every action in QuarkXPress involves the use of the tools in this palette, described in Figure 24.2. A tool is selected by clicking on it which determines what you can do with the mouse and the keyboard. The Tools palette can be accessed by clicking the **View** menu and then on **Show Tools**.

Item tool.
Enables you to move, group, ungroup, cut, copy, and paste items (text boxes, picture boxes, lines, and groups).

Content tool
Enables you to import, edit, cut, copy, paste, and modify box contents (text and pictures).

Rotation tool
Enables you to rotate items manually.

Zoom tool
Enables you to reduce or enlarge the view in your document window.

Text Box tool
Enables you to create text boxes.

Rectangular Picture Box tool
Enables you to create rectangular picture boxes.

Rounded-Corner Rectangle Picture Box tool
Enables you to create rectangular picture boxes with rounded corners.

Oval Picture Box tool
Enables you to create oval and circular picture boxes.

Polygon Picture Box tool
Enables you to create polygon picture boxes.

Orthogonal Line tool
Enables you to create horizontal and vertical lines.

Line tool
Enables you to create lines of any angle.

Linking tool
Enables you to create text chains to flow text from one text box to another.

Unlinking tool
Enables you to break links between text boxes.

FIGURE 24.2 The Tools palette.

The Two Most Important Tools: Item and Content

Item tool enables you to select, move, group, ungroup, modify (manipulate and edit certain ungrouped items only), cut, copy, paste, or delete items in a document. **Content** tool enables you to select, move, import, modify (manipulate and edit contents of any item, even though they may located within groups), cut, copy, paste, or clear the contents of items in your document.

If you need to change an attribute of an item inside a box, select that box with the **Content** tool. Choose the **Item** tool whenever you need to change the attributes of the box itself. The two tools' capabilities overlap in some areas. When you want to delete an item or group of items, select them with the **Item** tool and press the **Delete** key. Selecting with the **Content** tool and pressing the **Delete** key clears the contents of the item, not the item itself. An item selected with the **Content** tool can be deleted by pressing the keyboard equivalent of the **Delete** command, **Control+K**, or by choosing **Delete** from the **Item** menu.

Rotation and Zoom Tools

The **Rotation** tool lets you rotate any selected object around any origin point you create by clicking anywhere in the document window.

To rotate an item, it must first be active. The pointer's shape changes from the Arrow into the Rotation pointer when the pointer is over the page. Click to establish the rotation point, and the pointer changes into the Arrowhead. Crosshairs are displayed at the rotation point. Drag the Arrowhead pointer out of the crosshairs to rotate the active item. An outline of the item is displayed during rotation. By waiting until an item flashes after clicking on it and before rotating it, you can view the actual item rather than an outline. Release the mouse when you have placed the item. If the **Shift** key is held down while rotating an item, the movement is constrained to 45° increments. Release the **Shift** key during rotation to permit full movement.

If the Measurements palette is displayed (**View→Show Measurements**), the location of the item is updated on the palette during rotation. Fine adjustments to the angle of items to 0.001° can be made through the Measurements palette or **Modify (Item** menu).

The **Zoom** tool lets you magnify any portion of the page on the screen, working in two ways—by clicking on the **Zoom** tool and then on the item you want to enlarge or select or by clicking on the page with the tool by dragging the mouse diagonally across the area. You can enlarge any portion of the page up to 400% of its original size in increments you specify under the Tool Preferences dialog box. To zoom out, hold down the **Control** key and click. You can also change document view scale by entering a value directly in the **View Percent** field in the lower left corner of the document window, or by choosing one of the scales from the **View** menu.

Box Tools

The next five set of tools on the palette allow you to create new boxes in a document. The first creates new text boxes and is shown as a "boxed A." The next four, shown as "boxed Xs," create picture boxes. QuarkXpress allows you to create and manipulate text and images by means of these boxes in which you simply place text or images. These items, once created can be easily manipulated later with the **Item** and **Content** tools and by using the import features of the program with standard Cut, Copy, and Paste functions from the pull down menus.

Line Tools

The first of these creates new constrained horizontal or vertical lines regardless of the direction in which you drag the tool. The other tool draws lines at any angle that you specify. Lines drawn with either tool can be edited with the Measurements palette, **Style** menu, or the **Modify** command under the **Item** menu giving you access to changing the angle, weight, color, shade, or location of the lines.

Linking Tools

The last two tools control the flow of text from one text box to another, on the same page or across several pages.

Measurements Palette

You can edit values and use the controls in the Measurements Palette (**View→Show Measurements**) to modify items and their contents at any time. The fields and controls will vary according to whether a text box, a picture box, a line, or a group of items is active. The example shown in Figure 24.3 is for an active text box.

FIGURE 24.3 The Measurements palette.

Item information is displayed in the left half of the palette and content information is contained in the right half. Content information is only displayed when the Content tool is selected. Modify the location (**X,Y**), size (**W,H**), rotation, and number of columns (**Cols**) of an active text box. Modify leading, kerning/tracking, alignment (left, center, right, justify), font, size, and type style of text.

Document Layout Palette

The Document Layout palette (**View→Show Document Layout**) (Figure 24.4) provides a graphical method for inserting new pages in a document, and for moving or deleting existing pages. It enables you to

- Create, name, delete, arrange, and apply master pages;
- Insert, delete, and move document pages easily;
- Navigate through the document pages and master pages.

Master Pages. A Master page is a nonprinting page (which acts as a template for document pages) used to format document pages automatically. When you insert a new document page as you work, it will contain all the items specified on the master page on which it is based. Master pages frequently contain such items as headers, footers, page numbers, and other elements that are common to a number of document pages. You will only want to work with Master pages on large projects involving multiple sets of pages each with varying items such as margins, headers, footers, and text links.

Manipulating Pages. The Document Layout palette provides two views: one for working with document pages, and the other for working with master pages. The commands in the **Document/Masters** and **Apply** menus can be used with either document pages or master pages, depending on the view selected. The name of the first menu is **Document** when document page view is selected, and **Masters** when master page view is selected. You change the view by choosing **Show Master Pages/Show Document Pages** from the **Document/Masters** menu.

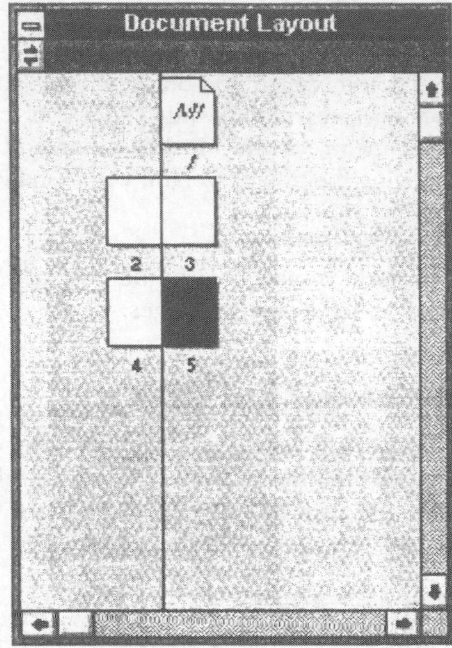

FIGURE 24.4 Document Layout palette.

In document page view, the icons displayed represent actual document pages. Each icon has a number below it indicating the page's absolute sequence number within the document. Within the icon is a user-definable indicator corresponding to the master page that is applied to that document page. No indicator is displayed if a blank master page is applied.

To insert a document page, choose an option from the **Insert** submenu (**Document** menu). Drag the pointer down into the page icon area and click when the pointer indicates the page will be placed where you want it. To insert multiple document pages, hold down the **Control** key as you choose an option from the **Insert** submenu (**Document** menu).

To select multiple document pages, click on the first icon in a range of document pages. **Shift+Click** on the last icon in the range. To select noncontiguous pages, hold down the **Control** key and click on individual icons.

To move document pages, select the pages you want to move, and drag them within the document page area of the palette. As you drag, the pointer changes to indicate where the pages will be placed. Release the mouse button when the pages are positioned where you want them. To delete document pages, select the pages you want to delete, and choose **Delete** from the **Document** menu. To scroll the document window to a specific master or document page, double-click on the desired page icon on the palette. The information on the icon is displayed in italic to indicate that page is currently displayed in the document window.

Library Palettes

A Library palette enables you to store and retrieve frequently used items in libraries. Library entries can include any page element you can create or import into QuarkXPress—logos, boilerplate text, graphic elements, etc. An entry can be a single line, picture box, text box, group, or multiple-selected items. To create a new library, choose **New→Library** (**File** menu) and the **New Library** dialog box is displayed. Click **Open** to select an existing library or type in a familiar name and click **Create** to open a blank library. Now you can build this library by dragging selected items from

FIGURE 24.5 Two sample Library palettes.

your document into the library; the dragging movement causes a copy of the selected item to be placed in the library.

You can open multiple libraries and drag selected items between them. Libraries are displayed as palettes in front of documents. When you place a copy of an item in a library, the item is displayed as a thumbnail. Each Library window contains a pull-down **Edit** menu that allows you to Cut, Copy, and Paste items to the Clipboard and other documents. You can also delete an item from the Library (**Edit→Delete**). Figure 24.5 displays two sample picture libraries that contain copies of picture boxes.

24.4 QUARKXPRESS BASICS

As the program is started the default floating tools and measurement palettes are displayed—your first choice is whether to work on an existing document (open a file), a template (a new document based on an old one), or create a new document (create a file).

Creating a New Document

To work on a new document, choose (**File→New→Document**) and the **New Document** dialog box (Fig. 24.6) is displayed. Specify the new document's page size, margin guide positions, the number of columns you want displayed on pages, and the space between columns (**Gutter Width**). You also have the option of creating a facing-page document with left- and right-facing pages, and of placing an automatic text box bounded by the margin on your document pages. When you create a new document, you automatically define its original master pages.

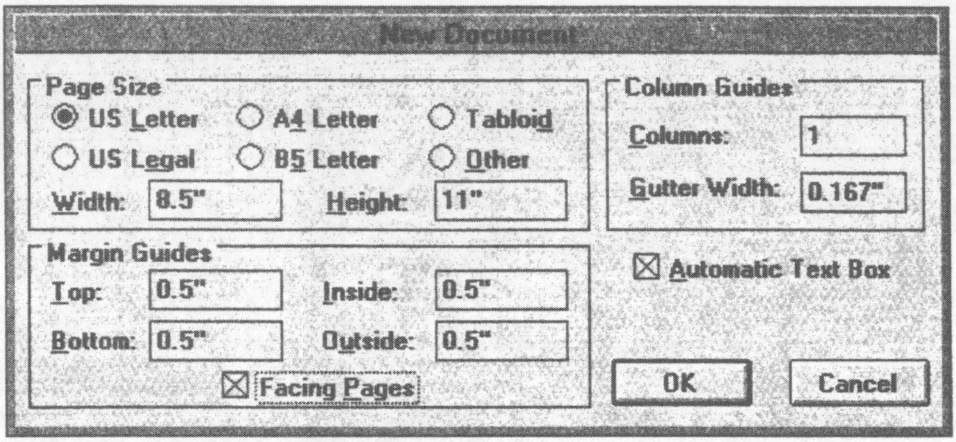

FIGURE 24.6 Dialog box for creating new documents.

Document Setup

To quickly change the page orientation, just select **Document Setup** from the **File** menu. Use Document Setup to change the page size of the active document or to switch between a single-sided and a facing-page document. When a new document is created, page margins are determined by the **Page Size** and **Margin Guides** specified in the **New Document** dialog box. If **Automatic Text Box** is checked, an automatic text box coinciding with the page margins is created on the first page of the document and on the original master page. When the **Page Size** is changed through **Document Setup**, any text box in the document whose dimensions coincide with the old margins is resized to match the new margins. You will be warned if changing to a smaller page size would cause existing items to no longer fit on the document pasteboard.

Preferences

After starting a new document or opening a template, you should check and change, if necessary, the document's preferences which are divided into four sets: **Application**, **General**, **Typographic**, and **Tools**. You access them all through the **Edit** menu, as shown in Figure 24.7.

Key preferences are highlighted below.

Application Preferences

Specify the colors you want the margin, ruler, and grid guides to appear in as well as the scrolling speed of the interface as you navigate from page to page. Settings made through Application Preferences are applied to all documents. Key settings are

- **Guide Colors**: displays a standard dialog box that enables you to specify the color of **Margin** guides, **Ruler** guides, or **Grid** lines.
- **Pasteboard Width**: defines the width of the pasteboard on either side of a page or spread as a percentage of the document width.
- **Reg. Marks Offset**: defines the distance from the edge of a page at which registration marks are printed when **Registration Marks** is checked in the **Print** dialog box.

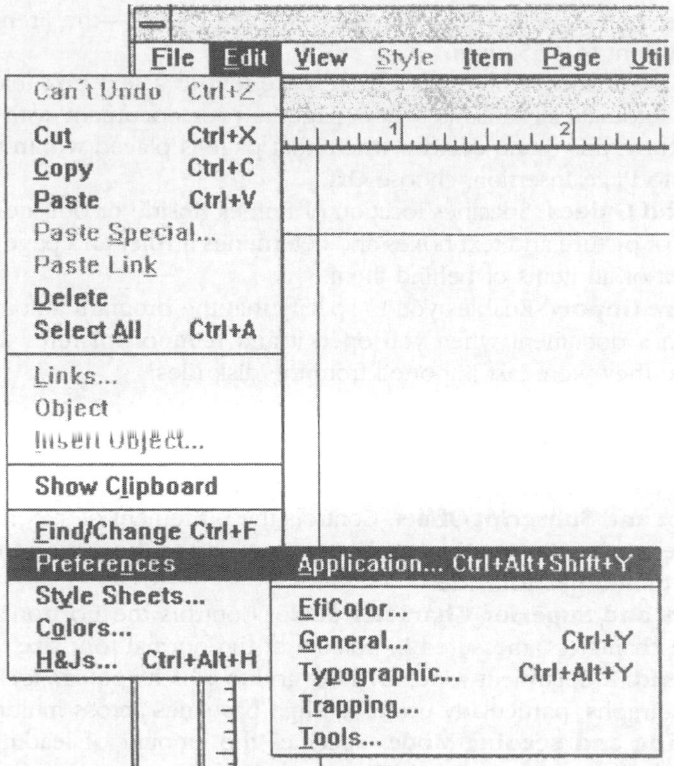

FIGURE 24.7 Setting program and document preferences.

- **Live Scroll**: updates the document view as you drag scroll boxes in document window scroll bars.
- **Auto Library Save**: causes libraries to be saved automatically whenever library entries are added.
- **Low Resolution TIFF**: determines whether the default resolution for importing TIFF pictures is half the dpi of your monitor or full resolution. Holding down the **Shift** key when importing a TIFF causes the opposite setting of Low Resolution TIFF to be used.
- **8-bit TIFF**: downgrades the display of 24-bit TIFFs to 8-bit.
- **256 Levels of Gray**: causes high resolution grayscale pictures to be displayed at full resolution on a monitor that is capable of such a display. Grayscale pictures are normally displayed using 16 levels.
- **Display DPI Value**: adjusts the display of your monitor so that an inch on-screen equals a real inch. Enter values and hold a ruler against the screen to determine the correct value.
- **Speed Scroll**: adjusts the rate at which the document window scroll bars function.

General Preferences

These preferences allow you to specify the defaults for QuarkXPress functions that are not Typography or tool related.

Horizontal and Vertical Measures. Specifies the measurement system on the ruler (**View→Show/Hide Rulers**) in your choice of units of measure—inches (1/8ths or

1/10ths), picas, points, millimeters, centimeters, even ciceros—the French typesetting measure equivalent to 4.552 mm.

Auto Page Insertion. When you import more text to a text box than it can hold, QuarkXPress automatically inserts a new page in your document with a text box to hold the overflow; this menu decides where that page is placed within the document. To disable Auto Page Insertion, choose **Off**.

Framing and Guides. Specifies location of frames (inside or outside) in relation to the perimeter of picture and text boxes and determines if ruler and page guides should appear in front of all items or behind them.

Auto Picture Import. Enables you to specify that the program automatically check the pictures in a document when you open it and reimport pictures that have been modified since they were last imported from the disk files.

Typographical Preferences

These settings allow you to control text flow and character placement and appearance.

Superscript and Subscript Offset. Controls the placement of superscript and subscript characters in relation to the baseline by entering appropriate values, measured in percent of the normal font size.

Small Caps and Superior Character Scale Controls the horizontal and vertical scale of these characters, measured in percent of the normal font size.

Baseline Grid. A document-wide, invisible grid to which you can lock the baselines of text in paragraphs, particularly useful to align baselines across multiple columns.

Auto Leading and Leading Mode. Specifies the amount of leading (the vertical spacing between lines of text) QuarkXPress uses in paragraphs that are "auto" leaded. Two modes of leading are available—**Typesetting** (line spacing measured from one baseline to another) or **Wordprocessing** (line spacing measured from the top of the ascent on one line of text to the top of the ascent on the line below it).

Auto Kern Above. Specifies the font size above which QuarkXPress uses the kerning tables built into most fonts to control intercharacter spacing.

Tool Preferences

Double-clicking on any tool displays the Tool Preferences dialog box (Fig. 24.8), which enables you to specify default settings for the Zoom and item creation tools (text boxes, picture boxes, and line tools). By clicking on the **Modify**, **Frame** and **Runaround** buttons you can change the appearance of each new box you draw. For text boxes, you can designate border style and thickness, text runaround style (how the text in this box will align or move when other items, like pictures, are placed nearby), and text inset attributes. For picture boxes, you can assign border style and thickness, text runaround style (how pictures in this box will cause other text to align within their boxes), and picture scale. Since lines do not have frames, the **Frame** button is dimmed when you try to modify it; however, you can designate whether a line drawn is automatically opaque or transparent using the **Runaround** mode button.

Ruler Guides

These nonprinting ruler lines can be created on your document pages to help you align items. Click within a ruler and drag onto the page. Drag them again to move, even to delete them by moving them off the page entirely by pressing the **Control** and **Shift** keys as you drag on the guide.

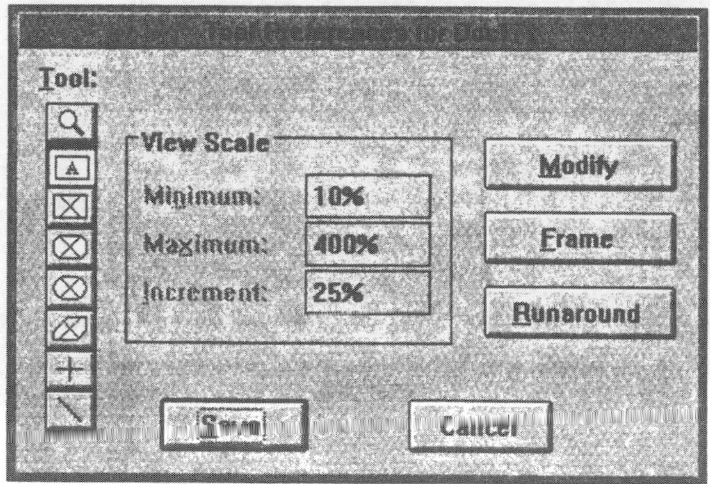

FIGURE 24.8 Tool Preferences dialog box.

Manipulating Items

QuarkXPress works with text boxes, picture boxes, and lines, each with its own tools in the palette, its own **Style** menu, and its own attributes. All three items are subject to a common set of commands in the **Edit** and **Item** menus.

Selecting Items. To select a single item, use the **Item** or **Content** tool and click on an item. Multiple items can only be selected by choosing the **Item** tool and then using **Shift+Click** to add additional items to the selection. Selected items appear with handles.

Cut, Copy, Delete, and Paste. Accessed from the **Edit** menu, select any item with the **Item** tool (or select them all with **Control+A**) to use these commands.

Duplicate. Accessed from the **Item** menu, make an immediate copy of the item instead of using **Copy** then **Paste** from the **Edit** menu.

Step and Repeat. In the **Item** menu, use it to make multiple copies of items in one operation. The **Repeat Count** field in the dialog box designates the number of copies.

Group and Ungroup. Also in the **Item** menu, allows you to convert a collection of several selected items into a single item that can itself be handled by other commands. This is particularly useful if you have spent time precisely positioning items and then have to move them—grouping items allows you to maintain the spatial relationship between items. Select the entire group with the **Content** tool and individual items with the **Item** tool.

Lock and Unlock. Prevent items from being changed (**Item** menu).

Send to Back, Bring to Front. Dictate the order of items that overlap to determine which items are completely displayed and which are obscured (**Item** menu).

Space/Align. Control the placement of active items in relation to one another. **Space/Align** (**Item** menu) is available when the **Item** tool is selected, and two or more items are active. You can space them a certain distance apart or distribute them evenly along a line, from the **Space/Align** dialog box shown in Figure 24.9.

24.5 WORD PROCESSING WITH QUARKXPRESS

QuarkXPress provides complete word processing capabilities that enable you to generate and edit text for your publications without having to use a stand-alone word processing program. Although it does not have all of the features you find in a program

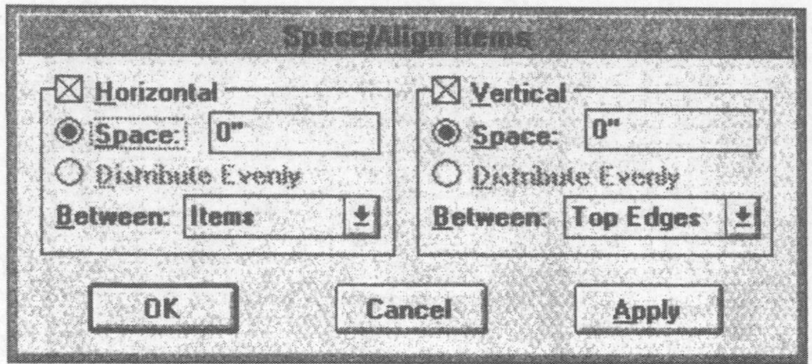

FIGURE 24.9 Space/Align dialog box.

dedicated to word processing, it has far more features than most other desktop publishing programs. You can create your own text in QuarkXPress or import text from other word processors.

Creating Text Boxes

Before you import text or enter text in QuarkXPress, you need a text box to work with. Open a document after setting up preferred rulers and guidelines needed to position the text box. Click on the text box creation tool, move the cursor to one corner of the desired location where you want the text displayed, click the mouse, and drag the cursor diagonally to the opposite corner of the new box position. Upon completion of the operation (when the mouse button is released), you will notice that the box frame is highlighted and contains handles, which shows that this box is selected. Now you can import text from a file on disk or from the Clipboard or type text directly in the box.

Importing and Exporting Text

You can import word processing documents from several other word processing applications, then edit and rearrange them to fit your layout. Complete translation of documents occurs when text is imported into QuarkXPress—the codes that define paragraph formats, boldface type, and tabs—are kept undisturbed. QuarkXPress also has the ability to save text in almost any word processing format, in case you need to move text back to the original program.

To import text, select the text box with the **Content** tool and move the cursor to the spot where the text is to be placed. Choose **Get Text . . .** from the **File** menu. A dialog similar to the one shown in Figure 24.10 will appear.

After finding the name of the text file you want and upon selecting it and clicking **OK**, the text from the file fills the text box. If there is too much to fit in the selected box, a small icon with an X in it appears in the bottom right corner of the text box.

Changing Text Style

The top half of the **Style** menu (Fig. 24.11) contains character attribute commands that enable you to change font, size, type style, color, shade, horizontal/vertical scale, kerning, tracking, and baseline shift position. Once you have created or imported your text, you can format it. There are four ways to set text attributes. In each case you start by

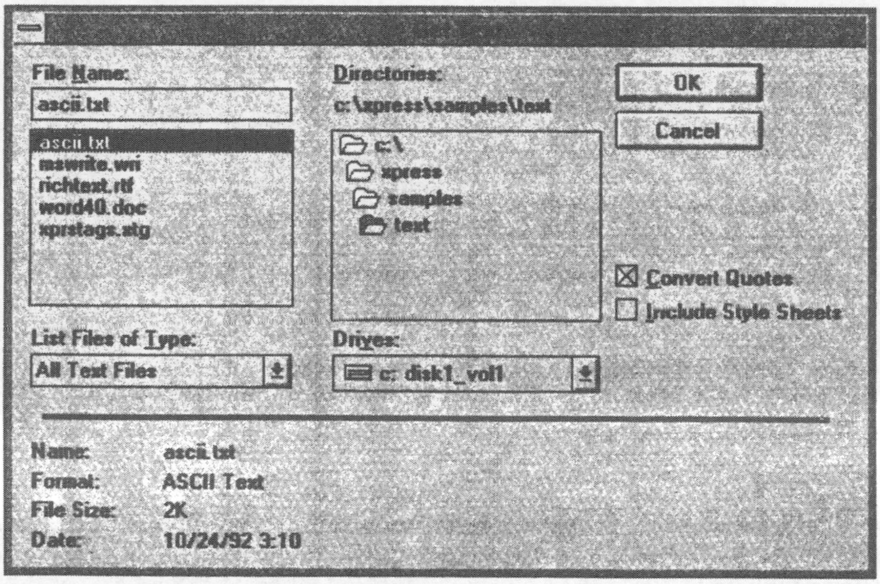

FIGURE 24.10 The Get Text dialog box.

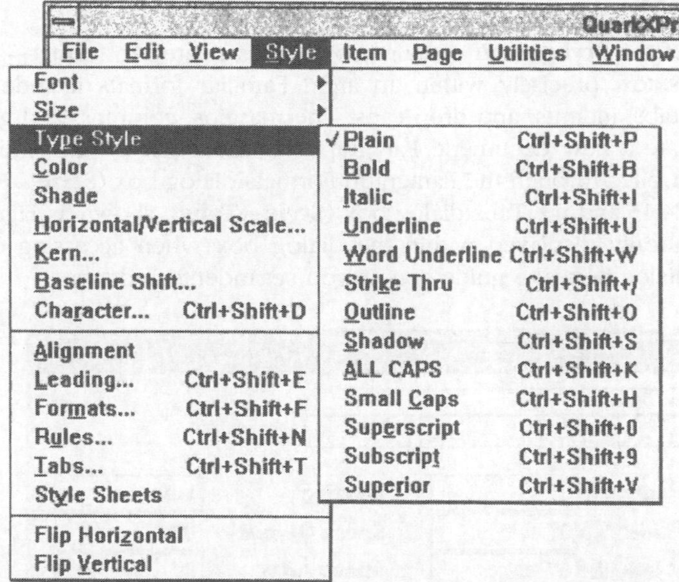

FIGURE 24.11 The Style menu.

highlighting the characters whose attributes you want to change by dragging the mouse across a section of text:

- The quickest way to apply a new format or style is by using the status section of the Measurements palette, as shown in Figure 24.3. Choose **Show Measurements** from the **View** menu, then click on the appropriate attribute in the palette.
- The **Style** menu (Fig. 24.11).
- Keyboard shortcuts accessed by pressing key combinations for a particular format, listed within the **Style** menu.

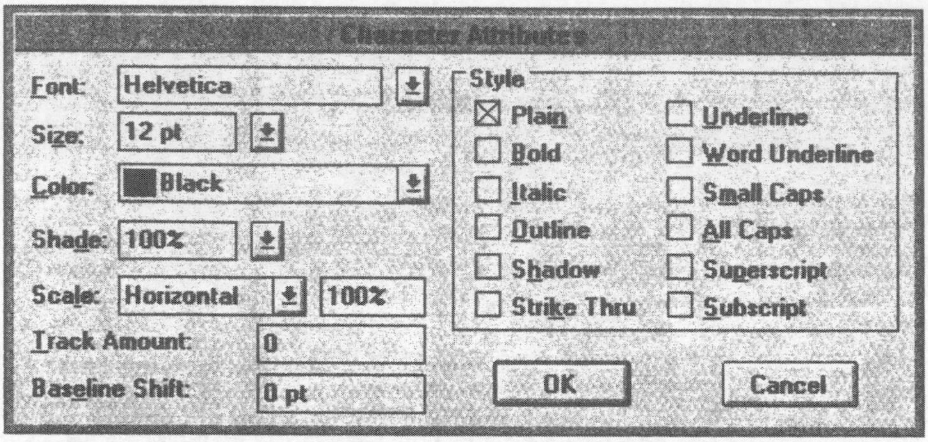

FIGURE 24.12 Character attributes.

- The **Character** attributes dialog box (Fig. 24.12), by choosing **Character . . .** from the **Style** menu—it shows all character attributes and is often used to apply multiple character attributes.

Formatting Text

The bottom half of the **Style** menu contains commands related to format—a position setting that places text precisely within an area. Familiar formats include margins, columns, gutters, tabs, indents, and drop caps. The margins, columns, and gutters are set when you create a new document. Paragraph formats such as tabs, indents, and drop caps are controlled through the Paragraph Formats dialog box (**Style→Formats**), shown as Figure 24.13 and the Tabs dialog box (**Style→Tabs**), shown as Figure 24.14. A ruler is automatically displayed within the dialog box when accessing either the Formats or Tabs dialog boxes to guide you as you set indents and tabs.

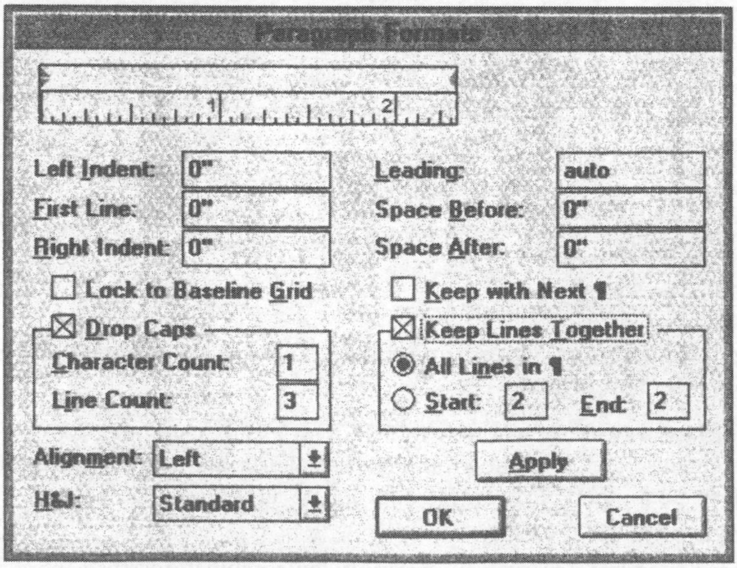

FIGURE 24.13 The Paragraph Formats dialog box.

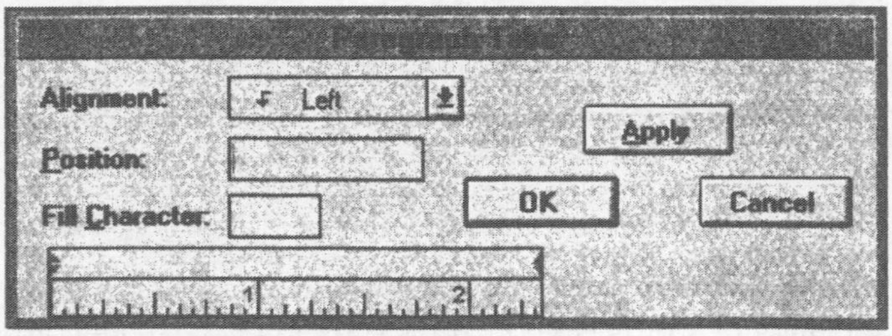

FIGURE 24.14 The Paragraph Tabs dialog box.

To use the Paragraph Formats dialog box first select a text box with the **Content** tool and then select the paragraphs you want to format. A paragraph will be affected if any of its characters are highlighted. Choose **Formats** from the **Style** menu and specify the distances you want to indent text from the left and right edges of the text. Specify the special indent for the first line of text and click **Apply** when you are done to see the results. Click on **OK** if satisfied with the results or **Cancel** to abandon the procedure.

To use the **Tabs** dialog box, first select a text box with the **Content** tool and then select the paragraphs for which you want to create tabs. A paragraph will be affected if any of its characters are highlighted. Choose **Tabs** from the **Style** menu and set your tab by choosing **Left**, **Center**, **Right**, or **Decimal** and an appropriate position number. Delete a single tab by clicking on it and dragging off the ruler or all tabs by holding down the **Ctrl** key and clicking on the ruler.

Finding or Replacing Text

QuarkXPress enables you to search for and selectively change specific text strings and attributes including Font, Size, and Style. Choosing **Find/Change** (**Edit** menu) displays the Find/Change dialog box (Fig. 24.15).

To search for and replace text, position the text insertion point where you want the search to begin, enter the text you want to find in the **Find what**: field and the replacement text in the **Change to**: field. Leave the **Change to**: field blank if you want to search for (but not change) specific text, or if you want to delete the text you find. Click on one of the following action buttons

- **Find Next**: begins the search, or finds the next occurrence of the specified search criteria. Holding down the **Control** key changes this button to **Find First**, which begins the search at the start of the document.
- **Change, then Find**: changes the highlighted text and finds the next occurrence of the search criteria.

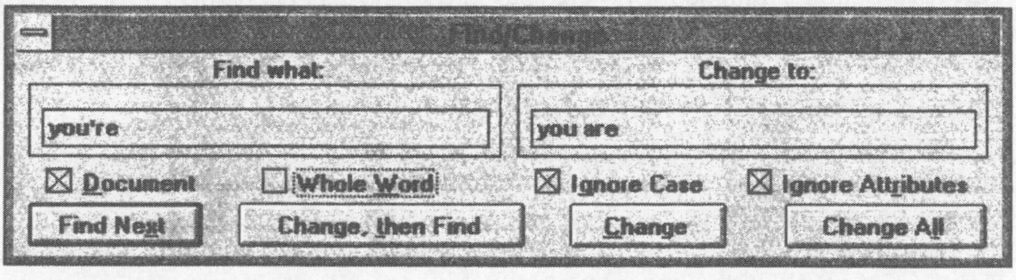

FIGURE 24.15 Find/Change dialog box.

- **Change**: changes the highlighted text.
- **Change All**: changes all occurrences of the search criteria. An alert is displayed showing the number of occurrences changed.

Spell Checking

Use **Check Spelling** (**Utilities** menu) to check spelling of a highlighted word or entire document. Spelling of words is verified by scanning a 120,000-word QuarkXPress dictionary. Choosing **Check Spelling** displays a submenu that contains three entries—Word, Story, and Document/Masters:

- **Word**: checks the word at the current text insertion point (or start of a highlighted range). Word is available when a text box is active.
- **Story**: checks all words in the current text chain. Story is available when a text box is active.
- **Document/Masters**: checks all stories on document or master pages. The menu entry changes depending on whether a document or master page is currently displayed in the document window.

Figure 24.16 shows the dialog box that is displayed after choosing **Check Spelling→Document/Masters** (**Utilities** menu):

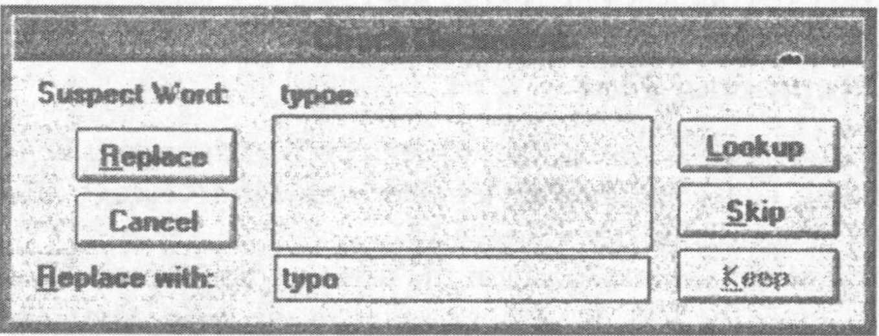

FIGURE 24.16 Check Document (for Spelling) dialog box.

If suspect words are found, the Check Document dialog box shows the first suspect word. To get a list of suggested spelling, click **Lookup**. To replace the suspect word with one in the list, select the word in the list and click **Replace**, or double-click on the replacement word. To leave the suspect word unchanged click **Skip**. To change the suspect word enter a replacement in the **Replace with**: field, or select an entry from the Lookup list, and click **Replace**. To add the word to a user dictionary, click **Keep** (this button is dimmed if no auxiliary dictionary has been specified). To abort the spelling check click **Cancel**.

Hyphenation and Justification (H&Js)

QuarkXPress lets you create and edit your own guides for any text. Hyphenation and Justification values can be set to automatically break words dependent on word-length or space from the column edges. These values may be set to control word spacing and character spacing defined as a percentage of word length. Use H&Js to make new hyphenation and justification specifications or to change existing specifications in a QuarkXPress document or for the QuarkXPress application itself.

To create a new H&J guide for a document, choose **H&J . . . (Edit** menu) and a dialog box is displayed that allows you to create a new specification or edit a selected specification. If you click **New** or **Edit**, the **Edit Hyphenation and Justification** dialog box (Fig. 24.17) appears.

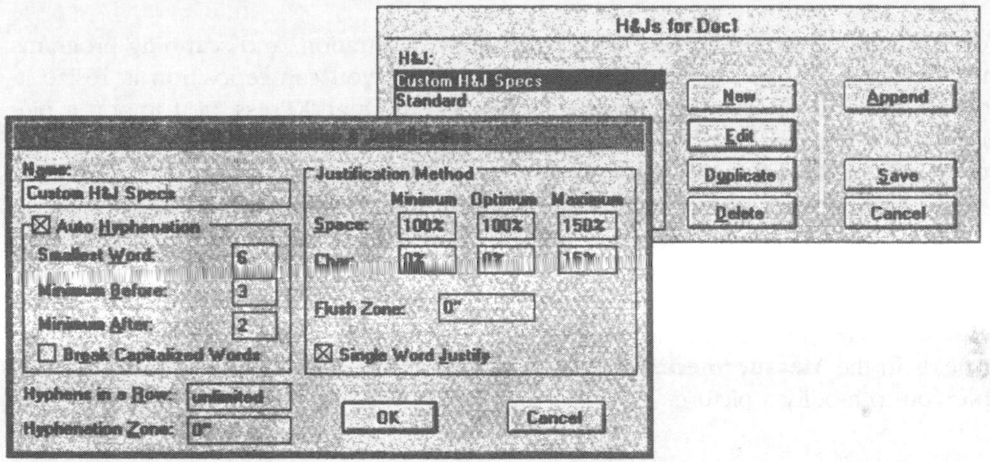

FIGURE 24.17 Edit Hyphenation & Justification dialog box.

Hyphenation values include the smallest word to hyphenate, minimum words before and after a hyphenation, and whether to hyphenate capitalized words. The maximum hyphens in a row setting designates the maximum number of consecutive lines that can end with a hyphenated word. The hyphenation zone is the area, measured from the right indent, where hyphenation can occur.

Justification values include the maximum (up to 500%), minimum (down to 0%), and optimum amounts of space to add between words, expressed as a percentage of the normal space for the font and size of the paragraph. Similar settings can be specified for the space to allocate between characters. The flush zone is measured from the right indent.

To define H&J specifications for a specific document choose **H&Js . . .** when the document is active. The H&J specifications you make are saved with and will be used with that document only. To define H&J specifications for the QuarkXPress application, choose **H&Js . . .** when there are no QuarkXPress documents open. Specifications that you make are saved with the QuarkXPress application itself and will be available when new documents are subsequently created (even after restarting the program).

24.6 GRAPHICS

The two basic graphical elements in QuarkXPress are pictures and lines. While pictures must be placed in picture boxes (just as text fits into text boxes), lines can be located anywhere in a document. QuarkXPress lets you overlay text, pictures, and lines, showing one through the other or wrapping text around graphics. QuarkXPress offers precise control over imported graphic images to within 1/1000 of a unit (pica, inch, cicero, or any other unit you specify). Rotation of images or the boxes within which they are contained can occur with a precision of within 0.001°, and borders, border styles, and weights (or thickness) to within 0.001 point.

Pictures

You can import pictures created with paint, draw, illustration, and scanning programs into QuarkXPress. Once a picture is in QuarkXPress, you can reposition it, resize it, and perform a host of stylistic modifications to it. In QuarkXPress, you import a picture into an active picture box. Any of the four box creation tools for creating picture boxes may be used from the Tools palette.

The Mover pointer is used to reposition picture boxes manually. Numerical repositioning occurs using the **Origin Across** and the **Origin Down** fields in **the Picture Box Specifications** dialog box (**Item→Modify**) or the **Measurements** palette. You can resize picture boxes manually using the Resizing pointer or numerically using the **Width** and **Height** fields in the **Picture Box Specifications** dialog box or the **W** and **H** fields in the **Measurements** palette. The **Item** menu also contains entries that enable you to modify a picture.

Creating Picture Boxes

To create a new picture box select one of the picture box creation tools first. When a creation tool is selected, the pointer's shape changes from the Arrow into the Item Creation pointer (a small set of crosshairs) when it is over the page. To create a picture box, move the pointer to the location where you want to place one of the corners of the box. Click, and drag the mouse to the diagonally opposite corner. Release the mouse button.

If the **Shift** key is held down during the creation process, the shape or orientation of the item is constrained. Rectangle or rounded-corner rectangle picture boxes are constrained to squares. Oval picture boxes are constrained to circles. The orientation of lines drawn with the Line tool is constrained to 0°, 45°, and 90°.

If the rulers are displayed (**View→Show Rulers**), the location of the pointer is tracked on the rulers before you press the mouse button to begin creation. As you drag, the content area of a picture box is tracked. This aids in the placement of the item. If the **Measurements** palette is displayed, the location of the pointer and the size of the item are displayed on the palette during item creation. Fine adjustments to item attributes to within 0.001 unit of any measurement system can be made through the **Measurements** palette or **Modify** (**Item** menu). After you use one of the creation tools, the palette reverts to either the **Item** or **Content** tool (whichever was last used). To keep a creation tool selected (when creating multiple items), hold down the **Alt** key when you select the tool.

Reshaping Boxes

Use **Box Shape** (**Item** menu), as shown in Figure 24.18, to change the shape of a box (or a text box) into a predefined shape or into a polygon. **Box Shape** is available when either a picture box or a test box is active. The box shape you have chosen is shown checked in the submenu.

Five predefined shapes can be chosen from the **Box Shape** submenu. The last entry in the menu converts a box into a polygon.

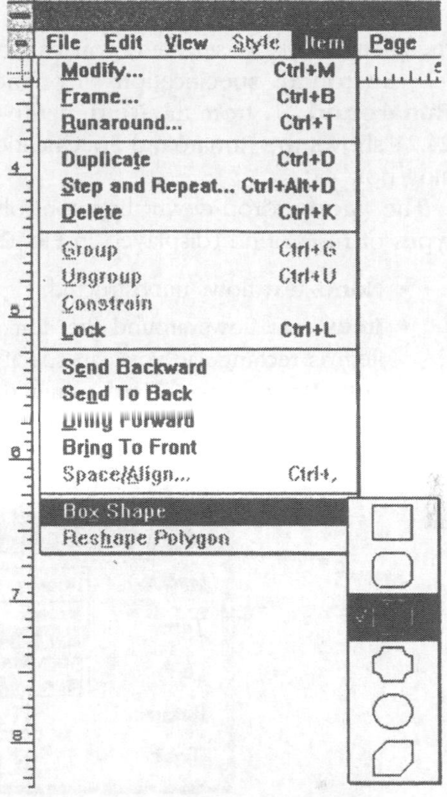

FIGURE 24.18 Box Shape menu.

Framing a Box

Use **Frame** (**Item** menu) to place a border around a picture box (or text box). The frame is placed inside or outside the box, depending on the setting of **Framing** in **General Preferences**. To place a frame around a box, choose the **Style**, **Width**, **Color**, and **Shade** for the frame, as shown in Figure 24.19. You can enter frame widths using any measurement system, but the values are displayed in points when you redisplay the **Frame Specifications** dialog box. A 12-point size sample of your selection is shown at the top of the dialog box. To remove a frame, enter 0 (zero) for the frame width.

Picture Box Runaround

The much acclaimed **Runaround** feature of QuarkXPress controls the flow of text behind an active item. Upon creating a picture box and placing it over a text box, the text automatically runs around it. The words and characters in the text move out of the way of an overlapping picture box. Figure 24.20 shows an example.

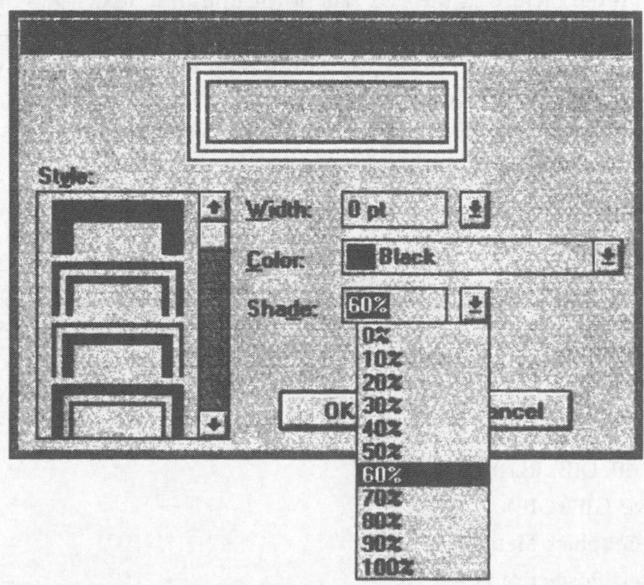

FIGURE 24.19 Frame Specifications dialog box.

There are several runaround options. While the picture box is selected, you can change its runaround specifications by choosing **Runaround . . .** from the **Item** menu. Figure 24.21 shows the Runaround Specifications dialog box

The **Mode** drop-down list controls four types of runaround (displayed in Fig. 24.22):

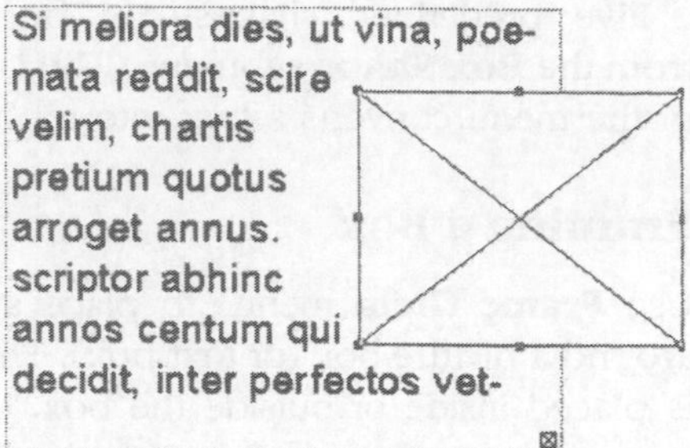

FIGURE 24.20 Example of Runaround—picture box overlaps a text box.

- **None**: text flows unobstructed.
- **Item**: text flows around the item. If the item is rectangular, you can specify **Top**, **Left**, **Bottom**, and **Right** values that offset the text from the item. If the item isn't rectangular, you can specify a uniform text outset.

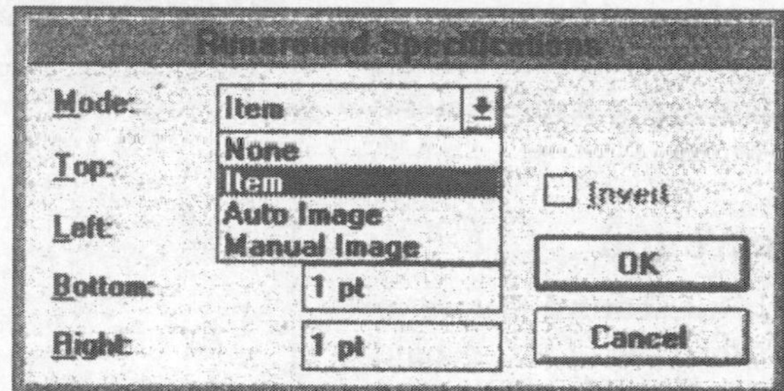

FIGURE 24.21 Runaround Specifications dialog box.

- **Auto Image**: (picture boxes only) text flows around the contours of the picture in the box (and not around the picture box itself). Text outset determines the offset of the text from the picture. If the box has no picture in it, text flows around the box.
- **Manual Image**: (picture boxes only) creates a runaround polygon in the shape of the picture in the box (or in the shape of the box, if there is no picture). This polygon can be customized to control the runaround of text behind the picture box. Text outset determines the offset of the runaround polygon from the picture. If **Invert** is checked, text will flow within the polygon.

Importing Images

QuarkXPress imports most popular file formats. Use **Get Picture** (**File** menu) to import pictures created outside QuarkXPress. Pictures can also be pasted into documents from the Clipboard. The **Content** tool must be currently selected in order to use **Get Picture**. If a picture box is active, the menu shows **Get Picture**. QuarkXPress imports pictures of the following types and extensions:

- Bitmap (BMP, DIB, RLE)
- Compuserve GIF (GIF)
- Computer Graphics Metafile (CGM)
- Encapsulated PostScript (EPS)
- HPGL (PLT)

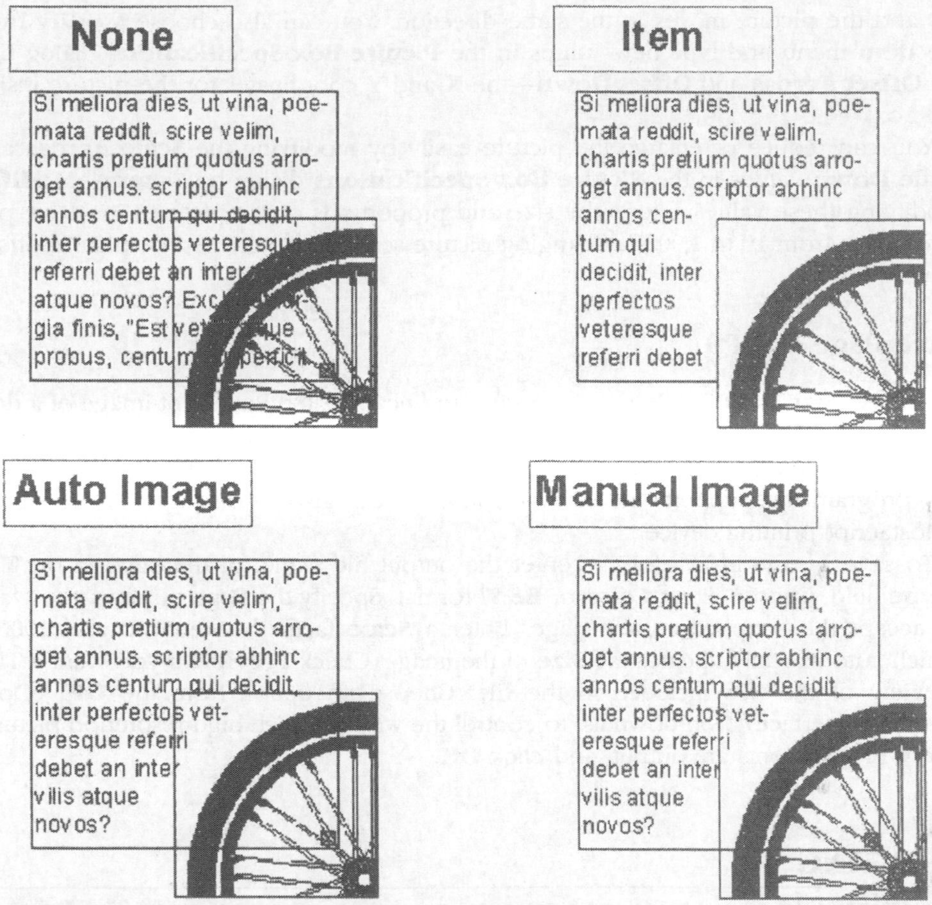

FIGURE 24.22 Examples of the four runaround modes.

- Macintosh PICT (PCT)
- Micrografx Designer (DRW)
- Paintbrush (PCX)
- Scitex (CT)
- TIFF (TIF)
- Windows Metafile (WMF)

To import a picture, select a name from the file list, then click **OK**, or double-click on the name. The percentage of the file that has been read in is displayed in the lower left corner of the document window. If the picture is on another disk, click on the **Drives** drop-down list and choose another drive from the list to see a list of files and directories on that drive.

Moving and Sizing Pictures Inside a Box

QuarkXPress imports the entire image into the selected picture box, regardless of the size of the box. You can move the picture inside the box by using the hand cursor that appears when you select the **Content** tool—just click and drag with the hand cursor and the picture moves in the same direction. You can also choose **Modify** from the **Item** menu and type new values in the **Picture Box Specifications** dialog box for **Offset Across** and **Offset Down**—the **X** and **Y** coordinates for the picture inside the box.

You can reduce or enlarge the picture easily by modifying the **Scale Across** and **Scale Down** values in the **Picture Box Specifications** dialog box (**Item→Modify**). Modifying these values adjusts the size and proportions of the pictures in active picture boxes from 10 to 1,000%. Changing picture scale does not alter the picture offset.

Save Page as EPS

Use **Save Page as EPS** (**File** menu) to save an Encapsulated PostScript image of a document page. The EPS file generated retains and reproduces all text layout and graphic elements of the original page. The image can be imported into and manipulated by any program that supports the EPS file format. It can also be directly downloaded to a PostScript printing device.

To save a page in EPS format, enter the output file name for the page in the **File Name** field. Choose either **Color** or **B&W** format. Specify the **Page** number you want, or accept the supplied current page. Enter a **Scale** factor between 10 and 1000%, which automatically updates the **Size** of the image. Check **Preview** if you want a TIFF preview of the EPS included in the file. Choose an option from the **OPI** (Open Prepress Interface) drop-down list to control the way in which high resolution pictures and OPI comments are output, and click **OK**.

24.7 PRINTING

Use **Print** (**File** menu) to print a document or portions of a document. The **Print** dialog box corresponds to the printer you are using and reflects the options available for your printer.

- **Copies**: the number of copies to print.
- **Pages**: specifies the page range you want to print (unless you want to print the entire document). If a master page is currently displayed in the document window, **Pages** is unavailable, and the current master page is printed.
- **Cover Page**: prints a cover page for the document.
- **Page Sequence**: **All Pages** prints all of your document; **Odd Pages** prints only the odd pages; **Even Pages** prints only the even pages.
- **Output**: **Normal** produces full standard output; **Rough** is faster but substitutes simple frames for complex frames and does not print pictures; **Low Resolution** prints pictures in low resolution.
- **Thumbnails**: (PostScript only) prints many small thumbnail images of the document's pages per sheet.
- **Back to Front**: prints the selected range of pages in reverse order. When more than one copy is requested, **Collate** will print successive copies of the complete range of pages, rather than a number of copies of each page in the range.
- **Spreads**: prints spreads of pages as continuous output.
- **Include Blank Pages**: prints pages containing no printable items.

- **Registration Marks**: places alignment marks on the edges of the output. These marks can be **Centered, Off Center**, or turned **Off**.
- **OPI**: (PostScript only) controls the way in which high resolution pictures and Open Prepress Interface comments are output.
- **Tiling**: used for printing large documents on smaller paper. The QuarkXPress manuals contain details on how to perform automatic and manual tiling.
- **Separation**: (PostScript only) prints an individual sheet for each color used in the document. When **On**, you can choose **All Plates** or a specific separation **plate** from the **Plate** drop-down list.
- **Print Colors as Grays**: (PostScript only) causes colors to be printed as corresponding shades of gray rather than 100% black. It is effective only when **Separation** is **Off** and you are printing to a black-and-white PostScript printer.
- **Setup**: displays QuarkXPress' Printer Setup dialog box.

The progress in downloading a job to a printer can be monitored by holding down the **Shift** key when you click **Print**. Information displayed includes the current page, separation plate, tile, and picture being processed.

DEFINING TERMS

Alignment: The four standard paragraph alignments: Left, Centered, Right, and Justified.

ASCII: An acronym for American Standard Code for Information Interchange. ASCII is an industry-standard, text-only file format. QuarkXPress can import and save text in the ASCII format.

Auto Leading: The value you enter in the Auto Leading field in the Typographic Preferences dialog box is applied to all paragraphs in a document for which auto leading has been specified. Line spacing in autoleaded paragraphs is relative to the largest character on a line.

Auto Page Insertion: The Auto Page Insertion drop-down list in the General Preferences dialog box (**Edit** menu) determines whether pages are inserted and where they are placed when you enter or import more text into a text box than the box can display.

Automatic Text Box: The text box on a master page and its corresponding document pages into which text flows when a new page is automatically inserted. On the master page, it is linked to the Intact Chain icon. The Automatic Text Box check box in the New dialog box (**File** menu) determines whether an automatic text box is placed on the first page of a new document and on the document's original master page.

Automatic Text Chain: The text chain defined by the automatic text box on a master page. An automatic text chain occurs when overflow text causes pages to be automatically inserted. A document can contain only one automatic text chain, but it can contain several manual text chains.

Auxiliary Dictionary: A user-defined dictionary used for checking spelling in documents that contain specialized vocabulary. When you check spelling, QuarkXPress uses its built-in 120,000-word dictionary (stored in XPRESS.DCT) and any auxiliary dictionary you have opened for the active document.

Background Color: The color applied to the background of a box. A background color is also called a box tint.

Baseline Grid: An invisible grid that underlies documents. When all paragraphs on a page are locked to the baseline grid, lines of text align from column to column and from box to box. You define the baseline grid in the Typographic Preferences dialog box (**Edit** menu). You can show the baseline grid by choosing **Show Baseline Grid** from the **View** menu.

Bitmap Graphic: A graphic image (picture) formed by a pattern of dots, generally of low resolution. Also called a *raster graphic*.

Bitmapped Font: A font made of bitmapped characters. Unlike scalable fonts, bitmap fonts require a font file for each font size and style combination. Screen font files are bitmapped fonts. Some printer fonts are also bitmapped fonts.

Bounding Box: A nonprinting rectangular box that fully encloses an item.

CGM: An acronym for Computer Graphics Metafile, a popular vector-based graphics file format supported by many applications.

Character Attribute: A specification applied to a character—Font, Size, Type Style, Color, Shade, Horizontal Scale, Kern, Track, and Baseline Shift.

Cicero: A unit of measurement in the Didot system, commonly used in Europe. A cicero is slightly larger than a pica and is equal to approximately 4.55 mm.

CMYK: An acronym for Cyan, Magenta, Yellow, and Black, the standard ink colors used in four-color printing. CMYK is a color model based on the subtractive color theory and is used by professional printers to reproduce color using offset lithography.

Column: A vertical division of a text box.

Constrain: The Constrain/Unconstrain command (**Item** menu) can be applied to a group whose rearmost box completely contains the other items in the group. Choosing Constrain prevents items contained entirely within the constraining box from being moved or enlarged beyond the sides of the box. These items are called constrained items.

Dots per Inch: A method of describing the resolution of printers and monitors (dpi).

EPS: Acronym for Encapsulated PostScript, a graphics format used to store high resolution pictures. EPS pictures created on a PC are often called EPSP pictures to distinguish them from EPSF files that were created on the Macintosh.

Export: To save text for use with other applications. Export and save text contained in text boxes via the **Save Text** command (**File** menu). You can save text in a number of formats that can be read by other programs.

Facing Pages: A facing-page document is one that has left and right pages that are bound in the gutter between the pages, such as a book, magazine, or newspaper. You create facing-page documents by checking Facing Pages in the New dialog box (**File** menu).

First Line Indent: The distance between the beginning of the first line of a paragraph and the Left Indent, specified in the Paragraph Formats dialog box (**Style** menu). A positive First Line indent value causes the first line of the paragraph to begin to the right of the left indent; a negative value combined with a left indent produces a hanging indent.

Font: A set of letters, numbers, punctuation marks, and symbols that share a unified design. The design is called a *typeface*. A group of related typefaces is called a *type family*.

Frame: A decorative border placed around a text box or a picture box. Frames are applied via the Frame command (**Item** menu).

Grayscale: Shades of gray that range from black to white. There are two types of gray: grayscale and true grays. Grayscales are shades of black. When you print separations, grayscale objects will print on the black plate. True grays, on the other hand, are colors composed of equal amounts of each CMYK color component. When you print separations, each CMYK plate will have a component of the gray object.

Group: A collection of items that have been combined via the Group command (**Item** menu) so that they act as a single item.

Gutter: The vertical space between adjacent columns or the area between the inside margins of facing-page documents.

H&J: The H&Js command (**Edit** menu) enables you to create and edit hyphenation and justification specifications that are used to control the way in which words are hyphenated in both justified and nonjustified paragraphs and the way in which space is added or subtracted between characters and words in justified paragraphs.

Hairline: A line that is 0.25 point wide. If you choose Hairline from the **Width** submenu (**Style** menu) to define the width of a line, the line will be printed 0.25 point wide on an imagesetter and 0.5 point wide on a laser printer.

Handles: Small, black squares displayed on active text boxes, picture boxes, and lines that are used to resize these items. Active polygon picture boxes have handles on their bounding boxes or vertices, depending on the setting of **Reshape Polygon** (**Item** menu). Active runaround polygons have handles on their vertices that you can drag to reshape them.

Hanging Indent: Created by positioning the first line of a paragraph to the left of all other lines in the paragraph; accomplished by specifying a Left Indent and a negative First Line Indent value in the Paragraph Formats dialog box (**Style** menu).

Horizontal/Vertical Scale: A command in the **Style** menu that enables you to compress or expand characters horizontally or vertically.

Import: To bring a picture or text file into a document. Import text into an active text box and a picture into an active picture box via the **Get Text** and **Get Picture** commands, respectively (**File** menu).

Import/Export Filter: A QuarkXPress translation file that enables the program to share text with other programs. QuarkXPress can exchange text in the following formats: Ami Pro, Microsoft Word, Microsoft Write, WordPerfect, XyWrite III Plus, RTF, XPress Tags, and ASCII.

Indent: The distance from the left (or right) sides of a text box (measured from the text inset) or column to the beginning (or end) of lines of text. Indents are specified with respect to the edges of columns, not the edges of a page.

Invisible Characters: Characters that can be displayed on-screen but do not print. The Tab, Return, and Space characters are examples. You can display invisible characters by choosing **Show Invisibles** from the **View** menu.

Item: There are three kinds of items in QuarkXPress: lines, text boxes, and picture boxes. Items can be combined into groups; a group is treated as a single item.

Item Creation Pointer: When you select an item creation tool to create a line or a box, the Arrow pointer changes to the Item Creation pointer when it is over the page or pasteboard.

Justification: (1) To horizontally justify a line of text by expanding or condensing the space between characters and words on lines of text to fill the width of a column so that the text has uniform (flush) left and right edges. (2) To vertically justify lines of text within a text box, adding space between paragraphs and/or lines so that the lines of text are spaced to fill the column from top to bottom.

Kerning: The adjustment of space between character pairs. QuarkXPress provides for automatic kerning, based on a font's built-in kerning table, and manual kerning, which enables you to adjust the space between selected character pairs.

Kerning Table: Kerning information is built into most scalable fonts. The values contained within a kerning table are applied to text during automatic kerning.

Knockout: In a color publication, a knockout is an area of a background color that is not printed on the background color separation plate. A knockout area is created by placing an object in front of a background color. The object is printed on the separation plate for the object color. Create knockouts and specify trapping values for background and object colors so that knockouts print properly. You also have the option to overprint a color on top of a background color.

Leading: In desktop publishing, leading has come to mean the amount of space occupied by a line of text measured from one baseline to the next baseline. A leading value includes the font size plus the space between lines. Leading has taken on the same meaning as the traditional typesetting term line space.

Library: A collection of QuarkXPress items. You create new libraries and open existing libraries via the Library command (**File** menu). An open library is displayed as a movable palette in front of all open documents. You can move items from document pages into an open library, from an open library onto document pages, and between open libraries.

Line: Draw horizontal and vertical lines with the **Orthogonal Line** tool; draw angled lines with the **Line** tool.

Link: Linking is a means of joining text boxes so that text automatically flows from one box to another. Links are formed using the **Linking** tool.

Master Items: Items on document pages that are automatically placed by the associated master page. Master items can be moved and modified like other page items.

Master Page: A nonprinting page used to automatically format (or reformat) document pages. A master page can contain such elements as headers, footers, page numbers, and other items that are repeated on multiple pages. A document can have up to 127 master pages.

Measurement System: Choose among seven measurement systems for displaying rulers and dialog box values: Inches, Decimal, Picas, Points, Millimeters, Centimeters, and Ciceros.

Metafile: A general term for graphics formats that describe a picture in mathematical terms as lines and arcs, rather than as dots (as bitmap graphics do). Windows Metafiles (WMF) and Computer Graphics Metafiles (CGM) are common metafile formats that QuarkXPress can import.

Multiple-Selected Items: When the **Item** tool is selected, you can activate more than one item at a time by holding down the **Shift** key while clicking on the items or by dragging out an enclosing box.

No Style: When applied to a paragraph, No Style detaches any associated style sheet without altering any of the text's character or paragraph attributes.

Normal Style Sheet: Automatically applied to paragraphs in newly created text boxes; its attributes determine the default text style.

Object Linking and Embedding (OLE): A method developed by Microsoft that enables Windows applications to share and manipulate data. For example, you can double-click on a picture box to launch the application that created the picture. Changes you make to the picture in the original application will also be made to the picture in QuarkXPress.

Overflow: An overflow occurs when a single unlinked text box or the last box in a text chain is not large enough to display all the text it contains. When this occurs, the overflow symbol is displayed in the lower right corner of the box.

Palette: A movable window that is always displayed in front of open documents. QuarkXPress has seven types of palettes: the Tools palette, the Measurements palette, the Document Layout palette, the Colors palette, the Trap Information palette, the Style Sheets palette, and Library palettes.

PANTONE Colors: Premixed ink colors that are often specified by graphic designers for spot color in multicolor print jobs. The QuarkXPress PANTONE Color Selector enables you to pick PANTONE Colors from the PANTONE matching system Catalog. A PANTONE Color can be specified as either a *spot color* or a *process color*.

Paragraph Format: A specification applied to a paragraph. QuarkXPress paragraph formats are Left Indent, First Line Indent, Right Indent, Leading, Space Before, Space After, Lock to Baseline Grid, Drop Caps, Keep with Next ¶, Keep Lines Together, Alignment, H&J, Rules, and Tabs. Paragraph formats are applied to a selected paragraph or a range of selected paragraphs via the commands in the lower portion (below the dividing line) of the **Style** menu.

Pasteboard: The nonprinting area that surrounds each page or multipage spread of a document.

Pica: A basic unit of typographic measurement. There are 6 picas in an inch and 12 points in a pica.

Picture Box: Pictures are imported into picture boxes. Picture boxes are created with any of the four Picture Box tools.

Point: A basic unit of typographic measurement. A point is approximately equal to 1/72 of an inch.

PostScript: A page description language developed by Adobe Systems, Inc. PostScript is used to describe fonts and graphics as well as the layout of pages.

Preset Defaults: The preprogrammed settings and values for specifications in QuarkXPress. They remain in use until they are changed by the user.

Printer Font: A bitmapped or scalable font that is resident in the printer or downloaded to the printer during printing. Computer fonts can have up to three components: a screen font for on-screen display, a printer font for printing, and a printer font metrics file that contains kerning and tracking information. Resident printer fonts are often called built-in fonts. Downloadable printer fonts are also called soft fonts.

Process Color: Any color (except Cyan, Magenta, Yellow, Black, White, and certain PANTONE colors) can be specified as a spot color or a process color. The way in which you specify color is significant for the printing of color separation plates. When separations are printed, all process colors on a page are broken down into their cyan, magenta, yellow, and black components, each of which is printed on its own separation plate. When combined during offset printing, the process colors can reproduce full-color page art.

Raster Graphic: A general term for graphics formats that store picture information as a series of dots rather than as mathematical descriptions of objects (as vector graphics do). Bitmaps and TIFFs are two common raster graphics.

Registration Mark: Symbols placed on camera-ready art and used to align overlaying plates. Print registration marks when **Registration Marks** is checked in the Print dialog box (**File** menu).

Resident Font: A font that is stored in a printer's memory. Also called built-in font.

Resolution: Refers to the degree of clarity of a monitor or a printer. Resolution is usually measured in dots per inch (dpi). The higher the resolution, the finer the detail of the screen or printed page.

RTF (Rich Text Format): A file exchange format that preserves information about the font, font size, and type style as well as style sheet information for those applications that support style sheets.

Rule: A line placed above or below a paragraph via the **Rules** command (**Style** menu). A rule is anchored to its associated paragraph and moves with the paragraph when text is reflowed.

Ruler Guides: Nonprinting lines used to align boxes and other items on a page. To obtain ruler guides, click on a ruler and drag a guide onto the page.

Ruler Origin: The zero point of the horizontal and vertical rulers; the intersection of the horizontal and vertical rulers displayed in a document window.

Ruler Origin Box: The ruler origin box is located in the upper left corner of a document window at the intersection of the horizontal and vertical rulers and enables you to reset or move the ruler origin.

Rulers: QuarkXPress displays a horizontal ruler along the top of the document window and a vertical ruler along the left edge of the document window when **Show Rulers** is selected from the **View** menu.

Runaround: The **Runaround** command (**Item** menu) enables you to control the way in which text flows with respect to items and pictures placed in front of the text. Also called *text wrap*.

Scalable Font: A font that is described mathematically. Scalable fonts print without jagged edges at all point sizes. Adobe Type 1, TrueType, and Bitstream Fontware Typefaces are common scalable font technologies.

Screen Font: A bitmap representation of a font that is used to display the characters onscreen.

Source Document: The document from which you copy items or pages when you are copying items or pages between documents. The document to which you copy items or pages is the target document.

Stacking Order: Refers to the position of a text box, picture box, or line relative to other boxes or lines in front of or behind it.

Style Sheet: A group of character attributes and paragraph formats that can be applied in one step to a paragraph or range of paragraphs via the Style Sheets command (**Style** menu).

Template: A preformatted document that is protected from overwriting and can be used repeatedly to create new documents.

Text Box: Text is entered into and imported into text boxes. Text boxes are created with the **Text Box** tool.

Text Chain: A set of linked text boxes through which text can flow.

Text Insertion Point: The place in a text box where newly entered or imported text is placed, indicated by the blinking Text Insertion "I" bar. You can establish the text insertion point when the **Content Tool** is selected by moving the I-beam pointer within an active text box and clicking the mouse.

Text Inset: The distance between the edge of a text box and the text within, as specified in the Text Inset field of the Text Box Specifications dialog box (**Item** menu).

Text Outset: This field in the Runaround Specifications dialog box (**Item** menu) enables you to specify the amount of space between a picture and the text that runs around it. Text Outset is available only when **Auto Image** or **Manual Image** is selected from the **Mode** drop-down list.

Thumbnail: A reduced view of a page. The **Thumbnails** command (**View** menu) displays a document as thumbnails and enables you to move pages within and between documents. When printing, you can check the Thumbnails check box in the Print dialog box (**File** menu) to print 1/8-size document pages.

TIFF (.TIF): Acronym for Tag Image File, a format used to store scanned images. TIFF pictures can be line art (black and white), grayscale, or color.

Tiling: The process by means of which a document larger than the paper it is to be printed on is broken into sections the size of the paper and then assembled manually. You can tile a document using the Tiling options in the Print dialog box (**File** menu).

Tracking: The adjustment of white space between selected characters and words. By specifying positive or negative tracking values in the Track Amount dialog box (**Style** menu) or the **Track** field in the Measurements palette, overall character spacing can be adjusted.

TRUMATCH Colors: TRUMATCH is a process color matching system for specifying predictable process colors. The TRUMATCH color system provides predictable four-color (CMYK) matching of more than 2,000 process colors.

Type Family: A group of related typefaces. For example, the Futura type family includes Futura, Futura Book, Futura Condensed, and Futura Extra Bold.

Typeface: A set of fonts that shares a unified design. For example, the Futura typeface includes Futura Bold, Futura Italic, and Futura Bold Italic.

Vertical Alignment: Controls the vertical placement of lines of text in a text box. The Text Box Specifications dialog box (**Item** menu) enables you to choose among four vertical alignments: Top, Centered, Bottom, and Justified.

WMF (Windows Metafile): A graphics format that QuarkXPress can import.

XPress Preferences: XPress Preferences (program defaults) are stored in the XPRESS.PRF file (located in the QuarkXPress program directory). This file contains a variety of information, including settings for Application, General, Typographic, and Tool Preferences; default style sheets, colors, and H&J specifications; and hyphenation exceptions. If the XPRESS.PRF file is not in the QuarkXPress program directory or in the Windows directory when you launch QuarkXPress, the program will create an XPRESS.PRF file.

XPress Tags: An option for saving text files (**File** menu) with complete QuarkXPress character-formatting information. Text saved in this format can be used only by QuarkXPress.

REFERENCES

QuarkXPress 3.0 User's Guide. Quark Corporation, Denver, Colorado.
QuarkXPress 3.0 Reference Manual. Quark Corporation, Denver, Colorado.
QuarkXPress 3.1 for Windows, On-Line Help. Quark Corporation, Denver, Colorado.

25

Typesetting with $\TeX$ and $\LaTeX$

Alan Hoenig
John Jay College, CUNY

25.1 BASIC CONCEPTS

Typesetting is the ability to place elements of a document on a page according to principles that make the pages as easy to read and understand as possible. It is surprisingly difficult to do that—a typesetter has to decide how best to break paragraphs into lines, how to hyphenate words, how much space to leave for footnotes, how to prepare indexes and the other detritus of scholarly publishing, and how to provide the optimum space between elements on the page (among many other things). The spacing issue is particularly critical for technical documents. Formulas make extensive use of arcane symbols that have different appearances and spacing depending on context. Consider, for example, how the position and size of the ordinary numeral "2" changes in

$$2x \qquad x^2 \qquad e^{-x^2}$$

and how the spacing surrounding minus sign changes in

$$-x \qquad \text{and} \qquad x - y.$$

Other symbols may change depending upon whether the expression appears in text $\int_0^{\frac{\pi}{2}} x dx$ or display mode:

$$\int_0^{\frac{\pi}{2}} x\,dx.$$

0-8493-2530-7/96/$0.00+$.50

Furthermore, if a computer system is going to control the typesetting, we expect more of it than from a mere human. We may expect, for example, to be able to label an equation in some logical way and then refer to it later by this label in our source document. It would be up to the typesetter to resolve these labels and references and replace the labels by properly formatted label numbers.

The TEX system has been freely available since the mid-1980s or so and accomplishes all of the above tasks (and more) in a particularly effective manner. TEX is the creation of Donald E. Knuth of Stanford University, who has placed all the source code for TEX in the public domain. The logo 'TEX' is related to the Greek root "τεχ" from whence come words like "technology." If pronounced properly, the face of your listener may become slightly moist (but no one complains if you say "tek").

The purpose of this chapter is to acquaint readers with the aspects of TEX and of the TEX cycle necessary to produce handsome papers and books. This chapter should *not* be regarded as a tutorial, since (like many other mature and sophisticated software systems) even lengthy books are not enough to do full justice to it. Nevertheless, simple documents should be well within the reach of any reader of this chapter. The reader should also know how to ask the questions necessary to gain further skills in TEX. Experienced readers may realize that some of the discussion, in the interest of conciseness, indulges in simplifications that are really little lies; these should be overlooked.

25.2 THE TEX PRODUCTION CYCLE

Why is "typesetting" not the same thing as "word processing"? Typically, a word processor allows editing of the document, but in its impatience to display the results immediately onscreen (most word processors are aggressively WYSIWYG in behavior), certain niceties are sacrificed. These niceties—fine control of spacing, word placement, hyphenation, and so forth—are never ignored in TEX. God, after all, is in the details.

It's useful to consider the TEX production cycle by comparing it with that of word processors. In a word processor, the program assists you in preparing the document, after which it is printed (two steps). TEX relies on three steps.

1. We use a *text editor* to prepare the *source document*—the document file that consists of the text and data of your document together with the TEX formatting commands. Let us suppose this file is called **`myfile.tex`**. (A programmer will recognize that the document source is analogous to the source code file for a program.)
2. We run **`myfile.tex`** through the TEX program. If all goes well, this generates a file in which the typesetting commands are made explicit using a generic printer description language independent of any particular printer; it is *device independent*. TEX names this file **`myfile.dvi`**. If there are any errors, we return to step 1 and correct them before continuing. (Again, programmers will recognize that this step compares with the compilation of a program source file. Experienced users of TEX often do refer to compiling a source document file.)
3. Finally, we need the assistance of a special **`device driver`** customized to the printer. It's the driver's task to translate the generic **`dvi`** commands into the form the printer understands.

The advantage of creating a **`dvi`** file is that we can print the document on any printer (at least, any printer for which device drivers exist) and may rest assured that the output is identical on each device (except for raster resolution).

25.3 MACROS; LOGICAL DOCUMENT DESIGN

TEX has been called an assembly language for typesetting. This means that there are plenty of primitive commands to control fine points, but these commands may not be entirely appropriate for creating a new section head or aligned equation. As a result, TEX has a rich and powerful macro creation facility. It is possible to string primitive commands and preexisting macro commands together to create new, custom typesetting commands.

Remember that the TEX production cycle means that we prepare a source file that is fed into TEX at a later point. This plus the nature of the macros means that TEX supports the notion of *logical document design*. We can embed components of the document by means of tags that can be defined or redefined depending upon context. One example suffices. Here is a theorem.

> **Theorem** There is no royal road to typesetting. Computer typesetting is a surprisingly complex task.

This was typeset by means of inserting **`\theorem There`** ... in the source document. It may happen that it is more appropriate to display that theorem as

> THEOREM *There is no royal road to typesetting. Computer typesetting is a surprisingly complex task.*

The same command string will accomplish this *provided* that only the macro definition of **`\theorem`** be changed. The implications are enormous—we can design our document so that it will properly printed for any set of particular formatting requirements provided only that we change the particular definitions of the macros. Many strategies exist for facilitating this change of definitions.

25.4 A FIRST TEX DOCUMENT

The "steps" for generating a TEX document are well-defined, but there are sufficiently idiosyncratic implementations of TEX floating around so that it may be necessary to adapt these procedures to a local adaptation. By the way, some readers may be interested in the TEX dialect called "LATEX"; as we will see, LATEX is still TEX, so these procedures follow for a LATEX document as well.

1. Use a *text editor* to create the *source document* for subsequent processing by TEX. The source document is the document file—text and typesetting commands. Take care not to use a word processor. These programs aim to do the formatting themselves, and tend to do so by inserting non-ASCII characters into the document file. Quite apart from the fact that TEX (or LATEX) needs no help with the typesetting, these binary characters will only confuse TEX. (If it is necessary to use a word processor, make sure to save the document in some way so as not to include the word processing formatting information.)

2. Run this source file through the TEX program. the simplest form of the command to do that is

   ```
   tex myfile
   ```

   where the source file has the name **`myfile.tex.`** LATEX users will use the command

```
tex &lplain myfile
```

(Unix users may have to enter the ampersand as **\&** or surround the arguments with backticks.)

As in any compilation process, TEX may uncover errors. (Warnings may be ignored, at least at this stage.) Return to step 1 to correct these errors, and rerun it through TEX. Repeat this process until all errors have been dealt with (or until there is enough of a document to print.)

The result of a successful TEX compilation is a new file with a `dvi` extension. In this example, we would have a new file **`myfile.dvi`**.

3. With the document in hand, it can be printed or previewed on screen. In each case, appropriate *device drivers* are necessary to properly render the document on screen or on paper.

As we see, the TEX process is actually a concerted action between several programs in addition to TEX—a text editor, a device driver, and a screen previewer. Many implementations of TEX may merge several or all of these into one integrated module.

25.5 THE TEX DOCUMENT: INPUT CONVENTIONS

Although it is not practical or possible to deal with all or even a completely useful subset of al TEX (or LATEX) commands in this article, it is possible to summarize the keyboard conventions that any TEX typesetting must adhere to.

Page Layout and TEX Defaults

One thing the beginning user need *not* worry about is the specifying of the page layout parameters for the document. Quantities such as the paragraph indentation, the offset of the printed page body from the top left corner, the vertical size of the text (which sets the top and bottom margins), the horizontal width of the text (which determines therefore the left and right margins), the amount of space between paragraphs, the amount of space preceding and following mathematical displays, and many, many more are given default values for us. Beginners need not worry about varying them, although the means exist for adjusting any and all of them.

White Space

TEX interprets white space in a document in a manner both similar yet dramatically different from the way a word processor does. Read the following very carefully.

Both TEX and any word processor agree that if we enter a space between two words, we will get a single space in the document. But TEX goes further.

TEX normally regards all white space as equivalent, where we include carriage returns, tabs, and of course spaces in this category. That means that **a␣b,** **a␣␣b,** and

```
a
b
```

generate the same output—"a b." The funny symbol '␣' represents the "space" character.

Furthermore, multiple spaces are generally equivalent to a single space. That means that **a␣b, a␣␣b, a␣␣␣b,** and even

```
a␣␣␣␣␣␣␣␣␣␣␣␣␣␣␣␣␣␣␣␣␣␣b
```

produce the same output. People coming to TEX from word processing find this freedom disorienting, but it provides great freedom to organize the source document in a logical way to better express the purpose of the proposed typesetting. Of course, multiple tabs are treated in the same way as multiple spaces, but see the next paragraph for an important exception with regard to carriage returns.

Important exception: We signal the end of one paragraph and the beginning of another by skipping a line in the source file; that is, we enter two hard carriage returns in a row. (But three or more consecutive carriage returns is still equivalent to a pair of carriage returns.) The following display

> This is a short paragraph.
>
> Here is a second paragraph, also quite short, but
> a bit longer than the first.

comes from typing

```
This is a short paragraph.

Here is a second . . .
```

Once in a while, spaces are special in that we do not want a line broken between two words or word groups. For example, in a discussion of Word War I, it would look silly if a line broke between "World War" and "I"; it would be too confusing to the reader. To guarantee that the line break will not happen at that point, we replace the space with the tilde character ~. If this *conditional space* is typeset in the middle of a line, it appears as a regular space. Aspiring TEX typists should develop the habit of typing things like **`King Henry~VIII, Dr.~Knuth`**, and **`pages~44-55`** to protect the manuscript from unwarranted and unwanted line breaks.

New users will be glad to know that TEX inserts a dab of extra space following punctuation. Look carefully at this sample:

I am what I am. What am I?

which we got by typing

```
I am what I am. What am I?
```

(but if you typed **`. . . I␣am.␣␣What . . .,`** TEX will not complain.) TEX assumes that lowercase letters followed by punctuation mark the end of a sentence, and it adds a bit of space accordingly. However, if an abbreviation (Dr., Mrs., etc.) occurs in a sentence, or if the final word of a sentence ends in a capital letter, TEX needs some assistance. Two special commands help out in these instances; we defer that discussion for a few paragraphs until we begin a discussion of TEX commands.

Characters

We generate most characters by simply entering the character in the source document. That is, we type

```
Oh! What a beautiful morning.
```

to get

Oh! What a beautiful morning.

On occasion, certain character pairs (or even longer sequences of consecutive characters) are replaced by special glyphs. For example, if we type "or" we get true quotes "and". With its special attention to details, TEX will replace certain character combinations such as **fi, fl,** and **ff** by the *ligatures* fi, fl, and ff (provided these ligatures are present in the font). Spanish punctuation is handled also by the ligature mechanism. To get ¿ or ¡, we type **'?'** or **'!'**

TEX does a similar thing with hyphens and dashes. We can type -, --, or --- to put-,–, or—in our documents. (And we will see later that the mathematical minus sign—yet a different dash—can be gotten using the hyphen character in mathematics mode.)

By the way, no user should *ever* put an explicit line-break hyphen in a word that has to be split at the end of a line. TEX's hyphenation algorithm takes care of such should a word break be necessary.

In summary, we type

```
"!'Oh, the selfish shell-fish---that lobster
mobster—tasted best when basted west!"
quoth Aaron while reading
pages~12--33 of his cookbook.
```

to typeset

"¡'Oh, the selfish shell-fish—that lobster mobster—tasted best when basted west!" quoth Aaron while reading pages 12–33 of his cookbook.

Units of Measurement

Since certain commands expect a unit of measurement, we present a few facts about measurement within the context of typesetting.

Very small movements on the page are visible to readers. Consequently, printers have become accustomed to dealing with small units of measurement. The basic such unit is the printer's point; there are a shade over 72 points to the inch (72.27, to be precise). A pica is another printer's measure; one pica equals 12 points. TEX also understands measurements expressed in inches and centimeters, as well as a few other arcane measures (see *The TEXbook* for a more complete discussion; a brief annotated bibliography concludes this chapter).

TEX expects units of measurement to be written using the abbreviations summarized in Table 25.1.

TABLE 25.1 Units and Their Abbreviations

Point	pt
Pica	pc
Inch	in
Millimeter	mm
Centimeter	cm

TEX Formatting Instructions and Commands

TEX is very good about applying default typesetting parameters to text, but there will be many times when you wish to actively control the printed appearance of your document by issuing commands to TEX. Since the entire file must contain ASCII characters only, TEX has decided to reserve the meanings of certain characters to itself. The tilde ~ is one such special character. To typeset an actual tilde ~ in the document, you must enter a short command to do so.

These characters

```
\ # $ % ^ & ~ { }
```

have special meanings to TEX. The backslash \ is TEX's *escape character*—it escapes the normal meaning of the following bit of text. This character generally begins all of TEX's commands. For example, to typeset an ampersand &, you would type **\&.** (Typesetting commands for some of these symbols are formed in the same way; see *The TEXbook* for details. All the symbols can be typeset, but for some, additional TEX expertise is needed.)

TEX can do àçëéñts as well. If you need them, check your local manual or *The TEXbook*.

We will discuss the special TEX characters bit by bit, but commands generally begin with the escape character. The escape character can be followed by one single non-letter, or by an arbitrary sequence of letters, terminated by a space or other non-letter. Examples of commands from the first category include **\&, \1, \$,** and \". Examples of the second category include **\TeX, \L, \noindent, \vskip,** and **\futurelet.** Note that TEX is case sensitive, so the command **\TeX** (which typesets the TEX logo) is different from the (nonsense) commands **\tex** and **\TEX.**

These rules lead to our first piece of TEXarcana. As part of TEX's digestive process, it is smart enough to know that when a nonletter follows an escape character (the backslash), the command name consists only of that single nonletter. Hence, anything following that command, such as a space, is typeset as you expect. To get the sequence, '& &', type **\& \&.**

The situation is subtly different when a command name follows the backslash. For now, TEX has no way a priori to know the length of the command name. It reads your file, and normally terminates the command name when it encounters a space, any nonletter, or a new command. Consequently the space following a command is "eaten up" by TEX—it serves not to introduce a space into the document but rather to delimit the command. But since at other times, spaces typed in the document file *do* generate a space, it is easy to see why newcomers are easily confused.

Anyway, to illustrate the point, suppose you wanted to typeset "TEX TEX." The way *not* to do it would be by entering **\TeX \TeX** into the source, for the interior space terminates the initial **\TeX** command and is therefore eaten alive. (**\TeX \TeX** typesets as TEXTEX.) Since multiple spaces count as a single space to TEX, the solution is *not* to insert additional spaces. The following list, which suggests several ways out of the impasse, also hones your beginning TEX skills.

1. Use the TEX command to explicitly generate an interword gobbet of space. This *control space* command is **\␣,** and so your source should like **\TeX\␣\TeX** if you use this method.

2. Terminate the command by inserting some command that does not print anything. The empty group **{}** is one such; thus, we could type **\TeX{} \TeX**.
3. Here is another way to use the grouping symbols. We may also type **{\TeX}, \TeX,** or **\TeX {\TeX}.**

See below for a brief discussion of groups and braces.

25.6 A FIRST TEX DOCUMENT

Creating the Document File

The purpose of this brief section is to hold the reader's hand in the course of producing a TEX document. The forgoing section should be sufficient to make clear what is the purpose of all the lines of this brief "story."

```
\nopagenumbers \parskip=1pc \parindent=15pt
Hello, world---this is \TeX!
And so the first of our tests ends.
\bye
```

Begin by typing this story *exactly* as it appears, making sure to use a text editor to ensure that the resulting file, which should be **myfile.tex** or something like that, is a *pure* ASCII file. (It is possible to use a word processor, but great care should be taken to make sure that the resulting file contains purely printable ASCII characters only.)

This file uses the conventions of so-called "plain" TEX. That is, we use the native TEX macros provided by the author of TEX, and not the macros and commands provided by LATEX and other macro packages.

Running the Document through TEX

To typeset this document, issue a command like

```
tex myfile
```

Someone will have had to make sure that the TEX executable file is visible to you (and, of course, that all of TEX has been properly installed). If all goes well, you should see TEX's banner, which looks something like this:

```
This is TeX ...
```

where the ellipsis ('...') represents other stuff. After a short while, you should see some more output to the screen culminating in something like the following:

```
(myfile.tex [1] )
Output written on myfile.dvi (1 page, 292 bytes).
```

If you get this message, great. You are ready to proceed to the next step. If you do not get this message, you have made an error in creating the file **myfile.tex.** It is a good idea to read the next subsection for some elementary procedures in dealing with TEX errors.

Errors

When TEX encounters what it thinks is an error, it stops. The console display shows line number and the context of what it thinks is the error. Here is a typical error display.

```
! Undefined control sequence.
1.2 Hello, world--this is \Tex
                              !

?
```

The line beginning with the bang ('!') let's you know what has confused TEX, and the next line shows the place where TEX thinks it has found the mistake. The line is split in two parts. The part up to and including the error **('1.2 Hello, world---this is \Tex')** is on one line, while the remaining portion of the line (here just '!') is printed below. The question mark is a prompt for action on your part.

In this discussion, we will mention only two of the several alternatives open.

1. You can simply hit the carriage return. TEX will do its best to carry on. Of course, serious errors lead to further errors downstream, and you should always be prepared see evermore mysterious errors occurring if you choose this alternative.
2. You can bail out at any time by typing **'x'** at the **'?'** prompt. TEX closes shop, and ships out any completed pages for the next step of the process.

Looking carefully at the display above makes clear what the problem is. We carelessly typed the command **'\Tex'** instead of **'\TeX'**—remember that TEX is case sensitive. Hitting **'x'** allows us to reedit the document file, make the correction, and re-TEX the document.

The subtlest and most insidious errors are those that look legal to TEX but nevertheless are not what you meant. We can say little else about these errors in this chapter.

Previewing and Printing

The successful result of running the document through TEX is a file **myfile.dvi**. We make use of two additional programs to actually paint the typeset images either onscreen or on paper. We might view the document via a command like

```
view myfile
```

and a similar command—something like

```
prt myfile
```

—will generate a printed sheet. (Check with your colleagues for the exact form of these commands.)

It is increasingly common to want to print TEX output on a PostScript™ printer. PostScript is another popular method for rendering print in a device independent manner. For those readers with access to a PostScript printer, it will be necessary to use a program that converts the **dvi** file to a **ps** file. This is easily done via a command like

```
dvips myfile
```

which produces **myfile.ps.** (The file **myfile.ps** correctly contains references either to PostScript outline fonts or the TEX bitmaps themselves.)

Further Experiments

Rerun this project, but make some slight changes to **myfile.tex** first. Some suggestions follow.

- Leave out the command **\nopagenumbers** and see what happens.
- Change some of the measurements, and see what difference (if any) results. Replace **'pc'** by **'pica'** and explain the results.
- Omit the final **\bye** and explain what happens. Try to figure out how to conclude the run successfully.

25.7 A BASIC TEX VOCABULARY

We introduce a basic core of commands that should suffice for typesetting simple letters and reports using the conventions of plain TEX. So far, we know several commands, namely **\TeX, \bye, \nopagenumbers,** and **\␣.**

Font Changing Commands; Introduction to Groups

We mentioned above that **{** and **}** are special to TEX. They serve to begin and groups for TEX. A *group* is a region in a document such that typesetting parameters set inside it are restored to the previous value when the group ends. Without doing anything special on your part, a TEX document is set in a default Roman typeface, similar to the text face you are currently reading. To typeset a word or phrase in **boldface**, surround the text with grouping symbols, and within the group, but prior to the first character in this group insert the boldface command **\bf.** Within the group, the prevailing typeface will be boldface, but the usual Roman face will be restored when the group is concluded. Thus, to get

a **bold** idea

type

```
a {\bf bold} idea
```

Other common font changing commands are **\it, \tt**, and **\sc**, which give *italic*, **typewriter-like**, and SMALL CAP type. Table 25.2 summarizes sample uses of these commands.

TABLE 25.2 Samples of Font Changes.

Type	To get
`a {\it remarkable\/} child`	a *remarkable* child
`{\bf Note} the following.`	**Note** the following.
`\TeX{} file {\tt myfile.tex} is . . .`	TeX file `myfile.tex` is . . .
`Knuth's {\sc Theorem}`	Knuth's THEOREM
`{\it I {\rm cannot} agree.`	*I* cannot *agree.*

The command **`\rm`** explicitly gives Roman type. The last example shows that groups within groups are also possible.

The use of italic type displays one subtlety. It often happens that the final character in an italic phrase leans so far to the right that it "invades the space" of its right Roman neighbor, even when they are separated by a space: a *tall* house. For that reason, it is a good idea to insert an *italic correction* to conclude every italic phrase (unless the last italic character is punctuation). The italic correction command is \/. Examine the following entries closely.

Naive way:	`a {\it bold} Texan`	a *bold* Texan
Preferred way:	`a {\it bold\/} Texan`	a *bold* Texan

We hope you agree that the "preferred way" is truly the preferred way.

Page Layout Commands

At the beginning of the document, insert commands that change global aspects of TeX's behavior. For example, Use **`\nopagenumbers`** to suppress pagination and use **`\parindent`** (followed by a legal TeX dimension) to control the amount by which each paragraph is indented. Here are some examples of the use of **`\parindent`**.

`\parindent12pt This . . .`	This is a short paragraph. It is amazing what good typesetting can do for morale.
`\parindent0in This . . .`	This is a short paragraph. It is amazing what good typesetting can do for morale.
`\parindent-0.25cm This . . .`	This is a short paragraph. It is amazing what good typesetting can do for morale.

The default page numbering (in the absence of **`\nopagenumbers`**) centers the page number at the bottom of the page.

A Few Other Commands

There are many ways to control the amount of white space in a document. Two explicit ways are via the commands **`\hskip`** and **`\vskip`**, each of which must be followed by legal dimension values. For example,

```
\hskip3in \vskip-2pc
```

are valid uses of these commands. The leading "h" and "v" refers to skipping done either horizontally or vertically. In the context of **\hskip**, positive and negative quantities refer to distances to the right and left. With regard to **\vskip**, the situation may be slightly counterintuitive: positive quantities go *down* the page, and negative quantities go up.

Several **. . . line** commands control placement of short passages of text on a line. We can center text, align it to the right, or align it on the left using commands **\centerline, \rightline**, or **\leftline**. If we type

```
\centerline{\bf The central idea}
```

we get

The central idea

(No, we do not normally get the box—it is here to emphasize the position of the text with respect to the entire line. In the same way, although we include boxes in the demonstration of the other **. . . line** commands, you will not see them in "real life".) Similarly, if we type

```
\leftline{\it Leaning to the left\ldots}
```

or

```
\rightline{I'm in the {\bf right}!}
```

we get

Leaning to the left . . .

or

I'm in the **right**!

We are allowed (and encouraged) to combine commands together in useful and imaginative ways, so if we wanted some text to left-align with a central vertical axis, we might type something like

```
\leftline{\hskip .5\hsize January 12, 1995}
```

to get

January 12, 1995

We use here a new $\TeX$ quantity— **\hsize**, which stores the width of the text on the page. We can adjust this value by placing a statement like

```
\hsize=41pc
```

at the beginning of our document.

We can go to a new page by using the command sequence

```
\vfill\eject
```

Normally, in extended passages of text, TEX does the page-braking for us.

Finally, the command **\null** is sometimes a handy way insert a typesetting command in a source file that produces no ink in the final document.

Challenge: A Simple Letter

Figure 25.1 shows a simple letter. The reader should try to figure out—and try it—how to achieve this before checking the answer in the Appendix.

Ms. Polly Darton
12345 Main Street
Alvarado, TN 54321

February 29, 1996

Dear Mr. Sofer,

I am very grateful for the effort you have taken in my behalf. True friends are true wealth—that's what this experience has taught me. Many thanks once again.

Sincerely yours,

Polly Darton

Figure 25.1 A simple letter in TEX

New Sections

The last of our layout commands controls the titling an beginning of a new section. The command is called

```
\beginsection
```

followed by the section title, which must *then be followed by a blank line!* Type

```
beginsection 1. An Address
Four-score and seven years ago, . . .
```

(note the blank line) to get

1. An Address
Four-score and seven years ago, . . .

`\beginsection` automatically suppresses indentation of the following paragraph.

Math

TEX's specialty, indeed its *raison d'être,* is its unsurpassed ability to typeset mathematical symbols and arcana. It is impossible to summarize a subset of commands that would be useful to more than a handful of readers. Instead, we will discuss just a few principles, and refer the interested reader to Chapters 16 through 19 in *The TEXbook.*

Single or double dollar signs delimit mathematical expressions, perhaps in memory of the great expense that used to be involved in the typesetting of mathematics. Math can either be typeset in text or in a special display mode. If we type

```
. . . we get $ x $.
```

then we get x. In contrast, we observe the displayed mode

$$x$$

if we type

```
. . . the displayed mode
$$ x $$
if we type . . .
```

Here, the only difference between the two modes appears to be the surrounding context, but in complicated expressions, TEX is careful to adjust the spacing and even the sizing of some symbols depending on the mode it is in.

Generally, most of the keys on the keyboard generate symbols you can type in a math expression. TEX is clever about inserting the proper spacing and such on its own.

`$x + y = -7$`	$x + y = -7$
`$ - 3(a - b) < 11u$`	$-3(a - b) < 11u$

In more involved cases, simple mathematical commands produce the proper symbols. The following examples give a taste.

`$2 + 2\times3\div4 = 2$`	$2 + 2 \times 3 \div 4 = 2$
`$\sqrt9=\pm3$`	$\sqrt{9} = \pm 3$
`$\int Cdx = Cx + \alpha$`	$\int Cdx = Cx + \alpha$
`$\Gamma(n + 1) = n\Gamma(n)$`	$\Gamma(n + 1) = n\Gamma(n)$
`$\ln(uv)=\ln u + \ln v$`	$\ln(uv) = \ln u + \ln v$

Of course, any of these examples can also be typed in display mode by replacing each single '$' by the pair '$$'. Note that typing things like **`$cos\theta$`** give *cos* θ. You need to type **`$\cos\theta$`** to get cos θ.

To understand some otherwise perplexing TEX phenomena, we have to understand something more about some commands, namely those that expect arguments. For example, the command **`\sqrt`** is one such. This is the command to typeset the square root symbol together with an overbar over some expression, and this expression is the *argument* of the **`\sqrt`** command. Normally, TEX assumes the next token to be the argument of the command, where (very loosely speaking) a *token* is either the next command or character in a file. In an expression like **`$\sqrt9$`**, the '9' is the token that **`\sqrt`** operates on. We get $\sqrt9$ from this. But if you type **`$\sqrt49$`** in your file, the single token following the command is "4", and you get $\sqrt4 9$, probably not what was wanted. Therefore, the author of TEX augmented the argument rule—if we surround an expression with grouping symbols (curly brackets), TEX regards the entire group as the argument.

In this instance, to get the more desirable $\sqrt{49}$, we would simply surround the the "49" with curly brackets, that is, we need to type

```
$\sqrt{49}$
```

to get $\sqrt{49}$.

The symbols ^ and _ control superscripting and subscripting. Type **`$e^x$`** and **`$H_2O$`** to get

$$e^x \quad \text{and} \quad H_2O.$$

The remarks above about bracketing material apply to these symbols whenever we need more than one character to appear above or below the line.

Input	Output
`$(x + 1)^2\equiv x^2 + 2x + 1$`	$2(x + 1)^2 \equiv x^2 + 2x + 1$
`$e^{-2x^2}$`	e^{-x2}
`${}_2T^3$`	$_2T^3$
`$R^{\alpha\beta\gamma}_{\delta\epsilon}$`	$R^{\alpha\beta\gamma}{}_{\delta\epsilon}$
`$\sqrt{a^2 + b^2}\neq a + b$`	$\sqrt{a^2 + b^2} \neq a + b$
`$\sum_{i = 1}^n i = n(n + 1)/2$`	$\sum_{i=1}^{n} i = n(n + 1)/2$
`\int_0^\infty e^{-r^2 x^2} = \sqrt\pi/(2r)$`	$\int_0^\infty e^{-r^2x^2} = \sqrt{\pi}/(2r)$
`$\sin \theta=\sqrt{1 - \cos^2\theta}$`	$\sin\theta = \sqrt{1 - \cos^2\theta}$

The grouping symbols are important also when constructing fractions (and related quantities) because it properly determines the scope of the **`\over`** command, which forms the core of a fraction construction. We are $\frac{1}{2}$-way there if we type **`$1\over2$`**, but if we type **`$x+(1\over2) y$`** to get $x + (\frac{1}{2})y$,, we get instead

$$\frac{x+(1}{2)y},$$

not what we need. We restrict the scope of **`\over`** with grouping symbols like so:

```
x + ({1\over2}) y
```

Study these examples.

`$a+{b\over c}+d$`	$a + \dfrac{b}{c} + d$
`$\tan A+\tan B={\sin(A+B)\over \cos A \cos B}$`	$\tan A + \tan B = \dfrac{\sin(A+B)}{\cos A \cos B}$
`${d(uv)\over dx}=u{dv\over dx}+v{du\over dx}$`	$\dfrac{d(uv)}{dx} = u\dfrac{dv}{dx} + v\dfrac{du}{dx}$
`${d\over dq}\int^q _p f(x)dx=f(q)$`	$\dfrac{d}{dq}\int_p^q f(x)dx = f(q)$

Delimiters

Delimiters often change sign to better enclose the expressions they surround. There are are severai ways of controlling the matching sizes of mathematical delimiters, which include **(), [], {}** (which you get by typing **\{and \}**), and several others.

`$|\kappa|=\sqrt{\kappa^2}$`	$\lvert\kappa\rvert = \sqrt{\kappa^2}$
`$\varphi(x+iy)$`	$\varphi(x+iy)$

If TEX does not make the delimiters grow in height the way you might like, the easiest way to fix that is with **\left** and **\right**. (But see *The TEXbook*, pp. 146 ff., for a few caveats.)

`$\tan{x\over2}=\sqrt{\left ({1 - \cos x\over1+\cos x}\right)}$`	$\tan\dfrac{x}{2} = \sqrt{\left(\dfrac{1-\cos x}{1+\cos x}\right)}$
`$\left[\phi=\sin^{-1}\left(a\over x\right),\ t k = {b\over a}\right]$`	$\left[\phi = \sin^{-1}\left(\dfrac{a}{x}\right),\ k = \dfrac{b}{a}\right]$

In the last example, the command **\␣** was necessary since TEX ignores keyboard spacing conventions when in math mode.

We have but scratched the surface in the foregoing discussion. (What, for example, of multilined equations, arrays, and such like?) The serious reader will need to delve into the relevant chapters of *The TEXbook*.

Second Challenge: A Brief Paper

The conscientious reader should pause now and attempt to create a plain TEX source file to reproduce the simple report shown in Figure 25.2. A document that is identical except for vertical spacing or font size is acceptable. One answer is given in the Appendix.

Macros and New Commands: An Introduction

As a software system, one of TEX's great strengths is its ability to accept new commands built out of sequences of pre-existing ones. These new commands are called *macros* and, once defined, macros can be components of other macros, and so on. Since TEX possesses decision-making commands, input/output commands, and the full panoply of commands associated with programming languages, it is fair to think of a macro as

Calculus in a Nut Shell
Center for Calculus Revisionism

Introduction
The curriculum of college calculus is too important to be left to instructors. It is crucial that the streamlining of the calculus course be done as soon as possible.

The Joy of Differential Calculus
Calculus is really as easy as π. We begin with a study of limits; how can we make sense of

$$\lim_{x \to x_0} \frac{f(x) - f(x_0)}{x - x_0}?$$

This limit, if it exists, is the derivative $f'(x)$.

Existential Integral Calculus
The definite integral of function $\int_a^b f(x)\,dx$ is related to the limit (if it exists) of

$$\sum_a^b \left[f_1(x) - f_2(x)\right]\Delta x.$$

It can further be shown .

Figure 25.2 A second challenge!

a small computer program. We will consider this subject only at the simplest level.

Each macro has a name and a string of commands associated with the macro. We match the name with the sequence of commands using the **`\def`** command preceding the new command name, followed by the sequence enclosed in curly brackets. For example, we can envision a personal **`\newchapter`** command, which finishes the old one, skips to a new page, and leaves a bit of space at the top for the title. One way to do this follows.

```
\def\newchapter{% begin definition
\vfill\eject % end old page
\null\vskip 1in % leave some space at top on new page
} % end macro definition
```

Following this **`\def`**`inition,` **`\newchapter`** is now a perfectly acceptable command to TEX. You may use it throughout your document, provided that its definition is included in the source document.

This example introduces one more TEX convention, which has so far escaped discussion. Any explanatory comments that we would like in our source file but would *not* like to typeset can be preceded by the percent symbol **`'%'`**, which renders it invisible to TEX. This symbol is good for one line only. Multiline comments need to be "percented" at each line. (Macros that will allow multiline comments do exist, and are surprisingly TEXnical and sophisticated.)

Macro Packages for TEX

It's possible to construct macros of such power and sophistication that they appear to redefine the actual syntax that TEX obeys. Several people have written collections of macros, *macro packages*, that aim to make TEX easier to use. Michael Spivak wrote the \AmSTEX package to make the mathematical power of TEX simpler. Leslie Lamporte wrote LaTeX to make general typesetting somewhat more straightforward, but also to make possible *logical document structure*. This means that commands identify not the appearance of the typesetting but the purpose of a section take precedence. The actual appearance of the typesetting is specified "behind the scenes" in style files, which assign the meaning of, say, a **\section** or **\title** command. An author specifies the particular style file once at the beginning of the document file, and can then concentrate on the structure of the article rather than its appearance. Then, when the article is rejected by the original journal, but accepted by a second, the only change that need be made is to the style file specification. The recompilation of the file is all that is needed to put the new style conventions into effect. LamSTeX is another macro package written by Spivak. Its aim is to combine the functionalities of both AmSTeX and LaTeX into one package.

LaTeX is by far the most widely used of the macro packages currently available, and a large collection of style files appropriate to many different journals and formats is freely available. Basics of LaTeX form the subject of the next section.

LaTeX was originally conceived and developed in the early and mid-1980s, and is beginning to show its age in the sense that many LaTeX users have become vocal at requesting features for a "new" LaTeX. As this chapter is written, efforts at creating an enhanced LaTeX (perhaps to be called LaTeX3) are currently underway. The discussion that follows adheres to that of the original, "old" LaTeX, but readers should be aware that a new, enhanced, and slightly different version of LaTeX—called LaTeX2$_\epsilon$—does now exist.

25.8 BASICS OF LaTeX

Although when you use LaTeX you are using TeX, so much of the syntax has been redefined and so many of the original commands of plain TeX have been redefined or disabled that it is important to get hold of the LaTeX manual or one of the several other primers to LaTeX that have appeared. It is also important to have read the foregoing material about plain TeX, since so many things (such as text input conventions, the concept of macro, and so on) apply to LaTeX as well.

To compile a LaTeX source file **mylatex.tex** issue a command like

```
tex &lplain mylatex
```

The **'&lplain'** indicates that TeX is to read a special format file containing all the LaTeX macros. A *format file* contains a large set of macros in a special binary format that TeX can ingest at high speed. (Somewhere on your system is a file **lplain.tex** that contains the ASCII version of these macros, should you be interested in close examination of them.) On many systems, the command **latex** will be an abbreviation for **tex &lplain.**

There are three lines that must appear in every LaTeX document:

```
\documentstyle{...}
...
\begin{document}
...
\end{document}
```

The ellipses (. . .) indicate other things that can appear. All text must appear between the **\begin{document}** and **\end{document}** commands. Additional declarations may appear between the **\documentstyle** and **\begin{document}** lines. Typical document styles might be "report," "article," or "book," or something quite specific like "ieee" or "apa," which refer to specific journals.

If you type, say, **\documentstyle{article}** your document will appear in some generic format appropriate for an article. The argument to **\documentstyle** gives the name of a *style file* that is responsible for declaring the meanings of commands that actually control the printed appearance of your document. In this case, LATEX expects to find the style file **article.sty** in the same place that all the other input files are located. Certain optional modifications may be possible. It may be possible to typeset the article in a larger type size or in a doublecolumn style. We invoke such changes by adding optional arguments to **\documentstyle.** We may say

```
\documentstyle[twocolumn, 12pt]{article}
```

but this will not work unless **article** is designed to accept these optional arguments. In all these cases, TEX will need to be able to find style files called **12pt.sty** and **twocolumn.sty** in some directory or folder containing all of TEX's other input files (in addition to **article.sty**). Material within the square brackets [ ] specifies optional arguments to LATEX.

You already know quite a bit of LATEX. The input conventions are the same as for TEX. The font changing commands and general syntax are also the same. Page layout parameters can be changed, but they often have different names in LATEX. Generally, though, the style files will be responsible for setting these parameters. Explicit white space is controlled by commands **\vspace{. . .}** and **\hspace{. . .}**, where explicit measurements take the place of the dots. These commands generate white space in a vertical or horizontal mode.

LATEX is organized around the concept of *environments,* which must explicitly **\begin** and **\end.** Other environments can be nested within outer environments. The document environment must occur in every document. We have time to consider only a few other environments in this chapter.

Text can be centered using the center environment. Explicit line breaks are indicated by the command \\. Thus

```
\begin{center}
Hey diddle diddle\\
the cat and the fiddle\\
the cow jumped over the moon.
\end{center}
```

produces

Hey diddle diddle
the cat and the fiddle
the cow jumped over the moon.

Notice the matching **\begin{. . .}** and **\end{. . .}** commands. In addition to the **\center** environments, there are **flushleft** and **flushright** environments, which work precisely as their names suggest.

(You now know enough LaTeX to generate the letter shown earlier. There is a general **letter** environment and style that does the same thing, but we will not discuss that here.)

Beginning a Paper; Sectioning Commands

Most large macro packages, including LaTeX, generate title page information in a way that may not be obvious. There are commands like **\author{ . . . }, \title{ . . . }**, and **\date{ . . . }**, which have to be filled in, but LaTeX will not typeset any of this information until **\maketitle** is issued; **\maketitle** must follow all the specifier commands *and* **\begin{document}**.

```
\author{Charles Dickens}
\date{The Best of Times}
\title{Tale of Two Cities}
...
\maketitle
```

generates something like

Tale of Two Cities

Charles Dickens
The Best of Times

but exact details of fonts and vertical spacing are left to the style file.

The numbered subsection title was typeset from

```
\subsection{Beginning a Paper; Sectioning Commands}
```

Notice that LaTeX handles the details of numbering. Notice too that TeX automatically updates the numbering scheme anytime a section is removed or added. In addition to **\subsection**, LaTeX provides **\chapter, \section, \subsubsection** and several others that we use in the same way.

TeX and LaTeX provide extensive mechanisms for the generation of bibliographies, indexes, tables of contents, lists of figures, insertion of figures and other "floating" objects, footnotes, and just about anything else possible, but we cannot discuss these topics here.

25.9 DISCURSIVE BIBLIOGRAPHY

Serious users of TeX should keep a copy of Donald Knuth's *The TeXbook* (Addison-Wesley, 1984) by their computer (if not by their bed). Make sure to acquire a very recent printing of the book, otherwise the recent changes to TeX will not be discussed. This volume is available either in hardback or paper covered formats.

Users of LaTeX will also require *LaTeX: A Document Preparation System* (Addison-Wesley, 1986) by Leslie Lamporte, the author of LaTeX. An updated version of this manual has been published in 1994. *The LaTeX Companion*, by Goossens, Mittelbach, and Samarin (Addison-Wesley, 1994), is also quite useful as it contains an extensive discussion of LaTeX styles along with the enhanced version of LaTeX alluded to above.

The TeXbook is actually only one volume of Knuth's five-volume series *Computers and Typesetting*. Other volumes in the series discuss METAFONT, Knuth's digital type-founding system he used to generate the Computer Modern fonts that accompany all installations of TeX and the Computer Modern fonts. The complete programs for both TeX and METAFONT, presented in a special narrative format, also form part of this important series.

Many excellent books (and ones not so excellent) on TeX have appeared in recent years. Beginners will get much out of Arvind Borde's two books *TeX* by Example and *Mathematical TeX by Example* (both published by Academic Press). Michael Doob's *Gentle Introduction to TeX* is of special interest because it has been distributed freely across the Internet superhighway. (Interested readers should check all good TeX archives, including **ftp.shsu.edu** and **ftp.tex.ac.uk**.) This manual has recently appeared in hardback format (Springer-Verlag, 1993). *TeX for the Impatient* by Paul Abrahams et al. (Addison-Wesley, 1990) is also helpful. It contains discussions of TeX organized around individual topics.

Finally, users at a high level of TeXpertise will want to consult Victor Eijkhout's *TeX by Topic* (Addison-Wesley, 1991) or Stephen von Bechtolheim's multivolume and idiosyncratic opus *TeX in Practice* (Springer-Verlag, 1994).

25.10 APPENDIX A: PLAIN TeX CODE FOR A SIMPLE LETTER

The following source will produce the simple letter shown in Figure 25.1.

```
\centerline{\bf Ms.\ Polly Darton}
\centerline{\it 12345 Main Street}
\centerline{\it Alvarado, TN 54321}

\vskip1cm

\leftline{\hskip .5\hsize January 12, 1996}

\vskip1pc
\noindent Dear Mr.\ Sofer,
\vskip 1pc

I am very grateful for the effort you have taken in my be-
half. True friends are true wealth—that's what this experi-
ence has taught me. Many thanks once again.
\vskip 2pc
\leftline{\hskip .5\hsize Sincerely yours,}

\vskip2pc
```

```
\leftline{\hskip .5\hsize Polly Darton}

\bye
```

It is fine if your units of vertical spacing differ from the ones shown here.

Simple—very simple—LaTeX code to generate the same text follows.

```
\documentstyle{article}
\begin{document}
\begin{center}
{\bf Ms.\ Polly Darton}\\
{\it 12345 Main Street\\
Alvarado, TN 54321}
\end{center}
\vspace{1cm}
\begin{flushleft}
\hspace{.5\textwidth}February 29, 1996
\end{flushleft}
\vspace{2pc}

\noindent Dear Mr.\ Sofer,

\vspace{1pc}

I am very grateful for the . . .

\vspace{2pc}

\begin{flushleft}
\hspace{.5\textwidth}Sincerely yours,\\

\vspace{2pc}

\hspace{.5\textwidth}Polly Darton
\end{flushleft}
\end{document}
```

Here is one solution to the challenge of Figure 25.2.

```
\centerline{\bf Calculus in a Nut Shell}
\medskip
\centerline{Center for Calculus Revisionism}
\bigskip
```

```
\beginsection 1. Introduction

The curriculum of college calculus is too important to be
left to instructors. It is crucial that the streamlining of
the calculus course be done as soon as possible.

\beginsection 2. The Joy of Differential Calculus

Calculus is really as easy as $\pi$. We begin with a study
of limits; how can we make sense of
$$
\lim_{x\rightarrow x_0} {f(x)-f(x_0)\over x-x_0}?
$$
This limit, if it exists, is the {\it derivative\/} $f'(x)$.

\beginsection 3. Existential Integral Calculus

The definite integral of a function $\int_a^b f(x)\,dx$ is
related to the limit (if it exists) of
$$
\sum_a^b [f_1(x)-f_2(x)]\Delta x.
$$
It can further be shown \dots
```

And here is simple LATEX code to do the same thing.

```
\document style{article]
\begin{document}
\begin{center}
{\bf Calculus in a Nut Shell}\\
\medskip
Center for Calculus Revisionism
\end{center}

\section{Introduction}

The curriculum of college calculus is too important to be
left to instructors. It is crucial that the streamlining of
the calculus course be done as soon as possible.

\section{The Joy of Differential Calculus}
```

```
Calculus is really as easy as $\pi$. We begin with a study
of limits; how can we make sense of
$$
\lim_{x\rightarrow x_0} {f(x)-f(x_0)\over x-x_0}?
$$
This limit, if it exists, is the {\it derivative\/} $f'(x)$.

\section{Existential Integral Calculus}

The definite integral of a function $\int_a^b f(x)\,dx$ is
related to the limit (if it exists) of
$$
\sum_a^b [f_1(x)-f_2(x)]\Delta x.
$$
It can further be shown \dots
\end{document}
```

26

Spreadsheets

Donald L. Byrkett
Miami University

Roberta Jaworski
Miami University

26.1 INTRODUCTION TO ELECTRONIC SPREADSHEETS

Engineers have traditionally needed to work with numbers and graphs. Prior to 1970 most of their work was done by hand, possibly with the aid of a slide rule or a computer for intensive numerical work. Worksheets were developed, calculations were performed, and graphs were drawn. As design parameters changed this process was repeated.

In the 1970s, this process was speeded up with the use of electronic calculators. Slide rules became obsolete. Calculations could be performed more rapidly and more accurately. Occasionally, engineers had to write computer programs to solve complex numerical problems but programming took a lot of time, turnaround was slow, and the results of the numerical calculations still needed to be organized and graphed by hand.

In the 1980s, electronic spreadsheets were developed. These electronic spreadsheet programs provided an easy way for engineers to organize and perform calculations. Electronic spreadsheets make it easy to vary design parameters and perform repetitive calculations instantly. These programs run on microcomputers so the engineer obtains immediate feedback. Originally, electronic spreadsheets were used primarily by business analysts, but they have improved greatly and provide many features that engineers can use:

- reports can be styled, organized and printed with little effort
- worksheet data can be graphed in a variety of ways and printed in a format that is ready for inclusion in a report
- trigonometric, logarithmic, and statistical functions and matrix operations are readily available to perform engineering calculations
- regression analysis is available
- equation solvers and optimizers are built in
- random numbers are available for simulation experiments

In addition, *no programming is necessary*. More and more, engineers are using electronic spreadsheet programs to perform functions they formerly performed by using their calculator or by writing computer programs.

0-8493-2530-7/96/$0.00+$.50

The remainder of this chapter describes three popular spreadsheet programs that run under the Microsoft Windows environment. Lotus 1-2-3 Release 4 is described first and in greater detail. The other two programs, Microsoft Excel 4.0 and Quattro Pro 6.0, are described in less detail. The focus in the latter two sections is to illustrate some of the differences between these products and 1-2-3.

26.2 LOTUS 1-2-3 RELEASE 4 FOR WINDOWS

Starting 1-2-3

From the Windows Program Manager, double click the 1-2-3 icon illustrated in Figure 26.1 to run 1-2-3 Release 4 for Windows. This action will bring up the 1-2-3 for Windows environment along with an empty worksheet as shown in Figure 26.2.

Worksheet Basics

Window operations: 1-2-3 for Windows is a typical Microsoft Windows application. The top row of the window provides a title bar, which includes the application name, the file currently being worked on (Untitled), a control box on the left, and the minimize and maximize buttons on the right. Just beneath the title bar is the main menu, which provides eight pull-down menus unique to this application. Notice that the traditional File menu is on the left and the Help menu is on the right.

FIGURE 26.1 Lotus 1-2-3 Release 4 icon.

SmartIcons: The fourth row provides a list of what Lotus Development Corporation calls SmartIcons. These SmartIcons are provided to automate certain menu selections. For example, instead of selecting the menu op-

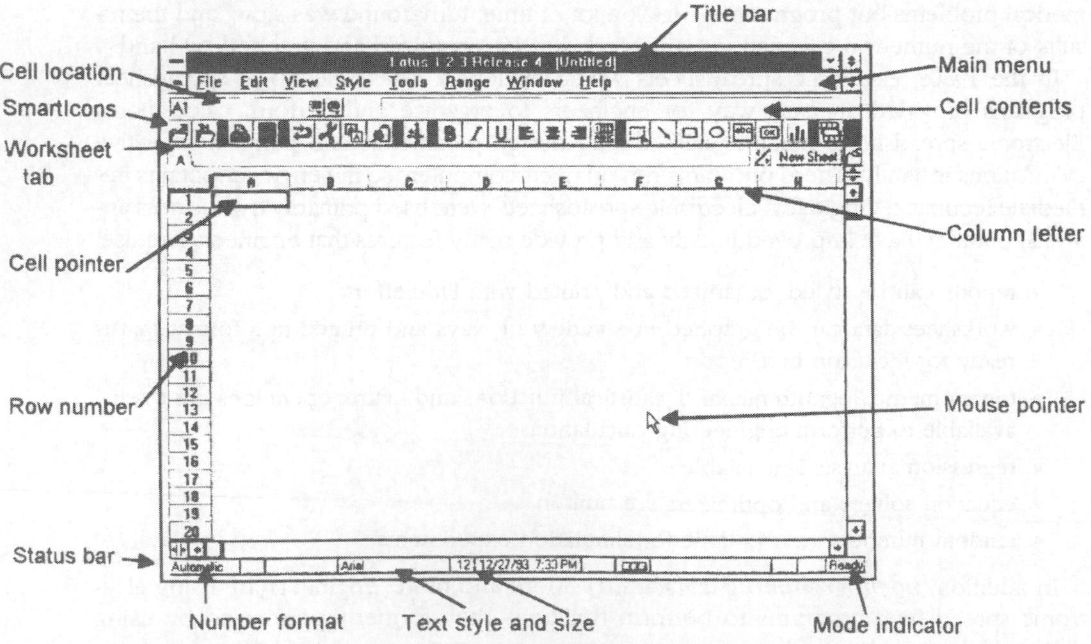

FIGURE 26.2 An empty 1-2-3 worksheet screen.

tion File Open (requiring two mouse clicks), the user can click the SmartIcon on the far left (one mouse click) to bring up the dialog box to open a file. The list of SmartIcons can be customized to meet individual needs. This discussion, however, will use menu selections, rather than SmartIcons, to perform all functions.

Worksheets: The remainder of the window is designed around the electronic spreadsheet or worksheet. The worksheet is basically an area of the screen divided into a series of rows separated by horizontal lines and a series of columns separated by vertical lines. By convention, the columns are assigned letters (A, B, . . ., Z, AA, AB, . . ., AZ, BA, . . .) and the rows are assigned numbers. Horizontal and vertical scroll bars are provided to access columns and rows that are not visible on the screen. A single spreadsheet file can be composed of multiple worksheets. The additional worksheets can be accessed by using the mouse pointer to click the appropriate worksheet tab. Initially, as in Figure 26.2, there is only one worksheet, and, hence, one tab. Additional worksheets can be obtained by clicking the button labeled New Sheet just below the SmartIcons on the right side of the screen.

Cells: At the intersection of each row and column is a cell. Cells are referred to by a cell address composed of the column letter followed by the row number (A1, E3, B9, and so on). One cell, designated by the cell pointer, is called the current cell. The location of the current cell and the contents in the current cell are displayed just beneath the main menu. In Figure 26.2, cell A1 is the location of the current cell and its content is empty. The current cell can be changed by using the mouse pointer to point to another cell and clicking the left mouse button. If the desired cell is not visible on the screen, the horizontal and vertical scroll bars must be used first to locate the cell.

To refer to a cell address on a different worksheet, precede the cell address with the name on the worksheet tab. For example, B:A1 refers to the first cell on worksheet B.

Ranges: A range is a rectangular group of cells. Ranges are referred to by a range address, which contains the cell addresses in opposite corners of the group separated by two periods (A1..F3 and A3..F1 refer to the cells in columns A–F and rows 1–3). The mouse pointer can be used to select a range of cells by pointing to a cell in one corner and dragging (hold the left mouse button down) the pointer to the cell in the opposite corner as illustrated in Figure 26.3. By moving the mouse pointer beyond the edge of the worksheet, it is possible to drag to a cell that does not initially appear on the screen. Note that the selected range address is highlighted on the worksheet and appears in the cell location window beneath the main menu. To deselect a range address, point to any cell on the worksheet and click the mouse.

It is also possible to refer to a three-dimensional range of cells by preceding each corner cell with the name on the worksheet tab. For example, A:A1..B:F3 refers to the cells including the first six columns and first three rows of the first two worksheets.

Specifying a range address in a dialog box: Many of the menu choices apply to a range of cells and will bring up a dialog box that requires you to specify a range address. The dialog box will typically include an edit box as illustrated in Figure 26.4.

There are basically three ways to specify this range address. The easiest way is to use the mouse to highlight the range address prior to bringing up the dialog box. If you do this, the highlighted range address will automatically be filled in the edit box when the dialog box is activated. A second method, if you forget to highlight the range address before bringing up the dialog box, is to simply click the mouse over the edit box, and type in the range address you want. Sometimes this is hard to do, however, since the dialog box may cover up the worksheet making it hard to see the appropriate range address. Finally, the third method is to click the icon to the right of the edit

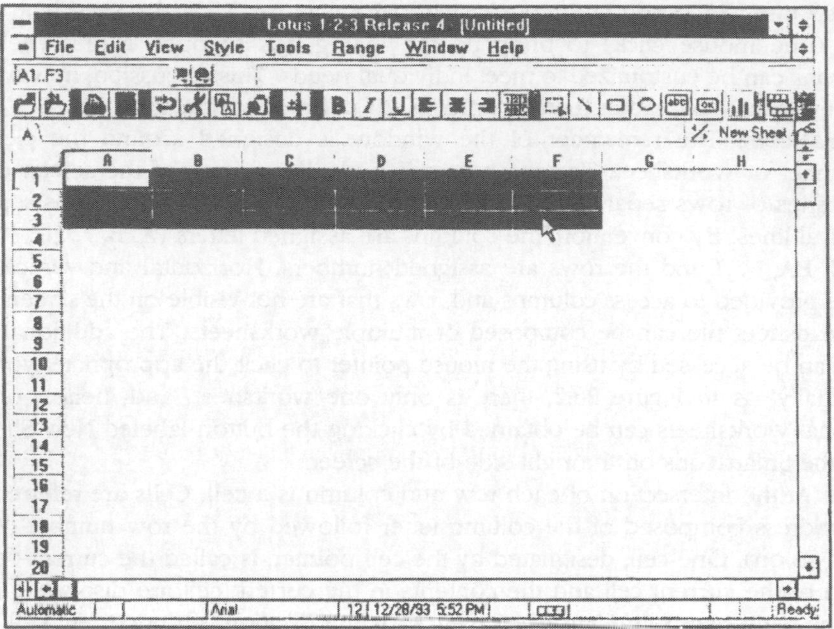

FIGURE 26.3 Pointing to a range address.

box. This will cause the dialog box to disappear allowing you to drag a special mouse pointer across the selected range. Once you have highlighted the desired range, the dialog box will reappear with the proper range address filled in.

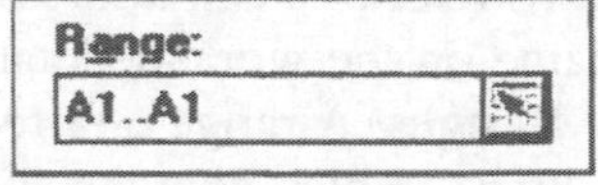

FIGURE 26.4 An edit box to specify a range address.

Menus: 1-2-3 for Windows provides seven pull down menus plus the Help menu. An overview of the functions provided in each menu category is provided below.

Menu	Functions
File	Opening, saving, and printing worksheet files
Edit	Inserting and deleting rows columns, and worksheets, copying cell contents, moving cell contents
View	Zooming in and out, setting viewer preferences
Style	Specifying number formats, fonts, visual enhancements, column widths, row widths, and setting worksheet defaults
Tools	Using special charting functions, drawing functions, spell checking, and database functions
Range	Filling a range in sequence, sorting, solving problems, and analyzing data
Window	Navigating between separate worksheet files in separate windows

In addition, when creating a chart, the Range menu is replaced by a Chart menu providing options for building the chart. The use of specific functions from these menus is described in subsequent sections.

Context-sensitive menus: Special context-sensitive pop up menus can be accessed by clicking the right mouse button. For example, if you are pointing to a row number, a column number, a cell, or a range address and click the right mouse button, a special menu will pop up on the screen to provide an appropriate set of options for the highlighted object. Like the SmartIcons, these pop-up menus provide a shortcut to common functions provided by the normal pull-down menus. The discussion that follows will emphasize the use of the regular menus, but many of the options can be accessed more quickly using the pop-up menus.

Creating Worksheets

Figure 26.5 illustrates a worksheet designed to calculate the distance (*d*) traveled by a projectile when launched from a cannon with a certain initial velocity (v) and a certain angle of elevation (θ). The equation for this calculation is provided below where *g* is the acceleration of gravity

$$d = (2 * v^2/g)\cos(\theta)\sin(\theta)$$

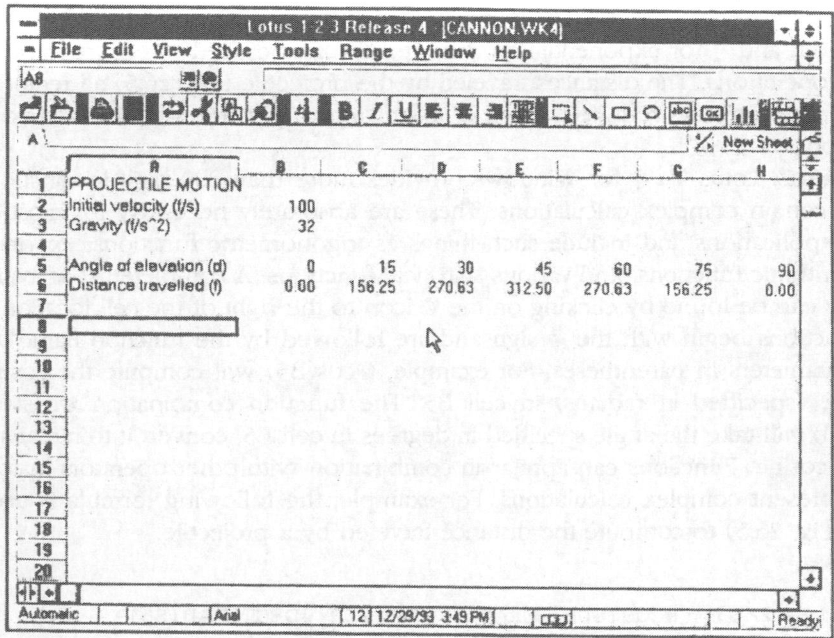

FIGURE 26.5 Projectile motion worksheet.

A worksheet is constructed by entering data into individual cells of the worksheet. There are three types of data that can be entered into cells: labels, numbers, and formulas.

Labels: A label is simply a text entry. All of the cells in column A (Fig. 26.5) contain labels. Labels are entered by positioning the cell pointer over the desired cell, typing the desired label, and pressing the Enter key. As a label is entered, the mode indicator will change from Ready to Label and the letters typed will be displayed on the worksheet and in the cell contents box.

Normally, as a label is entered, it will automatically be preceded by a single quote ('). This is called a label prefix character. This character indicates that the label is to be left justified in the cell. You can type other label prefix characters to position the label in the center (^) of the cell or along the right edge (") of the cell, All of the labels in Figure 26.5 are left justified.

Numbers: A number is a numeric constant. In Figure 26.5, the initial velocity (100), the acceleration of gravity (32), and the angles of elevation (0, 15, .., 90) are numeric constants. Numbers are entered in cells in the same way as labels. The only difference is that as a number is entered, the mode indicator will change from Ready to Value. Numbers are always right justified in the cell and can be used for arithmetic computation in formulas.

Formulas: A formula is an arithmetic expression to calculate the value in the given cell from values in other cells. For example, if the formula +B2*B3 is entered in cell C2 of Figure 26.5, the value 3200 (100*32) would be displayed in C2. Formulas are a powerful feature of spreadsheets because if you change one of the numbers in a worksheet all of the formulas depending on that number are recalculated automatically.

Formulas are entered in cells just like labels and numbers. Formulas should begin with an operator (such as a plus sign) or a parenthesis so that they can be distinguished from labels. When entering a formula, the mode indicator should change from Ready to Value. Formulas use + for addition, – for substraction, * for multiplication, / for division, and ^ for exponentiation. Parenthesis can be used to ensure the correct order of operations. The distances traveled by the projectile (Fig. 26.5) all require formulas. These formulas require the use of a function to perform the sine and consine calculations.

Functions: Lotus 1-2-3 for Windows provides more than 200 built-in functions to perform certain complex calculations. These are absolutely necessary for most engineering applications and include such things as trigonometric functions, exponential and logarithmic functions, and various statistical functions. A complete list of available functions can be found by clicking on the @ icon to the right of the cell location box.

All functions begin with the @ sign and are followed by the function name and a list of parameters in parentheses. For example, @cos(B5) will compute the cosine of the angle (specified in radians) in cell B5. The function combination @cos(@degtorad(B5)) will take the angle specified in degrees in cell B5, convert it to radians, and take the cosine. Functions can appear in combination with other operators in formulas to represent complex calculations. For example, the following formula is used in cell B6 (Fig. 26.5) to compute the distance traveled by a projectile.

```
(2 * B2^2/B3) * @COS(@DEGTORAD(B5)) * @SIN(@DEGTORAD(B5))
```

Similar formulas are contained in cells C6..H6.

Changing cell contents: There are two ways to change the label, number, or formula in a cell. One is to simply move the cell pointer to the given cell and type in the new contents. The new contents will replace the old. This is useful if you are entering a completely different label, number, or formula or the information in the cell is small.

The second way is to edit, or modify, what is currently in the cell. There are two ways to do this. One is to move the cell pointer to the given cell, and then click the mouse pointer in the cell contents box at the position you want to modify. Make the modifications desired using the normal editing keys and click the [✓] icon to the left of the cell contents box or press the Enter key. If you decide not to make the changes click the X or cancel icon or press the Escape key. Alternatively, you can double click over the cell to be modified and make the editing changes right on the worksheet.

What if analysis: Spreadsheets are a useful tool for performing what if analysis. The engineer can vary one of the design parameters on the worksheet and have all of the formula calculations performed again automatically. For example, in Figure 26.5, if the engineer changes the initial velocity in cell B2 to 200 feet/second, all of the distances traveled will be recalculated. Or, if the engineer changes the angle of elevation in cell B5 to 13 degrees, a new distance will be calculated. Typically, worksheets list the design parameters at the top and set up formulas to perform the related calculations. This makes the worksheet a useful tool for experimentation.

Opening, Saving, and Printing Worksheets

The File menu (Fig. 26.6) provides the standard group of menu options for opening a previously saved worksheet file, for saving the active worksheet file, and for printing the active worksheet.

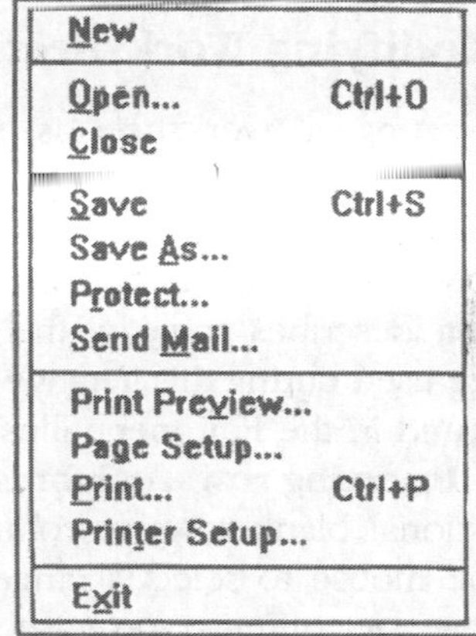

FIGURE 26.6 The File menu.

Opening a worksheet: Consistent with most Microsoft Windows applications, the New option creates a blank worksheet, Open displays a dialog box that allows you to retrieve a previously saved worksheet file from disk and display it on a worksheet for editing, and Close removes the active worksheet from memory. The Close option will prompt you to save the active worksheet if you have not already done so.

Saving a worksheet: Again, consistent with other Windows applications, the Save option copies the active worksheet to a previously named disk file (the filename appears in the title bar) and the Save As option displays a dialog box that allows you to specify the drive, directory, and filename to use for saving the active worksheet to disk. Lotus 1-2-3, Release 4 files are automatically saved with the filename extension WK4.

Printing a worksheet: The Print option displays the dialog box shown in Figure 26.7. This dialog box allows you to select the option to print the Current worksheet, All worksheets in the current worksheet file, or a Selected range address from the current worksheet file. Other options are available to print selected pages of the work-

FIGURE 26.7 The Print dialog box.

sheet or to print multiple copies. If you choose to print a selected range address, use one of the methods discussed earlier for specifying a range address in a dialog box. After making the desired choices, you can either click the Preview button to see how the printed report will look prior to the actual printing or click the OK button to commence printing.

Note that the four SmartIcons on the far left (Fig. 26.2) are for Opening a file, Saving a file, Printing, and Previewing, respectively. Clicking the appropriate SmartIcon is a shortcut for selecting the corresponding menu options.

Modifying Worksheets

Creating a worksheet is an iterative process. The process begins with a basic idea in mind, but usually requires continuous modifications and enhancements before the user is satisfied with the final results. This section describes some of the column modifications that are used during this iterative process. These features are found in the Edit menu illustrated in Figure 26.8.

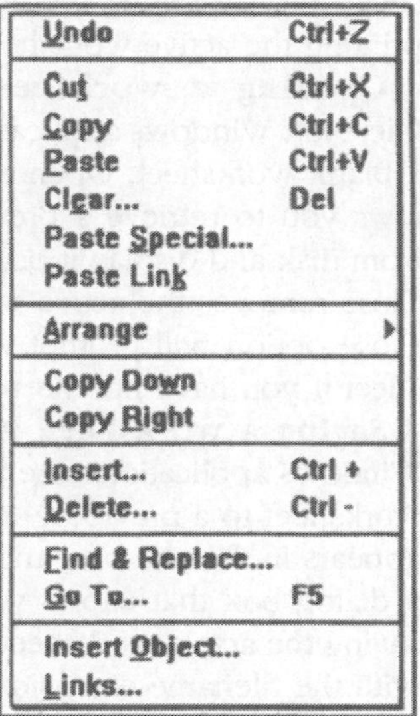

FIGURE 26.8 The Edit menu.

Inserting rows, columns, and sheets: To insert additional blank rows or columns into a worksheet, use the mouse to select a range address that includes the rows or columns you wish to insert. Then choose the Insert option from the Edit menu to obtain the dialog box in Figure 26.9. Select whether you want to insert blank rows, columns, or worksheets and click the OK button. Data (Labels, Numbers, or Formulas) that were previously in the selected rows, columns, or worksheets will be moved down, over, or back and blank space will be created. Any formulas referring to the previous data will be adjusted to reflect the new location. Labels, numbers, or formulas can now be entered in the new space. This is a handy modification when you forget to include a particular row or column of information and you want to insert it between existing rows or columns.

Deleting rows, columns, and sheets: To delete blank or filled rows or columns, use the mouse to select a range address that includes the rows or columns you wish to delete. Then choose the Delete option from the Edit menu to obtain the dialog box in Figure 26.10. Select whether you want to delete rows, columns, or worksheets and click the OK button. Data that were previously in the selected rows, columns, or worksheets will be deleted and information in the following rows, columns, or worksheet will be moved to fill in the empty space. Again, any formulas referring to the moved data will be adjusted to reflect the new location. Any formulas referring to the deleted data will contain the message ERR since the data is no longer there.

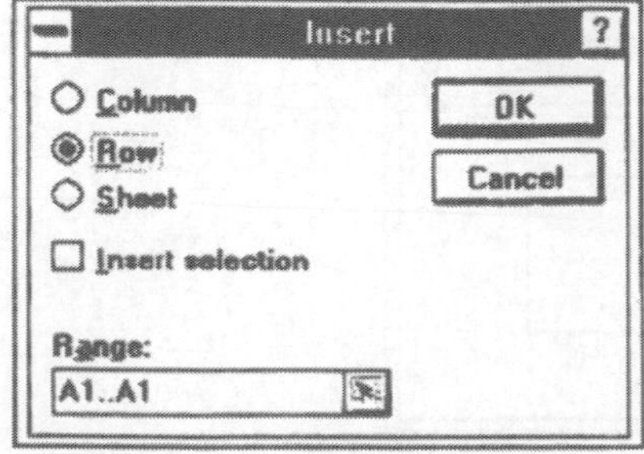

FIGURE 26.9 The Insert dialog box.

Cutting, Copying, and Pasting Worksheet Data

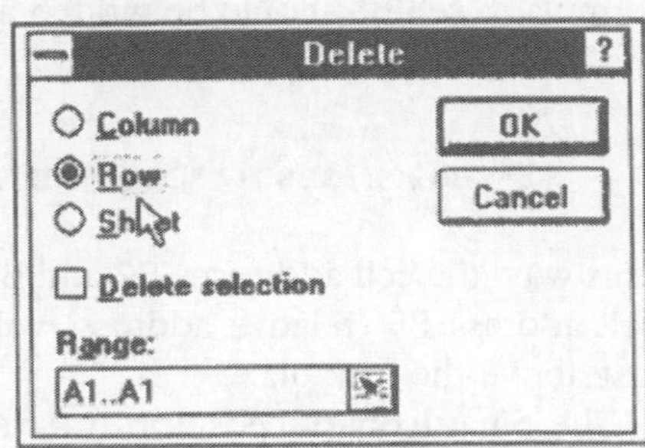

FIGURE 26.10 The Delete dialog box.

Working with worksheets is a lot like working with word processing documents, sometimes it is efficient to copy parts of the worksheet for faster data entry and sometimes it is useful to cut and paste to reorganize the arrangement of the worksheet. 1-2-3 provides the Edit options (Fig. 26.8) Cut, Copy, and Paste to do this.

These functions work just like a word processor. The process begins by dragging the mouse pointer across the range of cells you wish to cut or copy. This range will be highlighted. Then select the menu option Edit Cut or Edit Copy. The Cut option deletes the contents of all cells and places it on the clipboard (a temporary storage location in memory). The Copy option leaves the contents of the highlighted cells unchanged but copies the contents onto the clipboard. The next step is to move the cell pointer to the upper left corner of the range to receive the cut or copied cell contents. Then select Edit Paste to move the cell information saved on the clipboard to the worksheet at the specified location. Previous information in this range of cells will be replaced with the information from the clipboard. Edit Copy followed by Edit Paste is useful for entering duplicate or similar information quickly into the worksheet. Edit Cut followed by Edit Paste is useful for rearranging the worksheet.

The effect of a Cut and Paste and a Copy and Paste is fairly clear when dealing with labels and numbers. An exact copy is made in the new location. The way that formulas are copied, however, depends on the way cell addresses are specified in the formula. Suppose cell A1 contains one of the following formulas:

```
+A2*A3 or +$A$2*A3 or +$A$2*$A$3
```

All three do the same thing; they multiply the contents of cell A2 times A3 and place the result in cell A1. In the first formula, the cell addresses A2 and A3 are relative addresses. When a dollar sign precedes the column letter and the row number, the cell address becomes an absolute address. In the second formula, A2 is an absolute address and A3 is a relative address and, in the third formula, both A2 and A3 are absolute addresses. Whether a cell address is relative or absolute has nothing to do with the computation performed, but it has a big impact on how the formula is copied. Absolute cell address are copied exactly as written to another cell and relative cell addresses are adjusted during a copy operation so that they refer to the cell in the same relative position to the copied cell. If the above formulas were cut or copied and pasted into cell B1, they would appear as follows:

```
+B2*B3 or +$A$2*B3 or +$A$2*$A$3
```

In the first case, the relative cell addresses are adjusted so that the contents of the cells in the two rows beneath are multiplied together. In the other two cases, the absolute cell addresses are copied exactly.

To illustrate this idea, refer to the projectile motion worksheet in Figure 26.5. The formula in cell B6 should be written as follows to facilitate copying the formula to cells C6..H6:

```
(2*$B$2^2/$B$3)*@COS(@DEGTORAD(B5))*@SIN(@DEGTORAD(B5))
```

This way, the cell addresses B2 and B3 (absolute addresses) will be copied exactly and cell address B5 (relative address) will be adjusted in each case to refer to the angle just above the formula.

The SmartIcons in position 6 (scissors), 7 (two letter As), and 8 (jar of paste) of Figure 26.2 provide shortcuts for Edit Cut, Edit Copy, and Edit Paste, respectively.

Styling Worksheets

Once you have a worksheet that performs the correct calculations, you'll want to spruce it up so it's ready to include in a report. Options are available in the Styles menu (Fig. 26.11) for specifying the formatting of numbers, the fonts to use including sizing, boldfacing, and underlining, the use of shading and other ways to highlight certain parts of a worksheet, and the column widths and row heights.

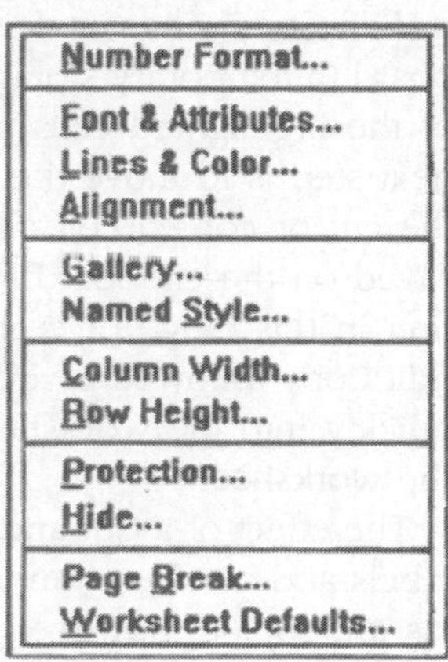

FIGURE 26.11 The Styles menu.

Adjusting column widths and row heights: Initially, all of the columns in a worksheet have the same width (9 characters) and a row height appropriate for the height of the font selected. Usually, it is necessary to adjust some of the column widths to accommodate the labels or other information in the cells. To make this adjustment, use the mouse to select a range address containing the columns you want to adjust and select the menu option Styles Column Width. This will bring up the column width dialog box illustrated in Figure 26.12.

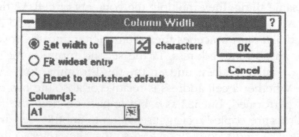

FIGURE 26.12 The Column Width dialog box.

You can choose to set the width to a specific number of columns using the small edit box at the top or you can choose to set the width to the widest entry in the given column. By selecting the columns to adjust prior to choosing this menu option, the appropriate range address will be filled in the Column(s) edit box. After making your selections, click the OK button to implement the desired action.

Similarly, the menu option Styles Row Height can be used to specify the row height of a group of rows. In this case, the row height can be either specified in points (where each point equals 1/72 of an inch) or can be determined automatically by 1-2-3 to fit the largest font used in the row.

In the projectile motion worksheet (Fig. 26.5), column A is set to a 17 character width and the remaining columns are set to 8 characters. The row height is set automatically to the height of the font.

Adjusting number formats: Numbers or the results of formula calculations are usually displayed using an Automatic format. This means that the format is set based on the way the data is entered. It is possible to use a variety of formats including Comma (3,113,456), Fixed decimal (14.2), Currency ($13.45), Percent (13.2%), various Date formats (January 3, 1994, 01/03/94, and so on), and others. These are specified by using the mouse to highlight the range of cells to format and selecting the menu option Styles Format. This action will bring up the Number Format dialog box illustrated in Figure 26.13.

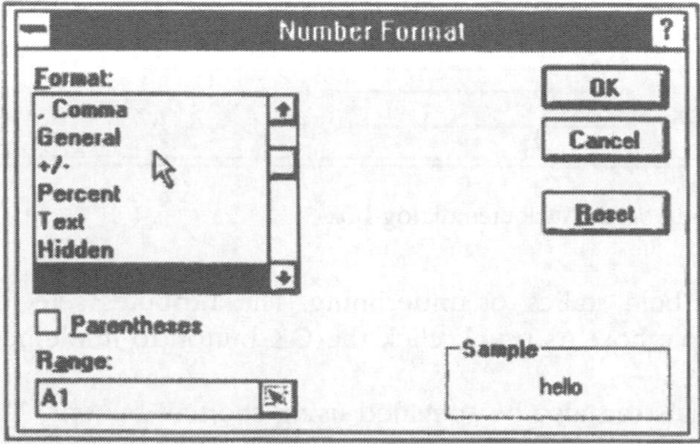

FIGURE 26.13 The Number Format dialog box.

Using this dialog box, you can select the number format desired from the list box and click the OK button. If you choose certain formats, an additional edit box will appear so that the number of decimal places to display can be specified. This format will be associated with each cell in the range address, thus any future number or formula entered in one of these cells will use the specified number format.

Another way to specify the number format is to use the special buttons in the lower left corner of the 1-2-3 window. Normally, these display the format and the number of decimal places used in the current cell. If you click one of these buttons, a list of available number formats and decimal places will appear. These buttons can be used to change the format or decimal places of the current cell or a currently selected range of cells.

In the projectile motion worksheet (Fig. 26.5), a Fixed format with two decimal places was used for cells in the range B6..H6. All of the remaining cells use the Automatic format.

Adjusting text formats: Labels and numbers are displayed using the default typeface and point size specified for the worksheet. Like number formats, the font and the font attributes can be easily adjusted. To change the text format of a range of cells, drag the mouse pointer across the desired range of cells and choose the menu option Styles Font & Attributes. This actions will display the Font & Attributes dialog box illustrated in Figure 26.14. You can choose the typeface from the list box on the left, the character size (in points) from the list box in the center, and the text

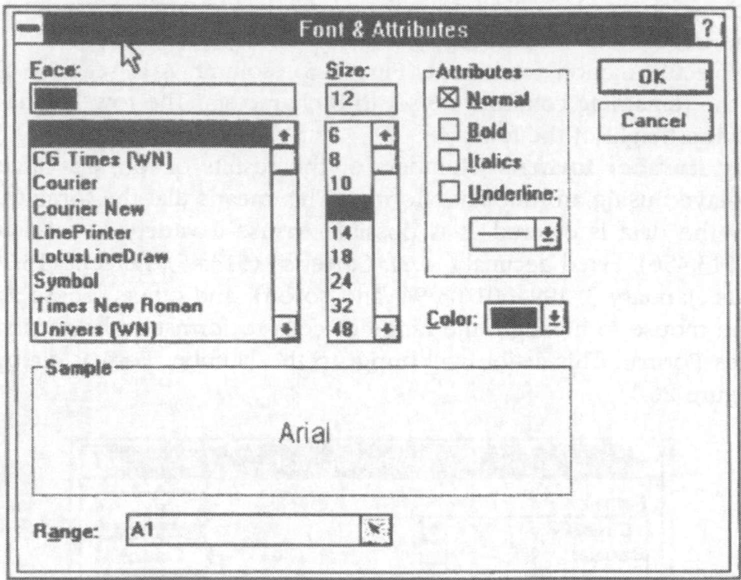

FIGURE 26.14 The Font & Attributes dialog box.

attributes such as bold, italics, or underlining. The attributes selected will be illustrated in the Sample box. As usual, click the OK button to implement the new text format.

These text formats can also be specified using shortcut buttons. The typeface and point size can be specified by clicking the special buttons at the bottom of the 1-2-3 window that normally display the typeface and point size of the text in the current cell. SmartIcons are available to specify bold (thick B), italics (slanted I), and underlining (U with underline).

Other styling options: The Lines & Color option in the Styles menu allows parts of the worksheet to be highlighted by using various fill options in the cells and/or by using lines or special designed styles to outline all or parts of the cell. The Alignment option is used to vary the alignment of labels (left, right, or center) in particular cells and the alignment of the entire worksheet on the page. The Gallery option provides a list of predefined highlighting combinations.

Worksheet defaults: You can use The Worksheet Defaults option in the Styles menu to define the default options for column width, row height, number format, text format, alignment, and fill patterns. A judicious selection of these worksheet defaults can save a lot of time in adjusting the style of individual parts of the worksheet.

Graphing

One of the most useful spreadsheet features for engineers is the ability to quickly enter data and produce a variety of graphs and analyses based on the data. For example, in Figure 26.15, the results of some tensile test experiments are recorded. The engineer has entered the elongation of a low carbon steel sample and an aluminum alloy sample at various loads, in kilopounds. The data in columns A, B, and C are composed of labels and numbers, there are no formulas. However, once the data are entered, it is

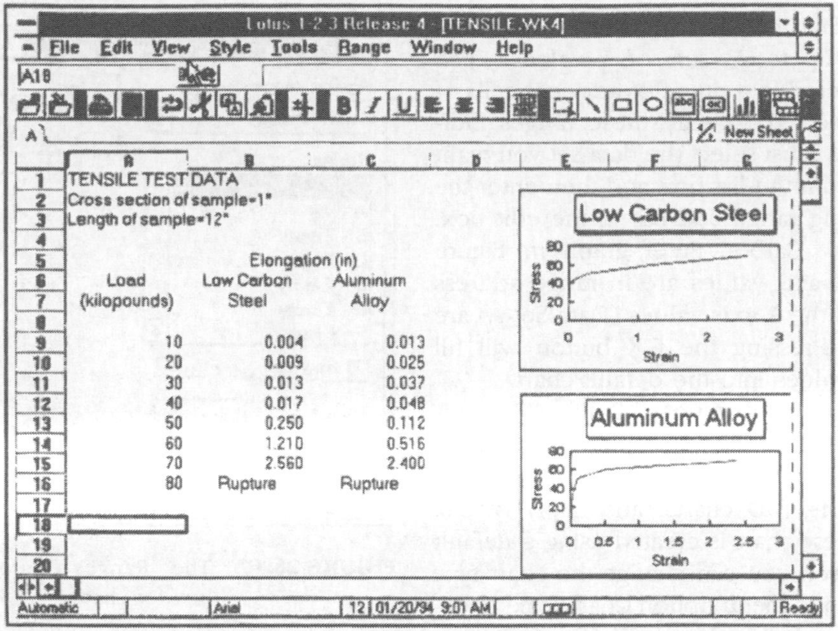

FIGURE 26.15 Data analysis worksheet.

easy to use 1-2-3 to produce the Stress–Strain curves at the right and to perform regression calculations to estimate the slope of the line before it yields. The slope of this line equals the modulus of elasticity.

Creating graphs: Suppose you have entered the data in columns A, B, and C of Figure 26.15 and would like to create the Low Carbon Steel graph. This can be accomplished by selecting the menu option Tools Chart (most spreadsheet packages refer to graphs as charts). The mouse pointer will change to a different shape and you'll be prompted to drag the pointer across the range of cells to contain the chart. Once this range has been highlighted, a default chart will be created in that area of the worksheet. The box around the chart will be active and the Range option on the main menu will be replaced by the Chart option as illustrated in Figure 26.16.

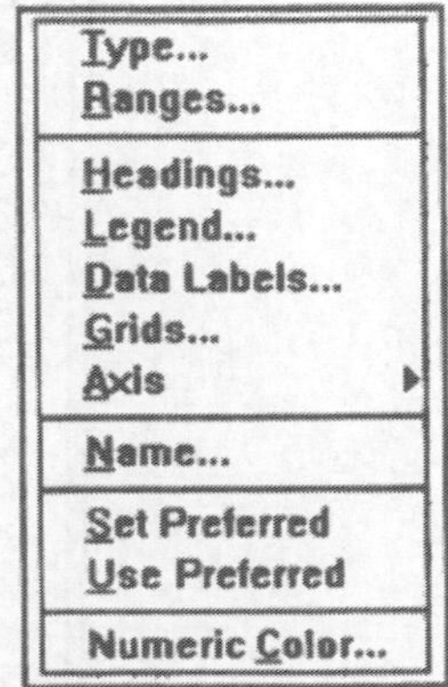

FIGURE 26.16 The Chart menu.

The Chart options allow you to specify the type, range address of the X and Y data, titles for the chart, axis labels, legends, and other enhancements to the chart. Clicking the mouse on a cell outside the chart area will deactivate the chart, cause the Chart menu to disappear, and cause the Range menu to return. By clicking on the chart again, the chart will become active and the Chart menu will return.

Specifying values to graph: If the data for the *X* axis is in the first column or row and the data for the *Y* axis is in the next column or row, you can specify the data automatically by highlighting the data before creating the chart. Otherwise, you need to do it after you create the graph by choosing the menu option Chart Ranges. Selecting this option will bring up the chart Ranges dialog box displayed in Figure 26.17.

A single chart may contain one set of *X* axis values and up to 23 sets of *Y* values. Each must be entered in a range of adjacent cells of the worksheet. To specify these ranges individually, you must select the data set you want to specify from the list box and then enter the corresponding range address in the edit box. For the Low Carbon Steel graph in Figure 26.15, the *X* axis values are in range address B9..B15 and the *Y* axis values (Data Set A) are in A9..A15. Pressing the OK button will fill these data values into the default chart.

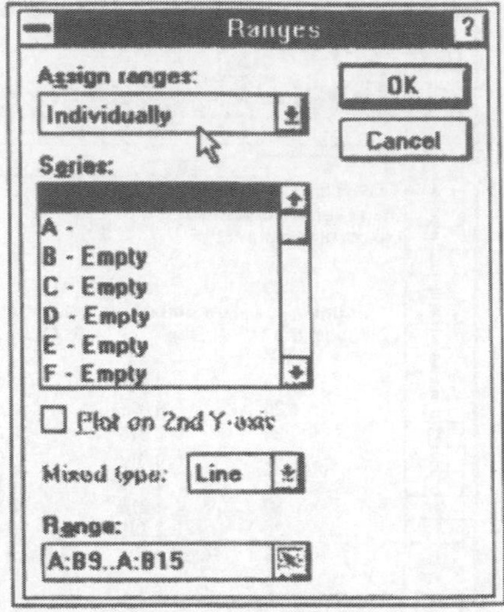

FIGURE 26.17 The Ranges dialog box for charts.

Specifying graph types: 1-2-3 for Windows provides a large variety of graphs including bar charts, pie charts, area charts, line graphs, stacked charts, 3-D charts, and many others. Initially, a new chart is created using a default type. However, you can change the chart type by choosing the menu option Chart Type to obtain the dialog box in Figure 26.18.

This dialog box allows you to select the type of chart you want and a specific style of the selected type. In Figure 26.18, the Bar type chart is selected and three styles, side by side bars, stacked bars, and connected stacked bars, are illustrated. Each type provides a variety of styles. In addition, each chart can be displayed with a vertical or horizontal orientation and can include the data values in a table with the chart.

For most of the charts, the *X* values serve as labels for the *Y* values. The only chart in which the *X* values are plotted a certain distance along the *X* axis is the *XY* type.

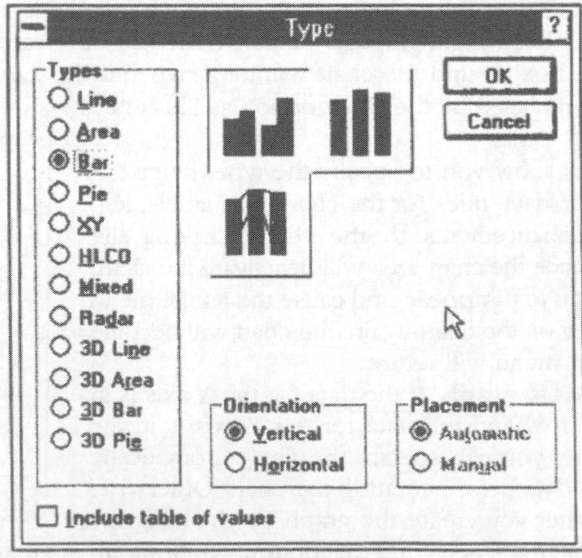

FIGURE 26.18 The Type dialog box for charts.

This is the type of chart that was used in Figure 26.15 and is a type that is useful for many engineering applications.

Enhancing graphs: The remainder of the Chart menu options are used to enhance the appearance of the graph. These can be accessed by clicking the appropriate menu choice on the Chart menu or by double-clicking on the appropriate item in the chart you are creating. The Headings option is used to establish a title for your chart. The Legend option is used to specify a legend to identify each data set when multiple data sets are used. The Grids option is used to draw horizontal and/or vertical lines on your chart and the Axis option is used to assign labels to each axis.

In the Low Carbon Steel graph in Figure 26.15, the heading is Low Carbon Steel, the *Y*-axis label is Stress, and the *X*-axis label is Strain. There is no legend and there are no grid lines on this graph.

Printing graphs: The graphs can be printed separately or as a part of the worksheet. To print the graph as a part of the worksheet, follow the usual procedure. Use the mouse to select the range address you want to print (which should include the chart) and select menu option File Print. If you just want to print the graph, click the mouse over the graph to make the graph window active and choose the File Print menu option. In this case, just the graph will print.

Analysis Tools

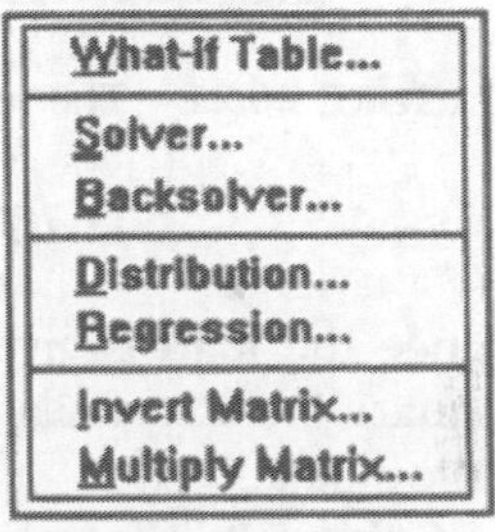

FIGURE 26.19 The Range Analyze menu.

1-2-3 for Windows provides several facilities for analyzing data and solving equations. These tools are accessed from the Range Analyze menu option. Selecting this menu option will pull up the cascading menu choices listed in Figure 26.19. Of particular importance to engineers are the Regression option to perform a linear regression analysis on a set of data, the Solver option to solve optimization problems, and the matrix options to multiply and invert matrices.

Regression analysis: Suppose you want to fit a regression line to estimate the modulus of elasticity for Low Carbon Steel (Fig. 26.15). This requires fitting a line to the Stress–Strain curve prior to the point of yield (loads of 10 to 40 kilopounds). Choose the menu option Range Analyze Regression. The dialog box in Figure 26.20 will appear.

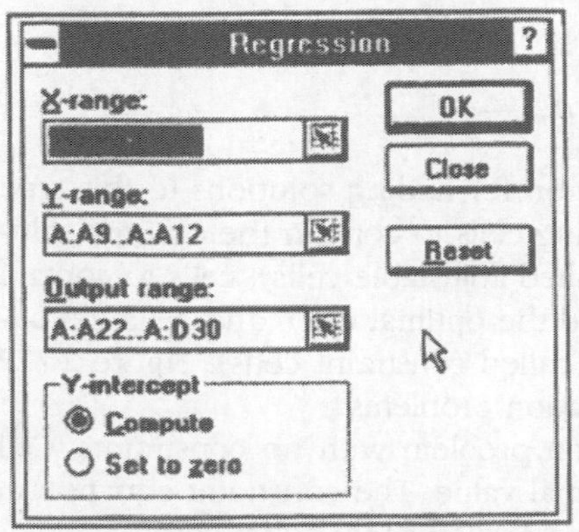

FIGURE 26.20 The Regression dialog box.

In the *X*-range edit box, you are to specify the range address containing the independent variables (B9..B12) and in the *Y*-range edit box you are to specify the range address for the dependent variable (A9..A12). The output range should provide the upper left corner of a range address to contain the regression analysis report (a minimum of nine rows and four columns are needed). Click the OK button and a report of the regression calculations will be provided, as illustrated in Figure 26.21.

According to this report, the line that best fits the data has the equation:

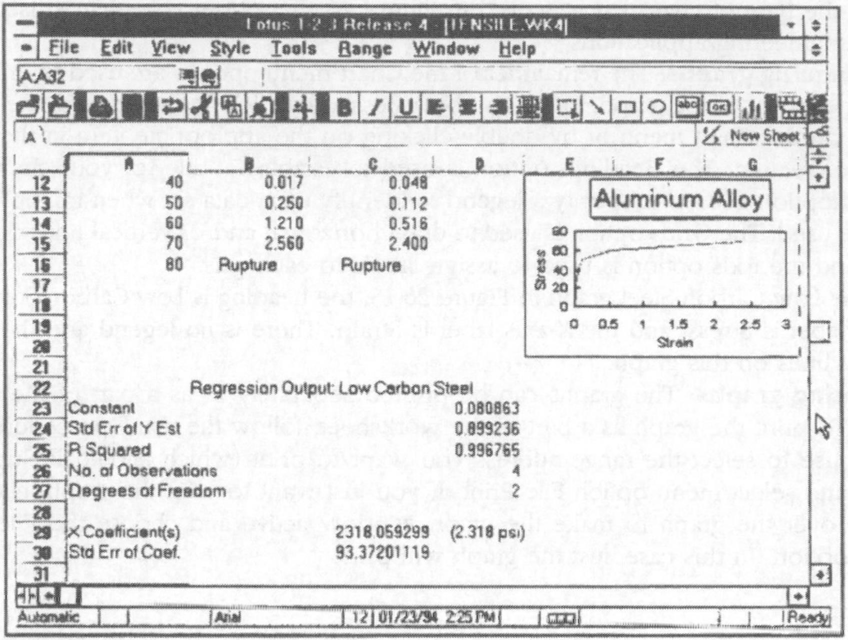

FIGURE 26.21 The regression analysis report.

$$\text{Load} = .080863 + 2318.059299^{*}\text{Elongation}$$

Since the load is given in kilopounds, the modulus of elasticity is 2.318 pounds per square inch. A similar process could have been followed to fit a line to the Aluminum Alloy data.

Solver: In general, an optimization problem has the following form:

$$\text{Maximize} : f(x_1, x_2, \ldots, x_n)$$

Subject to:

$$\begin{aligned} g_1(x_1, X_2, \ldots, X_n) &<= k_1 \\ \vdots \quad & \quad \vdots \\ g_m(x_1, X_2, \ldots, X_n) &<= k_m \end{aligned}$$

1-2-3 for Windows provides the Solver menu option for finding solutions to this type of problem. These problems are set up by defining cells to contain the optimal solution values for the variables x_1 to x_n (these are called adjustable cells), cells to contain the value of the objective function f (this is called the optimal cell), and cells to contain the constraint equations g_1 to g_m (these are called constraint cells). Figure 26.22 provides a worksheet with two example optimization problems.

The first example is a one variable optimization problem with no constraints. Cell B4 contains the adjustable value and B5 the optimal value. The constraint equation in cell B6 is a logical equation that requires that the first derivative equal zero.

```
+12*B4^2+18*B4-12=0
```

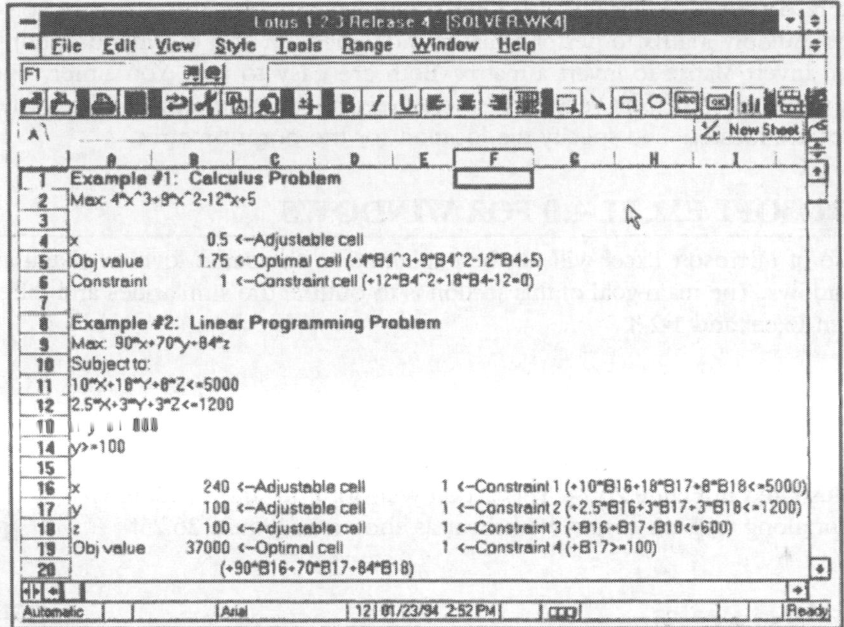

FIGURE 26.22 Two optimization problems.

This equation has the value 1 if it is true and the value 0 if it is false. The solver automatically adjusts the value in cell B4 in order to maximize the value in B5 while satisfying the constraint in B6.

The second example is a linear programming problem. In this case, there are three adjustable cells for the values *x*, *y*, and *z*, one optimal cell for the objective value, and four constraint cells. Again, the constraint cells contain logical conditions that are either true (value = 1) or false (value = 0).

To solve either of these equations, choose Range Analyze Solver. This will produce the dialog box in Figure 26.23. In this dialog box, you are to specify the range address of the adjustable cells (B16..B18 for example 2), the range address for the constraint cells (E16..E19 for example 2), and the range address of the optimal cell (B19 for example 2). You can specify a goal of maximization or minimization of the objective equation and multiple answers. When you click the OK button the optimization process will begin. After completion, another dialog box will appear to allow you to look at alternate solutions and to produce a variety of reports.

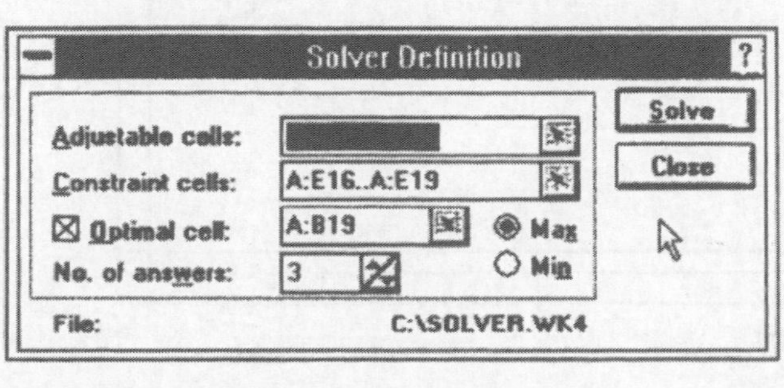

FIGURE 26.23 The Solver definition dialog box.

Matrix operations: Finally, 1-2-3 for Windows provides the menu option Range Analyze Multiply Matrix to perform matrix multiplication and the menu option Range Analyze Invert Matrix to invert a matrix. Both are easy to use. You simply enter the matrix values into adjacent cells of the worksheet, specify the range address where the matrices are located, and specify the location for the output matrix.

26.3 MICROSOFT EXCEL 4.0 FOR WINDOWS

The use of Microsoft Excel will not be explained in the same level of detail as 1-2-3 for Windows. The main goal of this section is to outline the similarities and differences between Excel and 1-2-3.

Starting Microsoft Excel

From the Windows Program Manager, double click the Microsoft Excel icon illustrated in Figure 26.24. This action will bring up the Excel environment along with an empty worksheet as shown in Figure 26.25.

FIGURE 26.24 Microsoft Excel icon.

Worksheet Basics

Window operations: Like 1-2-3, Excel for Windows is a typical windows application.

SmartIcons: Excel has a Toolbar analogous to the SmartIcons of 1-2-3. This is a row of special icons to automate certain menu selections. Besides the standard Toolbar there are eight specialized Toolbars dealing with commands for formatting, drawing, charts, utilities, Excel 3.0, and macros. Choose the Toolbars command under the Options menu to select other Toolbars for display. Excel Toolbars can also be customized.

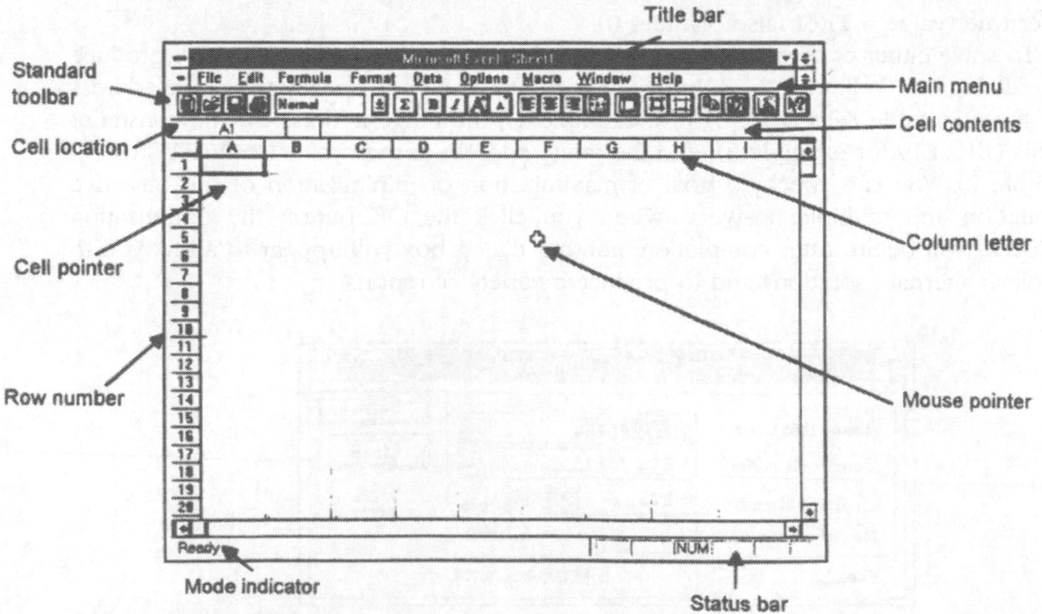

FIGURE 26.25 An empty Microsoft Excel worksheet screen.

Worksheets: Excel worksheets are set up basically the same as those of 1-2-3; however, Excel does not have the worksheet tab feature of 1-2-3. Worksheet documents can be grouped together under files called Workbooks, and an entire worksheet, or sections of a worksheet can be also be linked to other worksheets to maintain data integrity. These linking features are also available in 1-2-3 and Quattro Pro but will not be discussed in this chapter.

Cells: A cell pointer highlights the active cell. The bar beneath the Toolbar, called the Formula Bar, displays the location and contents of the active cell. Cells are addressed as in 1-2-3 and pointed to by clicking the mouse pointer over the desired cell.

Ranges: A range address in Excel is composed of the cell address in the upper left corner, a colon (:), and the cell address in the lower left corner (e.g., A1:F3). Select a range by pointing to one corner and dragging to the opposite corner as in 1-2-3. Ranges can be typed in by using the GoTo option in the Formula menu.

Specifying a range address in a dialog box: The basic approach to working with Excel is to first select the range to be worked on and to second choose the command to perform the desired actions. Many applications, however, have dialog boxes that allow the selected range to be edited.

Menus: Excel menus are organized somewhat differently from those of 1-2-3. The menu bar depends on what type of document is currently active. For instance, the menu bar for a chart is a little different from the menu bar for a worksheet. The standard menu items for worksheets are as follows:

Menu	Functions
File	Opening, closing, saving, and printing files
Edit	Cutting, copying, pasting, clearing, and changing data
Formula	Changing formulas, using names, selecting cells
Format	Specifying number formats, fonts, visual enhancements, column widths, row widths, and the overall format
Data	Sorting and various database functions
Options	Change settings for display, printing or calculations
Macro	Recording or running a macro
Window	Changing between separate workbook files in separate windows

Context-sensitive menus: Like 1-2-3, Excel provides a context-sensitive shortcut menu for various objects. These menus show the most frequently used commands for the selected item. This is accessed by clicking the right mouse button.

Creating Worksheets

The process for creating a worksheet in Excel is the same as that for 1-2-3. Text (labels in 1-2-3), numbers, and formulas are entered into individual cells.

Text: Text entries in Excel are left-aligned by default. Select the right-align or center icons in the Toolbar to change alignment.

Numbers: Excel, like 1-2-3 treats numeric entries as values and right justifies them in the cell. Numeric entries may contain the following characters: the digits (0–9), + – ( ) , / $ % . E e. Entries starting with $ are automatically formatted as currency; those ending with % are formatted as percentages.

Formulas: Formulas in Excel always begin with an equal sign (=). They use the same set of operators as 1-2-3.

Functions: Excel provides a large number of functions. An Excel function is defined by using the function name followed by a list of parameters in parentheses (no @ symbol as in 1-2-3). Many of the function names in Excel are the same as in 1-2-3, although some are different. A list of available functions can be found in the Paste Function option under the Formula menu. Functions can also be directly incorporated into a cell entry using this command. The Excel version of the formula in cell B6 (Fig. 26.5) follows:

```
=(2*$B$2^2/$B$3)*COS(RADIANS(B5))*SIN(RADIANS(B5))
```

Changing cell contents: In Excel, like 1-2-3, you can change the contents of a cell by retyping the entry or by editing the cell contents in the box beneath the Toolbar.

What if analysis: Excel can be used for the same type of what if analysis as 1-2-3.

Opening, Saving, and Printing Worksheets

Opening a worksheet: Excel like most other windows applications, contains the File menu with New, Open, and Close options for retrieving or starting a worksheet.

Saving a worksheet: The file menu option also contains the Save and Save As options for saving a worksheet in a disk file.

Printing a worksheet: The File menu option Print provides a dialog box so that you can print an entire worksheet or selected pages of a worksheet. To print a range that is not formed by entire pages, select the range and then use the Set Print Area option from the Options menu. Like 1-2-3, you can choose the Print Preview option to first review what is to be printed. Icons are available on the left side of the Toolbar to perform the opening, saving, and printing functions.

Modifying Worksheets

Inserting rows and columns: To insert additional space into the worksheet, use the mouse to select a range where you wish space inserted. Choose the Insert option from the Edit menu. This brings up a dialog box that allows you to choose to insert an entire row or column or to just shift the selected range down or to the right. As in 1-2-3, information previously in this space will be moved and formulas adjusted appropriately.

Deleting rows and columns: To delete, select the range you wish to delete and choose the Delete option from the Edit menu. As above, this will bring up a dialog box that will allow you to choose to delete the entire row or column or just the range selected with the option of shifting cells up or to the left. Make your choice and click to OK button. Any formulas referring to deleted data will place an error value in the cell. An error value begins with the number symbol (#) and is followed by a brief notation to describe the nature of the error.

Cutting, Copying and Pasting Worksheet Data

Cutting, copying, and pasting in Excel works like 1-2-3. These menu choices are under the Edit menu as in 1-2-3. Labels and numbers are copied exactly. In formulas, relative cell addresses are adjusted during a copy operation and absolute cell addresses are copied exactly. Absolute cell addresses are defined using a $ sign in front of the column letter and the row number as in 1-2-3.

Styling Worksheets

The styling options in Excel are available under the Format menu. To apply different styles to a cell or group of cells, first select the cells and then choose the appropriate option in the Format menu.

Adjusting column widths and row heights: Choosing the Column Width or Row Height options from the Format menu will bring up a dialog box that allows you to change these items. A convenient choice in the Column Width dialog box is Best Fit. Choosing Best Fit will automatically change the column width to fit the longest entry in the column. Adjustments can also be made by moving the mouse pointer to the dividing lines at the top of a column or the left of a row. Clicking and dragging with the mouse changes the size of the row or column.

Adjusting number formats: Choosing the Number option under the Format menu brings up a dialog box with many sample number formats. Move the mouse until the appropriate format is highlighted and click the OK button to execute the changes. You can easily add a custom format by just typing the format in the code box in the Number dialog box and clicking OK.

Adjusting text formats: Choosing the Alignment option under the Format menu brings up a dialog box that allows you to change horizontal or vertical alignment or orientation of text in cells. Choosing the Font option brings up a box that allows you to change font types and sizes for the selected cells. You can also choose the styles of bold, italics, underline, strikeout, outline, or shadow. The tool bar also has shortcut buttons for bold, italics, and for changing font type and size.

Other styling options: The option Border under Format allows you to add a border with various pattern options to the worksheet. The option Patterns provides a choice of patterns for cell shading. The option Style allows you to define or apply a previously defined style to a selected range.

AutoFormat: Choosing the AutoFormat option under the Format menu brings up a dialog box that allows you to select one of fourteen predefined styles to format an entire tabular worksheet. There is also an AutoFormat button on the toolbar.

Graphing

Creating graphs: A convenient way create a graph in Excel is to use the ChartWizard tool from the Toolbar. This is the second icon from the right on the Toolbar. To use this tool first select the range of data to be graphed. This can include cells containing text that will be used for labels. Next click the ChartWizard button and then select the range on the worksheet where you wish the graph to appear. Excel then produces a series of dialog boxes to lead you through the creation of the graph. The first dialog box (top of Fig. 26.26) asks you to verify or change the data range. The next box (bottom of Fig. 26.26) asks you to choose a general graph type (Area, Bar, Column, and so on); the next box (not shown) gives you choices of variations on the general type. Subsequent boxes allow you to add titles, legends, axes labels, and other styling features.

Another way to create a graph is to first select the range of data to be graphed and then choose the Chart toolbar from the Toolbars command on the Options menu. The

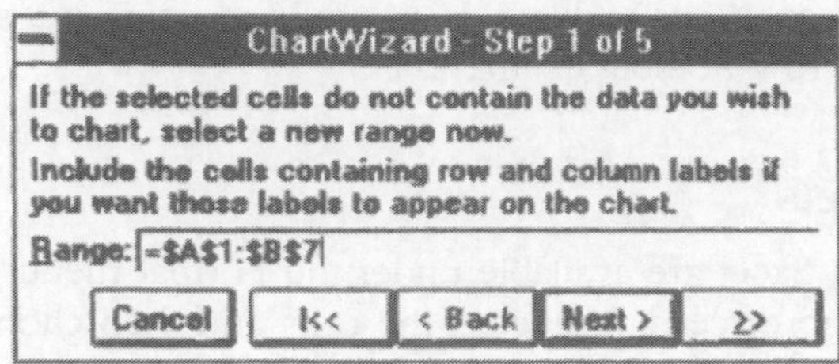

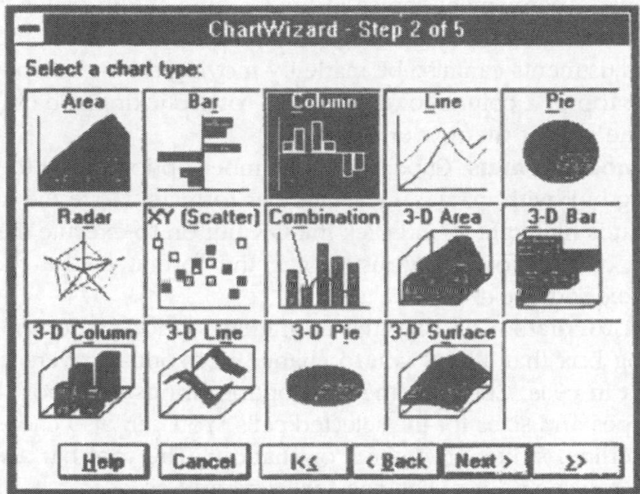

FIGURE 26.26 First two steps using the ChartWizard..

Chart toolbar has buttons for selecting the type of graph, for moving to the ChartWizard, for adding legends and text, and other options.

Enhancing graphs: Once a graph has been created in a worksheet, it can be selected by clicking on the graph. This will bring up a chart menu that contains various menus for editing and enhancing graphs.

Printing graphs: Graphs can be printed separately or as part of the worksheet. To print the graph separately, double-click on the graph to make it active and choose the option Print under the File menu.

Analysis Tools

Excel provides an Analysis ToolPak that provides a set of specialized tools for engineering applications and statistical analysis.

Regression analysis: Choose the Analysis Tools command under the Options menu to perform regression analysis calculations. As in 1-2-3, a dialog box is provided to enter the range address of the independent variables, the dependent variables, and the output report. After clicking the OK button, a report will be provided in the output range of the worksheet.

Solver: Excel also provides optimization capability. Like 1-2-3, the worksheet must be set up with cells for the variables being optimized and with a cell for the objective value; however, constraints are not represented in a single cell. Rather, constraints are specified in the Solver dialog box.

The worksheet should be set up as follows. The cell representing the function to be optimized should be a formula containing references to the cells containing the variables to be optimized. Then each constraint cell should contain a formula or cell reference representing the left side of the constraint. Once the worksheet is set up in this format, choose the option Solver under the Formula menu to bring up the Solver Parameters dialog box illustrated in Figure 26.27. The cells in this dialog box refer to Example 2 in Figure 26.22 with cells E16:E18 containing formulas representing the left side of the constraint inequalities.

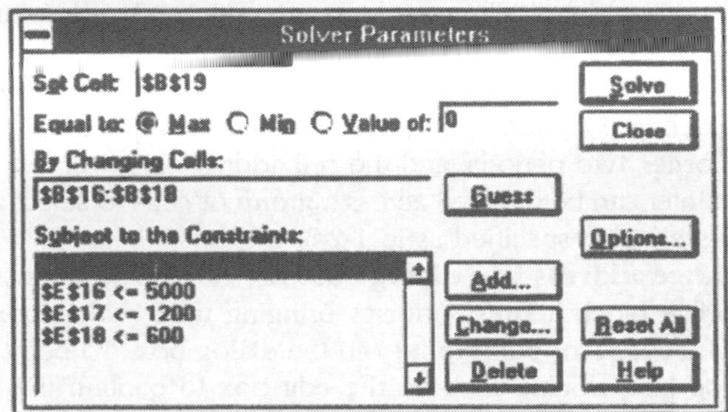

FIGURE 26.27 The Solver Parameters dialog box.

Next, fill in the address of the cell to be optimized in the Set Cell edit box and the range of the cells, which can vary in the By Changing Cells edit box. Choose the Add option to add constraints. This brings up the Add Constraint Dialog Box that allows you to enter the right hand side values and cell reference for the left side of each constraint. By clicking the Solve button, the values in the specified variable cells will be adjusted to solve the optimization problem.

Other analysis features of Excel: Excel provides a number of additional features of interest to engineers. These include analysis of variance, correlation analysis, exponential smoothing, Fourier transforms, moving averages, histograms, *F* tests, *T* tests, and *Z* tests.

26.4 BORLAND QUATTRO PRO 5.0 FOR WINDOWS

The use of Quattro Pro (QP) will not be explained in the same level of detail as 1-2-3 for Windows. The main goal of this abbreviated section is to outline the similarities and differences between QP and 1-2-3.

Starting Quattro Pro

From the Windows Program Manager, double-click the Quattro Pro icon illustrated in Figure 26.28. This action will bring up the QP environment along with an empty worksheet as shown in Figure 26.29.

Worksheet Basics

FIGURE 26.28 Quattro Pro icon.

Window operations: Like 1-2-3, QP is a typical windows application.

SmartIcons: QP provides the Speedbar, two rows of special icons to automate certain menu selections. The Speedbar can be customized to meet special needs.

Worksheets: QP uses a notebook metaphor to organize a series of worksheets as pages of the notebook. Each page is divided into lettered columns and numbered rows like 1-2-3. QP automatically starts with multiple notebook pages that can be accessed by clicking on the tabs along the bottom.

Cells: A cell pointer points to the current cell. Its location and contents are displayed in boxes beneath the Speedbar. Cells are addressed as in 1-2-3 and pointed to by clicking the mouse pointer over the desired cell.

Ranges: QP uses the term block, instead of range, to refer to a rectangular group of cells, but the blocks are identified by a block address composed of the cell address in the upper left corner, two periods, and the cell address in lower left corner as in 1-2-3. The mouse pointer can be dragged across a group of cells to select a block. Three-dimensional blocks are also specified as in 1-2-3.

Specifying a range address in a dialog box: As with 1-2-3, it is usually easiest to select the appropriate block address prior to bringing up a dialog box, but you can also type the block address or point to it from the dialog box. To point to a block address from a dialog box, double-click on the edit box to contain the block address. The dialog box will collapse leaving only a title bar on the screen. Next, drag across the desired cells and click the maximize button on the dialog box title bar. The dialog box will return with the appropriate block address filled in.

Menus: The QP menus are organized a little differently.

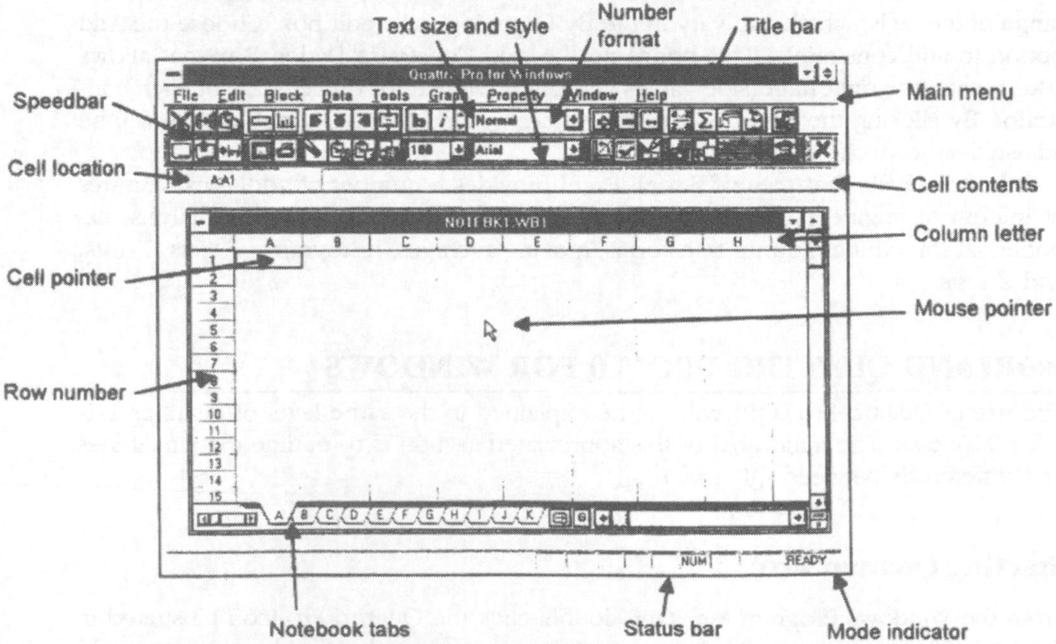

FIGURE 26.29 An empty Quattro Pro notebook screen.

Menu	Functions
File	Opening, saving, and printing notebook files
Edit	Cutting, copying, pasting, deleting, and changing data
Block	Moving blocks and pages, deleting columns, and deleting rows
Data	Sorting and various database functions
Tools	Defining macros, combining worksheets, analyzing data, and designing user interfaces
Graph	Defining graph types, data, and styling
Property	Defining properties of various notebook objects
Window	Navigating between separate notebook files in separate windows

Context-sensitive menus: Like 1-2-3, QP provides a context-sensitive pop-up menu for various objects such as a notebook page, the selected cell, or a selected block of cells. This is accessed by clicking the right mouse button. QP provides almost every conceivable function via these context sensitive menus.

Creating Worksheets

The process for creating a notebook in QP is identical with that for 1-2-3. Labels, numbers, and formulas are entered into individual cells.

Labels: In QP, like 1-2-3, text entries are treated as labels and the same label prefix characters are used to position the text entry in the cell.

Numbers: QP, like 1-2-3, treats numeric entries as values and right justifies them in the cell.

Formulas: Formulas in QP are entered in the same way as in 1-2-3 and use the same set of operators.

Functions: QP also provides a large number of functions. QP functions also begin with the @ sign, followed by the function name and a list of parameters in parentheses. Some of the function names are the same in both QP and 1-2-3 and some are different. Some function are available in one applications and not in the other. As an example, the formula in cell B6 (Fig. 26.5) would appear as follows in QP.

```
(2*$B$2^2/$B$3)*@COS(@RADIANS(B5))*@SIN(@RADIANS(B5))
```

Changing cell contents: In QP, like 1-2-3, you can change the contents of a cell by retyping or editing the cell contents in the box beneath the Speedbar. The option of editing the cell contents on the worksheet is not available in QP.

What if analysis: QP can be used for the same type of what if analysis as 1-2-3.

Opening, Saving, and Printing Worksheets

Opening a worksheet: QP, like 1-2-3 and most other windows applications, contains the File menu with New, Open, and Close options for retrieving or starting a worksheet.

Saving a worksheet: The File menu option also contains the Save and Save As options for saving a worksheet in a disk file.

Printing a worksheet: The File menu option Print provides a dialog box similar to 1-2-3 so that you can print an entire worksheet or a selected range of a worksheet. Like 1-2-3, you can choose to Preview the worksheet before the actual printing.

As with 1-2-3, icons are available on the second row of the Speedbar, on the left side, to perform the opening, saving, and printing functions.

Modifying Worksheets

Inserting rows, columns, and sheets: To insert additional space into the worksheet, use the mouse to select a range address that includes the rows, columns, or notebook pages you wish to insert and choose the menu option Block Insert. This will display a cascading menu with the choices Rows, Columns, and Pages. Make the selection you want, check the range address given in the dialog box, and click the OK button. As in 1-2-3, information previously in this space will be moved and formulas adjusted appropriately.

Deleting rows, columns, and sheets: To delete, select the rows, columns, or notebook pages you wish to delete and choose the menu option Block Delete. As above, a cascading menu will be provided so that you can specify whether to delete rows, columns, or pages. Make your choice and click the OK button. Any formulas referring to deleted data will display a zero as the cell value and the symbol ERR in the formula.

Cutting, Copying, and Pasting Worksheet Data

Cutting, copying, and pasting in QP works just like 1-2-3. These menu choices are under the Edit menu as in 1-2-3. Labels and numbers are copied exactly. In formulas, relative cell addresses are adjusted during a copy operation and absolute cell addresses are copied exactly. Absolute cell addresses are defined using a $ sign in front of the column letter and row number as in 1-2-3. Icons are available in the first row, left side, of the Speedbar to automate these actions.

Styling Worksheets

All of the styling options in QP are available by selecting the menu option Property Current Object. This action will bring up the dialog box in Figure 26.30. If a block of cells is selected, the block address will appear in the title bar and all actions you take will apply to the highlighted block of cells.

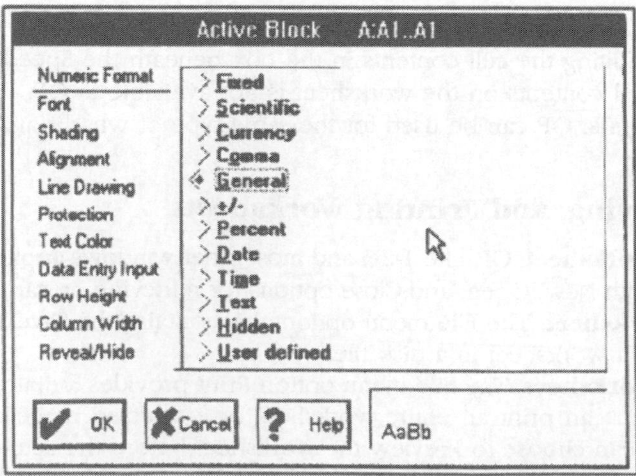

FIGURE 26.30 The Block Properties dialog box.

Adjusting column widths and row heights: To adjust the column width of the active block, click the Column Width button along the left side of the dialog box in Figure 26.30. This will cause a modification in the right half of the dialog box that will allow you to type in a column width or to automatically fit the column width to the largest entry in the column. Once the column width has been specified, you can change other block properties. Finally, select the OK button to execute the changes.

Similarly, the row height is adjusted by clicking the Row Height button and filling in the row height desired.

Adjusting number formats: Number formats can also be adjusted from the Block Properties dialog box (Fig. 26.30). Initially, this dialog box has the Numeric Format option selected and the various formats listed as illustrated in Figure 26.30. Simply select the format you want and click the OK button. Some options, such as Fixed, will produce additional edit boxes to specify the number of decimal places.

Another way to specify the numeric format is to use the list box under the Property menu option. By clicking the arrow to the right of this box, you can choose from a list of common formats. This provides a short cut, similar to the one available in 1-2-3, for specifying a numeric format.

Adjusting text formats: Text formats can also be adjusted from the Block Properties dialog box (Fig. 26.30). By clicking the Font button on the left side of the screen, you can specify the typeface, the point size, underlining, boldfacing, or italics. You can also specify these properties using the shortcut buttons (b, boldface; i, italics; arrows, point size) beneath the Graph menu option and the second list box (typeface) under the Property menu.

Other styling options: The Shading, Alignment, Line Drawing, and Text Color buttons on the left side of the Block Properties dialog box (Fig. 26.30) provide additional styling options similar to those provided with 1-2-3.

Worksheet defaults: Default styling options can be established for an entire page of the notebook or the entire notebook (all worksheets) by selecting the desired range and choosing the menu option Property Current Object to bring up the dialog box in Figure 26.30.

Graphing

Creating graphs: Creating graphs in QP is a little different than in 1-2-3. First, all graphs in QP are created in a separate window. If you recall, in 1-2-3 graphs were created in a specified range of the worksheet. Second, each graph has a name and an associated icon. These icons are displayed on the last page of the notebook (which is easily accessible by clicking the right arrow button at the bottom of the worksheet). To open a previously saved graph window, just double-click on the appropriate graph icon. A graph created in a separate window can later be moved to the worksheet by selecting the menu option Graph Insert.

To create a new graph in a separate window, choose the menu option Graph New. QP will display the Graph Series dialog box illustrated in Figure 26.31.

Specifying values to graph: Fill in the graph name and the edit boxes of the Graph Series dialog box (Fig. 26.31). The *X*-axis range should contain the data values to appear along the *X* axis and the 1st, 2nd, and other ranges should contain the numeric values to graph on the *Y* axis. To point to the desired range address, double-click on the appropriate edit box and highlight the desired range address. When the proper

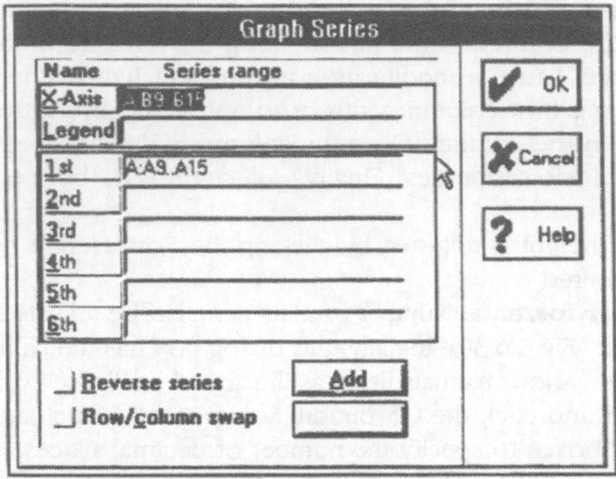

FIGURE 26.31 The Graph Series dialog box.

data have been highlighted, return to the dialog box and click the OK button. The graph will appear in a separate window.

Specifying graph types: Initially, the graph will appear as a bar graph. To select another graph type, choose the menu option Graph Type. A dialog box will allow you to select from a variety of graph types. Make your selection and click the OK button. The data will then be displayed using the newly specified graph format.

Enhancing graphs: Titles and axis labels can be added to the graph by selecting the menu option Graph Titles and filling in the appropriate edit boxes. General properties of the graph can be specified by choosing the menu options Property Graph Window, Property Graph Setup, or Property Graph Pane. The characteristics of individual objects on the graph can be controlled by clicking on the desired object and choosing the menu option Property Current Object. QP allows the user to control almost every aspect of the graph. In addition, drawing tools are provided to add annotations and other enhancements to the graphs.

Printing graphs: Graphs can be printed separately or as part of the worksheet. To print the graph separately, make the graph window active and choose the menu option File Print. To print the graph as part of the worksheet, insert the graph into the worksheet and print the worksheet using the File Print menu option.

Analysis Tools

Regression analysis: Choose the menu option Tools Advanced Math Regression to perform regression equation calculations. As in 1-2-3, a dialog box is provided to enter the range address of the independent variables, the dependent variables, and the output report. After clicking the OK button, a report identical to that in 1-2-3, will be provided in the output range of the worksheet.

Solver: QP also provides optimization capability. Like 1-2-3, the worksheet must be set up with cells for the variables being optimized, with a cell for the objective value, and with cells representing the constraints. In 1-2-3, the constraint equations are represented in a single cell by logical formulas. In QP, the constraint equations are represented in two cells, one containing the formula on the left side of the constraint and the other containing the constant on the right side of the constraint. Once the worksheet is set up properly, choose the menu option Tools Optimizer to bring up the dialog box in Fig. 26.32. Next, fill in the appropriate block addresses. The blocks specified in Figure 26.32 are designed to solve the optimization Example #2 in Figure 26.22 with the constraint equations modified so that the right side values of the constraints are given in F16..F19. By clicking the Solve button, the values in cells B16..B19 will be adjusted to solve the optimization problem.

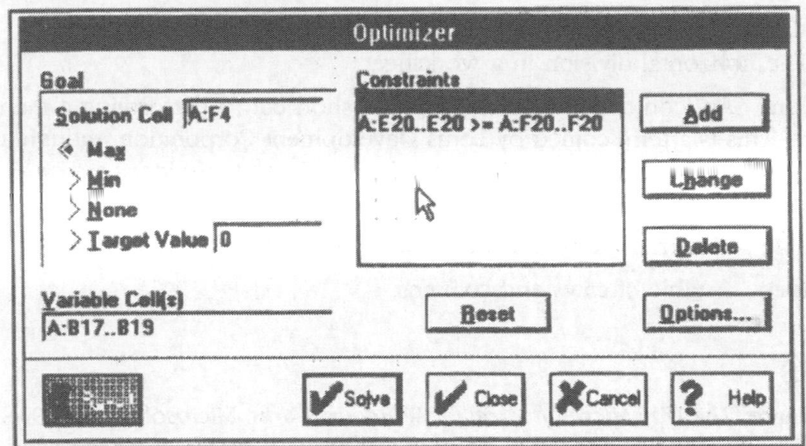

FIGURE 26.32 The Optimizer dialog box.

Matrix operations: QP, like 1-2-3, provides the menu option Tools Advanced Math Multiply and Tools Advanced Math Invert to perform matrix operations.

Other analysis features of QP: QP provides a number of additional features that would be of interest to engineers. These include exponential smoothing, Fourier transforms, histograms, moving averages, analysis of variance, correlation analysis, *F* tests, *T* tests, and *Z* tests.

DEFINING TERMS

Absolute Address: A cell address in which the column letter and/or row number is preceded by a $ sign. This address will be copied exactly during a copy operation.

Block: A term used by Quattro Pro to describe a rectangular group of cells.

Cell Pointer: The box that outlines the current cell.

Cell: The intersection of a row and a column in a worksheet.

Column: The vertical division in a worksheet.

Electronic Spreadsheet: A computer program used to create one or more tables of information organized into rows and columns.

Formula: An equation that specifies how the value in a cell is to be computed from the values in other cells.

Function: A predefined procedure to perform a certain calculation based on other values in the worksheet.

Label: A text entry in a cell.

Logical Formula: An equality or inequality that produces a true or false value in a cell.

Notebook: A term used by Quattro Pro to describe a group of worksheets.

Number: A numeric entry in a cell.

Range: A term used by 1-2-3 and Excel to describe a rectangular group of cells.

Relative Address: A normal cell address containing a column letter followed by a row number. This address will be adjusted during a copy operation.

Row: The horizontal division in a worksheet.

SmartIcon: An icon that can be clicked as a short cut to performing a menu selection. This is a term coined by Lotus Development Corporation, though all of the spreadsheet packages have this capability. In Excel, these icons are on Toolbar(s) and in Quattro Pro these icons are on Speedbar(s).

Workbook: A term used by Excel to describe a group of worksheet files.

Worksheet: A table of rows and columns.

REFERENCES

Brynly Clarke, *The Way Microsoft Excel for Windows Works*. Microsoft Press, 1994.

Douglas Hergert and Sheldon Dunn, *Understanding 1-2-3 Release 4 for Windows*.

Brian Underdahl, *Using Quattro Pro 5.0 for Windows*. Que Corporation, 1993.

User's Guide:*Borland Quattro Pro for Windows*, Version 5.0. Borland International, 1993.

User's Guide:*Lotus 1-2-3 Release 4 Spreadsheet for Windows*. Lotus Development Corporation, 1993.

User's Guide 1 and 2: *Microsoft Excel Spreadsheet with Business Graphics and Database*, Version 4.0. Microsoft Corporation, 1992.

27

dBASE IV

Dan Duricy
Miami University

DATABASES

Until the advent of the microcomputer and large storage capacity, fast and economical disk drives, databases were the province of the mainframe computer. With new developments in technology and easy-to-use software such as Ashton-Tate's dBASE IV or Microsoft's Access, databases are now available to any microcomputer user.

Databases can be used for many applications such as accounting and inventory systems that once required custom software. With a database language, these applications can be developed with much less effort and cost than would have been required to write the program in a traditional language such as COBOL, C, or Pascal.

You will remember that a spreadsheet is the electronic equivalent of the columnar pad used by accountants and financial analysts. In performing one of these tasks, you have organized numerical information into a series of rows and columns. You have then performed some sequence of mathematical operations on the various entries on the sheet.

There are two fundamental limitations to the spreadsheet:

- The calculations (formulas) are embedded in the spreadsheet.
- Adding more rows or columns to the spreadsheet often requires a complete overhaul of the spreadsheet design.

The fact that the program (calculations and formulas) is intermingled with the data and that the spreadsheet is finite in size leads to some severe limitations when more flex-

0-8493-2530-7/96/$0.00+$.50

ibility or large quantities of data must be managed. It is a simple process to separate the data from the calculations to form the more general idea of a database management system.

The essential idea of a database management system is that of a system of programs to group data together into a series of records, much like filing or ledger cards. A database management system allows you to add to the information in the database, alter it, delete it, and organize the database in different ways, giving you reports or information that you might require.

A database system offers many advantages over manual recordkeeping systems. Information needs to be stored in only one place. For instance, name and address information can be stored in one file along with account numbers. An accounting system can be organized separately and the account number information can be used to obtain the name and address information from the other database.

The means that if it is necessary to update name or address information, it needs to be updated only on one file. This leads to reduced effort and increased accuracy in maintaining files of information. With a database, information can be stored in a single place, or be given the appearance of having stored information only in a single place, even though it may be scattered about in a number of files.

Relational database systems such as dBASE IV and Access have the ability to work with more than one file at a time. This ability to tie or relate material from two or more files is what gives the relational database system its name.

27.1 INTRODUCTION

The goal of this section is to familiarize you with the beginnings of database theory as well as Database Management Systems in general. Users of computers have been adapting their business and personal processes to the ways of microcomputers through a variety of software packages. After using word processing and spreadsheets, the development of database applications becomes the focus for applying the power of the microcomputer to the numerous files maintained for business and personal use. Offices abound with files. These files are constantly created, maintained, and processed. The decision to place these files on a computer is made when the manual means of using files proves too slow, too error prone, and too labor intensive. The software that will manage these files is referred to as Database Management Systems (DBMS). In general, this software allows the creation of a computer file that is designed to hold the information as it appears in the manual file. The computer file can have records added, deleted, and edited. The computer file can be processed to produce query outputs and many different reports based upon the needs of the user.

27.2 ASSIST MODE

Also, we are going to walk through the range of functions that dBASE IV offers. dBASE was initially developed by Ashton-Tate Corporation and then was recently reworked by Borland. There are many different ways to manipulate files and information through dBASE. One way to work with dBASE is through the Assist mode. This is a simple and "user-friendly" designed system that uses a computerized desktop called the Control Center. From the Control Center (Fig. 27.1), the user has access to several different menus.

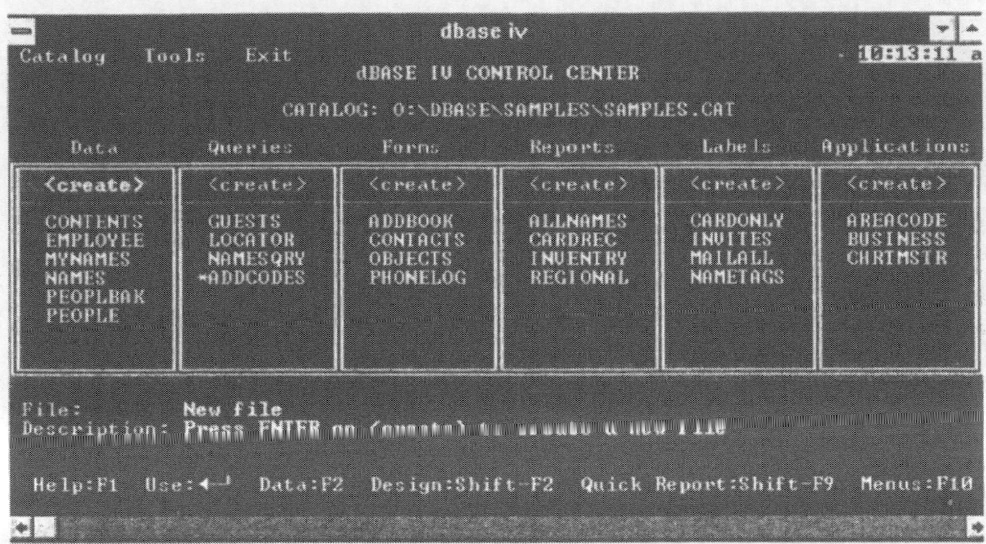

FIGURE 27.1 dBASE IV Control Center.

These menus allow users to do the following:

1. Design databases and organize them into catalogs.
2. Name fields.
3. Update records.
4. "Query" or search through databases.
5. Layout reports and mailing labels.
6. Access and create database applications.

27.3 DOT PROMPT

For those users that wish to delve more thoroughly into the full potential of dBASE IV, there is an interactive database interface called the Dot Prompt (Fig. 27.2). At this level, users can

1. Enter commands directly to dBASE.
2. Write Index files to put databases in order.
3. Make "ad hoc" reports and queries.

27.4 PROGRAMMING LANGUAGE

For those users that are willing to learn a new language, there is another more advanced level of interface that dBASE offers. This advanced level is that of structured programming (Fig. 27.3). By writing separate independent .PRG files, users can develop their own customized source code. This code is compatible with the XBASE family of programming languages. The XBASE standard is also used with Microsoft's FoxPro and Nantucket's Clipper. Using the dBASE IV programming language gives the user the ultimate in flexibility. Often users create their own versions of "user-friendly" menus. These menus offer a program that requires absolutely no prior knowledge of dBASE IV. Regardless of whether you are interested in simple data features or complex menu-driven programs, dBASE IV offers the power and precision to meet your needs for storing, sorting, manipulating, and using information.

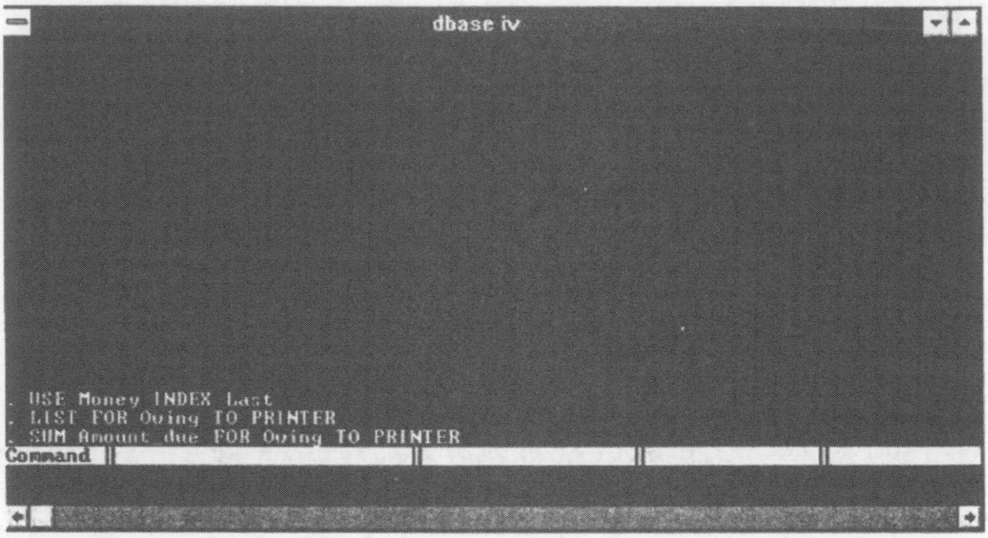

FIGURE 27.2 Dot Prompt.

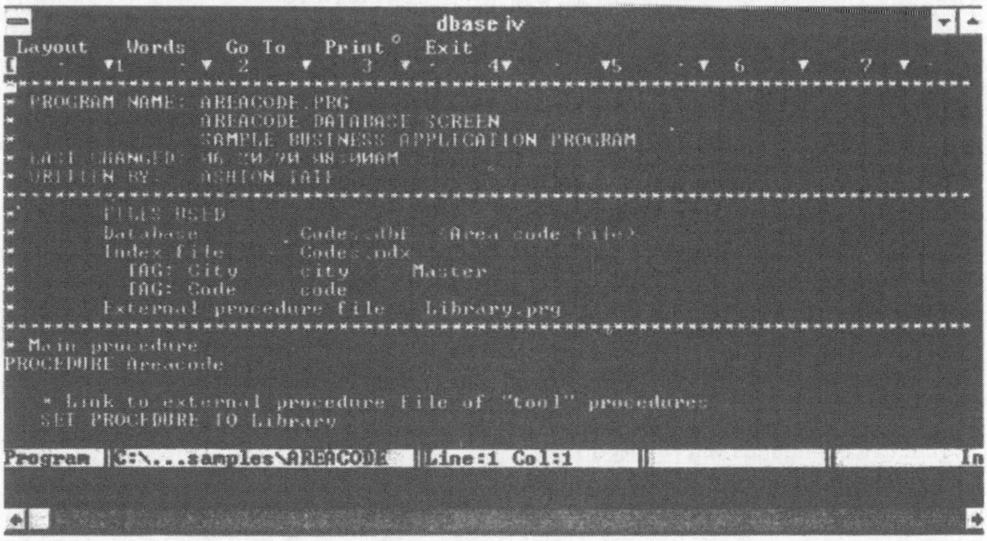

FIGURE 27.3 Structured programming.

27.5 STARTING REQUIREMENTS

To use dBASE IV, you will need to organize its files onto your hard drive. This first step is the installation process and is going to mean you will work with the DOS operating system in a limited way. The manual, "Getting Started With dBASE IV—DOS Version," sets out the steps to install dBASE IV on your system.

27.6 SYSTEM REQUIREMENTS

The system requirements indicate dBASE IV (DOS version) will run on computers based upon the Intel microcomputer processor, PC/XT and above, or 100% compatible clones and using DOS operating systems compatible with PC-DOS 2.0 or Compaq

DOS version 3.31. The files will occupy approximately 3.5 Megabytes on the hard disk and your computer must have a hard disk drive to run dBASE IV. The installed RAM must be 640 Kilobytes. This version of dBASE can use expanded and extended memory to improve its performance. according to the material provided by Ashton-Tate. This process (the single-user system) will include creating directories for dBASE files, tutorials, and work space. Make sure to use the program DBSETUP to configure dBASE properly for the particular aspects of your system. Your setup results will appear in a dBASE file called Config.db. This file contains startup information and sets the view you will see. The file could appear as follows:

```
**   dBASE IV Configuration File
*  Friday October 14, 199X
*
CATALOG              =
o:\dbase\samples\samples.cat
COLOR OF NORMAL      = W+/BG
COLOR OF HIGHLIGHT   = N/W
ECHO                 = ON
FILES                = 99
FULLPATH             = ON
PATH                 =
o:\dbase;o:\dbase\samples
PRINTER              = ON
PROMPT               = db4?
SQL                  = OFF
STATUS               = ON
```

27.7 RUNNING THE PROGRAM

Once everything is organized, you can call up the program by entering **dbase** at the DOS command prompt. Based upon your dBASE configuration file, you program will start at the Dot Prompt or go immediately into the Control Center Menu. To enter the menu system from the Dot Prompt you type **assist**. First you should see an opening screen, then the Control Center. The Control Center shows the default catalog and its DOS path. The example (Fig. 27.1) shows O:\DBASE\SAMPLES\SAMPLES.CAT or it will be whichever directory you set up during installation such as C:\dbase\work\catalog.cat.

27.8 WORKING ENVIRONMENT

The Control Center is like a screen telling you what computer file cabinet you are looking at. So a catalog is a group of files; you can create several different catalogs just like having different cabinets. You might have one catalog for inventory files and another for personnel files, or different catalogs for different corporate divisions. To better understand the many different features of dBASE IV, we can look at the traditional processes involved in bookkeeping and filing. dBASE was designed to automate traditional processes used in offices. Let us look at the largest unit of a theoretical database, the catalog. In a traditional office, think of where data are found. Before computer technology, most data were found in filing cabinets. Filing cabinets provide a common place for all of the companies files. Anyone with access to the filing cabinet can look through the different files and pick which ones to use.

27.9 CONTROL CENTER

The Control Center window is composed of six columns and three menu items. The first menu item is Catalog.

Catalog		Tools		Exit	
Data	Queries	Forms	Reports	Labels	Applications

Catalog

Highlighting this item and pressing return brings up the Catalog menu (Fig. 27.4).
From this location, you can use the following features:

1. Choose which catalog to use.
2. Change the name of one of your catalogs.
3. Make and change descriptions for catalog ("label your file cabinet").
4. Add and remove files to and from the catalog you are using.
5. Change the descriptions of your files within the catalog you are using.

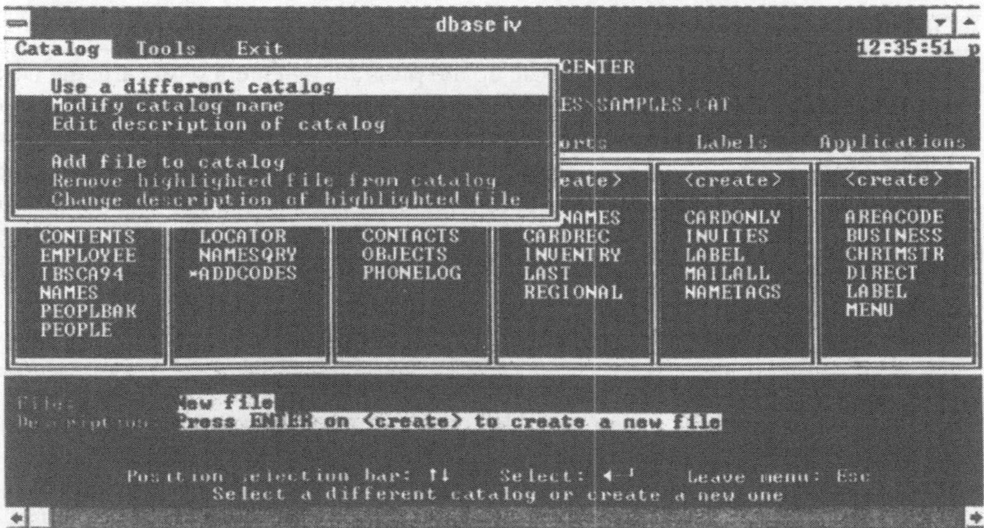

FIGURE 27.4 Catalog menu.

Tools

The Control Center offers another menu titled Tools (Fig. 27.5). This menu offers the users several ways to control the environment of dBASE IV. You can

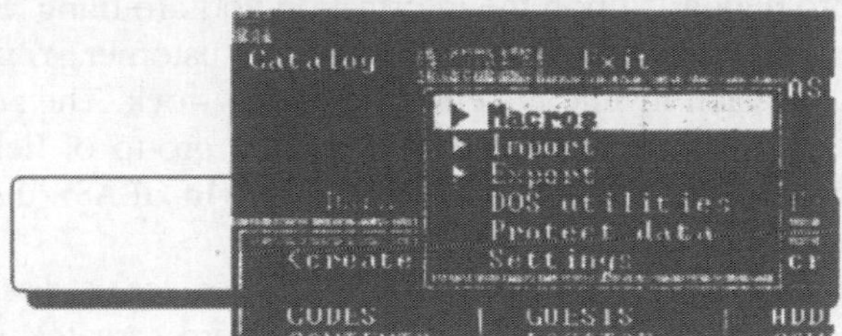

FIGURE 27.5 Tools menu.

1. Record keystrokes to form "macros" that save you time and repetitions.
2. Import files from other computerized databases.
3. Export dBASE IV files for use in other programs.
4. Protect your database for security purposes.

Exit

The final menu in the Control Center allows users to save their work and exit dBASE, returning to DOS.

27.10 ACTIVITIES

Now let us look at the Six Columns that correspond to the six major functions of dBASE.

```
Data    Queries    Forms    Reports    Labels    Applications
```

Data

The first column is the Data column; if there are database files in the active catalog they will be listed vertically in the Data column. Think of the traditional office; if the catalog is the file cabinet, then the Data column is a list of the files in the cabinet. The database file that is at the top of the list, above the line, is the one currently "in use." Any of the files can be put into use simply by highlighting the name and pressing enter. That file will go to the top of the list and will be activated. The description of your database is often called the schema. To set up the schema, you will need to define the characteristics of the file's records. A record is one part of a file. In a traditional office, the purchasing file would have many invoices. In this example, an invoice would be a record. Records are made up of many fields. For example, the invoice would have several pieces of information on it. Invoice-number, Invoice-date and part-number are all examples of typical fields that would make up each record. Although this might be a little confusing at first, take some time to learn the following basic structure of all databases: The Catalog is a holder for your database files. Grouping your databases (files) according to a particular area in a single catalog helps to organize your work. The catalog name will tell you what database files are available in that catalog. The files are the databases and each file consists of many records. The nature of a record depends upon the information you are filing. For example, an accounts receivable file would have a record for each customer. Your inventory file would have a record for each inventory item you keep in stock. The actual data about a record are described in one or more fields. A related group of fields makes up a record, and a group of related records makes up a file. In dBASE these related files can be organized into a Catalog.

Considering that you are just starting out with databases, let us discuss how to create a new file. Under the heading Data, there is the word ⟨create⟩. Highlight it and press enter. dBASE will move you to a new screen that will allow your to design the features of your new database. dBASE IV will now let you name each field in the file. Use names that have inherent meaning, that way fields are easy to keep track of. Next, you need to decide on a type for each field. There are six different fields types. Character fields have alphanumeric data like names and phone numbers. Logical fields are either true or false. Date fields are organized by day, month, and year. Numeric fields can be used in computations. Memo fields are used to make comments about data contents and Floating Point fields are used for very large numbers. The next part of the schema is the field width. This number denotes how many digits are allotted to each field. Look at the following sample schema:

Fieldname	Fieldtype	Width
FIRSTNAME	CHARACTER	20
LASTNAME	CHARACTER	20
ORDDATE	DATE	8
AMOUNT	NUMERIC	10

In the sample schema, each record would contain one value for each of the four fields. The screen for schema design has four menu items. The first menu item is the Layout. This menu allows you to save, edit, and print out database schemas. The next menu item is very useful for new databases, it is the Append function. Once your have designed your schema, you can highlight Append and go directly into entering the actual content of your database records. The Append function gives you the field names on the screen and you can enter the proper value for each field. The maximum number of records is based on the amount of memory on your individual system. Some databases end up with tens of thousands of different records. Once you have used the Append function to enter these records, you are going to need a way to organize all of this information into a useful order. The next menu item is called Organize. Organize uses index files to sort and order your data. An index file is a DOS file separate from your database. It contains information about how to organize your data in a particular way. For example, with this menu you could create a file called LAST.NDX that would allow you to use your data in alphabetical order according to last name. With traditional offices, data only come in one order. Whatever order the physical records are in, that is the only way to use the data. Computerized databases like dBASE IV allow you to use multiple indexes. For the same database you might also want to create FIRST.NDX, an index for order by first name. The Organize menu has the option to create, activate, remove, and modify indexes.

Queries

The final menu item on the schema design screen is a Goto menu. This menu has different commands that travel through the database, moving from the first to the last record, or go to record #45, etc. You can use this function to find certain records to edit or delete them.

Now that you have created a database and entered however many records you want, there are many different ways of using and accessing your data. Let us look at the next column of the Control Center, Queries. A query is like a question. I like to think of queries as talking to a superfast robot in a traditional office. I might ask the robot, "Give me all of the invoices for purchases above $200." A query is as if the robot instantly went through all of your files and brought you all of the records meeting your requirement. dBASE can make queries in 2 seconds that might take 2 hours in a traditional office system. The format that dBASE IV uses to make queries is called Query by Example or QBE. QBE allows you to look at all of the fields in your database, and then make conditions for searching according to your needs. QBE is very advanced because it uses a graphical approach. Older and less sophisticated database packages take a linguistic approach. To make a query on an older database, the user had to learn a complex Query language. But for our purposes let us look at making queries with QBE. Highlight ⟨create⟩, under the Query Column and press enter.

We are now at the Query Design screen. At the very top of the screen are the four menu items and below that is a horizontal list of the fields in your active database. Under each field you can enter the search condition. For example, under the field amount, you could enter ">5000." This would access all the records for which the invoice amount was greater than $5000. Queries can also be made for specific field values like =5000. One excellent feature of the Query Design screen is that you can construct queries that use fields from multiple databases. Using the Layout menu at the far left of the horizontal menu, you can add additional files to the Design screen. To do this the two files need to have a Key Field in common. The common key allows you to create a combination of the two files called a view. The layout menu lets you.

1. Add and remove files from a view.
2. Make a text description of a query.
3. Write a view into a new database (.dbf file).
4. Save the condition of a query to use it again later.

Forms

So far we have gone through the basics of creating a database and then searching through it. Now, as we look at the Forms Column we can begin to customize how you interact with dBASE IV. Everyone has different and unique needs in terms of working with data. The Third Column is the Forms Column; in this column you can "paint" the screen the way you want it to look when you are working with your database. Highlight the ⟨create⟩ under the Forms Column and press enter. This moves us to the Screen Generator. You can give your screen a title like "Invoice Data Entry screen." Using the Layout menu you can draw boxes and lines, add and delete text, and save your screen. The Fields menu gives you a list of all the fields in the active database. You can define where on the screen you want to put them. dBASE IV even checks to make sure the data are being entered according to editing and formatting rules. Forms are the way to format the way data appear on the screen as the data are being entered and edited.

Reports

The next column is a way to control the format of data when data are outputted. The next column is the Report Column. Highlight the ⟨create⟩ under the Report Column and press enter. Most reports associated with databases share some common features. At the top of the report is the page header. This information is found at the top of every page of the report. The corresponding feature at the end of each page is the page footer. The main section of the report, in which the actual data are presented, is called the detail lines. Most reports also have a section at the end of the last detail line; this part is called the report summary. The report summary often contains totals compiled throughout the detail lines. Another feature is the report introduction, which is a one time section at the beginning. The way that you can design each of these features is by using Report Bands. The Report Generation Screen shows each of the five bands. Only one band can be active at a time. To activate a particular band simply move the cursor to the band and enter whatever text or fields you like. Reports can consist of one through five bands depending on your needs. All reports have a detail band because that is what allows the data to be outputted. Like the Forms Column, the Report Column has a Fields menu that lists the fields in your active database and lets you place them on the detail band. Text is used in the detail band as well as the other bands to explain, label, and contextualize the data. One easy way to get started with reports is with the Layout menu. This menu contains several "Quick Layouts." These are reports that dBASE IV automatically prepares. Although they may not look exactly the way you would design them yourself, they will give you an excellent place to start. The three types of Quick Layouts are vertical column, form, and mailmerge. The Report Generation screen offers ways to

1. Add new fields to a report.
2. Add text to your report.
3. Remove and resize fields and text.
4. Print out your report design.

Labels

The next column of the Control Center is the Label Column, and it is another way to work with the way your data are output. Mailing labels themselves are a feature that make files worth computerizing. The traditional office used handwritten labels or typed on a roll of stickers. dBASE IV offers a quick and professional way to generate mailing labels from your database. To get to the Label screen, highlight ⟨create⟩ under the heading of the Label Column. Like the rest of the columns, this screen has menu items that help you create the type of design you want. At the center of the screen, there is space that represents a typical mailing label. Using the Field menu, you can chose fields from the active database and arrange them on the label. Another useful menu is Dimensions. The Dimensions menu lets you

1. Define the number of labels per page.
2. Set the number of lines between labels.
3. Work with Indentation, heights, and widths.

Using the other menus you can make editing changes like adding text and changing the size of fields on the labels. To use your new labels you are going to need to print them out. To do this you exit the Label screen and go back to the Control Center. Highlight the name of the label file you would like to print and press enter. dBASE IV asks you if you would like to print, and you can just press enter to continue. The print menu has many features; you can

1. Eject a blank page.
2. Generate sample labels.
3. View labels on the screen.
4. Save, change, and specify printer settings.

Applications

Now that we have explored creating databases, ordering data, creating forms, and reports and labels, we have reached the last column of the Control Center, the Applications Column. Applications are structured programs written in the Xbase programming language. They combine all of the different features of the first five columns into a program that requires no prior knowledge of dBASE IV. If you are familiar with current trends in systems analysis, the dBASE IV Applications Generator classifies as a fourth-generation computing language. This means that dBASE prompts the user for information about the desired program and then actually writes the code of the program itself. One good example of this feature is the Quick Application. dBASE will instantly create an predesigned standard application for you. Much like the Quick Layout in the Report Column, this application is easy to generate and very standard in design. The Applications Generator writes Xbase programs to run the necessary menus, searches, and printings. If you are willing to spend some time learning Xbase, then you will be able to edit these automatically generated programs and refine them to meet your unique needs. Programs give you a chance to organize menus that execute dBASE IV commands in the way that you frequently need them.

27.11 PROGRAMMING STRUCTURES

The Dot Prompt

Commands to manipulate your data can also be entered one by one rather than in the context of a program. This is the function of the Dot Prompt. At the Control Center, if you press escape and enter, dBASE will move you to the Dot Prompt. In this environment you can execute all of the features of the Control Center and more. You can do them quickly without being slowed down by all of the Control Center menus. The tradeoff is that the Dot Prompt offers little help and is not user friendly. The full power and scope of dBASE IV can be reached through the "dot prompt," which is an interactive command line interface. When commands are entered, the user receives an immediate response. This mode provides both speed and flexibility in processing database files. Once learned, the dot commands can be entered to provide the user with instant controlled access to create, query, update, output, or administer the database files. The command line may be up to 254 characters in length. Pressing enter executes the command. The commands may be entered in several forms—upper, lower, or mixed case. Many commands may be abbreviated entering only the first four letters. As you enter dot commands, they are saved in a memory buffer. You may go back to run a previous command, or to edit that command to perform a modified version of the original command. To review you actions, enter DISPLAY HISTORY or LIST HISTORY to see the entire contents of the command memory buffer.

To create a new database from the Dot Prompt, type CREATE and then the name of the new file. dBASE will move you to the Schema Design screen and you can proceed just like we did from the Control Center. To add records to your database, type APPEND at the Dot Prompt. This command will activate the input form and let you enter as many records as you like. To activate a database, enter USE and the name of the file at the Dot Prompt. When a database is activated its name and DOS path will appear below the Dot Prompt. You will also see the total number of records in the active database as well as the record that is currently active. To look at the record that is currently active, you can type EDIT at the Dot Prompt. You can use the PGup and PGdn to move through your database from there. Another way to edit your database from the Dot Prompt is with BROWSE. This command shows as many records as will fit on the screen. You can look through multiple records at the same time and make changes to one based on what you see in another. Another useful command is LIST. LIST takes all of the data in the file you are using and prints the data out on the screen. It lets you see the contents of the file. A subcommand of LIST is LIST STRUCTURE; this command shows you a printout on the screen of the schema of the active file. Another way to use the LIST commands is to get printouts. Just add TO PRINT to the end of the command and dBASE will send the contents of the screen to your printer. The SORT command takes the active database and puts it in the order of a key field, and then creates a second database out of that data. For example, you could type SORT ON FIRSTNAME TO FIRST. This would create a new file of records from the old file in order of firstname. You can order your present database without creating a new file through the use of indexes. Remember that index files are separate files that contain information on how to order you database files. You can type INDEX ON FIRSTNAME TO FIRST. This command will create an index that puts the database in order by firstname. To use that index file, type SET INDEX TO FIRST.

The Dot Prompt can direct you to the main screens that we used from the Control Center. Type CREATE REPORT and then the report filename to go to the Report Generation Screen. Type CREATE LABEL and then the filename to go to the Label screen. The same syntax is used for creating input forms.

The .PRG Files

The command lines you enter and execute one at a time may be grouped into a disk file. The disk file may be executed to carry out a long and complex set of steps automatically. The outputs and the system actions are the same as if you had entered the commands at the dot prompt. When executing the program file (.prg) dBASE creates an object code file that runs much faster. This object code file may be created directly from your source code by using the COMPILE command to generate a .dbo file. Many of your dBASE files, such as your existing report, label, or form files, can be accessed by the MODIFY command. This command can be used to create Xbase Programs by typing MODIFY COMMAND and then the name of your program. If the program already exists, you can edit, update, and change the program. For a new file, you use the modify editor, which opens to enter your commands, and then save the file with the .prg extension. Once you have written a program, it can be executed from the Dot Prompt by typing DO and then the name of the program.

For the dBASE user, these programs can be written to perform many repetitive utility tasks such as backing up your .dbf files or can be written to complete full, transaction-driven applications on your databases. Many dBASE IV users will print mailing labels on a regular basis. To run this application, the user would enter DO Label.prg while at the dot prompt. These few words represent a program that might appear in a .prg file:

```
*LABEL.PRG THIS MODULE PRINTS OUT MAILING
LABELS
USE ibdata94.dbf
SET ORDER to zip
LABEL form LABEL to PRINT
CLEAR
RETURN
```

These programming commands tell dBASE IV which database file to use; request that the file be referenced in zip code order, and then to print these labels according to a predetermined format. Many of the actions a user may apply to a database are entering new data into a database, listing the database by an index or a key field, printing the data to various output formats, and querying or searching the database for records that meet certain conditions. Rather than approach these uses as single issues, a dBASE program is often written that allows the user to select the function they wish to do by using a menu of choices. The menu program given allows a choice between printing labels or printing a listing of the records in the database.

```
*menu.prg
*this defines the main menu for the ibdata database manager
*
set status off
clear
DEFINE MENU Main MESSAGE "IBDATA DATABASE MANAGER—HIGHLIGHT
YOUR SELECTION AND PRESS ENTER"
DEFINE PAD Label OF Main PROMPT "PRINT MAILING LABELS" AT
6,33
DEFINE PAD direct OF Main PROMPT "PRINT IBDATA MEMBER DIREC-
TORY" AT 8,33
DEFINE PAD exit OF Main PROMPT "EXIT" AT 10,33
ON SELECTION PAD Label OF Main Do LABEL
ON SELECTION PAD direct OF Main Do DIRECT
ON SELECTION PAD exit OF Main DEACTIVATE MENU
ACTIVATE MENU MAIN
SET STATUS ON
RETURN
```

The command, DO menu.prg, causes a List of Choices to appear on the screen. The user selects which activity is wanted, and when the choice is entered, the program selects the procedure that completes the task.

Although very extensive programs can be developed and executed within the interactive dot prompt environment, many users will find that program development is easier using the ASSIST Application Generator. By using a menu option approach, the programmer can built a set of functions that provides a customized application for the user.

27.12 DBASE IV AND SQL

dBASE IV does support other programs to access and to query dBASE and other database files. Structured Query Language or SQL is an IBM language still used on many mainframe computers. SQL can be used to make queries in dBASE IV. If this feature is active, config.db will locate the SQL files. The Config.db file may record the location of SQL files by modifying the Config.db to read:

```
*
*dBASE IV Configuration File
*Monday, June 14, 199X
*
COMMAND              = ASSIST
SQLHOME              =
G:\dBASEIV\SQLHOME
SQLDATABASE          = SQLSAMPLES
SQL                  = ON
STATUS               = ON
```

Although SQL is an integral part of the dBase structure (command and programming language), SQL is considered an advanced topic for users. The .dbf files can be accessed and processed by both dBASE and SQL commands. SQL provides a link for users of mainframe and minicomputer database systems to access data in .dbf files. SQL broadens the application of dBASE by allowing users to utilize familiar, industry-standard commands.

As with dBASE IV, SQL lets you access information you have created in databases. These files may be on mainframe, minicomputer, or personal computers. Once you have access to these systems, SQL provides a well-organized, structured language for making queries, modifying records, and administering your database files.

The SQL command structure is designed to make it easy for the user to access computer data without having extensive computer knowledge. The SQL structure is based upon the table format placing records in row/column setups with the columns being the fields of the records. The principal method is to use the "query." The SELECT command provides interactive access to your database. To learn the name of a customer whose number is 421, and is in the CUSTADDR database, you could enter

```
SELECT Namecust              (field name)
    FROM Custaddr      (database table identification)
    WHERE Namecust = 421          (search condition)
```

The output might show this customer to be Greenville X. Smith. With knowledge of the field names and the database structure, SELECT queries can be created to show data in any number of informative sets. The database file may have been created in dBASE IV, or you may create the database with SQL's CREATE command.

Whichever relational database management system was used to establish your database, dBASE IV and/or SQL will give you powerful access to the information stored there.

27.13 CONCLUSION

We have looked at multiple ways to access the core set of features that dBASE IV offers. In general, this should have served as basic introduction to computerized database management systems. In today's fiercely competitive business environment, utilizing technology is become necessary just to stay in business. Good luck in developing you own hands-on competence with dBASE IV.

DEFINING TERMS

Application: A set of user tasks accomplished by a set of procedures accessed through one or more programs.

Append: The action of adding a new record to a database or having the computer action of adding additional file information to an already existing file.

Browse: The ability of a user to visually scan through the records of a database in a random manner; a feature of a database management system which provides users with a program to view the contents of database files.

Catalog: A method or technique of grouping together related files for ease of reference. Catalogs can contain different application files based upon the user's method of identification (sales catalog, inventory catalog, personnel catalog, etc.).

Command: A computer instruction which performs a specific task which can be combined with other commands to define a program or a procedure.

Control Center: A graphic form of interface providing the user with menu options for creating and using the files in the current catalog based upon: database, query, form, report, label, and application.

Database File (Relational): The .dbf file which contains the user's information in a table format for information processing purposes.

Dot Prompt: Prompt which permits the direct entry of commands for immediate execution.

Field: The smallest data unit which identifies a specific category of information found in a set of records in a given database file.

Form: A prescribed screen layout for viewing records on the monitor or for entering data into predefined areas setup up by the user or by using the default layout provided by the program.

Index: A reference file which indicates the order or sequence in which records will be presented.

Key Field: A field whose purpose is to separate records according to a unique identifier and is used to index and sort the records.

Program: A set of commands which accomplish a particular task or set of tasks.

Query: The act of seeking information from a database file by providing the set of conditions which determine which records are selected and presented.

Record: The defined information unit within a database consisting of fields which hold the data about each unit.

Report: The formatted soft or hard copy output of information extracted from a database file.

Structure: The list of fields in a database record which defines the logical and physical characteristics of each field.

Sort: The physical rearranging of the records in a database according to one or more key fields.

SQL: Structure Query Language—a method of communicating with a database which has become a standard and is supported by many database management system programs.

REFERENCES

dBASE IV Disk Pack. Ashton-Tate.
Getting Started with dBASE IV. Ashton-Tate.
Guide to dBASE IV. Ashton-Tate.
Introduction to the Dot Prompt. Ashton-Tate.
Learning dBASE IV. Ashton-Tate.
Using the Menu System. Ashton-Tate.
Using the dBASE IV Applications Generator. Ashton-Tate.
Quick Reference. Ashton-Tate.

28

Microsoft Access 2.0

Matthew Allen

28.1 DESCRIPTION

Microsoft Access 2.0 is a relational database authoring system that was designed to run under Microsoft Windows 3.1. With Access, one can create database systems that can include not only data, but also programs to compile and process those data. A database is "a comprehensive collection of related data organized for convenient access, generally in a computer" according to *The Random House Dictionary of the English Language*. Databases are generally used to facilitate the manipulation of data, e.g., adding, deleting, and modifying data quickly and easily. The data portion of an Access database is organized in tables of records. A classic example of a record is the individual cards of a rolodex. The table would be the rolodex itself. Other types of information often stored in a database include financial data, statistics, and experimental results.

Typically databases are used to store data (in tables) and retrieve data. The term associated with the search for information in a database is a "query." The result of a query is database records. Following our rolodex analogy, the result of a query of last names starting with the letter "T" would be all those cards in the T section of the rolodex.

28.2 RELATIONAL DATABASES

Databases are generally created to store related data in tables and query the data at a later time. Databases often have multiple tables with some relationship between them. For instance, a marketing database might have a table of companies and a table of contacts at those companies. There would be a one to many relationship between the records in the company table and the records in the contacts table as there may be more than one contact for each company. Access allows you to specify one-to-one or one-to-many relationships between tables in the database, which puts Access in the category of relational databases.

0-8493-2530-7/96/$0.00+$.50

Another concept associated with related databases is referential integrity. The idea behind referential integrity is to preserve the relationship between records in related tables of a database. For instance, in the marketing database example there is a table of companies and a table of contacts at those companies. If a company is deleted and the contacts associated with that company are not deleted then those contacts will remain in the database and not appear in any more queries that include company names. By enforcing referential integrity, Access will automatically report and prevent attempts to delete records that will violate such relationships.

28.3 MULTIUSER DATABASES

Access is a multiuser database. It is capable of allowing multiple users to add, modify, and delete records in the same table at the same time. However, users may not modify the same record at the same time. Users can access a record for "read," which allows the user to view the record, or for "write," which allows the user to modify or delete the record. Multiple users can access a record for "read" at the same time. Only one user can access a record for "write" at a time.

Access has the capability of administering permissions on objects within the database (e.g., tables, queries, forms, reports, macros, and modules) at the User and Group level. The following dialog box from Access lists the permissions a user may specify about objects within the database.

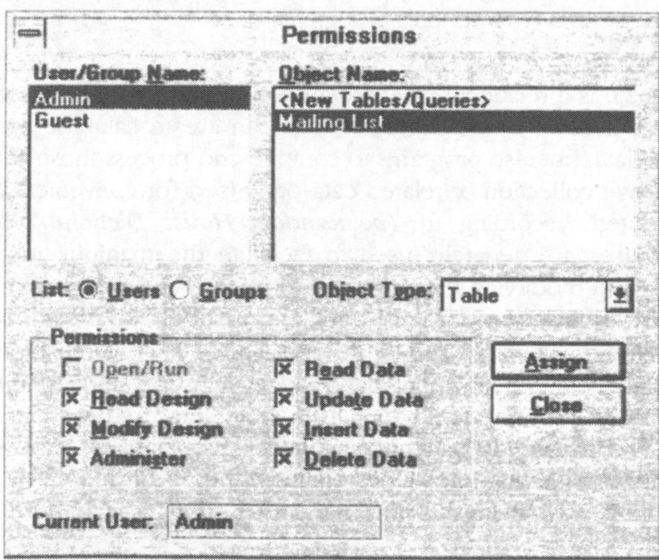

Users and Groups are defined within Access and bear no necessary relationship between users and groups on local area networks on which the database resides.

28.4 ACCESS DATABASES

The components of an Access database are tables, queries, forms, reports, macros, and modules. An access database can be made up of one or more of these components. Regardless of the number of tables, queries, etc. in the database, Access will always store the data and the structure of the database in two files, namely filename.MDB and filename.LDB.

28.5 CREATING A DATABASE

To create an Access database, open Access, then choose File, New Database. A dialog box will appear that will give you the opportunity to name the database and specify a path to designate the location where the database will be stored. The database can be stored on the local hard drive or a network drive.

Generally, the next logical step after creating a database is to create the database tables.

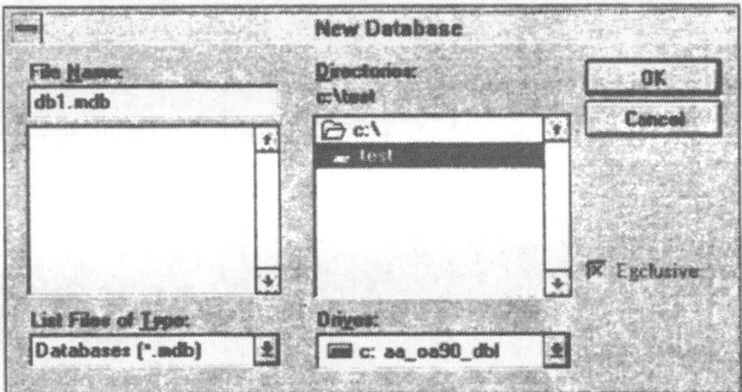

28.6 DESIGNING TABLES

A table is a series of zero or more records, which are in turn composed of one or more fields. Multiple tables can exist within an Access database. Access can also attach to tables from other databases including .dbf and SQL databases.

Tables are defined by specifying fields within the table. Fields can be of a number of types and can be defined as key fields or indexed fields or neither.

Key fields are unique for each record within the database, no record can have the same value in a keyed field. Indexed fields are useful for sorting and searching for records in a table. Using indexes can speed up the execution of queries dramatically. Indexes are defined as a combination of one or more fields and can be unique or nonunique depending on how they are defined. Key fields are considered indexes.

A table with nonunique indexes can be compared with a set of encyclopedias. To find a subject that starts with the letter "D" you first pick the encyclopedia for subjects starting with D. A quick glance at the shelve will allow you to find that book. However, there are multiple subjects in the book, which is why the index is not unique. An example of a unique index is the page numbers on the pages of this book. There is exactly one page number for each page, so finding any page is more or less a binary search through the book.

Creating Tables

Once you have decided what data your table will contain you are ready to create your table. Access provides an installable Table Wizard to make creating often used tables easy. However, to create a table "from scratch," from the File menu choose New then Table then New Table.

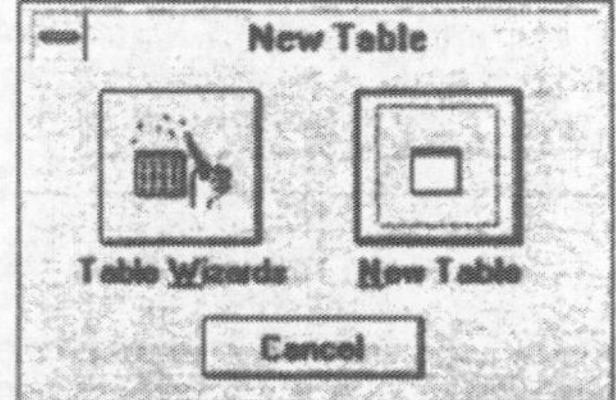

The following window will appear

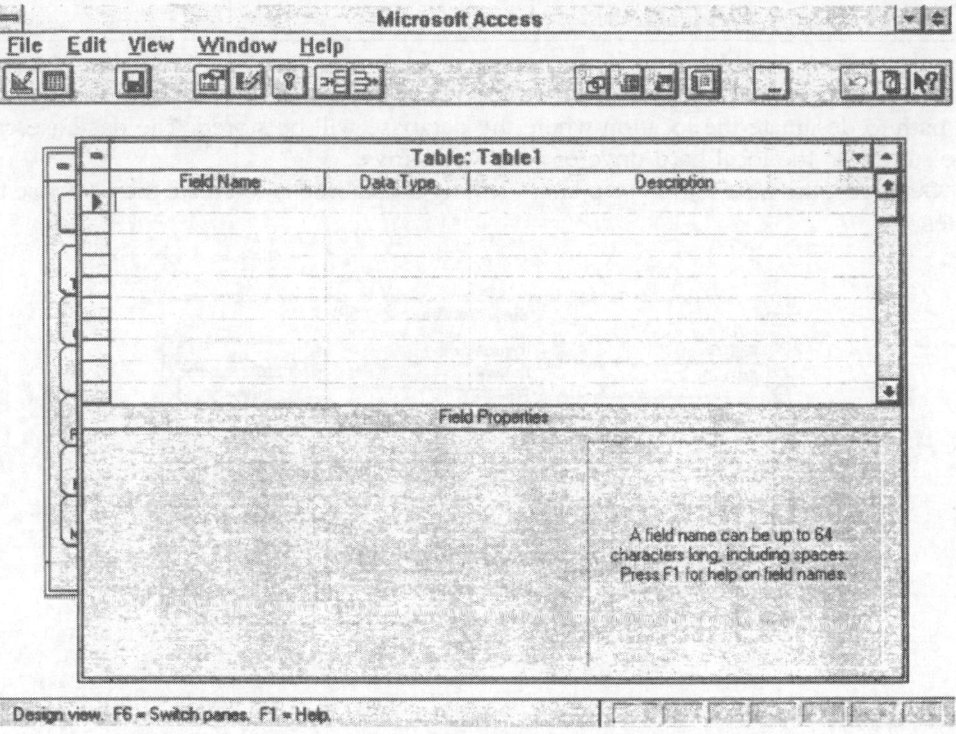

You can specify field names, types, and descriptions and choose which fields will be part of the key and which fields will be indexes.

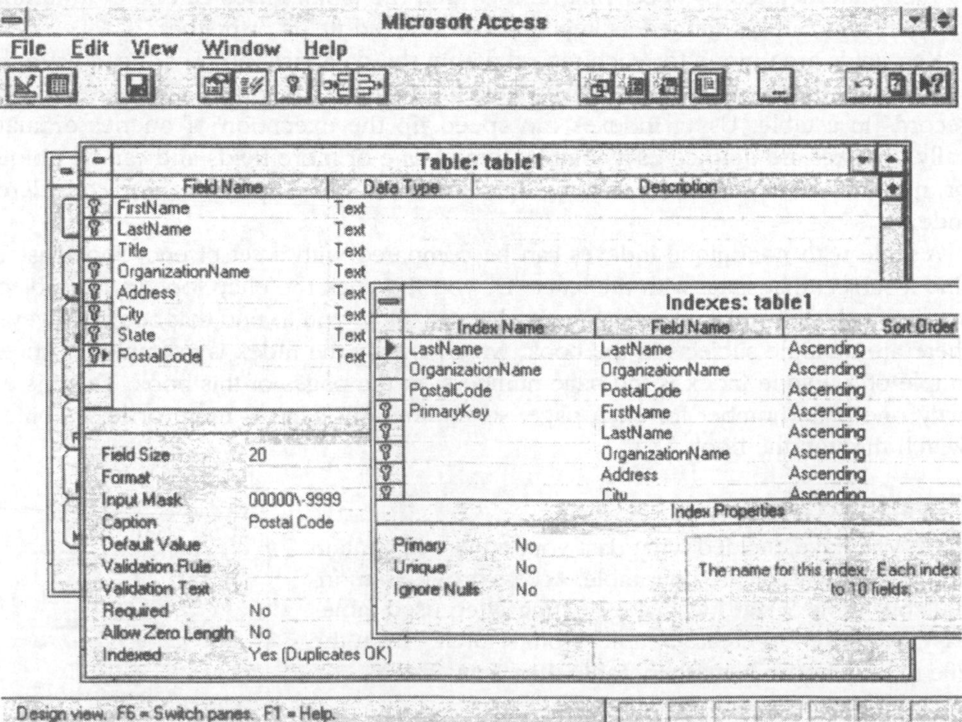

Relating Tables

Once you have defined two or more tables, it is possible to relate them. As mentioned earlier, relating tables is useful for establishing referential integrity. Another reason to relate tables is that Access will automatically use relationships previously specified on tables when two related tables are selected in the query editor. These relationships can be modified for each query, however. Defining queries will be covered later.

To relate tables, open the database window and click on the **Table** tab. Choose File, Edit, Relationships. Then choose the tables and/or queries you wish to include in the relationships. In this example two tables, companies and contacts, will be related. The relationship will allow multiple contacts to be related to each company.

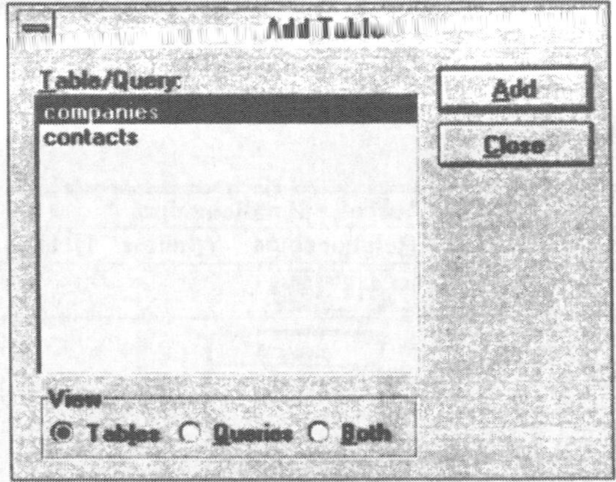

All selected tables will appear in the relationship layout window.

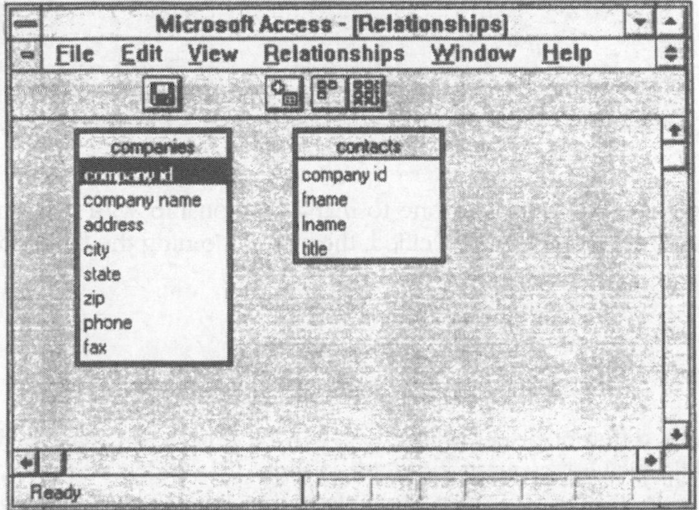

To define a relationship, drag a field from one table to a field in another. A dialog box will appear that will allow you to specify the type of relationship and choose other fields to relate between the two tables.

Relationships
Table/Query: Related Table/Query:
companies contacts
company id company id
Create
Cancel
Join Type...
Inherited Relationship
Enforce Referential Integrity
One To
One
Many
Cascade Update Related Fields
Cascade Delete Related Records

Choose Create to finalize the relationship between the tables.

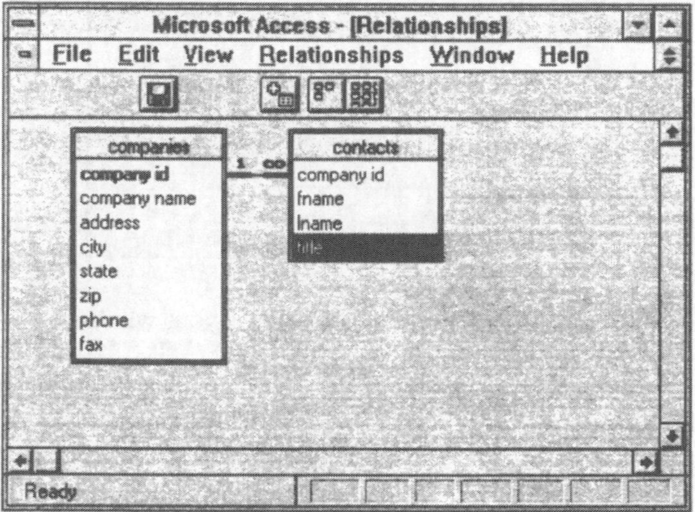

The symbol 1 ∞ implies a one to many relationship with referential integrity. When referential integrity is not specified, the symbol joining the fields looks like this: ⟶.

Data Entry

There are three different places where data can be entered in Access: while viewing the table, while viewing a query, or within a form. The quick and dirty way to enter data in to a table is simply to open the table and start typing!

To start entering data in a table open the database window and click on the Table tab. Then double-click on the table whose data you wish to modify. A window will appear that looks like a spreadsheet. The field names of the table are across the top. Each row is a record in the table.

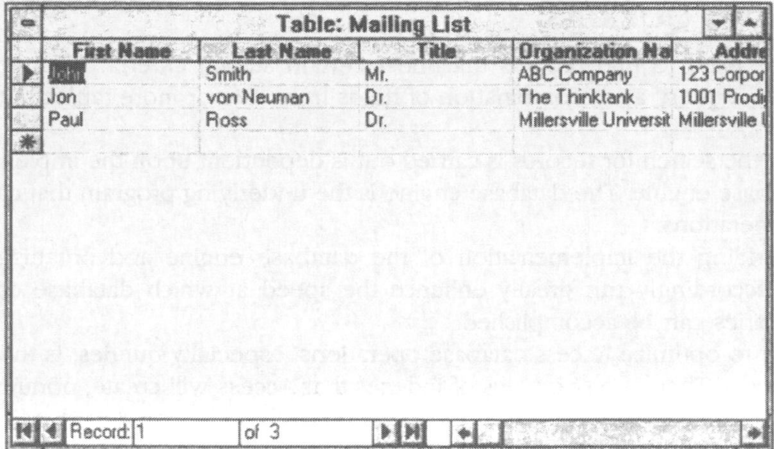

At the bottom of the table window there are several buttons that allow navigation through the records. To add data to the table, type data in the last record of the table. To modify a record, just click in the field you wish to modify. To delete a record, click on the record selector (the box next to the record) for the record you wish to delete and choose Edit, Delete.

Attaching External Tables

An Access database can directly reference data from a number of different databases including other Access databases. Adding tables from another database is accomplished by attaching to the table. To attach to a table from another database, open the database window and click on the **Table** tab, then choose File, Attach Table. A dialog box will appear that will allow you to choose the type of table to attach to.

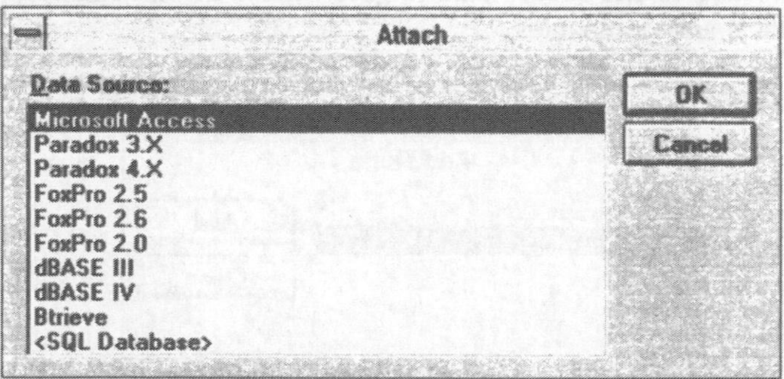

The attached table will be added to the table list. You cannot change the definition of an attached table (e.g., define key fields, add indexes) You can modify data in attached tables, however.

Copying and Deleting Tables

To copy a table in the database, open the database window and click on the **Table** tab. Next, highlight the table and choose Edit, Copy. Finally, choose Edit, Paste.

To delete a table, select the table and choose Edit, Delete.

28.7 DESIGNING QUERIES

Queries are used to find records that meet certain search criteria. The records resulting from a query are a combination of fields from one or more tables in the database.

The way the search for records is carried out is dependent upon the implementation of the database engine. The database engine is the underlying program that carries out database operations.

Understanding the implementation of the database engine and organizing one's databases accordingly can greatly enhance the speed at which database operations such as queries can be accomplished.

One way to optimize Access database operations, especially queries, is through the use of indexes. There are two types of indexes that Access will create, nonunique and unique indexes. Typically it is faster to search through tables for records with unique indexes.

Creating Queries

To create a query, you must first have one or more tables accessible in the database. Our goal in this example will be to generate a mailing list by selecting address information from a company table and linking that information with names from a contact table. Each contact will already have an association with a company in the company table.

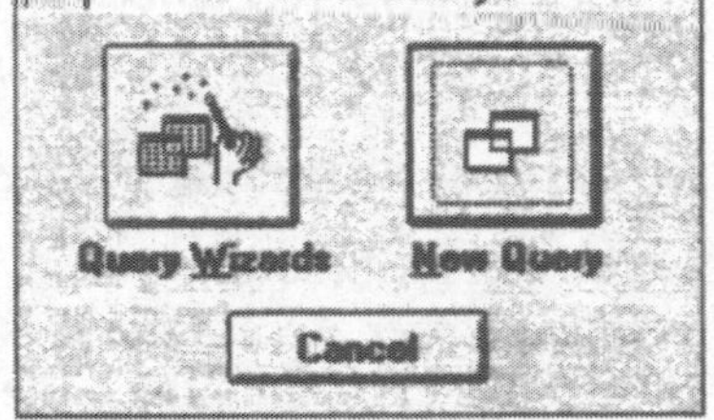

To create an Access query, choose File, New, Query. A dialog box will appear that allows you to build a new query "from scratch" or use the Query Wizard. Select New Query.

After selecting New Query, a dialog box will appear that is used to choose tables and/or queries to add to the source data for the query. In this example we will use two tables, contacts and companies.

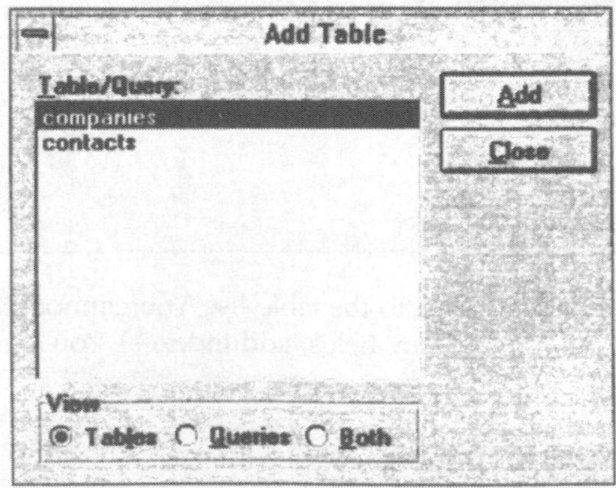

Once all tables and queries are added, select Close. A window will appear with all tables and queries that were selected.

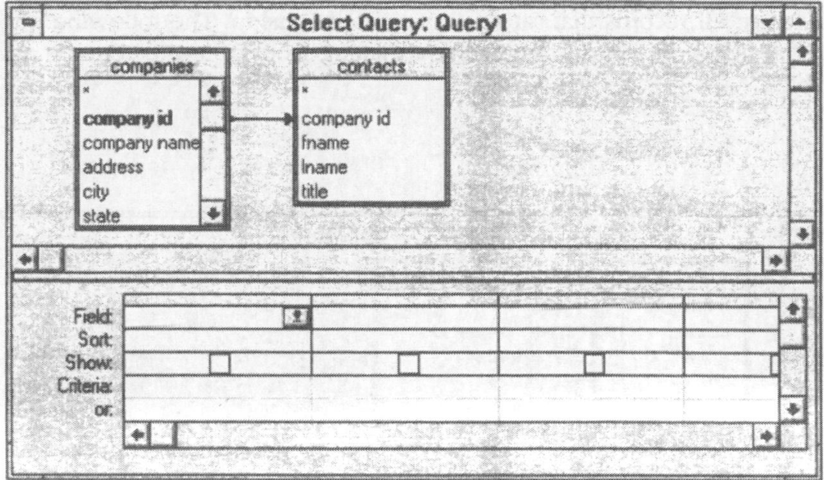

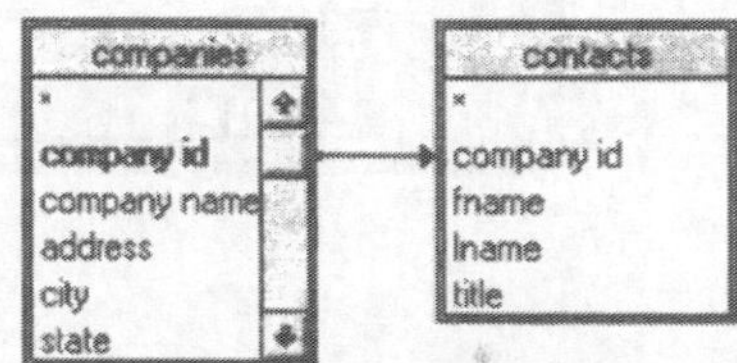

The next step is to join the queries as necessary. Because the contacts and companies tables used in this example were previously related, a one-to-many join relationship is already shown after adding the tables.

To modify the join relationship between the tables for this specific query, double-click on the middle of the join line between them. A dialog box will appear that allows you to specify the join properties.

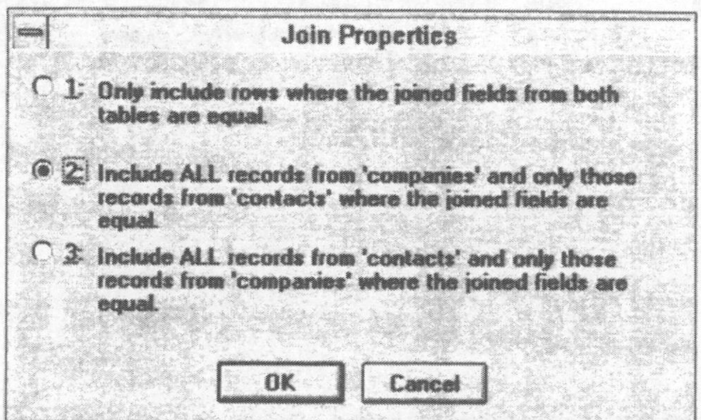

At the bottom of the query window there is an area for specifying what fields will appear as a result of the query and what the query criteria will be. Specifying no criteria will return all records that satisfy the join relationship. The following query will return all the fields necessary for creating a mailing label from the contacts and companies tables.

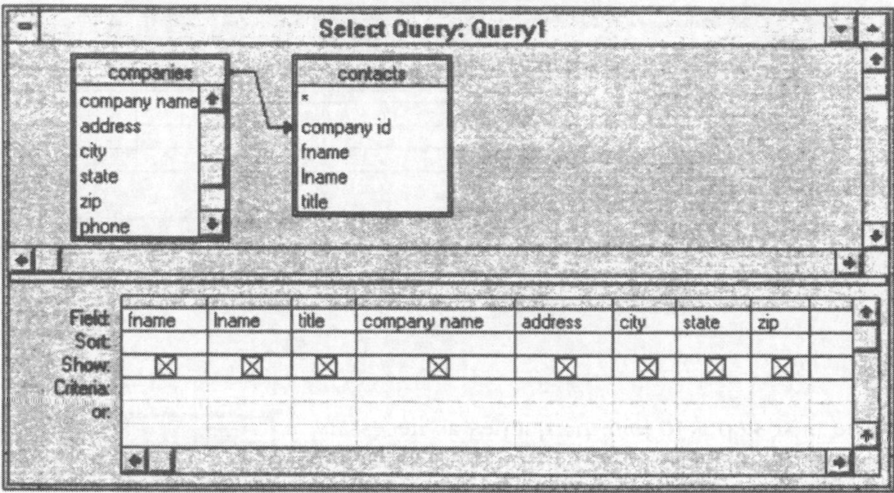

Choose View, Datasheet to see the data that result from the query.

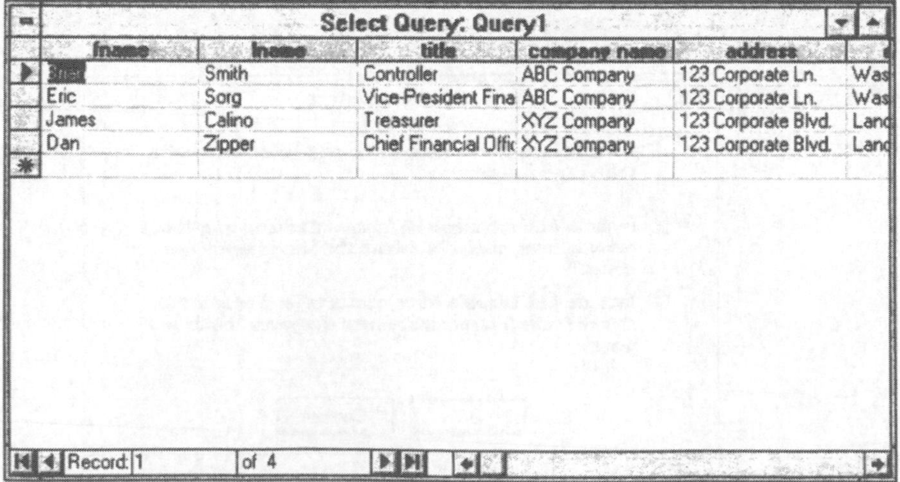

The following query will return records for a mailing label. The lname field in each record will start with the letter "S."

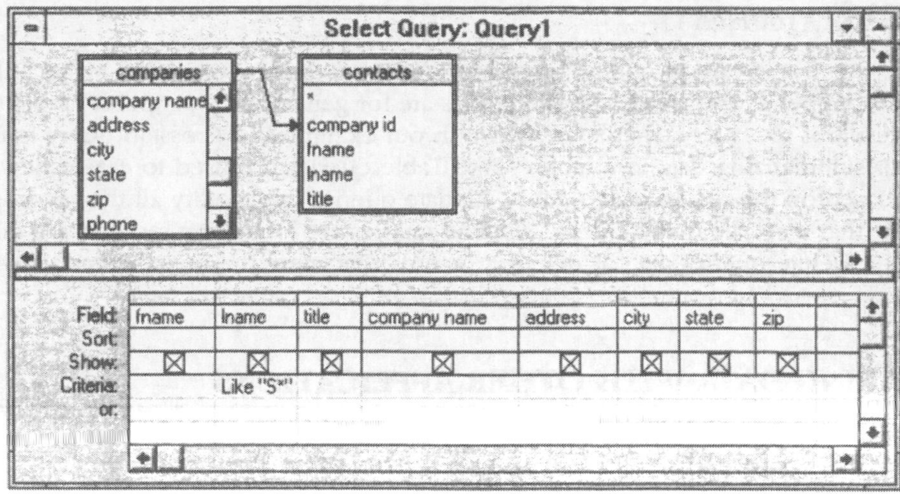

The query will return the following data:

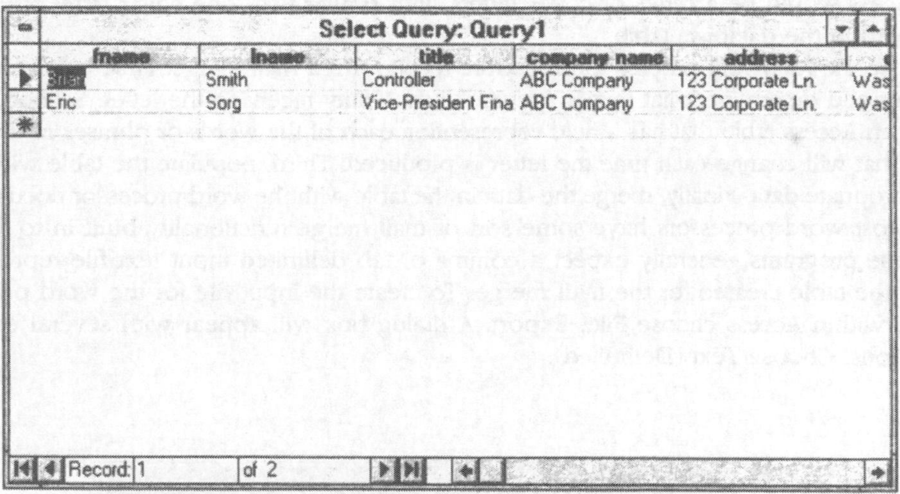

To view or modify the SQL code for your query, choose View, SQL. The code is generated automatically when you change the query design. Changing the SQL code will automatically update the query design as well.

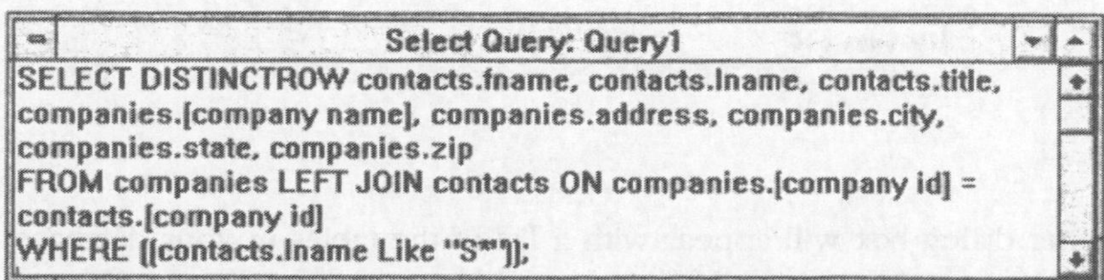

Types of Queries

There are five types of queries possible in Access: Select, Crosstab, Make Table, Update, Append, and Delete. Select queries are for generating a temporary table that is a result of the query criteria, as shown in our examples. A Crosstab query is useful for summarizing data in a table. Make Table queries are used to create new tables based on the results of the query. Update queries can modify all the records of a table that meet the query criteria. Append queries are for adding records to a table. Finally, Delete queries are for deleting records that meet the query criteria from a table.

28.8 SHARING DATA WITH OTHER APPLICATIONS

Merging Data with Word Processor Documents

A mail merge is a classic application for a database and a word processor. The goal of the merge is to create a report that is a word processor document. The inputs to this process could be a cover letter to accompany a resume and a database table containing a list of company addresses and human resource personnel. The output of this process would be a series of cover letters with similar text, one cover letter for each record in the database table.

Use the following process to procedure to perform a mail merge. First, compose the letter and determine what text is specific to the entity receiving the letter. Second, create an Access table that has a field representing each of the words or phrases in the letter that will change each time the letter is produced. Third, populate the table with the appropriate data. Finally, merge the data in the table with the word processor document.

Most word processors have some sort of mail merge functionality built in to them. These programs generally expect a comma or tab delimited input text file representing the table created for the mail merge. To create the input file for the word processor, within Access choose File, Export. A dialog box will appear with several export options. Choose Text (Delimited).

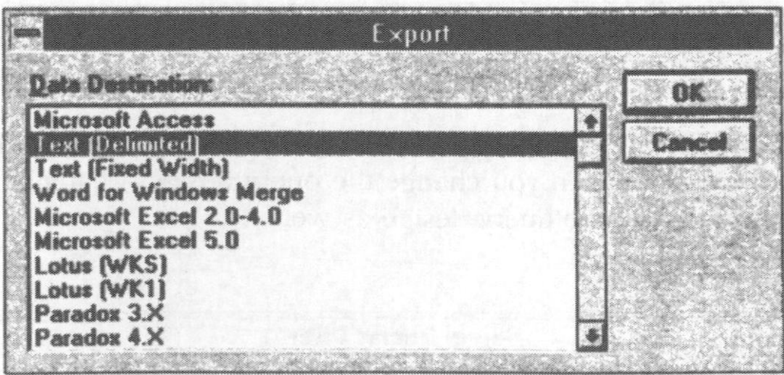

Another dialog box will appear with a list of the tables in your database. Choose the table you want to export (Mailing List in this case).

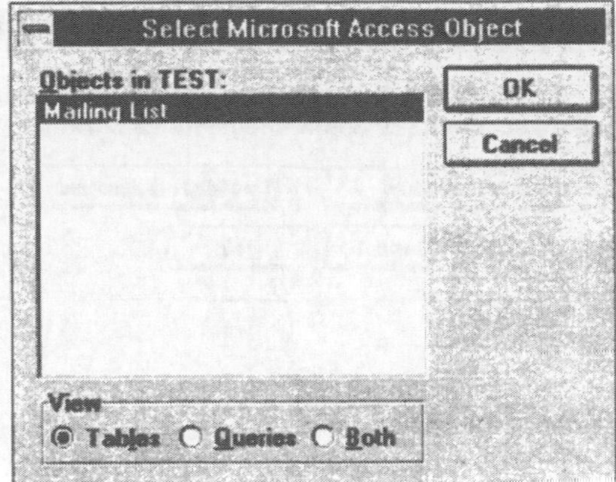

You will be prompted for a file name designation; after choosing the file name, check the box marked "Store Field Names in First Row."

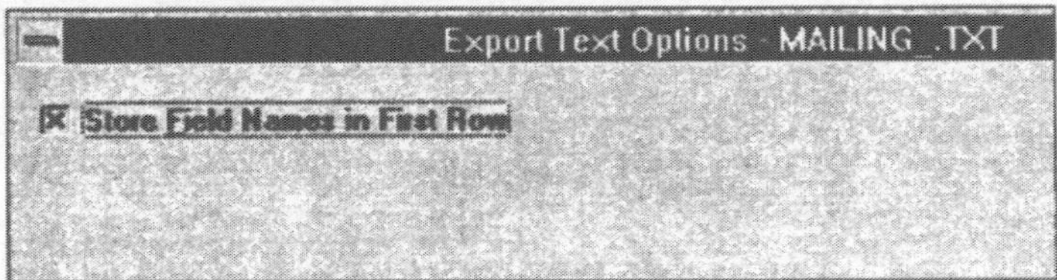

A text file will be created for use with the word processor. The first line of the text file will contain the field names from the Access table. The following lines will contain the data in the table.

The next section deals with merging Access table data with a word processor document under Microsoft Word 6.0.

Microsoft Word 6.0 Example

To create a Microsoft Word mergefile using an Access table and a Word document, open or compose the generic Word document then choose Tools, Mail Merge. Click "Create" on the Mail Merge Helper dialog and choose "Form Letters."

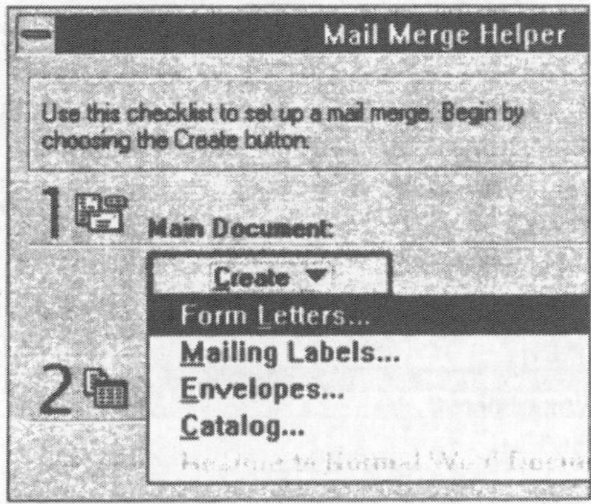

Now choose "Active Window."

Microsoft Word

To create the form letters, you can use the active document window Document3 or a new document window.

Active Window | New Main Document

Cancel | Help

Choose "Get Data" then "Open Data Source."

From the "List Files of Type" drop down menu choose "MS Access databases." Then open the Access database containing the source data table for the mail merge.

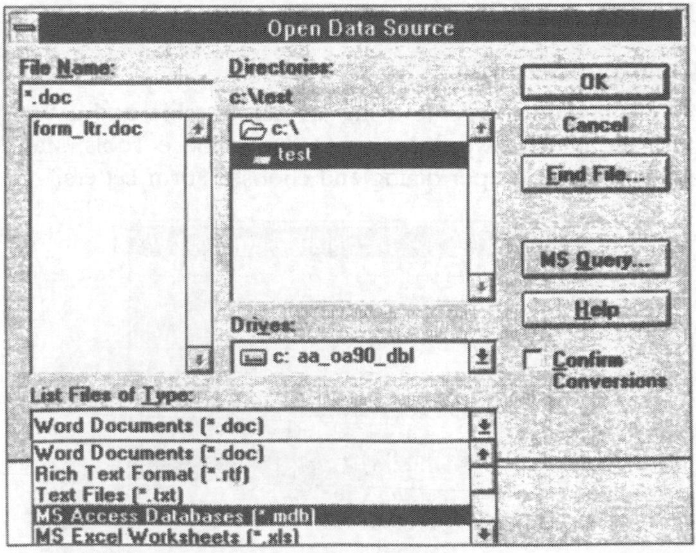

Now select the table containing the source data for the mail merge (Mailing List in this case).

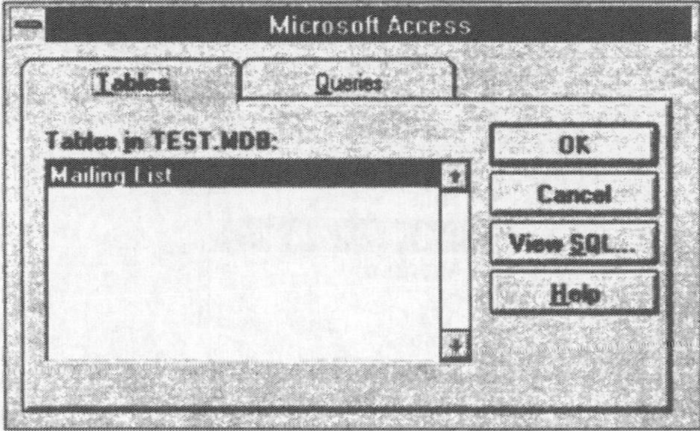

The next step is to add the fields from the table representing text in the source document to the document. Choose "Edit Main Document."

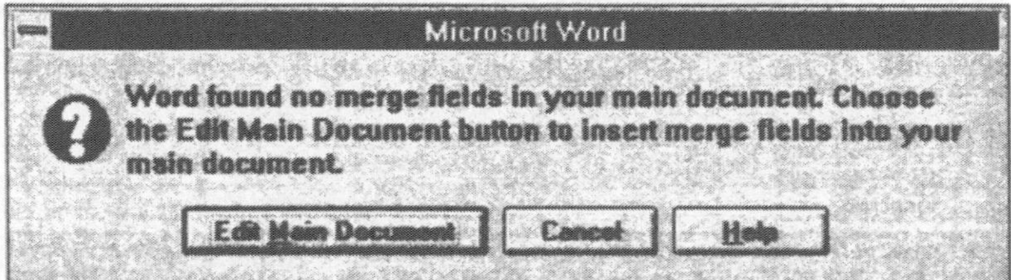

Select an area of the document represented by a field in the source data table, for instance the organization name.

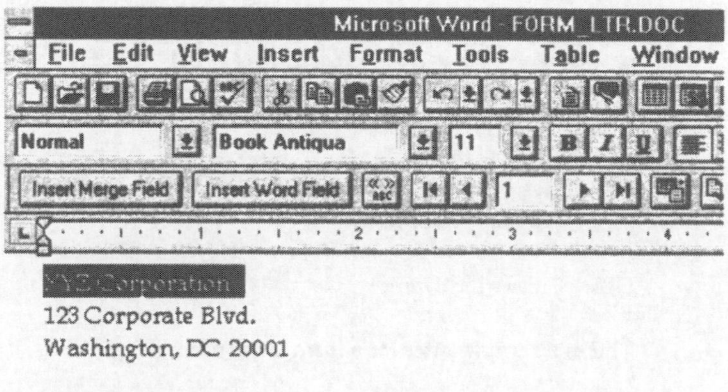

Choose "Insert Merge Field" and select the field representing the text to be replaced (Organization Name in this case).

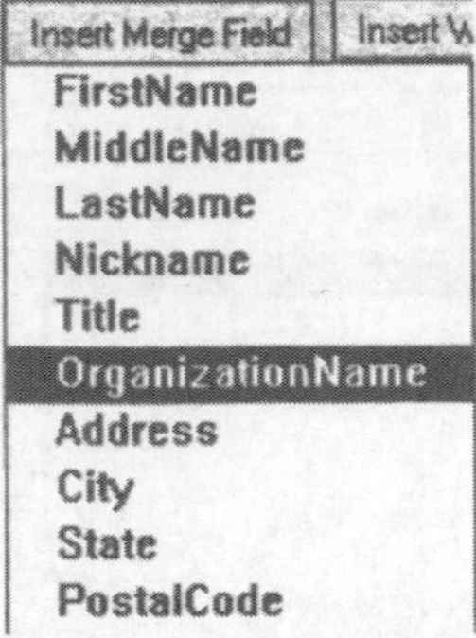

The text in the document will be replaced with the field representing that text.

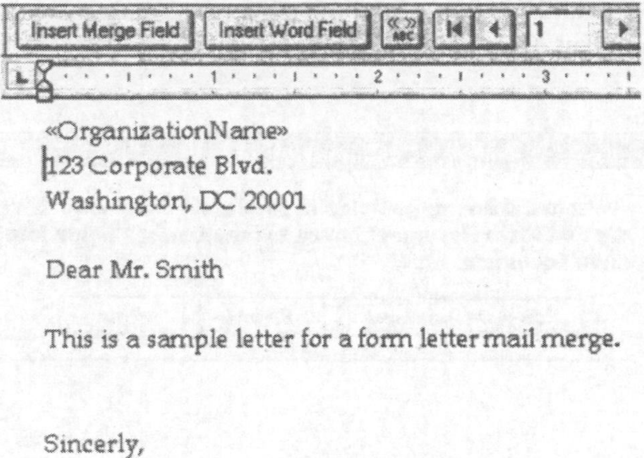

Continue the same operation until all text represented by fields in the Access source data table have been replaced.

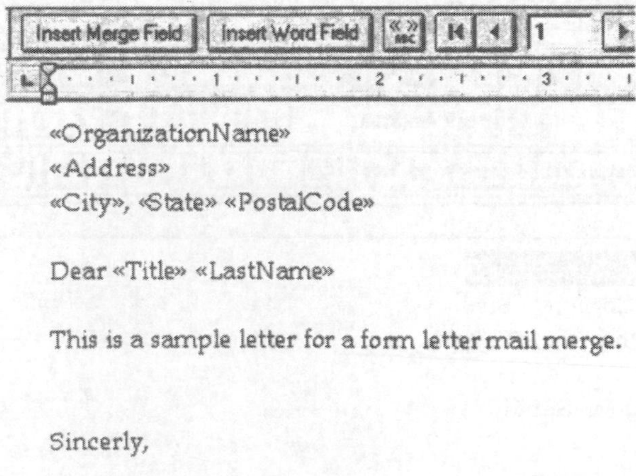

Click on the «» ABC button to see actual Access source table data in the document.

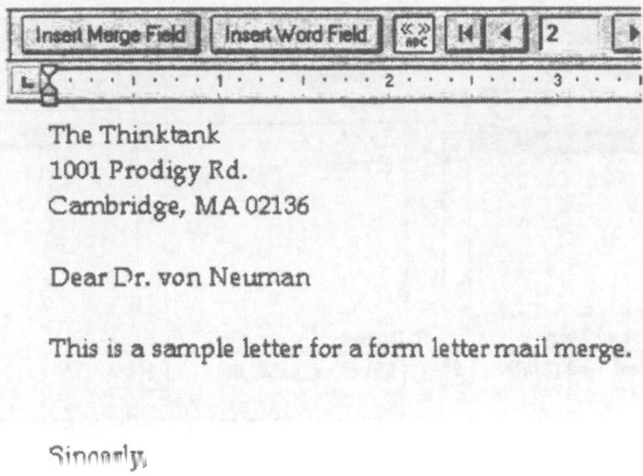

Insert Merge Field
Insert Word Field
The Thinktank
1001 Prodigy Rd.
Cambridge, MA 02136
Dear Dr. von Neuman
This is a sample letter for a form letter mail merge.
Sincerely,

To print the mail merged file, click on the button.

Importing Data

Access is capable of importing ANSI and PC-8 text and append that text to an existing table or create a new table. Each line of text will be represented by a new record.

Use the following procedure when importing text to an Access table. First, create a source text file such as the one below. The first line of the file should contain the field names for the Access table. Each field should be in quotes or single quotes if data will have spaces in the middle. Separate each field by a comma, tab, or space (quotes and commas in this case). Save the file with a .TXT file extension.

```
"FirstName","LastName","Title","OrganizationName"
"Paul","Ross","Dr.","Millersville University"
"John","Smith","Mr.","ABC Company"
"Jon","von Neuman","Dr.","The Thinktank"
```

Second, open or create an Access database to contain the text file data. Third, import the text file by choosing File, Import. Select "Text (Delimited)."

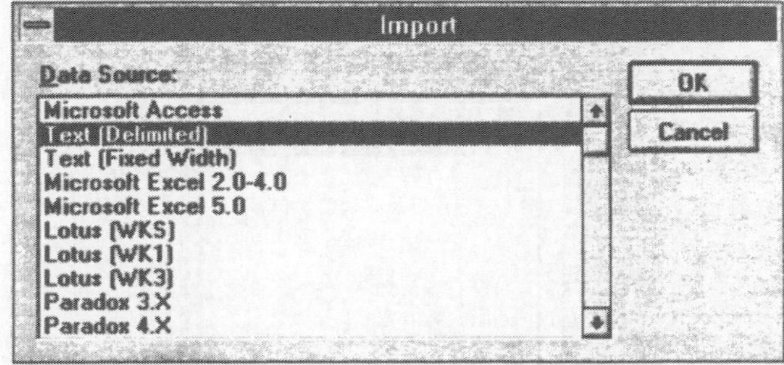

Fourth, open the text file containing the data you wish to add to an Access table.

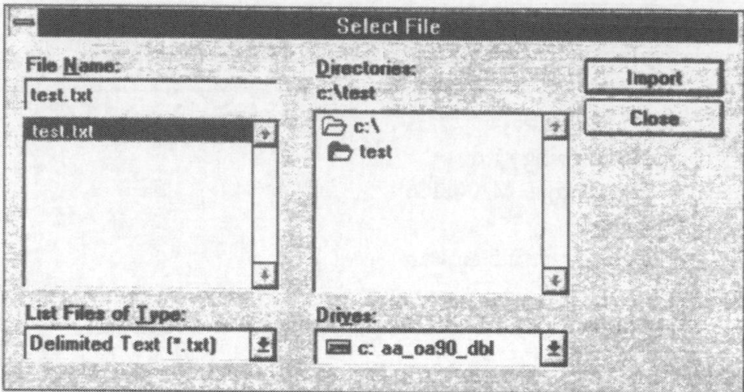

Finally, select the appropriate text import options. Choose "First Row Contains Field Names." Select "Create New Table" or select an existing table to append to. Choose a text delimiter and field separator; quotes and commas are the default.

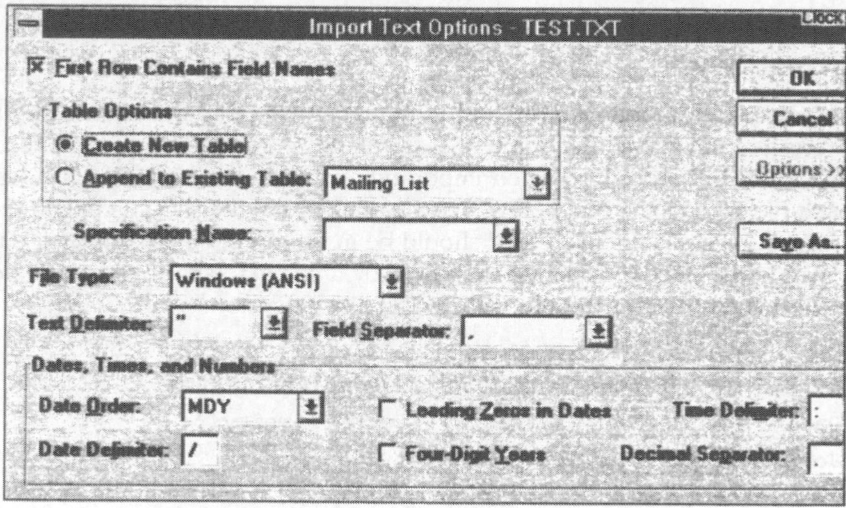

The new table will appear in the database table window.

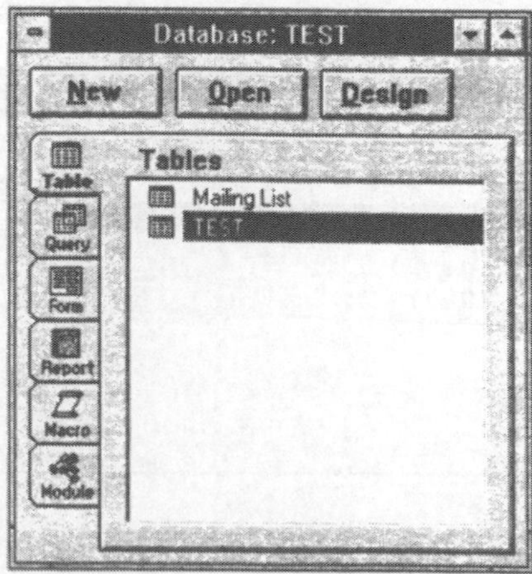

Microsoft Excel 5.0 Example

To import a Microsoft Excel spreadsheet into an Access table choose File, Import and select "Microsoft Excel 5.0."

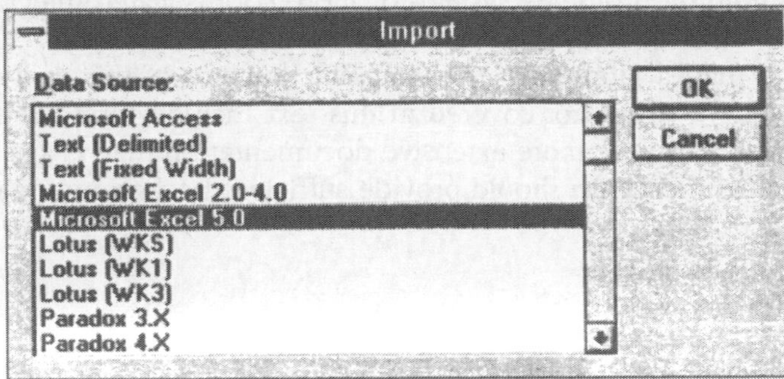

Open the source Excel document.

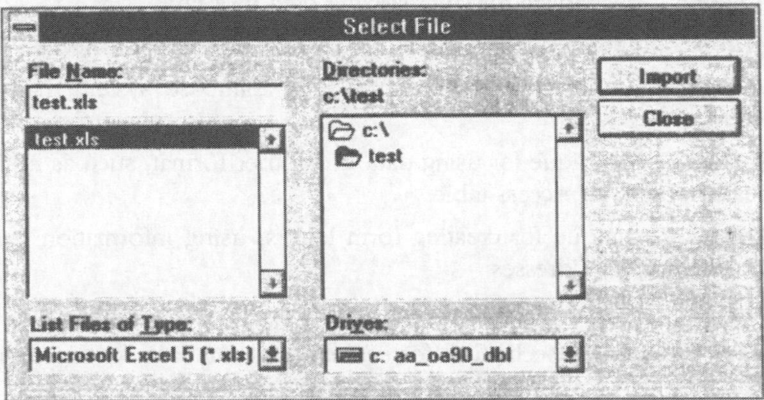

Choose "First Row Contains Field Namcs."

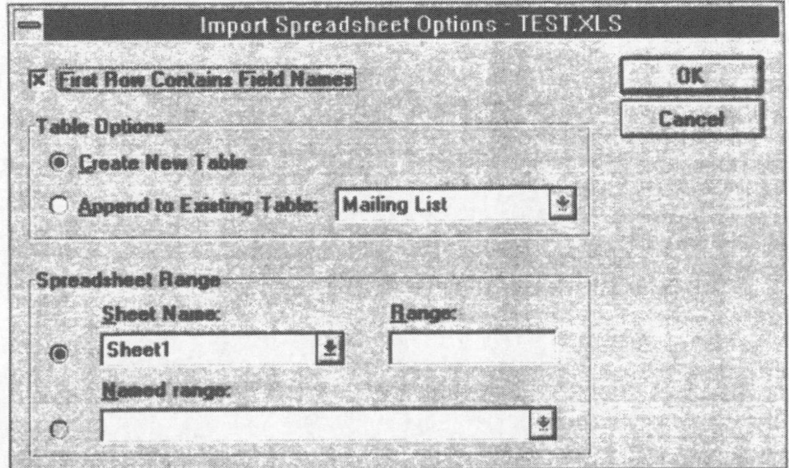

The new table will appear in the database table window.

28.9 SUMMARY

Microsoft Access is the perfect tool if you need to rapidly create a database from scratch or from existing data. Access also has functionality which makes it a great choice for designing a prototype system for more sophisticated integrated database applications.

We have only had a glimpse of the total functionality of this well-designed and intuitive application. Areas not covered in this text include forms, reports, macros, and modules which require more extensive documentation than is appropriate here. However, this documentation should provide sufficient background to explore these other areas.

DEFINING TERMS

The reader should consult the chapter on Windows for more details pertaining to the terms used in all Windows applications. Also, a number of the terms in Section 13.1 on dBASE IV apply to MicroSoft Access, as well.

Crosstab Query: Mechanism for summarizing data in a table.

Database Table: Tabular layout used as the model for MicroSoft Access.

External Table: A table created by some other database system than MicroSoft Access.

Importing Data: Technique for using data in another format, such as ASCII text, or a spreadsheet, in an Access table.

Merging Data: Technique for creating form letters, using information from Access for such things as addresses.

Query Window: Window used to create the query.

Query, Query Criteria: Mechanism for searching a database for the desired information.

Relating Tables, Relationship: Mechanism for tying two or more tables together.

29

Microsoft Works

Paul W. Ross
Millersville University

INTEGRATED PACKAGES

The modern research and industrial environment has changed considerably over the last 10 years. Word processors have replaced typewriters. Facsimile machines and fax modems are used to transmit graphical material to remote locations in minutes. Spreadsheets have replaced ruled pads and pocket calculators. Telecommunications systems link remote facilities to a main office or to the computer of your colleague in the next building.

Why has this happened? The answer is productivity: using your resources of time, people, and money to produce a better product, or service of better quality, in less time at lower cost. Let us consider a few of the typical tasks to be performed in a business. The majority of them are common to all businesses or organizations:

- Prepare memos, letters, and other correspondence.
- Print reports and prepare graphs and charts.
- Perform calculations for reports, and model decisions with "what if?" strategies.
- Schedule meetings and other business activities.
- Organize large amounts of data and use it for analysis and planning activities.
- Communicate with and pass computer-based information to other divisions of the organization, or to a central "mainframe" computer installation that handles organization-wide data.

Historically, the tasks of word processing, financial calculations, databases, graphics, and telecommunications have been done with separate software packages. Early microcomputers often had limited amounts of memory. Software was also expensive. As a consequence, it seemed reasonable first to buy a word processor to create basic business letters and reports. Next, buy a spreadsheet program for financial calculations and graphics, and finally a database system for maintaining data files. It was also useful to be able to communicate with another microcomputer or the corporate main-

0-8493-2530-7/96/$0.00+$.50

frame. This meant that companies had to buy an additional telecommunications package.

With separate systems or applications packages for word processing, spreadsheets, databases, and telecommunications, a number of serious problems arise, including the following:

- Separate training must be conducted for each computer application package. The commands for each application are liable to be very different from one package to the next. This can lead to user confusion when moving from application to application.
- It is often difficult to transfer data from one microcomputer application to another.
- Transfer data from one application to another is often difficult.
- The total cost of separate applications packages frequently exceeds the cost of a single integrated package, such as Works for Windows© or ClarisWorks©, to be discussed in this chapter.

An integrated system solves the problem users meet when employing stand-alone applications packages. These packages provide a uniform user interface for all applications within the package. The general structure of all commands in all applications is the same. This means that if a user is trained in one application, such as word processing, retraining problems in other applications are greatly reduced. Not only does the use of an integrated package reduce training costs, but the user is not confused by new philosophy in control commands or functions as well.

In integrated packages, data can be easily transferred from one application to another through a process known as *windowing*. This means that each application appears in a separate "window" on the user's screen. These windows can be viewed together, or one at a time.

Integrated packages such as Works for Windows or ClarisWorks (discussed in Chapter 30) provide a simple-to-use, inexpensive, yet effective software environment for a large number of common tasks.

29.1 INTRODUCTION TO WORKS FOR WINDOWS©

Microsoft Works for Windows© is an integrated software package for microcomputers. All of the standard applications, including word processing, spreadsheets, databases, graphics, and telecommunications, are available in a single program that allows a high degree of interaction between the various applications.

An integrated system such as Microsoft Works for Windows makes using microcomputer applications easier because it presents a unified design for commands and applications. The Works application package offers substantial benefits over conventional computer applications packages that have a separate program for each application. The benefits of Works derive from the reduced cost of a single package over many different packages. Further, with an integrated package, such as Works, there is a smaller amount of training required to learn how to use a single package versus many different packages. Finally, with Works, the different components can be easily used for the exchange of information, such as including a table from a spreadsheet in a word processing document.

In Works, data can easily be transferred from one application to another through windowing and the Windows Clipboard. This means that each application appears in a separate "window" on the user's screen. With an integrated package such as Works, you can begin a word processing document, then suspend it temporarily while you enter a spreadsheet to perform calculations. Information from one window can be captured and pasted into another Works application.

Files created by the different components of Works can be exchanged with many other commonly available applications packages. Works type packages, such as Works for Windows, do not have all the features such as Word for Windows©, Excel©, or other stand-alone packages. However, for many common day-to-day tasks, Works will produce an excellent level of functionality. It is especially useful for laptop computers, where the amount of disk storage or memory may be limited.

29.2 WORD PROCESSING IN WORKS FOR WINDOWS

The first step in using the Works for Windows word processing application is that of creating a file and entering text into it. Start Works for Windows. Click on the Word Processing button, unless you want to open an already existing file. If so, click on the Open Existing Document button and follow the directions to navigate to the desired file. Recently used files will be given in a list on the screen. Double-click on one of those, if you wish to use it.

When the word processing application loads, the screen will look like Figure 29.1 if you have started a new word processing document.

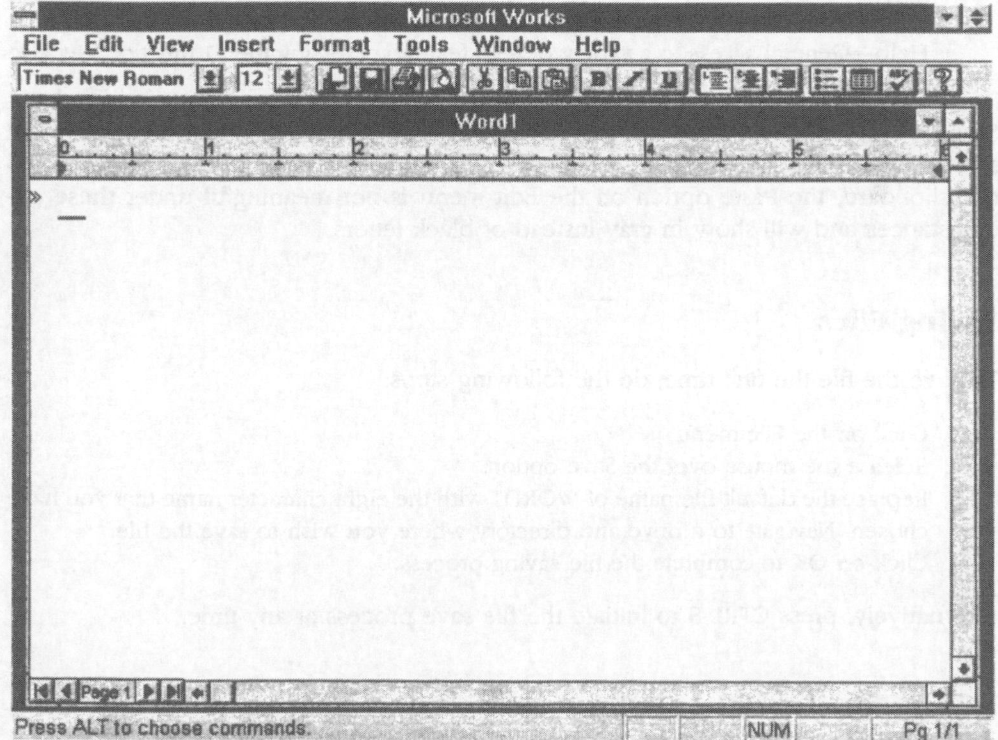

FIGURE 29.1 Initial empty word processing screen.

A default file name of WORD1 is assigned to the file, as shown at the top of the screen. When you save it the first time, you will be prompted to name the disk and directory where the file will be stored, and the name you wish to assign to your word processing file.

To enter text on the new word processing document, simply type it like any standard Windows-based word processor.

Use of the Menu System

Works for Windows is a menu-driven/mouse-driven system. The various features of Works for Windows are accessed through a series of menus at the top of the screen, either by selection with the Alt key or by clicking the mouse on the menu item.

The eight menus at the top of the screen perform the following functions:

- File—Create, save, open, and close files, exit from Works for Windows to Windows.
- Edit—Move, copy, delete, and restore deleted text.
- View—Headers and footers, document view format.
- Insert—Insert art, special characters, etc.
- Format—Appearance control with boldfacing, italics, centering, etc.
- Tools—Spell checker, thesaurus.
- Window—Access to the different Windows that you have open in Works for Windows, as well as features to change appearance and relationship between open Windows.
- Help—General access to a variety of help features. Pressing the F1 function key also gives you access to help in using Works for Windows.

Some menu items will often be "grayed." This means that a particular menu item will not be available at this time. For example, if nothing has been deleted or copied to the clipboard, the Paste option on the Edit menu is not meaningful under these circumstances and will show in gray instead of black letters.

Saving Files

To save the file the first time, do the following steps:

1. Click on the File menu.
2. Release the mouse over the Save option.
3. Replace the default file name of WORD1 with the eight character name that you have chosen. Navigate to a drive and directory where you wish to save the file.
4. Click on OK to complete the file saving process.

Alternatively, press CTRL-S to initiate the file save process at any time.

Moving the Insertion Point

Move the insertion point by using the up, down, left, and right cursor movement keys or click the mouse pointer at the location where you wish to place the insertion point.

For larger movements, such as an entire screen, the following keystrokes can be used:

Up one screen	Page Up
Down one screen	Page Down
To the end of a line	End
To the beginning of a line	Home
To the end of the document	CTRL-End
To the beginning of the document	CTRL-Home
To the beginning of word to right	CTRL-Right Arrow
To the beginning of word to left	CTRL-Left Arrow
To the top of the screen	CTRL-Page Up
To the bottom of the screen	CTRL-Page Down

Deleting and Inserting Text

There are two general situations that you must consider in editing or changing text with a word processor:

- Altering only a few characters (perhaps a spelling error).
- Altering whole words, lines, paragraphs, or the entire document.

To alter or delete text, take the following steps.

1. Position the insertion point under the first character to be altered with the mouse or cursor movement keys.
2. Press the Delete key to delete the character to the right of the insertion point. Press the Backspace key to delete to the left of the insertion point.
3. Press the Delete or Backspace key successively to delete any additional characters to the left or right of the insertion point.

Handle larger amounts of text by dragging the mouse over the desired area of text.

Setting Margins

To set the margins, use the mouse to move the small triangles on the ruler line at the top of the screen. Drag it to the desired position, and release the mouse button when it is at the desired location. The margin stops on the left end of the ruler line are different. You will see that instead of a single triangle, the triangle is broken down into two pieces. The upper one sets the margin for the first line in a paragraph. The lower triangle sets the left margin for succeeding lines.

Setting Tab Stops

To use the tabulation feature in Works for Windows, press the tab key. Initially, Works for Windows uses a set of predefined tab stops set for every 0.5 inch. To change the location of tab stops, double-click on the Ruler bar where you want a tab stop to appear. Follow the prompts to insert a tab at the desired location.

Centering and Aligning Text

1. Highlight the text (line or lines) using the mouse.
2. Click the mouse on the appropriate alignment symbol just above the ruler line.

Moving, Copying, and Deleting Blocks of Text

Highlight the desired text with the mouse. Go to the Edit menu and use one of the following options:

- Cut—This deletes the block of text (you can also use the Delete key, as before). Use CTRL-X.
- Copy—This copies the marked text to the Windows clipboard, which is available from any Windows application. You can then insert or paste it to another location in your document. Use CTRL-C.
- Paste—This moves the marked text that has been saved on the Windows clipboard to another location in your document. Use CTRL-V.

Copying Text from Another Document

Works for Windows produces a "Windowed" environment that allows you to have a combination of multiple spreadsheets, word processing documents, and databases running on your computer and be able to switch back and forth between the various Works for Windows applications.

To copy text between two Works for Windows word processing documents, do the following:

1. Open the first document, or create it from the File menu.
2. Open the second document from the File menu.
3. With the mouse, mark the text to be copied.
4. Press CTRL-C.
5. Select Window from the menu at the top of the screen. Select the target document.
6. You are now in the target document. Complete the copying process by positioning the insertion point to where you wish to place the text, and press CTRL-V, or select Paste from the edit menu.

Printing the Document in Works

Select Print from the Print menu at the top of the screen. You can get an idea of what your document will look like by selecting the Print Preview option. You can also set the number of copies in the Print dialog box.

The Spelling Checker

To activate the spelling checker, select Spell from the Tools menu at the top of the screen. If a word is not spelled correctly, follow these directions:

- Type in the correct word and press the Enter key. This replaces the incorrect word with the one you wish.
- Ask for a suggestion, and accept it.
- Change it as specified, for this or all occurrences.
- Ignore this word this time or now and in the future.
- Add it to the dictionary for future use, if it is spelled correctly.
- Indicate that you are done with the spelling checker by clicking on Cancel. This allows you to continue editing your document.

Finding Text

The procedure to locate or find text is as follows:

- Position the insertion point somewhere before the location where you expect to find the desired text. CTRL-Home will position you at the beginning of your word processing document.
- Choose Find in the Edit menu at the top of the screen.
- Type in the text you wish to find, either one word or a phrase. Press the Enter key to start the search, or click OK.
- Optionally, select the Match Case, if upper case and lower case letters are important in the search. Select Match Whole Word Only if you do not want the search to be satisfied for just a part of a word.

If the search is successful, the insertion point is automatically moved to the first occurrence of the target word or phrase and highlighted. If the word or phrase cannot be found, you will receive an error message telling you that no match is found.

Finding and Replacing Text

Perform the following steps:

- Place the insertion point at some point in the document before where you need to find and replace the text.
- Choose the Replace option in the Edit menu at the top of the screen.
- Type in the word or phrase to find.
- Press the tab key to move to the second line to enter the replacement word or phrase.
- Optionally, select either or both of the Match Case and Whole Word option.
- You now have three options: Replace, Replace All, or Cancel. Select the appropriate option.

Text Attributes and Appearance Issues

Text attributes refer to the font or design of type available and how it appears, such as boldface or italics. In addition, you can center text, make the left, right, or both margins justified, and create superscripts and subscripts for mathematical or chemical formulas.

The variety of fonts available to you depends upon the printer you have on your computer and the fonts available within Windows itself. Check what fonts are currently installed by pulling down the Fonts menu at the top of the Works word processing screen.

The text attributes such as boldfacing available in Works for Windows' word processes include plain text, italics, boldface, strikethrough, and underline.

- Select the desired text by dragging the mouse over the text.
- Select the desired attribute from the Format menu under the Fonts and Style option, or the B, I, or U buttons on the toolbar at the top of the screen.

You can perform justification and centering in four ways:

- Left—All lines are vertically aligned on the left side, but not the right side.
- Right—All lines are vertically aligned on the right side, but not on the left side.

- Justified—All lines are vertically aligned on both the left and right sides.
- Centered—Equal spaces on both the left and right.

Use the same strategy as boldfacing, but use the alignment icons on the toolbar.

29.3 SPREADSHEETS

What the word processor does for computing with words, the spreadsheet does for numbers. A spreadsheet is organized into rows and columns. Number, text, or formulas for calculations can be entered into a cell (the intersection of a row and column).

Start the Works spreadsheet system from the File menu (new file, then click on the spreadsheet icon).

The spreadsheet has the parts shown in Figure 29.2.

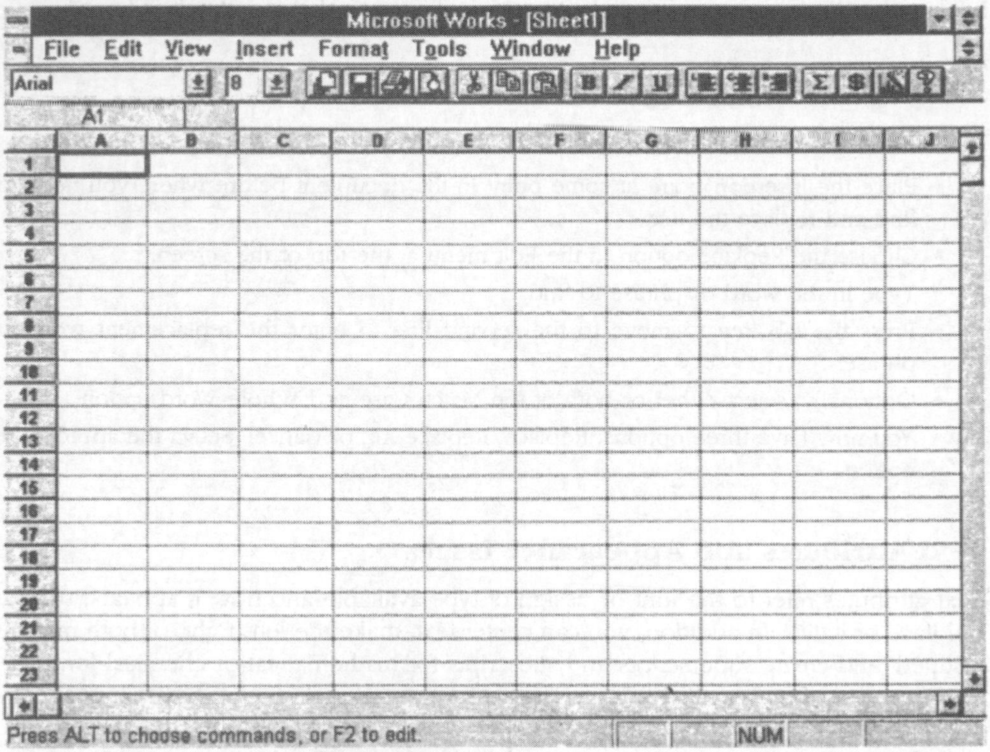

FIGURE 29.2 The Works spreadsheet screen.

There are 10 parts to the spreadsheet screen, each with an important function:

1. **The Menu Bar**—This is at the top of the screen, as in all Works applications, where all the commands available to you will be accessed. There are eight menus from which to select commands. Notice many are the same as other Works for Windows applications, and that several are identical.

2. **The Message Line**—This is always at the bottom of the screen in an area called the message line. This is where the spreadsheet sends you information as you work. The information can include hints, suggestions, warning messages, and brief command descriptions. The most useful of the messages are the brief command descriptions.
3. **The Status Box**—This box, in the upper left-hand corner of the screen, indicates where you are in the spreadsheet, in terms of the cell address (columns and rows).
4. **The Formula Bar**—This is below the toolbar. When information is entered into the spreadsheet, it appears on this line as it is typed.
5. **The Title Bar**—This contains the name of the current file.
6. **The Work Area**—The work area is where the actual numbers are shown as you work.
7. **Scroll Bars**—To move around quickly in a spreadsheet that is larger than can be shown on one screen you will use the scroll bars.
8. **Cells**—Cells are the intersection of a row and a column.
9. **Column and Row Headings**—At the beginning of each row and column you will find a label for each row and column. Columns are labeled with letters, while the rows are labeled with numbers.
10. **The Active Cell**—The cell that would contain data if you were to start typing information into the spreadsheet is the Active cell.

Column Widths

To alter the width of a column, place the mouse pointer at the boundary between columns. Drag the boundary to the right to increase the width of a column, to the left to decrease it.

Saving the Spreadsheet

The strategy for saving a spreadsheet is the same as that used for saving a Works word processing document. Once a spreadsheet has been saved the first time after being created, pressing CTRL-S will save it again.

Moving Around in the Works Spreadsheet

There are a number of keystrokes used to move quickly from one part of the spreadsheet to the other. They are as follows:

Move up one cell	Up Arrow
Move down one cell	Down Arrow
Left one cell	Left Arrow
Right one cell	Right Arrow
Move to the first row entry	Home
Move to the last row entry	End
Down one window	Page Down
Up one window	Page Up
Right one window	CTRL-Page Down
Left one window	CTRL-Page Up
To cell A1	CTRL-Home
To the last cell in the sheet	CTRL-End

Cell Formats

There are three types of information you can enter into a spreadsheet cell:

- Text—Alphabetic characters such as column headings or other words needed to label the spreadsheet.
- Numbers—Numeric characters that represent data.
- Formulas—The set of commands for executing a calculation.

Number Formats

Numbers in spreadsheets can be formatted in a variety of ways. Highlight the cell or cells to be formatted. Access the format menu, and select the desired formatting. The choices are as follows: General, Fixed, Currency, Comma, Percent, Exponential, Leading Zeros, Time/Date, and True/False. Select the appropriate formatting, as desired.

Font, Style, and Border

The font and style features are comparable to the corresponding features in Works for Windows' word processor application.

Spreadsheet Cell Ranges

So far we have talked about actions on the spreadsheet for a single cell. When we have talked about a cell we have used its cell reference with the letter for the column, followed by the row number. How do we refer to groups of cells?

Cell Ranges

A cell range is nothing more than a way of referring to a group of cells. To refer to the cells B3, B4, B5, B6, C3, C4, C5, C6, D3, D4, D5, and D6 we start with the first cell in the upper right corner of the box, followed by a colon, then the last cell in the lower right corner. If we are defining a single row it would be the left most cell, a colon, then the right most cell. There are several reasons for selecting ranges of cells in the Works for Windows spreadsheet:

- To format the numbers all the same. When the cell range is selected, number formats work upon the entire selection. In this way you can format large areas of a spreadsheet quickly.
- To format the justification: left, right, or centered or set styles, fonts, and borders.
- To refer to the area in a formula. You might want to compute the sum of a cell range, in which case you could type it into the formula, or insert it directly without typing, as we will show you later.

To select a group of cells, follow this procedure:

- Move the Active cell to where the upper left corner of the cell range will start. Do this most simply by clicking the mouse on the target cell.
- Drag the mouse to the lower right corner of the desired cell range, and release the mouse button.

You can then command some action on the cell range such as a number format or other function that allows for a range such as a formula.

Formulas and Simple Functions

To manipulate numerical data in the Works for Windows spreadsheet you will use formulas and simple functions. Formulas in your spreadsheet allow you to create a mathematical expression about some of the data and calculate a result. Since formulas are written with the cell references and not the actual values in the cells, we can change the data many times while the formulas recalculate automatically.

The mathematical operations that can be carried out are accomplished using symbols. These symbols act upon a reference to a cell. There are five basic operations denoted by symbols in the spreadsheet:

Symbol	Mathematical operation	Example
+	Addition	A2+B2
–	Subtraction	A2–B2
*	Multiplication	A2*B2
/	Division	A2/B2
^	Exponentiation	A2^B2

You can also use parentheses to indicate the order in which calculations will be performed.

Entering Formulas

There are four steps to entering a formula onto the Works for Windows spreadsheet:

- Move the Active cell to where you want to make the formula entry. This should be an empty cell unless you want to replace the data in the cell with the results of your formula.
- Type an equal sign to let the Works for Windows spreadsheet know what follows is a formula and not just a data entry.
- Indicate by typing the cell references and the operations to be performed. Instead of typing cell references, it is simpler and more accurate to click on the cell to be used as a reference. The cell reference will automatically be inserted in the formula.
- Press the Enter key or click on the check mark in the Formula bar, and the calculation will be performed and the result displayed in the cell.

Functions

Functions further assist you by making some shorthand expressions for the more common calculations you will be doing. For example, you can use a function to calculate the addition of cells B4, B5, B6, and B7 by typing

```
=SUM(B4:B7)
```

Works for Windows Spreadsheet Function

Example	Format
Average	AVG(P34:P45)
Integer part	INT(P34)
Logarithm	LOG(P34)
Maximum value	MAX(P24:Q49)
Minimum value	MIN(P24:Q49)
Square root	SQRT(P34)
Sum of numbers	SUM(P34:P45)

Other functions are available for financial, statistical, and scientific related calculations. There are a total of over 50 functions available to you in the Works for Windows spreadsheet.

Printing the Spreadsheet

Printing consists of two operations:

- Formatting the page
- Printing the spreadsheet

Formatting the page to print makes adjustments to the width and length of the printed area on the page; margins around the printed area; and placing headers, footers, and page numbers on the printout if needed.

To adjust the format of any of these parameters select Page Setup. Make changes as desired. Print the spreadsheet by clicking on the Printer icon on the toolbar, or the Print command from within the File menu.

Listing Spreadsheet Formulas for Documentation

As the spreadsheet becomes more complicated with formulas and data you will want to print some form of documentation. This documentation is a good way to check your formulas and to show others how the spreadsheet works. To show formulas instead of data or calculated results, select View Formulas from the View menu.

Copying from Spreadsheets to Word Processing Documents

Anyone who has ever typed in a table into a word processing document knows how much work it is to get the spacing correct, and have all of the decimal points line up correctly.

Since we can copy between a Works for Windows spreadsheet window and a Works word processing document, you have a way to avoid the tedious typing of tables. Perform the following steps:

1. Create a Works for Windows word processing document, if necessary, in one window, and the spreadsheet in a second window. You can use documents you already have or create new ones, as appropriate.
2. Highlight the source information from the spreadsheet window, using the mouse.
3. Press CTRL-C to copy the material to the clipboard.

4. Go to Window on the menu, and select the target window, which is the word processing window, in this case.
5. Move the cursor to the target location in the word processing document, and press CTRL-V.

Copying from Spreadsheet to Spreadsheet

Copying data between spreadsheets is also possible with Works for Windows. The steps for copying data from one spreadsheet to another are as follows:

1. Open your old spreadsheet that will be the source spreadsheet.
2. Open the new spreadsheet that is to be the recipient of data from the old spreadsheet.
3. With the mouse, highlight the material in the old (source) spreadsheet to be copied.
4. Press CTRL-C to copy the highlighted material to the clipboard.
5. Switch to the target (new) spreadsheet through the Window option at the top of your screen.
6. Position the cursor on the cell that will be the upper left-hand corner of the material to be copied.
7. Press CTRL-V to complete the copying process.

29.4 GRAPHING IN WORKS

There are five basic types of graphs available to you using Works for Windows. Variations on these basic graph types are available through extensive Gallery options. Each graph type and Gallery option offers a different visual format for presenting the data. The type you select for your purpose will depend upon:

- The form of the data, and its function.
- The type of presentation required.
- Which graph best represents the meaning of the data without misrepresentation.

You should experiment with the different types of graphs as you work through making charts from your data. This experimentation should help you to look at which type best presents the data in the most easily understood way. You should strive for clarity, completeness, and simplicity. The value of a graphic should add to, not detract from, the presentation of data. Keep in mind that Works can chart or graph only up to six variables at a time.

Works for Windows supports bar, line, pie, stacked line, X-Y (scatter), radar, and combination charts, as well as a variety of three-dimensional appearance enhancement features.

The strategy for creating a chart is simple.

1. Highlight the area on the spreadsheet you wish to chart.
2. Select Create New Chart from the Tools menu.
3. Follow the sequence of prompts in the dialog box.
4. Select a chart type from the pull-down list provided.
5. Select or change any additional parameters in the dialog box.
6. A sample of your chart will appear at the right side of the dialog box.
7. Click on OK when everything is as you wish. The graph appears on your screen.

Printing a Graph

Print the graph from the usual options in the File menu, or copy and paste the graph into a word processing document.

29.5 DATABASE TECHNIQUES

The database application in Works for Windows allows you to create a place to store data, organize it, maintain it, and make it accessible in the form of a report. A database in Works for Windows is made up of records, which are in turn made up of fields. Think of the database as very similar to a spreadsheet, only with just the data, but no complex calculations. Rows of a spreadsheet are equivalent to records in a database, and the intersection of a row and a column is a field in a database.

Although the database is similar to a spreadsheet, they are different in what they are used for. Spreadsheets are primarily used to make some kind of calculation, while a database is used to store and collect large amounts of information. Calculations can be performed later on the information in a database.

Designing and Creating the Database

The overall design of the database will be dependent upon

- The end use of the database (reports, archives).
- The kind of data that will be stored (numbers, text).
- The amount of data (a few addresses, thousands of records).
- What reports will be expected from the database.
- What data will be included in a database report.

A little planning up front will help to make the database more useful, easier to maintain, and easier to design with Works for Windows. The information you decide to put into your database is dependent upon what you need it for.

Although you will not be able to anticipate every way you could use the database, make sure you plan ahead as much as possible before you get started. It will save you some time and effort to design it correctly one time versus changing it every time you want a new report.

Let's make a database in Works for Windows. At the opening screen, select Create a New File, or respond to the standard startup screen by clicking on the database button.

When you have done this, the opening screen of the database application will appear in the Works for Windows workspace area, as shown in Figure 29.3.

Forms

The Works for Windows database is designed by defining a form. This form is similar to one you might use on paper to collect information. The Works for Windows database form is an on-screen representation of the columns of data that will be collected for our database. A completed form will constitute one record.

Each field of data on the Works for Windows database form consists of two parts:

- Titles—The name of the field, for example, Name, Address, City, State, ZIP code. You may have up to 256 fields in your database.

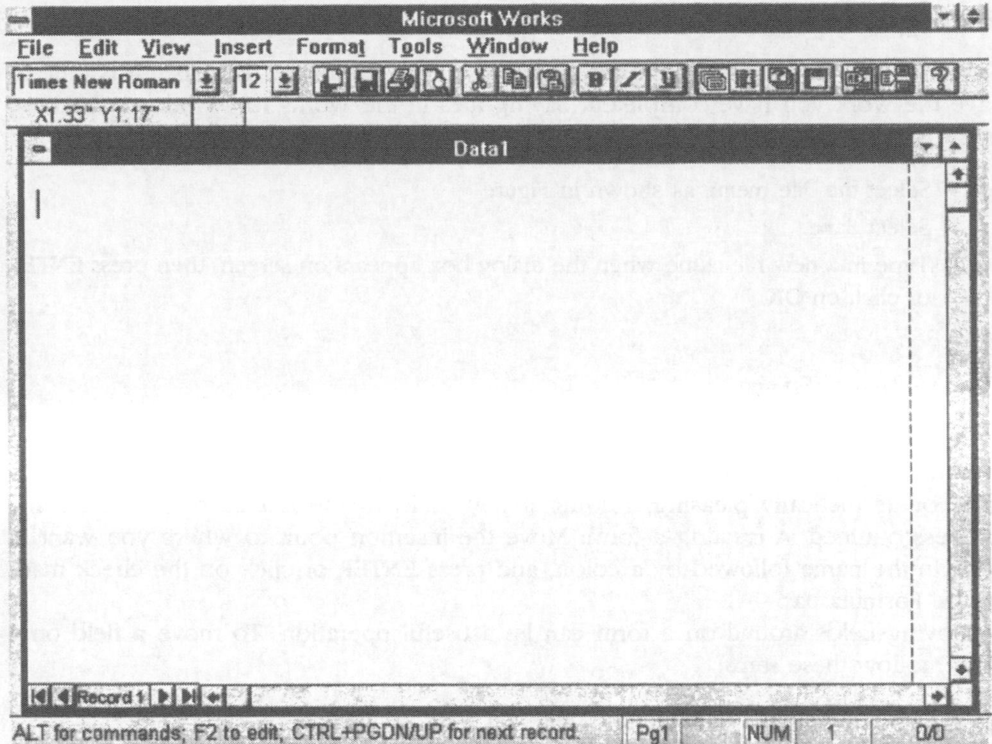

FIGURE 29.3 Initial database screen.

- Space to enter the data—The space should be long enough to account for the longest entry we might reasonably make. For example, the state field might only be two characters wide, while a street address might have to be 20 characters wide.

The form will also become the place where you fill in the data, just as you might on a paper form. Every Works for Windows database you create will have a single form.

To create the form follow these steps:

- Move the insertion point to where you want the field to appear on the form. As you do this try to think of how the whole form will appear.
- Next, type the name of the column, followed by a colon (:). The colon tells the Works for Windows database that the description we typed is a column name.
- After pressing the ENTER key or clicking on the check mark in the Formula bar (just like the Works for Windows spreadsheet application), a dialog box will appear asking for information about the column you are creating. This information includes Width and Height.

The width of the field is the total number of characters needed for each record entry in this column. The default value Works for Windows proposes is 20 characters. If this is acceptable leave it alone; if you want to change it, enter a new number in the space provided. The height entry is the number of rows high that the column will be. Usually the height will be one, which is also the default value.

The height of the field refers to how many lines high the field might be. If you want a field to have several lines in which you can enter data, increase the height.

Saving Your Work

Now that you have spent some time creating a form for a new database, it is time to save the work you have completed. Saving files in the Works for Windows database is the same as all the other Works for Windows applications. Perform these steps:

- Select the File menu, as shown in Figure
- Select Save.
- Type in a new file name when the dialog box appears on screen, then press ENTER or click on OK.

Changing a Form

You might need to change the form at some point to accommodate new data that may become important or to move the fields around within the form to make it more usable or aesthetically pleasing. Adding a new field to the database uses the same process outlined in creating a form: Move the insertion point to where you want it, type in the name followed by a colon, and press ENTER or click on the check mark in the Formula bar.

Moving fields around on a form can be a useful operation. To move a field on a form, follow these steps:

- Using the mouse, highlight the field you want to move.
- Drag the field to the new location with the mouse. The pointer of the mouse becomes a hand.
- Release the mouse button when the field is in the desired location.

The form is not a report, just a means of getting the data entered into the database. In a later section of this chapter, we will outline how to create a custom report.

Entering Data into the Database

Entering data into the database is as simple as moving from field to field and typing in the data. To move from field to field the Works for Windows database application uses the TAB and SHIFT-TAB keys. The TAB key moves the highlight from field to field in a forward direction, clockwise from top to bottom. The SHIFT-TAB key combination works in the reverse of the TAB key. It moves from field to field in a backward direction, counterclockwise from bottom to top. Continue entering your data until you have a complete the database.

Database Maintenance

Once your initial data have been entered into a database, the function of the application changes a bit. The main task changes from initial organization to maintenance. The primary database maintenance functions are:

- Editing—Changing entries already made in individual records.
- Adding Records—Appending records to your database.
- Deleting Records—Removing records from the database.

- Adding Fields—Adding another piece of data to be collected for each record.
- Deleting Fields—Removing pieces of a database that have been previously included.

Editing

Making edits to the database or particular records becomes necessary because the data have changed in some way. Changes can also include adding data into fields that were not entered when the rest of the data was placed into the database. To make a change follow these steps:

- Move to the record you want to edit. To do this quickly, use the key combinations CTRL-Page Up or CTRL-Page Down. CTRL-Page Up will move to the previous record in the database, while CTRL-Page Down moves to the next record.
- Move to the field you wish to edit. To move quickly between fields, use the TAB and SHIFT-TAB keys. TAB will move forward one field at a time, while SHIFT-TAB will move backward. Alternatively, click the mouse on the field you wish to edit.
- Finally, edit the record by clicking the mouse in the Formula bar. In the Formula bar you can edit the entry using the keyboard. When you have finished making the edit, press ENTER or click on the check box, and you will return to the record in the form with the new value in the field.

Adding Records

To add records at a later time:

- Press CTRL-Page Down until the last record passes from view or select Insert Record from the Edit menu.
- Press the TAB or SHIFT-TAB keys to move to the correct field (probably the first one).
- Enter the data for each field, pressing TAB to move to the next field. When you have entered the last field, press ENTER and you are done.

The only practical limit to the number of records you can have is a function of the disk space available to you to store the database file.

Deleting Records

You will find that from time to time it becomes necessary to remove entire records from the database. To permanently remove a record from the database, do the following:

- Move to the record in the database that needs to be discarded.
- Open the Edit menu, and select Delete Record/Field.

Adding Fields

Once the database is set up, it is inevitable that for some reason or another you will want to add a field. Accomplishing this is essentially the same as it was when you set up the original form. To create the form follow these steps:

- Move the insertion point to where you want the field to appear on the form with the mouse.

- Type the name of the field, followed by a colon (:).
- After pressing the ENTER key, complete the dialog box by entering the width and height of the field.

Deleting Fields

To remove a field from the database, follow these steps:

- Highlight the field you wish to delete.
- Select the Delete Field option from the Edit menu.
- A dialog box will then appear asking for your confirmation of deleting the field. Click on OK, if that is what you really want to do.

Once you have removed a field, this information is deleted from every record in the database.

Sorting and Searching a Database

In this section of the chapter we will discuss two important functions that will be very useful to you in using the database: sorting and searching. With these two functions, you will be able to increase the way in which you functionally view the data that have been entered into the database.

Sorting

Sorting a database rearranges the records in some logical order based on one of the fields. Sorting records will allow you to

- Organize information.
- Put records into an order that is important.
- View the data in an order other than how they were entered.

In the Works for Windows database application, rearranging is determined by alphabetic or numeric order. Alphabetic or numeric order can be either ascending order or descending order. To sort your Works for Windows database file follow these few steps:

- Open the database to be sorted if it is not already.
- Select Sort Records from the Select menu.
- When the dialog box comes on-screen, fill in the name of the field you want to sort the database by. Then indicate ascending or descending order by clicking the mouse on the appropriate button. The default is an ascending sort.

Note that at this point it is possible to sort by up to three fields. Filling in up to two more fields will sort your database by other fields within the first field.

- When you have completed the dialog box, click on OK or press ENTER, and the database will be sorted as you requested.

Searching

Often you will want to look at a particular record or set of records that meets some criteria. These criteria are usually thought of in terms of a question. In the Works for Windows database, applying a query, or question, to the database limits the records that are viewed. The records that are viewed after a query are those that satisfy the conditions defined in the search. In the List view of the data, all the records that satisfy the query show up on the screen in column format. In the Form view, records show up one at a time.

To create a query of a database in the Works for Windows database application, follow these steps:

- Open the database file you wish to work with.
- Select Query from the View menu. An apparently blank Form view appears.
- When you select this command, what looks like the Form view will come onto the screen. This is actually the place where you will write the query of the database.
- The query is entered into the form. Essentially, the data you enter onto the query form are used to find records that match your entry.
- When you have completed filling in the criteria in the form for which to query the database, switch to List view or Form view from the View menu. You can also click on the toolbar item that looks like a form (Form view) or list (List view). This will return to your previous view showing only the records that satisfy your conditions.

To return to looking at all the records in the database (not just the results of the query) select the Show All command from the Select menu.

Printing the Database

To print the view, select Print from the Print menu.

Defining and Printing Reports

A report is a more formal presentation of some or all of the data found in a database. The Works for Windows database reporting function will help you to

- Share data with others via paper printouts.
- Make presentations more effective.
- Bring together pertinent information, versus all fields and records.
- Provide a printout of your database or some selection of it.
- Combine the function of selecting, sorting, and summarizing information.
- Show a different aspect or perspective on the database information.

You can include or exclude many optional parts of a report:

- Introductory text.
- Sorted groups of data from your database.
- Column totals or formulas on fields, much like a spreadsheet.
- Formatting, including spaces, blank lines, page numbering, headers, and footers.

Where to Create the Report

After you have opened or created your database and have entered data you are ready to make it into a report. Works for Windows database reports are created using the New Report command, located in the Tools menu. When you have selected the command you will see the New Report dialog box. The New Report dialog box enables you to select which fields are to be included in the new report, and give the report a title. To do this, perform the following steps.

- Fill in the Report Title box with the desired text.
- Select one or more fields to be included in the report. The order you select them will determine their order in the report. Do this by clicking on the field name in the field list and clicking on the Add box. A list of fields to appear in the report will be given in the Fields in Report list. Do this for each field you wish to add to the report.
- To remove a field from the report, click on the field name in the Fields in Report list, and remove it by clicking the Remove button.
- When you have added all of the desired fields, and are satisfied, click on the OK button.

The Report Statistics dialog allows you to calculate such items as sums, averages, counts, and so forth, for any column of fields in the report. When all statistics have been determined, click on the OK button. The following statistical functions are available:

- SUM( )—The sum of a column.
- AVG( )—The average of a column.
- COUNT( )—The number of records.
- MAX( )—The largest number.
- MIN( )—The smallest number.
- STD( )—The statistical standard deviation.
- VAR( )—The statistical variance.

Works for Windows now automatically creates a report for you. To view the actual report, go to the File menu and select Print Preview. Select Zoom In for a closer look, and print the report, if you wish, at this time.

Understanding the Report Definition

Each part of the report that you set up will be made up of four parts:

- The Introductory report includes two lines for a title or other information to document what the report is about. The information will appear only on the first page of the report.
- The Introductory page includes two more lines following the introductory report. The first line is a field name; the second is blank and can be customized for your reporting purposes. This information appears at the top of every page.
- The third section is where the data show up.
- The fourth and final section is the Summary report area. This area is the final two lines of the report where summary information can be entered.

To add a column, follow these steps:

- Select the entire column next to the column of data where you want the new column to appear. Highlight the column by clicking at the top of the column with the mouse. The new column will move this column to the right by one.
- From the Edit menu, choose Insert Row/Column from the menu and the column will be added to your report definition.

You can also adjust the width of the columns by dragging the column boundaries with the mouse, as you did in the Works for Windows spreadsheet application. Likewise, you can format the entries using Format menu commands. These techniques are identical to those covered in the spreadsheet chapters.

To add the data from your database into the report in these new columns, you will enter information next to the record item label. All that is done to accomplish this is to enter the field name of the data item from your database next to the record label, and in the correct column. The correct column would be under the right label you created. In the example that follows, we have created an additional column at the left of the report.

Saving Your Work

Saving the report definition is the same as saving files in any of the other Works for Windows applications: Press CTRL-S, or select Save from the File menu. If this is the first time you have saved your report definition file you will of course be prompted for a file name. Once you have entered this name, press ENTER to save the file.

Printing and Viewing the Report

To view the report on-screen as it will print, go to the Edit menu, and select Print Preview. If it is as desired, select Print to print the report.

Selection and Queries for Reports

Selecting only certain records is useful because it narrows the amount of information you might want to work with. In very large database this is essential to the functionality of maintaining and using the information. Another reason is to simply ask questions (query) the database for information.

The Works for Windows database application allows you to look for specific entries, then examine them one at a time, or in the List view. You can also create reports using the Report view to output just the information that satisfies your query.

Searching for Records

In its simplest form, a query is a search for particular records in a database. To search for a particular record, open the Tools menu and select Create New Query. A dialog box will then prompt you for what it is you want to search for.

Queries and Reports

Once you have defined a query you should return to the List view to see the records that have been selected. To report a specific set of data that has been returned by a query, go into the View/Report after the query has been run. The report you create or use (from one that has been saved) will be run on the query that has just been completed.

Copying between Spreadsheets and Databases

As pointed out earlier in this book, you can view a database and a spreadsheet in Works as table-oriented constructs. In light of this, the columns in a spreadsheet correspond to fields in a database, and the rows of the spreadsheet correspond to the records of a database.

Again, using the general scheme that we have used before, follow these steps for moving data from a spreadsheet to a database:

1. Create or open a database file with fields to match the columns of the source spreadsheet.
2. Create or open the spreadsheet file that you will copy data from to the target database.
3. Mark the rows and columns in the source spreadsheet with the mouse.
4. Select Edit and Copy from the menu at the top of the screen, or press CTRL-C.
5. Use the Windows option on the menu at the top of the screen to switch to the database window.
6. Place the cursor on the first field of the first record where you wish to insert data. The database must be in List view mode (rows and columns).
7. Select Record/Field from the Edit menu, and press the ENTER key.

The data will be transferred from the spreadsheet to the database.

The process for moving data from a database to a spreadsheet is just the same as above, substituting spreadsheet for database and database for spreadsheet in the instructions. The Copy command is used to insert the information from the database into the spreadsheet.

With the foregoing capability, you can take any data from your database, convert it to spreadsheet format, and create a chart or graph for more complete analysis and charting or graphing of the data.

Copying from Word Processor Documents to Databases

It is easy to copy text from a Works word processing document into a database. For example, you may have information in a letter or other document that should be transferred to a database. The procedure to transfer data from a word processing document to a database is quite straightforward. Just follow these steps:

1. Open the database, or create it, if necessary. Make sure that the fields are of sufficient size and characters or numbers to fit the data that you will transfer from the word processing document. Make sure that the database is in "List view" mode (columns showing each field, rows showing each record).

2. Open the word processing document.
3. Follow each item in the word processing document by a TAB key press as a field separator. Each record goes on a separate line, terminated by pressing the ENTER key.
4. Highlight the lines of text to transfer to the database.
5. Go to the Edit menu and select Copy, or press CTRL-C.
6. Switch to the database window and position the cursor to the first field of the record where you will to copy the data.
7. Press CTRL-V or Paste from the Edit menu to insert the text.

The data from the word processing document are now transferred to the target record in the database.

29.6 MICROSOFT DRAW

Windows and Works for Windows form a handy graphical environment. Works for Windows contains a useful program for creating and editing drawings. This program is called Microsoft Draw. With this component of Works for Windows, you can create artwork to print directly or to include in your Works for Windows word processing documents. In the word processing application, select Drawing from the Insert menu. The screen shown in Figure 29.4 will appear.

For those readers who have used a drawing package, such as the Paint program in Windows, the Draw feature works much the same, but is more limited in its range of capabilities.

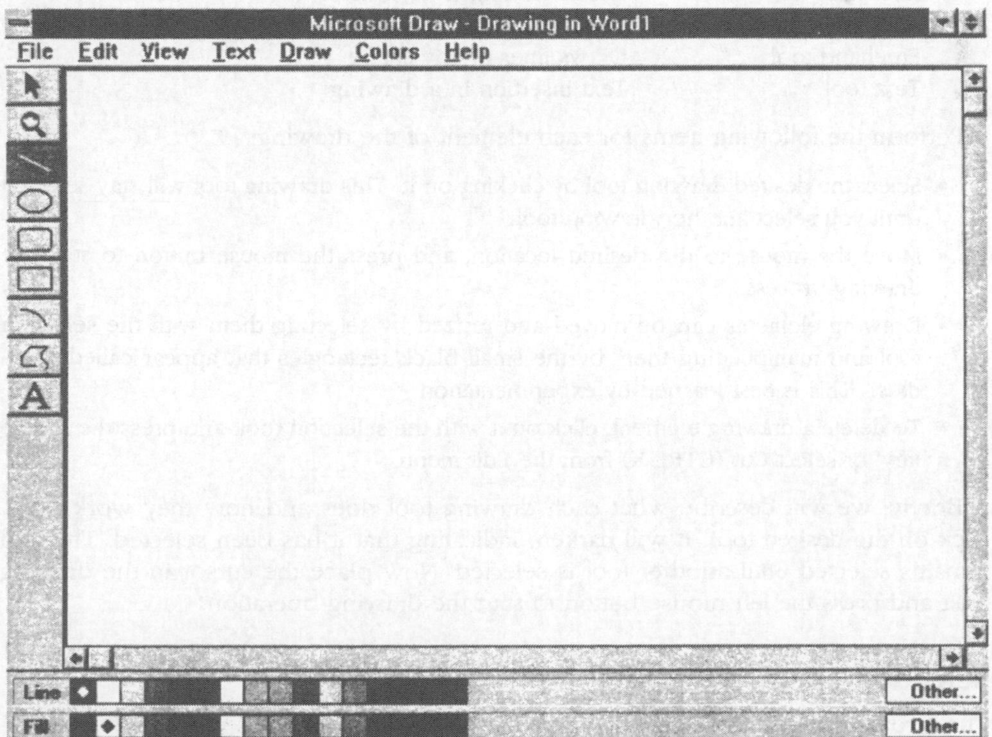

FIGURE 29.4 Microsoft Draw screen.

There are five things to notice on the screen in Figure 29.4:

1. The drawing tools on the left side. These are used to create lines, shapes, fill areas, cut, and paste drawing elements. These are the tools equivalent to what might be found on an artist's desk.
2. The color palette at the bottom of the screen. This is used to select the color for drawing lines, shapes, and filling areas.
3. The menus at the top of the screen. These are similar to all standard Windows applications for saving and loading files, editing, and setting program parameters.
4. The drawing area that constitutes the balance of the screen. This area is where your drawing will be created and edited.
5. The usual vertical and horizontal scroll bars to move around a large window.

The color palette at the bottom of the screen is used to select colors. To change the current line or fill color, simply click with the desired mouse button on any color on the palate areas. There are two areas: one for line colors and one for fill colors. We will discuss fill color later.

The tools palette contains the following tools, going from top to bottom on the tool palette:

Selection tool	Selects objects—the arrow symbol
Magnifying glass	To take a close look at a drawing area
Line Tool	Draws straight lines at any angle
Oval tool	Draws ovals (ellipses) or circles
Rounded-box tool	Draws rectangles or squares with rounded corners
Rectangle tool	Draws rectangles or squares
Arc tool	Draws 90° arcs of circles or ellipses
Freehand tool	Draws lines of any shape
Text tool	Text insertion in a drawing

Perform the following items for each element of the drawing:

- Select the desired drawing tool by clicking on it. This drawing tool will stay selected until you select another drawing tool.
- Move the mouse to the desired location, and press the mouse button to start the drawing process.
- Drawing elements can be moved and resized by selecting them with the selection tool and manipulating them by the small black rectangles that appear called "handles." This is best learned by experimentation.
- To delete a drawing element, click on it with the selection tool, and press the Delete key, or select Cut (CTRL-X) from the Edit menu.

Briefly, we will describe what each drawing tool does and how they work. First, click on the desired tool. It will darken, indicating that it has been selected. The tool remains selected until another tool is selected. Now place the cursor in the drawing area and press the left mouse button to start the drawing operation.

Selection Tool

The arrow symbol is the selection Tool. The selection tool allows you to select an object for editing. When you select an object, small black squares appear at the corners or ends of the drawing element selected. You can then drag the entire object to another location for exact positioning or change the shape of things like rectangles. Shape or size changing is done by dragging the handles, and positioning left, right, or center is done by using the justification feature as you do for text.

Magnifying Glass

This is used to expand the view, as if using a magnifying glass. Values of 200, 400, and 800% magnification can be obtained. This is equivalent to using the sizing options on the View menu at the top of the screen.

Line Tool

This is a simple straight line drawing tool. Press the mouse on the desired starting point. Drag the mouse to the end point and release the mouse button. A straight line will be drawn between the two points. You will notice as you move the end of the line that you can see where the line will fall. This is known as "rubber banding." When you release the mouse button, the end of the line is determined. Again, you can easily drag it into a final location by selecting it with the selection tool and positioning it with the mouse.

Circle/Ellipse, or Oval Tool

This tool is used to create circles or ellipses. Press the mouse button where you want the top of the circle or ellipse to be. Move the mouse to get the desired shape and size. Release the mouse button to anchor the circle or ellipse. You can think of the circle or ellipse as being inscribed in a box. Use the handles that we have previously discussed to obtain the exact position or shape you desire.

Rounded-Box Tool

This is used to create rectangles with rounded corners. Use it the same as the rectangle tool. This feature is handy for boxes on charts.

Rectangle Tool

Press the mouse button for the upper left corner of the rectangle. Drag the mouse to the point for the lower right corner of the rectangle and release it. An open rectangle is formed on the screen.

Arc Tool

This tool draws one quadrant of an ellipse or circle. The mouse position determines whether you have an elliptical arc or a circular arc.

Freehand Tool

This tool serves two purposes. If you use a selection of mouse clicks, you will obtain a series of connected straight line segments. If you hold the mouse button down, the cursor changes into the form of a pencil, and you can draw freehand lines.

Text Tool

One feature you will probably use frequently to place text in your drawings is the text tool. This technique can be used in figures to add captions and call-outs for clarifying some point in the text. If you want to use text in your drawing, perform the following steps:

- First, select the desired text attributes by dropping down the Text menu at the top of the screen. Make selections for font, text size, position, and appearance, such as bold or italics. These setting will remain in place until you change them.
- Select the text tool from the Toolbox by clicking on it with the mouse.
- Click the mouse on the place in your drawing where you want to position the text.
- Enter the text just as you would with the Works for Windows word processor.

Microsoft Draw Menu

The menus at the top of the Microsoft Draw screen are designed slightly differently than the menus in the other Works for Windows applications. This is primarily because the Drawing program is implemented only from within a Works for Windows application and not as a stand-alone program.

29.7 DATA COMMUNICATIONS

Works has a data communications capability very similar to the Windows Terminal program, included as part of Windows. You should read this article for further background in doing data communications in a Windows environment. With a telecommunications system consisting of a microcomputer, a modem, and the communications software provided with Works, it is possible to

- Send or receive information to or from another microcomputer.
- Receive or send information from or to a public database.
- Post and receive messages from a computer bulletin board that is a public database.
- Communicate with a central minicomputer or mainframe computer, using the Works communications software to emulate a standard computer terminal.

Settings

A number of steps are involved in the communications process:

- Establishing a telephone connection. This requires a modem and a telephone line that will access the remote computer through the dialed telephone network. The remote computer also has a modem to convert the signals from the telephone network back into a form acceptable to it.
- Obtaining access to the other computer by means of an account code and password (logging on).
- Emulation of the desired terminal for proper interaction with the other computer, if necessary. Some computers need a special terminal to utilize the features of the software on that computer. For example, a Digital Equipment Computer (DEC) needs a VT-100 terminal emulator to take advantage of many of the programs available on that computer.

- Optionally, passing files between the two computer systems.
- Disconnecting from the other system in an orderly manner (logging off) and breaking the telephone connection.

Some of these processes are associated with computer hardware, and others are functions of Works' telecommunications software. You will need a modem to use the Works communications capability. Works will automatically interrogate your modem for its connection port and baud rate. If you already know these parameters, go to the Settings menu item and set the various items, such as port number, baud rate, terminal, and transfer protocol.

You should next go to the Phone menu item and select Easy Connect or use the Telephone icon on the toolbar. Enter the telephone number for the data service or bulletin board you wish to use. You should also save this information with CTRL-S or Save on the file menu. Select a file name that reflects the name of the communications service. Click on OK to initiate the telephone call. On completion of the call, click on the button next to the telephone, if your data service does not automatically hang up your telephone. Also, you can disconnect by closing the Communications window.

Uploading and Downloading Files; Protocols

Computer bulletin boards and databases have the capability of receiving or sending entire files. This feature is especially useful if you want to exchange data or programs with another user of the bulletin board.

If you have a noisy telephone line that has static or heavy induced electrical noise from equipment such as motors, a transmitted file is liable to contain errors. If this file happens to be a computer program in machine language, the program may not run correctly.

There are a number of ways to pass files from one computer system to another without errors. These techniques are known as transmission protocols. The two most common ones used on microcomputers are called XMODEM and KERMIT. Works telecommunications system supports three versions of the XMODEM protocol, as well as the KERMIT protocol. See the article on KERMIT and XMODEM for a complete discussion of these topics.

Sending a File

First, we shall explain how to send a file to a remote computer. The steps are very simple.

- Connect to the remote computer, as was shown earlier in this chapter.
- Inform the remote computer that you are going to send it a file. The details will be specific to the remote computer.
- Select Send File from the Tools menu at the top of the screen.

Simply fill in the name of the file you wish to send. If it is not located in your current drive and directory, set the drive and directory to correspond to the location of the file, just as you did when you were setting the drive and directory for a word processing, spreadsheet file, or database file. Make sure to have previously set the appropriate protocol that is acceptable to the remote data service.

Receiving a File

Receiving a file is done with a process similar to sending a file. Just follow these steps:

- Connect to the remote computer, as was shown earlier in this chapter.
- Inform the remote computer that you are going to receive a file. The details will be specific to the remote computer.
- Select Receive the Tools menu at the top of the screen and click on OK. Make sure that you have previously selected the appropriate protocol, and named the drive and directory in which to store the received file.

DEFINING TERMS

Argument: The value(s) in a database or spreadsheet upon which the function acts.

Baud Rate: The speed at which data are transmitted and sent through a modem. The higher the baud rate, the faster the transmission.

Cell: The box space framed at the juncture of a column and row. The address of a cell is a combination of the column letter and the row number.

Cell Range: A group of adjacent cells in a spreadsheet. They may be selected by dragging from the first to the last, clicking on the upper left cell typing two periods, and clicking on the lower right cell, or clicking on the row or column label to select that row or column.

Cell Reference: The cell address used on a formula. Selecting by clicking rather than typing helps to eliminate mistakes.

Clipboard: A temporary file that stores information *cut* or *copied*. The information will stay there until something else is copied or cut or the file is closed. Only one set of information can be stored at a time.

Copy: The command found under the Edit menu that copies onto the clipboard the highlighted text, data, or graphic. The information remains in the original position and can be pasted to other locations.

Cut: The command found on the edit menu that removes highlighted text, data, or graphics from the document and transfers it to the clipboard. The *undo* command can reverse this action.

Drawing Tool: The tool in the toolbox that allows the user to enter the draw environment and draw lines or shapes.

Entry Bar: The line at the top, of a spreadsheet that identifies the cell location and shows the data being entered. Editing in a spreadsheet is done in the entry bar.

Find: A choice under the menu in a database that is used for searching for records with specific criteria.

Modem: The communication device that connects one computer to another over a network or telephone lines.

Paint Tool: The brush in the toolbox that allows you to enter the paint environment by creating a paint frame.

Paste: The action of copying the contents of the clipboard to the place selected by the cursor.

Record: The collection of data fields in a database that is relevant to one entry.

30

ClarisWorks

Judith W. Whitcomb

30.1 THE INTEGRATED PACKAGE

ClarisWorks (a trademark of Claris Corporation) is a package of six document types, text, draw, paint, spreadsheet, database, and communications, that can be integrated easily without closing windows. By drawing a frame with a text, spreadsheet, or paint tool in a document in one environment, you import into that frame the environment of the tool used and bring along all the tools that are active in that environment. The addition of a paint program in version 2.0 allows for much more sophisticated documents that are suitable for publishing. ClarisWorks also provides

- On line help that can be accessed under the question mark at the right end of the menu bar if you are operating under System 7.0 or later. With System 6, the file is found under the Apple on the left of the menu bar or can be accessed by *command?*. It is not necessary to exit the document in which you are working as you can leave *Help* open and refer to it as necessary. You can drag *Help* around to a convenient place on the screen by dragging on the horizontal bars as you do with other windows.
- Commands can be chosen from the menu and submenus with the mouse or can be keyboarded using the *shift*, *command*, *option key*, and a letter. Most of these shortcuts are intuitive with the letter being the first letter of the word involved, i.e., *command S* for *Save* and *command U* for *Underline*. In almost all cases there are keyboard and mouse alternatives. A third method of accessing commands is to use the shortcut icon menu found under file on the menu bar. Personal editing of those shortcuts can also be done in that file. One click on the appropriate icon will affect the highlighted text (see Fig. 30.1). The shortcuts and their icons are different in each application, but work in the same way.
- Easy importation of documents created under other Macintosh programs by just dragging the foreign file onto the ClarisWorks icon. In many cases ClarisWorks determines which program created the document and translates into a ClarisWorks file. Some translations require more input from the user.

0-8493-2530-7/96/$0.00+$.50

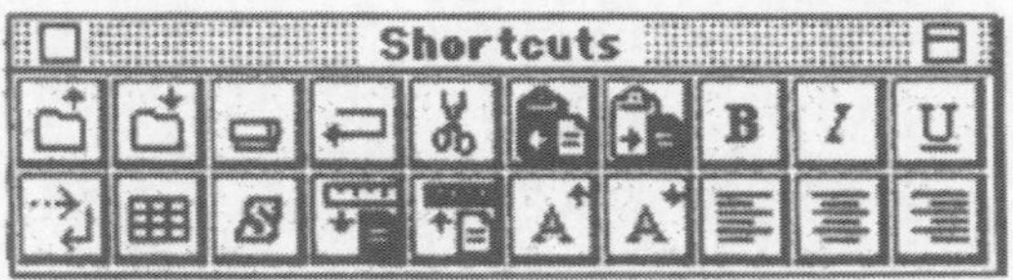

FIGURE 30.1 Shortcut icons.

ClarisWorks requires a Macintosh Plus or later model with an internal or external hard drive, 800 K floppy disk or a 1.4-Megabyte Superdrive disk drive. It needs 1Megabyte of memory under System 6 or 2Megabytes for any version of System 7 as it uses 600K of RAM. If System 7.0 or 7.0.1 is used, a system tuneup kit (available from Apple Computer, Inc.) should be ordered to make better use of available RAM.

30.2 WORD PROCESSING

To open a word processing document:

- Choose *new* from the File menu.
- In new document dialog box choose *word processing*.

ClarisWorks has automatic wrap and an easy access ruler to set tabs, margins, justification, and line spacing (Fig. 30.2). On the far right of the ruler is the columns bar, which allows a setting from one to nine columns. One click will rearrange the type into a progressively higher or lower number of columns on the monitor until you find the one that suits. The ruler can be toggled on and off with *show/hide* tools under View on the menu bar. The triangles on the right and left of the ruler set the margins. Margins and tabs are dragged to their positions and removed by dragging them back. Moving the margin marker when the cursor is before a paragraph will change the margin for that paragraph in already entered text. The upside down T is the automatic paragraph indentation. Pressing the return key starts a new paragraph. If block printing is desired, the paragraph indentation and the margin can be dragged together.

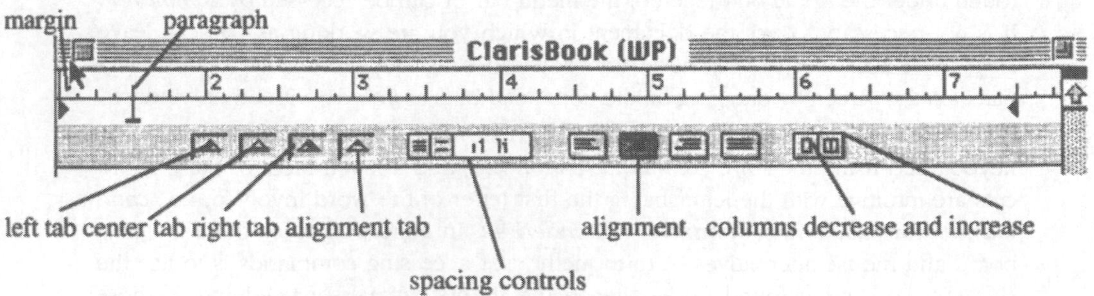

FIGURE 30.2 ClarisWorks ruler.

Text

To enter text:

- Set margin paragraph indent and tabs where needed. These are easily changed later if they are not correct.
- Choose justification from ruler. This may also be changed easily later.

- Choose style, font, and size by pulling down the proper heading in the menu.
- Keyboard in the desired text.
- To show invisible characters such as spaces, paragraphs, and tabs, choose the option under *preferences* in the Edit menu or toggle on and off by using command-semicolon.

To edit text for mistakes:

- Position cursor by using the arrows or dragging the mouse to a position one space to the right of that which is to be corrected.
- Press delete key once for each letter to be deleted to the left of the cursor.

or

- Highlight the desired text by dragging the mouse over it and typing directly over the highlighted area.

or

- Highlight the desired text and press the delete key.

or

- Select text by clicking the mouse twice to select a word, three times to select a line or four times to select a paragraph and press *delete*.
- To deselect text, click mouse anywhere outside the highlighted area.

To insert text:

- Position the blinking I-beam at the position for the insertion.
- Type the added text.

Note: One of the most useful options is the *Undo* under the Edit menu. The last typing sequence or text move can be undone and put back as a toggle (command-Z). Nothing is permanent until after the second sequence. This is great for those of us who are indecisive.

To move or copy text:

- Highlight the text using one of the techniques previously described.
- Under Edit on the menu choose *Cut*. (The text will not be lost, but will be transferred to the clipboard. If in doubt, highlight *Show clipboard* under Edit on the menu.)

or

- Under the Edit menu choose *Copy* if you wish the text to stay there and need a repeat somewhere else.
- Using the mouse, place the I-beam where the text is to be copied.
- Under Edit from the menu, choose *Paste*. The text will remain on the clipboard to be copied again until it is replaced by another cut-and-paste operation.

or

- Select the text.
- Move the I-beam with the mouse to the new location. (Do not click.)
- Press the command key and the option key together and click the mouse. (This bypasses the clipboard.)

Note: This will cut and paste the text once and will not copy.

To add an empty line:

- Press return.

To delete an empty line:

- Position I-beam at the beginning of the following line.
- Press *delete*.

To check spelling:

Note: The spellchecker (Fig. 30.3) works in all documents except the communication module. To check a communications document, copy it into a word processing document, check and copy back.

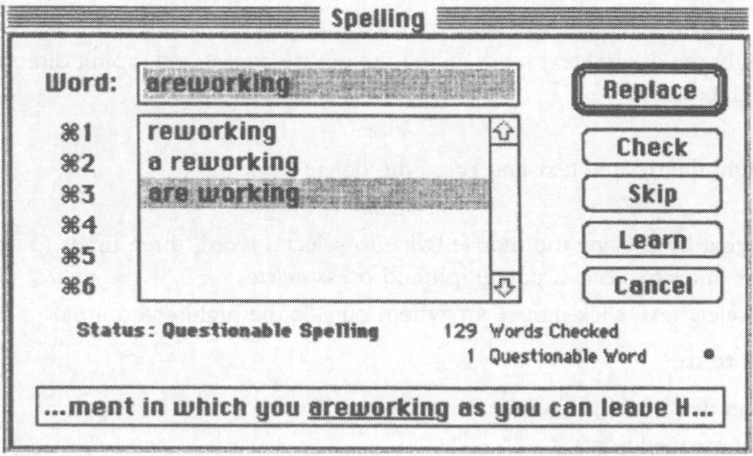

FIGURE 30.3 Spellchecker screen.

- Under the Edit menu choose *Spelling*.
- Choose *Check document*.
- When dialog box appears, click on lever at the right bottom to show the word in context.
- Click *Replace* or press return to choose the highlighted word.

or

- Choose *Check* to verify a corrected spelling and click.
- Choose *Skip* to ignore but not add to the dictionary.
- Choose *Learn* to add a new word to the user dictionary.

or

- Type the corrected spelling over the questioned word.
- Click *Replace* or press the return key.

Note: The checker checks the main dictionary as well as the user dictionary for spellings. Additional special dictionaries are available from Claris.

To use the thesaurus:

- Choose *Thesaurus* (Fig. 30.4) under *Spelling* in the Edit menu.
- Type in the unsatisfactory word. It will appear next to *Find*.
- Press return or click on *Lookup*.
- Choose a replacement word by clicking on it.
- Click on *Replace*.

or

- Highlight the synonym to be replaced as described earlier.
- Pull down *Thesaurus* under *Spelling* in the Edit menu.
- The designated word will automatically appear in *Find.*

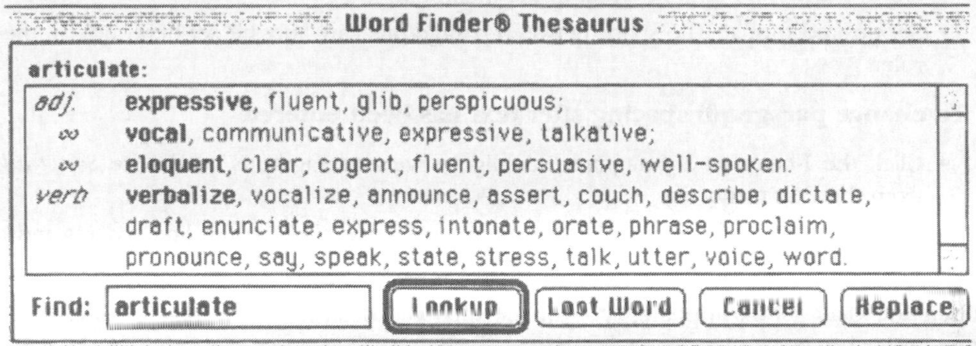

FIGURE 30.4 Thesaurus screen.

Formatting

To format a document into columns (Fig. 30.5):

- Click on the columns icon on the ruler to create multiple columns from a single column page (see Fig. 30.2).

or

- Choose *Columns* under the Format menu.
- In the columns dialog box choose the parameters desired and the number of columns. If unequal columns are going to be needed, click on *Unequal.*
- To resize a column, position the I-beam on either inner column margin while pressing the option key and drag the two-headed arrow with the mouse to position the column margin. This will increase or decrease the space between columns.
- To resize two columns at once, place the I-beam in the space between the columns, press the option key and drag with the mouse. This changes only the column size, not the space between them.
- Click on the column icon in the ruler to preview how the document would look with different numbers of columns.

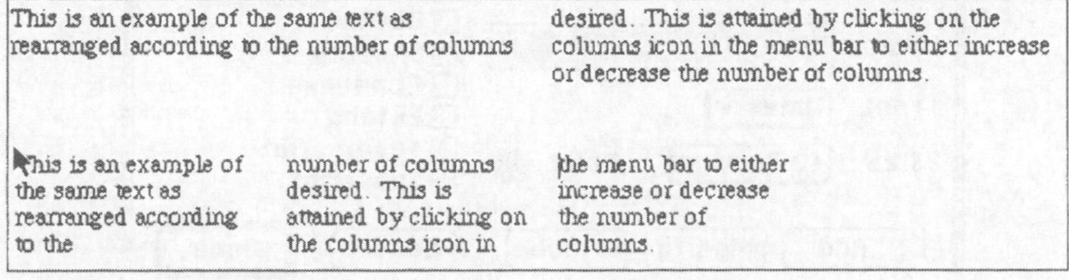

This is an example of the same text as rearranged according to the number of columns
desired. This is attained by clicking on the columns icon in the menu bar to either increase or decrease the number of columns.
This is an example of the same text as rearranged according to the
number of columns desired. This is attained by clicking on the columns icon in
the menu bar to either increase or decrease the number of columns.

FIGURE 30.5 Column arrangement.

To space paragraphs:

- Refer to Figure 30.2 under ClarisWorks Ruler and adjust indentations and margins. Paragraph alignment can only be set there.

or

- Go to *Paragraph* under Format in the menu bar.
- Set desired format. The space between paragraphs can only be set here, not on the ruler.

To change paragraph spacing after text has been entered:

- Click the I-beam in a paragraph, highlight several paragraphs or choose *Select all* from the Edit menu.
- Choose *Paragraph* from the Format menu or double click on the line spacing indicator on the ruler.

When the dialog box appears, type in the changes desired in the *Space before* and *Space after* boxes.

- Click *Apply* to preview.
- Click *OK* to make changes.
- Click *Cancel* to revert to former spacing.

The preset font, size, style, size and color may be determined at the outset by reading the dialog box in *Define styles* under the Style menu (Fig. 30.6). If these are changed in that box the new definitions will become the default settings.

- Highlight the text to be changed.
- Pull down the menu under Style, Size, or Font.
- Highlight the desired change. (A check will appear to signify that which is chosen.)
- The designated text will change format but remain highlighted.

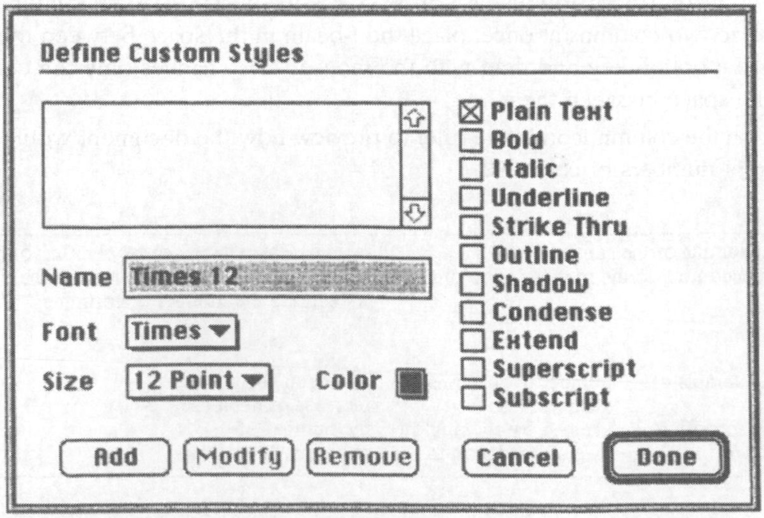

FIGURE 30.6 Fonts, Size, and Style screen.

- Click outside the highlighted area to deselect the text.
- If the change is not acceptable, pull down *Undo format* under Edit to return to the former format.

Note: Keyboard equivalents are listed to the right of the pulldown Style menu. For mouse devotees, the shortcut icons are also on the *Shortcuts* menu under File.

Page breaks are automatically formatted when the end of a page or column is reached. The next bit of text will go to the top of the next page or column. If an artificial page break is needed:

- Set the I-beam where the page break is to occur.
- Highlight *Insert break* from the Format menu.

or

- Set the I-beam where the page break is to occur.
- Press the enter key.

To remove a page break:

- Set the I-beam at the beginning of the line after the break.
- Press the delete key.

or

- If formatting characters are visible (toggle on/off using command-semicolon), select the formatting character (down arrow), and delete.

Adding Headers and Footers

ClarisWorks will automatically order consecutive pages and insert the date and time if needed. These are entered in headers or footers as preferred. A fixed page number or date can be added as text or by holding down the option key and choosing the applicable option under the edit menu. Footers and headers can be edited by using the standard text commands (Fig. 30.7).

To add footers and headers:

- Choose *Insert header* or *Insert footer* under the Format menu.
- When the gray outline box appears, type in the text.
- If you wish to enter the date and/or time highlight *Insert date, Insert time*, or *Insert page #* under the Edit menu. (If you wish the word *page* to appear before the number, you must type it.)
- To remove a footer or header, choose *Remove footer* or *Remove header* from the Format menu.

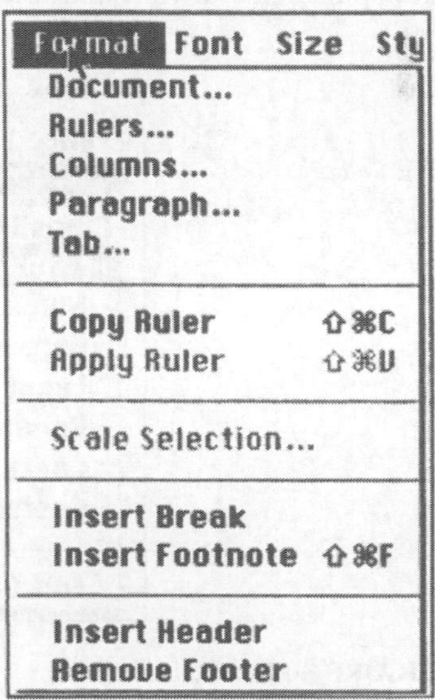

FIGURE 30.7 Format menu.

To add footnotes: Footnotes will be numbered automatically if that option is chosen in the preferences dialog box under the Edit menu. If you prefer to do it manually, you must turn off that option. ClarisWorks will automatically renumber footnotes if text is moved or if footnotes are added or deleted.

Note: See *ClarisWorks User's Guide*, p. 3–32, for inserting footnotes with custom characters.

- Click on the space to the right of where the footnote number is to appear.
- Under the Format menu choose *Insert footnote* or use command-shift-F. The proper number will appear and the screen will shift to the bottom of the page to a footnote panel.
- Type the footnote next to the number and edit as necessary using the Edit menu.
- Press enter to return to your place in the document.
- To remove footnote, position the I-beam one space to the right of the footnote number.
- Press the delete key.

or

- Highlight the number with the mouse.
- Press the delete key.

Outlining

Outlining is a new feature of Version 2.0. Topics and subtopics are indented at different levels and the outline can be expanded or collapsed as needed to create an outline that is easily changed to meet a specific purpose. A word processing document can be changed from standard to outline format by toggling on *Outline view* (Fig. 30.8) under Outline on the menu bar. This can be very useful for making overhead transparencies or for handouts in a group situation.

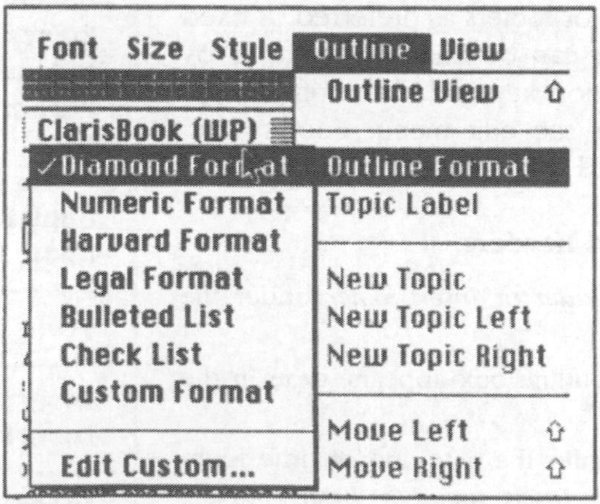

FIGURE 30.8 Outline menu.

To create an outline:

- Under Outline on the menu bar choose *Outline format* and drag to one of the six preformatted designs or choose *Custom format* to design your own.
- If custom is chosen, click on *Edit custom* to design a new format or alter an existing custom format.

To enter topics:

- Under Outline on the menu bar choose *New topic* or use the keyboard shortcut *command-return* (Fig. 30.9).
- Type the topic and the text. The text will wrap at the proper indentation for that level.
- To add an additional topic on the same level of indentation choose *New topic* under Outline on the menu bar or keyboard *command-return*.
- To enter a topic on an indentation to the left, go to Outline and choose *New topic left* or keyboard *command-l*.
- To enter a new topic to a right indentation, choose *New topic right* under Outline or keyboard *command-r*.

Outline View
✓Outline View ⇧⌘I
Outline Format ▶
Topic Label ▶
New Topic
New Topic Left ⌘L
New Topic Right ⌘R
Move Left ⇧⌘L
Move Right ⇧⌘R
Move Above ⇧⌘A
Move Below ⇧⌘B
Collapse
Expand
Expand To...
Raise Topic

FIGURE 30.9 Outline menu.

Moving Topics Already Entered to Different Places in the Outline

To move the topic label as well as all text and subtopics:

- Click once on the topic label. The topic text will be highlighted. Another click will change the cursor to a crossbar with arrows on the vertical bar.
- Drag text up or down, choose Outline from the menu and highlight *Move above* or *Move below*, or keyboard *command-a* or *command-b*.

Note: The text and the topic label cannot be dragged down to where no text has been before.

To move the topic without the subtopics:

- Click anywhere in the body of the text. The cursor will blink but the text will not be highlighted.
- Press the option key, choose Outline from the menu bar, and highlight the appropriate move.

XI. Do collection assessment in a few general and a few special areas
XII. Do curriculum mapping
A. first in science and social studies
B. ! Secondly in language arts
XIII. Do purchase planning using curriculum mapping with collection assessment
XIV. Include more non-print in purchases
A. electronic references geography materials
1. on-line databases
2. CD geography materials
3. CD art materials

FIGURE 30.10 Harvard Format Outline.

XI. **Do collection assessment in a few general and a few special areas**
XII. **Do curriculum mapping**
XIII. **Do purchase planning using curriculum mapping with collection assessment**
XIV. **Include more non-print in purchases**
XV. **Institute cooperative planning and execution of some social studies and science units with a view to expanding with more units yearly.**
XVI. **Develop system-wide professional library with copies of holdings distributed on disk.**

FIGURE 30.11 Collapsed Outline.

To change indentations:

- Click anywhere in the body of the text to be moved. The cursor will blink, but the text will not highlight.
- Choose Outline from the menu bar and highlight move left or move right. The shortcut is *shift-command-left* or *shift-command-right*.

Adding Art

Art can be added to a ClarisWorks text document by directly drawing or painting in the document or by importing art. There are two ways of importing illustrations to ClarisWorks documents. If the illustration is added as an inline character, that is, as if it were text, it can be moved, centered, and aligned with the standard text commands. If the picture is added as an object, it is subject to the commands of the draw or paint environments and does not respond to text commands. Quicktime movies can also be inserted into a document as art.

If the pointer tool is located in a text line when the art is copied, it will appear as inline text. The pointer must be located in a graphic Frame for the the import to be in the paint or draw mode.

To draw directly on a page:

- Under view in the menu bar open *Show tools*. The toolbox will appear on the screen left (Fig. 30.12).
- Click on the appropriate tool and draw.
- When the mouse is released, the palette will return to the pointer. To keep a drawing tool active, double clip or it and it will darken and stay active until you click on another tool. If the tool reverts to the **A**, you are in a text space.
- To erase, highlight pointer, click on object so handles show and press delete.

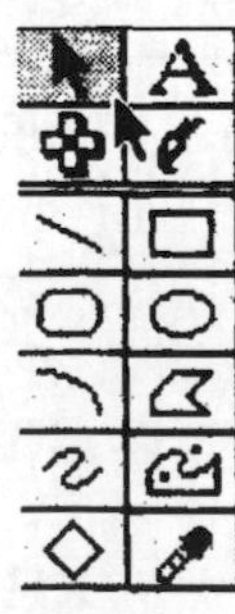

FIGURE 30.12 Toolbox.

To paint directly on the page within a frame:

- Choose *Show tools* under the View menu.
- Choose the paint tool (the paint brush) and draw a frame from the upper right to the lower left of the space you want to paint.
- See directions under the section on the paint program for further directions.

To paste a picture from the scrapbook or clipboard:

- Open a file with a picture. (If picture is a graphic, highlight with the pointer tool so the handles show. If picture is an inline character, highlight by dragging the mouse while in text mode.)
- Copy it onto the clipboard or into the scrapbook by choosing *Copy* under the Edit menu.
- In text document choose *Show tools* under View.
- Click on the draw tool.
- Position the pointer on the page or drag a paint frame.
- Choose *Paste* from the Edit menu.
- Position picture by dragging.

Layering and Wrapping

ClarisWorks allows the layering and wrapping of text in front of, behind, or around a graphic. Special effects can be obtained easily using these commands.

To layer a picture with text:

- Add a picture as an object to the text document.
- Click on picture to show handles.
- Drag picture to the preferred location.
- Choose *Move to back* or *Move backward* under the Arrange menu.

To wrap text around picture:

- Add a picture as an object to the text document.
- Click on the picture to show handles.
- Choose *Text wrap* under the Options menu.
- From the dialog box choose the icon that represents the way the text should wrap.
- To unwrap choose *None* in the dialog box.

or

- Choose *Undo format* under Edit.

To add a picture as an inline character:

- Copy a picture to the clipboard.
- Position the I-beam where the picture is to be positioned.
- Choose *Paste* from the Edit menu.

or

- Choose *Install* under File if picture is in a picture file.

To add tables to a text document: The most efficient way to add tables to a text document is to add a spreadsheet as an object. When you do that, all the spreadsheet commands are active. You can then move the whole image as a single unit for the best effect.

- Choose *Show tools* under the View menu.
- Highlight the spreadsheet tool (Fig. 30.13) in the toolbox by clicking once.

FIGURE 30.13 Spreadsheet tool.

- Using the the mouse, position the tool where you wish the upper left corner to be in the text document, and draw down diagonally to the bottom right (Fig. 30.14).
- Type in data using commands for a spreadsheet or copy tabulated data from another document. Be sure the spreadsheet tool is highlighted.
- To position the spreadsheet, click on the draw tool and click on the spreadsheet to make the handles visible.
- With the pointer anywhere except on the handles, drag the spreadsheet to the best position.
- To enlarge or shrink the spreadsheet, click on the spreadsheet to make the handles appear.
- Drag the handles with the mouse until the desired size is reached.

Note: See Spreadsheet for further explanation of spreadsheet commands.

	A	B	C
1	project expenses	January	February
2	Group A	$13650.00	$9072.00
3	Group B	$12405.00	$7925.00
4			
5			

FIGURE 30.14 Spreadsheet as an object.

Printing

After you have identified your printer through the Chooser (see your Mac manual), ClarisWorks will automatically select that printer unless you change the settings. The dialog boxes for each printer are slightly different but are essentially the same.

To print your document:

- Select *Page Setup* (Fig. 30.15) under the File menu.
- Choose the setup by clicking in the circle.
- Type in the reduce/enlarge number if you wish a change from 100 %.

Spreadsheets and other documents with charts and figures often need enlarging. This is also a help in making overhead transparencies.

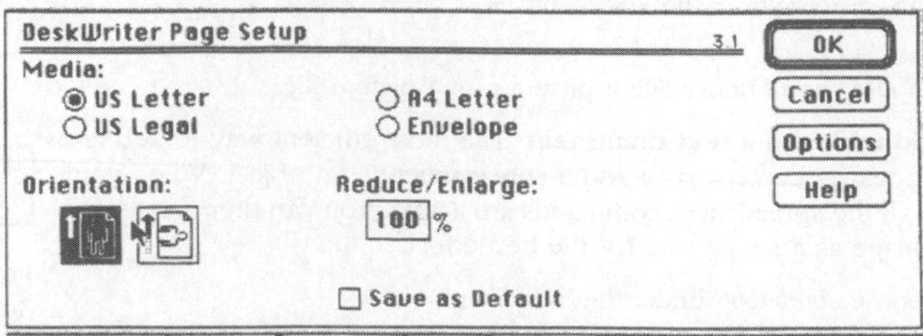

FIGURE 30.15 Page Setup.

- Click on *OK*.
- Open the document to be printed.
- Choose *Print* (Fig. 30.16) under the File menu or type command-P.
- Click on your choice to select quality.
- If you wish to print a portion of the document, click on From and type in the beginning and ending pages. The example in Fig. 30.16 will print only page 12.
- Click on *Preview* to see how your document will look. This is essential if you have changed the size percentage in the setup dialog box.
- Click on *OK* if document is ready or on *Cancel* if you wish to make changes.
- To stop printing, press command-period. If information has already spooled to the printer, that will be printed but no more will be sent.

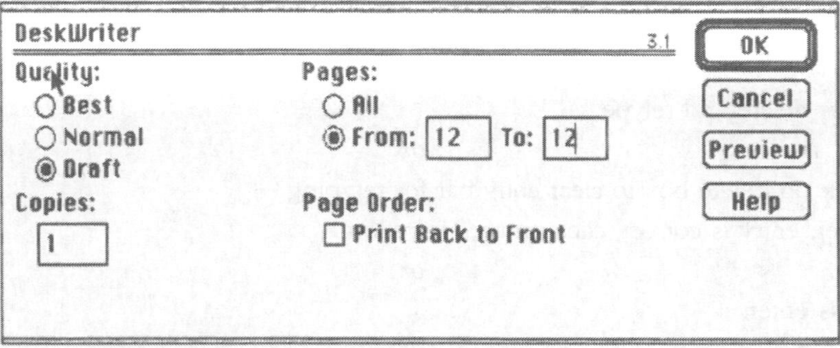

FIGURE 30.16 Printer menu.

30.3 THE SPREADSHEET

A spreadsheet can be copied into a word processing document and retain its spreadsheet environment (see Fig. 30.4). In ClarisWorks, the active cell is always outlined in a heavy border. Numbers, text, or formulas may be entered in a cell. Negative numbers must begin with a minus sign and text may be any combination of characters. For ClarisWorks to interpret a numeric entry as text, the entry must be preceded by an equal sign (=) and a quotation mark (") and followed by another quotation mark.

To open a spreadsheet:

- Choose *New* from the File menu.
- In new document dialog box choose *Spreadsheet* (Fig. 30.17).

To select active cell(s):

- Click on one cell to make it active.
- Drag over range of cells to highlight.
- Click on column heading to highlight column.
- Click on row heading to highlight row.

To enter data:

- Select active cell.
- Enter data. (It will appear on entry bar, not in active cell.)
- Correct mistakes by dragging across mistake on entry bar and retyping.

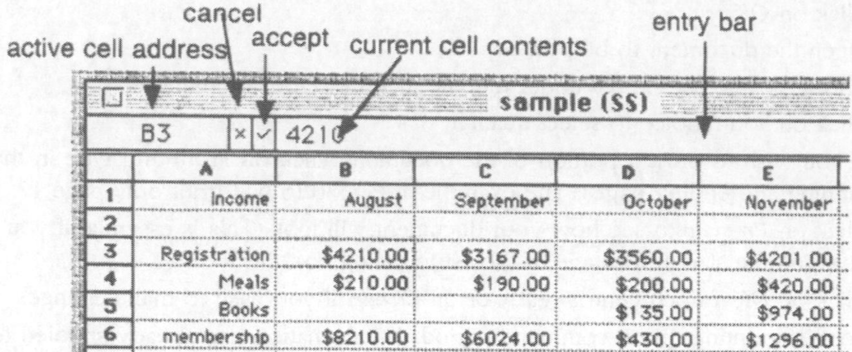

FIGURE 30.17 Spreadsheet.

or

- Press delete and retype.

or

- Click on delete box to clear entry bar for retyping.
- When entry is correct, click on accept button.

or

- Press enter.

or

- To accept and move cell selection press return (down), tab (right), shift return (up), or shift tab (left).

To move and copy data:

- To copy the contents of a cell, highlight the cell and choose *Copy* under the Edit menu (command-C).
- Select the destination cell and choose paste under the edit menu (command-V).

Note: If data fill more then the selected cell, the surrounding cells will take the data and previous contents of those cells will be deleted.

- To move data (Fig. 30.18), you can cut and paste.

or

- Choose *move* under the calculate menu.
- Type the address of the destination cell.
- Click on *OK*.

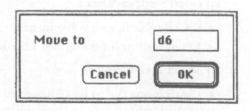

FIGURE 30.18 Move Menu box.

To enter formulas:

- Select a cell.
- Type an equal sign (=).
- Type the values, constants, operators, or function names that make up the formula.

or

- Paste the function by choosing *Paste function* (Fig. 30.19) from the Edit menu. This eliminates mistakes as well as being fast.

Note: Mathematical and logical functions are listed and explained in the on screen ClarisWorks Help document.

Using the example given in Figure 30.17, the function *sum* was pasted as shown in Figure 30.20.

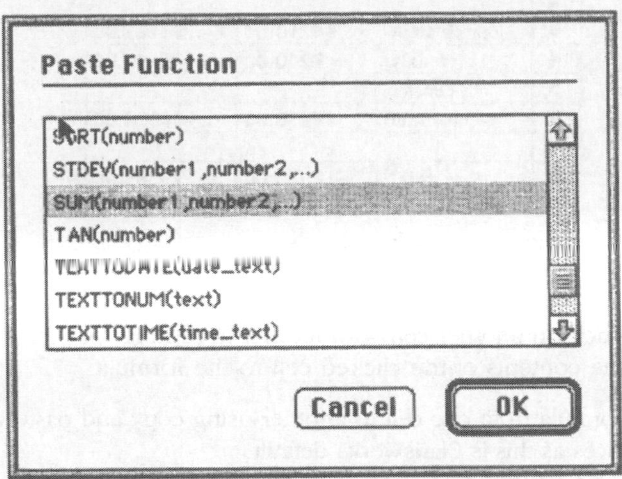

FIGURE 30.19 Paste function.

samp

C8 =SUM(C3..C6)

	A	B	C
1	Income	August	September
2			
3	Registration	$4210.00	$3167.00
4	Meals	$210.00	$190.00
5	Books		
6	membership	$8210.00	$6024.00
7			
8	total	$12630.00	$9381.00

FIGURE 30.20 Pasted function with dragged cells.

To complete a pasted function (Fig. 30.21):

- Click on the first cell to be entered as an argument and either drag to the end of a range.

or

- Click on first cell to be entered as part of the argument and add cells with additional clicks. (The default operator between separate clicks is +. If another operator is desired, it is necessary to replace the + with the other operator.

or

- Type in the desired numbers between the parentheses.
- Click on *accept* or press enter.

sampl

C8 =SUM(C3+C4+C5+C6)

	A	B	C
1	Income	August	September
2			
3	Registration	$4210.00	$3167.00
4	Meals	$210.00	$190.00
5	Books		
6	membership	$8210.00	$6024.00
7			
8	total	$12630.00	$9381.00

FIGURE 30.21 Pasted function with clicked cells.

Note: Do not click on another cell without first accepting the active formula cell as a click will add the contents of the clicked cell to the formula

- Copying a formula from one cell to another using copy and paste will result in relative references as this is ClarisWorks default.
- To create an absolute reference (Fig. 30.22), type a $ before each row designation and each column designation in the formula.

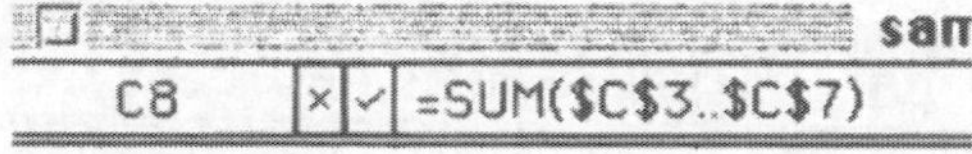

FIGURE 30.22 Absolute reference formula.

ClarisWorks will sort data or labels alphabetically or numerically throughout the spreadsheet or within a given range.

To sort data:

- Select by dragging the row or column to be sorted.
- Select sort from the calculate menu.
- Choose type of sort(s) desired.
- Click *OK*.

Warning: Sorting should be done before formulas are entered as only one row or column is affected so relativity to values in other columns is lost. If you discover this invalidates your data, choose *Undo sort* from the Edit menu. Paste mistakes can also be reversed by using the *Undo paste* from the Edit menu.

Formatting the Spreadsheet

A new spreadsheet is 40 columns wide and 500 rows deep. This can be changed by either inserting or deleting rows or columns under the calculate menu or using *Document* under the Format menu (Fig. 30.23).

To insert or delete rows and columns:

- Click to select on the row heading below where the new row is to go.
- From the Calculate menu, choose *Insert cells*.
- A new row appears above the one selected.
- If the new row is a mistake, choose *Undo insert* under the Edit menu.
- Click on the column heading to the right of where the new column is to go.
- The new column appears to the left of the selected column.

or

- Choose document from the Format menu.
- Type in the desired numbers of columns and rows.
- Click on *OK*.

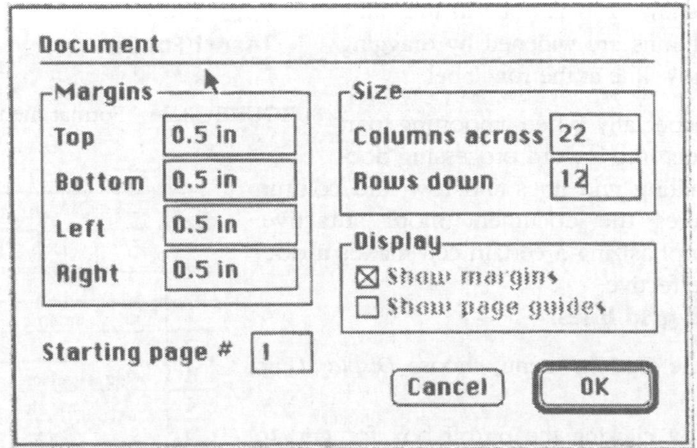

FIGURE 30.23 Adding or reducing columns.

Using the document method is advisable for shorter spreadsheets to make them more manageable from the beginning. Adding cells is convenient when you discover you need more as you work. When you add or delete, ClarisWorks renumbers and formulas are adjusted even if they contain absolute references.

To delete any rows or columns:

- Click on the row heading or column heading.
- Choose *Delete cells* from the Calculate menu.
- The row above or column to the left of the selected heading will be deleted.

To change column width:

- Click on the column line in the heading. A double pointed arrow appears.
- Drag the arrow to make the column narrower or wider.

or

- Choose *Column width* under the Format menu.
- Type in desired width. (All columns will be changed to this width.)

To set fonts, sizes, style, and text color:

- Highlight the active cell or range of cells by clicking or dragging.
- Under the Format menu (Fig. 30.24), choose the desired font, size, style, and text color and alignment.
- Choose the alignment that suits the entry. If the text is too long to fit in a cell and you do not want to widen the column, it is possible to wrap text by highlighting the cell and choosing wrap under the alignment portion of the format menu. To show wrapped text you must deepen the row to allow all the text to show. This is done in the same way columns are widened by dragging on the row line at the row label.

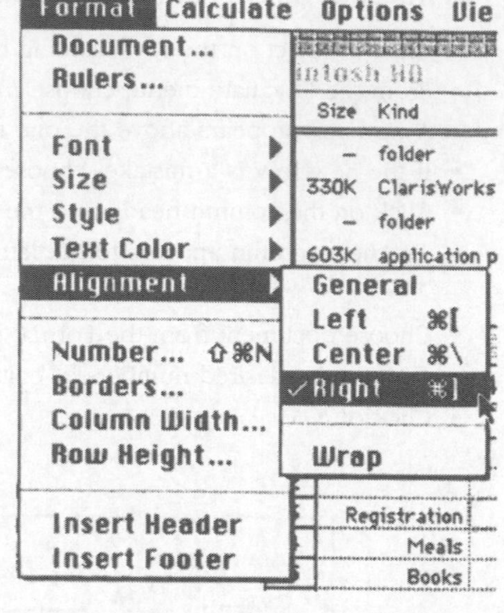

FIGURE 30.24 Format menu.

Sometimes, especially when importing part of a spreadsheet into a word processing document, eliminating grid lines and row and column headings makes the document more attractive. Sometimes emphasizing a certain cell makes a document more effective.

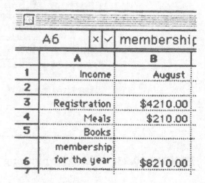

FIGURE 30.25 Wrapped text.

To turn off grid lines:

- Under the Options menu, choose *Display* (Fig. 30.26).
- Toggle by clicking the on/off box for grid to leave blank.
- Click *OK*.
- To border a cell, select a cell or drag a range of cells (Fig. 30.27).
- Highlight *Borders* on the Format menu.
- Choose Border for the desired side(s) or choose Outline.
- Click *OK*.

FIGURE 30.26 Display box.

Income	August	September	October	November
Registration	$4210.00	$3167.00	$3560.00	$4201.00
Meals	$210.00	$190.00	$200.00	$420.00
Books			$135.00	$974.00
membership for the year	$8210.00	$6024.00	$430.00	$1296.00

FIGURE 30.27 Grid and headings turned off with cell bordered.

To add headers, footers and page numbers:

- Choose *Insert header* or *Insert footer* from the Format menu.

The screen switches to the proper location for the header or footer and goes into text mode as a word processor.

- Type in the text desired.
- To eliminate the footer or header choose *Remove footer* or *Remove header* from the Format menu.
- For page numbering, choose *Insert page #* from the Edit menu. Paging will be automatic. If you want the word "page" to appear, you must type it in.

To manage a large spreadsheet you can lock the labels in place so that you can see them when scrolling off the page as you are working. There are two ways to do this.

To lock labels in place:

- Drag the horizontal and vertical pane controls (Figs. 30.28 and 30.29) to separate the labels from the data.
- Scroll through the data and the labels stay where they are.

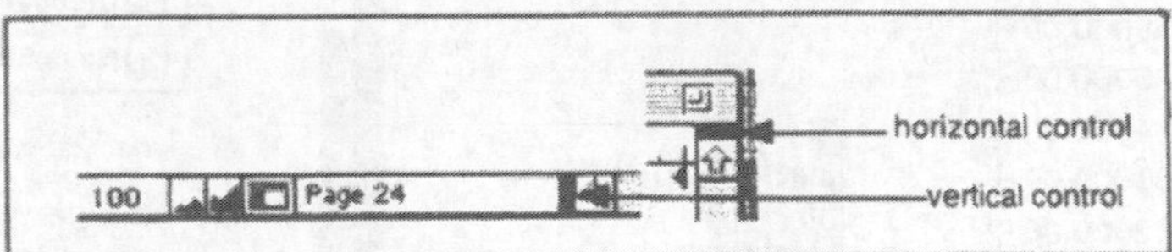

FIGURE 30.28 Pane controls.

or

- Select the intersection of the row and column you want to lock. (Usually it is A1.)
- Under the options menu choose *Lock title position*.

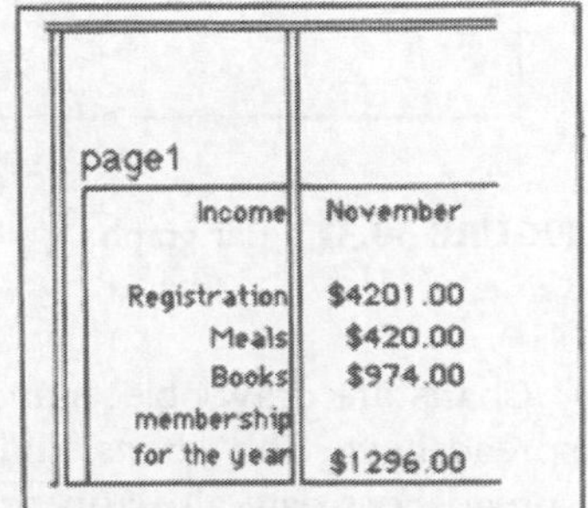

page1
Income
November
Registration
$4201.00
Meals
$420.00
Books
$974.00
membership
for the year
$1296.00

FIGURE 30.29 Spreadsheet divided into panes.

Spreadsheet Charting

ClarisWorks can easily turn your spreadsheet into a variety of kinds of charts (Figs. 30.30 and 30.31). This is a great feature to use for reports to enable readers to quickly wade through data or for overhead transparency projection to groups.

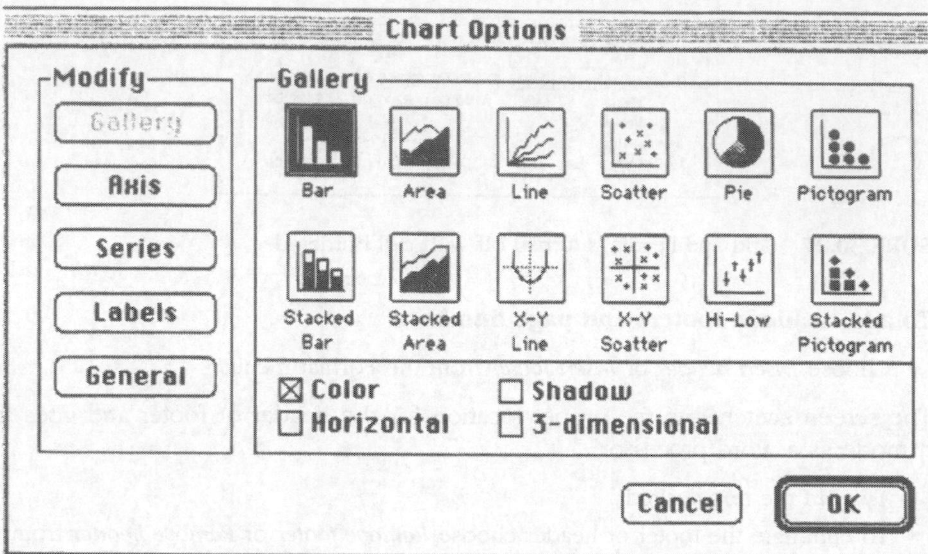

FIGURE 30.30 Graph dialog box.

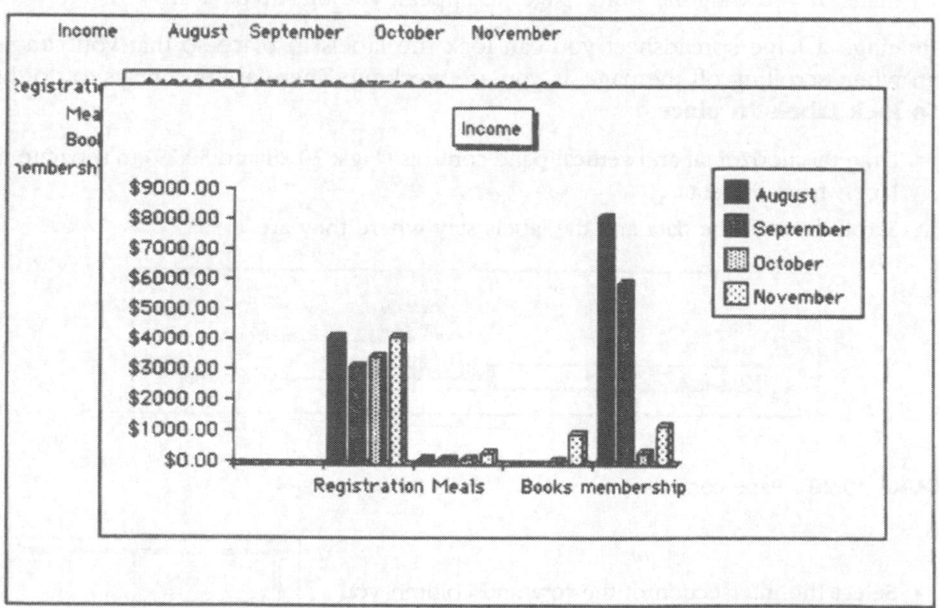

FIGURE 30.31 Bar graph.

Charts are draw objects and data can be modified only by returning to the original spreadsheet. The charts and spreadsheets are linked so that any changes to the spreadsheet data also appear on the chart. When a chart is copied to another document such as a word processing document, the link is broken and the chart must be recopied. Labels and explanations can be added to a chart using the draw tools after a chart is composed.

To make a chart:

- Select the data to be charted by clicking or dragging. It may be some or all of the data. If possible include any labels that you want to be on the chart.
- Choose *Make chart* from the Options menu.
- Choose a chart format in the *Gallery* portion.

The graph as a draw document is superimposed on the actual spreadsheet. To work on the spreadsheet you must pull the graph to one side. You can also alter the size of the graph by dragging on the handles of the frame.

You can make an unusual and effective presentation by changing the graph to a pictogram (Fig. 30.32)

The default is an arrow but a picture of your choice can be pasted from the clipboard.

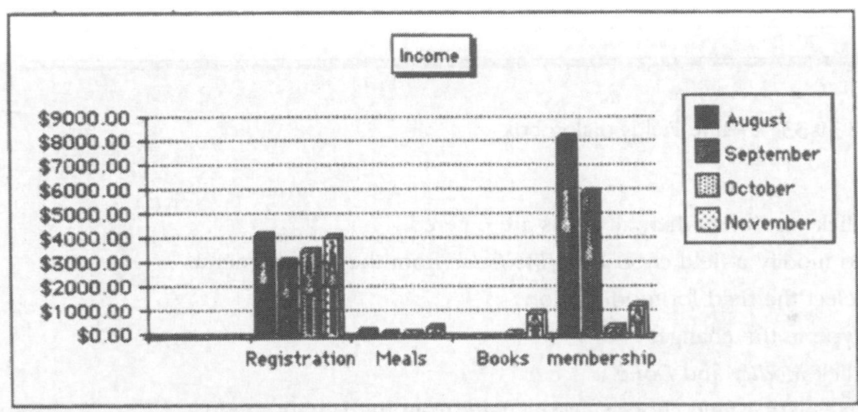

FIGURE 30.32 Pictogram.

When pasting a graph into another document, you can ungroup portions of it to enlarge or emphasize in other ways. The graph is treated as a draw document and as such will respond to the commands for that. (See section on draw documents.)

30.4 THE DATABASE

ClarisWorks supports three different modes that treat records in three different ways. You use *Browse* to enter, change, or view records, *Find* to select specific records according to named criteria, and *Layout* to tailor an existing layout or create a new layout.

To create fieldnames:

- Choose *New* from the File menu.
- Double-click on *Database* or click on *Database* and click on *OK*.
- The *Define fields* dialog box appears (Fig. 30.33).
- Type in the name of the field. Field names can be up to 63 characters long and may contain punctuation marks. (Correct in the same manner as in word processing.)
- Click on the proper type designation.
- Click *Create* or press the return key.
- Repeat for each field.

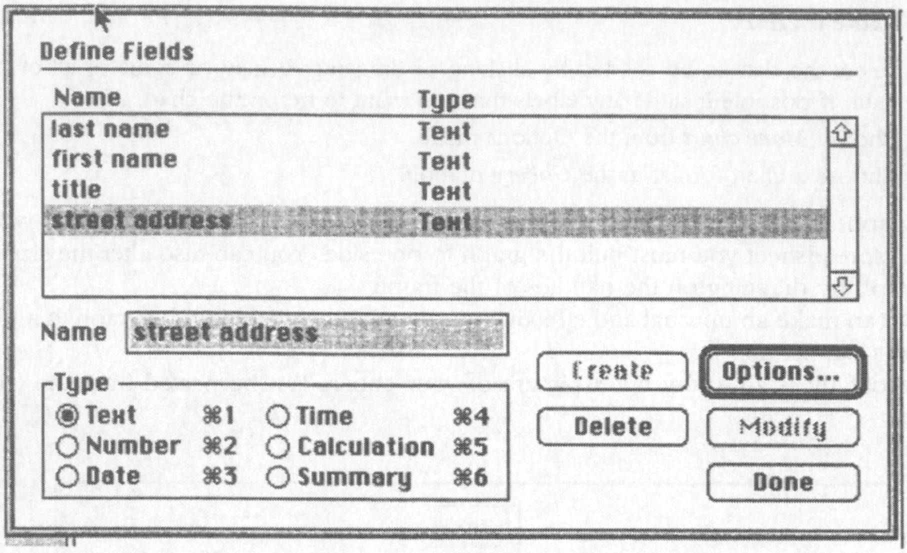

FIGURE 30.33 Define Fields dialog box.

- Click on *Done* when all fields are entered.
- To modify a field choose *Define fields* from the Layout menu.
- Select the field for modification.
- Type in the change.
- Click *Modify* and *Done*.
- To delete a field choose *Define fields* from the Layout menu
- Click on the field name.
- Click on the *Delete* button.

Warning: When you modify a field or field type after data has been entered, that field is changed throughout the database and some data may be affected. When you delete a field, the data in that field are lost. If those data are used in calculation or summary fields, those fields will be incorrect and *Badfield* will show up in the formula in the deleted fields place.

- When data for a record is entered (Fig. 30.34), choose *New record* under the Edit menu or type command R
- Click in the entry box next to the field name and type in data.
- Click in the next field or press tab.
- Shift-tab will take you to the previous field.
- Check ClarisWorks Help (command-?) for definitions and format for contents of the types of fields.
- You can change the order in which the tab moves to fields by choosing *Tab order* from the Layout menu (Fig. 30.35).
- Click to highlight the field in the left column that you would like to be selected first.
- Click on *Move*.
- Repeat in the order you want the tab to select fields for data entry

FIGURE 30.34 Data entry.

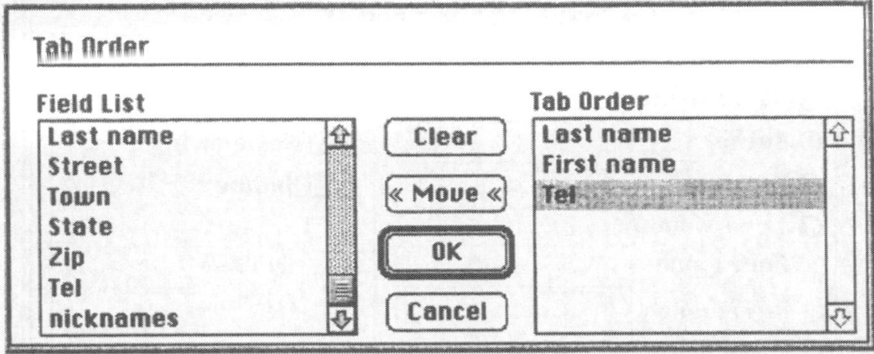

FIGURE 30.35 Changing tab order.

- Click on *OK*.
- You can copy data from one record to another be selecting the record (click outside the fields and the entire record is selected) and choosing *Copy* from the Edit menu. This copies the entire record and it will appear as a new record when pasted.
- You can also paste data from the clipboard. (This is handy if the same piece of information is to be added to many records.)
- When adding records in an already established database, use the same method.

After creating a database, ClarisWorks displays the fields in a *Standard layout*, in the order they were created.

To browse a database:

- Click or drag the vertical scroll bars.

or

- Click on the pages of the book icon in the upper left hand corner. (You can flip backward or forward as if it were a card file.)

or

- Drag the bookmark (the extension on the right side of the book icon) to move through the file. The current record number will change as you drag.

or

- Choose *Go to record* under the Organize menu, enter the record number, and press the return key.

Each type of field, except for the calculation and summary fields, allows the option of automatic entry (Fig. 30.36).

To enter data automatically:

- In Browse mode, click on the field to hold automated data.
- Click on *Define fields* under Layout menu.
- Click on *Data*.
- Type in the box the data to be entered.
- Each new record will automatically contain those data in the identified field.

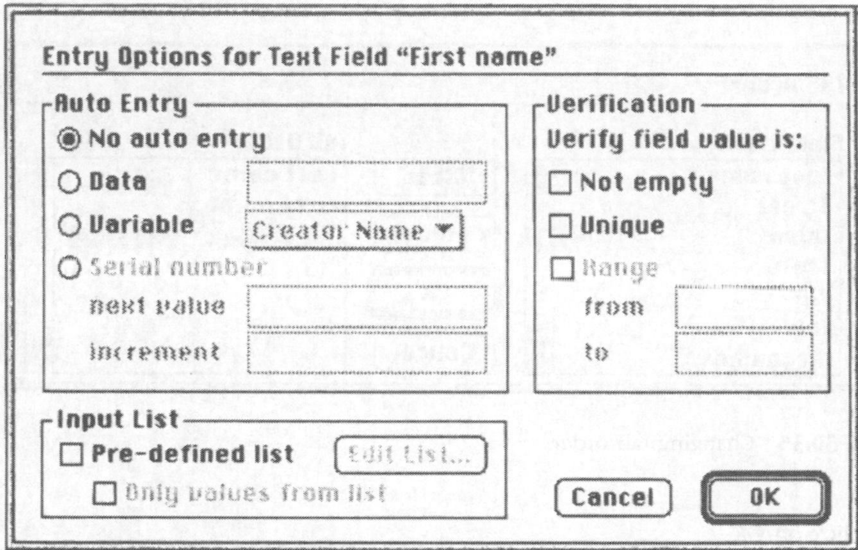

FIGURE 30.36 Entry options dialog box.

ClarisWorks allows you to view all records or a selected set of records. This is helpful when working with a subset for either entering or printing data.

To view selected records in the database:

- In Browse mode click outside the fields in any record to select it.

or

- Drag through a number of records to select them.
- Choose *Hide selected* from the Organize menu to view unselected records.

or

- Choose *Hide unselected* to view selected records.

Note: Be sure to click on *Show all records* under the Organize menu when you want to work with the full database again.

Finding and Sorting Records

Using the Find mode allows you to select records by using the data contained in them rather than clicking on the records themselves. This is much easier when records are separated in the database.

To use Find:

- Choose *Find* in the Layout menu (Fig. 30.37).
- Type the criterion for the find in the applicable space. In Figure 30.37 you might type a zip code or a state.
- Click on the All button on the left or press the return key.
- The book icon shows you how many records fulfill your criterion and how many records were searched.

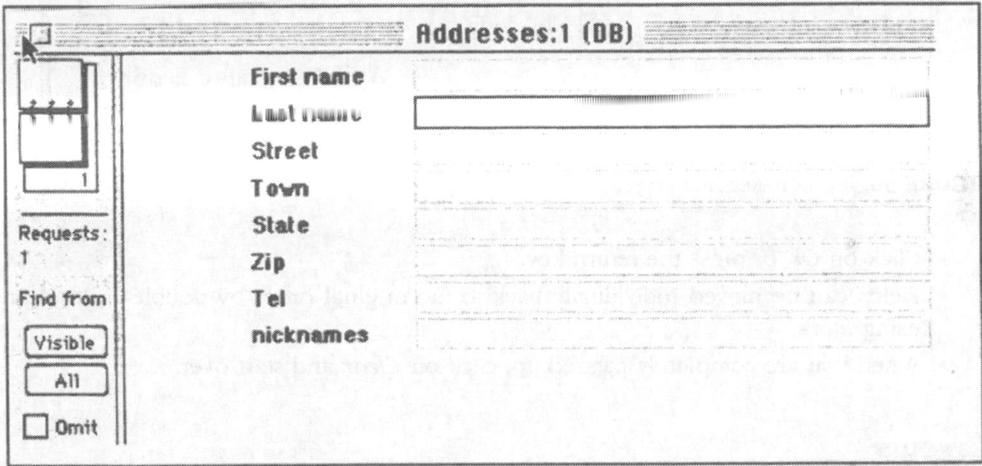

FIGURE 30.37 Find Mode screen.

You can search further within the selected subset by adding another criterion and clicking on Visible instead of All.

To do a multiple criteria find:

- Choose *Show all records* from the Organize menu.
- Click on *Find* in the Layout menu.
- Type the various criteria in the appropriate spaces in the find mode screen.
- Click on *All* or press the return key.

Note: The more criteria you specify, the fewer records will be found.

To sort records:

- Choose *Browse* under the Layout menu.
- Choose *Sort Records* from the Organize menu (Fig. 30.38).

In the same manner as you changed the tab selection order:

- Click on the first sort field to be moved.
- Click on *move* or double click on the field name.
- Continue clicking and moving until all fields are in the order in which you want to sort.
- Those fields not moved will remain in the original order.
- Click on the bar graph to set ascending (1–9, A–Z) or descending order.

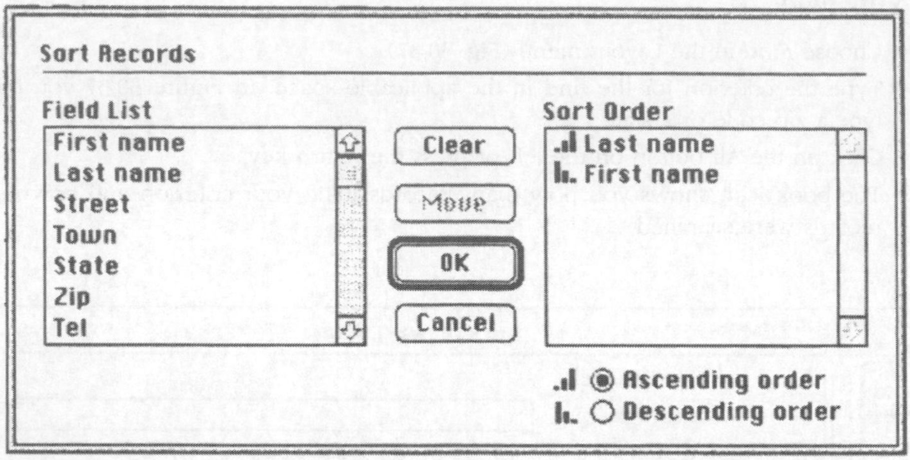

FIGURE 30.38 Sort Records screen.

- Click on *OK* or press the return key.
- Fields can be moved individually back to the original order by double-clicking or using *move*.
- When you are completely messed up, click on *Clear* and start over.

Layouts

The default layout in ClarisWorks is called the Standard Layout. For printing labels, filling in forms, importing organized information into word processing documents, and other purposes, ClarisWorks allows you to create customized layouts. The layout mode puts you automatically in a draw environment where you deal not with data, but with fields, text, and graphics. Each layout has fields, which contain the data seen in Browse; text, which includes the field names; and parts, which consist of the body, headers and footers, and summaries. You must be in Layout mode to modify a layout, but you can create a new one in Layout or Browse. Names of layouts are on the Layout menu in the order they were created. A check appears beside the one currently on the screen.

To make a new layout:

- Choose *New layout* from the Layout menu (Fig. 30.39).
- Type a name in the box that will help you remember why you created this layout.
- Click on type of layout needed.
- If you click on *Labels*, you may choose from an already customized list for Avery labels or you can make up the spacing for your own.
- If you choose *Columnar report*, a new screen will ask you to select which and in what order you want the fields. It is the same procedure used for selecting a subset or reordering tab movements.
- A *Blank* layout lets you start from scratch and add fields and graphics.
- A *Duplicate* layout lets you change some things in a special layout without changing the original.
- Click on *OK*.

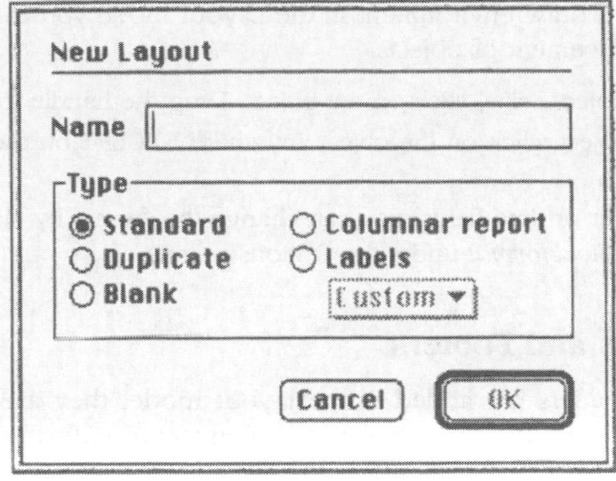

FIGURE 30.39 Layout choices.

For the different reports that you create, different parts of a layout will be needed.

To add parts to the layout:

- Choose *Insert part* from the Layout menu (Fig. 30.40).
- Choose the parts needed by clicking. If the part is already included in the layout, that option will not be available.
- To add fields choose *Insert field* from the Layout menu.

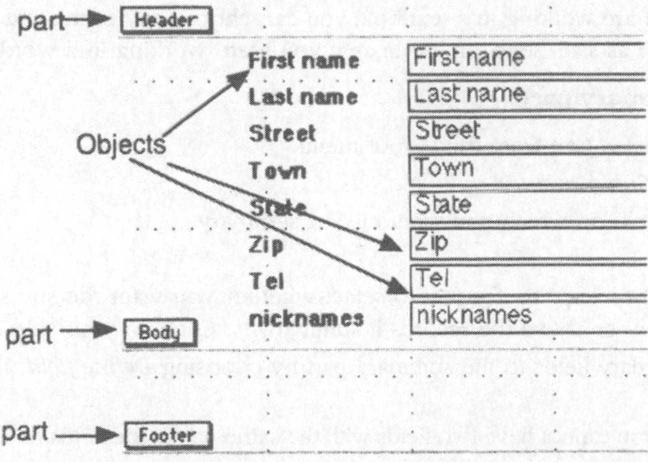

FIGURE 30.40 Parts of a layout.

Note: This does not add or delete fields in the database itself, only in the layout currently showing.

- To rearrange a new field, which is added to the bottom of the body when it is created, click on it to select it and drag it to where it belongs. (Remember you are in draw environment.)
- To delete a field, click on it to select it and press the delete key.

Because you are in a draw environment in the Layout mode you can change the size of a part and the placement of objects.

- To change a object's size, click on the object. Drag the handle to resize.
- To move an object, click on the object anywhere EXCEPT on the the handles. and drag the object.
- In date, number or time fields, you can change the format by clicking on the field and choosing *Field format* under the Options menu.

Adding Headers and Footers

When footers or headers are added in the layout mode, they are going to appear only when that particular layout is chosen. In columnar reports, the header with field names is automatically included. In other layouts, headers and footers are added by clicking on *Insert part* in the Layout menu and choosing *Header* or *Footer*. To add a header or footer to each page as part of the document, choose *Insert header* or *Insert footer* from the Format menu in the **Browse** mode. That text will appear in every layout.

To add headers or footers:

- In Layout menu choose *Insert part*.
- Choose *Header* or *Footer*.
- Shift-click on the field labels to drag them as a group where you want them.
- Click on the text tool and drag a text box.
- Type the header in the text box.
- Since you are working in a text box, you can choose any of the text formatting features such as size, style, and style as if you were working in a word processor.

To add summary parts:

- Choose *Insert part* from the Layout menu.
- Click on *Sub-summary*.
- Choose field on which you would like a summary.
- Click *OK*.
- A dialog box appears for you to select whether you want the sub-summary to appear below or above the records it summarizes. Click on your selection
- Add summary fields to the summary part by choosing *Define fields* from the Layout menu.
- Because you cannot have two fields with the same name, the fields have to be renamed if you want to include both a sub-summary and a grand summary (trailing summary).
- Type the field name in the box.
- Click on *Summary*. A dialog box appears (Fig. 30.41).
- Scroll the function list on the right and choose the appropriate function.
- Delete the contents of the function between the parentheses.
- Click on the field to be summarized. It will be added automatically inside the parentheses.
- Click *OK*.
- Repeat for a grand summary being sure to use a different name.

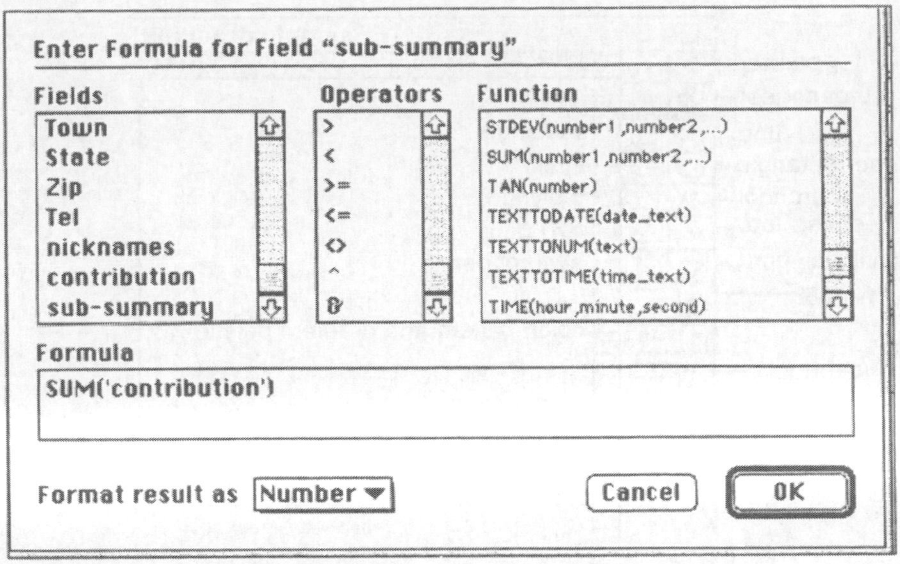

Enter Formula for Field "sub-summary"

Fields	Operators	Function
Town	>	STDEV(number1,number2,...)
State	<	SUM(number1,number2,...)
Zip	>=	TAN(number)
Tel	<=	TEXTTODATE(date_text)
nicknames	<>	TEXTTONUM(text)
contribution	^	TEXTTOTIME(time_text)
sub-summary	&	TIME(hour,minute,second)

Formula

SUM('contribution')

Format result as Number

Cancel OK

FIGURE 30.41 Formula for summary field.

30.5 DRAWING

The drawing program in ClarisWorks is operative as an environment in all programs except communication. When you draw you create an object. Objects can be enlarged, moved, layered, and pasted. The draw program allows documents to be enhanced easily to produce professional looking results.

To open a draw document:

- Choose *New* from the File menu.
- Double-click on Drawing.
- The grid lines help you to align objects. They are determined by the rulers, which are set by choosing *Rulers* under the Format menu. The rulers along the side and top margin show a dotted line to help you locate the exact position of the pointer.
- The toolbox appears on the left (see Fig. 30.42).

To draw an object:

- Click on the tool that will draw the desired object.
- Point to the place where you want the object to start, hold down the mouse button, and drag from the upper right to lower left of the object.
- Release the button and a solid line with "handles" will replace the dotted line that appears as you draw (Fig. 30.43).
- To keep a tool selected for multiple drawings, double-click on it. It will stay selected until you click on another tool. A single click allows the pointer tool to be reselected after you release the mouse button.
- To fill the object with a pattern, click on the fill tool while the object is selected. Click on the appropriate pattern. It will show in the space next to the fill icon and fill the object. As long as the fill tool and the object are selected, you can keep changing patterns until you find the one that's best suited (Fig. 30.44).

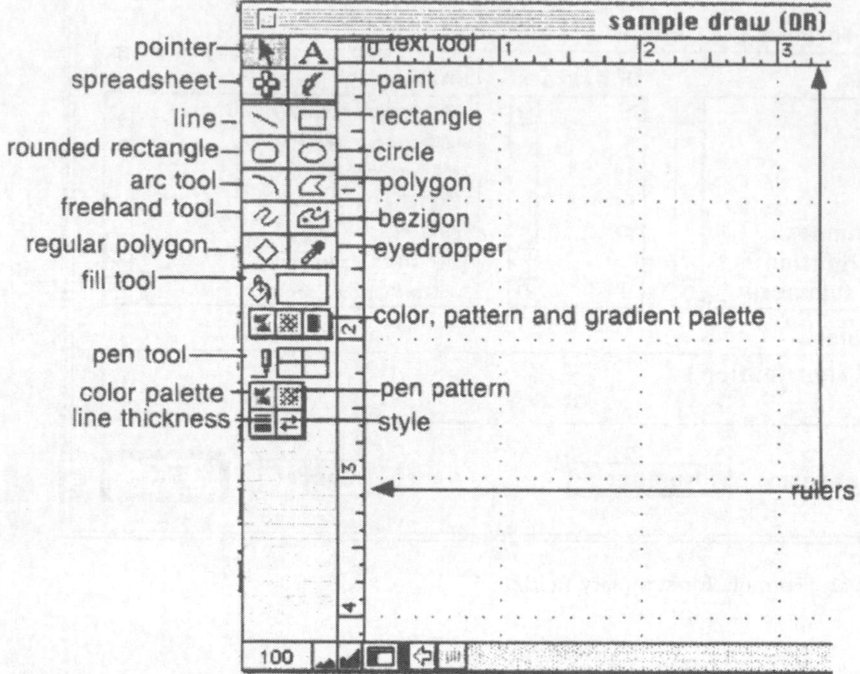

FIGURE 30.42 Draw window with Toolbox.

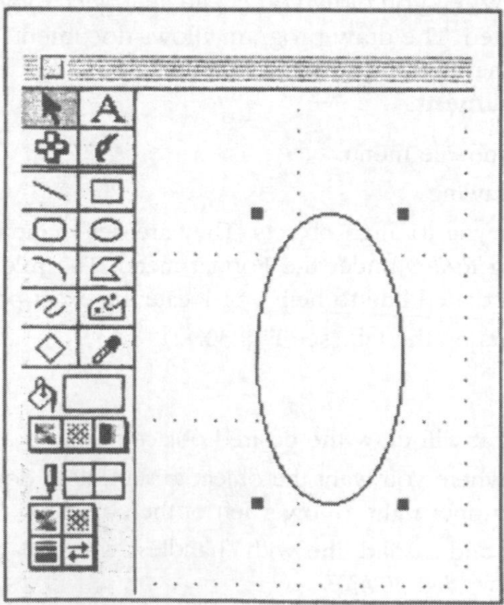

FIGURE 30.43 Object with handles.

To resize, move, or delete an object:

- Click on the object. Handles appear in the corners of the object space.
- Position the pointer on one of the handles and drag with the mouse to resize.
- Release the mouse when the desired shape and size are achieved.

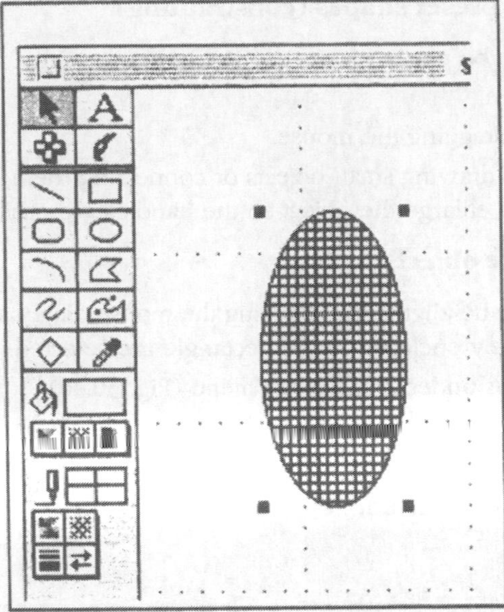

FIGURE 30.44 Filled object.

- To move an object on the page, click anywhere on the object EXCEPT on the handles.
- Drag with the mouse to the desired position on the page.
- If autogrid (the dotted grid lines) is on, the object will align to to the nearest grid point. If you want to place an object between grid points, choose *Trn autogrid off* under the Options menu.
- To delete an object, click on it to show the handles and press the *delete* key.

Polygons and bezigons are a bit different from the other draw shapes. You can set the polygons to close automatically by choosing that option under *Preferences* on the Edit menu. The polygon tool will give you straight lines with as many sides as needed. The bezigon tool creates polygons with curved lines between the click points (see Fig. 30.45).

To draw polygons and bezigons:

- Click where you want the first side to begin.
- Drag the dotted line with the mouse to the length wanted.
- Click and drag again for the next side.
- To end, double-click or close the polygon by clicking on the starting point.

ClarisWorks allows you to easily form a perfect square, circle, quarter circle, or 0, 45, and 90° angles without adjusting lines and angles. This is the easiest way to draw a straight line without those annoying jogs.

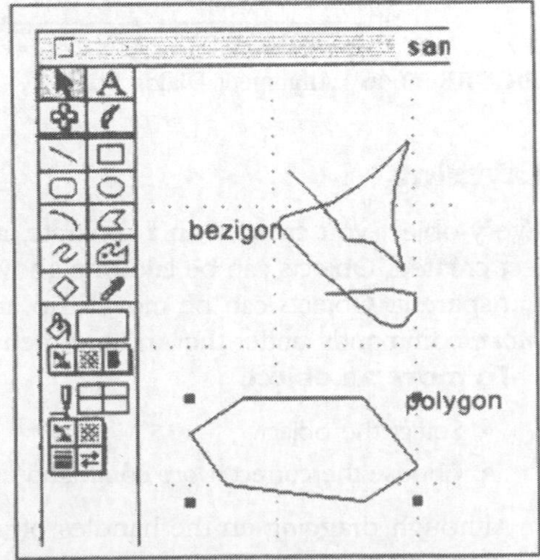

FIGURE 30.45 Polygon and bezigon.

To create automatic preset shapes (constraining):

- Click on the tool to be used.
- Hold shift key down.
- Begin drawing by dragging the mouse.
- If you have trouble drawing small objects or connecting them, use the zoom control on the lower left to enlarge the object so the handles are easily seen.

To align two or more objects:

- Select the objects to be aligned by dragging the mouse diagonally across the objects. All objects completely enclosed by the rectangle are selected.
- Choose *Align objects* under the Arrange menu (Fig. 30.46).
- Click on one button each side of the dialog box.
- The sample box will give you a preview of the alignment.
- Click on *OK* or press the return key.

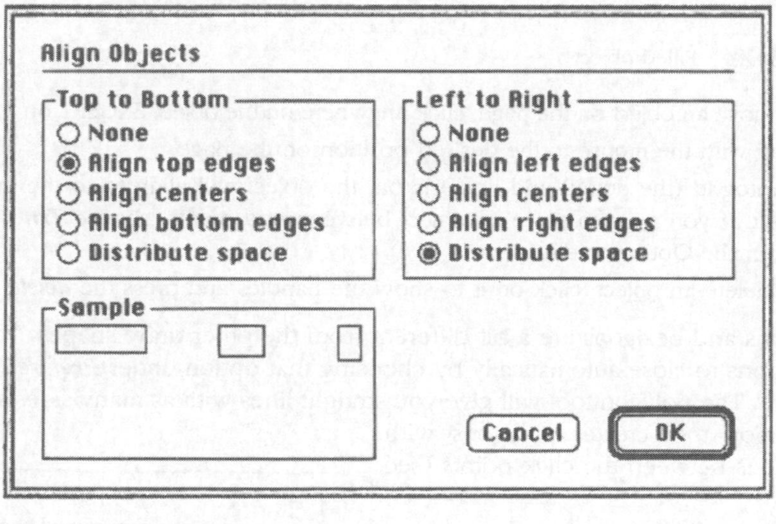

FIGURE 30.46 Alignment Dialog box.

Layering

Every object you create is in a separate layer with the bottom layer being the first object created. Objects can be laid one above the other and still be seen as the layers are transparent. Objects can be moved up and/or down through the layers by using the *move* commands under the Arrange menu.

To move an object:

- Select the object.
- Choose the correct *Move* command under the Arrange menu (Figs. 30.47 and 30.48).

Although dragging on the handles of an object will resize it, resizing precisely can be done by choosing *Object size* under the Options menu or by scaling the object (Fig. 30.49). You must know the exact size or position of the object.

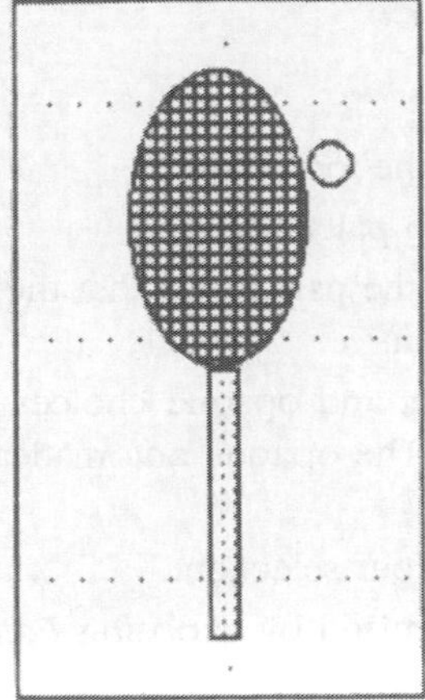

FIGURE 30.47
Racket to front.

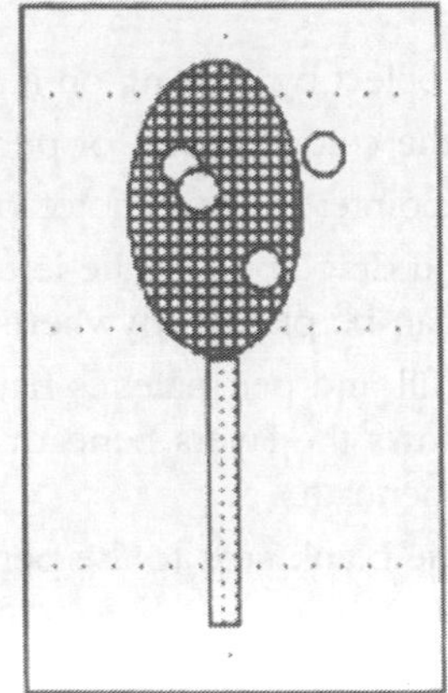

FIGURE 30.48
Racket to back.

To size an object from the menu:

- Select the object.
- Choose *Object size* under the Options menu.
- Click in the box(es) where measurement is to be changed.
- Press the *return* key.

FIGURE 30.49 Object size.

To scale an object:

- Select the object.
- Choose *Scale selection* from the Options menu.
- Choose the vertical or horizontal dimension box and enter a new value.
- Click *OK* or press the *return* key.
- To cancel a scale change, choose *Undo scale* under the Edit menu.

To change the orientation of an object:

- Select the object(s).
- Choose *Flip horizontal* or *Flip vertical* from the Arrange menu.

or

- Choose *Rotate* from the Arrange menu to rotate an object 90%.

Colors, Patterns and Pen Attributes

In Figure 30.42, the fill and pen indicators and palettes are identified. These tools have indicator squares next to them that show the current setting. You can make changes for the object selected or make default changes. If no object is selected when a change is made, the default is changed. If the change is made while an object is selected, only the fill of that object is changed.

To change a fill or pen selection for an object:

- Select an object by clicking on it.
- Click on the color, pattern, or pen options in the toolbox.
- Drag the pointer to your choice on the pop-up palettes.
- A click and drag up or to the left will tear off the palettes so that they will stay visible and can be placed anywhere on the screen.
- Both the fill and pen palettes have transparent and opaque choices. A transparent choice allows the layers beneath to be seen. The opaque automatically blocks out the layer beneath.
- The sample blank next to the pen will show your selection.
- Patterns and gradients can be edited or customized by choosing *Patterns* from the Options menu.
- The pen attributes have line width and style as well as color and pattern.
- The style palette gives the choice of an arrowhead on either or both ends.

The Master Page

The master page (Fig. 30.50) allows you to create a transparent layer that contains information that you want to appear on every page of a document. When you create a master page, that designation appears in the place where the page number usually appears at the bottom left.

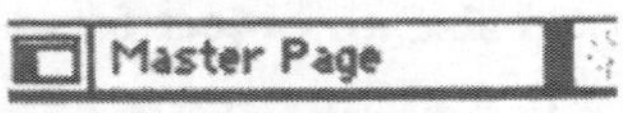

FIGURE 30.50
Master page indication.

To create a master page:

- Choose *Edit master page* from the Options menu.
- Enter data, text, or objects that will appear on each page.
- Exit the master page mode by choosing *Edit master page* again to toggle off. The check will disappear.

30.6 THE PAINTING PROGRAM

The painting program is different from the draw program in that it creates images rather than objects. The painting technique is made up of pixels that are turned on or off in color to create the image. To change an image you must change the pixels, not the outside form. Because the paint program takes up so much memory, only one-page documents can be created.

There are six new tools available in the paint program (Fig. 30.51):

- The **selection rectangle** selects portions of the document. Drag the rectangle over the part of the page you want selected. A double-click selects the entire page. The selection includes the white space around the objects. To select objects without the white space, press command while you double-click on the rectangle tool.
- The **lasso** draws a line around the object as you drag it to select the object without the white space. A double-click selects all the objects in the document without the white space.
- As you drag the **magic wand**, it changes every pixel in its path to match the first pixel selected. When the mouse is released the pointer reappears.

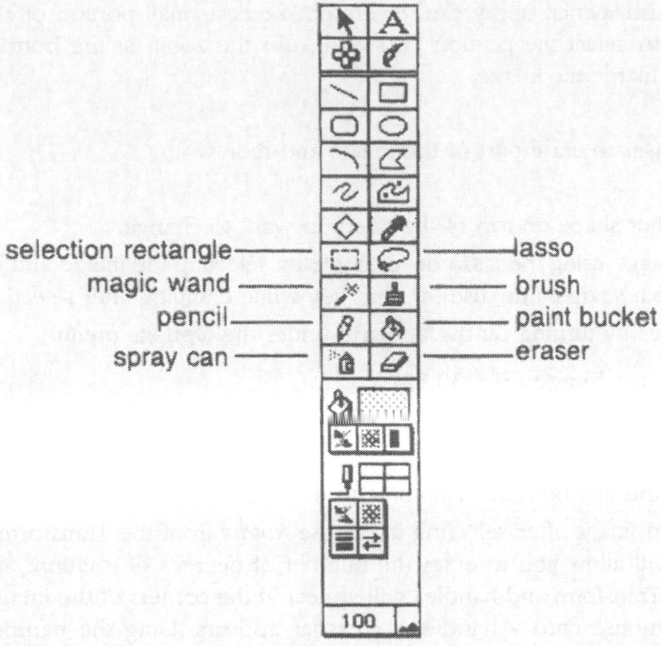

FIGURE 30.51 Paint tools.

- To paint freehand images, drag the **paint brush**. Choose the shape of the brush by double-clicking on the brush and choosing the shape from the dialog box that appears.
- The **pencil** creates a freehand thin line. The only option this tool has within the image is the color. The width and pattern options are applicable only to the outline.
- To fill an enclosed image with a color, pattern, or gradient, click on the **paint bucket**.
- The **spray can** creates a round spray of dots that you can drag around or spray in one place to fill in. The color, pattern, and gradient are options on the palettes. A double-click allows you to choose the dot size and the flow rate.
- The **eraser** clears anything over which it is dragged. A double-click clears the page

To create a paint document:

- Choose *New* under the File menu.
- Double click on *Painting* or click on *Painting* and *OK*.
- If you prefer the rulers to show, choose *show rulers* from the View menu.
- Choose a color, pattern, and/or gradient for the fill tool.
- Choose a line color, pattern, and width for the pen tool.
- Choose the opaque or transparent *Paint mode* from the Options menu.
- Click on the chosen tool to create an image.

To edit images:

- Select image(s) or parts of an image using the rectangle or lasso. A dotted line moves around the selected part.
- To deselect, select another tool.

- Use the brush, pencil, spray can, or eraser to edit a small portion of an image. If it is difficult to select the portion you want, use the zoom at the bottom left of the screen to enlarge the image.

or

- Use the eraser to erase part of the image and redraw.

or

- Draw another shape on top of the one you want to change.
- Move an image using the lasso or rectangle by selecting the image and dragging . To constrain to a straight line, use the shift key while dragging after selecting the image.
- Align images by turning on the autogrid under the Options menu.

To transform an image (Fig. 30.52):

- Because images are layered as you create them, you can replace an unwanted image by dragging it on top of the object you want. Choose *Pick up* from the Transform menu and the one beneath will come to the top.
- To rotate an image after selecting it, choose *Rotate* from the Transform menu. A dialog box will allow you to enter the number of degrees of rotation. Select *Free rotate* under Transform and handles will appear at the corners of the image. When you move the mouse onto a handle, a crossbar appears. Drag the handle to turn the image into whatever rotated position is desired.

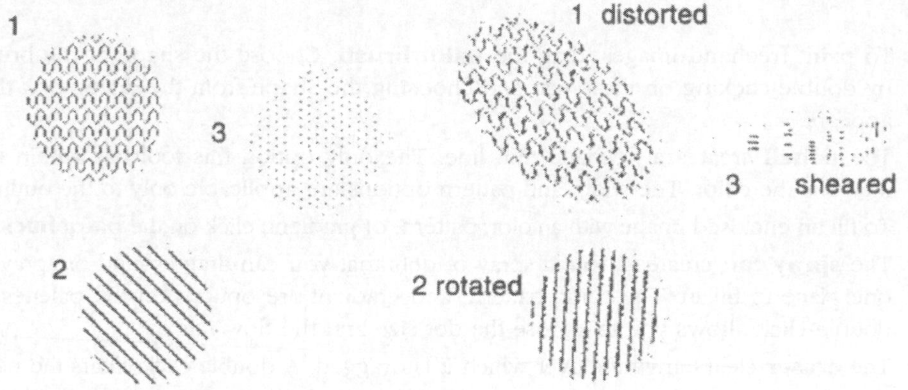

FIGURE 30.52 Transformed images.

To paint and fill:

- Fill a **blank** enclosed image with a fill color, pattern, or gradient with the paint bucket. The image MUST be enclosed or the color will "leak" to the adjoining areas until it finds itself enclosed.
- Choose the desired color and pattern from the paint bucket palettes.
- Click on the paint bucket tool.
- Click on the space to be filled.
- Fill an already colored image by choosing *Fill* from the Transform menu (Fig. 30.53) rather than using the paint bucket. You must reselect the area to be filled.
- Overlay one color on another for in-between colors by changing the color of the fill, selecting the image, and choosing *Tint* under the Transform menu.

- Make the image lighter or darker by selecting the image and clicking on Darker or Lighter under the Transform menu. Each time you click on that choice the image gets a little darker or lighter.

Note: With most choices that change images, you have the option of using *Undo* under the Edit menu. This gives you the ability to try out changes without losing your original image. Be sure you do not click in between your choice and the *Undo* because the *Undo* only works for the immediately previous action.

30.7 COMMUNICATIONS

ClarisWorks communications is a simple standard communications program. It has a phone book feature that will dial automatically when you select an entry using the phone tool. The communications program's biggest fault is that data that you enter cannot be edited or deleted. Therefore it is better to type in another environment and paste to the communications program. If you are copying a table of data, hold the *command* key down as you highlight the table and use *Copy table* from the Edit menu in the communications program to paste. This will preserve the tab stops and the receiver will get a table and not a randomly spaced line of data.

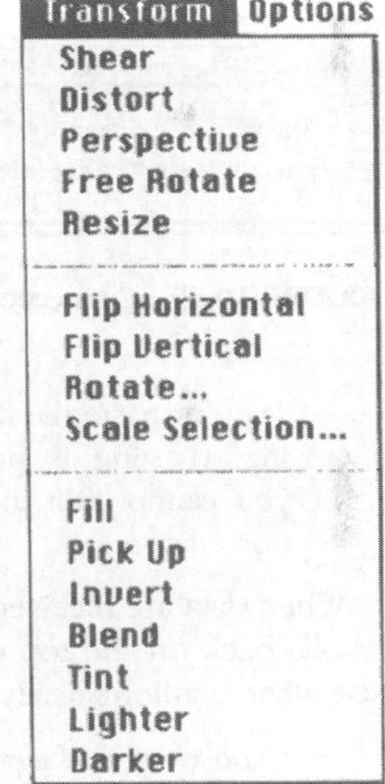

FIGURE 30.53 Transform menu.

Receiving and sending data:

- Choose *New* under the File menu. Figure 30.54 appears showing the phone tool with its pull-down menu. (Some names and numbers have already been entered into the phone book for this example.)
- Choose *Connections* under the Settings menu (Fig. 30.55).
- The default settings are for a Hayes-Compatible modem at 2400 Baud. These can be changed by choosing from the pop-up menus.
- The Apple Modem tool is the one used for regular telephone connections.
- Choose settings that match the computer with which you are communicating (Fig. 30.56).
- Enter the phone number by typing in or by choosing from the phone tool menu.

Note: To include a pause (possibly for an outside line) in the dialing, enter a comma as shown in Figure 30.55.

- Press the return key or click *OK* to save your settings.
- Turn on your modem after checking to be sure connections are secure.
- Click on *Open connection* under the Session menu.
- Wait for the message saying you are connected.
- Click on *OK* or press return to exit the dialog box.

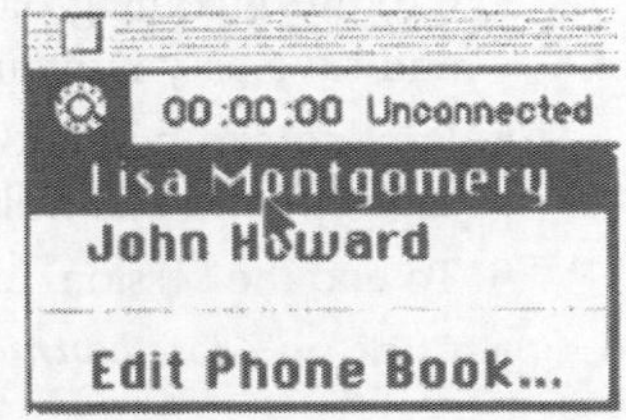

FIGURE 30.54 Phone Tool and menu.

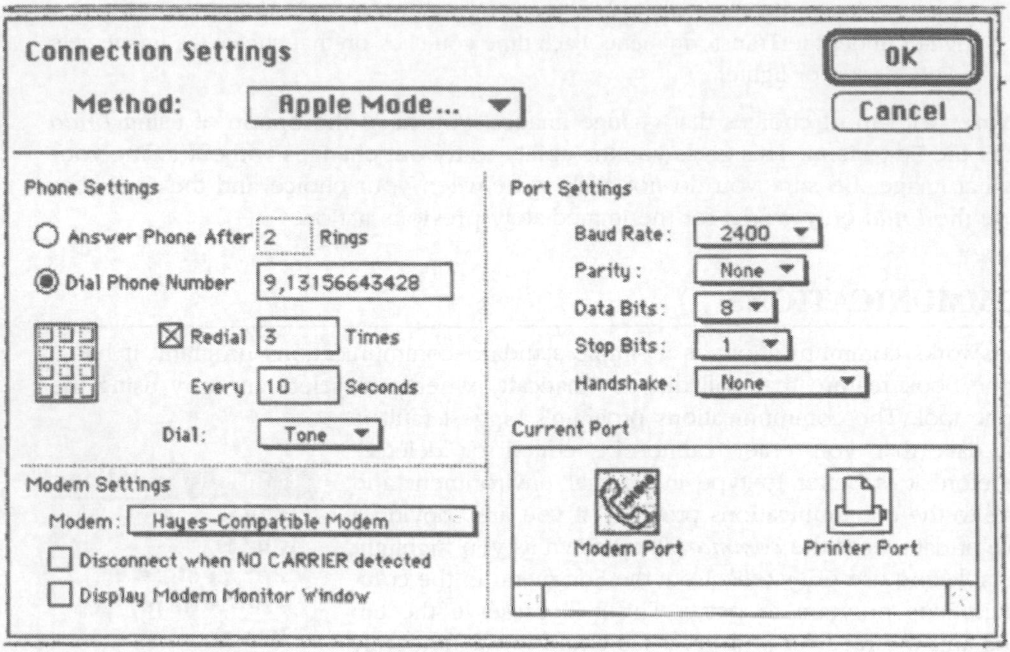

FIGURE 30.55 Connection settings.

- Type or paste the data you are sending. (Pasting is preferable because you cannot edit in the communications mode.)

When data are received, it automatically scrolls back off the top of the screen. To see what is automatically saved:

- Choose *Show scrollback* from the Settings menu.
- Choose *Hide scrollback* from the Settings menu to hide the scroll.

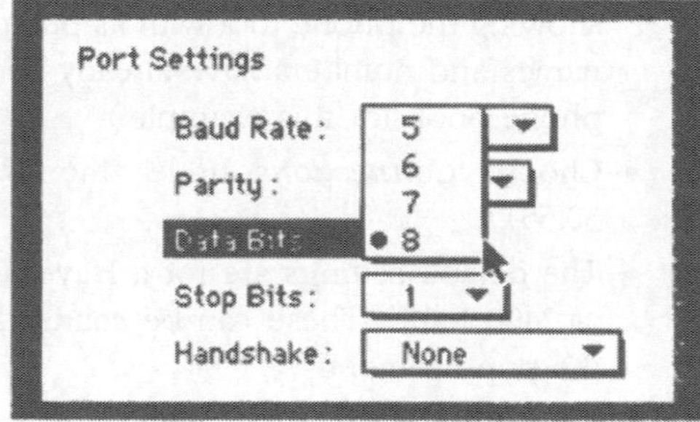

FIGURE 30.56 Port settings.

- You can select *Cut, Copy,* or *Paste* from the Edit menu or save the entire scrollback by choosing *Save lines off top* under the Session menu (Fig. 30.57).
- Clear the scrollback by choosing *Clear saved lines* from the Session menu.
- Capture incoming data by choosing *Capture to File* from the Session menu. (Name and save as you would any file. It can then be opened as a word processing file.)
- Printing can be done while you are online or from saved data. Usually printing later saves phone and online service charges.
- To end the session, use the online services's command to log off if you are using one.
- Click on *Close connection* under the session menu.

Receiving and sending files: You can send or receive files of all kinds while preserving the format. The receiving computer must be compatible with your file transfer settings.

To send a file:

- Check the file transfer options under *File transfer* under the Settings menu (Fig. 30.58). The XModem protocol does automatic checking and error correction. The other option is to send a text file.
- Leave other settings as ClarisWorks defaults.

Note: If the receiving computer requires you to make changes, check the *ClarisWorks Communication Handbook*.

- Follow the steps to open the session as described earlier.
- Choose *Send File* from the Session menu when the receiving computer answers.
- Click on the name of the file to be sent.
- Close the connection as earlier described.

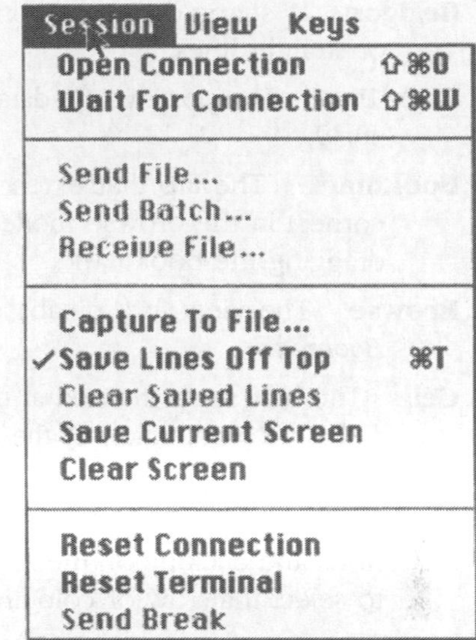

FIGURE 30.57 Session menu.

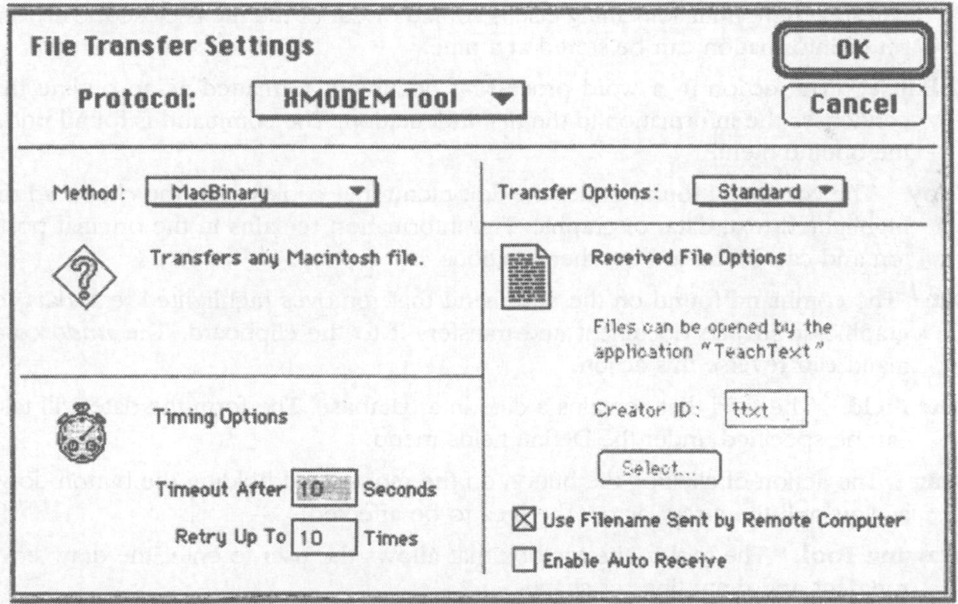

FIGURE 30.58 File Transfer settings.

DEFINING TERMS

Argument: The value(s) in a database or spreadsheet upon which the function acts.

Autogrid: A guide in the paint and draw environments that lines up objects and can create objects of an exact size. It is accessed under the Options menu.

Baud Rate: The speed at which data are transmitted and sent through a modem. The higher the baud rate, the faster the transmission.

Bezigon: A shape drawn by connecting lines through a series of points with curves or straight lines.

Body Part: The portion of a database layout that contains the record data. See Figure 30.39.

Bookmark: The tab that extends to the right of the record book (upper left hand corner) in the browse mode of a database. You can flip through the records by dragging the bookmark.

Browse: The view in a database where records are added, sorted, and found. See *Bookmark*.

Cell: The box space framed at the juncture of a column and row. The address of a cell is a combination of the column letter and the row number.

Cell Range: A group of adjacent cells in a spreadsheet. They may be selected by dragging from the first to the last, clicking on the upper left cell typing two periods and clicking on the lower right cell, or clicking on the row or column label to select that row or column.

Cell Reference: The cell address used on a formula. Selecting by clicking rather than typing helps to eliminate mistakes.

Clipboard: A temporary file that stores information *cut* or *copied*. The information will stay there until something else is copied or cut or the file is closed. Only one set of information can be stored at a time.

Collapse: The action in a word processing document formatted as an outline that condenses the information to the first indentation. The command is found under the outline menu.

Copy: The command found under the Edit menu that copies onto the clipboard the highlighted text, data, or graphic. The information remains in the original positon and can be pasted to other locations.

Cut: The command found on the edit menu that removes highlighted text, data, or graphics from the document and transfers it to the clipboard. The *undo* command can reverse this action.

Date Field: The field that contains a date in a database. The form the date will take can be specified under the Define fields menu.

Drag: The action of clicking the button on the mouse and holding the button down as you pull the arrow across the area to be affected.

Drawing Tool: The tool in the toolbox that allows the user to enter the draw environment and draw lines or shapes.

Entry Bar: The line at the top of a spreadsheet that identifies the cell location and shows the data being entered. Editing in a spreadsheet is done in the entry bar.

Environment: One of the six modes in which ClarisWorks operates.

Field Label: A text object in a database that contains the name of a field. See Figure 30.40.

Fill Tool: The tool that when highlighted can change the pattern and color of a selected object.

Find: A choice under the Layout menu in a database that is used for searching for records with specific criteria.

Footer: A feature of a page that is repeated automatically on every page of a document.

Footnote: A citation that gives added information to identified text in the body of a document. ClarisWorks automatically numbers footnotes consecutively when the option is chosen under the Format menu.

Formatting Characters: Characters that are invisible in a printed document but that are typed such as spaces, returns, and tabs. These characters can be made visible or invisible under Preferences under the Edit menu.

Frames: Graphic objects that can contain a spreadsheet, painting, drawing, or text. Frames allow a different environment to be introduced into a document.

Grand Summary Part: The field in a database that contains sub-summary fields.

Grouping: The process of selecting more than one object in a draw environment so that they can be treated as one. The option can be found under the Arrange menu.

Handles: The black squares that appear at the corners of a selected object in the draw environment. They are used to resize and reshape objects.

Header: A line that appears at the top of every page in a document. It is selected by choosing *Insert header* from the Format menu.

I-Beam Pointer: The indicator in a document that the text mode has been selected. It is used for text selection, cursor relocation, and moves with the mouse, and changes to an arrow when it leaves the page.

Image: A shape created of pixels that is created in the paint environment.

Layout: A depiction of the way a database will look if it is printed out Different layouts are used to display or emphasize different parts of a record or a selection of records.

Modem: The communication device that connects one computer to another over a network or telephone lines.

Paint Tool: The brush in the toolbox that allows you to enter the paint environment by creating a paint frame.

Panes: Split views, horizontal or vertical, of the same document created by dragging the Pane markers at the upper left or lower right of the screen.

Part: A section of the database such as the body, the footer, or the summaries that is used to organize data. See Figure 30.40.

Paste: The action of copying the contents of the clipboard to the place selected by the cursor or I-Beam.

Pixel: A single dot on the screen, which is the component unit of an image. It is an abbreviation of picture element.

Pointer: A short shafted arrow that is used to select a tool from the toolbox and make visible handles to resize or move an object. It changes to the appropriate shape when it moves into a document. Immediately after a tool is used, the selection will revert to the arrow unless the selected tool is clicked twice.

Record: The collection of data fields in a database that is relevant to one entry.

Record Book: The icon resembling a book that appears in the upper left corner of a database screen and shows the number of the current record.

Relative Reference: A reference to a cell whose location is identified by its relationship to other cells, not its column and row. A relative cell will change as the cells to which it relates are changed.

Selection: The identification of an object, image, text, data, or cell to be changed. Selection highlights text, makes visible handles for an object and outlines a spreadsheet cell.

Standard Layout: A database arrangement that shows all the fields in the order of creation.

Sub-summary: The part of a database layout that calculates the values of a selected field in selected records.

Trailing Grand Summary: The database part that is placed after the the information summarized by the part.

Zoom Controls: The small and large mountain icons that appear in the lower left of the screen that allow you to enlarge or reduce the view size of the document. The number to the left of the icon shows the percentage of the view size to the actual size.

REFERENCES

ClarisWorks Communications Handbook. Claris Corporation, Santa Clara, CA, 1993.

ClarisWorks User's Guide. Claris Corporation, Santa Clara, CA, 1993.

Shelley O'Haram, Catherine Morris, and Cyndie Shaffstall-Klofenstein, *Using ClarisWorks 2.1 for Macintosh*. QUE, Prentice-Hall, Englewood Cliffs, NJ, 1994.

Charles Rubin, *The Macintosh Bible Guide to ClarisWorks 2.1*. Peachpit Press, Berkeley, CA, 1994.

Barrie Sosinsky, *ClarisWorks Companion*. Hayden, Prentice-Hall, Englewood Cliffs, NJ, 1994.

31

Computer-Based Presentation Systems

T. M. Rajkumar
[illegible]

31.1 INTRODUCTION

A computer-based presentation consists of a set of computer visuals (slides) that you design and produce to deliver a message to an audience. The delivery can be via a computer or a traditional mechanism such as a slide or overhead. In this chapter, we will focus on delivery via a computer. The visuals can be pure text such as a list of bullets, a table (such as a spreadsheet data), a graphic chart such as a bar chart or pie chart, or a graphic such as a drawing, or a bitmap graphic such as a scanned logo. Presentation programs allow you to create/import the data, organize these visuals into a presentation, sort them, and provide transition effects. The presentation is stored in a file, that can be later edited. The presentation can be played back on a computer monitor or projected onto a screen using an LCD projector. You can also create speaker notes or make handouts for your audience using the presentation software. In addition, many presentation programs are currently providing the ability to incorporate audio, animation, and video into a presentation. In general, the following steps are used to create a computer-based presentation:

1. Create the slides you want to use.
2. Organize the slides in the order you want to deliver them in.
3. Add effects to provide transitions between the slides.
4. Add audio, video, or animation to provide special effects.
5. Present the slide show using a computer.

31.2 PAINT PROGRAMS

Paint programs try to replicate the traditional artistic media and processes inside the computer. Painting tools determine the color of each pixel they touch. The image shape and shade are manipulated by parameters of the tool such as brush type and geometry. Fill tools flood areas with solid colors, textures, or patterns.

0-8493-2530-7/96/$0.00+$.50

Paint programs work with bit mapped images. They can manipulate bits of information. The images do not scale or change resolutions well. Since the images are bit maps, they generally do not have the notion of layers of objects. Hence, if we lay a new graphic element on top of another, the information on the object underneath is lost. However, there is a new genre of paint programs making its appearance that allows some layering.

Microsoft PaintBrush is a paint program that comes standard with Windows and is found in the Accessories program group.

PaintBrush

When you start PaintBrush, you are taken to a screen with the following areas: Drawing area, Toolbox, Linesize Box, Palette (Fig. 31.1). The drawing area is where the image you draw is created. The linesize box allows you to specify the size of the line that you draw.

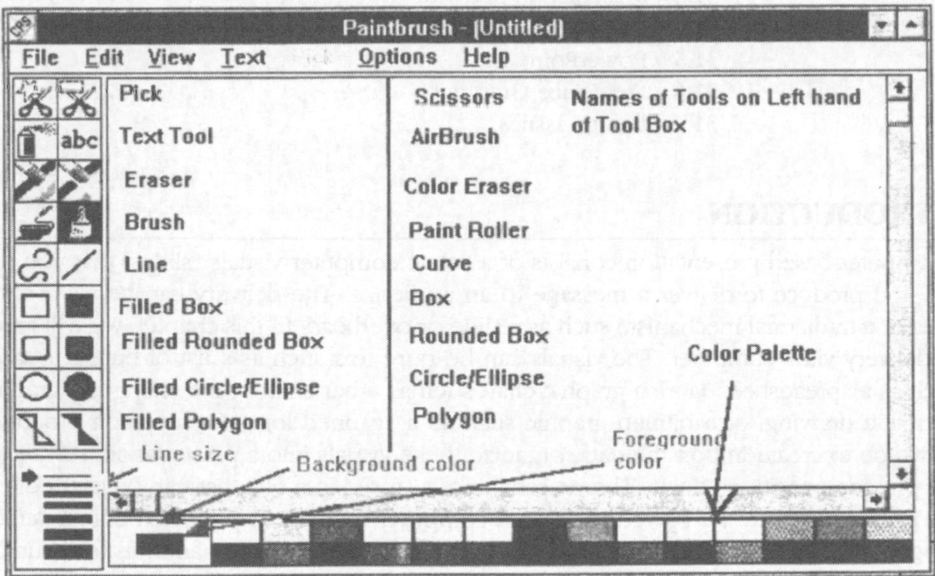

FIGURE 31.1 PaintBrush screen.

Palette

The palette specifies the foreground and background color that is being used at any point in time. The chosen colors are shown in the selected colors box.

1. To change the foreground color: click on the color in the color palette with the *left mouse button*.
2. To change the background color: click on the color palette with the *right mouse button*.

If you want to change the set of colors in the color palette, *double click on the color* in the color palette.

Paintbrush will allow you to edit the RGB color combination and create any palette you desire. Only the color that you are changing is reset. You can always go back to the original palette by clicking on reset within the edit color screen. If you like a certain set of colors on the palette, you can save them by choosing *Options, Save Colors* from the menu.

You can always set this to be the palette by choosing *Options, Get Colors*, and identifying the file you saved the palette to.

ToolBox

The ToolBox contains the tools shown on the left-hand side in Figure 31.1. These tools can be organized in the following broad categories: Drawing tools, Line Tools, Freehand tools, Text tools, Brush tools, and Editing Tools. Each of these is discussed next.

Drawing Tools

The Box, Ellipse tool, draws a box or ellipse in the foreground color.

1. *Click and drag* the mouse to draw the objects.
2. Hold the *shift key* while dragging with the mouse to constrain the object to a regular shape (square, circle).
3. The filled versions of these tools create a filled box or ellipse. The fill color is the foreground color, while the boundary is bordered with the background color.

The Polygon tool is used to draw a polygon.

1. Click with the *left mouse button to anchor.*
2. *Drag* to where the line should end and *click to end the line.*
3. *Drag and click* to create the rest of the sides.
4. When you are ready to close the polygon, *double-click the left mouse button* and the polygon will close.

Line Tools

The line tool allows you to draw a line.

1. Click with the *left mouse button* and *drag* to complete the line.
2. Hold the *shift key* while dragging to constrain the line to a straight line.

The curve tool allows you to draw free hand curves.

1. Click with the *left mouse button to start* the curve.
2. *Drag* to where you want the curve *to end. Release* the mouse button. This will show a straight line.
3. With the *left mouse button pushed, move the mouse* to choose the shape of the curve.
4. *Release the left mouse button* to finish the curve.

The thickness of the line and curve can be set by choosing the line width before drawing the line or curve.

Freehand Tools

This set of tools consists of the brush and airbrush. Brush draws freehand shapes with the foreground color. Choose the brush shape by double-clicking on the brush tool. Similar to the airbrush tool of the artist, the airbrush tool sprays dots of the foreground color when you click with the mouse and drag the tool. The density of the spray is determined by the speed with which you drag the tool with the mouse. By choosing a different line size, the diameter of the circle that is sprayed can be changed.

Text Tools

Text tools allow you to enter text in the drawing area. Prior to entering text

1. Set the font size, style (bold, underline, italics), shadow, or outline by choosing the appropriate option.
2. You can edit text while you are still typing by using the *backspace key*.

However, once you have selected some other tool, you cannot select or edit text. The text is simply stored as a bitmap.

Fill Tools

The fill tool in PaintBrush is the painter roller. The painter roller fills any enclosed space with the foreground color.

1. Just select the tool, *point* to the enclosed object, and *click*.

Editing Tools

The editing tools in PaintBrush are the Pick and the Scissors tool. These tools allow you to define an area, copy, cut, and paste to a different area of the screen.

1. Defining the area is done by *clicking* with the mouse and *dragging* the mouse.
2. Once you have picked an area, you can *cut, copy, paste* through options in the *Edit* menu similar to any other Windows program.
3. When you paste the item, after you have cut or copied the area, it appears on the top left-hand corner of your screen.
4. You can then *drag the pasted area to any location* on the screen.

Once you have defined an area, and picked the area, you can use the **Pick** menu to either invert colors, flip the area horizontally, vertically, or tilt the area. You can paste the picked area and shrink or grow it (change its size) when pasting by dragging the pasted object.

The pick tool allows you to select a rectangular area on the screen. The scissors tool *does not constrain you to select a rectangular area* on the screen. You can identify any shape by drawing freehand the area you want to select.

Eraser Tools

The eraser tools enable you to change the color of any selected object to the background color. Before performing any erase operation

1. *Set the foreground and background color* to the desired color.
2. It is helpful to use the Zoom option (*Ctrl+N*) when using either eraser or the color eraser.
3. *Define the area you want to zoom* to. Once you have zoomed in, you have control over each individual pixel in the screen. This is very useful during the erase process.
4. Press (*Ctrl+O*) to zoom back to the original size.

The color eraser tool is used to change the selected foreground color to the background color.

1. Select as the foreground color the color you want to change.
2. Select as the background color the new color.
3. Drag the color eraser tool over the objects that you want to change the foreground color to background.
4. Hold the shift key down while dragging to restrain the eraser tool to a straight line.

You can set the line width to control the thickness of the area affected. To change all occurrences of the color throughout the drawing *double-click* the color eraser tool.

The eraser tool is similar to the color eraser, and changes all colors to the background color when it is dragged over the object. Make sure that the background color of the object that you are trying to erase is the same as the background color you have selected.

Object Linking and Embedding

Paintbrush is an OLE server. This implies that any paint or bitmap object can be embedded in any other application that supports object linking.

PaintBrush Files

Paintbrush can store files in either the standard Windows bitmap format, or the PCX file format. It can store in 24-bit, 8-bit, or 4-bit color format. To decide on the format click on the *info box*, when you choose *File Save as*. This will let you know the number of distinct colors in your file. You can then choose the appropriate format. A 4-bit color format can store upto 16 colors, 8 bit—256 colors, and 24 bit is true color.

One of the uses of PaintBrush is to touch up a screen captured as a bitmap, or a scanned image such as a logo. Some scanners store images in a bitmap format such as TIFF that is different from either BMP or a PCX file. Use an image converter in such cases to store the image in BMP format. You can then use paintbrush's editing tools to clear up any errors, or add other enhancements to the image.

31.3 PRESENTATION SOFTWARE

Presentation software in general work with the slide metaphor. They help you generate and place text and graphics on a slide. Though these software can work with audio and video, they are best when used with simple text slides and data charts such as bar charts and pie charts. Many of these software include drawing capabilities, but do not match the sophisticated capabilities of special purpose drawing software. These software are highly useful to playback slides or make presentations using the computer. Animation capabilities are limited in these software. Harvard Graphics, Microsoft PowerPoint, and Lotus Freelance are industry leaders in this category of software.

31.4 HARVARD GRAPHICS

When you start Harvard Graphics (HG), you can either open an existing presentation or a new presentation. The presentation is the basic format with which HG works. A presentation is a sequence of slides. A slide is made up of text, graphics, or charts. You work with slides in one of three modes: slide editor, outline, or slide sorter mode in HG. With the default set up, when you open a presentation, HG places you in the slide editor mode.

Slide editor displays one slide at a time on the screen. You use this mode to enter data, format the data, and view a single slide.

Slide sorter displays all the slides in the presentation at once on the screen. It provides an overview of the entire presentation. You can use this view to reorder the slides in the presentation.

Outliner is useful only for text-based slides. It displays the structure of the presentation and is used to create an outline for a presentation. You can use this mode, if your presentation will basically contain text slides.

Chart

A slide consists of one or more charts. A chart is used to represent data either with text or graphics. HG supports text charts or data charts. HG provides support for various types of charts by providing templates for these charts. The chart is prepared by entering values for the various elements of the template.

Working with Text Charts

When you create a new presentation, HG asks for the chart/slide type you want to specify. For example, Figure 31.2 shows a slide for the text chart of type title. Within a title chart, multiple options exist with different formats. The preview option on the

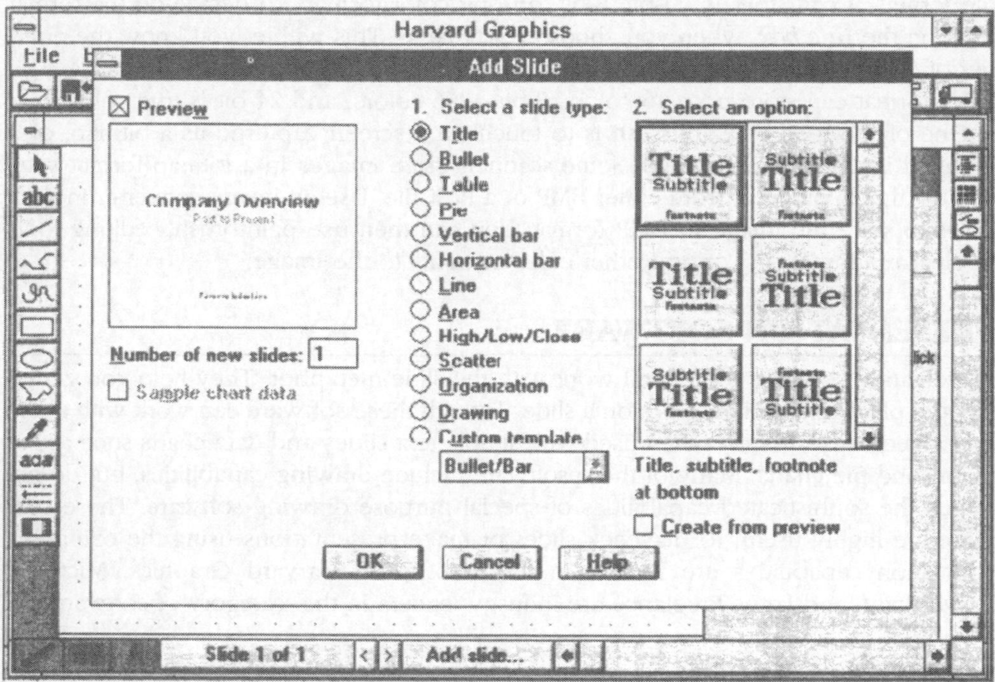

FIGURE 31.2 Title slide types in HG.

screen provides an overview of the slide with some sample data. Upon choosing the chart type, HG places you in the slide editor mode. HG presents a data form that you use to enter the data. For example,

1. Enter *Harvard Graphics* for the title.
2. Enter *your name* for the subtitle.
3. You can move between the regions by pressing *Tab*.
4. You can preview the slide by choosing *View, Preview*. Upon preview, you do not see the footnote that you see in the slide editor mode (Fig. 31.3). Regions that are not entered are not shown on preview. Once you have entered the text, you can click on OK to save the slide.

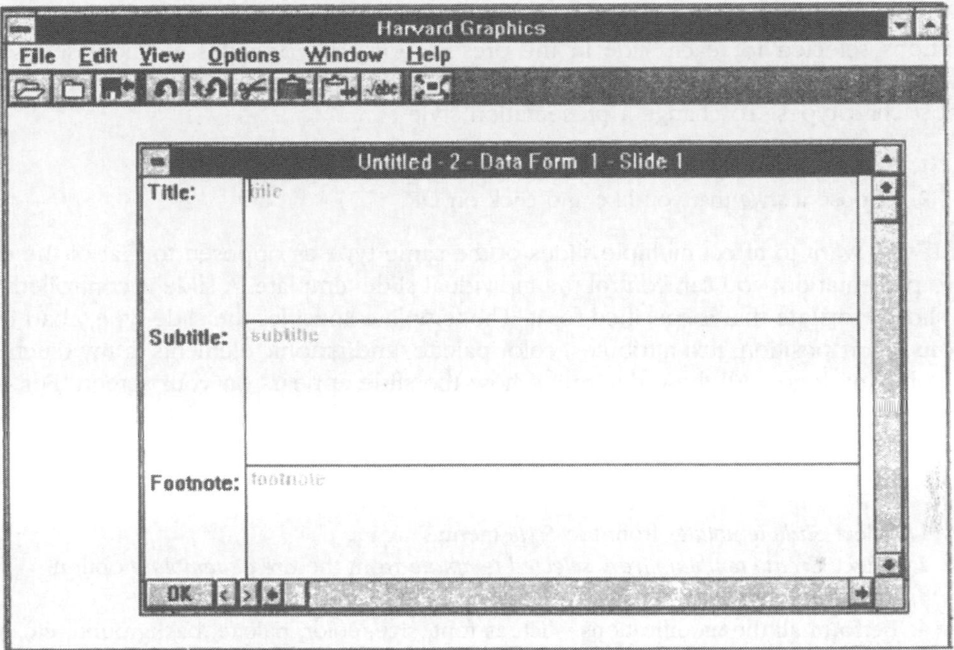

FIGURE 31.3 Slide editor for title chart.

Help

There are two ways of getting help in HG. Press *Shift+F1*, and obtain help. The second way is to use the HG advisor.

Advisor. The HG advisor is displayed in the slide editor. Two levels of advise are provided; Quick tips and Design Tips. Quick tips are useful to help you perform tasks. For example, if you want to add a footnote that you did not enter when you entered the data, the quick tip asks you to just double-click on the text. Design tips in contrast, contain suggestions to help improve the format and appearance of your chart. For example, design tip for a title chart informs you to not use all caps for the text.

1. To display design tips, click the *Design tips* button.
2. To display quick tips, click the *Quick tips* again.

Enhancing This Chart

Charts have enhancement options that you can make. If you want to change the font, color of the text, etc., the Quick tip provides the solution.

1. First edit the text by *double-clicking* on the text area. Select the text you want formatted differently by *highlighting* with the mouse.
2. Press the *right mouse button*.
3. Select the *new font, size, color* etc. in this option to reset the text.

This is the way to change individual text or element within a chart. However, if you want to change multiple elements on a chart, a better method is to choose a different style for the entire chart, especially when you are not satisfied with the colors or fonts.

A **presentation style** affects the slide background, and color palette, as well as the options selected for each slide in the presentation. HG provides a whole series of styles. These styles have been chosen by HG with good color, text, and options for most chart types. To change a presentation style

1. Select *select a style* from the *Style* menu.
2. Choose a *style* that you like and click on *OK*.

If you want to affect multiple slides of the same type as opposed to that of the entire presentation, you can control the individual slide template. A slide is controlled by a slide **template** that is specified for it. The template specifies the slide type, chart options, chart position, text attributes, color palette, and graphic elements if any (such as a company logo). All these determine how the slide appears on your screen. For example, the title slide we have created is controlled by the title template. You can modify the template, or create your own from the existing template and then apply it to all slides that you would like. To create your own template

1. Select *Slide templates* from the *Style* menu.
2. Select *Create template from selected template* from the *create template* option.
3. Select *Edit template*.
4. Perform all the modifications, such as font, size, color, palette, background, etc.
5. Save the template to slide by *clicking on back to the slide*.
6. Apply this template to any other slide with similar chart type for consistency.

In general you modify the template only when you want to be able to reuse a specific design without having to change the options for each similar chart. Once you have created your own custom template, whenever you add a slide, click on custom template, and select your template as option to pick from.

Bullet Charts

Generating other text charts is similar in nature. For example, to create a bullet chart with Harvard Graphics

1. Click on *Add Slide* button
2. From the list of slides, choose *Bullet*. Click on the *style* you like. Click on *OK*.
3. Enter the title.
4. Enter the bullet text in the bullet list. Each line gets entered as a separate text.
5. To create subbullets, press *Tab* before entering the text.
6. *Shift+Tab* takes you back to the main level.
7. Press *F2* to preview.
8. Click *OK* on completion.
9. To change the bullet attribute, select *Text Menu* and *Set Bullet Attribute*.
10. To save the file, press *Ctrl+S*.
11. Select *Edit, Check Spelling* to spell check the chart or presentation.
12. To Edit Text, *double-click* on the text. Make any changes you desire. *Click outside the region* to go back to the Slide Editor.

Build

A build allows you to reveal one bullet at a time, or highlight the current bullet, or both. To create a build of the bullet chart

1. Select *Edit, Autobuild.*
2. Click on *Use Autobuild, Build Type, OK.*
3. Press *Alt+F2* to view the screenshow from the current slide. The build is shown only in the screenshow, and not in the preview mode with F2.
4. Press *Enter* to view the next slide. Continue until you see all the bullets.

Table Charts

A table is a text chart with columnar data. It is very similar to a spreadsheet data.

1. Choose *Add, slide, Table, Choose the Style* of table desired, click on *OK.*
2. Enter the *title in the row below the column headings, A, B* etc.
3. Enter data in rows and columns.
4. You can use formulas such as *=sum(a1..b1),* or *=(a1+b1)* in column C1 to total that row. Other formulas are possible. Press *Shift+F1* to get help on formulas.
5. To adjust the column width, *click on the heading and drag* to the appropriate size.
6. You can import spreadsheet data from either Lotus or Excel by clicking on *Get Info.*
7. You can format numbers by selecting *Number Formats* from the *Chart Options* menu.
8. In general limit the number of rows and columns for computer-based presentations so that the audience at the back can read the data.

Working with Graphics Charts

Graphical charts are used to show data in a variety of formats such as pie charts, bar charts, or line charts. These are basically used to show relationships between data items. Before you create the charts make sure the charts convey the message that you want to convey.

For example, a pie chart is used to convey percentage or market share. Bar and column charts are used to compare several variables at one time or one variable at discrete points in time. A stepped bar is used to show frequency. Line charts are used to show changes over time of one or more data items. Area charts are used to see the parts as well as the whole over time. Make sure you use the appropriate chart for the message. We will discuss two chart types here. Regardless of the type of chart you make, the procedure is basically similar.

Pie Chart

Let us say you are manufacturing widgets, and your share of the widget market is 45%. Company B's share is 30%, and Company C's share is 25%. You also want to emphasize your widget share and show its breakdown by regional contribution. To create such a pie chart

TABLE 31.1 Appropriate Chart Types for Messages

Message	Chart Type
To show the proportion each item contributes or percentage	Pie
Change over time periods or trend	Line
Change over few time periods	Vertical bar
To rank different items at same point in time	Horizontal bar
Emphasize quantities of item, parts as well as whole	Area
To identify a pattern	Scatter
To show frequency	Stepped bar
To show high–low–opening–closing values of items (stocks)	Hi–Lo

1. Choose *Add Slide, Pie*, Choose the *style*, 3D Pie linked to a 3D Bar (last choice). Click on *OK*
2. The first pie labels and values go in the first two columns.
3. The columns' labels (breakdown of your regional contributions) are entered in the second pies labels and values.
4. Sample data are shown in Figure 31.4.
5. Click on OK. You should see Figure 31.5.
6. Choose *Chart Options* to vary the options for the pie, such as entering a legend (display a color coded legend that describes the pie slice), enter labels, or change the pie style.
7. If you want to change the entire style for the pie, click on *Change Chart Type, From Gallery, and pick the style you like.*
8. Press Ctrl-S to save the presentation

Bar Charts

Let us say you want to show production of widgets and bolts from January to June. Then we can choose the vertical bar chart.

1. Choose *Add Slide, Vertical Bar*. Choose the style *2D clustered* (first on list). Click on OK.
2. Enter a title, such as *Semi-Annual Production*.
3. Click on the *X-axis Labels*.
4. Click on *label Format—Month*.
5. Enter *Jan* for Start, and *June* as the end.
6. Click on *Series 1*, and enter *Widgets*.
7. Enter appropriate values for the widgets production under that column, for each row.

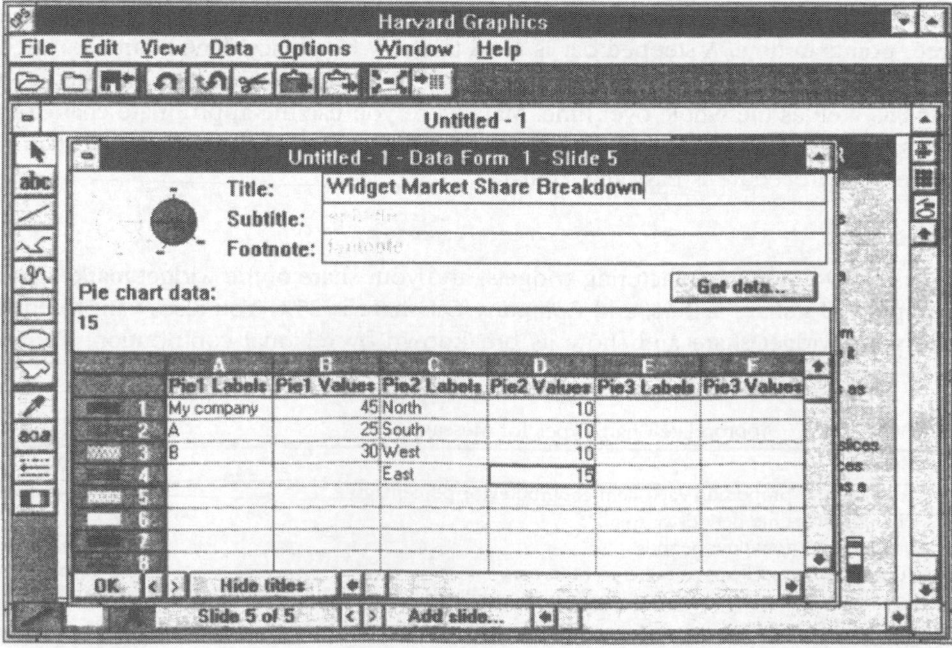

FIGURE 31.4 Pie Chart—Slide Editor.

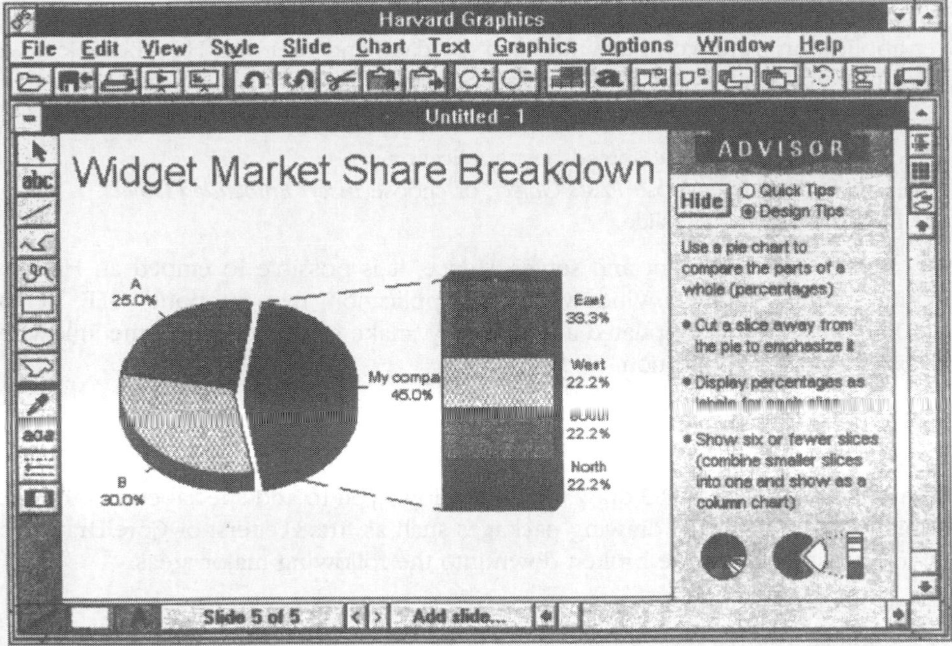

FIGURE 31.5 Pie Chart.

8. Click on *Series 2*, and enter *Bolts*.
9. Enter appropriate values for the bolts produced under that column.
10. Click on *OK* to finish the slide.
11. Press *F2* to view the slide.
12. Choose *Chart Options* to vary the options for Legend, series color, Axis to provide titles for the axes.
13. If the exact data values are needed, click on *Data Table Show* option from within *Chart Options*.
14. Once you are satisfied, press *Ctrl-S* to save the presentation.

After creating a chart, you may decide that a different chart type will better convey the message. For example, to change the chart type from a bar chart to an area chart, click on *Change Chart Type, From Gallery and pick the chart type* you like.

Importing Data

You can import spreadsheet data from either Lotus or Excel by clicking on *Get Info* within each chart. This creates a static link and imports the data.

If you need a more dynamic link, and need continous updates, use dynamic data embedding. In such a case HG will update the presentation when the source (spreadsheet) file changes. To create a DDE link

1. Both the spreadsheet and HG application must be *open*.
2. *Copy the data from* the spreadsheet application.
3. From within HG, choose *Paste Link*
4. Press *F2* to view the slide.

HG also supports object link and embedding (OLE). OLE allows HG to embed a spreadsheet object. The advantage with embedding is that you can modify the data from within HG by double-clicking on the object. This will take you to the spreadsheet application. After you have modified the data, and return to HG, the object will be instantly updated in HG. To embed an object,

1. Both the spreadsheet, and HG application must be *open*.
2. *Copy the data* from the spreadsheet application.
3. From within HG, choose *Paste Object*, or choose *Insert Embedded Object*.
4. Press *F2* to view the slide.

HG is both an OLE client and server. Hence, it is possible to embed an HG presentation within another Windows client application that supports OLE. If the OLE/DDE links are to be updated automatically, make sure you update the link when you exit the server application.

Creating a Drawing Chart

Visuals can be created using HG's drawing features, or can be enhanced using F/X (File exchange), which is HG draw tool, that allows you to add special effects. F/X has capability that is similar to drawing packages such as Arts&Letters, or CorelDraw. The drawing capabilities can be broken down into the following major areas:

1. Working with symbols.
2. Working with HG draw tools.
3. Working with HG F/X.

Working with Symbols

Symbols are a means of adding enhancements to a slide, by adding a visual dimension. These are basically clipart that can be imported for use inside your presentation. For example, if you are producing computers, you may like to add the computer symbol to your bar chart. To include a symbol

1. Click on the *truck icon* to open the symbol library.
2. A list of common symbols available in HG opens up.
3. To add a computer symbol, *double-click* on the computer.
4. *Select the computer* you like, and *drag it* to your slide.
5. You can resize and color the symbol within HG.
6. You can also import symbols that are either bitmaps (Tiff, PCX, BMP etc.) or metafiles (CGM, WMF) by choosing *File, Import*, specifying the file type and file. Bitmaps cannot be colored inside HG, but metafiles can be colored and resized.

Working with HG Draw Tools

HG comes with a minimal set of tools that allows you to draw on the screen, such as rectangle, ellipse, polygon, freehand, curves, etc. These are shown in Figure 31.6. You can draw any of the objects by clicking on the symbol, and clicking and dragging. For example, to draw a rectangle

1. Click on the *rectangle* symbol
2. *Click and drag* to draw the rectangle.
3. Chose the *Fill* tool to set the color of the fill.
4. Chose the *Line attributes* tool to change the thickness of the line.
5. You can work with other tools in a similar manner.
6. The eye dropper tool is useful if you want to copy a set of attributes from one object onto another. For example, the color, line thickness of a rectangle can be se-

lected and and dropped on to an ellipse object to set the color and line thickness for the ellipse.

7. HG places each object you create in its own layer. You can edit, resize, group, and align these objects by choosing the appropriate options from the menus.
8. HG has four levels of Undo and Redo. These are useful when drawing. Undo reverses an action. Redo cancels the most recent undo.
9. If you need to center your objects on your drawing, click on *Graphic, Center on slide.*

HG has the Group, Zoom, Align, Rotate, and Move objects to front/back icons at the top row of your screen. When you move the cursor on top of the icons the tools' name appears on your window's title bar. The align tool is useful to align objects so that the drawing is more attractive. The zoom tool is useful to zero in on a location and have closer control of your drawing. The rotate tool allows you to rotate an object by dragging or specifying an angle to rotate. The group tool allows you to group a set of objects so that you can manipulate them as a single object. The Move objects to front and back allow you to adjust the layer of the objects.

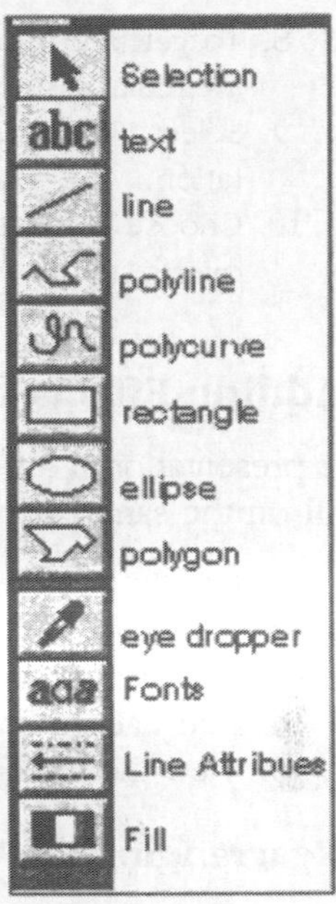

FIGURE 31.6
HG draw tools.

Once you have created a drawing you like, you can save the drawing as a symbol, or export it as a metafile for use with other programs. To create a symbol

1. Copy your drawing to the clipboard, with *Edit, Copy* after selecting your drawing.
2. Click on the *truck* (symbol) icon. Choose *cancel* to not open an existing file. A new file opens.
3. Paste the drawing with *Edit, Paste.* Enter the name for the symbol.
4. Save the symbol in a file with *File Save.* Give a file name.

Working with Harvard F/X

Harvard F/X is a stand alone drawing package that comes with HG for windows. It can be used to create additional drawings or provide special effects to your presentations. It provides some of the same tools found in sophisticated drawing packages such as Coreldraw. To start F/X, go to the Harvard Graphics group and *double-click on F/X.* to create a new drawing.

Since the capabilities of F/X are enormous, please refer to the F/X manual on how to use it. We shall look at using F/X only in terms of providing some special effects to the text charts such as a Title chart. In this format, we shall use F/X as an embedded object.

1. First create a *title chart* and enter a title such as *Special Effects Demo.*
2. Now select the words of the title and *cut it.*
3. Click on *Edit, Insert Embedded Object. Select Harvard F/X.*
4. Click on the *New Drawing* Button inside F/X.
5. *Edit, Paste* the selected title.
6. Select *Effects* Menu, choose *Quick F/X.*
7. Select *Shading, Simple Emboss* for a very simple embossed effect, You can choose other effects.

8. To get fancy, undo step 7, click on *circular copy*, and the text will be replicated in a circular fashion.
9. Select, *File, Exit, Update* from F/X and the object is pasted back on to your presentation.
10. Choose the special effects only for the Title slide. Most of the effects tend to be overdesign in presentations for other slides.

Adding Effects in Presentations

A presentation is an ordered sequence of slides. All the slides in your presentation are all on the same file. The following steps create a presentation:

1. Create the slides.
2. Rearrange the slides in the desired order.
3. Add transitional Effects.
4. Add Background, Style, and other effects.
5. View the presentation.

Rearranging the Order of the Slides

1. Choose *View, Slide sorter*, or click on the slide sorter icon on the right hand side of the screen. This view presents a thumbnail overview of your entire presentation. It is, therefore, very useful to look at the overall structure of your presentation in this view.
2. Now *select the slide* and *drag* to place it in the right order.
3. You can edit the slide by *double-clicking* on the slide.
4. You can preview any slide at its regular size by pressing F2.
5. You can delete, copy, and paste slides by choosing *Edit, Cut, Copy, Paste* commands.
6. You can copy slides from a different presentation by opening both the presentations side by side.

Transitional Effects

Effects are used to specify how a slide is drawn/erased from the screen. You can use a different transition for each slide or use the default transition effect. Effects that can be used include Wipe, Scroll, Blinds, etc. Wipe sweeps the chart vertically or horizontally of the screen. Scroll pushes the entire chart on or off the screen. Open—opens the screen like a curtain. Close—closes the screen like a curtain. Blinds—draws or erases an image in strips similar to the blinds on a window.

Please be sure that the effects you use are complementary, and do not overdo the effects. Some complementary transition effects that work well include the following:

Draw	Erase
Scroll left	Scroll right
Wipe down	Wipe up
Open horizontal	Close horizontal
Wipe right	Open horizontal

To create the transition effects

1. Choose *File, Screen Show*, and *Edit Screen Show Effects*.
2. Select the *Draw effect* you desire from the pull down list.

3. Specify the *Erase effect* similarly.
4. You can also specify the time for how long the slide should be on screen. Unless this is on an automated kiosk, the timing is not reliable, since it is dependent on the speed of computers.

Adding Style, Background and Other Effects

The presentation style is used to provide a consistent look to all the slides in your presentation. The style affects all the slides in your presentation. It affects the fonts, color palette, background, and the entire look of each slide. Hence, use it with care.

1. Choose *Style, Presentation Style,* and choose the style you like.
2. Use the *Edit Undo* command immediately if you do not like the new style.

To add a **background**:

1. Choose *View, Slide Editor, Edit Master Template.*
2. Choose *Style, Background.*
3. Choose *Solid, and a color such as blue* to provide a dark background.
4. You can choose a *pattern fill* if you like.
5. Click yes to change the master template.

Viewing the Screen Show

1. Display the screen show with *Ctrl+F2.*
2. To move to the next slide, press *Enter or click the right mouse button.*
3. To go back one slide, press *Backspace or click the left mouse button.*
4. You can use the *Home and End* keys to go to the first and last slides, respectively.
5. Use the *Esc* key to stop the presentation at any time.
6. While viewing the screenshow you may want to draw a circle around a figure for added emphasis. The chalk tool lets you point to and emphasize material on a slide during a presentation. To use the chalk position the pointer where you want to start drawing. Hold the mouse button and draw the circle. Release the mouse button on completion

Creating a HyperShow

In a regular screenshow, you move linearly from slide to slide. In a hypershow, you can move between slides in a nonlinear mode. To move nonlinearly, either keys such as function keys have to be programmed or a button used to specify the slide to move to. A button is an object on a slide that allows you to go to another slide in your presentation.

Programming Keys

Assume that you are used to the IBM mainframe world, and would like to use function keys F7, F8, and F3 to go to the next and previous slides and exit, respectively. To use keys to move between slides

1. Choose *File, Screenshow, Edit ScreenShow.*
2. Click on *Hypershow Links.*
3. Click on *Select a Key.* Scroll down and pick *F7.*
4. Click on the *Destination* and pick *Previous Slide.*
5. Similarly set up the rest of the function keys.

Buttons

1. Create an object that will be a button. Take any drawing object that you create and select it. Click on *Graphics, Button attributes.*
2. Give it a *name*, and select the *destination* you should go to (similar to function keys).
3. Within the screenshow if you click on the button, it will take you to the corresponding slide.

It is very useful to have buttons on your last slide for just the key slides in your presentation. Clicking on the buttons will take you to the corresponding key slide, and allow you to answer any questions that may come from the floor. You can also use Buttons to launch other applications. We shall see an example of this in the video player.

Adding Multimedia in Presentations

Multimedia is the inclusion of audio, animation, and video in presentations. Presentation software are slowly supporting the inclusion of multimedia in presentations if the computer has an audio card and appropriate video drivers. Including sound is the easiest.

Sound

Sound is added to set the tone or introduce a theme in a presentation. In Harvard Graphics it is easy to add sound to a slide. It supports the inclusion of Midi and wave files.

1. Click on *File, Screen Show, Edit Screen Show Effects, Enable Sound.*
2. *Select a slide* from the list of available slides and *Click on Add Sound.*
3. Select the *file* to be added.
4. You can set options for how long the sound file is to play such as To Completion (play entire sound file), Next Slide (until next slide is started), etc.

Video

HG comes with a video player that can play an AVI (Video for Windows) or an AVS (DVI—Digital Video Interactive) file. To play these files your computer must be set up with the appropriate hardware/drivers. The video player is a separate program and resides in the \spc\player subdirectory. The name of the file is **winvplay.exe**. To run video from within Harvard graphics presentation, the following needs to be done:

1. *Start Winvplay separately* from Program Manager.
2. Select the *video file* it has to play.
3. Choose *Settings* from the menu and specify whether it plays after going to the slide or before, etc.
4. *Save the settings* in a playback options file (with an extension .wvp—example test.wvp).
5. *Create a button* in HG screen show. Choose *Button Attributes* from the *Graphics menu.*
6. In the destination list for the button, *Select launch applications*, and enter *c:\spc\player\winvplay.exe* for the application file name and for the parameters *test.wvp*

7. On playback, when the user clicks on the button, the video file chosen in option 2 will play.

31.5 POWERPOINT

Similar to HG, Microsoft PowerPoint is another presentation software that has a Slide view, Slide sorter view, Slideshow view, Outline view, Notes view, and Handout view. You can create a presentation from scratch by designing each individual slide similar to HG, or use the auto contents wizard to help you create a presentation. The auto contents wizard has templates for standard presentation types. It prompts you to create the title slide and then provides an appropriate outline for the presentation that reflects the category you have chosen. You can type your own text over the outline the wizard provides to create the presentation. For example, if you want to create a presentation that reports on the status of an engineering project, you can choose the Auto-Content Wizard. To do so

1. *Double-click* on the PowerPoint icon to open it.
2. *Click on Auto-Content Wizard* and *OK*.
3. The Wizard welcomes you. Click on the *Next* button.
4. Enter the category you are going to talk about, such as *New Design*.
5. The Wizard provides various categories, and provides a broad outline on the left-hand side of the screen. Click on *Reporting Progress*.
6. Click on *Next* and *Finish*. PowerPoint creates the outline and puts you in the outline view.
7. PowerPoint also provides a cue card on the right-hand side of the screen. The cue cards, like Design tips in HG, provide directions on how to change and edit the presentation.
8. The outline and cue card view is shown in Figure 31.7.
9. You can follow the directions in the cue cards to change the outline to reflect details of your presentation. Simply type over the text in the outline. You can add additional text in any slide by just pressing enter, and typing. If you need help, just click on the appropriate button on the cue card.

Once you have entered all the text, you can preview the presentation by clicking on the slide-show icon at the bottom of your screen or clicking on View, SlideShow. If you are not satisfied you can change the look of your presentation using the Pick a Look Wizard. Click on *Format, Pick a Look Wizard*. This will take you through the steps of choosing a new look. This essentially allows you to change the template for the entire presentation. You can add additional slides to your presentation using the slide view (click on *View Slide*). This is detailed in the next subsection.

Creating Slides

Creating slides in PowerPoint is very similar to creating text slides in HG. To add a slide

1. Click on *Insert, New Slide*.
2. Click on the *layout* you desire.
3. Type the *title* and *text* to fill in the details of your slide.
4. If there is an icon for adding clipart in the layout, *double click on the clip art*, *select* the appropriate *category*, scroll to see the clipart, *select the picture*, and *click on OK* to add it to the slide.

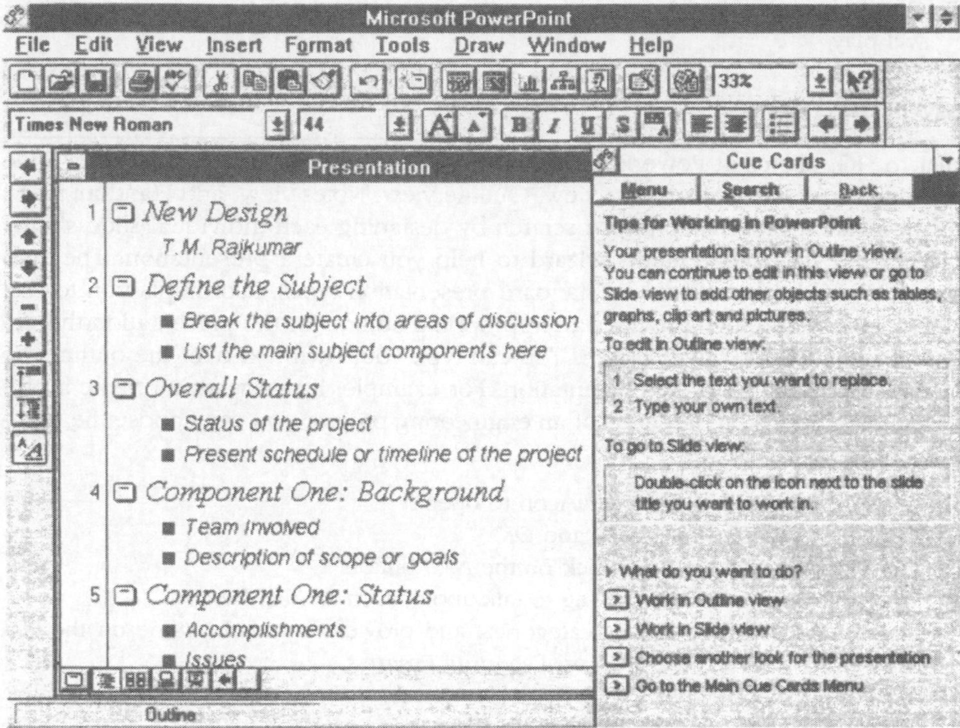

FIGURE 31.7 Progress report presentation from Auto-Content Wizard.

5. If there is an icon for graphs in the layout, *double-click on the graph icon* to create the chart.
   a. Change the data in the spreadsheet with your data.
   b. Change the chart type to the desired chart type by clicking on *Format, AutoFormat*. Choose the *type* from the *gallery* and select the desired *format*.
6. If there is an icon for creating an organization chart, *double-click on the organization chart* to add the organization chart.
   a. *Type the name and title* in the boxes.
   b. You can add a subordinate by *clicking on subordinate* and *clicking on the managers box*. Adding managers, co-workers etc. is similar
   c. Click on *File, Exit, and Return to Presentation*, to return to the slide. Click on *Yes* when PowerPoint asks for confirmation.
7. If there is an icon for a table, *double-click on the table icon*. This will take you to *Word* if Word is installed on your machine. Create the table in Word and it will be inserted in the presentation. If you do not have Word, you can create the table in any other word processor and cut and paste it or embed it as an object in Powerpoint.

Embedding Objects

PowerPoint is both an OLE server and client. You can embed a PowerPoint object in another Windows application that supports OLE or viceversa. To embed an object in PowerPoint:

1. Click on *Insert, Object.*
2. Select the application object you want to insert. For example, *Click on PaintBrush.* This will take you to PaintBrush. Create the PaintBrush object as you would normally create in PaintBrush. Click on *File, Update* and *Exit* to return to PowerPoint.
3. The bitmap in the slide is a PaintBrush object.

Other objects can be created and worked with in a similar manner. The organization chart, graph, and table icons in the slides view described in the previous subsection are essentially objects that are embedded. You can also add sound and movie files to the presentation as embedded objects. To create a movie object

1. Click on *Insert Object.*
2. Select *Media Clip* as the application object you want to insert. The Windows media player application is started.
3. Click on *File, Open* and select the movie you want to insert.
4. Click on *File, Update, and Exit* to return to PowerPoint.

The movie is played back when the slide containing the movie icon is reached. If desired the movie can be played back again by clicking on the movie icon. To create a sound object use Sound Recorder instead of media clip as the application object.

Adding Visuals

Visual drawings are created in PowerPoint using the different draw object icons that PowerPoint provides. PowerPoint like HG draws the objects in layers. The object attributes such as fill, line, and shadow, can be changed. These objects can be resized, rotated, and grouped as in any other drawing program. PowerPoint provides three toolbars with object icons that can be used for drawing. The drawing toolbar contains the standard icons such as ellipses, boxes, lines, text, and rotate. The drawing+ toolbar is made available when you choose *View, Toolbars, Drawing+*. The drawing+ toolbar provides icons that can change the fill color, alter line width, rotate, group a set of objects, or change the layer of an object. These are in general special tools. The AutoShapes toolbar is very useful, and is displayed by choosing *View, Toolbars, AutoShapes.* The AutoShapes toolbar contains icons for some of the most commonly used shapes such as rectangles, arrows, stars, and pentagons. To draw a standard shape

1. *Click on the shape* in the AutoShapes toolbar.
2. *Click and drag the shape* on your slide to draw it.
3. *Double click and type* to add text in the shape.

PowerPoint comes with a standard array of clipart. *Click on Insert, Clipart* to add clip art to your slide. You can also cut and paste from other Windows programs as bitmaps. PowerPoint can also import clipart in the windows metafile format.

Slide Shows

A slide show is started in PowerPoint by clicking on the Slideshow icon, or choosing *View, Slideshow.* You can set the transitions for the slide show in the slide sorter view. To add transitions:

1. Choose *View, Slide Sorter* to go to the slide sorter view.
2. Click on *Tools, Transition.*
3. *Select the transition* from the effects box.

4. Click on the speed by selecting *Slow, Medium, Fast*. In general, Fast is a good option as you do not want the transition to clutter the message present in the slide
5. Click on OK.

PowerPoint provides the ability to include more slides than you need to show in your presentation and hide the extra slides. Hidden slides are useful to show additional details in response to audience questions. To hide a slide during presentation, in the slide sorter view, click on *Tools, Hide Slide*. The slide is then hidden during presentation, unless you chose to show it by clicking on the icon in the lower right corner of the slide that precedes it.

The slideshow view, can be used to time and rehearse the presentation. To time the presentation

1. Choose *View, Slideshow* to go to the slide show view.
2. Click on *Rehearse New Timings*, and click on *Show* button.
3. Rehearse your presentation. Click anywhere to go to the next slide. As you move from one slide to the next, the time you spent rehearsing is added as the time of the slide. The next time you make the presentation, the slide will stay on the screen for the same amount of time.
4. Click on *Save*.

Slides can be annotated during presentation using the pencil icon available on the bottom right hand corner of the screen. Any annotation that is made is not permanent. Handouts and notes can be created using the appropriate views. You can print handouts, notes, and your slide show by choosing *File Print* from the main menu and choosing the appropriate options.

31.6 FREELANCE GRAPHICS

Lotus Freelance Graphics is another presentation graphics program that works similar to Harvard Graphics and provides similar views: page view, page sorter view, outline view, and screen show view. Creating a presentation in Freelance, is done in five steps.

1. Choose a look for the presentation. When you click on *File, New*, the choose a look option appears on the screen. Here, you choose a presentation style (smartmaster set in Freelance terms). The presentation style provides a consistent look for the entire presentation.
2. Create the slides (pages in Freelance terms). To create the slide, *choose a page layout*. The page layout shows how the slide or page is formatted.
3. *Fill in the blanks*. Once a page layout is selected, Freelance smartmasters provides prompts on the slide such as *Click here to add Title*. Typing the text, filling the blanks at appropriate locations, and following the instructions on the slide enable you to generate the slide. Freelance takes care of the design of the slide such as colors and fonts. All slides are created in a similar manner. If there is a problem, help is always available in Freelance by *clicking on the tips* button. Regular help is also available by pressing *F1*, and choosing *How Do I*, and the appropriate option for detailed help.
4. Once all the slides are created, you use the page-sorter view to get an overview of the presentation and reorganize the slides. Choose *View, Page-sorter* to go the page-sorter view.

5. The transition effects are provided in the Screen show. Choose *View, Screen Show, Edit Effect* to provide transition effects such as Fade and Blinds. Click on *Run show* to preview the presentation.

Creating Text Pages

Creating other text pages (slides) are similar. For example, to create a bullet slide

1. Click on *Page, New*.
2. Click on *Bullet List* as the page layout.
3. Fill in the appropriate blanks for each of the bullets. Press enter to create the next bullet.
4. Press *Tab* to create a sub bullet. Press *Shift+Tab* to go back one level.

To create a build of the bullet list, choose *Page, Build*. Freelance will automatically insert necessary pages to create the build. To create a bullet list with a symbol or chart, choose the appropriate page layout.

Freelance provides a table layout page to create a table easily.

1. Click on *Page, New*
2. Click on *Table* as the page layout.
3. Select the *table layout* from the table gallery that is presented.
4. Enter the *number of rows and columns* (include the title row as a regular row).
5. Click on OK.
6. Enter the data in the table. Use the arrow keys to go to adjacent cells in rows and columns.
7. *Highlight the title row* and press the *right mouse button*. Click on attributes, and emphasize the title row (bold or italics).

Freelance provides the organization tool icon, to create an organization chart.

1. Click on the *organization icon*.
2. Choose an organization *style* from the gallery that is displayed.
3. Enter the *name, title, and any comment* for each person on the organization chart.
4. Press *Tab* for a subordinate, and *Shift-tab* for a superior.

Creating Visuals

Freelance provides capabilities to add symbols, charts, and drawings that are similar to Harvard Graphics and PowerPoint. To add a symbol to a page,

1. Choose a page layout with a symbol.
2. Click on the *symbol*, and the gallery of symbols is shown.
3. *Select* the *symbol*. The symbol is sized and placed on the chart.

You can also add a symbol by clicking on the symbol icon (light bulb), selecting the symbol, and placing it on the page. You can resize the symbol to meet requirements.

Charts can be added in Freelance either by importing a lotus spreadsheet or by creating the chart from within Freelance. To create the chart,

1. Choose a page layout with a chart.
2. Click on the *chart*, and the chart gallery opens up. This lists all the available chart types and shows a variety of styles for each chart type.

3. Select a *chart type* and *style*. Click on *OK*. This opens up the chart data window.
4. Enter the *data, title, axis labels, and legends* for the chart. Click on *OK*.

The chart is drawn and placed on the page. You can also add a chart, by clicking on the chart icon, and proceeding from step 2 above. On completion of step 4, you can place and size the chart to meet requirements.

Freelance provides the basic drawing tools such as boxes, ellipses, and lines on the graphics toolbox. To draw any of these objects, all you need to do is to *select the icon, click, and drag*. Each object is then placed in a layer. You can change the layer, attributes such as color line thickness, rotate or flip the object by *selecting the object, clicking the right mouse button, and choosing the appropriate option*. You can group, align, or combine objects as in Harvard Graphics or PowerPoint. In addition to drawing these objects, you can import bitmaps and drawings created by other programs by using the *File, Import* command and choosing the appropriate file type.

Adding Multimedia

Freelance adds multimedia objects by using the OLE features of Windows. You can link or embed audio or video. Freelance provides the Lotus Sound and Lotus Media objects to add audio and video objects. In addition, you can use Lotus Sound to create, edit, and modify any wave files. To add a multimedia object

1. Click on *Edit, Insert Object*.
2. Select *Lotus Media*.
3. Select the type of *object*, wave files for sound files, Midi for music files, or movies for video files.
4. Select the *file* you want to play.
5. Click on *options* and set the options for playing such as once, continously, or when clicked. You can also set the play option by selecting the media object and clicking the right mouse button.

The play back of the multimedia object depends on the options you set during the screen show. Inserting multimedia as an object as above plays the multimedia object during the screen show under the control of the media player. If you need more control over the playback such as where the movie should appear on the screen, you can create a screen show button, and have the button control the playback during the screen show. To create a button

1. *Create a symbol* or any object that you want to act as the button.
2. *Select* the above object, and click on *View, Screen Show, Create/Edit Button*.
3. *Name* the button.
4. Click on *Play Multimedia Object*.
5. Click on *Browse Media*. The lotus media object pops up. Select the file type, file, and options for playback.
6. If you are playing back a movie, click on *Options*, and specify where the movie is to appear.
7. Click on OK to save.

Presentation

Screen shows in Freelance can be either linear or hypershows similar to the capability in Harvard Graphics. To set up a screen show as a hypershow, you need to:

1. Create a symbol or object to act as a button. For example, create an arrow.
2. Select the object, and click on *View, Screen Show, Create/Edit Button.*
3. Name the button (for example, Forward).
4. Click on *Jump to.* A list of the titles of each page, and other defaults such as Next and First appear. Select the page to go to, for example, *Next Page.*

You can also use the button to launch another application. This can be done by:

1. Choosing *Launch an application* in step 4 above, and
2. Clicking on *Browse*, and select the application to run.

You can create a set of speaker notes to assist with the presentation by clicking on *Page, Speaker Notes.* The speaker notes can be printed with *File, Print, Speaker Notes.* Handouts that can be made available to audiences can be printed with *File, Print, Handouts.*

Transition effects and timings for the display of each page on the screen can be set by choosing *Edit Effects* in the Screeen Show view. The screen show is run by clicking on the screen show icon, or clicking *View, Screen Show, Run.* You can highlight an area of the screen while the screen show is running by clicking with the left mouse button and dragging. These drawings are not saved with the presentation. The screenshow can be stopped at any time by pressing the *Esc* key.

31.7 DESIGN ISSUES

Before you make a presentation using computer-based software, it is worthwhile to consider the steps in making a presentation.

Define the Message

The focus of a presentation is the message. Understand what it is you are trying to convey to the audience, and then develop the rest of your presentation with the message. The message will vary with your goal as to what you want the audience to do after the presentation.

Understand the Audience

Audience analysis is important. This helps in identifying what terminologies you can use in your presentation. If it is a scientific or engineering audience, you can assume some basic terminologies. But if you are making a presentation to a marketing audience, you might have to define terms and adapt the presentation to the level of knowledge of the audience. The audience should influence your words and the visual charts you use.

Generate an Outline

If you can develop an outline of the presentation, it will help you organize the presentation in a logical manner. Most presentation software today come with an outlining tool. Use these to generate the outline. They will provide a quick overview of the entire presentation.

All presentations have a beginning, middle, and end. Tell the audience what you are going to tell them. Tell them; and tell them what you told them. Hence, your beginning should identify what you are going to talk about (the message). It must contain information as to why the message is important and how you will develop the argument.

The middle should contain convincing arguments for your message. Details of this are elaborated in the next step.

The ending should contain succinctly a review of the arguments for your message and conclude with the appropriate recommendation and actions that should take place. An effective presentation should lead the audience to the appropriate conclusion without your having to explicitly state it.

Choose Appropriate Visuals

Choosing the right chart type is very important in conveying the message. With the use of presentation software, it is possible to generate a wide variety of charts. But certain types of charts help get the message across quicker than others. Use the charts appropriately. For example, text and bullet charts help you preview and summarize information. Symbols, clipart, and drawings enhance or add relevance to the message. Graphic charts are used to show relationships between data. Irrespective of the type of chart used, the adage Keep it Simple and Stupid is true. Ensure that each chart has one major point and choose the appropriate chart that helps you convey that point. Keep the graphic design simple, and do not clutter the message with a lot of graphic design.

Develop Transitions

Transitions control how slides are drawn or erased from the screen. Presentation systems allow you to specify a variety of transition effects, such as wipes and fades. However, consistent transitions between slides can help keep the focus on the message and not on the presentation design.

Delivery Issues

Make sure you have the appropriate equipment to make the presentation. For example, make sure that the LCD projectors have the appropriate color capability (a minimum of 256 colors) and that the computer you are going to project from has the necessary speed and the appropriate playback software. If you include sound and video, ensure that the delivery mechanisms for these exist. Test your presentation with the appropriate delivery equipment to ensure that the presentation is fine. Always have a backup as a safeguard and keep a set of transparencies for important presentations.

In addition, if you are going to make handouts for your audience, make use of the software's capability to generate the handout. Make sure the format is readable by your audience. For presentations with lots of data, consider using an executive summary. Change the color scheme to a grayscale before you create the handout.

Practice

Practicing your presentation before you actually give it helps you make a professional presentation. It can resolve timing issues, delivery issues, remove any nervousness, and help you relax. It also is easier if you establish a rapport with the audience by maintaining eye contact, standing erect and relaxed, and sending appropriate nonverbal cues. Use the presentation you created to deliver a professional presentation. Evaluate any presentation you give, and use the feedback to improve on your presentation the next time.

General Guidelines

There are some general guidelines that are useful to making successful presentations.

Fonts

Keep the use of fonts minimal. If your delivery machine will be different from the authoring machine, limit the use of fonts to those that are supplied with Windows. For example, Arial and Times New Roman would be good choices. Unlike text in books that use serif fonts, sansserif fonts like Arial are better for computer-based presentations. Choose a large font size (18–24 pts) so that the audience can read them from the back. In general a single typeface at different sizes is sufficient for most presentations. Use bold or italics for emphasis.

Color

Colors can be used to affect the readability of the material and can enhance the message. You can use colors to differentiate (different elements of a bar chart in different color), associate (title is always yellow), or for emphasis (profits may be in a brigher color) of the material. Use color carefully as colors tend to evoke emotions. For example, red denotes loss and blue is cool. In general, a conservative choice of a dark blue background with white or yellow foregrounds works very well in most settings. Avoid any colors that clash, such as red on green. Also limit the colors you use in your charts to three or four. Choose the colors to complement your presentation. Most presentation software come with preset color schemes. These choices are chosen professionally since they work well.

Consistency

A common look and feel across the various screens in a presentation are important. It leads to the perception by viewers that there was unity and vision in the creation of the presentation. In general, they are percieved to be of higher quality by audiences. Use template and style mechanisms built into the software to provide consistency across the presentations.

DEFINING TERMS

Build: A build is a sequence of slides/charts that reveals more information gradually. For a bullet slide the first slide would reveal the first bullet. The second slide would reveal the second, and would have the first bullet dimmed. This process will continue until the entire series of slides has been shown.

Button: A button is an object in a slide that is assigned a specific action. The action is taken upon clicking the button.

Chart: Represents data with text or graphics.

Hypershow: A nonlinear screen show where the user can go to a particular slide by pressing a key or clicking on a button.

Page Layouts: Templates that you use for page formats, such as bulleted lists or pages with multiple charts.

Palette: A set of colors that complements each other and can be used in your presentation.

Scheme: The association of different colors in the palette to different elements in your slide. For example, the first color may be associated with the title, the second color for bullets, and the third color with the footnote.

SmartMaster Set: A collection of page layouts designed to have the same graphic style to ensure consistency among all the pages in a presentation.

Symbols: Symbols are clip arts that can be used by the presentation software. These come in libraries and are grouped in appropriate categories, for example, flags or people.

Template: A set of design choices that specifies how each graphic element of the slide appears. A template can be specified to provide consistency for the presentation.

Transition: This controls how slides are drawn on the screen and erased from the screen during the presentation.

REFERENCES

Bob Benedict, *Using Harvard Graphics 2 for Windows*. Que Publishing, Carmel, IN, 1993.

Using Harvard Graphics 2.0 for Windows. SPC Software Publishing, Mountain View, CA, 1994.

Edward R. Tufte, *The Visual Display of Quantitative Information*. Graphics Press, Cheshire, CT, 1983.

User Guide to MicroSoft Windows 3.1, Chapter 10 on PaintBrush. Microsoft Corporation, Redmond, WA, 1992.

User's Guide to MicroSoft PowerPoint 4.0. MicroSoft Corporation, Redmond, WA, 1994.

Using Freelance Graphics 2.0 for Windows. Lotus Corporation, Cambridge, MA, 1992.

32

Utilities and Miscellaneous Programs

Kevin Greiner
WARE, Inc.

Todd Leuthold
Todd Leuthold Consulting

Utilities are often viewed as mundane and unexciting programs. However, they are necessary for the effective and efficient operation of a computer system. Four such utilities for DOS-based systems are presented in this chapter.

- *Sorting*—Used to organize the information in files in a useful manner.
- *Disk Defragmentation Utilities*—Used to increase the efficiency of disk access by making files contiguous. After repeated additions and deletions, files may be scattered across the disk, slowing doing access to unacceptable levels.
- *PC Tools PRO*—An integrated package of disk maintenance utilities used to repair defective files, lost files, and accidentally deleted files, and perform a variety of other useful tasks.
- *Disk Backups*—"Blessed are the pessimists, for they have made backups" says it all. Backup utilities give you the capability of making archival copies of important files or complete systems. In the event of a disk malfunction or major system failure, having a complete backup of your critical files can get you back in operation in a matter of minutes, instead of hours.

32.1 THE MS-DOS SORT COMMAND

Kevin Greiner

The use and syntax of the MS-DOS SORT command go back to the command-line interfaces of the early years of computing. Although this command may appear to be of little relevance to users of modern computers, to overlook its power because of its interface would be a mistake. There are a number of situations where such a command is very useful.

The Sort command is an external DOS command. By definition, an external command is a separate executable file from the main operating system. On most MS-DOS computers, the Sort executable is stored in the DOS subdirectory, which should be referenced in your path. This way you can execute the Sort command from any subdirectory on your hard disk.

0-8493-2530-7/96/$0.00+$.50

Syntax

Sort works as a filter. When it is run, it reads input via redirection and provides output via redirection. If the input, output, or both are ommitted, the keyboard (CON) is the default input device and the screen is the default output device. Because sort is a filter, its input can be received through redirection. This feature allows you to send the output of a command like dir or find to the sort command. See the examples below.

A typical command line for sort might look like this: **`sort <mileage.txt> lpt1`**. This command sorts a file called **`mileage.txt`** and sends the output to a printer attached to LPT1. There are two switches that affect the operation of Sort. These switches immediately follow the Sort command but precede the input specification. They are listed in the following table.

Switch	Operation
/R	This switch reverses the order of the sort from ascending to descending
/+n	This switch specifies the beginning column of the sort key. The letter "n" is replaced with the column number. If none is specified, the entire line is used

Size Limitations

Sort is limited to 64 Kilobytes in total input data.

Example	Explanation
sort/r < test.dat	Sorts a file named test.dat in reverse order beginning with column 1 and sends output to the screen
sort 〈 things.txt 〉 ordered.out	Sorts a file named things.txt and writes the output to a file named ordered.out
sort < catch.prn | more	Sorts a file named **`catch.prn`** and sends the ouput to the screen, pausing after each screen of information
sort/+ 10 < data.txt	Sorts a file named **`data.txt`** beginning with column 10 and sends the output to the screen
find "Alexville" *.txt | sort | more	Searches all the files with an extension of txt in the current subdirectory and sorts the output, displaying it one screen at a time.

32.2 DISK DEFRAGMENTATION

Kevin Greiner

Introduction

Hard disk defragmentation is one of those subjects that borders on the fringe interest of most computer users. Generally, most users do not understand why defragmentation must be done or how it occurs. In simple terms, disk and file fragmentation occurs when a file is stored in two or more physically separate locations on a hard disk. As you might imagine, when the hard disk needs to fetch the file from two separate locations, a dely occurs. When enough of these delays occurs close enough together, the wait becomes discernible. The job of a disk or file defragmentation tool is to place these separated chunks of a file back together. Numerous utilities exist on the market for this purpose.

Much more detail regarding how and why disk defragmentation occurs underlies these few broad explanations, but a complete exploration of these topics is also beyond the scope of this article. Rather, the use of one specific disk defragmentation tool will be covered, the MS-DOS Defrag utility, which is shipped with each copy of MS-DOS after 6.0.

The disk defragmentation utility bundled with MS-DOS is a stripped-down version of Symantec's SpeedDisk utility. Defrag, as Microsoft titled this version, contains only the bare essentials to get the job done. This version runs in a DOS only environment. It is not designed to work in a Windows DOS box. Attempts to do this will result in a dialog box that informs you that Defrag cannot work in Windows. Do not try to circumvent this message. Damage to your hard disk and data files may occur. For a more advanced version, check out the SpeedDisk utility bundled with Norton Utilities. This utility can run in the background under Windows and provides many other advanced capabilities.

Use

The use of a disk deframentation tool varies widely. The key factor is the activity of your hard disk as it regards erasing files and adding new ones. Because of how MS-DOS handles erased files, as time passes, the files on your hard disk become broken into more and more chucks spread across the surface of your hard disk. In time, this reduces the operating efficiency of your daily tasks. Measurable increases in application response time are the result. For most users, I recommend using Defrag or a similar utility once or twice a month.

Because Defrag performs many low-level accesses to your computer's hard disk, it is best to run another utility first. This utility, called ScanDisk, tests the entire file system to ensure its integrity and safety. This program is run by typing "scandisk" at any DOS prompt. If ScanDisk finds any errors, it will display a dialog box that prompts you to correct the errors, or leave the hard disk in its current state. You should always allow ScanDisk to automatically fix these errors. Again, I strongly recommend running this utility before you run Defrag.

Because Defrag cannot run inside Windows, it must be operated from an MS-DOS environment. On most systems, you will be able to start Defrag by typing "defrag" at any DOS prompt (without the quotes). When Defrag begins, it scans your hard disk to determine the amount of disk fragmentation. Based upon percent of total files that are fragmented, it makes a recommendation on how to proceed. The two recommendations are documented below.

Unfragment Files Only

This type of disk defragmentation ensures only one thing: that the files on your hard disk are all contiguous. This means that all your files will be stored in one piece on your hard drive. Generally, this type of optimization requires only a short period of time. The trade-off is that with this type of optimization, disk defragmentation reoccurs much more rapidly. Gaps may be left between files which will accelerate the natural process of file fragmentation. On my 340-Megabyte hard disk with about 1,000 files, a Files Only defragmentation usually takes about 5–10 minutes.

Full Optimization

A complete disk defragmentation defragments the files, but also moves entire files from one place to another. This is done to ensure that there are no gaps between files. Although the tradeoff is that it takes much longer to complete this type of defragmentation, the benefit is usually worth the extra time. Running a complete disk defragmentation requires between 30 and 60 minutes on my 340-Megabyte hard disk.

Defrag Options

Defrag's options are quite limited when compared to third-party products. Basically, there are two sets of options. The first is the type of disk defragmentation, which we have already discussed. The second set of options regards the sorting of the file names. With the widespread use of Windows, this is less important today because many of the file lists in Windows are already sorted. If you spend much of your time in Windows, you can effectively ignore this set of options.

The file sorting options are covered below. Note that this set of options does not change where the files are physically located on the hard disk, but simply which files appear first in an unsorted directory listing from an MS-DOS prompt.

Sort criterion:

- Unsorted: The file order is not changed.
- Name: The files are ordered by name.
- Extension: The files are ordered by type. This is often a useful sort because many files types are indicated by extension.
- Date & Time: The files are ordered by the date first, then the time that they were last modified.
- Size: The files are ordered by size.

Sort order:

- Ascending: The above criterion begins at the lowest value and ascends to the highest.
- Descending: The above criterion begins at the highest value and descends to the lowest.

Command Line Options

A command line options is a word that is typed or placed after the name of the program. A common technique is to write a batch file that runs several of programs in a sequence, each with appropiate command line options to perform their tasks automatically. Defrag's options allow you to modify how it loads and operates. The following is a list of command line options for Defrag. Several examples follow of batch files that demonstrate this aspect of Defrag's operation.

Drive letter	This is the letter of the drive that you want to optimize. If you want to optimize several drives, place several letters here.
/F	Runs a Full Optimization regardless of how many or how few files actually need to be defragmented.
/U	Unfragments only files. This leaves space between the files. Read the previous explanation for more information.
/SN	Sorts the files in alphabetical order by Name. Add a - after the S to reverse the sorting order.

/SE	Sorts the files by Extension. Add a - after the S to reverse the sorting order.
/SD	Sorts the files by Date & Time. Add a - after the S to reverse the sorting order.
/SS	Sorts the files by Size. Add a - after the S to reverse the sorting order.
/B	This option reboots your computer after the optimization is complete. Some computers require this step to function normally.
/SKIPHIGH	This forces Defrag to use lower memory for its operations. Use of extended or upper memory is avoided. If you have problems running Defrag, attempt to use this option.
/LCD	If you are running Defrag on an LCD screen, use this option. Display colors are optimized for viewing on an LCD display.
/BW	This option forces Defrag to use only a black and white color scheme. This is useful is you are running a monochrome VGA monitor.
/G0	(Note this is G zero, not G oh) Defrag should automatically detect if your monitor and computer can correctly use advanced display techniques. If Defrag is incorrectly detecting your computer, use this option.
/H	Certain hidden files should not be moved while others may. If your hidden files may be moved, enable this option.

Sample Batch Files

Here are some sample batch files for use with Defrag along with explanations of what each one does.

```
ScanDisk
Defrag
```

This batch file simply runs the two programs listed, prompting the user for necessary choices along the way. As shown in the following example, each user prompt can be supplied with a command line parameter, allowing this batch file to run without user intervention.

```
ScanDisk /ALL /AUTOFIX /NOSUMMARY /NOSAVE /SURFACE
Defrag C: /F /H /SN
```

This batch file scans all the disks (hard disks and floppy disks) in my computer, automatically fixes errors, and does a surface scan. When the scan is complete, drive C: is completely defragmented, moving hidden files and sorting the file names alphabetically.

DEFINING TERMS

Command Line Options or Parameters: Words or options that are specified after a program to direct the program how to act or run. This technique is ofted used in batch files.

Defragmentation: The process of reorganizing a disks or files into contiguous pieces again.

Disk/File Fragmentation: When a file is stored in two or more physically separate locations on a hard disk is fragmented.

REFERENCES

The MS-DOS 6.x manual that came with your computer.

Any MS-DOS after-market book. Many cover this utility and its use in depth. Choose one that suits your use and comfort level.

32.3 PC TOOLS PRO

Todd Leuthold

PC Tools has an easy-to-use interface that the average user can become comfortable with in a short period of time. Use the Tab keys to move from window to window, and use the Enter key to select files and directories for specific operations. To access the menus, press and release the Alt key. The letters that are highlighted can be pressed, or you can use the cursoı keys to move the highlight bar to the desired choice. Press Enter to open that pull-down menu. F-keys at the bottom of the screen give quick access to the most-used features. If you do not like the options on the F-key menu, you can modify them to suit your needs. The colors are also configurable, though this will take some practice. There are many things that can be custom configured.

Configuration is done through the Configuration Menu option. Choose Define Function Keys . . . to change to other choices. You can also change other setting for the PC Shell window from this menu. Some options have the ability to be reset to the original defaults. However, you may wish to record the current settings, just in case you cannot reset them to their prior values.

Function Keys

Pressing the Function Keys by themselves will provide a shortcut to those options. You may also press the Alt, or Ctrl keys, along with a specific Function Key to access other options. Pressing the Alt or Ctrl key alone will change the options on the Function Key menu, showing you what is now available.

Pressing F1 from any screen will bring up help on the particular area that you are in. However, the help may be a bit hard to understand, or may not seem to apply to what you are trying to do. If it does not help, try going out to the main screen, then searching through the various help topics until you find what you are looking for.

Pressing F2 will bring up the Viewer. This is a file viewer, not a graphics viewer. You will be able to view various file types from this window, including binary, text, and special file formats, such as Word Perfect and dBASE files. When the Viewer is brought up, you have another set of Function Keys available, which provide short cuts to various options available in this portion of the program. These keys are not configurable. For the most part, the same keys will do the same job from window to window, except where there are options available that are not in other portions of the program.

Press F3 to Exit the PC Tools Shell. You will have the option to cancel and return to PC Tools, or to elect to save any changes to the environment that you may have made during this session. Clicking on OK (or pressing Enter) will exit the program.

Pressing F4 will deselect any files that you have selected up to this point. This is a good idea if you have selected a lot of files, then changed your mind. All files will be deselected, including the one that is currently highlighted, so be sure that this is what you want to do, as the operation proceeds without any further prompts.

The F5 key will copy the currently selected file (or files) to any other drive specified, provided that you have rights to the drive, and that the drive can be written to. For instance, you cannot copy a file to a read-only CD ROM drive. This operation makes a copy, rather than actually moving the file.

Using F6 will allow you to change the way that the files and related information are displayed. You can choose to display them by their name or extension information, by their attributes, or by their date or time. You can sort them in ascending or descending order. You can also have the list shown with no sorting at all. This option will show the file names just as they actually appear on the drive at the current time.

The F7 option will allow you to quickly find a file on any drive. You may also specify multiple drives to search. You can search for strings of characters in files, also. You can narrow your search to specific files, or search all files for a specified sequence of characters.

Using the F8 key will Zoom In on the currently active window. For instance, if you are currently working in the files window and you press F8, the screen will be filled with only the files window. Press the F8 again to exit this feature.

The F9 key will allow you to specify parameters to select a large number of files, or files that are widely separated. Just enter the appropriate information in the file name and extension windows, then choose Select to activate the selection process. All files that match the criteria will be selected, with a number placed before them in the file window.

The F10 key will provide access to the PC Tools menu. From here, you can run various other PC Tools programs, or other programs that you have previously set up for launching from within PC Tools. Just press and release the F10 key to access the menu. To return to the PC Tools Shell, press and release the F10 key again.

Using Alt with Function Keys

Using the Alt key, you can access another set of Function keys. These definitions follow. To use this function, hold down the Alt key while pressing the appropriate function key.

Alt-F1 will provide you with the System Information Pro description of the contents of memory at that time. It is a virtual map of the programs and drivers currently in memory. You can use this find out where programs are residing and where drivers sit after being loaded.

Alt-F2 shows you a map of the files on your various drives. You can see how the files are laid out and where they are located. You can also use this to determine whether or not files are fragmented, or broken into segments.

Alt-F3 will save any configuration changes that you have made during this session. You can also save your configuration when you exit. A third method is to use the menu at the top and choose Save Configuration under the Configure menu selection.

Alt-F4 performs an exit similar to that of Windows and DOSSHELL. You will be given a chance to change your mind before exiting. This is the same as using F3, or the Exit PC Tools Desktop option under File in the menu selection.

Alt-F5 does a disk copy from one drive to another, or to the same drive. It will also do a format, if necessary. Another feature is that it can copy from one size disk to another, with automatic detection.

Alt-F6 will reread the directory. This is useful if you have made changes by deleting, moving, or adding files. It is also used when changing floppy disks or removable media disks.

Alt-F7 will start the Directory Maintenance program. This is one of the best features of PC Tools Pro. It can do a great deal of management from one screen. You can exit the Directory Maintenance program and return to the Desktop Shell by pressing F3.

Alt-F8 starts the Disk Editor. You can directly change files from this program. It has a Hex editor, ASCII editor, and Binary editor. Be warned—this program can make changes to important files, making them unusable if errors are made. Do not use this program without the proper precautions!

Alt-F9 starts the Format program. This format is extremely good at its job. It is recommended above the DOS format. It will check a disk while formatting, skipping any areas that are even partially bad. This can prevent future data loss.

Alt-F10 will make a bootable system disk. Put a disk of the proper type in your A: drive and use this option to make a bootable disk. If you wish to make an emergency boot disk, use the Tools menu selection and choose Make Emergency Disk. This will copy all necessary files to the disk, as well as making it bootable.

Using Ctrl with Function Keys

The next set of function keys available are those that can be accessed with the Control Key. This is the key marked Ctrl on your keyboard. Press and hold this key, while selecting the desired Function key.

Ctrl-F1 will start the System Information program. This program will provide a wide range of information about your system. Among the information available is the speed of the CPU, amount of memory, size of hard drive, type and size of floppy drive, video card, and much more.

Ctrl-F2 will give additional information on the file that is currently highlighted in the Files windows. It provides the type of information not available with a normal directory scan. This information includes the number of storage clusters used by this file, and the starting cluster.

Ctrl-F3 will print the current file to an active, connected printer. You have several options with this selection—print as a text file, print using PC Tools Options, or dump the information in both Hex and ASCII (useful for binary files). The PC Tools Print Options allow you to control the number of li .es per page, whether or not to add headers and footers (you provide the information for these), and options such as page numbering, stopping between pages and ejecting the last page.

Ctrl-F4 allows you to rename the file currently highlighted. You can use the option to rename a number of different files (global), or just a single file. Use with caution. **Warning**—if a program needs this file, renaming it could make the program function in an unexpected manner!

Ctrl-F5 provides for moving one or more files. The directory pointers for the files are changed only if the move is made to the same drive as the original file. If moving to a different drive, the original file is actually copied to the new drive, then the old file is deleted.

Ctrl-F6 will allow you to sort your directory files. You have a wide range of options to choose from, including sorting by name, extension, date, etc., as well as whether to do it in Ascending or Descending order.

Ctrl-F7 will start the Undelete program. You can use this program to undelete any files that have been recently deleted. If you do not defragment your drive on a regular basis, you can also undelete files that have been deleted some time ago. The chances of successful deletion depend on the length of time the file has been deleted, as well as how much information was written to the disk since the deletion.

If you have installed the MIRROR or UNDELETE program (by loading from the AUTOEXEC.BAT file), with the Delete Tracking option, then files can be recovered easier. Mirror and Undelete will keep track of the clusters used by a deleted file. This information is used in the recovery of the file.

If you use the Delete Sentry option, retrieval is almost guaranteed. This is because the Delete Sentry will copy the file to an unused portion of the drive for a specified period of time. Remember, though, that this will take up storage space on your drive.

Note: These two options may be incompatible with some programs. Check the manual for your specific program if you are experiencing problems and you suspect that MIRROR or UNDELETE is the cause. You can also use the MIRROR /U, or UNDELETE /U option to unload MIRROR or UNDELETE from memory, if it was the last program loaded.

Ctrl-F8 will start the Hex Editor program. This program allows you to edit files in hexadecimal notation. Use caution when editing with the Hex editor. Your changes can be made permanent and you may make a file unusable.

Ctrl-F9 will select all files in the currently active directory. Use this option when you wish to make a change to all the files in a specific directory. You may want to do this for CP Backup, or to change the attributes of the files in a directory.

Ctrl-F10 deletes the currently highlighted file or files. You will be asked to confirm your decision. Use with caution, especially with the Select All option!

Menu

The following options are accessed through the various pull-down menus. You can access the menus by holding down the Alt key and pressing the highlighted letter of the menu choice that you wish. You can also press and release the Alt key and a menu option will be highlighted. Press Enter to choose this menu option, or use the cursor keys to select a different menu option. Press Enter to open that pull-down menu.

The first menu option is File. Selecting this option will give you a wide range of choices directly related to files. To execute one of these option, you can either move the highlight bar with your cursor keys and then press Enter, or press the highlighted letter for the option that you wish.

Selecting Open will run the highlighted program. It must be either an .EXE file, a .COM file, or a .BAT file to be opened. The Open command means the same thing as Run.

The next choice is Run DOS Command . . ., which will allow you to issue a command with parameters. Selecting this option will give you another screen where you can enter the command, along with any parameters that you need. When the command is completed (unless you choose the COMMAND option) you will be returned to the PC Tools Shell.

Print does the same thing as Ctrl-F3. It provides options for printing the highlighted file. Select your file first, then choose this option.

The option to Search Files . . . will allow you to search a list of files that you have preselected, for a specific string of characters. The search is case-insensitive, which means that you do not have to worry about capitalization.

The Quick View option will open the file in a viewer. The program will attempt to open the viewer that will give you the best display of the file. If the file is a text file, then the ASCII viewer is used. If it is a binary (program, etc.) file, then you will get the binary, or HEX editor. You can change the viewer being used by pressing F6. You will then be presented with other viewers to choose from.

Using Move . . . is much like using the Move option in DOS 6.x. You are asked to confirm that you actually want to move the file, since the source file (or files) will be deleted from their original position. If the destination is on the same drive and partition, then the file pointers are changed. In this case, the file is not deleted, so it cannot be undeleted, either.

If you are copying to another drive, then the file will first be copied to the destination, then the source will be deleted. You can use the Undelete functions in PC Tools to recover the file, or you can use the ones that come with DOS.

Using Copy . . . will do the same thing as Move, but the source file (or files) will not be deleted.

The Compare . . . option will allow you to compare two or more files in different locations. The files can be on the same drive or on another drive. You will be asked to specify the locations of the files and then the names of the files. This comparison is bit-by-bit, so letter case is important.

Using the Delete . . . feature is much the same as the other alternatives. This option will delete the highlighted file, if no other ones were previously selected. Again, you will be asked to confirm that you want to delete a file. However, you can undelete the file using the normal methods.

Rename will allow you to change the name of the specified file, or all files that have been selected. If you have several files selected, you cannot change all their names and extensions to the same thing. However, you can change either the file names, or the extensions, to a common name.

The Locate . . . option will allow you to enter parameters to search for a specific file, or set of files. You can search multiple drives and directories in the same search. You can also specify a type of file to search for, such as .ZIP files.

You can use the Selection option to either select all the files in the active directory, deselect all the files, or specify which files to select, using a filter. For example, you can specify all the .ZIP files, for moving to another drive or directory.

Using the Change option, you can make changes to selected or highlighted files. You have the option of using the text editor or the hex editor. You can also change the attributes of the file (or files) selected, or you can Wipe them. Wiping files will result in permanent deletion. This is accomplished by deleting the file and overwriting the deallocated clusters.

The Verify option will either verify that the file is OK (no crossed clusters or bad sectors, etc.), or it will provide you with information on the file, such as last time file was written to disk, date and time of file, name, and starting cluster.

The Secure option allows you to encrypt or decrypt a file, or set of files. You can encrypt an entire directory by selecting the files first, then using Secure. You will be asked to provide a master keyword in case you forget the password for a specific file. You can then encrypt files with various names or characters and save them. You can also use this option to safely send files to others that also have PC Tools 9 and give them a password that they can decrypt it with. It is strongly recommended that you do not give your master key password to anyone, and that you do not use it as a regular password on other systems.

The Compression option will allow you to compress or uncompress files. It is compatible with versions of PKZIP prior to 2.04x. You will have options to save directories and pathnames in the compressed libraries. You can also delete the original files to save room on your drive.

The final option under File is Exit PC Tools Desktop. This is the same as using F3 or Alt-F4. You will also have the option to save any configuration changes made during the current session. Be sure that you want to do this. If the changes (if any) that you have made were for this session only, then be sure that the Save Configuration check box is cleared (does not have an X in it). You can choose to cancel the exit by choosing Cancel.

The second Menu option is Disk. These options apply to operations that are normally done to an entire disk, rather than individual files. To access this menu option, press and hold the Alt key, while pressing the letter D. You will then get a drop-down window with several options to choose from.

The first option is Rename Volume . . . , which is the same as changing the label on a particular disk. You will have the option of entering the name that you would like to change to. You can use this option to enter label names that you cannot normally enter through the DOS Label or Format commands. You must have the drive that you wish to rename selected as the current drive.

The next option is Search Disk . . . , which will allow you to search the currently selected drive for a text string (case not considered), or in hex notation. The entire drive will be searched and if a match is found, you are given the option to continue your search, see the name of the file or directory with the file name, edit the file, or cancel the search..

The Fastcopy option allows you to copy floppy diskettes of any size, from 360K to 2.88 MB. You can also convert from one size to another. For example, you can copy a bootable 3.5-inch disk to a 5.25-inch disk, including the boot files, making the 5.25-inch disk bootable. You must follow directions completely, though, to get a successful copy. Note that you will not get an exact duplicate, since sector and track information is different between the two disk sizes. Also, Fastcopy will not copy empty sectors.

The Compare Disks . . . option allows you to compare the contents of an entire diskette with another diskette. If two diskettes have identical files, but are of a different format, you cannot compare them successfully, since the empty sectors will not match.

The Verify Disk option will verify the currently selected drive. It checks to see whether or not the disk is completely readable, including files, directories, and unused space. You must select the drive that you want to verify before using this option.

The Format option formats disks in a better manner than the DOS or Windows Format program. If a sector is in any way questionable, PC Format will mark the sector as bad, so that your data will not be saved there. Do not use this option from within Windows.

Make System Disk will format a diskette in your boot drive and put the system files on it. This will make the disk bootable. Note that all data on the disk will be lost. Be sure that you want to delete what is on the disk before you proceed.

Wipe Disk will delete everything on the specified drive. Once this option is run, you will NOT be able to recover anything on it. You can select the drive that you want to wipe after choosing this option.

Directory Maintenance . . . is an expandable choice that allows you to work with directories. Many of the things that you can do with files can also be done with directories, using this menu selection. You can Add, Rename, Prune and Graft (Move), or Delete a directory. You can also Expunge a directory, which, unlike Delete Directory, will erase all files and subdirectories contained in the directory specified. **Note**: you should use extreme caution when using this command, since ALL files will be deleted, including Read Only, Hidden, and System files. The DM (Directory Maintenance) Stand-Alone program will allow you to use a mouse in a graphical environment. You will see percentages that directories take up, as well as being able to move and delete directories. Again, please use caution when using this program.

Change Drive . . . will allow you to change to a different drive. You can also change to another drive by pressing and holding the Ctrl key and pressing the drive letter of the drive that you wish to change to. Clicking the mouse on the drive icon near the top of the screen will also change to that drive. You must have a valid disk in the drive before switching to it.

Re-Read Tree will reaccess the drive that is current. This is useful when changing disks and you need to read the new directory into memory. You may also wish to do this if you have made extensive changes to a directory or a set of files.

Sort Directory . . . will allow you to sort a directory listing in a number of ways, including ascending, descending, by name, extension, etc. The sorting is for the display, or you can write it to disk and make the sorting changes permanent.

Disk Info . . . will provide information on the currently active drive. It will show the number of clusters, total drive space, free drive space, file information and bad sectors. It will also show the configuration of the current drive, including heads, cylinders, and sectors. In the case of an IDE drive, these figures may be the translated geometry of the drive, rather than the actual configuration.

Disk Editor . . . will allow you to directly edit the sectors on your disk. Use with extreme caution, since your changes can be written to disk. Writing the wrong information, or changing a file, can make the file or the entire disk unreadable. You will be asked for confirmation before writing to control sectors. The editor will be in Hex mode when editing. You can either enter the hex values directly (on the left side of the screen), or enter characters, which will be translated into the hex values (on the right side of the screen). F3 will return you to the Desktop Shell.

The third menu option is Configure. This allows you to make changes to the configuration of various parts of the PC Tools Desktop Shell, as well as other features that you can run from the Shell. You can set things such as color, directory display, pull-down menus, and data protection. Following is a description of the commands available in this menu.

Load Pull-Downs . . . will allow you to load a description file that will change the contents or look of your pull-down menus. You can also edit the menu structures stored on disk. PC Tools comes with a variety of predesigned menus, but you can design your own.

The Edit Pull-downs . . . lets you actually create new pull-down menus, or edit old ones. You have a wide selection of choices and setups. When you finish your creation, you can save it with any valid name that you wish. You can stop creating/editing by choosing Exit (on the right side of the screen).

The Password . . . option will allow you to change your password, if any, or set a new password. This will be a global password that will be used by default. Remember your password. You may have to totally delete the program and reinstall if you forget!

Display Options allows you to change the general and specific features of the display, from what shows in the windows, to what colors to use. It also has file parameter settings. You can select the types of files to display. You can also switch to all upper case, or upper and lower case mixed.

You can use the *Tree option* to set up several different options for selecting and displaying the directory tree. You can also set several defaults from this screen. Be sure to save the current configuration after making changes here, or they will be discarded when you exit the program.

The Data Protection option will run the DOS version of Undelete. You cannot run this option in a DOS window from inside Windows. You can also set up options for using Wipe. These options include a fast method and the DOD method, which not only erases the file, but then writes specific characters over the file's old location to prevent recovery.

The Execution option in the Configuration Menu allows you to specify certain characteristics of running programs from the PC Tools Shell. The first option, Execution Passwords, will prevent people without the proper passwords from running your programs on your PC. The next option will allow you to elect to clear memory to run large programs (if necessary), or to leave the Shell in memory, thereby allowing you to run only small programs. This option is not available if you are running the Shell as a TSR. The last option allows you return to the Shell immediately after a program is done running, or to wait and give a prompt, before returning. Check this option to wait for a response before returning.

The DriveMap option allows you to access another computer's hard drive as another drive on your own computer. You must first have DRIVEMAP.EXE loaded as a TSR (loaded through the AUTOEXEC.BAT file). You must also be connected to another computer through either a network or a direct connection, such as serial ports.

The Confirmation . . . option allows you to set options to confirm on delete, replace, mouse operations (for when you are moving your mouse too fast), and operations concerning system files (files with the System attribute set). You can also set it up to not allow deletion of system files, to use date, time, and size when copying files, to hide system files, and to include all subdirectories (if any) in a drag and drop operation.

Define Function Keys . . . allows you to set the meanings of all the function keys. You can set the Control and Alt options from here, also. If you change your mind while making the changes, you can reset the keys to their original definitions.

Secure Options . . . is where you set the type of encryption, deletion, whether or not to use compression and expert mode. To set an option, select the option and press the spacebar to set or clear the box next to it. You will have to save your changes from the main menu, or they will be lost when you exit the Shell.

Startup Programs allows you to decide which programs to put in your AUTOEXEC.BAT and CONFIG. SYS files. You can choose various options for each program, as well. These program take up memory when loaded, so you may have to use a memory management program to load some of them into high memory (on a 386 or better computer).

Speed will allow you to configure some items that can speed up your computer. These options include Speed Search, which allows you to locate files quickly, Keyboard Speed . . . , which will load a keyboard buffer program that will also help fast typists to get a quicker response from the keyboard, Mouse Speed . . . , which allows you to set the speed of the mouse cursor when tracking on the screen, and PC

Cache Status, which shows the efficiency of the PC Cache TSR. (PC Cache is a TSR that takes the place of SMARTDRV. It is a disk caching program.)

Configure Editors allows you to set up specifics of the File Editor and Disk Editor. These files are used when doing work directly with files and sectors on the disk. Be sure to leave any safeguards in place to prevent accidental loss of data.

The final option will save any changes to the configuration setup. These changes will continue to be in force until you make other changes. You can also elect to save your changes when you exit from the Shell. Some configuration changes allow you to save those changes immediately. In these cases, you will not need to take this extra step.

The fourth menu option is Tools. These are other programs that are included with PC Tools. It is not necessary to install all the programs that comes with PC Tools, but it does help. With everything installed, you get the most functionality from the program. You can run all of these programs from this menu. Following is a brief description of each program and what it is used for.

Anti-Virus is a program that will detect and remove viruses (when possible). It also has options to protect files from infection (yes, this actually works!), to create checksums to verify integrity and protect from viruses that have not been recognized. It can even provide detection of stealth viruses!

CP Backup is a program that makes backing up and restoring your files a much more secure event. You can set it up in a number of different ways. You can also select a number of different types of backup options. This program can backup to floppy disks, removable hard drives, Bernoulli drives, and tape drives. **Note**: you will have to make some changes to your CONFIG. SYS and AUTOEXEC.BAT files. Some of these changes may not be well documented, or may not even be documented at all. If you have problems, you may need to call Central Point Software for help. However, the tech support staff there is quite good and can help with a wide range of problems.

Optimizer is a disk defragmenting program. It will safely organize the files on your hard drive in a number of different ways. You can sort your files by a number of different methods, you can just defragment your files (the quickest method), or you can defragment your files and fill all open clusters. You can also clear all empty sectors, making it easier to recover files deleted later. This option also eliminates "junk" from old files.

Build Emergency Disk will give you a safety boot disk with all the necessary files on it. If your drive ever fails to boot (due to virus, damage, or accidental deletions or formats), you can use this disk to recover from most problems. It does not guarantee that you will be able to fix everything, but most common problems can be remedied with this disk. I recommend using this option often, especially if you are changing things on a regular basis.

Unformat is normally used only to recover from an accidental format. It can (in very FEW cases) also be used to recover from an accidental use of the RECOVER.EXE program. You can use this program on floppy disks or hard drives. If you have accidentally formatted your hard drive, you can run this program from the original disks. If you have formatted with the /U switch, or if you have formatted with a version of DOS prior to 5.0, you may not be able to recover from a format operation. This is because versions of format before the one included with DOS 5.0 did not copy the system areas of the drive to a safe location. They also performed a very destructive format, which erased the file areas of a disk.

Diskfix is used to repair problems with the file structure, directory structure, and File Allocation Table (FAT). It can also provide some help in repairing bad sectors, or relocating sectors that are damaged. There is no guarantee that this will work, but it might help when other methods fail.

Filefix can repair damage to several different types of files, including database, spreadsheet and word processing. The repairs are successful in many cases, if the damage was solely due to software problems. If the disk surface itself is damaged, you may not be able to recover the file. Use DISKFIX for more options of repairing damage to your disk.

SI Pro is a program that provides extensive information about your computer. It can show you what is installed (in most cases), what different resources are being used, and how your memory and hard drive are configured. You can also use this program to check the speed of various components of your system, including CPU, hard drive, and video.

Maps will let you look at a map of a file on the disk, or a map of the entire disk itself. You can also get a description of what is in memory, how it is set up, and where it is located. Memory Info will give you a quick layout of what is in memory.

The Scheduler will allow you to set up certain events to take place at a specific time and day. You can set up events to take place daily, or you can specify an exact time and day. You can also set up long-term repetitive operations. This might be useful for setting up the CP Backup program.

Network Message can be configured to give a preselected message under certain circumstances on a network machine. This message can be set up to tell the user what extension to call if a problem occurs. It can also give information on what to do under certain conditions.

DOS Session will shell to DOS so that you can execute DOS programs. When you are done, you can type Exit to return to the Shell. You can set up the amount of memory and the conditions of returning in the Configuration Menu option.

The Window Menu has a number of choices for selecting what information is shown and in what manner. These options affect directory information, file information, and viewing information. The various options are described below.

You can have up to three different windows open on the screen at the same time. These are *Tree Window* (a display of the directory tree on the drive), *File Window* (which shows the files in the currently active directory), View Window (which shows the contents of a file), and the *Menu Window* (which shows the executable programs that you have set up in the PC Tools Shell). The Menu Window and View Window can be shown only if the other one is off, even if one of the other menus is off.

Single List Display will show one directory and one listing of the files in the active directory.

Dual List Display will show two directories and the files associated with each active directory tree. You can have two different directories active. You can also have information from two different drives displayed at the same time.

Menu Windows Only will show only the screen where your installed programs are listed. To return to the normal windows, press F10.

If you have a check next to *DOS Command Line,* you will have a line just above the Function Key list that you can use to execute DOS commands.

If you have a check next to *Function Key Line,* you will have the function key definitions listed on the last screen line. If you already know what they all do, you can turn this option off and gain another line for the display.

Background Mat will fill in any uncovered space with a blank area. This is to prevent distraction from the contents of the current DOS screen. If you like to see what is going on with this screen, you may wish to leave the mat off.

Hide All Windows will close all windows and show just the command line and the function keys. All other items are temporarily hidden. You can still press the Alt-Letter combinations to access the menus. When you do this, the upper portion of the screen returns. You can turn off the Hide All Windows option to return the screen to the normal display.

Compare Windows is functional only when you have a Dual List Display active. It will compare the contents of the two windows and report if they are identical. If there are differences, these will also be reported.

Move Window will do the same thing as putting the mouse cursor on the title bar of the desired window and holding down the left mouse button. You can drag the window to the desired location at that time. If you do not have a mouse, you can use the menu option and move the window with the keyboard.

Size Window will allow you to resize a specific window. You can also use the four-arrow square at the lowest right corner of each window to do this job.

Zoom Window will expand the currently active window to fill the entire window display area. When in this mode, selecting Zoom Window again will return the window to its previous size.

Reset Window Sizes will return the various windows to their default sizes.

The next Menu Option is Tree. This option deals directly with the way that the directory tree is displayed. The options are listed below.

Expand One Level will show the next level of the currently active directory (if any). You can also use the "+" key to accomplish this. Double-clicking on the directory with the mouse will do this, too.

Expand Branch will show all of the subdirectories of the currently active directory. You can use the asterisk key to do this, too.

Expand All will show all the subdirectories of all the directories on the currently active drive.

To collapse the tree again, choose Collapse. You can also use the "–" key to collapse the tree. To return to the normal view of the root directory and the first level of subdirectories, you can the press "+" or Expand One Level.

The final Menu Option available is Help. This option offers help in a number of different ways. These various methods of asking for help are described below.

Topics will give you a list of topics to search through. These are divided into different categories, from How To do something to Commands and options.

Index will give a listing of all the help available in PC Tools. You can scroll through the list, or press Page Down or Page Up to jump one screen full at a time. The items are divided into alphabetical order for more convenient browsing.

Keyboard offers help on the various options available from the keyboard. These items are listed in order and can be scrolled through using the cursor keys or the Page Up and Page Down keys.

Commands offers help on various commands available in PC Tools Shell. The commands are listed in menu order to facilitate finding the information that you want.

About . . . gives you brief information about the different parts of PC Tools Shell and its various additional programs. To get information on a specific item, find the word or phrase that is highlighted and press Enter, or click on it with your mouse cursor.

DOS Advice gives informative information on various errors and problem resolutions in DOS. It does not provide information for Windows operations.

All of the options in this menu can be accessed with the mouse. You can also use the scroll bar on the right of the screen to scroll through the help for the subject that you are looking for help on. If you do not know what to do in the screen, pressing F1 will provide more detailed help on the item that is currently highlighted.

You can also print the information on these screens by choosing Print (F6 with the keyboard). You can Go Back to the previous topic, or choose a different mode of help. There are also extensive manuals on-line for in-depth help on a subject or topic.

The next portion of the PC Tools Shell screen to cover is the small "–" box in the upper left corner of the screen. From this box, you can exit the Shell, get the version of the Shell that you are currently running, minimize or maximize the screen, and move, size, or restore the main program window.

Lastly, the drive icons just below the top menu are used to select a drive. To select a drive, move your mouse cursor to the desired drive icon and click your left button. If you select a floppy drive and there is no disk in the drive, you will cause an error that can be fixed only by putting a disk in the drive and closing the door (if necessary).

DEFINING TERMS

Allocate/Deallocate: To indicate that a cluster is/is not being used by a file or program.

ASCII: This is the normal characters and numbers that you can type directly from your keyboard. The ASCII character set for IBM-compatible computers includes 256 characters.

Attributes: Optional settings for a file. These settings can be used to protect a file from accidental erasure, to indicate that a file needs to be backed up, or to show that it is a system file (important for the operation of the system). You can also hide a file from a casual directory search. Many programs use these different attributes for different purposes.

Bad Sector: A storage segment that cannot reliably store data. This can be caused by physical damage or unreliable magnetic areas.

Binary: A file type that usually indicates a program, rather than a data file.

Bootable System Disk: A Bootable System disk is a disk that has the files necessary to start your computer. The disk must be of the same size as your A: drive.

Compress: This is an operation that examines a file or set of files and finds repetitive data. A compression program will then calculate a method of replicating that data, while taking a smaller amount of room to store that same data. Using this type of program may provide 50% or more in storage space savings.

Configure: To set up a program or option to your liking. This may involve choosing colors for various parts of the program, or telling the program where various things can be found on your computer. It may also involve telling the computer what hardware and software you may have in your computer.

Crossed Cluster: A cluster that is claimed by more than one program or file. This can be disastrous, as two files should never share the same space.

Decrypt: To put a file back to its normal state, to allow access by other people and programs.

Defragment: To sort the files on the drive into contiguous clusters. This helps the drive to work less to retrieve files, as well as making the storage more efficient.

This is much like sorting a deck of cards into suits and putting them in sequential order.

Directory: A listing of files on a disk.

Directory Tree: A listing of subdirectories. It resembles an upside down tree when displayed.

Double-Click: To click the left (or right) mouse button twice, quickly, without moving the mouse between clicks.

Drive: Usually indicates a hard drive, but may also indicate a floppy, or CD-ROM drive. A drive is where files and programs are stored.

Emergency Boot Disk: A disk that contains the files necessary to start your system, along with other necessary files. If your computer will no longer boot from the hard drive, you will need to use this disk to start up. You can then use the different programs on the disk to attempt to recover or repair the hard drive.

Encrypt: To change the normal file in such a manner that another person cannot easily read or access it.

Execute: To start a program or operation.

Exit: To stop using a program or option and return to your original location. If you exit a program, you will be returned to the operating system. If you exit an option or secondary program, you will be returned to the original program.

F-keys: F-keys are also called Function Keys. These keys can be found at the top of 101-key keyboards and along the left side, in a double row, on 88-key and specialized keyboards. These keys are used by many programs to indicate a choice or option.

Format: To prepare a disk for saving information.

Hex: A numbering method used within computers. It is set up on Base 16. This method uses the numbers 0 through 9, and the letters A through F. Hex notation is as follows: 0, 1, 2, 3, 4, 5, 6, 7, 8, 9, A, B, C, D, E, F. This means that a decimal (base 10) number of 26 would be shown as 1A and 16 would be shown as 10, in Hex notation.

Highlight Bar: Found in menus. A highlight bar is a portion of the screen that is shown in reverse video. If the screen has a white background and purple letters, the highlight bar may appear as a band of purple, with white letters. The highlight bar indicates the current choice or option.

Icon: A small picture that gives a basic idea of what the program does. Pointing to an icon with your mouse and double-clicking, or highlighting the program name with the keyboard and pressing the Return key, will usually run the program.

Interface: Method of showing information, requesting input from the user, and providing responses.

Menu: A list of options from which you can choose. Though an option may appear in a menu, there are times when it may not be available. Unavailable options may appear a different color (usually gray), or the program may produce an audible warning if you try to use that option. Some programs just will not do anything if you try to choose an unavailable option.

Navigate: To progress through a series of options to accomplish a specific task or set of tasks.

Parameters: Options that can be used when running a program or operation previously selected.

Read Only: An attribute of a file that indicates that it can be read and copied, but it cannot be modified.

Select/Deselect: Usually indicates that one or more items from a list have been chosen or unchosen. A selected item will normally be highlighted, have a check mark next to it, or have a number or some other character next to it, to indicate that this item has been selected. Deselect will unmark an item that had been selected.

Shell: A program from which other programs and utilities can be run. Usually, other operations can be carried out from here, also.

Storage Clusters: The area on a hard drive where a file is stored. A cluster consists of a number of sectors. The size of the partition determines the number of sectors per cluster. A sector holds 512 bytes. The minimum cluster size is one sector.

Strings: Strings are sequences of letters and/or numbers. In some cases, a string can also contain spaces and special characters. In other cases, a string may be limited to just letters, etc. The length of a string may also be limited.

Window: Area of the screen that is currently active. A window is showing its borders. When a window is active, other windows do not normally accept input, until you deactivate the currently active window and switch to the desired window. An active window is usually colored, or highlighted differently than the other portions of the screen.

32.4 MICROSOFT BACKUP FOR WINDOWS

Kevin Greiner

Starting

Since the earliest days of DOS, Microsoft has had at least a rudimentary type of backup. In the earliest days, one had to rely on floppy disks and the archive file attribute, which is turned on when a file is changed. Several versions later, Microsoft packaged its version of an automated backup system, called simply "backup" and "restore." For most users, these command-line utilities got little use. The data they stored were not compatible across the various versions of DOS and the were was not compressed.

Third-party vendors jumped to fill the gap with programs that provided these features and more. In time, tape drives grew to replace floppy drives as the backup medium of choice. Many tape drives came bundled with adequate software as well.

It was not until DOS 5.x that Microsoft improved the DOS backup utility by replacing it with a menu-driven one. In DOS 6.x, the standard backup utility is a Windows-based limited version of Symantec Backup. This chapter focuses on that version of the DOS backup program and how you might use it with your computer.

If your data storage needs are more than minimal, you need to seriously consider a tape drive, or some other storage mechanism that provides you with greater storage capacity than a floppy disk. On the other hand, if you are looking only for a simply way to periodically back up a few files, MS Backup is the tool for you.

Getting Started

When MS-DOS 6.x is installed on your system, it creates a Windows group titled "Microsoft Tools." This group includes an icon titled "Backup." This icon is shown in Figure 32.1.

Microsoft Backup is run by double-clicking on the icon labeled "Backup." The first time Backup is run, Backup will automatically configure itself for your system. You need a blank floppy for each floppy drive you plan on using to back up your files. Backup performs a small backup and reads it to verify that it can perform reliable backups on your system.

FIGURE 32.1 The Program Manager icon for Microsoft Backup.

When Backup is run you will see the window shown in Figure 32.2. To close Backup, double-click the system control box in the upper left corner, select Exit from the File menu, or type Alt-X on your keyboard.

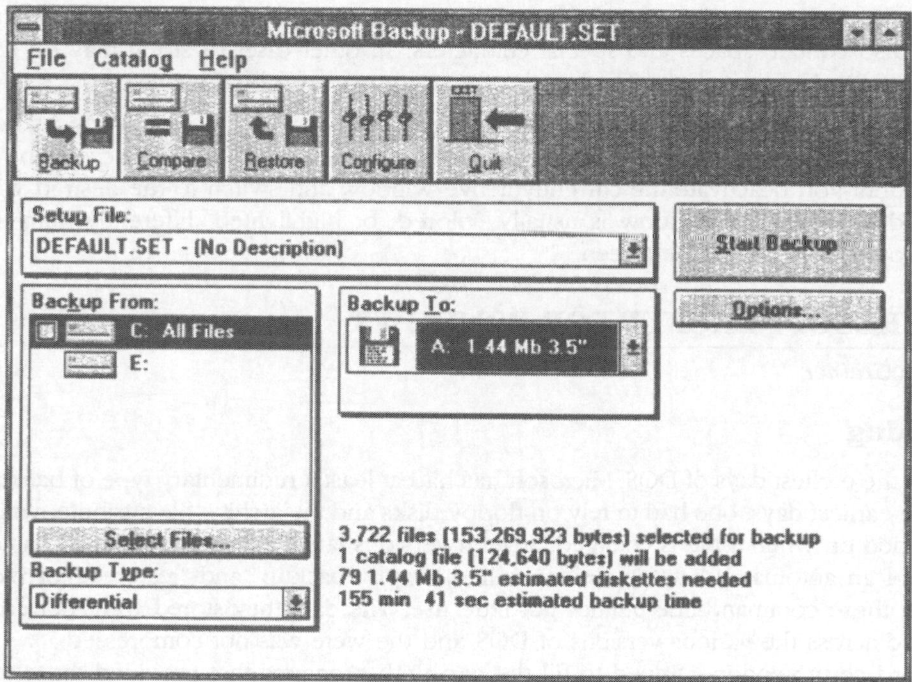

FIGURE 32.2 Backup mode.

The Microsoft Backup Application Window

The Microsoft Backup interface is somewhat different than many Windows programs. Backup uses a button bar across the top of the screen to provide access to four different areas of functionality. When you click on a specific button, Backup displays only the options necessary to use that specific function. The following paragraphs will explain each of the four areas, from left to right across the button bar.

Backup

Backup From. Choose the drive that you want to back up. Backup displays a list of all valid drives and allows you to choose one or more drives to back up. A label after the drive letter tells you if all files, some files, or no files on this drive are selected for backup.

Select Files. This button displays an expanded view containing each subdirectory and file from the drive selected in the "Backup From" dialog box above. From this list you can choose to include or exclude files based upon their names, attributes, and dates. You can also choose to back up files in certain subdirectories.

Backup To. This dialog box specifies the device where MS Backup will place the files to be backed up. This dialog box allows you to specify a floppy drive or a path. This might be a good option if you are connected to a network; you could then back up your local hard drive to the network drive. However, be sure to check with your network administrator before doing this.

When you choose a floppy drive, Backup displays in the bottom right of the screen approximately how many floppy disks you will need. If you are using data compression, the number of disks needed may actually be less than the number shown.

Backup Type. Three different types of backup exist: full, incremental, and differential. A full backup backs up every file selected, regardless of whether it was changed since the last backup. An incremental backup backs up only the files that were changed since the previous backup and marks all the files that are backed up. A differential backup backs up only files that were changed since the last backup and does not mark the files that are backed up. This allows you to daily back up files that you work with frequently, but may not change daily.

Options. Backup provides you with a number of options to customize how it works for you. This is the place to look to password protect your backups, turn on or off data compression, or have Backup automatically exit after performing a backup. Read the MS Backup online help for more information.

Setup File. If you find yourself choosing the same files and specifying the same options over and over again, read this section. Through the use of setup files, you can eliminate much of the repetitive nature of backups.

For example, if you wanted to perform a full backup weekly and an incremental backup daily, you would do it in the following way.

Weekly Backup:

1. Select the drive you want to back up and specify "all files."
2. Select the destination for the backup (floppy or path).
3. Select "Full" as the Backup Type.
4. Click on the File menu.
5. Click "Save Setup As."
6. Type "weekly" for the file name.
7. Type "Weekly Backup (Full)" for the description.
8. Click OK.

Daily Backup:

1. Select the drive you want to back up and specify "all files."
2. Select the destination for the backup (floppy or path).
3. Select "Incremental" as the Backup Type.
4. Click on the File menu.
5. Click "Save Setup As."
6. Type "daily" for the file name.

7. Type "Daily Backup (Incr)" for the description.
8. Click OK.

Wasn't that easy? Now when you run Backup to do your daily or weekly backup, just choose the one you want to run from the Backup Set list and click "Start Backup."

Compare

For most backups, you will want to verify that the files were backed up correctly. You can do this by running a Compare (Fig. 32.3). You can choose to compare the Backup against the same files, files on a different drive, or files in different directories. The other dialog boxes are similar in nature to those on the Backup screen. You can also save Compare setups in the same way.

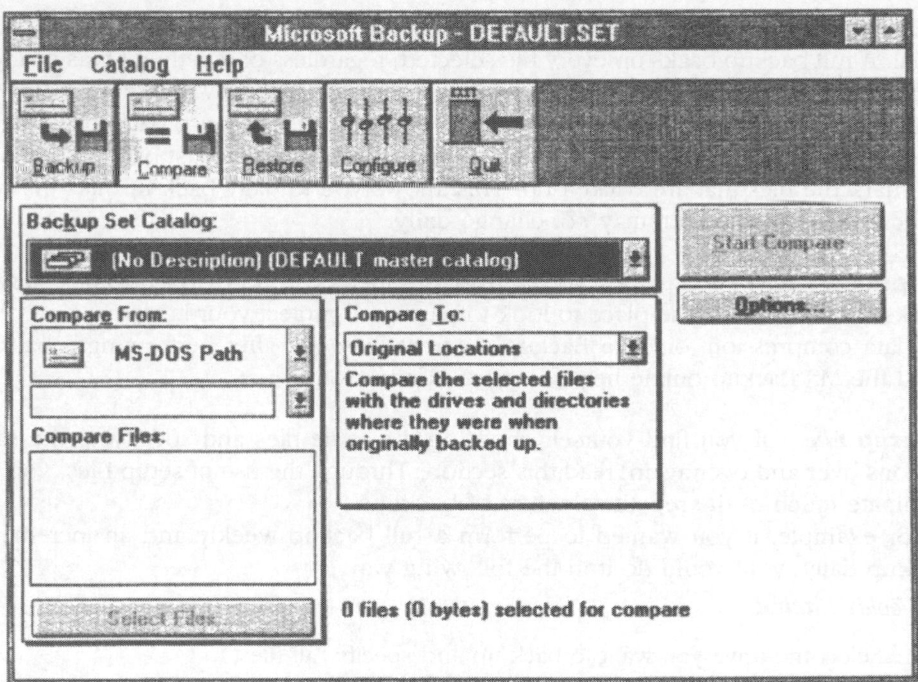

FIGURE 32.3 Compare Mode.

Restore

The purpose of a backup is to be able to restore the data if something goes wrong or you need a prior version. You can choose to restore the files from the backup to the same drive and directories they were backed up from, to a different drive, or to different directories. The other dialog boxes are similar in nature to those on the Backup screen. You can also save Restore setups in the same way (Fig. 32.4).

Configure

Auto Floppy Configure. This performs a test to determine if MS Backup can automatically detect when you change floppy disks during a backup, compare, or restore (Fig. 32.5).

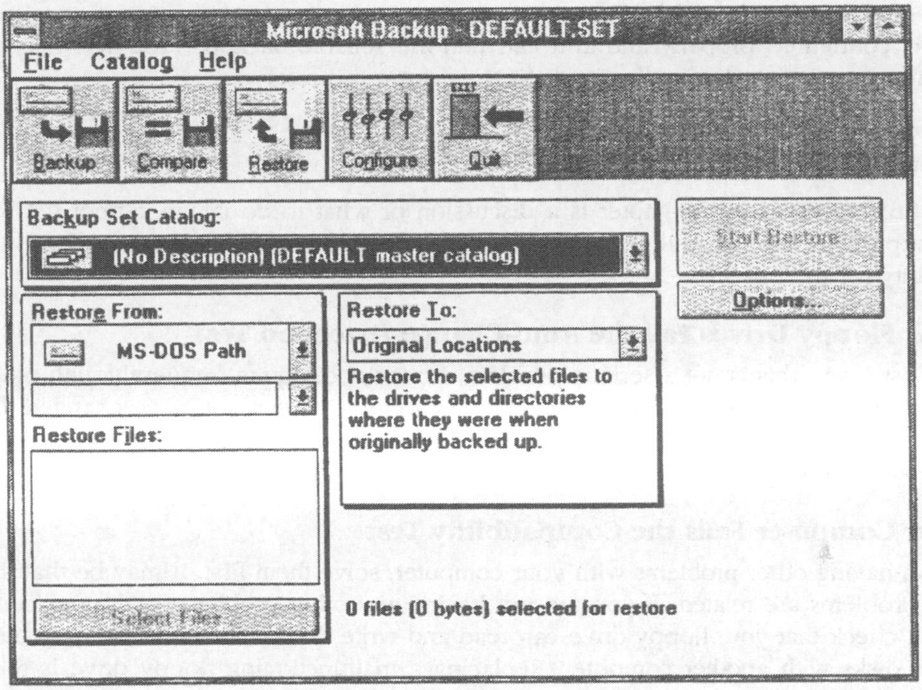

FIGURE 32.4 Restore mode.

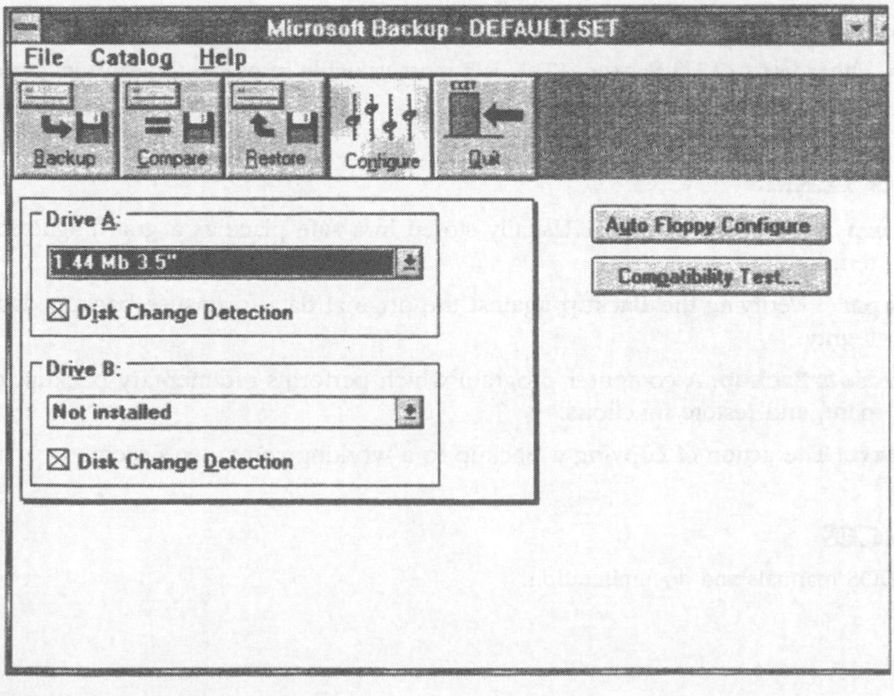

FIGURE 32.5 Configuration mode.

Compatibility Test. This runs an automatic mini-backup to determine if your system is configured properly and and can perform reliable backups. You will need a minimum to 2 high-density floppy disks to perform this test.

Troubleshooting

The final section in this chapter is a discussion of what to do if you cannot get MS Backup to work. In general, Backup is a simple program. If offers a great deal of functionality for the price.

Your Floppy Drives Fail the Auto-Change Detection Test

In most cases, this is not a serious problem. It may occur quite frequently with older machines. You will simply have to press a key to indicate that you have changed the floppy drive. If your computer fails this test, do not manually check the "Disk Change Detection" box.

Your Computer Fails the Compatibility Test

If you having other problems with your computer, solve them first. It may be that the two problems are related. If you are not having any other problems with your computer, check that your floppy drive can read and write disks accurately and exchange these disks with another computer. Replacing a malfunctioning floppy drive is relatively inexpensive with the time necessary to re-create lost data.

It Still Doesn't Work, Now What?

If you have tried everything you can think of to get a good backup, but without success, check your settings using the following sources:

- The Help available from within MS Backup.
- Other users of MS Backup. (This is a most valuable resource. Other users are often more than willing to aid a new user.)

DEFINING TERMS

Backup: A static copy of data. Usually stored in a safe place as a guard against system failure.

Compare: Verifying the Backup against the original data to ensure backup data integrity.

Microsoft Backup: A computer program which performs rudimentary backup, compare, and restore functions.

Restore: The action of copying a Backup to a working computer system.

REFERENCES

MS-DOS manuals and documentation.

33

ActionWorkflow

Action Technologies, Inc.
Alameda, California

The vertical corporation or organization—that time honored structure that is so often characterized by sluggish bureaucracy and many layers of management—is going the way of the dinosaur. Today's volatile and competitive marketplace is changing the rules, and companies are being forced to reinvent themselves along horizontal lines. Out of this sea of change in corporate organizational philosophy has emerged a new class of business software for workflow automation. As companies become more horizontal, their requirements for workflow automation focus more on managing and coordinating the flow of work for processes that span the entire organization. They need software that addresses the way people would ideally coordinate and collaborate with each other to get work done, consistently across all processes and on all levels in the organization.

This chapter and the next examine two such products, ActionWorkflow® from Action Technologies, Inc., and Lotus Notes® from Lotus Development Corporation.

33.1 REENGINEERING FOR CUSTOMER SATISFACTION

Successful strategies, such as total quality management (TQM), for improving the competitiveness of organizations agree on the essential qualities of an effective business process:

- The customer is central to the process.
- Workers understand the entire process.
- Workers are accountable for quality and are empowered to innovate.
- Improvement in quality is continuous.

ActionWorkflow is ideal for reengineering because of its focus on human interactions and customer satisfaction. ActionWorkflow assures that a workflow system produces tangible results in terms of greater customer satisfaction and increased job satisfaction for employees. ActionWorkflow accomplishes these results by

- Clarifying customer requirements
- Describing the human interactions required to meet customer requirements
- Organizing existing systems and technology to support the required human interactions.

0-8493-2530-7/96/$0.00+$.50

Why the Old Way May Not Work

Figure 33.1 illustrates three ways to look at the design of a process, depending on the resources that are being studied:

- Industrial engineering
- Information systems analysis
- Business process analysis

Materials Process	Information Process	Business Process
Industrial Engineering	Information Systems Analysis	Business Process Analysis
Mechanical Engineering	Input-Process-Output Model	ActionWorkflow Analysis
Time and Motion Studies	CASE Tools	ActionWorkflow System

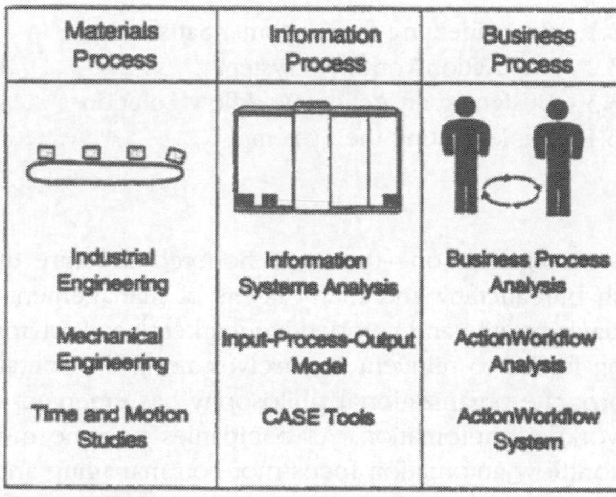

FIGURE 33.1 Three kinds of processes.

Industrial engineering is concerned with improving the manufacturing process and increasing the efficiency by which materials are moved within this process. Information systems analysis improves the availability and accuracy of information that is vital to a business. Both of these disciplines have provided tremendous benefits to organizations but the very success of these disciplines has led to diminishing returns. As the materials and information processes become more efficient, costs become more concentrated in the business process—the way people work together to achieve the goals of the organization.

Today, to achieve significant gains in overall productivity, organizations must improve the efficiency of the business process; but applying the approach of industrial engineering or information systems analysis to the business process does not ensure any improvement in the way people interact.

Business process analysis must improve the way that knowledge workers interact and collaborate to achieve greater customer satisfaction. ActionWorkflow provides the tools and methodology to analyze the business process and to integrate material and information resources so that they support this process. The ActionWorkflow methodology keeps the business process—the conversations and interactions between people—at the center of attention and prevents the analyst from slipping back into the more familiar viewpoint of the information or materials process.

33.2 THE ACTIONWORKFLOW SYSTEM

The Problem with Ordinary Workflow Systems

A different approach is required to analyze and design a business process because the information systems approach focuses on the input, processing, and output of information. Figure 33.2 illustrates a workflow system based on an input-process–output approach. This approach can provide an efficient materials and information process but leaves little room for workers to respond creatively when exceptions inevitably occur.

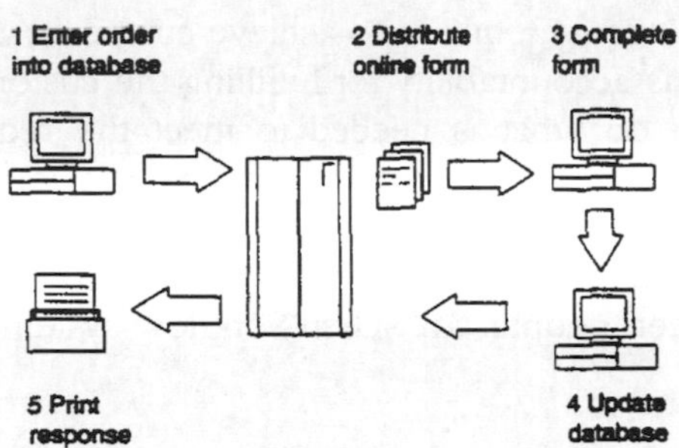

FIGURE 33.2 An ordinary workflow system.

This kind of workflow system focuses attention on how quickly a particular step is completed rather than encouraging workers to understand the goal of the entire process. An ordinary workflow system does not support workers' fulfilling their commitments—the system manages the movement of information and forms. The role of people in this process is basically to interact with the devices used for processing information and materials.

When a workflow system is broken into small steps to support efficient throughput, the resulting fragmentation makes it more difficult for an individual to take responsibility for quality or to innovate. Quality initiatives depend on encouraging workers to innovate and to provide suggestions for ongoing improvements. However, to change or improve a workflow system like this may involve stopping the entire system and making modifications to every step. Depending on how often changes occur, the time and energy spent modifying the system may eliminate most of the time or cost savings.

Information about this type of workflow system typically takes the form of historical statistics about throughput. Specific, real-time information that can reassure a customer, alert a manager, or guide a worker is not available. The kind of information that is available makes it easier to find out what happened than to discover what needs to happen next.

Figure 33.3 illustrates an information systems flow chart that might serve to create a workflow system like that in Figure 33.2.

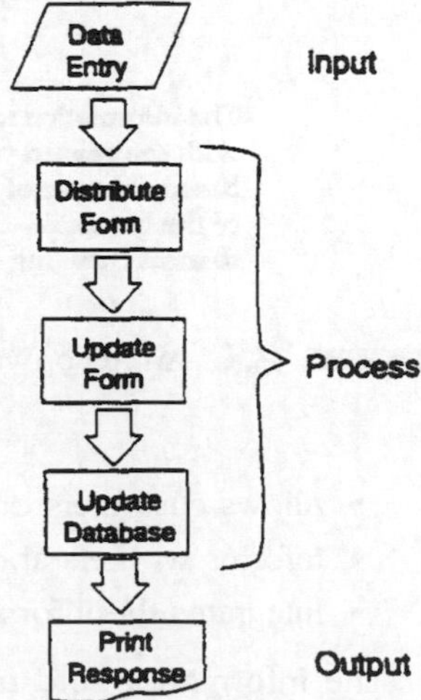

FIGURE 33.3 The input-process–output viewpoint.

A workflow system based on this model may encounter resistance from employees who feel that the system makes it more difficult for them to achieve accountability, to be innovative, or to know clearly what they need to do to achieve customer satisfaction. While the information processing viewpoint is useful for developing efficient information systems, another viewpoint is required to create a system that increases employee job satisfaction and fulfills customer requests.

Workflow Design with Action Workflow

In contrast to the previous example, an ActionWorkflow system is based on the network of personal commitments required to achieve customer satisfaction.

ActionWorkflow assigns accountability for fulfilling the customer's request while allowing the flexibility to do what is needed to meet the request. For this reason, ActionWorkflow provides the right viewpoint for trying to improve the efficiency or quality of a business process.

An ActionWorkflow system (Fig. 33.4) uses automation tools and technology to improve interactions between people. An ActionWorkflow system

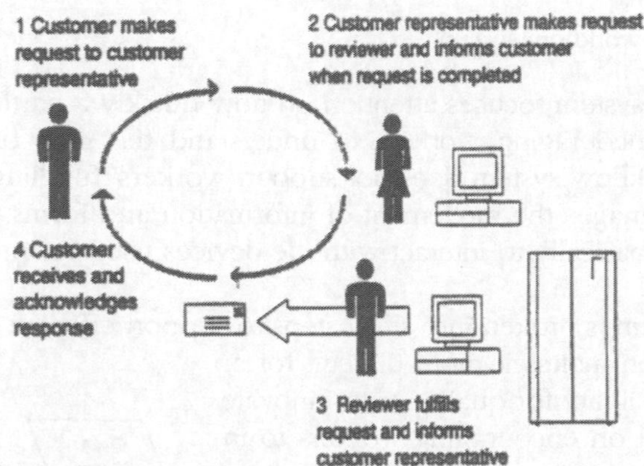

The information and materials process in an ActionWorkflow system can be managed with centralized processing systems or by participants using the microcomputer-based software of their choice. Because the workflow system is built on the structure of the business—commitments between people— it easily accommodates innovation to meet changing customer needs.

FIGURE 33.4 An ActionWorkflow system.

- Allows customers or managers to monitor the status of each transaction
- Informs workers about what needs to be done next
- Integrates the information resources needed to move the workflow to completion

The information and materials process in an ActionWorkflow system can be managed with centralized processing systems or by participants using the microcomputer-based software of their choice. Because it is built on the structure of the business—commitments between people—the workflow system easily accommodates innovation to meet changing customer needs.

The ActionWorkflow Methodology

The ActionWorkflow model, established through the research of Dr. Fernando Flores and Dr. Terry Winograd, ensures the necessary focus on worker accountability and customer satisfaction. U.S. patents 5208748 and 5216603 recognize their unique approach to business analysis. The ActionWorkflow methodology is a way of designing or reengineering a business process based on this model that focuses on the way people interact to achieve customer satisfaction.

The ActionWorkflow Loop

The ActionWorkflow loop (Fig. 33.5) describes interactions between participants in a business process as workflows occurring between customers and performers. Each Action workflow loop describes a complete interaction with a customer. During the process of completing interaction and receiving customer feedback, this way of organizing work empowers an individual to continually improve the quality of that interaction.

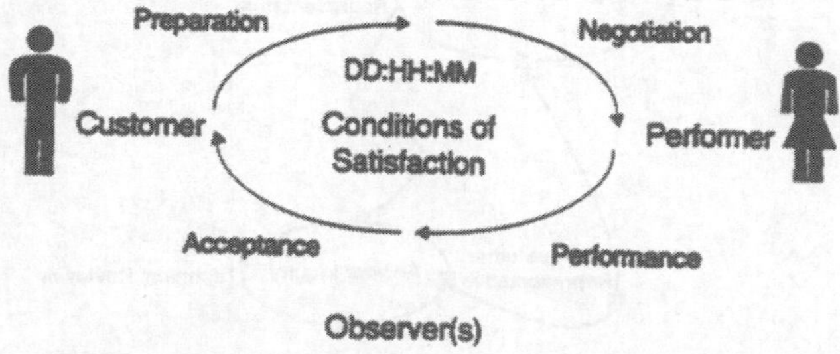

FIGURE 33.5 The ActionWorkflow loop.

In a workflow, either the customer makes a request of a performer or the performer makes an offer to provide a service to the customer. The conditions of satisfaction are the criteria used to determine if the customer is satisfied. The cycle time is the length of time required to complete the workflow. Cycle time, a critical part of the conditions of satisfaction, appears inside the upper part of the workflow loop.

The customer can be an external customer—someone outside the organization who is purchasing goods or services—or an internal customer—such as a worker's supervisor. The conditions of satisfaction can be explicit or implicit—that is, clearly stated or simply understood on the basis of background shared by customer and performer. The workflow structure also provides for an observer role.

Table 33.1 shows the four phases that occur in every workflow.

TABLE 33.1 Four Phases of a Workflow

Workflow Phase	What Happens
Preparation	The customer or performer proposes work to be done by the performer
Negotiation	The customer and performer come to agreement about the work to be performed
Performance	The performer performs the work and reports completion
Acceptance	The customer evaluates the work and declares satisfaction or dissatisfaction

A business or systems analyst can use this model to accurately describe the different interactions between people that occur in a business process. The resulting workflow map helps to define the goal of a business process and provides a very powerful tool for designing a system that

- Achieves greater customer satisfaction
- Provides workers with greater accountability and clarity
- Supplies the information required to manage the business process

ActionWorkflow Maps

An ActionWorkflow map, like the one shown in Figure 33.6, complements the information processing point of view. The map provides a description of the interactions between people, shows how these interactions support the goal of customer satisfaction, and guides the application developer to design information systems that support this goal.

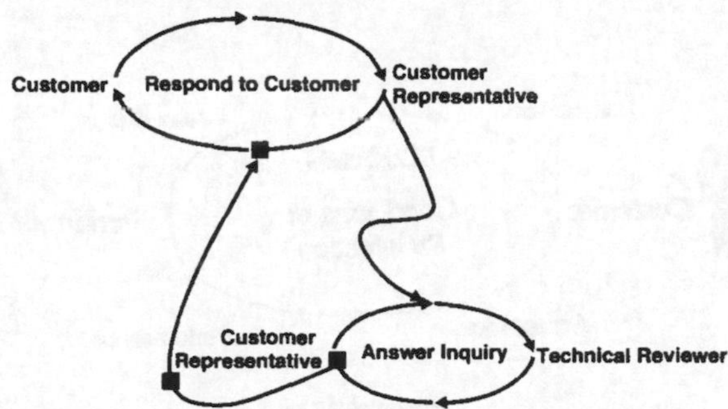

FIGURE 33.6 ActionWorkflow business process map.

Like any map, the purpose of a business process map is to allow people to get where they want to go. A good map is one that is easily read by different people and that helps them make the right choices to reach their goals.

ActionWorkflow mapping conventions ensure that the people implementing a workflow system—workflow participants, designers, and developers—agree about what needs to be done to produce customer satisfaction. This agreement reduces the miscommunication and miscoordination that have been too common in the past, and ensures that the resulting workflow system supports the client's business goals

Creating an ActionWorkflow map provides many opportunities to improve the business process before any coding begins. By analyzing the conversations that occur between people in a business process, this model helps to

- Identify key activities that must be completed to satisfy a customer
- Highlight opportunities to perform work in parallel and thereby reduce process cycle time
- Eliminate the performance of work that has not been requested by a customer
- Assign responsibility for work that has been requested but for which no performer is responsible
- Eliminate redoing work by clarifying the conditions of satisfaction
- Ensure that customers review and acknowledge work that has been completed

ActionWorkflow Software

ActionWorkflow software provides the tools and network services required to implement the design achieved through ActionWorkflow analysis, using existing applications and network resources. The ActionWorkflow System helps designers and developers create a workflow system to

- Automate repetitive tasks
- Remind people of the work they have to do next
- List the steps required to perform a task
- Provide access to the tools and information required to complete a task
- Allow customers, performers, and managers to monitor the status of specific transactions
- Empower individuals to be accountable and to receive credit for customer satisfaction
- Support the human interactions required to handle exceptions and ongoing change

These are the components of the ActionWorkflow System (Fig. 33.7):

- ActionWorkflow Analyst
- ActionWorkflow Application Builder
- ActionWorkflow Manager
- Action Workflow Client Library

Table 33.2 summarizes the role of each of these software products in the development of a workflow system.

The ActionWorkflow Analyst makes it easy to create business process maps to analyze and document the flow of work in an organization. The maps produced using the ActionWorkflow Analyst provide documentation and serve as input to the ActionWorkflow Application Builder.

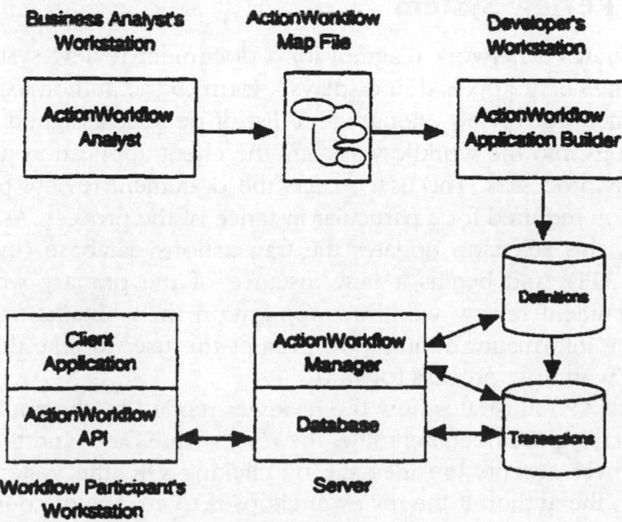

FIGURE 33.7 Action Workflow software components.

TABLE 33.2 Role of Each Software Product

Action Workflow Software Component	Development Phase
Action Workflow Analyst	Creating an Action Workflow map
Action Workflow Application Builder	Building ActionWorkflow definitions an generating the workflow-enable application
Action Workflow Client Library	Creating a client workflow application
Action Workflow Manager	Running and administering the workflow system

The Application Builder software lets the developer generate an application, including workflow definitions and application specific data, using the business process map. The Builder provides the same map drawing tools as the Analyst, so the developer can modify maps created with the Analyst. The Application Builder also provides tools for identifying the specific participants in a business process and for describing the forms and fields used to enter and display information to these participants. The Builder further defines the relationships between workflows, the interface with external systems, and the main functions of data that are specific to the business process.

The ActionWorkflow Manager is the workflow engine that allows the integration of existing hardware and software into a workflow system. Because each workflow is represented as a single transaction, the ActionWorkflow Manager can ensure process integrity, identifying any transactions that terminate abnormally. In conjunction with a third-party database server, the Manager maintains two principal databases:

- Definitions: The definitions database stores the descriptions of business processes created using the Application Builder.
- Transactions: The transactions database provides the information required for managing each repetition of a business process. This database also supplies application data used in the business process.

The ActionWorkflow Client Library provides an open interface that lets application developers create client workflow applications that can interact with the workflow system.

A Document Review System

Figure 33.8, illustrates a network diagram for a document review system. In the primary workflow the client application displays a form to the author, asking for the file name of a document for review, along with a list of persons assigned as reviewers.

When a user logs into the workflow system, the client application presents a list of available business processes. The user selects the document review process and enters the information required for a particular instance of the process. As shown by Step 1 in Figure 33.8, this selection updates the transactions database (by means of the ActionWorkflow API), and begins a new instance of the primary workflow for the process. The document review workflow application then obtains from the transactions database the information required to prompt the user to take the action necessary to carry the workflow process forward.

Step 2 in Figure 33.8 illustrates how the reviewer named by the author is prompted with a form displaying the document title, the review due date, and the options available. If the reviewer accepts the message by clicking OK, the system sends an acknowledgment to the author. If the reviewer chooses to change the due date or refuse the request, other workflows are initiated that allow negotiation of the date or selection of an alternate reviewer.

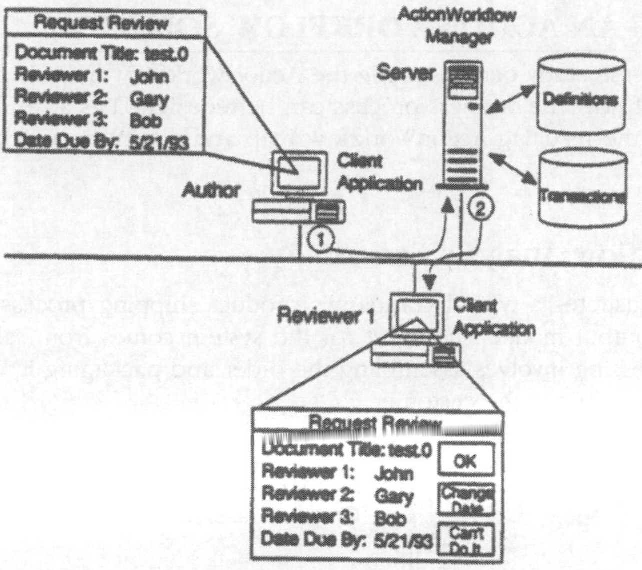

FIGURE 33.8 A Document review system.

Another form, shown in Figure 33.9, provides the author with a view of all documents currently under review, and shows the status of each document.

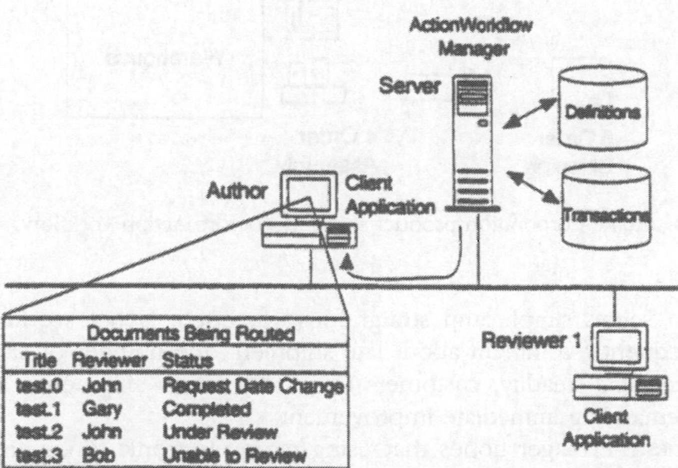

FIGURE 33.9 Monitoring workflow status.

Making Ongoing Improvements

Because the application is based on the ActionWorkflow structure, making incremental refinements is easy upon discovering breakdowns or areas for improvement. For instance, the application developer can add a workflow to the Performance Phase to ensure that reviewers notify the author when the review is complete. A workflow allowing the author to thank the reviewers for their effort can be added to the Acceptance Phase.

33.3 DESIGNING AN ACTIONWORKFLOW SOLUTION

Following is a case study demonstrating the ActionWorkflow analysis procedure, giving examples of how the business process can be redesigned as a result of the analysis, and shows the resulting ActionWorkflow map and workflow system for the business process.

ActionWorkflow Analysis Case Study

Figure 33.10 illustrates a typical company's product shipping process, based on an input-process–output model. The input for the system comes from sales representatives. The processing involves assembling the order and packaging it, and the output is sending the product to the customer.

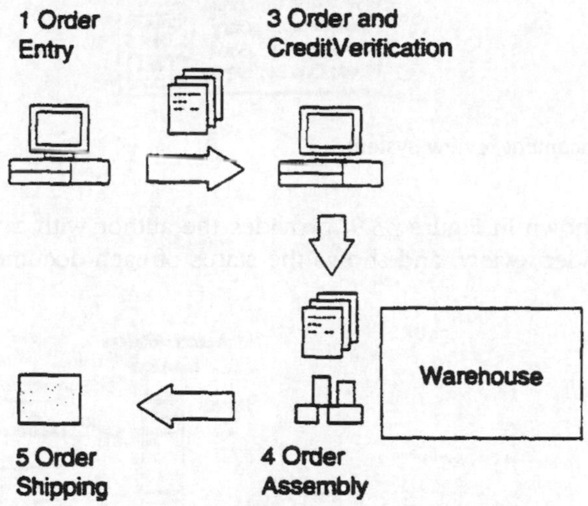

FIGURE 33.10 Acme Corporation product shipping (before ActionWorkflow).

The system seems simple and straightforward, but it is not working very well. Customers frequently complain about late shipment, the number of canceled orders has been increasing steadily, customers frequently return shipments, and top management is demanding immediate improvement.

Acme's general manager hopes that using more electronic forms and procedures will speed up the process and improve customer satisfaction. However, the analyst who has responsibility for implementing the workflow system recognizes that merely "automating the mess" will not achieve the desired results.

The problem with the current design is that no one in particular is responsible for the customer's order. Each worker has a part to play, but there is no coordination between these activities to ensure that the customer is satisfied with the service provided.

The ActionWorkflow model can help coordinate this activity because the ActionWorkflow model supports human beings' taking responsibility for the goal of the process. This model helps people interact effectively rather than asking them to fit into an inflexible information processing system.

Creating a Preliminary Map

Based on discussions with Acme's general manager and a review of the company's operations procedures, the analyst identified the workflows and the roles within each workflow (Fig. 33.11).

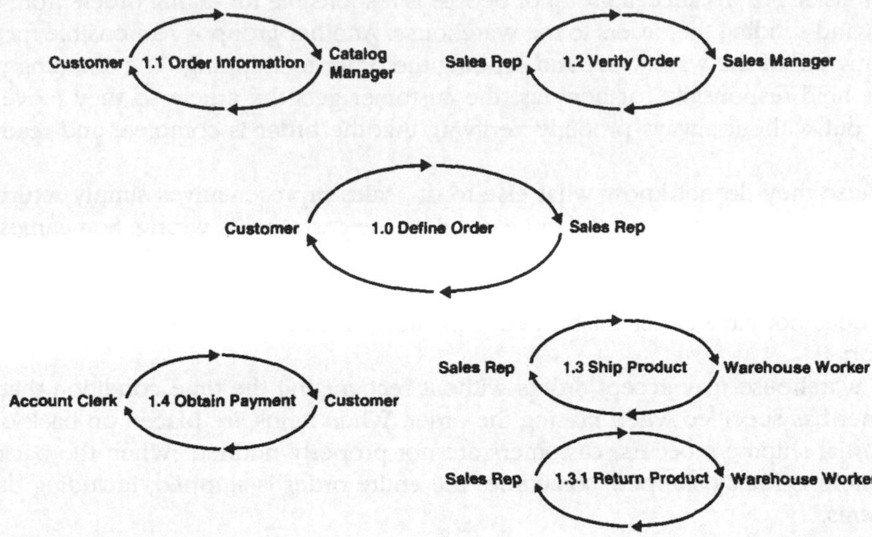

FIGURE 33.11 Preliminary workflow map.

The immediate effect of applying the ActionWorkflow methodology is to assign overall responsibility for customer satisfaction. The person in this role then becomes the customer for the performers in the secondary workflows. Recognizing that this person is the primary advocate for the customer clarifies the importance of this person's role in the process. Clearly the redesign of the process leads to greater responsibility (and status) for the sales representative.

The aim of the design process at this point is to clarify roles and conditions of satisfaction for each workflow. The design is based on the roles of customer, performer, and observer, rather than the identities of individual participants. The system provides the flexibility to assign identities to roles when the process is underway.

Interviewing Workflow Participants

Because there is a limited number of roles in this process, the analyst chooses to schedule interviews with an individual performing each role. These discussions reveal that although people are aware of problems that occur, no one feels empowered to take the initiative to meet the customer's needs.

The analyst discovers these breakdowns in the system:

- Customer requests for special handling are not being satisfied.
- Customers are not notified properly of a partial shipment because of items' being back-ordered.
- When back-ordered items are shipped, there is no proper record indicating which items were included in a partial shipment. The result is that sometimes duplicate items are sent and sometimes items are left out of the order.

- Customers cannot determine the status of the order when they call in.
- The price on the invoice does not always match the price quoted by the sales representative.

In the current system, responsibility for filling customer orders is divided into smaller tasks. For instance, a group of people is responsible for taking orders from customers and sending the orders to the warehouse. Another group is responsible for taking items out of the warehouse and sending the order to shipping. The shipping people are held responsible for how fast the customer gets the article so they move the orders out without always properly verifying that the order is complete and ready to send.

Because they do not know what else to do, sales representatives simply return an order form to the customer if the price given by the customer is wrong. Sometimes the representative accepts orders with no way to indicate the customer's stated time requirements. Sometimes a sales representative accepts an order on credit; but if the customer does not have credit established, accounting returns the order with a credit application.

The warehouse may accept orders without recognizing the time condition that the customer has specified when placing the order. When items are placed on back order and partial shipment occurs, customers are not properly notified. When the back-ordered item is finally shipped, sometimes the entire order is shipped, including duplicate items.

Mapping the Workflows

After completing the interviews, the analyst can understand some of the reasons for the breakdowns and problems that are occurring. Merely "force feeding" workers more items electronically probably will not provide the results that management wants. The analyst meets with the general manager and explains the problems that are occurring and how redesigning the system to provide participants with greater responsibility and autonomy will provide the basis for real improvements.

Figure 33.12 illustrates the map that emerges in the analyst's discussions with the manager. This map shows that a workflow system can reduce the number of roles involved in the process by providing an intelligent database system that helps individual participants perform a wider range of duties.

Creating Link Between Workflows

Once the roles and conditions of satisfaction for each workflow are established, the links between each workflow can be added. In this simple map, only one conditional link is required. The general manager is willing to grant greater authority to customer representatives, provided the manager is notified when customer requests fall outside certain parameters.

Evaluating the Map

The analyst in consultation with the manager and workflow participants now walks through the map to ensure that the breakdowns or bottlenecks that were discovered have been eliminated or dealt with to the satisfaction of the client.

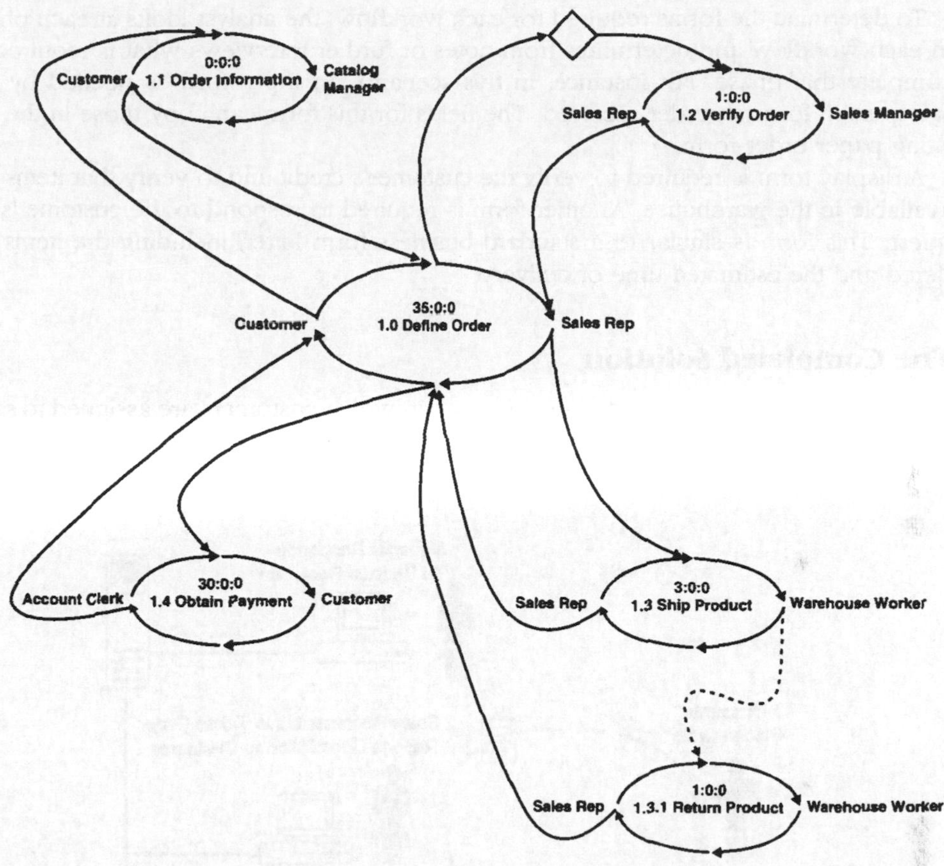

FIGURE 33.12 The completed workflow map.

In the new system, negotiation, declaring completion, and ensuring customer satisfaction are considered equally important. As a result, a sense of responsibility is created throughout the entire system. When accepting an order, the sales representatives can negotiate with the customer if the price or delivery time requested by the customer cannot be met. When exceptional situations arise, the representative knows where to go for help.

Integrating the Information Process

The last step in the process provides the application developer with the information required to actually build the workflow definitions and client workflow applications. This step can be completed by either the business analyst or the application developer or the two working together.

For each information process that supports a workflow, field names need to be assigned for two kinds of information:

- Information that the application must provide to the participant—display fields
- Information that the participant must provide for the workflow to proceed—data entry fields

To determine the forms required for each workflow, the analyst looks at each phase in each workflow and determines from notes or further interviews what is required to complete that phase. For instance, in this scenario, an input form is needed by the salesperson for each order received. The fields for this form can copy those in the existing paper order form.

A display form is required to verify the customer's credit and to verify that items are available in the warehouse. Another form is required to respond to the customer's request. This form is similar to a standard business form letter, including the items ordered and the estimated time of delivery.

The Completed Solution

Figure 33.13 illustrates the redesigned system, in which customers are assigned to sales representatives who have the overall responsibility for delivery of the product.

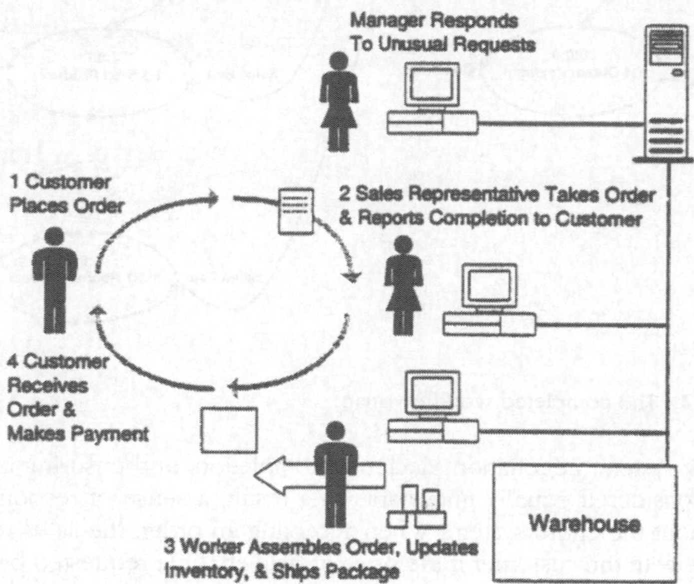

FIGURE 33.13 Acme's product delivery system.

Now, a representative who receives an order looks at inventory to verify that the customer's terms of delivery are acceptable. The representative's responsibility ends only after calling the customer to ensure that the item has been received and the customer is satisfied.

The customer representative is the customer for the workers who are responsible for assembling, packing, and shipping orders, and for maintaining warehouse inventory. The conditions of satisfaction that they must meet are "assemble and ship orders within 24 hours of their being received." For workers to achieve this goal, they must closely coordinate with their performers—workers who manufacture or order the merchandise in the warehouse.

33.4 IMPLEMENTING THE SYSTEM

There are three stages in implementing an ActionWorkflow system:

1. Creating the ActionWorkflow map
2. Building workflow definitions
3. Creating a client workflow application

The previous discussion described how Step 1 was completed. The following sections summarize how Step 2 and Step 3 might be performed in this case study.

Building the Workflow Definitions

The application developer opens the business process map using the Application Builder. The developer uses the annotations made on the map to determine where supporting information systems or forms are required. The developer then uses any existing forms along with interviews with participants to complete the definition of each entry or display field.

Using the Application Builder, the developer enters the application data fields to bind to each workflow and specifies the conditions for the process to proceed. The developer uses the Application Builder to generate the workflow application; this procedure writes the business process information to the definitions database. The ActionWorkflow Manager uses this information to make routing or other decisions (such as which acts and workflows trigger other workflows) and to determine available acts for each role in each workflow.

Creating the Client Workflow Application

The final phase is for the developer to create the client workflow applications that let the participant view the entry and display fields, as specified in the definitions database. Client applications typically have three kinds of interactions:

1. Starting: The application displays a list of processes from which the participant can choose.
2. Reviewing: The application displays a list of those workflows within the process for which the participant has a role.
3. Acting: The application displays forms for the participant to complete or review in the selected workflow.

After an instance of the workflow is started, the Action Workflow Manager creates a new transaction record from the definition record for the workflow in the definitions database.

The details of how the client workflow application is created depend on how the workflow system is implemented in hardware and software. The rest of this section describes two different implementations, using a Lotus Notes Workflow Application or a C-Language Workflow Application.

Lotus Notes Workflow Application

Figure 33.14 shows the implementation of the workflow system described in the scenario using a Lotus Notes Server and a Notes client.

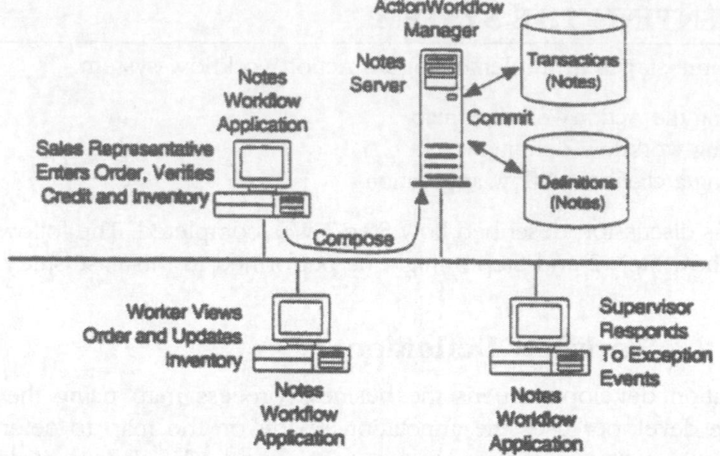

FIGURE 33.14 Lotus Notes application.

As described earlier, the ActionWorkflow Manager is the workflow engine that makes routing and other decisions based on information in the definitions database. These definitions, created by the Application Builder, are stored in the Lotus Notes database called "Definitions." Here is how the three parts of the client application are supplied:

1. Starting—from the Lotus Notes Compose option: This option provides the workflow participant with a list of forms to start different business processes.
2. Reviewing—from Lotus Notes View option: This option provides the participant with a list of workflows, shown with their different states of completion.
3. Acting—from Lotus Notes forms: These are the forms that are appropriate for the role played by the participant in the workflow selected.

The Lotus Notes version of the Action Workflow Builder generates template starting and acting forms in the transactions database. These template forms can be customized by a Lotus Notes programmer for greater ease of use.

The participant's completing a starting or acting form writes a new document to the transactions database. The Action Workflow Manager, which monitors all changes to the transactions database, determines from the user action and the definitions database the appropriate action to take next. This action might include generating an e-mail message to another participant or writing information from the definitions database to a document in the transactions database.

The Manager monitors open workflows and displays messages on the 0S/2 console as different events occur. If a workflow terminates abnormally, an error message is displayed. These messages are sent to the C language standard output, which can be redirected from the monitor to a log file for later review and analysis.

Workflow Applications with SQL Server

As shown in Figure 33.15, the same system can be implemented using client workflow applications written in C or Visual Basic, using the version of the ActionWorkflow System for SQL databases. As with the Notes version, these applications provide users with needed information and include data entry fields for entering information to

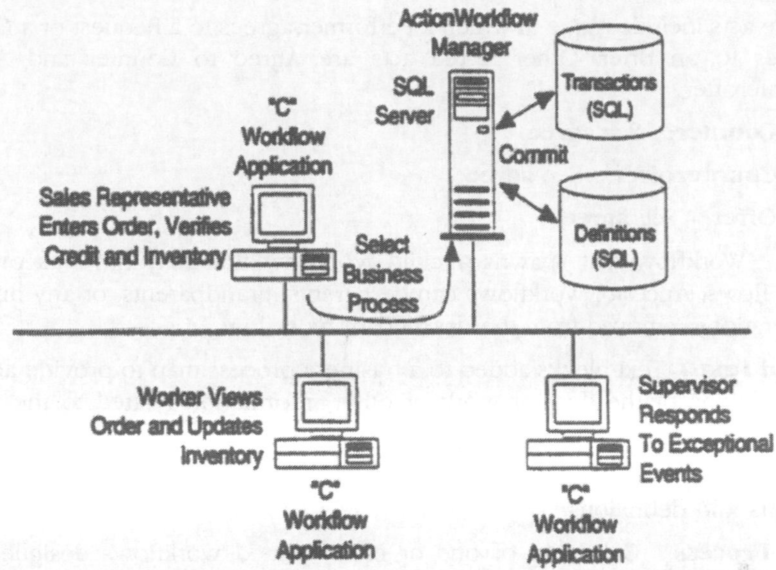

FIGURE 33.15 Workflow applications with SQL server.

make the workflow proceed. With the SOL version, the ActionWorkflow Manager interacts with the SQL server to make routing and other decisions based on information in the definitions database.

As with the Notes version, the definitions database is generated from the business process map created using the ActionWorkflow Analyst and the Application Builder. In the SQL version, however, the definitions are generated as a standard SQL database. For this version, the Action Workflow System provides a set of open APIs and sample applications that allows easy development of client workflow applications. Client applications can also use these APIs to query the ActionWorkflow Manager about open transactions.

DEFINING TERMS

Acceptance: Workflow phase in which the customer declares approval or disapproval of the work done; the lower-left quadrant of a workflow loop. Also the state of the workflow at the beginning of the Acceptance phase.

Act: Interaction between workflow participants that changes the state of the workflow. When participants in a workflow perform a set of acts, work is coordinated among the participants, and the workflow moves toward completion. Request, Agree, Report Completion, and Declare Satisfaction are examples of acts.

ActionWorkflow: Registered trademark of Action Technologies, Inc.; denotes the methodology and technology to analyze, redesign, and automate business processes.

Activate: Workflow act that starts a workflow. Activate is a system act rather than an act by a workflow participant. Activate places the workflow in the Preparation state.

Agree: Act that finalizes the negotiation between customer and performer regarding the conditions of satisfaction and moves the workflow to the Performance state.

Agree acts include Agree in which a Performer agrees to a Request or a Customer agrees to an offer. Other Agree acts are Agree to Counter and Agree to Counteroffer.

Agree to Counter: See **agree**.

Agree to Counteroffer: See **agree**.

Agree to Offer: See **agree**.

Ancestor: Workflow that spawns a child workflow that may have its own child workflows. Ancestor workflows can be parents, grandparents, or any number of generations removed from the descendant workflow.

Annotated Text: Text blocks added to a business process map to provide additional descriptions of the flow of work or other information related to the business process.

Author: Person responsible for creating the business process map and adding map objects and definitions.

Business Process: Collection of one or more related workflows designed to accomplish a specific purpose. Examples include offering a service, requesting a loan, filing an expense report, and making an order for a product. A map of a business process contains the collection of workflows and the links between them.

Cancel: Exception act that ends a workflow, moving it into the Canceled state. Cancel is an act that the customer can take at any moment of the workflow.

Child: Workflow that occurs during a phase or phases of a parent workflow. The child is triggered by a link from the parent workflow.

Collapse: Act of hiding child workflows of a parent workflow. A collapsed workflow is a parent workflow with hidden child workflows. Collapsed workflows are distinguished by thicker lines in the workflow loop.

Completion: Combination of the Negotiation and Minimum Performance phases. Values or cycle times set for Completion affect the Minimum Performance quadrant of a workflow because Minimum Performance is the difference between Completion and Negotiation.

Computed Phase: Phase that does not have a user-specified value or time but derives its value or time from the calculation of concurrent subprocesses.

Conditional Link: Specification of different courses of action whose occurrence depends on events that occur or conditions that exist (values of certain variables). The conditional link diamond marks the point in the path at which the mutually exclusive possibilities are defined.

Conditions of Satisfaction: Work that the customer and performer agree will be performed in an Action Workflow. Conditions of satisfaction always include, either implicitly or explicitly, a time by which the customer expects the work to be done.

Consistency: Adherence to the map rules that concern the logical and orderly connection of links to workflows and conditional links.

Cost: Monetary or numerical value assigned to a phase of a workflow. From this value, the total cost of the workflow, path, or business process can be computed.

Counter: Act taken by the customer while negotiating with the performer the conditions of satisfaction of the workflow. In a request workflow this act moves the workflow to the Negotiation state; in an offer workflow, this act moves the workflow to the Negotiation (countered) state.

Counteroffer: Act taken by the performer while negotiating with the customer the conditions of satisfaction of the workflow. In a request workflow this act moves the workflow to the Negotiation (countered) state; in an offer workflow, this act moves the workflow to the Negotiation state.

Customer: Person for whom the work is being done in a workflow; one of the two principal roles in an Action Workflow. In a request workflow, the customer makes the request; in an offer workflow, the performer makes the offer to the customer. In both cases, the workflow is not complete until the customer declares satisfaction.

Cycle Time: Time value assigned to a phase of a workflow. The cycle time for a workflow includes the time for each of the phases of the workflow. Cycle time is a statistic for measuring efficiency and effectiveness in achieving customer satisfaction.

Declare Satisfaction: Act by which the Customer notifies the Performer that the work performed is accepted and that the conditions of satisfaction have been met. This act ends the workflow and moves it into the Satisfied state.

Decline: Exception act taken by the performer in a request workflow. This act ends the workflow and moves it to the Declined state.

Decline Counter: Act taken by the performer in an offer workflow. This act moves the workflow to the Negotiation state.

Decline Counteroffer: Act taken by the customer in a request workflow. This act moves the workflow to the Negotiation state.

Decline Offer: Exception act taken by the customer in an offer workflow. This act ends the workflow and moves it to the Canceled state.

Decline to Accept: Act taken by the customer, in the acceptance state of a workflow, after the performer has reported completion. It indicates that the customer is not satisfied with the work done. This act moves the workflow back to the Performance state.

Descendant: A workflow that is the child workflow of a parent workflow that may have its own parent. Descendant workflows can be child, grandchild, or any number of generations removed from the ancestor workflow.

Error: Type of rule violation that occurs during consistency checking. If a map has an error it cannot pass consistency checking and cycle times and values cannot be computed. Errors also prevent application generation if the map is imported into the Application Builder.

Exception Act: Act that does not lead to the successful completion of a workflow. When a customer and performer cannot agree on the conditions of satisfaction, when the customer no longer needs the work performed, or when the performer wishes to cancel an agreement to perform the work, these circumstances can necessitate an exception act. Decline, Revoke, and Cancel are exception acts.

Informational: Type of rule violation that occurs during consistency checking. An informational warning is a recommendation that you should investigate a potential problem. Informational warnings do not cause maps to fail consistency checking and do not prevent computation of cycle times and values. Informational warnings do not indicate conditions that prevent application generation if the map is imported into the Application Builder.

Initiate: Act by which a customer or performer starts a workflow. Initiate is equivalent to a customer request or performer offer. An Initiate act moves the workflow into the Negotiation state.

Link: The line drawn connecting two workflows or connecting a workflow and a conditional link. Outgoing links from a workflow can be either act-based or state-based. In an act-based link, the taking of an act in the originating workflow causes the taking of an act in the linked workflow (or the workflow on the other side of a conditional link). In a state-based link, the arrival at a new state in the originating workflow causes the triggering of an act in the linked workflow. If either a state-based or act-based link is a return link to a parent workflow, the link may reach to the same state from which that path originated.

Map: Drawing that represents a business process and includes workflows, links, conditional links, and annotated text boxes.

Material Process: Management and movement of components and substances, for example, manufacturing products in an assembly line or the movement of paper through an office.

Minimum Performance: A period used for defining cycle time and value calculations. Minimum Performance is the difference between Completion and Negotiation.

Negotiation: Workflow phase during which the customer and performer negotiate the conditions of satisfaction and reach agreement about the work to be performed; upper-right quadrant of the workflow loop. Also the state of the workflow at the beginning of the Negotiation phase.

Negotiation (Countered): Workflow state during the Negotiation phase of a workflow in which the customer and performer are negotiating the conditions of satisfaction. Negotiation (Countered) occurs when the performer has counteroffered in a Request workflow or the customer has countered in an Offer workflow.

Object: Any item drawn in the map using the tool palette. Objects include workflows, links, conditional links, and annotated text boxes.

Observer: Workflow participant who is not required to take actions to move the process to completion. Observers are informed of the acts in a workflow and have access to information and data associated with the workflow.

Offer: Act by which the performer initiates a workflow by specifying the conditions of satisfaction to be completed by a specified time; workflow that begins with a performer's initiation.

Page Break: Dotted line that appears on a map and indicates the division between pages as they will print when the map is sent to the printer. The position of page breaks is determined by margin settings in the Page Setup dialog box and orientation and paper size settings in the Print Setup dialog box.

Parallel Workflows: Secondary workflows that are initiated simultaneously from the same parent workflow. These workflows are independent of each other. Parallel paths are paths that are initiated simultaneously from the same parent workflow.

Parent Workflow: Workflow that spawns child workflows that occur during a phase or phases of the parent.

Path: Flow of work through one or more child workflows of a parent workflow. A path begins with an outgoing link from the parent workflow, flows through all the connected workflows, and then usually ends with an incoming return link to the parent workflow. A path can be as simple as an outgoing link and a single child workflow. A path can also be very complex, including any number of serial workflows, each of which can have any number of child, or descendant workflows.

Performance: Phase in the workflow during which the performer completes the work that meets the customer's conditions of satisfaction; lower-right quadrant of the workflow loop. Also the state of the workflow at the beginning of the Performance phase.

Performer: Person responsible for completing the work and notifying the customer when the work is done; one of the principal roles in a workflow.

Phase: Characterization of the status of a workflow based on which acts have happened and which acts are permitted. There are four phases in a workflow: preparation, negotiation, minimum performance, and acceptance. *Phase also designates a quadrant of the workflow loop.*

Preparation: Workflow phase during which the customer drafts a request for some work to be done or the performer drafts an offer to do some work; the upper-left quadrant of a workflow loop. Also the state of the workflow at the beginning of the Preparation phase.

Primary Workflow: First workflow begun when a business process starts. The conditions of satisfaction of this workflow define the purpose of the business process. The primary workflow is the parent (or ancestor) of all other workflows in the business process (except when isolated workflows are intentionally used).

Reach: Connection of a link to a state in a target workflow. This term is distinct from "trigger," which is a connection to an act in a target workflow.

Report Completion: Act by which the performer notifies the customer that the conditions of satisfaction of the workflow have been met. This act moves the workflow into the Acceptance state.

Request: Act by which the customer initiates a workflow by specifying the conditions of satisfaction to be completed by a specified time; a workflow that begins with a customer's initiation.

Required Field: Input area in a dialog box. Required fields are distinguished by blue text in the field name. Failure to complete a required field will cause a warning during consistency checking.

Revoke: Exception act by which performer withdraws agreement to meet the conditions of satisfaction of the workflow. This act ends the workflow and moves it into the Revoked state.

Role: Part that a person plays in an ActionWorkflow; includes customer, performer, and observer; also the part that a person plays in a business process: Development Manager, East Coast Sales Manager, Secretary, Graphic Designer, etc.

Secondary Workflow: Child workflow.

Serial Workflow: Child workflows that are linked together so that they occur in sequential order. Each workflow in the series does not begin until the previous workflow is over. Serial workflows after the first in the series are triggered by a link from the previous workflow so they do not occur concurrently and do not have a parent–child relationship. Workflows in a series are siblings. Each workflow in the series is a child of the parent that spawned the first.

Sibling Workflow: Workflow that is a member of the same series in a set of serial workflows. The parent of the first workflow in a series is the parent of all siblings.

Specified: User-defined; not computed.

State: Condition of the workflow that reflects the progress of the work being done. A workflow arrives at a state as the result of an act.

Tool: Any of a set of drawing implements for creating map objects. The tool palette includes the pointer took, the workflow tool, the normal and exception flow link tools, and the conditional link tool.

Trigger: Connection of a link to an act in a target workflow or the initiation of that act. This term is distinct from *reach, which is a connection to a state in a target workflow.*

Value: Numeric amount assigned to a phase of a workflow. From this amount, the total value of the workflow, path, or business process can be computed.

Warning: Rule violation that occurs during consistency checking; a strong recommendation that you should investigate a potential problem. Warnings do not cause maps to fail consistency checking and do not prevent computation of cycle times and values. Warnings do not prevent application generation if the map is imported into the Application Builder.

Workflow: Unit of work in a business process, in which a customer and performer negotiate the conditions of satisfaction. In a successful workflow, the performer then carries out the work, reports that it is done, and the customer acknowledges satisfaction. An exception workflow occurs when the workflow does not complete successfully (for example, because it is canceled, or because the customer was not satisfied, or because the performer declined the work). Each workflow has four phases: preparation, negotiation, performance, and acceptance.

34

Lotus Notes

Debora Cole
Lotus Development Corporation

34.1 INTRODUCTION

Lotus Notes® is the industry-standard client/server platform for developing and deploying groupware applications. Groupware is a category of software that enables teams of people or workgroups to integrate their knowledge, work processes, and applications to achieve improved business effectiveness. It allows people across multiple computing platforms to access, track, share, and organize information in ways never before possible, even if they are only occasionally connected to a network. Groupware applications differ from traditional transaction processing applications in a number of ways:

- Unstructured information management. Much of corporate knowledge typically resides in unstructured format in documents that include text, image, voice, and video.
- Collaboration among users. Business information flows from one set of users to another in both defined and ad hoc patterns.
- Crossing traditional boundaries. Because qualitative information flows across departmental lines in many business processes, multilevel data security is essential to protect sensitive information without denying access to information to the appropriate users
- Asynchronous updates. Business process applications are typified by long transactions, in which information is "checked out" for use over a relatively long period of time. Meanwhile, other users may also be checking objects out and in.
- Remote users. Many business process applications include users at remote sites, on the road, or who work at different organizations (such as a customer or a business partner). These users are only occasionally connected to the system.

0-8493-2530-7/96/$0.00+$.50

A complete groupware environment includes a distributed document database, integral messaging services, multilevel security, and full application development facilities. It must support multiple network and operating systems, provide access to relational data, and support remote and mobile users.

This section describes the Notes™ application development environment, and shows how Notes is uniquely qualified to build groupware applications. It begins by distinguishing Notes from traditional relational databases. It then covers the tools for basic and advanced application development, explains how the environment is extensible, and discusses how the development process is managed.

34.2 COMPARING NOTES TO THE RELATIONAL DBMS MODEL

Developers familiar with relational database management systems (RDBMS) will find Notes fundamentally different from standard RDBMSs, but not altogether foreign. Much of the functionality found in Notes has analogous counterparts in an RDBMS. Each system's architecture reflects the functionality required by its target class of applications and users.

In many cases, Notes simply implements a particular database feature in a different manner than its RDBMS cousins to accomplish communications-centric workgroup tasks as opposed to data-centric, transaction processing tasks.

Object Store

At the heart of Notes is its Object Store. Notes can manage a variety of data types: numerical data, text, rich text, compound document objects, as well as image, voice, and video. While some relational databases have been extended to store this type of information, Notes is able to manage it more intelligently. Each object in a Notes database is self-describing. Each can be different from every other item. In a relational database, each item's structure, its record, is identical to every other item in the same table. Some RDBMSs support binary large objects (BLOBs), but these objects offer flexibility only within the scope of a single field. The flexibility that Notes provides is necessary if the system is to accommodate unstructured information.

Process Integrity

Notes' focus is on the work processes that surround unstructured information. By emphasizing process integrity, Notes is well suited to workflow management. Notes integrates its database with agents—programs that monitor the state of a business process as it changes in a shared database, treating each step in the process as a transaction. Notes manages the full lifecycle of the objects stored within it, unlike a relational database, which is more concerned with the current state of data. Tracking the lifecycle of objects through an entire business process is the focus of workgroup applications. Notes is able to support processes that require tracking of how information has moved or changed, including such details as who made the change, and when and why. In Notes, this functionality is built into every application.

Conversely, relational databases focus primarily on data integrity. They manage the structure and relationships of the tabular data they contain. SQL databases have some analogous process-oriented functionality in stored procedures and triggers. Instead of focusing on what happens to work over time, they generally are sophisticated scripts for maintaining tabular data relationships. In addition, RDBMSs do not track an object's lifecycle, but rather capture "snapshots" of business transactions (e.g., inventory has a current level, orders are received).

External Focus

Notes has a global directory that tracks all users and their roles, as well as the organizational hierarchy. This external focus on users is appropriate to Notes' role as a workgroup application environment in which the flow of documents and processes is necessarily concerned with who belongs to a workgroup, who has access to particular documents or parts of documents, and in what order they receive notifications. This is opposed to the RDBMS view of the world, which depends on a system catalog that tracks all the data elements in the system. Such an internal focus is more appropriate for transaction processing applications than for process-oriented applications.

Messaging-Based Applications

All Notes applications are inherently message-enabled, so that an application can notify users of events or send updates to other applications. Traditional RDBMSs can approximate this by the use of remote procedure calls or distributed updates, both of which require fully connected users.

Replication

Although some RDBMS vendors have begun to move toward replication-based solutions to augment their more traditional two-phase commit approach to computing, Notes' implementation of replication is more complete than simple database copying. First, Notes replication is bidirectional, so that changes made on both client and server databases can be propagated throughout the system. Notes can also perform partial replication according to specified parameters, as well as flag replication conflicts and inform the user of changes or additions to replicated copies. Like RDBMSs, Notes replication can take place from server to server, but it is also intended for use by individual users who are disconnected from their desktop.

In addition, Notes replication services can be used for application distribution. As soon as a user replicates with a server, the most recent version of a database/application is downloaded, ensuring that all users, regardless of their location, are using the same applications.

Distributed Applications

Moreover, Notes applications can refer to other Notes documents (or to RDBMS tables) found in other databases located on other servers. While some RDBMSs have provided support for distributed database management, they still require complex and restrictive locking systems, and are expensive to develop and maintain. Notes has an optimistic locking technique that supports more flexible data access.

Multiple Field Value Support

Notes fields support multiple values, as opposed to RDBMSs, which require a join to marry multiple values for a single field. This feature of Notes is useful when categorizing a document in multiple folders. In scenarios like this, the requirements of an RDBMS for normalized, or flattened, data would be unnecessarily restrictive.

Flexible Browsing

Notes provides the user with a flexible browsing capability. For example, if a parent document has multiple child documents, the user can elect to view just the parent, or the parent and all next-generation documents, or all the generations of documents related to the original parent. This would require three different queries in an RDBMS, the largest of which would significantly tax performance. More important, flexible browsing in Notes does not require that users make precise queries. Notes does not assume that users know exactly what they are searching for. They can use the outline-oriented, expand-collapse metaphor to easily browse through large volumes of information. SQL databases require a specific query, and they return a flat, tabular result set.

Security

SQL security, like its data, is based on uniform sets of data: tables or query results. Changes to access generally have to go through a database administrator who has access to the system catalog or server control center. Notes security, like its information, is far more flexible. Users can adjust access levels down to the field level without administrator intervention.

Application Framework

Notes represents a complete application framework, including database, development language, application templates, and full access to system-level services. This environment provides the developer with a platform for rapid application development. Although most traditional databases are accompanied by development tools, the developer often needs multiple tools that are not always integrated into an overall framework.

In Notes, the developer's environment is identical to the user's environment. This facilitates real-time collaboration between developers and users, and encourages users themselves to assume some development responsibility.

Taxonomy

Based upon some of the categories above, it should not be surprising that Notes uses a modified taxonomy from RDBMSs to refer to particular database features. The following table lists analogous terminology between Notes and traditional databases.

Lotus Notes	Traditional Database
View	Report
Form	Form, query
Rich text field	BLOB
Self-describing objects	Table, schema
Formula	Calculated field
Background macro	Stored procedure
Replication	Transaction, locking
Name and address book (global directory)	System catalog
Document-level access control	SQL grant (access to table, query)
Design templates	Repository

34.3 BASIC APPLICATION DESIGN AND DEVELOPMENT

A basic Notes application consists of three fundamental design elements: forms, fields, and views. With these three elements and a modicum of experience in using Notes in production, an end user can learn to develop a basic Notes application. More sophisticated applications require some of Notes more advanced features.

- Forms. The fundamental design element of Notes applications is the form. This is the medium in which information is entered and displayed. Most often the form is a container for a compound document created with users' desktop applications. Once the user has completed data entry, the form is then stored as a document in Notes.
- Fields. A form is made up of fields, such as author, date, summary, and document body. The developer can assign fields a particular data type: text, number, time, rich text. Some fields can be automatically populated, such as time and date using the system clock, or author name using the user ID. Other fields may require the user to choose from a list of keywords. Developers edit the attributes and properties of fields by responding to a series of dialog boxes.
- Views. Fields are critical to managing Notes databases because users view documents as sorted by particular fields. Views are designed by the developer to display lists of documents from any variety of perspectives. When Notes presents a view to the user, it lists field names as columns. This way, the user can choose to view all the documents by author, for example. Notes would then sort all the documents by the author field name and present them to the user, along with other field information (date, topic, etc.) listed in columns to the right. The developer designs how each of the columns will appear (column width, order, etc.) and how they will behave: by categorizing certain columns, developers can create the expand-collapse outline user interface that makes browsing easy.

Notes is shipped with a set of design templates. Notes is frequently used to develop applications such as status reports, customer tracking, document libraries, discussion databases, and correspondence. The included design templates are complete implementations of these applications that nondevelopers can use without modification as production applications, and which developers can use as a starting point to build more customized applications.

34.4 THE NOTES DEVELOPMENT PROCESS

Notes is a rapid application development environment. Notes offers developers an application framework that provides a skeleton of both an application's user interface as well as its functionality. All the tools needed to build the rest of a Notes application can be found in the Notes development environment. There is no need to compile and debug a Notes application. Once the developer specifies the forms, fields, and views, the Notes application is ready for production. This short development cycle allows Notes developers to work closely with users, providing for interactive development sessions. Users can see their specifications implemented in real time, offering suggestions and discovering new requirements.

Once developers have created a basic form with the required fields, they can build some sample documents, with direct user input, to see if the application meets the expectations of the users. Developers can modify their applications on the fly, changing the layout of forms, playing with fonts and column alignment, adding graphics, inserting tables, and importing images.

Developers can also add a set of useful views that will appear as menu items. Views serve as predefined queries that will satisfy most of the users' information viewing needs. The most common views are chronological by creation or modification date, or alphabetical by author or other key field such as department or product type. Users may request customized views, such as only documents created in the last 30 days, or a list of projects that have not been completed.

In this way, developers and many end users can create a basic Notes application. In most cases there is no need for scripted macros. More important, the developer does not have to be concerned with system level constructs. Once Notes has been installed on the network, it automatically handles all security, storage, replication, and connectivity issues.

34.5 NOTES SYSTEM-LEVEL SERVICES

Notes is unified platform that supports both application development and production. Its integral system-level services provide developers with a full range of development facilities without requiring that they acquire the skills of a systems integrator to piece together components that were not necessarily built to fit together.

Security

Developers do not have to delve into arcane security algorithms to build secure systems. In fact, in several instances, security issues are left to the user's discretion. In any case, defining security is generally a matter of selecting from a dialog box. Notes provides four classes of security that developers (and users) apply at various levels of granularity throughout an application.

- Authentication securely identifies users using the industry-standard X.500 hierarchical naming syntax. Authentication in Notes is bidirectional—users authenticate the identity of users using industry-standard X.500 hierarchical naming syntax. Authentication is used whenever a user and a server, or two servers, are communicating with each other. This security is established by the Notes administrator, who defines the list of authorized server users by user id within the server NOTES.INI file.
- Access control provides the ability to grant or deny access to shared databases, documents, views, forms, and fields. The application designer assigns each user or group of users an access level. When creating an access control list, the developer assigns each user or group of users a security level: depositor (can create documents

but cannot edit or delete them), reader (can read documents but cannot create, edit or delete them), author (can create, edit and delete his or her own documents), editor (can modify all documents), designer (can modify the structure of the database), and manager (has control over the access control list).

- Digital signatures ensure that a given message is from who it says it is from and verify that it has not been modified in transit. This is a user-to-user form of authentication, and does not require any design on the part of the developer.
- Encryption ciphers information using the RSA Public Key Cryptosystem, so that even if a message or document were accessed by the wrong individuals, it could not be understood. Encryption is available at three levels within the system. Users can encrypt individual messages (or parts of messages); the Notes administrator can use encryption at the network level, preventing lurkers from tapping into traffic on a LAN or dial-in line; and developers can build encryption directly into databases so that only specified users can read them.

Replication

As a client/server platform, Notes is inherently distributed. Notes servers contain databases, and users can make copies—or replicas—of those databases on their client desktops for use when they are not on the network. In other cases, they directly access the server databases, which may be replica copies from other servers. Notes makes database access a straightforward process both for users connected directly to a server through a LAN as well as for remote and mobile users.

Replication is the process of keeping multiple copies of a database in synchronization with each other. Users can make a replica of a database before disconnecting their portable device from the network. This allows users to continue to work on relevant documents while away from easy access to the server. Once the user is able to connect again by dialing in to the server, all changes made on the user's documents are replicated back onto the server database, and all changes that have been made on the server database are replicated to the user's portable. In those cases in which changes have occurred to the same documents on both ends of the "transaction," the user is sent a replication conflict notification.

Although the developer does not have to be concerned with the replication process itself, replication has two profound implications for the developer.

First, the developer no longer has to build applications that are constrained by connectivity issues. Unlike most RDBMS-based client/server systems that depend on continuous user access, pessimistic database locking techniques and two-phase commit architecture, Notes applications allow the developer to conceive and implement an information- and knowledge-sharing client/server system with considerably more flexibility. Concurrent access by multiple users does not present the same complexity in a Notes application.

Second, software distribution becomes a nonissue. Users that are connected to a Notes server, whether directly through a LAN or via remote communication, have access to all applications located on that server. Thus, any new database applications that have been implemented on the server are immediately "distributed" to all users. Moreover, as the developer makes changes during maintenance, each user automatically receives the new version of the application when the next database connection is made.

Object Location and Exchange

As mentioned, Notes documents can store a variety of data types. They can also store application objects, such as documents created with an altogether different application. Thus, a user can create a document with any Windows-based word processor, save it, and then embed it within a Notes document. The embedded object will be managed just as any other data type.

When other users view the embedded object, they can launch it—along with the relevant application—simply by double-clicking on it (provided that the original application exists on the viewer's computer). The viewer can also choose to save the document to his or her own desktop in the appropriate directory.

In addition to embedded objects, users can create links between associated Notes documents. When authoring a document that refers to another Notes document contained on the same or another Notes database, the user opens the related document and uses a menu command to create the link.

These examples of object management are initiated by the user and are handled by Notes itself. The database and application designer does not need to define potential or specific object relationships.

34.6 ADVANCED APPLICATION DESIGN AND DEVELOPMENT TOOLS

Using just forms, fields, and views, developers can easily build a basic Notes database application that allows users to share information and knowledge as captured in documents. Notes can support more sophisticated processes as well, including those that employ multiple Notes databases as well as non-Notes databases, drive a particular workflow process, or integrate common desktop systems. These applications use Notes' more advanced tools, such as formulas, @functions, and macros.

Formulas

Notes formulas are similar to mathematical formulas, consisting of variables, constants, and operators, although the outcomes are generally fundamentally different. For example, a developer might write a formula that formats text by capitalizing the first character of a word, or one that combines two or more independent fields to create a single field that is displayed to the user.

Although the main focus of Notes formulas is on document-oriented expressions, Notes does support basic arithmetic operators (addition, subtraction, multiplication, and division), so that a developer can combine fields that have a number data type with a function in order to derive a value (e.g., ProductPrice * 0.075 to calculate a sales tax). Notes works with integers and scientific notation. Because Notes works primarily with text-based data, most formulas will not need more complex mathematical calculations.

Developers can write customized formulas that perform a variety of functions.

- Calculate values for display in documents and views. Suppose a form contained two fields, FirstName and LastName. The developer could concatenate these two fields in a formula that defines how a column should appear in a view (e.g., first name followed by last name separated by a space).
- Validate and translate new values entered in documents.

- Select documents to include in a view. Formulas can include logical operators (such as <, >, =) that can be combined in conditional statements (@If...). If a particular field or fields conforms to the criteria set out in the formula, they may be included in a particular predefined view. For example, a special view might include only those customers who have made purchases in excess of $1,000 over the last year in order to develop a mailing list for a special event.
- Change, add, and delete values in documents. Related fields can be linked through a formula. For example, if in a Notes human resources application an employee changes from part-time to full-time, the vacation accrual field might change value to reflect this new status.
- Determine which form to use for displaying data in different situations.

@Functions

Notes is shipped with more than 100 predefined @functions that perform a variety of calculations with different data types. Most @functions return a value of text, number, time–date, or Boolean. @Functions fall into six categories.

- String @functions recognize and manipulate strings. For example, @Left searches a string from left to right and returns the leftmost characters of the string; @ProperCase converts the words in the string to proper name capitalization (first letter uppercase, all others lowercase).
- Mathematical @functions compute calculations using numeric values. @Log returns the log of any number; @Min finds and returns the smallest of the number of arguments specified.
- Time–date @functions are used for time–date generation and manipulation. @Created returns the time–date when the document was created; @Yesterday returns the time–date value that corresponds to yesterday's date.
- Logical @functions produce values based on the result of conditional statements. In many cases, the @If function is required to state the condition. @IsNumber evaluates whether the argument is a number or a number list, which can be useful in validating data types. @IsError checks to see if there is an error in a particular field name, and carries out the next action based on whether or not the error exists.
- View statistics @functions are used to view column formulas to maintain view statistics and hierarchy. @DocChildren indicates how many responses a document has. @IsExpandable indicates whether the current level of documents is expandable: it displays "+" if a level is expandable, "–" if it is not.
- Database and document statistics @functions calculate and display database and document information. @DocLength returns the approximate size of a document in bytes; @DbManager displays the name of users who currently have Manager access to the database.

Macros

Macros are a special set of @functions that, in addition to performing some type of calculation, performs some type of action based upon that calculation. There are five types of Notes macros.

- Filter macros update a series of documents in batch mode so that they do not have to be manually edited, updated, or saved. Filter macros can run in background mode. They are used to add, replace, or delete document fields (and their contents), rename document fields, and select documents for printing or other operations. Users must have designer access to create filter macros, which are stored as part of the database application.
- Search macros are a special kind of filter macro, which perform the same function as a filter macro but only on documents that have been found by a full text query. They are defined and activated as part of a query. Users must have Designer access to create search macros, which are stored as part of the database application.
- Execute-once macros perform one or more Notes menu commands on selected documents shown in a particular view (e.g., printing a series of selected documents). Users must have Designer access execute-once macros, which are stored as part of the database application.
- Buttons are macros that are activated when the user clicks on a button on the screen. Buttons work only on the document in which the button appears. They are used to simplify routine tasks such as saving or mailing documents. For example, in a workflow application, a button can be used for tasks such as approving a document and forwarding it to another person. Or, if a database manager has moved a database from one server to another, she or he might send a message to the database users, who can then push a button on the message that will automatically update their desktops. Any user with Author access can create a button, although the type of function performed may be restricted to higher access levels.
- SmartIcons® are macros that are activated when the user clicks on them. Unlike buttons, SmartIcon macros can be used within the context of the document (cut, paste, etc.), within a view, with a database, and with the Notes workspace itself. Any user can create a SmartIcon macro, which is stored as part of the user's workspace.

There are a few special Notes macros that help the developer create workflow process applications, modify the user interface, and make use of existing Notes resources and databases.

The @Command, which is usually found as the embedded logic in a button macro, executes a Notes menu command. It is used in conjunction with approximately 200 predefined Notes functions that handle database administration, management, and navigation; database, form, view, and workspace design; document and text editing; help system; mail routing and management.

- @Prompt macros display dialog boxes that prompt the user for input. The input can then be used as part of a formula or another macro.
- @DbLookup looks up specific values from documents or views in other Notes databases, even if those database are physically located on another server. This allows the developer to leverage existing resources. For example, by combining @DbLookup with a logical @function, one application can perform actions based upon the value of information in another application. Also, used in combination with a time-based macro, a document or a field can automatically be populated with values from another Notes database application.
- @DbColumn behaves similarly to @DbLookup except that it looks up a column of values in another Notes database.

Workflow and Agent Processing

The third layer monitors the process and keeps track of such workflow elements as users and their roles; time, for tracking and notifying; and activity, or what just happened and what is supposed to happen next.

In Notes, background macros comprise the critical third layer. Using conditional and time-based macros, Notes applications can advance a workflow process without any manual intervention by the user beyond entering information or signing (or clicking) an approve/disapprove field.

Full Text Index and Search

Notes allows users to make queries based upon the content value of rich text fields by providing a full text search engine. The engine can perform relevance-based searches, which prioritize selected documents by the number of times a keyword(s) is found or how closely two or more keywords are located to each other. For example, a search that includes the phrase "client/server" would likely return a large number of documents from a developer's database. If the user also included the word "asynchronous," Notes would indicate that some documents contained both phrases several times, and present them at the top of a browse list, whereas documents with fewer instances of either phrase would appear lower on the list and would be indicated with a differently highlighted background.

Version Management

Notes allows users to make changes to an original document and then save the revised version as a response to the original, so that both documents are saved. This feature can be turned on by clicking a box when the developer defines form attributes.

34.7 EXTENSIBILITY — EXPLOITING EXTERNAL RESOURCES

Notes in itself is a complete application development and production platform. Some developers will also want to include additional products and resources as part of their Notes application. Therefore, Notes maintains an open architecture through its published API. It also includes hooks into relational databases, desktop applications, and third-party messaging systems.

Notes API

The Lotus Notes API is used to construct standalone Notes application programs that run independently of Notes, but that use Notes services. The Notes API is a library of C-language subroutines and data structures that allow programmatic access to a variety of Notes facilities. API subroutines can be called from C programs that perform operations that cannot be performed from the Notes user interface.

With the Notes API, developers can add functionality to Notes applications, such as

- Create import/export processing utilities that are under the control of the Notes workstation user to extend support for other file formats and external data sources. For example, such a utility could be used to import non-Notes databases and use those data to create a Notes document, or to extract information from a Notes database for use with a specified word processor.

- Extend Notes software by writing menu add-ins that reside in the Notes workspace, such as a menu item for assigning a customer order number.
- Create Notes server add-in tasks that are executed and monitored by the same facilities that the Notes server uses to run many of its own subsystems. For example, such a program might periodically scan a phone message database and delete every message that is more than 6 months old.
- Raise and/or receive Notes events and act on them.

Notes/FX

Not every document that a business user works with is a Notes document. Users often work with spreadsheets, with stylized word processing documents, and with graphic presentations. Notes Field Exchange (Notes/FX™) allows developers to use common desktop applications as part of their toolset to build workgroup applications. Notes/FX is a tool to build integration between standalone desktop tools and Notes to extend common application functionality to the workgroup and to afford Notes services (security, replication, etc.) to desktop applications. Information contained in specified desktop application fields can be used as Notes fields so that users can manage documents based upon their content value. For example, a user can define a named range in a spreadsheet, and Notes will store and manage the named range and its value.

It is built on open, standards-based technology—OLE and the Notes API—so that any developer of a PC software product can integrate it with Notes.

Notes/FX works with the following Windows™ versions of Lotus products: 1-2-3® Release 4.0.1, Improv® Release 2.1 for Windows, Ami Pro® Release 3.01, and Freelance Graphics® Release 2.01.

- In 1-2-3, Notes F/X makes data contained in 1-2-3 cells or ranges available in Notes, not just as an embedded 1-2-3 worksheet, but as the contents of fields in Notes forms and views. As information published in a Notes field, the spreadsheet data can be used as input to Notes processes and applications. For example, summaries or "roll-ups" of a proposal or cost estimate can be displayed as part of a Notes view along with associated analyses, reports, charts, and work histories. Users have access to all information related to the project in one place, organized, and presented at progressive levels of detail. The spreadsheet data published in a Notes field can also be acted on by Notes agents and macros, making it possible to use data inside the spreadsheet to drive workflow applications.
- Because Notes/FX supports bidirectional data exchange, data in Notes fields can update a 1-2-3 worksheet. The source of the Notes data can range from data entered by a user to data accessed from a remote database. For example, a standard company expense report can be distributed by embedding it in a Notes form that includes links between the fields in the form and cells in the spreadsheet using Notes/FX. Whenever a user chooses "Compose/Expense Report" from the Notes Compose menu, Notes will launch the spreadsheet template while automatically adding the user's name, cost center, and travel authorization number to the template. When the user has completed the expense report, Notes can route the expense report to the person with the appropriate signing level based on the total expenses shown in the sheet.
- In Improv, as with 1-2-3, Notes/FX makes data contained in Improv ranges available in Notes. Any legal user-defined field name in Notes can be used to identify corre-

sponding ranges in an Improv spreadsheet for the purpose of exchanging text, number or time data.

- In Ami Pro, two sources of data are exchanged with Notes via Notes F/X. The first type of data is information that describes the Ami Pro document itself, which Notes uses to automatically categorize the document in a Notes View. These data includes subject, description, style sheet, file size, revisions, and date created and last modified. Data entered in Ami Pro Bookmarks are the second form of data exchanged with Notes. When used in conjunction with Notes/FX, Bookmarks designate text that can update a Notes form or be updated by Notes.
- As is the case for 1-2-3 above, Notes and Ami Pro form a potent development environment for creating workgroup applications. For example, an action management system can be implemented to improve communications and control within a project. Meeting minutes are recorded in an Ami Pro template launched from the "Compose/Minutes" menu command in the project's database in Notes. The project, committee name, agenda, and action item status are filled in automatically by Notes when the template is launched. While minutes of the meeting are recorded in expressive Ami Pro text, new action items are entered in a table that updates a Notes Action Item view when the minutes are "saved" to Notes as well as triggering mail messages to those people assigned actions.
- In Freelance Graphics, information that describes a presentation is published to Notes through Notes/FX. This information includes the subject, description, page titles, page count, size, and date created and last modified. The combination of Notes and Freelance Graphics can be used to create a presentation library that can be viewed by subject, author, date created, SmartMaster™, or even a user defined field. It also can be indexed so that the full text search capability of Notes can be used to find a particular slide in a library containing thousands of slides. Users can preview the presentations within Notes forms before launching Freelance Graphics. Freelance presentations can be launched from Notes in either edit mode or as screenshows that automatically play to completion and return the user to Notes.

DataLens

It is likely that many Notes applications will make use of information that is contained in non-Notes databases. DataLens®, a database gateway that was introduced for use with Lotus 1-2-3 in 1989, has been enabled for use with Notes as well.

Developers use the @DbLookup and @DbColumn functions with external databases just as they would with a Notes database, using DataLens as the API. In addition, @DbCommand is a DataLens-specific function that is used to invoke database-stored procedures or launch complex queries.

DataLens provides the following capabilities.

- Extract data from external tables into Notes fields, macros, buttons, and pop-ups.
- Accommodate the various data types used in different external databases. DataLens converts data types from external databases to Notes data types.
- Enable field-level lookups and keyword list generation optimized for single documents.

DataLens supports dBASE®, FoxPro®, Database Manager, Sybase® SQL Server, Informix-OnLine, Oracle®, Paradox®, and Microsoft® SQL Server, as well as Microsoft's ODBC drivers.

DBMS Integration

Through DataLens and such intrinsic Notes functions as @DBLookup, developers can integrate Notes with other DBMSs. Developers can also use a variety of other Notes/DBMS integration techniques, such as import/export operations that use intermediate files to transfer data between Notes and a DBMS, using Notes as a desktop launching point for DBMS applications, Notes database access from SQL-based tools, background/agent-based Notes/DBMS integration, and direct Notes server/DBMS integration. For a full description of these integration techniques, see the Lotus Notes/DBMS Integration Technical Brief.

34.8 MANAGING THE DEVELOPMENT PROCESS

As multiple Notes applications are put into production, developers will seek to leverage the work they or their co-workers have already invested. Notes provides facilities that help developers keep track of and share application designs.

Notes Data Dictionary and Design Templates

Because every Notes application is also a Notes database, it is easy to store application designs as a repository in a Notes database itself, which means it benefits from all the same attributes of any Notes database—security, replication, etc. As mentioned, Notes is shipped with a set of Design Templates, complete applications that can be used as is or that can be modified and extended to build custom applications. Any changes to templates or any completely new forms can be stored and reused by any other developer with the appropriate access rights. Moreover, the repository does not have to be centralized, although it would be a good idea to store template designs, along with all the relevant macros, in a single common database for use by developers.

This encourages departmental development, since any authorized departmental developer would have access to all the corporate designs. Of course, with departmental development, the question of consistency arises. How much freedom should a departmental developer have to make major changes to templates that reflect corporate standards? Notes addresses this issue from three perspectives:

- Developers can protect the design of an application. Once an application is developed and saved as a template, the designer can elect to close the design from further modification using Notes' Application Design Protection. Users who try to modify the design will see the design menu item disabled (grayed out).
- Departmental developers can modify standard templates. If the developer allows a template to remain open, departmental developers can make whatever modifications they wish. It is likely that these developers will not rewrite such standard fields as Customer, thereby preserving consistency across applications. Instead, developers will leverage the existing work, only adding new fields and formulas and perhaps deleting some unnecessary fields. Developers would also be able to combine elements from different applications to build custom departmental applications.
- Departmental applications can inherit changes. When a standard template is borrowed for use as the basis for a customized departmental application, all of its attributes are inherited in the copied template. If the original template is modified later on, all changes will be propagated to any application that used that template, ensuring consistency across applications.

- Departmental developers can choose when they make a copy of a template whether or not they wish to inherit changes made to the parent. If they choose not to inherit any changes, this decision can be reversed later with a simple click on a dialog box. Inheritance can be applied to a complete database/application, or can be applied at the field level.

Documentation

Every Notes application is self-documenting. Every application contains a policy document, which explains to the user (and to developers who may need to modify the design during maintenance) the overall purpose of the database and how it should be used.

In addition, Notes provides a new Design Menu option that displays all attributes of the application design in a single location. This feature displays all attributes of all forms (including attributes of all fields in each form), all views (including attributes of all columns), filters, and any other attributes that can be specified with the Notes user interface in a Notes database.

User as Developer

Design Templates imply that not only professional developers can build Notes applications. Users can modify application templates—or use them unmodified—with little training and no development experience. Moreover, users can create private views and personal macros (buttons and SmartIcons) to customize their own workspace. Thus, developers do not have to necessarily concern themselves with accommodating the myriad user preferences for screen layout, keystroke, and button shortcuts, and user interface design.

34.9 GROUPWARE AND TELEPHONY — PHONE NOTES

Two technologies upon which businesses depend today, the telephone and the computer, are converging to provide users with a new computing environment that combines the single most familiar and usable device for communication with the power of information management. By making the telephone a fully functional component of an overall computing infrastructure, developers can build applications that manage voice just like any other data type, extend application functionality to users who do not have access to a computer, and provide mobile workers with a convenient and alternative means of access to information.

Most implementations of computer telephony have been characterized by large startup costs, proprietary technology, and complex development environments, as well as their application-specific nature. Applications built on these systems are usually geared to computer developers and have focused on transaction processing, enterprise-wide legacy systems. In addition, existing interactive voice response (IVR) applications are not accessible to mainstream end users: they do not include a desktop PC component, and they cannot be modified easily because of their complex and proprietary technology.

Although a growing number of PC voice card vendors supply an important component required for computer telephony, most of them have not provided a development environment with which to build applications. The environments that do exist are highly proprietary and do not enjoy widespread use. On the other hand, as vendors and users grow more aware of the benefits of computer telephony integration, a handful of vendors have begun to explore the first steps of computer telephony, offering packaged applications that provide telephone access to some PC- and LAN-based office automation systems, including the ability to listen to electronic mail messages and schedule appointments via phone. Yet, to date, no vendor has delivered a low-cost, easy-to-use, complete development environment with which to create a full breadth of LAN-based IVR applications.

Lotus Development Corporation has extended Lotus Notes®, client/server platform for developing and deploying groupware applications, to include the telephone as a Notes client. Phone Notes™ is an innovative development environment that allows users to participate in Lotus Notes applications from any touchtone telephone. Phone Notes makes it possible to extend Notes applications to non-Notes users, and allows Notes users who do not have convenient access to a PC to connect to Notes from a ubiquitous client. Moreover, information captured over the telephone can be stored in a Notes document, which then uses the same integral Notes services as any other document. That is, voice messages can be replicated across a distributed workgroup and can become essential components of workflow applications.

With Phone Notes, Notes itself extends its unique "multimodal" application development environment. That is, Notes stands alone in its ability to integrate such diverse technologies as electronic messaging, distributed document database, pagers, and telephony in groupware and workflow applications

Support the Extended Enterprise

People use the telephone to conduct their daily business: they might call a retail outlet to see if a particular product is in stock or the movie theater to check which pictures are showing at what times, make reservations at a restaurant, or order clothes from a mail order catalog. In each case, the caller is inquiring into a knowledge base, and receives the answer from a human operator, a recorded message, or from an interactive voice response application.

It is a small leap, then, to imagine a caller making an inquiry and conducting business with Notes, a database designed to manage information in a variety of formats—including voice. If granted telephone access to a Notes database, a non-Notes user could navigate through a series of voice prompts to find information about available products, place an order, or inquire about the status of an existing order.

For example, a customer might call into a travel agent database that contains information about promotional travel packages. The caller would listen to a recorded voice stored in a Notes field, which describes the available product literature (cruises, tours, etc.). After the caller uses the telephone key pad to navigate to the desired product, the Notes application asks the caller to enter the telephone number of a nearby fax machine. When the caller hangs up, Notes automatically dials the fax machine and sends a fact sheet describing a 10-day bike tour in New England, for example, directly to the caller's site using the Notes fax gateway.

Another customer service application might provide a more customized level of interaction. For instance, a mortgage broker maintains information about borrowers and lenders, and makes this information available to mortgage applicants through Phone Notes. When prospective borrowers calls, they either enter a customer number or spell their last and first names through the telephone keypad. When the borrower's account is recognized, the Notes application can track the status of the lending process. Once the status is determined, Notes can use a combination of embedded voice and text-to-speech synthesis to provide the caller with current information about the account. For example, "Your application was approved by Assistant Manager Alberto Acevedo on Wednesday, April 6, and was forwarded to Vice President Lucy Harang. A final decision will be made by Friday, April 8." This application makes the customer an active, knowledgeable participant in the loan process, instead of a powerless bystander.

Make Notes Databases Available to Users Anytime, Anywhere

For many Notes users, Notes applications have become a critical part of their daily work life. They post inquiries on a discussion database and review answers and comments, check the status of products in development, scan news databases for timely information about important clients and competitors, and monitor customer accounts. Notes replication allows users who are occasionally disconnected from the network to work with current data with a laptop computer while traveling or working at a remote site.

Sometimes, however, these mobile workers do not have access to a laptop computer, and in these cases Phone Notes can provide the disconnected user with a ubiquitous, functional, and easy-to-use Notes client. By making the telephone a Notes client, Notes users are never disconnected from their business information.

A typical use of Phone Notes is to read electronic mail messages when users do not have access to a computer. Mobile users simply enter a phone ID and phone extension, and Phone Notes will read the name of the sender, the date/time received, and the message header for all new messages. The user can choose to read the entire message, delete the message, or reply to it. When replying to an e-mail, the user's voice message will be stored in a rich text field and mailed to the originator, who can either click on the voice object icon to hear it on a sound-capable (multimedia) PC or through a standard phone on the user's desktop.

Of course, as a Notes client, the telephone can be used to retrieve, create, modify, and delete documents in any application. The Phone Notes user can have the system read any document over the phone using text-to-voice synthesis: news reports, product development status reports, responses to documents—precisely the same information that the user would ordinarily read from a computer screen. In addition, the Phone Notes user can respond to document or create new documents that capture the user's voice instead of capturing the user's ideas as text. For example, a product manager attending a conference may have learned about a new competing product. If the manager had access to a laptop, he could describe the new product and any relevant details about delivery, price, or configuration in a Notes document, and then replicate the laptop database back to the Notes server so that the members of his workgroup could view it right away and begin responding with product development and marketing ideas. When deprived of a computer, the manager could use Phone Notes to relay the same information in a phone call, storing the voice object in a Notes document field on the server database. The team members could then open that document, click on the voice object icon, and listen to the manager's information.

A New Tool for Notes Developers

Phone Notes was designed as a tool for the Notes application developer. In essence, it is an extension of the Notes development environment that allows Notes developers to modify existing Notes applications by adding voice to them or to create altogether new applications that depend on telephony.

Unlike previous attempts at marrying telephony and computer technology, Phone Notes is easy to use. Any Notes developer will be able to use this forms-based development environment with little or no training. The Phone Notes command language includes only 17 telephony command forms, making it easy to learn and master. In addition, Phone Notes is packaged with a set of five sample applications—Sales Support Faxback, Customer Service Help Desk, Human Resource Benefits Selection, Notes Mail by Phone, and E-mail Page Summary—that developers can use immediately or customize. Moreover, Phone Notes does not require a significant investment in special purpose hardware and software.

Notes developers can use Phone Notes to extend existing Notes applications by adding telephony to them, and create an entirely new class of telephony and IVR Notes applications.

Extend an Existing Notes Application

One of the unique benefits of Notes applications is the ability to bring together suppliers, vendors, business partners, and customers into a unified, shared business process. For example, a manufacturing company might share a Notes application with its equipment service provider so that both parties can keep track of problems on a production line. When using traditional PC Notes clients, workers or a foreman on the line had to either report a problem to someone who could enter the information into a Notes database, which would automatically alert the equipment company, or leave the line and enter the information directly into a Notes workstation.

By using the telephone as a Notes client, however, the manufacturing company can eliminate steps in this process and ensure that accurate information is entered into the system. The foreman can now call into a Notes database and embed a voice message that completely describes the problem directly into a Notes document. The foreman or line worker does not have to leave the line and can actually observe the problem physically while "entering" the information. Notes can then prompt the caller to enter information about the level of urgency.

The rest of the process occurs as before, except that the equipment supplier now hears a voice description of the problem instead of a text-based description. That is, if the previous version of the application used Notes' workflow processing to alert the equipment supplier of an open troubleshooting document, then the Phone Notes extension would also be able to take advantage of these facilities. In this case, the equipment supplier would receive a document with an embedded voice message.

Notes has been used to build many human resource administration applications. Phone Notes developers can add a new dimension to a Notes HR application using telephony. The Human Resources department can automate its employee benefits selection process by allowing employees to call into a Notes database that contains the details of a benefits package (including such information as the cost of each option). Employees call into the database, supply an employee identification number using the keypad, and choose their options after navigating through a menu. As an employee makes each selection, Notes computes the total cost and provides the employee with a running tally. The Notes application can then confirm the employee's selection by sending an electronic mail message to the user with the information, or by printing out a hard copy document to be mailed to the employee's home.

Create Altogether New Applications

When using Phone Notes, callers do not leave information in a voice mail in-box that later has to be transcribed into a Notes or other database. Instead, information is now captured directly into a Notes document and stored and managed like other data. This not only gives developers the ability to add a new dimension to an existing Notes application, but also makes possible entirely new applications. One example of a new application is an "automated attendant" that helps a data center conduct lights out operations. In this Phone Notes application, a Notes client monitors the availability of a set of database servers by dialing into them on a scheduled basis (e.g., every 15 minutes). If the client is not able to make a connection, Notes automatically places a telephone call to the data center and leaves a prerecorded message, alerting the staff to the status of a particular server. The Phone Notes application can also page the appropriate equipment service administrator by using the Notes Pager Gateway. The administrator receiving the page would then be able to dial into the database and listen to the voice message. The administrator could also instruct Notes over the telephone to search for other reports regarding the same database server, and listen to those reports using text-to-speech synthesis.

Phone Notes also makes it possible to develop applications that carry out transactions over a telephone. For example, a Notes-based matriculation application allows students to dial into a Notes database, obtain information regarding class schedules, query Notes for class time and cost, and request complete course descriptions and prerequisites information through a fax back system. Once the student makes a course selection, the Phone Notes application asks the student to enter a confirmation number that it is able to read back to the student through text-to-speech synthesis to ensure accuracy. In addition, the same Phone Notes application could ask for payment for classes and request credit card information. Notes' security protects the caller's credit card number as if it had been entered through traditional keyboard entry.

User Proxy Support

Since a Notes ID is not available over a telephone, security in a Phone Notes application is handled differently from Notes LAN security. Phone Notes provides developers with the ability to assign "phone roles" to users in addition to their access level when they access Notes on a LAN. That is, a user may have manager level access to a database from a desktop Notes client, but only reader level access from a telephone. To ensure that access to a Notes database is secure, Phone Notes utilizes "password authentication." When user proxy support is implemented in a Phone Notes application, callers are required to supply a password and a phone ID before accessing a Notes database.

Open, Extensible Production Environment

Like Notes, Phone Notes is an open, extensible development environment. Phone Notes scripts are read by any Phone Notes-compatible voice processing engine, which are available from third-party vendors. These engines translate the Phone Notes command scripts into hardware-specific telephony commands to access Notes databases; create, delete or forward Notes documents; and read, write, or delete field entries. At present, Simpact Associates in San Diego, CA provides a Phone Notes Certified

Telephony Server, which also includes an appropriate voice card from Natural MicroSystems.

Summary

Notes provides developers with a complete application development environment that is specifically designed to build groupware applications. It meets the unique requirements of a successful groupware solution. It manages unstructured information, provides a mechanism for collaborative work, secures information for use across organizational boundaries, and equips users with the ability to work remotely.

Notes makes available to application developers an advanced distributed document database, an integral messing service, and multilevel security to build communications-centric applications that transparently support all major computing platforms. Its straightforward design metaphor and standard design templates make it a fully functional tool for end users, and its advanced programming facilities and extensible architecture make it an appropriate tool for professional application developers.

Phone Notes is an innovative development environment that allows developers to integrate Notes and telephony, extend Notes to new users, and provide current Notes users with a ubiquitous, easy-to-use Notes client. Using Phone Notes, developers can easily add value to existing Notes applications and create a new class of Notes applications.

DEFINING TERMS

Cross-Platform Software: Software that will work across different computer systems, such as Apple, Intel, and UNIX systems.

Groupware: Software enabling teams of people to integrate their knowledge, work processes, and applications to achieve improved business effectiveness.

Knowledge Base: The aggregation of information for a workgroup, kept as common computer files.

Messaging: Mechanism for exchanging information in a groupware environment.

SQL: System Query Language, a standardized language for making database queries.

Workgroup: A group of people sharing common goals, tasks, and information.

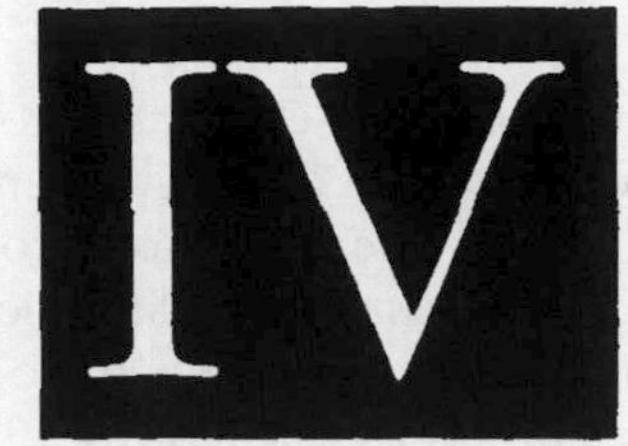

Mathematical and Statistical Software

Chapters 35 to 42 cover symbolic manipulation programs, also known (somewhat misleadingly) as computer algebra systems. Seven of the most popular systems are covered in detail: in alphabetical order, AXIOM, Derive, Macsyma, Maple, MathCad PLUS, Mathematica, and MATLAB. There are many other systems that we cannot cover for reasons of space, some of them quite good, particularly for specialized problems, or for interactive computation: a comprehensive list is given in Chapter 42.

Symbolic manipulation systems have been around for a while. The ACM's Special Interest Group on Symbolic and Algebraic Manipulation (SIGSAM) was formed in 1965, and by the early 1970s there were already several such systems available or under development (see D. Barton and J. P. Fitch, A review of algebraic manipulation programs and their applications, *Compute. J.* 15:362–381, 1972). But mostly, these early systems tended to be clumsy to program, being reminiscent of the language they themselves were written in, and the range of mathematics that they knew about was limited.

Even when a problem could be expressed in a natural way in these early systems, memory and speed limitations were such that only fairly small instances of the problem could be solved. To do any serious work one had to write a special-purpose program in (say) FORTRAN.

Of course, custom-written and optimized programs in a compiled language are still necessary in many scientific and engineering applications. But the boundary between computations that can be tackled in a casual manner and those requiring a skillful and determined programming effort has been pushed far indeed. The symbolic manipulation systems discussed here incorporate a large number of built-in mathematical functions, graphics ranging from adequate to excellent, and a programming language.

Each of them can be used in "calculator mode," that is, purely interactively, for problems ranging from simple arithmetic computations to (depending on the system) the solution of differential equations or sophisticated symbolic integration. But, in my view, it is the programming-language aspect of a symbolic manipulation system that ultimately gives it its power.

Partly because of that, my personal preference is for Mathematica, which has a relatively uniform design, a wide range of functional and pattern-matching capabilities, and versatile graphics. The importance of functional constructs and pattern matching cannot be overemphasized: many of the sorts of things that scientists and engineers like to do can be expressed much more briefly and cleanly (and often more efficiently as well) in a functional or declarative programming language than in a procedural one, such as C or FORTRAN.

The advantages of using a highly expressive language go beyond the fact that programming time and program size may be cut by an order of magnitude; they extend to the equally important issues of modularity, portability, robustness, and documentation.

There are several ongoing surveys that compare how well the various symbolic manipulation systems do when fed a menu of questions. (For instance, one by Michael Wester is available by anonymous FTP from math.unm.edu, in directory pub/cas.) These surveys are a start in comparing the relative worth of the systems, but they tend to ignore issues that cannot be easily quantified, such as ease of programming, adaptability, and extensibility.

They also sometimes consider as a bug what is merely a debatable design decision. Certainly none of the systems discussed here is bug-free, but most of them have become dramatically more reliable over the last few years. Unfortunately, even when the systems perform as specified, there are still many ways of running into trouble:

- Misleading names, such as "isprime" for a probabilistic primality testing routine in Maple, which until recently returned true for many composite numbers.
- Possibly unexpected but perfectly justifiable conventions, such as the choice of branch for fractional powers in Mathematica, which returns 0.5 + 0.866025i as the numerical value of $-1^{1/3}$.
- Numerical error; for example, when asked to numerically evaluate

$$\int_1^2 \frac{\sin(1000x)}{\sqrt{\sqrt{x}}}\,dx$$

with default options, some systems give an essentially random answer due to the oscillatory nature of the integrand (Mathematica gives a warning message and returns a good approximation, and the latest version of Maple gives the right value).

Most of these pitfalls can be avoided by checking answers independently (for instance, by plugging in known particular cases), and by reading the documentation carefully. This often represents a significant investment of time, but it will be time well spent. In any case, the software discussed in this chapter is not of the self-explanatory category; if you are impatient and try to use one of these systems for anything beyond basic arithmetic without having read the documentation, you will probably miss the best features of the system, repeatedly trip on its idiosyncrasies, and pretty much waste your time.

In fact, I recommend that you even go beyond the basic documentation, and study other books and publications for your system of choice (see the bibliographies for the individual sections). Another good source of information is the group sci.math.symbolic on the Internet.

Statistical software (Chapters 43 to 48) has the advantage over classical computational techniques in that the user can easily rerun computations to test both methodology and hypotheses with only minimal effort. Modern statistical software also contains features that allow for lucid and convenient presentation of results in a graphical form. This allows the researcher to easily explore relationships in their data that might otherwise go undiscovered.

35

AXIOM

Robert S. Sutor
IBM Corporation

35.1 INTRODUCTION

AXIOM,[1] a language, compiler, library, interactive shell, and hypertext help system for graphics and symbolic and numeric computation, is an integrated suite of applications for performing numeric and symbolic computations and creating two- and three-dimensional graphs. AXIOM is distinguished from the other major computer algebra systems by having an advanced object-oriented language (with compiler) and a library structured for maximum reuse and natural expression of algorithms. The interactive shell supports a subset of the compiled AXIOM language. The shell interprets user input and calls the appropriate library functions. The system documentation is available on-line in a hypertext help system. With one mouse click, the user can have an AXIOM expression (and all the expressions on which it depends) appearing in the documentation entered into an AXIOM workspace where it will be evaluated. Similarly, if a user clicks on a graph then the full graphics application becomes active and the graph can be interactively transformed. A library browser is fully integrated with the hypertext system.

35.2 THE INTERACTIVE SHELL

Most users of AXIOM spend the majority of their AXIOM time working in the *interactive shell.* This application accepts user input, parses, and analyzes it, calls library functions or issues error messages, and then prints the output. Several different forms of output are available, including FORTRAN and $\LaTeX$. The output shown below is the monospaced two-dimensional math output available on all AXIOM implementa-

[1]AXIOM and the AXIOM logo are trademarks of The Numerical Algorithms Group Limited. AXIOM was originally developed by the Mathematical Sciences Department of the Research Division of the IBM Corporation, Yorktown Heights, New York.

0-8493-2530-7/96/$0.00+$.50

tion platforms. Via the **)set** command, you have very flexible control over the messages and statistics displayed during and after a computation.

Here are some basic examples to familiarize you with AXIOM syntax. You can call functions accepting just one argument with or without parentheses around the argument.

```
factorial 30
  (1)  265252859812191058636308480000000
                                        Type: PositiveInteger
factor %
           26 14 7 4  2  2
  (2) 2  3   5 7 11 13 17  19  23  29
                                        Type: Factored Integer
```

A percent sign **(%)** is used to refer to the object returned by the previous computation. Note that the type of each returned object is displayed after the object. The type is an intrinsic property of the object and is the central idea that differentiates AXIOM from other computer algebra systems. The type of the second object above is *Factored Integer* and was created by the *factor* operation. Objects of this type can be added, subtracted, multiplied, have greatest common divisors and least common multiples computed, and have available most of the operations (with exactly the same names) available to objects of type *Integer*. *Factored Integer* objects also have operations available that are specific to them, for example, extraction of factors, number of factors, and so on. An object of type *Factored Integer* is not an object of type *Integer*, although it can be easily *coerced* into one. Inverting an integer usually results in a fraction. Similarly, you would not expect to get a *Factored Integer* object by inverting an object of that type.

```
1/%
                           1
  (3) ----------------------------
        26 14 7 4   2  2
       2  3   5 7 11 13 17  19  23  29
                                        Type: Fraction Factored Integer
```

The interactive shell analyzed the input and called the "/" operation found in *Fraction Factored Integer*. We did not need to specify anything special for this happen, the interactive shell did it for us. By observing what it does automatically, we can learn what to do when we want to override some of its default behavior (for example, when we need to select one of two operations having the same name and the same argument types, but defined in different *packages*).

```
f := % + 1/5
           1303 335743 1212658982820364533229 69
     (4) ------------------------------------
                   26 14 7 4  2  2
                  2  3   5 7 11 13 17 19 23 29
                                        Type: Fraction Factored Integer
```

This statement assigned the value of the sum to the variable **f**. The left-hand side of the sum was an object of type *Fraction Factored Integer* and the right-hand side was an object of type *Fraction Integer*. The interactive shell determined that the result should be of type *Fraction Factored Integer* and so had the right-hand side coerced into an object of that type and then called the library function "+" from *Fraction Factored Integer*. A coercion is an information-preserving transformation of an object of one type into an object of another.[2] For example, you can coerce a factored object into an unfactored one and vice versa. It might not be easy to do a coercion (for example, by factoring a large integer or a polynomial) but it will not fundamentally change the mathematical identity of the object.

A *conversion* is a transformation that may lose information. Creating an output representation of an object is a conversion because you cannot necessarily recreate the original object. For example, is **"7"** the output representation for "7 the integer," "7 the constant polynomial in the variables **x** and **y** with complex coefficients," or "20 mod 13"? Converting a rational number to floating point is an example of a conversion. In the next expression, "::" means "coerce or convert to the type."

```
f :: Float
    (5) 0.2
                                                  Type: Float
```

That does not look right! The type *Float* allows us to specify how many digits of precision are used, and the default is 20 digits. Let us change that to 50 digits.

```
digits (50)
    (6) 20
                                                  Type: PositiveInteger
f :: Float
    (7) 0.2000000000   0000000000   0000000000   0037699876
        2881590564
                                                  Type: Float
```

Now at least we get more information, but the answer is still not exact. Use *Float* only when necessary, either because you need a final result expressed that way or because the algorithm requires it (for example, in numerical integration). In these examples we used integers and floating point numbers that can get arbitrary large. For efficiency, the types *SingleInteger* and *DoubleFloat* provide objects with machine-level implementations and arithmetic. The latter type is often used in graphics computations.

In this section we have seen only a handful of the hundreds of types available from the AXIOM library. In the next section we will look at AXIOM's graphics facility. Subsequent sections will discuss types in more detail and will look at some of the mathematical and data structures available in the library. Beyond that, we will look at some of the other components that make up AXIOM.

[2]Mathematically speaking, a coercion is a isomorphism or an embedding. For example, *Integer* and *Factored Integer* are isomorphic and *Integer* can be thought of as being embedded in *Polynomial Integer* (although they are not actually implemented that way in AXIOM).

35.3 GRAPHICS

Almost all graph creation in AXIOM is done using the *draw* family of functions. The window on the screen containing the graph is called a *viewport*. A *draw* function creates both the viewport and the contained graph. It is possible to place multiple two-dimensional or multiple three-dimensional graphs in one viewport.

Two-Dimensional Graphics

AXIOM's two-dimensional graphics facility supports plotting of functions of one variable, parametric plane curves, and nonsingular plane algebraic curves. You can change the default plotting behavior in three ways. First, options given with the call to *draw* can be used to select features such as colors, coordinate system, and whether the graph is drawn to scale or stretched to fit the size of the viewport window. A full list of options is given in Jenks and Sutor (1992) and is available on-line in the AXIOM help system. Second, the return value from *draw* is a viewport object, and you can use it as an argument to functions that indicate whether units, points, lines, and axes should be drawn, and that set the size and position of the viewport window. Finally, if you click on the graph, then a control panel is displayed that allows you to interactively change many features of the plot. You can save the graph image in a proprietary AXIOM format for later display or in PostScript form.

The simplest two-dimensional version of *draw* takes an expression in one variable as the first argument and a range of domain values as the second argument. By default, Cartesian coordinates are assumed and adaptive refinement is used to determine how many points should be plotted.

In this example, the values of **x** are plotted on the horizontal axis and the corresponding values of **x*sin(x)** on the vertical axis. The values of **x** range from -2π to 2π.

```
draw( x*sin(x), x = -2*%pi..2*%pi )
```

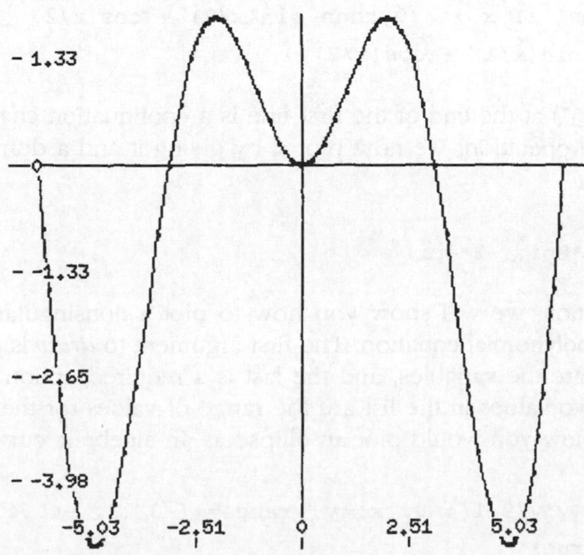

You can also plot curves given parametrically. The expression

```
draw( curve(t, t*sin(t)), t = -2*%pi..2*%pi )
```

produces the same graph as the last example. If the curve uses polar coordinates, add the option **coordinates==polar** in the call to *draw*.

```
draw( curve(t, t*sin(t)), t = -2*%pi..2*%pi, coordinates==polar)
```

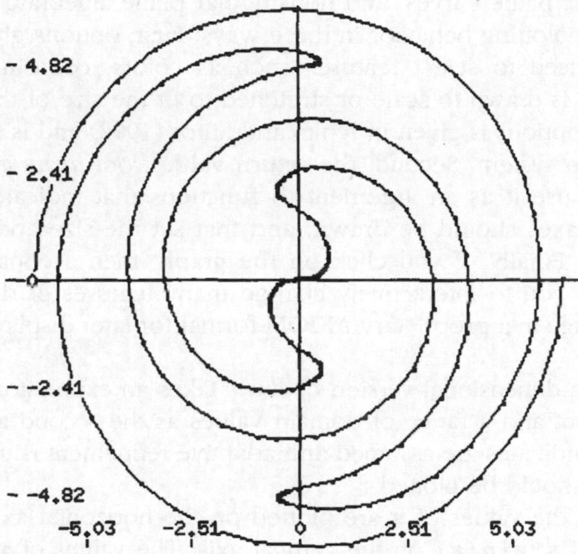

Since the interactive shell has a full programming language, you can define functions and then plot them. Define the function **f(x)** by

```
f(x) == abs( if x > = 0 then sin(x/2) + cos(x/2) _
    else - sin(x/2) + cos(x/2) )
```

The underscore ("_") at the end of the first line is a continuation character and *abs* is the absolute value operation. We now plot **f** by giving it and a domain range as the arguments to *draw.*

```
draw(f, -2*%pi..-2*%pi)
```

To end this section, we will show you how to plot a nonsingular plane algebraic curve given by a polynomial equation. The first argument to *draw* is the equation, the second and third are the variables, and the last is a required option list object called **"range"**. The two values in the list are the range of values for the first and second variables. This is how you would plot an ellipse as an algebraic curve.

```
draw(x*x/4+y*y/9-1 = 0, x, y, range==[-3..3, -4..4], _
toScale==true)
```

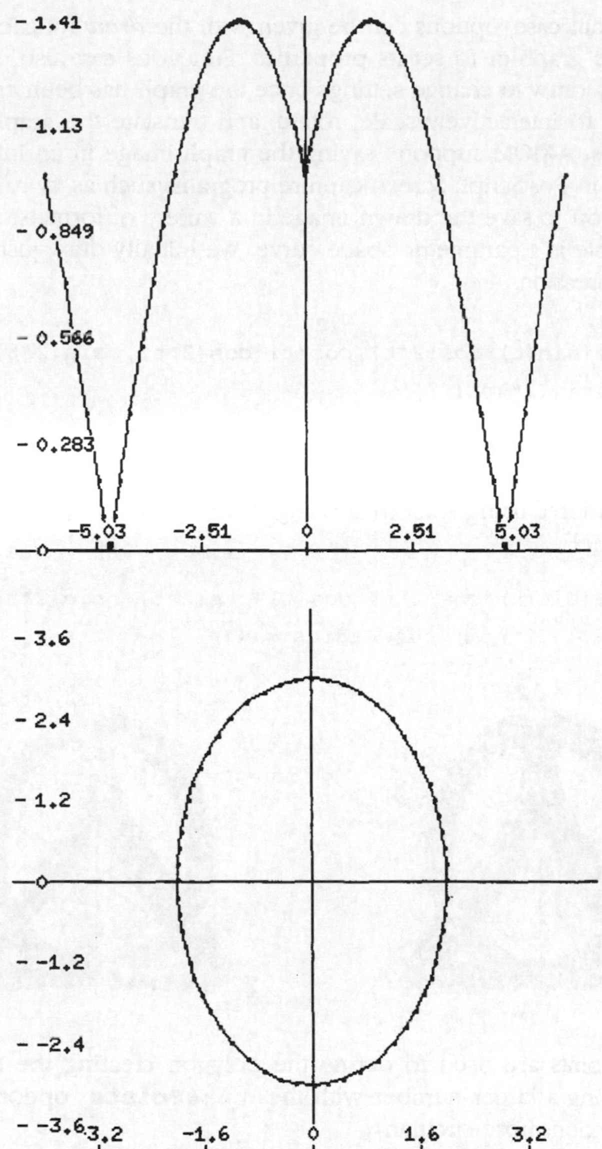

We used the **toScale** option so that the true shape of the ellipse would be obvious. Otherwise the units are scaled in each direction to fit the size of the viewport window.

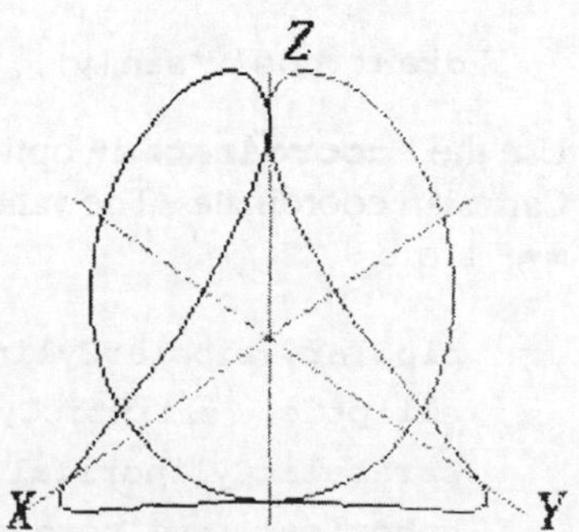

Three-Dimensional Graphics

AXIOM's three-dimensional graphics facility supports plotting functions of two real variables, parametrically defined surfaces and parametric space curves. Tubes can be drawn around parametric space curves. As in the two-dimensional case, options can be given with the *draw* function to change the appearance of the graph or to set its properties. Functions can also use the viewport object returned by draw to change settings once the graph has been created. A control panel allows you to interactively scale, rotate, and translate the graph and set colors and shading styles. AXIOM supports saving the graph image in an internal format for later viewing and in PostScript. Screen capture programs such as xv running under AIX or SunOS allow you to save the drawn image in a variety of formats.

Our first example is a parametric space curve. We initially draw just the curve using the following expression:

```
draw(curve(sin(t)*cos(2*t),cos(t)*cos(2*t), sin(2*t)), _
   t= -2*%pi..2*%pi)
```

Now we draw a tube around the curve. The view on the left is the tubular surface after the "Shade" and "Outline" options have been clicked in the control panel. The view on the right is rendered using smooth shading, which you get by clicking on "Smooth" in the control panel.

```
draw(curve(sin(t)*cos(2*t),cos(t)*cos(2*t), sin(2*t)), _
   t= -2*%pi..2*%pi, tubeRadius==.1)
```

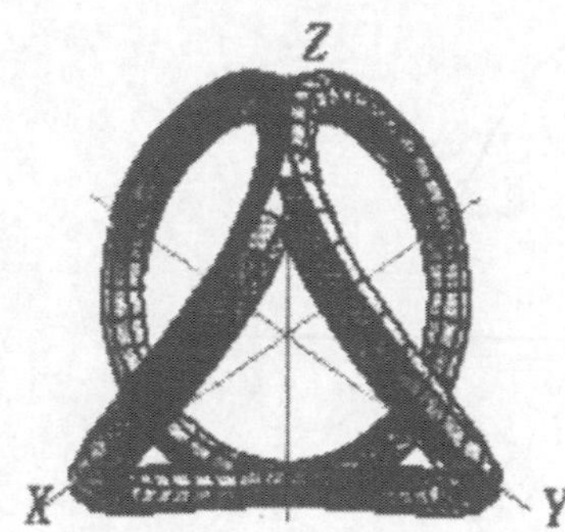

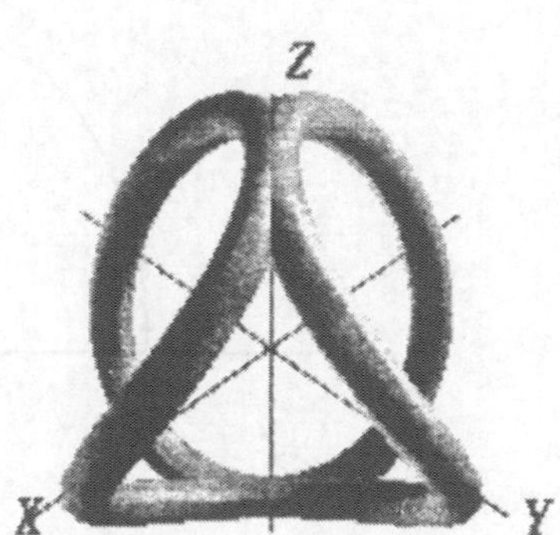

By default, six points are used to define the polygon creating the tube around the curve. By specifying a larger number with the **tubePoints** option, you can make the tube cross section more circular.

This is a plot of a function of two variables. We're including the **"shade"** style attribute in the call to *draw*. You get the outline by clicking on "Outline" on the control panel.

```
draw( cos(x*sin(y)), x=-4..4, y=-4..4, style=="shade")
```

Use the **"coordinates"** option to tell *draw* how to map your coordinate system into Cartesian coordinates. The valid coordinate system choices to follow **"coordinates =="** are

```
bipolar, bipolarCylindrical, cartesian, conical, cylindrical,
elliptic, ellipticCylindrical, oblateSpheroidal, parabolic,
parabolicCylindrical, paraboloidal, polar, prolateSpheroidal,
spherical, and toroidal
```

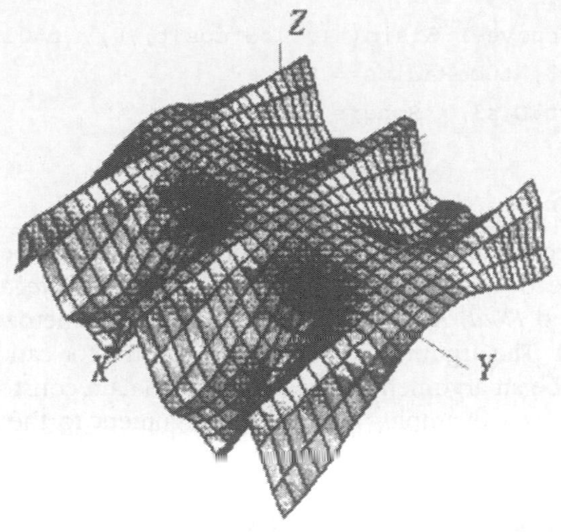

Three-dimensional graph objects are placed in 3-spaces and then displayed in viewport windows. When you use *draw*, neither the 3-space nor the viewport needs to be manually created. If you wish to display more than one graph object in a viewport, then first use *makeObject* to create a 3-space and place the first object in it. Subsequent calls to *makeObject* include the space object as an option. Other than the **"space"** option, *makeObject* accepts the same options as *draw*. Finally, call *makeViewport3D* with the 3-space and a window title to display the composed objects.

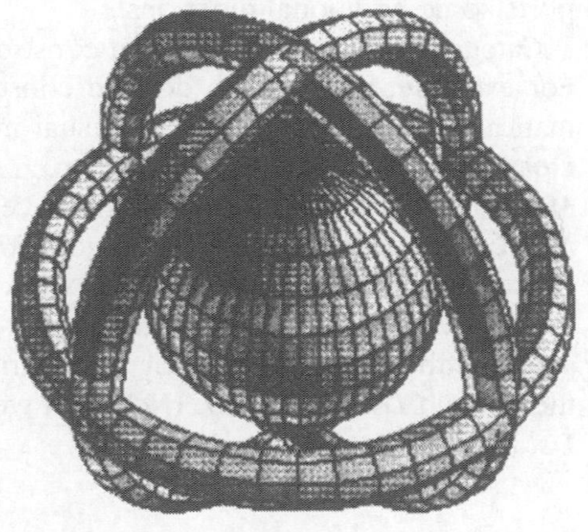

Here is an example of composing four three-dimensional objects in one viewport. The *makeObject* function is called once to create a sphere and then three times to make circular tubes wrapping around the sphere. The first argument in the first call to *makeObject* is an anonymous function: a function that is not given a name and only exists for the duration of the function call. In this case, it is a function of two variables that returns the constant 0.9, the radius of the sphere.

```
s3 := makeObject((u,v) +-> 0.9, 0..%pi, 0..2*%pi, _
    coordinates==spherical)
makeObject(curve(0, 1.5*sin(t), 1.5*cos(t)), t=0..2*%pi, _
    space==s3, tubeRadius==.15)
makeObject(curve(1.5*sin(t), 0, 1.5*cos(t)), t=0..2*%pi, _
    space==s3, tubeRadius==.15)
```

```
makeObject(curve(1.5*sin(t), 1.5*cos(t),0), t=0..2*%pi, _
    space==s3, tubeRadius==.15)
makeViewport3D(s3, "Sphere + 3")
```

35.4 AXIOM TYPES

A *domain* is a collection of objects and a set of functions that operate upon those objects. We have already seen several domains, including *Integer*, *Factored Integer*, *Fraction Integer*, and *Float*. A domain is created by a parametrized function called a *domain constructor*. The arguments to a domain constructor can be domains (for example, *Integer* can be an argument to the *Factored* domain constructor) or objects belonging to domains (for example, 7 can be an argument to the domain constructor *IntegerMod*). Domain constructors with no parameters can be defined and then written with the parentheses omitted (but you must be consistent: *Integer* is defined without the parentheses, so you must not include them).

Domains *export* the functions and constants that they allow to be seen by users. The collection of exports is called the *category* of the domain. Categories can also be defined independently and given names. A domain is said to *belong* to a category. For example, *Matrix R* belongs to the category *MatrixCategory R* but also specifically exports some additional functions.

Categories ensure that domain constructors are passed arguments that make sense. For example, because the domain constructor *Matrix* is intended to produce mathematical objects that support the usual matrix arithmetic, it is not possible for you to create in AXIOM the domain *Matrix List Integer*. Domains like *Matrix Integer* and *Matrix Complex Float* are valid, however. That is, the domain you substitute for *R* in *Matrix R* must belong to the category *Ring* (see below).

The interactive shell does much work to interpret your input, create objects of the intended domains, and then call library functions. In the following example, **m1** is assigned the value returned by the matrix function called with an object belonging to the domain *List List Integer*. (Note that we also use the terminology "and object of type *List List Integer*.")

```
m1 := matrix [[2, 4], [9, -7]]
```

$$(1)\quad \begin{bmatrix} 2 & 4 \\ 9 & -7 \end{bmatrix}$$

Type: *Matrix Integer*

Matrix Integer is a domain created by the *Matrix(R)* domain constructor, where *R* = *Integer*. The domain exports the usual arithmetic and linear algebra functions. In particular, the definition of *Matrix(R)* states that if multiplication in *R* is commutative, then a *determinant* operation is defined and exported. *Integer* does have commutative multiplication.

```
determinant m1
```

$$(2)\quad -50$$

Type: *Integer*

We now create another matrix, this time using more complicated entries.

```
m2 := matrix [[2*x, 1/2 - z], [9/4, x**6-7]]
          ┌            1┐
          |2x  - z + -  |
          |            2|
   (3)    |             |
          |9    6       |
          |- x  - 7     |
          └4            ┘
```

Type: *Matrix Polynomial Fraction Integer*

Some of the entries are polynomials with integer coefficients, one is a polynomial with a rational number constant term, and one is a rational number. The interactive shell determines that the list of lists that is the argument to *matrix* should have type *List List Polynomial Fraction Integer*. *Polynomial* is a domain constructor that is used to create multivariate polynomial domains where the variables names are not restricted.[3] The *matrix* function used here is *not* the one used in the integer example above. This function is from the *Matrix Polynomial Fraction Integer* domain; the other one is from the *Matrix Integer* domain. In this case, they use exactly the same definition.

Just as the *Matrix* domain conditionally exports the *determinant* function if multiplication in the entry domain is commutative, properties of the entry domain can be queried to select among alternative function definitions. For example, if the entry domain belongs to the category *Field*, then division by nonzero elements is supported and a Gaussian elimination algorithm can be chosen that uses division.

The interactive shell has extensive facilities for coercing objects of one type into objects of other types. Most of the time this happens automatically, as when the shell starts with the expression

```
[[2*x, 1/2 - z], [9/4, x**6-7]]
```

and ends with an expression of type *List List Polynomial Fraction Integer*. You can use coercion directly to change the form of an object for display purposes. Here is a matrix created from the one above where we express the entries as rational functions (instead of polynomials with rational number coefficients).

[3]The representation used by domains created by *Polynomial* is a recursive one where the variable names are part of the structure. If you wish to restrict the variable names to be in a given set, use, for example, *MultivariatePolynomial([x,y],Integer)*. If you wish to use a distributed representation for a Gröbner Basis calculation, for example, use either *DistributedMultivariatePolynomial* or *HomogeneousMultivariatePolynomial*. See Jenks and Sutor (1992) for more information and examples.

```
m2 :: Matrix Fraction Polynomial Integer
          ┌        - 2z + 1┐
          |2x    ---------|
          |          2     |
    (4)   |                |
          |9         6     |
          |-        x   - 7|
          └4               ┘
```

Type: *Matrix Fraction Polynomial Integer*

Coercion can be used to reformat an object so that a particular operation can be applied. The matrix **m2** can be broken into a sum of matrices with the variables extracted as factors. That is, we want to create a polynomial with matrix coefficients. From this polynomial we want to retrieve the leading coefficient.

```
m2 :: Polynomial Matrix Fraction Integer
```

Polynomial Matrix Fraction Integer is not a valid type.

That did not work! Categories came into play again, this time to prevent us from doing something incorrect. The *Matrix* domain constructor does not produce domains that belong to the category *Ring*.[4] Multiplication and addition of objects in the domain *Matrix Integer* are only defined when the dimensions are compatible. To have unrestricted arithmetic, we need to use square matrices of a given size.

```
m2 :: Polynomial SquareMatrix(2, Fraction Integer)
                                                          ┌     1┐
                                                          |0    -|
          ┌0   - 1┐     ┌0   0┐ 6      ┌2   0┐           |     2|
    (5)   |       |z +  |     |x   +  |     |x +         |      |
          └0     0┘     └0   1┘       └0   0┘           |9     |
                                                          |-  - 7|
                                                          └4     ┘
```

Type: *Polynomial SquareMatrix(2,Fraction Integer)*

```
leadingCoefficient %
          ┌0   - 1┐
    (6)   |       |
          └0     0┘
```

Type: *SquareMatrix(2,Fraction Integer)*

[4]*Ring* is the AXIOM category representing the mathematical notion of the same name. A ring supports unrestricted multiplication and addition and a distributive law. Addition is always commutative but multiplication need not be. The standard example of a ring is the set of integers. In AXIOM, *Integer* and *SquareMatrix(2, Integer)* both belong to the category *Ring*. The first has commutative multiplication, the second does not.

A special kind of domain is a *package*: a set of functions but no objects. The grouping together of functions into a package can be arbitrary, but it usually done to bring to one place functions with a common or complementary purpose. For example, in AXIOM there are packages that support integration, linear algebra, Gröbner basis computations, and graphics. Since packages are domains, they can be parametrized. For example, *ListFunctions2(A: SetCategory, B: SetCategory)* provides functions that apply to or return *List* domains with different entry domains. An important function implemented in this package is called *map* and it takes a function *f* from *A* to *B* and an object of type *List A* and returns an object of type *List B*. That is, it applies *f* to each element of type *A* in the first list and builds a new list containing objects of type *B*. The interactive shell uses such *map* functions to create coercions between list objects. In this case, *f* is a coercion from an object of type *A* to an object of type *B*. *AXIOM* automatically loads the domains and packages it needs.

```
l := [2/3, 4/5]
          2 4
   (1)   [-,-]
          3 5
                                   Type: List Fraction Integer
l :: List Polynomial Complex Fraction Integer
   Loading   /axiom/mnt/rios/algebra/COMPLEX.o   for   domain
   Complex
   Loading /axiom/mnt/rios/algebra/LIST2.o for package
         ListFunctions2
   Loading /axiom/mnt/rios/algebra/FLAGG2.o for package
         FiniteLinearAggregateFunctions2
   Loading/axiom/mnt/rios/algebra/FEVALAB-.o for domain
         FullyEvalableOver&
   Loading   /axiom/mnt/rios/algebra/BASTYPE-.o   for   domain
   BasicType&
   Loading /axiom/mnt/rios/algebra/COMPCAT-.o for domain
         ComplexCategory&
         2 4
   (2)  [-,-]
         3 5
                    Type: List Polynomial Complex Fraction Integer
```

If a domain is not a package, that is, there are objects belonging to the domain, the domain must define a *representation* for its objects. The representation is another domain and is frequently one built using the *List*, *Vector*, *Record*, or *Union* domain constructors. For example, *IntegerMod(n)* uses either *SingleInteger* or *Integer* as its representation, depending on the size of **n**. AXIOM therefore supports one form of inheritance via representation.

AXIOM uses the notion of category to define the interface a domain presents, but, unlike languages like C++, it does not affect the representation of the domain. One of the simplest categories in AXIOM is *SetCategory*: this category consists of six operations: an equality test, an inequality test, a function returning a sample element, a function that computes a hash code for an element, a function that produces a LaTeX rendering of an element in a string, and a function that produces the standard AXIOM output representation form of an element. One category can inherit from another: the category *SemiGroup* is *SetCategory* plus some basic arithmetic operations, *Monoid* is *SemiGroup* plus some operations, and so on. For a domain D to belong to a category C, C must appear somewhere in the domain's category ancestry.

AXIOM supports multiple inheritance via categories. For example, every finite field domain belongs to *FiniteFieldCategory* and this inherits from the four categories *FieldOfPrimeCharacteristic* (a finite field has characteristic $p > 0$, where p is a prime integer), *Finite* (a finite field has only a finite number of elements), *StepThrough* (it is possible to iterate across the elements in a finite field), and *DifferentialRing* (a finite field has a trivial differential operator).

Categories provide structure and safety to the AXIOM library. They ensure that domains are built using argument domains with the correct properties. Categories can also provide default definitions so that domain implementors need not recreate the standard function definitions (given 0 and a subtraction operation, the negation operator can be defined in a way that is independent of the actual representation used for a domain). More examples of categories will be shown in the next two sections. See Watt et al. (1994) for precise definitions of categories and domains, examples of how you can modify existing ones and create new ones, and a full exposition on the AXIOM language.

35.5 MATHEMATICAL FACILITIES

Like all of AXIOM's computational tools, its mathematical facilities are contained in domains and packages in the system library. Since these domains and packages are usually parameterized, the linear algebra routines, for example, will naturally extend to use new "matrix entry" domains as they are created.

We have already seen some of the number, polynomial, and matrix domains. The next section discusses data structures: domains used to hold a collection of objects. In this section we will show some of the mathematical structures that AXIOM supports and list some of its computational capabilities with those structures. This discussion is far from exhaustive. As always, the on-line documentation and the library itself are the best current references and sources of examples.

Numbers

AXIOM supports exact integer arithmetic and representations of integers in factored, based, and Roman forms. The domain *SingleInteger* uses machine-defined integer representation and arithmetic. Similarly, *DoubleFloat* corresponds to machine-defined floating point. *Float* implements floating point arithmetic using arbitrary-sized integers for the mantissa and exponent. The precision in digits can be set by the user. *Fraction* is used to create quotient fields of appropriate domains.[5] *Fraction Integer* implements exact rational numbers.

Complex creates complex "numbers" for any domain belonging to the category *CommutativeRing*. The usual operations such as *real*, *imag*, *conjugate*, and *norm* are all available. You can factor complex integers. The square root of –1 is displayed as **%i**.

[5]The domain must belong to the category *IntegralDomain*, just as in mathematics.

```
factor(2 :: Complex Integer)

   (1)  - %i (1 + %i^2)
```

Type: *Factored Complex Integer*

Quaternion, *CliffordAlgebra*, and *Octonion* create domains representing the mathematical structures of the same names.

Rational numbers can be expressed exactly as repeating decimals by coercing them to *DecimalExpansion*. Rational numbers can also be expressed in bases other than base 10. Continued and partial fraction expansions are also supported.

IntegerMod and *PrimeField* implement modular integral arithmetic. For coding theory and other applications, no fewer than nine domain constructors create finite field extensions, giving you unparalleled choice of representation for optimizing the efficiency of your computations.

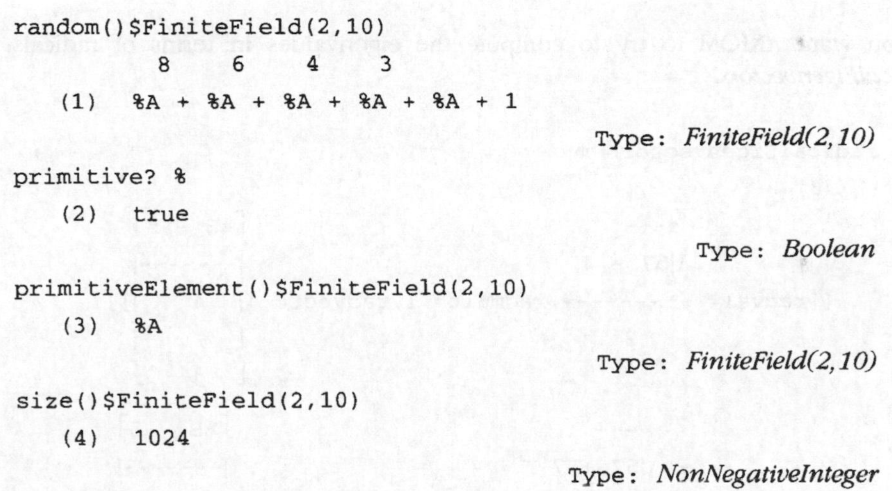

```
random()$FiniteField(2,10)
              8      6      4      3
   (1)   %A  + %A  + %A  + %A  + %A  + 1
```

Type: *FiniteField(2,10)*

```
primitive? %
   (2)   true
```

Type: *Boolean*

```
primitiveElement()$FiniteField(2,10)
   (3)   %A
```

Type: *FiniteField(2,10)*

```
size()$FiniteField(2,10)
   (4)   1024
```

Type: *NonNegativeInteger*

AlgebraicNumber and *Expression* allow you to express symbolic roots of polynomials. *Expression* is a general representation for symbolic function application, including trigonometric and hyperbolic functions. Simplification, evaluation, and conversion to floating point are all supported where appropriate. *Expression* also provides a powerful pattern matching facility.

Linear Algebra

AXIOM implements all the usual matrix and vector manipulation operations for the domain constructors *Matrix*, *SquareMatrix*, *RectangularMatrix*, and *Vector*. The null spaces and nullity can be computed and a matrix can be converted into row echelon form if the domain of the entries belongs to the category *Field*. For square matrices, *inverse*, *determinant*, and *permanent* operations are available.

The *eigenvectors* operation can be used to return a list of all eigenvectors and eigenvalues. For nonrational eigenvalues, the minimal polynomial is given. (**%A** is a polynomial variable name generated by AXIOM.)

```
m := matrix [[2, 3], [4, 5]]
        ┌2  3┐
   (1)  |    |
        └4  5┘
```

Type: *Matrix Integer*

```
eigenvectors m
                                  2
   (2) [[eigval= (%A | %A  - 7%A - 2),eigmult= 1,
                       ┌%A - 5┐
                       |------|
        eigvec= [|     4     |]]]
                       |      |
                       └  1   ┘
```

Type: *List Record(. . .)*

If you want AXIOM to try to compute the eigenvalues in terms of radicals, use *radicalEigenvectors*.

```
radicalEigenvectors m
   (3)
               +---                          ┌%B - 5┐
              \|57 + 7                        |------|
   [[radval= ---------,radmult= 1,radvect= [|    4   |]],
                 2                            |      |
                                              └  1   ┘

                +---                           ┌%B - 5┐
             - \|57 + 7                        |------|
    [radval= ----------,radmult= 1,radvect= [|    4   |]]]
                  2                            |      |
                                               └  1   ┘
```

Type: *List Record(. . .)*

To get approximate solutions, use either *realEigenvectors* or *complexEigenvectors*.

```
(4) ->realEigenvectors(m, 0.1)
   (4)
                                                ┌0.5703125┐
   [[outval= 7.28125,outmult= 1,outvect= [|             |]],
                                                └   1.0   ┘
                                                  ┌- 1.3203125┐
   [outval= - 0.28125,outmult= 1,outvect= [|                |]]]
                                                  └    1.0     ┘
```

Type: *List Record(. . .)*

Special Functions

The package *DoubleFloatSpecialFunctions* provides the following special functions for arguments of type *DoubleFloat* or *Complex DoubleFloat*: *Gamma*, *Beta*, *logGamma*, *digamma*, *polygamma*, *besselJ*, *besselY*, *besselI*, *besselK*, *airyAi*, *airyBi*, *hypergeometric0F1*.

Limits

AXIOM has a powerful limit facility. When the left and right limits differ, or one does not exist, the solution will reflect this.

```
limit(2 + x * log(x),x = 0)
    (1) [leftHandLimit= "failed",rightHandLimit= 2]
                                                    Type: Union(. . .)
```

Here is a limit of an expression involving an exponential singularity. The answer, **%e**, is the AXIOM input and output form for the mathematical constant *e*.

```
limit ( (x + 1)^(x + 1)/x^x - x^x/(x - 1)^(x - 1), _
x = %plusInfinity)
    (2) %e
                        Type: Union(OrderedCompletion Expression Integer, . . .)
```

Solutions of Equations

Use the *solve* operation for solving one or a system of linear or polynomial equations. For polynomial equations, solutions can expressed approximately or exactly, possibly in terms of roots of irreducible polynomials. The *radicalSolve* operation returns only those solutions it can express in terms of radicals. Use *complexSolve* if you want complex numeric approximations to the solutions.

Here are some examples involving one polynomial. For approximate solutions, the second argument is a numeric "tolerance" expressed either as a floating point or rational number.

```
pol := x^3 - 5
             3
    (1) x - 5
                                               Type: Polynomial Integer

solve(pol, .01)
    (2) [x = 1.70703125]
                                   Type: List Equation Polynomial Float

radicalSolve(pol, x)

                                  +----- +--     3+--      +----- +--     3+--
                      3+--       (\|- 1 \|3 - 1)\|5     (- \|- 1 \|3 - 1)\|5
    (3) [x= \|5 ,x= ------------------,x= --------------------]
                                    2                        2
                                   Type: List Equation Expression Integer
```

```
complexSolve(pol, 1/1000)

               7005          9402582047769      3033
     (4)  [x= -----, x= - --------------- - ----- %i,
               4096          10995116277760      2048

                  9702582047769      3033
           x= - --------------- + ----- %i]
                 10995116277760      2048
```

Type: *List Equation Polynomial Complex Fraction Integer*

Series

AXIOM power series are implemented using the *Stream* datatype and so can represent infinite series where the coefficients are computed only as necessary. Taylor, Laurent, and Puiseux series in one variable are all provided. You can perform arithmetic on two series to create a third. You can also create a new series via a function for the coefficients or a via coercion of an expression.

```
x := series 'x
   (1) x
```

Type: *UnivariatePuiseuxSeries(Expression Integer,x,0)*

```
s := 1 / (1 - 2*x**2)
                  2      4      6      8       10       11
   (2) 1 + 2x + 4x + 8x + 16x + 32x + O(x)
```

Type: *UnivariatePuiseuxSeries(Expression Integer,x,0)*

```
coefficient(s,30)
   (3) 32768
```

Type: *Expression Integer*

When you evaluate a series with a floating point number, the returned result is an infinite stream of the partial sums.

```
eval(s, .1)
   (4) [1.0,1.0,1.02,1.02,1.0204,1.0204,1.020408,1.020408,
        1.02040816,1.02040816, . . .]
```

Type: *Stream Expression Float*

Differentiation and Integration

AXIOM uses the *D* operator for differentiation. The first argument is the expression to be differentiated and the second is the differentiation variable.

```
D(sin(cos(x)/x), x)
```

```
                             cos (x)
        (- x sin(x) - cos(x))cos(------)
                                x
   (1)  ------------------------------
                        2
                       x
```

Type: *Expression Integer*

If present, the third argument is the order of differentiation. Here we take the fifth derivative with respect to **x**.

```
D(%e^(x^2), x, 5)
```

```
                5      3         ?
   (2)  (32x + 160x + 120x)%e
```

Type: *Expression Integer*

If the second and third arguments are lists, then we are computing partial derivatives. Here we compute the first derivative with respect to **x** and the second with respect to **y**.

```
D(sin(x*y), [x,y], [1,2])
```

```
                                  2
   (3)  - 2x sin(x y) - x y cos(x y)
```

Type: *Expression Integer*

The name of the family of functions for integration of real-valued functions is *integrate*. For real-valued elementary functions, AXIOM only returns a formal integral expression when it can prove that the integral is not an elementary function. If the values of the variables and parameters in the integrand may be complex, use *complexIntegrate*. If *integrate* is used and the signs of the parameters affect the integral, multiple solutions may be returned.

```
integrate( x^2/(x^2 - b), x)
```

```
                    +--   2
           +--   - 2x\|b + x + b
         \|b log(----------------) + 2x
                      2                     +----
                     x  - b       +----    \|- b
   (1) [------------------------,\|- b atan(------) + x]
                    2                          x
```

Type: *Union(List Expression Integer, . . .)*

```
complexIntegrate( x^2/(x^2 - b), x)
```

```
           +--       +--        +--          +--
       - \|b log(\|b + x) + \|b log(- \|b + x) + 2x
   (2)  -----------------------------------------
                          2
```

Type: *Expression Integer*

Solutions of Differential Equations

AXIOM's facility for solving differential equations in closed form can work with these kinds of problem data: linear ordinary differential equations and nonlinear first-order ordinary differential equations where integration is the only means necessary for finding the integrating factors. Some classes of systems of transcendental equations can also be solved.

```
y := operator 'y

   (1)   y
                                                          Type: BasicOperator

deq := D(y x, x) = y(x) / (3*x - y(x) * log y(x))

                                y(x)
   (2)   y'(x) = - ------------------
                     y(x)log(y(x)) - 3x
                                              Type: Equation Expression Integer
solve(deq, y, x)

          - 2y(x)log(y(x)) - y(x) + 4x
   (3)    ----------------------------
                          3
                     4y(x)
                                            Type: Union(Expression Integer, ...)
```

The *seriesSolve* operation can be used to obtain a series solution of a differential equation about a given point with given initial conditions.

35.6 DATA STRUCTURES

In AXIOM, a *data structure* is a domain that aggregates together one more objects, possibly of different types. In languages such as C++, data structures are often called *collection classes*. Data structures like *List Integer* are not mathematical objects, but some, like *SquareMatrix(3, Complex Float)* are. Most data structure domains belong to the category *Aggregate*. In the current version of AXIOM (Release 2.0), the following domain constructors create domains belonging to *Aggregate*:

ArrayStack	*Homogeneous-DirectProduct*	*OrderedDirectProduct*
AssociationList	*IndexedBits*	*PendantTree*
BalancedBinaryTree	*IndexedFlexibleArray*	*Point*
BinarySearchTree	*IndexedList*	*PrimitiveArray*
BinaryTournament	*IndexedMatrix*	*Queue*
BinaryTree	*IndexedOne-DimensionalArray*	*RectangularMatrix*
Bits	*IndexedString*	*Result*
CharacterClass	*IndexedTwo-DimensionalArray*	*Set*
DataList	*IndexedVector*	*SparseTable*

DenavitHartenberg-Matrix	*InnerIndexed-TwoDimensionalArray*	*SquareMatrix*
Dequeue	*InnerTable*	*Stack*
DirectProduct	*KeyedAccessFile*	*Stream*
DirectProduct MatrixModule	*Library*	*String*
DirectProductModule	*LieSquareMatrix*	*String Table*
EqTable	*List*	*Table*
FlexibleArray	*ListMultiDictionary*	*ThreeDimensional-Matrix*
GeneralSparseTable	*Matrix*	*Tree*
HashTable	*Multiset*	*TwoDimensionalArray*
Heap	*OneDimensionalArray*	*Vector*

Clearly there are too many domains to discuss here. We will note some important characteristics and then give some examples of the most commonly used data structures.

First note that many data structures are *homogeneous*: all contained objects have the same type. Those that are not homogeneous are usually collections of key-value pairs: *AssociationList, EqTable, GeneralSparseTable, HashTable*, and so on. AXIOM offers such a wealth of data structures so that you can choose the most efficient data representation to use in your application. These domains can be used directly in the interactive shell or for representations and intermediate datatypes in compiled domains.

Record is the most commonly used domain constructor for creating heterogeneous data structures. Although you can compose domain constructors in the interactive shell, you cannot define entirely new domain constructors (you need the system compiler, see Section 35.7 for more information). Here is how you might use *Record* and the interactive shell to test a representation for a domain you intend to write called *Molecule*.

First define *Molecule* as an alias for a domain whose objects are lists of records. Each record contains two components: an **"atom"** symbol entry for the atom abbreviation (such as H, C, O, Pb) and a **"count"** positive integer entry for the number of atoms of that kind that appear at that position in the list.[6]

```
Molecule := List Record( atom: Symbol, count: PositiveInteger)
   (1)  List Record(atom: Symbol,count: PositiveInteger)
                                                   Type: Domain
```

We next define two standard molecules. We put single quotes before the symbols to make sure they do not get evaluated. In this case we have not previously evaluated any statement such as **H := 2/3**, but it is good practice to quote the symbols in case you or someone else reuses your code in another situation. The output is the standard one for lists and records, but is not particularly appropriate for molecules.

[6]Note that the return type is *Domain*. Domains are objects and belong to the domain called *Domain*.

```
water : Molecule := [['H,2], ['O,1]]
   (2)  [[atom= H,count= 2],[atom= O,count= 1]]
                   Type: List Record(atom: Symbol,count: PositiveInteger)
carbonDioxide : Molecule := [['C,1], ['O,2]]
   (3)  [[atom= C,count= 1],[atom= O,count= 2]]
                   Type: List Record(atom: Symbol,count: PositiveInteger)
```

Let us write an interactive shell function to produce better output. The AXIOM domain *OutputForm* is the standard representation for expression output. *OutputForm* is not usually exposed, that is, its operations are not usually visible in the interactive shell. Among its operations are those with names "+," "–," "*," "/," and "^." Since almost everything can be converted to *OutputForm*, the interactive shell could easily get confused and produce an output representation instead of a polynomial when it encountered an expression like **7*x^4 + 2/3**. To tell the interactive shell "we want this operation to be gotten from *OutputForm*," we suffix the function call with "$OutputForm". This is called a package-call. For the sake of exposition, we've added line numbers to the function definition.

```
1.   showMolecule(m) ==
2.     o : OutputForm := empty()$OutputForm
3.     for r in m repeat
4.       t : OutputForm :=
5.        r.count > 1 => sub(r.atom, r.count)$OutputForm
6.        r.atom
7.     o := hconcat(o, t)$OutputForm
8.   o
                                                     Type: Void
```

This function definition uses indentation to indicate program structure and blocks. Line 1 indicates that we are defining a function called *showMolecule* and it accepts one argument which will be called **m** within the function definition. The interactive shell will determine the type of **m** and the return type of the function the first time we call the function. Our strategy is to define an empty object **o** of type *OutputForm* and then build it by horizontally concatenating new parts. Line 2 declares and defines **o**. We know that **m** will be a list of records. Line 3 starts an iteration across the records in **m**. Each one will successively be called **r**. Line 4 produces the term **t** that will be concatenated. Line 5 says that **t** includes the atom count if it is greater than 1, and line 6 says otherwise it is just the atom symbol. We access the parts of a record by using a dot (".") between the record variable name and the name of the record component. Thus **r.atom** and **r.count** access the atom and count components, respectively. "Dot" expressions can be used on the left-hand side of an assignment to set the record components. Line 7 horizontally concatenates the new term to right of the old output expression. Line 8 returns the final value.

Note that we don't actually call any functions to display the output. AXIOM automatically displays the result of an expression evaluation. We'll be able to see the desired output just by calling the function at the top-level of the interactive shell. We finish this example by doing just that for our two molecules.

```
showMolecule water
   Compiling function showMolecule with type
        List Record(atom: Symbol, count: PositiveInteger) -> OutputForm
   (5)  H₂O
```

Type: *OutputForm*

```
showMolecule carbonDioxide
   (6)  CO₂
```

Type: *OutputForm*

The *Union* domain constructor is syntactically similar to *Record.* However, instead of having several labeled components, objects in union domains have just one component at any given time. For example, *Union(d : DoubleFloat, f: Float)* objects contain either a *DoubleFloat* or a *Float* subobject. The case keyword is used with a component label to test for a particular kind of subobject. Once you know the kind of subobject the union object contains, use the dot operator (".") to extract the subobject.

```
union : Union( d: DoubleFloat, f: Float )
```

Type: *Void*

```
digits(40)
   (2)  20
```

Type: *PositiveInteger*

```
union := pi()$Float
   (3)  3.1415926535 8979323846 2643383279 502884197
```

Type: *Union(f: Float, . . .)*

```
union case d
   (4)  false
```

Type: *Boolean*

```
union case f
   (5)  true
```

Type: *Boolean*

```
union.f ^ 2
   (6)  9.8696044010 8935861883 4490999876 151135314
```

Type: *Float*

List and *Vector* are the most commonly used homogeneous data structures. Lists are most useful when you do not know in advance how many elements the structure will contain. Elements may be efficiently appended to the beginning of the list and *first* and *rest* can be used to iterate across the list.

```
l := [2,3,4,5]
   (1)  [2,3,4,5]
```

Type: *List PositiveInteger*

```
first l
   (2)  2
```

Type: *PositiveInteger*

```
rest l

   (3)   [3,4,5]

                                   Type: List PositiveInteger
```

The *cons* function appends an element to the beginning of a list and returns a new list. The new and old lists *share* structure: AXIOM does not make a copy of the old list. The second element of the new list is exactly the first element of the old list.

```
m := cons(1, l)
   (4)   [1,2,3,4,5]
                                   Type: List PositiveInteger
l
   (5)   [2,3,4,5]
                                   Type: List PositiveInteger
```

When you change an element in a list, that update is reflected in every list that shares that part of the structure.

```
m.3 := 99
   (6)  99
                                   Type: PositiveInteger
m
   (7)   [1,2,99,4,5]
                                   Type: List PositiveInteger
l
   (8)   [2,99,4,5]
                                   Type: List PositiveInteger
```

If you need to use a homogeneous aggregate of a fixed length, consider using a *Vector* object. Access to any one of its elements can be gotten in constant time, while a list object needs time proportional to the position of the element in the list. The *FlexibleArray* data structure is a good choice if you need constant access time but know that the size of the aggregate may change.

```
v : Vector Fraction Integer := [1/n for n in 1..6]

              1 1 1 1 1
   (1)    [1,-,-,-,-,-]
              2 3 4 5 6
                                   Type: Vector Fraction Integer
new(3, 0)$Vector(Integer)
   (2)    [0,0,0]
                                   Type: Vector Integer
```

Lists and vectors are inherently finite in length. What do you do if you need a linearly indexed collection of objects of potentially infinite size? Well, you cannot really store an infinite number of elements, but if you know how to compute the *n*+1th element given elements 1 through *n*, then you can use the *Stream* domain constructor. A stream can be used to hold an infinite sequence of numbers, such as the primes, or as a representation of a mathematical series.

Let us first define a stream of positive integers and then show how to create streams of primes in three different ways.

```
posints := [n for n in 1..]
   (1)  [1,2,3,4,5,6,7,8,9,10, . . .]
```
Type: *Stream PositiveInteger*

```
primes := [p for p in 1.. | prime? p]
   (2)  [2,3,5,7,11,13,17,19,23,29, . . .]
```
Type: *Stream PositiveInteger*

```
[p for p in posints | prime? p]
   (3)  [2,3,5,7,11,13,17,19,23,29, . . .]
```
Type: *Stream PositiveInteger*

```
select(prime?, posints)
   (4)  [2,3,5,7,11,13,17,19,23,29, . . .]
```
Type: *Stream PositiveInteger*

Given a stream, you can apply a function to every element in the stream. Given two streams, you can apply a function of two arguments to them element by element. In both cases you get new streams returned.

```
map(factor, posints)
   (5)  [1,2,3,2^2,5,2 3,7,2^3,3^2,2 5, . . .]
```
Type: *Stream Factored Integer*

```
map(_+, posints, primes)
   (6)  [3,5,8,11,16,19,24,27,32,39, . . .]
```
Type: *Stream PositiveInteger*

Elements in a stream are computed on demand. By default, AXIOM just displays the first 10 elements computed. To see them all, issue the following **)set** command.

```
)set streams showall on
primes
   (7)  [2,3,5,7,11,13,17,19,23,29, . . .]
```
Type: *Stream PositiveInteger*

If we ask for a prime later in the sequence, it and the preceding primes will be computed. Here we ask for the twentieth prime.

```
primes.20
     (8)  71
                                                    Type: PositiveInteger
  primes
     (9)
   [2,3,5,7,11,13,17,19,23,29,31,37,41,43,47,53,59,61,67,71, . . .]
                                             Type: Stream PositiveInteger
```

35.7 THE COMPILER, INSIDE AND OUTSIDE AXIOM

The AXIOM system compiler is used to create the library of categories, domains, and packages. The interactive shell can use these objects but they cannot be defined within it. The compiler can be invoked from within the interactive shell and the newly compiled objects loaded into it for testing or use.

Prior to Release 2.0 of AXIOM, the system compiler was an integral part of the runtime system. In Release 2.0, a new compiler written from scratch (in C) became the system compiler. Because it does not depend on executing within the AXIOM runtime system, the new compiler can be made available on platforms on which there now exists no AXIOM implementation. For availability information, please contact one of the NAG offices listed at the end of this AXIOM article.

The new system compiler supports many significant features that the older compiler did not. For example, functions that are "free," that is, not defined within a domain or package, can now be compiled and placed in the library. See Watt et al. (1994) for the definitive description of how to use the new compiler and how it differs from the old one.

Of great significance is that the new compiler can produce either Lisp code for use within AXIOM or C code to be used in stand-alone programs. As of this writing, the libraries used for these two applications are not identical. Work is underway to produce a common, layered library appropriate for both AXIOM work and independent applications.

35.8 HYPERTEXT HELP

One of the AXIOM component applications is HyperDoc, a hypertext help facility that contains the system documentation [most of what is in Jenks and Sutor (1992) and Watt et al. (1994)], additional examples, and a fill-in-the-blanks interface to the most commonly used AXIOM facilities. Two other facilities that use HyperDoc are the library browser and NAG Link. The on-line documentation is always the most up-to-date.

The underlying language for HyperDoc content is an extension of a subset of LaTeX. That is, not all LaTeX macros and environments are supported, but additional macros and environments to support hypertext, bitmaps, button controls, and the execution of AXIOM commands have been added. Users can write their own HyperDoc content and add it into the system. See the HyperDoc subtopic in the Reference topic gotten to from the HyperDoc top-level menu.

Interactive Examples

When you see an example AXIOM input in HyperDoc, you can either click on the box next to the expression to see the precomputed result or click on the expression itself to enter and evaluate the expression in an AXIOM window. Moreover, examples are stored with dependency information: if you click on an expression that requires definitions made in previous statements, HyperDoc will have AXIOM first evaluate the needed statements and then the expression you clicked.

If you click on a graph appearing in HyperDoc, the AXIOM graphics component displays the graph. You can then use the full interactive facilities of the graphics control panel to change the appearance of the graph. See Section 35.3 for additional information about AXIOM graphics.

The Browser

The AXIOM Browser is an application that uses HyperDoc as its display engine and the AXIOM runtime system as its database engine. On UNIX™ systems, HyperDoc and the AXIOM runtime system communicate via sockets. As users create queries by clicking on buttons and filling in fields, HyperDoc makes requests to the AXIOM runtime, which then executes the queries and returns display content in the LaTeX-like language HyperDoc processes.

The browser can retrieve and display the documentation and exports for each category, domain, and package in the AXIOM library. For each export, the browser can display its documentation and information about its implementation. Category ancestry can be listed for both domains and categories. Moreover, every client of a category can be shown. Searches using wildcards can be executed against the lists of domains, packages, categories, operations, and the documentation for each.

Clearly the browser is of great value to someone who is writing new AXIOM code and needs to know where to find and how to use elements of the system library. Even the most casual AXIOM user who does not care to deal much with notions such as domains can obtain significant knowledge by restricting his or her use of the browser to getting information about operations.

NAG Link

NAG Link is an interactive link between HyperDoc and the Numerical Algorithms Group Foundation Library. It allows you to call FORTRAN numerical routines using the HyperDoc interface or to call them directly from within an AXIOM workspace. AXIOM objects can be used as arguments to the numerical routines and the values returned are AXIOM objects. For efficient use of computing resources, computation can take place on another machine. You can use the browser for all AXIOM domains and operations supporting NAG Link. For those users who have seen the "chapter" organization of the NAG Foundation Library, the HyperDoc and browser organization of the NAG Link information will be quite familiar. For more information, please see the NAG Link topic available from the HyperDoc top-level menu.

Availability

AXIOM is marketed by the Numerical Algorithms Group (NAG). More information about the system can be obtained from the following sources:

NAG Inc., 1400 Opus Place, Suite 200, Downers Grove, IL 60515-5702, USA. Tel: 1-708-971-2337, Fax: 1-708-971-2706. World Wide Web: http://www.nag.co.uk:70/. Gopher: Host = www.nag.co.uk, Port=70.

NAG LTD, Wilkenson House, Jordan Hill Road, Oxford OX2 8DR, UK, Tel 44 865 511245, Fax 44 865 310139.

NAG GmbH, Schleissheimerstr. 5, D-8046 Garching bei Munchen, Deutschland, Tel: 49 89 3207395, Fax 49 89 3207396.

The preceding examples were produced on an IBM Corporation RISC System/6000 workstation running AIX 3.2.5. NAG also markets AXIOM for SUN workstations. Versions for other machine/operating system platforms (including Windows NT) will likely be available starting in 1995.

DEFINING TERMS

Category: A class of *domains*. A category is a collection of function signatures that is inherited by all other categories descended from the category and all domains belong to the category. Via inheritance, the collection of categories form a directed acyclic graph.

Coercion: Creating an object of one type from an object of another type without losing any information.

Conversion: Creating an object of one type from an object of another type, possibly losing some information (for example, conversion to an output representation).

Domain: A datatype: a collection of objects and functions with an associated *category*.

Interactive Shell: The AXIOM application that accepts user input, analyzes it, calls the appropriate library functions, and displays the output.

Package: A named, possibly parametrized, collection of functions. Alternatively, a *domain* with functions but no objects.

Signature: A function name, a list of arguments types, and the type of the return value. For example, *sin: Float -> Float* is a signature. Some signatures may contain argument names.

REFERENCES

Richard D. Jenks and Robert S. Sutor, *AXIOM: The Scientific Computation System*. Springer-Verlag, New York, 1992.

Stephen M. Watt and others. *The AXIOM Language Compiler User Guide*. Numerical Algorithms Group Ltd., Oxford, U.K., 1994.

36

DERIVE

Waldir L. Roque*
Universidade Federal do Rio Grande do Sul

36.1 INTRODUCTION

In the last three decades *Computer Algebra*, the field that deals with the development of techniques, algorithms, and creation of software for symbolic manipulation of mathematical expressions, formulas, and equations, has experienced enormous success. Currently, several softwares are available for computer algebra (see de Souza, Chapter 42, in this volume). With the new generation of processors and the high memory capability of the microcomputers, the use of computer algebra systems is no longer restricted to academia, but rather has spread to all kind of professionals who make use of mathematics as a tool for their job.

DERIVE[1] is a computer algebra system designed to run on PC microcomputers under DOS. Its executable file has only 246,302 bytes, which shows that with a little more free memory it can be run successfully. The current version of the system, called DERIVE XM, version 2.56, can profit from extended memory, while the previous versions were designed to run with only 640 kilobytes of main memory.

One of the nicest features of the computer algebra systems available is that the system can run interactively, providing the user with mathematical conversation simply and naturally. An addition feature of DERIVE is that it takes advantage of menu-oriented systems. Unfortunately, DERIVE is not as full a programming language for computer algebra as other systems, but rather is a highly sophisticated and powerful *computer algebra desk tool.*

*Email: roque@mat.ufrgs.br
[1]DERIVE is a trademark of The Soft Warehouse Inc., Honolulu, Hawaii, USA.

0-8493-2530-7/96/$0.00+$.50

In this chapter we provide an introduction to DERIVE's capabilities. There is no intention to show all of its strengths or to exhaustively explore the system to demonstrate the limits of its computational potentiality or its vulnerability. Rather, we intend to show how DERIVE can be useful to help professionals solve mathematical problems.

In this chapter the typewritten texts denote DERIVE menus or input commands, while the output are given in bidimensional mathematical notation similar to what is displayed by the system.

36.2 INTERACTING WITH DERIVE

The best and perhaps the easiest way to learn how to use a computer system is to apply the old, but rather efficient methodology of *try it by yourself.* In this regard, I will pretend that the reader has a running version of DERIVE XM.

To start our tour of DERIVE's capabilities, let us start loading DERIVE, invoking at the DOS prompt the executable file `derive.exe`. Soon after, the system exhibits its initial screen (see Fig. 36.1), which gives access to all high-level menu facilities.

```
                            D E R I V E   X M
                        A Mathematical Assistant

                              Version 2.56


                  Copyright (C) 1988 through 1993 by
                          Soft Warehouse, Inc.
                   3660 Waialae Avenue, Suite 304
                  Honolulu, Hawaii, 96816-3236, USA

If you have received this product as "shareware" or "freeware", you have an
unauthorized copy, because it is a violation of our copyright to distribute
DERIVE on a free trial basis.

To obtain a licensed copy, or if you know of any person or company distributing
DERIVE as shareware or freeware, please contact us at the above address or fax
(808) 735-1105.

                             Press H for help

COMMAND: Author Build Calculus Declare Expand Factor Help Jump soLve Manage
         Options Plot Quit Remove Simplify Transfer moVe Window approX
Enter option                                                       Derive XM
                                   Free:100%                         Algebra
```

FIGURE 36.1 DERIVE initialization window.

In the top part of the DERIVE initial screen, the user can see details about license policy and the text `Press H for help`. At the bottom part, after the status line, there appears the top-level menu's options, at the left corner the message line `Enter option`, which states that the system is available for executing commands, at the center the free function, and at the right corner the current window status, which is `algebra`.

The help menu is shown by pressing H. It provides access to instructions on several aspects of the system. Scrolling up and down the user gets an indication of how much can be done with DERIVE. To return to the initial screen press R.

On the initial screen, the highlight is on the top-level menu option `Author`. To scroll the highlight to the other menus the `Tab` and `Backspace` keys can be used. To invoke one of the menu's options just press `<enter>` when the desired option is highlighted. A quicker way to invoke a menu is to press the capital letter for the corresponding menu option chosen. This avoids spending extra time jumping the highlight.

DERIVE has three operational window modes: algebra, 2D plot, and 3D plot. The window status indicates which window is currently in action. These windows will be discussed in the sequel.

The `Free` function exhibits the percentage of free memory that is still available for execution according to the memory initially available to the system. To see the amount of initial memory that is available for the system, a new feature has been introduced in the DERIVE XM version. It can be found in the *hardware settings* through the `Help State` menu sequence.

A DERIVE session can be finished at any time by using the `Quit` command. This menu option is available at the top-level menu options for each one of the three operational windows.

Each top-level menu gives access to another submenu level, which may have yet another sublevel, and so on, until the final action is reached. The previous menu levels can be reached by pressing the `Esc` key.

One important feature of DERIVE is the possibility of turning to the DOS environment without quitting the session. As a result, the `Options Execute` menu sequence freezes the DERIVE session and any DOS command may be executed. In fact, we can use an editor to edit and/or see a file that has been output by the system. To return to DERIVE just enter the DOS command `exit`. Note: *do not use* any resident program within a frozen DERIVE session as it may cause an abnormal behavior.

36.3 ALGEBRA WINDOW

The algebra window provides an environment for executing symbolic and numeric computations. The main command option to pass on expressions to the algebra window is `Author`. However, the `Transfer` menu also allows you to do so, as will be seen below.

The Author Menu: Editing Expressions

When the `Author` menu is invoked, an editing environment is opened up where mathematical expressions can be edited. To see the commands of the editor, look at the help editing (`HE`) menu option. The commands are similar to the Wordstar text editor.

At the editing line any syntactically correct sequence of DERIVE data structure can be typed in. If any incorrect syntax occurs the system warns the user with a beep. If it was successful, the expression is sent to the algebra window above following the standard bidimensional mathematical notation. No computation happens until the user specifies what has to be done with the expression.

For instance, suppose we want to enter the expression 1/2 + 3/4 into the algebra window. Pressing `Author`, the system prompts you to enter the `AUTHOR expression:`

```
1/2 + 3/4
```

Pressing the <enter> key, it will appear in the algebra window as

```
    1   3
1:— + —
    2   4
```

This expression is highlighted. Once we have several expressions in the algebra window, the arrow keys are used to move the highlight up and down; the left and right arrows move the highlight horizontally. This is a good facility to copy past expressions to the `Author` line or to execute a command only on a subexpression.

To compute the result of the sum above, we should `Simplify` it. When this menu is invoked, it asks the user if the expression to be simplified is the one that has the number indicated. By default, the system always calls the number of the highlighted expression in the algebra window. By pressing <enter>, the result is

```
    5
2: —
    4
```

By pressing ^J (control J) soon after editing the expression, the result pops up on the algebra window, avoinding the call of `Simplify`.

As in the example above, DERIVE works with an exact arithmetic for integers and rationals. Now let us enter the expression $(x + a)^2$ on the algebra window.

```
3: (x + a)2
```

As a computer algebra system, DERIVE interprets both `x` and `a` as free variables. They have no attribute, standing by themselves.

By default, it can be changed by the user; the system is case insensitive. The case may be changed by the `Options Input` menu sequence, which provides the choice of setting `Mode: Character Word` and `Case: Insensitive Sensitive`.

The `Character` input mode allows undeclared variables to be formed with a single Latin or Greek alphabetical letter, while the `Word` input mode lets undeclared variables to be formed as a set of alphanumeric characters, as long as they start with a letter. One good point of the character input mode is that we may avoid inserting the infix multiplication operator `*`, as `xy` will be understood by the system as `x * y`. On the other hand, it can be inconvenient to use mnemonic undeclared variables, as for instance, the word `force` would be understood as `f*o*r*c*e`, which is not what we meant. This is solved by activating the word input mode.

The case sensitivity can be set by choosing either `Insensitive` or `Sensitive`. To avoid too much work, just set the system to `Option Input Character Insensitive`. This can be set up as an initial configuration of the system by changing the `DERIVE.INI` file.

The Transfer Menu: Moving Things

The `Transfer` menu allows the system to communicate with other devices, such as a printer or a drive. It has the following submenus sequence: `Load, Save, Merge, Clear, Demo, and Print`.

Loading Program Files

Expressions can also be sent to the algebra window through the use of the `Transfer Load` menu. This submenu provides the options DERIVE `State daTa Utility`. The DERIVE option allows the user to load into the system a file consisting of DERIVE'S commands. These files should have a filetype `.MTH`.

A numeric data file, which should have a filetype `.DAT`, is loaded with the `daTa` option, and finally the `Utility` is used to load the so-called utility files, which are files containing more complex mathematical function definitions. These files are provided with the system and have also a `.MTH` filetype.

Saving Results

In the `Author` environment the user has to write expressions according to DERIVE'S input syntax. That is a standard linear format of expressions including the character set, operators, etc. However, the output displayed in the algebra window is bidimensionally formated to be closer to what humans normally do with paper and pencil. Although this is visually more attractive, this bidimensional format is not readable by the system. To reload a file into the system, it has to be in the input format, i.e., in the `.MTH` format.

The `Save` submenu provides facilities to save expressions in different notations. It has the following options: DERIVE, `Basic, C, FORTRAN, Pascal, Options,` and `State`.

To save an expression of the algebra window in the DERIVE'S input format, the `Save` DERIVE menu sequence should be applied. This option asks the user to provide the name of the file and its location (drive and directory); the `.MTH` filetype is automatically inserted by the system. Long lines are automatically written on several lines having a tilde (~) as the connector.

The options `Basic`, `C`, `FORTRAN` and `Pascal` allow an expression from the algebra window to be saved in the respective language input format. For instance, the expression

$$ax^3 - x^2/b + (c - d)x - 1$$

is translated to Pascal through the `Transfer Save Pascal` to the expression `a*POW(x,3) - SQR(x)/b+(c-d)*x-1`.

The `Options` submenu prompts: `Range: All Some` and `Length: 79`. With `Range All` the whole algebra window will be saved, while with `Range Some` the systems ask for the range of lines to be saved. The `Length` specifies the column number to be wrapped.

The `State` submenu is quite useful as it allows the user to reset the initial configuration of the system. In other words, the current settings can be saved in the `DERIVE.INI` file. For instance, we can save the settings `Word` and `Insensitive` input modes and save them as the default configuration when the system initializes.

Appending Expression and Cleaning the Algebra Window

The `Transfer Merge` submenu is applied to append the contents of a `.MTH` file to the expression sequence in the current algebra window. It is important to use this command instead of load, as the latter would overwrite the algebra window, losing the previous expressions.

To erase all the expressions written in the algebra window, without quitting and starting a fresh session, use the `Transfer Clear` submenu. It is a very important command as it clears all constants, attributions, functions, variable domains, etc., that have been declared during that session. The `Remove` menu just removes the expressions from the screen without clearing them.

DERIVE provides the system distribution with some demonstration files. The `Transfer Demo` submenu is used to load a demonstration file, which can be either a system distribution file or a user prepared file. Demonstration files should have a filetype `.DMO`. Through this command the file being read is automatically simplified as with `Simplify`. This facility is very useful for training and education purposes (Zavin, 1992; Small, 1994).

Printing Graphics and Expressions

DERIVE has improved its facilities to print both expressions and graphics from the 2D and 3D plot windows.

The `Transfer Print` has the following subsubmenus: `Printer`, `File`, `Layout`, and `Options`.

The `Printer` has three options: `Expressions`, `Screen`, and `Window`. The `Expressions` command sends the highlighted expression in the algebra window to the printer. The `Screen` sends the printer the graphic image of the entire algebra window screen. A quick way to do the same action is just pressing `Shift-F10`.

The `Window` command has the options `All` and `Current`. The `All` option sets the `Transfer Print Printer` to send the entire graphic image of all opened windows to the printer, while the `Current` option sets it to send only the current activated window. The latter can be done quickly by pressing `Shift-F9`.

Sometimes it is necessary to have a portable file containing the information that was on the windows, particularly a graphic image of a plot. The `File` submenu provides this facility. It has the same command options as the `Printer` menu. However, its function is to generate either an ASCII file when `Expression` is activated or a tagged image file (TIF) for the other subcommands. TIF files can easily be loaded by several word processors and desktop publishing software systems.

`Layout` submenu provides options for setting the page layout, and the `Options` provides printer configuration commands.

Multiple Windows

DERIVE allows multiple window modes. Through the top-level `Window` menu the list of submenus can be reached to create and manipulate multiple windows. The `Window` submenu options are `Close`, `Designate`, `Flip`, `Goto`, `Next`, `Open`, `Previous`, and `Split`, which by their own may have some other command options.

The `Designate` submenu provides options to specify the kind of window into which the current algebra window should be transformed. Its command options are `2D-plot`, `3D-plot`, and `Algebra`. When either of these command options is invoked, DERIVE queries the user if the current highlighted expression in the algebra window should be abandoned or not. After that, it transforms the current window into the specified one.

When more than one window is simultaneously opened, a number flag indicates which one is in action. The submenus `Flip`, `Goto`, `Next`, and `Previous` are options to jump around the windows. `Close` closes the current window or the specified one and Open gives the same options as Designate with the difference that a new window will be opened in addition to the existing ones.

In one single screen we can have more than one window opened simultaneously. The Split submenu is used to specify a vertical or horizontal splitting of the current window. In Figure 36.6 we show a screen split into algebra and a 2D plot windows.

36.4 HANDS ON DERIVE

Arithmetic

One of the main features of the computer algebra systems is their exact arithmetic for integers and rationals and arbitrary precision arithmetic for reals, in contrast to numeric systems, which work with floating point numbers.

Let us show some simple examples. With Author enter the expressions and then Simplify. Simple calculations can be done in a straightforward manner.

```
    5(9-3)
1: -------
      2

2: 15

    108     33
3: ---- + ----
    29      4

    1389
4: -----
    116
```

To get the decimal representation of expression just make use of approX. This menu asks which expression is to be approximated. The system's default has six-digit precision. The result for 4: is then

```
5: 11.9741
```

The precision can be changed by using the menu sequence Options and Precision, which will prompt two alternative submenus Mode: Approximate, Exact, and Mixed, and Digits: 6.

Exact mode is the default. It is appropriate for doing exact calculations, which for heavy computations can be much more memory demanding. Approximate mode is used to do approximate calculations, with any number of digits of accuracy, which is somewhat limited only by the capacity of the hardware . Of course, large numbers of digits can exhaust the available memory. Mixed mode is a hybrid mode, where the irrational numbers are approximated, but rational numbers are not.

The irrational numbers pi (π) and the Neperian *e* can be approximated according to the user's desire. The identifier pi is reserved for DERIVE. It can be entered with Author in two ways: one is just typing the identifier or, alternatively, by pressing the Alt P keys simultaneously.

The irrational e is the base of the natural logarithm. It is entered with `Author` as `#e`, and appears on the algebra window as $\hat{e}$.

For instance, with `Digits` the values of $\sqrt{12}$, π, and e^2, with 18, 55, and 69 digits, respectively, are shown in Figure 36.2.

```
1:   √12

2:   3.46410161513775458

3:   π

4:   3.141592653589793238462643383279502884197169399375105820

       2
5:   ê

6:   7.38905609893065022723042746057500781318031557055184732408712782252257

COMMAND: Author Build Calculus Declare Expand Factor Help Jump soLve Manage
         Options Plot Quit Remove Simplify Transfer moVe Window approX
Compute time: 0.0 seconds                                          Derive XM
Approx(5)                          Free:100%                         Algebra
```

FIGURE 36.2 Large numbers: $\sqrt{12}$, π, and e^2 with 18, 55, and 69 digits, respectively.

Complex Numbers

DERIVE can do computations with complex numbers. The imaginary number $i = \sqrt{-1}$ is input with `Author` as `#i`. In the algebra window it appears as $\hat{\imath}$. The complex expressions `(1 + 2 #i)(3 - #i)` and `(-8) ^1/3` `Simplify` to:

1: $5 + 5\hat{\imath}$

2: $1 + \sqrt{3}\hat{\imath}$

The numbers may be written in different notations. Four notation styles are allowed in DERIVE. They can be set up through the menu `Options Notations`, which leads to the submenus `Styles: Decimal, Mixed, Rational,` and `Scientific,` and `Digits: 6`.

Algebraic Manipulation

The main feature of computer algebra systems is their power to manipulate mathematical expressions symbolically. DERIVE can manipulate quite complex algebraic expressions, simplify and collect terms, apply identities, expand polynomial products, find out the greatest common divisor of two polynomials, factorize expressions, etc.

Let us enter with `Author` the polynomial expression $(ax + b)^5$, where we assume that a, b are constants. To expand this expression the menu command `Expand` can be used. It prompts the user to enter the expression and then prompts to select the variable.

The expansion can be done in just one of the variables or to a sequence of variables. For instance, suppose we intend to expand the expression $(x + ay - b)^3$ in terms of x only, or y only. The results are shown in Figure 36.3. Notice that the term

$(ay - b)$ remains unexpanded. If the expansion is to be done for all variables, just hit the return key in the variable selection prompt.

In Figure 36.3, the expression 6: is the expansion of 5: with respect to y. The expansion would be a very long term if it is taken for x only, or for both x and y.

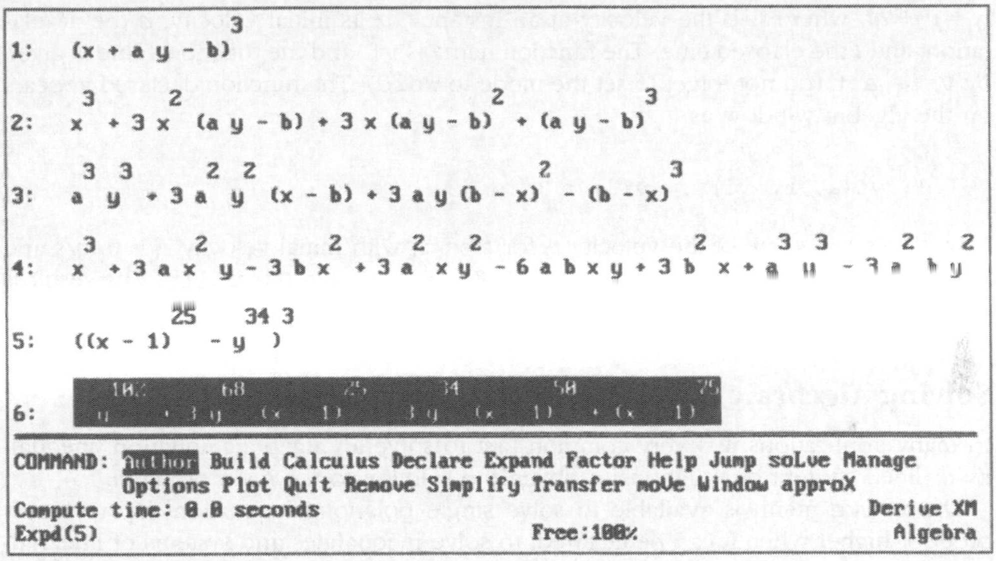

```
                3
1:  (x + a y - b)

     3   2          2           3
2:  x  + 3 x  (a y - b) + 3 x (a y - b)  + (a y - b)

     3 3      2 2           2             3
3:  a y  + 3 a  y  (x - b) + 3 a y (b - x)  - (b - x)

     3       2        2      2          2     3 3      2 2
4:  x  + 3 a x  y - 3 b x  + 3 a  x y  - 6 a b x y + 3 b  x + a  y  - 3 a  b y

             25    34 3
5:  ((x - 1)   - y  )

     102      68       25       34       50       75
6:  y    + 3 y   (x - 1)  - 3 y   (x - 1)  + (x - 1)

COMMAND: Author Build Calculus Declare Expand Factor Help Jump soLve Manage
         Options Plot Quit Remove Simplify Transfer moVe Window approX
Compute time: 0.0 seconds                                          Derive XM
Expd(5)                              Free:100%                       Algebra
```

FIGURE 36.3 Expanding expressions.

The Value of an Expression

To compute the value of an expression or to replace one variable or subexpression of a given expression, one can attribute values to the variables or parameters and then `Simplify` the intended expression to get the result. For instance, the value of the expression $(ax + b)^5$ for $x = 1$, $a = 2$, and $b = 3$ may be obtained by entering

```
1: a := 2
2: b := 3
3: x := 1
```

Here the symbol := means an attribution, or an assignment of a value to a variable. A variable with an attribute is no longer free; its value is the value assigned to it.

Assignment to variables can also be done by using the menu command sequence `Declare Variable Value`. It is possible to assign to a variable other characteristics such as `Positive`, `Nonnegative`, `Real`, `Complex`, or even an `Interval`.

With an attribute, the value of the expression is found if we `Simplify` it. To get a new value, either new assignments have to be done to the variables or, alternatively, the menu sequence `Manage Substitute` can be applied.

This menu sequence prompts for an expression and then asks for substitution to take place for each one of the variables involved in the expression. After that, the substituted values are shown on the algebra window. To compute the substitution just `Simplify` the new expression.

Notice that substitutions can be done from one variable to an expression, or from an expression to another. This feature allows you to compute a change of coordinates or composition of functions simply and quickly.

Defining Functions

In DERIVE functions can be created quite easily with the menu sequence `Declare Function`. It prompts the user to specify the function identification and then the function expression. Suppose we intend to write the velocity function of a particle, say $v_f = v_i + at$, where v_f is the velocity at an instant t, v_i its initial velocity, a the accelaration, and t the elapsed time. The function name is `vf`, and the function value is given by `vi + a t` (do not forget to set the mode to `Word`). The function declared appears on the algebra window as

```
1: vf(a, t, vi) := vi + a t
```

Now to get the value of the velocity v_f for $t = 5$ s, with initial velocity $v_i = 0$ m/s and accelaration $a = 9.8$ m/s^2, just write `vf(9.8, 5, 0)` and `Simplify`. The result is 49 m/s.

Solving Algebraic Equations and Linear Systems

In many applications it is very common first to solve an algebraic equation, inequality or linear system of equations to obtain the solution to the main problem.

The `soLve` menu is available to solve single polynomial equations up to fourth order or higher when it can be factored, to solve inequalities and systems of linear algebraic equations. This menu asks for an expression to be solved and then asks you to select the variables which the system has to be solved for.

To solve the simple linear equation: $ax + b = 7$ we can enter the equation using `Author` and then press `L`. The `soLve` menu will query which one of the identifiers a, b, or x should be considered as a variable. If x is chosen, the answer will be straightforward.

Alternatively, `soLve` can be seen as an operator. Therefore, with `Author` we can enter the following command:

```
1: solve(a x + b = 7, x)
```

By pressing `Ctrl J`, or `<enter>` then `Simplify`, the equation will be solved and the result displayed on the algebra window.

The above equation is a very simple one and so there is no real need to use a computer system to find the solution. Nevertheless, for more complicated relationships, expressing one variable in terms of others can be time-consuming and a tricky task.

Consider the equation $x^3 - 3/4x^2 + 9/16x - 27/64 = 0$. How long would you take to solve it by hand? Perhaps you would have to look up a table to remember the rules to find out the roots of a third-order algebraic equation. With DERIVE XM enter this equation and use the `soLve` menu; the three roots will appear in less than 1 second. Notice that the equation has one real solution and two complexes (see Fig. 36.4). The line `5:` in Figure 36.4 shows the operator `SOLVE` being applied to the same equation. Line `6:` gives the solution after `Simplify`. Their values can be better obtained with `approXimate`, which is shown at line `7:`.

A nice way to improve your understanding of the equation is to plot it. In the section on 2D plot you can see how to do it. The plotting is very useful to estimate the value of roots for transcendental equations, for instance.

```
       3    3  2    9      27
1:    x  - --- x  + --- x - ---
            4       16      64

           3
2:    x = ---
           4

             3 î
3:    x = - -----
              4

           3 î
4:    x = -----
             4

            ┌ 3    3  2    9       27   ┐
5:    SOLVE │x  - --- x  + --- x - ---, x│
            └      4       16      64   ┘

       ┌    3         3 î         3 î ┐
6:     │x = ---, x = - -----, x = ----- │
       └    4           4           4 ┘

7:     [x = 0.75, x = -0.75 î, x = 0.75 î]

COMMAND: Author Build Calculus Declare Expand Factor Help Jump soLve Manage
         Options Plot Quit Remove Simplify Transfer moVe Window approX
Compute time: 0.0 seconds                                          Derive XM
Approx(6)                                Free:100%                   Algebra
```

FIGURE 36.4 Solution of algebraic equations.

The operator soLve can also be used do solve inequalities.

2: solve(x^2 - 5 a < 7, x)

$$3 : \left[|x| < \frac{\left| 5a + 7 \right|^{3/2}}{5a + 7} \right]$$

DERIVE gives the answer in terms of absolute values. In fact, we can solve the following inequalities:

4: solve(|2 x - 1| > 3, x)
5: [x < -1, x > 2]
6: solve(|x^2 + 1| > 1,x)
7: [x/ = 0]
8: solve(|x^4 + 2| > 0,x)
9: [x = @1]

The cases above show the use of the abs operator to compute the ***absolute value*** of an expression. The solution number 7: is to be understood as [$x \neq 0$] and 9: shall be understood as an arbitrary number. When the equation or inequality is degenerate, the symbol @n is used by DERIVE to indicate an arbitrary number, with *n* 95 a sequential numbering label.

For systems of linear algebraic equations, the SOLVE menu can also be used. To specify the set of equations enter them in a square bracket with each equation separated by a comma. The systems $2ax + 3y = 5$, $x - b^3y = -3$, shall be entered as

```
11: [2 a x + 3 y = 5, x - b^3 y = -3]
```

SOLVE will ask the user to choose the set of variables to be solved for. The solution for x, y is then:

$$12: \left[x = \frac{5b^3 - 9}{2ab^3 + 3}, \quad y = \frac{5b^3 - 9}{b^3\ (2ab^3 + 3)} + \frac{3}{b^3} \right]$$

The operator SOLVE can be used instead. As its first argument just write the set of equations in square brackets and as the second argument, the variables in square brackets, as illustrated below:

```
13: solve([2 a x + 3 y = 5,x - b^3 y = -3],[x,y])
```

For nonlinear systems of algebraic equations DERIVE provides the utility file SOLVE.MTH, which has two operators with implementation of the Newton and fixed-point approximate methods.

Predefined Functions

The user can take advantage of predefined functions in DERIVE, such as the following:

exp(z), the exponential function. It can also be entered as #e^z, where z is a complex number.

`ln(z)`, the natural logarithm function. The `log(z,a)` represents the logarithm function of z to the base a.

`sin(z)`, the sine function. The angles are all in radians. If degrees are to be used, the identifier has to be attached to the number, as `60deg`, which will be understood as 60 degrees. If you `Simplify` an angle given in degrees, the result will be the equivalent value in radians.

`cos(z)`, the cosine function.

`tan(z)`, the tangent function. The tangent function is defined as `sin(z)/cos(z)`. Try to compute `tan(pi/2)`. The result obtained by DERIVE is $\frac{1}{0}$, which is not mathematically sound. Try to `Simplify` this result! DERIVE could provide a beep or have an intelligent way to avoid such mathematical indeterminacies. In fact, even the limit gives the same result.[2]

`cot(z)`, the cotangent function. The cotangent function is defined as `cos(z)/sin(z)`.

`sec(z)`, the secant function.

`csc(z)`, the cosecant function.

`sinh(z)`, the hyperbolic sine function of z.

`cosh(z)`, the hyperbolic cosine function of z.

`tanh(z)`, the hyperbolic tangent function of z.

`coth(z)`, the hyperbolic cotangent function of z.

`sech(z)`, the hyperbolic secant function of z.

[2]The new DERIVE version 3.02 provides the correct answer.

csch(z), the hyperbolic cosecant function of z.
sqrt(z), the (positive) square root function. This function provides the positive value of the square root.

The inverses of the trigonometric functions are given by asin(z), acos(z), atan(z), acot(z), asec(z), and acsc(z). The inverse functions provide the angles in radians. The operator atan(y,x) gives the angle between the vector $\mathbf{v} = (x, y)$ and the positive x-axis. Notice that the arguments for the operator are given in the opposite order as the vector coordinates.

The inverse of the hyperbolic functions are given by asinh(z), acosh(z), atanh(z), acoth(z), asech(z), and acsch(z). Some of the properties of the above functions are already assumed by the system. For instance,

$$e^z e^w = e^{z+w}, \quad (e^z)^\alpha = e^{\alpha z}$$

with α an integer, are valid. However, if the user intends to change these rules, the Manage menu provides submenu facilities for doing that. It has the following sumenus: Branch, Exponential, Logarithm, Ordering, Substitute, and Trigonometry.

The Manage Exponential gives the following options: Direction: Auto, Collect, and Expand. With Collect (the default) the left-hand sides of both relations given above are changed by their corresponding right-hand side. Expand does the opposite, applying the rules from the right to the left side. The Auto option assumes the first rule from the left to the right and the second rule from the right to the left side.

The logarithm properties $\ln(z) + \ln(w) = \ln(z + w)$, $\alpha \ln(z) = \ln(z^\alpha)$ can also be changed according to Manage Logarithm in the same way as for the exponential function. Also, trigonometric properties can be set up by adequate choice of the options from the submenu Manage Trigonometry, which are Direction: Auto, Collect, and Expand Toward: Auto, Sines, and Cosines.

An easy way to make a change of variables in an expression, either for a value, another variable, or expression, is to apply the Manage Substitute menu sequence, which takes the variables in the expression on the algebra window and asks for the new arguments. Note that this is a label substitution, being allowed to change a variable, say x, for an expression that contains x in itself.

Several additional functions and/or operators are predefined in DERIVE, extending its power. In the Help F menu, there is a list of functions and constants.

36.5 MANIPULATING WITH VECTORS AND MATRICES

Vectors and matrices are very common mathematical objects. In DERIVE they can be manipulated symbolically (Horbatsch, 1990) through theirs components.

Vectors

A vector is a mathematical object that has a magnitude and a direction. The usual representation of a vector in certain bases $\{\mathbf{e}_1, \mathbf{e}_2, \ldots, \mathbf{e}_n\}$ where n denotes the vector space dimension, is given by

$$\mathbf{v} = v_1\mathbf{e}_1 + v_2\mathbf{e}_2 + \cdots v_n\mathbf{e}_n$$

where the v_i, $i = 1, \ldots, n$, are the vector components.

A vector can be defined in DERIVE by using the menu command sequence `Declare` VECTOR. It prompts the user, first to specify the vector space dimension through the number of components and then by giving the vector components one by one.

Alternatively, a vector can be entered by using the `Author` menu. In this way a vector is a sequence of terms (the components) written in square brackets. For instance, `[a,b,c]` represents a vector in a three-dimensional vector space, whose components are *a, b, c*.

There is another facility to define a vector whose components are generated according to a rule. The VECTOR operator can be applied to easily generate such vectors. Its syntax is `VECTOR(F,X,M,N,S)`, where `F` is an expression, and `X` indicates the variable in the expression to be simplified from value `M` to `N` according to step size `S`.

As an example, suppose we intend to generate a vector **v** with 10 components given by 2^n, for n = 1,2,. . . 10. This is generated with VECTOR as

```
1: VECTOR(2^n , n, 1, 10)
2: [2, 4, 8, 16, 32, 64, 128, 256, 512, 1024]
```

When the step size is 1 it can be omitted.

The main operations defined for vectors are predefined in DERIVE. The magnitude (length) or Euclidean norm of a vector can be found applying the `ABS(v)` operator over a vector.

```
3: |(VECTOR(2^n , n, 1, 10)|
4: 10√13981
```

The dot and cross product are defined, respectively, as

```
5: [x,y,z] . [a,b,c]
6: ax + by + cz
7: cross([x,y,z], [a,b,c])
8: [cy − bz, az − cx, bx − ay]
```

Differential operators GRAD, DIV, CURL, and LAPLACIAN have also been defined in DERIVE to compute the gradient, the divergence, the curl, and the Laplacian.

```
9: GRAD(x^2 + y^2 + z^2)
10: [2x, 2y, 2z]
11: DIV([a x^2 , b y^2 , c z^2 ])
12: 2ax + 2by + 2cz
13: CURL([y x^2 , z y^2 , x z^2 ])
14: [-y^2,-z^2,-x^2]
15: LAPLACIAN(x^2 y z^2)
16: 2x^2y + 2yz^2
```

Matrices

A matrix can be defined by the menu sequence `Declare Matrix`. Initially, this menu sequence requests you to enter the number of rows and columns of the matrix and then each matrix element has to be specified. Observe that by default the number of rows and columns is set equal to 3 and each element is set initially equal to zero.

Alternatively, a matrix can be defined by entering with `Author` a vector where each one of its elements corresponds to a vector that represents a row of the matrix. For instance, `[[a,b,c],[d,e,f],[g,h,k]]` leads to

$$1: \begin{bmatrix} a & b & c \\ d & e & f \\ g & h & k \end{bmatrix}$$

Any element of a vector or matrix can be extracted with the operator `ELEMENT`. For a vector, its syntax is `ELEMENT(V,n)` where V is a vector and n corresponds to the nth element of the vector. For a matrix it is `ELEMENT(M,i,j)`, where i corresponds to the ith row and j to the jth column of the matrix M.

The transpose of a matrix is obtained when the operator " ` " is applied to it as follows: `[[a,b,c],[d,e,f],[g,h,k]]``.

The operators `TRACE` and `DET` are defined for square matrices to compute the trace and determinant, respectively. For instance, `TRACE([[a,b,c],[d,e,f],[g,h,k]])` and `DET([[a,0,0],[0,b,0],[0,0,c]])`, give, respectively

```
2: a + e + k
3: a b c
```

Eigenproblems

Many problems in engineering are eigenproblems. In other words, they are characterized by equations of the form $L[u] = \lambda u$, where L is an operator or transformation, λ is called an eigenvalue, and u an eigenvector. To solve such problems we need to know first the characteristic polynomial of the transformation u.

The predefined operator `CHARPOLY(M,k)` provides the characteristic polynomial of the matrix M in terms of a parameter k.

$$4: \text{CHARPOLY} \begin{bmatrix} a & 0 & 0 \\ 0 & b & 0 \\ 0 & 0 & c \end{bmatrix}$$

$$5: (a - k)(b - k)(c - k)$$

The operator `EIGENVALUES(M,k)` gives the eigenvalues of the matrix m in terms of k. For both operators, if the parameter k is omitted, the default parameter is w.

Additional features to deal with vectors and matrices can be found in the distribution file VECTOR.MTH.

36.6 CALCULUS WITH DERIVE

Differential calculus plays an important role in science and engineering. In fact, one of the first things that is taught to undergraduate students is the notions of limits, derivatives, and integrals.

The behavior of a single-valued function can be investigated by looking at how it behaves in the limits of its domain and near the singular points, finding out the regions where the function is increasing and/or decreasing, where it has its maximum and minimum values, finding out in which intervals the function changes its shape, etc. All this information can be obtained using the notions of limits and derivatives.

With DERIVE most of these calculations are done in a straightforward manner (Leinbach, 1991). The system provides facilities to do calculus through the top level `Calculus` menu. Once you press `C`, the system exhibits the set of submenus `Differentiate`, `Integrate`, `Limit`, `Product`, `Sum`, and `Taylor`.

Limit of Functions

Let us consider the function $y = e^{-1/x}$ as a working example. This function is well defined for all values of x, on the real line such that $x \neq 0$. To know the behavior of y when x approaches 0 we should take the limit of the function.

Entering the function with `Author`, the limit can be found using the sequence `C Limit`. The limit submenu provides the option to either write the expression or specify the entry number of the expression to be considered. By pressing the return key, the system queries which variable has to be the limit taken, and finally prompts the user to choose the limiting point and to specify whether the limit shall be taken from the left or right side, or both.

In our example, let us choose the option `Left`, `Right`, and finally `Both`. The results are obtained after simplifying the expressions.

1: $\lim_{x \to 0^-} e^{-1/x}$

2: ∞

3: $\lim_{x \to 0^+} e^{-1/x}$

4: 0

5: $\lim_{x \to 0} e^{-1/x}$

6: ?

As the limit from the left is different from the right, the limit of the function at 0 is not determined, as is pointed out by the question mark ?.

These limits just give information on the behavior of the function on the neighbor of $x = 0$. Now you can consider the limit for $x \to \pm\infty$. We can see with DERIVE that

7: $\lim_{x \to \infty} e^{-1/x}$

8: 1

An alternative way to get the limit is to use the `LIM` operator. In our example above the expression $\lim_{x \to \infty} e^{-1/x}$ should be entered with `Author` as `LIM(#e^(-1/x),x,0,1)`.

The last argument of the LIM operator assumes the value +1 when the limit is taken from the right and −1 when it should be taken from the left. If the limit is to be taken from both sides, the fourth argument can be omitted.

To know whether this function has a maximum or a minimum, we need to find out its critical points, i.e., the points that make the first derivative of the function vanished, $dy/dx = 0$.

Derivatives

The derivative of a function can be found using the `Calculus Differentiate` submenu, which prompts you to specify the expression; then the variable with respect to the derivative should be taken and finally the order of the derivative. The default is first-order derivative. Doing so for our function in the above example we get on the algebra window:

$$1:\ \frac{d}{dx}\, e^{-1/x}$$

With Simplify we immediately obtain:

$$2:\ \frac{e^{-1/x}}{x^2}$$

Alternatively, the `DIF` operator can be used to find the derivative of a function. Its syntax is `DIF(f,x,n)`, where *f* means the expression, *x* the variable, and *n* the order of the derivative. Thus, for instance, the second derivative of the function $e^{-1/x}$ is entered with `Author` as

```
DIF(#e^(-1/x), x, 2)
```

Pressing `Ctrl J` the result pops up:

$$3:\ \frac{e^{-1/x}(1-2x)}{x^4}$$

The user should keep in mind that `DIF` is a partial derivative operator. The basic rules of derivatives do apply automatically, such as the chain rule.

Integration

To find the primitive of a function is a quite challenging thing for first year students. With DERIVE it is very simple to find the definite integral of a class of functions involving polynomials, trigonometric functions, exponentials, logarithms, and some rational functions or expressions with square roots of other simple expressions.

It is beyond the scope of this text to discuss how the antiderivatives are actually computed. The reader may check the computer algebra literature on this subject to obtain a deeper understanding. The steps taken by DERIVE can be seen in Rich and Stoutemyer (1994).

Similarly to limits and derivatives, there are two options to integrate an expression. First, using the submenu `Integrate`, we should specify the expression or its sequence number, the variable of integration, and finally the lower and the upper limits. If the lower and upper limits are left blank, the integration will be considered an indefinite integration.

Alternatively, the operator `INT` can be used. Its syntax is `INT(f,x,a,b)`, where *f* is the integrand, *x* the variable of integration, *a* the lower limit, and *b* the upper limit. For the function given in the line `3:` above, we `Author INT(#e^(-1/x) (1-2x)/x^4, x)`.

Figure 36.5 shows on algebra window 1 the indefinite integration of the expression. Line `4:` is the derivative of expression `3:`. On algebra window 2 the definite integration is shown.

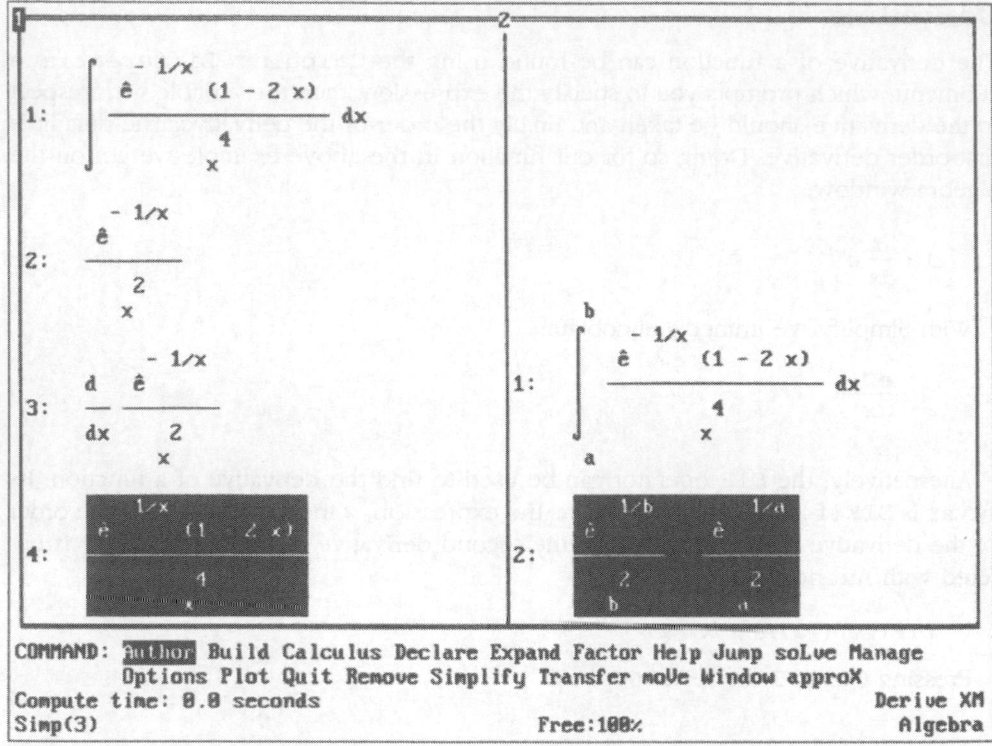

FIGURE 36.5 Integration with DERIVE.

Clearly, combinations with other operators can bring more flexibility to the system. Of course, users should know what they are doing and be alert in order to avoid mistakes.

Sum and Products

Closed form summation and products are done very easily with DERIVE using the `Calculus` submenu `Sum` and `Product`.

Suppose we need to know the result of the sum

$$1 + 2 + 4 + 8 + 16 + 32 + 64 + 128 + \cdots + 1024$$

Of course, entering with `Author` this summation and simplifying afterward, we get the result 2047. Nevertheless, it is quite tedious to type all the terms of the sum to get the result, particularly if we have a large number of terms.

It is easy to see that the summation can be written as $\sum_{i=0}^{10} 2^i$. Suppose that now $i = 100$. Shall we type all that to get the sum? Of course, not! DERIVE can do it quickly.

The `Sum` submenu prompts the user first to give the expression that has to be summed, then to indicate the running variable, and finally the range of the running variable, with its lower limit and upper limit.

Alternatively, the user can use the `Sum` operator, which has the syntax `SUM(f,i,a,b)`, where *f* means the sum expression, *i* the sum variable, and *a* and *b* correspond to the lower and upper limits, respectively. For our simple example, it is

SUM(2^i, i, 0, 100), which appears as

1 : $\sum_{i=0}^{10} 2^i$

Suppose that now the user is interested in computing the summation for $i = n$, with n an arbitrary integer. The closed form of this summation can be obtained just indicating the upper limit as n. Thus SUM(2^i, i, 0, n) gives

2 : $\sum_{i=0}^{n} 2^i$

Simplify the above expression and get the nice result:

3 : $2^{n+1} - 1$

To compute the value for $n = 100$ just apply Manage Substitute, which will ask the user to supply the value of the parameter n.

A very similar procedure should be taken to compute products. Either using the menu sequence Calculus Product or the operator Product(f,i,a,b). As an example, let us take the product $\pi_i^n\, i$. Enter PRODUCT(i, i, 1, n) to get

4 : $\pi_{i=7}^{n}$

Simplify the above expression and get the result:

5 : $n!$

Clearly, that is the factorial of a number n.

In both cases the upper limit can be infinity. For example,

6 : $\sum_{i=1}^{\infty} 1/i^2$

7 : $\dfrac{\pi^2}{6}$

This is known as Euler's sum.

Taylor Series

The Taylor series is a technique used to approximate a function by a polynomial expression. The Taylor series of a function $f(x)$ at the expansion point $x = a$ is given by

$$f(x) = f(a) + f'(a)(x-a) + \frac{f''(a)}{2}(x-a)^2 + \cdot\ \cdot\ \cdot + \frac{f^n(a)}{n!}(z-a)^n$$

where n means the order of the expansion.

The Taylor expansion of a function can be obtained using the Calculus Taylor submenu sequence, which prompts you to indicate the expression, then the expansion variable, and finally to specify the order and the expansion point.

Alternatively, the operator TAYLOR(f,x,a,n) can be used, where f is the function, x indicates the expansion variable, a the expansion point, and n the expansion order. As an example take the function sin x and determine the Taylor expansion of this function around $x = 0$ to order 5.

```
1: TAYLOR(sin(x), x, 0, 5)
```

$$2:\ \frac{x^5}{120} - \frac{x^3}{6} + x$$

A nice way to compare the function and its Taylor approximation is to plot both simultaneously.

36.7 2D AND 3D PLOTS

The plot windows provide an environment for 2D and 3D plotting of univariate and bivariate functions, respectively. From the algebra window the 2D and 3D plot windows will be automatically opened according to the function.

2D Plots

To display the graphic of the function x^2, enter it with `Author` and then ask `Plot`. In DERIVE XM the system will ask the user to set up where the plot window should be open. There are three options: `Beside`, `Under`, and `Overlay` With `Beside` the current screen will be vertically split into two windows, one for the current algebra window and the other for the 2D plot. The system provides the possibility for the user to specify the column to split.

The `Under` option is for horizontal splitting and `Overlay` is to open the plot window on the top of the current algebra window.

To 2D plot functions the highlight should be on the top of a univariate expression or in either form: $y = u$, where y is a variable and u is a univariate expression, or $y := u$, where here := denotes the assignment operator.

The top-level 2D plot menus are `Algebra`, `Center`, `Delete`, `Help`, `Move`, `Options`, `Plot`, `Quit`, `Scale`, `Ticks`, `Window`, and `Zoom`.

As examples see the plot of the function $y = x^4 - x^2$ in Figure 36.6. Notice that the scale is $x = 0.5$ and $y = 0.5$.

Polar Plots

The `Plot Options State` menu sequence provides the possibility to set the system to plot in either rectangular or polar coordinates. When `Polar` coordinates are selected, the system will query the user to set up the range of the angular coordinate. The default range is from –3.1415 to +3.1415 (from $[-\pi, \pi]$).

As an example, Figure 36.7 shows the polar plot of the functions $rose := 2\cos(5t)$, which corresponds to a rose with 5 petals (expression `1:`, 2D plot at window 2), and the butterfly curve (Fay, 1989) $b{:} = e^{\cos t} - 2\cos(4t) + \sin^5(t/12)$ (expression `2:`, 2D plot window 3).

Parametric Plot

Parametric plotting can be done with DERIVE for Cartesian or polar coordinate. The way it is done is very much like we have learned in undergraduate mathematics courses. DERIVE's syntax for a parametric plot is

```
[x(t),y(t)]
```

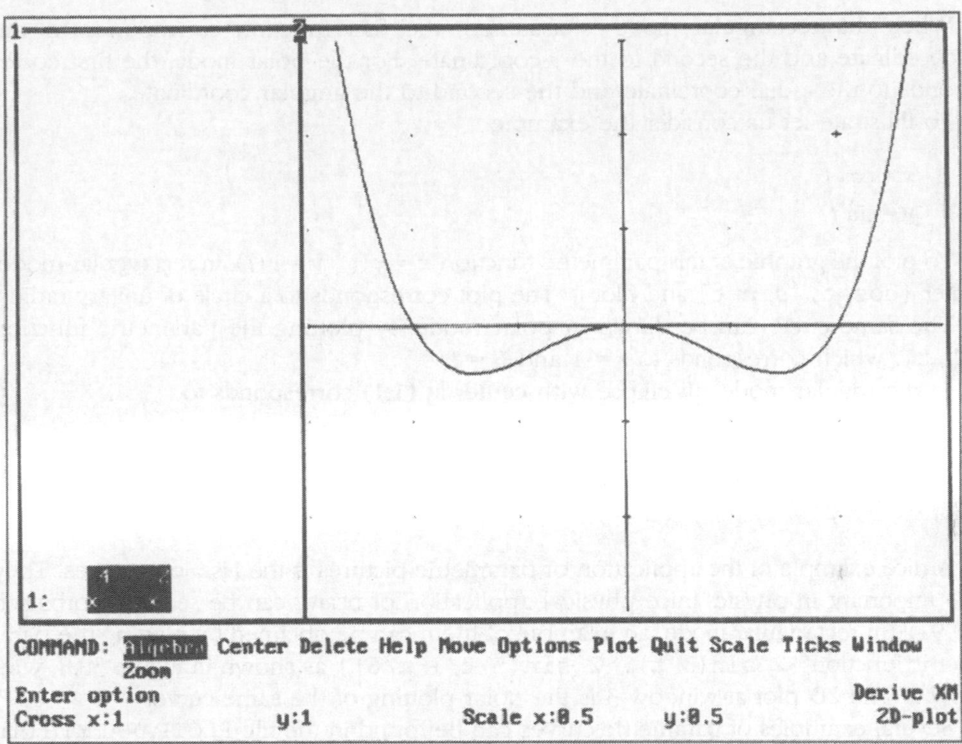

FIGURE 36.6 Algebra and 2D plot windows opened simultaneously.

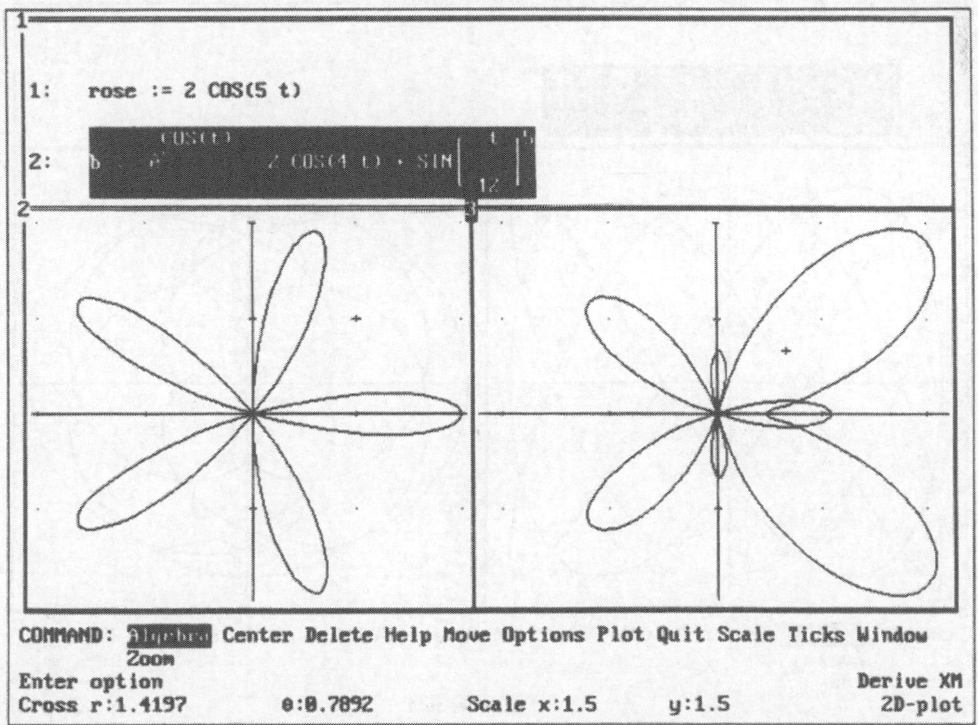

FIGURE 36.7 Polar plot of a rose with 5 petals and a butterfly.

When the rectangular mode is assumed, the first function corresponds to the x-coordinate and the second to the y-coordinate. For the polar mode, the first corresponds to the radial coordinate and the second to the angular coordinate.

To illustrate let us consider the example:

$$x = \cos t$$

$$y = \sin t$$

To plot the graphic of this parametric function $x = x(t)$, $y = y(t)$, in rectangular mode, enter `[cos t, sin t]` and plot it. The plot corresponds to a circle of unitary radius.

The same circle can be drawn in polar mode by plotting the parametric function `[1,t]`, which corresponds to $r = 1$ and $\theta = t$.

In rectangular mode, an ellipse with center at (1,1) corresponds to

```
        1
[1 + -- cos(t), 1 + sin(t)]
        2
```

A nice example of the application of parametric pictures is the Lissajous curves. They are important in physics [nice physical application of DERIVE can be seen in Horbatsch (1991)]. In rectangular mode, an example of them can be obtained by plotting the parametric function `[2 sin(3 t), 2 sin(5 t + π/6)]` as shown in Figure 38.8, window 2. The 2D plot at window 3 is the polar plotting of the same curve.

Several examples of parametric curves can be found in the file `PLOTPARA.MTH` that comes with the software.

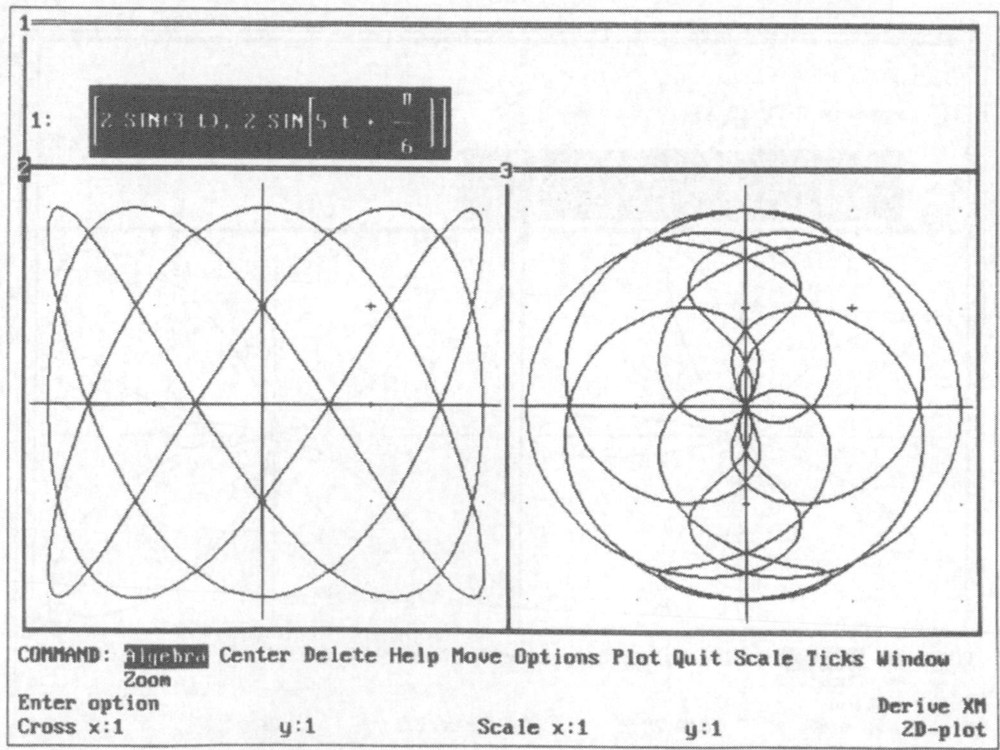

FIGURE 36.8 Lissajous curves.

3D Plots

Bivariate functions can be plotted in DERIVE using the 3D window. It is done by projecting the approximating grid of the surface embedded in a three-dimensional space. The top-level 3D plot menus are `Algebra, Center, Eye, Focal, Grids, Hide, Length, Options, Plot, Quit, Window,` and `Zoom`.

To plot the function $x^2 - y^2$ enter it with `Author`, open the 3D plot window, and press `Plot`. DERIVE will display the surface of a saddle (see Fig. 36.9). The axes are always x, y, z independently of the variables chosen, being x, y the independent variables and z the function value, i.e., the system always considers a function $z = u(x,y)$.

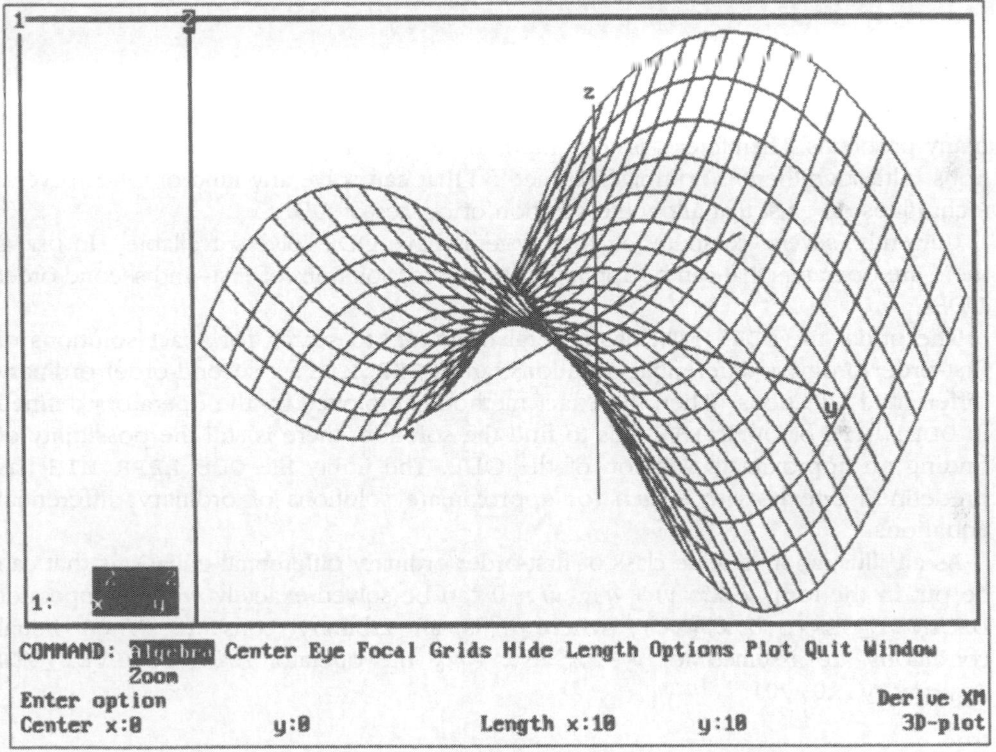

FIGURE 36.9 A saddle with grid size $x = 20$, $y = 20$.

To make the surface look smoother the size of the grid can be diminished by setting up a higher number of panels in the x- and y-axes, using the submenu `Grids`. Be aware that a large number of panels requires more memory to compute the surface. To display the surface without the presence of the axes, use the submenu `Options Axes` and set it to `No`.

The zoom facility for the 3D plot is similar to the one for the 2D plot. The shortkeys `F9` and `F10` can be used to zoom in or out the picture, respectively.

The surfaces displayed in the 3D plot window are encapsulated in an imaginary box with size equal to the maximum computed values of x, y, z. The length of this box is controlled by the submenu `Length`, which has a default value of `X:10 Y:10 Z:50`.

The imaginary box can be rotated by adjusting the submenus `Center`, `Eye`, and `Focal`. The lines that are hidden can be displayed by controlling the `Hide` submenu options to show them or not.

The graphics can be printed using the TIF facility referred to in the section on printing. In the file PLOT3D.MTH that comes with DERIVE there are several nice examples of functions for 3D plot.

36.8 MORE COMPLEX MATH: ADDITIONAL FACILITIES

What we have seen so far is still a small part of the full capability of DERIVE. Several operators are available in the utility files that are distributed with the software.

Solving ODEs with DERIVE

Ordinary differential equations (ODEs) play a very important role in the description of many problems. Therefore, it is very important to have techniques to solve such equations. Although there is no universal method that can solve any kind of ODE, several techniques do exist that allow the solution of classes of ODEs.

Currently, several computer algebra systems have ODE solvers available . In DERIVE there are operators that can be applied to search for solutions of first- and second-order ODEs.

The utility file `ODE1.MTH` has several operators to search for exact solutions of first-order ordinary differential equations, and `ODE2.MTH` for second-order ordinary differential equations. When the exact methods employed by the operators defined in `ODE1.MTH` or `ODE2.MTH` fails to find the solution, there is still the possibility of finding an approximate solution of the ODE. The utility file `ODE_APPR.MTH` has predefined operators to search for approximate solutions of ordinary differential equations.

As an illustration, a wide class of first-order ordinary differential equations that can be put in the form $y'p(x, y) + q(x, y) = 0$ can be solved exactely with the operator `DSOLVE1_GEN(p,q,x,y,c)`, where c is an arbitrary constant. When initial conditions are assumed for $y = y_0$ at $x = x_0$, the operator reads `DSOLVE1_GEN (p,q,x,y,x0,y0)`.

Statistics with DERIVE

For statistical computations a set of predefined functions is available in DERIVE. The main ones are `AVERAGE(a1,a2, . . . ,an)`, which provides the arithmetic average of the numbers $a_1, a_2, \ldots, a_n$; `RMS(a1,a2, . . . ,an)`, which gives the root mean square of the numbers a_1, $a2, \ldots, a_n$; `VAR(a1,a2, . . . ,an)`, which computes the variance of $a_1, a_2, \ldots, a_n$; `STDEV(a1,a2, . . . ,an)`, which computes the standard deviation of $a_1, \ldots, a_n$; and `FIT(x,M)`, which computes the least-square fit of an expression in the parameter x to a set of points given by the data matrix M.

A special utility file PROBABIL.MTH is available with probability operators.

Several other utility files are available to deal with numeric differentiation and integration, Pade rational approximation, hypergeometric functions, orthogonal polynomials, Zeta functions, elliptic integrals, solving recurrence equations, etc. (see Rich et al., 1992).

36.9 PROGRAMMING WITH DERIVE

Although DERIVE is very suitable for interactive computations much like a sophisticated algebraic calculator, it also has some facilities that provide some programming.

Its programming structure relies on a few functions defined for doing loops and deal with conditional statements.

Iterative processes can be resolved with the function ITERATES, which has the following syntax: ITERATES(F,x,a,n). Here F is an expression or function that has a dependence on the variable x, which is updated from the initial value $x = a$, stopping when the number of iterations n is reached. Notice that it is more appropriate here to use the Options Precision Approximate mode.

For instance, suppose that $F = \sin x$. The fixed point iteraction of F starting at $x = \pi/4$ to order $n = 5$ is given by

```
                         pi
1: ITERATES[sin(x), x, ---, 5]
                         4

2: [0.785398, 0.797106, 0.649636, 0.604897, 0.568677, 0.538518]
```

The answer is given by a vector, where each element corresponds to the value of the respective iteration. If the value of the last iteration is the only value that is needed, the function ITERATE provides just this value. Its syntax is similar to the ITERATES function.

Conditional expressions can be dealt with the IF function, which provides an additional programming facility to the system. Its syntax is

```
IF (test-clause, then-clause, else-clause, alterantive-clause)
```

A very nice feature of the conditional function is its application to plot partitions of a function. Suppose that we intend to plot the positive part of the function $f(x)= x^4$. It can be done by plotting the expression:

```
3: IF (x > 0, x^4, ?)
```

To plot of the regions where the function is increasing the derivative is used as shown in Figure 36.10. Window 2 shows the 2D plot of the function given in expression 1:, and window 3 plots the increasing intervals of the function, which is obtained just plotting expression 2: from algebra window 1.

The three logical operators AND, OR, and NOT can be combined with the relational operators =, /=, <, <=, >, >= to form a conditional clause, giving further strength to the programming capabilities of the system.

36.10 COMMENTS AND CONCLUSION

We have seen in this chapter that DERIVE XM is a powerful and rather simple menu-driven computer algebra system. There are versions of this system available to run on a PC-XT to PC-AT with 486 microprocessor under DOS.

DERIVE is a general purpose computer algebra system very suitable as a high level algebraic calculator. Its algebra and plot windows allow manipulations of algebraic expressions and visualization of 2D or 3D plots simultaneously.

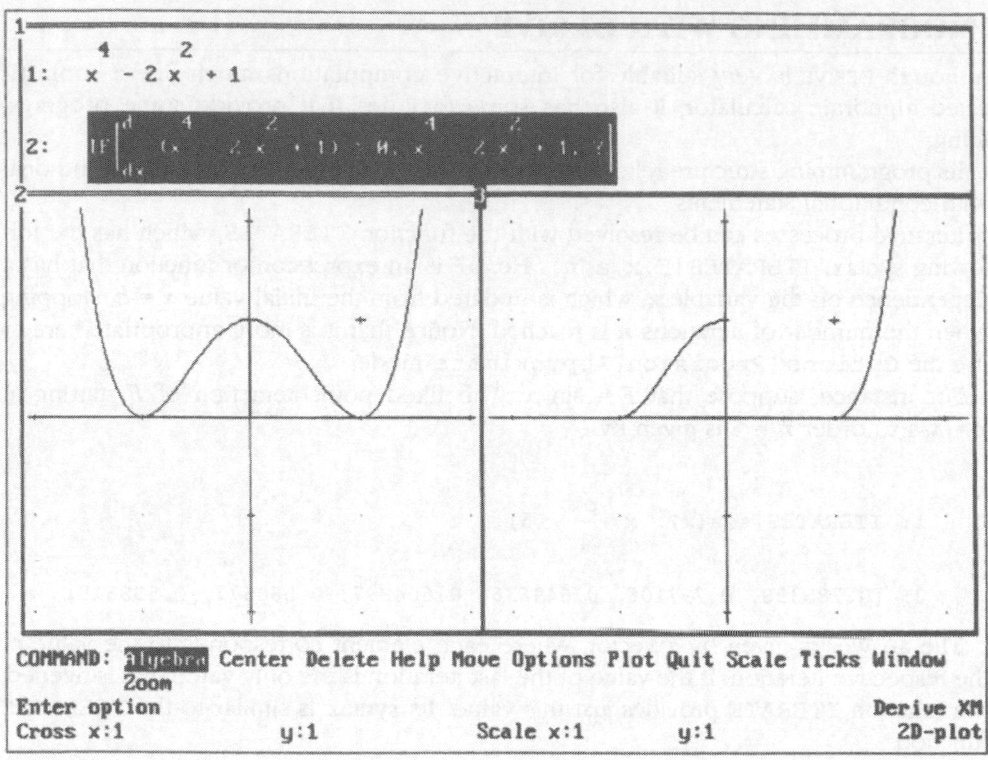

FIGURE 36.10 Partial plot of a function.

Today, the solution of many problems that would take too much time to do it by hand calculation pops up quickly with DERIVE. That is great! Thus, why shall an engineer or a biologist spend his or her time trying to solve tedious, lengthy, or even complicated calculations by hand if they can have the answer available interactively with a computer algebra system? Additionally, with plot facilities, graphics can be drawn to improve reasoning and understanding. Finally, the pictures can be printed easily.

Clearly, DERIVE is not the ultimate software for computer algebra. It needs improvements such as line length wrapping to let long expressions be displayed in multiple lines and let menu options be activated through a mouse, or even better a full version for Windows and an increase in its programming capability. Nevertheless, it is a powerful computer algebra system with low cost that is easy to learn and use.

36.11 FURTHER INFORMATION

There are a number of books and papers published about DERIVE and its applications where the reader can look for further details and applications of the system. Most of the titles can be found in DERIVE's Manual (Rich, 1992). An Electronic Bulletin Board and a Users Group have been created to give support, reply to users' queries and foster users' exchange of ideas and research. The Users Group produces a Newsletter that is distributed among the members.

To get information about DERIVE's license the reader should write to either of the following:

- Soft Warehouse, Inc.
  3660 Waialae Av., Suite 304
  Honolulu, HI 96816-3236, USA
  Fax: (808) 735-1105

- Soft Warehouse Europe GmB
  Schloss Hagenberg
  A-4232 Hagenberg, Austria

NOTE ADDED IN PROOF

Soon after this chapter was written, version 3.02 of DERIVE was released. This new version has several improvements and includes new graphical features.

REFERENCES

T. H. Fay, The butterfly curve. *Am. Math. Month*. 96(5):442–443, 1989.

M. Horbatsch, Teaching calculus and linear algebra with DERIVE. *Comput. Phys*. 5(6):656–660, 1990.

M. Horbatsch, Solving physics problems in the classroom with DERIVE 2.0. Technical Report, Dept. of Physics, York University, Toronto, Ontario, Canada, 1991.

L. C. Leinbach, *Calculus Laboratories Using* DERIVE. Wadsworth, Belmont, CA, 1991.

A. Rich, J. Rich, and D. Stoutemyer, DERIVE XM *User Manual, Version 2.5,* 5th ed. Soft Warehouse, Inc., Honolulu, Hawaii, 1992.

A. D. Rich and D. R. Stoutemyer, Inside DERIVE computer algebra system. *International* DERIVE *Journal*, 1(1):3–17, 1994.

D. Small (ed.), Computer Algebra Systems in Education NEWSLETTER, Dept. of Mathematical Sciences, U.S. Military Academy, West Point, NY, 1994.

A. K. Zavin, *Symbolic Computation in Undergraduate Mathematics Education*, MAA Notes 24. The Mathematical Association of America, 1992.

37

Macsyma

Richard Petti
Macsyma Inc.

37.1 A SHORT HISTORY OF MACSYMA®

The history of Macsyma is divided into three distinct stages.

Origins at M.I.T.: 1968 to 1982

The symbolic–numerical–graphical mathematics software program known as Macsyma was born as a research project in 1968 at the Massachusetts Institute of Technology. Predecessor research projects went back a further decade. By the early 1970s, Macsyma was being widely used for symbolic computation in research projects at M.I.T and the National Laboratories. Macsyma was one of the early applications that helped justify building the ARPA Net. Unix was first licensed outside Bell Laboratories to enable Macsyma to be ported to DEC VAXs thereby spawning the proliferation of Unix.

Macsyma development at M.I.T. had been largely funded by the U.S. Advanced Research Projects Agency. By 1980, Macsyma was easily the most advanced symbolic–numerical–graphical mathematics software in the world.

Commercialization by Symbolics, Inc.: 1982 to 1992

By 1980 the U.S. government had decreased funding for Macsyma and for symbolic mathematics software development in general. In 1982 M.I.T. licensed Macsyma to Symbolics, Inc., a workstation spin-off from M.I.T. Macsyma was so overpowering in its capabilities that it was assured market dominance for the first few years. By 1986 over 600 research papers and other publications had been published that referenced Macsyma.

In 1987, with Macsyma under new management, sales doubled, service customers quadrupled in one year, and the pace of software development increased. Macsyma began to incorporate packages developed outside the company: a Grobner basis pack-

0-8493-2530-7/96/$0.00+$.50

age,[1] GENTRAN,[2] a T_EX generator,[3] ODEFI,[4] and a new Poisson series package.[5] The Macsyma staff refined these contributions and added their own, including Taylor_solve, pattern matcher extensions, database extensions, special function libraries, and new O.D.E. perturbation and series solution methods.

As Symbolics' main business began to fail in 1987, Macsyma suffered further constraints. The Macsyma business declined with the rest of Symbolics between 1988 and 1992.

Revitalization by Macsyma Inc.: 1992 to the Present

In April of 1992 Macsyma Inc. acquired the Macsyma software business from Symbolics, Inc. In the fall of 1992 the company introduced a new PC Macsyma with revamped user interface and graphics. The April 1993 issue of *IEEE Spectrum* reviewed PC Macsyma, saying "Macsyma is a national treasure. . . . Users with heavy mathematics needs should insist on Macsyma."

In May 1993, the company shipped a new release of PC Macsyma that runs at three times the speed of the previous product, on the 1992 benchmark tests that appeared in *The Notices of the A.M.S.* and in *PC Magazine*. Macsyma solved more of the benchmark problems than any other product, and gave more reliable symbolic results even on some basic problems. Two independent studies in 1994 concluded that Macsyma was the most capable symbolic–numerical mathematics software system.[6,7]

In November 1994, Dr. Michael Wester of the University of New Mexico assembled the most comprehensive test ever devised for symbolic mathematics software. He selected 131 mathematics problems and attempted to solve them with six mathematics software products. We summarize his results with a numerical score as follows:

Result	*Numerical Score*
Satisfactory solution	1
Partially satisfactory solution	1/2
No solution	0
Partially incorrect result	–1/2
Incorrect result	–1

[1]Contributed by Gail Zachariaas.

[2]Contributed by Barbara Gates and Paul Wang. Gentran translates entire Macsyma programs to FORTRAN and C, including program flow control statements.

[3]Based on code contributed by Richard Faateman.

[4]Contributed by Roman Shtokhamer and B.F. Caviness. ODEFI strengthened Macsyma's already long lead in solving first-order ordinary differential equations symbolically.

[5]Contributed by Richard Fateman.

[6]Michael Wester(wester@amber.unm.edu) compared symbolic math capabilities of six math software systems on 123 carefully selected problems. This study is available by ftp at math.unm.edu. Change to directory pub/cas.

[7]The diploma project and thesis of Stefan Braun, Munich, Germany, was to carefully compare the mathematical capabilities of a range of mathematics software systems. Upon graduation, Mr. Braun started a business supporting Macsyma in Germany.

With this scoring scheme, Wester's results yield the following scores.

Axiom 1.2	Derive 2.0	Macsyma 419	Maple V.3	Mathematica 2.2	Reduce 3.5
59	73.5	108	91	87.5	49

Dr. Wester's report was made publicly available by ftp on the Internet in November. It has been published in *Computer Algebra Netherlands.*

User-contributed code continues to play an important role in evolution of new capabilities. For example, the ODELIN2[8] package for solving second-order linear ordinary differential equations (O.D.E.s) and user-contributed improvements[9] to Macsyma's inverse Laplace transform were incorporated in 1992 and 1993, respectively.

In mid-1993 the level of development effort was greatly expanded. By early 1994 the Unix workstation versions of Macsyma matched the level of the 1993 PC-Windows version, including mathematical improvements, formatted display of math expressions, on-line help systems, graphics, and most of the speed improvements in the PC product.

In the fall of 1994, the next generation of Macsyma began with the shipment of Macsyma 2.0 for PCs, which included vast improvements in user interface, graphics, linear algebra, numerical analysis, differential, equations and speed. The reception from independent reviewers was extraordinary.

- "While the Macsyma 2.0 symbolic math program's flashiest new feature is full fledged graphics animation, what users will appreciate most are its increased speed and formatting flexibility. . . . Speed is where Macsyma excels, especially in numeric matrix handling. . . . Macsyma is now the leader in the mathematical program marketplace."

  —*PC Magazine*, January 24, 1995 (USA)

- "Being a combination of mathematics program, graphics software and text processor, Macsyma offers a great deal for a very favorable price compared with competing products. For example, no comparable mathematics software can be found with such satisfying on-line help and such an abundance of tutorial demonstrations."

  —*MC Extra*, January 1995 (Germany)

- "A great improvement is the addition of executable notebooks. . . . Numerical commands are nearly as numerous as in specialized number-crunching programs like Matlab. . . . In its symbolic capabilities, Macsyma is the heavy artillery of the group: it has the most complete and arguably the most reliable capabilities of this type."

  "Macsyma is remarkably easy to use. It will load automatically any math packages it needs; competing product require users to pitch in manually. Macsyma has extensive on-line help, including sample uses of commands and demonstration files. . . . Users with heavy math needs should insist on Macsyma."

  —*IEEE Spectrum*, December 1994 (USA)

37.2 SELECTED MATHEMATICAL CAPABILITIES OF MACSYMA

We shall assume that users are familiar with the basic differentiation, integration, and equation-solving typical of symbolic–numerical mathematics software systems. The purpose of this section is to illustrate some of capabilities of Macsyma that enhance its

[8]Contributed by Shunro Watanabe. The ODELIN2 package raised Macsyma's success rate on the second-order linear O.D.E.s in the Kamke book on O.D.E.s from about 55 to 83% of all cases, and 96% of "relevant cases," as determined by Watanabe.

[9]Contributed by Michael Clarkson.

reliability and/or enable it to solve more problems than other systems. The discussion will focus on Macsyma's treatment of some quite elementary problems, rather than on solving unusual integrals and other advanced problems.

Many of the problems listed in this section are taken from the 1992 benchmarks published in the *Notices of the American Mathematical Society* and *PC Magazine*. Comparison of Macsyma's solutions with those of the other systems provide good measures of Macsyma's breadth, reliability, and ease of programming. Solutions using other software systems are not reproduced here.

Basic Arithmetic and Algebra

Arithmetic and Symbolic Procedures

Macsyma contains three types of floating point numbers (single floats, double floats, bigfloats), two types of integers (fixnums and bignums), rational numbers, and transcendental numbers. Macsyma combines the convenience and approximate nature of floating point numbers with exact symbolic arithmetic. Macsyma is responsible for making seamless transitions among these types of numbers (and among the types of algebraic expressions in the system) with a minimum of intervention by the user.

Macsyma has various methods for determining inequality information and integer properties of exact arithmetic expressions. The most advanced of these is an unproven heuristic algorithm, which is controlled by the option variable **`heuristic_precision_limit`**, which controls the number of digits used in the effort to evaluate inequality predicates. Below are two examples where Macsyma extracts inequality information from exact arithmetic expressions.

```
(c1) sqrt((%e-%pi)^2);
(d1)             π-e
```

```
(c1) block([heuristic_precision_limit:
false], csign(1+cos(355/113)));
(d1)             pnz
(c2) csign(1+cos(355/113));
(d2)             pos
```

Macsyma can compute in exact form a rather large collection of trig functions of special angles. The following example shows such an example, coupled with Macsyma's `algebraic` flag, which is used to clear radicals from the denominator of an expression.

```
(c1) 1/cot(%pi/24);
```

(d1) $\dfrac{1}{(\sqrt{2}+1)(\sqrt{3}+\sqrt{2})}$

```
(c2) ratsimp(d1),algebraic:true;
```

(d2) $(\sqrt{2}-1)\sqrt{3}+\sqrt{2}-2$

Symbolic Algebra

Since the 1970s Macsyma has had a database that enables users to specify information about mathematical expressions besides their symbolic form, and enables Macsyma to solve many problems correctly which other systems cannot. Some simple examples follow.

```
(c1) (-1)^n;                          (c1) sqrt(x^2);
(d1)        (-1)^n                    (d1)            |x|
(c2) declare(n,odd)$                  (c2) assume(x < 0)$
(c3) (-1)^n;                          (c3) sqrt(x^2);
(d3)          -1                      (d3)            -x
                                      (c4) forget(x < 0)$
(c4) remove(n,odd)$
                                      (c5) declare(z, complex)$
(c5) (-1)^n;                          (c4) sqrt (z^2);
(d5)        (-1)^n                    (d6)      sqrt(z^2)
```

```
(c1) integrate(x^n,x); is n+1         (c1) sin(n*(n+1)/2*%pi); . .
zero or nonzero? zero;
(d1)        log (x)                   (d1)            sin(πn(n + 1)/2)
(c2) assume(not equal(n,-1))$         (c2) declare(n,integer)$
(c3) integrate(x^n,x);                (c3) sin(n*(n+1)/2*%pi);
(d3)        x^(n+1)/(n + 1)           (d3)            0
```

Displayed results: (d1) $(-1)^n$; (d1) $|x|$; (d6) $\sqrt{z^2}$; (d1) $\sin\left(\frac{\pi n(n+1)}{2}\right)$; (d3) $\frac{x^{n+1}}{n+1}$.

Complex Algebra

In the example below, Macsyma computes the rectangular and polar form of a basic complex exponential expression. It appears to the author that no other major commercial mathematics software package can perform both of these computations with a few built-in commands.

```
(c1) (a+%i*b)^(c+%i*d);
```

(d1) $(ib + a)^{id + c}$

```
(c2) rectform(d1);
```

(d2) $$i\,(b^2 + a^2)^{c/2} e^{-\mathrm{atan2}(b,a)d} \sin\left(\frac{\log(b^2 + a^2)d}{2} + \mathrm{atan2}(b, a)c\right) + (b^2 + a^2)^{c/2} e^{-\mathrm{atan2}(b,a)d} \cos\left(\frac{\log(b^2 + a^2)d}{2} + \mathrm{atan2}(b, a)c\right)$$

```
(c3) polarform(d1);
```

(d3) $$(b^2 + a^2)^{c/2} e^{i\left(\frac{\log(b^2+a^2)d}{2} + \mathrm{atan2}(b,a)c\right) - \mathrm{atan2}(b,a)d}$$

A somewhat more intriguing problem in complex simplification is to ask Macsyma to find the arcsin of 2.

```
(c1) asin(2);

(d1)                                        asin (2)

(c2) rectform(d1);
```

$$(d2)\qquad \frac{\pi}{2} - i \log\left(\sqrt{3} + 2\right)$$

```
(c2) polarform(d1);
```

$$(d3)\qquad \sqrt{\log^2\left(\sqrt{3}+2\right) + \frac{\pi^2}{4}}\; e^{-i\,\mathrm{atan}\left(\frac{2\log\left(\sqrt{3}+2\right)}{\pi}\right)}$$

Solving Algebraic Equations

Recently an author of books about math software approached me about the complexity of Macsyma's solutions for simple algebraic equations. One of his examples was solving $x^3+x+1=0$.

```
(c1)    sol: solve(x^3+x+1=0,x)$
(c2)    pickapart(sol,1);
```

$$(e2)\qquad x = \left(\frac{\sqrt{31}}{6\sqrt{3}} - \frac{1}{2}\right)^{1/3}\left(-\frac{\sqrt{3}\,i}{2} - \frac{1}{2}\right) - \frac{\frac{\sqrt{3}\,i}{2} - \frac{1}{2}}{3\left(\frac{\sqrt{31}}{6\sqrt{3}} - \frac{1}{2}\right)^{1/3}}$$

$$(e3)\qquad x = \left(\frac{\sqrt{31}}{6\sqrt{3}} - \frac{1}{2}\right)^{1/3}\left(\frac{\sqrt{3}\,i}{2} - \frac{1}{2}\right) - \frac{-\frac{\sqrt{3}\,i}{2} - \frac{1}{2}}{3\left(\frac{\sqrt{31}}{6\sqrt{3}} - \frac{1}{2}\right)^{1/3}}$$

$$(e4)\qquad x = \left(\frac{\sqrt{31}}{6\sqrt{3}} - \frac{1}{2}\right)^{1/3} - \frac{1}{3\left(\frac{\sqrt{31}}{6\sqrt{3}} - \frac{1}{2}\right)^{1/3}}$$

```
(d4)                [e2,e3,e4]
```

This gentleman pointed out that Macsyma's solution to this problem is bulkier and harder to interpret than the solution returned by another, smaller symbolic math system. I apologized for this deficiency in Macsyma, and took the problem to the technical staff for resolution. Jeffrey Golden, who had 25 years of experience with Macsyma at that time, pointed out to me that Macsyma used to give the succinct solution provided by the other software product. In the late 1970s Macsyma was altered to give the current form of the solution, because this form includes the cube root of only one number (instead of the cube roots of several different numbers), which eliminates problems reconciling the branches of the cube root in subsequent algebraic simplifications and integration operations.

This simple example illustrates a generalization about Macsyma that is valid in many domains: Macsyma is designed with a bias for meeting the needs of the most demanding applications.

Basic Calculus Operations

Differential Calculus

Macsyma is able to utilize variable dependencies and values of derivatives that are declared by the user and that do not appear explicitly in symbolic expressions. These facilities are used below.

- In the example on the left, the user applies the chain rule to a function u(x,y,t), where both x and y depend on t, though these dependencies do not appear explicitly in the differentiation command line.
- The example on the right exhibits how to distinguish partial derivatives and total derivatives using the new differentiation capabilities.

Left:

(c1) depends(u, [x,y,t])$

(c2) diff(u,t);

(d2) $\frac{du}{dt}$

(c3) gradef(x,t,alpha)$

(c4) gradef(y,t,beta)$

(c5) diff(u,t);

(d5) $\beta \frac{du}{dy} + \alpha \frac{du}{dx} + \frac{du}{dt}$

Right:

(c1) (load(ndiff), newdiff:true)$

(c2) diff(u(x,y,t),t);

(d2) $u^{(0,0,1)}(x,y,t)$

(c3) gradef(x,t,alpha)$

(c4) gradef(y,t,beta)$

(c5) diff(u(x,y,t),t);

(d5) $\alpha u^{(1,0,0)}(x, y, t) + \beta u^{(0,1,0)}(x, y, t) + u^{(0,0,1)}(x, y, t)$

(c6) convert_to_de(d5);

(d6) $\beta \frac{d}{dy}(u(x, y, t)) + \alpha \frac{d}{dx}(u(x, y, t)) + \frac{d}{dt}(u(x, y, t))$

(c7) diff(u(xx,yy,t),t);

(d7) $u^{(0,0,1)}(xx, yy, t)$

(c8) u(x,y,t):=exp(-(x^2+y^2)/t)$

(c9) ev(d7);

(d9) $\frac{(yy^2 + xx^2)e^{-\frac{yy^2+xx^2}{t}}}{t^2}$

Integral Calculus

Macsyma takes special care that it does not make invalid assumptions that introduce errors into results in subtle ways.

- In the example on the left, Macsyma explicitly asks you about the exceptional value of n.
- In the example of an inverse Laplace transform on the right, Macsyma decides to employ the normal assumption for Laplace transforms without asking, and to point out to you that it had to make this assumption to get the result.

```
(c1) integrate(x^n,x);
is n+1 zero or nonzero?
zero;
(d1)          log (x)
(c2) assume(not equal(n,-1))$
(c3) integrate(x^n,x);
```

$$(d3)\qquad \frac{x^{n+1}}{n+1}$$

```
(c1) ilt(1,s,t);
Proviso: Assuming s > 0.
(d1)                     δ(t)
(c2) provisos
(d2)  [Proviso: Assuming s < 0. ]
```

Macsyma correctly evaluates the following double integral from Newtonian gravitational theory, due to use of the Macsyma database, and careful application of series expansions in the presence of absolute values.

```
(c1) expr: -r^2*sin(theta)/sqrt(x^2+r^2-2*x*r*cos(theta));
```

$$(d1)\qquad -\frac{r^2 \sin(\theta)}{\sqrt{x^2 - 2r\cos(\theta)x + r^2}}$$

```
(c2) integrate(integrate(expr,theta,0,%pi),r,0,1);
Is x positive, negative, or zero?
pos;
Is x-1 positive, negative, or zero?
pos;
```

$$(d2)\qquad -\frac{2}{3x}$$

```
If we tell Macsyma x-1 is negative, we get:
```

$$(d2)\qquad \frac{x^2}{3} - 1$$

```
Macsyma asks us to determine whether x < 0 or 0 < x < 1 or
1 < x. Macsyma can be told this information beforehand and
will not ask the questions.

(c3)  assume(x > 0,x < 1)$
(c4)  integrate(integrate(expr,theta,0,%pi),r,0,1);
```

$$(d4)\qquad \frac{x^2}{6} + \frac{x^2 - 6}{6}$$

New Features in 1994

Macsyma's symbolic integration capabilities continually grow stronger on an incremental basis. For example, Macsyma now solves all seven problems proposed in the paper by Schou and Broughan.[10]

New releases of Macsyma in 1994 also include a major step forward: a new implementation of the transcendental Risch algorithm for indefinite integration. The new Risch implementation can take successive field extensions of the field of rational functions, enabling it to solve many more problems than the older implementation. Macsyma's transcendental Risch implementation is now the fastest or nearly the fastest on the Bronstein benchmark integration problems (a family of integration problems of increasing difficulty). Figure 37.1 displays timings of several symbolic math software products solving the Bronstein problems.

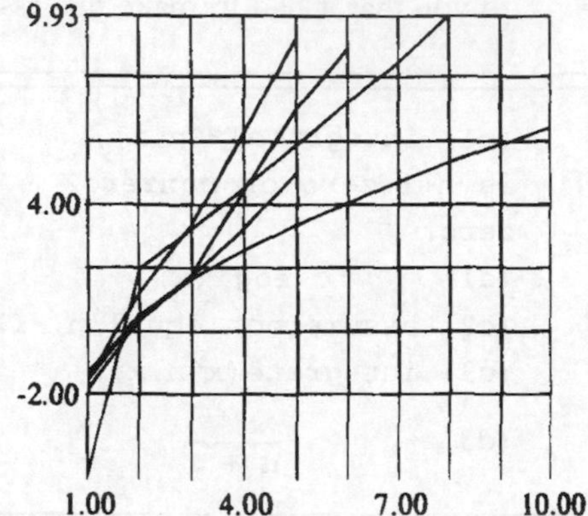

FIGURE 37.1 Log timings of Risch on problem Bronstein[n]. From top (slowest) to bottom (fastest):

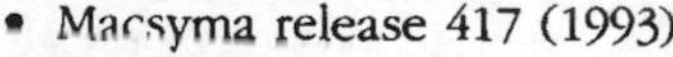

- Macsyma release 417 (1993)
- Competitor #1
- Competitor #2
- Competitor #3
- Macsyma release 2.0 (1994)

All timings are performed on the same PC except for competitor #3, which was computed indirectly from times on a different computer.

Differential Equations

Macsyma has the most extensive capability for exact symbolic solutions, perturbation solutions, series solutions and numerical solutions of O.D.E.s, including stiff systems of O.D.E.s. Macsyma includes a package for finding symmetries and symbolic solutions of partial differential equations. We shall give examples from all of these categories, except the numerical solution of systems of O.D.E.s.

Exact Solutions, First Order O.D.E.s

We start with an easy one, then a harder one. Symbols such as %c, %k1, %k2 are arbitrary integration constants. We also show how to insert initial conditions into a general solution.

```
(c1)   depends(y,x)$

(c2)   diff(y,x)=y*tan(x);

                          dy
(d2)                      --  =  tan (x)y
                          dx
```

[10]Schou and Broughan, *SIGSAM Bull.* 23(3):19–22, 1989.

```
(c3)  ode(d2,y,x);

(d3)  y = %c sec (x)

Insert initial conditions

(c4)  ic1(d3,x=x0,y=a);
```

(d4)
$$y = \frac{a \sec (x)}{\sec (x0)}$$

```
(c5)  diff(y,x)*x=(x*log(x^2/y)+2)*y;
```

(d5)
$$x \frac{dy}{dx} = \left(x\log\left(\frac{x^2}{y}\right) + 2\right)y$$

```
(c6)  ode(d5,y,x);
```

(d6)
$$\log\left(2\log (x) - \log (y)\right) + x - \log (2) = \%c$$

```
Insert initial conditions

(c7)  solve(ic1(d6,x=1,y=a),y);
```

(d7)
$$\left[y = x^2 e^{\log(a)e^{1-x}}\right]$$

In 1994, Macsyma added a command (ODE_IBC) for specifying initial and boundary conditions for O.D.E.s that combine values and derivatives of the dependent variable.

Exact Solutions, Second Order O.D.E.s

We start with a simple damped linear harmonic oscillator.

```
(c1)  depends(y,x)$

(c2)  m*diff(y,x,2)+b*diff(y,x)+k*y=0;
```

(d2)
$$m \frac{d^2y}{dx^2} + b \frac{dy}{dx} + ky = 0$$

```
(c3)  ode(d2,y,x);
Is 4 k m - b^2 positive, negative or zero?
pos;
```

(d3)
$$y = e^{-\frac{bx}{2m}} \left(\%k1 \sin\left(\frac{\sqrt{\frac{4k}{m} - \frac{b^2}{m^2}}\, x}{2}\right) + \%k2 \cos\left(\frac{\sqrt{\frac{4k}{m} - \frac{b^2}{m^2}}\, x}{2}\right)\right)$$

We can isolate the x-dependence in this solution to obtain a more compact form.

```
Isolate the x-dependency

(c4)    isolate(d3,x);
```

(e4) $$\frac{\sqrt{\frac{4k}{m} - \frac{b^2}{m^2}}}{2}$$

(d4) $$y = e^{-\frac{bx}{2m}}(\%k1 \sin(e4\ x) + \%k2 \cos(e4\ x))$$

```
        Specify initial amplitude A and initial velocity 0
(c5)    ic2(d4, x=0, y=a, diff(y,x)=0);
```

(d4) $$y = e^{-\frac{bx}{2m}}\left(\frac{ab \sin(e4\ x)}{2\ e4\ m} + a \cos(e4\ x)\right)$$

The new ODELIN2 package[11] solves a wide range of linear second order O.D.E.s. According to tests performed by the author, a precommercial version of the ODELIN2 package (before another generation of improvements by Macsyma Inc.) solved about 83% of the second order linear O.D.E.s in the book by Kamke.[12]

```
(c1)    'diff(y,x,2) + (1+(1-4*k^2)/(4*x^2))*y = 0;
```

(d1) $$\frac{d^2y}{dx^2} + \left(\frac{1 - 4k^2}{4x^2} + 1\right)y = 0$$

```
(c2)    ode(d1,y,x);
Is k an integer? (Answer e.g. yes; no; unknown; or help;.)
yes;
```

(d2) $$y = \sqrt{|x|}\,(\%k2\ \mathrm{bessel\_y}_k(|x|) + \%k1\ \mathrm{bessel\_j}_k(|x|))$$

```
We could have answered "no" to the question, and gotten
```

(d2) $$y = \sqrt{|x|}\,(\%k1\ \mathrm{bessel\_j}_k(|x|) + \%k2\ \mathrm{bessel\_j}_{-k}(|x|))$$

Approximate Symbolics Solutions to O.D.E.s

Macsyma helps symbolic and numerical analyses to complement one another. It provides many approximate symbolic solutions to algebraic and differential equations, plus symbolic utilities to support numerical analysis (such as finite difference generators, exceptionally strong FORTRAN and C code generators, and a companion finite element product, PDEase®, for solving P.D.E.'s numerically).

[11]Shunro Watanabe, An experiment toward a general quadrature for second order linear ordinary differential equations by symbolic computation. *EUROSAM 84 Proc.* 13–22, 1984.

[12]E. Kamke, *Differential Gleichungen-Lösungmethoden und Lösungen*. Chelsea, 1959.

Approximate symbolic solutions are primarily of two types.

- Series solutions give very accurate representation of exact solutions over a small interval around the center of expansion of the series; they are often extremely poor approximations to the exact solutions for larger times.
- Perturbation solutions accurately represent the asymptotic long-time behavior of the exact solutions; generally they do not capture the transient behavior of the solution.

First we give an example of a Taylor series solution of fourth order for a difficult equation with a derivative in an exponent. We specify initial conditions y(0)=a and dy/dx(0)=b.

```
(c1)  'diff(x,t,2)=x+%e^(c*t*'diff(x,t));
```

(d1)

$$\frac{d^2x}{dt^2} = e^{ct\frac{dx}{dt}} + x$$

```
(c2)  taylor_ode(eq[2],x,t,4,[0,a,b]);
```

(d2) /T/

$$\left[\left[x = a + bt + \frac{(a+1)t^2}{2} + \frac{(bc+b)t^3}{6} + \frac{(b^2c^2 + (2a+2)c + a + 1)t^4}{24} + \ldots\right]\right]$$

Macsyma implements several perturbation methods for O.D.E.s, the simplest of which is Lindstedt's method. Below we produce a solution of Duffing's equation for an almost-harmonic oscillator by Lindstedt's method. We specify initial conditions x(0)=a and dx/dt(0)=b. The initial conditions affect the form of the asymptotic solution, even though the perturbation method is not computing transient responses.

```
(c1)  duffing: 'diff(x,t,2)+x+e*x^3=0;
```

(d1)

$$\frac{d^2x}{dt^2} + ex^3 + x = 0$$

```
(c2)  lindstedt(duffing,e,1,[a,b]);
```

(d2)

$$\left[\left[\left[-\frac{\left((\sin(3\,\%tau) + 9\sin(\%tau))\,b^3 + (3\cos(3\,\%tau) + (21\sin(\%tau) - 3\sin(3\,\%tau))\,a^2b + (\cos(\%tau)\right.}{32}\right.\right.\right.$$

$$\left.\frac{\left.-3\cos(\%tau)) - \cos(3\,\%tau)\right)}{32} + \sin(\%tau)\,b + \cos(\%tau)\,a\right],$$

$$\left.\left.\%tau = \left(\frac{(3b^2 + 3a^2)\,e}{8} + 1\right)t\right]\right]$$

It is typical of nonlinear oscillators that the characteristic frequency $3/8\ (b^2 + a^2)e + 1$ is altered by terms that depend on the amplitude of oscillation.

Partial Differential Equations

Macsyma contains vector and tensor calculus packages that are useful for stating many physically important systems of partial differential equations (P.D.E.s) in a wide range of coordinate systems.

New Capabilities in 1994

- 1994 releases of Macsyma contain the PDELIE©[13] package for finding Lie symmetries and symbolic solutions of (systems of nonlinear) P.D.E.s.
- In January 1994, Macsyma Inc. began shipping a companion software product to Macsyma for solving systems of P.D.E.s by the finite element method. This product, PDEase®,[14] accepts a wide range of static and dynamic systems of P.D.E.s in two dimensions, including systems of mixed elliptic, parabolic, and hyperbolic type and eigenvalue problems. The input language requires only that you specify the equations with supporting definitions, the boundary shapes and boundary conditions, and which plots you want to produce. You need not specify the nodes or elements of the finite element grid; PDEase generates the grid, performs error analysis (using an approximation to the Sobolev error norm) to determine if the grid is fine enough, and refines or degrids as needed to meet error requirements efficiently. Macsyma can produce symbolic PDEase in string form that PDEase can read.
- Macsyma has facilities for automating the writing of systems of P.D.E.s in about 20 standard coordinate systems using vector calculus (a traditional Macsyma capability) or tensor calculus (capability enhanced in 1994).

Linear Algebra

Traditional Macsyma has a broad portfolio of linear algebra capabilities, including inversion and related capabilities, solving matrix equations, matrix normal forms (Jordan, QR, LU, LDU), matrix exponentiation, and multilinear algebra.[15]

New Capabilities in 1994

Starting with PC Macsyma 2.0 (and Macsyma 419 for workstations) in the summer of 1994, new versions of Macsyma include vastly expanded linear algebra capabilities. The product contains nearly all the functionality in MATLAB[16] release 3.0. The new features in these 1994 releases include:

- 180 new functions for linear algebra, mostly utility functions inspired by MATLAB. Important new algorithms include new normal forms (SVD, Schur, Hessenberg), and faster LU and QR decompositions.

[13]PDELIE copyright Trinity University and Peter Vafeades.

[14]PDEase is a registered trademark of SPDE, Inc.

[15]Since 1989, Macsyma has included a multilinear algebra package (ATENSOR) that performs computations in many kinds of associative algebras including universal tensor, Grassmann, Clifford, and symplectic algebras.

[16]MATLAB is a trademark of The Math Works, Inc.

- about a dozen new MATLAB-inspired language features for compact expression of linear algebra. Important improvements include:
  - left and right matrix division and multiple right-sides in linear equations,
    - — compact matrix notation (for matrix definitions, submatrix extraction, transposes, and adjoints), assigning values into submatrices and lists, threading many more functions over matrices, and
    - — more intelligence about symbolic identities [about, for example, idempotent and nilpotent operators, and that determinant[transpose(matrix)] = determinant(matrix)].
  - a MATLAB-to-Macsyma translator, which
    - — translates MATLAB batch command files into Macsyma command files, and
    - — enables users to type interactively unaltered MATLAB code directly to Macsyma.
  - vastly increased speed of numerical linear algebra. For example, PC Macsyma computes the determinant of a 10 × 10 floating point matrix about 1,000 times faster than did the last PC Macsyma from Symbolics in 1992.

Numerical Analysis

Most of numerical analysis can be viewed as consisting of six types of algorithms. Macsyma has extensive capabilities in all of these areas.

- Numerical integration and integral transforms

  New in 1994: The new QUADRATR package uses extrapolated Gaussian quadrature, the fifth and most general numerical quadrature method in Macsyma.
- Numerical solution of equations and critical point conditions

  New in 1994: The new NEWTON package for Newton iteration is faster and more robust. The MINFUNC package finds the minimum of a univariate function using Brent's method.
- Numerical solution of differential equations

  New in 1994: Macsyma has a solver for stiff systems of O.D.E.s, called ODE_STIFFSYS. The Runge–Kutta methods have also been expanded and made faster.

  Macsyma has a companion product, PDEase, that solves partial differential equations by the finite element method. Its most unique technical strengths are its adaptive gridding technology and its implementation of approximate Sobolev error norms to assure that solutions that are returned meet specified error conditions. These technical advances translate into ease of use, whereby the user is freed from most concerns about the definition and adequacy of the finite element discritization.
- Numerical linear algebra, including numerical eigenanalysis

  New in 1994: Massive improvements in numerical linear algebra are described in the section devoted to linear algebra.
- Numerical evaluation of special functions

  New in 1993–1994: Improvements were made to numerical evaluation of many kinds of Bessel functions, exponential integral functions, and the Riemann zeta function and its derivatives.
- Interpolation, extrapolation, fitting and statistical description of data

  New in 1993–1994: The new INTERPOL package interpolates numerical values from tabulated data in one or two dimensions using polynomials, rational functions, or splines. The improved LSQ package includes the Levenberg–Marquardt algorithm for nonlinear multivariate least squares fits of data with parameterized symbolic expressions. The improved STATS package computes descriptive statistics for univariate and multivariate data sets.

Graphics

Macsyma offers a full spectrum of two-dimensional and three-dimensional scientific graphics. This includes plotting functions, parametric curves and surfaces, vector fields, and surface contours.

In Figure 37.2 we demonstrate one of the more unusual graphical capabilities of Macsyma. You can place your viewpoint inside a plotted object and obtain a realistic perspective view from that location.

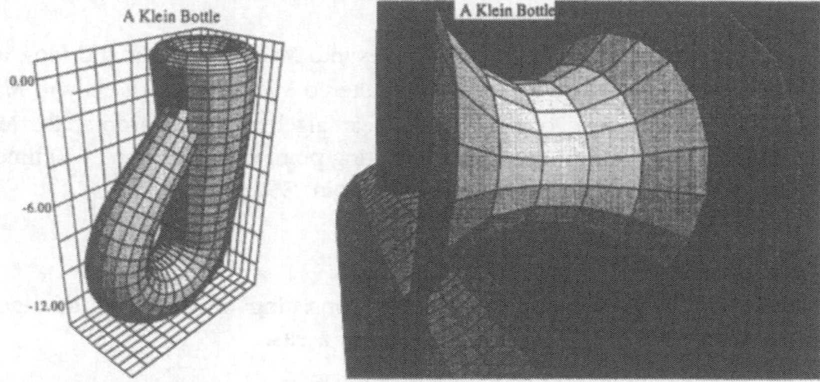

FIGURE 37.2 A plot of a Klein bottle, viewed from the outside, and one of many fascinating views from the inside.

New Graphics Features in 1994

The new features listed below appeared in the summer 1994 release of PC Macsyma. They are scheduled to appear in workstation versions in 1995.

- Plots appear in the notebook in vector graphic form and are editable in-place in notebooks.
- Plots can be animated to exhibit motion in two ways.
  - — Camera animation: the viewpoint moves along a specified trajectory.
  - — Data animation: the plotted data change with time.
- Users can clip plots using a slider control.
- Users can query for coordinates of a point with a mouse.
- Labeling of plots has been upgraded to publication quality.

The Macsyma demonstration file STRING1 derives the mathematical solution for a plucked guitar or violin string and animates the resulting motion. The solution is

$$u(x,t) = -\frac{\text{hp}\ \text{x1}^2 \displaystyle\sum_{m=1}^{\infty} \frac{\cos\left(\frac{\pi\, c\, m\, t}{\text{x1}}\right) \sin\left(\frac{\pi\, m\, t}{\text{x1}}\right) \sin\left(\frac{\pi\, m\, \text{xp}}{\text{x1}}\right)}{m^2}}{\pi^2\ \text{xp}\,(\text{xp} - \text{x1})}$$

where

xL	=	length of the string (0 <=x <= xL)
mu	=	mass per unit length of the string
tau	=	tension on string
u(x,t)	=	lateral displacement of the string
xp	=	location at which the string is plucked
xh	=	distance by which the string is displaced at the point xp

For a particular set of numerical values that corresponds to an "A" string, a still representation of the animated motion is shown in Figure 37.3. Macsyma can show the animated motion, store the animation in a notebook and replay it.

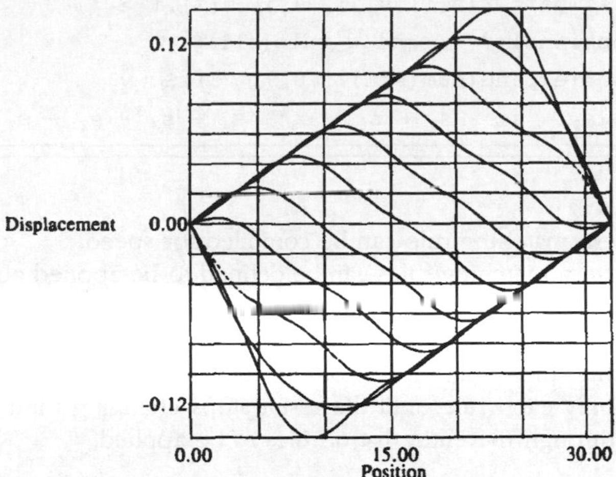

FIGURE 37.3 Still representation of animated motion.

Macsyma 2.0's new graphics viewer displays the great advances in user control and ease of use that is possible when fully object-oriented software technology is applied in the graphics domain. Users have access to an unprecedented range of plot attributes in a logically organized fashion. A typical 3D plot has over 200 attributes that are editable in-place in the notebook. Macsyma offers five graphics control dialogs (camera view, bounding box and axes, surfaces lines and points, titles and decorations, and graphics attributes) with custom controls for easy editing of plot attributes.

Programming Language

Macsyma contains a complete programming language. The language includes most common programming constructions, such as conditional branching ("if-then") statements, iteration loops, functions and procedure definitions, program blocks and localized variables, go-to and catch-throw mechanisms, many data types, and over 30 predefined type-testing predicates. Each version of Macsyma includes a compiler to convert user-written code into binary files.

Macsyma includes a collection of software development tools at the user level, including several varieties of debugger, function tracing, time metering (including reporting of time spent in specified functions), user settable break points, and setcheck capabilities (notification each time a specified variable is assigned a value).

Rule-Based Programming

The kernel of Macsyma is implemented with procedural programming software techniques. Rule-based pattern matching is used in various places in the system, and is available in several ways to users. An example follows of implementing the rules of a Clifford algebra over a 10-dimensional positive-definite inner product space. Macsyma is instructed to simplify $(s_0 + \ldots + s_9)$^5 in this Clifford algebra, where $\{s_i\}$ are a basis of the vector space that generates the algebra.

```
(c1)  matchdeclare([i,j],integerp)$
(c2)  tellsimpafter(s[i]^^2,1)$
(c3)  tellsimpafter(s[i].s[j],-s[j].s[i],i > j)
(c4)  /* optional */ compile_rule(all)$
(c5)  factor(expand(sum(s[i],i,0,9)^^5))$
(d5)  100 (s9 + s8 + s7 + s6 + s5 + s4 + s3 + s2 + s1 + s0)
```

Note that the pattern matching rule can be compiled for speed.

User-defined pattern matching rules can be defined to be applied automatically before or after simplification (as in the example above), or to be applied only when requested by the user. The user can specify in what order to apply several rules, whether to apply the rules from the bottom up the expression tree or the top down, and whether to apply each rule until the result stops changing then move the next rule, or to iterate through the entire list of rules to be applied.

Interfaces with Other Languages

All major symbolic-numerical math software products can translate mathematical expressions to FORTRAN and C. Macsyma and REDUCE each has the GENTRAN package for generation of FORTRAN and C code. This package translates not only mathematical expressions into the target language, but also function and subroutine definitions, program flow control statements (such as iteration loops and conditional branching statements), and data type information. In the following example, GENTRAN translates Macsyma program flow control statements into FORTRAN.

```
(c1)   gentran( for n:2 next n*2 thru 500 do s : s + n )$
         N = 2.0d0
25002  IF (N.GT.500.0d0) GOTO 25003
          S = S+N
          N = N*2.0d0
          GOTO 25002
25003  CONTINUE
(c2)   gentran( while f(x) > = 0 do x : x + 0.25 )$
25004  IF (.NOT.F(X).GE.0.0d0) GOTO 25005
          X = X + 0.25
          GOTO 25004
25005  CONTINUE
```

You can even mix Macsyma code with FORTRAN or C in one file. GENTRAN can perform the symbolic operations specified by the Macsyma code, and replace the Macsyma code with the resulting numerical code in the target language.

Macsyma also generates TeX output.

New Features in 1994

Macsyma can execute the MATLAB language. See the section on linear algebra for more information.

37.3 USER INTERFACE AND ON-LINE HELP SYSTEMS

According to *Personal Computer World* (1993), "[Macsyma's] enormity never compromises its ease of use." Macsyma's user interface and on-line help systems play a critical role in making this possible.

Fancy Display of Mathematical Expressions

Until 1992, all major symbolic mathematics software products displayed mathematical output expressions in traditional ASCII form in with a fixed width font. In 1992–1993 Macsyma pioneered the formatted display of mathematical expressions using graphical images of mathematical symbols (such as signs for integration, summation, products, radicals), Greek letters, and sizing and positioning of exponents.

The following example illustrates that formatted display of mathematical expressions is very useful, and not a cosmetic amenity.

```
(c1) p1: sqrt(ratexpand((sqrt(alpha)+beta)^6))$
(c2) p2: ratexpand((c+d)^11)$
(c3) p3: 'integrate(ratexpand((a+x)^12),x)$
(c4) p4: ratexpand((g+h)^13)$
(c5) fancy_display:false$
(c6) expr: p1/p2+p3/p4;
         /
         [   12          11        2 10       3 9        4 8        5 7
(d6) (I (x + 12 a x + 66 a x + 220 a x + 495 a x + 792 a x
         ]
         /

            6 6      7 5       8 4       9 3       10 2      11
 + 924 a x +792 a x + 495 a x +220 a x + 66 a x + 12 a x

    12         13          12        2 11       3 10       4 9
 + a ) dx)/(h + 13 g h + 78 g h + 286 g h + 715 g h

          5 8         6 7          7 6          8 5        9 4
 + 1287 g h + 1716 g h + 1716 g h + 1287 g h + 715 g h

         10 3       11 2       12      13                 6
 + 286 g h + 78 g h + 13 g h + g ) + sqrt(beta

                          5               4            3/2     3
 + 6 sqr(alpha) beta + 15 alpha beta + 20 alpha beta
```

```
           2           2              5/2                3      11        10
  + 15 alpha beta + 6 alpha beta + alpha )/(d + 11 c d

          2 9          3 8          4 7         5 6          6 5        7 4
  + 55 c d + 165 c d + 330 c d +462 c d + 462 c d+ 330 c d

            8 3        9 2         10       11
  + 165 c d + 55 c d + 11 c d + c
```

This expression is repeated below, using the new formatted display.

```
(c7)  fancy_display:true$
(c8)  expr;
```

(d8)
$$\frac{\displaystyle\int \left(\begin{array}{l} x^{12} + 12ax^{11} + 66a^2x^{10} + 220a^3x^9 + 495a^4x^8 \quad +792a^5x^7 + 924 \\ *a^6x^6 + 792a^7x^5 + 495a^8x^4 + 220a^9x^3 + 66a^{10}x^2 + 12a^{11}x + a^{12} \end{array} \right) dx}{\left(\begin{array}{l} h^{13} + 13gh^{12} + 78g^2h^{11} + 286g^3h^{10} + 715g^4h^9 + 1287g^5h^8 + 1716g^6h^7 \\ + 1716g^7h^6 + 1287g^8h^5 + 715g^9h^4 + 286g^{10}h^3 + 78g^{11}h^2 + 13g^{12}h + g^{13} \end{array} \right)} + \frac{\sqrt{\beta^6 + 6\sqrt{\alpha}\beta^5 + 15\alpha\beta^4 + 20\alpha^{3/2}\beta^3 + 15\alpha^2\beta^2 + 6\alpha^{5/2}\beta + \alpha^3}}{\left(\begin{array}{l} d^{11} + 11cd^{10} + 55c^2d^9 + 165c^3d^8 + 330c^4d^7 + 462c^5d^6 \\ + 462c^6d^5 + 330c^7d^4 + 165c^8d^3 + 55c^9d^2 + 11c^{10}d + c^{11} \end{array} \right)}$$

The main problem with the "ugly display" on line (d6) is that the user cannot easily identify the operator precedence: one of the plus signs is the primary operator of the expression, and the two division operators are the secondary operators. The formatted display, such as that on line (d8), enables the user to identify primary and secondary operators much more easily in a wide range of expressions.

Macsyma can echo input commands in formatted display, but does not accept two-dimensional input as of this writing (1995).

Document Processing

PC Macsyma can create Windows metafiles for exporting mathematical expressions and graphics to other applications. This enables users to create documents that combine fancy math expressions, text and graphics in their favorite Windows word processors, without learning how to operate another word processor.

New Features in 1994

PC Macsyma 2.0 has a "notebook interface" that combines formatted math display, in-place editable graphics, and full text processing in one reexecutable document. The sections can be rearranged and edited. Graphs can be edited in place in the notebook; for example, they can be rotated and decorated with labels without redrawing the plot with a Macsyma command. Text processing capabilities include full control of characters, paragraphs, tabs, margins, and so forth. The notebooks can be saved, reopened, and reexecuted.

Starting in early 1995, Macsyma notebooks include hypertext links that enable you to click on active words and phrases and jump to other locations in the same notebook or different notebooks. Macsyma notebooks also include a "navigator" dialog, which presents a compact summary of each section in a notebook. You can click on the summary line for any section and go to that section. These facilities make Macsyma notebooks the most convenient tools available for publishing large technical documents or lesson plans in notebook form, including documents which are spread over many notebooks.

On-Line Help Systems

Macsyma has a complete array of on-line help facilities.

- An interactive on-line primer, to introduce beginners to the system.
- Math topic browser menus with about 70 topics. These menus guide you through a taxonomy of mathematical capabilities in plain English, leading you to a small group of commands that addresses the area of your immediate interest.
- 2,000 hypertext topic descriptions with cross-references, plus about 100 additional in-depth descriptions of external packages and special topics.
- Over 600 executable examples of individual commands, and 200 executable demonstrations of applications and external packages that are included with Macsyma.
- 500 function templates for individual Macsyma commands. Each function template is a dialog box in which you receive help in composing your own command line using a built-in Macsyma function.

All of these facilities are accessible using mouse-and-menu interactive methods.

37.4 APPLICATIONS OF MACSYMA

Macsyma has been used in a wider variety of applications for a longer period of time than any other symbolic–numerical mathematics software package. By 1986, the number of papers and books referencing Macsyma exceeded 600, and Macsyma Inc. has not been able to maintain a complete listing of applications past that date.

Among the most ambitious applications of Macsyma at this time is a project to extend the PSPICE program for numerical simulation of analog electrical circuits to include symbolic circuit analysis. Feasibility studies and prototype software have been published by researchers at the University of Braunschweig in Germany. Work continues at the University of Kaiserslautern.[17]

[17]The principal researcher is Dr. Ralf Sommer of the University of Kaiserslautern.

Books about Macsyma (and PDEase)

The following books on Macsyma and its applications are available in 1995.

1. *Macsyma Mathematics and System Reference Manual*, 15th ed. Macsyma Inc., 1995. *Macsyma Graphics and User Interface Reference Manual*, 15th ed. Macsyma, Inc., 1995.

   A comprehensive reference for all the capabilities of Macsyma. A copy of the *Macsyma Reference Manual* comes with every full copy of Macsyma.
2. *Macsyma User's Guide*, 2nd ed. Macsyma Inc., 1995.

   A tutorial guide to performing basic operations in Macsyma. The *Macsyma User's Guide* comes with every full copy of Macsyma.
3. Chiang C. Mei, *Mathematical Analysis in Engineering*. Cambridge University Press, Cambridge, MA, 1994.

   Topics in engineering analysis, with one chapter introducing the use of Macsyma for perturbation computations.
4. Richard H. Rand, *Topics in Nonlinear Dynamics with Computer Algebra*. Gordon and Breach, 1994.

   Implements some perturbation theory methods for ordinary differential equations using Macsyma
5. Stefan Braun and Harald Häuser, *Macsyma Version 2—Systemulische und praxisnahe Einführing mit Anwendungsbeispielen* (*Macsyma 2.0—A Systematic and Practical Introduction with Applications*). Addison-Wesley, Reading, MA, 1994.
6. Barbara Heller, *Macsyma for Statisticians*. John Wiley, New York, 1991.

   An introduction to Macsyma with many exercises with answers in the fields of probability and statistics.
7. A. I. Beltzer, *Variational and Finite Element Methods*. Springer-Verlag, Berlin, 1990.

   Application of Macsyma to problems in variational calculus and finite element analysis.
8. Richard Rand and Dieter Armbruster, *Perturbation Methods, Bifurcation Theory and Computer Algebra*. Springer-Verlag, Berlin, 1987.

   Implements some perturbation theory methods for ordinary differential equations using Macsyma.
9. Gunnar Backstrom, *The Fields of Physics on the PC by Finite Element Analysis*. Studentlitteratur, 1994.

   Introduction to the use of PDEase to solve problems in electromagnetism, heat transfer, elasticity, and fluid mechanics, in Cartesian and polar coordinates.

Demonstration Applications Included with Macsyma

The 200 executable demonstration files included in Macsyma cover a wide range of elementary applications from physics, financial modeling, and probability and statistics. The applications demonstrations include the following:

Demo Name	Description
General physics	
DEMO(BALLISTICS);	Ballistics of cannon ball
DEMO(BALLISTIC2);	Ballistics problem, with wind
DEMO(CTENSOR1);	Elasticity, incompatibility tensor
DEMO(DYNAMICSYS);	Solve fifth order linear dynamic system

Demo Name	Description
DEMO(ITENSOR2);	Elastic strain, using indicial tensors
DEMO(OSCILLATOR);	General solution of linear oscillator
DEMO(STATMECH1);	Foundations of statistical mechanics
DEMO(STRING1);	Animated model of plucked guitar/violin string
DEMO(VECT_PDE);	Electrodynamics, P.D.E.s, finite differences
DEMO(TENS_PDE);	Fluid mechanics, P.D.E.s, finite differences
DEMO(UNITS);	Conversions of physical units of measure
Quantum mechanics	
DEMO(QMECH1);	Quantum mechanical equations in various coordinates
DEMO(FEYNBOUND);	Asymptotic behavior of Feynman integrals
General relativity	
DEMO(CTENSOR);	Schwarzschild solution, black hole
DEMO(CTENSOR2);	Robertson–Walker cosmological solution
DEMO(CTENSOR4);	Kerr–Newman solution: charge, rotating black hole
DEMO(ITENSOR3);	Weak field approx. to Einstein equations
DEMO(ITENSOR4);	Verify that div(Einstein) = 0
DEMO(ITENSOR6);	Verify that curvature invariant = 0
Economics and finance	
DEMOCASHFLOW);	Present values and discounted cash flows
DEMO(ECON1);	Microeconomic model of a consumer
DEMO(MORTAGE);	Present values and discounted cash flows
DEMO(OPTION_PRICE);	Black–Scholes option pricing
Medical diagnostics	
DEMO(BAYES);	Bayesian analysis of diagnostic tests

37.5 SYNOPSIS OF CAPABILITIES OF MACSYMA

Algebra

Arithmetic

- Exact arithmetic with integers, rational numbers, and transcendental numbers
- Floating point/arbitrary precision floating point numbers
- Complex arithmetic, numerical and symbolic
- Algebraic numbers and finite fields

Basic Algebra

- Simplification of rational functions, exponents, logs, radicals, various ways
- Factoring and expanding, various ways
- Partial fraction decomposition, continued fractions
- Greatest common divisors
- Simplification subject to constraints
- User-defined operators with user-defined properties
- Set theory computations and permutations
- Arithmetic of inequalities

Trigonometry

- Trigonometric and hyperbolic functions, inverses
- Trigonometric identities and simplifications, various
- Evaluation of trig functions as special angles
- Special efficient implementation of Poisson series

Matrix Algebra

- Add, multiply, transpose, rank, determinant, norm, inverse, LU inverse, Cholesky inverse, generalized inverse
- Triangularization, eigenvalues and eigenvectors, nullspace
- Gram–Schmidt orthogonalization
- *LDU*, *LU*, *QR*, Cholesky, singular value decompositions
- Jordan, Schur forms, matrix exponentiation
- Language features and matrix utilities approximate those in the numerical linear algebra product Matlab™ [18]

Solving Equations

- Exact solutions
  — Simultaneous multivariate linear and polynomial equations
  — Equations involving transcendental functions
  — Recurrence and difference equations
- Approximation Methods
  — Taylor series solutions
  — Least-squares solutions of overdetermined systems
- Numerical Solutions
  — Real and complex roots of polynomials
  — Bisection method and Newton's method

Sums and Products

- Sums over finite and infinite ranges
- Expressing indefinite summations in closed form
- Simplification of sums, many methods
- Products over finite and infinite ranges

Special Functions

- Orthogonal polynomials and related functions, Bessel, Neumann, Hankel, Bernoulli and Euler polynomials, zeta, Fibonacci, Airy functions
- Transcendental functions: error, polylogarithm, gamma and polygamma functions, exponential integrals, Jacobi elliptic functions
- Combinatorial functions: factorial, binomial, beta functions Riemann zeta (with simplification algorithms)

[18]Matlab is a trademark of The Math Works, Inc.

Calculus

Differential Calculus

- Differentiation and Limits
  - — Limits, including directional dependence
  - — Differentiation, partial differentiation of unknown functions
  - — Analytic optimization
  - — Euler–Lagrange equations; calculus of variations
  - — Operator algebra
- Taylor Series Methods
  - — Taylor and Laurent series, multivariate
  - — Reversion of Taylor series
  - — Pade approximants
  - — Infinite power series; expansions of many functions

Integral Calculus

- Indefinite Integration
  - — Pattern matching methods and tables
  - — Risch algorithm
  - — Integration by parts
  - — Trigonometric-exponential integrands, special methods
  - — Elliptic integrals (numerical only)
  - — Integrate derivatives of unknown functions
- Definite Integration
  - — Evaluation of indefinite integrals, including limits at endpoints and limits at infinity
  - — Contour integration
  - — Generalized gamma, exponential integrals
  - — Dirac delta functions and integrals
  - — Numerical integration: Romberg, Newton-Cotes, extrapolated Gaussian quadrature, Simpson, trapezoidal methods
- Integral Transforms
  - — Laplace transform and inverse, with transforms of special functions
  - — Fourier sine and cosine series
  - — Fourier transforms on finite, semiinfinite and infinite intervals
  - — Fast Fourier transform

Differential Equations

- Systems of Linear O.D.E.s
  - — Laplace transform method; transfer function matrix method
- First-Order O.D.E.s
  - — Linear, separable, exact, homogeneous, Ince, Prelle-Singer, Laplace transform methods; Bernoulli, Riccati equations; integrating factor
  - — Insertion of initial conditions
- Second-Order O.D.E.s
  - — Constant coefficient, exact, linear homogeneous, invariant constant, adjoint, factorized operator, Laplace transform methods; variation of parameters, free of x

or y, change of independent variable, Euler equation, trial list, series, hypergeometric and Whittaker solutions
  — Insertion of initial and boundary conditions
  — Solves 83% of linear second order O.D.E.s in Kamke's book
- Perturbation and Taylor Methods
  — Lindstedt's method, Method of averaging, Method of multiple scales, Taylor solutions of (systems of) O.D.E.s
- Numerical Solutions of O.D.E.s
  — Runge–Kutta for systems of O.D.E.s
  — Generation of finite difference equations
- Partial Differential Equations
  — Find Lie symmetries and solutions of systems of nonlinear P.D.E.s
  — Macsyma generates input to PDEase® for finite element analysis of systems of P.D.E.s (PDEase is available from Macsyma Inc.)

Integral Equations

- Equations of the First Kind
  Variable limit, fixed limit, Laplace transform methods, Abel's method, collocation, Taylor series, pseudo-Picard iteration
- Equations of the Second Kind
  Variable limit, fixed limit, Laplace transform methods, collocation, Taylor series, Fredholm–Carleman series, Neumann series

Statistics and Data Analysis

Statistics

- Univariate and multivariate descriptive statistics
- Probability densities and cumulative distributions, many

Data Analysis

- Nonlinear multivariate least squares fit of data (Levenberg–Marquardt algorithm)
- Polynomial and spline interpolation of tabulated data
- Dimensional analysis and units conversion

Vector and Tensor Analysis

Vector Calculus

- Dot, cross products, grad, div, curl, Laplacian operators
- Coordinate-invariant computations
- Transforms to specific coordinate systems, including 20 that are preprogrammed
- Many simplification options

Tensor Calculus

- Tensor summation, symmetry and contraction rules, simplification
- Covariant derivatives and curvature (various kinds)
- Exterior and Lie derivatives

- Geodesic coordinates
- Coordinate-invariant computations
- Transformation to arbitrary coordinates. Over 20 specific coordinate systems are preprogrammed
- Optional frame fields, affine torsion, nonmetricity

Tensor Algebras

- Universal, Grassmann, Clifford, Symmetric, and Symplectic

Exterior Calculus of Differential Forms

- Exterior multiplication, vector-form contraction
- Exterior derivative, Lie derivative

Graphics

Two-Dimensional Plots

- Plotting of functional relationships, parametric functions, vector fields
- Plotting lists of points; adaptive density plotting
- Contour plots

Three-Dimensional Plots

- Plotting of functional relationships
- Hidden line removal, change of perspective
- Plotting of parametric surfaces, vector fields
- Interactive query of coordinates with mouse[19]
- Color shows height or other user-selected function

Plot Utilities

- Camera animation and data animation of plots[19]
- User-controlled color graphics, axes, plot scale, plot labels
- Coordinate transformations (e.g., polar, log)
- Mouse/keyboard controlled translate, rotate, and zoom
- Superposing plots
- Postscript and other printer output

Utilities

Pattern Matching

- Rational function-based pattern matcher
- Simplifier based pattern matcher
- User-defined rules, auto and user-applied

[19]In 1994, available only in PC Macsyma 2.0.

Properties Data Base

- Assign mathematical properties and user-defined properties to symbols and use them in context

Foreign Language Interface

- FORTRAN and C code manufacturing
  - — Translate mathematical expressions
  - — Translate program control statements, (e.g., do-loops, if-then statements)
  - — Generate subroutine, function, and data type statements
  - — Optional segmentation and optimization of expressions
  - — Mix FORTRAN (or C) and Macsyma languages in template mode.
- Generate T_EX^{TM} output
- Translate MATLAB language to Macsyma language

Utilities Used with Numerical Analysis

- Horner's method to reduce floating point operations
- Find common subexpressions

Programming Tools

- Compile code user-written in Macsyma language
- Break points and setcheck
- Function tracing and timing
- Store/recall partial/total environment

User Interface

- Notebooks with combined math, text, graphics[19]
- Math display with Greek letters, drawn math symbols, half-height exponents, selectable fonts
- Export math and graphics using Windows metafiles
- Mouse-driven help menu system
- User formatting of floating point numbers
- Hypertext links within and between notebooks
- "Navigator" dialog presents compact summary of all notebook section. A mouse click takes you to any other section

Documentation

On-Line Documentation

- Math topic browser help menus
- Topic descriptions (2,000) with hypertext cross-references
- Executable examples of commands (700)

[19]In 1994, available only in PC Macsyma 2.0.

- Executable application demonstrations (200)
- Function templates (500)
- Interactive primer with six lesson scripts
- Find command from partial name

Hardcopy Documentation

- *Quick Reference Card*, 80 common commands
- *User's Guide*, with introductory tutorials and examples covering about 200 commands
- *Reference Manual*, covering 2,000 commands and the complete system
- Installation Guide and Release Notes
- *Macsyma Newsletter*

ACKNOWLEDGMENTS

I would like to acknowledge the invaluable assistance of the technical staff of Macsyma Inc. for reviewing and improving many aspects of this article. Any errors or omissions remain solely the responsibility of the author.

DEFINING TERMS

Algebraic: A Macsyma option that turns on machinery for automatic simplification of radicals and other algebraic numbers.

Assume and Declare: Macsyma commands for stating assumptions about inequality relationships or the type of a variable, function, operator, or other object.

Diff: A Macsyma command that returns the derivative of a function or expression.

Duffing's Equation: The equation of an almost-harmonic oscillator. It includes the cubic force term in the Taylor expansion of a general (antisymmetric) nonlinear force term.

Fancy_display: A Macsyma option that turns on or off advanced features in the 2D display of mathematical expressions.

Gentran: A Macsyma command that translates expressions and program flow control structures into FORTRAN or C.

Gradef: A Macsyma command that defines the gradient of a function without requiring that you specify the form of the function.

Ilt: A Macsyma command that returns the inverse Laplace transform a function or expression.

Isolate: A Macsyma command that reexpresses its input by compressing subexpressions that do not contain a specified variable, so you can clearly see the dependence on that variable.

Lindstedt: A Macsyma command that implements Lindstedt's method, and returns a perturbation solution to an ordinary differential equation.

Matchdeclare: A Macsyma command for declaring that a variable is a pattern matching variable that is used in definitions of pattern matching rules.

Ode: A Macsyma command that returns the symbolic solution of an ordinary differential equation.

PDEase: A companion product to Macsyma that solves systems of (nonlinear) partial differential equations by the finite element method.

Pickapart: A Macsyma command that pulls apart an expression into its component parts, to a specified level of operator structure.

Polarform: A Macsyma command that returns the polar form of a complex expression.

Ratsimp: A Macsyma command for general simplification of algebraic expressions.

Rectform: A Macsyma command that returns the rectangular form of a complex expression.

Rule-Based Pattern Matching: A method of transforming expressions wherein general rules are defined, and the order in which the rules are applied is determined by a general mechanism, not by explicit programming for each situation in which the rules are to be applied.

Stiff System of Differential Equations: A system of O.D.E.s is called "stiff" if it has eigenvalues whose magnitudes differ by two orders of magnitude or more.

Taylor_ode: A Macsyma command that returns an approximate Taylor series solution to an ordinary differential equation or system of equations.

Tellsimp and Tellsimpafter: Macsyma commands that define automatic pattern matching rules, which are applied either before or after the general simplifier is called.

Tex: A Macsyma command that translates mathematical expressions into the $\TeX$ document formatting language.

38

Maple

Frederick J. Wicklin
University of Minnesota

38.1 OVERVIEW

Algebraically complicated equations occur in all areas of science and engineering. In principle, standard manipulations will solve many of these equations, but in practice the required calculations are so unwieldy that producing the correct solution by hand is intractable or impossible. In such cases, the scientist can turn to *computer algebra* for assistance.

Maple is an example of a computer algebra package. It has the ability to work with floating point numbers, but its real power is as a symbolic manipulator. Maple would be an excellent choice for differentiating a complicated expression or for expanding $(x^2 - x + 1)^{13}$. Maple is probably *not* the best choice for finding eigenvalues of very large matrices or for numerically integrating a large system of differential equations (although it can perform these tasks).

Briefly, Maple allows users to:

- Symbolically manipulate quantities, formulas, and equations;
- Analytically solve integrals, derivatives, and simple differential equations;
- Graphically display curves and surfaces defined explicitly, implicitly, or parametrically;
- Numerically solve problems that are not amenable to a symbolic approach;
- Readily access thousands of predefined mathematical operations;
- View the algorithms underlying most predefined procedures;
- Easily define arbitrarily complicated functions and procedures.

In this article we use examples to illustrate Maple commands and syntax. The Maple program contains over 2500 mathematical functions and procedures, so this survey will describe only a small fraction of Maple's capabilities. The references at the end of this survey provide a more complete overview, as does Maple's on-line help facility.

0-8493-2530-7/96/$0.00+$.50

The version of Maple described here is Maple V Release 2, which is substantially enhanced from the first release of Maple V. In particular, Maple V Release 2 has a very nice user-interface, an excellent on-line help facility, and a variety of new functions for plotting curves and surfaces. For certain computer architectures, Release 2 also provides a "notebook" interface in which equations, graphics, and text may be combined into a single document. By using a notebook, you can write and print an entire technical document from within Maple.

Maple V Release 2 is available for most computers, including UNIX systems, Macintosh, DEC VMS, IBM VM/CMS, and PCs running DOS or Windows. Maple is commercial software and is distributed by Waterloo Maple Software, 450 Phillip St., Waterloo, Ontario, Canada N2L 5J2. Maple's name and trademark both reflect the Canadian origins of this program; it was written and developed primarily by the Symbolic Computing Group at Waterloo University.

38.2 ON-LINE HELP

When considering adopting a new software package, there are two important considerations:

- How easy is it to gain proficiency? In other words, how long will it take before this software helps you become more productive than you used to be?
- Once proficiency is gained, how easy is it to learn advanced features of the software?

Maple excels in both of these areas. Most of Maple's syntax is the same as mathematical writing, so that to enter the mathematical formula $e^{-2t}[A\cos(3t) + B\sin(3t)]$ we will type

```
> exp(-2*t)*(A*cos(3*t)+B*sin(3*t));
```

and Maple responds

```
exp(-2 t)  (A cos(3 t)+B sin(3 t))
```

Notice the semicolon that terminates the Maple command and tells Maple to echo its response. The variables **A** and **B** are *symbolic* and do not have specific values. Notice also that mathematical functions (e.g., **cos, sin**) do not require capital letters nor any special brackets to delimit their arguments. Maple names are usually short (which minimizes typing) and indicative of the functions they perform. For example, to factor a polynomial you use the **factor** command, to solve a set of equations you use the **solve** command, and to display the graph of a function you use **plot**. (There is a drawback to the use of common words to denote Maple commands: new users sometimes create a variable with the same name as a predefined command, thus inadvertently redefining the command!)

Of course, not all Maple names are intuitively obvious. Who would guess that **evalf** is the Maple function that gives numerical approximations? Fortunately, Maple's on-line help facility often comes to the rescue! Typing a question mark (**?**) followed by the name of a Maple command will produce and display a help file. For example, **?solve** will generate a detailed description about the syntax and usage of the solve command, and also refer the reader to 20 related topics such as **fsolve** and

linalg[linsolve]. Using **?** to guess a command is often fruitful: for example **?numerical** brings up the help page for **evalf!**

You can get general purpose information about Maple by typing **?help, ?intro, ?index,** or **?library.** A typical help window is shown in Figure 38.1. Incomplete requests are often met with useful responses:

```
> ?eigen
Try one of the following topics:
                    {eigenval, eigenvect, Eigenvals}
```

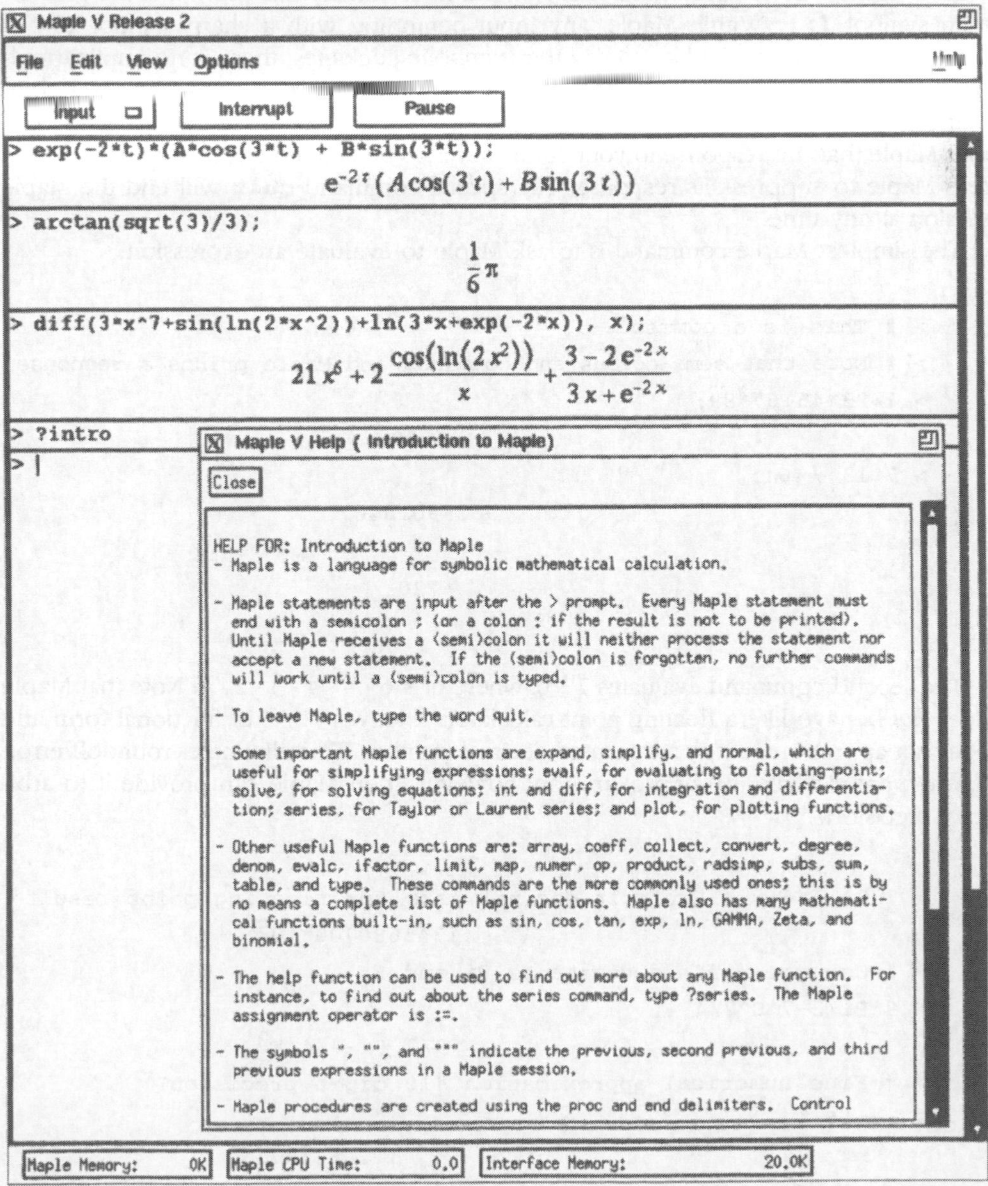

FIGURE 38.1 Help window for **?intro**.

In summary, Maple's on-line help facility makes it fairly straightforward to learn both Maple basics and advanced features. For those who prefer to learn new software by a "guided tour," Maple V Release 2 includes an on-line tutorial that can be initiated by typing **tutorial()**. The tutorial is broken into 14 sections, each section requiring between 5 and 15 minutes to complete.

38.3 ALGEBRAIC EXPRESSIONS AND MANIPULATIONS

In this section we will begin to present Maple's mathematical features. To help you follow the examples, we first mention a few words about Maple syntax. Maple provides a prompt at the beginning of each input line. Usually this prompt is the greater-than symbol **(>)**. Within Maple, any input beginning with a sharp symbol **(#)** is treated as a comment. Unlike some other symbolic packages, there is no implicit multiplication in Maple, so the mathematical expression $3x$ is entered as **3*x**. All commands must end with a semicolon or a colon. Ending a command with a semicolon tells Maple that the response to your command is to be echoed to the terminal; a colon tells Maple to suppress its response. The special command **quit** will end the Maple session at any time.

The simplest Maple command is to ask Maple to evaluate an expression:

```
> # This is a comment.
> # Note that semi-colons end commands and Maple prints a response
> 1+23*45+67*89;
                                    6999
> 7^13 / 6!;
                                 96889010407
                                 -----------
                                     720
```

The second command evaluates $7^{13}/6!$ where $6! = 6 \cdot 5 \cdot 4 \cdot 3 \cdot 2 \cdot 1$. Note that Maple does *not* behave like a floating point calculator: it leaves $7^{13}/6!$ in fractional form and does not approximate it as a truncated decimal number. This eliminates roundoff error. If you want a numerical approximation, however, then Maple can provide it to arbitrary precision.

```
> 7.0^13/6!; # Explicit decimal asks for floating point result
                                 .1345680700*10^9
> # Use pre-defined constants Pi and E
> 9*Pi/2-7*E^2/4;
                                    9/2 Pi-7/4 E^2
> # Find numerical approximation (10 digit precision)
> evalf(");
                                       1.20631877
> arctan(sqrt(3)/3);
                                  1/6 Pi
```

```
> # Approximate Pi to 20 digit precision
> evalf(Pi, 20);
                         3.1415926535897932385
```

Note that Maple knows about the constants π and e (denoted as **Pi** and **E** in Maple) where $e = 2.7182818\ldots$ is the base of the natural logarithm. Another commonly used constant is $i = \sqrt{-1}$, which Maple denotes as capital **I**. The double-quote symbol (") refers to the result of the immediately previous computation, whereas **evalf** (**eval**uate using **f**loating-point arithmetic) is a Maple function that returns a floating-point approximation of its argument. The **evalf** functions accepts an optional second parameter that specifies the number of digits to use during floating-point arithmetic.

Let's now assign and manipulate some algebraic expressions to see how Maple may be used as a desktop computational assistant.

```
> f := x^3-2*x^2+2*x-1;
                    f := x^3-2 x^2+2 x-1
> subs(x=2/3, f);
                          -7/27
```

In the above computation, we first assigned the polynomial expression $x^3 - 2x^2 + 2x - 1$ to the variable **f**. It should be noted that **f** is *not* a function; the expression f(2/3) does not make sense. (We give examples of Maple functions in Section 38.8: Programming Features.) Instead, to evaluate the polynomial **f** at $x = 2/3$, we use Maple's **subs** command to substitute **2/3** for every occurrence of **x**. Note also that there are two types of equal signs in Maple: the *assignment operator* **:=** and the *equation operator* **=**. The assignment operator takes a variable name (to the left of the **:=** sign) and assigns to it the value obtained by evaluating the expression to the right of the **:=** sign. In contrast, the equation operator is used to build mathematical equations, which then may be manipulated. The equation operator does not change the value of the variable to the left of the **=** sign.

Maple can also perform algebraic operations such as factoring polynomials. This can be extremely useful for finding the roots (i.e., *zeros*) of complicated expressions.

```
> g := 8*x^5-18*x^4*y-7*x^3*y^2+23*x^2*y^3-4*y^5;
                      5         4       3 2        2 3      5
              g := 8 x  - 18 x y - 7 x y  + 23 x y  - 4 y
> factor(g);
                                                2              2
          (x - 2 y) (2 x - y) (x + y) (4 x  - 3 y x - 2 y )
> # Find roots explicitly
> solve(g=0,x);
                                        1/2                  1/2
          -y, 2 y, 1/2 y, 3/8 y-1/8 41    y, 3/8 y+1/8 41    y
> # Symbolically solve a system of equations
> solve({a*x+b*y=0, y=r*sqrt(x)}, {x,y});
```

```
                          2              2 2
                        r b             b r
{y = 0, x = 0}, {y = - -----, x = -----}
                          a               2
                                         a
```

Even when Maple cannot factor or symbolically solve an expression, it is possible to find numerical roots of an expression, as the following example demonstrates.

```
> factor(x^5-x^3+2*x-1);

                     5    3
                    x  - x  + 2 x - 1

> # Maple was unable to factor. Find root numerically.
> fsolve(x^5-x^3+2*x-1);

                       .5603499931
```

38.4 CALCULUS

Maple can perform a multitude of calculus operations, including differentiation, integration, limits, manipulations of series, and the solution of simple differential equations. Using Maple, symbolic differentiation of complicated expressions becomes fast and error free.

```
> diff(3*x^7+sin(ln(2*x^2))+ln(3*x+exp(-2*x)), x);

                                   2
              6      cos(ln(2 x ))      3-2 exp(-2 x)
        21 x  + 2 ---------------- + ---------------
                          x           3 x+exp(-2 x)
```

In fact, Maple performs partial differentiation, so that **diff(f,x)** is mathematically equivalent to $\partial f/\partial x$; the second argument specifies the differentiation variable. You can take higher derivatives by stringing together variables at the end of the **diff** command, e.g., $\partial^2 f/\partial x\partial y$ is found by **diff(f,x,y):**

```
> g := x*y-6*sin(x)/y^2;

                                 sin(x)
                 g := x y - 6 ---------
                                   2
                                  y

> diff(g,x);

                         cos(x)
                    y - 6 ---------
                              2
                             y
```

```
> diff(g,x,y);
                          cos(x)
                 1 + 12 ---------
                             3
                            y
```

Even when the function being differentiated is unknown, Maple may be used to symbolically manipulate the resulting differential equation. For example, we can check that arbitrary scalar functions $f(x - ct)$ are solutions to the wave equation $\partial^2 f/\partial t^2 - c^2\partial^2 f/\partial x^2$.

```
> diff(f(x,t),t,t) = c^2*diff(f(x,t),x,x);
                 2                  /  2          \
                d                 2 | d           |
              ----- f(x, t) = c   |----- f(x, t)|
                 2                  |  2          |
                dt                  \ dx          /
> simplify( subs(f(x,t)=f(x-c*t),"));
                  (2)
                 D   (f)(x-c t) c^2 = D^(2) (f)(x-c t) c^2
> lhs(")-rhs("); # Both sides equal, thus f(x-ct) is soln.
                              0
```

The expression **f(x,t)** is one way to specify that **f** is an unknown function that depends on the variables **x** and **t**. Note that it is perfectly legal to nest Maple commands, so that the line **simplify(subs(. . .))** first makes the required substitutions, and then simplifies the result. (The **simplify** command is just one of the Maple commands to help you manipulate complicated expressions into "simpler" form. Other related commands are **normal, combine, expand**, and **collect.**) The command **lhs** (or **rhs**) returns the left-hand side (or right-hand side) of an equation.

Just as Maple excels at symbolic differentiation, Maple can also help you solve integrals. Instead of scanning long columns of tables for the closed form solution of complicated integrals, Maple can perform the computations for you.

```
> # Compute indefinite integral
> the_integral := int(1/(sin(a*x)*(cos(a*x)^2)), x);
                             1           ln(csc(a x)-cot(a x))
   the_integral := ---------- + ---------------------
                        a cos(a x)               a
> # Check the answer by differentiating the integral
> diff(the_integral, x);

   sin(a x)    -csc(a x) cot(a x) a-(-1-cot(a x)^2) a
   --------- + -----------------------------------------
                          a (csc(a x)-cot(a x))
   cos(a x)^2
```

```
> # Does this reduce to the original function?
> simplify(");

                              1
                     ------------------
                                      2
                     sin(a x)  cos(a x)
```

Definite integration is also implemented, and there are three different numerical algorithms to choose from. The limits of integration may be finite or infinite (by using the Maple constant **infinity**). As shown in the examples below, you can specify ranges in Maple by using two dots such as **0..infinity** or **a..b**. Sometimes Maple will express the results of integration in terms of special functions such as the gamma function, the error function, or elliptic functions.

```
> # Compute an improper definite integral
> int(cos(x)*exp(-x^2), x = 0..infinity);
                                          1/2
                                  1/2 Pi     exp (-1/4)
> # This next one doesn't have a closed-form solution
> int(sqrt(x)*exp(-x^2), x = 0..1);
                              1
                              /
                             |    1/2     2
                             |   x   exp(-x) dx
                             |
                             /
                             0
> # Maple was unable to evaluate analytically; find numeri-
  cal result
> evalf(");
                                     .4533919445
> # Another improper integral
> int(exp(-x^3), x = 0..infinity);

                                               1/2
                                         Pi 3
                                   2/9 ----------
                                       GAMMA (2/3)
> # Maple gave exact solution in terms of the GAMMA
> # function but perhaps we want a number.
> evalf(");
                                     .8929795121
```

When the integrands contain singularities, you should exercise extra care to ensure that the integral is solved correctly. Integration across a singularity is definitely a bad idea, and even Maple (which has only a few known bugs) can get confused with some improper integrals.

```
> # Be careful when integrating at or across singularities
> int(x/(x - a)^p, x = 0..1);
               (-p)                                (-p)
        (1-a)        (p a-a^2+1-p)           (-a)       a^2
      -------------------------------- + ------------------
             (-2+p)  (-1+p)                (-2+p)  (-1+p)
> # The above is only valid for certain values of a and p
> # For example, if a=1/2 and p=3/2 then
> int(x/(x-1/2)^(3/2), x=0..1);
                      1
                      /
                     |          x
                     |   ------------ dx
                     |             3/2
                    /  (x - 1/2)
                     0
> # Maple couldn't solve the integral. Try numerical.
> evalf(");
Error, (in evalf/int) unable to handle singularity

> # Reason: we are integrating across the singularity at x = 1/2
> int(x/(x-1/2)^(3/2), x=1/2..1);
                                  infinity
```

In the final Maple command above, we have evaluated a portion of the original integral and found that the improper integral is infinite.

In this last section, we will investigate the way that Maple treats and manipulates series, especially Taylor series. By default, Taylor series are computed up to sixth order, but the command **taylor** provides an optional third parameter that controls the order of the series.

```
> # Compute Taylor series about the point x = 0
> taylor(cos(x), x=0);
                                       2          4        6
                            1 - 1/2 x  + 1/24 x  + 0 (x )
> # Represent cos(2*Pi*x^3) as Taylor series around x = 1/2.
> taylor(cos(2*Pi*x^3), x = 1/2);
```

```
   1/2      1/2                      1/2    2     1/2        2
1/2 2  - 3/4 2  Pi (x - 1/2) + (-9/16 2  Pi  - 3/2 2  Pi) (x - 1/2)

          1/2           3                        1/2             3
  + (-1/2 2    (-9/16 Pi  + 2 Pi)  - 9/4 2    Pi2) (x - 1/2)

      /27  1/2  3          1/2 / 27     4         2\\         4
  + | — 2    Pi  + 1/2 2   | -- Pi  - 15/2 Pi || (x - 1/2)
      \16                  \128                  //

      /       1/2/      2   27    4\          1/2 /      3   81     \\
  + |1/2 2   |-6 Pi  + — Pi | -1/2 2    |-9 Pi  + -- Pi5 || (x - 1/2)5
      \         \             16     /              \         1280   //

  + O((x - 1/2)6)
> # Convert series to 6th order polynomial and
> # collect into polynomial in x; coefficients ordered by Pi
> taylor_poly := collect( convert(", polynom), [x,Pi]);

                      /  81  1/2 5   27 1/2 4           1/2 3        1/2 2\ 5
taylor_poly := |- — 2  Pi  + — 2  Pi  + 9/2 2  Pi  - 3 2  Pi | x
                      \ 2560          32                                   /

      / 81    1/2    5    513  1/2 4     153  1/2  3            1/2 2\  4
  + | — 2    Pi  - --- 2   Pi  - --- 2   Pi + 15/4 2   Pi | x
     \1024              256              16                             /

      /   81    1/2 5     243  1/2  4   261 1/2  3          1/2 2       1/2 \   3
  + |- —- 2   Pi  + --- 2    Pi  + — 2  Pi  - 9/4 2  Pi  - 2  Pi | x
     \ 1024              128             32                                 /

      / 81  1/2 5    459 1/2  4   225    1/2  3    15    1/2  2\   2
  + | — 2    Pi  - -- 2   Pi  - --- 2   Pi  + — 2   Pi | x
     \2048              512            64          16            /

      /  81    1/2 5    27    1/2  4    99   1/2  3             1/2     2\
  + |- — 2   Pi  + -- 2   Pi  + -- 2   Pi  - 3/16 2     Pi | x
     \ 8192              128            128                          /

      81   1/2  5    81   1/2  4           1/2  3          1/2              1/2
  + —- 2   Pi  - — 2   Pi  - 9/128 2   Pi  + 1/8 2   Pi + 1/2 2
     81920         4096
```

Although the polynomial above looks horribly complicated, Maple has no problem working with it—or even with polynomials with thousands of terms! Using Maple you can determine that the polynomial above is only a good approximation to the function $\cos(2\pi x^3)$ on the interval [0.4, 0.6] and that in order to have a Taylor polynomial which is a good approximation to $\cos(2\pi x^3)$ on the entire interval [0,1], you need to keep between 35 and 40 terms of the Taylor series for $\cos(2\pi x^3)$. To do such a calculation (correctly) by hand would be quite a challenge, but using Maple we can not only compute the Taylor polynomial to any order, but also determine that on [0,1] the maximum difference between $\cos(2\pi\ x^3)$ and the Taylor polynomial of order 40 centered at 1/2 occurs at $x = 1$, and the maximum difference is approximately given by

```
> approx := convert(taylor(cos(2*Pi*x^3), x=1/2, 40), polynom):
> evalf(subs(x=1, approx- cos(2*Pi*x^3)), 15);
                     -.2780425157*10^-5
```

38.5 GRAPHICS

Up to now we have looked at the way that Maple manipulates expressions and performs computations such as differentiation. These symbolic calculations are Maple's foremost strength. Yet, scientists and engineers should also consider Maple whenever they want to graph a function or view a two-dimensional surface. The power of Maple's color graphics cannot be adequately presented on a black and white piece of paper, but hopefully you can still get an idea of how Maple graphics can be useful.

The simplest graphics command is to **plot** a function of one variable over some domain:

```
> plot(sin(x^2)/x, x=0..2*Pi);
```

Figure 38.2 shows the result of this command under the X Window System. The `plot` command may be called with optional parameters that specify the display range, scal-

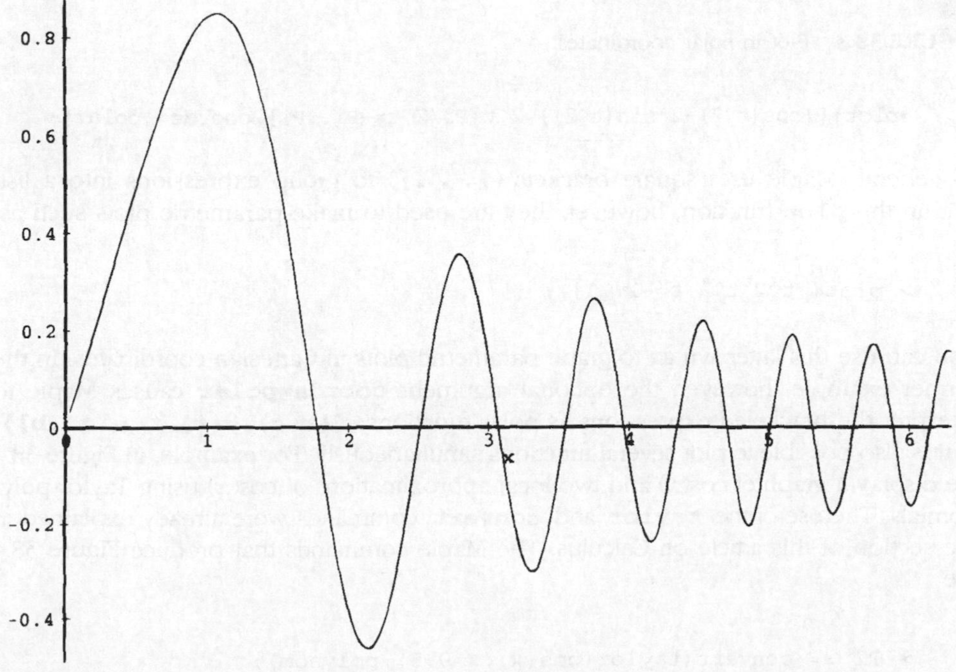

FIGURE 38.2 Plot of sin(x)/x.

ing information, color information, and number of evaluation points for the function. Another option, shown in Figure 38.3, is polar coordinates

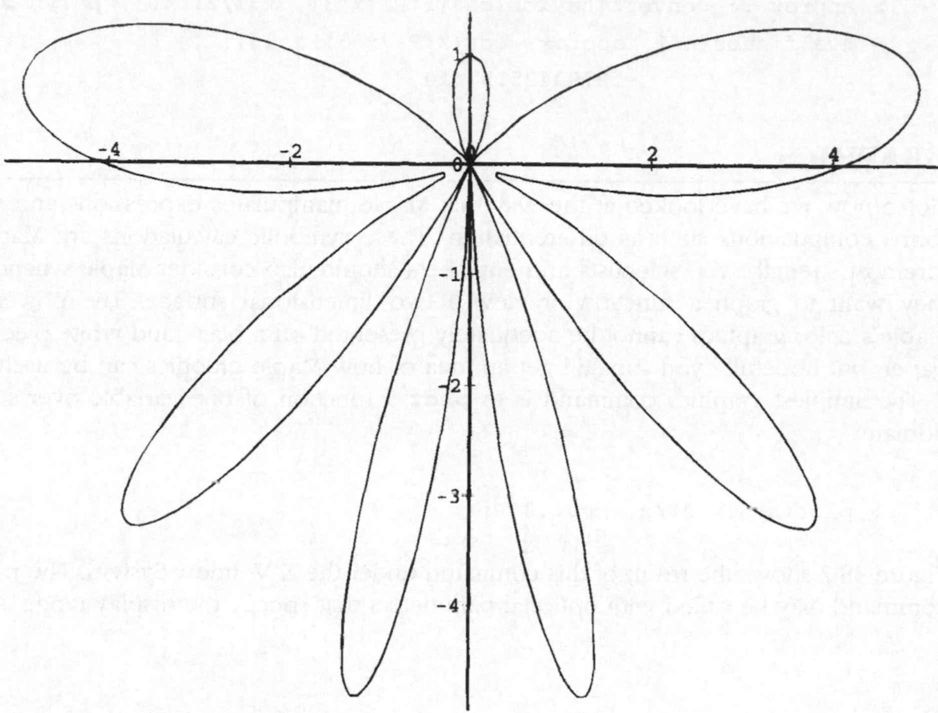

FIGURE 38.3 Plot in polar coordinates.

```
>plot([(cos(t^2)-2*sin(t^2))^2,t+Pi/2,t=-Pi..Pi],coords= polar);
```

In general, Maple uses square brackets (**[. . .]**) to group expressions into a list. Within the **plot** function, however, they are used to make parametric plots such as

```
> plot([t^2,t^3,t=-1..1]);
```

You can use this latter syntax to graph parametric plots in Cartesian coordinates. In the former example, however, the optional argument **coords=polar** causes Maple to interpret the bracketed expressions as polar functions **([*r*(*t*),θ(*t*),t = a..b])**.

It is also possible to plot several functions simultaneously. For example, in Figure 38.4 we display a graph of cos(x) and two local approximations of cos(x) using Taylor polynomials. The use of the **taylor** and **convert** commands were already explained in the section of this article on Calculus. The Maple commands that produce Figure 38.4 are

```
> T2 := convert(taylor(cos(x),x=0,3),polynom);
```

$$T2 := 1 - 1/2x^2$$

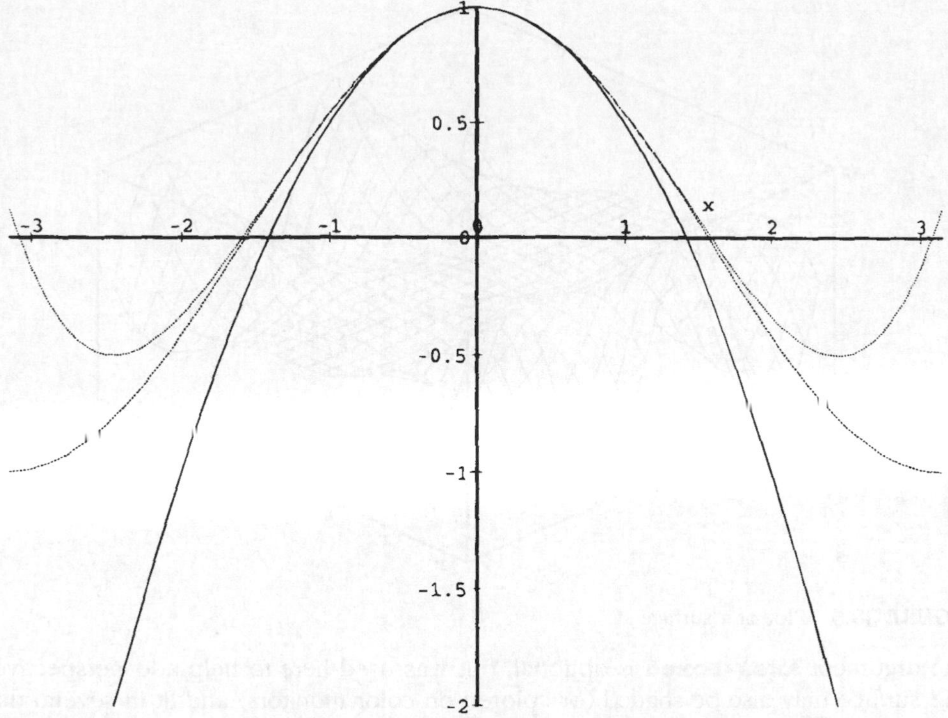

FIGURE 38.4 Plot of cos(x) and two approximations.

```
> T4 := convert(taylor(cos(x),x=0,5),polynom);

                              2        4
                T4 := 1 - 1/2x + 1/24x
> plot({cos(x),T2,T4}, x=-Pi..Pi, -2..1);
```

Braces (**{. . .}**) are used in Maple to collect objects into sets, so the plot command above instructs Maple to plot the set of functions containing **cos(x)**, **T2**, and **T4** over the domain $[-\pi,\pi]$, but to restrict display of the horizontal range to the interval $[-2,1]$. On a color monitor, Maple will display each of the three graphs in a different color.

It is also possible to plot lists of data points, geometric objects, and implicitly defined curves. Logarithmic and log–log plotting options are also available. Series of graphs may be animated using the **animate** procedure.

Until now we have dealt with two-dimensional graphics. But Maple also allows you to create and conveniently manipulate three-dimensional graphics, that is, curves or surfaces lying in three-dimensional space. All of the ideas presented so far for 2D graphics have 3D counterparts. Furthermore, Maple's interface to 3D graphics (seen in Fig. 38.5) makes it easy to change the direction from which you view a 3D object, so that you can interactively "walk around" an object to get a better feeling for "what's on the other side." Unfortunately, this interaction and manipulation of 3D objects are not easily conveyed through a static medium such as this article.

Figure 38.5 shows the result of the command

```
> plot3d(x^2*sin(x+2*y), x=-Pi..Pi, y =-Pi..Pi, axes=boxed);
```

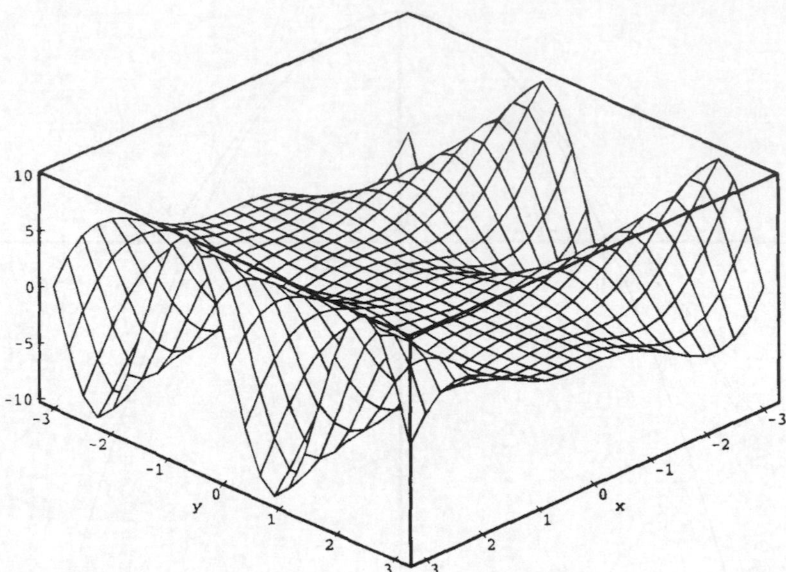

FIGURE 38.5 Plot of a surface.

The argument **axes=boxed** is optional, but was used here to help add perspective. The surface may also be shaded (or colored on color monitors) and lit in several different ways. Maple provides a variety of options that you can use to render the surface, the default being a rendering that uses a hidden line algorithm to indicate which portions of the surface are obstructed by other portions.

There are 3D routines that compute and display implicit surfaces, parametric surfaces, and contour plots. A 3D animator can show the evolution of surfaces, which is useful for displaying solutions of certain partial differential equations. Figure 38.6 shows a parametric plot of a portion of a periodic, self-intersecting surface. Note the use of **plot3d** options to control the placement and labels of the axes, the title, and (on color terminals) the coloration of the surface.

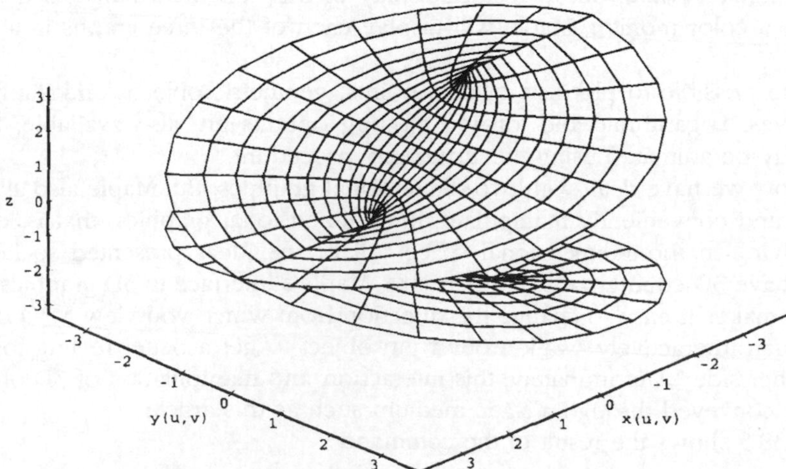

FIGURE 38.6 Plot of a parametric surface.

```
> plot3d([cos(u)*cosh(v), -sin(u)*sinh(v), u], u=-Pi..Pi,v=-2..2,
> axes=framed, color=v, title='Parametric Surface',
> labels=['x(u,v)', 'y(u,v)','v']);
```

It is a delightful feature of Maple that text and graphics can be printed with minimal effort. In particular, Maple can write its output in PostScript. The help file for **?interface** gives more details.

38.6 NUMERICS

Since Maple is designed to do symbolic mathematics, its numerical routines are less developed than its symbolic routines. We have already seen how the one command **evalf** can numerically evaluate complicated functions and integrals. In this example, we will present two additional numerical features of Maple: the numerical solution of ordinary differential equations (ODEs) and numerical linear algebra.

Although Maple can analytically solve many first- and second-order ordinary differential equations (including nonlinear equations of special forms such as Bernoulli, Riccati, and Bessel equations), these equations arise rarely in practice, so we refer the interested reader to the reference list at the end of this article for examples of Maple's ability. When approximate analytic solutions are desired, Maple can use series solutions to generate approximations to any order of accuracy.

```
> de := diff(x(t),t,t) - 2*diff(x(t),t) + x(t);
                  /  2      \
                  | d       |      / d      \
          de :=   | ----- x(t)| - 2 | ---- x(t)|  + x(t)
                  |   2     |      \ dt     /
                  | dt      |
                  \         /

> dsolve( de, x(t), series);
                                                             2
x(t) = x(0) + D(x)(0) t + (-1/2 x(0) + D(x)(0)) t

                                    3                                  4
     + (-1/3 x(0) + 1/2 D(x)(0)) t + (-1/8 x(0)+1/6 D(x)(0)) t

                                     5
     + (-1/30 x(0) + 1/24 D(x)(0)) t + O (t^6)
```

In the above example, Maple generates a series solution (to sixth order, by default) in the undetermined initial conditions **x(0)** and **D(x)(0)** [sometimes written mathematically as $x'(0)$ or $\dot{x}(0)$]. The operator **D** is similar to **diff** in that it denotes differentiation, but **D** operates on functions whereas **diff** operates on expressions. If you are only interested in solutions of **de** for short periods of time, then the series solution is adequate.

More often, however, scientists are interested in long term solutions of differential equations. For this, numerical solutions are often the better choice. Maple uses a Fehlberg fourth-fifth-order Runge–Kutta method to integrate ODEs. The user may set tolerances on the error of the computed solution. The following example is the Lorenz system of equations, which is famous for its "strange attractor" and chaotic trajectories. The Lorenz equations are a system of three ODEs given by

$$x' = 10(y - x)$$

$$y' = 28x - y - xz$$

$$z' = -8y3z + xy$$

We first define these equations

```
> lorenz := diff(x(t),t)=10*(y(t)-x(t)),
>           diff(y(t),t)=28*x(t)-y(t)-x(t)*z(t),
>           diff(z(t),t)=-8/3*z(t)+x(t)*y(t);
                    d
     lorenz := ---- x(t) = 10 y(t) - 10 x(t),
                    dt

  d                                              d
 ---- y(t) = 28 x(t) - y(t) - x(t) z(t), ---- z(t) = - 8/3 z(t) + x(t) y(t)
  dt                                             dt
```

The Lorenz system is an example of an initial boundary value problem, which means that we must specify the initial conditions $x(0)$,$y(0)$, and $z(0)$. For the sake of this example, we choose $x(0)= -1$,$y(0) = 0$, and $z(0) = 36$.

```
> IC := x(0)= -1,y(0) = 0,z(0) = 36;
                              IC := x(0) = -1, y(0) = 0, z(0) = 36
```

You can set up the numerical integration with these starting conditions by

```
> S := dsolve({lorenz,IC}, {x(t),y(t),z(t)},numeric);
S := proc(t)   'dsolve/numeric/result2'(t,269993936,[1,1,1])
  end
```

Note that the output of **dsolve** is a procedure that takes an argument, **t**. In particular, **S** is a function that will return a set of t,$x(t)$,$y(t)$,$z(t)$ values for any value of **t**. For example **S(1)** returns the x,y, and z positions at $t = 1$.

```
> S(1);
   {y(t) = -4.112767361, t = 1., x(t) = 7.399915314, z(t)=
     36.58585887}
```

By far the most useful way to view the numerical solution of an ODE is by graphing the solution. In Maple, a function called **odeplot** accomplishes this. The function is part of the **plots** package. (Packages are described in the next section.)

```
> with(plots): # Load names of functions in plots package
> odeplot(S, [x(t),y(t),z(t)], 0..20,
> numpoints = 600,axes = framed,orientation = [82,82]);
```

The above command instructs Maple to use the function **S** and at least 600 values of **t** in the range **0..20** to compute and display the solution **x(t), y(t), z(t)**. The **orientation** option determines the initial direction from which we view the image. The resulting plot is shown in Figure 38.7.

The second numerical aspect of Maple we will discuss is numerical linear algebra. Before using any linear algebra routines, you must load the names into memory by the command **with(linalg)**. Matrices are formed with the **matrix** command and vectors with the **vector** command. Matrix/vector or matrix/matrix multiplications are accomplished with the **multiply** command. Thus to multiply numerical matrices and vectors we might type

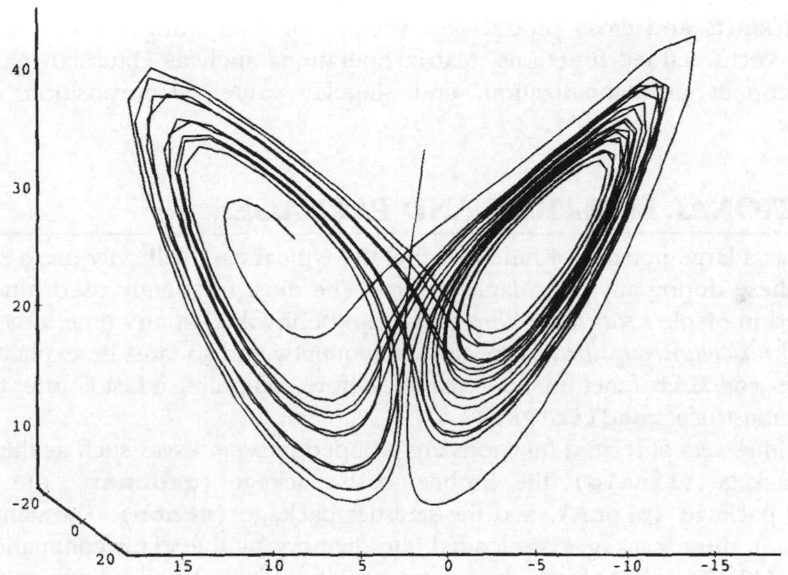

FIGURE 38.7 A numerical solution to the Lorenz system.

```
> with(linalg): # The warning messages are normal. Ignore them.
Warning: new definition for norm
Warning: new definition for trace
> A := matrix([[1.0,2,3],[4,5,6],[7,8,9]]); # Note floating point

                    [1.0 2 3]
                    [        ]
               A := [ 4  5 6]
                    [        ]
                    [ 7  8 9]

> b := vector([-1,1,3.0]); # Note floating point
               b := [ -1, 1, 3.0 ]
```

```
> c := multiply(A,b);
                        c := [ 10.0, 19.0, 28.0 ]
> eigenvects(A);

[-.2671993715*10^-9, 1, {[-.3734377923, .7468755784, -.3734377860]}],
[-1.116843982, 1, {[-.7616914013, -.0840865451, .5935183128]}],
[16.11684403, 1, {[.2719964252, .6159645983, .9599327726]}]
```

The command **eigenvects** gives both eigenvalues and eigenvectors for a matrix. The output is in the form [*eigenvalue, multiplicity, set of basis vectors for this eigenspace*]. In the above example, Maple reported the eigenvalues as the first entries in each row above. The true eigenvalues are $0, 15/2 \pm 3/2\sqrt{33}$, which are very close to the above numerical values. (Using the Maple command **det(A)** will convince you that **A** really is singular.)

Maple also provides routines for linear algebraic computations such as taking the inner products and cross products of vectors, or computing the divergence and curls of vector-valued functions. Matrix operations such as Gaussian elimination, Gram–Schmidt orthogonalization, and singular value decompositions are also available.

38.7 ADDITIONAL LIBRARIES AND PACKAGES

Maple has a large number of functions, but the typical user will only use a small fraction of these during any one Maple session. The most frequently used functions are contained in Maple's *standard library* and can be invoked at any time; others (in the so-called *miscellaneous library*) are used infrequently and so must be explicitly loaded using the **readlib** function. For example, before computing a fast Fourier transform, the user must use **readlib(FFT)**.

Sometimes sets of related functions are grouped into *packages* such as the linear algebra package **(linalg)**, the Grobner basis package **(grobner)**, the advanced graphics package **(plots)**, and the statistics package **(stats)**. The names of the functions in these packages are loaded into memory by the **with** command. For example, **with(linalg)** loads the names of linear algebra procedures such as **det** (compute the determinant of a matrix), **dotprod** (the inner product of two vectors), and **grad** (gradient of a vector-valued function). In fact, the single command **with(linalg)** notifies Maple of the existence of over 100 functions having to do with vectors, matrices, and linear transformations.

The number of Maple packages is growing. Not only are the Maple developers continuing to add features to Maple, but experienced users also contribute code to the so-called *share library* that is distributed with the Maple package.

38.8 PROGRAMMING FEATURES

Although Maple has hundreds of predefined functions, the Maple programming language makes it easy for users to define their own functions and procedures, and even to save them into a library. The simplest way to define your own function is to use "arrow notation." For example, if we want **parabola** to be the name of a Maple function that takes the variable x and returns the value $(-x^2/\pi^2) + 1$, then we can enter

```
> # define simple function using arrow notation
> parabola := x -> -1/Pi^2 * x^2 + 1;

                                                  2
                                                 x
                              parabola := x -> - --- + 1
                                                   2
                                                 Pi
> parabola(3);
                                   9
                                 - --- + 1
                                    2
                                  Pi
```

Note that by defining **parabola** to be a function, then **parabola(3)** makes sense (whereas it would not if **parabola** were an expression). The functions **map** and **unapply** convert functions into expressions and expressions into functions.

The arrow notation shown above is a shorthand method for defining a procedure. You can also define procedures explicitly by using a **proc..end** pair to delimit the scope of the procedure. Any symbols in parentheses following the word **proc** are used as parameters for that procedure. For example, suppose you daily use cylindrical coordinates in your work and want to define the operators for gradient, divergence, and curl in a cylindrical coordinate system. One way to do this is to define the following procedures:

```
> # Procedure to compute gradient in cylindrical coordinates
> cylgrad := proc(f)
>     [diff(f,r), 1/r*diff(f,theta), diff(f,z)]
> end:
> # Define divergence in cylindrical coordinates
> cyldiv := proc(V)
>     1/r*diff(r*V[1],r) + 1/r*diff(V[2],theta) + diff(V[3],z)
> end:
> # Define Laplacian in cylindrical coordinates
> cyllap := proc(V) cyldiv(cylgrad(V)) end:

> # Test the procedures
> cyldiv([r^2, r*theta, z^3/3]);
                                           2
                              3r + 1 + z

> cylgrad(r^2 + r*theta + theta*z);
                                     r + z
                        [2r + theta, -----, theta]
                                       r
```

```
> cyllap( r^2 + r*theta + theta*z );
                                        4r + theta
                                        ----------
                                            r
```

These procedures are simple enough that we could also use the "arrow notation" to define them. For example, it would be equivalent to define **cylgrad** by **cylgrad := f -> [diff(f,r), 1/r*diff(f,theta), diff(f,z)];**.

As a more complicated example of using procedures, suppose you use Fourier series in your work regularly. You might want to define your own Maple function **fourier_poly** which computes a Fourier approximation to a given periodic function on the interval $[-\pi, \pi]$. We will define **fourier_poly** as a procedure that takes three arguments: **f**, the name of a function (assumed periodic on $[-\pi, \pi]$), **x** the name of a variable, and **n** the power at which to truncate the Fourier series for **f**. One way to define **fourier_poly** is to enter the following Maple code:

```
> fourier_poly := proc(f,x,n)
> local a,b,k,t;
> a := array(0..n): b:= array(0..n); # Storage for coeffs
> for k from 0 to n do # Compute Fourier coeffs
>      a[k] := 1/Pi * int(f(t)*cos(k*t), t = -Pi..Pi):
>      b[k] := 1/Pi * int(f(t)*sin(k*t), t = -Pi..Pi):
> od:
> RETURN(a[0]/2 + sum('a[k]*cos(k*x) + b[k]*sin(k*x)', 'k' = 1..n) )
> end:
```

This procedure computes the Fourier coefficients defined by

$$a_k = \frac{1}{\pi}\int_{-\pi}^{\pi} f(t)\cos(kt)\,dt$$

and

$$b_k = \frac{1}{\pi}\int_{-\pi}^{\pi} f(t)\sin(kt)\,dt$$

and returns the Fourier polynomial

$$\frac{a_0}{2} + \sum_{k=0}^{n} a_k \cos(kx) + b_k \sin(kx)$$

Let us test the above procedure on the function parabola that we defined at the beginning of this section:

```
> FP := fourier_poly(parabola, y, 4);


                cos(y)    cos(2y)          cos(3y)          cos(4y)
FP := 2/3 4 ------ - -------- + 4/9 ------- - 1/4 -------
                  2         2                2                2
                Pi        Pi               Pi               Pi
```

This gives us the first few terms of the Fourier series for the function $x \to (-x^2/\pi^2) + 1$.

At the end of a Maple session you can save your procedures and definitions in a file (see **?save**). These definitions may be loaded into Maple the next time you want to use them. For example, suppose you save the above session in a file called **fourier_defs**. The next time you invoke Maple you can recover your definitions by issuing the command **read 'fourier_defs'**. See **?read** for more information.

Many commercial computer algebra packages contain a programming language, but a feature that makes Maple unique among its competitors is that Maple *permits the user to view a large portion of its source code.* This provides the Maple user with two distinct advantages: the user may check to see what algorithm underlies a certain Maple command, and the user may examine (presumably!) well-written Maple procedures. For the first advantage, suppose you know that the problem you want to solve is numerically unstable under algorithm X; you can check to see whether Maple uses algorithm X before trying to solve the problem. The second advantage is that Maple users can study the predefined Maple procedures to figure out how to write their own procedures.

For example, let us view the Maple code for the procedure **linalg[iszero]** that determines whether all entries of a matrix are zero.

```
> interface(verboseproc = 2 ); # Tell Maple to print procedure code
> print(linalg[iszero]);

proc(AA)
local A,m,n,i,j,typeAm,typeAv;
options 'Copyright 1991 by the University of Waterloo';
  if nargs <> 1 then ERROR('incorrect number of arguments') fi;
  typeAv := type(AA,'vector');
  typeAm := type(AA,'matrix');
  if not (typeAv or typeAm) then
    A := evalm(AA);
    typeAv := type(A,'vector');
    typeAm := type(A,'matrix')
  else A := AA
  fi;
  if typeAm then
    m := linalg['rowdim'](A);
    n := linalg['coldim'](A);
    for i to m do
     for j to n do if A[i,j] <> 0 then RETURN(false) fi od
    od;
    true
  elif typeAv then
    m := linalg['vectdim'](A);
    for i to m do if A[i] <> 0 then RETURN(false) fi od;
```

```
    true
  else ERROR('argument must be a matrix or a vector ')
  fi
end
```

Although this procedure may look formidable at first, it is not hard to interpret what Maple does. First it checks to see that the argument **AA** was a matrix or vector. (If it was not, then an error message is printed.) Then it checks each entry of the object and returns false if it finds any entry that is not zero. Otherwise, **true** is returned.

DEFINING TERMS

Computer Algebra: A software package that use symbolic computation.

Notebook: An interactive document that combines Maple commands and output with graphics and ordinary text.

Miscellaneous Library: A collection of lesser-used functions whose names are not initially known to Maple. Before using a function in the miscellaneous library, the function name must be loaded by executing a **readlib** command.

Package: A collection of related functions that can be loaded into Maple using the **with** command.

Standard Library: A collection of common functions whose names are always known to Maple.

Symbolic Computation: Mathematical expressions are manipulated as objects, without floating-point approximations.

Zeros: Given a function f, the zeros of f are the values of a variable x so that $f(x) = 0$.

REFERENCES

B. W. Char, K. O. Geddes, G. H. Gonnet, B. L. Leong, M. B. Monagan, and S. M. Watt, *Maple V Library Reference Manual,* first ed. Springer, New York, 1991.

B. W. Char, K. O. Geddes, G. H. Gonnet, B. L. Leong, M. B. Monagan, and S. M. Watt, *Maple V Language Reference Manual,* first ed. Springer, New York, 1991.

B. W. Char, K. O. Geddes, G. H. Gonnet, B. L. Leong, M. B. Monagan, and S. M. Watt, *First Leaves: A Tutorial Introduction,* first ed. Springer, New York, 1991.

André Heck, *Introduction to Maple*. Springer, New York, 1993.

T. Lee, G. P. Porciello, and S. Adams, *Engineering Applications of Maple V Release 2*. Waterloo Maple Software, 1993.

39

Mathcad PLUS

Faustino A. Lichauco
MathSoft, Inc.

39.1 INTRODUCTION

Think of a spreadsheet but without the cryptic math notation jammed between rows and columns. Add to that the features of a typical word-processor, the power to do symbolic math, and a suite of on-line Electronic Books. Combine all this in one integrated package and you can get a sense of what Mathcad is like.

Mathcad combines the live document interface of a spreadsheet with the WYSIWYG interface of a word processor. With Mathcad, you can typeset equations on the screen exactly the way you see them in a book. But Mathcad equations do more than look good on the screen. You can use them to actually do math.

Like a spreadsheet, as soon as you make a change anywhere in your document, Mathcad goes straight to work, updating results and redrawing graphs. With Mathcad, you can easily read data from a file and do mathematical chores ranging from adding up a column of numbers to evaluating integrals and derivatives, inverting matrices, solving differential equations and more. In fact, just about anything you can think of doing with math, you can do with Mathcad.

Like a word processor, Mathcad comes with a WYSIWYG interface, a spellchecker, headers and footers, and the ability to print what you see on the screen on any Windows supported printer. This, combined with its live document interface, makes Mathcad a good choice for producing up-to-date, publication quality engineering reports.

0-8493-2530-7/96/$0.00+$.50

39.2 A SIMPLE CALCULATION

Although Mathcad can perform sophisticated mathematics, you can just as easily use it as a simple calculator. To try your first calculation, follow these steps:

- Click anywhere in the document. You see a small crosshair. Anything you type appears at the crosshair.

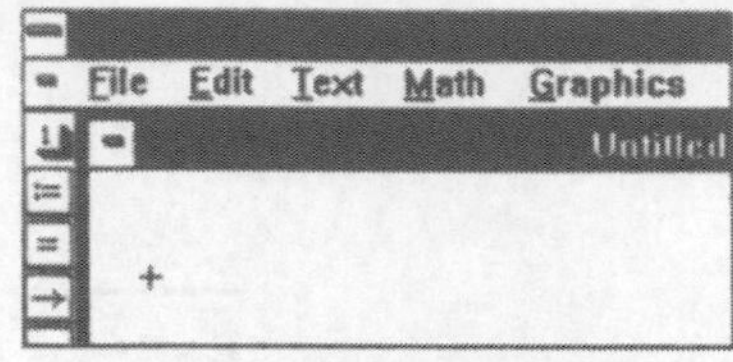

- Type **15-8/104.5=**. When you press the equals sign, Mathcad shows the result.

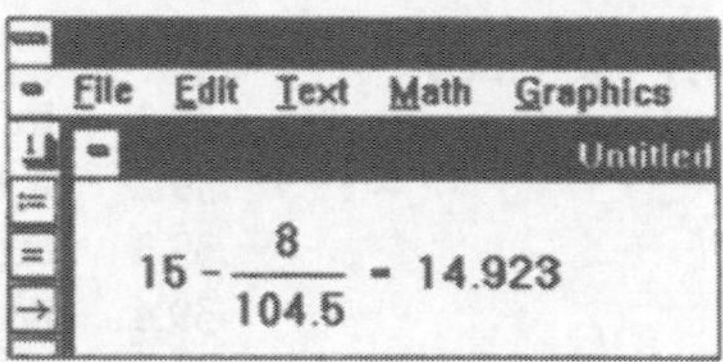

This calculation demonstrates the way Mathcad works:

- Mathcad shows equations as you might see them in a book, expanded fully in two dimensions. Mathcad sizes fraction bars, brackets, and other symbols to display equations the same way you would write them on paper.
- As soon as you type the equals sign, Mathcad returns the result.
- As you type each operator (in this case, – and /), Mathcad shows a small rectangle called a *placeholder*. Placeholders hold spaces open for numbers or expressions not yet typed. As soon as you type a number, it replaces the placeholder in the equation.

Once an equation is on the screen, you can edit it by clicking in the appropriate spot and typing new letters, digits, or operators. A symbol palette along the left side of the application window makes it easy to type various math operators.

39.3 MORE COMPLEX CALCULATIONS

Mathcad's power and versatility quickly become apparent once you begin to use *variables* and *functions*. By defining variables and functions, you can link equations together and use intermediate results in further calculations.

The following examples show how to define and use several variables.

Variables

To define a variable *t*, follow these steps:

- Type **t:** (the letter *t*, followed by a colon). Mathcad shows the colon as the definition symbol **:=**.

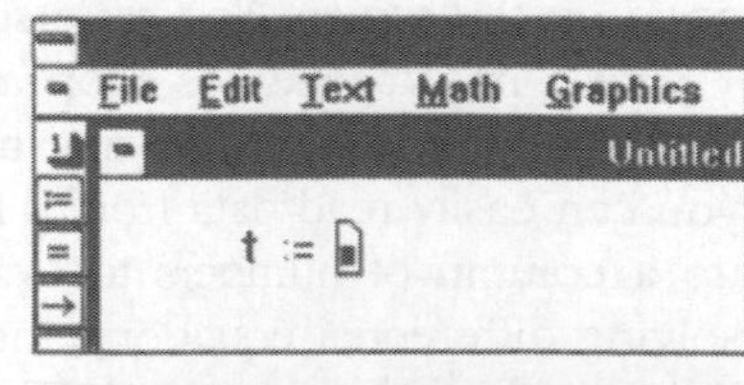

- Type **10** in the placeholder to complete the definition for *t*.

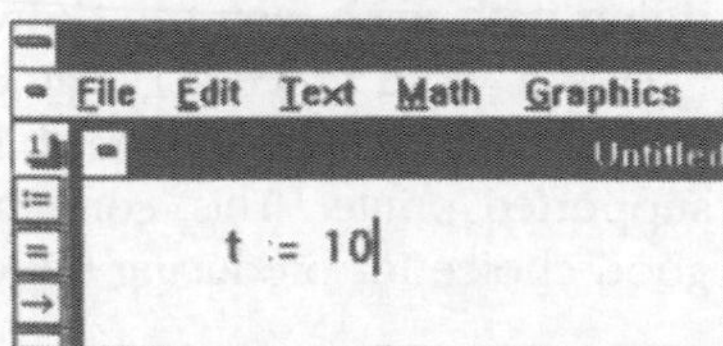

These steps show the form for typing any definition:

- Type the variable name to be defined.
- Type the colon key to insert the definition symbol.
- Type the value to be assigned to the variable. The value can be a single number, as in the example shown here, or a more complicated combination of numbers and previously defined variables.

Mathcad documents read from top to bottom and left to right. Once you have defined a variable like *t*, you can compute with it anywhere *below and to the right* of the equation that defines it.

- Now enter another definition. To define *acc* as –9.8, type: **acc:-9.8**.

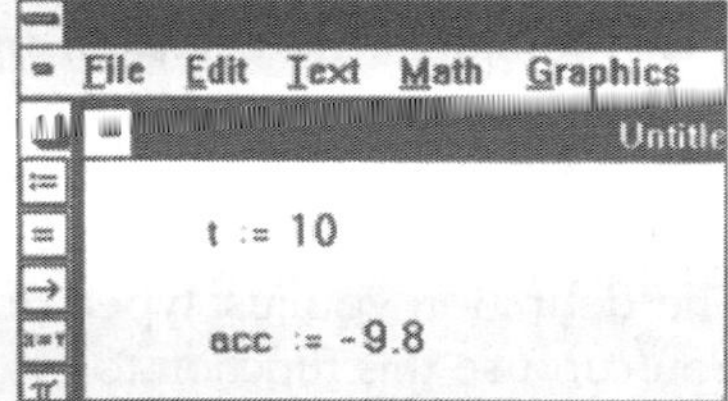

Now that the variables *acc* and *t* are defined, you can use them in other expressions.

- Below the previous definitions, type **acc/2[Space]*t^2**. The caret symbol **(^)** represents raising to a power, the asterisk **(*)** is multiplication, and the slash **(/)** represents division.

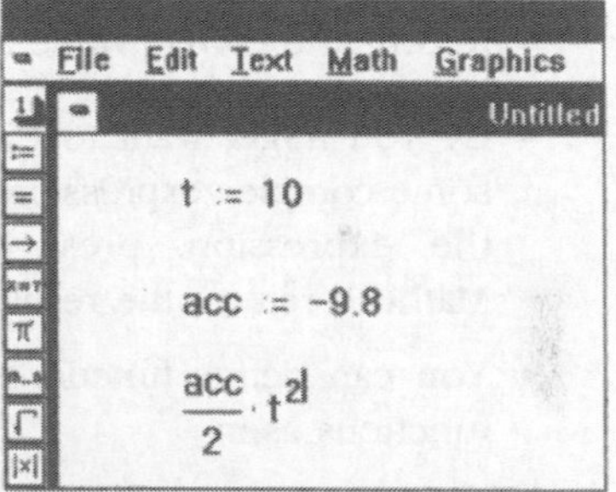

- Press the equals sign "=." The window now contains two *definitions*, which define variables, and one *evaluation*, which computes a result: the distance traveled by a falling body in time *t* with acceleration *acc*.

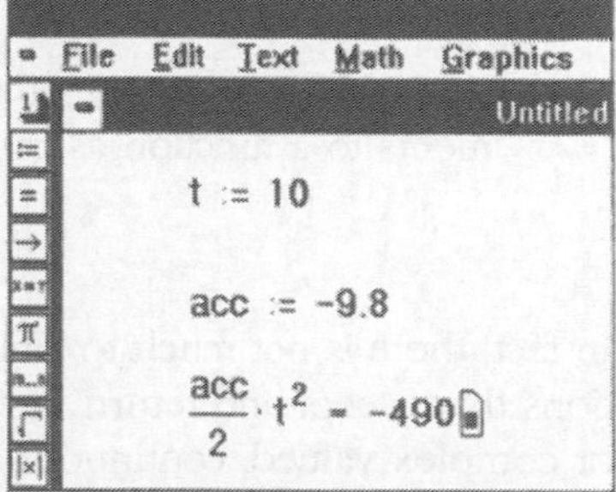

Like a spreadsheet, Mathcad updates results throughout the document as soon as you make changes. For example, if you click on the 10 on your screen and change it to some other number, Mathcad changes the result as soon as you click outside of the equation.

39.4 FUNCTIONS

Mathcad allows you to define and use functions in much the same way you define and use variables. The main difference between a variable and a function is that the value of a variable depends on how you define it, but the value of a function depends on how you define it *and* what arguments you supply when you finally use it.

User-Defined Functions

Here is how to use Mathcad to define the function:

$$d(t) = 1600 + \frac{acc}{2}t^2$$

Note how similar the Mathcad approach is to the way you would do it with pencil and paper.

- Start by typing **d(t):**

$d(t) :=$ ▮

- Complete the definition by typing the expression: **1600+acc/2[Space]*t^2**

$d(t) := 1600 + \frac{acc}{2}\cdot t^2$

The definition you just typed defines a function named *d* having a single argument, *t*. You can use this function to evaluate the above expression for different values of *t*. To do so, simply replace *t* with an appropriate number. For example:

- To evaluate d(12.5), type **d(12.5)=**. Mathcad returns the correct value as shown on the right.

$d(12.5) = 834.375$

- Or you might want to evaluate the function at some complex expression. To do so, just type in the expression, press the equal sign and Mathcad returns the result.

$d\left(\frac{3+\sqrt{\pi}}{6!}\right) = 1.6\cdot 10^3$

- You can define functions in terms of existing functions as in

$d(t) := 1600 + \frac{acc}{2}\cdot t^2$

$u(t) := d(t)^2 + 3\cdot d\left(\sqrt{t}+2\right)$

- And you can supply other functions as arguments to a function as in

$u\left(d\left(\frac{3+\sqrt{2}}{7}\right)\right) = 1.565\cdot 10^{14}$

In fact, there is not much to limit how you define and use functions. You can create functions that accept and return vectors and matrices as well as scalars. Functions can be real or complex valued, continuous or piecewise continuous. The function's definition can involve sums, derivatives, integrals, combinations of other functions, or boolean expressions.

Built-in Functions

Mathcad's built-in functions work just like user-defined functions except you never have to define them.

- For example, to evaluate the hyperbolic sine of an expression like 4 + 2*i* just type **sinh (4+2i) =**.

$\sinh(4 + 2i) = -11.357 + 24.831i$

Mathcad comes with an extensive collection of these built-in functions. They include common transcendental functions such as *sin* and *cos*, Bessel functions, log functions, various statistical and data analysis functions, functions for evaluating Fourier transforms and so on.

Users of Mathcad PLUS get an additional two dozen matrix functions for performing advanced matrix operations such as finding the pseudoinverse of a matrix, performing various types of decomposition, obtaining singular values, and evaluating several kinds of matrix norms.

External Functions

Mathcad's built-in functions would be difficult to define if they were not already built-in. For example, if you had to define atan(*x*) from scratch, you would have to use a truncated infinite series. Computation would be slow and sometimes inaccurate. Moreover, if you wanted to use this function in another document, you would have to define it all over again in that document.

Although Mathcad's built-in function list contains an extensive selection of functions, you may have functions you want to define that do not lend themselves to being defined in the conventional way.

If you are using Mathcad PLUS and you have a modicum of C programming expertise and a 32-bit compiler, you can program your own functions and compile them into DLL files. These functions will appear in Mathcad just as if they had been there all along. You can program your own customized error messages; the function will appear in a scrolling list of available functions; and you'll be able to use the function in any Mathcad document. The function you program will be indistinguishable from any of the Mathcad built-in functions. You can think of this as the equivalent, in software, of opening up your PC and adding extra boards to expand its capabilities.

If you are using Mathcad PLUS and you *do not* want to do any programming, you can still make use of Mathcad PLUS' external function interface. MathSoft provides a variety of *Function Packs*, each containing a collection of thoroughly documented and tested add-on functions organized by topic.

39.6 ADVANCED OPERATORS

Summations

You have already seen a few of Mathcad's basic operators: the four arithmetic operators, square roots, and exponents. Mathcad also comes with derivatives, summations, iterated products, and integrals. These work the same way as the more basic operators. You type a keystroke, fill in the placeholders around the operator, and press the equal sign to see a result. Here is an example of how you would evaluate a summation:

- Click on the summation button on the palette. A summation sign with four placeholders appears.

- In the placeholder to the left of the equal sign, type a variable name for the index of summation.

- In the placeholder to the right of the equal sign, type the lower limit of summation.
- In the single placeholder above the sigma, type the upper limit of summation.
- In the *remaining* placeholder, type the expression you want to sum. Usually, this expression will involve the index of summation.
- Now type the equal sign. Mathcad displays the result.

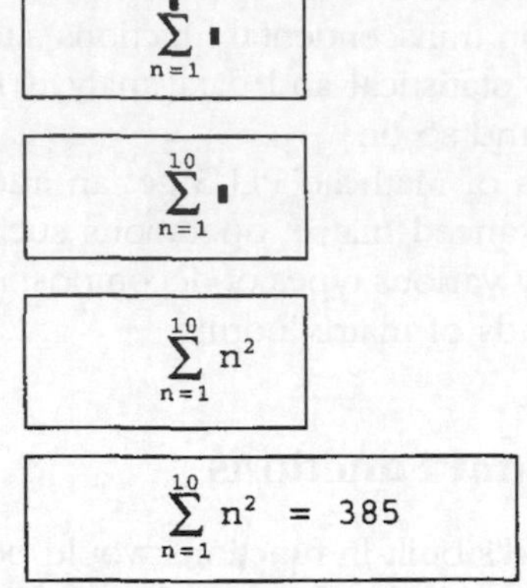

Integrals

As you might expect given their close mathematical relationship, integrals work exactly the same way as summations. You click the integral button on the palette, you fill in placeholders for the limits of integration, the integrand and the variable of integration. Then you press the equal sign to see the result.

Like sums, integrals can be nested together to form multiple integrals. Limits of integration are not restricted to numbers. They can be variables or functions. This allows you to solve more difficult problems like the ones below:

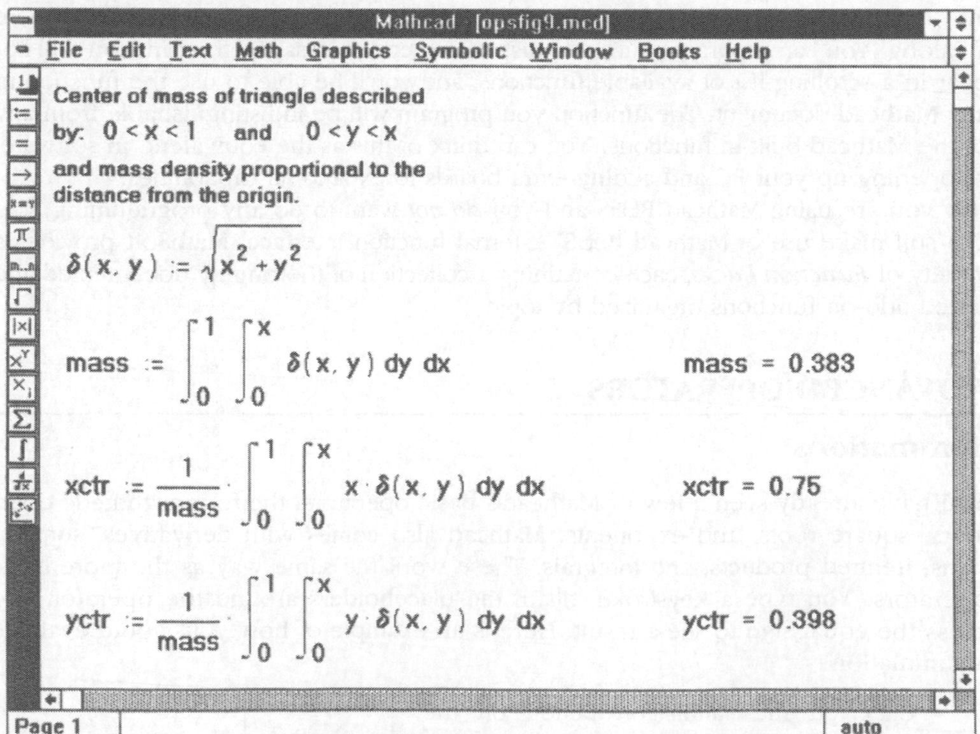

Derivatives

Mathcad has two derivative operators: one for a first derivative and another for higher order derivatives. You can use either operator for partial derivatives as well as total derivatives.

The first step in using a derivative is to define the point at which you want to evaluate the derivative. Once you have done this, the procedure is the same as that for integrals and summations. As an example, here is how you would evaluate the first derivative of x^3 with respect to x at the point $x = 2$:

- First define the point at which you want to evaluate the derivative. Type **x:2**.

  x := 2

- Click on the *d/dx* on the operator palette. A derivative operator appears with two placeholders.

  $\frac{d}{d\blacksquare}\blacksquare$

- Click on the bottom placeholder and type x. You are differentiating with respect to this variable.

  $\frac{d}{dx}\blacksquare$

- Click on the placeholder to the right of the *d/dx* and type **x^3**. This is the expression to be differentiated.

  $\frac{d}{dx}x^3$

- Press the equals sign to see the derivative of the expression at the indicated point.

  $\frac{d}{dx}x^3 = 12$

Complex Contour Integrals

By creatively combining these operators, you can easily perform very elaborate procedures. For example, to evaluate a contour integral along a specified path, you can parametrize the path and evaluate the integral shown in the following figure:

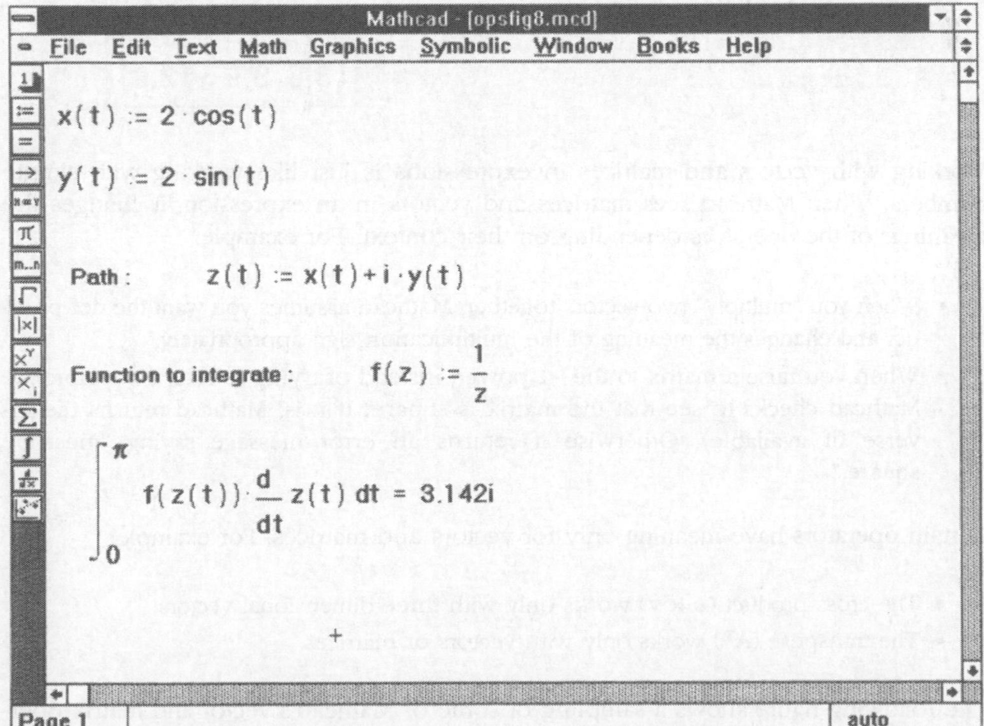

39.7 VECTORS AND MATRICES

Mathcad's WYSIWYG capabilities are especially useful when you begin to work with vectors and matrices. You create a vector or a matrix simply by filling in an array of empty placeholders. Here's an example of how to create a 2 × 3 matrix.

- Choose **Matrices** from the **Math** menu. The dialog box shown on the right appears.

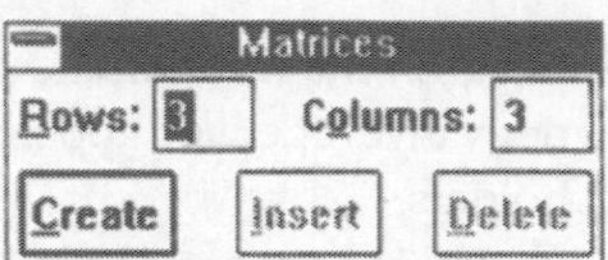

- Enter a number of rows and a number of columns in the appropriate boxes. In this example, there are two rows and three columns.

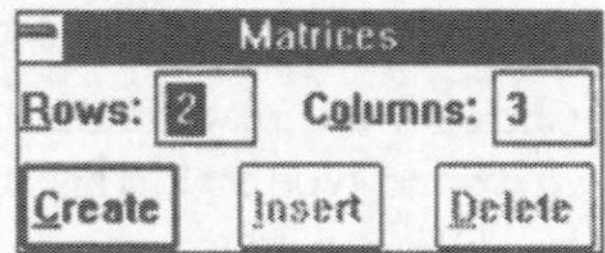

- Then click on "Create." Mathcad inserts a matrix of placeholders.

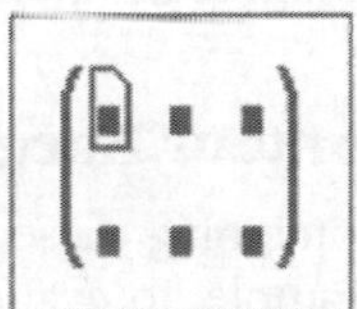

- Fill in the placeholders to complete the matrix.

$$\begin{pmatrix} 2 & 5 & 17 \\ 3.5 & 3.9 & -12.9 \end{pmatrix}$$

Working with vectors and matrices in expressions is just like working with simple numbers. When Mathcad sees matrices and vectors in an expression, it changes the meanings of the operators depending on their context. For example:

- When you "multiply" two vectors together, Mathcad assumes you want the dot product and changes the meaning of the multiplication sign appropriately.
- When you raise a matrix to the –1 power, instead of trying to return a reciprocal, Mathcad checks to see that the matrix is square. If it is, Mathcad returns the inverse (if available). Otherwise it returns an error message saying "must be square."

Certain operators have meaning only for vectors and matrices. For example:

- The cross product ($\mathbf{u} \times \mathbf{v}$) works only with three-dimensional vectors.
- The transpose ($\mathbf{A}^T$) works only with vectors or matrices.

The following figure shows a sampling of some of Mathcad's vector and matrix operators in actual use:

Mathcad - [&mo60vop.mcd]

File Edit Text Math Graphics Symbolic Window Books Help

Matrix M . . . Vectors v and w . . .

$$M := \begin{pmatrix} 0 & 1 & 2 \\ 3 & 0 & 2 \\ 5 & 3 & 1 \end{pmatrix} \quad v := \begin{pmatrix} 3+10 \\ 1-4 \\ 5\cdot 10 \end{pmatrix} \quad v = \begin{pmatrix} 13 \\ -3 \\ 50 \end{pmatrix} \quad w := 2\cdot v \quad w = \begin{pmatrix} 26 \\ -6 \\ 100 \end{pmatrix}$$

Sum . . . Determinant . . . Dot and Cross Product . . .

$$\Sigma v = 60 \quad |M| = 25 \quad v\cdot w = 5.356\cdot 10^{3} \quad v\times w = \begin{pmatrix} 0 \\ 0 \\ 0 \end{pmatrix}$$

Inverse . . .

$$M^{-1} = \begin{pmatrix} -0.24 & 0.2 & 0.08 \\ 0.28 & -0.4 & 0.24 \\ 0.36 & 0.2 & -0.12 \end{pmatrix}$$

Transpose . . .

$$w^{T} = (26 \ \ -6 \ \ 100)$$

$$M\cdot M^{-1} = \begin{pmatrix} 1 & 0 & 0 \\ 0 & 1 & 0 \\ 0 & 0 & 1 \end{pmatrix}$$

Solve linear system Mx=v with inverse . . .

$$x := M^{-1}\cdot v$$

$$x = \begin{pmatrix} 0.28 \\ 16.84 \\ -1.92 \end{pmatrix} \quad M\cdot x = \begin{pmatrix} 13 \\ -3 \\ 50 \end{pmatrix}$$

Page 1 auto

Mathcad 5.0 also provides numerous functions designed specifically for linear algebra. For example: identity(n) returns an $n \times n$ identity matrix; tr(**M**) returns the trace of **M**. The functions *eigenvals* and *eigenvec* can perform eigenanalysis as shown in the following figure:

Mathcad - [eigen.mcd]

File Edit Text Math Graphics Symbolic Window Books Help

Finding eigenvalues and eigenvectors of a real matrix

$$A := \begin{pmatrix} 1 & -7 & 6 \\ 3 & 0 & 10 \\ 2 & 5 & -1 \end{pmatrix} \quad c := \text{eigenvals}(A) \quad c = \begin{pmatrix} 3.805 + 1.194i \\ 3.805 - 1.194i \\ -7.609 \end{pmatrix}$$

To find the eigenvector corresponding to an eigenvalue, use *eigenvec*...

$$v := \text{eigenvec}(A, c_0) \quad v = \begin{pmatrix} 0.143 + 0.626i \\ 0.114 - 0.63i \\ 0.076 - 0.414i \end{pmatrix} \quad |v| = 1$$

Check...

$$A\cdot v = \begin{pmatrix} -0.203 + 2.554i \\ 1.188 - 2.261i \\ 0.783 - 1.484i \end{pmatrix} \quad c_0\cdot v = \begin{pmatrix} -0.203 + 2.554i \\ 1.188 - 2.261i \\ 0.783 - 1.484i \end{pmatrix}$$

Page 1 auto

Mathcad PLUS comes with a much larger repertoire of matrix functions including functions returning condition numbers, various norms and pseudoinverses, performing matrix decomposition and solving the generalized eigenvalue problem.

39.8 UNITS

Mathcad comes with a full complement of commonly used units ranging from common ones like *kg*, *joule*, and *newton* to the more unusual ones such as *slug* or *poise*. You use units the same way you use variables. To assign some combination of units to a number, you simply multiply the number by the appropriate combination of units. For example, to add 3 newtons to 2.3 force pounds you just create the following equation:

$$3\ \text{newton} + 2.3 \cdot \text{lbf} = 13.231 \cdot \text{kg} \cdot \text{m} \cdot \text{sec}^{-2}$$

Note that the result is displayed in terms of kilograms, meters and seconds. This is because by default, Mathcad displays results in terms of MKS units. To display the result in terms of another unit, say dynes, do the following:

- Click in the displayed result. You will see the *units placeholder*. Click on this placeholder to surround it with a selection box.

$$3\ \text{newton} + 2.3 \cdot \text{lbf} = 13.231 \cdot \text{kg} \cdot \text{m} \cdot \text{sec}^{-2}\ \blacksquare$$

- In the units placeholder, type **dyne**.

$$3\ \text{newton} + 2.3 \cdot \text{lbf} = 13.231 \cdot \text{kg} \cdot \text{m} \cdot \text{sec}^{-2} \cdot \text{dyne}$$

- Click outside the equation. Mathcad converts the result into dynes.

$$3\ \text{newton} + 2.3 \cdot \text{lbf} = 1.323 \cdot 10^{6} \cdot \text{dyne}$$

Naturally, you can define your own units in terms of existing ones. You do so just as if you were defining variables in terms of other variables. For example, if you routinely work with pounds per square foot, you could define and use this unit as shown in the following figure.

$$\text{psf} := \text{psi} \cdot \frac{\text{ft}^2}{\text{in}^2}$$

Working with units in Mathcad can help guard against errors. The following figure shows some potentially career-limiting mistakes that are easily caught as Mathcad routinely performs its dimension checking.

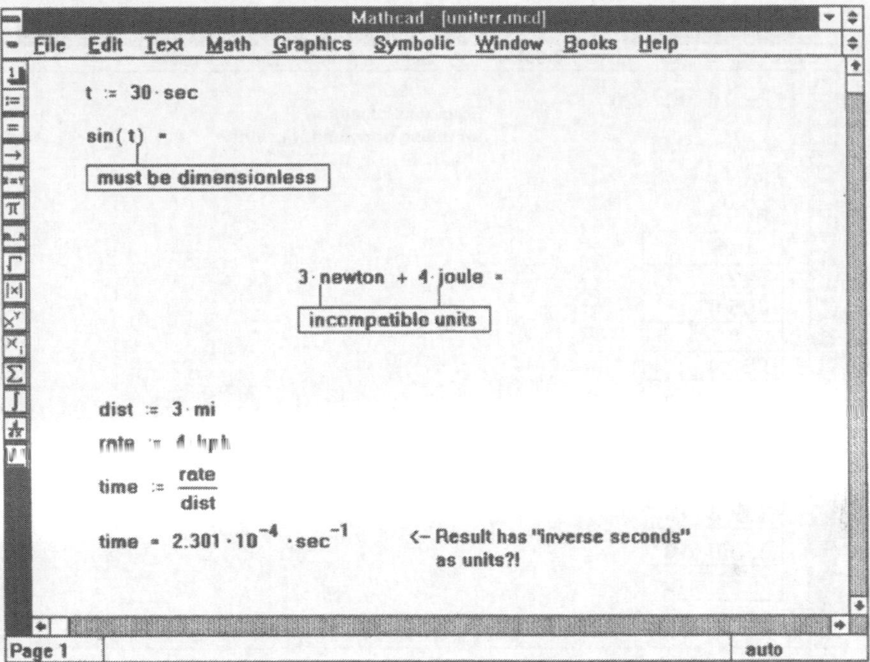

39.9 ITERATIVE CALCULATIONS

Mathcad can do repeated or iterative calculations as easily as it can perform individual calculations. Mathcad uses a special variable called a *range variable* to drive the iteration.

Range variables take on a range of values. For example, a single range variable could take on the values 1, 2, . . . 10 in succession. Whenever a range variable appears in a Mathcad equation, Mathcad evaluates the expression not just once, but once for each value of the range variable.

To define a range variable, you need to specify three things: its first value, its second value (Mathcad uses this to determine the step-size), and its last value. For example, the range variable *t* that goes from 10 to 20 in steps of 1 would be defined as follows:

- Begin as you would when you define any variable. Type **t:10**. `t := 10|`
- Type **,11**. This tells Mathcad that the next number in the range will be 11. `t := 10, 11|`
- Type **;20**. This tells Mathcad that the last number in the range will be 20. Mathcad shows the semicolon as a pair of dots. `t := 10, 11..20|`

The following example shows a range variable in action. Since *t* takes on eleven different values, there must also be eleven different answers. To display these answers, type the expression to be evaluated and press the equal sign. Mathcad displays the answers in a table as shown below.

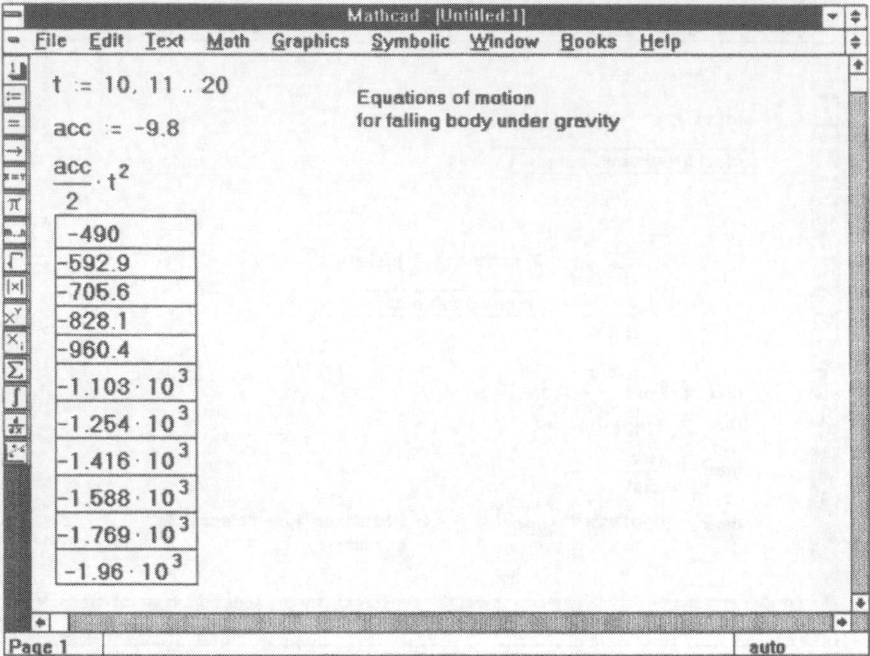

39.10 SOLVING EQUATIONS

When Mathcad solves a system of equations, it can display unknowns either in terms of numbers or in terms of other variables. If you want the unknowns displayed as numbers, the procedure depends on the type of equation you are solving:

- Systems of linear equations can be solved by matrix inversion.
- Systems of non-linear equations can be solved with a numerical algorithm embodied in a built-in function called *Find*.
- Systems of differential equations as well as a few partial differential equations can be solved using special built-in functions that come with Mathcad PLUS only.

If you want to display the unknowns in terms of the other variables in the equations, see the discussion on symbolic mathematics later in this chapter.

Linear Equations

For a system of linear equations, you can simply rewrite the system as a matrix equation ($\mathbf{A} \cdot \mathbf{x} = \mathbf{b}$) and invert the coefficient matrix ($\mathbf{x} = \mathbf{A}^{-1} \cdot \mathbf{b}$). An example illustrating this technique appears in the lower right of the figure on p. 907.

Nonlinear Equations

Mathcad uses an iterative method to converge to the solution of a system of equations. To solve a system of equations, Mathcad requires three things:

- Guess values to begin the iteration process,
- The equations themselves, and
- The variables you want to solve for.

The following example shows how you would solve two equations in two unknowns. Note that:

- The guess value determines which of the two solutions Mathcad returns.
- The equal sign you use for typing an equation is not the same one you use for displaying an answer. Here, you type **[Ctrl]=** rather than **=**.
- The equations you solve for must be confined to a band below *Given* and above *Find*. This band is called a *solve block*.

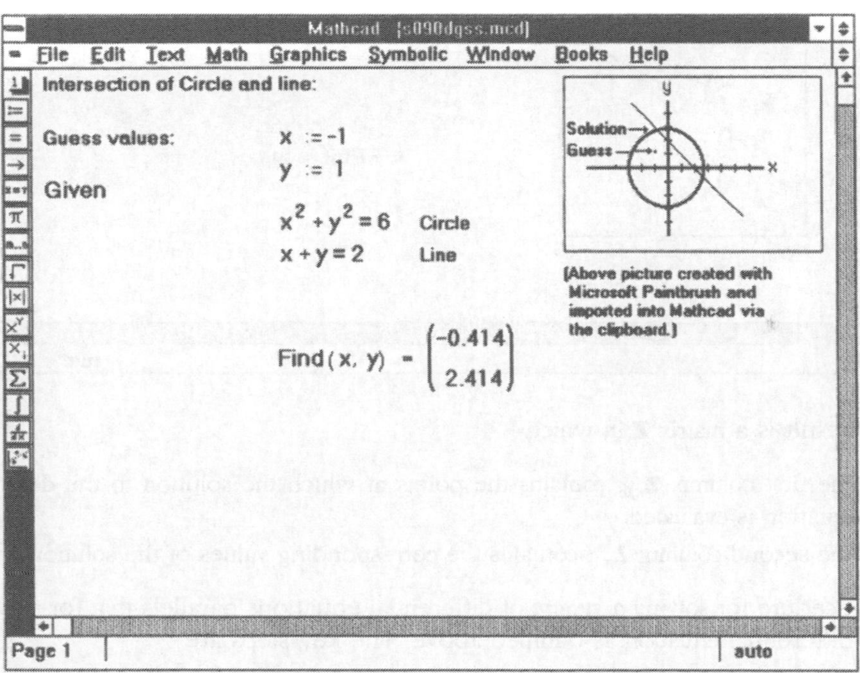

Differential Equations

Mathcad has a variety of functions for returning the solution to an ordinary differential equation. Each of these functions solves differential equations numerically. The result of each function call is a matrix containing the values of the function evaluated over a set of points. These functions differ in the particular algorithm each uses for solving differential equations. Despite these differences however, each of these functions requires you to specify at least three things:

- The initial conditions,
- A range of points over which you want the solution to be evaluated, and
- The differential equation itself, written in the particular form discussed in this chapter.

The following example shows how to solve a simple first order differential equation using, *rkfixed*, a function that uses the fourth-order Runge-Kutta method.

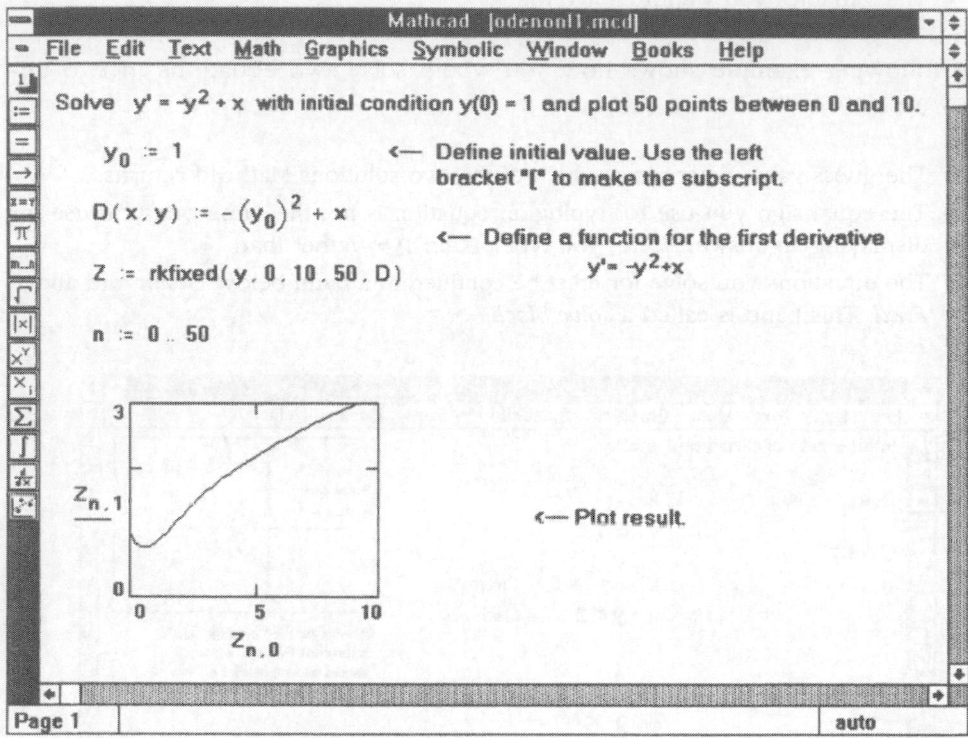

The result is a matrix **Z** in which

- The first column, $\mathbf{Z}_{n,0}$, contains the points at which the solution to the differential equation is evaluated.
- The second column, $\mathbf{Z}_{n,1}$, contains the corresponding values of the solution.

The procedure for solving a *system* of differential equations parallels that for solving a single differential equation as outlined above. The key steps are

- Define a vector containing the initial values of each unknown function.
- Define a vector-valued function containing the first derivatives of each of the unknown functions.
- Decide which points you want to evaluate the solutions at.
- Pass all this information into *rkfixed.*

The *rkfixed* function will return a matrix whose first column contains the points at which the solutions are evaluated and whose remaining columns contain the solution functions evaluated at the corresponding point. The following figure shows an example solving the equations:

$$x_0'(t) = \mu x_0(t) - x_1(t) - \left[x_0(t)^2 + x_1(t)^2\right]x_0(t)$$
$$x_1'(t) = \mu x_1(t) - x_0(t) - \left[x_0(t)^2 + x_1(t)^2\right]x_1(t)$$

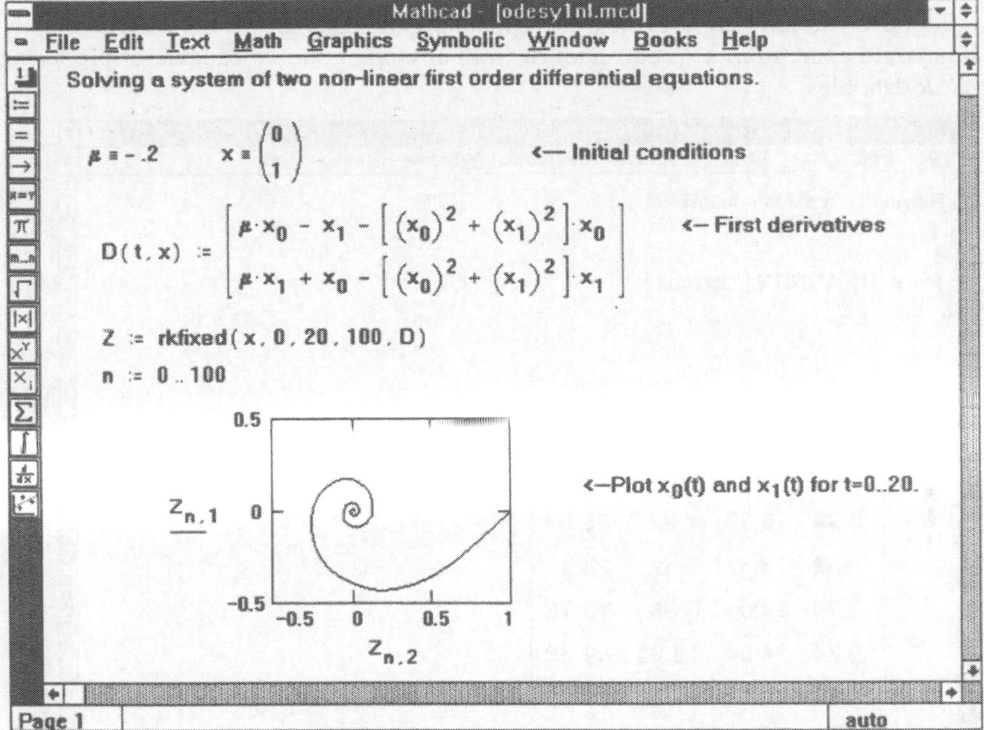

with initial conditions:

$$x_0(0) = 0 \text{ and } x_1(0) = 1$$

Mathcad comes with several functions similar to *rkfixed* that exploit specific properties of differential equations. For example, if you know you are working with stiff differential equations, you can use the Bulirsch-Stoer method for stiff systems as embodied in the function *stiffb*. You can even program your own differential equation solver. Mathcad has an external function interface through which functions you write yourself can behave exactly like Mathcad's own built-in functions.

39.11 DATA FILES

Mathcad's ability to perform data analysis is greatly enhanced by its ability to read and write ASCII data files. Using this feature, you can take a file containing numerical output from another application or from a measurement system and read the numbers directly into a Mathcad array.

Notepad - [spread.prn]

File Edit Search Help

```
0.25 0.91 0.72 1.8
0.77 2.18 1.63 4.4
2.74 8.08 5.46 16.63
3.94 16.33 7.92 33.63
3.82 13.82 7.52 28.5
2.79 8.09 5.58 16.73
6.84 24.04 13.86 49.63
4.68 12.68 9.76 26.06
```

Suppose you have an ASCII text file containing the data shown below. These numbers could come from a spreadsheet or from any other source capable of producing *ASCII* data files.

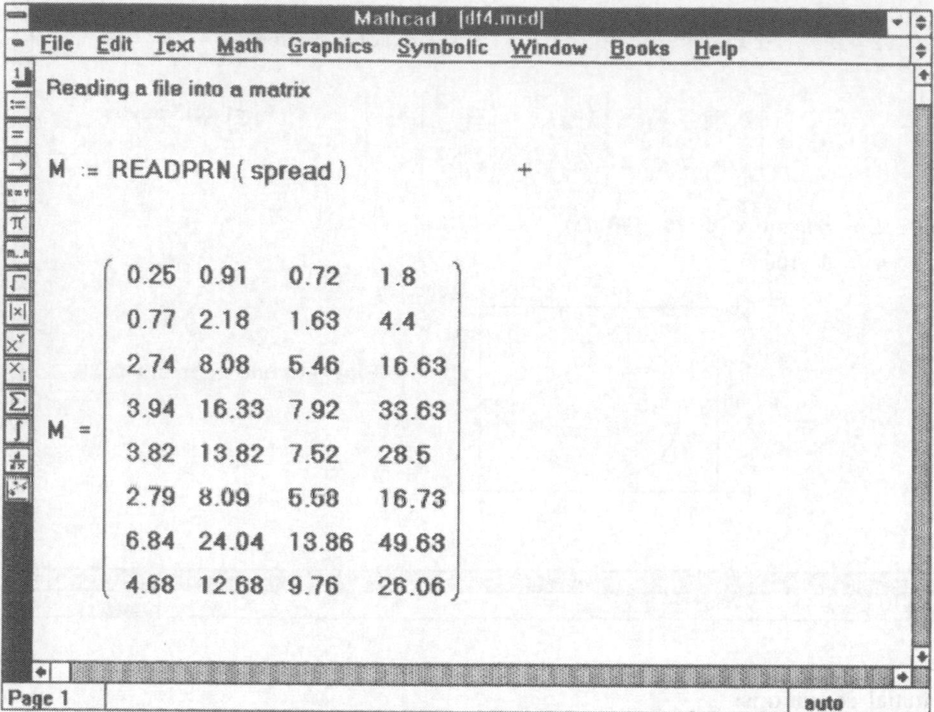

The following figure shows a *Mathcad* document that reads these numbers into a matrix.

Mathcad [df6.mcd]

File Edit Text Math Graphics Symbolic Window Books Help

$i := 0 .. 5$

$j := 0 .. 7$

$A_{i,j} := i \cdot \sin(j \cdot 5)$

$$A = \begin{pmatrix} 0 & 0 & 0 & 0 & 0 & 0 & 0 & 0 \\ 0 & -0.9589 & -0.544 & 0.6503 & 0.9129 & -0.1324 & -0.988 & -0.4282 \\ 0 & -1.9178 & -1.088 & 1.3006 & 1.8259 & -0.2647 & -1.9761 & -0.8564 \\ 0 & -2.8768 & -1.6321 & 1.9509 & 2.7388 & -0.3971 & -2.9641 & -1.2845 \\ 0 & -3.8357 & -2.1761 & 2.6012 & 3.6518 & -0.5294 & -3.9521 & -1.7127 \\ 0 & -4.7946 & -2.7201 & 3.2514 & 4.5647 & -0.6618 & -4.9402 & -2.1409 \end{pmatrix}$$

WRITEPRN(sinevals) := A

Page 1 auto

The *READPRN* function reads the entire data file and creates a matrix out of the data. Once the data are imported into Mathcad as a matrix, you can perform data analysis within Mathcad itself. This includes sorting elements, performing Fourier transforms, computing various statistics, interpolation, regression, plotting, or just about any other mathematical transformation.

You can also use Mathcad to write an array of numbers in ASCII format to a file. The following figure illustrates the use of the *WRITEPRN* function to do so.

When you calculate the document shown in the previous figure, Mathcad creates a data file containing the following numbers:

Notepad - sinevals.prn

File Edit Search Help

```
0        0       0       0       0        0       0       0
0  -0.9589  -0.544  0.6503  0.9129  -0.1324  -0.988 -0.4282
0   -1.918  -1.088   1.301   1.826  -0.2647  -1.976 -0.8564
0   -2.877  -1.632   1.951   2.739  -0.3971  -2.964  -1.285
0   -3.836  -2.176   2.601   3.652  -0.5294  -3.952  -1.713
0   -4.795   -2.72   3.251   4.565  -0.6618   -4.94  -2.141
```

39.12 ENTERING TEXT

So far, all you've learned to create in a Mathcad document is a "math region." Although math regions let you edit equations and perform a variety of calculations, they are not designed for annotating your work as you go along. To fill this need, Mathcad provides "text regions."

Text regions are small islands of text that you can drop anywhere in your document. When you are in a text region, Mathcad behaves very much like a typical word processor. You will be able to edit text and adjust its font characteristics, or spellcheck your text just as you would in a typical word processor.

To create a text region:

- Click in blank space anywhere in your document. You will see a small crosshair.
- Press " to tell Mathcad that you are about to enter some text. Mathcad changes the crosshair into a vertical line called the insertion point. A box surrounds the insertion point to indicate that you are now in a text region. This box grows as you enter text.
- Type **Equations of motion**

Mathcad shows the text in the document.

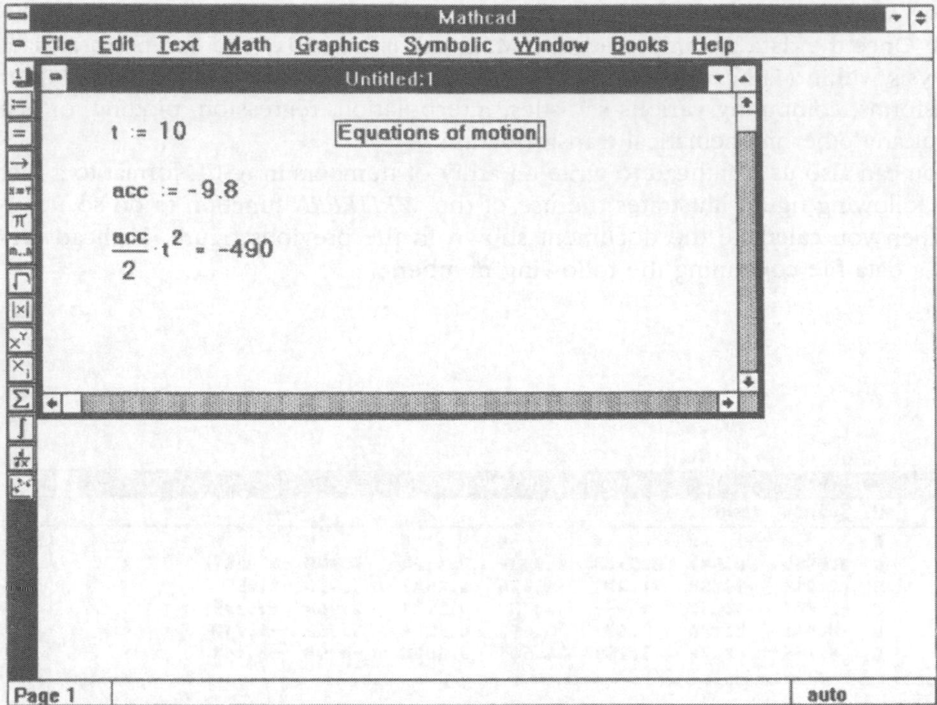

To enter a second line of text, just press [↵] and continue typing:

- Press [↵].
- Then type **`for falling body under gravity`**.
- Click in a different spot in the document to move out of the text region. The text box will disappear once you do so.

Once you have created a text region, you can move it or change its width. To do so, first press and hold down the mouse button and drag the mouse toward the text region. As you do so, a selection rectangle emerges. Drag the mouse until this selection rectangle encloses the text region, then let go of the mouse button.

- To change the width, move the pointer to the right edge of the text region and proceed as if you were resizing a window.
- To move the region, drag it as if you were moving a window.

39.13 GRAPHING

Once you have become accustomed to empty placeholders cueing you to supply a value, you are well on your way to knowing how to create plots. To make a plot, you start by creating a plot region. The choice of plot region depends on whether you plan to create a polar plot, a surface plot, a contour plot, or simply an *x-y* plot. Each plot region comes with several empty placeholders into which you place the quantities you want to plot.

The remainder of this section describes what to put into these placeholders to create the plot you want.

Creating *x-y* Plots

To create an *x-y* plot in Mathcad, click in blank space where you want the graph to appear and choose **Create X-Y Plot** from the **Graphics** menu. Mathcad creates a plot region with placeholders for the expressions to be graphed and additional placeholders for axis limits.

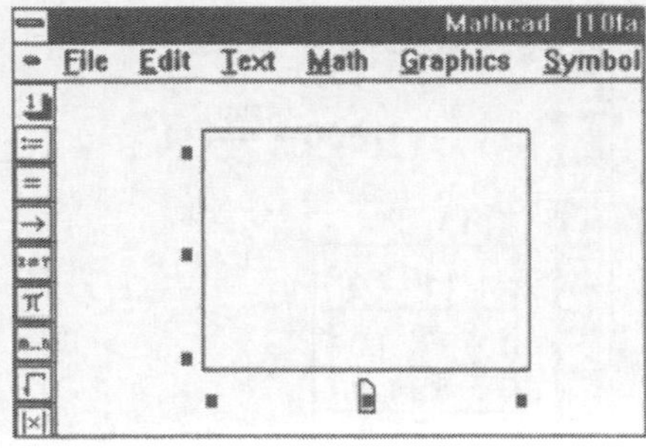

Graphs are driven by range variables: Mathcad will graph one point for each value of the range variable used in the graph. For example, here is how to create a graph of *d*(*t*) versus *t*, with one point for each value of *t*:

- Click below the equation for *d*(*t*) and choose **Create X-Y Plot** from the **Graphics** menu. Mathcad creates an empty graph.

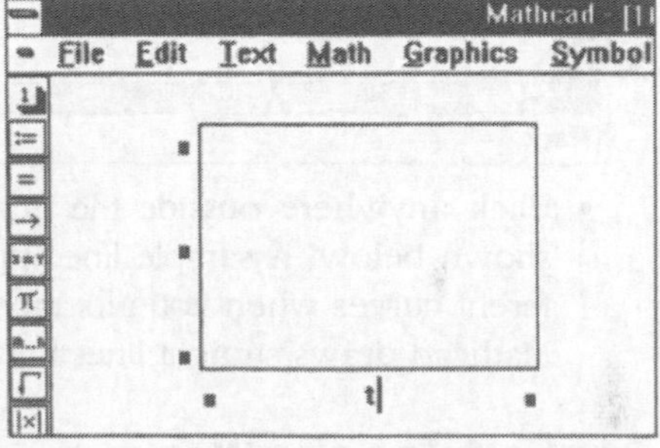

- The selection box should now be at the center of the bottom axis, on the *x*-axis placeholder. Type the variable name **t**. This tells Mathcad to graph *t* on this axis.

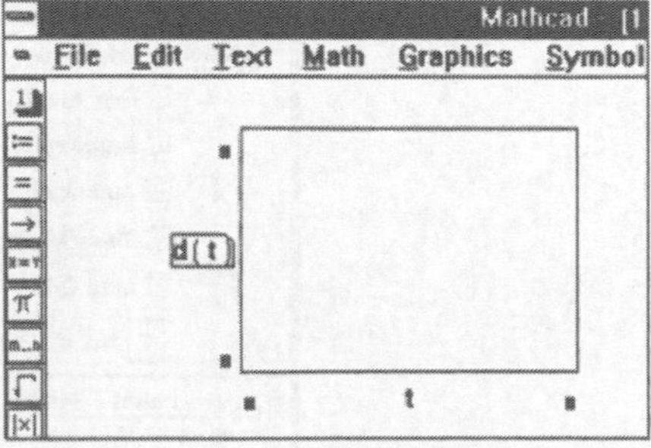

- Now click on the placeholder half-way up the left axis (the *y*-axis placeholder). Type **d(t)** here, to tell Mathcad to graph *d*(*t*) on this axis. The remaining placeholders are for *axis limits*—the high and low values for the axis. If you leave these blank, Mathcad automatically fills them when it creates the graph.

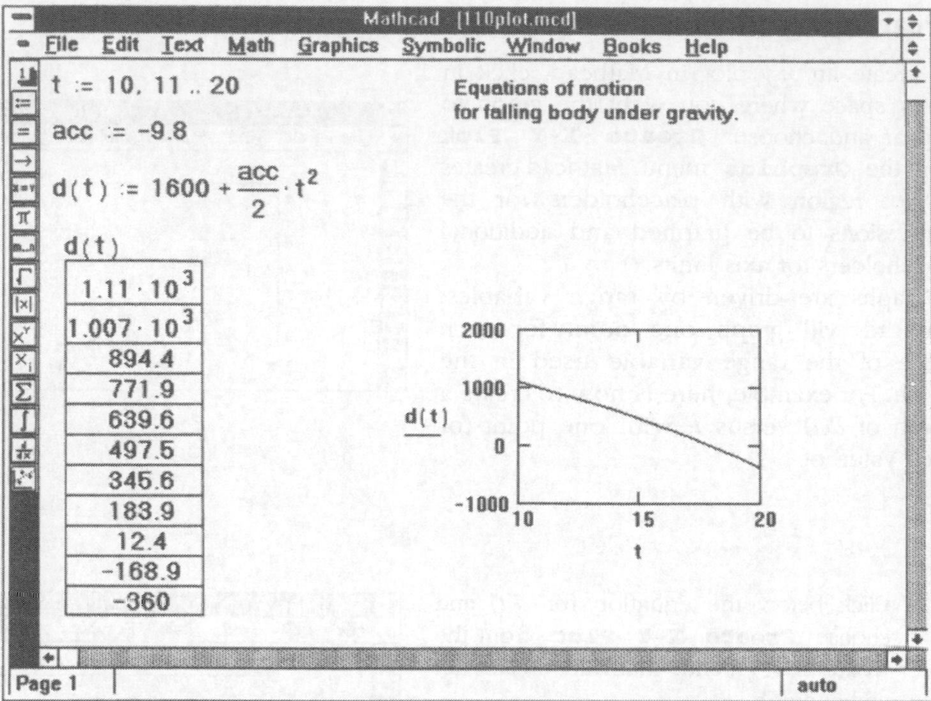

- Click anywhere outside the graph. Mathcad calculates and graphs the points as shown below. A sample line appears under the "*d(t)*" to help you identify the different curves when you plot more than one function. Unless you specify otherwise, Mathcad draws straight lines between the points and fills in the missing axis limits.

Customizing the Plot

The graph shown above still has the default characteristics: numbered linear axes, no grid lines, and points connected with solid lines. You can change these characteristics by *formatting* the graph.

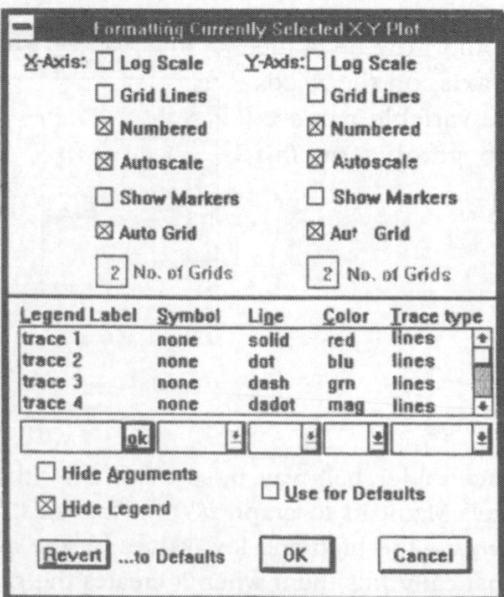

To format the graph, follow these steps:

- Click in the graph to select it. Mathcad puts a selection box around the graph.
- Choose **X-Y Plot Format** from the **X-Y Plot** menu.

The settings in this dialog box are divided into *x*-axis settings, *y*-axis settings, settings for the individual traces themselves, and settings that control the display of legends and axis labels. The following example shows how these settings can drastically alter the look of a particular graph.

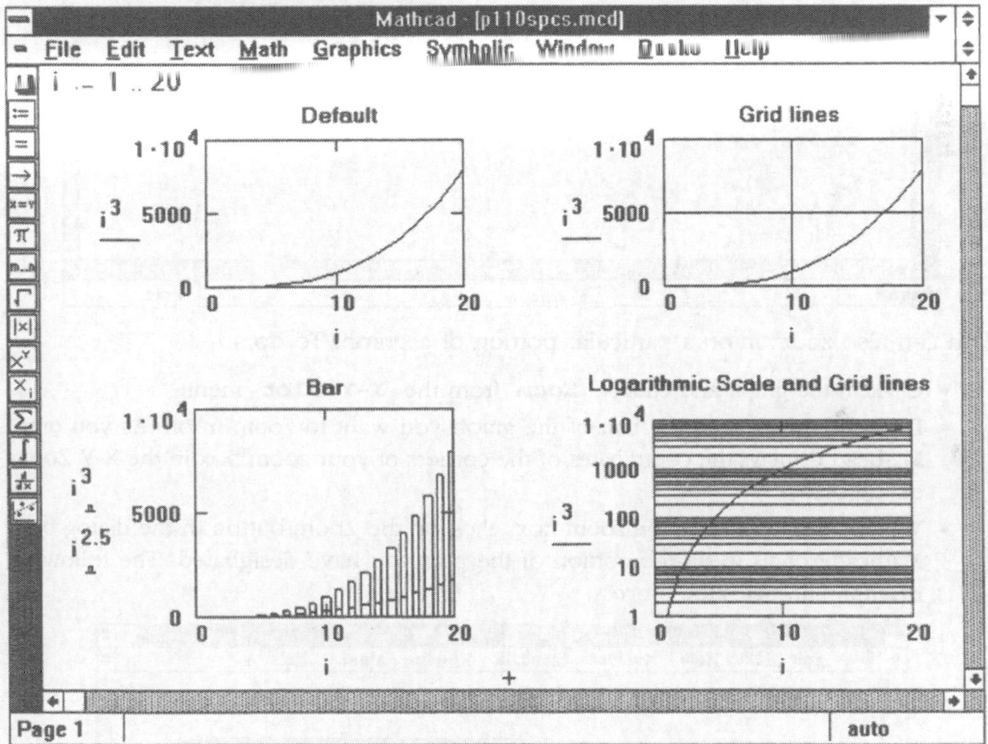

Plot Readouts

Although you can estimate the coordinates of a point by looking at the axes you may want to obtain those coordinates with greater precision. To do so

- Click on the graph and choose **Crosshair** from the the **X-Y Plot** menu. The X-Y Crosshair dialog box appears.
- In the graph region, click and drag the mouse over the points whose coordinates you want to see. A dotted crosshair follows your pointer as you drag it over the graph; the *x* and *y* values displayed in the dialog box change to reflect the current position of the crosshair.

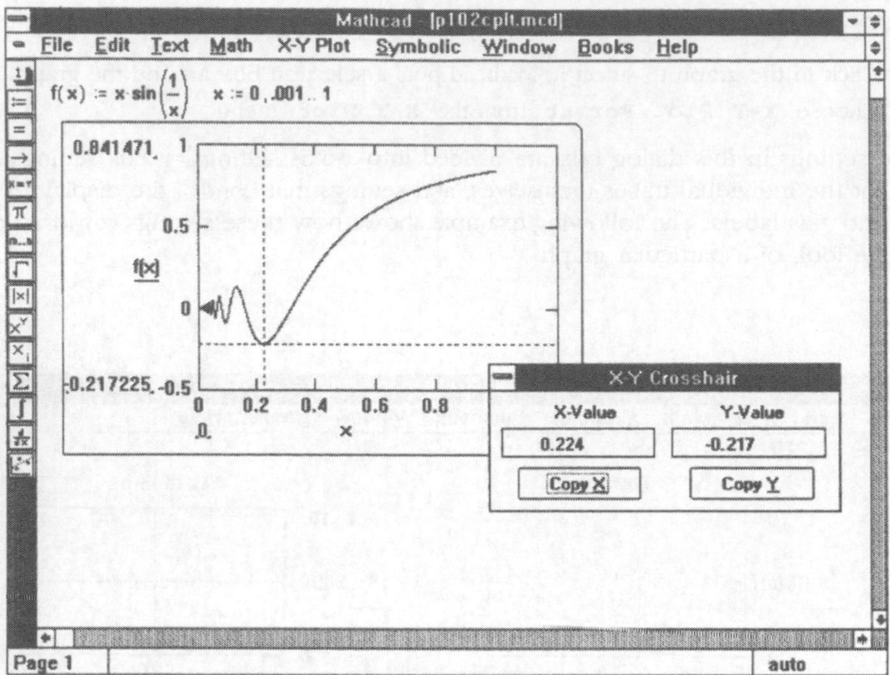

You can also zoom in on a particular portion of a graph. To do so

- Click in the graph and choose **Zoom** from the **X-Y Plot** menu.
- Drag the mouse over the part of the graph you want to zoom in on. As you drag, Mathcad displays the coordinates of the corners of your zoom box in the X-Y Zoom dialog box.
- When you have drawn the zoom box, click on the Zoom button in the dialog box. Mathcad zooms in on the portion of the plot you have designated. The following example illustrates the process.

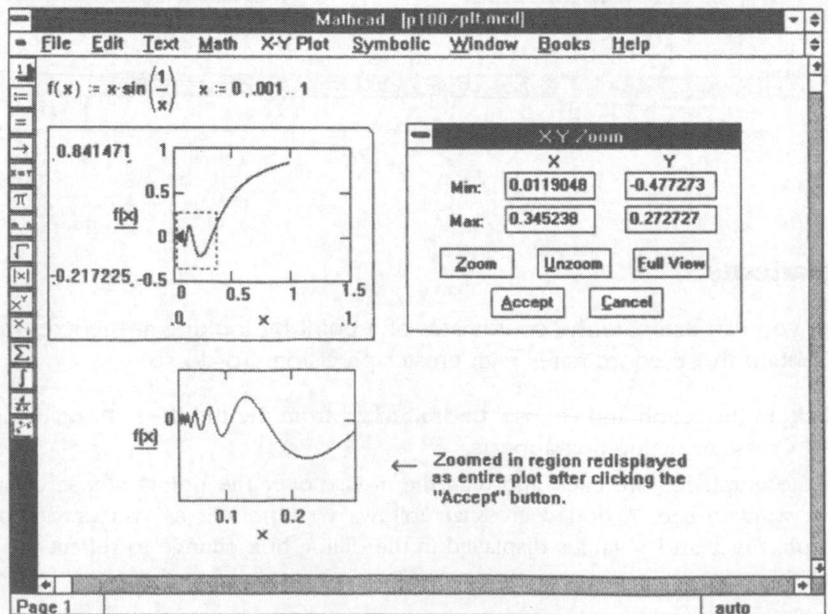

Creating Surface and Contour Plots

You can plot functions of the form $z = f(x,y)$ by doing the following:

- Create a matrix in which the elements correspond to the value of z and the row and column numbers correspond to values of x and y.
- Choose **Create Surface Plot** from the **Graphics** menu.
- Type the name of the matrix in the empty placeholder.

The following figure shows an example.

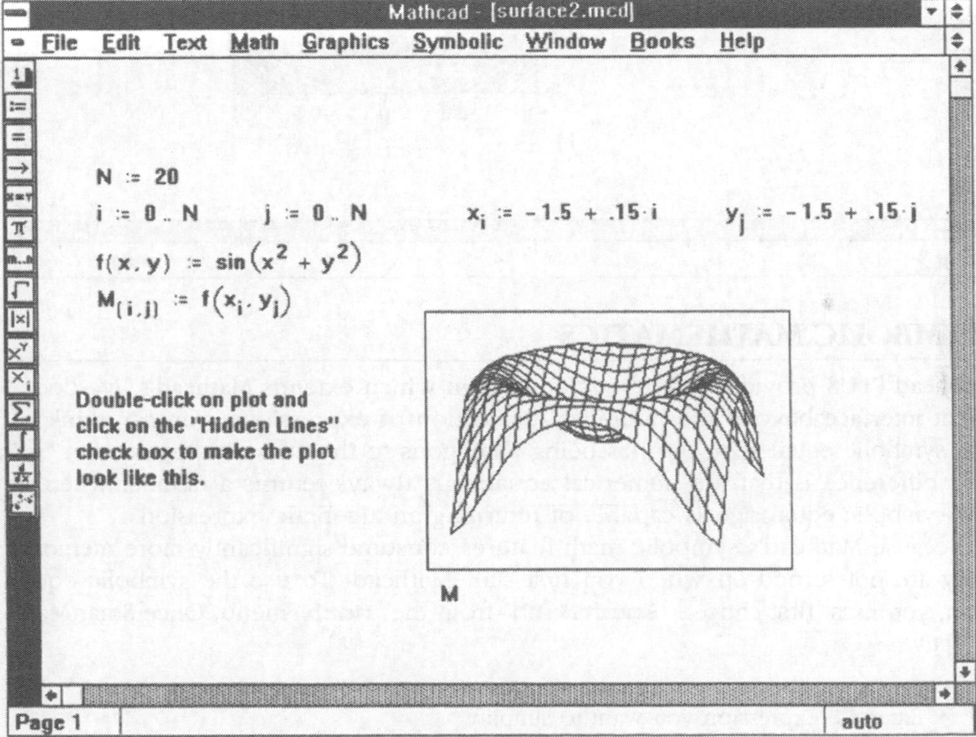

Contour plots are closely related to surface plots. You create them in exactly the same way as you create a surface plot except that you choose **Create Contour Plot** from the **Graphics** menu instead. In fact, you can switch back and forth between a surface plot and a contour plot of the same matrix by double-clicking on the plot to see the relevant dialog box and clicking the Change to Contour or Change to Surface button. The following figure shows a contour plot of the matrix shown earlier.

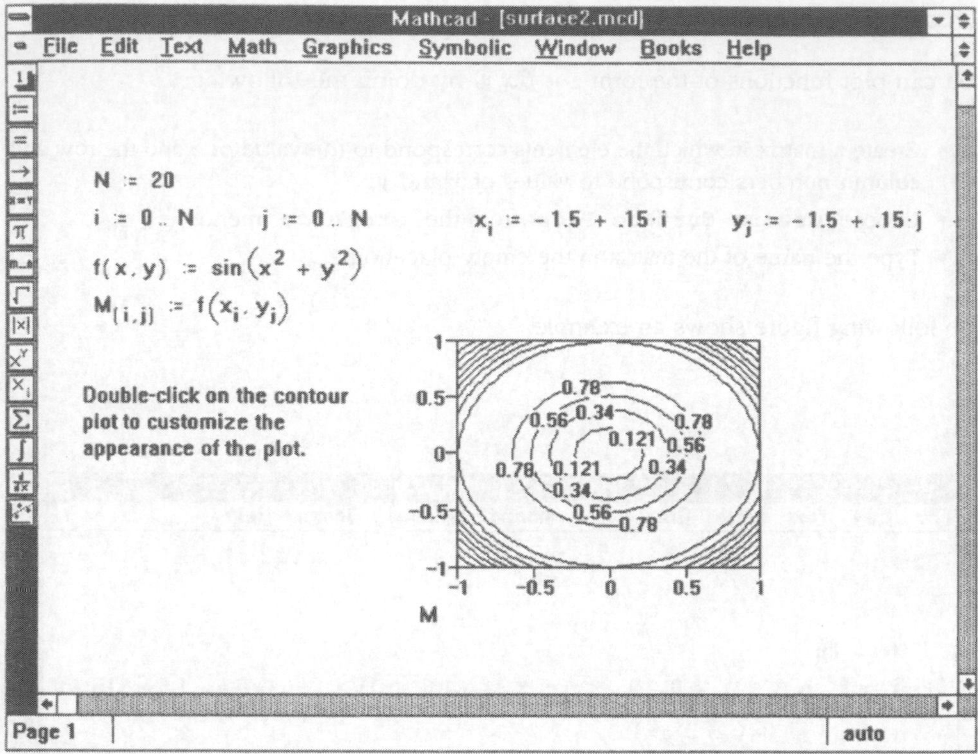

39.14 SYMBOLIC MATHEMATICS

Mathcad PLUS provides a symbolic equal sign which extends Mathcad's live document interface beyond the numerical evaluation of expressions. You can think of the symbolic equal sign, "→," as being analogous to the numerical equal sign "=." The difference is that the numerical equal sign always returns a number whereas the symbolic equal sign is capable of returning an algebraic expression.

Because Mathcad's symbolic math features consume significantly more memory, they are not turned on when you first start Mathcad. To use the symbolic equal sign, you must first choose **SmartMath** from the **Math** menu. Once SmartMath is active

- Enter the expression you want to simplify.
- Press **[Ctrl].** (the control key followed by a period). Mathcad displays an arrow, "→."

- Click outside the expression. Mathcad displays the transformed expression to the right of the arrow. If the expression cannot be simplified further, Mathcad simply repeats it to the right of the arrow.

The symbolic equal sign is a live operator just like any Mathcad operator. When you make a change anywhere above or to the left of it, Mathcad updates the result. The following figure illustrates the use of the symbolic equal sign.

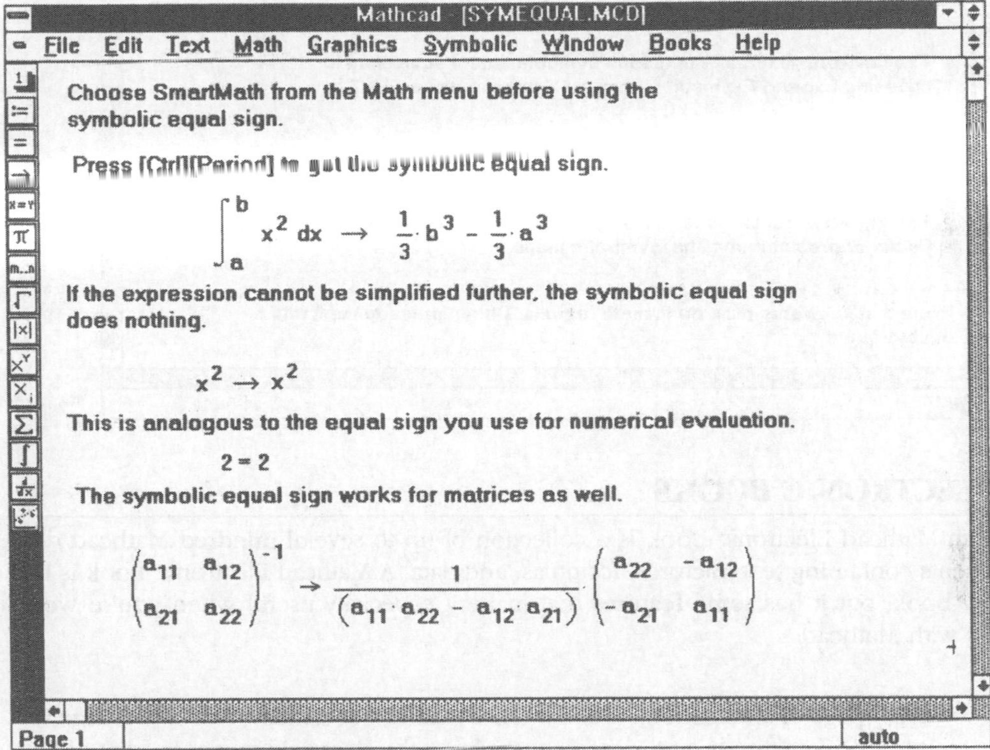

The symbolic equal sign attempts to "simplify" the expression to its left. Because "simplify" is inherently ambiguous, you may find that the result on the right is not exactly what you had in mind. To help you control what "simplify" means, Mathcad provides a half dozen keywords which, when used just before the symbolic equal sign, help to refine its meaning. The following example shows two of these keywords in action.

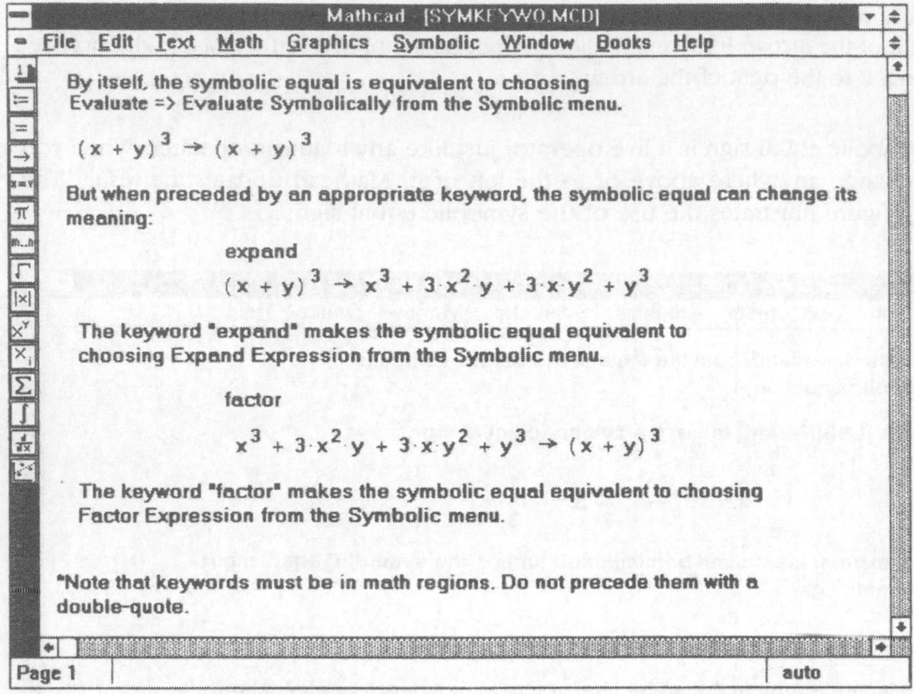

39.15 ELECTRONIC BOOKS

Each Mathcad Electronic Book is a collection of up to several hundred Mathcad documents containing text, pictures, formulas, and data. A Mathcad Electronic Book is like any book, but it has some features that make it especially useful when you're working with Mathcad:

- Every page of an Electronic Book is a live Mathcad document: You can change values, calculate results, and experiment right on the electronic "page" in front of you.
- If you have a second Mathcad document open, you can paste data and formulas from the Electronic Book into your working document simply by double-clicking on them, then going to the document in which you want to paste them and double-clicking again.
- Each Electronic Book has a table of contents and an extensive index. Double-clicking on a section title, index entry, or cross-reference automatically opens the appropriate section.
- A palette of Electronic Book controls lets you browse through the Electronic Book page by page or section by section, or jump directly to the table of contents or index.

An example of an Electronic Book is the *Desktop Reference*, which ships with every copy of Mathcad. The following figure shows an excerpt from this book alongside a Mathcad document making use of the information found in that book.

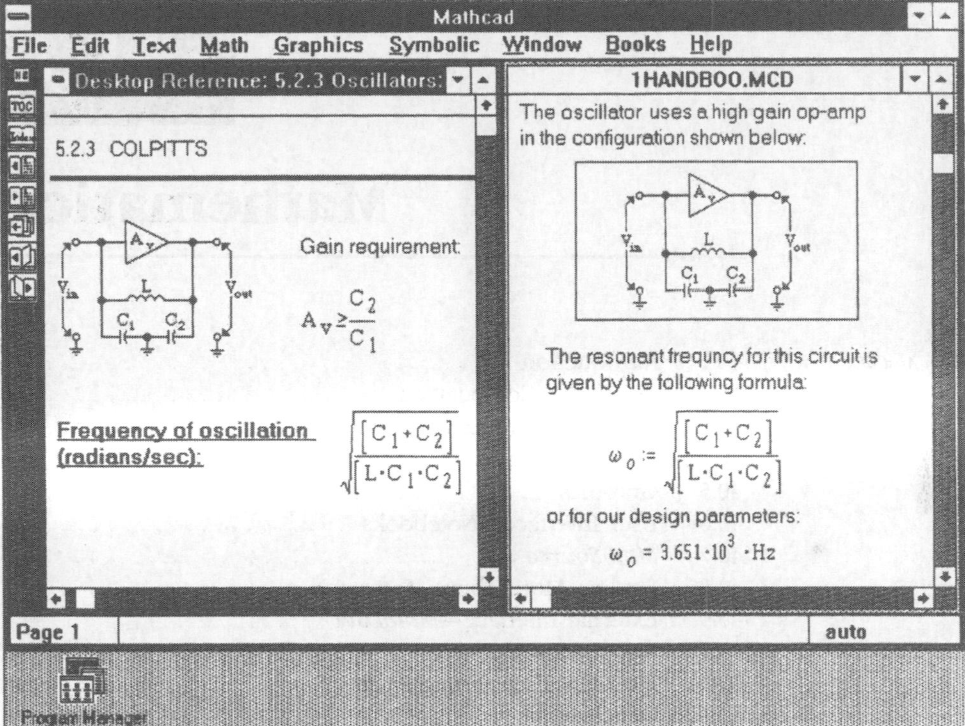

Finding Information in an Electronic Book

An Electronic Book, like a paper one, comes with a table of contents and an index. To see these, click on either the "TOC" button or the "Index" button on the palette at the left edge of the window. When you double-click on an entry in either the index or the table of contents, Mathcad jumps to the appropriate page.

In addition to the table of contents and the index, you can search for all occurrences of a particular word by choosing **Search Book** from the **Books** menu.

40

Mathematica

Paul C. Abbott
University of Western Australia

40.1 INTRODUCTION

Many scientists and engineers presently use FORTRAN, Pascal, or C extensively along with a range of special purpose subroutine libraries. *Mathematica* has been designed as a general, integrated, system for problem solving that includes numerical computation, graphics, symbolic or algebraic manipulation, and interprocess communication. The design goal was to produce a consistent environment that is easy to use and powerful, and, for a large range of calculations, can replace FORTRAN or C as the tool of choice for engineers and scientists.

Mathematica is also a high-level programming language that assists in the development of sophisticated custom applications. Its *Notebook* user-interface, with Notebooks portable between Microsoft Windows, Macintosh, NEXTSTEP, and Unix machines under the X Window system, provides an excellent documentation environment. Taken together, *Mathematica* is an ideal scientific and engineering problem-solving environment.

Throughout the following sections *Mathematica* input (output) will be in **bold** (plain) courier font, *i.e.*, **`input`** `(output)`.

40.2 NUMERICAL COMPUTATION

Mathematica covers the mathematical functions in standard books of tables (Abramowitz and Stegun, 1972) and many of the capabilities of standard numerical subroutine libraries such as IMSL, NAG, and Numerical Recipes.

0-8493-2530-7/96/$0.00+$.50

Arithmetic

You can do arithmetic just as on a calculator:

```
25325272342342342332798732 * 2342223423432342347239892324
59317446085037496585543465434876722977322673243733168
```

Unlike a calculator, you can get exact results:

```
3^100
515377520732011331036461129765621272702107522001
```

You can use **N** to compute results using any desired numerical precision. For example, the following computation used **60** significant digits:

```
N[Exp[Sqrt[163] Pi/3], 60]
640320.0000000006048637350490160394717418188185394757714858
```

Complex numbers are handled directly (**I** denotes $\sqrt{-1}$):

```
           40
(17 + 23 I)
17034685784040814898293226879810957705797710491744522469376 -
  5788921749816137954842206606934670288085790394882300313600 I
```

Special Functions

Mathematica knows the special functions used in engineering. For example, here is a particular value of the Bessel function of the first kind, $J_1(x)$:

```
BesselJ[1, 12.3]
-0.194259
```

The capabilities are integrated in such a fashion that you can combine several operations together. **FindRoot** uses Newton's method to locate roots of functions:

```
FindRoot[ BesselJ[1, x], {x, 3} ]
{x → 3.83171}
```

Importantly, the syntax is uniform so that graphically verifying this zero is straightforward:

```
Plot[ BesselJ[1, x], {x, 0, 10} ];
```

You can work with all the special functions to any desired precision in the complex plane.

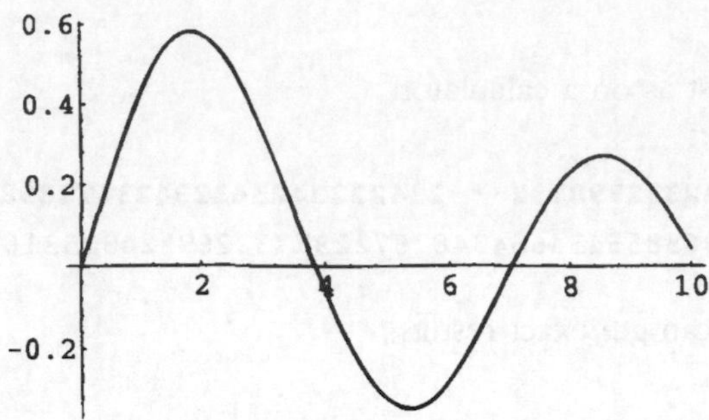


```
N[ BesselJ[0,   3 + I], 50]
-0.460492143882258459121757050410824590387075454129 -
  0.369565000014863578058657904039979146677576994696
```

Numerical Integration

Mathematica includes arbitrary precision numerical integration:

```
NIntegrate[ BesselJ[1,x], {x,0,5} ]
1.1776
```

Mathematica can also do contour integration: Consider the polygonal contour specified by

```
contour = {1, I, -1, -I, 1};
```

The numerical value of $(1/2\pi i)\int_0(e^z/z)dz$ around this contour is easily computed (**Chop** removes the small imaginary part):

```
Chop[N[NIntegrate[E^z/z, {z, 1, I, -1, -I, 1}]/(2 Pi I)]]
1.
```

and the contour visualized using **ListPlot** by computing the real **(Re)** and imaginary **(Im)** parts of each contour point:

```
ListPlot[Transpose[{Re[contour],
  Im[contour]}],   PlotJoined →
  True, AspectRatio → Automatic];
```

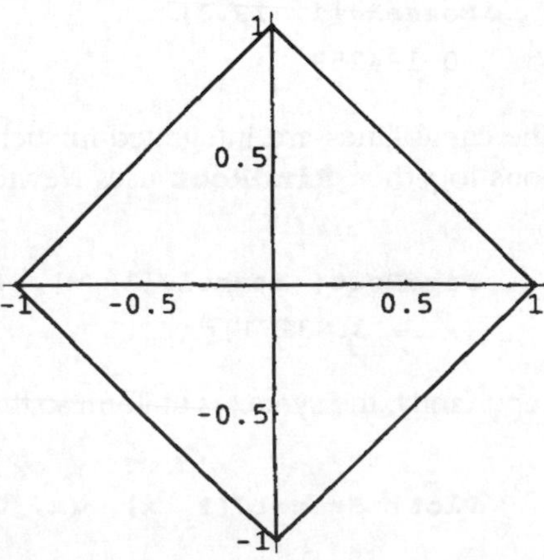

Compilation

Mathematica includes a compiler for numerical expressions that can dramatically speed up repetitive or iterative numerical computation. Compilation is used extensively internally for graphical and numerical functions such as **Plot** and **NIntegrate**.

Suppose that we are interested in the behavior of Newton's method for finding the complex roots of functions. For a general function, **f[z]**, the essence of Newton's method is the iteration of **z-f[z]/f'[z]**. For a particular function, say **z³ - 1**, the iteration reads:

```
z - f[z]/f'[z] /. f → Function[z, z^3 - 1] // Expand

  1      2 z
----- + ---
   2     3
3 z
```

Since **Compile** handles complex variables, compiling this iteration for a general complex argument is straightforward:

```
CubicRoot = Compile[{{w, _Complex}},
    Module[{z = w},
      While[Abs[z^3 - 1] >= 10^ (-10),
         z = 2/3 z + 1/(3z^2)]
    z]
    ];
```

As an example:

```
CubicRoot[0.2 + 0.3 I]
-0.5 - 0.866025 I
```

After computing the iterations on a 200 × 200 grid:

```
Table[CubicRoot[x + I y],
   {y, -0.8, 0.8, 0.008}, {x, -1.0, 0.6, 0.008}]; // Timing
{126.7 Second, Null}
```

which takes approximately 2 minutes on a DECstation 5000/200, we can visualize the result by coloring each initial value according to its destination—done here using the imaginary part of each iterated value:

```
ListDensityPlot[Im[points], Mesh → False,
   MeshRange → {{-1.0, 0.6},{-0.8, 0.8}}];
```

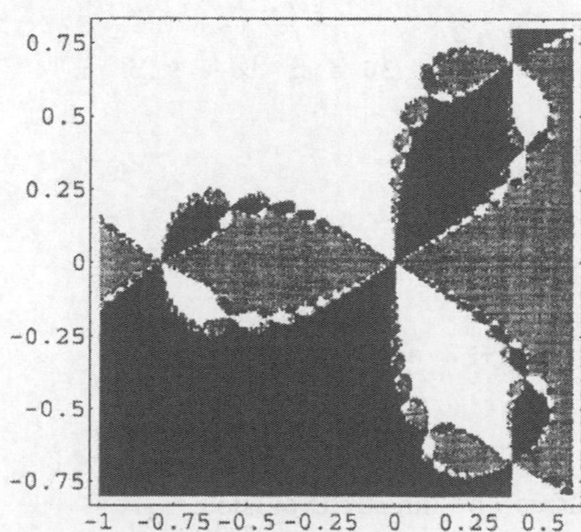

The fractal self-similar behavior is quite striking and indicates the sensitivity of the iteration to the choice of the initial value.

40.3 SYMBOLIC OR ALGEBRAIC COMPUTATION

Algebra

It is just as easy to work with symbols. Consider the expression

$$(a^2 - b^2)^3 + 3\ (x + 2)\ (a + b)^2\ ;$$

Note that a **;** at the end of a line suppresses intermediate output and that the immediately previous expression can always be referred to using **%**.

Raising this expression to the second power and expanding out all terms yields a reasonably complicated expression:

```
Expand[%^2]
```

$$36 a^4 + 12 a^8 + a^{12} + 144 a^3 b + 24 a^7 b + 216 a^2 b^2 -$$
$$24 a^6 b^2 - 6 a^{10} b^2 + 144 a b^3 - 72 a^5 b^3 + 36 b^4 + 15 a^8 b^4 +$$
$$72 a^3 b^5 + 24 a^7 b^5 - 20 a^6 b^6 - 24 a b^7 - 12 b^8 + 15 a^4 b^8 -$$
$$6 a^2 b^{10} + b^{12} + 36 a^4 x + 6 a^8 x + 144 a^3 b x + 12 a^7 b x +$$
$$216 a^2 b^2 x - 12 a^6 b^2 x + 144 a b^3 x - 36 a^5 b^3 x + 36 b^4 x +$$
$$36 a^3 b^5 x + 12 a^2 b^6 x - 12 a b^7 x - 6 b^8 x + 9 a^4 x^2 +$$
$$36 a^3 b x^2 + 54 a^2 b^2 x^2 + 36 a b^3 x^2 + 9 b^4 x^2$$

Factorizing this expression is immediate:

```
Factor[%]
```

$$(a + b)^4\ (6 + a^4 - 2 a^3 b + 2 a b^3 - b^4 + 3 x)^2$$

Calculus

Quite difficult definite integrals can be directly evaluated:

```
Integrate[Sin[x^p], {x, 0, Infinity}]

                        1    1
       Sqrt[Pi] Gamma[- + ---]
                        2   2 p
------------------------------------
                                    1
    1 1/2(2 p)     p Gamma[1 - ---]
2 (-)                              2 p
    4
```

Evaluating the following integral requires a $_1F_2$ hypergeometric function:

```
Integrate[Exp[-x] Sin[x^2], {x, 0, Infinity}]

        Pi       1                                  3   5        1
(Sqrt[--] Cos[-] - HypergeometricPFQ[{1}, {-, -}, -(--)] +
        2        4                                  4   4        64

         Pi       1
   Sqrt[--] Sin[-])/2
         2        4
```

Nevertheless, this result can be rapidly numerically evaluated to any desired precision:

```
N[%, 50]
0.27051358016221414425890085615784186281211951468968
```

Indefinite integrals are also straightforward:

```
Integrate[x^3 Sin[x]^2 Exp[x], x]

         x         x           x  2        x  3
(-3750 E + 3750 E  x - 1875 E  x + 625 E  x -

       x                  x                 x  2
   42 E  Cos[2 x] + 330 E  x Cos[2 x] - 225 E  x  Cos[2x] -

        x  3                  x                x
   125 E  x  Cos[2 x] - 144 E  Sin[2 x] + 60 E  x Sin[2 x] +

        x  2                 x  3
   300 E  x  Sin[2 x] - 250 E  x   Sin[2 x]) / 1250
```

Differentiating this expression yields:

```
D[%, x]

     x  3          x  3
625 E  x  - 625 E  x  Cos[2 x]
---------------------------------
              1250
```

One can **Simplify** this result using built in algebraic and trigonometric identities:

```
Simplify[%]

 x  3      2
E  x  Sin[x]
```

Series

Series manipulation tools include series expansion:

```
                                                   17
Sin[Tan[x]] - Tan[Sin[x]] + O[x]

   7       9         11        13             15
-x     29 x    1913 x    95 x     311148869 x          17
--- - ----- - ------- - ----- - ------------- + O[x]
30     756     75600     7392    54486432000
```

series inversion:

```
InverseSeries[E^Sin[x] + O[x]^5]

                           2           3           4
                   (-1 + x)    (-1 + x)    (-1 + x)              5
(-1 + x) - ---------- + -------- - --------- + O[-1 + x]
                      2           2           2
```

and the series expansion of arbitrary symbolic functions:

```
Simplify[1/p[x]^2 + O[x]^3]

                                        2                         2
    -2   2 p'[0] x     (3 p'[0]  - p[0] p"[0]) x          3
p[0]  - ---------  + ------------------------ + O[x]
                3                      4
           p[0]                    p[0]
```

Other tools include Padé approximation and general rational and minimax approximation.

Solving Equations

For the algebraic equation:

```
 3      2
x  - 7 x  + 3 x - 10 == 0;
```

the exact value for the real root reads:

```
x /. First[Solve[%, x]] // ComplexExpand

                  1/3                                1/3
7          40 2                   (767 + 3 Sqrt[36921])
- + ---------------------- + ------------------------
3                        1/3                   1/3
    3 (767 + Sqrt[36921])                  3 2
```

The general quintic equation is insoluble in terms of radicals. For example,

```
          5
Solve[x - x + 1 == 0, x]

                      5
{ToRules[Roots[-x + x == -1, x]]}
```

Direct numerical evaluation is still possible:

```
N[%, 20]
{{x → -1.1673039782614186843},
  {x → -0.18123244446987538390 - 1.0839541013177106684 I},
  {x → -0.18123244446987538390 + 1.0839541013177106684 I},
  {x → 0.76488443360058472603 - 0.35247154603172624932 I},
  {x → 0.76488443360058472603 + 0.35247154603172624932 I}}
```

Differential Equations

DSolve can symbolically solve a range of differential equations:

```
DSolve[x^3 y''[x] + x(2x-1) y'[x] + y[x] == 0, y[x], x]

                                                    1
            C[1] + E^(1/x) x C[2] - C[2] ExpIntegralEi[-]
                                                    x
{{y[x] → ----------------------------------------------}}
                           1/x
                          E    x
```

Here C[1] and C[2] are arbitrary constants.

NDSolve returns the numerical solution to systems of linear and nonlinear differential equations. For example:

```
NDSolve[{ y''[x] + (7 - x^2) y[x] == 0, y[0] == 0,
y[0] == 0, y'[0] == 1}, y, {x, -5, 5}]
{{y → InterpolatingFunction[{-5., 5.}, <>]}}
```

The result is returned as an *interpolating function*, which can be plotted immediately:

```
Plot[ y[x] /. %, {x, -5, 5}];
```

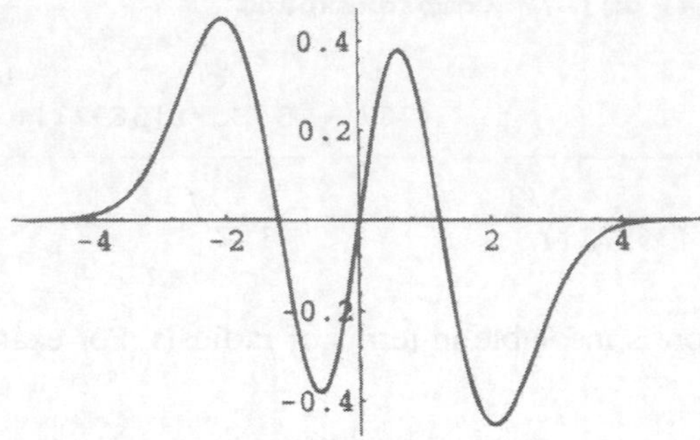

Matrices

Instead of using "Do loops," matrices in *Mathematica* are most easily produced using the **Table** command. For example,

```
mat = Table[ Random[], {100}, {100}];
```

produces a 100 × 100 matrix with entries that are random numbers uniformly distributed between 0 and 1. Arbitrary entries of **mat** can be examined. For example, here is the i=1, j=1 element:

```
mat[[1,1]]
0.0648696
```

and here are the first five entries of the second row:

```
mat[[2, Range[5]]]
{0.686831, 0.47071, 0.935435, 0.669428, 0.421835}
```

The relative size of entries in **mat** can be immediately visualized:

```
ListDensityPlot[mat, Mesh → False];
```

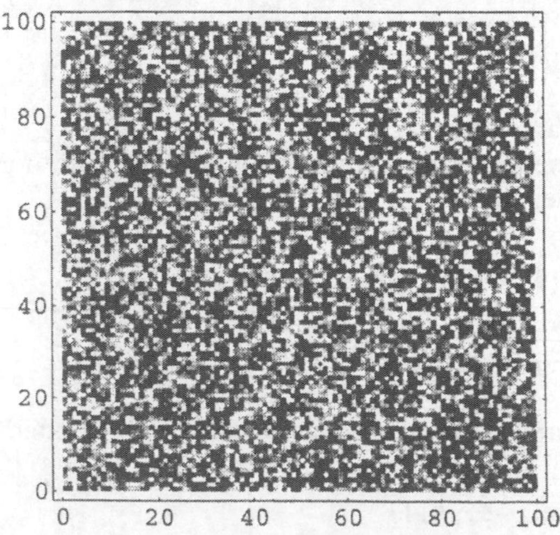

Standard matrix tools like determinant (**Det**), matrix inverse (**Inverse**), eigenvalues (**Eigenvalues**), eigenvectors (**Eigenvectors**), L-U decomposition (**LUDecomposition**), and singular value decomposition (**SingularValues**) are all built in. For example, the (complex) eigenvalues of **mat** are computed using

```
evals = Eigenvalues[mat];
```

Here is an Argand diagram of all the eigenvalues:

```
ListPlot[Transpose[{Re[evals],Im[evals]}],
    PlotRange → All,
    AspectRatio → Automatic,
    PlotStyle → PointSize[0.005]];
```

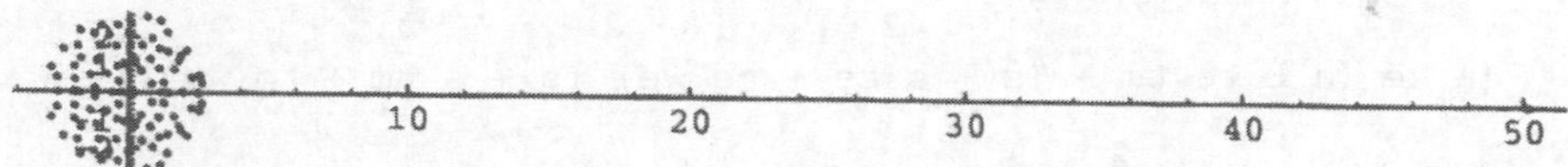

Note that the option **PlotRange → All** is required to reveal the existence of a real eigenvalue at a point corresponding to approximately half the size of the matrix.

Laplace Transform

Modeling a gravitational wave detector with a microwave resonant transducer as a mechanical oscillator coupled to an electrical one leads to two-coupled second-order differential equations:

```
eqv = V''[t] + 1/te V'[t] + we^2 V[t] == a X[t];
eqx = X''[t] + 1/tm X'[t] + wm^2 X[t] == b V[t];
```

After defining

```
LaplaceTransform[X_[t], t, s] = X[s];
```

and loading

```
<<Calculus`LaplaceTransform
```

the Laplace transforms read

```
lapeqv = LaplaceTransform[eqv, t, s]

             2            2         -V[0] + s V[s]
-(s V[0]) + s  V[s] + we  V[s] + -------------- - V'[0] == a X[s]
                                       te
```

```
lapeqx = LaplaceTransform[eqx, t, s]
                 2           2     -X[0] + s X[s]
-(s X[0]) + s X[s] + wm X[s] + ─────────────── - X'[0] == b V[s]
                                      tm
```

and can be solved immediately:

```
Solve[{lapeqx, lapeqv}, {X[s], V[s]}] // Simplify
               V[0] + s te V[0] + te V'[0]
{{X[s] -> -(────────────────────────────) -
                         a te

          2            2          2         2
   ((s + s te + te we ) ((s + s tm + tm wm )
      (V[0] + s te V[0] + te V'[0]) +
     a te (X[0] + s tm X[0] + tm X'[0]))) /

                              2          2        2
   (a te (a b te tm - (s + s te + te we) (s + s tm + tm wm^2))),

                 2
V[s] -> - (((s + s tm + tm wm^2) (V[0] + s te V[0] + te V'[0]) +
    a te (X[0] + s tm X[0] + tm X'[0])) /

                        2           2          2         2
   (a b te tm - (s + s te + te we) (s + s tm + tm wm)))}}
```

However, computing the inverse transform of this system is not simple. There is also a package for computing Fourier transforms.

40.4 PROGRAMMING

Mathematica also includes a high-level programming language. This makes it very useful for the development of sophisticated custom applications. Many different programming styles are incorporated into the programming language including ideas borrowed from FORTRAN, C, LISP, APL, and Prolog.

In this section the factorial function will be used to illustrate aspects of the language and the various programming styles available. Transformation and replacement rules, pattern-matching, and the object-oriented programming (OOP) idea of association are also briefly examined.

Built-in

The factorial function is, of course, built-in:

```
10!
3628800
```

Alternatively, one can use the **Gamma** function, which generalizes factorial to noninteger, negative, and complex values:

```
Gamma[11]
3628800
```

Procedural

A FORTRAN, Pascal, or C programmer will probably understand the following code without any explanation:

```
facProcedural[n_] :=
Block[{i = 1, s = 1},
    While[i < n,
        i++;
        s = s*i
      ];
    Return[s]
]
facProcedural[10]
3628800
```

The only syntactical subtleties are the use of **n_** (read as "n blank") to denote a generic variable **n** whose value is passed to the procedure and **:=** (delayed assignment), which prevents execution of the right-hand side until **n** is assigned a value.

Recursive

Because of the recursive nature of the factorial function, a recursive program may be more natural:

```
facRecursive[n_] := If[n == 1, 1, n facRecursive[n-1]]
facRecursive[10]
3628800
```

Functional

Since the factorial function can be defined by multiplying all numbers from 1 to n, one can use **Range**, say

```
Range[10]
{1, 2, 3, 4, 5, 6, 7, 8, 9, 10}
```

and then **Apply** the operation of **Times** to multiply these numbers:

```
Apply[Times, %]
3628800
```

This is easily converted into a procedure:

```
facFunctional[n_] := Apply[Times, Range[n]]
facFunctional[10]
3628800
```

Rule-Based

For many situations, *rule-based programming* is often the simplest and most elegant style. Specifying the initial condition:

```
facRule[0] = 1;
```

and the general recursive rule:

```
facRule[n_] := n facRule[n - 1]
```

Mathematica automatically applies the recursive definition:

```
facRule[10]
3628800
```

The advantage of this style is that it is often similar to the actual problem specification for a wide range of problems in science and engineering.

Dynamic Programming and Tracing Computations

As it stands, the above code for **facRule** is somewhat inefficient. Entering

```
?facRule
Global`facRule
facRule[0] = 1
facRule[n_] := n*facRule[n - 1]
```

one finds that, after computing **facRule[10]**, the intermediate values have not been saved. This can be addressed by using *dynamic programming*, i.e., saving the intermediate results. This is easily implemented:

```
facRule[n_] := facRule[n] = n facRule[n - 1]
```

Now, after computing the factorial of **3**:

```
facRule[3]
6
```

a check on **facRule** reveals that the intermediate values have been recorded:

```
?facRule
Global`facRule
```

```
facRule[0] = 1
facRule[1] = 1
facRule[2] = 2
facRule[3] = 6
facRule[n_] := facRule[n] = n*facRule[n - 1]
```

The next defect is that the code for **facRule** makes sense only for positive integer arguments. After removing the previous definition:

```
Remove[facRule]
```

the code can be made more rigorous by checking that the argument of the recursive rule is a positive integer:

```
facRule[0] = 1;
facRule[n_Integer?Positive] :=
    facRule[n] = n facRule[n - 1]
facRule[10]
3628800
```

Mathematica includes a **Trace** operation for tracing computations. This can be used to verify that intermediate factorial values are being saved and not recomputed. For example, when computing **facRule[12]**, only **facRule[11]** has to be computed as **facRule[10]** is already known:

```
Trace[facRule[12], facRule[_]]
{facRule[12], {{facRule[11], {{facRule[10]}}}}}
```

Transformation Rules, Pattern Matching, and Replacements

When simplifying expressions by hand one naturally tends to apply transformation rules as required. In addition, humans are excellent at pattern recognition. *Mathematica* attempts to include the basics of these concepts in a natural way.

Consider the following orthogonality integral that arises when computing Fourier series:

```
sincos = Integrate[Sin[n x] Cos[m x], {x, 0, 2Pi}]

     n
- (-------) + (m Cos[2 (m - n) Pi] + n Cos[2 (m - n) Pi] -
    2   2
   m  - n

                                                  2     2
m Cos[2 (m + n) Pi] + n Cos[2 (m + n) Pi]) / (2 (m  - n ))
```

At first, one may be surprised the result is not automatically simplified. However, one should note that *Mathematica* does not assume that **m** and **n** are integers (or even that they are real!). A human would recognize the occurrence of both the **Cos[2 (m + n) Pi]** and **Cos[2 (m - n) Pi]** terms in **sincos** and note that **Cos[2 n Pi]** is unity for *any* integral value of **n**. Since **n** and **m** are both integers then so is **(n + m)** and **(n - m)**. All this information is encapsulated in the following *replacement rule*:

```
sincos /. Cos[2 n_ Pi] → 1
0
```

Note that this replacement operation is literal, i.e., no care has been taken to check the limiting behavior of the expression. In fact, taking the limit as **m → n**, one finds

```
Limit[sincos, m → n]

              2
Sin[2 n Pi]
-----------
    2 n
```

Since **Sin[n Pi]** is zero for integral values of **n**, one finally obtains:

```
% /. Sin[n_ Pi] → 0
0
```

Association

Often, a definition or rule should be associated with a particular object. For example,

```
g/: g[a_] + g[b_] := g[a + b]
```

The rule is associated with **g** (instead of being associated with **+**) using the syntax **g/:**. This rule is applied repeatedly to expressions until they no longer change. For example,

```
g[a] + g[b] + g[c]
g[a + b + c]
```

This uses pattern matching (and the associativity property of **+**) since the rule above was only defined for the sum of two terms. **Trace** reveals something of the "inner workings" of *Mathematica*:

```
Trace[g[a] + g[b] + g[c]]
{g[a] + g[b] + g[c], g[c] + g[a + b], g[a + b] + g[c],
  g[a + b + c], {a + b + c, a + b + c}, g[a + b + c]}
```

40.5 GRAPHICS

Some examples of *Mathematica* graphics have already been presented in the sections above, including **Plot**, **DensityPlot**, and **ListDensityPlot**. These examples were chosen so as to illustrate that integrating numerical, symbolic, and graphical capabilities produces a system that is more powerful than the sum of its parts—during a computation, one is encouraged to do visualization because it is easy to do so.

Mathematica Graphics have been designed with a well-chosen set of standard defaults while still permitting the user to arbitrarily customize any setting.

The PostScript language was chosen for the intermediate representation of graphics for a number of reasons:

- PostScript is a de facto standard for graphics.
- PostScript can be interpreted by a wide range of displays and printers.
- PostScript is independent of the resolution of the output device.
- PostScript graphics contain only ASCII characters. This simplifies electronic document transfer between computers and operating systems.

Graphics Packages

Apart from the built-in capabilities there is a whole range of graphics packages distributed with *Mathematica*. After loading

```
<<Graphics`Master`
```

any required external graphics function is automatically loaded.

For example, plotting a function in decibels is possible using a log plot:

```
LogPlot[Abs[Sin[u]/u], {u, -4 Pi, 4 Pi},
PlotRange → {0.001, 1}, Frame → True,
GridLines → Automatic];
```

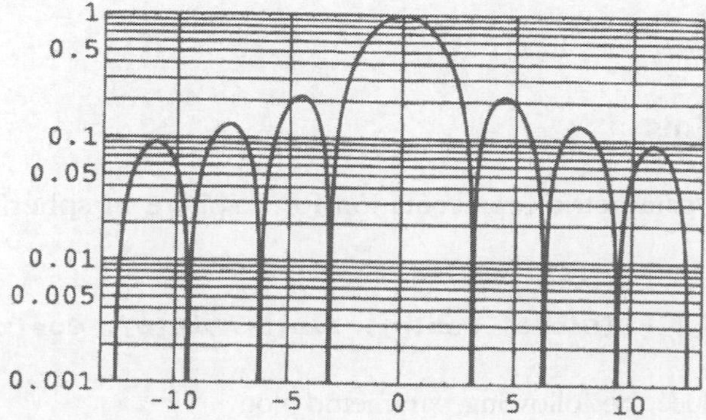

Contour Plots

The far-field diffraction pattern of a rectangular aperture twice as wide as it is high, with the z axis normal to the opening is easily visualized:

```
ContourPlot[Sin[20 Pi u]/(20 Pi u) *
  Sin[10 Pi v]/(10 Pi v), {u, -1,
  1}, {v, -1, 1}, PlotPoints → 40];
```

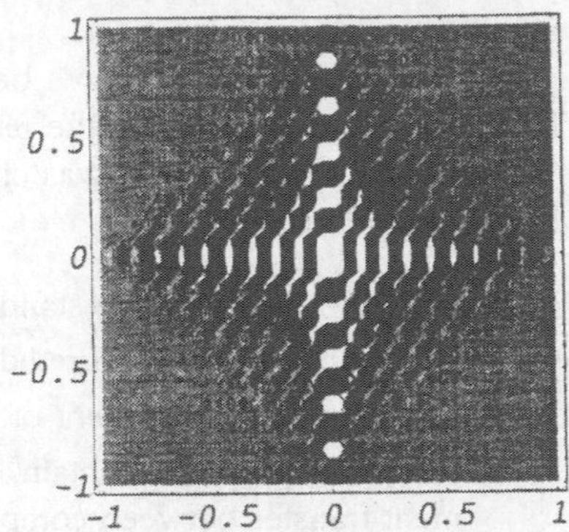

Three-Dimensional Plots

Here is a surface plot of the same function after zooming in near the origin:

```
Plot3D[ Sin[20 Pi u]/(20 Pi u) *
  Sin[10 Pi v]/(10 Pi v),
     {u, -0.2, 0.2},
     {v, -0.2, 0.2},
     PlotRange → All,
     PlotPoints → 35,
     Mesh → False];
```

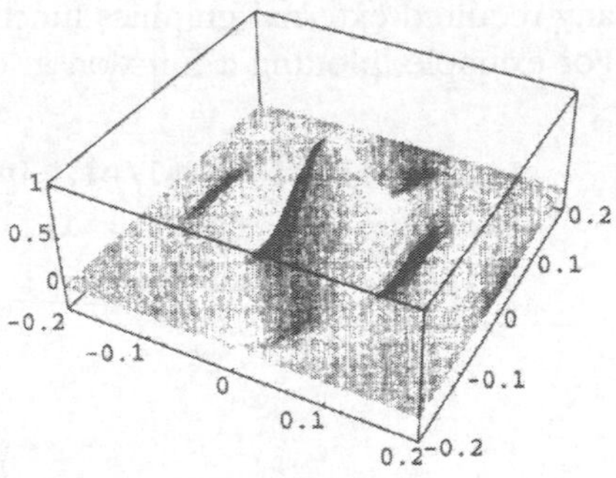

Parametric Plots

After defining a parametric representation for a sphere in spherical polar coordinates:

```
spc[t_, p_] = {Sin[t] Cos[p], Sin[t] Sin[p], Cos[t]};
```

it is easy to produce the following parametric plot:

```
ParametricPlot3D[
Evaluate[
{spc[t,p],
spc[t,4p/3]/2}],
   {t,0,Pi},
     {p,0,3Pi/2}];
```

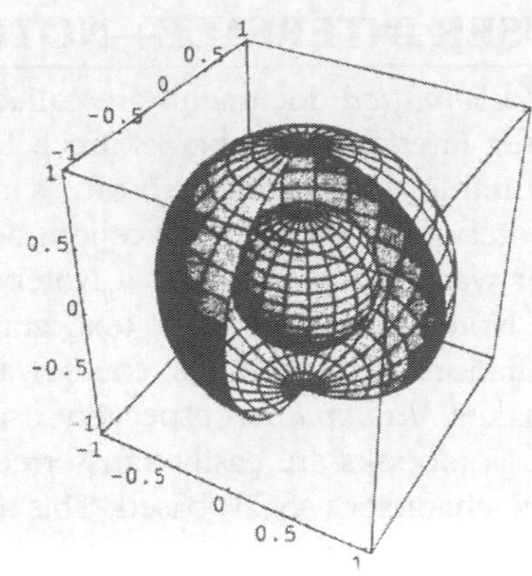

Animation and Arrays of Graphics

Visualization through animation is often useful. Any range of *Mathematica* graphics can be animated by the use of an appropriate menu command. When producing a set of graphics for a report or publication, say (output suppressed):

```
plots = Table[Plot[BesselJ[n,x],{x,0,10},
              PlotRange → {-0.4, 0.6}],{n,1,2}];
```

GraphicsArray, which displays a plot built up a number of subplots, is also very useful:

```
Show[GraphicsArray[plots]];
```

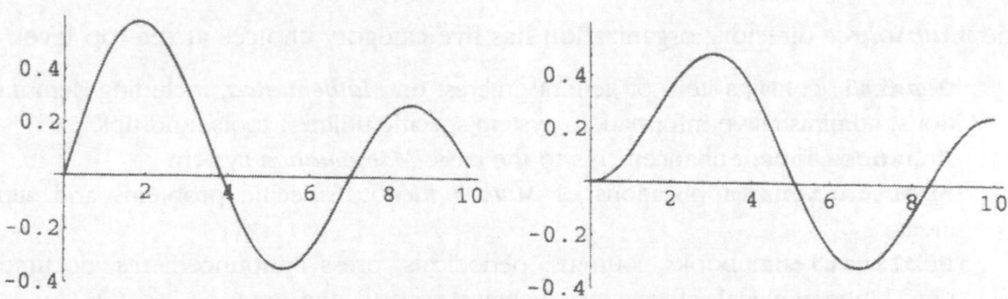

Other Graphics Formats

Mathematica graphics can be easily converted to a range of other formats including PostScript, bitmap, PICT, PICS, QuickTime, and Windows Metafile.

The package **Utilities `DXF`** converts *Mathematica* graphics to the DXF graphics interchange format, which can then be imported by AutoCAD, ray-tracers, and other three-dimensional rendering packages such as Ray Dream Designer. More generally, due to the high-level representation of *Mathematica* graphics, it is straightforward to convert to any number of graphics formats. See Sullivan (1991) for an example of the conversion to a format compatible with RenderMan and the **External Interface** section below for examples of the Live program.

40.6 USER INTERFACE—NOTEBOOKS

Mathematica documents are called Notebooks and these are the core of the graphical user interface. Notebooks are a form of interactive, electronic book. Since they are portable between Microsoft Windows, Macintosh, NEXTSTEP, and X Windows, Notebooks provide an excellent documentation environment. In fact, this entire chapter was itself prepared as a Notebook.

Notebooks can include text, graphics, animation, and sounds and are useful for calculations, presentations, courseware, and even books. It is the Notebook facility that makes *Mathematica* especially useful in education.

Notebooks are easily transferred between different operating systems because they are character (ASCII) based. This also simplifies electronic document transfer via ftp or electronic mail.

Although Notebooks are a powerful and friendly environment, conversion to other formats may be necessary. Notebooks can be saved as plain text, or RTF (rich text format) documents, which can then be directly imported by Microsoft Word (which was how this chapter was prepared). The program **`nb2tex`**, available from *MathSource*, automatically converts Notebooks to $\TeX$ format—the de facto standard for technical document preparation. Several books on *Mathematica* have been written as Notebooks and published after conversion to $\TeX$ format.

40.7 MATHSOURCE

MathSource is an online repository of *Mathematica* material including packages and Notebooks. Archives can be accessed using electronic mail, ftp, Gopher, and the World Wide Web.

Organization

The *MathSource* directory organization has five category choices at the top level:

`General`: contains item of general interest on *Mathematica*, including demonstrations, administrative information, system-specific utilities, tools, and tips.
`Enhancements`: enhancements to the basic *Mathematica* system.
`Applications`: applications of *Mathematica* to specific problems and subject areas.
`Publications`: books, journals, periodicals, press announcements, documentation, bibliographical information, technical reports, and notes.
`NumberedItems`: If you know a specific items number you can get it directly without moving through multiple directory levels.

Electronic Mail Access

For general information about *MathSource*, send an electronic mail message containing the line **`Help Intro`** to the *MathSource* server at **`mathsource@wri.com`**.

Electronic mail users can request self-extracting archives that results in simpler extraction and installation of packages and Notebooks.

Finding and requesting files is straightforward. For example, mailing a message to the *MathSource* server containing the line **`Find Elliptic`** will return a number of responses including

0203-421: Numerical Solution of an Elliptic PDE

This file can then be requested by sending a new message containing **Send 0203-421**.

ftp

MathSource is accessible via ftp at **mathsource.wri.com** (internet address **140.177.10.5**). Information on ftp access can be found in the file **/pub/README**.

MathSource Gopher

The *MathSource* Gopher accesses all material available via ftp and electronic mail. Additionally one can conduct a full-text search of the *MathSource* contents and link to other *Mathematica* FTP or Gopher sites around the world. To use this service, point your Gopher to **mathsource.wri.com, port 70** (Fig. 40.1).

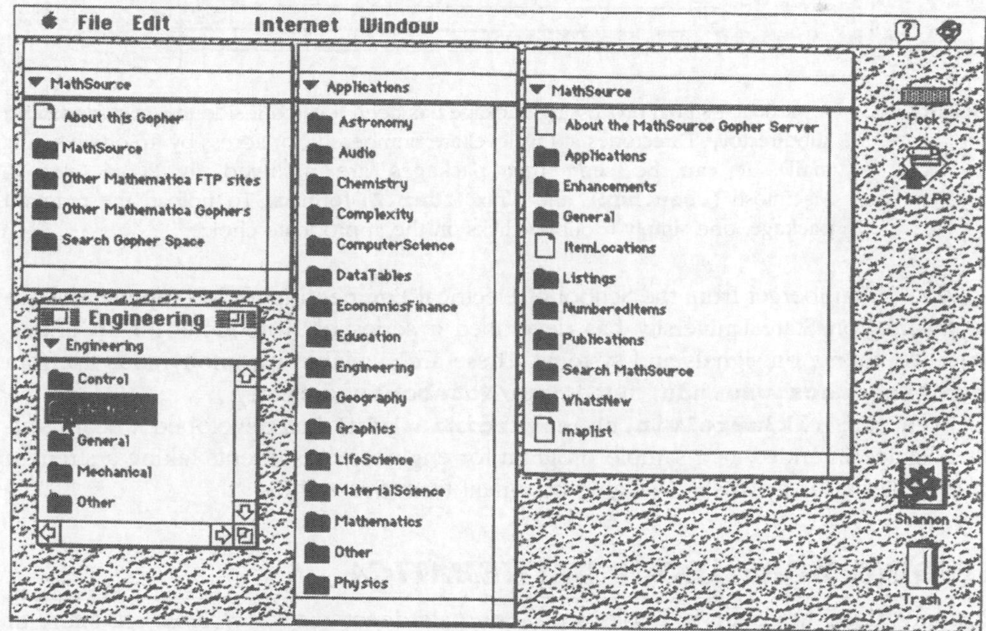

FIGURE 40.1 After making a gopher connection to **mathsource.wri.com**, the above hierarchy of windows is opened by (double)-clicking on **MathSource, Applications**, and **Engineering** in turn.

Some Relevant Packages

Brian Evans (Evans et al., 1990) from the Georgia Institute of Technology has developed a suite of Notebooks for Signal Processing that is available from *MathSource* (Fig. 40.2).

Malcolm Slaney from the Advanced Technology Group at Apple Computer has released Notebooks on cochlear modeling and digital filter design that might be of interest to signal processors (Slaney, 1990). These are available via anonymous ftp from **ftp.apple.com** in the **/pub/malcolm** directory.

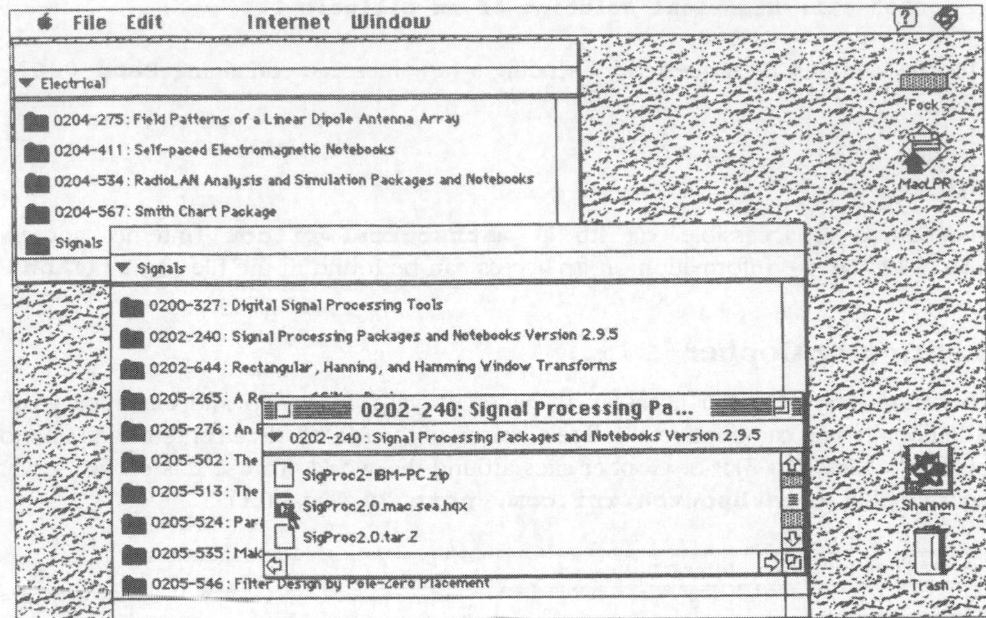

FIGURE 40.2 A particular signal processing package has been located inside the **Engineering** subdirectory. Directories at this level are numbered (for access by ftp or electronic mail). It can be seen that packages are archived in PC (**.zip**), Macintosh (**.sea.hqx**), and Unix (**.tar.Z**) formats. To upload the required package, one simply (double)-clicks on the appropriate choice.

Roberto Bamberger from the School of Electrical Engineering and Computer Science at Washington State University has developed a series of Notebooks for an undergraduate course on signals and systems. These are available via anonymous ftp from **yardbird.eecs.wsu.edu** in the **pub/Notebooks** directory.

Levent Kitis **(lk3a@kelvin.seas.virginia.edu)** has developed a beam statics package intended as a sample program for engineering students taking a strength of materials course. This is available by e-mail from the author.

40.8 FINDING MORE ABOUT *MATHEMATICA*

Presently there are over 50 books and 3 journals devoted to *Mathematica*. Many of these books include a disk or CD-ROM containing Notebooks in a range of electronic formats. An up to date list of *Mathematica* publications, packages, and Notebooks can be obtained from *MathSource*.

Electronic Newsgroups and Mailing Lists

Steve Christensen maintains the MathGroup mailing list. MathGroup is a forum devoted to *Mathematica* questions and answers, announcements of new packages and enhancements, and general discussion of *Mathematica*-related issues.

To subscribe, send email to **mathgroup-request@christensen.cybernetics.net**. All messages to MathGroup **(mathgroup@christensen.cybernetics.net)** are moderated.

The **sci.math.symbolic** newsgroup is a popular forum for discussion of all aspects of computer algebra software including general questions and answers, and comparisons between various computer algebra packages. The new **comp.soft-sys.math.mathematica** newsgroup focuses on *Mathematica*.

The *Mathematica Journal*

The *Mathematica Journal* commenced publication in 1990. Its mission is to provide a forum for the discussion of disparate quantitative sciences such as biology, economics, engineering, physics, and mathematics in a unified framework and to put relevant computational ideas and algorithms within reach of a much wider audience thus enabling greater cross-fertilization of ideas. It is hoped to achieve this by

- using the *Mathematica* language;
- including packages and Notebooks in its electronic supplement permitting readers to use and modify the algorithms presented with the articles;
- including the *MathReader* program with each supplement that can read Notebooks, show examples, and play animations and sounds.

Mathematica in Education

For the educator, the Notebook interface is the key to the usefulness of *Mathematica* as an authoring system. *Mathematica* is already popular for undergraduate courses in mathematics and physics (Brown et al., 1991; Crooke and Ratcliffe, 1991; Stroyan, 1992; Crandall, 1991; Vvedensky, 1992; Skeel and Keiper, 1993). In addition to these texts the *Mathematica in Education* journal is devoted entirely to educational issues and applications.

Mathematica World

Started in 1993, *Mathematica World* is an electronic journal published monthly and edited by Stephen Hunt **(smh@matilda.vut.edu.au)**. It consists of a number of linked packages and Notebooks distributed on Macintosh or DOS 3 1/2-inch disks. Sections include

- solutions to problems that have been posed over the previous month on MathGroup and internet network news groups;
- announcements of new packages and Notebooks;
- tutorials on Notebooks, front-end, kernel, and package functionality that extend the standard *Mathematica* documentation.

40.9 EXTERNAL INTERFACE—*MATHLINK*

MathLink (Wolfram Research, 1993) is a communication standard that allows programs to call or be called by *Mathematica*. *Mathematica* can interact with other programs exchanging data, numbers, strings, matrices, and graphics using *MathLink*: FORTRAN or C programs can be easily be connected to *Mathematica* using *MathLink*.

MathLink also works in the other direction: you can take an existing user interface and connect it to *Mathematica*—links to Excel, AVS (visualization software), HyperCard, Explorer (visual programming system), Transform, MATLAB, QuickTime, Live, and LabVIEW are already available.

Hierarchical Data Format

The Hierarchical Data Format (HDF) promulgated by the National Center for Supercomputing Applications in Urbana, Illinois is a portable and binary file format that can contain many different data types.

MathHDF (Janhunen and Stein, 1993) provides a *Mathematica*–HDF interface using *MathLink*. The present version of currently supports only the SDS (Scientific Data Set) data type.

MATLAB

The *MathLink* Symbolic Toolbox for MATLAB makes it possible to access symbolic and high-precision numerical capabilities from within MATLAB. MATLAB users can execute *Mathematica* programs or mathematical functions directly and matrices can be passed between the two programs.

The Symbolic Toolbox requires MATLAB 3.5 or 4.x and is available at no charge from *MathSource*.

HyperMath

HyperMath for the Macintosh interfaces HyperCard to *Mathematica*. The real-time nature of this application can rapidly generate a whole class of examples, which may provide insight into a particular problem (Fig. 40.3).

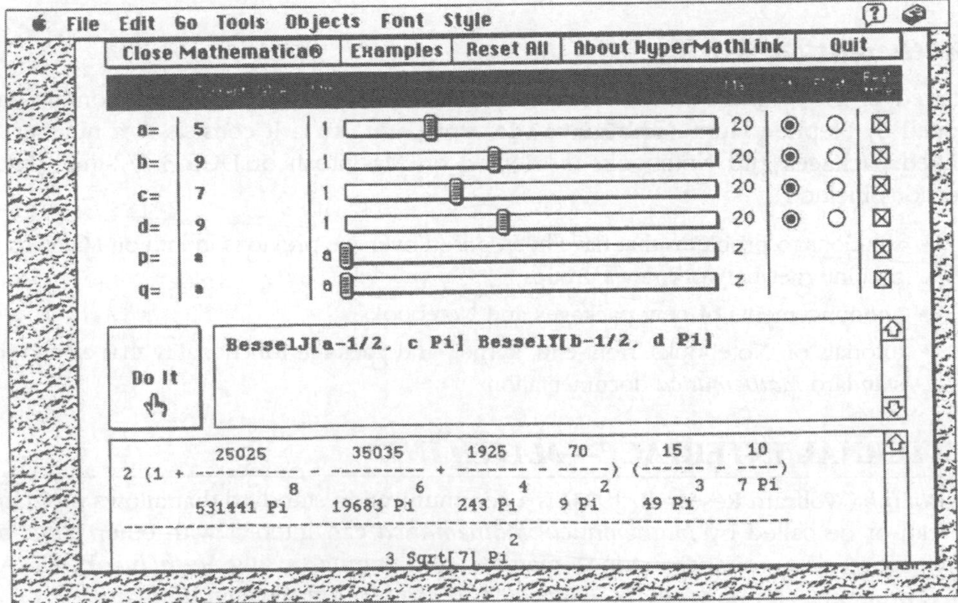

FIGURE 40.3 Computing **BesselJ[a - 1/2, c Pi] BesselY[b - 1/2, d Pi]** using HyperMath. The formula is entered into the middle box and the (real or integer) parameters **a, b, c, d, p, q**, can be changed by clicking and dragging on the appropriate slider. The answer appears in the output box.

Excel

MathLink for Microsoft Excel provides access to *Mathematica*'s numerical, symbolic, and programming capabilities from within Excel (Fig. 40.4). With *MathLink* for Excel you can

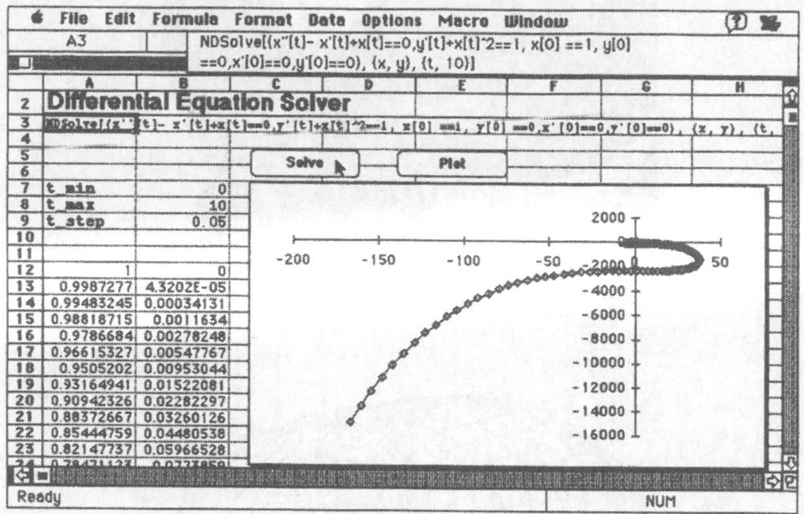

FIGURE 40.4 NDSOLVE.XLS demonstrates the numerical solution to a differential equation in Excel using *MathLink*. After entering the differential equation and boundary conditions, the user simply needs to press the **`Solve`** button. Values for the **`t_min, t_max,`** and **`t_step`** can be altered and pressing the **`Plot`** button will generate a set of points and a new plot.

- access *Mathematica*'s mathematical functions;
- use the *Mathematica* programming language to compute expressions that would be very complicated to do using Excel's macro language;
- send data sets from Excel to *Mathematica*, analyze them there, and return the results to the spreadsheet.

Note that you can connect Excel on a Macintosh or IBM compatible PC running Windows to a *Mathematica* kernel, which may be running on a completely different computer.

Live

Live, developed by True-D Software Limited, permits real-time 3D visualization of graphics. Live's capabilities integrate with, and complement those of, *Mathematica*. Three-dimensional *Mathematica* graphics can be interactively exported to Live using *MathLink*. You can do real-time animation, adjusting parameters such as rotation, zooming, and lighting in real time, convert to solid, wire frame, or scatter plot graphics, as well as alter surface color and surface properties, such as diffuse and specular reflectivity.

Figure 40.5 and 40.6 were generated by running Live on a PowerPC:

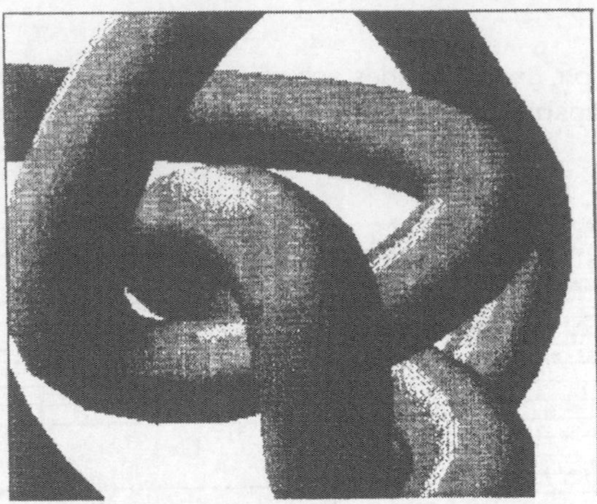

FIGURE 40.5 Frame from an animation of a knot.

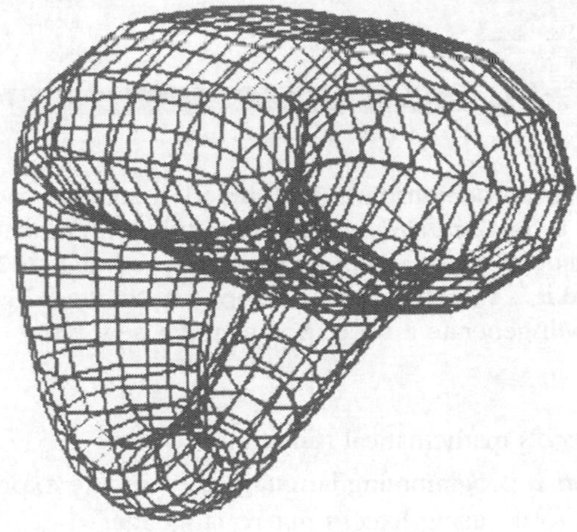

FIGURE 40.6 Wire-frame of a Klein bottle.

InterCall

InterCall is a *Mathematica* package that enables its user to interactively call external fortran or C code and external subroutine libraries (Fig. 40.7).

Subroutine libraries such as IMSL (IMSL, 1980) typically have many arguments. InterCall allows users to supply just a small number of arguments, with the rest being set to intelligent default values. InterCall includes a database of default settings for the routines in the IMSL and NAG libraries as well as for a number of public domain libraries such as LINPACK, MINPACK, ITPACK, and RKSUITE. All the routines in those, and other, libraries can be called in a simple manner—just as if they were normal *Mathematica* routines. Further information on InterCall and sample Notebooks are available from *MathSource*.

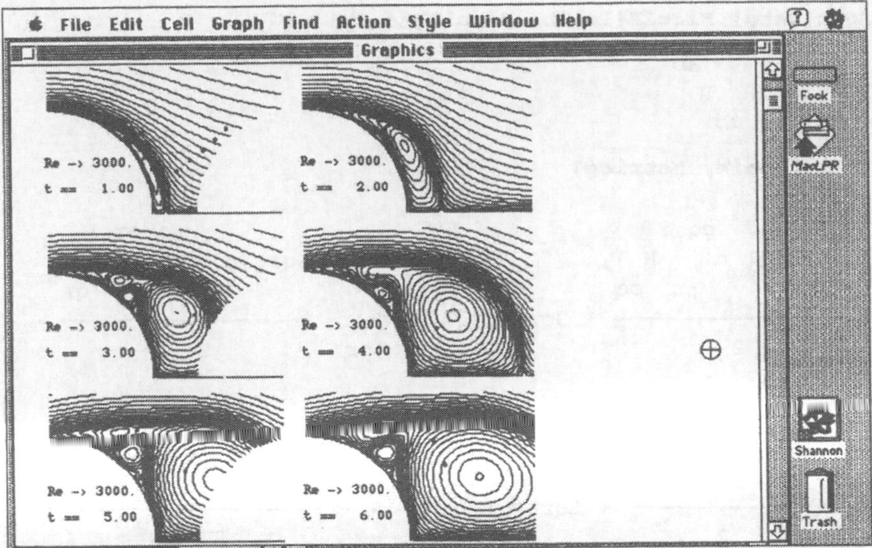

FIGURE 40.7 Unsteady flow (Re = 3000) around a cylinder obtained using InterCall. The motion was impulsively started at time $t = 0$. A wake forms downstream, and a small secondary vortex forms at about $t = 2.5$. Over a longer time-scale (about $t = 30$) the main wake becomes unstable, and a vortex is shed.

40.10 APPLICATION EXAMPLES

In this section a selection of application examples is presented. A wider range of applications can be found in the references.

Tensors

Tensor analysis has applications in a number of areas in science and engineering. Problems involving tensors often involve large numbers of terms and it makes sense to manipulate them using a computer algebra system.

MathTensor is a *Mathematica* package that provides functions and objects for both elementary and advanced tensor analysis (Christensen and Parker, 1990). MathTensor includes standard objects such as the Riemann and Ricci **(RicciR)** tensors along with functions including the covariant derivative, index commutation, raising and lowering of indices, and various differential forms operations.

As an example, after loading MathTensor using the command:

```
<<MathTensor`
```

one can compute variations with respect to the metric tensor of structures that are functions of the metric tensor, **Metricg**. For example,

```
Sqrt[Detg] RicciR[la,lb] RicciR[ua,ub]

                ab
Sqrt[g] R  R
          ab

Variation[%, Metricg]

              pq r                                          p qr
-(Sqrt[g] h       R )                            Sqrt[g] h     R
              ;r   pq                   p rq                  p;      qr
------------------------ + Sqrt[g] h     R  -  --------------------- -
          2                             q;   pr              2

             r  pq                                          p       qr
Sqrt[g] h        R                              Sqrt[g] h        R
          pq;r                              q  pr              p ;qr
---------------------- + Sqrt[g] h        R -  ---------------------- +
          2                              pq;r              2

          pq      rs
Sqrt[g] g   R  R   h
             rs    pq                        qr p          pr    q
------------------------ - Sqrt[g] R   R   h  - Sqrt[g] R   R h
          2                          pq     r             pq     r
```

Circuit Analysis

The principal advantage of a symbolic circuit analysis tool like Nodal (Riddle, 1990) over packages like SPICE is the integration of symbolic and numerical computations.

As an example, after loading the package:

```
<<Nodal2 `Nodal2`
```

here is the nodal description for a Sallen–Key active filter:

```
SKFilter = NodalNetwork[
    VoltageSource[{1,0},1],
    Capacitor[{1,2}, C],
    Capacitor[{2,3}, C],
    OpAmp[{3,4,4}],
    Resistor[{3,0}, 2 R],
    Capacitor[{4,5}, C],
    Resistor[{2,4}, R/2],
    Resistor[{5,0}, R]
];
```

This circuit can be analyzed symbolically and its transfer function reads

```
NodalAnalyze[ SKFilter, Result → V5, Frequency → Laplace]
```

```
         3  3
            C  s
------------------------------------
                     2  2
  -3     2 C s     2 C   s       3 3
R  +  ------ +  ------- + C s
          2          R
         R
```

which is recognizable as a Butterworth high pass filter. Finally, here is a **BodePlot** of this filter:

```
bd = NodalAnalyze[ SKFilter /. {C → 1, R → 1},
   Result → V5,
   Step → Frequency[10^Range[-2, 3.5, 0.25]]];
BodePlot[bd];
```

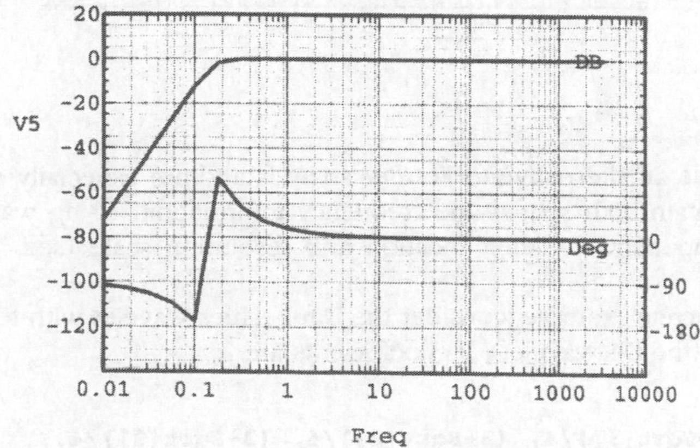

A Touch-Tone Phone

Mathematica includes the capability to visualize and play mathematically generated sounds. To aid in data analysis, lists of data can be played instead of plotted. Furthermore, sounds can be imported into Notebooks and digitized for further analysis. A neat example of the sound generation capabilities is to produce a telephone tone dialer (Cook, 1991):

A superposition of a pair of pure frequencies is easily generated:

```
PlayTouchTone[{f1_,  f2_}]  :=  Play[  Sin[2  Pi  f1  t]  +
  Sin[2 Pi f2 t], {t, 0, 0.1} ]
```

The list of touch-tone frequency pairs is:

```
TouchToneArray = Flatten[ Outer[List,{697, 770, 852, 941},
                                     {1209, 1336, 1477}], 1 ];
SetAttributes[TouchTone, Listable]
TouchTone[n_Integer /; 1<=n<=9]:=
     PlayTouchTone[TouchToneArray[[n]]]
TouchTone[0] := PlayTouchTone[TouchToneArray[[11]]]
TouchTone[n_String] := TouchTone[
     ToExpression[FromCharacterCode /@ ToCharacterCode[n]]]
```

For example, in Australia local directory assistance is reached by dialing

```
TouchTone["013"];
```

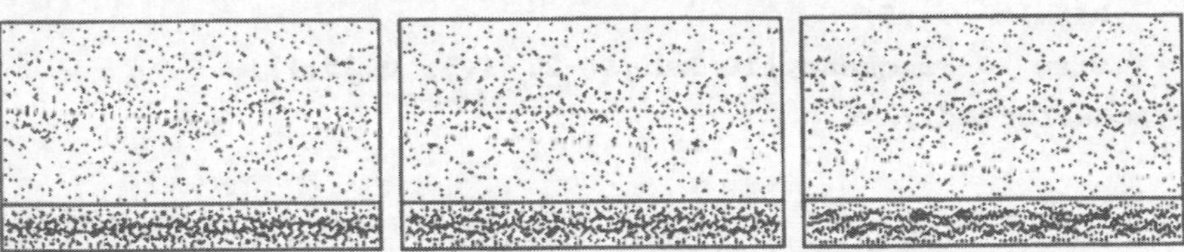

Wavelets

Wavelets provide an alternative to Fourier analysis and are especially appropriate for signals that contain both spatial and frequency information. Using wavelets one can sort signal components by their location and resolution scale (see, e.g., Newland, 1993).

To give a concrete example, consider the Daubechies wavelet with four nonzero filter coefficients (the D4 wavelet). Its coefficients are

```
c = {(1+Sqrt[3])/4, (3+Sqrt[3])/4, (3-Sqrt[3])/4,
     (1-Sqrt[3])/4};
```

The *scaling function*, $\phi(x)$, satisfies the dilation equation:

$$\phi(x) = c_0\phi(2x) + c_1\phi(2x - 1) + c_2\phi(2x - 2) + c_3\phi(2x - 3)$$

which can be implemented, in general, as

```
n := Length[c]
phi[x_] := phi[x] = c . Table[phi[2x-i+1],{i, n}] // Chop
```

along with the boundary condition:

$$\phi(x) = 0, \qquad x \leq 0 \qquad \text{or} \qquad x \geq n - 1$$

```
phi[x_ /; x <=0 || x >= n - 1] := 0
```

To start off the iterative procedure, consider the dilation equation at **x = 1** and **x = 2**. This yields a matrix eigenvalue equation:

$$\begin{pmatrix} c_1 & c_0 \\ c_3 & c_2 \end{pmatrix} \begin{pmatrix} \phi(1) \\ \phi(2) \end{pmatrix} = \begin{pmatrix} \phi(1) \\ \phi(2) \end{pmatrix}$$

Although we could use exact arithmetic, we proceed numerically:

```
c = N[c, 25]
{0.6830127018922193233818616,   1.183012701892219323381861 6,
  0.3169872981077806766181384, -0.1830127018922193233818616}
```

The matrix has the eigensystem:

```
{evals, evecs} =
  Eigensystem[{{c[[2]],c[[1]]},{c[[4]],c[[3]]}}]
{{1., 0.5}, {{0.9659258262890682867497 43,
  -0.2588190451025207623488988},
  {-0.7071067811865475244008 44, 0.7071067811865475244008 44}}}
```

Selecting the eigenvector corresponding to eigenvalue **1**:

```
{phi[1],phi[2]} = evecs[[1]]
{0.9659258262890682867497 43, -0.2588190451025207623488988}
```

The dilation equation can be used to compute the value of $\phi(x)$ at $x = k\,2^{-j}$ for all integers j and k (the *dyadic points*). We simplify these computations by rationalization of real arguments:

```
phi[x_Real] := phi[Rationalize[x]]
```

For example,

```
phi[1/2]
0.6597396084411710244750 83
```

Here is a plot of the scaling function (sampled only at the *dyadic points* through specification of the number of **PlotPoints** and the **PlotDivision**):

```
Plot[phi[x],
{x,0,3},
PlotPoints → 3 128 + 1,
PlotDivision → 1];
```

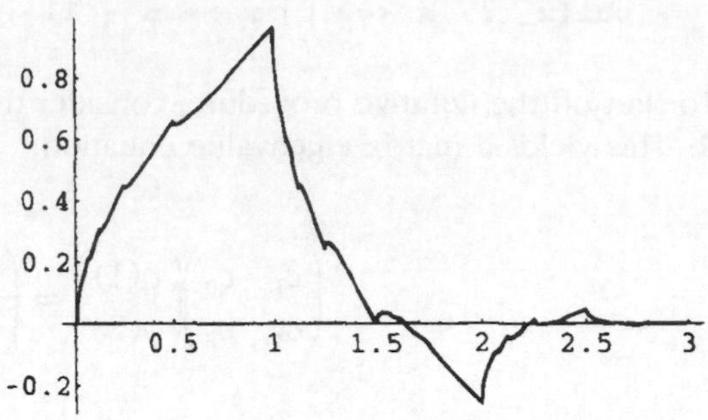

The D4 wavelet

$$W(x) = c_0\phi(2x-3) - c_1\phi(2x-2) + c_2\phi(2x-1) - c_3\phi(2x)$$

implemented generally as

```
w[x_] := w[x] = -Sum[c[[k]] (-1)^k phi[2x+k-n],{k,n}] // Chop
```

can be plotted directly:

```
Plot[w[x],
{x,0,3},
PlotPoints → 3 128 + 1,
PlotDivision → 1,
PlotRange → All];
```

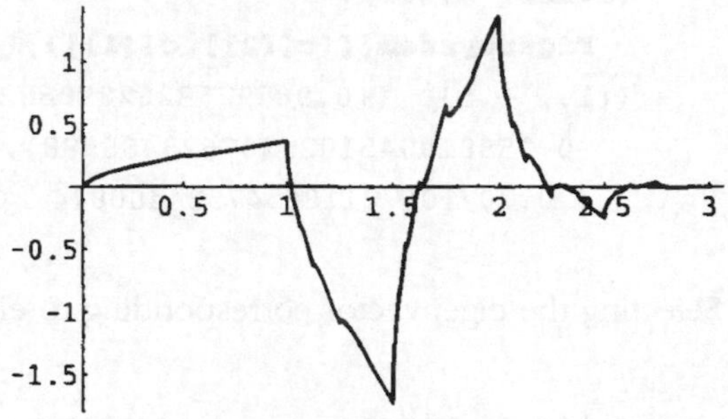

It is straightforward to generalize this analysis to wavelets with more than four nonzero filter coefficients.

Variational Methods

The Rayleigh–Ritz variational principle is a powerful technique for finding approximate solutions to self-adjoint differential operators. For example, the Laplace (Poisson), Helmholtz, and Schrödinger operators are all self-adjoint and applications range from quantum physics, wave propagation, stress or vibrational analyses, to evaluating capacitances or transmission line impedances, and antenna problems. Note also that discretization directly from the variational functional often leads to an effective numerical technique for solving these equations.

The package

```
<<Calculus 'VariationalMethods'
```

contains a number of functions useful for variational problems, e.g.,

```
?VariationalBound
VariationalBound[{f,g},u[x],{x,xmin,xmax},ut,{a,amin,amax},
   {b,bmin,bmax}, . . . ] determines the parameters
   {a,b,c . . . } of a trial function ut[x] that extremize Integrate
   [f,{x,xmin,xmax}]/Integrate[g,{x,xmin,xmax}]. It returns
   the extremal value of the functional and the optimal
   values of the parameters. The default range for a parame-
   ter is {-Infinity,Infinity}. VariationalBound[{f,g},u[x,y,
   . . . ],{{x,xmin,xmax},{y,ymin,ymax}, . . . },
   ut,{a,amin,amax},{b,bmin,bmax}, . . . ] does the same for
   more than one independent variable. If the argument g is
   absent the functional Integrate[f,{x,xmin,xmax}] is
   extremized.
```

The problem of the torsion of a rod of square cross section of side length a involves solving

$$\Delta u(x, y) = -1, \qquad u(\pm a, y) = u(x, \pm a) = 0$$

where u vanishes on the boundary and Δ is the two-dimensional Laplacian. The general Laplacian operator:

$$\Delta \equiv \nabla^2 = \sum_{i=1}^{n} \frac{\partial^2}{\partial x_i^2}$$

can be implemented as

```
Laplacian[exp_,var_] := Plus @@ (D[exp, {#,2}]& /@ var)
```

Note that the variational derivative of

```
f = 1/2 u[x,y] Laplacian[u[x,y],{x,y}] + u[x,y]

                      (0,2)            (2,0)
            u[x, y] (u      [x, y] + u     [x, y])
u[x, y] + -----------------------------------------
                              2
```

satisfies the differential equation

```
de = VariationalD[f, u[x,y], {x,y}] == 0

      (0,2)           (2,0)
1 + u      [x, y] + u      [x, y] == 0
```

After constructing a trial solution with free parameters satisfying the boundary conditions, say

```
ut[x_, y_] = (x^2-a^2)(y^2-a^2)(a1 + a2 (x^2+y^2));
```

one can extemize the functional and find the optimal parameter values:

```
VariationalBound[f, u[x,y],{{x,-a,a},{y,-a,a}},
  ut[x,y], {a1},{a2}]
             4          0.292193          0.0592283
{0.280786 a, {a1 -> ---------, a2 -> ----------}}
                          2                4
                         a                a
```

A more general trial solution, say

```
ut[x_, y_] = (x^2-a^2)(y^2-a^2)(a1 + a2 (x^2+y^2) +
  a3 (x^2 + y^2)^2 + a4 (x^2 + y^2)^3);
```

leads to an increased value for the functional:

```
VariationalBound[f, u[x,y], {{x,-a,a},{y,-a,a}},
  ut[x,y], {a1},{a2},{a3}, {a4}]
             4          0.294169          0.0551663 -        0.0262821
{0.281012 a, {a1 -> -------, a2 -> --------, a3 -> --------,
                          2                4                6
                         a                a                a

      0.0328708
a4 -> --------- }}
           8
          a
```

One can visualize the solution to see how well the equation is satisfied:

```
First[de] /. u -> ut /. {x -> x a, y -> y a} /. Last[%];
Plot3D[%, {x,-1,1},{y,-1,1}];
```

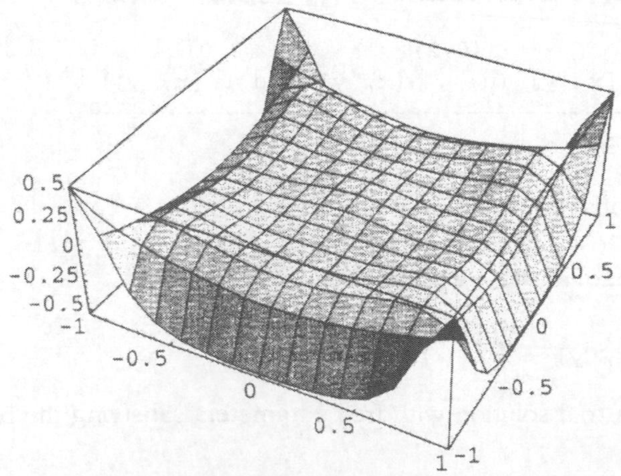

An exact solution would be identically zero over the whole domain.

Combinatorics

The package

```
<<DiscreteMath 'Combinatorica'
```

comprises over 230 functions in combinatorics and graph theory (Skiena, 1990). It includes functions for constructing graphs and other combinatorial objects such as permutations, subsets, partitions, and Young tableaux, computing invariants of these objects, and displaying them.

For example, the following random graph is expected to have half the number of edges of a complete graph:

```
ShowGraph[ RandomGraph[20, 0.5] ];
```

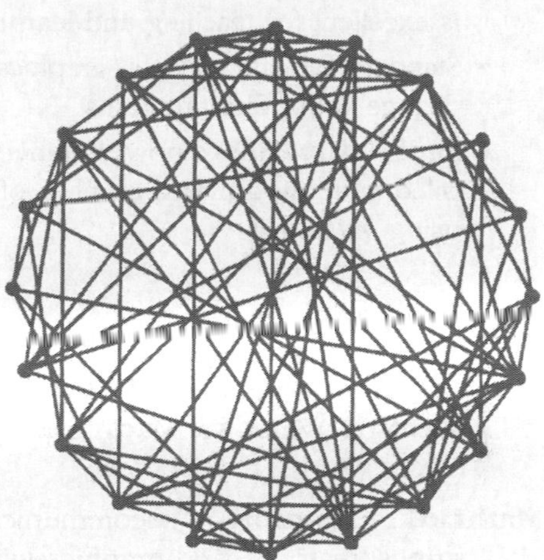

40.11 ELECTRICAL ENGINEERING PACK

The Electrical Engineering Pack available from Wolfram Research includes a number of specialized functions for specific electrical engineering in addition to solutions to problems in circuit analysis, transmission line theory, and antenna analysis. Specific areas covered include the following:

Circuit Analysis	Transmission Line Theory
Mesh circuit analysis	Symbolic transmission line parameters
Nodal circuit analysis	Analysis of reflections using animations
Laplace transforms	Suppression of line ringing by lossy lines
Circuit sensitivity analysis	Matrix techniques for microwave
Ideal diode and bipolar transistor models	Transmission line analysis
Chaotic behavior in a nonlinear circuit	Dispersion in uniform microstrip lines
Passive component noise	Smith chart design tools
Amplifier noise figure	Other Tools
Harmonic distortion	Bode plots
Crossover distortion	Nyquist plots
Clipping distortion	Nichols plots
Resistor selection	Root locus plots
Y to S parameter conversion	Reflections coefficient conversion
Antenna Analysis	
Antenna field patterns	
Moment methods for antenna analysis	

40.12 CONCLUSIONS

Mathematica includes a number of features that make it especially suitable for scientists and engineers:

- Calculator-like ease of use for simple operations. Interactive high-level computations do not require compilation and this makes for faster *problem solving* (as opposed to code execution time) than Fortran, C, or Pascal.

- Mathematical operations, including a very wide range of standard numerical functions and integrals, are built in.
- Symbolic computation gives exact rather than numerical results.
- The Notebook interface unifies numerics, graphics, and symbolic computations and is excellent for teaching and learning.
- Standard formats for data, graphics, and sound are supported permitting easy interchange with other programs.
- *MathLink* provides a powerful mechanism for linking to external programs and compiled code. A growing number of applications can be connected to *Mathematica* using *MathLink*.

DEFINING TERMS

Kernel: The *Kernel* is the *Mathematica* computational engine. It runs under DOS, Microsoft Windows, Macintosh, NEXTSTEP, VMS, and on a wide range of Unix machines.

MathLink: *MathLink* is a communication standard for exchanging data, numbers, strings, matrices, and graphics with external programs.

MathSource: *MathSource* is an online repository of *Mathematica* material including packages and Notebooks accessible using electronic mail, ftp, Gopher, or the World Wide Web.

Notebook: The *Notebook* is the *Mathematica* user-interface and documentation environment. Notebooks are portable between Microsoft Windows, Macintosh, NEXTSTEP, and X Windows.

REFERENCES

M. Abramowitz and I. Stegun, *Handbook of Mathematical Functions*. Dover, New York, 1972.

P. S. Aptaker, Ordinary differential equation models for eddy-currents using Fourier techniques, *Int. Conf. Comp. Electromagnet*. 26–29, 1991.

P. Boyland, *Guide to Standard Mathematica Packages*. Wolfram Research, Champaign, IL, 1993.

D. Brown, H. Porta, and J. Uhl, *Calculus & Mathematica*. Addison-Wesley, Redwood City, CA, 1991.

S. Christensen and L. Parker, MathTensor: A system for performing tensor analysis by computer. *Math. J.* 1(1):51–61, 1990.

D. K. Choi and S. Nomura, Application of symbolic computation to two-dimensional elasticity. *Comput. Struct.* 43:645–649, 1992.

M. Cook, Sound in *Mathematica*. In *1991 Mathematica Conference Tutorial Notes*. Wolfram Research, Champaign, IL, 1991.

R. Crandall, *Mathematica for the Sciences*. Addison-Wesley, Redwood City, CA, 1991.

P. Crooke and J. Ratcliffe, *A Guidebook to Calculus and Mathematica*. Wadsworth, 1991.

C. F. du Toit, Computation of integer-order Bessel functions of the first and second kind with complex arguments. *IEEE Antennas Propagation Mag.* 35:19–25, 1993.

B. L. Evans, J. H. McClellan, and W. B. McClure, Symbolic Z-transforms using DSP knowledge bases. *Int. Conf. Acoust. Speech Signal Process.* 3:1775–1778, 1990.

A. Gray, *Modern Differential Geometry of Curves and Surfaces*. CRC Press, Boca Raton, FL, 1993.

M. J. Hounslow and E. J. W. Wynn, Modelling particulate processes: full solutions and short cuts. *Comput. Chem. Eng.* 16:411–420, 1992.

IMSL, *IMSL Library Reference Manual*. IMSL Inc., Houston, TX, 1980.

N. I. Ioakimidis, Application of *Mathematica* to the direct solution of torsion problems by the energy method. *Comput. Struct.* 43:803–807, 1992.

P. Janhunen and D. Stein, MathHDF: *MathLink*-based distributed visualization between *Mathematica* and HDF files. *Comput. Phys.* 7(3):290–294, 1993.

R. H. Lance, Potential of symbolic computation in engineering education: The opportunity and the challenge. *Symbol. Comp. Their Impact Mechan.* 205:285–299, 1990.

M. C. Leu, Z. Ji, and Y. S. Wang, Studying robot kinematics and dynamics with the aid of *Mathematica*. *Int. J. Mech. Eng. Ed.* 19:213–228, 1991.

D. E. Newland, *Introduction to Random Vibrations, Spectral and Wavelet Analysis*. Longman, Essex, 1993:295–370.

A. K. Noor and C. M. Andersen, Hybrid analytical technique for the nonlinear analysis of curved beams. *Comput. Struct.* 43:823–830, 1992.

A. Riddle, A nodal circuit analysis program. *Mathematica J.* 1:62–68, 1990.

S. H. Rogers and A. B. Badiru, A fuzzy set theoretic framework for knowledge-based simulation. *Comput. Ind. Eng.* 25:119–122, 1993.

W. T. Shaw and T. Tigg, *Applied Mathematica, Getting Started, Getting it Done*. Addison-Wesley, Redwood City, CA, 1993.

R. Skeel and J. Keiper, *Elementary Numerical Computing with Mathematica*. McGraw-Hill, New York, 1993.

S. Skiena, *Implementing Discrete Mathematics: Combinatorics and Graph Theory with Mathematica*. Addison-Wesley, Redwood City, CA, 1990.

M. Slaney, Interactive signal processing documents. *IEEE ASSP Mag.* 7:8–20, 1990.

G. Sobelman, Computer algebra and fast algorithms. *Proc. Int. Conf. Acoust. Speech Signal Process.* 4:89–92, 1992.

K. Stroyan, *Calculus Using Mathematica* Academic Press, Sand Diego, 1992.

J. Sullivan, Generating and rendering four-dimensional polytopes. *Mathematica J.* 1(3):76–85, 1991.

H. Treat, Using *Mathematica* in support of LabVIEW: Power in the laboratory. *Northcon. Conf. Rec.* 353–8, 1990.

D. Vvedensky, *Partial Differential Equations with Mathematica*. Addison-Wesley, Reading, MA, 1992.

S. Wolfram, *Mathematica: A System for Doing Mathematics by Computer*. Addison-Wesley, Redwood City, CA, 1991.

Wolfram Research, *MathLink Reference Guide*. Wolfram Research, Champaign, IL, 1993.

He Xiaoyi, D. N. Ku, and J. E. Moore, Simple calculation of the velocity profiles for pulsatile flow in a blood vessel using *Mathematica*. *Ann. Biomed. Eng.* 21:45–49, 1993.

AVAILABILITY

Mathematica, *MathLink*, and *MathSource* are registered trademarks of Wolfram Research, Inc. *MathLink* is shipped as a C language library and accompanies *Mathematica*.

Mathematica is available from Wolfram Research, Inc., Champaign, Illinois, USA. Wolfram Research can be contacted on **+1 217 398 0700**.

General information is available from **info@wri.com** and **info-euro@wri.com** (Europe). For technical support contact **support@wri.com** or **support-euro@wri.com** (Europe).

TRADEMARKS

Unix is a registered trademark of AT&T.

X Window System is a trademark of the Massachusetts Institute of Technology.

Macintosh is a registered trademark of Apple Computer, Inc.

NeXT is a trademark of NeXT Computer, Inc.

Microsoft is a registered trademark and Windows is a trademark of Microsoft Corporation.

PostScript is a registered trademark of Adobe Systems, Inc.

AutoCAD and DXF are trademarks of Autodesk, Inc.

Nodal is a trademark of Macallan Consulting.

MathTensor is a trademark of MathSolutions, Inc.

Live is a trademark of True-D Software Limited (UK).

All other product names mentioned are trademarks of their producers.

41

MATLAB

Rudra Pratap
Cornell University

41.1 INTRODUCTION

What Is MATLAB?

MATLAB® is a software package for high performance numeric computation and visualization. It provides an interactive environment with hundreds of built-in functions for technical computation, graphics, and animation. Best of all, it also provides easy extensibility with its own high level programming language. The name MATLAB stands for MATrix LABoratory.

Figure 41.1 shows the main features and capabilities of MATLAB. Its built-in (those included in the basic MATLAB package) functions provide excellent tools for linear algebra computations, data analysis, signal processing, optimization, numerical solution of ODEs, quadrature, and many other scientific computations. There are numerous functions for 2-D and 3-D graphics as well as for animation. And for those, who cannot do without their Fortran or C codes, MATLAB even provides an external interface to run those programs from within MATLAB. Users, however, are not limited to the built-in functions; they can write their own functions in the MATLAB language. Once written, these functions behave just like the built-in functions. MATLAB's language is one of the easiest languages in which to program. There are also several optional "Toolboxes" available from the developers of MATLAB. These Toolboxes are collections of functions written for special applications such as symbolic computation, image processing, statistics, control system design, and neural networks. Some of these Toolboxes are introduced in this volume separately.

The basic building block of MATLAB is a matrix. The only data type, and that too you never need to declare, is a complex matrix. Vectors, scalars, real matrices, and integer matrices are all *automatically* handled as special cases of the basic data type. In addition, you almost never have to declare the dimensions of a matrix. MATLAB simply loves matrices and matrix operations. The built-in functions are optimized for vector operations, consequently, *vectorized* commands or codes run much faster in MATLAB.

0-8493-2530-7/96/$0.00+$.50

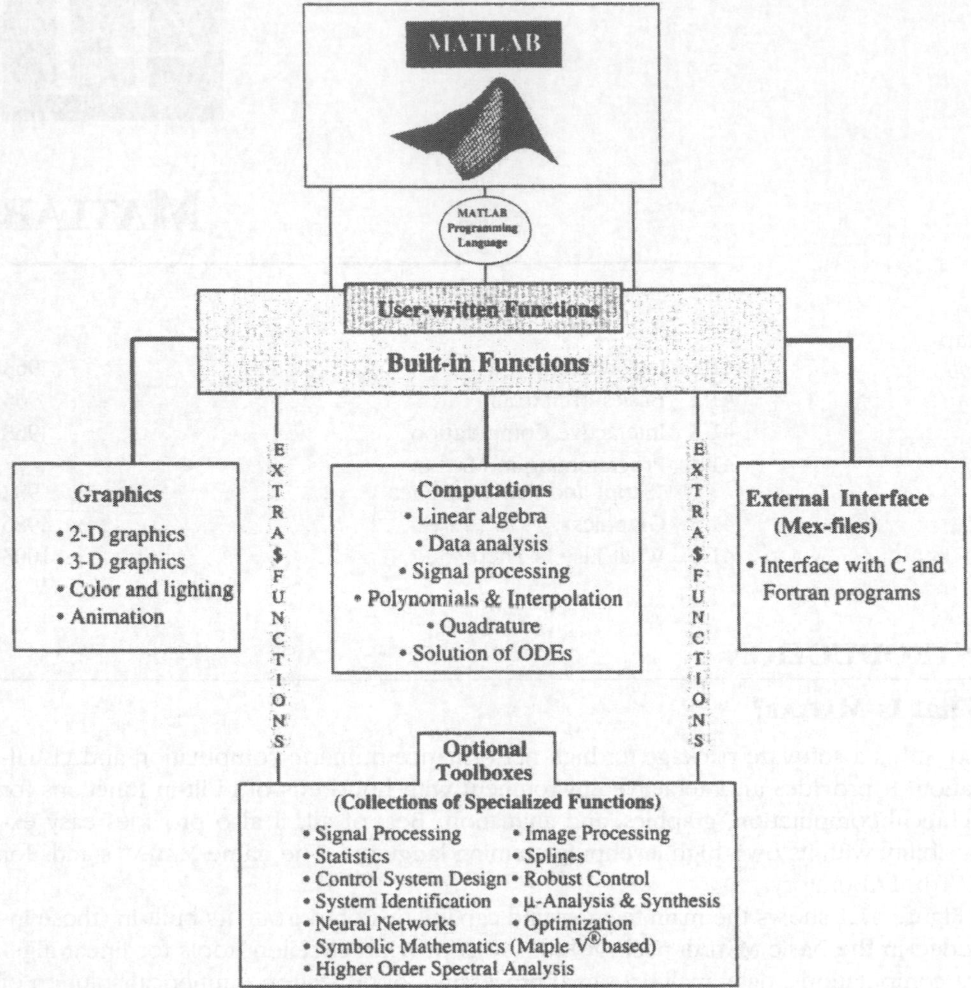

FIGURE 41.1 A schematic diagram of MATLAB's main features.

Will MATLAB Run on My Computer?

The most likely answer is yes because MATLAB supports almost every (useful) computational platform. Apart from PCs and Macintosh computers, appropriate versions are available for Sun SPARC stations, HP 9000 Series 700, IBM RS/6000, Silicon Graphics IRIS Series 4D, DEC RISC, DEC Alpha workstations, VAX computers, and even the CRAY Supercomputers.

Where Do I Get MATLAB?

MATLAB is a product of The MathWorks, Incorporated. Contact the company for product information and ordering at the following address: The MathWorks, Inc., 24 Prime Park Way, Natick, MA 01760. Phone: (508) 653-1415, Fax: (508) 653-2997, e-mail: info@mathworks.com.

How to Use This Article?

This article is intended to serve as an introduction to MATLAB. All features are discussed through examples. Following standard convention, all actual MATLAB commands or instructions are shown in `typed face`, place holders for variables or names in a command are shown in *italics*, and menu options and key names are shown in **bold face**. Actual examples carried out in MATLAB are shown in gray shaded boxes. Explanatory notes are added to the gray boxes within small white rectangles. If you do not have much patience or are not willing to read through the text, you are encouraged to just go through these boxed examples, preferably trying them out in MATLAB simultaneously. Most of the examples are designed such that you can (more or less) follow them without reading the entire text. All examples are system independent. We do, however, mention where appropriate, the differences in implementation of certain MATLAB features on different popular platforms.

41.2 BASICS OF MATLAB

Here we discuss some basic features and commands. To begin with let us look at the general structure of the MATLAB environment.

MATLAB Windows

On all UNIX systems, Macs, and PCs, MATLAB works through three basic windows that are shown in Figure 41.2 and discussed below.

1. **Command window:** This is the main window. It is characterized by the MATLAB prompt '>>'. When you launch the application program this is the window MATLAB puts you in. All commands, including those for running user-written programs, are typed in this window at the MATLAB prompt.
2. **Graphics window:** The output of all graphics commands typed in the command window are flushed to the graphics or *figure* window, a separate window with (default) black background color. The user can create as many figure windows as the system memory will allow.
3. **Edit window:** This is where you write, edit, create, and save your own programs in files called "*M-files*." You may use any text editor of your choice to carry out these tasks. On some systems such as Macs, MATLAB provides its own built-in editor. On other systems you can invoke the edit window by typing any standard file editing command that you normally use on your system. The command is typed at the MATLAB prompt following the special character `'!'`. The exclamation character prompts MATLAB to return the control temporarily to the local operating system, which executes the command following the `"!"` character. After the editing is completed, the control is returned to MATLAB. For example, on UNIX systems: typing `!vi myprogram.m` at the MATLAB prompt (and hitting the **return** key at the end) invokes the `vi` editor on the file "myprogram.m." Typing `!emacs myprogram.m` invokes the emacs editor instead on the same file.

Some Basic Features

- **On-line Help:** MATLAB provides on-line help for all built-in functions and its programming language constructs. See Using Built-in Functions and `help` for a description and use of the help facility.

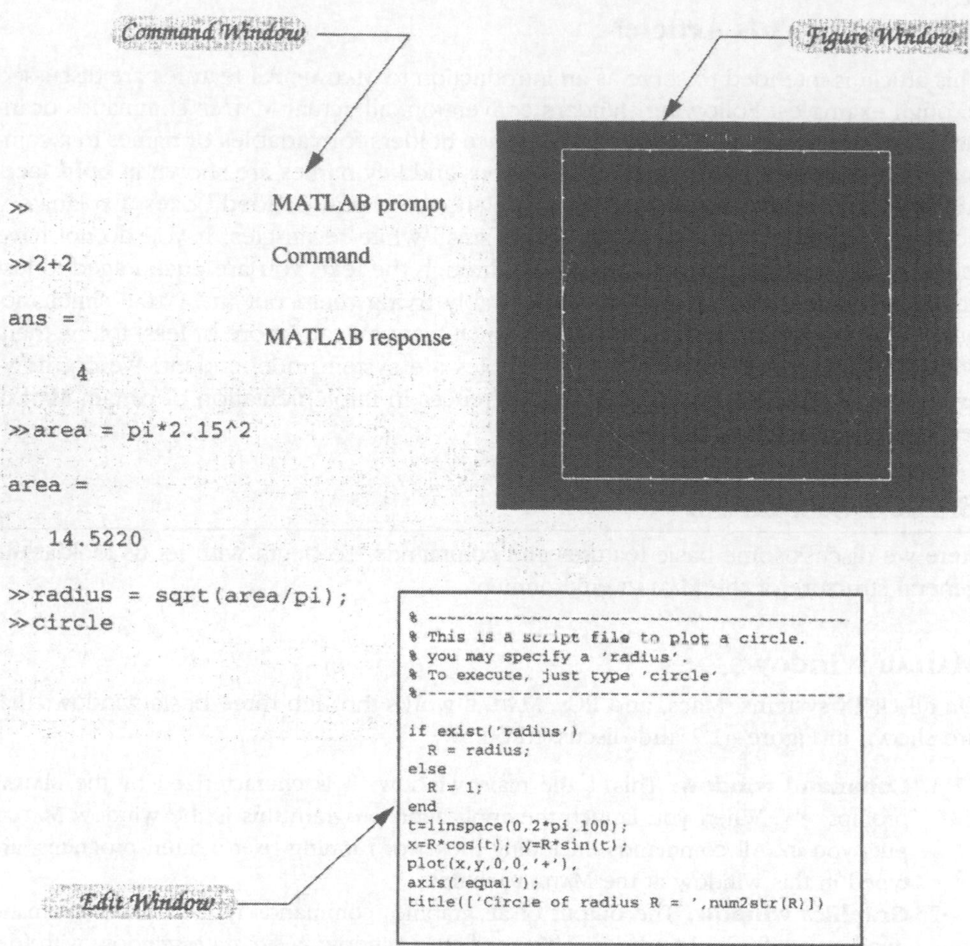

Command Window
Figure Window
>>
MATLAB prompt
>>2+2
Command
ans =
MATLAB response
4
>>area = pi*2.15^2
area =
14.5220
>>radius = sqrt(area/pi);
>>circle
% --
% This is a script file to plot a circle.
% you may specify a 'radius'.
% To execute, just type 'circle'
%---
if exist('radius')
R = radius;
else
R = 1;
end
t=linspace(0,2*pi,100);
x=R*cos(t); y=R*sin(t);
plot(x,y,0,0,'+'),
axis('equal');
title(['Circle of radius R = ',num2str(R)])
Edit Window

FIGURE 41.2 MATLAB environment consists of a command window, a figure window, and a platform-dependent edit window.

- **EXPO:** The MATLAB EXPO is a demonstration program that shows many features of MATLAB, including a graphical user interface totally implemented in MATLAB. It includes a number of examples worth browsing through. Type **`expo`** at the MATLAB prompt to invoke the EXPO program and follow the instructions on the screen.
- **Data Type:** There is only one data type in MATLAB and that is *complex matrix*. There is no need of any special declaration for real numbers. When a real number is entered as the value of a variable, MATLAB automatically sets the imaginary part to zero.
- **Dimensioning:** No dimension statements are required for vectors or arrays in MATLAB. Dimensioning is automatic. Dimensions of an existing matrix or a vector are obtained with **`size`** and **`length`** (for vectors only) commands.
- **Case Sensitivity:** MATLAB is case sensitive, that is, it differentiates between the lower and the upper case letters. Thus **`a`** and **`A`** are two different variables in MATLAB. Most of the MATLAB commands and built-in function calls are typed in lower case letters. It is possible, though not usually advisable, to change this sensitivity using **`casesen`** command.

- **Output Format:** While all computations inside MATLAB are done using double precision, the appearance of floating point numbers on the screen is controlled by the output format in use. There are 10 different screen output formats. The following table shows the printed value of 10π in 7 different formats.

```
format short        31.4159
format short e      3.1416e+01
format long         31.41592653589793
format long e       3.141592653589793e+01
format hex          403f6a7a2955385e
format rat          3550/113
format bank         31.42
```

The additional formats—**format compact** and **format loose**—control the spacing above and below the displayed numbers, and **format +** displays **+, -**, and blank for positive, negative, and zero numbers, respectively. The default is **format short**. Any format is invoked by typing **format** *type* on the command line.

File Types

There are three different types of files with which you will mostly work:

M-files: These are text files with ".m" extension to the file name. There are two types of these files: *script files* and *function files* (see Sections on Script Files and Function Files). Most of the programs that you write using MATLAB language are saved as m-files. All built-in functions in MATLAB are m-files, most of which reside on your computer in precompiled format. Some of the functions are provided with their source code in readable m-files, which can be copied and modified.

MAT-files: These are binary data files distinguished by the ".mat" extension to the file name. Mat-files are created by MATLAB when you save data with the **save** command. The data are written in a special format for MATLAB readability only. Mat-files may be loaded into MATLAB with the **load** command (see Saving and Loading Data for details).

MEX-files: MEX-files (filename with extension ".mex") are MATLAB callable Fortran and C programs. Use of these files requires some experience with MATLAB and a lot of patience.

Directory Structure

The most convenient place to save all user-created files is in a directory (or folder) immediately below the directory (or folder) in which the MATLAB application program is installed. This way all user-created files are automatically accessible to MATLAB. If you need to store the files somewhere else, you will have to specify the path to the files using the **path** command.

Some Basic Commands

A few of these like **help, expo**, and **size** are already mentioned above. Here are a few more that are frequently used.

help	lists topics on which help is available
help *topic*	provides help on *topic*
lookfor *string*	lists help topics containing *string*
demo	runs the MATLAB EXPO program
who	lists variables currently in the workspace
whos	lists variables currently in the workspace with their size
what	lists M-, MAT-, and MEX-files on the disk
clear	clears the workspace, all variables are removed
clear x y z	clears only variables x, y, and z
cd	change the current working directory
dir	lists contents of the current directory
^c (Control-c)	local abort, kills the current execution of a command or a program
computer	tells you the computer type; see on-line help
clock	gives you wall clock time
date	tells you the date
more	controls the paged output according to the screen size
flops	tells you how many floating point operations you have used so far
quit	quits MATLAB
exit	same as quit

41.3 INTERACTIVE COMPUTATION

In principle, one can do all calculations in MATLAB interactively by entering commands sequentially in the command window, although a *script file* (explained in Script Files) is perhaps a better choice for all computations which involve more than two or three steps. The interactive mode of computation, however, makes MATLAB an extremely powerful scientific calculator that puts hundreds of useful built-in functions for numerical and, now with the Symbolic Math Toolbox, algebraic calculations as well as sophisticated graphics on the finger tips of the user.

In this section, we introduce the reader to some of the MATLAB's built-in functions and capabilities through several examples of interactive computation. All commands in the interactive mode are entered at the MATLAB prompt ">>" in the command window. A **return** or **enter** at the end of the command makes MATLAB interpret and execute the command. Any line or part of a line starting with a **%** is ignored by MATLAB as a comment statement.

While executing commands in MATLAB, we have the option of seeing the result on the screen or suppressing the screen output. A semicolon (**;**) at the end of a command suppresses the screen output, although the command is carried out and the result is saved.

Since MATLAB derives most of its power from matrix computations and assumes every variable to be a matrix, we start with descriptions and examples of how to enter a matrix, index it, manipulate it, and do some useful calculations with it.

Matrices and Vectors: Input, Indexing, and Manipulation

Input

As mentioned before, the only data object recognized by MATLAB is a matrix. Vectors and scalars are taken as special cases of a matrix. A matrix is entered row-wise with elements of a row separated by spaces or commas and the rows separated by semicolons or carriage returns. The entire matrix must be enclosed within square brackets. Elements of the matrix may be real numbers, complex numbers, or valid MATLAB expressions. Thus **`A = [1 2; 5 9]`** and **`B = [2*x ln(x)+sin (y); 5i 3 + 2i]`** are valid matrix inputs that produce 2 × 2 matrices **`A`** and **`B`** with the specified elements. A vector is a special case of a matrix, with just one row or one column and is input the same way as a matrix. A scalar never needs brackets. Square brackets with no elements between them create a null vector (see Figure 41.3 for examples).

Indexing (or Subscripting)

Once a matrix exists, its elements are accessed by specifying their row and column indices. Thus **`A(i,j)`** in MATLAB refers to the element a_{ij} of matrix **`A`** where i is the row number and j is the column number of the element. This notation is fairly com-

```
>>A=[1 2 3; 4 5 6; 7 8 8]

A =

     1     2     3
     4     5     6
     7     8     8
```

Matrices are entered rowwise. Rows are separated by semicolon and columns are separated by a space or a comma.

```
>>A(2,3)

ans =

     6
```

Element A_{ij} of Matrix A is accessed by typing A(i,j).

```
>>A(3,3) = 9

A =

     1     2     3
     4     5     6
     7     8     9
```

Correcting any entry is easy through indexing.

```
>>B = A(2:3,1:3)

B =

     4     5     6
     7     8     9
```

Any submatrix of A is obtained using colon as a range specifier for row and column indices.

```
>>B = A(2:3,:)

B =

     4     5     6
     7     8     9
```

Note the use of (:) in the command. The colon by itself as a row or column index specifies all rows or columns of the matrix.

```
>>B(:,2)=[]

B =

     4     6
     7     9
```

A row or a column of a matrix is deleted by setting it to a null vector [ ].

FIGURE 41.3 Examples of matrix input and matrix index manipulation.

mon in many computational softwares and programming languages. MATLAB, however, provides a much higher level of index specification—it allows a range of row and column indices to be specified at the same time. Thus **`A(m:n,k:l)`** is a valid MATLAB statement that specifies rows *m* to *n* and columns *k* to *l* of matrix **`A`**. When the rows (or columns) to be specified range over all rows (or columns) of the matrix then use a colon as the row index. Thus **`A(:,5:20)`** gets all the elements in columns 5 through 20 of *each* row of matrix **`A`**. This feature makes matrix manipulation much easier and provides a way to take advantage of the vectorized nature of calculations in MATLAB (see Fig. 41.3 for examples).

Matrix Manipulation

As you can see from examples in Figure 41.3, correcting wrong entries of a matrix, extracting any part or submatrix of a matrix, or deleting or adding a row or a column to a matrix is fairly easy. Such manipulations are done with the smart indexing feature of MATLAB. By specifying vectors as the row and column indices of a matrix one can reference and modify any submatrix of a matrix. Thus if A is a 10 × 10 matrix, B is a 5 × 10 matrix, and *y* is a 20 elements long column vector, then **`A([1 3 6 9],:) = [B(1:3,:); y(1:10)']`** replaces rows 1, 3, and 6 of A by the first 3 rows of B and row 9 of A by the first 10 elements of *y*. In such manipulations, of course, it is imperative that the sizes of the submatrices to be manipulated be compatible. For example, in the above assignment, number of columns in A and B must be the same and total number of rows input on the right hand side must be the same as the number of rows specified on the left. A more sophisticated use of the indexing with vectors is to use 0–1 vectors (usually created by relational operations) to reference submatrices. For example, let **`v = [1 0 0 1 1]`** and **`Q`** be a 5 × 5 matrix. Then **`Q(v,:)`** picks out those rows of **`Q`** where **`v`** is nonzero, i.e., rows 1, 4 and 5. Finally, the all powerful colon can also be used to string all elements of a matrix **`A`** into a column vector **`b`** by the command **`b = A(:)`**.

To add to matrix generation and manipulation MATLAB provides many useful utility matrices. For example,

`eye(m,n)`	makes a matrix of size *m* by *n* with 1's on the main diagonal
`zeros(m,n)`	makes a *m* by *n* matrix of zeros
`ones(m,n)`	makes a *m* by *n* matrix of ones
`rand(m,n)`	makes a *m* by *n* matrix of random numbers
`randn(m,n)`	makes a *m* by *n* matrix of normally distributed random numbers
`diag(v)`	generates a diagonal matrix with vector **`v`** on the diagonal
`diag(A)`	extracts the diagonal of matrix **`A`** as a vector
`diag(A,1)`	extracts the first upper off-diagonal vector of matrix **`A`**.

The first four commands with a single argument, e.g., **`ones(m)`**, produce a square matrix of dimension **`m`**. A matrix can be built with many block matrices as well. See examples in Figure 41.4.

There is also a set of built-in special matrices such as **`hadamard, hankel, hilb, invhilb, kron, pascal, toeplitz, vander, magic`** that can be used to generate and manipulate special matrices. There are also special functions available to do useful manipulations:

`rot90`	rotates a matrix by 90°
`fliplr`	flips a matrix matrix from left to right

```
>>B = [ones(3) zeros(3,2); zeros(2,3) 4*eye(2)]

B =

   1    1    1    0    0
   1    1    1    0    0
   1    1    1    0    0
   0    0    0    4    0
   0    0    0    0    4
```

Create a matrix B using submatrices made up of elementary matrices: ones, zeros, and the identity matrix of the specified sizes.

```
>>diag(B)'

ans =

   1    1    1    4    4
```

This command pulls out the diagonal of B in a row vector. Without the transpose, the result will obvioulsy be a column vector.

```
>>diag(B,1)'

ans =

   1    1    0    0
```

The second argument of the command specifies the off-diagonal vector to be pulled out. Here we get the first upper off-diagonal vector. A negative value of the argument species the lower off-diagonal vectors.

FIGURE 41.4 Examples of matrix manipulation using utility matrices and functions.

flipud	flips a matrix matrix from up to down
tril	extracts the lower triangular part of a matrix
triu	extracts the upper triangular part of a matrix
reshape	changes the shape of a matrix.

Generating Vectors. Very often we need to generate a vector of numbers over a given range with a specified increment. The general command to do this in MATLAB is

v = *initial value:increment:final value*

Thus, **a = 0:10:100** produces a vector **a** with elements **[10 20 30 ... 100]**. The three values in the above assignment can also be valid MATLAB expressions, e.g., **a = sqrt(55*log(x)):x/1000:y** is a perfectly valid assignment if the scalars **x** and **y** already exist. If no increment is specified, MATLAB uses the default increment of 1. Thus, **u = 0:10** will create vector **u = [0 1 2 3 ... 10]**. As you may notice, no square brackets are required if a vector is generated this way, however, a vector assignment such as **u = [1:10 33:-2:19]** does require square brackets because of concatenation of the two vectors: **[1 2 3 ... 10]** and **[33 31 29 ... 19]**. Finally, we mention the use of two built-in functions used frequently to generate vectors:

linspace(a,b,n)	generates linearly spaced vector of length *n* from *a* to *b*, *Example:* **linspace (0,20,5)** generates **[0 5 10 15 20]**
logspace(a,b,n)	generates logarithmically spaced vector of length *n* from *a* to *b* *Example:* **logspace(0,3,4)** generates **[1 10 100 1000]**.

Matrix and Array Operations

Arithmetic Operations

For the people who have been programming in one of the conventional languages like Pascal, Fortran, or C, it is an absolute delight to be able to write a matrix product as **`C = A*B`** where **`A`** is an $m \times n$ matrix and **`B`** is an $n \times k$ matrix. MATLAB allows all arithmetic operations: +, –, *, /, and ^ (caret) to be carried out on matrices in straightforward ways as long as the operation makes sense mathematically and operands are compatible. Thus,

`A+B or A-B`	is valid if **`A`** and **`B`** are of the same size,
`A*B`	is valid if **`A`**'s number of columns equals **`B`**'s number of rows,
`A/B`	is tricky but valid and = $A \cdot B^{-1}$ for same size square matrices A and B,
`A^2 (=A*A)`	makes sense only if **`A`** is square.

In all the above commands if **`B`** is replaced by a scalar, say α, the arithmetic operations are still carried out. In this case, the command **`A`**+α adds α to each element of A, the command **`A`***α (or α***`A`**) multiplies each element of A by α and so on. Vectors, of course, are just treated as a single row or a column matrix and, therefore, a command such as **`w = u*v`** where u and v are same size vectors, say $m \times 1$, produces an error (because you cannot multiply an $m \times 1$ matrix with an $m \times 1$ matrix) while **`w = u*v'`** and **`w = u'*v`** execute correctly producing the inner and outer product of the two vectors, respectively (see examples in Fig. 41.5). So, how does one get products like $[u_1v_1\ u_2v_2\ u_3v_3 \ldots u_nv_n]$ from two vectors u and v? No, you do not have to use *DO* or *FOR* loops. You can do *array operation*—operations done on element-by-element basis. Element-by-element multiplication, division, and exponentiation between two matrices or vectors of the same size are done by preceding the corresponding arithmetic operators *, /, and ^ by a period (.). Thus **`u.*v, u./v,`** and **`u.^v`** produce $[u_1v_1\ u_2v_2\ u_3v_3 \ldots]$, $[u_1/v_1\ u_2/v_2\ u_3/v_3 \ldots]$, and $[u_1^{v_1}, u_2^{v_2}, u_3^{v_3}, \ldots]$ respectively. Clearly, there is a big difference between **`A^2`** and **`A.^2`** (see Fig. 41.5). Once again, scalars do enjoy a special status. While **`u./v`** or **`u.^v`** will produce an error if u and v are not the same size, **`1./v`** happily computes $[1/v_1\ 1/v_2\ 1/v_3 \ldots]$ and **`pi.^v`** gives $[\pi^{v_1}\ \pi^{v_2}\ \pi^{v_3} \ldots]$.

Relational Operations

There are six relational operators in MATLAB: <, <=, >, >=, ==, and ~=. The first four are obvious, "==" stands for "equal" and "~=" stands for "not equal." These operations result in a vector or matrix of the same size as the operands, with 1 where the relation is true and 0 where it is false. Thus, if **`x = [1 5 3 7]`** and **`y = [0 2 8 7]`** then

`k = x < y`	results in k = [0 0 1 0]	because $x_i < y_i$ for $i = 3$
`k = x <= y`	results in k = [0 0 1 1]	because $x \le y_i$ for $i = 3$ and 4
`k = x > y`	results in k = [1 1 0 0]	because $x_i > y_i$ for $i = 1$ and 2
`k = x >= y`	results in k = [1 1 0 1]	because $x_i \ge y_i$ for $i = 1, 2,$ and 4
`k = x == y`	results in k = [0 0 0 1]	because $x_i = y_i$ for $i = 4$ and
`k = x ~= y`	results in k = [1 1 1 0]	because $x_i \ne y_i$ for $i = 1, 2$ and 3.

Although these operations are usually used in conditional statements such as *if-then-else* to branch out to different cases, they can be used to do pretty sophisticated matrix manipulation. For example, **`u = v(v >= sin(pi/3))`** finds all elements of

```
>>A=[1 2 3; 4 5 6; 7 8 9];
>>x=A(1,:)'

x =

     1
     2
     3
```

Matrices are transposed using the single right quote character ('). Here x is the transpose of the first row of A.

```
>>x'*x

ans =

    14
```

Matrix or vector products are well defined between compatible pairs. A row vector (x') times a column vector (x) of the same length gives the inner product which is a scalar.

```
>>x*x'

ans =

     1     2     3
     2     4     6
     3     6     9
```

But a column vector times a row vector of the same length gives the outer product which is a matrix.

```
>>A*x

ans =

    14
    32
    50
```

Look how easy it is to multiply a vector with a matrix.

```
>>A^2

ans =

    30    36    42
    66    81    96
   102   126   150
```

You can even exponentiate a matrix if it is a square matrix. A^2 is simply A*A.

```
>>A.^2

ans =

     1     4     9
    16    25    36
    49    64    81
```

When a dot precedes the arithmatic operators *, ^, and /, MATLAB performs array operations (element by element operations). So, A.^2 produces a matrix with elements $(a_{ij})^2$.

FIGURE 41.5 Examples of matrix transpose, matrix multiplication, matrix exponentiation, and array exponentiation.

vector v such that $v_i \geq \sin \pi/3$ and stores them in vector u. Two or more of these operations can also be combined with the help of *logical operators* (described below).

Logical Operations

There are four logical operators: **&** (AND), | (OR), ~ (NOT), and **xor** (EXCLUSIVE OR). These operators work pretty much like the relational operators and produce vectors or matrices of the same size as the operands, with 1 where true and 0 where false.

Thus, for the two vectors $x = [1\ 5\ 3\ 7]$ and $y = [0\ 2\ 8\ 7]$, **`m = (x > y) & (x > 4)`** results in $m = [0\ 1\ 0\ 0]$. If you wish to see those elements of x that satisfy both the conditions, you may type **`x((x>y) & (x>4))`**.

There are also many built-in logical functions such as **`all, any, exist, find, finite`** that aid in matrix manipulation and execution of conditional statements.

Matrix Functions

We discussed the difference between the array exponentiation **`A.^2`** and the matrix exponentiation **`A^2`** above. There are some built-in functions that are truly matrix functions and they also have their array counterparts. The matrix functions are

`expm(A)`	finds the exponential of matrix A, e^A,
`logm(A)`	finds log(A) such that $A = e^{\log(A)}$,
`sqrtm(A)`	finds $\sqrt{A}$.

The array counterparts of these functions are **`exp, log`**, and **`sqrt`**, which operate on each element of the input matrix (see Fig. 41.6 for examples). The matrix exponential function **`expm`** also has some specialized variants **`expm1, expm2`**, and **`expm3`**. See on-line help or the *Reference Guide* (1993) for their proper usage. MATLAB also provides a general matrix function **`funm`** for evaluating true matrix functions.

```
>>A=[1 2; 3 4];
>>asqrt = sqrt(A)

asqrt =

    1.0000    1.4142
    1.7321    2.0000

>>Asqrt = sqrtm(A)

Asqrt =

   0.5537 + 0.4644i   0.8070 - 0.2124i
   1.2104 - 0.3186i   1.7641 + 0.1458i

>>exp_aij = exp(A)

exp_aij =

    2.7183    7.3891
   20.0855   54.5982

>>exp_A = expm(A)

exp_A =

   51.9690   74.7366
  112.1048  164.0738
```

sqrt is an array operation. It gives the square root of each element of A as is evident from the output here.

sqrtm, on the other hand, is a true matrix function, i.e., it computes $\sqrt{A}$. Thus `Asqrt*Asqrt = A`.

Similar to the example above, exp gives element by element exponential of the matrix whereas expm finds the true matrix exponential e^A. For info on other matrix functions type `help matfun`

FIGURE 41.6 Examples of differences between matrix functions and array functions.

Character Strings

All character strings are entered within two single right quote (') characters. MATLAB treats every string as a row vector with one element per character. For example, typing **`message = 'Leave me alone'`** creates a vector named **`message`** of size 1 × 14 (spaces in strings count as characters). Therefore, to create a column vector of text objects, each text string must have exactly the same number of characters. For example, **`names = ['John'; 'Ravi'; 'Mary'; 'Xiao']`** will create a column vector with each name per row, although, to MATLAB, the variable **`names`** will be a 4 × 4 matrix. Clearly, the command **`howdy = ['Hi'; 'Hello'; 'Namaste']`** will result in an error because each row has different length. Text strings of different lengths can be made to be of equal length by padding them with blanks. Thus the correct input for **howdy** will be **`howdy = ['Hi␣␣␣␣␣'; 'Hello␣␣'; 'Namaste']`** with each string being 7 characters long (␣ denotes a space).

Character strings can be manipulated just like matrices. Thus, typing **`c = [howdy(2,:) names(3,:)]`** produces **`Hello Mary`** as the output in variable **`c`**. This feature can be used along with number-to-string conversion functions, such as **`num2str`** and **`int2str`**, to create text objects containing dynamic values of variables. Such text objects are particularly useful in creating titles for figures and other graphics annotation commands (see Graphics). For example, say, you want to produce a few variance-study graphs with different values of the sample size *n*, an integer variable. Producing the title of the graph with the command

```
title(['Variance study with sample size n = ',int2str(n)])
```

writes titles with the current value of *n* printed in the title.

MATLAB provides a powerful function **`eval`** to *evaluate* text strings and execute them if they contain valid MATLAB commands. There are many usage of the **`eval`** function. See on-line help for more information.

Using Built-in Functions and help

MATLAB provides hundreds of built-in functions for numerical linear algebra, data analysis, Fourier transforms, data interpolation and curve fitting, root-finding, numerical solution of ordinary differential equations, numerical quadrature, sparse matrix calculations, and general purpose graphics. There is on-line help for all built-in functions. Although the help facility in MATLAB 4.x is improved over the previous versions, it could still use some help to serve the users better. With so many built-in functions and a *not-so-great* on-line help, it is important to know how to look for a function and how to use it.

If you know the exact name of a function, you may get help on it by typing **`help`** *functionname* on the command line. If you are not sure about the name of the function, use **`lookfor string`** to get a list of functions that use *string* in their description. Typing **`help`** by itself brings out a list of categories (see Fig. 41.7) that provides help on functions organized in functional groups. For example, typing **`help elfun`** gives a list of elementary math functions with a very brief description of each function. Further help can now be obtained on a function because the exact name of the function is now known. MATLAB'S **`help`** command is not forgiving of any typos or misspellings and hence you must know the exact command name.

```
>>help

HELP topics:

matlab:general      -  General purpose commands.
matlab:ops          -  Operators and special characters.
matlab:lang         -  Language constructs and debugging.
matlab:elmat        -  Elementary matrices and matrix manipulation.
matlab:specmat      -  Specialized matrices.
matlab:elfun        -  Elementary math functions.
matlab:specfun      -  Specialized math functions.
matlab:matfun       -  Matrix functions - numerical linear algebra.
matlab:datafun      -  Data analysis and Fourier transform functions.
matlab:polyfun      -  Polynomial and interpolation functions.
matlab:funfun       -  Function functions - nonlinear numerical methoc
matlab:sparfun      -  Sparse matrix functions.
matlab:plotxy       -  Two dimensional graphics.
matlab:plotxyz      -  Three dimensional graphics.
matlab:graphics     -  General purpose graphics functions.
matlab:color        -  Color control and lighting model functions.
matlab:sounds       -  Sound processing functions.
matlab:strfun       -  Character string functions.
matlab:iofun        -  Low-level file I/O functions.
matlab:demos        -  The MATLAB Expo and other demonstrations.
Toolbox:local       -  Local function library.
MATLAB 4.1:Extern     - (No table of contents file)
Macintosh HD:MATLAB 4.1 - (No table of contents file)

For more help on directory/topic, type "help topic".
```

FIGURE 41.7 MATLAB help facility.

On the other hand, **lookfor** is a friendlier command that lets you specify a string, a descriptive word about the function, on which you need help. The following example takes you through the process of looking for help on finding eigenvalues of a matrix, getting help on the exact function that serves the purpose, and using the function in the correct way to get the results.

Example 1: Finding Eigenvalues and Eigenvectors

Let us say that we are interested in finding out the eigenvalues and eigenvectors of a matrix *A* but we do not know what functions MATLAB provides to find the eigenvalues of a matrix. Since we do not know the exact name of the required function, our best bet to get help is to try **lookfor eigenvalue**. Figure 41.8 shows MATLAB's response to this command. It lists all functions that have the string "eigenvalue" either in their name or in their description. You can then browse through the list and pick the function that seems closest to your needs and ask for further help on it. In Figure 41.8, for example, we seek help on function **eig**. The on-line help on **eig** tells us what this function does and how to use it.

Thus, if we are interested in just the eigenvalues of a matrix A, we type **eig(A)** to get the eigenvalues, but if we want to find both the eigenvalues and the eigenvectors of A, we specify the output list explicitly and type **[eigvec, eigval] = eig(A)** (see Fig. 41.9). The names of the output or the input variables can, of course, be anything because these are dummy variables. Although obvious, we note that the list of the variables (input or output) must be in the same order as specified by the function.

```
>>lookfor eigenvalue

ROSSER A classic symmetric eigenvalue test problem.
WILKINSON Wilkinson's eigenvalue test matrix.
BALANCE Diagonal scaling to improve eigenvalue accuracy.
EIG   Eigenvalues and eigenvectors.
EXPM3 Matrix exponential via eigenvalues and eigenvectors.
POLYEIG Polynomial eigenvalue problem.
QZ    Generalized eigenvalues.
EIGMOVIE Symmetric eigenvalue movie.
EIGENSYS Symbolic matrix eigenvalues and eigenvectors.

>>help eig

 EIG  Eigenvalues and eigenvectors.
   EIG(X) is a vector containing the eigenvalues of a square
   matrix X.

  [V,D] = EIG(X) produces a diagonal matrix D of
   eigenvalues and a full matrix V whose columns are the
   corresponding eigenvectors so that X*V = V*D.

   [V,D] = EIG(X,'nobalance') performs the computation with
   balancing disabled, which sometimes gives more accurate
   results for certain problems with unusual scaling.

   Generalized eigenvalues and eigenvectors.

   EIG(A,B) is a vector containing the generalized
   eigenvalues of square matrices A and B.

   [V,D] = EIG(A,B) produces a diagonal matrix D of general-
   ized eigenvalues and a full matrix V whose columns are
   the corresponding eigenvectors so that A*V = B*V*D.
```

FIGURE 41.8 Example of use of on-line help.

Example 2: Solving a Linear System of Equations

One of the greatest strengths of MATLAB is its linear algebra package. For matrix analysis, eigenvalue analysis, and solution of linear systems, the functions provided and their ease of use are simply incredible.

There are many ways in which one can solve a system of linear algebraic equations in MATLAB. By typing **`help matfun`** to seek help on matrix functions, we get a list of functions used in linear algebra along with brief descriptive notes. Among the list is a sublist of functions under the heading **Linear equations**, designed for solving various linear systems. While matrix factorization functions such as **`lu, chol, qr`** can be used in appropriate cases to solve a system of equations, solving a simple system such as [**A**] {**x**} = {**b**} is as simple as typing **`x = A\b`**. As an example, Figure 41.10 shows the solution of the following system of equations.

$$\begin{bmatrix} 5 & -3 & 2 \\ -3 & 8 & 4 \\ 2 & 4 & -9 \end{bmatrix} \begin{Bmatrix} x_1 \\ x_2 \\ x_3 \end{Bmatrix} = \begin{Bmatrix} 10 \\ 20 \\ 9 \end{Bmatrix}$$

```
>>A = [5 -3 2; -3 8 4; 4 2 -9];
>>eig(A)

ans =

   4.4246
   9.7960
 -10.2206

>> [eigvec,eigval] = eig(A)

eigvec =

  -0.8706   -0.5375   -0.1725
  -0.3774    0.8429   -0.2382
  -0.3156   -0.0247    0.9558


eigval =

   4.4246         0         0
        0    9.7960         0
        0         0  -10.2206
```

Typing `eig(A)` without a list of outputs gives the eigenvalues of A in a column vector stored in the default output variable `ans`.

Specifying an explicit list of output variables `[eigvec,eigval]` gets the eigenvectors of A in the matrix `eigvec` and the eigenvalues of A on the diagonal of the matrix `eigval`.

FIGURE 41.9 Examples of how to use the function **eig** to find eigenvalues and eigenvectors of a matrix.

```
>>A = [5 -3 2; -3 8 4; 2 4 -9];   % Enter matrix A
>>b = [10; 20; 9];                 % Enter column vector B
>>x = A\b                          % Solve for x.

x =

   3.4442
   3.1982
   1.1868
```

The backslash (\) or the left division is used to solve a linear system of equations $[A]\{x\} = \{b\}$. For more information type: `help slash`.

FIGURE 41.10 Example of solution of a linear system of equations $[\mathbf{A}]\ \{\mathbf{x}\} = \{\mathbf{b}\}$.

Some Comments on Help Facility.

- MATLAB is case sensitive as we have pointed out earlier. All built-in functions in MATLAB use lowercase letters for their names, yet the **help** on any function lists the function in uppercase letters as is evident from Figures 41.7 and 41.8. For the first time user, it may cause some confusion if the user tries to type the command exactly as shown by the help facility, i.e., in uppercase, only to find an error message from MATLAB.
- On-line help can be obtained by typing help commands on the command line as we have described in the foregoing section. This method is applicable on all platforms that support MATLAB. In addition, on some computers such as Macintosh and IBM Compatibles with Windows, on-line help is also available on the menu bar. For example, on Macs, MATLAB HELP is available under Balloon Help in the top right corner of the menu bar. One can then scroll through the help files by selecting and clicking with the mouse.

- Although **lookfor** is a good command to find help if you are not sure about where to look for the desired functions, it has an annoying drawback—it takes only one string as an argument. So typing **lookfor linear equations** causes an error while both **lookfor linear** and **lookfor equations** do give you some help.

Saving and Loading Data

There are many ways of saving and loading data in MATLAB. The most direct way is to use the **save** and **load** commands that we describe here first. We also describe how to save a session or part of a session, including data and commands, using the **diary** command.

Saving into and Loading from the Binary Mat-files

The **save** command may be used to save either the entire workspace or a few selected variables in a file called a *MAT-file*. MAT-files are written in the binary format with full 16 bit precision. These files are recognized by the ".mat" extension. The data saved in these files can be loaded into the MATLAB workspace by the **load** command. Examples of proper usage of these commands are as follows.

save tubedata.mat x y stress	saves variables **x, y,** and **stress** in the file **tubedata.mat**.
save newdata rx ry rz	saves variables **rx, ry**, and **rz** in the file **newdata.mat**; here MATLAB automatically supplies the ".mat" extension to the file name.
load tubedata	loads the variables saved in the file **tubedata.mat**.
load	loads the variables saved in the default MAT-file **matlab.mat**.

Figure 41.11 shows an example session with **save** and **load** commands.

Recording a Session

An entire MATLAB session or a part of it can be recorded in a user-editable file using the **diary** command. A file name with any extension can be specified into which the session gets recorded. For example, typing **diary session1.out** opens a diary file named **session1.out**. Everything in the command window, including user-input, MATLAB output, and error messages, that follows the diary command gets recorded in the file **session1.out**. The recording is terminated by the command **diary off**. The same diary file may be opened later during the same session by typing **diary on**, which will append the subsequent part of the session into the same file **session1.out**. All figures in this article that show the commands typed in the command window and the consequent MATLAB output have been generated by first recording the session using the diary command and then adding the explanatory white boxes elsewhere. The diary files may be opened using any standard text editor and modified as wished.

Contents of a diary file may be loaded in the MATLAB workspace by converting the diary file into a *script file* (see Script Files for more details) and then executing the script file in MATLAB. To do so, we first edit the diary file to remove all unwanted lines (for example, MATLAB error messages), remove outputs if the command that

```
>>whos                                  Check what variables are in the work space
     Name          Size         Elements      Bytes     Density   Complex

        A        5 by 5             25         200        Full        No
 statevector     5 by 20           100         800        Full        No
     strain     20 by 1             20         160        Full        No
     stress     20 by 1             20         160        Full        No

Grand total is 165 elements using 1320 bytes

leaving 7070224 bytes of memory free.
                                              Save variables statevector,
>>save myheadache statevector stress strain   stress, and strain in the Mat-file
                                              named myheadache.mat

>>clear                                 Clear the entire workspace

>>whos                                  Check to see that the workspace is cleared

leaving 7045968 bytes of memory free.

>>load myheadache                       Load the file myheadache.mat

>>whos                                  Check to see if all variables saved in the
                                        Mat-file are loaded in the workspace.
     Name          Size         Elements      Bytes     Density   Complex

 statevector     5 by 20           100         800        Full        No
     strain     20 by 1             20         160        Full        No
     stress     20 by 1             20         160        Full        No

Grand total is 140 elements using 1120 bytes.....
```

FIGURE 41.11 Example of a session on saving and loading data using MAT-files.

generated the output exists, and then rename the file with an ".m" extension to make it a script file. Now, we can execute the file by typing the name of the file (without the ".m" extension) on the *command prompt*. This will execute all the commands saved in the diary file. Thus it is also possible to load the values of a variable that were written explicitly in a diary file. We would, of course, have to enclose the values in a square bracket if it is an array and assign it to a variable name. The values loaded this way will have only the precision explicitly written in the diary file. The number of digits of a variable recorded in the diary depends on the **format** in effect at the time (for a description of **format** see Some Basic Features). This method of saving and loading MATLAB generated data is not recommended. The diary is a good way of recording a session to be included in reports or to show someone how a particular set of calculations was done.

Plotting Simple Graphs

We close this section on interactive computation with an example of how to plot a simple graph in MATLAB and get a hardcopy of the graph.

As mentioned in the introduction, the plots in MATLAB appear in the graphics window. MATLAB provides very good facilities for both 2-D and 3-D graphics. The commands to produce simple plots are surprisingly simple. For complicated graphics and special effects there are a lot of built-in functions that empower the user to manipulate the graphics window in so many ways. Unfortunately, the more control you want the more complicated it gets. We describe the graphics facility in more detail later in this chapter.

The most direct command to produce a graph in 2-D is the **`plot`** command. If a variable **`ydata`** has n values corresponding to n values of variable **`xdata`**, then **`plot (xdata, ydata)`** produces a plot with **`xdata`** on the horizontal axis and **`ydata`** on the vertical axis. You can specify any number of pairs of vectors as the argument of the **`plot`** command to produce overlay plots. But more on this and much more on other aspects of plotting in the section on Graphics. Figure 41.12 shows an example of plotting a simple graph of $f(t) = e^{t/10} \sin(t)$, $0 \le t \le 20$. This function could also be plotted using **`fplot`**, a command for plotting functions of a single variable. The most important thing to remember about the **`plot`** command is that the vector arguments for the x-axis and the y-axis must be of the same length. Thus, in the command **`plot (x1,y1,x2,y2)`**, the vectors **`x1`**, **`y1`** must have the same length and **`x2`**, **`y2`** must have the same length while **`x1`**, **`x2`** may have different lengths.

```
>>x = 0: .1: 20;                                % create vector x

>>y = exp(0.1*x).*sin(x);                       % calculate y

>> plot(x,y)                                    % plot x vs. y

>> xlabel('Time (t) in Seconds')                % label x-axis

>> ylabel('The Response Amplitude in mm') % label y-axis

>> title('A Simple 2-D Plot')                   % put a title
```

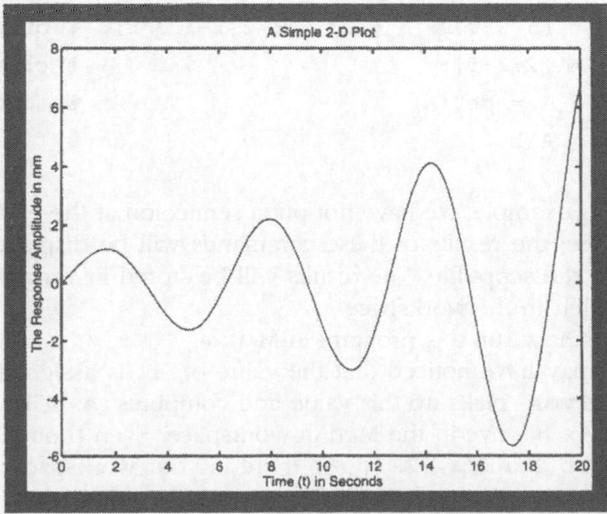

FIGURE 41.12 Example of a simple 2-D plot of function $f(t) = e^{t/10} \sin(t)$.

41.4 PROGRAMMING IN MATLAB: SCRIPT AND FUNCTION FILES

Script Files

A script file is an "M-file" with a set of valid MATLAB commands in it. A script file is executed by typing the name of the file (without the ".m" extension) on the command line. It is equivalent to typing all the commands stored in the script file, one by one, at the MATLAB prompt. Naturally, script files work on global variables, that is, variables currently present in the MATLAB workspace. Results obtained from executing script files are left in the workspace.

A script file may contain any number of commands including those that call built-in functions or functions written by you. This file is useful if you have to repeat a set of commands several times. Enough description, let us look at an example.

Example of a Script File

Let us write a script file to solve a linear system of equations given by

$$\begin{bmatrix} 5 & 2r & r \\ 3 & 6 & 2r-1 \\ 2 & r-1 & 3r \end{bmatrix} \begin{Bmatrix} x_1 \\ x_2 \\ x_3 \end{Bmatrix} = \begin{Bmatrix} 2 \\ 3 \\ 5 \end{Bmatrix} \tag{1}$$

$$\text{or} \qquad \mathbf{Ax} = \mathbf{b}$$

Clearly, **A** depends on the parameter r. We want to find the solution of the equation for various values of the parameter r. We also want to find, say, the determinant of matrix **A** in each case. Let us write a set of MATLAB commands to do the job and store these commands in a file called "solvex.m." How do you create this file, write the commands in it, and save the file, depends on which computer you are using.

```
%------This is the script file 'solvex.m'------
% It solves equation 1 for x and also calculates the det(A).
A = [5 2*r r; 3 6 2*r-1; 2 r-1 3*r]; % create matrix A
b = [2;3;5];                          % create vector b
det_A = det(A)                        % find the determinant
x = A\b                               % find x
```

In this example, we have not put a semicolon at the end of the last two commands. Therefore, the results of these commands will be displayed on the screen when we execute the script file. The results will be stored in variables **`det_A`** and **`x`**, which will be left in the workspace.

Let us now run this program in MATLAB.

You may have noticed that the value of **`r`** is assigned outside the script file and yet **`solvex`** picks up this value and computes **`A`**. This is because all the variables in a script file live in the MATLAB workspace. Even though the output of **`solvex`** is only **`det_A`** and **`x`**, **`A`** is also there in your workspace and you may see it by typing **`A`** at the MATLAB prompt. So, as you probably realize, if you want to do a big set of computations and finally want only a few outputs, a script file is not the right choice. What you need in such cases is a *function file*.

```
>>r = 1;                          % specify a value of r
>>solvex                          % execute the script file solvex.m

det_A =

        64
x =
    -0.0312
     0.2344
     1.6875
```

This is the output. The values of the variables det_A and x appear on the screen because there is no semicolon at the end of the corresponding lines in the script file.

Caution: Be careful with variable names while working with script files because all variables generated by a script file are left in the workspace unless you clear them. NEVER give a script file the same name as the name of a variable it computes.

Function Files

A function file is also an m-file, like a script file, except that the variables in a function file are all local. It is like a program or a subroutine in FORTRAN, a procedure in Pascal, and a function in C.

A function file begins with a function definition line, which has a well-defined list of inputs and outputs. Once you get to know MATLAB a little better, this is where you are likely to spend most of your time—in writing and refining your own function files. The syntax of the function definition line is as follows:

```
function [output variables] = functionname (input variables);
```

where the *functionname* must be the same as the *filename* (without the ".m" extension) in which the function is written.

Comments

- Comment lines start with a "%" sign and may be put anywhere. Anything after a % in a line is ignored by MATLAB as a comment.
- All comment lines immediately after the function definition line are displayed by MATLAB if **help** is sought on this function.
- A single output variable is not required to be enclosed in square brackets in the function definition line, but multiple output variables must be enclosed within [ ].
- Input variable names given in the function definition line are local to the function. So, other variable names or values may be used in the function call. The name of another function can also be passed as an input variable in which case the name must be passed as a character string (enclosed within two single right quotes).

These features will become clear in the following example.

Example of a Simple Function File

Let us write a function file to do the same thing that we did above using a script file. This time, we will make *r* an input to the function and *det_A* and *x* will be the output. Let us call this function **solvexf**. As a rule, it must be stored in a file called **solvexf.m**.

```
function [det_A, x] = solvexf (r);
%------------------------------------------------------
% This is the function file 'solvexf.m'
% To call this function, type:
%    [det_A,x] = solvexf(r);
% r is the input and det_A and x are output.
%------------------------------------------------------
A = [5 2*r r; 3 6 2*r-1; 2 r-1 3*r]; % create matrix A
b = [2;3;5];                          % create vector b
det_A = det(A);                       % find the determinant
x = A\b;                              % find x.
```

Now *r*, *x*, and *det_A* are all local variables. Therefore, any other variable names may be used in their places in the function call statement. Let us execute this function in MATLAB.

```
>> [detA, y] = solvexf(1);   % take r=1 and execute solvexf.m
>>detA                       % display the value of detA

ans =
     64

>>y                          % display the value of y

ans =
   -0.0312
    0.2344
    1.6875
```

Values of detA and y will be automatically displaced if the semicolon at the end of the function execution command is omitted.

After execution of a function, the only variables left in the workspace (by the function) will be the variables in the output list. Now we have more control over the input and the output. We may also put error checks and messages inside the function. For example, we could modify the function above to check if matrix A is empty or not and display an appropriate message before solving the system by changing the last line to

```
if isempty(A)                  % if matrix A is empty
  disp('Matrix A is empty');
else                           % if A is not empty
  x = A\b;                     % find x
end                            % end of if statement.
```

Modify the function file **solvexf.m** and run it with a value of *r* very close to 5 to see what happens.

A function file is the most important programming tool in MATLAB. It takes a while to be efficient at writing functions, but it takes much less time compared to any other language. You can learn the basics of the MATLAB language from the on-line help on **lang**.

Caution: A common mistake that causes many problems for beginners is to type **Function** (with upper case F) instead of **function** in the function definition line. MATLAB's poor error messages do not help a beginner much in identifying this error.

Language-Specific Features

We have already discussed numerous features of MATLAB language through many examples in the previous sections. You are advised to pay special attention to proper usage of punctuation marks, different delimiters, operators, especially the relational operators, and the use of a period (.) preceding the arithmetic operators to make them array operators. For control flow, MATLAB provides **for** and **while** loops and a **if-elseif-else** construct. All the three control flow statements must terminate with corresponding **end** statements. See on-line help for more details. Now we mention some specific features of the language.

Use of Comments to Create On-line Help

As we have already pointed out in the discussion on function files, all comment lines in the beginning (before any executable statement) of a script or a function file are used by MATLAB as on-line help on the particular file. This is a good facility that automatically creates on-line help for user written functions. It is a good idea to copy the function definition line without the word **function** among those first few comment lines so that the syntax of how to execute the function is displayed by the on-line help. The command **lookfor** looks for the argument string in the first commented line of m-files. Therefore, in keeping with the somewhat confusing convention of MATLAB's built-in functions, you should write the name of the script or function file in uppercase letters followed by a short description with keywords, as the first commented line. Thus the first line following the function definition line in the example function above should read **% SOLVEXF solves a 3 × 3 matrix equation with parameter r**.

Global Variables

It is possible to declare some variables to be globally accessible by all or some functions without passing the variables in the input list. This is done with the **global x y z** command where **x, y**, and **z** are the variables to be made global. This statement goes before any executable statement in all functions and scripts that need to access the values of the global variables. Be careful with the names of the global variables. It is generally a good idea to name such variables with long strings to avoid any unintentional match with other local variables.

Input–Output

MATLAB supports many standard C language file I/O functions to read and write formatted binary and text files. The functions supported include **fclose, fopen, fread, fwrite, fgetl, fgets, fprintf, fscanf, sprintf**, and **sscanf**. See on-line help or consult either the *Reference Guide* (1993) or a C language reference manual for more information.

Recursion

The MATLAB programming language supports recursion, i.e., a function can call itself during its execution. Thus all recursive algorithms can be directly implemented (what a break for Fortran users!).

41.5 GRAPHICS

MATLAB 4.x has very good built-in tools for visualization. From basic 2-D plots to fancy 3-D graphics with lighting and color-maps, a complete user-control of the graphics objects through *Handle Graphics*, tools for design of sophisticated graphics user-interface, and animation are now part of the standard MATLAB. What is special about MATLAB's graphics facility is its ease of use and expandability. Commands for most ordinary plotting are simple, easy to use, and somewhat intuitive. If you are not satisfied with what you get, you can control and manipulate virtually everything in the graphics window. This, however, requires an understanding of the Handle Graphics, a system of low-level functions to manipulate graphics objects. In this section we take you through the main features of MATLAB's graphics facilities.

Basic 2-D Plots

The most basic and perhaps the most useful command for producing a *simple* 2-D plot is

plot(*xvalues, yvalues, 'style-option'*)

where *xvalues* and *yvalues* are vectors containing the x- and y-coordinates of the points on the graph and the *style-option*, an optional argument, specifies the line or the point style (e.g., solid, dashed, dotted, o, or +) and the color of the line to be plotted. The two vectors *xvalues* and *yvalues* MUST have the same length. Unequal length of the two vectors is the most common source of error in the plot command. **plot** also works with a single vector argument, in which case the elements of the vector are plotted against their row or column index as appropriate. Thus, for two column vectors x and y each of length n,

plot(x,y)	plots y vs. x with a solid line (the default line style),
plot(x,y,'--')	plots y vs. x with a dashed line (more on this below),
plot(x)	plots the values in x against their row index.

Line Style Options

The *style-option* in the plot command is a character string which consists of 1, 2, or 3 characters that specify the color and/or the line style. There are eight color options: **y** (yellow), **m** (magenta), **c** (cyan), **r** (red), **g** (green), **b** (blue), **w** (white), and **k** (black); and nine line-style options: **-** (solid), **-** (dashed), **:** (dotted), **-.** (dash-dot), **.** (point), **o** (circle), **x** (x-mark), **+** (plus), and ***** (star). The *style-option* is made up of either the color option, the line-style option, or a combination of the two. Thus,

plot(x,y,'r')	plots y vs. x with a red solid line,
plot(x,y,':')	plots y vs. x with a dotted line,
plot(x,y,'b--')	plots y vs. x with a blue dashed line,
plot(x,y,'+')	plots y vs. x as unconnected discrete points marked by +.

When no style option is specified, MATLAB uses the default option, which is a yellow color solid line.

Labels, Title, and Other Text Objects

Plots may be annotated with **xlabel, ylabel, title**, and **text** commands. The first three commands take string arguments while the last one requires three arguments—**text** (*x-coordinate, y-coordinate, 'text'*) where the coordinate values are taken from the current plot. Thus,

xlabel('Pipe Length')	labels the *x*-axis with **Pipe Length**
ylabel('Fluid Pressure')	labels the *y*-axis with **Fluid Pressure**
title('Pressure Variation Study')	titles the plot with **Pressure Variation Study**,
text(2,6,'Note this dip')	writes **'Note this dip'** at the location (2.0,6.0) in the current plot coordinates.

We have already seen an example of **xlabel, ylabel, and title** in Figure 41.12. An example of **text** appears in Figure 41.13. The arguments of **text**(*x,y,'text'*) command may be vectors, in which case *x* and *y* must have the same length and *text* may be just one string or a vector of strings. If *text* is a vector then it must have the same length as *x* and, of course, like any other string vector, must have each element of the same length. A useful variant of the **text** command is **gtext**, which takes only string argument (a single string or a vector of strings) and lets the user specify the location of the text by clicking the mouse in the graphics window at the desired location.

Axis Control, Zoom-in, and Zoom-out

Once a plot is generated one can change the axes limits by the **axis** command. Typing **axis**([*xmin xmax ymin ymax*]) changes the current axes limits to the specified new values *xmin* and *xmax* for the *x*-axis and *ymin* and *ymax* for the *y*-axis. This command may thus be used to zoom-in on a particular section of the plot or to zoom-out. There are also some useful predefined string arguments for the **axis** command:

axis('equal')	sets equal scale on both axes
axis('square')	sets the default rectangular frame to a square
axis('normal')	resets the axis to default values
axis('axis')	freezes the current axes limits
axis('off')	removes the surrounding frame and the tick marks.

The **axis** command must come after the **plot** command to have the desired effect. Thus we could draw a unit circle by typing

```
t = linspace(0,2*pi,100);
plot(cos(t), sin(t))
axis('equal')
```

```
>>t=linspace(0,2*pi,100);                  % Generate vector x
>>y1=sin(t);  y2=t;                        % Calculate y1, y2, y3
>>y3=t-(t.^3)/6+(t.^5)/120;
>>plot(x,y1,x,y2,'--',x,y3,':')            % Plot (x,y1) with solid line
                                           %- (x,y2) with dahed line and
                                           %- (x,y3) with dotted line
>>axis([0 5 -1 5])                         % Zoom-in with new axis limits
>>xlabel('t')                              % Put x-label
>>ylabel('Approximations of sin(t)')% Put y-label
>>title('Fun with sin(t)')                 % Put title
>>text(3.5,0,'sin(t)')                     % Write 'sin(t)' at point (3.5,0)
>>gtext('Linear approximation')
>>gtext('First 3 terms')
>>gtext('in Taylor series')
```

gtext writes the specified string at a location clicked with the mouse in the graphics window. So after hitting return at the end of gtext command go to the graphics window and click a location.

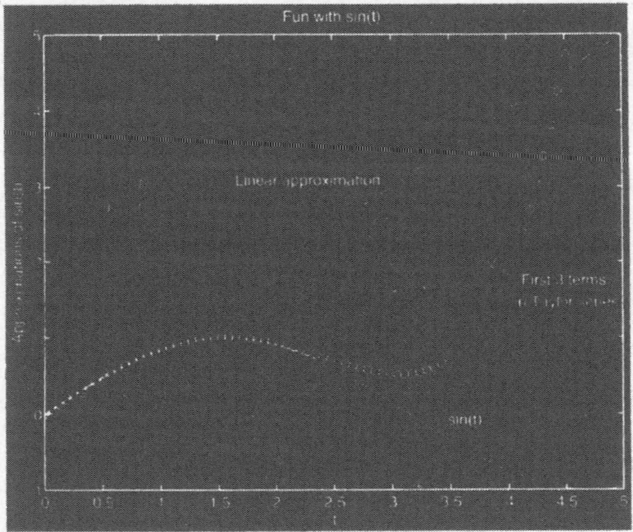

FIGURE 41.13 Example of an overlay plot along with examples of **xlabel, ylabel, title, axis, text, gtext** commands. The three lines plotted are $y_1 = \sin t$, $y_2 = t$, and $y_3 = t - (t^3/3!) + (t^5/5!)$.

Overlay Plots

There are three different ways of generating overlay plots in MATLAB: using **plot, hold**, or **line** commands.

Using the* plot *Command. If the entire set of data is available, **plot** command with multiple arguments may be used to generate an overlay plot. For example, if we have three sets of data—(x1,y1), (x2,y2), and (x3,y3), the command **plot(x1,y1, x2,y2, ':', x3,y3, 'o')** plots (x1,y1) with a solid line, (x2,y2) with a dotted line, and (x3,y3) as unconnected points marked by small circles ('o') on the same graph (see Fig. 41.13 for example). Note that the vectors (xi,yi) must have the same length pairwise. If the length of all vectors is the same then it may be more convenient to make a matrix of **x** vectors and a matrix of **y** vectors and then use the two matrices as the argument of the **plot** command. For example, if x1, y1, x2, y2, x3, and y3 are all column vectors of length *n* then typing **X=[x1 x2 x3]; Y=[y1 y2 y3]; plot(X,Y)** produces a plot with three lines drawn in different colors. When **plot** command is used with matrix arguments, each column of the second argument matrix is plotted against the corresponding column of the first argument matrix.

***Using the* hold *Command*.** Another way of making overlay plots is with the **hold** command. Invoking **hold on** at any point during a session freezes the current plot in the graphics window. All subsequent plots generated by the **plot** command are simply added on to the existing plot. The following script file shows how to generate the same plot as in Figure 41.13 by using the **hold** command.

```
% ----- Script file to generate an overlay plot with
  hold command
x=linspace(0,2*pi,100);      % Generate vector x
y1=sin(x);                   % Calculate y1
plot(x,y1)                   % Plot (x,y1) with solid line
hold on                      % Invoke hold for overlay plots
y2=x; plot(x,y2,'—')         % Plot (x,y2) with dashed line
y3=t-(t.^3)/6+(t.^5)/120;    % Calculate y3
plot(x,y3,'o')               % Plot (x,y3) as pts. marked
                             % by 'o'
axis([0 5 -1 5])             % Zoom-in with new axis limits
hold off                     % Clear hold command
```

The **hold** command is more useful for overlay plots when the entire data set to be plotted is not available at the same time. You should use this command if you want to keep adding plots as the data become available. For example, if a set of calculations done in a **for** loop generates vectors *x* and *y* at the end of each loop and you would like to plot them on the same graph, **hold** may be a convenient way to do it.

***Using the* line *Command*.** The **line** is a low-level graphics command that is used by the **plot** command to generate lines. Once a plot exists in the graphics window, additional lines may be added by using the **line** command directly. The **line** command takes a pair of vectors (or a triplet in 3-D) followed by *parameter name/parameter value* pairs as arguments. This command simply adds lines to the existing axes. For example, the overlay plot created by the above script file could also be created without using the **hold on** command by replacing the last two **plot** commands in the script file by **line (x,y2,'linestyle','—')** and **line(x,y2,'linestyle','o')**, respectively.

Specialized 2D Plots

There are many specialized graphics functions for 2-D plotting:

bar	creates a bar graph,
comet	makes animated 2-D plot,
compass	creates arrow graph for complex numbers,
contour	makes contour plots,

errorbar	plots a graph and puts error bars,
fill	draws filled polygons of specified color,
fplot	plots a function of a single variable,
hist	makes histograms,
loglog	creates plot with log scale on both *x*- and *y*-axes,
polar	plots curves in polar coordinates,
quiver	plots vector fields,
rose	makes angled histograms,
semilogx	makes semilog plot with log scale on the *x*-axis,
semilogy	makes semilog plot with log scale on the *y*-axis,
stem	plots a stem graph,
stair	plots a stair graph.

For more information on these functions see on-line help. There are also some other functions such as **line, legend**, and **grid**, which add to the 2-D plotting. We close this section with examples of **fill, polar**, and **stem** plots. We plot the curve $r^2 = \sin 5t$, $0 \leq t \leq 2$ using **fill** and **polar**, and the curve $f(t) = e^{-t/5} \sin t$ using the **stem** plot over the same range. The following script file generates the plots shown in Figure 41.14.

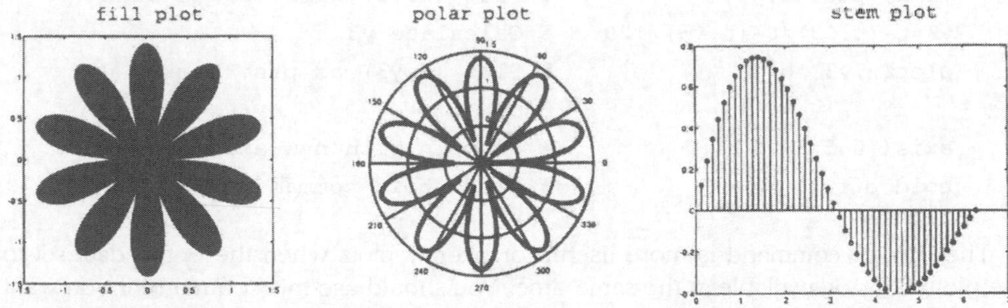

FIGURE 41.14 Examples of specialized 2-D plot commands **fill, polar**, and **stem**.

```
% ---- Script file showing examples of fill, polar, and
  stem plots
t=linspace(0,2*pi,200);         % Generate t from 0 to 2*pi
r=sqrt(abs(2*sin(5*t)));        % Calculate r
x=r.*cos(t); y=r.*sin(t);       % Calculate cartesian
                                % coordinates x and y
fill(x,y,'k'),axis('square')    % Draw a fill graph and
                                % fill the area with
                                % black color. Use square
                                % axes.
figure(2)                       % Initialize a new figure
                                % window, #2
polar(t,r)                      % Make a polar plot in the
                                % new window
f=exp(.2*t).*sin(t);            % Calculate f(t)
```

```
figure(3)              % Initialize a third figure window
stem(t,y)              % Make a stem plot of f(t) vs t.
```

Subplot, a Useful Graphics Layout

If you want to plot a few things and place the plots side by side (not overlay), you may use the **subplot** command to design your layout. The subplot command with three integer arguments, **subplot(m,n,p)**, divides the graphics window into $m \times n$ subwindows and puts the plot generated by the next plotting command into the pth subwindow where the subwindows are counted rowwise. Thus, the command **subplot(2,2,3), plot(x,y)** divides the graphics window into four subwindows and plots **y** vs. **x** in the third subwindow, which is the first subwindow in the second row (see Figure 41.15).

3-D Plots

MATLAB 4.x provides extensive facilities for visualization of 3-D data. In fact, the built-in *colormaps* may be used to represent the 4th dimension. The facilities provided include built-in functions for plotting space-curves, wire-frame objects, surfaces, and shaded surfaces, generating contours automatically, specifying light sources, interpolating colors and shading, and even displaying images. Typing **help plotxyz** in the command window gives a list of functions available for general 3-D graphics. A list of commonly used functions follows.

plot3	plots curves in space,
comet3	makes animated 3-D line plot,
fill3	draws filled 3-D polygons,
contour3	makes 3-D contour plots,
mesh	draws 3-D mesh surfaces (wire-frame),
meshc	draws 3-D mesh surface along with contours,
meshz	draws 3-D mesh surface with curtain plot of reference planes,
surf	creates 3-D surface plots,
surfc	creates 3-D surface plots along with contours,
surfl	creates 3-D surface plots with specified light source,
slice	draws a volumetric surface with slices,
cylinder	generates a cylinder,
sphere	generates a sphere.

Among these functions, **plot3** and **comet3** are the 3-D analogues of **plot** and **comet** commands mentioned in the 2-D graphics section. The general syntax for the **plot3** command is **plot3**(*x*, *y*, *z*, '*style option*'), which plots a curve in 3-D space with the specified line style. The argument list can be repeated to make overlay plots just the same way as in the **plot** command.

Plots in 3-D may be annotated with functions already mentioned for 2-D plots—**xlabel, ylabel, title, text, gtext, grid**, etc. along with an obvious addition—**zlabel**. The **grid** command in 3-D makes the 3-D appearance of the plots better, especially for curves in space (see Fig. 41.15, for example).

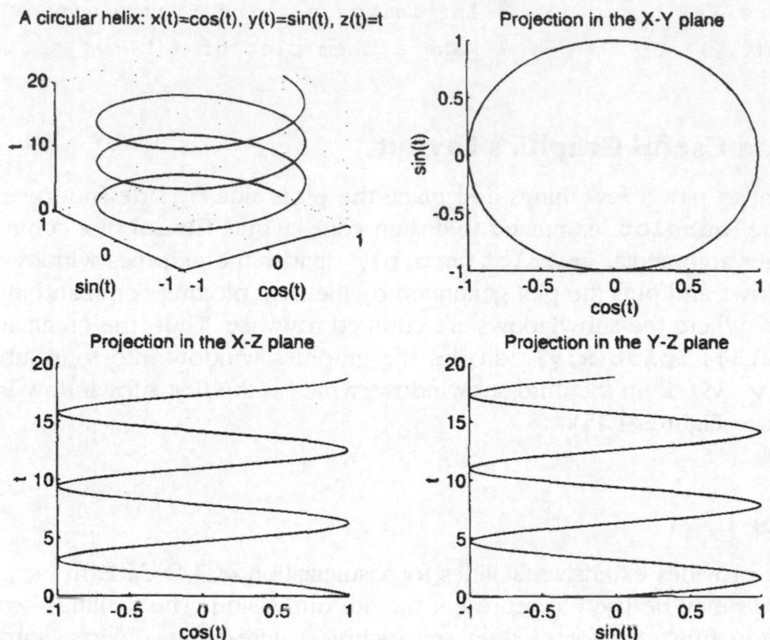

FIGURE 41.15 Examples of **plot3, subplot**, and **view**. Note how putting a 3-D grid in the background helps in the 3-D appearance of the space curve. Although the three 2-D pictures could be made using the plot command, this example illustrates the use of viewing angles.

View

The viewing angle of the observer is specified by the command **view** (*azimuth, elevation*) where *azimuth* in degrees specifies the horizontal rotation from the *y*-axis, measured positive counterclockwise, and elevation in degrees specifies the vertical angle measured positive above the *xy*-plane. The default values for these angles are –37.5° and 30°, respectively. By specifying appropriate values of these angles one can plot the projections of a three-dimensional object on different 2-D planes. For example, the command **view(90,0)** puts the viewer on the positive *x*-axis looking straight on the *yz*-plane and thus produces a 2-D projection of the object on the *yz*-plane. Figure 41.15 shows the projections obtained by specifying view angles. The script file used to generate the data, plot the curves, and obtain the different projected views is listed below.

```
% ------ Script file with examples of 'subplot' and 'view'
  commands
clg                             % Clear any previous graph
t=linspace(0,6*pi,100);         % Generate vector t with 100
                                % equally spaced points between
                                % 0 and 6*pi
x=cos(t); y=sin(t); z=t;        % Calculate x, y, and z
subplot(2,2,1)                  % Divide the graphics window
                                % into 4
                                % subwindows and plot in the
                                % first window
```

```
plot3(x,y,z), grid                 % Plot the space curve in 3-D and
                                   % put grid
xlabel('cos(t)'), ylabel('sin(t)'), zlabel('t')
title ('A circular helix: x(t)=cos(t), y(t)=sin(t), z(t)=t')

subplot(2,2,2),
plot3(x,y,z), view(0,90), % View along the z-axis from above
xlabel('cos(t)'), ylabel ('sin(t)'), zlabel('t')
title('Projection in the X-Y plane')

subplot(2,2,3),
plot3(x,y,z), view(0,0), % View along the y-axis
xlabel('cos(t)'),ylabel('sin(t)'),zlabel('t')
title('Projection in the X-Z plane')

subplot(2,2,4),
plot3(x,y,z), view(90,0), % View along the x-axis
xlabel('cos(t)'),ylabel('sin(t)'),zlabel('t')
title('Projection in the Y-Z plane')
```

Mesh and Surface Plots

The functions for plotting meshes and surfaces **mesh** and **surf** and their various variants **meshz, meshc, surfc, surfl** take multiple optional arguments, the most basic form being **mesh**(Z) or **surf**(Z), where Z represents a matrix. Usually surfaces are represented by the values of z-coordinates sampled on a grid of (x,y) values. A very useful function for generating such a grid is **meshgrid**, which transforms a domain specified by two vectors x of length m and y of length n into an $m \times n$ grid of points. The values of z-coordinates of a surface or the values of a function of two variables may then be calculated on the grid points and the surface or the function be plotted using the **mesh** or **surf** command. Typing **mesh(x,y,Z)** or **surf(x,y,Z)** results in the surface plot of Z with x- and y-values shown on the x- and y-axes if x and y are vectors. The following script file should serve as an example of how to use **meshgrid** and **mesh** command. Here we try to plot the surface

$$z = \frac{xy(x^2 - y^2)}{x^2 + y^2}, \qquad -3 \le x \le 3, \qquad -3 \le y \le 3$$

by computing the values of z over a 50×50 grid on the specified domain. The results of the two plot commands are shown in Figure 41.16.

```
%------------------------------------------------------------
% Script file to generate and plot the surface
%     z = xy(x^2-y^2)/(x^2+y^2)
% using meshgrid and mesh commands.
%------------------------------------------------------------
x=linspace(-3,3,50); y=x;               % Generate 50 element
                                        % long vectors x and y
```

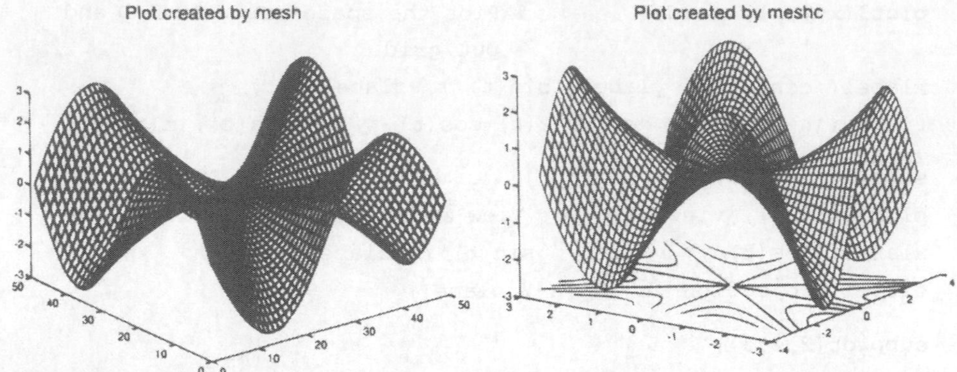

FIGURE 41.16 3-D surface plots created by **mesh** and **meshc** commands. The second plot uses a different viewing angle to show the center of the contour lines. Note that the surfaces do not show hidden lines (this is the default setting that can be changed with the **hidden** command).

```
[X,Y]=meshgrid(x,y);                  % Create a grid over the
                                      % specified domain
Z=X.*Y.*(X.^2-Y.^2)./(X.^2+Y.^2);     % Calculate Z at each
                                      % grid point (Z is 50
                                      % by 50)
mesh(x,y,Z)                           % Make a wire-frame
                                      % surface plot of Z and
                                      % use x and y values on
                                      % the x and y-axes.
figure(2)                             % Open a new figure
                                      % window
meshc(x,y,Z),view(-55,20)             % Plot the same surface
                                      % along with contours
                                      % and show the view
                                      % from the specified
                                      % angles.
```

While surfaces created by **mesh** or its variants have a wire-frame appearance, surfaces created by the **surf** command or its variants produce true surface-like appearance, particularly when used with the **shading** command. There are three kinds of shading available—**shading flat** produces simple flat shading, **shading interp** produces more dramatic interpolated shading, and **shading faceted**, the default shading, shows shaded facets of the surface. Both **mesh** and **surf** can plot parametric surfaces with color scaling to indicate a fourth dimension. This is accomplished by giving four matrix arguments to these commands, e.g., **surf(X,Y,Z,C)** where X, Y, and Z are matrices representing a surface in parametric form and C is the matrix indicating color scaling. The command **surfl** can be used to produce special effects with a specified location of a light source as well as control of light reflectance. See on line help on **surfl** for more information. We close this section with an example of a surface plot showing the deformed configuration of a cylinder under axial loading. The plot is shown in Figure 41.17. The data for this plot were generated by Finite Element analysis outside MATLAB and then imported in MATLAB.

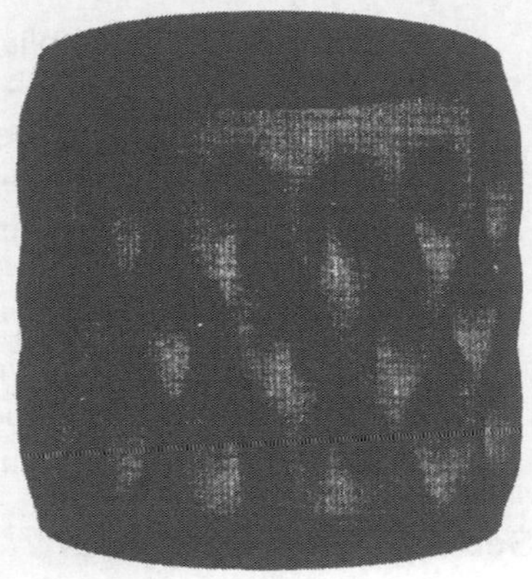

FIGURE 41.17 Example of surface plot generated by **surf** with **shading interp**. Figure printed from Dr. Chris Wohlever's Ph.D. thesis with the permission of the author.

Handle Graphics

The graphics facility in MATLAB 4.x is based on object-oriented architecture. Everything that exists in the graphics window, including the window itself, is a graphic object that has certain properties associated with it. MATLAB provides a handle for each object through which the user can get access to the individual objects and their properties and change any property of the object without affecting other properties or objects. Thus the user gets a complete control over graphics objects. This entire system of object oriented graphics and its user controllability is referred to as "Handle Graphics." Here we briefly discuss this system and its usage, but we urge the more interested reader to consult the *User's Guide* (1993) for more details.

The Object Hierarchy

Graphics objects follow a hierarchy of parent-child relationship. The following tree diagram shows the hierarchy:

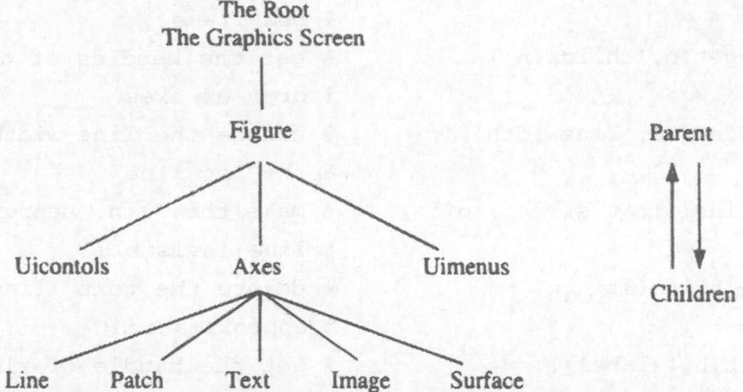

It is important to know this structure for two reasons: (1) it shows you which objects will be affected if you change a default property value at a particular level and (2) it tells you at which level you can query for handles of which objects.

Getting Object Handles

Object handles are unique identifiers associated with each graphics object. These handles have a floating point representation. Handles are created at the time of creation of the object by the graphics function such as **`plot(x,y)`**, **`contour(z)`**, **`line(z1, z2)`**, and **`text(xc, yc, 'Look at this')`**. There are two ways of getting hold of handles:

1. By creating handles explicitly at the object creation level commands. For example, **hl = plot(x, y, 'r-')** and **hxl = xlabel('Angle')**, which return the handles of the line and the x-label in **hl** and **hxl**, respectively.
2. By using explicit handle returning functions—**gcf** and **gca. gcf** return the handle of the current figure and **gca** returns the handle of the current axes. Handles of other objects, in turn, may be obtained with **get** command. For example, **hlines = get(gca, 'children')** returns the handles of all children of the current axes in a column vector **hlines**. The function **get** is used to get a property value of an object, specified by its handle in the command form **get**(*handle, 'PropertyName'*). See on-line help on **get** for more information.

Setting Property Values

Once the handle of an object is known, the list of its properties and their current values can be seen by the command **get**(*handle*), the list of properties that can be set can be seen by the command **set**(*handle*), and any set property can be changed by the command **set**(*handle, 'PropertyName", PropertyValue'*), where *PropertyValue* may be a character string or a number. If the *PropertyValue* is a string then it must be enclosed within single quotes.

As an example, we take Figure 41.13 and use the aforementioned Handle Graphics features to modify the figure. The following script file is used to change the plot in Figure 41.13 to the one shown in Figure 41.18.

```
h=gca;                              % get the handle of the
                                    % current axes
set(h,'box','off');                 % throw away the enclosing
                                    % box frame
hline=get(h,'children');            % get the handles of chil-
                                    % dren of axes
set(hline(7),'linewidth',4)         % change the line width of
                                    % the 1st line
set(hline(6),'visible','off')       % make the 'lin. approx'
                                    % line invisible
delete(hline(3))                    % delete the text 'linear
                                    % approximation'
hxl=get(h,'xlabel');                % get the handle of xlabel
set(hxl,'string','t (angle)')       % change the text of xlabel
set(hxl,'fontname','times')         % change the font of xlabel
set(hxl,'fontsize', ...
  20,'fontweight','bold')           % change the fontsize &
                                    % weight%
```

In the above script file the reader may perhaps be confused about the use of handle **hline**. The command **hline = get(h, 'children')** above gets the handles of all children of the current axes (specified by handle h) in the column vector **hline**, which has six elements—three handles for the three lines and three handles for xlabel, ylabel, and title.

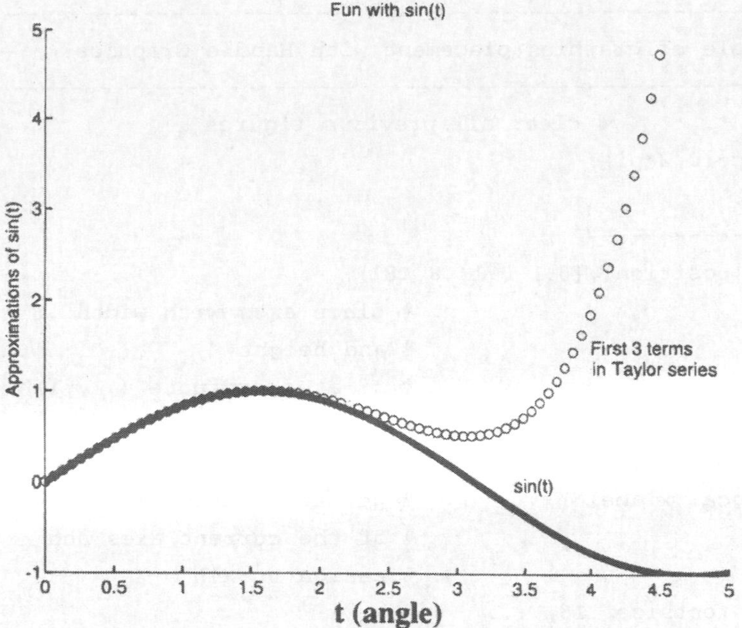

FIGURE 41.18 Example of manipulation of a figure with Handle Graphics. This figure is a result of executing the above script file after generating Figure 41.13.

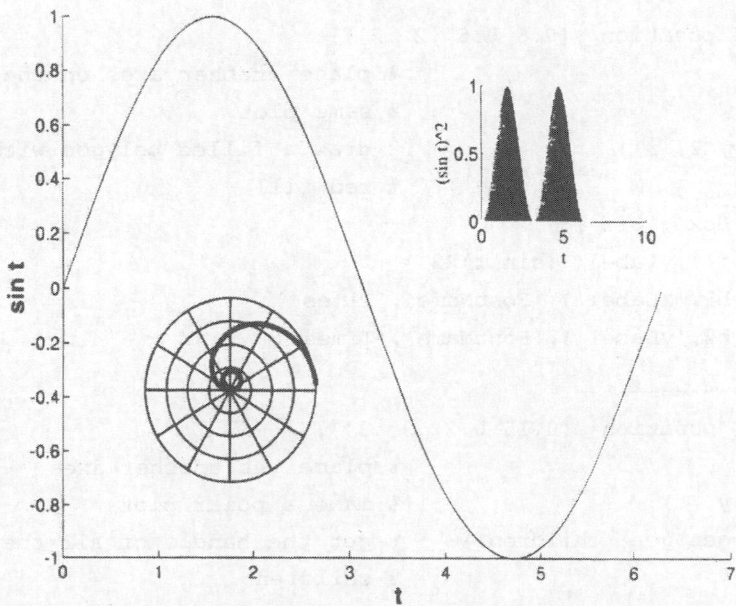

FIGURE 41.19 Example of manipulation of the Figure Window with Handle Graphics. Virtually anything in the figure window, including the placement of axes, can be manipulated with Handle Graphics.

```
%-----------------------------------------------------------
%   Example of graphics placement with Handle Graphics
%-----------------------------------------------------------
clf                  % clear all previous figures
t=linspace(0,2*pi);
y=sin(t);
%------------------
h1=axes('position',[0.1 0.1 .8 .8]);
                               % place axes with width .8
                               % and height
                               % .8 at coordinate (.1,.1)
plot(t,y),xlabel('t'),ylabel('sin t')
set(h1,'Box','Off');           % Turn the enclosing box off
xhl=get(gca,'xlabel');         % get the handle of 'xlabel'
                               % of the current axes and
                               % assign to xlh
set(xhl,'fontsize',16, ...
  'fontweight','bold')         % change attributes of
                               % 'xlabel'
yhl=get(gca,'ylabel');         % do the same with 'ylabel'
set(yhl,'fontsize',16,'fontweight','bold')
%------------------

h2=axes('position',[0.6 0.6 .2 .2]);
                               % place another axes on the
                               % same plot
fill(t,y,^2,'r')               % draw a filled polygon with
                               % red fill
set(h2,'Box','Off');
xlabel('t'),ylabel('(sin t)^2')
set(get(h2,'XLabel'),'FontName','Times')
set(get(h2,'yLabel'),'FontName','Times')
%------------------
h3=axes('position',[0.15 0.2 .3 .3]);
                               % place yet another axes
polar(t,y,/t);                 % make a polar plot
polarch=get(gca,'children');   % get the handle of all the
                               % children
                               % of the current axes
set(polarch(1),'linewidth',3)  % set the line width of the
                               % first child
                               % which is actually the line
```

```
                                     % we plotted
for i=1:length(polarch)              % let us clear the clutter
                                     % due to text
    if strcmp(get(polarch(i),'type'),'text')
                                     % look for all 'text'
                                     % children
        delete(polarch(1))           % delete all text children
    end
end
%------------------
```

Saving and Printing Graphs

The simplest way to get a hardcopy of a graph is to type **print** in the command window after the graph appears in the figure window. The **print** command sends the graph in the current figure window to the default printer in an appropriate form. On PCs (running windows) and Macs you could, alternatively, activate the figure window (bring to the front by clicking on it) and then select **print** from the file menu.

The figure can also be saved into a specified file as a PostScript file or encapsulated PostScript file (EPSF) either for black and white printers or color printers. The PostScript supported includes both Level 1 and Level 2 PostScript. The command to save graphics to a file has the form

print -d*devicetype* -*options filename*

where *devicetype* for PostScript printers can be one of the following: **ps** (black and white PostScript), **psc** (color PostScript), **ps2** (Level2 BW PostScript), **psc2** (Level 2 color PostScript), **eps** (black and white EPSF), **epsc** (color EPSF), **eps2** (Level 2 black and white EPSF), and **epsc2** (Level 2 color EPSF). For example, **print -deps sineplot** saves the current figure in the Encapsulated PostScript file **sineplot.eps**. The ".eps" extension is automatically generated by MATLAB.

In addition to the PostScript devices, MATLAB supports a number of other printer devices on UNIX and PC systems. There are device options available for HP Laser Jet, Desk Jet, and Paint Jet printers, DEC LN03 printer, Epson printers, and other types of printers. See on-line help on **print** to check the available devices and options.

Other than printer devices, MATLAB can also generate a graphics file in the Adobe Illustrator format by specifying **-dill** in the -**d***devicetype* field of the **print** command. This option is extremely useful if you want to dress-up or modify the figure in a way that is either not possible or extremely difficult to do in MATLAB. Of course, you must have access to Adobe Illustrator to be able to open and edit the saved graphs. Figure 41.20 shows an example of a graph generated in MATLAB and then modified in Adobe Illustrator. The most annoying aspect of MATLAB 4.x's graphics is the lack of simple facilities to write subscripts and superscripts and mix fonts in the labels. There is no way of producing something like sin ($\lambda^2_{critical}$) as the label on the axes without playing with the PostScript file (MathWorks says they will incorporate such features in Version 5.x)!

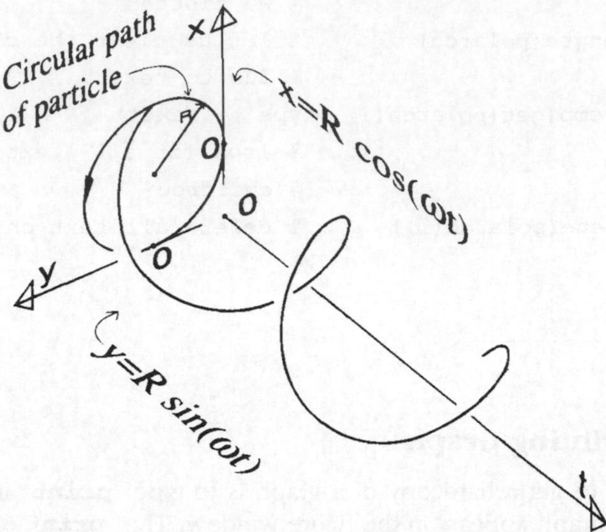

FIGURE 41.20 Example of a figure generated in MATLAB, saved in the Illustrator format and then modified in Adobe Illustrator. The rotation and shearing of texts were done in Illustrator (figure generated by Prof. Andy Ruina of Cornell University).

The figures in MATLAB appear on a black background on the screen. When the figures are saved in a file or printed on paper, the colors are reversed. If you wish to obtain a hardcopy with black backgound and lines drawn in white, type the command **`set(gcf, 'inverthardcopy', 'off')`** before sending the print command. We have used this command to produce a few figures in this article.

Animation

We all know the visual impact of animation. If you have many data representing a function or a system at several time sequences, you may wish to take advantage of MATLAB's capability to animate your data.

There are three types of facilities for animation in MATLAB.

1. **Comet plot:** This is the simplest and the most restricted facility to display a 2-D or 3-D line graph as an animated plot. The command **`comet(x,y)`** plots the data in vectors **`x`** and **`y`** with a comet moving through the data points. The trail of the comet traces a line connecting the data points. So, rather than having the entire plot appear on the screen at once, you can see the graph "being plotted." This facility may be useful in visualizing trajectories in a phase plane. For example, see the built-in demo on the Lorenz attractor.
2. **Movies**: If you have a sequence of plots that you would like to animate, use the built-in **`movie`** facility. The basic idea is to store each figure as a frame of the movie, with each frame stored as a column vector of a big matrix, say M, and then

to play the frames on the screen with the command **movie(M)**. A frame is stored in a column vector using the command **getframe**. For efficient storage you should first initialize the matrix M. The built-in command **moviein** is provided precisely for this initialization, although you can do it yourself too. An example script file to make a movie may look like this:

```
%---- skeleton of a script file to generate and play a movie
------
nframes = 36;                    % number of frames in the the
                                 % movie
Frames = moviein(nframes);       % initialize the matrix
                                 % 'Frames'
for i = 1:nframes
  :                              % you may have calculations
                                 % here to generate data
  :
  x = . . . .;
  y = . . . .;
  plot(x,y)                      % you may use any plotting
                                 % function
  Frames(:,i) = getframe;        % store the current figure
                                 % as a frame
end
movie(Frames,5)                  % play the movie stored in
                                 % frames 5 times
```

You can also specify the speed (frames/second) at which you want to play the movie (the actual speed will eventually depend on your CPU) by typing **movie(Frames**, *m, fps*), which plays the movie, stored in **Frames**, *m* times at the rate *fps* frames per second.

3. **Handle Graphics:** Another way, and perhaps the most versatile way, of creating animation is to use the Handle Graphics facilities. The basic idea here is to plot an object on the screen, get its handle, use the handle to change the desired properties of the object (most likely its 'xdata' and 'ydata' values), and replot the object over a selected sequence of times. There are two important things to know to be able to create animation using Handle Graphics:

- The command **drawnow**, which flushes the graphics output to the screen without waiting for the control to return to MATLAB. (Due to the lack of space, we urge the reader to see the on-line help on **drawnow** to understand its working.)
- The object property **'erasemode'**, which can be set to **'normal'**, **'background'**, **'none'**, or **'xor'** to control the appearance of the object when the graphics screen is redrawn. For example, if a script file containing the following lines is executed

```
h1 = plot(x1,y1,'erasemode','none');
```

```
h2 = plot(x2,y2,'erasemode','xor');
:
newx1= . . .; newy1= . . .; newx2= . . .; newy2= . . .;
:
set(h1,'data',newx1,'ydata',newy1);
set(h2,'xdata',newx2,'ydata',newy2);
:
```

then the first **set** command draws the first object with the new x-data and y-data but the same object drawn before remains on the screen while the second **set** command redraws the second object with new data and also erases the object drawn with the previous data **x2** and **y2**. Thus it is possible to keep some objects fixed on the screen while some other objects change with each pass of a control flow.

The following example script file shows how to animate the motion of a bar pendulum governed by the ODE: $\ddot{\theta} + \sin\theta = 0$.

```
%----- script file to animate a bar pendulum ------------
clf                             % clear figure and stuff
data=[0 0; -1.5 0];             % coordinates of endpoints
                                % of the bar
phi=0;                          % initial orientation
R=[cos(phi) -sin(phi); +sin(phi) cos(phi)];
                                % rotation matrix
data=R*data;
axis([-2 2 -2 2])               % set axis limits axis limits
('equal')
%-----define the objects called bar, hinge, and path.
bar=line('xdata',data(1,:),'ydata',data(2,:), ...
   'linewidth'>,3,'erase','xor');
hinge=line('xdata',0,'ydata',0,'linestyle''>o',
   'markersize',[10]);
path=line('xdata',[],'ydata',[],'linestyle','.',
   'erasemode','none');
theta=pi-pi/1000;               % initial angle
thetadot=0;                     % initial angular speed
dt=.2; tfinal=50; t=0;          % time step, initial
                                % and final time
%------Euler's method for numerical integration
while(t<tfinal);
   t=t+dt;
```

```
    theta=theta + thetadot*dt;
    thetadot=thetadot -sin(theta)*dt;
    R=[cos(theta) (-sin(theta)); sin(theta) cos(theta)];
    datanew= R*data;
    %---- change the property values of the objects:
      path and bar.
    set(path,'xdata', datanew(1,1), 'ydata', datanew(2,1) );
    set(bar,'xdata',datanew(1,:),'ydata',datanew(2,:) );
    drawnow;
  end
```

41.6 WHAT ELSE IS THERE?

There are mainly three other significant facilities in MATLAB that we are unable to discuss here because of lack of space.

Debugging Tools

The new versions of MATLAB (version 4.x) support a built-in debugger that consists of 10 commands—**dbclear, dbcont, dbdown, dbquit, dbstack, dbstatus, dbstep, dbstop, dbtype**, and **dbup**, to help you debug your MATLAB programs. See the on-line help under **lang** for details of these commands.

External Interface

If you wish to dynamically link your FORTRAN or C programs to MATLAB functions so that they can communicate and exchange data, you need to learn about *MEX*-files. Consult the *External Interface Guide* (1993) to learn about these files. The process of developing MEX-files is fairly complicated and highly system dependent. You should perhaps first consider a nondynamic link with your external programs through standard ASCII data files.

Graphics User Interface

Now it is also possible to design your own Graphical User Interface (GUI) with menus, buttons, and slider controls in MATLAB. This facility is very useful if you are developing an application package to be used by others. You can now build many visual "user friendly" features in your application. For more information consult MATLAB's *Building a Graphical User Interface* (1993).

DEFINING TERMS

The reader should consult Chapter 2 on Windows for more details pertaining to the terms used in all Windows applications. All pertinent terms for this material are defined in Chapter 55 on circuit and controls simulation.

REFERENCES

MATLAB User's Guide. The MathWorks, Inc., Natick, MA, 1993.
MATLAB Reference Guide. The MathWorks, Inc., Natick, MA, 1993.
MATLAB External Interface Guide. The MathWorks, Inc., Natick, MA, 1993.
MATLAB Building a Graphical User Interface. The MathWorks, Inc., Natick, MA, 1993.
MATLAB New Features Guide. The MathWorks, Inc., Natick, MA, 1993.

42

Computer Algebra Systems

Paulo Ney de Souza
University of California, Berkeley

This is the list of currently developed and distributed software for symbolic mathematical applications. It was started a few years ago by Booker Bense of the University of California San Diego Supercomputer Center. He is responsible for the layout and for the idea of collecting this database. Most of the information here is obtained from the developers and we try to keep it up to date, a not so easy task. Any errors in the list are mine and I would like to hear about them.

No information is supplied on systems no longer being supported such as SAINT, FORMAC, ALPAK, ALTRAN, MATHLAB, SIN, SAC, CAMAL, ScratchPad, MuMath, TRIGMAN, ANALITIK, SMP, CCALC, or Cayley.

A more thorough description and detailed information for each of the systems is available by gopher or anonymous-FTP in the machine

```
math.berkeley.edu
```

in the directory pub/Symbolic_Soft. An updated version of this file can also be obtained from the same directory. No particular recommendation is made for any of the systems listed. If you want the commercial packages prices contact the developers.

If you do not have access to FTP you can get the files by e-mail using the BITFTP server at Princeton University (and at other locations). To find out more about this, send a one-line mail message containing the word HELP to one of the following addresses:

```
bitftp@pucc.BITNET
bitftp@pucc.princeton.edu
BITFTP@PLEARN.BITNET
bitftp@plearn.edu.pl
ftpmail@decwrl.dec.com
```

0-8493-2530-7/96/$0.00+$.50

Help will be sent to you in the form of electronic mail.

Programs are listed in alphabetical order, inside each class, general purpose systems first.

42.1 GENERAL PURPOSE

AXIOM

Type: Commercial

Machines: SUN Sparc, IBM RS 6000s and other IBM mainframe platforms.

Contact: USA:
ryan@nag.com
NAG Inc
1400 Opus Place, Suite 200
Downers Grove, IL 60515-5702

Phone: 708-971-2337

Fax: 708-971-2706

Rest of the world:
valerie@nag.co.uk
NAG Ltd
Wilkinson House
Jordan Hill Rd.
Oxford OX2 8DR, United Kingdom
Phone: [+44] 865-51-1245
Fax: [+44] 865-31-0139
or
axiom@watson.ibm.com

Version: 1.2

Comments: General purpose. Object oriented with multiple inheritance based on algebraic concepts. Powerful type-inferencing techniques to minimize the need for type declarations. Hypertext browser and on-line documentation with source code available for all library functions. High level interactive language and powerful graphics capability. Design goal: unlimited extensibility without degradation in performance or usability. Release 2 will provide ability to create stand-alone packages and links to Nag FORTRAN Libraries of numerical and statistical routines and a compiler for AXIOM programs.

DERIVE

Type: Commercial

Machines: Runs on PCs and HP 95s.

Contact: Soft Warehouse, Inc.
3660 Harding Avenue, Suite 304
Honolulu, Hawaii 96816-3236
Phone: +1 808 734 5801
Fax: +1 808 735 1105

Version: 3.0

Comments: Very robust, gets problems that other larger programs fail on. Low cost. Runs on the tiny palmtops like HP 95-100 and on meager machines like the PC XT.

FLAC

Machines: IBM PCs (DOS).
Contact: Victor L. Kistlerov, Institute for Control Sciences, Profsoyuznaya 65, Moscow, USSR
Comments: Functional language.

FORM

Type: Version 1 is public domain, Version 2 is commercial
Machines: MS-DOS, AtariSt, Amiga, Mac, Sun3, SunSparc, Apollo, NeXT, VAX/VMS, VAX/Ultrix, DECStation, HP, IBM Risc, IBM 3090, MIPS, Alliant, Gould, and SGI.
Contact: t68@nikhef.nl (Jos Vermaseren)
Binary versions of version 1 are available
by anonymous FTP from nikhef.nikhef.nl (192.16.199.1)
Version 2 is commercially distributed by
CAN, Kruislaan 419
1098 VA Amsterdam, The Netherlands
form@can.nl
Phone: [+31] 20 560-8400
Fax: [+31] 20 560-8448
Version: 1 and 2.2.a
Comments: General purpose , designed for BIG problems, batch-like interface.

GNU-calc

Type: GNU copyleft
Machines: Where Emacs runs.
Contact: Free Software Foundation
Comments: It runs inside GNU Emacs and is written entirely in Emacs Lisp. It does the usual things: arbitrary precision integer, real, and complex arithmetic (all written in Lisp), scientific functions, symbolic algebra and calculus, matrices, graphics, etc. and can display expressions with square root signs and integrals by drawing them on the screen with ASCII characters. It comes with a well-written 600 page on-line manual. You can FTP it from any GNU site.

JACAL

Type: Gnu CopyLeft
Machines: Needs Scheme ver. R4RS.
Contact: Aubrey Jaffer
84 Pleasant St.

Wakefield MA 01880
jaffer@martigny.ai.mit.edu

Version: 1a4

Comments: An IBM PC version on floppy for $99 is available from Aubrey Jaffer, 84 Pleasant St., Wakefield, MA 01880, USA. There is a Scheme implementation available from the same sites as JACAL that will run it on just about any machine with a C compiler (Amiga, Atari-ST, MacOS, MS-DOS, OS/2, NOS/VE, Unicos, VMS, Unix, and similar systems). FTP sites: altdorf.ai.mit.edu:archive/scm/jacal1a4.tar.gz prep.ai.mit.edu:pub/gnu/jacal/jacal1a4.tar.gz nexus.yorku.ca:pub/scheme/new/jacal1a4.tar.gz ftp.maths.tcd.ie:pub/bosullvn/jacal/jacal1a4.tar.gz JACAL is a symbolic mathematics system for the simplification and manipulation of equations and single and multiple valued algebraic expressions constructed of numbers, variables, radicals, and algebraic functions, differential, and holonomic functions. In addition, vectors and matrices of the above objects are included.

Macsyma

Macsyma was developed at MIT and has spun-off a series of different versions that run on specific machines. The full list consists of Macsyma, DOE-Macsyma, Maxima, Aljabr, Paramacs, and Vaxima.

1. Macsyma

Type: Commercial

Machines: PC386/387 and 486 (no SXs) PCs. SPARC Solaris, SPARC SunOS, HP9000 RISC, and IBM RS/6000; the SGI IRIX release will ship soon.

Contact: Macsyma Inc.
20 Academy St.
Arlington, MA 02174-6436
info-macsyma@macsyma.com
Phone: 800-MACSYMA
Fax: 617-646-3161

Version: Version 2 for PCs and 419 for Unix Workstations.

Comments: General purpose, many diverse capabilities, one of the oldest around. Includes proprietary improvements from Symbolics and Macsyma Inc. Descendant of MIT's Macsyma. Recent major enhancements include new capabilities in ODEs, Laplace transforms, integrations, inequalities, linear algebra and partial differential equations.

2. DOE-Macsyma:

Type: Distribution fee only

Machines: GigaMos, Symbolics, and TI Explorer Lisp machines. The NIL version runs on Vaxes using the VMS system. The public domain Franz Lisp version, runs on Unix machines, including Suns and Vaxes using Unix.

Contact: ESTSC—Energy Science & Technology Software Center
P. O. Box 1020

Oak Ridge, TN 37831-1020
Phone: 615-576-2606

Comments: Help with DOE-Macsyma; general and specific help with issues such as support, new versions, etc: lph@paradigm.com Leo Harten
Paradigm Associates, Inc.
29 Putnam Avenue, Suite 6
Cambridge, MA 02139
Phone: 617-492-6079

3. Maxima

Type: License for a fee. Get license from ESTSC before download.
Machines: Unix workstations (Sun, MIPS, HP, PCs) and PC-DOS (beta).
Contact: Bill Schelter
wfs@math.utexas.edu
Version: 4.155
Comments: General purpose—MIT Macsyma family. Common Lisp implementation by William F. Schelter, based on Kyoto Common Lisp. Modified version of DOE-Macsyma available to ESTSC (DOE) sites. Get the license from ESTSC (Phone: 615-576-2606) and then download the software from DOS: math.utexas.edu:pub/beta-max.zip or UNIX: rascal.ics.utexas.edu:pub/maxima-4-155.tar.Z Currently their charge for 1 machine license is $165 to universities. Site licenses are also available.

4. Aljabr

Type: Commercial
Machines: Macs with 4 Megabytes of RAM.
Contact: Fort Pond Research,
15 Fort Pond Road
Acton, MA 01720 USA aljabr@fpr.com
Phone: 508-263-9692
Version: 1.0
Comments: MIT Macsyma family descendant, uses Franz LISP.

5. Paramacs

Type: Commercial
Machines: VAX-VMS, Sun3, Sun4 (SGI and Macs on the works).
Contact: Paradigm Associates, Inc.
29 Putnam Avenue, Suite 6
Cambridge, MA 02139
lph@paradigm.com
Phone: 617-492-6079
Version: 1
Comments: Improved SHARE library and enhanced ODE solver (over DOE-Macsyma) and animated graphics. Maintenance, phone services, and site licensing available.

6. Vaxima

Type: Distribution fee only
Machines: VAX-Unix
Contact: ESTSC (see DOE-Macsyma above)
Comments: General purpose—MIT Macsyma family descendant. Includes source and binaries with assembler for Macsyma and Franz Lisp Opus 38.

Maple

Type: Commercial
Machines: Unix workstations (DEC, HP, IBM, Intergraph, MIPS, Sun, SGI) 386/486/586 (DOS, Windows 3.1 and SCO Unix), Mac (System 6.0.7+, PRO, At Ease, A/UX), NeXT, Amiga, Atari, VAX/VMS, Convex, Sequent, VM/CMS, Fujitsu, and Cray's. Under preparation: Windows NT, DEC Alpha (OSF/1, OpenVMS), NT PowerPC, NEXTSTEP Intel 486 and OS/2.
Contact: Waterloo Maple Software
450 Phillip Street, Waterloo, Ontario, Canada N2L 5J2
info@maplesoft.on.ca, support@maplesoft.on.ca
Phone: 519-747-2373 and 800-267-6583
Fax: 519-747-5284
Version: 5 release 3
Comments: General purpose, source available for most routines. Graphic and animation. On-screen and printable real-math notation. Worksheet interface on Amiga, Macintosh, NeXT, Unix X, Windows, PC Windows, and DEC VMS. Demo programs available for DOS, Windows 3.1, and Macintosh. The demo of the program for PC-DOS can be obtained from anonymous FTP at wuarchive.wustl.edu:/edu/math/msdos/modern.algebra/maplev.zip Student versions available for MAC and PCs. A share library contains many additional Maple routines, packages, and application worksheets. It is distributed with the product, and is also available through anonymous ftp from
daisy.uwaterloo.ca 129.97.140.58
neptune.inf.ethz.ch 129.132.101.33
ftp.inria.fr 128.93.2.54
canb.can.nl 192.16.187.2
ftp.maplesoft.on.ca
Maple's symbolic technology is incorporated into MathCAD (MathSoft, Inc.), MATLAB (The MathWORKS, Inc.), and PV-WAVE (Visual Numerics, Inc).

MAS

Type: Anonymous FTP
Machines: Atari ST (TDI and SPC Modula-2 compilers), IBM PC/AT (M2SDS and Topspeed Modula-2 compilers), and Commodore Amiga (M2AMIGA compiler).
Contact: H. Kredel
Computer Algebra Group
University of Passau, Germany
Version: 0.60
Comments: MAS is an experimental computer algebra system combining imperative programming facilities with algebraic specification capabilities for design and study of algebraic algorithms. MAS is available via anonymous FTP from alice.fmi.uni-passau.de = 123.231.10.1

MATLAB Symbolic Math Toolbox

Type: Commercial

Machines: Microsoft Windows, Sun SPARC, HP 9000/700, DEC, IBM RS/6000, Silicon Graphics. Under preparation: Macintosh.

Contact: The MathWorks, Inc.
24 Prime Park Way Natick,
MA 01760-1500
info@mathworks.com
Phone: 508-653-1415
Fax: 508-653-2997

Version: 1.0

Comments: General purpose. Based on the Maple V computational kernel, with a set of MATLAB M-files designed to make symbolic computation more easy. The MATLAB Symbolic Math Toolbox includes the computational functions in the Maple kernel and the linear algebra package. An Extended Symbolic Math Toolbox is also available that supports Maple procedure execution and includes other Maple packages for statistics, Grobner basis, combinatorial functions, number theory, Euclidean geometry, Lie symmetries, etc.

Mathematica

Type: Commercial

Machines: Macintosh; Power Macintosh; Microsoft Windows; Windows NT; MS-DOS; OS/2; Sun SPARC; Sun Solaris; HP; DEC Alpha OSF/1, OpenVMS AXP, RISC, VAX/VMS; Hitachi 3050RX; IBM RISC; SGI; NEC PC; NEC EWS; NEXTSTEP; CONVEX, and others.

Contact: Wolfram Research, Inc.
100 Trade Center Drive, Champaign, IL 61820-7237
info@wri.com
Phone: 1-800-441-MATH and 217-398-0700
Fax: 217-398-0747
Wolfram Research Europe Ltd.
10 Blenheim Office Park, Lower Road, Long Hanborough
Oxfordshire OX8 8LN, United Kingdom
Phone: +44-(0)1993-883400 Fax: +44-(0)1993-883800
email: info-euro@wri.com
Wolfram Research Asia Ltd.
Izumi Building, 3-2-15 Misaki-cho, Chiyoda-ku Tokyo 101, Japan
info-asia@wri.com
Phone: +81-(0)3-526-05069
Fax: +81-(0)3-526-0509

Version: 2.2

Comments: General purpose, Notebook interface on Windows, NEXTSTEP, Mac, and most Unix platforms (Sun SPARC and Solaris, Silicon Graphics, HP, DEC RISC, IBM Risc System/6000) nice graphics. Packages include MathTensor for Tensors and NCAlgebra for Non-

Commutative Algebra and Combinatorica for Graph Theory. Specialty packages for Finance and Electrical Eng. are sold by WRI. Student version available for MAC and PCs. MathLink libraries included for communicating with external programs (not MS-DOS version).

Mock-Mma

Type:	Anonymous FTP
Machines:	Anywhere running Common LISP.
Contact:	Richard Fateman Department of Computer Science University of California Berkeley, CA 94720 fateman@cs.berkeley.edu
Version:	1.5
Comments:	It is a framework of a parser, display, polynomial manipulator, and a few other pieces, that allow one to program other pieces of a common-lisp systems for computer algebra. The pieces supplied are roughly consistent with the conventions adopted by the Mathematica system. Available through anonymous FTP from from peoplesparc.berkeley.edu:/pub directory.

MuPAD

Type:	Anonymous FTP (free, but not PD; registration required)
Machines:	Sun SPARCstation (under SunOS and Solaris), IBM RS6000, Silicon Graphics, PC (under LINUX), and Apple Macintosh. Prereleased versions exist for 80386-based MS-DOS systems and Sequent Symmetry.
Contact:	MuPAD-Distribution Fachbereich 17 Mathematik-Informatik University of Paderborn Warburger Str. 100 D 4790 Paderborn, Germany MuPAD-distribution@uni-paderborn.de Phone: +49-5251-602633
Version:	1.2
Comments:	General purpose, source available for library routines, graphics support, source code debugger, on-line hypertext help system. Obtained by anonymous FTP at athene.uni-paderborn.de:unix/MuPAD

Reduce

Type:	Commercial
Machines:	Acorn Archimedes, Apple Macintosh, Atari 1040ST and Mega, CDC Cyber 910 and 4000 series, Convex C100, C200, and C300 series, Cray X-MP, Y-MP, and C90, Data General AViiON series, DEC Alpha series running Windows NT, OSF-1 or Open VMS,

DECStation series 2000, 3000, and 5000, DEC VAX running VAX/VMS or Ultrix, Fujitsu M Mainframe Unix series, Fujitsu 2400 series running UXP/M, HP 9000/300, 400, 700, and 800 series, IBM-compatible 286-based PCs running MS-DOS, IBM-compatible 386-based PCs and above running MS-DOS, Windows 3, OS/2, Windows NT, SCO-Unix, Interactive Unix, Solaris and Linux, IBM RISC System/6000, ICL mainframes running VME, ICL DRS6000, NEC EWS 4800 series, NEC PC-9800 series, NEXTstations, Siemens S400.40 series running UXP/M, Silicon Graphics Iris, Indigo, and Onyx machines, Sony NEWS, Sun 3, 4, Sparcstations, and compatibles, Thinking Machines CM5, generic ANSI-C and Common Lisp based versions.

Contact: Anthony C. Hearn
RAND
1700 Main Street
P. O. Box 2138
Santa Monica CA 90407-2138
reduce@rand.org
Phone: 310-393-0411, Ext. 7681
Fax: 310-393-4818

Version: 3.5

Comments: Interactive system and programming language for general purpose symbolic and numeric calculations. Specific application packages for various fields. Complete source available. Demos for MS DOS, Windows, and Linux available from ftp.zib-berlin.de:pub/reduce/demo. Demos for Macintosh and MS DOS available from ftp.bath.ac.uk:pub/jpff/REDUCE. More complete information is available from URL http://www.rrz.uni-koeln.de/REDUCE/ or by sending the e-mail message "send info-package" to reduce-netlib@rand.org.

SACLIB

Type: Anonymous FTP, registration required

Machines: Sun SPARC, Ultrix, Sequent, Generic Unix, and Amiga.

Contact: saclib@risc.uni-linz.ac.at
SACLIB Maintenance
Research Institute for Symbolic Computation
Johannes Kepler University
4020 Linz, Austria

Version: 1.1

Comments: SACLIB is a library of C programs for computer algebra derived from the SAC2 system, and incorporating many improvements. It contains programs for list processing, infinite precision arithmetic (integer, rational and modular), operations on multivariate polynomials, polynomial real root isolation and refinement, and operations with real algebraic numbers and polynomials having algebraic number coefficients. Improved and extended versions of SACLIB, and several SACLIB application packages, are already in advanced stages of development. These include arbitrary precision floating point and interval arithmetic, improved greatest common divisor and factorization algorithms for polynomials, isolation and refinement of complex roots of polynomials, Groebner basis computation, and quantifier elimination.

SENAC

Type: Commercial

Machines: Microsoft Windows (PCs), Unix workstations (DEC, HP, IBM, Sun, SGI), Convex, and Cray. For other implementations please contact us.

Contact: senac@ulcc.ac.uk, senac@waikato.ac.uk
For Europe:
Minaz Punjani
University of London Comp Centre
20 Guilford Street
London WC1N 1DZ
England
Tel: [+44] 71 405 8400
Fax: [+44] 71 242 1845
For rest of the world:
Kevin. A. Broughan
Mathematical Soft Project
University of Waikato
Private Bag 3105
Hamilton, New Zealand
Tel: [+64] 7 856 2889
Fax: [+64] 7 838 4155

Version: 6.0

Comments: General purpose environment for numeric and algebraic computing. An interactive computer algebra host language, an interactive library including Numerical Recipes, graphics library with postscript output, a fully automated symbolic numeric interface to the NAG library, a symbolic graphic interface to the NAG Graphics library. Specialties in large scale optimization and finite element modeling.

Schoonschip

Type: Anonymous FTP

Machines: Amiga, Atari, Sun3, NeXT, MAC II (680x0 machines).

Contact: M. G. J. Veltman or David N. Williams
Physics Department
University of Michigan
Ann Arbor MI 48109-1120
veltman@umiphys.bitnet or veltman@mich.physics.lsa.umich.edu
david.n.williams@um.cc.umich.edu, dwilliams@umiphys.bitnet or 75706.3124@compuserve.com

Version:	91
Comments:	Developed at CERN by Veltman and Strubbe in the 1960s to tackle problems in quantum electrodynamics on the CDC 6000 machine. Written in 600x0 assembler, it is relatively bug free, very fast, capable, and efficient in its use of machine resources. Available from the server "archive.umich.edu" in the directory "physics/schip."

Theorist

Type:	Commercial
Machines:	Macs.
Contact:	Prescience Corp 939 Howard St #333 San Francisco, CA 94103 prescien@well.sf.ca.us Phone: 415-543-2252 Fax: 415-882-0530
Version:	2.53
Comments:	General purpose, good graphics. If you like the Mac interface you will love this. Fixed precision (19 digits). Runs on small Macintoshes, with at least 1 Meg of memory.

Weyl

Type:	Anonymous FTP
Machines:	Needs Common LISP.
Contact:	rz@cs.cornell.edu (Richard Zippel)
Version:	4.275
Comments:	Intended to be incorporated in larger, more specialized systems. Available from ftp.cs.cornell.edu:/pub/Weyl.

42.2 GROUP THEORY

CHEVIE

Type:	Anonymous FTP
Machines:	All Unix workstations, ATARI ST, IBM PC, and MAC (3.1 only).
Contact:	Meinolf Geck and Goetz Pfeiffer Lehrstuhl D für Mathematik RWTH Templergraben 64 D-52062 Aachen, Germany and Gerhard Hiss, Frank Luebeck, and Gunter Malle IWR der Universitaet Heidelberg Im Neuenheimer Feld 368 D-69120 Heidelberg, Germany
Version:	

Comments: CHEVIE is a computer algebra package for symbolic calculations with generic character tables of groups of Lie type (including Green functions), Weyl groups, and Hecke algebras. CHEVIE is accompanied by a library of generic character tables, tables of Green functions, tables of unipotent characters, and character tables of Hecke algebras. It can be obtained via anonymous ftp through samson.math.rwth-aachen.de (137.226.152.6) or ftp.iwr.uni-heidelberg.de (129.206.104.40) in the directory pub/chevie.

GAP

Type: Anonymous FTP (free, but not PD; basically GNU copyleft)
Machines: All Unix workstations, IBM PC (under DOS, Windows, OS/2), MAC (under MPW), Atari ST (under TOS), VAX (under VMS).
Contact: gap@math.rwth-aachen.de
Martin Schoenert
Lehrstuhl D für Mathematik
RWTH
52056 Aachen, Germany
Version: 3.4.1
Comments: GAP is a system mainly for computational group theory. It consists of a small kernel implementing the language, a large library of functions implementing the algorithms, a large library of important groups and character tables, and a 1000 pages manual in LaTeX (also available online). It is available via 'ftp' from ftp.math.rwth-aachen.de, math.ucla.edu, pell.anu.edu.au, ftp.cc.umanitoba.ca(MAC).
Interface to Mathematica available from ftp.math.rwth-aachen.de:/pub/incoming.

LiE

Type: Commercial
Machines: Unix workstations (Sun, DEC, SGI, IBM), NeXT, PCs, Atari, and Macs.
Contact: CAN
Kruislaan 419
1098 VA Amsterdam, The Netherlands
lie@can.nl
Phone: [+31] 20 560-8400
Fax: [+31] 20 560-8448
Version: 2
Comments: Lie group computations

Schur

Type: Commercial
Machines: 386 (with coprocessor) and higher under DOS, Sun3, SPARC, NeXT, RS/6000 (AIX 3.2), HP 9000, SGI, and DEC Alpha series.

Versions for Macintosh, PowerPC, Pentium, PC Windows, and others under development.

Contact: Dr. Steven M. Christensen
P.O. Box 16175
Chapel Hill, NC 27516
stevec@wri.com
Phone/Fax: 919-967-9853

Version: 5.1

Comments: Schur is an interactive program for calculating properties of Lie groups and symmetric functions.

42.3 ALGEBRA AND NUMBER THEORY

Bergman

Type: Anonymous FTP-GNU type license

Machines: Sun SPARC with PSL (Portable Standard Lisp) 3.4.

Contact: Prof. Joergen Backelin
Department of Mathematics
Stockholm University
106 91 Stockholm, Sweden
joeb@matematik.su.se

Version: 0.8—experimental

Comments: Bergman is a Groebner basis calculation program with homogeneous input (although there are some possibilities to homogenize/dehomogenize an arbitrary commutative input), by polynomials in the commutative or the noncommutative polynomial ring over the rationals or a finite prime field, in a finite number of variables. References in Jan-Erik Roos, *J. Pure Appl. Algebra* 91:255–315, 1994 (Bergman relevant pages are 300–302).
Available from matematik.su.se:pub/src/bergman.tar.Z.

CoCoA

Type: Anonymous FTP

Machines: Macs; PCs on the works.

Contact: Gianfranco Niesi
Dipartimento di Matematica
Via L. B. Alberti, 4
I-16134 Genova, Italy
cocoa@igecuniv.bitnet

Version: 1.5

Comments: Computations in commutative algebra. Available on the anonymous FTP at ftp.dm.unipi.it:/pub/alpi-cocoa/cocoa or diskettes (2) sent in a self-addressed to the author.

FELIX

Type: Anonymous FTP

Machines: PC MS-DOS.

Contact:	Joachim Apel and Uwe Klaus Universität Leipzig Augustusplatz 10–11 O-7010 Leipzig, Germany apel@informatik.uni-leipzig.dbp.de klaus@informatik.uni-leizig.dbp.de
Version:	2.5
Comments:	FELIX is designed for computations in and with algebraic structures. The structures implemented so far are commutative polynomial rings, noncommutative algebras, as well as modules over them. Available by anonymous FTP from canb.can.nl in the directory pub/felix.

Galois

Type:	Commercial
Machines:	IBM-PC DOS.
Contact:	CIFEG Inc. Kalkgruberweg 26 A-4040 Linz, Austria
Comments:	Algebra and number theory microcomputer written by R. Lidl, R. W. Matthews, and R. Wells from the University of Tasmania in Turbo Pascal v3.0.

GANITH

Type:	Version 1 is public domain. Current Version 3 is commercial
Machines:	Unix Workstations under X-11 R5 (SUN, SGI, HP, ..).
Contact:	Chandrajit Bajaj Department of Computer Science Purdue University West Lafayette, IN 47907 bajaj@cs.purdue.edu
Comments:	GANITH is an algebraic geometry toolkit, for the computation and visualization of solutions to systems of algebraic equations. Example applications of this for geometric modeling and computer graphics are algebraic curve and surface display, curve–curve intersections, surface–surface intersections, global and local parameterizations, implicitization. GANITH also incorporates techniques for interpolation and least-squares approximation (multivariate data fitting) with algebraic curves and surfaces. Version 1 of GANITH is available from anonymous FTP at ftp.cs.purdue.edu in the file /pub/shastra/ganith-sun.tar.Z. The latest release is being sold by the Technology Transfer Division of Purdue University for $250 (binary) and $1000 (source code).

GRB

Type:	Anonymous FTP
Machines:	Unix workstations and PC DOS, specific executables for DOS running on 8086, 80286, and 80386, available upon request.

Contact: E.L. Green
Virginia Polytechnic Institute and State University
Blacksburg, VA 24061
green@math.vt.edu

Version: 1.0

Comments: GRB provides a computational basis to do algebraic and homological manipulations on algebras and modules. Functionality currently in the package includes computation of a basis for the algebra, reduced Groebner basis for the ideal of relations, the Cartan matrix and determinant for finite dimensional algebras, construction of minimal projective resolutions and coefficients of the Hilbert and Poincare' series. Available from calvin.math.vt.edu in the directory: pub/green.

KAN

Type: Anonymous FTP

Machines: Sun SPARC station.

Contact: Nobuki Takayama
Department of Mathematics
Kobe University
Kobe, Japan
kan@math.s.kobe-u.ac.jp

Version: 1.930528

Comments: Kan is a system for doing algebraic analysis by computer based on computations of Groebner bases. It does computations in the rings of polynomials, differential operators, difference operators, and q-difference operators. Available by anonymous FTP from gradient.scitec.kobe-u.ac.jp (133.30.90.19) in the directory public/Kan.

KANT

Type: Anonymous FTP

Machines: Any machine with a C-compiler; tested on Siemens 7.580-S, Atari, Apollo, Sun3.

Contact: KANT Group
Prof. Dr. M. E. Pohst/Dr. Johannes Graf v. Schmettow
Mathematisches Institut
Heinrich-Heine-Universität
Universitätsstr. 1
D-4000 Düsseldorf 1, Germany pohst@ze8.rz.uni-duesseldorf.de or JUENTGEN@ze8.rz.uni-duesseldorf.de (Max Juentgen).

Version: 2 Rev. 2

Comments: KANT (Computational Algebraic Number Theory) is a subroutine package for algorithms from geometry of numbers and algebraic number theory. There are two versions of KANT: Ver. 1, written in Ansi-FORTRAN 77, has been superseded by Ver. 2, built for the Cayley Platform and written in Ansi-C. Available from the site: ftp.math.uni-sb.de(134.96.32.23):pub/kant/V2R2.tar.z or clio.rz.uni-duesseldorf.- de(134.99.128.3):kant/V2R2.tar.Z.

Macaulay

Type: Anonymous FTP
Machines: Sun, DEC, 386-PC, Mac, binary Mac versions available.
Contact: Dave Bayer (Barnard College)
Department of Mathematics
Columbia University
New York, NY 10027
bayer@cunixa.cc.columbia.edu
Phone: 212-854-2643, 864-4235
Mike Stillman
Department of Mathematics
Cornell University
Ithaca, NY 14853
mike@mssun7.msi.cornell.edu
Phone: 607-255-7240, 277-1835
Version: 3.0
Comments: A system for computing in algebraic geometry and commutative algebra (polynomial rings over finite fields). Available from anonymous FTP from math.harvard.edu (128.103.28.10). IBM-PC versions (2.1 and 3.0) are available from hopf.math.purdue.edu.

Magma

Type: Cost recovery
Machines: Sun3, Sun4, Sun10 under SunOS 4.x and Soalris 2.x DECStation under Ultrix, DEC Alpha under OSF/1, SGI, HP9000/7000, IBM RS6000 under AIX, Apollo M680x0 based machines, DN10000, and Mac running A/UX 2.01.
Contact: The Secretary
Computational Algebra Group
School of Mathematics
University of Sydney
NSW 2006
Australia
magma@maths.su.oz.au
Phone: +61 2 692 3338
Fax: +61 2 692 4534
Version: 1.01–2
Comments: The system is designed to support computation in algebra, number theory, geometry, and algebraic combinatorics. It has an advanced functional programming language with many novel features designed for concise and efficient specification of algebraic algorithms. The kernel has (coded in) the fundamental algorithms for ring theory (polynomial rings, matrix rings, integer rings), field theory (general algebraic number fields—KANT V.2, finite fields, real and complex fields), module theory, group theory (fp groups, permutation groups, soluble groups, and matrix groups), and algebraic combinatorics (coding theory and graph theory). For more information see: http://www.maths.usyd.edu.au:8000/comp/magma/Overview.html.

Numbers

Type: Free but not Public Domain, registration required
Machines: PC-DOS.
Contact: Ivo Duentsch
Rechenzentrum
Universitat Osnabrueck
Postfach 4469
W 4500 Osnabrueck, Germany
duentsch@dosuni1.bitnet
Phone: [49] 541-969-2346
Fax: [49] 541-969-2470
Version: 202c
Comments: Numbers is a calculator for number theory. It performs various routines in elementary number theory, some of which are also usable in algebra or combinatorics. Available in the anonymous FTP in ftp.rz.Uni-Osnabrueck.de in the directory /pub/msdos/math.

PARI

Type: Anonymous FTP
Machines: Most Unix workstations, NeXT, PC386, and Mac. Special 64-bit version for the Dec Alpha machines.
Contact: pari@ceremab.u-bordeaux.fr
Version: 1.38
Comments: Number theoretical computations, source available, key routines are in assembler, ASCII, and Xwindows graphics. Complete package for algebraic number theory computations including handling of ideals, prime ideals, prime ideal factorization, p-adic factorization, etc. Unix and Amiga versions available from megrez.ceremab.u-bordeaux.fr (147.210.16.17), math.ucla.edu, and ftp.inria.fr. PC-DOS version available from anonymous FTP at wuarchive.wustl.edu:/edu/math/msdos/modern.algebra/pari386 or math.ucla.edu (128.97.64.16) in the directory /pub/pari.

SIMATH

Type: Anonymous FTP
Machines: Sun 3/140 under SunOS 3.4; Sun 3/80 under SunOS 4.0.3; Sun SPARCstation under SunOS 4.1.1; Apollo DN 3000 under SR10.1; Apollo DN 4500 under SR10.1; Apollo DN 10000 under SR10.3; Siemens MX2 under SINIX V2.0; Siemens MX2+ under SINIX V5.2, Siemens MX300 under SINIX V5.2, and hopefully other 32-bit UNIX machines.
Contact: SIMATH-Gruppe
Lehrstuhl Prof. Dr. H.G. Zimmer
FB 9 Mathematik
Universitaet des Saarlandes

	D-W-6600 Saarbruecken, Germany simath@math.uni-sb.de, Phone: [0681]302-2206.
Version:	3.7
Comments:	SIMATH is written in C, contains an interactive calculator (simcalc) and many C-functions over algebraic structures as arbitrary long integers, rational numbers, floating point numbers, polynomials, Galoisfields, matrices, elliptic curves, algebraic number fields, modular integers, finite fields, Groebner basis, etc. Version 3.7 contains a handbook written in English. Available from: ftp.math.uni-sb.de (134.96.32.23) or ftp.math.orst.edu (128.193.80.160). You have to get one extra file by an e-mail-Server, see the installation hints.

UBASIC

Type:	Anonymous FTP
Machines:	IBM PCs and compatibles.
Contact:	Yuji Kida, kida@rkmath.rikkyo.ac.jp
Version:	8.65
Comments:	BASIC-like environment for number theory. In the collection of programs written for it one can find MALM (Collection of UBASIC Number Theory Programs) by Donald E. G. Malm (Department of Mathematical Sciences, Oakland University, Rochester, MI 48309-4401, malm@vela.acs.oakland.edu, Phone: 313-370-3425), which includes Baillie-Wagstaff Lucas pseudoprime test, algorithm for Chinese remainder, elliptic curve method to factorize n, Fermat's method of factoring, general periodic continued fraction to quadratic routine, evaluates Carmichael's function and D. H. Lehmer's method of solving $x^2 = q \pmod p$. Other programs include PPMPQS [Prime factorization program for numbers over 80 digits (ubmpqs32.zip)], which comes with the distribution, and many others as well. Available by anonymous FTP from shape.mps.ohio-state.edu:/pub/msdos/ubasic.

42.4 TENSOR ANALYSIS

GRTensorII

Type:	Anonymous FTP
Machines:	All computers that run Maple V, release 3.
Contact:	Kayll Lake and Peter Musgrave Department of Physics Queens university Kingston, Ontario, Canada K7L 3N6 lake@bill.phy.queensu.ca
Version:	1.0 (beta)
Comments:	GRTensorII is a package for component calculations of tensors, basis/tetrads, and NP tetrads. Available by anonymous ftp from

astro.queensu.ca:pub/grtensor or http://astro.queensu.ca/~grtensor/GRHome.html.
Mathematica version under development.

MathTensor

Type:	Commercial
Machines:	All computers that run Mathematica 1.2, 2.0, or greater.
Contact:	MathSolutions, Inc. P. O. Box 16175 Chapel Hill, NC 27516 mathtensor@wri.com Phone and Fax: 919-967-9853
Version:	2.1.5
Comments:	Elementary and advanced tensor analysis. Adds more than 250 functions and objects to Mathematica.

Ricci

Type:	Anonymous FTP from otter.stanford.edu
Machines:	All computers that run Mathematica 2.0 or greater.
Contact:	Jack Lee Department of Mathematics University of Washington Seattle, WA 98195 lee@math.washington.edu Fax: 206-543-0397
Version:	1.0
Comments:	Mathematica package for doing tensor computations in differential geometry, including coordinate-free tensor operations as well as indexed tensors with dummy indices and the summation convention.

STENSOR

Type:	Symbolic contribution of $250
Machines:	IBM 3090, VAX DECStation, Sun, Apollos, Orion, Atari and Amiga, IBM 386, 486, and MacIntosh in the works.
Contact:	Lars Hornfeldt Physics Department University of Stockholm Vanadisvagen 9 S-113 46 Stockholm, Sweden lh@vand.physto.se, Phone [46] 8-837114
Version:	2.31
Comments:	System for tensor calculus and noncommutative algebra. It includes the systems SHEEP/CLASSI.

42.5 DIFFERENTIAL EQUATIONS

DELiA

Type:	Informal distribution
Machines:	IBM PC's (DOS), MS-Windows version in the works.
Contact:	IBB Trading Co. 715 Oceanview Ave. Brooklyn NY 11235 or A. V. Bocharov Wolfram Research 100 Trade Center Dr. Urbana-Champaign, IL 61820-7237 alexei@wri.com
Comments:	Differential equation computations.

LIE

Type:	Public Domain
Machines:	PC-DOS
Contact:	A. K. Head CSIRO Division of Materials Science and Technology Melbourne Australia or Locked Bag 33, Clayton, Vic 3168, Australia head@rivett.mst.csiro.au Phone: 03-542-2861 Telex: AA 32945 Fax: 03-544-1128
Version:	4.3
Comments:	LIE is a program written in the MuMath language (not a package) for Lie analysis of differential equations. Available from anonymous FTP at wuarchive.wustl.edu: /edu/math/msdos/adv.diff.equations/ lie43.

ODE

Type:	Commercial and Public Domain versions
Machines:	IBM PCs (DOS).
Contact:	Keith Briggs Department of Applied Mathematics University of Adelaide, South Australia 5005 kbriggs@maths.adelaide.edu.au Phone: (08) 303 5577 Fax: (08) 232 5670 or PO Box 75 Heidelberg West Australia 3081
Version:	2.6

Comments: Differential equation computations, easy-to-use menus with mouse support. Support for CGA, EGA, and VGA with outputs in LaTeX and PostScript for graphics. For the commercial version contact the author; the public Domain version is available at wuarchive.wustl.edu:edu/math/msdos/diff.equations/ode 26.zip.

42.6 PC SHAREWARE

Calculus

Type: Shareware
Machines: PC-DOS with EGA.
Contact: Byoung Keum
Department of Mathematics
University of Illinois
Urbana, IL 61801
Version: 9.0
Comments: Program for calculus and differential equations. It has symbolic differentiation and integration (simple functions), graphs. Very unstable program—no reason to use it, except for price (suggested registration fee is $30.00). Available from anonymous FTP at wuarchive.wustl.edu: /edu/math/msdos/calculus/calc.arc.

CLA

Type: Anonymous FTP
Machines: PC-DOS.
Contact: Lenimar Nunes de Andrade
UFPB-CCEN-Dep. de Matemática
58.059 João Pessoa, PB Brazil
ccendm03@brufpb.bitnet
Version: 2.0
Comments: A linear or matrix algebra package that computes rank, determinant, row-reduced echelon form, Jordan canonical form, characteristic equation, eigenvalues, etc. of a matrix. Available from anonymous FTP at wuarchive.wustl.edu:/edu/math/msdos/linear.algebra/cla20.zip.

Mercury

Type: Shareware
Machines: PC-DOS.
Contact: Roger Schlafly
Real Software
PO Box 1680
Soquel, CA 95073
76646.323@compuserve.com
TTDX08A@prodigy.com
Phone or Fax: 408-476-3550

Version: 2.06

Comments: Limited in symbolic capabilities, but is extremely adept at numerically solving equations and produces publication quality graphical output. This used to be Borland's Eureka!, but when Borland abandoned it, its original author started selling it as shareware under the name Mercury. Available from anonymous FTP at wuarchive.wustl.edu:/edu/math/ msdos/calculus/mrcry206.zip.

PFSA

Type: Public Domain

Machines: PC-DOS and a Unix noninteractive version.

Contact: Don Stevens
Courant Institute
251 Mercer St. New York, NY 10012
stevens@cims.nyu.edu
Phone: 212-998-3275

Version: 5.46

Comments: Written in FORTRAN, it is very fast but is limited to polynomial algebra and calculus. The DOS version is available from the anonymous FTP at wuarchive.wustl.edu:/edu/math/msdos /modern.algebra/vol546.zip and the Unix version at math.berkeley.edu:pub/PFSA.

SymbMath

Type: Shareware, student and advanced versions

Machines: IBM PC.

Contact: Dr Weiguang Huang
Department of Analytical Chemistry
University of New South Wales
Kensington NSW 2033 Australia
w.huang@unsw.edu.au or s9300078@cumulus.csd.unsw.oz.au
Phone: [61] 2-697-4643
Fax: [61] 2-662-2835

Version: 3.1

Comments: Runs on plain (640k) DOS machines (8086). An expert system that can learn from the user. More capable versions are available by mailorder from the author. Also available by anonymous FTP from math.bekerley.edu: /pub/Software/SymbMath/sm31a.zip oak.oakland.edu: /pub/msdos/calculator/sm31a.zip garbo.uwasa.fi: /pc/math/sm31a.zip wuarchive.wustl.edu: /mirrors/msdos/calculator/sm31a.zip or by sending the following e-mail: /pdget mail pd:<msdos.calculator>sm31a.zip to listserv@vm1.nodak.edu, or listserv@ndsuvm1.bitnet.

(X)PLORE

Type: Anonymous FTP

Machines: PC-DOS, Windows.

Contact: David Meredith
Department of Mathematics
San Francisco State University
San Francisco, CA 94132
meredith@sfsuvax1.sfsu.edu

Version: 4.0

Comments: Formerly called CCALC. Well-integrated graphics with (numerical) matrix manipulation routines. Intended for calculus students. Prentice-Hall sells this with a book (ISBN 0-13-014226-3 or by calling 800-947-7700), but it is also available (without the manual but with a comprehensive help system) by anonymous FTP from wuarchive.wustl.edu: /edu/math/msdos/calculus/cc4-9206.zip. The author can provide users with a version that uses the coprocessor (80x87), and contains a few bug fixes. You can ftp the most recent version from sutro.sfsu.edu:math/.

42.7 THEOREM PROVERS

IMPS

Type: Anonymous FTP

Machines: SUN SPARCstation under OpenWindows or X11, with 24 Megabytes of memory.

Contact: imps-request@linus.mitre.org

Version: 1.0

Comments: IMPS is an interactive mathematical reasoning system and theorem prover developed at The MITRE Corporation. It is intended to provide organizational and computational support for the traditional techniques of mathematical reasoning. The system consists of a database of mathematics (represented as a network of axiomatic theories linked by theory interpretations) and a collection of tools for exploring, applying, extending, and communicating the mathematics in the database. It is available from math.harvard.edu:imps.

43

Minitab

William C. Rinaman
Le Moyne College

Minitab is a widely used statistical package. Versions of it run on DOS, Microsoft Windows, Macintosh, VMS, and Unix. It provides a wide variety of procedures for data analysis. It allows the user to enter, edit, and manipulate data. The data analysis features include exploratory data analysis, basic statistics, regression analysis, analysis of variance, multivariate analysis, nonparametric statistics, time series analysis, simulations, professional graphics, and quality control and improvement. In addition, Minitab has a macro capability that permits users to program commands to carry out procedures that are not part of Minitab.

The user interface for the Minitab system varies depending on the platform on which it is run. However, the essential features—the worksheet structure and commands—are the same for all environments. This means that a user who is familiar with Minitab under, say, VMS should encounter little difficulty using the program in Microsoft Windows. This exposition will discuss only the elements of Minitab that are common to all environments.

43.1 WORKSHEET

The Minitab system is centered on what is termed the *worksheet*. It consists of a number of columns denoted C1, C2, C3, . . ., a similar number of constants denoted K1, K2, K3, . . . and a number of matrices denoted M1, M2, M3. . . . The number of these depends on the platform on which Minitab is installed and available memory. For example, Minitab Release 8 for DOS has a default worksheet size of 16,714 data elements consisting of 100 columns, 100 constants, and 15 matrices. Release 9 for Windows has a default worksheet size of 100,000 data elements that are allocated to 1000 columns, 1000 constants, and 100 matrices. In all versions the highest numbered constant has a default value of π, the next highest numbered constant has a default value of e, and the third highest numbered constant has a default value of '*', for use as a missing data indicator. Columns, constants, and matrices are referred to in one of two ways. First you may use their numbers, such as C12. Second, you may assign each a name and use their names when referencing them.

0-8493-2530-7/96/$0.00+$.50

43.2 COMMANDS

All Minitab commands have a common structure. This is independent of the interface. Menu-driven versions allow the user to generate and execute commands by making menu selections. On systems without a menu interface, commands are entered by typing them at the Minitab prompt. A command always begins with a command name. Minitab requires only the first four letters of the command name, although the full command name is recommended to enhance readability. Minitab commands are not case sensitive. The command name is usually followed by a list of *arguments.* The arguments specify the appropriate columns, constants, or matrices that are to be used. In addition, many Minitab commands have subcommands that have a structure like that of commands.

Commands are described in both Help and the documentation according to the following structure. Command descriptions use the special symbols: C, K, E, M, FILENAME, and square brackets. Any command description that uses C means that you must replace C by a specific column, using either the number (e.g., C12) or name, enclosed in single quotes (e.g., 'TEMPERATURE'). If a command description uses a K you must replace K by either a specific number (e.g., 9.8) or a stored constant (e.g., K9 or its name). An instance of E indicates that you replace E by either a column, a number or a stored constant. Replace M by a matrix (e.g., M4 or its name). Any extra text used to explain the command is written in lower case. Any argument surrounded by square brackets ([ ]) is optional and may be omitted. The command is entered on a single line with the available subcommands listed below it. In general, if optional arguments and subcommands are not used, then Minitab uses built-in default settings. Some commands such as STOP, which exits the user from the Minitab system, have no arguments. Here are some examples.

1. The description of the TTEST command is given as follows

   ```
   TTEST [of mu = K] on data in C,..,C
   ALTERNATIVE = K
   ```

   This command performs a single-sample *t* test on the columns listed. The particular value of the null hypothesis mean is indicated as being optional. A detailed description of this command will show that the default value for mu is 0. In addition, the fact that "of mu =" and "on data in" are shown in lower case means that they are extra text that may be included to enhance the readability of the command, omitted or altered. The command has one possible subcommand that specifies the alternative hypothesis. The default test is a two-tailed *t* test. ALTERNATIVE = 1 gives an upper tail test, and ALTERNATIVE = –1 gives a lower tail test. The fact that a subcommand is being used is indicated by terminating the main command line with a semicolon. Upon typing a semicolon and entering that line the Minitab system prompt MTB> will change to a SUBC> prompt indicating that a subcommand is expected. Each subcommand should be terminated with a semicolon except for the final one, which is terminated with a period. Thus, the command sequence

   ```
   MTB> TTEST of mu = 50 on data C1,C4,C7;
   SUBC> ALTERNATIVE = 1.
   ```

will conduct three separate *t* tests on the data in columns C1, C4, and C7. Each test will test the null hypothesis that the mean is 50 against the alternative hypothesis that the mean is greater than 50. A terse version of this command could look like the following:

```
MTB> TTES 50 C1,C4,C7;
SUBC< ALTE 1.
```

2. The sample mean is computed using the MEAN command. Its description is as follows:

```
MEAN of the values in C [put into K]
```

This will compute the sample mean of the values in a column. If no constant is specified for storage, the result will be printed on the output device, typically the terminal screen. Thus,

```
MEAN C2 K3
```

will compute the mean of the data in column C2 and store the value in K3. The command

```
MEAN C2
```

will compute the mean of the data in C2 and display the result on the screen.

3. The FILENAME parameter shows up in commands that read or store worksheets, data, commands, and Minitab sessions. For example, the READ command, which can copy data from an ASCII file into a Minitab worksheet, is described as follows:

```
READ data [from 'FILENAME'] into C, . . ., C
```

This means that the optional FILENAME parameter should be replaced by a specific file name, enclosed in single quotes. To read data from the file TENSILE.DAT into columns C1 through C4 you would use the following.

```
READ 'TENSILE.DAT' C1,C2,C3,C4
```

Files and devices will be discussed in more detail later.

43.3 USING HELP

Minitab has a help system that explains how to use all of the built-in commands and explains other aspects of the system. Accessing it depends on the environment. For example, in Release 8 for DOS one types *help commands* at the Minitab prompt to obtain a listing of the categories of commands as shown below.

```
MTB > help commands
To get a list of the Minitab commands in one of the cate-
gories below, type HELP COMMANDS followed by the appropriate
number, for example, HELP COMMANDS 1 for General Information.
1 General Information            12 Time Series
2 Input and Output of Data       13 Statistical Process Control
3 Editing and Manipulating       14 Exploratory Data Analysis
  data                           15 Distributions & Random Data
4 Arithmetic                     16 Sorting
5 Plotting Data                  17 Matrices
6 Basic Statistics               18 Miscellaneous
7 Regression                     19 Stored Commands and Loops
8 Analysis of Variance           20 Design of Experiments
9 Multivariate Analysis          21 QC Macros
10 Nonparametrics                22 How Commands are
11 Tables                           Explained in Help
```

To see what commands are in a given category you simply enter ***help commands k***, where ***k*** is the number of the category of interest. For example, to see what commands are available for plotting you would proceed as follows.

```
MTB > help commands 5
Plotting Data

Standard graphics commands:

 HISTOGRAM        displays a histogram
 STEM-AND-LEAF    does a stem-and-leaf plot
 DOTPLOT          displays a dotplot
 BOXPLOT          does a boxplot
 PLOT             plots y versus x
 MPLOT            plots several variables all on the same axes
 LPLOT            plots y versus x, using letters for group
                     membership
 TPLOT            does a pseudo three-dimensional plot
 TSPLOT           plots a time series
 MTSPLOT          plots several time series, all on the same
                     axes
 GRID             creates a grid of points for a contour plot
 CONTOUR          does a contour plot
 WIDTH            specifies the width of plots
 HEIGHT           specifies the height of plots

Statistical Process Control Charts:

 XBARCHART        MACHART         NPCHART
 RCHART           EWMACHART       CCHART
 SCHART           MRCHART         UCHART
 ICHART           PCHART
```

For high-resolution graphics add a G before the command name, for example, GHISTOGRAM, GPLOT, or GXBARCHART. GOPTIONS. . . . select graphics device and set options. To produce high-resolution graphics, you must first select the appropriate screen, printer, and/or plotter drivers using the MSETUP procedure. See the *Reference Manual*, PC Version for details. Note, a graphics monitor is required to execute high-resolution graphics commands.

By default, Minitab produces graphics output suitable for printers. To produce plotter-style output, type

```
MTB > GOPTIONS;
SUBC> PLOTTER.
```

before entering your graphics command. See Help on GOPTIONS for more information.

This command gives a very brief description of what the commands in a category do. To find out exactly what the command description is, to obtain information on how Minitab processes the data for a command, and to find out what subcommands are available, you *help command* for that command. For example, the information provided for the REGRESS command is shown below.

```
MTB > help regress
REGRESS C on K predictors C, . . ., C
```

Subcommands:

```
NOCONSTANT      RMATRIX        TRESIDUALS     DW
WEIGHTS         HI             COOKD          PURE
MSE             FITS           DFITS          XLOF
COEFFICIENTS    SRESIDUALS     PREDICT        TOLERANCE
XPXINV          RESIDUALS      VIF
```

Fits the regression equation y = b0 + b1*X1 + b2*X2 + . . . + bk*Xk to data in selected response and predictor variables. The first column given on the command line corresponds to the response variable. The number of predictors is given next, followed by columns containing the predictors.

Use the subcommands SRESIDUALS and FITS to store the standardized residuals and fits. The standardized residuals are ei/stdev (ei), where ei is the residual and stdev(ei) = SQRT(MSE - Var(Yhati)). The fitted value for i-th observation is

```
Yhati = b0 + b1X1 + · · · + bkXk.
```

Note: In previous versions of Minitab, the syntax for REGRESS was

```
REGRESS C on K predictors C . . . C [put stand. residuals in
   C [fits in C]]
```

That is, you could store the standardized residuals and fits on the REGRESS command line rather than with the SRES and FITS subcommands. This older syntax still works as long as you do not also use the SRES and FITS subcommands.

To control the amount of printed output, use the command BRIEF.

Missing Data

All observations that contain one or more missing values (either in the dependent or one or more of the independent variables) are not used in any of the regression calculations, with two exceptions: If Yi is missing but all predictors are present (or if case i has a weight = 0 and all predictors are present), then Yhati is calculated, and hi is calculated as xi*(INV(X'X))*xi', where xi is the row vector of predictors for the i-th observation and X is the design matrix with the i-th observation deleted.

Ill-Conditioned Data

See the *Minitab Reference Manual* for a discussion on the handling of ill-conditioned data. Ill-conditioned data refers to cases where some predictors are highly correlated with other predictors, or when a predictor variable has a small coefficient of variation.

The computational method used for regression is Givens transformations using Linpack routines. The method is described in Chapter 10 of the Linpack *User's Guide*. See the *Minitab Reference Manual* for reference listings.

To obtain a description of a subcommand you type *help command subcommand* at the Minitab prompt. For example, the information on the RMATRIX subcommand in REGRESS would be obtained as follows.

```
MTB > help regress rmatrix
RMATRIX put into M
```

Stores the R matrix of the QR decomposition. Suppose X is the n by p design matrix (n = number of observations, p = number of coefficients) of the regression. Then there is an n by p orthogonal matrix Q (thus, Q'Q = I, the pxp identity matrix) and a p by p upper triangular matrix R such that X = QR. This upper triangular matrix is stored in M.

43.4 ENTERING AND SAVING DATA

The Minitab commands READ, SET, INSERT, END, NAME, and RETRIEVE are used for placing data in the worksheet. NAME is used to assign names to columns, constants, and matrices. This is an aid in making Minitab sessions more readable. END is used to signal the end of data in the SET and READ commands.

Data may be entered directly with the SET, READ, and INSERT commands. These commands also facilitate entering data from already existing ASCII files. The RETRIEVE command enters the contents of a previously saved worksheet into the current worksheet. If no extension to the file name is given Minitab will retrieve the file with the specified FILENAME and a default extension of .MTW. If the PORTABLE subcommand is used, the default extension is .MTP. The PORTABLE format for Minitab worksheet files is used for files that are saved using one version of Minitab that are to be read using another.

To illustrate data entry, placing data into columns C1, C2, and C3 using the READ command is accomplished as follows.

```
MTB> read C1,C2,C3
DATA> 20 70 53
DATA> 25 88 66
DATA> 30 65 52
DATA> 35 72 63
DATA> 40 64 66
DATA> end
```

Here, each line corresponds to a row. That is, 20, 70, and 53 will be entered in row 1 for columns C1, C2, and C3, respectively. These data could have been entered by using the SET command three times, once for each column. This would look like the following.

```
MTB> SET C1
DATA> 20 25 30 35 40
DATA> end
MTB> SET C2
DATA> 70 88 65 72 64
DATA> end
MTB> SET C3
DATA> 53 66 52 63 66
DATA> end
```

In addition, the READ command using a FILENAME would read these data from a file that contained the following.

```
20 70 53
25 88 66
30 65 52
35 72 63
40 64 66
```

If the file were named MYDATA.DAT the command would look like the following.

```
MTB> read 'mydata.dat' C1,C2,C3
```

It should be noted that data in the READ, SET, and INSERT commands may be separated by spaces, as shown, or by commas. The READ, SET, and INSERT commands allow the data in ASCII files to have a format that is described by FORTRAN format codes. In many environments, such as Release 8 for DOS, Release 9 for Windows, and the Macintosh, data may be placed in columns by typing them directly in a data window that is navigated in a manner like using a spreadsheet.

The contents of any columns, constants, or matrices may be displayed on the terminal screen using the PRINT command. PRINT does not save data nor does it send output to a printer. Printing may be displayed according to FORTRAN format codes specified by the user.

Data may be saved using the WRITE or SAVE commands. WRITE saves data in an ASCII file. The output of WRITE may be formatted using FORTRAN format codes. The current contents of a Minitab worksheet are saved using the SAVE command. The resulting file will contain the contents of all nonempty columns, constants, and matrices. A worksheet file created by the SAVE command will write a file with a default extension .MTW. If the PORTABLE subcommand is used the default extension is .MTP.

Minitab provides two methods of recording what takes place during a session. A copy of all commands you type and output seen on the screen is made by using the OUTFILE command. The general form is as follows.

```
MTB> OUTFILE 'FILENAME'
```

This will produce an ASCII file of what takes place in your session until the command NOOUTFILE is issued. High-resolution graphics and data that are entered are not saved by OUTFILE. The default extension for OUTFILE files is .LIS.

A copy of all commands you type and data you enter but no responses by Minitab can be maintained in an ASCII file by using the JOURNAL command. The general form is as follows.

```
MTB> JOURNAL 'FILENAME'
```

The default extension for journal files is .MTJ. Recording is stopped by issuing the command NOJOURNAL. This feature is useful in creating and editing Minitab macros.

43.5 EDITING AND MANIPULATING DATA

There are a wide variety of commands that facilitate making changes to data stored in a worksheet. We shall give a brief description of some of these that are most generally useful. Use Help to find out about commands not mentioned here. We begin with some editing commands. Here we are assuming that changes are being made to data that have been entered in a worksheet by any of the methods discussed above.

Data may be edited using the LET, DELETE, ERASE, and COPY commands. ERASE will delete the contents of any columns or constants that are specified. The form of the command is

```
MTB> ERASE E, . . ., E
```

DELETE allows the user to erase the contents of specified rows of indicated columns. This command looks like the following.

```
MTB> DELETE rows K, . . ., K of columns C, . . ., C
```

Individual entries in a worksheet may be altered by using the LET command. This command may also be used for transforming data as we shall see later. For editing purposes the LET command looks like the following.

```
MTB> LET C (K) = K
```

The first K specifies which row in column C is to have the second value of K entered. The LET command may be used to place values in constants. In those environments with a data window the contents of worksheet columns are most easily altered by going to the data window and making changes as you would in a spreadsheet.

Data in the columns of a worksheet may be manipulated using the COPY, CODE, CONVERT, STACK, UNSTACK, or CONCATENATE commands. We shall give a brief overview of these. The COPY command is highly flexible in that it permits the user to copy entire columns or portions of columns to other columns. In addition, a list of constants may be copied to a column. CODE is designed to let the user assign values to ranges of numbers in a column. CONVERT allows the user to take character data and convert them to numerical data and vice versa. STACK and UNSTACK are designed to reconfigure data that are classified in separate groups. STACK will combine grouped data from separate columns in a single column, and UNSTACK does the reverse. CONCATENATE is designed to combine a number of columns of alphabetic data into a single column.

Numerical data may be transformed in a variety of ways. There are a number of built-in functions such as absolute value, sign, exponential functions, logarithms, trigonometric functions, inverse trigonometric functions for processing data. Logical operators AND, OR, and NOT are also available. There are also a large number of statistical functions such as MEAN, MEDIAN, STDEV, SSQ, SORT, and RANK available. For a complete list of these functions consult either Help, the *Minitab Handbook* or the *Minitab Reference Manual*. These functions may be combined with the LET command to alter data in columns or constants. For example, the natural logarithm of the data in column C1 can be stored in column C10 as follows.

```
MTB> LET C10 = LOGE (C1)
```

These functions may be combined with the standard FORTRAN operators for addition, subtraction, multiplication, division, and exponentiation to produce more complex transformations. Thus, to compute the standardized values of data in column C1 and place them in column C11 we would proceed as follows.

```
MTB> LET C11 = (C1 - MEAN(C1))/STDEV(C1)
```

In addition, there are separate procedures ADD, SUBTRACT, MULTIPLY, DIVIDE, and RAISE available for performing addition, subtraction, multiplication, division, and exponentiation, respectively, on data in columns, constants, and matrices. Consult the references for details on using them.

43.6 STATISTICAL PROCEDURES

Once data have been entered into a worksheet (and, if necessary, manipulated and transformed), you are now ready to perform statistical analyses. It is not possible to give detailed descriptions of every statistical procedure in the space allotted. We shall, instead, give a list of command names that are in each category and provide a short indication of what each does. There is a small difference in the availability of some commands among different versions of Minitab. Consult the *Reference Manual* for your version to make sure a command you wish to use is supported. Examples of the output from selected commands will also be shown.

Basic Statistics

DESCRIBE	Computes standard descriptive statistics of data in a column
ZINTERVAL	Constructs a confidence interval of the mean for data with a known variance
ZTEST	Performs a single sample z-test on the mean
TINTERVAL	Constructs a confidence interval of the mean using Student's t test
TTEST	Performs a single sample t test on the mean
TWOSAMPLE	A two sample t test and confidence interval with one sample per column
TWOT	Like TWOSAMPLE but data in one column and subscripts in another
CORRELATION	Computes correlations and creates a correlation matrix
COVARIANCE	Computes covariances and creates a covariance matrix
CENTER	Centers and scales data. Can create standardized observations

Plots

HISTOGRAM	Produces a histogram of data in a column
STEM-AND-LEAF	Generates a stem-and-leaf plot of data in a column
DOTPLOT	Does a dotplot of data in a column
BOXPLOT	Produces a box-and-whisker plot of data in a column
PLOT	Plots data in one column versus that in another
MPLOT	Shows several variables on the same axes
LPLOT	Like PLOT but uses letters for group membership
TPLOT	Produces a pseudo three-dimensional plot. Uses symbols for z value
GRID	Creates a grid for a contour plot
CONTOUR	Generates a contour plot

Plot names with a G appended at the start of the name will produce high-resolution graphics.

Regression

REGRESS	Performs linear and polynomial regression
STEPWISE	Performs stepwise regression using forward selection or backward elimination
BREGRESS	Performs best subsets regression using the maximum R-squared criterion
RREGRESS	Performs robust and rank-based regression

Analysis of Variance

AOVONEWAY	Performs one way analysis of variance, with one group per column
ONEWAYAOV	Like AOVONEWAY but all data in one column and subscripts in another
TWOWAYAOV	Performs a balanced two-way analysis of variance
ANOVA	Does analysis of balanced mixed models with crossed and nested factors

ANCOVA	Analyzes orthogonal designs with fixed effects
GLM	Fits the general linear model. Includes unbalanced designs

Multivariate Analysis

PCA	Performs principal component analysis
DISCRIMINANT	Performs linear and quadratic discriminant analysis
FACTOR	Does factor analysis

Nonparametrics

RUNS	Performs a runs test for randomness
STEST	Conducts a single sample sign test on the median
SINTERVAL	Constructs a confidence interval for the median based on the sign test
WTEST	Performs a one-sample Wilcoxon signed rank test on the median
WINTERVAL	Constructs a confidence interval for the median based on the Wilcoxon test
MANN-WHITNEY	Performs a two-sample Mann–Whitney–Wilcoxon rank sum test and constructs a confidence interval
KRUSKAL-WALLIS	Performs the Kruskal–Wallis test of equality of k medians
MOOD	Performs the Mood median test
FRIEDMAN	Performs the Friedman test for randomized blocks
WALSH	Computes Walsh averages of all pairs
WDIFF	Computes all pairwise differences
WSLOPE	Computes all pairwise slopes

Contingency Tables

TABLE	Constructs contingency tables and tables of statistics for associated variables
TALLY	Computes one-way tables of counts and percents
CHISQUARE	Performs a chi-square analysis for a contingency table

Time Series

TSPLOT	Constructs a time series plot
MTSPLOT	Plots several time series
ACF	Computes the autocorrelation function for a time series
PACF	Computes the partial autocorrelation function for a time series
CCF	Computes a cross-correlation function
DIFFERENCE	Differences a time series
LAG	Lags a time series
ARIMA	Fits a Box-Jenkins ARIMA model to a time series

Statistical Process Control Charts

XBARCHART	Produces a chart of sample means
MACHART	A moving average chart

NPCHART	A number of nonconformities chart
RCHART	Sample range chart
EWMACHART	Exponentially weighted moving average chart
CCHART	Poisson counts of nonconformities chart
SCHART	Standard deviations chart
MRCHART	Moving ranges chart
UCHART	Poisson counts per unit for nonconformities chart
ICHART	Chart of individual observations
PCHART	Proportion of nonconformities chart

High-resolution versions of these chart are obtained by prepending each command with the G.

Exploratory Data Analysis

STEM-AND-LEAF	Produces a stem-and-leaf display of data in a column
BOXPLOT	Generates box-and-whisker plots of data
GBOXPLOT	Creates a high-resolution graphics version of BOXPLOT
LVALS	Produces a letter-value display
CPLOT	Creates a condensed scatterplot
RLINE	Fits an outlier resistant line to data
RSMOOTH	Smooths data (usually a time series)
CTABLE	Creates a two-way table in a coded form
MPOLISH	Analyzes a two-way layout using median Polish
ROOTOGRAM	Creates a suspended rootogram

Probability Distribution and Random Data

RANDOM	Generates random observations from a variety of distributions
PD	Evaluates the probability functions for discrete distributions and the probability density functions for continuous distribution
CDF	Evaluates the cumulative distribution function for a variety of distributions
INVCDF	Calculates the inverse cumulative distribution function for a variety of distributions
SAMPLE	Takes a random sample, with or without replacement, of specified rows

Design of Experiments

FFDESIGN	Creates full and fractional factorial two level designs
PBDESIGN	Creates Blackett–Burman designs
FFACTORIAL	Fits full and fractional 2-level designs, both orthogonal and nonorthogonal

Quality Control Macros

ANOM	Conducts one-way and two-way analysis of means
CAPA	Generates a process capability histogram and statistics
CUSUM	Plots cumulative sum control charts
PARETO	Plots a pareto diagram

RSDESIGN	Creates Central Composite designs for 2 through 6 factors and Box–Behnken designs for 3 through 6 factors
RSMODEL	Fits a quadratic model to designs created by RSDESIGN

We now give some examples of the use of these commands.

Example 1: Confidence Intervals

The description of the TINTERVAL procedure is as follows.

```
TINTERVAL [K% confidence] for C, . . ., C
```

The lifetimes of 101 aluminum strips, in thousands of cycles, subjected to alternating stress have been stored in column C1. We wish to construct a 99% confidence interval for the mean lifetime. The Minitab session would look as follows.

```
MTB > Read 'A:\ALUM.DAT' c1
Entering data from file: A:\ALUM.DAT
  101 rows read.
MTB > TInterval 99.0 C1
        N      MEAN     STDEV   SE MEAN   99.0 PERCENT C.I.
C1    101    1400.9     391.3      38.9   (1298.6, 1503.2)
```

Example 2: Regression Analysis

The general form for the REGRESS command is shown in the section describing Minitab help. We illustrate its use in the following session. The dependent variable, Y, is stored in column C1, and the independent variables, X_1, X_2, and X_3 are stored in columns C2, C3, and C4, respectively. The residuals will be stored in column C10, and the predicted values for Y will be stored in column C11. Additional residual analysis will be done by generating a histogram of the residuals, plotting the residuals against the predicted values, and creating a normal probability plot of the residuals. These will use the HISTOGRAM, PLOT, and NSCORES commands.

```
MTB > read c1-c4
DATA> 20 2 1 1
DATA> 25 3 2 4
DATA> 10 1 4 3
DATA> 12 2 8 6
DATA> 4 1 8 4
DATA> 24 2 4 4
DATA> end
 6 ROWS READ
MTB > name c1 = 'Y' C2 = 'X_1' C3 = 'X_2' C4 = 'X_3'
MTB > regress 'Y' 3 'X_1' 'X_2' 'X_3' C10 C11
```

The regression equation is

$$Y = 11.7 + 4.25\ X\_1 - 2.45\ X\_2 + 2.01\ X\_3$$

Predictor	Coef	Stdev	*t*-ratio	*p*
Constant	11.662	9.904	1.18	0.360
X_1	4.250	5.680	0.75	0.532
X_2	−2.445	2.169	−1.13	0.377
X_3	2.014	3.424	0.59	0.616
s = 4.638	R-sq = 87.9%	R-sq(adj) = 69.9%		

Analysis of Variance

SOURCE	DF	SS	MS	*F*	*p*
Regression	3	313.82	104.61	4.86	0.175
Error	2	43.02	21.51		
Total	5	356.83			

SOURCE	DF	SEQ SS
X_1	1	254.13
X_2	1	52.25
X_3	1	7.44

```
MTB > print c10, c11
```

ROW	C10	C11
1	0.15947	19.7305
2	−0.96685	27.5762
3	−1.32456	12.1719
4	−0.23939	12.6816
5	−0.15947	4.4043
6	1.39823	18.4355

```
Histogram of Residuals
MTB > ghistogram c10
```

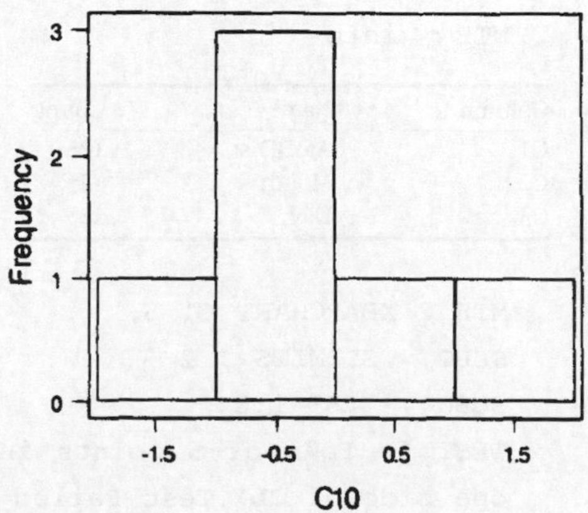

Normal Plot of Residuals

```
MTB > nscores c10 c12
MTB > GPlot C12 C10;
SUBC > Symbol.
```

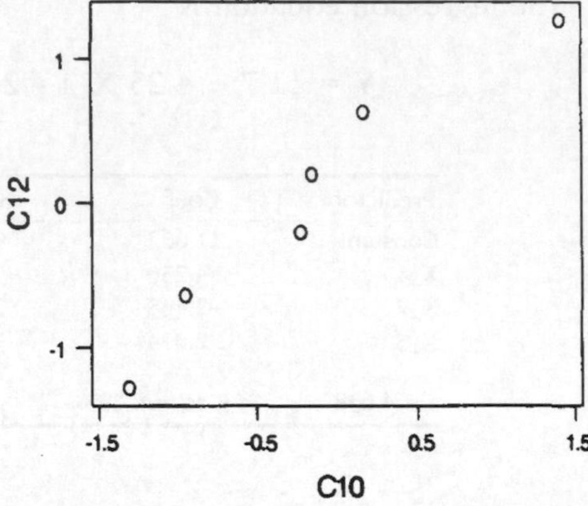

Residual Plot

```
MTB > GPlot C10 C11;
SUBC> Symbol.
```

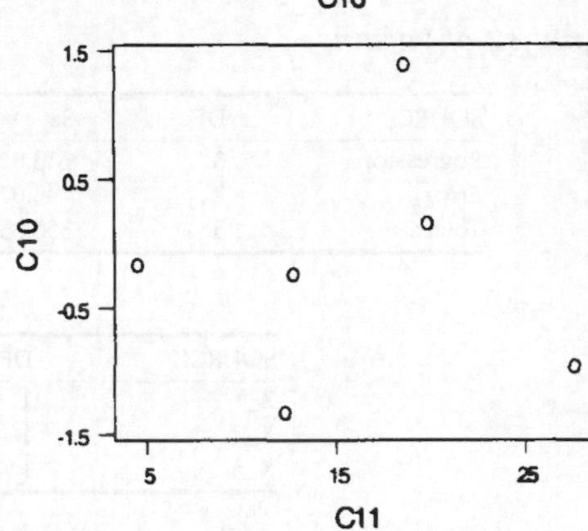

Example 3: Quality Control Chart

As a final example we show an $\bar{X}$ chart. The data are the crankshaft worksheet that is included with Minitab. Means are computed for 5 consecutive observations, limits of 1, 2, and 3 sigma, and 8 tests of the data are computed.

```
MTB > Retrieve 'D:\MTBWIN\ DATA\CRANKSH.MTW'
Retrieving worksheet from file: D:\MTBWIN\DATA\
  CRANKSH. MTW
Worksheet was saved on 2/25/1993
MTB > info
```

Column	Name	Count
C1	AtoBDist	125
C2	Month	125
C3	Day	125

```
MTB > XBARCHART C1 5;
SUBC > SLIMITS 1 2 3;
SUBC > TEST 1:8.
TEST 6. Four of 5 points in a row in zone B or beyond (on
one side of CL).Test Failed at points: 5
```

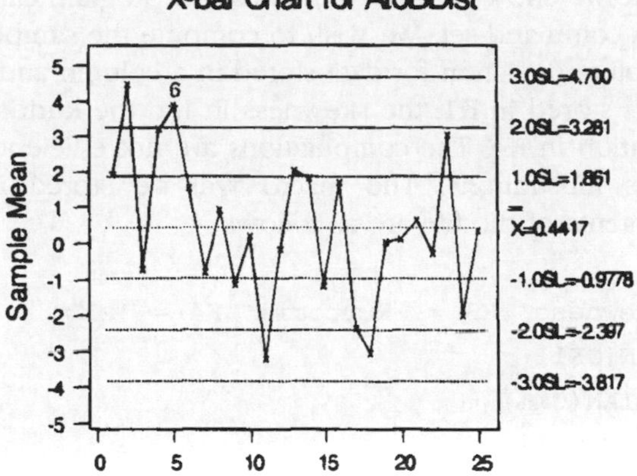

43.7 MACROS

Minitab provides two methods for creating and executing macros. Generally speaking, macros are files of Minitab commands that are executed in sequence when called. Macros are useful for the following.

1. Storing sequences of commands that you use frequently to avoid much retyping.
2. Repeating blocks of commands many times for doing simulations.
3. Looping through the columns of a worksheet to perform the same calculations on each.
4. Using Minitab as a programming language to perform calculations not provided as built-in commands.

In all versions of Minitab macros stored in a file can be run by using the EXECUTE command. The default extension for macro to be run using this command is .MTB. The general form for this command is as follows.

```
MTB> EXECUTE 'FILENAME' [K times]
```

The default value for K is 1. Macro files can be created by using an external text editor, the JOURNAL command or the STORE command. As a simple example, suppose you wished to compute the mean for columns C1 through C10 and store them in column C20. This example illustrates a helpful feature of Minitab that is often called the CK capability. It permits column indices to be stored in constants. Changing the value stored in that constant allows a macro to access columns in sequence for looping. Matrices can also be referred to in this manner. We could use, for example, MK1 to loop through matrices according to the value stored in K1. The contents of a macro file named MEANS.MTB file could look as follows.

```
LET C20(K1) = MEAN(CK1)
LET K1 = K1 + 1
```

This file would then be called as follows.

```
MTB> LET K1 = 1
MTB> EXECUTE 'MEANS' 10
```

As a second example we show how to use Minitab to program calculations that are not part of the built-in command set. We wish to compute the sample skewness, kurtosis, and median absolute deviation for data stored in a column and print the results. The column number is stored in K1, the skewness in K2, the kurtosis in K3, and the median absolute deviation in K4. The computations are not efficient numerically, but are to serve only as illustrations. The macro will be stored in a file named NUMER.MTB. The contents of the file are as follows.

```
NAME K2 = 'Skewness' K3 = 'Kurtosis' K4 = 'MAD'
LET K10 = MEAN(CK1)
LET K11 = MEDIAN(CK1)
LET K12 = STDEV(CK1)
LET C40 = (CK1 - K10)**3/(K12**3)
LET 'Skewness' = SUM(C40)/COUNT(CK1)
LET C41 = C40*(CK1 - K10)/K12
LET 'Kurtosis' = SUM(C41)/COUNT(CK1)
LET C42 = ABS(CK1 - K11)
LET 'MAD' = MEDIAN(C42)
PRINT 'Skewness'
PRINT 'Kurtosis'
PRINT 'MAD'
```

These statistics would be computed for data stored in column C5 as follows.

```
MTB> LET K1 = 5
MTB> EXECUTE 'NUMER'
```

Minitab macros can perform conditional execution of commands, have a variable number of arguments, and be run interactively. See the *Minitab Reference Manual* for your implementation for details.

Later releases of Minitab provide a second means of running macros. In these, macros are classified as either global or local. Global macros are much like the examples just presented. They can access only columns, constants, and matrices in your worksheet. The general structure of a global macro is as follows.

GMACRO
template
body of macro
ENDMACRO

The template is simply the name given to the macro. The default extension for such files is .MAC. The macro given in the previous would be written as a global macro as follows.

```
GMACRO
NUMER
NAME K2 = 'Skewness' K3 = 'Kurtosis' K4 = 'MAD'
```

```
LET K10 = MEAN(CK1)
LET K11 = MEDIAN(CK1)
LET K12 = STDEV(CK1)
LET C40 = (CK1 - K10)**3/(K12**3)
LET 'Skewness' = SUM(C40)/COUNT(CK1)
LET C41 = C40*(CK1 - K10)/K12
LET 'Kurtosis' = SUM(C41)/COUNT(CK1)
LET C42 = ABS(CK1 - K11)
LET 'MAD' = MEDIAN(C42)
PRINT 'Skewness'
PRINT 'Kurtosis'
PRINT 'MAD'
ENDMACRO
```

This macro would be called to process data in column C5 as follows.

```
MTB> LET K1 = 5
TB> %NUMER
```

Local macros allow you to write commands that can be made to look like built-in Minitab commands. Local macros can have arguments and subcommands. A macro file can contain more than one macro. In such a case the first macro in the file is the main macro and those that follow are considered to be subcommand macros. Local macros create a *local worksheet* that contains results known only to the macro. The local worksheet disappears upon completion of the local macro. Local worksheet entries can be saved, however, by using the WRITE command within the macro. The local worksheet means that you need not fill your worksheet with intermediate values. The general structure of a local macro is as follows.

MACRO
template
declaration statements
DEFAULT
body of macro
ENDMACRO

The template in this case gives the name of the macro, the names of macro subcommands, and arguments whose values are passed to and from the macro. All variables used in a local macro must be described in the declaration statements. DEFAULT allows the user to assign default values to constants that are used in subcommand calls. If a subcommand is not used then the default value is used. As an example of a local macro without subcommands we rewrite the preceding example as follows.

```
MACRO
NUMER X SKEWNESS KURTOSIS MAD
#
# NUMER takes data in a column, X, and computes and returns
```

```
# the sample SKEWNESS, sample KURTOSIS, and sample
# median absolute deviation (MAD).
#
MCOLUMN X T1 T2 T3
MCONSTANT N XBAR MED SDEV SKEWNESS KURTOSIS MAD
LET XBAR = MEAN(X)
LET MED = MEDIAN(X)
LET SDEV = STDEV(X)
LET N = COUNT(X)
LET T1 = (X - XBAR)**3/(SDEV**3)
LET SKEWNESS = SUM(T1)/N
LET T2 = T1*(X - XBAR)/SDEV
LET KURTOSIS = SUM(T2)/N
LET T3 = ABS(X - MED)
LET MAD = MEDIAN(T3)
ENDMACRO
```

This macro would compute the desired statistics for data in column C5 and store the results in K1, K2, and K3 as follows.

```
MTB> %NUMER C5 K1 K2 K3
```

A number of macros are supplied with Minitab. Consult these to see examples of more complex macros, including the implementation of subcommands.

DEFINING TERMS

Argument: A value, column, column name, constant name, or matrix name that is passed to a command.

Command: A named procedure that is built into Minitab.

Local Worksheet: Columns, constants, or matrices that are used within a macro for storage of intermediate results. The contents are not accessible outside the macro.

Macro: A stored file of Minitab instructions. May be global or local.

Subcommand: A command within a command to invoke options within that command.

Worksheet: Columns, constants, and matrices where data and results are stored for processing or output.

REFERENCES

Minitab Graphics Manual. Minitab, Inc., State College, PA.

Minitab QC Manual. Minitab, Inc., State College, PA.

Minitab Reference Manual. Minitab, Inc., State College, PA.

B. F. Ryan, B. L. Joiner, and T. A. Ryan, Jr., *Minitab Handbook*. PWS-Kent Publishing Company, Boston, MA, 1992.

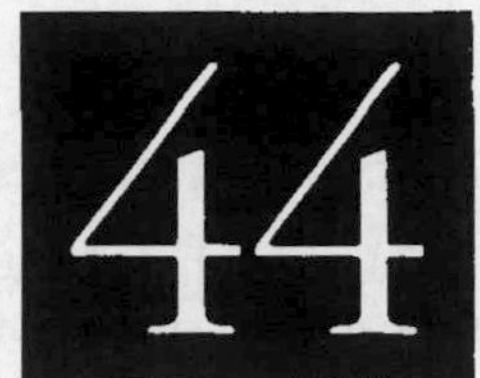

SAS

Timothy C. Krehbiel
Miami University

44.1 SAS SYSTEM

The SAS System is an integrated system of computer software products designed to allow for the easy analysis of data. The SAS system includes many components that perform many functions including database management, report writing, graphics, operations research, applications development, and statistical analysis. SAS, which stands for Statistical Analysis System, is foremost a software package devoted to statistical analysis.

This chapter discusses the statistical analysis features of the SAS system most useful to engineers. These features include descriptive statistics, graphical analysis, control charts, and the design and analysis of experiments. To perform such analyses, programs using the SAS programming language must be written and executed. The SAS System products needed to perform basic statistical analyses are Base SAS and SAS/STAT. A wider and more sophisticated range of analyses can be used if SAS/QC and SAS/GRAPH are available. A short description of these four products are as follows:

- Base SAS. Base SAS is the foundation of the entire SAS system and is a required part of all systems. It provides data access, management, analysis, and presentation.
- SAS/STAT. One of the most complete statistical analysis software products in the world. It includes descriptive statistics, regression analysis, multivariate statistics, and much more.
- SAS/GRAPH. SAS/GRAPH allows you to construct high-resolution graphics of reports, graphs, and figures produced by SAS. Work produced by SAS/STAT, SAS/QC, and other SAS products is greatly enhanced with very little extra effort on the programmer's part.
- SAS/QC. SAS/QC is a set of statistical tools important in quality improvement techniques. These tools include the design and analysis of experiments and statistical control charts.

0-8493-2530-7/96/$0.00+$.50

Macintosh Users

JMP (pronounced "jump"), JMP Serve, and JMP Design are interactive user-friendly statistical analysis software for the Macintosh. JMP is icon-oriented and provides the user with an impressive selection of graphical and numerical statistical procedures. JMP is now available for PC's running Windows. This chapter does not discuss JMP software.

Other SAS System Products

The SAS system is an extensive collection of software products designed to meet the needs of today's varied and complex computing environments. The number and diversity of products available from the SAS Institute are changing rapidly. All SAS System products are available in English. In addition, some products are available in a variety of other languages.

44.2 BASIC OPERATIONS

SAS can be used on many different platforms and in many different operating systems. Although each operating system and platform has a few unique features and commands, the vast majority of operations are the same regardless of the computing environment.

To perform statistical analysis in SAS, programs using the SAS programming language must be written. Novice programmers should refer to *Introducing the SAS System*. In most cases, the commands in the programming language are identical across environments. Commands that import or export data, however, are likely to differ. Also, the ways in which you create, store, and execute the commands are often unique to the environment in which you are operating. Consult your local SAS consultant or host system documentation for details.

There are three types of files in the SAS programming language:

- SAS program files.
- Log files.
- Listing files.

After a SAS program file is executed, the SAS program creates two output files, the log file and the listing file. The log file is a written logbook of everything that occurs when the program file is sent to the cpu. Any errors occurring during execution will be recorded in the log file. The listing file contains the reports, data analysis, and graphs requested by the input file.

Within the different environments themselves, there are several methods of running SAS programs:

- Batch. In batch mode SAS programs are written using an external editor and then submitted to the SAS compiler. After completion of the program the output generated by SAS is stored in separate files.
- Display Manager Mode. In most PC environments, and some mainframe environments, the Display Manager System (DMS) is the prevalent method of executing SAS. In DMS your screen is split into three windows, one each for the SAS program file, the log file, and the listing file. The SAS editor is used to edit the program files written in the program file window.
- Interactive Line Mode. In interactive mode, SAS statements are written and executed one at a time. Interactive mode is very cumbersome when writing long or complex programs.

In each of these methods of operations, SAS programs are written and executed. If working in a mainframe environment, SAS is typically used in a batch mode and the three files are accessed separately. For example, when using CMS the program, log, and listing files are three separate files all having the same file name with file types, SAS, SASLOG, and LISTING, respectively. Check with your local SAS consultant or host system documentation for the options available to you.

SAS Program Files

SAS program files contain the SAS programming commands. SAS program files can be broken into three parts:

- Titles, footnotes, and options.
- Data section.
- Procedures section.

Each of these parts is made up of individual SAS commands. The data and procedures sections are required in all programs, while the titles, footnotes, and options section is not.

Titles and Footnotes

Title and footnote commands are not required by any SAS program, but effective use of them will greatly enhance your programs and overall productivity. The TITLE command creates one or more title lines on each page of the listing file:

```
TITLE1 "This is the first heading line";
TITLE2 "This is the second heading";
TITLE3 "This is the third";
```

Notice that all the commands end with a semicolon. The semicolon indicates to SAS that the end of the command has occurred. Commands in SAS can be included one per line, several per line, or a command can occupy two or more lines. It is the semicolon that signifies the end of a command. If a semicolon is mistakenly left off the end of a command, SAS believes that the next command is part of the previous command and will almost always cause an error. One of the most common mistakes made by novice SAS users is to omit the semicolon at the end of a command.

The syntax used for the footnote command is the same as for the title command illustrated above. To change a footnote or heading line, use a second command to redefine the text previously entered. To remove a heading use "" as the new heading. All headings after the one set to "" will be removed as well. For example, if the following title lines are added to the ones presented above, then there will now be one heading line per page, reading "New Title Line 1". A footnote has also been added to each page:

```
TITLE1 "New Title Line 1";
TITLE2 "";
FOOTNOTE "This Appears at the Bottom of the Page";
```

Options

Options are system specific commands that allow you to override the default values of your SAS system. For example, when using SAS in CMS or VMS, the default line length is 132 characters. If you wish to print to legal or letter size paper, the line size needs to be changed. Also, a line length of 132 characters is not easily displayed on most monitors, thereby making browsing of SAS output files very cumbersome. Likewise, if printing on legal size paper, the page length would need to be changed from the default value of 66 to 84. The following options command allows for printing on legal size paper and suppresses the date, which is typically printed at the top of every page:

```
OPTIONS LS-72 PS-84 NODATE;
```

If using the SAS Display Manager System, you can specify many of the system options in the OPTIONS window. To invoke the options window enter OPTIONS on a command line.

44.3 ENTERING DATA

Before performing analyses in SAS, you must enter the data into a SAS data set. Data sets contain one or more observations. Each observation contains one or more variables. SAS variable names are 1–8 characters in length and must begin with a letter.

There are several ways of entering data, each requiring a different form of the DATA command. There are three possible ways of entering data:

- Data are included in SAS program file. This is useful if you have a very small data set that is not already stored on disk.
- Reading an external ASCII file. SAS can read any data stored in ASCII format. This is the most practical way of entering data into SAS for it allows you to enter data already available on your system. When the integrity of the data is important (which is almost always), the data can be entered once and then stored in read-only format. This provides you with reasonable assurance that the data file remains intact and correct. If the data are currently being used or stored in another application package, the data can be exported to an ASCII file and then imported into SAS.
- Reading previously constructed SAS Data Sets. If the data have currently been entered into a SAS data set and saved as such, the SAS data set can be recalled. If large data sets are being used, this is by far the fastest and most efficient way of running subsequent SAS programs.

Data Are Included in SAS Program File

The first line of the data section of your SAS program begins with a data statement. The remaining lines depend on the current application. If the data are not currently stored on your computer system, an easy way to enter the data is to make the data part of the program itself:

```
DATA EXAMPLE1;
INPUT SHIFT $ DIAMETER;
```

```
CARDS;
A 250.12
A 250.05
B     .
B 249.99
```

In the above example the first statement begins the construction of a data set named "EXAMPLE1". In the second line, you are telling SAS that the data set has two variables, shift and diameter. The $ tells SAS that the variable SHIFT is a qualitative variable and therefore can contain alphanumeric characters. There is no $ following DIAMETER, which defines this variable as strictly numeric. The CARDS statement informs SAS that the data will follow directly, with one row of data per observation. In this example there are four observations. The columns of data must be separated by at least one blank space when entered in this way. The diameter of the third observation is unknown. The period, ".", indicates a missing value to SAS.

Reading an External ASCII File

The most common way to enter data into a SAS data set is through the use of an INFILE command. The INFILE command tells SAS that the data are located in a separate file:

```
DATA EXAMPLE2;
INFILE 'a:exp2.dat';
INPUT MACHINE $ PRESSURE TEMP;
```

The above commands, written for a PC version of SAS, indicate that the data set has three variables (MACHINE, PRESSURE, and TEMP) and are located in the ASCII file a:exp2.dat. PRESSURE and TEMP are numeric variables and MACHINE is a qualitative variable. This type of format requires that the data file have three columns of data separated by at least one blank space. Also, any missing values must be indicated with periods, ".".

To conserve disk space, data are often stored without spaces between variables. Furthermore, it is sometimes convenient to only use a few of the many variables contained in a stored data set. Moreover, it is possible that individual observations may occupy more than one line of the data file. In any of these cases, formatted input statements can be used:

```
DATA EXAMPLE3;
INFILE 'a:exp3.dat';
INPUT MACHINE $ 1-2 PRESSURE 3-10 TEMP 30-35 #2 HEAT 1-5;
```

In the above commands each observation occupies 2 rows. The values for MACHINE are found in columns 1–2, PRESSURE is in columns 3–10, and TEMP in 30–35, and all three of these variables are in the same row. The values for the variable HEAT are in columns 1–5 but in the next row of the ASCII file. The second observation occupies lines three and four of the file "a:exp3.dat," the third observation lines five and six, etc.

FILEREF Commands for CMS and VMS

If you are not using SAS on a PC, then the infile statement requires that you use a fileref command. For example, if running VMS a FILENAME command is needed before the data section. In CMS a FILEDEF statement is needed. In the following lines the data are stored in the CMS file EXP4 DATA A. Once the CMS command is given, the fileref FILE4 can be used anywhere in the SAS program:

```
CMS FILEDEF FILE4 DISK EXP4 DATA A;
DATA EXAMPLE4;
INFILE FILE4;
INPUT MACHINE $ PRESSURE TEMP;
```

Saving a SAS Data Set

SAS automatically saves data sets being used in the current program. At the completion of the program, however, the data sets are discarded. If you have sufficient disk spacc, the data sets can be saved for use in later programs. To save a SAS data set permanently, you need to enter a two-level data set name in the data section that created the data set. Since the exact form of the command depends on the operating system, you may need to consult your local SAS consultant or host system documentation. The following lines save a data set in CMS. After completion of the program, the CMS file EXP8 SASDATA A contains the SAS data set.

```
DATA SASDATA.EXP8;
```

If large data sets are needed, reading the raw data into the SAS data set can be quite time consuming. Also, if a significant amount of data manipulation is needed, these steps can be timely. In these cases, it is much better to save the SAS data set and recall it at a later date. A small program used only for reading the raw data and saving the SAS data set can be executed first. Then, when the data set is needed in future SAS programs, the data statement simply refers to the previously constructed file.

Reading Previously Constructed SAS Data Sets

To recall a previously constructed SAS data set, a DATA statement must be used. In the DATA statement, the two-level SAS data set name must be used. There are two methods of doing this:

- SET. Recalls a previous SAS data set.
- MERGE. Merges two or more data sets into one.

The following lines recall a CMS SAS data set created earlier. The data have been stored in an encrypted format in the file EXP8 SASDATA A. The new data set is called NEW:

```
DATA NEW;
SET SASDATA.EXP8;
```

Two data sets can be easily merged into one via a merge command. The following commands merge the data from SAS data sets EXAMPLE1 and EXAMPLE5 into a new data set known as EXAMPLE6:

```
DATA EXAMPLE6;
MERGE EXAMPLE1 EXAMPLE5;
```

Data Issues in LANs and Client-Server Environments

SAS system software is quickly growing to support data access and integrity in local area networks (LANs) and in client-server environments. Some of the software products currently available are as follows:

- SAS/ACCESS. Provides an easy interface to many different databases and files including MVS, CMS, VSE, Open VMS VAX, Open VMS AXP, Solaris 1.0, HP-UX, AIX for RS/6000, OS/2, Windows, and Windows NT.
- SAS/CONNECT. Allows for cooperative and distributed processing.
- SAS/FSP. Allows for customized data entry and retrieval.
- SAS/SHARE. Provides concurrent access to data among several users.

44.4 MANIPULATING DATA IN A DATA STATEMENT

The data statement can be used to manipulate data already stored in a SAS data set. The following lines modify an existing SAS data set, EXAMPLE4, and create a new data set, EXAMPLE5. EXAMPLE5 contains all the variables in EXAMPLE4, plus two new variables as well. Variable PRES2 is equal to pressure squared, and TEMPF is the fahrenheit temperature calculated from the celsius temperature:

```
DATA EXAMPLE5;
SET EXAMPLE4;
PRES2 - PRESSURE * PRESSURE;
TEMPF - (TEMP * 1.8) + 32;
```

IF, THEN, ELSE Statements

Data can be manipulated using IF, THEN, and ELSE statements. These statements can be used individually or nested. In the following example, a new variable called CASE is constructed. CASE=1 when the pressure is greater than 50 and the temperature is greater than 100. CASE=2 when the pressure is greater than 50 and the temperature is less than or equal to 100. Whenever the pressure is less than or equal to 50, CASE=3:

```
IF PRESSURE > 50 THEN
     IF TEMPF > 100 THEN CASE=1;
     ELSE CASE=2;
ELSE CASE=3;
```

SAS Functions

The SAS programming language contains many functions that allow for ease of programming commonly performed operations. A SAS function returns a single value from an algebraic or system computation. These operations cover a wide range of operations including random numbers, statistics, probability quantiles, trigonometric, state and zip codes, and truncation. A complete listing of SAS Functions can be found in *SAS Language: Reference*. The following illustrate the absolute value, square root, and maximum value functions:

```
DATA EXAMPLE7;
INFILE 'a:exp10.dat';
INPUT A B;
C = ABS(A);
D = SQRT(B);
E = MAX(B);
```

44.5 BASIC STATISTICAL PROCEDURES

SAS procedures, known as PROCs, are SAS instructions that allow for a certain type of statistical or mathematical analysis. Base SAS software along with SAS/STAT allow you to perform many commonly performed statistical analysis techniques. For instance, PROC REG is a large set of routines that allows you to perform many types of regression analysis.

All procedure sections begin with a PROC statement. A PROC statement has the following form:

```
PROC proc_name DATA=data_set_name;
```

Proc_name is the procedure name to be used. The procedure will be performed on data in the SAS data set named data_set_name. If the DATA portion of the statement is omitted, SAS will use the last data set used or created in the program.

Procedures usually require additional statements to define exactly what analysis needs to be performed. For example, when performing an analysis of variance using the general linear models procedure, PROC GLM, a CLASS statement identifying the classification variables and a MODEL statement must also be included. The following SAS program performs a two-way analysis of variance on a previously created SAS data set called TOOLLIFE. The data for this program are found in Table 44.1. The data are taken with permission from Montgomery (1991a:239).

```
TITLE "Two-way ANOVA using Tool Life Data";
DATA TOOLLIFE;
INPUT ANGLE $ SPEED $ Y;
CARDS;
L L -2
L L -1
L M -3
```

TABLE 44.1 Data for Tool Life Experiment[a]

Tool Angle (degrees)	Cutting Speed (in/min)		
	125	150	175
15	−2	−3	2
	−1	0	3
20	0	1	4
	2	3	6
25	−1	5	0
	0	6	−1

[a]Adapted from Douglas C. Montgomery, *Design and Analysis of Experiments*, 3rd ed. John Wiley, New York, 1991. Reprinted by permission of John Wiley & Sons, Inc.

```
L M 0
L H 2
L H 3
M L 0
M L 2
M M 1
M M 3
M H 4
M H 6
H L -1
H L 0
H M 5
H M 6
H H -1
PROC GLM DATA=TOOLLIFE;
CLASS ANGLE SPEED;
MODEL Y=ANGLE SPEED ANGLE*SPEED;
```

Many options are available for all SAS procedures. When no options are used, as is the case above, the most general type of statistical analysis is performed. Options allow for a large variety of analyses often with added detail and diagnostics. In the following program, the option SOLUTION computes the least squares estimates to the two-way ANOVA model:

```
PROC GLM DATA=TOOLLIFE;
CLASS ANGLE SPEED;
MODEL Y=ANGLE SPEED ANGLE*SPEED / SOLUTION;
```

Commonly Used Procedures

The SAS/STAT software has over 35 different procedures. Many other procedures can be obtained with additional SAS system products. A few of the more commonly used SAS procedures available with SAS/STAT are as follows:

PRINT	Prints lists
SORT	Sorts data sets
REG	Performs regression analysis
ANOVA	Performs analysis of variance
MEANS	Analyzes means
CHART	Produces charts
PLOT	Produces plots
FREQ	Performs analysis on frequency data
CORR	Performs correlation analysis
FACTOR	Performs factor analysis
CLUSTER	Performs cluster analysis

For more information, consult the *SAS/STAT User's Guide* and the *SAS Procedures Guide.*

44.6 SAS/GRAPH SOFTWARE

SAS/GRAPH software is a computer graphics system for producing high-quality graphics on CRTs and printing devices. SAS/GRAPH includes many SAS procedures that allow you to produce color graphs, plots, slides, maps, chart, etc. The addition of SAS/GRAPH also allows you to enhance the output of other SAS system products, e.g., the control charts constructed using PROC SHEWHART in Figure 44.2.

SAS/GRAPH is a device-intelligent system that supports a full range of graphics devices. A list of SAS/GRAPH device drivers can be found in the *SAS/GRAPH Software: Reference*. SAS programs using SAS/GRAPH must identify the graphics device to be used in a GOPTIONS command. The general form of the command is as follows:

```
GOPTIONS DEVICE=device_name;
```

Consult your local SAS consultant or host system documentation for details regarding which device_name options are available to you.

Example Using PROC G3D

The following example is adapted (with permission) from Montgomery (1991a:239–245). An experiment is presented that investigates the life time of a cutting tool based on the cutting angle and cutting speed of the tool. After performing a regression analysis on the data in Table 44.1, the following response surface was estimated:

$$
\begin{aligned}
Y = {} & 2.667 + 0.700^{*} (\text{angle} - 20) - 0.0267^{*} ((\text{angle} - 20)^{**}2) \\
& + 0.0532^{*} (\text{speed} - 150) - 0.008^{*} (\text{angle} - 20)^{*} (\text{speed} - 150) \\
& - 0.00128^{*} (\text{angle} - 20)^{*} ((\text{speed} - 150)^{**}2) \\
& - 0.000128^{*} ((\text{angle} - 20)^{**}2)^{*}((\text{speed} - 150)^{**}2)
\end{aligned}
$$

The following SAS code produces the response surface found in Figure 44.1. The device statement directs the output to a Tektronics terminal emulator in full screen mode. (The plot was then written to an HPGL file.) The data section creates a grid of data points for all combinations of angle and speed for angles 15 to 25 (by 1) and speed 125 to 175 (by 5). After these points are calculated, PROC G3D is used to construct the three-dimensional response surface found in Figure 44.1.

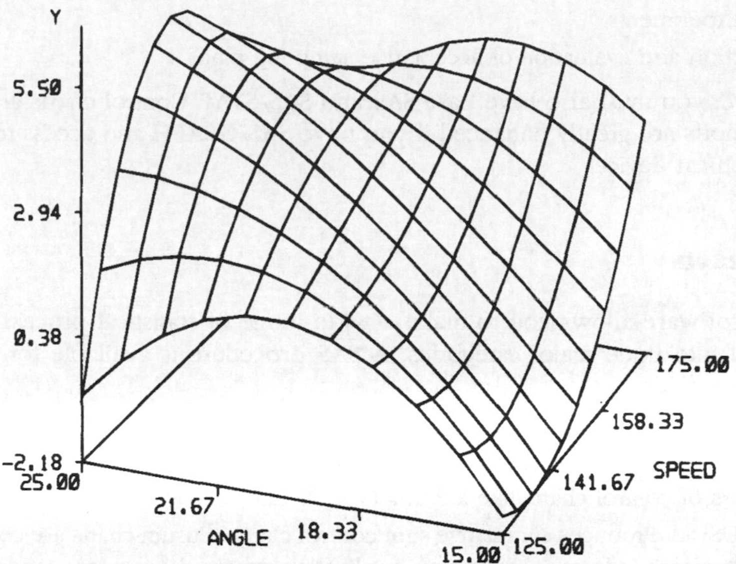

FIGURE 44.1 Response surface for tool life experiment. Adapted from Douglas C. Montgomery, *Design and Analysis of Experiments*, 3rd ed., by John Wiley, New York, Reprinted by permission of John Wiley & Sons, Inc.

```
GOPTIONS DEVICE=TEK4010;
DATA TOOLLIFE;
DO ANGLE=15 TO 25 BY 1;
  DO SPEED=125 TO 175 BY 5;
    Y = 2.667 + 0.700*(ANGLE - 20)-0.0267*((ANGLE - 20)**2)
        + 0. 0532*(SPEED - 150) - 0.008*(ANGLE - 20)*(SPEED
        - 150) -0.00128*(ANGLE - 20)*((SPEED - 150)**2) -
        0.000128*((ANGLE - 28)**2)*((SPEED - 150)**2);
    OUTPUT;
  END;
END;
PROC G3D;
PLOT ANGLE*SPEED = y;
```

44.7 SAS/QC SOFTWARE

SAS/QC Software is a collection of procedures and functions that aids in quality improvement activities. The statistical tools included in SAS/QC can help you perform the following tasks:

- Process stability analysis.
- Process capability analysis.

- Design experiments.
- Construction and evaluation of acceptance sampling plans.

To run SAS/QC, you must also have Base SAS and SAS/STAT. Control charts and other graphical methods are greatly enhanced if you have SAS/GRAPH and access to an appropriate graphical device.

Control Charts

The SAS/QC software allows you to make a wide range of statistical process control charts that fall into three major categories. A SAS procedure is available for each of these three categories.

- PROC SHEWHART. Produces a wide variety of statistical process control charts that can be accredited to the work of Walter P. Shewhart. These include the most common types of control charts like a P chart, or $\bar{X}$ and R charts.
- PROC CUSUM. Produces cumulative sum control charts. Cusum charts are commonly used in process industries, or in any application where the process tends to drift gradually and does not typically suffer from quick dramatic shifts. PROC CUSUM can produce two-sided charts using a V-mask scheme, or one-sided charts using a decision interval scheme.
- PROC MACONTROL. Produces moving average control charts. Moving average control charts, like cusum charts, are often used when the process tends to drift gradually instead of experiencing quick shifts. PROC MACONTROL can produce charts using uniformly weighted moving averages (MA charts), or exponentially weighted moving averages (EWMA charts).

PROC SHEWHART

As of release 6, PROC SHEWHART allows you to make 13 different types of statistical process control charts. A wide range of options is available as well. The data may be arranged in a variety of ways. The easiest arrangement is to create a data set containing two variables, one that contains the subgroup number and one that contains the quality variable of interest.

Assuming that you have a data set as described in the previous paragraph, the general form of the procedure section is as follows:

```
PROC SHEWHART DATA=data_set_name;
chart_type quality_variable*subgroup / options;
```

Data_set_name is a previously created SAS data set, quality_variable is the SAS variable containing the measurements, and subgroup is a SAS variable identifying which subgroup each measurement comes from. The chart_type identifies the chart or charts to be created. Some of the commands produce a single chart (e.g., the nonconformities chart) while some commands produce two charts displayed on top of each other (e.g., the $\bar{X}$ and R chart displayed in Fig. 44.2).

Some of the control charts available are listed below. Consult the *SAS/QC User's Guide* for details.

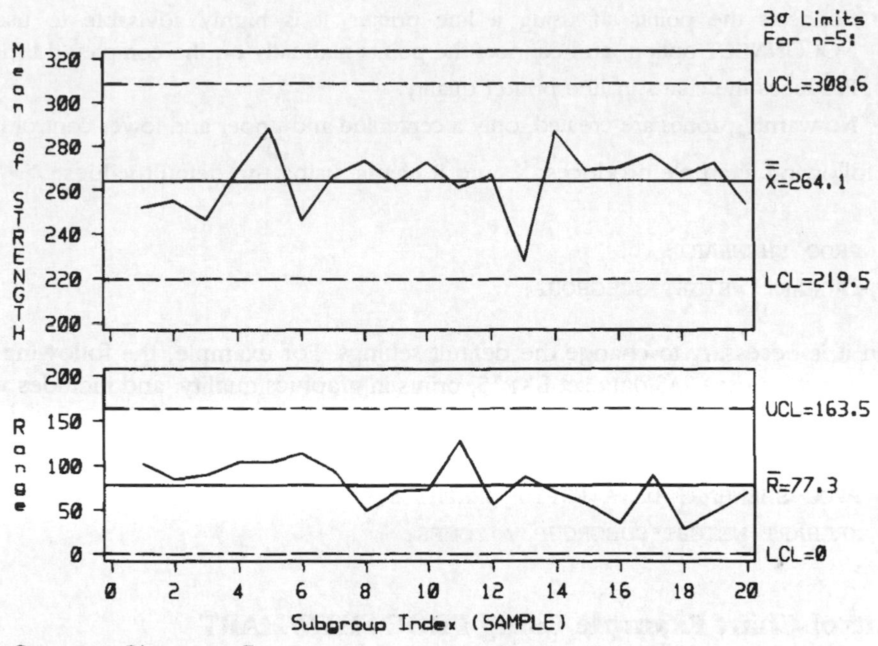

FIGURE 44.2 $\bar{X}$ and R charts using PROC Shewhart. Adapted from Douglas C. Montgomery, *Introduction to Statistical Quality Control*, 2nd ed. by John Wiley, New York, 1991. Reprinted by permission of John Wiley & Sons, Inc.

BOXCHART	$\bar{X}$ or median charts overlaid with box-and-whisker charts for each rational subgroup
CCHART	Number of nonconformities
IRCHART	Individuals and moving range charts
MRCHART	Median and range charts
NPCHART	Number of nonconforming items (equal subgroup size)
PCHART	Proportion of nonconforming items
UCHART	Number of nonconforming items (unequal subgroup size)
XRCHART	$\bar{X}$ and range charts
XSCHART	$\bar{X}$ and standard deviations chart

Default Options in PROC SHEWHART

All of the charts listed above have multiple options available to them. Many of the options, and their defaulted values, are the same for all the charts. If no options are given, then PROC SHEWHART produces statistical process control charts using the following details:

- The latest SAS data set is used.
- The control limits are estimated using the data given.
- Three sigma control limits are used.

- Connects the points. If using a line printer, it is highly advisable to use the NOCONNECT option, and connect the points manually on the completed listing.
- Produces the charts in line printer quality.
- No warning zones are created, only a centerline and upper and lower control limits.

The following example produces $\overline{X}$ and R charts, using the default values.

```
PROC SHEWHART;
XRCHART WEIGHT*SUBGROUP;
```

Often it is necessary to change the default settings. For example, the following procedure section uses SAS data set EXP15, prints in graphics quality, and includes warning zones.

```
PROC SHEWHART DATA=EXP15 GRAPHICS;
XRCHART WEIGHT*SUBGROUP / ZONES;
```

Control Chart Example Using PROC SHEWHART

The following example is adapted with permission from Montgomery (1991b:386–387). The SAS data set SOFTDRINK contains the bottle-strength of softdrink bottles listed in Table 44.2. The $\overline{X}$ and R chart constructed from this example is given in Figure 44.2.

```
PROC SHEWHART DATA=SOFTDRINK GRAPHICS;
XRCHART STRENGTH*SAMPLE;
```

TABLE 44.2 Bottle-Strength Data[a]

Sample			Data		
1	265	205	263	307	220
2	268	260	234	299	215
3	197	286	274	243	231
4	267	281	265	214	318
5	346	317	242	258	276
6	300	208	187	264	271
7	280	242	260	321	228
8	250	299	258	267	293
9	265	254	281	294	223
10	260	308	235	283	277
11	200	235	246	328	296
12	276	264	269	235	290
13	221	176	248	263	231
14	334	280	265	272	283
15	265	262	271	245	301
16	280	274	253	287	258
17	261	248	260	274	337
18	250	278	254	274	275
19	278	250	265	270	298
20	257	210	280	269	251

[a]Adapted from Douglas C. Montgomery, *Introduction to Statistical Quality Control*, 2nd ed. John Wiley, New York, 1991. Reprinted by permission of John Wiley & Sons, Inc.

Capability

PROC CAPABILITY is a collection of statistical tools to assess a stable process's capability of meeting product specifications. Many helpful graphical tools such as probability plots, histograms, Q-Q plots, and P-P plots can be constructed. All the charts are greatly enhanced if you have SAS/GRAPH and invoke the GRAPHICS option. The numerical measures include descriptive statistics, process capability indices and tolerance, prediction, and confidence intervals.

Designing Fractional Factorial Experiments

SAS/QC software contains four features to help you design and analyze fractional factorial experiments:

- PROC FACTEX. Generates standard symmetrical and asymmetrical designs. Some of the tasks performed by this interactive procedure are to print design points, examine the alias structure, randomize the design order, and decode the factor levels to values of your choice.
- PROC OPTEX. Generates nonstandard designs using A-optimal or D-optimal criterion. Can be used to augment existing or standard designs.
- ADX Macro System. Generates all the designs in FACTEX and OPTEX as well as Plackett-Burman, Box-Wilson, McLean-Anderson and other mixture designs. ADX can also perform power-transformations and analyze all the designs generated.
- ADX Menu System. Provides the same design generation features as in the ADX Macro System using a user-friendly menu interface. The ADX menu system also provides the user with a large number of analysis tools. To use the ADX menu system on a personal computer you must have expanded memory capability. Version 4.0 of LIM EMS is recommended.

Acceptance Sampling Plans

The construction and evaluation of double-sampling lot acceptance plans can be eased with the SAS/QC software. The following four functions are available:

- AOQ2 Returns the average outgoing quality.
- ASN2 Returns the average sample number.
- ATI2 Returns the average total inspection.
- PROBACC2 Returns the acceptance probability.

Consult the *SAS/QC User's Guide* for details concerning these functions, as well as details concerning control charts, process capability, and the design of experiments.

44.8 FINDING ERRORS IN SAS CODE

The SAS log file contains information regarding the latest execution or attempted execution of a SAS file. All SAS commands resulting in an error are identified in this file. When using the display management system the log file is found in one of the three windows. If you executed SAS in batch mode, the log file will be contained in a separate file with the same file name as the SAS program but with a different extension. When operating SAS interactively, all errors will be displayed on the screen as they occur and a log file is not generated.

Any SAS statement that cannot be completed results in an error. The SAS log identifies the statement that could not be completed. For example, in the following section of a log file an error occurred in line 36 of the SAS program. The mistake here is obvious: The procedure REG was mistakenly written REGR:

```
36          PROC REGR;
ERROR: Procedure REGR not found.
```

In the above example the SAS log identified exactly where to look. In some cases, however, the actual change that needs to be made is several lines above where the ERROR message is written. In other words, the ERROR message is printed under the line in which SAS could not complete, not necessarily the line that contains the coding error.

In the following example, the coding error (a missing semicolon) occurs in line 33 but the execution error occurs in line 34:

```
32          PROC REGR;
33          MODEL Y=X1 X2 / P
34          PROC CORR;
            202 202
NOTE: The previous statement has been deleted.
ERROR: 202-322: The option or parameter is not recognized.
```

The underlined code identifies where the execution halted. The missing semicolon in line 33 inadvertently makes line 34 an extension of the command in line 33. Therefore SAS believes that PROC and CORR are options to the model statement, not a new statement.

Warning statements are issued by SAS in cases where actual execution is not halted but atypical events or operations are observed. In the following example, a quotation mark is missing in line 29. The regression output requested from lines 30–31 is ignored due to the previous mistake. The SAS log, however, does not report this as an error since no attempt was made to execute lines 30–31. The SAS log issues a WARNING statement before the coding error actual occurs:

```
WARNING: The current word or quoted string has become more
than 200 characters long. You may have unbalanced quotation
marks.
29          TITLE1 "Regression Example;
30          PROC REG;
31          MODEL Y=X1 X2 / P;
```

Anytime your SAS log issues a WARNING statement, it is highly advisable to critically examine the output to ensure that it contains all the procedures and options requested.

General Suggestions for Debugging SAS Code

You should read the ERROR message very carefully. Remember that SAS cannot tell you exactly where your coding mistakes are, only where the program failed to execute properly. Scan backward the code preceding the error message, keeping in mind the wording and nature of the error message.

Always look for SAS commands with missing semicolons. It is highly recommended that you use only one command per line, therefore all lines of your program will end in semicolons and scanning for missing semicolons is quick and easy.

Try correcting the first error first. Once an error in a SAS program occurs, multiple subsequent errors may occur. Consider the case where you have an error in a data section and the SAS data set is not created. All procedures that use this data set will result in an error. If multiple errors are present in the SAS log and their source is not quickly identifiable, correct as many as you can and resubmit the program.

Read carefully the NOTES and WARNINGS in the log file. The program may not be producing the desired output without actually causing an execution error. You might be able to locate these problems by carefully considering the messages marked as NOTES and WARNINGS.

44.9 SAS MACROS

The SAS Macro facility is a tool for customizing and extending Base SAS software. Advanced statistical or analytical routines that are not part of your SAS system can be written as macros and then called from within a SAS program. These macros can be exceptionally useful for iterative procedures.

The %MACRO command signals the beginning of a macro definition. The general form of the command is as follows:

```
%MACRO macro_name;
```

Macro_name is a user-defined name of eight characters or less. The %MACRO command is then followed by the SAS commands that make up the macro. The %MEND command is then used to signal the end of a macro definition. To invoke the macro, issue the SAS command %macro_name.

The following macro performs iterative reweighted least-squares regression to a binary response variable. The initial data set includes the binary-dependent variable Y, an independent variable X, and weights W (which are initially set to 0.25). To perform an iteration, the command %RWBINARY is issued. To perform multiple iterations, multiple %RWBINARY commands need to be issued:

```
%MACRO RWBINARY;
      PROC GLM;
      WEIGHTS W;
      MODEL Y-X / P;
      OUTPUT OUT=NEW P=PHAT;
      DATA NEW2;
      SET NEW;
      W=PHAT*(1-PHAT);
      DROP PHAT;
%MEND;
```

The SAS macro facility includes macro program statements, macro functions, macro variables, and data step interfaces. It also includes an autocall facility that allows you to invoke a macro without having to define the macro in the same program. Base SAS software also comes with a small library of sample macros. For more information you should contact your local SAS consultant, your local host documentation, or the *SAS Guide to Macro Processing*.

44.10 OTHER SAS SYSTEM PRODUCTS

The SAS system is a collection of integrated software products designed to meet data analysis needs in a wide variety of computing environments. The number and diversity of products available from the SAS Institute are changing rapidly. All SAS system products are available in English. In addition, some are available in Spanish, Japanese, German, Italian, French, Danish, Finnish, Norwegian, and Swedish.

Some of the SAS system products currently available are discussed below. For more information on these products and new releases, contact the SAS Institute.

- SAS/ACCESS. Provides transparent interfaces to many different databases and files including MVS, CMS, VSE, Open VMS VAX, Open VMS AXP, Solaris 1.0, HP-UX, AIX for RS/6000, OS/2, Windows, and Windows NT.
- SAS/ASSIST. A menu-driven task-oriented interface designed for novice or infrequent users of the SAS system.
- SAS/AF. A collection of applications development facilities.
- SAS/CALC. An electronic spreadsheet.
- SAS/CONNECT. System software that allows cooperative and distributed processing.
- SAS/CPE. System software that performs computer performance evaluation, capacity planning, and network performance management.
- SAS/EIS. A menu-driven development tool for Executive Information Systems.
- SAS/ENGLISH. A portable natural language interface for querying and reporting.
- SAS/ETS. A collection of SAS procedures commonly used in time series analysis, forecasting, and econometrics.
- SAS/FSP. A data base package allowing customized data entry, retrieval, letter writing, mail merges, and report generation.
- SAS/GIS. A collection of procedures for the numerical and graphical analysis of spatial data.
- SAS/IMAGE. A software package for processing digitized and scanned images.
- SAS/IML. A matrix programming language.
- SAS/INSIGHT. An interactive high-end graphics program.
- SAS/LAB. Interactive guidance-driven analysis program for the novice or infrequent SAS user.
- SAS/NVISION. A three-dimensional high-end graphics package.
- SAS/OR. A mathematical modeling package for linear programming, operations research and project management.
- SAS/PH-Clinical. A collection of SAS procedures and data review system designed especially for the pharmaceutical and biotechnology industries.
- SAS/SHARE. Network software providing concurrent access to data.
- SAS/TOOLKIT. A collection of programming tools for user-written extensions.
- SAS/TUTOR. A library of online training courses aimed at novice SAS users.

DEFINING TERMS

Batch Mode: SAS programs are written in an external editor and processed as a single unit.

Data Set: File created by a SAS data command that stores data in encrypted form readable by SAS. Each row represents a single observation and each column a different variable.

Default Values: Instructions used by SAS to perform specific operations when no options statements are used.

Display Management System (DMS): Windowing facility for SAS.

Fractional Factorial Experiments: Designed experiments where only a few of the many treatment combinations are being used.

Function: A SAS programming command representing a set of instructions used to calculate frequently used mathematical or statistical operations.

JMP (pronounced "jump"): Interactive Icon-oriented version of the SAS system available for Macintosh and Windows.

Local Host Documentation: Software manuals containing the SAS documentation that specifically applies to the computer system and operating system you are using.

Log File: A file containing messages concerning the execution of the program, including any errors that have occurred.

Listing File: A file containing the requested reports, tables, graphs, plots, etc.

Macro: A portion of a SAS program that can be invoked at a later time with a single command.

Observation: A row in a SAS data set that contains the value for each of the variables.

Options: Methods of performing SAS operations different than that specified by the default value.

Procedure or PROC: Built-in SAS programs that allow you to manipulate and analyze data.

Program File: A file containing the SAS programming commands.

Subgroup: A sample taken from a process.

Variable: A column in a SAS data set.

REFERENCES

Introducing the SAS System, *Version 6*. SAS Institute Inc., Cary, NC, 1990.

Douglas C. Montgomery, *Design and Analysis of Experiments*, 3rd ed. John Wiley, New York, 1991a.

Douglas C. Montgomery, *Introduction to Statistical Quality Control*, 2nd ed. John Wiley, New York, 1991b.

SAS Guide to Macro Processing, *Version 6*. SAS Institute Inc., Cary, NC, 1990.

SAS Language: Reference, *Version 6*. SAS Institute Inc., Cary, NC, 1990.

SAS Procedures Guide, *Version 6*. SAS Institute Inc., Cary, NC, 1990.

SAS/GRAPH Software: Reference, *Version 6*. SAS Institute Inc., Cary, NC, 1990.

SAS/QC Software: Reference, *Version 6*. SAS Institute Inc., Cary, NC, 1990.

SAS/STAT User's Guide, *Version 6*. *Volume 1 and Volume 2*. SAS Institute Inc., Cary, NC, 1990.

For further information contact SAS Institute Inc., SAS Campus Drive, Cary, NC 27513.

45

BMDP

William C. Rinaman
Le Moyne College

The BMDP system consists of a number of programs that provide a powerful, flexible capability to perform statistical analysis. It has versions that run on a wide range of mainframe computers, DOS-based personal computers, personal computers running Microsoft Windows, and UNIX workstations. The heart of the system consists of 44 separate statistical programs. To facilitate the use of these programs the VAX and all PC and UNIX versions provide menu-driven interfaces. These interfaces have a slightly different appearance in each version, but they all perform essentially the same tasks. Each of these provides the following:

- A MENTOR utility that facilitates automatic generation of BMDP commands.
- A built-in editor for creating or revising BMDP command files.
- A data import/export utility. The formats supported depend on the platform.

In addition, some versions provide the following as part of the graphical interface:

- Context-sensitive HELP.
- A high-resolution PLOT program.
- A DATA ENTRY (DE) utility that lets users enter data using a spreadsheet-like editor.

This discussion will center mainly on the part of BMDP that is common to all environments. That is, we shall focus on describing the BMDP statistical programs, the format of BMDP command files, and running BMDP programs. We shall briefly discuss MENTOR and PLOT. The discussion will focus on the latest version of BMDP, Release 7.0. If you have an earlier release, consult the BMDP manual to determine any differences between it and Release 7.0.

45.1 STATISTICAL PROGRAMS

BMDP programs are referred to using a two-character identifier. For most programs this consists of a number followed by a letter. The remaining cases use two letter designators. The letter in the second position is usually indicative of the general class of

0-8493-2530-7/96/$0.00+$.50

statistical analysis performed by that program. The designators for the major blocks of programs are as follows:

- D—Data Description
- F—Frequency Tables
- L—Survival Analysis
- M—Multivariate Analysis
- R—Regression Analysis
- S—Nonparametric Statistics
- T—Time Series Analysis
- V—Analysis of Variance

In addition, there are some specialized programs that do not fit in the above categories. What follows is a listing of the available programs with a brief description of some of the features of each. Consult the *BMDP Software Manual* for complete descriptions.

CA—Correspondence Analysis

- Simple correspondence
- Multiple correspondence
- Two-dimensional plots of row and column profiles
- Plots of rows and/or columns in one dimension
- Face-plane projections for three dimensions

1D—Simple Data Description

- Descriptive statistics for all cases or within subgroups
- Location of missing data and out of range values

2D—Detailed Data Description, Including Frequencies

- Frequency and cumulative percentages for each distinct value of a variable
- Univariate statistics, including skewness, kurtosis, and robust estimates of location.
- Histograms
- Normality testing using the Shapiro–Wilks *W* test

3D—*t* Tests

- One-sample, two-sample, and matched pairs *t* tests
- Test for equality of variances for two groups using the Levene test
- Group histograms
- Trimmed *t* test
- Nonparametric tests
- Hotelling's T^2 and Mahalonobis' D^2 for multivariate data
- Interactive capability to manipulate analysis and plot data

5D—Histograms and Univariate Plots

- Histograms
- Cumulative frequency plots
- Cumulative histograms
- Normal plots

- Half-normal plots

6D—Bivariate Plots

- Scatter plots showing simple linear regression line
- Correlation
- Simple linear regression analysis

7D—One- and Two-Way ANOVA with Data Screening

- Within groups histograms
- One- and two-way ANOVA
- Levene test of equality of variances
- Brown–Forsythe and Welch tests for cases with equality of variances not assumed
- Trimmed means and ANOVA based on trimmed data
- Box–Cox plots
- Post hoc comparisons using Bonferroni, Tukey, Scheffe, and Dunnett procedures
- Duncan and Newman–Keuls multiple range tests
- User–defined contrasts on group menas

8D—Correlations with Missing Data

- Correlation matrix for cases with complete data or with missing values
- Matrices of pairwise frequencies, means, and variances
- *t* tests based on incomplete data pattern

9D—Multiway Description of Groups

- Miniplots of cell means with symbols identifying groups, variables, or repeated measures data
- Means and standard deviations of cells defined by grouping variables
- Marginal means and standard deviations of models defined by grouping variables

LE—Maximum Likelihood Estimation

- General maximum likelihood estimates for a user-specified function
- Estimates of functions of parameters
- Likelihood profile plots of parameters with confidence intervals
- Plot of estimated density function

4F—Frequency Tables—Measures of Association and the Log-Linear Model
Two-way tables:

- Tests of independence (chi-square, likelihood ratio, Fisher's exact, Cramer's *V*, Yule's *Q*, etc.)
- Measures of prediction (McNemar's test of symmetry, Kappa test of reliability, row-wise and column-wise test of symmetry, test of linear trend across ordered proportions)

  Multiway tables:

- Log-linear models
- Stepwise models
- Tests of marginal and partial association

- Tests of fit, parameter estimation
- Stepwise identification of unusual cells and strata

1L—Life Tables and Survivor Function

- Life-table and product-limit estimates of the survival distributions
- Survival curves for all subjects or subsets of subjects
- Mantel–Cox, Tarone–Ware, Breslow, and Peto–Prentice tests and associated trend tests for equality of curves between groups
- Plots of cumulative survival function and hazard function
- Stratified analysis
- Brookmeyer–Crowley confidence intervals for survival time

2L—Survival Analysis with Covariates

- Cox proportional hazard models with built-in risk functions
- Accelerated failure time models (exponential, Weibull, log-logistic, and lognormal distributions)
- Optional stepwise variable selection
- User-specified time-dependent covariates
- Competing risk analysis
- Plots of cumulative survival function and log cumulative hazard function
- Stratified analysis
- Hypothesis tests for joint significance of subsets of regression coefficients
- Diagnostic plots (Cox–Snell residuals, standardized residuals vs. each covariate, etc.)

AM—Description and Estimation of Missing Data

- Display of pattern of missing values
- Estimated covariance and correlation matrix
- Five options for replacing missing values
- *t* test and variance ratio results based on pattern of incomplete data
- Asymptotic standard errors, *t* statistics, and confidence intervals for means
- Rod Little's chi-square test for MCAR

KM—K-Means Clustering of Cases

- Four standardization methods
- ANOVA, cluster profiles, covariances, and correlations
- Crosstabulation of user-selected variables versus final cluster membership

1M—Cluster Analysis of Variables

- Four measures of association for clustering variables
- Three criteria for combining clusters
- Cluster tree
- Shaded correlation matrix display

2M—Cluster Analysis of Cases

- Eleven distance measures for clustering of cases
- Three linkage algorithms (single, centroid, *k*-th nearest neighbor)

- Cluster tree
- Shaded distance matrix

3M—Block Clustering

- Clustering of categorical data into blocks of cases by variables
- Block symbol diagram to describe submatrices identified

4M—Factor Analysis

- Four methods of initial factor extraction
- Correlation matrix in sorted, shaded form
- Several methods of factor rotation
- Squared multiple correlations, sorted factor loadings, factor score coefficients, Mahalanobis distances
- Scree plot
- Goodness-of-fit chi-square for maximum likelihood factor analysis
- Cronbach's alpha
- Second-order factors computed from the factor correlation matrix

5M—Quadratic Discriminant Function Analysis

- Linear and quadratic classification functions
- Uniform, proportions, and user-specified prior probabilities
- Test of homogeneity of within-groups variance
- Pairwise test of equality of group means
- Pairwise Mahalanobis distances between group means
- Posterior probabilities and coefficients for linear and quadratic discriminant analysis

6M—Canonical Correlation Analysis

- Canonical correlation analysis for two sets of variables
- Bartlett's test for significance of remaining eigenvalues
- Canonical variable loadings, scores, bivariate plots

7M—Stepwise Discriminant Analysis

- Stepwise linear classification functions
- Built-in cross-validation procedure
- User-defined contrasts for selection of variables
- Jackknife validation procedure
- Multivariate tests (Wilks' lambda, Pillia's trace, Hotelling–Lawley trace, Roy's maximum root)
- Posterior probabilities and Mahalanobis distance of each case from group mean
- Standardized canonical variable coefficients
- Average squared canonical correlation
- Plot of canonical variables

8M—Boolean Factor Analysis

- Boolean factor analysis of dichotomous data

- Program- or user-specified initial estimates of loading matrix

9M—Linear Scores for Preference Pairs

- Stepwise construction of one linear score per case
- Coefficients based on judgments of experts, comparing several pairs of cases, *t* values, plots

AR—Derivative-Free Nonlinear Regression

- Estimates of parameters of nonlinear function using pseudo-Gauss–Newton algorithm
- Uses seven built-in functions or user-defined functions
- Estimates of user-specified functions of the parameters and their standard deviations
- Models defined by system of differential equations
- Linear inequality constraints on the parameters
- Plots of predicted values, observed values, and residuals
- Serial correlation, runs test, and pseudo R^2 for residuals

LR—Stepwise Logistic Regression

- Binary-dependent variable, categorical and interval-scaled independent variables
- Stepwise MLE of coefficients based on maximum likelihood ratio or asymptotic covariance estimate
- Log-likelihood and three goodness-of-fit statistics
- Histograms, classification tables, predicted probabilities scatter plots
- ROC curve

PR—Polychotomous Logistic Regression

- Generalized version of LR that accommodates multinomial and ordinal dependent variables
- Stepwise MLE of coefficients based on maximum likelihood ratio or asymptotic covariance estimate
- Log-likelihood and goodness-of-fit statistics

1R—Linear Regression by Groups

- Least-squares regression
- Test for equality of regression lines
- Standardized and unstandardized regression coefficients with *t* tests
- Durbin–Watson statistics

2R–Stepwise Regression

- Forward selection and backward elimination alone or in combination
- User may specify sets of variables to enter in one step
- Standardized and unstandardized regression coefficients, tolerance, *F*-to-enter, *F*-to remove, partial correlations, R^2, adjusted R^2
- 21 diagnostic statistics, including the standardized residual, and diagonal elements of the "hat" matrix
- Durbin–Watson statistics and serial correlation
- Case plots of diagnostics

3R—Nonlinear Regression

- Estimates of parameters of nonlinear functions using the Gauss–Newton algorithm
- Uses seven built-in functions or user-defined functions. Can use CDFs in user-defined functions
- Estimates of user-specified functions of the parameters along with their standard deviations
- Exact linear contrasts on the parameters
- Plots of predicted values, observed values, and residuals
- Sum of squares profile plots of parameters, with confidence intervals
- Durbin–Watson statistics

4R—Regression on Principal Components and Ridge Regression

- Stepwise regression using principal components
- Eigenvectors, cumulative proportion of σ^2 explained, R^2, regression coefficients, residual sum of squares, plots

5R—Polynomial Regression

- Regression using orthogonal polynomials
- Goodness-of-fit statistics, plots
- Table of predicted values and residuals at the specified polynomial degree

6R—Partial Correlation and Multiple Regression

- Partial correlation of a set of variables after removing the linear effects of a second set of variables
- R^2 of independent variables with dependent and other independent variables
- Shaded correlation matrix, partial correlations

9R—All Possible Subsets Regression

- Estimated regression equations for 10 "best" subsets of each subset size using Furnival–Wilson algorithm
- Three criteria to define "best"—Mallow's C_p, R^2, and adjusted R^2 across different subset sizes
- Residual analysis, Cook's distance, Durbin–Watson statistics, serial correlation

3S—Nonparametric Statistics

- Sign test and Wilcoxon signed-rank test
- Mann–Whitney rank sum test and Kruskal–Wallis one-way ANOVA
- Friedman two-way ANOVA and Kendall's coefficient of concordance
- Multiple comparisons for the Kruskal–Wallis and Friedman tests
- Kendall and Spearman rank correlations
- Kolmogorov–Smirnov test

1T—Univariate and Bivariate Spectral Analysis

- Frequency domain approach
- Graphical and numeric displays of periodograms and spectra of single or pairs of time series

- Filtering and recoloring
- Estimates of degree of coherence and regression relation between two time series in different frequency bands
- Smoothing by removing seasonal means or linear trends, or by constructing filters
- Confidence intervals about spectral estimates

2T—Box–Jenkins Time Series Analysis

- Parametric time domain models—ARIMA, intervention, and transfer function models
- Selection, estimation, and testing of models including residual analysis
- Forecasting of future observations

1V—One-Way Analysis of Covariance

- ANOVA or ANCOVA table with tests of zero slope and equality of slopes
- Pairwise *t* tests for adjusted group means
- Regression coefficients, standard errors, *t* values
- Within-group slope for each covariate
- Scatter plots and correlations
- User may specify contrasts across group means or adjusted group means

2V—Analysis of Variance and Covariance with Repeated Measures

- Factorial designs, repeated measures designs, or a combination of both
- Orthogonal decomposition of trial factors—linear, quadratic, and cubic effects
- Adjusted cell means with standard errors, marginals, and plots. Confidence intervals for adjusted cell means
- Regression coefficients
- User may specify contrasts for factorial and repeated measures factors
- Greenhouse–Geisser and Huynh–Feldt adjustments for repeated measures
- Bonferroni confidence limits for repeated measures

3V—General Mixed Model ANOVA

- Maximum likelihood and restricted maximum likelihood for fixed and random effects models
- Estimates of parameters, asymptotic standard errors, *t* statistics
- Log-likelihoods and likelihood ratio test

4V—Univariate and Multivariate ANOVA and ANCOVA Including Repeated Measures

- Multifactor designs and multivariate ANOVA
- Repeated measures, splot plot, and changeover designs
- User may specify contrasts and tests of simple effects
- Analysis with empty cells
- Generates eigenvalues, eigenvectors, and contrast matrices
- Greenhouse–Geisser and Huynh–Feldt adjustments
- Multivariate test statistics include Wilks lambda, Roy's largest root, and Hotelling–Lawley trace

5V—Unbalanced Repeated Measures Models with Structured Covariance Matrices

- Fits models to incomplete or unbalanced data using maximum likelihood or restricted maximum likelihood
- User may specify the covariance structure
- Covariates may be fixed or time-dependent
- Regression or ANOVA type contrasts
- Linear combinations of parameters may be estimated and tested with user-specified Wald tests
- Asymptotic Wald-type chi-square tests for regression model terms
- Imputed values of missing responses
- Predicted means, standard errors of predicted means, residuals and their standard errors, standardized residuals

8V—Mixed Model ANOVA-Equal Cell Sizes

- Nested, crossed, fixed-effects, random-effects, or mixed models
- Expected mean squares
- Estimates of variance components

In addition to these statistical programs BMDP provides a data manipulation capability in its DM (Data Manager) program. We shall not discuss it here. The interested reader should consult the *BMDP Software Manual.* BMDP also makes provisions for users to include FORTRAN subroutines to transform data or to perform computations not available in BMDP. This feature is not available in all implementations. Consult the BMDP manuals for further details.

45.2 RUNNING BMDP PROGRAMS

All BMDP programs are executed by referring to a list of instructions contained in an ASCII file that is created by using a text editor, the BMDP editor, or, for certain BMDP programs, the MENTOR utility. The default extension for instruction files is INP. We give an example of such a file. The data represent a soil erosion study that measured the moisture content of soil in a forest after logging. The treatments were to leave no trees standing after cutting, leave a small number of tress, and to leave a moderate number of trees. The purpose of the following instruction file is to generate summary statistics using program 1D. The data are stored in a file named SOIL.DAT, and the instruction file is called SOILDESC.INP.

```
/ INPUT FILE = 'D:\DYNAMIC\SOIL.DAT'
        FORMAT = FREE.
        VARIABLES = 2.
/ VARIABLE NAMES = Moisture, Method.
/ GROUP VARIABLE = Method.
        CODES(Method) = 1, 2, 3.
        NAMES(Method) = Clear, Few, Moderate.
/ PRINT DATA.
/ END
```

The contents of a BMDP instruction file are structured similar to English. That is, instructions are divided into *paragraphs*, and the paragraphs are composed of *sentences*. All BMDP programs require that a basic set of paragraphs be provided. We shall discuss these in more detail later. Each paragraph begins with a slash (/), which is followed by the paragraph name. Although it is not necessary, it is good practice to start each paragraph on a new line. Next the sentences for that paragraph are given, each terminated by a period (.). Again, it is recommended, but not required, that each sentence begin on a new line. We briefly explain the paragraphs given in the above instruction file.

- The INPUT paragraph contains three sentences. The FILE = sentence tells the program the name of the ASCII file containing the data. The VARIABLES = sentence tells the program how many variables per case (line) are to be read. The FORMAT sentence specifies how the data are to be read.
- The VARIABLE paragraph contains a single sentence that specifies the names to be assigned to the two variables.
- The GROUP paragraph contains three sentences. The VARIABLE sentence tells the program which input variable is to be used to determine group membership. The CODE sentence gives the specific values of the grouping variable that are to be used. The NAMES sentence gives the group names that correspond to each value of the grouping variable.
- The PRINT paragraph contains a single sentence. The DATA sentence instructs the program to show the raw data as a part of the final printout.
- The END paragraph contains no sentences, but it is required by all programs to indicate the termination of the instruction file.

All programs use the INPUT, VARIABLE, and END paragraphs.

Program 1D can be run with this instruction file in a number of ways. The instructions can be created and run directly from the MENTOR utility. For BMDP installations with a menu-driven user interface the instruction file can be created separately and run by selection of the RUN option from the interface menu. On systems with a command line interface a program is run from the system prompt. On a VAX operating under VMS the command line would look like the following:

```
$BMDPRUN 1D IN=soildesc.inp OUT=soildesc.out
```

It is also possible to run BMDP programs interactively. We shall not discuss this method here. Interested readers should consult the documentation for BMDP appropriate for their system. In all cases, upon completion of the BMDP program a file named SOILDESC.OUT will be created that looks like the following. This particular run was conducted using BMDP/DYNAMIC on an IBM-compatible personal computer.

```
BMDP Instruction File  : D:\DYNAMIC\SOILDESC.INP

BMDP Program Output File: D:\DYNAMIC\SOILDESC.OUT

BMDP1D—SIMPLE DATA DESCRIPTION

Release: 7.0    (BMDP/DYNAMIC)    Date: 02/22/94 at 13:44:46

Site: 160941500B
        rinaman
```

```
/ INPUT FILE = 'D:\DYNAMIC\soil.dat'
FORMAT = FREE.
VARIABLES = 2.
/ VARIABLE NAMES = Moisture, Method.
/ GROUP VARIABLE = Method.
        CODES(Method) = 1, 2, 3.
        NAMES(Method) = Clear, Few, Moderate.
/ PRINT DATA.
/ END
NUMBER OF CASES READ          16
GROUPING VARIABLE  Method
                    CATEGORY   FREQUENCY
                    Clear          5
                    Few            6
                    Moderate       5
DESCRIPTIVE STATISTICS OF DATA
VARIABLE  TOTAL   STANDARD ST.ERR COEFF SMALLEST    LARGEST
NO. NAME FREQ. MEAN DEV. OF MEAN OF VAR VALUEZ-SCR VALUEZ-SCR
1 Moisture 16 1.8031 .65070 .16267 .36087 1.0100-1.22 3.2100 2.16
PRINT SUMMARY STATISTICS OVER ALL CASES AND BROKEN DOWN BY
INDIVIDUAL CATEGORY ON Method
CASE    1   2
NO.  Moisture Method
1          1.41 Clear
2          1.27 Clear
3          1.19 Clear
4          1.37 Clear
5          1.52 Clear
6          1.53 Few
7          1.81 Few
8          1.24 Few
9          1.01 Few
10         2.19 Few
11         1.34 Few
12         2.43 Moderate
13         3.21 Moderate
14         2.65 Moderate
15         2.01 Moderate
16         2.67 Moderate
VARIABLE GROUPING TOTAL   STANDARD       SMALLEST  LARGEST
NO. NAME VAR/LEVL FREQ. MEAN  DEVIATION VALUE Z-SC VALUE Z-SC
```

```
1 Moisture         16  1.8031 .65070 1.0100 -1.22  3.2100 2.16
     Method
     Clear          5  1.3520 .12736 1.1900 -1.27  1.5200 1.32
     Few            6  1.5200 .42521 1.0100 -1.20  2.1900 1.58
     Moderate       5  2.5940 .43483 2.0100 -1.34  3.2100 1.42
NUMBER OF INTEGER WORDS USED IN PRECEDING PROBLEM 634
END OF INSTRUCTIONS
PROGRAM TERMINATED
```

This sample printout illustrates features that are produced by all BMDP programs. These include the following:

- A header showing which program is being run along with other information regarding the version of BMDP that was used.
- A listing of the BMDP instructions you submitted.
- Information showing how BMDP interpreted the instructions, including default settings that were used.
- A list of variable names along with a sequence number that BMDP assigned to each.
- The format used to read the data along with additional interpretation of the BMDP instructions.
- The number of cases read by the program.
- Grouping information showing how BMDP interpreted the instructions in your GROUP paragraph.
- Descriptive statistics for variables not used for grouping.
- A printout of the data in response to your PRINT command. ID is the case number.
- Descriptive statistics grouped according to method. This section is specific to program 1D.
- Information about the CPU time and problem size. Error messages show up here if the program fails to run properly.

45.3 GUIDELINES FOR WRITING BMDP INSTRUCTIONS

The instructions in SOILDESC.INP were written following the general guidelines listed below. These rules apply to sentences in all paragraphs.

- The paragraph name, such as VARIABLE, comes first. Paragraphs are separated by a slash (/). It is recommended that you precede each paragraph name with the slash, and begin a new paragraph on a new line.
- Except for the END paragraph, and, where otherwise indicated in the program description, paragraphs may be given in any order. END indicates termination of the problem or subproblem. Note from the example that END contains no sentences and, hence, should not be terminated with a period (.).
- Additional instructions are given in the sentences. One form of sentence has a "keyword" followed by IS, ARE, or =, followed by a list of items. IS, ARE, and = are interchangeable. The one that seems more natural should be used. The other form of sentence consists of a single "keyword." The DATA sentence in the PRINT paragraph given above is an example of such a sentence.

- Every sentence must be terminated with a period (.) or a semicolon (;). Sentences within a paragraph may be in any order.
- Names or values in a list are separated by commas (,).
- Names can be no longer than eight characters. If a name contains characters other than letters or digits it must be enclosed by single quotes.
- Instructions are not case sensitive.

45.4 PROGRAM DESCRIPTIONS

Both the *BMDP Software Manual* and the *BMDP User's Digest* give descriptions of each of the programs. They show which paragraphs are necessary and which are optional. Within each paragraph the required sentences are indicated, and any default settings are shown. Symbols are used in these descriptions that have the following meanings.

■ Required paragraph or command
▲ Frequently used instruction
✓ Default option
● Assigned value is retained for multiple problems

Sentences that are marked with a check (✓) as defaults will be performed unless that sentence name is placed in the paragraph preceded by the keyword NO. That is, if a paragraph description indicates that a CHISQUARE sentence will be executed by default, it can be suppressed only if that paragraph contains the sentence NO CHISQUARE.

Program descriptions in the *BMDP Manual* are organized in the following manner. Each one begins with a brief description of what the program does. This is followed with some examples showing how the program is used on the data sets that are provided with BMDP. Then each paragraph that is appropriate for that program is listed with descriptions of the sentences in each. Next the recommended order for paragraphs is given. Finally a summary table for commands that are specific to that program is given. The *BMDP User's Digest* omits the examples and presents terse descriptions of the instructions that are specific to that program. By the term "instructions that are specific to that program" we mean the following. Some paragraphs, such as INPUT, are required by all programs. In such cases there are a number of sentences that are common to all programs in that paragraph. Individual programs may also provide for the use of additional sentences in that paragraph. It is these additional sentences that are discussed in the program descriptions. The sentences that are common to all programs are described elsewhere.

We shall give a brief introduction to those paragraphs that are common to all BMDP programs. After this we shall give as examples the descriptions of BMDP programs 2R and 2T. The format we shall use is similar to that of the *BMDP User's Digest*. Our introduction will show only those sentences that are required or are frequently used options. Consult the BMDP documentation for complete information on these sentences as well as those not discussed here. The convention used in giving the paragraph and sentence names is that the portion in capital letters is all that BMDP requires for proper interpretation. Use of the full name enhances readability.

■ / INPut

This paragraph is required. It describes the data.

■ VARiables=#.

Required. Specifies the number of variables per case. This sentence is ommited if data are coming from a BMDP file.

■ **FORMat = FREE, STREAM, SLASH, 'c',** *or* **BINARY.** *(one only)*
Required. Specifies the way data are entered. Again, this sentence is omitted if data are coming from a BMDP file.

FREE	Variables are separated by one or more blank spaces or a comma. One case per record. May have only the number of variables per record as are specified in the **VAR** sentence. Can handle exponential notation.
STREAM	Allows more than one case per record.
SLASH	Like **STREAM** but cases are separated by a slash (/).
'c'	For FORTRAN-like specifications. Replace **c** with the appropriate FORTRAN format specifiers. Use this if the number of variables per record differs from that given in the **VAR** sentence.
BINARY	For unformatted data.

▲ **FILE = 'name'** *or* **UNIT = #.**
Specifies the location of the file. On VAX, UNIX and PC systems use the **FILE** sentence and replace **name** with the name of your data file. Use **UNIT** on IBM mainframes. If this sentence is omitted BMDP assumes that the data come immediately after the **END** paragraph. This sentence is required if the data come from a BMDP file. If data come from a BMDP file, then a **CODE** sentence is required.

The **INPUT** paragraph has optional sentences that allow the user, among other things, to create a title that is placed on each page of output, limit the number of cases read, assign a missing value flag, and set the maximum number of input errors that are allowed before aborting the program.

▲ **/VARiable**
This paragraph describes the characteristics of the variables.

▲ **NAMEs = list.**
Assigns names to variables. Names are associated to variables in the order the names are given in the list unless tabbing is used. For example, variables 1, 2, and 5 would be named A, B, and C, respectively, as follows:

```
NAMES ARE A, B, (5)C.
```

Names may be 1 to 8 characters in length. If the name begins with a number or contains blanks or symbols, then it must be enclosed by apostrophes.

▲ **USE = list.**
List the variables which are to be used in the analysis. Default is all.
The **VARIABLE** paragraph has optional sentences that permit the user to specify missing value codes for variables, permissible ranges for variables, specify variables to be case labels, how to interpret blanks, etc.

▲ **/GROUP** *or* **CATEGory**
This paragraph is used to classify cases into categories for tables, ANOVA, plots, etc.

▲ **VARiable = list.**

This sentence lists the variables that are to be used to define groupings. **CODES** or **CUTPOINTS** are required if a variable has more than 10 distinct values.

▲ **CODEs(j) = # list.**

Used for defining groups according to values of a discrete variable. The number list contains values of the variable *j*, where *j* is the variable name or number. Only those cases with values in the number list will be analyzed.

▲ **CUTPoints(j) = # list.**

Used for defining groups according to values of a continuous variable. Numbers in the list are the upper limits of intervals for the variable *j*, where *j* is the name or number of a variable. For example, if a list contains *a* and *b*, then three groups will be defined. The groups will be those cases with values less than or equal to *a*, those cases with values greater than *a* and less than or equal to *b*, and those cases with values greater than *b*.

▲ **NAMEs(j) = list.**

Defines names for categories in the order defined by the **CODES** or **CUTPOINTS** sentence for the variable *j*, where *j* is the name or number of a grouping variable.

/ TRANSform

This paragraph permits the user to compute functions of values of variables, select cases, edit and recode values, and generate random numbers. In addition to the usual arithmetic operators (+,–,*,/,**,mod), BMDP provides an extensive range of mathematical functions, data functions, summary statistical functions, and cumulative distribution functions for transforming data. For example, the data on soil moisture content could be converted to standardized values as follows:

```
Z = (MOISTURE - MEAN (MOISTURE))/SD(MOISTURE).
```

Standardized values will be computed for all valid cases. The new variable will be appended to the data set after the last variable. Subsequent variables created in a TRANSFORM paragraph are appended to the data set in this manner in the order in which they are computed in the TRANSFORM paragraph. BMDP also provides logical operators so that conditional transformations may be computed. In addition, there are sentences that permit users to select cases for analysis, to perform linear interpolations for replacing missing values, and to generate random numbers from uniform and normal distributions. Consult the BMDP documentation for detailed information on transformations.

/ PRINT

This paragraph controls the amount and format of program output. The LINESIZE and PAGESIZE sentences allow the user to control printing for various paper sizes. Other sentences are used to control the amount of output that is generated. Consult the BMDP manuals for details on these features.

/ SAVE

This paragraph is required if the user desires to save raw data, transformed data, and analysis results in a raw data file or BMDP file for later use.

■ FILE = 'name'. *or* UNIT = #.
This sentence is required to identify the system file or unit where data are to be saved. Use the FILE sentence on VAX, UNIX, and PC systems. Replace **name** with the file name where that data are to be stored. Use UNIT on IBM mainframes.

▲ CODE = name.
This sentence is required to name a BMDP file. The name must be 1 to 8 characters in length. This sentence is not to be used when creating a raw data file.

▲ NEW.
This sentence is to be used only when saving to a BMDP file. It tells BMDP to create a new file or to write over the contents of an existing file. If this sentence is omitted, then the data will be appended at the end of the existing file having the specified code name.

Other sentences in this paragraph permit the user to print the data according to a specified format, print a subset of the variables, print a subset of the cases, specify the content of the file if results other than data are to be stored, or to create a BMDP file that can be read by all implementations of BMDP.

/ CONTrol
This paragraph, if used, must precede all other BMDP instructions and be followed by an END paragraph. Sentences in this paragraph include those that let the user specify the maximum width of BMDP instructions, control how BMDP checks for errors in instructions, and specify the location of BMDP instructions and macros.

■ / END
This sentence is required. It indicates the termination of BMDP instructions for a problem.

45.5 EXAMPLES

We now give two examples of using BMDP programs. We shall show the order of instructions as given in the BMDP documentation, omitting that part that is appropriate for running BMDP interactively. Next the desired analysis will be described. The output generated by the BMDP instructions will not be included. Run these instruction files on your own system to create output files.

The first example illustrates stepwise regression using program 2R. The order of instructions is as follows.

■ / INPUT
▲ / VARIABLE
/ TRANSFORM
/ SAVE
■ / REGRESS—repeat for subproblems
/ PRINT
/ PLOT
■ / END
data

BMDP provides data on air pollution in 60 U.S. cities. The file contains the names of the cities, average annual rainfall, median educational level, population density, percentage nonwhites living in the city, relative pollution potential of oxides of nitrogen, relative pollution potential of sulfur dioxide, and age-adjusted mortality rates. We wish to find a good set of variables other than name for predicting the mortality rate. Forward selection will be used with the *F*-value being the criterion for inclusion. These are the default settings for this program. For this reason the only sentence necessary in the REGRESS paragraph is that declaring which variable is dependent. All other variables except for those declared to be LABELs in the VARIABLE paragraph are considered independent. For regression diagnostics we wish to plot the residuals and a normal plot of the residuals. The sentences for this are in the PLOT paragraph. In addition to the default case of printing ANOVA tables, results, summary tables of coefficients, R, R^2, and F to enter values at each step, we wish to see a correlation matrix and a summary table of partial correlations. The sentences in the PRINT paragraph accomplish this. The LINESIZE sentence sets the width of printed output. The default for noninteractive runs is 132.

```
/ INPUT    FILE = 'AIRPOLL.DAT'.
           FORMAT = FREE.
           VARIABLES = 9.
/ VARIABLE NAMES = Name1, Name2, Rain, Educ, Pop_Den,
                   Non_whit, Nox, SO2, Mort.
           LABELS = Name1, Name2.
/ REGRESS DEPENDENT = Mort.
/ PLOT     RESIDUAL.
           NORMAL.
/PRINT     LINESIZE = 80.
           CORRELATION.
           PARTIAL.
/ END
```

The second example illustrates time series analysis using program 2T. The order of instructions is as follows.

```
■ / INPUT
  / VARIABLE
  / TRANSFORM
  / PRINT
  / SAVE
■ / END
  data
  PRINT . . . /
  SAVE . . . /
  TPLOT . . . /
  BLOCK . . . /
  ACF . . . /
```

```
PACF . . . /
CCF . . . /
DIFFERENCE . . . /
FILTER . . . /
ARIMA . . . /
INDEP . . . /
INDEP . . . /
ESTIMATION . . . /
PSIWEIGHT . . . /
FORECAST . . . /
CHECK . . . /
ERASE . . . /
END /
FINISH /
```

The paragraphs between the data and the FINISH paragraph may be repeated for subproblems.

BMDP provides data on the operations of a furnace. It contains 296 observations on two variables—feed rate of a mixture of gases (feed) and carbon dioxide concentration in the output (CO2). We wish to perform time series analysis on the carbon dioxide concentration. In particular, we wish to compute the autocorrelation function for lags up to 25, the partial autocorrelation function for lags up to 25, and fit an autoregressive model with AR polynomial powers of 1 and 2 plus a mean term. In addition, we wish to estimate the parameters of the model and compute their correlations. The residuals will be stored in a variable RESCO2. Upon completion of the analysis the model will be erased. The contents of a BMDP instruction file to do this is as follows:

```
/ INPUT FILE = 'FURNACE.DAT'.
    VARIABLES ARE 2.
    FORMAT IS FREE.
/VARIABLE NAMES ARE FEED,CO2.
/END
ACF VAR IS CO2.
    MAXLAG IS 25./
PACF VAR IS CO2.
    MAXLAG IS 25./
ARIMA VAR IS CO2.
    ARORDERS ARE (1,2)'.
    CONSTANT./
ESTIMATION RESID IS RESCO2.
    PCOR./
ERASE MODEL.
/END
/FINISH
```

45.6 USING MENTOR

The MENTOR utility is available on PC and VAX implementations of BMDP. It is a menu-driven program that generates BMDP instruction files and then executes them. It supports the most commonly used options in frequently used programs. The BMDP programs it presently supports are 1D, 2D, 3D, 6D, 7D, 4F, 1L, 2L, 1R, 2R, LR, and 2V. For new BMDP users it provides an easy way to generate instruction files. For users with more experience MENTOR can be used to generate commands for common paragraphs, such as INPUT and VARIABLE. These instructions can then be brought up in the BMDP editor or any other editor where the remaining paragraphs can be written. Using MENTOR proceeds in the following manner. The menus guide you through each of these steps.

1. Select MENTOR from the main BMDP menu.
2. Select the desired BMDP program from the MENTOR main menu.
3. Specify the contents of the INPUT paragraph. The user is led through this with submenus.
4. Specify the contents of the VARIABLE paragraph. You can create variable names, declare which variables are to be used, which variables are to serve as case labels, missing values, and acceptable ranges for variables.
5. Specify groups of cases where appropriate. Groups can be defined using either CODES or CUTPOINTS.
6. Specify the analyses to be done. This part varies according to the BMDP program that is being used.
7. Save the instruction file, run the analysis, and specify the output file.

MENTOR also allows the user to use an instruction file as a basis for creating other sets of instructions for analysis. These can then be run from MENTOR. The interested reader should consult the *BMDP User's Guide* for details on these features.

45.7 USING PLOT

PLOT is a utility that is available on PC and VAX/VMS versions of BMDP. It allows the user to do the following:

- Produce high-resolution graphics without running any BMDP programs.
- Interactively alter plot settings, list coordinates, and save plots.
- Retrieve existing plot files for viewing or editing.

PLOT files have a default extension of .PLT. PLOT will produce line graphs, bar charts, pie charts, scatter diagrams, box plots, error bar charts, normal probability plots, quantile plots, quantile–quantile plots, and histograms. PLOT is menu driven. To use it select PLOT from the BMDP main menu or type BMDPLT at the operating system prompt. The user will then be asked to specify a file for analysis. This file can be a BMDP data file, a raw data file, or a previously saved PLOT file. Next, the type of plot is specified, variables to be plotted are selected, and plot options are chosen. Following this the requested plot will be displayed on the screen. An example of a histogram of the sulfur dioxide data from the air pollution data set is shown in Figure 45.1.

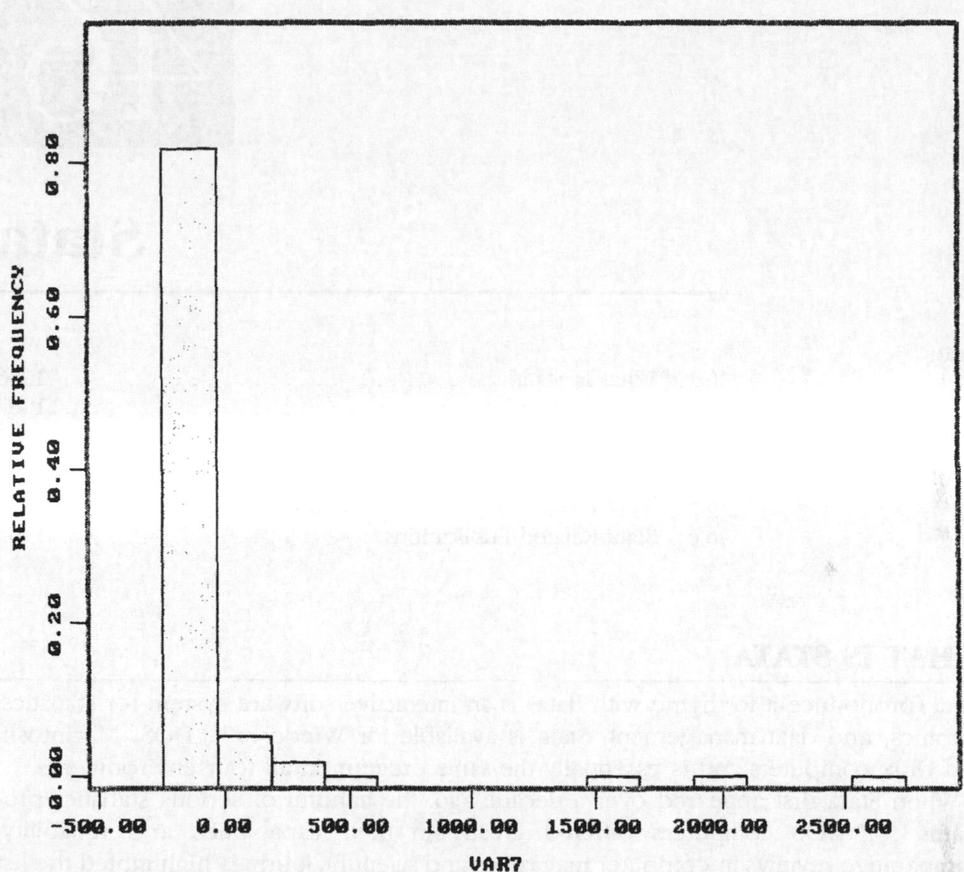

FIGURE 45.1 Histogram of the sulfur dioxide data from the air pollution data set.

DEFINING TERMS

Instruction: A sentence in a BMDP paragraph.

Paragraph: A grouping of BMDP instructions that accomplishes a set of tasks while running a program.

Problem: The statistical analysis to be performed using a BMDP program on a specified data file.

Program: The basic BMDP unit. There are 44 programs, each of which performs a different type of statistical analysis.

Sentence: A BMDP instruction to specify tasks to be performed.

Subproblems: Different analyses that use the same BMDP program and the same data set.

REFERENCES

BMDP Software Manual, Vols. 1, 2, and 3. BMDP Statistical Software, Inc., Los Angeles, CA.

BMDP/DYNAMIC User's Guide. BMDP Statistical Software, Inc., Los Angeles, CA.

BMDP/GRAPHIC User's Guide. BMDP Statistical Software, Inc., Los Angeles, CA.

BMDP User's Digest. BMDP Statistical Software, Inc., Los Angeles, CA.

46

Stata

Sean Becketti
Stata Corporation

46.1 WHAT IS STATA?

Stata (pronounce it to rhyme with data) is an interactive software system for statistics, graphics, and data management. Stata is available for Windows™, DOS, Macintosh, and Unix computers and is essentially the same program in all four environments.

When Stata first appeared over a decade ago, the handful of serious statistics programs for DOS computers differed greatly in their capabilities and reliability. Comparative reviews in computer magazines and scientific journals highlighted the list of statistics calculated by each program, data set size limitations and program speeds, and performance on standardized tests of numerical accuracy. Those days are gone. There are now many high-quality, full-featured statistics packages available, including both programs written specifically for PCs and workstations and classic mainframe programs repackaged for the workstation environment.

How can you distinguish between high-quality statistical programs that offer essentially the same menu of statistics? The answer lies in the path that links your ideas and questions to the statistical answers the program calculates: the **interface**.

Scientific and engineering questions are complex and abstract:

- Can this treatment extend the lives of cancer patients?
- Which of these designs produces the most reliable engine?
- Why are divorce rates higher in some states than in others?

It takes work to answer questions like these, even after data have been collected and stored in the computer. You must

- check the data for accuracy and outliers;
- examine the distributions of the variables and consider appropriate transformations;
- estimate models, plot residuals, and calculate model diagnostics;
- reformulate and reestimate;
- construct out-of-sample forecasts;

0-8493-2530-7/96/$0.00+$.50

and so on. You cannot tell whether a program makes it easy to do all this by reading a list of the statistics a program can calculate. The only way to find out whether a program's interface works for you is to try it.

So let us try Stata.

46.2 A SAMPLE SESSION

Stata comes with example data sets that are used in on-line tutorials and in the *Reference Manual* to illustrate Stata's commands. These data sets cover a variety of topics: census data on birth, death, marriage, and divorce rates; survival times of cancer patients; failure times of equipment; and so on. Here we use data on various 1978 automobile models. (These data were originally published in Chambers et al., 1983.)

We begin by bringing the auto data into memory and describing the contents of the data set.

```
. use auto
(1978 Automobile Data)
. describe
Contains data from auto.dta
Obs:    74 (max= 32766)                1978 Automobile Data
Vars:   12 (max= 99)
Width:  44 (max= 200)
  1. make      str18  %18s             Make and Model
  2. price     int    %8.0g            Price
  3. mpg       int    %8.0g            Mileage (mpg)
  4. rep78     int    %8.0g            Repair Record 1978
  5. hdroom    float  %6.1f            Headroom (in.)
  6. trunk     int    %8.0g            Trunk space (cu. ft.)
  7. weight    int    %8.0g            Weight (lbs.)
  8. length    int    %8.0g            Length (in.)
  9. turn      int    %8.0g            Turn Circle (ft.)
 10. displ     int    %8.0g            Displacement (cu. in.)
 11 gratio     float  %6.2f            Gear Ratio
 12. foreign   int    %8.0g   forlbl   Car type
Sorted by: foreign
```

Stata's prompt is the period (.). Stata operates on data stored in RAM. We load the automobile data into RAM by typing **use auto**. This command looks for a file named **auto.dta** (by default, the **.dta** extension identifies the file as a Stata-format data set), copies its contents into memory, and displays a *data label* (**1978 Automobile Data**) that has been attached to the data set.

Next we use the **describe** command to display information about this data set. The first line of the description indicates that the data came from the disk file **auto.dta**. The following lines reveal that there are 74 observations on 12 variables in this data set. Apparently this data set contains 12 pieces of information on each of 74 different automobiles. The data set label is displayed to the right of the information on the number of observations in the data set. The maximum numbers of observations and variables reflect the amount of RAM on this computer and can easily be adjusted. The information on **Width** is related to the number of bytes allocated to an individual observation and is rarely used.

Next are 12 numbered lines describing each of the 12 variables in the data set. Each line reports five bits of information about a variable. First is the *variable name*. The first variable names are **make**, **price**, . . ., and **foreign**. Second is the *data type* of the variable. **make** is an 18-character string variable (**str18**); **price** an integer (**int**); **hdroom**, a floating-point number (**float**); etc. These data types will be familiar to C programmers. Stata uses three additional data types not represented in this data set: **long** (long integer), **double** (double precision floating point), and **byte** (an integer stored as a single byte).

The third item in each row is the *format*, which determines how Stata displays the values of each variable. Formats can be changed at will. The fourth item—which is blank except for **foreign**—is the name of the *value label* associated with each variable. Statistical data sets frequently contain variables that take on a few distinct variables. In a medical study, for example, a variable called **sex** might be coded **1** for females and **0** for males. Stata allows you to define value labels—paired lists of values and labels—and to display the labels (**Female** and **Male**) rather than the values (**1** and **0**). In this case the value label **forlbl** pairs **0** with **Domestic** and **1** with **Foreign**.

The fifth, and final, item in each row is the *variable label*, a brief description of the contents of each variable. The variable labels in this data set tell us, for instance, that the mysteriously named **rep78** contains the automobile's 1978 repair record and the equally indecipherable **gratio** the gear ratio.

The final line of this description indicates that the observations in the automobile data set have been sorted according to the values of the **foreign** variable.

These first two commands already reveal a bit of Stata's style. Commands are terse, imperative "sentences." They begin with an ordinary English verb (**use, describe**) that indicates what Stata is to do.

Stata's responses are also terse. For example, the **use** command displays only the data label when it loads a data set into memory. The **use** command could display the full data set description produced by the **describe** command, but it does not. The **describe** command could display more text explaining the interpretation of each of the items displayed (data types, formats, etc.), but does not. This additional information might be helpful the first few times you use Stata, but these two commands are used so frequently that this extra information would rapidly become irritating clutter.

We use the **list** command to list the first five observations of the variables **make, price, mpg,** and **foreign**.

```
. list make price mpg foreign in 1/5
                make       price      mpg      foreign
 1.      AMC Concord        4099       22     Domestic
 2.        AMC Pacer        4749       17     Domestic
 3.       AMC Spirit        3799       22     Domestic
 4.    Buick Century        4816       20     Domestic
 5.    Buick Electra        7827       15     Domestic
```

Those 1978 prices look pretty good—an AMC Spirit for only $3,799 (plus tax, license, and dealer prep). The mileage, however, seems low by current standards. The first five observations are all domestic cars (notice the value label **"Domestic"** hiding the numeric code **0**), but the **describe** command indicated that the data are sorted by **foreign**, so the imports must appear in the latter part of the data set.

How many imports are covered in this data set?

```
. tabulate foreign

Car type |     Freq.     Percent       Cum.
---------+-----------------------------------
Domestic |       52       70.27      70.27
 Foreign |       22       29.73     100.00
---------+-----------------------------------
   Total |       74      100.00
```

The **tabulate** command counts the number of times each value of a variable occurs. In this data set, 52 of the 74 observations contain information on domestic cars, while the remaining 22 contain information on foreign cars.

Let us turn back to price and mileage. What were the average price and mileage among the 1978 models? The **summarize** command provides summary statistics:

```
. summarize price mpg

Variable |     Obs        Mean    Std. Dev.       Min        Max
---------+-----------------------------------------------------
   price |      74    6165.257    2949.496      3291      15906
     mpg |      74     21.2973    5.785503        12         41
```

The average price was $6,165, so that AMC Pacer was a relatively inexpensive car. It was not the least expensive, though; the minimum price was $3,291 and the maximum was $15,906. Mileage ranged from 12 to 41 miles per gallon with an average of 21.3.

In 1978, were imports generally more expensive or less expensive than domestic cars, and how did their mileage compare?

```
. by foreign: sum pri mpg

-> foreign = Domestic
Variable |     Obs        Mean    Std. Dev.       Min        Max
---------+-----------------------------------------------------
   price |      52    6072.423    3097.104      3291      15906
     mpg |      52    19.82692    4.743297        12         34

-> foreign = Foreign
Variable |     Obs        Mean    Std. Dev.       Min        Max
---------+-----------------------------------------------------
   price |      22    6384.682    2621.915      3748      12990
     mpg |      22    24.77273    6.611187        14         41
```

There are a couple of new features here. First is the **by foreign**: prefix, which tells Stata to execute the command independently for each value of the variable **foreign**. In this example, **foreign** takes on only two values. Data sets can be sorted by multiple variables each with many values, so the *by prefix* can produce a large amount of output. In a cross-country medical study, for instance, we might execute a statistical command **by condition country sex year**:.

The second new feature is abbreviation. The **summarize** command has been shortened to **sum** and the **price** variable has been shortened to **pri**. Stata accepts abbreviations almost everywhere: command names, variable name, option names. Abbreviations are required to be unique. In this data set, **price** can be abbreviated all the way down to **p**, but if we added a variable called **problem**, the minimum abbreviation would be **pri**.

Stata's output shows that the average foreign car was 5% ($313) more expensive than the average domestic car in 1978 and got 25% (5 MPG) better mileage. Are these differences statistically significant? If we randomly selected another 1978 import not included in these data, could we reasonably expect it to cost more and get better mileage than the average domestic car? One way to approach this question is to calculate *t* tests of the differences in means for each variable separately. (Hotelling's *T*-squared test, also available in Stata, can be used to test the differences for both variables simultaneously.)

```
. ttest price, by(foreign)

Variable |      Obs          Mean     Std. Dev.
---------+---------------------------------------
       0 |       52      6072.423     3097.104
       1 |       22      6384.682     2621.915
---------+---------------------------------------
combined |       74      6165.257     2949.496

       Ho: mean(x) = mean(y) (assuming equal variances)
                   t = -0.41 with 72 d.f.
               Pr > |t| = 0.6802
```

The difference in average price is not significant: the *t* statistic is –0.41 with a *p* value of 0.68.

```
. ttest mpg, by(foreign) unequal

Variable |     Obs         Mean     Std. Dev.
---------+-------------------------------------
       0 |      52     19.82692     4.743297
       1 |      22     24.77273     6.611187
---------+-------------------------------------
combined |      74     21.2973

     Ho: mean(x) = mean(y) (assuming unequal variances)
                t = -3.18 with 31 d.f.
            Pr > |t| = 0.0033
```

The difference in average mileage is highly significant: the *p* value is less than 0.01.

These commands introduce another element of Stata's syntax: options. In Stata, the comma indicates the end of the body of the command and the beginning of the options. Both **ttest** commands included the **by(foreign)** option, indicating that the means were to be compared for the groups defined by the values of **foreign**. In the second **ttest** command, the **unequal** option was added. This option specifies that the variance of **mpg** is allowed to be different for domestic and foreign cars. We could test this hypothesis as well using Stata's **sdtest** command, but let us move on.

Imagine that we believe that a car's price is determined by its objective qualities: its mileage, its size, and so on. We can explore this hypothesis quickly and easily using Stata.

One of the first steps in any analysis is to graph the data in a variety of ways. One way is shown in Figure 46.1.

```
. graph price mpg
```

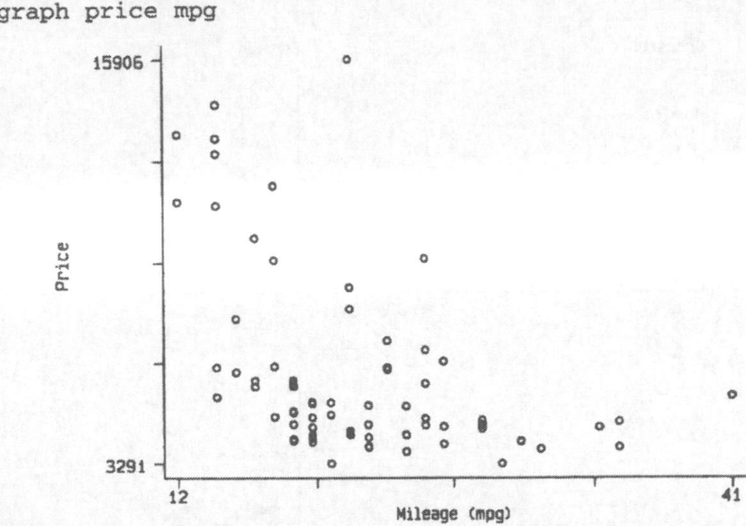

FIGURE 46.1

Stata's **graph** command produces many different types of statistical graphs. But the workhorse of statistical graphs—and the default when more than one variable is specified—is the scatterplot. The command **graph price mpg** produced the scatterplot of auto price against mileage displayed above. This scatterplot shows that expensive cars had relatively low mileage in 1978. Perhaps the expensive cars were high-performance imports. We can highlight the imported cars in this graph as shown in Figure 46.2.

```
. hilite price mpg, hilite(foreign)
```

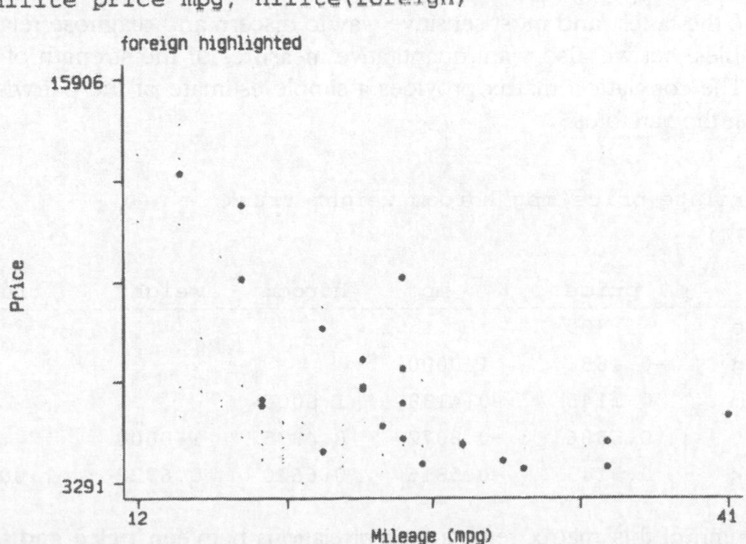

FIGURE 46.2

In this scatterplot, the points corresponding to imports are highlighted, that is, they are plotted with circles, while the others are plotted with small dots. There is no apparent difference between the domestic and foreign cars in this graph. Another way of displaying this information is shown in Figure 46.3.

```
. graph price mpg, by(foreign) total
```

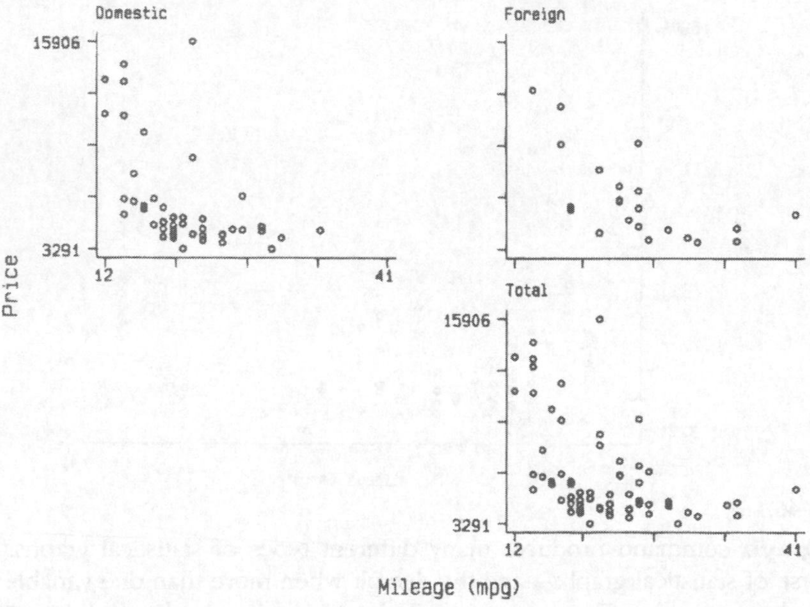

FIGURE 46.3

This command produces three scatterplots on one screen: the first includes only domestic cars; the second includes only foreign cars; and a third includes all the cars, that is, it replicates Figure 46.1.

Graphs are the fastest and most sensitive way to discern and diagnose relationships among variables, but we also want quantitative measures of the strength of these relationships. The correlation matrix provides a simple estimate of the pairwise associations among the variables.

```
. correlate price mpg hdroom weight trunk
(obs=74)

         |   price      mpg      hdroom    weight     trunk
---------+--------------------------------------------------
   price |   1.000
     mpg | -0.4686    1.0000
  hdroom |  0.1145   -0.4138    1.0000
  weight |  0.5386   -0.8072    0.4835    1.0000
   trunk |  0.3143   -0.5816    0.6620    0.6722    1.0000
```

The first column of this matrix reports the correlations between **price** and four other variables, including **mpg**. There appears to be a moderate negative association between price and mileage. Remember though that correlation measures the linear association, but the graph suggests a possibly nonlinear relationship between **price** and **mpg**. There is some evidence of pairwise positive associations between price and headroom, weight, and trunk size.

As you might expect, mileage, headroom, weight, and trunk size are more highly correlated with each other than they are with price. It is possible that one of these variables is the primary determinant of a car's price, and the correlations between price and the other variables reflect the indirect effect of the primary determinant.

Multiple regression can help to disentangle the relationships between a single dependent variable and a set of explanatory variables.

```
. regress price mpg hdroom weight trunk foreign


  Source |       SS       df       MS                  Number of obs =      74
---------+------------------------------              F(5,  68)     =   15.07
   Model |  333779438     5  66755887.5               Prob > F      =  0.0000
Residual |  301285958    68  4430675.86               R-squared     =  0.5256
---------+------------------------------              Adj R-squared =  0.4907
   Total |  635065396    73  8699525.97               Root MSE      =  2104.9

------------------------------------------------------------------------------
   price |      Coef.   Std. Err.       t     P>|t|     [95% Conf. Interval]
---------+--------------------------------------------------------------------
     mpg |    14.4324    73.55266     0.196   0.845     -132.3396    161.2044
  hdroom |  -615.6944    390.0197    -1.579   0.119     -1393.967    162.5776
  weight |    3.78137     .677797     5.579   0.000      2.428848    5.133893
   trunk |  -11.80202    91.57615    -0.129   0.898     -194.5394    170.9353
 foreign |   3654.777     677.486     5.395   0.000      2302.875    5006.679
   _cons |  -4641.084    3394.703    -1.367   0.176     -11415.11     2132.94
------------------------------------------------------------------------------
```

According to this regression, the weight of the car is a statistically significant predictor of a car's price. A one pound increase in the weight of the car is estimated to increase the price of a car by \$3.78, and the *t* statistic for this coefficient is 5.579. However, after the effect of weight is accounted for, none of **mpg, hdroom**, and **trunk** appears to significantly improve predictions of **price**.

We have added **foreign** as a dummy variable to this regression. In contradiction to the *t* test above—which found no difference in the average price of imports and domestic cars—the coefficient on **foreign** is highly significant. The *t* statistic is 5.395. The price of an import is estimated to be \$3,655 higher than an otherwise identical domestic car. Perhaps imports in 1978 were smaller, and hence lighter than domestic cars, on average. As a consequence, the depressing price effect of foreign cars' low weight may have obscured the foreign price premium and, thus, may account for the failure of the *t* test to detect this premium.

Before pursuing this conjecture further, we need to assess the reliability of this regression. We should examine graphs of the regression residuals and calculate diagnostic statistics. First, let us see whether the residuals are normally distributed. Nonnormal residuals can affect the coefficient estimates and can invalidate hypothesis tests.

The first step is to calculate the residuals. Stata's **predict** command calculates fitted values, residuals, and other regression-related measures. The command

```
. predict resid, residual
```

calculates the residuals and stores them in a new variable called **resid**. The **residual** option overrides the default action of **predict**, which is to calculate fitted values.

There are a variety of tests for normality, and most of the commonly used tests are available in Stata. In a debate published in the *Stata Technical Bulletin*—a journal we discuss below—these tests were compared, and the Shapiro–Francia test was found to give reliable results in a variety of settings. To calculate the Shapiro–Francia test of normality for these residuals, we type

```
. sfrancia resid

                  Shapiro-Francia W' test for normal data

Variable | Obs       W'        V'       z      Pr > z
---------+---------------------------------------------
   resid |  74   0.90836    6.486   3.553   0.00019
```

This test decisively rejects the null hypothesis that the regression residuals are distributed normally. Perhaps the data contain one or more outliers. Perhaps the relationship between price and the explanatory variables is nonlinear (recall the scatterplot of **price** and **mpg**). Perhaps an important variable was omitted from the regression. Price takes on only positive variables; perhaps a log or Box–Cox transformation is needed prior to the regression.

All these possibilities and more can easily be examined with Stata. However it is time for us to move on.

46.3 THE SECRET TO STATA'S INTERFACE

In the introduction, we emphasized the importance of a program's interface. In the sample session above, Stata's interface allowed us to explore the automobile data quickly and easily. On a low-budget 486 computer, the session above took less than 6 seconds when executed as a *do-file*, Stata's version of a batch or script file.

Particularly in an exploratory session like the sample above, the direction of exploration is difficult to forecast. As each new aspect of the data is revealed, another set of questions arises. With Stata's command interface, you move without delay from command to command, from idea to idea. There are no branching menus to descend to find the command you want. More importantly, there are no branching menus to reascend when you change your strategy in response to new information.

The secret to Stata's interface is its command syntax. It is simple, flexible, powerful, and consistent. It is easy to learn—the brief sample session already introduced most of its elements. Once you learn this simple syntax, you can tap all of Stata's features.

There are nine basic syntactic elements—nine parts of speech—in Stata. These elements function the same way in every Stata command. Once you learn these nine things, you "speak" Stata.

We can illustrate these nine elements as we examine another one of Stata's sample data sets. This one contains state-level data from the 1980 Census.

1. *Command names* are usually verbs that describe the action Stata is to take (**summarize, correlate, regress, predict**, etc.). Often the command name is all that is required: **list**, by itself, lists all the observations on all the variables in the data set; **regress**, by itself, redisplays the most recent regression; **help**, by itself, displays information about the **help** command. Another example is

```
summarize
```

which displays summary statistics for all the variables in the data set.

2. Referring to *filenames* does not fit quite as neatly into the general scheme as the other eight elements, so let us get it out of the way early. To bring a disk data set into memory you **use** it. For example

```
. use census
(1980 Census data by state)
```

reads a disk file of data from the 1980 Census into memory. The command to save a copy of the data set in memory to disk is **save**. Other commands that refer to filenames typically precede the filename with the word **using** or **saving**. For instance,

```
. merge using census
```

merges the current data set with the Census data on disk.

3. The *variable list* indicates the variables to which the command is to be applied. It always appears immediately after the command name:

```
. describe pop divorce
  3.  pop       long   %10.0g          Population
 12.  divorce   long   %10.0g          Number of divorces

. summarize pop divorce

Variable | Obs       Mean    Std. Dev.       Min        Max
---------+-------------------------------------------------
     pop |  50    4518149     4715038     401851   2.37e+07
 divorce |  50   23679.44    25094.01       2142     133541
```

4. An = *expression* is used to evaluate expressions. For example, this data set contains information on the number of divorces in each of the 50 United States in 1980. The divorce rate per 100,000 population is an easier measure to analyze:

```
. generate dvcrate = 100000*divorce/pop

. summarize dvcrate
```

```
Variable | Obs       Mean    Std. Dev.        Min        Max
---------+-------------------------------------------------
dvcrate  |  50   566.4144    224.7315   294.3553   1729.184
```

The average state had a divorce rate of a bit over 0.5% (566 per 100,000) in 1980. Nevada had the highest rate (1,729 per 100,000). Stata has a full complement of statistical, mathematical, string, and special functions that can be used freely in expressions.

5. The **in** *range* modifier restricts the range of observations on which the command acts. For instance,

```
. summarize dvcrate in 5/20

Variable | Obs        Mean    Std. Dev.        Min        Max
---------+-------------------------------------------------
dvcrate  |  16    535.1574    121.3409   389.1725   734.4205
```

calculates summary statistics for the divorce rate for the fifth through the twentieth observation.

6. A more powerful and flexible tool for restricting the range of observations is the **if** *expression* modifier. We can use the **if** *expression* modifier to eliminate the most urbanized states from the analysis:

```
. generate pcturban = popurban/pop

. summarize dvcrate if pcturban<.85

Variable | Obs        Mean    Std. Dev.        Min       Max
---------+-------------------------------------------------
dvcrate  |  45    551.3139    153.0665   294.3553     875.2
```

You can see from the reduction in the maximum divorce rate (1,729 per 100,000 to 875 per 100,000) that this restriction has eliminated Nevada from the sample. Note, however, that the **if** and **in** modifiers leave the data set unchanged. They restrict the sample only for the duration of the command in which they appear.

7. Many statistical calculations can and should be weighted—so many, that weights are one of Stata's nine syntactic elements. Stata can accommodate four different kinds of weights, where appropriate, and the weighting formulas for each command are spelled out explicitly in the *Stata Reference Manual*.

The 50 states have widely different populations. It may make more sense to calculate weighted summary statistics for the divorce rate where population is used as a frequency weight, that is, as a measure of the true number of observations underlying each observation in the data set.

```
. summarize dvcrate if pcturban<.85 [fweight = pop]

Variable |     Obs       Mean    Std. Dev.        Min      Max
---------+---------------------------------------------------
dvcrate  | 1.9e+08   520.7829    146.7395   294.3553    875.2
```

8. *Options* modify the command, adding or subtracting calculations, changing the way results are displayed, etc. We can use the **detail** option to get more detailed summary statistics for the divorce rate:

```
. summarize dvcrate if pcturban<.85 [fweight = pop], detail

                           dvcrate
-------------------------------------------------------------
      Percentiles      Smallest
 1%      294.3553      294.3553
 5%      294.3553      311.5371
10%      352.9545      328.1667       Obs            192162409
25%      406.8216      352.9545       Sum of Wgt.    192162409

50%      406.3595                     Mean            520.7829
                        Largest       Std. Dev.       146.7395
75%      680.3549      800.2186
90%       728.677      800.7827       Variance        21532.47
95%      734.4205      852.5057       Skewness        .1905231
99%      800.7827         875.2       Kurtosis        1.819456
```

9. The **by** *prefix*: allows you to execute commands independently for subgroups of the data set. For example, we can see if the divorce rate differs across regions of the United States:

```
. sort region

. by region: summarize dvcrate if pcturban<.85 [fweight = pop]

-> region=      NE
Variable |     Obs        Mean   Std. Dev.        Min        Max
---------+-------------------------------------------------------
 dvcrate |  4.1e+07    348.666   64.80059   294.3553   570.7086

-> region=  N Cntrl
Variable |     Obs        Mean   Std. Dev.        Min        Max
---------+-------------------------------------------------------
 dvcrate |  5.9e+07   496.0956   98.60465   328.1667    728.677

-> region=     South
Variable |     Obs        Mean   Std. Dev.        Min        Max
---------+-------------------------------------------------------
 dvcrate |  7.5e+07   594.3946   124.1046   389.1725   800.7827

-> region=      West
Variable |     Obs        Mean   Std. Dev.        Min        Max
---------+-------------------------------------------------------
 dvcrate |  1.8e+07   688.6759   72.28129   534.0043      875.2
```

Stata's syntax is simple and consistent. As a consequence, Stata commands are like Lego™ blocks. Each command is simple and similar in structure to the others, but these simple commands can be combined in unlimited ways. In fact, one hallmark of Stata is that command extensions that logically should work, usually do. For example, here is another natural way to see the variation in divorce rates across regions:

```
. tabulate region if pcturban<.85 [fweight = pop], summa-
rize(dvcrate)

  Census  |              Summary of dvcrate
  region  |       Mean     Std. Dev.          Freq.        Obs.
----------+--------------------------------------------------------
      NE  | 348.66602    64.800588       40823306     40823305
 N Cntrl  | 496.09559    98.604651       58865670     58865670
   South  | 594.39455    124.10465       74734029     74734029
    West  | 688.67589    72.281286       17739404     17739404
----------+--------------------------------------------------------
   Total  | 520.78291    146.73946      1.922e+08    192162409
```

Here, **summarize** is an option to the **tabulate** command. Are these differences statistically significant? Stata's **oneway** command estimates one-way layouts. (The **anova** command estimates general analysis-of-variance models.)

```
. oneway dvcrate region [aweight=pop], tabulate scheffe

  Census  |              Summary of dvcrate
  region  |       Mean        Std. Dev.         Freq.     Obs.
----------+---------------------------------------------------
      NE  |   353.59315     63.711961       49135283        9
 N Cntrl  |   496.09559      102.9892       58865670       12
   South  |   594.39455     128.17473       74734029       16
    West  |   634.63561     177.12263       43172490       13
----------+---------------------------------------------------
   Total  |   524.09598      157.9369      2.259e+08       50

                       Analysis of Variance
   Source        SS            df      MS          F      Prob > F
-----------------------------------------------------------------
Between groups   524866.801     3   174955.60    11.54    0.0000
Within groups    697392.386    46   15160.704
-----------------------------------------------------------------
   Total         1222259.19    49   24944.065

Bartlett's test for equal variances:   chi2(3)    = 5.1205
                                       Prob>chi2  = 0.163
```

```
                 Comparison of dvcrate by Census region
                               (Scheffe)
Row Mean-|
Col Mean |       NE      N Cntrl      South
---------+---------------------------------
 N Cntrl |  142.502
         |    0.090
         |
   South |  240.801      98.299
         |    0.000       0.239
         |
    West |  281.042      138.54     40.2411
         |    0.000       0.061       0.857
```

Now **tabulate** is an option to the **oneway** command. And **scheffe** calculates the Scheffe test for multiple comparisons. **oneway** also offers the Bonferroni and Sidak tests. We have changed the interpretation of population weights from frequency weights to analytic weights, a more appropriate choice when estimating a linear model for a rate such as the divorce rate. These results suggest that the divorce rate is significantly higher in the South and West than in the Northeast.

46.4 STATA'S FEATURES

Stata's interface provides convenient and flexible access to an impressive array of features. The introductory examples above display Stata's more elementary features, but Stata is known for its advanced statistical commands. Several techniques have appeared in Stata before they were published in scientific journals. Stata also produces a multitude of graphics tailored for the statistician—graphs that can display thousands of observations intelligibly, in contrast to business-style charts that fill a page with four or five bits of data. Stata supports its statistical and graphical capabilities with data management features that are second to none.

Statistics

Stata's statistics command offer a full range of standard statistical techniques along with advanced techniques, some just now becoming familiar to researchers. The bulleted paragraphs below highlight some, but definitely not all, of Stata's statistical features.

- *Fundamental techniques* offered by Stata include complete summary statistics; correlation and covariance matrices; one-way and cross-tabulations; *t* tests for means; confidence intervals for means, proportions, and counts; tests for the equality of variances; and tests for normality.
- A wide variety of *nonparametric* and *semiparametric techniques* are available: the exact binomial probability test; Spearman's and Kendall's rank correlations; run tests, sign and rank tests, and nonparametric tests for trends in ordered data; Kolmogorov–Smirnov tests for the equality of distributions; Wilcoxon (Mann–Whitney) and Kruskal–Wallis tests; and quantile regression models.

- Commands for estimating *linear models* include regression; one-way layouts and general ANOVA; and multivariate regression. Stata can calculate in- and out-of-sample forecasts, test hypotheses about the estimates, and estimate linear models subject to constraints. Diagnostic statistics and plots are provided and several measures of outliers and influential observations are available. *Nonlinear models* can also be estimated.
- *Qualitative, categorical*, and other *limited dependent variable models* are easily estimated in Stata. Qualitative dependent variable models include CUSUM plots and tests for binary variables, maximum-likelihood logistic regression (logit) and probit, grouped logit and probit, conditional (stratified) logit, and multinomial logit. Categorical dependent variable models include ordered logit and probit, and a host of tests for cross-tabulated variables. Limited dependent variable models include censored normal regression and Tobit models.
- *Multivariate techniques* include Cronbach's alpha, canonical correlation, principal components and factor analysis, Hotelling's *T*-squared generalized means test, and multivariate and seemingly unrelated regression models.
- Techniques for *robust* and *exploratory data analysis* include Huber's robust variance estimates, stem-and-leaf displays, letter values, ladder-of-powers calculations for transforming a variable to normality, symmetry plots, quantile plots, normal quantile and probability plots, bivariate quantile plots, locally weighted smoothing including lowess smoothing, nonlinear compound smoothing, and robust regression.
- *Epidemiology, biostatistics*, and *survival analysis* are represented by life tables, the Cox proportional hazards model, Weibull and exponential regression models, Poisson regression, binary, conditional, and multinomial logistic regression, and a host of tables for epidemiologists including tables for incidence rates and for cohort study, case-control, and matched case-control data.
- Charts for *statistical quality control* include *c*-charts and *p*-charts (charts for the number and fraction of nonconforming units), *X*-charts and *R*-charts (charts for the mean and range of repeated measurements), and standard error-bar charts.

Graphics

Graphs have always played an important role in statistics and data analysis, but the advent of inexpensive computers combined with improvements in display technology touched off an explosion of new statistical graphics in recent decades. Compare the leading statistical programs used on university mainframes a generation ago to the leading programs used on PCs and workstations today, and you will be struck by the rapid increase in the quantity and quality of statistical graphs. Stata is an outstanding example of the modern generation of statistical software.

Stata offers eight basic graph styles, and they all appear in Figure 46.4. All these graphs were produced using the **graph** command.

- A *histogram* appears in the upper left of Figure 46.4. The **graph** command produces a histogram by default if only one variable name is specified. This particular histogram is overlaid with a normal curve—added with the **normal** option—fitted to the data underlying the histogram. The **bins( )** option can be used to control the number of bars in the histogram.
- The graph in the middle of the top row is a *two-way scatterplot*, usually called simply a scatterplot. The two-way scatterplot is the default when two or more variables

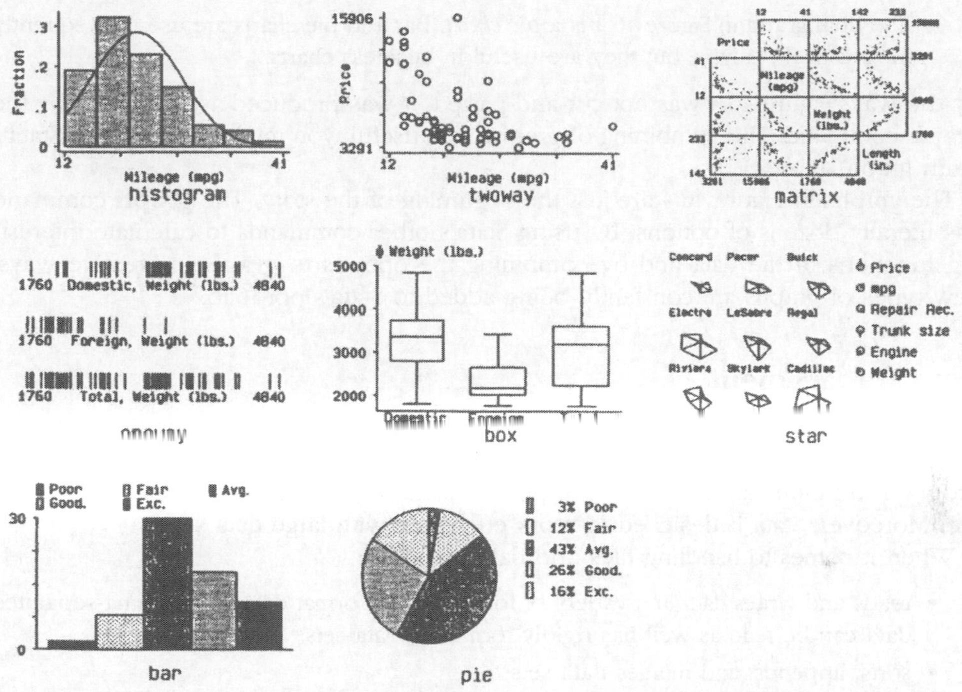

FIGURE 46.4

are specified. The last variable in the list defines the horizontal axis. All the other variables are scattered against the last variable using a variety of plotting symbols. The symbols can be connected simply with line segments or in more complex ways.

- The graph in the upper-right corner of Figure 46.4 is a *scatterplot matrix*. This very useful graph is a generalization of the correlation matrix. Replace each of the correlations with the two-way scatterplot of the appropriate pair of variables, and you have a scatterplot matrix.
- The left-most graph in the middle row may look like the bar code used on packaged groceries, but it is, in fact, a *one-way scatterplot*. Because only one variable is being scattered, there is only one axis, the horizontal one. Each observation is plotted with a short vertical line. This graph is useful for comparing densities and for detecting outliers. This one-way scatterplot displays the distribution of automobile weights, first for domestic cars, next for foreign cars, and finally for all cars. As we suspected in our sample session, the foreign cars tend to be much lighter than the domestic cars.
- The center of Figure 46.4 is occupied by a *box-and-whisker plot*, also called a boxplot. Developed initially by J. W. Tukey (1977), the box portion of the box-and-whisker plot displays the 25th percentile, the median, and the 75th percentile of a group of observations. The whiskers display a measure of the dispersion of the observations.
- The right-most graph in the middle row is a *star chart*. Each star represents a different observation—a different car in this example. The length of each ray of the star is proportional to the value of a particular variable. Star charts are used to detect qualitative differences in multivariate data. This example displays information on the price, mileage, repair record, trunk size, engine size, and weight of nine cars.
- The graph in the lower-left corner is a conventional *bar chart*. Stata can produce charts with multiple side-by-side bars, stacked bars, even bars with negative heights.

- The final graph in Figure 46.4 is a *pie chart*. Bar and pie charts are used infrequently in statistical analysis, but they are useful in business charts.

By the way, Figure 46.4 was not cut-and-pasted; it was produced as you see it by the **graph** command. This combining of graphs is so useful, you might think of it as Stata's ninth graph style.

The graphs in Figure 46.4 are just the beginning of the story. The **graph** command has literally dozens of options. By using Stata's other commands to calculate interesting functions of the data and by combining the options to **graph** in creative ways, new types of graphs are constantly being added to Stata's portfolio.

Data Management

The reality of data analysis and statistics is that far more time is spent getting data into the computer and into shape than in analyzing data and estimating models. While no program can eliminate this work completely, Stata makes it about as easy as is possible. Moreover, Stata is designed to work efficiently with large data sets.

When it comes to handling files and data sets, Stata

- reads and writes data in a variety of formats. Free format data and comma-separated data can be read as well has rigidly formatted data sets;
- sorts, appends, and merges data sets;
- transposes data sets, interchanging observations and variables;
- reshapes panel data sets, from a "wide" representation—where the same concept is measured by several variables, one for each level of a grouping variable—to a "long" representation—where the measurements of a variable are stored in different observations for each level of the grouping variable—and back again;
- draws random samples from a data set.

After data have been loaded, sorted, transposed, sampled, and so on, they typically need to be transformed in other ways. When it comes to transforming data, Stata

- generates new variables that can be complicated functions—including string functions—of other variables;
- replaces the values of existing variables and renames the variables;
- recodes variables, creating and converting missing values as needed;
- eliminates variables and/or observations;
- changes the order of the variables in a data set;
- changes the data types (**float, int, byte**) of variables;
- compresses a data set so all variables are in the smallest possible data type while preserving full precision;
- collapses a data set into a smaller data set of summary statistics (means, standard deviations, etc.) of the original variables;
- performs date calculations and conversions;
- generates random numbers;
- compares files and variables for similarities and differences;
- labels data sets, variables, and values of variables and controls the format used to display data;
- composes codebooks describing the contents of a data set in detail.

46.5 STATA PROGRAMS AND STATA'S MATRIX LANGUAGE

No statistics program is ever finished. New statistical techniques are constantly being introduced. As a consequence, program developers are continually producing newer and better versions of their programs. But producing a new version of a program takes time. Researchers do not want to wait for a new version of their favorite program; they want to use the newest techniques right away.

Adding New Commands to Stata: ado-files

With Stata, there is no need to wait. Stata's programming commands make it possible for you easily to add new commands to Stata. These new commands can do whatever you want—they can calculate new statistics, display new types of graphs, update and maintain your data sets, and so on. More importantly, these new commands are indistinguishable from ordinary Stata commands. They use Stata's full syntax. They take advantage of Stata's error-handling features. They can use the results of previous Stata commands and programs, and they can pass their results to other Stata commands and programs.

At the simplest level, Stata offers *do-files*. These are ordinary ASCII files of Stata commands that can be executed like a batch file or shell script. For example, suppose you store the following commands in a file named **myjob.do**:

```
use auto
tabulate foreign
regress price mpg hdroom weight trunk foreign
```

Then, while in Stata, if you type **"do myjob"**, these three commands will execute in sequence.

Stata programs are called *ado-files*, which is mnemonic for automatically-loaded *do*-files. However, ado-files are much more powerful than do-files. Stata's ado-files are used just like Stata commands. For example, if you wrote an ado-file and called it **myado.ado**, you could use it just by typing **myado**. In addition, your ado-file can use Stata's syntax, if you wish. At your option, Stata will parse the variable lists, **if** and **in** modifiers, weights, and options for you. This facility frees you from the chore of parsing program input and allows you to concentrate on your calculations. Stata also provides additional commands specifically for use in Stata programs. These commands allow your Stata program to create temporary variables and files, to loop, to restore the user's original data set if an error is detected, and much more.

Stata's ado-files make it easy to extend Stata's capabilities as rapidly as new statistical techniques are introduced. In fact, many of Stata's official commands are really ado-files, and some of these were originally provided by Stata users who did not want to wait for the next version of Stata for some new feature.

Stata's Matrix Language

Stata contains a complete matrix language. The matrix commands can be used interactively, but they are used primarily in Stata programs. The matrix language enables you to create matrices, obtain copies of matrices calculated by Stata's statistical commands, and perform matrix calculations. Inverses can be calculated as can eigenvalues and eigenvectors and the singular value decomposition. There are matrix facilities for constrained estimation as well.

The matrix language rounds out Stata's programming capabilities. With the matrix language, virtually any statistic can be calculated.

46.6 STATA-RELATED PUBLICATIONS

The *Stata Technical Bulletin*

Stata has attracted a loyal and enthusiastic following. The *Stata Technical Bulletin* (STB) is a bimonthly journal that provides a forum for Stata users. Users contribute articles to the STB covering every aspect both of Stata and of statistical computing generally. Some articles take the form of questions; a user may seek clarification of some issue or may ask if other users have developed programs for some application. Some articles provide the answers to these questions. The STB also publishes announcements by Stata Corporation and by other companies and persons involved with Stata.

The typical article in the STB describes one or more Stata programs contributed by the article's authors. Among the hundreds of programs published in the STB so far are programs for

- extracting Gaussian components from mixed distributions;
- calculating kernel density estimators;
- documenting variable dates;
- graphing arbitrary functions, including recursive functions;
- estimating regression switching models;
- bootstrapping standard errors of new or unanalyzed estimators;
- displaying 3-D graphs;
- estimating generalized linear models;
- solving equations by Ridder's method;
- calculating U.S. marginal income tax rates.

These programs form a library of useful Stata routines. In addition, the editor of the STB includes a complete library of time series programs, updated bimonthly, in each issue. Now in its fourth volume, the STB has grown so much that each issue comes with a computerized index to all previous issues. Completed volumes are reprinted as softcover books for easier reference.

Books That Use Stata

Stata is popular with teachers as well as researchers. Many professors assign students to use Stata in their courses. Some textbook authors have used Stata in the examples in their books. Some of these authors supply diskettes with sample data sets in Stata format and even with a student version of the Stata program. A selection of these books includes

- *Statistics with Stata 3* by Lawrence Hamilton provides an introduction both to statistics and to Stata.
- *Regression with Graphics* by Lawrence Hamilton explains regression and emphasizes the role of graphics in regression analysis.

- *StataQuest* by Ted Anagnoson and Richard DeLeon explains the new menu-driven version of Stata for undergraduate teaching.
- *Principles of Biostatistics* by Marcello Pagano and Kimberlee Gauvreau is an introduction to biostatistics for students of the health sciences.

REFERENCES

J. T. Anagnoson and R. DeLeon, *StataQuest*. Duxbury Press, Belmont, CA, 1994.

J. M. Chambers, W. S. Cleveland, B. Kleiner, and P. A. Tukey, *Graphical Methods for Data Analysis*. Wadsworth International Group, Belmont, CA, 1983.

W. S. Cleveland, *The Elements of Graphing Data*. Wadsworth Advanced Books and Software, Monterey, CA, 1985.

L. Hamilton, *Regression with Graphics*. Brooks/Cole, Pacific Grove, CA, 1992.

L. Hamilton, *Statistics with State 3*. Duxbury Press, Belmont, CA, 1993.

M. Pagano, and K. Gauvreau, *Principles of Biostatistics*. Duxbury Press, Belmont, CA, 1993.

J. W. Tukey, *Exploratory Data Analysis*. Addison-Wesley, Reading, MA, 1977.

47

S-PLUS

Robert P. Treder
MathSoft, Inc.

S-PLUS is a graphical data analysis and mathematical computing system and an object-oriented programming language developed by MathSoft, Inc., StatSci Division. It is an enhanced superset of the S language developed at AT&T Bell Laboratories by Richard Becker, John Chambers, and Allan Wilks, described in *The New S Language* (Becker et al., 1988), and is the result of over 16 years of research and development at Bell Labs, plus over 7 years of development at Statistical Sciences (now MathSoft, StatSci division). S-PLUS is a general toolbox of methods for statisticians, applied mathematicians, and scientific researchers. S-PLUS emphasizes

- dynamic, interactive, and presentation graphics,
- exploratory data-analytic methods,
- statistical methods,
- mathematical computations, and
- extensibility through programming and interfaces to other languages.

An important feature of S-PLUS is that it is an interpreted language that promotes exploratory, data-driven, graphically oriented data analysis. The purpose of this chapter is to give you a brief overview of the S-PLUS system. Because of the limitation of space, this is not intended to be comprehensive but merely to touch on some of the highlights of the S-PLUS language and data analysis system.

47.1 THE COMPUTING ENVIRONMENT

S-PLUS is best suited for computers that run a windowing system such as the X Window System on Unix workstations and MS-Windows on PCs. Window systems create ideal computing environments for data analysis. In a typical data analysis session you may have separate windows for

0-8493-2530-7/96/$0.00+$.50

- typing S-PLUS expressions
- reading help files
- displaying graphics
- interacting with the operating system.

Figure 47.1 displays a typical S-PLUS session on a Sun workstation. Starting in the upper left corner and proceeding counterclockwise are the S-PLUS expression window, three help file windows, a Unix command window, and the S-PLUS graphics window.

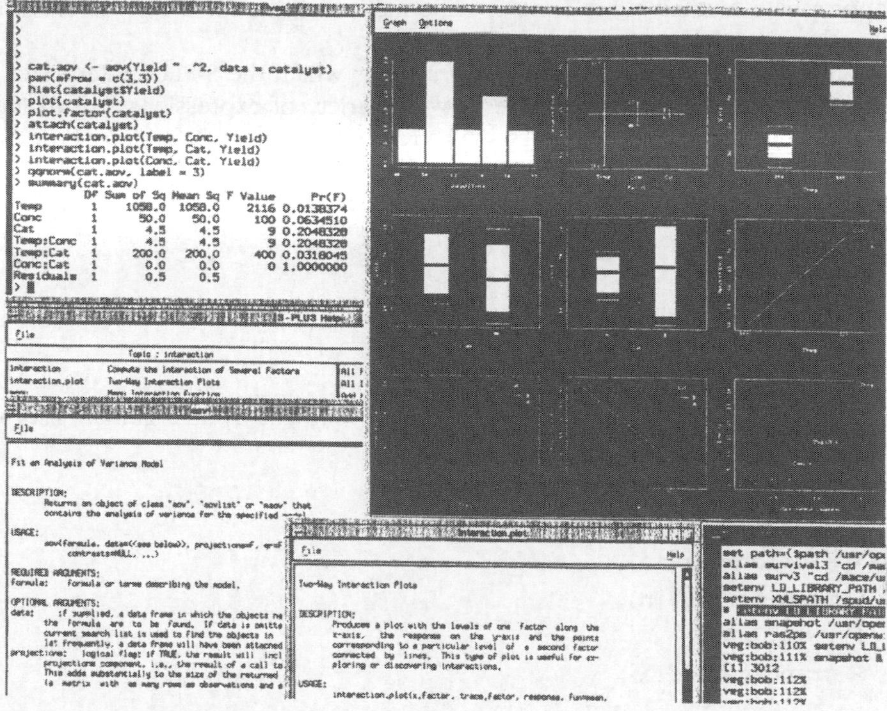

FIGURE 47.1 A typical S-PLUS computing environment including S-PLUS expression window, three help file windows, a Unix command window, and a graphics window.

47.2 LANGUAGE BASICS

S-PLUS is an *interpreted* language, in which individual language expressions are read and then immediately executed. The S-PLUS interpreter, which carries out the actions specified by the S-PLUS expressions, is always interposed between the S-PLUS expressions being evaluated and the CPU of the machine running S-PLUS. This contrasts with compiled languages for which a compiler translates the source code to machine language, and the program, once compiled, runs independently of the compiler and any additional translation. The great advantage of an interpreted language is that it allows for *incremental* and, consequently, more flexible analysis. Analysis becomes more data driven, rather than procedure driven.

S-PLUS is a *functional* language. The expressions you type are function calls with data passed as arguments to those calls and data returned by those calls. A purely functional language passes data only as arguments to functions and as values returned from evaluated function calls. A function does not modify its arguments, but only *copies* of the arguments. Complex functions may require many copies of the original data. S-PLUS is not a pure functional language, but it does make extensive use of the functional programming paradigm. Consequently, the larger the data set analyzed, the more RAM and virtual memory required.

S-PLUS is an *object-oriented* programming language, characterized by a set of *generic functions* for which specific *methods* are defined to operate on corresponding *classes of* objects. A generic function can be called with any type of object, and the appropriate method for the class of this object is automatically used; the generic function embodies a dispatch mechanism that recognizes the class of the object and finds the appropriate method for it. If no specific method exists for the given class, a default method is used.

You interact with S-PLUS by typing *expressions*, which the S-PLUS interpreter evaluates and executes. S-PLUS recognizes a wide variety of expressions, but in interactive use the most common are *names*, which return the current definition of the named data object, and *function calls*, which carry out a specified computation. Typing the name of a built-in S-PLUS function, for example, shows the definition of the function:

```
> sqrt
function(x)
x^0.5
```

A function call is usually typed as a function name followed by an argument list (which may be empty) enclosed in parentheses:

```
> motif()
> plot(corn.rain)
> mean(corn.rain)
[1] 10.78421
> graphics.off()
```

All S-PLUS expressions return a value. Normally, this return value is automatically printed. Some functions, however, such as **motif, graphics.off, plot,** and **q** are called primarily for their side effects, such as starting or closing a graphics device, plotting points, or ending an S-PLUS session, and, consequently, have the automatic printing of their values suppressed.

In the examples above, the **>** character beginning each line is the default S-PLUS prompt indicating that S-PLUS is ready to interpret expressions.

Saving and Deleting Objects

Objects created in S-PLUS are saved by assigning the object a name. For example,

```
> a <- 7
```

saves the value 7 in an object named **a**. Typing the name **a**, displays the stored value:

```
> a
[1] 7
```

Assignments typed at the S-PLUS prompt are *permanent*; objects created in this way endure from session to session, until removed. To delete permanent objects, you need to explicitly remove them by using the **rm** function. You can examine all the names of all the objects you have saved in your S-PLUS session by calling the **objects** function.

Operators

Operators are functions for performing mathematical or logical operations on one or two arguments, and are most often used in the convenient *infix* form, that is, between the two arguments. Familiar examples are **+,-, *,** and **/**. S-PLUS uses a precedence ordering of its operators similar to other programming languages and the precedence may be overridden with parentheses or curly braces {} also similar to other languages. A typical mathematical calculation looks like the following:

```
> 7 + 5 - 8^2 / 19 * 2
[1] 5.263158
```

Mathematical operations in S-PLUS are *vectorized*, meaning that they act on whole data objects at once, element by element. Calculations are performed in double precision, and numeric values are stored in double precision unless otherwise specified. S-PLUS performs its floating-point calculation in accordance with IEEE conventions on those machines that support IEEE arithmetic. This includes proper handling of **Inf** and **NaN** values.

To illustrate the notion of vectorized arithmetic, suppose we have the following vectors **x** and **y** :

```
> x
[1] 0.0 0.5 1.0 1.5 2.0 2.5 3.0 3.5 4.0 4.5 5.0
> y
[1] 1 2 3 4 5 6 7 8 9 10 11
```

To add the two vectors element-by-element, simply type the usual arithmetic expression:

```
> x + y
[1] 1.0 2.5 4.0 5.5 7.0 8.5 10.0 11.5 13.0 14.5 16.0
```

Subscripting

One of the most commonly used operators is the *subscript* operator **[**, used to extract subsets of an S-PLUS data object. The syntax for subscripting is *object* [*subscript*]. Here *object* can be any S-PLUS object and *subscript* typically takes one of the following forms:

- Positive integers corresponding to the position in the data object of the desired subset. For example, the `letters` data set consists of the 26 lowercase letters. We can pick the third letter using a positive integer subscript as follows:

```
> letters[3]
[1] "c"
```

- Negative integers corresponding to the position in the data object of points to be *excluded*:

```
> letters[-3]
 [1] "a" "b" "d" "e" "f" "g" "h" "i" "j" "k" "l" "m" "n"
[14] "o" "p" "q" "r" "s" "t" "u" "v" "w" "x" "y" "z"
```

- Logical values; true values correspond to the points in the desired subset, false values correspond to excluded points:

```
> i <- 1:26
> letters[i < 13]
 [1] "a" "b" "c" "d" "e" "f" "g" "h" "i" "j" "k" "l"
```

The colon operator (**:**) in the first expression above generates the positive integers 1 through 26.

Subscripting is *extremely* important in making efficient use of S-PLUS because it emphasizes treating data objects as whole entities, rather than as collections of individual observations.

47.3 DATA OBJECTS

Every S-PLUS expression is interpreted by the S-PLUS evaluator and returns a value. The return value is a *data object* that can be assigned a name. Data objects are composed of any number of *elements*, which in simple data objects correspond to individual data points and in more complicated objects may consist of whole data objects. Simple elements are literal expressions of four *modes*: **logical**, the values **T** and **F**; **numeric**, floating-point real numbers such as **-4, 4.52,** and **6.02e23; complex**, complex numbers such as **3 + 1.23i; character**, and character strings such as "**Alabama**".

Data Types

The fundamental data type is the *vector*, a one-dimensional array of simple elements. Other data types are constructed with vectors as starting points but with additional defining *attributes*. The most commonly used data types in S-PLUS are as follows:

vector	A vector is a set of elements with the same mode in a specified order.
matrix	A matrix is a two-dimensional array of elements of the same mode.
factor	A factor is a vector of categorical data.
time series	A vector (or matrix or data frame) of elements with observation times associated with the elements (or rows of a matrix or data frame).
data frame	A data frame is a two-dimensional array whose columns may represent data of different modes.
list	A list is a set of components that can be any of the possible object types.

47.4 GRAPHICS

Graphics are central to the S-PLUS philosophy of looking at your data visually as a first and last step in any data analysis. With its broad range of built-in graphics functions and its programmability, S-PLUS lets you look at your data from many views.

S-PLUS has functions for interactive and dynamic graphics, presentation graphics, controlling graphics devices and printing, and page layout. The presentation graphics functions are divided into *high-level* and and *low-level* plotting functions. The high-level functions produce an entirely new plot, while low-level functions add some feature to an existing plot. You create customized plots by combining calls to high-level and low-level graphics functions. Graphics device drivers are activated and controlled through a set of functions designed for managing multiple devices. A hardcopy of a graph is obtained either through point-and-click operations or by calling the **printgraph** function on UNIX workstations or the **win.printer** function on PCs.

Dynamic Graphics

For data with three or more variables, methods of graphical visualization are often highly interactive. One useful technique for evaluating multivariate data sets is to graph *pairs* of variables in an array called a *scatterplot matrix*. Scatterplot matrices are even more useful if you can interactively highlight a point in one scatter plot and see the same case highlighted in the remaining scatter plots of the matrix. Such highlighting is called *brushing*. Another interactive technique is to create a three-dimensional scatterplot from any three variables and then *spin* the plot to observe structure among the variables. The **brush** and **spin** functions implement brushing and 3D spinning in S-PLUS.

Figure 47.2 displays a session using **brush** on a matrix, **cpac** , of measurements of chemical sensor data. The scatterplot matrix in the lower left and the spin plot in the upper right give a quick assessment of how linearly related the variables are. Brushing enhances our ability to see multidimensional relationships including interactions and locate interesting subpopulations including outliers or extreme values. The call that produced Figure 47.2 is

```
> brush(cpac, hist = T)
```

The developers of S-PLUS are currently developing DATAVIEWER, which extends **brush** and **spin**. It manages the interaction of a *variety* of graphical displays using a "linked views" paradigm similar to the way **brush** works. In this paradigm, multiple views of one or more data sets are linked so that DATAVIEWER always knows exactly how to update all views after every possible operation. One possible operation is brushing, but others include adding and deleting points or changing the color of the points. DATAVIEWER provides a variety of graphical displays including histograms, scatterplots, scatterplot matrices, images, 3-D spinning plots, and spreadsheets. As an example of the dynamic nature of these methods, scatterplots may display fitted curves that are automatically updated as points are added or deleted.

High-Level Graphics Functions

A high-level (presentation) graphics function produces an entire new graphic, including a bounding box, axes, and axes labels by default. It controls the clearing of the "screen" or advancing to the next frame when multiple plots are placed on a page.

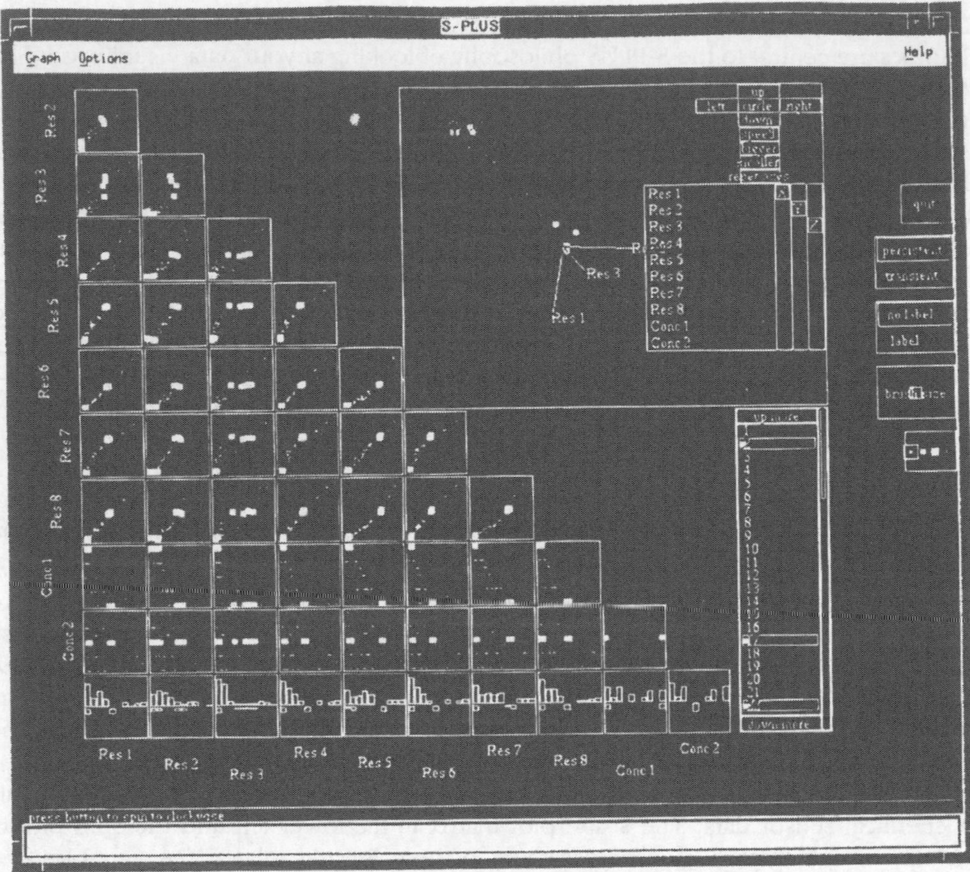

FIGURE 47.2 The brushing of chemical sensor data.

The generic plotting function is **`plot`**. When supplied two vectors of equal length, **`plot`** produces a scatterplot of the two variables. To plot a mathematical function, first create a vector of values spanning the domain of the function, evaluate the desired function at these points, and then plot the ordered pair of values connecting them with lines. For example, to plot the sine curve from $-\pi$ to π shown in Figure 47.3:

```
> x <- seq(-pi, pi, len=100)
> plot(x, sin(x), type="l")
```

The type argument specifies a plot type "`l`" to connect the points with lines.

Table 47.1 summarizes the commonly used high-level plotting functions, organized according to the type of data they can be used to represent.

Low-Level Graphics Functions

Add to existing plots by using low-level graphics functions, which do not clear the device screen, but simply add to what is already displayed. Points are plotted in the coordinate system already established by the previous high-level plotting function.

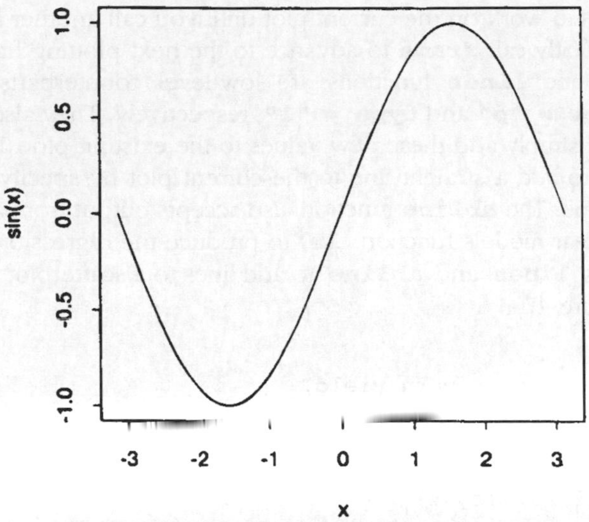

FIGURE 47.3 Sine curve created with `plot`.

TABLE 47.1 High-Level Graphics Functions

Univariate Data	
barplot	Bar plot or bar chart
hist	Histogram
dotchart	Dot chart
pie	Pie chart
Bivariate Data	
plot	Scatterplot
boxplot	Side-by-side box plots
qqnorm	Normal probability plot for one sample
qqplot	Quantile-quantile plot for two samples
plot.surv.fit	Survival curve plot
shewhart	Shewhart quality control chart
cusum	Cusum quality control chart
Three-Dimensional Plots	
contour	Contour plot
persp	Perspective or mesh plot
image	Image plot
Multivariate Data	
faces	Chernoff faces
matplot	Matplot
pairs	Pairwise scatterplots
stars	Starplots
symbols	Plot symbols
Time Series Data	
tsplot	Univariate and multivariate time series plot
acf	Autocorrelation function plot
spectrum	Spectrum plot
Dynamic Graphics	
brush	Linked scatterplot matrix for highlighting with mouse
spin	Rotatable 3-D point cloud

S-PLUS continues to work on the current plot until you call another high-level plotting function, or explicitly call **frame** to advance to the next plotting frame.

The **points** and **lines** functions are low-level counterparts to **plot**, corresponding to **type = "p"** and **type = "l"**, respectively. They also require two vectors as input and simply add these new values to the existing plot. The abline function allows you to add a straight line to the current plot by specifying values for the intercept and slope. The **abline** function also accepts output from a, regression function (e.g., the linear models function, **lm**) to produce the regression line. The following example uses **lines** and **abline** to add lines to a scatterplot of the **corn** data, as shown in Figure 47.4.

```
> plot(corn.rain, corn.yield)
> abline(lm(corn.yield - corn.rain))
> lines(lowess(corn.rain, corn.yield), lty=2)
> abline(2, 3, lty=4)
> legend(11, 25, c("regression line", "lowess smoothed line",
+ "some kind of trend"), lty=c(1,2,4))
```

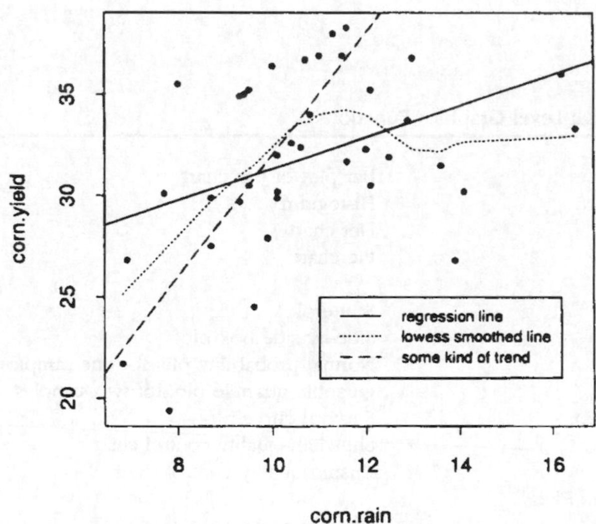

FIGURE 47.4 Adding lines to plots.

Interactive Graphics

Often, you may want to build upon an existing plot in an interactive way. For example, you may want to identify individual points in a plot and label them for future reference. Or you may want to add some text or a legend, or overlay some new data. Different S-PLUS functions facilitate interactively adding information to your plots.

The **identify** and **locator** functions are interactive functions that require you to click with the mouse in the graphics window. The **identify** function returns the indices of the observations on which you click, in addition to printing them on the graph. The **locator** function does not add anything to the plot, but returns the (x,y)

coordinates where you click. The output of **locator** can be used for input to graphics functions that require an (x,y) location for placement such as **text** for placing text strings in a plot or **legend** for adding a legend box to a plot.

47.5 S-PLUS PROGRAMMING BASICS

S-PLUS is not simply a programming language, like C or Fortran. It is much more: a full-function calculator, a statistical package, a mathematical problem solver, a graphics tool, and so on. You can use S-PLUS productively and effectively without ever writing a program in the S-PLUS language. However, most users begin programming in S-PLUS almost subconsciously—defining functions to streamline repetitive computations, avoid typing mistakes in multiline expressions, or simply keep a record of a sequence of commands for future use.

Writing S-PLUS Functions

You can create functions by modifying existing S-PLUS functions or writing original functions in the S-PLUS language, either at the command line or using one of the editing functions provided to do so.

The general syntax for defining a function at the command line is

```
> name <- function (arguments) { body }
```

Specify arguments to your function in the parentheses, separated by commas. You can supply default values for some or all of the arguments by following the name of the argument with an **=** sign and a value. Arguments without a specified default value are usually required arguments to the function.

Enclose the S-PLUS expressions that form the body of your function in **{}**. The value returned by a function is the value of the last evaluated expression in the body of the function. All S-PLUS functions return a single S-PLUS object.

Objects defined within the body of a function are *local* to that function, existing in memory only for the duration of the execution of the function. Local objects do not interfere with objects of the same name in other nested function calls nor in data directories containing objects for your S-PLUS session.

As an example, consider the the plot of the sine function in Figure 47.3. You can encapsulate this idea into a new function that would plot any function of a single variable. The function might look like the following:

```
> fcn.plot
function(fcn, min.x, max.x, nx=100, type = "l", . . .)
{
    x < seq(min.x, max.x, length.out=nx)
    y <- fcn(x)
    plot(x, y, type = type, . . .)
}
```

The arguments **fcn, min.x, max.x,** and **nx** are the function being plotted, the minimum and maximum x values over which the function is plotted, and the number of *x* values at which the function is evaluated prior to plotting it. The ... is a special argument that allows passing an unspecified number of additional arguments. These additional arguments are passed on to the **plot** function so they can be used for specifying, for example, a main title, axis labels, line type, or line color. The plot of the sine curve in Figure 47.3 is now more easily produced.

```
> fcn.plot(sin, -pi, pi)
```

The **fcn.plot** function can now take *any* function of a single variable as its first argument. You can, for example, plot a quadratic polynomial as follows:

```
> fcn.plot(function(x) {1 + 2*x + x^2}, -5, 5)
```

Here, the quadratic function is actually defined at the time **fcn.plot** is called.

Iteration and Flow of Control

S-PLUS expressions are normally evaluated sequentially. Groups of expressions can be collected within curly braces **{ }**; such groups are treated as a single S-PLUS expression, but within that expression evaluation proceeds sequentially.

You can override the normal flow of control with various looping and control constructions. The most commonly used are **if-else** expressions and **for** loops described as follows:

if (*cond*) *expr* **else** *condexpr* Evaluates *cond*; if true, evaluates *expr*. If false, evaluates *condexpr*.

for (*name* in *expr1*) *expr2* Evaluates *expr2* once for each *name* in *expr1*.

Other important functions for flow of control include: **ifelse** a vectorized version of the **if-else** construct above, **while** for continual looping as long as a condition is satisfied, **repeat** for indefinite repetition, and switch for evaluating optional expressions. Additionally, **break, next, return,** and **stop,** respectively, terminate a loop, jump to the next iteration of a loop, terminate a function and return a value, and terminate a function and print an error message.

Object-Oriented Programming

S-PLUS is based on an object-oriented language, characterized by a set of *generic functions* for which specific *methods* are defined to operate on corresponding *classes* of objects. A generic function can be called with any type of object, and the appropriate method for the class of this object is automatically used. You can easily recognize a generic function by its definition, as all generic functions in S-PLUS are simply calls to the **UseMethod** dispatch mechanism. The **plot** function, for instance, is a generic function:

```
> plot
function(x, . . .)
UseMethod("plot")
```

The dispatch mechanism recognizes the class of an object through its **"class"** attribute and calls the appropriate method. The class attribute is a character vector, the first element of which defines the class of the object. Any other strings in the vector indicate the classes that the object *inherits* from. Whenever there is not a specific method defined for the class of an object called in a generic function, S-PLUS searches for methods for the classes from which the object inherits, in order, before using the default method. For example, ordered factors are of class **"ordered"**, but inherit from the class **"factor"**. There is not a plotting function for ordered factors but there is one for factors that is used when we call **plot** with an ordered factor as the argument.

47.6 PROBABILITY AND STATISTICS

S-PLUS provides functions for doing many kinds of statistical analyses including hypothesis tests, goodness of fit tests, contingency tables, design of experiments, analysis of variance, linear and nonlinear regression, parametric and nonparametric regression, survival analysis, time series analysis, quality control, and multivariate techniques such as principal components analysis, factor analysis, cluster analysis, and discriminant analysis.

This section begins with a description of the S-PLUS functions for calculating values from theoretical probability distributions, followed by overviews of the functions that produce summary statistics and perform hypothesis tests. Then we introduce the syntax for creating formulas to represent statistical models, with a brief regression example.

Probability Distributions

S-PLUS includes functions for calculating values from a variety of probability distributions. There are four functions for each available distribution:

d*distname*(**x**) Computes density values.
p*distname*(**q**) Computes probabilities from the cumulative distribution function (cdf).
q*distname*(**p**) Computes quantiles from the inverse cdf.
r*distname*(**n**) Generates random numbers.

The **ppoints** function creates a vector of evenly spaced values between 0 and 1 (probabilities), and is useful for creating plots of distributions. The following example draws the theoretical density curve on top of a histogram of data (Fig. 47.5).

```
> sample.data <- rbeta(100,2,9)
> hist(sample.data, den=-1, prob=T)
> p <- ppoints(100)
> lines(qbeta(p,2,9), dbeta (qbeta(p,2,9),2,9))
> title(main="Data from Beta(2,9) Distribution")
```

The distributions represented in S-PLUS are

```
beta            binomial     Cauchy         chi-squared
exponential     F            gamma          geometric
hypergeometric  log-normal   logistic       negative binomial
normal          Poisson      stable         Student's t
uniform         Weibull      Wilcoxon rank
                               sum
```

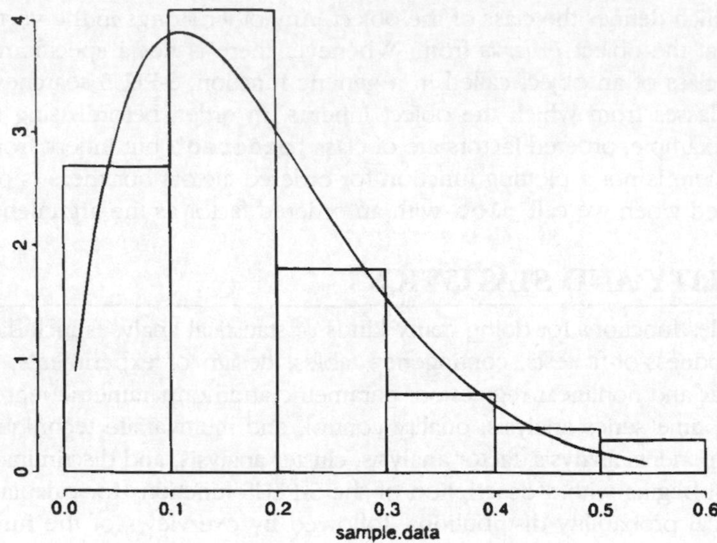

FIGURE 47.5 Histogram with theoretical density curve added.

Use the **sample** function to take a random sample from a set. For example, a sample of size 10 from the integers ranging from 100 to 199 is produced as follows:

```
> sample(100:199, 10)
[1] 157 148 115 118 121 126 173 164 189 149
```

Summary Statistics

S-PLUS includes most functions for calculating the standard summary statistics for a data set, and also functions that compute various robust estimators of location and scale (Table 47.2). The **summary** function is a generic function, providing appropriate summaries for different types of objects. For example, **summary** returns a summary for an object of class **"lm"** obtained from fitting a linear model that includes the table of estimated coefficients, their standard errors, and *t* values, along with other information. The default summary for a numeric vector is shown by the following example:

```
> summary(stack.loss)
Min.  1st  Qu.  Median   Mean 3rd Qu.   Max.
   7        11      15  17.52      19     42
```

Statistical Tests

S-PLUS contains several functions for doing classical hypothesis testing. Each of the functions listed below returns an object of class **"htest"** that prints out in a format describing the results of the test.

TABLE 47.2 Statistical Summary Functions

	Classical Estimators
mean	Arithmetic mean
median	Median
var	Variance of vector, covariance matrix of matrix
cor	Correlation between matrices or vectors
quantile	Empirical quantiles
summary	Summary statistics of vector or other object
	Robust Estimators
location.m	M-estimate of location
mad	Mean absolute deviation estimate of variance
scale.a	Bisquare-A estimate of variance
scale.tau	Huber's τ estimate of variance
robloc	M-estimate of location and Huber estimate of variance
cov.mve	Robust location and variance estimates for multivariate data

t.test	Student's t tests: one-sample, two-sample, paired, equal, and unequal variance.
wilcox.test	Wilcoxon rank sum and signed rank sum tests.
var.test	F test to compare two variances.
kruskal.test	Kruskal–Wallis rank sum test for a one-way design.
friedman.test	Friedman rank sum test for an unreplicated blocked design.
cor.test	Test for zero correlation; includes Pearson's, Kendall's τ, and Spearman's ρ.
binom.test	Exact binomial test for a single proportion.
prop.test	Pearson's chi-square test for equality of proportions.
chisq.test	Pearson's chi-square test for two-dimensional contingency table.
fisher.test	Fisher's exact test for a two-dimensional contingency table.
mcnemar.test	McNemar's chi-square test for a two-dimensional contingency table.
mantelhaen.test	Mantel–Haenszel chi-square test for a three-dimensional contingency table.

The following example illustrates how to use **t.test** to perform a two-sample t test to detect a difference in means. This example uses two random samples generated from $N(0, 1)$ and $N(1, 1)$ distributions.

```
> x <- rnorm(10)
> y <- rnorm(5, mean=1)
> t.test(x,y)

                    Standard Two-Sample t-Test

data: x and y
t = -1.4312, df = 13, p-value = 0.176
alternative hypothesis: true difference in means is not equal
   to 0
95 percent confidence interval:
```

```
-1.7254080 0.3502894
sample estimates:
mean of x mean of y
-0.4269014 0.2606579
```

Statistical Models

Many of the statistical modeling functions in S-PLUS follow a unified modeling paradigm, in which the input data are represented as a data frame and the model to be fit is represented as a *formula*. Some of the modeling functions that use formulas are listed below:

crosstabs	Create a multiway contingency table from a collection of factors.
aov, manova	Fit univariate and multivariate analysis of variance models.
lm	Fit a linear regression model.
glm	Fit a generalized linear model.
gam	Fit a generalized additive model.
loess	Fit a local regression model.
tree	Fit a classification or regression tree model.
nls, ms	Fit parametric nonlinear models.
factanal	Fit a factor analysis model.
princomp	Performs principal components analysis.

In a formula, you specify the response variable first, followed by a tilde (~) and the terms to be included in the model. Table 47.3 contains some examples of formulas and what they mean.

Additional functions are available for fitting survival models, analyzing time series, and charting quality control data.

Survival Analysis

surv.fit	Fit a Kaplan–Meier survival model.
coxreg	Fit a Cox proportional hazards model.
agreg	Fit the Andersen–Gill extension to the Cox model.

Time Series Analysis

ar	Fit a univariate or multivariate autoregressive model.
arima.mle	Fit an ARIMA model.
spectrum	Estimate the spectrum for a time series.

Quality Control Charting

shewhart	Produce Shewhart charts (xbar, s, R, p, np, u, and c).
cusum	Produce cusum charts (xbar, s, p, np, u, and c).

TABLE 47.3 Examples of Formulas

Formula	Meaning
Yield ~ Temp + Conc	**Yield** is modeled as depending on **Temp** and **Conc**
Yield ~ Temp + Conc + Temp:Conc or **Yield ~ Temp * Conc**	**Yield** is modeled as depending on **Temp, Conc**, and the interaction between these two variables
log(Mileage) ~ Weight + poly(HP, 2)	The log of **Mileage** is modeled as depending on **Weight** and a second-order polynomial in **HP**

Regression

Linear least-squares regression is performed in S-PLUS using the linear models function **lm**. Logistic regression and Poisson regression are two examples of models that are usually fit using the generalized linear models function **glm** and the generalized additive models function **gam**. S-PLus functions for doing robust regression include **l1fit** for L_1 regression, **rreg** for M-estimate regression, and **ltsreg** least trimmed squares regression (includes least median squares regression).

A brief example shows the interplay between different modeling techniques. The data frame **ethanol** contains data from a study of nitrogen oxide and dioxide emissions (**NOx**) in the exhaust of a small piston engine. Design parameters are the compression ratio (**C**) and the equivalence ratio (**E**), a measure of the fuel to air mixture. The goal is to model **NOx** as a function of the two design parameters **C** and **E**.

Start by looking at statistical summaries of the data and pairwise scatterplots.

```
> summary(ethanol)
           NOx                   C                  E
Min.     :0.370    Min.     : 7.500    Min.     :0.5350
1st Qu.  :0.953    1st Qu.  : 8.625    1st Qu.  :0.7618
Median   :1.754    Median   :12.000    Median   :0.9320
Mean     :1.957    Mean     :12.030    Mean     :0.9265
3rd Qu.  :3.003    3rd Qu.  :15.000    3rd Qu.  :1.1100
Max.     :4.028    Max.     :18.0000   Max.     :1.2320

> pairs(ethanol)
```

Figure 47.6 shows the scatterplot matrix resulting from the call to **pairs**. From Figure 47.6 it is clear that there is a strong relationship between **NOx** and **E** that appears quadratic. However, it is difficult to tell much about the relationship between **NOx** and **C**.

An alternative plot, the conditioning plot, produced by the function **coplot** is more revealing. First create nine intervals for **E** that have a 25% data overlap and then produce a separate plot of **NOx** versus **C** for each interval. The **panel.smooth** function plots the data points in each **E** interval and then adds a local regression smooth (loess smooth) fit.

```
> attach (ethanol)
> E.intervals <- co.intervals(E, number = 9, overlap = 1/4)
> coplot(NOx ~ C | E, given = E.intervals, panel =
+ function(x,y) panel.smooth(x,y, degree = 1, span = 1))
```

The resulting plot (Figure 47.7) shows a linear relationship between **NOx** and **C** conditional on **E**. In fact, the slope appears to increase and then decrease with increasing values of **E** indicating an interaction between the two design parameters.

The conditioning plot is read as follows. The top box is simply a legend to explain which data points are represented in the plots in the bottom panels. In this example,

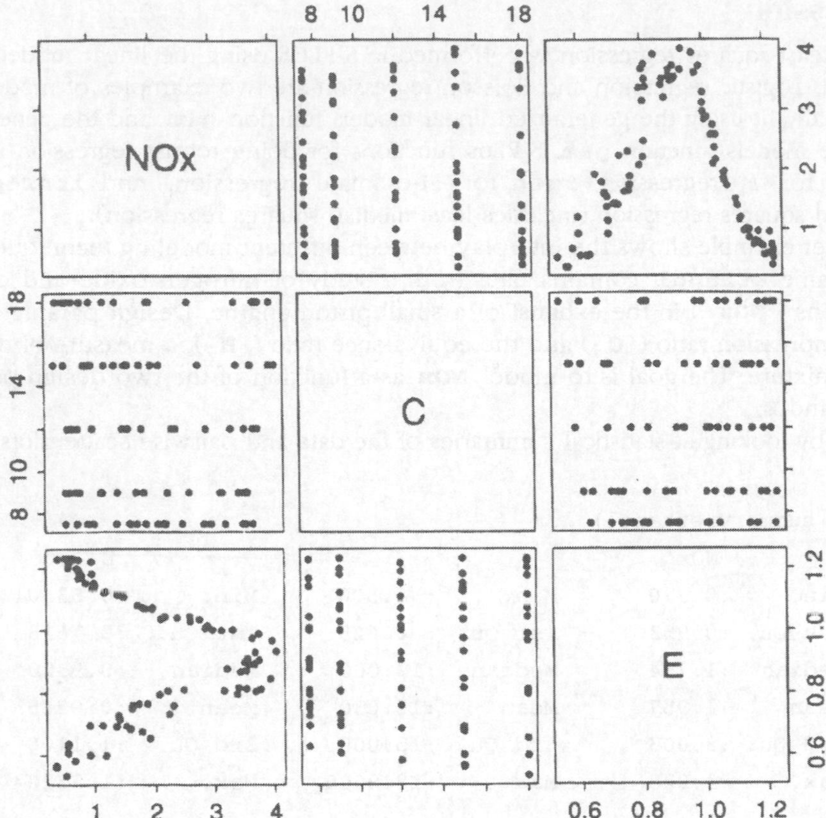

FIGURE 47.6 Scatterplot matrix of the **ethanol** data frame.

we have three rows of panels, so the legend box at the top is separated into three sections by dotted lines. The three solid lines below the lowest dotted line correspond to the bottom row of plots and indicate the intervals of **E** used to produce each plot from left to right. The two other rows of plots correspond to the other two sections of the legend panel.

A similar plot (not shown) for **NOx** versus **E** conditional on **C** can be produced but adds no new information to the scatterplot matrix in Figure 47.6.

Now fit a model based on the information we have gathered so far. Start with a linear model, fitting a second-degree polynomial to **E**.

```
> mod1 <- lm(NOx ~ C + poly(E, 2), data = ethanol)
> summary(mod1)

Call: lm(formula = NOx ~ C + poly(E, 2), data = ethanol)
Residuals:
    Min             1Q       Median         3Q        Max
 -0.823        -0.4063      0.03611     0.4061     0.9681
Coefficients:
```

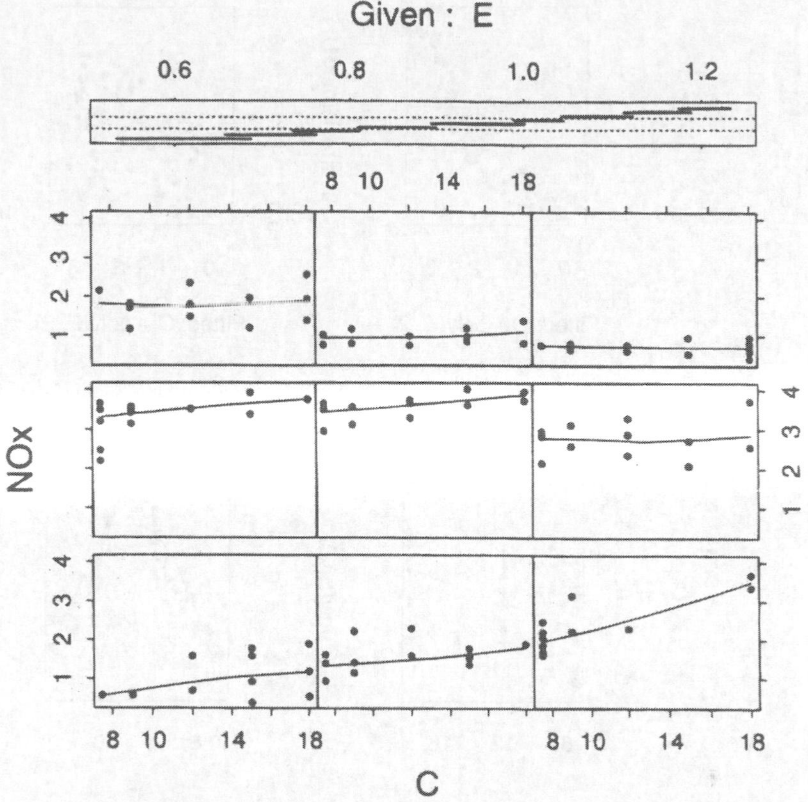

FIGURE 47.7 Coplot of NO_x versus **C** for nine overlapping intervals of **E** .

```
               Value Std. Error   t value Pr(>|t|)
(Intercept)   1.1932     0.1728    6.9065   0.0000
C             0.0635     0.0137    4.6347   0.0000
poly(E, 2)1  -1.2247     0.4851   -2.5247   0.0135
poly(E, 2)2  -9.8693     0.5015  -19.6792   0.0000

Residual standard error: 0.484 on 84 degrees of freedom
Multiple R-Squared: 0.8237
F-statistic:  130.8  on  3  and 84 degrees of freedom, the p-
  value is 0
```

It is clear that **C** and linear and quadratic terms for **E** are significant in the model, but do they fit the data well? A sequence of plots (Figure 47.8) helps assess the fit.

```
> par(mfrow =c(2,2))
> plot(mod1)
> plot.gam(mod1, resid = T, rug = F)
```

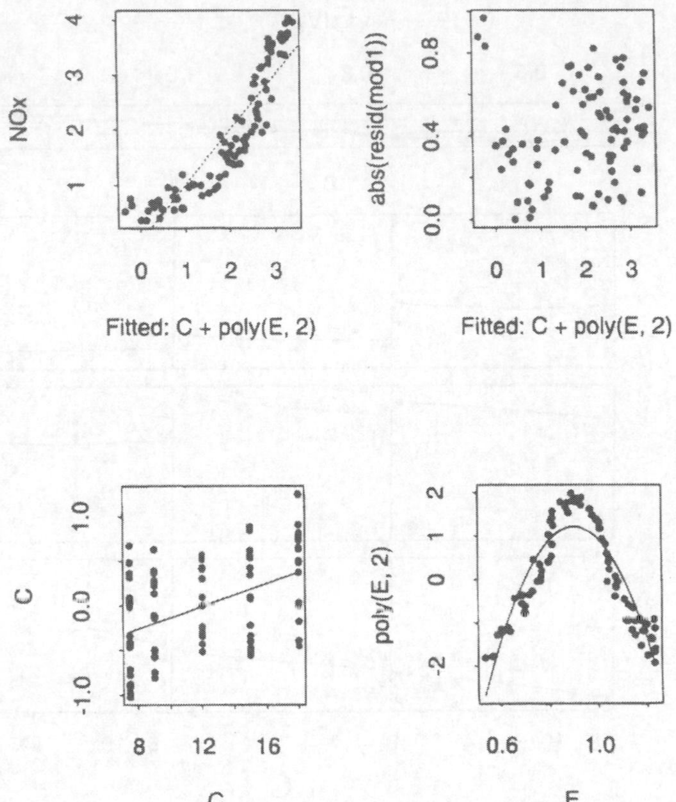

FIGURE 47.8 Diagnostic plots for **mod1**.

The top row of Figure 47.8 plots the observed response versus fitted values and the absolute residual versus fitted values, from left to right. This is the default plot for a linear model. The bottom row of Figure 47.8 plots the fitted curve with respect to each variable versus each variable with partial residuals overlaid. While the fit with respect to **C** looks pretty good, it is clear that a quadratic does not model the relationship with respect to **E** very well.

Now try using a local regression smoothing operation to fit the relationship with respect to **E**. To do so, use the generalized additive modeling function **gam** and the loess smoothing operator **lo**, specifying the degree of the local regression to be two for local quadratic fits.

```
> mod2 <- gam(NOx - C + lo(E, degree = 2), data = ethanol)
> plot(mod2, resid = T, rug = F)
```

The resulting fitted curves with partial residuals (Figure 47.9) shows a substantial improvement in the fit, as does an ANOVA comparing the models.

```
> anova(mod1, mod2, test = "F")
Analysis of Variance Table

Response: NOx
```

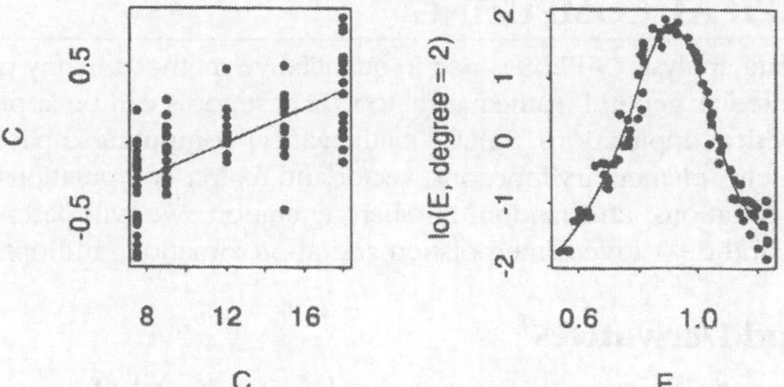

FIGURE 47.9 Diagnostic plots for mod2.

```
                 Terms Res. Df     RSS    Test       Df Sum of Sq  F-Val Pr(F)
1       C+poly(E, 2) 84.000 19.678
2 C+lo(E, degree=2) 80.118   5.168 1 vs. 2 3.882      14.511 57.959     0
```

Now, fit an additional model that includes interaction between **C** and **E** in a nonparametric, data-driven way. If, in the gam model, both **C** and **E** are included in the loess smooth, the resulting fit incorporates any interaction between two variables. To make this third model comparable with **mod2** cut the default span of 1/2 for the loess smoother to 1/4.

```
> mod3 <- gam(NOx~ lo(C, E, span = 1/4, degree = 2), data =
  ethanol)
```

```
> anova(mod2, mod3, test = "F")
Analysis of Deviance Table

Response: NOx

                  Terms Res. Df Res.Dev    Test     Df Sum of S F-Val Pr(F)
1      C+lo(E, deg=2)80.118   5.168
2 lo(C,E,span=1/4, deg=2)62.194   1.740 1 vs. 2 17.924    3.427  6.834    0
```

Thus, there is significant interaction between **C** and **E** . However, the fitted relationship between **NOx** and C used local *quadratic* fits that is probably overfitting in light of Figure 47.7. The local regression model fitting function **loess** is more flexible than the local regression smoothing operator **lo**. With **loess**, the fit for **C** can be locally linear or even conditionally parametric and linear while fitting **E** with local quadratic fits. Fitting such a model with **loess** maintains the good fit of **mod3** but reduces the (equivalent) number of parameters estimated from about 18 to about 9 making it a better model.

47.7 MATHEMATICAL COMPUTING

Designed for data analysis, S-PLUS is rich in quantitative methods, many of which have been implemented as general mathematical tools. These tools can be applied to a wide variety of numerical applications. S-PLUS mathematical computing capabilities include complex arithmetic, elementary functions, vector and matrix computations, solving systems of linear equations, and random number generation. We will describe in greater detail integrals and derivatives, interpolation and approximation, and optimization.

Integrals and Derivatives

Use the **integrate** function to compute the definite integral of a real-valued function over a given interval. The **integrate** function returns a list, of which the first two components are the integral and the absolute error:

```
> integrate(sin, 0, pi)[1:2]
$integral:
[1] 2
$abs.error:
[1] 2.220446e-14
```

To perform symbolic differentiation, use the D function.

```
> D(expression(3*x^2),"x")
3 * (2 * x)
> D(expression(exp(x^2)),"x")
exp(x^2) * (2 * x)
> D(expression(log(y)), "y")
1/y
```

Interpolation and Approximation

S-PLUS has a variety of functions for interpolation and approximation, most of them developed to aid in fitting curves and lines to data. However, they are sufficiently general to have wide application in mathematical settings.

Linear Interpolation

To find interpolated values in S-PLUS, use the **approx** function. You provide a vector of x values and a vector of associated y values, and (optionally) a vector of different x values at which you want interpolated values. S-PLUS returns a list of x values and the associated y values:

```
> approx(1:10, (1:10)^2, xout = 1.5:3.5)
$x:
[1] 1.5 2.5 3.5
$y:
[1] 2.5 6.5 12.5
```

A more specialized interpolation function, **interp**, can be used to generate input for the three-dimensional plotting functions **image, contour,** and **persp**. The **interp** function interpolates the value of the *z* variable onto an evenly spaced grid of the *x* and *y* variables.

Cubic Spline Approximation

Splines approximate a function with a set of polynomials defined on subintervals. A cubic spline is a collection of polynomials of degree less than or equal to 3 such that the second derivatives agree at the "knots" (the points where the polynomials hook together), i.e., the spline has a continuous second derivative.

When interpolating a number of points, a spline can be a much better solution than a polynomial interpolation, since the polynomial can oscillate wildly to hit all of the points (polynomials fit the data globally while splines fit the data locally).

Using the **spline** function to obtain a cubic spline approximation to the quadratic function above does better than simple linear interpolation. Here we do a linear interpolation on the cubic spline approximation because the **spline** function returns estimates for *y* at an evenly spaced set of **x**'s.

```
> approx(spline(1:10, (1:10)^2), xout = 1.5:3.5)
$x:
[1] 1.5 2.5 3.5

$y:
[1] 2.359119 6.255975 12.283648
```

The exact values are

```
> (1.5:3.5)^2
[1] 2.25 6.25 12.25
```

Convex Hull

To obtain the convex hull of a planar set of points, use the **chull** function, which returns the indices of the points defining the hull:

```
> ch <- chull(corn.rain, corn.yield)
> ch
[1] 35 12 5 24 33 31 13 26
```

A plot of the convex hull is now created by doing

```
> plot(corn.rain, corn.yield)
> polygon(corn.rain[ch], corn.yield[ch], density = 0)
```

The **density = 0** argument to **polygon** simply prevents the inside of the hull from being filled in so you can see the data points.

The **peel** option allows you to peel off the convex hull, take the convex hull of the remaining points, peel off *that* hull, and so on, until either all points are assigned to a hull or a user-specified limit is reached.

Optimization

S-PLUS has several functions for finding roots of equations and local maxima and minima of functions, as shown in Table 47.4. As a simple example, obtain the roots of the quadratic polynomial $(x - 3)(x - 2) = 6 - 5x + x^2$ as follows:

```
> polyroot(c(6, -5, 1))
[1] 2+0i 3+0i
```

47.8 ADD-ON MODULES

MathSoft, StatSci Division, is currently shipping add-on modules to S-PLUS. Three of them are briefly described here.

S+DOX

S+DOX is an S-PLUS module of industrial design software. It extends and enriches the ANOVA modeling already in S-PLUS, to provide graphics and analysis techniques for industrial experimentation. The aim of industrial experimental design is to improve the quality of a product by experimenting with the settings of factors that affect the product. Typically, cost constrains the number of observations that can be collected. This module provides graphical and analytic tools for fractional factorial, mixed fraction, response surface, and robust designed experiments. S+ DOX is distinguished by a visual emphasis throughout: displays are customized to factorial and response surface methods for both exploratory analysis and display of model and inference results.

S-PLUS for ARC/INFO

S-PLUS for ARC/INFO provides a bidirectional data bridge that allows ARC/INFO users to access the full data analysis capabilities of S-PLUS, using ARC/INFO AML commands. The built-in interface translates ARC/INFO coverage attribute tables, GRID raster data, and INFO files to S-PLUS format. Data can be mapped to ARC/INFO format, then transferred to S-PLUS for analysis, and the statistical results transferred back to ARC/INFO for enhanced geographic and spatial analysis.

TABLE 47.4 Optimization functions

Function	Explanation
polyroot	Finds the roots of a complex polynomial equation
uniroot	Finds the root of a univariate real-valued function in a user-supplied interval
peaks	Finds local maxima in a set of discrete points
optimize	Finds a local maximum or minimum of a univariate function within a user-supplied interval
ms	Finds a local minimum of a multivariate function (expects a formula object and a data frame)
nlminb	Finds a local minimum of a multivariate function, subject to bound-constrained parameters
nls	Finds a local minimum of the sums of squares of one or more multivariate functions (expects a formula object and a data frame)
nlregb	Finds a local minimum of the sums of squares of one or more multivariate functions, subject to bound-constrained parameters
nnls	Finds least-squares solution subject to the constraint that the coefficients be nonnegative

S+INTERFACE

S+INTERFACE is a programmers toolkit, available on the UNIX environment, for integrating S-PLUS into other applications and for building a menu front-end to S-PLUS functions. It consists of two distinct parts, S+CASIM and S+MENU. S+CASIM is the C Application/S-PLUS Interface Manager, a set of callable C routines you can use to incorporate S-PLUS functionality into a new or existing C application. S+MENU is the S-PLUS Menu Builder, a set of S-PLUS functions for building a customized Motif-based graphical user interface to S-PLUS.

ACKNOWLEDGMENTS

This manuscript has borrowed heavily from the S-PLUS documentation set produced by MathSoft Inc., StatSci Division. The documentation team is lead by Richard Calaway. James Schimert made important contributions to content and organization of this chapter.

DEFINING TERMS

Brushing: An dynamic graphical procedure in which points selected with a mouse in one plot of a scatterplot matrix cause the same cases to be highlighted in the other plots of the matrix.

Point-Cloud Rotation: The dynamic spinning of a 3-D scatterplot. The spinning is controlled with a mouse.

Conditioning Plot: Plot of one variable versus another conditional on the values of a third variable.

High-Level Graphics Function: An S-PLUS function producing an entirely new graph, including axes, labels, and a box (if appropriate) around the plot.

Low-Level Graphics Function: An S-PLUS function for adding detail such as lines, points, and text to an existing plot.

Mode: The mode of an S-PLUS data object. Vectors have a single mode that is one of `logical`, `numeric`, `complex`, or `character`

REFERENCES

R.A. Becker, J.M. Chambers, and A.R. Wilks, *The New S Language*. Wadsworth & Brooks/Cole, Pacific Grove, CA, 1988.

S-PLUS User's Manual: Version 3.2. MathSoft, Inc., StatSci Division, Seattle, WA.

S-PLUS Programmers's Manual: Version 3.2. MathSoft, Inc., StatSci Division, Seattle, WA.

S-PLUS Guide to Statistical and Mathematical Analysis: Version 3.2. MathSoft, Inc., StatSci Division, Seattle, WA.

FURTHER INFORMATION

J.M. Chambers and T.J. Hastie, (Eds.), *Statistical Models in S.* Wadsworth & Brooks/Cole, Pacific Grove, CA, 1992.

P. Spector, *An Introduction to S and S-PLUS.* Duxbury Press, Belmont, CA, 1994.

S-PLUS, A Crash Course in S-PLUS: Version 3.2. MathSoft, Inc., StatSci Division, Seattle, WA.

48

WinSTAT

Robert K. Fitch
Greulich Software

48.1 INTRODUCTION

The ability to apply statistical methods in evaluating research results closely paralleled the development of electronic computers. Many of the calculations involved in these methods include the use of large-scale matrix manipulations that were hardly thinkable before the introduction of computers. With the advent of the PC, the makers of statistics software shifted their focus correspondingly, and soon it was possible to do "personal" statistical research without the need to access a university computation center.

Development of WinSTAT began in the late 1980s. It was clear by this time that Microsoft Windows had been accepted as the predominant operating system for PCs. There were, at this stage, a number of statistics programs running under DOS, many of which had been ported from mainframe programs by companies that had been in the statistics software business for decades.

Because they were starting from scratch, the programmers of WinSTAT were able to work with modern C++ compilers right from the start, using class definitions made specifically for the development of Windows applications. At the end of 1991, version 1.0 was released, and it was in fact the first statistics program to reach the Windows market. Obviously, at this early stage, it could not offer the complete realm of statistics methods available from the established programs.

The present (1994) version 3.0 of WinSTAT has expanded to offer most methods needed by the average user of a statistics package. In addition, it has many graphics capabilities. The graphics may be edited interactively with mouse control and are of presentation quality.

48.2 REFERENCE MANUAL AND HELP SYSTEM

The reference manual contains descriptions for all of WinSTAT's commands. The product comes with several sample files, and the manual includes an example from the sample files to illustrate each command. As expected of a Windows application, the Help system is context-sensitive and includes information about all commands and the corresponding dialog boxes. In addition, a small expert system leads the novice user to the proper statistical method by a series of questions and answers.

0-8493-2530-7/96/$0.00+$.50

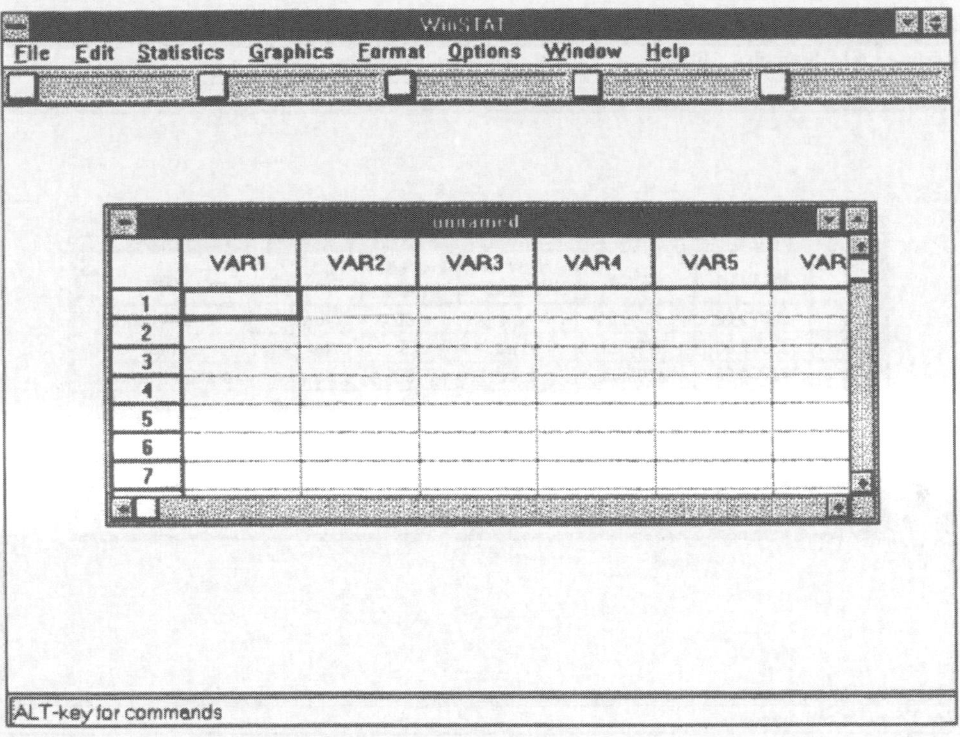

FIGURE 48.1 A new worksheet has been opened.

48.3 THE WinSTAT WORKSHEET

The WinSTAT worksheet is the basis for entering and editing raw data. Of course, data may also be copied from other programs to the worksheet using the Windows clipboard.

As can be seen in Figure 48.1, care was taken in designing the worksheet to make it as similar as possible to popular spreadsheet programs. This helps the user feel at home almost immediately. The worksheet is a two-dimensional data matrix. Along the top of the table appear the names of the observed variables. In a new worksheet, they are simply labeled consecutively VAR1, VAR2, etc. The numbers down the left-hand side of the table stand for the cases. Thus, for each case (e.g., object, person), the observed value of each of the variables may be entered. A tool bar is also visible just below the menus. The buttons on the tool bar can be freely programmed with whichever functions the users needs most frequently.

The cursor key functions, as well as all editing functions (Copy, Paste, etc.) are implemented as in their spreadsheet counterparts. Figure 48.2 shows a file after the variable names have been defined and data have been entered.

The user may choose any font for the worksheet. Column widths may also be set as desired by simply dragging with the mouse. Note that data entries may be numeric or text. In addition, date and time formats are allowed. An interesting short-cut is available when entering text variables. Suppose, for example, the variable "Sex" can contain either the word "male" or the word "female." Once these text values have been entered, it is enough to enter "m" or "f" for each of the remaining cases of that variable. The program automatically finishes each entry correctly.

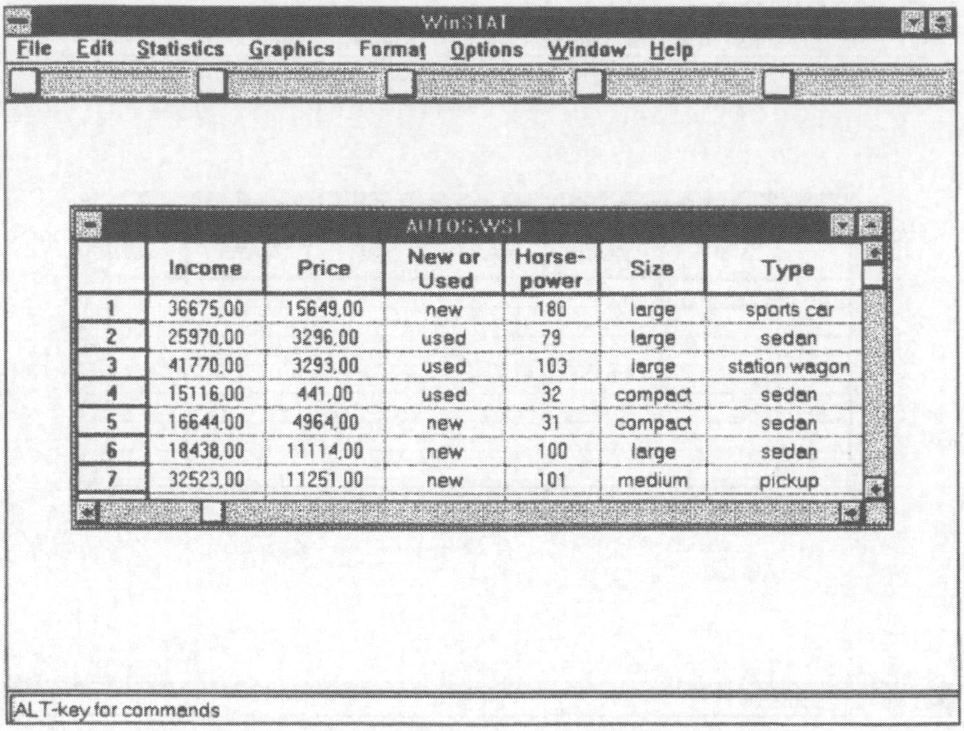

WinSTAT

File Edit Statistics Graphics Format Options Window Help

AUTOS.WST

	Income	Price	New or Used	Horse-power	Size	Type
1	36675.00	15649.00	new	180	large	sports car
2	25970.00	3296.00	used	79	large	sedan
3	41770.00	3293.00	used	103	large	station wagon
4	15116.00	441.00	used	32	compact	sedan
5	16644.00	4964.00	new	31	compact	sedan
6	18438.00	11114.00	new	100	large	sedan
7	32523.00	11251.00	new	101	medium	pickup

ALT-key for commands

FIGURE 48.2 A completed worksheet.

Importing Foreign Data

WinSTAT can read the following file types for data input:

```
Excel (.xls), dBase (.dbf), 1-2-3 (.wks, .wk1, .wk3),
Quattro Pro (.wkq), ASCII (.txt)
```

Sort, Select, and Transformation

In preparation for statistical analysis, it is often necessary to perform various manipulations on the data at hand. The worksheet cases may be sorted, using up to four sorting variables, each in either ascending or descending order.

Furthermore, a subset of the data file may be selected using conditions applied to any number of variables, combined with logical operators. For example,

```
("Income" > 30000) and ("Type" = 'Sports car')
```

Finally, a new variable may be defined as a mathematical function or transformation of existing variables. Examples would be

```
("fahrenheit" -32) * 5/9
LOG "age"
```

An entire range of mathematical functions is available, as well as well as the Z-transformation, Box–Cox transformation, and random normal distribution. In addition, an IF-function is provided for transformations dependent on input values, for example,

```
IF("age" < 18; 1; 2)
```

This returns 1 if the value found in variable "age" is less than 18, otherwise 2. The IF-function may be nested for more complex dependencies.

48.4 STATISTICS CAPABILITIES

In the following sections, each of WinSTAT's statistics commands is briefly described.

Descriptive Statistics

This command yields the basic characteristics for any number of variables. To give the reader a first look at the way a WinSTAT user directs control of the program via dialog boxes, we include Figure 48.3 showing the Descriptive Statistics dialog box:

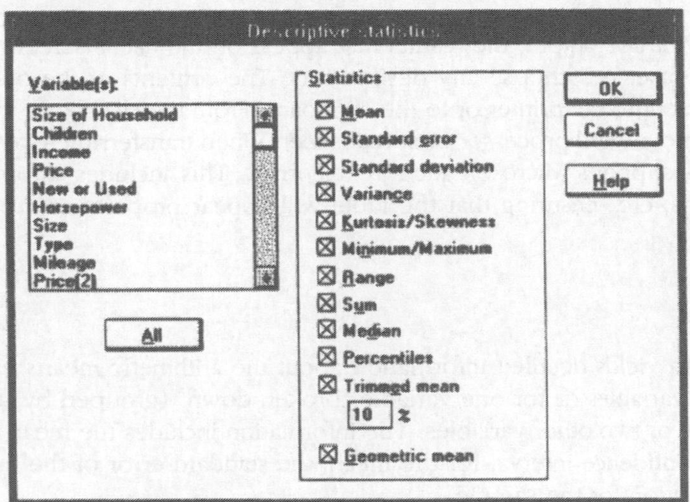

FIGURE 48.3 The Descriptive Statistics dialog box.

Using the list of variables at the left-hand side of the dialog box, any number of variables may be chosen for which the statistics are desired. The actual statistics to be output are then selected from the list on the right. Note the "Help" button. As expected of a Windows program, almost all dialog boxes include this button, so that the less-experienced user can receive on-line, context-sensitive help regarding the operation at hand.

Frequencies

This command displays frequency information for any number of variables. The output tells you the number of cases in which each value or range of values of the given variable appears. To give the reader a first look at the way WinSTAT outputs its results, we include a sample output for the Frequencies command:

Variable: Price

Value	Frequen-cies	Percent	Percent valid	Percent cumulative
<0.00 .. 2499.99>	9	12.5000	12.5000	12.5000
<2500.00 .. 4999.99>*	22	30.5556	30.5556	43.0556
<5000.00 .. 7499.99>	10	13.8889	13.8889	56.9444
<7500.00 .. 9999.99>	11	15.2778	15.2778	72.2222
<10000.00 .. 12499.99>	10	13.8889	13.8889	86.1111
<12500.00 .. 14999.99>	2	2.77778	2.77778	88.8889
<15000.00 .. 17499.99>	2	2.77778	2.77778	91.6667
<17500.00 .. 19999.99>	3	4.16667	4.16667	95.8333
<20000.00 .. 22499.99>	1	1.38889	1.38889	97.2222
<22500.00 .. 24999.99>	1	1.38889	1.38889	98.6111
<25000.00 .. 27499.99>	0	0.00000	0.00000	98.6111
<27500.00 .. 27904.00>	1	1.38889	1.38889	100.000

As in all WinSTAT output, the results first appear on the user's screen in a separate window. The user can choose any desired font. The contents of the output window may then be sent to the printer or to the clipboard. From the clipboard, the results can be inserted into a word processor or spreadsheet. When transferring to a word processor, WinSTAT supports Microsoft Rich Text Format. This includes information about font, tab stops, etc., ensuring that the table will appear properly in the target document.

Means

This command yields detailed information about the arithmetic means either for several separate variables or for one variable "broken down" (grouped by) the values or classes of one or two other variables. The information includes the mean of each sample itself, a confidence interval for the mean, the standard error of the mean, and the standard deviation for the sample.

Tests of Randomness

It is sometimes of interest to investigate whether the values of a variable, *in the order in which they are recorded*, display some sort of significant behavior. For example, do the values rise or fall with increasing case number (trend), or do they exhibit periodicity? Two tests of randomness are included, the test based on Turning Points and the Rank Test. The resulting **significance** values indicate whether the variable may be regarded as random or not.

Cross-tabulation

This command produces a cross-tabulation (bivariate frequency distribution) table for a pair of variables. It is used to look for dependencies between **nominal-level** variables. The statistics included in the output are the true and expected frequencies for each cell of the table, the cell chi-square values, the total chi-square and its corresponding significance (*p* value), as well the as the contingency coefficient and Cramer's *V* for the table. If the table is 2 by 2, Fisher's exact test is computed and displayed.

Chi-Square Test

This command compares measured frequencies with expected frequencies for a single variable, and is best explained with an example. Suppose we throw a die a number of times and record the number of occurrences for each possible outcome, 1 to 6. The expected frequencies would be equal for each possibility. The chi-square test compares the actual frequencies with these expected frequencies, and determines the significance of the differences. A high significance (low *p* value) indicates that we can assume the variable is behaving other than expected.

Kolmogorov–Smirnov Test

This command compares the distribution of a given variable with a normal (Gaussian) distribution. Since many statistical methods require that the variables be distributed normally, the user will want to support that claim by using this test on the variables involved. If it cannot be demonstrated that a given variable is distributed normally, WinSTAT provides transformations (in particular, the Box–Cox transformation) that in many cases suffice to transform the nonnormal variable into a normal one. This new variable may then be used for subsequent calculations.

t Test for Independent Samples

The values for a given dependent variable (measured at the **interval level** and assumed to have a normal distribution) are placed into two groups by a second, independent variable. The independent variable must be dichotomous, i.e., contain only 2 possible values. The arithmetic mean of the dependent variable will generally be different for the two different groups. The *t* test determines to what extent the difference is significant. An *F* test is automatically performed to determine if the variances of the two groups may be considered equal (homogeneous) or not.

t Test for Dependent (Correlated) Samples

A pair of variables is analyzed. The variables must be measured at the interval level. Each must belong to "the same object" and must measure "the same thing" under different circumstances. For example, one might want to investigate a subject's weight before a diet and after the diet. For each case, the difference of the two variables is computed. These differences are assumed to have a normal distribution. Then, the mean and standard deviation of these differences over the entire sample are computed. The *t* test determines to what extent the differences are significant.

U Test (Mann–Whitney Test)

This test is similar to the *t* test for independent samples, in that the values for a selected dependent variable are placed into two groups by a second, independent, dichotomous variable. The dependent variable need only be measured at the **ordinal level**, and this is the main difference between the *t* test and the *U* test.

All cases of the dependent variable are first sorted into ascending order. Each value is then replaced by its position after sorting (its *rank*). Thus, the lowest value becomes 1, the next lowest becomes 2, etc. These rank values are then separated into two groups according to the independent variable. Each group has its own mean, this being the mean rank of the original values, and the *U* test determines to what extent the difference in mean rank is significant.

The output includes the mean rank and the number of cases for each group. The *U* value is a statistical measure of the difference in mean rank. This value is then translated to a *Z* value, which is the abscissa of the equivalent point on the Gaussian normal curve. The *Z* value is used to compute the significances.

Wilcoxon Test

This test is similar to the *t* test for dependent samples in that, for each test, a pair of variables is analyzed. Each variable must belong to "the same object" and must measure "the same thing" under different circumstances. The variables need be measured only at the ordinal level, and this is the main difference between the *t* test and the Wilcoxon test.

For each case, the difference between the two variables is computed. The absolute values of these differences are then sorted to get a rank order. Finally, the mean rank of the negative differences is compared to the mean rank of the positive differences. The Wilcoxon test determines to what extent the difference in mean rank is significant.

The number of negative and positive differences is displayed, as well as the corresponding sum of ranks. This value is then translated to a *Z* value, which is the abscissa of the equivalent point on the Gaussian normal curve. The *Z* value is used to compute the significances.

Analysis of Variance

WinSTAT can perform a univariate one-factor or two-factor analysis of variance. This means that a single variable (measured at the interval level and assumed to have a normal distribution) is grouped according to the values of one or two independent variables. The arithmetic mean of the dependent variable will generally be different for the different groups. The analysis of variance determines to what extent the difference is significant. The analysis of variance is valid only if the groups can be assumed to have equal variances. A *Bartlett test* is automatically performed to test this assumption.

One-Factor ANOVA (Analysis of Variance)

If the one-factor ANOVA determines that differences in the means exist, it is often of interest to ask which of the group means differ significantly from which other group means. Or put another way, which subsets of groups can be built whose members have means that do not differ significantly from each other. The answer to this question is provided by a set of methods called "multiple comparisons," also known as "a posteriori" or "post hoc" tests.

All of the methods available display the same general pattern: A range statistic is calculated for each possible subset size, and if the greatest difference in means within a subset is less than the calculated range, then the subset is considered to contain members that do not differ significantly from each other. Various theories regarding the calculation of the ranges are reflected in the different methods.

The methods supported by WinSTAT are LSD (Least Significant Difference), Bonferroni, Scheffé, Tukey, S-N-K (Student–Newman–Keuls), Duncan, and REGW (Ryan, Einot, Gabriel, Welsch).

Two-Factor ANOVA

The goal of a two-factor (or two-way) ANOVA is to look at the effect of one variable after controlling for the effect of another variable. For instance, one may want to compare the average mileage of the cars from different manufacturers after controlling for the number of cylinders.

If the first factor groups the dependent variable into m groups and the second factor groups the dependent variable into n groups, then an $m \times n$ table is created. Each cell contains all cases with a given value-pair for the two factors, and the arithmetic mean is calculated for all cases of the dependent variable within each cell. If each cell contains the same number of cases, then the experiment is called ***balanced***. This, of course, will usually be the case only if the experiment was designed with this in mind (e.g., randomized block design).

WinSTAT uses the *Method of Unweighted Means*, which assumes that the data are fairly well balanced. To be more precise, the method is valid only if there is not more than a *twofold variation* in the number of observations in the row–column combinations.

Repeated Measures ANOVA

This analysis examines a given set of variables. These variables must be measured at the interval level and have a normal distribution. Each variable must belong to "the same object" and must measure "the same thing" under different circumstances, for example, the cholesterol levels of patients at several different time periods after a given treatment. The similarity to the t test for dependent samples is evident, this being an extension to the n-variable case. The repeated measures analysis of variance determines to what extent the differences among the variables are significant.

Kruskal–Wallis Test

This test is similar to a one-factor analysis of variance in that the values for a given dependent variable are placed into groups according to a given independent variable. The dependent variable need be measured only at the ordinal level, and this is the main difference between the Kruskal–Wallis test and the one-factor analysis of variance.

All cases of the dependent variable are first sorted into ascending order. Each value is then replaced by its position after sorting (its *rank*). Thus, the lowest value becomes 1, the next lowest becomes 2, etc. These rank values are then separated into groups according to the independent variable. Each group has its own mean, this being the mean rank of the original values, and the Kruskal–Wallis test determines to what extent the difference in mean rank is significant.

Friedman Test

This test is similar to a repeated measures analysis of variance, in that several variables are examined, each of which must belong to "the same object" and measure "the same thing" under different circumstances. The variables need be measured only at the ordinal level, and this is the main difference between repeated measures and the Friedman test. An example of such variables could be the ratings of three similar products, each case representing the ratings of one expert.

For each case, the values of the given variables are inspected. The variable with the lowest value has its value replaced by 1, the second-lowest variable becomes 2, etc. After all cases have been handled, the mean of each variable is computed, this being its mean rank. The Friedman test determines to what extent the differences in mean rank are significant.

Pearson Correlation

The Pearson correlation is a measure of the linear dependency between two variables, which must be measured at the interval level. The results include the correlation value itself as well as an indication of its significance.

Spearman Rank Correlation

The Spearman rank correlation is also an indication of the dependency between two variables. The variables need be measured only at the ordinal level.

Kendall's Tau

This coefficient is an alternative to the Spearman rank correlation as an indication of the dependency between to ordinal variables.

Partial Correlation

The partial correlation is a measure of the linear dependency between two variables, where the influence of a third variable is "partialled out." If one suspects that the correlation results of two variables are being clouded because they are both related to a third variable, this command can be used to find out. As with Pearson correlation, the variables must be measured at the interval level. An example from one of the sample files may illustrate the principle.

Calculating first the Pearson correlation between the number of children in the families sampled and the amount of money paid for the family car, the result is –0.226. Repeating the calculation, but partialling out the family income, the correlation coefficient jumps to –0.547 and becomes much more significant. One now has reason to believe that, *for a given income*, an increasing number of children has a negative effect on the amount of money spent on the family car. The effect was not as obvious before partialling out the family income, because family income itself plays a signicant role in the number of children *and* the amount spent on the car.

Cross-Correlation

The cross-correlation is closely related to *Time Series Analysis*, and may best be described by an example. Suppose we take temperature readings every *x* minutes in two neighboring rooms. One of the rooms is heated and cooled directly and the adjacent room indirectly through its proximity to the first room. Obviously, the two temperature variables will be highly correlated, but we would expect the temperature in the second room to lag behind the temperature in the primary room by some number of time units. That is, the correlation would be even higher if we could shift the variables in time with respect to one another. This is exactly what a cross-correlation accomplishes. It shifts the variables one unit (case) at a time forward and backward and calculates the resulting Pearson correlation coefficient, up to a predetermined number of *lags*.

Simple Regression

This procedure look for simple equations relating a given dependent variable to a single independent variable. Various classes of equations (linear, power, exponential, logarithmic, and hyperbolic) may be taken into consideration. The results of the analysis appear as a table in which the best-fit coefficients for each of the equation types is given. The regression coefficient for each equation is also listed, giving an indication of the goodness-of-fit. The best curve found can also be plotted against the data, as in Figure 48.4.

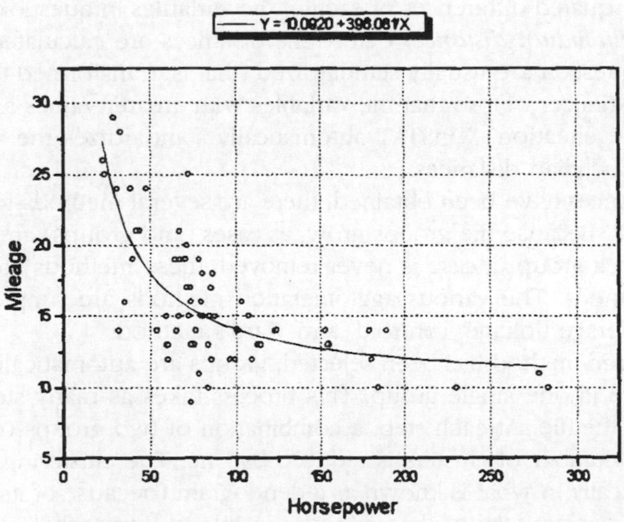

FIGURE 48.4 A best-fit simple regression curve.

WinSTAT also allows the calculation of the residuals, that is, the differences between the actual data points and the values as yielded by the best-fit curve. Various tests can be perfomed on the residuals that may lead to more information about the quality of the best-fit function.

Polynomial Regression

This procedure looks for the best polynomial equation relating a given dependent variable to a single independent variable. The degree of the polynomial to be taken into consideration may be freely chosen. For each term, the calculated coefficient is displayed as well as a confidence interval for the coefficient. As in simple regression, the polynomial function can be plotted against the data.

Multiple Linear Regression

This procedure looks for the best linear equation relating a given dependent variable to any number of independent variables. It is also possible to choose among several methods of stepwise regression, in which the program decides on its own, according to certain criteria, which of the independent variables are to be included in the regression equation. The possible choices are stepwise, forward, backward, or maximum *R*-square.

Cluster Analysis

Cluster analysis refers to methods which attempt to group cases in such a manner that the members of each group are, in some sense, "close" to one another. Several variables may be chosen for the analysis, and the differences in these variables between two cases determine the "distance" between the two cases.

The most common method of calculating the distance is, for the two cases in question, to sum the squared differences of each of the variables in question. This is known as the *squared Euclidian distance*. Before the distances are calculated, the values of all variables in question are usually standardized (that is, transformed to variables with mean = 0 and variance = 1), so that the variables with greater values and variances do not dominate the equation. WinSTAT automatically standardizes the variables before calculating the Euclidian distances.

Once the distances have been obtained, there are several methods for collecting the cases into groups. Because the groups grow, as cases (and groups) are combined, and because, once in a group, a case is never removed, these methods are known as *agglomeration* methods. The various agglomeration methods are simple linkage, complete linkage, average linkage, centroid, and Ward's method.

Once the desired method has been selected, groups are automatically agglomerated until all cases are in one single group. This process takes as many steps as there are cases present in the file. At each step, a combination of two groups (clusters) occurs, and so the total number of clusters is reduced by one. The clustering process can be displayed graphically in what is known as a dendogram (because of its tree-like form). The dendogram can be used to determine a sensible point at which to stop the clustering process. One looks for a step that combines two clusters that are significantly farther apart than the clusters that were combined in previous steps.

Once the user has decided on the number of clusters to keep, the *Cluster Separation* command may be used to assign cases to respective groups. This command automatically adds a new variable to the worksheet. The new variable indicates, for each case, the group to which that case belongs. One may then have a look at the clusters created by graphing a scatterplot and grouping the data according to the newly created variable (Fig. 48.5).

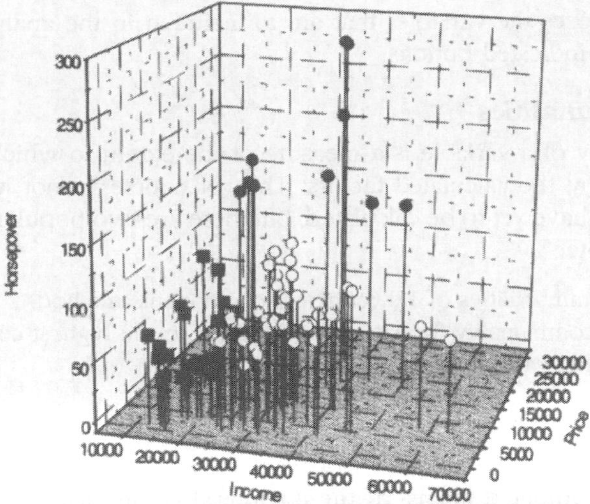

FIGURE 48.5 Scatterplot with clusters.

Factor Analysis

Factor analysis is used to look for basic, independent dimensions underlying the existing variables. One tries to find the smallest set of such dimensions that nevertheless explains the variables to a sufficient degree. There are many options to consider when preparing a factor analysis and often a satisfactory solution will be found only by repeating the analysis a number of times with different parameters. The user should have a general idea ahead of time which variables are to be explained and the approximate number of factors that may lie behind these variables.

Factor analysis is a good example of one of the more complex capabilities of WinSTAT, and so we demonstrate it here in more detail. The Factor Analysis dialog box appears in Figure 48.6.

Factor analysis

Variables:
Size of Household
Children
Income
Price
New or Used
Horsepower
Size
Type
Mileage
Price[2]

All

Factor extraction
15 Maximum number of factors
1.0 Minimum eigenvalue per factor

OK
Cancel
Help

Calculated communalities
☒ Repeat analysis with new communalities
until residues are
smaller than: 0.05
or 10 repetitions

Assumed communalities
○ 1.0
◉ Highest correlation

Rotation
☒ Kaiser normal
◉ Varimax
○ Quartimax
○ No rotation

Output
☒ Table of eigenvalues
☐ Unrotated factor loadings
☒ Sort

FIGURE 48.6 The Factor Analysis dialog box.

First, one chooses the variables that are to included in the analysis. Then, one selects among the indicated options:

Initial Communalities

The communality of a variable is a measure of the extent to which that variable can be determined by the calculated factors. This, of course, is not known in advance, since the factors have yet to be calculated, but there are two popular methods for making a first estimate:

1. Assume communalities of 1.0 (principal components method).
2. Assume a communality for each variable equal to the highest correlation coefficient involving that variable (principal axis method).

Revised Communalities

The factor analysis, when complete, delivers values for the actual communalities as yielded by the computed factors. If the computed communalities do not correspond well with the estimated communalities, the user may wish to repeat the analysis with the revised values, since the initial assumption was apparently weak. This process can be continued until a given correspondence level is reached. A standard acceptance level is a difference no greater than 0.05 between the estimated values and the computed values, but there is no guarantee that this level can be reached with a restricted number of factors.

Factor Extraction

A decision must be made as to how many factors are to be extracted. One standard method is to have the program continue extracting factors as long as the calculated eigenvalue of the factor is greater than 1.0. This is equivalent to saying that each factor must contain at least as much information as any single variable. Of course, any other minimum eigenvalue may be specified. One may limit the maximum number of factors extracted in this process by filling in the appropriate field. Alternately, a predetermined number of factors may be extracted. In general, the desired number of factors will also depend on how well the calculated factor loadings can be interpreted. Thus, several analyses can be performed, each with a different number of factors, and a subjective decision reached as to which is best.

Rotation

One goal of factor analysis is to be able to interpret the factors as they relate to groups of variables. Since the initial calculation of the factors is in a more or less random orientation with respect to the variables, it is desirable to manipulate the factor axes in such a way that the majority of variables lie close to one of the axes. This process is called factor rotation, and the two most popular methods are **Varimax** and **Quartimax**. Before rotation, the Kaiser normalization may be applied to the factor loadings. This procedure ensures that the variables with large communalities are given greater weight during the rotation. Finally, the table of rotated factor loadings may be sorted, making it easier to recognize which variables are grouped with which factor.

Output

Output of the table of eigenvalues and output of the unrotated factor loadings may be suppressed, if desired.

In the following example, variables relating to the sampled families and their cars were selected. The *Highest correlation* option was chosen for the initial communalities, and the *Revised commualities* option was selected. Also, the output of the unrotated factor loadings was suppressed.

```
Valid cases: 72

ANALYSIS 1 EIGENVALUES:

                              Communal.    Communal.
                              assumed      calculated

Size of Household             0,96877      0,94547
Children                      0,96877      0,93476
Income                        0,68760      0,64626
Price                         0,79565      0,80862
New or Used                   0,70064      0,34325
Horsepower                    0,79565      0,80691
Size                          0,62759      0,53406
Type                          0,70008      0,49499
Mileage                       0,62288      0,50538

                              Variance     Percent
Factor      Eigenvalue        percent      cumulative
  *1        3,37929           37,5476      37,5476
  *2        2,64045           29,3383      66,8859
   3        0,75781           8,42011      75,3061
   4        0,32935           3,65946      78,9655
   5        0,15617           1,73529      80,7008
   6        0,14868           1,65206      82,3529
   7        6,07045e-02       0,67449      83,0274
   8        1,68733e-02       0,18748      83,2148
   9        1,19448e-02       0,13272      83,3475

ANALYSIS 2 EIGENVALUES:

                              Communal.    Communal.
                              assumed      calculated

Size of Household             0,94547      0,95088
Children                      0,93476      0,93195
Income                        0,64626      0,64076
Price                         0,80862      0,82705
New or Used                   0,34325      0,26988
Horsepower                    0,80691      0,83179
```

```
Size                                    0,53406      0,51616
Type                                    0,49499      0,44932
Mileage                                 0,50538      0,47866

                                        Variance     Percent
Factor          Eigenvalue              percent      cumulative

 *1             3,30942                 36,7713      36,7713
 *2             2,58707                 28,7452      65,5165
  3             0,54650                 6,07222      71,5887
  4             0,28708                 3,18984      74,7786
  5             0,26452                 2,93912      77,7177
  6             0,20114                 2,23498      79,9527
  7             4,89672e-02             0,54408      80,4968
  8             2,79151e-02             0,31016      80,8069
  9             4,08794e-03             4,54216e-02  80,8523

ANALYSIS 3 EIGENVALUES:

                                        Communal.    Communal.
                                        assumed      calculated

Size of Household                       0,95088      0,95662
Children                                0,93195      0,93152
Income                                  0,64076      0,63967
Price                                   0,82705      0,83631
New or Used                             0,26988      0,25693
Horsepower                              0,83179      0,84568
Size                                    0,51616      0,51163
Type                                    0,44932      0,43825
Mileage                                 0,47866      0,47121

                                        Variance     Percent
Factor          Eigenvalue              percent      cumulative

 *1             3,30400                 36,7111      36,7111
 *2             2,58384                 28,7093      65,4205
  3             0,51283                 5,69821      71,1187
  4             0,31905                 3,54507      74,6637
  5             0,29000                 3,22227      77,8860
  6             0,17574                 1,95267      79,8387
  7             5,04243e-02             0,56027      80,3989
  8             2,31390e-02             0,25710      80,6561
  9             2,69060e-03             2,98956e-02  80,6859
```

VARIMAX FACTOR LOADINGS:

	Factors 1	2	Commu-nality
Horsepower	0,91857	-0,04357	0,84568
Price	0,8956	-0,18492	0,83631
Size	0,66517	0,263	0,51163
Type	0,662	0,00177	0,43825
Mileage	-0,62888	-0,27515	0,47121
New or Used	-0,40214	0,30856	0,25693
Size of Household	-0,03277	0,97752	0,95662
Children	-0,09759	0,9602	0,93151
Income	0,4062	0,68897	0,63967
Sum of squares	3,25943	2,62841	5,88784
Percent of variance	36,2159	29,2046	65,4205

In the output of eigenvalues, an asterisk is placed next to the factors that have been accepted according to the options chosen. In this case, they are the factors with an eigenvalue of at least 1.0.

Note that the factor analysis had to be repeated three times before the residuals of the communalities became acceptable. Two factors were extracted, and the rotated factor loadings yield the following groups of variables:

Factor 1 (Car)	Horsepower
	Price
	Size
	Type
	Mileage (negative)
	New or Used (negative)
Factor 2 (Family Situation)	Size of Household
	Children
	Income

In all, 65% of the variance of all selected variables can be explained by these two factors. WinStat can also produce a 2-D or 3-D graphical display of the factor loadings, as in Figure 48.7.

Discriminant Analysis

In discriminant analysis, it is assumed that the data at hand are grouped according to a given nominal-level variable. Then, any number of further variables are examined to see to what extent they can be used to "predict" the value of the grouping variable. The similarity to multiple regression is apparent, the difference being that the variable to be "predicted" in multiple regression is interval-level.

Once the analysis has been performed, the derived equations can be used to classify cases that have not yet been assigned to groups. A classic example is the bond market. First, discriminant analysis is performed using bonds of known quality (**AAA, AA**, . . . , **C**), relating these ratings to various financial indicators of the respective com-

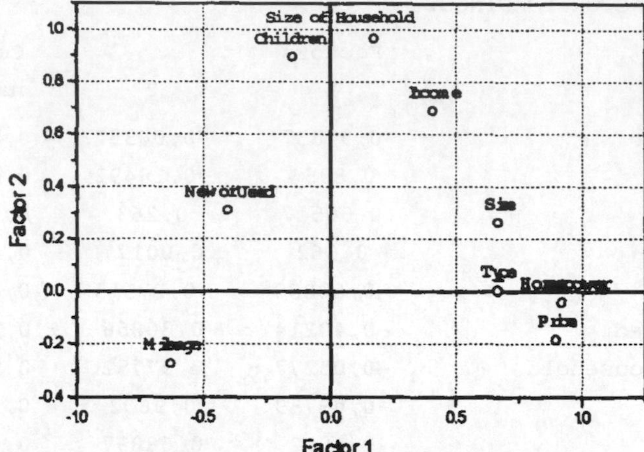

FIGURE 48.7 A plot of factor loadings.

panies. Following this, the financial indicators of previously unrated companies can be fed into the equations to get bond ratings for those companies.

The results include the eigenvalues of the discriminant functions, the canonical correlation, Wilks' lambda, the standardized coefficients of the discriminant functions, the values of the discriminant functions at the group centroids, and the Mahalanobis distances between the groups.

Kaplan–Meier Survival Analysis

Survival analysis is used in medical research to plot life expectencies for patients undergoing different types of treatment. It is also used in industrial quality control to analyze time-to-fail data. A *life table* is calculated, based on the data at hand. This table presents the probability of individual survival throughout the entire time span of the experiment. The data may also be displayed graphically as a survival probability curve. WinSTAT also allows the comparison of two survival curves, indicating the significance of the differences.

Time Series Analysis

The goal of time series analysis is to predict future values of a given variable by looking at its behavior in the past. Obviously, this is possible only if the variable displays patterns that can be recognized by the mathematical methods at one's disposal. The analysis itself is rather complex and involves several steps.

In the first step, the variable undergoes appropriate transformations to turn it into a *stationary variable.* A stationary variable is one that has constant variance and displays no trend or periodicity. WinSTAT offers several tools and transformations to create such a variable. Among the tools is the *periodogram*, which is a graph of the frequencies contained within the variable (using Fourier analysis).

In a second step, various functions are used to try to model the data of the stationary variable at hand. These functions are collectively called the *Box–Jenkins model*, and include an autoregressive (AR) and a moving average (MA) part. Again, WinSTAT provides tools to decide which parts are necessary for appropriate modeling of the data. The most important of these are graphs of the *autocorrelation function* and *partial autocorrelation function* of the stationary variable.

In a third step, the program calculates the best-fit coefficients for all Box–Jenkins models that are being taken into consideration. The user may decide on the best model by evaluating the resulting goodness-of-fit statistics.

As a last step, future values of the original variable can be calculated and plotted by the program, using the model decided upon by the user. This is called *forecasting*.

48.5 GRAPHICS CAPABILITIES

WinSTAT produces many kinds of graphics. Some have been used as illustrations in the above section on statistics capabilites. All graphics may be edited extensively by the user in an interactive manner. There are three basic methods of making changes:

1. The mouse: Users may double-click on any element (axis, text, legend, data point) to open a dialog box that allows them to change the characteristics of that element. The mouse can drag any free-standing text to another position. A single-click on an axis opens a size box, allowing the absolute size of the entire plot to be changed.
2. The graphics tool box: Selecting the appropriate tool from the tool box allows the user to add new elements. Text can be entered at any position and rotation, using any of the available fonts. Arrows, circles, and rectangles may be added at will, with a large number of color and shadow effects. There is also a magnifying glass to zoom in on plot details.
3. The menus: Whenever a graphics window is active, the menu bar displays commands relating only to graphics windows. Using these commands, one can change characteristics not directly accessible by the mouse.

Three-dimensional plots include an additional tool box, allowing one to rotate the plot in space and change the perspective.

Histogram

A histogram is a display of the number of cases per class of a given variable. Figure 48.8 shows the histogram of just one variable.

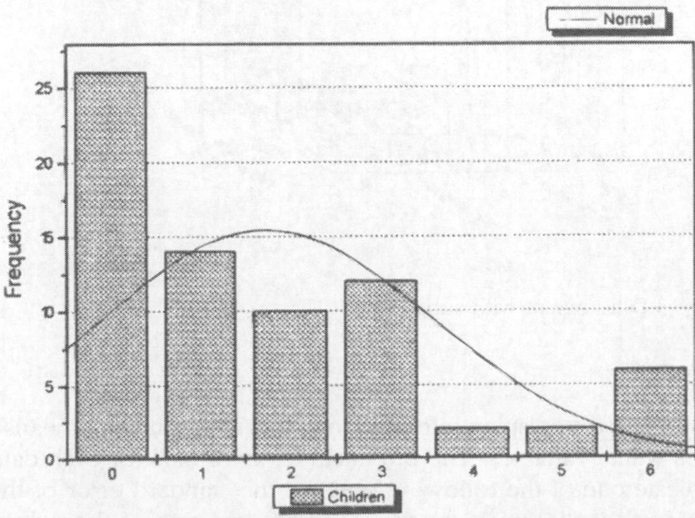

FIGURE 48.8 Histogram with normal curve.

If a single variable is specified, the program automatically superimposes the best-fit normal curve for the data. Figure 48.9 shows an example with two variables.

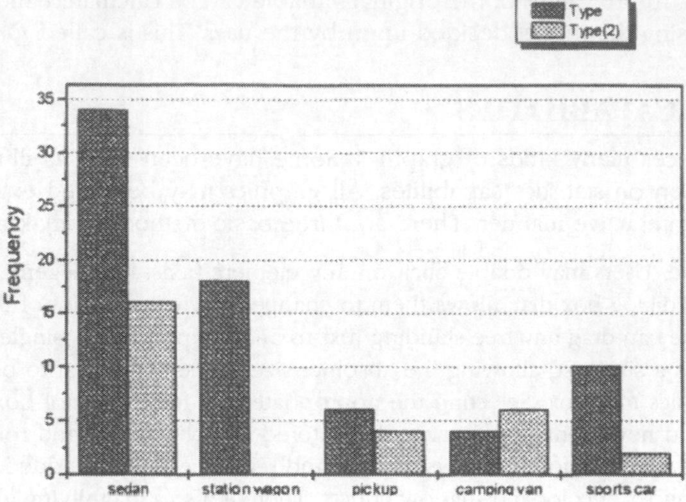

FIGURE 48.9 Histogram with two variables.

All histograms may be plotted in 3-D (Fig. 48.10).

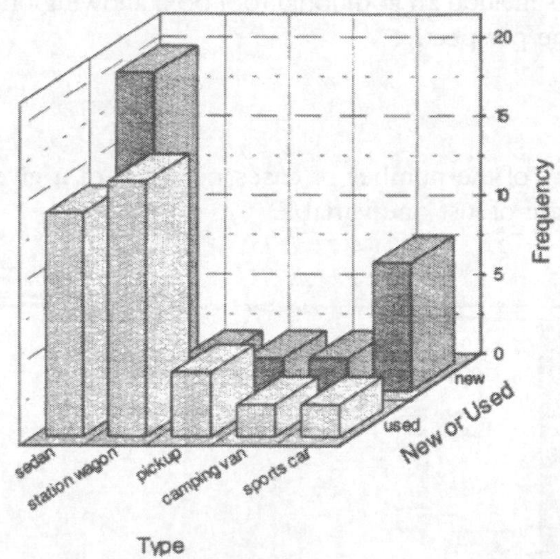

FIGURE 48.10 3-D histogram with subgroups.

Plot of Means

This command is used to display information about the arithmetic means of variables and subgroups within variables. The plot includes error bars for each data point. The error bars indicate one of the following options: the standard error of the mean, the confidence interval of the mean, or the standard deviation of the subgroup. Figure 48.11 shows the mean price of various kinds of cars.

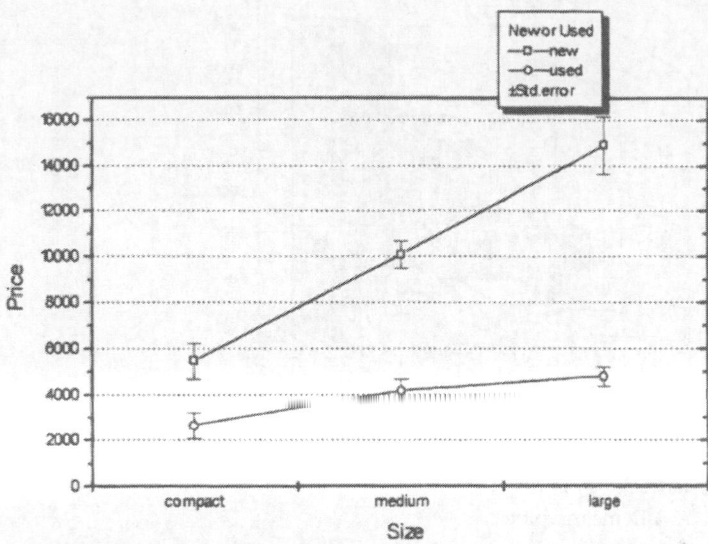

FIGURE 48.11 A means plot of subgroups.

Various visual effects can be chosen. Choosing *Column/Bar*, for example, results in the plot shown in Figure 48.12.

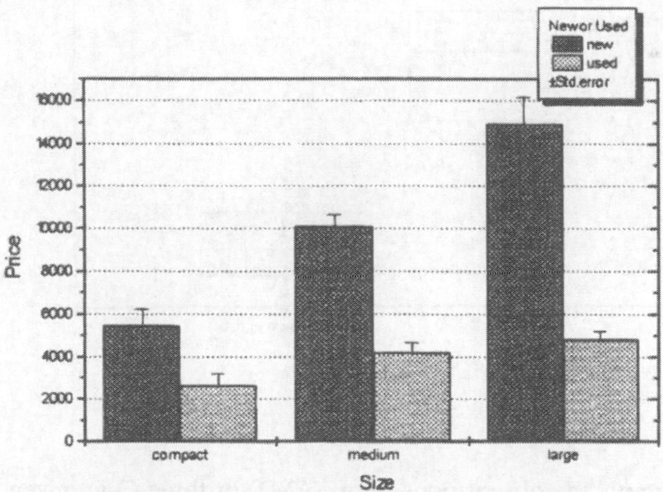

FIGURE 48.12 A means plot as columns.

All means plots may be created in 3-D (Fig. 48.13).

Box-and-Whisker Plot

This plot yields information regarding percentiles instead of means. Figure 48.14 displays the box plots of two variables. The line within the box represents the median of the given variable. The bottom and top edges represent the 25th and 75th percentiles. In other words, 50% of the data fall within the box, and 25% each above and below. The "whiskers" extend to the 5th and 95th percentiles.

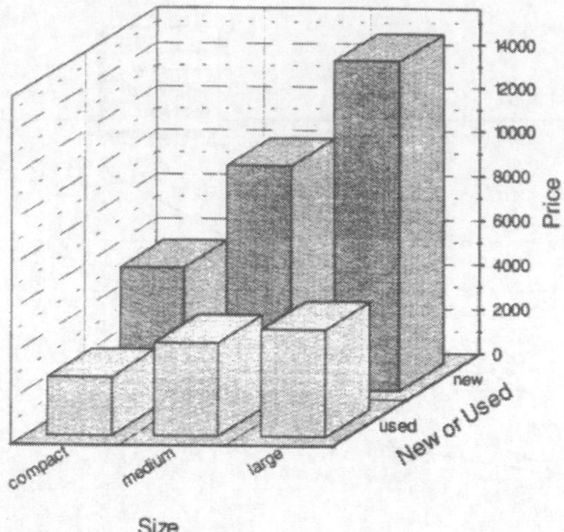

FIGURE 48.13 A 3-D means plot.

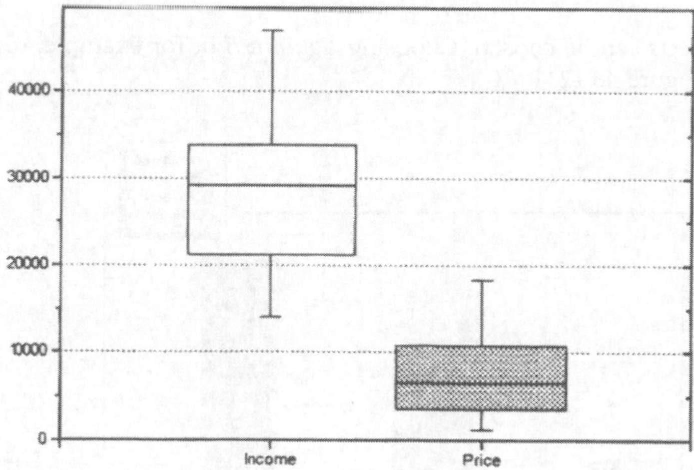

FIGURE 48.14 A box and whisker plot.

Scatterplot

In this type of graph, the observations of two (2-D) or three (3-D) given variables are plotted case for case against each other. In addition, the plotted points may be grouped according to the values of an additional, grouping variable. Figure 48.15 shows an example of a scatterplot with a grouping variable.

Figure 48.16 shows an example of a 3-D scatterplot.

Cumulative Frequency Plot

This plot displays the cumulative distribution curve (in percent) for a single variable. In addition, it superimposes the best-fit normal distribution curve, so that one can make a visual comparison. After selecting the desired variable, the plot appears as shown in Figure 48.17.

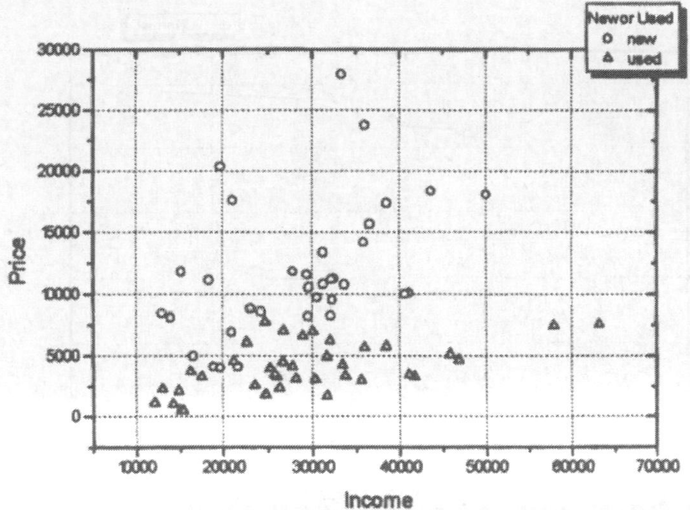

FIGURE 48.15 A 2-D scatterplot with groups.

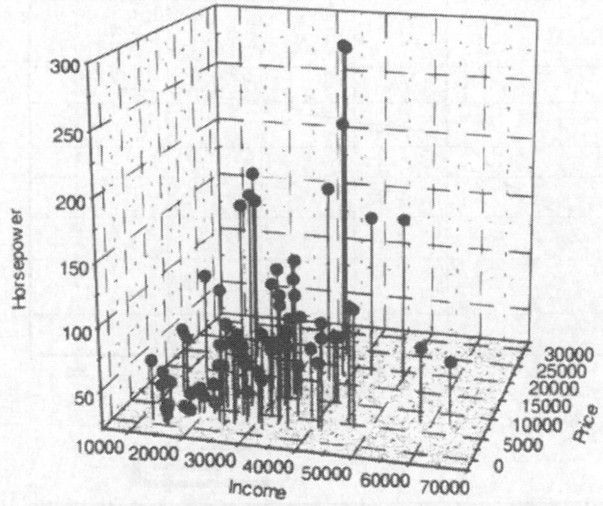

FIGURE 48.16 A 3-D scatterplot.

The *Y*-axis may be transformed to create a *Probability Plot*. The transformation is such that a normal distribution will plot as a straight line (Fig. 48.18).

Time Series Plot

The time series curve is simply a plot of the variable values (*Y*-axis) against the case numbers (*X*-axis). It is assumed that the case numbers reflect the time sequence for a variable that has been measured regularly (Fig. 48.19).

Quality Control Charts

WinSTAT offers several types of charts for use in quality control. Each type of chart takes data from one variable, which are then subdivided into groups of equal size and evaluated. The nature of the data, group size, and evaluation differs from one chart

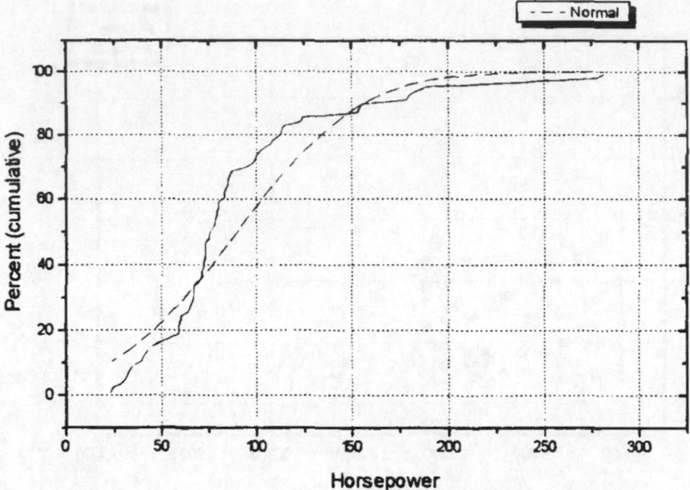

FIGURE 48.17 Cumulative plot with normal distribution.

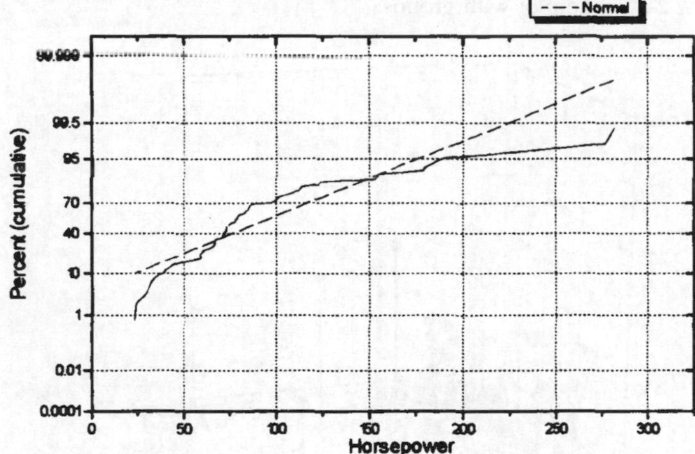

FIGURE 48.18 Probability plot.

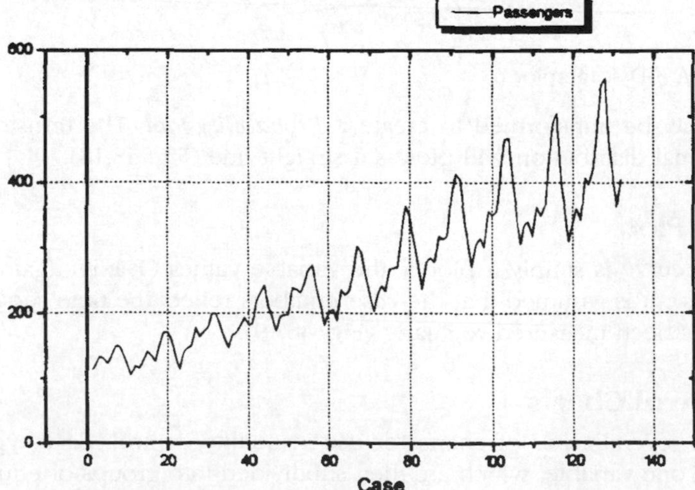

FIGURE 48.19 Time series plot.

type to the next. The chart types supported are X-bar chart, R chart, S chart, p chart, np chart, c chart, and u chart. Figure 48.20 shows an example of an X-bar chart. In the illustration, there are 25 data points, each one showing the average diameter of the 5 piston rings in the corresponding sample. The labels LCL and UCL refer to *lower control line* and *upper control line*. The corresponding lines indicate limits outside of which the process is considered to be out of control. The limits themselves are determined by a value *sigma* set by the user.

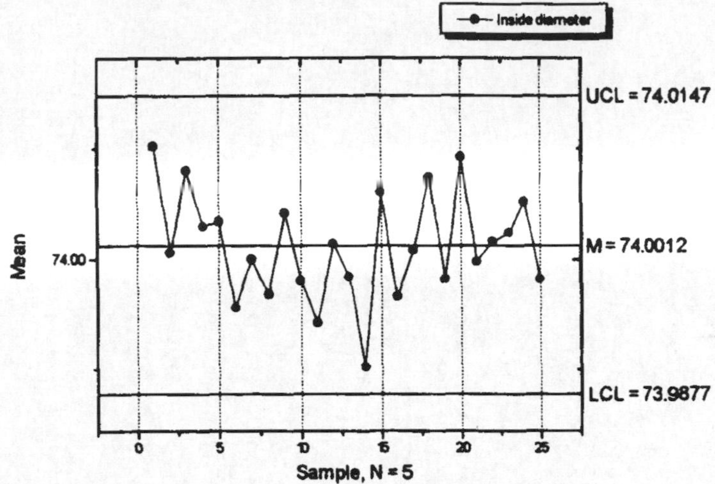

FIGURE 48.20 X-bar chart.

DEFINING TERMS

Interval Level: The variable measures some linear property. An example would be height. The difference between 65 in. and 70 in. is the same as the difference between 70 in. and 75 in.

Nominal Level: The data represent codes for nonnumeric properties, for example, occupation (1 = barber, 2 = engineer, etc.).

Ordinal Level: The variable indicates a judgment where linearity cannot be assumed. For example, consider school grades A through F (coded 1 through 6). While it is true that A is a higher grade than B and B is a higher grade than C, it cannot be claimed that the difference from A to B the same is as the difference from B to C.

Significance: This refers to the overall strength of the findings. For example, a difference in the arithmetic mean of a variable between two groups will be more significant the larger the difference is and the greater the number of subjects in the groups. The significance is usually expressed as a p value. This is the probability that the observed difference would occur by chance alone when in fact the two groups come from the same population. Thus, a low p value indicates that it is unlikely that the two groups come from the same population. Instead, it is likely that there is an inherent difference between the two groups. In other words, a low p value indicates significant findings.

REFERENCE

Reference Manual, WinSTAT 3.0. Kalmia Co Inc., Cambridge, MA.

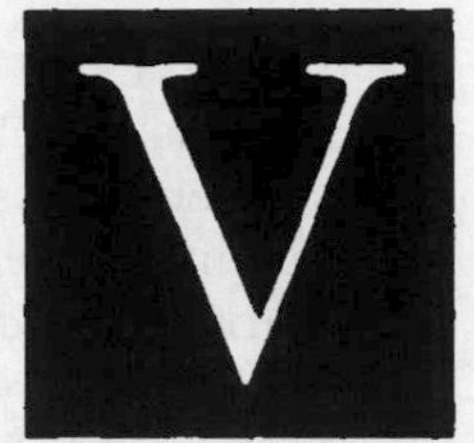

Engineering Tools

Today, computers have a very major role in every area of electrical engineering and artificial intelligence. The software environments presented in Chapters 49 to 56 constitute very useful tools for a wide spectrum of fields.

SPICE, perhaps the most popular simulation environment for electrical circuits and networks, has gone through numerous stages of maturity; today every sophomore electrical engineering student is unable to do without it in his or her network courses.

Software environments MATRIX$_x$, CADACS, and MATLAB all have a common goal—they treat analysis and design of control systems. MATLAB, or MATrix LABoratory, was originally intended for applied and numerical mathematics with special emphasis on matrix theory, but today several computer-assisted control systems programs are based on MATLAB. MATRIX$_x$, which stemmed originally from MATLAB, is arguably the most elaborate of those used in industry and research organizations around the world. MATLAB is now perhaps the most common tool for control systems, primarily due to its availability and relatively low price. Both MATRIX$_x$ and MATLAB are accompanied by powerful graphics interfaces such as SYSTEM BUILD or SIMULINK, which make the treatment of even nonlinear or user-defined algorithms easier.

CADACS is arguably the most comprehensive non-MATLAB-based software environment for control systems. It is intended for more serious users with practical and/or industrial use to follow. For example, CADACS is in use in many German manufacturing processes today, but it does have a steep learning curve.

FULDEK, or FUzzy Logic DEvelopment Kit, is a simulation environment for fuzzy logic control. It is a user-friendly program designed for the average user as well as for the sophisticated users at the professional level in real-time control applications. FULDEK has four versions: Demo, Starter Kit, Standard, and Professional, in order of complexity and cost. It generates codes in FORTRAN, BASIC, C, MATLAB, and Ada. Within the realm of artificial intelligence the MATLAB's Neural Networks toolbox is also a very useful program. It can be used in neural networks and intelligent (neural) control.

The last two software environments, Vision and Khoros, are programs that target upper-class engineering students and professionals in the fields of vision, image, and signal processing. Khoros is available on Unix-based systems and has a wide spectrum of users around the world.

Finite element analysis systems are tools for the mechanical engineer for the analysis of structures or the solution of differential equations. Algor and ANSYS, two such tools, are discussed in Chapters 57 and 58.

CAD/CAM systems are to the engineer, draftsman, and designer what the word processor is to the writer. Chapter 59 covers COGO, a coordinate geometry system designed primarily for civil engineers. With COGO, field survey information is integrated for various mathematical tool to enter data into a computer-aided design system. SmartCAM, discussed in Chapter 60, is a CAD/CAM package used primarily by mechanical engineers for design, portrayal, and fabrication of complex machined parts.

49

PSpice® Circuit Analysis and Schematic Capture

Dan Moore
Rose-Hulman Institute of Technology

49.1 INTRODUCTION

The early days of analyzing electrical and electronic circuits using large "mainframe" computers was anything but simple. The need to streamline the process and provide a more "friendly" user interface led to the development of various circuit simulators. The *SPICE2* simulator developed at the University of California at Berkeley around the mid-1970s has become the "basis" for many of the simulators on the market today. The *SPICE2* software was developed using public funds and, as such, is termed "public domain software," therefore allowing free access to U.S. citizens. The software, however, was developed for the specific needs of the research community at U.C. Berkeley and was not intended to be supported like commercial software through routine updates and consulting services. The recognition of the large market for this type of circuit simulator software led to the development of various commercial packages that were based on *SPICE2*. These packages offered extensive consulting support, updates, and improvements designed to meet the needs of the industrial and academic communities. The early commercial programs, while easier to use than the original "canned" analysis programs, still required large mainframe type computers and included HSPICE, IG-SPICE, and I-SPICE[1]. The advent of the personal computer (PC) in the mid-1980s led to the development of a number of simulators that could run on the PC. These PC-based programs, like their mainframe counterparts, are based on the *SPICE2* software and continue the "alphabet soup" naming convention. The more common ones include PSpice®, IS-SPICE, AccuSim, Spectre, and SPICE-Plus[1].

PSpice® by MicroSim will be the basis for this presentation on PC-based circuit analysis tools. The software was originally introduced in the mid-1980s and has been

0-8493-2530-7/96/$0.00+$.50

continuously upgraded and improved since that time. Evaluation version 6.0 of "The Design Center®" including Schematic Capture for use with Microsoft Windows 3.1 was used to develop the circuits and graphs included in this brief tutorial. MicroSim currently offers both full and evaluation versions of the PSpice® software that can run on IBM-compatible PCs, Sun and HP9000 Workstations, and Macintosh computers (MicroSim has indicated that Macintosh application support will not be included in future versions). The evaluation version of "The Design Center®" includes PSpice® for mixed analog/digital analysis, Schematics Capture for circuit development, and Probe for graphical displays of the analysis results. A Filter Synthesis package is also available in both full and evaluation versions that assists in the design of both active and passive filters. Information on obtaining the various packages can be obtained from MicroSim directly by calling (800) 245-3022 or writing to MicroSim Corporation, 20 Fairbanks, Irvine, CA 92718.

49.2 THE DESIGN CENTER® OVERVIEW

The "heart" of the analysis package is the actual PSpice® program. The program will allow inputs from a number of circuit drawing packages as well as a text file describing the circuit or circuits to be analyzed. The program checks for major errors, such as open circuits, improper input statements, or analysis statements, but *does not* check for "valid" circuit element values or connections before running the simulation. You are responsible for checking the circuit description and reviewing the output data to see if it "appears" correct. The analysis is done using various "standard" device equations and numerical analysis techniques. You have the choice, in many instances, to select the level of the device model used for the analysis along with the component tolerances. The results of the simulation are automatically included in an output text file and, if a ".Probe" statement is included in the circuit description, in a "*.dat" file.

The first and most important item to remember is that PSpice® is an *analysis* tool and not a *design* tool. That is, the primary function of the software is to perform an analysis of a specific circuit in a batch type process and not in an interactive process. The speed, flexibility, and analysis capability of the software make it a very strong support tool for circuit design, but the user must have the necessary background to perform the actual design along with the knowledge of what changes to make based on the analysis results.

Two additional primary tools for use with PSpice® are the Schematic Capture program for actually "breadboarding" the circuit on the computer and the Probe program for graphical displays of the analysis results. While neither of these programs is required, they provide a very powerful set of tools for automatic development of the circuit description file and a visual method for use in interpreting the results.

The filter synthesis package, unlike PSpice® is primarily a design tool. The package is a stand alone DOS-based application that can interface with PSpice® or provide results directly without additional analysis packages.

All the PC-based PSpice® packages require a minimum 386-based processor, the related math coprocessor, and some extended memory. The amount of extended memory depends on the specific application with 8 Megabytes recommended for the complete package including PSpice®, Probe, and Schematic Capture. The evaluation versions of the software are basically identical to the full versions, but are limited in the circuit size, complexity, and/or number of active components that can be included. The evaluation versions also include a small subset of the device models and vendor part specifications available with the full versions.

49.3 CHAPTER ORGANIZATION

The presentation will be divided into four main sections. Section 49.4 will discuss the primary operation and analysis functions that can be performed using PSpice®. This section will include some sample circuit file descriptions along with the required specification formats that are used to provide input to PSpice®. Section 49.5 will cover the Schematics Capture program including some discussion of common "pitfalls." Section 49.6 will provide an overview of Probe, the graphical post processor, and will include representative outputs from the circuit files provided in the PSpice® and Schematics Capture sections. Section 49.7 will provide a general discussion of the filter synthesis package.

The information presented in this chapter is intended to provide a starting point for the use of the circuit analysis package. The material, while not all inclusive, should give the reader a solid foundation for performing routine analysis. A list of references that can be used for more detailed information is provided at end of the chapter.

49.4 PSPICE® ANALYSIS

Microsim's "The Design Center®" is based on the PSpice® circuit analysis package. The program requires an input circuit description file that includes the circuit elements with their related parameters and connections and the types of analysis to be performed. The input file may include more then one analysis type for the circuit under test and may also include more then one circuit to be analyzed if each circuit description is separated by a ".end" line and NO BLANK or REMARK lines are placed between the ".end" and the next circuit title line. A sample input file is shown in List 1. This file listing includes several different circuit descriptions, analysis types, and library references and may be used as a reference for the discussion to follow.

Input File Specifications

The circuit description file must be a standard ASCII (text) file and must not include any hidden or extended characters. The actual file may be obtained using any text editor or word processor as long as the file is saved as a "text only" file. The file may also be produced indirectly using different circuit generation programs such as Schematics Capture or other industry programs that generate a circuit connection listing. The **FIRST LINE** in the file, or **FIRST LINE** after the ".end" statement, *must* be blank or contain the optional title of the circuit to be analyzed. Any library references should generally be included immediately after the first line, but can occur anywhere in the file as long they appear before the referenced device description. The remaining order is not important. The ".end" statement is optional unless more then one circuit description is included in the complete file, in which case it (.end) must occur between the circuit descriptions. The sample file shown in List 1 includes several "blank" or "remark" lines that are started with an "*." These lines are ignored by the PSpice® program as is any information in a line after the ";." PSpice® will allow any valid file name for the circuit description but uses a ".cir" as a default extension. The output file defaults to the input file name with a "*.out" extension and the data file for Probe has a default extension of "*.dat." While the file extensions are not required, they allow much easier access to the related information.

Component Descriptions

While the order in which the circuit elements and analysis types are listed is not important, the order of the individual description and command lines is important. The first item in the line must include a letter indicative of the element type. The standard element types are fairly straightforward, such as "R" for resistor and "C" for capacitor. The problem arises when you are including an element such as a GaAs FET, which is designated with a "B" or "K" for a transformer core. Dependent sources likewise are designated with letters "E, F, G, and H" for Voltage Controlled Voltage, Current Controlled Current, Voltage Controlled Current, and Current Controlled Voltage Sources, respectively. The user is referred to the reference sources included at the end of the chapter for more detailed information on other special designations. The part type letter is followed immediately, without any blank spaces, with additional numbers and/or letters to uniquely identify each element in the circuit. The part name is followed by a blank space and a node identification. The node may be identified by either letters or numbers. The node identifications do not have to be all numbers or letters and do not have to be continuous, i.e., nodes 0, 3, 6, A, D are completely acceptable as long as each node has a least two connections. The only exception to the above statement relates to the fact that PSpice® basically performs nodal analysis and, as such, requires a reference node. *One node must* be identified as the reference node and labeled as node "0." All node connections for each element/component are separated from each other by a blank space. The order of the connection is basically unimportant for passive elements but not for active or complex elements. An Op-Amp, for example, must have the nodes listed in the order of noninverting, inverting, positive supply, negative supply, and output. The connection information is followed by parameter descriptions. Again these descriptions may be a single value, a string of numbers, or names of reference files or statements. A review of the circuits listed in the sample file (List 1) will illustrate some of the different types of items mentioned.

There are several important items to note relating to the actual listings. PSpice® is not case sensitive and therefore the elements R5 and r5 would not result in two separate elements. The lack of case sensitivity also causes problems in the dimensional designators. The standard engineering prefixes are acceptable with two exceptions; U is used to designate micro (10^{-6}) and Meg to designate mega (10^{6}). The element value may also be expressed as E3 instead of using k. The actual element units, Ohms, Farads, etc. are not needed and may, on occasion, actually cause problems. The following two lines might be thought to represent the same size elements, but they do, in fact, represent elements differing in size by 15 orders of magnitude!

```
C1 2 3 25uf
C2 4 5 25E-6F
```

a 25 micro Farad capacitor as interpreted by PSpice®

a 25×10^{-21} Farad capacitor as interpreted by PSpice®

The primary reason for this problem relates to the fact that f represents femto (10^{-15}) and PSpice® uses the *first* letter following a numerical value as the size multiplier if appropriate and ignores all others. In the first case the first letter is "u," which represents micro, and the f is ignored. In the second case the first letter following the numerical value is f, which is interpreted as 10^{-15}.

Source Descriptions

The basic format for source descriptions follows the same rules as the other circuit elements but may contain additional parts. A voltage or current source may, for example, contain a DC value, an AC value, and a time varying value, each of which is used differently by PSpice®. The AC value is used when an AC analysis is performed, the time varying description is used when a transient analysis is performed, but the two items *do not* have to be equal. The AC analysis may be performed using a value of 5 V while the transient analysis may use a value of 250 mV. The voltage description for this might look like.

```
VS 1 0 DC 0.0 AC 5 SIN(0 250m 100 0 0 90)
```

where the sin () values represent the offset, peak amp, freq, delay, damping, and phase, respectively. Another source that would be used only for a transient analysis might omit the DC and AC portion and look like the following pulse source:

```
VA 2 0 PULSE(0V -5.2V 1us 0.1us 0.1us 1us 2.2us)
```

where the values represent the minimum, maximum, start delay, rise, fall, pulse width, and period, respectively. The important point in identifying the sources is the type of analysis to be performed. You may also indicate a DC value of "0" if a sweep of the values will be performed since the actual value used in the analysis is taken from the analysis command line and not the source description line.

Dependent sources require additional node connections along with the gain or related factor. They may involve simple linear relationships between the controlling and controlled source or complex polynomial relationships.

Analysis Commands

There are several types of analysis that can be performed on a circuit using PSpice®. You may basically assume that any analysis you require can, either directly or indirectly, be obtained. The more common types are included in the file description in List 1. A brief comment on these will be included below.

`.DC` This analysis performs either single or nested (double) sweep of a source, model parameter, element value, temperature, or global parameter. The sweep may be linear, logarithmic (by decades or octaves), or from a list. A couple of examples are shown below:

```
.DC VY 23.5 24.5 100mV; sweeps VY from 23.5 V to 24.5
V in 100mv steps
.DC VCE1 0 15 500mV IB1 0 500UA 50UA; nested sweep of
VCE and IB.
```

`.AC` This analysis is used to examine the small signal response of a circuit over a range of frequencies with the sweep being of the same type as indicated for the .DC analysis. All circuit voltages and currents are analyzed at each frequency. The source(s) must include an AC value in the description in addition to any time varying description. A couple of examples are shown below:

```
.AC lin 50 90 110; linear sweep of 50 points, start
freq, stop freq
.AC dec 25 10 10MEG; Decade sweep (25 points/dec) from
10 Hz - 10 MHz.
```

`.four` This will calculate the Fourier series components in conjunction with a transient analysis. The individual component values are included in the output file but are not available for viewing in Probe. The analysis defaults to the first 10 components.

`.temp` This command sets the temperature(s) at which the analysis is done. The analysis is repeated for each temperature listed in the command.

`.tran` This is referred to as the transient analysis command, but it does more then determine the typical transient response to a change in the input. The command actually performs a time-based analysis of the circuit from t = 0 to the final time indicated in the command line. All nodal voltages and element currents are included in the analysis. A couple of examples are shown below:

```
.tran 50ns 300us; print step and analysis duration
.TRAN 1m 150m 50m 250u; print step, duration, print
delay, step ceiling.
```

`.watch` This command allows the calculated values of up to three separate variables to be displayed on the screen while the analysis in being performed. Only one analysis type (AC, DC, TRAN) can be specified per .watch line but more then one .watch line may be included in the file, unlike the .ac, .dc, .temp, .tran command lines.

There are several additional analysis functions that can be performed, such as noise and distortion, but they are beyond the scope of this material.

Output Commands

There are basically two types of output from a PSpice® analysis. All analyzed circuits will have a *.out file generated that contains a circuit description, a listing of the parameters used in the analysis, currents through all sources, and node voltages for each nonreference node. In addition to the standard items indicated, the user may specify additional information to be included in the output file by using a ".print" and/or ".plot" command. If the user included a ".Probe" command line in the circuit file, an additional *.dat file is generated that contains all calculated variable values for the various analyses that were performed. The user does have the opportunity to limit the information stored in the *.dat file by specifying the specific variables of interest on the ".probe" command line.

`.Print` This command must be followed by the analysis type (.AC, .DC, etc.) and the variables of interest such as V(3) and I(R3). The requested information will be included in tabular form in the *.out file. There is no limit as to the number of variables listed per line, but only one analysis type may be included on any one ".print" command line. Additional ".print" lines may be included for additional analysis types. For example:

```
.print DC v(1) v(2) v(3) v(1,2) v(2,3)
.print AC VP(1) VM(1) IP(VS) V(4) IP(L1) VP(1,2)
IP(C1)
```

where the VP represents the phase and V and VM represent the magnitude.

`.Plot` This command basically follows the same format as the ".print" statement but the output is stored in the *.out file in a manner such that when the file is displayed or printed, the variable values form an apparent graph. There is a limit of 8 variables on any one ".plot" line, but there is no limit on the number of ".plot" lines that can be included in the circuit file.

`.Probe` When this command is included in the circuit file, the calculated values of the node voltages and element/component currents are included in the *.dat file for subsequent use with the Probe processor. The material may also be stored in a text type file for importing into other software that can use the "Common Simulation Data Format" structure of the text file.

Starting the Analysis

The analysis is performed by invoking the PSpice® pspice.exe program and indicating the name of the input file. The program first checks the file for major errors such as open nodes, invalid model parameters, or similar items. The error checks do not include invalid circuit construction or a check of "reasonable" values used throughout the circuit. The progress of the analysis is displayed while the analysis is being performed by indicating the type of analysis occurring, the start and end limits, and the "current" value being used. The display will indicate the completion of the analysis and, depending on the manner in which the program was started, will start Probe automatically or wait for the user to invoke it. Remember that no *.dat file for use by Probe will be generated unless a .probe statement is included in the circuit description file.

List 1

```
DC Circuit Analysis
*Simple Nodal Analysis Example
*
*Voltage Source
VS 1 0 24V
*
*Resistor elements
R1 1 2 2k
R2 2 0 30k
R3 2 3 5k
R4 3 0 10k
*
*Analysis
.DC VS 24 24 1
.options nopage
*
*Output File Information
.print DC v(1) v(2) v(3)
  v(1,2) v(2,3)
.print DC I(r1) I(r2) I(r3)
  I(r4) I(vs)
*
.End
Diode Temperature Sweep
*Temperature Sweep Example
*
```

```
AC Circuit
* RLC example
*
*Voltage Source
VS 1 0 DC 0.0 AC 50 SIN(0 50 100
 0 0 90)
*
*Resistor elements
R1 3 0 150
*
*Capacitor and Inductor Values
C1 1 2 25uF
L1 4 0 250mH
*
*Voltage Sources for Current
 measurements
VX 2 3 0.0
VY 3 4 0.0
*
*Analysis
.TRAN 500U 400m 360m
.AC    lin    50    90    110
*
*Output File Information
.print AC VP(1) IP(VS) VP(4)
IP(L1) VP(1,2)
.print AC V(1) I(VS) V(4) I(L1)
 V(1,2) I(C1)
```

```
*Voltage Source
VS 1 0 DC 0.0
*
*Voltage Sources for Current
 measurements
V1 1 2 0.0
*
*Resistor elements
R1 2 3 10mOhm
*
*Diode
D1 3 0 MDNOM
.model MDNOM D
 (IS = 1pA N = 1.5 BV = 8V)
*
*Analysis
.temp -100 0 100 200
.DC VS -10 1 50mV
*
.Probe
.End
```

```
*
.Probe
.End
```

```
Circuit Analysis for Diode
*Voltage Source
VY 1 0 24V
*
*Resistors
R1 1 2 3kOhm
R2 2 0 1.5kOhm
R3 2 3 1kOhm
*
*Diode Connection
D1 3 0 dnom
.MODEL dnom D (IS = 25pA N = 1.8
 BV = 8V)
*
*Analysis
.DC VY 23.5 24.5 100mV
*
*Output File Information
.print DC V(1) V(2) V(3) I(D1)
.Probe
.End
```

```
Clamping Circuit
*Diode Analysis Problem
*
*Input Source
VS 1 0 0 AC 4 sin(0 4 60)
*
*Diodes
D1 2 3 Dnom
.model Dnom d
 (IS = 1pA BV = 50V N = 1)
*
*Circuit Elements
R1 3 4 2kOhm
C1 1 2 50nF
VX 4 0 1V
*
```

```
ECL Logic Circuit (DC sweep)
*Logic Circuit Example
*
*Input Sources
VEE 1 0 -5.2
VA 2 0 0.0
*
*Resistor elements
R1 3 0 300ohm
R2 5 0 300ohm
R3 4 1 1.5kohm
R4 6 1 1kohm
R5 6 0 295ohm
R6 8 1 1.5kohm
R7 9 1 1.5kohm
```

```
*Analysis Routine
.TRAN 1m 150m 0m 250u
*
.Probe
.End
ECL Logic Circuit (Pulsed input)
*Logic Circuit Example
*
*Input Sources
VEE 1 0 -5.2
VA 2 0 PULSE(0V -5.2V 1us 0.1us
 0.1us + 1us 2.2us)
*
*Resistor elements
R1 3 0 300ohm
R2 5 0 300ohm
R3 4 1 1.5kohm
R4 6 1 1kohm
R5 6 0 295ohm
R6 8 1 1.5kohm
R7 9 1 1.5kohm
*
*Transistors
 (Collector-Base-Emitter)
Q1 3 2 4 MQNOM
Q2 5 6 4 MQNOM
Q3 0 5 9 MQNOM
Q4 0 3 8 MQNOM
.model MQNOM NPN (BF = 130)
*
*Analysis Routine
.tran 10ns 6u
.Probe
.End
*
*Transistors (Collector-Base-
 Emitter)
Q1 3 2 4 MQNOM
Q2 5 6 4 MQNOM
Q3 0 5 9 MQNOM
Q4 0 3 8 MQNOM
.model MQNOM NPN (BF = 130)
*
*Analysis Routine
.DC LIN VA -5.2 0 100m
.Probe
.End
BJT Transistor Curves (with LIB
 function)
*Transistor Curve Generation
*
*Input Sources
VC1 1 0 15V
IB1 0 2 DC 0.0
VCE1 3 0 DC 0.0
*
*BJT Devices*
.lib "C:\msimev60\lib\eval.lib"
Q1 3 2 0 Q2N2222
*
*Resistors
RC1 1 3 500
*
*Analysis Routine
.DC VCE1 0 15 500mV IB1 0 500UA
50UA
.Probe
.End
```

```
BJT Transistor Curves
 (with user defined values)
*Transistor Curve Generation
*
*Input Sources
VCC 10 0 15V
IB2 0 20 DC 0.0
VCE2 30 0 DC 0.0
*
*BJT Devices*
Q2 30 20 0 MQNOM
.model MQNOM NPN
 (BF = 256 CJE = 25pF
+ CJC = 8pF)
*
*Resistors
RC2 10 30 500
*
*Analysis Routine
.DC VCE2 0 15 500mV
 IB2 0 500UA 50UA
.Probe
.End
```

```
Fourier Analysis Circuit
*Voltage Source
*
V_V1 2 0 DC 5 AC 5
+ PULSE 0 5 0 1ns 1ns 25us 50us
*
*Resistor elements
R_R1 1 2 1k
R_R2 1 0 1k
*
*Analysis
.tran 50ns 300us
.four 20k 10 v(R_R2)
.End
```

```
CE Transistor circuit
*Single Stage Amplifier Example
*
*Input Sources
VS 1 0 DC 0.0 AC 75e-3 sin(0 75m
   10k)
VCC 4 0 12.0
*
*Resistor elements
RS 1 2 1Kohm
R1 4 3 225K
R2 3 0 47K
RC 4 5 5.1K
RE 6 0 1K
RL 7 0 2K
*
*Capacitor Values
C1 2 3 3.3UFd
C2 51 7 3.3UF
C3 6 0 47UF
*
*Voltage Sources for Current
measurements
VB 3 31 0.0
VC 5 51 0.0
VE 61 6 0.0
*
*Transistor (Collector-Base-
 Emitter)
Q1 51 31 61 MQNOM
.model MQNOM NPN (BF = 130
 CJE = 25pF
+ CJC = 8pF)
*
*Analysis Routine
.AC dec 25 10 10MEG
.tran 1u 200u 100u
.Probe
.End
```

49.5 SCHEMATICS CAPTURE

The Schematics Capture program requires at least 4 Megabytes of extended RAM with 8 megabits as the recommended minimum when it is used in conjunction with the other "The Design Center®" programs. The program provides a graphical user interface to the PSpice® analysis program and is only available within Microsoft Windows 3.1. A *.cir file for use by PSpice® is automatically generated by Schematics Capture.

Schematics Capture allows you to "breadboard" the circuit of interest using standard circuit symbols and conventions. The process involves the selection of "parts" from various libraries included with "The Design Center®" or user-specific libraries. The parts listing includes actual vendor and/or generic semiconductor and IC components ranging from simple diodes, BJTs, and FETs to complex circuits such as Op-Amps, Timers, and Programmable Logic Arrays. The parts can be selected using either a mouse (preferred method) or the keyboard.

Sample circuit diagrams are shown in Figures 49.1 and 49.2. The first circuit diagram (Fig. 49.1) involves UA741 Op-Amps, resistors, DC voltage sources, and ammeters. This diagram also includes several analysis-based components such as the printer symbol and the "viewpoint" for observing DC voltages at specific points within the circuit and printing them to the *.out file. The second diagram includes an FET, AC and DC

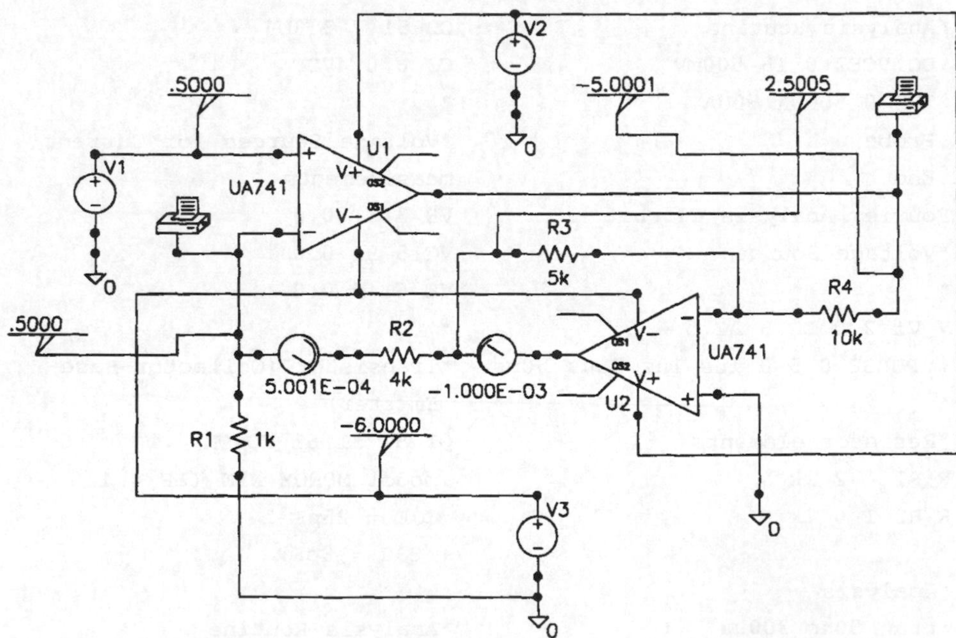

FIGURE 49.1 Schematics Capture Circuit Op-Amp diagram.

voltage sources, resistors, capacitors, and voltage and current "markers." These simple diagrams include only a small subset of the parts that can be used and do not involve any digital components or mixed analog/digital circuits. They are included to provide a reference for the following discussion.

The schematics capture program automatically increments the part identification as similar parts are added to the circuit to avoid duplication of part names. The user has the option to relabel any of the components after they have been placed. The selected

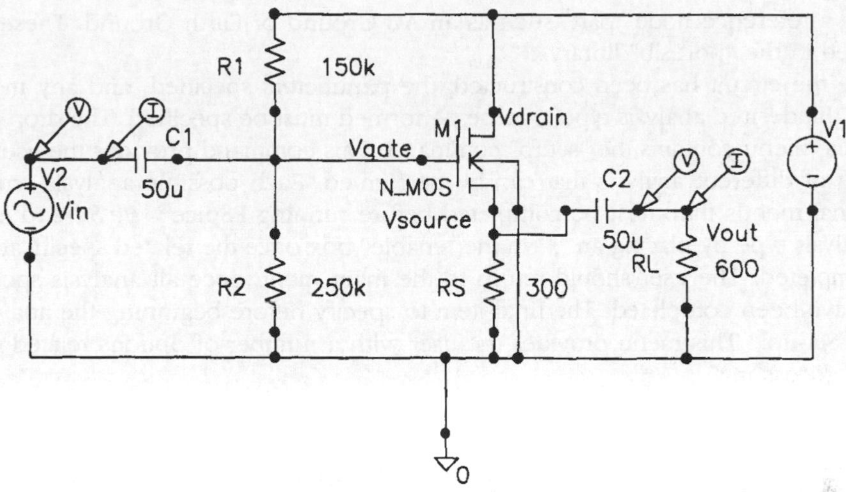

FIGURE 49.2 Schematics Capture MOSFET Amplifier diagram.

parts are all placed with default values such as 1 kΩ for resistors and 1 nF for capacitors. As with the part identification, these values can be changed by the user after the parts have been placed. The more complex parts, such as transistors, power supplies, and similar components, will have a dialog box for the user to determine the default values and to adjust those necessary for the specific circuit. In the case of vendor supplied parts, such as the UA741 Op-Amp, the part "default" values match the nominal specification sheet values. The user cannot generally change these vendor-supplied values. Users do have the option of "generating" their own part with specific values when the existing libraries do not meet the needs of the circuit. The process of "generating" a part is complex and beyond the scope of this section. The user may, however, decide to edit the "breakout.lib" file and supply the desired parameters to one of the models listed. The user can then use the edited device in the circuit diagram. The same process can be done on the other *.lib files but it must be done with care to avoid "losing" the vendor-supplied device parameters.

The process of providing the values for the elements and/or components in the circuit follows the same rules as outlined in the PSpice® component description section. The user is again cautioned as to the potential pitfalls associated with the engineering multiplier prefixes and identification labels for the various components as discussed previously. Once the parts have been placed, they must be connected or "wired" together. The parts are connected by using the "wire" tool under the "draw" menu heading. The actual connections are shown by a "solder" dot. The "wires" can cross each other without problems unless a "solder" dot appears where there should not be a connection. When this occurs the user should delete the affected line, select "redraw" (under the "view" menu) to bring back the circuit before the deletion, and then redraw the wire using a different path or being careful not to stop the drawing on the crossover.

A couple strong words of caution! Be sure to start the "wires" as close to the element or component as possible. The "leads" on the parts should not be thought of as extended wires. The user MUST start the "wires" well before the end of the element/component lead to ensure a proper connection. A good check of the procedure is the connection for series connected parts. A correctly drawn "wire" will have *two* "solder" dots between the series parts not just one. The user must also remember to include a reference node "part" such as an AC Ground or Earth Ground. These parts are listed in the "port.slb" library.

Once the circuit has been constructed, the parameters specified, and any markers added, the desired analysis type(s) to be performed must be specified. The drop down "analysis" menu contains the "setup" command. This command provides the user with a menu of different analysis that can be performed. Each possible analysis type has additional menus that must be completed before running PSpice®. *Be Sure to Enable* the analysis type by placing an "x" in the "enable" box once the related specification(s) are completed. The user should return to the main menu once all analysis specifications have been completed. The final item to specify before beginning the analysis is "Probe Set-up." This menu provides the user with a number of options related to the Probe processor.

The final step in the analysis is to select "simulate" from the "analysis" menu. This will start the process of generating the netlist and related circuit description files and run PSpice®. Once the analysis is completed, the user can observe the results in the *.out file or by using Probe.

49.6 PROBE GRAPHICAL DISPLAYS

The Probe program allows the user to display the results of the various circuit simulations graphically. The user has control over both the X axis and the Y axis in terms of scale type, variable type, and range. The display defaults to the appropriate X axis for the analysis type, such as time for a ".tran" analysis and frequency for a ".ac" analysis, but the user can change the default settings. The user may also specify additional Y axis using the same X axis on a single display and/or specify additional "plots" (X and Y axis pair) be added to the single display. The variable traces appear on screen in different colors (on a color monitor) but will, if a hard copy is desired, print with different markers automatically added if a noncolor printer is attached to the computer. Figures 49.3, 49.4, 49.5, and 49.6 illustrate several output plots associated with the example files and circuits provided under the PSpice® and Schematics Capture sections.

The Probe *.dat file contains data for all node voltages and element/component currents unless a smaller subset was selected in the analysis file. The display variables are, however, not limited to just the variables themselves. The user can "modify" them using many mathematical formulas in order to generate a more meaningful graph. The gain of an amplifier (Fig. 49.3) can be graphed in both the ratio of the input and output variables and the dB gain. The input impedance can also be displayed for a specific circuit as shown in Figure 49.4. The graphs may be used to display a family of curves typical for multiple terminal elements. Figure 49.5 shows a set of characteristic curves for the 2N2222 BJT transistor. This graph was made possible by using a nested DC sweep in the actual PSpice® analysis. Figure 49.6 shows the actual small signal, time-based currents and voltages for a single stage MOSFET amplifier. While the frequency sweeps associated with the ".AC" analysis might give a nice flat response over a wide frequency range, the time response associated with the ."tran" analysis provides a measure of the quality of the amplifier. The relative amount of distortion can also be seen graphically using the time based, ".tran" analysis.

The graph(s) can be printed directly from Probe if a printer is connected to the computer. The print function can be performed on printers ranging from a "text" only printer to a full featured color graphics printer. The amount of resolution and display quality is, however, a function of the printer; a highly detailed display on the monitor will only

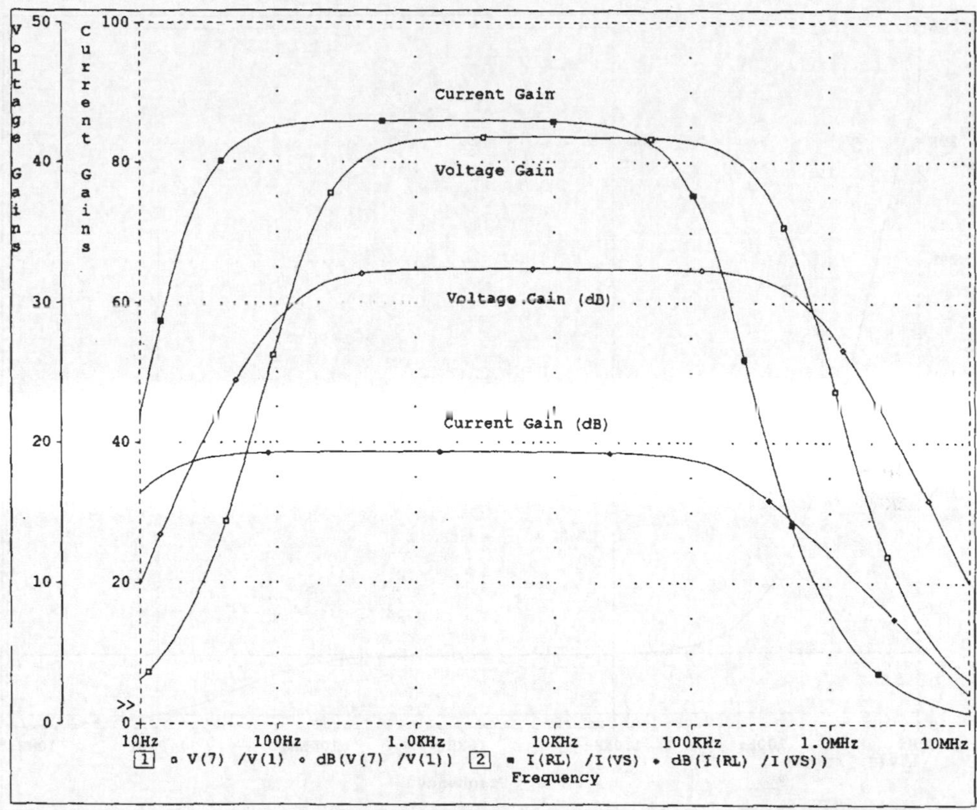

FIGURE 49.3 Probe display of BJT amplifier gains.

appear on the hard copy at the resolution of the printer. The user can specify a number of the parameters associated with the printout by selecting the page setup menu from within the "File" main menu and then selecting the related submenus. A "standard" set of page parameters is included with the software, but the user can individualize the printouts and save the desired parameters as the new defaults. The graphs included in this section were generated on an HP Laser Jet IIP with a Postscript cartridge.

As can be seen in the four figures, the Probe processor is a very powerful tool for use in understanding the circuit analysis results and has many more features then can be included here. The user is referred to the reference manuals from MicroSim as listed in the reference section for detailed information on the range of functions available.

49.7 FILTER SYNTHESIS

A Filter Synthesis Software package is also available from MicroSim but not included with "The Design Center®." The software is a DOS-based package that, as the name implies, provides the user with synthesis capability. The package also provides related analysis through a graphical plot of the filter response. The user can have the package generate a ".cir" file for use at a later time by PSpice® as well.

The package is menu driven and "walks" the user through the design process by "requesting" the filter parameters. Filter types include both active and passive and

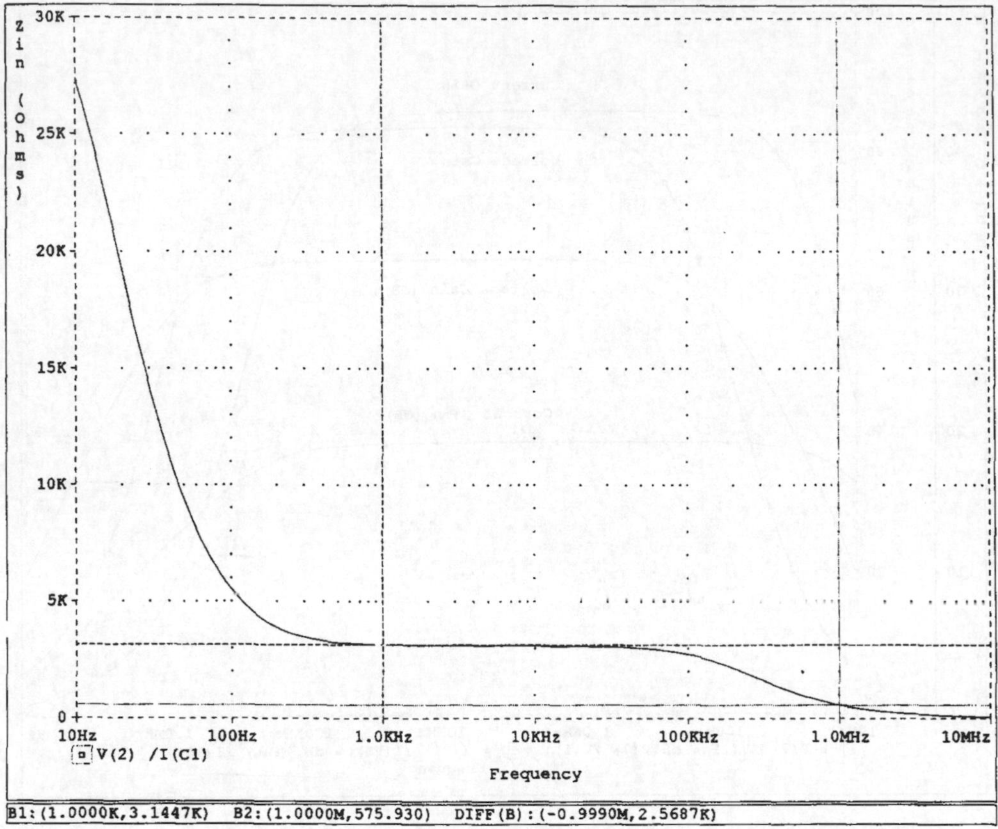

FIGURE 49.4 Probe display of BJT amplifier input impedance.

highpass, lowpass, bandpass, and bandstop. The filter implementations include RC Biquads, LC Ladders, and several Commercial IC filters. The transfer approximation or filter "type" include Butterworth, Chebyshev, Inverse Chebyshev, Elliptic, and Bessel. The user completes the initial specifications and can then make adjustments based on the order of the filter, the approximation used, or implementation.

The specifications for the filter elements are calculated once the user has completed the design parameter section. The actual circuit diagram and response curves are also available to the user. The software, as indicated earlier, can then be used to creat an input file for a more detailed PSpice® analysis such as input and output impedances, power consumption, and individual voltages and currents.

The evaluation version of the Filter Synthesis package provides all the features of the full version package in terms of plots, design, and implementation. The evaluation package is, however, limited to third order filters from a standpoint of generating a netlist for further analysis in PSpice®, saving the implementation, or editing the final design. "The Filter Synthesis User's Guide," available from MicroSim, provides a good summary of filter design process, the different types of implementations, and the basic procedure to follow in specifying the filter parameters.

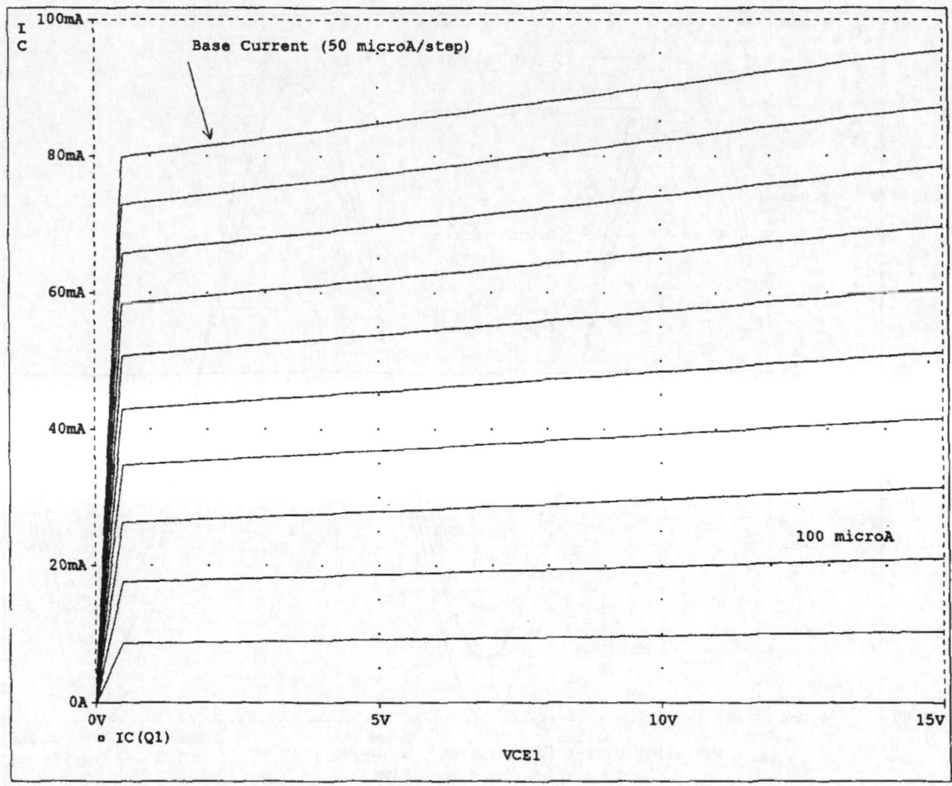

FIGURE 49.5 Probe display of BJT characteristic curves.

49.8 CONCLUSION

This PSpice® Circuit Analysis and Schematic Capture material, while brief in nature, should provide the reader with a solid basis in using the various PC and Workstation-based circuit analysis software packages available from MicroSim. There are also several other commercial packages available on the market, each of which is broadly based on the *SPICE2* simulator developed at the University of California at Berkeley, which are not discussed in this section. The reader may contact the manufacturers directly for information on their specific packages and needs. The evaluation versions of PSpice® and related software provided by MicroSim have all the features of the full versions but with a part number or circuit size limitation. The actual software discussed can be obtained from MicroSim directly and then copied freely by colleagues and/or students. The cost of the packages ranges from approximately $150 for the fully documented packages to "free" without any manuals or user guides. MicroSim will supply information about specific prices and available evaluation and full packages upon request.

The software described is very powerful and only a few of the capabilities have been included in this material. The reader is encouraged to obtain a copy of the software and "experiment" with various circuits to gain a better understanding of the analysis process and capabilities.

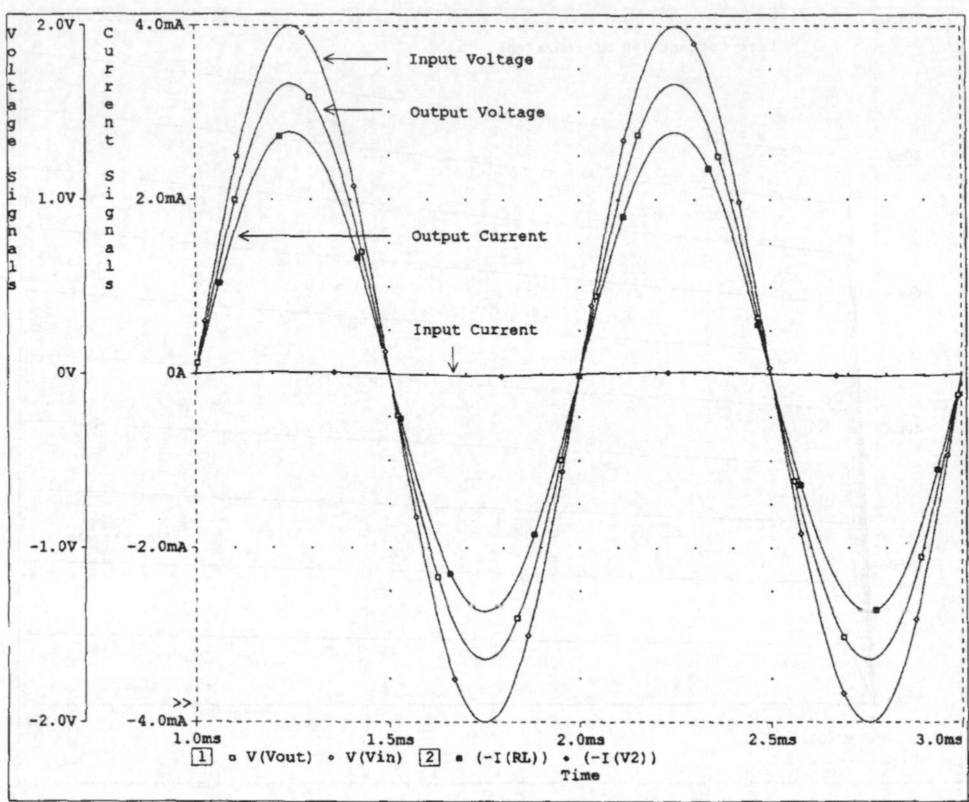

FIGURE 49.6 Probe display of MOSFET amplifier voltages and currents.

DEFINING TERMS

Bandpass Filter: Circuit that effectively "allows" signals over a range of frequencies to travel from the input of the circuit to the output while "blocking" both high and low frequencies outside the desired range of frequencies.

Bandstop Filter: Circuit that effectively "blocks" signals over a range of frequencies from traveling from the input of the circuit to the output while "allowing" both high and low frequencies outside the desired range to travel from input to output.

Breadboarding: The process of temporarily connecting the various circuit components to electrically check the circuit performance.

dB: Decibel identifier used in relation to output/input ratios. Power gain in dB is defined as 10 log (P_{out}/P_{in}) while voltage and Current gains in dB are defined as 20 $\log(V_{out}/V_{in})$ or 20 log (I_{out}/I_{in}).

Highpass Filter: Circuit that effectively "allows" high-frequency signals to travel from the input of the circuit to the output while "blocking" low frequencies.

Input Impedance: The value of the impedance appearing between (looking into) the two terminals identified as the input of the circuit.

Lowpass Filter: Circuit that effectively "allows" low-frequency signals to travel from the input of the circuit to the output while "blocking" high frequencies.

Netlist: A file that contains the listing of the circuit components and how they are connected to each other.

Nodal Analysis: A circuit analysis procedure based on Kirchhoff's current law (KCL) where the currents entering each node are summed. One equation is developed for each node not selected as the reference node.

Node: A junction of two or more components.

Open Circuit: A circuit connection that does not allow current to flow through it.

Output Impedance: The value of the impedance appearing between (looking into) the two terminals identified as the output of the circuit

Reference Node: A node chosen to be used as a "0 V" location for subsequent nodal analysis. The actual node does not need to be connected to "0 V" or other physical ground.

Schematics: The diagram of a circuit showing the symbols for the components used and the connections necessary to construct the circuit. A type of circuit roadmap.

Short Circuit: A circuit connection that does not allow a voltage to develop across it.

REFERENCES

John Keown, *PSpice® and Circuit Analysis*, 2nd ed. Merrill, New York, 1993.

The Design Center®, Circuit Analysis Reference Manual. MicroSim, Irvine, CA 1994.

The Design Center®, Circuit Analysis User's Guide. MicroSim, Irvine, CA 1994.

The Design Center®, Schematic Capture User's Guide. MicroSim, Irvine, CA 1994.

The Design Center®, Filter Synthesis User's Guide. MicroSim, Irvine, CA 1994.

Paul W. Tuinenga, *SPICE A Guide to Circuit Simulation & Analysis Using PSpice™*, 2nd ed. Prentice-Hall, Englewood Cliffs, NJ, 1992.

Andrei Vladimerescu *The SPICE Book*, John Wiley, New York, 1994.

50

Control and Embedded System Design with the $MATRIX_x$ Product Family

Alistair Adams
Tamara Artim
Martin Chorich
Ashish Gupta
Integrated Systems, Inc.

The best way to view the $MATRIX_x$[1] Product Family is as an application development methodology that begins with conceptual system design and results in a completed software program. The methodology combines embedded and control system modeling, simulation, code generation, and documentation in an integrated collection of software and hardware products. The tool set has proven invaluable in a number of roles—application development, simulation, production testing, system analysis and modeling, and general purpose research and development. Although originally intended to support control system design, the $MATRIX_x$ Product Family also plays an extensive role in general-purpose real-time embedded systems development, reflecting the fact that almost all embedded systems involve control processes of one kind or another.

50.1 $MATRIX_X$ PRODUCT FAMILY DESIGN PROCESS

The $MATRIX_x$ Product Family offers a top-down, visually based design process that sharply contrasts with the traditional bottom-up, programming-oriented approach to control and embedded systems development. In the bottom-up programming paradigm, software engineers write software that aims at achieving application performance objectives. The problem is that while software engineers may have considerable expertise in programming languages, operating systems, and computer science in general, they often lack background in how a particular class of application is supposed to operate, or the basic engineering practice underlying, for example, automotive, process control, or aerodynamic systems.

[1]AutoCode, DocumentIt, $MATRIX_x$, pSOS+, pSOSystem, RealSim, SystemBuild, and Xmath are trademarks or registered trademarks of Integrated Systems, Inc. Other product names are property of their respective owners. For more information contact Integrated Systems, Inc. at 408-980-1500.

0-8493-2530-7/96/$0.00+$.50

The MATRIX$_x$ Product Family enables incremental, interactive top-down design that empowers application designers to take applications from initial concept to final implementation. The key decisions determining finished application performance are made at the front end of the development process—during system modeling and simulation. By shuttling between modeling, simulation, and analysis, designers can take a design through many iterations before proceeding to code generation and field testing. This contrasts with traditional development where developers had to take an application all the way to field testing before they could evaluate performance and adjust design parameters. Design processes with short iterative cycles also give developers room to experiment with alternative application architectures and design parameters without committing to expensive hardware prototyping.

The block diagram modeling method allows designs to build up from pretested subsystems. Each subsystem can be defined, simulated, and tested before integration into a larger design. This "build-a-little, test-a-little" development process fosters concurrent engineering practice and means subsystems may be reused many times in an application or in future design efforts.

It is also important to note that users can move freely from tool to tool in the MATRIX$_x$ Product Family. This allows design errors to be discovered and corrected immediately, contrasting with traditional development processes where errors are uncovered late in the development process—during debugging or application testing—requiring a long trip back to the early stages of development.

The MATRIX$_x$ Product Family also speeds software development by automating many, previously manual, time-consuming activities. The product family's AutoCode® software automatically generates C or Ada source code from system models. Similarly, the DocumentIt™ package extracts information from system models, formatting it into a variety of documentation formats.

50.2 MATRIX$_X$ PRODUCT FAMILY LINE-UP

The MATRIX$_x$ Product Family (Fig. 50.1) consists of five separate packages, each handling separate phases of the application development process:

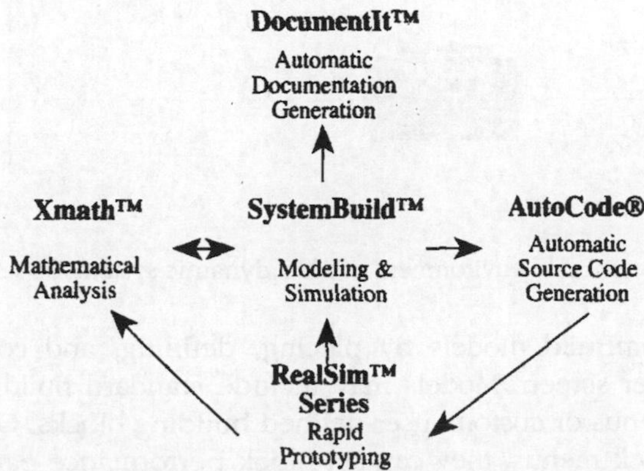

FIGURE 50.1 The MATRIX$_x$ Product Family offers an iterative control and embedded system design process combining modeling and simulation, system analysis, automatic code generation, rapid prototyping, and documentation extraction.

- **SystemBuild**™ modeling and simulation software.
- **Xmath**™ mathematical analysis and visualization software.
- **AutoCode** automatic C and Ada code generator.
- **DocumentIt** automatic documentation capture and generation software.
- **RealSim Series**™ prototyping and test software and hardware environment.

SystemBuild Modeling and Simulation Environment

SystemBuild software is a visual, block diagram-based system modeling and simulation tool for control and embedded systems design. In the SystemBuild environment, users can model applications as sets of connected block diagrams, with each individual block diagram representing the operations of a system or subsystem. The product's simulation tools let users simulate models with interactive, animated simulations enabling users to run through various "what if" scenarios and tests of the application's behavior (Fig. 50.2).

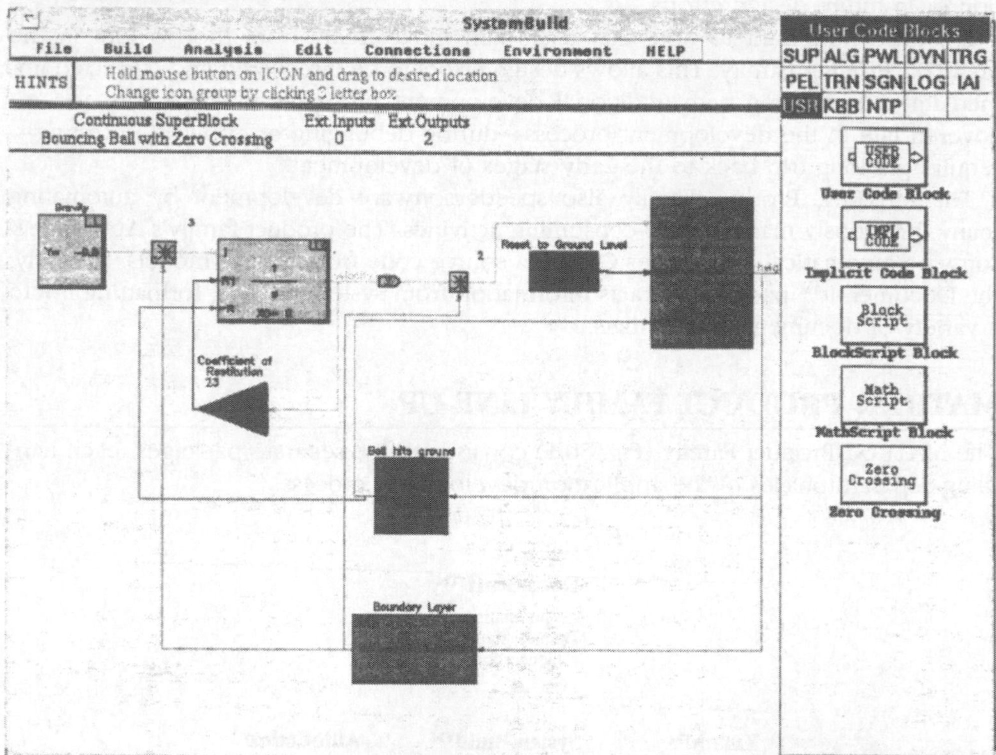

FIGURE 50.2 The SystemBuild environment models dynamic systems as sets of block diagrams.

Users create SystemBuild models by placing, defining, and connecting building blocks on a computer screen. Models may include standard building blocks chosen from SystemBuild menus or custom, user-defined building blocks. Once users chooses a block from the block menus, they can set block performance parameters in the dialog form. Building block types available under the SystemBuild environment enable modeling of continuous, discrete, hybrid, and asynchronous systems. In addition, a connection editor dialog form assists the user in connecting a block's inputs and outputs to other blocks, signal sources, or to the outside world.

The product supplies many standard block choices, including

- Algebraic blocks
- Dynamic systems
- Nonlinear systems
- Trigonometric systems
- Coordinate transformations
- Logic functions
- Signal generators
- Multidimensional interpolation
- Interactive animation icons
- Custom, user defined blocks
- Hierarchical SuperBlocks

Custom, user-defined building blocks may originate from Xmath-based mathematical analyses (written in the Xmath MathScript programming language), be developed with the SystemBuild environment's BlockScript language, or be defined in C or FORTRAN code. The BlockScript language is an intuitive, easy to learn scripting language specially tailored to support SystemBuild-based modeling and simulation. Custom blocks written in BlockScript fully support AutoCode-based code generation and are fully supported in the SystemBuild environment. The SystemBuild environment also accepts custom User Code Blocks which are blocks written in the C or FORTRAN programming languages.

The SystemBuild environment's hierarchical architecture fosters subsystem-based, build-a-little, test-a-little design cycles, enables system and subsystem reusability, and reduces the screen clutter of complex models. Users can group related building blocks into SuperBlocks. Although a SuperBlock may appear as a single entity in a model, it may contain many individual building blocks or SuperBlocks. In addition, SystemBuild supports unlimited levels of hierarchy.

Users make connections between model objects (blocks and SuperBlocks) through a point-and-click Connection Editor. The Connection Editor maintains lists of building block inputs and outputs, and displays block signal labels, data type, and channel numbers in a dialog box. Once the user defines connections, SystemBuild automatically lays out signal routings on the computer display (Fig. 50.3).

Specialized Building Block Libraries

To expand the scope of the SystemBuild product beyond standard classical and modern control engineering, the environment also supports several optional block libraries for specialized modeling. These include

- **RT/Expert Module** for developing knowledge-based expert system applications.
- **RT/Fuzzy Module** for incorporating fuzzy logic-based control regimes. The module enables users to establish and evaluate block behavior rules, define block classes and generate certain outputs from fuzzy logic processes.
- **State Transition Diagrams Module** (Fig. 50.4) offering a visual environment for modeling finite state-machines. Users can incorporate state machine analysis into SystemBuild models by defining states as bubbles connected to other bubbles. Sets

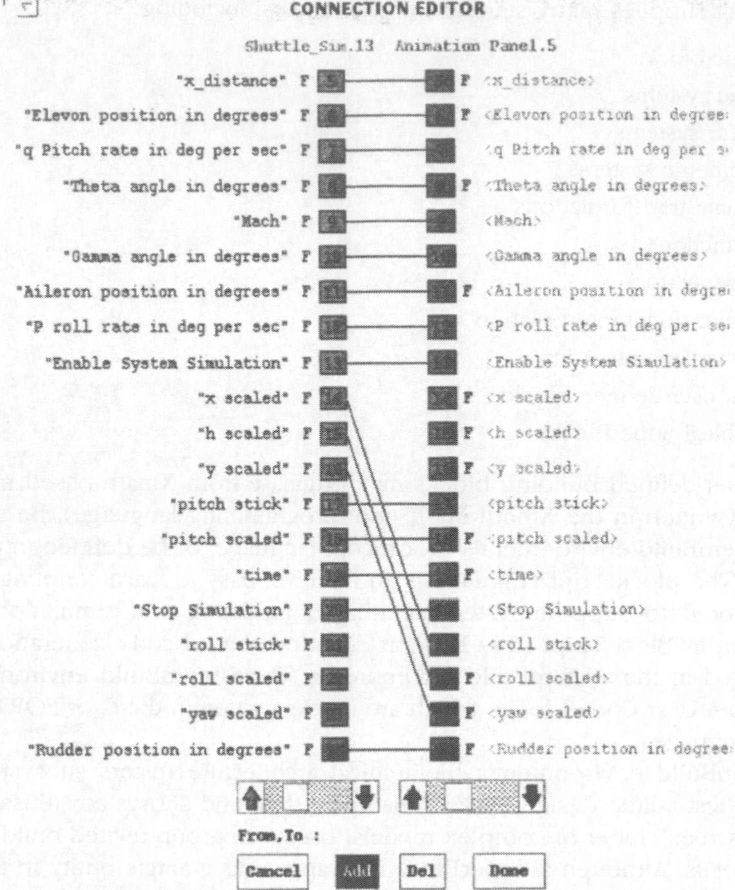

FIGURE 50.3 The SystemBuild Connection Editor provides a point and click mechanism to simplify connecting SystemBuild building block objects.

of bubbles may be gathered into SuperBubbles much in the same way SystemBuild SuperBlocks may contain complex subsystems. The State Transition Diagrams Module is fully supported by the AutoCode automatic code generator as well.

System Simulation Features

In addition to model building, SystemBuild software also offers powerful simulation facilities to test the behavior of models created in the SystemBuild environment. Simulations not only show system or subsystem behavior, but generate performance data that can be used to adjust SystemBuild model parameters and structure. Integrating modeling and simulation in a common environment allows users to shuttle between the two processes, rapidly working out kinks and optimizing model performance. Simulation data may be easily imported into the MATRIX$_x$ Xmath environment for detailed analysis and visualization.

The SystemBuild environment supports three simulation modes—Noninteractive, Interactive, and Animated Interactive. Noninteractive simulations run without user intervention, usually in background mode on a workstation or PC. This allows other work to proceed as a simulation runs.

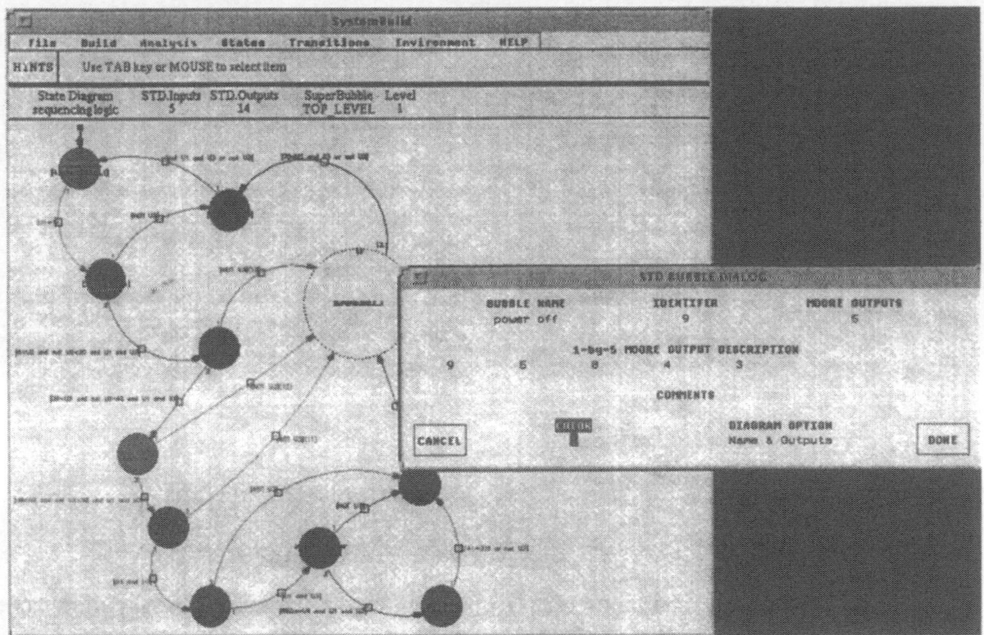

FIGURE 50.4 The SystemBuild State Transition Diagrams Module extends SystemBuild diagramming to state machine system design.

Interactive simulations provide real-time displays of system activities and permit user intervention. In the SystemBuild environment, users create Interactive simulations by connecting simulation icons (screen graphics of gauges, meters, switches, status lights, etc.) to model blocks and SuperBlocks. As the simulation runs, users can intervene in simulation operations—often mimicking a real-world end user—or add disturbances to the system to validate the model under realistic operating conditions. Again, data generated by simulations can be used to adjust models, either directly or after further analysis in the Xmath environment.

The optional SystemBuild Interactive Animation Module enhances the software's Interactive Animation features by providing additional animation tools to create richer and more customized interactive simulation environments. The Module includes a predefined Icon Library of screen graphics for strip charts, bar charts, indicator graphs, LED displays, numeric displays, controls such as sliders and switches, and icons for pumps, valves, and pipes. The software also includes a simulation control panel editor for designing interactive displays. The Interactive Animation Module is particularly valuable in creating displays for use as application control panels, user training systems or permanent simulations (Fig. 50.5).

The SystemBuild environment enjoys a close relationship with the Xmath analysis tool. Most significantly, users can operate both Xmath software and SystemBuild software simultaneously, allowing a simulation to run while users edit SystemBuild-based models. As noted earlier, Xmath software acts as the main visual analysis environment for SystemBuild and is used to formulate building block parameters, aid system identification, and evaluate simulation results.

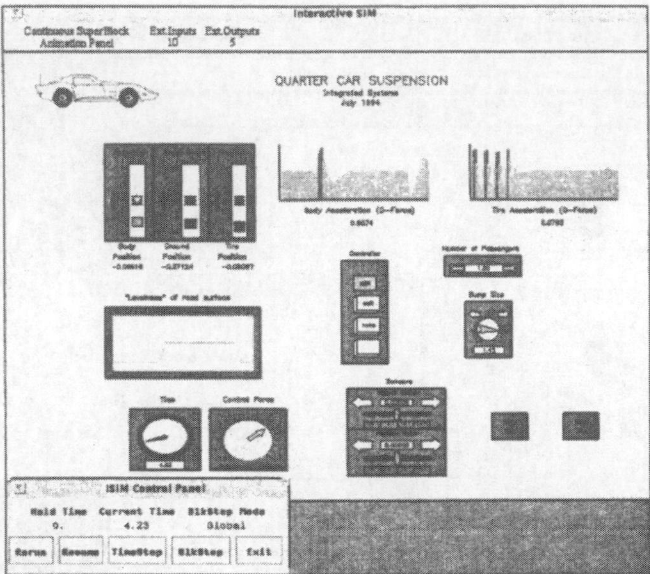

FIGURE 50.5 Interactive simulations offer a powerful way to visualize system performance early in the design cycle.

Xmath Analysis Software

The Xmath environment offers a powerful set of mathematically based analysis and visualization tools focused on supporting control and embedded systems design. Major pillars of the software include powerful, easy to use graphics, object-oriented design, a wealth of predefined "plug and play" mathematical functions, a scripting language, and several special purpose optional modules.

Principal uses for Xmath software include analyzing and characterizing phenomena to be modeled and simulated in the SystemBuild environment. The product lets users apply widely used engineering analysis techniques to system design such as Bode plots, root locus, and matrix mathematics. With the Interactive System Identification Module the product enables users to identify model parameters for SystemBuild models. In addition, the software can analyze data generated by SystemBuild-based simulations and application prototyping test runs conducted in the RealSim Series environment. Results of these analyses can help the user debug and adjust block diagram models in the context of the MATRIX$_x$ design methodology.

Xmath Software Features

Xmath software emphasizes visualizing the results of mathematical calculations. The product features vivid two- and three-dimensional full color graphics with many options to help users alter, annotate, and customize graphics. Users can rotate, scale, and manipulate graphics using either mouse-and-menu-driven interactive user interface or keyboard entry.

The Xmath software's analysis tools focus on solving system development problems. The product offers a selection of over 200 predefined functions and operators that go beyond conventional matrix mathematics and linear algebra tools. Primary predefined function classifications include trigonometric functions, matrix decomposition and matrix algebra, basic mathematical functions, system modeling and analysis, and filter design tools. The software also offers full abilities to write and solve user-defined equations and algorithms.

The software's MathScript programming language simplifies custom algorithm development. Easy to learn and powerful to use, the MathScript language describes algorithms clearly, compactly, and succinctly. To aid in debugging the algorithm, the MathScript interactive debugger includes features to single step through algorithms and watch variables.

Algorithms written in the MathScript language (Fig. 50.6) can be incorporated into the SystemBuild environment using the MathScript Block feature. However, the AutoCode automatic code generator is not supported for MathScript Blocks. Alternately, algorithms written in the BlockScript language can be prototyped in the Xmath environment. Although related to the MathScript language, BlockScript custom building blocks are supported by the AutoCode automatic code generator.

Xmath software's object-oriented design adds convenience and cuts engineering workload (Fig. 50.7). Treating algorithms, graphics, and data parameters as objects makes for easier data management. Plots and graphics can be stored as objects, avoiding the need for recalculation when retrieved. Also, users can specify alternative data sets for algorithm objects rather than tying algorithms and data together. The ability to overload data operators allows more compact representation of expressions.

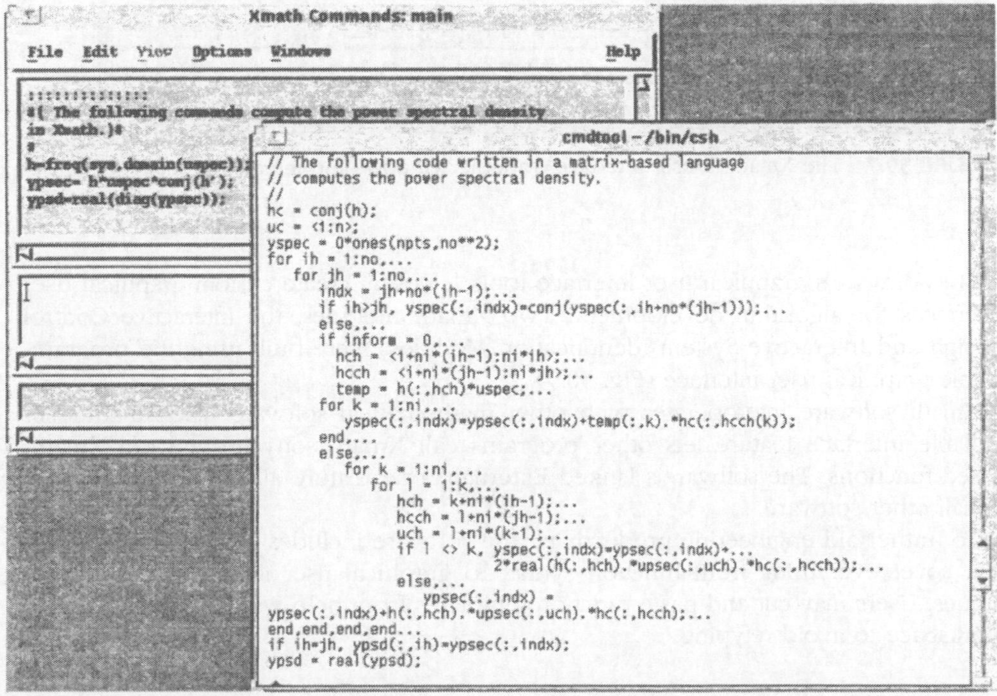

FIGURE 50.6 The object-oriented MathScript language (background) simplifies the process of describing complex algorithms compared to the older matrix-based scripting languages (foreground).

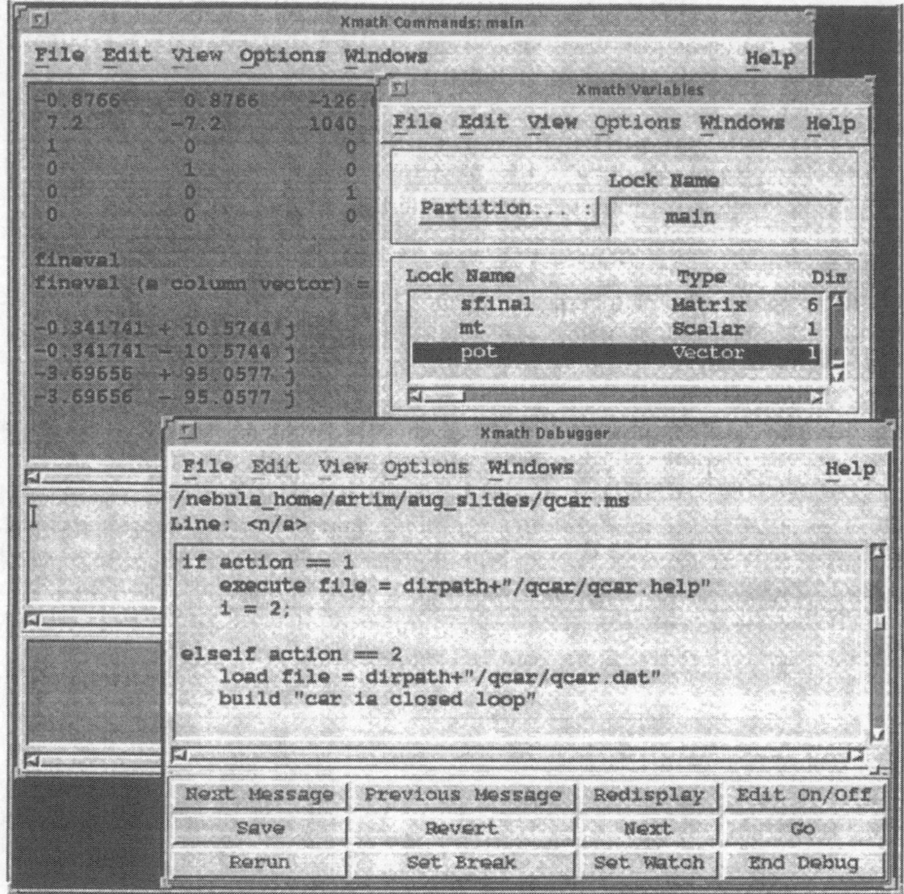

FIGURE 50.7 The Xmath, a user interface, variable window, and interactive debugger.

The software's graphical user interface tools let users create custom graphical user interfaces for algorithm development. Two Xmath modules, the Interactive Control Design and Interactive System Identification Modules, were built using the programmable graphical user interface (Fig. 50.9).

Xmath software interoperates with other mathematical software. The Xmath User Callable Interface feature lets other programs call Xmath software and run Xmath-based functions. The software's Linked External Code feature allows Xmath software to call other software.

To further aid engineering productivity, the software includes on-line help facilities that cover everything from function syntax to graphical user interface design techniques. Users may cut and paste function templates from help windows into the main workspace to avoid retyping.

Xmath Optional Modules

For specialized analysis, the Xmath environment supports a growing list of optional modules. Currently available choices include:

- **Control Design Module**—A library of functions pertaining to classical and modern control and dynamic system analysis.
- **Interactive Control Design Module**—This innovative tool helps users find parameters for controller design by manipulating Xmath-generated plots describing system specifications. For example, by tugging at the slope of a curve with a mouse in a graphic window, users can change the data underlying a graphic (Fig. 50.8).

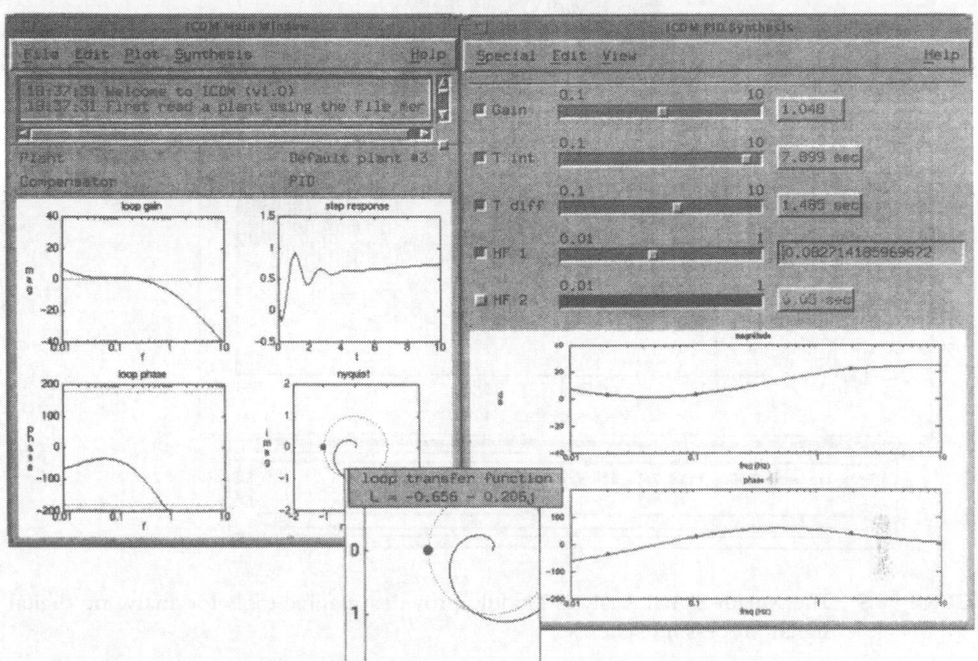

FIGURE 50.8 The Interactive Control Design Module lets users manipulate algorithm parameters by adjusting their graphic representations.

- **Interactive System Identification Module**—Helps users develop mathematical descriptions of observed phenomena interactively. Users can then use system descriptions in system Xmath-based analyses and SystemBuild-based models and simulations.
- **Model Reduction Module**—Helps users reduce otherwise elaborate controllers into less complex, lower order designs.
- **Robust Control Module**—For robustness analysis and synthesis of mission critical and multiinput, multioutput (MIMO) control systems.
- **Optimization Module**—Provides tools to analyze and improve the performance of system models using specialized linear, quadratic, and nonlinear programming techniques.
- **Signal Analysis Module**—This provides a functional toolbox used to analyze signal processing and communications systems (Fig. 50.9).
- **Xμ Module**—This option adds functions enabling analysis and synthesis of linear robust control systems.

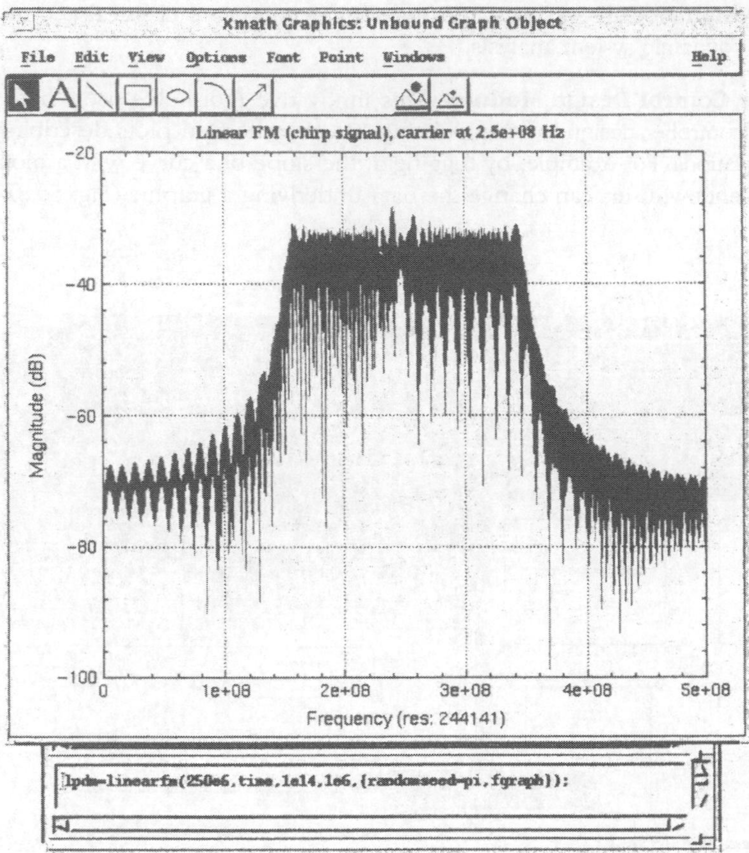

FIGURE 50.9 The Xmath Signal Analysis Module provides graphic tools for analyzing digital signal processing systems.

AutoCode Automatic Code Generation Software

The AutoCode tool automatically generates C or Ada source code from SystemBuild-based system models. The product produces code in minutes, a fraction of the time required for hand-coding software. Resulting C or Ada code structures correspond to SystemBuild-based models. SystemBuild-based block labels become names of variables in the source code, making it easy to trace relationships between model elements and the automatically generated code.

Automating code generation not only saves considerable time, but significantly changes the development process. Coding no longer stands as the main activity in product development, but a stepping stone between design and implementation. The iterative development process provided by the MATRIX$_x$ Product Family lets users test an application in a prototyping environment, make changes in the model's design environment, and almost instantly proceed back to prototyping without waiting for extensive recoding. Automatic code generation also changes the role of implementation level software engineers, freeing them to focus on activities such as optimizing the code and integrating automatically generated code with software from other sources (Fig. 50.10).

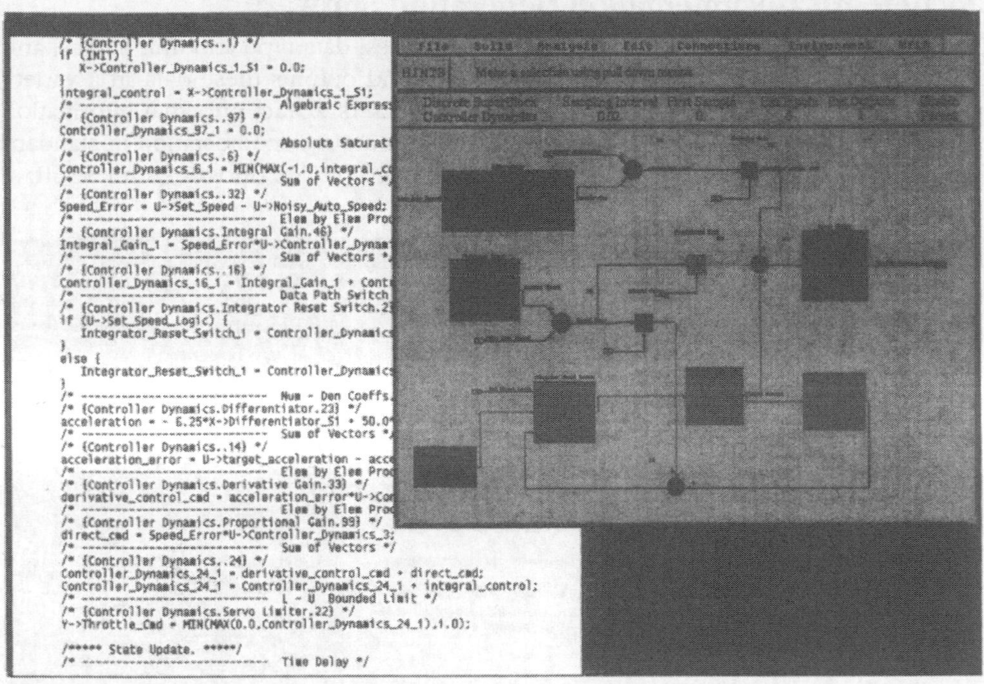

FIGURE 50.10 The AutoCode product automatically converts SystemBuild-based block diagram modes into C or Ada language source code.

Automatically generated code contains comments and annotations indicating block names, variable values, and other information. Well commented code improves traceability, thus simplifying code maintenance. Users can maintain an application at the visual modeling level, rather than manually modifying and revising program code. Automatic code generation also makes it easy to switch an application from one language to another.

To customize code generation options, the AutoCode product includes a Template Programming Language (tpl). Template programming allows users to control the structure of automatically generated applications including tasking, I/O management, scheduling algorithms, and targeting code to run on specific operating systems.

The source code generated by AutoCode software is platform-independent and compiles to run on any microprocessor supported by a C or Ada language compiler. In the context of the MATRIX$_x$ environment, this allows users to test an application using a prototyping computer featuring, for example, an Intel i860 microprocessor; but run the production application on another, completely unrelated microprocessor architecture. AutoCode-generated software to run in multiprocessor systems.

AutoCode software template functions allow users to tailor code to run on discrete, continuous, hybrid discrete/continuous, multirate, multiprocessor, or event-driven systems. Users may also use templates to add hardware-specific code modules such as initialization routines, interrupt handlers, and device drivers. For real-time applications, users can invoke an a rate-monotonic scheduler or automatically generate code compatible with the pSOS+™ priority-driven multitasking real-time operating system.

DocumentIt Documentation Generation Software

The DocumentIt tool (Fig. 50.11) extracts signal names, data type and unit names, and comments attached to SystemBuild-based models and outputs these data in a variety of documentation formats. Adjustable templates let users extract only that information needed for a particular document. The product can output documentation in standard formats such as FrameMaker, Interleaf, Microsoft Word, HTML (Mosaic), and ASCII.

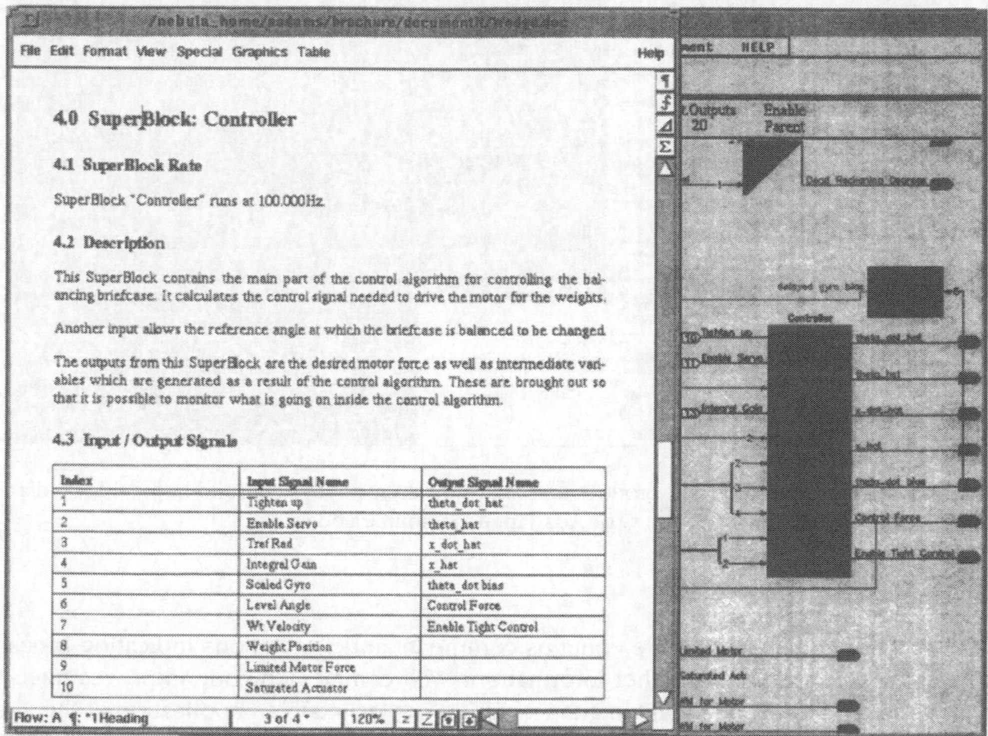

Index	Input Signal Name	Output Signal Name
1	Tighten up	theta_dot_hat
2	Enable Servo	theta_hat
3	Tref Rad	x_dot_hat
4	Integral Gain	x_hat
5	Scaled Gyro	theta_dot bias
6	Level Angle	Control Force
7	Wt Velocity	Enable Tight Control
8	Weight Position	
9	Limited Motor Force	
10	Saturated Actuator	

FIGURE 50.11 DocumentIt software extracts information pertaining to SystemBuild-based models and outputs it in a number of standard or user-definable formats.

Users can define outputs using the same template language used by the AutoCode software tpl feature. In addition to standard templates to generate documents compatible with FrameMaker, Microsoft Word, Interleaf, etc., users can generate documents conforming to other publishing and word processing software by inserting the target package's markup commands into a DocumentIt software output template.

Automated documentation generation improves the application development process in a number of ways. Documentation can become an ongoing activity synchronized with application development. Using the DocumentIt product, the same engineers who design a system can automatically generate documentation concurrently with development. Managers can use this incremental documentation process to monitor work progress. Improved documentation of subsystems makes integration of larger scale systems easier. It enables a team to follow design creation, changes, and reworks as they occur. The SystemBuild and DocumentIt tools can also be used to support developing project proposals that include detailed descriptions of how a design will work. The tool also can generate material for inclusion in progress reports during the course of development.

RealSim Series Application Prototyping, Simulation, and Testing Environment

The RealSim Series products offer an integrated software and hardware environment for application prototyping, simulation, and test. Architecturally, the RealSim Series divides into software and hardware components. RealSim Series software consists of a set of development tools running on PC or workstation host computers, and run-time software that ensures the smooth running of MATRIX$_x$-developed applications on the various RealSim Series computing platforms. RealSim Series computers include bench-top and portable computers configured to run MATRIX$_x$-developed applications, control real-world hardware, and collect data on application performance.

The main uses for the RealSim Series products entail application prototyping, real-time simulation, hardware-in-the-loop testing and production testing. In application prototyping, RealSim Series computers run software designed and developed with other MATRIX$_x$ software tools connected to real-world hardware. As the application runs, the RealSim Series computer can collect data about application performance. Performance data can be analyzed in the Xmath environment and used to adjust SystemBuild models. In addition, users can interact with applications as they run through custom graphical user interfaces created with RealSim Series Host Tools and the SystemBuild Interactive Animation module.

Users can prototype applications even if portions of the run-time hardware are not available for testing. The RealSim Series environment can simulate the operations of missing hardware while the rest of the application runs in real life. For example, a team designing an active suspension system for a car can simulate the operation of antilock brakes even if the brakes are not installed in the test vehicle or jig. Then, when the hardware is complete, the actual hardware can be interfaced and run together with the remaining simulated components. This is referred to as hardware-in-the-loop testing.

The same RealSim Series tools used to create user interfaces for application prototyping can develop graphic simulation environments. RealSim Series simulations differ from those created to support SystemBuild-based application design in that they can involve multiscreen displays and run in real time on dedicated, high-performance RealSim Series computers. Simulations may be permanently installed as user training systems, user control stations, or application diagnostic and management systems.

RealSim Series environments also have been used in application production run testing. In production testing, users program RealSim Series computers to test subsystems and subject the product to simulated real-world conditions.

RealSim Series Software Environment

The RealSim Series software environment consists of development environment Host Tools. These help users set up device driver and I/O connections, graphical user interfaces, and run-time Target Support Tools that mate MATRIX$_x$-developed application to specific computing platforms. Major elements of the RealSim Series Host Tools environment include

- **Hardware Connections Editor.** This point-and-click tool guides the user in connecting SystemBuild-based model block inputs and outputs to application hardware device drivers. Users also may apply scaling factors and offsets to signals, assign device channel numbers, and leave connected devices in safe states when the application stops running (Fig. 50.12).

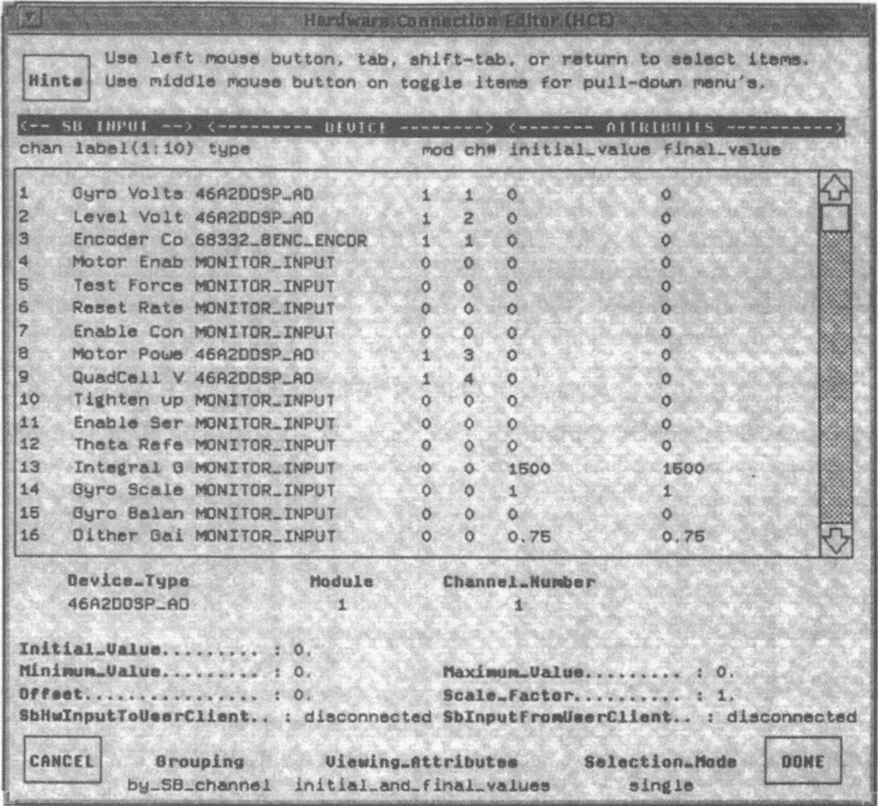

FIGURE 50.12 The RealSim Series connections editor feature expedites connecting MATRIX$_x$-developed applications to real-word hardware device drivers.

- **Interactive Animation Builder.** This tool lets users link screen graphics and application building blocks to drive real-time, interactive displays of application behavior. Available graphics include gauges, status lights, strip charts, sliders, buttons, switches, etc. Users can build hierarchical or linked displays, in which clicking on a process icon, for example, brings up another window showing less commonly observed phenomena. Users also can define alarm conditions that activate special displays if the application crosses a performance threshold. Users may also create multiscreen interactive application displays, overcoming the space limitations of single-monitor displays.
- **Data Acquisition Editor.** This enables users to define data sets to be collected in test runs and their individual sampling rates. Data to be collected may be transferred to disk for later storage and/or drive real-time status displays.

RealSim Series Run-Time Target Support tools include

- **Real-Time Monitor.** Provides a user interface to the RealSim Series computer for navigating through the MATRIX$_X$ product family, downloading the application, running the Interactive Animation display, and performing data acquisition.
- **Application Builder.** This is an automated compile and link function that builds the executable for the target RealSim Series computer.
- **Complete Host Run-Time Packages.** Users may order RealSim Series run-time software separately from host tools. This lets developers create turn-key RealSim

Series-based products for sale to end-users who may have no need for the RealSim Series development tool environment. Typical RealSim Series-based end-user products include process control environments, simulation systems, production test equipment, etc.

RealSim Series Computer Platforms

The RealSim Series concept offers users a choice of two ranges of prototyping computers specially configured to run MATRIX$_x$-RealSim Series-based applications. Currently available RealSim Series computers include the AC-100 Model 3, a multiprocessing-capable Multibus II benchtop and portable computer, and the AC-100 Model C30, a compact unit based on the Texas Instruments TMS320C30 microprocessor-based computer. In addition, a recently announced software package allows users to configure Intel i386/i486/Pentium microprocessor-based PC-compatible platforms into Ada language application prototyping computers.

RealSim Series AC-100 Model 3 Computers

The RealSim Series AC-100 Model 3 computer is a real-time multiprocessing computer designed to support demanding application prototyping, simulation, and testing applications. Built on a Multibus-II backplane, benchtop versions of the AC-100 Model 3 computer can contain up to 11 Intel i860 or i486 microprocessor-based CPU boards, delivering up to 400 Megaflops of performance. Intel i860 versions of the Model 3 computer run C language programs and i486 version run Ada language applications.

The computers can be configured to accept a wide variety of Multibus II or VMEbus-compatible I/O boards along with an Ethernet controller handling communications between the AC-100 computer and a host workstation. Intelligent I/O controllers relieve CPU boards of managing I/O functions and networking.

The RealSim Series AC-100 Model 3 portable computer enables prototyping to occur in moving vehicles such as aircraft, motor vehicles, process control environments, or other locations outside of development labs. Portable versions of the Model 3 can accommodate up to two processor boards and run off either AC or DC current.

RealSim Series AC-100 Model C30/C40 Real-Time Computer

The RealSim Series AC-100 Model C30/C40 real-time computer incorporates a Texas Instruments TMS320C30-based and TMS320C40-based PC/AT bus compatible board integrated into a standard PC chassis. The AC-100 Model C30/C40 processor board interfaces to a separate DSPLINK I/O bus operating at 10 Megabytes/second. The DSPLINK bus bypasses the slower PC bus and enables users to configure the computer to support high-performance data acquisition and peripheral control. The Texas Instruments TMS320C30 processor provides a peak throughput of 16.7 MIPS and 33 MFLOPS. The Texas Instruments TMS320C40 processor board can carry up to 4 C40 DSP processors providing up to 200 MFLOPS in a single board.

The product accepts standard Industry Pack modules that can add functions such as analog and digital I/O, ADC, and DAC conversion, timers, and communications. The AC-100 Model C30/C40 computer may be ordered in desktop, portable, or ruggedized in-vehicle computer versions.

RealSim Series Ada-PC Package

The RealSim Series Ada-PC Package is a run-time software environment that allows standard Intel i386/486/Pentium microprocessor PCs to run Ada language MATRIX$_x$ applications. Although Ada is the main language of this system, users may connect applications to device drivers written in either the C or Ada languages. The Ada–PC package is compatible with RealSim Series host tools, allowing easy setup and implementation of MATRIX$_x$ applications to run in PC-based prototyping environments.

50.3 CONCLUSION: THE MATRIX$_X$ PRODUCT FAMILY AND BEYOND

The MATRIX$_x$ Product Family has been used successfully to design and develop high-performance real-time, control applications in a wide variety of industries including aerospace, automotive, industrial process control, computer peripherals, and instrumentation. Recently, the product line made a solid connection to the world of general purpose real-time embedded systems development by adding a new feature allowing automatic generation of C language code running under the pSOS+ real-time operating system. The AutoCode-to-pSOS+ software link joins a powerful set of design tools for a widely used off-the shelf real-time operating system and delivers a complete design through implementation solution for embedded systems development.

The MATRIX$_x$ product line remains a work in progress, with new features, capabilities, and software set to join the product line over the foreseeable future. The future, however, seems clear. More control and embedded systems developers will abandon programming-focused bottom up system development and turn to top-down, concept-to-implementation methodologies such as those offered by the MATRIX$_x$ Product Family-pSOSystem™ environments.

DEFINING TERMS

AutoCode®: C and Ada software generator.

Block Diagram Modeling: Visual development strategy using graphical pretested subsystems.

DocumentIt®: Documentation generator.

RealSim® Series: Prototyping and test software and hardware development package.

SystemBuild®: Modeling and simulation system.

Xmath®: Mathematical analysis and visualization software.

51

CADACS[1] for System Analysis, Synthesis, and Real-Time Control

Chr. Schmid
Ruhr-Bochum University, Germany

M. Jamshidi[2]
University of New Mexico

51.1 INTRODUCTION

In recent years, a growing number of software packages for Computer-Aided Control System Design (CACSD) have become available. With varying degrees of sophistication, all these packages aid the control system engineer with efficient numerical analysis and graphical data representation. Only a few CACSD systems, however, go beyond theoretical control system analysis to offer support for the task of implementing control laws in hardware and software: interfacing to the real world for data acquisition, control operation, and experimental design verification.

To develop microprocessor-based industrial control devices, special real-time environments are needed. They may be housed by real-time workstations for CACSD, which can be connected directly to the plant to be controlled. Besides performing off-line control analysis and design, a real-time-based CACSD system must support on-line identification of the plant dynamics, as well as evaluation of prototype control systems. Furthermore, it is advantageous to have a software environment available, where real-time code can be generated and tested on the same computer or downloaded to a process control component. The automatic source-code generation may be one task in such a system, as currently followed by MATRIX_x (see Chapter 50). But there are additional requirements for a real-time control environment, such as monitoring, supervisory control, or scheduling. This contribution deals with a unified software frame for such tasks that is independent of the hardware or operating system and is relatively sim-

[1]CADACS can be obtained through CAD Laboatory, University of New Mexico. FAX 505-291-0013 or e-mail through wac@eece.unm.edu.

[2]This work was supported, in part, by CNRS-LAAS, Toulouse, France, where this author was on leave from September 1, 1994 to August 1, 1995.

0-8493-2530-7/96/$0.00+$.50

ple. Such a frame is contained in the real-time suite of the KEDDC system. The frame has been extended by many projects and is still being improved. In this chapter, the main interest is focused to the implementation of the real-time suite on PC-based systems, where the entire power of KEDDC is packed into one low-cost portable system that can be taken to the plant to be analyzed and controlled. We call such a system CADACS-PC, which is the mature implementation on PC-type computers.

Sections 51.2 and 51.3 start with a short general description of CADACS, where we focus our interest to the location of the real-time suite in the scope and structure of CADACS. Section 51.4 comprises details about the real-time suite and its different implementations. Examples are offered in Section 51.5 and, finally, Section 51.6 provides concluding remarks for the chapter.

51.2 CAPABILITIES OF CADACS-PC

Scope

CADACS-PC is an integrated software system that supports all phases of control engineering: (1) system identification (on-line and off-line), (2) control system analysis, design, and simulation, (3) real-time implementation of control algorithms (including adaptive control), and (4) documentation of design analysis and real-time control operation.

CADACS-PC supports the analysis of multivariable dynamic systems in the frequency domain, state space, and time domain. Figure 51.1 shows where one can find the subsystem that performs the real-time implementation.

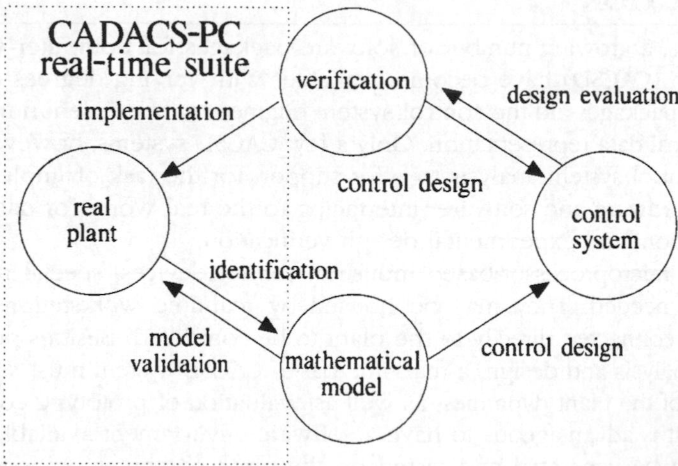

FIGURE 51.1 CADACS-PC real-time suite location within the simplified development cycle of control system analysis and design.

User Interface

CADACS-PC is interactively controlled by an efficient command-driven dialogue featuring "learning" defaults, which adjust to the last user input. Menu, help, and on-line manual facilities provide a friendly environment for the novice, without impeding the expert. In a workstation environment, CADACS-PC supports window techniques, thus enabling the engineer to concurrently work on different tasks. Versatile graphics, using a graphics working-sheet approach, complement the numerical capabilities. Consequently, the user environment actually resembles an engineer's desk.

A hard copy of the user dialogue and graphic output to screen or file in DTP format provides complete documentation of the work performed with CADACS-PC. Sessions do not require any planning or programming and can be interrupted at any time, as well as during real-time operation. All data can be stored on file for future reference to build a project database.

Analysis and Design Tools

CADACS-PC supports a wide variety of control engineering methods. It offers tools for systematic design strategies and helps the engineer to combine intuition with the results of high-accuracy numerical solutions. In addition to modern analysis and design concepts, classical methods are also implemented and may be used in combination with modern control techniques.

Frequently used tasks, such as transformations to different representational forms, database definitions, or basic calculations, are grouped together into several interactive modules, the so-called "managers," which together form the core system of CADACS-PC. In this contribution we will also become acquainted with two managers: the Documentation Manager DOKMG as part of the real-time suite and the Graphics Manager GRMGR, from the non-real-time facilities.

There are about 300 commands implemented in the core system that play the role of elementary utility functions. The core system is augmented by individual modules for particular tasks and methods:

- *System identification*:
  - –approximation in time domain
  - –approximation in frequency domain
  - –correlation and spectral analysis
  - –parameter estimation by various methods for different model structures
  - –model reduction in time and frequency domain
  - –parameter estimation for nonlinear systems, including structure selection
  - –harmonic analysis suite
- *Control System design*:
  - –continuous- and discrete-time compensators
  - –optimized PID-type controllers
  - –finite-settling-time methods
  - –various pseudocompensator approaches
  - –multiobjective parameter optimization
  - –state-space controllers (LQ, LQG)
  - –full- and reduced-order observer design
  - –Inverse–Nyquist–Array technique and systematic compensator design
  - –servo compensator design using PMDs
  - –open- and closed-loop system reduction
- *Adaptive control*:
  - –pole assignment or LQ control
  - –model-reference adaptive control

51.3 COMPONENTS OF CADACS-PC

Figure 51.2 portrays the block diagram of the main components of CADACS and their relation to the applications environment. The most important components consist of a group of *interactive programs* for the management of dynamic systems, systems analysis, controller synthesis, simulation, and signal processing. A unified *database* for system models allows a smooth exchange of data between the different programs and between different groups of a development or research project. Due to the wide range of methods, the system may be used for industrial planning of plants, subsequent analysis of measured data, modeling of dynamic systems, and design of control systems. The CADACS concept is not tailored for a specific control engineering use; on the contrary, it is able to cover, due to its universality, the entire range of control engineering methods. This fact is based on the open structure of the system concept, which is supported by a *subroutine library*, consisting of more the 1000 routines, and by comfortable tools. The program developer is able to change existing parts of programs or add new ones. In a research-oriented environment, this feature enjoys a particularly useful application: the *real-time suite* allows the experimenter to implement the controller just developed, using the real-time workstation, in such a way that all possibilities of the interactive part of CADACS are available in parallel to the real-time operation. In the following section we shall focus our interest on the real-time suite and its implementation.

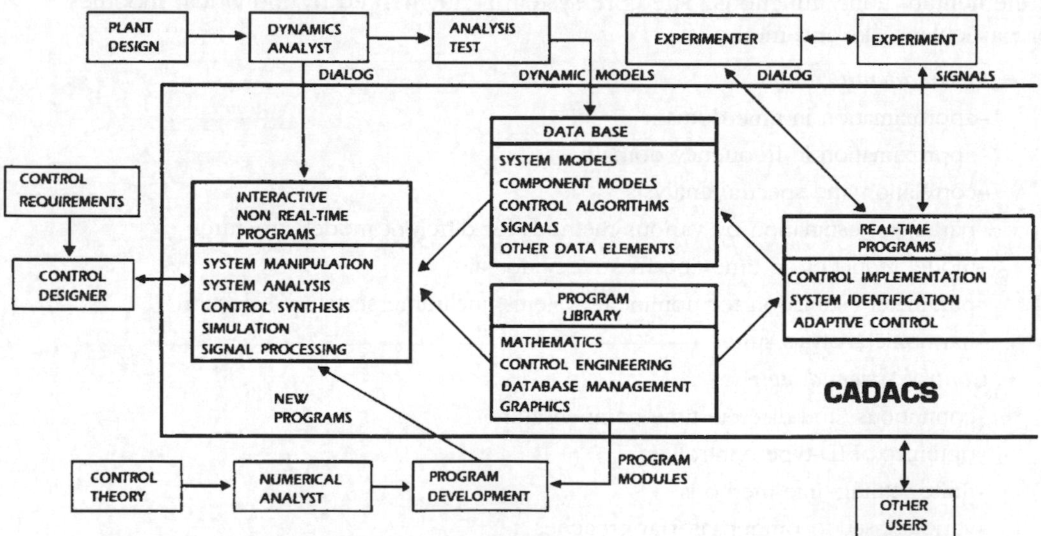

FIGURE 51.2 CADACS applications environment.

51.4 CADACS-PC REAL-TIME SUITE

CADACS supports the real-time implementation of all control schemes that are designed using CADACS. The multipurpose real-time suite serves as a development tool in basic studies or pilot experiments. Parts of the suite can be exported to other environments to run in automation systems or DDC substations in automation networks. Therefore, the real-time suite must show a highly modular structure. CADACS provides a high degree of portability, which results from an interface of CADACS to the operating system and to hardware-dependent functions. Generally some kind of basic routines build an interface to the world outside of the programs and modules. In addition, as the suite is an open system and not designed to be used by real-time system specialists, the usage must be as simple as possible. Thus a set of libraries with very powerful real-time entries is defined. In the following, a detailed study of all the components shall be presented.

General Concept

Figure 51.3 shows a block diagram similar to Figure 51.2 being, from the real-time suite point of view, more detailed and elaborated. The rectangular boxes identify one or several tasks. Names of interface libraries can be found in the ellipses within the boxes and the lines representing the information flow between the components. All parts within the dotted box belong to the real-time suite itself, which can be connected to the non-real-time facilities, as indicated below this box.

A command-driven *class monitor*, selected by the user within the start-up procedure of the real-time suite, is responsible for one class of real-time algorithms, e.g., for parameter adaptive controllers using parameter estimation methods or for a state-feedback controller. The class monitor serves as an intelligent interface between the user and the related *real-time task*, which itself runs on the same or other hardware. The user can specify all the data necessary to implement the controller and to run the real-time task. The user can change or list parameters and can call components from the database. In a research environment, the class monitor shows full support of all features suitable for the researcher. In an application environment, the class monitor is designed for a special application and serves as an interface to technicians of a special branch. For stand-alone purposes, it may also be implemented together with a teachbox suitable for technicians. The communication between the class monitor and the real-time task is performed by a shared data section where all the real-time relevant control data reside. This shared data section may be handled by the operating system, by the real-time monitor, or by the real-time task itself. This depends on the capabilities of the underlying operating system.

The real-time task contains all problem-oriented code to realize a controller or any code that runs in real time. It can be include, for instance, a single controller or a set of controllers of the same class that will be dispatched by it.

The *Real-Time Monitor* (RTM) connects the real-time suite to the operating system and to the time-base clock. If the operating system supports real-time operation (e.g., iRMX), the RTM is implemented as an interface to the real-time operating system. If real-time or even multitasking operation is not supported by the operating system (e.g., MSDOS), or if there is no operating system (e.g., DDC devices), the RTM is implemented as a real-time server, supporting all features necessary for real-time operation including multitasking.

To monitor the real-time response, the *Documentation Manager* (DOKMG) is linked between the real-time task and the *Graphics Manager* (GRMGR). Recording of signals and internal data and storing them in the database can be performed simultaneously with other CADACS activities through the DOKMG facility. Thus, the DOKMG serves as an interface between the real-time task and the user. The main KEDDC *Monitor* links the class monitor to all non-real-time facilities. Therefore, the entire analysis and design programs are available during real-time operation. When the real-time suite is implemented in a DDC substation, the class monitor and the non-real-time facilities are

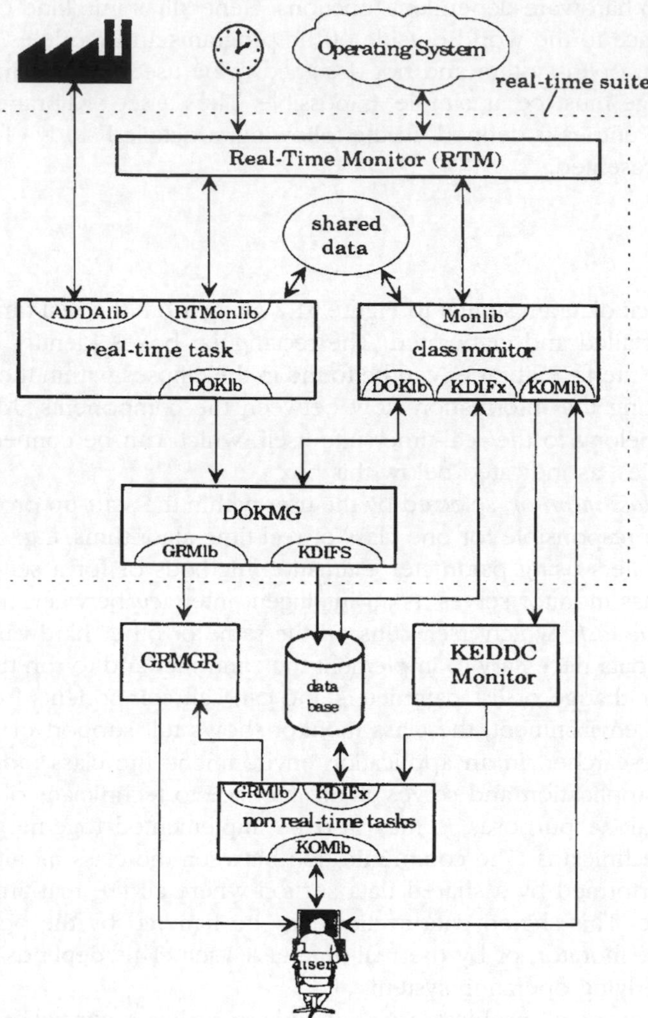

FIGURE 51.3 Components and basic function of the CADACS real-time suite.

often installed in a separate real-time workstation attached to this substation. After the implementation, the workstation can be removed.

Using these facilities, the user can both monitor and store the real-time behavior of any external or internal signal. At the same time or at a later date the user can evaluate and present results. It is possible to interlace a redesign version of a control scheme into the running start-up or monitoring phase of a controller. For example, actual plant input and output signals of an adaptive closed-loop system can be fed to an off-line identification task. The result could be the model basis for a redesign of an adaptive controller that might be tested with a simulation tool with the same specifications as the actual real-time task. After extensive verifications, the new controller can be called by the class monitor to replace the actual one. Figure 51.4 shows a graphics screen copy of such a session, where graphical results from several analysis and synthesis tasks are displayed simultaneously with real-time results.

User Libraries

To support programming, different libraries with a powerful set of entries are defined. The names and the functionality within the real-time suite can be taken from Figure 51.3. These libraries include

RTMonlib	Interface that integrates a real-time task into the real-time suite by sending requests to the RTM.
Monlib	Interface that integrates a class monitor into the real-time suite by sending requests to the RTM.
ADDAlib	Process interface library that connects to the plant.
DOKlb	Interface to the DOKMG for monitoring and recording signals.
GRMlb	Interface to the GRMGR for sending objects to be displayed in graphical form.
KDIFx	Database interface for control engineering objects (x=S for time signals).
KOMlb	Nongraphical user I/O and communication interface.

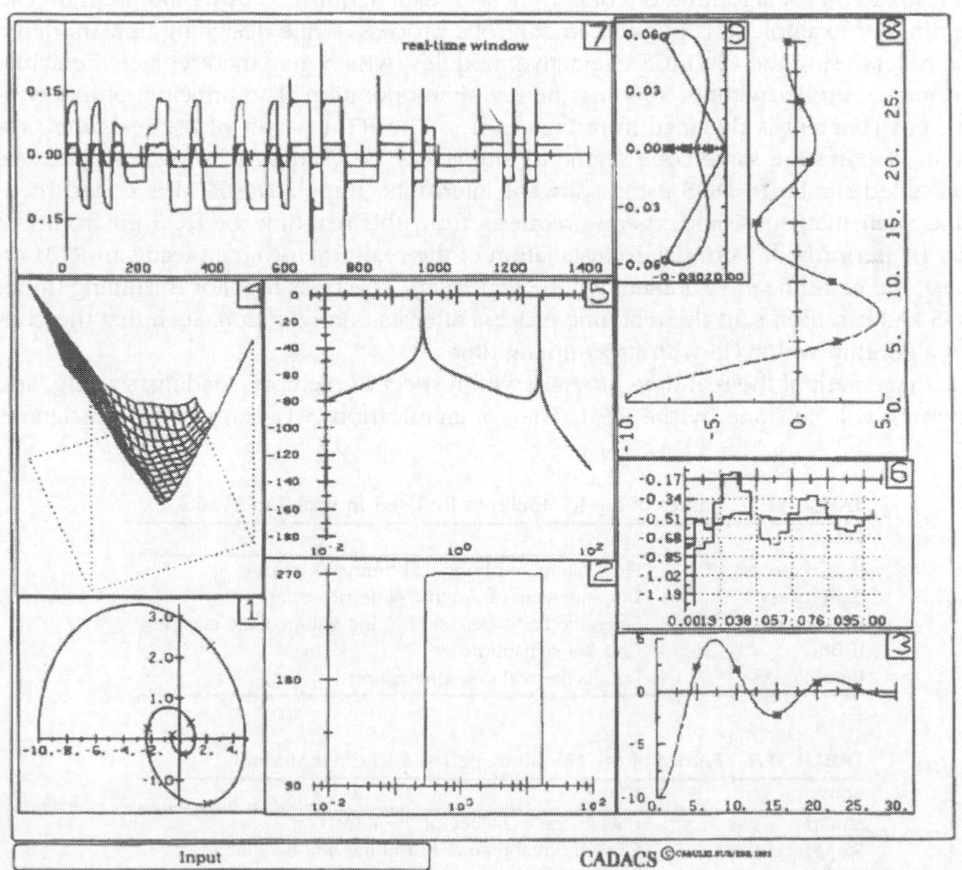

FIGURE 51.4 Hardcopy example of a graphic user screen during real-time operation.

The most important entries of the first two libraries are listed in Tables 51.1 and 51.2, respectively. Using these libraries, it is not necessary to code any basic real-time calls to an operating system or to the RTM. The usage of this libraries shall be demonstrated later in the DOS real-time toolbox implementation in the section on the real-time toolbox.

MS-DOS Implementation

The MS-DOS operating system (henceforth called DOS) is not capable of multitasking, which means that it is not possible to start quasiparallel tasks. Further, neither the basic I/O system (BIOS) nor DOS itself is coded reentrant, such that the simultaneous usage of system request from two different applications is not allowed.

These are reasons for generating a simple real-time environment with an RTM, which does not access DOS or BIOS for performing its task of managing real-time programs. Further, no hardware device already managed by the BIOS, as, for example, the timer/counter 8253 that is needed to generate system time, timeouts, refresh, and sound, will be used (as far as possible) to perform real-time tasks beneath normal DOS processes.

The real-time task is not operated as a standard DOS application, which means that beneath a control algorithm or concurrent with data acquisition any DOS program can be run. For example, it is possible to control a process, while designing new modified controllers using the CADACS interactive modules, which may then replace the actual running controller without aborting the real-time operation. The principle of this two-task environment is deduced from Figure 51.3. The RTM is part of the real-time program (it shares the same code segment), but it may be used by all programs, because it is called similar to DOS using software interrupts (traps). The RTM is coded reentrant, such that quasisimultaneous requests from the real-time or DOS environment may be performed. During auto installation of the real-time program using an RTM request, the real-time environment will be generated. The class monitor is running under DOS and can then start the real-time process after its configuration, such that the control algorithm will cycle with its sampling time.

Management of the real-time program with respect to memory, loading, starting, and stopping will be done by the RTM. The communication between real-time programs

TABLE 51.1 Entries of the RTMonlib to Be Used in Real-Time Tasks

Entry	Usage
BeginProcess	First statement of any real-time procedure
EndProcess	Last statement of any real-time procedure
StopProcess	Stops the real-time task; routine will become inactive
If Brk	Check a semaphore
InstallProcess	Installs the real-time procedure

TABLE 51.2 Entries of the Monlib to Be Used in Class Monitors

Entry	Usage
MoOpn	Checks the existence of the RTM
StopProcess	Stops the real-time task; routine will become inactive
SetBrk	Sets a semaphore
StartProcess	Starts a real-time task
GetSharedCommon	Get the shared data section address
StateUserTask	Display the state of the real-time task to the user

and their associated class monitors is performed using a shared memory, which is part of the real-time program. A pointer to this area of memory is provided after installation of the real-time program by the RTM. Figure 51.5 shows the typical partitioning of memory for loaded or active real-time programs on an IBM-PC under DOS.

address		
A0000H	resident command line interpreter	
	free memory	DOS environment
	class monitor	
	real-time task	real-time environment
	RTM	
	DOKMG	
	KEDDC Monitor	
	GRMGR	resident programs
	drivers	
	DOS	
	BIOS	
00400H		

FIGURE 51.5 Partitioning of memory for loaded real-time application.

The actual configuration offers the opportunity to schedule a real-time task with high priority using a cyclic sampling time greater than 1 millisecond. Processor time is not divided using a time slice between the real-time program and DOS programs, because a controller has to cycle with a sampling time specified during its synthesis. The user can assume that the controller will not be interrupted by normal DOS applications. To guarantee that user interaction is also possible in case of a large computational burden of a control algorithm (i.e., the operating time of the control algorithm is greater than the sampling time), the DOS environment gets at least 1 millisecond computation time per sampling cycle. But the user may also code routines with higher priority (keyboard interrupt, timer interrupt) to influence or to supervise the control task.

The Real-Time Toolbox

The programming on a PC under DOS is simplified by a toolbox for Turbo Pascal or Turbo C. It is not necessary for the user to code basic calls, because the two user libraries RTMonlib and Monlib are realized in the toolbox.

The usage of this toolbox shall be demonstrated for a simple proportional controller. A typical real-time program is given in the small Pascal source code segment shown in Figure 51.6. The real-time program consists of a main program that allocates the common area of memory (shared common, real-time common) as a global variable in form of a record, to guarantee a contiguous memory, which may be simply accessed using a pointer. Further, it initializes its variables according to its problem and installs the real-time procedure using the monitor-request InstallProcess (see Tables 51.1 and 51.2). This procedure call will not return to the caller, but will make the real-time program resident and returns to the DOS command line interpreter. The parameters for InstallProcess are the data segment together with segment and offset address of the real-time procedure and the shared common.

```
PROGRAM PDDCA;
USES
  RTMonlib, {Monitor library}
  ADDALib;  {Library for A/D-D/A conversion}
TYPE
  RealTimeCommonType = RECORD
    Gain : Double;
    SetPoint   :  DOUBLE;
    ADChannel :  INTEGER;
    DAChannel :  INTEGER;
  END;

VAR
  RealTimeCommon   :  RealTimeCommonType;

PROCEDURE PController; INTERRUPT;
BEGIN { PController }
  BeginProcess;
  WITH RealTimeCommon DO
  BEGIN
    DAReal(DAChannel,Gain*(SetPoint-ADReal(ADChannel));
  END; { WITH }
  EndProcess;
END; { PController }

BEGIN { main program }
  Gain := 0.0; ADChannel := 0; DAChannel := 0; SetPoint:=0
  InstallProcess(DSeg,Ofs(PController),Seg(PController),
                Ofs(RealTimeCommon),Seg(RealTimeCommon));
END. { main program }
```

FIGURE 51.6 Real-time program example for a proportional controller.

The real-time procedure, which will be cyclically called by the scheduler, must be declared an INTERRUPT procedure, due to the scheduler's calling mechanism. The first statement of this real-time procedure shall be the call to BeginProcess; here, for example, will be the processor interrupt flag set to guarantee interruptability of the control task. This is very important for processor time management within the real-time monitor, because all counters have to be actualized using the basic timer tick.

The common memory area has to be declared in the form of a record, to guarantee a contiguous allocation of this memory block. The supervisory program uses a pointer variable to the same type of record to access this area of memory. The WITH statement allows a simple notation for accessing this structure. The realtime procedure shall finish by invocation of the procedure EndProcess. Herein parameters for internal management of monitor data will be set.

The problem of proportional control solved here is trivial, but even the most complex adaptive control algorithm does not need any further elaboration with respect to real-time implementation. The skeleton of the controller will remain the same, only the problem oriented part will change.

The programmer should initialize all variables of the shared common within the real-time program. The class monitor shall not do any basic unrequested initialization, but has to adapt its data management to the real-time task when it is started. This renders possible a stand-alone operation of the control algorithm. The supervisory program has to be run only if changes in the control algorithm are necessary.

The class monitor program gains access to the functions of the RTM. For the usage of RTM requests, a noninteractive class monitor example is given in Figure 51.7. This trivial example sets the P controller parameters and starts the controller in the PDDCA real-time program of Figure 51.6. First a test on existence of the RTM is made using the subprogram MoOpn. Then the pointer to the shared common is requested and the common memory is moved into a buffer variable. Controller gain and numbers of D/A- A/D-channels are set using the buffer variables. After moving the buffer to the common memory, the controller is started using a sampling time of 100 milliseconds. Prior to finishing the class monitor, the state of the real-time task is displayed.

Such a toolbox implementation is most suitable for basic controller studies of laboratory-scale processes, where during the development new software parts must be generated. This environment proved to be suitable for developing adaptive controllers running at a later time in DDC substations in automation communication networks.

The real-time toolbox contains examples of different kinds of real-time tasks. They are realized both as source-code templates and as turn-key ready programs according to Figure 51.8.

Special Implementations

An important factor in the selection of a hardware system is software and hardware compatibility with future generations of microprocessors. For a fullscale professional implementation, the INTEL 3xx microcomputer and the INTEL RMX operating system were selected as the basis for the implementation. This system fulfills the requirements of having a real-time multitasking operating system. It recognizes a large variety of multibus boards and peripheral devices. An advantage of the system is the ability to configure it in many different ways, for example, with different and also multiple CPU boards. Several options are supported for the process interface.

For low-cost systems, an IBM-compatible personal computer running under MSDOS can be prepared to run a small set of real-time tasks concurrently with other programs, as shown in the previous sections. This environment is suitable for developing adaptive controllers running later in DDC substations. In the first stage of development, the adaptive controller runs locally on the same computer. Later, the control software will be downloaded to the target processor of a DDC substation and the operation will be supervised by the development system in the same way as running on the development system. Finally, the software will be burned into the EPROM format to be plugged into the DDC substation for permanent use. Substations of this kind can be temporarily linked with an IBM-compatible personal computer, where the CADACS system is running together with the real-time suite on the PC. For fast real-time monitoring, the real-time suite on the PC is connected to the substation via fast communication lines. For stand-alone purposes, a teachbox may be attached. Using the keypad of this small device, the controller can be configured with the aid of an LCD-based alphanumeric

```
PROGRAM PDDCS;
USES Monlib;
CONST
  RealTimeProgramName = 'PDDCA';
TYPE
  RealTimeCommonType = RECORD
    GAIN : DOUBLE;
    ADChannel : INTEGER;
    DAChannel : INTEGER;
  END;

VAR
  RealTimeCommonPtr : ^RealTimeCommonType;
  RealTimeCommonBuffer : RealTimeCommonType;
  Error : INTEGER;

BEGIN { supervisory program }
  IF NOT MoOpn THEN
  BEGIN
    WRITELN('Monitor not installed ');
    WRITELN('Install using RP ',RealTimeProgramName);
    HALT(1);
  END;
  GetSharedCommon(Error,RealTimeCommonPtr);
  IF Error<>0 THEN
  BEGIN
    WRITELN('Error ',Error,' when getting common'); HALT(Error);
  END;
  RealTimeCommonBuffer := RealTimeCommonPtr^;
  WITH RealTimeCommonBuffer DO
  BEGIN
    Gain := 2.0; ADChannel := 1; DAChannel := 1;
  END;
  RealTimeCommonPtr^ := RealTimeCommonBuffer;
  StartProcess(Error,0.1,RealTimeProgramName);
  IF Error<>0 THEN
  BEGIN
    WRITELN('Error ',Error,' when starting real-time program');
    HALT(Error);
  END;
  StateUserTask;
END. { class monitor }
```

FIGURE 51.7 Class monitor program example for a proportional controller.

and graphic display. The entire control device is built in an industrial rack mountable form, where the flat hand-held teachbox can be plugged into the module frame.

Keuchel et al. (1991) presented a real-time toolbox implementation of adaptive controllers to a muitilevel control engineering framework, where different levels of networking are used. Figure 51.9 shows such a process automation network where the real-time suite of CADACS-PC has been integrated into such a network. The software has been taken directly from the real-time toolbox and connected to a supervisor via a shared data section. Signal recording and remote controlling of the plant can be performed locally using the built-in DOKMG and GRMGR, and remotely via a process automation executive server, which is connected to a separate computer. This process automation execu-

class monitor/real-time task name	usage
DATST/DAT	data acquisition with optional stimulating signal
DDCST/DDC	decentralized MIMO controller with feedback and feedforward dynamics
ZSTEW/SFEDB	MIMO state-feedback controller with optional observer
APPCS/APPCA	decentralized adaptive MIMO controller using pole-placement or LQ control laws
MRACS/MRACA	decentralized MIMO model-reference adaptive control

FIGURE 51.8 Real-time program templates of the real-time toolbox.

tive computer may be connected to different servers that are distributed in a factory or big building. The system has been applied for climate control in an office building. The local class monitor within the real-time suite is designed for the porters work in the building. For the expert, the process control system is linked to other CADACS systems via SNA, where signal history data can be evaluated using CADACS tools.

51.5 EXAMPLES

In the following, a CACSD example will be given that comprises the full cycle of activities from Figure 51.1, and that also includes real-time experiments. The final results

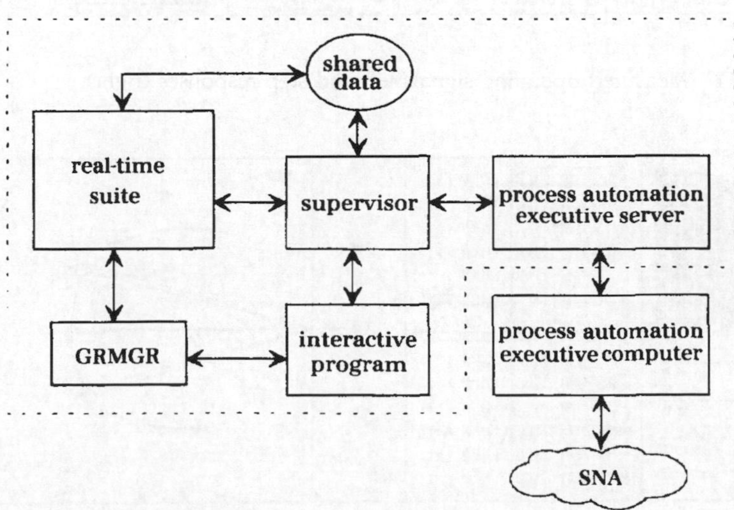

FIGURE 51.9 The real-time suite in a process automation network.

of the analysis and design steps will be presented in graphical form including short explanations.

The example is an air conditioning system according to Figure 51.10 where the temperature and the humidity must be controlled. In this example there was no model available. An IBM-PC equipped with some data acquisition boards and the real-time toolbox data acquisition template has been used to measure the operating signals over a time of 7200 seconds. Figure 51.11 shows the monitored operating signals (left). In diagram 1 the input signals are displayed. The temperature can be seen in diagram 2 and the humidity in 3. In the right part of Figure 51.11, the measured step responses are ordered according to the position of the elements in a transfer matrix. From the information from signals, different MIMO models have been estimated by several methods. Figure 51.12 shows a comparative plot of the time and frequency responses of the temperature submodel. Measured and model responses are compared.

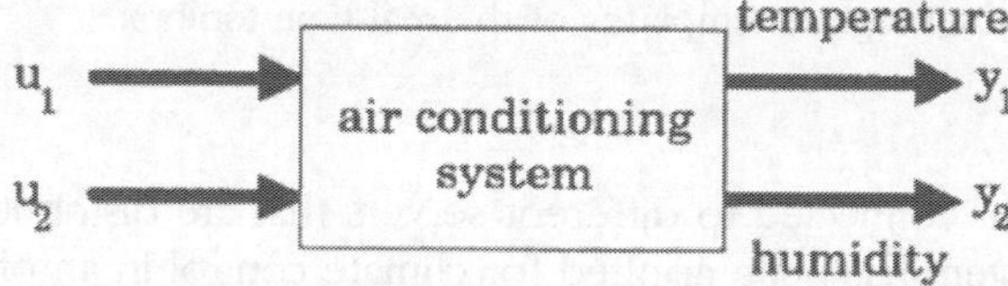

FIGURE 51.10 Input-output configuration of the air conditioning system.

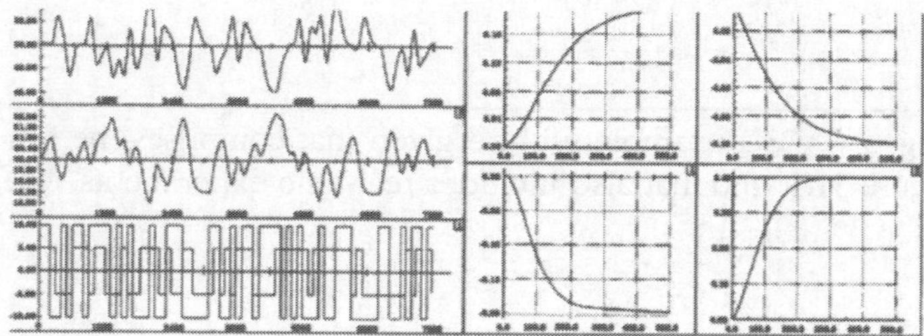

FIGURE 51.11 Measured operating signals left and step responses (right).

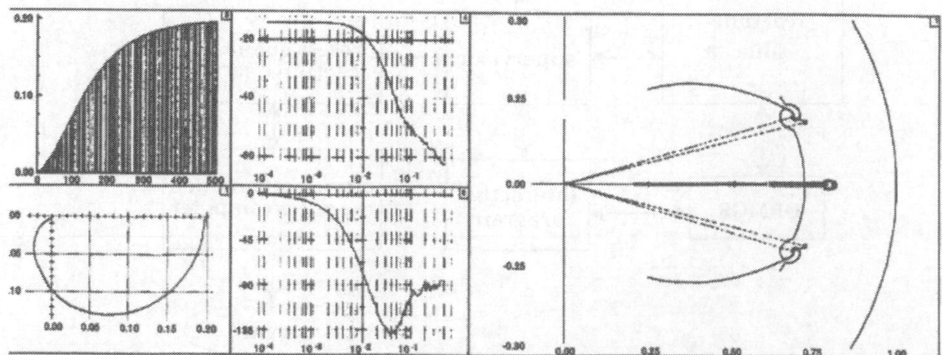

FIGURE 51.12 Comparative plot of the time and frequency responses (left) and root-locus diagram of an LQ regulator (right).

Simulation studies with single-loop PID control and with multivariable LQ control verify the controller design. This will be shown in the real-time results from Figure 51.13. Diagram 1 contains the manipulated and diagram 2 the controlled signals.

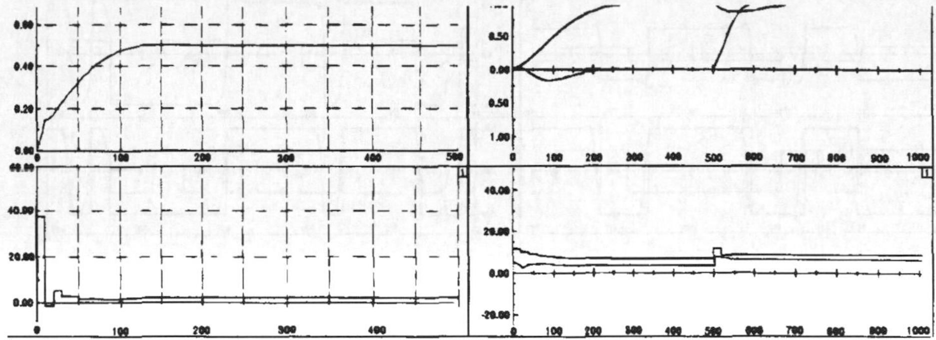

FIGURE 51.13 Setpoint behavior for single-loop temperature PID control (left) and for multivariable LQ control (right).

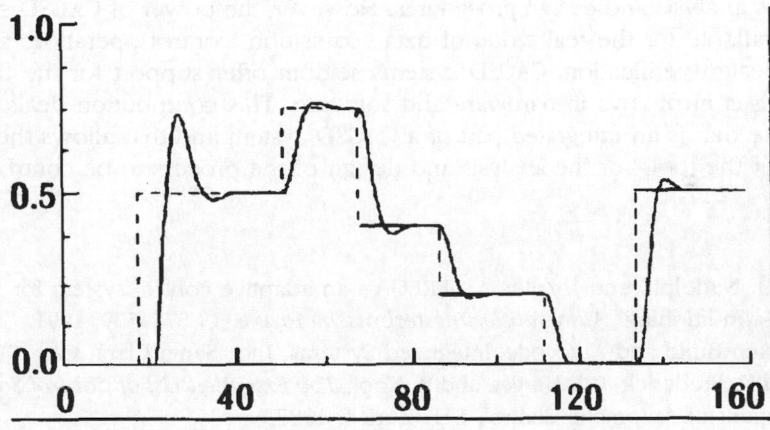

FIGURE 51.14 Temperature behavior during adaptive setup operation.

To test adaptive control, the multivariable control scheme has been made adaptive. Figure 51.14 shows the temperature behavior during the adaptive set-up operation. The temperature setpoint has been changed several times. During the first change of the setpoint, an overshoot in temperature can be recognized. But then the changes in temperature will become smooth. This experiment has been continued for several days. Figure 51.14 (left) shows a section from history data, where the air flow is constant. Diagram 1 contains setpoint and controlled variable data (temperature and humidity), diagram 2 the manipulated variables, and diagram 3 the behavior of some estimated plant parameters. A phase of changes in the air flow is shown in the diagrams on the right-hand side of Figure 51.15 where the changes in dynamics can be seen from the parameters. After a short identification period, the closed-loop behavior will be as smooth as in the case of constant air flow.

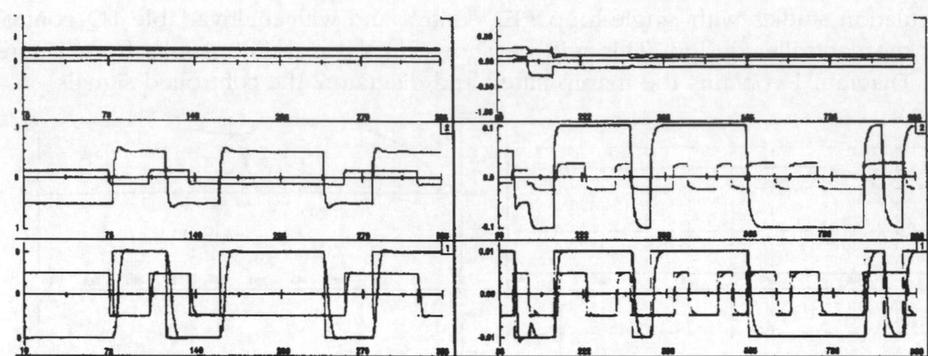

FIGURE 51.15 Adaptive control during constant air flow (left) and during air flow changes (right).

51.6 CONCLUSIONS

Industrial process automation systems are generally not supporting highly sophisticated process analysis and design procedures. However, the power of CACSD systems is seldom available for the real world of data acquisition, control operation, and experimental design verification. CACSD systems seldom offer support for the tasks of implementing control laws in hardware and software. This contribution deals with a real-time suite that is an integrated part of a CACSD system and that allows the direct connection of the tools for the analysis and design of the process to be controlled.

REFERENCES

U. Keuchel, U. Nadolph, and J. Ziller, ASTEREGA—an adaptive control system for heating and air-conditioning. *Automatisierungstechnische Praxis 33* 571–579, 1991.

MATRIX$_x$, SystemBuild and Autocode. Integrated Systems, Inc., Santa Clara, CA, 1990.

M. Rimvall, D.K. Frederick, C.J. Herget, and R. Kool, *The Extended List of Control Software*. Department of Automatic Control, ETH Zurich, 1987.

M.E. Stieber and Chr. Schmid, A real-time workstation for computer-aided design and implementation of control systems. *Proceedings of the IEEE CSS 3rd Symposium on Computer-Aided Control System Design-CACSD*, IEEE, New York, 1991, 79–84.

H. Unbehauen, P. Du, and U. Keuchel, Application of a digital adaptive controller to a hydraulic system. Conference Publication No. 285, *IEE Int. Conf. Control 88*, Oxford, 177–182, 1988.

FURTHER INFORMATION

Chr. Schmid, KEDDC—a computer-aided analysis and design package for control systems. In M. Jamshidi and C.J. Herget (eds.), *Advances in Computer-Aided Control Engineering*. North-Holland, Amsterdam, 1985, 159–180.

Chr. Schmid, Adaptive controllers: Design packages. In M.G. Singh (ed.), *System & Control Encyclopaedia*, suppl. Vol. I. Pergamon Press, New York, 1990, 1–4.

M. Jamshidi, M. Tarokh, and B. Shafai, *Computer-Aided Analysis and Design of Linear Control Systems*. Prentice-Hall, Englewood Cliffs, NJ, 1992, 425–434.

52

The Fuzzy Logic Development Kit for IBM PCs and Compatibles

Mark E. Dreier
Bell Helicopter Textron, Inc.

George A. Cunningham
New Mexico Institute of Mining and Technology

52.1 INTRODUCTION

The Fuzzy Logic Development Kit, FULDEK©,[1] is a fast, user-friendly method to develop **fuzzy logic** control laws. FULDEK runs on personal computers using the WINDOWS™[2] operating system. It has a graphical editor for membership functions and rules, and provides a method to test the control rules in a **closed-loop simulation**. It also features automatic rule generation and fuzzy code generation. Experienced fuzzy logic practitioners as well as novices benefit from the intuitive ordering of menus and functions in FULDEK. This chapter describes key features of FULDEK by presenting a simple example of its operation.

Bell Helicopter Textron, Inc. (BHTI) designed the FUzzy Logic DEvelopment Kit (FULDEK) to explore fuzzy logic, and to serve as a development tool for applications in manufacture and control. The current version of FULDEK provides several important tools that make Fuzzy Rule Base (FRB) development intuitive and easy. Furthermore, Fuzzy Logic Controllers (FLC) are easy to design with fuzzy code gener-

[1]FULDEK, Copyright 1994, Bell Helicopter Textron, Inc.
[2]WINDOWS is a registered trademark of Microsoft Corporation.

0-8493-2530-7/96/$0.00+$.50

ators that are a part of the FULDEK professional suite. Other user-friendly capabilities are listed below.

1. FULDEK operates in the WINDOWS environment, with or without a mouse.
2. FULDEK provides six types of **membership functions**, and also allows the user to define functions.
3. FULDEK exhibits rule interaction using text and three graphic modes.
4. FULDEK provides shortcuts to reduce editing time and testing.
5. FULDEK generates **fuzzy inference rules** from observed data in a one-step process.
6. FULDEK links the user's fuzzy inference rules to a linear model of the plant to be controlled, then performs a time domain simulation. The simulation, including phase plane analysis, helps the user assess the effect of the FLC.
7. FULDEK generates fuzzy rule base code in four languages—Ada, BASIC, C, and FORTRAN. The code, in the form of subroutines or procedures, can be compiled and linked to a user-written driver to become part of an embedded controller.

This chapter describes the structure of FULDEK by presenting a simple control problem—balancing an inverted pendulum. The inverted pendulum has become the favorite first problem to solve because it is easy to describe, the control law admits an analytic solution, and the fuzzy rules are intuitive and easy to express. Before the pendulum problem is described and its FLC solution is given, a brief survey of FULDEK is presented.

52.2 OVERVIEW

FULDEK is divided into two major sections, the *EDITOR* option and the *RUN* option. The EDITOR option lets the user manipulate files, and edit fuzzy variables, membership functions, rules, and ASCII files. The user may also select **defuzzification** methods, **composition** methods, and scaling factors for the fuzzy variables. A new feature in version 3.1 of FULDEK automatically generates fuzzy inference rules from observed data, a capability that is useful in parameter estimation.

The RUN option exercises the rules in three different formats. The first format is a step-by-step examination as one input variable is swept through its **universe of discourse** while the other input variables, except feedback variables, are held at user-defined values. The output of this slice through the rule hyperplane is a list of the fuzzy variables and their values.

The second RUN option format is a continuous sweep of one or two input variables through their universes of discourse. The result is a graphical display of a line or surface in the rule hyperplane. The graphics, which are device-independent, come in three useful styles—X-Y, two dimensional contour, and three dimensional surface.

The third format, which is most useful for those interested in evaluating the performance of the rules, is a link to a linear model (plant) expressed in **state space**. In control parlance, this format "closes the loop" between the rule base and the plant, affording the user the opportunity to see the behavior of the plant as it is controlled by the fuzzy rules. The key features of FULDEK are illustrated using the classic inverted pendulum problem. The state space formulation of the inverted pendulum problem is described in the next section.

52.3 THE INVERTED PENDULUM

The linearized differential equation of motion of the simple inverted pendulum is

$$d^2\theta/dt^2 - \omega^2 * \theta = Q/I_p$$

where Q is the torque applied at the base of the pendulum, I_p is the rotational inertia of the pendulum about its base, $\omega^2 = g/l$, g is the acceleration of gravity, l is the length of the pendulum, and θ represents the angle of the pendulum measured from vertical. The positive sense of θ and its time derivatives is arbitrarily assigned clockwise rotation. This differential equation is undamped and unstable, meaning that as soon as the pendulum begins to tip over, it will continue to tip rather than try to right itself. FULDEK requires the differential equations of motion in state space format. The state space model is

$$\begin{aligned} dx/dt &= A * x + B * u \\ y &= C * x + D * u \end{aligned} \quad (1)$$

where

$$x = \begin{Bmatrix} \theta \\ d\theta/dt \end{Bmatrix}, \quad u = \{Q/I_p\}, \quad A = \begin{bmatrix} 0 & 1 \\ \omega^2 & 0 \end{bmatrix},$$

$$B = \begin{bmatrix} 0 \\ 1 \end{bmatrix}, \quad C = \begin{bmatrix} 1 & 0 \\ 0 & 1 \end{bmatrix}, \quad D = \begin{bmatrix} 0 \\ 0 \end{bmatrix}$$

The vectors x, u, and y are called the state vector, control vector, and observability vector, respectively. The matrix A is called the dynamic matrix, B is called the control power matrix, C is the state observation matrix, and D is the control (sometimes called feed forward) observation matrix. Often, the C matrix is simply the unity matrix and D is null, though with some systems this is not true. (The FULDEK user's manual describes state space modeling in greater detail.) Once the problem is in state space form, names for the variables in the state vector can be assigned. Names are used in place of symbols in FULDEK. Thus, THETA, THETAD, and THETADD refer to Θ and its first and second time derivatives, and TORQUE refers to Q/I_p. The editor option in FULDEK can now be invoked.

52.4 THE EDITOR OPTION

The EDITOR option is loaded at execution start and presents a menu bar. Clicking with the mouse or depressing a short cut key activates all of the menu items. The major menu items are described below.

52.5 THE FILES MENU

The *Files* menu item provides access to all ASCII files that FULDEK uses. The basic FULDEK file is the Fuzzy Rule Base or FRB. The FRB contains all the information that describes the fuzzy variables, the membership functions, and the rules. This file is cryptic, so a Verbose Rule Base or VRB is also available. The VRB describes the fuzzy rule base using easy to read English. However, this file is for the convenience of the user only, it is not a substitute for the FRB. FULDEK also has a built in full screen editor that is useful when the user wishes to modify an ASCII file. All files are accessed with a file-finder familiar to WINDOWS users. From this menu, users can also setup and access the printer and get information about the maximum allowable sizes and capabilities of FULDEK.

52.6 THE EDIT MENU

The *Edit* menu item comprises operations that describe fuzzy variables, membership functions, fuzzy rules, and a feature that enables the user to extract knowledge from observed data. The user may also select the defuzzification and composition methods.

The *Names* operation declares fuzzy variables to be INPUT, OUTPUT or BOTH. Each fuzzy variable has left and right physical values that form the boundaries of its universe of discourse. The INPUT and OUTPUT classifications are self-explanatory. The BOTH classification permits a fuzzy variable to appear as a feedback variable or a chaining variable. For instance, the variable V is a feedback variable in the following rule:

If V is FAST and RETRO-ROCKET is ON Then V is MEDIUM

In the next two rules, Y is a chaining variable:

If X1 is BIG and X2 is BIG Then Y is SMALL
IF Y is SMALL and X3 is BIG Then Z is VERY SMALL

FULDEK does not have forward or backward chaining inference engines as used in expert systems. The chaining example above would need two steps to determine that Z is VERY SMALL if X1, X2, and X3 are all BIG.

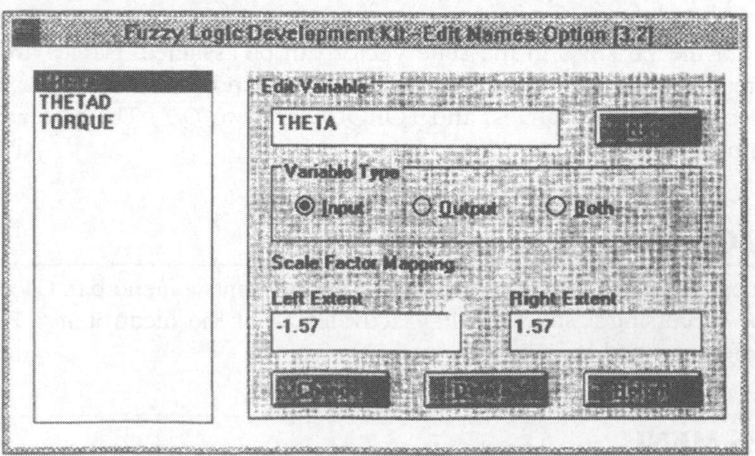

FIGURE 52.1. Defining the attributes of a variable in FULDEK.

The *Names* operation also scales and maps fuzzy variables by using the left and right physical values. Scaling and mapping reduces a real variable with physical values to a scaled, nondimensional value extending from –1 to +1, i.e., the normalized fuzzy variable has a universe of discourse that extends from –1 to +1. A typical variable definition is shown in Figure 52.1.

The *Membership* operation describes and builds fuzzy membership functions. Fuzzy membership functions describe the degree of belonging that a variable has in a given set. Fuzzy membership functions come in six standard types, plus a user-defined type in FULDEK. The standard types are singleton, rectangular, triangular, trapezoidal, Gaussian distribution, and sigmoidal. The *AutoGen* feature quickly builds and distributes these functions, lessening the user's editing effort. The user may select either linear or nonlinear distribution of the function centers. Alternatively, the user may manually build a function with up to nine x–y pairs. An instruction bar at the bottom of the screen guides the user through the process. In Figure 52.2, the membership function TTH_POS1 is shown. The user sees the numerical values that define the function, and a picture of the function with a simple click of the mouse.

The *Defaults* operation lets the user select the composition method and the defuzzification method. Composition refers to the way that the value of a compound antecedent is calculated. The standard conjunction method uses the MIN function, and the standard disjunction method uses the MAX function. FULDEK also offers multiplication as a substitute for MIN, and limited addition as a substitute for MAX. Defuzzification refers to the manner in which conclusions from competing rules are combined to arrive at a best answer. Kosko (1992) describes three methods:

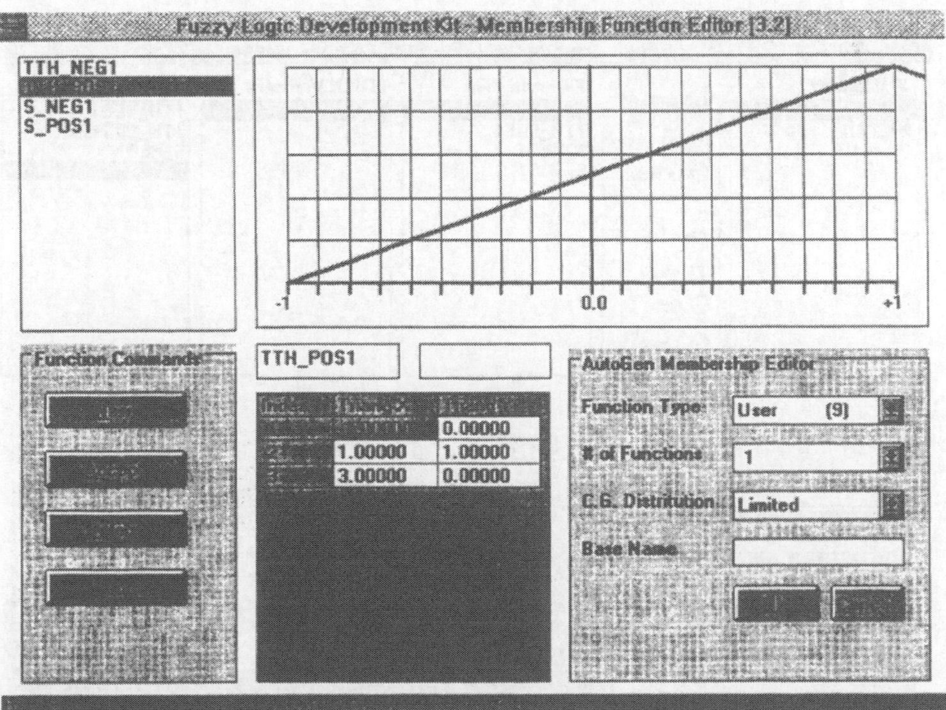

FIGURE 52.2. Examination of a membership function.

Correlation Minimum Inference, Correlation Product Inference and Max-Product (Winner-Take-All) Inference. FULDEK offers all three methods.

The *Rules* operation works intuitively, enabling the user to write rules in a natural manner. FULDEK uses a point-and-click rule editor and has an instruction bar at the bottom of the page that gives the user instant hints on how to write a rule. Start with the antecedents, click a fuzzy variable, click the IS or IS NOT function, and then click a membership function. If more than one antecedent clause is required, simply click AND, or OR, and build the second antecedent clause in the same manner. Up to four antecedent clauses may be used. The conclusions are built in a similar manner. FULDEK allows multiple conclusions (up to three) in a rule. In many cases, this reduces the number of rules a user must write. Just before accepting a rule into the rule base, the rule building page looks something like Figure 52.3.

Before leaving the *Edit* menu, some of the smart features of the editor deserve mention. If the user wishes to remove a fuzzy variable from the FRB, FULDEK checks the rules for that variable. If the variable is found, FULDEK warns the user and then offers three options. The first is simply to stop the deletion process. The second is to continue with the deletion process by discarding the entire rule that has the variable. The third is to continue with the deletion by proposing a new rule which does not contain the variable, if possible. See Figure 52.4 for an example of this feature. The same sort of check takes place when the user tries to delete a membership function. This option reduces the entry and editing time substantially. Another time saver is the rule editor. If it is determined that a particular rule has a useless antecedent clause or a useless conclusion clause, then the user simply displays that rule in the rule grid with the *find, next*, or *previous* commands, selects the clause, and presses the Delete key. The selected clause is removed, and the lower clauses move up to fill the gap.

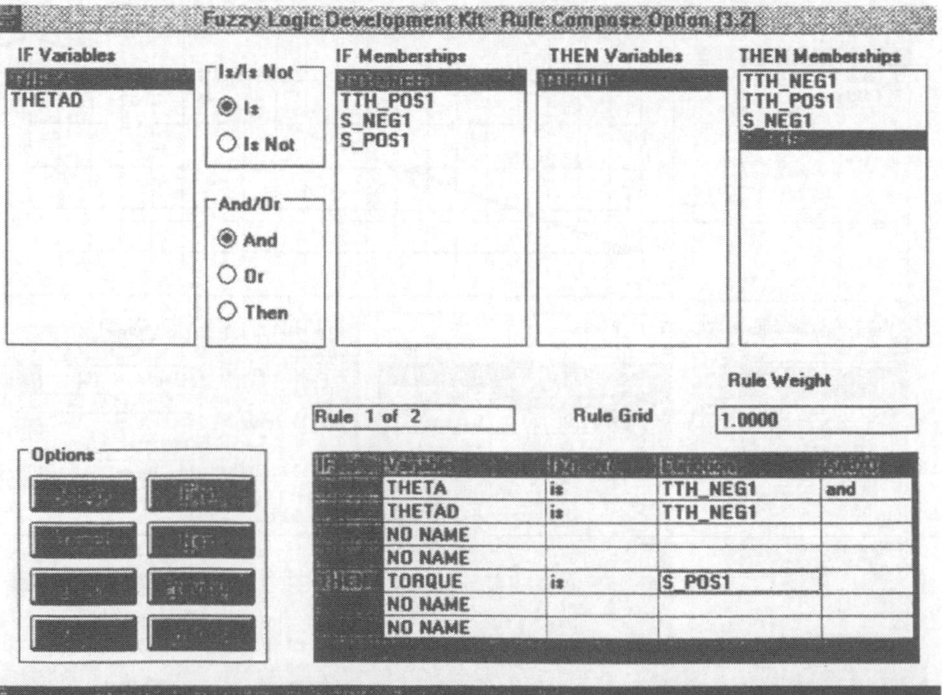

FIGURE 52.3. Writing a rule with the rule editor.

Rule 1		Rule 2	
If	THETA is POS and THETAD is POS	IF	THETA is NEG and THETAD is NEG
Then	TORQUE is NEG_Q	Then	TORQUE is POS_Q

Fuzzy Logic Development Kit - Main Menu [3.2]

File Edit Run Palette

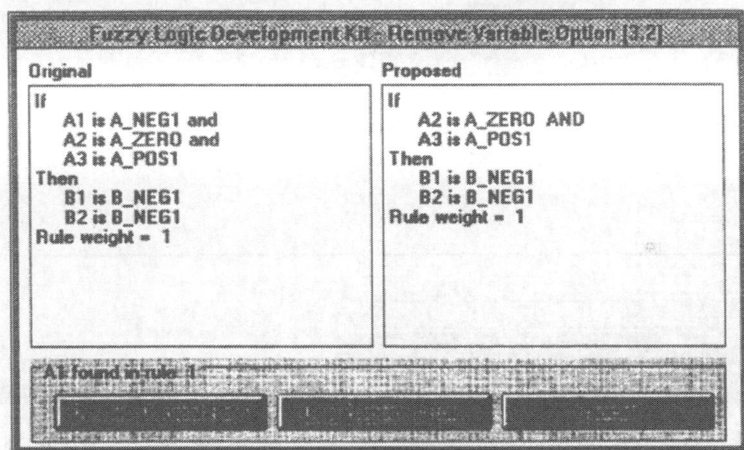

FIGURE 52.4. FULDEK proposes a new rule when variable A1 is deleted.

Once the rules are complete, FULDEK is ready to test the rules. The first two tests are available on the editor screen. They examine the universe of discourse by variable and by rule to look for obvious gaps in the rules. If the user is satisfied, then the RUN option is called for in-depth testing.

52.7 THE RUN OPTION

The RUN Option allows the user to exercise the fuzzy rule base in a number of ways including STEP, MAP, and ABCDE Sim. Each of these options is illustrated by executing a set of rules that regulates an inverted pendulum. We will use a two-rule inverted pendulum controller:

While it may seem that two rules are missing, it is shown in Cunningham (1994) that by clever definition of the linguistic terms POS and NEG, these two rules are sufficient. Let us examine the options available to us on the RUN option screen.

The *Step* option steps through the rules, sweeping one fuzzy variable while holding all others at initial values set by the user. The exception to this is those variables that are typed as BOTH; they vary as the rules progress. The output lists all fuzzy variables and their scaled and unscaled values for a particular instance of the sweep variable.

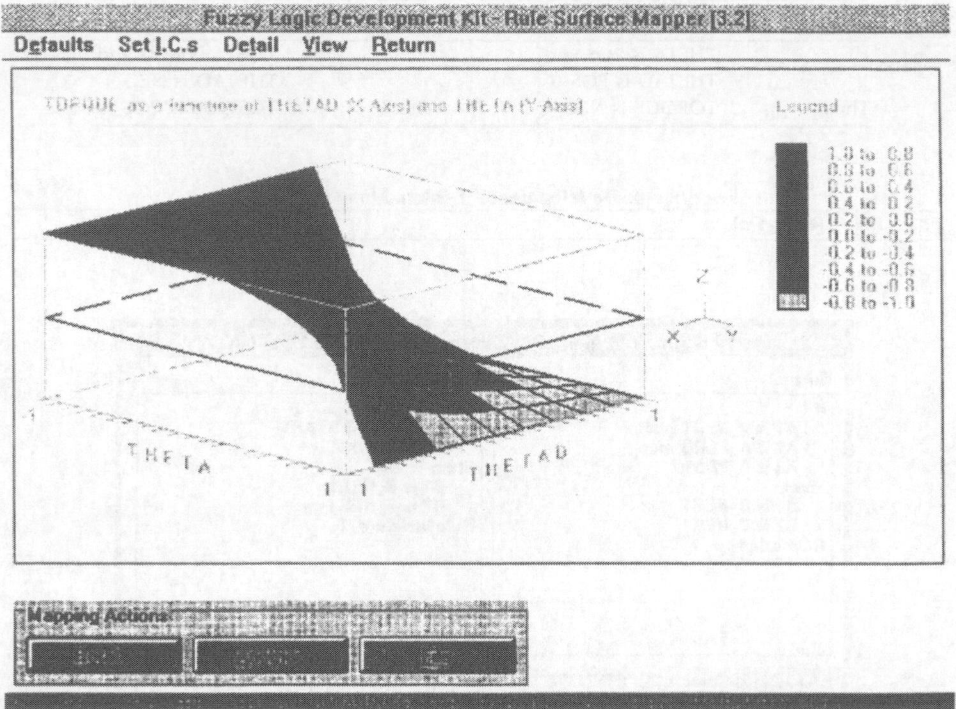

FIGURE 52.5. The rule surface for a two-rule controller of an inverted pendulum is examined.

The *Map* option offers three graphical formats to examine rule interaction. These formats (views) are X-Y, Contour, and 3-D. With any of these options, the user selects one or two fuzzy variables to sweep, and one fuzzy variable to examine as output from the rule base. All other variables, except those typed as BOTH, are held at user defined initial values. The user may also select the detail in the view, either coarse or fine. The two rule controller generates the 3-D surface in Figure 52.5. In some cases, the surface is presented edge on. To see the surface from different perspectives, a ROTATE button turns the surface 90° counterclockwise about the vertical axis with each click. The MAP option provides a way to detect unintentional discontinuities in the rules, gaps in the influence of a variable on the outcome, and a succinct summary of the control logic over the range of each variable. The RUN option tools above provide excellent examination of the fuzzy rules in a standalone sense. They are not actually controlling anything.

To explore the behavior of the rules in a "closed-loop simulation", FULDEK is equipped with a link tool that ties the fuzzy rule base to a dynamic simulation model expressed in state space. The *ABCDE Sim* operation performs this link. As described above, the state space model for the inverted pendulum is

$$\begin{aligned} dx/dt &= A * x + B * u \\ y &= C * x + D * u \end{aligned} \tag{2}$$

where x is now a column vector with θ and $d\theta/dt$ and u is a column with Q/I_p. To this model FULDEK adds two additional matrices and another vector. The matrices are B_e and D_e and represent the control power matrix and feedforward matrix for an external input or disturbance vector called u_e. Thus, the FULDEK state space model is

$$dx/dt = A * x + B * u + B_e * u_e$$
$$y = C * x + D * u + D_e * u_e \quad (3)$$

FULDEK can express up to 20 first-order differential equations in this way.

In the closed-loop simulation, the time varying inputs to the fuzzy rules are the outputs of the dynamic plant, and the inputs to the dynamic plant are the outputs of the fuzzy rule base. The details of the link are beyond the scope of this chapter. Briefly though, the user merely matches a fuzzy variable name with an element in the any of the vectors described above. An interactive screen lets the user create these links easily.

The state space model is stored in an ASCII file. The external inputs are defined with a second time history file. The user selects the integration step size and end time, and the variables to be plotted as a function of time. The user may also select regular time history simulation in X versus Time format, or a phase plane analysis that plots two states against each other as a function of time.

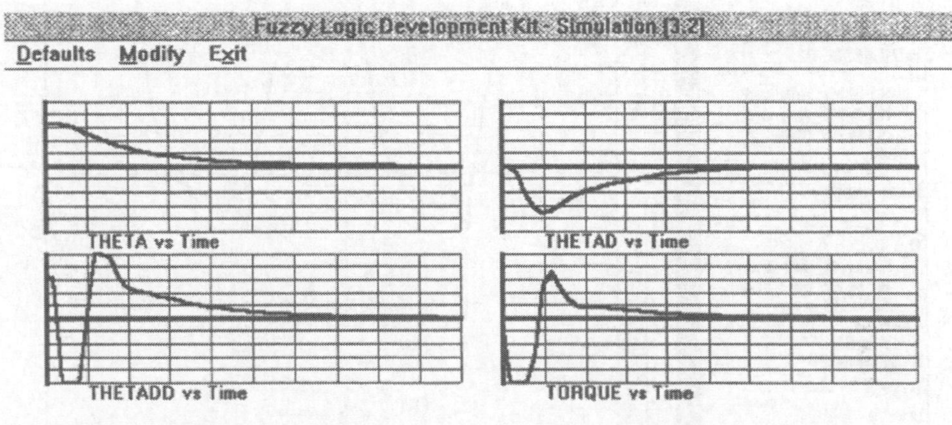

Time runs from 0 to 1.02

FIGURE 52.6. A time history of an inverted pendulum controlled by a two-rule fuzzy controller.

As can be seen in Figure 52.6, the fuzzy rule base does a very good job of regulating the inverted pendulum. It drives the pendulum (the THETA trace on the upper left grid) from an initial displacement of 1 radian up to vertical without overshoot.

52.8 ADDITIONAL FEATURES IN FULDEK

Two additional features of FULDEK warrant description. The first is an automatic rule generation option. The second is an automatic code generation option.

The *AutoRule* option generates fuzzy inference rules from observed data. That is, given a collection of inputs and an output, FULDEK will generate a series of fuzzy inference rules that emulate the collection. In one sense, this is a knowledge extraction process for static data. However, if the collection of data is a time history of multiple inputs and a single output (MISO), then this is the first step in a parameter identification process. The *AutoRule* option is described in Dreier (1994) and an example of the output is shown in Figure 52.7.

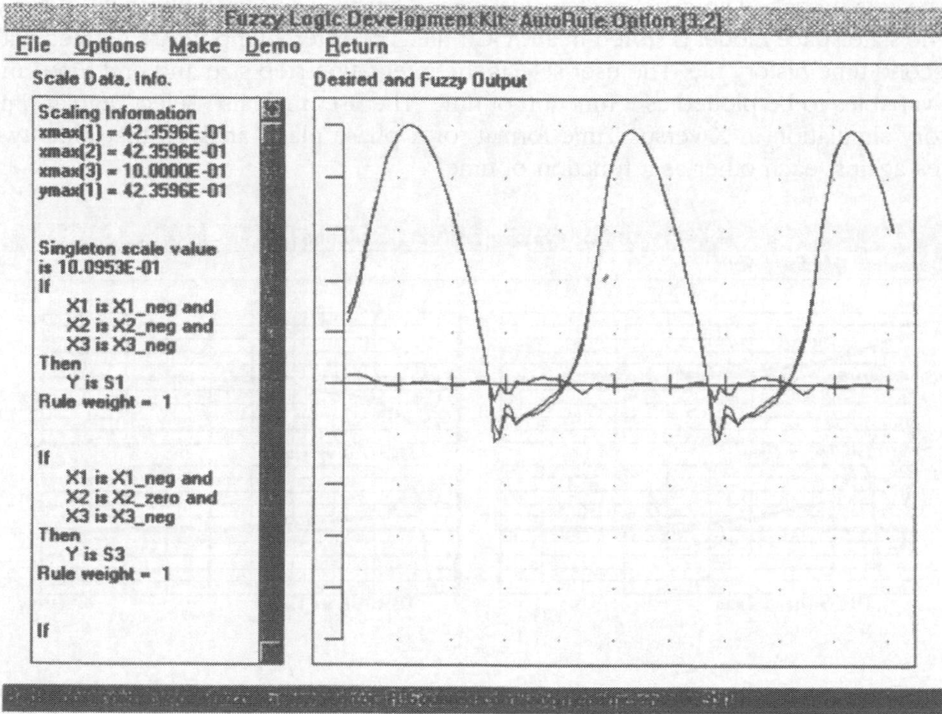

FIGURE 52.7. Rules generated from observed data using the AutoRule option.

The other option is a postprocessor program that reads the fuzzy rule base that FULDEK generates, and writes fuzzy logic controller code in Ada, BASIC, C, or FORTRAN. This is a very useful option when building an embedded controller.

52.9 CONCLUSION

The FULDEK program is a user-friendly program that runs under the WINDOWS environment. Its screens and operations are laid out in an intuitive manner, and it provides many tools to write and test fuzzy logic inference rules. The FULDEK program has matured and will continue to mature, with the primary goal to make fuzzy logic affordable for the beginner and yet comprehensive for the experienced user. Ways to use fuzzy logic are limited only by imagination; FULDEK is a good tool for getting there.

DEFINING TERMS

Closed Loop Simulation: A mathematical model of a system and its controller working together. The output of the system is the input to the controller, and the output of the controller is the input to the system, thus a closed loop is formed.

Composition: The method used to combine the individual strengths of antecedent clauses to arrive at an overall value for a rule. In classic fuzzy logic, the AND operator (conjunction) uses the minimum of two antecedent clause values, and the OR operation (disjunction) uses the maximum. In recent years, multiplication has also been used for conjunction and limited addition or numerical average has been used for disjunction.

Defuzzification: The method to obtain a "crisp" value for an output variable from fuzzy inference rules. Defuzzification resolves conflicts between rules by weighting the membership function for the conclusion variable of each rule, and then combining the conclusion variable values. The weighting value is the rule value, and the combination method can be a centroid calculation or simply a winner-take-all selection.

Fuzzification: The process of determining to what degree a variable belongs to a fuzzy set defined by a membership function. If the variable lies outside of the shoulders of a function, its fuzzified value is zero, otherwise, it takes a value between 0 and 1, where 1 indicates maximal belonging to a fuzzy set.

Fuzzy Logic: A multivalued or multivalent logic that tolerates imprecision and inexact knowledge. Fuzzy logic admits shades of gray, and linguistic concepts such as "maybe," "somewhat," etc. Boolean logic with its TRUE or FALSE only is a special limiting case of fuzzy logic.

Fuzzy Inference Rules: Linguistic expression of rules, a generalized Modus Ponens construction. Given the statement "A is TRUE implies B is TRUE," and if A is indeed true, then B will be true. This is an example of strict Modus Ponens. If, however, A is only partially true, then B would be false in strict Modus Ponens, but may be partially true in generalized Modus Ponens.

Membership Functions: A function that defines a region in which a real variable may be classified (fuzzified). They cover ranges of real variables and specify the amount a real variable belongs to that range. The functions are usually simple in shape and analytic to ease computational burden. Favorite functions are triangles, trapezoids, and Gaussian distribution curves.

State Space: A representation of one or many higher order differential equations in the form of many simultaneous first-order differential equations. Each equation represents a state of the system being modeled.

Universe of Discourse: The entire range (universe) that a variable may take or speak in (discourse). By defining a universe of discourse, an implicit limit has been placed on the values a real variable may take.

REFERENCES

George A. Cunningham, Stability and performance of two rule fuzzy systems. *J. Intelligent Fuzzy Sys.*, in press.

Mark E. Dreier, A fast, non-iterative method to generate fuzzy inference rules from observed data. *J. Intelligent Fuzzy Sys.*, 3(2):181–185, 1995.

Bart Kosko, *Neural Networks and Fuzzy Systems—A Dynamical Systems Approach to Machine Intelligence*. Prentice-Hall, Englewood Cliffs, NJ, 1992.

FOR FURTHER INFORMATION

For more information on FULDEK, including a price list, contact TSIE, Inc., P. O. Box 14155, Albuquerque, New Mexico 87191-4155. Phone (505) 298-5817, FAX (505) 291-0013.

A good overview of fuzzy logic software, hardware, and applications is presented in *Fuzzy Logic and Control—Software and Hardware Applications* by Mohammad Jamshidi, Nader Vadiee, and Timothy J. Ross. The authors present the fundamentals of fuzzy logic in lay terms, and present classroom problems with real world application.

A complete description of the capabilities and features of FULDEK is found in the user's manual, *Learning to Use the Bell Helicopter Textron Incorporated Fuzzy Logic Development Kit (FULDEK)* by Mark E. Dreier. The manual discusses the use of FULDEK, but also provides an introductory chapter on fuzzy logic, a detailed chapter on automated rule generation, and an important analytical demonstration of the equivalence between a modern suboptimal controller and a fuzzy logic controller.

53

Khoros

Gregory W. Donohoe
University of New Mexico

53.1 THE KHOROS ENVIRONMENT

Khoros is widely used in scientific circles for image and signal processing, image analysis, and pattern recognition. Available free by anonymous FTP, Khoros runs on Unix workstations under XWindows. This chapter discusses Khoros Version 1.0, Patch 5. The major features are

- Over 250 routines for image and signal processing and pattern recognition
- The Cantata visual programming language
- A common date file format
- A machine-independent system that operates over networks
- A set of application generator tools for extending the system

In addition to libraries for digital image and signal processing and pattern recognition, Khoros has utilities for

- Interactive image editing (**editimage**)
- Two- and three-dimensional plotting (**xprism2, xprism3**)
- Image sequence animation (**animate**)

The Khoros functions are stand-alone programs called *vroutines* or *xvroutines*. Routines that perform similar functions or serve similar application areas are grouped into collections called *toolboxes*. Example toolboxes are matrix arithmetic and grayscale morphology. File conversion routines import and export many common file formats, including graphics files like TIFF and mathematical formats like MATLAB. Khoros provides an interface to many of the public-domain Linpack linear algebra routines.

Khoros was created to serve as a common user and development environment to encourage research teams to share their work. The principal design goals are *flexibility, portability*, and *extensibility*: the user should not be constrained by limits built into the system. Source code for the entire system comes with the distribution. Application

0-8493-2530-7/96/$0.00+$.50

generation tools generate the boilerplate code required for file and error handling and for generating the graphical user interfaces, shielding the developer from the vagaries of xlib, and greatly simplifying the development of Khoros-compatible software. As a result, numerous research organizations have developed application-specific programs and toolboxes. Many of these have been contributed to the public domain, and are available by anonymous FTP along with Khoros itself.

53.2 DATA REPRESENTATION

Most Khoros routines use the Visualization/Image File Format (VIFF). Each file consists of a 1024-byte *header*, an optional *data area*, and zero or more *maps*.

All VIFF data are stored as a matrix of up to three dimensions, organized as one or more rows, columns, and bands (Fig. 53.1). If the file represents an image, the matrix elements are pixels. A file containing a single scalar value is a special case, with one row, one column, and one band. A monochromatic image might have 512 columns, 480 rows, and one band. A full-color image would have three bands, one each for the red, green, and blue component. An image sequence containing 10 frames, or snapshots in time, would have 10 bands. Khoros follows the image processing raster convention: the horizontal axis units (columns) increase to the right, and the vertical axis units increase downward. Row, column, and band indices start at 0, so a file with dimension M × N × P has column indices from 0 to M–1, etc.

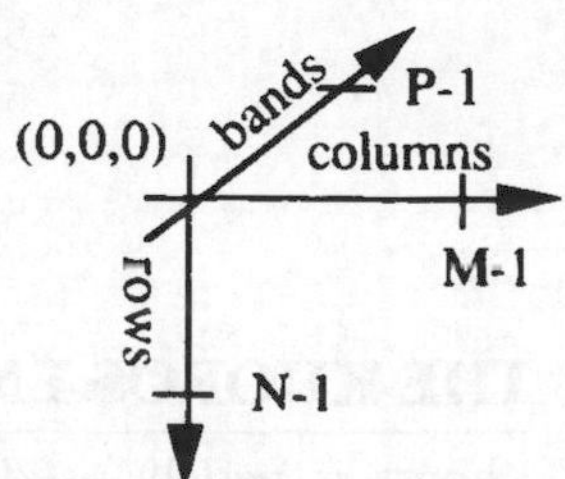

FIGURE 53.1 VIFF file indexing conventions.

The scalar elements that make up the VIFF data area can take most of the common computer data storage types: bit, byte, short integer (usually two bytes), long integer (usually four bytes), floating point, double precision, and complex. In addition to the standard types, there are "meta types" that have special meaning to Khoros programs. One is Byte Binary, in which the foreground regions of a binary image are stored as unsigned bytes with the value 255, and the background region, 0. Another meta type is Labeled Integer. Region labeling is a common step in image analysis. An image is segmented into regions that have meaning to an application. So that subsequent steps can identify the regions, each connected region is assigned a unique numerical value. The background might have value 0, one labeled foreground region value 1, another 2, and so on. These concepts are illustrated in Figure 53.2.

The VIFF *header* contains descriptive information like the number of rows, columns, and bands, data storage type (byte, float), machine type (Sun, Decstation), number of maps, etc.

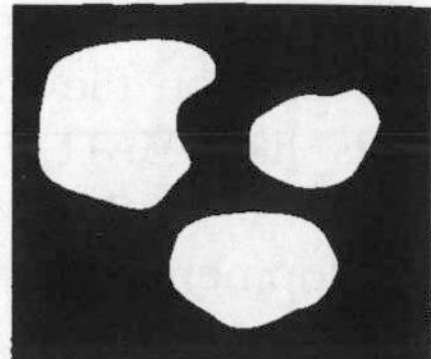

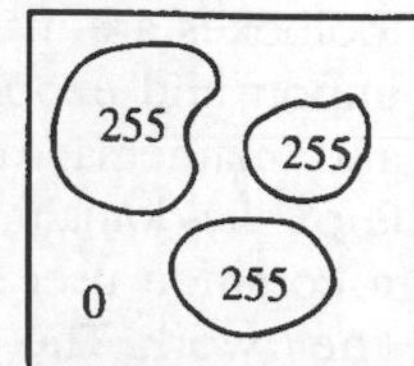

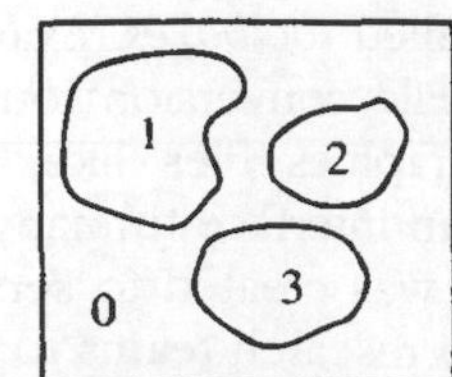

FIGURE 53.2 (a) Segmented image with background and three foreground "blobs." (b) Byte binary format: background = 0, foreground = 255. (c) Labeled integer: background = 0, each connected foreground blob has a unique grayscale value.

Maps, also called lookup tables or LUTs, are commonly used to create pseudocolor and 8-bit-encoded full color images. A pseudocolor image has one 8-bit data area and three maps, one for each of the red, green, and blue display channels of a workstation screen. When a pseudocolor image is displayed, each 8-bit datum serves as an index into each of three arrays; the resulting pixel has a mixture of red, green, and blue, as specified by the map. Khoros supports full-color (24 bit) images, but most workstations are limited to 8 bit color. An RGB image is encoded into an 8-bit image and three maps in a scheme sometimes called *indexed color*. When such an image is encoded in a VIFF file, the maps serve to decode the 8-bit image so the workstation can display something akin to true color. In Khoros, the map concept has been extended, and maps can be used to store user-specified information, such as calibration data. See *Digital Image Processing* by R.C. Gonzalez and R.E. Woods, Addison-Wesley, Readings, MA, 1992, for more on image processing and display.

Figure 53.3 illustrates the four possible VIFF file arrangements. Khoros provides tools for inserting, extracting, and modifying data and maps, making for a very flexible environment for image manipulation.

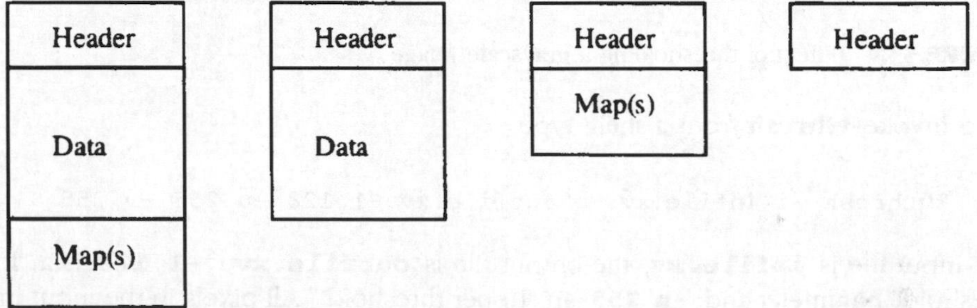

FIGURE 53.3 Four possible VIFF file combinations.

53.3 LIBRARY ROUTINES

The Khoros application library programs are divided into *vroutines* and *xvroutines*. Xvroutines, like the **editimage** interactive image editor, contain a built-in graphical user interface. Patterned after the Unix text filters, vroutines normally accept one or more VIFF files as input, and output one or more VIFF files. Some example vroutines are

- Image arithmetic. Add, subtract, multiply, or divide two image. Apply a point operation to each pixel: take the log, the square, the square root, etc.
- Image manipulation. Rotate, transpose, shrink, expand, or warp an image. Extract a subimage and insert it into another image. Pad an image.
- Histogram manipulation and contrast enhancement. Color manipulation.
- Spatial filters. Convolution filters, edge detection, mathematical morphology.
- Discrete Fourier transform and frequency domain filters. Inverse and Wiener filters.
- Image labeling and segmentation. Extracting image features like texture properties or fractal dimension.
- Statistical classification.

Khoros routines can be run from the Unix shell, or within the Cantata visual language (Section 53.4). When run from the shell, all arguments are preceded by a minus sign and a tag. Thus, **-i file1** indicates that **file1** is an input file, while **-o file2** designates an output file called **file2**. We will illustrate with a thresholding program called **vthresh**, which converts a grayscale image into a byte binary image having only two values, black and white, illustrated in Figure 53.4. The lefthand picture is an 8-bit grayscale image. The right-hand image has been thresholded; the white pixels correspond to grayscale values greater than 128 in the left-hand image.

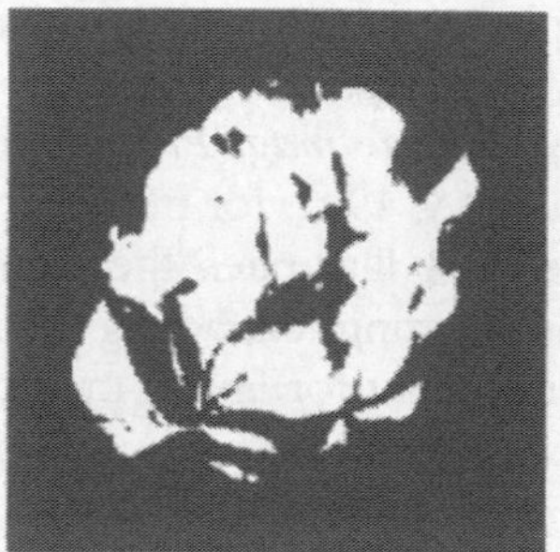

FIGURE 53.4 Effect of thresholding a grayscale image.

To invoke **vthresh** from a shell, type

```
%vthresh -i infile.xv -o outfile.xv -l 128 -u 255 -v 255
```

The input file is **infile.xv**, the output file is **outfile.xv. -l 128** is a "lower threshold" parameter and **-u 255** an "upper threshold." All pixels in the input image with gray levels between these two numbers are given the value 255, or white, specified by **-v.** All others are replaced by gray value 0, or black.

The order in which the arguments appear doesn't matter; we could just as well have typed

```
%vthresh -u 255 -o outfile -v 255 -i infile -l 128
```

Each vroutine has its own set of parameters. To learn what these are for **vthresh**, type

```
%vthresh -U
```

or simply

```
%vthresh
```

To be prompted for input parameters one by one, type

```
%vthresh -P
```

For a manual-page description of **vthresh**, type

```
%vman vthresh
```

53.4 THE CANTATA VISUAL PROGRAMMING LANGUAGE

Cantata serves as a graphical substitute for the Unix shell. Based on a data flow program model, Cantata uses a combination of menus and process glyphs (icons) to construct an application.

To run Cantata, type

```
%cantata &
```

in a terminal window. Cantata takes all its input from the XWindows interface; placing it in the background with "&" frees up the window for other work.

When Cantata starts, you will see a workspace, shown in Figure 53.5a. The rectangular objects inside the workspace are icons or "glyphs." (When you first run Cantata, there are no glyphs: the workspace is blank.) Each glyph represents a program or Unix process, usually a vroutine. The glyph marked **input** acts a source of data, such as an image. **put_update** is an image display program; the displayed image changes when the input changes. **vthresh** is the thresholding routine we used in the previous example.

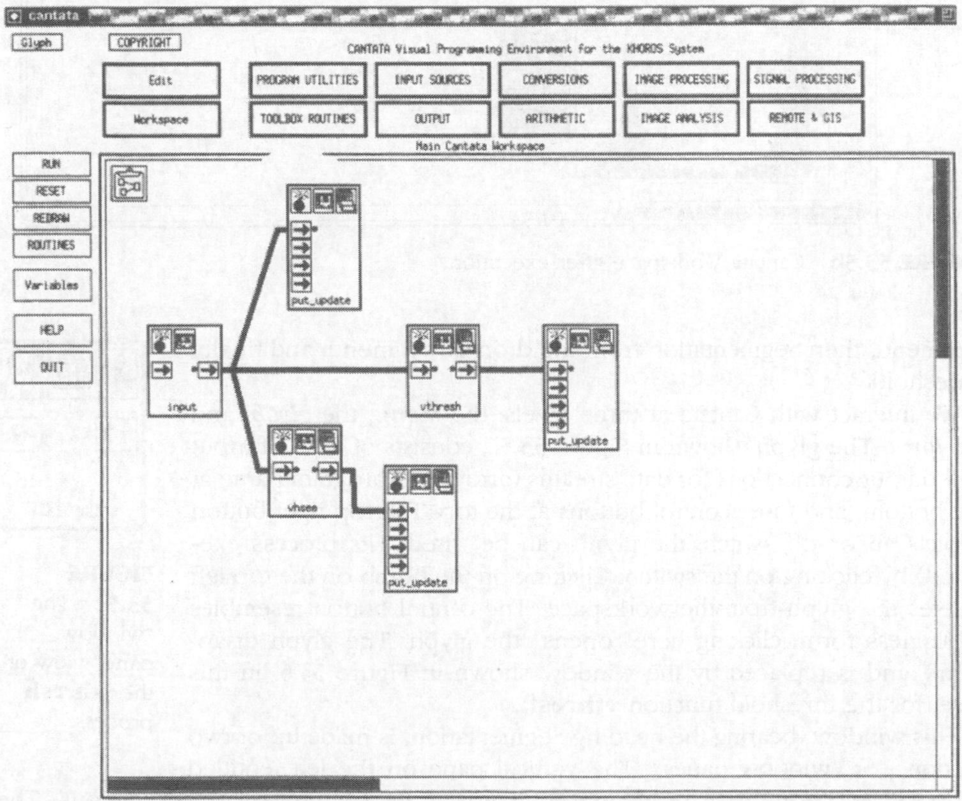

FIGURE 53.5a Cantata Workspace before execution.

The lines connecting the glyphs carry both data and process status information. The buttons arrayed on the left select workspace actions, like RUN and RESET. The two rows of menu buttons along the top produce drop-down menus that are used to construct programs. To get a **vthresh** glyph, select IMAGE ANALYSIS from the

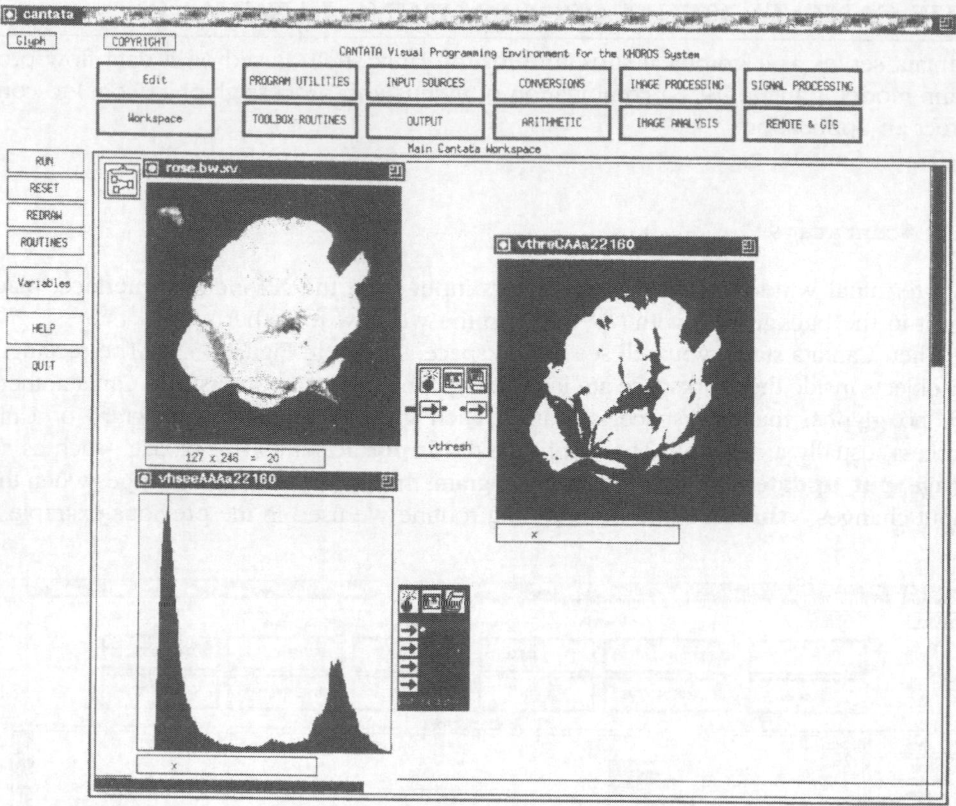

FIGURE 53.5b Cantata Workspace after execution.

top menu, then Segmentation from the drop-down menu, and finally Threshold.

We interact with Cantata at three levels: the *menu*, the *glyph*, and the *form*. The glyph, shown in Figure 53.5c, consists of a set of input and output connections for data streams (arrows), a program name at the bottom, and three control buttons at the top. The top right button depicts an on-off switch; the glyph can be "fired" (its process executed) by clicking on the switch. Clicking on the bomb on the top left deletes the glyph from the workspace. The central button resembles a business form; clicking here "opens" the glyph. The glyph disappears, and is replaced by the window shown in Figure 53.6, in this case, for the threshold function **vthresh**.

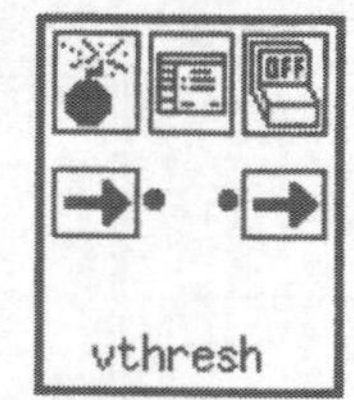

FIGURE 53.5c The "window pane" view of the **vthresh** process.

This window, bearing the heading Segmentation, is made up of two sections, or "window panes." The vertical pane on the left, entitled "Choose Selection," is a **guide pane**. It lists a set of segmentation functions. The Threshold function is highlighted, indicating that (1) the Threshold glyph is chosen, and (2) the pane on the right is the user interface for the Threshold function. The Threshold pane lists the name of the input file (rose.bw.xv) and an output file, which has been generated by Cantata. There are three numerical user inputs: Lower Threshold Level, Upper Threshold Level, and Nonzero Pixel Value; each is preceded by a small square. The Lower Threshold Level and Upper Threshold Level squares are

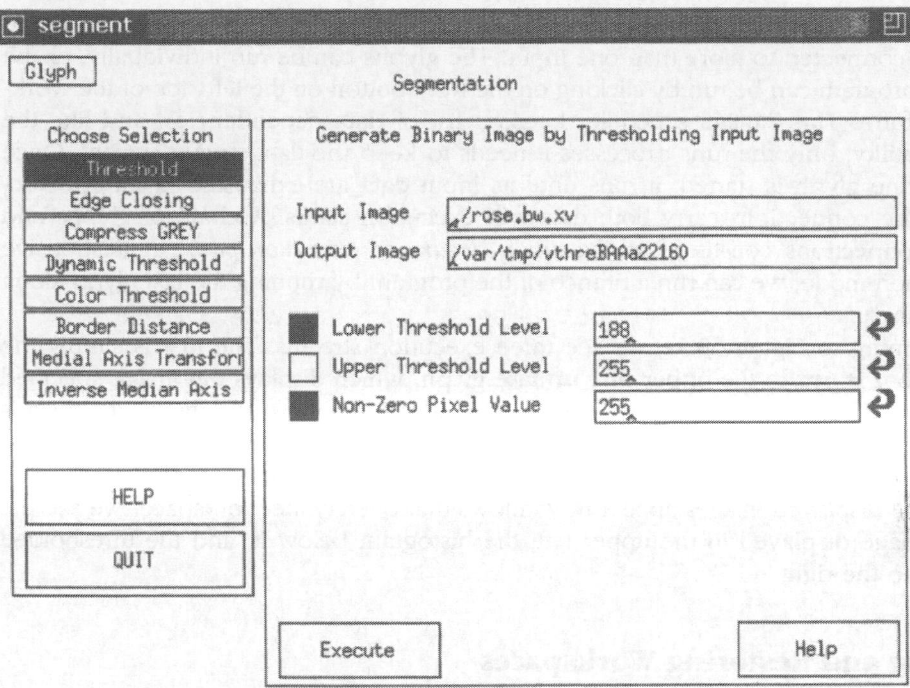

FIGURE 53.6 The "windowpane" view of the **vthresh** process.

darkened; this indicates that the parameters are selected, or active. (Optional inputs are normally inactive, but may be activated by clicking in the corresponding square.) The pane now describes a simple threshold: pixel values below the threshold (188) become white (255), and those below, black (0). Had the Upper Threshold Level been selected, and an upper level entered (say, 210), we would have a windowing function: pixels between 188 and 210 would become white, and those outside this range, black.

The stylized return arrows at the end of the text boxes indicates that these selections are **active**. This has two consequences. First, if Cantata is in "Responsive Execution" mode, as soon as the user hits return in this window, Cantata reruns the Threshold process and all processes "downstream" from it. The second consequence is a side effect: if you type in a new numerical value, this is not made available to the program until the you type return. (In unadorned text boxes, changes are registered without the carriage return.)

The button labeled Execute runs the program, just as if we had click on the ON/OFF switch on the glyph. Like a shell, Cantata searches the user's execution path for the first occurrence of an executable file whose name matches the glyph name. The QUIT button on the guide pane has the same effect as clicking on the bomb: the glyph is deleted. *Every level has a HELP button*. As we shall see, when users write their own routines, Khoros encourages them to create their own HELP facilities. This is a tremendous boon when members of a research team contribute code to a common library, at least to the extent that users/programmers are conscientious about entering useful text in the HELP section.

To construct a Cantata program, select the desired processes (glyphs) from the menu and place them on the workspace. To connect the output of one glyph to the input of the next, click on the output arrow of the source glyph, then on the input arrow of the destination. A thick line appears to indicate the connection. An output can be connected to more than one input. The glyphs can be run individually, or the entire program can be run by clicking on the RUN button on the left side of the workspace form. The Cantata scheduler keeps a list of data dependencies, and like the make utility, only the runs processes it needs to keep the data stream current. Once a vroutine glyph is started, it runs until its input data are exhausted or an error occurs. The connections carry both data and execution status. A chain of glyphs and their connections constitute an *execution stream* or *execution path*. In Responsive Execution mode, we can run a branch of the program by running the last glyph along the branch.

Returning to Figure 53.5a, we see three execution streams. The first is simply the path from **input** to the upper **put_update** glyph, which displays the image specified in **input**. The second is the path to **vthresh** and a second display glyph. The third is the path from **input** to **vhsee** and **put_update**. **vhsee** computes the grayscale histogram of the image for display. Figure 53.5b shows the workspace after it has been run. The displayed images appear in windows that overlay the workspace. We see the rose image displayed in the upper left, the histogram below it, and the thresholded image to the right.

Saving and Restoring Workspaces

If we view Cantata as a graphical shell, then glyphs and connections in the workspace constitute a shell script. If we think of Cantata as a visual language, then the workspace is a program. To save a workspace to a file, select Workspace from the Cantata menu, then File Utilities; use the browser, or type in a file name. The workspace is compactly encoded as a text file, which Cantata can reload and execute. As a text file, the workspace can be edited, annotated, and e-mailed to colleagues. This is a handy way to share work over the network.

Expressions, Variables, and Program Control

Like a text-based shell, Cantata supports "shell variables," which can be used in mathematical expressions and in program control. Cantata's built-in parser evaluates expressions and assigns values to variables before they are used. Virtually everywhere you can type in a numerical parameter, you can use an expression instead. Relational expressions, like (x<y), evaluate to TRUE (nonzero) or FALSE (zero). This lets us use them in control structures like loops, IF-THEN-ELSE branches, and procedures.

Branches

IF-THEN-ELSE branches are implemented with the branch glyph shown in Figure 53.7. The input connection to the glyph is an execution stream, and the two outputs represent possible execution paths. Within the glyph, an expression is tested. If the expression evaluates to TRUE (nonzero), the upper path is executed; if the expression evaluates to FALSE (zero), the lower path is taken. Unlike a text-based language, the convergence of the two execution paths must be made explicit with a Merge glyph.

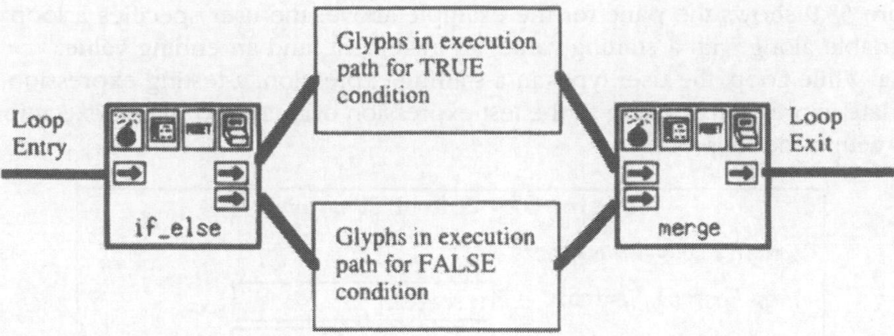

FIGURE 53.7 Conditional branch.

Loops

Cantata supports *counting loops*, analogous to the Fortran "do loop" or the "for loop" in C, and *while loops*. Externally, the two kinds loops are similar; they differ in the way the controlling conditions are specified. We illustrate loop construction with the counting loop below. The rose image is corrupted by impulse noise; to clean up the image, we wish to apply a median filter five times. We select a Counting Loop glyph from PROGRAM UTILITIES in the main menu. The noisy rose image is connected to the upper input arrow, which is the loop entry point. The body of the loop consists of the median filter (**vhmed**) glyph, connected between the lower exit arrow and the lower entrance arrow. The loop exits through the upper output arrow, which is connected to the **put_update** image display glyph. The Counting Loop is configured to execute the loop five times. The filtered image is shown in the upper right (Fig. 53.8).

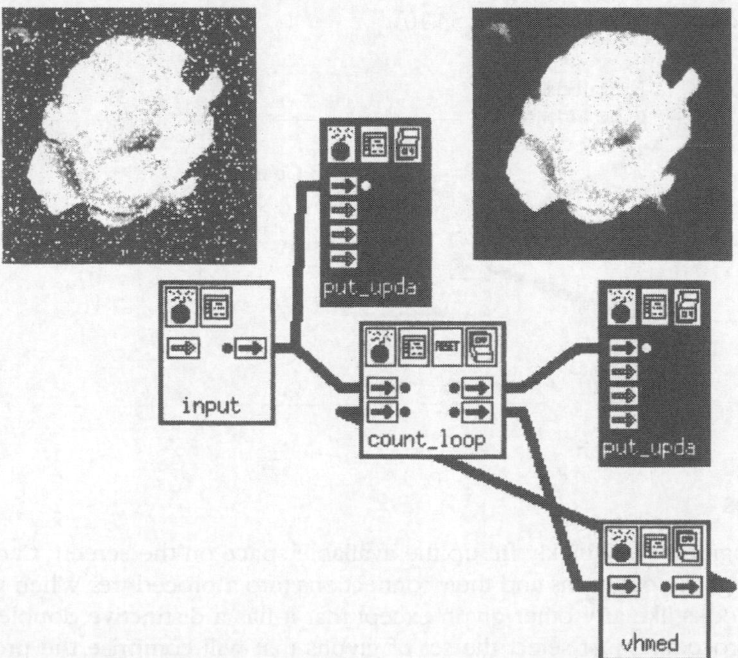

FIGURE 53.8 Removing impulse noise with repetitive applications of a median filter in a counting loop.

Figure 53.9 shows the pane for the example above: the user specifies a loop control variable along with a starting value, an increment, and an ending value.

For a While Loop, the user types in a starting expression, a testing expression, and an update expression. As long as the test expression evaluates to TRUE, execution remains within the loop body.

Counting-Loop Control Structure

Counting-Loop Parameters:

Loop Control Variable	iteration
Initial Value	1
Increment Value	1
Final Value	5

FIGURE 53.9 Part of the pane for the Counting Loop glyph.

Process Synchronization: the Trigger Glyph

A complex Cantata workspace resembles a tree or a forest, with execution paths bifurcating (branching out) as we more from input to output. Sometimes the execution path in one branch of this tree depends on the results of a computation in another branch. In this case, we want to block execution in the dependent branch until the controlling process has completed. The Trigger Glyph allows us to make this dependency explicit to the Cantata scheduler, and thus synchronize the processes. The top input of the trigger glyph is the execution path to be blocked; the bottom input is the controlling process (Fig. 53.10).

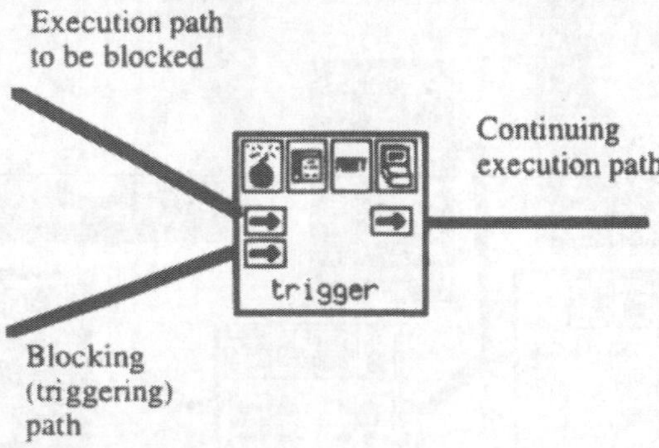

FIGURE 53.10 Trigger glyph.

Procedures

Graphical programs can quickly fill up the available space on the screen. Cantata lets us combine a group of glyphs and their connections into a procedure. When glyphed, a procedure looks like any other glyph except that it has a distinctive double border. To create a procedure, just select the set of glyphs that will comprise the procedure, and from the EDIT menu, select Procedure. Give the procedure a name, and the selected glyphs will collapse into one.

In Figure 53.11 the glyphs **vconvert**, **vscale**, and **voffset** serve to enhance an image for display. **Vconvert** converts the data type to INTEGER (to avoid overflow), **vscale** multiplies each pixel by a constant to increase contrast, and **voffset** adds a positive or negative constant to each pixel, adjusting the brightness. All three have been selected with a mouse click, indicated by their heavy borders.

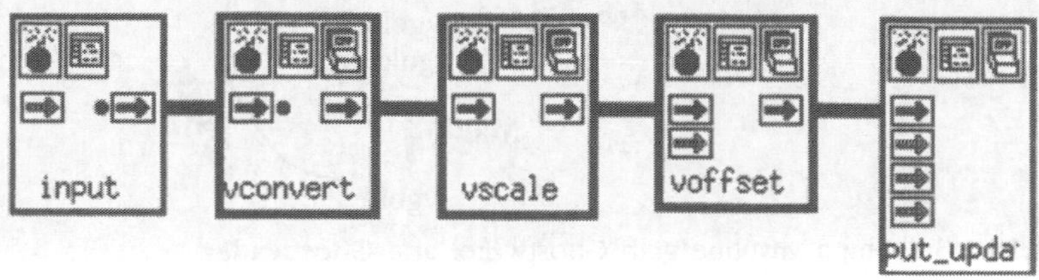

FIGURE 53.11 Selected glyphs to be combined in a procedure.

In Figure 53.12 the three selected glyphs have been combined into a procedure called scale.

When a procedure is opened, we see a second workspace just like the original. We can connect glyphs from the parent workspace directly do those in the procedure. Procedures may be nested.

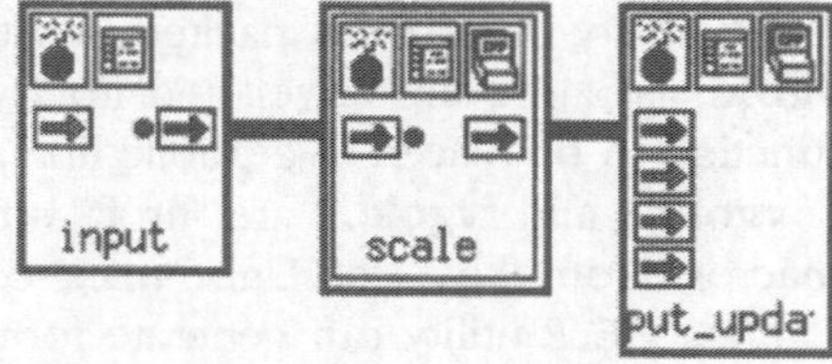

FIGURE 53.12 A procedure.

53.5 EXTENDING KHOROS WITH CUSTOM ROUTINES AND TOOLBOXES

Khoros provides tools for user interface specification and code generation. The vroutines, whose graphical user interface must be interpreted by Cantata, are easier to create and are illustrated here. Xvroutines, with their standalone graphical interface and added flexibility, are produced with a similar set of tools. Once created, custom programs can be run immediately from a shell or from Cantata. For maximum convenience, they should be installed in a toolbox; this makes them available through the Cantata menu. We can only sketch the procedures here; refer to the Khoros *Programmer's Manual* (Khoros Manual Volume 2) for details.

Creating a vroutine

We shall generate a hypothetical vroutine called **vgold**, and install it in a toolbox called **metals**. The figure below illustrates the process for creating **vgold**. Using the Khoros tool **Composer** and text editors, the user creates two source files, **vgold.pane** and **vgold.prog. vgold.pane** is a compact encoding of the user interface: it specifies the number and types of user parameters; whether these are optional or required, default values, if any; and defines mutually exclusive "toggle groups." **vgold.pane** also specifies what will happen when the user clicks the RUN or HELP buttons. **vgold.prog** contains the editing history, help information, and code fragments (in C) that determine what **vgold** will do. From these files, the Ghostwriter tool creates a host of files as shown in Figure 53.13. The Imake utility reads the Imakefile and generates a

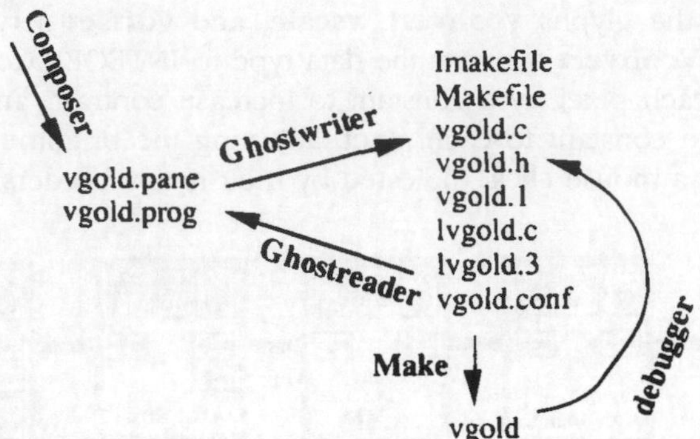

FIGURE 53.13 Building a vroutine with Ghostwriter and Ghostreader.

Makefile.[1] **Ghostwriter** creates header file **vgold.h**, and **vgold.c**, a driver program that manages the UIS, file I/O, error checking, etc. **lvgold.c** is a "library" function: the functionality of **vgold** is packed into this routine, which is called as a C function by **vgold**. Isloating the driver and library routines int separate files allows the **lvgold** function to be placed in a public library so that other programs can use it.

vgold.1 and **lvgold.3** are "nroff" source files written with the manual page (-man) macros. From these, the Unix "man" command, the Khoros **vman** command, and the Cantata HELP utility can generate formatted help pages. (For more on nroff and the format of "man" pages, see Dale Dougherty and Tim O'Reilly, *Unix Text Processing*, O'Reilly & Associates, 1987.) **vgold.conf** is a configuration file used when installing **vgold** in a toolbox.

The Imakefile and Makefile reflect the current environment; just type "make" to build the **vgold** executable file. **vgold** can be tested from the shell, or invoked from the Cantata menu: select WORKSPACE | File Utilities | UIS file, and specify file "vgold.pane." If there are run-time errors, use a debugger and an editor to fix the offending file (**vgold.c** or **lvgold.c**). When **vgold** is working, all is well except that **vgold.pane** and **vgold.prog** no longer reflect the code in the C sources. Run **Ghostreader** to regenerate **vgold.pane** and **vgold.prog** from the other files.

The **.pane** and **.prog** files now completely specify program **vgold**. To share **vgold** with a friend, e-mail these two files. Your friend then places them in a clean directory, removes the e-mail headers, and runs **Ghostwriter** and **make**.

Installing a Toolbox

Twenty years ago, a typical computer "program" was completely specified by a source file and a compiled executable. Today, we may need half a dozen files to run a program. We think of this collection of files as a "program object," which resides in a toolbox directory tree, shown in Figure 53.14.

The directory **src** contains source code. **repos** (for "repository") contains graphical support files (those with .form and .pane extensions) necessary for the graphical user

[1]For a description of the Unix "make" utility, see Andrew Oram and Steve Talbott, *Managing Projects with Make,* O'Reilly & Associates, city 1991. The "imake" utility is distributed with XWindows. A "Makefile maker," imake, extends the make utility. See Paul Dubois, *Software Portability with imake*, O'Reilly & Associates, city 1993.

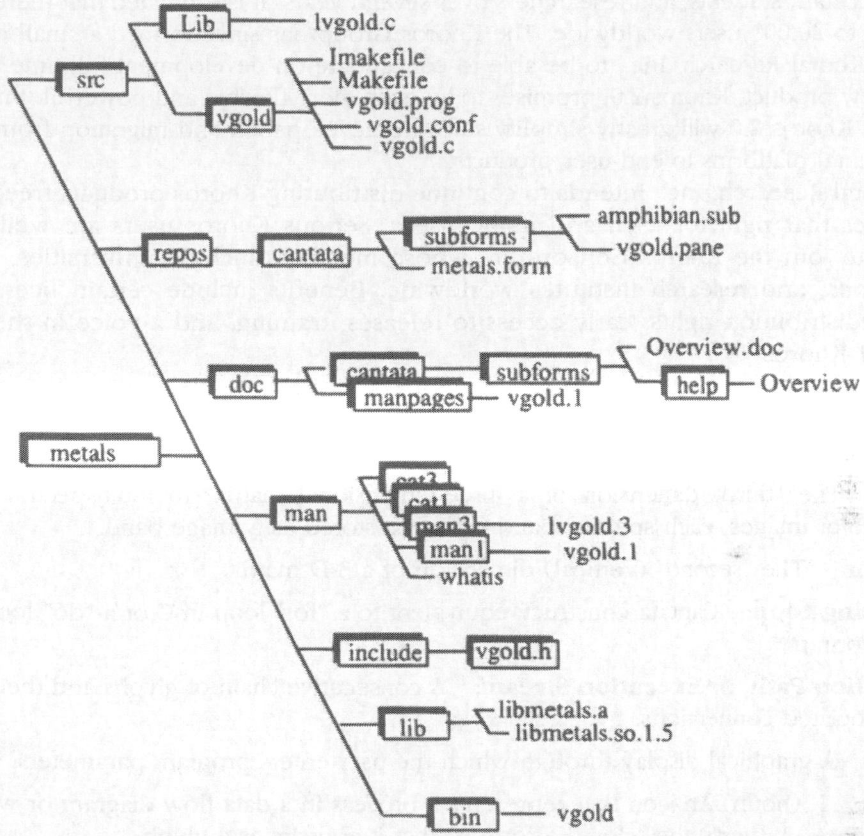

FIGURE 53.14 The **metals** toolbox directory structure, with the **vgold** program installed.

interface. The **doc** directory contains formatted help text; the **man** directory, text for the **vman** command. The **include** directory has global include files. **lib** contains both the compiled library functions for the metals toolbox (**libmetals.a**) and the shared libraries supported on some systems (**libmetals.so.1.5**). Finally, the **bin** directory has the executable file.

The Khoros tool **kraftsman** creates the directories for the toolbox. Program **kinstall** installs individual programs (**vgold**, **vsilver**, etc.). When a toolbox is created or a routine added, some of the graphical support files and documentation files must be edited to reflect the change. To execute programs in this toolbox, you have to define some environment variables and add . . . metals/bin to your search path. If you use shared libraries, the environment LD_LIBRARY_PATH must include . . . /metals/libmetals.so.1.5.

Developing an xvroutine is much like developing a vroutine, using many of the same tools, and a few new ones. See the Khoros *Programmer's Manual* for details.

53.6 SUMMARY

Khoros 1.5 grew out of a project conceived and directed by Dr. John Rasure at the University of New Mexico. Khoros 1.5 is the result of the cumulative efforts of development staff, students, and researchers over several years. It is estimated that there are 10,000 to 20,000 users worldwide. The Khoros Group has since formed a small company, Khoral Research, Inc., to be able to concentrate on development full-time. The first new product, Khoros 2.0, promises to be even more flexible and powerful. In particular, Khoros 2.0 will greatly simplify software development and migration from experimental platforms to end-user products.

Khoral Research, Inc., intends to continue distributing Khoros products free, but reserves that right for itself and its licensees. Serious Khoros users are well advised to join the Khoros Consortium, whose members include universities, corporations, and research institutes worldwide. Benefits include certain licensing and redistribution rights, early access to releases, training, and a voice in the future of Khoros.

DEFINING TERMS

Band: The "third" dimension of a three-dimensional matrix. In multispectral and color images, each spectral band (color) is stored as a image band.

Column: The "second" (vertical) dimension of a 3-D matrix.

Counting Loop: Cantata construct, equivalent to a "for" loop in C or a "do" loop in Fortran.

Execution Path, or Execution Stream: A consecutive chain of glyphs and their associated connections.

Form: A graphical display through which the user enters program parameters.

Glyph: 1. (noun) An icon that represents a process in a data flow diagram or workspace; 2. (verb) the "close" a form so that it appears as a glyph.

Header: Part of a VIFF file that carries descriptive information about the file.

Library Routine: A vroutine normally consists of a driver, which handles I/O and error checking, and a library routine, which encapsulates the functionality of the program. For example, the fast Fourier transform routine vfft calls a library routine, lvfft, which actually computes the transform.

Open: (a glyph). To click on the central control button of the glyph, converting it into a *form* into which parameters can be entered.

Pane: 1. A region in a display window. Normally contains the user interface for one routine. 2. File suffix: the .pane file encodes the user interface specification which Cantata interprets.

Map: A lookup table stored as part of a VIFF file. Often used by display programs to "decode" 8-bit-encoded color.

Vroutine: A Khoros routine that does not have a graphical user interface except when run under Cantata. Often written as "filters," accepting one or more input data streams and producing one or more output streams.

XVroutine: A Khoros routine that has a built-in graphical user interface.

Workspace: (1) the Cantata graphical interface, with menu selections, control buttons, and glyphs. (2) a program that runs under Cantata, compactly encoded as a text file.

FURTHER INFORMATION

The definitive source on the details of Khoros is the three-volume set of Khoros manuals. Volume 1 is the *User's Manual*, and Volume 2 the *Programmer's Manual*. Volume 3 is simply the collected "man" pages for the routines, and so is more easily accessed on-line. For information about purchasing Khoros on a CD ROM, to purchase manuals, or to inquire about membership in the Khoros Consortium, write to

Khoral Research Inc.
6001 Indian School Rd N.E., Suite 200
Albuquerque, NM 87110 USA
Telephone: 505 837 6500
e-mail: khoros@khoros.unm.edu

Internet is the preferred mechanism for Khoros information: support is available on a volunteer basis over the Internet. The Khoros software, its documentation, and related information are available by anonymous FTP from ftp.khoros.unm.edu. Use the user name **anonymous**, and your email address as the password. Change directories to **pub/khoros**, and read the README files. You will find a set of Frequently Asked Question (FAQ) files, sample data files and workspaces, contributed routines, and contributed toolboxes.

To be placed on the Khoros mailing list, mail to khoros-request@khoros.unm.edu. To ask a specific question, mail to khoros@khoros.unm.edu. To reach this author, e-mail to donohoe@eece.unm.edu.

54

Computer Vision

Ali Zilouchian
Florida Atlantic University

54.1 INTRODUCTION

Even with three decades of research efforts, machine vision has not yet achieved the power and the flexibility of human vision. Nevertheless, computer vision outperforms human vision for well-performed tasks in controlled environments. Machine Vision has been a forcing function for new development of computer technology, primarily because it is computationally intensive. It requires enormous amounts of computational resources as well as high bandwidth inputs and outputs. As the machine vision technology finds its way into various medical, industrial, and military applications, the development of powerful image processing software packages is critical for these applications oriented problems.

The chapter is organized as follows. In Section 54.2, the fundamentals as well as a generic set of image processing functions are described. A list of commercial packages is provided in Section 54.3. In Section 54.4 the *Image Analyst* package is briefly discussed to provide a sample of commercial software packages. Finally, in Section 54.5, the implementation of two machine vision applications is presented.

54.2 FUNDAMENTALS OF COMPUTER VISION

Digital Image

When a picture is digitized, a sampling process is used to extract a discrete set of numbers or samples. These sets of samples can be represented as an array of integers. The elements of a digital picture array are called picture elements or **pixels**.

A simple method to segment a gray image is to use an **intensity threshold**. The intensity threshold partitions the image into two types of regions: subject regions, where the intensity values of the pixels are within the threshold range, and background regions, where the intensity values of the pixels are outside the threshold range.

Most of the available image processing functions for commercial packages can be categorized into the following groups:

0-8493-2530-7/96/$0.00+$.50

1. Hardware initializations
2. Image acquisition
3. Image I/O and image display
4. Look-up tables
5. Binary image manipulations
6. Interactive tools
7. Image enhancement
8. Feature extractions
9. Object recognitions
10. Camera calibration

- **Hardware Initialization**. Certain systems require some hardware initialization functions to be executed to reset the Hardware before use; other systems do not. Hardware initialization includes tasks such as defining addresses for registers and frame grabbers and configuration of a frame buffer to certain dimension. These functions are system dependent and varied based on processing modes of image processors.
- **Image Acquisition**. Image acquisition functions generally allow the user to select the camera and capture and store the image into **frame buffer**, once or repeatedly. Some software packages even provide functions to select gain, average several images acquired by the camera, and board synchronization to system clock.
- **Image I/O**. These functions are provided basically to store and retrieve images to and from the disk.
- **Image Display**. These functions provide a window to display on the monitor a live image from the camera or a stored image in one of the frame buffers.
- **Look-Up Tables (LUT)**. Functions in this group provide some means to change the appearance or intensity of an image by selecting different LUT or by changing the mapping values in the LUT. These include the selection of threshold for an image as well as gray selection levels.
- **Frame Buffer Manipulators**. These functions perform a series of operations between the frame buffers. A wide range of functions should be expected for systems that are supporting more than two frame buffers. These operations include addition, subtraction, minimum, maximum, division, multiplication, AND, OR, XOR, and NOT.
- **Image Enhancement**. To enhance a given image, a number of filtering operations can be performed using mean or median filters on the **grayscale** or binary image. The enhancement functions usually cover a wide range of symmetrical and directional grayscale image enhancement and filtering functions. They consist of **erosion** and **dilation** operations often known as region growing and shrinking functions. Various packages provide different enhancement operations.
- **Feature Extractions**. Most features are based on the size of the object or its shape. The most obvious feature based on size is the area of the object: this is simply the number of pixels comprising the object multiplied by the area of a single pixel. The length and the width of an object also describe its size. The features such as shape numbers that encode the shape of an object are usually very useful for the purposed of classification. All commercial image processing package have a number of these futures for image identification and classification.
- **Classification and Object Recognition**. The final stage of pattern recognition is the classification and recognition of the objects on the basis of the set of the features extracted from an image. As an example, consider a pattern recognition application that requires the distinction between nuts and bolts on a convey belt. Assuming that we

can segment these objects adequately, the circularly and maximum dimension futures can be considered for object classification. Thus, we can distinguish the nuts and bolts based on these extracted futures from the images. Various image-processing packages normally include a number of algorithms for object classification and recognition.

- **Camera Calibration**. While various vision functions can obtain the futures of the image, certain vision applications often need the real world position of the corresponding physical object. Many computer vision packages provide conversion functions between image coordinate and spatial position within a chosen plane. For these functions to work correctly, the vision system must first be calibrated, thereby establishing the viewing function for the system. If the system uses several viewpoints, calibration must be performed from every one.

Digital Image Processing and Analysis Algorithms

In general, digital image processing can be thought of as a transformation that takes an image into an image, i.e., it starts with an image and produces an enhanced image. Normalized **correlation**, threshholding, and **morphology** are examples of image processing algorithms. On the other hand, image analysis is a transformation of an image into something other than an image, i.e., it produces some information representing the image. **Connectivity** and **Hough** transformation are examples of image analysis algorithms. In this section four different algorithms related to image processing and analysis are briefly described.

Edge Detection

We define an edge as the boundary between two regions with relatively distinct gray level properties. Basically, the idea underlying most edge detection techniques is the computation of a local derivative operator, which measures the rate of change at the transitional boundary areas. Therefore, these derivative-based operations enhance the image by estimating its gradient function and then signal that an edge is present if the gradient value is greater than some defined threshold.

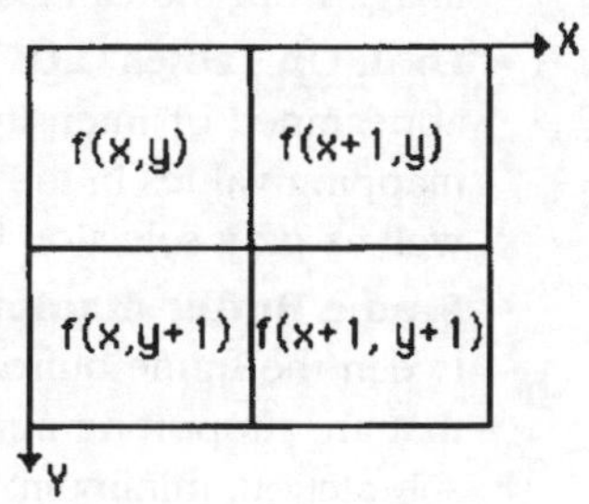

FIGURE 54.1 A 2 × 2 neighborhood of a pixel.

If we confine ourselves to a discrete domain of digital images then the derivatives become simple differences. For example, the first derivative of the two-dimensional function in Figure 54.1 in the x direction is simply

$$f(x+1, y) - f(x, y) \quad (1)$$

Similarly, the first derivative of the two-dimensional function in the y direction is simply

$$f(x, y+1) - f(x, y) \quad (2)$$

Various operators can be utilized to obtain the magnitude of the gradient $g(x,y)$ at an image point $f(x,y)$. For example, the Robert cross operator is given as

$$g(x, y) = \sqrt{\left\{\left[f(x, y) - f(x+1, y+1)\right]^2 + \left[f(x, y+1) - f(x+1, y)\right]^2\right\}} \quad (3)$$

We can also estimate the gradient by combining the differencing neighbor with the local average, which provides a Sobel operation estimation. For details about various edge detection operations see image processing textbooks (Ballard and Brown, 1982; Gonzalez and Sefa-baksh, 1994; Gonzalez and Wintz, 1994; Vernon, 1991).

Edge detection is a very popular algorithm for image processing as well as image segmentation of a given picture. It is very fast and relatively robust, depending upon the utilized operation.

Connectivity (Blob Analysis)

The connectivity analysis (blob) algorithm presents one of the most useful and popular image analysis algorithms. The connectivity analysis routine finds all the individual connected blobs (components) from an image. It reports the centers, sizes, orientations, and moments for all the blobs. It is a popular algorithm due to its simplicity and applicability to various image processing applications. In Figure 54.2 an arbitrarily shape component is shown. The area, centroid, moments, and orientation of each component can be computed accordingly. The connectivity algorithm is

- fast
- available in most commercial packages
- provides adequate information in many cases
- sensitive to noise.

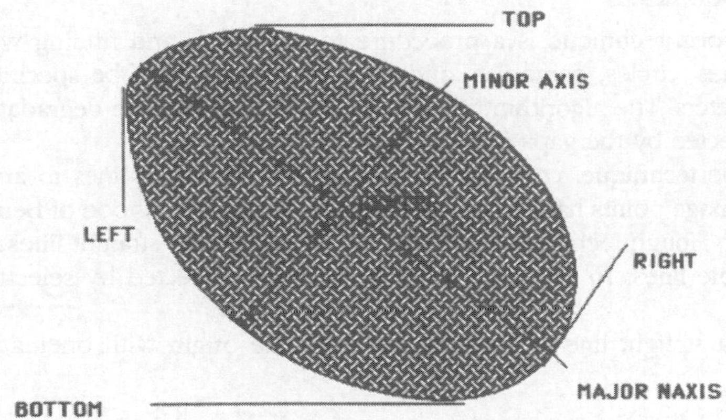

FIGURE 54.2 An arbitrary shape component.

Normalized Correlation

Given a digitized image $S(x,y)$ of size $M \times N$, we wish to determine if it contains in a region $T(x,y)$ of size $J \times K$ where $J < M$ and $K < N$ as shown in Figure 54.3.

A method often used for the solution of this problem is to perform a simple cross-correlation between $S(x,y)$ and $T(x,y)$, which is a sum of products of pixels values:

$$R(m,n) = \sum \sum S(x,y)T(x-m, y-n) \tag{4}$$

where $m = 1,2,3, \ldots, M-1$, $n = 0,1,2,3, \ldots, N-1$, and the summation is taken over the image region where $S(x,y)$ is defined.

Unfortunately this type of correlation does not factor out **brightness** and contract differences between the image and **template**. Thus, it is sensitive to **contrast** and illumination variation. Normalized correlation, on the other hand, provides a more suitable measure of image similarity. For detail descriptions of correlation and normalized correlation refer to classical image processing textbooks (Rosenfeld and Kab, 1992; Ballard and Brown, 1982; Gonzalez and Sefabaksh, 1994; Gonzalez and Wintz, 1994; Vernon 1991).

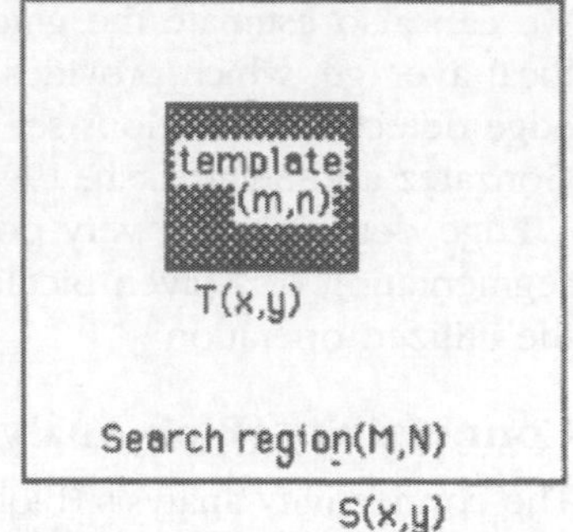

FIGURE 54.3 Digitized image $S(x,y)$ of size $M \times N$.

This algorithm has the following characteristics:

- Computationally expensive
- Correlation coefficients are independent of linear gray level changes in either template or search region.
- Computation time primarily depends on the size of the template and the search region.
- High tolerance for image degradation and noise.
- Some commercial packages cannot handle the rotation and scaling changes.
- Subpixel resolution is available in some packages.

Hough Transformation

The Hough transform technique is a procedure for detecting and finding within an image, straight lines, circles, parabolas, and other curves that can be specified by a number of parameters. The algorithm has a high tolerance for image degradation and is relatively unaffected by the gaps in the curves.

To introduce the technique, consider the problem of detecting lines in an image. Assume that the image points have been selected with a high likelihood of being a linear boundary. The Hough technique organizes these points into straight lines by considering all possible lines. In sequence, straight lines are detected by selection of a threshold.

In Figure 54.4 a straight line at a distance P from the origin with orientation θ is shown.

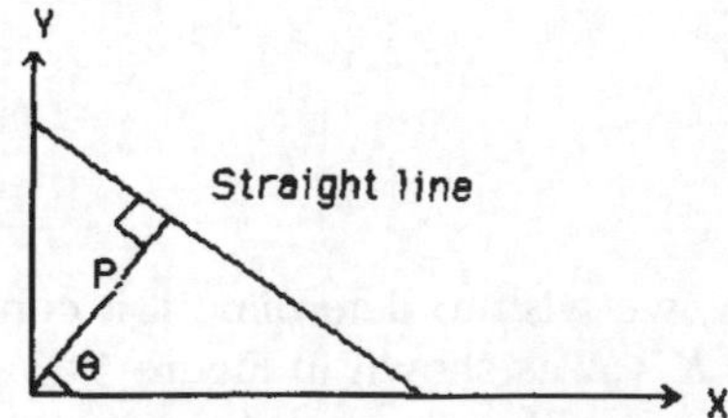

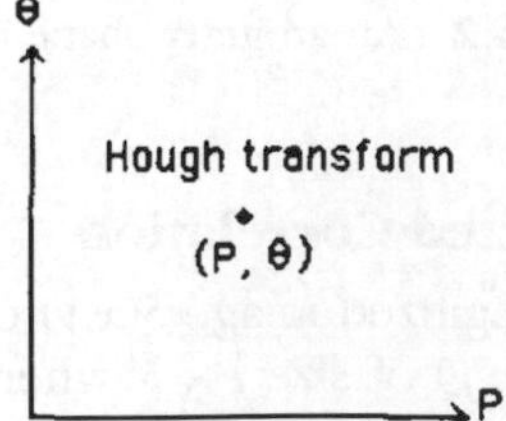

FIGURE 54.4 Straight line and Hough transform.

The Hough transform of this line is just a point in the (P,θ) plane, that is, all the points on the line map to a single line. This fact is utilized to detect straight lines in the set of boundary lines. The techniques can be generalized for the other shapes such as circles and ellipsoids.

The algorithm is

- Robust to noise
- Sensitive to quantization noise
- Computationally expensive

The algorithm may perform

- Line Hough transform
- Circle Hough transform
- Ellipsoid Hough transform
- Generalized Hough transform based on an LUT.

54.3 COMMERCIAL PACKAGES

A list of commercial machine vision packages is provided in Table 54.1. The author or publisher does not make any representation with respect to the quality, readiness, fitness, or applications of the products. The list does not imply the endorsement of any product, but supplies limited guidelines and information regarding some machine vision packages. Customers and/or other interested parties requiring information regarding particular products are urged to contact the relevant company to obtain references, documents, or demonstrations.

54.4 IMAGE ANALYST SOFTWARE PACKAGE

Introduction

In this section, one of the computer vision packages, namely *Image Analyst,* is discussed in detail. Image analyst is a Mac II package aimed at scientific and industrial users who need to extract quantitative data from video images. It is designed for various tasks such as robot guidance, parts inspections in assembly automation lines, computing the number and size of cells in images acquired by video cameras attached to microscopes, or enhancing distances in radiographs.

Image Analyst offers a standard Mac graphic interface that allows users to quickly apply the latest image processing algorithms to images without the complexities of traditional programming. Application setup is as simple as click and drag from pull down menus. The user can experiment with various image processing techniques to find the proper algorithms for a particular application. Images can be filtered and manipulated either interactively or by running "sequences" containing as of commends which have been previously entered and saved.

For users who wish to develop their own processing and analysis algorithms, a high level, interactive language called *MacRail* is available that allows you to add routine into Image Analyst software. The MacRail development environment provides a complete interpreter and program editor for developing custom applications.

The User

It is assumed that the user has some experience with the Macintosh. If you do not have any experience, it is recommended that you look at relevant chapters in this handbook or read the Macintosh manual. You need only one or two hours of experience to be ready to learn about the Image Analyst package.

TABLE 54.1 Commercial Softwares

Package	Company	Address	Phone	Platform
ALACRON	Alacron	Nashua, NH	(603)-891-2750	PC/AT
ARIES	Skylord design	Bethesda, MD	(301)-530-7533	PC/AT
C-Image	Foster Findlay	Newcastle, UK	(091)273-1111	PC/AT
AURORA	Data Translation	Marlboro, MA	(508) 481-3700	PC/AT
COLO TROL	YSC	Youngstown, OH	(216) 758-1240	PC/AT
FREELANCE	Sight System	Newbury, UK	44(0)635-529121	PC/AT
IMAGE REP/ETC	Engineering Data Systems	Hancock, MI	(906) 482-4872	PC/AT
GEOGRAF	Geoocomp	Concord, MA	(508) 369-8304	PC/AT, OS/2
IMAGE ACTION PLUS	Imaging Tech. Inc.	Woburn, MA	(617) 275-2700	PC/AT
IMAGE SCALE PLUS	Electrical Imag. Inc.	Delray Beach, FL	(407) 243-7947	PC/AT
Image-pro/ETC.	Media Cybernetic	Silver Spring, MD	(301) 495-3305	PC/AT
ISIS	Vision Harvest	Hatch, NM	(505)-267-1014	PC/AT
JAVA	Jandel Scientific	Corte Madera, CA	(305) 924-8640	PC/AT
MICROCOMP	SM	Atlanta, GA	(404) 956-0343	PC/AT
MSHELL	Applied Coherent	Reston, VA	(703) 860-9689	PC/AT
NDS1000	Nester Inc.	Providence, RI	(401) 333-9640	PC/AT, Unix
OASIS	Sight System	Newbury, UK	44(0) 635-529121	PC/AT
OPTICS NERVE SY	Umist	Manchester, UK	061-236-3311	PC/AT
OPTIMA	Opto-Tech	Meersburg, Germany	075332-9623	PC/AT
OPTIMAS	Optimas Inc.	Edmonds, WA	(206)-775-3640	PC/AT
PC ALBUM	PCM Image Software	Vienna, VA	(703)356-1260	PC/AT
SCI IMAGE ANALYS	Earth View	Ottawa, Canada	(613)-727-5853	PC/AT
PICTURE PUBLI.	Astral Devel. Co.	Londonderry, NH	(603) 432-6800	PC/AT
UNIVISION	Univision Tech.	BURLINGTON, MA	(617)-221-6777	PC/,SUN,PS/2
SHAPVIEW IMAGE	Aut. Visual Inspec.	Santa Clara, CA	(408) 296-4947	PC/AT
TCL IMAGE	Myltihouse TSI	Amsterdam, Netherlands	31-20-82-10-81	PC,OS/2,Unix
VISILOG	Noesis Vision Inc.	Ville Saint, Canada	(514) 345-1400	PC/AT, Unix
BIOSCAN	Optimas	Edmonds, WA	(800) 653-7226	OS/2
DT-IRIS	Data Translation	Marlboro, MA	(508) 481-3700	PC/AT, Unix
LYNXOS	Lynx	Los Gatos, CA	(408)-370-2233	Unix
ICOMETRICS	Everst Tech. Inc.	Houston, TX	(713) 975-8866	Unix
VISUAL WORKBE.	Paragon Imaging	Lowell, MA	(508) 441-2212	Unix
BLACK & WHITE	Computer Photo	Goettingen, Germany	49-551-7700708	Mac
EDITWORX	Spe. Computer Sy	Santa Rosa, CA	(707) 539-3212	Mac
IMAGE	NIH	Bethesda, MD	(313)-487-4650	Mac
ENHANCE	Microfrontier	Des Moines, IA	(800) 388-8109	Mac
IMAGE ANALYST	Automatix	Billerica, MA	(508) 667-7900	Mac
IPLAB	Signal Analytics	Vienna, VA	(703) 281-3277	Mac
ULTIMAGE	GTFS INC.	Santa Rosa, CA	(703) 579-1733	Mac
VCS	Vision Dynamic Ltd.	Herts, UK	(44) 442-216088	Mac

It is assumed that the user is familiar with the following basic techniques throughout the rest of this chapter:

- Clicking
- Double-clicking
- Command-clicking
- Shift-clicking
- Option-dragging
- Pulling down menus
- Moving, sizing, and scrolling windows
- Selecting and editing text

Hardware Requirements

The minimum system configuration needed to run the software is

Computer:	Mac II series/Power PC
Disk drive:	20-Megabyte hard disk drive
Memory:	5-Megabytes of RAM
Operator interface:	Keyboard and mouse
Video source:	Standard RS-170, NTSC, CCIR, PAL video Camera, VCR. **Frame Grabber**: AM2000 for RS-170 or NTSC interlaced, 60 Hz. Digitizes 640 (horizontal) by 480 (vertical) pixels by 256 shades of gray; up to 4 video sources can be connected.

Software

You must have the Video Expansion Kit installed on Video card in your MacII. This enables you to use the 256 bit mode for the images. To see if your MacII has the Video Expansion Kit installed

1. Turn on your MacII.
2. When the desktop appears, select ***Control Panel*** from the Apple menu.
3. Click on ***Monitors***.
4. If you are using a color monitor, select ***Color***.
5. If you are using a black and white monitor, select ***Black & White***.
6. Select ***2 5 6*** under the number of colors and gray levels to be displayed.
7. Click on the ***Control Panel*** close box (upper left corner of control panel).

If your MacII monitor control panel does not have a 256-bit mode, you must install the Video Expansion Kit. This option consists of 8 memory chips that are installed on the video card. Follow the instructions provided with your Video Expansion Kit.

Image Analyst Installation

1. Make a new folder on the hard disk.
2. Copy the contents of software disks to this new folder.
3. Double-click on each of these files in the new folder.
4. Delete all the archive file.
5. If you are loading Image Analyst for the first time, you must personalize it by entering your name and your company name in the sign-on screen.

Software Specifications

Image Analyst can be utilized to interrogate or extract quantitative data using either the interactive or automatic sequencing modes. Images, statistical data, raw data, and tabulated results can be output to print applications, such as Microsoft Excel. With the Image Analysis/MacRail source vision, results can also be compared to pass/fail criteria or trending analysis with decisions automatically generating external signals. The use of the software with MacRail is common for industrial product inspection. Examples of the measurement that are available with image analysis include

- Point to point
- Line histogram
- Area histogram angle
- Area
- Blob dimensions
- Counts
- Blob statistic
- Major axis
- Centroid

Interactive Image Analyst

For various scientific and industrial tasks, the interactive tool palette and enhancement mode are very useful. Generally these images have been previously acquired by other devices (camera and frame grabber) and can be accessed by Image Analyst when in **PICT** or **TIFF** format. Information may be extracted through subjective interpretation of the video monitor. Measurement of data points can be collected interactively. A calibration algorithm converts the provided data to the real-world dimension.

Image processing information comes from user-defined Region of Interest (**ROI**). An ROI can be as small as a single pixel or can be selected as a whole image. Position as well as the size of ROIs are defined graphically by dragging the mouse.

In Figure 54.5 various stages as well as functions of Image Analyst package is shown.

The software uses the standard Mac interface consisting of icons, windows, and pull-down menus. The tool Palette allows the user to utilize various interactive commands and can be found under the ***File*** menu. Select ***Show Tools*** from the ***File*** menu to activate Master Palette as shown in Figure 54.6.

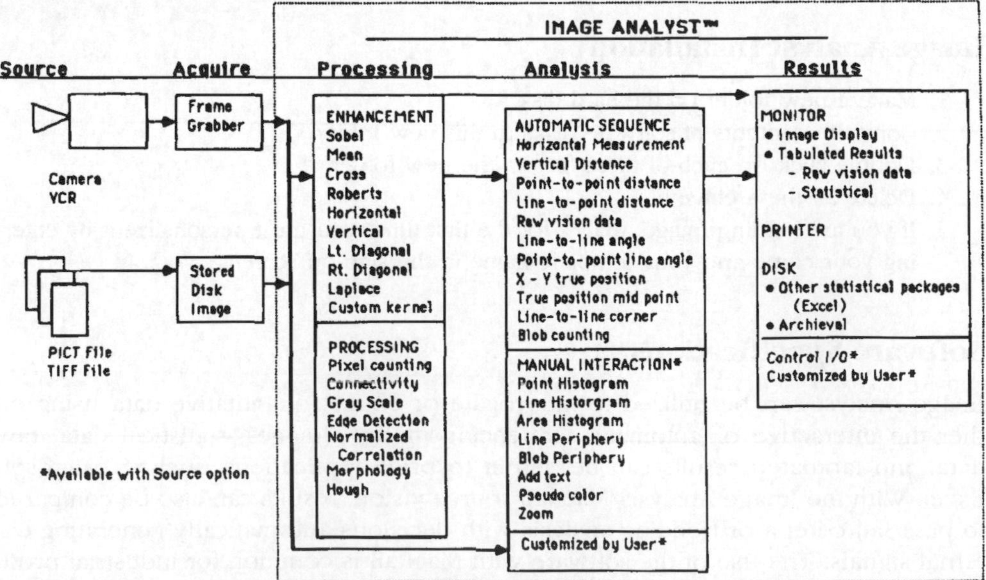

FIGURE 54.5 Schematics of Image Analyst package.

The Master Palette is divided into modal tools in the top row and nonmodal tools below. Individual tools are activated by clicking on the desired tool icon. Two tools **Zoom** and **Set Aside**, do not have a corresponding Palette. The **Histogram** tool operates on the current ROI, whereas other tools operate on the entire display buffer window. The position of all the tools in palette is saved in the MacRail Preference file when the palette is closed. The next time the tool palette is activated, it will be in the same position as it was when it was last closed.

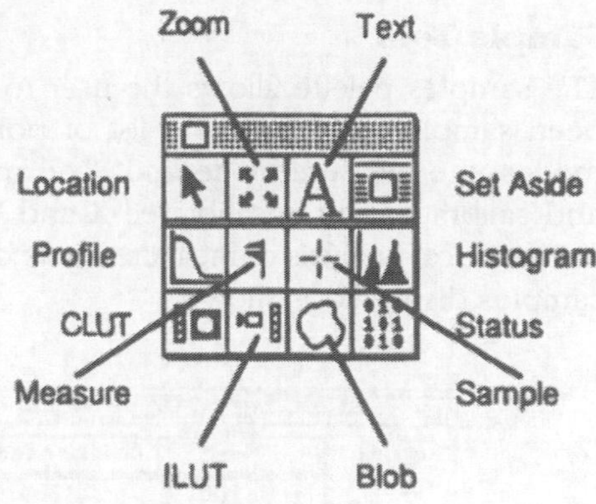

FIGURE 54.6 Menu for Master Palette.

Location Tool

The location tool is utilized to display the current cursor pixel, calibrated coordinates, and gray value under the cursor. Figure 54.7 shows the parameters related to this tool.

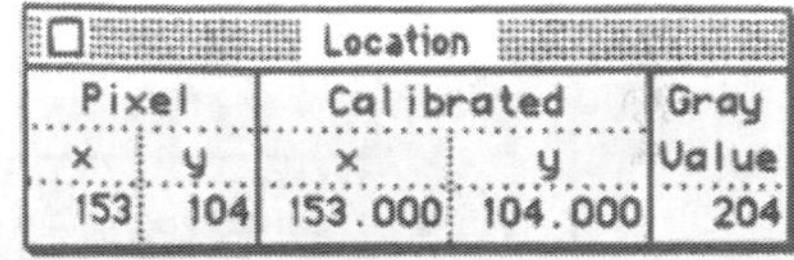

Location

Pixel		Calibrated		Gray
x	y	x	y	Value
153	104	153.000	104.000	204

FIGURE 54.7 Display of location tool.

Zoom Tool

The Zoom tool can be used to zoom in on the image. When this tool is activated, the cursor will change to zoom in cursor, indicating its activation mode. If you click anywhere within the active window, the image inside the window will be zoomed in by the factor of two. To exist from zoom mode, select another tool from the Master Palette.

Text Tool

This tool allow you to attach your desire text to the image inside the active window. The text tool allows for the modification of font, size, style, and color of any character in a text. The text can be stored into the off screen buffer to become a permanent part of the image when the image is stored. Text objects can also be resized or removed. In addition, multiple text objects can be created using this tool.

Set Aside Tool

This tool allows the user to hide all the tool palettes on the screen except for the master palette. Once the tools have been removed, they can be reactivated by clicking on any icon on the master palette.

Profile Tool

The profile tool allows the user to view the pixel gray value along a line. The length and position of the line are adjustable. To move the profile line, place the cursor over the line until the cursor changes to the line drag cursor. The line can be change by simultaneously dragging and clicking until the line is in the desired position. The endpoints of the profile line can be also be changed using this tool.

Measure Tool

The location tool can be used to measure distance. It is like a ruler. The length and position of the line are user adjustable. Utilization of the measure tool in conjunction with the zoom tool allows for accurate distance measurement.

Sample Tool

The samples palette allows the user to list measurement and pixel values that have been sampled. It can save a list of points to another file. The sample tool has two modes, the measure mode and Pixel mode. With single pixel points, the gray value and calibrated and uncalibrated X and Y values are reported. This tool is very useful for statistical analysis of measurements data. In Figure 54.8 a sample palette with pixel samples displayed is shown.

Samples

Pixel	Calibrated	ΣPixel	ΣCalibrated
200.0	200.0	200.0	200.0
170.5	170.5	370.4	370.4
126.0	126.0	496.4	496.4
115.3	115.3	611.7	611.7
118.1	118.1	729.8	729.8
92.8	92.8	822.6	822.6
72.4	72.4	895.0	895.0

7 Items
0 Selected

	Pixel	Calibrated
Avg	127.9	127.9
Std Dev	43.9	43.9
Median	118.1	118.1

Delete Append Save As...

FIGURE 54.8 Display of a sample palette.

Histogram Tool

The Histogram palette displays the histogram of the active ROI of the image. If you move the cursor across the histogram, the gray as well as the histogram values are updated. The user can select the threshold value by dragging the threshold knob with the cursor. The vertical axis of the histogram is automatically scaled. However, the user can change the vertical axis maximum value by double-clicking on the vertical axis label. Figure 54.9 shows a typical histogram for an ROI.

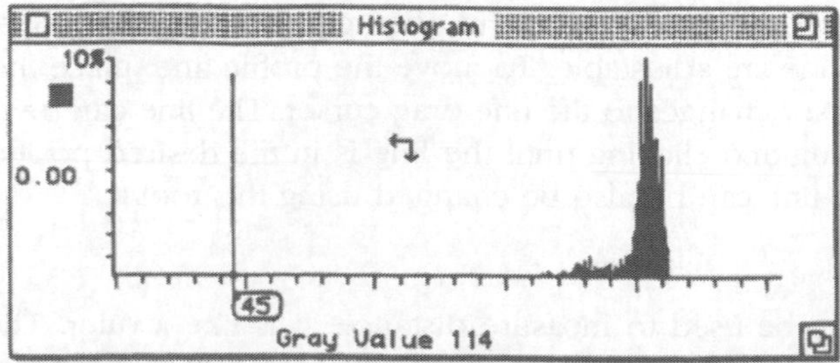

FIGURE 54.9 Display of a D histogram palette.

CLUT Tool

The CLUT tool is used to modify the color look-up table by the active buffer window. The CLUT can be selectively modified. Colors in the palette can be selected by clicking and dragging within the colored region. The current selection can be extended by pressing the shift key. The selected color can be changed by clicking on of the button on the plate as shown in Figure 54.10.

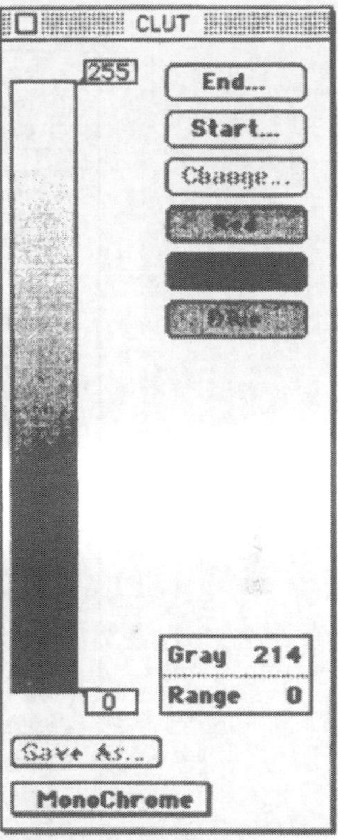

FIGURE 54.10 Display of CLUT tool.

ILUT

The Input Look-Up Table (ILUT) tool can be used to change the gray value of each pixel within an image before the image is displayed or processed. Possible icons on ILUT configurations include normal, inverse, bipolar threshold, posterize, contrast, enhance, and binary threshold. When you click on one of these icons, the ILUT will change to the selected predefind ILUT, using the built in settings. If you would like to modify the predefind ILUTs, press the commend key while clicking. A dialog will be displayed, through which the predefined ILUTs can be modified and stored for subsequent use.

BLOB Tool

The blob palette is utilized to list various parameters from the connectivity algorithm for a given image. The Blob tool has two possible modes, ***Features*** and ***Statistics***. Theses modes are entered by ***clicking*** into the proper radio button in the ***blob Palette***.

In the ***feature mode***, the blob tool displays the ID and feature values of currently selected blob. Blobs are chosen by clicking on the desired blob in the window. The features list is arranged to display up to 22 features. A sample of this mode is shown in Figure 54.11.

In the ***statistics mode***, the blob tool displays the feature statistics for the blob. Such statistical parameters include average, standard deviation, median, maximum, and minimum. The list scroll bar can be used to display other blob features for both features and statistics modes of the blob palette.

Status Tool

The status palette displays the ID, size, and zoom of the active buffer, the amount of available memory, the date, and the time.

Using Image Analyst

Image analysis consists of extracting useful information about objects by examining the gray scale values of pixels in a buffer, using various vision processing methods. Some of these methods have been briefly described in the beginning of this chapter.

In following sections, several available algorithms in Image Analyst package are presented. Once you are familiar with various algorithms, you can create the desired ROI and set up the desired measurements for a particular application. The Image Analyst provides the following algorithms.

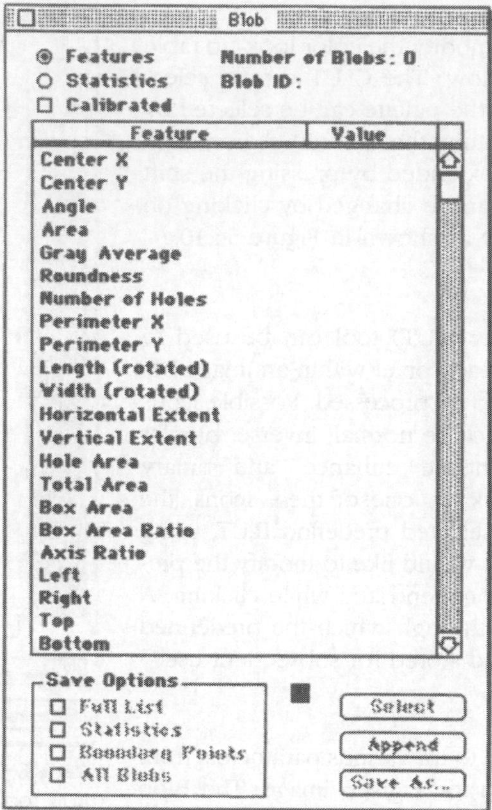

FIGURE 54.11 Menu for blob tool.

Area (Pixel) Counting

Area counting is a fast way to calculate the area of a part. This process simply counts the number of pixels above or below the threshold in the current ROI.

Click on the ***Area Count*** button after you select ***Choose Processing*** from the ***ROI*** menu (Fig. 54.12). The area count of the feature contained in ROI is automatically calculated. You can choose either ***Dark*** or ***Light Pixels***, depending on the future of interest. This algorithm is primarily used for processing ROIs that contain only one future. If the image contains more than one feature, individual ROIs can be set around each feature to be measured.

Connectivity Analysis

Blob analysis identifies objects from the background in the image and provides a variety of geometric features of the objects.

Select ***Choose Processing*** from ***ROI*** menu. Click on the connectivity button. Choose ***Dark*** or ***Light Object***, depending on the color of the object or part. The blob algorithm uses two main processes: analysis and processing. In the image analysis phase the algorithm determines which groups of pixels are "connected" to each other. Connected pixels are of the same color. During the image processing phase, each part and hole is processed. Subsequently features such as area, perimeter, centroid, and location as well as number of holes are calculated. These features are useful for either blob measurement or to distinguish one blob from others.

ROI: Untitled ROI-3
Processing Type
Area Count
Enhancement
Connectivity
Correlation
Edge
Gray Scale
Morphology
Hough
Attributes
Dynamic Locator
Cancel
OK

FIGURE 54.12 Menu for area counting.

Edge Detection

Edge detection evaluates the locations of the edges that separate objects from backgrounds in a given image or ROI, (Fig. 54.13). Various steps for execution of this algorithm is as follows:

1. Select ***Choose processing*** from the ***ROI*** and click on the ***Edge*** button.
2. Select***Configure Parameters*** from the ***ROI*** menu.
3. Select either ***Edge*** or ***line***.
4. Choose a ***scanning direction*** by clicking on the appreciate arrow.
5. Select ***Dark to light*** or ***Light to dark*** color.
6. Choose ***scanning mode***.
7. Choose ***Min.*** and ***Max.*** gradient levels.
8. Select a ***processing type*** base on the information you desire from the image.

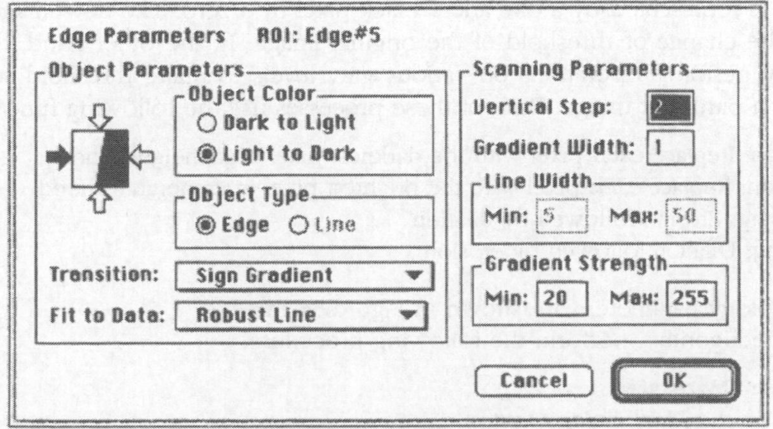

FIGURE 54.13 Menu for edge recognition parameters.

The algorithm is based on the gradient method described in Section 54.2 of this chapter. As pointed out in gradient-based detection, a point is located where the greatest contrast changes occur. The user can place both minimum as well as maximum limits on the gradient. Theses feature help edge detection in the presence of noise.

Image Enhancement

Image enhancement is a useful algorithm that provides a sharper and more suitable image for further analysis, improved feature extraction, and object recognition. The algorithm uses several operators to enhance the image. These operations include Sobel, Mean, Median, Cross, Robert, Horizontal, Vertical, Left Diagonal, and Right Diagonal. For descriptions of these operations refer to the user's manual.

For this algorithm, select ***Choose Processing*** from the ***ROI*** menu and click on ***Enhancement***. To select an appropriate enhancement operator, select ***Configure Parameters***. Then click on the button of the enhancement algorithm to be used for your provided image or selected ROI of the image.

Selecting the appropriate enhancement operator depends on the characteristics of the image that needs to be enhanced. For example, ***Sobel*** operates on the entire image, while ***Horizontal*** operates only on the horizontal pixels, as its name implies.

Normalized Correlation

A brief description and applicability of this algorithm have been presented in a previous section of this chapter. It is a very effective technique and a robust method to compare an image with an ideal image using this algorithm.

To set up normalized correlation processing, select ***Choose processing*** and click on ***Correlation***. Click on **OK**. The ***Correlation Parameters*** dialog box appears. These parameters include size, interval, and coarseness. In addition, the are two windows used by the algorithm, ***Search window*** and ***Template window***. These windows are shown in Figure 54.3.

Morphology

This algorithm exposes the geometric relationships or structure in the image. The morphological algorithm consists of various processes for binary images as well as gray scale images. Binary morphology works on two gray scale levels, black and white. A white pixel is represented by a one and a black pixel by a zero. The new image is obtained by the change of threshold of the original image. In the other hand, the gray morphology performs operations on various gray levels to make threshold selection suitable for a particular image. Both of these processes use the following functions:

Erosion: Replace each pixel with the darkness pixel in its neighborhood
Dilation: Replace each pixel with the brightest pixel in its neighborhood
Opening: Erosion followed by dilation
Closing: Dilation followed by erosion

The morphology parameters are shown in Figure 54.14.

To use the algorithm perform the following procedure:

1. Choose ***Processing***
2. Click on the ***Morphology*** button
3. Select a ***proper structure*** on the button of the dialog box
4. Select ***threshold***

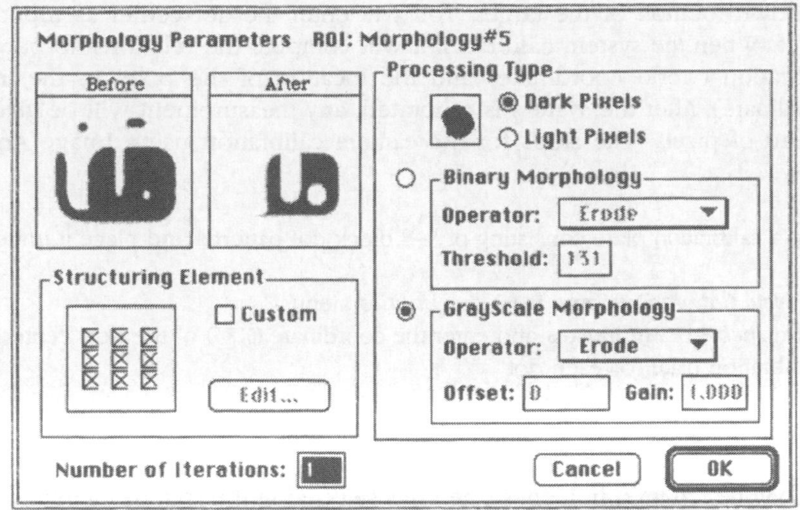

FIGURE 54.14 Menu for morphology parameters.

5. Choose your desired operation either from the ***Binary*** or ***Gray scale*** table on the dialog box. After each morphology operation the new image will appear on your computer screen. You cannot recover the previous image after the operation is performed.

Hough Transform

The theoretical background as well as description of this algorithm were presented in Section 54.2 of this chapter. The only available operations for this algorithm using the Image Analyst package are lines and circles operations. The general Hough transform cannot be performed using the Image Analysis package.

This algorithm provides robust feature extraction based on a Scan subroutine and robust approaches such as least squares for line or circle selection.

To use the Hough transform perform the following steps:

1. Select ***Choose processing*** from the ***ROI*** menu
2. Click in the ***Hough*** button
3. Select the ***Kernel size*** (3 × 3, 5 × 5, 7 × 7) button
4. Choose the ***Points per kernel*** (up to 4) button
5. Select the ***Gradient threshold*** button
6. Select ***Pick features*** from the ***ROI*** menu
7. Choose the ***shape*** (***line*** or ***circle***)
8. Click on the ***Fine tune*** button
9. Choose ***line*** (or ***circle***) parameters
10. Click on the ***OK*** button to see the best line (circle) on your selected ROI.

Calibration

The above algorithms will report all the parameters of an object in pixels, not inches or millimeters. To obtain measurement in these units for an object, you must first calibrate the camera. Calibration involves putting a reference or calibration target of known dimension in front of the camera. The calibration target contains an arrays of dots in a known location of the target. You will enter the dot center as input into the computer. When the system calibrates, it will compute the relationship between the dots' location (world coordinate) and the location of the pixel in the image (image coordinate). After the system is calibrated, any measurement will be given in inches instead of pixels. The steps for the camera calibration using Image Analyst are as follow:

1. Obtain a calibration plate consisting of 5–8 block dot patterns and place it under the camera
2. Select *New Calibration Setup* from the *Utilities* menu
3. Click on *Inches* or *Millimeters* and enter the coordinate (X,Y) of the dots center from the calibration plate for each dot
4. Click on *OK*
5. Run the *Calibration* routine from the *Utilities* menu.

The images will be reported in inches (the units selected in the routine) when the calibration routine is used. You can save the calibration sequences by selecting ***Save*** from the File menu.

54.5 APPLICATIONS

Although there are many applications for computer vision in various engineering and scientific disciplines, it is difficult to get started. This is because there is no single, generic machine vision software and technology. In this section, we present only two applications of the Image Analyst software: the off-line automatic sequence mode of an image and on-line automated inspection of a printed board circuit (PBC).

Off-Line Automatic Sequence Mode

In this mode, the multiple measurement functions as well as image enhancement routine can be sequenced and executed as a single step for an object. The steps for defining an automatic sequence are shown in Figure 54.15.

The technique is particularly useful for repetitive analysis of collected images. It consists of the following steps:

Step 1: Take a picture using camera and frame grabber or select an image from memory.
Step 2: Choose *New ROI* from the ROI menu.
Step 3: Select *Choose processing*.
Step 4: Choose the proper image processing *algorithm* that appears in the dialog box. These algorithms have been discussed in previous sections in detail.
Step 5: Click on *configure parameter* from the ROI menu.
Step 6: Choose appropriate *features* needed to make the required measurement.
Step 7: *Label* the sequence and *save* to the disk. Each time you would like to run the sequence, you simply load the sequence name and select *run*.
Step 8: The feature data will be displayed (as well as statistical summary if applicable) on the screen.

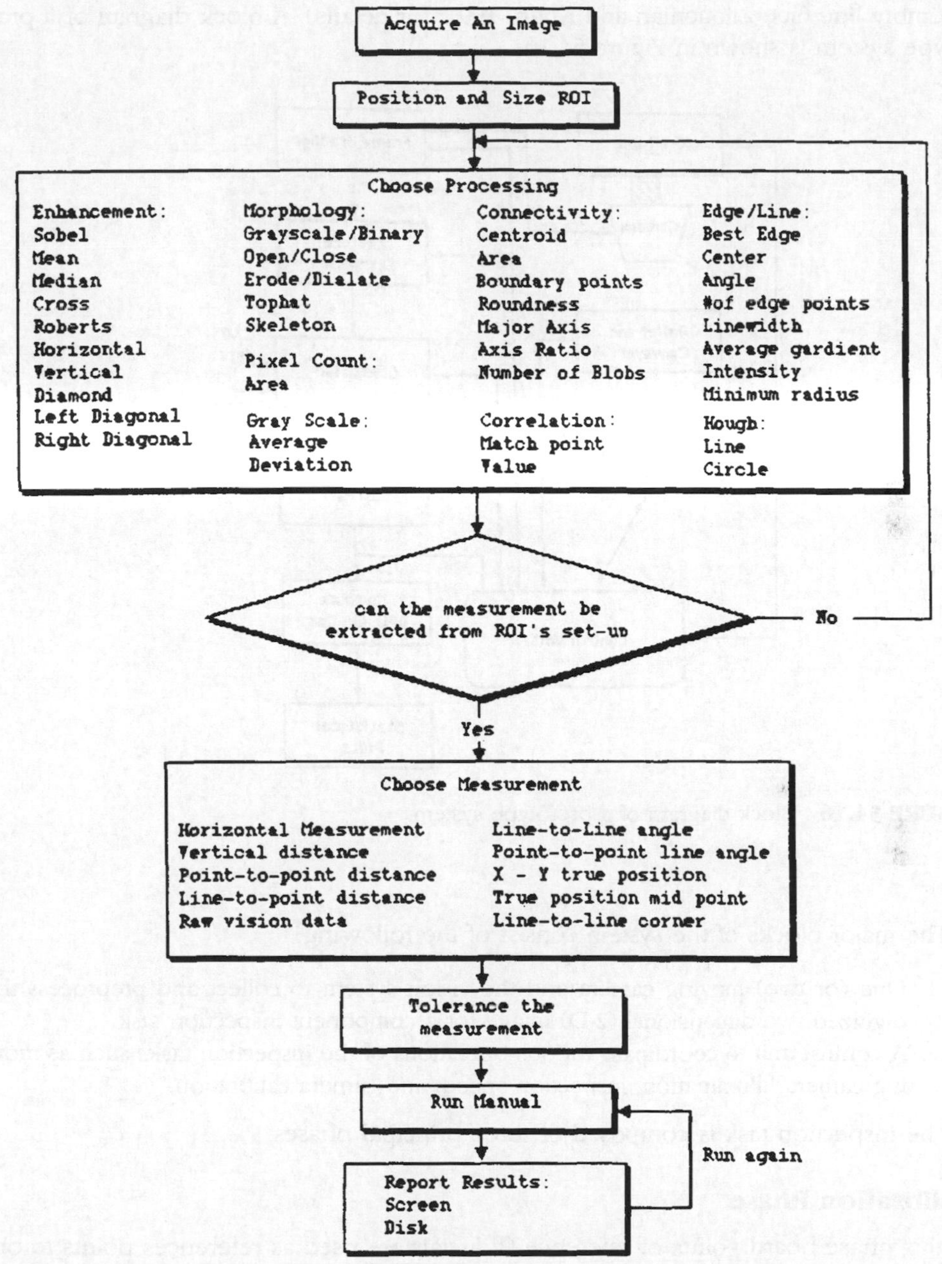

FIGURE 54.15 Flow chart for automatic sequencing.

On-Line Printed Board Circuit (PBC) Inspection

Image analyst (or other vision packages) can be utilized for on-line inspections of various objects in manufacturing. As an example, the automatic inspection of a printed circuit board is briefly presented. The task includes the position and orientation of various electronic components on PCB during the production processes in an automated assembly line (see Zilouchian and Khan, 1991, for details). A block diagram of a prototype system is shown in Figure 54.16.

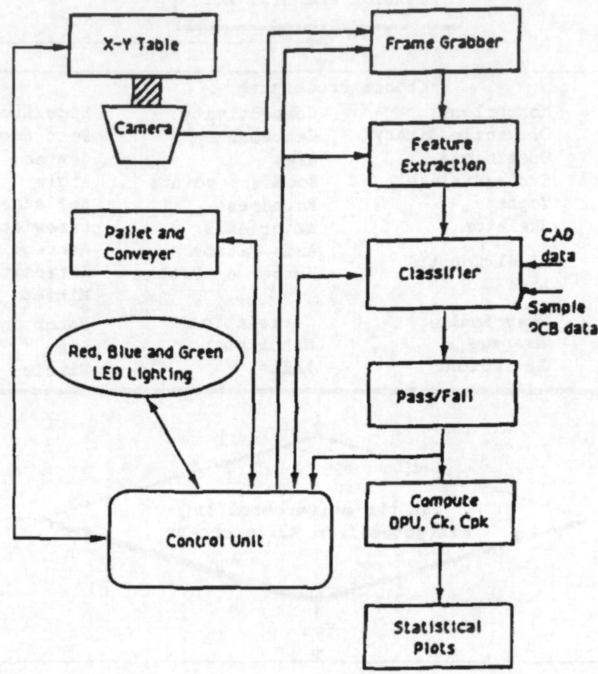

FIGURE 54.16 Block diagram of a prototype system.

The major blocks of the system consist of the following:

1. One (or two) moving camera and the vision system to collect and preprocess the digitized two-dimensional (2-D) image for a component inspection task.
2. A control unit to coordinate various operations of the inspection tasks such as moving camera, illumination, inspection speed, and camera calibration.

The inspection task is composed of three principal phases.

Calibration Phase

In this phase board points of reference (fiducial) are used as references points to obtain the camera parameters. The X-Y table is moved a number of positions to the required data to calculate the transformation matrix between the world coordinate (table) and the image coordinate.

Teaching Phase

In this phase selection of teaching points for X-Y table movement as well as a collection of image samples of defected free PCB is performed. If CAD data for the required inspected components are available, such information can be utilized and stored in the memory of the computer for future extraction. A library of various component functions might then be developed for complete board inspection.

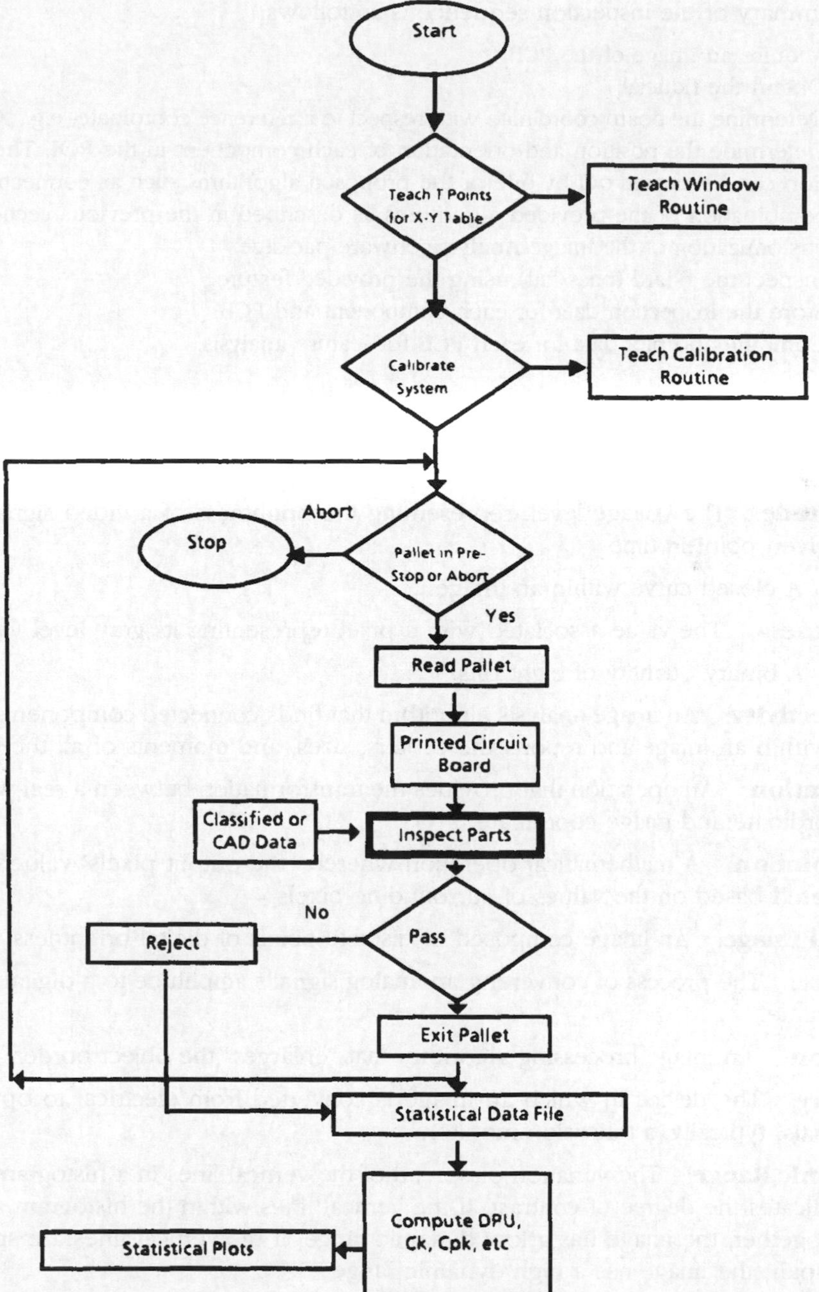

FIGURE 54.17 Flow chart for implementation of prototype component inspection station.

On-Line Inspection Phase

In this phase the image features are extracted and calculated for each component according to the rules provided by the classifier. A decision is then made as to whether the part would pass inspection.

A summary of the inspection sequence is as follows:

1. Acquire an image of the PCB.
2. Obtain the fiducial.
3. Determine the board coordinate with respect to a reference coordinate, e.g., X-Y table.
4. Determine the position and orientation of each component in the ROI. The inspection can be carried out by one of the proposed algorithms such as connectivity or a combination of the provided algorithms as described in the previous section using customization of the Image Analyst software package.
5. Inspect the board (pass/fail) using the provided feature.
6. Store the inspection data for each component and PCB.
7. Compute and plot data for each PCB for feature analysis.

DEFINING TERMS

Amplitude: The voltage level, representing the brightness of a video signal at any given point in time.

Blob: A closed curve within an image.

Brightness: The value associated with a pixel representing its gray level value.

Byte: A binary quantity of eight bits.

Connectivity: An image analysis algorithm that finds connected components (blobs) within an image and reports the centers, sizes, and moments of all the blobs.

Calibration: An operation that provides the transformation between a real-world coordinate and image coordinate.

Convolution: A mathematical operation whereby the output pixels' value is calculated based on the values of surrounding pixels.

Digital Image: An image composed of discrete pixels of digital brightness values.

Digitize: The process of converting an analog signal's amplitude to a digital (binary) value.

Dilation: An image processing algorithm that "enlarges" the object border.

Display: The device in which an image is converted from electrical to optical signals; typically, a television monitor.

Dynamic Range: The variation or spread of the vertical lines in a histogram that indicates the degree of contrast. If the vertical lines within the histogram are close together, the image has a low dynamic range; if the vertical lines are spread far apart, the image has a high dynamic range.

Edge Detection: Any operation that detect edges within an image.

Enhancement: An operation that changes the image into a more cleared one.

Erosion: An image processing algorithm that "shrinks" the object border.

Fiducial: Reference point on the corner of a printed circuit board.

Filter: An operation that changes the spatial and intensity characteristics of an image.

Frame: The total number of lines of scan that represents an image on a display device.

Frame Buffer: A high-speed memory designed to store one or more images and allow simultaneous video display and CPU access.

Frame Grabber: An interface card that digitizes a captured picture by camera.

Gradient Operation: An edge enhancement operation possessing a directional quality.

Gray Level: The brightness value assigned to a pixel. A value may range from black, through the grays, to white.

Gray Scale: Variations in the luminance value of "white" light, from black to white.

Histogram: A graphic representation of the grayscale value contained in an image that effectively measures the color/gray values in an image. The horizontal axis represents the gray levels and the vertical axis displays the number of pixels within each gray level.

Hough Transformation: A technique to find lines, circles, ellipses, and other curves within an image.

Morphology: An image processing algorithm that includes erosion and dilation for image enhancement.

PICT: Special format for image saving in frame buffer.

Pixel: The smallest unit in a digital image. A pixel is addressed by its horizontal and vertical coordinate indicating its location within an image.

Resolution: The number of bits of accuracy or number of gray levels that can be represented in a pixel.

ROI: Region of interest within an image.

TIFF: Special format for image saving in a frame buffer.

Sample: A discrete pixel of analog brightness. A sample is subsequently quantized, yielding a pixel of digital brightness.

Spatial: Pertaining to the two-dimensional nature of an image.

Zoom: To enlarge or reduce the size of a displayed image on a pixel-by-pixel basis.

REFERENCES

D.H. Ballard and C. M. Brown, *Computer Vision*. Prentice-Hall, Englewood Cliffs, NJ, 1982.

R.C. Gonzalez and R. Sefa-baksh, Computer Vision techniques for industrial inspection and robotic control: A tutorial overview. In *Tutorial on Robotic Control*. C. Lee, R.

Gonzalez, and Fu (eds.). IEEE Computer Society Press, New York, 1994, 300–324.

R. Gonzalez and P. Wintz, *Digital Image Processing*, 3rd ed. Addison-Wesley, Reading, MA, 1994.

B.K.P. Horn, *Robot Vision*. MIT Press, Cambridge, MA, 1986.

A. Rosenfeld and A. Kak, *Digital Picture Processing*, Vols. one and two, 2nd ed. Academic Press, Orlando, FL, 1982.

D. Vernon, *Machine Vision: Automatic Visual Inspection and Robot Vision*. Prentice-Hall, Englewood Cliffs, NJ, 1991.

A. Zillouchian and P. Kahn, *Implementation of Different Pattern Matching Algorithm for Component Inspection*. Motorola Internal Report, Boyton Beach, FL, 1991.

FOR FURTHER INFORMATION

Image Processing Handbook. Data Translation Inc., Marlboro, MA.
User's Guide: Image-Scale Plus. Electronic Imagery, Inc., Delray, FL.
User's Manual: Image Analyst 8.1. Automatrix Inc., Billerica, MA.
User's Guide: Image-Action Plus. Imaging Technology Inc., Woburn, MA.
User's Source Handbook. Data Translation Inc., Marlboro, MA.

55

MATLAB in Systems and Controls

John S. Bay
Virginia Polytechnic Institute and State University

55.1 INTRODUCTION

As discussed in Section 18.7 of this book, MATLAB®,[1] by the MathWorks, Inc., is a general-purpose mathematics package. It can be used as an interpreted, command-line processor, somewhat like a sophisticated desktop calculator. However, it is also a complete language, running .M files within the MATLAB environment or as called from other .M files. These features make it ideal for both quick computations and extensive simulations, analysis, and software development.

Because MATLAB's variables are all equivalent to arrays or matrices, it is important as a computational tool in linear systems analysis and, consequently, as a tool for the control systems community. Although there are similarly structured competing programs available, MATLAB is quickly becoming a de facto standard among systems and controls educators.

In fact, most of MATLAB's numerical analysis routines were written by control systems researchers and practitioners, using the most accurate and robust numerical routines available. In combination with the specialized toolboxes, such as the Control System Toolbox, the Signal Processing Toolbox, the System Identification Toolbox, and several others, MATLAB contains many features that enable a control systems designer or researcher to perform in a single, one-line command, a sophisticated operation that would take entire subroutines in any other language.

The discussion in this section will be a basic overview of those features of MATLAB most important for signal processing and control systems. It is at an introductory level and a great many powerful and sophisticated commands are neglected. All of the com-

[1]MATLAB and SIMULINK are registered trademarks of The MathWorks, Inc., 24 Prime Park Way, Natick, MA 01760-1500, (508)653-1415, info@mathworks.com, http://www.mathworks.com.

0-8493-2530-7/96/$0.00+$.50

mands are fully explained and documented in the MATLAB toolbox manuals. These manuals also contain handy references to the original source of the underlying algorithms and related literature. They also often contain brief but informative tutorials that can be valuable references in lieu of an available text.

55.2 SIGNAL AND SYSTEM REPRESENTATION IN MATLAB

The two fundamental requirements necessary for an efficient systems/controls analysis and simulation package are the convenient representation of **signals** and of **systems**.

Signal Representation

Except for software that allows symbolic representation of functions, MATLAB is like most other programs in that a signal is entered as a series of samples. These may be evenly or unevenly spaced in time (or some other independent variable of choice). For example, if we wanted to construct a series of samples from the function:

$$y(t) = \sin(2\pi * 60t) + 2 * \sin(2\pi * 100t) \tag{1}$$

(the sum of a 60 and a 100 Hz sine wave), evenly spaced over the time interval $t \in [0,1]$ seconds, with a 1000 Hz sampling rate, we could execute the following command sequence:

```
>> t = 0:.001:1;
>> y=sin(2*pi*60*t)+2*sin(2*pi*100*t);
```

After execution, **y** and **t** are both row vectors of 1001 elements (1000 time steps plus the initial point at $t = 0$), and each sample of the vector **y** is taken at the corresponding element of vector **t**. One of the most convenient features of MATLAB is its implicit ability to process vector data. In this case, we can create the vector **y** by inserting the vector **t** in a single command line, rather than having to loop through each element of vector **t** and computing samples of **y** one at a time. If **t** were transposed to a column, then **y** would inherit this column shape from **t**.

Note that in the example above, MATLAB understands that the string ' **pi** ' stands for 3.14159 . . . [actually, it is computed as 4 arctan(1)]. It also understands that the symbol **i** (and **j**) stands for the "imaginary" value of $\sqrt{-1}$. Complex numbers are entered in the natural manner, for example

```
>> z = 3 + 4*i
```

or

```
>> z = 3 + 4*j
```
(preferred by electrical engineers because i is the symbol for electric current)

or equivalently,

```
>> z = 5*exp(i*theta)
```
where **theta** is some representation of arctan(4/3).

Thus, complex numbers are handled as naturally as real ones.

For laboratory applications where real data is required, the **load** command can be used to read flat ASCII files, either comma- or space-delimited.

System Representation

Some software packages intended for signal and system analysis constrain the user to one specific system representation, for example, either state-variable matrices or transfer function coefficients. Most MATLAB commands that use system representations are flexible and can use systems representations in several formats. These may be a set of state-variable matrices, the coefficients of a transfer function (either in pole-zero or rational polynomial form), the coefficients of a differential equation (which are the often the same as the coefficients of a transfer function), or as a partial-fraction expansion of a transfer function.

Furthermore, the systems can be interpreted as operators in discrete time or continuous time, usually by equivalent commands in either domain. For example, the two commands **step** and **dstep** will return the step responses of a system interpreted, respectively, as a continuous-time or discrete-time system. As we shall see, it is possible also to convert a system representation back and forth between continuous and discrete time. In the following discussion, the continuous-time terminology will be used throughout. However, most operations apply equivalently to both time domains.

Differential/Difference Equation Coefficients

Perhaps the most fundamental way to represent a linear, time invariant (LTI) system is with a differential equation. Consider the LTI differential equation:

$$a_3 \frac{d^3 y(t)}{dt^3} + a_2 \frac{d^2 y(t)}{dt^2} + a_1 \frac{dy(t)}{dt} + a_0 y(t) = b_2 \frac{d^2 u(t)}{dt^2} + b_1 \frac{du(t)}{dt} + b_0 u(t) \quad (2)$$

The input-output behavior of such a system can be represented by its transfer function in *rational polynomial* form:

$$H(s) = \frac{Y(s)}{U(s)} = \frac{b_2 s^2 + b_1 s + b_0}{a_3 s^3 + a_2 s^2 + a_1 s + a_0}$$

These representations are given to MATLAB in identical ways: by an array giving the coefficients of the numerator of the transfer function and one giving the denominator. That is, we could enter

```
>> num = [b2 b1 b0];
>> den = [a3 a2 a1 a0];
```

(usually, the a_3 coefficient is factored from the denominator and absorbed into the numerator coefficient b_2). As we shall see soon, there are simple ways to combine such blocks into composite diagrams, and to have MATLAB compute the equivalent overall transfer function. For now, we are interested in equivalent ways to look at this single system.

First, if we wish simply to examine the roots of the numerator or denominator polynomials, we can convert the arrays into lists of roots:

```
>> poles = roots(den)
```

This command will return a variable **poles** which is a list of the roots of the polynomial called **den**. Doing the same thing with the numerator to create a vector **zeros**, and associating an overall gain value k will put the system in *pole-zero-gain* form:

$$H(s) = k \frac{\left[s - zeros(1)\right]\left[s - zeros(2)\right]}{\left[s - poles(1)\right]\left[s - poles(2)\right]\left[s - poles(3)\right]} \tag{3}$$

This form is conceptually convenient, but most MATLAB commands do not operate on it directly.

Another useful form that explicitly displays the poles is the partial fraction expansion (pfe). Like the pole-zero-gain form, MATLAB's commands do not directly accept pfe's, but they can be computed with the command **residue(num, den)**. For example, the partial fraction expansion

$$H(s) = \frac{s-1}{s^2 + 8s + 15} = \frac{3}{s+5} + \frac{-2}{s+3} \tag{4}$$

would be generated by the command sequence

```
>> num = [1 -1];
>> den = [1 8 15];
>> [r, p] = residue(num, den);
```

The result is a pair of arrays, r and p. Array r, [3 –2], and array p = [–5 –3]. Note the correspondence between r and the residues of the poles (numerators of the fractions above) and between p and the corresponding poles themselves.

State-Space Representations

The different frequency-domain representations above are really variations on the same basic entity: the transfer function. The transfer function is convenient for use in block diagrams and when it is only the input–output relationship of a system that is of interest.

In so-called *modern* control systems, the **state-variable** representation is used. State variables contain information about the internal dynamics of a system, and thus provide some flexibility in the selection of performance criteria when designing a control system. State variables are considered vectors in the linear vector space referred to as the *state space* (for a complete explanation and definition of the concept of state-variables, one can refer to any of a great many texts, e.g., Kailath, 1980).

Conversions between Representations

To convert a transfer function in rational polynomial form into a state variable form, MATLAB provides the simple command **tf2ss** ("transfer function to state-space"). The reverse process, naturally, is **tf2ss** . For example, the rational polynomial $H(s)$ from Eq. (4) above can be converted to state space as follows:

```
>> [a, b, c, d] = tf2ss(num, den);
```

This command will return the four matrices that give the state-variable representation

$$\begin{aligned} \dot{x} &= Ax + Bu \\ y &= Cx + Du \end{aligned} \tag{5}$$

In this example, returning[2]

$$A = \begin{bmatrix} -8 & -15 \\ 1 & 0 \end{bmatrix} \qquad B = \begin{bmatrix} 1 \\ 0 \end{bmatrix} \qquad C = \begin{bmatrix} 1 & -1 \end{bmatrix} \qquad D = 0 \tag{6}$$

Or, if a system is given in discrete-time by its rational polynomial Z-domain transfer function, the same command will give matrices representing

$$\begin{aligned} x(k+1) &= A_d x(k) + B_d u(k) \\ y(k) &= C_d x(k) + D_d u(k) \end{aligned} \tag{7}$$

where k is the number of our discrete time step.

Continuous—Discrete Time Conversion

Readers familiar with discrete-time systems will know that the state-space and transfer function representations given for the same system will look entirely different depending on whether the system is interpreted in continuous time or discrete time. The discrete-time representation of a system will change as the sampling period T_s changes and, in fact, this sampling period can have a significant effect on the stability of a system.

MATLAB's Control System Toolbox (to be discussed in more detail later) will convert a continuous-time system that is given in state-variable form into its discrete-time equivalent with the **c2d** command ("continuous to discrete") as follows:

```
>> [ad, bd] = c2d(a, b, Ts);
```

In this expression, **a** and **b** are the A and B matrices from the state-space representation (5), and **ad** and **bd** are the matrices in (7). Note that the C and D matrices are the same in each case, because they simply provide the algebraic combination of state variables and inputs to construct a system's output. They do not affect the *dynamics*, i.e., the time differential or difference relationships.

As any student of discrete-time systems will have noticed, the **c2d** command discussed above can be performed analytically in a number of ways. As given, **c2d** uses the zero-order hold equivalent on the system inputs to discretize the system. This method approximates the continuous input as one that is held constant over each sampling period. There are a number of other ways to discretize a system, including the first-order hold, Tustin's approximation (with or without prewarping), and matched pole-zero (Franklin and Powell 1980). All of the techniques can be performed by MATLAB with a variation on the **c2d** command, the **c2dm** ("continuous to discrete using *method*"). Its syntax is

```
>> [ad, bd, cd, dd] = c2dm(a, b, c, d, Ts, 'method');
```

[2]Note that MATLAB is by default case-sensitive (although this sensitivity can be turned off by issuing a `casesen` command). The presentation in this book is *not* case-sensitive, so that our variable *a* represents the matrix A.

or

```
>> [numd, dend] = c2dm(num, den, Ts, 'method');
```

where 'method' is a string (in single-quotes), that is one of the following: *zoh* (same conversion as **c2d**), *foh* (first-order hold), *tustin*, *prewarp*, or *matched* (pole-zero matched). Note that this command allows transfer function arguments and requires and returns discretized versions of the C and D matrices also. This is because the Tustin and pole-zero matching methods convert a system by considering its continuous/discrete equivalent behavior in frequency domain. Thus, the entire input-output behavior is of concern, and the C and D matrices are important.

Displaying System Representations

MATLAB has always been, until its recent release of a Symbolic Math Toolbox in conjunction with the symbolic program Maple, a *numerical* math package. When manipulating transfer function and state-space representations of system, however, it is convenient to be able to display systems not just as arrays of polynomial coefficients or matrices in a state space. The **printsys** command (also in the Control Systems Toolbox) offers some assistance in this regard. For the $H(s)$ above,

```
>> printsys(num, den)
```

will return

```
num/den =

              s - 1
        ---------------
        s^2 + 8 s + 15
```

For a state-space system, printsys allows the user to give each state-variable a label, such as position and velocity. For example, if one defines a system with the following commands:

```
>> a = [0 1; -2 -4]; b = [0; 1]; c = [1 0]; d = 0;
>> xlabel = ['position velocity'];
>> ylabel = 'displacement';
>> input = 'force';
```

then the following **printsys** command and response will result:

```
>> printsys (a, b, c, d, input, ylabel, xlabel)
a =
                    position              velocity
   position                0               1.00000
   velocity         -2.00000              -4.00000
```

```
     b =
                        force
         position           0
         velocity     1.00000
     c =
                     position          velocity
         displacement 1.00000                 0
     d =
                        force
         displacement       0
```

The **printsys** command can be used with no labels also, in which case the state variables, inputs, and outputs are given the default labels **x1, x2, . . . ; u1, . . . ; y1, . . . ,** etc.

55.3 DIFFERENTIAL EQUATION SOLVERS

One of the most widely needed numerical methods in all of science and engineering is the ordinary differential equation (ODE) solver. Although it is possible to compute closed-form solutions for most linear differential equations with given initial values, some linear and most nonlinear differential equations are difficult to solve. Numerical differential equation solving methods such as the Runge–Kutta algorithm treat differential equations the same regardless of the linearity or time-invariance. That is, the numerical methods consider a generalized expression

$$\frac{dx(t)}{dt} = f(x,t) \tag{8}$$

where x and f are vectors of the same dimension. If a numerical solution is acceptable, this property of numerical methods often makes numerical ODE solvers more convenient than analytical techniques, even for simple linear systems.

Basic Differential Equation Solvers

MATLAB's basic configuration (i.e., without the special toolboxes or SIMULINK) offers two ODE solving commands: **ode23** and **ode45**. These each use the adaptive step-size Runge–Kutta algorithm, of second and third, or fourth and fifth order, respectively. These commands use the syntax

```
>> [t, x] = ode45('derivs', t0, tf, x0, tol, trace);
```

where

t0	initial simulation time
tf	final simulation time
x0	initial state condition
tol	(optional) desired accuracy of the solution
trace	(optional) flag to print intermediate results

The string 'derivs' in the command line above provides the name of a separately constructed M-file that contains the *dynamics* of the system. It is written as a MATLAB *function* and returns the vector of state derivatives as computed from the expression for the differential equation (8).

For example, the solution to the differential equation

$$\dot{x} = -2x + \sin(x) \tag{9}$$

could be generated by using the ODE23 or ODE45 command with desired initial time and condition and a final time, for example

```
>> [t, x] = ode45('derivs', 0, 10, 1);
```

where the string *derivs* specifies a predefined file called DERIVS.M that has the structure

```
function xdot = derivs(t, x)
xdot = -2*x + sin(x);
```

If the differential equation to be simulated is of higher dimension, it should first be expressed in terms of a set of state variables. The DERIVS.M function would then accept a vector argument for **x** and in turn would compute and return a vector of derivatives to the calling **ODE45** command. If such a multidimensional system were also linear, the derivs.m function would be equivalent to evaluating the equation $\acute{A} = Ax + Bu$ [where the input function $u(t)$ is either generated inside the function routine, or declared 'global' so that the function could access it]. There are also a number of solution options in addition to the **odeXX** command available in the Control System Toolbox. Some of these will be discussed in a subsequent section devoted to this toolbox.

The **odeXX** commands return two outputs, **t** (the vector of time samples), and **x**, and vector solution to (8). Note that it is especially important when examining the returned solution that it be plotted against this time variable **t** [that is, using the command **plot(t, x)**], and not simply versus the sample number. Because variable step sizes are used, the elements of **t** are unevenly spaced, so that **x** would appear distorted if plotted against any regularly spaced independent variable.

The MATLAB command **odedemo** provides a complete example of the basic differential equation simulating capabilities.

55.4 BASIC TOOLBOXES

There is now a large and still growing library of toolboxes available for use with MATLAB. For the purposes of the systems/control and signal processing engineer, the minimum requirement would be the Signal Processing Toolbox and the Control System Toolbox. These two toolboxes will be introduced here. Others, some containing advanced routines, including the Robust Control Toolbox, the Nonlinear Control Design Toolbox, the System Identification Toolbox, and the Optimization Toolbox, are invaluable tools to controls designers, but are too specialized for discussion here.

Signal Processing Toolbox

The term "signal processing" refers to the study of signals and the systems through which they pass. Of utmost concern to a signal processing engineer is the efficient and concise characterization of the signal data themselves, and the properties and response of the systems, or *filters*. MATLAB provides a powerful set of commands to assist in signal analysis and filter design, as well as in the filtering process itself.

Fourier Analysis

An indispensable tool in signal processing is Fourier analysis, in which signals and responses to filters are described not in the time domain, but according to the amplitude and phases of the sinusoids into which all real continuous signals can be expanded. These sinusoids of various frequencies then become the independent variables in the representation, and the analysis is performed in *frequency domain*. Frequency is mathematically convenient because the response of a system can be expressed by a simple multiplication of its frequency-domain representation with the frequency-domain (e.g., transfer-function) representation of the system.

Fast Fourier Transforms. For general signals, the first step in frequency-domain analysis is determining its Fourier transform. Fourier transforms can be written for common, simple signals with a closed-form expression, but numerically, a Fourier transform is simply a list (perhaps infinitely long) of the amplitudes and phases of the sinusoidal components of a signal. For a signal consisting of a finite data set, the fast Fourier transform (FFT) has been developed for this purpose. It is most efficient on data sets whose length is a power of two, but can be altered to apply to any length data set.

Given a set of signal data points **x** evenly spaced in the independent variable, the FFT of this signal can be found with MATLAB's **fft** command:

```
>> y = fft(x);
```

This command will perform a so-called "mixed-radix" FFT on a data set whose length is not a power of 2, or a radix-2 FFT otherwise.

```
>> y = fft(x, n);
```

forces the application of the radix-2 transform, where an insufficient number of data points are "padded" to a power of 2 with trailing zeros.

The **fft** command returns complex vector variable **y**, the magnitude and phase of which represent the harmonic components (and the DC component) of the signal. This vector includes the "negative" frequency components of the signal that appear as "foldover" components beyond one-half of the sampling frequency (sampling at the **Nyquist frequency**).

The power spectral density, which is used to determine the relative signal power at different frequencies and thus examine the relative magnitudes of signals and embedded noise, is easily calculated by multiplying the signal's **fft** by its own complex conjugate and normalizing by the array length of the signal:

```
>> Py = y .* conj(y)/length(y);
```

Note here that the .* operator is element-by-element array multiplication.

If the sampling interval were, say, 1000 Hz (i.e., the time signal **x** was evenly spaced 0.001 second apart), then this would be the Nyquist frequency for signals with bandwidth of half this, or 500 Hz. If the input signal **x** were 1024 elements long, then the frequency axis of the resulting fft and spectral density would be 513 (1024/2 + 1 for the DC component) elements long, and evenly spaced between 0 and 500 Hz. Thus, a useful frequency axis for plotting is easily generated by the command

```
>> f = 500*(0:512)/512;
```

Example. We first generate a signal with two sinusoidal components and some additive noise:

```
>> t = 0:.001:1;           % time axis: sampling at 1000Hz
>> x = sin(2*pi*60*t) + 2*sin(2*pi*100*t); %define signal
>> x = x + 0.5*randn(size(t));   %add a Gaussian noise
                                 %vector of same length as
                                 %time vector;
```

Now we compute the 1024 point FFT and the power spectral density:

```
>> y = fft(x, 1024);
>> Py = y .* conj(y)/length(y);
```

Now if we generate the frequency axis as above, the spectral density can be plotted versus these frequencies:

```
>> Py = Py(1:513);         %discard foldover (reflected)
                           %portion of spectral density
>> plot(f, Py);
>> title('Spectral Density');xlabel('frequency (Hz)');
```

The result is shown in Figure 55.1.

Filter Design

With suitable signal analysis as afforded by the Fourier tools provided in MATLAB, we may need to design a particular filter to alter the *spectral* (i.e., sinusoidally decomposed) content of the signal. The Signal Processing Toolbox provides a wide array of filter design tools, including some predefined standard filters and some tools for designing filters with custom characteristics. The toolbox makes it a simple matter to design **IIR** and **FIR filters**, either from frequency response specifications (cutoff frequencies, maximum ripple, e.g.) or from user-supplied data points. Filters can be designed for lowpass, bandpass, high pass, or band-reject characteristics, and stored signals can be filtered with a single command. Detailed discussion of these capabilities is beyond the scope of this chapter.

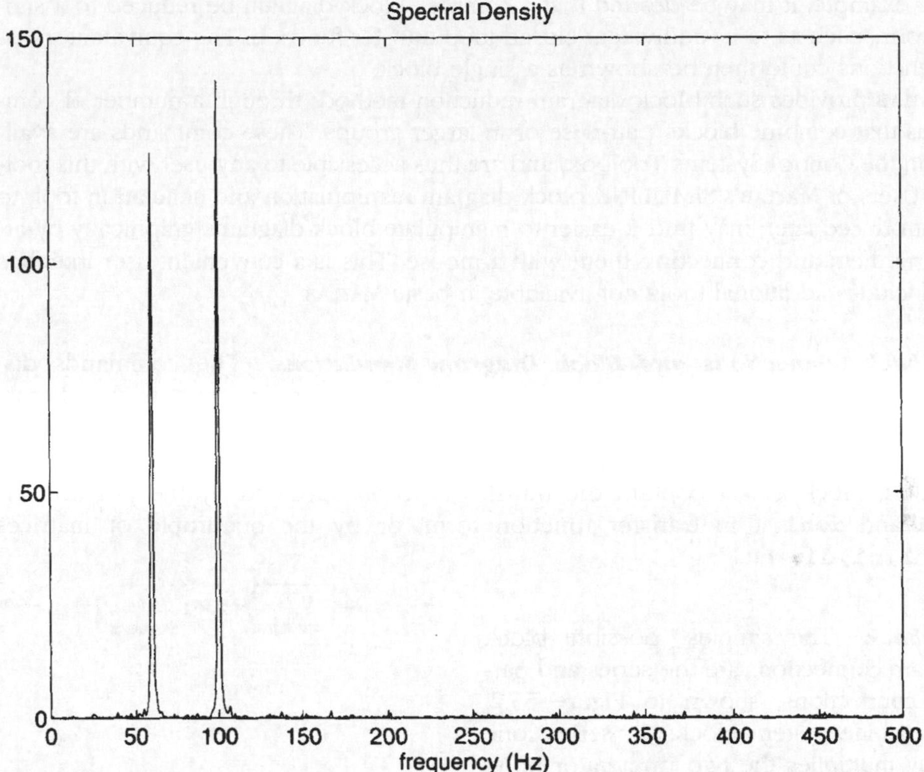

FIGURE 55.1 Spectral density of signal with 60 Hz and 100 Hz components.

Control System Toolbox

MATLAB tools for control system designers range from simple stability and damping tests all the way up to advanced nonlinear control and H_∞ methods. Like the filter design methods discussed in the previous subsections, control system tools also come in two flavors: discrete time and continuous time, although in many cases the distinction is unnecessary for the computations at hand. For example, when computing the root locus or eigenvalues of a system, it makes no difference whether the system is in discrete or continuous time; only the interpretation of the results matters.

In this section, we introduce some of the basic control system tools, suitable for the beginning student. Advanced tools are contained in the more advanced toolboxes, and, as always, the toolbox manuals do an excellent job giving a brief tutorial and explanation of the commands.

Frequency Domain

Block Diagram Systems Representation. We have already discussed the several ways in which we can represent a particular transfer function in MATLAB. Control system analysis usually involves a further level of complexity in system representation by requiring interconnected networks of separate transfer functions in a *block diagram*. Complex block diagrams may be reduced to simpler forms with only a few blocks, which then facilitates easier stability and performance analysis.

For example, it may be desired that a complex block diagram be reduced to a simpler form, such as to its equivalent closed-loop transfer function. The equivalent transfer functions could then be shown as a single block.

MATLAB provides such block diagram reduction methods through a number of commands that combine blocks pair-wise or in larger groups. These commands are available in the Control Systems Toolbox, and are thus accessible to any user with this toolbox. Users of MATLAB's SIMULINK block diagram manipulation and simulation tool, to be introduced later, may find it easier to manipulate block diagrams graphically by arranging them and connecting them with a mouse. This is a convenient user interface that includes additional tools not available in basic MATLAB.

Feedback Connections and Block Diagram Simulations. The commands discussed below each apply to particular type of block interconnection. In this discussion, we will assume that individual blocks contain transfer functions specified either as rational polynomials or state-variable matrices. That is, by our previous notation, block 1 will contain the transfer function specified by the polynomials **num1** and **den1** if in transfer function form, or by the quadruple of matrices **a1,b1,c1,d1,** etc.

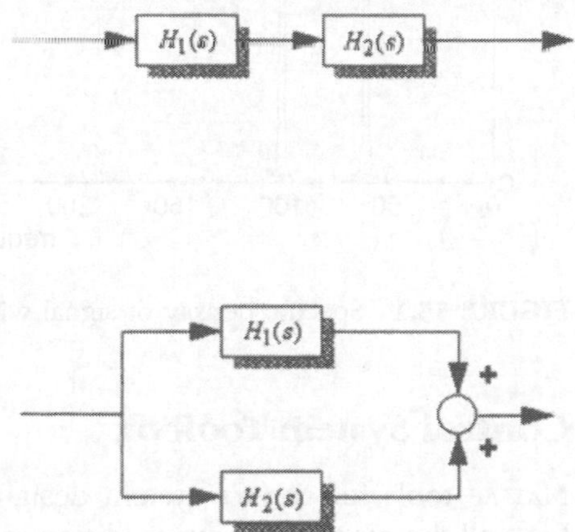

FIGURE 55.2 Series (top) and parallel block connections.

Feedback. The simplest possible block diagram connections are the series and parallel connections, shown in Figure 55.2. To combine system blocks in series, one simply multiplies the two numerator polynomials together and the two denominator polynomials together. Polynomial multiplication is conceptually accomplished via the convolution command **conv(a,b)**, where a and b are the coefficient arrays of the polynomials to be multiplied. The result is an array giving the product polynomial. MATLAB simplifies this process by providing the series command. Using

```
>> [num, den] = series(num1, den1, num2, den2);
```

or

```
>> [a, b, c, d] = series(a1, b1, c1, d1, a2, b2, c2, d2);
```

the resulting transfer function or state-space representation is the multiplicative product of the two original systems.

The parallel connection requires the *addition* of two transfer functions (or state-space representations). System addition in a parallel connection is performed with the command

```
>> [a, b, c, d] = parallel(a1, b1, c1, d1, a2, b2, c2, d2)
```

or

```
>> [num, den] = parallel(num1, den1, num2, den2)
```

Perhaps the most common block diagram simplification in practical control systems analysis is the feedback connection. In this configuration, a forward path transfer function is connected to a feedback path transfer function as shown in Figure 55.3. When it is desired that the equivalent closed-loop transfer function be computed, the **feedback** command can be used. The forms

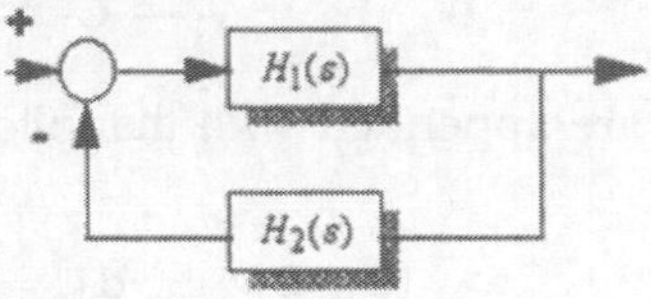

FIGURE 55.3 Basic feedback connections.

```
>> [a, b, c, d] = feedback(a1, b1, c1, d1, a2, b2, c2, d2)
```

or

```
>> [num, den] = feedback(num1, den1, num2, den2)
```

can be used to find closed-loop equivalents. Analytically, the closed-loop equivalent is

$$H_{eq}(s) = \frac{num(s)}{den(s)} = \frac{Y(s)}{U(s)} = \frac{H_1(s)}{1 + H_1(s)H_2(s)} \tag{10}$$

but this expression is inconvenient to compute by direct manipulation of the polynomials involved. Not only would it require polynomial multiplication, but it also requires a common denominator for simplification.

A similar command, **cloop**, works exactly like **feedback**, but takes only a single system as its argument, because it assumes unity feedback.. That is, if the transfer function in the feedback loop is $H_2(s) = 1$, **cloop(num1, den1)** will find the closed-loop equivalent. In the sense that the basic connection is the same as in the feedback case, **cloop** is a special case of feedback.

More general and complex block diagrams can be reduced using the basic block connections commands above. One would simply reduce a block diagram stage by stage, slowly collapsing it to its equivalent single block. This iterative method is often taught to engineering students as an exercise in block diagram reduction techniques. However, there is a single-step process available in MATLAB that can be used on complex systems with, potentially, many inputs and outputs.

Reduction of Complex Block Diagrams. The MATLAB Control Systems Toolbox commands **connect** and **blkbuild** are used in conjunction to specify the interconnections and reduce, respectively, large block diagrams. These are powerful tools that can be used to reduce almost any block diagram, even those that contain multivariable system blocks (i.e., more than one input and/or output). Because of this generality and flexibility, the use of these commands is somewhat complex.

The command **blkbuild** is a generalization of a simpler command called **append**. **Append** operates on two state-space representations, and produces a single, composite, "augmented" state space system that includes the dynamics of both components, but performs no algebraic combinations like the parallel or series connections. For example, if the two state space systems

$$\begin{aligned} \dot{x}_1 &= A_1 x_1 + B_1 u_1 \\ y_1 &= C_1 x_1 + D_1 u_1 \end{aligned} \quad \text{and} \quad \begin{aligned} \dot{x}_2 &= A_2 x_2 + B_2 u_2 \\ y_2 &= C_2 x_2 + D_2 u_2 \end{aligned} \tag{11}$$

are appended with the following command:

```
>> [a, b, c, d] = append(a1, b1, c1, d1, a2, b2, c2, d2);
```

the resulting matrices A, B, C, and D, will represent the augmented system:

$$\begin{aligned} \begin{bmatrix} \dot{x}_1 \\ \dot{x}_2 \end{bmatrix} &= \begin{bmatrix} A_1 & 0 \\ 0 & A_2 \end{bmatrix} \begin{bmatrix} x_1 \\ x_2 \end{bmatrix} + \begin{bmatrix} B_1 & 0 \\ 0 & B_2 \end{bmatrix} \begin{bmatrix} u_1 \\ u_2 \end{bmatrix} \\ \begin{bmatrix} y_1 \\ y_2 \end{bmatrix} &= \begin{bmatrix} C_1 & 0 \\ 0 & C_2 \end{bmatrix} \begin{bmatrix} x_1 \\ x_2 \end{bmatrix} + \begin{bmatrix} D_1 & 0 \\ 0 & D_2 \end{bmatrix} \begin{bmatrix} u_1 \\ u_2 \end{bmatrix} \end{aligned} \tag{12}$$

Now if we had, say, five such systems that we wanted to append into one, we would have to repeatedly call **append** with two systems as arguments until all subsystems were grouped together. Instead, with the standard notation for system matrices (a1, a2, etc.), we could simply execute the commands:

```
>> nblocks = 5;
>> blkbuild;
```

This would produce the desired conglomeration of five subsystems into one as in Eq. (12). MATLAB automatically detects the systems numbered according to the above convention and produces four composite matrices A, B, C, and D. Note that the composite system is in state variable form, although the individual component subsystems need not be. They may occur in any combination of state-space or transfer function descriptions. **Blkbuild** will automatically convert any transfer function representations into state-space matrices before combining them.

Whenever a large block diagram is to be combined into an equivalent single system, the subsystems must first be combined, either with **append** or **blkbuild**. Then to perform the actual reduction as specified by the block diagram interconnections, the connect command is used. It is necessary to tell connect which block inputs and outputs are connected together, essentially creating a description of the block diagram as a list of the connections.

Use of the connect command in conjunction with **blkbuild** is best illustrated with an example. Consider the example block diagram given in Figure 55.4. This system contains three component subsystems, two given as transfer functions (*num1*, *den1* and *num3*, *den3*) and one as a state-space quadruple (*a2*, *b2*, *c2*, *d2*). With all of the necessary polynomials and matrices defined, one would first use **blkbuild** to create the composite state-space description:

```
>> nblocks = 3;
>> blkbuild;
```

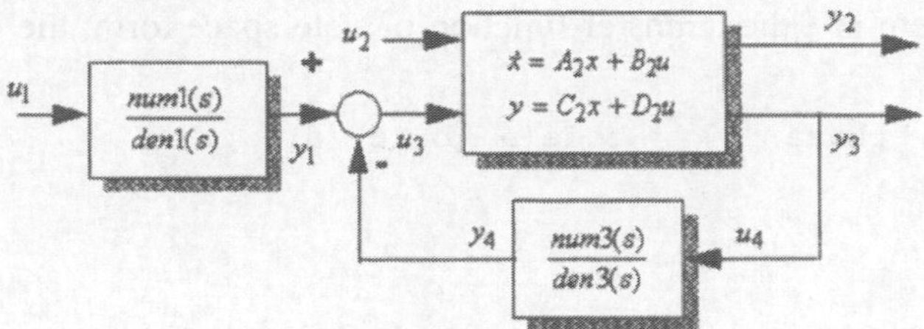

FIGURE 55.4 Complex block diagram with multiple inputs and outputs and a mixture of transfer function and state-space representations.

To use **connect**, we create a matrix that will tell **connect** which inputs and outputs are connected together. Suppose we call this matrix Q. Matrix Q's first column will contain the numbers designating system inputs into which other blocks feed. Subsequent elements of each row in Q specify those composite system outputs that feed into those inputs. For the diagram of Figure 55.4, the Q matrix would be

$$Q = \begin{bmatrix} 3 & 1 & -4 \\ 4 & 3 & 0 \end{bmatrix} \tag{13}$$

The first row of Q thus implies that outputs 1 and the negative of output 4 feed into input 3. The second row says that only output 3 feeds into input 4.

We also provide **connect** with knowledge of which inputs and outputs constitute the inputs and outputs to and from the outside world (as opposed to those that have connections only to other blocks). In this case, we do this by defining the arrays

```
>> inputs  =  [1 2];
>> outputs  =  [2 3];
```

We can now call **connect**:

```
>> [ac, bc, cc, dc] = connect(a, b, c, d, Q, inputs, outputs);
```

which will return the combined, equivalent system (*ac, bc, cc, dc*). Of course, this could in turn be converted to transfer function form with **ss2tf**, if desired.

Bode, Nyquist, and Root Locus Plots

We now turn to the design tools common to introductory control systems. Historically, classical control systems (using transfer functions) have employed graphical design aids. These include the frequency-response plot, or **Bode plot**. Bode plots provide two graphs, one of the magnitude response of a transfer function on a semilog scale, and one of the phase; both given versus frequency. A similar plot can be formed by evaluating the transfer function over a closed path in the complex s-plane, one that encompasses the entire right-half plane (called the *Nyquist contour*). The result is called a **Nyquist plot**. Both types of plots are used for similar purposes, but certain stability characteristics of systems are interpreted from these plots in different ways.

Given a system in either transfer function or state space form, the commands

```
>> [mag, phase, w] = bode(a, b, c, d);
```

or

```
>> [mag, phase, w] = bode(num, den);
```

will both return the same result: an array of points giving the magnitude of the response of the system to a unit sinusoid (*mag*), and an array giving the corresponding phase shift (*phase*). These points are calculated at the frequencies given in the returned array of frequencies (*w*). Bode thus generates its own frequency axis, but one could be supplied by including a desired frequency array as a final argument. Also, if a system is represented in state variables and it happens to have more than one input, the response to a particular input, say the ith one, can be returned by specifying **bode(a, b, c, d, ui, w)**.

The Nyquist plot is similarly generated, except that it returns not the magnitude, phase, and frequency, but the real part, imaginary part, and frequency of the response instead, all evaluated on the Nyquist contour. That is, the commands necessary to produce a Nyquist plot are either:

```
>> [re, im, w] = nyquist(a, b, c, d);
```

or

```
>> [re, im, w] = nyquist(num, den);
```

The same options apply for **nyquist** as for **bode**, i.e., we can use one of many inputs of a state-space system, or we can provide our own frequency axis.

If either of these commands is executed without the left-hand arguments, e.g.,

```
>> bode(a, b, c, d);
```

then MATLAB proceeds to create the plots themselves (with the magnitude in dB an the phase in degrees), rather than simply creating the data arrays. As executed before, the *mag*, *phase*, etc., variables are created, but the user would have to then create the graph with the plot command. For example, the lines below will produce the Bode plot shown in Figure 55.5, and the Nyquist plot in Figure 55.6.

```
>> num = [1 30];
>> den = [1 15 100];
>> bode(num,den); title('Bode plot')
>> nyquist(num,den); title('Nyquist plot')
```

Given the *mag*, *phase*, and *w* variables, stability analysis can be performed by examination of the graphs (see Kuo, 1991), or through use of the **margin** command. The command

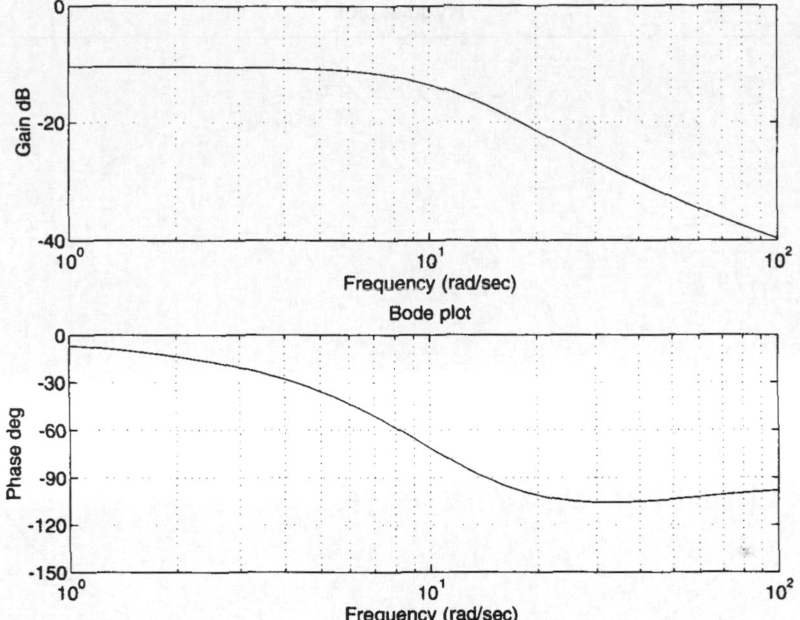

FIGURE 55.5 Bode plot of a transfer function with a zero at $s = +1$, and poles at $s = -3, -5$.

```
>>[Gm, Pm, Wcg, Wcp] = margin(mag, phase, w);
```

will return the so-called *gain margin*, *phase margin*, *gain crossover frequency*, and *phase crossover frequency* of the system. Approximate values for these quantities can be read from an accurately produced graph, but this command is a convenient way to quickly get their exact values. These four values give a measure of the relative stability of a system, i.e., how much excess gain (or phase) can be tolerated for a stable system to become unstable, and thus are critical pieces of information for a control system designer.

Root Locus Plots. Another popular s-plane design tool is the root-locus plot. The root-locus plot is an s-plane graph that shows the loci of all possible closed-loop system poles as a parameter, such as a gain, is varied throughout its range, usually taken to be zero to infinity.

In the case of a control system represented in the standard configuration shown in Figure 55.3, with forward and feedback transfer functions, a multiplicative gain K may be inserted next to the forward transfer function $H_1(s)$. In that case, the expression for the closed-loop transfer function is

$$T(s) = \frac{KH_1(s)}{1 + KH_1(s)H_2(s)} \tag{14}$$

in which case the closed-loop characteristic polynomial is given by the numerator of the expression $1 + KH_1(s)H_2(s)$; that is, the poles of the system occur at all s such that $1 + KH_1(s)H_2(s) = 0$. If the control system of interest is not in this standard form, for example, when the parameter K is not a multiplicative gain in the forward path, it will always be possible to algebraically manipulate the expression for the characteristic polynomial so that it resembles this form. That is, it will always be possible to find an

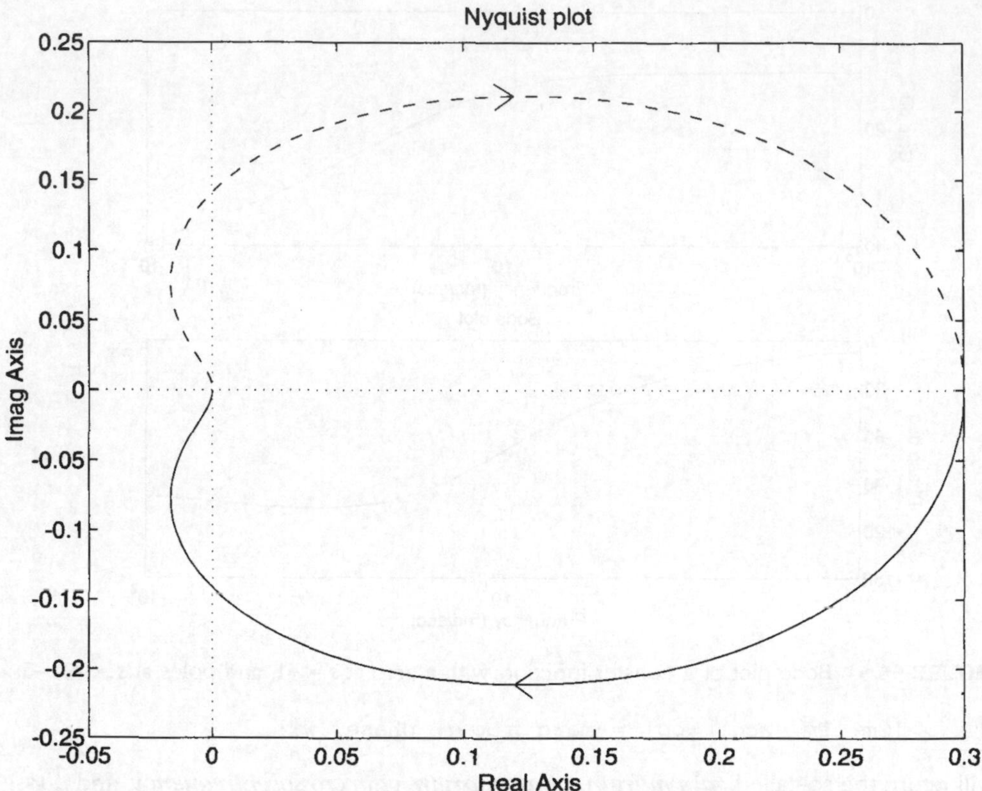

FIGURE 55.6 Nyquist plot of same transfer function whose Bode plot is given in Figure 55.5. Arrows show the direction of increasing frequency.

expression for the denominator of the closed loop transfer function in the form

$$1 + K\,\frac{num(s)}{den(s)} = 0 \tag{15}$$

In this form, it is the polynomials *num* and *den* that are important in constructing a root-locus plot.

To generate a root-locus plot, these two polynomials, *num*(*s*) and *den*(*s*), are entered as always: as arrays of coefficients. One then simply executes the MATLAB command

```
>> rlocus(num,den);
```

MATLAB will then plot the root locus graph, using a different color for each branch (there will, of course, be one branch for every closed loop pole). For example, the following lines

```
>> num = [1 4];
>> den = [1 6 -40 0];
>> rlocus(num, den); title('Root Locus Plot');
```

will construct the plot shown in Figure 55.7.

In the root-locus plots, the x's are placed where the "open-loop" poles are, which is at the roots of the polynomial *den*(*s*). The o's are at the "open-loop" zeros, or the

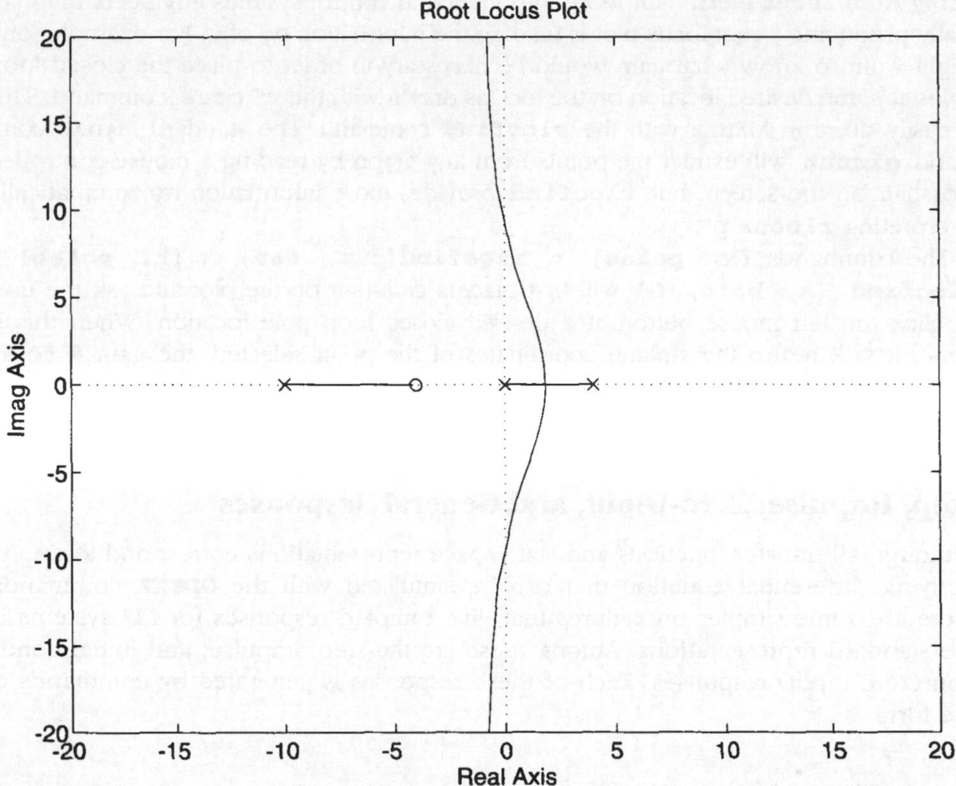

FIGURE 55.7 Root locus plot showing the loci of closed-loop poles versus a change in loop gain.

roots of polynomial *num(s)*. The root loci will start (where $K = 0$) at these open-loop poles, and travel toward the open-loop zeros [which may be at infinity (off the graph) for strictly proper open-loop transfer functions] as $K \to \infty$. [One can plot *just* these open-loop pole and zero locations with the command **`pzmap (num, den)`**.]

Of course, the root locus can also be generated by supplying the **`rlocus`** command with the four state-space matrices of a system. One can also provide a gain array *k* if desired, specifying only those gains for which the root locus is to be plotted. **`rlocus`** also allows left-hand arguments, **`[r, k] = rlocus (a, b, c, d)`**, which will return a matrix *r* of closed-loop poles, and the corresponding gain vector *k* for each.

The construction of the root locus plot is identical for discrete and continuous-time systems, but, of course, the interpretation is different. In continuous time, the closed-loop poles must be in the left-half s-plane for stability, whereas in discrete time, they must be inside the unit circle. Selection of poles is thus facilitated by knowing, in each domain, where the lines of constant damping ζ and constant natural frequency ω_n are. This is facilitated by the **`sgrid`** and **`zgrid`** commands. Each of these commands, when executed after a root locus plot is generated, will add to the plot lines of constant ζ and constant ω_n. Of course, these lines differ in each domain, hence the need for two different commands in the *s*- or *z*-plane.

Using Root Locus Plots. Of more importance to control systems engineers than actually producing a root locus plot is reading the information *off* of it. For example, one might want to know what gain would be necessary in order to place the closed loop poles at some desired location on the loci, as drawn with the **rlocus** command. This is easily done in MATLAB with the **rlocfind** command. The standard MATLAB command **ginput** will extract the points from any graph by reading a mouse-controlled crosshair on the screen, but **rlocfind** provides more information by automatically interpreting **rlocus** plots.

The commands **[k, poles] = rlocfind(num, den)** or **[k, poles] = rlocfind (a, b, c, d)** will first place a crosshair on the plot and ask the user to click the left mouse button at a desired closed loop pole location. When this is done, it will return the *s*-plane coordinates of the point selected, the gain *K* corresponding to that point, and the complete list of that and all the other poles that will result if the gain *K* is chosen.

Step, Impulse, Zero-Input, and General Responses

Although all transfer functions and state-space representations correspond to an underlying differential equation that can be simulated with the **ODEXX** commands, there are some simpler procedures that give sampled responses for LTI systems in the standard representations. Among these are the step, impulse, and initial condition (zero-input) responses. Each of these responses is generated by commands of the form

```
>> [y, x, t] = step(a, b, c, d);
```

This command produces a vector of outputs, state-variables, and time points giving the step response of the system (as usual, a transfer-function could also be given, an input number can be specified for multivariable systems, and a final argument can provide a user-supplied time scale). They are plotted automatically if the left-hand arguments are omitted. **Step** can be replaced with **impulse** to give the impulse response.

Using the command

```
>> [y, x, t] = initial(a, b, c, d, x0);
```

an initial-condition response is obtained. This command takes state-space system matrices only.

For more general responses, **lsim** can be used. In its most general form, **lsim** is executed as follows:

```
>> [y, x] = lsim(a, b, c, d, u, t, x0);
```

which gives the output *y* and state vector *x* corresponding to time vector *t*. These can be plotted using the usual plotting commands. As usual, the state-space matrices above can be replaced by **num** and **den**, representing a transfer function.

If the time response of a discrete time system is desired, MATLAB provides the equivalent commands **dstep, dimpulse,** and **dlsim**.

Time Domain

Many of the analysis techniques discussed above, for example, the frequency response plots and root locus plots, can be generated from either transfer function or state-space descriptions. The design techniques, though, are performed from frequency-domain analysis, and are regarded as *classical*. *Modern* control includes many powerful design techniques that work directly in time domain, with state-variables.

In the next section, we will introduce some of MATLAB's many modern control system design techniques.

State-Space Representation

Earlier we discussed the conversion of system descriptions back and forth between transfer functions and state-variables using the **ss2tf** and **tf2ss** commands. Because state-space descriptions are not unique for a given system, but are related by equivalence, or *similarity* **transformations**, MATLAB also provides a command to convert between one state-space representation and another.

To do this, the user must supply the similarity transformation matrix T (really a change of basis) converting state vector representations x into some other representation $\hat{x}$, which converts a standard state variable description given by matrices A, B, C, and D, to one of the form

$$\dot{\hat{x}} = TAT^{-1}\hat{x} + TBu$$

$$y = CT^{-1}\hat{x} + Du$$

The command to do this is, as one might guess, is

```
>> (at, bt, ct, dt,) = ss2ss(a, b, c, d, T).
```

The change of basis matrix T can be generated in many useful ways, for example by using the matrix of eigenvectors of A (found with the **[v, d] = eig(a)** command), which will diagonalize the system, or by using other canonical transformations (Kailath, 1980). Canonical transformations may be generated using MATLAB's **canon (a, b, c, d, 'type')** command, where type may be the strings *modal* or *companion*. Controllable or observable staircase forms can be generated with the `ctrbf` and `obsvf` commands. MATLAB also provides a command **T = balance(a)**, which will find a matrix T that results in a "balanced" realization.

Controllability and Observability

Before designing a controller for state-variable systems, the first step is often the analysis of the system's **controllability** and **observability**. These properties indicate the possibilities, respectively, of arbitrarily placing the pole locations through state-feedback, or of estimating the plant's state by processing the input and output vectors.

Controllability and observability for an nth order system are determined by the rank of the matrices

$$P = \begin{bmatrix} B & AB & A^2B & \cdots & A^{n-1}B \end{bmatrix} \tag{16}$$

and

$$Q = \begin{bmatrix} C \\ CA \\ C^2A \\ \vdots \\ C^{n-1}A \end{bmatrix} \tag{17}$$

If these matrices have "full" rank, i.e., rank n, then the system is controllable and observable. While the rank of a matrix is easily determined with the **rank(P)** command, the controllability and observability matrices themselves are produced with the one-line commands

```
>> P = ctrb(a, b);
>> Q = obsv(a, c);
```

State Feedback

For a system that is completely controllable, all poles can be "placed" by feeding back the state vector into the inputs, $u = -Kx$. When this is done, simple analysis shows that the equivalent closed loop "A-matrix" becomes A-BK. Thus, the eigenvalues, and hence, the closed loop poles, are affected by the feedback gain matrix K, which is determined by the designer for a desired response.

If the designer has a list of desired closed loop eigenvalues in an array, say **p**, MATLAB will compute the necessary gain matrix K with one of two commands, **place** or **acker**. These commands are called in the same way, for example,

```
>> k = place(a, b, p);
```

which, of course, returns the required K matrix. The closed loop eigenvalues can then be checked with the command **eig(a-b*k)**.

It should be noted here that the MATLAB manual warns against using the **acker** command, which uses the numerically "unreliable" Ackermann's formula to place poles. On the other hand, the **place** command refuses to place poles with multiplicity greater than the number of inputs, something which is done often in practice. With either of these commands, it is recommended that the results be thoroughly checked for accuracy and conditioning before using.

Optimal Control and Estimation

Rather than choosing a feedback gain matrix to place the eigenvalues in a particular place, it is often desired to place them such that the corresponding response minimizes a chosen cost criterion. This is the domain of "optimal" control (Kirk, 1970).

MATLAB provides a complete suite of optimal control techniques, many of which reside in the advanced toolboxes, such as the Robust Control Toolbox. The most basic form of optimal control is the LQR, or **Linear Quadratic Regulator**. The LQR controller gives a state-feedback matrix K that results in the system minimizing the infinite-time cost function

$$J = \int_0^\infty \left(x^T Q x + u^T R u\right) dt \tag{18}$$

where Q and R are designer-chosen weighting matrices. (MATLAB can automatically handle "cross-terms," too.)

Called as

```
>> [k, s, e] = lqr(a, b, q, r);
```

the **lqr** command returns the state-feedback matrix K, the matrix solution S to the associated algebraic Riccati equation:

$$0 = SA + A^T S - SBR^{-1}B^T S + Q$$

(matrix K is computed from S by $K = R^{-1}B^TS$, see Kirk, 1970), and E, the set of resulting closed-loop eigenvalues.

By way of explanation, this control that minimizes cost function J is justified as follows: The term x^TQx in the integral above provides a weighted penalty for magnitude of the state variables as the system response evolves (Q must be symmetric and positive semidefinite), the term u^TRu gives a similar penalty for the magnitude of the required inputs (R must be symmetric and positive-definite). If *output* weighting is desired, the **lqry** command is available.

Analogous procedures are followed when designing an **observer** for a linear system. In this case, the intent is to design a system that produces an estimated state vector $\hat{x}$ that can be used to generate estimated state feedback $u = -K\hat{x}$ when the full state-vector is unavailable for measurement.

MATLAB provides the **lqe** (Linear Quadratic Estimator) command for this purpose. The **lqe** command is general enough to allow systems with process and measurement noise,

$$\dot{x} = Ax + Bu + Gw$$
$$y = Cx + Du + v \tag{19}$$

where G is a constant coefficient matrix, and the noise statistics

$$E[w] = E[v] = 0, \qquad E[ww^T] = Q, \qquad E[vv^T] = R, \tag{20}$$

are known (a variation on the **lqe** command allows correlated noise, $E[wv^T] = N \neq 0$). In this case with noise included, the LQE observer is equivalent to the **Kalman filter** (Lewis, 1986). Thus, Kalman filtering and observer design is performed with a single command.

The command invoked by the line

```
>> [l, p, e] = lqe(a, g, c, q, r);
```

returns the Kalman (observer) gain matrix L, the solution P to the associated Riccati equation (which is also the error covariance of the Kalman filter's state estimate), and the system's closed loop eigenvalues E. The observer thus designed has the form

$$\dot{\hat{x}} = A\hat{x} + Bu + L(y - C\hat{x} - Du) \tag{21}$$

Controller Design. Thus, the two one-line commands **lqr** and **lqe** (**dlqr** and **dlqe** in discrete-time) are sufficient to compute the necessary gains required for any standard "LQG" control problems. When complete, the overall closed-loop system, together with its state-feedback and observer, can be represented with a single representation through the **reg** command.

That is, when the feedback and estimator gains K and L have been found, the system consists of two component subsystems,

$$\begin{aligned} \dot{x} &= Ax + Bu \\ y &= Cx + Du \end{aligned} \quad \text{and} \quad \begin{aligned} \dot{\hat{x}} &= [A - BK - LC + LDK]\hat{x} + Ly \\ u &= K\hat{x} \end{aligned} \tag{22}$$

As with the **append** command, these systems may be combined to form a single representation, this time using the **reg** command (**dreg** for discrete time). As given by the command

```
>> [ac, bc, cc, dc] = reg(a, b, c, d, k, l);
```

the **reg** command produces a composite system assuming all inputs are control inputs and all outputs are sensor outputs (*known* inputs can also be provided by specifying them in a separate argument of the **reg** command.

55.5 SUMMARY

The reader is reminded once again that this discussion has focused only on the basic ingredients of MATLAB and its toolboxes necessary for systems and controls analysis and design. A great many commands and, in most cases, the complete capabilities of some commands, have been omitted. An observant MATLAB user will notice that topics such as Lyapunov equations, Nichols plots, singular-value analysis, Householder transformations, Grammians, etc., are all missing from the above discussion. They are all, however, along with many more, provided by MATLAB with single commands.

It should be apparent that MATLAB provides a systems/controls engineer with single commands that would require hours of programming in any other language. A real flavor for the comprehensiveness of MATLAB can be gained by simply reading the list of commands as printed in the manuals. The MATLAB manuals conveniently provide these lists as an index to the many M-files available, each grouped according to functionality.

55.6 ADVANCED TOOLBOXES

Of course since we have not had the room here to completely cover all of MATLAB's systems/controls capabilities even for the basic M-files and the Signal Processing and Control Systems Toolboxes, we cannot hope to do justice to the other toolboxes. Briefly, they are (at the time of writing)

Symbolic Math
Extended Symbolic Math
Image Processing
Neural Network
Statistics
Nonlinear Control Design
System Identification
Robust Control
μ-Analysis and Synthesis
Spline

Hi-Spec
MMLE3 Identification
Chemometrics
Model Predictive Control, and
Frequency Domain

In addition, because toolboxes are simply collections of M-files for particular applications, there are third-party toolboxes available in diverse areas, such as Robotics (Sun Teq).

55.7 SIMULINK

Like the advanced toolboxes, it is not possible to discuss SIMULINK in any detail here. SIMULINK is MATLAB's graphical block diagram manipulation package that allows users to assemble complex systems by simply "drag-and-drop"-ing the component blocks into a simulation workspace. Blocks can be connected with a mouse, and input and output blocks attached to provide system inputs (even from a real-time source such as a data acquisition card), and outputs (as data files, workspace variables, or one of many types of displays).

Many standard blocks are available in the block libraries, which are simply dragged into the workspace and customized so that they contain the desired values, transfer functions, state-space matrices, initial conditions, or other parameters. For example, a *small* sample of the libraries and the blocks they contain is given below:

- Sources
  step functions
  clocks
  workspace variables
  data files
  signal generators (sine, square, sawtooth, and noise)
  constants
- Sinks
  oscilloscopes
  .mat data files
  workspace variables
- Discrete
  delays
  transfer functions
  state-space systems
- Linear
  gains
  summing junctions
  integrators
  transfer functions
  state-space systems
- Nonlinear
  multipliers
  arbitrary functions
  backlash
  saturation
  dead zone

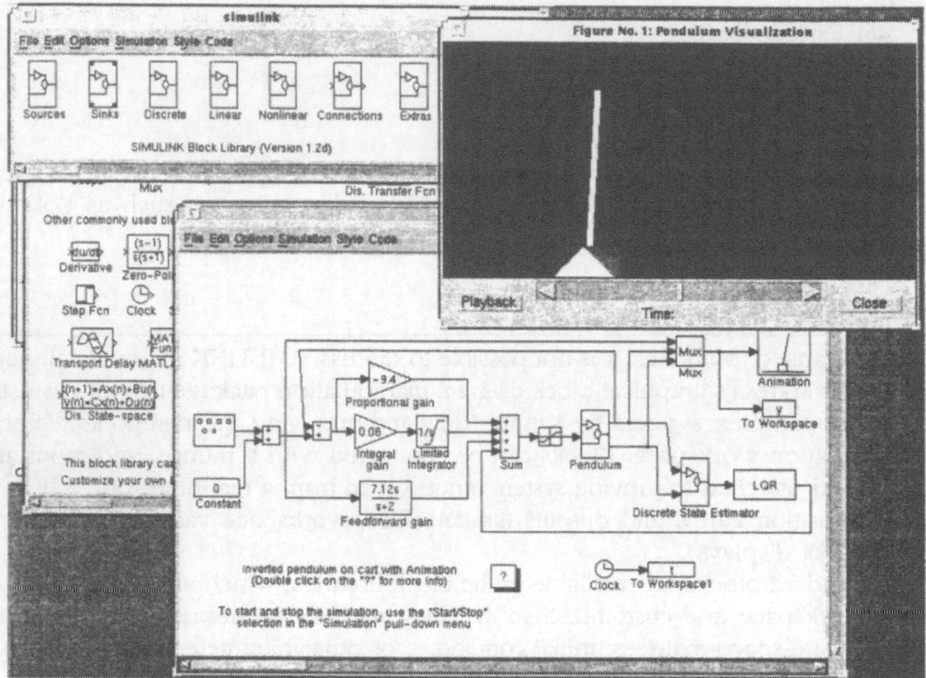

FIGURE 55.8 Screen-dump of SIMULINK simulation.

- Connections
  multivariable line multiplexers and demultiplexers
- Displays
  oscilloscopes
  graph windows
  power spectral density displays
  spectrum analyzers
  x-y displays
- Filters
  analog lowpass, highpass, bandpass, and bandreject
  discrete lowpass, highpass, bandpass, bandreject
- Controllers
  PID, with or without anti-wind-up
  matrix gain
  Kalman filter

While a complete discussion of SIMULINK would justify an entire chapter of this book, an appreciation for its ease of use and simple operation can be gained from the screen-dump shown in Figure 55.8.

DEFINING TERMS

Bode Plot: A set of two graphs, the first giving the magnitude of a system's response (on a log-log scale) versus the frequency of the applied sinusoidal input, and the second a plot of the phase shifts on the same frequency scale.

Controllability: The property of a linear system that admits a finite control input to force the state to reach equilibrium in a specified time, or, equivalently, allows one to arbitrarily place all of the eigenvalues with static state feedback.

FIR Filter: (Finite-duration Impulse Response filter) A digital filter that has "all zeros," i.e., one that provides as output as a moving average of the applied input sequence.

IIR Filter: (Infinite-duration Impulse Response filter) A digital filter that has poles and may have zeros also.

Kalman Filter: An optimal linear filter that provides the best linear estimate of a system's state, given its inputs, measured outputs, and knowledge of the noise statistics in the process and in the sensors.

Linear Quadratic Regulator: A control system design that produces a linear state feedback matrix resulting in the minimization of an integrated cost function consisting of quadratic functions of the state and input.

Nyquist Frequency: The minimum frequency necessary to sample a signal to unambiguously reconstruct it; proven to be two times the bandwidth, or highest frequency present, in the signal.

Nyquist Plot: A frequency-response plot similar to a Bode plot, but with the complex response of a linear system graphed on the complex plane, as the value of s varies along a closed path in its complex s-plane. The frequency then becomes a parameter along the curve.

Observability: The property of a linear system that enables one to construct estimates of the state by observing the input and output signals.

Observer: A linear system constructed to provide an estimate of another system, whose internal state is not available, but whose input and output sequences are.

Similarity Transformation: A full-rank matrix transformation on the collective matrices of a state-space representation that accomplishes a change of basis of the underlying vector space.

State Variable: An element of the state vector that constitutes a complete set of information about the dynamics of a linear system, such that given an initial condition and the system model, the state variables suffice to determine the state at any future time.

REFERENCES

G. F. Franklin and J. D. Powell, *Digital Control of Dynamic Systems*. Addison-Wesley, Reading, MA, 1980.

A. Grace; et al., *Control System Toolbox* (manual). The MathWorks, Inc., Natick, MA.

T. A. Kailath, *Linear Systems*. Prentice-Hall, Englewood Cliffs, NJ, 1980.

D. E. Kirk, *Optimal Control Theory*. Prentice-Hall, Englewood Cliffs, NJ, 1970.

B. C. Kuo, *Automatic Control Systems*. Prentice-Hall, Englewood Cliffs, NJ, 1991.

B. C. Kuo, and D. C. Hanselman, *MATLAB Tools for Control System Analysis and Design*. Prentice-Hall, Englewood Cliffs, NJ, 1994.

N. E. Leonard and W. S. Levine, *Using MATLAB to Analyze and Design Control Systems*. Benjamin/Cummings, Redwood City, CA, 1992.

F. L. Lewis, *Optimal Estimation with an Introduction to Stochastic Control Theory*. John Wiley, New York, 1986.

J. N. Little and L. Shure, *Signal Processing Toolbox* (manual). The MathWorks, Inc., Natick, MA.

MATLAB (manual), The MathWorks, Inc., Natick, MA.

K. Ogata, *Solving Control Engineering Problems with MATLAB*. Prentice-Hall, Englewood Cliffs, NJ, 1994.

R. D. Strum and D. E. Kirk, *Contemporary Linear Systems Using MATLAB*. PWS Publishing, Boston, MA, 1994.

Sun Teq Engineering, Inc., Albuquerque, NM.

FOR FURTHER INFORMATION

By far the most informative single source on MATLAB's capabilities are the MATLAB manuals. As we have mentioned, they contain concise tutorials and references on the theory behind each command and references to the source of each command's numerical algorithm (Little, 1986; *MATLAB Manual*, 1990; Grace et al., 1990).

In addition, several books are now available that directly target the systems/controls student and educator: texts or companion texts that incorporate MATLAB command and .M-file examples into the lessons. Among these are Kuo and Hanselman (1994), Strum and Kirk (1994), Ogata (1994), and Leonard and Levine (1992).

56

Neural Networks with MATLAB

Salahalddin Abusalah
Florida Atlantic University

Ali Zilouchian
Florida Atlantic University

56.1 INTRODUCTION

For many decades it has been a goal of science and engineering to develop intelligent machines with a large number of simple elements. References to this subject can be found even in the scientific literature of the nineteenth century. During the 1940s, researchers desiring to duplicate the function of the human brain developed simple hardware (and later software) models of biological neurons and their interconnection system. McCulloch and Pitts (1943) published the first systematic study of the artificial neural network (NN). Four years latter McCulloch and Pitts (1947) explored network paradigms for pattern recognition using a single layer perceptron. However, from 1960 to 1980, due to severe restrictions on what a perceptron can represent, neural network research went into near eclipse. The recent discovery of training methods for multilayer methods has, more than any other factor, been responsible for the recent resurgence of interest in the neural network.

In general, neural networks are mathematical models of various brain activities. They exploit the massively parallel local processing, and distributed representation properties that are believed to exist in the brain. The primary intent of neural networks is to explore and produce human information processing tasks such as speech, vision, knowledge processing, and motor-control. In addition, neural networks can be used for data compression, near-optimal solutions, pattern matching, system identification, and function approximation.

The attempt to organize human information processing tasks highlights the classical comparison between information processing capabilities of the human and computer. The computer can multiply large numbers at a fast speed, yet it is not capable of understanding unconstrained speech. On the other hand, humans understand speech, yet lack the ability to compute the square root of a prime number without the aid of a pencil and paper or a calculator. The differences between these two opposite capabilities can be traced to the processing methods each employs. Digital computers rely

0-8493-2530-7/96/$0.00+$.50

bilities can be traced to the processing methods each employs. Digital computers rely upon algorithm-based programs that operate serially, are controlled by a complex central processing unit (CPU), and store information at the particular locations in memory. In the other hand, the brain relies on highly distributed representations and transformations that operate in parallel, have distributed control through billions of highly interconnected neurons or processing elements (PE), and appear to store information in variable straight connections called synapses.

Neural network theory evolved from many disciplines including psychology, mathematics, physics, engineering, computer science, philosophy, biology, and linguistics. It is evident from this diverse listing that neural network technology represents a unique topic among the sciences working towards a common goal: Building intelligent systems. It is equally evident from the listing that an accurate and complete descriptions of neural networks are impossible.

The neural network toolbox using MATLAB is one of the first and most popular commercial programs available for neural networks in the market. It is a useful tool for the simulations of various neural network algorithms. It can assist the user regarding various theoretical as well as implementation issues related to neural network architectures.

This chapter is organized as follows. In Section 56.2 the fundamentals of neural networks are presented. In Section 56.3 perceptron architecture using the neural network toolbox as well as a number of examples are provided. In Section 56.4 the well-known backpropagation algorithm is discussed. Section 56.5 is devoted to the Hopfield network. Finally, in Section 56.6, various applications of neural networks are discussed.

56.2 FUNDAMENTAL OF NEURAL NETWORKS

Models of Neurons

It is convenient to visualize neurons as arranged in a number of layers. Neurons in the same layer usually behave the same. The arrangement of neurons into layers and connection patterns within and between layers is called the network architecture. The key factors in determining the behavior of a neuron are its activation function and the pattern of weighted connection over which it sends and receives signals. The basic operation of an artificial neuron involves summing its weighted input signals and applying an activation function to the single output of the summing point. A neuron can be classified based upon the number of inputs and outputs as follows.

Single Input Neuron

The basic architecture of the neural networks consists of a single input unit and a single output unit as shown in Figure 56.1. The input X is transmitted through a connection that multiplies its strength by weight W, to form the product $X * W$. The product $X * W$ is the only argument of the transfer function F, which produces the output y.

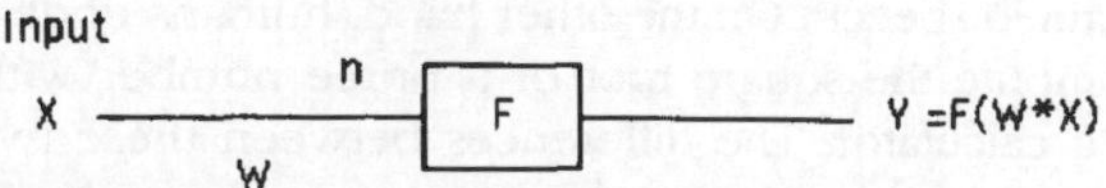

FIGURE 56.1 Neural network architecture.

Multiple Inputs Neuron

A neuron with multiple inputs is shown in Figure 56.2. An individual input $X(i)$, weighted by elements $W(1,j)$ of a matrix W, is summed to form the weighted inputs. R is the number of inputs.

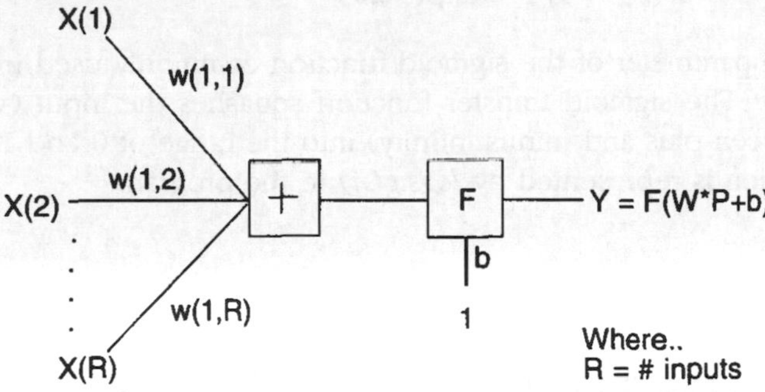

FIGURE 56.2 Neuron with multiple inputs.

In neuron network toolbox inputs as well as outputs values are represented as arrays. The array of the inputs $X(j)$, and the array weights $W(1,j)$, can be represented by the following vectors:

$$W(j) = [w(1,1)\ w(1,2) \ldots\ w(1,j)]$$

$$X(j) = \begin{bmatrix} x(1) \\ x(2) \\ \vdots \\ x(j) \end{bmatrix}$$

The output of the neuron is the sum of weighted inputs and the bias b.

Activation Functions

An activation function defines the output of a neuron in terms of the activity levels at its inputs. Three types of activation functions are available on MATLAB toolbox.

Threshold Function

$$F(v) = \begin{cases} 1 & \text{if } v \geq 0 \\ 0 & \text{else} \end{cases}$$

This activation function is represented in the neural network toolbox by ***hardlim(n)***. The hard limit transfer function limits the output of the neuron to either 0 or 1.

Piecewise Linear Function

$$F(v) = \begin{cases} 1 & \text{if} \quad v \geq 0 \\ v & \text{if} \quad -1/2 > v \geq 1/2 \\ 0 & \text{if} \quad v \leq -1/2 \end{cases}$$

This transfer function has an output equal to its input plus the bias. It is represented by *purelin(n)*.

Sigmoid Functions (S-Shaped Curve)

$$F(v) = 1/1 + \exp(-av)$$

where a is the slope parameter of the sigmoid function commonly used in the back-propagation network. The sigmoid transfer function squashes the input (which may have any value between plus and minus infinity) into the range of 0 and 1.

The sigmoid function is represented by *logsig(n)* in the program.

Examples

Example 1

Consider the neuron with a single input $X = 1$, weight $W = -1$, and a sigmoid function. The output of this network can be computed as follows:

$$Y = \text{logsig}(X \cdot W) = \text{logsig}[1 \cdot (-1)] = 0.2689$$

If a single neuron has a bias value, it can be considered as a weight except with a constant value of 1. In general, the output of a given neuron is the sum of a weighted input $X * W$ and bias b.

Example 2

Consider the single input neuron with an input $X = 1$, weight $W = -1$, bias $b = 1$, and a linear transfer function. The output Y can be obtained as follows:

$$Y = \text{purelin}(w \cdot x + b) = \text{purelin}(-1 \cdot 1 + 1) = -1$$

Example 3

Consider the network that consists of four inputs where

$$X = \begin{bmatrix} 1 \\ 2 \\ 5 \\ 8 \end{bmatrix}, \qquad W = \begin{bmatrix} 1 & 1 & -1 & -2 \end{bmatrix}$$

The output of the network is given as

$$R = w \cdot x = 14$$

$$y = \text{logsig}(R) = 1.52^{-8}$$

Batching

In general, a neural network can be represented by a number of inputs, hidden layers, and outputs. The outputs are calculated and, subsequently, an algorithm is applied to determine the weight element changes. However, you may want to apply Q input vector simultaneously and get the network response to each of them. Such a batching operation often is more efficient. The inputs and outputs can be represented by matrices called P and Y, which can be written in the following forms:

$$P = \begin{bmatrix} p(1,1) & p(1,2) & \cdots & P(1,Q) \\ p(2,1) & p(2,2) & \cdots & P(2,Q) \\ \vdots & \vdots & & \vdots \\ p(X,1) & p(X,2) & \cdots & P(X,Q) \end{bmatrix}$$

$$Y = \begin{bmatrix} y(1,1) & y(1,2) & \cdots & Y(1,Q) \\ y(2,1) & y(2,2) & \cdots & Y(2,Q) \\ \vdots & \vdots & & \vdots \\ y(S,1) & y(S,2) & \cdots & Y(S,Q) \end{bmatrix}$$

The network will produce an output vector $y(:,j)$ for each input vector $p(:,j)$. Figure 56.3 represents a batch Q inputs vectors and R elements applied to the network of S neurons.

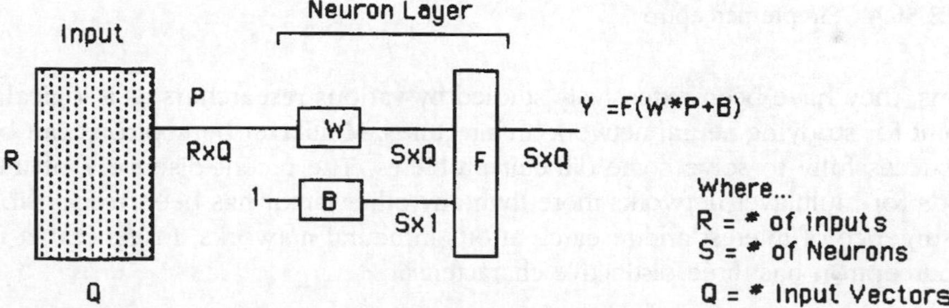

FIGURE 56.3 Network of S neurons.

The following instruction represents the calculation of log-sigmoid transfer function for Q dimension Y and P vectors with weight matrix W, and bias vector B:

$$Y = \text{logsig}(W \cdot P, B)$$

The above notation allows an easy and natural extension from one dimension to multidimensional input and output vectors.

56.3 PERCEPTRON

The perceptron is the simplest form of a neural network used for classification. It consists of a single neuron with adjustable synaptic weights and a threshold. A single-layer perceptron is limited to performing pattern classification with only two separate classes. By expanding the output layer of a single perceptron to include more than one neuron, we may obtain the classification of the data for more than two set of regions classes. A simple perceptron is shown in Figure 56.4.

Each external input is weighted with appropriate W, and the sum of the weighted inputs is sent through a hard limit transfer function. The hard limit transfer function provides a perceptron with the ability to classify the input vector into two regions. The hard limit transfer function return either –1 or 1.

However, it can be shown that there are severe restrictions on what a single-layer perceptron can represent, and hence what it can learn. Despite the limitations of per-

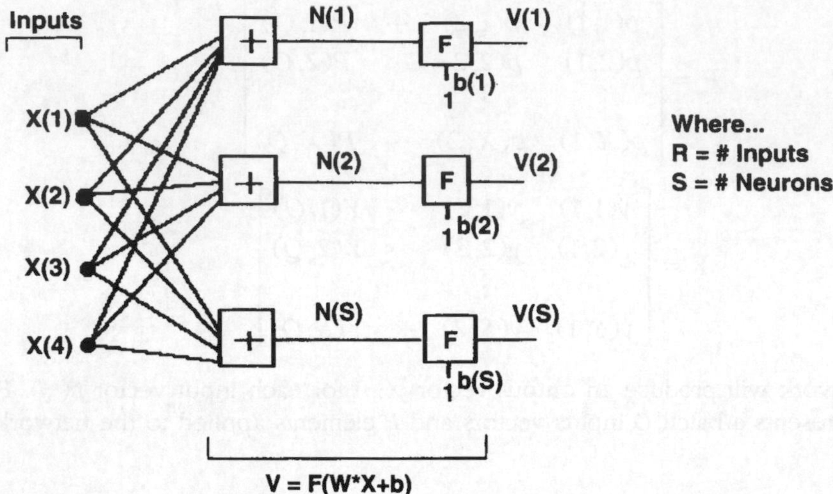

FIGURE 56.4 Simple perceptron.

ceptrons, they have been extensively studied by various researchers as a logical starting point for studying neural network architectures. Multilayer perceptrons can be applied successfully to solve some difficult problems. The recent discovery of training methods for multilayer networks more than any other factor has been responsible for the resurgence of interest and research efforts in neural networks. In general, a multilayer perceptron has three distinctive characteristics:

- The model of each neuron in the network includes a nonlinearity element at the output end.
- The network contains one or more layers of hidden neurons that are not part of the input or output of the network. The hidden neurons enable the network to learn complex tasks by extracting progressively more meaningful features from the input patterns.
- The network exhibits a high degree of connectivity, determined by the synapses of the network.

Solving the perceptron problem using the MATLAB toolbox is useful due to its simplicity for regional data classification. We can clearly observe how different regions can be separated, and how the training and the learning algorithms should be implemented. The simulation of perceptron consists of three phases: initialization, learning, and training.

Initialization

The MATLAB function for the initialization is *rad.* This function used to initialize the weights and bias elements to small positive and negative values. For simplicity we set the weights and bias to zero. For example, a perceptron with 4 input and 8 neurons can be initialized as follows:

$$R = 4;\ S = 8;$$
$$[W, B] = \text{rand}(S, R)$$

Learning Rules

The learning rules can be utilized to converge on finite time if a solution exists. The perceptron network is trained to respond to each input vector with a corresponding target output. The function *learnp* is used to calculate the changes in the weight matrix elements. Bias weights are updated only for those patterns that do not produce the correct value of the output. For example, the following function can be used to calculate the changes in the weights and bias:

$$[dW, dB] = \text{learnp}(P, Y, T)$$

In addition, the function *learnp* can be used to batch and train the network simultaneously by representing different values of the inputs vectors.

In neural network toolbox, one method to represent a batch of inputs vector P for perceptron and updates its weights W and biases B is as follows

$$Y = \text{hardlim}(WP, B)$$
$$[dW, dB] = \text{learnp}(P, A, T)$$
$$W = W + dW$$
$$B = B + dB$$

Training

The major steps in the training phase can be summarized as follow:

1. Presentation step: present the inputs and calculate the network outputs.
2. Checking step: check to see if each output vector is equal to the target vector associated with the given input vector.
3. Learning step: Adjust weights and biases accordingly using perceptron learning rule.

The following codes summarize the above steps:

```
% PRESENTATION PHASE:
A = hardlim(W*P,B);
for epoch = 1 : max_epoch
% CHECKING PHASE:
    if all (A = = 1)
            epoch = epoch-1;
            break
    end
% LEARNING PHASE:
    [dW,dB] = learnp(P,A,T);
    W = W + dW;
    B = B + dB;
% PRESENTATION PHASE:
    A = hardlim(W*P,B);
end;
```

If a network does not perform successfully, it can be trained further by calling *trainp* function. The code to batch train perceptrons is

```
Tp = [ disp-freq. max.-epoch]
[ W,B,epoch] = trainp(W,B,P,T,TP)
```

See *User's Guide* (1994) for details.

56.4 BACKPROPAGATION ALGORITHM

The backpropagation network is an extension of the perceptron structure involving addition of a nonlinear function on the output of each neuron, and the use of multiple adaptable layers. The term is often used to refer to a hierarchical network architecture that uses the backpropagation update algorithm for updating the weights of each layer based on the error presented at the network output. A backpropagation network always consists of at least three layers; an input layer, a hidden layer, and an output layer. For some applications more than one hidden layer is used. In a backpropagation network there are no interconnection between neurons in the same layer. However, each neuron on a layer provides an input to each neuron on the next layer, e.g., each input neuron will send its output to every neuron of the hidden layer. The neuron used in the backpropagation is a sigmoidal function, i.e., a function that is continuous, monotonically increasing, continuously differentiable, and asymptotically approaches fixed finite values as the input approaches plus or minus infinity.

The algorithm consists of two phases: a training phase and recall phase. In the training phase the two distinct operations include the feedforward computation and the update of the weights based upon the error of the network output. Generally, the weights between the layers are initially selected as small random values. During the recall phase only the feedforward computations consisting of a single calculation take place. The feedforward process involves presenting an input pattern to the input neurons that pass the values onto the hidden layer. Typically each neuron in the network will use the same activation function and threshold value. The various stages for backpropagation algorithm are as follows.

Initialization

The choice of initial weights influences whether the net reaches a global minimum of error. The initial weight matrix for the backpropagation network is created with random elements between –1 and 1. The update of the weight between two units depends on both the derivative of the upper unit's activation function and the activation of the lower unit. The values for the initial weights must not be too large, otherwise the initial input signals to each hidden or output layers will fall into the regions where the derivative of the sigmoid function has a very small value. However, if the initial weights are too small, the net input to a hidden or output unit will be close to zero. In MATLAB toolbox the initialization code is presented as follows:

```
[R, Q] = size(p);
S1 = m;
[s2,Q] = size(T);
[W1,B1] = rads(S1,R);
[W2,B2] = rands(S2,S1);
```

Learning Rules

The backpropagation learning rule is used to train nonlinear, multilayer networks to perform function approximation, pattern association, and pattern classification. The backpropgation learning rule can be used to adjust the weights and biases of network to minimize the sum square error of the network.

The following rules represent the learning procedures for the backpropagation algorithm:

deltalin(A,E) returns the delta vector for the an output layer of a linear neuron with an output vector *A* and error vector *E*.
deltalin(A,D,W) returns the delta vector for a hidden layer of linear neurons with an output vector *A*, preceding a layer with a delta vector *D* and weight matrix *W*.
deltalog (A,E) returns the deltas for an output layer of log-sigmoid neurons.
deltatan (A,D,W) returns the deltas for a hidden layer of tan-sigmoid neurons.

The changes to be made in a layer's weights *W* and a bias vector *B* are calculated using that layer's delta vector *D*, and the layer's input vector *P*. *[dw,dB]* = learnbp(*p,d*, 1*r*) function returns both a weight change matrix and a bias change vector for a layer.

Training

The training process consists of the following steps:

1. Presentation phase
2. Checking phase
3. Backpropagation phase
4. Learning phase.

The deltails of the above steps are wrapped up in the training function *trainbp*. For two layer networks *trainbp* can be written as allows:

```
[w1,B1,w2,B2,epochs,TR] = trainbp(W1,B1,'F1',W2,B2,'F2'P,TP)
```

Example 5

Consider the following network consists of three inputs, two hidden layers, and two outputs (Fig. 56.5). Obtain the final weights by using the backpropagation algorithm.

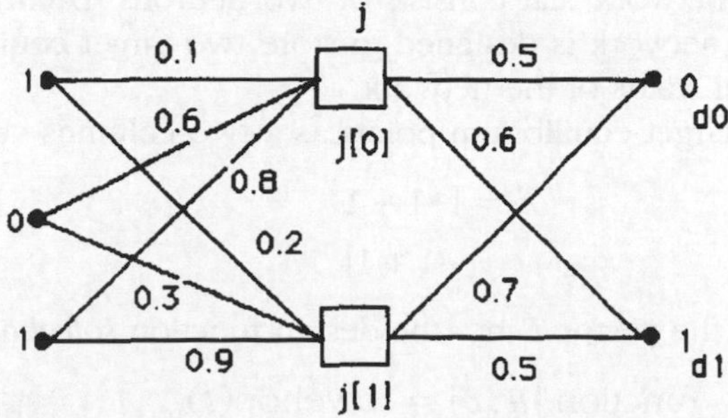

FIGURE 56.5 Neural network.

```
% Define one 3-element input vectors
p = [1 0 1]';
% Define the associated one 2-element targets.
T = [0 1]';
% Initialize weight and bias
b1 = [0;0];
b2 = [0;0];
w1 = [0.1 0.6 0.8;0.2 0.3 0.9];
w2 = [0.5 0.6;0.7 0.5];
% Training parameters
Tp = [1 100 0.001 1];
% Training network
[w1,b1,w2,b2,te,tr]=trainbp(w1,b1,'logsig',w2,b2,'logsig',p,t,tp)
```

The final weights are:

$$w1 = \begin{bmatrix} 0.246 & 0.600 & 0.946 \\ 0.291 & 0.300 & 0.991 \end{bmatrix}$$

$$w2 = \begin{bmatrix} -0.634 & -0.563 \\ 1.266 & 1.083 \end{bmatrix}$$

56.5 THE HOPFIELD NETWORK

The analog Hopfield network is a famous neural network. The main objective of this analog circuit is to design a network that stores a specific set of equilibrium points such that when initial conditions are provided the circuit will eventually come to rest. The architecture of the circuit can be described by the network shown in Figure 56.6.

Testing the network can be performed with a number of inputs that are presented as initial conditions to the network. After all initial conditions are given, the network produces an output. The output will feed back to the input of the network. The following example demonstrates the construction of this network using the MATLAB toolbox.

Example 5

Consider a Hopfield network that consists of two neurons. Each neuron has two inputs and a bias. The network is designed to store two target equilibrium points (Fig. 56.7). Obtain the final status of the network.

Step #1: Store the target equilibrium points as a two columns vector T.

$$T = [+1 - 1$$
$$-1 + 1]$$

Step #2: Substitute the vector T into the design function *solvehop*

$$\text{function } [w, b] = \text{solvehop}(t)$$

Solvehop returns sets of weights and bias for each neuron. The results are

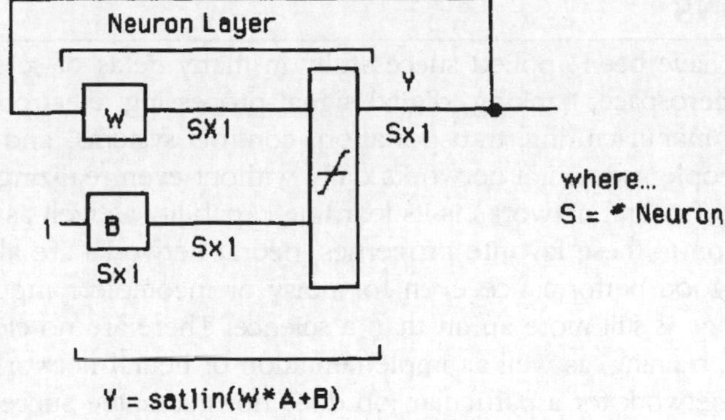

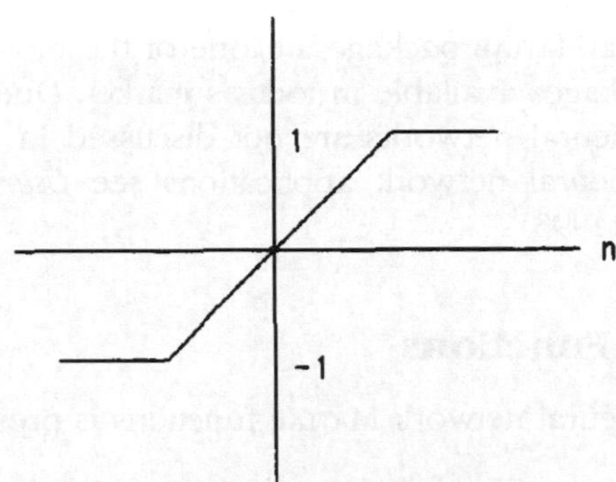

FIGURE 56.6 Neural network showing circuit architecture.

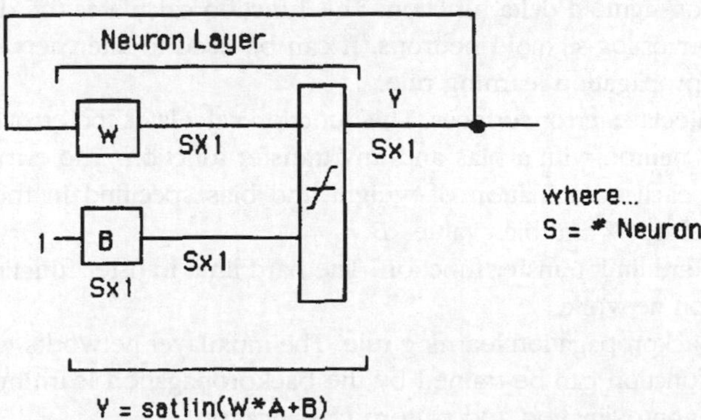

FIGURE 56.7 Hopfield network.

$$W = [0.6925 \ -0.4694; \ -0.4694 \ 0.6925]$$

$$B = 1.0^{-16} * [0.6900; \ 0.6900]$$

Step #3 Confirm the desired and actual output vectors in the network. For this particular example the solution will converge after three cycles. The network has stored the equilibrium points.

56.6 APPLICATIONS

Neural networks have been applied successfully in many fields of science and engineering such as aerospace, banking, digital signal processing, electronics, insurance, robotics, speech, manufacturing, transportation, controls systems, and medical engineering. Many people use neural networks daily without even realizing it. One of the main advantages of neural networks is its learning capability as well as adaptation capability. In addition to these favorite properties, neural networks are also robust, that is, they provide good performance even for noisy or incomplete input data. Neural network technology is still more an art than a science. There are no clearcut rules regarding selection, training, as well as implementation of neural networks. Even selection of the right network for a particular job does not guarantee success. Neural networks conform to the data on which they train. If the data are not accurate or sufficient, the result will not necessarily be acceptable. Despite these drawbacks, neural networks are being used from Wall Street to the Silicon Valley. In general, many designers concur that a commercially developed package is the way to get a project started. Neural networks with a MATLAB package are one of the most comprehensive and advanced commercial packages available in today's market. Due to space limitations various applications of neural networks are not discussed in this chapter. For more information regarding neural network applications see *User's Guide* (1994), Miller (1990), and Wasserman (1993).

MATLAB Neural Network Functions

In this section a summary of neural network MATLAB functions is presented:

deltalin: Linear delta function. This function calculates the derivatives of error (or deltas) for a layer of linear neurons. It is used to alter a network weights with the backpropagation network learning rule.

deltalog: Log-sigmoid delta function. This function calculates the derivative of error for a layer of log-sigmoid neurons. It can be used to alter network weights with the backpropagation learning rule.

errsurf: Calculates error surfaces. This function calculates the error surface for a single-input neuron with a bias and any transfer function. The error surface is evaluated at each combination of weight and bias specified in the row vectors of weight values W and bias values B.

hardlim: Hard limit transfer function. The hard limit transfer function is used by the perceptron network.

learnbp: Backpropagation learning rule. The multilayer networks with differentiable transfer function can be trained by the backpropagation learning rule to perform function approximation and pattern classification.

learnk: Kohonen learning rule. The Kohonen rule allows a neuron to store the input vector by adjusting the weights of a neuron to equal its current input vector.

learnp: Perceptron learning rule. The preceptron learning rule adjusts the weights and biases of a layer to produce correct classification of the network's input vectors.

learnwh: Widrow–Hoff (delta) learning rule. This function is used in the Widrow–Hoff learning rule. The Widrow–Hoff learning rule can be used to train a layer of linear neurons to linearly approximate a function or perform pattern association.

logsig: Log-sigmoid transfer function. This an activation transfer function. This activation function is always used in the backpropagation networks. The log-sigmoid function is a differentiable function that makes it suitable for neurons being trained with backproagation.

maxlinlr: Maximum linear learning rate. This function calculates the maximum stable learning rate for training linear networks with the Widrow–Hoff rule. This function is used to generate weight matrices and bias vectors with positive and negative value.

satlin: Saturated linear transfer function. This is an activation transfer function. The saturated linear transfer function returns its input if the input is greater than –1 and less than 1. This function is used by the Hopfield network.

simhop: Simulates a Hopfield network. This a simulation function for a Hopfield network.

solvehop: Designs Hopfield networks. This function returns a weight matrix and bias vector that result in a Hopfield network with stable output vectors given by the matrix of target vector.

sumsqr: Sum of squared matrix elements. This function is used to calculate the sums squared elements of a matrix.

tansig: Tangent-sigmoid transfer function. This is an activation function. The tan-sigmoid is a fully differentiable function that makes it suitable for neurons being trained with backpropagation.

trainbp: Trains network with backpropagation. This function is used in the backpropagation training for feedforward networks.

trainp: trains hard limit layer with perceptron rule. Trains a layer of hard limit neurons to classify input vectors as desired using the perceptron learning rule.

DEFINING TERMS

Architecture: A description of the number of layers in a neural network, each layer's transfer function, the number of neurons per layer, and the connections between layers.

Backpropagation Learning Rule: A learning rule in which weights and biases are adjusted by error derivative (delta) vectors backpropagated through the network.

Batch: A matrix of inputs (or target) vectors applied to the network simultaneously.

Bias: A neuron parameter that is summed with the neuron's weighted inputs and passed through the neuron's transfer function to generate the neuron's output.

Connection: A oneway link between neurons in a network.

Delta rule: The Widrow–Hoff rule.

Epoch: The presentation of the set of training vectors to a network and the calculation of new weights and biases.

Feedback Network: A network with connections from a layers output to the layer's input. The feedback connection may direct or pass through several layers.

Feedforward network: A layered network in which each layer receives inputs only from previous layers.

Function Approximation: The task performed by a network trained to respond to inputs with an approximation of a desired function.

Hard Limit Transfer Function: A transfer that maps inputs greater than or equal to 0 to 1, and all the values to 0.

Hidden Layer: A layer of a network that neither receives input from outside the network nor sends outputs outside the network.

Input layer: A layer of neurons receiving input directly from outside the network.

Input space: The range of all possible input vectors.

Input vector: A vector presented to the network.

Input weight vector: The row vector of weights going to a neuron.

Star Learning Rule: A learning rule that trains a neuron's weight vector to take on the values of the current input vector. Changes in the weights are proppational to the neuron's output.

Learning: The process by which the weights and biases are adjusted to achieve somedesired network behavior.

Learning Rate: A training parameter that controls the size of weight and bias change during learning.

Log-Sigmoid Transfer Function: This is an activation function. The log-sigmoid is a differentiable function, which makes it suitable for neurons being trained with backpropagation.

Output Layer: A layer whose output is passed to the world outside the network.

Output Weight Vector: The column vector of weights coming from a neuron or input.

Pattern: A vector.

Pattern Recognition: The task performed by a network trained to respond to an input vector.

Perceptron: A single layer network with a hard limit transfer function. The network is trained with the perceptron learning rule.

Perceptron Learning Rule: A learning rule for training single layer hard limit networks.

Saturated Linear Transfer Function: A function that is linear in the interval (1,+1) and saturates outside this interval to –1 to +1

Sum Squared Error: The sum of squared differences between the network targets and actual outputs for a given input vector or set of vectors.

Supervised Learning: A learning process in which changes in a network's weights and biases are due to the intervention of any external teacher.

Target Vector: The desired output vector for a given input vector.

Training Vector: An input and/or target vector used to train a network.

Transfer Function: The function that maps a neuron's (or layer's) input to its output.

Weight Matrix: A matrix containing connection strengths from a layer's inputs to its neurons. The element $w(i,j)$ of a weight matrix w refers to the connection strength from input j to neuron i.

REFERENCES

L. Fausett. *Fundamental of Neural Networks*. Prentice-Hall, Englewood Cliffs, NJ, 1994.

S. Haykin. *Neural Networks*. Macmillan, New York, 1994.

W.W. McCulloch, and W. Pitts. A logical calculus of the idea imminent in nervous activity. *Bull. Math. Biophys*. 115–133, 1943.

W. Pitts, and W.W. McCulloch. How we know universals. *Bull. Math. Biophys*.. 127–147, 1947.

P. Simpson, R.S. Sutton, and P.J. Werbos. *Neural Networks for Control*. MIT Press, Cambridge, MA, 1990.

User's Guide: Neural Network Toolbox. The Math Works Inc., 1994.

P. Wasserman. *Neural Computing, Theory and Practice*. VNR, New York, 1989.

57

Algor: Finite Element Modeling and Engineering Analysis Software

W. Charles Paulsen
Algor, Inc.

57.1 DESIGN DRIVES MANUFACTURING

One can, if pressed, divide the business world into two sectors, the service industry or the manufacturing industry. The role of the service industry needs little explanation. In the world of manufacturing the efficient production of goods is the primary objective. The goods or products always start from an idea, as shown in Figure 57.1. From the time the idea is conceived to the time the product is shipped many engineering disciplines and methodologies are brought to play. The process cycles (iterates) between design intent, design capture, and analysis. In the background are considerations in regards to manufacturability, cost, esthetics, environment, and safety. Finally, with respect to the global economy all companies are in, design optimization is now a factor.

The psychology of how the human mind initiates a design, even though it is the first step in the process, is probably the least understood, at least quantitatively. The rest of the process is fairly well defined, and can be taught, and computer aids to the process abound. Capturing the design intent was, at one time, done with pencil and paper (affectionately called drafting), but the universal trend is to use a computer to aid the process, thus, Computer-Aided Design or CAD for short. The methodologies, or more accurately the mathematical underpinnings, of CAD fall in three categories. The first uses wire frame entities to describe the design.

Wire Frame CAD

Wire frame entities are just what the name implies, a collection of entities that "frames" the design, as if they were made from wires. The wires are, of course, more accurately described as members of the conics, that is, lines, arcs, circles, and ellipses, all, of

0-8493-2530-7/96/$0.00+$.50

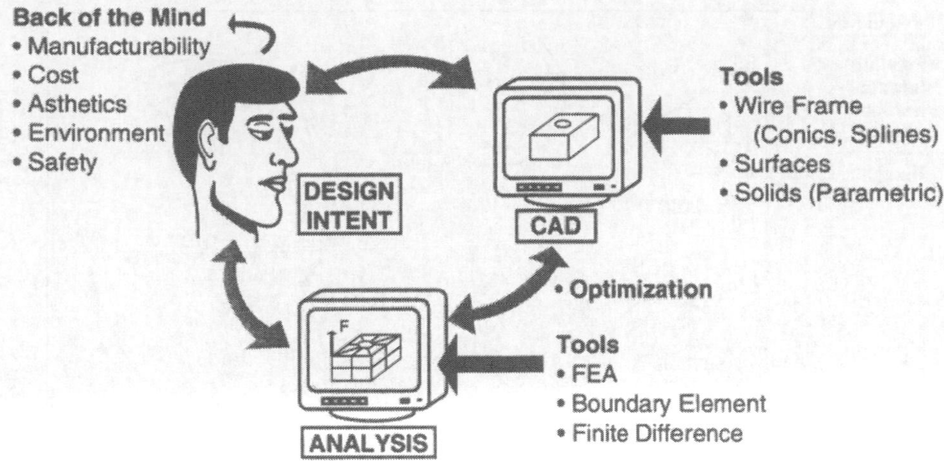

FIGURE 57.1 The Big 3 + 1 in design.

course, capable of being implemented in 3-D space. Wire frame CAD systems now must include a set of spline entities of be complete and/or competitive. Splines are like stiff rope that can be curved in space. In the early days designers used long wooden pieces of wood (called lathes) and bent the strips of wood to describe the curve. Ship designers often performed this task on the top floor, or loft, of the factory and came to be known as loftsmen. Lofting, or the laying down of splines, is now done on a computer screen. The first spline invented was the patched cubic spline. The first "free form" spline was the Bezier spline and is still popular. The principal deficiency of the Bezier spline was its inability to accept local changes. Any changes (or edits) performed anywhere on the spline changed the whole spline. This is analogous to having a clay sculptor change the nose of a bust only to have the ear change! A more robust spline, the B-Spline (B stand for basis or, more accurately, the blending function-basis), allows for local editing. The newest member of the spline family, a superset of the B-Spline called the nonuniform rational B-Spline, usually referred to as the NURB-Spline (or NURBS for short), introduced two more features. First the NURB introduced a second parameter, the weight parameter, that allows control points to alter their "attraction" to the curve, this ability is essential if one wants to describe a conic (e.g., ellipse) curve exactly. The second feature allows for nonuniform knot spacing and is used by only a handful of experts. Algor's modeling system incorporates all the latest CAD spline entities and allows for complete control of NURB entities as shown in Figure 57.2.

The exchange of wire frame entities between different CAD systems is often done by two popular specifications, the DXF format (invented by the creators of AutoCAD) and the IGES format, created by the then U. S. National Bureau of Standards. Compliance to IGES is the format by which CAD companies are judged.

Surfaces

A surface CAD entity can be thought of as a flag allowing for virtually unlimited surfaces shapes as the wind blows. Or else it can be thought of as Saran wrap used for covering the tops of refrigerator containers. It is the last paradigm that mathematicians

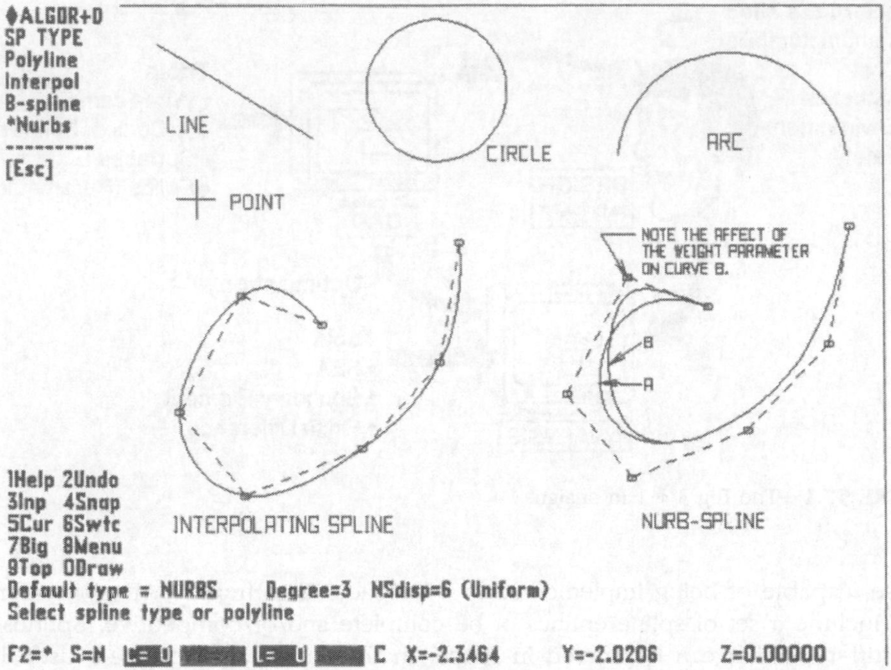

FIGURE 57.2 Algor provides for all CAD wire frame entities including NURB-Splines with full control.

utilize in constructing so called B-rep solids, mainly to start with a wire frame model and stretch "Saran Wrap™" over the wires to create a patch of surfaces that totally encloses the volume. This hollow but totally surface-bounded watertight (mathematicians call this a manifold) entity is a valid description of a solid. There are basically two kinds of surfaces, analytical and freeform. Analytical surfaces are formed by applying three straightforward operations: translation sweeping (extruding), rotational sweeping, and helical sweeping. Freeform surfaces are nothing more than two parameter extensions of splines. They are created by patching or applying operations similar to those described above. Surfaces created by the operation method include ruled (or lofted), tabulated (translational sweeping along a curve), sweeping (translating and rotating along a curve), and bi-rail sweeping. Patching is completely analogous to patching a hole in a pair of blue jeans. Patches, for the most part, are bounded by basic wire frame entities (lines, arcs, or splines in any combination). These four entities are "covered" (or patched) mathematically by one of several formulations. The first historically was the bilinear surface, followed by the cubic, Bezier, B-spline surface, and NURB surface. The NURB surface is by far the most popular. Algor has a wide assortment of traditional tools for making these patches: the 2-entity, 3-entity, and 4-entity (the classic patcher). Algor has also invented some unique tools such as the Multi and Global patchers. The Global patch is particularly powerful in that it can take any wire frame model and automatically patch the outer surface only; it has the intelligence not to make internal patches or cover up any holes. See Figure 57.3 for some typical surfaces as created in Supersurf Algor's surface modeler. The real benefit of Supersurf is that once a surface is created it can automatically be finite element analysis (FEA) meshed with any mesh density (uniform, nonuniform, and gradient) desired.

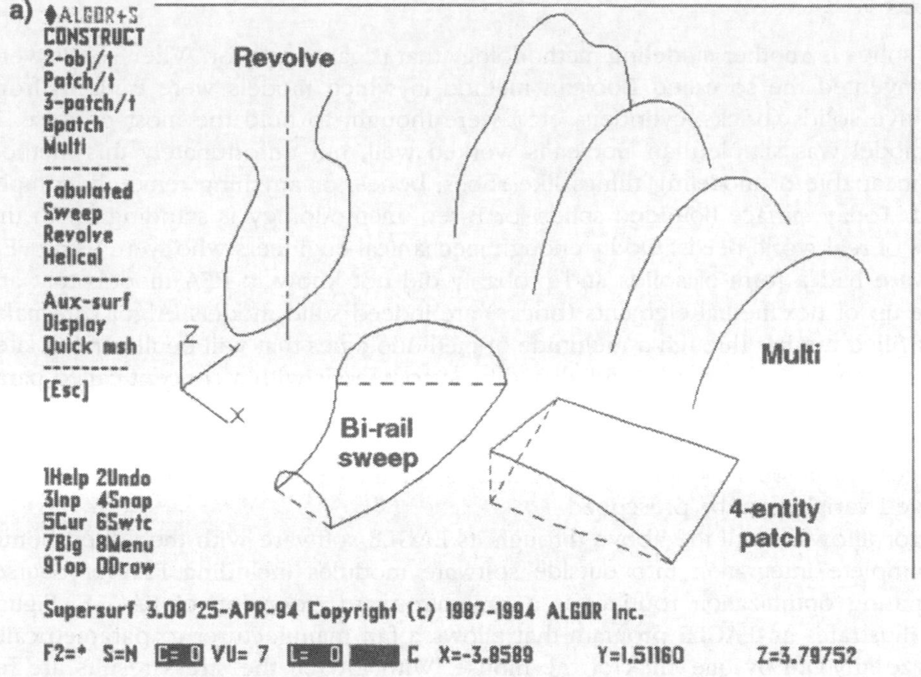

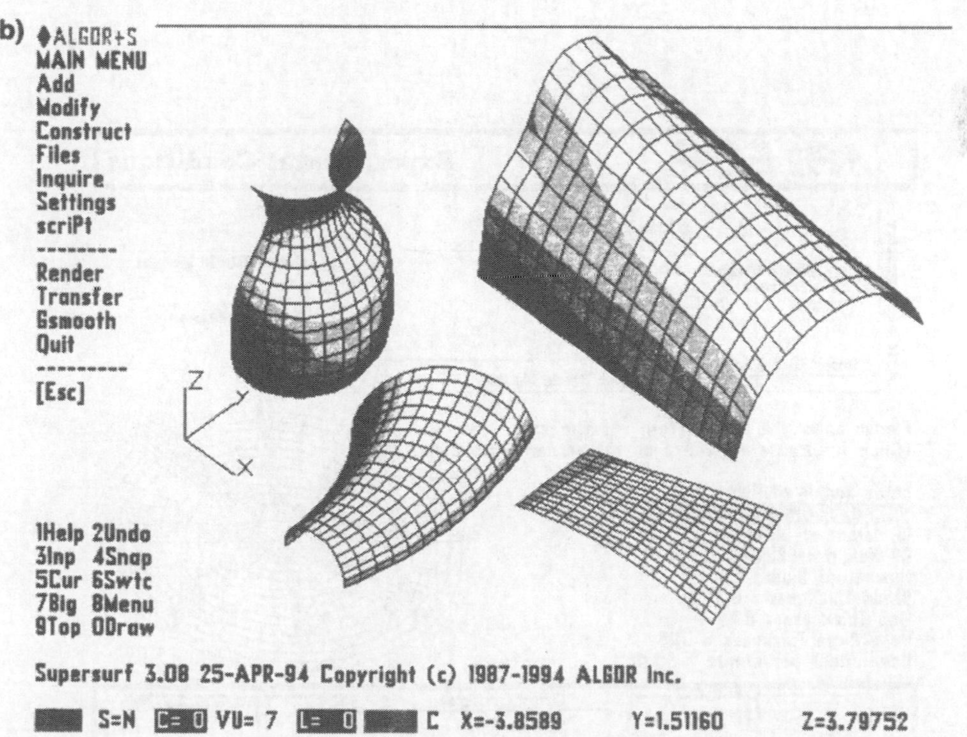

FIGURE 57.3 (a) Algor provides for a large assortment of surface building tools. Revolve, bi-rail sweep, multi, and the classical 4-entity patch are shown. (b) The resulting NURB-Spline surfaces with a FEA mesh automatically applied.

Solids

CAD solids is another modeling methodology that is gaining favor. When solids were first invented the so called Boolean method in which models were built-up from primitive solids (bricks, cylinders, etc.) were thought to hold the most promise. If the model was simple than Booleans worked well, but unfortunately this method was incapable of modeling things like shoes, bones, or anything remotely complicated. Today surface bounded solids, or B-rep, methodology is standing up to the rigors of real world needs. Oddly enough mechanical engineers who were using FEA software had a form of solids and probably did not know it! FEA models that are made up of hexahedral elements (bricks) are indeed solid models. Algor can make brick-filled models through a multitude of methodologies that will be illustrated later in this chapter. Today solids tend also to be synonymous with a concept called parametric or variational mathematics, a relatively recent concept. Parametrics attaches variables to dimensions in the model. If one changes the dimension the model changes. Variational modeling simply allows for mathematical dependence among selected variables to be prescribed.

Algor allows for all the above through its EAGLE software with the added bonus of complete integration into outside software modules including FEA of course. Integrating optimization routines is a straightforward extension of EAGLE. Figure 57.4 illustrates an EAGLE program that allows a fan manufacturer to parametrically analyze any fan by one click of the mouse. With EAGLE the stress results are returned.

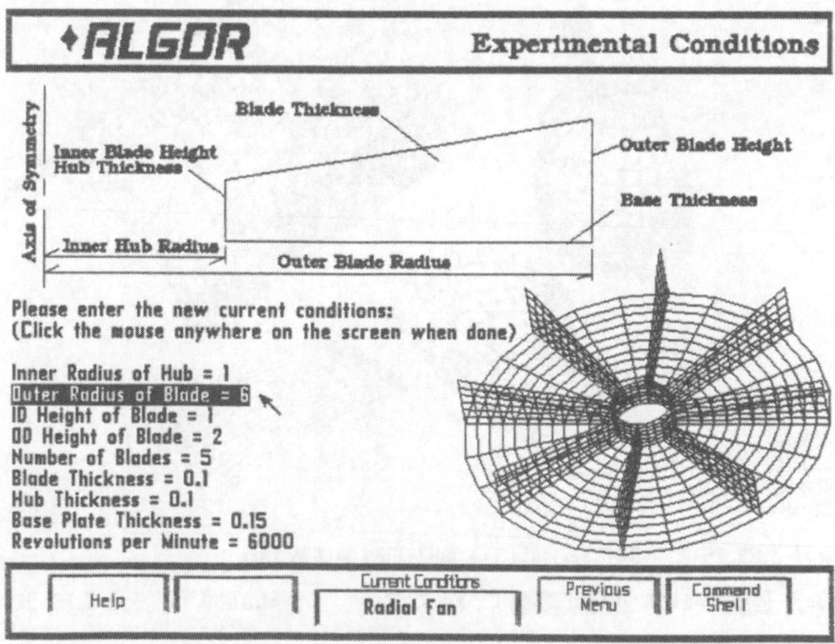

FIGURE 57.4 The EAGLE language allows for parametric and variational control for both design and analysis.

57.2 ANALYSIS IN SUPPORT OF DESIGN

Congratulations, the design has been captured and well documented in your CAD system. It even looks great. The model shop was able to effortlessly machine all the parts simply by "IGESing" over the CAD drawing into the CAM software. One minor detail has been overlooked: will the design work! More specifically, will the design stand-up to the rigors of the environment to which it will be subjected? To be totally candid, a very large number of designs are turned into product and shipped out the door without consideration of the above. It is the role of physical testing and/or design analysis to answer this all important final question.

Analysis Is the Key

Although physical testing may be the final experiment to ensure that the design is acceptable, it is unacceptable during the iterative design phase. It is the role of analysis to predict what physical testing would reveal. Although the word analysis refers to all physical phenomena, fortunately the world of practical design reduces to just a few disciplines. Probably the most important overall discipline is stress analysis, or, more accurately, deflection analysis. Simply stated stress analysis reveals whether a design will or will not support its loads. Other disciplines are heat transfer, where one is concerned about temperature gradients and heat flows. Fluid mechanics is the science of liquid and/or gas flow where one is concerned about velocities, pressures, and turbulence. A superset of stress analysis is vibration analysis, where instead of steady loads vibratory loads such as earthquakes, unbalanced rotating parts, and others give rise to potentially destructive resonance phenomena.

Field Problems

Mathematically the above problems can be characterized as field problems. Field problems mean that the primary unknown, displacement, temperature, velocity, etc., is simply a function of the design's x, y, and z coordinates. If loads, or more precisely the external influences, can be viewed as the input, then the responses (the field variable) are the output. Analysis in this context then is viewed as an input output system as shown in Figure 57.5. Figure 57.5 is unfortunately deceptively simple, for finding the unknowns analytically is nearly impossible. Why is that? The problem lies with our knowledge of solving partial differential equations.

Analysis Methodologies

Let us take a step back to basics. Every physical phenomena that designers encounter can always be described by the phenomena's partial differential equation (PDQ) that governs how the primary unknown (the field variable) behaves. Table 57.1 lists the equations for heat transfer, stress, and fluid flow, three of the key phenomena in many design situations. Solve for the unknown and you are done. How does one solve for the unknown?

- Solve the PDQ directly. Obviously this is the most logical and direct way. The problem is that humans can solve these equations only for very simple situations. Most normal design problems simply cannot be solved. Until mathematicians figure out the "secret" universal methodology one can only stare fondly at the PDQ.

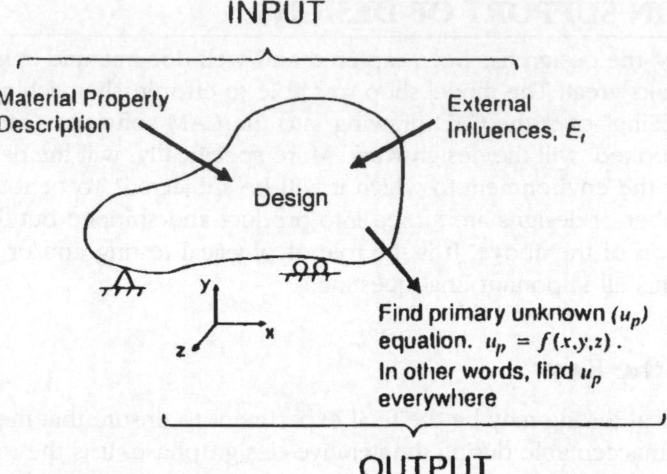

FIGURE 57.5 The ultimate goal of all field problems is to find the unknown field variable as a function of the geometry, the material properties, and the external influences.

- Solve the PDQ numerically. Numerically means to get a table of numbers for the unknown and it almost universally employs the use of a computer. Mathematicians tend to reject this method but engineers gladly embrace it. Unfortunately the methodology, officially called the finite difference method, is plagued by the same deficiency as above, the arguable inability to solve complex problems.
- Solve the PDQ numerically using finite elements. This numerical method is relatively new (1960 time frame) in that instead of breaking up the equation into pieces one divides the design into pieces, thus the name finite elements. The finite element method (FEA) has no limitations.

57.3 THE KING: FINITE ELEMENT ANALYSIS

Civil engineers, in the early 1960s, were the first to use finite element analysis, only then is was called matrix methods of structures. Dividing up a structure into elements was quite natural, for the structures being analyzed were indeed natural elements, mainly beams and trusses bolted and riveted together. The essence of the method is best shown in Figure 57.6. The actual structure is mathematically transformed into a series of "spring-like" elements. The mathematical relation that relates the force to displacement for a single spring is known as Hoo! e's law:

$$f = kd$$

where

$$f = \text{force}$$
$$d = \text{stretch or displacement} \tag{1}$$
$$k = \text{spring stiffness}$$

If you isolate one single spring one can see that the external influences are the forces at the nodes, and response to the force is the defection. Recalling Figure 57.5, this is the input–output for a single element.

TABLE 57.1 The Principal Governing Equations and Unknowns for Stress, Heat, and Fluid Mechanics

Application	Partial Differential Equations (2-D Versions Shown): Conservation or Balance	Partial Differential Equations (2-D Versions Shown): Partial Differential Equations	Unknowns to Be Solved for: Primary	Unknowns to Be Solved for: Secondary
Heat transfer 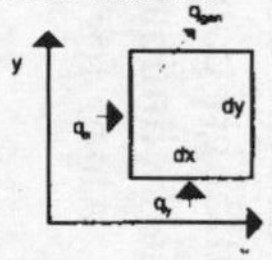 *Properties*: k = conductivity ρ = mass density c = specific heat	Conservation of energy $\text{Heat}_{in} + \text{Heat}_{generated} = \text{Heat}_{out} + \text{Energy}_{stored}$	$\frac{\partial^2 T}{\partial x^2} + \frac{\partial^2 T}{\partial y^2} + \frac{\dot{q}\text{gen}}{k} = \frac{\rho c}{k}\frac{\partial T}{\partial t}$	Temperature T	Heat flux $q'' = -k\frac{dT}{dx}$
Solid mechanics 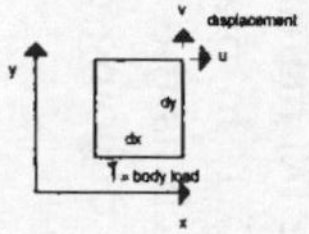 *Properties*: ν = Poisson's ration E = Young's modulus	Equilibrium (balance of forces) Sum of forces in x and y directions = 0	$\frac{E}{1-\nu^2}\frac{\partial^2 u}{\partial x^2} + \frac{E}{2(1-\nu)}\frac{\partial^2 v}{\partial x \partial y} + \frac{E}{2(1+\nu)}\frac{\partial^2 u}{\partial y^2} = -f_x$ $\frac{E}{1-\nu^2}\frac{\partial^2 v}{\partial y^2} + \frac{E}{2(1-\nu)}\frac{\partial^2 u}{\partial x \partial y} + \frac{E}{2(1+\nu)}\frac{\partial^2 v}{\partial x^2} = -f_y$	Displacement u	Stress $\sigma = E\frac{\partial u}{\partial x}$
Fluid Mechanics 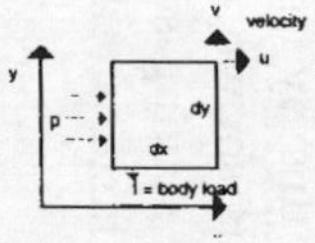 *Properties*: μ = viscosity ρ = mass density p = pressure	Conservation of momentum (Newton's second law) Net force = Rate of change of momentum $\text{Rate of momentum}_{in} + \text{Rate of momentum}_{gen} = \text{Rate of momentum}_{out} + \text{Net forces}$	$\rho\left[\frac{\partial u}{\partial t} + u\frac{\partial u}{\partial x} + v\frac{\partial u}{\partial y}\right] - \mu\left[\frac{\partial^2 u}{\partial x^2} + \frac{\partial^2 u}{\partial y^2}\right] + \frac{\partial v}{\partial x} = f_x$ $\rho\left[\frac{\partial v}{\partial t} + u\frac{\partial v}{\partial x} + v\frac{\partial v}{\partial y}\right] - \mu\left[\frac{\partial^2 v}{\partial x^2} + \frac{\partial^2 v}{\partial y^2}\right] + \frac{\partial v}{\partial y} = f_y$	Velocity u	Shear stress $\tau = \mu\frac{du}{dy}$

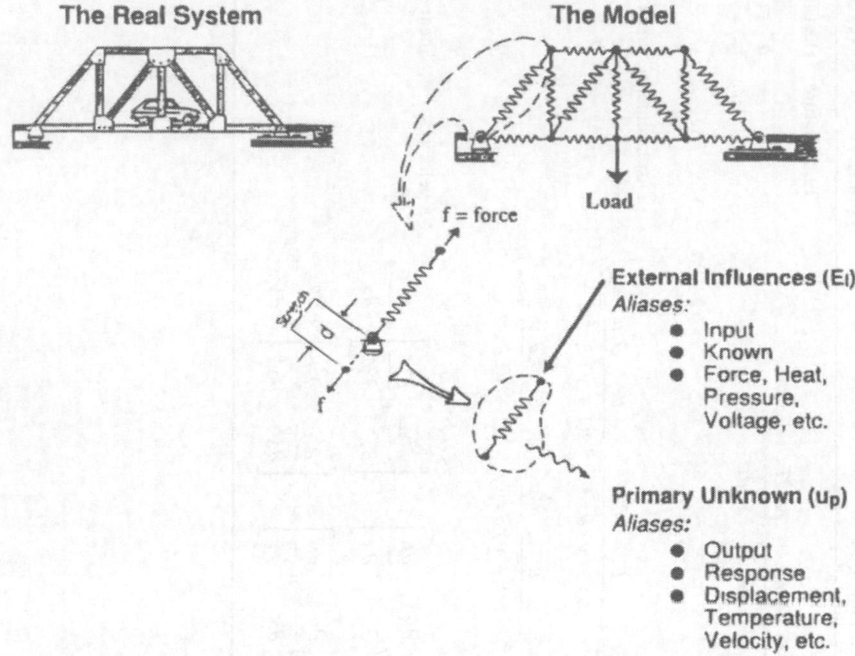

FIGURE 57.6 The essence of FEA.

Putting the Pieces Together

The spring element is a good start. A bridge structure is modeled as an assembly of spring-like elements. More exactly they are made up of truss elements that behave like the spring and have a mathematical formula exactly like the simple spring. The 2-D truss element allows for displacements in two directions (for the 3-D case we have three directions), u and v at each node. Therefore forces are related to displacements through a 4×4 matrix as shown in Figure 57.7. As elements get more complicated so do their stiffness matrices, as will be discussed later. For many elements finding the stiffness matrices is as simple as invoking Newton's equilibrium law. For more complex elements these matrices are done through so-called numerical integration techniques on a computer. Algor has a rich assortment of elements, as shown in Figure 57.8, that allows for the "building up" of virtually any design.

Building the Master

Once an element is selected the total system or master equation is assembled. The assembly is made by inserting the element matrix into the proper row and column of the master as shown in Figure 57.7. Finally the master equation is solved for the primary unknown, the displacement vector for the case of the bridge.

$$\{d\} = [K]^{-1}\{F\} \tag{2}$$

Symbolically we show this by inverting the $[K]$ matrix. Stresses are then calculated from the displacements as shown in Table 57.1. In practice though companies such as Algor use a variety of solving schemes such as the Gaussian elimination method. Research into better and faster methods is an ongoing project. Algor for instance was

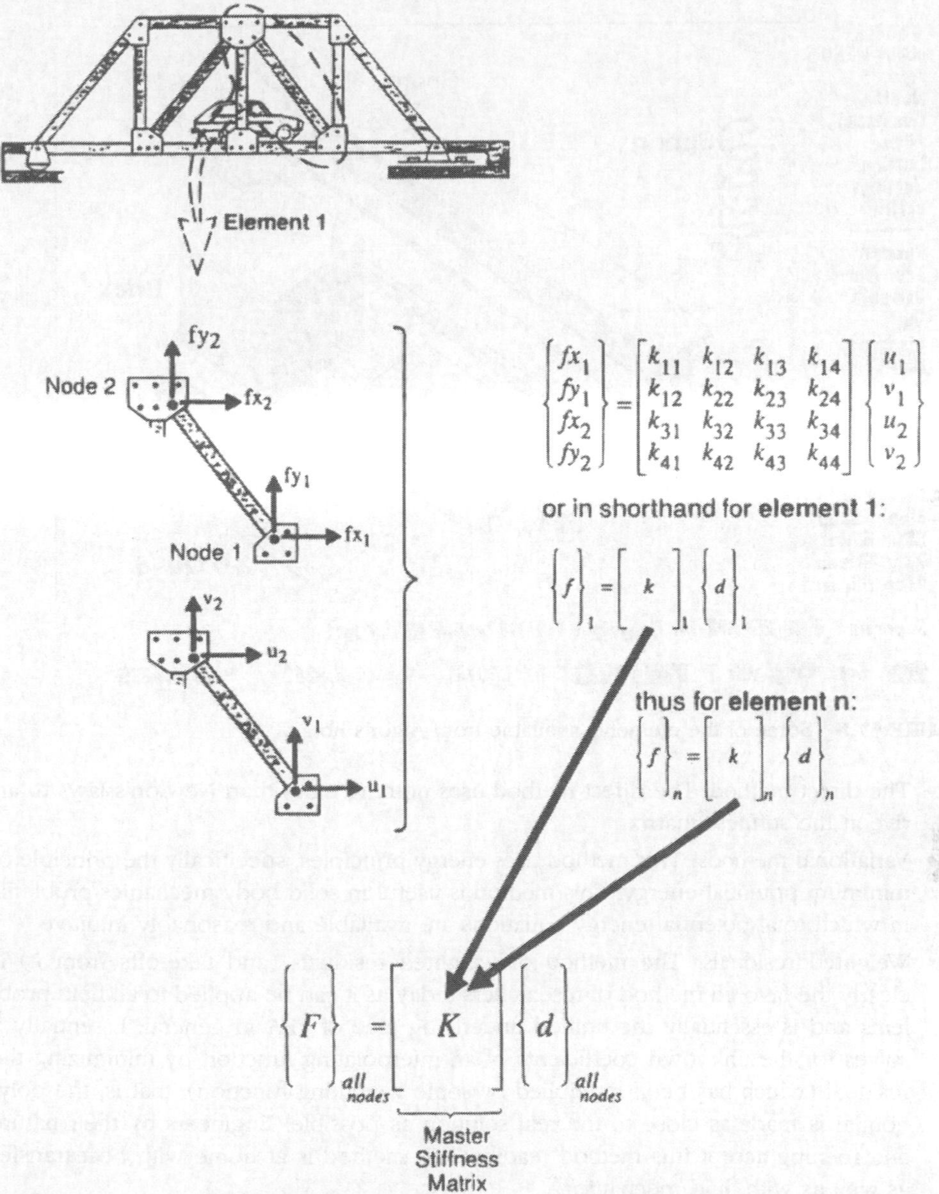

FIGURE 57.7 The truss element is used to model the complete structure.

able to develop a proprietary storage compression algorithm that allowed engineers to solve virtually unlimited size problems even on a PC computer. Algor, in no small part, has been instrumental in turning the PC into a viable engineering workstation.

More about the Element Stiffness Matrix

It should be clear now that the key to FEA analysis is in finding the element stiffness matrix. Once found, the process is the same for all elements. Recent advances have extended the FEA methodology from truss structures into solid body mechanics areas, heat transfer, electrostatics, magnetics, and fluid mechanics. There are essentially three mathematical techniques used today for developing these matrices.

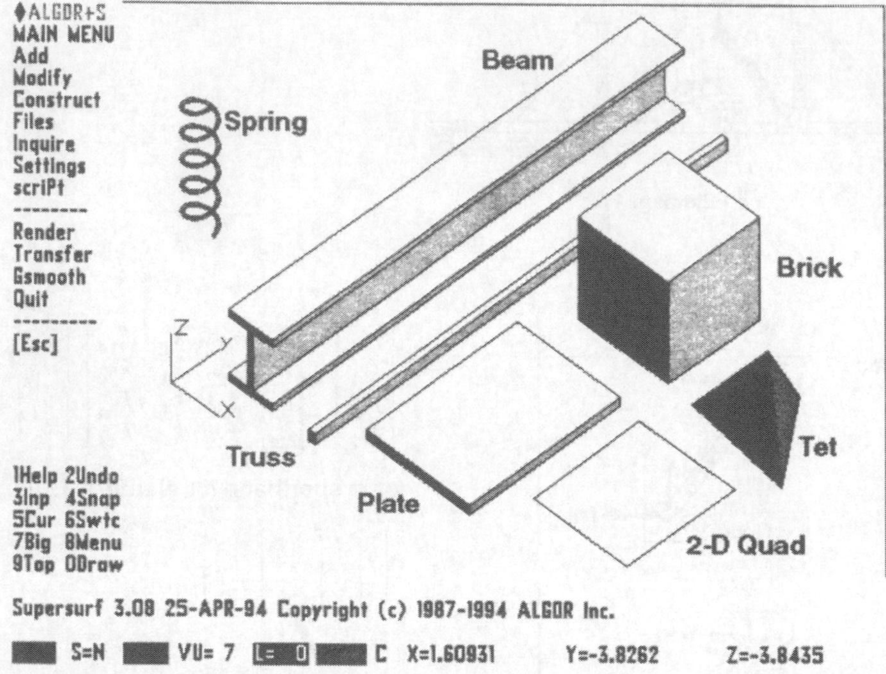

FIGURE 57.8 Some of the elements available from Algor's library.

- The direct method. The direct method uses nothing more than Newton's laws to arrive at the stiffness matrix.
- Variational methods. This method uses energy principles, specifically the principle of minimum potential energy. This method is useful in solid body mechanics problems in which total potential energy equations are available and reasonably intuitive.
- Weighted residuals. The method of weighted residuals (and take-offs from it) is clearly the favored method of researchers today as it can be applied to all field problems and is essentially the unified underlying glue of FEA in general. Essentially it solves for the unknown coefficients of an interpolating function by minimizing the residual (which has been multiplied by some weighting function), that is, the polynomial is made as close to the real solution as possible. Engineers by their nature and training accept this method readily. This method is at home with heat transfer as well as with fluid mechanics.

Interpolating Functions

As mentioned, the key to the method of weighted residuals, and thus FEA in general, is that it starts by approximating the unknown (displacement, temperature, etc.) as a polynomial interpolating function. One soon discovers that the higher order the polynomial the better the results. On the other hand a polynomial can not be any larger than can be "supported" by the elements nodes. In Figure 57.9 it can be seen that a tetrahedral element can support 12 coefficients, which translates to a polynomial of the first degree. What is really significant is the fact (from the theory of elasticity) that the strain, which is the first derivative of displacement, is constant throughout the element. The hexahedron (often called bricks), on the other hand, can support a second-order

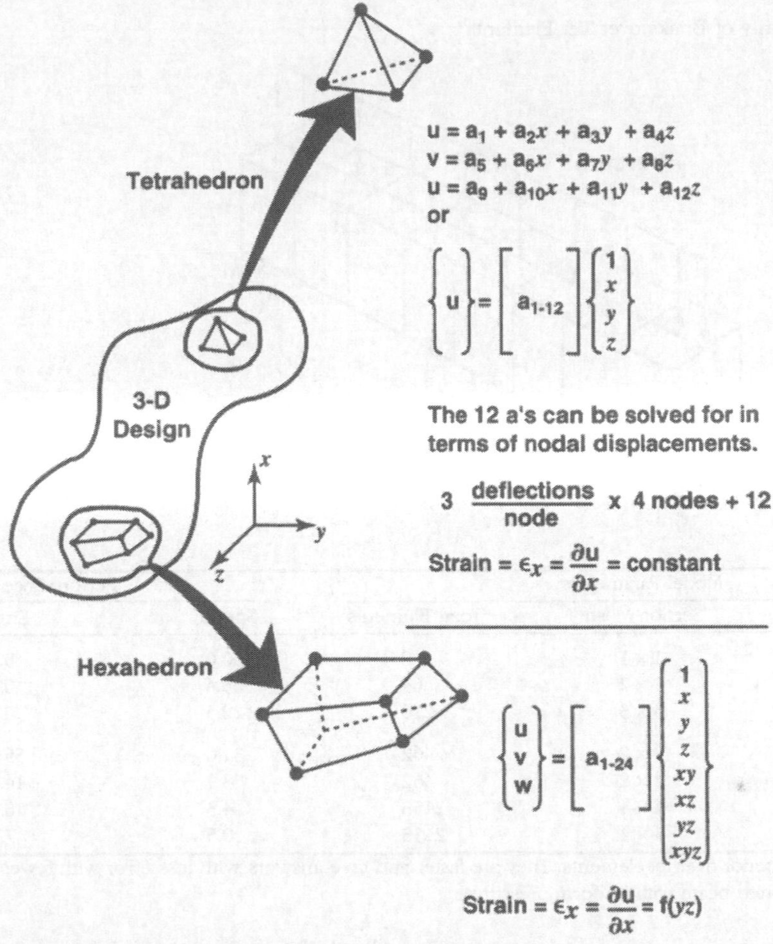

Tetrahedron
u = a1 + a2x + a3y + a4z
v = a5 + a6x + a7y + a8z
u = a9 + a10x + a11y + a12z
or
{u} = [a1-12] {1 x y z}
3-D
Design
The 12 a's can be solved for in
terms of nodal displacements.
3 deflections/node x 4 nodes + 12
Strain = εx = ∂u/∂x = constant
Hexahedron
{u v w} = [a1-24] {1 x y z xy xz yz xyz}
Strain = εx = ∂u/∂x = f(yz)

FIGURE 57.9 Three-dimensional element interpolation.

polynomial, which gives rise to a linear variation of strain. So what? It turns out that a linear strain gradient is vastly superior to a constant strain element. Studies run at Algor's Research Center revealed (see Table 57.2) that for a simply supported beam model the brick outperformed the tetrahedral element by an order of magnitude. It is for this reason that Algor has initiated a professional crusade informing engineers of the importance of making models made up of bricks. It bears repeating: bricks are vastly superior to tets.

57.4 MODELING IS EVERYTHING

FEA has matured to the point that the theoretical underpinnings are so well understood and well documented that the shift in research has transferred to modeling. As previously mentioned, one's design needs to be broken up into elements; a sampling of Algor elements (from a much larger library) is shown in Figure 57.8. Not too long ago making a finite element model was a serious impediment to the growth of FEA. It was not uncommon for engineers to literally spend months to make complex models. Since the day it opened for business in 1976 Algor has dedicated itself to making modeling easier and easier. Today Algor has a family of "plug and play" modules that

TABLE 57.2 Superiority of Bricks over Tet Elements[a]

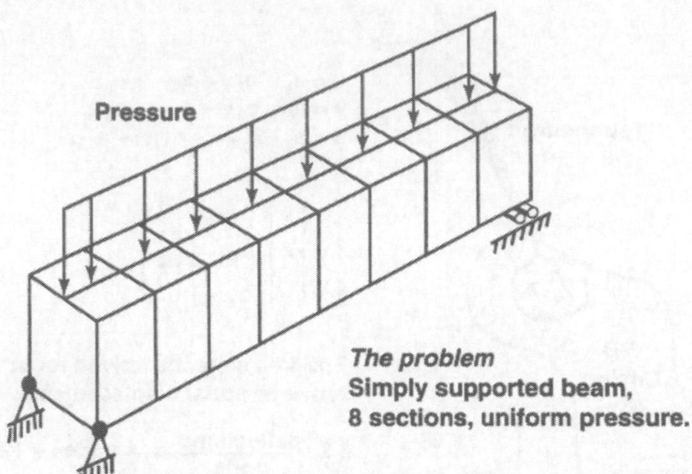

	Model Parameters		Performance	
Element Type	Section Mesh	Total Elements	Speed	Error
Brick	1 × 1	8	100.0	3.7
Brick	2 × 2	64	65.4	0.8
Brick	3 × 3	216	22.5	0.1
Tet	2 × 2	462	31.0	55.0
Tet-10	2 × 2	462	3.1	14.6
Tet	4 × 4	2158	4.3	16.0
Tet-10	4 × 4	2158	0.5	7.7

[a] Bricks are vastly superior over tet elements. They are faster and give answers with less error with fewer elements. Model: Simply supported beam with uniform pressure.

can address virtually any design scenario. All of the modules can be controlled by Algor's unique push button Roadmap system as shown in Figure 57.10. The Roadmap system can effortlessly (simply click on the appropriate button) lead both beginners and experts through the Algor family of products.

The Unique Algor Modeling Methodology

In 1984 Algor took the world by surprise when it introduced the world's first CAD Decoder. The Decoder was, and still is, unique in that it could take a CAD drawing (AutoCAD, and Cadkey were the first) and turn it into a FEA model ready for solving with Algor's processors. Results were displayed in beautiful color images. All of this was achievable on PC computers. Because most of the CAD systems did not have the commands that engineers needed, Algor introduced the first CAD system designed from the ground up to be an FEA modeler. Today, Superdraw II is still the key module in what has grown into a family of products. Superdraw II uses a familiar tree structure menu system, as shown in Figure 57.11, and runs identically in DOS, Unix, and Windows 95/NT environments. The real significance of Algor's invention is that designers could start with their CAD system, complete the design, and then transfer to Algor for analysis. Many companies like having one source document. Algor has expanded this paradigm to the point that the software has been designed to coexist, at several entry points, in any corporate environment.

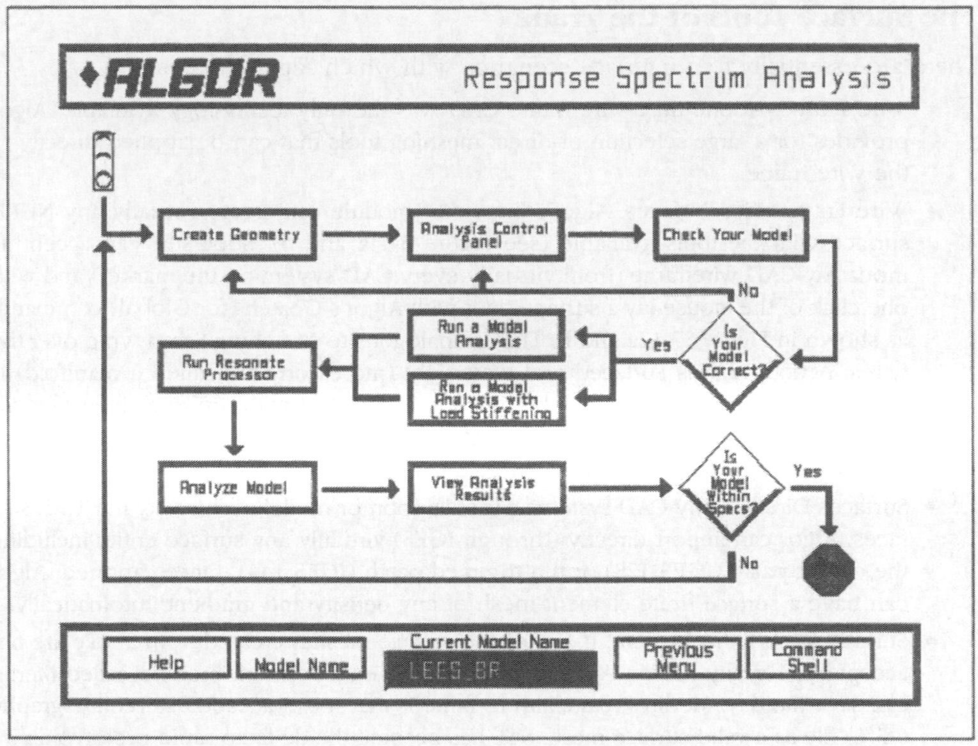

FIGURE 57.10 One of several interconnected Roadmap screens.

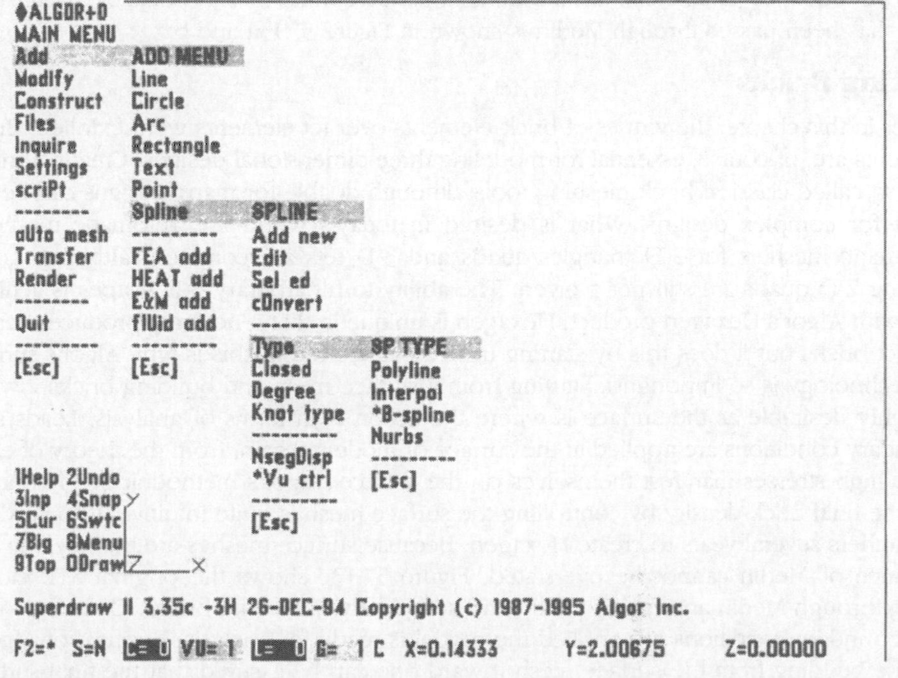

FIGURE 57.11 All Algor graphic modules use a tree structure menu system.

The Surface Tools of the Trade

There are essentially four interface scenarios with which Algor can deal.

- Wire Frame. At one time wire frame CAD was the only technology available. Algor provides for a large selection of direct meshing tools that can be applied directly to the wire frame.
- Wire Frame with Surfaces. Algor's Supersurf module can create virtually any NURB surface with the tools available (see Figure 57.3a and b). Supersurf can accept as input any CAD wireframe (from virtually every CAD system on the market) and with one click of the mouse lay a surface on it with Algor's Gpatch (for Global) command, as shown in Figure 57.12a and b. This is analogous to stretching Saran wrap over the wires. A body that is surfaced and watertight (more correctly called a manifold) is one form of a solid (other "solids" that Algor can deal with are explained below). Once the design has been surfaced a surface mesh of any density (with gradients if desired) can be "paved" onto the surface.
- Surfaces Direct. Many CAD systems either support or model exclusively in NURB surfaces. Algor can import directly (through IGES) virtually any surface entity including the classic patch (IGES 128) or the trimmed patch (IGES 144). Once captured, Algor can pave a surface finite element mesh of any density and gradient automatically.
- Surface Mesh Enhancement. If any of the surface meshes created from above are not acceptable for any reason Algor's surface mesh enhancement module called Merlin can save the day. Merlin's real claim to fame is that it can accept a stereolithography (.STL) file as a valid surface mesh. STL has become the de facto rapid prototyping interface standard. STL's highly irregular triangular surface "mesh" was created only to make rapid prototype parts and was never intended for FEA usage. Algor developers taught Merlin to accept and then enhance (through either default settings or with user-defined parameters) this mesh into a perfect surface mesh. A Cadkey STL file that has been passed through Merlin is shown in Figure 57.13a and b.

Making Bricks

Earlier in this chapter the virtues of brick elements over tet elements were extolled. Brick elements are, of course, essential for modeling three-dimensional designs. Creating bricks with so called classical brick meshing tools although doable for many designs can be tedious for complex designs. What is desired in today's world are automatic meshers. Automatic meshers for 2-D triangles/quads and 3-D tets are common, although good looking 2-D quads are still not a given. The ability to fill arbitrary 3-D shapes is achievable with Algor's Hexagen product. Hexagen is unique in that it not only produces nearly perfect bricks but it does this by starting from a surface mesh (this is why Algor's surfacing technology is so important). Starting from a surface mesh and building bricks inward is highly desirable as the surface is where the action is in terms of analysis. Loads and boundary conditions are applied at the surface of models. Again, from the theory of elasticity, high stresses manifest themselves on the surface. Algor's methodology of controlling the final brick density by controlling the surface mesh is quite intuitive. It took Algor researchers several years to create Hexagen. Because surface meshes are the key, the importance of Merlin cannot be overstated. Figure 57.13c shows the original STL model going through Merlin and finally totally and completely filled with bricks. Once the loads and boundary conditions are applied the model is ready for analysis. Because of Algor's unique building from the surface mesh inward one can be assured that the final surface mesh will not be altered, a definite bonus if you want to apply the loads and boundary conditions before processing through Hexagen.

a)

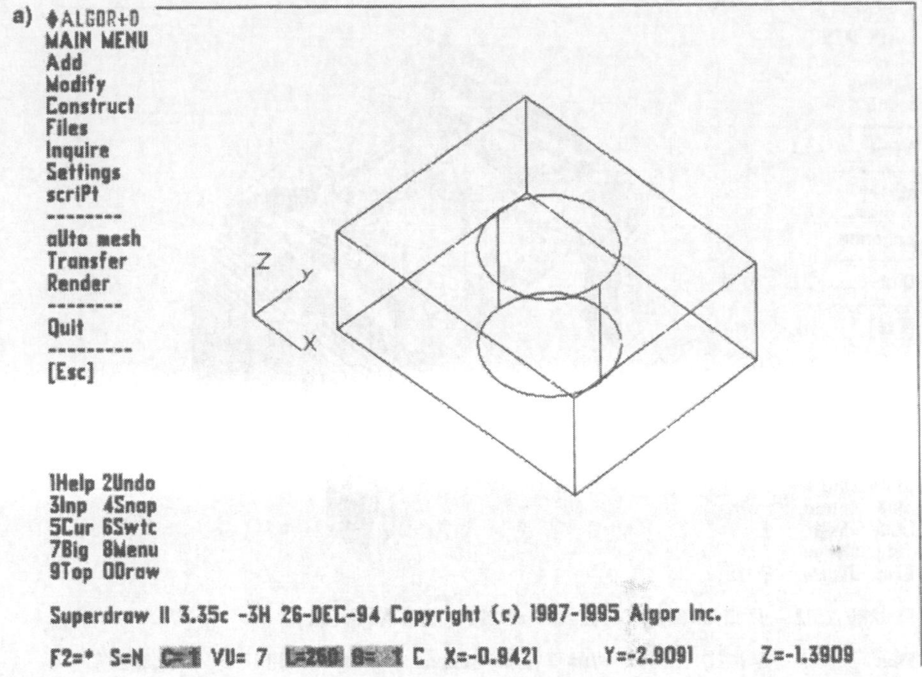

b)

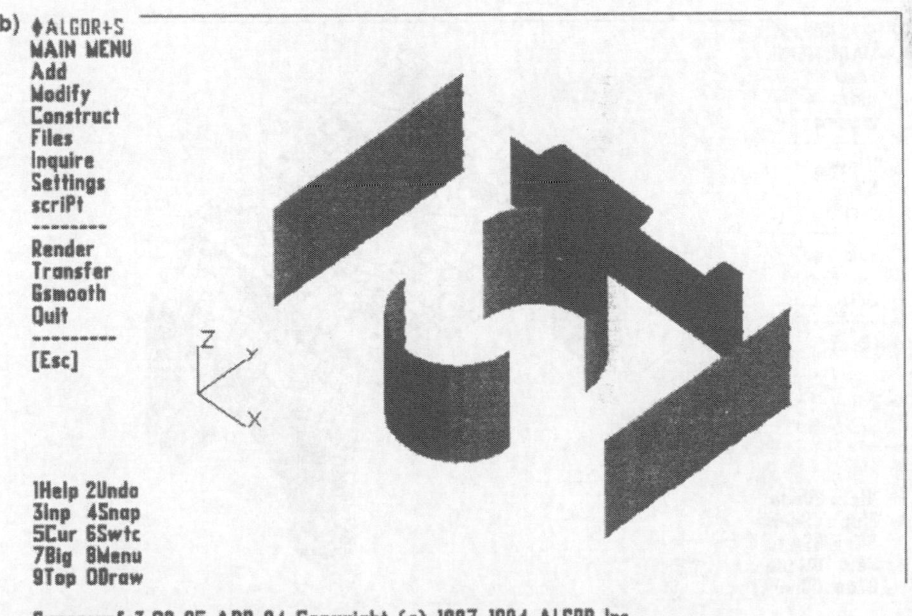

FIGURE 57.12 (a) Algor can import wire frame models from virtually any CAD system. (b) With one click of the mouse Supersurf can lay a NURB Surface over the wire frame. The model has been exploded for better viewing.

a)

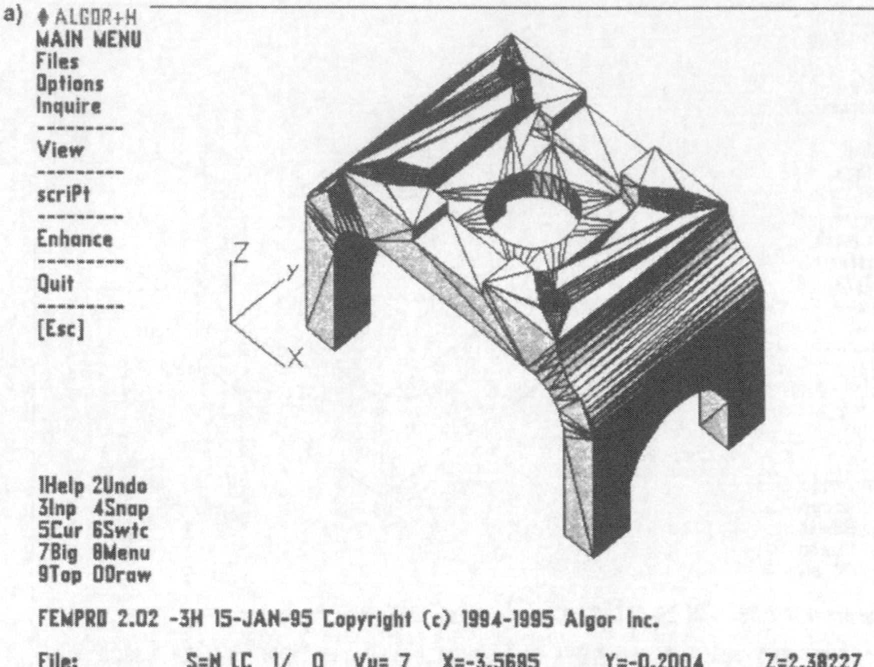

b)

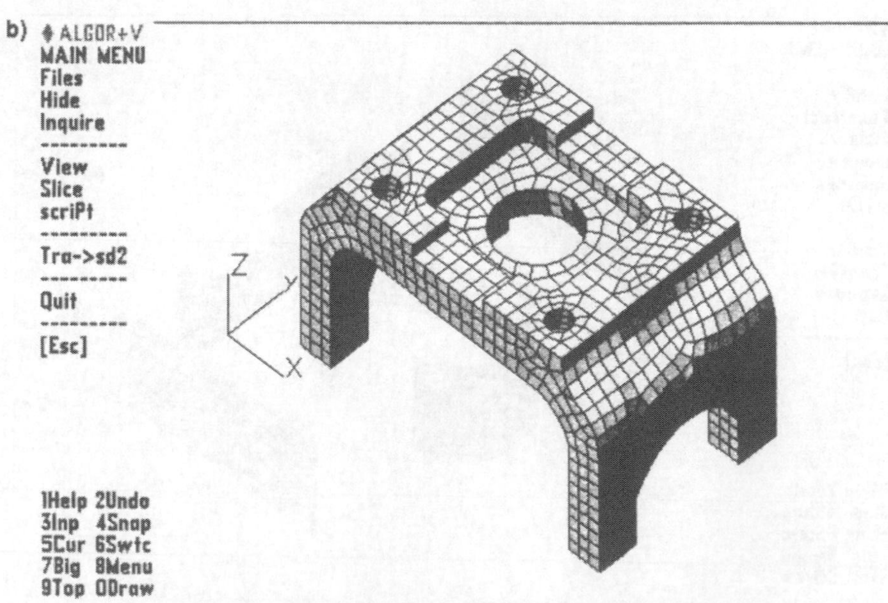

FIGURE 57.13 (a) A .STL file from Cadkey. (b) Merlin's mesh enhancement capabilities turns the .STL file into a very nice surface mesh.

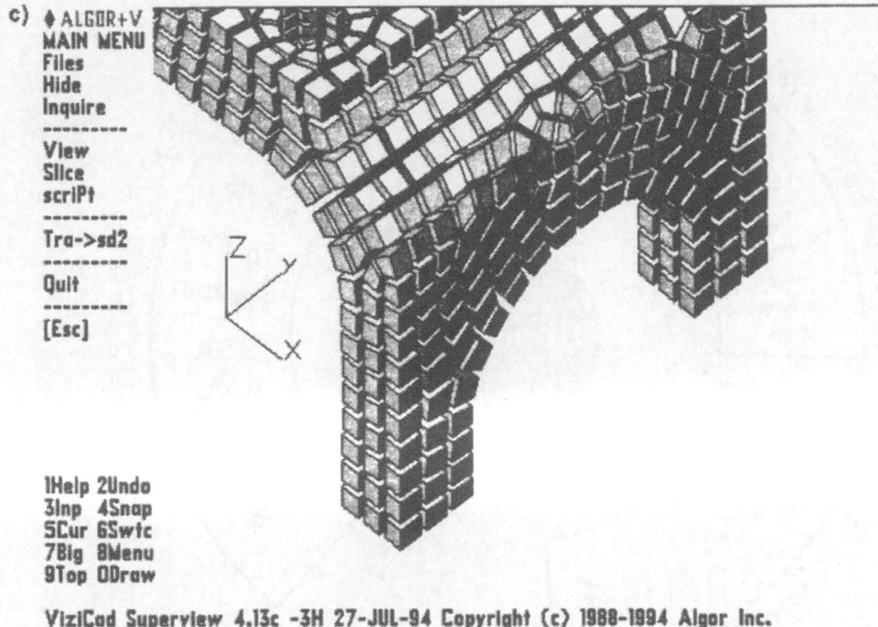

FIGURE 57.13 (*Continued*) (c) A zoomed-in view of the final bricked model.

57.5 THE ULTIMATE INTERFACE: HOUDINI

Their are many CAD systems for designers to choose from plus a large selection of FEA systems. With state-of-the-art modeling tools in hand the developers at Algor set out to devise the ultimate interface to address their myriad of CAD/FEA interface combinations. The product is Houdini. Houdini was created for engineers who utilize CAD solid modelers and was invented to save time and money by providing an easy way to integrate FEA into the design process. A productive synergy between CAD and FEA has developed because of Houdini. Houdini provides for more than 100 different CAD to FEA interface options (see Fig. 57.14). Houdini accepts models created in virtually all the popular CAD solid modeling systems, including Pro/ENGINEER, Unigraphics, SDRC, Aries, Catia, AutoCAD, Cadkey, IBM, and Intergraph. Houdini will also accept finite element (FE) models from those CAD solid modelers that produce them. CAD solid modelers usually create FE models by generating a tetrahedral mesh from the inside out. In addition, tetrahedral or brick FE solid models or plate/shell models from other FE programs may be used. A new Menupak user interface (similar to Roadmaps) puts all this power within a few mouse clicks.

Multilevel Interface Scenarios

Depending on the type and quality level of the final mesh desired, you may choose to operate on one of six Houdini levels:

- Level 1. Like other FE programs Houdini enables you to process 4- or 10-node tetrahedral FE models from CAD solid modelers "as-is" for a fast, low-level analysis. Unlike the others, Houdini provides for direct, "as-is" processing of FE models created in one system, by FEA solvers in another system, such as Algor.

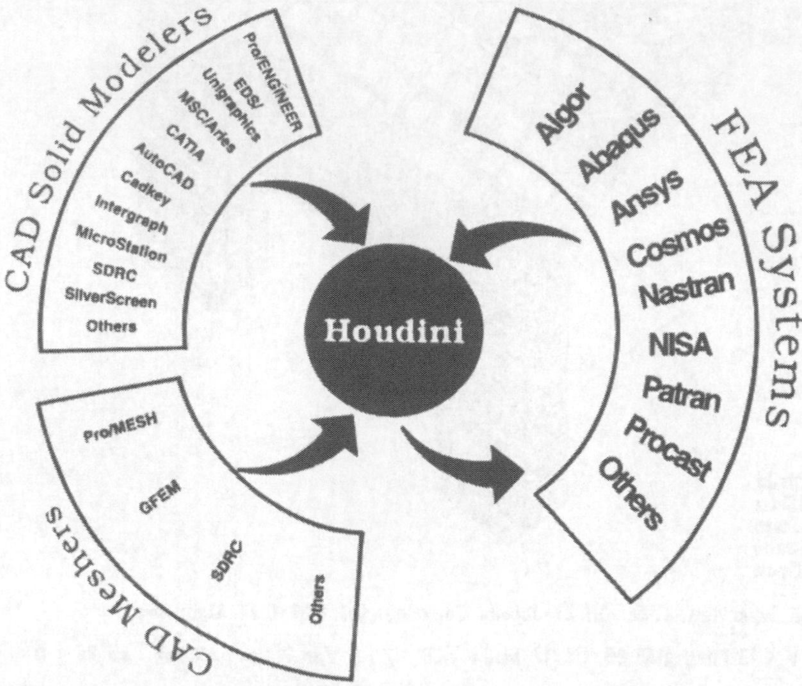

FIGURE 57.14 Houdini adds value to FEA systems and/or CAD solid modelers.

Level 1 is rarely recommended, as most engineers prefer to take advantage of Houdini's proprietary technology to increase accuracy by using one or more of the following higher-levels:

- Level 2. Use Algor's Hypergen to automatically create a new 4- or 10-node solid tetrahedral mesh from the surface inward. Like its big brother Hexagen, accuracy is improved by using the surface mesh as the starting point.
- Level 3. Increase the quality of the solid model by creating an improved surface mesh, then creating a more accurate tetrahedral mesh using Hypergen.
- Level 4. Convert the improved surface mesh into quadrilateral elements, then use Hexagen to create an even more accurate 8-node, solid brick model from the surface in.
- Level 5. Directly access Algor's exclusive Merlin Mesh Enhancement Technology to automatically produce an engineering quality quadrilateral surface mesh prior to using Hexagen. This scenario can improve brick models from an FE program or mesh generator.
- Level 6. Gain maximum control and highest accuracy by using Advance Merlin functions such as feature line creation and suppression, mesh specifications, adherence to features, local refinement, etc. in the creation of the final meshed model.

Unlimited Analysis Options

Houdini-produced models can be analyzed using Algor's full range of FEA processors, including linear or nonlinear stress, vibration and natural frequencies (mode shapes), heat transfer, fluid flow, composite material elements, and electrostatics. In addition, an extender module, called FEMOUT, enables models from Houdini to be processed by other finite element analysis packages, including Nastran, Abacus, Ansys, Rasna, Cosmos, NISA, and others.

58

ANSYS: Finite Element Analysis

Timothy R. Trainer
ANSYS, Inc..

58.1 WHAT IS FINITE ELEMENT ANALYSIS?

Before the advent of finite element analysis (FEA) technology, the only design/analysis tools available to engineers for accurately predicting the behavior of a component or system were scientific and mathematical equations, applied to simplified systems, and testing models. The difficulty the analytic (mathematical) approach caused for design engineers and analysts rested in the fact that they still had to run the component or system through a whole battery of tests to verify its calculated behavior, resulting, in many cases, in repetitive prototype building and testing. The end result was often a slow, resource-intensive design process. If only there were a way to see how a design behaves before the prototype was built.

Enter FEA. Numerous analytic procedures have been used by engineers over the years. However, the advent of FEA coupled with the rapid advances in computer technology have made it possible to simulate responses of designs on the computer and only test the final product. In some instances, in which tests are impractical or impossible, FEA is the preferred procedure for evaluating and "testing" the design.

Based on the structural principles of equilibrium and compatibility (and the comparable heat flow balance for heat transfer), FEA utilizes influence coefficients (stiffness or flexibility) to mathematically represent the response of a design to a force, displacement, or other loading. The influence coefficients are generated for each individual building block, or element, in an FEA model. The entire design is represented by the assemblage of all the individual elements. The design response is given as the combination of all influence coefficients. Therefore, complex designs can be represented by an assemblage of simple elements.

To use a simple analogy, the engineer uses an FEA program with its elements much like a child would use Legos®[1] to build a system. The earliest FEA programs used discrete elements (bars and beams), which were more like Tinkertoy® stick models. Current technology allows the use of more general continuum (solid) representation.

[1]All brand and product names are trademarks or registered trademarks of their respective companies.

0-8493-2530-7/96/$0.00+$.50

The actual elements even respond very much like a Lego block by transmitting forces through the connecting points between blocks. Elements are "connected" at discrete locations called nodes. These nodes are most commonly found on the corners of elements and along the edges between corners.

The development of FEA software permitted an engineer to test a design before it was built. By assessing the effect of a force on the individual elements and by graphically displaying deflections, strains, or stresses, one can determine and "see" the effect of that force on the whole. Thus, FEA programs became powerful engineering tools and are the most widely used engineering software packages for design simulations.

Today, FEA programs analyze a variety of physical characteristics of computer-generated models. And although the Lego analogy is a simple means for understanding the process, FEA programs actually employ a mathematical technique known as the finite element method (FEM). This technique is used for solving differential equations that govern physical phenomena as they apply to simulated geometries.

Utilizing the principles of matrix algebra, virtual work, and variational calculus, FEM creates a set of linear or nonlinear equations representing the response of sections or "elements" within a model. FEM allows an engineer to individually determine the impact of physical forces on each of numerous elements (equations), which are meshed or linked together to simulate a component or system, thus characterizing the response of the entire model. In short, FEM is a numerical discretization technique for differential equations.

Originating in the field of structural mechanics to predict equilibrium stresses and displacements, FEA has evolved to include a broad spectrum of engineering applications, ranging from structural, thermal, and kinematic analyses to magnetic field, fluid flow, coupled-field, acoustic, and piezoelectric analyses.

The advantage of using a computer program to conduct design analysis and simulation is the ability to visualize the responses through such techniques as contouring data, isosurface plots, section plots, and animation. An engineer can also use the computer to see the effect of a physical force, whether a load, force, temperature change, or movement. Without the aid of a computer, interpreting the vast quantities of numbers generated by the FEA would be extremely difficult and virtually impossible to visualize.

58.2 THE ANSYS® FINITE ELEMENT ANALYSIS PROGRAM

The ANSYS FEA program is one of the most widely used engineering design analysis software packages in the world, with installations at thousands of commercial sites and universities. Originally developed in the early 1970s for use by the power generation industry in heat transfer and linear structural analyses, ANSYS FEA has evolved into a large-scale, general-purpose FEA software program, offering diverse engineering disciplines a wide range of analysis capabilities, including

- Structural static analysis
- Structural dynamic analysis
- Structural buckling analysis
- Structural nonlinearities
- Static and dynamic kinematic analysis
- Thermal analysis
- Magnetic field analysis
- Fluid flow and acoustic analysis

- Coupled-field analysis
- Piezoelectric analysis
- Design optimization

The ANSYS program can help reduce design and manufacturing costs substantially by accelerating the design–prototype–testing process. It allows engineers to construct computer models of structures or machine components, apply operating loads and other forces, and evaluate physical responses, such as stress levels and temperature distribution. Analysis with the ANSYS program can detect potential design defects or determine the optimum design geometry before a prototype is produced, obviating the need for expensive multiple prototype-building, testing, and rebuilding, as well as overdesign or product retrofits for underdesign.

ANSYS FEA is most effective when used at the conceptual design stage but also provides benefits later on in the manufacturing process to verify final design before prototyping, as shown in Figure 58.1.

An important attribute of any FEA program is its quality and reliability. The ANSYS program was the first FEA software package to develop a quality assurance error report and an error correction system, incorporating vigorous verification testing. Since then, additional quality assurance efforts—including regression checks on error corrections, tests for modules, libraries, and elements, and graphics and FE meshing tests—have been instituted for the program. The ANSYS Program is the only FEA package to achieve certification under ISO 9001, an international quality standard.

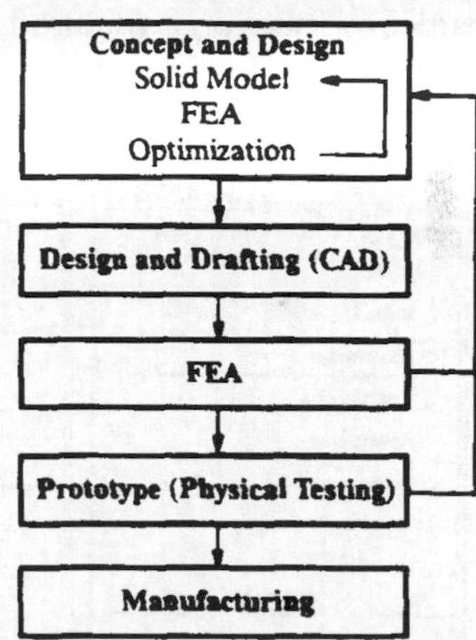

FIGURE 58.1 The ANSYS program is most effective when used at the conceptual design stage, but can also be used later in the manufacturing process to verify final design, as this diagram illustrates.

Program Overview

As a result of its extensive and complex capabilities, the ANSYS program is appropriate for use at all levels of proficiency, from beginners to experts, because of its organization and user-friendly menu system.

User Interface

The ANSYS program utilizes a comprehensive graphical user interface based on the X Windows® environment overlaid with the Open Software Foundation (OSF)/Motif® Standard. The X Window format was developed as an industry standard and is not dependent on any particular computer hardware. OSF/Motif is a flexible software system layered on top of the X Windows environment to create individual scroll bars and menus. The two standards work together to graphically emulate the organization of paper on a desk for windows on the computer screen. Menu windows can be stacked, partially obscured, staggered, filed, discarded, added, etc.

Because its user interface is based on a common standard, the ANSYS program provides the user with easy, interactive access to program functions, commands, documentation, and reference material. An intuitive menu system has been developed for the X Windows environment to help the user navigate through the ANSYS program. Users can input data through a mouse, a keyboard, or a combination of the two.

Within the user interface, there are four general methods for instructing the ANSYS program:

- Window menus
- Dialog boxes
- Toolbar
- Direct input of commands

Window menus are groupings of related commands and functions for operating the ANSYS program. Window areas include a utility menu, main menu, input menu, graphics window, output window, and a toolbar. The user can access these windows, which can be moved or hidden, with a mouse at any point in the process.

Dialog Boxes are windows that present the user with choices for completing operations or specifying settings. These boxes pop up on the screen whenever input for a particular function is required by the user (Fig. 58.2). Selecting the "OK" command will complete the function and dismiss the dialog box (or menu). Choosing "apply" will complete the function but leave the dialog box displayed.

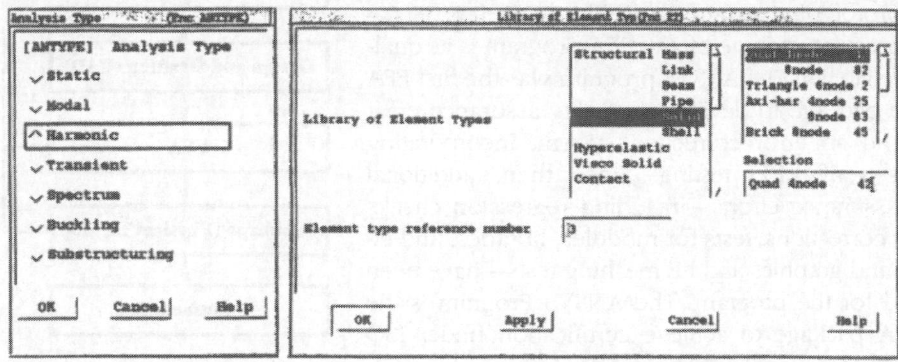

FIGURE 58.2 Dialog boxes help the user navigate through the problem by appearing whenever input for a particular function is required by the user.

Because of its wide range of configurability, the *toolbar* represents a very efficient means for inputting commands for the ANSYS program. It provides the user with the capability to create abbreviations for and have immediate access to commonly used, or "hot," commands (Fig. 58.3). The toolbar can accommodate up to 200 abbreviations, and each abbreviation can be used to represent a subset of up to 200 abbreviations. Thus, if desired, the toolbar can be used to designate or execute all of the program's commands (about 1000).

Regardless of how they are specified, commands are ultimately used to supply all data and control all program functions. The user interface is designed to make com-

ANSYS Toolbar

SAVE_DB	ISO	VECTOR	SQRWIN	POSTBAR
RESUM_DB	FRONT	RASTER	FULLWIN	ZOOM
QUIT	NOERASE	HIDDEN	TOOLBAR	NODES
REPLOT	ERASE	ZBUFFER	SYS_EDIT	LINES

FIGURE 58.3 The toolbar allows users to create abbreviations for and have immediate access to commonly used commands.

mand selection and execution an easy, intuitive process, through the use of menus, dialog boxes, and the toolbar. However, those users who are familiar with ANSYS commands can opt to input commands directly via the keyboard.

Once executed, a command is listed in a session log file by the program. This log file is accessible through the program's output window, permitting the user to review a list of commands in the event of an error or to save a list of commands as a file for batch processing.

Processors

ANSYS functions are organized into groups called processors. The program features the following primary processors:

- The *preprocessor* is used to create a finite element model and specify options needed for a subsequent solution.
- The *solution processor* is used to apply loads, forces, and boundary conditions and determine model response.
- The *postprocessor* is used to retrieve and examine solution results, evaluate how the model performed, and perform additional calculations.

In addition to the three main processors, the program features several auxiliary processors, such as the design optimizer.

Database

The ANSYS program uses a single, centralized database for storage of all model data and solution results (Fig. 58.4). As a result, data can be written to the database while using one processor and can be accessed, as needed, in all other processors.

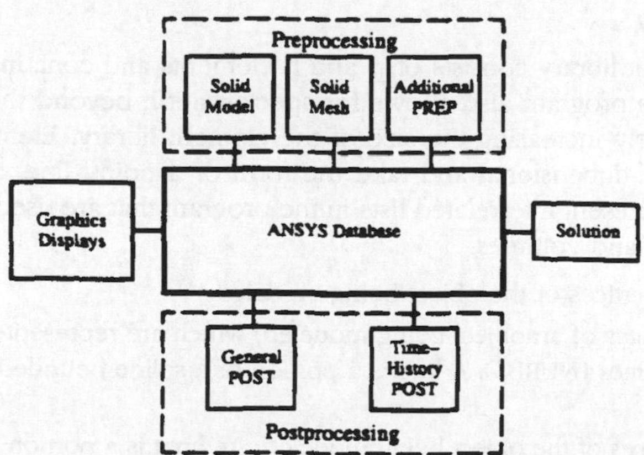

FIGURE 58.4 This diagram illustrates the centralized nature of the ANSYS database.

File Format

Files are used when necessary to pass data from one part of the program to another, store the program database, and save program output. The program writes files in either coded (i.e., can be easily read or edited) or binary format. By default, the ANSYS program writes any binary files using an external format (IEEE Standard), which permits transportability across different hardware systems.

Graphics Capabilities

Full interactive graphics are an integral part of the ANSYS program. Graphics are important for verifying preprocessing data and reviewing solution results in postprocessing. ANSYS graphics features include

- Animation
- Up to 256 colors
- Boundary condition displays
- Color contour displays of results
- Graphs of results versus time
- Display manipulation (viewing direction, zoom, magnification, rotation)
- Multiple display windows
- Overlays
- X-Y data display
- Outline display
- Three-dimensional volume visualization
- Annotation capability
- Translucency
- Composite material layer display
- Distorted ratio displays
- Submodeling
- Cross-sections
- Multivariate display

Element Library

The ANSYS element library consists of p and h elements and contains more than 100 element types. The program also allows for specialization beyond these element formulations, effectively increasing the size of the element library. Elements are two-dimensional or three-dimensional and take the form of a point, line, area, or volume. These elements represent interrelated lists in the program that are used to describe vertices, edges, faces, and volumes.

- **Points**: the vertices of the object being modeled.
- **Lines**: the edges of an object being modeled, which are represented as nonuniform rational B-splines (NURBS). A line is a portion of a spline bounded on both ends by a keypoint.
- **Areas**: the faces of the object being modeled. An area is a portion of a surface completely bound by three or more lines.
- **Volumes**: the volumes of the object being modeled. A volume is a portion of a solid that is completely bound by four or more areas.

Nodes are used to form the geometric shape and function of an element. Both linear and quadratic elements are available. Quadratic elements, which are represented through midside nodes, offer higher accuracy for a given element mesh. However, linear elements generally include extra shape functions to improve their accuracy.

The user has a great degree of control over these elements. Midside nodes of any element edge can generally be deleted. Most three-dimensional brick elements can be degenerated to prisms or tetrahedrons. And most two-dimensional quadrilateral elements can be degenerated to triangles.

Most elements allow appropriate element loadings, such as pressure, temperature, and convection. These loadings are applied to the element, which then calculates the corresponding load vector terms. Inertia loads, such as gravity, are available for most elements as applicable. Nodal loads, such as forces, temperature, and displacements, are allowed for all elements as appropriate.

An alternative method for applying loads to an element is to use surface effect elements, which represent special loads such as surface tension and foundation stiffness.

Engineering Applications (ANSYS-Specific)

As a result of its extensive capabilities, the ANSYS program can be used for numerous types of engineering analysis and offers both linear, when the response varies proportionately to the forces, and nonlinear analyses. Detailed descriptions of ANSYS engineering applications follow.

Structural Static Analysis

The ANSYS program can be used to determine displacements, stresses, strains and forces that occur in a structure or component as a result of applied loads (Fig. 58.5). Most mechanical and structural engineers are familiar with static analysis and probably have solved numerous static problems using classical analysis methods or equations from engineering handbooks. The ANSYS program solves these problems by applying numerical techniques to traditional engineering concepts. The following equation governs static analysis in the ANSYS program:

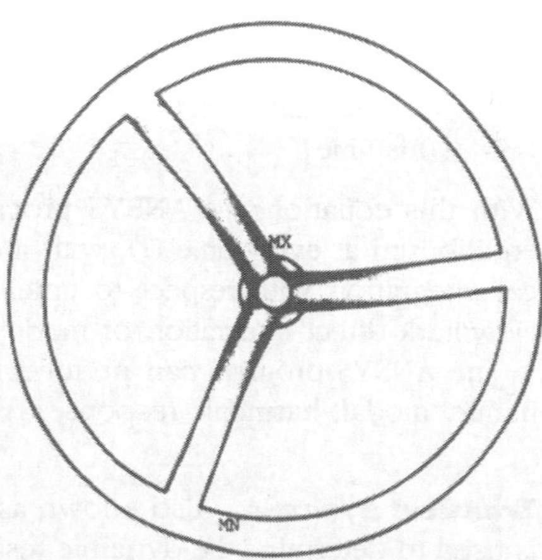

FIGURE 58.5 Equivalent (von Mises) stress contours of a pulley under belt loading are shown. The darkest portions of the stress contour in the hub and spokes indicate areas of high stress.

$$[K]\{u\} = \{F\} \tag{1}$$

$[K]$ is the structural stiffness and $\{u\}$ is the displacement vector. The force vector $\{F\}$ can include concentrated forces, thermal loads, pressures, and inertia loads. Inertia relief calculations, in which accelerations are necessary to counterbalance the applied loads, can also be performed.

Structural Dynamic Analysis

Determining the effects of time-varying loads on a structure or component, taking damping and inertia effects into account, is the function of the ANSYS program's structural dynamic analysis capability. This feature can be used to analyze loads that are characterized by

- Alternating forces (rotating machinery)
- Suddenly applied forces (impact or explosion)

- Random forces (earthquake)
- Any other transient forces, such as moving loads on a bridge

Dynamic analyses in the ANSYS program are based on the following general equation for motion for a finite element system:

$$[M]\{\ddot{u}\}+[C]\{\dot{u}\}+[K]\{u\}=\{F(t)\} \tag{2}$$

where

[M] is the mass matrix
[C] is the damping matrix
[K] is the stiffness matrix
$\{\ddot{u}\}$ is the nodal acceleration vector
{Â} is the nodal velocity vector
{u} is the nodal displacement vector
{F} is the load vector
(t) is time

With this equation, the ANSYS program determines the value of {u}, which satisfies equilibrium at every time (t), with inertia and damping effects included. The numerical integration with respect to time, when required, is accomplished through either Newmark direct integration or mode superposition.

The ANSYS program can perform the five types of dynamic analyses: transient dynamic, modal, harmonic response, response spectrum, and random vibration.

Transient Dynamic. Also known as time-history analysis, transient dynamic analysis is used to determine the dynamic response of a structure subjected to time-dependent loads. Solutions can be obtained using one of three methods based on the general equation of motion for dynamic analysis: *the full transient dynamic method*, the *reduced method*, and *mode superposition*.

Of the three methods, *full transient dynamic* is the most general and most powerful. Because this method uses the full mass ([M]), damping ([C]) and stiffness ([K]) matrices, it has full nonlinear capability and may include plasticity, creep, large deflection, large strain, stress stiffening, and nonlinear elements (such as contact surfaces).

Full transient dynamic analysis with the ANSYS program can accommodate any type of structural load, including nodal forces and imposed displacements, element loads (such as pressures and temperatures), and inertia loads (such as gravity and rotational velocities and accelerations). This method can also model the kinematic behavior of flexible structures. A combination of nodal couplings and specialized elements can be used to represent hinges, universal joints, rigid or elastic links, hydraulic cylinders, and other features found in many flexible mechanical systems.

With the *full transient dynamic analysis method*, which features an automatic variable integration time-stepping option, the solution of the equation of motion is based on the Newmark direct integration scheme in conjunction with the Newton–Raphson method (to account for nonlinear effects).

When nonlinear effects are assumed to be negligible, the user can take advantage of the speed of either the *reduced* or *mode superposition transient dynamic methods*, both of which assume linear behavior. Although general nonlinearities are not included in either of these methods, a special gap condition is available that can be used for impact problems.

With *reduced transient dynamic analysis*, the most commonly used form of transient dynamic capability in the ANSYS program, [*M*], [*C*], and [*K*] matrices are assumed to be linear. These matrices are condensed (through Guyan reduction) and expressed in terms of dynamic, or master, degrees of freedom. The Newmark direct integration method is used to solve the equations of motion, and constant time steps are employed. Loading may include nodal forces, imposed displacements, and gravity.

Solutions in the *reduced method* are achieved through a two-step process. Nodal displacements at the master degrees of freedom are solved first. If determining strains, reaction forces, stresses, etc., is necessary, an expansion pass (an optional second step) can be performed to expand the solution at desired time points to the full degree of freedom set.

The *mode superposition method* is similar to the *reduced method* with the following difference: the solution adds individual mode responses from a modal analysis to calculate a structure's total response. This modal analysis must be performed prior to any other solution step. The modal analysis may be based on reduced matrices (through Guyan reduction) or full matrices (through full subspace iteration).

Determining which method of transient dynamic analysis is most appropriate will depend on the application involved and the needs of the user.

Modal Modal analysis is used to extract the natural frequencies and mode shapes of a structure (Fig. 58.6). Such an analysis is important in preparation for any dynamic analysis because identifying a structure's fundamental modes and frequencies can help to characterize its dynamic response. For modal analysis, the ANSYS program assumes free (unforced), damped, or undamped vibration, as described by the following equation of motion:

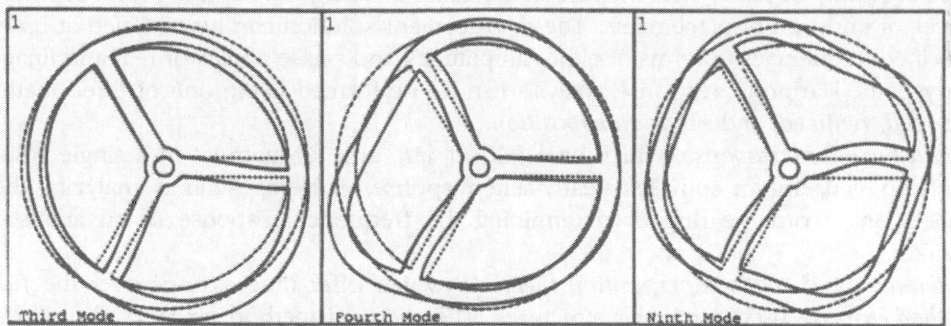

FIGURE 58.6 A modal analysis of the pulley is performed to determine the third, fourth, and ninth modes. The broken lines represent the undisplaced shape.

$$[M]\{\ddot{u}\}+[C]\{\dot{u}\}+[K]\{u\}=0 \tag{3}$$

The equation is recast as an eigenvalue problem. For undamped cases, which are most common for modal analysis, the damping term, $[C]\{\hat{A}\}$, is ignored, and the equation reduces to

$$([K]-\omega^2[M])\{\bar{u}\}=0 \tag{4}$$

In this equation, ω^2 (the square of natural frequencies) represents the eigenvalues, and $\{\ddot{u}\}$ (the mode shapes, which do not change with time) represents the eigenvectors. For damped cases, the equation of motion reduces to

$$([K]+i\omega[C]-\omega^2[M])\{\bar{u}\}=0 \tag{5}$$

Four methods of eigenvalue extraction are available for modal analysis: reduced (Householder–Bisection–Inverse iteration), full subspace, unsymmetric, and damped. The reduced method uses reduced matrices, while the remaining methods use full matrices. The unsymmetric method is used when the stiffness and/or mass matrices are unsymmetric, such as occurs in acoustical fluid–structure interaction analyses. The damped method is for situations where damping cannot be ignored, such as rotor dynamic applications. Both the unsymmetric and damped methods are based on the Lanczos algorithm.

Modal analysis is useful for any application in which the natural frequencies of a structure are of interest.

Harmonic Response. This type of dynamic analysis is used to determine the steady-state response of a linear structure to a sinusoidally varying, forcing function. It is useful for studying the effects of load conditions that vary harmonically with time, such as those experienced by the housings, mountings, and foundations of rotating machinery.

Harmonic response analysis is governed by a special form of the general equation for motion, in which the forcing function $\{F(t)\}$ is a known function of time varying sinusoidally with a known amplitude F_0 at a known frequency ω (and phase angle, ϕ):

$$\{F(t)\}+F_0\left[\cos(\omega t+\phi)+i\sin(\omega t+\phi)\right] \tag{6}$$

The displacements are assumed to vary sinusoidally at the same frequency ω but are not necessarily in phase with the forcing function. Loading can be in the form of nodal forces or imposed displacements. The displacement solution can be obtained at user-specified frequencies in terms of either amplitudes and phase angles or real and imaginary parts. Harmonic response analysis can be performed using one of three methods: *full, reduced*, and *mode superposition*.

Full harmonic response, which uses full [*K*], [*M*], and [*C*] matrices in a single solution step, is useful for complex steady-state response problems, such as analyzing the stresses in a rotor bearing or determining the frequency response of an acoustic speaker.

Reduced and *mode superposition methods*, which offer time savings over the *full method*, can be used for linear structures. The reduced method employs Guyan reduction to condense the [*K*], [*M*], and [*C*] matrices through a two-step procedure: a reduced solution and an expansion pass. The mode superposition method may be based on reduced (through Guyan reduction) or full matrices and requires a modal analysis prior to additional solution steps.

Response Spectrum. A response spectrum analysis is used to determine the response of a structure to shock loading conditions, such as a seismic analysis, a study of the effects of earthquakes on structures (piping, towers, bridges, etc.). It uses the results of a modal analysis with a known spectrum to calculate actual displacements and stresses that occur in the structure at each of its natural frequencies.

The response spectrum data are supplied as a response-versus-frequency function and can take four different forms: the displacement, velocity, acceleration, and force spectra. The user can select a single-response spectrum or a series of spectra at different damping ratios for a single- or multipoint analysis.

When a response spectrum analysis is performed, actual structural displacements are calculated for each mode. The total response is obtained by combining all modes through one of the following methods: Wilson-CQC, Ten-Percent, Double-Sum, Square-Root-of-Sum-of-Squares, or a user-defined method.

Random Vibration. This type of spectrum analysis is used to study the response of a structure to random excitations, such as those generated by jet or rocket engines. The spectrum utilized is a power-spectral-density (PSD)-versus-frequency curve, a statistical measure of random excitation.

The PSD spectrum can be input in terms of displacement, velocity, acceleration, or force with the user specifying single- or multipoint analysis. Normal (Gaussian) distribution of the PSD is assumed, and the ANSYS program calculates a response that is normally distributed.

Regardless of the type of PSD used, three sets of solution quantities are available: the displacement solution (displacements, stresses, strains, and forces), the velocity solution (velocities, stress velocities, force velocities, etc.), and the acceleration solution (accelerations, stress accelerations, force accelerations, etc.).

Random vibration analysis is especially useful in the aerospace industry, where components must be designed to withstand the effects of flight conditions.

Structural Buckling Analysis

This capability is used to determine the load level at which a structure becomes unstable or whether or not a structure remains stable at a particular load level. Structural buckling analysis is important for ensuring the stability of any load-carrying structure, such as a bridge or a tower. The ANSYS program can perform both linear (eigenvalue) and nonlinear buckling analyses.

Linear Buckling. Linear buckling accounts for stress stiffness effects in situations in which compressive stresses tend to lessen a structure's ability to resist lateral loads. As compressive stresses increase, resistance to lateral forces decreases. And at some load level, this negative stress stiffening overcomes the linear structural stiffness, causing the structure to buckle.

The ANSYS program uses an eigenvalue formulation to perform linear buckling analysis, determining the scaling factors (eigenvalues) for the stress stiffness matrix, which offset the structural stiffness matrix through the following equation:

$$([K]) - \lambda[S])\{u\} = 0 \tag{7}$$

where

$[K]$ is the structural stiffness matrix
$[S]$ is the stress stiffness matrix
λ are eigenvalues representing the scale factors
$\{u\}$ is the eigenvector representing the buckled shape

The point at which buckling occurs, where the two paths of the force-deflection curve intersect, is called the bifurcation point. Once the bifurcation point is exceeded, the structure will either buckle or continue to take on load in an unstable state.

Linear buckling cannot account for any nonlinearities or structural imperfections. When present, which usually is the case, these factors would cause the buckling load to be lower than the analysis results.

Nonlinear Buckling. This capability is a more accurate means for determining buckling loads with the ANSYS program because it can account for initial imperfections and nonlinearities that exist in real structures. The program updates the orientation of a structure's elements in a large deflection analysis through an incremental Newton–Raphson method for any given equilibrium iteration:

$$[K]_{i-1}\{\Delta u\}_i = \{F\} - \{F^{el}\}_{i-1} \tag{8}$$

where

$[K]_{i-1}$ is the stiffness matrix from the previous iteration
$\{\Delta u\}_i$ is the incremental displacement vector, $\{u\}_i = \{u\}_{i-1} + \{\Delta u\}_i$
$\{u\}_i$ is the displacement vector at the current iteration
$\{F\}$ is the applied force vector
$\{F^{el}\}_{i-1}$ is the elastic force vector based on displacements for iteration $(i - 1)$

The ANSYS program performs nonlinear buckling analysis by monitoring Δu through the iterative process. Typically, in a large deflection analysis, the change in displacements between equilibrium iterations will decrease as the structure converges to a stable configuration. If the structure is loaded beyond its stability limit, however, Δu will increase from iteration to iteration; that is, the solution diverges. The limit (buckling) load is the load level at which the solution begins to diverge.

Another application for nonlinear buckling analysis is the *snap-through analysis.* Many types of structures will reach a second stable state after buckling if the load continues to increase (Fig. 58.7). One example is a shallow arch pinned at each end with a downward load applied at its apex. The arch will begin to deflect downward as the force increases until it reaches its buckling point and can no longer resist the applied load. It will then snap through, inverting its shape, and begin to resist the load once more.

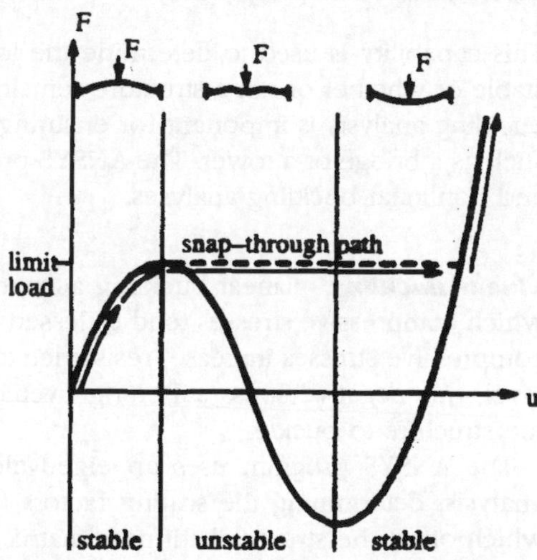

FIGURE 58.7 Nonlinear buckling analyses, such as the snap-through behavior of a shallow arch under displacement loading, can be done with the large deflection capabilities of the ANSYS program.

Structural Nonlinearities

Structural nonlinearities cause the response of a structure or component to vary disproportionately with applied forces. In reality, all structures are nonlinear in nature. But nonlinearities do not always exist in a structure to the degree that they have a significant effect on an analysis.

The ANSYS program solves both static and transient nonlinear problems. Static analyses are performed by subdividing the load into a series of incremental load steps and performing a succession of linear approximations (known as equilibrium iterations) at each step to obtain equilibrium. Similarly, nonlinear transient analyses are broken into a series of time-varying load increments with equilibrium iterations at each step. Transient analyses can also include the integration over time of inertial effects.

Using the Newton–Raphson method (Fig. 58.8), a series of linear approximations converges to the actual nonlinear solution based on the following equation:

$$[K]_{i-1}\{\Delta u\}_i = \{F\} - \{F^{el}\}_{i-1} \quad (9)$$

where

$[K^T]_{i-1}$ is the tangent stiffness matrix based on the deformed geometry from the $(i-1)$ iteration

$\{\Delta u\}_1$ is the incremental displacement vector, $\{\Delta u\}_i = \{u\}_i - \{u\}_{i-1}$

$\{u\}_i$ is the displacement vector at the current iteration

$\{F^A\}$ is the applied force vector

$\{F^{NR}\}_{i-1}$ is the Newton–Raphson restoring load based on displacement for iteration $(i-1)$

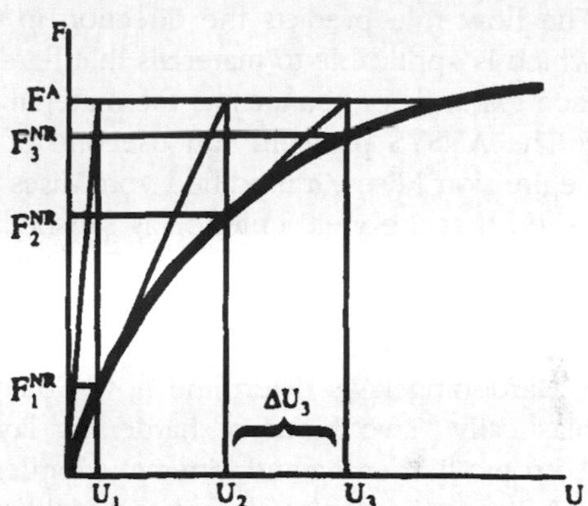

FIGURE 58.8 The ANSYS program uses the Newton–Raphson method to calculate nonlinear analysis results. As indicated by the force deflection (F–U) curve shown here, the solution is approximated first, then refined in a linear, iterative scheme until nonlinear convergence is reached.

The user can control both the subdivision of the load and the maximum number of equilibrium iterations at each substep. Equilibrium iterations will continue until convergence is achieved or the maximum iteration limit is reached.

In many nonlinear static analyses, the loading must be applied in increments, or ramped, to obtain an accurate solution. The ANSYS program features an automatic load-stepping capability, which will stagger the load order in increments to obtain accurate and convergent solutions.

In nonlinear transient analyses, the dynamic equilibrium equations are solved through the Newmark time integration method. The ANSYS program features an automatic time-stepping capability, which, depending on the response frequency and the degree of nonlinearities, increases or decreases the integration time step.

In addition to automatic load- and time-stepping, the program provides other convergence enhancement features, such as prediction, bisection, and adaptive descent. Prediction activates a linear predictor on the degree of freedom solution at the beginning of each substep, while bisection and adaptive descent cause a solution to back up and restart if the solution is detected to be proceeding inaccurately.

For both static and transient analyses, the ANSYS program can represent many different types of nonlinear effects, whether material, geometric, or element-related.

Material Nonlinearities. A material nonlinearity exists when stress is not proportional to strain. The ANSYS program can simulate various types of nonlinear material behavior. Plasticity, multilinear elasticity, and hyperelasticity are characterized by a nonlinear stress–strain relationship. Viscoplasticity, creep, swelling, and viscoelasticity are behaviors in which strain may depend on other factors, such as time, temperature, neutron flux levels, or stress.

To fully account for plastic material behavior in an analysis, the program considers the yield criterion, the flow rule, and the hardening law. The yield criterion measures the three-dimensional stress state by computing a single-valued equivalent stress, which is compared against the yield strength to determine when the material will yield. The flow rule predicts the direction in which strain will occur. The hardening law, which is applicable to materials that harden under strain, describes how the yield surface expands or changes as the material strains.

The ANSYS program can use one of three criteria to predict when yielding will begin: von Mises, a modified von Mises (Hill), and Drucker–Prager. Once it is established that the yield criterion is satisfied, the flow rule determines the direction and magnitude of the plastic strain. The flow rule is associative for all yield criteria in the ANSYS program except for Drucker–Prager, in which the flow rule can be associative or nonassociative.

Hardening laws determine how a material yield surface is changed as it deforms plastically. Two kinds of hardening laws are represented in the ANSYS program: isotropic hardening and kinematic hardening.

Isotropic hardening describes a yield surface that expands the same in all directions and implies that an increase in tensile yield strength due to hardening results in an equal increase in compressive yield strength.

Kinematic hardening predicts an increase in tensile yield strength and produces a corresponding decrease in compressive yield strength known as the Bauschinger effect.

A particular combination of yield criterion, flow rule, and hardening law describes a unique plasticity behavior in the ANSYS program, which can model these behaviors:

- *Classical bilinear kinematic hardening* describes general metallic materials that are considered to be bilinear, having one elastic and one plastic slope. It is applicable to most common, initially isotropic engineering metals in the small strain region.
- *Multilinear kinematic hardening* also describes metallic materials, but is more applicable for materials having stress–strain curves with more than two slopes. It uses the overlay or Besseling model to characterize complex multilinear behavior by combining simple stress–strain responses.
- *Bilinear isotropic hardening* describes general metallic materials that are considered to be bilinear. It is applicable to isotropic materials and is preferable to kinematic hardening at higher strains.
- *Multilinear isotropic hardening* describes general strain-hardening materials, especially in conjunction with large strain.
- *Anisotropic behavior* describes materials that behave differently in tension and compression or that have different behaviors in different directions. By applying isotropic hardening to anisotropic material, this option can represent the effects of work hardening.
- *Drucker–Prager* describes granular materials such as rock, concrete, or soil. It simulates the increase in yield strength that is produced by an increase in confinement pressure (hydrostatic stress).
- *The Anand model* describes the behavior of metals at elevated temperature, although it may also apply at lower temperatures. It is an isotropic, rate-dependent, strain-hardening model with input through material parameters rather than in the form of stress–strain curves.

- *User-defined models* may also be incorporated to define virtually any nonlinear material behavior. The user-programmed FORTRAN subroutine is linked with the ANSYS program and is accessed in a manner similar to the other plasticity options.

In addition to these plasticity behaviors, the ANSYS program offers several other specialized material behaviors.

- *Multilinear elasticity* is a conservative type of nonlinear stress–strain relationship in which all strains are recovered after the load is removed.
- *Hyperelasticity* represents the large strain behavior of very nearly incompressible and rubber-like materials. Elastic or rubber-like materials may be modeled using a nine-term Mooney–Rivlin function. A Blatz–Ko function is available for compressible foam-type polyurethane rubber materials. A user-programmable feature allows for customized material functions.
- *Creep*, a time-dependent, stress–strain relationship, accounts for additional nonlinear strain under a constant load or reduced stress under a constant displacement (stress relaxation). Libraries of creep–strain equations are built into the ANSYS program for primary, secondary, and irradiation-induced creep. User-defined functions are also available.
- *Viscoplasticity* is a combination of plasticity and creep. Primary applications include metal-forming processes, such as rolling and deep drawing, which involve large plastic strains and displacements with small elastic strains.
- *Swelling* is another time-dependent, stress–strain relationship that can be simulated in the ANSYS program. The swelling option represents the three-dimensional expansion of a material resulting from neutron bombardment and is useful in the analysis of materials exposed to highly radioactive environments. This option can also be used to account for other volumetric changes, such as the sintering of powdered metal or the curing of concrete.
- *Viscoelasticity* is an elastic time-dependent stress–strain relationship that characterizes viscously flowing materials, such as heated gas. The material behavior is represented by a series of Maxwell models that allow for both shear modulus and bulk modulus relaxation with respect to time and temperature.

Geometric Nonlinearities. Geometric nonlinearities occur when the displacements of a structure significantly change its stiffness. The ANSYS program accounts for several types of geometric nonlinear effects.

- *Large strain* geometric nonlinearities account for the large localized deformations that occur as a structure deflects. The program accounts for large strain by adjusting element shapes to reflect the changing geometry.
- *Large deflection* represents a change in global structural stiffness resulting from a change in element spatial orientation as the structure deflects (Fig. 58.9). The program accounts for large deflection by updating the element orientations as the structure deflects. Another program capability for large deflection analysis is the simulation of follower loads, which always act normal to the structure's elements and are described as element pressures.
- *Stress stiffening* (also known as geometric, initial stress, incremental, or differential stiffening) accounts for changes in structural stiffness due to the stress state. It represents the coupling between the in-plane and transverse deflections in a structure. The program's stress stiffening analysis option can be used for any structure, but is most appropriate for structures that are weak in bending resistance, such as pres-

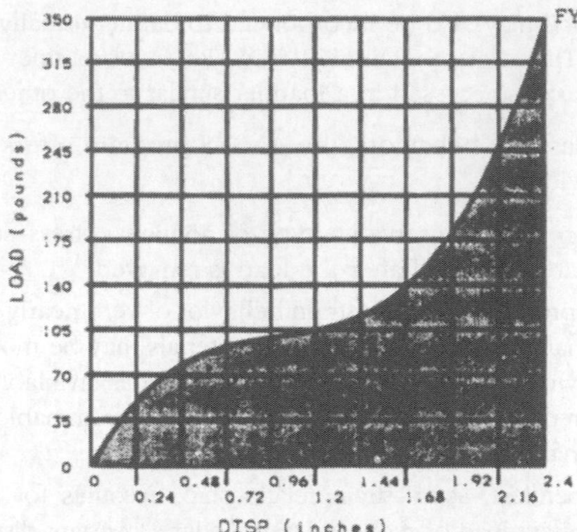

FIGURE 58.9 A nonlinear, large deflection analysis is required to determine this force-deflection curve of a two-strut shallow arch.

surized membranes or turbine blades rotating at high speed. The program can also determine stress stiffening effects in what are otherwise linear problems.

- *Spin softening* in rotating bodies models a decrease in stiffness due to the deflections of a body, such as a turbine blade, in the plane of rotation. This capability is usually used together with stress stiffening in analyses of spinning bodies.

Element Nonlinearities. Nonlinear elements exhibit their own nonlinear behavior independent from other elements. This behavior is typically characterized by an abrupt change in stiffness due to a change in status (such as a contact surface element changing from open to closed). Element nonlinearities provide various capabilities that are not normally possible with global nonlinearities. The ANSYS element library includes several nonlinear elements.

- *General contact surface elements* are general surface-to-surface contact elements that can include significant sliding and transmission of loads between surfaces. Elastic or rigid coulomb friction may be specified between surfaces. The element may be closed and sliding, closed and sticking, or open.
- *Interface elements* represent point-to-point contact with limited sliding or point-to-ground contact with significant sliding. The element may be closed and sliding, closed and sticking, or open.
- The *reinforced solid element* represents concrete, rock, or composites with up to three different sets of directional reinforcing material. The solid portion of the element is capable of crushing, cracking, deforming plastically, and creeping, while the reinforcing materials in the element can include plastic deformation and creep behavior.
- The *nonlinear damper element* represents a longitudinal or torsional spring-damper with nonlinear damping response. The element's nonlinearity is a continuous function that is evaluated at each iteration.
- The *nonlinear spring element* represents a varying-stiffness rate spring with conservative or nonconservative response. The element's force deflection curve is specified by the user with up to 40 linear segments.

- The *tension-only/compression-only spar element* is a bilinear element used to represent a cable (tension-only) or a gap (compression-only). The element can be tensioned or slacked for the cable option or compressed or opened for the gap option.
- The *shell with wrinkle option element* represents a membrane shell that collapses or wrinkles under compression. The element can be tensioned in both directions, collapsed in one direction, or collapsed in both directions.
- The *combination element* is a single element that has combined mass, damping, gap, spring, and slider effects. This element has a lock-up option, which prevents the gap from opening once the gap has closed; and a breakaway option that prevents the interface from closing once it has opened.
- The *control element* is a powerful element with mass, damper, and slider effects. It is used to remotely control portions of a structure under predetermined conditions through binary (on/off) controls or controls defined by quadratic functions. The element can represent mechanical snubbers, friction clutches, thermostats, relief valves, electrical switches, etc.

An additional nonlinear element capability of the ANSYS program is the element birth and death option, allowing the user to activate or deactivate the contribution of an element to the matrices during the solutions phase. It can be used to simulate the addition or removal of material (e.g., excavation and fabrication), the interaction of moving parts (e.g., chain and sprocket interaction), or any application in which an element's contribution to the solution depends on its location.

Static and Dynamic Kinematic Analysis

Kinematics is the branch of mechanics that deals with motion in the abstract without reference to force or mass. Kinematic motion can be described as either rigid-body, which assumes that the flexibility of moving structural members will have a negligible effect on the solution, or flexible-body, which accounts for local deformations that occur in a structure as it moves. The flexible-body approach to kinematic motion is more appropriate for real-world applications (Fig. 58.10).

The ANSYS program can analyze large, three-dimensional motions of flexible bodies as part of its large deflection analysis capabilities. This feature is useful when the cumulative effects of motion play a critical role.

ANSYS elements have been formulated to allow unlimited spatial motion in both two-dimensional and three-dimensional space. This capability combined with the Newton–Raphson solution method allows the ANSYS program to track element movement and accurately determine motion occurring throughout a model.

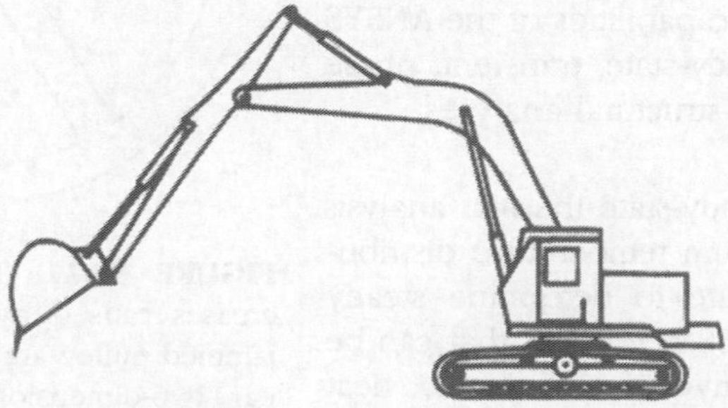

FIGURE 58.10 The ANSYS flexible-body kinematics capabilities can analyze complicated, linked structures, such as this model of an excavator.

A specific element in the ANSYS program, called the three-dimensional revolute joint element, was developed exclusively for analyses involving three-dimensional flexible-body kinematics. This element represents a hinge or pin joint and is used to connect two parts of a model. It can represent a variety of effects, including joint flexibility or stiffness, friction, damping, and certain control features. But the revolute joint element's most important attribute is the ability of its axis to translate and rotate as the linkage moves.

The user can determine the behavior of the joint's movement by inputting specifications for friction torque, preload torque, rotational viscous friction, interference rotation, joint flexibility, two differential rotation limits or "stops," and feedback control instructions. For example, the user can specify the behavior of an element if the upper or lower rotational stop is reached: should it lock in place or bounce off the stop?

Another element that is useful in kinematic analyses is the linear actuator element, which was developed to model linkage members that both rotate and experience change in length, such as an hydraulic cylinder.

All of these features enable the ANSYS program to realistically evaluate the dynamics of complex motions through space and determine the resulting stresses, strains, and deflections that occur in the structure.

Thermal Analysis

The ANSYS program can analyze three different methods of head transfer: conduction, convection (both free and forced), and radiation. Heat transfer can be accounted for through steady-state or transient, linear, or nonlinear analysis. The governing equation for heat transfer in finite element system is

$$[C]\{\bar{T}\} + [K]\{T\} = Q \tag{10}$$

where

[*C*] is the specific heat matrix
{*T*} is the time derivative of the nodal temperature
[*K*] is the effective thermal conductivity matrix
{*T*} is the nodal temperature vector
{*Q*} is the effective nodal heat flow rate vector

The thermal analysis capabilities of the ANSYS program include steady-state, transient, phase change, and thermal–structural analyses.

Steady-State. A steady-state thermal analysis predicts the equilibrium temperature distribution within a structure to determine steady heat flow rates (Fig. 58.11). Loads that can be applied include convection surfaces, heat fluxes, heat flow rates, heat generation rates, and specified temperatures. The analysis can be either linear or nonlinear.

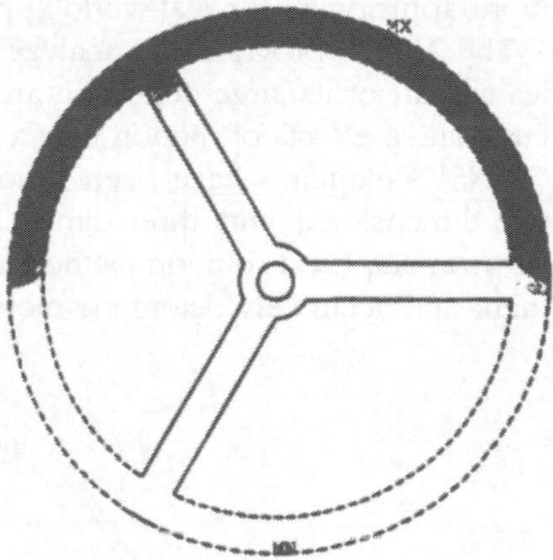

FIGURE 58.11 The temperature increases caused by belt slippage on a jammed pulley are calculated in this linear two-dimensional thermal analysis. The darkest portion of the temperature contour indicates the region of maximum temperature.

A linear steady-state heat transfer analysis does not consider thermal mass (specific heat) effects or temperature-dependent material properties. This method of analysis is appropriate for conduction and linear convection heat transfer.

In a nonlinear steady-state heat transfer analysis, time-dependent (thermal mass) effects are not considered. However, material properties, including convection film coefficients, may vary with changes in temperature, and radiation effects may be present.

Radiation can be defined in one of three ways for thermal analyses in the ANSYS program. The radiation link element represents radiation between two points. Surface effect elements are useful for describing radiation between a surface and a point. A radiation matrix generator is also available for analyses involving several surfaces, both hidden and partially hidden, that are receiving and emitting radiation. In general, the heat flow rate for radiation is a nonlinear process.

For both linear and nonlinear steady-state heat transfer analyses, the solution data are in the form of nodal temperatures and heat flow rates. These data are used in the postprocessing phase to produce displays of temperature contours, called isotherms, throughout the model. Other display options are available for more specific information, such as the thermal gradient and thermal flux at nodes and element centroids, and the heat flow rate across convection surfaces.

Transient. A transient thermal analysis is used to determine the temperature distribution in a structure as a function of time or to predict the rates of heat transfer or heat storage in a system. This type of analysis can be either linear or nonlinear. Loads and nonlinearities are the same as those described for a steady-state analysis. Specific heat, which is input as a material property, is used to account for heat storage effects.

A transient thermal analysis integrates the governing equation with respect to time through the Crank–Nicholson/Euler theta integration method, in which the equation is solved at discrete time points within the transient. The difference between any two time points is known as the integration time step, which is specified by the user and can be automatically controlled with the time-stepping feature.

Once the program completes the analysis, isotherms can be viewed as well as temperature-versus-time graphs and other data output obtained for specific points in the model.

Phase Change. A phase change analysis is merely a special case of a transient thermal analysis that accounts for the solidification or melting of a material in the heat transfer process. It is useful in numerous applications, such as continuous metal-casting or solar storage systems.

The energy released or absorbed when the phase change occurs (latent heat) must be accounted for in a phase change analysis. The ANSYS program accomplishes this by defining the enthalpy of the material as a function of temperature.

The results of the analysis can be viewed on a "solid-liquid" contour display that can be created by narrowing the displayed temperature range to that of the phase change region. The ANSYS animation capability allows the user to view a series of these contours at different times to visualize phase change propagation through the model.

Thermal-Structural. The program's thermal–structural analysis capabilities allow the input of the results of a thermal analysis into a structural analysis, which is useful for determining the effects of temperature distributions on the structural response of a model. A temperature load can be applied by itself or in conjunction with other mechanical loads.

A thermal–structural analysis can be performed through two methods. A thermal analysis can be done first, and the results transferred into a structural analysis. Or both types of analyses can be completed simultaneously by using the ANSYS programs coupled-field elements. With the simultaneous solution process, it is possible to couple complex heat transfer and structural problems, such as transient thermal and nonlinear dynamic analysis.

Magnetic Field Analysis

The ANSYS program's magnetic capabilities can be used to analyze the different aspects of magnetic fields, including inductance, flux density, flux lines, forces, power loss, and other related phenomena. Capabilities such as these are useful in analyzing devices ranging from solenoids, actuators, and motors to permanent magnet devices and transformers.

Magnetic analysis can be performed by the ANSYS program on either static or time-varying two-dimensional planar, axisymmetric, and three-dimensional magnetic fields. ANSYS finite element formulations for magnetic analyses are derived from Maxwell's equations for magnetic fields through the introduction of scalar or vector potential and consideration of their constitutive relationships.

The user can choose either CGS or MKS units for magnetic analysis as well as the Jacobian Conjugate Gradient solver, which can be very useful for magnetic field analysis because it provides a faster solution to potential field problems having symmetric, sparse, positive, definite matrices.

Two-dimensional and three-dimensional infinite boundary matrices eliminate the need to model large portions of the infinite medium surrounding the electromagnetic device (e.g., air), resulting in smaller models and less demand on computer resources. Virtual work force calculations are available for all magnetic elements.

A major advantage of using the full ANSYS program for magnetic analysis is the ability to conduct coupled-field analysis, in which the field loads of a magnetic analysis can be automatically coupled to structural and thermal elements.

Static Magnetic Fields. Static magnetic field analysis can be performed in two or three dimensions for linear and nonlinear analyses (Fig. 58.12). Current conductors and permanent magnets can be modeled as sources. Conductors can be modeled with elements or described by bar, arc, or coil primitives, or electromagnetic field coupling. Other items that can be modeled include saturable irons and nonmagnetic materials.

The ANSYS program provides for a variety of linear and nonlinear magnetic material representations, including isotropic or othrotropic linear permeability, material B-H curves, and permanent magnet demagnetization curves. The program can be used to view flux lines, flux density, and field intensity as well as to perform calculations for force, torque, source input energy, terminal voltage, and other parameters.

Time-Varying Magnetic Fields. Two types of time-varying magnetic field analyses are available in the ANSYS program: alternating current (AC) (or harmonic) and transient. AC, or harmonic, magnetic field analysis solves for complex vector potentials, flux density, and field intensity, while transient magnetic field analysis solves for time-varying (real) vector potentials, flux density, and field intensity. Both types of analyses can be performed for two-dimensional planar and axisymmetric or three-dimensional analyses.

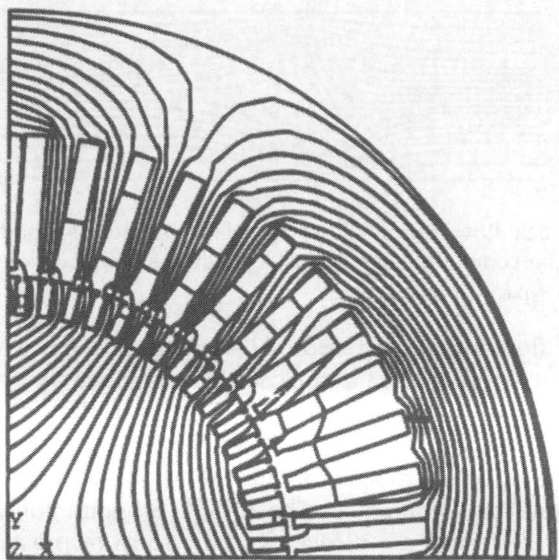

FIGURE 58.12 The ANSYS program is used to calculate magnetic flux lines in a typical induction motor.

Time-varying analysis is used to calculate the effects of eddy currents in a system. By using the coupled-field capabilities of the ANSYS program, the user can allow for skin effects in a two-dimensional, time-varying analysis.

Once an analysis is complete, the program allows the user to calculate source impedance, power loss, eddy currents, eddy current density, stored magnetic energy, magnetic forces, and other field effects, either for each element or for the total system. In a transient analysis, these calculations can be made at any point in time.

Fluid Flow and Acoustic Analysis

The ANSYS program's fluid capabilities allow for the study of flow or pressure wave characteristics of a liquid or gas within a given system. Several types of analyses are available.

FLOTRAN®. The complete integration of the FLOTRAN program, a general-purpose computational fluid dynamics (CFD) package utilizing the finite element method, is the basis for the ANSYS CFD capabilities.

A CFD analysis is conducted to determine the flow characteristics of a fluid, such as pressure drop, velocity distribution, direction of flow, lift and draft forces, and heating or cooling effects (Fig. 58.13). Loadings that can be prescribed in a CFD analysis include velocities, pressures, temperatures, turbulent kinetic energy, turbulent dissipation rate, convection, heat fluxes, and heat generation rates. With FLOTRAN integration, the ANSYS program can be used to solve fluid flow problems involving laminar or turbulent flow, incompressible or compressible fluids, forced, free, or mixed convection, heat conduction, conjugate heat transfer, and distributed flow resistance through porous media.

In addition to the fluid flow capabilities provided by FLOTRAN, the ANSYS program can be used for pipe flow, potential flow, seepage, free convection or mass transport flow, and acoustics analysis. Another option is structural analysis of contained fluid displacement.

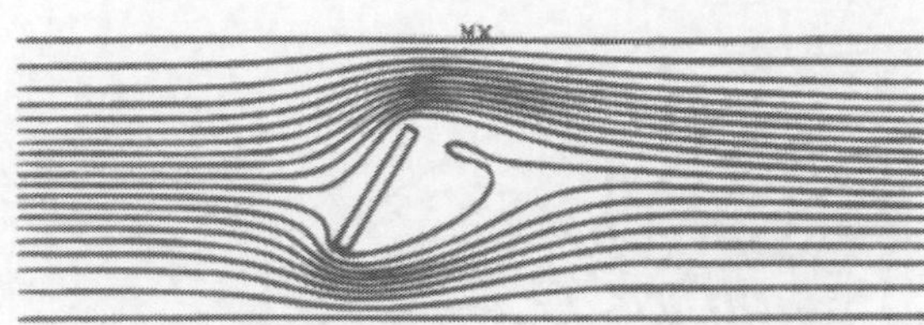

FIGURE 58.13 Particle flow tracing allows the trajectory path of a particle or many particles to be computed and displayed, as in this coupled fluid-structural analysis of flow past a hinged plate.

Pipe Flow. Pipe flow analysis is used to determine pressures, velocities, and heat-exchange characteristics of a fluid in a closed system, such as a water supply piping network. It is applicable for any system with a constant flow rate of an incompressible fluid.

Pipe flow problems are nonlinear because the conductivity matrix changes with variations in the pressure differential. The ANSYS program solves for flow rate and temperature gradient. Results are displayed in the form of pressures, flow rates, and temperature distribution.

By using the thermal-fluid pipe element with the three-dimensional surface effect element, the user can simulate a fluid mass flow around the exterior of a structure, such as air passing over a rotating turbine blade, and determine convection heat transfer effects.

Potential Flow. Potential flow analysis is used to determine the velocity, pressure, and streamline patterns of incompressible fluid in a system. It is useful for predicting the primary flow characteristics of an ideal or inviscid fluid at a steady flow rate, such as air passing through a ventilation duct.

Results of this type of analysis, which can be performed in two or three dimensions, can be displayed as a streamline pattern, velocity contour, velocity head, or pressure field, allowing for the visualization of fluid behavior as it flows over or around an obstruction.

Seepage. The ANSYS program can also analyze the flow of liquid or gas through a porous, permeable medium. This capability is useful for applications such as seepage of water through brick, sandstone, or filters.

After a solution is obtained, results can be viewed in terms of overall pressure gradient, component pressure gradients in global directions, mass flow rate versus time, fluid velocity magnitude, and component fluid velocities in global directions.

Acoustics. ANSYS acoustic capabilities are used to study the propagation of sound pressure waves in a contained fluid medium or to analyze the dynamics of a structure submerged in fluid. Typical acoustic analyses using the ANSYS program include a determination of the frequency response of an audio speaker, a study of sound distribution in a concert hall, and an evaluation of the damping effects of water on a vibrating ship hull.

Acoustic analysis is possible because of specialized two- and three-dimensional fluid elements that were designed specifically for this purpose. These elements are used to represent the fluid medium and the fluid–structure interface in the finite element model, which assumes small changes in density.

Results of the analysis are displayed as pressure contours or structural deflections.

Coupled-Field Analysis

In designing components under the influence of thermal, structural, fluid, electrical, or magnetic fields, there is often a need to examine the coupled influence of these fields. Examples of the usefulness of a coupled-field analysis capability are a current-carrying conductor with temperature-dependent resistivity, and a transmission line, on which a determination of eddy currents and skin effects of coupled electromagnetic fields is desirable, or a slot-embedded conductor in an electric apparatus, for which an analysis of eddy currents and skin effects in coupled electromagnetic fields is necessary.

In the ANSYS program, field coupling can be achieved either directly, through the use of specialized coupled-field elements in a single analysis, or indirectly, through sequential field analyses.

Multiple degrees of freedom (spanning several fields) at each node in the direct method allow for continual cross-communication between the analytical disciplines involved. Sequential analysis is not necessary because the coupling is built into the governing equations, through the element matrices or the element load vectors. An application requiring the direct method is a current-carrying conductor with temperature-dependent resistivity.

In the indirect method, the results of the first field analysis are applied as loads for the second analysis through a single ANSYS command. This method is preferable and more efficient in many cases, such as a thermal structural analysis in which the thermal problem is nonlinear and transient, while the stress analysis is static.

The ANSYS program provides coupled-field analysis for the following types of interaction:

- Thermal–stress
- Magnetic–thermal
- Magnetic–structural
- Fluid flow–thermal
- Fluid flow–structural
- Electromagnetic
- Piezoelectric coupling

Most of these interactions can be modeled by either the direct or indirect coupling methods, except piezoelectric and electromagnetic skin-effect analyses, for which the direct coupling method must be used.

Piezoelectric Analysis

The ANSYS program's piezoelectric capabilities are used to analyze the response of a three-dimensional structure to an AC, DC, or arbitrary time-varying electrical or mechanical load. This capability is useful for analyzing components such as transducers, oscillators, resonators, microphones, and other electromechanical devices.

Four types of analysis are available for determining piezoelectric response:

- *Static analysis* for determining deflection, potential, electric field, electric flux density, and stress distribution.
- *Modal analysis* for determining natural frequencies and mode shapes.
- *Harmonic response analysis* for determining system response to harmonic loads (current, voltage, forces, etc.), including electrical admittance, impedance, electro-

mechanical couplings, deflections, electric field, electric flux density, and stress distribution.

- *Transient response analysis* for determining system response to arbitrary time-varying loads (current, voltage, forces, etc.), including electrical admittance, impedance, electromechanical couplings, deflections, electric field, electric flux density, and stress distribution.

A piezoelectric structure can be modeled using three ANSYS coupled-field solid elements, which allow for a variety of linear material property data input, such as complete 6×3 piezoelectric constants, isotropic, orthotropic, or anisotropic elastic stiffness or compliance constants, and diagonal real dielectric constants.

Using the Program

Completing an engineering analysis with the ANSYS program is a three-phase process, comprising preprocessing, solution, and postprocessing, with distinct functions within each phase.

Preprocessing

The preprocessing phase of the ANSYS program is the point at which data needed to perform the analysis are input by the user. Activities driven by the user in the preprocessing phase include the selection of coordinate systems and element types, the definition of real constants, material properties, and coupling and constraint equations, the creation and meshing of solid models, and the manipulation of nodes and elements.

A run statistics module can calculate the expected wavefront, which is related to the number of elements and their connectivity. The wavefront is an indication of solution efficiency, file sizes, and memory requirements for solutions during preprocessing. (The ANSYS program automatically conditions the matrices to reduce the wavefront and provide shorter solution times in the solution phase.)

The program uses coordinate systems to locate geometry in space, to specify degree of freedom directions at nodes, to define material property directions, and to change graphic displays and listings. Available coordinate systems, all of which can be located anywhere in space and in any orientation, include Cartesian, cylindrical, spherical, elliptical, and toroidal types.

All data input during the preprocessing phase becomes part of the centralized ANSYS database, which is organized in tables of coordinate systems, element types, material properties, keypoints, nodes, loads, etc. The user has the ability to define these tables and activate them later.

Model data are selected on, among others, the following criteria: geometric locations, solid modeling entities, element types, material types, and node and element numbers.

The user can also conveniently select model data by dividing the model into components, which are groups of geometric entities defined by the user for clarity or logical organization (Fig. 58.14). These components can be displayed in different colors to clearly differentiate them in a complex model. Models can be generated directly, built with the solid modeling technique or created with a combination of the two.

Solid Modeling. This ANSYS capability enables the user to work directly with the geometry of the model without concern for the specific entities (nodes and elements) of the finite element model. The program separates the definition of geometry and

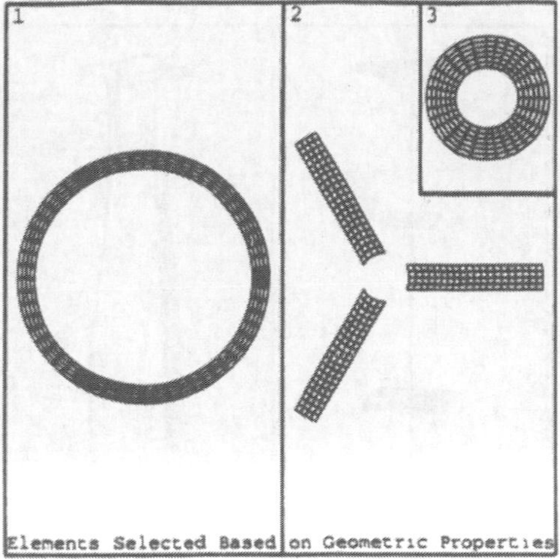

FIGURE 58.14 Portions of the database can be selected to construct different segments of a model as indicated with the separate displays of these pulley components.

boundary conditions from the creation of the finite element mesh to facilitate model generation. This also allows for the importing of models produced on a CAD system through the use of an Initial Graphics Exchange Standard (IGES) file. The ANSYS program automatically meshes the model after the geometry and boundary conditions are described by the user. Solid modeling can be performed in either a top-down or bottom-up manner.

In top-down modeling the user specifies only the highest order entities of a model. This method involves the use of geometric primitives, commonly used solid modeling shapes, such as spheres and prisms, which can be created with a single command. Thus, a user can define a volume primitive, and the program automatically defines associated areas, lines, and keypoints.

In bottom-up modeling, the user builds the model from the lowest order entity up, defining keypoints, associated lines, areas, and volumes.

Both solid modeling methods can be freely combined in any model. Regardless of which solid modeling technique is employed, the user can use Boolean algebraic operations, such as intersection, subtraction, and union, to combine sets of data that "sculpt" a solid model. This capability can save considerable amounts of time and effort in building complex solid models.

Because the solid model representation is NURBS-based in the ANSYS program, users can take advantage of a surface construction technique known as "skinning" or "lofting" (Fig. 58.15). With this technique, the user can define a set of cross sections in the model, including curves, and instruct the program to automatically generate a surface that fits through those cross-sections. This skinning technique can help facilitate modeling of complex shapes.

The ANSYS interactive graphics-based menu system provides useful model generation tools, such as mouse picking, rubber banding, and working planes. With a mouse or other pointing device, the user can define or retrieve portions of a model by selecting their locations at a point defined by cursor position on the display screen. With a working plane, the user can quickly and easily locate or select two-dimensional model entities from a three-dimensional model (Fig. 58.16).

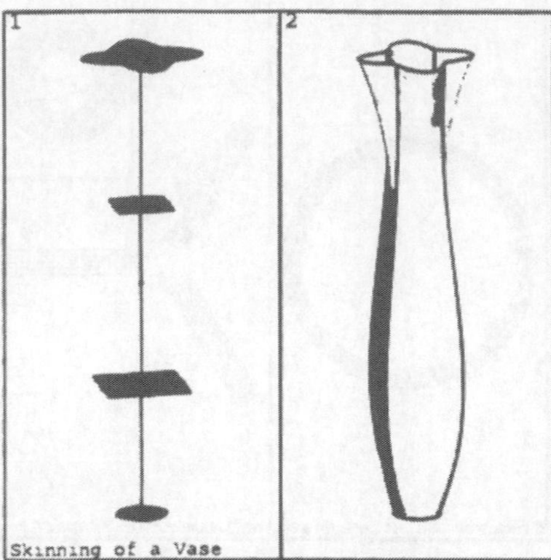

FIGURE 58.15 The ANSYS solid modeler makes complicated shapes, like this vase, easy to model and mesh through skinning, generating the exterior surface onto defined cross sections.

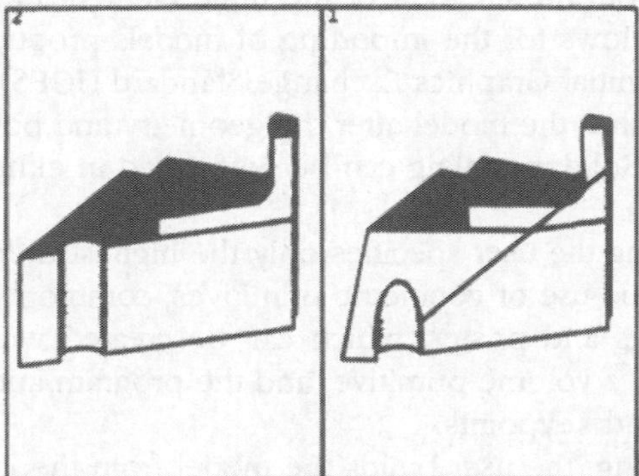

FIGURE 58.16 ANSYS graphics allow sophisticated manipulation of model data. The display on the left shows a 1/8-symmetry bearing mount model. The use of a clipping plane, the results of which are shown in the clipped display on the right, enable the user to view stress contours and other load effects within a solid three-dimensional model.

Once the solid model is created, the finite element model or mesh, comprising nodes and elements, is generated through mapped, free, or adaptive meshing.

- *Mapped meshing* requires the user to select appropriate element attributes and meshing controls to create a mesh of only quadrilateral or brick elements.
- *Free meshing* does not restrict the solid model to any special meshing requirements. In this case, ANSYS meshing algorithms automatically generate a best-fit pattern of nodes and elements. For two-dimensional models, free meshing can utilize a combination of quadrilateral and triangular elements or all triangular elements. For three-dimensional models, all tetrahedral elements must be used.
- *Adaptive meshing* involves instructing the program to automatically generate a mesh following the creation of a solid model with boundary conditions. This program will

perform the analysis on the mesh, evaluate the mesh discretization error, and resize the mesh through a series of solutions until the measured error drops below a user-defined value (or until a user-defined limit on the number of solutions is reached).

The ANSYS program permits the user to modify a meshed solid model, changing nodes and element attributes. For models with repetitive features, the user can model and mesh a pattern region of the model and then generate copies of that meshed region to create a whole.

Once the solid model has been meshed, the program automatically provides solid model cross-reference checking to ensure the validity of any modifications made by the user to the meshed model. Such checking prevents the user from incorrectly deleting or otherwise contaminating solid model and finite element model data.

Direct Generation of Models. With this method, models are defined in ANSYS preprocessing by specifying the location of every node and the size, shape, and connectivity of every element. The user can conveniently copy, select, and scale a given pattern of nodes or elements to facilitate this process.

Nodes are used to locate elements in space and do not have to be numbered consecutively or in sequence relative to their geometric position. Elements define the connectivity of the model.

Direct generation of nodes and elements is appropriate for line models and small models with regular geometry. For large, complex models, the ANSYS program's solid modeling approach is preferable.

Solution

After the model has been built in the preprocessing phase, or imported via ANSYS interfaces with leading computer-assisted design (CAD) programs, the solution or results of the analysis are obtained in the solution phase. It is at this point that the user specifies analysis type, analysis options, load data, and load step options, and then initiates the finite element solution.

Selecting the analysis type—structural, thermal, magnetic field, electric field, fluid, and coupled-field—activates the appropriate governing equations for solving the problem. Each analysis category includes additional options, such as static or dynamic. The user can select more options, such as Newton–Raphson options, to solve nonlinear equations, to further define the specific analysis to be run.

Specified load data and constraints, including degree of freedom constraints, point loads, surface loads, body loads, and inertia loads, define the boundary conditions on the finite element model. Each configuration of loads is called a load step. Any analysis may consist of one or more load steps.

The user can use additional solution phase features to change material properties and real constants, reactivate and deactivate elements (birth and death option), specify master degrees of freedom (MDOFs), and define gap conditions.

Once these specifications are made and the solution is executed, the program will solve the governing equations and compute the results for the selected analysis. This is the computationally intensive part of an ANSYS analysis and requires no user interaction.

All ANSYS analyses are based on classical engineering concepts that can be formulated into matrix equations, through proven numerical techniques, for completing an analysis using the finite element method. The model to be analyzed is represented by a mathematical model, consisting of discrete regions (elements) connected at a finite number of points (nodes).

The primary unknowns in an analysis are the degrees of freedom for each node in the finite element model. Degrees of freedom may include displacements, rotations, temperatures, pressures, voltages, or magnetic potentials, and are defined by the elements attached to the node. Corresponding to the degrees of freedom, stiffness (or conductivity), mass, and damping (or specific heat) matrices are generated as appropriate for each element in the model. These matrices are then assembled to form sets of simultaneous equations that can be processed by the program's solver.

Postprocessing

The postprocessing phase of the ANSYS program displays the results of the analysis—displacements, temperatures, stresses, strains, velocities, heat flows, etc.—that was set up in the preprocessing phase and computed in the solution phase. Solution results can be output as graphic displays and/or tabular reports.

The amount and type of data available are controlled by the type of analysis that was performed and the options that were defined by the user in the solution phase. Results can be viewed graphically through either the general postprocessor or a time-history results processor.

The *general postprocessor* can be used to view the results of any type of analysis. These data can be selected, sorted, algebraically manipulated, combined with data from another analysis, listed, or graphically displayed. A set of select commands allows portions of the data to be flagged for specific operations.

Once the desired postprocessing data have been obtained, they can be displayed in many graphic forms. Contour displays, which are depicted in the form of lines, color bands, or isosurfaces (surfaces of constant value in three dimensions), show how a result varies over the model. Other available graphic displays include a vector display, a path plot, and a particle flow trace.

The *time-history results postprocessor* enables an engineer to select items, such as nodal displacements, stresses, or reaction forces, and examine them over a period of time or substep history of an analysis. These results can be viewed as graph plots or tabular listings.

Both postprocessors allow the user to perform mathematical and algebraic calculations on results data.

Specialized Functions

The ANSYS program features several specialized functions that either make an analysis easier to perform, saving computer time and resources, or produce analysis results that are more meaningful and useful to design engineers and analysts.

Substructuring

The program's extensive substructuring capabilities can improve solution run times or increase modeling efficiency by reducing a group or set of elements to an equivalent, single, independent element. This technique is based on matrix condensation and is used to form a superelement. The stiffness (or conductivity), and, if necessary, the mass (or specific heat) and damping matrices are reduced to MDOFs, forming the superelement.

The user can generate superelements with any element type or combination of element types. The only restriction on this process is that superelements are assumed to be linear and are treated as such, even if nonlinear elements were used in their creation. Multiple superelements, as well as superelements within superelements, can be defined. Superelements are most commonly used to separate or isolate certain portions of a model from the rest of the structure or to simplify repetitive areas of a model. Substructuring can be used duplicate like areas of a model quickly, saving time and computer resources (Fig. 58.17).

Other beneficial uses for superelements include separating a linear portion of a structure from nonlinear portions, having several users independently model sections of a structure that are then brought together to form a model, modeling components that cannot vary by "fixing" their design, separating stress and displacement calculations in a single pass, and more efficiently solving flexible kinematic analyses because of the superelement's large rotation capability.

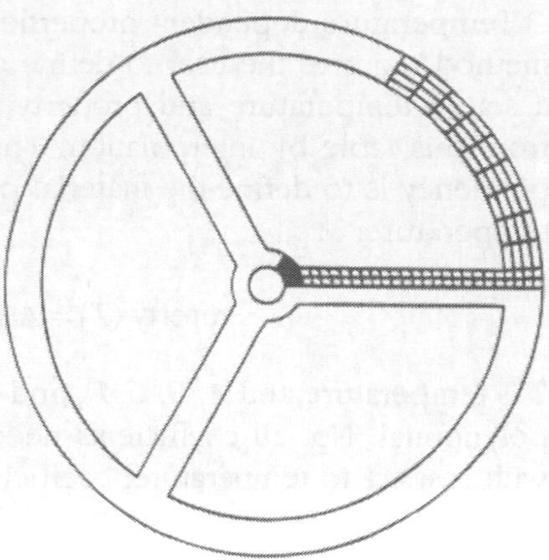

FIGURE 58.17 The use of substructuring for repeated geometry can significantly reduce modeling effort and computer time. A 60-degree sector of the pulley shown above was formed into a single-matrix superelement, then repeated to form the entire model.

Submodeling

The ANSYS program's submodeling capability is similar to substructuring except the subportion of the model is intentionally separated from the rest of the model for more detailed analysis. This modeling method can be more efficient because it allows for pinpoint analysis of portions of a model (Fig. 58.18). A preliminary analysis can be done first on the whole structure using a coarse element mesh. Then the user can analyze finely meshed submodels to obtain more accurate information about a particular area or areas of the structure without increasing the complexity of the entire model.

Additional advantages of using the program's submodeling capability include eliminating the need for complicated mesh transitions from fine to coarse regions in a model, studying the effect of local geometric changes of alternate designs without reanalysis of the entire model, reanalyzing areas of concern, such as high-stress regions, without prior knowledge of where these areas are located, eliminating the need to initially include small geometric details, such as holes, fillets, etc., because they are included later through submodeling, and creating solid element submodels from shell element coarse models.

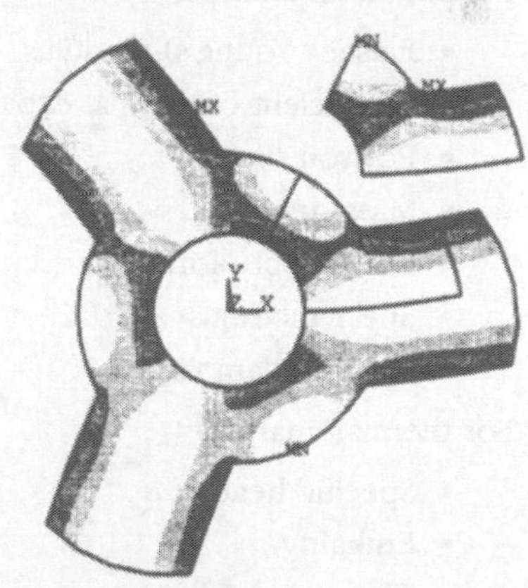

FIGURE 58.18 The relatively coarse finite element model of a pulley hub and spoke junction, shown on the left, is overlaid by a submodel. The results of an analysis of the same submodel, with more accurate result contours, is shown in the postprocessing display on the right.

Material Properties

Any material property can easily be defined in the ANSYS program as isotropic and constant with respect to temperature. Most material properties can also be defined as orthotropic and temperature-dependent.

Temperature-dependent properties are defined through one of two methods. One method requires the user to define a property-versus-temperature table by inputting a set of temperature and property data points. Property values are then obtained from this table by interpolation. The second method of specifying temperature dependency is to define the material property as a fourth order polynomial function of temperature:

$$\text{Property}\ (T) = A + B(T) = C(T)^2 + D(T)^3 + E(T)^4 \tag{11}$$

T is temperature and A, B, C, D, and E are input values representing coefficients of the polynomial. Not all coefficients need to be defined. For properties that are constant with respect to temperature, coefficients B through E are zero. If this method of data input is used, the property curve is converted by the program to a temperature table, similar to the one constructed directly in the first method.

With either method, the data can be written to a file to create a material property library, allowing the information to be used for future analyses or by other users.

Values for orthotropic materials are specified for the X, Y, and Z directions in the element or global coordinate system. If the property is defined in the X direction only, the Y and Z values default to the X direction values and represent isotropic materials. For some structural and piezoelectric materials, a special constitutive matrix input can be used to represent anisotropic behavior.

Composite materials can be modeled by means of special multilayer shell and solid elements, allowing for the stacking of isotropic or orthotropic material layers with varying thicknesses and material orientations.

The ANSYS program features the following linear material properties for each analysis type:

For structural analyses:

- Elastic (Young's) modulus
- Coefficient of thermal expansion and reference temperature
- Poisson's ratio
- Mass density
- Coefficient of friction
- Shear modulus
- Material damping

For thermal analyses:

- Specific heat
- Enthalpy
- Thermal conductivity
- Convection (film) coefficient
- Emissivity

For fluid analyses:

- Viscosity

For electric analyses:

- Resistivity
- Permittivity

For magnetic analyses:

- Material B-H curve
- Permanent magnet B-H curve
- Relative permeability
- Permanent magnet coercive force

For piezoelectric analyses:

- Piezoelectric matrix
- Elastic stiffness matrix
- Dielectric matrix

ANSYS Parametric Design Language (APDL)

In addition to the normal analysis procedure—defining the model and its loading, obtaining a solution, and interpreting results—using the finite element method, users can take steps to conserve resources by using ANSYS Parametric Design Language (APDL). Using the normal procedure would require the running of a whole analysis each time the design is altered.

APDL provides the ability to automate or "streamline" the analysis cycle by using program input to make decisions based on specified functions, variables, and selected analysis criteria (Fig. 58.19). APDL expands ANSYS capabilities beyond the realm of traditional finite element analysis and into more advanced operations, including sensitivity studies, parametric modeling from parts libraries, innovative design changes, and design optimization.

APDL gives the user the ability to create a highly sophisticated controlling scheme that will maximize the program's efficiency for a particular area of application. This control stems from several features within the APDL—parameters, array parameters, branching and looping, repeat functions and abbreviations, macros, and user routines—that can be used together or separately as desired.

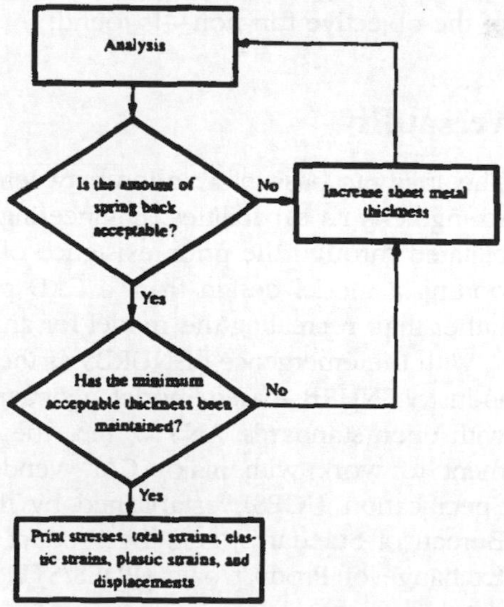

FIGURE 58.19 The design of a stamped sheet metal part might require decisions based on calculated quantities. An efficient analysis of the stamping process is possible with the ANSYS Parametric Design Language (APDL).

Design Optimization

The ANSYS program includes a design optimization capability, a computer technique that generates a series of finite element designs to improve upon the original design. The user defines the criteria and bounds of the design and sets up the model parametrically. The optimization routine then controls and executes the design cycle, determining which new values to supply for the parameters to be used in each trial design.

ANSYS design optimization is not limited to just cost or weight but allows the user to optimize virtually any aspect of the design, including shape, stress, natural frequencies, temperature, magnetic potential, or discrete quantities. The design optimization process is described in terms of design variables, state variables, and an objective function.

Design variables represent those input parameters of a design that are subject to change, such as length, radius, and fillet radius. Design variables are not only geometric parameters but may also be descriptors, such as materials, locations of load, or locations of constraints. The user specifies minimum and maximum limits for each design variable.

State variables are response parameters of the model used to evaluate the design based on criteria specified by the user. Typical state variables include stresses, deflections, temperatures, or natural frequencies. Again, the user specifies upper and lower limits for each state variable, representing the engineering criteria that determine the feasibility of the design.

The *objective function* is the single variable that characterizes the design and represents the function to be minimized. Any quantity that can be expressed as an ANSYS parameter, including user-defined formulas, can be defined as an objective function. Typical objective functions include total weight, cost, volume of material, and various performance criteria.

The user specifies the parametric input data for the initial design, sets design and state variables (with limits), and identifies the objective function. The optimization routine selects new values for design variables, analyzes the resulting design, and evaluates the design against the state variables. The program then uses the results of the evaluation to repeat the sequence until the objective function is minimized through the sequential unconstrained minimization technique (SUMT). Once the minimum objective function for a particular set of design variables is reached, the program selects a new set of design variables for the next trial design. The process continues until an optimum design—the design variables that provide the highest degree of minimization of the objective function—is found.

Versatility

The ability to pass information between the ANSYS program and many CAD packages strengthens its capabilities. Engineering analysis with the ANSYS program may be facilitated through the prior existence of a model. Efficiency can be improved by importing a model design from a CAD program into the ANSYS program for analysis, rather than recreating the model for analysis.

With the emergence of NURBS as the recognized mathematical standard in the CAE industry, NURBs-based model geometry can be transferred among many programs with open standards. ANSYS, Inc., the developer of ANSYS, has an ongoing commitment to work with major CAD vendors to pursue the Initial Graphics Exchange Specification (IGES), established by the U.S. Department of Commerce's National Bureau of Standards, and the Product Data Exchange Specification/Standard for the Exchange of Product data (PDES/STEP) as industry standards for transferring model data.

That commitment has led to the integration of the ANSYS program with leading CAD programs. For example, ANSYS/ProFEA™ is the integration of the ANSYS program with the Parametric Technology Corporation's Pro/ENGINEER® and Pro/Mesh™ packages. UG ANSYS® Design Optimization is the marriage of the ANSYS program with the GFEMPlus™ and Unigraphics® packages offered by Electronic Data Systems (EDS)/Unigraphics. ANSYS/AutoFEA™ embeds ANSYS within the AutoCAD® environment.

Other CAD packages that can interface with the ANSYS program's auxiliary translators, through the transfer of geometry via an IGES file, include ANVIL™, AutoCAD®, AUTO-TROL®, Bravo®, CADAM®, CADKEY®, GRAFTEK® Finite Element Modeler, MSC/NASTRAN®, PATRAN®, CV MEDUSA™, Pro/ENGINEER, and VDAFS.

Many programs have their own translators to the ANSYS program. These include MSC/ARIES™, Bravo FEM™, CADDS®, FEM, CALMA®, CAEDS, CATIA®, EUCLID-IS, I-DEAS™, Supertab®, Intergraph®, PATRAN, Pro/MESH, and Unigraphics.

Program Support

A variety of support services are available to ANSYS licensees, ranging from a telephone hotline to ANSYS training courses around the globe. These services are provided by SASI and its network of ANSYS Support Distributors (ASDs). The hotline, available 40 hours per week, provides users with immediate assistance, during regular working hours, on using the ANSYS program.

Training

A complete series of ANSYS training programs is available. The series begins with a three-day introductory seminar that provides a comprehensive overview of ANSYS capabilities and operational techniques. Advanced seminars offer more in-depth study of ANSYS applications, including the study of dynamics, heat transfer, solid modeling, nonlinearities, and substructures. Seminars on specialized topics, such as design optimization, undersea structures, magnetics, and user elements, are also regularly presented.

Documentation

A variety of on-line documentation is available to assist the ANSYS user, including the following publications:

- *Getting Started*: An introductory manual containing a short overview of the ANSYS user interface and a step-by-step example.
- *User's Manual*: A four-volume set that provides complete program description and data input information. The set comprises a Procedures Manual, Command Manual, Element Manual, and Theory Manual.
- *Verification Manual*: More than 200 examples of various engineering analyses. Each example includes a description, model sketch, ANSYS input, and a comparison of ANSYS output with the theoretical solution.
- *User Guides*: In-depth discussions of specific features, such as fracture mechanics or composite structures elements. Tutorials include theory, methodology, command explanations, and examples.
- *Command Reference Guide*: A pocket-size listing of all ANSYS commands and the arguments associated with each.

Program Life Cycle

As with any computer software, the ANSYS program is continually being improved and revised to increase its usefulness to design engineers and analysts. As revisions to the program, which incorporate the strengths of prior releases, are made, additional documentation, support, and training are provided.

Product Listing/Hardware Requirements

In addition to offering a complete range of diverse capabilities within the full ANSYS program, ANSYS also offers these capabilities as specialized versions of the program to meet definite application needs and hardware requirements. The full ANSYS program runs on all classes of computer hardware, including workstations, mainframes, and 386 and 486 personal computers. Other ANSYS-derived programs include

ANSYS/ED—designed for educational and instructional needs.

ANSYS/LinearPlus—designed for linear structural (limited nonlinear) analyses on personal computers and workstations.

ANSYS/Thermal—designed for thermal analyses on personal computers and workstations.

ANSYS/Emag—designed for magnetic fields analyses on personal computers.

ANSYS/Fluid—designed for analyses of computational fluid dynamics (CFD) using the FLOTRAN program.

ANSYS/ProFEA—A CAD/FEA program that incorporates the model-building strengths of the Pro/ENGINEER program, the meshing attributes of Pro/MESH, and the analysis and optimization prowess of the ANSYS program into a single package.

ANSYS/AutoFEA—A CAD/FEA program embedded within the AutoCAD program.

Computers Supported by the ANSYS Program

- Apollo® (DN 3000, 3500, 4000, 4500, 10000)
- Control Data Corp.® (Cyber® 800, 930, 990x, and 990-Vector Version)
- CONVEX® (C Series)
- Cray® Research, Inc. (CRAY-2™, CRAY Y-MP®, CRAY Y-MP C90™)
- Digital Equipment Corp. (all VAX models, DECstation™ and DECsystem™ Series)
- Fujitsu (VP 2000 Series)
- Hewlett-Packard® (HP9000/3XX, 4XX, 7XX, and 8XX Series)
- IBM® (9000, 30nn series and compatibles; RISC System/6000™, PS/2™ Model 70, 80, and Intel386™; i486™ compatible PCs)
- Intergraph® (Clipper® Series)
- Prime® (50 Series, 55 Series)
- Silicon Graphics™ (Iris 4D™ and Power Series, Indigo™, Crimson™, Onyx™, Challenge™, Power Challenge™, Silicon Graphics Compatibles/CDC® Cyber 910/500 and Prime-PXCL™ 5500 Series)
- Stardent™ (Stardent 30nn)
- Sun™ (Sun-3™ Series, Sun-4™ and SPARC® Series, Computervision® CADDstation®)

58.3 CRITICAL ISSUES FOR USERS

Design engineers and analysts should evaluate several critical issues before selecting which FEA package to use. FEA software is diverse and no two packages are the same. Before deciding on a particular program, consider the following:

Support: Is assistance in using the program readily available?

Training: Does the company marketing the software offer training in using the program?

Quality: Does the company conduct testing to ensure the program's accuracy, issue error reports, or have an error correction system? Is the company certified under ISO 9001?

Documentation: Is the program adequately documented in terms of user's manuals, user guides, command listings, and verification manuals?

Technology: Does the program incorporate the latest advances in finite element and graphics technology and offer continuous updates to this technology.

Program Life Cycle: Is the program improved, updated, and revised on a regular basis?

Standards: Does the program meet internal and external standards, such as the IEEE Standard for transportability across hardware systems?

Interface Technology: Does the program accommodate IGES and/or PDES/STEP transfer of data or have interfaces with leading CAD programs?

If the answers to those questions are affirmative, you probably are considering a quality FEA program on which you can count.

DEFINING TERMS

Anisotropic: Having physical properties, such as conductivity, speed of transmission of light, etc., that vary according to the direction in which they are measured.

Area: A portion of a surface completely bound by three or more lines; a face of an object being modeled.

Bifurcation Point: The point at which buckling occurs; where the two paths of a force-deflection curve intersect.

Boundary Conditions: The specific load data and constraints that are applied against a finite element model to conduct an analysis.

Coupled-Field Analysis: An analysis that examines the impact of the application of two different physical phenomena, such as temperature and stress.

Design Optimization: A computer technique for identifying the best design of a component or system, based on optimizing or minimizing a single variable.

Elements: The building blocks of finite element analysis, elements are actually mathematical representations—polynomials—that are used to compute the influence of forces on a model.

FEA (Finite Element Analysis): A computational technique for analyzing the physical, magnetic, etc., characteristics of a component or object that utilizes the finite element method.

FEM (Finite Element Method): A mathematical technique, drawing from matrix algebra and variational calculus, that is used to solve differential equations governing physical phenomena as they apply to simulated geometries.

FLOTRAN: A computational flow dynamics (CFD) program that is integrated within the ANSYS program.

Isosurfaces: Like surfaces of constant value.

Isotherm: A color contour display differentiating portions of a mass by temperature.

Isotropic: Having physical properties, such as conductivity and elasticity, that are the same regardless of the direction of measurement.

Iteration: A computational pass.

Kinematics: The branch of mechanics that deals with motion in the abstract without reference to force or mass.

Linear: An engineering analysis in which response varies proportionately to forces.

Line: A portion of a spline bounded on both ends by a keypoint; an edge of an object being modeled.

Load Step: An increment at which an engineering variable, such as force, temperature, and time, etc., is applied to a model.

Material Nonlinearity: An engineering analysis in which stress is not proportional to strain.

Node: A coordinate in space that is one of many used to form the geometric shape and function of an element.

Nonlinear: An engineering analysis in which the response varies disproportionately to the forces.

Orthotropic: Having a tendency to grow in a vertical direction or position or having different properties in the *X* and *Y* directions.

Points (Keypoints): Vertices of an object being modeled.

Postprocessor: The portion of the ANSYS program used to retrieve and examine solution results, evaluate how a model performed, and perform additional calculations.

Preprocessor: The portion of the ANSYS program used to create a finite element model and specify options for a subsequent solution.

Skinning: A technique that facilities modeling by applying an external geometry to user-defined cross sections.

Solution Processor: The portion of the ANSYS program used to apply loads, forces, and boundary conditions and determine model response.

Superelement: A combination of elements and/or element types.

Volume: A portion of a solid that is completely bound by four or more areas; a volume of the object being modeled.

Wavefront: For the frontal solver technique, an indication of the number of elements and their connectivity in a model, which is related to solution efficiency, file sizes, and memory requirements.

REFERENCES

Technical Description of Capabilities, ANSYS Revision 5.0. Swanson Analysis Systems, Inc., Houston, PA.

Tim Trainer, *A Powerful Tool That's Easier to Use*: ANSYS Revision 5.1 To Feature Standardized Graphical User Interface. *ANSYS News* (Second Issue) 1994.

FURTHER INFORMATION

Command Reference Guide (pocket-size). Swanson Analysis Systems, Inc., Houston, PA.

Getting Started (an introductory ANSYS manual). Swanson Analysis Systems, Inc., Houston, PA.

User Guides (tutorials). Swanson Analysis Systems, Inc., Houston, PA.

User's Manual (four-volume set including Procedures, Command, Element, and Theory manuals), Revision 5.1. Swanson Analysis Systems, Inc., Houston, PA.

Verification Manual (more than 200 analysis examples). Swanson Analysis Systems, Inc., Houston, PA.

59

COGO

Osama Ettouney
Miami University

59.1 WHAT IS COGO?

COGO, an acronym for COordinate GeOmetry, was originated by C.L. Miller in 1959 and his staff at MIT as a part of the Integrated Civil Engineering System project (Harlow, 1991; *COGO Command Descriptions*, 1900; Miller and Lin, 1900). Originally, COGO was developed to teach a way of thinking more than a way of solving geometric or surveying problems. More recently, COGO has been modified, customized, and extended at a number of computer centers and software companies (as will be shown later in Section 59.4) to solve geometric problems in control surveys, highway design, right-of-way surveys, interchange design, bridge geometry, subdivision work, land surveying, and construction layout.

COGO's primary role is to construct design models for proposed facilities and to relate these to its real world environment (Harlow, 1991; Miller and Lin, 1900). The program accepts descriptions of coordinate geometry problems and allows the user to measure and calculate objects that can be referenced on *x–y* coordinates. COGO uses the basic building blocks of geometry which include objects—points, curves, chains, spirals, polygons—and operations—locate, intersect, and project.

There are four basic operations in COGO. Two of these are expressed in Cartesian coordinates: locate and inverse; and the other two in curvilinear coordinates: locate and project. For example, in Cartesian coordinates, a user can locate a new point from a known point, given the distance and direction. In the inverse case of Cartesian coordinates, a user can find the distance and direction between two given points; it is the basic dimensioning tool of COGO and it is the inverse of locate. In the curvilinear-locate operation the user can locate a point relative to a stored chain (curvilinear object), given the station along and offset distance from the chain. The project-curvilinear operation is the inverse of the curvilinear locate: it computes the station and offset of a known point projected onto a stored chain.

The principal aim of using a COGO software system is to enable the user to calculate and locate coordinates of points on a site by utilizing data from aerial and field surveys, graphical data, and design specification. After locating these points, the user can utilize them as control points to create boundary lines, arcs, and dimensions for the given job. Control surveys provide the reference points for tying this to the real

0-8493-2530-7/96/$0.00+$.50

world, and the output may be represented in a tabular format, plots, and graphical layouts (Harlow, 1991; Anderson, 1991).

COGO is used by civil engineers, surveyors, and CAD and geographic information system (GIS) technicians; each group uses it for a different application. For example, civil engineers may use COGO as a mathematical modeling tool that allows them to enter their designs into a CAD system. A surveyor, on the other hand, may use COGO to complete the qualification of the field survey or to plot the results as a schematic drawing or a map. For a CAD operator, COGO may be used as one of many graphics tools to speed up the process of creating drawings. This is also true in the GIS environment: to use COGO for database management and to create the base maps that are useful in a GIS analysis.

59.2 COGO AND CAD

In civil engineering and surveying firms, CAD has addressed, traditionally, drafting far more than design (LeMay, 1991). In these firms, large or small, much of the design work used to be done on drafting boards and then reentered into the computer for checking and final drafting (LeMay, 1991). Now, there are many, fully integrated automatic systems that allow field-to-finish electronic design and drafting. Also, some CAD systems can produce 3-D rendering of sites or merge proposed designs with captured video images of the existing land (LeMay, 1991). In addition, CAD systems can be interfaced with electronic data collection to speed project flow from the initial topography to final stakeout, which allows a surveyor to create topography maps in the field.

The advent of digital terrain modeling (DTM), which was originally used to create contour maps and to calculate earth work, has added new features and applications to civil engineering software. It allowed the use of numeric values in lieu of elevations, which means that contour maps representing such diverse attributes as slope intensity, depth of cut or fill, property value, runoff coefficients, or topsoil thickness can be created with a simple contouring package (LeMay, 1991). GIS is another important technology with exciting applications (Morgan and Fraley, 1993) in civil engineering and surveying. This technology grew from the increasing availability of accurate electronic mapping and the power of automated information management systems (LeMay, 1991).

The customizing features and user-modification tools built into CAD packages, such as internal programming languages, make third-party software products such as COGO very user friendly and quite popular among civil engineers and surveyors. These features allow the user to execute COGO commands from within CAD software, and make the required calculations with the benefit of having the results drawn instantly on the CAD screen. Since this process is performed inside a CAD software package, all editing and plotting functions are available to the CAD operator. Also, COGO software packages make it possible for the user to predefine CAD/COGO drawing object properties, such as screen color, and pen and layer assignments (Anderson, 1991).

With the advance in programming access within CAD software, users of COGO software can semiautomatically draw boundaries from legal descriptions and plats, reduce and adjust field notes of almost any description, design and layout semicircular and spiral curves, and automatically design a number of given tasks, such as right-of-way intersections. Also, the software creates automatically topographic maps with elevation contour lines drawn at any interval and output field stakeout data. In addition, the user can connect an electronic data collector (field book) directly to the CAD workstation with the COGO software loaded and automatically down-load raw field data through RS-232 communications (or through long distance communication if both the data collectors and the CAD/COGO workstations are equipped with modems) (Anderson, 1991).

This technology has been utilized, in 1989, by the City of Tacoma, when it started developing its third generation geographical system (Baldwin and Wheeler, 1992). This project is sponsored by the City's Engineering Groups: Light Division, Water Division, and Public Works. These groups required highly detailed, accurate maps for preliminary engineering and infrastructure mapping that could be accessed from the City's existing 76 AutoCAD workstations. To achieve this goal, the engineers developed a highly accurate base map that was constructed using GPS. Also, they were able to perform in-house conversion of the local land records by using COGO. This enabled them to form a legal base, thus creating of planimetric maps with two foot contours from aerial photography. From AutoCAD the maps are ported to Arc/Info and the existing Arc/Info parcels are adjusted to the more accurate base.

59.3 HOW TO SELECT A COGO COMPUTER SOFTWARE?

To select a COGO software, you need to consider specific characteristics that will make it useful for your application and computer system. At present, there are many makers of commercial COGO computer software (Meyer, 1991). To help you select among these, we considered several characteristics that will cover most of your needs. In Section 59.4, we will discuss in more details these characteristics with 19 commercial COGO packages.

The Type of Platform Used

This is related to the package's capability for integration with other existing computer software systems, especially CAD. Some systems are stand-alone, self-contained packages while others can communicate with many commercial CAD software, especially AutoCAD. Also, the packages vary in the types of computers that they use to run, such as Macintosh, PC compatible, MS-DOS systems, Intergraph Clipper, Unix, and VAX-VMS, as well as the supporting network system.

The Input Format

This is concerned with the format wherein the information stored in the data collectors is down-loaded to the COGO system. Most data collectors can down-load an ASCII file composed of point numbers, northings, eastings, and elevations; and most also provide their own file formats for storing other types of information in addition to special codes for data such as temperature and instrument height. Almost all COGO packages can accept straight ASCII files from the data collectors and many can interpret this information as well. Some packages will also include the communication facilities (both software and cables) necessary to make a direct connection with the data collector rather than relying on the user to figure out how to make this connection (Meyer, 1991).

The Output Format

Some COGO packages will export DXF or IGES files that can be imported into a CAD package; others will simply provide HP-GL output to a pen plotter. Many new COGO packages run within a CAD package, such as AutoCAD, which facilitates your ability to see the data as they are laid down in file, make connections and modifications, and take advantages of features in the CAD package itself, such as area and volume calculations.

Maximum Number of Points

Such is the information found in an ASCII file down-loaded from the data collector to the COGO software; it indicates the capacity or the maximum number of data that the system can handle. (It also relates to the COGO system's computer memory capacity.)

2-D and 3-D Capabilities

This is an important feature, especially if the software is to be incorporated with a 2-D or 3-D CAD package. All the available COGO software is capable of working with 2-D CAD packages, but some are not yet capable of working fully with 3-D CAD packages.

Plan Check and Data Correction

This is the system's flexibility in allowing the user to edit and check the accuracy of the collected data. Many of the products can output an audit file after applying data correction routines so that quick comparisons can be made with the original data.

Search and Query Routines

The software is able to aid the user in locating and identifying specific information within the generated drawing. This can be found in some COGO softwares that highlight data within a drawing; it is made possible, especially, for the COGO software that work within a CAD package.

Other Supporting Information

The packages enables the user to generate useful information, such as cross sections and profile drawings, and calculate areas and volumes for excavation purposes. It also aids in generating separate CAD layers for coordinate grids, profile annotations, and contour drawings.

59.4 COGO SOFTWARE PRODUCT SPECIFICATION

In this section, we review 19 different COGO software packages that are available in the market. We hope that the review will become a source guide for you to select the appropriate software for your application. In addition, please see also *The Windows Sources Catalog* (1994), which is a comprehensive Windows-Source Catalog for other products that utilize Windows.

ARC/INFO COGO

Company. Environmental Systems Research Institute (ESRI), Inc., 380 New York St., Redlands, CA 92373-8100; Ph: 909-793-2853, ext. 1375; FAX: 909-793-5953.

Product Summary. A set of software tools that creates and analyzes spatial data based on the principles of COGO. These tools support the functions performed by land surveyors and civil engineers in a host of design and mapping applications such as the layout of subdivisions, roads, and related facilities.

Part of a family of software packages, built by ESRI, that works with the ARC/INFO Geographic Information System software. Tools used for engineering and survey procedures can enter COGO-generated data into an ARC/INFO database. Data in an ARC/INFO database can be used with COGO for engineering and surveying descriptions. The ability to share data between ARC/INFO and COGO provides a link between Geographic Information Systems technology and the analytical capabilities of coordinate geometry.

Tools allow users to set coordinates for a point, delete points, intersect lines, define traverses, calculate inverse bearings, report lot closure characteristics, and perform other data analysis techniques. It supports a variety of applications including reduction and analysis of field survey data, entry of land records, preparation of Records of Survey, alignment calculations for roads and utilities, and subdivision design and mapping.

Other Information. Release date 1993; application: Geographic Information Systems; compatibility: DEC VAX, DECstation/VMS, ULTRIX, DG/UX, HP-UX, Sun/SunOS, Solaris, Silicon Graphics Iris/IRIX, and IBM/AIX; minimum RAM required: 32 MB; disk storage required: 2–18 MB; additional hardware and software required: ESRI's ARC/INFO; Network compatibility: Token-Ring, Ethernet, DECnet, and TCP/IP; and source language: FORTRAN 77 and C.

Architectural Power Tools

Company. Caricato Systems, 405 W. College Ave, Ste. C, Santa Rosa, CA 95401; Ph: 707-576-1140.

Product Summary. AutoCAD architectural system provides parametric commands, integrated 3-D routines, 3-D symbol libraries, industry standard keynoting, and flexible layering system. Also, the system includes hatch pattern and automatic elevation generators, movie generator, COGO routines, unlimited scheduling capabilities, full-screen text editor, source code, and dynamic space reporting.

Other Information. The manufacture-suggested list price $795; release date 1985; application: civil engineering and architecture; compatibility: PC-MS/DOS, Apple Macintosh, Sun SPARCstation, and SunOS; disk storage required: 2 MB; source language: AutoLISP and C.

AutoCOGO

Company. A/E Microsystems, Inc., 10134 Mosteller Lane, West Chester, OH 45069; Ph: 513-772-6700; FAX: 513-672-6191.

Product Summary. COGO program for interactive use with AutoCAD. It includes routines for locating points, inquiry commands for distances, bearings, azimuths, locating lines and intersection points, dividing lines, curves, and curve design. Also, it can be used for basic and advanced civil design, profiling, cross-sectioning and cut, fill area functions.

Other Information. The manufacture-suggested list price $995; release date 1985; application: civil engineering and architecture; compatibility: PC-MS/DOS; minimum RAM required: 640 KB, and disk storage required: 4 MB; additional hardware and software required: AutoCAD; source language: AutoLISP and C.

Civil Design Series-Plan & Profile, COGO & Development, and Construction Details

Company. LANDCADD International, Inc., 7388 S. Revere Pkwy., Bldg. 900, Ste. 901–902, Englewood, CO 80112-3942; Ph: 800-876-LAND; 303-799-3600; FAX: 303-799-3696.

Product Summary. Plan & Profile module allows users to stay within safety guidelines when engineering vertical curves. COGO & Development module aids in engineering layout and design of any type of land planning project including subdivisions, roads, parking lots, golf courses, and commercial sites. Construction Details module provides library of civil construction details including drainage, erosion control, soil stabilization, bumper stops, paving, parking, and utilities.

Other Information. Release date 1991; application: civil engineering and architecture; compatibility: PC-MS/DOS, Sun and SunOS, Apple macintosh, IBM RS/6000/AIX, and Windows 3.X; minimum RAM required: 4 MB, and disk storage required: 663 KB; additional hardware and software required: Novell Netware and Windows 3.X; network compatibility: Novell; source language: AutoLISP and C.

COGO

Company. Softdesk, Inc., 7 Liberty Hill Rd., Henniker, NH 03242; Ph: 603-428-3199; FAX: 603-428-7901.

Product Summary. Provides coordinate geometry tasks for wide range of civil engineering and surveying work. Operates from inside AutoCAD. Allows user to set points, draw lines, calculate areas, perform closures, create special line types, label metes and bounds, and produce information for legal descriptions. Includes point protection and access features, enhanced layer and selection set manipulation, and extended coordinate entry modes including point number and northern and eastern coordinates.

Other Information. The manufacture-suggested list price: $595; application: civil engineering and architecture; compatibility: PC-MS/DOS, Sun SPARCstation, and SunOS; additional hardware and software required: AutoCAD.

COGO

Company. Eagle Point, 4131 Westmark Dr., Dubuque, IA 52002-2627; Ph: 800-678-6565; 319-556-8392; FAX: 319-556-5321.

Product Summary. COGO provides a set of functionality to reduce field notes, enter and query drawing information, generate formatted reports, and make design revisions. Industry-standard COGO routines are used to input data, create and query drawings, and design cul-de-sacs and intersections. Uses the mathematical concepts of coordinate geometry to place simple or multiple data points within a drawing. Based on the points entered, the system draws lines, curves, alignments, spirals, cul-de-sacs, and areas. Intersections of lines can be determined and data from elements or points in the drawing can be queried.

Allows the user to place points in the drawing, draw lines by point number, draw feature lines, place curves using multiple methods, place points by station and offset, and create right-of-way lines. Three-dimensional nodes can be created with attached symbology and descriptions. Traverses can be performed for boundaries as well as alignments and can contain station equations, lines, circular curves, and clothoid spirals. Includes full support for point protection and provides grid-to-ground or ground-to-grid conversion, intersection trimming, and location of prints using a bearing–bearing–station intersection. The user can also traverse through combining spirals.

Other Information. Application: civil engineering and architecture; compatibility: PC-MS/DOS, AT&T UNIX System V; source language: C. Available in MicroStation PC, MicroStation Clipper, and AutoCAD versions.

COGOMASTER III and DEEDMASTER and TRIG-n-COGO

Company. Guenzi Surveys, Box 127, Glendive, MT 59330; Ph. 406-365-3527; sold through Electronic Supply Co., Box 20556, Billings, MT 59104; Ph: 406-252-2197.

Product Summary. COGOMASTER III, DEEDMASTER, and TRIG-n-COGO provide many features such as traverse, inverse, curve calculation, modifying lines and curves, automatic area calculations, stake-out, bearing–bearing, bearing–distance, distance–distance, compass rule adjust, change decimal output, automatic point numbering, import, export and merge coordinates, screen plot, automatic plot update, print plots, export plot as pictures, and reference bearings and distances.

Other Information. The manufacture-suggested list price: $545; application: surveying, education, engineering, architecture, and realtors; compatibility: Macintosh; minimum RAM reouired: 2 MB.

COGO-PC Plus

Company. Research Engineers, Inc., 1570 N. Batavia St., Orange, CA 92667; Ph: 800-FOR-RESE; 714-974-2500; FAX: 714-974-4771.

Product Summary. Coordinate geometry program that provides more than 280 COGO commands. Features Speed Menus, on-line help screens, and special assist mode with data promoting.

Other Information. The manufacture-suggested list price: $1,895; release date: 1993; application: civil engineering and architecture; compatibility: PC-MS/DOS; minimum RAM required: 640 KB; source language: FORTRAN.

Entry 2000

Company. Informed Management Environment, Inc., subsidiary of Debratex, Inc., 3605 Katy Frwy., Ste. 200, Houston, TX 77007; Ph: 713-869-4630; FAX: 713-869-3172.

Product Summary. PC-based desktop AM, FM, and GIS software. Provides table-driven symbology, attribute database map annotation, topologically structured data, polygon creation, interactive thematic mapping, Helmert coordinate transformation, COGO and traverse and projection conversions.

Other Information. The manufacture-suggested list price: $1,150–$2,000; release date: 1992; application: Geographic Information Systems; compatibility: PC-MS/DOS; minimum RAM required: 640 KB; disk storage required: 20 MB; additional hardware and software required: GTCO or comp. digitizer; network compatibility: Novell and Ethernet; source language: Pascal. It is also available for Windows.

GWN COGO

Company. GWN Systems, Inc., 11133 124th St., Ste. 200, Edmonton, Alberta, Canada T5M OJ2; Ph: 403-452-0090; FAX: 403-453-5207.

Product Summary. Provides mapping and design capabilities for professional surveyors. Allows users to enter, edit, and adjust survey data according to standard geometric routines. Includes automatic plotting with annotation, curve and spiral design, area calculations, alignments, audit trail generation, reporting and batch processing.

Allows for graphic selection of command inputs such as points, angles, and distances. Supports MicroStation commands for planning and zooming. Users can move between MicroStation and COGO environments for cartographic positioning or manual placement of text and digitizing.

Plots a variety of graphic types, providing user control of color, style, weight, font, and text height. Allows for different annotation types such as distance annotation being displayed as legal distances as input from plans of survey, not the adjusted distances. Supports generalized plotting routines.

Produces reports such as coordinate listings, inverses, traverse adjustments, layout ties, station offsets, describe alignment, tangent offset, chord offset, deflection angle, and area calculations. Supports least-squares and compass rules for traverse adjustments for up to 300 leg traverses. Includes horizontal alignment design commands and curve stakeout and stationing.

Performs in 2-D or 3-D. Converts input data as necessary and can apply any conversion such as chains to feet to distances before coordinate calculation. Applies a combined scale factor based on position and elevation to input distances.

Other Information. The manufacture-suggested list price: $1,650; release date 1988; application: civil engineering and architecture; compatibility: PC-MS/DOS, Sun SPARCstation, and SunOS; minimum RAM required: 4 MB; disk storage required: 2 MB; additional hardware and software required: Integraph's MicroStation or AutoCAD.

InFoCAD and InFoCAD for Windows

Company. Digital Matrix Services, Inc., 3191 Coral Way, Ste. 900, Miami, FL 33145; Ph: 305-445-6100; FAX: 305-442-1823.

Product Summary. Geographic information system provides an intuitive user interface and application development tools; integrates the worlds of graphics, databases, images, and analysis within a vector environment whose data structure is founded on coordinate geometry. All graphical information is stored in one or more of 10 independent tables that reference each other and collectively make up a drawing library. The COGO point table stores COGO point numbers and their x, y, and z coordinates. The Segment table stores unique segment numbers and the COGO point numbers that make up a graphic representation. The Figure table stores figure names as well as the COGO point numbers that make up figures. Data administration allows users to share access to all information simultaneously, and enables users to analyze and process large and complex data sets.

The system includes a computer-aided design and drafting module that is completely integrated and provides COGO and 3-D functions. The layer and External Library managers provide users with the ability to create and manage unlimited continuous base-maps of information. The map projection utility allows the user to define up to 15 projection systems from 24 different projections. Also available projections include geographic, state plane, lambert conformal conic, polar stereographic, equidistant conic, orthographic, sinusoidal, mercator, ployconic, and others.

Other Information. The manufacture-suggested list price $7,500–$12,500; release date 1993; application: Geographic Information Systems; compatibility: Sun SPARCstation, SunOS, IBM RS/6000, AIX, DEC, ULTRIX, DG/UX, HP, HP-UX, SCO UNIX, and windows NT; minimum RAM required: 24 MB, and disk storage required: 80 MB; network compatibility: TCP/IP and Ethernet; source language: FORTRAN and C.

LANDCADD Light

Company. LANDCADD International, Inc., 7388 S. Revere Pkwy., Bldg. 900, Ste. 901–902, Englewood, CO 80112-3942; Ph: 800-876-LAND; 303-799-3600; FAX: 303-799-3696.

Product Summary. Land planning package. Provides solutions for site planning, landscape and irrigation design for civil engineers, architects, and contractors. Provides routines to lay out property lines, generates parking lots, COGO routines, shadow calculations, planting plans, and more than 350 symbols of sports facilities, trees, cars, and site amenities

Other Information. The manufacture-suggested price: $795; release date: 1990; application: civil engineering and architecture; compatibility: PC-MS/DOS, Sun SPARCstation and SunOS, Apple macintosh, IBM RS/6000/AIX, and Windows 3.X; minimum RAM required: 4 MB, and disk storage required: 682 KB; additional hardware and software required: Novell Netware, Windows 3.X, and AutoCAD; network compatibility: Novell; source language: AutoLISP and C.

LANDesign

Company. Compuneering, Inc., 113 McCabe Crescent, Thornhill, ON, CD L4J 2S6; Ph: 905-738-4601; FAX: 905-738-5207.

Product Summary. It is used for land surveyors and provides graphics ability to set and move points, lines, and curves by pointing and clicking. All conventional COGO features are supported: import and export of point data; Data Collector module by directly down-loading from data collectors by Lietz, Topcon, Geodimeter, and others; Export Module allows export in DXF, Claris CAD, MacDraw II, CGDEF, and FlexiScript formats; and intersections and field traverse editing and adjusting. Additional features include deed plotting, radial stakeout, slideshots, and area calculation.

Other Information. The manufacture-suggested list price $595; release date 1988; application: civil engineering and architecture; compatibility: Apple macintosh; minimum RAM required: 1 MB, and disk storage required: 800 KB; source language: Pascal.

M-COGO and V-COGO Family of Products

Company. Engineering Desktop, 1776 N. State St., Ste. 200, Orem, UT 84057; Ph: 801-225-3133; FAX: 801-225-8333.

Product Summary. M-COGO and V-COGO are the cornerstone of the system. M-COGO provides civil engineers, land surveyors, architects, and mapping professionals with productivity advantages of GUI for site planning and development needs; and V-COGO is a CAD coordinate geometry for VersaCAD that provides standard coordinate geometry routines including northing and easting, elevation, object drawing, annotation, data storage, map-check, painting, ASCII test file import and export, and area calculation applications, and automatically assigns point numbers.

Other available modules include *M-Collect*, which provides M-COGO and MicroStation 4.0 with the ability to receive ASCII data generated by electronic field book, and handles output from popular data collectors and data stored in AASHTO format. Also, it allows coordinates, point numbers, elevations, and descriptions to be automatically drawn to design file. *M-Contour*, which provides contour mapping capabilities inside of MicroStation PC V.4.0. Automatically generates, smooths, and annotates major and minor contour lines. Creates 3-D digital terrain model of contour area.

M-Design, which is integrated into M-COGO and adds advanced coordinate geometry site design and layout functionality. Suited for companies involved in design of right-of-ways including highways and railroads and subdivisions. Features spiral curve entry and annotation, street intersections, semicircular arc, compound and reverse curves, and other features. *M-Field*, which is a professional surveying software for use with MicroStation PC V.4.0 and M-COGO. Designed for office that employs field land-surveying crews. Includes least-squares adjustments, radial stakeout reports, and other features.

V-Field, which is an add-on module to V-COGO. Includes open-ended and closed-loop field traverse reduction. Provides traverse adjustment using compass rule or least-squares. Calculates and prints radial stakeout reports. Includes audit trail feature and double-precision accuracy.

Other Information. The manufacture-suggested list price for M-COGO is $995, for M-Collect is $995, for M-Contour is $1,495, for M-Design is $995, for M-Field is $995, for V-COGO is $495, and for V-Field is $795; release date 1991/92; application: computer-aided design, civil engineering and architecture; compatibility: PC-MS/DOS and Apple Macintosh; source language: C. Runs inside of MicroStation PC V.4.0.

Profiles

Company. Eagle Point, 4131 Westmark Dr., Dubuque, IA 52002-2627; Ph: 800-678-6565; 319-556-8392; FAX: 319-556-5321.

Product Summary. Provides for the vertical alignment design of a project. A profile is a 2-D curve that represents a slice or cross section of the real world. The *X*-axis in a profile space represents the horizontal distance along a path traced across the earth's surface. The *Y*-axis represents the elevation along the path. A profile Coordinate System can be developed for a drawing created using Surface Modeling, site Design, COGO, or Roadcalc. The user can use the tangent, curve, annotate, and extract options to design and create a complete vertical curve alignment. The user can report coordinate data for a tangent line, curve, point, or structure using the inlet, manhole, pipe, pipe structure, and annotate functions; the user can create a complete storm sewer, sanitary sewer, or water main alignment.

Other Information. Application: civil engineering and architecture; compatibility: PC-MS/DOS, AT&T UNIX System V; source language: C. Available in MicroStation PC, MicroStation Clipper, and AutoCAD versions.

Roadworks

Company. Intergraph Corp., One Madison Industrial Park, Huntsville, AL 35894-0001; Ph: 800-345-4856, 205-730-2000; FAX: 205-730-2108.

Product Summary. Highway engineering application. Provides design capabilities including multilayer digital terrain modeling and COGO features. Includes alignment layout and design, automated criteria-driven highway design, independent ditch, right-of-way, and material-controlled slopes. Provides simultaneous subgrade design and multiple methods of earthworks computations. Compatible with Intergraph's In Roads.

Other Information. The manufacture-suggested list price: $5,000; release date: 1993; application: civil engineering and architecture; compatibility: PC-MS/DOS, AT&T UNIX System V, and Microsoft Windows NT.

SOKKIA

Company. Sokkia Corp., 9111 Barton, P.O. Box 2934, Overland Park, KS 66201; Ph: 913-492-4900; FAX: 913-492-0188.

Product Summary. Sokkia has a comprehensive line of interactive, modular programs. MAP is the cornerstone of Sokkia Software; it contains the common database required for all other modules (with the exception of LINK and COMMS Plus). MAP automatically reduces raw data to coordinates in the database and generates maps, including line joining and symbol generation. It includes data communication, editing capabilities, automatic map generation, complete CAD facilities, full two-way AutoCAD DXF transfer, customized borders and title blocks, and plotting and grid layout libraries.

The other modules that can be interfaced with MAP include *CALC*, a coordinate geometry module; *CONTOUR* uses the the Triangular Irregular Network (TIN) method of calculation to create digital terrain models from existing data; *GPSMAP* adds kinematic Global Positioning System (GPS) data to the types of survey data that MAP can read; *ROADING* includes a series of comprehensive programs to allow engineers to design and produce construction drawings for a wide variety of roading and corridor-type applications.

In addition to the above, Sokkia software provides *EARTHWORK*, which is a bundled package that contains two modules that are available separately but designed to work in tandem for best results, *VOLUMES* (to calculate earth volumes), and *PROFILES* (to plot surface profiles and cross sections). Other stand-alone packages include *LINK*, which provides quick communication between an SDR Electronic Field Book and non-Sokkia software programs. *COMMS Plus* is a software designed for transmitting data files between a computer and a data collector, either a total station or an Electronic Field Book.

Other Information. The manufacture-suggested list price for MAP/LINK is about $2,000, and each other item is an additional cost that varies from about $200 to $1,795; there are also combined kits of MAP, ROADING, and EARTHWORKS that cost between $1,000 and $5,000; applications include field data reduction and automated map generation, coordinate geometry and solar observation calculations, digital terrain modeling, contouring and 3-D perspectives; calculate volumes and profiles, as well as cross sections, plan, design, plot, and report roadways, provide kinematic GPS application support, digitize existing map information, and multiple user support in local area network environment; compatibility: IBM-compatible personal computers, PCXT, 486; RAM required: 640K–4 MB; disk storage required: 20–100 MB; additional hardware and software required: CGA and VGA Monitors, Mouse; network compatibility: Novell.

Survey

Company. Softdesk, Inc., 7 Liberty Hill Rd., Henniker, NH 03242; Ph: 603-428-3199; FAX: 603-428-7901.

Product Summary. Combines variety of survey-related capabilities inside AutoCAD. Provides user-defined command line COGO, least-squares adjustment, geodetic coordinate transformations, and default screen setup of all variables. Provides sun and starshot calculations, traverse and sideshop entry through spreadsheet input, data collector down-loads, and batch mode input.

Other Information. The manufacture-suggested list price: $1,295; application: civil engineering and architecture; compatibility: PC-MS/DOS; additional hardware and software required: AutoCAD.

Tech-Mac

Company. Technical Advisors, Inc., 5918 Lilley Rd., Ste. 1, Canton, MI 48187; Ph: 800-521-9958, 313-981-3430; FAX: 313-981-5228.

Product Summary. Surveying and graphics package. Includes text editor, batch command file input, performs COGO computations, point values with elevations and labels, legal description writer, backup facility and control of printer output, and numerous graphics features for plotting.

Other Information. The manufacture-suggested list price: $595; release date: 1993; application: civil engineering and architecture; compatibility: PC-MS/DOS; minimum RAM required: 640 KB; disk storage required: 10 MB; additional hardware and software required: math coprocessor; and source language: C.

DEFINING TERMS

COGO: COordinate GeOmetry.

CAD: Computer-aided design.

GUI: Graphical-user interface.

DXF: Drawing-interchange file.

DOS: Disk-operating system.

RAM: Random-access memory.

ROM: Read-only memory.

ASCII: American Standard Code for Information Interchange.

REFERENCES

M.E. Anderson, Third-party COGO expands. *MicroCAD News* 6(5):39–42, 1991.

P.E. Baldwin and G.L. Wheeler, Developing a third generation engineering accurate GIS using AutoCAD and Arc/Info. In *URISA Proceedings of the Annual Conference of the Urban and Regional Information Systems Association*, Vol. 1, D.M. Freund (ed.). Urban & Regional Information Systems Association, Washington, D.C., 1992, 138–145.

COGO Command Descriptions. Computing Services, the University of Alberta, Edmonton, Alberta, Canada, T6G 2H1.

C. Harlow, The many faces of COGO. *MicroCAD News* 6:33–35, 1991.

B. LeMay, Field-to-finish design. *MicroCAD News* 6(5):37–38, 1991.

A. Meyer, A survey of COGO software. *MicroCAD News* 6(5):43–51, 1991.

C.L. Miller and S. Lin, *The COGO Book.* CLMSystems, Inc., Tampa, FL.

R.A. Morgan and M. Fraley, The use of COGO/GIS in the Route 58 Corridor Development Study: A case study. *In a Microcomputer in Transportation, Proceedings of the 4th International Conference*. J. Chow et al. (eds.). American Society of Civil Engineers, New York, 1993, 48–59.

The Windows Sources Catalog, Vol. 2, No. 1, pp. 337. Ziff-Davis Publ. Co., 1994.

SmartCAM

Osama Ettouney
Miami University

60.1 THE CAM ENVIRONMENT

Digital computers are the key to computer-aided manufacturing (CAM) as well as computer-aided design (CAD). The objectives are to use computers to automate and control all manufacturing operations to increase productivity, reduce cost, and achieve high-quality products. Computers have been used for more than 30 years in continuous and discrete manufacturing operations. Applications included (Bedworth, 1991): product testing and quality control, foundry control, numerical control equipment interface, nuclear reactor control and monitoring, utility plants start up and control, and automobile assembly lines. Computer-aided manufacturing is not just a computer software that is used to generate codes for numerical-control (NC) machines or to simulate the operation; it is a philosophy of operation that requires complete understanding of the capabilities and functions of the manufacturing process to be automated and controlled (Bedworth, 1991; Chang et al., 1991; Amirouche, 1993).

To integrate computers into any manufacturing process the engineering, design team needs to learn and identify the short- and long-term goals from automating the process (Bedworth, 1991; Chang et al., 1991; Amirouche, 1993; Thacker, 1989; Urbaniak, 1986; Kuttner, 1986; Jackson and Jones, 1987).

Examples for short-term goals are:

- using CAM software to generate codes for NC machines that will be integrated in the production of a specific family of parts (for more details see Puttre, 1992a; Noaker, 1993; Kramer, 1990).
- integrating the CAM software with an existing CAD software. Tying CAD and CAM systems together will enable you to take a product from design conception to manufacturing, which eliminates many problems, such as overdesign. This tying will allow you to apply the concepts of design for manufacturing (DFM), or making it right the first time, and concurrent engineering (Bedworth, 1991; Chang et al., 1991; Amirouche, 1993; Thacker, 1989; Kramer, 1990; Stoll, 1988; MicroCAD News Staff, 1988; Puttre, 1992b).

Examples for long term goals are:

- expanding the CAM capabilities and NC machines to more production cells in your production line. This may be the beginning of implementing such concepts as cel-

0-8493-2530-7/96/$0.00+$.50

lular manufacturing, flexible-machining cells, and flexible-manufacturing systems (Bedworth, 1991; Chang et al., 1991; Amirouche, 1993; Thacker, 1989; Urbaniak, 1986; Kuttner, 1986; Jackson and Jones, 1987; Shrensker, 1987; Romanini, 1987; Spur et al., 1985).

- integrating the CAM system into an overall computer-integrated manufacturing systems (CIMS) philosophy that utilize computers not only in automating manufacturing processes, but also in breaking down the walls among the different departments of the company, such as business functions, human resources, management, and engineering. The idea is to enable these departments to communicate effectively and continuously among themselves to produce quality, inexpensive products. The CIMS integration philosophy of conducting business has been used in many companies (such as IBM and Ford), and the philosophy is considered essential for the successful competition in the international market (for more details see Bedworth, 1991; Amirouche, 1993; Thacker, 1989; Shrensker, 1987; Romanini, 1987).

After learning and identifying your short- and long-term goals, you need to learn about the different elements that constitute a CAM system. The following is a brief discussion of those essential elements that will help you familiarizing yourself with the CAM environment.

The Different Elements of CAM

- Computer-aided manufacturing (CAM) software. This is used to convert a CAD drawing (or its own drawing) into an NC code. In this chapter, we will use this term to define both the software and the big picture that includes computers, machines, and methods.
- Computer-numerical control (CNC). Machining operations that include tool positions, the desired cutting, and lubrication are accomplished by numerical control commands generated and controlled by the computers (Bedworth, 1991; Chang et al., 1991; Amirouche, 1993; Kramer, 1990; Stoll, 1988; Puttre, 1992b; Spur et al., 1985; Coleman, 1990).
- Direct Numerical Control (DNC). The use of one mainframe computer to control several CNC machines in a time-shared fashion (Bedworth, 1991; Chang et al., 1991; Amirouche, 1993).
- Robotics. Another type of CNC machines used for loading and unloading parts, assembly, and painting operations (Bedworth, 1991; Change et al., 1991; Amirouche, 1993; Urbaniak, 1986; Shrensker, 1987; Romanini, 1987; Spur et al., 1985).
- Flexible-machining cells (FMC). Groupings of machines, equipments, and operations to produce discrete families of parts (Bedworth, 1991; Chang et al., 1991; Amirouche, 1993).
- Flexible-manufacturing systems (FMS). Groupings of FMC and the production of more than five different families of parts (Bedworth, 1991; Chang et al., 1991; Amirouche, 1993; Shrensker, 1987; Romanini, 1987; Spur et al., 1985).

In this chapter, we are concerned primarily with the CAM software. There are many makers of commercial computer software for these types of operations (for excellent reviews and backgrounds about these different software, see Puttre, 1992a,b; Noaker, 1993; Kramer, 1990; MicroCAD News Staffs, 1988; Coleman, 1990; CAM Software Update, 1993; Watkins, 1987). To select among these, you need to consider the issues discussed above as well as the specifics about the CAM software that you will use. In the following sections, we will discuss one of those softwares: SmartCAM, which is among the top two software packages available in today's market.

60.2 THE WORKPLACE ENVIRONMENT OF SmartCAM

SmartCAM is used to perform complex machining routines for mills, lathes, punch presses, and profiles. SmartCAM V7.0 works with MS DOS 3.3 or later systems, 100% IBM-compatible 80286 computer, 640 KB RAM, with 40 MB hard disk drive, 4 MB memory and 80387 math processor or higher, mouse or digitizer, EGA (VGA or better) monitor, and SmartCAM hardware lock installed on a parallel port.

In SmartCAM, you can interact easily with the software's different features by using a Graphical User Interface (GUI) environment. The GUI promotes capability, and coupled with a pointing device such as a mouse, makes it easier for you to interact with SmartCAM. However, you also have the option of entering commands and manipulating information with the keyboard.

When SmartCAM is installed on your computer, the program creates several subdirectories that store program, work, and documentation files. These subdirectories allow the different files of SmartCAM to be stored and retrieved efficiently and in a consistent manner. The basic subdirectories, and the resident files and extensions, in SmartCAM are

- \SM6—This directory stores program files, hardware setup files, and batch files. These have the following extensions:
  - .EXE—Executable program files
  - .SU—Hardware setup files
  - .BAT—Batch files
- \SM6\DATA—All job plans are stored in the DATA subdirectory, unless specified otherwise by you. The following extension system is used in this subdirectory:
  - .JSF—Job plan files
  - .MAC—Macro routine files
- \SM6\AMDATA—All shape files and code files are stored in the Advanced Machining DATA subdirectory, unless specified otherwise by you. In addition, advanced machining job plan files can also reside in this subdirectory. The following extension system is used in this subdirectory:
  - .SH2—2-D shape files
  - .SH3—3-D shape files
  - .—Code files, no default extension
- \SM6\AMSMF—All code generator files and some notes files are stored in the Advanced Machining SMF subdirectory. These have the following extensions:
  - .SMF—Machine define files
  - .TMP—Template files that determines the format of the output code
  - .NTS—Notes files

- \SM6\DEV—All hardware support files are stored here. These have the following extensions:
  - .DEV—Device configuration files
  - .SU—Hardware setup files
- \CGI—All device driver files used to configure your computer system, and those loaded during the installation process are stored in this subdirectory. The extension for these files is *.SYS.

The Main Menu of SmartCAM V7.0 is shown in Figure 60.1. In general, the program contains the following functions: Job Plan, Applications, Edit Plus, Communicate, Design_System, CAM-Connection, Accessories, and Install. These functions are used to create the appropriate NC code for machining a part. Basically, there are seven steps to achieve that. A brief discussion of these steps and the functions are presented below.

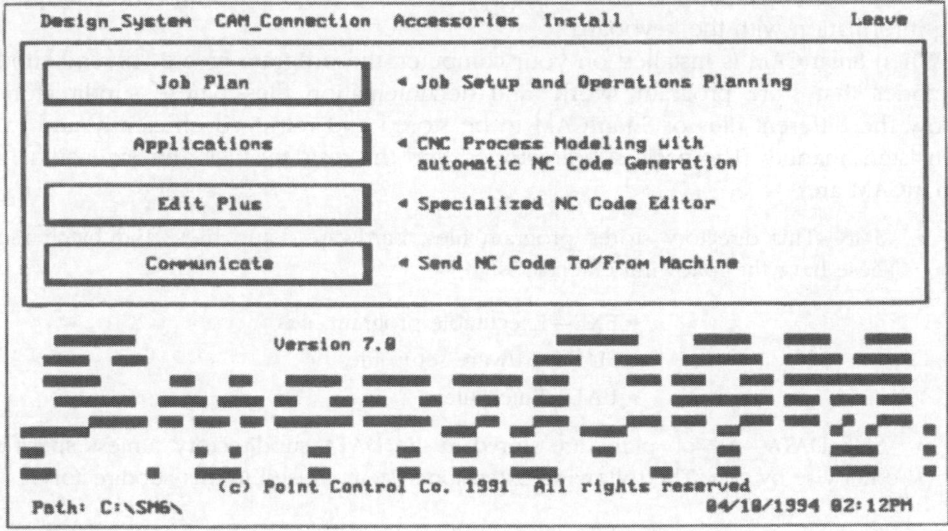

FIGURE 60.1 The Main Menu in SmartCAM.

60.3 THE SEVEN STEPS TO CREATE AND EXECUTE AN NC CODE

Step One: Machine Define

Machine define can be found by using the pull down menu in **Accessories** (Fig. 60.2). In this step, the program allows you to adjust the parameters of the machining operation that will be used, such as the type of unit system: Metric or British, and the type of scaling: incremental or absolute. (There are more than 150 different parameters available for you to adjust for their own specific machine and operation.)

Step Two: Job Plan

Here, the program allows you to select tool types and sizes, cutting parameters, feeds, speeds, offset values, and types of machines to be used, such as mills or lathes. In addition, you plan the sequence of operations, select the unit system, specify the CAD

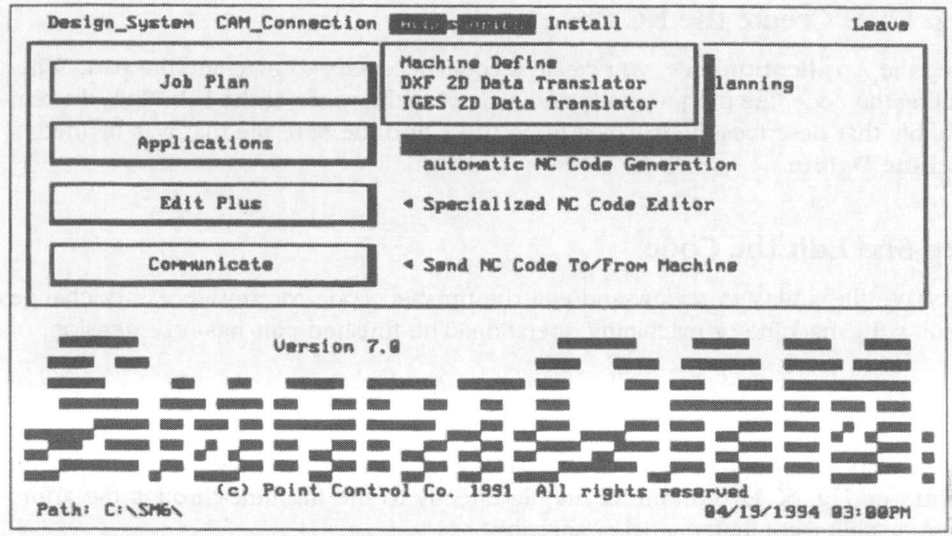

FIGURE 60.2 The Machine Define File in SmartCAM.

layers to be transferred, if a CAD file is to be used later, and provide comments to other users. The extension of this file is *.JSF.

Step Three: Convert CAD to CAM

SmartCAM is capable of using the geometry, layers, and blocks from existing CAD files or creating new models. In this step, you utilize the **DXF CAM Connection** function, V5.3, to convert CAD-prepared drawing files saved in a **DXF** format into **SmartCAM** CNC Process Model files; SmartCAM takes the CAD file's geometry (entities) and changes it into a tool path by assigning tool properties to it. These properties are controlled by the **Job Plan** that you select. Geometry that does not represent a tool path, such as part blank, clamps, or fixtures on different layers, is recognized by SmartCAM and assigned the same layers within the CNC Process Model. **CAM Connection** can also reverse this process and convert data from SmartCAM model's file into a CAD drawing file.

Step Four: Create a Model

Here, you model the part either by adjusting the converted CAD model or by creating it using SmartCAM's graphic capabilities. These are found in the **Applications** file (Fig. 60.1). You can also simulate the machining or cutting operation, and can then prepare the code for the machine. The model that is created by SmartCAM integrates tool path sequence and manufacturing properties with part geometry, and it is an actual manufacturing database. In this section you has the ability to edit or create a model, examine the different elements (operation steps) of the model, include text, view actual cutting process, and modify sequence of operation. The generated shape file has the extension *.SH3 for three-dimensional machining.

Step Five: Create the NC Code

Using the **Applications** file, you create a code necessary to machine the part. When creating the code, the program asks you to provide the name of the **Job Plan**, the template file that describes the machine to be used, and the **SMF** file that was defined by **Machine Define**.

Step Six: Edit the Code

You have the ability to review and edit the finished code for any necessary changes to suit your machine or machining operation. The finished part has no extension.

Step Seven: Communicate with the Machine

You specify the appropriate machine (using the **Communicate** file in the **Main Menu**, see Fig. 60.1) and transfer the file directly to the machine through the appropriate machine–computer serial connection.

60.4 A MILLING EXAMPLE USING SMARTCAM

In this section, we will describe in some details an example using SmartCAM to prepare a milling operation. Thus, if you are not familiar with the software, you will able be to get an idea of the capabilities of SmartCAM to prepare and execute an NC code. The example is taken from *SmartCAM User Manuals* (1993) and is shown in Figure 60.3. We will use the seven steps discussed above to prepare and execute this example. First, we must run the SmartCAM program by typing **SC** from the SmartCAM directory: **SM6>SC**. Next, the **Main Menu** appears (Fig. 60.1) and you can access any of the different files and information management options using the interactive, pull-down menus.

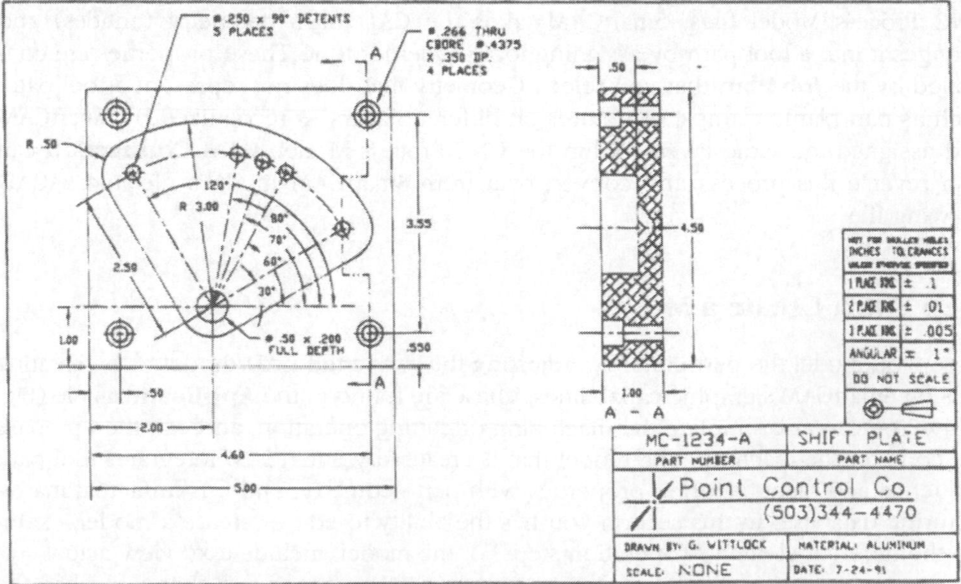

FIGURE 60.3 The part that will be modeled using SmartCAM.

Step One: Machine Define

If you need to change any of the parameters in the program to accommodate your needs, choose **Accessories** from the pull-down menu. Open the file, scan through the requirements, and adjust accordingly. After finishing this step, close the file and you will be back to the **Main Menu**.

Step Two: Job Plan

- To select the job plan file, click on the job plan icon or type [J] in the **Main Menu** screen. You may start a new job plan (select **New** option) or modify an existing job plan.
- To modify an existing job plan, press the **[F1]** key to view a list of the files currently available; before reading the file, the screen will show you the **Drive, C:**, the SmartCAM **Main Directory, SM6**, and options of subdirectories to select from (Fig. 60.4). The path SmartCAM uses, **\AMDATA or \DATA**, appears at the top of the screen (Fig. 60.5).

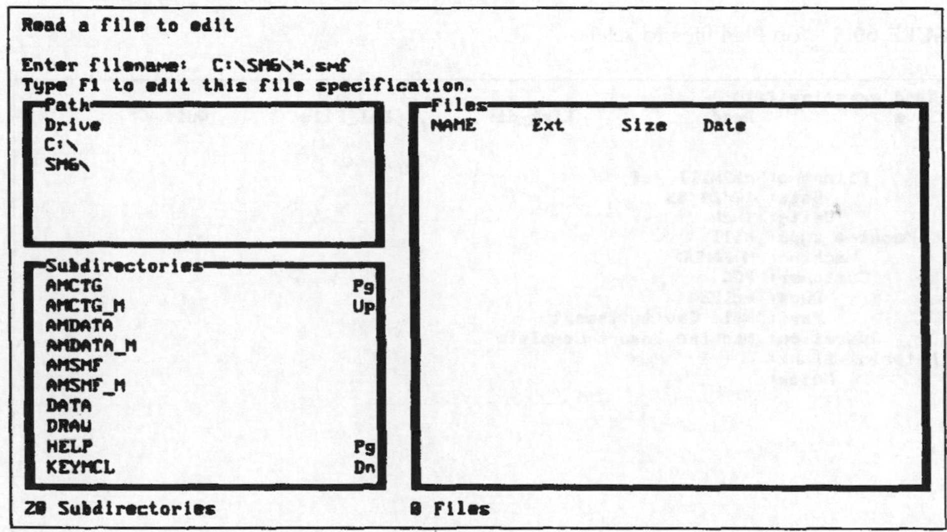

FIGURE 60.4 The subdirectories in SmartCAM.

- To start a new job plan, select **New** and you will enter first into **Job Info** (Fig. 60.6). Here, you will provide the following information: File name, Date, Units, Machine Type (this is automatically selected from the previous step), Machine (the file name of the correct machine defines file without the extension), Customer, Dwg#, Part, operation, and Material Blank.
- After finishing the data section, press [**Q**] or [**Esc**] and then press [**T**] to enter the tooling selection section. Select edit and start typing the tools and other information that you use; an example is shown in Figure 60.7. The following is a brief description of this information:

  TL#—this is important if you have an automatic tool changer, otherwise it is just the number of tools that you use in the operation.

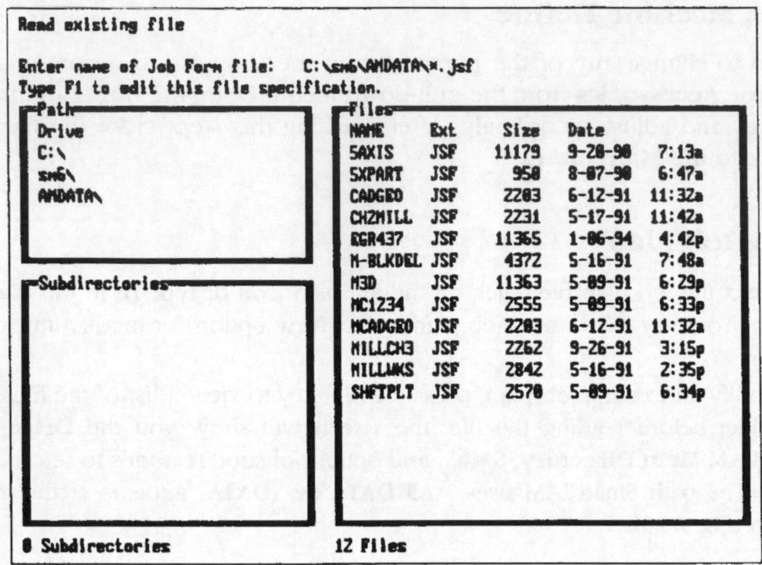

FIGURE 60.5 Job Plan files to select.

```
Read existing file
Save            Read            List_dir        Del_File        Quit

       Filename: ch2mill.jsf
           Date: 10/20/89
          Units: Inch
   Machine type: Mill
        Machine: M_FANUC
       Customer: PCC
           Dwg#: mc1234
           Part: Mold Cavity Insert
      Operation: Machine Insert Complete
 Material Blank:
          Notes:
```

FIGURE 60.6 The screen for the Job Info of the Job Plan.

Type—the type of tool is defined by selecting a number from the menu in the tooling section on the top section of the screen (Fig. 60.6). For example press [5] for an **End Mill**, press [**Enter**], and you will see the tool name shown in place and you are advanced to the next step.

ID#—optional tool number; you can skip this one.

CAD—this selects a layer of the CAD file to assign this tool to (in case you are using a CAD file), but you can skip this part if you are not using a CAD file.

DOFF—this part does not pertain to all controls. It specifies the number of the diameter offset register in the CNC control (not the actual value of the diameter offset); skip this one for now.

```
Edit - change data for existing tool
Edit          Insert          Delete          Move          Quit

TL# Type       ID#    CAD Doff Loff  Diamtr Length Ang/Rd Speed       Feed
  1 Drill      SHFT_1                0.7650 4.0000 118.00   374 RPM   5.24  IPM
 Comment: Drill Pilot Hole for Pocketing

  2 End Mill   SHFT_2                0.7500 1.0000   0.00   382 RPM   3.06  IPM
 Comment: Rough Mill Pocket

  4 Spot DrillSHFT_4                 0.2500 0.5000  90.00  1146 RPM   9.17  IPM
 Comment: Spot drill all Holes

  5 Drill      SHFT_5                0.5000 2.0000 118.00   573 RPM   4.58  IPM
 Comment: Drill Pivot Hole

  6 Drill      SHFT_6                0.2660 2.0000 118.00  1077 RPM   8.62  IPM
 Comment: Drill 4 Corner holes Through

  7 C'Bore     SHFT_7                0.4375 1.2500          655 RPM   5.24  IPM
 Comment: Counter Bore 4 Places .35 Deep
```

FIGURE 60.7 The tooling section of the Job Plan.

LOFF—this specifies the number of the length offset register in the CNC control (not the actual length offset value), and may not pertain to all controls; skip this one for now.

Diamtr—selects the diameter of the tool, either as a fraction or decimal value. For our example, you should enter 0.50.

Length—this is the tool's length as defined by the flute length, maximum cutting depth of an insert, or the length from the end to the first full thread of a tab. For our example, enter 0.75.

Ang/Rd—this section is for tool point angle and will be automatically skipped if you are using a ball mill, tap, reamer, and bore. Tool point angles are defined as follows: Spot Drill includes angle of drill point; Drill same as spot drill; End Mill angle need not be input or can be 180 or 0; face Mill lead angle of the cutter; and Draft Mill draft angle of the cutter, enter 0.00.

Speed—RPM, may be determined using the speed and feed calculations, or you can input the desired value of 573.

Feed—IPM, also can be input through the system's speed and feed calculations, enter 4.58.

- press [**Esc**], then press [**Q**] to quit, and press [**L**] to leave the job plan. SmartCAM indicates that you need to save your changes. Press [**Y**] to save them [chose any name you like or **SHFTPLT.JSF** to be consistent with the SmartCAM title (SmartCAM User Manuals, 1993)]. The SmartCAM **Main Menu** reappears and you can start the modeling process.

Step Three: Convert CAD to CAM

SmartCAM is capable of using geometry from virtually any CAD system. **CAM Connection** (Fig. 60.1) converts CAD files into information that SmartCAM can use to create the geometry for a CNC process model. With the proper settings included in the

Job Plan, SmartCAM automatically assigns tools to the specified layers and resequences the geometry into an acceptable tool path profile. To achieve an acceptable tool path, use the **CAM Connection** function to convert a CAD file into a SmartCAM file. In this example, we are not going to do that, instead we will create the model using SmartCAM Shape capabilities in the **Application** File.

Step Four: Create the Model

To start creating the model shown in Figure 60.3, you will need to complete the following steps:

- Select **Applications** from the SmartCAM **Main Menu**. Select **Advanced Machining—Milling** from the *Applications* menu (Fig. 60.1). When you enter the **Milling** Applications, SmartCAM displays a menu bar at the top of the screen with labeled sections (Fig. 60.8). Each of these sections is a pull-down menu with its own set of options. When you select one, the available tool boxes, dialogue boxes, or submenus appear grouped according to function. Selecting one of these options is similar to pulling a toolbox out of a tool crib.

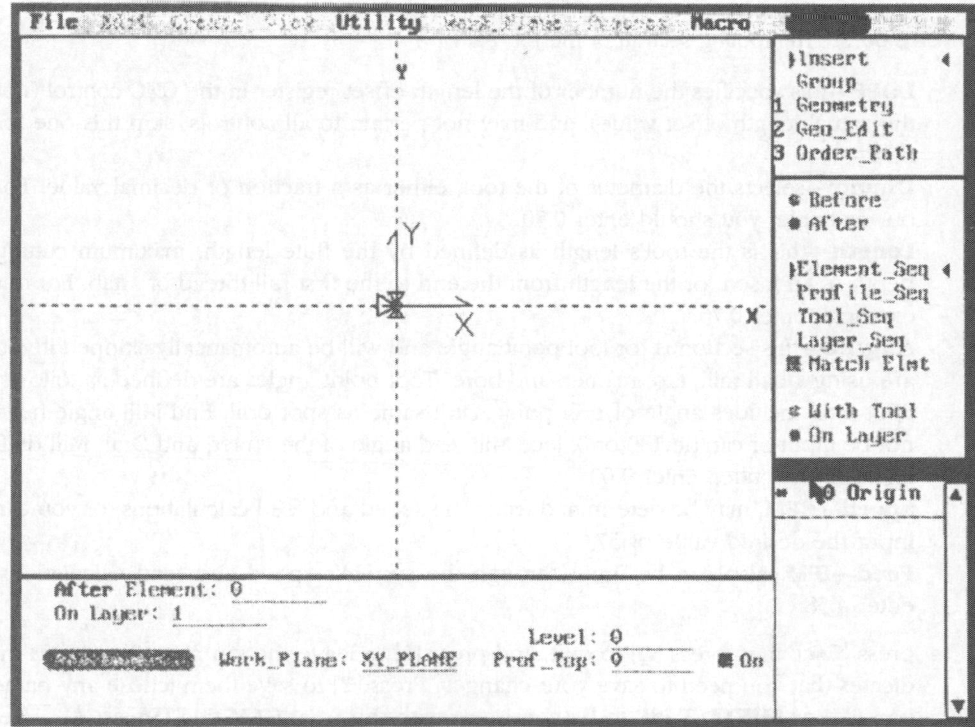

FIGURE 60.8 The SmartCAM screen after entering the Milling Applications.

- Select **File** from the top menu bar by moving the pointer until it points to **File**, and then click the left mouse button to "open," or pull down, the menu. Select **New** and the dialogue box appears asking for a **Job Plan** file. Now use the **File Select** button to choose the appropriate **Job Plan**, and the **File Select** dialogue box appears. Position the pointer over the bottom scroll bar arrow (on the right of the file list) and hold down the left mouse button until the **Job Plan** file called "**SHFTPLT.JSF**" appears. Click the left mouse button to select the file, click on the **Accept** action button, and SmartCAM reads the **SHFTPLT.JSF** file.

- Before you work on the model, you need to specify the sequence of operations and the properties of the part that you will machine. Select **On Layer** or **With Tool** (Fig. 60.8); your choice determines whether new geometry generates machine codes or defines nontool path elements, such as clamps and hold downs.
- To create the part of Figure 60.3, you need to define a part blank. In SmartCAM, and to speed up the modeling process, it is a good practice to create special files that include commonly used geometry such as machining-center tables, fixtures, part blanks, and other information typically used in CNC operations. To demonstrate this idea, you will use an already defined part blank with nothing else in the file. Select **Read** from the **File** Menu; use **File Select** to read in the **SHFTPLTB.SH3** file (Fig. 60.9).

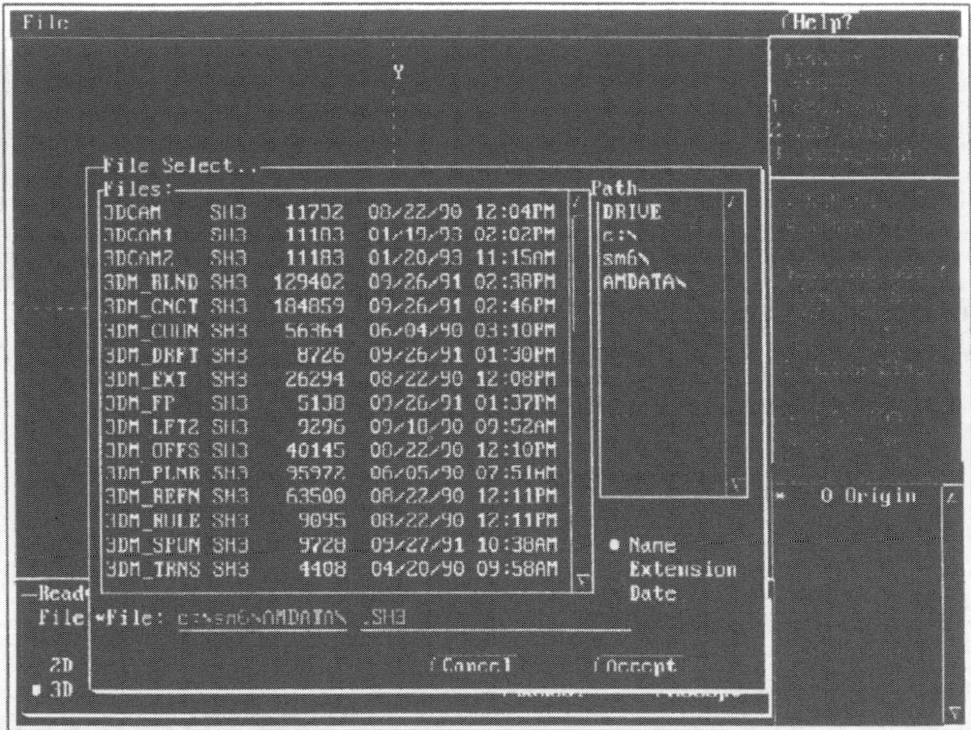

FIGURE 60.9 The SmartCAM screen to select a file in the Milling Applications.

- This is a part blank (Fig. 60.10) that can be used for many different designs; the only thing that you need to do is to adjust its given dimensions to suit your new design.
- To view the isometric shape of part blank, select **Get_View** from the **View** menu and pick ISO (isometric) (Fig. 60.11).
- In Figure 60.3, you see that there are four holes to be drilled in the corners of the part. You need to tell SmartCAM when and how to cut these holes. To do that, call up the tool by selecting the **Insert** toolbox (Fig. 60.10), then select **After** and **Element-Seq**. The **Match Elmt** should be **Off** to avoid using the same properties of the preceding element, which is on a layer. Click on **With Tool** in the control

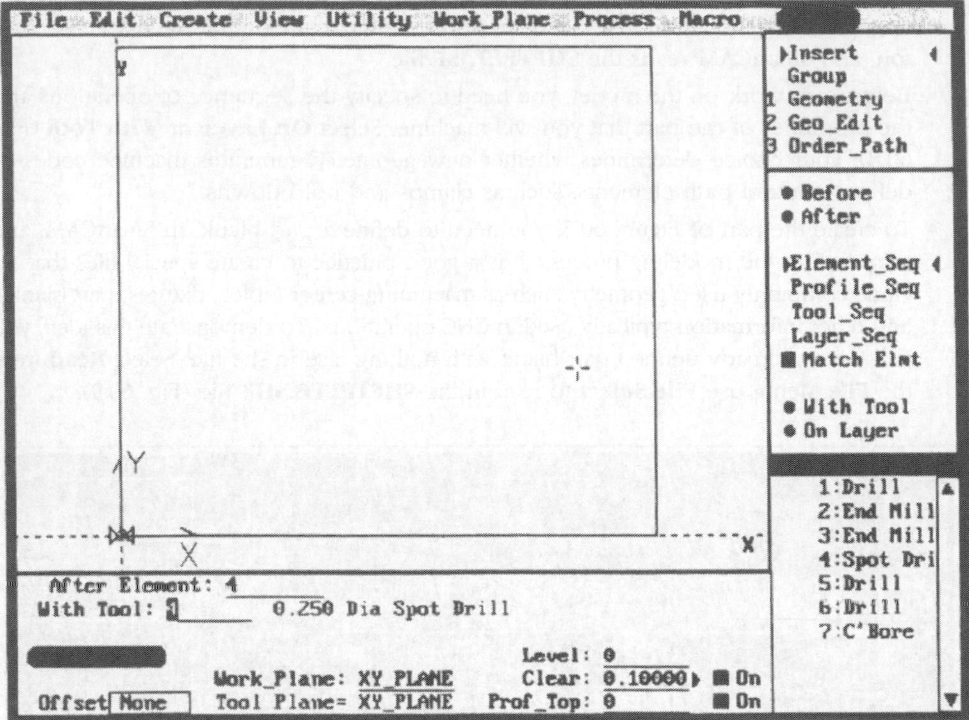

FIGURE 60.10 The part blank named SHFTPLTB.SH3.

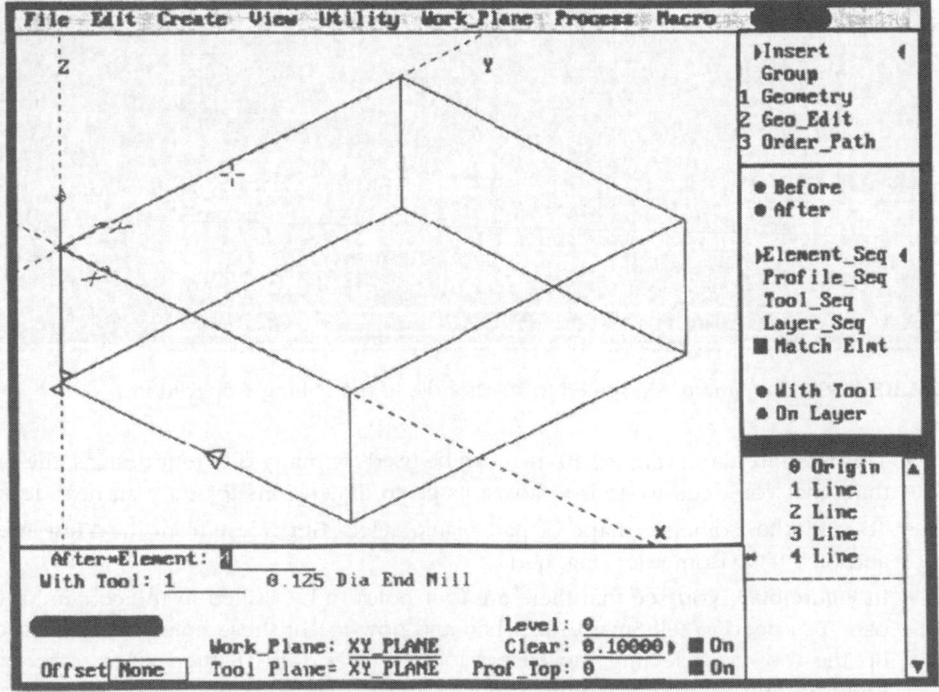

FIGURE 60.11 The isometric view of the part blank.

panel and pick tool 4 (a **0.25 [8.00] Spot Drill**) from the list view, set the **Offset** at None for Drilling, and set **Clear** to **0.1 [2.0]** above the part.

- Now, enter tool path geometry to tell SmartCAM **where** to drill the holes. From the **Create** toolbox (Fig. 60.10), select **Geometry**, select the Hole tool, enter a **Spot-Dia** of **0.25 [8.00]**; SmartCAM automatically calculates and enters tip depth (use **View** to observe this using **Top View**). There are different methods to enter coordinate values, and following is a demonstration of four of these methods.
- To specify the lower left hole, click on the X input field for **Hole Point** (Fig. 60.12), and enter **0.5 [15.00]**. Enter **0.55 [15.00]** for **Y**, and notice the hole appears on the screen. For the upper left hole, click on **X** in the control panel, and snap to the first hole to pick up the X value. For Y, enter **0.55 + 3.55 [15.00 + 85.00]** (calculated input), and again, a hole appears at the top of the screen.

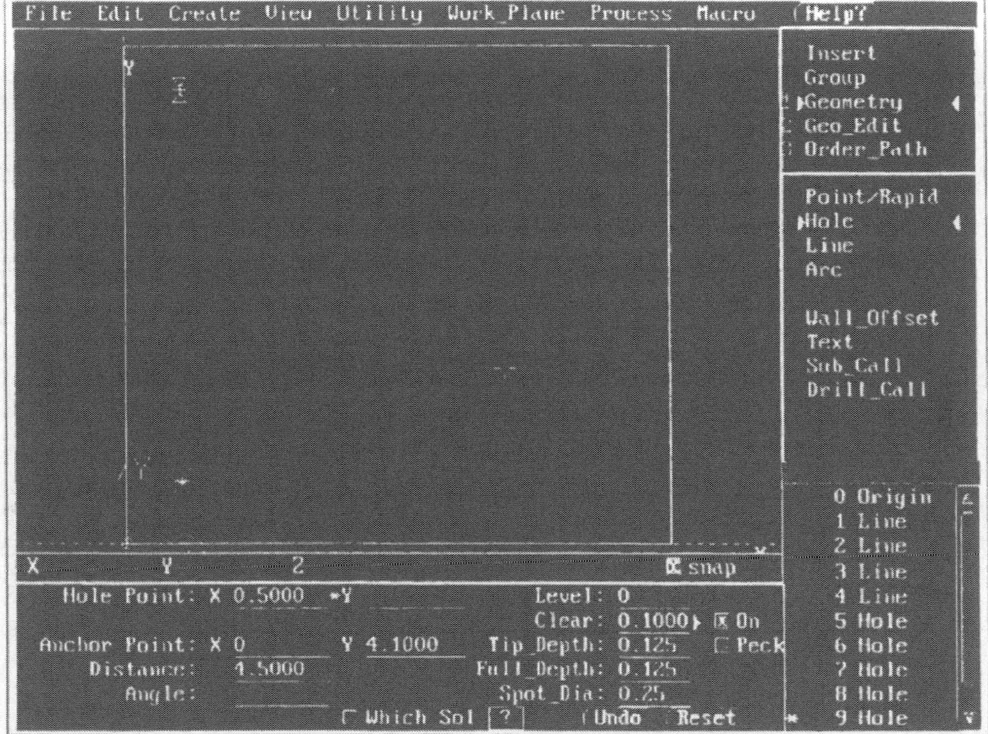

FIGURE 60.12 The screen in SmartCAM that shows Hole Point operation.

- To enter the upper right hole, use the mouse. Click on **Hole Point** so you can enter both coordinates at the same time. Turn Snap off, watch the Read-out Line, and when you have X = **4.6 [115.00]** and Y = **4.1 [100.00]**, click the mouse to enter the information.
- For the last hole, turn **Snap On** so that you can use existing holes for the X and Y values. Click on the X of the **Hole Point** and snap to the upper right hole for the X value; then, click on the Y of the **Hole Point** and snap on the lower left hole for the Y value, and hit return.
- After locating the holes, the next step is to perform the drilling and counter bore operations. **After Tool 4** appears in the control panel, use the **Insert** command, and

then click on **With_Tool**. Find the **0.266 [7.00]** Dia Drill in the list views (**Tool 6**) and select it; the other settings remain the same.

- In the **Geometry** toolbox, enter **1.0 [25.00]** in the **Full_Depth** option in the control panel. For optimal tool path, click on **Hole Point** in the control panel, then snap to the lower right hole (the new drilling operation appears there), and the **Hole Point** trigger is active (note the asterisk). Then snap to the upper right, the upper left, and the lower left holes, in that order.
- Now, you need to change to the counter bore tool. Select **Insert-After-Tool 6**, click on **With Tool** in the control panel, and select **Tool 7** from the list view. Enter **0.35 [10.00]** as the **Full Depth** in the Geometry toolbox. Click on **Hole Point** for the trigger, and enter the four holes in tool path order. Again, start with the lower left, move to the upper left, and so on, to optimize tool path.
- To view the model, select **Get_View** from the **View** menu and pick ISO (isometric). Pick **View-Show_Path**, set the speed to 5, and click on **Start**. To check vertical clearance for the rapid moves, select **Get_View-Right**, and watch each of the tools move at the specified Z clearance plane and drill to the indicated depth.
- With **Show_Path**, you can verify exactly what will happen at the machine. If these holes were all that the part required, you could go directly to Process Code, generate the machine code, and you would be finished. But in this example, there are several operations to be performed.
- From Figure 60.3, you see that the drawing indicates that additional spot drill holes need to be machined; however, we will leave that as an exercise for you to do. The completed job should look like the one shown in Figure 60.13.

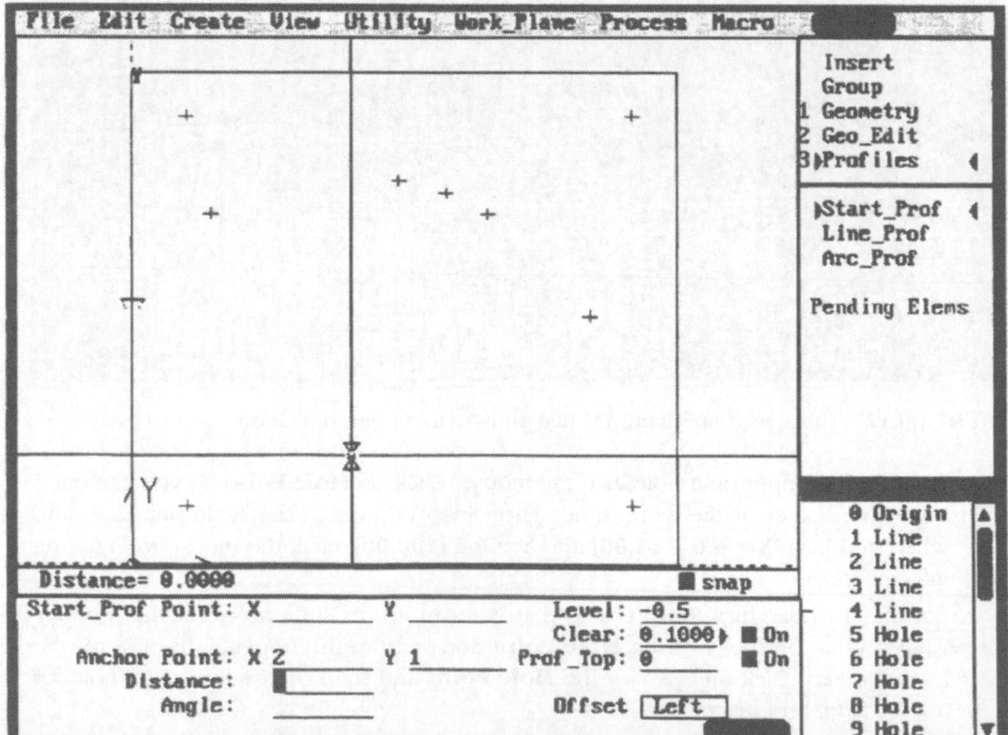

FIGURE 60.13 The part after locating the five spot drilling holes.

- The next step is to cut the pocket, as shown in Figure 60.3. In SmartCAM, there are two methods to cut the pocket and to enter geometry—free-form and profile. Both feature unique solution logic to make the process faster and easier for you. Free-form geometry is what you have been using to enter holes, but you can also use it to construct lines, arcs, and points (rapid moves). We will discuss this pocket operation using profile.
- To mill the pocket before drilling any holes, select **Insert Before Tool 4 With Tool 3** (a **0.500 [12.00]** Dia. End Mill) for finishing; select **Offset Left**, with **Level –0.5 [–10.00]** for the depth of the pocket; select **Profiles** from the **Create** pull-down menu; and pick **Start_Prof**.
- Use the lower **0.5 [12.00]** Dia. Hole to determine where the profile starts, and progress counterclockwise around the pocket. Click on **Anchor Point**, and snap to the hole center (Fig. 60.13).
- Enter a **Distance** of **0.5 [15.00]**, the radius of the tangent arc, and for **Angle** value, enter a calculated input of **30–90**. Select **Line_Prof**, press the **Advance** button, and then change to **Arc_Prof**. Select **Advance**, since the end point of this line is not known. Select a **Tangent** arc, **CCW**, with a **Radius** of **0.5 [15.00]**, snap to the upper right spot drill hole for the arc center, and SmartCAM solves for the line (Fig. 60.14).

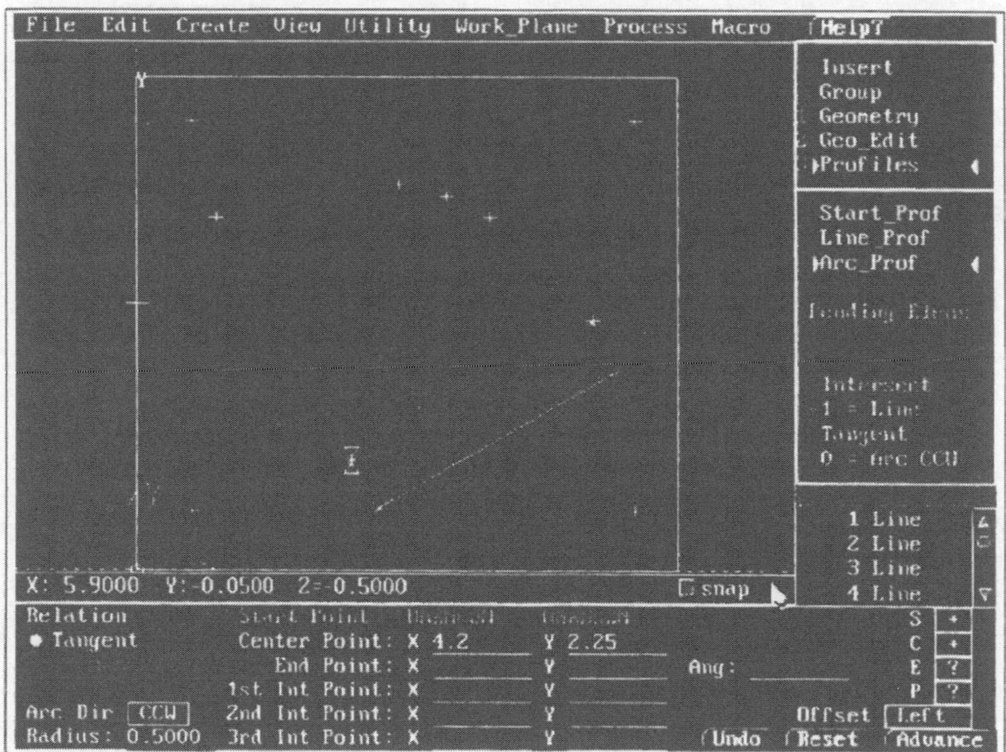

FIGURE 60.14 Solving pending elements.

- To place the large arc at the top, select **Advance**, then select **Tangent** relationship, **CCW**, and enter a **Radius** of **3.00 [75.00]**. Since you do not know the center point or an intermediate point, press **Advance** and move to the next arc.

- Enter a **Radius** of **0.5 [15.00]** and snap to the center of the upper left spot drill hole for the arc center point. You will see a possible solution and a Which Solution dialogue box. Click on **Other** to look at different solutions, and then **Accept** the one shown in Figure 60.15.

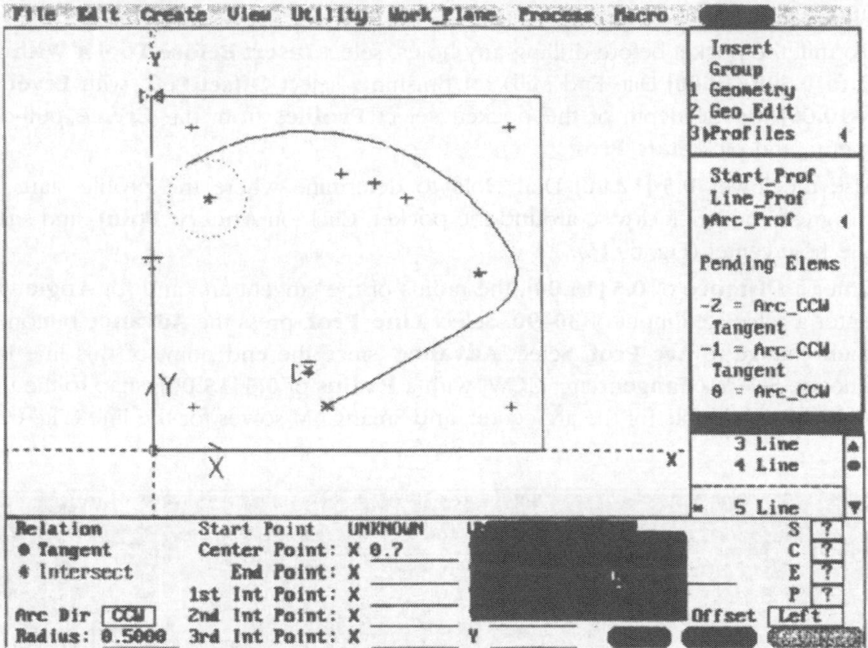

FIGURE 60.15 Completing the large Arc.

- Select **Advance** and then **Line_Prof**; press the **Advance** button, select **Arc_Prof**, set **Relation** to **Tangent**, and enter a **0.5 [15.00]** radius. Again, since you do not know the line's end point, press **Advance** to start defining the arc, and let SmartCAM solve for the line later.
- For a center point, snap to the center of the lower hole. SmartCAM solves for a line tangent to the two arcs, so close the profile by entering the arc **End Point**. To do this snap to the start of profile, pick the proper solution, and **Accept**. Redraw the screen by pressing **[F8]** to remove the extra lines, which allows you to see a clean profile. Then, you need to add the roughing operation.
- To rough out the pocket, select **Rough** from the **Process** menu, choose **Pocket**, and enter the following settings: Pocket: Spiral: Outside Boundary: Snap to finish profile; Finish Amt: **0.00 [0.00]**; Width of Cut: **0.25 [6.0]**; First Pass Depth: **–0.10 [–2.5]**; Depth of Cut: **0.10 [2.5]**; Final Depth: **–0.50 [–10.00]**; and **Ramp Angle: 45**. Since you want to rough the pocket before finishing, you will need to rearrange the sequence and insert a tool.
- Select **Insert** and check the sequence location **Before Tool 3**. Select **With Tool—Tool 2**, return to **Process—Rough—Pocket**, select **Spiral**, and snap to the pocket boundary. Leave a finish amount of **0.05 [1.00]**, and use the defaults on the other settings. Finally, select **Go** (Fig. 60.16).
- You may use View-**Show_Path** to view the actual tool motion, and **Show_Path** with ISO view and animated 3-D tools (Fig. 60.17). Press [**Esc**] to stop.

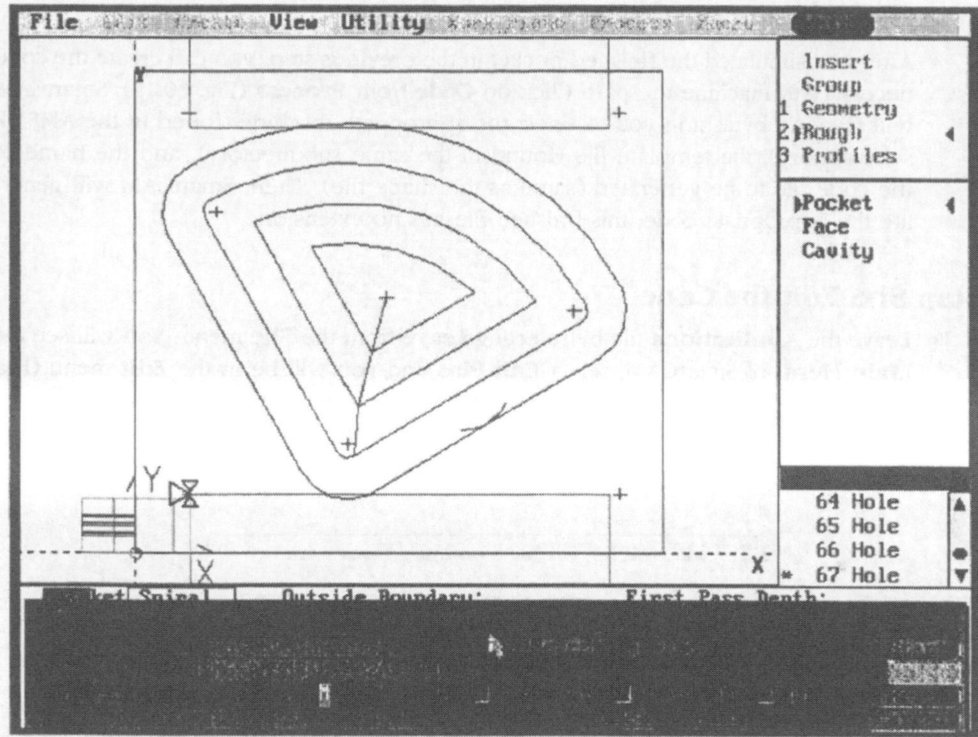

FIGURE 60.16 Completed roughing passes.

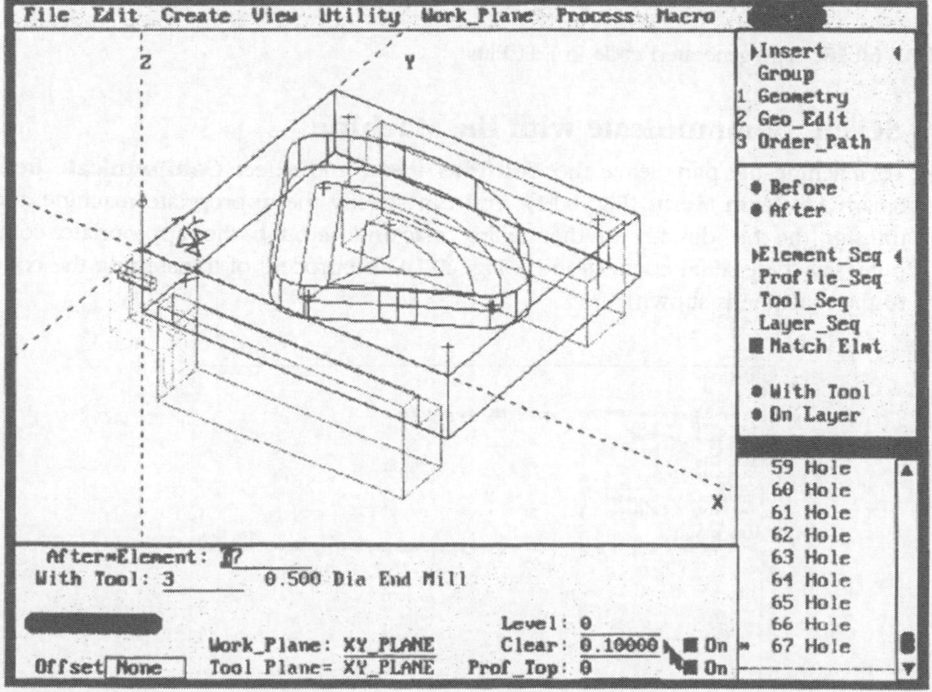

FIGURE 60.17 Isometric view of the generated Pocket.

Step Five: Create the Code

- After you simulated the finished pocket in the previous step, you can create the code necessary to machine the part. Click on **Code** from **Process** (Fig. 60.17). SmartCAM will respond by asking you to select the appropriate machine (found in the AMSMF subdirectory), the template file (found in the same subdirectory), and the name of the code file to be generated (same as the shape file). Then, SmartCAM will generate the appropriate code; this finished file has no extension.

Step Six: Edit the Code

- Leave the **Applications** file by selecting **Leave** from the **File** menu. You will see the **Main Menu** of SmartCAM, select **Edit Plus** and you will be in the **Edit** menu (Fig. 60.1). Use **Read** to bring in the code that you just generated and you will see the generated code (Fig. 60.18). Use **Quit** from the top menu, and click on **Edit** to review and modify the generated code if necessary.

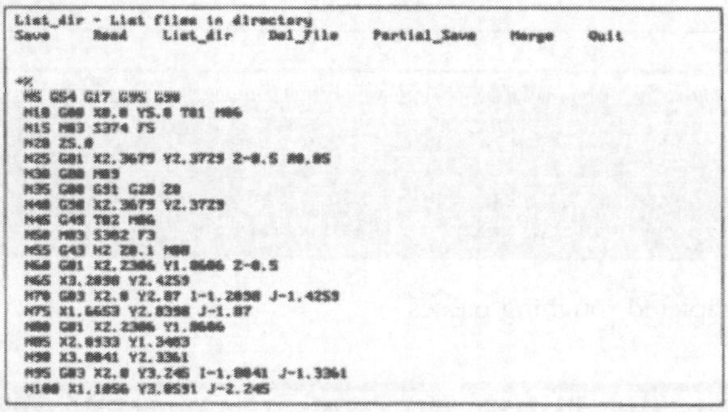

```
List_dir - List files in directory
Save     Read     List_dir     Del_file     Partial_Save     Merge     Quit

%
N5 G54 G17 G95 G90
N10 G00 X0.0 Y5.0 T01 M06
N15 M03 S374 F5
N20 Z5.0
N25 G01 X2.3679 Y2.3729 Z-0.5 R0.05
N30 G00 M89
N35 G00 G91 G28 Z0
N40 G90 X2.3679 Y2.3729
N45 G49 T02 M06
N50 M03 S302 F3
N55 G43 H2 Z0.1 M08
N60 G01 X2.2306 Y1.8606 Z-0.5
N65 X3.2090 Y2.4259
N70 G03 X2.0 Y2.87 I-1.2090 J-1.4259
N75 X1.6653 Y2.8398 J-1.87
N80 G01 X2.2306 Y1.8606
N85 X2.8933 Y1.3403
N90 X3.0041 Y2.3361
N95 G03 X2.0 Y3.245 I-1.0041 J-1.3361
N100 X1.1056 Y3.0591 J-2.245
```

FIGURE 60.18 The generated code in Edit Plus.

Step Seven: Communicate with the Machine

- To machine the part, leave the **Edit Plus** menu and select **Communicate** from SmartCAM **Main Menu** (Fig. 60.1). You can specify the appropriate machine and transfer the file directly to the milling machine through the appropriate computer–machine serial connection (in Fig. 60.19, the process of transferring the code to the machine is shown).

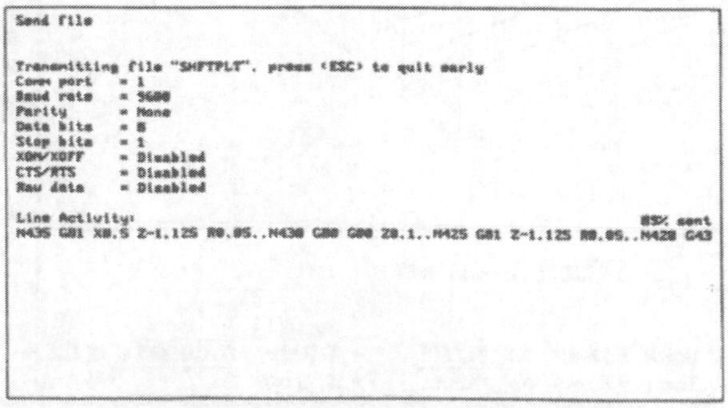

```
Send file

Transmitting file "SHFTPLT". press <ESC> to quit early
Comm port     = 1
Baud rate     = 9600
Parity        = None
Data bits     = 8
Stop bits     = 1
XON/XOFF      = Disabled
CTS/RTS       = Disabled
Raw data      = Disabled

Line Activity:                                                       85% sent
N435 G01 X0.5 Z-1.125 R0.05..N430 G00 G00 Z0.1..N425 G01 Z-1.125 R0.05..N420 G43
```

FIGURE 60.19 Transferring the generated code to the Milling Machine.

DEFINING TERMS

CAD: Computer-aided design.

CAM: Computer-aided manufacturing.

DFM: Design for manufacturing.

NC: Numerical control.

CNC: Computer-numerical control.

CIMS: Computer-integrated manufacturing systems.

DNC: Direct-numerical control.

FMS: Flexible-manufacturing systems.

FMC: Flexible-machining cell.

GUI: Graphical-user interface.

REFERENCES

F.M.L. Amirouche, *Computer-Aided Design and Manufacturing*. Prentice Hall, Englewood Cliffs, NJ, 1993.

D.B. Bedworth, *Computer-Integrated Design and Manufacturing*. McGraw-Hill, New York, 1991.

CAM Software Update, *Manufacturing Eng*. 111(1):28, 1993.

T. Chang, R.A. Wysk, and H. Wang, *Computer-Aided Manufacturing*. Prentice-Hall, Englewood Cliffs, NJ, 1991.

J.R. Coleman, PCs make it with MEs. *Manufacturing Eng. Part I*. 50–55, 1990; and Revisited, 32–36, 1990.

R.H. Jackson and A.W.T. Jones, An architecture for decision making in the factory of the future. *Interfaces* 17(6):15, 1987.

B. Kramer, NC code generators. *MicroCAD News* 5(5):39–41, 1990.

B. Kuttner, Approaching the factory of the future. In *Automated Assembly*, 2nd. ed. J.D. Lane (ed.) SME, 1986, 417–420.

MicroCAD News Staff, Pairing CAD and CAM systems for manufacturing. *MicroCAD News* 31–35, 1988.

P.M. Noaker, Job shops enter the information age. *Manufacturing Eng*. 111(2):27–31, 1993.

M. Puttre, Computer-aided manufacturing: Sculping parts from stored patterns. *Mech. Eng*. 114(4):66–70, 1992a.

M. Puttre, Advances in CAM signal: A manufacturing renaissance. *Mech. Eng*. 60–63, 1992b.

S. Romanini, The importance of R-D in computer-integrated manufacturing. In *Flexible-Manufacturing Systems: Proceedings of the 6th International Conference*, Italy, 4–6 November. G.F. Micheletti (ed.). 1987, 121–125.

W.L. Shrensker, A brief history of CIM. In *A Program for CIM Implementation*, 2nd ed. *A Project of the CASA/SME Technical Council*. SME, 1987, 163–164.

SmartCAM V7.0 User Manuals, Point Control Co., P. O. Box 2709, Eugene, OR, 1993.

G. Spur, G. Selinger, and B. Viehweger, Cell concepts for flexible automated manufacturing. In *CAD/CAM: Integration and Innovation*, 1st. ed. K. Taraman (ed.). SME, 1985, 197–213.

H.W. Stoll, Design for manufacturing. *Manufacturing Eng*. 100(1):67–73, 1988.

R.M. Thacker, *A New CIM Model*, A Blueprint for the Computer-Integrated Manufacturing Enterprise. SME, 1989.

D.F. Urbaniak, The unattended factory FANUC's new flexibility automated manufacturing plant using industrial robots. In *Automated Assembly*, 2nd. ed. J.D. Lane (ed.). SME, 1986, 417–420.

C. Watkins, Competitive manufacturing: AutoCAD integrated with SmartCAM. *CADENCE* 2(10):54–58, 1987.

Data Communications and Networking

This introduction explains the basics of data communications and networking. Our ever increasing ability to transmit data stems from (1) insights into the characteristics of electrical energy (which we express as concepts) and (2) techniques that exploit these characteristics. Each concept is presented as a brief sentence that names and describes it. To help the reader build a mental picture, the concepts are grouped logically.

Readers who master this terminology (thus building the mental picture) and its associated techniques should understand enough not only to be able to use simple data communications equipment (a personal computer loaded with a standard communications software program, for example), but also to solve simple problems, to see new ways to employ data communications, and to carry on informed discussions with data communications experts.

Data communications and networks have existed for over a century and are continually evolving or being replaced. The telegraph, the first electronic communications device, transmitted data using discrete signals in two states: a dot and a dash. Telephones came later and became more widely used because they transmitted the human voice as a continuous range of signals between two limits. (That is, the telephone used continuous or analog signals.) Now computers with their two state, discrete signals (one and zero) have become widespread, but the only communications system that can serve most of them (the telephone system) still uses some analog signals. The answer has been the modem, which converts analog to digital (two-state) signals and digital to analog. The telephone companies have found digital communications so efficient that they have replaced their analog circuits between their local exchanges with digital circuits leaving analog telephone signals in use only between individual telephones and local signals.

Thus, while analog signals (and modems) are widely used in data communications, digital circuits are being introduced wherever it is economically feasible to do so. Readers who grew up with and now use modems and analog telephone circuits should be aware that they may not need them in a few years. It is, therefore, not enough to know how to use today's data communications and software: you will need to understand the theory on which tomorrow's software and hardware will be built.

BASIC TERMINOLOGY

This section, the first step in presenting the mental picture of data telecommunications and networking, lists the most basic concepts. Terms in **boldface** are the basic vocabulary we are trying to convey.

Telecommunications is passing information over distance. **Data communications** is passing information coded for machine use. Data are represented by **binary values** (1 or 0). Binary values work well for numbers. To represent other things (letters, sounds, and graphics, for example) codes are used. ASCII is the most widely known code. Each binary value is called a **bit**. Bits are strung together to form binary numbers. To form ASCII code, eight bits are grouped together to form a **byte**. Each unique pattern of bits represents a different character (e.g., in ASCII, 1000001 represents the letter A).

Multimedia communications conveys information in multiple forms: audio, video, data, graphics, etc.

A **Network** is a collection of senders and receivers ("nodes") that can communicate to or through each other or through "links." There are two categories of networks: **local area networks (LAN)** and **wide area networks (WAN)**. A LAN connects computers in, at most, several buildings that are close to each other. A WAN connects more widely separated computers. Computers that are connected by a LAN or WAN can be formed into a **distributed system**, a group of computers that can operate together sharing data and software capabilities.

The arrangement of computers in networks has been categorized as **network topologies**. The more common topologies are star, ring, tree, and bus (a bus is a data channel that has multiple input/output taps—in essence a piece of wire with multiple connections).

Networks can set up to distribute data in one of two ways: **broadcast**—data are transmitted to more than one recipient at a time and **switched**—data are transmitted to a specific station. (This may and often does require that transmissions be relayed through a series of intermediate nodes.)

The **medium** is anything that can pass information. The medium connects a *sender* to a *receiver* enabling them to communicate. Every medium introduces errors in the data to a greater or lesser degree.

Duplexity describes the capability for data flow. In a **simplex** system, communications can flow in one direction only. In a **half duplex** system, communications can flow in both directions but only one way at a time. And in a **full duplex**, communications can flow in both directions simultaneously.

The terminology presented so far has built a model of a telecommunications, data communications, and the networks over which these activities are conducted. We now turn to the concepts that describe how data are passed between computers.

Bandwidth is the range of frequencies that can pass through a medium calculated as the difference between the highest and lowest frequencies that can be transmitted. This is important because the amount of information a signal can hold is proportional to the signals bandwidth.

Bits per second (bps) describes the rate at which bits are passed over a telecommunications circuit, for example, 2400 bps. **Bytes per second** is the rate in bps divided by eight.

The **signaling rate (Baud)** is the rate at which electronic signals are *sampled* each second. As we will see, this can but need not be equal to the bps.

Data can be transmitted faster if each character follows another without time between them. For this to happen, the sender and the receiver must share common, precise timing. When this happens, the transmission is **synchronous**. For high-speed (and expensive) circuits, the solution often is to let the telephone company provide the clock or "timing." Low cost, low speed circuits use **asynchronous** transmissions. Instead of using a clock, "start" and "stop" bits are included in transmission with a gap between characters. A start bit tells the receiving computer that a data transmission is starting and, therefore, that the sampling should start.

Because computers work with 1s and 0s, to process continuous or "analog" signals they must sample them to turn them into numbers. The rate at which this sampling occurs is the **sampling rate**. The information a sampled or incoming signal can convey is limited by the frequency with which it is sampled. To capture all information, it must be sampled at the *Nyquist rate*, i.e., at twice the highest analog frequency of the signal. For example, a 300-kHz signal must be sampled 600,000 times a second for all the incoming information to be captured.

In synchronous transmissions, the sending source must provide a whole block of information at once. This block is called a **frame** and is identified by **flag** characters, the beginning and end of the frame. A common flag is the byte (eight bits, an **octet**) 01111110. When the receiver senses this pattern, the receiver knows were to find the incoming data bits without start and stop bits.

A **session** is a continuous period of data exchange among two or more communicating elements on a network.

The **components of electronic signals** are **amplitude, period, frequency**, and **phase**. In the formula $f(t) = A \sin(2\pi ft + \phi)$, A is the amplitude, f is the frequency, and ϕ is the phase.

There are two types of current: **alternating (AC)** and **direct (DC)**. In AC, current flows in first one direction and then the other. Each change of the direction of flow of the electricity is called a cycle. The number of **hertz** is the number of cycles per second.

We now turn to the terminology used to categorize the equipment for data communications. The **channel** is the medium (e.g., telephone line) over which data are transmitted. A **host computer** is the physical device (computer) in which the data to be transmitted over the channel are located. The host computer or other user equipment connected to the communications circuits is the **data terminal equipment (DTE)**. This may or may not be the host computer. Sometimes a special computer is provided for this purpose. Such a computer is called a **front end**. The DTE connects to **data circuit terminating equipment (DCE). Modems**, which convert the digital signals of computers to and from the analog signals used by telephones, are examples of DCE. (The name Modem is a contraction of modulator–demodulator, the functions performed by a modem.) If the computer's digital signals are to be transmitted as digital signals, a **digital service unit/channel service unit (DSU/CSU)** converts a computer's digital signals to the digital signals transmitted over a communications circuit that uses digital signals. A DSU/CSU transmits and receives data at 56 Kbps or faster rates.

A **multiplexer** is used to combine the signals from several inputs and transmit them over one line. This can increases the amount of data that can be transmitted over available communications circuits.

An **internetworking unit** (also called a **bridge** or **router**) connects two or more networks.

STANDARDIZATION ORGANIZATIONS

For data communications to work, myriad specific details must be in order. To facilitate the standardization of all these details, various agreements called protocols have been developed, commented on, and published as standards. These agreements cover a wide variety of topics, some of which will be mentioned later. Agencies that publish these agreements or protocols include the International Telecommunications Union (ITU-T, previously known as CCITT), the International Standards Organization (ISO), the American National Standards Institute (ANSI), the Electronic Industries Association (EIA), the National Communications System (NCS), the National Institute of Standards and Technology (NIST), and the Internet Engineering Task Force (IETF).

TRANSMISSION MEDIA

We now turn to the media over which data are transmitted: wire, coaxial cable, microwave, satellite, and optical fiber. Information is passed over mediums as bits in the form of "state values." For current this can be as a + or – (flow of charge in "current"). For optical fiber, this can be the presence or absence of light.

Wire is frequently used in twisted pairs, sometimes "shielded" with foil. With wire, the signal is **attenuated** (made weaker) with distance, however, the larger the wire, the less the signal is attenuated. Twisted pairs can handle speeds up to 10 Mbps, however, the lower the speed the further the signal goes without being so degraded as to be useless. To reduce effects of "noise," **balanced** lines can be used.

Coaxial cable is a cable that contains two conducting elements. One is an inner core of wire surrounded by an insulator. The other is an outer wire (usually made of braided metal). Both inner core and outer wire share a common axis, thus the name *co*axial. Coaxial cable are noise resistant, can transmit frequencies to tens of MHz, and handle data rates to hundreds of Mbps.

Microwaves are waves that have a relatively short wavelength (and therefore relatively high frequency), at frequencies of hundreds of millions or billions of cycles per second measured in megahertz or gigahertz. Microwaves are line-of-sight, that is need antenna that can "see" the antenna at other end. This means microwave antennas generally are located on high points. Microwaves can carry a lot of data, tens of Mbps. They require a license and maintenance and are easily monitored by third parties.

Satellites are microwave transponders (repeaters) in geostationary orbit (22,300 miles above equator). They have high data rates, a broadcast capability, the same cost and delay (about 1/4 second) for sites anywhere in satellite "footprint," and a significant bit error rate—often one in 10^4 or 10^5 (often represented as 10^{-4} or 10^{-5}). For some satellite links sun interferes for 10 minutes per day for two 5-day periods a year.

Optical fiber passes light through very thin glass fibers and transmits data at a high rate. For example, OC3 optical fiber passes 155 Mbps. Optical fiber is immune to electrical interference, cannot produce sparks, has low attenuation (repeaters are placed every 20 miles in the best case, compared to the 2.8 miles used on copper T1 lines), and are hard for third parties to tap because there is no radiating signal. Optical fibers are also very thin and very robust, and have a very low error rate: one in 10^9 (or 10^{-9}) or better. There are two versions, multimode and single mode. The latter carries data at higher rates. Optical fiber is used in high-speed LANs and the carrier backbone of commercial telecommunications operations including submarine cables.

Modems and DSU/CSUs are used to connect computers to wire, coaxial, or microwave circuits: modems for analog circuits, DSU/CSUs for digital circuits. The DSU/CSU may also include a **line driver** to send the digital signal over copper wires (up to a few miles). If external modems (not built into a computer) are used, **interfaces** connect the computers and the modems. The **EIA-232-D ("RS232")** and **EIA-449 ("RS449")** are widely used interfaces. Interfaces derive their names from the protocols that describe their specific wiring plans.

PLACING DATA ON THE SIGNAL

Data inside a computer are represented by on and off, 1s and 0s. If the data are to be passed digitally to another computer, the first computer need only send a string of bits in the form of 1s and 0s (current or no current) to the other computer. If, however, the channel between the two computers is an analog circuit, the answer is not so simple because there is a minimum and a maximum (not an on and off) and an infinite number of values between them. The phrase **"modulating the signal"** describes the process by which digital data are represented by an analog signal.

The answer lies in three aspects of a signal that were discussed earlier: amplitude, phase, and frequency. To place digital data on an analog signal using amplitude, high and low amplitude values are established. If a zero is to be transmitted, a low amplitude is transmitted for a specific period. If a one is to be transmitted next, the signal is increased to a high value and held there for a short period. At the receiving modem, the signals are monitored by the periodic sampling mentioned above. Values in the range of the high amplitude value are evaluated as representing a one and values of the low amplitude are evaluated as representing a zero. Similar use is made of the phase and frequency components of the signals. By combining these, a particular signal can convey more than one bit. It has been this ability to combine these components that has enabled modem makers to keep producing faster and faster modems.

ERRORS: THEIR SOURCES AND HANDLING

As has already been noted, errors will be introduced into some data when they are communicated. If the error is in a 25,000 word paper, it may be unimportant, but if the error changes a $100 payment to a $1000, it will be important. The following introduces error sources and handling with a focus on the information that will help readers determine what they need to do about error handling.

Extraneous signals that distort data come under the heading of noise. Types of noise include **White noise**—happens randomly due to the physics of electricity; **Electrical**

noise—caused by high-powered electrical sources near signal, e.g., arc welder; **Transients**—short duration changes in circuit (e.g., impulse noise and dropouts); **Cross talk**—interference from other channels in the same transmission medium; **Echoes**—signal reflected from the far end of circuit. Also substandard performance from communications components may distort the signal and cause errors. This happens mostly to analog signals.

The **signal-to-noise ratio** is used to state the relationship between a signal and the noise in a channel. Because the signal is usually much larger than the noise, the ratio is often scaled down logarithmically and stated in **decibels**, defined as 10 log(signal power ratio).

Error management methods include sending test signals, measuring the results, and taking appropriate corrective action; doing loop-back tests, tests that send a signal to the far end of the circuit, which is then sent back to the beginning of the circuit; using error detecting codes; keeping statistics on errors (most routers do this today); and using **forward error correction**, the sending of redundant information so the receiving station can make a good guess on what should have been sent if a possible error is discovered.

Digital signals are much less prone to errors than are analog signals.

NETWORK BASICS: SWITCHING AND PACKETS

Data communications requires a means to connect computers, a **connection strategy**. A **direct connection** (e.g., leased line) is the simplest, most expensive, and least flexible approach because there is no switching. **Switching**, the establishment of short-term connections between communicating elements, is the alternative to a direct connection. **Circuit switching** is making and maintaining a connection between two nodes *until one party terminates the connection*. **Packet switching** breaks messages into pieces or **packets**, which are relayed among **packet switches** or **routers** from source to destination. Two methods are usually used to route packets. In the **virtual circuit** approach, a route is established and all the packets for a message are transmitted over that route, i.e., through the same sequence of packet switches. In the **datagram** approach, each packet is transmitted independently. The packet switches are responsible for routing from source to destination, and routes change over time. This approach is different from circuit switching because the route is not reserved exclusively for traffic between the sending and receiving nodes. **Public packet switching** offers packet switching communications for sale as a service, often tied to the Internet.

A **packet** is a string of bits that has been structured for transmission over a network. A packet usually contains only part of a file to be transmitted because files are often divided between many packets before being transmitted and then reassembled at the receiving station. When a packet is built, the information to be transmitted has added to it information required, for example, to route the message, to reassemble the pieces of the message, and to check for errors.

REFERENCES

Uyless Black, *Data Networks: Concepts, Theory and Practice*. Prentice-Hall, Englewood Cliffs, NJ, 1989

Douglas Comer, *Internetworking with TCP/IP*. Prentice-Hall, Englewood Cliffs, NJ, 1991

William Stallings, *Data and Computer Communications*. Macmillan, New York, 1994

61

Modems and Interfaces

J. Mark Pullen
George Mason University
David Jensen
George Mason University

61.1 MODEMS

Modems exist to permit digital signals to be passed over circuits intended for analog voice signals, which means the data rate is inherently low. Because analog paths always involve an extra set of interfaces and often use older communications technologies that are prone to noise, modem links tend to have higher error rates. But because they are inexpensive and readily available, voice grade circuits with modems are widely used where their performance is sufficient to the application.

61.2 PHYSICAL CONFIGURATION

The most obvious choice in selecting a modem is the choice of internal or external location. This choice normally has little impact on performance (an exception is noted below), but it is important in assembling a computer configuration. Internal modems make for a more compact configuration, present fewer difficulties if the equipment must be moved, and normally have direct access to the computer's internal bus, which means there is no impediment to their performance; also they are generally less expensive as they can draw power from the computer's internal power supply. External modems have the advantage of flexibility: they connect to the computer externally (usually through a serial port), thus either computer or modem may be upgraded independently. However, external modems, just as internal ones, require communications software (see Chapter 63), configured with commands recognized by the modem. (In this regard the Hayes command set has become a defacto standard that is generally available in communications software.) If performance is an issue, it is advisable to check the data rate on the interface between external modem and computer.

61.3 MODEM TRANSMISSION RATE

The transmission rate of a modem (sometimes called its "speed") is the data rate in bits per second it is capable of supporting. Modems are normally used for asynchronous transmission in which each character has its own start and stop bits; accounting for this overhead the data rate in bytes per second will be about one-tenth

0-8493-2530-7/96/$0.00+$.50

TABLE 61.1 Common Serial Link Options

Type	Interface Unit	Data rate (bits per second)	Interfaces
Low-speed analog	Modem	300 to 1200	RS-232D or internal bus
High-speed analog	Modem	2400 to 28.8K	RS-232D or internal bus
Leased digital (carrier provided or microwave)	DSU/CSU	56K or 1.544M (NorthAmerica) 64K or 2.048M (Europe)	RS-449
Private digital (on premises)	Line driver	56K to 2.048M	RS-449
ISDN	NT1+TA	Two channels at 56K or 64K	RS-449 or internal bus
Broadband ISDN	ATM interface card	155M and higher	ATM to internal bus

the bits-per-second transmission rate. Older modems had operating speeds of 150, 300, 600, or 1200 bits per second, using amplitude or frequency shift keying (modulation). These rates should be considered obsolete for new purchases, but the older modems generally will interoperate with new, higher-speed modems and need not be replaced unless higher performance is required.

The sampling rate ("baud rate") on a modem link used with a voice circuit is normally not greater than 1200 samples per second (1200 baud). Newer modems, taking advantage of the fact that multiple aspects (amplitude, frequency, phase) of the analog signal can be modulated, are able to achieve data rates of 2400, 4800, 9600, and 19,200 bits per second with high-quality analog circuits. In addition, the latest modems include compression processing that takes advantage of the fact that most data streams have information density much lower than the raw bits-per-second presented. These modems operate at nominal speeds of 14,400 and 28,800 bits per second but can actually support data rates much higher when the information rate of the data stream is low, for example, when the data contains long strings of similar patterns (such as blanks or zeros).

Facsimile or "fax" modems are compatible with international standards for facsimile transmission, for example, "group 3," which offer a particular range of data rates and a specific compression protocols. If your data modem is also fax compatible, you will be able to use facsimile software such as BitFax or Delwina WinFax to send and receive fax documents, treating them just as fax machines do—as bits on a page which are black or white. (In other words, just because you can receive a fax and display it on your computer does not guarantee the computer will capture the fax contents in a form it can manipulate.)

61.4 SERIAL INTERFACES

A modem has two interfaces: one connects to a telephone line (typically an RJ-11 modular phone connector) and the other connects to the computer. The phone side uses an analog signal that is capable of a continuous range of values and was originally intended to carry signals derived from human voice by a microphone. The computer side may be any of several standard interfaces, but it is sure to be digital, which means that each wire between the modem and computer can only be in one of two states ("on" or "off") at any given time. The reason the modem is necessary is to convert digital signals to a form that can pass over an analog circuit.

Internal modems generally plug into the computer's internal bus, which links its internal components at very high information transfer rates. Therefore internal modems are generally limited only by the information capacity of the telephone line. External modems must use a digital interface, most often Electronic Industries Association (EIA) RS-232D or EIA RS-449.

RS-232D (usually just called "RS-232") is used for lower data rates. Because most modems operate at relatively low data rates by computer standards, they use RS-232. The RS-232 standard defines up to 25 wires at the interface, but these are mostly not used, so serial port jacks with 9 and with 25 pins are both common. RS-232 is intended to be used with asynchronous data communication, where each character consists of 7 or 8 information bits plus "start" and "stop" bits. Because the start bit defines where the character starts, no other timing information is needed. This is the arrangement normally used to connect an external modem.

RS-449 is used for higher data rates, up to 2.048 megabits per second. The RS-449 standard assumes synchronous data. This means one side of the interface provides a highly regular timing signal called the "clock"; the rate of change of the clock signal determines the transfer rate of the data in bits per second. This sophisticated arrangement is not normally used with modems, but it does apply to the next step up in data communications: digital transmission.

Digital transmission service is increasingly available from local communications carriers ("the phone company"). In general the installed communications plant in industrial countries is at its core digital, with analog voice service to the customer premises. Using this digital capacity directly, rather than through a modem, provides higher data rates and lower error rates. To do this a different sort of interface box is required; rather than a modem a "channel service unit/data service unit" (CSU/DSU) is needed to establish correct digital signal levels and format between the computer and the phone line. The CSU/DSU provides synchronous data communications and therefore uses an RS-449 interface (or a variation such as RS-422 or V.35).

Most carriers offer two data rates. The lower one uses the phone system capacity normally allocated to a single voice connection, and provides a data rate of 56 kilobits per second (64 kilobits per second in Europe and over some wide-area services), called a "DS0." The faster rate, called "DS1," has the capacity to carry 24 voice channels and provides 1.544 megabits per second (2.048 megabits per second, equivalent to 32 voice channels in Europe). The nominal rate may be derated by up to 10% by the carrier's timing and connection management scheme (in general the last 10% can be had by insisting in the contract, but with the negative result that the carrier can no longer monitor the circuit to ensure its functioning; this is probably unwise).

The DS1 data rate is becoming broadly available in the United States, typically at a cost equal to that of three DS0 circuits (although this is highly variable among markets). Many Internet service providers now routinely use this rate, which is generally called "T1," its name in the carriers' internal digital hierarchy. Higher rates than this are hard to get and quite expensive. T2 (four times T1) is available only from microwave systems installed on the customer premises; T3 can be leased from telephone carriers and brought in over fiber optic connections (and sometimes copper from the phone company's equipment rack to your computer).

All of this assumes you are at the mercy of the telephone carrier, or perhaps a microwave service provider. If you have access to the entire physical data path, you may choose to circumvent the carrier and run your own wire. Doing this will require yet another interface box, called a "line driver," that maintains timing between the ends of the data link and compensates for signal lost over long runs (up to three miles is possible now over twisted wire pairs originally intended for telephone connections). The line driver will also have a synchronous interface, probably RS-449, and may be integrated with a CSU/DSU. For higher rates than DS1 you will probably have to install a fiber optic cable.

61.5 ISDN

To complete the topic of transmission we must consider the Integrated Services Digital Network (ISDN) that is widely available in Europe and is becoming available slowly in the United States. With ISDN the carrier brings a digital signal all the way to the customer's terminating equipment (usually a digital telephone set or computer; the telephone part is interoperable with the regular analog voice system). In its most flexible form, this can be used interchangeably by data and digital voice; however, the carrier may charge different rates for the two and treat them differently in his plant, so at least in the United States we have found it advisable to pick one of the two. In fact, with high-quality analog voice service readily available, the real attraction of ISDN for most users is that it represents a digital (and therefore higher-rate, lower-error) path that is available for a much smaller monthly charge than a DS0 circuit and is billed by actual connect time, much as an analog voice call.

ISDN is available in two forms. By far the most common is the "basic rate interface" (BRI), which provides two DS0 channels (56 or 64 kilobits per second, depending on the carrier), called "B" channels, plus a channel for "signaling" (dialing) called the "D" channel. The more exotic form is a "primary rate interface" (PRI) that provides 23 B and one D channels over a T1 link. This arrangement offers great flexibility; terminating equipment manufacturers are only beginning to take advantage. For example, Internet routers that have recently become available will sense the offered data load and dial up as many B channels as needed from a PRI. If the application offers occasional periods of high traffic separated by a long period of little or no traffic, this arrangement can be much less expensive than a leased T1 circuit.

Another service known as "broadband ISDN" (BISDN) is now becoming available in a few large-city markets. The base technology of the so-called "information superhighway," BISDN will in the future make data rates up to gigabits per second available, at least between major communications markets. Based on the Synchronous Optical Network (SONET) transmission standard and Asynchronous Transfer Mode (ATM) switching standard, BISDN will be able to offer a mixture of voice, video, data, and graphics that is truly integrated. SONET links are specified "optical channel" (OC) units, which are multiples of 51.8 megabits per second, thus an OC3 is 155 megabits per second. Common sizes are OC3, OC12, OC24, and OC48, the last two being over a gigabit per second! BISDN is introduced here only for future planning as it is not generally available today.

To use an ISDN BRI the customer must have an "NT1" adapter, which performs the DSU/CSU functions and provides the dialing interface. In Europe the NT1 is provided by the carrier, but in the United States the customer buys it. For computer applications a Terminal Adapter (TA) is also required; as with modems, TAs may be internal to the equipment, in fact they are available on plug-in cards for IBM PC-compatible systems. To use an ISDN PRI a more sophisticated box, the "NT2," is needed. ISDN offers clear economic benefits, but the author's advice (based on experience) is to plan on a long install period and "learning curve" as both carrier personnel and equipment vendors are still quite inexperienced with ISDN.

62

Local Area Networks

J. Mark Pullen
George Mason University
David Jensen
George Mason University

62.1 COMMON LAN CHARACTERISTICS

Local Area Networks (LANs) have the following characteristics:

- **Limited area:** Most LANs cover distances from a few feet to at most a few miles. The most common LAN technology today, twisted pair Ethernet, is typically limited to 100 meters of wire between networked device and LAN hub.
- **High speed:** Most LANs support data rates of 1M to 100M bits per second. Typical is 10M bits per second burst capacity (which does not equate to 10M bits per second sustained transmission, see below).
- **Private operation:** LANs are normally owned and operated by the user, and reside entirely on the user's premises.
- **Data mode:** Today most LANs are used only for data transmission. This is beginning to change as multimedia systems push real-time voice and video applications onto the LAN.
- **Broadcast mode:** The fundamental mode of operation is that when one station sends a data frame, all other stations receive it. Each station is responsible to listen to all the data frames and select those addressed to that particular station. This implies a need for a control mechanism to ensure that only one station sends at a time. The particular control mechanism is part of the LAN standard (see IEEE Standards below). It also facilitates multicast (sending to a group address, see Section 64.3).
- **Baseband or broadband:** Baseband LANs transmit digital signals directly; broadband LANs modulate a carrier, typical frequency in the range 5 to 400 MHz, which permits transmission using cable TV technology.
- **Physical address:** Each station in a LAN must have a unique address. The general solution to this is that every LAN adapter is built with a different address, 6 bytes long. Address assignment is coordinated between the manufacturer and Xerox Corporation, the creator of Ethernet. Each data frame carries the address of the destination station. Exchange of addresses at LAN startup is a responsibility of the network software in the stations.

0-8493-2530-7/96/$0.00+$.50

62.2 LAN TECHNOLOGIES

LANs are available using several transmission technologies:

- **Coaxial cable:** The oldest LAN technology, coaxial cable (thick, with taps cut through the braid, or thin, with taps made using coaxial BNC "tee" connectors).
- **Twisted pair:** Although the signal will not travel as far across twisted pairs, most buildings have them already installed for telephone. By installing a LAN hub in the telephone closet it is often possible to avoid rewiring for the LAN.
- **Optical fiber:** High-speed LANs (100M bits per second) typically use fiber transmission. This allows longer distance between stations with very low error rates. However, a new "100 megabit Ethernet" technology now in development may diminish the role of fiber.
- **Infrared:** Wireless LANs are available that link all units within a room by infrared transmission similar to that used in remote controls for consumer products. These typically use a hub-repeater mounted on the ceiling. The technology is appealing for wandering stations such as laptop computers, and to save rewiring.
- **Radio frequency (RF):** This form of wireless has the advantage that it passes through walls. The radio transmission is typically of the spread spectrum type that is insensitive to other local uses of radio spectrum and transmits at very low power. The typical range is up to 1500 feet with burst data rates of 10M bits per second.

To use a LAN you will need the transmission medium (wire, fiber, etc.), any hubs required by the LAN standard, and a LAN adapter for each station. Like serial link interfaces, LAN adapters come either in an external form that connects to a port on the computer, and in the form of circuit boards to be inserted into the computer. Prices range from around $40 for high-volume products such as the IBM Personal Computer and its many clones, to over $1000 for FDDI adapters.

62.3 LAN PROTOCOLS

There are two basic approaches to coordinating a LAN. Pure broadcast LANs share the transmission medium (wire, infrared or RF), thus their topology is logically that of a bus (although some of them are organized around hubs to simplify management and growth). All stations listen to the medium ("ether") and send only when it is idle. Because there is no form of central coordination, two or more stations may start sending at or near the same instant. This requires stations to sense the medium while transmitting; if they sense a collision they stop transmitting, wait a random interval so they are unlikely to send again at the same time, then send their frames again. The amount of backoff is increased when the LAN is heavily loaded, as sensed by frequent collisions. This arrangement is called "carrier sense multiple access with collision detection" or CSMA/CD. It has the characteristic that its average transmission capacity is somewhat less than half of the peak burst capacity, because collisions limit its capacity under heavy loading. If only one station is sending, the average capacity can approach the peak capacity.

The other basic approach to coordinating a LAN is use of a "token," a bit pattern that indicates the LAN is either "free" or "busy." The topology may be a bus, but is more commonly a loop (ring). The token is a unique code that is held by only one station at a time. Only the station holding the token can send; it does this by converting the free token to a busy token and sending it forward, followed by a data frame. The busy token is changed back to free either by the receiving station or by the sender when it comes back around the loop. There is a rule for passing the token, the most common rule being that a station may send only one frame and then must pass the token on. A station with no data to send passes the token to the next station, thus when the LAN is idle the token circulates continuously. The stations "elect" one of themselves by a distributed algorithm, to initiate the token and deal with the two error cases that can arise: loss of the token and multiple tokens in circulation. This sort of LAN goes by the names "token ring" and "token bus," depending on the physical topology. Rings larger than a few stations are generally organized around hubs for management purposes, where the medium (wire or fiber) passes through the hub between each station. In this way the hub forms a natural test point.

Even though CSMA/CD and token bus/ring are quite reliable, conditions can arise when frames are lost. Therefore they are used with higher level protocols (e.g., TCP, see section 26.1) that manage reliable delivery by accounting for transmission between sender and receiver and retransmitting any lost segments. The LAN protocol is referred to as the "medium access control" (MAC) sublayer of the data link control protocol, and fits between the physical medium and wide-area network protocols such as X.25 or TCP/IP.

62.4 LAN STANDARDS

The Institute of Electrical and Electronics Engineers, Inc. (IEEE) has a standards group that maintains LAN standards, which fall in the IEEE 802 series. Some of the more important LAN standards are listed below.

- IEEE 802.1 defines the software interface to the 802 series.
- IEEE 802.2 defines a generic Data Link Control (DLC) protocol for the series, specifying frame format, automatic retransmission procedures, and error-checking code.
- IEEE 802.3 is "Ethernet" or CSMA/CD.
- IEEE 802.4 is Token Bus.
- IEEE 802.5 is Token Ring.

There are more IEEE 802 series protocols, and the number continues to grow. Not all of them deal with LANs, for example, 802.6 is the Dual-Queue, Dual-Bus metropolitan area network (MAN) intended for larger distances such as building complexes or urban areas.

The most popular of the IEEE standards is 802.3. It now comes in several forms, for example, "10BASE5" with 10M bit per second data rate, baseband transmission, and up to 500 meters of coaxial cable; "10BASE2" or "Cheapernet" with a 200 meter limit; and "10BASET," which uses twisted pairs and hubs but otherwise complies with 802.3.

High-speed LAN standards are

- Fiber-Distributed Data Interface (FDDI), a token ring system.
- Asynchronous Transfer Mode (ATM), the Broadband ISDN data format (see Chapter 23), used as a LAN technology, with ATM switches in place of hubs.
- 100M bit per second CSMA/CD, now available based on the "100BASEX" draft specification of IEEE 802.3.

63

Telecommunications Software

Kevin Greiner
WARE, Inc.

Paul W. Ross
Millersville University

It is frequently necessary to access a central computer from a remote site, or pass data files over the dialed telephone network. There are two issues involved:

- Emulation of a computer terminal, and management of the modem and dialed network.
- Passing files across the dialed network to or from the remote location to the central computer.

The emulation of a computer terminal is done with a terminal emulation program. These programs will emulate a variety of commonly available terminal types. Two such common programs are described: Procomm, and the Windows Terminal program. The later, the Windows Terminal program, is provided with the Windows software system. Procomm is typical of many terminal programs on the market.

File passing is anther critical issue. Large files, especially ones containing executable machine code, often cannot be passed through systems that expect only ASCII text characters. Systems known as protocols are necessary, both to ensure the correctness of the transmission, or to convert the machine language code into characters that can be passed correctly by a variety of systems. Two common protocols, KERMIT and XMODEM, are described.

63.1 PROCOMM PLUS FOR WINDOWS

Kevin Greiner

Starting Procomm Plus

The most popular communications program for DOS is Procomm Plus. The Windows version of this package is now available. Although one can use the Windows Terminal program to access remote computer systems, Terminal is limited in many respects. Although the basic function is the same, Procomm Plus for Windows addresses many of the shortcomings of Terminal and is a viable choice for flexible, powerful PC communications.

0-8493-2530-7/96/$0.00+$.50

Rather than dealing with the nuts and bolts of communications, this chapter focuses on Procomm, its unique feature set, and how to use it on your computer. If you are new to modems, COM ports, communication programs, you should consult a book or acquaintance that can explain these areas to your satisfaction. At the same time, an understanding of every communications term is not essenntial to benefit from a communications program such as Procomm. Many of the complex configuration options default to appropiate choices for nearly every user. Only in a few unusual cases should you find yourself changing more than a few of these settings.

When Procomm Plus for Windows is installed, it creates a group or folder in the Program Manager called "PCPLUS/Win." If this label is beneath a minimized icon, double-click that icon to show the contents of this folder.

FIGURE 63.1 The Program Manager icon for Procomm Plus.

The Procomm Plus program is run by double-clicking on the icon labeled "PCPLUS/Win." This is the icon in the upper left hand corner of the folder. When Procomm is run you will see the window shown in Figure 63.1. To close Procomm, double-click the system control box in the upper left corner, select Exit from the File menu, or type Alt-X on your keyboard.

The Procomm Plus Application Window

The Procomm Plus window (Fig. 63.2) is basically similar in function to all other windows you have seen in Windows. Of course, additional functionality has been added to suit unique communication purposes such as dialing another computer and sending files. The five main parts of the Procomm Plus window are listed and described below:

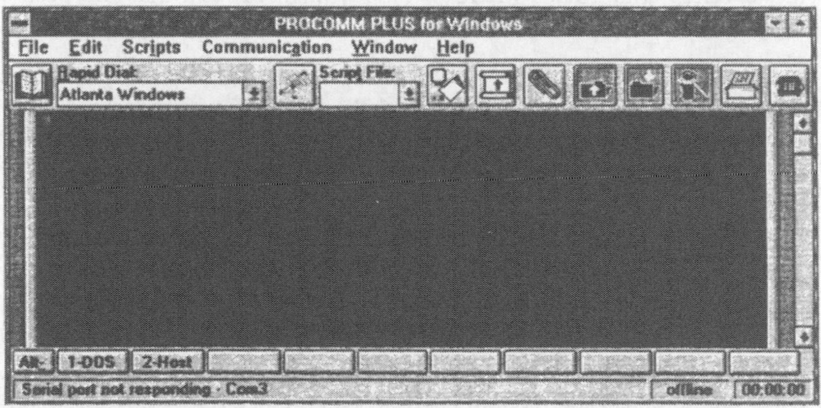

FIGURE 63.2 The Procomm Plus application window.

- Title Bar—The Title Bar appears at the very top of the application window. The name of the program (PROCOMM PLUS for Windows) appears centered in the title bar.
- Menu Bar—The menu bar is immediately below the title bar. The menu bar provides you with access to all the features of Procomm Plus. In addition, many of these features can be accessed with function keys or the toolbar (see below).
- Tool Bar—The tool bar follows the menu bar as you move down the screen. This area contains buttons that provide short-cuts for often used actions in Procomm. You choose a specific button by clicking on it with the mouse.

- Text Area—The next area down the screeen is the text area. All text received from the host computer is displayed in this area. You can use the mouse to select text for cutting and pasting, as you expect from a Windows application.
- Scroll Bars—The scroll bars appear at the right and bottom of the Procomm Plus window. The purpose of these scroll bars is to allow you to review text that has scrolled off the screen during your current session. Text is not saved from session to session. Using the mouse, click on the arrow that points in the direction you wish to go.

Selecting and Copying Text in the Window

Text that appears in the Text Area can be selected, copied, and pasted into other applications. You can also paste that same text or text from other applications into the text window. The functions used for selecting, copying, and pasting are similar to those used in other Windows applications.

- Select Text—Place the mouse cursor at the beginning of the text you want to copy. While holding down the left mouse button, drag the mouse over the text you want to select. You will see the text change color. You can copy one complete screen at a time or the complete scroll-back buffer (see Scroll Bars above) at once. From the **Edit** menu, choose **Screen To** or **Scrollback Buffer to**.
- Copy Text—After the text has been selected, choose **Copy** from the **Edit** menu or press Ctrl-Insert.
- Paste Text—Text that has been selected and copied can be pasted into the text area of the windows or into another Windows application. To paste, select **Paste** from the **Edit** menu or press Shift-Insert on the keyboard. Note that you can only paste into the text area at the place where you are typing. Any text that you paste is immediately sent to the remote computer.

Hardware Requirements and the Remote Computer

Procomm Plus or any terminal program needs several items to communicate with other computers:

- A modem—A modem is a computing device that allows your computer to send a digital signal over analog phone lines to a remote computer. The remote computer also uses a modem to translate the signal you sent into understandable commands. The modem should also be Hayes compatible (most PC modems are). It does not matter if the modem is internal or external to your computer.
- A telephone line—The modem needs to be connected to a telephone line. The line need not be used only by the modem. It could be the phone line into your home or apartment. Just be sure when you are using the modem no one picks up or attempts to use any of the other extensions. At best, you will see garbage on your screen or your modem will hang up the phone.
- An available serial port—Most personal computers have several serial communications ports available. Often one is used for the mouse; the other should be available for your modem.

A null modem cable can take the place of a modem and phone line. The null modem cable stretches between two computers, providing communication between them. The only limitation to a null modem cable is that the computers must be fairly close to one another. Null modem cables are really an option only for two computers that are located in the same room.

The Remote Computer

The key to successful communication between computers is getting them to talk in terms that both understand. Procomm Plus on your end is your computer's way of establishing communications. The communication capabilities of the computer at the other end of the line is important because you need to know how it expects the dialog to occur.

Think of the process as two people speaking different languages and needing an interpreter. The function of the interpreter is the same as the Procomm program. Procomm will send information that is understandable to the remote computer, and will interpret signals sent to you so that your computer can understand. Communication does not occur until both sides of the link are able to understand what the other is trying to articulate.

There are seemingly an overwhelming number of possible options that might be used to communicate with another computer. Although there are some loose standards, it is very important to find out the exact requirements of the host computer. Defining these details up front will save you a great deal of time configuring and establishing a successful communication link. The next section of this chapter will cover the exact specifications you will need to gather about the computer to which you want to connect. The requirements can usually be obtained from either a manual for the service, or from the host computer's operator. If you absolutely cannot find out the proper settings, try the default. The worst that can happen is that you will connect, but not be able to communicate, and the host will terminate the call. In this case, attempt other settings and retry.

Procomm Setup

Communication settings for different host computers vary widely. In Procomm, you create named settings that can be saved and used over and over again (Fig. 63.3). Usually, you need to research which particular settings are necessary to communicate with a specific host computer. When you are asked to choose a specific setting that may not apply to every host you call, choose the one that you will use most often. I will cover host-specific settings in Section 63.6.

Many of the settings can be saved using a name to identify that specific set of parameters. Create a named setting for each different communication scenario you use on a regular basis. Later, in the Dialing Directory, you will identify which host uses which named settings.

To configure communication settings in Procomm for Windows, select the Setup choice on the Window menu, or with the keyboard, press Alt-S.

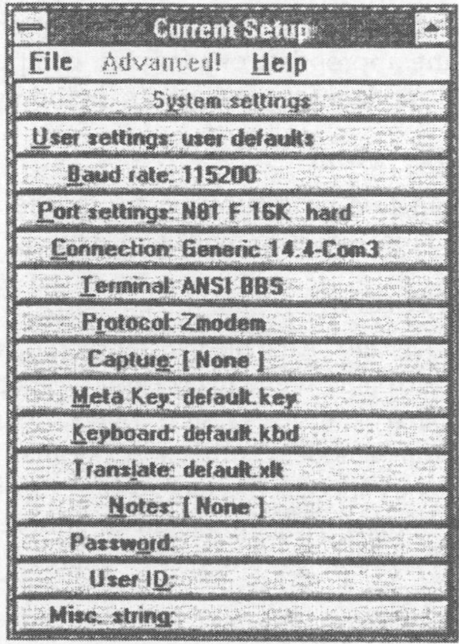

FIGURE 63.3 The Setup menu.

System Settings

All the System Settings should be correct for most communications tasks. See your Procomm manual for specifics on these settings.

User Defaults: User Settings

This button provides two functions rolled into one. Double-click the button to change various settings such as download and upload paths and add new named setting collections. Click the button once to choose a previously named setting as your current configuration.

Baud Rate

Set the highest baud rate that your modem supports. In general, with a high-speed modem, this number should be double the highest baud rate.

Specifically, the baud rate is a measurement of the number of bits per second at which the two computers will communicate. All modern modems have syncronization built in that allows the two modems to automatically determine the highest speed at which they can communicate. A general rule is to set the baud rate at one or two settings higher than your modem¸s baud rate. This allows the computer to send data in high-speed batches to the modem. For example, if you are using a 14.4 modem, set the baud rate to 19,200 or 38,400.

Port Settings

In this dialog box, choose the settings that correspond to the host computers that you will be using most often. The default configuration will be correct for most users.

Connection

This button allows you to choose the default modem type and change a large number of settings associated with a specific modem. If you need to enter custom initialization strings, you will use this dialog box.

Single-click the connection button to see currently installed modem types and select the appropriate one for your computer and modem. Double-click the button to set up a modem for your specific brand and model modem.

Terminal

Single-click this button to choose the Terminal type for the host you access most. Double-click to modify an existing Terminal emulation or create a new one.

Protocol

Over the years, file transfer methods and protocols have greatly increased in speed and reliability. While many early protocols could transfer text only, new protocols combine binary transfer, crash recovery, adjustable packet lengths, and automatic error correction. Zmodem is the most popular modem transfer protocol for good reason. It combines speed, efficiency, robust error detection and correction, and wide availability.

Single-click the button to choose from the list of available protocols. Double-click the button to modify the settings for the currently selected protocol.

Capture

Single-clicking allows you to append to or overwrite an existing capture file. Double-click to change the way in which Procomm writes incoming characters to this file.

Meta Key

Meta keys are the large buttons arranged in a row across the bottom of the Procomm screen. These buttons specify actions to be performed when they are clicked with the mouse or accessed with the keyboard. You can customize these keys to perform special tasks such as inserting text, running a program, or executing a Procomm script.

Single-click this key to choose from the list of available meta key files. Double-click to modify the current selection and set up your own keys.

Keyboard

Using this button, you can change the mapping of the keyboard to suit your needs. You can also name and save custom keyboard maps.

Translate

Each host computer to which you connect assigns sequences of characters to indicate specific keys. Should your host computer require, you can change the translation of sequences of characters from one system to another.

Notes

This button allows you to create files (with an editor or word processor of your choice) that contain typed notes.

Password, User Id, Misc. string

These three fields contain any test that you type into them. They are accessible via the Procomm script language. See the Procomm manual for more information.

The Dialing Directory

The Dialing Directory is where you enter information about each host you call. Procomm saves this information, allowing you to quickly reconnect to the host again. This section covers host-specific settings.

To configure communication settings in Procomm for Windows, select the Dialing Directory choice on the Window menu, or with the keyboard, press Alt-D.

The Dialing Directory (Fig. 63.4) is arranged in a grid. Each row contains information about a different host. The columns contain specific settings that you may need

PROCOMM PLUS Dialing Directory - PW.DIR
File Edit Dial Help

	Name	Number	Baud	Port	Co
1	Atlanta Windows	1-404-516-0048	default	N81 F 16K hard	default co
2	LABB v.32bis	394-2882	default	N81 F 16K	default co
3	Semware	1-404-641-8968	default	N81 F 16K hard	default co
4	Parsons Consulting	1-312-752-1258	default	N81 F 16K	default co
5	LABB Line 1	394-1357	default	N81 F 16K	default co
6	Warlock's Cave	394-3654	default	N81 F 16K	default co
7	DataLine BBS	295-7048	default	N81 F 16K	default co
8	MS BBS	1-206-936-6735	default	N81 F 16K	default co

1 entry selected
Dial
Manual Dial
Stats ...
Advanced

FIGURE 63.4 The Dialing Directory.

to change, depending upon the host's communications characteristics. The default for each column comes from the General Settings covered in the previous section.

Name

This is the name of the host computer that you arbitrarily assign. It needs to be meaningful only to you. This is the name that you use to identify the specific host to which you want to connect.

Number

This is the phone number as you would dial it. You may enter the number using parentheses, hyphens, or spaces if you prefer. Procomm will use only the numerical digits.

Port

The port button allows you to change very specific characteristics of your communications hardware. Only three of these settings need to be modified with any frequency. The three are Data, Parity, and Stop. To connect to bulletin boards, set Data to 8, Parity to None, and Stop to 1. To use CompuServe, set Data to 7, Parity to Even, and Stop to 1. These settings do vary from host to host, so be sure these three are correctly set for your specific host.

Half duplex means that only received characters are shown on your screen. When you type a character, the host must "echo" it back to you for it to be displayed. If the characters that you type are not appearing on the screen, change this setting to full duplex. If the characters you type are appearing twice, change this setting to half.

The com buffer refers to how much memory your computer sets aside to buffer the flow of data in and out of your computer. Generally, the higher the connection speed, the larger the buffer should be set.

The remaining settings are highly technical and need to be changed only in rare cases. Consult your Procomm documentation for more information regarding these settings.

The rest of the columns allow you to choose named settings that you configured from the Settings menu. If you need to use one that is not there, you need to enter the Setup menu by choosing Setup from the Window menu or pressing Alt-S. Then add the named setting that you need.

On-Line Timesavers

Connecting to other computer systems and navigating through the host computer's commands can mean many repetitive keystrokes. To aid your effiency, Procomm allows you to program a set of keystrokes and more complex commands into buttons called meta keys. These buttons appear at the bottom of the screen by default. If they are currently not displayed, you can see them by selecting the Meta Keys choice on the Window menu.

To configure the meta keys, select the Setup choice on the Window menu or type Alt-S, then double-click the meta keys button. You will see a dialog box that allows you to assign four different commands to F1 through F10 by using the Alt, Alt Shift, Alt Control, and Alt Shift Control modifiers for a total of 40 different commands (Fig. 63.5).

For each function key assignment, you can select text, a script, or a program. For text, type the actual text in the Contents box. When you click on this button or type this keystroke, the text you enter here will be entered as if you have typed it. Script refers to a set of Procomm commands that enables you to perform complex communication tasks. Enter the name and path of the script you want to run in the Contents area. Read the Procomm documentation for more information. The Program selection

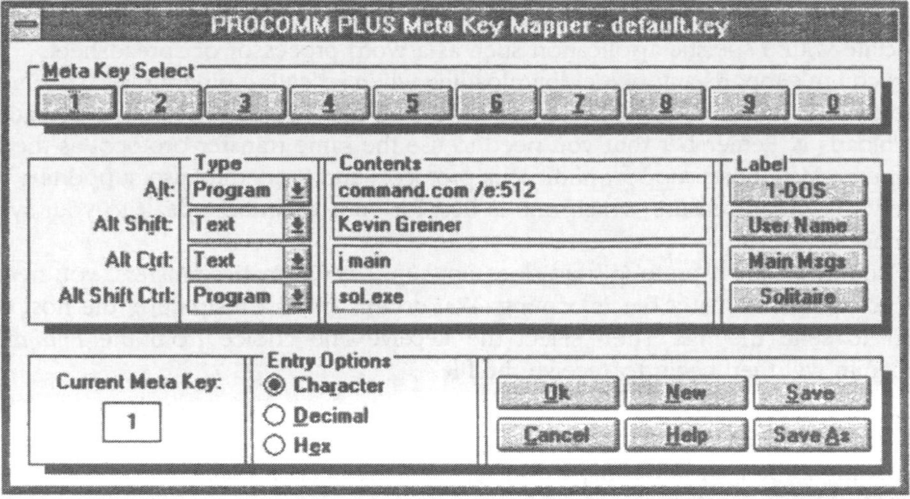

FIGURE 63.5 Meta Key setup.

allows you to run another application. Enter the name and path of the program you want to run in the Contents area. This would be useful if you liked to play a game while waiting for a file to finish transferring.

Connecting to a Host

Procomm provides several ways to dial a host. The first is to choose the host you want to connect with from the Rapid Dial list (Fig. 63.6). This list is placed on the left side of the button bar. Procomm immediately begins to dial the number and connect to the specified host. As an alternative, you can display the Dialing Directory, select the host, and click on the Dial button.

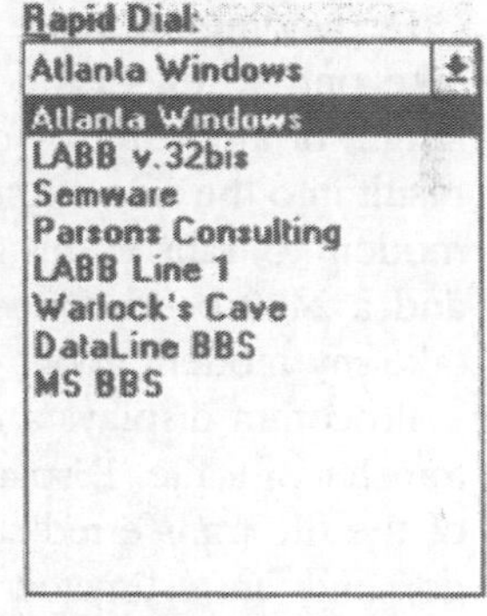

FIGURE 63.6 Rapid Dial list.

Capture Files

After you are connected, there are several activities which you might find useful or necessary. First, you may want to keep a complete record of what you did while you were online. This is referred to as a "log" of your current session. You can instruct Procomm to capture every incoming character and write it to a text file or a printer. To initiate a capture file, you first need to be connected to a host. Then choose Capture File from the File menu or press Alt-F1. You can then choose various options that control how data are written to the file.

Downloading Files

One of the primary reasons that modems are so popular is that they allow the transfer of both text and binary files from remote locations. The transfer method that Procomm uses is dependent on the file's type.

If the file you want to move is an ASCII text only file, use the ASCII transfer. These files contain only printable characters and a few simple formatting characters such as tabs and carriage returns.

Binary files are, quite simply, all other files. Usually, these are files that you would associate with a specific application such as a word processor or spreadsheet.

Procomm supports automatic downloading when receiving binary files. That is, you simply need to instruct the host computer to send the file and Procomm automatically downloads it. Remember that you need to use the same transfer protocol as the host computer. Most hosts support numerous protocols and Procomm also supports a wide variety. If you are unsure which one is best for your situation, speak to your system administrator.

If you are using a protocol that does not support automatic transfers, you need to instruct Procomm that a file is coming. You do this by first instructing the host computer to send the file. Then select the Receive File choice from the File menu. Procomm will then begin to receive the file.

Uploading Files

Uploading a file always begins with preparing the host to receive it. At a minimum, many hosts require that you choose a file name for the uploaded file. You may also need to supply additional information about the file's location, size, purpose, or a description.

After you are done preparing the host to receive the file, you need to instruct Procomm to send a file. You do this by selecting Send File from the File menu. You will see a dialog box prompting you for the name of the file to send. You can either type the complete name of the file in the text box or traverse through the directory structure of your disk to locate the file you want to send. When you press Enter or click the OK button, Procomm will begin to send the file.

Depending upon the size of the file and the speed of your modem, the transfer may take only a few seconds, or it may require several hours. You can approximate the length of a transfer by dividing the speed of your connecction by 10 and dividing the result into the size of the file. The result is the number of seconds that it will take your modem to transfer this file. For example, if I had a file that was 10,000 bytes in length and a 2400 baud connection I would divide 10,000 by 240. The result is that it will take my modem about 42 seconds to transfer this file.

Procomm displays a dialog box that shows you numerous numbers relating to the transfer of a file. Displayed are the file name, the total length of the file, the amount of the file transferred so far, the amount of time taken so far, the amount of time to go, and an estimated completion percent. If you are running Windows in 386 Enhanced mode, you can minimize Procomm or switch away from it to use another application while transferring a file. The file transfer proceeds in the background. If you have an old or slow processor, or a small amount of free memory, Procomm may detect and correct a few errors, but this is a small price to pay for being able to perform two different jobs at once! (Warning: If you are not using an error-correcting protocol, do not transfer files in the background due to the inability of Procomm to detect errors. Such protocols include YModem-G.)

Troubleshooting

The final section in this chapter is a discussion of what to do if you cannot get Procomm to work. Communications between computer systems must follow some rigid standards for the link to be successful. Because the exchange is defined so specifically, there is little room for any variation in the proper settings. Be patient in setting up the link; sometimes you will need to experiment with a few or even many settings to get the right combination that works.

It may be useful for you to read the following section even if you made a successful connection the first time. You will gain additional insight on where things might go wrong when setting up to connect with other computers.

The Modem Does Not Dial

The most common problem with a modem interacting with a computer using a program like Procomm is that it does not work at all.

If this is happening to you, the first place to look is the port you assigned to the modem. For example, if you selected COM1, try another COM port.

Another possibility is that Procomm is set up to use an incorrect modem type. When this happens, the wrong commands are sent to the modem. To be sure that you are using the correct modem type, refer to the original manufacturer's documentation.

After Connecting to the Host Garbage Appears on the Screen

If you successfully connect to the remote system, but the screen becomes cluttered with characters that make no sense or no characters appear, check the following items:

- Baud rate
- Data bits
- Stop bits
- Parity

These settings should be checked against the host computer's specifications. If one of the specifications is not given in the documentation or unclearly written, some experimentation may yield successful results. A very good source of help when troubleshooting is others who have successfully connected to the same host. You can compare their settings to your own even if you are using different communications programs. The above settings are referred to by the same names no matter which program they might be using.

File Transfers Fail after Repeated Attempts

Both ASCII and binary transfers fail most often because of a poor quality transmission line. This has nothing to do with your system or the host, but rather with the noise or static on the phone line. When this happens during a voice call, we are usually able to carry a normal conversation amidst the static, but modems are unable to do this as successfully as humans. The best solution for this problem is to logout of the remote system, hang up, and try again. If the problem was indeed the phone line, you should have better results.

If you continually experience noisy lines and lost transmissions, faultly phone wiring in your building may be the culprit. You can verify if this is the case by using a direct line from the phone company's feed into your building. If this solves the problem, have a qualified person verify the problem and fix it. If this fails to clear up the problem, continue to look elsewhere.

If your file transfer never seems to get started, check that Procomm and the host are using the same file transfer protocol. Many mainframe or Unix-based systems provide the public domain protocol, Kermit, at a minimum. Most PC-based host systems use Xmodem as a default, but offer a wide range of other options.

File transfers may also fail or be greatly slowed because your computer is too busy with other applications. Although during a transfer with Procomm you can switch to another application or program, doing so may slow down the transmission to the point of failure. If you suspect this is the problem, do not run another application during a file transfer to test your hypothesis.

It Still Does Not Work, Now What?

If you have tried everything you can think of to get a good connection, but without success, check your settings using the following sources:

- The Help available from within Procomm
- Your modem's documentation
- Documentation for the host computer
- Other users of the host computer (This is a most valuable resource. Other users are often more than willing to aid newcomers.)
- A communications textbook that explains each setting in depth (probably only as a last resort)

REFERENCE

Procomm Plus for Windows User Guide.

63.2 XMODEM AND KERMIT FILE TRANSFER PROTOCOLS

Paul W. Ross

Protocols

When transferring files between computers using the telephone system, there is always the chance that electrical noise will result in data transmission errors. To ensure proper transfer of files it is necessary to detect data transmission errors and to retransmit any data that contain errors. Most people think that asynchronous parity error detection provides that capability. It does not. Parity error detection may tell the user when a data transfer error has occurred, but it is up to the user to retransmit the data to correct errors. The problem is that parity error detection is not performed by most communication packages. Also, parity error detection systems are not robust in the presence of high error levels, or "bursts" of errors that typically occur on telephone lines.

To ensure "error-free" data transfer, the user needs a protocol file transfer technique. The general strategy is to divide the file into a number of blocks of convenient length, typically 100 to 200 bytes. Each block is then "encapsulated" with error control information. Typically, additional information is added to indicate the beginning and ends of the block, block sequences, block length, and some sort of checksum to ensure data integrity.

Two common protocols often encountered are the XMODEM and Kermit protocols. A protocol is a set of rules and conventions that applies to a specific area of communication and that allows participants to communicate properly regardless of the hardware or software being used. The protocol file transfer is a set of rules for transferring files that specify a set of ASCII handshaking characters, and the sequence of handshaking required to perform file transfer functions.

Protocol handshaking signals allow communication software to transfer text, data, and machine code files, and to perform more sophisticated error-checking. The handicap in using protocol file transfer techniques is that the computers on both ends of the communications link must use compatible software. In addition, there is the overhead (near 50% under certain circumstances) due to the handshaking and error detection characters that must be transmitted and processed.

XMODEM

The XMODEM protocol was developed some years ago by Ward Christiansen. XMODEM is one specific file transfer protocol that has become a common standard in personal communications because of its widespread use on bulletin boards and because of its inclusion in personal computer communication packages. Its primary use is file transfer between IBM PCs, or functional equivalents; it is also used for file transfer between bulletin boards and PCs.

The original XMODEM protocol was designed for the reliable transfer of a single file. Additional modifications and enhancements, in the form of the YMODEM and ZMODEM protocols, increase the block size, which is acceptable if the line error rate is small. Further, the enhanced versions of XMODEM allow the transmission of multiple files without user intervention, a great convenience when many files must be transmitted that have similar file names or extensions.

XMODEM does not begin the transfer of data until the receiving computer signals the transmitting computer that it is ready to receive data. The ASCII Negative Acknowledge (NAK) character is used for this signal and is sent to the transmitting computer *each* 10 seconds until the file transfer begins. If the file transfer does not begin after 10 NAKs are sent, the process has to be restarted manually.

After a NAK is received, the transmitting computer uses a Start of Header (SOH) character and two block numbers (a true block number followed by a 1s complement of the number) to signal the start of a 128-byte block of data to be transferred. The balance of the block is sent, followed by an error-checking checksum. The checksum is calculated by adding the ASCII values of each character in the 128 character block; the sum is then divided by 255 and the remainder is retained as the checksum. After each block of data is transferred, the receiving computer computes its own checksum and compares the result to the checksum received from the transmitting computer.

If the two values are the same, the receiving computer sends an Acknowledge (ACK) character to tell the receiver to send the next block in sequence. If the two values are not the same, the receiving computer sends the transmitter an NAK to request a retransmission of the last block.

This retransmission process is repeated until the block of data is properly received, or until 10 attempts have been made to transmit the block. If the communications link is noisy, resulting in improper block transmission after 10 attempts, the file transfer is aborted. XMODEM uses two block numbers at the start of each block to be sure the same block is not transmitted twice because of a handshake character loss during the transfer. The receiving computer checks the transmitted block to be sure that it is the one requested and blocks that are retransmitted by mistake are thrown away. When all data have been successfully transmitted, the transmitting computer sends the receiver an End of Transmission (EOT) character to indicate the end of file.

The XMODEM protocol offers several advantages over a simple ASCII file transfer without a protocol.

- The XMODEM protocol is in the public domain, which makes it readily available for software designers to incorporate into a communications package.
- The protocol is easy to implement using high level languages such as BASIC, C, or Pascal.
- The protocol allows a user to transfer non-ASCII 8-bit data files (i.e., COM and EXE files) between microcomputers. The end of a file is based on file size. Handshake signals are used to indicate the end of a file instead of relying on an end of file marker character (control-Z) to conclude a file transfer.

XMODEM error checking is superior to normal asynchronous parity error checking. The parity method of error checking is typically 95% effective if the software on the receiving end checks for parity errors. XMODEM error checking is 99.6% effective, and the software on the receiving end must check for errors. Parity errors detected also do not result in automatic retransmission of the bad data; XMODEM-detected errors result in data retransmission until no errors are detected or until 10 retransmissions have been attempted.

Figure 63.7 illustrates the sequence of events in the transfer of a file by the XMODEM protocol for a single file.

The Kermit Protocol

A slightly different situation occurs when it is necessary to transfer a file between a minicomputer or mainframe system and a PC. In this case, it is entirely possible that the mainframe will malfunction in transferring eight-bit characters or ASCII control characters. Further, mainframe to PC transfers frequently involve the transfer of multiple files. This implies that a protocol such as XMODEM may be unsatisfactory. The Kermit protocol, developed at Columbia University, addresses this problem. Kermit is available on a wide variety of computers, and is in the public domain. Kermit is also included as a transfer protocol in most communications packages.

To implement the transfer of multiple files and accommodate a variety of computers, the design of Kermit differs substantially from XMODEM. Kermit must address the following major issues:

- Packet length: some computer systems cannot handle long packets. In the interest of efficiency, as long a packet size as possible should be implemented.
- Some characters cannot be handled by one of the computers. Typically, these are any eight-bit characters, or the ASCII control characters. The ASCII control characters, such as CTRL-D, might cause the communications session to terminate. Kermit handles this with a sequence of "escape" sequences to avoid transmitting nonprinting ASCII characters, such as the ASCII control characters.
- A difference in file naming conventions between computers. For example, MS-DOS allows an eight-character name and three-character file extension, while IBM's VM/CMS allows for eight-character names with an eight-character file extension. The lowest common denominator of file naming must be implemented.

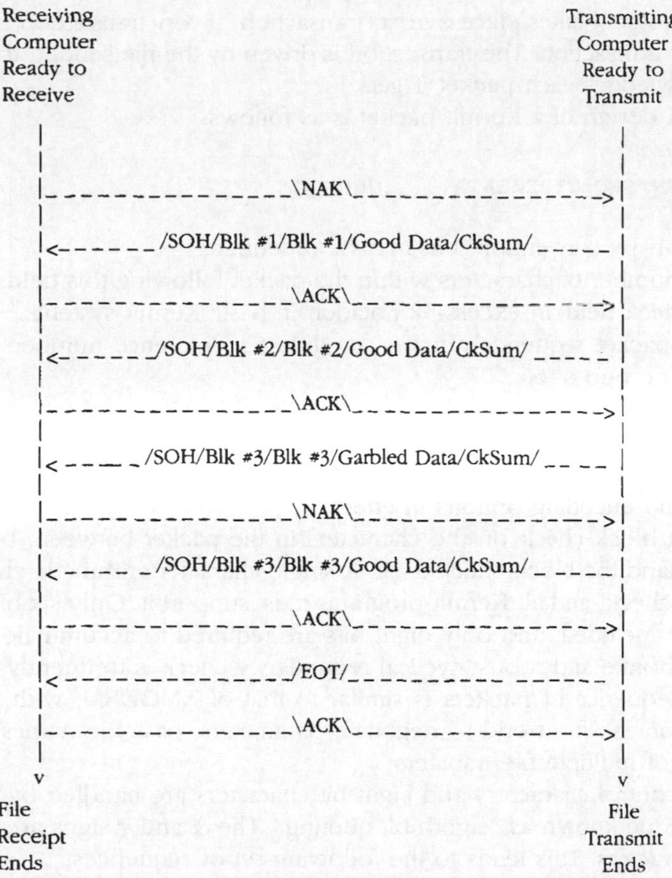

FIGURE 63.7 File transfer by the XMODEM protocol.

The general strategy of ACK/NAK and retries is done in Kermit as is done in XMODEM. However, Kermit uses a variety of packet structures and a more elaborate handshaking sequence to negotiate between the sending and receiving systems. The various packet types are as follows:

- S Send initiation. Parameters are exchanged.
- F File header, which gives the name of the file to come.
- D File data.
- Z End of file.
- B Break transmission, used for end of transaction.
- Y Acknowledgment.
- N Negative Acknowledgment.
- E Fatal error.

The basic protocol takes place over a "transaction." Every transaction is independent of every other transaction. The transaction is driven by the file sender; the file receiver simply acknowledges each packet it gets.

The general design of a Kermit packet is as follows:

```
/SOH/LEN/SEQ/TYPE/DATA...../CHECK/
```

SOH is ASCII 1, by convention, but may be redefined.

LEN is the number of characters within the packet following this field, up to and including the check field, in excess-32 notation in basic Kermit systems.

SEQ is the packet sequence number, modulo 64. Sequence numbers wrap around to 0 after each group of 64.

TYPE is the packet type given in the previous table. Other values are allowed. For a complete description of the Kermit system, see da Cruz (1987).

DATA is the contents of the data packet, encoded or interpreted according to the packet type and encoding options in effect.

CHECK is a block check on the characters in the packet between, but not including, the mark and the block check itself. A single-character arithmetic checksum is the normal block check, and all Kermit programs must support it. Only six bits of the arithmetic sum are included, and only eight bits are required to accumulate and calculate it. A more elaborate and robust cyclical redundancy check is frequently implemented.

The basic sequence of packets is similar to that of XMODEM, with the additional handshaking on the front end to negotiate parameters, pass file names, and manage the sequence of multiple file transfers.

The ASCII control characters and eight bit characters are handled by an escape sequence technique known as "eighth bit quoting." The & and # signs are typically used as escape characters. This leads to the following set of sequences:

If printable ASCII text is being sent, the only overhead incurred is packet encapsulation code. If a binary file is being sent, the efficiency of the transfer process is decreased by approximately 50%. The use of a compression program to compress the data before

Character	Prefixed	
A	A	
meta-A	&A	(A with eighth bit set)
Ctrl-A	#A	
Ctrl-meta-A	&#A	
Rubout	#?	
Meta-rubout	&#?	
#	##	
Meta-#	&##	
&	#&	
Meta-&	&#&	

transmission (such as any of the Lempel–Zev–Welsh programs) can often improve the transmission efficiency to that comparable to a plain ASCII text file.

A Kermit Session

As an example of transferring a file from a mainframe to or from a PC, the following transcription of a session is given. The mainframe is a DEC VAX. Kermit is available on most systems, or may be obtained in source format for local implementation from

Columbia University. The Kermit implementation on the mainframe has an extensive help system.

```
Ready for Login Procedure
Username: PW_ROSS
Password:                (Password does not show)
Logged on to Millersville University's uVAX-3600, OpenVMS VAX
V5.5-2
                AUTHORIZED USE ONLY
    Last interactive login on Friday, 1-JUL-1994 06:45
  PW_ROSS logged in at 1-JUL-1994 08:23:45.25 on VTA7:
  Process Name: PW_ROSS Process I D: 000000A7
$ kermit
VMS Kermit-32 version 3.3.125
Default terminal for transfers is: _VTA7:
Kermit-32>help            (ask for general help)
   Information available:

ASCII_Set BYE     CONNECT Control_Chars EXIT   FINISH
GET       HELP    LOCAL   LOG  Logicals LOGOUT Notes
PUSH      QUIT    RECEIVE REMOTE   SEND SERVER   SET
SHOW      Startup STATUS  TAKE TRANSMIT

Kermit-32>help send (ask for help on SEND)

SEND

The SEND command will allow the user to send a file(s) to the
other Kermit. The SEND command will allow file wild-card pro-
cessing as is found in VAX/VMS. If Kermit-32 is running in
remote mode, the file will be sent on the controlling termi-
nal line after waiting the number of seconds specified by the
SET DELAY command. This gives the user time to escape back to
the other Kermit and issue a receive command. If Kermit-32 is
running in local mode, the file will be sent immediately on
the terminal line specified by the SET LINE command.

Kermit-32>SEND file-spec

Where file-spec is any valid VAX/VMS file specification.

Kermit-32>send toc.txt              (send file toc.txt)
Kermit-32>q                         (terminate Kermit)
```

```
$ kermit
VMS Kermit-32 version 3.3.125
Default terminal for transfers is: _VTA7:
Kermit-32>help receive              (ask for help on RECEIVE)

RECEIVE

The RECEIVE command is used to put Kermit-32 into remote
mode waiting for a single file transfer transaction. This
is most useful if the other Kermit does not support local
server commands. If no file specification is given, Kermit-
32 will use whatever file specification is supplied by the
other Kermit (suitably altered to conform to VAX/VMS stan-
dards). If a file specification is given, Kermit-32 will
use that file specification instead of that supplied by the
other Kermit. This is most useful when the file name on the
other system is such that it does not map well into a
VAX/VMS file specification. Note that if the other Kermit
sends more than one file, the same name will be used for
all of them. Only the version numbers will be different.
Therefore, it is best to use a file-specification on this
command only when transferring a single file. The format of
the command is:

        Kermit-32>RECEIVE

    or

        Kermit-32>RECEIVE file-spec

Where file-spec is any valid VAX/VMS file specification.

Kermit-32>receive (set Kermit to receive)

(Transmitting system specifies file to be transmitted, and
starts transfer of a file called DEFAULT.CAP)

Kermit-32>q                       (terminate Kermit)
$ dir *.cap                       (check for file transferred)

Directory SYS$PUB1:[FACULTY.PWROSS]

DEFAULT.CAP;1

Total of 1 file.
```

By the use of wildcards (* for any character sequence, ? for any single character), a sequence of files may be sent or received.

REFERENCE

Frank da Cruz, Kermit, A File Transfer Protocol. Digital Press, Maynard, MA, 1987.

63.3 WINDOWS TERMINAL PROGRAM

Paul W. Ross

The Windows Terminal program, shipped with Windows, allows you to use your computer to communicate with other systems. Terminal is the software that is needed to control the connection and allow you to create the correct communication parameters so that the link is possible. As the name implies, your computer becomes a remote terminal to the computer to which you are connected.

The Terminal program is found in the accessories group in the program manager. Double-click on the icon to start the program. The Terminal Program icon is shown in Figure 63.8.

In most cases, the default selections will be correct. There are quite a few selections available for customization, most of which never need to be adjusted.

FIGURE 63.8 The Terminal Program icon.

When the program is run you will see the window shown in Figure 63.9.

The Terminal Application Window

The Terminal Application window is very similar to all the other windows you have been using in Windows 3.1. There are five main parts to the Terminal window, many of which should now be familiar to you:

- **Title bar**—The Title Bar appears at the very top of the Application window. The name of the program (Terminal) and the files you are using appear on the bar.

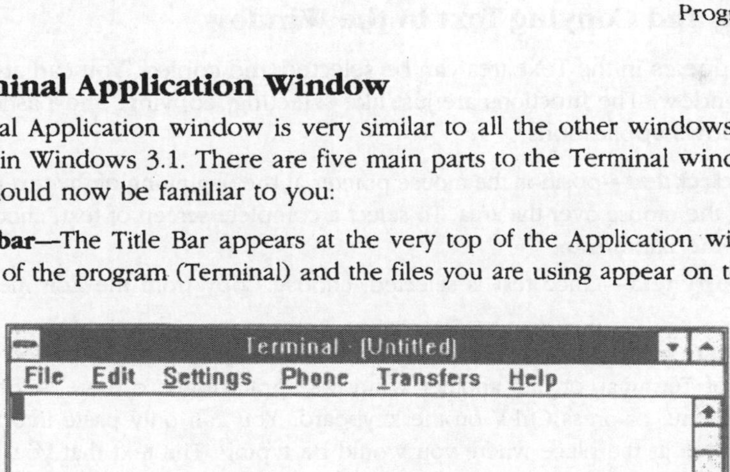

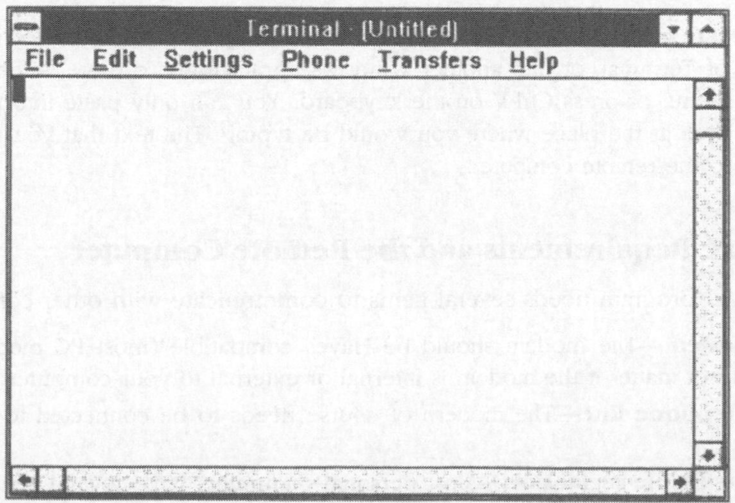

FIGURE 63.9 The Terminal Application window.

- **Menu bar**—The menu bar follows the title bar as you move down the menu. As with all the other Windows applications, the Menu bar is where you enter commands to the Terminal Program. In Terminal there are six menu selections including Help.
- **Text area**—The area directly below the menus is where you will see the data coming from the remote computer. This area will also show any commands the Terminal program sends to the modem, and any messages the modem sends to Terminal. You cannot type directly into this area unless you are in communication with another computer.
- **Function keys**—The area just below the Text area shows what looks like 10 buttons across the bottom of the window. We will discuss how you might use these later in the chapter. Their function is to allow you to program strokes that you would send to the remote computer, and access them by clicking on the area with the mouse. Initially, these will not appear on your screen; you have to turn them on from the menus for them you appear in the Application window.
- **Scroll bars**—The scroll bars appear at the right and bottom portions of the Terminal Application window. The purpose of the scroll bars in terminal is to move forward, backward, left, or right for text that appears on the screen from the remote computer. These become very useful if you want to see a history of text that has been made during the connection.

Selecting and Copying Text in the Window

Text that appears in the Text area can be selected and copied. You can also paste text into the window. The functions are just like selecting, copying, and pasting in all the other Windows applications.

- **To select text**—position the mouse pointer at the beginning of the text to be copied. Drag the mouse over the area. To select a complete screen of text, choose Select All from the Edit menu.
- **To copy text**—Once text is selected, choose Copy from the Edit menu, or press Ctrl-C.
- **To paste text**—Text that has been selected, and copied can be pasted into the Text area of Terminal, or into another Windows application. To paste, select it from the Edit menu, or press Ctrl-V on the keyboard. You can only paste into the Terminal Text area at the place where you would be typing. The text that you paste is then sent to the remote computer.

Hardware Requirements and the Remote Computer

The terminal program needs several items to communicate with other computers:

- **A modem**—The modem should be Hayes compatible (most PC modems are). It does not matter if the modem is internal or external to your computer.
- **A telephone line**—The modern of course needs to be connected to a telephone line.
- **An available serial port**—Most personal computers have several serial communications ports available. Often one is used for the mouse. The other should be available for your modem.

The Remote Computer

The key to successful communication between computers is getting them to talk in terms that both understand. The Terminal program on your end is your computer's way of establishing communications. The communication capabilities of the computer at the other end of the line are important because you need to know how it expects the dialog to occur.

There appears to be an overwhelming number of possible options that might be used to communicate with another computer. There are some loose standards, but it is important to find out the exact requirements of the host computer.

Terminal Configuration

The terminal configuration or settings are the key to a successful communication between computers. Settings for different computers can be saved into files and used repeatedly. In the Terminal program you configure the communications settings from within the Settings menu, shown in Figure 63.10.

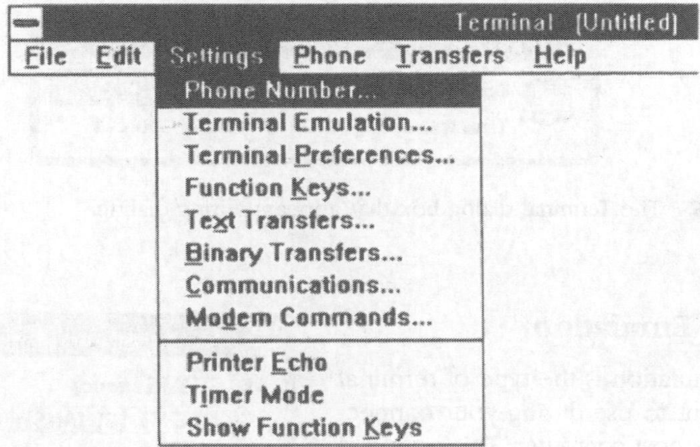

FIGURE 63.10 The Terminal Settings menu.

The Phone Number

The first setting Terminal needs to know is the telephone number of the remote computer. To enter the necessary information, select Phone Number from the Settings menu (Fig. 63.11).

Phone Number
Dial: OK
Timeout If Not Connected In 30 Seconds Cancel
☐ Redial After Timing Out ☐ Signal When Connected

FIGURE 63.11 The Phone Number dialog box.

Enter the following information:

- **Dial**—The text box labeled dial is the place where the actual telephone number of the remote computer is entered.
- **Time-out if not connected in 00 seconds**—Time-out is the amount of time terminal will wait until it assumes a connection cannot be made. The default is 30 seconds. It may be necessary to increase the time for long distance or other connections requiring additional time. During the connection process you will see the Time-out count down from the number you set. You cannot set the Time-out less than 30 seconds.
- **Redial after timing out**—Terminal will continue to try until you either connect, or cancel the command. To cancel the redial, click on the Cancel button when the redial Dialog Box appears. The number of times that a redial is attempted will appear in the dialog box (Fig. 63.12).
- **Signal when connected**—Selecting this option will cause Terminal to notify you when the connection is made by sounding a "beep."

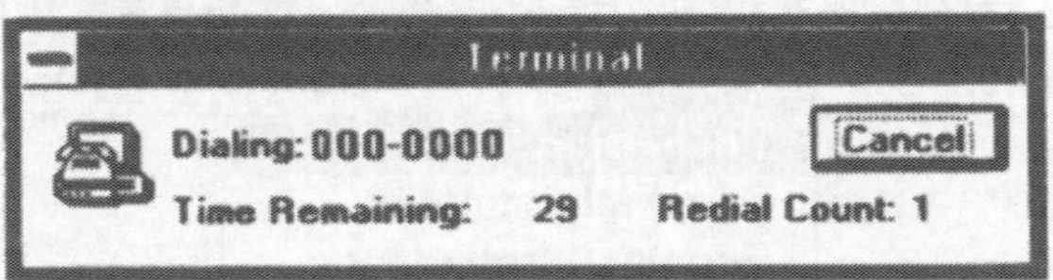

FIGURE 63.12 The Terminal dialog box that appears during dialing.

Terminal Emulation

Terminal Emulation is the type of terminal that you want to use during your connection with the host computer. This option is dependent upon the requirements of the host. Terminal can emulate three types of terminals during the connection, as shown in Figure 63.13.

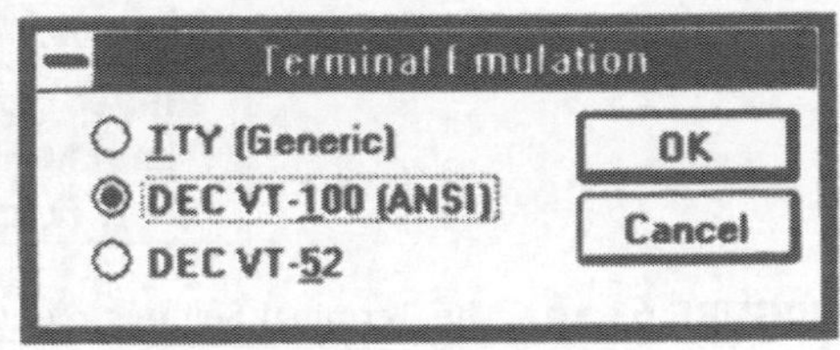

FIGURE 63.13 The Terminal Emulation dialog box.

- TTY (Generic)
- DEC VT-100 (ANSI)
- DEC VT-52

These three emulations are the most prevalent terminals connected to mainframe computers. Selection of the correct emulation will allow Terminal to send the keys you press and any special formatting to the remote computer. Additional setting such as Line Wrap, Local Echo, Sound, Carriage Returns, Line Feeds, Terminal Screen Width, Type of Cursor, Terminal Font, and County translation can be set from the Terminal Preferences selection from the settings menu. It is recommenced that these settings be adjusted only to meet any special needs the computer system you are connecting to requires. In most cases the default settings will be satisfactory.

Communications Settings

Following the selection of a phone number and terminal type, continue with the communication settings. The selections you make in this area are based upon the requirements of the remote computer and the speed capabilities of your modem.

If you are not sure about a particular setting, leave the choice at the default value. A few of the settings are not commonly documented because it is assumed you will be using what is most common. Always investigate as many *settings as possible*. This will better the chances at a successful link with a remote computer.

There are eight settings available when you select Communications from the Settings menu. The dialog box to make these selections is shown in Figure 63.14.

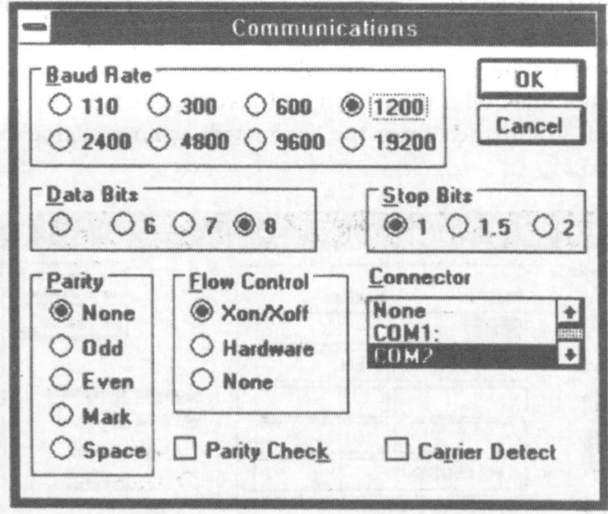

FIGURE 63.14 The Communications Settings dialog box.

The first two selections we want to discuss are Baud Rate and Connector. These two items are really the only ones you should have a firm understanding of in setting up the parameters. The other options are important, but simply need to be made based upon the requirements of the computer you want to connect.

The Baud Rate offers eight numeric choices from 110 to 19200. The Baud Rate setting is the speed that your modem should try to connect with the host computer. The higher the number, the faster the two computers can communicate. The most common baud rate is 2400. If you are in doubt, select that option.

The Connector option is the communications port where your modem is attached. The available choices in the scroll box are COM1, COM2, COM3, and COM4. When your modem was installed on the system, one of these ports was selected. Make sure this setting reflects that choice. In Windows the most likely situation is that the modem is located at COM2. Your mouse probably uses COM1. While these settings are the most common, remember that it is possible that your system is different. A little experimentation will find the answer if you do not know. Selecting the wrong or nonexistent COM port will result in an error message.

The first host-specific selection is the Data Bits. The five available choices are 4, 5, 6, 7, and 8. Eight is the most common, and the default selection.

Stop Bits. The most common selection is the default of one.

Parity is your next selection, and refers to data checking that is performed during the communications process. Parity checking is available only if you selected Data Bits of seven or less..

Flow Control deals with the method the communications process will use to control the speed at which data are sent. This is the method one computer uses *to tell the other to slow down the transmission rate.* The most common method, and default is Xon/Xoff.

Modem Commands

The Modem Commands selection from the Settings menu allows for the customization of commands to be sent to your modem. For example, using a Hayes-compatible modem, the command to get the modem to dial on a touch tone phone line is ATDT.

Your modem may require a special command be sent to the modem to dial, hang-up, etc. This option is where the commands can be customized if needed. Read the documentation that comes with your modem for further details.

Selecting Modem Commands from the menu will open a dialog box like the one shown in Figure 63.15.

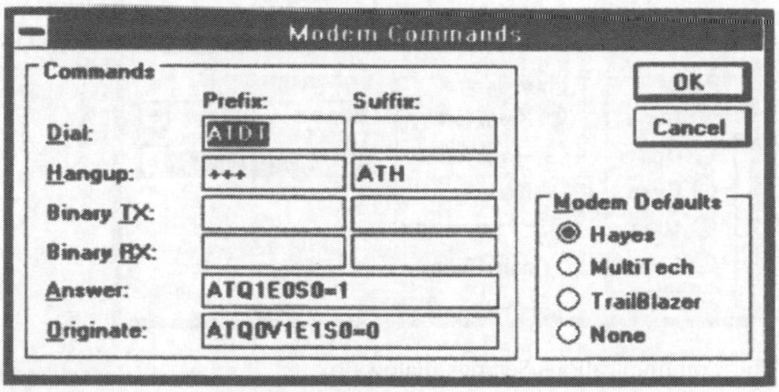

FIGURE 63.15 The Modem Command dialog box.

The most important selections in this dialog box are the Modem Defaults Radio button selections. The four modem types shown on the screen will determine the different commands that need to be entered in the command area. Once you select the type of modem you will be using, you can customize the commands by typing in the edit boxes next to the commands.

Saving Your Selections

Because different hosts require different settings and phone numbers, Terminal allows you to save your selections in a file. If you connect to several different hosts with Terminal, you will find it more convenient to save the particular communications settings and phone numbers rather than select them each time you want to connect.

Connecting and Terminating Host Connections

Making the connection involves commanding the modem to dial, logging in, using the service, then disconnecting.

To command the modem to dial, select Dial from the Phone menu. A dialog box appears on the screen, showing the progress of the attempted connection. This is shown in Figure 63.16.

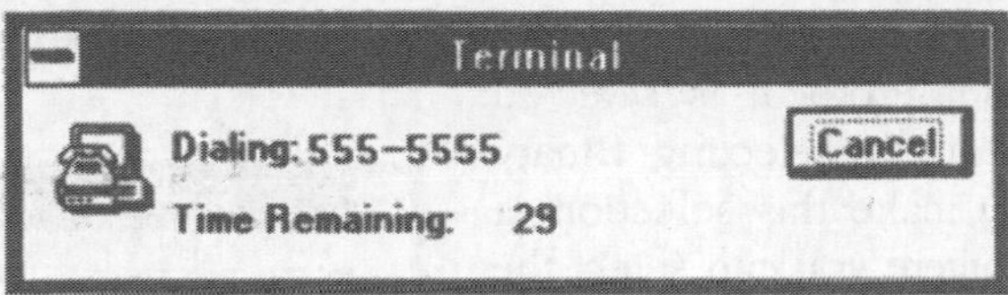

FIGURE 63.16 The Terminal dialog box showing the connection status.

In the dialog box, during the connection process you will notice a few items:

- The phone number being dialed by the modem
- The time remaining. This is the Time-out number entered in the Phone Number dialog box under Settings.
- The Cancel button. Click on this to immediately stop the modem and abort the call.

In the text area of the Application window you will notice that the commands we selected for the modem appear on the screen. Upon connection you will see a notification from the modem that a connection has been made.

Once you have established the connection, you can proceed to conduct your business on the host computer just as if you were a directly connected terminal. Typically this involves logging in, checking mail, running programs, or transferring files to your system or to the host. When you have completed the work on the host computer, log out.

After logging out, you should send the modem the Hang-up command from the Phone menu.

File Transfers and Printing

File transfers involve moving a file from your system to the host, or vice versa. In this way we can share data and programs between computer systems.

Text and Binary File Transfers

There are two types of file transfers available to you in the Windows Terminal program: Text and Binary. The choice between the two is dependent upon the type of file being moved from one computer to the other.

If the file you want to move is an ASCII text only file, use the Text transfer method. These files contain only printable characters, and a few formatting codes such as carriage returns.

Binary files are: files that contain unprintable characters. Usually these files are programs or special files associated with an application. For example, if you wanted to transfer a file from your spreadsheet you would use the binary transfer method because the file contains much more than the raw ASCII data.

Binary File Transfer Protocols

Because of the incompatibilities between most dissimilar computer systems, a set of rules for the file transfer must be established. These rules, or protocols, control the way in which the data are sent from one computer to the other. Windows supports two of the most popular file transfer protocols: XMODEM and Kermit.

The Binary Transfer Protocol is selected from the Setting menu by selecting Binary Transfers. When you make this selection, a dialog box appears where you can select the protocol you want to use. Click on the Radio button next to your choice. In the sample dialog box shown in Figure 63.17, XMODEM has been selected as the Binary File Transfer Protocol.

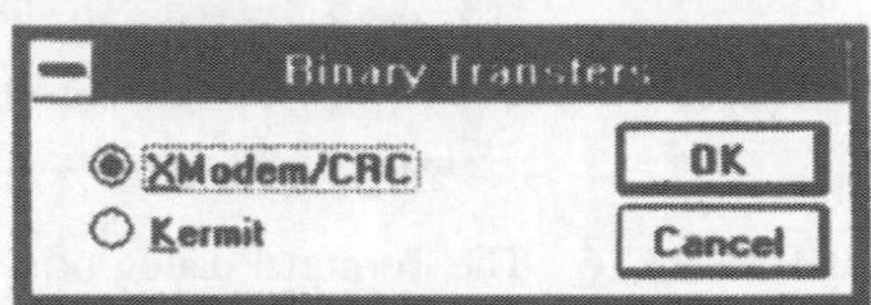

FIGURE 63.17 The Binary Transfer dialog box.

Text Transfer Settings

Flow control options determine the way in which the data are transferred from one system to the other: Standard Control, one Character at a time, or one line at a time. The Standard Control is the same as the Flow Control you selected in *the Communications dialog box*. It is also the default *value*. The reason for including the other two is to offer alternate methods if the Standard Flow Control does not seem to work. The alternative methods are also slower because of the reduced amount of data that are sent at a given time from one system to the other.

The Word Wrap option is a way to control the length of the line sent to the remote system.

The Text Transfer Settings are made by selecting Text Transfers from the settings menu. When you select the option, it initiates a dialog box to make the settings (Fig. 63.18).

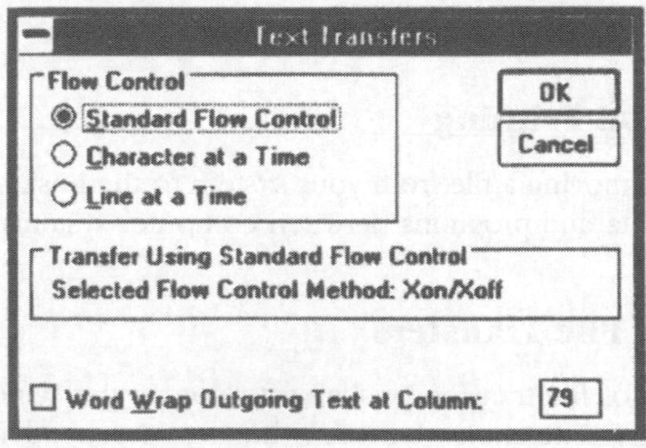

FIGURE 63.18 The Text Transfer dialog box.

Sending, Receiving, Stopping, and Pausing the Transfer

The method used to start either the receipt or sending of a file is entirely dependent upon the host's software. Usually the method is well documented and consists of selecting an option from a menu of commands. Moving a file always begins with telling the host that you are starting the transfer.

Once you have sent the host the command that the file is to be transferred you can begin the process on your end with the commands in the Transfer menu shown in Figure 63.19.

The selection of a command on the Transfer menu depends upon the answer to two questions:

- Are you transferring a binary or text file?
- Are you sending or receiving files?

If the files are being sent to the remote computer, use either the Send Binary or Send Text commands from the Transfer menu. Once you select the proper send command, a dialog box will prompt you for the name of the file on your computer that you would like to send (Figs. 63.20 and 63.21).

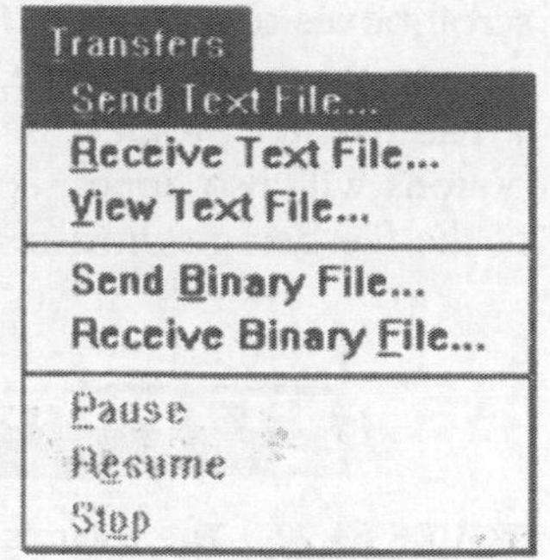

FIGURE 63.19 The Transfer menu.

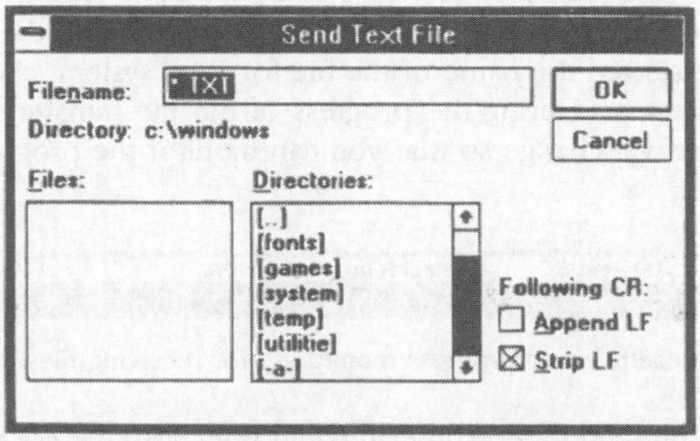

FIGURE 63.20 The Send Text File dialog box.

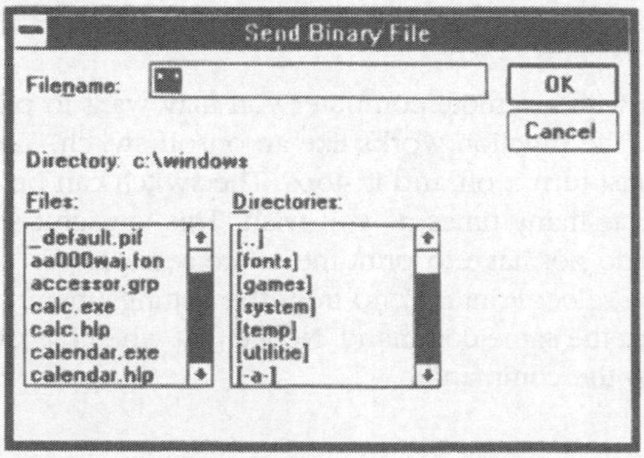

FIGURE 63.21 The Send Binary File Dialog Box.

In either of the Send commands you will be prompted for the file name. Type the file name into the Edit Box, and Click on OK. The Terminal program will begin the transfer using the options you selected.

When you enter the name of the file to transfer, be sure to include the drive letter and path name. If you are unsure of the name of the file, use the file and directory scroll boxes to locate the file. It may be advisable to search for the file in this manner so that the exact drive, path, and file name are entered into the text box.

File transfer progress is shown at the bottom of the Application window. Control buttons will also appear so that you can stop and pause the transmission. The name of the file being sent also appears at the bottom of the screen (Fig. 63.22).

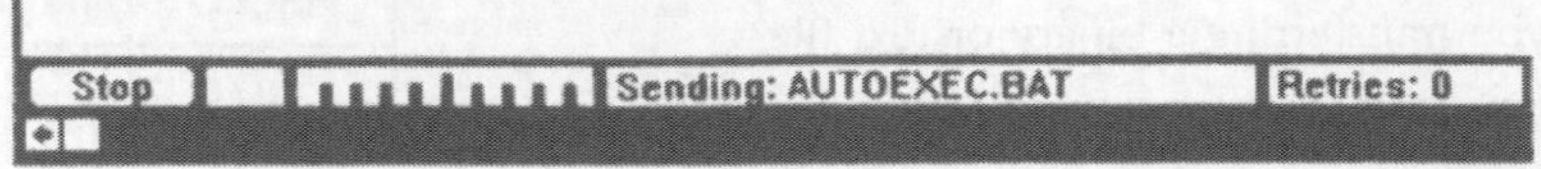

FIGURE 63.22 The controls and progress monitoring for sending files from your computer.

Receiving files is similar to sending them. First, initiate the command to send the file on the host. To receive a file, select either the Receive Text or Receive Binary commands from the Transfer menu. Again, as with the Send commands, a dialog box prompting for the file name will appear. This time, type in the name of the file that will be received on your system. Be sure to specify the drive and path along with the name of the file. In most instances, the name of the file is the same as the one on the host computer.

Once you have selected the name of the file for your system, click on OK and the file will begin to be sent. During the progress of the file transfer the bottom of the Application window will change so that you can monitor the progress (Fig. 63.23).

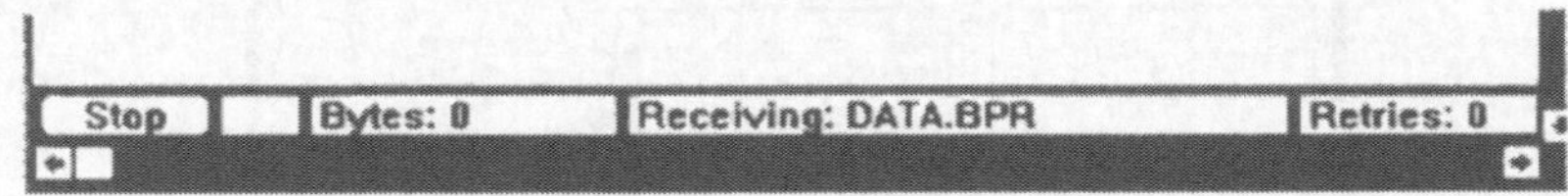

FIGURE 63.23 The controls and progress monitoring for receiving files from the host.

When files have finished transferring (in either direction) the commands at the bottom of the screen will disappear, and you will resume the session with the host.

Printing

During the a session with a remote computer you may want to print the text that appears on the screen. The function works like an on/off switch. You turn on the function and the text prints; turn it off and it stops. The switch can be thrown at any time during your session, as many times as you wish. This is useful so that you get only what you want, and do not have to print the entire session.

To start the printer, select Printer Echo from the Settings menu. When you want to turn it off, again select the same command. Notice that when the printer is on, a check mark appears next to the command.

DEFINING TERMS

The reader should consult the chapter on Windows for more details pertaining to the terms used in all Windows applications.

Baud Rate: Data signaling rate. *Not* the same as data transfer rate (BPS—bits per second)

Flow Control: Process by which two communications systems can control data flow so that one will not fall behind the other, or get ahead of the other system.

MODEM: Acronym for modulator–demodulator. Device used to connect a computer to a telephone or other communications service.

Protocol: A file or data transfer method that ensures the integrity of the transmission. Typical ones for PCs are XMODEM, YMODEM, Kermit, etc.

Terminal Emulation: The capability of the software to function as a terminal such as a DEC VT-340, etc.

REFERENCES

Microsoft Windows Reference Manual and Terminal Help Screens.

64

Internetworking

James Griffioen
University of Kentucky

J. Mark Pullen
George Mason University

Satish Chaliki
Keane, Inc.

In recent years we have witnessed a plethora of technological changes that have drastically altered our view of the role computers play in the workplace. One such change has been in the area of computer networks. Computer users now view the computer as an instantaneous medium for communication among users and also as a window to a wide world of information. Electronic mail, bulletin boards, audio broadcasts, electronic literature, whiteboards, conference calls, and many other services are now in widespread use. The so-called "information superhighway" is rapidly becoming a reality.

However, the wide-scale network communication we have all become accustomed to did not appear overnight, but is, in fact, the fruit of many years of research and experimentation. The "magic" that allows us to create large-scale communication networks is the concept of ***internetworking***. Internetworking refers to the process of connecting many independent networks together into a single logical communication network.

Because the real world consists of users with a wide range of communication requirements, network vendors have developed a wide variety of network technologies, each having a different set of characteristics. For example, some technologies provide reliable data transmission while others provide unreliable (best-effort) data transmission. As a result, every organization or group judiciously chooses a network architecture that achieves the organization's communication needs. Unfortunately, each individual network architecture is incapable of communicating directly with other types of network technologies. Consequently communication between organizations with different underlying network technologies becomes impossible. Moreover, at the time internetworking was first being developed, no single hardware technology was capable of scaling to a nation-wide network. Even today with technologies such as broadband ISDN (e.g., ATM), few would venture to say that such technologies could scale up to the size of to a national or international network like the current Internet has.

To solve the interorganization communication problem, a new technology called internetworking was developed. The fundamental idea behind internetworking is that

0-8493-2530-7/96/$0.00+$.50

internetwork communication can occur only if all networks appear the same and can be connected into a single seemless network. Clearly, getting vendors to agree on a single hardware technology was (and is) unthinkable and undesirable. To achieve a single network technology, the internet designers decided to develop a new "logical network technology" that could be built on any underlying hardware technology. Thus, all network technologies could be made to look alike and could be easily connected to allow communication between otherwise independent organizations and differing network hardware.

64.1 TCP/IP AND THE INTERNET

James Griffioen

One of the chief objectives of early network designers was the ability to internetwork multiple independent heterogeneous networks into a single communication fabric that would allow full connectivity and communication between otherwise autonomous organizations. DARPA (the Defense Advanced Research Projects Agency as it was called) was particular interested in developing an internetwork technology as the basis for a large-scale national network that would connect government labs and universities. Among other things, DARPA's internetworking research efforts resulted in two enormous contributions: the *TCP/IP protocol suite* and the *Internet*.

The *TCP/IP protocol suite* is a large set of communication and routing protocol standards that dictate how computers communicate. Although the TCP/IP protocol suite consists of many different protocols, the entire suite of protocol standards is typically referred to by the suite's two primary protocol standards: the *TCP protocol* and the *IP protocol*. The IP protocol defines a logical or virtual network and achieves the goal of internetworking by allowing machines on different networks to communicate. The TCP protocol uses the IP protocol to transfer messages between machines. In addition, TCP provides features such as reliability and flow control that increase the scalability of communication to large networks consisting of thousands of independent networks that span large geographical areas. The TCP/IP protocol suite has been thoroughly tested, runs across almost any underlying hardware network architecture, and can be used to connect as many networks as one will ever want to connect. Many corporations have now adopted TCP/IP as the standard for their corporate networks. Consequently, new TCP/IP internets are being created daily. Some of these private internets are very large and span large geographical distances.

The most powerful illustration of the robustness of the TCP/IP protocol suite is the existence of the *Internet*. Originally called the DARPA Internet and more recently called the NSF Internet, it is now simply called the Internet, with a capital "I" to distinguish it from corporate or private internets (lowercase "i"). The Internet started as a small research network with only a few research institutions and government agencies scattered across the United States. However, as TCP/IP became more robust and demand for wide-scale communication increased, the Internet has grown into a monstrous network connecting hundreds of thousands of computers across the United States, Canada, Europe, and beyond (Fig. 64.1). Computer access to the Internet is now quite common. Hundreds of universities, colleges, high schools, government agencies, corporations, small businesses, etc. now connect to the Internet. E-mail addresses are now a standard entry on many a business card. Moreover, the number of services provided by the Internet and the wide range of information available on the

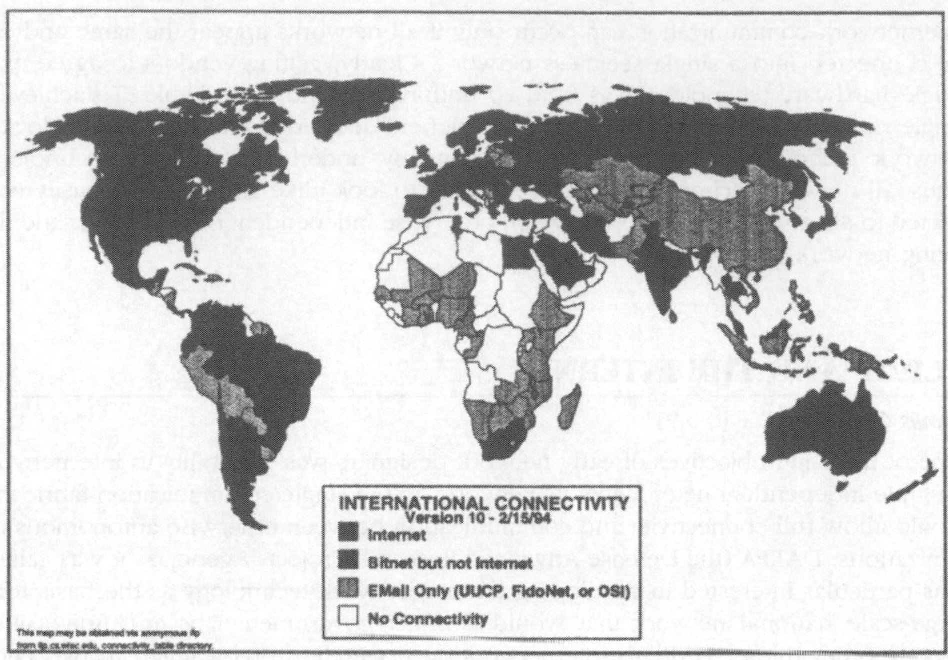

FIGURE 64.1 The Internet is the largest existing internet, providing connectivity from almost any point in the world to almost any other point in the world. Coverage of other major networks such as Bitnet and UUCP are also shown. Copyright © 1994 Larry Landweber and the Internet Society. Reprinted with permission.

Internet are staggering. As a result, the term "information superhighway" has been coined, likening the Internet to the U.S. Interstate Highway system with information rather than goods being transported.

In the following sections we describe the TCP/IP protocol suite and the way TCP/IP internets work. We begin by describing the layered design used in the TCP/IP protocol suite. Given the overall design philosophy, we examine three of the suites major protocols: IP, TCP, and UDP.

Layering

The TCP/IP protocol suite, like many other protocol suites, is designed in a layered fashion. Layering divides up the task of communication into a set of services organized into a hierarchy where the services at each level of the hierarchy depend on services below it in the hierarchy. Such a hierarchy is called the *protocol stack*. A protocol stack "loosely" defines the services at each level of the hierarchy. Thus a protocol suite may implement multiple protocols at any given level. All the protocols at a given level provide the basic services required at that level but otherwise differ in some fundamental way.

The TCP/IP protocol suite employs a four-layer model (as opposed to the other prominent layering model, ISO/OSI's seven-layer model). The four service layers provided by TCP/IP are shown in Figure 64.2.

The *network interface layer* is responsible for sending and receiving packets over the underlying network technology. The *internet layer* provides internet host-to-host communication that involves routing and fragmentation. The *transport layer* provides

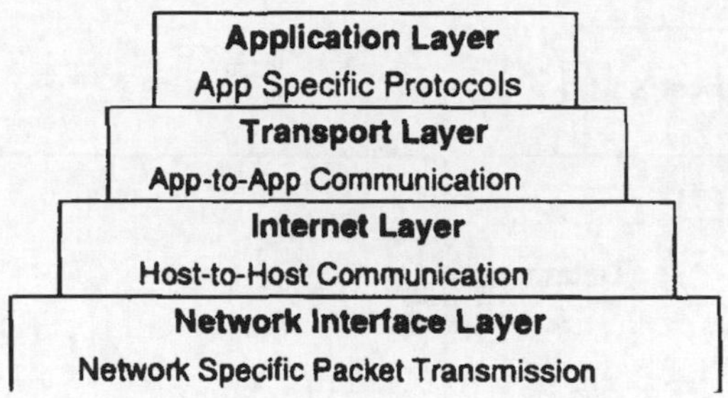

FIGURE 64.2 The four conceptual levels of service provided by the TCP/IP protocol stack.

application-to-application communication and may optionally provide services such as reliable and sequenced delivery or flow control. The *application layer* provides application specific services. In the case of the TCP/IP protocol suite, the TCP protocol occurs at the transport level, while the IP protocol occurs at the internet level.

The three lower layers of the TCP/IP protocol stack are typically implemented inside the operating system kernel. Because the application layer is application dependent, it is generally implemented by the application. Consequently, the border that divides the operating system from the application usually occurs between the transport and the application layer. Thus, the interface that the operating system exports to applications is that of a transport service.

In the following sections we will introduce the major components of the TCP/IP protocol suite. The entire TCP/IP protocol suite contains many different protocols, some rather complex, some still in development, and some already discarded as outdated by newer protocols. A complete description of the suite is far beyond the scope of this book. Consequently, the following sections simply provide a high-level overview of various features and characteristics of the TCP/IP protocol suite. We will begin with the internet layer and work our way up, showing how each layer builds on the services of the layers below it.

IP: The Internet Protocol

The Internet Protocol (IP) is perhaps the most important protocol in the TCP/IP protocol suite. The IP protocol resides at the internet layer of the protocol stack. IP provides the support required to internetwork independent networks into a single logical network, allowing any machine on the logical network to send or receive messages to or from any other machine on the logical network.

The Virtual Network

The major goal of the IP protocol is to create the illusion of a single logical network spanning multiple organizations and their physical nets. To connect two or more independent networks, IP requires that a machine called a *gateway* be placed between the networks. The gateway is responsible for relaying packets from one network to the other.

To illustrate the role of gateway machines, consider the internet shown in Figure 64.3. Two gateway machines are used to connect three different types of physical networks (an Ethernet style network, a ring style network, and a star-based network technology). Both gateway machines have two network interfaces—one on each network to which they connect.

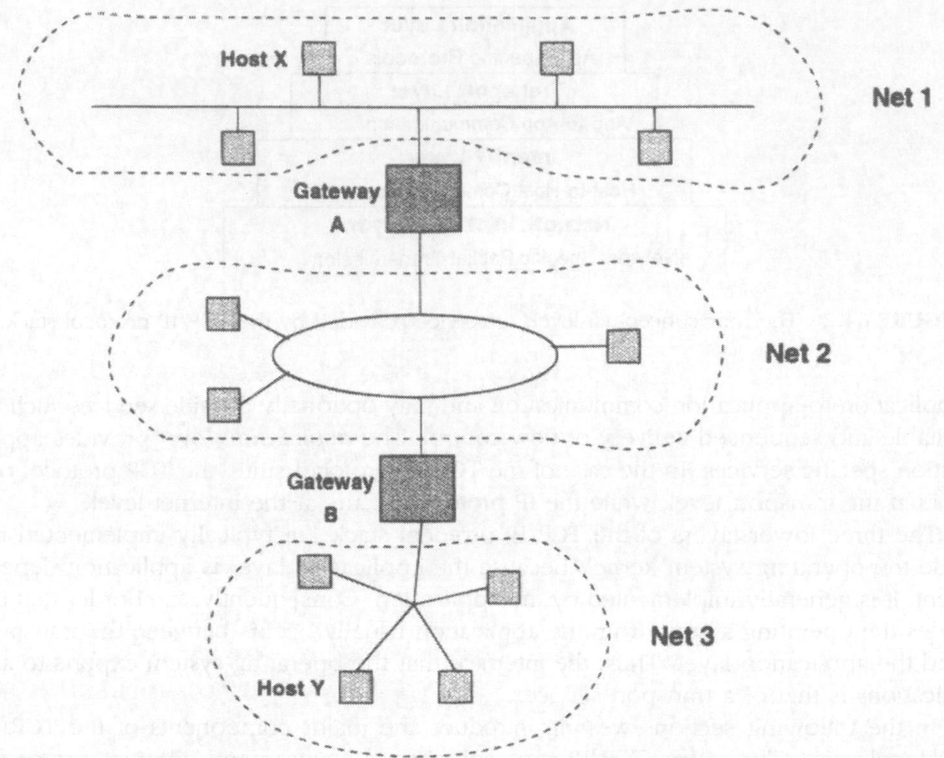

FIGURE 64.3 Two gateways interconnecting three independent networks. Note that the resulting internet links three completely different physical network technologies together. As a result, any machine on any one of the three networks can communicate with any other machine on any of the other networks.

Assume that host X on network 1 wants to send a message to host Y on network 3. For a message to get from X to Y, it will have to pass through gateway A. Consequently, X begins by sending the message to gateway A, which is on X's network. Gateway A knows that host Y is on network 3 and forwards the message onto gateway B. Gateway B is connected to the same network as host Y and thus forwards the message directly to Y. By relaying messages from one network to another, gateways allow communication from any host on the internet to any other host.

Encapsulation and Fragmentation

Recall that the interconnected physical networks may have drastically different architectures. Consequently, the format of the data coming into a gateway may not be anything like the format of the outgoing data that must be forwarded onto the next physical network.

Each gateway machine must be capable of communicating over all network architectures to which the gateway attaches. For example, Gateway B in Figure 64.3 is connected to a token ring network architecture and a star network architecture and thus must be able to send and receive messages over both network architectures. At first glance, one might think that the gateway must be able to translate incoming data into the format of the outgoing link. However, gateways use the concept of *encapsulation* rather than translation to forward data from one network architecture to another. When a host wishes to send an IP message, the host encapsulates the entire IP message, including the receiver's address and various other pieces of information, in the data portion of the underlying network packet structure. Thus, each IP message is completely independent from the underlying network technology over which the message is sent. When a message arrives at a gateway, the gateway simply extracts the encapsulated IP message from the incoming network packet and then encapsulates the message in an outgoing network packet.

By encapsulating the entire IP messages in the data portion of the underlying network architecture's packets, an IP message can flow across any network architecture assuming the network's maximum packet size or maximum transfer unit (MTU) is larger than the IP message. To ensure complete connectivity, IP would have to know and limit IP message sizes based on the smallest MTU of the physical networks that comprise the internet; this is a severe drawback. Instead, IP allows very large messages (up to 65,535 octets in length) regardless of the MTU sizes of the underlying networks. IP uses the concept of *fragmentation* to provide this service. When sending an IP message, the sending host or gateway first breaks the message into pieces called *fragments* that fit within the MTU of the underlying physical network. On the receiving end, the destination machine reassembles incoming fragments to produce the original IP message. IP's fragmentation capability ensures that any network technology can be integrated into an internet.

The Communication Service

The IP protocol provides a very basic communication service and relies on the transport layer to provide additional functionality. Simply stated, IP provides an unreliable connectionless datagram service.

IP is unreliable in the sense that it does not guarantee that a message will ever arrive. Instead, IP provides best-effort delivery, which means IP will do its best to get the message there, but if an error arises, IP will not correct the error.

Datagram service implies that IP transmits only self-contained messages. A datagram is much like a postal letter. Each datagram has an address stating where to deliver the datagram and a data portion that contains the message. Each message is independent of other messages and may be directed at any host. The data portion of each datagram is delivered "as-is" to the transport layer protocol. Datagrams vary in size with the maximum size being 65,535 octets.

IP provides connectionless data delivery. Unlike a telephone connection where an imaginary point-to-point link is established at the time of a call and all subsequent data automatically flow across that link, IP provides connectionless delivery. Connectionless delivery means that every datagram presented to IP for delivery must contain a destination address. Thus each message can be independently directed at any host, rather than all going to a single predefined host.

Rather than providing features such as connections and reliability, at the internet layer, the TCP/IP protocol suite provides such features at the transport layer of the protocol stack. In some sense, the communication services provided by IP are minimalistic for an internet layer protocol. However, because IP provides a bare-bones service, IP has relatively low overhead and does not force advanced features on applications that do not require such features.

IP Addresses

For hosts to communicate over the logical network created by IP, each host must have a unique logical address called an *IP address*. To ensure that no two hosts are assigned the same IP address, a single person or organization is typically responsible for assigning IP addresses to hosts. In the case of the Internet, unique IP addresses must be obtained from the Network Information Center (NIC).

To simplify the task of routing, the IP protocol assigns a unique number to each physical network. The hosts on each physical network are then assigned numbers such that no two hosts on the same physical network have the same number. The network number combined with the host number are sufficient to uniquely identify any machine on an internet. Thus, an IP address is a structured 32-bit value consisting of a network identifier and a host identifier. Four different IP address structures have been defined to support different size networks. Class A addresses are for networks having a large number of hosts (2^{24}). Not many networks of this size are expected, so an internet may have only 2^7 such networks. Class B addresses are for networks with a substantial, but not extremely large, number of hosts (2^{16}). An internet may have 2^{14} such networks. Class C addresses are for networks with relatively few hosts (2^8). However, an internet may have many class C networks (2^{21}). Class D addresses are reserved for hosts belonging to a group of hosts. Figure 64.4 shows the structure of each IP address class.

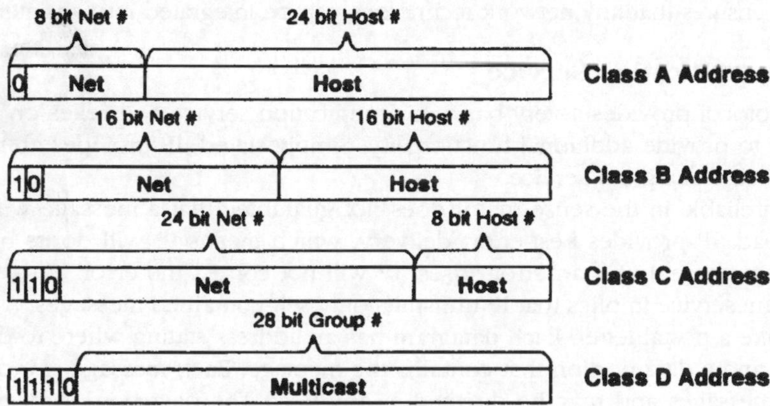

FIGURE 64.4 The components of an IP address. The bits shown on the left are fixed and used to determine the class to which the address belongs.

By convention IP addresses are written (in human-readable format) as a series of four numbers separated by periods (dots). Each number represents one octet of the IP address. For example, the 32-bit IP address $80A38601_{hex}$ would be written 128.163.134.1 and represents a Class B address (because the highest two bits are 10) with network number $32931_{decimal}$ and host number $34305_{decimal}$.

IP Packet Format

Every IP datagram sent across the network contains an IP header and a data area. The IP header does not have a fixed length. Instead, the length of the header depends on the number of IP options that are used. Figure 64.5 illustrates the format of an IP datagram. The following briefly describes the meaning of each field in the header [see Postel (1981a) or Comer (1991) for a detailed description]:

Version Number: Identifies the version of the IP protocol used.
Header Length: The length of the IP packet header (in 32-bit words).

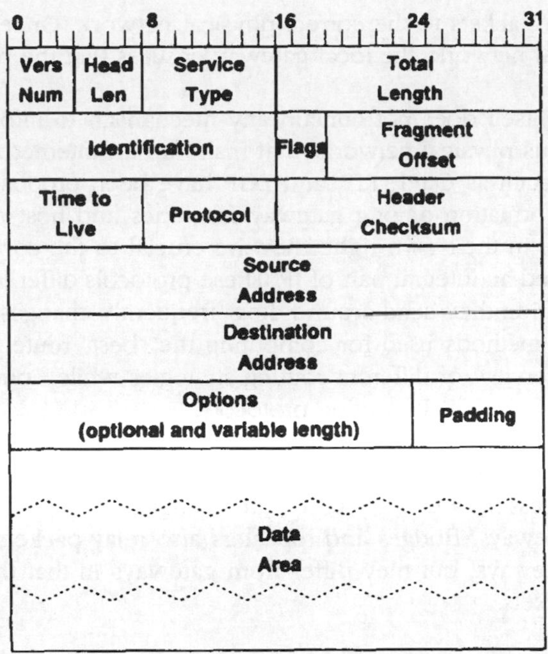

FIGURE 64.5 The IP packet format. The header length depends on the options used.

Service Type: How to handle the packet to achieve the desired transmission quality.
Total Length: Size of the header and data (in octets).
Identification: Uniquely identifies the IP datagram to which the fragment belongs.
Flags: To control the fragmentation process.
Fragment Offset: Specifies the location of the this packet's data relative to the beginning of the IP datagram (in 8 octet units).
Time to Live: How long the packet may wander around in the network. Prevents packets from remaining in the network indefinitely.
Protocol: The higher-level protocol used in the data portion of the packet.
Header Checksum: The checksum of the header portion only (assumes the header checksum field is zero).
Source Address: The IP address of the sender.
Destination Address: The IP address of the receiver.
Options: A variable length list of IP options. Useful for controlling aspects such as routing. Padding fills out the options to a 32-bit boundary.
Data Area: Contains the IP message.

IP Routing

Gateway machines play a crucial role in an IP internet by routing and forwarding IP packets toward their destination. Whenever a gateway receives an IP packet, the gateway must choose an appropriate link over which to forward the packet. Consequently, each gateway must know the "best" way to get to every host in the internet. Instead of storing huge tables containing every **host** in the internet, each gateway maintains a table that tells how to get to every **network** in the internet. Because IP addresses are structured as a pair (network id, host id), gateways use the network id portion to route packets to the correct physical network. Once a packet arrives at the correct physical network, the local gateway ensures that the packet arrives at the correct host.

The IP protocol itself does not contain any mechanism to inform gateways about routes to the various physical networks that make up an internet. Instead, a wide variety of protocols such as OSPF, RIP, and EGP have been proposed and used to exchange routing information among gateway machines and host machines. Although these are protocols in their own right, they are crucial to the correct operation of IP and often considered an integral part of IP. These protocols differ in such things as the amount of information they send around, how frequently changes are propagated, or the definition and methods used for computing the "best" route. Tanenbaum (1988) provides a nice overview of different routing strategies while Comer (1991) discusses the specifics of various TCP/IP routing protocols.

Over the years a variety of names have been given to gateways in different contexts. As a result, the terms *gateway*, *router*, *PSN* (packet switching node), and *IMP* (interface message processor) are often used interchangeably. Note that *bridges* and *repeaters* are not gateways. Bridges and repeaters also relay packets from one network to another like gateways, but they differ from gateways in that they do not perform any routing of packets.

ICMP

The ICMP protocol (Internet Control Message Protocol) is an integral (and required) part of the IP protocol. ICMP is responsible for the transmission of error and control messages. ICMP message never reach a user or a program, but instead are sent to the IP protocol component of an operating system. When the IP protocol software of an operating system receives an ICMP message, it modifies the way it is behaving to correct the error indicated by the ICMP message, or it responds to a control request message.

ICMP error messages include such things as "destination unreachable," indicating the packet could not be delivered, "source quench," indicating congestion, "redirect," indicating an incorrect route was selected, or "time exceeded," indicating the time-to-live timer has expired. In addition, control messages such as an "echo request," to determine whether a machine is up, or an "address mask request," to obtain the subnet mask, may be issued to find out information about the current state of other hosts.

Host Names

To make host identification more palatable to users, the TCP/IP protocol suite provides a mechanism for assigning "ASCII-readable" names to hosts rather than requiring users to deal with IP addresses. TCP/IP internets optionally include a translation system called the Domain Name System that maps ASCII hostnames (called full hostnames or domain names) onto IP addresses. The set of servers that comprises the Domain Name System maintains a set of distributed tables that contains (domain name, IP address) pairs. Given a domain name, these servers work together to locate the IP address associated with the specified domain name. The servers communicate via the DOMAIN protocol[1] to locate the desired mapping if the mapping is not known locally.

[1]RFCs describing the TCP/IP protocol suite and the Internet are available via anonymous ftp from nic.ddn.mil.

Although the domain name system is highly flexible, the Internet has adopted a convention where machine names typically have one of the following formats:

hostname.deptname.instname.insttype
or
hostname.deptname.instname.city.province.country.

hostname: the local machine name.
deptname: The department name or abbreviation.
instname: The institution or organization's name or abbreviation.
insttype: The type of institution or organization.
city.province.country: The name or abbreviation of the city, province or state, and country where the institution resides.

It should be noted that although most names fall into one of these two formats, not all names contain all the fields listed in the format. Example domain names include **violin.dcs.uky.edu, snoopy.cnri.reston.va.us, or sophia.inria.fr.**

TCP: Transmission Control Protocol

Recall that the IP protocol provides host to host communication across an internet. The TCP protocol builds on the functionality provided by the IP protocol to support application-to-application communication between applications executing on different machines anywhere in an internet. In addition, TCP defines a generic communication abstraction/service called a *stream* that is conceptually easy to understand and use.

The Stream Abstraction

Rather than requiring applications to package data up into individual messages for transmission, TCP supports the abstraction of a *byte stream*. Conceptually, a TCP byte stream is like a dedicated wire connecting two applications. Any amount of data (bytes) can be sent over the wire (byte stream) and will be delivered to the application on the other end of the wire. We often call this imaginary connection a *virtual circuit* or *virtual connection*. Applications do not need to worry about creating or formatting messages or packets for transmission. Instead, an application simply sends data directly over the connection and/or reads data directly from the connection. Each byte stream is bidirectional and can carry information in both directions simultaneously. One major advantage is that TCP connections behave just like files. Applications read and write data to/from a connection just like they would a file.

Streams and Connections

The previous section described the basic stream abstraction. However, there is still an aspect of the stream abstraction that needs to be defined. In particular, there must be a mechanism for initiating and establishing a stream connection.

Before any communication can occur between two applications executing on a different (or possibly the same) host, a *TCP connection* must be established. To establish a connection between the applications, one of the applications must assume an active role (initiating the connection) while the other application assumes a passive role (waiting for the active application to connect to it). The active application is often referred to as the *client* application while the passive application is referred to as a *server* application.

This model of communication implies that the active application can somehow uniquely identify the passive application with which it wants to communicate. That is, when two applications wish to communicate, they must (1) locate one another, and (2) establish a unique TCP connection.

The TCP protocol supports the abstraction of *ports*. Ports are particularly helpful when applications need to locate one another. Each port is uniquely identified by a *TCP port number*. In some sense, a port is like a telephone. To call someone you only need to know their telephone number. Similarly, if you know an application's port number, you can connect to and communicate with the application. When a client application wants to talk to a server, the client establishes a connection to the TCP port that represents the server.

Some TCP port numbers have been reserved for common applications and the services they offer. Examples services include time-of-day (port 13), file transfer (port 21), telnet (port 23), and mail (port 25). Such servers listen to the appropriate TCP port, passively waiting for clients to connect. The remaining unreserved TCP port numbers are available for user-defined servers. Such servers dynamically obtain an unused port that clients can then connect to.

The server's TCP port clearly identifies one end of the connection. However, streams are bidirectional. To identify the other end of a connection clients, like servers, obtain a TCP port to serve as their endpoint of the connection. Thus, each TCP stream is uniquely identified by the two ports that represent the endpoints. Because client applications actively initiate TCP connections, they typically do not need a well-known TCP port number and thus choose an arbitrary (unused) TCP port number as their port. When a client connects to a server, the client tells the server the TCP port number that it will be using. Consequently, both hosts know both endpoints of the connection.

Because streams are identified by two endpoints, a server application may accept multiple connections through the same TCP port. Thus it is possible for a server to communicate over two or more TCP connections (through the same TCP port) at the same time (Fig. 64.6). As a result, mail servers and other servers can simultaneously serve multiple clients via connections through a single well-known TCP port.

Buffering

Recall that TCP builds on the functionality of the IP protocol. However, IP provides datagram, not byte steam, support. To create the stream abstraction, TCP breaks up the data stream into sections that can be transmitted in an IP packet. Because bytes may be written to the stream one at a time, TCP typically buffers the data either for a certain amount of time or until enough data have been written onto the stream so that a sufficiently large IP packet can be constructed. On the receiving end, TCP extracts the stream data from the incoming IP packets and delivers (only) the data to the application. Consequently, writing data over a TCP connection does not imply that the data will be sent immediately. Although this may delay some data, it results in much better utilization of the network, and ultimately better TCP performance.

Reliable Delivery

Recall that IP provides unreliable datagram delivery. Not only does TCP build a new abstraction (streams) on top of the IP datagram service, but it also provides reliable, ordered, delivery of data. That is, TCP guarantees that all bytes written to a TCP stream

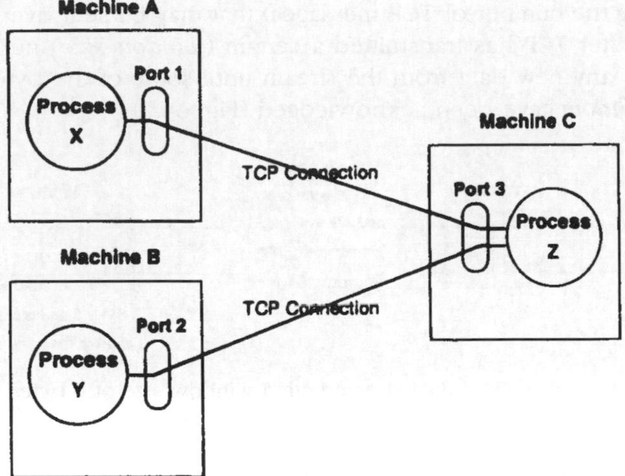

FIGURE 64.6 Server Z simultaneously serves two clients via two TCP connections through port 3.

will be delivered to the receiver.[2] Moreover, all bytes will be delivered in the order they were sent.

To ensure reliable delivery, TCP uses *acknowledgments*, *timeouts*, and *retransmissions*. Whenever TCP transmits a packet across an internet, it saves the message until the destination replies with an acknowledgment message saying that the packet was correctly received. If an acknowledgment is not received within some *timeout period*, TCP assumes the message was lost, transmits the message again, restarts the timer, and again waits for an acknowledgment message.

However, TCP cannot be certain that the message was lost. In fact, the message may have been delayed but not lost. In this case the destination will receive duplicate messages (i.e., two copies of the message). For the receiver to identify duplicate messages, the sender stamps each TCP message with a unique *sequence number*. If the receiver receives two (or more) messages with the same sequence number, it knows the second message is a duplicate and does not need to be delivered to the application. Every connection has two sequence numbers, one for each direction of the bidirectional connection. Each time a host sends a new TCP message, the host increments the connection's sequence number and tags the message with the new number.

Flow Control

Unlike protocols that use a stop-and-wait flow control mechanism (i.e., a new message cannot be sent until the previous message has been acknowledged), TCP allows multiple messages to be sent before receiving any acknowledgments. This parallelism significantly improves the attainable throughput. However, the increase in traffic may also lead to an increase in lost or dropped packets because hosts and gateways are more likely to become congested. Consequently, TCP limits the number of bytes (ef-

[2]TCP cannot really guarantee delivery (especially if the hardware fails), but it will continue to try until it determines that an unrecoverable error has occurred or until it has tried long enough and has concluded that the data cannot be delivered.

fectively limiting the number of TCP messages) that may be sent over a connection at any one time. After TCP has transmitted a certain (*window size*) number of bytes, it will not transfer any new data from the stream until some of the bytes in the current transmission *window* have been acknowledged (Fig. 64.7).

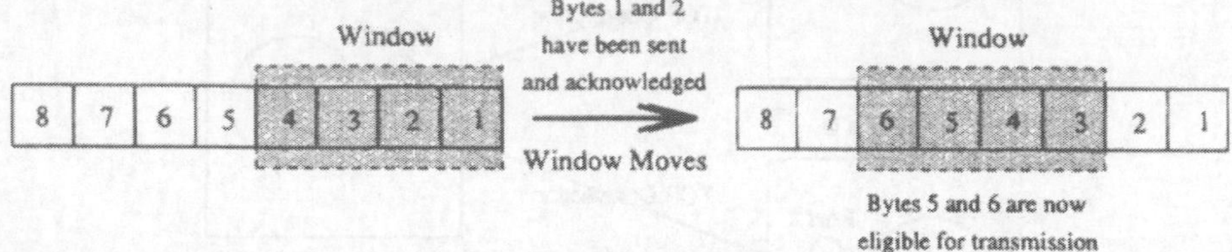

FIGURE 64.7 An outgoing TCP data stream with a window size of 4 bytes. All 4 bytes in the window shown on the left may be transmitted immediately. As acknowledgments arrive, the window "slides" left to allow new bytes to be sent, as shown by the window on the right after bytes 1 and 2 have been acknowledged.

To remedy the problem of congestion that occurs suddenly, TCP provides a mechanism where the sending and receiving machines can dynamically adjust the current window size. When a receiver sends an acknowledgment message to the sender, it includes a window advertisement that indicates how the window size should be adjusted. By dynamically adjusting the window size, hosts can adjust for congestion as it arises.

TCP Packet Format

The TCP packet format is shown in Figure 64.8. Like IP packets, TCP packets consist of a (variable size) header region and a data region. The size of the header region depends on the TCP options selected. The following briefly describes the meaning of

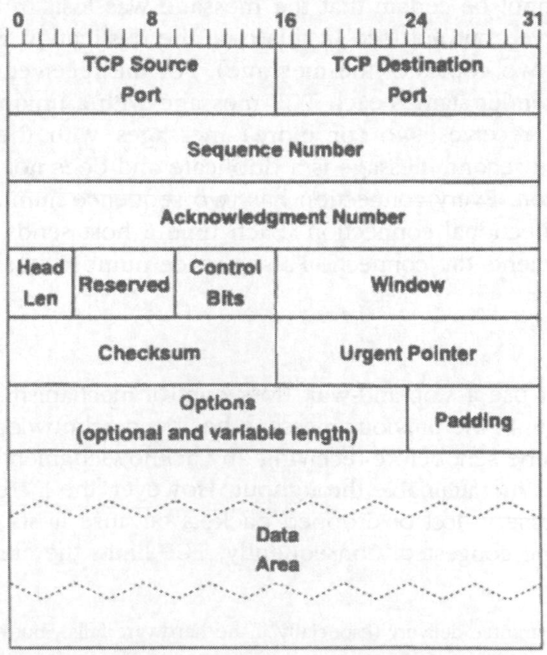

FIGURE 64.8 The format of a TCP packet.

each field in the TCP header [see Postel (1981b) or Comer (1991) for a detailed description]:

TCP Source Port: The sender's TCP port number.
TCP Destination Port: The receiver's TCP port number.
Sequence Number: The packet's sequence number (based on the sequence number associated with packets flowing from the source to the destination).
Acknowledgment Number: The sequence number of the next octet expected in the opposite direction.
Header Length: The length of the TCP header region (also called the data offset) in 32-bit words.
Reserved Bits: Six bits reserved for future use.
Control Bits: These bits tell how to interpret various fields in the packet header (e.g., urgent pointer valid) or convey control information (e.g., no more data from sender).
Window: The window advertisement.
Checksum: The TCP checksum of header and data.
Urgent Pointer: Used to specify where urgent data (out-of-band data) ends.
Options: A variable length list of TCP options. Padding fills out the options to a 32-bit boundary.
Data Area: Contains the TCP message.

UDP: User Datagram Protocol

The User Datagram Protocol (UDP) provides an alternative to the transport-level service provided by TCP. Like TCP, UDP is a transport-level protocol and builds on the service provided by the IP protocol. However, unlike TCP, UDP does not provide reliability, flow control, or a connection-oriented stream abstraction. Instead, UDP provides an unreliable application-to-application datagram delivery service. In short, UDP enhances the IP protocol only in its ability to direct messages to a specific application executing on a host.

However, UDP provides sufficient service for a wide variety of applications. UDP is often used when the duration of communication is very short (e.g., a single message). Because there is no connection setup or tear-down, UDP exhibits lower overhead than TCP, which can result in better performance when sending many short messages. UDP also can make use of IP's 1-to-many (multicast) or 1-to-all (broadcast) communication facilities if the IP implementation supports them. Of course reliability must be implemented by the application if reliability is desired.

The UDP packet format is shown in Figure 64.9. Unlike TCP, UDP packets have a fixed size (eight octet) header region. However UDP does have a variable size data region, limited only by the maximum size of an IP packet. The following briefly describes the meaning of each field in the UDP header [see Postel (1980) or Comer (1991) for a detailed description]:

UDP Source Port: The sender's UDP port number.
UDP Destination Port: The receiver's UDP port number.
Length: The length of the UDP header and data region.
Checksum: The UDP checksum of the UDP psuedo-header. The pseudo-header is comprised of information taken from the IP header, the UDP header, and the data area.
Data Area: contains the TCP message.

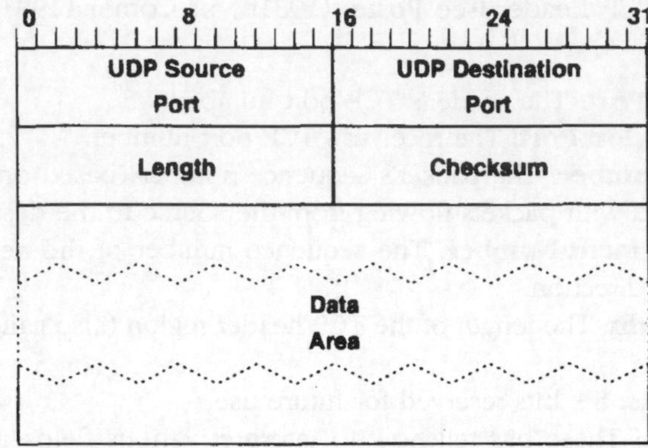

FIGURE 64.9 The format of a UDP packet.

The Application Layer

The application layer of the TCP/IP protocol suite consists of a wide variety of protocols designed to support specific tasks. The vast majority of applications build on the functionality provided by either the TCP or UDP protocols. Table 64.1 lists several example application-level protocols. Note that some application-level protocols build on services provided by other application-level protocols. For example, the NFS protocol builds on the services of the RPC protocol, which builds on the services of UDP protocol.

TABLE 64.1 Example Application-Level Protocols and Their Base Protocols

Application-Level Protocol	Typical Base Protocols(s)	Application/ Purpose
TELNET	TCP	Remote login service
FTP (File Transfer Protocol)	TELNET TCP	Interactive file transfer
SMTP (Simple Mail Transfer Protocol)	TCP	Electronic mail exchange
RPC (Remote Procedure Call)	UDP	Remote procedure invocation
NFS (Network File System)	RPC UDP	Remote file system service
FINGER	TCP	Remote user information
NTP (Network Time Protocol)	UDP	Remote time service
NNTP (Network News Transfer Protocol)	TCP	Remote news/bboard
SNMP (Simple Network Management Protocol)	ASN.1 UDP	Network management access

Application Layer Interface to the Transport Layer

Up to this point, we have defined only the services available at the transport layer of the TCP/IP protocol suite; we have not defined an interface to these services. Unfortunately, there is no standard that defines the interface to the services provided by the transport layer. Consequently, vendors of systems with support for TCP/IP have been free to implement whatever interface they desire. The disadvantage is that an application written for one type of interface cannot be compiled or run directly on a system with a different interface to the same set of services.

Although no standard exists, one could argue that the Unix *socket* abstraction has emerged as the dominant interface (although not necessarily the preferred interface!). The Unix socket abstraction allows applications to control the type of communication used (e.g., stream vs. datagram) or the desired local/remote port used but otherwise provides an interface very similar to the Unix file system interface. In fact, sockets can be read from or written to using the Unix's normal file system read and write routines. The socket abstraction and its implementation are described in more detail in Leffler et al. (1989) and specifics can be found in the Unix Programmer's *Reference Manual*.

Navigating the Internet

The Internet has been nicknamed "the information superhighway" because of the vast amount of information available and the rapidly increasing number of services available via the Internet. A wide variety of application programs have been developed that help users access these services and information. Example applications include telnet, ftp, gopher, archie, wais, worldwideweb, mosaic, and the list goes on. Several books have been written that describe these applications in great detail. Dern (1994) provides a nice introduction to the Internet for new users.

We will briefly describe one of the most popular applications for retrieving information in the Internet; the *FTP program*. FTP is an acronym for the File Transport Protocol, one of the protocols in the TCP/IP protocol suite. FTP is primarily used to transfer files to or from another machine and comes standard on most Unix systems. Various versions of FTP are also readily available for PC systems.

Normally a user must have access to (i.e., an account on) both machines to use FTP. However, a feature of most FTP servers, known as *anonymous FTP*, allows any user in the Internet to access publicly available files stored at the server.

To use anonymous FTP, issue the command "ftp machinename," where machinename is the name of the machine from which you wish to retrieve files. You will be asked for a login ID. Type "anonymous" as the login ID (most systems will also accept "ftp" as the login ID). You will then be prompted for a password. It is polite to type your e-mail address as the password so that the owners of the remote machine know who has obtained their files. However, most servers will accept anything as the password.

At this point you have access to the remote system. You can then use a series of "ls" and "cd" commands to move around the server's directory structure looking for the files you wish to retrieve. Once you have located the file you want, you can use the "get" command to copy the file from the remote machine to your machine. Typing "help" will give you a complete list of FTP's commands.

On-Line Information

In the References we present a brief list of articles that describe the TCP/IP protocol suite and the Internet in much greater detail. However, two of these references deserve special attention.

While the TCP/IP protocol suite was under development, a series of papers and proposals called RFCs (Request For Comments) was generated and placed on-line. These RFCs cover a wide variety of internet topics. They also vary in seriousness, with some being quite comical. However, they do contain the specifications for various TCP/IP protocol suite standards and proposed standards. New RFCs continue to appear and all RFCs are still available on-line. RFCs can be obtained via anonymous ftp from nic.ddn.mil.

A second source of information is a set of on-line reports and notes that summarize the workings of the IETF/IESG (which stands for the Internet Engineering Task Force and the Internet Engineering Steering Group). These groups are responsible for future Internet developments. The documents they produce can be found at various locations via anonymous ftp. One such location is nis.nsf.net.

DEFINING TERMS

FTP: File Transfer Protocol: used to copy files between hosts.

Host: A computer.

Internet: Internet is the term used to refer to the current worldwide network often referred to as the "information superhighway."

Internetworking: Building a large virtual (logical) network out of many small independent local area networks so that all machines can communicate with all other machines despite the fact that they reside on different physical networks.

IP: The Internet Protocol is a communication protocol that provides the illusion of an Internet by allowing hosts on different physical networks to communicate as if they were on the same network.

LAN: Local Area Network. A relatively small network confined to a small geographic area (e.g., an office, business, or college) and serving a small number of machines (at most a few hundred).

Network Address: The address used by the network to identify and locate a computer.

Packet: A structured message consisting of a header (usually containing the address) and a body (the data to be sent). Conceptually a packet is similar to a postal letter.

Routing: The task of determining a route or path through the network that a packet must travel to get from the sender to the receiver.

TCP: Transmission Control Protocol: provides a reliable byte-stream connection between two processes on different hosts.

Telnet: A protocol used to Login to other machines on the network.

UDP: User Datagram Protocol: provides simple unreliable packet delivery between processes.

WAN: Wide Area Network: A large network connecting machines scattered across a large geographic area (e.g., the United States or the world) usually supporting many hosts (hundreds of thousands or more).

REFERENCES

Douglas Comer, *Internetworking with TCP/IP: Principles, Protocols, and Architecture*, Vol. 1, 2nd ed., Prentice-Hall, Englewood-Cliffs, NJ, 1991.

Douglas Comer and David Stevens *Internetworking with TCP/IP: Design, Implementation, and Internals*, Vol. 2. Prentice-Hall, Englewood-Cliffs, NJ, 1991.

Daniel Dern, *The Internet Guide for New Users*. McGraw-Hill, New York, 1994.
Craig Hunt, *TCP/IP Network Administration*. O'Reilly and Associates, 1992.
Ed Krol, *The Whole Internet: User's Guide and Catalog*. O'Reilly and Associates, 1994.
Leffler, McKusick, Karels, and Quarterman, *The Design and Implementation of the 4.3BSD UNIX Operating System*. Addison-Wesley, Reading, MA, 1989.
J. Postel, RFC 768, *User Datagram Protocol*, 1980.
J. Postel, RFC 791, *Internet Protocol*, 1981a.
J. Postel, RFC 793, *Transmission Control Protocol*, 1981b.
Andrew Tanenbaum, *Computer Networks*, 2nd ed. Prentice-Hall, Englewood Cliffs, NJ, 1988.

64.2 Unix, UUCP, AND INTERNET ROUTERS

J. Mark Pullen

Unix

The Unix operating system was originally developed by AT&T Bell Laboratories, expanded by the University of California at Berkeley, and later commercialized by AT&T and others. It provides a full suite of powerful capabilities in an open system that is intended to be tailorable and expandable. As an open, multiplatform system it has proved quite popular and has been adopted by Sun Microsystems, IBM, Hewlett-Packard, and Silicon Graphics Incorporated among others.

UUCP

The Unix to Unix Copy Program (UUCP) is a utility for copying files between UNIX systems. It provides a capability to copy files both within and between systems based on a simple command. It easier to invoke but less powerful than FTP (in general, a lot of systems files must be restricting from UUCP copying for security reasons). Because UUCP is also the basis for the electronic mail in Unix, the term UUCP is often associated with e-mail.

Internet Routers

The Internet works because all of its interconnected networks are able to route Internet Protocol (IP) packets toward their destinations, based on the address in the packet header. This function is performed by "routers," computers that interconnect networks (or subnets of a single network). The routers exchange accessibility ("routing") data among themselves using standard protocols, for example, the Routing Information Protocol (RIP). To be connected to the Internet, a LAN must have at least one router connection. Routing is a system function and is not normally under the direct control of users. Routers are really just computers dedicated to this function. Some vendors specialize in router hardware and software, for example, Cisco Data Systems, Inc. and Bay Networks, Inc.

The TCP/IP protocol suite owes much of its early growth to the fact that it was built into Berkeley Unix and therefore became widely available in academic and research settings. As a result of this legacy Unix systems tend to have exceptionally flexible features for dealing with TCP/IP. For example there are programs (Unix standard "routed" and Cornell University's "gated") that will cause a Unix platform to function as an IP packet router and network gateway. Because Unix supports multiprogramming, this function can be transparent to the user of the system, thus a typical workstation that supports Unix can provide a router function at the cost of an extra interface for each connected network or subnet. The router is run as a "daemon" or independent system process that is activated by incoming packets.

Router software is also available for other platforms. A notable example is the KA9Q software created by the amateur ("ham") radio community, which runs on IBM PC-compatible systems. Contact bdale@col.hp.com for information.

64.3 INTERNET MULTICAST AND MULTIMEDIA

J. Mark Pullen and Satish Chaliki

The Internet has recently gained a new capability: transmission of multimedia (voice, video, graphics) information in real time, in addition to the previously existing data transfer function. What this means is that suitably equipped workstations can capture and communicate information for their human users. This capability has a variety of uses, from capturing experimental data and functioning as a "picture phone" for small group conferences, to distributing lectures and even programs from the space shuttle.

To use the multimedia Internet capability you will need a suitably equipped workstation (a Sun or SGI, plus video and sound capture if you intend to originate those media), plus software that is available over the Internet. To get started, FTP to venera.isi.edu, cd to mbone, and get faq.txt.

The MBONE works by using special "multicast" (group delivery) addresses on the Internet. A special router "mrouted" that runs on Sun and SGI platforms is required. Where multicast routing is not available, the MBONE creates a unicast "tunnel." This is problematic in that it can lead to multiple copies of MBONE packets crossing the same link, which is contrary to the basic purpose of multicast delivery. As a result careful coordination with Internet service providers is required when joining the MBONE.

Multimedia Workstations and Software

The MBONE is only the most visible part of the multimedia revolution in computing that is beginning to be widely seen in networked systems. Multimedia is not a single technology but rather a class of technologies and applications that spans two information types or more. Multimedia is an interdisciplinary, application-oriented technology that capitalizes on the multisensory nature of humans and the ability of computers to store, manipulate, and convey nonnumerical information such as video, graphics, and audio in addition to numerical and textual information. Multimedia has the intrinsic goal of improving the bandwidth and effectiveness of computer-to-human communication and, ultimately, human-to-human communication. Applications range from internetted flight simulators, to collaborative learning systems, to business support services such as presentations and teleconferencing.

Communication technology will be critical in enabling multimedia to migrate from dedicated desktop systems to more efficient distributed systems, moving toward a future where most educational institutions and corporate employees become users of multimedia applications. But multimedia data stretch the state of the art in network technology. The most constraining problem is the limitation of bandwidth in many networks. Networks supporting multimedia applications operate delivering from as little as 56 Kbps of information to a user to as much as 155 Mbps, 622 Mbps, and even gigabit-per-second rates. New protocols are being developed to deal with the need to dedicate some fraction of network capacity to real-time, multimedia traffic.

The synergistic confluence of the PC and workstation, the video, and the file server makes multimedia possible. Broadband networking is another catalytic force driving multimedia. Multimedia requires the integration of storage, communications, and presentation mechanisms for diverse information types in a single technology/platform so that the information can be simultaneously manipulated and displayed.

Software plays a key role in multimedia applications—from its most visible manifestation in workstation windowing systems, graphical user interfaces, and actual applications to more endogenous utilization in data compression, data storage and retrieval, and network synchronization systems. Multimedia information consumes large amounts of storage space; playback of full-motion digital video requires real-time decompression and high input/output bandwidth. Decompression systems include a set of hardware and software tools that enables integrated multimedia information types to be brought to the desktop.

Desktop Video and Networked Applications

A goal of multimedia developers is to bring two-way, real-time video to the desktop. Relatively inexpensive add-on systems for PCs, including a video camera and a digitizer/compressor adjunct, can be purchased for about $2000. In one application, multimedia conferencing enables designers in remote locations to review and/or work cooperatively on the same project by incorporating text, graphics, audio, visual, and tactile capabilities. A typical workstation-based multimedia conferencing system provides

- Desk-to-desk video conferencing with local and remote users;
- Shared-screen work space (to collaborate on documents, designs, software, presentations, and proposals);
- High-speed file transfer, and
- Call-management features that simplify setting up and controlling conference calls (including participants and PCs).

At the system level, multimedia spans the following functions: capture, storage, retrieval, presentation, transfer, and group sessioning. The first four functions support multimedia, while the last two support networked multimedia.

Capture deals with collecting and transforming external signals pertinent to multimedia, typically analog voice, sound, and video, into a form that can be utilized by a computer system. Capture devices include scanners, video cameras, off-air receivers, facsimile devices, microphones, musical instrument digital interface devices (MIDI), and optical character recognition (OCR) systems.

Storage entails the appropriate hardware needed to retain multimedia information in a form so that a user can quickly gain access to it. Such hardware typically consists of optical storage systems. It also includes videotape players and video disc players. The issue of managing time-stamped data becomes important here.

Retrieval involves sophisticated database access/navigation software and communication (if the information is archived).

Presentation entails the delivery of the information to the user at the logical level over the human–computer interface. The information is physically delivered by the display and speakers. The preferred avenues for object-oriented interfaces are graphical user interface tools.

Transfer refers to the ability to prepare multimedia messages for delivery to a remote recipient. Communications is an important support technology for this.

Group sessioning (connectivity) refers to the logical ability to bring two or more parties together to allow them to participate in a multimedia session with minimum complexity, minimum protocol conversion, and minimum multistep multicommand sequences. Communications, once again, is a key support technology. Another term used is "group-to-group conferencing."

At the resource level, an MM-system consists of five elements:

- A workstation with multimedia hardware for audio and video (either integral to the system or added on), possibly including capture elements such as camera;
- An operating system with a GUI;
- Storage (typically in the form of CD-ROM drives);
- Multimedia compression/decompression and other processing elements and multimedia transmission hardware;
- Software-development tools allowing the manipulation of multiple information types.

Multimedia Standards

On the technical side, there are many issues that affect the implementation of a multimedia system. The ever-changing technology inevitably has no standards at the leading edge, for example, relating to compression and decompression. Consistent control protocols, data/information formats, and interworking methods between various domains of the system are desirable for these multifaceted multimedia applications.

Compression algorithms are critical to multimedia. Video digitization enables the user to store video segments on a hard drive or network server in addition to videodisks or videocassettes. It also enables the user to transmit the information over a medium-capacity network.

The embedded generation of CD-ROM drives can deliver data at a rate of 150 Kbps; broadcast video delivers information at a rate of 30 fps. These constraints define the requirements of the multimedia system to support full-motion video with current equipment. Allocating the 150 Kbps to 30 frames implies that current technology dictates an encoding of 5 Kilobits per frame. A frame is typically digitized at a resolution of 512 × 480 pixels with a color palette of 24 bits per pixel. This results in about 740 Kilobits of information per frame. Here the compression algorithms come to compress the 740 Kilobits to 5 Kilobits, that is by a factor of 150.

Some applications may require real-time compression and decompression, for example, high quality video conferencing. Currently a variety of compression standards exist. The three most common compression standards are ITU-T Recommendation H.261 (also known as p × 64), the Joint Photographic Experts Group (JPEG) standard, and the Motion Picture Experts Group (MPEG) standard.

Digital Video Interactive (DVI) methods allow storage compression and real-time decompression for presentation purposes. DVI uses two proprietary image processing chips that compress images up to about 200 times. MPEG uses a modified interframe coding technique also used by DVI. Chip sets supporting one or more standard are now available and sell for around $300. CD-ROMS can store as much as 680 MB of information in a compact, portable optical medium and are available at around $2000. A 5.25-inch rewritable optical disc drive costs around $3000. The 5.35-inch discs can store as much as 1 Gigabyte. The discs cost around $100. However, one drawback of this technology is the read/write speed. It typically takes 19 to 60 seconds to write a 3-MB file and 16 to 33 seconds to read the same file. Techniques used to improve the speed include using RAM caches, turning off the verify cycle, using split-head optics, and increasing the spindle speed.

Presentation Platforms

Any system that manipulates more than one media can be considered a multimedia platform. To support presentation, the hardware platform needs to include input devices (cameras, keyboard, microphones), MIDI devices, output devices, and the processor itself in addition to storage. The major multimedia platforms on the market include

- Intel-based PCs with Microsoft Windows,
- IBM's PS/2 with OS/2,
- Apple Computer's Macintosh with System 7, and
- RISC (reduced-instruction-set computer) workstations with Unix and the X Window System.

The platforms identified above also require extensions to the hereto-available operating system to support the management of time-variable information objects such as graphics and animation, digital video, and sound. The more well-known extensions are for Intel-based PCs, Video for Windows; for Macintoshes, QuickTime; for IBM PS/2s, OS/2 Multimedia Presentation Manager/2; and, for Unix, Continuous Media Extension to X Window. These extensions can be purchased for $25 to $200.

PC Platforms

The PC meeting the requirements expounded by the Multimedia PC Marketing Council can be called a "Multimedia PC" in the strict sense of the term. Those requirements are

- A 386SX (16 MHz or better) or a 486SX processor,
- 2 MB RAM,
- VGA graphics (4 bits/16 colors or 8 bits/256 colors),
- A sound card (8 bits),
- A pulse-code modulated (PCM) (digital) audio channel,
- A CD-ROM drive with audio capability,
- A 30 MB hard drive, and
- Support for standard peripherals (mouse, printer, LAN).

386SX-based systems have difficulties running the more-demanding multimedia applications, particularly those entailing quality video. Hence, faster systems (486DX) are preferred.

Macintosh Platforms

QuickTime is a multimedia extension to Apple's Operating System 7.0 and higher. It enables developers to integrate video, audio, and animation into applications such as video conferencing and CD-ROM-driven magazines and learning systems. While PCs support frame rates between 6 and 10 fps, a Macintosh workstation can support 8–12 fps (system like Apple's Quadra has power to support 15–30 fps). Audio encoding is based on 8–16 bits. Sound is already part of Macintosh, so no additional equipment is required. QuickTime can be installed on any Macintosh with 2 MB RAM.

Unix Platforms

The Unix-based workstation is shedding its image as a purely scientific and engineering-based system with development in multimedia technology. A Unix workstation can be made multimedia ready by adding video capture cards, video image compression/decompression cards, and a CD-ROM drive, all usually available from third parties. Typically audio support is included with the workstation. Workstations cost from $6000 to $60,000. Systems selling for $8000 to $15,000 provide reasonable features for an end-user multimedia system.

Transfer: Networked Multimedia

Networked multimedia applications are still in the minority. There are three key problems that have held back its development:

1. The amount of bandwidth needed to transmit multimedia streams (for example, uncompressed quality video requires 270 Mbps, although compression can reduce this to a fair degree);
2. The refinements needed to the client–server paradigm to enable remote procedure calls (RPC) to work effectively in the multimedia environment; and
3. The network sophistication needed in terms of effective and expeditious signaling.

Two categories of networked multimedia applications are currently available. The first are those on LANs that support simple animation and small bitmaps stored on a file server. Some LAN-based video conferencing has emerged, but presentation quality leaves a lot to be desired. The second category consists of high-end sophisticated multimedia applications that use full-motion video and high-fidelity audio.

Some of the current network limitations—particularly for multiparty connectivity—include

- A conference bridge must be employed, and an individual/group must act as a hub;
- A hub must stay on the line as long as the conference lasts;
- Users typically have to establish separate connections for each stream; and
- The network treats individual streams as completely independent and unrelated "calls," making synchronization difficult.

Group Sessioning

The communication challenges involved in supporting multimedia fall into the following categories:

1. Sufficient bandwidth to transport voice, data, and video.
2. Stringent QOS in regard to end-to-end delay, delay variation, and loss.
3. Synchronized delivery of these high-capacity streams.
4. Complex user–network signaling interactions, such as the supporting multipoint sessions, and allowing remote users to join or leave the conference without action or permission from the "controller," user, or node.

Setting up a multimedia conference automatically requires a sequence of many signaling events. Therefore the goal should be to make the process of setting up such a conference as transparent to the user as possible. New approaches are under development to support the capability of originating, modifying, and terminating sessions in a networked multimedia environment. It is becoming necessary to rethink the protocols for call setup.

CU-Seeme

A number of desktop-based video conferencing systems are under development. We describe here one that is representative and has been adopted widely. Cornell University's Information Technology organization (CIT) has developed a Macintosh videoconferencing program called CU-SeeMe. It displays 4-bit grayscale windows at either 320 × 240 or half that diameter, 160 × 120, and includes audio. CU-SeeMe in version 0.40 provides a one–one connection or, by use of a "reflector," a one-many, several-to-several, or several-to-many conference. Each participant can decide whether to be a sender, a receiver, or both. Receiving requires only a Macintosh with a screen capable of displaying 16 grays and a connection to the Internet. Sending requires the same plus a SuperMac VideoSpigot board, a camera, QuickTime, and SpigotVDIG extensions added to the system folder.

The software is freely available via anonymous ftp from gated.cornell.edu in the directory /pub/video as CU-SeeMe0.40.bin. This file is README.CU-SeeMe.txt. There is also a choice of VDIG files needed for use with the SuperMac VideoSpigot frame grabber board. The executable and VDIG files are stored in a MacBinary II format.

CU-SeeMe specifications to RECEIVE video are

- Macintosh platform only with a 68020 processor or higher;
- System 7 and higher operating system (it "may" run on system 6.07 and above);
- Ability to set your monitor to 16 grayscale—an IP network connection;
- MacTCP; and
- CU-SeeMe0.40 file (file size is approximately 28K).

Specifications to SEND video are

- The specifications to receive video mentioned above;
- Video Spigot hardware (street price is approximately $380);
- Camera with NTSC 1vpp output (like a camcorder) and RCA cable;
- QuickTime installed (requires approximately 2 Megabytes of memory); and
- SpigotVDIG QuickTime component (driver) on disk (approximately 300 Kilobytes).

MacTCP is available on the same ftp server, in two versions. The license from Apple allowed the developers to distribute it only with their own software (CU-SeeMe), and not for use with anything else. You must obtain MacTCP another way to use it with other applications. Use v1.1.1 with System 7.1. There are two new spigot VDIGs on the server; if you have QuickTime 1.5, you need the one called VDIG 1.5b18. Otherwise the one called SpigotVDIG is good. The third one is an old one, which should be tried only if all else fails.

65

Novell Networks

Rodney Hocutt
Rubicon Technologies

65.1 NETWORKING BASICS

This chapter introduces the concept of computer networking. Computer networks have, in recent years, come to be considered a bare essential in the corporate world where the need to share information and peripherals has become paramount not only to a companies success, but to its very survival.

At the most basic level, a network is a means by which both computer peripherals and information may be shared. Computer networks fall into one of two categories, Local Area Networks (LANs) and Wide Area Networks (WANs). LANs and WANs are differentiated from one another by the fact that a LAN serves a specific, limited geographic area, such as an individual department or organization. By contrast, WANs are large networks comprised of smaller LANs and are not limited by geography. The primary constituents of a LAN are PCs and other peripherals such as printers, modems, and gateways to other systems such as minis and mainframes. These components are connected (usually physically) by a cabling system such as fiber optic cable, coaxial cable, twisted-pair cable, or a host of other means. The cable acts as a data path between different computers (or peripherals) on the network. Finally, there is a Network Interface Card (NIC) that provides the physical connection between the computer (or other components) and the cable system.

The Novell NetWare Environment

While all networked computers share the same basics listed above, network operating system (NOS) vendors are free to implement their network environment as they see fit. This often leads to incompatibilities between different NOS vendors and creates quite a challenge for network administrators. This section will talk exclusively about the NetWare environment.

In the NetWare environment, the NIC connects a client PC to the cabling system, which in turn is connected to a file server. The file server (running NetWare version 3.xx) enables users to share information with specified users of the network. Client

0-8493-2530-7/96/$0.00+$.50

computers run two programs that are jointly called the NetWare *shell* that enables users (client computers) to access shared resources on the file servers as though they were physical hard drives on their own PCs.

65.2 THE NETWARE SHELL

Novell NetWare utilizes a shell that is responsible for performing several basic tasks involved in initializing the network environment prior to an end users starting to work. Some of these tasks include loading the network drivers and connecting the workstation to the file server. The NetWare shell consists of two programs: IPX and NETx. The first program, IPX, is an acronym for Internetwork Packet eXchange and is responsible for controlling the network interface card (NIC). Due to the plethora of NICs available today, each IPX is written using drivers specific to the NIC with which it is to be used. This helps ensure there is protocol commonality throughout the entire network.

The second program, NETx (where the x represents the version of DOS that the client computer is running) is responsible for handling all communications between the client computer and the IPX program. Once the NetWare shell has been loaded, the user will be able to use ordinary DOS commands on shared resources at the file server as well as access additional network-specific commands to perform additional functions, such as assigning security to shared resources or files.

When the network shell has been loaded, the user will be at the network login drive. For illustration purposes, we will use the first network drive as drive **F:**. Please note that this is network dependent and may vary from network to network as well as from workstation to workstation. From the **F:\LOGIN>** prompt, type the command **LOGIN <username>** . The network will then prompt you for a password. After entering your password, the network will reference a data base to determine if your account/password combination is valid. If so, you will be granted access to the system.

65.3 DRIVE MAPPINGS

When a user logs on to a Novell network, the file server performs several tasks that enable the user to work. The first of these is to establish drive mappings. A mapped drive is a logical letter that represents a specific directory (or subdirectory) on a server. NetWare permits up to 26 mapped drives. Additionally, NetWare allows specified drive mappings to be used as search drives. This is equivalent to the PATH statement in DOS in that any drive mapping that is listed as a search drive, NetWare will search if a requested file or executable program is not found in the current directory. Search drives can only be assigned the letters K through Z (inclusive). By default, Novell establishes several drive mappings whenever a user logs onto the file server. The first mapped drive is the drive that represents the **SYS:PUBLIC** directory. **SYS:PUBLIC** is the directory that contains general NetWare utilities and programs for regular network users. Additional drive mappings may be set up by the system administrator based on the needs of an individual or a group. By establishing a drive map to the SYS:PUBLIC directory, every user has access to a host of NetWare utilities.

65.4 NETWARE UTILITIES

In NetWare 3.xx, there are many utilities available for both end users as well as network administrators. Below is a description of available utilities for both users and administrators.

User

In the NetWare environment, an end user of a network has access and control over only those things that pertain to his or her account. Exceptions to this are instances where the Network Administrator has subsumed the users rights (such as determining whether users may change their password or not). Once logged into the file server (and granted access to the **SYS:PUBLIC** directory), there are a number of NetWare utilities that are available to a general network user. A subset of these utilities is listed below.

- *Filer:* Filer is a general purpose file maintenance utility that is used to control volume, directory, and file information as well as providing the ability to change file and directory security. All network users may execute Filer, but some options are restricted to the Network Administrator. To access Filer, type **<FILER>** at the DOS prompt.
- *Syscon:* Syscon (System Console) is a utility that is used primarily by Network Administrators, but general network users may find it useful for making changes to their specific account, such as changing a password or altering access rights.
- *Send:* At times, users may wish to communicate with other users on the network in other locations. This is made possible by means of the Send utility. The Send command will transmit a single line of text across the network to a specified user (if he or she is logged onto the network). The format for Send is

```
SEND "message" TO USER.
```

- *Salvage:* The Salvage utility allows a user to recover any/all files that have been recently deleted. To run Salvage, type **<SALVAGE>** from the DOS prompt. From the available files menu, select the files that you wish to reclaim. Note that NO files may be salvaged following a **PURGE** command.
- *Setpass:* The Setpass command allows a user to change his or her password at any time, if this right has been granted by the System Administrator. To run SetPass, type **<SETPASS>** from the DOS prompt. The system will ask for your current password. Once validated, you will be asked for your new password and asked to enter it again for verification purposes.
- *Rights:* The Rights utility allows a user to display their current rights for a given directory. Each user is assigned, or may assign to other users, specific rights to public or private directories. To display the current rights mask, type **<RIGHTS>** at the DOS prompt. The Rights command also provides an explanation of all NetWare rights.
- *Grant:* To allow the sharing of data files etc., users may allow specific individuals (or groups of individuals) limited or unlimited rights to a specified directory in their account. To grant another user or group of users rights to a specific directory, change to that directory and type

```
<GRANT [rights] TO [user|group]>.
```

Note that rights may be granted only on a directory by directory basis.

- *Revoke:* The Revoke utility allows a user to revoke some or all of the rights previously granted to an individual or group. To revoke a users rights from a specific subdirectory, first change to that directory and type

```
<REVOKE [rights] FROM [user|group]>.
```

- *PConsole:* PConsole is a utility that allows users to control print queues and jobs. To run PConsole, type **<PCONSOLE>** from the DOS prompt.

Administrator

A Network Administrator has unlimited access to the network as well as every user or group of users. The bulk of network administration is done through the SYSCON utility. Some of the most common uses of the SYSCON utility are listed below:

- *Group Information:*

  In large networks, such as those running several file servers or many users, it is often useful to "collect" users with similar jobs or from the same department into groups. Once placed into a group, the administrator can work with the single group entity and have the modifications affect every user assigned to that group. This approach has two distinct benefits in that

  1. it gives a group of users rights to the same directories
  2. it permits the administrator to grant a limited subset of control rights to other users called **workgroup-managers**.

  To create or modify groups:

  1. Start Syscon by typing **<SYSCON>** at the DOS prompt.
  2. From the main menu, select **Group Information**.
  3. Add a new group by pressing the **<insert>** key and enter the name of the group that you wish to add. This name is then added to the group list. To delete a group, highlight it and press the **<delete>** key.
  4. To change a group name, highlight it and press **F3**. You are then prompted to revise the group name and the group list is updated.

- *User Information:* The User Information option in Syscon is where most user-specific modification takes place. When the User Information option is selected, a list of all network users is displayed. From this list, highlight the specific user that you wish to modify and press **<ENTER>** . This brings up a windows with the following options:

  Account Restrictions
  Change Password
  Full Name
  Groups Belonged To
  Login Script
  Other Information
  Security Equivalencies
  Station Restrictions
  Time Restrictions
  Trustee Directory Assignments

- *Account Superintendent Restrictions:* This option allows the network administrator to specify such user-specific items such as

  Password requirements
  Account expiration date
  Number of permitted concurrent connections
  Grace logins
  Disk space limitations on the file server

- *Change Password:* This option enables the network administrator to create and/or alter a users password requirements at any time.

- *Full Name:* This selection simply displays the full name of the user.
- *Groups Belonged to:* This allows an administrator to assign a user to one or more groups to make administering the network easier.
- *Login Script:* This options allows for the execution of one or more internal (NetWare) or external (DOS) programs to be run, batch style, every time the user logs into the network.
- *Other Information:* This option displays user-specific information, such as

  Date of last login
  File server operator(?)
  File server disk space restrictions
  Amount of disk space in use
  User ID.

- *Security Equivalencies:* This option displays a list of users and groups that the specific user is security equivalent to. This means that the user has all the rights of the users and groups that he or she is equivalent to, in addition to the rights that have been explicitly granted to him or her as a user.
- *Station Restrictions:* This selection allows the administrator to restrict the physical computers that the user may login to the network from. If this list is empty, the user may login from any computer on the network. To add a restriction, press **`<Insert>`** and enter the network address of the restricted computer(s).
- *Time Restrictions:* This option allows the administrator to regulate the time of day or night that a user may login to the network.
- *Trustee Directory Assignments:* This option allows an administrator to create or modify trustees for users and groups. When a user (or group of users) is given rights to access a directory or subdirectory on a server, they become a trustee of that directory. The specific level of access to the directory is controlled by granting or withholding a set of eight (8) rights. These rights are as follows:

  `Supervisory(S):` This right is specific to NetWare 386. Any user who possesses the Supervisory right essentially has all other rights to the specified directory as well as all its subdirectories. Additionally, possession of the Supervisory right prohibits removing any rights from directories below the specified directory. This right should not be granted unless absolutely necessary.

  `Read(R):` This right permits a user to read the contents of any existing files within the specified directory.

  `Write(W):` This right allows a user to add to or modify the contents of any existing files within the specified directory, but does not allow them to create an entirely new file.

  `Create(C):` This right enables a user to create a new file in the specified directory. Additionally, the Create right allows a user to create new subdirectories within the specified directory.

  `Erase(E):` The Erase right allows a user to delete a file in the specified directory.

  `Access Control(A):` The Access Control right empowers a trustee to grant rights to other users as well as allowing for the modification of a directories rights mask. With this right, a trustee may grant or revoke from other users, any right or combination of rights with the exception of the Supervisory right. The supervisory right may not be granted to any other users.

File Scan(F): The File Scan right gives a user a listing of all files and/or subdirectories in the specified directory. This is the NetWare equivalent to the DOS DIRectory command.

Modify(M): The Modify right allows a trustee to rename a file or a subdirectory, or to change the file attributes of any file within the specified directory.

65.5 BUILDING A NOVELL NETWORK

You will find that the actual administration of the completed network will be far more time consuming than building the network in the first place. Since every network environment is different, this is meant to be a general guide to creating and administering a "typical" Novell network. Prior to installing NetWare 3.xx on your file server, make certain that all required hardware is properly installed and recognized by the computer.

Requirements

Memory

A NetWare file server requires a minimum of 4 MB of RAM. However, this minimum can be misleading. As server memory is one of the primary factors in how well or how poorly your file server performs, it is very important to ensure that the server has enough memory to function well, and perhaps even 4 more MB above that. The optimum amount of RAM for the file server is contingent on the number of users or nodes on the network and the functions that the server is going to perform. Some of the functions that have the largest impact on a server's memory requirements are the following:

1. If you wish to run PSERVER.NLM (an NLM that allows for remote printing across the network), add a minimum of 256K to 512K of RAM to the base configuration.
2. If your server is going to support the TCP/IP protocol suite, add a minimum 256K of RAM to the base configuration.
3. If the server is going to support Macintosh clients, add a minimum of 256K plus an additional 128K for each Macintosh print queue you define.

To be safe, always user more RAM than your calculations indicate are necessary. Also note that while NetWare 3.11 can address 4 GB of memory, only the first 16 MB can be addressed automatically. To address any memory above 16 MB, it is necessary to use the REGISTER utility. Additionally, if you install more than 16 MB of RAM in an ISA or EISA file server, it is advisable **not** to use any 8 or 16 bit cards that use bus-mastering or Direct Memory Addressing to avoid memory conflicts.

Disk Drive Requirements

Currently, several computer manufacturers supply special RAID (Redundant Array of Inexpensive Disks) arrays for NetWare environments that provide outstanding performance and fault tolerance. These arrays typically feature hot swappable drives, tremendous throughput, and a high degree of data redundancy for fault tolerance. However, if such an array is not required for your environment then the burden of choosing drives is left up to you. NetWare supports most of the most common drive types. These include

1. Any drive type supported by the ROM-BIOS such as MFM, RLL, ESDI, and IDE drives.
2. Drives supported by DCBs (Disk Coprocessor Board). A DCB is an intelligent SCSI host adapter developed specifically for use with Novell NetWare.
3. Drives that are supported via third party drivers. Most SCSI drives fall into this category.

Although any of the above drive types are supported by Novell, it is highly recommended that SCSI drives are used in a file server. MFM and RLL drives are old technology and are very slow compared to SCSI, ESDI, and IDE drives so they tend to slow down the entire network. ESDI drives, while offering adequate performance, are generally considered to be old technology and are therefore not supported as readily or as easily as IDE and SCSI drives.

IDE drives, while offering excellent performance and storage capacities, are not capable of supporting some of NetWare's advanced drive features such as **Mirroring** and **Duplexing**.

Disk Mirroring. Disk mirroring allows two drives (or pairs of drives) attached to the same controller (or host adapter) to contain the exact same information. Since the drives work in tandem, mirroring provides a great level of security in case of a drive failure. If the primary drives fails for some reason, the secondary drive will continue to function.

Disk Duplexing. Disk duplexing is superior to mirroring because it mirrors not only the drives, but the entire disk channel components as well. A disk channel consists of the drive controller (or host bus adapter), the disk drive(s), and all cables etc. Duplexing provides a much higher degree of data redundancy and fault tolerance than mirroring. Additionally, duplexing can dramatically improve the throughput of the network because it allows the file server to perform split seeks on read operations. In a situation where one disk channel is busy performing a task, the file server can read requested information from an alternate channel, thus enabling a user to get their information without waiting. Duplexing is most commonly performed with SCSI drives. One reason for this is because while SCSI and IDE drives are considered to be the latest in drive technology, IDE drives cannot be duplexed. This is because IDE drives, when connected to the same controller, are configured in a master/slave fashion. The master drive controls the slave drive and performs all diagnostics for both drives. Therefore, if the master drive crashes, both drives crash since the slave drive cannot function independently of the master. Also, if the slave drive crashes, the master drive will continue looking for it until it times out. It is for these reasons that SCSI drives are the drive type of choice for NetWare environments.

Disk Drive Preparation

1. A NetWare 3.x file server boots from a DOS partition, so the first step is to create a bootable DOS partition on the first hard drive connected to the system. This is accomplished by booting with a DOS disk in drive A: and running the <FDISK> program. It is recommended that a DOS partition of 5 MB is created. This allows sufficient space for the necessary NetWare files while supplying the most space to the NetWare volumes. Once having created the DOS partition (and marking it ACTIVE), use the DOS Format command to format the partition by typing

<FORMAT C: /S>

(the /S parameter copies DOS system files to the hard drive to make it bootable).

2. Once the partition has been formatted, create a NetWare directory by typing

   <MD NETWARE>.

   Next, copy the SYSTEM-1 and SYSTEM-2 diskettes to the NetWare directory, as well as any third party disk drivers and LAN drivers.

3. From the NetWare directory, type

   <SERVER>.

   Server is the NetWare command processor. At this point, you will be prompted for the file server name and an IPX internal network number. Both the file server name and IPX number must be different from any other file server on the network, or any other remote file server. The file server name can be up to 47 characters (alphanumeric), and the IPX number must be a hexadecimal number within the range 1 to FFFFFFFE. Once this information has been entered, the file server should display the system prompt (:).

4. Load the disk driver that corresponds to the drive types in the server. For most AT type drives (MFM, RLL, IDE), this will be the ISADISK.DSK driver. If you are using third party drives, the vendor will have supplied a disk driver that should be copied into the NetWare directory and loaded in place of ISADISK in the following examples. For AT type drives, load the driver by typing

   <LOAD ISADISK>

   At this point, you will be prompted to enter the parameters for the I/O port address and the interrupt parameter for the network interface card. In most instances, you can accept the Novell default. If your card has settings other than the default, you will need to know them before you can continue.

5. Next, load the Install module. This is done by typing

   <LOAD INSTALL>

   after which the Install Options menu will appear.

6. From the **Install Options** menu, select **Disk Options** to display the Available Disk Options menu. From this menu, select **Partition Tables**. This will display all available drives recognized by NetWare. Select the first drive (since the drives are zero relative, the first device will be device #0 and not device #1).
7. From the Partition Options menu, select **Create NetWare Partition**. NetWare will create a partition that spans the entire amount of free space left on the drive. NetWare will also reserve two percent of the partition as a **Hot Fix Redirection** area.

Hot Fix. The hot fix redirection area is NetWare's utility for fixing disk errors on the fly. When data are written to the hard disk, NetWare rereads the data and compares them to the original data that were written. If the two are not the same, NetWare will write the data to the hot fix redirection area and mark the current disk block as "bad," thus preventing any more data being written there. NetWare allocates two percent of the drive by default, but you may adjust the size if necessary. There are usually two reasons that you may want to change the hot fix:

- In a large LAN with many users, it is advisable to increase the hot fix to ensure that there is sufficient space in the hot fix to handle any unseen bad disk blocks.
- If you are mirroring or duplexing your drives, you may have to adjust the hot fix to make the data areas on each drive the same. In mirrored or duplexed drives, the drives may be of different physical sizes, but the data area (the logical size) must be the same for each drive pair. Note that the hot fix redirection area is created during the install procedure and cannot be changed without regenerating the file server. Be sure to allocate sufficient space for the hot fix.

If you have more than one drive in the server return to the **Available Disk Options** menu and select **Partition Tables** to repeat the above process for the remainder of your drives.

8. If you are mirroring or duplexing drives, select the **Mirroring** option from the **Available Disk Options** menu. This will display a list of available NetWare partitions. To mirror one partition with another (they both must be of the same size), select the desired partition and press **<ENTER>** . This will display a list of **Mirrored NetWare Partitions**. If the selected partition is not mirrored, its status will be *In Sync*. Next, press the key **<ENTER>** to display a list of remaining partitions. Select the partition to serve as the back half of the mirrored pair and press **<ENTER>** Both partitions will now appear as *In Sync* in the **Mirrored NetWare Partitions** list. Repeat this process for each mirrored pair.
9. Once having partitioned the drives, it is necessary to create the NetWare *Volumes*. A volume is nothing more than a fixed amount of disk space. Under NetWare 3.x, a volume can

   - be a single drive
   - be a partial drive
   - span multiple drives

   By default, NetWare creates a single volume (SYS:) that spans the entire drive. For most small networks, this is sufficient. However, there are situations where it is advisable to create multiple volumes on a drive (or set of drives):

   - If you are going to support multiple operating systems on the network. If you are going to have name spaces for non-DOS files (i.e., Macintosh, OS/2 and NFS), it is recommended that multiple volumes be used.
   - If you are doing software development across the network, it is advisable to have a special volume for the development team. This is useful in situations where a bug in a program causes drive irregularities. By isolating any unproven software to a separate volume, any problems that arise will not affect the general users of the network.

   To create the volume(s)

   - Select the **Volume Options** from the **Installation Options** menu and press **<ENTER>**.

- Display the **New Volume Information** menu by pressing **<INSERT>**.
- Specify the volume parameters:

Volume Name: The name that the volume will be addressed as. By default, NetWare creates a **SYS**: volume, and requires that the SYS: volume be present.
Block Size: The size of each block on the physical disk. By default, NetWare uses a 4K block size, but if the volume is going to store primarily large files, the block size can be increased to 8, 16, 32, or 64K. To change the block size, press **<INSERT>** and select the desired block size from the menu.
Initial Segment Size: This is the number of blocks that the volume contains. If there is only one volume on the drive, the Initial Segment Size will be the entire partition. If there will be multiple volumes on the drive, this number must be decreased.
Volume Size: This entry is calculated automatically by the Install program. It represents the number of disk blocks times the segment size.
Status: This field indicates if the volume is mounted or not (mounting a volume makes it available to NetWare). This field cannot be altered within the **New Volume Information** screen because the volume must exist before it can be mounted. When finished entering the necessary information, press **<ESCAPE>** to return to the previous menu. The system will prompt you with *Create Volume?* Enter **Yes** and repeat the above process for each volume that you want to make. If you want to have a single volume span multiple physical drives, add a **Volume Segment** to the volume as follows:

- Choose **Volume Options** from the **Installation Options** menu.
- From the Volumes list, select the volume that you wish to add the drive to and press **<ENTER>** to display the **Volume Information** menu.
- From the **Volume Segments** option, press **<ENTER>** to display the Volume Segments screen.
- Press the **<INSERT>** key to display any available volume segments. If a **Free Space Available** list is displayed, choose the segment that you wish to add and press **<ENTER>** If a **Free Space Available** menu does not appear, the **New Volume Segment Size** screen will appear.
- Enter the number of blocks that you wish to add to the volume (press **<ENTER>** to add the entire drive segment) and then press **<ENTER>**.
- When the system prompts you for *Adding New Segment To volume?*, choose **Yes** and press **<ENTER>**.

10. Next, mount all volumes that were created by the above procedures.
    - From the **Volumes** menu, select a volume and press **<ENTER>** to display the **Volume Information** menu.
    - Go to the **Status** field and press **<ENTER>**.
    - Select the **Mount Volume** option and press **<ENTER>**.

    Repeat this process for each volume you wish to define.

File Upload

The longest part of generating a file server is uploading the files. To do this

1. Return to the **Installation Options** menu and select **System Options**.
2. Select the **Copy System and Public Files** option. As files are copied to the server, you will be prompted to insert disks as necessary. To start copying after switching disks, press **<ESC>**.

When all the files have been copied to the server, the system will respond with the message *File Upload Complete*.

Protocol

11. Next, the LAN driver for your Network Interface Card (NIC) must be loaded and then bound to the IPX protocol. This driver should be in either the NetWare directory on the DOS partition of the hard drive, or in the SYS:SYSTEM directory of the file server (if the driver is included with NetWare 3.x). To load the LAN driver

    - Switch to the Command prompt by pressing **<ALT><ESC>**
    - Type `LOAD (lan_driver_name)` **<ENTER>** where `lan_driver_name` is the name of the LAN driver for your NIC. For example, to load the LAN driver for a Novell-Eagle NE1000 card, type

    ```
    LOAD NE1000 <ENTER>
    ```

    To load the LAN driver from the DOS partition (or a floppy drive), type in the full path to the driver. For example, if the driver resides in a **NETWARE** directory on a floppy disk in drive A:, the command would be

    ```
    LOAD A:\NETWARE\NE1000 <ENTER>
    ```

    - If the system finds the necessary LAN driver, it will prompt you for an I/O port and an Interrupt Request line (IRQ) (and possibly other parameters depending on the NIC). The system will display initial values for both settings that usually represent defaults for the NIC.

    Note: The default IRQ for most NICs is IRQ:3. This is also the IRQ that is used by the COMM:2 serial port. If you have a second serial port on your computer, you will develop IRQ conflicts. If this situation occurs, choose another IRQ for the NIC.

    - Next, the IPX must be bound to the LAN driver. This is done by typing

    ```
    BIND IPX TO (lan_driver_name) <ENTER>
    ```

    The system will respond by prompting you for a network number. This is a hexadecimal address of the network attached to the file server and must be in the range of 1 to FFFFFFFE.

12. Next, it is necessary to create two files to automate the NetWare bootstrap functions (loading the LAN drivers, BINDing the IPX etc.). To do this

    - Switch back to the Install Screen by pressing <ALT><ESC>.
    - Select `Create STARTUP.NCF File` from the **Available System Options** menu.
    - Make sure the path points to `C:\NETWARE\STARTUP.NCF` and press <ENTER>.
    - The `STARTUP.NCF` file should contain the command

```
Load ISADISK port = xxx int = x
```

where `port = xxx` is the I/O port address and `int = x` is the interrupt that you entered in the previous step.

- Press <ESC> and save the file.
- Select `Create AUTOEXEC.NCF File` from the **Available System Options** menu. The `Autoexec.NCF` file should contain the following information:
- `—file server name *******`
- `—ipx internal net *******`
- `—load (lan_driver_name) port = xxx int = xxx frame = ETH-ERNET_802.3`
- `—bind IPX to (lan_driver_name) net = xxx`
- `—mount all` (only if multiple volumes)
- —Press <ESC> and save the file

13. Finally, edit the `Autoexec.bat` file on the DOS partition so that there is no memory management software running and the path points to the NetWare directory. The `autoexec.bat` should look similar to the following:

```
path = c:\;c:\netware; server
```

The `Server.exe` program loads the NetWare command processor and executes the commands located in the **`STARTUP.NCF`** and **`AUTOEXEC.NCF`** files. Once the file server is booted, you can remove the DOS command processor from the system memory to increase the NetWare disk cache. This is done by typing

```
<REMOVE DOS>
```

from the: prompt

65.6 NETWORK PRINTING

One of the most troublesome matters in a network environment is that of network printing. This is so for several reasons:

1. A large majority of the DOS applications that are popular today are not network aware. That is to say that they cannot send print jobs directly to a network print queue.
2. There are literally hundreds of printers that most DOS applications can use. Very often, each printer requires its own printer control codes. In a large network with many applications, this can be quite a formidable problem.
3. Finally, while the printing capabilities included in NetWare are adequate for most environments, the utilities for printer administration and printing control are awkward to work with.

To be able to access shared printers across the network, it is necessary to create a PRINT SERVER. A print server under NetWare 3.x allows up to 16 printers to be shared across the network. Of these 16 printers, only 5 may reside on the file server and the rest must reside on workstations.

Generating a Print Server

Generate a NetWare print server as follows:

1. Load the PCONSOLE program from the DOS prompt by typing <PCONSOLE>. This will display an **Available Options Menu**.

2. From the **Available Options Menu**, select **Print Server Information** to display an initially empty list of print servers.
3. Press **<INSERT>** and enter the name of the new print server. For each printer to be added to the network, it is necessary to identify or *define* the printer to NetWare. Define each network printer as follows:

   - From the **Print Server Configuration** menu, select the **Printer Configuration** option. This will display a list of available printer (0–15) that will initially be labeled "not installed." Add the new printer by selecting printer number 0 and press **<ENTER>**.
   - The Printer # Configuration menu allows you to specify the logical and physical properties of your printer. The following options are available:

   Name: Up to 47 characters to identify the printer.
   Type: This option may be one of the following:
   Parallel (LPT 1,2,3): A parallel printer physically connected to the print server.
   Serial (COM 1,2,3,4): A serial printer physically connected to the print server.
   Remote Parallel (LPT 1,2,3): A parallel printer physically connected to a workstation.
   Remote Serial (COM 1,2,3,4): A serial printer physically connected to a workstation.
   Remote Other/Unknown: For use with remote printers controlled by RPRINTER at a workstation.
   Defined Elsewhere: For use with printers defined on another print server.
   Use Interrupts: The use of interrupts on dedicated print servers tends to improve performance but may sometimes cause conflicts if there are several other interrupt driven devices in the server. Note that if you are going to use interrupts, you must know the printer port configuration(s).
   IRQ: The interrupt value that corresponds to the selected LPT port that your printer is attached to.
   Buffer Size in K: Sets the size of the print buffer. If you experience printer pauses during the printing of active jobs, it will help to increase this value.
   Serial Port Parameters: For printer attached to RS232 ports, it is necessary to specify the communication parameters. Refer to your printer documentation for the correct parameters if you are unsure of them.

Once having created the print server and defined the printers, it is necessary to create the print queues that each printer will spool to. Create print queues as follows:

1. From the menu, select Print Queue Information to display a list of available print queues. This list will initially be empty.
2. Press the **<INSERT>** key to create the new queue. This will display a New Print Queue Name box. Enter the name of the print queue, keeping in mind that a short, easy to use name is by far the best.
3. Next, the print server must be attached to the newly created queue. To attach a print server to a queue, select Currently Attached Servers from the Print Queue Information menu and press **<ENTER>**.
4. From the system colon prompt (:), type the command **LOAD PSERVER** . To automatically have NetWare load **PSERVER** and attach all print servers and queues at boot time, edit the system **AUTOEXEC.NCF** file to add the **LOAD PSERVER** command. When the file server is booted, you will be prompted for the name of the print server.

65.7 NETWARE 4.x

The next generation of NetWare, version 4.x is designed around organizations with large internetworks with over 10 file servers. Networks of this size have a unique set of administrative problems such as the need for tighter security, as well as the need to provide easy access to shared resources. NetWare 4.x addresses these needs by implementing a global directory facility. With a network of many file servers and shared resources, it can be difficult to determine what server hosts what printers and shared directories. NetWare 4.x implements all file servers, printers, users, and user groups as objects in a database that is replicated across multiple servers. In the NetWare 4 environment, users do not log onto a specific server, but rather to the network itself. During the login process, users are validated for access to all network objects that they have been granted rights to, thus allowing users to access shared resources without knowing where, or on what server they reside.

DEFINING TERMS

Administrator: Individual in charge of maintaining the network, server, etc.

Binding: Connection of software to hardware.

Client: Workstation connected to the network.

Disk Duplexing: Similar to mirroring, but involves both the disks and support hardware, such as disk drive cards, cables. Will be much more fault tolerant than mirroring.

Disk Mirroring: Allows two drives or pairs of drives to contain the exact same information.

Drive Mapping: Mechanism in Novell NetWare used to make files on server appear as drives to user.

IDE (Integrated Disk Electronics): Common disk drive system for microcomputers. See SCSI for contrast.

IPX: Component of system that handles communications between clients and server.

LAN: Local Area Network.

MFM, RLL: Early disk drive systems for microcomputers. See IDE and SCSI.

NIC: Network Interface Card.

NLM: Network Load Module.

NOS: Network Operating System (Novell NetWare, in this case).

Partition: Section of disk drive storage area used by DOS or Novell NetWare to store programs and data. A physical drive may have one or more logical drives, or partitions.

Print Server: Network component allowing a single printer to be accessed in common by all users on the network.

RAID: Redundant Array of Inexpensive Disks.

Rights: Mechanism for controlling access to information on the server for a single user or group of users.

SCSI Drive: Small Computer System Interface; drive system often used in networked microcomputer environments.

Server: Dedicated computer used to provide network file services.

User: See client (also, a four-letter word!).

WAN: Wide Area Network.

66

Windows for Workgroups

Paul W. Ross
Millersville University

66.1 PEER-TO-PEER NETWORKS

Windows for Workgroups®, from Microsoft, is an example of a peer-to-peer network, in contrast to the Novell Netware® system, which is an example of a server-based system. In a peer-to-peer network, each user may share files, clipboards, directories, or disk drives with their fellow users on the network. In a server-based system, all sharable files reside on one or more dedicated (usually) server systems. For small organizations, a system such as Windows for Workgroups is especially useful. In addition, a Novell session may be run concurrently with the Windows for Workgroups session. This means that files may be shared among users (peer-to-peer) or may be accessed from the server (server-based).

Windows for Workgroups appears quite similar to Windows. However, Windows for Workgroups has a number of major extensions and features to enhance its performance as a peer-to-peer networking environment. These are as follows:

- File sharing
- Disk and directory sharing
- Electronic mail
- Personal scheduling
- Sharing the Clipboard
- Device sharing (i.e., printers, modems)

In addition, Windows for Workgroups maintains an extensive Help system, comparable to that for Windows 3.1. A Windows 3.1 user should find the transition from

0-8493-2530-7/96/$0.00+$.50

Windows 3.1 to Windows for Workgroups quite simple. The systems appear the same to the user except for the enhancements and extensions previously mentioned.

66.2 LOGGING ONTO THE WINDOWS FOR WORKGROUPS SYSTEM

The first thing that is obviously different about Windows for Workgroups appears when you first start your system. You will be asked to log on to the system with your user name, and a password. The following screen will appear, as shown in Figure 66.1.

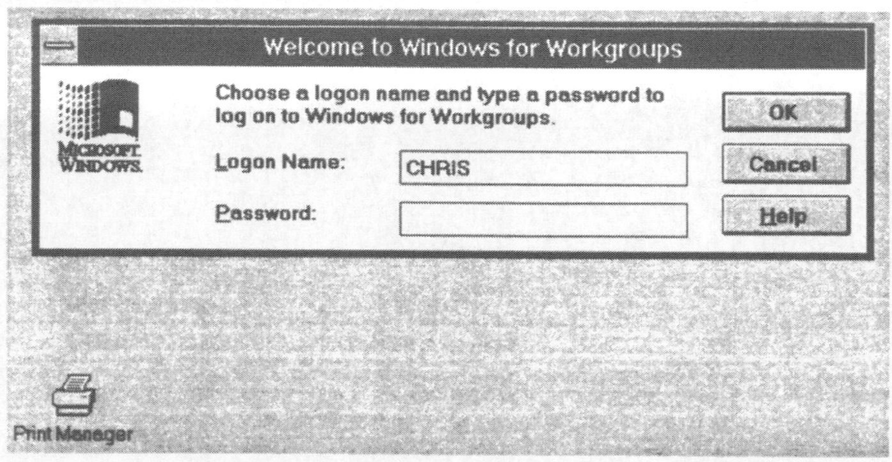

FIGURE 66.1 Windows for Workgroups logon screen.

Simply enter your logon name and password. Click on OK when you have entered them, or press the **Enter** key. When the system is initially installed, you will be prompted for your user name and password. Also, when the system is initially installed, the system will prompt the first user for assignment of a "Postoffice Manager," which is the system administrator function in Windows for Workgroups. The Postoffice Manager can set and reset passwords, and perform various system tasks. For more discussion of this topic, refer to the Windows for Workgroups manuals.

66.3 ELECTRONIC MAIL

An electronic mail capability gives you an easy way to send and receive messages with other people on your network using Windows for Workgroups. If you invoke the mail capability by double-clicking on the Mail icon, or invoking Mail from the Files Manager, you will be presented with the following screen, as shown in Figure 66.2. Enter information for the destination user, copies (optional), and subject (optional). When done, click on Send. Address files and attachments may also be included.

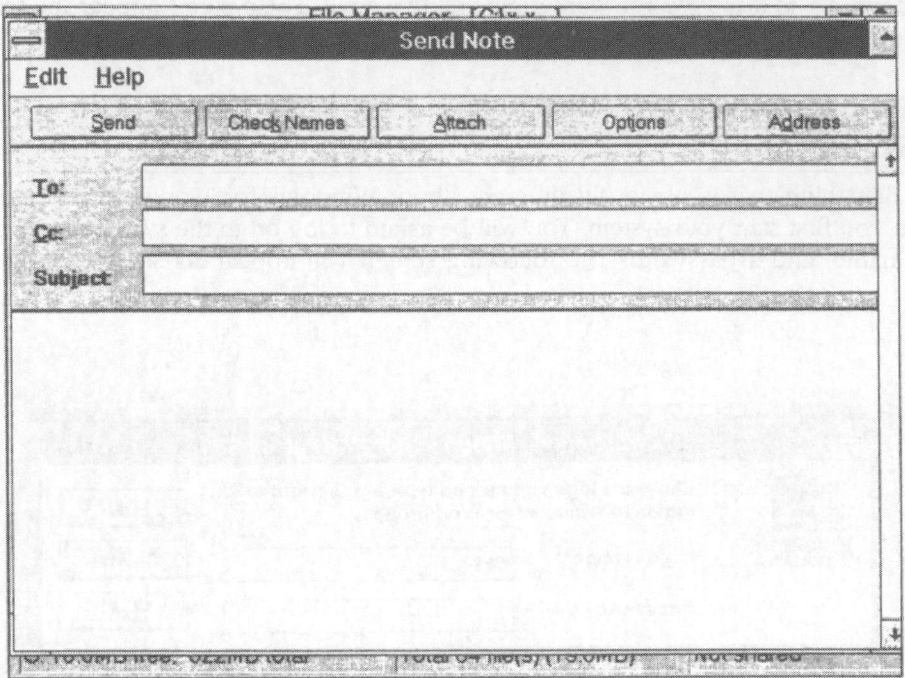

FIGURE 66.2 Windows for Workgroups electronic mail screen.

66.4 THE PRINT MANAGER

The Print Manager for Windows for Workgroups works slightly differently than the Print Manager found in Windows 3.1. The Print Manager serves not only to control the spooling queue, but to allow sharing of printing facilities. A typical Windows for Workgroups Print Manager screen is shown in Figure 66.3. With the Print Manger, you may designate printers as sharable to other users on the network, and select the printer that you wish to use for your printing. In a typical situation, this might mean that an expensive high-resolution laser printer could be shared between a number of users. A user might also have a slower and less expensive local Inkjet printer that is not shared. Other than being able to designate sharing, the Print Manger in Windows for Workgroups is essentially the same in functionality as the Print Manger in Windows 3.1.

66.5 THE SCHEDULING SYSTEM

Windows for Workgroups supports a powerful personal scheduling system. If desired, the schedules of various users may be shared, with lockout capabilities for private information. This would allow the secretary for a group to schedule joint activities, or for users of the system to check each other's schedules for mutually available times.

A typical schedule screen is shown in Figure 66.4.

As you can see in Figure 66.4, an appointment book feature occupies the left-hand side of the screen. An area for notes, and a current calendar appears on the right-hand

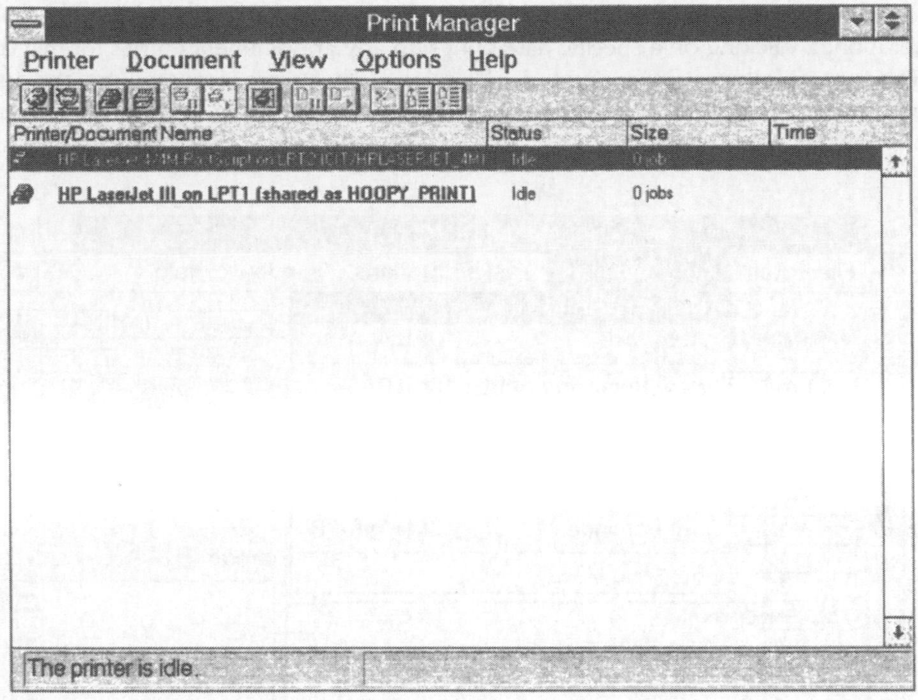

FIGURE 66.3 Windows for Workgroups Print Manager.

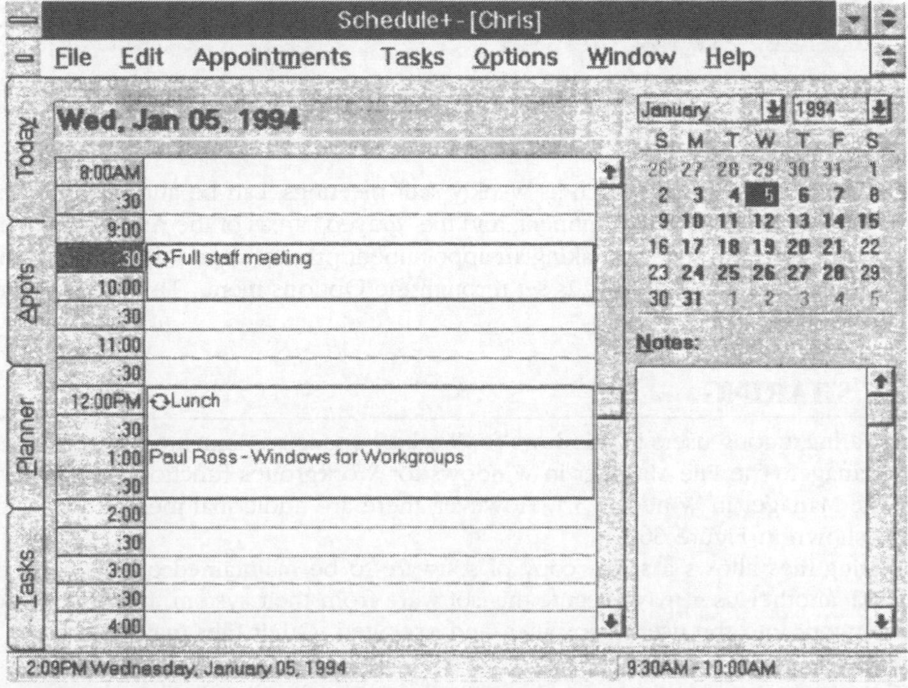

FIGURE 66.4 Windows for Workgroups scheduling system.

side of the screen. The calendar can be scrolled forward and backward in time for another month. Clicking on a specific date brings up any appointment entries for that date.

The system allows for a number of options. These are accessed by the usual Windows menu system. The Appointments menu, as shown in Figure 66.5, contains the following useful items.

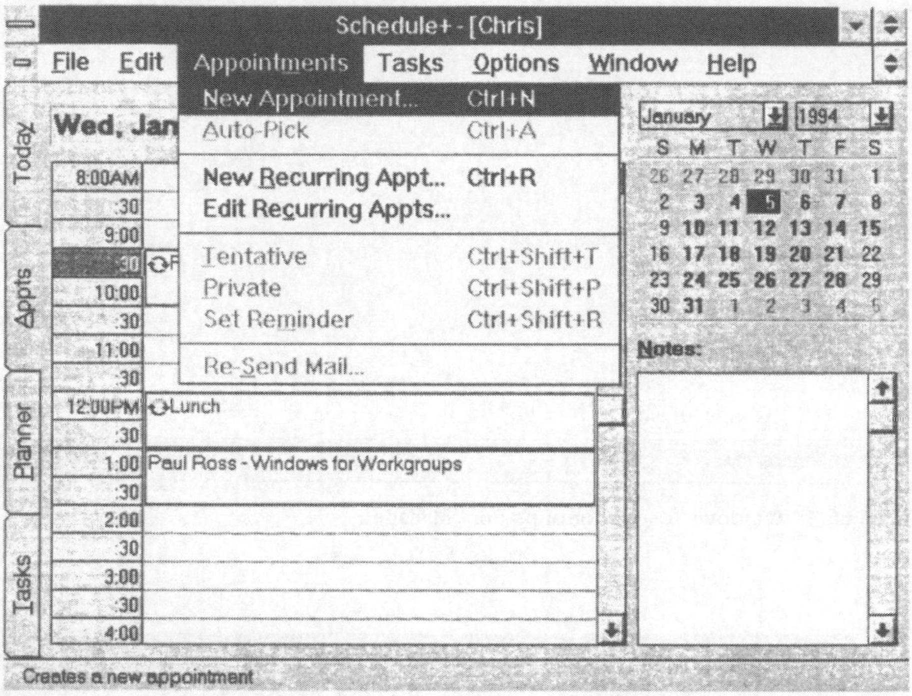

FIGURE 66.5 Windows for Workgroups scheduling system Appointment menu.

Recurring appointments, such as weekly staff meetings, can be automatically scheduled. Click on a specific appointment, and the "grayed" areas of the Appointment menu will be made active, such as making an appointment private (cannot be read by others).

Schedule access and sharing is set through the Options menu. This menu is shown in Figure 66.6.

66.6 FILE SHARING

File sharing among users in Windows for Workgroups is implemented by means of the File Manager. The File Manager in Windows for Workgroups functions essentially like the File Manager in Windows 3.1. However, there are additional menu items, such as those shown in Figure 66.7.

Sharing files allows a single copy of software to be maintained on the network. If desired, another user may execute this software from their system. It is loaded across the network into the user's computer, and executed. Other files may be shared, such as those in a joint project.

The mechanism for sharing a file is simple. Select the file, files, or directory to be shared. Select the Share As option, and when prompted, as shown in Figure 66.8, des-

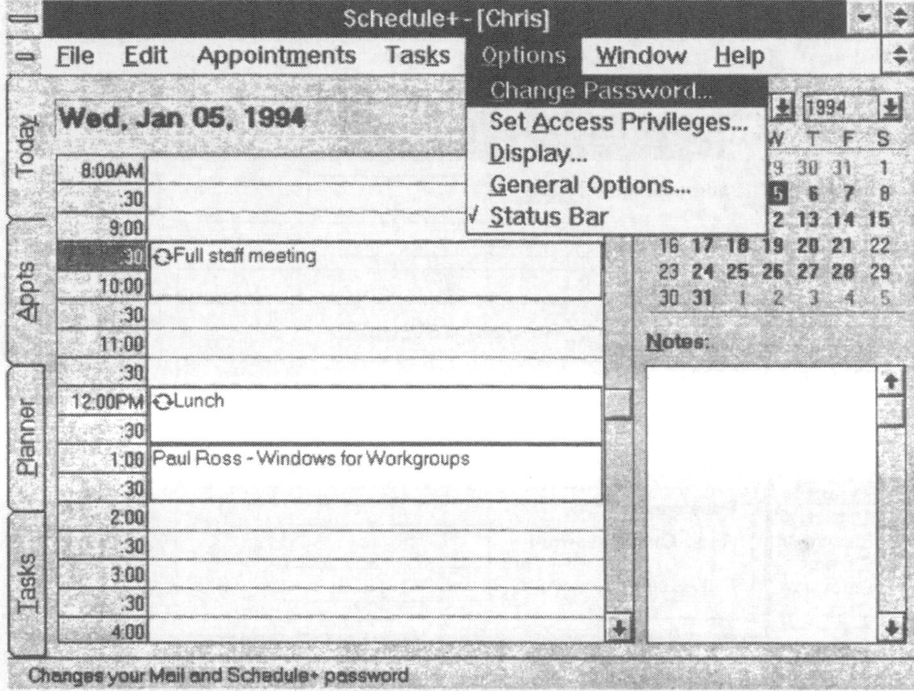

FIGURE 66.6 Windows for Workgroups scheduling system Options menu.

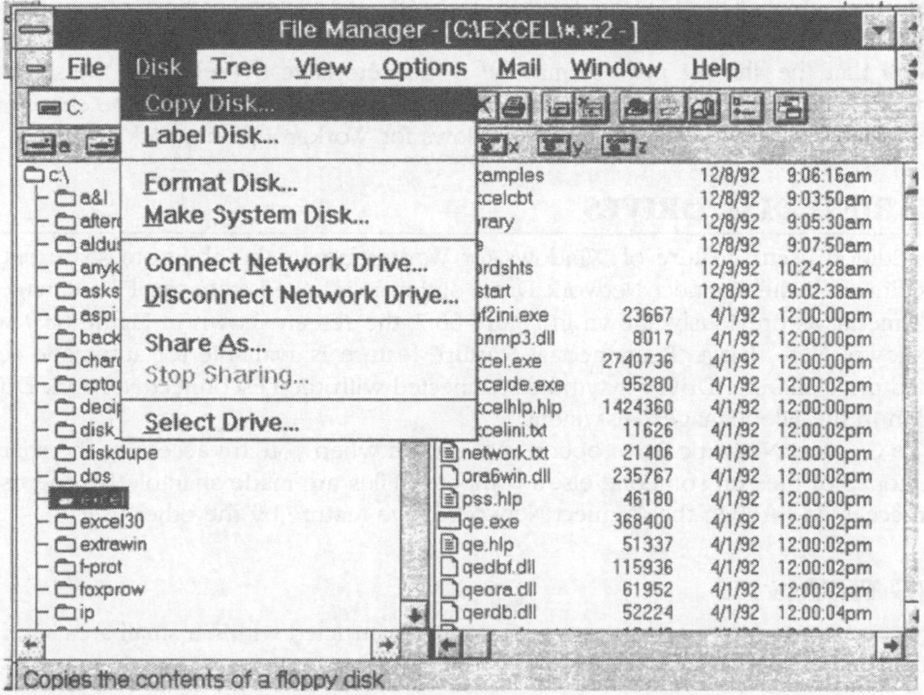

FIGURE 66.7 Windows for Workgroups File Manager Disk menu.

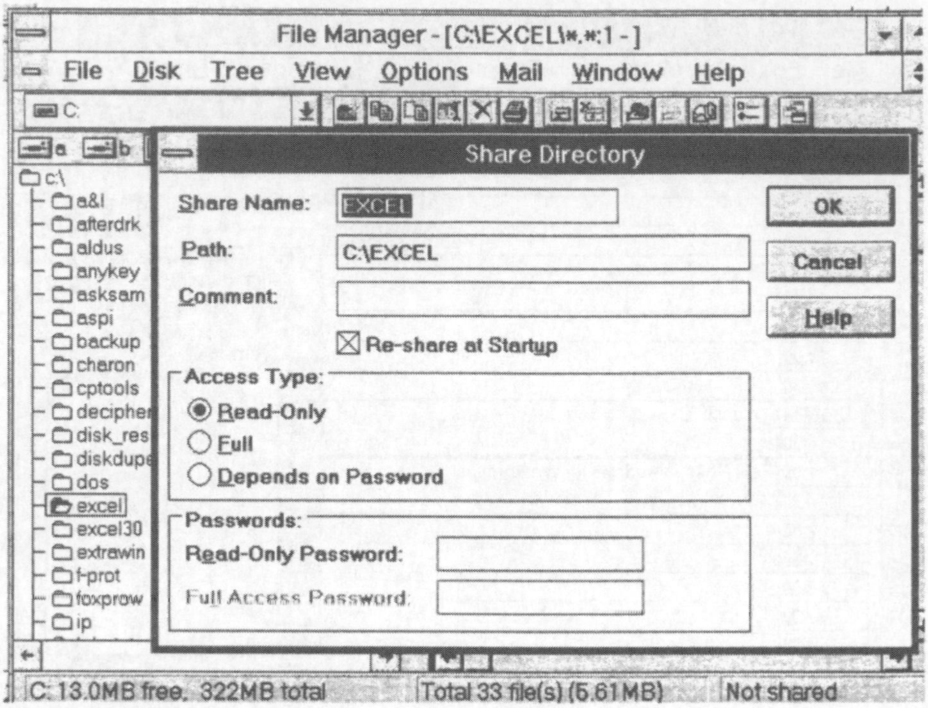

FIGURE 66.8 Windows for Workgroups file sharing directory.

ignate the name of the files or directory to be shared. Selecting the Unshare option terminates sharing of designated files or directories.

Note that the sharing process may be made automatic by selecting "Re-share at Startup." This is useful when a given piece of software or file needs to be shared on a regular basis between users on the Windows for Workgroups network.

66.7 SHARING DISK DRIVES

An addition useful feature of Windows for Workgroups is the ability to share entire disk drives. If the Connect Network Drive option is invoked from the File Manager's Disk menu, as previously shown in Figure 66.7, the screen shown in Figure 66.9 will be shown. Note that a "Reconnect at Startup" feature is available for automatic continued drive sharing. Drives may be disconnected with the Disconnect Network Drive option in the File Manager Disk menu.

The Connect Network Drive option is also used when you are accessing the shared directories or files on someone else's computer. Files are made sharable by one user, then accessed through the Connect Network Drive feature by the other user.

DEFINING TERMS

LAN: Local Area Network. A network usually constructed within a small area such as a building or groups of buildings.

Peer-to-Peer Network: Computer network where all nodes on the network have equal rights for sharing files and other information. Contrast with File Server networks.

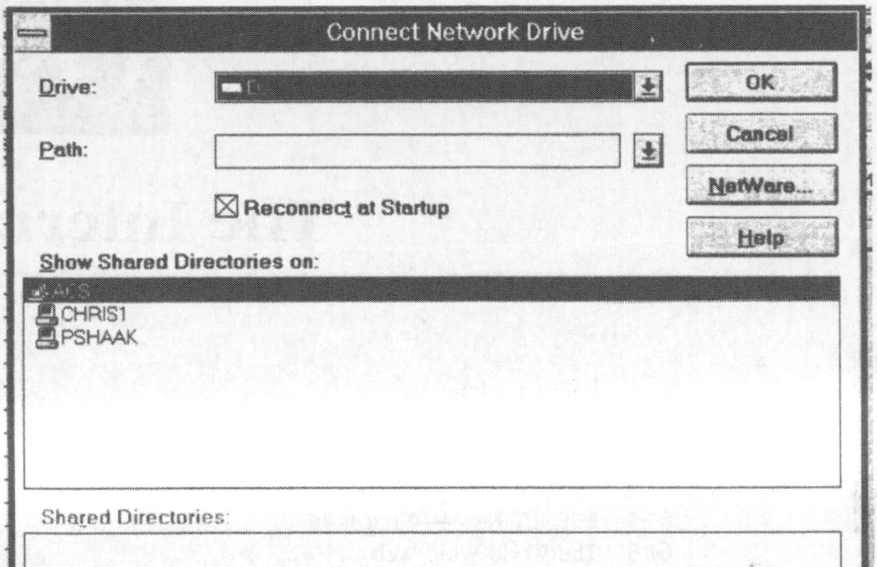

FIGURE 66.9 Connect Network Drive dialog box.

REFERENCES

Read the various .WRI (Windows Write format) files found in the Windows for Workgroups directory. These files explain many of the features of the .INI files and customization features.

FURTHER INFORMATION

Reference manuals for Windows for Workgroups from Microsoft, Inc.

67

The Internet

Laurie A. Knox
Millersville University

Paul W. Ross
Millersville University

The Internet is a super network of universities, research sites, commercial organizations, and government agencies that allows information exchange and sharing of data and resources. The predecessor of the Internet was ARPANET, a Department of Defense computer network created in 1969. When the Department of Defense moved its military applications to a new network (MILNET) in the mid 1980s, researchers—and eventually the National Science Foundation—took over the physical network, and the Internet was born. Today, the Internet connects over 30,000 smaller computer networks, over one million individual (host) computers, and over 5 million users.

67.1 INTERNET ADDRESSES

Internet addresses fall into two categories: all-numeric "IP" addresses (e.g., 192.206.29.16) or the more-common alphanumeric fully qualified domain name or FQDN (e.g., cs.millersv.edu). Note that these are not personal e-mail addresses—they are the addresses of computers or networks on the Internet. To specify a particular user's Internet address, the individual's userid is prefixed to a domain name (e.g., postmast@cs.millersv.edu); this is discussed later.

You must specify either an IP address or domain name when using the Internet commands PING, TELNET, or FTP. The domain name has three or four parts, separated by periods. The last part specifies the type of organization that owns the host computer. The most common are

com—commercial organization
edu—college or university
gov—government agency
mil—military site
net—network organizations
org—miscellaneous private organization

The middle portion denotes the particular site. The first part is the actual computer; if the domain name has four parts, the second part usually indicates a network.

0-8493-2530-7/96/$0.00+$.50

To specify the Internet address of an individual user, you give their userid followed by an at sign (@), then the domain name of the host computer they are on. For example, the postmaster account (POSTMAST) on cs at Millersville University has the Internet address postmast@cs.millersv.edu.

67.2 COMMON INTERNET UTILITIES AND FUNCTIONS

TELNET

TELNET is a powerful Internet command that allows remote logon to another computer ("remote-host") on the Internet. TELNET is similar to dialing into another computer, except that you are spared the expense and hassle of a modem, communication software, and a long-distance phone call. You must have a valid login id and (usually) a password to use a remote computer. Some computers feature "generic" userids (e.g., GOPHER, LIBCAT, INFORMU) that allow you to logon (often without supplying a password) and access general information files via a series of menus. Typically, you would initiate a TELNET connections as follows. Note that in many UNIX-based systems, commands may be case-sensitive.

```
TELNET <remote-host domain name>
```

FTP

FTP (or anonymous FTP) stands for File Transfer Protocol. This command is similar to TELNET; you start by logging into a remote system. However, you are then provided with an environment in which you can transfer files between systems—usually from the remote host to your account. To start FTP, enter the FTP command, followed by the Internet address (or IP number) of the remote host:

```
FTP <remote host domain name>
```

For example, to FTP to the University of Illinois, type

```
FTP vmd.cso.uiuc.edu
```

You should see a message "Connecting to..," then you will be asked for your login id. For most systems, this is the word anonymous. Type ANONYMOUS and press <Enter>. You will then be asked for a password—this is usually your Internet address; type it in and press <Enter>. You should now see a message telling you that you are logged in, along with any general information messages. At this point, you can enter other FTP commands. Some of these commands are as follows:

DIR	lists files in current directory on remote host
CD <new dir>	change directory to new dir on remote host
GET <filename>	transfer file from remote host to your account
PUT <filename>	transfer file from your account to remote host (only with proper authorization!)
MGET <*.ext>	transfer all files of type *.ext from remote host to your account

MPUT <*.ext>	transfer all files of type *.ext from your account to remote host (if authorized)
QUIT	exit FTP and logoff remote host

Getting Directory Listings

- dir provides a more complete listing of files
- 1s provides a simple listing of files

Maneuvering through an FTP Site

- Look for a file named Index or README at the highest directory level; these files usually provide information as to the type of files located at the FTP site.
- Use the **cd** (change directory) command to go "downward" through directories.
- Use the **cd..** command to go "upward" through directories.
- Use the **pwd** command to see what directory you are currently viewing.

Using the GET Command

- Use the command **get** to obtain one file:

```
GET filename-to-get local-filename
```

- The name of the filename-to-get should be indicated as it appears at the FTP site, no matter how files are named in your local environment.
- The name of the local-filename should be in a format that your computer can handle.
- The **mget** command may be used to obtain multiple files.

Different File Types: ASCII and Binary

To change between modes:

- Use the command **binary** to set the file type to binary.
- Use the command **ascii** to set the file type to ASCII ("text").
- Default file type is ASCII.

How to Identify Files at FTP Sites

File Extension	Operating System	Binary or ASCII?	Program to Translate
com, exe	DOS	binary	executable file
doc	any	ASCII	text file
gif	any	Binary	GIF viewer program
hqx	Macintosh	ASCII	BinHex 4.0
pit	Macintosh	ASCII	PackIt 3.13
ps	any	ASCII	print to PostScript printer
sit	Macintosh	Binary	Stuffit
tar	Unix	Binary	tar (Unix command)
txt	any	ASCII	text file
uu	Unix	ASCII	uudecode
wp	DOS	Binary	WordPerfect file
z	DOS	Binary	Pack/Unpack
Z	DOS	Binary	uncompress
zip	DOS	Binary	PKZIP/PKUNZIP

Modes to Use for File Transfer

File Type	Mode to Use for Transfer
Text file	ASCII
Spreadsheet	Binary
Database file	May be Binary or ASCII
Word processor file	May be Binary or ASCII
Program source code	ASCII
Electronic mail messages	ASCII
Backup file	Binary
"Compressed" file	Binary
"Uuencoded" file	ASCII
Executable file	Binary
Postscript file	ASCII

FTP Etiquette

- Try to access remote sites after hours. FTP requires some computer processing time, so some FTP sites may restrict hours for anonymous FTP access. Also, if many people are accessing a popular site, response time may be slower.
- FTP access is a **privilege** for Internet users. Host sites provide information voluntarily.
- Do not remain connected to sites after you have obtained the information you need. After accessing an FTP site, disconnect as soon as possible to free up access for other Internet users.
- If you are not sure as to what files you want at an FTP site, look for an INDEX or README file. Use the get command to obtain that file locally, and disconnect from the FTP host to view the directory listing and determine the files you want to obtain.

PING

The PING command (like TELNET and FTP) operates on an Internet host computer (as opposed to MAIL or NOTE, which use an individual user's Internet address). The purpose of the PING command is to determine whether or not a specified host computer is operating and connected to the Internet. A typical use of the PING command would be

```
PING acad.fandm.edu
```

If the remote computer is up and connected, you might see a message similar to the following:

```
Ping #1 response took 0.154 seconds. Successes so far 1.
```

On the other hand, if you see a "time-out" message (e.g., "Ping #1 timed out"), either the remote ("host") computer is not up, or the connection to it has failed. If you see an "unknown host" message, the domain name you specified does not exist.

Electronic Mail

The mail facility allows you to compose and send a message in the form of a file to another user on the Internet. To use it, you must know the Internet address of your intended recipient (e.g. jsmith@um.psu.edu) The command to invoke mail varies with the system you are using. Typical mail systems are MAIL on the DEC VAX, and mail, elm, or pine on UNIX systems. They vary, and their usage is system-specific.

67.3 SUBSCRIBING TO MAILING LISTS AND ACCESSING LISTSERVERS

What Is LISTSERV?

- Provides a means for groups of computer users with similar interests to communicate among themselves.
- Subscribing is similar to subscribing to a magazine.
- Information is automatically delivered to you via electronic mail. You may respond to messages you receive, which creates an interactive environment.

How to Communicate with LISTSERV Servers

- Commands are sent to LISTSERV hosts via e-mail.
- LISTSERV hosts usually provide access to a wide range of interest groups.
- Send informational commands to LISTSERV@host.
- Send mail to be shared with your interest group to Interest-group@host.

Using BITnet/Internet Gateways

- LISTSERV servers are mainly located on BITnet; to communicate with them via the Internet, you need to use a BITnet/Internet Gateway when sending electronic mail.

Example: You receive information about a discussion group called GRANOLA (a vegetarian discussion list). You are provided with the following information:

```
GRANOLA@VTVM2
```

This is a BITnet address (such an address may also be followed by .BITNET). An Internet address has more than one name after the @ sign, and will end in .EDU, .COM, .ORG, etc.

Subscribing to a Discussion Group

- Send the SUBSCRIBE command to the LISTSERV:

```
SUBSCRIBE list-name your-full-name
```

- Providing your name to the LISTSERV is not always necessary, but it will identify you by name to the LISTSERV.
- You need to indicate the list-name (i.e., in the above example, GRANOLA) because many LISTSERV hosts handle more than one discussion group.
- You will receive a message back from the LISTSERV as to the status of your subscription. Some Listservers require you to respond back within 24 to 48 hours to confirm your subscription; otherwise, your subscription will not take effect.

Controlling the Mail You Receive from a Discussion Group

- Use the SET command to control the information you receive:

```
SET list-name Mail    (default)
```

List mail is distributed to you as mail.

```
SET list-name DIGESTS
```

Provides you with a digest of mail messages sent to the list. Digests may contain messages from a day, a week, or a month, depending on how the list is handled. Mail messages in the digest are complete.

```
SET list-name INDEX
```

Provides an index of mail messages sent to the list (by date, time, subject, sender's name, and length of message). You may select and retrieve any of the messages by sending a message to the Listserver.

```
SET list-name NOMAIL
```

Will suspend the mail you receive from a list; your subscription will be retained. **Very** useful for suspending your mail from the list when you know you will not have a chance to check your mail (holidays, vacations, etc.)

Obtaining Files from a Listserver

- To obtain a list of files available at a LISTSERV host:
- INDEX filelist
- Files are stored under the LISTSERV host in filelists, which are normally directly associated with the name of discussion groups.
- To obtain a file from a LISTSERV host:

```
GET filename filetype filelist
```

- You obtain the filename and filetype from information received via the INDEX command, or information contained in indexes sent to you (if you have indicated SET list-name INDEX).
- Indicating a filelist is not necessary, but it saves time. If you do not indicate a filelist, the LISTSERV host will search all of its archived files for the file you have indicated.

Unsubscribing from a Discussion List

- Send the UNSUBSCRIBE command to the LISTSERV:

```
UNSUBSCRIBE list-name
```

- You will receive confirmation of your removal from the discussion group.

Additional Commands to Send to LISTSERV

- LIST—Provides a listing of all discussion groups handled by the particular host to which you send this request.
- REVIEW list-name—Provides information about the mailing list you indicate: who is authorized to review or join the list, whether the list is archived, and a listing of members of the discussion group. (You may not receive a list of all members of a discussion group; members may use the command SET list-name CONCEAL to have their name concealed from such a general listing.)

Tips to Remember When Using Lists

- Learn how to use your electronic mail system before you become actively involved with a list.
- Save your subscription information! This includes any information you received when you first subscribed.
- Once you join a list, access your mail on a regular basis; if you cannot access your mail, change your options for how you receive list information. If you go on vacation, send the SET NOMAIL command to the Listserver.
- If you send a message to a list requesting information, include your e-mail address in the message so others may respond to you directly. You may then summarize the information you received from others and post it to the list.
- Be careful when you use the REPLY command when responding to a message from the list—make sure your message is being directed properly.

List Etiquette

- Keep your messages relevant to the list topic.
- Avoid using all caps. IT HAS THE EFFECT OF SHOUTING!
- When you respond to a message, clearly reference the message to which you are responding. Using the > character before a message indicates text from a previous message. Some mail systems insert this automatically.

- When responding to a message, try to avoid including entire previous messages in your replies. List recipients have already seen the original message, and such duplication usually creates very long messages.
- Avoid flaming (sending negative messages, especially directed at individuals). Beware of any strong, emotional, angry, or sarcastic comments.
- Always include a meaningful subject line; many people use this initial information to determine whether they want to even read a message.
- Have patience with people who post messages to the list that were intended to be sent to the LISTSERV. Rather than "flaming" such a message, note the e-mail address of the person sending the improperly directed message and (time permitting) send that person a direct message informing them of the proper procedure.

67.4 USING SEARCH TOOLS

There are a number of tools available to aid your search for information on the Internet.

Veronica

- Searches Gopher sites (over 500) by document *title*.
- Prompts you for keywords
- Multiple keywords may be used, separated by *and*, *or*, *not*
- Wildcard character may be used—*
- From the list of documents matching your keyword(s), you can retrieve as with any Gopher menu.

WAIS Wide Area Information Servers

- Search tool for databases on the Internet.
- Steps:
  1. Select a database to search.
  2. Form a query by providing keywords.
  3. Documents are listed, ranked according to number of matches to keyword(s).
  4. To retrieve a document, select it from the list.

Archie

One of the questions that comes up about Internet is "Where do I telnet/ftp to get a certain type of information?" To help address this concern, the Archie database was developed. This resides at several Internet sites, including Rutgers University. The Archie system can usually be accessed through the GOPHER service. See the section on GOPHER for details.

You may also access Archie by telnetting to a site offering Archie services, such as Rutgers University:

```
telnet archie.rutgers.edu
login with the username "archie"
```

You will now be in the Archie database. Here are some useful commands:

about	what does Archie do?
help	lists all commands—good for beginners
list	lists Internet sites in database (over 1000!)
find *word*	searches for the keyword provided and returns matching filenames with site names (where each file is located)
prog	searches database for a file
site	lists files at an archive site
whatis	searches for a keyword in the database
exit	quits Archie

Archie permits scanning of over 1000 anonymous FTP sites, worldwide. It provides a list of FTP site addresses and files, sorted by host address.

Finding People and Computers

There are two common mechanisms for finding people and computers:

Whois

- Directory service to network users
- Main database is maintained by IRC
- Will address *known* databases (will not look at *all* databases at *all* sites)
- telnet rs.internic.net
- type whois at the prompt

Netfind

- Known as a "white pages" tool
- Provide a name and keywords indicating where a person works
- TELNET to the location closest to where your target may be (USA, Canada, etc.)
- telnet ds.internic.net (example of one site to use)
- login: netfind

Gopher

Gopher is a menu-driven information system that is installed at several Internet sites.

To access GOPHER, typically type the command "gopher." Be patient—the remote system is working when you see messages like "Connecting . . .," "Reading . . .," or "System. . . ." Once you have successfully accessed Gopher, you will see a menu of information appear. Simply follow the directions on the screen.

Veronica

Veronica is an "Archie for gophers"—that is, a mechanism for searching (by keywords) titles of gopher items. The result of a search is a menu that can then be used to select a particular item.

Veronica can be accessed under Gopher. When you choose an area of "gopherspace" to search, you will be prompted for index keywords for which to search. When matches are found to your search, you will be provided with a menu to access the information.

67.5 USENET NEWS—WHAT IS IT?

- Collections of articles (submissions by people reading Usenet news), organized by subject into "news groups."
- Provides a forum for discussion and exchange of ideas.
- Over 3500 news groups exist on a variety of diverse subjects.

The major categories of Usenet Newsgroups are

comp	Computer science and related topics
news	News network and news software
rec	Hobbies, recreational activities, the arts
sci	Scientific research and applications
talk	Forums for debates on controversial topics
misc	Miscellaneous discussions (may fit in several categories)

Alternative Newsgroups

alt	Discussions of "alternative ways of looking at things"
bionet	Interest groups for biologists
bit	BITNET listserv discussion groups
biz	Discussions related to business
de	Discussions in German, various topics
ieee	Discussions related to the IEEE (Institute of Electronic and Electrical Engineers)
gnu	Discussions related to the Free Software Foundations (FSF) and the GNU project
k12	Discussions related to teachers and students through high school (K–12)
rec	Recreational interests

You either need a news reader to read newsgroups, or, alternatively, may be able to find a reader through your GOPHER service.

67.6 THE WORLD WIDE WEB

WWW stands for "World Wide Web." The WWW project, started by CERN (the European Laboratory for Particle Physics), seeks to build a distributed hypermedia system.

The World Wide Web is the vision of programs that can understand the numerous different information-retrieval protocols (FTP, TELNET, NNTP, WAIS, gopher, etc.) in use on the Internet today as well as the data formats of those protocols (ASCII, GIF, Postscript, DVI, TeXinfo, etc.) and provide a single consistent user-interface to them all. In addition, these programs would understand a new protocol (HTTP) and a new data format (HTML) both geared toward hypermedia.

The advantage of hypertext is that in a hypertext document, if you want more information about a particular subject mentioned, you can usually "just click on it" to read further detail. In fact, documents can be and often are linked to other documents by completely different authors—much like footnoting, but you can get the referenced document instantly!

Web "Browsers"

To access the web, you run a browser program. The browser reads documents, and can fetch documents from other sources. Information providers set up hypermedia servers from which browsers can get documents.

The documents that the browsers display are hypertext documents. Hypertext is text with pointers to other text. The browsers let you deal with the pointers in a transparent way—select the pointer, and you are presented with the text that is pointed to.

Hypermedia is a superset of hypertext—it is any medium with pointers to other media. This means that browsers might not display just plain text files, but might display images or sound or animations. (Browsers usually need "companion applications" to help display images and sound.)

Browsers that are available now include

Plain Text Browsers	Graphical Interface Browsers
Lynx	Mosaic (for IBM PC, Macintosh, UNIX)
www	Mac Web (Mosaic)
	Cello (IBM PC)
	NetScape (various platforms)

What Is Available on the Web?

- anything served through gopher
- anything served through WAIS
- anything on an FTP site
- anything on Usenet
- anything accessible through telnet
- anything in hytelnet
- anything in hyper-g
- anything in techinfo
- anything in texinfo
- anything in the form of man pages
- hypertext documents

What Is HTML?

Documents on the World Wide Web are written in a simple "markup language" called HTML, which stands for Hypertext Markup Language.

What Is a URL?

URL stands for "Uniform Resource Locator." It is a draft standard for specifying an object on the Internet, such as a file or newsgroup.

URLs look like this: (file: and ftp: URLs are synonymous.)

- **`file://wuarchive.wustl.edu/mirrors/msdos/graphics/gifkit.zip`**
- **`ftp://wuarchive.wustl.edu/mirrors`**

- **`http://info.cern.ch:80/default.html`**
- **`news:alt.hypertext`**
- **`telnet://dra.com`**

The first part of the URL, before the colon, specifies the *access method*. The part of the URL after the colon is interpreted specific to the access method. In general, two slashes after the colon indicate a machine name (machine:port is also valid).

When you are told to "check out this URL," what to do next depends on your browser; please check the help for your particular browser.

How Does WWW Compare to Gopher?

While both of these information presentation systems are client-server based, they differ in terms of their model of data. In gopher, data are either a menu, a document, an index, or a telnet connection. In WWW, everything is a (possibly) hypertext document that may be searchable.

In practice, this means that WWW can represent the gopher (a menu is a list of links, a gopher document is a hypertext document without links, searches are the same, telnet sessions are the same) as well as providing extra functionality.

How Do I Find Out What's New on the Web?

What's New With NCSA Mosaic

```
(URL is http://www.ncsa.uiuc.edu/SDG/Software/Mosaic/Docs/
    whats-new.html)
```

Carries announcements of new servers on the web and also of new web-related tools. This should be in your hot list if you are not using Mosaic (which can access it directly through the help menu).

Where Is the Subject Catalog of the Web?

There are several. There is no mechanism inherent in the web that forces the creation of a single catalog (although there is work underway on automatic mechanisms to catalog web sites).

The best-known (and first) catalog: The WWW Virtual Library

```
(URL is http://info.cern.ch/hypertext/DataSources/bySubject/
    Overview.html)
```

Maintained by CERN. The Virtual Library is a good place to find resources on a particular subject, and has separate maintainers for many subject areas. Another good source is ALIWEB:

```
(URL is http://web.nexor.co.uk/aliweb/doc/aliweb.html).
```

How Can I Search Through ALL Web Sites?

Several people have written robots that create indexes of web sites. Here are a few such automatic indexes you can search:

- WebCrawler

```
(URL is http://www.biotech.washington.edu/WebQuery.html)
```

Builds an impressively complete index; on the other hand, since it indexes the content of documents, it may find many links that are not exactly what you had in mind. However, it does a good job of sorting the documents it finds according to how closely they match your search.

- World Wide Web Worm

```
(URL is http://www.cs.colorado.edu/home/mcbryan/WWWW.html)
```

Builds its index based on page titles and URL contents only. This is somewhat less inclusive, but pages it finds are more likely to be an exact match with your needs.

- Lycos

```
(URL is http://fuzine.mt.cs.cmu.edu/mlm/lycos-home.html)
```

Another web-indexing robot, which includes the ability to submit the URLs of your own documents by hand, ensuring that they are available for searching.

Web Sites to Try

The Whitehouse:	**http://www.whitehouse.gov/**
Spencer Gifts Home Page	**http://www.btg.com/spencer/**
Star Trek: Generations	**http://generations.viacom.com/**
The Awesome List	**http://www.clark.net/pub/journal-ism/awesome.html (messy, but has many entries)**
The Daily Planet (weather)	**http://www.atmos.uiuc.edu**

DEFINING TERMS

Archie: Archival search systems, searches FTP sites.

Electronic Mail: The Internet equivalent of memos.

FTP: File Transfer Protocol; used for transferring files between sites on the Internet.

Gopher: Menu-driven information searching system.

HTML: Hypertext Markup Language; used in creating Mosaic pages.

Internet Addresses: A set of numbers or symbols identifying you and your computer site on the Internet.

Listservers: System to exchange information with users with same interests.

Mosaic: A graphical browser for WWW.

PING: A utility to check if a computer site is currently connected to the Internet.

TELNET: A terminal program used to interact with another site on the Internet.

URL: Uniform Resource Locator; a standard (draft) for specifying an object on the Internet, such as a file or newsgroup.

Veronica: Software to search Gopher sites.

WAIS: Wide Area Information Servers; search tool for databases on the Internet.

Whois: The system for locating computers and users.

WWW: The World Wide Web; a distributed hypermedia system, using text or graphics browsers, such as Mosaic (graphics) or Lynx and www (text).

REFERENCES

Bernard Aboba, *The Online User's Encyclopedia: Bulletin Boards and Beyond*. Addison-Wesley, Reading, MA, 1994.

Eric Braun, *The Complete Internet Directory*. Fawcett, New York, 1993.

Daniel P. Dern, *The Internet Guide for New Users*. McGraw-Hill, New York, 1994.

Joshua Eddings, *How the Internet Works*. Ziff-Davis Press, Emeryville, CA, 1994.

Bennett Falk, *The Internet Roadmap*. SYBEX, San Francisco, 1994.

Sharon Fisher, *Riding the Internet Highway*. New Riders, Carmel, IN, 1993.

Harley Hahn and Rick Stout, *The Internet Yellow Pages*. Osborne McGraw-Hill, Berkeley, CA, 1994.

Brendan Kehoe, *Zen and the Art of the Internet: A Beginner's Guide*, 3rd ed. Prentice-Hall, Englewood Cliffs, NJ, 1994.

Ed Krol, *The Whole Internet: User's Guide & Catalog*, 2nd ed. O'Reilly & Associates, Sebastapol, CA, 1994.

John S. Quarterman and Carl-Mitchell Smoot, *The Internet Connection, System Connectivity and Configuration*. Addison-Wesley, Reading, MA, 1994.

Richard Smith and Mark Gibbs, *Navigating the Internet*. SAMS, Carmel, IN, 1993.

Appendices

The following appendices are included in this handbook. They contain information that you will find useful in your day-to-day activities.

- Appendix A: ASCII Codes

  The codes for ASCII text and control characters.

- Appendix B: RS-232 Ports and Pinouts

  Quick reference material that is handy when you need to either connect RS-232 equipment, or debug serial interfaces on communications and computer equipment.

- Appendix C: Centronics Parallel Printer Ports: The Parallel Interface

  The pinouts and signal information for the common Centronics connector. This is useful for testing and debugging parallel interfaces on printers and computers.

- Appendix D: ANSI Escape Sequences

  ANSI terminals accept a special sequence of control characters to move the cursor, set color options if available, and perform other useful functions. This appendix contains the necessary escape sequences for you to include in programs and other software for implementation of these features.

Appendix A

ASCII Codes

Ctl	Decimal	Hex	Value	Decimal	Hex	Value
^@	000	00	NUL	041	29	)
^A	001	01	SOH	042	2A	*
^B	002	02	STX	043	2B	+
^C	003	03	ETX	044	2C	,
^D	004	04	EOT	045	2D	–
^E	005	05	ENQ	046	2E	.
^F	006	06	ACK	047	2F	/
^G	007	07	BEL	048	30	0
^H	008	08	BS	049	31	1
^I	009	09	HT	050	32	2
^J	010	0A	LF	051	33	3
^K	011	0B	VT	052	34	4
^L	012	0C	FF	053	35	5
^M	013	0D	CR	054	36	6
^N	014	0E	SO	055	37	7
^O	015	0F	SI	056	38	8
^P	016	10	DLE	057	39	9
^Q	017	11	DC1	058	3A	:
^R	018	12	DC2	059	3B	;
^S	019	13	DC3	060	3C	<
^T	020	14	DC4	061	3D	=
^U	021	15	NAK	062	3E	>
^V	022	16	SYN	063	3F	?
^W	023	17	ETB	064	40	@
^X	024	18	CAN	065	41	A
^Y	025	19	EM	066	42	B
^Z	026	1A	SUB	067	43	C
^[	027	1B	ESC	068	44	D
^\	028	1C	FS	069	45	E
^]	029	1D	GS	070	46	F
^^	030	1E	RS	071	47	G
^_	031	1F	US	072	48	H
	032	20	(space)	073	49	I
	033	21	!	074	4A	J
	034	22	"	075	4B	K
	035	23	#	076	4C	L
	036	24	$	077	4D	M
	037	25	%	078	4E	N
	038	26	&	079	4F	O
	039	27	'	080	50	P
	040	28	(	081	51	Q
	082	52	R	104	68	h
	083	53	S	105	69	i
	084	54	T	106	6A	j
	085	55	U	107	6B	k

Decimal	Hex	Value	Decimal	Hex	Value
086	56	V	108	6C	l
087	57	W	109	6D	m
088	58	X	110	6E	n
089	59	Y	111	6F	o
090	5A	Z	112	70	p
091	5B	[	113	71	q
092	5C	\	114	72	r
093	5D	]	115	73	s
094	5E	^	116	74	t
095	5F	_	117	75	u
096	60	’	118	76	v
097	61	a	119	77	w
098	62	b	120	78	x
099	63	c	121	79	y
100	64	d	122	7A	z
101	65	e	123	7B	{
102	66	f	124	7C	|
103	67	g	125	7D	}
			126	7E	~
			127	7F	DEL

Appendix B

RS-232 Ports and Pinouts

The following are the standard circuits and their definitions for the RS-232 interface.

For the purposes of the RS-232 standard, a circuit is defined as a continuous wire from one device to the other. There are 25 circuits in the full specification, less than half of which are at all likely to be found in a given interface. There is a certain amount of confusion associated with the names of these circuits, partly because there are three different naming conventions (common name, EIA circuit name, and CCITT circuit name). The table below lists all three names, along with the circuit number (which is also the connector pin with which that circuit is normally associated on both ends). Note that the signal names are from the viewpoint of the DTE (e.g., Transmit Data are data being sent by the DTE, but received by the DCE).

PIN	NAME	EIA	CCITT	DTE DCE	FUNCTION
1	CG	AA	101	—	Chassis Ground
2	TD	BA	103	→	Transmit Data
3	RD	BB	104	←	Receive Data
4	RTS	CA	105	→	Request To Send
5	CTS	CB	106	←	Clear To Send
6	DSR	CC	107	←	Data Set Ready
7	SG	AB	102	—	Signal Ground
8	DCD	CF	109	←	Data Carrier Detect
9*				←	Pos. Test Voltage
10*				←	Neg. Test Voltage
11					(usually not used)
12+	SCDC	SCF	122	←	Sec. Data Car. Detect
13+	SCTS	SCB	121	←	Sec. Clear To Send
14+	STD	SBA	118	→	Sec. Transmit Data
15#	TC	DB	114	←	Transmit Clock
16+	SRD	SBB	119	←	Sec. Receive Data
17#	RC	DD	115	←	Receive Clock
18					(not usually used)
19+	SRTS	SCA	120	→	Sec. Request To Send
20	DTR	CD	108.2	→	Data Terminal Ready
21*	SQ	CG	110	←	Signal Quality
22	RI	CE	125	←	Ring Indicator
23*		CH	111	→	Data Rate Selector
		CI	112	←	Data Rate Selector
24*	XTC	DA	113	→	Ext. Transmit Clock
25*				→	Busy

In the foregoing, the character following the pin number has the following meaning:

* rarely used
+ used only if secondary channel is implemented
\# used only on synchronous interfaces

The direction of the arrow indicates which end (DTE or DCE) originates each signal, except for the ground lines (——). For example, circuit 2 (TD) is originated by the DTE, and received by the DCE. Certain of the above circuits (11, 14, 16, and 18) are used only by or in a different way by some 208A modems.

A secondary channel is sometimes used to provide a very slow (5 to 10 bits per second) path for return information (such as ACK or NAK characters) on a primarily half duplex channel. If the modem used supports this feature, it is possible for the receiver to accept or reject a message without having to turn the line around, a process that usually takes 100 to 200 milliseconds.

On the above circuits, all voltages are with respect to the Signal Ground (SG) line. The following conventions are used:

Voltage	Signal	Logic	Control
+3 to +25	SPACE	0	On
-3 to -25	MARK	1	Off

Note that the voltage values are inverted from the logic values (i.e., the more positive logic value corresponds to the more negative voltage). A logic 0 corresponds to the signal name being true. For example, if the DTR line is at logic 0, that is, in the +3 to +25 V range, then the Data Terminal IS Ready.

The following criteria apply to the electrical characteristics of each of the above lines:

1. The magnitude of an open circuit voltage shall not exceed 25 V.
2. The driver shall be able to sustain a short circuit to any other wire in the cable without damage to itself or to the other equipment. The short circuit current shall not exceed 0.5 A.
3. Signals shall be considered in the MARK (logic 1) state when the voltage is more negative than –3 V with respect to the Signal Ground. Signals shall be considered in the SPACE (logic 0) state when the voltage is more positive that 3 V with respect to the Signal Ground. The range between –3 V and 3 V is defined as the transition region, within which the signal state is not defined.
4. The load impedance shall have a DC resistance of less than 7000 Ω when measured with an applied voltage of from 3 to 25 V, but more than 3000 Ω when measured with a voltage of less than 25 V.
5. When the terminator load resistance meets the requirements of Rule 4 above, and the terminator open circuit voltage is 0 V, the magnitude of the potential of that circuit with respect to Signal Ground will be in the 5 to 15 V range.
6. The driver shall assert a voltage between –5 and –15 V relative to the signal ground to represent a MARK signal condition. The driver shall assert a voltage between 5 and 15 V relative to the Signal Ground to represent a SPACE signal condition. Note that this rule in conjunction with Rule 3 above allows for 2 V of noise margin. In practice, –12 and 12 V are typically used.
7. The driver shall change the output voltage at a rate not exceeding 30 V/μsec. The time required for the signal to pass through the –3 to +3 V transition region shall not exceed 1 msec, or 4% of a bit time, whichever is smaller.
8. The shunt capacitance of the terminator shall not exceed 2500 pF, including the capacitance of the cable. Note that when using standard cable with 40 to 50 pF per foot capacitance, this limits the cable length to no more than 50 feet. Lower capacitance cable allows longer runs.

9. The impedance of the driver circuit under power-off conditions shall be greater than 300 Ω.

The following are definitions of the most common circuits.

1. **CG: Chassis Ground**
   This circuit (also called Frame Ground) is a mechanism to ensure that the chasses of the two devices are at the same potential, to prevent electrical shock to the operator. This circuit is not used as the reference for any of the other voltages. This circuit is optional. If it is used, care should be taken not to set up ground loops.
2. **TD: Transmit Data**
   This circuit is the path whereby serial data are sent from the DTE to the DCE. This circuit must be present if data are to travel in that direction at any time.
3. **RD: Receive Data**
   This circuit is the path whereby serial data are sent from the DCE to the DTE. This circuit must be present if data are to travel in that direction at any time.
4. **RTS: Request to Send**
   This circuit is the signal that indicates that the DTE wishes to send data to the DCE. Note that no such line is available for the opposite direction, hence the DTE must always be ready to accept data. In normal operation, the RTS line will be OFF (logic 1/MARK). Once the DTE has data to send, and has determined that the channel is not busy, it will set RTS to ON (logic 0/SPACE), and await an ON condition on CTS from the DCE. Then, it may begin sending. Once the DTE is through sending, it will reset RTS to OFF (logic 1/MARK). On a full-duplex or simplex channel, this signal may be set to ON once at initialization and left in that state. Note that some DCEs must have an incoming RTS to transmit (although this is not strictly according to the standard). In this case, this signal must either be brought across from the DTE, or provided by a wraparound (e.g., from DSR) locally at the DCE end of the cable.
5. **CTS: Clear to Send**
   This circuit is the signal that indicates that the DCE is ready to accept data from the DTE. In normal operation, the CTS line will be in the OFF state. When the DTE asserts RTS, the DCE will do whatever is necessary to allow data to be sent (e.g., a modem would raise carrier, and wait until it stabilized). At this time, the DCE would set CTS to the ON state, which would then allow the DTE to send data. When the RTS from the DTE returns to the OFF state, the DCE releases the channel (i.e., a modem would drop carrier), and then set CTS back to the OFF state. A typical DTE must have an incoming CTS before it can transmit. This signal must either be brought over from the DCE or provided by a wraparound (e.g., from DTR) locally at the DTE end of the cable.
6. **DSR: Data Set Ready**
   This circuit is the signal that informs the DTE that the DCE is operational. It is normally set to the ON state by the DCE on power-up and left there. Note that a typical DTE must have an incoming DSR to function normally. This line must either be brought over from the DCE, or provided by a wraparound (e.g., from DTR) locally at the DTE end of the cable. On the DCE end of the interface, this signal is almost always present, and may be wrapped back around (to DTR and/or RTS) to satisfy required signals whose normal function is not required.
7. **SG: Signal Ground**
   This circuit is the ground to which all other voltages are relative. It must be present in any RS-232 interface.

8. **DCD: Data Carrier Detect**
   This circuit is the signal whereby the DCE informs the DTE that it has an incoming carrier. It may be used by the DTE to determine if the channel is idle, so that the DTE can request it with RTS. Note that some DTEs must have an incoming DCD before they will operate. In this case, this signal must either be brought over from the DCE, or provided by a wraparound (i.e., from DTR) locally at the DTE end of the cable.
9. **TC: Transmit Clock**
   This circuit provides the clock for the transmitter section of a synchronous DTE. It may or may not be running at the same rate as the receiver clock. This circuit must be present on synchronous interfaces.
10. **RC: Receiver Clock**
    This circuit provides the clock for the receiver section of a synchronous DTE. It may not be running at the same rate as the transmitter clock. Note that both TC and RC are sourced by the DCE. This circuit must be present on synchronous interfaces.
11. **DTR: Data Terminal Ready**
    This circuit provides the signal that informs the DCE that the DTE is alive and well. It is normally set to the ON state by the DTE at power-up and left there. Note that a typical DCE must have an incoming DTR before it will function normally. This signal must either be brought over from the DTE, or provided by a wraparound (i.e., from DSR) locally at the DCE end of the cable. On the DTE side of the interface, this signal is almost always present. It may be wrapped back around to other circuits (e.g., DSR, CTS, and/or DCD) to satisfy required hand-shaking signals if their normal function is not required.

In an asynchronous channel, both ends provide their own internal timing. As long as they are within 5% of each other, it is sufficient for them to agree when the bits occur within a single character. In this case, no timing information need be sent over the interface between the two devices. In a synchronous channel, however, both ends must agree when the bits occur over possibly thousands of characters. In this case, both devices must use the same clocks. The transmitter and receiver may be running at different rates. Note also that both clocks are provided by the DCE. When a synchronous terminal is tied into a synchronous port on a computer through two synchronous modems, and the terminal is transmitting, the terminal's modem supplies the Transmit Clock. This clock is brought directly out to the terminal at its end, encodes the clock with the data, and sends it to the computer's modem. The computer's modem recovers the clock and brings it out as the Receive Clock to the computer. When the computer is transmitting, the same thing happens in the other direction. Hence, whichever modem is transmitting must supply the clock for that direction, but on each end, the DCE device supplies both clocks to the DTE device.

All the above applies to interfacing a DTE device to a DCE device. To interface two DTE devices, it is usually sufficient to provide a null modem cable, in which the pairs (TD, RD), (RTS, CTS), and (DTR, DSR) have been reversed. Hence, the TD of one DTE is connected to the RD of the other DTE, and vice versa.

It may be necessary to wrap various of the hand-shaking lines back around from the DTR on each end to have both ends work. Similarly, two DCE devices can be interfaced to each other. An RS-232 break-out box is particularly useful in solving interfacing problems. This is a device that is inserted between the DTE and DCE. First, it allows you to monitor the state of the various hand-shaking lines (light on = signal ON/logic 0), and watch the serial data flicker on TD and/or RD. Second, it allows you to break the connection on one or more of the lines (with dip-switches), and make any kind of cross-connections and/or wraparounds (with jumper wires). Using this, it is fairly easy to determine which line(s) are not functioning as required, and quickly build a prototype of a cable that will serve to interface the two devices. At this point, the break-out box can be removed and a real cable built that performs the same function. Care should be taken with this type of device to connect the correct end of it to the DTE device, or the lights and switches do not correspond to the actual signals.

An alternative abbreviated version of the RS-232 interface, often call the "AT 9-pin RS-232 adapter," is connected as follows:

Pin	Function
1	DCD-Data Carrier detect
2	RD-Serial input
3	TD-Serial output
4	DTR-Data terminal ready
5	CNG-Signal ground
6	DSR-Data set ready
7	RTS-Request to send
8	CTS-Clear to send
9	RI-Ring indicator

Appendix C

Centronics Parallel Printer Ports: The Parallel Interface

This appendix describes the parallel interface used by many computers. It is frequently known as the "Centronics" interface, for the first generally available printer to use the interface. Like many things in the computer field, it has become an industry standard for the majority of printers using a parallel interface, primarily the IBM PC or functionally compatible machines. This parallel interface is also frequently used on many Unix machines. Another common interface is the RS-232 serial interface. See Appendix A for a complete description of this serial interface.

The connector pin assignments and a description of the interface signals are given below. The parallel interface is represented by a 36-pin female connector on the printer. It is connected to a mating 36-pin connector on the computer, or through a 25-pin to 36-pin connector. This later connection is generally used on IBM PC or functionally equivalent machines.

Signal Pin	Return Pin	Signal	Direction	Description
1	19	'STROBE	In	'STROBE pulse to read data in. Pulse width must be more than 0.5 μsec at the receiving terminal.
2	20	DATA 1	In	These signals represent information of the first to eighth bits of parallel data, respectively. Each signal is at HIGH level when data are logical 1 and LOW when they are logical 0.
3	21	DATA 2	In	
4	22	DATA 3	In	
5	23	DATA 4	In	
6	24	DATA 5	In	
7	25	DATA 6	In	
8	26	DATA 7	In	
9	27	DATA 8	In	
10	28	'ACKNLG	Out	Approximately 12-μsec pulse. LOW indicates that data have been received and that the printer is ready to accept more data.

Signal Pin	Return Pin	Signal	Direction	Description
11	29	BUSY	Out	A HIGH signal indicates that the printer cannot receive data. The signal goes HIGH in the following cases: 1. During data entry. 2. During printing. 3. When off-line. 4. During the printer-error state.
12	30	PE	Out	Always LOW.
13				Pulled to +5 v through a 3.3K Ω resistance.
14		'AUTOFEED 'XT	In	When this signal is LOW, the paper is automatically fed 1 line after printing (the signal level can often be fixed to this level with a DIP switch on the printer).
15		NC		Unused.
16		OV		Logic ground level.
17				Chassis Ground Printer's chassis ground, which is isolated from the logic ground.
18		NC		Unused.
19–30		GND		Twisted-pair return signal ground level.
31	'INIT	IN		When this level becomes LOW, the printer controller is reset to its initial state and the print buffer is cleared. This level is usually HIGH; its pulse width must be more than 50 μsec at the receiving terminal.
32	ERROR	OUT		This level becomes LOW when the printer is in 1. Off-line. 2. Error state.
33	GND			Same as for pins 19-30.
34	NC			Unused.
35				Pulled up to +5 V through 3.3K Ω resistance.
36	'SLCT	IN		Unused.

Notes

1. The column "Direction" refers to the direction of signal flow as viewed from the printer.

2. "Return" demonstrates the twisted-pair return, to be connected at signal ground level. For the interface wiring, be sure to use a twisted-pair cable for each signal and to complete the connection on the return side. To prevent noise, these cables should be shielded and connected to the chassis of the host computer or the printer.

3. All interface conditions are based on TTL level. Both the rise and fall times of each signal must be less than 0.2 μsec.

4. Data transfer must be carried out by ignoring 'ACKNLG or BUSY signal. Data transfer to the printer can be carried out only after receipt of the 'ACKNLG signal, or when the level of the BUSY signal is LOW.

Appendix D

ANSI Escape Sequences

An ANSI escape sequence is a series of characters beginning with an escape character that can be used to define functions to an ANSI compatible terminal. This appendix explains how the ANSI escape sequences are defined. Please note the following:

1. The default value is ued when no explicit value or a value of zero is specified.
2. Pn represents a "numeric parameter." This is a decimal number specified with ASCII digits.
3. Ps represents "selective parameter." This is any decimal number that is used to select a subfunction. Multiple subfunctions may be selected by separating the parameters with simicolons.

The folloiwng escape sequences affect the cursor position on the screen.

CUP - Cursor position

```
ESC [ Pl ; Pc h
```

HVP - Horizontal and Vertical Position

```
ESC [ Pl ; Pc f
```

CUP and HVP move the cursor to the position specified by the parameters. The first parameter specifies the line number, and the second parameter specifies the column number. The default value is 1. When no parameters are specified, the cursor s moved to the home position.

CUP - Cursor UP

```
ESC [ Pn A
```

This sequence moves the cursor up one line without changing columns. The value of Pn determines the number of lines moved The default value of Pn is 1. The CUP sequence is ignored if the cursor is already on the top line.

CUD - Cursor Down

```
ESC [ Pn B
```

This sequence moves the cursor down one line without changing columns. The value of Pn determines the number of lines moved. The deault value for Pn is 1. The CUD sequence is ignored if the cursor is already on the bottom line.

CUF - Cursor Forward

```
ESC [ Pn C
```

The CUR sequence moves the cursor forward one column without changing lines. The value of Pn determines the number of columns moved. The default value of Pn is 1. The CUF sequence is ignored if the cursor is already in the far right column.

CUB - Cursor Backward

```
ESC [ Pn D
```

This escape sequence moves the cursor back one column without changing lines. The value of Pn determines the nunber of columns moved the defgault value for Pn is 1. The CUB sequence is ignored if the cursor is already in the far left column.

DSR - Îevice Status Report

```
ESC [ 6 n
```

The terminal will output a CPR sequence (see below) on receipt of the DS‰ escape sequence.

CPR - Cursor Position Report

```
ESC [ Pn ; Pn R
```

The CPR sequence reports current cursor position via standard input. The first parameter secifies the current line and the second parameter specifies the current column.

SCP - Save Cursor Position

```
ESC [ s
```

The current cursor position is saved. This cursor position can be restored with the RCP sequence (see below).

RCP - Restore Cursor Position

```
ESC [ u
```

This sequence restores the cursor position to the value it had when the system received the SCP sequence.

The following escape sequences affect erase functions.

ED - Erase Display

```
ESC [ 2 J
```

The ED sequence erases the screen, and the cursor goes to the home position.

EL - Erase line

```
ESC K
```

This sequence erases from the cursor to the end of the line (including the cursor position).

The following escape sequences affect sreen graphics.

SGR - Set Graphics Rendition

```
ESC [ Ps ; ... ; ps m
```

The SGR escape sequence invokes the graphics functions specified by the parameter(s) describd below. The graphics functions remain until the next occurrence of an SGR escape sequence.

Parameter	Parameter Function
0	All Attributes off
1	Bold on
4	Underscore on (monochrome displays only)
5	Blink on
7	Reverse Video on
8	Concealed on
30	Black foreground
31	Red foreground
32	Green foreground
33	Yellow foreground
34	Blue foreground
35	Magenta foreground
36	Cyan foreground
37	White foreground
40	Black background
41	Red backgound
42	Green background
43	Yellow background
44	Blud background
45	Magenta background
46	Cyan background
47	White background

Index

B

C

F

G

H

M

N

P

R

S

T

U

V

W

X

Y

Z

Notes